Psychiatric Care of the Medical Patient

Psychiatric Care of the Medical Patient

EDITED BY

ALAN STOUDEMIRE, M.D.
Professor of Psychiatry
Emory University School of Medicine
Atlanta, Georgia

BARRY S. FOGEL, M.D.
Professor of Psychiatry and Human Behavior
Brown University
Providence, Rhode Island

New York Oxford
OXFORD UNIVERSITY PRESS
1993

Oxford University Press

Oxford New York Toronto
Delhi Bombay Calcutta Madras Karachi
Kuala Lumpur Singapore Hong Kong Tokyo
Nairobi Dar es Salaam Cape Town
Melbourne Auckland Madrid

and associated companies in
Berlin Ibadan

Published by Oxford University Press, Inc.,
200 Madison Avenue, New York, New York, 10016

Oxford is a registered trademark of Oxford University Press

Library of Congress Cataloging-in-Publication Data

Psychiatric Care of the Medical Patient /
edited by Alan Stoudemire and Barry S. Fogel.
p. cm.
Includes bibliographical references and index.
ISBN 0-19-506477-1
1. Psychiatry. 2. Sick—Mental health. I. Stoudemire, Alan.
II. Fogel, Barry S.
[DNLM: 1. Mental Disorders—complications. 2. Mental Disorders—
etiology. 3. Psychophysiologic Disorders. WM 100 P957]
RC454.4.P735 1993 616.89—dc20 DNLM/DLC 92-16193

9 8 7 6 5 4 3 2 1

Printed in the United States of America
on acid-free paper

This text is dedicated to the memory of George A. Stoudemire and Daniel Fogel

Introduction

This text entitled *Psychiatric Care of the Medical Patient* represents a revised and expanded version of our original volume entitled *Principles of Medical Psychiatry.* In this current volume, the editors continue to pursue the goal of developing a comprehensive reference text on the treatment of psychiatric disorders in the medically ill. When we planned the first edition, published in 1987, there was no major comprehensive reference text on this topic that provided information on special diagnostic problems, psychotropic medication use in various medically ill populations, and special issues of psychodynamic psychotherapy for medically ill patients. Moreover, at that time the identity of consultation-liaison psychiatry was somewhat muddled. We aimed to help define modern consultation-liaison psychiatry by defining its knowledge base in our text.

In the first edition we proposed the term "medical psychiatry" to designate the subspecialty of psychiatry concerned with the medically ill, believing that the new name was potentially clearer than "consultation-liaison psychiatry." For many reasons, including the concern of some psychiatrists that we implied the rest of psychiatry was not "medical," the term was not widely accepted. In our view, an accurate and acceptable title to designate the subspecialty involved in the diagnosis and treatment of psychiatric disorders in the medically ill remains to be adopted.

The need not just to designate the name of the subspecialty but also to define the specialized skills and knowledge it comprises has taken on an added significance because of the subspecialization movement in American psychiatry. The subspecialties of geriatric psychiatry and addiction psychiatry are now formally recognized, and efforts are well under way to achieve subspecialty status for consultation-liaison psychiatry as well, spearheaded by the Academy of Psychosomatic Medicine. We hope that the knowledge base for this field has been encapsulated here.

Since the first edition, psychiatric research has made substantial progress in elucidating the biological bases of the major mental illnesses. The importance of biological factors in mental illness is now broadly accepted by both mental health professionals and the general public. Nonetheless, we have little doubt that clinical psychiatry will remain *psycho*biological in both theory and practice. We have continued to emphasize the value of both psychological and biological perspectives in this volume and have attempted to integrate psychodynamic, behavioral, and biological concepts in our approach to diagnosis and treatment.

In keeping with our aim of comprehensiveness, we have covered several areas not included in the first edition. We have added chapters on group therapy and family therapy with the medically ill to complement the chapter on individual medical psychotherapy. The chapter on behavioral medicine is now augmented by specific material on interventions to treat obesity and nicotine dependence, two of the most important behavioral risk factors influencing general health. The chapter on medical-legal issues has been supplemented by material on legal issues of dementia patients. Finally, several additional medical and surgical specialties are covered, including dermatology, hematology, ophthalmology, and otolaryngology. Topics in plastic surgery and urology—concerned with burns and sexual dysfunction, respectively—are also included, and there is a special chapter on sexual rehabilitation of spinal cord trauma patients.

This volume contains a new chapter on neurological diseases, and updated chapters on neurodiagnostic assessment and common traumatic brain injury but does not attempt to cover completely the burgeoning field of neuropsychiatry. Although we do believe that psychiatrists who work with the medically ill should be acquainted with the literature of neuropsychiatry, we see neuropsychiatry as a field that is distinct from, though overlapping with, medical psychiatry.

This volume omits our earlier work on the organization and administration of medical-psychiatric units. Since the publication of the first edition, many new medical psychiatric units have opened, and information about them has become widely available. For this reason, we believe it has become less important to discuss medical-psychiatric units in this book, although we believe that such units fill an important gap in the health care system.

The clinical need for physicians with expertise in integrating psychiatric and medical care has never been greater. Progress in acute care medicine has made chronic disease, disability, and unhealthy behavior stand out in sharp relief as major unresolved public health problems. Those who assess health care and those who finance it have become increasingly concerned with the ultimate outcomes of health care, including well-being and functioning in everyday life. This perspective, which is increasingly in the mainstream of medicine, has led to the realization, supported by research, that the diagnosis and treatment of concurrent psychiatric problems can significantly improve the functional outcomes of medical care. Psychiatrists who can consistently improve medical outcomes, in this broad sense, are thus in ever-increasing demand.

This text, together with the editors' concurrent series of updated volumes on *Medical-Psychiatric Practice*, aim to consolidate the literature on the psychiatric care of the medically ill and to make this information available to clinicians in a readily accessible and useful form. In so doing, we hope to encourage the growth and the effectiveness of this subspecialty of psychiatry and to promote excellent and comprehensive care of patients with combined medical and psychiatric disorders.

Atlanta A.S.
Providence B.S.F.

Acknowledgments

Building on the foundation of the first edition of this text, the editors have attempted to assemble a definitive reference source for the treatment of psychiatric disorders in the medically ill. The effort, almost encyclopedic in scope, was made possible by the superb contributions of our chapter contributors, who shared our ambitious goals for this volume. We would like to thank them for their hard work, dedication, and outstanding contributions.

As with our previous texts, we would also like to thank our assistants, Ms. Lynda Mathews and Ms. Rita St. Pierre, for their splendidly efficient efforts, patience, and dedication. Without their support, this text would never have been completed.

Contents

SECTION IV SPECIFIC DISEASES AND MEDICAL SUBSPECIALTIES

SECTION V SURGICAL SUBSPECIALTIES AND TRAUMA

SECTION VI BEHAVIORAL MEDICINE

SECTION VII MEDICAL-LEGAL ISSUES

Contributors

Gene G. Abel, M.D.
Professor of Clinical Psychiatry
Emory University School of Medicine
Atlanta, Georgia

David B. Abrams, Ph.D.
Professor of Psychiatry and Human Behavior
Brown University School of Medicine
Director, Division of Behavioral Medicine
The Miriam Hospital
Providence, Rhode Island

Theodore J. Anfinson, M.D.
Chief Resident, Department of Internal Medicine
Fellow Associate, Division of General Medicine
Research Associate, Department of Psychiatry
University of Iowa Hospitals and Clinics
Iowa City, Iowa

Drue H. Barrett, Ph.D.
Clinical Instructor of Psychiatry
Morehouse School of Medicine
Atlanta, Georgia

Lawson F. Bernstein, M.D.
Assistant Professor of Psychiatry
University of Pittsburgh School of Medicine
Pittsburgh, Pennsylvania

Hugh F. Biller, M.D.
Professor and Chair, Department of Otolaryngology
Mount Sinai School of Medicine
New York, New York

David J. Blake, M.D., Ph.D.
Clinical Associate Professor of Psychiatry
Bowman Gray School of Medicine
Winston-Salem, North Carolina

Donald R. Bodner, M.D.
Associate Professor of Urology
Case Western Reserve University School of Medicine
Cleveland, Ohio

Harold Bronheim, M.D.
Assistant Clinical Professor of Psychiatry
Mount Sinai School of Medicine
New York, New York

Frank W. Brown, M.D.
Assistant Professor of Psychiatry
Emory University School of Medicine
Atlanta, Georgia

Richard A. Brown, Ph.D.
Clinical Assistant Professor of Psychiatry and Human Behavior
Brown University School of Medicine
Providence, Rhode Island

Harold J. Bursztajn, M.D.
Co-Director, Program in Psychiatry and the Law
Associate Clinical Professor of Psychiatry
Harvard Medical School
Cambridge, Massachusetts

Matthew M. Clark, Ph.D.
Clinical Assistant Professor of Psychiatry and Human Behavior
Brown University School of Medicine
Providence, Rhode Island

C. Edward Coffey, M.D.
Professor of Psychiatry (Neuropsychiatry) and Medicine (Neurology)
Medical College of Pennsylvania, Allegheny
Clinical Director, Allegheny Neuropsychiatric Institute
Oakdale, Pennsylvania

Steven A. Cohen-Cole, M.D.
Associate Professor of Psychiatry
Emory University School of Medicine
Atlanta, Georgia

Christopher C. Colenda, M.D., M.P.H.
Associate Professor of Psychiatry
Associate in Public Health Sciences
Bowman Gray School of Medicine
Winston-Salem, North Carolina

Barry J. Coyne, Ph.D.
Associate Specialist, School of Social Work
University of Hawaii, Manoa
Honolulu, Hawaii

Lucy Davidson, M.D., Ed.S.
Clinical Associate Professor of Psychiatry
Emory University School of Medicine
Atlanta, Georgia

Rodney J. S. Deaton, M.D., J.D.
Staff Psychiatrist
Gallahue Mental Health Center
Indianapolis, Indiana

Sherry N. Dubester, M.D.
Assistant Faculty Member, Department of Medicine
National Jewish Center for Immunology and Respiratory Medicine
Assistant Professor of Psychiatry
University of Colorado School of Medicine
Denver, Colorado

Karen M. Emmons, Ph.D.
Assistant Professor of Psychiatry and Human Behavior
Brown University School of Medicine
Providence, Rhode Island

Steven A. Epstein, M.D.
Assistant Professor of Psychiatry
Georgetown University School of Medicine
Director, Consultation-Liaison Psychiatry
Georgetown University Medical Center
Washington, D.C.

Steven Robert Erle, M.D.
Medical Director, Dual Diagnosis Treatment Unit
Fair Oaks Hospital
Summit, New Jersey

Peter J. Fagan, Ph.D.
Associate Professor of Medical Psychology
Department of Psychiatry and Behavioral Sciences
Johns Hopkins School of Medicine
Baltimore, Maryland

David Faust, Ph.D.
Professor of Psychology
University of Rhode Island School of Medicine
Kingston, Rhode Island

Barry S. Fogel, M.D.
Professor of Psychiatry and Human Behavior
Brown University School of Medicine
Providence, Rhode Island

David G. Folks, M.D.
Professor and Vice Chair, Department of Psychiatry
University of Alabama, at Birmingham School of Medicine
Birmingham, Alabama

David F. Gardner, M.D.
Professor of Medicine
Medical College of Virginia
Richmond, Virginia

Richard J. Goldberg, M.D.
Professor of Medicine and Psychiatry
Brown University School of Medicine
Psychiatrist-in-Chief
Rhode Island Hospital and Womens & Infants' Hospital
Providence, Rhode Island

Richard L. Goldberg, M.D.
Professor and Chair, Department of Psychiatry
Georgetown University School of Medicine
Washington, D.C.

Michael G. Goldstein, M.D.
Assistant Professor of Psychiatry and Human Behavior
Brown University School of Medicine
Providence, Rhode Island

William V. Good, M.D.
Assistant Professor of Opthalmology
University of California
San Francisco, California

Stephen A. Green, M.D.
Clinical Professor of Psychiatry
Georgetown University School of Medicine
Washington, D.C.

C. Thomas Gualtieri, M.D.
Medical Director, North Carolina Neuropsychiatry and Rebound, Inc.
Chapel Hill, North Carolina

Lawrence R. Gulley, M.D.
Assistant Professor of Psychiatry
Emory University School of Medicine
Atlanta, Georgia

Madhulika A. Gupta, M.D.
Clinical Assistant Professor of Psychiatry
University of Michigan School of Medicine
Ann Arbor, Michigan

Scott D. Haltzman, M.D.
Clinical Assistant Professor of Psychiatry and Behavioral Medicine
Brown University School of Medicine
Providence, Rhode Island

Jimmie Holland, M.D.
Professor
Department of Psychiatry
Cornell University Medical College
New York, New York

Carl A. Houck, M.D.
Assistant Professor of Psychiatry
University of Alabama at Birmingham School of Medicine
Birmingham, Alabama

Robert M. House, M.D.
Director, Consultation-Liaison Psychiatry
Colorado Psychiatric Hospital
Associate Professor of Psychiatry
University of Colorado School of Medicine
Denver, Colorado

Jane Jacobs, Ed.D.
Associate Research Professor
Department of Psychiatry and Behavioral Sciences
George Washington University School of Medicine
Washington, D.C.

Jonathan P. Jarow, M.D.
Assistant Professor of Urology
Bowman Gray School of Medicine
Winston-Salem, North Carolina

Roger G. Kathol, M.D.
Professor of Psychiatry and Internal Medicine
University of Iowa College of Medicine
Iowa City, Iowa

Gundy B. Knos, M.D.
Assistant Professor of Anesthesiology
Emory University School of Medicine
Atlanta, Georgia

Susan G. Kornstein, M.D.
Assistant Professor of Psychiatry
Medical College of Virginia
Richmond, Virginia

Lynna M. Lesko, M.D., Ph.D.
Assistant Professor of Psychiatry
Cornell University Medical College
New York, New York

James L. Levenson, M.D.
Associate Professor of Psychiatry and Medicine
Chairman, Division of Consultation-Liaison Psychiatry
Medical College of Virginia, Virginia Commonwealth University
Richmond, Virginia

Norman B. Levy, M.D.
Professor of Psychiatry, Medicine and Surgery
Director, Liaison Psychiatry Division
New York Medical College
Valhalla, New York

Mary Jane Massie, M.D.
Associate Attending Psychiatrist
Memorial Sloan-Kettering Cancer Center
New York, New York

J. Stephen McDaniel, M.D.
Assistant Professor of Psychiatry
Emory University School of Medicine
Atlanta, Georgia

Mary Eileen McNamara, M.D.
Assistant Professor of Psychiatry
Brown University School of Medicine
Director of Ambulatory EEG
Rhode Island Hospital
Providence, Rhode Island

Michael G. Moran, M.D.
Director, Adult Psychosocial Medicine
National Jewish Center for Immunology and Respiratory Medicine
Associate Professor of Psychiatry
University of Colorado School of Medicine
Denver, Colorado

Rebecca R. Neal, M.D., M.S.W.
Assistant in Psychiatry, Massachusetts General Hospital
Instructor in Psychiatry
Harvard Medical School
Cambridge, Massachusetts

Raymond Niaura, Ph.D.
Assistant Professor of Psychiatry and Human Behavior
Brown University School of Medicine
Providence, Rhode Island

William H. Overman, J.D.
Attorney, Private Practice
William H. Overman, P.C.
Atlanta, Georgia

Vincent Pera, Jr., M.D.
Clinical Instructor in Medicine
Brown University School of Medicine
Providence, Rhode Island

Janice Petersen, M.D.
Assistant Professor of Psychiatry
University of Colorado School of Medicine
Denver, Colorado

Russell K. Portenoy, M.D.
Associate Professor of Neurology
Cornell University Medical College
New York, New York

Donn A. Posner, Ph.D.
Clinical Assistant Professor of Psychiatry
Brown University School of Medicine
Providence, Rhode Island

Scott L. Rauch, M.D.
Instructor in Psychiatry, Harvard Medical School
Clinical Research Fellow, Massachusetts General Hospital
Boston, Massachusetts

Quentin R. Regestein, M.D.
Associate Professor of Psychiatry
Harvard Medical School
Boston, Massachusetts

Anne Marie Riether, M.D.
Clinical Assistant Professor of Psychiatry
Emory University School of Medicine
Atlanta, Georgia

Joanne L. Rouleau, Ph.D.
Department of Psychology, University of Montreal
Montreal, Quebec, Canada

Laurie Ruggiero, Ph.D.
Adjunct Assistant Professor (Research) of Psychiatry and Human Behavior, Brown University School of Medicine
Assistant Professor (Research) of Psychology
University of Rhode Island School of Medicine
Kingston, Rhode Island

Chester W. Schmidt, Jr., M.D.
Chairman, Department of Psychiatry
Johns Hopkins School of Medicine
Francis Scott Key Medical Center
Baltimore, Maryland

Elisabeth Jeanne Shakin, M.D.
Associate Director, Consultation-Liaison Psychiatry
Assistant Professor, Jefferson Medical College
Philadelphia, Pennsylvania

Andrew Edmund Slaby, M.D., Ph.D., M.P.H.
Clinical Professor of Psychiatry
New York University College of Medicine
New York, New York

David Spiegel, M.D.
Professor of Psychiatry and Behavioral Sciences
Stanford University School of Medicine
Stanford, California

James L. Spira, Ph.D., M.P.H.
Fellow, Department of Psychiatry and Behavioral Sciences
Stanford University School of Medicine
Stanford, California

Alan Stoudemire, M.D.
Professor of Psychiatry
Emory University School of Medicine
Atlanta, Georgia

James J. Strain, M.D.
Professor of Psychiatry
Mount Sinai School of Medicine
Director, Behavioral Medicine and Consultation Psychiatry
Mount Sinai Medical Center
New York, New York

Paul Summergrad, M.D.
Director, Inpatient Psychiatric Service
Massachusetts General Hospital
Assistant Professor of Psychiatry
Harvard Medical School
Boston, Massachusetts

Yung-Fong Sung, M.D.
Director of Anesthesiology, The Emory Clinic
Associate Professor of Anesthesiology
Emory University School of Medicine
Atlanta, Georgia

Robert M. Swift, M.D., Ph.D.
Associate Professor of Psychiatry and Human Behavior
Brown University School of Medicine
Providence, Rhode Island

Troy L. Thompson II, M.D.
The Daniel Lieberman Professor and Chair, Department of Psychiatry
Jefferson Medical College
Philadelphia, Pennsylvania

Wendy L. Thompson, M.D.
Clinical Associate Professor of Psychiatry
Jefferson Medical College
Philadelphia, Pennsylvania

Richard D. Weiner, M.D., Ph.D.
Associate Professor of Psychiatry
Duke University School of Medicine
Director of ECT Program, Duke University Medical Center
Durham, North Carolina

Thomas N. Wise, M.D.
Professor and Vice Chair, Department of Psychiatry
Georgetown University School of Medicine
Washington, D.C.

I Psychotherapeutic Principles

1 Principles of medical psychotherapy

STEPHEN A. GREEN, M.D.

The interaction between many physicians and patients remains an impersonal, if not mechanistic, encounter. Symptomatic individuals present themselves for diagnosis and treatment with the expectation that they will be cured, or at least significantly relieved of their particular suffering. Because of this cause-and-effect mindset, the *intellectual* activity of problem solving has become the basis of medical practice for many clinicians. Obviously, considerable empathy and concern are involved in this process, because the specific problems under consideration affect the physical and mental health of individuals. However, treating illness is generally considered an exercise of determining etiology, prescribing the indicated therapeutic regimen, and meticulously monitoring the treatment response and course of the illness. This approach is based on the biomedical model, a parochial standard that compromises patient care by embracing the inaccurate notion of a mind-body dichotomy, and perpetuating the reductionistic notion that complex phenomena ultimately are explained by a single principle (Engel, 1977).

Given the evolution of medicine into a highly specialized world that utilizes increasingly sophisticated—and impersonal—technology, a major impact of the biomedical model has been to dissect the patient into more and more composite parts. It isolates individuals from their daily environment of psychosocial supports and stressors, thereby promoting "the physician's preoccupation with the body and disease and the corresponding neglect of the patient as a person" (Engel, 1980). It fails to consider "how the patient behaves and what he reports about himself and his life" because "it does not include the patient and his attributes as a person, a human being" (Engel, 1980). In sum, the biomedical model suggests that treating a patient's specific organic pathology is synonymous with treating the patient, and ignores the fundamental truth that all illness simultaneously affects one's mind and body.

Optimal medical treatment requires a comprehensive appreciation of the interplay between patients' organic pathology, their intrapsychic life, and the positive and negative impact of the social environment. The biopsychosocial model advocated by Engel (1980) places the individual within two hierarchies: biologic and psychosocial (Figure 1-1). These two hierarchies are in a dynamic equilibrium, since every system within each hierarchy is interrelated with every other system. Consequently, disturbances at any level can alter any of the other systems; through feedback controls they in turn may modify the system originally affected.

The biopsychosocial model highlights the link between mind and body and consequently enhances understanding of such diverse processes as endocrinologic function (Haskett & Rose, 1981; Irwin, Daniels, & Weiner, 1987; Kiely, 1974; Mason, 1975; Whybrow & Silberfarb, 1974), immunologic response (Amkraut & Solomon, 1974; Dorian et al., 1986; Kennedy, Kiecolt-Glaser, & Glaser, 1988; Kiecolt-Glaser et al., 1984; Locke et al., 1984; Schleifer et al., 1984; Stein, 1981; Stein, Keller, & Schleifer, 1985), stress response (Cannon, 1920, 1932; Selye, 1950), pathogenesis of disease (Rogers, Dubey, & Reich, 1979; Schindler, 1985; Wolff, Wolf, & Hare, 1950), the course of illness (Engel, 1980; Reiser, 1975), and mechanisms of death (Engel, 1968, 1971; Greene, Goldstein, & Moss, 1972; Lown, Verrier, & Rabinowitz, 1977; Reich et al., 1981). It demonstrates that disturbances in physiologic functioning influence, and are in turn affected by, one's psychologic functioning, and consequently emphasizes the clinical reality that optimal medical treatment requires constant assessment of an individual's affective response to illness. An abnormal response to illness requires comprehensive study of the patient's "illness dynamics" (Green, 1985)—that is, the varied psychosocial factors affecting, and affected by, the patient's organic pathology. This information is then incorporated into an individualized psychotherapeutic approach. The following pages discuss diagnosis and treatment of the abnormal emotional states that can afflict patients with medical or surgical illness.

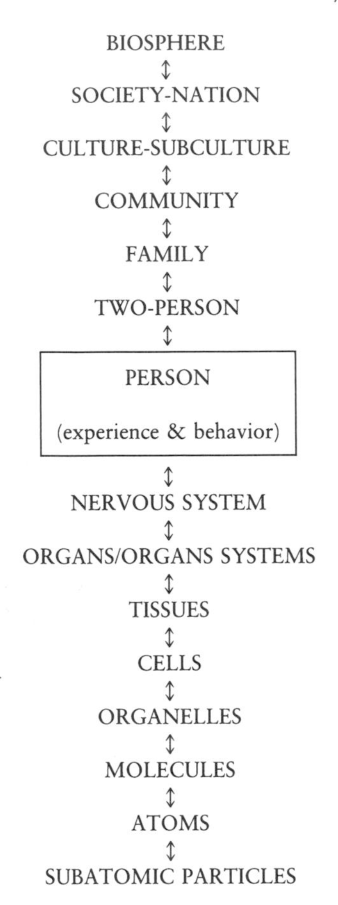

FIG. 1–1. Hierarchy of natural systems. (Reprinted with permission from Engel G [1980]. The clinical application of the biopsychosocial model. Am J Psychiatry 137:535–544.)

RESPONSES TO ILLNESS

Impaired health, whether it takes the form of a mild upper respiratory infection or a life-threatening cerebral vascular accident, is a universal condition that precipitates a predictable emotional response. Illness often is experienced as a *loss*—specifically, the loss of health—because it diminishes one's level of functioning. This is most obvious when serious ailments permanently and profoundly change an individual's physical status, as with the severe restrictions caused by blindness. However, even a limited illness can precipitate considerable feelings of loss because of its symbolic significance to the patient. In addition to challenging omnipotent wishes—highlighting one's vulnerability to physical ailments whose onset, course, and response to treatment are often unexplainable and unpredictable—an episode of ill health conveys specific *symbolic meaning* that derives from the previously noted highly personalized interplay between the patient's physical pathology, intrapsychic life, and social environment. This determines how, and to what extent, the patient reacts affectively to a particular illness and the subsequent loss of health.

Whether illness causes concrete physical restrictions or imposes implied or anticipated limitations, it promotes the same psychologic response—*a grief reaction during which a patient mourns the loss of the previous state of health.* The process is identical to bereavement. Illness obliges patients to acknowledge and confront a diminished level of functioning, causing them to progress through a series of feeling states—denial, anxiety, anger, depression, helplessness—before resolving the various emotions precipitated by the loss. This is most obvious during a medical emergency, such as acute appendicitis. The initial pain cause by an inflamed appendix, which may be quite severe, often is dismissed as indigestion or the effects of a virus. However, other feelings supplant the early denial as the significance of the symptoms becomes clearer. Anxiety emerges as a patient ruminates about the morbidity (and possible mortality) of surgery, as well as the need to abdicate considerable responsibility for well-being and comfort to anonymous caretakers. This anxiety can persist during the recuperative period, focused on such issues as the restrictions and length of convalescence. Emerging resentments, however, usually predominate in this stage, during which the patient discharges anger on family members, friends, and even medical personnel who provide succor and support. These feelings eventually give way to a period of helplessness and overt depression, characterized by affective, behavioral, and cognitive changes, when one is most aware of the losses brought on by a surgical emergency that has relegated all other activities (e.g., professional and social responsibilities) to positions of secondary importance. Most patients emerge from this phase progressively, reestablishing an emotional equilibrium by working through all of the feelings precipitated by illness. This requires appropriate assessment of the specific limitations caused by the disease, as well as a general acknowledgment of one's vulnerabilities and ultimate mortality.

An abnormal emotional response to illness occurs when one is unable to effectively grieve the losses caused by ill health. For a variety of reasons idiosyncratic to the particular individual during a particular period of life, the patient may be unable to recognize, experience, and put into proper perspective some or all of the feelings precipitated by illness. That individual may completely deny all emotions, or more commonly, experience a specific affect to the relative exclusion of all others. In effect, the patient becomes mired in some phase of the grief process, preoccupied with the feelings characteristic of that stage. This forestalls attainment of an emotional resolution regarding the loss of health, a situation that can evolve into an emotional state similar to pathologic grief whose symptomatic expression may take the form of an affective disturbance (e.g., excessive anxiety and/or depression), a behavioral disturbance (e.g., compulsive eating or substance abuse), or impaired object relationships (Brown & Stoudemire, 1983). Subsequent effects on somatic functioning—which may aggravate the initial organic pathology—derive from mind-body interactions. Because of this interdependence, *one's affective resolution of the loss of health is essential for physical, as well as psychologic, well-being.* Absent this grief work, a pathologic illness response emerges that is characterized by a psychobiologic disequilibrium generally more disabling than that caused by the initial illness. The most frequent abnormal psychologic reactions to illness include denial, anxiety, anger, depression, and dependency.

The Denial Response

Denial, the refusal to perceive or accept external reality, is a common ego mechanism that enables one to completely ablate distressing cognitions and/or affects. If absolute, it can be considered a primitive behavior, classified by Vaillant (1977) as a psychotic defense mechanism. It has some adaptive purpose, such as enabling an individual to temporarily ignore the overwhelming impact of a stressor that might otherwise threaten psychologic integrity; consequently, denial is often invoked by emotionally healthy individuals when specific circumstances limit the effectiveness of higher-level ego functioning. This is exemplified by the belief that one will survive in battle, and reflected in better short-term psychologic adjustment following a myocardial infarction (Hackett & Cassem, 1974).

Denial becomes maladaptive when it is so excessive that it prevents accurate assessment and acceptance of the realities of life. At this point it is exclusively defensive, providing illusory comfort by prompting actions (or ensuring inactivity) that promote a false sense of well-being—a frequent reaction of medical and surgical patients whose response to illness is actually a nonresponse. Denial may be present throughout the course of illness, from the moment of onset, during the period of initial treatment and convalescence, and extending into the rehabilitative phase. The long-standing cardiac patient who attributes cramping in his left arm to "overwork," or the woman who decides to have her breast mass evaluated "in a few months," exemplifies the acute denial of illness. Noncompliance with one's therapeutic regimen often reflects continued denial. This can apply to the period of initial hospitalization, as illustrated by the patient who refuses to remain at bed rest (Reichard, 1964), as well as to chronic care, as seen in the individual with diabetes mellitus who repeatedly ignores dietary restrictions. It may also be observed in the terminal patient who adamantly refuses to acknowledge impending death. A patient's unconscious rejection of illness, invoked to maintain emotional stability, can complicate accurate diagnosis, interfere with definitive treatment, and, consequently, aggravate organic pathology.

The Anxiety Response

Signal anxiety, the adaptive expression of this affect, alerts one to danger and subsequently prompts purposeful, self-protective behavior. The onset of illness is always accompanied by heightened anxiety, as the patient questions the extent and degree of the particular ailment. Common concerns, such as whether or not the symptoms signal the beginning of a fatal illness, usually give way to speculation about the diagnostic workup, course of treatment, and ultimate prognosis. Patients may wonder how they will be perceived by others, and how they will regard themselves, if disfigured or chronically debilitated. They ruminate about the toll of the illness on family relationships, professional responsibilities, friendships, and day-to-day functioning. In short, they attempt to assess the impact of illness on the overall quality of life in the near and distant future. Although a difficult and disconcerting exercise, this usually is an adaptive response to illness, spurring them to seek timely medical attention and adhere to prescribed treatments.

Illness can also precipitate a pathologic anxiety that causes the patient undue concern about physical status. As a consequence, patients become hypersensitive to all aspects of care, from diagnostic procedures to therapeutic interventions—a preoccupation that has diverse detrimental effects. A new sign or symptom may take on excessive significance, interfering with the patient's ability to provide an accurate medical history, significantly augmenting the apprehension accompanying routine examination and treatment, and generally distracting that individual from the pleasures and responsibilities of daily life. If this evolves into chronic anxiety, the patient becomes burdened with an unpleasant affective state in addition to the discomfort of the physical pathology. Moreover, continued demands by the patient for reassurance from family, friends, and health care professionals may progressively alienate those persons, ultimately isolating the patient from fundamental sources of emotional support. In this fashion pathologic anxiety can compromise every phase of medical treatment, and decrease the patient's potential for returning to optimal functioning.

The Anger Response

Patients who demonstrate anger as the predominant response to loss of health express the affect in the same way one grieves the death of a loved one—focusing feelings of frustration, resentment, and overt hostility in several directions. They may rail at the gods with a diffuse, globally expressed anger, cursing their bad luck and wondering, "Why me?" Alternatively, they may direct that anger toward themselves; the individual who herniates a lumbar disc while moving furniture holds himself most responsible for the painful disruption of his life. Anger is also focused on family members and friends, whose various behaviors often are blamed as contributors to the development or exacerbation of an illness; a patient may view the demands of a spouse as the primary cause of her peptic ulcer disease, hypertension, or migraine headaches. Patients may judge members of their social network as insensitive to their plight, or resent family members because of greater dependence on them. Patients may even blame ancestors for a defective gene pool that predisposes them to a particular illness.

Health care personnel also frequently bear the brunt of a patient's hostility. In addition to providing succor and support, caretakers exert considerable control, which often takes the form of deprivation and discomfort. They subject individuals to unpleasant diagnostic tests, restrict diets, limit activity levels, and prescribe medications that cause unpleasant, sometimes dangerous, side effects. Moreover, practitioners frequently fail to cure disease, and often bear grim news about an individual's welfare and mortality. For these reasons, they are prime targets for patients' angry feelings.

The hallmark of the anger response is conflict. Instead of working through feelings fostered by the illness, the patient engages in a variety of struggles with the people in his life. These may be obvious, such as debate surrounding the refusal to submit to a diagnostic procedure or consent to rehabilitative therapy. The impotent pleas of family members urging reconsideration, as well as the frustration of health providers, reflect the patient's considerable hostility. Struggles may also be covert. These can take the form of passive aggression, such as ignoring dietary restrictions or irresponsibility in adhering to a medication regimen, behaviors that afford the patient the pleasure of secret control over caretakers. Displacement is an additional means of disguising one's anger about the loss of health; for example, the recurrence of old marital disputes permits discharge of the affect without directly identifying illness as its root cause.

In sum, the anger response promotes a generalized noncompliance with the treatment, progressively transforming the patient's supportive alliances—including those with caretakers—into adversarial relationships. In addition to detracting from the level of medical care, this reaction often isolates patients from a support system that could help them negotiate the stressful life situation of ill health.

The Depression Response

Individuals who react to physical illness with depression manifest affective, cognitive, and behavioral changes characteristic of a clinical depression (Klerman, 1981; Rodin & Voshart, 1986). These vary in degree, depending on factors idiosyncratic to the patient and the particular disease process; consequently, the intensity of the depression response may range from a relatively mild adjustment disorder to a major depressive episode. The clinical presentation encompasses four general areas of objective and subjective signs and symptoms.

First, alterations of affect occur. This usually presents as a sustained lowering of mood accompanied by pronounced sadness, tearfulness, and anhedonia, although affective changes may also include irritability, agitation, and hypomania. Next, the physiologic sequelae of depression are manifested in wide-ranging somatic symptoms that can affect any organ system (Lindemann, 1944). Third, the depression response causes changes in patients' usual patterns of behavior beyond those resulting from restrictions of the illness state. This often takes the form of a generalized withdrawal—physically and psychologically—from their environment. Mounting preoccupation with the loss of health causes patients to invest less energy in family matters, social relationships, and professional responsibilities. As this isolation grows, the world becomes increasingly joyless. Finally, as with all depression, a prominent aspect of this illness response is diminished self-esteem. The effects of a patient's physical impairment, aggravated by a subsequent mood disturbance, raise considerable fear and doubt about the prospect of regaining the premorbid state of health. This negativism fosters self-criticism characterized by feelings of weakness, self-reproach, and worthlessness.

This illness response affects medical treatment in a variety of ways. Psychophysiologic symptoms of depression complicate the diagnosis of an organic illness (e.g., when attempting to differentiate between a dementia and pseudodementia), making ongoing evaluation and treatment a more difficult task. Depression impedes the immunologic response (Schindler, 1985; Schleifer et al., 1984), with obvious adverse effects on the patient's ability to battle an illness, and can become so severe that it poses a greater threat to the patient than does the physical pathology. Neurovegetative symptoms, such as extreme anorexia, may progress to the point of passive self-destructive behavior; purposeful suicidal actions also occur (Slaby & Glicksman, 1985). Finally, the characteristic withdrawal of the depression response progressively isolates patients from the usual supports of their environment. This subverts the therapeutic alliance with various health care personnel, alienates the patient from family and friends, and may become so extreme that the individual essentially abandons the will to live (Engel, 1968).

The Dependency Response

Illness limits autonomy because of factors internal to the patient (e.g., restrictions caused by specific symptoms, the individual's general preoccupation with the disease), as well as external issues (e.g., limitations imposed by the structured hospital routine, the requirements of a therapeutic regimen). The adaptive aspect of this regression permits the patient to revert to more passive behaviors characteristic of early developmental phases. This allows caretakers to take a necessary degree of control of the treatment, and simultaneously provides the patient with the comfort and familiarity of diminished responsibility. (In his discussion of regression, Freud [1924] made the analogy to the actions of a migrating population under siege; it falls back to previously strengthened positions, revitalizes its forces, and then proceeds with renewed vigor). The defensive aspect of regression helps patients contend with the distressing emotions precipitated by illness, by assigning to others concern and accountability for their welfare.

Although regression helps patients maintain emotional equilibrium during periods of illness, it can interfere with medical treatment if their dependency becomes too extensive or intense. This is addressed in Parsons' (1951) description of the dynamics of the *sick role*, which emphasizes the collaborative aspect of medical treatment. During the early phase of illness, individuals' impairment is legitimized by health care personnel, who identify them as patients and, consequently, exempt them from various societal responsibilities. The patient's dependency on caretakers is acknowledged and accepted; however, with improvement the individual is expected to assume increasing levels of independent functioning. This phase of illness, a collaborative endeavor that Parsons termed the "common task" between patient and provider, highlights the fact that individuals bear definite responsibility for important aspects of their health care, which range from adherence to a prescribed diet to active participation in a long-term rehabilitation program. The last phase of the sick role, as outlined by Parsons, has particular relevance to the dependency response. Patients may find the ministrations of others so gratifying that they are disinclined to strive toward the autonomy of premorbid life. They abdicate increasing responsibility to family, friends, and health care personnel, and savor the nurturing pleasures of dependency. Unfortunately, this decline to greater levels of helplessness compromises patients' physical and emotional well-being.

Excessive dependency often is manifested as noncompliance with medical treatment. This may cause an individual to continue harmful habits (e.g., smoking), to be irresponsible with medications, or to allow known symptoms to get worse before seeking medical attention. Whether these motivations are purposeful or unconscious, they yield the same results—prolonged dependency on health care personnel, who are obliged to provide the care necessitated by the patient's neglect. In addition to possibly aggravating the patient's physical status, pathologic dependency often subverts supportive relationships. The prevailing dynamics within a family can be considerably disrupted by a patient's continued regression, altering established roles in a manner that undermines the emotional welfare of all concerned. In a similar fashion, relationships with professionals may suffer, and medical personnel may come to view the regressed individual as a "hateful patient" (Groves, 1978), a countertransference response often triggered by increasingly annoying dependent demands. All these behaviors—which reflect the individual's evolution into a professional patient more concerned with being cared for than with discharging the responsibilities of adult life—may be overt or covert. It is particularly important for health care providers to appreciate a dependency response that might be disguised by an individual's seeming compliance. The longer the patient's wish to remain ill goes unrecognized, the likelier it is that the collaborative doctor-patient relationship will steadily degenerate into an adversarial association harmful to both.

ILLNESS DYNAMICS

Everyone lives within a distinctive context of supports and stressors that influence day-to-day functioning and interpersonal relationships. Illness dynamics refer to those diverse factors as they affect an individual's response to a specific disease at a particular time in life (Green,

1985). They derive from conscious and unconscious determinants that converge in the mind, causing the patient to perceive, assess, and defend against the loss of health in a highly subjective manner (Table 1-1). Patients' illness dynamics cause them to evaluate all illness-related information in a fashion that reflects particular values, needs, wishes, and fears, thereby shaping a distinctive perception of ill health. This idiosyncratic biopsychosocial profile may help them accommodate to the emotions precipitated by loss of health or, alternatively, impede that effort.

Illness fosters regression when it "cuts off [one's] major source of gratification, or enhances intrapsychic conflicts by facilitating emergence of repudiated wishes, weakening ego mechanisms of defense or signifying punishment for realistic or neurotic guilt" (Lipowski, 1975). This can have widespread impact on the total organism as a result of diffuse mind-body interactions. Reiser (1975) explored the role of the brain as a clearinghouse for the diverse emotional and physiologic reactions, discussing how it "orchestrates, integrates, and at points transduces across the biologic, psychologic, and the social realms," thereby translating idiosyncratic symbolic meanings into physiologic changes throughout the body that may promote the onset of illness, aggravate its clinical course, or help maintain a healthy homeostasis. He described a "transactional continuity extending from subcellular metabolic processes throughout the body via the brain to the social environment" that suggests how "major life experiences, such as bereavement, can influence even the capacity to sustain the life process itself." This thesis is supported by a considerable body of laboratory and clinical data that includes investigation of total body processes, such as immunologic response, and more discrete physiologic activities, such as cardiovascular functioning (Henry, 1975).

TABLE 1–1. *Illness Dynamics*

Biologic
Nature, severity, and time course of disease
Affected organ system, body part, or body function
Baseline physiologic functioning and physical resilience
Psychologic
Maturity of ego functioning and object relationships
Personality type
Stage in the life cycle
Interpersonal aspects of the therapeutic relationship (e.g., countertransference of health care providers)
Past psychiatric history
Effect of medical history on attitudes toward treatment (e.g., postoperative complications)
Social
Dynamics of family relationships
Family attitudes toward illness
Level of interpersonal functioning (e.g., educational and occupational achievements; ability to form and maintain friendships)
Cultural attitudes

Illness dynamics also affect the doctor-patient relationship, as well as sources of support in an individual's social network. According to Lipowski (1975) the process of evaluating the state of ill health

> influences the patient's perceptions and mood as well as the content and form of what he communicates concerning his illness, how, to whom, and when. It affects his decision to seek or delay medical consultation, his degree of compliance with medical advice and management, as well as his relationship with his family, employers, health professionals and other concerned people.

These interactions significantly influence an individual's ability to contend with illness. Engel has discussed their day-to-day impact on clinical care in both theoretical (1977) and practical (1980) terms. His illustration of the clinical application of the biopychosocial model relates how treatment of a middle-aged man suffering an acute myocardial infarction was compromised by inattention to his characterologic structure and style of interpersonal relationships. Heightened anxiety, which precipitated neurogenic responses that initiated or augmented a potentially fatal arrhythmia, might have been averted had the patient's caretakers been familiar with his illness dynamics.

In sum, *illness dynamics derive from the interplay between the singular components of one's biologic, psychologic, and social existence, transforming an episode of ill health into a highly subjective experience. By shaping its meaning in this fashion, they can convert the same disease process into seemingly different illnesses in two individuals requiring identical diagnostic workup, acute treatment, and ongoing care.* Because illness dynamics span an enormous and complex range—given the breadth of psychosocial factors influencing and influenced by a specific organic pathology—combinations of these many biologic, psychologic, and social issues cause patients to define their particular ailments in highly idiosyncratic terms. When these individualistic reactions negatively affect the day-to-day treatment of an illness, they set the stage for a maladaptive illness response. As discussed below, medical psychotherapy is the indicated treatment for these clinical situations.

MEDICAL PSYCHOTHERAPY

Reasoned attention by medical personnel to the heightened emotions precipitated by an individual's ill health can prevent the onset of an abnormal illness response.

However, all therapeutic measures must be carefully tailored to each patient, based on a comprehensive understanding of that person's illness and illness dynamics. Interventions designed to initiate the grief work accompanying the loss of health are required for some individuals, but contraindicated in patients obviously overwhelmed by their current affective state. The latter group benefits from measures that support adaptive ego functioning, such as supplying an obsessional patient with considerable information about his day-to-day condition. Unfortunately, many patients do not respond to this aspect of primary medical care, and become increasingly encumbered by the distressing feelings and harmful behaviors that characterize a particular pathologic reaction to illness. Continued negative impact on their physical and emotional well-being signals the need for medical psychotherapy, a specialized form of psychotherapeutic intervention. As discussed by Goldberg and Green (1985), medical psychotherapy is based on "the communicated understanding between physician and patient concerning the biologic, psychologic and social aspects" of an individual's illness. The treatment, which can be utilized for crisis situations (e.g., acute onset of illness) or chronic conditions (e.g., persisting illness), can take a supportive or introspective approach.

Despite the correlation between grief work and the usual emotional reaction to illness, helping the patient work through feelings about the loss of health is sometimes contraindicated. Such insight-oriented therapy rests on the patient's ability to tolerate the heightened anxiety attending exploratory work, and certain patients have characteristic personality features that render intensive, anxiety-provoking psychotherapy ineffective and/or harmful. Clinicians' expert knowledge concerning the substantive and temporal course of grief work is counterproductive if it prompts them to doggedly confront the defensive structure and maladaptive behaviors of psychologically immature individuals. Attempts to pressure patients through the various stages of grief are more likely to foster a clinical regression.

Many individuals afflicted with one of the traditional "psychosomatic" ailments fall into this category (Sifneos, 1972–1973). Attempts to help them resolve the powerful emotions accompanying illness are often frustrated by their diminished ability or motivation for self-examination or a predisposition to hypochondriasis, factors that significantly undermine meaningful attempts to explore and understand affective states. Because alexithymia limits an individual's ability to communicate in an abstract, symbolic manner, it, too, is often a serious impediment to grieving the loss of health. Alexithymic patients have great difficulty exploring inner feelings because of their fundamentally concrete style of perceiving, evaluating, and discussing events (Nemiah & Sifneos, 1970; Sifneos, 1972). This deficit is often reflected in impoverished object relations, including interactions with health care providers (Krystal, 1979), sometimes provoking a negative countertransference characterized by frustration and boredom (Taylor, 1977, 1984). For these reasons, supportive techniques usually are the most therapeutic approach for alexithymic patients (Freyberger, 1977).

There also are clinical situations in which patients' psychologic distress is so extreme that psychotherapeutic interventions alone fail to ameliorate their level of dysfunction. When individuals experience severe affective turmoil, or exhibit a pathologic denial that inhibits the emotional and/or intellectual appreciation of an illness, pharmacotherapy is indicated. Psychoactive medications may then be employed as the primary treatment modality (e.g., prescribing major tranquilizers for delirium or a postoperative psychosis), or adjunctive to psychotherapy (e.g., utilizing minor tranquilizers to control the anticipatory anxiety associated with activities that frequently exacerbate an individual's angina).

When treating medical and surgical patients unable to contend with the emotions precipitated by illness, the clinician's most important decision concerns a clear delineation of therapeutic goals. The physician must choose between commending and encouraging the patient's usual style of emotional functioning and pursuing a more analytic path. The former approach, which attempts to foster an atmosphere of support and reassurance that can then be exploited to enhance the clinician's positive influence, may or may not include the use of psychoactive medications. The latter course attempts to help the patient work through the conflictual feelings precipitated by the illness. Correct treatment of an abnormal illness response thus requires the physician to choose between a supportive, anxiety-suppressing approach, sometimes in conjunction with pharmacotherapy, and an introspective, anxiety-provoking stance as the appropriate form of medical psychotherapy for a particular patient. The clinical data used to make this determination are discussed below, followed by a review of the general characteristics of these distinctive therapeutic strategies. Clinical material explicating the two types of medical psychotherapy is presented elsewhere (Green, 1985).

Determining the Therapeutic Approach

The decision as to whether patients suffering from an abnormal illness response require introspective investigation of their emotions (anxiety-provoking treatment) or are more in need of reassurance and support (anxiety-suppressing treatment) is founded on the systematic evaluation of objective clinical findings. The most fundamental of these is an accurate multiaxial psychiatric diagnosis, particularly as it applies to major mental illness. Psychosis, for example, demands intensive support in the form of psychosocial and pharmacologic interventions. This applies to all patients, whether the cause be organic (e.g., due to hypo- or hyperthyroidism), or reactive due to acute regression in an individual suffering from a chronic mental illness. Similarly, medically ill patients suffering from major depression require initial supportive interventions, and usually somatic treatments, in order to alleviate their symptoms (Cohen-Cole & Stoudemire, 1987; Rifkin et al., 1985).

When patients do not exhibit evidence of major Axis I pathology, their level of ego functioning and maturity of object relationships are the most useful criteria for determining the type of psychotherapy they require. Other important parameters parallel the more rigid selection criteria for brief psychotherapy proposed by Sifneos (1972). All these factors help establish what Karasu and Skodol (1980) called a "psychotherapy diagnosis," a judgment concerning "the psychotherapeutic approach by which the patient is likely to derive maximal benefit," which is not exclusively determined by diagnostic criteria.

General measures of a patient's ego strength include the level of sexual development and adjustment, intellectual skills, educational and occupational achievements, and degree of autonomy and ability to assume responsibility. These criteria, which broadly assess psychologic strength and weakness, measure an individual's ability to pursue self-selected goals while contending with exceptional stresses of life, such as ill health.

A more detailed evaluation of ego functioning, discussed by Vaillant in his excellent study of basic adaptational styles (1977), derives from a determination of prevailing ego mechanisms. Anxiety-provoking work is definitely contraindicated in an individual who relies heavily on psychotic mechanisms (e.g., denial, delusional projection, and distortion). For example, a middle-aged diabetic woman who responds to the recent amputation of her lower leg by acting as if the limb were still intact is ill-suited for introspective work. Such an individual requires considerable ego support, provided in the form of psychopharmacologic treatment and consistent reality testing. Although immature mechanisms (e.g., schizoid withdrawal, projection, passive-dependent and passive-aggressive behavior, acting out, and hypochondriasis) may not cause such severe regression, they typify a personality structure that is generally unresponsive to an anxiety-provoking approach because of limitations on the capacity to modulate affect and control impulses. Conversely, the presence of neurotic mechanisms (e.g., intellectualization, repression, displacement, reaction formation, and dissociation) and the more sophisticated mechanisms (e.g., suppression, humor, anticipation, altruism, and sublimation) reflect flexibility and resilience of ego functioning sufficient to tolerate the heightened anxiety of emotional self-scrutiny.

The quality of an individual's object relationships provides a general measure of interpersonal functioning, another important barometer of the ability to tolerate anxiety. This information helps define the patient's motivation and ability to be involved in a productive and gratifying manner with others, important data when assessing the capacity for the emotional intensity and mutuality of a therapeutic relationship. The pattern of the patient's interaction with family, friends, and professional colleagues may yield a picture of marked passivity, characterized by dependency on most of the people in that individual's life. Other maladaptive modes of relating, such as fight-flight behavior or a series of short-lived superficial encounters involving fluctuating idealization and unreasonable denigration of a partner, may reflect the extreme anxiety emanating from unresolved issues of separation and individuation. These patterns of object relationships suggest the necessity of a supportive psychotherapeutic approach, one that would be supplemented by pharmacotherapy in more extreme instances, such as the inability to differentiate self from nonself. Conversely, patients' interpersonal relationships may reflect the importance that they ascribe to independent functioning, as well as their ability to abdicate autonomy when they recognize the need to rely on the support of others. Also, the history of at least one meaningful relationship reflects basic trust and a genuine emotional involvement, important criteria for introspective therapy.

Other factors signaling the appropriateness of an anxiety-provoking approach include the patient's willingness to engage in self-scrutiny and desire for emotional change as opposed to symptomatic relief. These are predicated on a requisite degree of intelligence and psychological-mindedness, often evidenced in individuals' insights concerning their illness. These factors ensure a workable transference relationship, one with sufficient observing ego for the patient to appreciate, and then modulate, the intense affects stirred by any distorted perceptions. Attention to the therapeutic relationship can supply other indicators for introspective work, including the patient's realistic expectations concerning the expertise of the therapist and the anticipated goals of treatment, the patient's ability to be affectively and cognitively involved in psychotherapy, and the patient's facility to relate to the therapist throughout encouraging and disheartening phases of the medical illness and psychiatric treatment. An individual's responsiveness to trial interpretations provides data concerning each of these indicators for introspective work. Malan (1976), Sifneos (1972), and Davanloo (1980) detailed the diagnostic value of these interventions. Viederman (1984) discussed their therapeutic worth; communicating back to patients an empathic and accurate understanding of their plight enhances their sense of trust and, consequently, commitment to the psychotherapy.

Anxiety-Suppressing Psychotherapy

Ever since Freud (1919) praised the "gold" of psychoanalytic interpretations, "supportive" psychotherapy has been considered an inferior form of treatment compared to insight-oriented work. This negative, and unfair, reputation derives from a variety of factors, ranging from the long-standing belief that characterologic change is the primary goal of psychotherapy to the frustration of practitioners who struggle to successfully implement this taxing treatment modality (Wallace, 1983). The relative inattention afforded anxiety-suppressing psychotherapy over the years reflects this bias, and may explain why it still remains a somewhat nebulous form of treatment. Divergent conceptualizations confuse basic therapeutic issues, such as the distinction between supportive therapy and supportive relationships, and, consequently, anxiety-suppressing therapy has frequently been explicated via "negative definitions" that basically declare what it is not (Buckley, 1986). Winston, Pinsker, and McCullough's (1986) excellent review of the literature significantly clarifies the fundamental goals and techniques of this treatment modality, which highlights its relevance to medical psychotherapy.

Anxiety-suppressing psychotherapy requires the clinician to be active and supportive in ways that contain the patient's anxiety within limits that were acceptable premorbidly. The overall treatment is focused on symptomatic relief, as opposed to structural intrapsychic change, an approach based on concrete, sometimes controlling, therapeutic directives. The major goal of treatment is to dissipate the powerful emotions that negatively affect the patient's psychologic well-being. The patient is not guided toward discovery of unconscious motivations and conflicts that crystallized into an abnormal illness response. Rather, the patient is offered consistent encouragement and support by therapists who wish to maximize their influence by exploiting a working alliance solidified by trust and cooperation (Werman, 1984). The therapeutic maneuvers most often used to achieve this end are *suggestion*, *manipulation*, and *abreaction*, as described by Bibring (1954). When treating medical patients, the clinician attempts to curtail emotional dysfunction precipitated by illness by altering factors external to the individual, such as effecting changes in their living situation or prescribing psychoactive medication.

Because anxiety-suppressing psychotherapy rests on the clinician's ability to influence and guide the patient, eliciting and maintaining a positive transference is a primary goal of treatment. Careful attention to unconscious issues embodied in an individual's psychodynamics and illness dynamics identifies doctor-patient interactions that may be used by the therapist, whose inherent power (Frank, 1961) can be enhanced in a manner that parallels Alexander and French's studied manipulation of the transference (1946). Techniques used to foster the positive transference, summarized by Winston, Pinsker, and McCullough (1986), emanate from clinicians' basic posture of presenting themselves as real, objective, and analytic. They offer advice and reassurance, (e.g., opinions concerning the patient's response to a given therapeutic regimen), as well as praise when warranted, (e.g., when individuals engage in meaningful discussion about their illness with family members or progress in a rehabilitative program). They provide auxiliary ego to the patient, which may take the form of educating patients about the facts of their illness, identifying and justifying patients' affective and behavioral responses to the sick role, explaining changes in patients' interactions with friends and family members, and aiding in problem-solving for life issues caused by ill health. Therapists communicate all this in a personable style, more conversational than clinical, which may reveal some of their own feelings and values. The negative transference is addressed only when it directly interferes with treatment, and is interpreted in a circumscribed manner so as to overcome the patient's resistance without mobilizing intense affects that could be potentially disorganizing.

Supporting and enhancing a patient's defensive structure is another major goal of anxiety-suppressing psychotherapy. The therapist must identify and maintain those mechanisms that provide the basis for an individual's ego functioning. This knowledge is useful in helping to reinforce the patient's reality testing, particularly if the patient is observed to utilize considerable projection and distortion. It also highlights important defenses that should not be challenged by the clinician. Primitive mechanisms can be adaptive, such as the denial following a myocardial infarction (Hackett & Cassem, 1974); lacking them, a patient may become overwhelmed by emotional distress and/or act out in a self-destructive fashion. In addition to preserving a patient's fundamental defenses, the therapist should also attempt to discover and mobilize healthier ego functioning. For example, the reality testing of an individual with diabetes mellitus preoccupied with "slipping into coma," despite excellent control of the illness, might be enhanced by working with his obsessional defenses. Repeated review of the patient's insulin needs and blood sugar levels could afford a greater sense of control over the illness and, consequently, the emotions it stirs.

Attending to an individual's diminished self-esteem is the third goal of supportive psychotherapy. When stressed by life events that limit the ability to cope with a variety of intensely painful affects, patients often become demoralized and begin to doubt their self-worth. The medical patient needs protection from the isolation attending the sick role, and, consequently, benefits significantly from interventions that bolster self-esteem. The therapist's active interest and attention, which communicate concern and respect, can offset feelings of helplessness and hopelessness. Specifically, the clinician is reassuring (e.g., validating the normality of a patient's feeling state) and offers advice (e.g., recommending actions that have helped others plagued by the same illness). These active interventions provide support and educate patients in a way that legitimizes their affective distress. They additionally convey an empathic view of the patient's status, which serves to bolster self-esteem through the more passive process of identification with the therapist's values.

Two related aspects of supportive psychotherapy that are particularly relevant to treatment of an abnormal illness response are its attention to a specific focus and the structure of the therapeutic environment. The benign neglect afforded unconscious material (e.g., fantasies, dreams, and transference distortions) helps to contain the patient's level of anxiety, which permits the patient to concentrate psychic energies on the here and now. Treatment of an individual in the throes of an anxiety response, for example, would address worries concerning the potentially debilitating effects of the illness, and would not explore the dynamic roots of his considerable separation anxiety. This focus is mirrored in the structure of the therapeutic interaction, which is predominantly characterized by an active give-and-take between the two participants, as opposed to the open-ended communication of introspective work and its characteristic intervals of silence. This helps the patient concentrate fully on concrete treatment goals.

Anxiety-Provoking Psychotherapy

Insight-oriented psychotherapy seeks to promote psychologic maturation by exploiting the turmoil of an emotional upheaval. The fundamental therapeutic task is to help patients acknowledge, bear, and put into perspective painful feelings that adversely affect their lives (Semrad, 1969), a conceptualization of the treatment that highlights its relationship to the grief process. The patient is first made aware of previously unrecognized emotions that have been kept out of consciousness by various defensive maneuvers. The patient then experiences those affects within the therapeutic environment, as well as in the context of important relationships. Finally, by recognizing the intensity, diversity, and ambivalence of his feelings, the patient achieves some degree of conflict resolution.

Several disciples of the analytic school have tailored the anxiety-provoking model into a variety of brief psychotherapies well suited for the treatment of an abnormal illness response (Alexander & French, 1946; Davanloo, 1980; Malan, 1976; Mann, 1973; Sifneos, 1972). Despite conceptual and technical differences, all these approaches share several characteristics: (1) persistent pursuit of unconscious material (instinctual drives, as well as ego defenses); (2) active, confronting interventions by the therapist; and (3) attention to a central focus that helps limit the degree of clinical regression. The goal of such treatment is "working through," a therapeutic process that guides the patient toward insight and self-understanding so as to effect significant and lasting behavioral change (Greenson, 1967). Except for time limitations, and the fact that the focus of treatment is circumscribed (e.g., confined to the patient's emotional reaction to physical illness), the process of these therapies is identical to long-term, insight-oriented psychodynamic work.

The fundamental task of anxiety-provoking therapy is studied attention to the interrelationship between two general areas of an individual's life: first, libidinal impulses, the anxiety they provoke, and the defenses utilized to contain them; and second, the lifelong pattern of object relationships that characterize one's historical past, current existence, and the transference relationship. Elucidating the dynamic interaction between these two triads provides patients with meaningful insight that helps liberate them from long-standing maladaptive patterns. Although introspective therapy continually challenges defenses that contain unacknowledged affects, it periodically employs supportive techniques when the level of therapeutic tension threatens the treatment. Consequently, all the technical maneuvers discussed by Bibring (1954) come into play, although *clarification* and *interpretation* are most often used.

Combined Treatment: Psychotherapy and Pharmacotherapy

Despite increasing acceptance of the biopsychosocial model, the Cartesian notion of mind-body duality still influences the practice of medicine. As a result, psychiatry continues to suffer from a polarization between biologic and psychologic orientation, which influences how mental health personnel relate to patients—objectively, as diseased organs, or subjectively, as disturbed individuals (Docherty et al., 1977). This has significant implications for treatment (Muskin, 1990). Biologists, who view mental illness as distinctive manifestations of abnormal neurochemical processes, argue that medications are the mainstay of all psychiatric treatment. Psychoanalysts, in contrast, traditionally have believed that pharmacotherapy negatively affects the therapeutic process—for example, by diminishing the patient's level of anxiety and, consequently, motivation, or by facilitating state-dependent insight that dissipates if drugs are withdrawn. Ostow (1962) maintained that drugs should only be used in psychoanalysis and psychotherapy when "they are essential to protect the patient or to protect the treatment."

In an attempt to reconcile these divergent therapeutic philosophies, the interactive effects of psychotherapeutic and biologic treatments have been intensively studied during the past two decades. Klerman (1975) demonstrated that the patient's level of affective distress influences the accessibility to psychosocial interventions, an argument for the use of medications to facilitate and/or optimize psychotherapy. Subsequent investigation disproved long-standing hypotheses concerning negative interactions between medications and psychotherapy (Rounsaville, Klerman, & Weissman, 1981), and further suggested the benefits of combining these therapeutic approaches (Luborsky, Singer, & Luborsky, 1976).

Practitioners more readily accept that psychosocial and biologic interventions are complementary, and that combined therapy is usually superior to the exclusive use of one or the other treatment. The combined approach has been conceptualized as a means of tailoring treatment to the specific state and trait issues affecting an individual (Extein & Bowers, 1979). Van Praag (1979) described it as the combination of "tablets and talking," the former directed toward neurochemically mediated issues (e.g., vegetative symptoms of depression; mania) and the latter targeting psychosocial issues (e.g., intrapsychic and interpersonal events). Given the broad biopsychosocial determinants of patients' illness dynamics, this philosophy of combined treatment is a logical, pragmatic, and effective method for implementing medical psychotherapy.

When trait-related issues form the basis of an individual's pathologic illness response, then medical psychotherapy alone, either as supportive or introspective work, is the indicated treatment. However, pronounced state-related symptoms often predominate in a patient's illness response. Up to 20% of general medical inpatients have an anxiety disorder (Strain, Leibowitz, & Klein, 1981), panic disorder is present up to seven times more often in patients in medical settings than in the general population (Rosenbaum & Pollack, 1987), and depression is present in up to 30% of medical inpatients (Rodin & Voshart, 1986). When these conditions are pronounced, the practitioner must rely on somatic therapies to alleviate affective turmoil. This, in turn, facilitates the therapeutic alliance (Klerman, 1976). Unless severe agitation or psychomotor retardation is effectively controlled, patients have difficulty working collaboratively with caretakers and, consequently, derive limited benefit from varied psychotherapeutic modalities (e.g., individual, group, or milieu therapy). This is obviously true to a greater degree in patients with pathologic denial of illness, who suffer from delusional thinking, possibly with an associated psychotic transference.

Relief from state-related symptoms permits subsequent psychotherapy, which may be anxiety-provoking (e.g., in patients capable of proceeding with the grief work associated with the loss of health once they feel their painful emotions can be effectively controlled with medications) or anxiety-suppressing (e.g., in those individuals whose idiosyncratic illness dynamics predispose to the need for ongoing psychotherapeutic support). There are several schools of brief psychotherapy within each of these broad categories, and consequently, there is debate as to the therapeutic efficacy of each particular approach. Burke, White, and Havens (1979), for example, argued for the utility of matching a particular patient with a specific form of brief therapy based on the level of that patient's psychosexual development. Utilizing an Ericksonian framework (1963), they suggested that Mann's (1973) time-limited, existential technique is most effective for passive-dependent individuals who have not resolved the adolescent conflict of "identity vs. role confusion," whereas patients with a more oedipal orientation, struggling with issues of "intimacy vs. isolation," are better served by the more confronting analytic approaches of Malan (1976) or Sifneos (1972). The latter approaches emphasize psychodynamic insight as opposed to Mann's focus on mastery of separation. Ursano and Hales (1986), in contrast, believe that the spectrum of psychotherapies differ more in types of interventions than in their goals or identified problems areas, based on the substantial overlap in selection criteria, technique, duration, and goals of treatment of Malan (1976), Sifneos (1972), Mann (1973), and Davanloo (1980). They also underscored how the distinction between brief psychodynamic psychotherapy, interpersonal psychotherapy, and cognitive techniques has more to do with perspective and emphasis of specific aspects of treatment than with fundamental differences in the treatment approach. For example, much of the cognitive therapist's work, helping an individual identify faulty perceptions that evolve into maladaptive defenses, is quite similar to the clarification and interpretation of defenses in brief psychodynamic psychotherapy.

In general, there are more similarities than differences among the brief therapies. As Bennett (1984) suggested, the clinician should refrain from seeking "perfectionism," an absolute fit between a patient and a particular form of brief treatment. It is the focused, short-term *technique* that is most important to medical psychotherapy; whether this is employed as an anxiety-suppressing or anxiety-provoking treatment depends on clinical issues idiosyncratic to the patient.

Combined therapy, then, can be viewed as a two-stage treatment strategy. Somatic therapies are employed during the acute phase of a major affective episode, and relief from that distress allows the shift to the psychotherapeutic treatment of dynamic causation (Klerman, 1976). For in-depth discussion of the psychopharmacology of the medically ill, the reader is referred to Chapter 9 of this volume, as well as to Bidder (1981), Cohen-Cole and Stoudemire (1987), Massie and Holland (1984), and Strain, Liebowitz, and Klein (1981). However, some general points concerning combined therapy should be noted here.

First, practitioners must be particularly wary of the side effects of psychoactive medications. Some, which are little more than an annoyance in fundamentally healthy patients, can seriously complicate treatment of the medically ill and can sometimes be life threatening. For example, the anticholinergic impact of tricyclic antidepressants may precipitate delirium in geriatric patients, and the use of beta-blockers for anticipatory anxiety can induce bronchospasm in an individual with bronchial asthma. When practicing combined medical psychotherapy, "choosing the safest side effect profile for a given patient is usually the starting point of choosing an appropriate agent" (Cohen-Cole & Stoudemire, 1987).

Second, because many patients who require medical psychotherapy are already taking a variety of medications, the clinician must be particularly mindful of drug-drug interactions when prescribing psychoactive substances. The impact of these effects ranges from altering blood levels of medications (e.g., tricyclic antidepressants can increase prothrombin time in a patient on warfarin) to precipitating dangerous blood pressure changes (e.g., prescribing monoamine oxidase inhibitors to an individual being treated for bronchial asthma with sympathomimetic agents).

Third, a particular medication within a general class of psychoactive medications may be indicated for specific medical conditions. For example, because the 2-keto-benzodiazepines (chlordiazepoxide, diazepam, flurazepam) are metabolized via oxidation in the liver, their clinical effects are subject to the status of hepatic enzymes. Liver disease or concurrent medications (e.g., oral contraceptives) that inhibit liver enzymes will slow degradation of these agents. The 3-hydroxybenzodiazepines (lorazepam, oxazepam, temazepam) are metabolized via conjugation, a less vulnerable process than oxidation, and are therefore a better choice of anxiolytic for individuals with compromised liver function (see Chapter 9).

THE ART OF MEDICAL PSYCHOTHERAPY

The type of medical psychotherapy utilized when treating an individual's abnormal illness response generally conforms to either the anxiety-suppressing or the anxiety-provoking approach. However, by focusing on the theoretical underpinnings and practice of these two treatment modalities, the preceding description neglects some realities of the clinical environment. The actual implementation of supportive and introspective interventions, within the context of a comprehensive treatment plan, requires flexibility and pragmatism on the part of the clinician. The task, which is fundamentally guided by constant attention to the patient's illness dynamics, may be facilitated by the following observations.

First, although a specific type of medical psychotherapy is predominant in the treatment of a patient's abnormal illness response, *it rarely occurs in pure form because of an existing interrelationship between anxiety-suppressing and anxiety-provoking techniques.* Psychoanalysis, for example, places a premium on therapeutic neutrality and abstinence. However, it has always acknowledged the importance of supportive maneuvers, most notably as they pertain to the working alliance (Freud, 1913; Greenson, 1967). Alternatively, Winston, Pinsker, and McCullough (1986) discussed "a continuum of supportive therapies" that incorporate expressive (anxiety-provoking) work to a degree determined by the patient's level of ego strength. Pine (1986) emphasized this point, describing techniques for achieving "interpretive content" in the context of supportive therapy. He outlined specific maneuvers that enable individuals "generally characterized by a fragility of defense" to tolerate this anxiety-provoking intervention. These observations do not blur the distinction between the supportive and introspective approaches as much as they indicate a positive interaction between aspects of each approach, such as the need to interpret negative transference feelings that arise during anxiety-suppressing therapy. In Pine's words, "supportive and insight therapies are not *counterposed*, in some *opposition* to one another, but are *counterpoised*, in some *balance* with one another." The degree to which this occurs in the treatment of an individual's abnormal illness response is guided by the particulars of his illness dynamics.

Second, the psychiatrist involved with the diagnosis and treatment of psychiatric disorders in the medically ill must constantly deal with the interaction between psychic and physical events, which may be overt (e.g., delirium) or subtle (e.g., a masked depression). These problems are mind-body in origin, and their treatment frequently utilizes biologic interventions in conjunction with psychotherapy. The use of psychoactive medications is often an integral aspect of medical psychotherapy; theoretically this can be conceptualized as an anxiety-suppressing technique, because medications help provide structure and support that the patient is lacking because of the physical and/or psychologic impact of illness. Pharmacotherapy provides acute relief of symptoms, and, in certain individuals, continued remission via maintenance medication. When the treatment goals of an abnormal illness response are concerned with interpersonal relationships, family issues, or occupational and social adjustment, then treatment shifts to predominantly talking psychotherapy. This would focus on the working through of the feelings associated with loss of health, if the individual's illness dynamics permitted grief work, or on continued anxiety-suppressing psychotherapy, if ongoing support is indicated.

Third, although anxiety-suppressing and anxiety-provoking therapies are founded on a psychodynamic understanding of the patient, each of these treatment modalities has distinctive cognitive and behavioral attributes. For example, clarification and interpretation of defenses and unconscious material remains the foundation of insight-oriented work; however, behavioral interventions (e.g., limit-setting maneuvers) complement the exploratory process. In addition, supportive work can occur within a clinical framework completely apart from psychodynamic psychotherapy. Moos and Schaefer (1984), for example, utilized a cognitive approach based on crisis theory when treating patients unable to cope with the impact of physical illness. They outlined two sets of treatment goals for these individuals: illness-related tasks (e.g., dealing with pain and incapacitation, dealing with the hospital environment, and developing adequate relationships with heath care staff) and general tasks (e.g., preserving a reasonable emotional balance, preserving a satisfactory self-image and maintaining a sense of competence and mastery, sustaining relationships with family and friends, and preparing for an uncertain future). They also described the major types of coping skills employed to deal with these various tasks. The treatment includes fundamental interventions of supportive therapy, such as education, reassurance, and abreaction, which are effected via cognitive, problem-solving techniques. Anxiety-suppressing therapy can also take the form of behavioral interventions, such as relaxation protocols and desensitization (e.g., to treat excessive anxiety) or positive and negative reinforcement (e.g., to deal with excessive regression).

Fourth, although supportive and insight-oriented therapies have been discussed in terms of individual treatment, medical psychotherapy is effected in a variety of clinical formats. An individual's abnormal illness response can be treated in the context of family therapy, couples work, or group psychotherapy, each of which may be anxiety suppressing or anxiety provoking depending on the prevailing illness dynamics. When the psychologic well-being of a family is adversely affected by the illness of one of its members (Binger et al., 1969; Borden, 1962; Livsey, 1972), supportive work can facilitate the family's accommodation to the illness, and may actually be part of the individual's treatment (Rosman, Minuchin, & Liebman, 1975). Couples work is indicated when an individual's abnormal illness response causes increasing distance in the relationship and threatens its viability. The psychological health of each member, their dynamic interaction, and the particular losses incurred by the patient's ill health (e.g., decreased sexual function) determine whether a supportive or an introspective intervention is indicated.

The same decision applies to group therapy. Insight-oriented work has traditionally been used in the treatment of classic "psychosomatic" illnesses (Stein, 1971), and has been effective in helping patients adjust to terminal illness (Spiegel, 1979; Yalom & Greaves, 1977). However, several attributes of the group process (e.g., diffusion of the transference and greater interpersonal contact) bolster impaired ego functioning, and consequently favor its use as a supportive form of medical psychotherapy. Group members benefit from information sharing (Bilodeau & Hackett, 1971) and a generalized acceptance of emotional expression (Adsett & Bruhn, 1968; Stein, 1971), as well as the mutual support derived from discussing common medical experiences, such as amputation, myocardial infarction, or the impact of a mastectomy or colostomy (Slaby & Glicksman, 1985). As Moos and Schaefer (1984) pointed out, "a patient may deny or minimize the seriousness of a crisis while talking to a family member, seek relevant information about prognosis from a physician, [and] request reassurance and emotional support from a friend." Because an individual's illness response can take on such complexity, these varied therapeutic modalities are often used in combination to effect a medical psychotherapy that targets the patient's distinctive needs. Group psychotherapy with the medically ill is discussed in detail in Chapter 3.

Finally, supportive and introspective treatments have been discussed within the context of a single episode of ill health. However, most illness is chronic, and its elongated time course has implications for the practice of medical psychotherapy, both for the patient and for the clinician. The onset of a long-term illness may so overwhelm an individual that initially he can only tolerate an anxiety-suppressing approach; however, this does not suggest that the patient will require support throughout the clinical course. Slaby and Glicksman (1985) observed that coping mechanisms appropriate to one stage of life-threatening illness are not necessarily appropriate at later stages. Similarly, although some patients repeatedly require anxiety-suppressing work during the prolonged course of a disease, others might regress with that approach, since supportive treatment may help reestablish their emotional equilibrium in a way that fosters a sense of mastery. This can protect against the recurrence of an abnormal illness response during acute clinical exacerbations and/or subsequent episodes of ill health, as well as generally enhance adaptive ego function via a positive ripple effect. This parallels the work of crisis intervention, as described by Caplan (1964), which promotes the acquisition of new coping mechanisms, increased self-esteem, and diminished fear about the recurrence of a particular stressor. The same accommodation may derive from patients' assessment of their illness over time, which can replace feared fantasies with the reality of symptomatic relief, loving support from family and friends, and a level of functioning that affords some sense of control over the illness. This should minimize the emotional turmoil accompanying future progression of the disease, and may even facilitate subsequent anxiety-provoking work designed to help patients grieve for their additional loss of health.

There are also implications for the clinician when long-term illness necessitates ongoing medical psychotherapy, particularly when the patient fails to positively accommodate to the loss of health. That circumstance can readily promote a negative countertransference, which is often related to patients' character pathology (Groves, 1978; Kahana & Bibring, 1964; Lipsitt, 1970) or to specific clinical issues such as hypochondriasis (Adler, 1981; Brown & Vaillant, 1981). In these instances health care providers may become increasingly frustrated and resentful, as they respond to patients' overt and covert negativism by feeling a corresponding helplessness and hopelessness. When this type of countertransference remains unrecognized, clinicians are more likely to act out their anger, as opposed to using it constructively. Sternbach (1974) described how this destructive cycle sadly produces a multitude of "crocks" and "quacks" whose fundamental antagonism promotes a progressive dissolution of the therapeutic alliance.

SUMMARY

All medical illness produces emotional responses that form an integral part of the disease process. Although many patients successfully accommodate to their particular impairment, others suffer abnormal illness responses because they cannot negotiate the affective turmoil precipitated by ill health. Medical personnel who ignore these psychologic reactions provide suboptimal, if not detrimental, treatment. Effective medical care is founded on an understanding of what a particular disease means to a particular patient at a particular time in life. This requires a full understanding of the patient's illness dynamics; they define a distinctive biopsychosocial formulation that helps primary care providers identify the specific psychologic needs to be addressed during the course of treatment. If this approach fails to abort an abnormal illness response, mental health personnel can offer more focused attention via medical psychotherapy. The form of the treatment also is based on patients' illness dynamics, which determine their ability to tolerate anxiety, providing information concerning the maturity of their ego functioning and object relationships. The psychotherapeutic approach may be supplemented with pharmacotherapy when the patient's affective distress is extreme. This comprehensive biopsychosocial approach is the only acceptable medical model if "one is to treat the whole patient and not merely characterize the nature of an illness and impede biological deterioration" (Slaby & Glicksman, 1985).

REFERENCES

Adler G (1981). The physician and the hypochondriacal patient. N Engl J Med 304(23):1394–1396.

Adsett C, & Bruhn J (1968). Short-term group psychotherapy for myocardial infarction patients and their wives. Can Med Assoc J 99:577–581.

Alexander F, & French T (1946). Psychoanalytic therapy. New York: Ronald Press.

Amkraut A, & Solomon G (1974). From the symbolic stimulus to the pathophysiologic response: Immune mechanisms. Int J Psychiatry 5:541–563.

BENNETT M (1984). Brief psychotherapy in adult development. Psychother Theory Res Pract 21:171–177.

BIBRING E (1954). Psychoanalysis and the dynamic psychotherapies. J Am Psychoanal Assoc 2:745–770.

BIDDER TG (1981). Electroconvulsive therapy in the medically ill: Resistances and possibilities. Int J Psychiatry Med 14:99–108.

BILODEAU C, & HACKETT T (1971). Issues raised in a group setting by patients recovering from myocardial infarction. Am J Psychiatry 128:73–78.

BINGER C, ALBLIN A, FEUERSTEIN R, ET AL. (1969). Childhood leukemia: Emotional impact of patient and family. N Engl J Med 280:414–418.

BORDEN W (1962). Psychological aspects of a stroke: Patient and family. Ann Intern Med 57:689–692.

BROWN H, & VAILLANT G (1981). Hypochondriasis. Arch Gen Psychiatry 141:723–726.

BROWN J, & STOUDEMIRE G (1983). Normal and pathological grief. JAMA 250(3):378–382.

BUCKLEY P (1986). Supportive psychotherapy: A neglected treatment. Psychiatr Ann 16(9):515–521.

BURKE J, WHITE H, & HAVENS L (1979). Which short-term therapy? Arch Gen Psychiatry 36:177–187.

CANNON W (1920). Bodily changes in pain, hunger, fear and rage. New York: Appleton-Century-Crofts.

CANNON W (1932). The wisdom of the body. New York: WW Norton.

CAPLAN G (1964). Principles of preventive psychiatry. New York: Basic Books.

COHEN-COLE S, & STOUDEMIRE A (1987). Major depression and physical illness. Psychiatr Clin North Am 10(1):1–17.

DAVANLOO H (Ed.) (1980). Short term dynamic psychotherapy. New York: Aronson.

DOCHERTY J, MARDER S, VAN KAMMEN D, ET AL. (1977). Psychotherapy and pharmacotherapy: Conceptual issues. Am J Psychiatry 134:529–533.

DORIAN B, GARFINKEL P, KEYSTONE E, ET AL. (March, 1986). Stress, immunity, and illness, presented at the annual meeting of the American Psychosomatic Society, Baltimore.

ENGEL G (1968). A life-setting conductive to illness: The giving-up–given-up complex. Ann Intern Med 69:293–298.

ENGEL G (1971). Sudden and rapid death during psychological stress. Ann Intern Med 74:771–782.

ENGEL G (1977). The need for a new medical model: A challenge for biomedicine. Science 196:129–136.

ENGEL G (1980). The clinical application of the biopsychosocial model. Am J Psychiatry 137:535–544.

ERICKSON E (1963). Childhood and society. New York: WW Norton.

EXTEIN I, & BOWERS M (1979). State and trait in psychiatric practice. Am J Psychiatry 136:690–693.

FRANK J (1961). Persuasion and healing. Baltimore: Johns Hopkins University Press.

FREUD S (1913). On beginning the treatment. In J STRACHEY (Trans. and Ed.), The standard edition of the complete psychological works of Sigmund Freud (Vol. 12; pp. 121–144). London: Hogarth Press, 1955.

FREUD S (1919). Lines of advance in psychoanalytic therapy. In J STRACHEY (Trans. and Ed.), The standard edition of the complete psychological works of Sigmund Freud (Vol. 17; pp. 157–168). London: Hogarth Press, 1955.

FREUD S (1924). Aspects of development and regression: Etiology. In J RIVIERE (Ed.), A general introduction to psychoanalysis (Ch. 22, pp. 348–366). New York: Pocket Books, 1975.

FREYBERGER H (1977). Supportive psychotherapeutic techniques in primary and secondary alexithymia. Psychother Psychosom 28:337–342.

GOLDBERG R, & GREEN S (1985). Medical psychotherapy. Am Fam Physician 31(1):173–178.

GREEN S (1985). Mind and body: The psychology of physical illness. Washington DC: American Psychiatric Press, Inc.

GREENE W, GOLDSTEIN S, & MOSS A (1972). Psychosocial aspects of sudden death. Arch Intern Med 129:725–731.

GREENSON RR (1967). The technique and practice of psychoanalysis (Vol. I). New York: International Universities Press, Inc.

GROVES J (1978). Taking care of the hateful patient. N Engl J Med 298:883–887.

HACKETT T, & CASSEM N (1974). Development of a quantitative rating scale to assess denial. J Psychosom Res 18:93–100.

HASKETT R, & ROSE R (1981). Neuroendocrine disorders and psychopathology. Psychiatr Clin North Am 4(2):239–252.

HENRY J (1975). The induction of acute and chronic cardiovascular disease in animals by psychosocial stimulation. Int J Psychiatry Med 6:147–158.

IRWIN M, DANIELS M, & WEINER H (1987). Immune and neuroendocrine changes during bereavement. Psychiatr Clin North Am 10(3):449–466.

KAHANA R, & BIBRING G (1964). Personality types in medical management. In N ZINBERG (Ed.), Psychiatry and medical practice in a general hospital (pp. 108–123). New York: International Universities Press, Inc.

KARASU T, & SKODOL A (1980). VIth axis for DSM-III: Psychodynamic evaluation. Am J Psychiatry 137:607–610.

KENNEDY S, KIECOLT-GLASER J, & GLASER R (1988). Immunological consequences of acute and chronic stressors: Mediating roles of interpersonal relationships. Br J Psychol 61:77.

KIECOLT-GLASER J, GARNER W, SPEICHER C, ET AL. (1984). Psychosocial modifiers of immunocompetency in medical students. Psychosom Med 46:7–14.

KIELY W (1974). From the symbolic stimulus to the pathophysiological response: neurophysiological mechanisms. Int J Psychiatry Med 5:517–529.

KLERMAN G (1975). Combining drugs and psychotherapy in the treatment of depression. In M GREENBLATT (Ed.), Drugs in combination with other therapies (Ch. 13, pp. 213–228). New York: Grune & Stratton.

KLERMAN G (1976). Combining drugs and psychotherapy in the treatment of depression. In J COLE J, A SCHATZBERG, & S FRAZIER (Eds.), Depression: Biology, psychodynamics and treatment (Ch. 13, pp. 213–227). New York: Plenum.

KLERMAN G (1981). Depression in the medically ill. Psychiatr Clin North Am 4(2):301–318.

KRYSTAL H (1979). Alexithymia and psychotherapy. Am J Psychother 33:17–31.

LINDEMANN E (1944). Symptomatology and management of acute grief. Am J Psychiatry 101:141–146.

LIPOWSKI Z (1975). Psychiatry of somatic disease: Epidemiology, pathogenesis, classification. Compr Psychiatry 16:105–124.

LIPSITT D (1970). Medical and psychological characteristics of "crocks." Psychiatry Med 1:15–25.

LIVSEY C (1972). Physical illness and family dynamics. Adv Psychosom Med 8:237–251.

LOCKE S, KRAUS L, LESERMAN J, ET AL. (1984). Life change, stress, psychiatric symptoms, and natural killer cell activity. Psychosom Med 46:441–453.

LOWN B, VERRIER R, & RABINOWITZ S (1977). Neural and psychologic mechanisms and the problem of sudden cardiac death. Am J Cardiol 39:890–902.

LUBORSKY L, SINGER B, & LUBORSKY L (1976). Comparative studies of psychotherapies. In R SPITZER & D KLEIN (Eds.), Evaluation of psychological therapies (Ch. 1, pp. 3–22). Baltimore: John Hopkins University Press.

MALAN D (1976). The frontier of brief psychiatry. New York: Plenum.

MANN J (1973). Time-limited psychotherapy. Cambridge, England: Harvard University Press.

MASON J (1975). Clinical psychophysiology: Psychoendocrine mechanisms. In M REISER (Ed.), American handbook of psychiatry (Vol. 4; pp. 553–582). New York: Basic Books.

MASSIE MJ, & HOLLAND JC (1984). Diagnosis and treatment of depression in the cancer patient. J Clin Psychiatry 45:25–28.

MOOS R, & SCHAEFER J (1984). The crisis of physical illness: an overview and conceptual approach. In R MOOS (Ed.), Coping with physical illness. 2: New perspectives (Ch. 1, pp. 3–26). New York: Plenum.

MUSKIN P (1990). The combined use of psychotherapy and pharmacotherapy in the medical setting. Psychiatr Clin North Am 13(2):341–353.

NEMIAH J, & SIFNEOS P (1970). Psychosomatic illness: A problem of communication. Psychother Psychosom 18:154–160.

OSTOW M (1962). Drugs in psychoanalysis and psychotherapy. New York: Basic Books.

PARSONS T (1951). The social system. Glencoe, IL: The Free Press.

PINE F (1986). Supportive psychotherapy: A psychoanalytic perspective. Psychiatr Ann 16(9):526–529.

REICH P, DESILVA R, LOWN B, ET AL. (1981). Acute psychological disturbances preceding life-threatening ventricular arrhythmias. JAMA 246:233–235.

REICHARD J (1964). Teaching principles of medical psychology to medical house officers: Methods and problems. In N ZINBERG (Ed.), Psychiatry and medical practice in a general hospital (pp. 169–204). New York: International Universities Press.

REISER M (1975). Changing theoretical concepts in psychosomatic medicine. In M REISER (Ed.), American handbook of psychiatry (Vol. 4; pp. 477–500). New York: Basic Books.

RIFKIN A, REARDON G, SIRIS S, ET AL. (1985). Trimipramine in physical illness with depression. J Clin Psychiatry 46:4–8.

RODIN G, & VOSHART K (1986). Depression in the medically ill: Overview. Am J Psychiatry 143:696–705.

ROGERS M, DUBEY D, & REICH P (1979). The influence of the psyche and the brain on immunity and disease susceptibility: A critical review. Psychosom Med 41:147–164.

ROSENBAUM J, & POLLACK M (1987). Anxiety. In T HACKETT, N CASSEM, & M LITTLETON (Eds.), Massachusetts General Hospital handbook of general hospital psychiatry (2nd ed., Ch. 9, pp. 154–183). Littleton, MA: PSG Publishing Company.

ROSMAN B, MINUCHIN S, & LEIBMAN R (1975). Family lunch session: An introduction to family therapy in anorexia nervosa. Am J Orthopsychiatry 45:846–853.

ROUNSAVILLE B, KLERMAN G, & WEISSMAN M (1981). Do psychotherapy and pharmacotherapy of depression conflict? Arch Gen Psychiatry 38:24–29.

SCHINDLER B (1985). Stress, affective disorders, and immune function. Med Clin North Am 69:585–597.

SCHLEIFER S, KELLER S, MYERSON A, ET AL. (1984). Lymphocyte function in major depressive disorder. Arch Gen Psychiatry 41:484–486.

SELYE H (1950). Physiology and pathology of exposure to stress. Montreal, Acta Press.

SEMRAD E (1969). A clinical formulation of the psychoses. In E SEMRAD & D VAN BUSKIRK (Eds.), Teaching psychotherapy of psychotic patients (Ch. 2, pp. 17–30). New York: Grune & Stratton.

SIFNEOS P (1972). Short-term psychotherapy and emotional crisis. Cambridge, MA: Harvard University Press.

SIFNEOS P (1972–73). Is dynamic psychotherapy contraindicated for a large number of patients with psychosomatic diseases? Psychother Psychosom 21:133–136.

SIFNEOS P (1973). The prevalence of "alexithymic" characteristics in psychosomatic patients. Psychother Psychosom 22:255–262.

SLABY A, & GLICKSMAN A (1985). Adapting to life-threatening illness. New York: Praeger.

SPIEGEL D (1979). Psychological support for women with metastatic carcinoma. Psychosomatics 20:780–787.

STEIN A (1971). Group therapy with psychosomatically ill patients. In H KAPLAN & B SADOCK (Eds.), Comprehensive group psychotherapy (Ch. 4, pp. 581–601). Baltimore: Williams & Wilkins.

STEIN M (1981). A biopsychosocial approach to immune function and medical disorders. Psychiatr Clin North Am 4:203–222.

STEIN M, KELLER S, & SCHLEIFER S (1985). Stress and immunomodulation: The role of depression and neuroendocrine function. J Immunol 135(suppl):827–833.

STERNBACH R (1974). Varieties of pain games. Adv Neurol 4:423–430.

STRAIN J, LEIBOWITZ M, & KLEIN D (1981). Anxiety and panic attacks in the medically ill. Psychiatr Clin North Am 4:333–350.

TAYLOR G (1977). Alexithymia and the countertransference. Psychother Psychosom 28:141–147.

TAYLOR G (1984). Alexithymia: concept, measurement, and implications for treatment. Am J Psychiatry 141:725–732.

URSANO R, & HALES R (1986). A review of brief individual psychotherapies. Am J Psychiatry 143:1507–1517.

VAILLANT G (1977). Adaptation to life. Boston: Little, Brown.

VAN PRAAG H (1979). Tablets and talking: A spurious contrast in psychiatry. Compr Psychiatry 20:502–510.

VIEDERMAN M (1984). The active dynamic interview and the supportive relationship. Compr Psychiatry 25:145–157.

WALLACE E (1983). Dynamic psychiatry in theory and practice. Philadelphia: Lea & Febiger.

WERMAN D (1984). The practice of supportive psychotherapy. New York: Brunner/Mazel.

WHYBROW P, & SILBERFARB P (1974). Neuroendocrine mediating mechanisms: From the symbolic stimulus to the physiological response. Int J Psychiatry Med 5:531–539.

WINSTON A, PINSKER H, & MCCULLOUGH L (1986). A review of supportive psychotherapy. Hosp Community Psychiatry 37(11):1105–1114.

WOLFF H, WOLF S, & HARE S (Eds.) (1950). Life stress and bodily disease. Baltimore: Williams & Wilkins.

YALOM I, & GREAVES C (1977). Group therapy with the terminally ill. Am J Psychiatry 134:396–400.

2 Family therapy in the context of chronic medical illness

JANE JACOBS, ED.D.

Technical advances in the treatment of serious medical illness and trauma in the last 20 years have dramatically increased the number of adults who live for years or even decades with a chronic illness or disability. The great majority of these individuals live with their immediate families. We have come to understand that these medical miracles can result in prolonged burdens for family members, who often settle into a life organized around illness demands.

The psychiatrist often is in a position to provide help for families with caregiving responsibilities, either by intervening directly or by acting as a consultant to a specialized medical service. This chapter describes the major problems facing families who must care for a chronically ill relative, provides an assessment model, and outlines a method for selecting a family intervention when specific help is needed.

COMMON PROBLEMS OF FAMILIES FACING CHRONIC ILLNESS

The impact of chronic illness on families varies tremendously. Evaluations of caregivers' psychologic status in relation to a variety of medical conditions reveal that more than half have a "healthy" range of functioning (Gallagher et al., 1989; Rabins et al., 1990). However, the rest experience anxiety, depression, anger, or somatic complaints.

One striking element of family stress in the context of medical illness is that the severity of the illness alone does not seem to be a strong predictor of caregiver functioning. Rather, one must understand what type of challenge a particular illness constitutes in the context of a particular family environment. Certain areas of family life have been shown to be especially vulnerable to disruption by chronic illness, while being essential to the family's success in coping with these disorders. These areas include five dimensions of functioning within the family as well as two areas in which families interact with important outside systems (Table 2-1).

Intrafamilial Problems

Boundary Regulation

Boundary regulation involves the appropriate management of space and privacy among individuals in the family, between generations, and between the family and the outside world. All three boundaries are vulnerable to disruption in the presence of a chronic disease. Individual family members may feel they have no right to private time or recreational activities when the patient requires care or other family members need emotional support. Spending time with friends, at school activities, or privately in one's room may appear selfish under the circumstances.

Generational boundaries may suddenly become more fluid; the partner of the ill parent, for example, may feel justified in confiding his frustration and fear to his children. Extrafamilial boundaries may become more rigid, particularly if the illness carries some stigma with it. The extent and quality of the family's social support network has been shown repeatedly to influence the family's ability to cope successfully with the illness (Ell et al., 1988; Rabins et al., 1990).

Role Allocation

Familiar family roles must change substantially in the face of a chronic disease. Families organized around a particular role structure, such as a traditional gender division, may have difficulties adapting their preexisting values to the requirements of the illness; for example, they may not be able to accept an increased domestic role for the husband or a primary breadwinning role for the wife. Children may realistically be needed to assist in the care of a disabled relative; families must learn to strike a delicate balance between overburdening children and giving them appropriate caregiving tasks that will heighten their sense of self-esteem.

In the case of episodic illnesses, families must learn to shift back and forth between "ill phase" and "well phase" role arrangements. This repeated cycling requires that family members have a fluid, situation-based view of one another's roles and responsibilities. Finally, the patients themselves must take on a meaningful role in their own care to the fullest possible extent; patients' assumption of maximum responsibility in their own care is associated with the most positive illness outcome and the highest levels of family adaptation to the disease (Evans et al., 1987).

Problem Solving

A family with an effective problem-solving approach is able to identify the need to address a problem without having to have a crisis; family members are able to identify the problems areas, open the discussion to entirely novel ideas, come up with a concrete new strategy, and subsequently evaluate it (Falloon et al., 1984). Empirical evidence reveals a strong relationship between effective family problem solving and favorable illness course (Evans et al., 1987) as well as successful family adaptation to the disease (Lewis et al., 1989). However, families with a chronically ill member tend to back off from discussing controversial issues because they fear that disagreements may trigger an illness episode (Gonzalez, Steinglass, & Reiss, 1987).

Problem-solving strategies that work adequately in the absence of adverse conditions often are badly strained in the presence of a chronic illness. Family members must be able to assess accurately the nature and severity of the disorder. They must overcome the tendency to deny its presence, yet they must also absorb the shock of the news so that they may determine realistic measures that can be taken to minimize the illness impact.

TABLE 2–1. *Common Family Problems With Medical Illness*

Intrafamilial problems
Boundary regulation
Role allocation
Problem solving
Communication
Family beliefs
Family interface problems
The family and the medical care team
Financial problems

Families that are unable to mobilize themselves for an initial appraisal risk poor outcome in the management of the illness (Waltz et al., 1988).

The negotiation and resolution of illness management requires a level of explicit discussion that may be alien to some families. Families that cannot talk frankly about illness demands or make direct requests of one another or of extended family members and friends risk heightened depression and anxiety in family members over an extended period of time (Gonzalez, Steinglass, & Reiss, 1987; Venters, 1981).

Communication

The most common communication problem in a family dealing with a chronic illness is the gradual restriction of affect among family members in the interest of protecting the person with the illness. Particularly in cases in which the illness is life threatening, family members fear that strong expressions of feelings may exacerbate the medical condition and perhaps even cause the person's death (Gonzalez, Steinglass, & Reiss, 1987; Koocher & O'Malley, 1981).

The unspoken rules that govern the family's expression of affect may shift in important ways; if one family member was responsible for the expression of feelings in the family, that individual may now abdicate the role if he is the patient or the primary caregiver. Conversely, the family may develop an implicit belief that only the patient is now entitled to the free expression of feelings. If no new rules supercede the old ones, there will be an unsettling impasse in the family's implicit arrangements for intimate exchanges.

A supportive marital relationship has a buffering effect on depression, which occurs frequently in adults caring for a chronically ill relative (Brown, Harris, & Bifulco, 1986). Strong marital communication has been shown to mediate the level of depression experienced by patients suffering from a variety of chronic diseases, including cardiac illness, breast cancer, and diabetes (Waltz et al., 1988). In the presence of successful marital communication, higher illness demands do not threaten the successful implementation of family problem-solving methods (Lewis et al., 1989).

Confusion about how much assistance and support the patient needs is a major source of tension between marital partners. In family discussion groups, patients frequently report that their spouses are either too negligent or too solicitous; the well spouse usually assumes that the patient needs more attention than is the case, consequently sacrificing important nonillness priorities to meet these needs (Gonzalez, Steinglass, & Reiss, 1987).

The result is a situation in which one partner feels like a martyr while the other feels misunderstood.

Family Beliefs

In recent years there has been greater appreciation of the role that belief systems play in the successful management of chronic illness (Jacobs, 1989). Family members' initial appraisal of their own efficacy in meeting the challenge of the illness is a powerful predictor of family adaptation (Waltz et al., 1988). Blaming of patients by family members or by the patients themselves for bringing on the illness is known to be associated with increased patient stress and symptomatology (Brown, Birley, & Wing, 1972; Jacobs, 1991).

Many families have a set of organizing values growing from their family history, religious convictions, moral precepts, or individualistic beliefs. These values, often an implicit basis for the marriage, are reshaped in an increasingly collective vein over time and are expressed in the family's important stories, traditions, and priorities. In moments of crisis these beliefs can provide the family with an immediate rationale for a particular coping strategy. Families who endow the arrival of the illness with a special meaning of a spiritual or philosophical nature or who view it as a vehicle through which to express the strengths of the family bring greater resilience to the demands of the situation; this is particularly true when these meanings are developed in the context of a preexisting set of family beliefs (Venters, 1981).

At times, however, the very nature of the crisis can challenge the underlying tenets of the family's belief system. Family members' belief in a particular moral order or in the efficacy of certain forms of behavior may be profoundly violated by a cataclysmic event such as a severe illness. When a family cannot reconcile the adversity it must face with the values that have sustained the family, or when the family does not have a transcendent belief system through which to experience the illness, a profound crisis may ensue.

Family Interface Problems

Problems between Family and Medical Care Team

In the course of managing the long-term care of a patient with serious medical problems, the immediate family and the medical team form an enduring caregiving system; the intensity of these attachments and their effect on the course of the illness are frequently underestimated (Reiss & DeNour, 1989). Small perturbations in these relationships occur in response to phasic changes in the illness, or to membership reconfigurations within the family or the medical staff.

There are two common problems that arise within the caregiving system. The first concerns the *priorities* of the caregiving team. The medical providers, who are focused on the patient's care, emphasize everything the family can do to promote the recovery of the patient, while the family members, who become dependent on the medical providers, want to comply with their advice. This exclusive focus on the patient fails to take nonillness family priorities into account and reinforces the family's tendency to reorganize around the illness.

The second problem arises from the difficulty experienced by family members and medical caregivers in *accepting the patient's illness* and, by definition, the limits of what they can do. Both groups, faced with initial diagnosis, with relapses, or with deterioration, must work through a variety of feelings that may include anger, guilt, and grief. Family members may avoid dealing with their feelings by not hearing information provided by the medical care team, by maintaining unrealistic expectations of the medical team, and by harboring resentments toward them for perceived mistakes and insensitivities. Physicians may be narcissistically wounded by their inability to prevent enduring illness or disability; they may unwittingly contribute to the problem by not talking to family members frankly about their relative's prognosis in a way that permits the expression of disappointment by all concerned.

Financial Problems

Financial burden is one of the greatest sources of distress for families dealing with chronic medical illness. A financial crisis usually arises from one of two sources: uncovered medical and rehabilitation expenses and loss of income from curtailed spousal employment (Hobbs, Perrin, & Ireys, 1985).

The pattern of reimbursement by third-party payers to families with a chronically ill member is of special interest to the medically oriented psychiatrist. Public and private reimbursement entities typically pay for hospitalization and emergency services but rarely pay for maintenance functions such as transportation to specialty care facilities, special equipment needed at home, child care, social services, and respite care. Such reimbursement policies reinforce the family's tendency to organize around the illness by making heroic caregiving strategies necessary.

The lack of reimbursed maintenance services leaves family members with difficult financial and moral choices about the proper allocation of family resources. The primary caregiver, usually the spouse, must decide whether to cut back on paid employment to care for the ill partner or to remain in the work force to pay for private support services. The difficulty of making for such morally complex decisions exacerbates the level of distress in the family. Rigid organizational features in the family may block a creative approach to this dilemma.

FAMILY ASSESSMENT

Although some families would be challenged by any illness, other families with considerable strengths have difficulties with a particular aspect of their relative's condition. These families struggle with the "fit" between the particular demands of the illness and their values and operating style. Therefore it is useful to consider the particular characteristics of the illness that impact on family functioning, as well as the strengths and limitations of the family itself (Table 2-2).

Illness Phase

Assessment of families dealing with a chronic disease includes two steps: evaluation of illness phase and features and evaluation of the family's response to the illness. *Illness phase* is important to understand because, from the standpoint of family management, each phase requires a very different response from family members. The *acute phase* of a chronic disease consists of the initial period of diagnosis when medical activities are directed toward controlling symptoms, preventing further progression of the disease, and in some cases inducing a remission. The primary tasks for the family dealing with the acute phase of an illness are to incorporate accurate information about the disease as quickly as possible and to mobilize family resources to deal with the ensuing medical demands. The arrival of a serious illness constitutes a significant family crisis; it is adaptive, therefore, for family members to make the illness their first priority and to delegate their time and their financial and emotional resources to dealing aggressively with it. Clinicians should realize that a temporary narrowing of the family's focus is a phase-specific necessity for adequate resolution of the crisis.

Assessment of family response to an acute episode of serious illness should focus on family members' ability to comprehend the nature of the illness and to appropriately allocate family resources for efficacious management of the medical condition. Specific elements include: (1) the family's ability to obtain and retain critical information about the disorder; (2) the parents' ability to explain the disorder to the children, grandparents, and other key family members; (3) the family's ability to mobilize extended family and friends for emotional and pragmatic support; (4) the family members' ability to deal effectively with anxiety and depression about the illness; and (5) the family members' ability to reallocate family resources on a temporary basis to deal with the illness.

The *chronic phase* begins after initial treatments have controlled the disease to the extent possible. For many illnesses the chronic phase involves minimizing exacerbations through medications, diet, restrictions on patient activity, or specialized treatments. For other illnesses, chronic care involves providing social supports and rehabilitative interventions to mitigate disability from a stable or deteriorating medical condition. The caregiving tasks during the chronic phase may be great or small, but by a certain point they are known.

It is critical during the chronic phase of illness that family members find ways to routinize and contain illness management and find ways to prevent the illness from completely dominating family life. A family's successful adaptation to the chronic phase involves protecting important family routines, priorities, and traditions that preceded the illness.

TABLE 2–2. *Assessment Steps*

Illness characteristics
Illness phase
Acute
Chronic
Specific illness features
Unpredictability
Need for intensive monitoring
Uncertain prognosis
Extensive disability
Stigma and disfigurement
Family response
Fit between family style and demands of the illness
Reallocation of family roles
Preservation of preillness activities and traditions
Development progress of family members
Connection of illness management to long-standing family problems

Impact of Illness on Family

A promising development of recent years is a consideration of the types of illness characteristics that are most likely to threaten family functioning. Stein and Jessop (1989) provided an empirical rationale for this approach when they demonstrated that there are greater differences in specific psychologic impact on family members *within* traditional diagnostic categories than *between* categories. They made a cogent case for abandoning individual diagnoses and looking for commonalities among disease categories in considering illness impact on families. Rolland (1984) and Jacobs (1991) have also proposed grouping illnesses by risk factors that are meaningful to the families who must cope with them. This method makes it possible to classify disorders in terms of the *type of challenge* they pose to family life.

A group of illness characteristics has emerged in the empirical literature that appear to contain those risk factors that are most challenging to family life. These are the factors that families themselves describe as the most toxic (Gonzalez, Steinglass, & Reiss, 1989) as well as those that appear most strongly predictive of family functioning in the context of serious illness (Evans et al., 1987). A specific illness may contain two or more of these factors. The following five illness characteristics are recommended for use in assessing the *nature* of the illness challenge to families.

Degree of Predictability

The degree to which family members can anticipate acute episodes of the illness has a significant impact on the extent to which they can plan short- and long-term activities. Much of family coherence is grounded in the *planning and carrying out of meaningful events*, such as vacations, attendance at children's school or sports events, and bedtime storytelling (Wolin, Bennett, & Jacobs, 1988). The loss of predictability triggers an unusual degree of anxiety in some families because earlier disruptions in the family history, such as traumatic personal losses or repeated changes of circumstance, have left the family vulnerable to such unexpected events (Jacobs, 1991). Examples of illnesses that present high levels of unpredictability include lupus, bipolar disorder, chronic fatigue syndrome, and sickle cell anemia.

Degree of Disability

The extent of a patient's disability has major implications for *role allocations* in the family. To care for a pervasively disabled person, either the family must purchase services or family members must provide services themselves. The wife or mother in the family is often the individual singled out for the lion's share of this task (Hobbs, Perrin, & Ireys, 1985). This may mean that she must relinquish her full- or part-time paid employment, thus depleting family financial resources.

Individuals who act as the primary caregiver report significant levels of burden (Gubman & Tessler, 1987), although it is widely believed that women are reluctant to admit to negative feelings in relation to the care of a loved one (Rabins et al., 1990). Parents who share the responsibility for the care of a child with a demanding disorder, such as spina bifida, relate their stress specifically to the everyday demands of the illness as opposed to other possible sources of stress (Goldberg et al., 1990). Other examples of disorders with pervasive disability include multiple sclerosis, stroke, and trauma.

Stigma

Disorders carrying significant social stigma primarily affect *boundary regulation* in the family. Family members are reluctant to share information about the condition with friends, neighbors, extended family, school personnel, or clergy. Both children and adults are hesitant to invite friends to their home. Opportunities for concrete financial assistance (e.g., from insurance companies) may be lost. Infrequent access to the outside world resulting from collective shame has a tendency to exacerbate boundary problems within the family or self-defeating family myths; restrictive coalitions and distortions in the family's perceptions of its capacities are further reinforced by the family's self-imposed isolation (Gubman & Tessler, 1987; Walker, 1991). Examples of illnesses with considerable stigma include acquired immunodeficiency syndrome (AIDS), major psychiatric disorders, alcoholism, physical anomalies, and mental retardation.

Degree of Monitoring

The extent to which blood sugar levels, physical activity levels, or medication must be supervised has profound effects on *autonomous functioning* in the family. This becomes particularly important when the patient is an adolescent or a noncompliant adult. The family must struggle with the degree of regulation it should provide and the patient with the degree of regulation he or she should accept. Examples of disabilities that may require extensive monitoring include renal disease, diabetes, psychoses, and Alzheimer's disease.

Certainty of Prognosis

Ambiguity about the ultimate fate of the patient, particularly when the stakes are very high, has a major impact on *risk-taking behavior* and *long-term planning* in the family. Studies of parents with children with a variety of chronic illnesses (Jessop & Stein, 1985), as well as those specifically dealing with cancer (Koocher & O'Malley, 1981) or a congenital heart defect (Goldberg et al., 1990), demonstrated that uncertainty was a major cause of stress. Healthy family fights and the necessary airing of controversial subjects are truncated as family members live in fear of triggering a relapse or sudden death (Gonzalez, Steinglass, & Reiss, 1989). Family members are hesitant to express fears of their relative's demise and to discuss future phases of the family when the patient may be institutionalized or deceased (Walker, 1991). Examples of disorders with uncertain prognoses include many forms of cancer, cardiac disease, and renal disease.

Family Response

Family response to chronic illness is a product of the "fit" between the particular demands of the disorder and the family's characteristic belief systems and behavioral strategies. A family with a tradition of self-sacrifice, for example, will readily adapt to the exigencies of a pervasively disabling illness such as renal disease or stroke; in this case, however the clinician will need to assess whether family members can allow themselves to set limits on their devotion to the patient's needs. By contrast, a family with a strict generational hierarchy will have difficulty permitting the children to assume roles that were previously the exclusive realm of the ill parent or the beleaguered spouse.

The clinician can question the patient or family about its functioning in the five areas outlined in the previous section. Special attention should be paid to signs of *constriction* in the family's ability to operate flexibly. One sign is a history of solutions that are repeated over and over despite their ineffectiveness. Another is an inability to determine how family members really feel about the illness circumstances. Several research tools can provide a useful adjunct to a clinical interview in eliciting these signs (Table 2-3). The Family Assessment Device (Epstein, Baldwin, & Bishop, 1983) covers many of the elements in the five areas described above.

A concrete behavioral assessment of the family's routines will shed light on important changes in roles and responsibilities and on family members' ability to balance illness and nonillness activities. The physician should ask about the family's success in preserving preillness activities, pleasures, and relationships in the face of the illness demands. A review in concrete behavioral terms of a typical day in the family's life before and after the appearance of the illness helps to clarify the nature of any changes.

Family rituals and routines, such as dinnertime, vacations, and holidays, are concrete, accessible markers for assessing the extent to which the family has been able to maintain certain traditions in the face of the illness. The Ritual Interview (Wolin, Bennett, & Jacobs, 1988), originally developed to assess the protection of routines and rituals in alcoholic families, can easily be adapted to assess the preservation of family routines in families with a physically ill member.

TABLE 2–3. *Assessment Questionnaires and Interviews*

Instrument	Type	Administration Time	Target Area
Psychosocial Adaptation to Illness Scale (Derogatis, 1986)	Semistructured interview	30 minutes	Patient/family adjustment
Ritual Interview (Wolin, Bennett, & Jacobs, 1988)	Semistructured interview	45 minutes	Family ritual protection
Family Assessment Device (Epstein, Baldwin, & Bishop, 1983)	Questionnaire	30 minutes	Family problem solving/ communication
Genogram (McGoldrick & Gerson, 1984)	Interview	30 minutes	Intergenerational process regarding illness
Card Sort Procedure (Reiss, 1981)	Laboratory observation	45 minutes	Shared perceptions

Chronic illness can seriously compromise the developmental progression of family members as they become preoccupied with illness responsibilities (Jacobs, 1991). It is possible for the physician to make note of the life cycle stage of the family and to briefly to assess whether preoccupation with the illness has prevented the family from successfully negotiating the tasks of their current phase. In summary the major life cycle phases and developmental tasks are:

1. A *new couple* must differentiate themselves from their respective families of origin and establish their own values and traditions. A partner's illness may pull the well spouse back toward the family of origin, or may distort the initial patterns established between the new partners.
2. A *family with young children* must focus on the developmental needs of children without losing intimacy as a couple. A chronic illness can drain energy from this very frenetic period, usually leaving the well partner feeling depleted and depressed. Suppression of conflict may establish a dangerous degree of overcompliance in the children.
3. A *family with adolescents* must allow teenagers more autonomy while still providing a stable environ-

ment to which they can retreat. Parents must find substitute interests through their work and personal lives to replace the close supervision of children that occupied them earlier. The arrival of an illness may pull adolescent children back into the house physically or emotionally, and may interfere with their entertaining friends at home.

4. In the *launching phase* parents must permit children to make their own school, work, or marriage decisions while providing guidance as requested. Children must struggle with adult identity and commitments to an intimate relationship and career track. The concrete and psychologic responsibilities of attending to a disabled family member may interfere with the young adult's level of comfort in leaving home, and may result in the parent's conveying to the young adult that he is needed at home.

5. In the *retirement phase* parents must find meaning in life that is not centered around paid work while adult children must balance responsibilities to parents with the pleasures and responsibilities of their own families. The illness can ruin lifelong retirement plans or pull adult children away from their own family responsibilities to attend to their disabled parents.

Establishing whether illness management has become entangled in long-term family problems requires relatively detailed interviewing; it involves systematic questioning about the beliefs family members have about the illness. Beliefs about what brought on the illness, how it will develop, and what should be done to manage it often have their roots in earlier generations. Assumptions are made and courses of action played out without family members' conscious awareness of the patterns they are repeating. A particularly tragic death as the result of an illness may have devastated an earlier generation, and this experience may now cast a pall over the current family's ordeal despite a more hopeful prognosis for the present illness. Many health professionals are now trained in the art of gathering genogram information; "problem-centered genograms," or genograms that specifically explore the family's process and history regarding illness management, are particularly relevant for this purpose. A short practical book has been published that outlines the steps involved in gathering genogram information (McGoldrick & Gerson, 1985); a computer program has even been developed for those who interview large numbers of families or who are using the procedure for research.

FAMILY INTERVENTIONS

Clinical theory and methods in the treatment of families with a medically ill member have changed substantially in recent years. Earlier theoretical models of families and medical illness emphasized the pathogenic role of the family in the development and severity of the medical condition (Coyne & Anderson, 1988; Minuchin, Rosman, & Baker, 1978). The best known was that of the "psychosomatic family"; this model asserted that particular family processes predispose a target family member to produce a somatic response to stress.

The classic work articulating the psychosomatic family model was by Minuchin and his group, who studied families of children with brittle diabetes, anorexia, and intractable asthma (Minuchin, Rosman, & Baker, 1978). In an influential series of experiments with families of children with insulin-dependent diabetes mellitus, Minuchin studied children's arousal levels as they entered the room while their parents were having a disagreement. He observed that, in certain families, the more aroused parent would calm down and the child would become more aroused (as measured by free fatty acid [FFA] levels). Because elevated FFA levels are associated with diabetic ketoacidosis, Minuchin concluded that problems in the management of conflict within the family were responsible for the emergence or exacerbation of diabetes in the child. Minuchin and his group also conducted family therapy trials with families of anorexic children; they reported that the intervention, which helped keep the child out of parental conflicts, alleviated symptoms successfully (Minuchin, Rosman, & Baker, 1978).

Although Minuchin has never published key methodologic details pertaining to the FFA experiments and has not systematically tested the family intervention with other illnesses, the generic concept of the pathogenic family remains popular among family therapists. The limitation of the concept lies in its implication that such gross somatic phenomena as acute asthmatic episodes can be generated by family dysfunction alone.

A more comprehensive biopsychosocial model has emerged in the past 10 years as a generic treatment approach for families with a medically ill member. This model conceptualizes medical illness as an external stressor to which the family members respond with varying degrees of competence. The family is not seen as inherently pathogenic; however, preexisting patterns of behavior in the family will influence the degree to which the new challenge is mastered. The critical difference is that preexisting biologic realities of the illness are accepted as "givens" with which the family must cope.

The biopsychosocial model, exemplified by the empirically tested success of a set of structured family interventions in the management of schizophrenia (Anderson, Reiss, & Hogarty, 1986; Falloon, Boyd, &

McGill, 1984), focuses on educational and behavioral interventions that promote successful management of the illness. The illness remains the focus of attention as families learn both to promote recovery and to pursue other family priorities. Family therapists concerned with the treatment of medical disorders increasingly are adopting this approach. Interest has focused around identifying generic issues that arise in families in response to serious illness as well as issues that are specific to particular illness conditions.

SUPPORT, EDUCATION, OR THERAPY?

Families having difficulty coping with a medically ill member may come to the attention of a medical practitioner in several ways. The most common are:

1. *Problems in care delivery.* Persistent medical compliance problems, lack of improvement in the medical condition despite apparent compliance, and frequent, inappropriate or adversarial contacts with the medical staff signal family dysfunction.
2. *Symptomatology in the patient.* The medical staff may hear about or observe signs of depression or anxiety in the patient, such as irritability, work problems, weight loss, sleep problems, or dysphoria.
3. *Symptomatology in the family.* Persistent disagreements or misunderstandings between spouses regarding medical care, frequent family illness, spouse's sleeplessness or weight changes, and problems with children are probable signs of family distress.

Depending on the degree of dysfunction, the medically oriented psychiatrist may consider three levels of intervention for families having difficulty with their medically ill child. They include: (1) educational and supportive intervention by members of the medical staff, (2) time-limited psychoeducational intervention, and (3) family therapy.

Educational and Supportive Interventions by Medical Staff

Families commonly experience transient distress regarding three specific illness events: initial diagnosis (often the greatest crisis period), relapse, and the time at which the ceiling of recovery is reached in a chronic condition. Families with a medically ill member are far more likely to accept psychosocial intervention from members of their relative's own medical team than from external specialists. The psychiatrist can play a consultative role in helping physicians, nurses, and social work staff to provide supportive interventions that are effective with families experiencing transient distress. The most common transient problems are:

1. *Inadequate understanding of the illness and its implications* resulting from excessive anxiety that has prevented family members from integrating basic medical information. A careful review of the factual information may be sufficient to correct this problem; consulting psychiatrists can also be helpful to the medical team by explaining the risks and limitations of providing detailed information to families immediately after hearing a distressing diagnosis.
2. *Grief and anger* regarding the relative's loss of normal functioning. Supportive, nonjudgmental listening may facilitate the expression of these feelings and ease family members through this phase.
3. *Failure to reallocate or supplement family resources* to accommodate the new illness requirements. Information about financial or nursing assistance, or suggestions about the reallocation of duties among immediate and extended family, will be helpful.

If the problems fall into any of these categories, one or two consultations with a physician, nurse, or social worker should be adequate to address these issues.

Time-Limited Psychoeducational Intervention

At an intermediate level, one may use a brief family-focused educational intervention aimed at management of a specific illness or mastery of generic illness-related issues (see Table 2-3). Nurses, social workers, or rehabilitation therapists can be trained to provide this type of intervention.

An example of a brief structured intervention is the eight-session multiple family discussion group (MFDG), developed at the Center for Family Research, George Washington University Medical Center, which focuses on generic issues facing families with a medically ill member (Gonzalez, Steinglass, & Reiss, 1989). Through a highly structured format, two discussion leaders guide a group of four to five families through a series of discussions that highlight the typical challenges facing families with a chronically ill member. The leaders provide the families with useful principles and metaphors with which to articulate their experience, while the families are encouraged to call on their own experiences to help other families with illness-related problems.

The theoretical framework for the MFDG is based on the notion of the family's reorganization around the illness. Group leaders describe the illness as an unwanted intruder on family life, a "2-year-old terrorist" who shows up uninvited to a family meal and interrupts conversations and knocks over glasses. The family with such a "terrorist" in its midst must become self-conscious about its valued traditions and plans, striving to maintain them despite the disruption of the illness. Successful adaptation to chronic illness involves a two-fold process: accepting the reality of the illness (making a place for the illness) and protecting family traditions in the face of the illness (putting the illness in its place).

The MFDG consists of three structured components that systematically provide participating families with an opportunity to explore how these principles apply to their family, as well as to help other families to address these questions. Group leaders follow a detailed treatment manual that gives instructions for conducting each session (Gonzalez, Steinglass, & Reiss, 1987).

Each of the three components of the MFDG employs a "group-within-a-group" format. The first stage is called the "educational component"; in this component the "patient" family members (including children) are asked to sit in a circle and, with the guidance of a group leader, to talk with each other about living with a chronic illness in a family. The "nonpatient" family members are asked to listen quietly and later to respond to what they have heard. In the second session, the process is reversed, with patient members listening to a conversation between nonpatient members from every family. The purpose of this exercise is to help family members to think systematically about the impact of the illness—that is, to think about the characteristic *roles* played by patients and family members in the face of serious medical illness. The third session consists of a didactic meeting in which the leaders present the concept of family reorganization around the illness and the principle of maintaining important family traditions and priorities; family members are urged to think of examples of the illness intruding on the family's important priorities.

In the "family issues" component, each family has a turn to describe in detail an important family issue that has emerged in relation to the illness. In this component the individual family is the group-within-a-group; the purpose of the exercise is to help each family to articulate how the illness has disrupted an important aspect of family life. Group members then add additional observations to the formulation, and offer suggestions for addressing each family's dilemma. An underlying principle of these discussions is to identify the valued features of family life that have been disrupted because of the illness; group members help other families to recapture these core elements while adapting to the reality of the illness.

An example of a family issue is that of a couple in their 60s in which the wife suffered from multiple sclerosis. Throughout their marriage they had been active in their church, their children's school, and many other neighborhood institutions. They always participated in these activities together, proudly developing the identity of "pillars of the community." When the wife's disability coincided with the husband's retirement, the husband could not bring himself to participate in any of the community activities without his wife, resulting in isolation of both partners from the outside world. Over the years, the couple had developed important ideas about "sharing their experiences," which had led to their doing everything together; the husband venturing off without his wife would have meant a betrayal of that valued tradition. The leader helped the couple to reevaluate this tradition in view of the wife's disability. The group members gently helped the couple to recontract so that the husband could participate in outside events and then come home and "share the experience" with his wife.

In the "affective component," families explore the ways in which the illness has disrupted the affective rules by which the family operates. In this case, the leaders select the most psychologically minded family member for the group-within-a-group to discuss this somewhat sophisticated concept. A common pattern that emerges in this discussion is one in which the family member who usually expresses important feelings for the family has become a caregiver for the patient and fears that emotional upheavals may trigger a relapse. Critical expressions of conflict go underground as the family tries not to rock the boat. The remaining family members, now quite outspoken, advise each other on ways to protect the free expression of feeling.

INTERVENTIONS FOR FAMILIES WITH CHRONIC ILLNESS

There have been few controlled trials of family therapy with medically ill children (Table 2-4). While the outcomes described with the use of structural family therapy with brittle diabetes, intractable asthma, and anorexia nervosa were impressive (Minuchin et al., 1978), this study had no control group and lacked critical methodologic details. In one of the few randomized trials of family therapy with a somatic illness (Lask & Matthew, 1979), children with moderate to severe asthma whose families developed coping strategies for wheezing behavior had less daily wheezing and lower residual volume than did children of a group of families who received no psychosocial intervention.

The existing literature supports the idea that family interventions that provide highly specific behavioral interventions targeted toward management of illness behavior (Anderson et al., 1986; Brownell, 1983; Clark,

TABLE 2–4. *Interventions for Families With Chronic Illness*

Reference	Illness	Intervention	Study Type
Morisky et al. (1983)	Cardiovascular disease	Educational counseling	
Clark, Feldman, & Evans (1981)	Asthma	Family education	Randomized control ($n = 300$)
Gonzalez et al. (1989)	Generic	Multiple family psycho-education	Randomized control ($n = 31$)
Lask & Matthew (1979)	Asthma	Family therapy	Randomized control ($n = 33$)
Minuchin et al. (1978)	Diabetes, asthma	Family therapy	No control ($n = 48$)

Feldman, & Evans, 1981; Lask & Matthew, 1979) can reduce illness symptoms, and that intervention focusing on the protection of nonillness family priorities (Gonzalez, Steinglass, & Reiss, 1989) can reduce family stress. The absence of rigorous empirical evidence of the efficacy of more open-ended, process-oriented family therapy with this population should be considered not a definitive statement of the superiority of focused behavioral interventions, but rather a reflection of the fact that the latter type of intervention has been tested more comprehensively.

Therapeutic Principles

Family therapy is best used in cases in which a time-limited psychoeducational intervention has revealed issues needing further work, and in cases in which the family itself sees the importance of pursuing particular issues in greater depth. These are usually cases in which the illness has become embedded in a persistent dysfunctional pattern that is impervious to education or support alone. Based on the success of the behaviorally focused models, family therapy interventions would best be based on the following principles.

Illness-Related Definition of the Problem

Although the presenting problem is typically reframed as a broader systems issue in traditional family therapy, families facing chronic illness often experience this conceptual leap as blaming and irrelevant. The therapist is advised to use the assessment framework described earlier in the chapter to: (1) assess problematic behavioral sequences related to the illness, (2) assess the impact of the illness on core family beliefs and traditions, and (3) undertake a brief review of the family's experience with illness in the previous two generations. These systematic inquiries should lead to a formulation of the problem that is directly linked to the presence of the illness in the family.

Treatment Objective Grounded in Successful Management of the Illness

Even if illness management problems are reflections of more general family problems, the illness is usually the major family preoccupation at the time of referral; an initial successful experience in managing the illness will provide family members with confidence in the therapy process as well as a greater capacity to grapple with more complex emotional issues. The overall objective can be conceptualized as a successful balance between attending to illness requirements and maintaining nonillness family priorities. To address these issues meaningfully, the family must tackle the difficult question of "how much is enough" to give to the illness.

Sequence of Intervention from Focused to Broad

Particularly for families who are not psychologically minded, an initial focused intervention that alleviates family burden is usually most effective. The most common interventions in this category include: (1) involving the more peripheral family members in the care of the patient; (2) delegating limited responsibilities to the children; and (3) giving increased responsibility to the patient for his own care.

In families experiencing an excessive degree of burden, the enactment of caregiving responsibilities is often tied to a dysfunctional family pattern. Frequently, the primary caregiver has taken on full responsibility for the patient's recovery process, while other family members, including the patient, have abrogated responsibility for the patient's care. The therapist must help the family reallocate as much care responsibility as possible to the person with the illness, explaining the benefits to all family members and holding each family member responsible for his role in the reallocation.

The second phase of intervention addresses the larger implications of the illness for the family, in particular the family's recapturing of its own priorities and its balancing of illness and nonillness activities over the long term. Based on a careful assessment of illness phase and characteristics and family response, the therapist helps the family members to put the illness in perspective in the context of their current and future plans. The therapist helps family members to recall preillness practices and priorities that provided them with a meaningful collective identity.

When families are unable to take this step, it is usually because the management of the illness has become entangled in a long-standing, often multigenerational, family problem. For example, a physician referred a man with diabetes and his wife for treatment because the wife seemed increasingly depressed and unwilling to prepare food that conformed to her husband's dietary requirements. As the only son in a prominent family, the husband had long felt entitled to special treatment, and the development of the diabetes had given him a fertile arena in which to express his demands. The wife was the daughter of an abusive alcoholic who had forced her to serve him beer every night until he passed out. She had succumbed to many of her husband's demands over the years, but the meal requirements related to the diabetes became so intense and so reminiscent of her experience with her father that she gradually became unable to continue with her husband's care. The therapist helped the couple to identify the long-standing pattern of demands and compliance that permeated both of their histories. The husband was gradually able to moderate his expectations, while the wife began to learn to distinguish unreasonable demands from legitimate needs related to the illness. Ultimately the therapist's goal was to help them separate the illness management process from the dysfunctional patterns in their respective families.

SUMMARY

Although there are few controlled clinical trials of family therapy interventions in cases of chronic medical illness, interest in addressing family dilemmas in regard to these conditions has increased considerably in recent years. Studies of sources of family stress and preliminary clinical trials have provided us with valuable information about the aspects of chronic illness and family life that are most crucial to successful family adaptation.

This chapter has outlined the most common problems families experience in living with a chronically ill member. It has provided a method for the medically oriented psychiatrist to assess the extent of a family's problems with the illness by considering key characteristics of the illness itself as well as the family response. Finally, a set of interventions have been described, together with criteria that would lead to the selection of one over the other.

REFERENCES

Anderson C, Reiss D, & Hogarty G (1986). Schizophrenia and the family. New York: Guilford Press.

Brown G, Birley J, & Wing J (1972). Influence of family life on the course of schizophrenic disorders: A replication. Br J Psychiatry 121:241–258.

Brown G, Harris T, & Bifulco A (1986). Long-term effects of early loss of parent. In M Rutter, C Izard, & P Read (Eds.), Depression in young people. New York: Guilford Press.

Brownell K, & Wadden T (1986). Behavior therapy for obesity: Modern approaches and better results. In K Brownell & J Foreyt (Eds.), Handbook of eating disorders. New York: Basic Books.

Campbell T (1986). Family's impact on health: A critical review. Fam Systems Med 4:135–328.

Carter E, & McGoldrick M (1987). Family therapy and the life cycle. New York: WW Norton.

Clark N, Feldman C, & Evans D (1981). The effectiveness of education for family management of asthma in children: A preliminary report. Health Educ Q 8:166–174.

Coyne J, & Anderson B (1988). The "psychosomatic family" model reconsidered: Diabetes in context. J Marital Fam Ther 14:113–123.

Derogatis L (1986). Psychosocial adaptation to illness scale. Psychosomatics 30:77–91.

Ell K, Nishimoto R, Mantell J, et al. (1988). Longitudinal analysis of psychological adaptation among family members of patients with cancer. J Psychosom Res 32:429–438.

Epstein N, Baldwin L, & Bishop S (1983). The McMaster Family Assessment Device. J Marital Fam Ther 9:171–180.

Evans R, Bishop D, Matlock A, et al. (1987). Prestroke family interaction as a predictor of stroke outcome. Arch Phys Med Rehabil 68:508–512.

Falloon I, Boyd J, & McGill C (1984). Family care of schizophrenia. New York: Guilford Press.

Gallagher D, Rose J, Rivera P, et al. (1989). Prevalence of depression in family caregivers. Gerontological Society of America, 29(4):449–456.

Goldberg S, Morris P, Simmons R, et al. (1990). Chronic illness in infancy and parenting stress: A comparison of three groups of parents. J Pediatr Psychol 15:347–358.

Gonzalez S, Steinglass P, & Reiss D (1986). Family-centered interventions for chronically disabled: The eight-session multiple family discussion group program (Treatment manual). Washington, DC: George Washington University Rehabilitation and Research Center, Center for Family Research.

Gonzalez S, Steinglass P, & Reiss D (1987). Family-centered interventions for people with chronic disabilities. Advances in psychosocial rehabilitation (pre-publication report on rehabilitation research at the George Washington University Medical Center).

Gonzalez S, Steinglass P, & Reiss D (1989). Putting the illness in its place. Fam Process 28:69–87.

Gubman G, & Tessler R (1987). The impact of mental illness on families: Concepts and priorities. J Fam Issues 8:226–245.

Hobbs N, Perrin J, & Ireys H (1985). Chronically ill children and their families: Problems, prospects, and proposals from the Vanderbilt Study. New York: Jossey-Bass.

Jacobs J (1989). Family resilience in the context of chronic medical illness. Plenary presentation at the annual meeting of the American Family Therapy Association.

Jacobs J (1991). Family therapy with families with a chronically ill member. In F Herz-Brown (Ed.), Reweaving the family tapestry. New York: WW Norton.

Jessop D, & Stein R (1985).Uncertainty and its relation to psychological and social correlates of chronic illness in children. Soc Sci Med 10:993–999.

Koocher G, & O'Malley J (1981). The Damocles syndrome: Psychosocial consequences of surviving childhood cancer. New York: McGraw-Hill.

LASK B, & MATTHEW D (1979). Childhood asthma: A controlled trial of family psychotherapy. Arch Dis Child 54:116–119.

LEWIS F, WOODS N, HOUGH E, ET AL. (1989). The family's functioning with chronic illness in the mother: The spouses's perspective. Social Sci Med 11:1261–1269.

MCGOLDRICK M, & GERSON R (1985). Genograms in family therapy. New York: WW Norton.

MINUCHIN S, ROSMAN L, & BAKER L (1978). Psychosomatic families. Cambridge MA: Harvard University Press.

MORISKY D, LEVINE D, GREEN L, ET AL. (1983). Five year blood pressure control and mortality following health education for hypertensive patients. Am J Public Health 73:153–172.

RABINS P, FITTING M, EASTHAM J, ET AL. (1990). The emotional impact of caring for the chronically ill. Psychosomatics 31:331–335.

REISS D (1981). The family's construction of reality. Cambridge, MA: Harvard University Press.

REISS D, & DENOUR A (1989). The family and medical team in chronic illness: A transactional and developmental perspective. In C RAMSEY (Ed.), Family systems in medicine. New York: Guilford Press.

ROLLAND J (1984). Toward a psychosocial typology of chronic illness. Fam Systems Med 2:245–262.

STEIN R, & JESSOP D (1989). What diagnosis does not tell: The case for a noncategorical approach to chronic illness in childhood. Social Sci Med 29:769–778.

VENTERS M (1981). Familial coping with chronic and severe childhood illness: The case of cystic fibrosis. Social Sci Med 15A:289–297.

WALKER G (1991). In the midst of winter: Systemic therapy with individuals, couples and families with AIDS infection. New York: WW Norton.

WALTZ M, BADURA B, PFAFF H, ET AL. (1988). Marriage and the psychological consequence of a heart attack: A longitudinal study of adaptation of chronic illness after 3 years. Social Sci Med 27:149–158.

WOLIN S, BENNETT L, & JACOBS J (1988). Assessing family rituals. In E IMBER-BLACK, J ROBERTS, & R WHITING (Eds.), Rituals in families and family therapy. New York: WW Norton.

3 Group psychotherapy of the medically ill

JAMES L. SPIRA, PH.D., M.P.H., AND
DAVID SPIEGEL, M.D.

Group therapy has been used with almost every type of psychiatric or psychologic problem, and it is now being considered as an adjunct to regular medical care of the medically ill. This may be due to the recent surge in studies linking psychosocial issues to health outcome, or to new studies demonstrating health improvement for patients undergoing treatment that includes group psychotherapy. Furthermore, the relatively low cost of group therapy makes it an attractive psychosocial treatment in an era of cost containment. Despite the increased use of group therapy for the medically ill, relatively little has been written about this application of group therapy, nor are there clearly outlined protocols for its implementation. This chapter examines the circumstances under which group psychotherapy for the medically ill should be considered as a part of medical treatment. In the first section, the advantages, types, and uses of group therapy are reviewed. The second section discusses the influence on health status of psychosocial factors that can be modified through group therapy, and the effects of various types of therapeutic intervention on health outcome. Finally, in the third section, specific protocols for group therapy with the medically ill are described. Different approaches are delineated for various types and degrees of illness.

WHY GROUP PSYCHOTHERAPY?

There are times when group therapy should be considered the primary form of psychotherapeutic treatment. First, group therapy is up to four times more affordable than individual psychotherapy (Hellman et al., 1990; Yalom & Yalom, 1990). This may enable patients to benefit from psychotherapy they could not otherwise afford. Interpersonal problems, behavioral skills deficits, and educational needs often can be addressed just as easily in a group setting as in individual treatment. Specifically, interpersonal, as opposed to intrapsychic, difficulties often can be most clearly recognized, explored, and worked through during group therapy (Yalom, 1985). Although developing a transference to an individual therapist can be of use for working with intrapsychic patterns that have interpersonal ramifications, group therapy offers patients the advantages of: (1) directly learning to notice and give appropriate feedback to other members in the group; (2) deriving benefits from others' paths to mastery of social skills; and (3) learning to recognize a variety of social patterns and develop interpersonal skills that are often more natural and generalizable than those that commonly occur in interaction with only one person. Being part of a group can also afford patients a sense of community (being understood by and understanding others, receiving help from and being able to help others) in a way that they might have been unable to achieve either in individual therapy or in their lives outside of therapy. Thus, patients can benefit from a number of interactions with others in the group, and for many, this group interaction can be a model for interacting more successfully in other relationships outside the group (Yalom, 1985).

Types of Group Therapy (Table 3-1)

Most therapeutic groups have an *interpersonal-existential* orientation (Yalom, 1980, 1985). Emphasis is placed on the relationship between individuals in the group and in the present moment. Therapists do not attempt to "lead" the groups in terms of establishing a set protocol to be followed, but instead respond to the situation at hand. The therapist's role is to encourage patient expression or interaction between patients, and occasionally make "meta-statements" about the group process: where the group seems to be focusing now and how it appears to be developing.

Interpersonal-psychodynamic approaches (Adler, 1949; Benezra, 1990) attempt to help patients recognize how early childhood patterns can be recognized as con-

TABLE 3–1. *Approaches to Group Therapy: Typical Problems, Methods, Goals, and Relevance for the Medically Ill*

THERAPIST-LED GROUPS

Interpersonal-existential
- Problem: fears about existing fully in the moment are revealed by inauthentic relationships
- Method: helping patients face fears of death occurring in this very moment and place
- Goals: being able to live more fully in the moment in relationship with others
- Relevance: useful for patients who experience extreme distress due to their illness
- Drawbacks: specific coping skills may not be addressed

Interpersonal-psychodynamic
- Problem: early patterns affect present perceptions and actions
- Method: address problems arising during group interaction
- Goals: extend understanding to past problems and future actions
- Relevance: may be useful for chronic medical problems that have a behavioral etiology
- Drawbacks: for the severely ill, exploring historic patterns may not be relevant

Interpersonal-systems
- Problem: *a priori* or unique structures operating between individuals maintain psychosocial dysfunction
- Method: reveal implicit and constraining "rules" of communication in the system
- Goals: find alternative ways of relating so that each member operates at fullest potential
- Relevance: useful for patients whose family system exacerbates the illness
- Drawbacks: emphasis on the system alone may not address needed coping skills

Cognitive Restructuring
- Problem: unconscious perceptions, emotions, and actions operate habitually
- Method: through group interaction, patient recognizes these habit patterns
- Goals: recognition leads to increased control over disruptive habit patterns
- Relevance: helps patients shift from feeling helpless to gaining mastery over their actions
- Drawbacks: for the severely ill, support in the moment may be more important than recognizing patterns

Behavioral Medicine
- Problem: lack of skills, resources, and information keep patient from overcoming stressor
- Method: patients learn specific skills, gaining mastery over their response to the stressor
- Goals: patients learn to control specific behavioral problem
- Relevance: useful for specific medical problems that have a behavioral etiology or reaction
- Drawbacks: not as valuable for general and existential problems associated with severe illnesses

Supportive-Expressive Therapy
- Problem: sufficient support is lacking to allow discharge and exploration of emotional distress
- Method: carefully led groups allow supportive relationship needed to openly explore crisis
- Goals: patients can explore their problems more fully, reduce distress, and feel less isolated
- Relevance: beneficial when patients are undergoing existential crises such as illness and loss
- Drawbacks: less useful for specific behavioral problems leading to medical illness

LEADERLESS GROUPS

- Problem: persons who assemble to offer each other support for:
  - —recuperation from earlier tragedy (e.g., abuse)
  - —assistance in overcoming some behavior-related problems (e.g., drug abuse)
  - —examination of common interests (e.g., men's groups)
- Method: taking turns stating one's experience, discussing common problems, or a combination
- Goals: feel less isolated, receive emotional support from others in similar situation
- Relevance: can offer much needed support to those who feel isolated from others
- Drawbacks: for the severely ill, leaderless groups can raise anxiety

tributing to current interactions in the group. This approach almost always combines individual psychotherapy with the group process. In fact, the groups frequently are comprised of the individual patients seen by the group leader (Yalom, 1989).

Interpersonal-systems approaches, as described in texts on structural family therapy (Minuchin, 1974), assume that certain *a priori* dynamics operate between family members. Some clinicians follow a deductive methodology, wherein the *a priori* relationships are believed to be a standardized factor within any family, requiring the therapist to help describe these relationships to the family members so that they can learn to modify them. Alternatively, some therapists follow an inductive approach to the analysis of a group system, attempting to discover along with the group members how the members have defined themselves and others in their self-constructed system. In either case, efforts are made to not only understand the underlying assumptions that members of the system make about their own and others' roles, but also to help increase the flexibility of the system. This approach can extend beyond typical familial relationships to applications to the workplace, couples therapy, or any other close social network.

Cognitive restructuring is a problem-oriented approach to therapy (Beck et al., 1979) that examines common irrational or limited patterns of handling stresses. In its application to group psychotherapy, "little emphasis is placed on understanding the interaction between patient and therapist" (Hollon & Shaw, 1979, p. 336). Rather, attention is focused more on the individual problems of group members and less on the interaction between the members. Each patient takes turns focusing on a particular problem on which they wish to work, and others have an opportunity to comment on the problem. This approach is useful because it offers patients the opportunity to discuss their perceptions and in turn receive feedback from others in a controlled environment (Flores, 1988; Fram, 1990).

Group therapy for patients with medical conditions (the *behavioral medicine* approach) often focuses on developing skills or providing education to help patients with specific illness-related problems. While many of the above models for group therapy emphasize general patterns of interpersonal behavior, groups involving medical patients frequently focus on skills designed to assist with illness-specific behavioral disturbances.

The degree of structure and leadership in group therapy varies widely. Some groups function without any professional leadership. These are often highly structured, as in Alcoholics Anonymous (AA), but may have relatively little designated organization, as in some "women's" and "men's" groups. Some *"leaderless" groups* have a designated coordinator whose responsibility it is to organize the meeting times and locations, and occasionally to invite expert speakers (Spiegel, 1981). Such groups are not psychotherapy groups, in that they are not optimal for exploring problem habit patterns or learning new coping strategies. Nevertheless, they can be useful in providing social support and can help to bolster self-efficacy (Wasserman & Danforth, 1988).

Most groups use a combination of elements from each of these methods. An example of such a combined approach is *supportive-expressive therapy* (Spiegel, Bloom, & Yalom, 1981; Spiegel & Spira, 1992; Spiegel & Yalom, 1978), which combines an interpersonal-existential approach with group support and specific coping skills for use with the seriously ill. The focus here is on creating a supportive environment in which patients can express how they are feeling and what they are thinking. Patients are encouraged, without direct confrontation, to share their sorrow, fears, hopes, and other intimate feelings they may not be able to share with others outside the group. Open expression of this range of emotion is associated with improved quality of life (Spiegel, Bloom, & Yalom, 1981; Spiegel & Glafkides, 1983) and survival (Spiegel et al., 1989). Patients begin to care for other group members, offering support to others as well as deriving support from them. Therapists help to maintain the group as warm and supportive, and to encourage members to stick to personal, concrete, affective issues. The focus is not only on the here and now, but also on developing important social relationships and realistic future goals and plans. Unlike rules laid down in other therapy groups forbidding contact among group members outside the meetings, members of these groups are encouraged to see one another outside of the formal meeting times. In addition, specific coping skills are taught to the group so that pain and other problems common to the illness can be more effectively tolerated. Group exercises also are occasionally utilized in order to help members address difficult issues that would otherwise be too stressful to consider directly (Spiegel, Bloom, & Gottheil, 1983; Spiegel & Spiegel, 1985). This approach is covered in more detail below (see "Protocols for Group Therapy for the Medically Ill").

Structure of Groups

The number of patients in a group, the frequency of meetings, and the makeup of the group depend on such factors as the therapeutic orientation, the problems to be dealt with, and the therapeutic goals. Behaviorally oriented groups that focus on education may consist of 20 or more patients who meet once or twice to receive useful information or to be introduced to specific coping strategies. Alternatively, groups can meet for a year or more to help members master a set of skills (Ornish et al., 1990; Powell & Thoresen, 1987).

The more specific the goal, the more structured and time limited the meetings. In highly focused groups, each meeting has a specific agenda to be covered and has certain skills to be learned. Other types of groups that attempt to deal with deeper personality patterns that affect many current aspects of the patients' life may meet indefinitely. Instead of planning what is to be covered in advance, each session generates its own topic of concern. The therapists are responsive instead of directive, reflective rather than preemptive and didactic.

BENEFITS OF PSYCHOSOCIAL INTERVENTION FOR HEALTH-RELATED PROBLEMS

Correlations of Psychosocial Factors with Health Outcomes

Why examine group psychotherapy for health-related issues? First of all, there is a growing body of evidence showing that a variety of psychosocial factors influence both quality and quantity of life. Group psychotherapy has the potential to assist the patient in every psychosocial domain that affects psychologic and health outcomes. The most frequently explored relationships are outlined in Table 3-2.

Psychosocial variables that have been found to negatively affect the incidence, progression, and mortality

TABLE 3–2. *Combinations of Psychosocial Variables and Outcome Measures Common in Research*

Psychosocial Variables	Outcomes Examined	Typical Measures Used
Past traumas and stresses	Psychiatric symptoms	SCID[1]
Trait personality; self-image; self-efficacy	Current mood	POMS;[2] CESD[3] WOC[4]
Current or past mood	Self-efficacy	Stanford Cancer SES[5]
Coping strategies	Psychologic adjustment to illness	MAC;[6] CECS;[7] CLC;[8] WOC[9]
trait styles over situations	Psychophysical coping (pain, sleep, etc.)	(various)
state-dependent coping processes	Treatment compliance	(various)
Social relatedness	Physiologic measures	NK; CD4/8: cortisol[10];
social network	Incidence of disease	other primary diseases;
social support	Progression of disease	co-morbid diseases[11]
family functioning	Mortality	recurrence; staging; etc. all causes; disease-specific

[1]Structured Clinical Interview Device; Spitzer & Williams
[2]Profile of Mood States; McNair, Lorr, & Drappelman (1971).
[3]Center for Epidemiology Scale of Depression; NIMH
[4]Ways of Coping Scale; Folkman & Lazarus (1980); see also Moos & Moos (1988).
[5]Stanford Cancer Self-efficacy Scale; Joss, Spira, & Spiegel (unpublished)
[6]Mental Adjustment To Cancer Scale; Greer & Watson (1985).
[7]Cortauld Emotional Control Scale; Greer & Morris (1975).
[8]Cancer Locus of Control, Watson, Greer, Van Pryun, et al. (1990).
[9] Ways of Coping Scale; Folkman & Lazarus (1980)
[10]Natural killer cell function and number; T-helper and suppressor cell ratio; cortisol, and other functional and frequency measures of immunity and neuro-endocrine measures.
[11]Initial diagnosis for a specific illness; or utilization of health services for other concerns.

of many illnesses include prior stressful life events (Holmes & Masuda, 1967; Holmes & Rahe, 1967; Rabkin & Struening, 1976) and the amount and intensity of daily stresses (Kanner et al., 1981; Lazarus & Folkman, 1987). A maladaptive current attitude toward one's illness, with nonacceptance of the fact of the illness, coupled with a sense of helplessness, and inability or unwillingness to express one's feelings about having the illness, is associated with relatively poor prognosis (Greer & Watson, 1985; Heltz & Templeton, 1990; Wallston & Wallston, 1982). Poor social support has been associated with poor health outcomes (Berkman & Syme, 1979; Cohen, Sherrodd, & Clark, 1986; Cohen et al., 1985; House, Landis, & Umberson, 1988). Group therapy, which can facilitate coping skills and establish supportive relationships, might be expected to improve health outcomes.

Several common goals of group therapy can have a positive effect on health outcomes. These include being able to openly discuss and express one's problems (Kneier & Temoshok, 1984; Naor et al., 1983; Temoshok et al., 1985); developing enhanced control over the course of one's illness (Watson et al., 1990; Taylor, 1984); and using appropriate coping mechanisms (Greer & Watson, 1985). Realigning social relationships, such as being able to focus on positive relationships and learning to reject unwanted favors, may also be relevant to more favorable outcomes (Kneier & Temoshok, 1984; Solomon & Temoshok, 1987).

General personality characteristics, such as a tendency to suppress anger, or depressive traits, have not been consistently shown to be risk factors for specific medical illness (Cassileth et al., 1985; Jamison, Burnish, & Walston, 1987; Plumb & Holland, 1981; Zonderman, Costa, & McCrae, 1989). Thus, emphasizing change in such personality characteristics in group therapy may not be desirable if better health outcomes are the goal. Specific coping skills, however, such as being able to openly express one's feelings and thoughts regarding the illness, do appear to act consistently as predictors of health outcomes (Greer & Watson, 1985; Naor et al., 1983; Watson et al., 1990). It is time consuming, extremely difficult, and of questionable value to attempt to alter basic personality characteristics during relatively time-limited group therapy for a medically ill population, whereas emphasizing supportive and expressive relationships is relatively easier and considerably more relevant to patients' health outcome. Table 3-3 describes a trend toward an emphasis on coping and social support that has been emerging in the psychosocial literature on cancer and heart disease.

Group Therapy Interventions and Health Outcomes: Research Results

Although the research on psychosocial aspects of health far exceeds that on psychosocial intervention, new evidence supports the possibility that therapy can influence a change in health outcome. Many of these interventions have focused on *education* for (1) prevention of developing illness (acquired immunodeficiency syndrome [AIDS], cancer, bereavement); (2) modifying

TABLE 3–3. *Success of Studies Finding Psychosocial Variables Related to Cancer Incidence, Progression, and Survival*

Variable	
Trait Personality	LEAST CONSISTENT
Intrapersonal	↓
anger suppression	
trait depression	
general sense of control or helplessness	
Interpersonal	
emotional expression/control	
social compliance/assertiveness	
openness or inability to discuss illness	
Situational Coping and Adjusting to Cancer	
Mental adjustment to cancer	
Pain, sleeplessness, anxiety	
Cancer locus of control	
Self-efficacy in coping	
Social Relatedness	
Network	
Support	
perceived emotional connections	
functional support	↓
Family functioning	MOST CONSISTENT

behavior-related illnesses; or (3) teaching coping strategies for improving quality of life for patients with a current illness and also for their family members. Other approaches have utilized more traditional *cognitive* or *structural* group therapy in order to help alter personal behaviors or family structures that interfere with health or recovery from illness. Rather than remove individuals from their stressful context, therapy is directed toward helping to identify the patterns that contribute to creating or maintaining the health problem and finding more productive solutions. Finally, *supportive-expressive* therapies help to reduce a sense of isolation and helplessness that stems from serious disability or life-threatening illness, through encouraging open expression by patients of thoughts and feelings about their illness or disability in order to live life more fully. Frequently, group therapy for the medically ill employs a combination of these approaches, although usually one aspect is predominant. The most common approaches are outlined in Table 3-4.

Presented below are some approaches that have been empirically studied (comparing treatment with control groups), in which both the treatments used and the patient populations examined have been well defined, and that appear to have the best "fit" between therapeutic style and the patients' particular needs. Although many of these approaches do not have the direct intention of improving health outcome, some reports are beginning to emerge of how physical health, as well as quality of life, can be improved. Health benefits have included improved physiologic functioning, retarded progression of disease, and increased survival.

TABLE 3–4. *Methods, Target Populations, and Primary Goals in Group Therapy for the Medically Ill*

Method	Populations	Goal
Education		
Information	At risk (AIDS; heart disease)	Prevention of illness
Behavioral skills	Psychological exacerbation (psychological problems causing health problems: eating disorders, heart disease)	Modifying behavior-mediated illness
Coping strategies and health-related skills	Chronically ill (diabetes, Alzheimer's, stroke, asthma, cancer)	Improved quality of life through better coping with illness; improved level of health through health strategies or treatment compliance
Cognitive therapy and family therapy	Psychosocial exacerbation (psychosocial dysfunction affecting health: anorexia, asthma, recovery from heart disease, latent stress-related viruses, ulcers)	Helping alter personal or family behaviors that interfere with health or recovery
Supportive-expressive	Terminally ill (cancer; AIDS) Severely impaired (multiple sclerosis, burn patients)	Reduce sense of isolation and helplessness; encourage open expression of thoughts and feelings; confront fact of illness and death or deformity in order to live more fully

Education

Education aimed at preventing illness usually takes the form of providing information and life-style counseling. It is usually applied to persons at risk for specific illnesses, such as atherosclerotic heart disease (Ornish et al., 1990; Powell & Thoresen, 1987) or AIDS (Coates et al., 1989; Perry et al., 1991). Usually solicited from physician referral or public health screenings, such "information-oriented" groups have produced mixed results. One negative example can be found in the failure of the massive MR FIT study to significantly reduce high-risk behaviors of subjects in an educational condition compared to controls who were equally at risk for myocardial infarction (Multiple Risk Factor Intervention Trial Research Group, 1982). In this large multisite trial, classes were given on smoking cessation, dietary considerations, and exercise. The study concluded that those who received education about changing their health habits did no better than controls who were aware of their risk and volunteered to be in the educational groups. However, both of these groups were able to change their health behaviors in order to reduce future myocardial infarction significantly more than a second control group who did not volunteer to participate in the educational groups. These results were widely viewed as indicating that personal motivation rather than educational intervention is more important in altering health behaviors. In contrast, educational approaches that have been found to have the most impact on altering health-related behaviors have been those designed to be sensitive to the individual needs, motivations, and life context of the persons in the group (Powell & Thoresen, 1987). Timing of the intervention can also be important. Research by Schneiderman et al. (1990) showed that stress management interventions prior to diagnosis helped to buffer the psychologic distress of subsequently receiving a diagnosis of positive human immunodeficiency virus (HIV) infection, and also modestly elevated the values of some immune parameters in HIV-seropositive individuals.

When education for change in health behavior is integrated with specific coping strategies, the results are generally more positive than when information is merely meted out to a group. In a series of studies with geriatric patients, Rodin (1980) found that patients offered education and counseling lived longer on average than their peers who were not randomized to this type of intervention. Richardson (Levine et al., 1987; Richardson et al., 1990a) studied patients with hematologic malignancies. The patients were randomized to intensive home visiting intervention that included the patient and a family member in educational programs to increase compliance in taking allopurinol, a medication intended to reduce potential side effects of chemotherapy but that does not directly influence survival. They found a significant increase both in compliance with allopurinol and survival for patients who received health counseling. Perhaps increased compliance with allopurinol was reflective of increased compliance in other areas of health care, yet increased survival was not found to be merely due to increased compliance alone. Rather, the nature of the interaction between counselor, patient, and family appears to have an additional effect on survival.

While home visits sometimes are preferable for the elderly or disabled, it is usually far more cost effective to have patients come to a clinic or hospital setting. In a prospective randomized study with patients suffering from so-called psychosomatic complaints, Hellman et al. (1990) compared information alone to information plus training in specific coping strategies in a group therapy format. The group therapy that combined education and coping strategies significantly reduced visits to the hospital, as well as reducing discomfort from physical and psychologic symptoms. Health counseling and coping skills training groups are particularly beneficial when problem-related behavior can be eliminated or modified in a relatively short amount of time, when the patients are highly motivated, and when the problem is limited to specific behaviors in a patient's life.

Behavioral Skills and Cognitive Restructuring

Intervention for altering behaviors that cause or contribute to an illness are perhaps the most frequent of all group interventions for health purposes. Because of the nature of the clearly identifiable behaviors and effects on health, this area is also the most frequently studied. Such intervention often is conducted in individual sessions, yet because of the educational methods used and the psychosocial character of the behavioral problems dealt with in many of the interventions, group therapy can be especially beneficial. Group therapy has been found helpful in treating drug dependency (Brown & Yalom, 1977), posttraumatic stress disorder (Marmar et al., 1988), eating disorders (Telch et al., 1990), and hypertension (Nunes, Frank, & Kornfeld, 1987). Matano and Yalom (1991) developed the Stanford University Drug and Alcohol Treatment Program around the principle that most drug treatment programs emphasize personal expression without sufficient interpersonal development. Their interactional approach focuses on the here and now, how patients relate to each other in the group, and the common patterns that arise during the meetings. This approach is distinct from typical AA meetings, and has been found to be a valuable adjunct to recovery.

Strategic problem-focused therapists direct their attention to ineffective coping strategies. This is evident in therapy that focuses on the covert strategies used by families that have an ill member (Coyne, 1982). In a study by Hellman et al. (1990), group treatment that emphasized either cognitive restructuring, stress-reduction strategies, information only, or a combination of these was compared. After 6 weeks of 90-minute weekly sessions, the cognitive restructuring groups were significantly more effective than the stress-reduction groups, while both together reduced the number of hospital visits by half compared to the groups that received information alone. In another study emphasizing cognitive restructuring, families of stroke victims were treated to modify the family's overprotective and overly solicitous behavior, allowing patients to begin to become more outgoing and to exercise regularly (Coyne, 1982).

Patients with serious health problems can benefit from group therapy that allows for expression and exploration of deeply held feelings, yet such exploration becomes difficult when the subject is experiencing extreme pain or fear. Therefore, many therapy groups include specific coping strategies that can help subjects to adjust to the onset of serious illness and help them to deal with specific aspects of their condition. In one such study, cancer patients were treated with standard oncologic treatment alone, standard treatment plus group therapy, or standard treatment plus group therapy along with self-hypnosis as part of every session (Spiegel, 1985). Meeting weekly for 1 year, group therapy consisted of discussions of fears about dying and strategies for patients to maintain control over their lives and the management of their illness, grieving over the loss of group members who had died, and establishing realistic goals for the remainder of their lives with friends and family. Whereas the control group had a large increase in pain and suffering over the course of the year, the group receiving therapy only had a slight increase in pain over the course of the year, and the patients receiving self-hypnosis along with group therapy reported a *decrease* in pain sensation and suffering.

Supportive-Expressive Therapy

The beneficial nature of group therapy can extend far beyond assisting with specific coping skills to help deal with stress or pain. In several studies, breast cancer patients receiving group therapy have been found to have reduced mood disturbance over the course of treatment compared to nontreated controls. These patients were significantly less depressed, fatigued, confused, and phobic than the control patients, and used better coping responses (Cain et al., 1986; Capone et al., 1980; Ferlic, Goldman, & Kennedy, 1979; Gustafson & Whitman, 1978; Spiegel, Bloom, & Gottheil, 1983; Spiegel, Bloom, & Yalom, 1981; Turns, 1988; Wood et al., 1978). Recent studies with HIV-positive patients have shown similarly that group interventions increase social support (Kelly et al., 1987) reduce depression and anxiety, and increase active coping (Fawzy et al., 1990a). Typically, these studies also employed some therapy for families as well as offering behavioral skills to assist in coping with the illness. This approach is explored in more depth on p. 39 "Protocols for Group Therapy for the Medically Ill."

Family Therapy

Family therapy can help to improve the adjustment to illness (Cohen & Wellisch, 1978). Spiegel, Bloom, and Gottheil (1983) found that family functioning was related to mood in cancer patients. Furthermore, improved family functioning resulting from group therapy for both patients and their spouses was related to an improvement in mood. These studies have generally shown more consistently positive results than have individual interventions for cancer patients (Gorden et al., 1980), which have been most useful for those with depression (Maguire, 1988).

Besides providing an opportunity for the acquisition of specific coping skills, or for improving relationships among family members, group therapy has the additional powerful advantage of providing for beneficial modeling (Bandura, Lanchard, & Ritter, 1969). In one study (Anderson & Masur, 1989), seeing another patient go through cardiac catheterization and coronary angiography reduced anxiety even more than did information or coping skills, thus allowing patients to undergo the procedure with far fewer complaints of nausea, fear, pain, and other verbally and nonverbally demonstrated stress reactions.

Changing Health Outcome with Group Therapy

Some group therapies intervene with the direct intention of altering physical health. Other therapeutic interventions are designed to improve the quality of life for patients, with the surprising outcome of influencing quantity of life as well. Such groups usually are oriented toward illnesses in which the attitude or behavior is clearly related to the health of the patient. If a patient can behave in such a way as to exacerbate a problem, then presumably intervening to modify this behavior will result in reduced risk. There are three types of psychosocial interventions that have been found to influence health of patients: home visits, short-term therapy, and long-term therapy.

Home visits have tremendous potential benefit for certain types of patients since the psychosocial intervention can be individually tailored to meet the needs of the patient in ways that will work for them in the context of their lives. This approach has worked well for the elderly (Rodin, 1980, 1986); for newborns in inner cities, where nurses instruct families on infant care in their own homes (City of Berkeley Infant Home-Care Project, 1980); and in the medically ill (Levine et al., 1987; Richardson et al., 1990b).

Short-term group therapy also has been shown to be effective. Combining coping skills, cognitive restructuring, and group support, Fawzy et al. (1990a, 1990b) conducted a prospective longitudinal study of post-surgical patients with malignant melanoma with good prognosis. They studied the effects of a structured, 6-week intervention consisting of health education, relaxation training, problem-solving skills, psychologic support for distress, and use of effective active cognitive and behavioral coping methods. At the 6-month follow-up, the patients showed less depression, fatigue, confusion, and total mood disturbance as well as higher vigor (using the Profile of Mood States; McNair, Lorr, & Drappleman, 1971). Patients also were able to shift from a passive to an active coping strategy (Moos & Moos, 1988). In addition, while no physiologic changes between treatment and control patients could be detected at the 6-week assessment, 6 months after therapy began (and 18 weeks after the group therapy ended), treatment patients showed significantly better immune function (e.g., natural killer cell function) than control patients. This suggests that even a brief intervention can have long-lasting effects.

Longer term group therapy combining coping skills, cognitive restructuring, and group support has proven most effective for the severely ill. Powell and Thoresen (1987) reported positive effects of group training of patients with so-called *type A* personalities on the course of myocardial disease. Individuals with type A personalities have been thought to lack motivation to pursue sufficient mental and physical recuperation from their driving life-style, compared to those with *type B* personalities (Friedman & Rosenman, 1974). However, recent work by Thoresen and others (Fontana et al., 1989; Powell & Thoresen, 1987; Thoresen et al., 1985) have shown that it is not the type A life-style that leads to increased recurrence and mortality from myocardial infarction. Rather, the inability to sustain a loving, supportive relationship may be the causal behavioral factor among type A individuals who have recurrent infarctions. In fact, Thoresen has preliminary evidence that type A patients who are able to both offer and receive emotional support have one fourth the risk of infarction of type A patients who are more socially isolated (Thoresen & Powell, 1992). These emotionally healthy type A patients even appeared to have a better prognosis than did patients with type B personalities.

Thoresen's therapy groups emphasized cognitive restructuring of members' life-style and relationships. Therapists led patients through a series of structured exercises that helped them to examine the way they viewed their life, their family and friends, and their future. Much like alcoholics in an AA meeting, patients were helped to acknowledge their problem and gain control over their behaviors in a variety of contexts. Specific homework assignments aimed at modifying their health-related behaviors were given out at each class, helping patients to notice and control their dietary, exercise, and work patterns. Much of patients' aggressive and antisocial behavior was seen in terms of low self-esteem. Therapy was thus directed toward helping to raise their self-esteem through improving their understanding and perceptions of control, assertiveness, and recuperation. Clearly, such significant life changes require time and effort. The patients in this study were involved in therapy for a minimum of 3 years, and were encouraged to return to groups whenever they liked.

In a very different type of long-term intervention, Ornish and associates (1990) showed that an intensive and multifaceted treatment was able to achieve significant improvement in cardiac health. Treatment subjects committed themselves to a rigorous program that began with a week-long retreat. Patients stopped smoking, ate a vegetarian diet, relaxed for at least an hour a day, and walked at least three times a week. Twice a week, they also attended 4-hour group sessions that involved group support, hatha yoga, and education. The control group was asked to continue conventional care, which included advice on how to improve their health-related behaviors (diet, exercise, quitting smoking). Coronary artery lesions were analyzed by quantitative coronary angiography. After 1 year, the diameter of stenotic vessels had improved in 82% of the treatment subjects compared to 42% of the controls. In both groups, those who changed their life-style most dramatically improved the most. The intensive group approach was able to assist patients in sticking to a healthy regimen.

These studies demonstrate that patients with heart disease can learn new skills that may help them to live a longer life. Patients with very poor prognoses, however, may not have the strength to engage in such rigorous therapies or may not live long enough to make use of what is learned in the groups. Instead, these patients can benefit from an approach that helps them to improve the quality of their present life. This benefit can be accomplished by a combination of therapy that encourages patients to openly address and express their fears, gain emotional support from others in similar circumstances, and learn better methods of coping with daily stresses associated with their illness or treatment.

A *supportive-expressive* approach of this kind was taken by Spiegel et al. (1989). Eighty-six women with metastatic carcinoma of the breast were randomized to 1 year of standard oncologic care plus group therapy or to standard care alone. Even though the treatment and control groups were not different in terms of disease parameters or treatment, the 50 women undergoing psychosocial treatment lived an average of 18 months

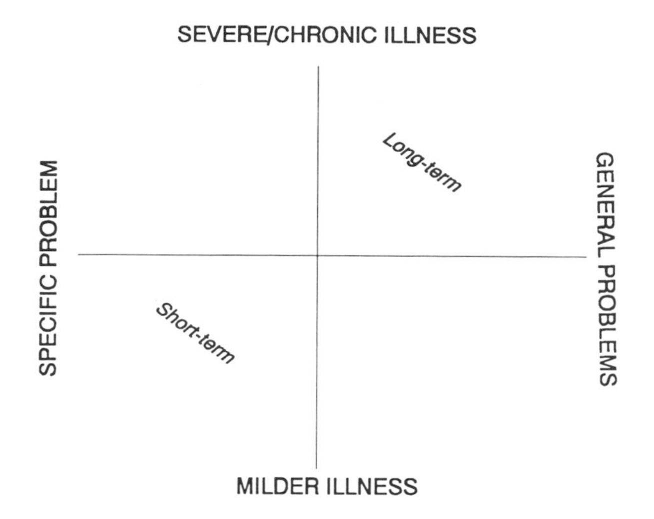

FIG. 3–1. Less serious and more specific problems are associated with briefer therapies. More serious and fundamental problems benefit from longer therapy.

longer than did 36 control patients. Four years after the study began, all of the control patients had died, whereas one third of the treatment sample were still alive. In addition, the quality of life in those surviving longer was significantly improved as a result of group therapy, as indicated by ratings of pain, mood, and family functioning (Spiegel, Bloom & Yalom, 1981) (Figure 3-1).

While some studies on the role of psychosocial intervention in improving health outcome were negative (Fox, 1983; Morganstern et al., 1984), several have demonstrated significantly better health among cancer patients randomly assigned to psychosocial support. Thus, there is increasing evidence that group support for the medically ill has positive physical as well as psychologic effects. That such interventions are effective is less surprising than the fact that they are not more widely employed, especially given their relatively low cost. Few persons are adequately prepared to deal with the effects of serious medical illness, and most can benefit from the combination of emotional support and learning that is available in groups.

In the next section, specific protocols for working with the medically ill are described. A supportive-expressive approach is examined in some depth as a model for therapy groups for the severely ill.

PROTOCOLS FOR GROUP THERAPY FOR THE MEDICALLY ILL

Patients who come to group therapy because of medical illness are different in many ways from those who seek group therapy primarily for psychiatric disturbances. They are often faced with a crisis that would be difficult for any individual to handle. In addition, there are comorbid factors, such as depression and anxiety, that blur the line between psychopathology and normal stress due to physical illnesses. The distinction becomes even less clear when a person's pathologic behavior directly leads to medical illness (eating disorders, drug abuse, etc.). In some cases (e.g., type A behavior), the psychologic characteristics are not usually perceived as pathologic until the disease becomes apparent. Therefore, the structure of groups for the medically ill must take into consideration the special characteristics and problems of the population to be treated.

The issues dealt with in such groups depend on (1) the motivation of the patients, (2) the goals of the therapists, and (3) prospects for change, given patients' medical prognoses. The further away the illness in time and severity, the lower the motivation to change but the greater the opportunity to make significant changes if desired. The more severe and current the illness, the greater the motivation to attend and benefit from the groups, but the less time and energy there may be to spend on making major changes and integrating them into one's life.

Population of the Group

The question often arises as to whether the groups should be homogeneous in terms of diagnosis, prognosis, sex, age, and the like. Homogeneity is less important for groups that meet for brief educational interventions or to acquire coping skills for specific, temporary ailments. Greater homogeneity is recommended for long-term, interactive groups for patients with life-threatening illness. For example, cancer groups most often are comprised of same-sex patients who have similar type and staging of the disease (e.g., women with recurrent breast cancer).

Occasionally mixed-stage groups meet (e.g., groups of patients with AIDS, AIDS-related complex [ARC], or HIV seropositivity) when there is likelihood of parallel progression of the illness among group members, or when there are other aspects that are common (gay men). Mixed groups have the benefit of modeling—seeing how well those who are more progressed in the illness handle problems likely to occur to less progressed members of the group. In addition, patients at all levels of health and illness can find value in supporting each other, since each has something of value to offer to others. However, it is common to find some discomfort in healthier members of the group, who feel they may not end up as ill as other members. Also, there occasionally arise among sicker members feelings of isolation from those who are "not like *me*," who have a different diagnosis or different prognosis and therefore different concerns and experiences.

Preventive Medicine

Preventive medicine group meetings primarily are concerned with information exchange. For classes and meetings regarding AIDS prevention, breast cancer, heart disease, and the like, often a community meeting or a series of lectures/discussions at a clinic or hospital is sufficient. Even so, leaving sufficient time for discussion and personal expression can increase the benefits of the gathering.

For persons "at risk" for illnesses such as heart disease, breast cancer, and AIDS, a useful approach is a series of meetings at which risk factors can be described and health promotion skills can be learned. As important as information and good health behaviors can be for persons "at risk," therapists should not underestimate the value of patients having an opportunity to meet with others in a similar situation to exchange stories, fears, and hopes. As the research reviewed above points out, the gathering together of persons with similar health concerns appears to be at least as effective as health education and skills training in influencing patients' behavior.

Salient issues addressed in such groups go beyond medical information, and may include the importance of supportive relationships with open communication; personality patterns that are related to the illness; health behaviors that directly bear on the illness; and coping strategies that enable the individual to best deal with stress, from daily hassles to major events (such as the diagnosis of the illness). These issues can be raised as points for awareness and discussion at a single meeting, which is often sufficient for those tentatively at risk (a man whose father died of a heart attack; a woman whose mother had breast cancer). Alternatively, these issues can be addressed in more depth in ongoing group therapy for those at greatest risk (HIV-positive but asymptomatic individuals; men who have survived a first heart attack; women who have had a malignant breast tumor removed and are now in remission).

Many psychotherapists might not place a high priority on conducting groups for persons who are not yet ill but for whom there is some increased risk of becoming ill. Certainly the motivation and intensity in such groups is generally less than in groups of those who are currently ill. Yet if prevention-oriented groups were more common, the need to conduct groups for those already ill might be reduced.

Groups for the Mildly Ill

Groups for the mildly ill with relatively good prognoses will have goals distinct from groups of persons at high risk for major illness and also from groups of patients with more pessimistic prognoses. Depending on the patient population and the structure of the group therapy, the goals for this type of group might include helping patients:

1. Gain better control over behaviors that might contribute to their health in general (moderate type A behavior, improve diet, exercise more, learn to relax in order to sleep better at night, etc.) and the illness in particular (fear might interfere with compliance, a poor diet might exacerbate high blood pressure, etc.).
2. Learn coping strategies to adjust to the conditions brought about by the illness or treatments received.
3. Receive social support, which lessens isolation and enables patients to express concerns openly and honestly to each other, and thus to themselves.
4. (if longer term therapy is available or necessary) Examine personality characteristics that may affect health, so that patients can recognize actions that lead to worse prognosis and moderate these activities.

For most patients with an illness of moderate intensity, that is heavily influenced by personal behavior and attitude, or that is severe yet has a good prognosis for recovery, a combination of educational and supportive-expressive methods is most valuable (Figure 3-2).

Groups for the Severely Ill

Patients with severe illnesses and poor prognoses require mainly supportive-expressive therapy with some

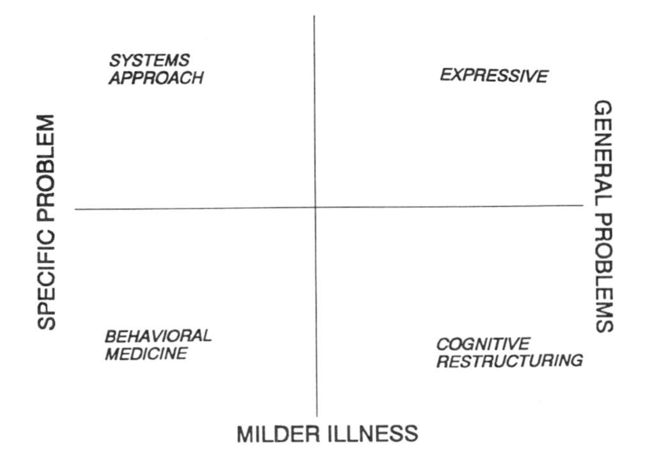

FIG. 3–2. Types of therapy by seriousness and specificity of problem presented by patients.

behavioral coping skills (for pain and anxiety). This population of patients may not have the time or energy to benefit from in-depth analysis and education regarding personality factors that may be influencing the course of their illness or interfering with their quality of life. Therapy for these patients must center around their current life situation. Clarification and expression of how they are feeling in the present are crucial so that they can live their life to their fullest in the time remaining to them. In order to accomplish this, a supportive-expressive approach to therapy is recommended as the primary psychosocial intervention for patients with terminal illness. This should include learning specific skills to help cope with the intense feelings associated with facing one's illness as well as means of handling pain and stress involved with the illness and its treatments.

Supportive-Expressive Group Therapy for Women with Metastatic Breast Cancer: A Model for Group Therapy to Reduce Long-Term Illness Morbidity and Mortality

The following is a description of a specific protocol that has been found to be extremely effective in working with patients who have a poor medical prognosis. However, it is an excellent approach to utilize as part of the overall treatment for groups who are moderately ill as well.

In general, less severe illnesses that have a strong behavioral component may benefit from a more direct cognitive restructuring approach, examining the behaviors and personality patterns that interfere with recovery (Figure 3-2). However, the more severe the illness, the more important it will be to offer support for emotional expression and strategies for coping with the problems of adjusting to the illness. This model has been successfully applied to women with metastatic breast cancer (Spiegel et al., 1989), and is also being applied to a variety of groups with different forms of cancer, AIDS, and other serious illnesses. Typically, these groups are led by cotherapists and meet weekly for at least 1 year, with each meeting lasting 90 minutes. Family and friends of the patients are also encouraged to attend separate meetings.

Goals of Therapy

Supportive-expressive group therapy offers a safe and supportive environment in which to (1) admit distressing feelings and thoughts to oneself, (2) express their feelings and thoughts to others, (3) support others by responding to their thoughts and feelings, and (4) examine issues that would otherwise be difficult to examine. Supportive-expressive therapy does not attempt to alter basic personality characteristics or to educate patients so that they can change their lives in the future. Rather, the emphasis is on how to live most fully in the present and the near future, and on personal expressions and relationships. This approach to therapy is ideally suited for those who are seriously ill and may not have a distant future. Thus, this style of therapy helps patients to feel less isolated, enhance expression of feelings and thoughts, and make changes that will affect their present life. In particular, this type of group would foster and encourage:

- Increased openness to feelings and thoughts, wherein expression of affect is encouraged and nothing is beyond discussion.
- An atmosphere of acceptance and a suspension of direct confrontation.
- Building of a cohesive, respectful group wherein every member can contribute to the extent he likes.
- Emphasis on the present time and the near future, versus past problems.
- Developing relationships outside of the groups instead of forbidding contact outside the group.
- Dealing with existential issues of death and dying.
- Coping with the illness rather than attempting to correct long-standing problems or change personality.
- Consideration of personality issues to the extent that (1) they interfere with group process; (2) they interfere with medical treatment (e.g., working with the physician); and (3) dealing with them helps to address unfinished business that is of crucial concern to the patient.
- Free expression of affect, both negative and positive, mild and intense.

Content of Therapy

Rather than guide the conversations, therapists allow patients to discuss whatever is of most concern to them. The meetings begin by "taking stock" of who is present and who is absent. In this way, all members feel they are valued. There may be one dominant topic over several weeks, especially at the death of a group member, or there may be several different topics raised in any meeting. The following is a list of the major topics that have been most frequently discussed in several groups of women with recurrent breast cancer (Spiegel & Glafkides, 1983; Spira, 1991).

- Small Talk—especially at the beginning and end of a meeting

- Information Gathering and Sharing—who is present or absent
- Medical Status and Treatment—new treatments or diagnoses patients have received
- Doctor-Patient Relationship—difficulties and suggestions for improved communication
- Family and Social Network—difficulties in communicating with family and friends
- Therapy Group Issues—problems or aspects of the group that need attention
- Illness-Related Coping Skills—strategies to improve coping with various problems
- Illness, Dying, and Death-Related Issues and Feelings—emotions surrounding the illness
- Life Values and Perspectives—current and new values that are developed by patients
- Self-Image—negative and positive changes that occur as the illness progresses

Medical status and treatment, doctor-patient relationships, economic hardships, and the like are discussed more often at the beginning of therapy, since patients have these topics in common and there is relatively little affect involved. As therapy progresses, patients will begin discussing increasingly personal issues associated with negative affect. Topics usually center around one or two main issues in any given meeting, yet it is common that, following an emotionally intense period, the discussions will turn to more informational and less affective topics. Here, the therapist can gently guide the group back to the underlying issues at hand. This is especially important following the hospitalization or death of a group member. Such a turn of events raises personal fears in each group member. Some members may attempt to distance themselves from the event by focusing on how they are different from that group member, whereas other members may become depressed by focusing on how they are similar. How the group deals with the crisis in therapy will to a large extent determine how well they are able to handle their own future crises. Being able to address these underlying issues openly and honestly in the groups helps ensure that patients will be able to confront their own issues with more clarity in the future.

Yet beyond the topics that are discussed in the meetings, there is an underlying sense of bonding with others in the group. Patients frequently report that the group becomes a central part of their lives. For example, they may schedule treatments to occur just after a group meeting, so that they can discuss their fears and receive support from other members before treatment. The results of the procedure will then be the first topic discussed at the next meeting. Typically, patients say that "Six months ago, I would have been terrified by the prospect of the procedure. Now I am concerned, but able to handle it better. I can even sleep the night before and relax much more during the procedure." When a patient becomes too ill to attend a group, others call to find out how the patient is doing. Ill patients are visited in their homes or hospital. Funerals are also attended by many group members. Bonding among group members occurs rapidly, since they share not only a common disease and associated problems but also a process of sharing intimate feelings.

Therapeutic Facilitation

There are several aspects of supportive-expressive therapy that must be specifically addressed in groups for the medically ill. These fall under the headings of social support, psychotherapeutic interventions, existential considerations, family communications, and specific coping skills.

Social Support. As discussed above, groups can provide a powerful counterweight to the social isolation often experienced by patients. The very condition that separates them from others is the ticket of admission to the group. The expression of deep fears and concerns not only deprives them of some of their dread but normalizes them. The loss of friends is seen as a consequence of the illness, not some personal failure, when the same has happened to numerous other members of the group. However, losses must be taken seriously. Failure to grieve a dying or recently deceased member will arouse considerable anxiety in the group, since it conveys to others the message that they, too, will slip away unnoticed. Yet expression, while important, is just one aspect of social support. It is equally important for group members to be able to offer support as well. Being able to help another person gives patients a sense of importance and value, and reaching out to another helps to end feelings of isolation. After a while, members of these groups come to feel like experts in living. The very tragedy of the illness and their coping with it now gives them something they can offer others. This hard-won expertise becomes an asset to others, and they can observe its effect. Members in a cancer support group remind one another: "I thought of what you said all week long—I didn't feel so alone." This has been referred to as the "helper-therapy principle" (Spiegel, 1981). A genuine increase in self-esteem comes from tangibly helping someone else in the same position.

Psychotherapeutic Interventions. Leading groups with the medically ill involves helping patients with both problem-focused and emotion-focused coping (Moos & Moos, 1988). The threat of separation from loved ones is often reinforced by social isolation. Fear of illness and the confrontation with personal vulnerability that contact with someone who is ill represents often lead to avoidance. This makes patients feel not only isolated but already a bit dead, removed from the flow of life. Practical problems encountered include financial worries, difficulties in communicating with physicians, knowing what to tell children about the illness, and physical symptoms such as pain and energy loss.

The therapist attempts to lead the group away from small talk or technical information about cancer or cancer treatments to more personal emotional issues. The therapists can be most beneficial when they engage the group members to be more active in this process, rather than overtly taking control of the group and guiding them directly. Techniques include asking open-ended questions of the group or a group member and referring a personal statement made by one group member to the other group members. This approach helps the group to gain enhanced control over their own lives in the midst of otherwise uncontrollable illness.

Therapists assist patients in discussing these specific issues by guiding patients from expressions that are external and abstract ("No one understands how frightening it is to have cancer") to statements that are personal and concrete ("I wake up at three in the morning, feeling like I have an elephant on my chest, I am so scared"). Therapists also encourage ventilation of the strong emotions associated with these thoughts. In addition, although personal expression is beneficial for patients, therapists encourage patients to respond directly to others' needs. Patients also come to actively reflect on the process of the group meetings and to choose the direction they wish to pursue. In fact, the therapist can facilitate this by occasionally asking the group to reflect on what they have accomplished in the meetings up to this point and, looking back from the perspective of a year from now, what they would wish they would have been able to have achieved. By taking charge of the direction of the meetings, the therapist promotes in patients a more active participation in their health care and greater decision making in their lives.

Therapists employ a great deal of verbal and nonverbal rapport, showing an understanding of the emotional and situational state of the patients. From this basis, patients are gently encouraged to further explore and express their feelings, while the therapist respects the fact that each patient will have a different therapeutic readiness or threshold. Proceeding from this basis, independent of what topic is discussed, therapists encourage:

1. Patient autonomy, choice, and control.
2. Clarification of problems, ideal and attainable goals worth pursuing, and personal and external resources that can be identified to help attain these goals.
3. Continuity of meetings through encouraging constant attendance, checking in with each member at each meeting, and continuing themes from previous meetings.
4. Development of group cohesion through personal modeling of supportive dialogue and encouraging widespread involvement and interaction among members. Cohesion can be further developed by: (1) giving support to a patient, (2) commenting positively on support given by others, and (3) realizing that the therapist need not solve *every* problem (by first giving the patient and then the group every opportunity to solve problems).

The following example can serve to illustrate how this approach to therapy for the medically ill differs from the usual approach to group psychotherapy for those with primarily psychologic dysfunction. If a patient in typical group psychotherapy states that she feels isolated from her family and friends, a common response would be to discuss how she is similarly splitting herself off from the group, and how this pattern plays out in various aspects of her life. By becoming conscious of this unconscious pattern, the patient might be able to recognize this tendency in her behavior outside as well as inside the group, and begin to act differently. In supportive-expressive therapy for the severely ill, patients may not have years ahead of them in which to implement such restructuring. Instead, when a patient makes a statement that she feels isolated from friends or family, the therapist can redirect her *experientially* to others in the group: "Lisa, do you feel comforted by anyone in the group?", or "Lisa, are you able to connect with anyone here, now?", or "Mary, how do you feel about Lisa feeling isolated? You were just saying how much you enjoyed her in the group." Thus, rather than abstracting the concept of Lisa's isolation for reflection in order to facilitate long-term change, the group offers an opportunity to help ameliorate such feelings of isolation immediately. If this allows for longer term changes, that is a welcome additional benefit.

Existential Considerations. Emotional issues ventilated in such groups include fears of dying and death, growing disability, and loss of highly valued social roles, physical abilities, and plans for the future. People seem to have inherent mechanisms to avoid issues of death and dying. Yet when one receives a diagnosis such as recurrent cancer, coupled with the changes experienced in one's body and self-image, the fact of one's vulnerability becomes difficult to avoid. Faced with the imminent loss of one's stability and possibly one's life, a constant sense of dread is not uncommon. Paradoxically, directly confronting such losses can make fears, even of dying, more bearable by giving them definable dimensions. Dying can be treated as a series of problems: loss of physical control, management of pain, the need for a living will, and so forth, each of which can be addressed. One metastatic breast cancer patient described her experience in the group as follows:

> What I felt being in the group was like looking down into the Grand Canyon—I don't like heights. I knew that falling down would be the end, but I felt better about myself because I could look. I can't say I feel serene about it, but I can look. (Spiegel, 1985)

Thus, by *detoxifying dying*, patients are able to begin to appreciate the life they have, able to live each moment more fully as if it may be their last—indeed, for some it may be. The significance of various social relationships and commitments on a patient's time and energy begin to be reassessed, with the patient letting go of those aspects of life that seem less important and giving more of herself to those aspects of life that are seen to be most meaningful for the future remaining to her.

There are some exercises that can assist in confronting such existential issues (Bugental, 1973-4, 1978, 1988). When patients in the group raise the issue of the confusion and stress involved with the inevitable change in self-image, or when the therapist feels that the group could use some gentle assistance in that direction, taking time to explicitly examine self-images and values can be beneficial. For example, group members can write out a list of the 10 most central characteristics that define themselves. These could be things patients do or have, people they are involved with, body image, or qualities they have. Attributes should be as specific and concrete as possible, yet can be positive or negative. When this is finished, the list can then be ranked from most central to least central defining attributes. Then, beginning with the least centrally defining item, patients can take a minute to cross off that item from the list, and consider what their lives would be like without that aspect of themselves. Continuing until each item is crossed off the list, it is useful to finally consider what remains, if anything. For some patients, what remains is their breath, blood, bones, or warm body sensation. For others, what remains are the sounds or light around them. For still others, the group remains. At that point, patients can consider several qualities they would *like* to have and activities they would like to do in the next year.

Such exercises can help patients better confront and accept the changes that they are undergoing. Patients can better appreciate the habitual ways they have been viewing themselves, acknowledge a more central sense of self as more superficial attributes begin to change, and commit to activities that can offer renewed value in their lives. While these type of exercises can be of assistance in facilitating the discussion in the group and self-exploration in each member, they should be used judiciously. Directed activities can be overly confrontive for those not yet ready to address the issues, or they could appear like a game, distracting patients from important interpersonal issues at hand. Used with sensitivity, they can stimulate continued exploration and increased authenticity.

Family Communication. An important function of group therapy is the discussion of family problems and hearing how others cope with family difficulties. Positive communication and a feeling of emotional support in one's family is important for health and well-being (Spiegel, Bloom, & Gottheil, 1983). It is thus useful for families of patients to meet with one another. Following much the same approach to therapy as in the patient groups, family meetings can occur as infrequently as once a month or as frequently as the patient meetings (once a week). Meetings with families and friends of the patients without the patients being present allows those close to the patient to express their own problems, which are often as numerous as those of the patient. If those close to the patient can begin to confront their own anxieties about taking care of the patient and potentially losing the patient, channels of communication between patient and family that were previously closed or clouded will begin to open and clarify. The caretaker also needs to be cared for as well, and family meetings offer the opportunity for this to occur. Moreover, this is not only good therapy for the patients, in that it helps to improve their family environment, but is also good preventive medicine for the caretakers, who are at increased risk of morbidity and mortality in the years following the death of a loved one (Rynerson, 1986, 1990).

Family meetings for the terminally ill help the patient and family by promoting a sense of emotional support and open communication, rather than attempting to fundamentally restructure family dynamics. In fact, such restructuring sometimes can be more disruptive than helpful to the patient (see Chapter 2).

Coping Strategies. Groups also are excellent vehicles for teaching specific problem-solving strategies. Patients can learn from fellow group members alternative ways of handling a variety of stresses in their lives. Sometimes patients get specific ideas for improving their communication with their physician or for handling side-effects of medication. Other times, merely seeing that others have gotten through a difficult situation raises patients' confidence that they, too, can go through it.

Meditation, relaxation, and hypnosis are helpful to patients experiencing extreme arousal. These techniques initially may offer some temporary respite from a sense of being overwhelmed by uncontrollable negative emotions. Such techniques help patients cope with specific symptoms such as pain and nausea, or with anxiety over upcoming events such as surgery or tests. These techniques can be taught in a group format and are quickly and easily learned by most patients.

The best *stress reduction exercises* are those that can be practiced sitting in a chair, lying down in bed, or on a gurney awaiting a procedure. Patients are asked to notice the breath flowing naturally and effortlessly into and out of different parts of their body. Patients can learn to "relax with the breath out for a moment at the bottom of the exhale, before the next natural inhale flows in." Such exercises have the potential to immediately reduce sympathetic arousal. Exercises can employ imagery to recall a past or fantasized moment of comfort and safety. Relaxation techniques that use movement (progressively tensing and relaxing of muscles, raising and dropping fingers with each breath, etc.) are an alternative for patients who find it difficult to be still.

It is possible for patients to reduce the reactive component to pain. Since patients experience different levels of pain at any given meeting, patients should be offered a range of pain control *techniques*. As a general rule, overwhelming pain is best dealt with by distraction, helping patients to focus on something outside their body or on a different sensation in their body. For pain that is difficult to deal with but not overwhelming, patients can be instructed to focus on a pleasant feeling in their body (or a pleasant image or sound) and allow this pleasant sensation to merge with the unpleasant sensation of pain. For pain that is bothersome but tolerable, it is useful to go directly to the sensation, allowing the breath (or some pleasant visual image or sound) to go to the center of the sensation, helping it to become more like the breath (or image or sound). Exercises that allow patients to quickly and easily manage stress and pain offer tangible benefits that help to increase their confidence in the benefit of the group, increasing enthusiasm for pursuing other group concerns.

Self-hypnosis can be easily taught to a group of patients. It has been found to be extremely effective in reducing pain, relaxing somatic arousal, and reducing mood disturbance (Spira & Spiegel, 1992). The basic principles of medical hypnosis are discussed elsewhere (Wain, in press).

Cautions and Considerations

Utilizing Coping Skills in Groups. Despite the value of the coping skills described above, their appropriateness for different types of groups must be considered. Experiential self-help coping skills may be the primary focus of short-term groups that meet for members to learn how to cope with a temporary or specific problem. Examples of this could include an educational group led by social workers in a surgery unit to teach patients better ways to handle the period immediately before and after the surgery, or to assist groups of patients experiencing anticipatory nausea before oncologic treatment. Short-term groups with this sort of primary focus also can assist patients with specific somatic complaints related to stress. Such groups can also serve as a useful adjunct to other types of group therapy for the medically ill. Supportive-expressive principles are also important to consider, because positive interactions among patients with somewhat similar complaints can be facilitated in the ways described above.

However, when long-term groups of patients with serious life-threatening illnesses are conducted, then overuse of these techniques can detract from the interpersonal issues at hand. Patients will have a tendency to want to discuss their use of the techniques rather than the underlying emotional issues. Such exercises can also be inappropriately used as a way of "rescuing" patients who are experiencing negative emotions. Attempting to buffer patients against these difficult emotions can also unfortunately deprive patients of the benefits of coming to terms with their feelings. More usefully, such exercises can be a means to further explore the feelings that arise or exist just below the surface. Used at the right time, they may allow negative emotions to surface in a way that enables the patients to better explore and handle what arises. Conducting such exercises at the end of a group meeting circumvents many of the problems while preserving the benefits.

Additional Cautions for Group Therapy. Any method that has the power to assist also has the potential to disrupt. There are instances of studies of psychotherapy groups for the medically ill that had to be abandoned in midtreatment because the treatment subjects were faring worse than the untreated controls. In one instance of patients with HIV disease, unstructured support meetings (at which the therapist was mostly a coordinator and observer) were held to discuss issues of mutual concern. After several months, the level of anxiety was so high in the groups that they were considered detrimental to the patients and stopped. This example of unstructured meetings of patients with severe illness offers a clear contrast to therapist-led groups, in which an explicit method of confronting difficult issues and techniques to better handle them are offered to patients.

Another example of an unsuccessful group intervention was one that ceased because group therapy patients (with cancer) were progressing in disease more rapidly than closely matched controls. Therapists in the experimental group used a cognitive restructuring approach that directly confronted patients on issues of denial and repression, and that encouraged interpersonal assertiveness. Apparently, this confrontational approach was overly challenging, arousing patients without the support needed to adequately handle the fears and worries that arose.

The two studies just described were abandoned because of therapist concerns for the patients and remain unpublished. However, they serve as excellent examples of extremes that should be avoided for the medically ill. In contrast to unstructured or overly confrontive groups is the supportive-expressive approach coupled with training in coping strategies to handle excess arousal. It is a safe and effective way of working with the great majority of persons experiencing emotional distress as a result of medical illness.

There are several researchers who have investigated the initial phase of acute crises in a wide range of patients, including initial diagnosis of HIV-related disease (Folkman, Cohen, & Coates, manuscript in preparation), newly diagnosed cancer (Levy et al., 1987; Temoshok, 1987), and rape (Miller, Brody, & Summerton, 1988). These investigators reported that, on a variety of physical and psychologic measures, initial repression of feelings appears to *help* patients while later repression hurts. This may very well be true where sufficient psychosocial support is not present to buffer the initial trauma of the illness. However, if psychosocial support is immediately offered to patients, the initial phase of repression will be shorter lasting and less disruptive.

Generalization of This Method to Other Forms of Illness. The issues facing members of a recurrent cancer group have much in common with those who have other life-threatening illnesses, and also much in common with patients facing less severe prognoses. For instance, 50% of those with first-occurrence breast or colon cancers have recurrences, following which the average survival time is about 2 years. Patients in groups with ambiguous prognoses will naturally hold out hope that they will be among the fortunate percentage that do not have recurrences, and this may lead them to evade important issues that those with poor prognoses cannot easily avoid. Yet, as is evident from research examining psychosocial correlates of disease progression, those who do not consider such issues openly, affectively, and interpersonally may be at increased risk. Similarly, groups of patients who have undergone a severe trauma must consider issues of self-image, self-worth, and interpersonal relationships. Comparable issues are found among patients whose behavioral patterns are associated with disease incidence and progression. Because of the similarities of concerns for all patients with medical illnesses, supportive-expressive group therapy can be of significant value to the entire range of medically ill patients. Of course, just as any good individual therapist needs to adjust to meet the specific needs and context of the patients, so too must the group therapist adjust the protocol to meet the needs of the specific group.

Just as the issues considered in supportive-expressive group therapy are of universal concern in any therapy group, the ways that patients react to medical illness have much in common. Every person must subtly and gradually adjust in order to interact successfully with a continually changing environment. When medical illness occurs, such subtlety is no longer possible. Modern industrial societies have a way of bracketing illness and death away from the normal experience of most persons. For many, some of the most central issues of our lives, such as self-image, life goals, and the value of our lives in light of our impending death, can be postponed indefinitely. It often takes a crisis before we are able to rally all our resources to deal with out lives as authentically as possible. Yet even in times of crisis, when such matters are bound to arise, they can arouse such distress that related thoughts are avoided and concomitant feelings repressed. Therefore, serious medical illness befalling oneself or a loved one has the *potential* to mobilize a person to consider these issues immediately and fully, if the person is given the *opportunity* to do so. By providing a safe and supportive environment, group psychotherapy can enable patients to confront these issues, allowing them to live more fully in the face of illness.

Establishing and Implementing Group Psychotherapy for the Medically Ill

As evidence mounts in the scientific literature as to the efficacy of group therapy for the medically ill, health professionals are increasingly interested in providing this valuable service for their patients. As news of these scientific findings filter into the media, patients themselves are becoming interested in joining such groups. Careful consideration of the many factors involved in organizing a group will help ensure a therapy that will prove most effective for the patients. Some of these factors are outlined in Table 3-5.

Whether groups are held in large or small institutions or in private practice may determine the size of the population from which they are drawn, therefore constraining the type of group recruited. Groups of 8 to 12 patients work well, yet finding a group of this size that is entirely homogeneous is not always an easy task. Ideally, groups should be recruited that are fairly homogeneous in terms of disease type, severity, and patient distress. When extreme disparities exist in disease severity, patients usually defer their own concerns to those with more "pressing" problems. When the groups are heterogeneous with regard to disease type, splits often emerge between group members. In addition, the therapeutic issues often vary considerably from one disease type to another, even though the prognosis may be similar. When homogeneity across all of these factors is impractical because of the population available, care must be given in deciding what type of heterogeneity is acceptable.

Most groups meet weekly, with each meeting lasting 60 to 120 minutes. However, groups also can begin with a weekend or week-long retreat and then continue weekly for some period. Some groups have family members meeting separately once per week or month to facilitate patient support and change. These meetings either can be for family members alone or can be attended by patients. It is important to appreciate that family members have their own special concerns and problems, and that by dealing with their concerns they can improve family support for the patient.

In planning for a new group, training of therapists must not be overlooked. Co-therapists are useful for several reasons. First, it is difficult to find one therapist who has sufficient experience in group dynamics, psychotherapeutic methods, medical illness, and specific coping strategies, all of which are useful in group therapy for the medically ill. Cotherapists usually can be found who share all the requisite skills between them. In addition, cotherapists can notice persons or problems that escape notice by a single therapist. Specialized training for therapists is also helpful. Observing groups in progress, coleading groups with an experienced therapist, and observing videotaped sessions all are useful to the aspiring group therapist. Treatment manuals have been written to assist with training (Spira, 1991).

Payment issues are also a common problem. Since group therapy is usually not considered part of a standard treatment regimen for most medical illnesses, most group therapy is billed under mental health codes, and often is poorly reimbursed. Fortunately, group therapy sessions are less costly than other medical procedures or individual psychotherapy sessions. Often those whose benefits run out can be allowed to continue in the groups free of charge, or for reduced fees, since overhead is covered by other patients. Attending the group without paying is a different issue than for the usual psychotherapy groups, where payment is often synonymous with commitment. Patients with medical illnesses are among the most seriously committed patients one ever sees. When confronted with loss of health or life, defensive resistance is at a minimum and patients are eager to change and grow.

TABLE 3–5. *Steps in Determining the Type and Structure of Groups*

1. Determine patient population
type of illness (cancer; gynecologic cancer; ovarian cancer)
disease severity (life-threatening; metastatic cancer; hospice)
level of patient distress (mildly concerned, severe distress)
sex (mixed, single sex)
Attempt to obtain greatest homogeneity of factors
Consider heterogeneity when necessary for a proper group size
2. Determine goals of therapy and primary emphasis
3. Determine structure of therapy:
Duration of therapy
Frequency of meetings
Family involvement
4. Determine billing for services
5. Train therapists
6. Recruit group
7. Evaluate group and modify for future groups

SUMMARY

Certainly some patients have the ability to handle a crisis well on their own. Some may even discover that they have grown richer in some ways from working through the tragedy of illness. Yet those who have difficulty expressing their thoughts and feelings to themselves and others, as well as those who experience anxiety as a result of their illness, can benefit from supportive psychotherapy. Individual psychotherapy offers an advantage over group therapy in that individual patterns that lead to or interfere with adjusting to the illness can be examined and potentially modified. However, many persons who become ill do not necessarily have personality disorders that, if corrected, will cure the illness. Rather, most of the medically ill are psychologically competent, yet faced with a situation that would be distressing to anyone. Group psychotherapy has the advantage of acknowledging every member's intrapersonal and interpersonal resources that can be used to cope with the stress of the illness. Universal issues that the illness brings forth can be more readily understood when discussed among those who share similar experiences. Moreover, group therapy is more affordable and therefore can be continued for a longer period for many who cannot afford long-term individual therapy. The efficacy and practical nature of group therapy for persons with medical illness augers that this treatment modality will increasingly become a fundamental adjunct to medical care for many disorders.

REFERENCES

Adler A (1949). Understanding human nature. New York: Permabooks.

Anderson K, & Masur F (1989). Psychological preparation for cardiac catheterization. Heart Lung 18:154–163.

Beck AT, Rush AJ, Shaw BF, et al. (1979). Cognitive therapy of depression. New York: Guilford Press.

Benezra EE (1990). Psychodynamic group therapy: A multiple treatment approach for private practice. Psychiatr Ann 20(17):375–378.

Berkman LF, & Syme SL (1979). Social networks, host resistance, and mortality: A nine-year follow-up study of Alameda County residents. Am J Epidemiol 109:186–204.

Brown S, & Yalom I (1977). Group therapy with alcoholics. J Stud Alcohol 38:426–456.

Bugental J (1973–4). Confronting the existential meaning of "my death" through group exercises. Interpersonal Dev 4:148–163.

Bugental J (1978). Psychotherapy and process: The fundamentals of an existential-humanistic approach. New York: Random House.

Bugental J (1988). The search for existential identity. San Francisco: Jossey-Bass.

Cain EN, Kohorn EI, Quinlan DM, et al. (1986). Psychosocial benefits of a cancer support group. Cancer 1:183–198.

Capone MA, Good RS, Westie KS, et al. (1980). Psychosocial rehabilitation of gynecologic oncology patients. Arch Phys Med Rehabil 61:128–132.

Cassileth BR, Lusk EJ, Miller DS, et al. (1985). Psychosocial correlates of survival in advanced malignant disease? N Engl J Med 312:1551–1555.

City of Berkeley Infant Home-Care Project (1980). Berkeley, CA: Department of Health and Welfare.

Coates TJ, McKusick L, Kuno R, et al. (1989). Stress reduction training changed number of sexual partners but not immune function in men with HIV. Am J Public Health 79:885–887.

Cohen M, & Wellisch D (1978). Living in limbo: Psychosocial intervention in families with a cancer patient. Am J Psychother 32:561–571.

Cohen S, Mermelstein R, Kamarck T, et al. (1985). Measuring the functional components of social support. In IG Sarason & BR Sarason (Eds.), Social support: Theory, research and application. The Hague, Netherlands: Martinus Nijhoff.

Cohen S, Sherrodd D, & Clark M (1986). Social skills and the stress-protective role of social support. J Pers Soc Psychol 50:963–973.

Corsini RJ, & Rosenberg B (1955). Mechanisms of group psychotherapy processes and dynamics. J Abnorm Soc Psychol 51:406–411.

Coyne JC (1982). Stress, coping, and illness: A transactional perspective. In T Millon, C Creen, & R Meagher (Eds.), Handbook of health care clinical psychology. New York: Plenum.

Fawzy F, Cousins N, Fawzy N, et al. (1990a). A structured psychiatric intervention for cancer patients: I. Changes over time in methods of coping and affective disturbance. Arch Gen Psychiatry 47:720–725.

Fawzy F, Kemeny M, Fawzy W, et al. (1990b). A structured psychiatric intervention for cancer patients: II. Changes over time in immunological measures. Arch Gen Psychiatry 47:729–735.

Ferlic M, Goldman A, & Kennedy BJ (1979). Group counseling in adult patients with advanced cancer. Cancer 43:760–766.

Flores PJ (1988). Group psychotherapy with addicted populations. New York: WW Norton.

Folkman S, Cohen F, & Coates T (manuscript in preparation). Coping with HIV disease. San Francisco: University of California School of Medicine.

Folkman S, & Lazarus RS (1980). An analysis of coping in a middle-aged sample. J Health Social Behavior 21:219–239.

Fontana AF, Kerns RD, Blatt SJ, et al. (1989). Cynical distrust and search for self-worth. J Psychosom Res 33:449–456.

Fox BH (1983). Current theory of psychogenic effects on cancer incidence and prognosis. J Psychosoc Oncol 1:17–31.

Fram DH (1990). Group methods in the treatment of substance abusers. Psychiatr Ann 20(17):385–388.

Friedman M, & Rosenman R (1974). Type-A behavior and your heart. New York: Alfred A. Knopf.

Gordon WA, Freidenbergs I, Diller L, et al. (1980). Efficacy of psychosocial intervention with cancer patients. J Consult Clin Psychol 48:743–759.

Greer S, & Watson M (1985). Towards a psychobiological model of cancer: Psychological considerations. Soc Sci Med 20:773–777.

Grossman HY, Brink S, & Hauser S (1982). Self-efficacy, psychosocial adjustment and metabolic control in adolescents with diabetes. Diabetes 45(suppl):64–68.

Gustafson J, & Whitman H (1978). Towards a balanced social environment on the oncology service. Soc Psychiatry 13:147–152.

Hellman CJ, Budd M, Borysenko J, et al. (1990). A study of the effectiveness of two group behavioral medicine interventions for patients with psychosomatic complaints. Behav Med 16(4):165–173.

Helz JW, & Templeton B (1990). Evidence of the role of psychosocial factors in diabetes mellitus: A review. Am J Psychiatry 147:1275–1282.

Hollon SD, & Shaw BF (1979). Group cognitive therapy for depressed patients. In AT Beck, AJ Rush, & BF Shaw, et al. (Eds.), Cognitive therapy of depression (Ch. 16, pp. 328–353). New York: Guilford Press.

Holmes R, & Masuda M (1967). Life change and illness susceptibility. In BS Dohrenwend & BP Dohrenwend (Eds.), Stressful life events: Their nature and effects. New York: John Wiley & Sons.

Holmes R, & Rahe R (1967). The social readjustment rating scale. J Psychosom Res 11:213–218.

House JS, Landis KR, & Umberson D (1988). Social relationships and health. Science 241:540–544.

Jamison RN, Burish TG, & Walston KA (1987). Psychogenic factors in predicting survival of breast cancer patients. J Clin Oncol 5:768–772.

Kanner A, Coyne J, Schaefer C, et al. (1981). Comparisons of two modes of stress measurement: Daily hassles and uplifts versus major life events. J Behav Med 4:1–39.

Kelly JA, St Lawrence JS, Hood HV, et al. (1987). Behavioral interventions to reduce AIDS risk activities. Unpublished manuscript.

Kneier AW, & Temoshok L (1984). Repressive coping reactions in patients with malignant melanoma as compared to cardiovascular disease patients. J Psychosom Res 28:145–155.

Lazarus R, & Folkman S (1987). Transactional theory and research on emotions and coping. Eur J Pers 1:141–169.

LEVINE AM, RICHARDSON JL, FARKS G, ET AL. (1987). Compliance with oral drug therapy in patients with hematologic malignancy. J Clin Oncol 5:1469–1476.

LEVY S, HERBERMAN R, LIPPMAN M, ET AL. (1987). Correlation of stress factors with sustained compression of natural killer cell activity and predicted prognosis in patients with breast cancer. J Clin Oncol 5:348–353.

MAGUIRE GP (1988). Psychological morbidity among cancer patients—who needs help? Oncology 2(6):55–64.

MARMAR CR, HOROWITZ MJ, WEISS DS, ET AL. (1988). A controlled trial of brief psychotherapy and mutual-help group treatment of conjugal bereavement. Am J Psychiatry 145:203–209.

MATANO RA, & YALOM ID (1991). Approaches to chemical dependency: Chemical dependency and interactive group therapy: A synthesis. Int J Group Psychother 41:269–293.

MCNAIR PM, LORR M, & DRAPPELMAN L (1971). POMS manual (pp. 24–25). San Diego: Education and Industrial Testing Services.

MILLER S, BRODY D, & SUMMERTON J (1988). Styles of coping with threat: Implications for health. J Pers Soc Psychol 54:345–353.

MINUCHIN S (1974). Families and family therapy. Cambridge, MA: Harvard University Press.

MOOS RH, & MOOS BS (1986). Family Environment Scale manual (2nd ed.). Palo Alto, CA: Consulting Psychologists Press.

MOOS RH, & MOOS BS (1988). Life Stressors and Social Resources Inventory: Preliminary manual. Palo Alto, CA: Social Ecology Laboratory, Stanford University and Veterans Administration Medical Center.

Multiple Risk Factor Intervention Trial Research Group (1982). Risk factor changes and mortality results. JAMA 248:1465–1477.

NAOR S, ASSAEL M, PECHT M, ET AL. (1983). Correlation between emotional reaction to loss of an unborn child and lymphocyte response to mitogenic stimulation in women. Isr J Psychiatry Relat Sci 20:231–239.

NUNES EV, FRANK KA, & KORNFELD DS (1987). Psychologic treatment for the type A behavior pattern and for coronary heart disease: A meta-analysis of the literature. Psychosom Med 48:159–173.

ORNISH D, BROWN S, SCHERWITZ L, ET AL. (1990). Can lifestyle changes influence coronary heart disease? The Lifestyle Heart Trial. Lancet 2:129–133.

PERRY S, FISHMAN B, JACOBSBERG L, ET AL. (1991). Effectiveness of psychoeducational interventions in reducing emotional distress after human immunodeficiency virus antibody testing. Arch Gen Psychiatry 48:143–147.

PLUMB M, & HOLLAND J (1981). Comparative studies of psychological function in patients with advanced cancer—II. Psychosom Med 43:243–254.

POWELL L, & THORESEN C (1987). Modifying the type A behavior pattern: A small group treatment approach. In J BLUMENTHAL & D MCKEE (Eds.), Applications in behavioral medicine and health psychology: A clinician's source book. Sarasota, FL: Professional Resource Exchange.

RABKIN JG, & STRUENING EL (1976). Life events, stress, and illness. Science 194:1013–1020.

RICHARDSON JL, SHELTON DR, KRAILO M, ET AL. (1990a). The effect of compliance with treatment on survival among patients with hematologic malignancies. J Clin Oncol 8:356–364.

RICHARDSON JL, ZARNEGAR Z, BISNO B, ET AL. (1990b). Psychosocial status at initiation of cancer treatment and survival. J Psychosom Res 34:189–201.

RODIN J (1980). Managing the stress of aging: The role of control and coping. In LEVINE S & URSIN H (Eds.), Coping and health (pp. 171–202). New York: Plenum.

RODIN J (1986). Health, control, and aging. In BALTES MM & BALTES PB (Eds.), Aging and the psychology of control. Hillsdale, NJ: Lawrence Earlbaum Associates.

RYNERSON EK (1986). Psychological effects of unnatural dying on bereavement. Psychiatr Ann 16(5):272–275.

RYNERSON EK (1990). Pathological Bereavement. Psychiatr Ann 20(6):294–334.

SCHNEIDERMAN N, ANTONI M, LAPERRIERE A, ET AL. (1990). Stress management, immunity, and HIV-1. Presented at the First International Congress of Behavioral Medicine, University of Uppsala, Sweden.

SOLOMON GF, & TEMOSHOK L (1987). A psychoneuroimmunologic perspective on AIDS research: Questions, preliminary findings, and suggestions. J Appl Soc Psychol 17:286–308.

SPIEGEL D (1981). Group counseling in cancer. In Proceedings of the National Conference on Human Values and Cancer: Psychological, social and ethical issues. New York: American Cancer Society.

SPIEGEL D (1985). The use of hypnosis in controlling cancer pain. CA 35:221–231.

SPIEGEL D, BLOOM JR, & GOTTHEIL E (1983). Family environment of patients with metastatic carcinoma. J Psychosoc Oncol 1:33–44.

SPIEGEL D, BLOOM J, KRAEMER HC, ET AL. (1989). The beneficial effect of psychosocial treatment on survival of metastatic breast cancer patients: A randomized prospective outcome study. Lancet 12:888–891.

SPIEGEL D, BLOOM J, & YALOM I (1981). Group support for patients with metastatic cancer. Arch Gen Psychiatry 38:527–533.

SPIEGEL D, & GLAFKIDES MS (1983). Effects of group confrontation with death and dying. Int J Group Psychother 33:433–447.

SPIEGEL D, & SPIRA J (1992). Hypnosis for psychiatric disorders. In D DUNNER (Ed.), Current psychiatric therapy. Philadelphia: WB Saunders.

SPIEGEL D, & YALOM ID (1978). A support group for dying patients. Int J Group Psychother 28:233–245.

SPIEGEL H, & SPIEGEL D (1985). Trance and treatment: Clinical uses of hypnosis. New York: Basic Books.

SPIRA J (1991). Treatment manual for women with recurrent breast cancer. Presented at the 12th Annual Meeting of the Society For Behavioral Medicine, Washington, DC.

SPIRA J, & SPIEGEL D (in press). Hypnosis and related techniques in cancer pain management. In D TURK & C FELDMAN (Eds.), Cancer pain control: Non-invasive approaches. New York: Haworth Press.

TAYLOR S (1984). Attributions, beliefs about control, and adjustment to breast cancer. J Pers Soc Psychol 46:489–502.

TELCH C, AGRAS WS, ROSSITER E, ET AL. (1990). Group cognitive-behavioral treatment for the nonpurging bulimic: An initial evaluation. J Consult Clin Psychol 58:629–635.

TEMOSHOK L (1987). Personality, coping style, emotion and cancer: Towards an integrative model. Cancer Surv 6:545–567.

TEMOSHOK L, HELLER B, SAGEBIEL RW, ET AL. (1985). The relationship of psychosocial factors to prognostic indicators in cutaneous malignant melanoma. J Psychosom Res 29:139–153.

THORESEN CE, FRIEDMAN M, POWELL L, ET AL. (1985). Altering the type-A behavior pattern in postinfarction patients. J Cardiopulm Rehabil 5:258–266.

THORESEN CE, & POWELL LH (1992). Type A behavior pattern: New perspectives on theory, assessment & intervention. J Consult Clin Psychology 60:595–604.

TURNS DM (1988). Psychosocial factors. In DONEGAN WL, & SPRATT JS (Eds.), Cancer of the breast (pp. 728–738). Philadelphia: WB Saunders.

WAIN HJ (in press). Hypnosis in medical disorders. In A STOUDEMIRE & BS FOGEL (Eds.), Medical psychiatric practice (Vol. 2). Washington, DC: American Psychiatric Press, Inc.

WALLSTON KA, & WALLSTON BS (1982). Who is responsible for your health?: The construct of health locus of control. In G SANDERS & J SULS (Eds.), Social psychology of health & illness (pp. 160–170). Hillsdale, NJ: Earlbaum.

WASSERMAN H, & DANFORTH H (1988). The human bond: Support groups and mutual aid. New York: Springer-Verlag.

WATSON M, GREER S, PRUYN J, ET AL. (1990). Locus of control and mental adjustment to cancer. Psychol Rep 66:39–48.

WEINBERGER DA, SCHWARTZ GE, & DAVIDSON RJ (1979). Low-anxious, high-anxious, and repressive coping styles: Psychometric patterns and behavioral and physiological responses to stress. J Abnorm Psychol 88:369–380.

WOOD PE, MILLIGAN I, CHRIST D, ET AL. (1978). Group counseling for cancer patients in a community hospital. Psychosomatics 19:555–561.

YALOM ID (1980). Existential psychotherapy. New York: Basic Books.

YALOM ID (1985). The theory and practice of group psychotherapy (3rd ed.). New York: Basic Books.

YALOM ID (1989). Love's executioner. New York: Basic Books.

YALOM VJ, & YALOM ID (1990). Brief interactive group psychotherapy. Psychiatr Ann 20(17):362–367.

ZONDERMAN AB, COSTA P, & MCCRAE RR (1989). Depression as a risk for cancer morbidity and mortality in a nationally representative sample. JAMA 262:1191–1195.

SUGGESTED READINGS

ALLEN MG (1990). Group psychotherapy—past, present, and future. Psychiatr Ann 20(17):358–361.

BANDURA A, LANCHARD EB, & RITTER B (1969). Relative efficacy of desensitization and modeling approaches for including behavioral, affective, and attitudinal changes. J Pers Soc Psychol 13:173–199.

BANDURA A, & WALTERS R (1963). Social learning and personality development. New York: Holt, Rinehart, & Winston.

BERGER MM (1990). Combined group therapy: Treating patients with both individual and group therapy. Psychiatr Ann 20(17):379–384.

DEROGATIS LR, ABELOFF MD, & MELISARTOS N (1979). Psychological coping mechanisms and survival time in metastatic breast cancer. JAMA 242:1504–1508.

DEW MA, RAGNI MV, & NIMORWICZ P (1990). Infection with human immunodeficiency virus and vulnerability to psychiatric distress. Arch Gen Psychiatry 47:737–744.

ELL K, NISHIMOTO R, MANTELL JE, ET AL. (1988). A longitudinal analysis of psychological adaptation among survivors of cancer. Fam Systems Med 6:335–348.

EPSTEIN NB, & VLOK LA (1981). Research on the results of psychotherapy: A summary of evidence. Am J Psych 138:1027–1035.

FREUD S (1921). Group psychology and the development of the ego. New York: Bantam.

GLASER R, KENNEDY S, LAFUSE W, ET AL. (1990). Psychological stress-induced modulation of interleukin 2 receptor gene expression and interleukin 2 production in peripheral blood leukocytes. Arch Gen Psychiatry 47:707–712.

GREER S, & MORRIS T (1975). Psychological attributes of women who develop breast cancer: A controlled study. J Psychosom Res 19:147–153.

GREER S, MORRIS T, & PETTINGALE KW (1979). Psychological response to breast cancer: Effect on outcome. Lancet 2:785–787.

GURMAN AJ, & KNISKERN DP (1981). Family therapy outcome research: Knowns and unknowns. In AS GURMAN & DP KNISKERN (Eds.), Handbook of family therapy. New York: Brunner/Mazel.

HISLOP GT, WAXLER NE, COLDMAN AJ, ET AL. (1987). The prognostic significance of psychosocial factors in women with breast cancer. J Chronic Dis 40:729–735.

HOROWITZ MJ, WILNER N, MARMAR C, ET AL. (1980). Pathological grief and the activation of latent self-images. Am J Psychiatry 137:1157–1162.

HOUSE JS, ROBBINS C, & METZNER HL (1982). The association of social relationships and activities with mortality: Prospective evidence from the Tecumseh Community Health Study. Am J Epidemiol 116:123–140.

IRWIN M, CALDWELL C, SMITH TL, ET AL. (1990). Major depressive disorder, alcoholism, and reduced natural killer cell cytotoxicity: Role of severity of depressive symptoms and alcohol consumption. Arch Gen Psychiatry 47:713–719.

JACOBSON AM, HAUSER ST, & WERTLIEB JI (1986). Psychological adjustment of children with recently diagnosed diabetes mellitus. Diabetes Care 9:323–329.

KOMRO K, & DESIDERATO O (1990). Stress, repressive coping style, and immune response. Presented at the 11th Annual Meeting of the Society of Behavioral Medicine, Chicago.

KRANT MJ, & JOHNSTON L (1977-8). Family members' perceptions of communications in late stage cancer. Int J Psychiatry Med 8:203–217.

MAY PR (1976). Rational treatment for an irrational disorder: What does the schizophrenic patient need? Am J Psychiatry 133:1008–1012.

MINUCHIN S, ROSMAN BL, & BAKER L (1978). Psychosomatic families. Cambridge, MA: Harvard University Press.

MORENO JL (Ed.) (1966). International handbook of group psychotherapy. New York: Philosophical Library Inc.

MORGANSTERN H, GELLERT GA, WALTER SD, ET AL. (1984). The impact of a psychosocial support program on survival with breast cancer: The importance of selection bias in program evaluation. J Chronic Dis 37:273–282.

MORRIS T, GREER S, & PETTINGALE K (1981). Patterns of expression of anger and their psychological correlates in women with breast cancer. J Psychosom Res 25:111–118.

PARKES CM (1972). The cost of commitment. In Bereavement: Studies of grief in adult life (2nd Am. Ed.). Madison, CT: International Universities Press.

PARKES CM (1985). Bereavement. Br J Psychiatry 146:11–17.

RAPHAEL B (1977). Preventive intervention with the recently bereaved. Arch Gen Psychiatry 34:1450–1454.

RAPHAEL B (1983). The anatomy of bereavement. London, Hutchinson.

REES W, & LUTKINS S (1967). Mortality of bereavement. Br Med J 4:13–16.

SCHLEIFER SV, KELLER SE, MEYERSON AT, ET AL. (1984). Lymphocyte function in major depressive disorder. Arch Gen Psychiatry 41:484–486.

SOLOMON GF (1981). Emotional and personality factors in the onset and course of autoimmune disease, particularly rheumatoid arthritis. In ADER R (Ed.), Psychoneuroimmunology. New York: Academic Press.

SOLOMON GF (1985). The emerging field of psychoneuroimmunology. Adv 2(1):6–19.

STABLER B, SURWIT RS, LANE JD, ET AL. (1987). Type A behavior pattern and blood glucose control in diabetic children. Psychosom Med 49:313–316.

SULLIVAN HS (1953). The interpersonal theory of psychiatry. New York: WW Norton.

TURNER RJ (1981). Social support as a contingency in psychological well-being. J Health Soc Behav 22:357–367.

WEINER H (1977). Psychobiology and human disease. New York: Elsevier.

II General Principles of Diagnosis and Treatment

4 Assessment of depression and grief reactions in the medically ill

STEVEN A. COHEN-COLE, M.D., FRANK W. BROWN, M.D.,
AND J. STEPHEN McDANIEL, M.D.

This chapter reviews the wide range of depressive syndromes that physicians may encounter in medical and surgical patients, suggests an approach to the diagnosis of major depression in medical contexts, and offers a conceptual approach to treatment for each of the different syndromes.

Five different depressive diagnoses are described in the DSM-III-R (American Psychiatric Association, 1987): major depression, organic mood disorder, adjustment disorder with depressed mood, dysthymic disorder, and dementia with depression. With the exception of dementia with depression, each of these conditions is described in detail. One non-DSM-III-R concept, covert ("masked") depression, is discussed separately. Grief reactions in medical patients can complicate the assessment of depressive disorders and are also reviewed in detail. Proposed changes in diagnostic terminology for DSM-IV are discussed at the end of this chapter.

TERMINOLOGY: ARE THERE "GOOD REASONS" FOR DEPRESSION?

Before each of these different syndromes is described, the use of the term "depression" must be clarified. The word can be used in one of three different ways: as a *synonym* for the affect of sadness; as a *symptom* of an illness; and as one of several disease *syndromes*. Unfortunately, practitioners rarely distinguish these various meanings when they use the term. For example, physicians often say, "Mrs. Jones is depressed, but she has good reasons to be depressed." The many layers of meaning embedded in this phrase are rarely clear. Does the statement indicate that the patient suffers from a major depression or simply from a normal and expected sadness associated with severe physical illness? Either interpretation could be acceptable. In practice, the phrase usually reflects diagnostic uncertainty between the two poles of meaning. Because the word "depression" has so many overlapping meanings, the statement "Mrs. Jones has good reasons to be depressed" can never be disproved. It presents the form of a clinical, scientific statement but, since it cannot be disproved, it is confusing, impressionistic, and nonscientific.

When the phrase is used to indicate that a patient has understandable reasons to be sad, it is tautologic because the severe illness and sadness are seen as inextricably linked by the speaker. When the statement is used, however, to indicate that a patient has "good reasons" to have a major depression, the phrase is empirically incorrect. Used in this way, it implies that everyone with such a physical condition will suffer from major depression. This is not supported by research, which demonstrates that less than half of any medically ill population, including terminal cancer patients, meet clinical criteria for major depressive illness (Bukberg, Penman, & Holland, 1984; Kathol et al., 1990; Rodin & Voshart, 1986; Stoudemire, 1985; Wells et al., 1989a).

In addition, such statements about the "understandability" of major depression are dangerous in that they tend to explain away a serious psychiatric disorder that is eminently treatable. When clinicians say that patients have "good reasons" to be depressed, they often withhold specific antidepressant therapies, believing that depressive symptoms will typically abate with stabilization or improvement in the underlying physical disorder. Such patients often do not receive independent or aggressive psychiatric treatment for their major depression. This omission is a serious error, since patients with physical illness and major depression usually respond to the standard psychiatric treatments for depression and often remain depressed if not treated adequately

(Maguire et al., 1985; Rifkin et al., 1985). Controlled outcome studies of the treatment of major depression in specific medical and neurologic populations have been reported. Table 4-1 indicates that six out of nine such studies demonstrate a benefit for standard antidepressants over placebo.

To help with this conceptual and diagnostic muddle, clinicians should avoid the use of the word "depression" as a synonym for sadness. The terms "depressed" and "depression" should be reserved for distinct clinical entities. Different terms, such as "sad" or "demoralized," should be used to describe the affect associated with physical illness. Distinct psychiatric syndromes can then be distinguished from more nonspecific affects, and appropriate treatment plans can be developed for each particular depressive syndrome.

EPIDEMIOLOGY OF MAJOR DEPRESSION IN THE MEDICALLY ILL

Table 4-2 summarizes 35 studies culled from an international literature search for the prevalence of major depression in general medical patients and in specific medical illnesses during the period from 1960 to 1990. Only studies using structured or semi-structured interviews for diagnosis were included. Overall prevalence ranged from 0% to 54%; specifically, prevalence of major depression in general medical outpatients ranged from 2.2% to 14.6% (average of 7.7%); in cancer patients, from 6% to 39% (average of 31%); in patients after myocardial infarction, from 18% to 19%; in patients with Parkinson's disease, from 21% to 37% (average of 29%); and in patients after stroke, from 30% to 50% (average of 38%).

Because of the different methodologies employed in these studies (e.g., type of interview, diagnostic criteria, population studied, sample size), it is not appropriate to consolidate their results to establish one true prevalence rate or range. The averages listed above should be used only as rough approximations to any true prevalence rate. The most reasonable conclusion from these diverse studies should be the recognition that patients with medical illnesses have increased rates of major depression when compared to community studies of nonpatients, which demonstrate prevalences ranging from 2% to 4% (Robins et al., 1984; Weissman, Myers, & Thompson, 1981). The one exception to this finding of increased depression in the medically ill (Howell et al., 1981) must be viewed with caution because of the very small sample size of 22. Furthermore, the community studies cited utilized a 6-month prevalence figure, whereas

TABLE 4–1. *Controlled Outcome Studies in the Treatment of Depression in Medical-Neurologic Populations*

Disease	Study	Duration (weeks)	Study Size	Treatment	Outcome
Diabetic neuropathy	Turkington (1980)	12	59	Imipramine/amitriptyline/diazepam	Imipramine and amitriptyline groups improved compared to diazepam group
Heart disease	Veith et al. (1982)	4	24	Imipramine/doxepin/placebo	Imipramine and doxepin groups improved compared to placebo group ($p < 0.001$)
Stroke	Lipsey et al. (1984)	6	34	Nortriptyline/placebo	Nortriptyline group improved compared to placebo group ($p < 0.006$)
Cancer	Costa, Mogos, & Toma (1985)	4	73	Mianserin/placebo	Mianserin group improved compared to placebo group ($p < 0.01$)
Medical illness	Rifkin et al. (1985)	6	42	Trimipramine/placebo	Trimipramine group improved compared to placebo group ($p < 0.06$)
Epilepsy	Robertson & Trimble (1985)	6	39	Amitriptyline/nomifensine/placebo	Significant difference not found
COPD[a]	Light et al. (1986)	6	12	Doxepin/placebo	Significant difference not found
COPD	Borson et al. (1990)	12	36	Nortriptyline/placebo	Nortriptyline group improved compared to placebo group ($p = 0.003$)
Alzheimer's disease	Reifler et al. (1989)	8	24	Imipramine/placebo	Depressed and nondepressed patients all responded to imipramine *and* to placebo; imipramine might have impaired cognitive function

[a]Chronic obstructive pulmonary disease.

TABLE 4–2. *Prevalence Rates For Depression in Medical Illness*

Setting or Disease	Study	Depression Measure[a]	Prevalence Rate (%)
Outpatient	Hoeper et al. (1979)	SADS	5.8
Outpatient	Schulberg et al. (1985)	DIS	9
Outpatient	Magni, Schifano, & de Leo (1986)	DSM-III/SSI	8
Outpatient	Barret et al. (1988)	SADS	2.2
Outpatient	Blacker & Clare (1987)	SADS	4.8
Outpatient	Feldman et al. (1987)	PSE	14.6
Outpatient	Von Kopff et al. (1987)	DIS	5
Inpatient	Koenig, Meador, & Cohen (1988)	SADS	12
Pancreatic cancer	Joffe et al. (1986)	SADS	32
Cancer	Plumb & Holland (1981)	SSI	32.5 mild 24 moderate 20 severe
Cancer	Derogatis et al. (1983)	DSM-III/SSI	6
Cancer	Bukberg, Penman, & Holland (1984)	DSM-III/SSI	18 moderate 24 severe
Cancer	Lloyd et al. (1984)	SPI	37.5[c]
Cancer	Evans et al. (1986)	DSM-III/SSI	23
Cancer	Devlen et al. (1987)	PSE	39[c]
Cancer	Hardman, Maguire, & Crowther (1989)	SPI	23
Cancer	Kathol et al. (1990)	SADS	38
Myocardial infarction	Lloyd & Cawley (1983)	SPI	19[d]
Myocardial infarction	Schleifer et al. (1989)	SADS	18[e] 15[f]
Rheumatoid arthritis	Murphy, Creed, & Jayson (1988)	PAS	12.5
Parkinson disease	Mayeux et al. (1986)	DSM-III/SSI	25
Parkinson disease	Santamaria, Tolosa, & Valles (1986)	DSM-III/SSI	32[g]
Parkinson disease	Starkstein et al. (1989)	PSE	37[h] 10[i]
Parkinson disease	Starkstein et al. (1990)	PSE	21
Multiple sclerosis	Minden, Orav, & Reich (1987)	SADS	54
Stroke	Robinson, Starr, & Price (1984)	PSE	34[j]
Stroke	Dam, Pederson, & Ahlgren (1989)	RDC/SSI	30
Stroke	Eastwood et al. (1989)	SADS	50
Stroke	Federoff et al. (1991)	PSE	22
Spinal cord injury	Howell et al. (1981)	SADS	0[k]
Alzheimer's disease	Rovner et al. (1989)	SADS	17
End-stage renal disease	Smith, Hong, & Robson (1985)	SADS[b]	5
Diabetes	Popkin (1988)	DIS	11

[a]SADS, Schedule for Affective Disorders and Schizophrenia; DIS, Diagnostic Interview Schedule; SSI, Semi-structured Interview; PSE, Present State Examination; SPI, Standardized Psychiatric Interview; PAS, Psychiatric Assessment Schedule; RDC, Research Diagnostic Criteria.
[b]Structured interview based on SADS to generate DSM-III diagnostic criteria.
[c]Major depression and/or anxiety.
[d]One year after myocardial infarction.
[e]Immediately after myocardial infarction.
[f]Three to 4 months after myocardial infarction.
[g]Major depression and dysthymia.
[h]Age at onset <55.
[i]Age at onset >55.
[j]Six-month prevalence.
[k]No major depressive episodes; 22% prevalence of minor depression; 22 patients studied.

most of the studies in the medically ill population utilized cross-sectional point prevalence. Thus, the differences found would be even greater had the studies in the medically ill used 6-month prevalence figures.

The explanation for the differences in observed prevalence rates among different illness groups are to date unexplained, but have probably been influenced by several factors, including the population studied (inpatient, outpatient, community sample, cultural orientation), type and severity of medical illness, and sample size. We believe that the most potent factor is the type and severity of medical illness. As Table 4-2 indicates, the highest prevalence rates, in general, seem to be for patients with the most serious illnesses: cancer, myocardial infarction, Parkinson's disease, and stroke.

When screening instruments such as the Beck Depression Inventory and Zung Depression Scale are utilized, prevalence rates of depressive symptoms have been shown to vary from 2% to 70% for medically ill patients (Bieliauskas & Glantz, 1989; Brown et al., 1988; Fava et al., 1982; Katon et al., 1986; Lansky et al., 1985; Lykouras et al., 1989; Massie & Holland, 1984; Schwab et al., 1967; Wade, Legh-Smith, & Hewer, 1987). When these instruments are utilized to study the epidemiology of depression in medical patients, researchers have chosen varying cutoff points to dichotomize patients into categories of "depressed" and "not depressed." Despite suggestive research demonstrating a preponderence of discriminating "psychologic" symptoms as opposed to "somatic" profiles among severely depressed medical patients (Clark, Cavenaugh, & Gibbons, 1983), very few efforts have attempted to validate results from screening instruments against the "gold standard" of a diagnostic interview.

Two important recent studies yielded somewhat divergent results concerning the validity of screening instruments. Kathol and colleagues' recent work (1990) in cancer patients demonstrated that the Beck Depression Inventory performs very poorly as a diagnostic instrument, whether or not the scale is divided into psychologic and/or somatic subscales. In contrast, the work of Parikh et al. (1988) suggested that a cutoff score of 16 on the Center for Epidemiological Studies' Depression Scale (CES-D) was found to be highly sensitive (86%) and specific (90%), with a positive predictive value of 80% for the diagnosis of poststroke major depression. Without more convincing data, however, clinical use of depression scales should be reserved for screening purposes and measurement of change, but should not be used alone for purposes of diagnosis. Thus, hundreds of studies in the literature professing to estimate the prevalence of depression in medical populations by the use of screening instruments alone must be viewed with considerable skepticism. Such studies should more correctly be interpreted as indicating levels of symptoms, which may or may not be related to the prevalence of the specific syndrome, major depression.

MAJOR DEPRESSION

Major depression is the most important depressive syndrome for the psychiatrist to recognize and treat in the medical setting because of its high prevalence and because many patients with this illness will improve significantly with appropriate biologic or psychologic therapy, or both. Unfortunately, however, the diagnosis is frequently overlooked, and many treatable patients never receive appropriate psychiatric treatment. Several studies indicate that about 30% to 50% of major depression is missed in the medical setting (Cavenaugh & Kennedy, 1986; Cohen-Cole et al., 1982; Knights & Folstein, 1977; Schulberg et al., 1985; Thompson, Stoudemire, & Mitchell, 1983; Wells et al., 1989a).

The assessment of major depression in the medical setting, however, is problematic. First, since some sad feelings are an acknowledged part of the experience of physical illness, what are the boundaries between these normal sad feelings and the syndrome of major depression? Second, many of the symptoms of physical illness (e.g., decreased energy and anorexia) are identical to symptoms of depression, and the clinician may find it difficult to determine the etiology of the symptoms encountered. These two issues are considered in detail, with reference to the research literature, in order to provide some guidelines to clinicians caring for depressed patients in the medical-surgical context.

Major depression, according to the DSM-III-R, requires a patient to demonstrate five of the nine symptoms listed in Table 4-3 nearly every day during the same 2-week period of time. These must represent a change from previous functioning. Symptoms clearly due to a physical condition should not be included. One of the five must be either a persistent depressed mood (feeling sad, blue, down in the dumps) or pervasive anhedonia (the loss of interest or pleasure in all or almost all activities).

Dysphoric Mood: Neither Necessary nor Sufficient for Diagnosis

Note that dysphoric mood (i.e., sadness), no matter how severe, is not sufficient by itself for the diagnosis of major depression. The patient must also demonstrate some of the other symptoms. Many, perhaps most, patients with severe physical illness will experience varying degrees and persistence of sadness, but the full syndrome of major depression does not necessarily follow. The presence of a dysphoric mood, therefore, may prop-

TABLE 4–3. *Symptoms of Major Depression*

A. *General symptoms*[a]
   1. Depressed mood
   2. Anhedonia (markedly diminished interest or pleasure in all, or almost all, activities)

B. *Physical symptoms*
   3. Insomnia or hypersomnia
   4. Significant weight or appetite change (increase or decrease)
   5. Psychomotor agitation or retardation
   6. Fatigue or loss of energy

C. *Psychologic symptoms*
   7. Feelings of worthlessness or excessive or inappropriate guilt
   8. Diminished ability to think or concentrate, or indecisiveness
   9. Recurrent thoughts of death or suicidal ideation

[a]At least one of these is required for the diagnosis of a major depressive episode.

erly be used as a clue for a possible major depression, but the two are not synonymous.

Besides being insufficient for the diagnosis of major depression, dysphoria is not even necessary for the diagnosis. Paradoxically, a patient need not be sad to have a major depression. According to the DSM-III-R, a patient need only present with anhedonia and four more symptoms to meet the criteria for major depression. This fact is important clinically because many patients with physical illness will not report feelings of sadness, yet they may meet these other criteria and respond well to treatment. Thus, the question of whether or not a patient's dysphoria is a "normal" reaction to the stress of the physical illness is not the key issue in diagnosing major depression. It is far more important to determine whether or not other symptoms of depression are present.

Four Approaches to the Evaluation of Depressive Symptoms in the Medically Ill

The determination of whether or not the symptoms of depression are present in a medically ill patient is exceedingly complex because some of the apparent symptoms of major depression could be accounted for by the medical illness itself. Should the fatigue, sleep impairment, psychomotor retardation, and anorexia associated with many physical illnesses be interpreted as symptoms of major depression as well? Unfortunately, the literature supplies no clear answer. The research that supported the DSM-III-R concept of major depression was conducted on patients without significant physical illnesses (J. Williams, personal communication, 1986), and there are not yet sufficient studies of medically ill patients to indicate the reliability or validity of different approaches to diagnosis. Chandler and Gerndt (1988) have suggested from their research that the majority of the increases in somatic symptoms in depressed patients may be due not to depressive psychopathology but rather to their increased number of medical problems and increased age. Wise et al. (1985-6) have suggested that the actual symptoms of a medical illness, rather than the diagnosis of a medical illness, may increase the dysphoric moods of patients.

To date, there have been four major approaches to diagnosing major depression in the physically ill. These approaches are here termed "inclusive," "etiologic," "substitutive," and "exclusive." Each is described and discussed separately.

Inclusive Approach

The inclusive approach, used by Rifkin and colleagues, is the simplest. Using the Schedule for Affective Disorders and Schizophrenia (SADS) and the Research Diagnostic Criteria (RDC), this group studied physically ill patients and counted symptoms of depression, whether or not they might be attributable to a primary physical problem (Rifkin et al., 1985). Low energy, if present, was thus considered a symptom of depression whether or not the patient or the physician believed that it was a "normal" response to the physical illness. The approach is conceptually clear and "clean." It is phenomenologic, in keeping with the spirit of DSM-III-R; that is, diagnostic decisions are made on the evaluation of observable phenomena and do not require etiologic inferences. This approach tends to maximize interrater reliability in diagnosis, but because so many symptoms of physical illness are also those of depression, it could lead to overdiagnosis by including false-positives (i.e., considering as depressed people who are only physically ill). Its sensitivity should thus be high, but its specificity relatively lower.

Two recent studies, in fact, have examined this question in detail. In cancer patients, Kathol and colleagues (1990) found that the "inclusive" approach to diagnosis led to a prevalence rate of 38% (58 of 152 patients), which fell to 30% (45 of 152) when symptoms that were considered to "have a definite relationship [to] the physical condition" were excluded. Thus, this study suggests that 8% of patients diagnosed with an inclusive approach might not be so diagnosed using another system.

In another recent report, Federoff et al. (1991) used inclusive criteria to determine that 24% of 205 stroke patients met DSM-III-R criteria for major depression. "Tightening" the criteria to discount up to two symptoms that could be attributed to strokes themselves led to a changed diagnosis in only five patients, thus decreasing the prevalence by 1.5% to 22% overall.

These two studies indicate somewhat conflicting results. The findings of Federoff and colleagues seem to indicate that an inclusive approach would not change overall prevalence rates significantly, whereas Kathol's group demonstrated that the inclusive approach yielded results somewhat different (8%) from a "tighter" approach. Further studies in more populations are needed to clarify this important issue.

Etiologic Approach

An etiologic approach, advocated by Spitzer, Williams, and Gibbon (1984) and the developers of the DSM-III-R, requires that diagnosticians count a symptom toward the diagnosis of depression only if it is not "clearly due to a physical illness." This is the decision rule for the Structured Clinical Interview for DSM-III-R (SCID) (Spitzer, Endicott, & Robins, 1978) as well as for the

Diagnostic Interview Schedule (DIS) (Robins et al., 1981). A study of depression in Parkinson disease patients utilized this approach (Mayeux et al., 1986). Unfortunately, the report did not specify how these distinctions were actually made clinically. Since this approach requires inferences of causality from unclear criteria, it seems its reliability might be quite low.

An important group of studies in stroke patients utilized a similar approach (Lipsey et al., 1984; Robinson, Lipsey, & Price, 1985; Robinson, Starr, & Price, 1984). Researchers in those studies used the Present State Examination but made dysphoria a necessary requirement for a diagnosis of major depression (anhedonia was not an acceptable alternative). Their decision rules, moreover, allowed them to count a physical symptom such as fatigue toward the diagnosis of depression only if the patient or the physician believed the symptom was in excess of what might be caused by the stroke (R.G. Robinson, personal communication, 1986). Although the authors reported high interrater reliability using this approach, it is uncertain whether other psychiatrists could make these same distinctions reliably. The authors were probably able to achieve this reliability because of their close partnership in their research endeavors and the ability to specify decision rules to guide their diagnostic thinking.

It is possible that some psychiatrists working in very specific medical settings (like Robinson and Lipsey's group with stroke patients) might become familiar enough with one specific disease process to be able to reliably differentiate between "normal" symptoms of a physical illness (e.g., fatigue) and "excessive" symptoms, which might be more properly attributed to a depressive syndrome. This determination, however, would be nearly impossible for the general psychiatrist to make in a reliable manner. In addition, the validity of the approach also needs to be established through outcome studies, to determine if major depression diagnosed in this manner has a clinical course different from the outcome for patients who have some symptoms of depression (e.g., sadness) but do not meet criteria for major depression.

Substitutive Approach

The substitutive approach suggests changing the criteria for the diagnosis of depression in the medically ill. From a factor-analytic study (latent trait analysis) using the Beck Depression Inventory in large numbers of medical patients, Cavenaugh, Clark, and Gibbons found that decreased energy was a poor discriminator of depression and that "indecisiveness" was a relatively good one (Cavenaugh, Clark, & Gibbons, 1983; Clark, Cavenaugh, & Gibbons, 1983). They suggested that psychiatrists incorporate this finding into their clinical work. Although weakened by the lack of clinical interviews and by reliance on an instrument that emphasizes cognitive symptoms, this research is promising in its attempt to reconsider the diagnostic criteria for major depression in the medically ill.

Similarly, Endicott (1984) suggested substituting other symptoms of depression when DSM criteria are problematic. For example, if it cannot be determined whether a patient's low energy is due to physical disease or depression, she suggested substituting consideration of other depressive symptoms, such as "brooding, self-pity, or pessimism." Endicott does not indicate the source of these other criteria, and it is unclear whether her suggested list would lead to over- or underdiagnosis of depression.

In the only published study to date utilizing this approach, Kathol et al. (1990) used the Endicott criteria (except for emotional reactivity) and found prevalence rates quite similar to those found using an inclusive approach (36% compared to 38%, respectively). Only 2 of 152 patients were diagnosed as depressed by the Endicott criteria who did not also meet the inclusive criteria for major depression. However, when "tighter" criteria were utilized (i.e., excluding symptoms that may be caused by cancer), 11 patients were found to be depressed by Endicott criteria who were "not depressed" by tighter criteria.

Exclusive Approach

Research on depression in cancer patients by Holland's group at the Sloan-Kettering Cancer Institute represents the exclusive approach (Bukberg, Penman, & Holland, 1984; Plumb & Holland, 1981). This group simply eliminated anorexia and fatigue from the list of nine depressive criteria, and required four of the remaining seven symptoms for a diagnosis of major depression. This, like the inclusive approach, is relatively "clean" because it specifies how to apply the criteria. It leads to the reverse problem, however, in that it is harder for patients to meet the restricted criteria. Thus, some depressed patients presenting primarily with vegetative symptoms might be missed, leading to false-negatives. Specificity is probably high at the cost of a loss of sensitivity. The exclusion of symptoms that may be caused by the physical illness led to a reduction in prevalence from 38% to 30% in Kathol et al.'s study (1990), but only a 1.5% reduction, from 24% to 22%, in Federoff et al.'s study (1991).

Clinical Evaluation

All four diagnostic approaches have specific advantages and disadvantages depending on the particular research or clinical objectives involved. These advantages and disadvantages are summarized in Table 4-4. For research purposes, the best model to follow is that of the Sloan-Kettering group, which maximizes specificity (i.e., confidence that the true disorder is really present when diagnosed). This approach ensures the most homogeneous depressed group possible, with the fewest extraneous or confounding variables, thereby increasing the clinical and statistical significance of the research data.

For the clinician, however, maximizing sensitivity (i.e., detecting all possible disorders within a population) is the most important first step in the proper evaluation of the patient. Although the clinical goal will always be to maximize sensitivity and specificity at the same time, this is not always possible. Depression is generally underdiagnosed in the medically ill (Schulberg et al., 1985), and the syndrome can be long lasting and debilitating (Maguire, 1984). Furthermore, even when depression is recognized by primary care physicians, the patients are often treated inadequately with either anxiolytic medications or subtherapeutic doses of antidepressants (Keller, Klerman, & Lavor, 1982). Because of these findings, psychiatrists should lean in the direction of ensuring that depressions are recognized and treated adequately.

For clinical practice, therefore, the best approach seems to be the *inclusive* one, which preserves the phenomenologic spirit of the DSM-III-R. Using this guideline, clinicians diagnose disorders based on the presence or absence of signs and symptoms, without reliance on inferences of intrapsychic dynamics or complex etiologic sequences. Thus, the initial approach to diagnosis should count all relevant symptoms if there is any reason to believe that they may be part of a depressive syndrome. Even if the etiology of a particular symptom is questionable (i.e., the symptom may result from both a physical and a psychiatric process), we suggest that it be counted toward those required for the diagnosis. Fatigue, anorexia, and psychomotor changes should be evaluated in the same way. It is worth noting that, even if all four physical symptoms of major depression are present in an individual with physical illness, the patient will still need to demonstrate significant and persistent depressed mood or anhedonia to meet the criteria for major depression.

This inclusive approach may lead to some false-positive diagnoses, but such errors are preferable to the alternative: lack of treatment for patients who need it. In support of this position, Cassem's review (1990) specifically endorsed the "inclusive" diagnostic approach we had proposed in an earlier publication (Cohen-Cole & Stoudemire, 1987). Data available to date indicate that the risk of overdiagnosis is small—between 1.5% and 8% (Federoff et al., 1991; Kathol et al., 1990). This seems a small risk to take compared with the risk of neglecting treatment in patients suffering from depression. Depression in the context of physical illness is associated with increased morbidity, length of hos-

TABLE 4–4. *Approaches to the Diagnosis of Depression in the Medically Ill*

Approach	Key Features	Strengths	Weaknesses	Comments
Inclusive	Count all possible symptoms, regardless of whether or not they could be caused by an underlying physical disorder	Simple, phenomenologic, high reliability, high sensitivity	Possible overinclusiveness, possible high number of false positives (low specificity); validity uncertain	Recommended for clinical use, modified case by case with clinical judgment
Etiologic (DSM-III-R)	Do not count a symptom toward the diagnosis of depression if it is "clearly due to a physical condition"	Theoretically sound	In practice, difficulty in attributing etiology of symptoms in a reliable way; validity uncertain	Theoretically sound, but practically difficult to implement
Exclusive	Exclude fatigue and anorexia from the list of depressive symptoms	Lowers the possibility of confounding variables contributing to diagnoses; higher specificity and low number of false positives	Harder to reach diagnosis; possibly excessive number of false negatives (low sensitivity); may deny treatment to those who could benefit	Good for research; too exclusive for clinical use
Substitutive	Substitute cognitive symptoms such as brooding or indecision for fatigue and anorexia	Reasonable conceptually	Unclear	Conceptually reasonable, but little experience to date

pital stay, and overall disability (Wells et al., 1989b). Furthermore, some of the false-positive patients identified by an inclusive approach will have some other milder form of depressive syndrome (e.g., minor depression) requiring attention, and possibly treatment (either psychosocial or pharmacologic), of its own. Wells and colleagues (1989b) and Broadhead and colleagues (1990) have convincingly shown that symptom complexes not meeting the threshold for major depression are still associated with significant social impairment and disability. Quitkin and colleagues (1990) have demonstrated that minor depression may still respond to antidepressant medications.

In these complex situations of mixed physical and psychiatric symptoms, we suggest that clinicians pay particular attention and lend more relative weight to psychologic symptoms and anhedonia. Most physically ill patients retain variable affective responses to their environment, some of them pleasurable. Unless they are in continuous pain or have impaired consciousness, physically ill patients usually can respond to humor, to intimacy from family or friends, or to gestures of support from the staff. That is, they will usually be able to report that they get some degree of pleasure from or have some interest in activities around them. They may also smile or laugh at jokes. The patient with major depression, however, may be unable to respond with significant interest, pleasure, or humor to any event.

Questions that can be especially helpful in this regard are: "What are you able to enjoy these days?", "What interests you now?", "When did you last have a good time?", and "Have you been able to laugh at anything lately?" Patients who report pervasive loss of interest or pleasure for at least 2 weeks should be considered likely candidates for the diagnosis of major depression. Questions regarding the presence of other psychologic symptoms of depression also can be helpful in problematic circumstances. Patients can be asked about their ability to concentrate, their sense of self-esteem or guilt, and their thoughts about death. As part of the mental status evaluation, tests of concentration and short-term memory should be utilized as well. Since major depressive illness has been associated with deficits in these areas, the clinician may find that the presence of these cognitive problems may aid in the diagnosis of major depression (Stoudemire et al., 1987).

The Multidimensional Approach to Diagnostic Assessment

The clinician applying DSM-III-R criteria to medically ill patients should remember that the criteria for major depression (as for many other psychiatric illnesses) were established in large part by consensus of expert panel members and have not yet been validated by external standards, such as natural history, family history, response to treatment, and biologic markers (Carroll, 1984). Furthermore, the criteria are cross-sectional in nature and, even under the best of circumstances, are subject to problems of reliability. Therefore, the thorough evaluation of the depressed patient requires a multidimensional approach that includes the following elements:

1. Clinical signs and symptoms (descriptive, cross-sectional)
2. Personal history of depression
3. Previous response to treatment
4. Family history of depression (genetic vulnerability)
5. Response to treatment—current episode

Possible biologic markers have not proven to be clinically useful to date. These include the dexamethasone suppression test, the thyrotropin-releasing hormone stimulation test, sleep electroencephalography (EEG), platelet tritiated imipramine binding, platelet monoamine oxidase level, and levels of cerebrospinal fluid neuropeptides and/or neurotransmitters.

The personal and family histories of the patient are of key importance. A prior personal or family history of major depression increases the likelihood of a current episode (Carroll, 1984). Previous response to treatment (of the patient or a family member) is useful information to obtain because successful past treatments can predict similar responses with the same treatments.

Treatment of Major Depression

The diagnostic distinctions in the DSM-III-R and the multidimensional factors described above are important because they have implications for treatment. A patient receiving a diagnosis of major depression usually should be treated aggressively with a balanced combination of biologic and psychologic therapies. In treatment planning, it may be helpful to point to current research suggesting that psychotherapy alone (either cognitive or interpersonal) may be as effective as antidepressant medication alone for the treatment of some ambulatory patients with major depression (Elkin et al., 1989; Rush & Jarrett, 1986). Combination treatment, however, is probably the most efficacious. These findings are of importance because some physical illnesses or their treatments present relative contraindications to the use of biologic psychiatric treatments (i.e., antidepressants or electroconvulsive therapy [ECT]). Although safe and effective somatic therapies can almost always be selected (see Chapter 9), psychotherapy of the depressed medically ill patient should be an important adjunct to the somatic treatments, and it can be the primary treatment in patients with milder forms of major depression, especially those without profound vegetative symptoms (which may be predictive of response to biologic interventions).

ORGANIC MOOD DISORDERS

The essential feature of an organic mood syndrome is a disturbance in mood, resembling a major depressive episode, that is due to a specific organic factor. The diagnosis is not made if the disturbance occurs in a clouded state of consciousness (as in delirium), if it is accompanied by a significant loss of intellectual abilities (as in dementia), or if there are persistent delusions or hallucinations (as in organic hallucinosis or organic delusional syndromes). A number of medical illnesses and medications have been linked to depressive syndromes, ranging from endocrine conditions (e.g., hypothyroidism and hypoparathyroidism) and cancer of the brain or pancreas to medications such as beta-blockers, alphamethyldopa, diazepam, chemotherapeutic agents, reserpine, and steroids (Holland, Fasanello, & Ohnuma, 1974; Maricle, Kinzie, & Lewinsohn, 1988; Patten, 1990; Stiefel, Breitbart, & Holland, 1989; Stoudemire, 1987; Weddington, 1982; Young, 1982).

Some authors have provided a virtual medical textbook of drugs and medical conditions "associated" with depression. Stoudemire (1985) cited literature revealing more than 200 such causes, but quite correctly urged caution in interpreting these catalog-style lists, in that one must understand what is meant by "depression" in each situation. For example, vague reports of low energy with a particular drug may be interpreted as a symptom of depression. Furthermore, even a group of such symptoms may not meet clinical criteria for major depression. There are no prospective controlled studies showing an association of any medication with a clinical diagnosis of major depression. One study with propranolol failed to reveal the expected association (Stoudemire et al., 1984). One case-control study, however, recently reported a higher prevalence of antidepressant use among hypertensive patients treated with beta-blockers than among patients treated with other hypertension medication (Avorn, Everitt, & Weiss, 1986). In light of these problems, the medications and illnesses mentioned in Table 4-5 are those that have been most commonly reported to be associated with depressive problems. Table 4-5 represents a selective, but not comprehensive, review.

TABLE 4–5. *Medications and Illnesses Most Commonly Associated With Depression*

Medical illnesses
- Carcinoid syndrome
- Carcinoma (pancreatic)
- Cerebrovascular disease (stroke)
- Collagen-vascular disease (systemic lupus erythematosus)
- Endocrinopathies (Cushing's syndrome, Addison's disease, hypoglycemia, hyper- and hypocalcemia, hyper- and hypothyroidism)
- Lymphoma
- Parkinson's disease
- Pernicious anemia (B_{12} deficiency)
- Viral illness (hepatitis, mononucleosis, influenza, Epstein-Barr virus)

Medications
- Antihypertensives (reserpine, methyldopa, propranolol)
- Barbiturates
- Cimetidine
- Corticosteroids
- Indomethacin
- Levodopa
- Psychostimulants (amphetamine and cocaine in the postwithdrawal phase)
- Analgesics (codeine, oxycodeine)
- Chemotherapeutic agents (vincristine, vinblastine, procarbazine, L-asparaginase, amphotericin B, interferon)
- Diazepam

Diagnosis of Organic Mood Disorder

The DSM-III-R criteria for this diagnosis are "loose," in that no length of time is required and only two of the nine associated symptoms of major depression are required. There are many problems with the DSM-III-R concept of organic mood disorder, the primary one centering on the question of causality. For example, most clinicians would agree that a depression following a leg fracture is not an organic mood disorder but that one associated with hypothyroidism is an organic mood disorder. However, what about depression in the context of Parkinson's disease or Huntington's disease? Many such illnesses are on the boundary between the physical and the psychiatric, and the etiologic issues involved are of sufficient complexity to make us seriously question the overall usefulness of the concept of organic mood disorder. In addition, the loose criteria contribute even further to the lack of usefulness of the category. Despite these problems, it is important for the clinician to always consider occult physical illnesses or medication effects that might be contributing to a patient's depressive symptoms.

Fogel (1990) has explored the issue of the distinction between major depression and organic mood disorder in detail, pointing to potentially insolvable problems with reliability and validity of the organic mood disorder diagnosis. Because the DSM-III-R diagnosis of organic mood disorder is so phenomenologically heterogeneous, Fogel recommended eliminating this category from Axis I and diagnosing major depression when a patient's phenomenologic presentation meets diagnostic criteria. Any organic factors that may contribute to the Axis I diagnosis could be denoted by an additional symbol and recorded on Axis III. This approach would decrease unreliable inferences of organic etiology on Axis I and lead to an overall greater reliability. We agree with these suggestions. Proposed terminology changes for DSM-IV are discussed in an addendum to this chapter.

Treatment of Organic Mood Disorder

The treatment of a putative organic mood syndrome should focus first on the medical treatment of the underlying physical condition or on the removal of the offending drug. At times, however, the physical illness cannot be treated or the drug cannot be removed. In such situations, if the condition meets the clinical criteria for a diagnosis of major depression, clinicians should cautiously initiate the same biologic and psychologic treatments they might use for major depression without an organic etiologic factor. In other cases, even after the illness has been successfully treated or the drug removed, the depressive syndrome remains (as if a depressive process had begun and now is maintained by its own momentum). Such cases should also be treated as major depression in their own right.

ADJUSTMENT DISORDER WITH DEPRESSED MOOD

The essential feature of an adjustment disorder is the occurrence of a maladaptive reaction to an identifiable stressor. The only criteria for the diagnosis are the presence of the maladaptive or "excessive" reaction and a dysphoric mood. An adjustment disorder, for example, might occur after a patient learned he was paralyzed from an automobile accident or a stroke. Grief and demoralization would be expected as normal, but withdrawal associated with hopeless feelings and refusal to participate in potentially effective rehabilitation efforts would be considered maladaptive.

Treatment of adjustment disorders is primarily psychologic and social (see Chapter 1). The focus should be on nonspecific support, education, and psychotherapeutic clarification of the individual's particular conflicts in the context of the physical illness. If patients develop the symptoms of major depression, they should receive biologic treatment as well.

In clinical practice, it is worth noting that many patients with minor depression who do not meet all the criteria for major depression (subthreshold depression) may have significant disability (Broadhead et al., 1990; Wells et al., 1989b) and also may respond to antidepressant medication (Quitkin et al., 1990). In our clinical experience in the medically ill, this seems to be particularly true for patients experiencing sleep disruption or pain associated with their physical problems. A trial of a sedating antidepressant medication (such as nortriptyline or trazodone) is often very helpful not only in achieving better sleep, but also in improving general mood. Benefit often occurs at relatively low doses, such as 25 to 50 mg of a tricyclic every night at bedtime. For most patients, these antidepressants are more effective and safer than hypnotic medications. Such low doses of antidepressants, however, are usually inadequate for alleviating a more serious major depression.

DYSTHYMIC DISORDER AND DEPRESSIVE PERSONALITY

Patients with dysthymic disorders have experienced depressive symptoms most of the time during the preceding 2 years. Many such patients become even more dysphoric when they experience physical illnesses. If the symptoms meet the criteria for major depression, these patients should be treated just as any other patient with major depression, with a combination of biologic and psychologic therapies. When they do not meet the criteria for a major depressive diagnosis, the treatment focus rests, as in the adjustment disorders, on psychosocial interventions.

Akiskal et al. (1980) have studied dysthymic patients and postulated that two distinct subgroups make up this population: a "subaffective" dysthymic group and a group with "character spectrum" disorder. The so-called subaffective dysthymic patients were distinguished by their sleep EEGs, their personal and family histories, and their good response to medication. This distinction is difficult to make cross-sectionally on clinical criteria, but it points to the possibility of antidepressant efficacy in groups other than those with major depression. When psychotherapeutic approaches are making little headway in dysthymic patients, we suggest an empirical trial of antidepressants. In general, less sedating agents tend to be better tolerated by dysthymic patients. Chapter 9 reviews the options for antidepressant drug therapy.

Many patients with dysthymic disorder will eventually develop a major depressive disorder (Keller, Klerman, & Lavor, 1982). This condition has been termed "double depression," that is, major depression superimposed on a dysthymic disorder. Such a condition should be treated like any other major depression.

COVERT, MASKED, OR SOMATIZED DEPRESSION

Many depressed patients present to their physicians with primary complaints related to bodily symptoms—for example, pain, fatigue, or "spells." It is likely, in fact, that the presentations of depression in primary care settings are routinely masked by other, more prominent complaints. Chronic abdominal pain and headaches are extremely common manifestations of depression, as is diarrhea, insomnia, or general nervousness. Primary care physicians unaware of these common depressive equivalents will fail to recognize and properly treat a serious psychiatric disturbance.

Such patients often vigorously deny being depressed and adamantly refuse to see psychiatrists. If they do visit a psychiatrist's office after strong encouragement from a primary physician, they often are angry because they feel their primary physician does not believe they are suffering physically, and they feel humiliated because their physician believes there is something wrong with their minds.

Such patients can present diagnostic and psychotherapeutic challenges. Successful management usually begins with the communication of an empathic understanding of the reasons for the patient's anger, frustration, and disillusionment with the medical profession. Examples of useful intervention are statements like "I can understand why you are so upset about the referral to a psychiatrist. You're frustrated that your doctor can't find any explanation for your pain and you think the doctor doesn't believe you really hurt. And to make matters worse, you're insulted that your doctor has referred you to a psychiatrist."

Once the anger and frustration have been acknowledged, an explanation can be given of how the psychiatrist might be able to help. An open-minded biopsychosocial approach, emphasizing some of the uncertainties of medical practice, usually is effective. The psychiatrist can point out that many physical problems are never fully resolved, but that the patient has been thoroughly evaluated medically and that no ominous illnesses have been found. The discomfort, distress, and disability may be profound, however. The psychiatrist can point out that many patients with physical illnesses, such as diabetes and cardiac disease, get depressed or have difficulty coping. Psychiatric intervention often is helpful to them. Similarly, psychiatric evaluation may be of assistance with this patient's particular problem.

The most important differential diagnosis in such cases is between masked depression and somatoform disorders. There is a great deal of overlapping symptomatology in the two disorders. Many depressed patients have a wide variety of unexplained physical complaints, and many patients with chronic somatoform disorders (e.g., somatization disorder) have episodes of major depression. Recent studies have shown that 55% percent of primary care patients with somatization disorder have concurrent major depression as assessed by the DIS (Brown, Golding, & Smith, 1990).

Masked depression is not described in the DSM but awareness of the problem can often lead to accurate diagnosis and successful treatment. Even if patients deny depressed mood, they may admit to adhedonia. They may say they are unable to experience pleasure because of the physical problem. Four other symptoms of depression often are present. In such cases, the diagnosis can be made relatively easily, and the response to treatment usually is gratifying.

Such patients often respond best initially to a rather biologic medical model of depression. They prefer to think of the illness as a biologic one, with chemical "imbalances" that are corrected with psychotropic drugs, the way insulin helps a diabetic. This rather oversimplified model can be endorsed temporarily to develop a therapeutic alliance with psychologically resistant or unsophisticated patients, but eventual expansion to a biopsychosocial approach usually is possible and is therapeutically useful. Diabetes is a good analogy because the psychiatrist can discuss the relevance of psychosocial issues to diabetes management, making the point that psychosocial issues can contribute to depressive problems in diabetes and psychosocial intervention is an important adjunct to the treatment of depression in such patients.

BEREAVEMENT, GRIEF, AND DEPRESSION IN MEDICAL PATIENTS

Bereavement is the reaction to the loss of a loved person by death (Clayton, 1990). It is a ubiquitous human experience that most individuals face several times in their lives. Other terms sometimes used interchangeably are grief (the emotional and psychologic reaction to any loss not limited to death) and mourning (the social expression of bereavement, often formalized by cultural practices). Many patients with medical illnesses suffer grief reactions related to losses they experience secondary to their illness: loss of job, function, social relationships, finances, self-esteem, and the like. Estimates of 1-year incidence rates of bereavement in the general population range from 5% to 9% (Frost & Clayton, 1977; Imboden, Canter, & Cluff, 1963). When medical patients are bereaved or suffering from grief related to their illnesses, the question of diagnosis and treatment of a superimposed major depression can become problematic. This final section of the chapter reviews the current state of knowledge of this complex topic.

Phases of Bereavement

In order to determine the similarities and differences between major depression and bereavement, the phases of bereavement must be understood. A number of investigators have outlined various stages of bereavement (Brown & Stoudemire, 1983; Clayton, Desmarais, & Winokur, 1968; Parkes, 1970) and, although their terminologies differ, their general outline of emotional reactions is very similar. Clayton et al. referred to the first stage as "numbness"—a period of disbelief that can last a few hours, a few days, or a few weeks. During this phase, the bereaved person is generally dazed and functions automatically. Most of what is said and done is poorly remembered, and anxiety symptoms may be prominent.

The numbness phase is followed by a second stage that Clayton et al. referred to as "depression." This phase has been further divided by Parkes (1970) into two parts: (1) yearning and protest, and (2) disorganization. The depressive phase lasts from a few weeks after bereavement up to 1 year. Many depressive symptoms are commonly present; however, irritability and restlessness are the most prominent ones. Hostility is common, and is almost always directed at the deceased (Van Erdewegh & Clayton, 1988). Many people begin to recover from this second phase of depression by 4 to 6 months, although some continue to display some symptoms at 1 year.

The third phase, referred to by Clayton et al. as "recovery," is acceptance of the death of the loved one with return to the level of functioning established before the death. This may manifest itself as a willingness to look for new relationships and new roles. Some persons may react by merely strengthening a role that they filled prior to the death. Generally, there is cultivation of memories of the deceased. Some degree of emotional involvement with the deceased is common and should not be viewed as pathologic, even it it continues for years (Goin, Burgoyne, & Goin, 1979).

Some lessons on phases of bereavement can be taken from the early work of Bowlby (1969). His exposure to children separated from their parents during the Second World War provided the impetus for his work on attachment theory. He has suggested that responses to separation can be systematized. Protest, despair, and grief have been repeatedly observed on separation. If these attachment behaviors become prolonged, denial of need ensues, accompanied by an increased likelihood of emotional sequelae. Like the phases of bereavement, the phases of attachment behavior on separation require successful resolution to lower the risk of psychiatric morbidity.

Differential Diagnosis of Major Depression from Bereavement

Because depressive symptoms are so clearly associated with bereavement, decisions concerning the diagnosis of and the intervention and treatment for depression can be difficult. As with its requirement that major depression should not be diagnosed when physical symptoms of depression may result from a physical illness, so does the DSM-III-R depart from phenomenologic diagnosis when depressive symptoms result from bereavement. The DSM-III-R, in fact, specifically permits a "full depressive syndrome" in uncomplicated bereavement, but suggests that preoccupation with worthlessness, prolonged and marked functional impairment, and psychomotor retardation may indicate bereavement complicated by major depression.

We take exception to this for the same reason that we proposed using the "inclusive" approach to diagnosis for depression in the medically ill and for removing the category of organic mood disorder. In general, we propose that major depression should be diagnosed whenever the phenomenologic signs and symptoms are present to meet DSM-III-R criteria, regardless of etiology (i.e., whether or not they occur in the context of bereavement or physical illness). The occurrence of bereavement or of a grief reaction related to medical illness can be noted on Axis IV, but should not obviate diagnosis on Axis I.

In an effort to clarify this diagnostic issue, Clayton and Darvish (1979) studied 171 men and women at 1 month and 1 year after the deaths of their spouses. A full spectrum of depressive symptoms was present at 1 month. By 1 year, most of the vegetative symptoms had remitted; however, the frequency of some symptoms remained high (sleep disturbance, 48%; restlessness, 48%; low mood, 41%). Somatic symptoms generally improved within the first year of bereavement, whereas psychologic symptoms such as death wishes, hopelessness, worthlessness, and feelings of anger persisted.

Some symptoms of depression are rare in bereavement or present differently in bereaved individuals. Psychomotor retardation is a rare finding, and should be considered pathologic. Although guilt is frequently noted, it is not the ruminative guilt of major depression. It is usually a guilt of omission about things not done surrounding the events of death. Morbid cultivation of the deceased, or "mummification," is not normally observed. Rather, the usual response of the bereaved is to save significant mementos, or perhaps preserve certain rooms or possessions. Hallucinations have been noted during the process of normal bereavement; however, delusions should be considered abnormal.

Suicidal thoughts are uncommon in the bereaved; however, thoughts of death are frequently noted. Actual thoughts of killing oneself should always be considered pathologic. Some populations have been shown to have significantly higher rates of suicidal ideation: (1) young men during the first month after the deaths of their spouses (Clayton & Darvish, 1979), (2) parents of deceased children (Clayton, Desmarais, & Winokur, 1968), and (3) relatives of victims of violent or sudden death (Vargas, Loya, & Hodde-Vargas, 1989).

In her comparison of bereaved individuals with a community sample of controls, Clayton (Clayton, 1986; Clayton & Darvish, 1979) found a markedly increased use of alcohol, tranquilizers, and hypnotics in the bereaved group. Clearly, this is a cofactor in the psychiatric morbidity associated with bereavement.

Time Course

Generally, the time course of bereavement is one of gradual improvement. Although sleep may remain disturbed throughout the first year, most bereaved individuals have regained their appetites and begun to regain lost weight by the end of the fourth month. Similarly, sexual interest returns about this same time. Mood remains intermittently low with exacerbations on significant days such as anniversaries (Clayton, 1990). By the end of the first year, most bereaved persons can discuss the lost loved one with equanimity. Those who cannot should be considered to have an abnormal response (Bornstein & Clayton, 1972).

Pathologic Outcomes

Specific aspects of the depressive responses in bereavement have been outlined above. Grief reactions can have the same clinical presentation and course as bereavement. Those persons who grieve such losses as physical health and impaired body functioning may have a presentation indistinguishable from bereavement. A number of authors (Brown & Stoudemire, 1983; Osterweis, Solomon, & Green, 1984) have summarized the pathologic outcomes of grief as follows:

1. *Prolonged or chronic grief*, defined as persistent grieving without diminution and intensity despite the passage of time.
2. *Absent grief*, defined as the "fending off" of threatening emotions that are too painful to experience.
3. *Delayed grief*, defined as a long period (months to years) of absence of grief after which grief-like symptoms emerge.

Predictors of Outcome

Predicting outcome in bereavement is a difficult endeavor because previous studies have used various outcome measures: number of hospitalizations, remarriage, sudden death, and continued presence of depressive symptoms, to name a few. Clayton (1990) has found the best predictor of the development of a chronic depressive syndrome to be poor physical or mental health before the loss. Although any psychiatric disorder can make a bereaved individual more vulnerable, those persons with a history of addictive disorders are at a substantially increased risk of having a poor outcome secondary to the increased substance use associated with bereavement. Studies of bereavement indicate that those persons most disturbed in the early stages were most disturbed 1 year later. Prolonged, severe grief accounted for the poorest outcome. There was also considerable agreement that if the bereaved perceive their social network as unsupportive for grieving, they are at an increased risk of a poor outcome (Middleton & Raphael, 1987). Age, sex, race, social support, finances, religious involvement, relationship to the deceased, and degree of closeness to the deceased have not been shown to predict long-term outcome, although lack of social support, low income, and young age do predict poor immediate response. Bowlby (1980), reviewing studies of pathologic grief, enumerated several risk factors. These are listed in Table 4-6.

Treatment Considerations

Just as Engel (1961) asked whether grief was a disease three decades ago, we continue to ponder this question today. It is not known whether or not the majority of bereaved persons would benefit from psychiatric treatment (either biologic or psychosocial). There are, however, numerous psychosocial and supportive treatments for the bereaved, mostly built on the idea that expression of affect is beneficial. Alexander (1988) identified

TABLE 4–6. *Risk Factors For Pathologic Grief*

Deceased is a parent or parent substitute
Sudden death
False or inadequate information at the time of the death
Difficulties in the relationship with the deceased just before the death
Death by suicide
Multiple losses
Lack of social supports
Anxious or ambivalent attachment to the deceased
Compulsive caregiving as a life pattern of the bereaved person prior to the death
Excessive dependence or counterdependence
History of early major losses

the following interventions: "regrief" therapy, guided imagery, guided mourning, carthersis, brief dynamic psychotherapy, and music therapy.

The positive role of supportive groups for the treatment of grief has now been well established, particularly with respect to the grief associated with the diagnosis of cancer. In a landmark study (Spiegel, Bloom, & Yalom, 1981), it was shown that weekly supportive group meetings for women with metastatic breast carcinoma resulted in improved mood, coping strategies, and self-esteem when compared to a nontreatment group. Interestingly, when these patients were evaluated at 10-year follow-up, the mean survival time of those patients who had received the 1 year of supportive group intervention was twice as long as those patients who had been randomized to no psychosocial intervention (Spiegel et al., 1989). Recently, Fawzy et al. (1990) examined the effects of short-term (6 weeks) structured group intervention in patients with malignant melanoma. These investigations have found that short-term psychiatric intervention effectively reduced psychologic distress and enhanced longer term effective coping.

For those individuals who develop the full syndrome of major depression during bereavement or grief (and this may be the majority), treatment can consist of any of the therapies designed to provide psychosocial intervention or, if the depression is severe enough, biologic treatment can be used. Such treatment could include combinations of interpersonal therapy, cognitive therapy, antidepressant treatment, or ECT.

Some clinicians have questioned the use of antidepressant therapy in bereaved persons. Our own experience and some recent data, however, indicate that depressive syndromes (especially when associated with several "vegetative" symptoms) in bereaved persons respond to traditional antidepressant treatment. In their pilot study of depressed, bereaved spouses, Jacobs, Nelson, & Zisook (1987) found that 7 of 10 patients had moderate to marked improvement after 4 weeks of treatment with desipramine. More systematic, controlled studies with larger samples are needed to further evaluate this important clinical issue.

Although there are no current data on the need to prophylactically treat certain patients with pharmacotherapy in the context of grief reactions, those patients at high risk of pathologic outcomes should be closely monitored. Those patients with poor physical or mental health prior to loss may be particularly vulnerable.

ADDENDUM: DSM-IV and Depression

The DSM-IV *Options Book* (Task Force on DSM-IV, *DSM-IV Options Book*, American Psychiatric Association, 1991) in which proposed changes in DSM-III-R criteria are discussed, suggested several changes for the assessment of depression in the medically ill. It is important to note, however, that these proposed changes are preliminary and may not be included in the final version of DSM-IV.

First, the criteria for major depression have not substantially changed, although the DSM-III-R caveat "do not include symptoms that are clearly due to a physical condition" has been taken out. This change would greatly simplify the diagnosis of major depression in the medically ill and seems to support the recommendation for an "inclusive" approach that is discussed in this chapter. However, clinicians are instructed not to diagnose major depression if the disorder is due to a "secondary" mood disorder (see below). Unfortunately, there are not clear guidelines for clinicians to follow in this regard and the attempt to distinguish major depression from secondary mood disorder may lead to problems of reliability and validity.

The second important proposal for DSM-IV is to change the name of "Organic mood disorder" to "Secondary mood disorder". This change is in keeping with an increased understanding of the physical basis of many psychiatric conditions, previously considered only "functional." According to the *Options Book*, a secondary mood disorder requires a depressed mood or anhedonia that is etiologically judged to be contributory to the mood. Furthermore, a secondary mood disorder can be of the major depressive type (i.e., meeting criteria for major depression) or simply with depressive features (i.e., not meeting full criteria for major depression).

Lastly, the *Options Book* suggests the new category of "minor depression." This proposal represents an attempt to recognize that many patients suffer from depressive syndromes that may warrant treatment interventions but whose problems do not meet the full threshold for diagnosis of major depression. While this seems to be an important development, this new category of minor depression will present special diagnostic challenges for patients with medical illness. The *Options Book* does not provide clinicians with any clear directions concerning the extent to which patients with mild depressive syndromes in the context of medical illness should receive this diagnosis of minor depression, the diagnosis of adjustment disorder with depressed mood, or the diagnosis of secondary depression, with depressive features.

The final version of DSM-IV may resolve some of the dilemmas discussed above although historically, the DSM system for the diagnosis of depression has not been standardized for use in medically ill populations. Readers should consult the final version of DSM-IV when it is published.

REFERENCES

Akiskal H, Rosenthal T, Haykal R, et al. (1980). Characterological depressions: Clinical and sleep EEG findings separating subaffective dysthymias from character spectrum disorders. Arch Gen Psychiatry 37:777–783.

Alexander DA (1988). Bereavement and the management of grief. Br J Psychiatry 153:860–864.

American Psychiatric Association (1987). Diagnostic and statistical manual of mental disorders (3rd ed., rev.). Washington, DC: American Psychiatric Press, Inc.

Avorn J, Everitt DE, & Weiss S (1986). Increased antidepressant use in patients prescribed beta-blockers. JAMA 255:357–360.

Barret JE, Barret JA, Oxman TE, et al. (1988). The prevalence of psychiatric disorders in a primary care practice. Arch Gen Psychiatry 45:1100–1106.

Bieliauskas LA, & Glantz RH (1989). Depression type in Parkinson disease. J Clin Exp Neuropsychol 11:597–604.

Blacker CVR, & Clare AW (1987). Depression in primary care. Presented at "Mental Disorders in General Health Care Settings," Seattle, Washington.

Bornstein PE, & Clayton PJ (1972). The anniversary reaction. J Dis Nerv System 33:470–472.

Borson S, McDonald GJ, Gayle TC, et al. (1992). Improvement in mood, physical symptoms, and function with nortriptyline for depression in patients with chronic obstructive pulmonary disease. Psychosomatics 33:190–201.

Bowlby J (1969). Attachment and loss. Vol. 1: Attachment. New York: Basic Books.

Bowlby J (1980). Loss: Sadness and depression. New York: Basic Books.

Broadhead WE, Blazer DG, George LK, et al. (1990). Depression, disability days, and days lost from work in a prospective epidemiologic survey. JAMA 264:2524–2528.

Brown FW, Golding JM, & Smith GR (1990). Psychiatric comorbidity in primary care somatization disorder. Psychosom Med 52:445–451.

Brown JT, & Stoudemire GA (1983). Normal and pathological grief. JAMA 250:378–382.

Brown RG, MacCarthy B, Gotham AM, et al. (1988). Depression and disability in Parkinson's disease: A follow-up of 132 cases. Psychol Med 18:49–55.

Bukberg J, Penman D, & Holland JC (1984). Depression in hospitalized cancer patients. Psychosom Med 46:199–212.

Carroll BJ (1984). Problems with diagnostic criteria for depression. J Clin Psychiatry 45:14–18.

Cassem EH (1990). Depression and anxiety secondary to medical illness. Psychiatr Clin North Am 13:597–612.

Cavenaugh S, Clark D, & Gibbons R (1983). Diagnosing depression in the hospitalized medically ill. Psychosomatics 24:809–815.

Cavenaugh S, & Kennedy S (1986). A successful training program for medical residents. Gen Hosp Psychiatry 8:73–80.

Chandler JD, & Gerndt J (1988). Somatization, depression, and medical illness in psychiatric inpatients. Acta Psychiatr Scand 77:67–73.

Clark D, Cavenaugh S, & Gibbons R (1983). Core symptoms of depression in medical and psychiatric patients. J Nerv Ment Disord 171:705–713.

Clayton PJ (1986). Bereavement and its relation to clinical depression. In H Hippius, GL Klerman, & N Matussek (Eds.), New results in depression research (pp. 59–69). Berlin: Springer-Verlag.

Clayton PJ (1990). Bereavement and depression. J Clin Psychiatry 51(7)(suppl):34–40.

Clayton PJ, & Darvish HS (1979). Course of depressive symptoms following the course of bereavement. In JE Barrett, RM Rose, & GL Klerman (Eds.), Stress and mental disorder (pp. 121–139). New York: Raven Press.

Clayton PJ, Desmarais L, & Winokur G (1968). A study of normal bereavement. Am J Psychiatry 125:168–178.

Cohen-Cole SA, Bird J, Freeman A, et al. (1982). Psychiatry for internists: A study of needs. J Operational Psychiatry 13:100–105.

Cohen-Cole SA, & Stoudemire A (1987). Major depression and physical illness: Special considerations in diagnosis and biologic treatment. Psychiatr Clin North Am 10:1–17.

Costa D, Mogos I, & Toma T (1985). Efficacy and safety of mianserin in the treatment of depression of women with cancer. Acta Psychiatr Scand 72(suppl 320):85–92.

Dam H, Pedersen HE, & Ahlgren P (1989). Depression among patients with stroke. Acta Psychiatr Scand 80:118–124.

Derogatis LR, Morrow GR, Fetting J, et al. (1983). The prevalence of psychiatric disorders among cancer patients. JAMA 249:751–757.

Devlen J, Maguire P, Phillips P, et al. (1987). Psychological problems associated with diagnosis and treatment of lymphomas. I: Retrospective study; II: Prospective study. Br Med J 295:953–955.

Eastwood MR, Rifat SL, Nobbs H, et al. (1989). Mood disorder following cerebrovascular accident. Br J Psychiatry 154:195–200.

Elkin I, Shea T, Watkins J, et al. (1989). NIMH Treatment of Depression Collaborative Research Program: General effectiveness of treatments. Arch Gen Psychiatry 46:971–982.

Endicott J (1984). Measurement of depression in patients with cancer. Cancer 53:2243–2248.

Engel G (1961). Is grief a disease? Psychosom Med 23:18–23.

Evans DL, McCartney CF, Nemeroff CB, et al. (1986). Depression in women treated for gynecological cancer: Clinical and neuroendocrine assessment. Am J Psychiatry 143:447–452.

Fava GA, Pilowsky I, Pierfederici A, et al. (1982). Depressive symptoms and abnormal illness behavior in general hospital patients. Gen Hosp Psychiatry 4:171–178.

Fawzy FI, Cousins N, Fawzy NW, et al. (1990). A structured psychiatric intervention in cancer patients: Changes over time in methods of coping and affective disturbance. Arch Gen Psychiatry 47:720–725.

Federoff JP, Starkstein SE, Parikh RM, et al. (1991). Are depressive symptoms non-specific in patients with acute stroke? Am J Psychiatry 148:1172–1176.

FELDMAN E, MAYOU R, HAWTON K, ET AL. (1987). Psychiatric disorder in medical inpatients. Q J Med 63:405–412.

FOGEL BS (1990). Major depression versus organic mood disorder: A questionable distinction. J Clin Psychiatry 51:53–56.

FROST NR, & CLAYTON PJ (1977). Bereavement and psychiatric hospitalization. Arch Gen Psychiatry 34:1172–1175.

GOIN MK, BURGOYNE RW, & GOIN JM (1979). Timeless attachment to a dead relative. Am J Psychiatry 136:988–989.

HARDMAN A, MAGUIRE P, & CROWTHER D (1989). The recognition of psychiatric morbidity on a medical oncology unit. J Psychosom Res 33:235–239.

HOEPER EW, NYCZ GR, CLEARY PD, ET AL. (1979). Estimated prevalence of RDC mental disorder in primary care. Int J Ment Health 8:6–15.

HOLLAND JC, FASANELLO S, & OHNUMA T (1974). Psychiatric symptoms associated with L-asparaginase administration. J Psychiatry Res 10:105–113.

HOWELL T, FULLERTON DT, HARVEY RF, ET AL. (1981). Depression in spinal cord injured patients. Paraplegia 19:284–288.

IMBODEN JB, CANTER A, & CLUFF L (1963). Separation experience and health records in a group of normal adults. Psychosom Med 25:433–440.

JACOBS SC, NELSON JC, & ZISOOK S (1987). Treating depressions of bereavement with antidepressants: A pilot study. Psychiatr Clin North Am 10:501–511.

JOFFE RT, RUBINOW DR, DENICOFF KD, ET AL. (1986). Depression and carcinoma of the pancreas. Gen Hosp Psychiatry 8:241–245.

KATHOL RG, MUTGI A, WILLIAM SJ, ET AL. (1990). Major depression diagnosed by DSM-III, DSM-III-R, RDC, and Endicott criteria in patients with cancer. Am J Psychiatry 147:1021–1024.

KATON W, BERG AO, ROBINS AJ, ET AL. (1986). Depression: Medical utilization and somatization. West J Med 144:564–568.

KELLER M, KLERMAN G, & LAVORI P (1982). Treatment received by depressed patients. JAMA 248:1848–1855.

KNIGHTS E, & FOLSTEIN M (1977). Unsuspected emotional and cognitive disturbance in medical patients. Ann Intern Med 87:723–724.

KOENIG HG, MEADOR KG, & COHEN HJ (1988). Depression in elderly hospitalized patients with medical illness. Arch Intern Med 148:1929–1936.

LANSKY SB, LIST MA, HERRMANN CA, ET AL. (1985). Absence of major depressive disorder in female cancer patients. J Clin Oncol 3:1553–1560.

LIGHT RW, MERRILL EJ, DESPARS J, ET AL. (1986). Doxepin treatment of depressed patients with chronic obstructive pulmonary disease. Arch Intern Med 146:1377–1382.

LIPSEY IR, ROBINSON RG, PEARLSON GD, ET AL. (1984). Nortriptyline treatment of post-stroke depression: A double-blind study. Lancet 1:297–300.

LLOYD GG, & CAWLEY RH (1983). Distress or illness: A study of psychological symptoms after myocardial infarction. Br J Psychiatry 142:120–125.

LLOYD GG, PARKER AC, LUDLAM CA, ET AL. (1984). Emotional impact of diagnosis and early treatment of lymphoma. J Psychosom Res 28:157–162.

LYKOURAS E, IOANNIDIS C, VOULGARI A, ET AL. (1989). Depression among general hospital patients in Greece. Acta Psychiatr Scand 79:148–152.

MAGNI G, SCHIFANO F, & DELEO D (1986). Assessment of depression in an elderly medical population. J Affective Disord 11:121–124.

MAGUIRE P (1984). The recognition and treatment of affective disorder in cancer patients. Int Rev Appl Psychol 33:479–491.

MAGUIRE P, HOPWOOD P, TARRIER N, ET AL. (1985). Treatment of depression in cancer patients. Acta Psychiatr Scand 72:81–84.

MARICLE RA, KINZIE JD, & LEWINSOHN P (1988). Medication-associated depression: A two and one-half year follow-up of a community sample. Int J Psychiatry Med 18:283–292.

MASSIE MJ, & HOLLAND JC (1984). Diagnosis and treatment of depression in the cancer patient. J Clin Psychiatry 45(sec 2):25–29.

MAYEUX R, STERN Y, WILLIAMS J, ET AL. (1986). Clinical and biochemical features of depression in Parkinson's disease. Am J Psychiatry 143:756–759.

MIDDLETON W, & RAPHAEL B (1987). Bereavement: State of the art and state of the science. Psychiatr Clin North Am 10:329–343.

MINDEN SL, ORAV J, & REICH P (1987). Depression in multiple sclerosis. Gen Hosp Psychiatry 9:426–434.

MURPHY S, CREED F, & JAYSON MIV (1988). Psychiatric disorder and illness behavior in rheumatoid arthritis. Br J Rheumatol 27:357–363.

OSTERWEIS M, SOLOMON F, & GREEN M (1984). Adult reactions to bereavement. In M OSTERWEIS, F SOLOMON, & M GREEN (Eds.), Bereavement: Reactions, consequences, and care (Ch. 3, pp. 47–71). Washington, DC: National Academy Press.

PARIKH RM, EDEN DT, PRICE TR, ET AL. (1988). The sensitivity and specificity of the Center for Epidemiological Studies Depression Scale in screening for post-stroke depression. Int J Psychiatry Med 18:169–181.

PARKES CM (1970). The first year of bereavement: A longitudinal study of the reaction of London widows to the deaths of their husbands. Psychiatry 33:444–467.

PATTEN SB (1990). Propranolol and depression: Evidence from the anithypertensive trials. Can J Psychiatry 35:257–259.

PLUMB M, & HOLLAND J (1981). Comparative studies of psychological function in patients with advanced cancer: 2. Interviewer rated current and past psychological symptoms. Psychosom Med 43:243–254.

POPKIN MK (1988). Prevalence of major depression, simple phobia and other psychiatric disorders in patients with long-standing type I diabetes mellitus. Arch Gen Psychiatry 45:64–68.

QUITKIN FM, STEWART JW, MORRISON W, ET AL. (1990). Antidepressants for mild depressives. Presented at the Annual Meeting of the American Psychiatric Association, New York.

REIFLER BV, TEVI L, RASKIND M, ET AL. (1989). Double-blind trial of imipramine in Alzheimer's disease patients with and without depression. Am J Psychiatry 146:45–49.

RIFKIN A, REARDON G, SIRIS S, ET AL. (1985). Trimipramine in physical illness with depression. J Clin Psychiatry 46:4–8.

ROBERTSON MN, & TRIMBLE MR (1985). The treatment of depression in patients with epilepsy. J Affective Disord 9:127–131.

ROBINS LN, HELZER JE, CROUGHAN J, ET AL. (1981). National Institute of Mental Health Diagnostic Interview Schedule. Arch Gen Psychiatry 38:381–389.

ROBINS LN, HELZER JE, WEISSMAN M, ET AL. (1984). Lifetime prevalence of specific psychiatric disorders in three sites. Arch Gen Psychiatry 41:949–958.

ROBINSON RG, LIPSEY JR, & PRICE TR (1985). Diagnosis and clinical management of post-stroke depression. Psychosomatics 26:769–778.

ROBINSON RG, STARR LB, & PRICE TR (1984). A two year longitudinal study of mood disorder following stroke: Prevalence and duration at six month follow-up. Br J Psychiatry 144:256–262.

RODIN G, & VOSHART K (1986). Depression in the medically ill: An overview. Am J Psychiatry 143:696–705.

ROVNER B, BROADHEAD J, SPENCER M, ET AL. (1989). Depression and Alzheimer's disease. Am J Psychiatry 146:350–355.

RUSH AJ, & JARRETT RB (1986). Psychotherapeutic approaches to depression. In RN MICHELS, & J CAVENAR (Eds.), Psychiatry (Vol. 1; sec 2, p. 65). Philadelphia: JB Lippincott.

SANTAMARIA J, TOLOSA E, & VALLES A (1986). Parkinson's disease with depression: A possible subgroup of idiopathic Parkinsonism. Neurology 36:1130–1133.

SCHLEIFER SJ, MACARI-HINSON MM, KAHN M, ET AL. (1989). The nature and course of depression following myocardial infarction. Arch Intern Med 149:1785–1789.

SCHULBERG HC, SAUL M, MCCLELLAND MN, ET AL. (1985). Assessing depression in primary medical and psychiatric practices. Arch Gen Psychiatry 42:1164–1170.

SCHWAB JJ, BIALOW M, BROWN JM, ET AL. (1967). Diagnosing depression in medical inpatients. Ann Intern Med 67:695–707.

SMITH MD, HONG BA, & ROBSON AM (1985). Diagnosis of depression in patients with end-stage renal disease. Am J Med 79:160–166.

SPIEGEL D, BLOOM JR, KRAEMER H, ET AL. (1989). The beneficial effect of psychosocial treatment on survival of metastatic breast cancer patients: A randomized prospective outcome study. Lancet 2:888–891.

SPIEGEL D, BLOOM JR, & YALOM I (1981). Group support for patients with metastatic cancer. Arch Gen Psychiatry 38:527–533.

SPITZER RL, ENDICOTT J, & ROBINS E (1978). Research diagnostic criteria. Arch Gen Psychiatry 35:773–782.

SPITZER R, WILLIAMS J, GIBBONS M, ET AL. (1990). Structured Clinical Interview for DSM-III-R (SCID). Users guide for the Structured Clinical Interview for DSM-III-R. Washington, DC: American Psychiatric Press, Inc.

STARKSTEIN SE, BERTHIER ML, BOLDUC PL, ET AL. (1989). Depression in patients with early versus late onset of Parkinson's disease. Neurology 39:1441–1445.

STARKSTEIN SE, PREZIOSI TJ, BOLDUC PL, ET AL. (1990). Depression in Parkinson's disease. J Nerv Ment Dis 178:27–31.

STIEFEL FC, BREITBART WS, & HOLLAND JC (1989). Corticosteroids in cancer: Neuropsychiatric complications. Cancer Invest 7:479–491.

STOUDEMIRE A (1985). Depression in the medically ill. In J CAVENAR (Ed.), Psychiatry (Vol. 2; sec. 99, pp. 1–8). Philadelphia: JB Lippincott.

STOUDEMIRE A (1987). Selected organic brain syndromes. In R HALES & S YUDOFSKY (Eds.), Textbook of neuropsychiatry (Ch. 7, pp. 125–139). Washington, DC: American Psychiatric Press, Inc.

STOUDEMIRE A, BROWN JT, HARRIS R, ET AL. (1984). Propranolol and depression: A reevaluation based on a pilot clinical trial. Psychiatr Med 2:211–218.

STOUDEMIRE A, HILL C, KAPLAN W, ET AL. (1987). Clinical issues in the assessment of dementia and depression in the elderly. Psychiatr Med 6:40–52.

THOMPSON TL, STOUDEMIRE A, & MITCHELL WE (1983). Underrecognition of patients' psychosocial distress in a university hospital medical clinic. Am J Psychiatry 140:158–161.

TURKINGTON RW (1980). Depression masquerading as diabetic neuropathy. JAMA 243:1147–1150.

VAN ERDEWEGH M, & CLAYTON P (1988). Bereavement. In R MICHELS, JO CAVENAR, AM COOPER, ET AL. (Eds.), Psychiatry (rev. ed.) (Ch. 67, pp. 1–11). Philadelphia: JB Lippincott.

VARGAS LA, LOYA F, & HODDE-VARGAS J (1989). Exploring the multidimensional aspects of grief reactions. Am J Psychiatry 146:1484–1488.

VEITH RC, RASKIND MR, CALDWELL JH, ET AL. (1982). Cardiovascular effects of tricyclic antidepressants in depressed patients with chronic heart disease. N Engl J Med 306:954–959.

VON KOPFF M, SHAPIRO S, BURKE JD, ET AL. (1987). Anxiety and depression in a primary care clinic: Comparison of Diagnostic Interview Schedule, General Health Questionnaire, and practitioner assessments. Arch Gen Psychiatry 44:152–156.

WADE DT, LEGH-SMITH J, & HEWER RA (1987). Depressed mood after stroke: A community study of its frequency. Br J Psychiatry 151:200–205.

WEDDINGTON WW (1982). Delirium and depression associated with amphotericin B. Psychosomatics 23:1076–1078.

WEISSMAN MM, MYERS JK, & THOMPSON WD (1981). Depression and its treatment in a US urban community 1975–1976. Arch Gen Psychiatry 38:417–421.

WELLS KB, HAYS RD, BURNAM A, ET AL. (1989a). Detection of depressive disorders for patients receiving prepaid or fee-for-service care. JAMA 262:3298–3302.

WELLS KB, STEWART A, HAYS R, ET AL. (1989b). The functioning and well-being of depressed patients: Results from the medical outcomes study. JAMA 262:914–919.

WISE TN, MANN LS, PUSCHECK E, ET AL. (1985-6). Factors affecting anxiety and depression in psychiatric consultation patients. Int J Psychiatry Med 15:177–184.

YOUNG DF (1982). Neurological complications of cancer chemotherapy. In A SILVERSTEIN (Ed.), Neurological complications of therapy: Selected topics (Ch. 3, pp. 57–113). Mount Kisco, NY: Futura Publishing Co.

5 Suicide and aggression in the medical setting

LUCY DAVIDSON, M.D., ED.S.

No form of treatment is effective with a dead patient.

R.S. Mintz (1971)

Violence, whether self-inflicted (suicidal) or interpersonal, assails our therapeutic efforts. Destructive patients endanger more than the immediate physical well-being of themselves and others. Even if only a trivial physical injury occurs, the therapeutic collaboration among patient, physician, and other health care professionals is disrupted. The patient may break off treatment precipitously, or the physician may unconsciously retaliate in ways that drive the patient away. Witnessing or hearing about suicidal or assaultive behavior may worsen other patients' illnesses and erode confidence in the physician's ability to heal and protect.

Medical staff who have been assaulted may be psychologically disabled by it and may have extraordinary trouble caring for other patients or recovering from their own physical injuries. A patient's suicide or suicide attempt exposes the physician's own fears of inadequacy. When these conflicts are rekindled, our ability to appropriately attend to patient needs suffers. The dread of malpractice litigation for assault or wrongful death can be debilitating in itself.

Physicians clearly are motivated to prevent violence and these harmful outcomes. Moreover, the opportunity for physician-patient contact should make suicidal behavior and assaults in the medical setting especially preventable. This chapter identifies characteristics of patients at high risk for violence and situations in the medical setting in which violence is likely to occur, and it describes techniques for interviewing, stabilization, and support to reduce the likelihood of violence. Frequently, psychiatric consultation is initiated after an assaultive or potentially violent confrontation. The intervention and prevention strategies discussed are designed to restore the patient and staff to a therapeutic relationship as rapidly as possible. Although self-directed and interpersonal violence are closely related, each topic is discussed separately, beginning with suicide.

SUICIDAL PATIENTS

Assessing the Suicidal Patient

A suicide is a self-inflicted, intentional death. A suicide attempt is a nonlethal self-inflicted act that has as its intended purpose death or the appearance of willingness to die.

How does the psychiatrist assess suicide potential? As Bakwin (1957) said, "the sole approach to the . . . suicide problem lies in recognizing beforehand the susceptible individuals and in their proper management." Four general categories of data are weighed in assessing suicidality: (1) personal risk factors, (2) environmental and situational factors, (3) psychodynamic meaning, and (4) information from the mental status examination (Table 5-1).

Epidemiology and Risk Factors

Over 30,000 Americans complete suicide annually (National Center for Health Statistics, 1991). Suicide has become the eighth leading cause of death among all ages and the second leading cause among youth. Epidemiologic data help profile high-risk groups. White males have the highest rates of any race or sex group and account for 70% of all suicide deaths in the United States. Suicide rates decrease in the following order among race and sex groups: white males (70%), white females (22%), black and other males (6%), and black and other females (2%). Suicide rates vary markedly among native American tribes. The ratio of male to female suicides is about 3:1 overall, increasing to 5:1 for youth (Centers for Disease Control, 1985).

The proportion of suicides among youths 15 to 24 years old has increased in the last four decades. Male suicide rates are now bimodally distributed, with peaks in the young adult years and in old age. Rates for females peak in midlife. The lifetime suicide risk among persons with mood disorders, schizophrenia, or substance abuse is increased. The lifetime risk is estimated

TABLE 5–1. *Conditions Predisposing to Completed Suicide*

Expression of intent to die	Explicit, verbal Implicit or indirect (e.g., hopelessness, suicide plan, rehearsing fatal behavior)
Demographic factors	White, male 15–24 or over 65, unmarried, urban, western U.S. residence
Psychiatric factors	Affective disorder, schizophrenia, anxiety disorder, or substance abuse; especially one of these Axis I comorbid with Axis II Psychodynamic conflict or fantasy that would be resolved via suicide
Situational factors	Recent loss: tangible, interpersonal, narcissistic Another recent suicide Ready access to lethal means, especially a gun Impaired judgment and insight (e.g., intoxicated patient) Few social supports Recovery from depressive episode or schizophrenic break
Historic factors	Previous suicide attempt or threat Abused or neglected Violent toward self or others Family member suicide
Physical factors	Serious or chronic illness, especially with central nervous system effects Serotonergic system dysfunction

at 10% for schizophrenics and at 15% for persons with mood disorders (Miles, 1977). Among affectively disordered patients, those who commit suicide show these differentiating clinical features: hopelessness, loss of pleasure or interest, and loss of reactivity. Suicide risk may be lower in bipolar than in unipolar patients and is not associated with the presence of psychosis during episodes of illness (Black, Winokur, & Nasrallah, 1988). Patients who complete suicide have fewer preceding episodes of mood disorder than others with the same diagnosis (Fawcett et al., 1987). This suggests that suicide may occur fairly early in the course of mood disorders.

In contrast to the situation for patients with mood disorders, suicide is a relatively late complication of alcoholism (Robbins, 1981). Murphy and Wetzel (1990) calculated a mean of 19 years' active alcoholism preceding completed suicide. Medical examiner data show that nearly 20% of suicides are alcohol abusers (by physical evidence) and 25% of suicides have a positive blood alcohol concentration at death (Riddick & Luke, 1978). Although alcoholics comprise approximately 25% of suicides, the lifetime suicide prevalence among alcoholics is 2% to 3.4%, with completed suicide being largely determined by comorbid psychiatric disorders (Murphy & Wetzel, 1990).

Alcoholics who attempt suicide begin heavy drinking at an earlier age, have an earlier onset of alcohol-related problems, and are more often comorbid for other psychiatric disorders than alcoholic nonattempters (Roy et al., 1990). Alcoholics who committed suicide have been compared with living alcoholics, highlighting factors that increased the alcoholics' susceptibility to suicide. More of the alcoholic suicides were divorced or widowed. A history of previous suicide attempts was obtained for 67% of the alcoholic suicides versus 10% of the controls. When compared with nonalcoholic suicides, more alcoholic suicides had made an overt suicide threat, and more had seen a psychiatrist or other doctor in the week before death (Barraclough et al., 1974).

Of alcoholic patients with bipolar disorder, 80% had attempted suicide versus 13% of nonalcoholic bipolar patients (Johnson & Hunt, 1979). Of persons admitted for alcoholism treatment, 27% had made a previous suicide attempt and 12% had current suicidal ideation (Beck, Steer, & McElroy, 1982). In another series, 15% of alcoholics who died from all causes had a history of suicide attempt (Choi, 1975).

Patients with panic disorder or panic attacks have elevated rates of suicide attempts. Estimates based on Epidemiologic Catchment Area (ECA) study data indicate that 20% of persons with panic disorder and 12% of those with panic attacks have a lifetime history of suicide attempts (Weissman et al., 1989). The frequent comorbidity of these disorders with depression, alcoholism, and other drug abuse further increases the likelihood of suicidal behaviors (Johnson, Weissman, & Klerman, 1990; Markowitz et al., 1989; Vollrath & Angst, 1989).

Patients with comorbid personality disorders show more severe initial Axis I illness, poorer response to anxiety treatment, and more residual symptoms and social maladjustment at follow-up (Noyes et al., 1990; Reich & Green, 1991). Mixed anxiety and depression was more often associated with borderline personality disorder compared with the other Axis II disorders (Alnaes & Torgersen, 1990). Furthermore, coexisting personality disorder was an important predictor of relapse after ending pharmacologic treatment (Mavissakalian, 1990).

The typical suicidal patient is often thought of as a depressed, elderly white male who abuses alcohol, is in poor health, and has experienced a recent emotional loss. These risk factors for older adult suicides are not, however, prototypes for youth suicides. Depression, while still an important risk factor, is less common among youth suicides. Youth suicides are often impulsive acts, and the suicidal youth often fits into the diagnostic spectrum of conduct disorders (Shafii et al., 1985). Adult alcoholics meeting criteria for antisocial personality disorder who had attempted suicide were significantly more likely to have shown conduct disorder symptoms before age 15 than those with the same adult diagnoses who had never attempted suicide (Whitters et al., 1987). Substance-abusing college students are at greater risk for suicidal ideation and suicide attempts. Children and adolescents with prior history of abuse or neglect are more likely to attempt suicide (Deykin, Alpert, & McNamarra, 1985; Rosenthal & Rosenthal, 1984). Clusters of youth suicides have also occurred, superimposing a milieu for behavioral contagion on otherwise susceptible youth (Davidson et al., 1989).

A family history of suicide increases risk, probably through both genetic and modeling factors (Egeland & Sussex, 1985). Neurochemical studies have demonstrated biochemical and structural abnormalities associated with suicide (van Praag, Plutchik, & Apter, 1990). Serotonergic neurotransmission dysfunction is associated with completed suicide and suicide attempts by violent methods (Arango et al., 1990; Arora & Meltzer, 1989; Jones et al., 1990). Other lines of research have indicated that alpha$_1$-noradrenergic receptors (Gross-Isseroff, Dillon, & Fieldust, 1990) and parahippocampal and temporal lobe structures (Altshuler et al., 1990) may also be aberrant.

Persons who have ever attempted suicide are also at increased risk. Married persons are less susceptible to suicide than are the never married, widowed, or divorced. Suicides are more common in urban than rural areas, and rates peak in the western United States. Gunshot is the most frequent method of suicide in this country for men and women of all ages (Centers for Disease Control, 1985).

None of these epidemiologic factors represents suicidal *intent*, the essential variable in predicting suicidality. Nonetheless, the presence in a patient of many of these risk factors for suicide should heighten the clinician's sensitivity to predicting, detecting, and managing possible suicidal intent.

Environmental and Situational Factors

Environmental and situational factors offer clues to suicidal intent. Most suicidal persons express their intent to die either directly or indirectly. The myth of "talkers" versus "doers" provides false reassurance at the expense of lives. Suicidal persons may tell others, "You'd be better off without me," or "It just doesn't seem worth going on." The term "suicide threat" connotes manipulation and histrionic confrontation. The apparent threat may, however, be a very matter-of-fact communication: "I'm going to kill myself."

More indirect indications of intent to die include expressions of hopelessness and intolerable pain (physical or emotional). Patients may give away cherished possessions or put things in order and make wills at a time out of context for their life situation and medical prognosis. They may rehearse fatal behavior in a tentative way, imagining what it would be like to hang or suffocate themselves. They may contact significant people to say their last good-byes (Rosenberg et al., 1988).

Some time periods are high-risk intervals for suicide. The recovery period from a major depressive episode or schizophrenic break juxtaposes the increasing energy and organization needed to carry out a fatal act with increasing insight and despair over the effects of illness. Bereavement is also a period of heightened risk, especially if the decedent committed suicide. The term "suicide contagion" has been used to describe a series of suicides in which one death appears to influence the others, probably through identification and modeling. Reported clusters of such suicides, especially among youth, recently have received widespread media attention. The time period surrounding an ongoing cluster of these suicides or around intense media attention to a real or fictional suicide story constitutes a high-risk period for susceptible young individuals (Davidson & Gould, 1989).

Family and other social supports may be a protective factor, and their absence may be an indication of increasing isolation, alienation, and withdrawal leading to suicide. Self-destructive feelings are transient, and even completed suicides represent a fatal outcome of an ambivalent state. The availability of social supports may provide auxiliary emotional resources to bridge the time period of intense suicidality. The family interview complements the patient interview in the management of the suicidal patient. Families can provide the information patients edit from their interviews with the doctor. Willing family members can also store and administer medications, remove dangerous materials from the household, transport patients to appointments, and convey by their interest and involvement that the patient's situation is not hopeless.

More disturbed families may communicate that the patient is "expendable." The patient may have already exhausted the family's resources, and the "burned out" family may present him to the hospital to be relieved of their burden. Even well-meaning families often have unrealistic hopes and expectations of psychiatric treatment. The patient and family may benefit from plain talk about the time course of psychiatric treatment and its limitations in affecting behavior. This tactful disillusionment is particularly relevant for a chronically suicidal patient's family, which wishes that the psychiatrist could assume responsibility for the patient's life.

Psychodynamic Factors

In considering psychodynamic factors, the psychiatrist examines the meaning of suicide in the light of the patient's ego strengths, patterns of defense, and fantasies. No single paradigm accurately reconstructs the intrapsychic function of suicidal impulses and behavior for all patients. Nonetheless, loss is a predominant theme. Loss may be either tangible (the death of a pet) or intangible (an injury to self-esteem). The loss may be actual or threatened. Understanding the immediate loss in its personal context, the patient's defensive strategies in coping with loss, and the function of loss in the patient's fantasies is useful in determining probable outcomes for the patient and possible interventions. A suicide attempt and its sequelae may, in effect, make some restoration of or restitution for loss, and this can be a major determinant in assessing what's next for the patient.

Classically, suicide is viewed as retroflexed rage—taking revenge on an internalized bad object (Menninger, 1933). The suicide may be an effort to punish significant others or to express hostility toward others that would be socially prohibited. Some suicides and attempts are most usefully viewed as a cry for help. For example, persons who cannot tolerate dependency needs or cannot express needs for affection may enact these needs self-destructively. Other people can never be totally responsible for preventing a patient's suicide, and whatever intervention is planned must reestablish the patient's internal impulse control. Understanding the intrapsychic meaning of suicide for the particular patient is central in this transition.

The Mental Status Examination in Suicide Assessment

The mental status examination identifies psychotic patients whose reality testing is impaired, perhaps even to the extent of responding to self-destructive command hallucinations or delusions. Typically, however, data from the mental status examination are used to assess more subtle impairments of insight and judgment. Patients may present a rational, intellectualized discourse on the logic of committing suicide or the reasons they would never complete suicide. Their conclusions, however, may reflect as much impairment of insight and judgment as a schizophrenic's distorted thinking. Ability to establish rapport with the interviewer also contributes to the psychiatrist's impression that the patient could contact the physician if feeling overwhelmed by suicidal impulses, avail himself of proffered supports, or keep a follow-up appointment. Clinical judgment about the presence of treatable psychiatric illnesses, such as depression or anxiety disorders, that predispose to suicide is grounded in the mental status examination as well as the history. Murphy has documented physicians' errors of omission in neglecting to examine patients for depression (Murphy, 1975, 1983).

The Chronically Suicidal Patient

Chronic self-destructiveness and suicidal behaviors frequently occur among patients with borderline personality organization whose day-to-day functioning is impulsive and whose desperate attempts to control others may be enacted with increasingly lethal props. For these patients, the risk of suicide is heightened throughout an episode of depression or dysthymia rather than more specifically during the recovery phase, as in patients with Axis I disorders only. "Patients who simultaneously display general impulsivity, dishonesty, chronic self-mutilating tendencies, alcohol and/or drug abuse, and a profound interpersonal aloofness or emotional unavailability may develop suicidal behavior at any time" (Kernberg, 1984).

Episodes of self-mutilation may relieve tension associated with feelings of derealization and clearly not demonstrate conscious suicidal intent. However, self-mutilative behavior increases the risk of subsequent suicide. The likelihood of suicide may be unappreciated by clinicians who must repetitively provide emergency medical care when the patient has slash or burn injuries. Similarly, cumulative suicide attempts increase the likelihood of completed suicide (Shearer et al., 1988), but the actual increased risk may not be appreciated if the clinician views such behaviors exclusively as indications of manipulative threat or "acting out." Repetitive suicidal behavior may foster a deceptive sense of assurance that the patient will remain an attempter, not a completer.

Litman (1989) estimated that 20% to 25% of chronically suicidal patients eventually complete suicide. Although such patients may repeatedly present in crisis, crisis intervention alone is ineffective in preventing their suicides because the suicidal behaviors represent a self-destructive life-style, which needs to be transformed. Litman's experience in working with chronically suicidal patients reinforces the value of a team approach to treatment, with flexible combinations of medications and psychotherapy. As with other severe, chronic conditions, "sometimes the treating physician has to be satisfied, not with a cure, but with an effort to substantially improve the quality of life for the patient, and to postpone death" (Litman, 1989).

Some chronically suicidal patients who function at the borderline level, including those diagnosed with borderline, narcissistic, and schizoid personality disorders, can benefit from a trial of low-dose neuroleptic medication. The pharmacologic goal is mood and impulse stabalization and decreased frequency of psychotic symptoms.

Clinical management of these patients includes immediate hospitalization of those with a major mood disorder or who are psychotic. The patient's bland denial of suicidal intent and indifferent or devaluing attitude toward the therapist heighten the risk of suicide; the patient's communication is dishonest, either consciously or through splitting and denial. Kernberg's (1984) treatment recommendations include the following:

1. Confronting patients with their behavior patterns and how their behavior affects others.
2. Refusing to carry out treatment under circumstances that would cripple it, such as the patient's refusal of integral components (e.g., regular sessions, medication, family involvement).
3. Continuing open and direct communication, which does not ignore the angry and manipulative aspects of the patient's life-style and through which the therapist's interpretations consistently explore suicidality.
4. Examining and limiting secondary gains of suicidal behavior, especially because the patient may attempt to dominate and control the family and the therapist.
5. Communicating to the patient that the therapist would be saddened if the patient killed himself, but that the therapist's life would not be shattered.
6. Communicating the patient's chronic suicide risk to the family and realistically limiting the therapist's role in changing such behavior.

Suicidal Patients in the Emergency Room and on the Medical-Surgical Ward

The most pressing consideration in the evaluation of the suicidal patient in the emergency room is to protect the patient from immediate self-harm. The patient should not be left alone, even momentarily, until the interviewer is convinced that the patient is not immediately suicidal (Bassuk, 1984). The interviewing area should not offer any easily accessible means of suicide, such as breakable windows, cigarette lighters, cleaning or pharmaceutical supplies, electric cords and window sashes, curtain or closet rods, or plastic trash can liners (Litman & Farberow, 1970).

A great deal of time and effort, including clever mnemonics, has been devoted to suicide screening questionnaires that might be rapidly and universally administered. None have demonstrated adequate sensitivity and specificity in the short-term prediction of suicide (Eddy, Wolpert, & Rosenberg, 1989). Their utility lies in increasing the likelihood that the clinician will ask about suicidal ideation and behaviors and other known risk factors. The sum of the positive risk factors (even if weighted) does not equal assessment of suicidal risk.

In assessing potentially suicidal patients, the psychiatrist should not abruptly ask, "Are you suicidal?" Patients who feel that things are hopeless, wish that they were dead, and yet have no concrete plan may answer "No," precluding a thorough assessment of their suicidal ideation. Patients often do not categorize themselves as suicidal because they are not planning an immediate attempt. General inquiries such as, "How badly do you feel?" allow a natural progression to questions about hopelessness, wishing to be dead, wishing to harm oneself, wishing to die, and any details of plans and opportunities (Sletten & Barton, 1979). If the interviewer proceeds in an empathic, nonjudgmental way, the patient is more likely to be able to confide details of suicidal thoughts or plans. This approach provides data for the psychiatrist's clinical judgment instead of asking the patient to render a summary judgment of his suicidality.

Patient characteristics that make outpatient management reasonable are evidence of satisfactory impulse control, absence of psychosis or intoxication, absence of a specific plan and readily accessible means, presence of social supports, and ability to establish rapport with the interviewer. If the suicidal patient is not being admitted, the interviewer should set up a definite follow-up appointment with a treatment provider. Nonspecific treatment referrals to the community mental health clinic or instructions to contact a psychiatrist are ineffectual. Calling the patient back to check on arrangements made increases the likelihood of compliance.

Among those at increased risk of suicide in the hospital are patients admitted for injuries sustained in attempted suicide. The percentage of suicide attempters who kill themselves while hospitalized for their attempt is 0.2 (Glickman, 1980). Hospital policy usually provides that a psychiatrist be called to evaluate persons seen in the emergency room or admitted for intentional self-inflicted injuries. Determining whether the attempter should be discharged to outpatient care or transferred to the psychiatry service when medically cleared is often a pressing decision.

A request to see a recently extubated self-poisoning patient, who is demanding to go home and whose doctor needs an intensive care unit bed for the next admission forces the issue. If the patient is no longer appropriate for the medical-surgical service and the psychiatric evaluation is not yet thorough enough to determine present suicidality, then psychiatric hospitalization is imperative. A patient who has ingested a large dose of a drug with a long half-life, such as diazepam, may be stable metabolically and out of danger of respiratory depression long before the drug's psychotropic effects have dissipated. The massive anxiety that precipitated the attempt may thus be masked. A period of psychiatric hospitalization for evaluation also allows time to meet with family members and significant others to provide a more reliable assessment of the patient's support systems.

Attempters who survive construct a rationale for their actions (Kiev, 1975). The physician who uncritically accepts this post hoc construction may mistakenly trivialize the attempt, consider it a "gesture," and dismiss serious suicidal behavior. The survivor's embarrassed denial of suicidal intent (e.g., "I just wanted to get some sleep") cannot be taken in lieu of a more thoughtful assessment of suicidality. Understanding the meaning of the attempt in its intrapsychic and interpersonal contexts provides a much sounder basis for assessing the patient's capacity to survive as an outpatient. In many instances this assessment may not be possible in the time and milieu available on the medical-surgical ward.

For some patients, transfer to the psychiatric service may be adventitiously therapeutic. The transfer itself may convey the seriousness of the patient's actions to the patient and family in ways that underscore the need for continued treatment or that mobilize support systems. Patients may channel their rage toward the transferring physician or staff so that the vector of that rage is redirected from self-destructive injury to verbally expressed anger.

Many suicide attempters have been in some form of psychotherapy. Inpatient psychiatric hospitalization may be indicated to evaluate that therapy, its relationship to the attempt, and the configuration of therapies best suited to the patient's needs after discharge. Discharging patients directly from the medical-surgical ward because they are in ongoing psychiatric treatment is a tempting option, but the patient in therapy who has acted out self-destructive impulses has at least temporarily exceeded the patient-therapist capacity for containing and working on those impulses. Analyzing these limitations of the patient, the therapist, and their interaction while the patient is hospitalized can allow their therapy to be restructured or permit different therapeutic arrangements to be made. Conversely, patients with excessively punitive superegos may project their guilty, bad feelings onto the therapist and insist on an immediate disposition to another clinician. Grappling with the internal conflict may prevent the patient from reenacting the self-directed riddance (suicide attempt) through an other-directed riddance (dismissing the therapist).

Some facilities offer the option of admitting suicidal patients to a medical-psychiatric unit, which allows more intensive intervention than liaison consultation on a medical-surgical unit can provide. Patients most appropriately managed on the medical-psychiatric unit are motivated for inpatient treatment and combine an organic physical disorder with an Axis I disorder. Patients with severe personality disorders tend to do poorly on the medical-psychiatric unit because the milieu is usually not set up to control their limit testing and acting out behaviors.

With intensely suicidal patients, no environmental arrangements are sufficient deterrents. The patient must be under constant observation, even while toileting. Although environmental controls alone cannot prevent suicide, attention to these details reduces risk. The physician may leave orders for "suicide precautions," yet patients can hang themselves with intravenous tubing or jump out the window. The trash can may have a plastic liner suitable for asphyxiation. The meal tray may arrive with metal cutlery. A roommate may have antiseptics, mouthwash, bandage scissors, shampoo, cologne, or a cigarette lighter on the bedside table. Patients may "cheek" their medications and stockpile a lethal dose. The clinician is always less inventive than the determinedly suicidal patient. None of us is infallible in deciding that a patient is not immediately suicidal, but it is folly to ever believe that the patient cannot kill himself.

Aggressive diagnosis and treatment of underlying psychiatric disorders, environmental safety, and constant observation are the keys to managing the suicidal patient in the general hospital. Electroconvulsive therapy should receive primary consideration for the severely depressed and imminently suicidal patient. Necessary medications can be given in elixir form or by injection, both to ensure that the patient takes them and to prevent stockpiling. The psychiatric unit generally provides a structurally safer environment, with Plexiglas windows, breakaway curtain rods, and locked storage of sharp and toxic materials. Even on the psychiatric unit, however, the imminently suicidal patient requires constant, one-to-one nursing.

Less intensive suicide prevention measures work on the basis of restricting access to immediately lethal methods and providing frequent staff contact to serve as auxiliary ego controls and to frequently reassess the patient's condition. This arrangement conveys that the hospital takes the patient's suicidal ideation seriously and will assist the patient in controlling suicidal impulses. It assumes that the patient is ambivalent enough about suicide to collaborate in these arrangements.

Treatment of the Suicidal Outpatient

Most persons with suicidal ideation are treated as outpatients. Pharmacotherapy of their underlying psychiatric disorders, however, may involve medications, such as tricyclic antidepressants, that are a readily lethal means of suicide. Amitriptyline is the most frequently ingested medication in suicide by overdose in the United States (Davidson, 1985), whereas doxepin, for instance, leads in Finland (Vuori et al., 1989). Differences among rates of fatal self-poisonings with antidepressants reflect both how widely prescribed each medicine is and its inherent toxicity. The relative toxicity of amitriptyline, dothiepin, doxepin, and desipramine is greater than that of trazodone, clomipramine, lofepramine, or mianserin (Beaumont, 1989; Henry, 1989; Montgomery, Baldwin, & Green, 1989). Newer agents such as fluoxetine have a much wider safety margin (Borys et al., 1989; James & Lippmann, 1991). Limiting tricyclics to 1,200 mg total (or a week's supply if less) in a nonrefillable prescription may help reduce deaths from antidepressant overdose, although stockpiling of medications by the patient can subvert this precaution. Limiting prescription size should not be confused with limiting the dose. Homeopathic doses of antidepressants prescribed by overly tentative physicians increase the risk of suicide by leaving the patient's illness pharmacologically undertreated.

Recently, Teicher, Glod, and Cole (1990) reported intense suicidal preoccupation among some patients treated with fluoxetine; they attributed these suicidal thoughts to the patients' antidepressant medication. The Church of Scientology subsequently capitalized on this adverse publicity as part of their antipsychiatry campaign, as did predatory plaintiff attorneys and litigious patients. Much patient distress and apprehension about medication has been generated by these reports. Teicher, Glod, and Cole's original report overlooked earlier published work showing a rare worsening of depressive episodes with consequent suicidality while on other antidepressant medications (Damluji & Ferguson, 1988). Subsequently, a chart review study by psychiatrists blind to the study hypotheses demonstrated no significant difference among various antidepressant regimens and the rates of new-onset suicidal ideation (Fava & Rosenbaum, 1991). A comprehensive review of studies of treatment-emergent suicidal ideation by Mann and Kapur (1991) reached an identical conclusion. The new onset of suicidal ideation was more likely to occur among psychotic patients, those with borderline personality disorder, and those with worsening depression. Clinicians must always be alert to the rare possibility of new-onset suicidal ideation among patients treated with antidepressants, but the relatively low toxicity of fluoxetine in overdose makes it an extremely safe antidepressant.

Sedative-hypnotic and anxiolytic drugs can also be lethal, especially combined with alcohol. Diazepam is the third most frequently ingested medication in suicide by overdose (Davidson, 1985). Secondary depressions in anxiety disorders and mixed anxiety-depressive symptomatology are characterized by enormous psychic pain and perturbation, which must be rapidly addressed to prevent suicide. Aggressive but carefully monitored therapy with benzodiazepines can provide symptomatic relief during the time it takes antidepressant medication to reach effectiveness (Fawcett, 1990).

The shift from prescribing barbiturate and respiratory suppressant sedative-hypnotics to prescribing benzodiazepines was associated with a steady decline in overdose suicide rates (Whitlock, 1975). The household pharmacy can be reduced by appropriately treating mood disorders and psychoses specifically, rather than with symptomatic medications such as sedative-hypnotics and anxiolytics, and by recommending disposal of old medications. A reliable family member can administer medications to noncompliant or unreliable patients, and frequent office visits should be scheduled to assess patient response and suicidality. Partial responders or nonresponders to antidepressant medications can be asked to bring unconsumed medication to the doctor's office for disposal when prescription of a different medicine is anticipated (Farmer & Pinder, 1989).

Some clinicians have advocated using suicide prevention contracts as a means of reducing the likelihood of suicidal behavior without unnecessarily restricting the patient's autonomy and freedom of movement (Assey, 1985). The clinician communicates the wish for the patient to live and negotiates specific arrangements that will help the patient manage self-destructive urges. The patient then verbally affirms the contract specifics, including the time frame, and promises to honor the contract. Suicide prevention contracts have no empirically demonstrated effectiveness. Nonetheless, the process of negotiating with the clinician and verbally affirming the contract with the caregiver may be regarded as a mode of supportive psychotherapy that is helpful to some patients. Additionally, some clinicians may find the process useful in ensuring that they develop a clear and specific plan with the patient for dealing with exacerbations of self-destructive urges. Most suicidal patients need concrete, written information about how and under what circumstances to call the clinician (Jamison, 1988). Otherwise confusion, guilt, and hopelessness can make answering services, call coverage schedules, and pagers seem like impossible hurdles.

Medical-Surgical Patients at High Risk

Depression precipitated by major medical illness can affect up to 20% of severely ill medical-surgical patients and more often represents their first lifetime psychiatric illness (Winokur, Black, & Nasrallah, 1988). (See also Chapter 4). Outpatients with concurrent psychiatric and physical illnesses who attempt suicide are also more likely to use violent methods (e.g., gunshot) rather than drug overdose (Kontaxakis et al., 1988). Neurologic disorders, such as stroke, temporal lobe epilepsy, and multiple sclerosis, that directly impact neuropsychiatric functioning have higher rates of depression than in the general population (Kontaxakis et al., 1988; Silver, Hales, & Yudofsky, 1990).

Human immunodeficiency virus (HIV) infection and acquired immunodeficiency syndrome (AIDS) may differ in their impact on suicidal ideation and behaviors. Persons seeking HIV testing have higher lifetime rates of mood disorders before notification of test results (Perry et al., 1990), putting them at increased risk of suicide on that basis alone. Measures of suicidal ideation, however, decrease from before to after HIV testing for both seropositive and seronegative respondents (Perry, Jacobsberg, & Fishman, 1990). As a catastrophic illness producing neurogenic neurobehavioral changes (Grant & Atkinson, 1990), AIDS is likely to be more strongly associated with suicide than other major illnesses without direct central nervous system effects. Data from New York City demonstrate a 36-fold increase in the relative risk of suicide among men ages 20 to 59 with AIDS (Marzuk et al., 1988). Suicide attempts with zidovudine among patients with HIV infection (Terragna et al., 1990) may represent this illnesses' equivalent of suicide attempts via antidepressant medications among mood-disordered patients.

Cancer patients generally have been regarded as a group at higher than average risk for suicide. Although patients hear the diagnosis of cancer as less of a death knell now than it was in the past, cancer is still fearfully associated with pain, disfigurement, and loss of function. In the vulnerable patient, any of these may precipitate suicide. The physician's impression that it is normal to be depressed if one has cancer ("Well, I'd be depressed too if I had . . .") may leave treatment-responsive major depression untreated and thereby increase the likelihood of suicide (see also Chapter 24).

The risk of suicide among cancer patients has been calculated from the Finnish Cancer Registry. Suicide was 1.3 times higher among male and 1.9 times higher among female cancer patients than in the general population. Patients with nonlocalized cancers had twice the risk of suicide as those with localized cancers, and gastrointestional cancer patients had the highest relative risk. Those patients who underwent traditional surgical or radiation therapy were not at increased risk of suicide, but the suicide risk among patients receiving other or no specific treatment was significantly elevated (Louhivuori & Hakama, 1979). This latter group included patients receiving steroid therapy as well as those who committed suicide before any treatment could be initiated. Subsequent analyses with the Swedish Cancer-Environment Register (Allebeck, Bolund, & Ringback, 1989) confirmed an increased rate of suicide among cancer patients, with highest rates during the first year after cancer diagnosis. These data showed statistically significant differences among suicide rates by tumor site, with increases noted for gastrointestinal tumors for males and lung/upper airway tumors for both sexes.

These findings are consistent with the hypothesis that cancer patients with better prognoses may be less likely to commit suicide. Even terminally ill cancer patients, however, have been reported *not* to wish to die unless they are also clinically depressed (Brown et al., 1986). Being suicidal or wishing for an early death was associated with the presence of clinical depression in the patients studied, not with the severity of their physical illnesses, since all were terminally ill, were aware of their prognosis, and had severe pain, disfigurement, or disability. Feeling suicidal is thus not inevitably part of the natural course of terminal illness.

Retrospectively comparing cancer patients who have committed suicide with nonsuicidal hospital controls highlights some of the situational variables besides being depressed that may cue the physician to suicidality (Farberow, Shneidman, & Leonard, 1963). The suicidal patients had been overly involved in their treatment, with behaviors perceived as controlling, complaining, and demanding. They were less tolerant of pain and more likely to feel that their physical and emotional resources were depleted. These needy and insistent patients can exhaust the responsiveness of their caregivers. Yet, such patients are acutely sensitive to withdrawal of contact or interest by others. Transfer to a private room to die exemplifies the type of environmental change that may be perceived as rejection and may culminate in suicide.

Farberow, Shneidman, and Leonard (1970) emphasized the power of suicidal intent over physical frailty or motor limitations. A patient's deteriorating behavioral pattern or more overt suicidal communications must be taken seriously and not discounted because of the patient's debilitated physical state. In Farberow et al.'s series, one partially paralyzed patient was immobilized in a Stryker frame and still committed suicide by self-immolation after dousing himself with lighter fluid. Clearly, when suicidal intent is strong, one cannot assume that physical limitations will prevent a suicide attempt.

The insistent, complaining, and controlling cancer patients who committed suicide were in a treatment setting that radically opposed their lifelong character structures. Once such potentially suicidal patients are identified, therapy can be initiated that (1) treats underlying depression or other concurrent psychiatric disorder, (2) reestablishes social and emotional supports, and (3) reestablishes the patient's sense of control. For some terminal patients, psychotherapy provides an arena for actively confronting the personal psychodynamic meaning of their cancer and thus reestablishing emotional potency despite physical deterioration (Leigh, 1974). The psychiatrist's acknowledgment that suicide remains an option for the patient, although not condoning it or accepting it as "rational," may help some patients regain their feelings of autonomy sufficiently to work toward other goals. Reestablishing purpose for oneself takes many forms; for instance, teaching the doctor about the feelings and experiences of approaching death.

Hemodialysis patients are another group at increased risk for suicide because of the characteristics of their illness. Their suicide rates are relatively high. Including those whose suicides are effected by discontinuation of dialysis or intentionally fatal breaches of the treatment regimen, rates are up to 100 times that of the general population (Cloonan, Gatrell, & Cushner, 1990). About 5% of dialysis patients commit suicide. Suicidal behavior is more common among patients attending dialysis centers than those on home dialysis. Excess mortality from causes other than overt suicide has been correlated with higher scores on the Beck Depression Inventory, because depression can predispose to death through dietary neglect, carelessness with vascular access sites, and inattention to complex medication regimens (Shulman, Price, & Spinelli, 1989). Long-term dialysis survivors are characterized by high levels of denial, but denial combined with overlapping somatic complaints may make depression difficult to recognize in the early stages of dialysis (Burton et al., 1986).

Reasons for suicide among these patients are legion. The quality of life on dialysis is significantly impaired. Losses associated with this change in physical health are major threats, such as job loss and social isolation (Dorpat, Anderson, & Ripley, 1968). The loss with transplant failure may even exceed that in initiating dialysis and can precipitate severe depression. As with the suicidal cancer patient, the chronic hemodialysis patient may act as though the locus of control were primarily external (Goldstein & Reznikoff, 1971). Death is an ever-present possibility. Those for whom the fearfulness and anxiety attendant on death are intolerable may defend against death through reaction formation and, paradoxically, commit suicide (Dorpat, Anderson, & Ripley, 1968). Ready access to lethal methods also contributes to the dialysis patient's likelihood of suicide. Disconnecting a shunt, severing the arteriovenous fistula, or inducing hyperkalemia through food binges are quite common. These methods particular to dialysis may also be overly determined because they symbolically express some of the patient's conflicts.

Dialysis patients without psychologic problems attendant on the illness are so rare that recommendations for suicide prevention in this special population emphasize the need for universal, ongoing psychologic support (Abram, Moore, & Westervelt, 1971; Haenel, Brunner, & Battegay, 1980). Although crises occur at nodal points in dialysis treatment, the severity of ongoing stresses for these patients is so great that attempts to initiate treatment amid crises are much less likely to succeed. Self-destructive impulses are too readily enacted, and the sum of previous losses rekindled by the current crisis may be too overwhelming. Psychiatric aspects of chronic renal failure and dialysis are discussed in more detail by Levy in Chapter 27.

Other medical-surgical patients at high risk for suicide are those in alcohol withdrawal delirium (delirium tremens), particularly if they are inadequately sedated (Kellher et al., 1985). Additionally, patients with severe personality disorders may make suicide attempts when the organic basis of their medical symptoms is challenged (Reich & Kelly, 1976). Patients with respiratory diseases have also been overrepresented among hospital suicides (Baker, 1984). Whether their relative anoxia, progressive incapacity or some other factor accounts for their increased suicide risk is unknown. Ongoing evaluation of medical-surgical patients' emotional status can detect potentially suicidal patients who did not suggest psychiatric concern on admission but have developed intense reactions to pain, unbridled fears of death, or clouded consciousness impairing reason (Pollack, 1957).

Countertransference to the Suicidal Patient

Acutely and chronically suicidal patients mobilize countertransference responses that can exacerbate their self-destructiveness. The frustration, peril, and feelings of helplessness that their treatment can entail can evoke hatred from the therapist. This countertransference hatred is projected and then felt as dread that the patient will commit suicide (Maltsberger & Buie, 1974). Another countertransference expression is the therapist's feeling of relief when a difficult, chronically suicidal patient drops out of treatment, whereas the patient experiences dropping out with increasing feelings of hopelessness and doom (Litman, 1989). The therapist may issue an ultimatum to terminate treatment if the acting-out patient repeats provocative behaviors. When the forbidden behavior is repeated, the therapist may unconsciously express countertransference aversion in dismissing the patient, thereby also shifting responsibility for treatment failure. Suicidal danger is intensified by the personality-disordered patient's lack of social supports and internal emotional resources when deprived of the therapist's ongoing relatedness (Maltsberger & Buie, 1990).

Excessive reliance on clinical intuition may distort professional judgment, since countertransference influences are part of the same preconscious processes as intuition. For instance, the therapist may ward off empathically experiencing a profoundly disturbed patient's worthlessness, aloneness, or despair. Without this resonance, the therapist may "intuitively" feel that the patient is not suicidal. This countertransference avoidance can also lead to too-literal reliance on the mental status examination, with hazardous nonrecognition of lethal suicidal preoccupation in patients who are not overtly depressed (Maltsberger & Buie, 1990).

Projective identification from suicidal patients with severe personality disorders may evoke countertransference responses that ostensibly return the split-off part to the patient. This hasty return, however, is experienced by the patient not as a restored part of himself, but as a sadistic attack from an external object (the therapist), which may further provoke suicidal behavior. The patient may die to kill the loathsome introject or in masochistic surrender to (or attack on) the intensely valued therapist.

A psychiatrist's self-image as a physician is challenged by suicidal patients who do not respond to therapeutic interventions. If feelings of personal worth are overly attached to outcome rather than the ability to exercise one's best professional skills, the psychiatrist will feel increasingly unhappy and inadequate in working with suicidal patients (Maltsberger & Buie, 1974). The patient's sadism fuels lethal behavior when the patient perceives that the therapist has established an image as a "model of omnipotent goodness" (Zee, 1972).

INTERPERSONAL VIOLENCE IN THE MEDICAL-SURGICAL SETTING

Interpersonal violence is a major public health problem in the United States. Perpetrators of violence outside the hospital have illnesses and injuries that make them our patients, too. Hospitalization is always stressful and represents a setting in which physical and verbal assaults may erupt. If we can extrapolate from psychiatric patients, the best predictor of hospital violence is a recent history of violence in the community (Binder & McNiel, 1990; McNeil, Binder, & Greenfield, 1988). Thus, taking a careful history of recent aggression or fear-inducing behavior increases one's predictive ability. Predicting near-term violence is much more accurate than postdischarge or posttreatment predictions.

Instances of interpersonal violence among hospitalized medical and surgical patients are relatively uncommon but of great consequence. Like self-destructive patients, those who are violent toward others are overrepresented among certain diagnostic groups and in particular clinical contexts. Most typical is a male substance-abusing patient admitted for a complication of his drug use, such as cellulitis or endocarditis. This patient may chafe at hospital rules, have escalating demands for pain medication, and lose control of his aggressive impulses. When violent patients were compared with nonviolent controls, the violent patients had longer hospital stays. The most likely time for violence to erupt was between 7:00 p.m. and 7:00 a.m. Prior documented threats were uncommon, but patients who later became violent had been perceived as uncooperative (Ochitill & Kreiger, 1982).

To maintain their dependency, drug abusers must be rule breakers, so the usual hospital rules and regulations represent major sources of conflict. The hospital is an alien environment, and particular difficulties may arise at the end of visiting hours if the substance abuser does not want his friends to leave. Challenging the reality of the patient's pain can provoke violence. Substance abusers are often undermedicated for pain, especially if the physician calculates that a maintenance dose of methadone will also block pain or that the half-life of pain medications will be as long for these patients as for nonabusers. Drugs are the way that these patients treat their emotional needs, and the hospital situation exacerbates their problems while preventing access to the customary remedy. Avoiding unnecessary confrontation and maintaining a sufficient pain control regimen reduces the likelihood of violence among substance abusers (see Chapters 8 and 16).

Patients may be precipitously violent during confusional states. Regulating sensory stimuli, treating the underlying medical condition, and preserving sleep are helpful. Cognitively-impaired or brain-injured patients may increasingly rely on one or two staff members to organize tasks of daily living and attend to their needs. When those caregivers are away for extended periods or busy with others, patients can feel increasingly anxious and helpless, leading to fight-flight responses (Armstrong, 1978). Delirium from acute alcohol withdrawal is a medical condition that may be unsuspected in the trauma or surgical patient. The postoperative course may begin benignly, but increasing agitation and irritability can later explode in violence.

Among general hospital patients, those with identified psychiatric disorders are infrequently violent. This may reflect the benefit of more frequent requests for psychiatric consultation early in the hospital stay. Assaultiveness among psychotic inpatients correlates with hallucinatory behavior, conceptual disorganization, and unusual thought content as measured by the Brief Psychiatric Rating Scale. In this population, overtly hostile behavior is a less robust predictor than these indicators of thought disorder (Yesavage et al., 1981).

Hospitalization affords little privacy. Diagnostic procedures and room arrangements put patients physically closer to others than many would choose to be. Kinzel (1970) has identified a "body-buffer zone," which is the unpeopled space around an individual that is perceived necessary for emotional safety. This perimeter is larger for violent persons and is configured with greater distance behind than in front. The paranoid male, fearful of homosexual assault, may develop unbearable anxiety in crowded hospital situations and may explode aggressively. Hospital boredom may also be intolerable to those character-disordered patients who feel empty when they are unstimulated. Verbal explosions, assaults, and self-destructive behaviors are pathologic responses that disperse the emptiness and reestablish a sense of reality and focus (Kalogerakis, 1971).

Rating scales have been developed to document and quantify verbal and physical aggression (Kay, Wolkenfeld, & Murrill, 1988; Yudofsky et al., 1986). These are not predictors of subsequent aggression, but are useful tools in monitoring hospital aggression and pharmacologic, behavioral, or environmental intervention efforts with individual patients. Quantifying the observed behaviors allows a more objective assessment of intervention effectiveness.

Emergency Evaluation of and Response to Interpersonal Violence

Patients who present with fears of becoming violent require immediate evaluation. Many physicians are reluctant to take a history that inquires specifically about violence and available weapons, especially guns. However, "patients who experience violent impulses desperately want help in curbing such urges. Violent patients are terrified of losing control and welcome therapeutic efforts that restore a sense of control and prevent them from acting on their urges" (Lion & Pasternack, 1973).

A patient who is pacing in the waiting room or sitting on the edge of the chair gripping the armrests, who is hypervigilant and easily startled, and who cannot respond to a quiet offer of food or the invitation to talk requires immediate external controls (Lion, Bach-y-Rita, & Ervin, 1969). If the patient will take medication by mouth, anxiolytics or oral-concentrate neuroleptics should be given. Injections create their own problems with fears of homosexual penetration or bodily assault. When necessary, however, intramuscular neuroleptics, lorazepam, or amobarbital sodium rapidly help the patient shift away from fight-flight responses.

Medicating the Violent Patient

Rapid tranquilization requires close observation of the patient so that the medication can be titrated to an appropriate end point. Patients whose extreme agitation is relieved as quickly as possible without somnolence are less likely to feel that they have been chemically abused. Obtunding the patient is not the same as relieving anxiety and agitation sufficiently for the patient to talk, rest, or sleep. Most cases of psychotic agitation and paranoid intoxication, such as with phencyclidine, cocaine, or amphetamines, respond to neuroleptics (e.g., haloperidol, 5 to 10 mg oral concentrate or intramuscular injection every 30 minutes). Chlorpromazine, 50 to 100 mg, has been widely used but is more sedating and more likely to be dose limited by orthostatic hypotension. Alternatives for patients with a history or symptoms of neuroleptic malignant syndrome or other contraindications to neuroleptics are lorazepam, 1 to 2 mg orally or intramuscularly every 30 minutes, or amobarbital sodium, 250 mg intramuscularly, repeated as needed.

Specific medications are useful for intoxications or withdrawals that present with agitation, irritability, and combativeness (see Chapter 8 for details).

Interviewing the Potentially Violent Patient

The interviewer should sit closest to, but never blocking, the exit. Other staff or security personnel within view, but not menacingly close, can feel protective to the patient and the clinician. An interview area that is free of distractions (especially any other obstreperous or disorganized patient) and yet not isolated is conductive to a productive interview.

The interview should begin with neutral questions rather than a direct opening inquiry about the precipitants for the patient's presentation. As the interview progresses, the physician does need to broach the particulars of the patient's impulses, past actions, immediate plans, and opportunity to carry out these plans. The psychiatrist may acknowledge the patient's anger and also indicate that the patient can be assisted in controlling aggressive urges. One goal of the interview is to "convert physical agitation and belligerence into verbal catharsis" (Lion, 1985).

The patient may produce a weapon during the interview. The physician can offer to have the weapon stored, indicating that the patient may have brought it to the hospital to be relieved of the possibility of losing control and using it. The patient should be instructed to lay the weapon down and told that the physician will then take it. Attempting to take a gun directly from the patient may cause the patient to pull the trigger reflexively.

If the decision is made to hospitalize the patient, it should be presented as a nonnegotiable event but with thoughtful conveyance of its purpose. Patients who appear resistant to hospitalization frequently are relieved by it. Those whom the psychiatrist believes might adamantly resist should be informed of hospitalization at the moment of transfer, to prevent escalating anxiety and loss of control during their wait. Sufficient staff should be assembled so that it is indisputably clear that the staff represent an overwhelming force and resistance would be futile.

Advance planning for the rare instance in which forcible restraint is necessary encompasses policy and logistic decisions. Included are designations of permissible techniques, appropriate circumstances, and the chain of command. Controlling imminent violence is not the time for participatory democracy. Established guidelines should specify who will decide on the necessity of restraint, who will direct the staff in implementation, and what roles will be filled in carrying out this decision. A reporting system for events in which forcible restraint is necessary and an internal review process for examining what transpired in each instance are part of the unit plan (Mental Health and Behavioral Science Service Program Guide, 1983).

On the hospital ward, the individual staff member may need to decide whether to intercede immediately with a violent patient before other help can arrive. Damage to property can usually wait; our obligation to intercede in violence directed toward other patients is more imperative. As reinforcements arrive, someone should be assigned to move other patients out of the area. Sufficient staff and unarmed security personnel should be assembled to convey to even an irrational patient that resistance is impossible. The patient should be told clearly what to do (e.g., to lie face down on the bed). Violent patients requiring physical restraint should be told that restraint is to prevent them from injuring themselves or others and to allow time for them to regain control of their impulses, and that restraint will be discontinued as they are able to do that. Efforts to direct both the staff and the patient away from viewing restraint as retaliation or punishment make subsequent therapeutic interventions much easier.

Five staff members are the minimum necessary to restrain a patient—one to control each limb and one the head. Agitated patients can bang their heads and may bite. Grasping limbs close to the sockets reduces the likelihood of adventitious fracture or dislocation (Penningroth, 1975). Rapid-acting neuroleptic or anxiolytic medication is administered immediately, barring rare contravening metabolic situations. Persons in three- or four-point restraints need to be under constant observation. The immobilized patient's potential for aspiration, asphyxiation, and other hazards makes 15-minute checks useful only for discovering a warm body, not preserving life. HIV positive patients present obvious risks in such situations.

Ethical and Legal Concerns

Treatment of suicidal or violent patients raises ethical and legal concerns, including involuntary hospitalization and nonconsensual treatment, confidentially, and the protection of potential victims. The psychiatrist assesses the risk of danger to the patient or to others, and assessment that the risk is substantial and that it is due to a mental disorder usually constitutes the legal basis for commitment. In some instances, the patient willing to be hospitalized but too psychotic to understand the nature of the treatment contract may be better served by the review and documentation process of involuntary hospitalization (Skodol, 1984).

No matter how compelling the likelihood of injury, violence per se is not synonymous with psychiatric disorder (Skodol, 1984). When the bases of the patient's violence are social or environmental, not psychiatric, the evaluating psychiatrist should say so and decline custodial responsibility for those to whom no treatment can be offered. In assessing potential for violence, the psychiatrist relies on the patient's past history of violence, his current level of functioning, and the risk that accrues if the patient belongs diagnostically to a group with a significantly higher likelihood of violence. The ability of mental health professionals to predict individual violence over the long term does not exceed 50%, however, and is even less reliable in the short term (National Academy of Sciences, 1978). How risk was assessed should be carefully documented, along with a consultative opinion from another professional if the case is ambiguous.

Tarasoff v Board of Regents (1976) illuminated the issue of injured third parties in forensic psychiatry (Dix, 1985). The psychiatrist's ethical duty to exercise reasonable professional care in identifying patients who pose substantial risk of harming others extends to exercising reasonable professional care in protecting those at risk. Warning those at risk is one way, but not the only way, to exercise that professional responsibility. Continuing the treatment itself may constitute reasonable professional care, but should warning individuals or notifying law enforcement authorities be necessary, information will be disclosed that the patient revealed in confidence. At a minimum, the psychiatrist must inform the patient of the need to disclose confidential information and must tell the patient what information will be disclosed to whom (Dix, 1985). Numerous lawsuits filed since the original *Tarasoff* decision continue to further define the psychotherapist's potential liability for patients' destructive behavior. Courts are inconsistent from state to state, but are more often limiting the therapist's duty to protect by informing to instances of overt threats and identifiable victims (Appelbaum, 1988; Menninger, 1989).

Nonconsensual treatment of suicidal or interpersonally violent patients may require increasingly intrusive interventions, including restraint, seclusion, and involuntary medication. Their justification rests on the determination that the risk of serious harm is substantial and imminent and that less intrusive methods have not been or are not likely to be effective. Periodic room searches or personal searches may be necessary if the patient appears to be secreting potentially dangerous items. Detailing the behavior or condition that necessitated the treatment, the way the treatment decision was reached, the persons notified, and the outcomes of periodic reassessment to determine the patient's well-being and need for continued intervention will minimize the risk of litigation (Dix, 1985; Fogel, Mills, & Landen, 1986).

Negligent release cases have been brought in instances in which patients treated for aggressive or suicidal behavior or committed for treatment on the basis of dangerousness have harmed someone or themselves after release from the hospital. No national specific standard of care for release has been drafted. Thus, clinical staff must draft hospital or system policies in collaboration with legal consultants. Without following such a preexisting policy, the individual clinician is less able to demonstrate that he was diligent in making release decisions. Explicitly documenting violence or suicidal assessment and annotating information from social history, progress notes, mental status examinations, and other hospital records help demonstrate that the release decision was made systematically. Independent review and videotaped exit interviews are further provisions for documenting that the patient's potential for subsequent injury was carefully addressed (Poythress, 1990).

Treatment Issues for Aggressive Patients

More long-range treatments of violence are based on establishing the etiology of the patient's aggressive behavior. Outbursts resulting from temporal lobe epilepsy have a different therapeutic approach than violence as a reaction formation to feelings of passivity and weakness. The patient should be taught to recognize premonitory signs and physical sensations of impending aggression. For example, pathologic intoxication may encompass a repetitive sequence of loosening inhibitions with alcohol, projection of closeness as a homosexual advance, autonomic nervous system responses to that, and assault. The patient certainly need not share the psychiatrist's psychodynamic formulation but can learn to recognize repetitive situational factors and proprioceptive responses. Patients are also taught to verbalize anger rather than handling it behaviorally and are taught assertiveness, which can reduce the buildup of anger. Working with patients so that they can elaborate fantasies of the probable consequences of their acts is designed to interpose insight and judgment between impulse and action.

Countertransference responses to violent patients are universal. Denial is most typical. It leads the psychiatrist to miss taking a history of violence: criminal history, driving history, weapons ownership, and history of injury to others. With seductive but menacing patients, the psychiatrist may identify with the aggressor. Inappropriate attempts may be made to foster a positive transference; if this is based on avoiding discussion of aggressive impulses, it will not reduce the patient's potential for violence. Violent patients can make the physician feel angry and helpless. These feelings can lead to preoccupying fantasies that the patient will do something heinous for which the physician will be held responsible. The physician's anger toward such frustrating patients may be projected as excessive fears and ruminations about the patient's dangerousness and likelihood of assaulting the physician. The patient's patricidal or infanticidal impulses may touch similar unconscious impulses in the physician and generate the same sort of obsessive ruminations about the patient's dangerousness. Withdrawal from violent patients by physician and staff removes the patients from therapeutic human contact and actually increases the chance of violence. It conveys to patients the sense that they are truly uncontrollable (Lion, Madden, & Christopher, 1976; Lion & Pasternack, 1973).

SUMMARY

Neither suicide nor interpersonal violence constitutes a psychiatric diagnosis. Both are behavioral outcomes of complex social, emotional, biochemical, and environmental factors. The psychiatrist can, however, have a major role in reducing the likelihood of these destructive acts among medical and surgical patients. This role includes:

1. Carefully assessing each patient's potential for suicide and interpersonal violence.
2. Recognizing high-risk patients by their personal and situational risk factors, psychiatric diagnosis, and their psychodynamic issues.
3. Incorporating environmental controls, family and social supports, and protective staffing.
4. Aggressively treating underlying psychiatric disorders.

AUDIOVISUAL RESOURCES

Teen Suicide and Depression
Ridgeview Institute
3995 South Cobb Drive
Smyrna, GA 30080
1-800-345-9775

Suicide
Bureau of Audio/Visual Instruction
University of Wisconsin Extension
1327 University Avenue
Madison, WI 53706

REFERENCES

Abram HS, Moore GL, & Westervelt FB (1971). Suicidal behavior in chronic dialysis patients. Am J Psychiatry 127:1199–1204.

Allebeck P, Bolund C, & Ringback G (1989). Increased suicide rate in cancer patients: A cohort study based on the Swedish Cancer-Environment register. J Clin Epidemiol 42:611–616.

Alnaes R, & Torgersen S (1990). DSM-III personality disorders among patients with major depression, anxiety disorders, and mixed conditions. J. Nerv Ment Dis 178:693–698.

Altschuler LL, Casanova MF, Goldberg TE, et al. (1990). The hippocampus and parahippocampus in schizophrenic, suicide, and control brains. Arch Gen Psychiatry 47:1029–1034.

Appelbaum PS (1988). The new preventive detention: Psychiatry's problematic responsibility for the control of violence. Am J Psychiatry 145:779–785.

Arango V, Ernsberger P, Marzuk PM, et al. (1990). Autoradiographic demonstration of increased serotonin 5-HT_2 and β-adrenergic receptor binding sites in the brain of suicide victims. Arch Gen Psychiatry 47:1038–1047.

Armstrong B (1978). Handling the violent patient in the hospital. Hosp Community Psychiatry 29:463–467.

Arora RC, & Meltzer HY (1989). Serotonergic measures in the brains of suicide victims: 5-HT_2 binding sites in the frontal cortex of suicide victims and control subjects. Am J Psychiatry 146:730–736.

Assey J (1985). The suicide prevention contract. Perspect Psychiatr Care 23:99–103.

Baker JE (1984). Monitoring of suicidal behavior among patients in the VA health care system. Psychiatr Ann 14:272–275.

Bakwin H (1957). Suicide in children and adolescents. J Pediatr 50:749–769.

Barraclough B, Bunch J, Nelson B, et al. (1974). A hundred cases of suicide: Clinical aspects. Br J Psychiatry 125:355–373.

Bassuk EL (1984). Emergency care of suicidal patients. In EL Bassuk & AW Birk (Eds.), Emergency psychiatry: Concepts, methods, and practices (Ch. 2, pp. 97–125). New York: Plenum.

Beaumont G (1989). The toxicity of antidepressants. Br J Psychiatry 154:454–458.

Beck AT, Steer RA, & McElroy MG (1982). Relationships of hopelessness, depression, and previous suicide attempts to suicidal ideation in alcoholics. J Stud Alcohol 43:1042–1046.

Binder RL, & McNeil DE (1990). The relationship of gender to violent behavior in acutely disturbed psychiatric patients. J Clin Psychiatry 51:110–114.

Black DW, Winokur G, & Nasrallah A (1988). Effect of psychosis on suicide risk in 1,593 patients with unipolar and bipolar affective disorders. Am J Psychiatry 145:849–852.

Borys DJ, Setzer SC, Ling LJ, et al. (1989). The effect of fluoxetine in the overdose patient. Vet Hum Toxicol 31:364.

Brown JH, Henteleff P, Barakat S, et al. (1986). Is it normal for terminally ill patients to desire death? Am J Psychiatry 143:208–211.

Burton HJ, Kline SA, Lindsay RM, et al. (1986). The relationship of depression to survival in chronic renal failure. Psychosom Med 48:261–269.

Centers for Disease Control (1985). Suicide surveillance. Atlanta, GA.

Choi SY (1975). Death in young alcoholics. J Stud Alcohol 36:1224–1229.

Cloonan CC, Gatrell CB, & Cushner HM (1990). Emergencies in continuous dialysis patients: Diagnosis and management. Am J Emerg Med 8:134–148.

Damluji NF, & Ferguson JM (1988). Paradoxical worsening of depressive symptomatology caused by antidepressants. J Clin Psychopharmacol 8:347–349.

Davidson L (1985). Suicide and suicide attempts by the nonmedical use of drugs. MMWR 34:570–571.

Davidson L, & Gould M (1989). Contagion as a risk factor for youth suicide. In L Davidson & Linnoila M (Eds.), Report of the Secretary's Task Force on Youth Suicide. 2: Risk factors for youth suicide (pp. 88–109). DHHS Publ. No. (ADM) 89-1622, Washington, DC: U. S. Government Printing Office.

Davidson LE, Rosenberg ML, Mercy JA, et al. (1989). An epidemiologic study of risk factors in two teenage suicide clusters. JAMA 262:2687–2692.

Deykin EY, Alpert JJ, & McNamarra JJ (1985). A pilot study of the effects of exposure to child abuse or neglect on adolescent suicidal behavior. Am J Psychiatry 142:1299–1303.

DIX GE (1985). Legal and ethical issues in the treatment and handling of violent behavior. In LH ROTH (Ed.), Clinical treatment of the violent person (Ch. 9, pp. 178–206). Rockville, MD: National Institute of Mental Health.

DORPAT TL, ANDERSON WF, & RIPLEY HS (1968). The relationship of physical illness to suicide. In HL RESNIK (Ed.), Suicidal behaviors: Diagnosis and management (Ch. 15, pp. 209–219). Boston: Little, Brown.

EDDY DM, WOLPERT RL, & ROSENBERG ML (1989). Estimating the effectiveness of interventions to prevent youth suicides: A report to the Secretary's Task Force on Youth Suicide. In ML ROSENBERG, & K BAER (Eds.), Report of the Secretary's Task Force on Youth Suicide (Vol. 4; pp. 37–81). DHHS Publ. No. (ADM)89-1624. Washington, DC: U. S. Government Printing Office.

EGELAND JA, & SUSSEX JN (1985). Suicide and family loading for affective disorders. JAMA 254:915–918.

FARBEROW NL, SHNEIDMAN ES, & LEONARD CV (1963). Suicide among general medical and surgical hospital patients with malignant neoplasms. Department of Medicine and Surgery Medical Bulletin (Veterans Administration) 9:1–11.

FARBEROW NL, SHNEIDMAN ES, & LEONARD CV (1970). Suicide among patients with malignant neoplasms. In ES SHNEIDMAN, NL FARBEROW, & RE LITMAN (Eds.), The psychology of suicide (Ch. 19, pp. 325–344). New York: Science House.

FARMER RDT, & PINDER RM (1989). Why do fatal overdose rates vary between antidepressants? Acta Psychiatr Scand 80(suppl. 354):25–35.

FAVA M, & ROSENBAUM J (1991). Suicidality and fluoxetine: Is there a relationship? J Clin Psychiatry 52:108–111.

FAWCETT J (1990). Targeting treatment in patients with mixed symptoms of anxiety and depression. J Clin Psychiatry 51(suppl):40–43.

FAWCETT J, SCHEFTNER W, CLARK D, ET AL. (1987). Clinical predictors of suicide in patients with major affective disorders: A controlled prospective study. Am J Psychiatry 144:35–40.

FOGEL BS, MILLS MJ, & LANDEN JE (1986). Legal aspects of the treatment of delirium. Hosp Community Psychiatry 37:154–158.

GLICKMAN LS (1980). Psychiatric consultation in the general hospital. New York: Marcel Dekker.

GOLDSTEIN AM, & REZNIKOFF M (1971). Suicide in chronic hemodialysis patients from an external locus of control framework. Am J Psychiatry 127:1204–1207.

GRANT I, & ATKINSON JH (1990). Neurogenic and psychogenic behavioral correlates of HIV infection. In BH WAKSMAN (Ed.), Immunologic mechanisms in neurologic and psychiatric disease (pp. 291–304). New York: Raven Press.

GROSS-ISSEROFF R, DILLON KA, & FIELDUST SJ (1990). Autoradiographic analysis of α_1-noradrenergic receptors in the human brain postmortem—effect of suicide. Arch Gen Psychiatry 47:1049–1053.

HAENEL T, BRUNNER F, & BATTEGAY R (1980). Renal dialysis and suicide: Occurrence in Switzerland and in Europe. Compr Psychiatry 21:140–145.

HENRY JA (1989). A fatal toxicity index for antidepressant poisoning. Acta Psychiatr Scand 80(suppl. 354):37–45.

JAMES WA, & LIPPMANN S (1991). Bupropion: Overview and prescribing guidelines in depression. South Med J 84:222–224.

JAMISON KR (1988). Suicide prevention in depressed women. J Clin Psychiatry 49(suppl):42–45.

JOHNSON GF, & HUNT G (1979). Suicidal behavior in bipolar manic-depressive patients and their families. Compr Psychiatry 20:159–164.

JOHNSON J, WEISSMAN MM, & KLERMAN GL (1990). Panic disorder, comorbidity, and suicide attempts. Arch Gen Psychiatry 47:805–808.

JONES JS, STANLEY B, MANN JJ, ET AL. (1990). CSF 5-HIAA and HVA concentrations in elderly depressed patients who attempted suicide. Am J Psychiatry 147:1225–1227.

KALOGERAKIS MG (1971). The assaultive psychiatric patient. Psychiatr Q 45:372–381.

KAY SR, WOLKENFELD F, & MURRILL LM (1988). Profiles of aggression among psychiatric patients—I. Nature and prevalence. J Nerv Ment Dis 176:539–546.

KELLHER CH, BEST CL, ROBERTS JM, ET AL. (1985). Self-destructive behavior in hospitalized medical and surgical patients. Psychiatr Clin North Am 8:279–289.

KERNBERG OK (1984). Severe personality disorders: Psychotherapeutic strategies. New Haven, CT: Yale University Press.

KIEV A (1975). Psychotherapeutic strategies in the management of depressed and suicidal patients. Am J Psychiatry 29:345–354.

KINZEL AF (1970). Body-buffer zone in violent prisoners. Am J Psychiatry 127:99–104.

KONTAXAKIS VP, CHRISTODOULOU GN, MAVREAS VG, ET AL. (1988). Attempted suicide in psychiatric outpatients with concurrent physical illness. Psychother Psychosom 50:201–206.

LEIGH H (1974). Psychotherapy of a suicidal, terminal cancer patient. Int J Psychiatry Med 5:173–182.

LION JR (1985). Clinical assessment of violent patients. In LH ROTH (Ed.), Clinical treatment of the violent person (Ch. 1, pp. 1–19). Rockville, MD: NIMH.

LION JR, BACH-Y-RITA G, ERVIN FR (1969). Violent patients in the emergency room. Am J Psychiatry 125:120–125.

LION JR, MADDEN DJ, & CHRISTOPHER RL (1976). A violence clinic: Three years' experience. Am J Psychiatry 133:432–435.

LION JR, & PASTERNACK SA (1973). Countertransference reactions to violent patients. Am J Psychiatry 130:207–210.

LITMAN RE (1989). Long-term treatment of chronically suicidal patients. Bull Menninger Clin 53:215–228.

LITMAN RE, & FARBEROW NL (1970). Suicide prevention in hospitals. In ES SHNEIDMAN, NL FARBEROW, & RE LITMAN (Eds.), The psychology of suicide (Ch. 28, pp. 461–473). New York: Science House.

LOUHIVUORI KA, & HAKAMA M (1979). Risk of suicide among cancer patients. Am J Epidemiol 109:59–65.

MALTSBERGER JT, & BUIE D (1974). Countertransference hate in the treatment of suicidal patients. Arch Gen Psychiatry 30:625–633.

MALTSBERGER JT, & BUIE DH (1990). Common errors in the management of suicidal patients. In D JACOBS & HN BROWN (Eds.), Suicide: Understanding and responding (Ch. 15, pp. 285–294). Madison, CT: International Universities Press, Inc.

MARKOWITZ JS, WEISSMAN MM, OUELLETTE R, ET AL. (1989). Quality of life in panic disorder. Arch Gen Psychiatry 46:984–992.

MARZUK PM, TIERNEY H, TARDIFF K, ET AL. (1988). Increased risk of suicide in persons with AIDS. JAMA 259:1333–1337.

MAVISSAKALIAN M (1990). The relationship between panic disorder/agoraphobia and personality disorders. Psychiatr Clin North Am 13:661–684.

MCNIEL DE, BINDER RL, & GREENFIELD TK (1988). Predictors of violence in civilly committed acute psychiatric patients. Am J Psychiatry 145:965–970.

MENNINGER KA (1933). Psychoanalytic aspects of suicide. Int J Psychoanal 14:376–390.

MENNINGER WW (1989). The impact of litigation and court decisions on clinical practice. Bull Menninger Clin 53:203–214.

Mental Health and Behavioral Sciences Service Program guide: Management of the violent and suicidal patient (1983). Department of Medicine and Surgery Manual M-2, Part X, G-15. Washington, DC: Veterans Administration.

MILES CP (1977). Conditions predisposing to suicide: A review. J Nerv Ment Dis 164:231–246.

MINTZ RS (1971). Basic considerations in the psychotherapy of the depressed suicidal patient. Am J Psychother 25:56–73.

MONTGOMERY SA, BALDWIN D, & GREEN M (1989). Why do amitriptyline and dothiepin appear to be so dangerous in overdose? Acta Psychiatr Scand 80(suppl. 354):47–53.

MURPHY GE (1975). The physician's responsibility for suicide: II. Errors of omission. Ann Intern Med 82:305–309.

MURPHY GE (1983). On suicide prediction and prevention. Arch Gen Psychiatry 40:343–344.

MURPHY GE, & WETZEL RD (1990). The lifetime risk of suicide in alcoholism. Arch Gen Psychiatry 47:383–392.

National Academy of Sciences (1978). Deferrence and incapacitation: Estimating the effects of criminal sanctions on crime rates. Washington, DC.

National Center for Health Statistics (1991). Births, marriages, divorces, and deaths for October 1990. Monthly Vital Statistics Report 39, no. 10. Hyattsville, MD: U.S. Public Health Service.

NOYES R, REICH J, CHRISTIANSEN J, ET AL. (1990). Outcome of panic disorder. Arch Gen Psychiatry 47:809–818.

OCHITILL HN, & KRIEGER M (1982). Violent behavior among hospitalized medical and surgical patients. South Med J 75:151–155.

PENNINGROTH PE (1975). Control of violence in a mental health setting. Am J Nurs 75:606–609.

PERRY S, JACOBSBERG L, & FISHMAN B (1990). Suicidal ideation and HIV testing. JAMA 263:679–682.

PERRY S, JACOBSBERG LB, FISHMAN B, ET AL. (1990). Psychiatric diagnosis before serological testing for the human immunodeficiency virus. Am J Psychiatry 147:89–93.

POLLACK S (1957). Suicide in the general hospital. In ES SHNEIDMAN & NL FARBEROW (Eds.), Clues to suicide (Ch. 15, pp. 152–163). New York: McGraw-Hill.

POYTHRESS NG (1990). Avoiding negligent release: Contemporary clinical and risk management strategies. Am J Psychiatry 147:994–997.

REICH JH, & GREEN AI (1991). Effect of personality disorders on outcome of treatment. J Nerv Ment Dis 179:74–82.

REICH P, & KELLY MJ (1976). Suicide attempts by hospitalized medical and surgical patients. N Engl J Med 294:298–301.

RIDDICK L, & LUKE JL (1978). Alcohol-associated deaths in the District of Columbia: A postmortem study. J Forensic Sci 23:493–502.

ROBBINS J (1981). The final months. New York: Oxford University Press.

ROSENBERG ML, DAVIDSON LE, SMITH JC, ET AL. (1988). Operational criteria for the determination of suicide. J Forensic Sci 33:1445–1456.

ROSENTHAL PA, & ROSENTHAL S (1984). Suicidal behavior by preschool children. Am J Psychiatry 141:520–525.

ROY A, LAMPARSKI D, DEJONG J, ET AL. (1990). Characteristics of alcoholics who attempt suicide. Am J Psychiatry 147:761–765.

SHAFII M, CARRIGAN S, WHITTINGHILL JR, ET AL. (1985). Psychological autopsy of completed suicide in children and adolescents. Am J Psychiatry 142:1061–1064.

SHEARER SL, PETERS CP, QUAYTMAN MS, ET AL. (1988). Intent and lethality of suicide attempts among female borderline inpatients. Am J Psychiatry 145:1424–1427.

SHULMAN R, PRICE JD, & SPINELLI J (1989). Biopsychosocial aspects of long-term survival on end-stage renal failure therapy. Psychol Med 19:945–954.

SILVER JM, HALES RE, & YUDOFSKY SC (1990). Psychopharmacology of depression in neurologic disorders. J Clin Psychiatry 51(suppl):33–39.

SKODOL AE (1984). Emergency management of potentially violent patients. In EL BASSUK & AW BIRK (Eds.), Emergency psychiatry: Concepts, methods, and practices (Ch. 6, pp. 83–96). New York: Plenum.

SLETTEN IW, BARTON JL (1979). Suicidal patients in the emergency room: A guide for evaluation and disposition. Hosp Community Psychiatry 30:407–411.

TEICHER MH, GLOD C, & COLE JO (1990). Emergence of intense suicidal preoccupation during fluoxetine treatment. Am J Psychiatry 147:207–210.

TERRAGNA A, MAZZARELLO G, ANSELMO M, ET AL. (1990). Suicidal attempts with zidovudine [letter]. AIDS 4:88.

VAN PRAAG HM, PLUTCHIK R, & APTER A (Eds.) (1990). Violence and suicidality: Perspectives in clinical and psychobiological research. New York: Brunner/Mazel.

VOLLRATH M, & ANGST J (1989). Outcome of panic and depression in a seven-year follow-up: Results of the Zurich study. Acta Psychiatr Scand 80:591–596.

VUORI E, PENTTILA A, KLAUKKA T, ET AL. (1989). Fatal poisonings with antidepressants in Finland 1985–1987. Acta Psychiatr Scand 80(suppl 354):55–60.

WEISSMAN MM, KLERMAN GL, MARKOWITZ JS, ET AL. (1989). Suicidal ideation and suicide attempts in panic disorder and attacks. N Engl J Med 321:1209–1214.

WHITLOCK FA (1975). Suicide in Brisbane, 1956–1973: The drug-death epidemic. Med J Aust 1:737–743.

WHITTERS AC, CADORET RJ, TROUGHTON E, ET AL. (1987). Suicide attempts in antisocial alcoholics. J Nerv Ment Dis 175:624–626.

WINOKUR G, BLACK DW, & NASRALLAH A (1988). Depressions secondary to other psychiatric disorders and medical illnesses. Am J Psychiatry 145:233–237.

YESAVAGE JA, WERNER PD, BECKER J, ET AL. (1981). Inpatient evaluation of aggression in psychiatric patients. J Nerv Ment Dis 169:299–302.

YUDOFSKY SC, SILVER JM, JACKSON W, ET AL. (1986). The overt aggression scale for the objective rating of verbal and physical aggression. Am J Psychiatry 143:35–39.

ZEE H (1972). Blindspots in recognizing serious suicidal intentions. Bull Menninger Clin 36:551–555.

6 Anxiety in the medically ill

RICHARD J. GOLDBERG, M.D., AND
DONN A. POSNER, PH.D.

It is always an error to dismiss anxiety by saying, "Wouldn't you be anxious if you were in that situation?" This chapter presents an approach to the evaluation and treatment of anxiety in medical patients that addresses three biopsychosocial dimensions. First, clinicians should look for a possible underlying medical basis for the anxiety. Second, they must obtain a personal and psychiatric history and determine whether the patient's anxiety is part of some underlying psychiatric disorder, such as panic disorder, affective disorder, or a personality disorder. The third step is to assess the anxiety as a psychologic adjustment response.

MEDICAL DISORDERS THAT PRESENT AS ANXIETY: ORGANIC ANXIETY SYNDROMES

Medical disorders can produce anxiety directly or indirectly. Some medical disorders have specific effects on neurotransmitter systems or on neuroanatomic sites associated directly with the production of anxiety. Two examples of such direct causes are tumors of the temporal lobe (Dietch, 1984) and hyperthyroidism (Hall, 1983; MacCrimmon et al., 1979). Other medical disorders produce anxiety through autonomic arousal, which the patient interprets as a psychologic state. Medical disorders that can present with anxiety as a prominent symptom are listed in Table 6-1.

Toxic and Withdrawal Effects of Drugs

The toxic and withdrawal effects of drugs are a common cause of anxiety in medical patients (Abramowicz, 1989). If there is any doubt about what drugs the patient has been taking, a toxicology screen should be ordered. Such screening is especially useful when the patient is suspected of surreptitious drug use.

Caffeine

Caffeine is one of the most widely used psychotropic drugs in the United States (Abelson & Fishburne, 1976), found in many beverages and drug combinations. The approximate amount of caffeine in a cup of coffee is 150 mg. Although sensitivity varies, symptoms of caffeinism may occur at doses of only 200 mg (Victor, Lubetsky, & Greden, 1981). Caffeine increases norepinephrine levels in the plasma and urine (Robertson et al., 1978), inhibits phosphodiesterase breakdown of cyclic adenosine monophosphate in the central nervous system (CNS), and sensitizes central catecholamine receptors, particularly for dopamine (Waldeck, 1975). Caffeine can precipitate an actual panic attack in patients with panic disorder (Charney, Heninger, & Jatlow, 1985). In chronic users, caffeine abstinence should also be considered as a possible source of intermittent anxiety (White et al., 1980).

Cocaine

Symptoms of anxiety, irritability, tremulousness, fatigue, or depression may appear soon after the initial euphoriant effects of cocaine (Resnick, Kestenbaum, & Schwartz, 1976; Abramowicz, 1986a). Cocaine has been reported to induce panic attacks in susceptible individuals and actually to precipitate panic disorder, which continues autonomously even after the cocaine use is discontinued (Aronson & Craig, 1986). Detectable amounts of cocaine are found in the urine or plasma only for a few hours after use, although one of the metabolites, benzoylecgonine, can be detected in a urine sample for as long as 48 hours after use (Wilkerson et al., 1980).

Alcohol

Since 30% to 50% of hospitalized medical patients have a problem related to alcohol use (Moore, 1985), alcohol withdrawal (Lerner & Fallon, 1985) is a frequent etiology for agitation and anxiety in this population. Minor abstinence syndrome, which has its modal onset about 24 hours after cessation of drinking, presents with anxiety, tremulousness (the shakes), and insomnia. The major concern is whether such symptoms are the harbinger

TABLE 6–1. *Medical Conditions Associated with Anxiety*

Cardiovascular conditions	Neurologic conditions
Angina pectoris	Akathisia
Arrhythmia	Encephalopathy
Congestive heart failure	Mass lesion
Hypovolemia	Postconcussion syndrome
Myocardial infarction	Seizure disorder
Valvular disease	Vertigo
Endocrine conditions	Peptic ulcer disease
Carcinoid	Respiratory conditions
Hyperadrenalism	Asthma
Hypercalcemia	Chronic obstructive pulmonary disease
Hyperthyroidism	Pneumothorax
Hypocalcemia	Pulmonary edema
Hypothyroidism	Pulmonary embolism
Pheochromocytoma	Immunologic conditions
Metabolic conditions	Anaphylaxis
Hyperkalemia	Systemic lupus erythematosus
Hyperthermia	
Hypoglycemia	
Hyponatremia	
Hypoxia	
Porphyria	

of alcohol withdrawal delirium (delirium tremens), which has a modal onset 72 hours following cessation or significant reduction of drinking but can occur as early as 24 hours or up to 7 days later. In major abstinence, anxiety and tremulousness are accompanied by symptoms of autonomic arousal, and should be considered a serious medical problem that requires supervised medical management (Sellers & Kalant, 1976). The classic signs and symptoms of sedative withdrawal can be masked by the concurrent use of other medications. For example, mydriasis may be absent if the patient is on narcotics, and tachycardia may be masked by beta-blockers. Because of the association of Wernicke-Korsakoff syndrome with alcohol use (Reuler, Girard, & Cooney, 1985), such patients should always be given thiamine. The treatment of alcohol withdrawal is covered in Chapter 8.

Neuroleptics

It is estimated that as many as 40% to 50% of patients on neuroleptics (and other dopamine receptor blockers, such as metoclopramide) have akathisia (Ratey & Salzman, 1984), a sense of internal restlessness often described as anxiety. Akathisia may be treated by lowering the neuroleptic dose. While akathisia may respond to anticholinergic agents, especially if there is concurrent dyskinesia, it is often more effective to use low-dose beta-blockers (e.g., propranolol 10 mg tid). Benzodiazepines such as lorazepam, 0.5 mg tid, or clonazepam, 0.5 mg qd, offer another treatment option (Fleischhacker, Roth, & Kane, 1990; Wells et al., 1991).

Antidepressants

Akathisia is a common side effect of fluoxetine (Lipinski et al., 1989), characterized by motor restlessness and anxiety that is indistinguishable from neuroleptic-induced anxiety. This side effect appears to be similar to the jitteriness associated with both amoxapine and the tricyclic antidepressants (Nierenberg & Cole, 1991; Pohl et al., 1988). Such jitteriness (including increased anxiety and insomnia) can occur even at low doses. Agitation, insomnia, and tremor are also among the most common side effects of bupropion (Bryant, Guernsey, & Ingrim, 1983; Gardner, 1983). Tolerance usually develops with continuation of treatment. As with neuroleptic-induced akathisia, beta-adrenergic blockers or benzodiazepines may be helpful in reducing or abolishing the symptoms, although dose reduction is often necessary.

Opiate Withdrawal

Narcotic abstinence is always accompanied by anxiety. Iatrogenic withdrawal (and recurrent pain) may occur when patients are changed from parenteral to oral narcotics without consideration of differences in oral-parenteral ratios.

Other Drugs

Alpha-adrenergic stimulants such as pseudoephedrine are often included in over-the-counter decongestants. These agents are closely related to amphetamines and can produce symptoms of anxiety, restlessness, irritability, and insomnia (Weiner, 1980).

Bronchodilators, which chemically resemble the catecholamines (such as isoproterenol and albuterol), have peripheral effects that include increased heart rate and blood pressure that contribute to feelings of anxiety. Systemic effects, reported even with metered-dose inhalers (Harris, 1985), include symptoms of anxiety, restlessness, nervousness, tremor, irritability, insomnia, and emotional lability.

Theophylline, a methylxanthine related chemically to caffeine, is capable of producing powerful cardiovascular effects with tachycardia, along with nervousness and anxiety states (Jacobs, Senior, & Kessler, 1976).

Calcium channel blockers have been reported to produce neuropsychiatric symptoms that include anxiety, tremulousness, jitteriness, and sleep disturbance (Bela & Raftery, 1980; Mueller & Chahine, 1981; Rinkenberger et al., 1980; Singh, Ellrodt, & Peter, 1978).

Other drugs reported to produce anxiety as a side effect include amphetamines and similar anorexic agents, antihistamines, baclofen, cycloserine, indomethacin, oxymetazoline, and quinacrine (Abramowicz, 1989).

Other Medical Disorders

Hypoglycemia

A review of the relationship of hypoglycemia to anxiety, along with its diagnosis, may be found in Chapter 29.

Complex Partial Seizures

Anxiety is the most common ictal emotion associated with temporal lobe epilepsy (Weil, 1959). The phenomenology of panic attacks and temporal lobe epilepsy can be difficult to distinguish, especially when the electroencephalogram (EEG) does not support the diagnosis of temporal lobe epilepsy. McNamara and Fogel (1990) reported on five cases of patients with panic attacks with associated paroxysmal emotional, autonomic, or psychosensory symptoms. The EEG findings, although compatible with interictal temporal lobe epilepsy, were nonspecific. However, the cases responded well to anticonvulsant therapy. Therefore, it seems warranted to consider the possibility of an anticonvulsant-responsive panic disorder in patients whose panic symptoms fail to respond to standard therapies, as well as in those with other risk factors for epilepsy. The likelihood of complex partial seizures is increased if the patient has a history of head injury, febrile convulsions in childhood, birth trauma, or encephalitis, or a personal or family history of seizures.

The characteristic abnormality in temporal lobe seizures is the anterior temporal spike focus; however, in the waking state at least one half of patients have normal EEGs (Gibbs & Gibbs, 1952). A sleep record does increase the percentage of abnormal EEGs in epileptic patients, although the percentage of false-negatives remains about 30 to 40. The value of nasopharyngeal and sphenoidal leads is uncertain, although the use of continuous ambulatory EEG monitoring (Kristensen & Sendrup, 1978; Lieb et al., 1976; McNamara & Fogel, 1990) can increase the diagnostic yield (see Chapter 20). Because of the high incidence of false-negatives, the EEG is not a definitive test, and the diagnosis of temporal lobe seizure disorder often must be made on clinical grounds and tested by an empiric trial of anticonvulsants.

Angina Pectoris

Angina may present as anxiety with episodes of dyspnea and palpitations accompanied by only mild chest discomfort. When such episodes are precipitated by exercise or emotional stress, angina should be suspected and cardiac evaluation considered, especially in patients over 40 or those with a cardiac history. However, when such evaluations are done, panic disorder is found to be a common cause of chest pain in patients with negative cardiac test results (Katon, 1990).

Cardiac Arrhythmias

Because cardiac arrhythmias may produce symptoms mistaken for anxiety, pulse regularity should always be checked during an episode of anxiety. If the diagnosis of arrhythmia is suspected, Holter monitoring may be helpful in confirming it. For those patients with arrhythmias, a host of medical conditions that are also associated with anxiety symptoms need to be considered, such as hyperthyroidism, caffeinism, and nicotine abuse (Lynch et al., 1977).

Recurrent Pulmonary Emboli

Recurrent pulmonary emboli can present as repeated episodes of acute anxiety associated with hyperventilation and dyspnea (Ferrer, 1968). Arterial blood gases may reveal decreased oxygen during an episode, although the confirmatory test of most use is the ventilation-perfusion lung scan. This diagnosis should be considered in those patients who have some predisposition, such as hyperviscosity, prolonged bed rest, peripheral venous thrombosis, or recent pelvic surgery.

Hyperthyroidism

Hyperthyroidism usually presents with symptoms of nervousness, palpitations, diaphoresis, heat intolerance, and diarrhea. Signs include tachycardia, tremor, weight loss, and hot, moist skin. The contribution of hyperthyroidism to anxiety, along with its diagnosis and treatment, are comprehensively reviewed in Chapters 7 and 29.

Pheochromocytoma

Pheochromocytoma is a rare disorder involving catecholamine secretion by a tumor of the renal medulla. The output of catecholamines may be episodic or continuous, producing acute or chronic symptoms of anxiety, often accompanied by headache and flushing (Lishman, 1978). Hypertension is usually present during acute episodes, and 60% to 80% of patients have sustained hypertension. The most accurate diagnostic test for detection involves plasma catecholamine assay (Bravo et al., 1979). Further details of the evaluation can be found in Chapter 7. In one study of 17 patients with active pheochromocytoma (Starkman et al., 1985), none experienced panic attacks and only three had prominent anxiety symptoms.

Mitral Valve Prolapse

Although some studies have shown that patients with panic disorders have a higher incidence of mitral valve prolapse (MVP) than psychiatrically normal people (Kantor, Zitrin, & Zeldis, 1980; Pariser, Pinta, & Jones, 1978; Venkatesh et al., 1980), other studies fail to document this increased prevalence (Kathol et al., 1980; Shear et al., 1984), possibly because of differences in the stringency of criteria used for the diagnosis of MVP. About half of patients with MVP at some time complain of palpitations, but continuous cardiac monitoring of such patients often reveals no relationship between the complaint of palpitations and any form of cardiac rhythm disturbance (Devereux et al., 1976; Shear et al., 1984). Furthermore, a study of the prevalence of anxiety disorder in patients with MVP found no differences between patients and controls, casting further doubt on the etiologic role of MVP in panic attacks (Mazza et al., 1986). Although there continues to be support for some relationship between MVP and panic attacks (Liberthson et al., 1986), there is no evidence that MVP causes panic attacks. A more extensive discussion of MVP may be found in Chapter 23.

Hyperventilation Syndrome

Hyperventilation syndrome (HVS) has an estimated prevalence of 10% in a general medical clinic (Rice, 1950), 5.8% in a gastroenterology practice (McKell & Sullivan, 1947), and 5% in a neurology clinic practice (Pincus & Tucker, 1985). Anxiety is a cardinal symptom along with a variety of other medical symptoms, including faintness, visual disturbances, nausea, vertigo, headache, palpitations, dyspnea, diaphoresis, and paresthesias. The diagnosis of HVS can be made on the basis of the patient's response to overbreathing (breathing by mouth for up to 3 minutes or until dizzy). If the symptoms in question are entirely reproduced, without an alternative explanation by physical examination, medical history, or laboratory tests, the diagnosis can be established.

Hyperventilation syndrome probably represents a form of panic disorder in most cases, although there are no study data to confirm this. Hyperventilation leads to excessive elimination of carbon dioxide, acute respiratory alkalosis, and cerebral arterial constriction. In 240 seconds of overbreathing, cerebral blood flow can be reduced by 40% (Plum & Posner, 1972), causing EEG slowing (Gotoh, Meyer, & Takagi, 1965). Muscular tension is heightened by a decreased ionization of calcium associated with the increase in pH (Neill & Hattenhauer, 1975). Hyperventilation also causes increased coronary artery resistance and can cause chest discomfort that is difficult to distinguish from angina pectoris (Evans & Lum, 1977), along with nonspecific downward depression of the ST segment and T wave flattening (Christensen, 1946). Unlike ischemic ST changes, those caused by hyperventilation usually appear early during exercise and tend to disappear as exercise continues (McHenry et al., 1970).

Postconcussion Syndrome

Cerebral concussion is classically regarded as a disorder that produces no irreversible anatomic lesions. Clinically, concussion results in an instantaneous diminution of function or loss of consciousness followed by rapid and complete recovery. The episode may be surrounded by a sphere of amnesia, with about one tenth of the total as retrograde amnesia (Parkinson, 1977). For most patients, the symptoms abate after a few weeks to a few months. However, about 20% of concussions lead to a syndrome involving medically unexplained symptoms along with anxiety, impairment of sleep and appetite, irritability, lightheadedness, headaches, and poor concentration (Leigh, 1979).

Symptoms that follow concussion may be a direct result of neuronal damage (Oppenheimer, 1968) and alteration of cerebral blood flow, producing regional and generalized abnormalities correlating with the presence of psychologic symptoms (Taylor & Bell, 1966) and impaired information processing (Gronwall & Wrightson, 1974). Mild head trauma (1- to 2-minute loss of consciousness without external signs of trauma) (Wortsman et al., 1980) has been reported to lead to an increase in catecholamines lasting for about 4 months.

Nonspecific Medical Causes

Many medical disorders may be associated with a broad array of psychiatric symptoms, including anxiety as a nonspecific reaction to delirium, although not as a prominent feature. Such disorders include Cushing's syndrome (Lishman, 1978), hypomagnesemia (Hall & Joffe, 1973), hyponatremia (Gehi et al., 1981), renal failure (Marshall, 1979), hypoparathyroidism (Denko & Kaelbling, 1962) and other electrolyte imbalances (Webb & Gehi, 1981). The diagnostic workup for an organic cause of anxiety must be guided by the clinical review of systems and a complete survey of current and recent drugs and medications, and known concurrent medical conditions. In an otherwise healthy patient with significant sustained anxiety, thyroid screening alone might be sufficient. At the other end of the spectrum (e.g., new-onset sustained anxiety in the elderly), a comprehensive metabolic and neurologic assessment is needed.

CONCURRENT PSYCHIATRIC DISORDERS IN MEDICAL PATIENTS

The second step in evaluating the presentation of anxiety in a medical or surgical patient is to evaluate for the possibility of underlying psychiatric disorder. From 4% to 14% of medical patients have an anxiety disorder (Barrett et al., 1988; Strain, Liebowitz & Klein, 1981; Wells, Golding, & Burnam, 1989), with higher rates in selected subgroups such as those with Parkinson's disease (Stein et al., 1990) and high utilizers of medical care (Katon et al., 1990). The clinician should inquire about previous personal and family psychiatric history. Medical patients may present with anxiety as a result of generalized anxiety disorder, relapsing schizophrenia, or borderline personality. Anxiety is often a significant component of a mood disorder. Somatoform disorder can present as continuous anxiety focused on physical symptoms (Kaplan, Lipkin, & Gordon, 1988; Warwick & Salkovskis, 1990). Finally, dementia can impair a patient's coping responses and thereby lead to anxiety.

PSYCHOSOCIAL ISSUES AND ANXIETY IN MEDICAL PATIENTS

No evaluation of the anxious medical patient is complete without an inquiry into the psychosocial dimension of the patient's experience. The psychosocial dimension encompasses the intrapsychic meanings that patients attach to experiences and their behaviorally conditioned responses. Although physicians often assume that the distress associated with illness is accounted for by the physical morbidity, it is now appreciated that a significant component of the distress associated with cancer, for example, is due to issues involving psychosocial adjustment (Goldberg & Cullen, 1985). It is often therapeutic to help the patient identify emotional problem areas that are being transformed into symptoms and to help contain the anxiety by putting it into words. The following sections discuss four major areas worth reviewing in terms of understanding potential psychologic sources of anxiety.

Alienation

Social support plays an important role in maintaining mental and physical health. Anxiety over abandonment and isolation can be even more important than fear associated with the disease itself. The physician should identify the patient's major social supports and assess their involvement. Whenever possible, the physician should meet with patient and key supports together, to observe their interaction and facilitate better sharing and communication. The patient's intrapsychic concern about abandonment is sometimes best dealt with by a concrete intervention, such as arranging homemaker or visiting nurse assistance.

As disease progresses, patients may become anxious because of a growing sense of distance from the physician. Patients nearing the end of an intensive treatment program are noted to experience an increase in anxiety rather than a sense of relief (Mastrovito, 1972; Peck & Boland, 1977), fearing that the detachment will jeopardize their survival. Maintenance of a regular contact and communication of concern are valuable positive interventions, although to the action-oriented physician such visits may seem like doing very little.

Loss of Control

For many patients, the loss of control inherent in illness may be the crucial factor underlying anxiety symptoms. The consultant must be creative in identifying areas in which the patient can exercise some control without jeopardizing medical treatment. Intellectual mastery is another means of reasserting control for some patients. Sharing information in a way appropriate to the patient's personality style should therefore be considered an important means of reducing anxiety, although physicians often worry that patients may become more upset by hearing about their diagnosis (Goldberg, 1983).

Physical Damage

Threatened or real loss of bodily integrity or a body part can trigger profound anxiety, along with insomnia, anorexia, and difficulty in concentrating. Patients who view themselves as less than whole may withdraw from relationships because of anxiety over rejection. Before surgery, patients can become panicked and feel they have a poor chance of survival (Bard & Sutherland, 1977). It is always important to explore what patients have heard from others about the procedure they are awaiting, because misconceptions can arise from stories about a friend or relative who did poorly in a similar situation (Baudry & Weiner, 1968). Preoperative review of potential misconceptions may even decrease certain surgical complications (Egbert et al., 1964).

With anxious patients who ask "Am I going to be all right?", there is a tendency to reassure prematurely rather than to explore difficult feelings. Some preoperative anxiety, however, appears to be helpful since it stimulates realistic planning. Patients who are extraordinarily calm may be masking fears that place them at higher risk for not coping with later inevitable events (Sutherland et al., 1977).

Death

The issue of dying may or may not be brought up directly by the patient; it may emerge instead through some related symptoms of anxiety. As long as patients with life-threatening illness function adaptively, elements of denial play an important role in continued function (Dimsdale & Hackett, 1982). The natural course of progressive illness, however, usually challenges the patient to slowly adjust denial mechanisms to the emerging medical reality (Weisman, 1979). Recurrence of disease often erodes initial optimism and creates a situation in which the patient for the first time deals with issues of dying. Signs of physician willingness to engage the patient on these issues are important in allaying fears, which otherwise would remain unexpressed. Dealing in some way with their own sense of mortality probably is important for clinicians to be effective in dealing with patients' anxiety about death.

PHARMACOLOGIC MANAGEMENT

Benzodiazepines

The best indication for using benzodiazepines is when a time-limited cause of anxiety can be identified, as in the case of the patient awaiting a cardiac catheterization. Benzodiazepines are also helpful in treating primary anxiety disorders, and adjunctively for patients with organic anxiety syndromes. (See Chapter 9.)

Pharmacokinetics

All the benzodiazepines are well absorbed orally and reach peak blood levels after a single oral dose in times varying from 1 to 6 hours. The differences in time to reach peak plasma level largely reflect differences in gastrointestinal absorption. As metabolites reach a steady state, however, initial differences related to absorption and distribution disappear, and differences related to metabolism become prominent. Chlordiazepoxide (Greenblatt et al., 1974) and diazepam (Greenblatt & Koch-Weser, 1976) are poorly and unpredictably absorbed from intramuscular (IM) sites, whereas lorazepam and midazolam have the distinct property of prompt and reliable IM absorption (Greenblatt et al., 1982). The metabolic fate of various benzodiazepines may be simplified by appreciating that they fall into two classes—long acting and short acting (see Table 6-2).

Drug accumulation, with potential impairment of cognitive and motor performance, is a special risk to be kept in mind with long-acting agents, especially for older patients and those with liver impairment (Greenblatt et al., 1980; Hoyoumpa, 1978). Despite the long half-lives and accumulation of active substances, chronic use of these drugs usually does not lead to oversedation. The sedative and antianxiety effects of benzodiazepines appear to be distinct, and, as a steady state is reached, the CNS seems to adapt to the nonspecific sedative effects (Johnson & Chernik, 1982). However, there is an age-related increase in the sensitivity of elderly individuals to the central depressant effects of long-acting benzodiazepines (Pomara et al., 1985).

Drug Interactions

Benzodiazepines have relatively few significant adverse drug interactions. Their major drug interaction, augmentation of other CNS depressants, can be controlled by adjusting dosage downward. There are a few reports of benzodiazepines increasing phenytoin levels (Vajda, Prineas, & Lovell, 1971) and decreasing prothrombin times for patients on warfarin (Taylor, 1967). Concomitant use of cimetidine alters the pharmacokinetics of diazepam but without clinically relevant effects (Greenblatt et al., 1984). It has been reported that diazepam increases digoxin binding to plasma proteins, resulting in increased digoxin serum levels (Castillo-Ferando, Garci, & Carmona, 1980). Long-term use of low-dose estrogen-containing oral contraceptives greatly increases the elimination half-life of diazepam (Abernethy et al., 1982). There is a report that erythromycin significantly inhibits the metabolism of triazolam (Phillips, Antal, & Smith, 1986). Because absorption of benzodiazepines may be impaired under conditions of high pH (over 7.4), antacid preparations should not be given concomitantly. Concurrent use of anticholinergic agents, which delay gastric emptying, also can impair bioavailability.

TABLE 6–2. *Pharmacokinetic Summary Comparison of Benzodiazepines*

Drug Given	Peak Plasma Level (Hours)	Mean (Range) Elimination Half-Life (Hours)	Active Metabolites	Mean (Range) Elimination Half-Life (Hours)	Approximate Dose Equivalent (mg)
Alprazolam (Xanax)	1–2	11(6–16)	Alphahydroxyalprazolam	6	0.5
Chlordiazepoxide (Librium)	0.5–4	10(5–30)	Desmethylchlordiazepoxide Demoxepam Desmethyldiazepam Oxazepam	(24–96) (14–95) 73(30–100) 7(5–15)	10
Clonazepam (Klonopin)	1–4	23(18–50)	None		0.25
Clorazepate (Tranxene)	1–2	*	Desmethyldiazepam Oxazepam	73(30–100) 7(5–15)	7.5
Diazepam (Valium)	1–2	43(20–70)	Desmethyldiazepam Oxazepam	73(30–100) 7(5–15)	5
Estazolam (ProSom)	2	14(10–24)	2 metabolites with low concentrations and potencies	10–14	1.0
Flurazepam (Dalmane)	0.5–2	*	*N*-desalkylflurazepam	74(36–120)	5
Lorazepam (Ativan)	1–2 PO (20 min IM)	14(10–25) 14	None		1.0
Midazolam	5 min IV	68 min	None		For sedation 1-2 mg IV up to 0.15 mg/kg
Oxazepam (Serax)	2	7(5–15)	None		15
Prazepam (Centrax)	6	*	Desmethyldiazepam Oxazepam	73(30–100) 7(5–15)	
Quazepam (Doral)	1–3	39	2-oxoquazepam *N*-desalkylflurazepam	39 74(36–120)	
Temazepam (Restoril)	1–1.5	13(8–20)	None		15
Triazolam (Halcion)	1–2 (0.25 sl)	2 3(1.5–5.5)	None		0.25

Note: Drugs marked with (*) are prodrugs; all CNS effects are due to their active metabolites.

Side Effects

The most common adverse effects of the benzodiazepines involve CNS depression: muscle weakness, ataxia, dysarthria, vertigo, somnolence, and confusion. These side effects can be a major problem, especially for the medically ill patient who may be weak from prolonged bed rest or already impaired by other CNS illness. Older patients are especially susceptible to psychomotor impairment and falls (Ray, Griffin, & Downey, 1989; Tinetti, Speechley, & Ginter, 1988).

There is some question whether or not benzodiazepines may stimulate some people (Hall & Joffe, 1972) or release hostility and rage reactions (Karch, 1979). Such instances are infrequent (Dietch & Jennings, 1988). Disinhibition seems most common in patients with preexisting personality disorders, substance abuse, or underlying organic disorders. However, benzodiazepines can produce or increase depressive symptomatology (Greenblatt & Shader, 1974; Ryan et al., 1968).

Van der Kropf (1979) first reported finding depersonalization, anxiety, and paranoia in subjects with chronic insomnia treated with triazolam. Benzodiazepines impair memory function in two ways. The first is an acute anterograde amnestic effect, usually after intravenous use. However, this effect is also being reported with therapeutic oral doses of the high-potency, short half-life benzodiazepines, especially if taken with alcohol (Healy et al., 1983; Scharf et al., 1984; Shader et al., 1986; Wolkowitz et al., 1987). The second type of memory impairment involves recall during chronic use. Because benzodiazepines interfere with memory consolidation, users, especially the elderly (Nikaido et al., 1987), may have impaired long-term recall (Angus & Romney, 1984; Lucki, Rickels, & Geller, 1986).

Large doses of benzodiazepines produce only minor

changes in cardiovascular function even in patients with underlying cardiac disease (Rao et al., 1973). In patients without pulmonary disease, changes in tidal volume and response to elevated pCO_2 are barely detectable (Lakshminarayan et al., 1976). In a single-blind study, 25 mg of diazepam per day actually improved the breathlessness of patients with chronic airflow obstruction associated with emphysema (Mitchells-Heggs et al., 1980). The respiratory depressant effects of benzodiazepines appear to be most marked in patients with carbon dioxide retention. Benzodiazepines should not be given to patients with clinically significant pulmonary disease who might be retaining carbon dioxide before measuring arterial blood gases. Anxious carbon dioxide retainers may be treated with buspirone or low-dose neuroleptics, which do not alter respiratory drive. Intravenous benzodiazepines given concurrently with parenteral narcotics produce significant respiratory depression of a greater degree than is found with opiates alone, especially in patients with some pulmonary impairment (Cohen, Finn, & Steen, 1969).

Toxicity and Dependence

Benzodiazepines are relatively nonlethal in overdose. In the medical literature there have been fewer than a dozen suicides by diazepam ingestion alone (Finkel, McCloskey, & Goodman, 1979). Benzodiazepines are, however, often used in combination with other sedative drugs and ethanol in fatal overdoses.

Overall, the risks of overuse, dependence, and addiction are low considering the widespread use of benzodiazepines (Rifkin et al., 1989; Uhlenhuth et al., 1988). However, when used regularly for at least several months, addiction can occur. The potential for addiction may be greater for alcoholic patients (Ciraulo, Sands, & Shader, 1988).

Withdrawal

Symptoms of benzodiazepine withdrawal may include anxiety, insomnia, dizziness, headache, anorexia, hypotension, hyperthermia, neuromuscular irritability, tinnitus, blurred vision, shakiness, and psychosis. Higher doses taken over longer durations create a greater risk of moderate to severe withdrawal (Hollister, Motzenbecker, & Degnan, 1961). There are reports of abstinence phenomena beginning at 5 to 7 days and lasting from 2 to 4 weeks after cessation of usual therapeutic doses of diazepam (Busto et al., 1986; Pevnick, Jasinski, & Haertzen, 1978; Winokur et al., 1980), and of seizures occurring in association with the discontinuation of moderate doses of lorazepam (de la Fuenta et al., 1980) and triazolam (Tien & Gujavarty, 1985). One advantage of long-acting benzodiazepines is that they tend to self-taper if discontinued. Although abrupt discontinuation of long-acting benzodiazepines is usually not dangerous, it can lead to a persistent state of heightened anxiety. The short-acting benzodiazepines are associated with a greater prevalence and severity of withdrawal reactions, because their plasma concentrations decline more rapidly following discontinuation. Withdrawal symptoms can be minimized by tapering rather than abruptly discontinuing the drug.

During withdrawal, autonomic symptoms can be relieved by a beta-adrenergic blocker (Abernethy, Greenblatt, & Shader, 1981), and possibly by carbamazepine as well (Malcolm et al., 1989; Ries et al., 1989). It should be kept in mind that withdrawal symptoms from alprazolam and triazolam may not be fully covered by other benzodiazepines (Schneider, Syapin, & Pawluczyk, 1987; Zipursky, Baker, & Zimmer, 1985). In these cases, substitution with clonazepam may be an effective alternative (Albeck, 1987; Patterson, 1988).

Use during Pregnancy

Although benzodiazepine use during the first trimester of pregnancy has been suspected of causing an increased risk of cleft palate (Calabrese & Gulledge, 1985) in addition to dysmorphism (Laegreid et al., 1987), others have not confirmed this association (Rosenberg et al., 1983). The use of benzodiazepines late in pregnancy or during nursing has been associated with "floppy infant syndrome" (Gilberg, 1977), and there are two case reports—one of severe congenital malfunction possibly associated with use of the clorazepate dipotassium in the first trimester (Patel & Patel, 1980) and another of intrauterine growth retardation and possible withdrawal symptoms in a neonate exposed to diazepam during several months of gestation (Backes & Cadero, 1980). Some of the mothers involved in these case studies were using other drugs, and other reviews indicate no increased risk of congenital abnormalities from benzodiazepines alone (Cohen, Heller, & Rosenbaum, 1989). However, such cases serve to heighten awareness of and concern for the judicious use of medication during pregnancy. If one is compelled to use medication to treat severe anxiety or agitated depression in pregnancy, tricyclic antidepressants and diazepam both appear relatively safe and preferable to neuroleptics (Calabrese & Gulledge, 1985). (See Chapter 28.)

Midazolam

Midazolam is a parenteral benzodiazepine recently marketed in the United States for intravenous sedation for short diagnostic or endoscopic procedures and sedation associated with general anesthesia. Its onset of action is within 15 minutes, reaching peak activity within 30 minutes, with a duration of action of 1 to 2 hours. One advantage of midazolam over diazepam is that it is formulated in an aqueous solution; therefore, it does not cause local irritation after IM or intravenous (IV) injection.

Like other benzodiazepines, midazolam can cause respiratory depression and lead to apnea, especially when combined with narcotics. Hypotension has also been reported. Anterograde amnesia may persist for 1 or 2 hours after injection; therefore, postoperative patients need to be given instructions in writing. The approximate dose used for intravenous sedation for a diagnostic procedure is an initial dose of 1 or 2 mg, up to 0.15 mg/kg. Dosage should be reduced in patients on concomitant narcotics (Abramowicz, 1986b).

Buspirone

Buspirone is a nonbenzodiazepine anxiolytic that has an anxiolytic effect without the sedative or cognitive side effects associated with the benzodiazepines. Its mechanism of action is thought to involve its function as a partial agonist at the serotonin 1A receptor (Eison & Eison, 1984; Eison & Temple, 1986).

Table 6-3 summarizes the pertinent clinical issues involving buspirone, in comparison with the benzodiazepines. To begin, buspirone does not have acute effects. In fact, it usually takes at least 7 to 10 days (and often 4 weeks) to have an effect. Therefore, it is best used for patients with chronic generalized anxiety.

Aside from generalized anxiety disorder itself, medical patients with chronic symptoms related to autonomic arousal may benefit from buspirone. These groups would include, for example, patients with irritable bowel syndrome, peptic ulcer disease, asthma, and chronic obstructive pulmonary disease (COPD) with anxiety-related dyspnea. Patients such as those with epilepsy, whose symptoms are increased by anxiety, may benefit from anxiety reduction with buspirone.

TABLE 6–3. *Comparison of Benzodiazepines (BZ) and Buspirone (BUS)*

	BZ	BUS
Anxiolytic effect	+	+
Psychomotor impairment	+	–
Cognitive impairment	+	–
Sedation	+	–
Addiction potential	+	–
Sedative augmentation	+	–
Respiratory depression	+	–
Withdrawal	+	–
Muscle relaxation	+	–
Anticonvulsant	+	–

There are, in addition, several niches for buspirone that emerge from its clinical profile. Buspirone does not impair respiratory drive as the benzodiazepines do, and in fact may be somewhat of a respiratory stimulant (Garner et al., 1989; Rapoport & Mendelson, 1989). Therefore, it is safe to treat anxious carbon dioxide-retaining patients with buspirone, whereas it would not be safe to do so with a benzodiazepine. Buspirone also does not augment the sedative potential of other CNS sedatives. Therefore, anxious patients on other sedatives (such as anticonvulsants, antihistamines, and narcotic analgesics) may be able to benefit from anxiolytic treatment without the additive sedative effects associated with the benzodiazepines. Unlike the benzodiazepines, buspirone lacks muscle relaxant, anticonvulsant, and hypnotic activities, and does not block the withdrawal syndrome associated with CNS sedatives (Schweizer & Rickels, 1986).

Pharmacokinetics

Buspirone is rapidly and completely absorbed from the gastrointestinal tract, with extensive first-pass hepatic metabolism. Taking the drug with food appears to decrease its rate of absorption and its first-pass metabolism, making more drug available. Peak plasma level is reached in about 1 hour. Average elimination half-life is 2.5 hours. About 65% is eliminated by the kidneys, mostly in a metabolized form; 35% undergoes fecal elimination. There have not been any adverse reports about the effects of liver or renal impairment on drug effects.

Side Effects

The major side effects include dizziness, headache, and nervousness. Buspirone is notably less sedative than the benzodiazepines, which is one of its primary advantages (Cohn & Wilcox, 1986; Newton et al., 1986). In fact, buspirone causes no more drowsiness than placebo and does not impair psychomotor performance skills, such as driving. There have been no deaths reported as a result of buspirone overdose, out of more than 375 cases. Major sequelae of overdose have included nausea and vomiting, dizziness, drowsiness, and miosis. No withdrawal syndrome has been reported following the abrupt discontinuation of its use.

Buspirone appears to have no abuse potential (Griffith, Jasinski, & McKinney, 1986), lacking euphoric properties and actually having some dysphoric properties with repeated, excessive use. This property, coupled with the fact that it does not potentiate the sedative effects of alcohol (or other CNS sedatives), seems to make buspirone a good drug for the treatment of anxiety in alcoholic patients (Kastenholz & Crismon, 1984; Meyer, 1986).

Drug Interactions

There are few reports of drug interactions in humans. Use of buspirone with monoamine oxidase (MAO) inhibitors is considered contraindicated because of potential hypertensive reactions. In addition, there are reports that its serotonin-augmenting effects may be synergistic with fluoxetine, which may be an advantage leading to increased drug response for obsessive-compulsive disorder (Markovitz, Stagno, & Calabrese, 1990). However, fluoxetine may impair the anxiolytic response to buspirone (Bodkin & Teicher, 1989). Finally, albeit without increased extrapyramidal effects buspirone can increase haloperidol levels.

Clinical Use

Buspirone has been demonstrated to have clinical anxiolytic efficacy comparable to that of the benzodiazepines (Cohn et al., 1986; Rickels et al., 1982; Schuckit, 1984). However, a dose of 5 mg tid often is inadequate. More patients show a response to 10 mg tid but some cannot start this high without developing tinnitus, lightheadedness, or aggravation of anxiety. From 10 tid, dosage should be raised by 10-mg/day increments every 7 to 10 days as needed up to a maximum of 60 mg/day. Since the anxiolytic effects of buspirone may take several weeks to occur, the patient may become prematurely discouraged about the potential effectiveness of this medication. An additional issue, the significance of which is yet to be determined, is that some patients who have been on benzodiazepines appear not to report comparable anxiolytic benefit from buspirone (Schweizer, Rickels, & Lucki, 1986). However, buspirone may be helpful in concurrent use with benzodiazepines in treating panic disorder and helping prevent relapse after withdrawal of benzodiazepines.

Buspirone also appears to have a clinically relevant antiaggression effect, which may be helpful in populations with different types of brain damage, including the episodic hyperarousal that accompanies head injury (Levine, 1988), mental retardation (Ratey et al., 1989), dementia (Colenda, 1988), and attention deficit disorder (Balon, 1990). In these patients, a lower initial dose (no more than 5 mg tid) is recommended because some patients with brain damage can show an agitation response at a higher initial dose.

Beta-Adrenergic Blocking Agents

Autonomic symptoms associated with anxiety (such as palpitations and tremulousness) are mediated by beta-adrenergic sympathetic activity. Beta-adrenergic blocking agents, such as propranolol, have been demonstrated to antagonize both the somatic (Granvill-Grossman & Turner, 1966; Tyrer & Lader, 1974) and the emotional (Kathol et al., 1980) symptoms of anxiety.

It has been suggested that propranolol has a unique role for patients with acute situational distress, such as public performance anxiety, in which the psychomotor intellectual impairment produced by benzodiazepines is not desirable. Forty milligrams of propranolol given 90 minutes before performance was shown to have a positive effect in decreasing performance anxiety in a group of musicians (James et al., 1978).

Evaluation of the effectiveness of beta-blockers in anxiety has become more complex with the recognition that these agents (especially propranolol) enter the CNS and can produce direct neurologic and behavioral effects. Although weakness and depression are the most common neuropsychiatric symptoms associated with propranolol, patients may also have insomnia, vivid nightmares, hypnogogic hallucinations, or toxic psychosis, even at relatively low doses (Fraser & Carr, 1976; Gershon et al., 1979). Such mental changes tend to reverse within 48 hours after discontinuation of propranolol. Nadolol and atenolol are beta-blockers that do not cross the blood-brain barrier to any significant extent but also seem to improve some components of anxiety. It may be, therefore, that the central effects are exerted by some kind of neurohumoral feedback to the CNS rather than by a direct action. Finally, pindolol (with intrinsic sympathomimetic activity) and labetalol (with $alpha_1$-blocking activity) are beta-blockers that do not tend to lower pulse rate. However, their antianxiety effects remain relatively unproven.

Issues in Prescribing

Propranolol is almost completely absorbed following oral administration. It undergoes extensive first-pass metabolism in the liver, and variation in this component results in as much as 20-fold variability in plasma concentration among individuals on comparable doses. The half-life is initially about 3 hours, and it may increase to 4 hours during chronic use. It is 90% to 95% bound to plasma protein, which also may contribute to its variabilities in plasma concentrations. Propranolol is almost completely metabolized in the liver before urinary excretion. Nadolol is poorly absorbed from the gastrointestinal tract (Dreyfuss et al., 1979) and is excreted largely by the kidney in unchanged form (Frishman, 1981). The elimination half-life is 14 to 24 hours, increasing dramatically in patients with renal dysfunction. Contraindications to the use of beta-blockers include bradycardia, atrioventricular block, and congestive heart failure. Because of their effects on bronchial smooth muscle, beta-blockers are relatively contraindicated in patients with bronchospastic disease or COPD. They should be used cautiously in diabetics because they can mask clinical signs of hypoglycemia and interference with glycogenolysis during hypoglycemia.

Beta-receptor blockade has little effect on the normal heart at rest, although there is some decrease in heart rate, cardiac output, and blood pressure. During exercise and anxiety, however, sympathetic responses may be significantly blocked. Maximum exercise tolerance may therefore be decreased in otherwise normal patients. Serious cardiac depression is uncommon, but heart failure may develop slowly or suddenly, expecially if the heart is compromised by intrinsic disease or by other drugs, such as digitalis.

Other Anxiolytics

There are certain situations in which neuroleptics can be effectively used to treat anxiety, especially for short periods of time so the risk of tardive dyskinesia is much less of an issue (Rickels, 1983; Fann, Lake & Majors, 1974). These situations include acute anxiety in patients with carbon dioxide retention, in whom suppression of the hypoxic respiratory drive by benzodiazepines should be avoided; anxiety that represents an incipient psychotic disorganization, as in some borderline personality-disordered patients under stress; schizophrenia; organic anxiety syndrome with disorganized cognition or behavior, as in the severely anxious patient with racing thoughts precipitated by steroids; and anxiety in the context of significant delirium. Some of the more sedative neuroleptics, such as thioridazine and perphenazine, are more anxiolytic than high potency agents such as haloperidol. However, no neuroleptic maintains a Food and Drug Administration indication for treating anxiety alone.

Antihistamines such as hydroxyzine and diphenhydramine are sometimes prescribed to treat anxiety. They have no specific anxiolytic properties (Rickels et al., 1970), although patients may feel less anxious on them because of their sedative effects. There is little rationale to support their use, since they have a number of unwanted side effects, such as anticholinergic properties, especially if used in repeated doses. Diphenhydramine is less effective than benzodiazepines for sleep induction (Rickels et al., 1983). The only situation involving anxiety in which antihistamines might have a special use would be when anxiety is the accompaniment of an allergic response for which their antihistaminic properties are of primary value.

Finally, antidepressant drugs have been noted to be of value in the treatment of anxiety. For many years, the tricyclics and MAO inhibitors have been recommended for the treatment of panic attacks (Klein, 1982; Pohl, Berchou, & Rainey, 1982; Sheehan, Ballenger, & Jacobsen, 1980; Zitrin, Klein, & Woerner, 1978). Both groups of drugs are also effective in patients with mixed anxiety and depression (Paykel et al., 1982; Rickels et al., 1974), especially in so-called atypical depressions, which are characterized by high levels of anxiety (Robinson et al., 1973). There is also evidence that the tricyclics may be effective in the treatment of some patients with chronic anxiety alone (Lipman et al., 1981). In a double-blind, placebo-controlled study (even eliminating patients with panic-phobic syndromes) of the treatment of anxiety with imipramine, chlordiazepoxide, or placebo, antianxiety effects of imipramine were superior to those of the others by the second treatment week, becoming clearly more significant thereafter, independent of baseline levels of depression and anxiety (Kahn et al., 1986).

SELF-REGULATION METHODS

It is now well established that in a relatively brief period of time most patients can be taught techniques that induce relaxation. These techniques should be considered appropriate:

1. For patients with generalized anxiety.
2. As a component of a desensitization therapy for specific anxiety-producing situations.
3. When stress seems to be a major factor in initiating or maintaining other medical symptoms.

Muscle Relaxation Therapies

Learning to specifically sense and control muscle tension is a widely utilized and effective method for anxiety reduction. The Jacobson method of progressive relaxation depends on systematically tensing and relaxing muscle groups starting with the feet and eventually involving the entire body (Jacobson, 1938). Of course, it is important to discuss the intent of the procedure with the patient beforehand and to elicit any specific questions, misconceptions, or concerns the patient might have. A comfortable, quiet setting without interruptions is important. Although tape-recorded or printed instructions can be used, the initial session with the physician can be an important factor in establishing a positive alliance for future work. Muscle relaxation procedures may be more successful in introverted than in extroverted personality types (Stoudemire, 1972). Furthermore, while participating in biofeedback, some extroverted patients have actually shown an increase in anxiety, along with dysphoric symptoms such as fear of losing control (Leboeuf, 1977).

Electromyographic Biofeedback

The Jacobson method was devised to induce relaxation by heightening awareness of muscle tension. Electromyographic (EMG) biofeedback goes one step further by providing the person with precise information about the electrical potentials of selected muscle groups. In EMG biofeedback treatment of anxiety the frontalis muscle is usually selected for monitoring, because the frontalis EMG activity level is regarded as an index of overall physiologic arousal. The biceps brachialis has also been used to monitor general muscle relaxation (deVries et al., 1977). Using data that show no correlation between EMG activity reductions and decreased anxiety levels, Raskin, Bali, and Peeke (1980) have challenged the assumption that a profound degree of muscle relaxation is necessary for achieving anxiety relief when using self-regulatory therapies.

Meditation Techniques and the Relaxation Response

Meditation has been shown to produce a physiologic state of restful alertness that is different from sleeping or waking (Wallace, 1970). Physiologic findings during meditation are generally opposite to those encountered in an anxiety patient. After reviewing a range of meditative practices, Benson concluded that within the apparent multiplicity of practices reside common features that could serve as the basis for a nonsectarian form of practice capable of inducing a state he called the relaxation response (Benson, Beary, & Carol, 1974). The technique for eliciting the relaxation response consists of four basic elements:

1. *A mental device*. There needs to be some constant stimulus—for example, a sound, a word (such as "one"), or a phrase repeated silently or audibly. Fixed gazing at an object is a suitable alternative. The purpose of this procedure is to focus attention away from the continuous flow of sensory distractions and intellectual preoccupations.
2. *A passive attitude*. During the aural or visual practice, distracting thoughts are to be disregarded. One should not be concerned with performance standards. When lapses are recognized, the practitioner should patiently return to the mental device without self-criticism or concern about success or failure.
3. *Decreased muscle tone*. The subject should be in a comfortable position to minimize any muscular strain or tension.
4. *A quiet environment*. A quiet environment with decreased stimuli should be chosen. A quiet room where there is no concern about unexpected interruptions is usually suitable. Most techniques instruct the practitioner to close the eyes.

The relaxation response has achieved wide popular recognition and in many places is emerging as part of the armamentarium of the primary care physician as a technique to manage anxiety and a wide array of stress-related disorders (Goldberg, 1982a, 1982b).

Behavioral Techniques Used with Anxiety Disorder Patients and the Medically Ill

Since behavior therapy is an effective treatment for anxiety, especially as an adjunct to the use of medication, a physician can consider making a referral for such therapy from the outset of treatment. Such referrals become even more important in cases in which the patient is hesitant to take medication or when medication alone is not enough or is completely contraindicated. Although the cost of behavior therapy will vary among professionals, it is a highly structured and usually time-limited form of treatment and, therefore, quite cost effective.

One area in which behavioral techniques have proven quite successful is in the management of blood and injury phobias. Blood phobias are unique in that, unlike other phobias, they do not result in the typical sympathetic arousal, tachycardia, dyspnea, and paresthesias. Rather, blood phobias usually result in an initial rise in heart rate that is often quickly followed by vasovagal bradycardia, decrease in blood pressure, and eventually syncope. It is estimated that approximately 3.1% of the normal population experience severe blood phobia (Agras, Sylvester, & Oliveau, 1969). Such a phobia can represent a severe threat if it results in avoidance of life-saving medical procedures such as insulin injections, surgery, and blood transfusions.

A variety of behavioral treatments for blood and injury phobia have been elaborated in case study reports. Of all these treatments, the most frequently reported have been in vivo exposure to needles and blood (Leitenberg et al., 1970) and exposure to films of venipuncture, surgery, and injury in an emergency room (Marks et al., 1977). However, a variety of other behavioral treatments, such as systematic desensitization (Cohn, Kron, & Brady, 1976), modeling (Fryrear & Wearer, 1970), implosion (Ollendick & Gruen, 1972), and relaxation techniques (Ost et al., 1984), have been used with varying degrees of success. An applied tension procedure (Ost et al., 1989) involves having subjects tense all the muscles in their arms, chest, and legs until a feeling of warmth is felt in the face. After 20 to 30 seconds the subjects are required to release tension until they feel they have returned to pretension level. This cycle of tensing and letting go is repeated five times in a given session and the subject is required to practice this procedure five times a day. Since this procedure could be taught in half the time of relaxation (five sessions as opposed to 10), the authors concluded that applied tension should be the treatment of choice for blood phobias.

Another area in which behavioral techniques such as progressive relaxation and biofeedback have been shown to be beneficial has been with patients who suffer from asthma and COPD (Davis et al., 1973; Hock et al., 1978). Relaxation techniques are particularly helpful in the treatment of COPD because they are relatively easy to teach and because they do not have the respiratory depressant side effects that various anxiolytic medications may have (Renfroe, 1988). The need for anxiety reduction in COPD patients is clear given that dyspnea, a common symptom of COPD, often leads the patient to become more anxious, which results in increased muscle tension and increased need for oxygen. As this happens, dyspnea becomes magnified and a vicious cycle of increasing muscle tension, anxiety, and dyspnea ensues. One case report has shown biofeedback and progressive muscle relaxation to aid in weaning an anxious COPD patient from a ventilator (Acosta, 1988).

Finally, the relaxation response has been demonstrated to be effective in reducing stress and psychophysiologic reactivity in post-myocardial infarction patients (Gatchel, Gaffney, & Smith, 1986) and reducing preoperative anxiety in ambulatory surgery patients (Domar, Noe, & Benson, 1987). However, the literature on the use of behavioral techniques and the relaxation response in other medically ill patients is sparse, and further research is needed.

Contraindications to Relaxation Therapy

There seem to be few contraindications to the use of relaxation methods. People with organic mental disorders may have difficulty maintaining attention, and psychotic patients are unable to control intrusive thoughts. One other possible contraindication to the use of relaxation training would be suspicion of panic disorder. At least one case study report has shown significant increases in heart rate, temperature, and frontalis EMG activity in two individuals during a relaxation induction exercise (Cohen, Barlow, & Blanchard, 1985). Furthermore, systematic study of 15 panic disorder patients revealed greater increases in symptoms similar to those experienced in natural attacks, such as chest pain, dizziness, paresthesias, and derealization, while listening to a relaxation tape (Adler, Craske, & Barlow, 1987). Studies with large samples have shown that anywhere from 17% to 31% of subjects with generalized anxiety disorder will experience increases in restlessness, muscle tension, and general anxiety when listening to relaxation tapes (Braith, McCullough, & Bush, 1988; Heide & Borkovec, 1983; Ley, 1988). In these studies, relaxation-induced anxiety was associated with measures indicating a high internal locus of control and a high fear of losing control.

REFERENCES

ABELSON HI, & FISHBURNE PM (1976). Nonmedical use of psychoactive substances: 1975–1976. Princeton, NJ: Response Analysis Corporation.

ABERNETHY DR, GREENBLATT DJ, DIVOLL M, ET AL. (1982). Impairment of diazepam metabolism by low dose estrogen-containing oral contraceptive steroids. N Engl J Med 306:791–792.

ABERNETHY DR, GREENBLATT DJ, & SHADER RI (1981). Treatment of diazepam withdrawal syndrome with propranolol. Ann Intern Med 94:354–355.

ABRAMOWICZ M (1986a). Crack. Med Lett 28:69–70.

ABRAMOWICZ M (Ed.). (1986b). Midazolam. Med Lett 28:73–76.

ABRAMOWICZ M (1989). Drugs that cause psychiatric symptoms. Med Lett 31:113–118.

ACOSTA F (1988). Biofeedback and progressive relaxation in weaning the anxious patient from the ventilator: A brief report. Heart Lung 17:299–301.

ADLER CM, CRASKE MG, & BARLOW DH (1987). The use of modified relaxation in the experimental induction of anxiety and panic. Presented at the annual meeting of the Association for the Advancement of Behavior Therapy, Boston.

AGRAS S, SILVESTER D, & OLIVEAU D (1969). Epidemiology of common fears and phobias. Compr Psychiatry 10:151–156.

ALBECK JH (1987). Withdrawal and detoxification from benzodiazepine dependence: A potential role for clonazepam. J Clin Psychiatry 48:10S.

ANGUS WR, & ROMNEY DM (1984). The effect of diazepam on patients' memory. J Clin Psychopharmacol 4:203–206.

ARONSON TA, & CRAIG TJ (1986). Cocaine precipitation of panic disorder. Am J Psychiatry 143:643–645.

BACKES CR, & CADERO L (1980). Withdrawal symptoms in the neonate from presumptive intrauterine exposure to diazepam: Report of a case. J Am Osteopath Assoc 79:584–585.

BALON R (1990). Buspirone for attention deficit hyperactivity disorder? J Clin Psychopharmacol 10:77.

BARD M, & SUTHERLAND AM (1977). Adaptation to radical mastectomy. In The psychological impact of cancer (Ch. 3, pp. 55–71). New York: American Cancer Society.

BARRETT JE, BARRETT JA, OXMAN TE, ET AL. (1988). The prevalence of psychiatric disorders in a primary care practice. Arch Gen Psychiatry 45:1100–1106.

BAUDRY F, & WEINER A (1968). Preoperative preparation of the surgical patient. Surgery 63:885–889.

BELA SV, RAFTERY EF (1980). The role of verapamil in chronic stable angina: A controlled study with computerized multistage treadmill exercise. Lancet 1:841–844.

BENSON H, BEARY JF, & CAROL MP (1974). The relaxation response. Psychiatry 37:37–46.

BODKIN JA, & TEICHER MH (1989). Fluoxetine may antagonize the anxiolytic action of buspirone. J Clin Psychopharmacol 9:150.

BRAITH JA, MCCULLOUGH JP, & BUSH JP (1988). Relaxation-induced anxiety in a subclinical sample of chronically anxious subjects. J Behav Ther Exp Psychiatry 19:193–198.

BRAVO EL, TARAZI RC, GIFFORD RW, ET AL. (1979). Circulating and urinary catecholamines in pheochromocytoma. N Eng J Med 301:682–686.

BRYANT SG, GUERNSEY BG, & INGRIM NB (1983). Review of bupropion. Clin Pharmacy 2:525–537.

BUSTO U, SILLERS EM, NARANJO CA, ET AL. (1986). Withdrawal reaction after long-term therapeutic use of benzodiazepine. N Engl J Med 315:854–859.

CALABRESE JR, & GULLEDGE AD (1985). Psychotropics during pregnancy and lactation: A review. Psychosomatics 26:413–426.

CASTILLO-FERRANDO JR, GARCI M, & CARMONA J (1980). Digoxin levels and diazepam. Lancet 2:368.

CHARNEY DS, HENINGER GR, & JATLOW PI (1985). Increased anxiogenic effects of caffeine in panic disorders. Arch Gen Psychiatry 423:233–243.

CHRISTENSEN B (1946). Studies on hyperventilation: II. Electrocardiographic changes in normal man during voluntary hyperventilation. J Clin Invest 24:880.

CIRAULO DA, SANDS BF, & SHADER RI (1988). Critical review of liability for benzodiazepines: Abuse among alcoholics. Am J Psychiatry 145:1501–1506.

COHEN AS, BARLOW DH, & BLANCHARD EB (1985). The psychophysiology of relaxation associated panic attacks. J Abnorm Psychol 94:96–101.

COHEN LS, HELLER VL, & ROSENBAUM JF (1989). Treatment guidelines for psychotropic drug use in pregnancy. Psychosomatics 30:25–33.

COHEN RB, FINN H, & STEEN SM (1969). Effect of diazepam and meperidine, alone and in combination, on the respiratory response to carbon dioxide. Anesth Analg 48:353–355.

COHN CK, KRON RA, & BRADY JP (1976). A case study of blood-illness-injury phobia treated behaviorally. J Nerv Ment Dis 162:65–68.

COHN JB, BOWDEN CL, FISHER JG, ET AL. (1986). Double-blind comparison of buspirone and clorazepate in anxious outpatients. Am J Med 80(suppl 3B):10–16.

COHN JB, & WILCOX CS (1986). Low-sedation potential of buspirone compared with alprazolam and lorazepam in the treatment of anxious patients: A double-blind study. J Clin Psychiatry 47:409–412.

COLENDA CC (1988). Buspirone in treatment of agitated demented patient. Lancet 1:1169.

DAVIS MH, SUANDERS DR, CREER TH, ET AL. (1973). Relaxation training facilitated by biofeedback apparatus as a supplemental treatment in bronchial asthma. J Psychosom Res 17:121–128.

DE LA FUENTA JR, ROSENBAUM AH, MARTIN HR, ET AL. (1980). Lorazepam-related withdrawal seizures. Mayo Clin Proc 55:190–192.

DENKO JD, & KAELBLING R (1962). The psychiatric aspects of hypoparathyroidism. Acta Psychiatr Scand 38(suppl 164):1–70.

DEVEREUX RB, PERLOFF JK, REICHEK N, ET AL. (1976). Mitral valve prolapse. Circulation 54:3–14.

DEVRIES HA, BURKE RK, HOPPER RT, ET AL. (1977). Efficacy of EMG biofeedback in relaxation training: A controlled study. Am J Phys Med 56(2):75–81.

DIETCH JT (1984). Cerebral tumor presenting with panic attacks. Psychosomatics 25:861–863.

DIETCH JT, & JENNINGS RK (1988). Aggressive dyscontrol in patients treated with benzodiazepines. J Clin Psychiatry 49:184–188.

DIMSDALE JE, & HACKETT TP (1982). Effect of denial on cardiac health and psychological assessment. Am J Psychiatry 139:1477–1480.

DOMAR AD, NOE JM, & BENSON H (1987). The preoperative use of the relaxation response with ambulatory surgery patients. J Hum Stress 13:101–107.

DREYFUSS J, GRIFFITH DL, SINGHVI SM, ET AL. (1979). Pharmacokinetics of nadolol, a beta-receptor antagonist: Administration of therapeutic single and multiple-dosage regimens to hypertensive patients. J Clin Pharmacol 19:712–720.

EGBERT LD, BATTIT GE, WELCH CE, ET AL. (1964). Reduction of post-operative pain by encouragement and instruction of patients: A study of doctor-patient rapport. N Engl J Med 270:825–827.

EISON AS, & TEMPLE DL JR (1986). Buspirone: Review of its pharmacology and current perspectives on its mechanism of action. Am J Med 80(suppl 3B):1–9.

EISON MS, & EISON AS (1984). Buspirone as a midbrain modulator: Anxiolysis unrelated to traditional benzodiazepine mechanisms. Drug Dev Res 4:109–119.

EVANS DW, & LUM LC (1977). Hyperventilation: An important cause of pseudoangina. Lancet 1:155–157.

FANN WE, LAKE RC, & MAJORS LF (1974). Thioridazine in neurotic, anxious, and depressed patients. Psychosomatics 15:117–121.

FERRER M (1968). Mistaken psychiatric referral of occult serious cardiovascular disease. Arch Gen Psychiatry 18:112–113.

FINKEL BS, MCCLOSKEY KL, & GOODMAN LS (1979). Diazepam and drug-associated deaths. JAMA 242:429–434.

FLEISCHHACKER WW, ROTH SD, & KANE JM (1990). The pharmacologic treatment of neuroleptic-induced akathisia. J Clin Psychopharmacol 10:12–21.

FRASER HS, & CARR AC (1976). Propranolol psychosis. Br J Psychiatry 129:508–509.

FRISHMAN (1981). β-Adrenergic antagonists: New drugs and new indications. N Engl J Med 305:500–505.

FRYREAR JL, & WERNER S (1970). Treatment of a phobia by use of video taped modeling procedure: A case study. Behav Ther 1:391–394.

GARDNER EA (1983). Long-term preventive care in depression: The use of bupropion in patients intolerant of other antidepressants. J Clin Psychiatry 44:157–162.

GARNER SJ, ELDRIDGE FL, WAGNER PG, ET AL. (1989). Buspirone, an anxiolytic drug that stimulates respiration. Am Rev Respir Dis 139:946–950.

GATCHEL RJ, GAFFNEY FA, & SMITH JE (1986). Comparative efficacy of behavioral stress management versus propranolol in reducing psychophysiological reactivity in post-myocardial infarction patients. J Behav Med 9:503–513.

GEHI MM, ROSENTHAL RH, FIZETTE NB, ET AL. (1981). Psychiatric manifestations of hyponatremia. Psychosomatics 22:739–743.

GERSHON ES, GOLDSTEIN RE, MOSS AJ, ET AL. (1979). Psychosis with ordinary doses of propranolol. Ann Intern Med 90:938–940.

GIBBS FA, & GIBBS EC (1952). Atlas of electroencephalography (Vol. 2). Cambridge, MA: Addison-Wesley.

GILBERG C (1977). "Floppy infant syndrome" and maternal diazepam. Lancet 2:224.

GOLDBERG RJ (1982a). Anxiety: A guide to biobehavioral diagnosis and therapy for physicians and mental health clinicians. New York: Free Press.

GOLDBERG RJ (1982b). Anxiety reduction by self-regulation: Theory, practice and evaluation. Ann Intern Med 96:483–487.

GOLDBERG RJ (1983). Personality types and personality disorders. In H LEIGH (Ed.), Psychiatry in the practice of medicine (Ch. 4, pp. 37–56). Menlo Park: Addison-Wesley.

GOLDBERG RJ, & CULLEN LO (1985). Factors important to psychosocial adjustment to cancer: A review of the evidence. Soc Sci Med 20:803–807.

GOTOH F, MEYER JS, & TAKAGI Y (1965). Cerebral effects of hyperventilation in man. Arch Neurol 12:410.

GRANVILLE-GROSSMAN KL, & TURNER P (1966). The effect of propranolol on anxiety. Lancet 1:788–790.

GREENBLATT DJ, ABERNETHY DR, MORSE DS, ET AL. (1984). Clinical importance of the interaction of diazepam and cimetidine. N Engl J Med 310:1639–1643.

GREENBLATT DJ, ALLEN MD, HARMATZ MJ, ET AL. (1980). Diazepam disposition determinants. Clin Pharmacol Ther 27:301–312.

GREENBLATT DJ, DIVOLL M, HARMATZ JS, ET AL. (1982). Pharmacokinetic comparison of sublingual lorazepam with intravenous, intramuscular and oral lorazepam. J Pharm Sci 71:248–252.

GREENBLATT DJ, & KOCH-WESER J (1976). Intramuscular injection of drugs. N Engl J Med 295:542–546.

GREENBLATT DJ, & SHADER RI (1974a). Benzodiazepines in clinical practice. New York: Raven Press.

GREENBLATT DJ, SHADER RI, KOCH-WESER J, ET AL. (1974). Slow absorption of intramuscular chlordiazepoxide. N Engl J Med 291:1116–1118.

GRIFFITH JD, JASINSKI DR, & MCKINNEY GR (1986). Investigation of the abuse liability of buspirone in alcohol-dependent patients. Am J Med 80(suppl 3B):30–35.

GRONWALL D, & WRIGHTSON P (1974). Delayed recovery of intellectual function after minor head injury. Lancet 2:605–609.

HALL RCW (1983). Psychiatric effects on thyroid hormone disturbance. Psychosomatics 24:7–18.

HALL RCW, & JOFFE JR (1972). Aberrant response to diazepam: A new syndrome. Am J Psychiatry 126P:738–742.

HALL RCW, & JOFFE JR (1973). Hypomagnesemia: Physical and psychiatric symptoms. JAMA 224:1749–1751.

HARRIS MC (1985). The use and abuse of pocket nebulizers in the treatment of asthma. Postgrad Med 23:170–173.

HEALEY M, PICKENS R, MEISCH R, ET AL. (1983). Effects of clorazepate, diazepam, lorazepam, and placebo on human memory. J Clin Psychiatry 44:436–439.

HEIDE FJ, & BORKOVEC TD (1983). Relaxation-induced anxiety: Paradoxical anxiety enhancement due to relaxation training. J Consult Clin Psychol 51:171–182.

HOCK RA, RODGERS CH, REDD C, ET AL. (1978). Medical-psychological interventions in male asthmatic children: An evaluation of psychological change. Psychosom Med 40:210–215.

HOLLISTER LE, MOTZENBECKER FP, & DEGNAN RO (1961). Withdrawal reactions from chlordiazepoxide (Librium). Psychopharmacologia 2:63–68.

HOYUMPA AM (1978). Disposition and elimination of minor tranquilizers in the aged and in patients with liver disease. South Med J 71:23–28.

JACOBS MA, SENIOR RM, & KESSLER G (1976). Clinical experience with theophylline: Relationship between dosage, serum concentration, and toxicity. JAMA 235:1983–1986.

JACOBSON E (1938). Progressive relaxation (2nd ed.). Chicago: University of Chicago Press.

JAMES IM, PEARSON RM, GRIFFITH DNW, ET AL. (1978). Reducing the somatic manifestations of anxiety by beta-blockage: A study of stage fright. J Psychosom Res 22:327–337.

JOHNSON LC, & CHERNIK DA (1982). Sedative-hypnotics and human performance. Psychopharmacology 76:101–113.

KAHN RJ, MCNAIR DM, LIPMAN RS, ET AL. (1986). Imipramine and chlordiazepoxide in depressive and anxiety disorders: II. Efficacy in anxious outpatients. Arch Gen Psychiatry 43:79–85.

KANTOR JS, ZITRIN CM, & ZELDIS SM (1980). Mitral valve prolapse syndrome in agoraphobic patients. Am J Psychiatry 137:467–469.

KAPLAN C, LIPKIN M, & GORDON GH (1988). Somatization in primary care: Patients with unexplained and vexing medical complaints. J Gen Intern Med 3:177–190.

KARCH FE (1979). Rage reaction associated with clorazepate dipotassium. Ann Intern Med 91:61–62.

KASTENHOLZ KV, & CRISMON ML (1984). Buspirone, a novel nonbenzodiazepine anxiolytic. Clin Pharm 3:600–607.

KATHOL RG, NOYES R JR, SLYMEN DJ, ET AL. (1980). Propranolol in chronic anxiety disorders. A controlled study. Arch Gen Psychiatry 37:1361–1365.

KATON WJ (1990). Chest pain, cardiac disease, and panic disorder. J Clin Psychiatry 51:27–30.

KATON W, VONKORFF M, LIN E, ET AL. (1990). Distressed high utilizers of medical care. DSM-III-R diagnoses and treatment needs. Gen Hosp Psychiatry 12:355–362.

KLEIN DF (1982). Medication in the treatment of panic attacks and phobic states. Psychopharmacol Bull 18:85–90.

KRISTENSEN O, & SENDRUP EH (1978). Sphenoidal electrodes. Acta Neurol Scand 58:157–166.

LAEGREID L, OLEGARD R, WAHLSTROM J, ET AL. (1987). Abnormalities in children exposed to benzodiazepines in utero. Lancet 1:108–109.

LAKSHMINARAYAN MD, SAHN SA, HUDSON LD, ET AL. (1976). Effect of diazepam on ventilatory responses. Clin Pharmacol Ther 20:178–183.

LEBOEUF A (1977). The effects of EMG feedback training on state anxiety in introverts and extroverts. J Clin Psychol 33:251–253.

LEIGH D (1979). Psychiatric aspects of head injury. Psychiatry Digest 40:21–32.

LEITENBERG H, WINCZE JP, BUTZ RA, ET AL. (1970). Comparison of the effects of instructions and reinforcement in the treatment of a neurotic avoidance response: A single case experiment. J Behav Ther Exp Psychiatry 1:53–58.

LERNER WD, & FALLON HJ (1985). The alcohol withdrawal syndrome. N Engl J Med 313:951–952.

LEVINE AM (1988). Buspirone and agitation in head injury. Brain Injury 2:165–167.

LEY R (1988). Panic attacks during relaxation and relaxation-induced anxiety: A hyperventilation interpretation. J Behav Ther Exp Psychiatry 19:253–259.

LIBERTHSON R, SHEEHAN DV, KING ME, ET AL. (1986). The prevalence of mitral valve prolapse in patients with panic disorders. Am J Psychiatry 143:511–515.

LIEB JP, WALSH GO, BABB TL, ET AL. (1976). A comparison of EEG seizure patterns recorded with surface and depth electrodes in patients with temporal lobe epilepsy. Epilepsia 17:137–160.

LIPINSKI JF, MALLYA G, ZIMMERMAN P, ET AL. (1989). Fluoxetine-induced akathisia: Clinical and theoretical implications. J Clin Psychiatry 50:339–342.

LIPMAN RS, COVI L, DOWNING RW, ET AL. (1981). Pharmacotherapy of anxiety and depression. Psychopharmacol Bull 17:91–103.

Lishman WA (1987). Endocrine disorders and metabolic disorders. In Organic psychiatry (Ch. 11, pp. 428–485). London: Blackwell.

Lucki I, Rickels K, & Geller AM (1986). Chronic use of benzodiazepines and psychomotor and cognitive test performance. Psychopharmacology 89:S55.

Lynch JJ, Paskewitz DA, Gimbel KS, et al. (1977). Psychological aspects of cardiac arrhythmia. Am Heart J 93:645–657.

MacCrimmon DJ, Wallace JE, Goldberg WM, et al. (1979). Emotional disturbance and cognitive deficits in hyperthyroidism. Psychosom Med 41:331–340.

Malcolm R, Ballenger JC, Sturgis ET, et al. (1989). Double-blind controlled trial comparing carbamazepine to oxazepam treatment of alcohol withdrawal. Am J Psychiatry 146:617–621.

Markovitz PJ, Stagno SJ, & Calabrese JR (1990). Buspirone augmentation of fluoxetine in obsessive-compulsive disorder. Am J Psychiatry 147:798–800.

Marks IM, Hallam RS, Connolly J, et al. (1977). Nursing in behavioral psychotherapy: An advanced role for nurses. London: Royal College of Nursing.

Marshall JR (1979). Neuropsychiatric aspects of renal failure. J Clin Psychiatry 40:81–85.

Mastrovito RC (1972). Symposium: Emotional considerations in cancer and stroke. NY State J Med 72:2874–2877.

Mazza DL, Martin D, Spacavento L, et al. (1986). Prevalence of anxiety disorders in patients with mitral valve prolapse. Am J Psychiatry 143:349–352.

McHenry PL, Cogan OJ, Elliott WC, et al. (1970). False-positive ECG response to exercise secondary to hyperventilation: Cineangiographic correlation. Am Heart J 79:683–687.

McKell TE, & Sullivan AJ (1947). The hyperventilation syndrome in gastroenterology. Gastroenterology 9:6–16.

McNamara ME, & Fogel BS (1990). Anticonvulsant-responsive panic attacks with temporal lobe EEG abnormalities. J Neuropsychiatry 2:193–196.

Meyer RE (1986). Anxiolytics and the alcoholic patient. J Stud Alcohol 47:269–273.

Mitchells-Heggs P, Murphy K, Minty K, et al. (1980). Diazepam in the treatment of dyspnea in the "pink puffer" syndrome. Q J Med 49:9–20.

Moore RA (1985). The prevalence of alcoholism in medical and surgical patients. In MA Schuckit & AE Slaby (Eds.), Alcohol patterns and problems (Ch. 8, pp. 247–265). New Brunswick, NJ: Rutgers University Press.

Mueller HS, & Chahine RA (1981). Interim report of multicenter double-blind placebo-controlled studies of nifedipine in chronic stable angina. Am J Med 71:645–657.

Neill WA, & Hattenhauer M (1975). Impairment of myocardial O_2 supply due to hyperventilation. Circulation 52:854–858.

Newton RE, Marunycz JD, Alderdice MT, et al. (1986). Review of the side-effect profile of buspirone. Am J Med 80(suppl 3B):17–21.

Nierenberg A, & Cole JO (1991). Antidepressant adverse drug reactions. J Clin Psychiatry 52:40–47.

Nikaido AM, Ellinwood EH, Heatherly D, et al. (1987). Differential CNS effects of diazepam in elderly adults. Pharmacol Biochem Behav 27:273–281.

Ollendick TH, & Gruen GE (1972). Treatment of a bodily injury phobia with implosive therapy. J Consult Clin Psychol 38:389–393.

Oppenheimer RD (1968). Microscopic lesions in the brain following head injury. Neurol Neurosurg Psychiatry 3:299–306.

Ost LG, Lindahl IL, Sterner U, et al. (1984). Exposure in vivo vs. applied relaxation in the treatment of blood phobia. Behav Res Ther 22:205–216.

Ost LG, Sterner U, & Fellenius J (1989). Applied tension, applied relaxation and the combination in treatment of blood phobia. Behav Res Ther 27:109–121.

Pariser SF, Pinta ER, & Jones BA (1978). Mitral valve prolapse syndrome and anxiety neurosis/panic disorder. Am J Psychiatry 135:246–247.

Parkinson D (1977). Concussion. Mayo Clin Proc 52:492–496.

Patel DA, & Patel AR (1980). Clorazepate and congenital malformations. JAMA 244:135–136.

Patterson J (1988). Alprazolam dependency: Use of clonazepam for withdrawal. South Med J 81:830–831.

Paykel ES, Rowman PR, Parker RR, et al. (1982). Response to phenelzine and amitriptyline in subtypes of outpatient depression. Arch Gen Psychiatry 39:1041–1049.

Peck A, & Boland J (1977). Emotional reactions to radiation treatment. Cancer 40:180–184.

Pevnick JS, Jasinski DR, & Haertzen CA (1978). Abrupt withdrawal from therapeutically administered diazepam. Arch Gen Psychiatry 35:995–998.

Phillips JP, Antal EJ, & Smith RB (1986). A pharmacokinetic drug interaction between erythmycin and triazolam. J Clin Psychopharmacol 6:297–302.

Pincus JH, & Tucker GJ (1985). Behavioral neurology (3rd ed.). New York: Oxford University Press.

Plum F, & Posner JB (1972). Diagnosis of stupor and coma (2nd ed.). Contemporary Neurology Series. Philadelphia: FA Davis.

Pohl R, Berchou R, & Rainey JM (1982). Tricyclic antidepressants and monoamine oxidase inhibitors in the treatment of agoraphobia. J Clin Psychopharmacol 2:399–407.

Pohl R, Yeragani VK, Balon R, et al. (1988). The jitteriness syndrome in panic disorder patients treated with antidepressants. J Clin Psychiatry 49:100–104.

Pomara N, Stanley B, Block R, et al. (1985). Increased sensitivity of the elderly to the central depressant effects of diazepam. J Clin Psychiatry 46:185–187.

Rao S, Sherbaniuk RW, Prasad K, et al. (1973). Cardiopulmonary effects of diazepam. Clin Pharmacol Ther 14:182–189.

Rapoport DM, & Mendelson WH (1989). Buspirone: A new respiratory stimulant [abstract]. Am Rev Respir Dis 139:A625.

Raskin M, Bali LR, & Peeke HV (1980). Muscle biofeedback and transcendental meditation. Arch Gen Psychiatry 37:93–97.

Ratey JJ, & Salzman C (1984). Recognizing and managing akathisia. Hosp Community Psychiatry 35:975–977.

Ratey JJ, Sovner R, Mikkelsen E, et al. (1989). Buspirone therapy for maladaptive behavior and anxiety in developmentally disabled persons. J Clin Psychiatry 509:382–384.

Ray WA, Griffin MR, & Downey W (1989). Benzodiazepines of long and short elimination half-life and the risk of hip fracture. JAMA 262:3303–3307.

Renfroe KL (1988). Effect of progressive relaxation on dyspnea and state anxiety in patients with chronic obstructive pulmonary disease. Heart Lung 17:408–413.

Resnick RB, Kestenbaum RS, & Schwartz LK (1976). Acute systemic effects of cocaine in man: A controlled study by intranasal and intravenous routes. Science 195:696–698.

Reuler JB, Girard DE, & Cooney TG (1985). Waernicke's encephalopathy. N Engl J Med 312:1035–1039.

Rice RL (1950). Symptom patterns of the hyperventilation syndrome. Am J Med 8:691–700.

Rickels K (1983). Nonbenzodiazepine anxiolytics: Clinical usefulness. J Clin Psychiatry 44(11 sec 2):38–43.

Rickels K, Csanalosi I, Chung HR, et al. (1974). Amitriptyline in anxious-depressed outpatients: A controlled study. Am J Psychiatry 131:25–30.

Rickels K, Gordon PE, Zamostein BB, et al. (1970). Hydroxyzine and chlordiazepoxide in anxious neurotic outpatients: A collaborative controlled study. Comp Psychiatry 11:457–474.

Rickels K, Morris RJ, Newman H, et al. (1983). Diphenhydramine in insomniac family practice patients: A double-blind study. J Clin Pharmacol 23:235–242.

Rickels K, Weisman K, Norstad N, et al. (1982). Buspirone and diazepam in anxiety: A controlled study. J Clin Psychiatry 43(12, sec 2): 81–86.

Ries RK, Roy-Byrne PP, Ward NG, et al. (1989). Carbamazepine treatment for benzodiazepine withdrawal. Am J Psychiatry 146:536–537.

Rifkin A, Doddi S, Karajgi B, et al. (1989). Benzodiazepine use and abuse by patients at outpatient clinics. Am J Psychiatry 146:1331–1332.

Rinkenberger RL, Psystowsky EN, Heger JJ, et al. (1980). Effect of intravenous and chronic oral verapamil administration in patients with supraventricular arrhythmias. Circulation 62:996–1010.

Robertson D, Frolich JC, Carr RK, et al. (1978). Effects of caffeine on plasma renin activity, catecholamines and blood pressure. N Engl J Med 298:181–186.

Robinson DS, Nies A, Ravaris CL, et al. (1973). The monoamine oxidase inhibitor, phenelzine, in the treatment of depressive-anxiety states: A controlled clinical trial. Arch Gen Psychiatry 29:407–413.

Rosenberg L, Mitchell AA, Parsells JL, et al. (1983). Lack of relation of oral clefts to diazepam use during pregnancy. N Engl J Med 309:1282–1285.

Ryan HF, Merrill FB, Scott GE, et al. (1968). Increase in suicidal thoughts and tendencies: Association with diazepam therapy. JAMA 203:1137–1139.

Scharf MB, Khosla N, Brocker N, et al. (1984). Differential amnestic properties of short and long-acting benzodiazepines. J Clin Psychiatry 45:51–53.

Schneider LS, Syapin PJ, & Pawluczyk S (1987). Seizures following triazolam withdrawal despite benzodiazepine treatment. J Clin Psychiatry 48:418–419.

Schuckit MA (1984). Clinical studies of buspirone. Psychopathology 17(suppl 3):61–68.

Schweizer E, & Rickels K (1986). Failure of buspirone to manage benzodiazepine withdrawal. Am J Psychiatry 143:1590–1592.

Schweizer E, Rickels K, & Lucki I (1986). Resistance to the antianxiety effect of buspirone in patients with a history of benzodiazepine use. N Engl J Med 314:719–720.

Sellers EM, & Kalant H (1976). Alcohol intoxication and withdrawal. N Engl J Med 294:757–762.

Shader RI, Dreyfuss D, Gerrein JR, et al. (1986). Sedative effects and impaired learning and recall following single oral doses of lorazepam. Clin Pharmacol Ther 39:526–529.

Shear MK, Devereux RB, Dramer-Fox R, et al. (1984). Panic patients with a low prevalence of mitral valve prolapse. Am J Psychiatry 141:302–303.

Sheehan DV, Ballenger J, & Jacobsen G (1980). Treatment of endogenous anxiety with phobic, hysterical and hypochondriacal symptoms. Arch Gen Psychiatry 37:51–59.

Singh BN, Ellrodt G, & Peter CT (1978). Verapamil: A review of its pharmacological properties and therapeutic use. Drugs 15:169–197.

Starkman MN, Zelnik TC, Nesse RM, et al. (1985). Anxiety in patients with pheochromocytomas. Arch Intern Med 145:248–252.

Stein MB, Heuser IJ, Juncos JL, et al. (1990). Anxiety disorders in patients with Parkinson's disease. Am J Psychiatry 147:217–220.

Strain JJ, Liebowitz MR, & Klein DF (1981). Anxiety and panic attacks in the medically ill. Psychiatr Clin North Am 4:333–350.

Stoudemire J (1972). Effects of muscle relaxation training on state and trait anxiety in introverts and extroverts. J Pers Soc Psychol 24:273–275.

Sutherland AM, Orbach CE, Duk RB, et al. (1977). Adaptation to the dry colostomy; preliminary report and summary of findings. In The psychological impact of cancer (Ch. 1, pp. 1–16). New York: American Cancer Society.

Taylor AR, & Bell TK (1966). Slowing of cerebral circulation after concussional head injury: A controlled trial. Lancet 2:178–180.

Taylor PJ (1967). Hemorrhage while on anticoagulant therapy precipitated by drug interaction. Arizona Med 24:697–699.

Tien AY, & Gujavarty KS (1985). Seizure following withdrawal from triazolam [letter to the editor]. Am J Psychiatry 142:1516–1517.

Tinetti ME, Speechley M, & Ginter SF (1988). Risk factors for falls among elderly persons living in the community. N Engl J Med 319:1701–1707.

Tyrer PJ, & Lader MH (1974). Response to propranolol and diazepam in somatic anxiety. Br Med J 2:14–16.

Uhlenhuth EH, DeWit H, Balter MB, et al. (1988). Risks and benefits of long-term benzodiazepine use. J Clin Psychopharmacol 8:161–167.

Vajda FJE, Prineas RJ, & Lovell RPH (1971). Interaction between phenytoin and the benzodiazepines. Br Med J 1:346.

Van der Kropf C (1979). Reactions to triazolam. Lancet 2:526.

Venkatesh A, Pauls DL, Crowe R, et al. (1980). Mitral valve prolapse in anxiety neurosis (panic disorder). Am Heart J 100:302–305.

Victor BS, Lubetsky M, & Greden JF (1981). Somatic manifestations of caffeinism. J Clin Psychiatry 42:185–188.

Waldeck B (1975). Effect of caffeine on locomotor activity in central catecholamine mechanisms: A study with special reference to drug interaction. Acta Pharm Toxicolog 36:1–23.

Wallace RK (1970). Physiological effects of transcendental meditation. Science 167:1751–1754.

Warwick HMC, & Salkovskis PM (1990). Hypochondriasis. Behav Res Ther 28:105–117.

Webb WL, & Gehi M (1981). Electrolyte and fluid imbalance: Neuropsychiatric manifestations. Psychosomatics 22:199–203.

Weil AA (1959). Ictal emotions occurring in temporal lobe dysfunction. Arch Neurol 1:87–97.

Weiner N (1980). Norephedrine, ephedrine, and sympathomimetic amines. In AG Goodman, LS Goodman, & A Gilman (Eds.), Pharmacological basis of therapeutics (P. 163). New York: Macmillan.

Weisman A (1979). Coping with cancer. New York: McGraw-Hill.

Wells BG, Cold JA, Marken PA, et al. (1991). A placebo-controlled trial of Nadolol in the treatment of neuroleptic-induced akathisia. J Clin Psychiatry. 52:255–260.

Wells KB, Golding JM, & Burnam MA (1989). Affective, substance use, and anxiety disorders in persons with arthritis, diabetes, heart disease, high blood pressure, or chronic lung conditions. Gen Hosp Psychiatry 11:320–327.

White BC, Lincoln CA, Pearce NW, et al. (1980). Anxiety and muscle tension as consequences of caffeine withdrawal. Science 209:1547–1548.

Wilkerson P, Van Dyke C, Jatlow P, et al. (1980). Intranasal and oral cocaine kinetics. Clin Pharmacol Ther 27:386–394.

Winokur A, Rickels K, Greenblatt DJ, et al. (1980). Withdrawal reaction from long-term low-dosage administration of diazepam. Arch Gen Psychiatry 37:101–105.

Wolkowitz OM, Weingartner H, Thompson K, et al. (1987).

Diazepam-induced amnesia: A neuropharmacological model of an "organic amnestic syndrome." Am J Psychiatry 144:25–29.

Wortsman J, Burns G, Van Beek AL, et al. (1980). Hyperadrenergic state after trauma to the neuroaxis. JAMA 243:1459–1460.

Zipursky RB, Baker RW, & Zimmer B (1985). Alprazolam withdrawal delirium unresponsive to diazepam: Case report. J Clin Psychiatry 46:344–345.

Zitrin CM, Klein DF, & Woerner MG (1978). Behavior therapy, supportive psychotherapy, imipramine and phobias. Arch Gen Psychiatry 35:307–316.

7 Laboratory and neuroendocrine assessment in medical-psychiatric patients

THEODORE J. ANFINSON, M.D., AND
ROGER G. KATHOL, M.D.

The use of the clinical laboratory by psychiatrists has increased dramatically in recent years. Economic, technologic, medical-legal, and research issues have fueled this increased reliance on laboratory data. However, laboratory evaluations should not be seen as a substitute for the traditional history and physical assessment (Hampton et al., 1975). Historic information must include a thorough review of systems and collateral histories obtained from the patient's family or acquaintances, whenever possible. Physical examination must include a complete neurologic evaluation with mental status examination. However, some form of laboratory assessment, guided by the history and physical findings, is usually necessary in the evaluation of the often confusing, atypical, and complex patients who present to the medically-oriented psychiatrist.

Topics in laboratory medicine of concern to the psychiatrist include the utility of screening evaluations; the integration of historic, physical examination, and laboratory data; the critical appraisal of tests related to medical conditions commonly encountered by psychiatrists; and endocrine and neuroendocrine testing in psychiatric patients.

SCREENING LABORATORY EVALUATION

Routine laboratory screening of psychiatric patients is a common clinical practice which deserves scrutiny. The rationale for screening includes the detection of physical disease presenting with psychiatric symptoms, physical disease resulting from psychiatric disorder, coincidental medical disease, and side effects of therapy (Thomas, 1979). Criteria such as acceptability, simplicity, reliability, validity, the appropriateness of the test for the population screened, and the cost of the test should be considered when deciding on screening tests. In addition, the diseases for which screening is performed should be treatable, serious or potentially serious, and relatively prevalent, and have a better prognosis with early diagnosis (Frankenburg & Camp, 1975).

Pertinent statistical concepts include sensitivity, specificity, positive predictive value, and disease prevalence. Analytic imprecision, biologic variability, including regression to the mean (Sackett, 1973), and statistical assumptions regarding the reference or "normal" range are common sources of error in diagnostic classification (Cebul & Beck, 1987). In addition, the prevalence of a particular disease has a dramatic and often underappreciated effect on the predictive value of a given test (Fleeson & Wenk, 1970). These concepts need to be considered when making recommendations regarding screening tests.

Several studies have demonstrated the limited utility and high cost of extensive ambulatory and preadmission screening in the evaluation of general medical patients (Cebul & Beck, 1987; Durbridge et al., 1976; Korvin, Pearce, & Stanley, 1975). However, the data are less compelling concerning such screening in psychiatric patients. Epidemiologic studies evaluating the rate of medical illness in patients presenting with psychiatric complaints differ widely in their methodology and in the populations studied (Table 7-1). However, many of the conclusions reached are very similar: that physical illness is present in psychiatric patients far more frequently than is often appreciated, and that the pre-

TABLE 7–1. *Studies Evaluating the Prevalence of Medical Disease in Psychiatric Populations*

Study	N	Rate (%)	Related to Psychiatric Diagnosis	Previously Undiagnosed
Philips (1937)	164	45	24	*
Marshall (1949)	*	44	*	*
Herrige (1960)	209	50	*	*
Davies (1965)	36	58	42	*
Maguire (1968)	200	33	*	49
Johnson (1968)	250	60	12	80
Eastwood, Mindham, and Tennent (1970)	100	40	*	16
Koranyi (1972)	100	49	20	71
Burke (1972)	202	43	*	*
Eastwood (1975)	124	("high")*	*	*
Burke (1978)	133	50	*	*
Hall et al. (1978)	658	*	9	46
Koranyi (1979)	2090	43	18	46
Hall et al. (1981)	100	80	46	80
Barnes et al. (1983)	147	26	*	13
Colgan and Philpot (1985)	167	20	*	*
Dolan and Mushlin (1985)	250	4	*	*
Ferguson and Dudleston (1986)	650	17	*	*
Roca, Breakey, and Fischer (1987)	42	93	*	46
Sox et al. (1989)	509	39	*	12
Koran et al. (1989)	529	34	6	12

*No data given.
Adapted from Koranyi (1980, p. 889), with updates.

senting psychiatric complaints are often caused or exacerbated by the medical illness discovered at presentation. The rate of physical illness in these studies of psychiatric patients ranged from 20% to 80%. The rate of physical illness believed to be related in some way to the patients' psychopathology ranged from 9% to 46%.

These investigators noted that the medical disorders often were missed at the time of the initial evaluation. Many of these patients had been referred by primary care physicians and had had physical examinations and laboratory assessments prior to referral. Rates of previously undiagnosed physical illness ranged from 12% to 80%.

Studies specifically evaluating the utility of laboratory screening psychiatric patients are summarized in Tables 7-2 and 7-3. The retrospective data are fairly consistent regarding the relatively low frequency of clinically significant laboratory abnormalities. Of note is the frequent finding that physicians often fail to pursue abnormal results. Considerable disparity exists among the prospective data, which generally can be attributed to the population studied and other methodologic issues. In general, outpatients have a significantly lower rate of active medical disease, and thus lower rates of laboratory abnormalities. The elderly, patients of lower socioeconomic status, and state hospital patients have the highest rates of abnormal laboratory results (Hall et al., 1980; Hall et al., 1981; Koran et al., 1989; Roca, Breakey, & Fischer, 1987). The likelihood of finding an abnormal result increases with the pretest probability of an abnormality (i.e., clinical suspicion). The data emphasize that abnormal results are predicted by the history, review of systems, and/or physical examination (Barnes et al., 1983; Ferguson & Dudleston, 1986; Hall et al., 1978; Kolman, 1984). Historic predictors of laboratory abnormalities included patients with evidence of alcohol or drug excess, disorientation, self-neglect, or age greater than 65 years. "Organic psychosis" and alcohol/drug dependence are psychiatric diagnoses associated with higher rate of abnormalities (Ferguson & Dudleston, 1986).

Several limited screening profiles have been recommended, but again are variable and population dependent. The serum glucose, electrolytes, blood urea nitrogen (BUN), and creatinine levels and urinalysis are the most common tests included in such a profile (Kolman, 1984). Serum calcium level, liver function tests, syphilis serology, and vitamin B_{12} and folate levels are less frequently abnormal (Colgan & Philpot, 1985; Dolan & Mushlin, 1985; White & Barraclough, 1989).

Extensive testing batteries are often cumbersome, inappropriate for a given population, and expensive. Furthermore, important clinical findings are infrequently revealed and are often ignored. Physicians often have difficulty interpreting the results of such batteries (Casscells, Schoenberger, & Graboys, 1978; McNeil, Keeler, & Adelstein, 1975), and the diseases that they are designed to detect often have such a low prevalence that the false-positive rate is unacceptably high (Fleeson &

TABLE 7–2. *Retrospective Studies Evaluating the Utility of Screening Laboratory Assessment in Psychiatric Patients*[a]

Study/ Population	Abnormal Results (%)	Clinically Significant Results (%)	Conclusions & Recommendations	Comments
Thomas (1979)/613 acute adult inpatients	13	10	Small percentage of laboratory tests lead to new diagnoses or changes in management; when abnormal results occur, the majority lead to no further investigation	27% of patients received no screening tests; 61% of abnormal tests were not followed up; selective ordering of tests resulted in higher yield
Colgan and Philpot (1985)/168 consecutive geriatric admissions	20	4	CBC, BUN, electrolytes, folate and midstream urine recommended; other tests should be ordered selectively	Tests ordered selectively (such as ECG, EEG) resulted in a higher yield of positive results
Dolan and Mushlin (1985)/250 adult inpatients	4	1.8	CBC, thyroxine, calcium, aspartate aminotransferase, alkaline phosphatase, urinalysis, syphilis serology recommended	True-positive rate in patients without physical signs/ symptoms, 0.06%; organic mental disorder associated with 4.0% true-positive abnormal test rate
Gabel and Hsu (1986)/ 100 adolescent inpatients	*	*	No definite evidence of undetected medical illness discovered through screening tests in adolescent population	No diagnosis changed from "functional" to "organic" on basis of lab tests; many insignificant findings noted
Whitle and Barraclough (1989)/ 1007 admissions (719 patients), adult; only 19 patients >65 years of age	10.2	0.8	Thyroid function tests in affectively ill females with personal or family history of thyroid disease; urinalysis on all females; white blood cell count in unexplained unremitting episodes of psychiatric illness; syphilis serology when history suggests exposure	40% received no laboratory testing; only 19 patients >65 years

[a]Investigations in all studies included complete blood count (CBC), electrolytes, blood urea nitrogen (BUN), glucose level, liver function tests, urinalysis, vitamin B_{12} and folate levels, syphilis serology, electrocardiogram (ECG), electroencephalogram (EEG), and skull roentgenograms. Many tests were ordered broadly (CBC, serum chemistries); others were performed more selectively and, therefore, were not truly screening tests. All studies revealed that the yield of a particular test improved with clinical suspicion.

*Data not presented in fashion to allow inclusion in table.

Wenk, 1970). There is little evidence that *extensive* routine screening batteries contribute more to management in general psychiatric patients than they do in general medical populations.

Patients with physical disease resulting from their psychiatric disorder appear to benefit most from laboratory screening. These patients have a high prevalence of physical illness resulting from neglect and complications of the underlying disorder. Therefore, patients with historic evidence of diminished self-care, alcohol or substance abuse, and secondary mental disorders merit more thorough evaluation.

The finding that physicians often are unable to prospectively predict abnormal laboratory findings based on information gained from history and physical examination cannot be ignored and suggests that a limited screening program may be valid (Kolman, 1985). However, it is impossible to determine from the data the precise components of this battery, which would vary with the population.

An algorithmic approach to detecting physical disease in psychiatric patients has been advocated and may represent a compromise between the conflicting demands of cost effectiveness and diagnostic power. This approach identified 90% of patients with active medical disease at a cost of $156.00 per detected case, compared to the

TABLE 7–3. *Prospective Studies Evaluating the Utility of Screening Laboratory Assessment in Psychiatric Patients*

Study/ Population	Investigations Performed[a]	Results/Conclusions	Comments
Hall et al. (1978)/ N = 658 suburban outpatients	H/P, ROS, MSE, CBC, complete serum chemistry, ECG, syphilis serology	3% of symptom-free patients had lab evidence of disease compared to 60% of symptom-positive patients	Organic predictors included visual hallucinations, illusions, and distortions
Hall et al. (1981)/ N = 100 inpatients under mental health warrant	H/P, ROS, MSE, additional detailed neurologic exam, CBC, UA, complete serum chemistry, vitamin B_{12} and folate levels, ECG, sleep-deprived EEG	High percentage of state hospital patients have medical comorbidity not apparent on presentation; complete workup utilizing all tests in study should be included in the workup of all psychiatric inpatients	Population highly selected; patients at high risk for concomitant physical illness; data not presented in a manner allowing determination of relative contribution of history, physical, and lab exams to diagnosis; causal link between physical illness and psychiatric diagnosis questionable in many cases
Barnes et al. (1983)/N = 147 outpatients	H/P, ROS, CBC, complete serum chemistry, T_3, T_4, VDRL, UA, ECG, CXR	ROS, PE, laboratory revealed 26, 31, & 19 abnormalities, respectively	ROS, blood pressure, fasting blood glucose, and UA would have detected 75% of all new diagnoses or those requiring treatment
Kolman (1984)/ N = 68 geriatric inpatients	H/P, ROS, CBC, complete serum chemistry, vitamin B_{12} and folate levels, T_4, ESR, VDRL, ECG, CXR, skull roentgenogram	23% of all tests abnormal; 1.4% of abnormals revealed diagnoses not apparent on H/P, ROS, PE	Most useful tests: UA, glucose, vitamin B_{12}, BUN, ECG, CXR; 74% of abnormal labs not followed up
Willett and King (1977)/N = 636 adult inpatients	CBC, VDRL, BUN, complete serum chemistry, UA	Overall abnormality rate, 5.9%; rate of new diagnoses made on the basis of laboratory results, 2.2–8.5%	61% of unexpected abnormals not followed up (unexpected abnormals identified retrospectively)
Ferguson and Dudleston (1986)/N = 650 acute adult inpatients	Naturalistic design; variety of batteries employed	Overall abnormality rate 17%; highest rate of abnormals in patients >65, or those with history of alcohol/drug abuse, disorientation, self-neglect, or abnormal physical exam	57% of abnormals not followed up

[a]Abbreviations: H/P, history and physical; ROS, review of systems; MSE, mental status examination; CBC, complete blood count; ECG, electrocardiogram; UA, urinalysis; EEG, electroencephalogram; T_3, triiodothyronine; T_4, thyroxine; VDRL, syphilis serology; CXR, chest roentgenogram; PE, physical examination; ESR, erythrocyte sedimentation rate.

method then employed in the mental health system, which identified 59% of patients at a cost of $230.00 per case (Sox et al., 1989). Laboratory components of the algorithm included a completed blood count; levels of sodium, potassium, calcium, albumin, aspartate aminotransferase (AST), cholesterol, vitamin B_{12}, thyroxine (T_4), and free T_4; and a urine dipstick. More prospective data are needed to evaluate this algorithm in other settings.

A relatively high prevalence of hypothyroidism (up to 3.9%), regardless of symptoms, in women over 50 justifies the use of thyroid screening in this population. Conversely, the low prevalence of thyroid dysfunction in men and in younger women (<1%) supports the recommendation that thyroid screening in these groups should be reserved for those patients with two or more signs of hypothyroidism (Bauer, Halpern, & Schriger, 1991). Clinical use of thyroid function tests is discussed in more detail later in this chapter.

Although frequently performed, chest radiography is generally not a useful screening tool in psychiatric patients because abnormal results are usually predicted by clinical information (Hughes & Barraclough, 1980; Liston et al., 1979). This agrees with data in the general medical population (Tape & Mushlin, 1986).

In summary, widespread use of extensive screening batteries consisting of CBC, complete blood chemistry analysis, erythrocyte sedimentation rate (ESR), urinalysis, levels of vitamin B_{12}, and folate, electroencephalogram (EEG), electrocardiogram (ECG), and chest roentgenogram is not indicated in the majority of psychiatric patients. Such investigations result in many abnormal findings, most of which are clinically insignificant and do not affect patient management and outcome. Most abnormal results can be predicted by information obtained from a careful history, review of systems, and physical examination. *Certain populations appear to benefit from more extensive evaluation, including those older than 65 years of age or of low socioeconomic status, state hospital patients, patients with drug and alcohol histories, and patients with evidence of disorientation, self-neglect, or cognitive impairment disorders. These patients should be evaluated more aggressively. The few tests that have unequivocal merit as broader screening tests in asymptomatic patients include serum glucose, BUN, and creatinine levels and urinalysis. In addition, females older than 50 may benefit from screening for occult thyroid disease.* Finally, patients on psychotropic medications should be monitored for side effects of their particular therapy.

INTEGRATION OF LABORATORY, HISTORY, AND PHYSICAL FINDINGS

Eating Disorders

Patients with anorexia nervosa and bulimia represent a profoundly challenging management problem, in that they present with a host of medical complications in nearly every organ system (Brotman, Rigotti, & Herzog, 1985; Hall et al., 1989; Mitchell et al., 1987). The spectrum of laboratory abnormalities encountered in anorexic and bulimic patients is presented in Table 7-4. The more serious and dramatic metabolic changes usually take place as a result of purging. Up to 40% of bulimics have significant medical complications (Hall et al., 1989). These complications result from behaviors such as surreptitious vomiting, syrup of ipecac abuse, and laxative and diuretic abuse. Given the potentially fatal complications that can arise, the screening laboratory evaluations listed in Table 7-5 are recommended for all patients who meet criteria for anorexia or bulimia.

TABLE 7–4. *Laboratory Abnormalities in Anorexia Nervosa and Bulimia*

Hematologic
Anemia
Leukopenia
Thrombocytopenia
Coagulopathies
Renal/Electrolyte
Hyponatremia
Hypokalemia
Metabolic acidosis
Metabolic alkalosis
Azotemia
Diabetes insipidus
Endocrine
Hypothalamic amenorrhea
Euthyroid sick syndrome
Osteoporosis
Gastrointestinal
Pancreatitis
Steatohepatitis
Nonspecific evaluations of liver function tests
Cardiovascular
Nonspecific ECG changes, including nonspecific ST-T wave changes, low voltage
Bradycardia and other dysrhythmias

Adapted from Brotman, Rigotti, and Herzog (1985, p. 266).

TABLE 7–5. *Recommended Screening Laboratory Evaluation in Anorexia and Bulimia*

CBC with white blood cell differential
Serum glucose
Serum electrolytes
BUN, creatinine
Calcium, magnesium, phosphate
Serum osmolality
Liver function tests
Serum amylase
Urinalysis with urine osmolality and electrolytes
ECG
Bone density determination
T_4, TSH

Specific clinical situations may warrant additional laboratory investigation. Although the anemia in anorexia is of multifactorial etiology, there is usually a prominent nutritional component; serum iron, total iron-binding capacity (TIBC), ferritin, vitamin B_{12}, and folate levels may be required. When a history of ipecac use is given or suspected, a careful cardiovascular assessment is necessary to evaluate for the presence of an ipecac-induced cardiomyopathy. Appropriate additional laboratory studies in that setting would include a serum creatine kinase (CK) level with isoenzymes, chest roentgenogram, ECG, and possible echocardiogram, if clinical symptoms suggested heart failure. The CK MB fraction is elevated in settings of active cardiac muscle injury, whereas an elevated CK MM fraction may reflect an ipecac-induced skeletal myopathy. Electrocardiographic findings associated with ipecac cardiotoxicity include nonspecific ST-T-wave changes and P-R and Q-T interval prolongations. Rhythm disturbances include sinus tachycardia, supraventricular tachycardia, ventricular tachycardia, and ventricular fibrillation (Manno & Manno, 1977).

The clinical laboratory can be particularly helpful in the outpatient management and follow-up of eating disorder patients. These patients are often deceptive and reluctant to give an accurate history when questioned about restrictive or purging behaviors. The laboratory can be used to discover the presence of acting purging behaviors when they are not apparent from the history or physical examination.

The serum amylase level frequently becomes elevated in the context of repeated vomiting; this amylase is of salivary and pancreatic origin (Gwirtsman et al., 1986; Humphries et al., 1987). Elevated serum amylase levels appear to separate bulimic anorexics from restrictor anorexics, with a two- to fourfold increase in serum amylase accompanying binging-purging episodes. Therefore, serial amylase determinations are recommended in monitoring surreptitious purging behavior (Gwirstman et al., 1989). Urinary screens for laxatives (phenolphthalein, bisacodyl) and diuretic ingestion are available and are useful when the history or serum electrolyte levels suggest abuse of these substances. Laxative compounds can be detected 18 to 32 hours after a single oral dose using high-performance thin-layer chromatography techniques (deWolff, deHaas, & Verweij, 1981). Diuretics generally are analyzed utilizing liquid chromatography, which has the sensitivity to detect most compounds 24 hours after a single dose (Fullinfaw, Bury, & Moulds, 1987). Table 7-6 illustrates typical electrolyte patterns seen in various forms of purging, specifically vomiting and diuretic and laxative abuse.

Alcohol-Related Disorders

Evaluation of the Intoxicated Patient

When a patient enters the emergency room in an intoxicated state, the clinician must determine the patient's degree of tolerance to alcohol in order to anticipate withdrawal and the need for prophylaxis. This is best done by obtaining a serum alcohol level and considering the patient's clinical state in the context of a given level. Table 7-7 illustrates the correlation between the blood alcohol level (BAL) and signs and symptoms of ethanol intoxication in *nontolerant* subjects. It is not uncommon to encounter an alcoholic patient with a blood alcohol concentration greater than 0.25% (250 mg/dl) with signs of intoxication minimally present or absent.

However, it is often not possible to obtain the results of an ethanol determination in a timely fashion. An alternative method involves the calculation of the osmolar gap, which can be determined from laboratory data readily obtained within minutes to hours (Glasser et al., 1973). The osmolar gap is determined by subtracting the calculated serum osmolality from the measured osmolality as reported by the laboratory:

$$\text{Osmolar gap} = \text{measured osmolality} - [2\,Na^{+} + (\text{glucose}/20) + (\text{BUN}/3)]$$

The concentration of a given susbstance can be estimated by multiplying the osmolar gap by the molecular weight of the substance divided by 10 (to convert to milligrams/deciliter). The molecular weight of ethanol is 46. The ethanol concentration can then be estimated as follows:

$$\text{Ethanol concentration (mg/dl)} = \text{osmolar gap} \times 4.6$$

It must be noted, however, that substances other than ethanol can contribute to an elevated serum osmolality. These substances include, but are not limited to, methanol, isopropanol, ethylene glycol, acetone, beta-hydroxybutyrate, and acetaldehyde. Commonly prescribed psychotropic agents are generally highly protein

TABLE 7–6. *Use of Serum and Urine Indices in the Differential Diagnosis of Purging Behaviors*

	Vomiting	Laxatives	Diuretics
Serum			
Na^{+}	N,I,D[a]	N,I	N,D
K^{+}	D	D	D
Cl^{-}	D	N,I	D
HCO_3^{-} (CO_2)	I	N,D	I
Urine			
pH	I	D	I
Na^{+}	D	D	I
K^{+}	I	I	I
Cl^{-}	D	N	I

[a]N, normal; I, increased; D, decreased.
Adapted from Brotman, Rigotti, and Herzog (1985, p. 266).

TABLE 7–7. *Blood Alcohol Levels and Associated Neuropsychiatric Signs in Nontolerant Subjects*

Blood Alcohol Level	Neuropsychiatric Signs
0.05% (50 mg/dl)	Impaired attention, truncal swaying while standing
0.10% (100 mg/dl)	Impaired short-term memory and arithmetic calculations; Romberg test reveals staggering with eyes closed
0.15% (150 mg/dl)	Slurred speech, latency of response; Romberg test reveals staggering with eyes open
0.20% (200 mg/dl)	Stupor
0.25% (250 mg/dl)	Anesthesia
0.35–0.50% (350–500 mg/dl)	Abolishment of respiration

Adapted from Barry (1979, p. 515) and Glasser et al. (1973).

bound or are present in small concentrations and so contribute little to the serum osmolality. Table 7-8 illustrates the utility of the osmolar gap in a variety of clinical situations (Glasser et al., 1973).

Other priorities include determining the presence of gastrointestinal (GI) bleeding, pancreatitis, alcoholic hepatitis, cirrhosis, and Wernicke's encephalopathy. Determination of orthostatic vital signs, the hematocrit, and the presence of occult blood in the stool are essential in evaluating GI bleeding in the alcoholic. The presence of significant orthostatic hypotension in the face of an intoxicated patient with guaiac-positive stool certainly suggests that a significant GI bleed may be present. A normal hematocrit is reassuring but is no guarantee that significant blood loss has not occurred, because the hematocrit nadir following acute blood loss manifests only after rehydration (Adamson & Hillman, 1968).

Pancreatitis generally presents with severe abdominal pain that may mimic a surgical abdomen. However, the characteristic boring pain of pancreatitis can be absent in the intoxicated patient. Pancreatitis can be identified by the presence of elevated serum amylase and lipase levels, with the latter being more specific for the disorder. In general, alcoholic pancreatitis carries a mortality rate of less than 5%, although patients are likely to have recurrent episodes. However, once pancreatitis is identified, a number of prognostic variables must be assessed (Table 7-9). The presence of zero to two of these factors is indicative of mild pancreatitis, with a mortality of 3%. The presence of three to five factors suggests moderate pancreatitis, and six or more factors suggest severe pancreatitis, with a mortality of 30% to 60% (Barkin & Garrido, 1986, p. 255; Ranson et al., 1974). Imaging with either ultrasound or computerized tomography (CT) can help rule out gallstone pancreatitis and identify the presence of a phlegmon or pseudocyst.

Another urgent consideration in the evaluation of the intoxicated patient is the potential for thiamine deficiency and its sequela, the Wernicke-Korsakoff syndrome. Clinical signs of Wernicke's encephalopathy include confusion, ataxia, and disorders of extraocular movement. Frequently, the clinical approach is to administer at the time of presentation 50 to 100 mg of thiamine intramuscularly (IM) to all patients suspected of having alcoholism, those with altered consciousness, or those considered to be at risk for nutritional deficiency. If documentation of the thiamine-deficient state is desired, this can be performed by measurement of erythrocyte transketolase activity before and after the addition of thiamine pyrophosphate (TPP). The addition of TPP results in optimal enzyme activity, and this increase in activity is known as the TPP effect (McCormick, 1987). The TPP effect, however, is eliminated within 2 to 4 hours of the administration of 50 mg thiamine IM (McCormick, 1987), so samples should be drawn before or concurrent with thiamine administration.

TABLE 7–8. *Commonly Encountered Substances with Potential To Create an Elevated Osmolar Gap in Settings of Intoxication or Overdose*

Substance	MW	LD_{50} (mg/dl)	Change in osmolarity at LD_{50}
Ethanol	46	350	80
Isopropanol	60	340 (toxic)	60
Methanol	32	80	27
Acetone	58	55	10
Ethyl ether	26	180	70

Adapted from Glasser et al. (1973, p. 698).

TABLE 7–9. *Laboratory Variables Used for Assessing Prognosis in Pancreatitis*[a]

Admission/Diagnosis	Within Initial 48 Hours
Age >55 years	Hematocrit decrease >10%
White blood cell count >16,000 cells/mm³	BUN increase >5 mg/dl
Glucose >200 mg/dl	Serum calcium <8 mg/dl
Lactate dehydrogenase >700 U	PaO_2 <65 mm Hg
AST >250	Base deficit >4 mEq/liter
	Estimated fluid sequestration >6 liters

[a]0–2 factors, mild pancreatitis (3% mortality); 3–5 factors, moderate pancreatitis; ≥6 factors, severe pancreatitis (mortality 30–60%).
Adapted from Barkin and Garrido (1986, p. 255).

Evaluation of the Chronic Alcoholic

When approaching the patient with a known or suspected chronic alcohol problem, anticipation of the complications of chronic liver disease—including cirrhosis, ascites, variceal bleeding, hepatic encephalopathy, and coagulopathy—is necessary, in addition to the employment of the acute interventions mentioned above. Table 7-10 illustrates the recommended laboratory profile for evaluationg the patient with chronic alcohol problems.

Hepatic Encephalopathy

Hepatic encephalopathy is a complex disorder generally characterized by a rapid onset and prominent disturbances in consciousness ranging from mild delirium with disorientation, inattention, indifference, and agitation to stupor and coma. The psychiatric symptoms are ac-

TABLE 7–10. *Laboratory Diagnosis of the Chronic Alcoholic*

Test	Diagnosis/Problem
CBC with white blood cell differential	Anemia, macrocytosis, thrombocytopenia
Sodium, potassium	Common electrolyte disturbances
Albumin, prothrombin time	Hepatic synthetic function
Transaminases, bilirubin	Degree of hepatocellular damage
Stool guaiac	Presence of occult GI blood loss
Amylase, lipase	Evaluation of abdominal pain, pancreatitis
Ammonia level	Hepatic encephalopathy
Vitamin B_{12}, folate, magnesium, phosphate levels	Nutritional deficiencies

companied by an irregular tremor and asterixis; other neurologic findings are uncommon, except in stuporous or comatose patients (Adams & Salam-Adams, 1986). Repeated episodes can lead to chronic hepatocerebral degeneration (Gilbertstad et al., 1980).

The etiology of the cerebral disturbance is not fully understood; however, the presence of some degree of portal-systemic shunting appears to be a necessary component of the disorder. The action of intestinal flora on nitrogenous compounds appears to play an integral role as well (Bircher et al., 1966; James et al., 1979; Weber, Fresard, & Lally, 1982). The most prevalent neurochemical theory concerning hepatic encephalopathy is the ammonia hypothesis. Ammonia, derived from enteric bacterial degradation of proteins and normally metabolized to urea and renally excreted, accumulates because of portal-systemic shunting and impaired hepatic function. The serum level of ammonia correlates roughly with the degree of encephalopathy, and methods to enhance its elimination result in improvement (Bircher et al., 1966; Weber, Fresard, & Lally, 1982; Weber et al., 1985). Studies investigating the correlation between encephalopathy and ammonia, however, have yielded conflicting results. Furthermore, substantial differences exist between arterial and venous samples, with ammonia from muscle metabolism contributing to the latter. A fasting arterial specimen is preferred (Pappas & Jones, 1983) and should be obtained whenever possible; however, in one study elevated venous ammonia levels correctly identified 15 of 17 cirrhotics with a history of encephalopathy (Ansley et al., 1978). The fasting ammonia level should not be viewed in isolation; it is best interpreted in light of other clinical information (e.g., the severity of the disturbance in consciousness and the presence and severity of asterixis) (Conn & Lieberthal, 1978).

Other biochemical factors have been implicated, including mercaptans, a reduced ratio of branched chain to aromatic amino acids, "false transmitters," and gamma-aminobutyric acid (GABA) (Fischer & Baldessarini, 1975; Jones et al., 1989; McClain, Zieve, & Diozake, 1980; Morgan, Milsom, & Sherlock, 1978). The finding of triphasic waves on the EEG was at one time believed to be specific for hepatic encephalopathy. However, recent data have revealed this finding to be relatively nonspecific (Weissenborn et al., 1990).

Diagnosis of Early or Occult Alcoholism

The diagnosis of early or occult alcoholism has received much attention in the literature. A number of investigators have attempted to identify laboratory markers for alcoholism. Many studies have been disappointing because laboratory measures proved to be no more sensitive or specific than screening instruments such as the Michigan Alcoholism Screening Test (MAST) or CAGE questionnaires (Bernadt et al., 1982; Kristensen & Trell, 1982).

Most studies have focused on the evaluation of mean corpuscular volume (MCV) or liver function tests and revealed a great heterogeneity of end organ damage among alcoholics, thus limiting the utility of liver enzymes in the diagnosis of early or occult alcoholism. Markers such as MCV or gamma-glutamyl transferase (GGT) have low predictive power when used as screening tests (Chick, Kreitman, & Plant, 1981). Measurements of AST (or serum glutamic oxaloacetic transaminase [SGOT]) and alanine aminotransferase (ALT, or serum glutamic pyruvic transaminase [SGPT]) have also been utilized in the laboratory detection and follow-up of alcoholism. Elevations in these enzymes reflect subacute or chronic exposure and are rare in the acute setting (Devgun et al., 1985). Depending on the population studied, a high percentage of alcoholic patients have normal transaminase levels, thereby limiting the use of this enzyme in the detection of occult alcholism.

However, the pattern of transaminase elevation may be helpful in that the AST/ALT (SGOT/SGPT) ratio is an important discriminator in differentiating between various types of liver disease (DeRitis, Coltorti, & Giusti, 1972). An AST/ALT ratio of greater than 2:1 is highly suggestive of alcoholic hepatitis or cirrhosis (Cohen & Kaplan, 1979). The elevated AST/ALT ratio reflects diminished hepatic ALT activity and is in part related to depletion of pyridoxal 5-phosphate (Matloff, Selinger, & Kaplan, 1980; Diehl et al., 1984).

The determination of AST isoenzymes can also be used to differentiate alcoholic liver damage from other acute hepatic processes. Total AST activity consists of two contributing isoenzymes, the cytoplasmic and mitochondrial fractions. In normal subjects, mitochondrial AST contributes little to serum activity (Rej, 1978, 1980). In liver disease, regardless of etiology, both isoenzymes are elevated, whereas in alcoholic liver disease the mitochondrial fraction is elevated disproportionately. Elevated mitochondrial AST levels are noted in alcoholics even in the presence of normal liver function as measured routinely by the laboratory. The ratio of mitochondrial AST to total AST activity is much more sensitive and specific than previously discussed markers for alcohol use, but has limited availability (Ink et al., 1989; Nalpas et al., 1984, 1986).

While examining serum and cerebrospinal fluid (CSF) proteins in patients with a variety of neurologic diseases, it was noted that patients with alcoholic cerebellar ataxia often had an abnormal transferrin variant present (Stibler, Borg, & Allgulander, 1979; Stibler & Kjellin, 1976). Later investigations revealed this abnormal transferrin to be present in the serum of a large percentage of alcoholics without ataxia, and that this abnormality appeared to result from the displacement of sialic acid from the native transferrin (Stibler & Borg, 1981). Methods of quantifying this desialotransferrin were soon developed and refined (Stibler, Sydow, & Borg, 1980; Storey et al., 1985, 1987). It has remarkably high sensitivity and specificity for identifying excessive drinking and monitoring abstinence (Gjerde et al., 1988); however, widespread clinical use of this test has not yet occurred.

Other putative tests for occult alcoholism include serum methanol, glutamate dehydrogenase, acetaldehyde protein adducts, dolichols, and the alpha-amino-*n*-butyric acid/leucine ratio (Leggett, Powell, & Halliday, 1989; Pullarkat & Raguthu, 1985; Roine et al., 1989).

It is clear that most of the widely available laboratory tests for the detection of excessive drinking lack sufficient sensitivity, specificity, or positive predictive value to be considered superior to a careful history. The historic information may be enhanced by laboratory data in some situations, and these laboratory data can be followed over time to monitor success of treatment. Alternative laboratory measurements have been developed that, although more sensitive and specific for identifying heavy alcohol consumption, are limited by either procedural complexity or relative unavailability (Table 7-11). However, as these techniques are simplified or become more available, they may prove extremely valuable in the early identification of the alcoholic patient.

TABLE 7–11. *Laboratory Markers for Occult Alcohol Use*

Test	Availability	Sensitivity	Specificity
MCV	Wide	40–54%	Low
GGT	Wide	34–85%	50%
AST	Wide	24–70%	Low
AST/ALT	Wide	Moderate	Moderate
Mitochondrial AST/AST	Limited	85–100%	82%
Desialo-transferrin	Very limited	81–100%	91–100%

Reprinted by permission from Leggett BA, Powell LW, & Halliday JW (1989). Laboratory markers of alcoholism. Dig Dis 7:127.

Substance Abuse Disorders

When a substance abuse disorder is suspected, a clinical laboratory evaluation is essential for definitive diagnosis, because the presenting history, signs, and symptoms are often nonspecific or misleading. The clinical laboratory results can be equally misleading, however, without adequate knowledge of the types of procedures available and their relative strengths and weaknesses. Furthermore, considerable variability exists between individual laboratories, thus mandating an awareness on the part of the clinician concerning the specific procedures performed at a given institution.

Analytic techniques include thin-layer chromatography (TLC), high-performance liquid chromatography (HPLC), immunoassays (including radioimmunoassay [RIA], gas chromatography (GC), and GC linked to mass spectrometry (GC-MS). Samples that are presumptively positive by one technique must be confirmed by a different analytic technique. Urine usually is used as the biologic sample, because it is easily obtained and often contains drug concentrations several times that of serum. Several commercial assays are commonly employed, including Abuscreen (Roche), EMIT (Syva), TDx (Abbott), and Toxilab (Gold & Dackis, 1986; Saxon et al., 1988). Table 7-12 illustrates the relevant data concerning these laboratory determinations. Chapter 8 discusses drug testing in more detail.

Delirium

Delirium is associated with considerable morbidity and mortality; therefore, accurate and rapid diagnosis of its etiology is extremely important. The most common causes of delirium include drug or alcohol intoxication or withdrawal, toxic encephalopathy caused by endogenous substances, electrolyte disturbances, hypoxia, systemic infection, head trauma, and seizure disorders. When a diagnosis of delirium is made, the presenting history may reveal its etiology. However, many cases of delirium are multifactorial, and the morbidity increases with the number of contributing factors (Francis, Martin, & Kapoor, 1990). Therefore, careful, complete metabolic assessment is warranted. Table 7-13 illustrates the typical screening laboratories for the evaluation of delir-

TABLE 7–12. *Detection Limits For Urine Drug Testing*

Drug	Dose (mg)	Detection Time	Screening Frequency (times/week)[a]
Amphetamines	30	1–120 hr	1–2
Barbiturates			
Short acting	100	4.5 days	1–2
Long acting	30	6–24 days	1–2
Benzodiazepines			
Diazepam	10	7 days	1
Triazolam	0.5	24 hr	2–3
Cocaine (benzoylecgonine)	250	8–48 hr	2–3
Methadone	40	7.5–56 hr	2–3
Methaqualone	150	60 hr	1–2
Morphine-opiates (intravenous)	10	84 hr	1
Tetrahydrocannabinol metabolites	Daily use	6–81 days	Monthly
	Weekly use	7–34 days	

[a]Refers to optimal frequency of screening to detect surreptitious drug use while participating in treatment program.
Reprinted by permission from Saxon AJ, Calsyn DA, Haver VA, et al. (1988). Clinical evaluation and use of urine screening for drug abuse. West J Med 149:297.

TABLE 7–13. *Laboratory Evaluation in Patients with Delirium of Undetermined Etiology*

General Tests	Optional Tests
CBC with white blood cell differential	EEG (serial EEGs)
Urinalysis	Syphilis serology
Serum electrolytes	Cobalamin (B_{12})
Calcium, magnesium	Folate
BUN/creatinine	Head CT/magnetic resonance imaging
Serum glucose	Lumbar puncture
Liver function tests	Urine drug screen
Thyroid-stimulating hormone, free T_4	Arterial ammonia level
Arterial blood gases	CK
Blood cultures	
Chest roentgenogram	
ECG	

ium. When the etiology is evident from clinical assessment, laboratory investigations can be limited, based on clinical judgment. (See also Chapter 19.)

COMMONLY ENCOUNTERED MEDICAL CONDITIONS AND RELATED TESTS

Neuroleptic Malignant Syndrome

Laboratory determinations in neuroleptic malignant syndrome are primarily directed at ruling out other disorders and identification of complications. The cardinal features include hyperthermia, rigidity, autonomic instability, and delirium (Caroff, 1980). Complications of the disorder include rhabdomyolysis with or without myoglobinuric renal failure, deep venous thrombosis, pulmonary emboli, myocardial infarction, shock, sepsis, and disseminated intravascular coagulation (DIC) (Caroff, 1980; Eiser, Neff, & Slifkin, 1982; Pearlman, 1986). Several laboratory abnormalities have been associated with the disorder as well, including leukocytosis; elevations in CK, hepatic transaminases, and BUN; hypercalcemia, hypocalcemia, hypophosphatemia, hyponatremia, hypernatremia, and hypoferrinemia; proteinuria; and myoglobinuria (Caroff, 1980; Eiser, Neff, & Slifkin, 1982; Pearlman, 1986; Rosebush & Stewart, 1989). Creatine kinase isoenzymes may be necessary to rule out myocardial damage, if suspected. Interpretation of CK values is discussed later in the chapter (see section "Miscellaneous Tests") and in Table 7-22, below.

Disorders of Sodium and Water Balance

Hyponatremia

Hyponatremia is one of the most common electrolyte disturbances in hospitalized patients. Neuropsychiatric manifestations of hyponatremia can include agitation, anxiety, somnolence, apathy, delirium, tonic-clonic seizures, and coma. The development of hyponatremia has particular importance in psychiatric patients because of the frequent occurrence of primary polydipsia in this population.

Hyponatremia is a derangement in sodium and water balance manifested by an absolute or relative free water excess; a true sodium-deficient state may or may not be present. Treatment methods are directed at limiting the ingestion of free water and/or facilitating its excretion. The clinical history, supplemented by a systematic, algorithmic approach to hyponatremia, prevents misdiagnosis in most cases (Figure 7-1) (Narins et al., 1982).

The first step in the evaluation of hyponatremia utilizes the serum osmolality to help determine if the low

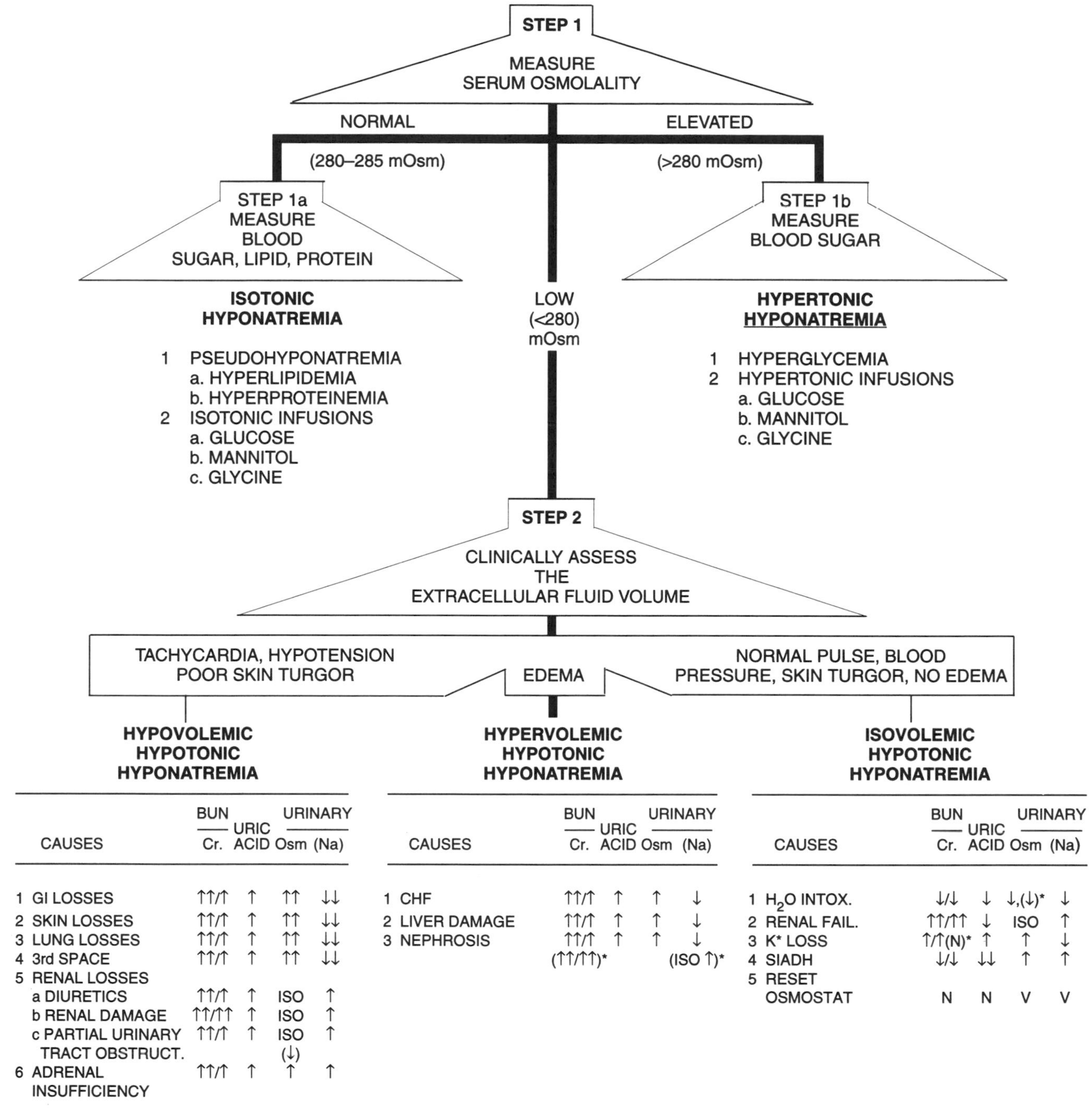

CAUSES	BUN/Cr.	URIC ACID	URINARY Osm	URINARY (Na)
1 GI LOSSES	↑↑/↑	↑	↑↑	↓↓
2 SKIN LOSSES	↑↑/↑	↑	↑↑	↓↓
3 LUNG LOSSES	↑↑/↑	↑	↑↑	↓↓
4 3rd SPACE	↑↑/↑	↑	↑↑	↓↓
5 RENAL LOSSES				
a DIURETICS	↑↑/↑	↑	ISO	↑
b RENAL DAMAGE	↑↑/↑↑	↑	ISO	↑
c PARTIAL URINARY TRACT OBSTRUCT.	↑↑/↑	↑	ISO (↓)	↑
6 ADRENAL INSUFFICIENCY	↑↑/↑	↑	↑	↑

CAUSES	BUN/Cr.	URIC ACID	URINARY Osm	URINARY (Na)
1 CHF	↑↑/↑	↑	↑	↓
2 LIVER DAMAGE	↑↑/↑	↑	↑	↓
3 NEPHROSIS	↑↑/↑ (↑↑/↑↑)*	↑	↑ (ISO ↑)*	↓

CAUSES	BUN/Cr.	URIC ACID	URINARY Osm	URINARY (Na)
1 H_2O INTOX.	↓/↓	↓	↓,(↓)*	↓
2 RENAL FAIL.	↑↑/↑↑	↓	ISO	↑
3 K* LOSS	↑/↑(N)*	↑	↑	↓
4 SIADH	↓/↓	↓↓	↑	↑
5 RESET OSMOSTAT	N	N	V	V

FIG. 7–1. Diagnostic approach to hyponatremia. Iso = isotonic; N = normal; V = variable. (Reprinted by permission from Narins RG, Jones ER, Stom MC, et al. [1982]. Diagnostic strategies in disorders of fluid, electrolyte and acid-base homeostasis. Am J Med 72:498.)

sodium is a real or artifactual finding. Hyponatremia may thus be divided into hyperosmolar, isotonic, and hypoosmolar types.

Hyperosmolar Hyponatremia. Hyperosmolar hyponatremia is often caused by hyperglycemia or the infusion of hypertonic solutions, such as glucose, mannitol, or glycine. This can be thought of as a dilutional phenomenon, wherein the serum sodium concentration falls following the influx of free water into the extracellular fluid space as a result of the osmotic pull of a particular solute. A reasonable estimate of the "true" sodium concentration in dilutional hyponatremia resulting from hyperglycemia is obtained by the following calculation:

$$\text{True Na}^+ = \left[\left(\frac{\text{Measured glucose} - 100^*}{100}\right) \times 1.6\right] + \text{measured Na}^+$$

(*Represents normal glucose in mg/dl.)

A typical scenario is that of diabetic ketoacidosis. The patient may present with a blood glucose level of 700 mg/dl and a serum sodium concentration of 125 mEq/liter. The sodium concentration to be expected, were the hyperglycemia not present, would be approximately 135 mEq/liter. It must be understood that this manipulation renders only an approximation of the sodium-water balance and should not be relied on as absolute. Hyperosmolar hyponatremia must be distinguished from hypoosmolar hyponatremia because they require different treatment. Treatment of hyperosmolar hyponatremia involves treatment of the underlying condition (hyperglycemia), or simply noting that the laboratory finding is a result of treatment (mannitol infusion). Treating the low sodium value by fluid restriction could potentially be disastrous in cases of hyperosmolar hyponatremia.

Isotonic Hyponatremia. Isotonic hyponatremia may be, in fact, a pseudohyponatremia secondary to hyperlipidemia or hyperproteinemia or the result of infusions of isotonic solutions containing glucose, mannitol, or glycine.

Hypoosmolar Hyponatremia. Hypoosmolar hyponatremia is the most common type encountered and is associated with the greatest morbidity because of the fluid shifts that occur, resulting in cerebral edema. The assessment of hypoosmolar hyponatremia is begun by determining the volume status of the patient. The patients may be hypovolemic, as manifested by tachycardia, orthostatic hypotension, dry mucous membranes, or poor skin turgor. Alternatively, they may be hypervolemia as manifested by peripheral, presacral, or pulmonary edema. Other patients present with no apparent derangement in their volume status.

Medical conditions resulting in hypovolemic hypotonic hyponatremia reflect a loss of sodium and free water. These losses can be via the GI tract, lung, skin, or renal routes. Adrenal insufficiency can also lead to hyponatremia in the setting of volume depletion and is accompanied by hyperkalemia and metabolic acidosis.

Hypervolemic hypotonic hyponatremia becomes manifest in the settings of congestive heart failure, cirrhosis, and nephrosis.

Isovolemic hypotonic hyponatremia is associated with water intoxication (as in the setting of primary polydipsia), renal insufficiency with iatrogenic administration of hypotonic fluids, potassium depletion, and syndrome of inappropriate secretion of antidiuretic hormone (SIADH), and is found in instances in which patients appear to have a reset osmostat (lowered serum osmolality while preserving the ability to suppress antidiuretic hormone [ADH] release with water loading). Tuberculosis and cirrhosis appear to be associated with the latter circumstance (deFronzo, Goldberg, & Agus, 1976).

Syndrome of Inappropriate Secretion of Antidiuretic Hormone. Hyponatremia in the setting of normal clinical volume status should alert the clinician to the possibility of the presence of SIADH. The classic features of the disorder include hypoosmolar, hyponatremia, continued renal sodium excretion absence of hypovolemia as assessed clinically, excretion of urine that is less than maximally dilute, and normal renal and adrenal function (Bartter & Schwartz, 1967).

In the development of the disorder, patients initially retain water, which is manifested by a weight gain of 3 to 4 kg. This is followed by an increase in glomerular filtration rate and subsequent enhanced sodium excretion. Edema is rarely manifested in SIADH. Urinary sodium values are usually greater than 20 mEq/liter and often are greater than 40 mEq/liter. Urine osmolality is higher than that which would be expected given the degree of serum hypoosmolality. Often the urine osmolality is greater than that of serum; it is in this setting that the diagnosis is most clear. The serum uric acid level is often low in SIADH secondary to decreased reabsorption (Beck, 1979).

Several pathophysiologic states have been associated with SIADH (Table 7-14). Medications can also cause hyponatremia by a mechanism that appears to be SIADH. Table 7-15 lists the medications that have been associated with hyponatremia. Antipsychotic agents and tricyclic antidepressants have been reported to cause SIADH. There are, however, no prospective studies that support this hypothesis. Since a significant number of patients with psychiatric disease have hyponatremia secondary to primary polydipsia, the role of SIADH remains to be clarified. This issue is particularly compelling since the hyponatremia in patients with primary polydipsia does not worsen when treated with antipsychotic or antidepressant medications.

TABLE 7–14. *Causes of SIADH*

Neoplasms:	Lung (80% small cell), pancreas, duodenum, lymphoma, ureter, Ewing's sarcoma, prostate
Pulmonary:	Pneumonia, active tuberculosis, abscess, asthma, intermittent positive pressure breathing
CNS:	Infection, trauma, neoplasm, vascular, degenerative diseases, psychosis
Cardiac:	Post mitral commissurotomy, atrial tachycardia
Metabolic:	Myxedema, hypoadrenalism (primary and secondary), acute porphyria
Medications:	See Table 7–15

Reprinted by permission from Narins RG, Jones ER, Stom MC, et al. (1982). Diagnostic strategies in disorders of fluid, electrolyte and acid-base homeostasis. Am J Med 72:498.

TABLE 7–15. *Drugs Associated with Hyponatremia*

Psychotropic agents[a]	Hypoglycemic agents
Thioridazine	Chlorpropamide
Thiothixine	Tolbutamide
Fluphenazine	Sedatives/analgesics
Amitriptyline	Barbiturates
Desipramine	Morphine
Fluoxetine	Miscellaneous
Carbamazepine	Indomethacin
Diuretics	Nicotine
Antineoplastic agents	Clofibrate
Cyclophosphamide	Acetaminophen
Vincristine	

[a]Causal link debatable (see text).
Adapted from Narins et al. (1982), with updates.

Primary Polydipsia

Primary polydipsia is a relatively commonly encountered clinical entity in psychiatric practice. The term "psychogenic" polydipsia is misleading and is not used in this discussion since polydipsia related to purely psychogenic factors (e.g., response to delusions or hallucinations) is relatively uncommon. Furthermore, there is a substantial body of data that suggests that patients with polydipsic behavior have dysregulation of ADH and thirst mechanisms (Goldman, Luchins, & Robertson, 1988). This is referred to as primary polydipsia to distinguish it from secondary polydipsic states wherein increased water intake results from a physiologic need, such as volume depletion or hyperosmolarity. Diabetes mellitus and diabetes insipidus are clinical states associated with secondary polydipsia.

Diagnoses that have been associated with primary polydipsia include schizophrenia, psychotic depression, dementia, organic delusional disorder, alcoholism, neuroses, and hysteria (Chinn, 1974). Other associations include accidental water intoxication (Anastassiades et al., 1983) and intractable hiccups (Cronin, 1987). Primary polydipsia also occurs in some individuals without psychiatric illness (Zerbe & Robertson, 1981).

Clinical manifestations resulting from hyponatremia secondary to water intoxication range from syndromes involving agitation, psychotic exacerbation, delirium, stupor, seizures, coma, and, ultimately, to death. Hypocalcemia, rhabdomyolysis, congestive heart failure, atonic bowel, megalocystis, hydronephrosis, and renal failure may also occur in the setting of primary polydipsia (Blum & Friedland, 1983; Cronin, 1987; Vieweg et al., 1987). It also can be asymptomatic.

The incidence and etiology of primary polydipsia are unknown. Incidence estimates in hospitalized patients range from 6.6% utilizing a behavioral history to 17% utilizing hyposthenuria as a marker for polydipsia (Blum, Tempey, & Lynch, 1983; Jose & Perez-Cruet, 1979). Etiologic theories include the idiopathic and intermittent development of SIADH (Brown, Kocsis, & Cohen, 1983; Dubovsky et al., 1973; Hariprasad, Eisinger, & Nadler, 1980; Raskind, Orenstein, & Christopher, 1975) and the use of psychotropic medications (Ajlouni, Kern, & Tures, 1974; Luzecky, 1974; Matuk & Kalyanaraman, 1977). Rarely, polydipsic behavior occurs in response to delusions or as a manifestation of a thought disorder (Alexander, Crow, & Hamilton, 1973; Rendell, McGrane, & Cuesta, 1978; Rosenbaum, Rothman, & Murray, 1979). Chronic hyponatremia may result in a loss of the medullary concentration gradient, limiting the kidney's ability to excrete a dilute urine (Blum & Friedland, 1983). Hyperdopaminergic central nervous system (CNS) activity has also been implicated, with denervation supersensitivity occurring as a result of neuroleptic agents (Smith & Clark, 1980). Nicotine is known to stimulate an antidiuresis; thus polydipsic patients who smoke are often more hyponatremic than patients who do not (Allon et al., 1990; Blum, 1984; Chin et al., 1976). Recent data suggest that patients with primary polydipsia have impaired excretion of a water load in spite of adequate suppression of vasopressin, a reduced osmotic threshold for the release of ADH, and altered osmoregulation of water intake (Goldman, Luchins, & Robertson, 1988). The defect in urinary dilution may result from an enhanced sensitivity of the distal tubule to ADH, with increased ADH receptor number and/or affinity postulated as mechanisms (Goldman, Luchins, & Robertson, 1988; Zerbe & Robertson, 1987).

Table 7-16 illustrates how the laboratory can be used to differentiate between pure water intoxication, SIADH, and diabetes insipidus when confronted with the polydipsic patient. Primary polydipsia may present with pure water intoxication and/or SIADH.

TABLE 7–16. *Use of Serum and Urine Indices in the Differential Diagnosis of Polydispic States*

Clinical Situation	Serum Na+[a]	Serum Osmolarity	Urine Na+	Urine Osmolarity
Water intoxication	N, D	N, D	<20[b]	D[b]
SIADH	N, D	N, D	>20[c]	I[d]
Diabetes insipidus	N, I	N, I	<20	D

[a]N, normal; D, decreased; I, increased.
[b]In chronic polydipsic states; in the acute psychiatric presentation of hyponatremia, urine sodium and osmolality are consistent with SIADH (Hariprasad, Eisinger, & Nadler, 1980).
[c]Often >40 mEq/liter.
[d]Less than maximally dilute.

Management of primary polydipsia involves fluid restriction. This can be very difficult even in an inpatient setting, necessitating behaviorial measures such as shutting off water on the ward or in other patients' rooms, and locking bathrooms to prevent surreptitious drinking from the shower or toilet. Pharmacologic treatment of primary polydipsia has been attempted with demeclocycline and propranolol (Nixon, Rothman, & Chin, 1982; Shevitz et al., 1980). Angiotensin-converting enzyme (ACE) inhibitors are of theoretical interest but result in increased polydipsia, if anything (Kathol et al., 1986).

Hypernatremia

Hypernatremic patients may also present with an altered mental status. The approach to their assessment is illustrated in Figure 7-2. The most common clinical scenario encountered by the medical psychiatrist involves hypernatremia secondary to lithium-induced nephrogenic diabetes insipidus.

Lithium-Induced Nephrogenic Diabetes Insipidus

Polyuria and polydipsia occur frequently among patients treated with lithium, with up to 70% of patients excreting more than 2 liters/day; as many as 37% of patients have daily urinary volumes of greater than 3 liters (Borton, Gaviria, & Batlle, 1987; Vestergaard, Amdisen, & Schou, 1980). The mechanism of this polyuria is related to impaired concentrating ability secondary to impairment of ADH-sensitive adenylate cyclase systems in the collecting tubule (Christiansen et al., 1985; Dousa, 1974).

Definitive diagnosis of nephrogenic diabetes insipidus depends on the results of a water deprivation test. Patients are deprived of water for approximately 2 to 24 hours, with a serum osmolality of greater than 285 and/or clinical symptoms serving as end points of testing. Basal and stimulated urine osmolality determinations are obtained. The patient is then administered 5 units of aqueous vasopressin subcutaneously (SQ) and another urine osmolality is obtained after 1 hour (Wass & Besser, 1989, pp. 499–500). Figure 7-3 illustrates the pattern of response seen in normal patients and in those with either central or nephrogenic diabetes insipidus (Narins et al., 1982). Plasma ADH levels can be obtained as well to interpret equivocal responses, but results are not usually available for weeks. Potential complications of the procedure include severe volume depletion, hypotension, and hypernatremia. The test is often difficult to administer to severely psychotic patients.

An alternative method involves measuring urine osmolality 3 hours following the administration of intranasal 1-desamino-8-D-arginine vasopressin (DDAVP). The DDAVP test has the advantages of being easier to administer to psychotic patients and potentially less hazardous. Results of this test correlate well with those of the water deprivation test (Asplund, Wahlin, & Rapp, 1979). Patients with severe lithium-induced nephrogenic diabetes insipidus will have a negligible rise in urine osmolality, whereas those with milder degrees of tubular concentrating impairment, or those with a component of central diabetes insipidus, will have intermediate increases in urine osmolality. Normal subjects will exhibit a highly concentrated urine (>800 mOsm/liter) following the administration of DDAVP.

Nutritional and Metabolic Disorders

Vitamin B_{12} (Cobalamin) Deficiency

An association between vitamin B_{12} deficiency and psychiatric disturbance has long been recognized, with affective, anxiety, psychotic, delirious, and dementing syndromes having been described (Edwin et al., 1965; Holmes, 1956; Schulman, 1967b, 1967c; Woltmann, 1919); however, studies using current criteria have not been done. Rates of psychiatric disturbance reported are from 35% to 82%. Early prospective data, however, suggest that the frequency of B_{12} deficiency is no higher in psychiatric patients than in the general population (Shulman, 1967a). Treatment of the B_{12}-deficient state

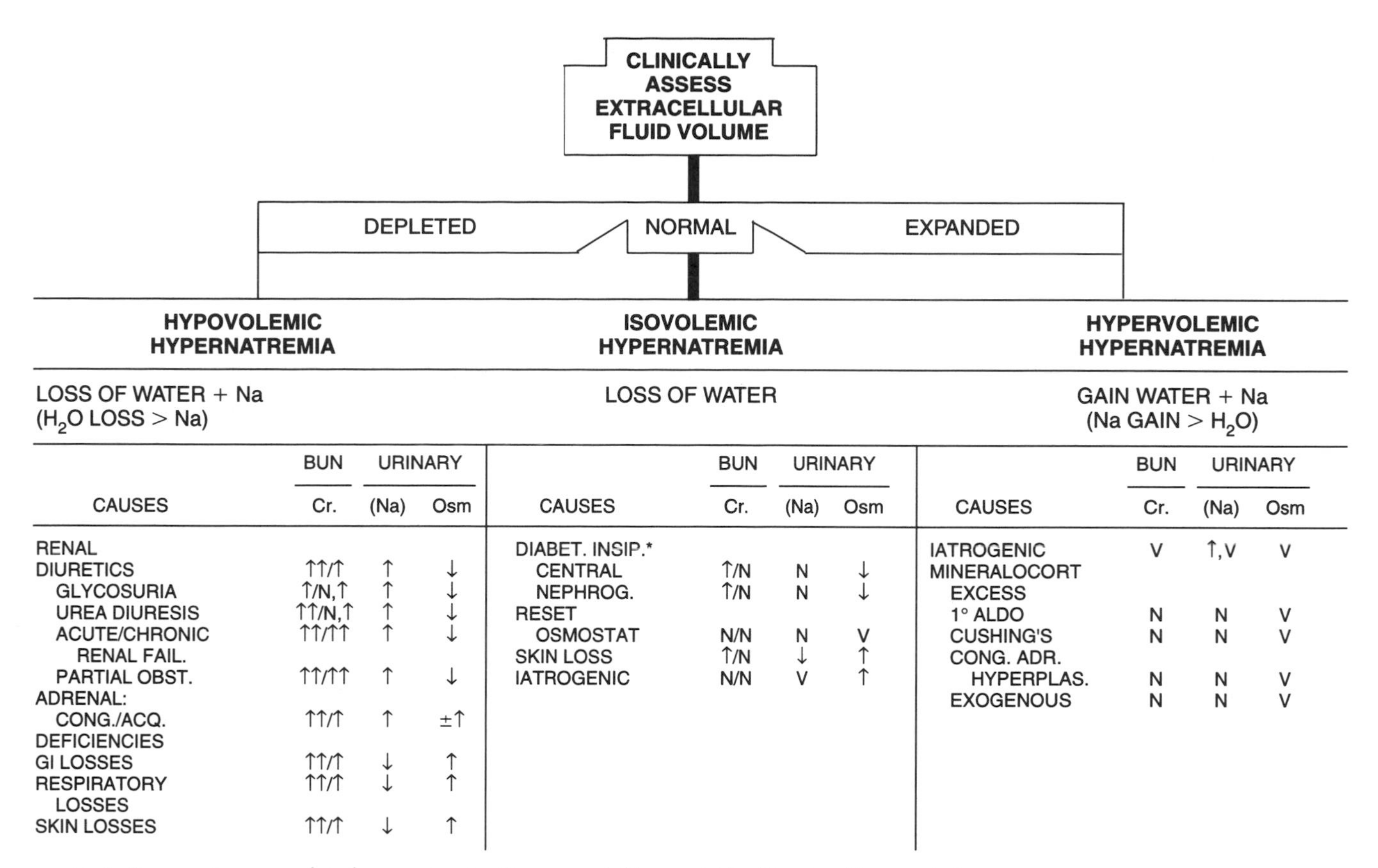

FIG. 7–2. Diagnostic approach to hypernatremia. N = normal; V = variable. (Reprinted by permission from Narins RG, Jones ER, Stom MC, et al. [1982]. Diagnostic strategies in disorders of fluid, electrolyte and acid-based homeostasis. Am J Med 72:501.)

results in clinical improvement of the psychiatric symptoms in many patients (Edwin et al., 1965; Hunter et al., 1967; Smith, 1960). Both B_{12} and folate deficiencies are associated with the hematologic findings of anemia, with macroovalocytes and hypersegmented neutrophils on the peripheral blood smear. Neurologic manifestations include paresthesias, sensory loss, particularly in the vibratory and proprioceptive modalities, and ataxia.

It has long been noted that psychiatric and neurologic manifestations may occur in the absence of or preceding the hematologic findings (Hunter & Matthews, 1965; Langdon, 1905; Lindenbaum et al., 1988; Strachan & Henderson, 1965). Reliance on the routine peripheral blood smear as a screening tool for B_{12}/folate deficiencies is relatively insensitive (Carmel, 1988), and the megaloblastosis can be masked by other hematologic diseases (Carmel, Weiner, & Johnson, 1987; Green et al., 1982; Spivak, 1982). In suspected cases, the blood smear should be carefully reviewed and a serum B_{12} level obtained.

Proper interpretation of the laboratory data necessitates a brief review of the physiology of B_{12} absorption and transport (Herbert, 1987). Ingested B_{12} is bound to intrinsic factor (IF), which is produced by the parietal cell in the gastric antum. This B_{12}-IF complex is absorbed in the distal ileum. Impaired absorption results from an absence or deficiency of IF, ileal disease, or competition for utilization by intestinal bacterial overgrowth. Once absorbed, the B_{12} is bound to specific serum proteins, transcobalamin II (TC II) and the haptocorrins (TC I and TC III) (Jacob, Baker, & Herbert, 1980). The B_{12}–serum protein complex is referred to as either holotranscobalamin II or holohaptocorrin depending on the binding protein involved. Only holotranscobalamin II is biologically active, but both complexes are measured in immunoassays of B_{12} levels. This discrepancy is one source of false-negatives in blood tests for B_{12} deficiency. There is also a second, less efficient mechanism of absorption that is independent of IF (Doscherholmen & Hagen, 1957).

Elevations in levels of serum methylmalonic acid and total homocysteine (metabolites in B_{12}-dependent enzyme systems) occur earlier than the clinical manifestations and are more sensitive than the serum B_{12} level

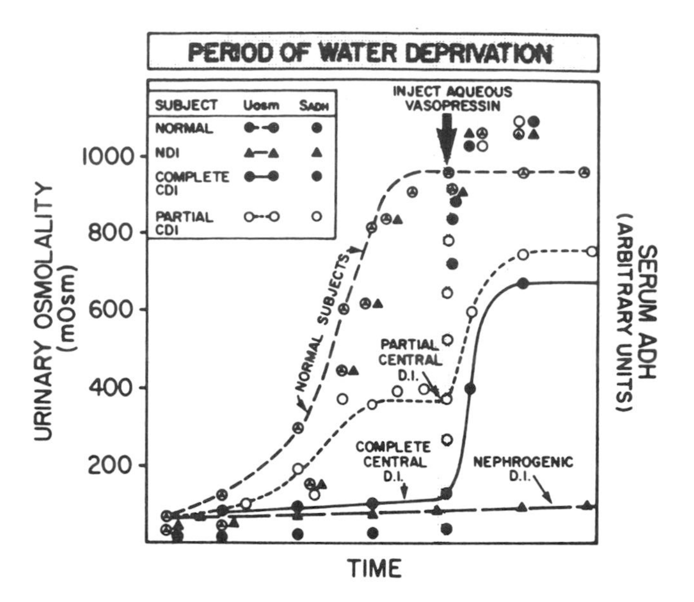

FIG. 7–3. Water deprivation test, revealing the response of normal subjects, patients with complete or partial central diabetes insipidus, and patients with nephrogenic diabetes insipidus. (Reprinted by permission from Narins RG, Jones ER, Stom MC, et al. [1982]. Diagnostic strategies in disorders of fluid, electrolyte and acid-base homeostasis. Am J Med 72:503.)

(Herbert, 1987; Stabler et al., 1986, 1988). Other investigators have suggested that these tests are clinically useful techniques in the evaluation of occult cobalamin deficiency, and advocated their more widespread use (Lindenbaum et al., 1988). However, critical review of their data (Herbert, 1988) reveals that all but 2 of 40 patients had low serum B_{12} levels, obviating the need for confirmatory tests in these patients. Furthermore, 38% of the patients had an elevated MCV, and a careful review of the patients' blood smears revealed the presence of macroovalocytes and/or hypersegmented neutrophils. There were only two normal peripheral blood smears, both of these in patients with low serum B_{12} levels. Therefore, measurement of methylmalonic acid and homocysteine levels should be reserved for those patients with a low-normal B_{12} level and a strong suspicion of B_{12} deficiency, since the combination of a serum B_{12} level and a carefully reviewed peripheral blood smear reveals most cases of cobalamin deficiency. Depletion of holotranscobalamin II may be the earliest and most sensitive indicator of occult B_{12} deficiency, but its role in diagnosing early B_{12} deficiency has yet to be determined (Herzlich & Herbert, 1988). (see also Chapter 31).

Schilling Test. Once a diagnosis of B_{12} deficiency has been made, the Schilling test is performed to differentiate between a dietary deficiency of B_{12}, impaired absorption of B_{12} resulting from absent IF, and the presence of intestinal disease interfering with absorption (Schilling, 1953). In stage I, the patient ingests B_{12} radiolabeled with cobalt-57 (^{57}Co). An injection of 1000 μg of unlabeled B_{12} follows after 2 hours. This unlabeled B_{12} saturates tissue receptors and displaces any bound radiolabeled B_{12}. A 24-hour urine collection is obtained and the amount of radiolabeled B_{12} is measured. Excreted fractions less than 10% of the oral dose are considered abnormal. Patients with diets deficient in B_{12} have a normal stage I Schilling test, and proceeding to stage II or III is not necessary in this case.

If stage I is abnormal, a second stage is undertaken after 5 to 7 days in which the patient ingests B_{12} and IF to differentiate between pernicious anemia and ileal disease. The second stage usually is normal in pernicious anemia, because replacement of the missing IF normalizes the absorption of B_{12}. However, if the second stage also reveals impaired B_{12} absorption, the defect exists in the intestinal mucosa. Situations that result in abnormal first- and second-stage Schilling tests include sprue, inflammatory bowel disease, bacterial overgrowth, and the presence of antibodies to intrinsic factor. However, impaired absorption of B_{12}-IF complex has been described in occasional cases of pernicious anemia; it reverses with continued B_{12} therapy.

A third stage of the Schilling test may be performed after the administration of antibiotics in patients with abnormal first- and second-stage tests to rule out the possibility of bacterial overgrowth causing impaired B_{12} absorption. Typically, these patients have a history of gastrointestinal pathology or surgery, such as a Billroth II gastroenterostomy.

Renal disease and inaccurate specimen collection may result in falsely low urinary B_{12} levels (Rath, McCurdy, & Duffy, 1957); determination of serum B_{12} values or a combination of serum and urine values has been advocated by some (Armstrong & Woodliff, 1970; Chanarin & Waters, 1974). A one-stage Schilling test has been devised wherein the patient simultaneously ingests ^{58}Co-labeled B_{12} and ^{57}Co-labeled B_{12} bound to IF. However, this procedure is fraught with misleading results (Fairbanks, Wahner, & Phyliky, 1983).

Intrinsic Factor Antibodies. Antibodies to IF are present in up to 76% of patients with pernicious anemia (Chanarin, 1979, p. 363; Rothenberg, Kantha, & Ficarra, 1971). Two types of IF antibodies exist; type I (blocking antibody) prevents B_{12} from binding to IF, and type II (binding antibody) binds to the B_{12}-IF complex and prevents its absorption (Roitt, Donaich, & Shapland 1964). Type I antibodies appear to be more common than type II and neither appear to be present in patients with folate deficiency (Rothenberg, Kantha, & Ficarra, 1971).

Occasionally, an empiric trial of B_{12} can be employed in situations in which the clinical data are confusing (e.g., a patient with neurologic signs, an equivocal B_{12} level, and a normal Schilling test). Intrinsic factor antibodies, if present, may also be helpful in confirming the presence of pernicious anemia if previous injections of B_{12} have rendered either the serum B_{12} level or the Schilling test normal. If available, ancillary tests such as the determination of serum methylmalonic acid and total homocysteine may assist in diagnosis as well in patients with low-normal serum B_{12} levels in whom B_{12} deficiency is strongly suspected. (The diagnosis of B_{12} deficiency is discussed further in Chapter 31.)

Treatment. Treatment of B_{12} deficiency in the United States has traditionally involved the use of parenteral vitamin B_{12}, although oral administration of B_{12} has been employed with success in Europe for decades (Berlin, Berlin, & Brante, 1968; Hathcock & Troendle, 1991; Lederle, 1991). Absorption occurs independently of IF and results in resolution of the deficient state with negligible toxicity noted. A daily dose of 1,000 μg oral B_{12} has been recommended (Berlin, Berlin, & Brante, 1968; Berlin et al., 1978).

Folate Deficiency

Folate deficiency commonly accompanies cobalamin deficiency and has similar hematologic sequelae. In clinical practice, vitamin B_{12} and folate levels are often obtained simultaneously. Folate-deficient patients often have a history of dietary inadequacy, alcoholism, or anticonvulsant exposure. Relative folate deficiency may be seen in pregnant women and in dialysis patients. However, isolated folate deficiency is less frequently associated with neuropsychiatric sequelae.

Tests for diagnosing folate deficiency include serum and red blood cell (RBC) folate levels. A low serum folate level reflects a state of negative folate balance for at least 2 weeks prior to the laboratory determination. Folate depletion follows later and is revealed by the presence of a diminished RBC folate level. Thus, in recently hospitalized patients who may have been given folate, the RBC folate is a more reliable marker for folate deficiency. The first hematologic manifestation is the presence of neutrophil hypersegmentation and larger-than-normal reticulocytes and platelets, followed by the development of macroovalocytes with an elevated MCV and anemia.

Hepatolenticular Degeneration (Wilson's Disease)

Wilson's disease is an autosomal recessive disorder characterized by abnormal accumulation of copper resulting in hepatic cirrhosis, degeneration of the basal ganglia, a variety of neuropsychiatric syndromes, and, less frequently, hemolytic anemia (Brewer & Yuzbasiyan-Burkan, 1989). The prevalence of the disorder is 1:20,000 to 1:40,000, with most cases presenting between the ages of 12 and 20. Wilson noted in his original series that 8 of 12 patients had significant mental symptoms (Wilson, 1912). A recent retrospective review of 195 cases confirms that the prevalence of mental abnormalities is greater than 50%. Furthermore, 20% of these patients had seen a psychiatrist before the diagnosis of Wilson's disease was made (Dening & Berrios, 1989). Psychiatric manifestations can include depression, mania, anxiety, psychosis, confusion, cognitive impairment, and emotional lability (Inose, 1968; Lishman, 1987). Psychotic states frequently are described as characteristic of Wilson disease; however, careful review suggests they are an infrequent finding (Beard, 1959; Dening & Berrios, 1989). Cognitive impairment often improves with treatment of the disorder (Goldstein et al., 1968; Scheinberg, Sternlieb, & Richman, 1968; Sternlieb & Scheinberg, 1964). Psychiatric morbidity is strongly correlated with the neurologic manifestations, which include micrographia, dysarthria, tremor, dystonia, choreoathetoid movements, and ataxia. A slit-lamp examination may identify Kayser-Fleischer rings; these are present more often in patients presenting with neuropsychiatric signs than in those presenting with hepatic dysfunction (Brewer & Yuzbasiyan-Gurkan, 1989; Dening & Barrios, 1989).

Screening procedures include the serum ceruloplasmin and 24-hour urinary copper levels. Ninety percent of Wilson's disease patients have a very low ceruloplasmin level of 1 to 5 mg/dl (normal levels are 25 to 35 mg/dl); the remaining 10% have low-normal or less markedly low levels. Ten percent of patients heterozygous for Wilson disease also have low and occasionally very low levels of ceruloplasmin. Low ceruloplasmin can also be caused by copper deficiency, rare inherited hypoceruloplasminemic disorders, and liver disease of such severity that protein synthesis is markedly impaired. Twenty-four–hour urinary copper levels are elevated in Wilson disease, often to greater than 100 μg (normal values are 20 to 50 μg/24 hr); however, these levels must be interpreted with caution. Contamination or insensitive assays may produce misleading results; cirrhosis, in and of itself, can cause elevated urinary copper levels (Brewer & Yuzbasiyan-Gurkan, 1989).

Other diagnostic tests include measurement of ^{64}Cu

incorporated into ceruloplasmin after oral or intravenous administration and liver biopsy to determine hepatic copper levels. Table 7-17 illustrates the copper-related tests for Wilson's disease. Patients who should be screened for Wilson's disease include: (1) those with a family history suggestive of Wilson's disease, 2) those with unexplained liver disease plus psychiatric symptoms, and 3) young patients with signs of basal ganglia or frontal lobe disease not explained by other causes.

The Porphyrias

For a complete discussion of the interpretation of laboratory data in the diagnosis of the acute porphyrias, the reader is referred to Chapter 31.

ENDOCRINE AND NEUROENDOCRINE TESTING

Hypoglycemia

Comatose and delirious patients, those with adrenergic and neuroglycopenic symptoms on fasting, those with diabetes mellitus, and those with alcoholism or cirrhosis all deserve evaluation for hypoglycemia. Symptoms of hypoglycemia can be divided into adrenergic and neuroglycopenic phenomena (Table 7-18).

Factitious hypoglycemia due to exogenous insulin administration is suggested by the presence of elevated insulin antibodies, hypoglycemia, and low C-peptide levels (Horwitz, 1989; Scarlett et al., 1977). (C-peptide is co-secreted with endogenous insulin but is not part of commercial insulin preparations.)

"Hypoglycemia" as a self-diagnosed symptom and the problem of reactive hypoglycemia are addressed in detail in Chapter 29.

Thyroid Function Tests

Before discussing the use of the various thyroid function tests in psychiatric practice, a brief review of basic thyroid psysiology is warranted. Principal stimulatory hormones include thyrotropin-releasing hormone (TRH) from the hypothalamus and thyroid-stimulating hormone (TSH) from the pituitary. Dopamine and somatostatin exert an inhibitory effect on the pituitary, and T_4 and tri-iodothyronine (T_3) exert inhibitory feedback on both the pituitary and the hypothalamus. The thyroid releases thyroid hormone in the form of T_4 and T_3, with the latter being more physiologically active. The majority of T_3 is produced by peripheral conversion of T_4 in tissues.

Abnormalities in thyroid function are associated with a variety of psychiatric syndromes, including anxiety, mood disorders, psychosis, delirium, and dementia (Jellinek, 1962; Lahey, 1931; Taylor, 1975; Whybrow, Prange, & Treadway, 1969). Psychotropic drugs such as lithium, carbamazepine, and phenytoin have effects on thyroid function as well. In addition, primary psychiatric conditions have been associated with abnormalities in thyroid function tests, which may ultimately have diagnostic and prognostic importance. Table 7-19 summarizes the results of thyroid function tests in a variety of disease states.

Thyroid-Stimulating Hormone Assay

A highly sensitive radioimmunometric assay for TSH is now widely available that is capable of distinguishing hypothyroidism from euthyroid and hyperthyroid states. Formerly, less sensitive TSH assays were unable to distinguish the very low levels of TSH found in hyperthyroidism from the normal levels found in the euthyroid

TABLE 7–17. *Copper-related Variables Useful for the Diagnosis of Wilson's Disease*

	Normal	Symptomatic of Wilson's Disease	Presymptomatic of Wilson's Disease
Serum ceruloplasmin (mg/dl)	25–35	Typically 1–5; 10% have intermediate or low-normal values (10% of carriers also have low values)	Same
24-hour urine copper (μg)	20–50	Over 100	Normal to 50
Incorporation of ^{64}Cu into ceruloplasmin at 24 hours			
Oral: ratio to initial peak	0.8–1.2	0.1–0.3	Same
Intravenous: % of dose	Over 4	Less than 1	Same
Hepatic copper (μg/g dry weight)	20–50	Over 250	? Over 150

Reprinted by permission from Brewer GJ, & Yuzbasiyan-Gurkan V (1989). Wilson's disease: An update, with emphasis on new approaches to treatment. Dig Dis 7:182.

TABLE 7–18. *Symptoms of Hypoglycemia*

Adrenergic	Neuroglycopenic
Anxiety	Fatigue
Palpitations	Mental confusion
Sweating	Blurred vision
Irritability	Headache
Hunger	Amnesia
Tremor	Dizziness
	Fainting spells
	Primitive movements
	Muscle spasms
	Seizures
	Deep coma
	Bradycardia
	Hyporeflexia

Reprinted by permission from Hale F, Margen S, & Rabak D (1981). Postprandial hypoglycemia and "psychological" symptoms. Biol Psychiatry 17:126.

state. This "supersensitive" TSH assay has been advocated as *the best single screening test* for the evaluation of thyroid function. Figure 7-4 illustrates the recommended diagnostic strategy using this test (Klee & Hay, 1986).

Assessment of Circulating Thyroid Hormone

The hormones T_4 and T_3 exist both in the free form (<1%) and bound to thyroxine-binding blobulin (TBG), thyroxine-binding prealbumin (TBPA), and albumin. Changes in the total T_4 (TT_4) concentration may reflect changes in the serum level of TBG and may not represent thyroidal illness. Factors that affect TBG levels are listed in Table 7-20. Methods for determining the level of biologically active free T_4 involve either indirect estimations based on the degree of TBG binding or more direct measurements of the free hormone.

Triiodothyronine Resin Uptake and Free Thyroxine Index

The triiodothyronine resin uptake test (T_3RU or RT_3U) is performed by incubating serum with radioactive iodine–labeled T_3. Labeled hormone binds either to the anion exchange resin of the assay or to TBG in the sample of the patient's serum. The amount of radioactivity adsorbed onto the resin is inversely proportional to unoccupied binding sites on the sample of TBG. The resin uptake is thus increased in the presence of excess thyroid hormone or decreased TBG. Conversely, the uptake is decreased when the level of unsaturated TBG is high, secondary to diminished thyroid hormone or increased TBG. Estimations of free T_4 are usually determined utilizing the free thyroxine index (FTI), a calculation based on the TT_4 and T^3RU tests:

$$\text{FTI} = TT_4 \times \left(\frac{T_3\text{RU patient}}{T_3\text{RU standard}}\right)$$

Determination of Free Thyroxine

Formerly, accurate determination of free T_4 was accomplished by the equilibrium dialysis method; however, the cumbersome nature of the technique did not lend itself to routine clinical use. More recently, a number of radioimmunoassays have been developed to determine the fT_4 level. Although some authors suggest that these tests are not superior to measurement of the FTI or dialyzable T_4 level (Chopra et al., 1980; Keptein et al., 1981), they are replacing the FTI as a screening test in many centers.

Thyrotropin-Releasing Hormone Stimulation Test

The TRH stimulation test is a provocative test originally designed to detect hyperthyroidism before the availability

TABLE 7–19. *Thyroid Function Tests*

	Thyroid Function Test[a]								
Disorder	TSH	TT_4	fT_4	FTI	T_3RU	T_3	rT_3	TRH Stim	Antibody +
Hyperthyroidism	D	I	I	I	I	I	I	flat	+/−
T_3 toxicosis	D	N	N	N		I	—	flat	+/−
Hypothyroidism									
Grade I (overt)	I	D	D	D	D	N/D	D	I	+/−
Grade II	I	N	N	N	N	N	—	I	+/−
Grade III	N	N	N	N	N	N	—	I	+/−
Grade IV	N	N	N	N	N	N	—	N	+
Euthyroid sick syndrome									
Low T_3	N	N	N	N	N	D	N/I	var	+/−
Low T_3-T_4	N	D	D	D	N/I	D	I	var	+/−
High T_4	N	I	N/I	N/I		N	—	var	+/−

[a]Abbreviations: TT_4, total thyroxine; fT_4, free thyroxine; FTI, free thyroxine index; T_3RU, triiodothyronine resin uptake; rT_3, reverse T_3; TRH stim, response to TSH to TRH stimulation; antibody +, presence of antithyroid antibodies; D, decreased, I, increased, N, normal.

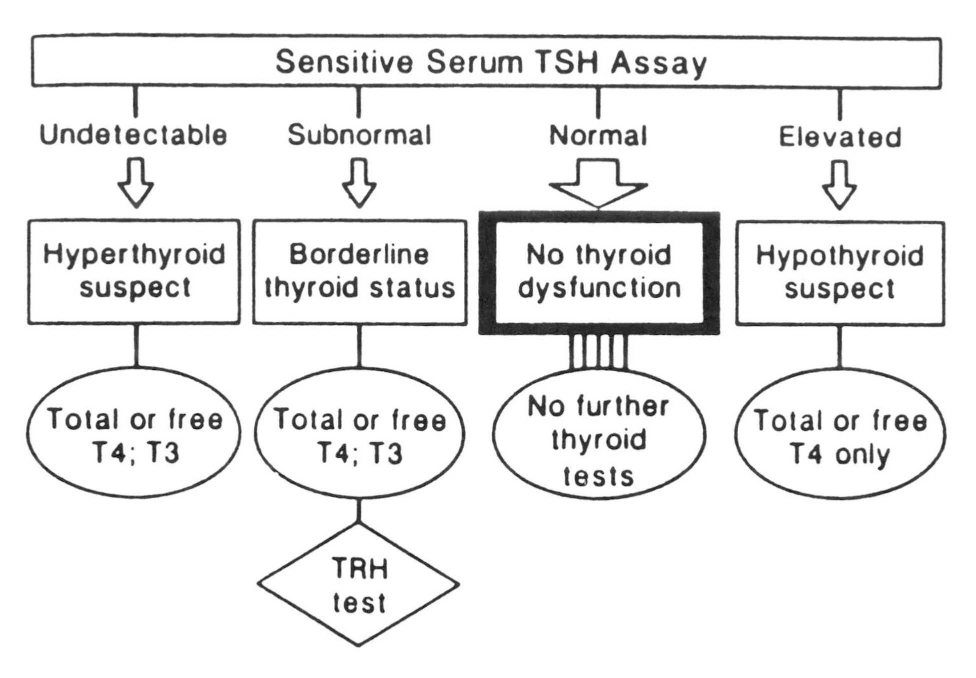

FIG. 7–4. Diagnostic strategy for use of the highly sensitive TSH assay as a cost-effective screening tool. (Reprinted by permission from Klee GC, & Hay ID [1986]. Assessment of sensitive thyrotropin assays for an expanded role in thyroid function testing: Proposed criteria for analytic performance and clinical utility. J Clin Endocrinol Metab 64:461.)

TABLE 7–20. *Factors Influencing TBG Levels*

Increased TBG	Decreased TBG
Inherited TBG excess	Inherited TBG deficiency
Estrogens	Chronic liver disease
Liver disease	Nephrosis
Acute intermittent porphyria	Renal dialysis
5-Fluorouracil	Protein-losing enteropathy
Clofibrate	Androgens
Methadone and heroin	L-Asparaginase
	Glucocorticoids

Adapted from Ladenson (1985).

the sensitive TSH assay. Hyperthyroid patients exhibit a blunted or flat TSH response to TRH, whereas patients with primary hypothyroidism exhibit an exaggerated response, in addition to an elevated baseline TSH (Haigler, Hershman, & Pittman, 1972; Snyder & Utiger, 1972). Patients with early hypothyroidism may also have an exaggerated response before elevations in basal TSH become apparent. Shortly after the development of the TRH stimulation test, investigators began evaluating the response in a variety of psychiatric conditions, including mood disorders, anorexia nervosa, schizophrenia, borderline personality disorder, and alcoholism (Garbutt et al., 1983; Gold et al., 1981; Kirkegaard et al., 1978; Loosen, 1985; Loosen, Prange, & Wilson, 1979; Targum SD, Greenberg RD, & Harmon RL, 1984).

Suggested clinical application of the TRH stimulation test in depressed patients has included prediction of treatment response and relapse (Loosen, 1985; Pottash, Gold, & Extein, 1986). However, the data in depressed patients are mixed, with some suggesting a blunted TSH response to TRH (Loosen, 1985; Loosen & Prange, 1982), some suggesting the response is normal (Coppen et al., 1980; Karlberg, Kjellman, & Kagedol, 1978), and some suggesting that there is an exaggerated response (greater than five- to eightfold increase in TSH above baseline) (Brambilla et al., 1978; Nemeroff & Evans, 1989). This inconsistency, combined with the substantial cost of the test, severely limits its utility; therefore, *it is not recommended for clinical use in the assessment of depression or other psychiatric disorders.*

Thyroid Autoantibodies

Antibodies to thyroglobulin and thyroid cell microsomal proteins are commonly present in serum of patients with thyroid disease. Antimitochondrial antibodies are present in approximately 95% of patients with Hashimoto thyroiditis and 85% of patients with Graves disease, and antithyroglobulin antibodies are present in approximately 60% of Hashimoto's thyroiditis patients and 30% of patients with Graves disease. Positive tests can be seen in up to 10% of the population, especially among women and the elderly, and may respresent subclinical autoimmune thyroid disease. The clinical utility of antithyroid antibodies lies in the evaluation of patients with atypical manifestations of autoimmune thyroid disease, such as isolated ophthalmopathy or dermopathy (Refetoff, 1989, p. 609).

Thyroid Function in Nonthyroidal Illness (Euthyroid Sick Syndrome)

A number of disturbances in thyroid function have been noted in patients with nonthyroidal illness (Chopra et al., 1983; Nicoloff, 1989; Wartofsky & Burman, 1982). Most common among these is the low T_3 syndrome. Other abnormalities can include either depressed or elevated T_4 and an elevated reverse T_3. Thyroid-stimulating hormone and fT_4 levels are usually normal or only mildly abnormal in the euthyroid sick syndrome and are useful in distinguishing nonthyroidal from thyroidal illness (Table 7-19). Patients with alcohol dependence and anorexia nervosa exhibit manifestations of the euthyroid sick syndrome.

The reverse T_3 level is usually elevated in the euthyroid sick syndrome; however, its clinical utility is limited to those situations associated with alterations in serum T_3 and T_4 levels when systemic or thyroid abnormalities are not readily apparent (Refetoff, 1989, p. 604). When T_3 and T_4 levels are low, TSH is normal, and reverse T_3 is elevated, the diagnosis of euthyroidism, rather than hypothyroidism, would be made.

An elevated T_4 level is noted in 7% to 18% of acute psychiatric admissions, but this elevation almost resolves within 2 weeks of admission (Levy et al., 1981; Spratt et al, 1982). Diagnoses most associated with this increase in T_4 were schizophrenia, major affective syndromes, and amphetamine and phencyclidine ingestion (Morley & Shafer, 1982; Morley et al., 1980; Spratt et al., 1982). The probable mechanism is transient TSH hypersecretion; patients rarely have intrinsic thyroid disease. A normal or elevated TSH rules out a hyperactive thyroid as the etiology of the high T_4 level.

Grades of Hypothyroidism

As methods for measuring thyroid function have become more refined, a number of subtle gradations in thyroid dysfunction have been identified. Grade I, or overt, hypothyroidism is defined by a low FTI (or fT_4) and an elevated TSH, along with clinical symptoms. Grade II hypothyroidism is associated with normal T_4 levels but elevations in TSH. Both Grade I and Grade II hypothyroidism exhibit an exaggerated TSH response to TRH. Grade III hypothyroidism is identified by an exaggerated TSH response in the setting of normal circulating T_4 and basal TSH levels. Antithyroid antibodies are variably present or absent in all three grades of hypothyroidism (Evered el al., 1973; Wenzel et al., 1974). Symptomless thyroiditis, characterized by the presence of antithyroid antibodies in the serum but normal circulating basal TSH and T_4 levels and normal TSH response to TRH, may represent a prodromal grade of hypothyroidism (grade IV) (Gordin & Lamberg, 1981; Haggerty et al., 1990; Tunbridge et al., 1981). Treatment implications are discussed in Chapter 29.

Medication Effects on Thyroid Function

Alteration in the level of TBG is the most common mechanism for drug-induced changes in thyroid function (Table 7-20). Other mechanisms include inhibition of the organification of iodine and release of thyroid hormone, the induction of hepatic enzymes, and stimulation of antithyroid antibodies (Emerson, Dyson, & Utiger, 1973; Ramsay, 1985).

The spectrum of lithium-induced thyroid dysfunction includes hypothyroidism, hyperthyroidism, goiter, and the development of antithyroid antibodies. Lithium-induced hypothyroidism results primarily from impaired release of hormone, with impaired organification of iodine playing a lesser role (Berens et al., 1970). Rates of grade I (overt) hypothyroidism induced by lithium range from 0 to 23% (Lazarus, 1986). Age greater than 40, female sex, the presence of thyroid autoantibodies, and rapidly cycling bipolar affective disorder have been implicated as possible risk factors (Cho et al., 1979; Emerson, Dyson, & Utiger, 1973; Lazarus, 1986). The rate of grades II and III hypothyroidism is approximately 50% (Lazarus, 1986; Salata & Klein, 1987). Elevations in TSH may be transient (Maarbjerg, Vestergaard, & Schou, 1987), may persist at low levels without symptoms, or may progress to overt hypothyroidism. Patients treated with lithium should have baseline thyroid function tests performed, with follow-up TSH levels at 3 months, 6 months, 1 year, and annually thereafter. The majority of clinically significant thyroid function abnormalities occur in the first 4 years of treatment (Lazarus, 1986). The cost-effectiveness of yearly screening beyond this period has not been determined.

Carbamazepine, phenytoin, and phenobarbital are associated with reduced levels of T_3 and T_4, resulting from displacement of these hormones from TBG and from hepatic enzyme induction, resulting in increased clearance. However, TSH levels generally remain normal and overt hypothyroidism rarely occurs with these agents (Heyma et al., 1977; Larsen et al., 1970; Smith & Surks, 1984; Wenzel, 1981). However, because carbamazepine and phenytoin can depress TSH levels, it is possible for patients on these drugs to be hypothyroid with a normal TSH level. On occasion, empiric trials of thyroxine therapy are useful in such patients who have signs and symptoms suggesting hypothyroidism.

Dexamethasone Suppression Test

The dexamethasone suppression test (DST) is the most widely investigated laboratory test in psychiatry. Its use as a research tool has fueled an information explosion in biologic psychiatry. Appropriate clinical use of the DST is quite limited, however. The reader is referred to several reviews for complete discussion of the DST (Arana, Baldessarini, & Ornsteen, 1985; Baldessarini & Arana, 1985; Braddock, 1986; Carroll, 1982, 1985, 1986; Insel & Goodwin, 1983; Shapiro, Lehman & Greenfield, 1983).

The DST was originally developed for the evaluation of Cushing's syndrome (Liddle, 1960). Its use in the investigation of affective disorders was based on the concept that dysregulation of the hypothalamic-pituitary-adrenal axis could be studied as a potential window for hypothalamic and limbic dysfunction in these disorders (Carroll, Martin, & Davies, 1968; Stokes, 1966). The complex regulation of this system is illustrated in Figure 7-5 (Meador-Woodruff & Greden, 1988).

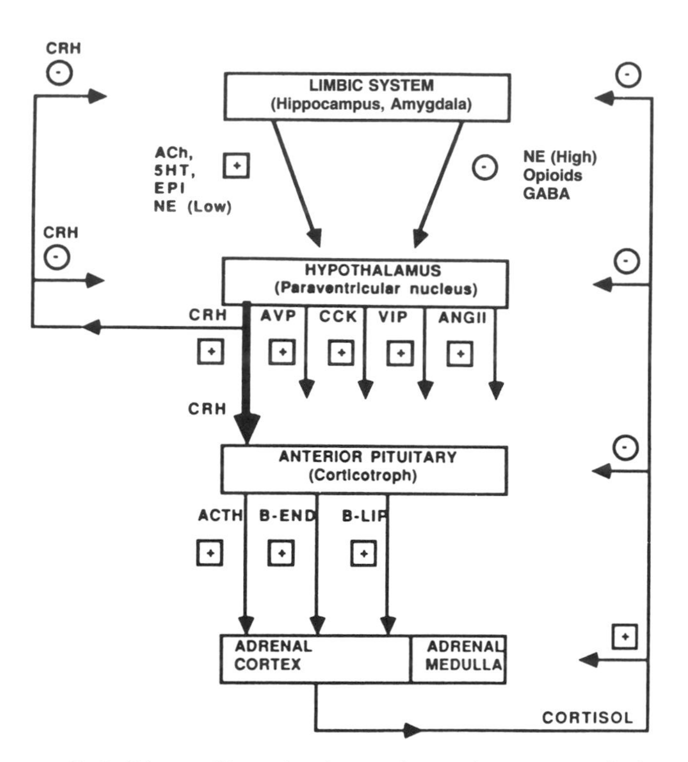

FIG. 7–5. Diagram illustrating the complex regulatory systems in the hypothalamic-pituitary-adrenal axis. (Reprinted by permission from Meador-Woodruff JH, & Greden JF [1988]. Effects of psychotropic medications on hypothalamic-pituitary-adrenal regulation. Endocrinol Metab Clin North Am 17:226.)

Test Protocol

Variations in procedure exist in the literature; however, the following represents one of the most widely used protocols. On day 1, baseline plasma cortisol is obtained (optional), usually at 4:00 p.m., followed by the administration of 1 mg dexamethasone at 11:00 p.m. On day 2, plasma cortisol samples are obtained at 8:00 a.m., 4:00 p.m., and/or 11:00 p.m. Reducing the number of postdexamethasone samples reduces the test's sensitivity while increasing its specificity. A plasma cortisol level greater than 5 μg/dl (50 ng/ml or 138 pmol/ml) obtained between 9 and 24 hours after dexamethasone administration is considered a positive result (nonsuppression) (American Psychiatric Association [APA] Task Force, 1987). Some investigators have employed different criteria for nonsuppression, based on the analytic method used in their laboratory.

Several medical, psychiatric, and pharmacologic factors have been associated with spurious DST results and must be considered in the interpretation of results (Table 7-21) (Carroll, 1986). The sensitivity of the DST for major depression is modest, ranging from 30% to 70%, with higher rates of nonsuppression in melancholic depressives tested with two or three plasma cortisol samples. Thus, it has little utility as a laboratory screening test for major depression, and in any case would be inferior to clinical examination in establishing a diagnosis (APA Task Force, 1987). Early reports of specificity noted rates of over 90% (Carroll, Curtis, & Mendels, 1976a, 1976b). However, the comparison group consisted of normal controls. In studies using other psychiatric patients as controls, the specificity drops to approximately 80% (APA Task Force, 1987). Studies using medically ill, demented, or elderly controls show even lower specificity.

The initial DST suppressor status does not significantly contribute to the clinical diagnosis of major depressive disorder, nor does it consistently predict response to tricyclic antidepressants or electroconvulsive therapy (ECT) therapy (Gitlin & Gerner, 1986). Results from pooled data revealed that DST nonsuppressors had a 76% antidepressant response rate, compared to 64% of DST suppressors (APA Task Force, 1987). However, persistent nonsuppression even after apparent clinical improvement may be associated with a higher relapse rate. Arana and Mossman (1988) noted a poor clinical outcome in 19% of depressed nonsuppressors who converted to and maintained normal suppression

TABLE 7–21. *Factors That Interfere with the DST*

False-positive results	
Drugs:	Phenytoin; barbiturates; carbamazepine; meprobamate; glutethimide; methyprylon; methaqualone; heavy alcohol use; ?recent withdrawal of TCA's or benzodiazepines; ?reserpine
Endocrine:	Cushing's disease or syndrome; pregnancy; high-dose estrogen replacement; diabetes mellitus; ?hypothyroidism
Medical:	Major medical disorders, i.e., serious infections; uncontrolled CHF; hypertension; advanced renal or hepatic disease; cancer
Metabolic:	Recent withdrawal of alcohol (2 weeks); dehydration; fever; nausea, vomiting; low body weight (anorexia nervosa, malnutrition); recent weight loss (rigid dieting)
Neurologic:	Alzheimer's disease; multi-infarct dementia; high intracranial pressure; ?brain tumor; ?temporal lobe disease; ?status post stroke
Other:	Rapid metabolism of dexamethasone; ?ECT or generalized seizure on post dexamethasone day; ?unstable circadian rhythm (shift workers, airline travelers); acute physical pain (multiple venipuncture attempts); noncompliance with dexamethasone administration; recent admission to hospital
False-negative results	
Drugs:	Synthetic corticosteroids; indomethacin; high-dose cyproheptadine; ?high-dose benzodiazepines; ?tricyclic antidepressants; L-tryptophan
Endocrine:	Hypopituitarism; Addison's disease; slow metabolism of dexamethasone

Reprinted by permission from Carroll BJ (1986). Informed use of the dexamethasone suppression test. J Clin Psychiatry 47(1 suppl):10–12; copyright 1986, Physicians Postgraduate Press.

status. However, 77% of those who maintained or reconverted to nonsuppressor status during follow-up had a poor or even fatal outcome.

The DST has provided insight into the regulation of the hypothalamic-pituitary-adrenal axis in a variety of disease states. However, it still must be regarded largely as a research tool. Clinical decisions should depend on standard methods of assessment and not be swayed by the results of the test. The encouraging results regarding its prognostic role would need to be confirmed by other studies before the DST could be used to supplement clinical judgment regarding prognosis and maintenance therapy requirements.

Prolactin

Hyperprolactinemia is the most common primary hypothalamic-pituitary disorder, with prolactinomas accounting for over 40% of pituitary tumors (Faglia et al., 1980). Signs and symptoms of hyperprolactinemia include galactorrhea, amenorrhea, impotence, and infertility. Twenty-five percent of women with secondary amenorrhea have hyperprolactinemia (Franks et al., 1975). Hostility and aggression are psychiatric symptoms that have been attributed to elevated prolactin levels (Kellner et al., 1984), but there are conflicting data regarding this (Steiner et al., 1984). The presence of headache and visual field deficits suggests a prolactinoma. Stress, exercise, pregnancy, breast manipulation, and coitus are physiologic stimuli for prolactin release. Dopamine is the primary inhibitory stimulus. Pathologic causes of hyperprolactinemia include hypothyroidism, pituitary micro- and macroadenomas, and craniopharyngiomas (Kletzky, 1984). Neuroleptics, monoamine oxidase (MAO) inhibitors, estrogens, and methyldopa are medications associated with prolactin elevation (Sherman et al., 1984). While prolactin levels are not part of screening batteries, they should be strongly considered in the evaluation of female psychiatric patients with irregular menses and male psychiatric patients with impotence or loss of libido.

Neuroleptic-induced hyperprolactinemia is the most common kind encountered by most psychiatrists. Prolactin levels peak within 1 to 2 hours after a single intramuscular injection of neuroleptics, and gradually fall over the next 6 hours. The doses required for a peak response are much smaller than those used in clinical practice (Gruen et al., 1978; Rubin et al., 1976). Schizophrenic patients exhibit persistent 3.2-fold to 3.8-fold elevations in serum prolactin 72 hours following initiation of neuroleptic therapy. Prolactin levels return to baseline within 48 to 96 hours following discontinuation of treatment (Meltzer & Fang, 1976). Chronic treatment may result in tolerance to the prolactin-elevating effects of neuroleptics (Brown & Laughren, 1981). Galactorrhea, amenorrhea, and impotence secondary to neuroleptic-induced hyperprolactinemia can be successfully treated with bromocriptine, without an exacerbation of psychotic symptoms (Matsuoka et al., 1986). Clozapine is associated with only mild and transient elevations in serum prolactin (Meltzer et al., 1979) and has been suggested as an alternative to conventional neuroleptics in patients with neuroleptic-induced gynecomastia (Uehinger & Baumann, 1991).

The serum prolactin level can be used as a diagnostic tool to help differentiate generalized seizures from pseudoseizures. Transient elevations in serum prolactin frequently occur following generalized seizures (Abbott, Browning, & Davidson, 1980; Bye, Nunn, & Wilson, 1985) and complex partial seizures with bilateral medial temporal involvement (Sperling et al., 1986), but rarely in association with pseudoseizures (Collins, Lanigan, & Callaghan, 1983; Laxer, Mullhooly, & Howell, 1985; Yerby et al., 1987). This prolactin peak occurs 15 to

20 minutes following the seizure and returns to baseline within 1 to 3 hours. Prolactin values two to three times normal are considered supportive of a diagnosis of true seizure activity by most (Yerby et al., 1987).

Anabolic Steroids

Anabolic steroid use should be considered in a psychiatric patient with a history of weight lifting, bodybuilding, or highly competitive sport activity. Testicular atrophy often is present on physical examination. Psychiatric symptoms include irritability, aggression, depression, delirium, and psychosis (Annitto & Layman, 1980; Campbell, 1987; Pope & Katz, 1988). Patients often are quite frank about their androgen use, but may be reluctant to discuss it because of the potential for adverse effects on their athletic careers.

Analysis of anabolic steroids is most commonly performed using GC/MS. It is possible to use HPLC or RIA, but both of these methods lack the necessary sensitivity and specificity to confirm a presumptively positive sample (Chung et al., 1990). A testosterone/epitestosterone ratio of greater than 6:1 has been advocated as a marker for exogenous testosterone administration by the International Olympic Committee (Park et al., 1990). Beta–human chorionic gonadotropin (HCG) is often used by athletes in an attempt to increase the endogenous production of testosterone, thereby reducing the testosterone/epitestosterone ratio. Quantitative measurement of beta-HCG is performed by immunoassay techniques.

For a discussion of male hypogonadism, the reader is referred to Chapter 29.

MISCELLANEOUS TESTS

Creatine Kinase

The determination of serum CK activity is important in several clinical situations involving muscle injury, most notably rhabdomyolysis, myocardial infarction, and a variety of muscle disorders such as muscular dystrophy and polymyositis. Rhabdomyolysis is a syndrome of striated muscle destruction associated with the release of muscle cell contents into the serum. Alcohol and substance abuse disorders are the most frequent conditions associated with the development of rhabdomyolysis (Gabow, Kaehny, & Kelleher, 1982). In addition, rhabdomyolysis is a frequent complication of neuroleptic malignant syndrome (Caroff, 1980; Eiser, Neff, & Slifkin, 1982), and may occur in other catatonic states as well. Muscle pain and swelling are the most frequent clinical manifestations; however, up to 50% of patients are asymptomatic (Gabow, Kaehny, & Kelleher, 1982). In its most extreme form, tissue destruction and edema can result in compartment syndromes, leading to massive muscle destruction. Renal failure is a frequent complication, as a result of myoglobin-induced glomerular damage.

Creatine kinase activity is greatest in striated muscle, brain, and cardiac tissue and is a dimer composed of two subunits: brain-associated (B) and muscle-associated (M). Thus, three forms of the enzyme exist. Fraction BB (CK-1) exists primarily in brain, but is also present in other tissues such as prostate, placenta, and bladder. Fraction MB (CK-2) exists primarily in cardiac tissue, with a small amount (<5%) produced by skeletal muscle. Fraction MM (CK-3) is present in skeletal and cardiac muscle. Myocardial infarction is associated with an elevated MB fraction (>6%) (Moss, Henderson, & Kachmar, 1987). Elevated MB fractions are also observed in chronic hemodialysis patients (Ma et al., 1981) and in marathon runners (Kielblock et al., 1979; Olivier et al., 1978); however, these elevations do not appear to be indicative of myocardial damage.

Elevated CK levels (MM fraction) can be seen in response to physical exercise (Olerud, Homer, & Carroll, 1976; Priest, Oei, & Moorehead, 1982; Shumate, Brooke, & Carroll, 1979), generalized tonic-clonic seizures (Chesson, Kasarskis, & Small, 1983; Glotzner, 1979; Os & Lyngdal, 1989; Wyllie et al., 1985), intramuscular injections (Armstrong, Lloyd, & Balazs, 1976; Attar & Mata, 1971; Sidell, Culver & Kamindkis, 1974), and surgery (Klein, Shell, & Sobel, 1973). Psychotic patients often present with elevated CK levels independent of the above-noted factors (Meltzer, 1976); furthermore, the use of restraints is also associated with an increase in CK activity (Goode et al., 1977). Racial differences also exist, with black subjects exhibiting higher CK values than either Hispanics or Caucasians (Black, Quallich, & Gareleck, 1986).

Creatine kinase activity is the most useful biochemical marker of rhabdomyolysis. Most investigators have defined rhabdomyolysis by elevations in CK level of five times normal or greater (Gabow, Kaehny, & Kelleher, 1982; Ward, 1988). Urine myoglobin may be revealed by heme dipstick (orthotoluidine reaction) in the absence of RBCs in the urine. Significant hematuria, however, prohibits a diagnosis of myoglobinuria by dipstick. Moreover, a negative dipstick does not preclude a diagnosis of rhabdomyolysis (Gabow, Kaehny, & Kelleher, 1982). If heavy urinary pigment is present, a concomitant serum sample will be tinged pink if the pigment is hemoglobin but will be clear yellow in the presence of myoglobin. Other associated laboratory findings include hyperkalemia, hyperphosphatemia, decreased BUN-to-creatinine ratio, and proteinuria by dipstick (Gabow, Kaehny, & Kelleher, 1982). Factors associated with the development of renal failure in rhabdomyolysis include a CK level greater than 16,000 IU/liter, elevations in serum potassium and phosphorus levels, and the presence of volume depletion and/or sepsis (Ward, 1988). Table 7-22 illustrates the interpretation of CK values in a variety of clinical settings.

When a significant elevation of CK (greater than five times normal) is encountered in a patient on neuroleptics, a vigorous search for its etiology is indicated. Creatine kinase and the physical examination should be monitored until CK returns to normal or its elevation is satisfactorily explained. If physical signs of neuroleptic malignant syndrome develop, the neuroleptic should be discontinued. Discontinuation of the neuroleptic in the absence of other signs of neuroleptic malignant syndrome is not always necessary, and should depend on an individualized risk-benefit assessment of neuroleptic therapy. The risk of renal failure is only significant with CK values greater than 70 times normal.

Pheochromocytoma

Anxiety often is cited as one of the most common symptoms encountered in pheochromocytoma. Other cardinal symptoms include headache, diaphoresis, palpitations, tremulousness, abdominal or chest pain, nausea, vomiting and weakness. Prominent signs include sustained or paroxysmal hypertension, orthostatic hypotension, hyperhidrosis, hypertensive retinopathy, pallor (very rarely flushing), Raynaud's phenomenon, and livedo reticularis (Manger, Gifford, & Hoffman, 1985). Although anxiety symptoms are frequently encountered, full syndromal pictures resembling panic disorder or generalized anxiety disorder are relatively uncommon (Starkman et al., 1985). Anxiety symptoms accompanied by headache, significant blood pressure abnormalities, and diaphoresis should alert the clinician to the possibility of the presence of pheochromocytoma. However, given the relative rarity of the syndrome in

TABLE 7–22. *Interpretation of Creatine Kinase Values*

Clinical Situation	Magnitude of Peak	Time Course of Peak and Resolution	Isoenzyme	Comments
Psychotic state	1–50-fold	Noted on presentation; duration 1–2 days, occasionally >1 month	MM	Approximately 50% frequency (Meltzer, 1976)
Electroconvulsive therapy	Small	Not applicable	MM	Succinylcholine may elevate MM; early sample draw time may have led to false-negatives (Taylor, von Witt, & Fry, 1981)
Neuroleptic malignant syndrome	Variable, 1–150-fold	Variable	MM	May progress to rhabdomyolysis and acute renal failure
Rhabdomyolysis	≥5-fold	Generally decreases by 30%/day	MM	Risk of renal failure increases when CK >16,000 (>70-fold increase)
Tonic-clonic seizures	8–20-fold; up to 150-fold in alcohol-related seizures		MM	Alcohol-related seizures associated with most significant rise; repeated seizures (>4) associated with higher CK
Pseudoseizures	No change			
Intramuscular injections	Usually 4-fold, occasionally greater	12 hr to peak, 3 days to resolution	MM	Frequency 7–50%; appears to correlate with volume, osmolarity of injection
Exercise	Variable, usually 4–8-fold	12–24 hr to peak, 3–5 days to resolution	MM	Higher peaks in untrained vs. trained & in males; MB fraction noted with severe exertion
Myocardial infarction	1–4-fold	8–24 hr to peak, 2–4 days to resolution	MB >6% total	
Surgery	Up to 18-fold		MM	

hypertensive populations, routine screening of anxious, hypertensive patients for pheochromocytoma is not recommended. A search for pheochromocytoma should be reserved for those atypical situations wherein the clinician is struck by the dramatic nature of the patient's autonomic symptoms, or in which conventional antihypertensive and antianxiety treatments have failed to alleviate symptoms.

The diagnosis of pheochromocytoma is dependent on the demonstration of biochemical evidence for catecholamine hypersecretion. The most commonly employed tests used for this purpose are the determination of urinary catecholamines, vanillylmandelic acid (VMA), and metanephrines. These tests have the advantage of being readily available and fairly sensitive and specific (Bravo & Gifford, 1984; Bravo et al., 1979). Urinary metanephrines have a sensitivity of 79% to 98% (Bravo et al., 1979; Gitlow, Mendelowitz, & Bertani, 1970; Remine et al., 1974) and are more sensitive than either VMA or catecholamines (Bravo & Gifford, 1984). Criticism of urinary determinations centers around the inconvenience and difficulty of obtaining an accurate collection. Medications and other conditions can elevate urinary catecholamine levels and MAO inhibitors can spuriously decrease urinary VMA levels (Atuk, 1983). Some assays for VMA are sensitive to the presence of vanillin and phenolic acids, necessitating a special diet involving the elimination of chocolate, coffee, bananas, citrus fruits, and foods containing vanilla (Chattoraj & Watts, 1987; Shapiro & Fig, 1989). Table 7-23 illustrates the clinical situations other than pheochromocytoma associated with elevated urinary catecholamines.

TABLE 7–23. *Clinical Situations Associated with Elevated Urinary Catecholamines*

Medications
Tricyclic antidepressants, especially desipramine
Trifluoperazine
Methyldopa
Clonidine withdrawal
Theophylline
Aminophylline
Levodopa
Ephedrine
Nitroprusside
Other conditions
Guillain-Barré syndrome
Porphyria
Carcinoid syndrome
Neuroblastoma
Astrocytoma
Acute psychosis
Glomus jugulare tumor
Benign recurrent intrahepatic cholestasis

Adapted from Atuk (1983, p. 194).

Because of difficulty with urinary sample collection and the risk of false-positive results, plasma catecholamine levels have been advocated as the initial diagnostic procedure of choice (Bravo & Gifford, 1984; Bravo et al., 1979). The sample collection procedure involves having a fasting patient rest comfortably for approximately 30 minutes before the procedure. A large-bore intravenous needle is inserted 20 minutes beforehand to avoid elevations in catecholamines secondary to pain and apprehension. Elevated values (>950 pg/ml) are considered specific, and the test has an estimated sensitivity of 94%. Values of plasma catecholamines above 2,000 pg/ml are considered pathognomonic of pheochromocytoma (Bravo & Gifford, 1984). However, plasma catecholamines may be normal in patients with pheochromocytoma who are normotensive at the time of sampling (Jones et al., 1980; Plouin et al., 1981). In addition, others have suggested that the procedure for obtaining plasma samples is at least as cumbersome as the method for urinary collection (Nicoll & Gerard, 1985).

Provocative tests for pheochromocytoma include the clonidine suppression tests and the glucagon stimulation test. Clonidine is a centrally acting alpha$_2$-agonist that reduces plasma catecholamines in normal subjects. The clonidine suppression test is based on the fact that pheochromocytomas secrete catecholamines autonomously and that, in the presence of such tumors, lowering of plasma catecholamines does not occur after clonidine administration. A failure of catecholamine suppression is considered diagnostic of pheochromocytoma (Bravo et al., 1981). A modification of the technique involving overnight collections of urinary catecholamines has been suggested as an alternative to the standard clonidine suppression test (MacDougall et al., 1988). The glucagon stimulation test is a potentially dangerous provocative test often useful in the diagnosis of paroxysmally secreting tumors. It should be performed only by those familiar with the procedure and its complications. It is reviewed elsewhere (Bravo & Gifford, 1984).

Once elevated catecholamine secretion has been demonstrated, the next step in the diagnosis of pheochromocytoma is tumor localization. Computerized tomography is currently the principal method of localization. Other localization studies include magnetic resonance imaging, ^{131}I-*meta*-iodobenzylguandine (^{131}I-MIBG) scintigraphy, angiography, selective venous sampling for catecholamines, and ^{99m}Tc-labeled methylene diphosphonate bone scan to localize metastatic lesions (Shapiro & Fig, 1989).

REFERENCES

Abbott, RJ, Browning MCK, & Davidson DLW (1980). Serum prolactin and cortisol concentrations after grand mal seizures. J Neurol Neurosurg Psychiatry 43:163–167.

Adams RD, & Salam-Adams M (1986). Acquired hepatocerebral syndromes. In PJ Vinken, GW Bruyn, & GL Klawans (Eds.), Handbook of clinical neurology. 5: Extrapyramidal disorders (Ch. 11, pp. 213–221). 49 Amsterdam: Elsevier.

Adamson J, & Hillman RS (1968). Blood volume and plasma protein replacement following acute blood loss in normal man. JAMA 205:609–612.

Ajlouni K, Kern MW, & Tures JF (1974). Thiothixene-induced hyponatremia. Arch Intern Med 134:1103–1105.

Alexander ER, Crow TJ, & Hamilton SM (1973). Water intoxication in relation to acute psychotic disorder. Br Med J 1:89.

Allon M, Allen HM, Deck LV, et al. (1990). Role of cigarette use in hyponatremia in schizophrenic patients. Am J Psychiatry 147:1075–1077.

American Psychiatric Association Task Force on Laboratory Tests in Psychiatry (1978). The dexamethasone suppression test: An overview of its current status in psychiatry. Am J Psychiatry 144:1253–1262.

Anastassiades E, Wilson R, Stewart JSW, et al. (1983). Fatal brain oedema due to accidental water intoxication. Br Med J 287:1181–1182.

Annitto WJ, & Layman WA (1980). Anabolic steroids and acute schizophrenic episode. J Clin Psychiatry 41:143–144.

Ansley JD, Isaacs JW, Rikkers LF, et al. (1978). Quantitative tests of nitrogen metabolism in cirrhosis: Relation to other manifestations of liver disease. Gastroenterology 75:570–579.

Arana GW, Baldessarini R, & Ornsteen M (1985). The dexamethasone suppression test for diagnosis and prognosis in psychiatry: Commentary and review. Arch Gen Psychiatry 42:1193–1204.

Arana GW, & Mossman D (1988). The dexamethasone suppression test and depression—approaches to the use of a laboratory test in psychiatry. Endocrinol Metab Clin North Am 17:21–39.

Armstrong B, Lloyd BL, & Balazs NDH (1976). Changes in plasma enzyme concentrations following intramuscular injections and gastroscopy. Aust NZ J Med 6:548–551.

Armstrong BK, & Woodliff HJ (1970). Studies on the ^{57}Co vitamin B_{12} plasma level absorption test. J Clin Pathol 23:569–571.

Asplund K, Wahlin A, & Rapp W (1979). D.D.A.V.P. test in assessment of renal function during lithium therapy. Lancet 1:491.

Attar AM, & Mata C (1971). Increased levels of creatine phosphokinase after intramuscular injections. Med Ann Dist Columbia 15:92–93.

Atuk NO (1983). Pheochromocytoma: Diagnosis, localization, and treatment. Hosp Pract 4:187–202.

Baldessarini R, & Arana GW (1985). Does the dexamethasone suppression test have clinical utility in psychiatry? J Clin Psychiatry 46:25–29.

Barkin JS, & Garrido JA (1986). Management of acute pancreatitis. In LB Gardner (Ed.), Acute internal medicine (Ch. 14, pp. 253–263). New York: Elsevier Science Publishing Co.

Barnes RF, Mason JC, Greer C, et al. (1983). Medical illness in chronic psychiatric outpatients. Gen Hosp Psychiatry 5:191–195.

Barry H III (1979). Behavioral manifestations of ethanol intoxication and physical dependence. In E Majchrowisz & EP Noble (Eds.), Biochemistry and pharmacology of ethanol (Vol. 2; Ch. 26, pp. 511–529). New York: Plenum.

Bartter FC, & Schwartz WB (1967). The syndrome of inappropriate secretion of antidiuretic hormone. Am J Med 42:790–801.

Bauer MS, Halpern L, & Schriger D (1991). Screening ambulatory depressives for causative medical illnesses: The example of thyroid function screening. Report for the Agency for Health Care Policy and Research (AHCPR).

Beard AW (1959). The association of hepatolenticular degeneration with schizophrenia. Acta Psychiatr Scand 34:411–428.

Beck LH (1979). Hypouricemia in the syndrome of inappropriate secretion of antidiuretic hormone. N Engl J Med 301:528–530.

Berens SC, Bernstein RS, Robbins J, et al. (1970). Antithyroid effects of lithium. J Clin Invest 49:1357–1367.

Berlin H, Berlin R, & Brante G (1968). Oral treatment of pernicious anemia with high doses of vitamin B_{12} without intrinsic factor. Acta Med Scand 184:247–258.

Berlin R, Berlin H, Brante G, et al. (1978). Vitamin B_{12} body stores during oral and parenteral treatment of pernicious anaemia. Acta Med Scand 204:81–84.

Bernadt MW, Mumford J, Taylor C, et al. (1982). Comparison of questionnaire and laboratory tests in the detection of excessive drinking and alcoholism. Lancet 1:325–328.

Bircher J, Muller J, Guggenheim P, et al. (1966). Treatment of chronic portal systemic encephalopathy with lactulose. Lancet 1:890–893.

Black HR, Quallich H, & Gareleck CB (1986). Racial differences in serum creatine kinase levels. Am J Med 81:479–487.

Blum A (1984). The possible role of tobacco cigarette smoking in hyponatremia of long-term psychiatric patients. JAMA 252:2864–2865.

Blum A, & Friedland GW (1983). Urinary tract abnormalities due to chronic psychogenic polydipsia. Am J Psychiatry 140:915–916.

Blum A, Tempey F, & Lynch W (1983). Somatic findings in patients with psychogenic polydipsia. J Clin Psychiatry 44:55–56.

Borton R, Gaviria M, & Batlle DC (1987). Prevalence, pathogenesis and treatment of renal dysfunction associated with chronic lithium therapy. Am J Kidney Dis 10:329–345.

Braddock L (1986). The dexamethasone suppression test: Fact and artefact. Br J Psychiatry 148:363–374.

Brambilla F, Smeraldi E, Sacchetti E, et al. (1978). Deranged anterior pituitary responsiveness to hypothalamic hormones in depressed patients. Arch Gen Psychiatry 35:1231–1238.

Bravo EL, & Gifford RW (1984). Pheochromocytoma: Diagnosis, localization and management. N Engl J Med 311:1298–1303.

Bravo EL, Tarazi RC, Fouad FM, et al. (1981). Clonidine suppression test. A useful aid in the diagnosis of pheochromocytoma. N Engl J Med 305:623–626.

Bravo EL, Tarazi RC, Gifford RW, et al. (1979). Circulating and urinary catecholamines in pheochromocytoma. Diagnostic and pathophysiologic implications. N Engl J Med 301:682–686.

Brewer GJ, & Yuzbasiyan-Gurkan V (1989). Wilson's disease: An update, with emphasis on new approaches to treatment. Dig Dis 7:178–193.

Brotman AW, Rigotti N, & Herzog D (1985). Medical complications of eating disorders: Outpatient evaluation and management. Compr Psychiatry 26:258–272.

Brown RP, Kocsis JH, & Cohen SK (1983). Delusional depression and inappropriate antidiuretic hormone secretion. Biol Psychiatry 18:1059–1063.

Brown WA, & Laughren TP (1981). Tolerance to the prolactin-elevating effect of neuroleptics. Psychiatry Res 5:317–322.

Burke AW (1972). Physical illness in psychiatric hospitals in Jamaica. Br J Psychiatry 121:321–322.

Burke AW (1978). Physical disorder among day hospital patients. Br J Psychiatry 133:22–26.

BYE AME, NUNN KP, & WILSON J (1985). Prolactin and seizure activity. Arch Dis Child 60:848–851.

CAMPBELL IA (1987). Aggressive psychosis in AIDS patient on high-dose steroids. Lancet 2:750–751.

CARMEL R (1988). Pernicious anemia—the expected findings of very low serum cobalamin levels, anemia, and macrocytosis are often lacking. Arch Intern Med 148:1712–1714.

CARMEL R, WEINER JM, & JOHNSON CS (1987). Iron deficiency occurs frequently in patients with pernicious anemia. JAMA 257:1081–1083.

CAROFF SN (1980). The neuroleptic malignant syndrome. J Clin Psychiatry 41:79–83.

CARROLL BJ (1982). The dexamethasone suppression test for melancholia. Br J Psychiatry 140:292–304.

CARROLL BJ (1985). Dexamethasone suppression test: A review of contemporary confusion. J Clin Psychiatry 46(2, part 2):13–24.

CARROLL BJ (1986). Informed use of the dexamethasone suppression test. J Clin Psychiatry 47(1, suppl):10–12.

CARROLL BJ, CURTIS GC, & MENDELS J (1976a). Neuroendocrine regulation in depression, I: Limbic system—adrenocortical dysfunction. Arch Gen Psychiatry 33:1039–1044.

CARROLL BJ, CURTIS GC, & MENDELS J (1976b). Neuroendocrine regulation in depression, II: Discrimination of depressed from non-depressed patients. Arch Gen Psychiatry 33:1051–1058.

CARROLL BJ, MARTIN FIR, & DAVIES BM (1968). Resistance to suppression by dexamethasone of plasma 11-OHCS levels in severe depressive illness. Br Med J 3:285–287.

CASSCELLS W, SCHOENBERGER A, & GRABOYS TB (1978). Interpretation by physicians of clinical laboratory results. N Engl J Med 299:999–1001.

CEBUL RD, & BECK JR (1987). Biochemical profiles—applications in ambulatory screening and preadmission testing of adults. Ann Intern Med 106:403–413.

CHANARIN I (1979). The megaloblastic anemias (2nd ed.). Oxford, England: Blackwell Scientific Publications.

CHANARIN I, & WATERS DAW (1974). Failed Schilling tests. Scand J Haematol 12:245–248.

CHATTORAJ SC, & WATTS NB (1987). Endocrinology. In NW TIETZ (Ed.), Fundamentals of clinical chemistry (3rd ed., Ch. 18, p. 604). Philadelphia: WB Saunders.

CHESSON AL, KASARSKIS EJ, & SMALL VW (1983). Postictal elevation of serum creatine kinase level. Arch Neurol 40:315–317.

CHICK J, KREITMAN N, & PLANT M (1981). Mean cell volume and gamma-glutamyl-transpeptidase as markers of drinking in working men. Lancet 1:1249–1251.

CHIN WW, COOPER DS, CRAPO L, ET AL. (1976). Water intoxication caused by smoking in a compulsive water drinker [abstract]. Clin Res 24:625A.

CHINN TA (1974). Compulsive water drinking—a review of the literature and an additional case. J Nerv Ment Dis 158:78–80.

CHO JT, BONE S, DUNNER DL, ET AL. (1979). The effect of lithium treatment on thyroid function in patients with primary affective disorder. Am J Psychiatry 136:115–116.

CHOPRA IJ, HERSHMAN JM, PARDRIDGE WM, ET AL. (1983). Thyroid function in nonthyroidal illnesses. Ann Intern Med 98:946–957.

CHOPRA IJ, VAN HERLE AJ, TECO GNC, ET AL. (1980). Serum free thyroxine in thyroidal and nonthyroidal illnesses: A comparison of measurements by radioimmunoassay, equilibrium dialysis, and free thyroxine index. J Clin Endocrinol Metab 51:135–143.

CHRISTIANSEN S, KUSANO E, YUSUFI ANK, ET AL. (1985). Pathogenesis of nephrogenic diabetes insipidus due to chronic administration of lithium in rats. J Clin Invest 75:1869–1879.

CHUNG BC, CHOO HYP, KIM TW, ET AL. (1990). Analysis of anabolic steroids using GC/MS with selected ion monitoring. J Anal Toxicol 14:91–101.

COHEN JA, & KAPLAN MM (1979). The SGOT/SGPT ratio—an indicator of alcoholic liver disease. Am J Dig Dis 24:835–838.

COLGAN J, & PHILPOT M (1985). The routine use of investigations in elderly psychiatric patients. Age Aging 14:163–167.

COLLINS WCJ, LANIGAN O, & CALLAGHAN N (1983). Plasma prolactin concentrations following epileptic and pseudoseizures. J Neurol Neurosurg Psychiatry 46:505–508.

CONN HO, & LIEBERTHAL MM (1978). The hepatic coma syndromes and lactulose. Baltimore: Williams & Wilkins.

COPPEN A, RAMA RAO VA, BISHOP M, ET AL. (1980). Neuroendocrine studies in affective disorders. J Affective Disord 2:311–320.

CRONIN RE (1987). Psychogenic polydipsia with hyponatremia: Report of eleven cases. Am J Kidney Dis 9:410–416.

DAVIES WD (1965). Physical illness in psychiatric outpatients. Br J Psychiatry 111:27–37.

DEFRONZO RA, GOLDBERG M, & AGUS ZS (1976). Normal diluting capacity in hyponatremic patients: Reset osmostat or a variant of SIADH. Ann Intern Med 84:538–542.

DENING TR, & BERRIOS GE (1989). Wilson's disease—psychiatric symptoms in 195 cases. Arch Gen Psychiatry 46:1126–1134.

DERITIS F, COLTORTI M, & GIUSTI G (1972). Serum-transaminase activities in liver disease. Lancet 1:685–687.

DEVGUN MS, DUNBAR JA, HAGART J, ET AL. (1985). Effects of acute and varying amounts of alcohol consumption on alkaline phosphatase, aspartate transaminase, and gamma-glutamyltransferase. Alcoholism Clin Exp Res 9:235–237.

DEWOLFF FA, DEHAAS EJM, & VERWEIJ M (1981). A screening method for establishing laxative abuse. Clin Chem 27:914–917.

DIEHL AM, POTTER J, BOITNOTT J, ET AL. (1984). Relationship between pyridoxal 5′-phosphate deficiency and aminotransferase levels in alcoholic hepatitis. Gastroenterology 86:632–636.

DOLAN JG, & MUSHLIN AI (1985). Routine laboratory testing for medical disorders in psychiatric inpatients. Arch Intern Med 145:2085–2088.

DOSCHERHOLMEN A, & HAGEN PS (1957). A dual mechanism of vitamin B_{12} plasma absorption. J Clin Invest 36:1551–1557.

DOUSA TP (1974). Interaction of lithium with vasopressin-sensitive cyclic AMP system of human renal medulla. Endocrinology 95:1359–1366.

DUBOVSKY SL, GRABON S, BERL T, ET AL. (1973). Syndrome of inappropriate secretion of antidiuretic hormone with exacerbated psychosis. Ann Intern Med 79:551–554.

DURBRIDGE TC, EDWARDS F, EDWARDS RG, ET AL. (1976). An evaluation of multiphasic screening on admission to hospital: Precis of a report to the National Health and Medical Research Council. Med J Aust 1:703–705.

EASTWOOD MR (1975). The relation between physical and mental illness. Toronto: University of Toronto Press.

EASTWOOD MR, MINDHAM RHS, & TENNENT TG (1970). The physical status of psychiatric emergencies. Br J Psychiatry 116:545–550.

EDWIN E, HOLTEN K, NORMUM KR, ET AL. (1965). Vitamin B_{12} hypovitaminosis in mental diseases. Acta Med Scand 177:689–699.

EISER AR, NEFF MS, & SLIFKIN RF (1982). Acute myoglobinuric renal failure—a consequence of the neuroleptic malignant syndrome. Arch Intern Med 142:601–603.

EMERSON CH, DYSON WL, & UTIGER RD (1973). Serum thyrotropin and thyroxine concentrations in patients receiving lithium carbonate. J Clin Endocrinol Metab 36:338–346.

EVERED DC, ORMSTON JB, SMITH PA, ET AL. (1973). Grades of hypothyroidism. Br Med J 1:657–662.

FAGLIA G, PARACCHI A, FERRARI C, ET AL. (1980). Use of neuroactive drugs and hypothalamic regulatory hormones in the diagnosis of hyperprolactinemic states. In E MULLER (Ed.), Neuroactive drugs in endocrinology (Vol. 9; pp. 262–278). Amsterdam: Elsevier.

FAIRBANKS VF, WAHNER HW, & PHYLIKY RL (1983). Tests for pernicious anemia: The "Schilling test." Mayo Clin Proc 58:541–544.

FERGUSON B, & DUDLESTON K (1986). Detection of physical disorder in newly admitted psychiatric patients. Acta Psychiatr Scand 74:485–489.

FISCHER JE, & BALDESSARINI RJ (1975). Pathogenesis and treatment of hepatic coma. In H POPPER & F SCHAFFNER (Eds.), Progress in liver diseases (pp. 363–397). New York: Grune & Stratton.

FLEESON WP, & WENK RE (1970). Pitfalls of mass chemical screening. Postgrad Med 48:59–64.

FRANCIS J, MARTIN D, & KAPOOR WN (1990). A prospective study of delirium in hospitalized elderly. JAMA 263:1097–1101.

FRANKENBURG W, & CAMP, B (1975). Criteria in screening test selection. In Pediatric screening tests (Ch. 2, pp. 23–37). Springfield, IL: Charles C Thomas.

FRANKS S, MURRAY MAF, JEQUIER AM, ET AL. (1975). Incidence and significance of hyperprolactinemia in women with amenorrhea. Clin Endocrinol 4:597–607.

FULLINFAW RO, BURY RW, & MOULDS RFW (1987). Liquid chromatographic screening of diuretics in urine. J Chromatogr 415:347–356.

GABEL S, & HSU G (1986). Routine laboratory tests in adolescent psychiatric inpatients: Their value in making psychiatric diagnoses and in detecting medical disorders. J Am Child Psychiatry 25:113–119.

GABOW PA, KAEHNY WD, & KELLEHER SP (1982). The spectrum of rhabdomyolysis. Medicine 61:141–152.

GARBUTT JC, LOOSEN PT, TIPERMAS A, ET AL. (1983). The TRH test in borderline personality disorder. Psychiatry Res 9:107–113.

GILBERTSTADT SJ, GILBERTSTADT H, ZIEVE L, ET AL. (1980). Psychomotor performance defects in patients without overt encephalopathy. Arch Intern Med 140:519–521.

GITLIN MJ, & GERNER RH (1986). The dexamethasone suppression test and response to somatic treatment: A review. J Clin Psychiatry 47:16–21.

GITLOW SE, MENDELOWITZ M, & BERTANI LM (1970). The biochemical techniques for detecting and establishing the presence of a pheochromocytoma: A review of ten years' experience. Am J Cardiol 26:270–279.

GJERDE H, JOHNSEN J, BJORNEBOE A, ET AL. (1988). A comparison of serum carbohydrate-deficient transferrin with other biological markers of excessive drinking. Scand J Clin Lab Invest 48:1–6.

GLASSER L, STERNGLANZ PD, COMBIE J, ET AL. (1973). Serum osmolality and its applicability to drug overdose. Am J Clin Pathol 60:695–699.

GLOTZNER PFL (1979). Creatine kinase in serum after generalized seizures. Arch Neurol 36:661.

GOLD MS, & DACKIS CA (1986). Role of the laboratory in the evaluation of suspected drug abuse. J Clin Psychiatry 47(1, suppl):17–23.

GOLD MS, POTTASH AL, EXTEIN I, ET AL. (1981). The TRH test in the diagnosis of major and minor depression. Psychoneuroendocrinology 6:159–169.

GOLDMAN MB, LUCHINS DJ, & ROBERTSON GL (1988). Mechanisms of altered water metabolism in psychotic patients with polydipsia and hyponatremia. N Engl J Med 318:397–403.

GOLDSTEIN NP, EWERT JC, RANDALL RV, ET AL. (1968). Psychiatric aspects of Wilson's disease (hepatolenticular degeneration): Results of psychometric tests during long-term therapy. Am J Psychiatry 124:1555–1561.

GOODE DJ, WEINBERG DH, MAZURA TA, ET AL. (1977). Effect of limb restraints on serum creatine phosphokinase activity in normal volunteers. Biol Psychiatry 12:743–755.

GORDIN A, & LAMBERG BA (1981). Spontaneous hypothyroidism in symptomless autoimmune thyroiditis: A long term follow up study. Clin Endocrinol 53:537–543.

GREEN R, KUHL W, JACOBSON R, ET AL. (1982). Masking of macrocytosis by alpha thalassemia in blacks with pernicious anemia. N Engl J Med 307:1322–1325.

GRUEN PH, SACHAR EJ, LANGER G, ET AL. (1978). Prolactin responses to neuroleptics in normal and schizophrenic subjects. Arch Gen Psychiatry 35:108–116.

GWIRSTMAN HE, KAYE WH, GEORGE DT, ET AL. (1989). Hyperamylasemia and its relationship to binge-purge episodes: Development of a clinically relevant laboratory test. J Clin Psychiatry 50:196–204.

GWIRSTMAN HE, YAGER J, GILLARD BK, ET AL. (1986). Serum amylase and its isoenzymes in normal weight bulimia. Int J Eating Disord 5:355–361.

HAGGERTY JJ, GARBUTT JC, EVANS DL, ET AL. (1990). Subclinical hypothyroidism: A review of neuropsychiatric aspects. Int J Psychiatry Med 20:193–208.

HAIGLER ED JR, HERSHMAN JM, & PITTMAN JA JR (1972). Response to orally administered synthetic thyrotropin-releasing hormone in man. J Clin Endocrinol Metab 35:631–635.

HALL RCW, GARDNER ER, POPKIN MK, ET AL. (1981). Unrecognized physical illness prompting psychiatric admission: A prospective study. Am J Psychiatry 138:629–635.

HALL RCW, GARDNER ER, STICKNEY SK, ET AL. (1980). Physical illness manifesting as psychiatric disease II. Analysis of a state hospital inpatient population. Arch Gen Psychiatry 37:989–995.

HALL RCW, HOFFMAN RS, BERESFORD TP, ET AL. (1989). Physical illness encountered in patients with eating disorders. Psychosomatics 30:174–191.

HALL RCW, POPKIN MK, DEVAUL RA, ET AL. (1978). Physical illness presenting as psychiatric disease. Arch Gen Psychiatry 35:1315–1320.

HAMPTON JR, HARRISON MJG, MITCHELL JRA, ET AL. (1975). Relative contributions of history-taking, physical examination, and laboratory investigation to diagnosis and management of medical outpatients. Br Med J 2:486–489.

HARIPRASAD MK, EISINGER RP, & NADLER IM (1980). Hyponatremia in psychogenic polydipsia. Arch Intern Med 140:1639–1642.

HATHCOCK JM, & TROENDLE GJ (1991). Oral cobalamin for treatment of pernicious anemia. JAMA 265:96–97.

HERBERT V (1987). The 1986 Herman Award Lecture. Nutrition science as a continually unfolding story: The folate and vitamin B_{12} paradigm. Am J Clin Nutr 46:387–402.

HERBERT V (1988). Cobalamin deficiency and neuropsychiatric disorders [letter]. N Engl J Med 319:1733–1734.

HERRIGE CF (1960). Physical disorders in psychiatric illness: A study of 209 consecutive admissions. Lancet 2:949–951.

HERZLICH B, & HERBERT V (1988). Depletion of serum holotranscobalamin II—an early sign of negative vitamin B_{12} balance. Lab Invest 58:332–337.

HEYMA P, LARKINS RG, PERRY-KEENE D, ET AL. (1977). Thyroid hormone levels and protein binding in patients on long term diphenylhydantoin treatment. Clin Endocrinol 6:369–376.

HOLMES JM (1956). Cerebral manifestations of vitamin-B_{12} deficiency. Br Med J 2:1394–1398.

HORWITZ DL (1989). Factitious and artifactual hypoglycemia. Endocrinol Metab Clin North Am 18:203–210.

HUGHES J, & BARRACLOUGH BM (1980). Value of routine chest radiography of psychiatric patients. Br Med J 281:1461–1462.

HUMPHRIES LL, ADAMS LJ, ECKFELDT JH, ET AL. (1987). Hyperamylasemia in patients with eating disorders. Ann Intern Med 106:50–52.

HUNTER R, JONES M, JONES TG, ET AL. (1967). Serum B_{12} and folate concentrations in mental patients. Br J Psychiatry 113:1291–1295.

HUNTER R, & MATTHEWS DM (1965). Mental symptoms in vitamin B_{12} deficiency. Lancet 2:738.

INK O, BOUTRON A, HANNY P, ET AL. (1989). Place du taux serique de l'aspartate aminotransferase mitochondriale comme marquer d'intoxication chez le cirrhotique alcoolique. Presse Med 18:111–114.

INOSE T (1968). Neuropsychiatric manifestations in Wilson's disease: Attacks of disturbance of consciousness. Birth Defects 4(2):74–76.

INSEL TR, & GOODWIN FK (1983). The dexamethasone suppression test: Promises and problems of diagnostic laboratory tests in psychiatry. Hosp Community Psychiatry 34:1131–1138.

JACOB E, BAKER SJ, & HERBERT V (1980). Vitamin B_{12} binding proteins. Physiol Rev 60:918–960.

JAMES JH, ZIPARO V, JEPPSON B, ET AL. (1979). Hyperammonemia, plasma amino acid imbalance, and blood-brain amino acid transport: A unified theory of portal systemic encephalopathy. Lancet 2:772–775.

JELLINEK EH (1962). Fits, faints, coma, and dementia in myxoedema. Lancet 2:1010–1012.

JOHNSON DAW (1968). The evaluation of routine physical examination in psychiatric cases. Practitioner 200:686–691.

JONES DH, REID JL, HAMILTON CA, ET AL. (1980). The biochemical diagnosis, localization and follow-up of phaeochromocytoma: The role of plasma and urinary catecholamine measurements. Q J Med 49:341–361.

JONES EA, SKOLNICK P, GAMMAL SH, ET AL. (1989). The gamma-aminobutyric acid A ($GABA_A$) receptor complex and hepatic encephalopathy. Ann Intern Med 110:532–546.

JOSE CJ, & PEREZ-CRUET J (1979). Incidence and morbidity of self-induced water intoxication in state mental hospital patients. Am J Psychiatry 13:221–222.

KARLBERG BE, KJELLMAN BF, & KAGEDOL B (1978). Treatment of endogenous depression with oral thyrotropin. Acta Psychiatr Scand 58:389–400.

KATHOL RG, WILCOX JA, TURNER RD, ET AL. (1986). Pharmacologic approaches to psychogenic polydipsia: Case reports. Prog Neuropsychopharmacol Biol Psychiatry 10:95–100.

KELLNER R, BUCKMAN MT, FAVA GA, ET AL. (1984). Hyperprolactinemia, distress, and hostility. Am J Psychiatry 141:759–763.

KEPTEIN EM, MACINTYRE SS, WEINER JM, ET AL. (1981). Free thyroxine estimates in nonthyroidal illness: Comparison of eight methods. J Clin Endocrinol Metab 52:1073–1077.

KIELBLOCK AJ, MANJOO M, BOOYENS J, ET AL. (1979). Creatine phosphokinase and lactate dehydrogenase levels after ultra long distance running. S Afr Med J 55:1061–1064.

KIRKEGAARD C, BJORUM N, COHN D, ET AL. (1978). TRH stimulation test in manic depressive disease. Arch Gen Psychiatry 35:1017–1023.

KLEE GG, & HAY ID (1986). Assessment of sensitive thyrotropin assays for an expanded role in thyroid function testing: Proposed criteria for analytic performance and clinical utility. J Clin Endocrinol Metab 64:461–471.

KLEIN MS, SHELL WE, & SOBEL BE (1973). Serum creatine phosphokinase (CPK) isoenzymes after intramuscular injections, surgery, and myocardial infarction. Experimental and clinical studies. Cardiovasc Res 7:412–418.

KLETZKY OA (1984). Diagnostic approaches to hyperprolactinemic states. Semin Reprod Endocrinol 2:23–30.

KOLMAN PBR (1984). The value of laboratory investigations of elderly psychiatric patients. J Clin Psychiatry 45:112–116.

KOLMAN PBR (1985). Predicting the results of routine laboratory tests in elderly psychiatric patients admitted to hospital. J Clin Psychiatry 46:532–534.

KORAN LM, SOX HC, MARTON KI, ET AL. (1989). Medical evaluation of psychiatric patients. I. Results in a state mental health system. Arch Gen Psychiatry 46:733–740.

KORANYI EK (1972). Physical health and illness in a psychiatric outpatient department population. Can Psychiatr Assoc J 17(suppl):109–116.

KORANYI EK (1979). Morbidity and rate of undiagnosed physical illness in a psychiatric clinic population. Arch Gen Psychiatry 36:414–419.

KORANYI EK (1980). Somatic illness in psychiatric patients. Psychosomatics 21:887–891.

KORVIN CC, PEARCE RH, & STANLEY J (1975). Admission screening: Clinical benefits. Ann Intern Med 83:197–203.

KRISTENSON H, & TRELL E (1982). Indicators of alcohol consumption: Comparisons between a questionnaire (Mm-MAST), interviews and serum gamma-glutamyl transferase (GGT) in a health survey of middle-aged males. Br J Addict 77:297–304.

LADENSON PW (1985). Diseases of the thyroid gland. Clin Endocrinol Metab 14:145–173.

LAHEY FH (1931). Nonactivated (apathetic) type of hypothyroidism. N Engl J Med 204:747–748.

LANGDON FW (1905). Nervous and mental manifestations of prepernicious anemia. JAMA 45:1635–1642.

LARSEN PR, ATKINSON AJ JR, WELLMAN HN, ET AL. (1970). The effect of diphenylhydantoin on thyroxine metabolism in man. J Clin Invest 49:1266–1279.

LAXER KD, MULLHOOLY JP, & HOWELL B (1985). Prolactin changes after seizures classified by EEG monitoring. Neurology 35:31–35.

LAZARUS JH (1986). Endocrine and metabolic effects of lithium (pp. 99–124). New York: Plenum.

LEDERLE FA (1991). Oral cobalamin for pernicious anemia. Medicines' best kept secret? JAMA 265:94–95.

LEGGETT BA, POWELL LW, & HALLIDAY JW (1989). Laboratory markers of alcoholism. Dig Dis 7:125–134.

LEVY RP, JENSEN JB, LAUS VG, ET AL. (1981). Serum thyroid hormone abnormalities in psychiatric disease. Metabolism 30:1060–1064.

LIDDLE GW (1960). Test of pituitary-adrenal suppression in the diagnosis of Cushing's syndrome. J Clin Endocrinol Metab 20:1539–1560.

LINDENBAUM J, HEALTON EB, SAVAGE DG, ET AL. (1988). Neuropsychiatric disorders caused by cobalamin deficiency in the absence of anemia or macrocytosis. N Engl J Med 318:1720–1728.

LISHMAN AW (1987). Organic psychiatry. The psychological consequences of cerebral disorder. London: Blackwell Scientific Publications.

LISTON EH, GERNER RH, ROBERTSON AG, ET AL. (1979). Routine thoracic radiography for psychiatric inpatients. Hosp Community Psychiatry 30:474–476.

LOOSEN PT (1985). The TRH-induced TSH response in psychiatric patients: A possible neuroendocrine marker. Psychoneuroendocrinology 10:237–260.

LOOSEN PT, & PRANGE AJ JR (1982). Serum thyrotropin response to thyrotropin-releasing hormone in psychiatric patients: A review. Am J Psychiatry 139:405–416.

LOOSEN PT, PRANGE AJ JR, & WILSON IC (1979). TRH (protirelin) in depressed alcoholic men: Behavioral changes and endocrine responses. Arch Gen Psychiatry 36:540–547.

LUZECKY MH (1974). The syndrome of inappropriate secretion of antidiuretic hormone associated with amitriptyline administration. South Med J 67:495–497.

MA KW, BROWN DC, STEELE BW, ET AL. (1981). Serum creatine kinase MB isoenzyme activity in long-term hemodialysis patients. Arch Intern Med 141:164–166.

MAARBJERG K, VESTERGAARD P, & SCHOU M (1987). Changes in serum thyroxine (T4) and serum thyroid stimulating hormone (TSH) during prolonged lithium treatment. Acta Psychiatr Scand 75:217–221.

MACDOUGALL IC, ISLES CG, STEWART H, ET AL. (1988). Overnight clonidine suppression test in the diagnosis and exclusion of pheochromocytoma. Am J Med 84:993–1000.

MAGUIRE GP, & GRANVILLE-GROSSMAN KL (1968). Physical illness in psychiatric patients. Br Med J 115:1365–1369.

MANGER WM, GIFFORD RW JR, & HOFFMAN BB (1985). Pheochromocytoma: A clinical and experimental overview. Curr Probl Cancer 9:1–67.

MANNO BR, & MANNO JE (1977). Toxicology of Ipecac: A review. Clin Toxicol 10:221–242.

MARSHALL H (1949). Incidence of physical disorders among psychiatric inpatients. Br Med J 2:468–470.

MATLOFF DS, SELINGER MJ, & KAPLAN MM (1980). Hepatic transaminase activity in alcoholic liver disease. Gastroenterology 78:1389–1392.

MATSUOKA I, NAKAI T, MIYAKE M, ET AL. (1986). Effects of bromocriptine on neuroleptic-induced amenorrhea, galactorrhea and impotence. Jpn J Psychiatry Neurol 40:639–646.

MATUK F, & KALYANARAMAN K (1977). Syndrome of inappropriate secretion of antidiuretic hormone in patients treated with psychotherapeutic drugs. Arch Neurol 34:374–375.

MCCLAIN CJ, ZIEVE L, & DIOZAKE WM (1980). Blood methanethiol in alcoholic liver diseases with and without subacute encephalopathy. Gut 21:318–328.

MCCORMICK DB (1987). Vitamins. In NW TIETZ (Ed.), Fundamentals of clinical chemistry (3rd ed., Ch. 16, pp. 505–506). Philadelphia: WB Saunders.

MCNEIL BJ, KEELER E, & ADELSTEIN SJ (1975). Primer on certain elements of medical decision making. N Engl J Med 293:211–215.

MEADOR-WOODRUFF JH, & GREDEN JF (1988). Effects of psychotropic medications on hypothalamic-pituitary-adrenal regulation. Endocrinol Metab Clin North Am 17:225–234.

MELTZER HY (1976). Neuromuscular dysfunction in schizophrenia. Schizophr Bull 2:106–135.

MELTZER HY, & FANG VS (1976). The effect of neuroleptics on serum prolactin in schizophrenic patients. Arch Gen Psychiatry 33:279–286.

MELTZER HY, GOODE DJ, SCHYVE PM, ET AL. (1979). Effect of clozapine in human serum prolactin levels. Am J Psychiatry 136:1550–1555.

MITCHELL JE, SEIM HC, COLON E, ET AL. (1987). Medical complications and medical management of bulimia. Ann Intern Med 107:71–77.

MORGAN MY, MILSOM, JP, & SHERLOCK S (1978). Plasma ratio of valine, leucine and isoleucine to phenylalanine and tyrosine in liver disease. Gut 19:1068–1073.

MORLEY JE, & SHAFER RB (1982). Thyroid function screening in new psychiatric admissions. Arch Intern Med 142:591–593.

MORLEY JE, SHAFER RB, ELSON MK, ET AL. (1980). Amphetamine-induced hyperthyroxinemia. Ann Intern Med 93:707–709.

MOSS DW, HENDERSON AR, & KACHMAR JF (1987). Enzymes. In NW TIETZ (Ed.), Fundamentals of clinical chemistry (3rd ed., Ch. 12, pp. 374–375). Philadelphia: WB Saunders.

NALPAS B, VASSAULT A, CHARPIN S, ET AL. (1986). Serum mitochondrial aspartate aminotransferase as a marker of chronic alcoholism: Diagnostic value and interpretation in a liver unit. Hepatology 6:608–614.

NALPAS B, VASSAULT A, LE GUILLOU A, ET AL. (1984). Serum activity of mitochondrial aspartate aminotransferase: A sensitive marker of alcoholism with or without alcoholic hepatitis. Hepatology 4:893–896.

NARINS RG, JONES ER, STOM MC, ET AL. (1982). Diagnostic strategies in disorders of fluid, electrolyte and acid-base homeostasis. Am J Med 72:496–520.

NEMEROFF CB, & EVANS DL (1989). Thyrotropin-releasing hormone (TRH) the thyroid axis and affective disorder. Ann NY Acad Sci 553:304–310.

NICOLL CD, & GERARD SK (1985). Diagnosis of pheochromocytoma [letter]. N Engl J Med 312:721.

NICOLOFF JT (1989). Thyroid function in nonthyroidal disease. In L DEGROOT (Ed.), Endocrinology (2nd ed., Ch. 42, pp. 640–645). Philadelphia: WB Saunders.

NIXON RA, ROTHMAN JS, & CHIN W (1982). Demeclocycline in the prophylaxis of self-induced water intoxication. Am J Psychiatry 139:828–830.

OLERUD JE, HOMER LD, & CARROLL HW (1976). Incidence of acute exertional rhabdomyolysis. Serum myoglobin and enzyme levels as indicators of muscle injury. Arch Intern Med 136:692–697.

OLIVIER LR, DE WAAL A, RETIEF FJ, ET AL. (1978). Electrocardiographic and biochemical studies on marathon runners. S Afr Med J 53:783–787.

OS I, & LYNGDAL PT (1989). General convulsions and rhabdomyolysis. Acta Neurol Scand 79:246–248.

PAPPAS SC, & JONES EA (1983). Methods for assessing hepatic encephalopathy. Semin Liv Dis 3:298–307.

PARK J, PARK S, LHO D, ET AL. (1990). Drug testing at the 10th Asian Games and the 24th Seoul Olympic Games. J Anal Toxicol 14:66–72.

PEARLMAN CA (1986). Neuroleptic malignant syndrome: A review of the literature. J Clin Psychopharmacol 6:257–273.

PHILLIPS RJ (1937). Physical disorder in 164 consecutive admissions to a mental hospital: The incidence and significance. Br Med J 2:363–366.

PLOUIN PJ, DUCLOS JM, MENARD J, ET AL. (1981). Biochemical tests for diagnosis of pheochromocytoma: Urinary versus plasma determinations. Br Med J 282:853–854.

POPE HG, & KATZ DL (1988). Affective and psychotic symptoms associated with anabolic steroid use. Am J Psychiatry 145:487–490.

POTTASH ALC, GOLD MS, & EXTEIN I (1986). The use of the clinical laboratory. In LI SEDERER (Ed.), Inpatient psychiatry—diagnosis and treatment (2nd ed., Ch. 9, pp. 197–217). Baltimore: Williams & Wilkins.

PRIEST JB, OEI TO, & MOOREHEAD WR (1982). Exercise-induced changes in common laboratory tests. Am J Clin Pathol 77:285–289.

PULLARKAT RK, & RAGUTHU S (1985). Elevated urinary dolichol levels in chronic alcoholics. Alcoholism Clin Exp Res 9:28–30.

RAMSAY I (1985). Drug and non-thyroid induced changes in thyroid function tests. Postgrad Med J 61:375–377.

RANSON JHC, RIFKIND KM, ROSES F, ET AL. (1974). Prognostic signs and the role of operative management in acute pancreatitis. Surg Gynecol Obstet 139:69–81.

RASKIND MA, ORENSTEIN H, & CHRISTOPHER TG (1975). Acute psychosis, increased water ingestion, and inappropriate antidiuretic hormone secretion. Am J Psychiatry 132:907–910.

RATH CE, MCCURDY PR, & DUFFY BJ (1957). Effect of renal disease on the Schilling test. N Engl J Med 256:111–114.

REFETOFF S (1989). Thyroid function tests and effects of drugs on thyroid function. In LJ DeGroot (Ed.), Endocrinology. (Ch. 41, pp. 590–639). Philadelphia: WB Saunders.

Rej R (1978). Aspartate aminotransferase activity and isoenzyme proportions in human liver tissues. Clin Chem 24:1971–1979.

Rej R (1980). An immunochemical procedure for determination of mitochondrial aspartate aminotransferase in human serum. Clin Chem 26:1692–1700.

Remine WH, Chong GC, Van Heerden JA, et al. (1974). Current management of pheochromocytoma. Ann Surg 179:740–748.

Rendell M, McGrane D, & Cuesta M (1978). Fatal compulsive water drinking. JAMA 240:2557–2559.

Roca RP, Breakey WR, & Fischer PJ (1987). Medical care of chronic psychiatric outpatients. Hosp Community Psychiatry 38:741–745.

Roine RP, Eriksson JP, Ylikahri R, et al. (1989). Methanol as a marker of alcohol abuse. Alcoholism Clin Exp Res 13:172–175.

Roitt IM, Donaich D, & Shapland C (1964). Intrinsic factor autoantibodies. Lancet 2:469–470.

Rosebush P, & Stewart T (1989). A prospective analysis of 24 episodes of neuroleptic malignant syndrome. Am J Psychiatry 146:717–725.

Rosenbaum JF, Rothman JS, & Murray GB (1979). Psychosis and water intoxication. J Clin Psychiatry 40:287–291.

Rothenberg SP, Kantha KRK, & Ficarra A (1971). Autoantibodies to intrinsic factor: Their determination and clinical usefulness. J Lab Clin Med 77:476–484.

Rubin RT, Poland RE, O'Connor D, et al. (1976). Selective neuroendocrine effects of low dose haloperidol in normal adult men. Psychopharmacology 47:135–140.

Sackett D (1973). The usefulness of laboratory tests in health screening programs. Clin Chem 19:366–372.

Salata R, & Klein I (1987). Effects of lithium on the endocrine system: A review. J Lab Clin Med 110:130–136.

Saxon AJ, Calsyn DA, Haver VA, et al. (1988). Clinical evaluation and use of urine screening for drug abuse. West J Med 149:296–303.

Scarlett JA, Mako ME, Rubenstein AH, et al. (1977). Factitious hypoglycemia. Diagnosis by measurement of serum C-peptide immunoreactivity and insulin-binding antibodies. N Engl J Med 297:1029–1032.

Scheinberg IH, Sternleib I, & Richman J (1968). Psychiatric manifestations in patients with Wilson's disease. Birth Defects 4:85–87.

Schilling RF (1953). Intrinsic factor studies. II. The effect of gastric juice on the urinary excretion of radioactivity after the oral administration of radioactive vitamin B_{12}. J Lab Clin Med 48:860–866.

Shapiro B, & Fig LM (1989). Management of pheochromocytoma. Endocrinol Metab Clin North Am 18:443–479.

Shapiro MF, Lehman AF, & Greenfield S (1983). Biases in the laboratory diagnosis of depression in medical practice. Arch Intern Med 143:2085–2088.

Sherman L, Fisher A, Klass E, et al. (1984). Pharmacologic causes of hyperprolactinemia. Semin Reprod Endocrinol 2:31–34.

Shevitz SA, Jamieson RC, Petrie WM, et al. (1980). Compulsive water drinking treated with high dose propranolol. J Nerv Ment Dis 168:246–248.

Shulman R (1967a). Psychiatric aspects of pernicious anemia: A prospective controlled investigation. Br Med J 3:266–270.

Shulman R (1967b). A survey of vitamin B_{12} in an elderly psychiatric population. Br J Psychiatry 113:241–251.

Shulman R (1967c). Vitamin B_{12} deficiency and psychiatric illness. Br J Psychiatry 113:252–256.

Shumate JB, Brooke MH, & Carroll JE (1979). Increased serum creatine kinase after exercise: A sex linked phenomenon. Neurology 29:902–904.

Sidell FR, Culver DL, & Kamindkis A (1974). Serum creatine phosphokinase activity after intramuscular injection. The effect of dose, concentration, and volume. JAMA 229:1894–1897.

Smith ADM (1960). Megaloblastic madness. Br Med J 2:1840–1845.

Smith PJ, & Surks MI (1984). Multiple Effects of 5,5'- diphenylhydantoin on the thyroid system. Endocr Rev 5:514–524.

Smith WO, & Clark ML (1980). Self-induced water intoxication in schizophrenic patients. Am J Psychiatry 137:1055–1060.

Snyder PJ, & Utiger RD (1972). Response to thyrotropin releasing hormone (TRH) in normal man. J Clin Endocrinol Metab 54:635–637.

Sox HC, Koran LM, Sox CH, et al. (1989). A medical algorithm for detecting physical disease in psychiatric patients. Hosp Community Psychiatry 40:1270–1276.

Sperling MR, Pritchard PB III, Engel J Jr, et al. (1986). Prolactin in partial epilepsy: An indicator of limbic seizures. Ann Neurol 20:716–722.

Spivak JL (1982). Masked megaloblastic anemia. Arch Intern Med 142:2111–2114.

Spratt DI, Pont A, Miller MB, et al. (1982). Hyperthyroxinemia in patients with acute psychiatric disorders. Am J Med 73:41–48.

Stabler SP, Marcell PD, Podell ER, et al. (1986). Assay of methylmalonic acid in the serum of patients with cobalamin deficiency using capillary gas chromatography-mass spectrometry. J Clin Invest 77:1606–1612.

Stabler SP, Marcell PD, Podell ER, et al. (1988). Elevation of total homocysteine in the serum of patients with cobalamin or folate deficiency detected by capillary gas chromatography-mass spectrometry. J Clin Invest 81:466–474.

Starkman MN, Zelnik TC, Nesse RM, et al. (1985). Anxiety in patients with pheochromocytomas. Arch Gen Psychiatry 145:248–252.

Steiner M, Haskett RF, Carroll BJ, et al. (1984). Plasma prolactin and severe premenstrual tension. Psychoneuroendocrinology 9:29–35.

Sternlieb I, & Scheinberg H (1964). Penicillamine therapy for hepatolenticular degeneration. JAMA 189:748–754.

Stibler H, & Borg S (1981). Evidence of a reduced sialic acid content in serum transferrin in male alcoholics. Alcoholism Clin Exp Res 5:545–549.

Stibler H, Borg S, & Allguander C (1979). Clinical significance of abnormal heterogeneity of transferrin in relation to alcohol consumption. Acta Med Scand 206:275–281.

Stibler H, & Kjellin KG (1976). Isoelectric focusing and electrophoresis of the CSF protein in tremor of different origins. J Neurol Sci 30:269–278.

Stibler H, Sydow O, & Borg S (1980). Quantitative estimation of abnormal microheterogeneity of serum transferrin in alcoholics. Pharmacol Biochem Behav 13:47–51.

Stokes PE (1966). Pituitary suppression in psychiatric patients. In 48th Meeting Abstracts. Chicago: The Endocrine Society (USA).

Storey EL, Anderson GJ, Mack U, et al. (1987). Desialated transferrin as a serological marker of chronic excessive alcohol ingestion. Lancet 1:1292–1294.

Storey EL, Mack U, Powell LW, et al. (1985). Use of chromatofocusing to detect a transferrin variant in serum of alcoholic subjects. Clin Chem 31:1543–1545.

Strachan RW, & Henderson JG (1965). Psychiatric syndromes due to vitaminosis B_{12} with normal blood and marrow. Q J Med 34:303–307.

TAPE TG, & MUSHLIN AI (1986). The utility of routine chest radiographs. Ann Intern Med 104:663–670.

TARGUM SD, GREENBERG RD, & HARMON RL (1984). Thyroid hormone and the TRH stimulation test in refractory depression. J Clin Psychiatry 45:345–346.

TAYLOR JW (1975). Depression in thyrotoxicosis. Am J Psychiatry 132:552–553.

TAYLOR RJ, VON WITT RJ, & FRY AH (1981). Serum creatine phosphokinase activity in psychiatric patients receiving electroconvulsive therapy. J Clin Psychiatry 42:103–105.

THOMAS CJ (1979). The use of screening investigations in psychiatry. Br J Psychiatry 135:67–72.

TURNBRIDGE WMG, BREWIS M, FRENCH JM, ET AL. (1981). Natural history of autoimmune thyroiditis. Br Med J (Clin Res) 282:258–262.

UEHINGER C, & BAUMANN P (1991). Clozapine as an alternative treatment for neuroleptic-induced gynecomastia. Am J Psychiatry 148:392–393.

VESTERGAARD P, AMDISEN A, & SCHOU M (1980). Clinically significant side effects of lithium treatment. A survey of 237 patients in long-term treatment. Acta Psychiatr Scand 62:193–200.

VIEWEG WVR, DAVID JJ, ROWE WT, ET AL. (1987). Hypocalcemia: An additional complication of the syndrome of self-induced water intoxication and psychosis. Psychiatr Med 4:291–297.

WARD MM (1988). Factors predictive of acute renal failure in rhabdomyolysis. Arch Intern Med 148:1553–1557.

WARTOFSKY L, & BURMAN KD (1982). Alterations in thyroid function in patients with systemic illness: The "euthyroid sick syndrome." Endocr Rev 3:164–217.

WASS JAH, & BESSER GM (1989). Tests of pituitary function. In LJ DEGROOT (Ed.), Endocrinology. (Ch. 41, pp. 492–502). Philadelphia: WB Saunders.

WEBER FL, FRESARD KM, & LALLY BR (1982). Effects of lactulose and neomycin on urea metabolism in cirrhotic subjects. Gastroenterology 82:213–217.

WEBER FL, MINCO D, FRESARD KM, ET AL. (1985). Effects of vegetable diets on nitrogen metabolism in cirrhotic subjects. Gastroenterology 89:538–544.

WEISSENBORN K, SCHOLZ M, HINRICHS H, ET AL. (1990). Neurophysiological assessment of early hepatic encephalopathy. Electroencephalogr Clin Neurophysiol 75:289–295.

WENZEL KW (1981). Pharmacological interference with in vitro tests of thyroid function. Metabolism 30:717–732.

WENZEL KW, MEINHOLD M, RAFFENBERG, ET AL. (1974). Classification of hypothyroidism in evaluating patients after radioiodine therapy by serum cholesterol, T3 uptake, total T4, F-T4 index, total T3, basal TSH, and TRH test. Eur J Clin Invest 4:141–148.

WHITE AJ, & BARRACLOUGH B (1989). Benefits and problems of routine laboratory investigations in adult psychiatric admissions. Br J Psychiatry 155:65–72.

WHYBROW PC, PRANGE AJ JR, & TREADWAY CR (1969). Mental changes accompanying thyroid gland dysfunction. Arch Gen Psychiatry 20:48–63.

WILLETT AB, & KING T (1977). Implementation of laboratory screening procedures on a short term psychiatric inpatient unit. Dis Nerv Syst 38:867–870.

WILSON SAK (1912). Progressive lenticular degeneration: A familial nervous disease associated with cirrhosis of the liver. Brain 34:295–509.

WOLTMANN HW (1919). The nervous symptoms in pernicious anemia: An analysis of one hundred and fifty cases. Am J Med Sci 3:400–409.

WYLLIE E. LEUDERS H, PIPPENGER C, ET AL. (1985). Postictal serum creatine kinase in the diagnosis of seizure disorders. Arch Neurol 42:123–126.

YERBY MS, VANBELLE G, FRIEL PN, ET AL. (1987). Serum prolactins in the diagnosis of epilepsy: Sensitivity, specificity, and predictive value. Neurology 37:1224–1226.

ZERBE RL, & ROBERTSON GL (1981). A comparison of plasma vasopressin measurements with a standard indirect test in the differential diagnosis of polyuria. N Engl J Med 305:1539–1546.

ZERBE RL, & Robertson GL (1987). Osmotic and non-osmotic regulation of thirst and vasopressin secretion. In MH MAXWELL, CR KLEEMAN, & RG NARINS (Eds.), Clinical Disorders of fluid and electrolyte metabolism (4th ed., Ch. 4, pp. 61–78). New York: McGraw-Hill.

8 Alcohol and drug abuse in the medical setting

ROBERT M. SWIFT, M.D., PH.D.

The alcohol- or drug-abusing patient presents a special challenge to clinicians who treat medically ill patients. Substance abuse is found in up to one-third of patients presenting to general hospitals, and at least as many patients in ambulatory settings. The presence of chemical dependency may be etiologically related to presenting medical or surgical problems, and it usually complicates the medical and surgical treatment of other illnesses.

Despite the high prevalence of psychoactive substance use among patients, physicians are ill prepared to identify substance abuse or dependence and to treat these conditions. The sources of these deficiencies include poor training in the identification and treatment of substance abuse and dependence, a lack of knowledge about treatment resources, lack of confidence in treatment efficacy, and negative attitudes toward substance abusers as patients (Clark, 1981; Geller et al., 1989; Holden, 1985).

Effective treatment of substance abuse and dependence requires knowledge about therapies for the acute management of intoxicated patients or patients undergoing withdrawal, and knowledge about options for long-term treatment and rehabilitation. This chapter provides basic information on the identification and treatment of substance abuse and dependence in medical patients.

CRITERIA FOR DIAGNOSIS OF SUBSTANCE ABUSE

The description of substance use disorders is complicated by ambiguity in the words used to describe psychoactive substance use. Words such as tolerance, dependence, withdrawal, abuse, and addiction are often confused. *Tolerance* is a pharmacologic concept describing the need for a larger dose of a drug to achieve the same effect after repeated drug use. *Withdrawal* describes a physiologic state that follows cessation or reduction in amount of drug used. *Dependence* refers both to a condition in which a drug-specific withdrawal state follows cessation or reduction in drug dose and to continued drug use despite adverse consequences. *Addiction* describes a repertoire of pathologic behaviors that serve to maintain drug use.

The concept of substance use disorders is also complicated by basic questions as to whether use of psychoactive substances constitutes a medical or a moral condition. In addressing these issues, organizations such as the World Health Organization (WHO) and the American Psychiatric Association (APA) consider problem-causing use of psychoactive substances to be a medical disorder with defined diagnostic criteria.

The acute and chronic effects of psychoactive substances are classified under two major headings. Psychoactive Substance-Induced Organic Mental Disorders describe direct effects of the drug on the central nervous system (CNS), such as intoxication, withdrawal, delirium, delusions, or mood changes. Psychoactive Substance Use Disorders (Dependence and Abuse) describe behavioral symptoms and maladaptive behaviors resulting from the acute or chronic effects of the drug. DSM-III-R widened the concept of dependence to include the association of substance use with uncontrolled use or with use in spite of adverse medical, behavioral, or social consequences.

Eleven distinct classes of psychoactive substances were designated by the DSM-III-R: alcohol, amphetamines (and similarly acting sympathomimetics); caffeine; cannabis; cocaine; hallucinogens; inhalants; nicotine; opioids; phencyclidine (or similarly acting arylcyclohexylamines); and sedatives, hypnotics, or anxiolytics. Each class is associated with both an organic (secondary) mental disorder and a substance use disorder. Under the category of Psychoactive Substance Abuse disorders, 10 of these classes (all but nicotine) are associated with abuse and dependence; only dependence is defined for nicotine. Polysubstance abuse or dependence is defined for individuals using three or more classes of substances.

Psychoactive substance *dependence* is defined by the presence of at least three of the following, persisting for at least 1 month or occurring repeatedly over a longer period:

1. The substance is taken in larger amounts over a longer period of time than intended.
2. There is a persistent desire, or one or more unsuccessful attempts to cut down or to control substance use.
3. Much time is spent in getting the substance, taking the substance, or recovering from its effects.
4. There are intoxication or withdrawal symptoms during major role obligations or when substance use is hazardous.
5. Social, occupational, or recreational activities are reduced or given up because of substance use.
6. Substance use continues despite knowledge of social, occupational, psychologic, or physical problems.
7. Presence of marked tolerance to the substance.
8. Occurrence of characteristic withdrawal symptoms (may not apply to cannabis or hallucinogens).
9. Substance use to relieve or avoid withdrawal symptoms.

Substance "abuse" is a residual category that describes patterns of drug use that do not meet the criteria for dependence. Psychoactive Substance Abuse is a pattern of substance use of at least 1 month's duration that causes impairment in social or occupational functioning, in the presence of a psychologic or physical problem, or in situations in which use of the substance is physically hazardous, such as driving while intoxicated. Relatively minor changes in this nosology are planned for DSM-IV.

EVALUATING SUBSTANCE ABUSE IN PATIENTS

Although some patients present with substance abuse or its sequelae as a chief complaint, many patients present with other medical or surgical problems and only later reveal a substance use disorder through physical or laboratory findings, or incidental discovery. Many patients are reluctant to report the extent of their drug and alcohol use to physicians or other health care providers. Often patients deny the extent of their substance use to themselves. Obviously, patients experiencing an altered mental status as a result of intoxication or withdrawal may be incapable of providing an accurate history. In situations in which a patient is unable or unwilling to give a history of substance use, it is important to obtain additional history from family or acquaintances of the patient. However, family members may collude in the denial of substance use. It is also important to examine pill bottles or other medications in the patient's possession.

The primary task of the physician evaluating a patient for substance abuse or dependence is to establish an effective therapeutic relationship. In the context of this relationship, the physician should conduct a detailed alcohol and drug history, conduct a physical and mental status examination, order and interpret necessary laboratory tests, and meet with family or significant others to obtain additional information and to involve them in evaluation and treatment.

While obtaining information about the patient's alcohol and drug use, most clinicians routinely ask quantity and frequency questions about psychoactive substances, such as "how much?" and "how often?" Unfortunately, these questions are notoriously unreliable in their sensitivity for detecting substance abuse. A more effective interview method focuses on whether the patient has experienced negative consequences from use of psychoactive substances, has poor control of use, or has received criticism from others about the substance use.

Several well-validated, formalized interviews have been developed that discriminate alcoholism using these criteria. The most reliable screen, with a sensitivity of 90% to 98%, is the Michigan Alcohol Screening Test (MAST), a 25-item scale that identifies abnormal drinking through its social and behavioral consequences (Selzer, 1971). A shortened 10-item test, the Brief MAST, has similar efficacy in the diagnosis of alcoholism. The CAGE questionnaire (Ewing, 1984), is a simple, highly sensitive, four-item test that uses the letters *C*, *A*, *G*, and *E* as a mnemonic for the following questions about alcohol use:

Have you ever felt the need to Cut down on drinking?
Have you ever felt Annoyed by criticisms of drinking?
Have you ever had Guilty feelings about drinking?
Have you ever taken a morning Eye opener?

An affirmative answer on more than one question is considered suspicious for alcohol abuse. A recent study suggests that the CAGE may be a better predictor of alcoholism than any combination of laboratory tests (Beresford et al., 1990).

Although less documentation exists regarding the optimal interview for the assessment of the drug-abusing patient, the same considerations apply; it is more effective to ask about the behavioral consequences of drug abuse, rather than questions about quantity and frequency of use.

Other "red flags" in the patient history that should increase suspicion about psychoactive substance abuse include: divorce, problems at work (frequent job changes, tardiness, absenteeism, work-related injuries), injuries (falls, auto accidents, fights), arrests, driving while intoxicated, leisure activities involving drugs or alcohol, and financial problems. Having an alcohol- or drug-abusing biologic parent or spouse or a concomitant psychiatric disorder increases the risk for problem substance abuse.

Physical Examination

The physical examination of the patient provides important information about the presence of substance abuse and its medical complications. The presence of signs of repeated trauma, especially to the head, strongly suggests substance abuse (Skinner et al., 1984). Other physical stigma of substance abuse include track marks of intravenous drug abuse, a necrotic nasal septum from cocaine abuse, peripheral neuropathy from solvent inhalation, signs of liver disease or gastrointestinal bleeding from alcoholism and needle-acquired hepatitis B, and signs and symptoms of acquired immunodeficiency syndrome (AIDS) or human immunodeficiency virus (HIV)–related illnesses (Stein, 1990).

Mental Status Examination

Each patient should receive a mental status examination to assess cognitive, neurologic, or psychiatric impairments. Psychiatric disorders such as mood disorders, anxiety disorders, and personality disorders are known to frequently coexist with substance abuse disorders (Kosten & Kleber, 1988; Mirin et al., 1988). Cognitive mental status testing should be particularly detailed, because alcohol and drug abusers can have significant deficits in memory, concentration, and abstract reasoning, yet pass simple bedside tests of orientation, calculation, and immediate memory. Formal neuropsychologic testing has utility for detecting subtle deficits in attention, cognition, and performance and can better identify the type and localization of cerebral dysfunction (Berg, Franzen, & Wedding, 1987).

Laboratory Screening

Abnormal results on laboratory testing provide an important adjunct for confirming the diagnosis of substance abuse, but are not highly reliable or specific. In heavy users of alcohol, laboratory tests such as mean corpuscular volume (MCV) and liver function tests such as aspartate aminotransferase (AST) and gamma-glutamyl transferase (GGT) may be abnormal in a high percentage of patients. However, medical illnesses such as liver disease and nutritional deficiencies may produce similar abnormal results. A British study comparing the use of questionnaires and abnormal laboratory testing results for the detection of alcoholism found MCV, AST, and GGT to be somewhat poor discriminators of alcohol abuse (Bernadt et al., 1982). In the past decade, several promising biochemical tests have been proposed to detect alcohol use, including tests of carbohydrate-deficient glycoproteins (especially transferrin), serum and tissue acetaldehyde adducts, apolipoprotein II, high-density lipoprotein cholesterol, red blood cell aldehyde dehydrogenase, serum acetate, 2,3-butanediol, 5-hydroxytryptophol, and 5-hydroxytryptamine (Schellenberg et al., 1989; Takase et al., 1985). However, measurement of biochemical markers is still experimental and requires specialized, and often expensive, methodology. Laboratory assessment of alcohol dependence is discussed in detail in Chapter 7.

Toxicologic Screening

Serum and urine toxicologic screens have an important role in the assessment and treatment of patients with substance use disorders. However, it is important that such testing be properly conducted, and that the results of the testing be properly interpreted. Informed consent should be obtained for all drug testing. As with all laboratory tests, both false-positive and false-negative results may be obtained; the test result may be affected by methods of sample collection and accuracy of the laboratory (Hansen, Caudhill, & Boone, 1985). To minimize collection errors, all samples should be obtained under direct observation. Both serum and urine should be obtained, because substances may be differentially distributed in body fluids. Positive test results should be confirmed by a second test on the same sample using a different analytic method, because chemically similar compounds in foods or medications can mimic illicit drugs in some drug analyses. For example, use of sympathomimetic agents to treat asthma may yield a positive urine test for amphetamines. A positive test suggests past use of a psychoactive substance, but it may not indicate the extent of use, when it occurred, or whether there was behavioral impairment because of the use. Details of laboratory testing in drug users and in psychiatric patients are discussed in recent reviews (Hawks & Chiang, 1986; Schwartz, 1988; Swift, Griffiths, & Camara, 1991), as well as in Chapter 7 of this text.

TREATMENT OF SUBSTANCE

General Considerations

The treatment of the substance-abusing patient comprises several components. The primary task of the clinician is to establish an effective therapeutic relationship. The clinician should assess his own attitudes toward substance abuse to avoid negative countertransference reactions. Optimal treatment of substance abuse and dependence requires knowledge about therapies for the acute management of intoxicated or withdrawing patients, as well as knowledge regarding long-term treatment and rehabilitation. A treatment plan should be developed that is practical, economic, and based on well-established principles. The clinician should engage both the patient and the family or significant others in the treatment process.

The objectives of short-term treatment include: (1) establishment of a drug- or alcohol-free state; (2) relief of distress and discomfort resulting from intoxication or withdrawal; (3) treatment of medical or psychiatric complications of substance intoxication, substance withdrawal, or dependence; and (4) preparation for and referral for longer term treatment or rehabilitation.

The objectives for longer term treatment or rehabilitation include: (1) maintenance of the alcohol- or drug-free state and (2) psychologic, family, and vocational interventions to ensure its persistence. Changes in living situation, work situation, or friendships may be necessary, to decrease availability of drugs and to reduce peer pressure to use drugs. Halfway houses, therapeutic communities, and other residential treatment situations may be useful in this regard. Individual and group psychotherapy can be useful for understanding the role of the drug in the individual's life, improving self-esteem, and providing alternative methods of relieving psychosocial distress. Treatment of underlying psychiatric or medical illness may reduce the impetus for self-medication. Self-help groups, such as Alcoholics Anonymous (AA), Narcotics Anonymous (NA), and Al-Anon provide effective treatment, education, emotional support, and hope to substance users and their families (Emrick, 1987).

Most patients presenting for treatment do so in the context of a family that is also experiencing dysfunction. It is important for the clinician to be aware of dysfunctional family dynamics, and the denial, defensiveness, and hostility often present in family members. Family members need education and emotional and social support. Organizations such as Al-Anon and Alateen also may provide meaningful education and support for spouses and family members. It is important to involve family members in the patient's treatment as much as possible, and to recommend treatment for other family members, when appropriate.

Alcohol

It is estimated that as many as 7% to 10% of adult Americans have "alcoholism," defined as a "repetitive, but inconsistent and sometimes unpredictable loss of control of drinking which produces symptoms of serious dysfunction or disability" (Clark, 1981). While alcoholism exists in 20% to 50% of hospital admissions, it is diagnosed in fewer than 5% of the cases (Holden, 1985; Lewis & Gorden, 1983). In an American Medical Association–sponsored poll of physicians, 71% believed they were either not competent enough or were too ambivalent to treat alcoholic patients correctly (Kennedy, 1985).

Treatment of the alcoholic patient consists of two phases: detoxification and rehabilitation. Some patients may present to a facility with relatively uncomplicated alcohol intoxication, but others can have significant medical or psychiatric problems.

The treatment of alcohol intoxication is essentially supportive, to support vital functions, maintain physiologic homeostasis, and prevent behavioral problems. Patients using alcohol *always* should be medicated with thiamine and other B vitamin supplements to prevent the development of Wernicke-Korsakoff syndrome. This frequently underdiagnosed condition is classically characterized by ocular disturbances (nystagmus and sixth cranial nerve ophthalmoplegia), ataxia, and mental status changes, although most patients do not have the full triad of signs. The etiology is thiamine deficiency. Its presence should be considered a medical emergency, because delay in treatment diminishes its reversibility. Although low doses of oral thiamine probably are effective, most patients receive 25 to 100 mg intramuscular thiamine daily for 3 days to ensure adequate dosing. In patients with florid neurologic symptoms or signs, saving minutes may be critical, and thiamine should be administered intravenously. Alcoholics also are frequently hypomagnesemic. Magnesium levels should be determined and deficits replaced with intramuscular magnesium sulfate. Low magnesium can intensify withdrawal symptoms and predispose to seizures.

Physical dependence on alcohol and the alcohol withdrawal syndrome are due to compensatory CNS changes in response to a chronically administered depressant substance (ethanol). The withdrawal syndrome that follows the cessation of chronic alcohol intake results from increased neuronal activity in the CNS and peripheral nervous system. The consequences of this process can be minimal, or can include autonomic hyperactivity, seizures, delirium, and general physiologic dysregulation (Gross, Lewis, & Hastey, 1974). Studies done before the 1930s suggested that the mortality of untreated alcohol withdrawal was a high as 50%, although current sources quote mortality figures of 5% to 15% (Lewis & Gomolin, 1982). More recent studies suggest that approximately 5% of alcohol-dependent individuals undergoing detoxification will develop severe withdrawal delirium (delirium tremens) and, of those, approximately 10% to 15% will die. For those patients who do not develop delirium tremens, detoxification from alcohol still can be associated with a variety of serious medical complications, including seizures, pneumonia or other infections, myocardial infarction, cardiac arrhythmias, and electrolyte disturbances, and less serious physiologic derangements, including hypertension, tachycardia, tremors, agitation, and insomnia (McIntosh, 1982).

The clinician should be aware that signs and symptoms of withdrawal can obscure an underlying illness. For example, fever and change in mental status associated with withdrawal may coexist with an infection of the CNS, which requires a lumbar puncture for diagnosis. Treatment of the alcohol withdrawal syndrome includes correction of physiologic abnormalities, hydration, nutritional support, and pharmacologic therapy for the increased activity of the nervous system (Sellers & Kalant, 1976).

Pharmacologic Treatment of Ethanol Withdrawal

Recent studies on the physiology of alcohol dependence and withdrawal suggest that the CNS effects of alcohol may be mediated through modification of neurotransmitter signal-transduction mechanisms involving inhibitory gamma-aminobutyric acid (GABA) receptors and excitatory *n*-methyl-*d*-aspartate (NMDA) receptors (Hoffman et al., 1990). Alcohol modifies the binding of GABA to its receptors and augments the electrophysiologic and behavioral effects of GABA (Hunt, 1983). Alcohol also appears to have effects on the binding of other sedative drugs to the GABA receptor–chloride channel complex, presumably by dissolving in the membrane and altering its fluidity (Seeman, 1972; Skolnick et al., 1981). In the case of NMDA receptors, acute, low doses of ethanol cause NMDA receptor inhibition and chronic ethanol intake results in an increase in the number of these excitatory receptors. Alcohol also affects other neurotransmitters, such as serotonin, dopamine, norepinephrine, and adenosine.

It has long been known that readministration of depressant substances markedly attenuates the signs and symptoms of withdrawal, and greatly decreases the medical morbidity and mortality. Historically, a variety of pharmacologic agents has been noted to reduce the signs and symptoms of the withdrawal syndrome, including chloral derivatives, paraldehyde, barbiturates, antihistamines, neuroleptics, antidepressants, lithium, adrenergic blocking agents, and benzodiazepines (Golbert et al., 1967; Palestine & Alatorre, 1976; Sellers et al., 1980). Indeed, almost any CNS depressant substance may have efficacy. The reader is referred to two excellent recent review articles on the pharmacologic treatment of alcohol withdrawal and alcoholism (Liskow & Goodwin, 1987; Litten & Allen, 1991).

Benzodiazepines. Today, benzodiazepine derivatives are the treatment of choice, and their efficacy is well established by double-blind, controlled studies (Sellers et al., 1983). Benzodiazepines are superior to other agents because of their low toxicity and anticonvulsant effects. Recent studies suggest that benzodiazepines are optimally used when the dose is titrated according to a withdrawal severity scale such as the CIWA-A Scale (Foy, March, & Drinkwater, 1988). Several methods have been described for titrating medication dosage to symptoms. The benzodiazepine loading method appears to have utility in most patients (Sellers et al., 1983). The advantages of this method include matching dose of medication to an individual patient's tolerance and the avoidance of cumulative pharmacokinetics. During this procedure, patients receive an initial oral or intravenous dose of a long half-life benzodiazepine (10 to 20 mg diazepam or 50 to 100 mg chlordiazepoxide) that repeats every hour until the patient is sedated, develops nystagmus, or has a significant decrease in withdrawal signs and symptoms as shown by a reduced withdrawal scale score.

The vast majority of patients treated in this manner respond within several hours with a marked reduction in signs and symptoms. Many patients require no additional medication for the duration of their detoxification, presumably because of the long half-life of the drugs. Occasionally, patients may require additional doses of medication after several days to suppress emergent symptoms. During the period of benzodiazepine loading, patients must be closely observed to avoid undermedication or overmedication. In particular, close attention must be paid to patients with respiratory, cardiovascular, or hepatic disease. In patients receiving adrenergic blocking drugs, some signs of withdrawal, such as hypertension, tachycardia, and tremor, may be obscured.

Other Agents Used for Withdrawal. Other pharmacologic agents, such as beta-adrenergic blocking drugs, anticonvulsants, and antipsychotics, often are administered to control withdrawal symptoms. Carbamazepine appears effective in reducing most signs and symptoms of alcohol withdrawal (Malcolm et al., 1989). Patients initially receive 400 to 600 mg of carbamazepine daily in divided doses, with the dose subsequently adjusted to achieve therapeutic blood levels. Beta-blocking drugs such as propranolol and atenolol have been used as primary agents in the treatment of alcohol withdrawal (Sellers & Kalant, 1976). However, beta-blockers are most effective in reduction of peripheral autonomic signs of withdrawal, and less so for CNS signs such as delirium. Adrenergic blocking drugs are particularly useful for controlling tachycardia and hypertension in patients with coronary disease.

Controversy exists regarding the use of anticonvulsant medications to control withdrawal seizures. Single seizures do not require anticonvulsant treatment; for patients with multiple seizures during withdrawal, or who have a chronic seizure disorder, anticonvulsants such as phenytoin may be useful. Given that many alcoholics are noncompliant with treatment, the erratic use of anticonvulsants on an outpatient basis may actually worsen a seizure problem (Hillbom & Hjelm-Jager, 1984). Neuroleptics such as haloperidol are useful for the treatment of hallucinosis and paranoid symptoms.

Nonpharmacologic Treatment Methods

Recently, the widespread use of pharmacologic agents in detoxification from alcohol has been called into question. Alcohol treatment facilities have been experimenting with "social setting detoxification," a nondrug method. This procedure relies on the extensive use of peer and group support and usually occurs in nonmedical settings. This method seems effective in reducing withdrawal signs and symptoms without an increased incidence of medical complications. There have been some questions about the possible bias in selection of healthier patients for detoxification in these nonmedical settings. However, Shaw et al. (1981) have shown that, even within a medical setting, most patients respond to "supportive care" and do not require pharmacologic intervention. In a double-blind study comparing parenteral diazepam treatment with placebo, over half the patients receiving a placebo injection responded with a marked attenuation of withdrawal symptoms within 5 hours (Sellers et al., 1983). British practitioners routinely detoxify most alcohol-dependent patients as outpatients, with minimal use of psychotropic drugs (Whitfield et al., 1978).

Thus, the available evidence suggests that many, and perhaps even most, alcohol-dependent individuals may be detoxified without the use of sedatives or other psychotropic drugs. However, there exists a subgroup of patients who apparently do require careful monitoring and pharmacologic treatment during the withdrawal process. A history of delirium tremens or seizures or the presence of medical or psychiatric comorbidities increases the need for closer medical supervision. However, few data exist for the clinician to predict which alcohol-dependent patients have an absolute requirement for pharmacologic treatment. The existence of predictive data would optimize clinical care for patients, and would reduce the cost of care, because patients who did not require drug administration within a hospital setting could be detoxified out of hospital at lower cost.

During and after detoxification, many alcoholic patients will appear to suffer from major depression. Some of these mood disorders are directly related to alcohol and will resolve within 4 weeks of detoxification. Major depressive symptoms persisting beyond this time should be considered for treatment with antidepressants or electroconvulsive therapy (ECT). Patients previously known to have recurrent mood disorders should be treated immediately after detoxification if mood symptoms are severe.

Long-Term Treatment

The goals of long-term treatment include maintaining a state of abstinence from alcohol and psychologic, family, and social interventions to maintain this recovery. These goals are best achieved through the patient's participation in a comprehensive treatment program, beginning after discharge from the acute care setting. Nevertheless, there are some aspects of long-term treatment that can be initiated during an acute hospital stay.

Alcoholics Anonymous is an independent organization, founded by a physician alcoholic in 1939. Its only goal is to help individuals maintain a state of total abstinence from alcohol and other addictive substances, through group and individual interactions between alcoholics at various stages of recovery. Self-help programs such as AA are helpful for many patients, but there is a paucity of objective outcome data on the actual long-term efficacy of such groups (Emrick, 1974, 1987). Most detoxification and rehabilitation programs encourage liberal attendance at AA meetings by their patients. General hospitals often are used as meeting sites by local AA groups, and medical and surgical inpatients easily can attend these meetings.

Drugs to Reduce Alcohol Consumption

Disulfiram (Antabuse) is an inhibitor of the enzyme acetaldehyde dehydrogenase, and is used as an adjunctive treatment in selected alcoholics. If alcohol is consumed in the presence of this drug, the toxic metabolite acetaldehyde accumulates, producing tachycardia, flushing of skin, dyspnea, nausea, and vomiting. The presence of this unpleasant reaction provides a deterrent to the consumption of alcohol (Keventus & Major, 1979). Recently, a few professional groups have questioned whether the toxicity of disulfiram precludes its therapeutic use under any circumstances.

Patients started on disulfiram must be informed about the dangers of even small amounts of alcohol. Alcohol present in foods, shaving lotion, mouthwashes, or over-the-counter medications can produce a disulfiram reaction. A usual dose of disulfiram is 250 to 500 mg once daily. Disulfiram may have interactions with other medications, notably anticoagulants and phenytoin. It is contraindicated in patients with liver disease.

Antidepressant medications such as imipramine, desipramine, and amitriptyline have been reported to be effective in reducing alcohol consumption (Mason & Kocsis, 1991), although the mechanism by which this occurs is not known. Serotonin reuptake blocker antidepressants such as zimelidine and fluoxetine show efficacy in reducing alcohol use in nondepressed, heavy drinkers (Naranjo et al., 1990). In several studies, lithium has been found to reduce alcohol consumption and to block the intoxicating effects of alcohol (Fawcett et al., 1987; Sellers et al., 1980). However, lithium appears to be most effective in alcoholics with coexisting depression (Dorus et al., 1989). Anxiolytic medications such as buspirone have been reported to reduce relapse in anxious alcoholics (Kranzler & Meyer, 1989).

The appropriate role for psychotropics in promoting alcohol abstinence is not yet settled. Practically, one should certainly treat concurrent Axis I disorders such as major depression, panic disorder, or bipolar disorder when they occur together with alcohol dependence. The author further believes that psychotropic trials are appropriate in recovering alcoholics with subsyndromal but clinically significant anxiety, depression, or mood instability. The author's drugs of choice are buspirone for anxiety, serotonin reuptake blockers for depression, and lithium for mood instability. These drugs would be continued if they helped the target symptoms without undue side effects; improved chances of abstinence would be hoped for as a corollary benefit.

Amphetamines and Similarly Acting Sympathomimetic Amines

The amphetamines are a group of drugs structurally related to the catecholamine neurotransmitters norepinephrine, epinephrine, and dopamine. Amphetamines release catecholamines from nerve endings and are catecholamine agonists at receptors in the peripheral autonomic and central nervous systems. Intoxication with stimulants such as amphetamines, methylphenidate, or other sympathomimetics produces a clinical picture of sympathetic and behavioral hyperactivity. An "amphetamine psychosis" with manifestations of agitation, paranoia, delusions, and hallucinosis can follow chronic use of these drugs (Ellinwood, 1969). Antipsychotic medication such as haloperidol is useful in the treatment of stimulant psychoses; however, such patients may frequently require psychiatric hospitalization. Severe hypertension is seen in overdose, and may be treated with alpha-adrenergic blockers such as phentolamine.

Chronic users of amphetamines use escalating doses of the drug for periods of several days to weeks, followed by a period of abstinence. A paranoid psychosis that is diagnostically similar to schizophrenia can occur with chronic use and persist following cessation of stimulant use. A withdrawal syndrome with physiologic dysfunction does not follow abstinence from amphetamines; however, marked dysphoria, fatigue, and restlessness may occur. Stimulant users also may suffer from underlying psychiatric disorders such as mood disorders and attention deficit disorders. Thus, all stimulant abusers should receive a comprehensive psychiatric evaluation.

The use of over-the-counter sympathomimetic amines such as ephedrine and phenylpropanolamine has increased dramatically in the past decade (Dietz, 1981). These medications, marketed as appetite suppressants, decongestants, and bronchodilators, have moderate stimulant activity. Signs of intoxication are similar to those produced by amphetamines, although there tends to be less CNS stimulation and greater autonomic effects. Hypertensive crises, cerebral bleeds, and vasculitis have resulted from the use of these drugs, as they have from the use of amphetamines.

Barbiturates or Other Sedative-Hypnotics and Anxiolytics

Sedative medications are a major source of adverse drug interactions and drug emergencies, including overdose (Gottschalk et al., 1979), yet they are among the most prescribed drugs, and are routinely used for their anxiolytic and hypnotic effects. Medications in this group include barbiturates, benzodiazepines, chloral derivatives, ethchlorvynol, glutethimide, meprobamate, and methaqualone. Patients may obtain these medications illicitly from the street or from physicians who unwittingly (or purposely) may be contributing to abuse or dependence. The medications in this group derived their pharmacologic activity by affecting the chloride channel–GABA receptor complex in the brain (Seeman, 1972). Specific binding sites exist for benzodiazepines, barbiturates, and other drugs that, when occupied by drug, increase hyperpolarization (and inhibition) of neurons

(Costa & Guidotti, 1979; Schulz & MacDonald, 1981; Skolnick et al., 1981).

As with alcohol treatment, treatment of the sedative abuser occurs in two stages, detoxification and long-term rehabilitation. The withdrawal syndrome that follows cessation of sedative drug use may be severe, including seizures, cardiac arrhythmias, and death. The need for detoxification depends on the duration and amount of sedative drug abuse, which can be estimated by means of a *pentobarbital challenge test* (Wesson & Smith, 1977; Wikler, 1968). Pentobarbital 200 mg is administered orally and the patient is observed 1 hour later. The patient's condition after the test dose will range from no effect to sleep. If the patient develops drowsiness or nystagmus on a 200-mg dose, the patient is not barbiturate dependent. If 200 mg lacks effect, the dose should be repeated hourly until nystagmus or drowsiness develop. The total dosage administered at this end point approximates the patient's daily barbiturate habit and may be used as a starting point for detoxification. The barbiturate dose should be tapered over 10 days, with approximately a 10% reduction each day. Alternatively, the patient can be loaded with a long-acting benzodiazepine such as diazepam or clonazepam in a dose sufficient to produce mild drowsiness and nystagmus. This medication can then be tapered over 7 to 10 days (Patterson, 1990).

An alternative treatment increasingly used for sedative detoxification is carbamazepine (Klein et al., 1986). The medication is administered until blood levels are in the range for effective anticonvulsant activity, maintained for up to 2 weeks, and then tapered and discontinued.

Following detoxification, the patient should be engaged in long-term treatment. Treatments should be individualized to the patient, but may include residential drug-free programs, outpatient counseling, and self-help groups such as AA or NA.

Caffeine

Caffeine, and the related methylxanthines theophylline and theobromine, are ubiquitous drugs in our society. These agents occur in coffee, tea, cola, and other carbonated drinks, and are consumed by more than 80% of the population (Dews, 1982). Caffeine is present in chocolate and in many prescribed and over-the-counter medications, including stimulants (No-Doz), and appetite suppressants (Dexatrim), analgesics (Anacin, APC tablets), and cold and sinus preparations (Dristan, Contac).

Central nervous system effects of caffeine include psychomotor stimulation, increased attention and concentration, and suppression of the need for sleep. Even at low or moderate doses, caffeine can exacerbate symptoms of anxiety disorders and may increase requirements for neuroleptic or sedative medications (Charney, Henninger, & Jatlow, 1985). At high doses and in sensitive individuals, methylxanthines can produce tolerance and can cause behavioral symptoms of tremor, insomnia, jitteriness, and agitation. In moderate to heavy users, a withdrawal syndrome characterized by lethargy, hypersomnia, irritability, and severe headache follows cessation of use. Clinically significant caffeine withdrawal symptoms are commonly observed in even low to moderate users of caffeine (Hughes et al., 1991) and can occur with reduced caffeine intake during a medical or psychiatric hospitalization. The signs and symptoms of caffeine intoxication or caffeine withdrawal can complicate medical or psychiatric treatment by increasing patient distress and by leading to an unnecessary workup for other disorders.

Methylxanthines produce physiologic effects through actions at the cellular level. They produce cardiac stimulation, diuresis, bronchodilation, and CNS stimulation through several mechanisms. They inhibit the enzyme cyclic adenosine monophosphate phosphodiesterase and increase intracellular levels of this second messenger, thereby augmenting the action of many hormones and neurotransmitters, such as norepinephrine. They also have a direct inhibitory effect on adenosine receptors and may have other neurotransmitter effects as well.

Treatment of caffeine dependence limits consumption of caffeine-containing foods, medications, and beverages. Beverages such as coffee or cola may be substituted with decaffeinated forms. Often, patients are unaware of the extent of their caffeine consumption and of the caffeine content of consumables. They require education about the caffeine content of these substances. Withdrawal symptoms such as headache and lethargy can be treated with a caffeine taper or symptomatically with analgesics and rest.

Cocaine and Other Stimulants

The usage of cocaine has undergone a dramatic increase in the last two decades. The Haight-Ashbury Clinic reported a rise in cocaine use from less than 1% of patients in 1970 to greater than 6% in 1982 (Gay, 1982). Based on the 1985 National Survey on Drug Abuse, 22 million Americans tried cocaine at least once and 12 million used it during the preceding year. During the mid-1970s the pattern of cocaine use changed from intranasal "snorting" of cocaine powder to smoking or intravenous injection of the more potent cocaine "freebase." Freebase cocaine is now widely available in a product called "crack," which is extremely potent, inexpensive, and easily distributed. Crack is self-administered by smoking, usually by adding a small piece to a burning cigarette and inhaling the vapor. Because crack is so inexpensive and freely available, it has greatly increased the pool of cocaine users.

Cocaine has major physiologic and behavioral effects (Gawin & Ellinwood, 1988). First, it is a local anesthetic of high potency, the only naturally occurring local anesthetic. It blocks the initiation and propagation of nerve impulses by affecting the sodium conductance of nerve cell membranes (Seeman, 1972). Second, it is a potent sympathomimetic agent that potentiates the actions of catecholamines in the autonomic nervous system, producing tachycardia and hypertension. It is a potent vasoconstrictor. Third, it is a potent stimulant of the CNS, potentiating the action of central catecholamine neurotransmitters, norepinephrine and dopamine. Its effects include increased arousal, euphoria, excitement, and motor activation. This may progress to agitation, irritability, apprehension, and paranoia at high doses.

Cocaine intoxication produces elation, euphoria, excitement, pressured speech, restlessness, stereotyped movements, and bruxism. Physiologic signs of sympathetic stimulation are present, including tachycardia, mydriasis, and sweating. With chronic use, paranoia, suspiciousness, and frank psychotic symptoms may occur. Overdosage of cocaine produces hyperpyrexia, hyperreflexia, and seizures, which may progress to coma and respiratory arrest. Propranolol and haloperidol administration have been reported as useful in overdosage (Rappolt, Gay, & Inaba, 1977). Cocaine also has major deleterious effects on pregnancy, producing low-birth-weight infants, abruptio placentae, and behavioral abnormalities in the newborn (Chasnoff et al., 1985).

The plasma half-life of cocaine following oral, nasal, or intravenous administration is approximately 1 to 2 hours, which correlates with its behavioral effects (Van Dyke et al., 1978). With the decline in plasma levels, most users experience a period of dysphoria or "crash," which often leads to additional cocaine use within a short period. The dysphoria of the "crash" is intensified and prolonged following repeated use. However, some investigators have questioned whether a distinct cocaine abstinence syndrome actually exists (Satel et al., 1991).

Treatment

The optimal treatment of the chronic cocaine user is still not established. While cessation of cocaine use is not followed by a physiologic withdrawal syndrome of the magnitude of that seen with opioids or alcohol, the dysphoria, depression, and drug craving that follow chronic cocaine use are often intense and make abstinence difficult. Psychotherapy, group therapy, and behavior modification are useful in maintaining abstinence (Rounsaville, Gawin, & Kleber, 1985).

A variety of pharmacologic agents have shown promise as adjunctive treatments. Several reports have shown efficacy of antidepressant agents such as imipramine, desipramine, lithium, or trazodone in reducing cocaine craving and usage (Gawin & Kleber, 1984; Rosecan, 1983; Tennant & Rawson, 1983). The doses of medication used were similar to those used for antidepressant therapy. Carbamazepine has recently been reported as a useful adjunct for treatment (Halikas et al., 1991). The postsynaptic dopamine agonists bromocriptine and amantadine also may have usefulness in cocaine treatment, because they appear to block cocaine craving (Dackis & Gold, 1985; Giannini et al., 1989). The doses of bromocriptine (2.5 mg two to four times daily) and amantadine (100 mg twice daily) are usual therapeutic doses. However, not all researchers report effectiveness of these agents (Teller & Devenyi, 1988). The partial opioid agonist buprenorphine is reported to reduce cocaine use in opioid-dependent cocaine users (Kosten, Kleber, & Morgan, 1989).

Many psychiatric and drug hospitals now offer short-term inpatient treatment of the cocaine user, providing intensive psychologic treatment and drug education in a drug-free environment. For recidivists, long-term residential drug-free programs, including therapeutic communities, may be efficacious. Self-help groups such as NA may be useful both as a primary treatment modality for cocaine dependence and as an adjunct to other treatment.

Certain psychiatric disorders such as depression and attention-deficit hyperactivity disorder may be more common in cocaine users than in the general population, although good epidemiologic data are lacking. Recognition and treatment of these underlying disorders may be necessary to stop cocaine use. However, it would be wise to avoid the use of amphetamines in former cocaine abusers, except under closely monitored conditions. In addition, many cocaine users also use alcohol or other drugs, particularly sedatives and heroin, and may require treatment for abuse of these substances as well.

Cannabis

Cannabis sativa, also called marijuana or hemp, is a plant indigenous to India but now grown worldwide. The leaves, flowers, and seeds of the plant contain many biologically active compounds, the most important of which are the lipophilic cannabinoids, especially delta-9-tetrahydrocannabinol (THC). The biologically active substances are administered by smoking or ingesting dried plant parts (marijuana, bhang, ganga), the resin from the plant (hashish), or extracts of the resin (THC or hash oil). After inhalation or ingestion, THC rapidly enters the CNS. It has a biphasic elimination with a short initial half-life (1 to 2 hours) reflecting redistribution and a second half-life of days to weeks. Tetrahydrocannabinol is hydroxylated and excreted in bile and urine.

In spite of its illicit status, much of the American population has used marijuana. In 1982, 64% of young adults (18 to 25) had used marijuana at least once. Although its use is declining, millions of individuals continue to use marijuana regularly. Marijuana is frequently used simultaneously with other psychoactive substances, such as hallucinogens and opiates, which are mixed with the marijuana and smoked together.

Cannabis intoxication is characterized by tachycardia, muscle relaxation, euphoria, and a sense of well-being. Time sense is altered and emotional lability, particularly inappropriate laughter, can be seen. Performance on psychomotor tasks, including driving, is impaired (Klonoff, 1974). Marijuana has antiemetic effects and reduces intraocular pressure, and has been used medically for these effects. Occasionally, with high doses of the drug, depersonalization, paranoia, and anxiety reactions occur. Although tolerance to the effects of cannabis occurs with chronic use, cessation of use does not produce significant withdrawal phenomena. Chronic use of cannabis has been associated with an apathetic, amotivational state that improves on discontinuation of the drug (Gersten, 1980).

Treatment

Treatment of cannabis dependence is similar to treatment of other drug dependencies. As part of the initial assessment, all patients should undergo complete psychiatric and medical examinations. Short-term goals should focus on reducing or stopping cannabis use and interventions to ensure compliance. Inpatient treatment may be necessary to achieve an abstinent state. Since many patients with cannabis dependence are adolescents or young adults, involvement of the family in assessment and treatment is important.

Long-term treatment should involve behavioral and psychologic interventions to maintain an abstinent state. Often, a change in social situation is necessary to decrease drug availability and reduce peer pressure to use drugs. Individual and group psychotherapy may be useful for understanding the role of the drug in the individual's life, improving self-esteem, and providing alternate methods of relieving psychosocial distress. Self-help groups such as NA can provide group and individual support.

Inhalants

Inhalants are volatile organic compounds that are inhaled for their psychotropic effects. Substances in this class include organic solvents such as gasoline, toluene, and ethyl ether; fluorocarbons; and volatile nitrates, including nitrous oxide and butyl nitrate. Inhalants are ubiquitous and readily available in most households and places of employment. At low doses, inhalants produce mood changes and ataxia; at high doses they can produce dissociative states and hallucinosis. Dangers of organic solvent use include suffocation and organ damage, especially hepatotoxicity and neurotoxicity in the CNS and peripheral nervous system (Watson, 1982). Cardiac arrhythmias and sudden death may occur. Inhaled nitrates can produce hypotension and methemoglobinemia.

The typical user of inhalants is adolescent and male. According to the National Survey on Drug Abuse (1985) 9.1% of 12- to 17-year-olds and 12.8% of 18- to 25-year-olds have tried an inhalant at least once.

Optimal treatment of the inhalant user is not well established. Since most users are adolescents, treatment must involve the family. Long-term residential treatment may be helpful in the treatment of heavy users.

Nicotine

Nicotine is an alkaloid drug present in the leaves of the tobacco plant, *Nicotiana tabacum*. The plant has been used for centuries by Native Americans in ceremonies and rituals and as a medicinal herb. Since its discovery by Europeans, tobacco use has spread worldwide, and today nicotine is the most prevalent psychoactive drug in use. Over 50 million persons in the United States are daily users of cigarettes, with another 10 million using another form of tobacco. Since the publication of the Surgeon General's Report on Smoking and Health in 1964, there has been a gradual decline in the percentage of Americans who smoke. Most of this decline has occurred in men. The numbers of young women who smoke, and the use of other tobacco products such as smokeless tobacco, has increased. The morbidity and mortality resulting from use of nicotine are extensive and include an increase in cardiovascular and respiratory disease and in cancers, particularly of the lung and oropharynx. Many deleterious effects of tobacco are due not to nicotine but to other toxic and carcinogenic compounds present in tobacco extract or smoke.

To maximize the absorption of nicotine, tobacco products are usually smoked in pipe tobacco, cigars, or cigarettes or instilled intranasally or intraorally as snuff or "smokeless tobacco." Following absorption from the lungs or buccal mucosa, nicotine levels peak rapidly and then decline, with a half-life of 30 to 60 minutes.

Nicotine has several effects on the peripheral autonomic and central nervous systems. It is an agonist at "nicotinic" cholinergic receptor sites and stimulates autonomic ganglia in the parasympathetic and sympathetic nervous systems, producing salivation, increased gastric motility and acid secretion, and increased catecholamine release. In the CNS, nicotine acts as a mild psychomotor stimulant, producing increased alertness, increased attention and concentration, and appetite suppression. The fact that tobacco use can prevent weight gain makes the drug attractive, particularly to young women.

Repeated use of nicotine produces tolerance and dependence. The degree of dependence is considerable, because over 70% of dependent individuals relapse within 1 year of stopping use. Cessation of nicotine use in dependent individuals is followed by a withdrawal syndrome characterized by increased irritability, decreased attention and concentration, and an intense craving for and preoccupation with nicotine. Frequently, appetite and food consumption increase and a significant weight gain occurs. Withdrawal symptoms may begin within several hours of cessation of use or reduction in dosage, and typically last about a week. Craving and weight gain may persist for weeks, however.

The treatment of nicotine dependence is discussed in great detail in Chapter 42.

Opioids

Opioid abuse and dependence is a significant sociologic and medical problem in the United States, with an estimated opioid addict population of greater than 500,000. These patients are frequent users of medical and surgical services because of the multiple medical sequelae of intravenous drug use and its associated life-style.

Current diagnostic criteria for opioid abuse require the presence of a pattern of pathologic use, an inability to stop use, and frequent intoxication or overdose, with impairment in social or occupational functioning and a duration of at least 1 month. Requirements for a diagnosis of opioid dependence include the presence of tolerance or evidence of opioid withdrawal.

Opiate drugs have effects on many organ systems. Their action is due to stimulation of receptors for endogenous hormones—enkephalins, endorphins, and dynorphins. Recent evidence suggests that there exist at least four distinct opioid receptors, which are designated by the Greek letters mu, kappa, sigma, and delta (Jaffe & Martin, 1990). Drugs that act primarily through mu receptor effects include heroin, morphine, and methadone; such drugs produce analgesia, euphoria, and respiratory depression. Drugs that are mediated through the kappa receptor include the so-called mixed agonist-antagonists, butorphanol and pentazocine, which produce analgesia but less respiratory depression. The sigma receptor appears identical with the receptor for the hallucinogenic drug phencyclidine. The delta receptor appears to bind endogenous opioid peptides. At high doses, opioid drugs may lose receptor specificity and have agonist or antagonist properties at multiple receptor subtypes.

Opiate overdose is a life-threatening emergency, and should be suspected in any patient who presents with coma and respiratory suppression. Although miotic pupils are usually present, they are nondiagnostic and may not appear with ingestion of mixed agonist-antagonists. Other effects of intoxication include hypotension, seizures, and pulmonary edema. Treatment of suspected overdosage includes emergency support of respiration and cardiovascular functions. Parenteral administration of the opioid antagonist naloxone, 0.4 to 0.8 mg is of both diagnostic and therapeutic value (Martin, 1976). Although naloxone will rapidly reverse the effects of opioids, including coma and respiratory suppression, it does not reverse CNS depression caused by other drugs, such as alcohol or sedative-hypnotics. Naloxone will precipitate withdrawal in any patient who is dependent on opioids, causing the patient whose life was just saved to be most ungrateful.

Detoxification Regimens

The opioid withdrawal syndrome is unpleasant but rarely life threatening. It is characterized by the presence of increased sympathetic nervous system activity coupled with gastrointestinal symptoms of nausea, vomiting, cramps, and diarrhea. Patients also may report myalgias and arthralgias. There is increased restlessness, increased anxiety, insomnia, and an intense craving for opioids.

Many opioid-dependent patients are detoxified during their hospital stay, because current federal law allows narcotic maintenance with methadone only under the auspices of an approved methadone maintenance program. As a result of the expansion of funding for drug abuse treatment, most methadone programs have short or no waiting lists for new patients. If a patient is accepted into methadone treatment, methadone can be started in the hospital and be continued after discharge.

For patients who are detoxified while in the hospital, two methods are commonly used. Most in-hospital opioid detoxifications are performed by substitution of the abused opioid with methadone and then gradually decreasing the dose of methadone over a period of up to 21 days, as specified by federal law (Fultz & Senay, 1975). Although a gradually decreasing dose of any opioid could be used for detoxifications, in practice methadone is preferable because of its long half-life and once-daily oral administration. Initially, patients should be given 10 to 20 mg methadone orally every 2 to 4 hours until withdrawal symptoms are suppressed. The total daily dose received is typically 20 to 40 mg for heroin addicts. This dose is then decreased by approximately 10% daily. For those patients who are unable to receive oral medications, the same dose of intramuscular methadone may be administered twice daily in divided doses.

The alpha$_2$-adrenergic agonist clonidine hydrochloride suppresses many autonomic signs and symptoms of opioid withdrawal. Clonidine acts at presynaptic noradrenergic nerve endings in the locus ceruleus of the brain, and blocks the adrenergic discharge produced by opioid withdrawal (Aghajanian, 1976; Gold, Redmond, & Kleber, 1979). Clonidine has been reported as effective for primarily suppressing the autonomic signs of opioid withdrawal following discontinuation of opioids in dependent patients (Charney et al., 1981; Gold et al., 1979; Washton & Resnick, 1981).

Clonidine detoxification is performed as follows. On the day before beginning clonidine, the usual dose of opioid is received. On day 1, the opioid is stopped completely and instead, clonidine is given at a dose of 0.1 mg every 8 hours. From day 2 to day 4 the dose of clonidine is gradually increased to suppress withdrawal signs and symptoms, but without allowing blood pressure to decrease below 80 mm Hg systolic and 60 mm Hg diastolic. Typically, for most patients, a dose of 0.6 to 1.2 mg clonidine is required by day 4, but the dose will depend on the quantity of opioid used. This dose continues until day 7 for patients using short-acting opioids, such as heroin, morphine, or meperidine, and until day 10 to 12 for those using longer acting methadone. The dose of clonidine is then reduced by 0.2 to 0.3 mg per day until discontinued. Clonidine detoxification has been performed in an outpatient setting without significant morbidity. However, outpatients should be monitored closely and blood pressure be determined at least once daily.

Clonidine can cause orthostatic hypotension, and patients should be cautioned to stand up slowly, and to lie down if they become dizzy or light-headed. Clonidine causes sedation, and the patient should be advised to avoid activities that require alertness, such as driving. The drug should be used with caution in patients with hypotension and those receiving other antihypertensive medications. Most patients also describe dry mouth, which may be annoying but is otherwise harmless. Withdrawal symptoms that are not significantly ameliorated by clonidine include drug craving, insomnia, and arthralgias and myalgias. Insomnia is best treated with a short-acting hypnotic such as chloral hydrate. Muscle and joint pains may respond to acetaminophen or an antiinflammatory agent such as ibuprofen.

Methadone Maintenance

Since its introduction in 1965, methadone maintenance has become a major modality of long-term treatment of opioid abuse and dependence (Dole & Nyswander, 1965). Currently, over 85,000 individuals are maintained on methadone in the United States. Because of recent increased state and federal funding, the number of patients who may be accommodated in methadone maintenance programs has increased to satisfy demand. Although some programs may have waiting lists for treatment, patients who are pregnant or who have significant problems, such as renal failure, heart disease, or AIDS, usually are accepted directly without a waiting period. Senay (1985) provides an excellent recent review of the theory and practice of methadone maintenance.

Patients who are receiving methadone maintenance should have their usual daily dose of methadone continued in the hospital. However, it is important to maintain frequent communication with the methadone program, particularly regarding changes in methadone dosage. If pain medication is necessary, patients should receive additional short- to intermediate-acting opioids, such as meperidine or oxycodone, *besides* their usual dose of methadone, rather than increasing the methadone dose. The use of narcotic analgesics other than methadone keeps separate the idea of opioids for analgesia and opioids for maintenance and does not change the dose of methadone as determined by the methadone maintenance program. Because methadone has made these patients tolerant to the effects of opioids, additional pain medication may be required beyond that required by most patients. Mixed agonist-antagonist drugs such as pentazocine and butorphanol should be avoided, because they will precipitate withdrawal in opioid-dependent individuals.

Other Treatment Methods

Alternatives to methadone maintenance include opioid antagonist therapy with naltrexone or being drug free. Naltrexone is a long-acting orally active opioid antagonist that, when taken regularly, entirely blocks the euphoric, analgesic, and sedative properties of opioids (Resnick et al., 1980). It is administered either daily at a dose of 50 mg or three times weekly at doses of 100 mg or 150 mg. The drug is most effective in highly motivated individuals with good social supports, and appears less helpful for heroin addicts. Although naltrexone may be prescribed by any physician, it is most effective when part of comprehensive rehabilitative efforts. Because of reports of hepatitis with naltrexone use, it is recommended to avoid its use in patients with liver disease and to monitor liver function tests periodically.

Nonpharmacologic and behavioral treatment modalities are quite efficacious in the treatment of the opioid abuser. Programs may differ in their lengths of stay, their intensity, and their theoretical orientation. Long-term residential treatment may be most useful for the chronic opioid abuser who requires a change in lifestyle, with vocational and psychologic rehabilitation. Attending NA is helpful for many patients.

Hallucinogens and Phencyclidine

Many drugs are used for their hallucinogenic or psychotomimetic effects. These include psychedelics such as lysergic acid diethylamide (LSD), mescaline, psilocybin, and dimethyltryptamine (DOT); hallucinogenic amphetamines such as methylenedioxyamphetamine (MDA) and methylenedioxymethamphetamine (MDMA, or "ecstasy"); phencyclidine (PCP) and similarly acting arylcyclohexylamines; and anticholinergics, such as scopolamine. All cause a state of intoxication characterized by hallucinosis, affective changes, and delusional states. The mechanism of action of hallucinogens is not well understood and varies according to the drug. Hallucinogens of the LSD and amphetamine class are thought to act on dopaminergic and/or serotonergic brain systems. Hallucinogens related to PCP probably act at sigma "opioid" receptors, which are part of the NMDA receptor complex. Anticholinergics act at muscarinic cholinergic receptors.

Phencyclidine intoxication has several definitive features (Young, Lawson, & Gacocn, 1987). Patients often present with violence, directed either at the self or at others. Eye signs, including vertical and horizontal nystagmus, are often present, and myoclonus and ataxia are frequent. Autonomic instability, with hypertension and tachycardia, is common.

The differential diagnosis of hallucinogen- or PCP-induced psychosis includes schizophrenia, bipolar disorder, delusional disorder, secondary mental disorders due to primary CNS disease, and other toxic ingestions. Psychoses, including those that are drug induced, may produce an analgesic state, and the clinician must be aware of any coexisting medical problems such as injuries or abdominal emergencies, which may be obscured.

Treatment of the psychotic state induced by hallucinogens includes supportive measures to prevent patients from harming themselves or others, maintaining cardiovascular and respiratory functions, and amelioration of agitation and psychotic symptoms. Often agitation and psychosis respond to decreased sensory stimulation and verbal reassurance; still, patients usually require sedation with benzodiazepines or high-potency neuroleptics. Severe tachycardia and hypertension, if present, can be treated with propranolol. Most cases of hallucinogen intoxication are short lived (several hours), although prolonged drug-induced psychoses may occur. This is particularly common with use of PCP, which may produce a prolonged psychosis lasting 2 to 7 days (Walker, Yesavage & Tinklenberg, 1981). In addition, it is believed that hallucinogenic drug use can precipitate psychotic illnesses in certain individuals predisposed to the development of such illnesses (Bowers & Swigar, 1983). If psychosis persists beyond 2 weeks after hallucinogen ingestion, it should be regarded as another primary psychiatric illness.

REFERENCES

AGHAJANIAN GK (1976). Tolerance of locus ceruleus neurons to morphine and inhibition of withdrawal response by clonidine. Nature 276:186–188.

American Psychiatric Association (1987). Diagnostic and statistical manual of mental disorders (3rd ed., rev.). Washington, DC: American Psychiatric Press, Inc.

BERESFORD TP, BLOW KC, HILL E, ET AL. (1990). Clinical practice: Comparison of CAGE questionnaire and computer assisted laboratory profiles in screening for covert alcoholism. Lancet 336:482–485.

BERG R, FRANZEN M, & WEDDING D (1987). Screening for brain impairment. New York: Springer-Verlag.

BERNADT MW, TAYLOR C, MUMFORD J, ET AL. (1982). Comparison of questionnaire and laboratory tests in the detection of excessive drinking and alcoholism. Lancet 1:325–328.

BOWERS MB, & SWIGAR ME (1983). Vulnerability to psychosis associated with hallucinogen use. Psychiatry Res 9:91–97.

CHARNEY DS, HENNINGER GR, & JATLOW PI (1985). Increased anxiogenic effects of caffeine in panic disorders. Arch Gen Psychiatry 42:233–243.

Charney DS, Sternberg DE, Kleber HD, et al. (1981). Clinical use of clonidine in abrupt withdrawal from methadone. Arch Gen Psychiatry 38:1273–1278.

Chasnoff IJ, Burns WJ, Schnoll SH, et al. (1985). Cocaine use in pregnancy. N Engl J Med 313:666–669.

Clark WD (1981). Alcoholism: Blocks to diagnosis and treatment. Am J Med 71:275–285.

Costa E, & Guidotti A (1979). Molecular mechanisms in the receptor action of benzodiazepines. Annu Rev Pharmacol Toxicol 19:531–545.

Dackis CA, & Gold M (1985). Bromocriptine as treatment of cocaine abuse. Lancet 1:1151.

Dews PB (1982). Caffeine. Annu Rev Nutr 2:323–341.

Dietz AJ (1981). Amphetamine-like reactions to phenylpropanolamine. JAMA 245:601–602.

Dole VP, & Nyswander M (1965). A medical treatment for diacetylmorphine (heroin) addiction: Clinical trial with methadone hydrochloride. JAMA 193:646–650.

Dorus W, Ostrow DG, Anton R, et al. (1989). Lithium treatment of depressed and non-depressed alcoholics. JAMA 262:1646–1652.

Ellinwood EH (1969). Amphetamine psychosis: A multidimensional process. Semin Psychiatry 1:208–226.

Emrick C (1974). A review of psychologically oriented treatment of alcoholism. Q J Stud Alcohol 38:1004–1031.

Emrick CD (1987). Alcoholics Anonymous: Affiliation processes and effectiveness as treatment. Alcoholism 11:416–423.

Ewing JA (1984). Detecting alcoholism: The CAGE questionnaire. JAMA 252:1905–1907.

Fawcett J, Clark DC, Aagesen CA, et al. (1987). A double-blind, placebo controlled trial of lithium carbonate therapy for alcoholism. Arch Gen Psychiatry 44:448–456.

Foy A, March S, & Drinkwater V (1988). Use of an objective clinical scale in the assessment and management of alcohol withdrawal in a large general hospital. Alcoholism Clin Exp Res 12:360–364.

Fultz JM Jr, & Senay EC (1975). Guidelines for the management of hospitalized narcotics addicts. Ann Intern Med 82:815–818.

Gawin FH, & Ellinwood EH Jr (1988). Cocaine and other stimulants. Actions, abuse and treatment. N Engl J Med 318:1173–1182.

Gawin FH, & Kleber HD (1984). Cocaine abuse treatment: Open trial with desiprimine and lithium carbonate. Arch Gen Psychiatry 41:903–909.

Gay GR (1982). Clinical management of acute and chronic cocaine poisoning. Ann Emerg Med 11:562–572.

Geller G, Levine DM, Mamom JA, et al. (1989). Knowledge, attitudes and reported practices of medical students and housestaff regarding the diagnosis and treatment of alcoholism. JAMA 261:3115–3120.

Gersten SP (1980). Long-term adverse effects of brief marihuana usage. J Clin Psychiatry 41:60.

Giannini AJ, Folts DJ, Feather JN, et al. (1989). Bromocriptine and amantadine in cocaine detoxification. Psychiatry Res 29:11–16.

Golbert TM, Sanz CJ, Rose HD, et al. (1967). Comparative evaluation of treatments of alcohol withdrawal syndromes. JAMA 201:113–116.

Gold MS, Redmond DE Jr, & Kleber HD (1979). Noradrenergic hyperactivity in opiate withdrawal suppressed by clonidine. Am J Psychiatry 136:100–102.

Gold MS, Pottash AC, Sweeney DR, et al. (1979). Opiate withdrawal using clonidine: A safe, effective and rapid nonopiate treatment. JAMA 234:343–346.

Gottschalk L, McGuire F, Heiser J, et al. (1979). Drug abuse deaths in nine cities: A survey report. NIDA Research Monograph No. 29. Washington, DC: U.S. Government Printing Office.

Gross M, Lewis E, & Hastey J (1974). Acute alcohol withdrawal syndrome. In Kissin B, Begleiter H (Eds.), The biology of alcoholism (Vol. 3). New York: Plenum.

Halikas JA, Crosby RD, Carlson GA, et al. (1991). Cocaine reduction in unmotivated crack users using carbamazepine versus placebo in a short-term, double-blind crossover design. Clin Pharmacol Ther 59:81–95.

Hansen HJ, Caudhill SP, & Boone DJ (1985). Crisis in drug therapy: Results of the CDC blind study. JAMA 253:2382–2387.

Harvey SC (1980). Hypnotics and sedatives. In Gilman AG, Goodman LS, & Gilman A (Eds.), The pharmacological basis of therapeutics (6th ed., Ch. 17, pp. 339–375). New York: Macmillan.

Hawks RL, & Chiang CN (1986). Urine testing for drugs of abuse. NIDA Research Monograph No. 73. Washington, DC: U.S. Government Printing Office.

Hillbom ME, & Hjelm-Jager M (1984). Should alcohol withdrawal seizures be treated with anti-epileptic drugs? Acta Neurol Scand 69:39–42.

Hoffman PL, Rabe CS, Grant KA, et al. (1990). Ethanol and the NMDA receptor. Alcohol 7:229–231.

Holden C (1985). The neglected disease in medical education. Science 229:741–742.

Hughes JR, Higgins ST, Bickel WK, et al. (1991). Caffeine self-administration, withdrawal and adverse effects among coffee drinkers. Arch Gen Psychiatry 48:611–617.

Hunt WA (1983). The effect of ethanol on GABAergic transmission. Neurosci Biobehav Rev 7:87–95.

Jaffe JH, & Martin WR (1990). Opioid analgesics and antagonists. In Gilman AC, Goodman LS, Rall TW, et al. (Eds.), The pharmacological basis of therapeutics (8th ed., Ch. 21, pp. 485–521). New York: Macmillan.

Kennedy W (1985). Chemical dependency: A treatable disease. Ohio State Med J 71:77–79.

Keventus J, & Major LF (1979). Disulfiram in the treatment of alcoholism. Q J Stud Alcohol 40:428–446.

Klein E, Uhde T, & Post RM (1986). Preliminary evidence for the utility of carabamazepine in alprazolam withdrawal. Am J Psychiatry 143:235–236.

Klonoff H (1974). Marihuana and driving in real-life situations. Science 186:317–324.

Kosten TR, & Kleber HD (1988). Differential diagnosis of psychiatric comorbidity in substance abusers. J Subst Abuse Treat 5:201–206.

Kosten TR, Kleber HD, & Morgan C (1989). Treatment of cocaine abuse with buprenorphine. Biol Psychiatry 26:637–639.

Kranzler H, & Meyer R (1989). An open trial of buspirone in alcoholics. J Psychopharmacol 9:379–380.

Lewis DC, & Gomolin IH (1982). Emergency treatment of drug and alcohol intoxication and withdrawal. Brown University Program in Alcoholism and Drug Abuse Medical Monograph II. Providence, RI: Brown University Press.

Lewis D, & Gordon A (1983). Alcoholism and the general hospital: The Roger Williams intervention program. Bull NY Acad Med 59:181–197.

Liskow BI, & Goodwin DW (1987). Pharmacological treatment of alcohol intoxication, withdrawal and dependence: A critical review. J Stud Alcohol 48:356–370.

Litten RZ, & Allen JP (1991). Pharmacotherapies for alcoholism: Promising agents and clinical issues. Alcoholism Clin Exp Res 15:620–633.

Malcolm R, Ballenger JC, Sturgis ET, et al. (1989). Double

blind controlled trial comparing carbamazepine to oxazepam treatment of alcohol withdrawal. Am J Psychiatry 146:617–621.

MARTIN WR (1976). Naloxone. Ann Intern Med 85:765–768.

MASON BJ, & KOCSIS JH (1991). Desipramine treatment of alcoholism. Psychopharmacol Bull 27:155–161.

MCINTOSH I (1982). Alcohol-related disabilities in general hospital patients: A critical assessment of the evidence. Int J Addict 17:609–639.

MIRIN SM, WEISS RD, MICHAEL J, ET AL. (1988). Psychopathology in substance abusers: Diagnosis and treatment. Am J Drug Alcohol Abuse 14:139–157.

NARANJO CA, KADLAC KE, SANHUEZA P, ET AL. (1990). Fluoxetine differentially alters alcohol intake and other consumatory behaviors in problem drinkers. Clin Pharmacol Ther 47:490–498.

National Survey on Drug Abuse (1981). Rockville, MD: Institute of Drug Abuse.

PALESTINE ML, & ALATORRE E (1976). Control of acute alcoholic withdrawal symptoms: A comparative study of haloperidol and chlordiazepoxide. Curr Ther Res 20:289–299.

PATTERSON JF (1990). Withdrawal from alprazolam using clonazepam: Clinical observations. J Clin Psychiatry 51(5, suppl):47–49.

RAPPOLT RT, GAY GR, & INABA D (1977). Propranolol: A specific antagonist to cocaine. Clin Toxicol 10:265–271.

RESNICK RB, SCHUYTEN-RESNICK E, & WASHTON AM (1980). Assessment of narcotic antagonists in the treatment of opioid dependence. Annu Rev Pharmacol Toxicol 20:463–474.

ROSECAN JS (1983). The psychopharmacologic treatment of cocaine addiction. Presented at the annual meeting of the VIIth World Congress of Psychiatry, Vienna.

ROUNSAVILLE BJ, GAWIN FH, & KLEBER HD (1985). Interpersonal psychotherapy adapted for ambulatory cocaine users. Am J Drug Alcohol Abuse 11:171.

SATEL SL, PRICE LH, PALUMBO JM, ET AL. (1991). Clinical phenomenology and neurobiology of cocaine abstinence: A prospective inpatient study. Am J Psychiatry 148:1712–1716.

SCHELLENBERG F, BENARD JY, LEGOFF AM, ET AL. (1989). Evaluation of carbohydrate-deficient transferrin compared with Tf index and other markers of alcohol abuse. Alcoholism Clin Exp Res 13:605–610.

SCHULZ DW, & MACDONALD RL (1981). Barbiturate enhancement of GABA-mediated inhibition and activation of chloride channel conductance: Correlation with anticonvulsant and anesthetic actions. Brain Res 209:177–188.

SCHWARTZ RH (1988). Urine testing in the detection of drugs of abuse. Arch Intern Med 148:2407–2412.

SEEMAN P (1972). Membrane effects of anesthetics and tranquilizers. Pharmacol Rev 24:583–655.

SELLERS EM, COOPER SD, ZILM DH, ET AL. (1980). Lithium treatment during alcohol withdrawal. Clin Pharmacol Ther 20:199–206.

SELLERS EM, & KALANT H (1976). Drug therapy: Alcohol intoxication and withdrawal. N Engl J Med 294:757–762.

SELLERS EM, NARANGO CA, HARRISON M, ET AL. (1983). Diazepam loading: Simplified treatment for alcohol withdrawal. Clin Pharmacol Ther 6:822.

SELZER ML (1971). The Michigan Alcoholism Screening Test: The quest for a new diagnostic instrument. Am J Psychiatry 127:89–94.

SENAY EC (1985). Methadone maintenance treatment. Int J Addict 20:803–821.

SHAW JM, KOLESAR GS, SELLERS EM, ET AL. (1981). Development of optimal treatment tactics for alcohol withdrawal: Assessment and effectiveness of supportive care. J Clin Psychopharmacol 1:382–387.

SKINNER HA, HOLT S, SCHULLER R, ET AL. (1984). Identification of alcohol abuse using laboratory tests and a history of trauma. Ann Intern Med 101:847–851.

SKOLNICK P, MONCADA V, BARKER J, ET AL. (1981). Pentobarbital: Dual action to increase brain benzodiazepine receptor affinity. Science 211:1448–1450.

STEIN M (1990). Medical consequences of intravenous drug abuse. J Gen Intern Med 5:249–257.

SWIFT RM, GRIFFITHS W, & CAMARA P (1991). Special technical considerations in laboratory testing for illicit drugs. In A STOUDEMIRE & BS FOGEL (Eds.), Medical psychiatric practice (Vol. 1; Ch. 4, pp. 145–161). Washington, DC: American Psychiatric Press, Inc.

TAKASE S, TAKADA A, TSUTSUMI M, ET AL. (1985). Biochemical markers of chronic alcohol. Alcohol 2:405–410.

TELLER DW, & DEVENYI P (1988). Bromocriptine in cocaine withdrawal—does it work? Int J Addict 23:1197–1205.

TENNANT FSG, & RAWSON RA (1983). Cocaine and amphetamine dependence treated with desipramine. In HARRIS L (Ed.), Problems of drug dependence. NIDA Monograph Series No. 43 (pp. 351–355). Washington, DC: U.S. Government Printing Office.

VAN DYKE C, JATLOW P, UNGERER J, ET AL. (1978). Oral cocaine: Plasma concentration and central effects. Science 200:211–213.

WALKER S, YESAVAGE JA, & TINKLENBERG JR (1981). Acute phencyclidine (PCP) intoxication. Quantitative urine levels and clinical management. Am J Psychiatry 138:674–675.

WASHTON AM, & RESNICK RB (1981). Clonidine for opiate detoxification: Outpatient clinical trials. J Clin Psychiatry 43:39–41.

WATSON JM (1982). Solvent abuse: Presentation and clinical diagnosis. Hum Toxicol 1:249–256.

WESSON DR, & SMITH DE (1977). A new method for the treatment of barbiturate dependence. JAMA 231:294–295.

WHITFIELD CL, THOMPSON G, LAMB A, ET AL. (1978). Detoxification of 1024 alcoholic patients without psychoactive drugs. JAMA 239:1409–1410.

WIKLER A (1968). Diagnosis and treatment of drug dependence of the barbiturate type. Am J Psychiatry 125:758–765.

YOUNG T, LAWSON GW, & GACOCN CB (1987). Clinical aspects of phencyclidine (PCP). Int J Addict 22:1–15.

9 Psychopharmacology in the medical patient

ALAN STOUDEMIRE, M.D., BARRY S. FOGEL, M.D.,
LAWRENCE R. GULLEY, M.D., AND
MICHAEL G. MORAN, M.D.

The decision to use psychotropic agents in patients with combined medical and psychiatric illness requires careful risk-benefit assessment for a number of reasons. Psychotropic drugs can interact with underlying medical illness, causing serious complications, as when tricyclic antidepressants exacerbate heart block in patients with cardiovascular disease. Metabolic abnormalities associated with physical illness can increase the chances of drug toxicity because of altered pharmacokinetics, as when lithium is used in patients with renal insufficiency. Since medical patients are likely to be taking other nonpsychotropic medications, the chance of a clinically significant drug interaction is increased. Finally, elderly medically ill patients are at higher risk for adverse central nervous system (CNS) effects as a result of their altered pharmacodynamic responses to psychotropic agents. This chapter gives an overview of the basic psychopharmacologic principles that should be considered in selecting and using psychotropic agents in the medically ill. Individual specialty chapters in this volume discuss psychopharmacologic considerations in greater detail for specific patient populations.

CYCLIC ANTIDEPRESSANTS

Cardiovascular Complications

One of the primary concerns in using cyclic antidepressants (CyADs) in medically ill patients is the possible precipitation or exacerbation of cardiovascular complications. (In this chapter, the abbreviation CyADs refers to all unicyclic, bicyclic, tricyclic and tetracyclic antidepressants, excepting the monoamine oxidase inhibitors. TCAs refer to tricyclic antidepressants.) Absolute cardiovascular contraindications to the use of CyADs, however, are few and, outside of the known high-risk groups, the risks are relatively small. For example, the Boston Collaborative Drug Surveillance Program found no evidence that tricyclics caused arrhythmias or sudden death (Boston Collaborative Drug Surveillance Program, 1972). While CyADs are relatively safe when used appropriately, they can cause serious toxicity in excessive dosage, in deliberate overdosage, or when used injudiciously in patients at high risk for specific side effects.

Excessive fear of complications in patient groups who are not actually at high risk for severe cardiovascular complications may explain why seriously depressed medically ill patients may not receive antidepressant treatment, or receive treatment at inadequate doses. In fact, the majority of patients with cardiovascular disease can be safely treated with CyADs if appropriate consideration is given to careful pretreatment evaluation and drug monitoring.

Conduction Effects

Standard tricyclic antidepressants (TCAs) such as imipramine and amitriptyline have quinidine-like effects on the electrocardiogram (ECG). Tricyclic antidepressants can increase the P-R interval, QRS duration, and Q-Tc time, and can cause T wave flattening. Clinically significant lengthening of the P-R, QRS, and Q-T intervals may imply excessive plasma levels (Smith, Chojnacki, & Hu, 1980). Direct relationships have been demonstrated between TCA serum concentrations and slowing of intracardiac conduction under steady state conditions (Kantor et al., 1978; Preskorn, 1989; Preskorn

Portions of this chapter have been adapted, with the publisher's permission, from STOUDEMIRE A, FOGEL BS, & GULLEY LR (1991). Psychopharmacology in the medically ill. In A STOUDEMIRE & BS FOGEL (Eds.), Medical psychiatric practice (Vol. 1; Ch. 2, pp. 29–97). Washington, DC: American Psychiatric Press, Inc.

et al., 1983; Rudorfer & Young 1980; Veith et al., 1980). The clinical implication of this relationship is that monitoring of TCA serum levels may be helpful in preventing clinically significant cardiac conduction delays in vulnerable patients.

By virtue of their quinidine-like effects, the TCAs have type 1 antiarrhythmic properties, inhibiting fast sodium channels and thus prolonging the refractory period of the action potential of the cardiac conduction system. This effect tends to suppress ectopic pacemakers that are thought to cause atrial flutter, atrial fibrillation, ventricular tachycardia, and premature ventricular contractions (PVCs). Thus, when they are used in depressed patients with cardiovascular disease, TCAs can suppress PVCs (Bigger et al., 1977).

Clinically relevant deleterious effects on conduction time at therapeutic dosage levels are observed almost exclusively in patients with preexisting, usually advanced, conduction problems such as atrioventricular (AV) nodal block (Preskorn, 1989). Even in patients with first-degree block, quinidine-like actions of tricyclics at therapeutic doses usually are minor and need not impede treatment. For example, Roose and colleagues compared 150 depressed patients with normal ECGs and 41 depressed patients with first-degree AV and/or bundle-branch block treated for depression with imipramine or nortriptyline. The likelihood of second-degree AV block developing during treatment was greater in patients who had preexisting bundle-branch block (defined as a QRS interval greater than 0.11 seconds) than it was in patients with normal pretreatment ECGs (9% versus 0.7%). However, more than 90% of patients with preexisting conduction disease did *not* develop second-degree heart block (Roose et al., 1987a).

Certain types of heart block present particularly high risks with TCAs. In patients with preexisting bundle-branch block (especially those with second-degree heart block), dissociative (third-degree) AV heart block may develop. If type 1 antiarrhythmic medications (quinidine, disopyramide, procainamide) are concurrently administered with tricyclics, additive effects on conduction are possible (Levenson, 1985). If such a combination (TCAs plus type 1 antiarrhythmics) is considered in patients prone to heart block, treatment usually should be undertaken with serial ECGs every 3 to 4 days until steady state TCA levels are reached, in order to monitor cardiac conduction parameters.

Certain calcium channel blockers may prolong AV conduction and at least theoretically could interact with TCAs in this regard. Diltiazem and verapamil both may slow AV conduction; nifedipine, in contrast, generally does not affect AV conduction and this calcium channel blocker would at least theoretically be preferred if TCA/calcium channel blocker combination therapy were being considered in a patient with cardiac conduction disease.

Recent evidence also indicates that terfenadine, a new nonsedating antihistamine, can cause dangerous prolongation of the Q-Tc interval and fatal arrythmias when used in combination with erythromycin or ketoconazole, which interact with terfenadine to raise its plasma level. It is not as yet known if clinically significant interactions occur with tricyclics. Until more data are available, clinicians should consider potential cardiac effects when prescribing tricyclics with terfenadine.

Right bundle-branch block may be seen as a part of underlying cardiovascular disease but is not in itself a contraindication for CyAD treatment; left bundle-branch block usually implies some degree of ischemic or hypertensive heart disease. When treating patients with these conduction defects, the psychiatrist and cardiologist should jointly establish a schedule for monitoring cardiac effects when CyAD treatment is pursued. In our experience, asymptomatic patients with right bundle-branch block, isolated left anterior fascicular block, and left posterior fascicular block usually can be treated safely with TCAs if dosage is increased gradually and ECGs obtained following each dosage increase (Stoudemire & Atkinson, 1988).

Patients with *symptomatic* conduction defects usually should begin drug therapy in the hospital. Patients with bifascicular or trifascicular block are at high risk for bradyarrhythmias and Stokes-Adams attacks. These patients, and patients with chronic bifascicular block associated with syncopal attacks, should not be treated with tricyclics unless a pacemaker is implanted first. Alternatives to be considered would be antidepressants that have low or no quinidine-like side effects, such as fluoxetine, sertraline, bupropion, or MAOIs.

Although asymptomatic first-degree heart block, right bundle-branch block, and focal fascicular blocks are not absolute contraindications to tricyclics, these same defects contraindicate tricyclic therapy if they are associated with syncopal attacks. When they are, patients usually require pacemaker insertion first if TCAs are to be used. The presence of symptoms is crucial because they suggest that a higher degree of block occurs intermittently.

Prolonged Q-T Syndromes. One of the quinidine-like effects of CyADs on the electrocardiogram is a prolongation of the Q-T interval. Significant prolongation may be associated with an increased risk of ventricular tachycardia and ventricular fibrillation, particularly in patients with congenital or acquired heart disease. The Q-T interval ordinarily varies with heart rate, so that guidelines for CyAD use are best based on the corrected Q-T interval (Q-Tc), which is defined as the actual Q-T interval divided by the square root of the R-R interval. The upper limit of normal for the Q-Tc is 0.42 second for men and 0.43 second for women (Lipman, Dunn, & Massie, 1984).

Patients who have had myocardial infarction (MI) with persistently prolonged Q-T intervals are at relatively higher risk for subsequent fatal ventricular fibrillation. Since the standard tricyclics can potentially further prolong the Q-Tc interval, ECGs following each dosage increase should be obtained if Q-Tc lengthening is a possibility. An upper bound for the Q-Tc should be established in consultation with the cardiologist. A Q-Tc interval of 0.44 second may be used as a guideline.

Other Effects on Cardiac Conduction. Reports of doxepin's superior safety margin in cardiac disease in various studies have been appropriately criticized because doxepin blood levels may have been subtherapeutic (Burrows, Vohra, & Dumovic 1977; Luchins, 1983). For example, Roose et al. (1991a) cited recently completed research showing that the cardiovascular effects of doxepin are comparable to those of other TCAs.

Trazodone initially appeared not to promote cardiac arrhythmias in animals and in humans free of heart disease (Himmelhoch, 1981). A few rare case reports, however, have described complete heart block (Rausch, Pavlinac, & Newman, 1984), aggravation of ventricular arrhythmias (Janowsky, Curtis, & Zisook, 1983), and development of first-degree heart block (Irwin & Spar, 1983). Nevertheless, cardiovascular complications from trazodone are extremely rare, even in overdoses (Richelson, 1984).

Despite the safety of CyADs in the vast majority of patients without preexisting cardiac disease, malignant ventricular arrhythmias rarely have been reported as a complication of CyAD treatment (Fowler, McCall, & Chou, 1976; Krikler & Curry, 1976). Two groups of patients at risk are patients with congenital long Q-T syndrome and patients who develop undue Q-T interval prolongation during antidepressant treatment (Flugelman, Tal, & Pollack, 1985).

The first group is identified by a pre-treatment ECG. The second group is identified by a follow-up ECG on CyAD treatment. While follow-up ECGs clearly are indicated in patients with significant baseline abnormalities, it is not clear which patients with normal baseline ECGs and no cardiac history need follow-up ECGs to screen for drug-induced Q-T prolongation. At present, we advise follow-up ECGs in patients with borderline elevated Q-Tc intervals at baseline, and in patients with a family history of sudden cardiac death, as well as those with definite cardiac histories. A Q-Tc of 0.44 second is regarded by some cardiologists as an acceptable upper limit of Q-T prolongation (Schwartz & Wolf, 1978). We would recommend, however, the setting of a criterion for each individual case in consultation with a cardiologist. Cyclic antidepressants would be reduced or discontinued if the Q-Tc exceeded the selected criterion, or a less cardiotoxic agent would be substituted.

Therefore, in evaluating cardiac patients with depression, particularly patients with a history of cardiac arrhythmias, special attention should be given to considering whether or not a quinidine-like effect would aggravate heart block or increase vulnerability to ventricular tachycardia–ventricular fibrillation. However, the toxic effects of tricyclics in inducing life-threatening arrhythmias have been documented primarily in overdose situations and not at therapeutic blood levels (Fricchione & Vlay, 1986).

Wolff-Parkinson-White Syndrome

Depressed patients with the Wolff-Parkinson-White (WPW) syndrome require special consideration in considering the use of cyclic antidepressants. The WPW syndrome is characterized by the presence of a short P-R interval and widened QRS complex, associated with paroxysmal tachycardia. The condition is caused by an accessory pathway between the atrium and ventricle that allows atrial impulses to prematurely activate the ventricular tissue, short-circuiting the normal AV conduction system. Arrhythmias associated with WPW syndrome have been described as deriving from "circus movement"—that is, a reentrant tachycardia usually based on ventriculoatrial retrograde conduction over the accessory pathway and antegrade conduction through the AV node. Electrocardiographic findings reveal a prolonged QRS duration, a shortened P-R interval, and a delta wave (slowing of the upstroke component of the QRS complex).

In some patients with WPW syndrome, atrial flutter-fibrillation may occur. If quinidine-like drugs such as the TCAs are administered, the atrial rate may slow during atrial fibrillation-flutter to the extent that antegrade conduction via the anomalous accessory pathway will predominate, sometimes leading to ventricular tachycardia or ventricular fibrillation (Sellers et al., 1977).

Patients WPW syndrome who have a short (less than 0.27 second) refractory period are at higher risk for life-threatening ventricular tachycardia if they develop atrial fibrillation (Wellens & Durrer, 1974). Patients with a short refractory period can be identified by a special cardiology procedure known as the procainamide infusion test, in which patients with WPW are monitored with an ECG during an intravenous infusion of procainamide (a type IA quinidine-like antiarrthythmic). If the delta wave does not dissipate and the QRS duration shorten during this infusion, this indicates the presence of a short refractory period and vulnerability to atrial fibrillation (Vieweg et al., 1988; Wellens et al., 1982). All patients with WPW syndrome should be evaluated by a cardiologist prior to treatment with a CyAD, particularly a tricyclic with quinidine-like properties, to assess if a quinidine effect would put them at high risk for arrhythmias.

Congestive Heart Failure

Cyclic antidepressants are relatively safe in the majority of patients with adequately treated cardiac heart failure unless the patient has symptomatic orthostatic hypotension or markedly impaired ejection fraction. In elderly patients (mean age 70) with evidence of preexisting left ventricular dysfunction, imipramine (mean daily dose 223 mg/day, mean plasma level 338 ng/ml) had little or no effect on cardiac ejection fraction as measured by first-pass radionuclide angiography (Glassman, Johnson & Giardina, 1983). These findings are consistent with those of Veith, Raskind, and Caldwell (1982) that demonstrated the safety of imipramine and doxepin in patients with chronic heart disease. Nortriptyline also does not significantly affect ejection fraction in patients with stable congestive heart failure, and may be less likely than imipramine to induce orthostatic hypotension (see later in this section) (Roose, Glassman, & Giardina, 1986). Nortriptyline would thus be a reasonable first-choice TCA in a patient with stable congestive heart failure. The use of bupropion, fluoxetine, and sertraline in patients with cardiovascular disease is discussed later.

Myocardial Infarction

In patients with uncomplicated MI, institution or reinstitution of cyclic antidepressant therapy is relatively noncontroversial after a 6-week waiting period; in the absence of complications such as heart block or orthostatic hypotension, CyADs can be begun sooner. If the depression is serious, however, earlier treatment may be justifiable. Data are lacking to determine a lower time limit for restarting tricyclics following uncomplicated MI in patients who have previously done well on them. Dialogue between the psychiatrist and the cardiologist should address the patient's particular psychiatric and cardiac vulnerabilities.

The presence of secondary complications of MI, such as heart failure, arrhythmias, orthostatic hypotension, and cardiac conduction abnormalities—and the potential influence of CyADs on cardiac rhythm, conduction, and blood pressure, and their potential interactions with other drugs—are important in assessing the relative safety of CyADs in the post-MI phase (Stoudemire & Fogel, 1987). It should be noted that there has been no convincing demonstration through prospective study of increased morbidity or mortality associated with the use of TCAs by post-MI patients (Smith, Chojnacki, & Hu, 1980).

Several recent reports have called into question the routine practice of discontinuing TCAs abruptly after MI. First, the antiarrhythmic efficacy of the TCAs continues to be confirmed, although their use as first-line antiarrhythmics is limited by side effects (CAPS, 1988). Moreover, cardiac arrhythmias have been observed to emerge after TCAs are discontinued, particularly if the withdrawal is abrupt (Regan, Margolin, & Mathew, 1989).

The decision to continue or discontinue TCAs after a cardiac event must be considered on an individual basis, dependent on multiple medical factors, such as the presence of heart block, orthostatic hypotension, and concurrent arrhythmias. Moreover, the complications caused by abrupt tricyclic withdrawal or exacerbation or relapse of depression in the post-MI period also should be considered. Hence, the use of antidepressants in the peri- and post-MI period always must be based on a risk-benefit assessment done in consultation with the patient's cardiologist.

Results of animal experimentation further suggest that tricyclics may actually promote myocardial reperfusion by increasing collateral supply to the ischemic heart (Manoach et al., 1986, 1987, 1989). In experimentally induced MI in cats, administration of a TCA either before or following experimental coronary artery occlusion increased retrograde perfusion of the distal component of the occluded artery and significantly decreased the volume of ischemic myocardium (Manoach et al., 1986). While these results cannot necessarily be generalized to human clinical situations, they further support the relative safety of tricyclics in patients with recent myocardial infarction (Manoach et al., 1989) and make an argument for their continuation, unless there are specific complications related to or aggravated by the tricyclic.

Orthostatic Hypotension

The cardiovascular effect of TCAs most often leading to discontinuation of treatment is orthostatic hypotension. Elderly patients treated with TCAs have a 60% increase in the risk of hip fracture as compared to matched controls, with the highest risk period for falls being the first 90 days of treatment (Ray, Griffin, & Malcolm, 1991). In a study by Glassman and associates, almost half of the study patients had to discontinue the drug because of orthostatic hypotension (Glassman, Bigger, & Giardina, 1979). Impairment of left ventricular function with or without hypotension is a significant risk factor for the development of tricyclic-induced orthostatic hypotension (Glassman, Johnson, & Giardina, 1986). Bundle-branch block may be another risk factor for orthostatic hypotension during treatment (Roose, Glassman, & Giardina, 1986). Glassman, Bigger, and Giardina (1979) reported no consistent relationship between the daily dose or plasma level of imipramine and subsequent development of orthostatic hypotension. A similar lack of predictive correlation between plasma levels and subsequent orthostatic hypotension has also been observed with nortriptyline (Smith, Chojnacki, & Hu, 1980).

Among the TCAs, nortriptyline is less likely than imipramine and amitriptyline to produce orthostatic hypotension at therapeutically equivalent dosages. Nortriptyline causes little orthostatic hypotension in patients with compensated left ventricular impairment and does not appear to significantly alter the ejection fraction when used at therapeutic serum levels (Roose, Glassman, & Giardina, 1986).

Bupropion appears to have minimal effects on the cardiovascular system and does not appear to cause orthostatic hypertension (Roose et al., 1987b) and may even raise blood pressure. Fluoxetine and sertraline appear to have little or no effect on blood pressure (Cooper 1988).

Anticholinergic Effects

Particularly in elderly patients, drugs with high cholinergic receptor antagonism are relatively poorly tolerated, with common complications being constipation and delirium. In men with benign prostatic hypertrophy, urinary retention may be a major problem, necessitating surgical correction before tricyclics can be tolerated. In patients with diabetic gastroparesis, the anticholinergic side effects can exacerbate problems with delayed gastric motility. Precipitation of narrow-angle glaucoma "crises" may occur, although patients with narrow-angle glaucoma can be safely treated with TCAs if they have first been treated ophthalmologically. Open-angle glaucoma, which is more common than the narrow-angle type, is not exacerbated by tricyclics (Lieberman & Stoudemire, 1987).

In patients prone to the anticholinergic side effects, drugs with lesser anticholinergic effect should be chosen. Drugs of choice among the older agents would be desipramine and trazodone. The newer agents bupropion, sertraline, and fluoxetine have almost no anticholinergic effects. Amitriptyline is the most anticholinergic antidepressant and should not be used in such patients.

While the use of trazodone does not appear to produce anticholinergic effects, it may cause priapism or bladder outlet obstruction, and it frequently is excessively sedating in therapeutic doses (300 to 600 mg/day). (The occurrence of priapism with trazodone is often in the first 28 days of treatment at doses that may be lower than 150 mg/day.) The relative side effect profiles of the currently available antidepressants are presented in Table 9–1 (Stoudemire, Fogel, & Gulley, 1991).

Effects in Patients with Renal Failure

Clinicians have noted that patients with renal failure and those on dialysis appear to be somewhat more sensitive to the side effects of TCAs, although the reasons for this phenomenon are not entirely clear (Levy, 1987). Part of the answer may lie in the accumulation of hydroxylated tricyclic metabolites, which have been found to be markedly elevated in renal failure and dialysis patients (Dawling et al., 1982; Lieberman et al., 1985). Although it appears that the hydroxylated metabolites of tricyclics are higher in patients with renal disease as compared to normal control patients, serum levels of the parent tricyclic compounds (amitriptyline and nortriptyline) also have been found to be somewhat higher in dialysis patients compared to control subjects after oral doses (Lieberman et al., 1985). Hence, although there are no data suggesting the need for routine measurement of the hydroxylated metabolites of TCAs in patients with chronic renal failure or those on dialysis, the likelihood that these metabolites contribute to side-effect hypersensitivity argues for more conservative titration of doses in this patient population.

Effects on Seizure Disorders

Although the occurrence of seizures in tricyclic overdose has led to the clinical maxim that tricyclics lower seizure thresholds and should be used with "caution" in patients with seizure disorders, these agents in animal models actually have suppressant effects on some forms of seizure activity within their therapeutic dose ranges (Clifford, Rutherford, & Hicks, 1985). Vernier (1961) demonstrated that amitriptyline and imipramine were comparable to phenytoin as anticonvulsants in maximal electroshock seizures in mice. Both human and animal reports vary in their conclusions as to the effects of CyADs on the seizure threshold (Edwards, Long, & Sedgwick, 1986). Antidepressants may actually reduce

TABLE 9–1. *Side Effect Profiles of the CyADs*[a]

	Effect on Serotonin Reuptake	Effect on Norepinephrine Reuptake	Sedating Effect	Anti-cholinergic Effect	Orthostatic Effect	Dose Range (mg)[e]
Amitriptyline[b]	+ + + +	+ +	+ + + +	+ + + +	+ + + +	75–300
Imipramine[b]	+ + + +	+ +	+ + +	+ + +	+ + + +	75–300
Nortriptyline	+ + +	+ + +	+ +	+ +	+	40–150
Protriptyline	+ + +	+ + + +	+	+ + +	+	10–60
Trazodone	+ + +	+	+ + +	+[c]	+ +	200–600
Desipramine	+ + +	+ + + +	+	+	+ +	75–300
Amoxapine[d]	+ +	+ + +	+ +	+ +	+ +	75–600
Maprotiline	+	+ +	+ +	+	+ +	150–200
Doxepin	+ + +	+ +	+ + +	+ +	+ +	75–300
Trimipramine[d]	+	+	+ +	+ +	+ +	50–300
Fluoxetine	+ + + +	–	–	–	–	20–60
Sertraline	+ + + +	–	–	–	–	50–200
Bupropion	–	–	–	+	–	150–450

[a]Relative potencies (some ratings are approximated) based partly on affinities of these agents for brain receptors in competitive binding studies (Richelson, 1982): – none; +, slight or indeterminate; + +, moderate; + + +, marked; + + + +, pronounced.
[b]Available in injectable form.
[c]Most in vivo and clinical studies report the absence of anticholinergic effects (or difference from placebo). There have been case reports, however, of apparent anticholinergic effects.
[d]Amoxapine and trimipramine have dopamine receptor blocking activity.
[e]Dose ranges are for treatment of major depression in healthy young adults. Lower doses may be appropriate for other therapeutic uses, and for use in the elderly and medically ill.
Adapted by permission from Stoudemire, A, Fogel BS, & Gulley LR (1991). Psychopharmacology in the medically ill. In A Stoudemire & BS Fogel (Eds.), Medical psychiatric practice (Vol. I; Ch. 2, p. 31). Washington, DC: American Psychiatric Press, Inc.

after-discharge duration of electrically kindled hippocampal seizures in rats (Clifford, Rutherford, & Hicks, 1985). Of the drugs studied, amitriptyline, imipramine, desipramine, and maprotiline reduced behavioral seizures but bupropion and trazodone did not. Another study, using a somewhat different animal model, showed proconvulsant effects for imipramine and amitriptyline but anticonvulsant effects for protriptyline and trimipramine (Luchins, Oliver & Wyatt, 1984). Not surprisingly, clinical effects of CyADs on seizure frequently in known epileptics are variable (Edwards, Long, & Sedgwick, 1986). The implication of these studies for the clinical treatment of depression in patients with seizure disorders is discussed in Chapter 20.

Maprotiline, a tetracyclic agent, has a record of producing seizures in nonepileptic patients, usually with doses above 200 mg/day and with rapid loading doses. Among the tricyclic agents, amitriptyline appears most likely to aggravate seizures in humans (Edwards, Long, & Sedgwick, 1986). When epileptic patients develop seizures on CyADs, pharmacokinetic interactions between antidepressants and anticonvulsants often are the cause (Fogel, 1988). Bupropion is relatively contraindicated in patients with epilepsy.

When CyADs are given to treated epileptic patients, anticonvulsant levels should be periodically checked and dosages adjusted to maintain the level in the therapeutic range because of possible drug interactions with anticonvulsants (Table 9–2) (Stoudemire, Moran, & Fogel, 1990). Patients with known brain damage, such as head trauma victims, should receive an electroencephalogram (EEG) prior to antidepressant therapy, and those with significant paroxysmal EEG abnormalities should receive prophylactic anticonvulsants. Because bupropion carries greater risk of seizures than other CyADs, this drug generally should be avoided in patients with epilepsy as well as in other high-risk patients, such as those with a history of head trauma or with focal or paroxysmal EEG abnormalities (Davidson, 1989). Prophylactic anticonvulsants are advisable when bupropion is used in high-risk patients.

A recent retrospective review of a group of 68 patients with traumatic brain injury and seizures suggested that treatment with TCAs (predominantly amitriptyline and protriptyline) led to an increase of frequency in seizure activity in this population. Fourteen patients had seizures after initiation of TCA treatment, and seven of these were already on anticonvulsants prior to TCAs. Although it might be predicted that patients with a history of traumatic brain injury would be a somewhat higher risk for seizures with TCAs, these findings (although retrospective and uncontrolled) suggest that concurrent anticonvulsant therapy does not necessarily guarantee protection from the onset or exacerbation of seizures in this high-risk population (Wroblewski et al., 1990).

At present, there appears to be no absolute contraindication to CyAD therapy for patients with seizures, provided that (1) adequate levels of anticonvulsant drugs

TABLE 9–2. *Reported Drug Interactions with CyADs*

Medication	Interactive Effect
Type IA Antiarrhythmics (quinidine, procainamide)	May prolong cardiac conduction time
Phenothiazines	May prolong Q-T interval and raise CyAD levels
Reserpine Guanethidine Clonidine	May decrease antihypertensive effect
Prazosin and other alpha-adrenergic blocking agents	Potentiate hypotensive effect
Parenteral Sympathomimetic pressor amines (e.g., epinephrine, norepinephrine, phenylephrine)	May cause slight increases in blood pressure
Disulfiram Methylphenidate Cimetidine	Raise CyAD levels
Warfarin	Fluoxetine may increase prothrombin time
Oral contraceptives Ethanol Barbiturates Phenytoin	May lower CyAD levels
Anticholinergic agents	TCA may potentiate side effects
Carbamazepine	Additive cardiotoxicity possible Lower tricyclic levels
Propafenone (Type IC Anti-arrhythmic)	May elevate tricyclic levels
Digitoxin	Fluoxetine may displace digitoxin from protein binding sites and increase bioactive levels of digitoxin (converse is also true)
Cyclosporine	Fluoxetine may elevate cyclosporine levels

Reprinted by permission from Stoudemire A, Moran MG, & Fogel BS (1990). Psychotropic drug use in the medically ill, Part I. Psychosomatics, 31:382.

are maintained, (2) seizures are controlled, and (3) the CyADs are discontinued if they exacerbate seizures in a particular patient despite maintenance of therapeutic anticonvulsant blood levels. For patients without seizures but with known brain damage, a pretreatment EEG is recommended. If the EEG shows definite paroxysmal features, an anticonvulsant such as phenytoin, carbamazepine, or valproate should be started prior to initiating CyAD therapy.

Other Clinically Significant Side Effects

While the side effects of the CyADs are of most frequent concern in elderly patients and patients with preexisting cardiac disease, disease in other organ systems also is relevant to any choice and dosage. The CyADs are metabolized in the liver and, theoretically, their half-life could be increased in patients with liver disease. However, this hypothesis is poorly supported by systematic data. It is reasonable to predict slower metabolism of CyADs in patients with liver disease and to reduce the standard CyAD doses by one-half to two-thirds in patients with severe liver disease until stable drug levels can be obtained or tolerance to side effects is established.

In patients with asthma and chronic obstructive pulmonary disease, the anticholinergic properties of TCAs may increase drying of bronchial secretions. Of much more concern is the possibility of the precipitation of acute asthmatic attacks in asthmatic patients with tartrazine sensitivity, since certain antidepressant preparations may still contain tartrazine as coloring component of FD&C yellow dye No. 5. Urticaria has been reported to occur even in patients treated with tartrazine-free desipramine (Bajwa & Asnis, 1991). The implication of the yellow dye tartrazine contained in antidepressant compounds as a cause of allergic reactions continues to be noted (Pohl et al., 1987). While tartrazine has been removed from most trade name antidepressant dosage forms (e.g., Tofranil and Norpramin), it may still be found in some generic preparations of tricyclics. Patients who develop symptoms of urticaria, bronchospasm, nonthrombocytopenic purpura, angioedema, rhinitis, or anaphylaxis should be considered as possibly reacting to this dye, and patients with a history of aspirin sensitivity should not receive tartrazine-containing compounds at all.

A few final miscellaneous points regarding the CyADs will be noted; they are further discussed in subsequent chapters. Amoxapine has the potential for extrapyramidal side effects because of dopamine receptor blocking activity and should be avoided in elderly patients prone to parkinsonism and in patients with preexisting Parkinson's disease. Tardive dyskinesia is a possibility with prolonged use of amoxapine, and patients receiving amoxapine should be informed of the risk and closely monitored for early signs of involuntary movements. Nortriptyline and trazodone have received special study in stroke patients and were found to be efficacious for poststroke patients (see Chapter 20), but their antidepressant effect in these patients is probably shared by all the CyADs (Lipsey, Robinson, & Pearlson, 1984). Because of its potent antihistaminic effects, doxepin may be the preferred drug in patients with peptic ulcer disease, gastritis, or skin allergy. Doxepin and trimipramine have been used experimentally with some success as primary treatments for peptic ulcers in nondepressed patients (Hoff, Ruud, & Tornder, 1981; Moshal & Kahn, 1981; Rees, Gilbert, & Katon 1984).

Antidepressants that primarily block serotonin reuptake (fluoxetine, sertraline, paroxetine) may be preferred on theoretical grounds in depressed patients with Parkinson's disease, because of the findings of low 5-hydroxyindoleacetic acid metabolites in the cerebrospinal fluid (CSF) found in these patients (Mayeux, Stern, & Williams, 1986). No particular antidepressant, however, has established clinical superiority in the treatment of depression in Parkinson's disease. Cyclic antidepressants may be used in patients with myasthenia gravis, since it is the nicotinic, not the muscarinic, receptor that is affected in this disorder (Cohen-Cole & Stoudemire, 1987).

Dosage and Plasma Levels of Tricyclic Antidepressants

An essential principle for the safe use of antidepressant medication in elderly and medically ill patients is careful titration of dosage. Starting doses of traditional tricyclic drugs such as imipramine, desipramine, and nortriptyline should be low (10 or 25 mg) in patients with orthostatic hypotension or with other predictable vulnerabilities to side effects. Doses may be increased every 2 to 4 days as tolerated. Assessment for expected side effects (e.g., hypotension or cardiac arrhythmia) should be carried out after each dosage increase. Peak plasma levels for most tricyclic antidepressants are reached 2 to 4 hours after oral ingestion, and because of the long half-life of most of the standard antidepressants, the dose may be given once daily in the evening, which can facilitate both sleep induction and compliance. Trazodone generally should be given immediately after meals or with a light snack because it can be a gastric irritant, because it is better absorbed if given with food, and because it can cause dizziness if taken on an empty stomach. If a particular patient experiences severe sedation, however, qhs dosage may be preferable despite these considerations.

Regarding specific tricyclics, consensus on therapeutic levels is best for the secondary amine tricyclics nortriptyline and desipramine (American Psychiatric Association, 1985). For nortriptyline a therapeutic range of 50 to 150 ng/ml is suggested, with higher levels associated not only with more side effects but also with poorer antidepressant effect. For desipramine, Nelson, Jatlow, and Quenlan (1984) have shown that raising levels over 125 ng/ml can convert nonresponders to responders.

It appears that, at least with nortriptyline, age-related effects on metabolism are minimal and other alterations in pharmacokinetics or pharmacodynamic factors, or concurrent organ disease, are more important considerations medically ill patients (Katz et al., 1989). Hence, according to Katz et al., "The most important differences between young and old are in pharmacodynamics rather than pharmacokinetics" (1989, p. 234). Interindividual variation in TCA metabolism is great. Nomograms based on TCA serum levels 24 hours after a 25-mg test dose have been developed to facilitate predicting oral doses to achieve therapeutic serum levels with nortriptyline (Cooper & Simpson, 1978).

In evaluating drug levels, the clinician should strive to obtain levels under uniform conditions. A standard procedure is to draw blood to determine levels in the morning, approximately 12 hours after the last dose of medication.

Regarding antidepressant serum levels and doses in elderly patients, desipramine and nortriptyline are the best studied. In depressed elderly patients the steady state level and half-life of desipramine appear to be approximately the same as in younger and middle-aged individuals (Antal et al., 1982; Cutler & Narang, 1984; Cutler, Zavadil, & Eisdorfer, 1981). Results for nortriptyline conflict as to whether or not half-lives are prolonged in the elderly. One report cites an average therapeutic dose of 40 mg/day in elderly patients (Dawling, Crome, & Braithwaite, 1980), and another cites 30 mg/day (Dawling et al., 1981).

Hydroxylated Metabolites of Tricyclic Antidepressants

Several recent studies have assessed the contribution of the hydroxylated metabolites of nortriptyline and desipramine to the development of side effects and to clinical outcome (Nelson, Atillasoy, & Mazure, 1988; Nelson, Mazure, & Jatlow, 1988; Nordin, Bertilsson, & Siwers, 1987). Although therapeutic serum levels for both drugs are reasonably well established (50 to 150 ng/ml for nortriptyline and >120 ng/ml for desipramine), less is known about the effects of their hydroxylated metabolites and whether or not the measurement of those metabolites is helpful in predicting either efficacy or toxicity.

Several groups have attempted to determine whether levels of these hydroxylated isomers are of clinical importance in predicting cardiotoxicity (Georgotas et al., 1987; McCue et al., 1989a; Schneider et al., 1988; Young et al., 1984, 1985). Although it is well established that nortriptyline can affect cardiac conduction, it has not been clear to what extent the two isomeric forms of 10-hydroxynortriptyline (*cis* and *trans*) contribute to this effect. In fact, investigations to date of the relationship of 10–hydroxynortriptyline isomers to cardiotoxic effects have produced conflicting results (McCue et al., 1989a).

The clinical significance of hydroxylated TCA metabolites is not fully known, and currently available data do not support routine measurement of the 10-hydroxy isomeric metabolites of nortriptyline (McCue et al., 1989b). Nevertheless, the possibility of elevated levels of hydroxylated metabolites of nortriptyline, especially in elderly patients, may partially explain toxicity that develops at apparently therapeutic nortriptyline serum levels (McCue et al., 1989b; Schneider et al., 1988). Hydroxymetabolites also deserve consideration as variables in treatment outcome studies of nortriptyline in depressed elderly or medically ill patients.

The hydroxymetabolites of imipramine and desmethylimipramine do possess quinidine-like type 1A antiarrhythmic properties (Muir, Strauch, & Schaal, 1982; Roose, Dalack, & Woodring, 1991; Wilkerson, 1978). Heart rate and P-R, QRS, and Q-Tc intervals appear to be affected by both desipramine and hydroxydesipramine levels (Stern et al., 1991). Both 1-hydroxydesipramine and desipramine serum levels predict prolonged intracardiac conduction, consistent with results of Kutcher and colleagues noting 2-hydroxydesmethylimipramine involvement in ECG changes in elderly patients (Kutcher et al., 1986). Elderly patients are likely to have slightly higher levels of hydroxymetabolites of desipramine than younger patients, although the true clinical significance of this observation, as with nortriptyline, is not fully known (Nelson, Mazure, & Jatlow, 1988). For example, 2-hydroxydesipramine, the principal hydroxymetabolite of desipramine, was examined in one group of elderly patients (Kutchner et al., 1986). Prolongation of P-R intervals and increased Q-Tc intervals were significantly correlated with steady state 2-hydroxydesipramine concentrations but not with desipramine concentrations alone. In other reports, the adverse effects of desipramine were not significantly correlated with levels of hydroxydesipramine (Nelson et al., 1982). As with nortriptyline, routine measurement of hydroxydesipramine in cardiac patients is not supported by the literature to date (Stern et al., 1991). However, hydroxydesipramine levels may be useful in understanding toxicity when it occurs in renal failure patients (Potter, Rudorder, & Lane, 1984).

A major practical implication of the hydroxymetabolite literature is that serum levels of tricyclics may not necessarily suffice to monitor for cardiac toxicity. Follow-up ECGs are needed in addition to serum levels in monitoring patients at high risk for cardiac toxicity from CyADs.

"SEROTONERGIC" ANTIDEPRESSANTS

Fluoxetine

Fluoxetine, a serotonin reuptake blocker, has very little affinity for muscarinic, histaminic, dopaminergic, serotonergic (5-HT_1 or 5-HT_2), and $alpha_1$ or $alpha_2$ noradrenergic receptors in vitro (Stark, Fuller, & Wong, 1985). Its low affinity for other neurotransmitter systems probably accounts for its very low incidence of anticholinergic, antihistaminic, sedative, and hypotensive effects. In most studies, the side effects that have been most frequently reported are anxiety, insomnia, anorexia, and nausea (Feighner et al., 1988). Recently, however, Lipinski et al. (1989) suggested that the "anxiety, nervousness and insomnia" that affect 10% to 15% of fluoxetine-treated patients actually represent a form of akathisia, a side effect that can limit treatment even if benzodiazepines or beta-blockers are coadministered to attenuate it.

Although fluoxetine causes insomnia in some patients, a substantial number of patients (10% to 20%) experience drowsiness or sedation (Feighner & Cohn, 1985). The sedation may be continuous, or most pronounced in the late afternoon, a pattern also seen with MAOIs.

Fluoxetine appears to be benign in respect to the heart, and no deaths have been reported from overdoses with fluoxetine used as a single agent (Cooper, 1988). In a comprehensive examination of the effects of fluoxetine on cardiac conduction, Fisch (1985) examined a total of 1506 ECGs from 753 patients treated with amitriptyline (54), doxepin (56), imipramine (165), fluoxetine (312), and placebo (166). No appreciable changes in the ECG were seen with the use of fluoxetine. This observation is confirmed by fluoxetine's lack of cardiotoxicity in overdose.

Accumulating sporadic case reports, however, report some cardiac side effects of fluoxetine, although cause-and-effect relationships are difficult to assess because of the anecdotal nature of these observations. Case reports of bradycardia (with syncope) and of atrial fibrillation associated with fluoxetine have been published (Buff et al., 1991; Ellison, Milofsky, & Ely, 1990; Feder, 1991). As of 1990, according to reports made directly to Eli Lilly, 10 cases of atrial fibrillation, 2 cases of atrial flutter, 23 cases of bradycardia, and 2 cases of AV block occurred out of an estimated 1,273,000 patients exposed (Buff et al., 1991). The cause of these purported cardiac side effects is not clear, and the drug continues to appear remarkably benign even in patients with significant degrees of cardiovascular illness.

Several drug interactions are relevant. Fluoxetine has been reported to prolong the half-life of diazepam and reduce its clearance, resulting in higher plasma concentrations of diazepam and a reduction in the rate of formation of the active metabolite *N*-desmethyldiazepam. However, there is no evidence that this effect has any major clinical significance as judged by performance on tests of psychomotor abilities (Lemberger et al., 1988). A few other agents have been studied specifically for pharmacokinetic interactions with fluoxetine; no effect on plasma half-life has been observed for chlorothiazide or tolbutamide (Lemberger et al., 1985).

When fluoxetine is coadministered with other drugs that are highly protein-bound, displacement effects may occur that result in more unbound bioactive drug being available. Such effects are possible with warfarin and digitoxin (but much less so with digoxin). There is no strong evidence that such effects are large enough to require dosage adjustment of either warfarin, digitoxin, or fluoxetine in such situations when they are coadministered with each other. However, rechecking prothrombin times or drug levels after instituting fluoxetine is a reasonable precaution until more definitive information is available.

Fluoxetine's metabolism does appear to be affected by cirrhosis since, in stable alcoholic cirrhosis, the elimination of fluoxetine is substantially reduced. The mean half-life of fluoxetine was 6.6 days in middle-aged patients with cirrhosis, compared to 2.2 days in age-matched volunteer controls. Formation of fluoxetine's principal metabolite, norfluoxetine, was delayed (Schenker et al., 1988). These data are compatible with previous studies in which other low-clearance drugs metabolized predominantly by demethylation have shown impaired elimination in liver disease (Secor & Schenker, 1987; Williams & Mamelok, 1980). It has been recommended that the doses of fluoxetine be reduced by as much as 50% in patients with liver disease, which can be accomplished practically by giving the drug on an every-other-day (or Monday-Wednesday-Friday) schedule. Because the half-life may be even more prolonged in patients with unstable or decompensating liver disease, even lower doses may need to be given. Even under normal circumstances in healthy patients, the elimination half-life of fluoxetine is 2 to 3 days, and that of norfluoxetine is 7 to 9 days. (However, there is a great deal of interindividual variation. Furthermore, chronic high-dose therapy of 80 mg/day has been associated with half-lives of 8 days for fluoxetine and 19 days for norfluoxetine [Pato & Murphy, 1991]). Thus, less-than-daily dosage is rational, and frequently necessary, particularly in elderly patients. The availability of fluoxetine in liquid form has rendered more dosing flexibility for patients needing lower daily doses.

It is now well established that fluoxetine may increase the serum levels of tricyclics and trazodone, and if fluoxetine is added to the drug regimen of a patient with an established tricyclic level, abrupt elevation in tricyclic levels may result and cause tricyclic toxicity. Tricyclic overdoses in patients taking fluoxetine can lead to very prolonged elevation of tricyclic levels (Rosenstein et al., 1991). Fluoxetine inhibits the cytochrome P-450 system (Cavanaugh, 1990) and may elevate plasma levels of any medication that is demethylated, hydroxylated, dealkylated, or sulfoxidated—which would include most psychotropics (but not lithium) (Fabre, Scharf, & Itil, 1991). Carbamazepine levels are known to rise significantly with coadministration of fluoxetine, with a 27% increase in the area under the concentration-time curve when 20 mg of fluoxetine was added to stable carbamazepine doses in a volunteer sample (Grimsley et al., 1991). Details of fluoxetine interactions with other psychotropics as well as other medications can be found elsewhere (Ciraulo & Shader, 1990a, 1990b).

Fluoxetine infrequently provokes seizures, with the incidence of this side effect being probably less than 0.2% and similar to reported rates for seizure activity with tricyclics (Cooper, 1988; Weber, 1989). When using fluoxetine in treating known epileptics, care should be taken to monitor anticonvulsant levels frequently, because the frequency and magnitude of pharmacokinetic interactions between fluoxetine and anticonvulsants are not fully known.

Steiner and Fontaine (1986) reported a toxic reaction, a hyperserotonergic state, when fluoxetine was coadministered with L-tryptophan. When L-tryptophan was added to an existing regimen of fluoxetine in five patients with obsessive-compulsive disorder, the patients developed severe agitation, restlessness, insomnia, nausea, and abdominal cramps that resolved with discontinuation of L-tryptophan (Steiner & Fontaine, 1986). Similar and more severe reactions have resulted from combining MAOIs with fluoxetine (see below).

The syndrome of inappropriate secretion of antidiuretic hormone (SIADH) has been reported as an side effect of fluoxetine, particularly in the elderly (Cohen, Mahelsky & Adler, 1990; Hwang & Magraw, 1989; Vishwanath et al., 1991). SIADH has also been observed occasionally and unpredictably both with tricyclics and with neuroleptics.

The combination of fluoxetine and lithium has been reported to cause toxicity when either agent was added to a stable dose of the other (Noveske, Hahn, & Flynn, 1989; Salama & Shafey, 1989); symptoms were similar to those of lithium toxicity. The interaction was not due to toxic lithium levels; in one of the reported cases the lithium level was actually decreased, and in the other it was only slightly elevated. This interaction is likely to be very rare.

Skin rashes occur in approximately 3% of fluoxetine-treated patients (Stark & Hardison, 1985). The usual fluoxetine rash is a nonspecific maculopapular eruption typical of drug hypersensitivity; drug discontinuation is indicated when it occurs. Exfoliative dermatitis has occurred with use of fluoxetine, although it is rare. Fluoxetine has been associated with anorgasmia in both women and men, with the preponderance of reports affecting women. The plateau phase of the sexual response cycle was reached in almost all cases but orgasm was inhibited. Fluoxetine also has been reported to cause extrapyramidal effects, including parkinsonian symptoms, akathisia, and reversible dyskinesias (Fallon & Liebowitz, 1991).

Sertraline

Sertraline is a new "serotonergic" antidepressant with marked similarities to fluoxetine. As compared to fluoxetine, however, pharmacokinetic differences exist in that its elimination half-life is approximately 24 hours and steady state levels are reached more rapidly than with fluoxetine. The drug has selective effects on serotonin reuptake similar to those of fluoxetine, and thus appears to have minimal effects on pulse rate and blood pressure and minimal or no anticholinergic effects. There do not appear to be clinically significant affects on cardiac conduction, at least in patients with normal pretreatment ECGs. There is some evidence that sertraline can cause sinus bradycardia in some patients (Fisch, 1991). In all likelihood its cardiovascular side effect profile will be as benign as that of fluoxetine, and data from its use in relatively medically healthy adults suggest a very safe cardiac profile.

Sertraline, like fluoxetine, theoretically may increase prothrombin time by competing with warfarin for protein binding sites, leaving more unbound bioactive warfarin; such effects are believed to be of minor clinical significance (Wilner et al., 1991). Sertraline also has been shown to have no clinically significant effects on the plasma concentration or renal clearance of digoxin (Forster et al., 1991). Sertraline has been reported to rarely cause reversible elevations in aspartate aminotransferase (AST), alanine aminotransferase (ALT) and alkaline phosphatase levels (Cohn et al., 1990). Sertraline interacts minimally with the cytochrome P-450 system in the liver and is much less likely than fluoxetine to affect serum levels of other medications. Sertraline's half-life is also shorter, approximately 24 hours.

Paroxetine

Paroxetine is another potent and selective inhibitor of neuronal reuptake of serotonin. It has a mean terminal elimination half-life of 24 hours. Like fluoxetine and sertraline, it does not appear to have clinically significant effects on the ECG. In doses of 30 to 40 mg/day, it also does not appear to alter heart rate or blood pressure.

Maximal plasma concentrations of paroxetine tend to increase with deteriorating renal function, but elimination half-life is prolonged only for patients with severe renal impairment. In patients with hepatic disease there is some evidence of reduced clearance, and patients in this population should be started on doses in the lower end of the therapeutic range. Experience with this drug in patients with significant medical illness remains quite limited at this time, but it appears to have the same range of side effects as fluoxetine and sertraline (Dechant & Clissold, 1991).

BUPROPION

Bupropion has a half-life of approximately 10 hours, and multiple daily doses are recommended. The drug has some structural similarities to amphetamine and diethylpropion, a sympathomimetic agent. Although bupropion has an activating effect that can be beneficial in some patients, it can induce agitation or provoke psychotic episodes (Golden et al., 1985). The properties of bupropion of greatest interest to psychiatrists working with medical patients are its lack of cardiovascular toxicity and its association with seizures.

Cardiovascular Disease and Bupropion

Bupropion has an excellent side effect profile (Roose et al., 1991b). It has minimal anticholinergic properties and has minimal effects on histamine and $alpha_1$ and $alpha_2$ adrenergic receptors. Its relative lack of sedative, anticholinergic, antihistaminic, and anti-alpha-adrenergic effects would make it particularly attractive for medically ill patients if its efficacy were conclusively demonstrated to be equal to that of more established agents.

Early studies demonstrated that bupropion had little significant effect on blood pressure, heart rate, or the ECG, and that it was not associated with orthostatic hypotension (Branconnier et al., 1983a, 1983b; Burroughs Wellcome Company, 1989; Chouinard, Annable, & Langlois, 1981; Farid et al., 1983; Preskorn & Othmer, 1984; Roose et al., 1987b; Wenger, Bustrack, & Cohn, 1982; Wenger, Cohn, & Bustrack, 1983; Wenger & Stern, 1983; Zung, 1983). Bupropion-treated patients have nevertheless reported symptoms of syncope, dizziness, and fainting. Syncope has been reported in 1.2% of bupropion-treated patients versus 0.5% of placebo-treated patients (Bronconnier et al., 1983b). Occasional patients develop *hyper*tension on bupropion.

The previously cited studies did not specifically focus on patients with concurrent advanced cardiovascular disease. However, Roose and associates (1987b) compared the cardiovascular effects of bupropion to imipramine in a small series of depressed patients with congestive heart failure. In a randomized, double-blind, crossover trial 10 depressed patients with left ventricular failure received bupropion (mean dose 445 mg/day) or imipramine (mean dose 197 mg/day). Neither bupropion nor imipramine adversely affected ejection fraction or other indices of left ventricular function. Half of the patients treated with imipramine, however, developed severe orthostatic hypotension. Bupropion, in contrast, was well-tolerated and did not cause orthostatic hypotension. Bupropion had no effect on the ECG. However, since experience with this drug in patients with preexisting heart block is limited, repeating an ECG after dosage increases in such patients would be a reasonable precaution until more systematic studies are reported.

A recent report (Roose et al., 1991b) of 36 depressed patients with cardiac disease has confirmed bupropion's general safety in cardiac patients, indicating that: (1) pulse rate was not changed; (2) supine blood pressure sometimes was elevated; (3) orthostatic hypotension was observed in only 1 of 36 study patients; (4) no significant effects were observed on cardiac conduction, nor were higher degrees of AV block induced in patients with preexisting bundle-branch block; (5) preexisting ventricular arrhythmias were not exacerbated; and (6) no effect was observed on ejection function in patients with impaired left ventricular function. In two patients, however, the drug had to be discontinued because of exacerbation of baseline hypertension.

Association with Seizures

Davidson (1989) critically reviewed the prevalence of seizures with bupropion by comparing bupropion with other available antidepressants. According to his 1989 review, there were at that time 37 known reports of tonic-clonic seizures in patients receiving bupropion. Patients treated for bulimia have shown the highest incidence of associated seizures (5.4%). In patients receiving daily doses of bupropion of no more than 450 mg/day, the risk of seizures was 0.35%. This incidence was compared with estimates of the incidence of seizures with other antidepressants. The summary estimate of the incidence of seizures in nonepileptic depressed patients on long-term (average 28 months) tricyclic treatment at an average dose of 150 mg/day (range 50 to 300 mg) was 0.5% (Davidson, 1989; Lowry & Dunner, 1980).

A seizure incidence of 0.35% to 0.40% with bupropion in the therapeutic dose range is not meaningfully higher than the estimated incidence for tricyclics at the higher end of the dosage spectrum (Johnston et al., 1991). However, bupropion appears riskier than tricyclics for patients with underlying risk factors for epilepsy, and riskier than tricyclics at the typical doses used in the medically ill. Risk factors for seizures with bupropion (and other antidepressants) include active withdrawal from alcohol or benzodiazepines; a history of epilepsy; concurrent medications, such as neuroleptics, known to lower the seizure threshold; and a history of head trauma.

Even patients with risk factors for seizures on bupropion may still receive a bupropion trial if they give informed consent. In this high-risk situation, a relatively higher estimate of risk (e.g., 1% to 5%) should be offered to the patient, and anticonvulsant prophylaxis with phenytoin, carbamazepine, or valproate should be considered, especially in patients with a definite history of seizures or with epileptiform EEGs.

Theoretically, bupropion might be relatively contraindicated in the treatment of depression in alcoholics because of their increased risk of seizures (as a result of head trauma, withdrawal, etc.). Furthermore, bupropion undergoes a first-pass effect influenced by alcoholic liver disease, which can alter both pharmacokinetics and dose-response relationships. Although prolongation of the half-life of bupropion metabolites in patients with a history of alcoholism may be only minimal, the drug should be used with lower initial doses and carefully titrated, if it is used at all in alcoholic patients (DeVane et al., 1990).

NEUROLEPTICS

The appropriate use of neuroleptics in the medically ill and geriatric populations requires considerable knowledge and care on the part of the physician. In addition to their use in psychiatric patients who have concurrent medical illnesses and in medically ill patients with secondary psychiatric symptoms, neuroleptics are widely used in geriatric patients to control such symptoms as agitation, wandering, behavioral dyscontrol, and confusion. In fact, neuroleptics are the third most utilized class of drugs in nursing homes in the United States (Phillipson et al., 1990). The extensive use of these drugs in a group of patients who are especially susceptible to their untoward side effects calls for careful attention to treatment strategies and risk-benefit assessment. We recommend adhering to the following principles in the use of neuroleptics in the medically ill:

1. Whenever possible, identify and treat the underlying illness rather than superficial symptoms; for example, treat the etiology of a patient's delirium or cognitive impairment instead of merely controlling secondary symptoms of agitation or hallucinations. This is especially important since, in the medically ill, neuroleptics can produce improvement in overt psychiatric symptoms while underlying medical illness progresses.
2. Avoid irrational polypharmacy and be cognizant of the potential for drug interactions, and of pharmacokinetic and pharmacodynamic reasons for an increased risk of adverse effects. For example, decreased albumin in malnourished patients implies increased free fractions of protein-bound drugs (Jenike, 1988).
3. Identify specific target symptoms and carefully evaluate the treatment response objectively. Whereas agitation and behavioral symptoms may respond promptly, symptoms such as hallucinations and delusions may require weeks to respond.
4. Use the minimum effective dosage. There is growing evidence of a "therapeutic window" in neuroleptic treatment. High neuroleptic dosages are associated with increased side effects and do not necessarily increase antipsychotic efficacy (Kane, 1990). Even when higher doses give somewhat better control of positive psychotic symptoms, the patient's overall function may be worse because of extrapyramidal side effects such as bradykinesia, apathy, akathisia, or rigidity. When treating geriatric patients, starting doses as low as 0.5 to 1 mg of haloperidol often are sufficient.
5. Since available neuroleptics, when corrected for dosage, are generally equally efficacious (with the possible exception of clozapine, which may offer increased efficacy in refractory schizophrenia), neuroleptic choice should be made on the basis of the side effect profile. When possible, desirable side effects such as sedation in patients with disturbed sleep, or H_2-antihistaminic activity in patients with peptic ulcer disease, should be exploited.

The major classes of standard neuroleptics include phenothiazines (such as chlorpromazine, trifluoperazine, perphenazine, fluphenazine, and thioridazine), butyrophenones (such as haloperidol), thioxanthenes (such as thiothixene), dihydroindolones (such as molidone), and dibenzoxazepines (such as loxapine). Clozapine is an atypical neuroleptic that is chemically related to loxapine but has a distinctive pharmacologic profile (see below).

All neuroleptics are well absorbed from the gastrointestinal tract. Hypoacidity (as in antacid therapy) and increased gastric emptying may result in increased absorption, whereas increased gastric acidity and delayed gastric emptying will result in decreased absorption, although such effects are of doubtful clinical significance. Neuroleptics are highly lipophilic and highly protein bound. Thus, as with cyclic antidepressants, dialysis is not useful in overdose. Neuroleptics are metabolized primarily via hepatic oxidation pathways. Portal hypertension or congestive failure can decrease first-pass metabolism and increase the activity of these drugs. Generally speaking, the serum half-lives of neuroleptics are greater than 24 hours (Silver & Simpson, 1988), although chlorpromazine has a fairly short half-life of about 8 hours (Lader, 1989).

Common drug interactions with neuroleptics are listed in Table 9-3, p. 172 (Stoudemire, Moran, & Fogel, 1991).

Low-Potency Neuroleptics

Neuroleptics may be broadly classified into two major groups: low-potency agents such as chlorpromazine and thioridazine, and high-potency agents such as haloperidol and fluphenazine. Low-potency neuroleptics have a side effect profile that includes alpha-adrenergic receptor blockade resulting in a risk of orthostatic hypotension, histaminic blockade resulting in sedation, and anticholinergic effects that can cause dry mouth, constipation, blurred vision, urinary retention, and sinus tachycardia.

In the elderly, the central anticholinergic actions of these medications are especially likely to increase cognitive and memory deficits and place the patient at increased risk for central anticholinergic toxicity and delirium. In particular, patients with Alzheimer's disease are more susceptible to these deleterious side effects.

In addition, the low-potency neuroleptics are more likely to be associated with miscellaneous dermatologic and systemic side effects such as photosensitivity and cholestatic jaundice. The former has about a 3% incidence in the elderly. Neuroleptic-induced cholestatic jaundice is almost always benign and is thought to be an allergic response since it is frequently accompanied by rash, fever, and eosinophilia (Lader, 1989). Its incidence appears to be decreasing for unknown reasons. At higher doses (600 to 800 mg/day), thioridazine carries a small risk of pigmentary retinopathy; doses over 800 mg/day are contraindicated because of an unacceptable risk of this side effect. It is also the neuroleptic most likely to be associated with male sexual dysfunction (e.g., retrograde ejaculation, impotence).

Low-potency phenothiazines are also more likely to have quinidine-like side effects on myocardial conduction. This may result in P-R and Q-T interval prolongation and in flattening of the T wave on the ECG. These effects rarely are clinically significant unless the patient is also receiving a type 1 antiarrhythmic or tricyclic antidepressant, or has a preexisting heart block or prolonged Q-T syndrome (Stoudemire, Moran, & Fogel, 1991). Thioridazine appears to have the most pronounced quinidine-like effects of the neuroleptics and should not be used in patients with heart block.

Although all neuroleptics can lower the seizure threshold, the low-potency phenothiazine chlorpromazine is strongly associated with seizures, and should be avoided in epileptics and patients at high risk for seizures. Hematologic side effects such as agranulocytosis, transient leukopenia, and thrombocytopenia can occur with any neuroleptic, but have a higher incidence with low-potency neuroleptics (Balon & Berchou, 1986). Common and reversible endocrinologic side effects of all neuroleptics include gynecomastia, amenorrhea, and galactorhea.

In general, because of their propensity to cause orthostatic hypotension and sedation, low-potency neuroleptics should be avoided in patients with acute medical illness. This is especially true in delirious patients, in whom monitoring the sensorium is of critical importance, and in patients with cardiovascular or cerebrovascular disease, in whom an episode of hypotension could have catastrophic effects.

The use of low-potency agents in patients with chronic and stable medical illness, particularly patients with Alzheimer's disease, is reasonable in the special situation in which a relatively small dose of a low-potency agent (e.g., 10 to 25 mg of thioridazine) gives dramatic relief of target symptoms. At very low dosages, the hypotensive, sedative, and anticholinergic effects of low-potency agents may be minimal or easily tolerated. If low doses of low-potency agents do not control symptoms, switching to a high-potency agent usually is safer than raising the dose of the low-potency agent.

High-Potency Neuroleptics

High-potency neuroleptics, with haloperidol as the prototypic agent, are less likely to induce orthostatic hypotension and sedation, but are more likely to be associated with extrapyramidal side effects (EPS). Acute EPS syndromes include acute dystonias, parkinsonian symptoms, bradykinesia, akathisia, and the "rabbit syndrome," an unusual rhythmic perioral tremor. In addition, high-potency neuroleptics are more commonly associated with neuroleptic-induced catatonia and neuroleptic malignant syndrome.

Acute Dystonias

The risk for acute dystonia is greatest in young men. Dystonia may be manifested by torticollis, retrocollis, tongue protrusion, opisthotonos, facial grimacing, or oculogyric crisis. It occurs in approximately 2.5% of patients receiving neuroleptics (Lader, 1989). The symptoms usually respond to 25 to 50 mg of diphenhydramine or 1 to 2 mg of benztropine intramuscularly or intravenously, followed by oral doses of diphenhydramine 25 to 50 mg tid or benztropine 1 to 2 mg tid. Trihexyphenidyl also is effective. If patients are unable to tolerate anticholinergic agents, amantadine, a weak dopamine agonist with milder anticholinergic effects, can be used instead; 100 mg PO bid is the usual dose (Stoudemire & Fogel, 1987).

Parkinsonian Symptoms

Parkinsonian side effects are particularly prevalent and problematic in elderly patients, and can limit the use of high-potency neuroleptics in this population. Neuroleptic-induced parkinsonism can be clinically indistinguishable from idiopathic Parkinson's disease, with masked facies, bradykinesia, rigidity, pill-rolling tremor that is greater at rest, and festinating gait.

However, parkinsonian side effects from neuroleptics comprise a continuum, and functionally significant bradykinesia, rigidity, or gait disturbance can occur without a prominent tremor or the typical "cogwheeling" on examination of the limbs. The finding of rigidity in the neck is a useful confirmatory sign on physical examination when a partial parkinsonian syndrome is suspected. The occurrence of an incomplete parkinsonian syndrome should trigger the same management approach as a full syndrome, if the symptoms and signs are associated with either subjective distress or impairment in daily activities. Management strategies include reducing the dosage of neuroleptic, switching to a lower-potency neuroleptic, and prescribing concurrent antiparkinsonian agents.

The choice of antiparkinsonian drug should depend on the specific EPS as well as medical risk factors. Anticholinergic drugs, such as trihexyphenidyl and benztropine, are rapidly effective for tremor and dystonia and can be given preventively, but cause a full range of peripheral and central anticholinergic side effects. Patients with constipation, urinary retention, or vulnerability to confusion may tolerate them poorly. Amantadine, a dopamine agonist with mild anticholinergic effects, has fewer physical side effects, although it occasionally can cause hallucinations, confusion, and other mental symptoms. It may be more effective for rigidity and akinesia than the anticholinergics. It is not available parenterally. Finally, severe and disabling drug-induced parkinsonism can be treated with major dopamine agonist drugs, such as levodopa-carbidopa or bromocriptine. Because of the risk of mental side effects of these drugs, they would generally be used only if motor disability outweighed mental disability. Neuroleptics can also unmask idiopathic Parkinson's disease, and, even if antipsychotics are discontinued, up to 11% of patients may show persistent symptoms (Sakauye, 1990).

Akathisia

Akathisia (literally, "unable to remain seated") is an intense, unpleasant subjective sense of inner restlessness and associated motor restlessness. The most common estimate of the prevalence in neuroleptic-treated patients is approximately 20% (Adler et al., 1989). Unlike neuroleptic-induced parkinsonian symptoms, akathisia typically does not respond well to antiparkinsonian therapy, but may respond to beta-adrenergic blockade. Beyond neuroleptic dose reduction, when feasible, propranolol, 30 to 120 mg/day in divided doses, may be effective in most patients, although concerns regarding hypotension and bradycardia in the medically ill may limit this option. It is important to exclude other organic etiologies of intense anxiety states, such as hypoxia, drug withdrawal states, and metabolic abnormalities. For a review of the phenomenology and treatment of akathisia, see Adler et al. (1989).

"Tardive" Disorders

In addition to acute and subacute neuroleptic-induced EPS syndromes, there is a spectrum of belated or "tardive" disorders including tardive dyskinesia, tardive dystonia, and tardive akathisia. Tardive dyskinesia refers to a late-onset movement disorder that is characterized by purposeless repetitive movements that most commonly involve the oral-facial buccal musculature but can include distal tremors and choreiform movement, vermiform movements of the tongue, and writhing movements of the trunk. The most consistently reported risk factors are increased age, female gender, and increased duration of neuroleptic exposure. A recent controlled study of diabetic patients treated with neuroleptics suggests that patients with impaired glucose control may also be at increased risk of developing tardive dyskinesia. Ganzini et al. (1991) studied 38 neuroleptic treated diabetics and compared the incidence of tardive dyskinesia to that in 38 nondiabetic controls. Seventy-nine percent of the diabetics, as opposed to 53% of the nondiabetic patients, developed tardive dyskinesia.

The incidence of tardive dyskinesia is approximately 4% per year of neuroleptic exposure, at least for the first 5 or 6 years, after which incidence may plateau (Kane, 1989). Unfortunately, there is no safe and effective treatment for tardive dyskinesia. Treatment regimens involving presynaptic dopamine depletion via medications such as reserpine and tetrabenazine have met with minimal success. The clinician should therefore emphasize prevention of the syndrome when possible by minimizing dosage of neuroleptics, especially in patient groups at increased risk, and by considering nonneuroleptic alternative therapies when feasible.

Tardive forms of akathisia and dystonia resemble the acute forms clinically, but arise or aggravate after neuroleptics are reduced or discontinued. They can occur with or without associated oral-facial dyskinesia. While their incidence and prevalence are not known, they appear to be less common than typical tardive dyskinesia (Burke, 1992).

Neuroleptic Malignant Syndrome

Neuroleptic malignant syndrome (NMS) is an idiosyncratic and potentially fatal reaction to neuroleptic medications with the following essential features: hyperthermia (98% of cases), muscular rigidity (97% of cases) with attendant elevations in creatine kinase and myoglobinuria, autonomic dysfunction, and altered consciousness (97% of cases) (Caroff et al., 1991). Reported estimates of the syndrome vary from 0.02% to 2.44%, with a mean estimate of 0.67%, according to Keck, McElroy, and Pope (1991). The time frame of the clinical course of the syndrome is somewhat controversial, since previously reported fulminant onset and high mortality rates may have been due to late recognition of the reaction. Some authors suggest that a more appropriate name for the disorder would be extrapyramidal syndrome with fever, in recognition of a spectrum of clinical severity from prodromal to life-threatening symptoms (Sewell & Jeste, 1991). According to a review of 115 cases by Addonizio, Susman, and Roth (1987), NMS typically evolved within 3 days of initial symptoms, and most cases involved high-potency neuroleptics. Symptoms resolved within 13 days of onset of symptoms in nondepot neuroleptic cases.

Putative risk factors for development of NMS include preexisting medical and neurologic illness, dehydration, a previous history of NMS, recent withdrawal of dopaminergic drugs such as amantadine or levodopa-carbidopa, and intramuscular injections. In addition to supportive measures such as intravenous fluids, antipyretics, and intensive monitoring of electrolytes, creatine kinase, and urine myoglobin, specific pharmacologic treatment with the dopamine agonist bromocriptine and the muscle relaxant dantrolene is recommended. The initial recommended dose of bromocriptine is 5 mg PO tid (2.5 mg PO tid in elderly patients). If the patient is unable to take PO medications, then 1 to 2 mg/kg dantrolene as an initial dose should be given intravenously followed by qid dosing if the patient shows a response (Sewell & Jeste, 1991). When feasible, bromocriptine can be started concurrently via nasogastric tube. Differential diagnosis includes CNS infections such as viral encephalitis; structural lesions such as tumors, CNS abscesses, and cerebrovascular accidents; status epilepticus; thyrotoxicosis; pheochromocytoma; drug-related hyperthermia; anesthetic-related malignant hyperthermia; and lethal catatonia (Caroff et al., 1991).

In patients with comorbid medical and psychiatric illness, and in elderly patients who may have preexisting cognitive impairment or motor signs, the patient's known medical condition sometimes provides an apparent explanation for some of the signs of NMS (Addonizio, 1991), making the differential diagnosis particularly difficult. Since untreated NMS is frequently fatal and the toxicity of both bromocriptine and dantrolene is relatively low, we recommend a bias in favor of treating possible cases of NMS, unless an alternate diagnosis appears much more likely.

Reinstituting neuroleptic therapy in a patient who has a previous history of NMS and yet requires further antipsychotic treatment is a clinical dilemma associated with understandable anxiety on the part of the physician. There is a consensus that low-potency neuroleptics should be utilized and the patient should have a drug-free interval of at least 2 weeks (Rosebush, Stewart, & Gelenberg, 1989). Since clozapine has a minimal EPS profile, it was initially believed to be a relatively safe choice for rechallenging patients with a previous history of NMS (Stoudemire & Clayton, 1989). However, there have been recent reports of well-documented clozapine-induced NMS (Anderson & Powers, 1991; Miller, Sharafuddin, & Kathol, 1991).

CLOZAPINE

Clozapine is a novel antipsychotic with a structure that is chemically related to the dibenzoxazepine neuroleptic loxapine. In contrast to loxapine, however, clozapine has greater $alpha_1$- and $alpha_2$-adrenergic, 5-HT_2, and H_1-blocking potency and stronger muscarinic anticholinergic properties. In addition, its D_1 and D_2 receptor affinity is relatively weak when compared with that of traditional neuroleptics (Kane et al., 1988), and its dopamine receptor binding is more pronounced in mesolimbic than in striatal regions (Mattes, 1989). Clozapine is quickly absorbed, and peak plasma concentration is achieved after an average of 3 hours (Cheng et al., 1988). It is 94% protein bound, has a half-life of approximately 16 hours, and is completely metabolized via demethylation and oxidation prior to excretion (Lieberman, Kane, & Johns, 1989). Clozapine does not cause hyperprolactinemia. It is considered to be approximately 1.5 to 2 times as potent clinically as chlorpromazine (Baldessarini & Frankenberg, 1991). Little is known about drug interactions with clozapine, although there are reports of phenytoin causing a decrease in clozapine plasma levels (Miller, 1991) and cimetidine increasing clozapine levels by inhibiting the cytochrome P-450 oxidative system (Szymanski et al., 1991).

While most studies in Europe and the United States clearly support clozapine's clinical efficacy as an antipsychotic, the greatest obstacle to its widespread usage lies in its distinctive side effect profile, primarily its potential for myelosuppression. Agranulocytosis, defined as a white blood count less than 2,000 cells/mm^3 and a polymorphonuclear leukocyte count of less than 500 cells/mm^3, occurs in 1% to 2% of patients treated with clozapine. This potentially fatal side effect is reversible if detected early. A total of 12 deaths worldwide as a result of agranulocytosis were reported from 1984 to 1987. Most cases of clozapine-induced agranulocytosis occur within the first 6 months of therapy, and, if the medication is discontinued, blood counts generally normalize within 7 to 21 days (Lieberman, Kane, & Johns, 1989).

Many of clozapine's other side effects are more predictable and result from its antihistaminic, anticholinergic, and antiadrenergic activity. In one recent multicenter study, the most frequent adverse side effects were sedation (21%), tachycardia (17%), constipation (16%), dizziness (14%), hypotension (13%), and hypersalivation (13%) (Kane et al., 1988). Although tremor, akathisia, and rigidity have been reported with clozapine treatment, these are uncommon (approximately 5%) and generally mild. Moreover, the drug does not appear to be associated with masked facies, acute dystonic reactions, and parkinsonian gait abnormalities. Indeed, improvement of both parkinsonian tremor and psychotic features has been reported with clozapine (Ostergaard & Dupont, 1988).

Friedman and Lannon reported in 1989 on the successful treatment with clozapine of psychotic symptoms in six patients with Parkinson's disease. In patients who exhibited psychotic symptoms such as auditory hallucinations and paranoia and who ranged in age from 52 to 78 years, they found that clozapine improved psychiatric symptoms without worsening motor manifestations of Parkinson's disease such as rigidity and gait disturbance. In two patients, parkinsonian symptoms actually improved. They noted that the likely explanation is that clozapine has relatively low affinity for striatal dopamine receptors. Its affinity for the D_2 receptor in the caudate nucleus is only 1/50th that of haloperidol and 1/10th that of chlorpromazine. The patients in this case series did not develop agranulocytosis, and the major side effect encountered was hypersalivation. Psychotic symptoms were controlled with relatively low doses, ranging from 25 or 50 mg/day up to 275 mg/day (Friedman & Lannon, 1989). Another case report (Roberts, Dean, & Stoudemire, 1989) showed similar symptomatic improvement in hallucinosis and paranoia in a 64-year-old woman with Parkinson's disease. When treated with clozapine 25 mg PO qhs, not only did mental symptoms improve but the patient's parkinsonian symptoms improved to the point at which specific parkinsonian pharmacotherapy (carbidopa, levodopa, and trihexyphenidyl) could be discontinued. These reports have been confirmed more recently by several other papers (Bernardi & DelZampa, 1990; Kahn et al., 1991; Pfeiffer et al., 1990).

Patients with Parkinson's disease may experience psychotic symptoms related either to antiparkinsonian medications or to underlying dementia. Because conventional neuroleptics typically exacerbate the motor symptoms of Parkinson's disease, patients with Parkinson's disease and psychosis always have been a difficult group to manage. Clozapine offers such patients a rational alternative to conventional neuroleptics.

Clozapine lowers the seizure threshold, with the incidence of seizures increasing with dosage: 1% at dosages below 300 mg/day, 2.7% at dosages of 300 to 599 mg/day, and 4.4% at higher dosages of 600 to 900 mg/day. The risk of seizures with clozapine is cumulative over time, reaching 10% at 3.8 years of treatment (Devinsky, Honigfeld, & Patin, 1991). Clozapine is contraindicated in patients with myeloproliferative disorders or granulocytopenia, and should not be used with other medications known to have myelosuppressive effects (e.g., carbamazepine).

If clozapine is psychiatrically indicated in a patient with epilepsy, the patient should be switched to an anticonvulsant other than carbamazepine. To minimize the risk of seizure, anticonvulsant levels should be monitored frequently during initiation of clozapine and kept at the higher end of the therapeutic range. Some authors favor concurrent use of an anticonvulsant such as phenytoin or valproate if clozapine doses in excess of 500 mg are utilized, even in nonepileptic patients (Baldessarini & Frankenberg, 1991).

It is recommended that clozapine treatment be instituted at low dosages (25 to 50 mg on day 1) because of its potential to induce marked sedation and hypotension. In elderly patients a starting dose of 12.5 mg BID is recommended. The typical therapeutic range for schizophrenia in physically healthy patients is 300 to 500 mg/day, with a maximum of 900 mg/day. Dosing should be on a bid or tid schedule (Lieberman, Kane, & Johns, 1989).

Fever up to 103°F may occur when starting clozapine and may persist with fluctuation for 4 to 6 weeks. Fever should be managed supportively with aspirin or acetaminophen and is not in itself a reason to discontinue the drug. However, care should be taken to ensure that patients are adequately hydrated and to frequently reassess patients for symptoms suggesting an infectious cause for the fever.

Food and Drug Administration (FDA) labeling recommends that clozapine be reserved for patients who have not responded to two standard neuroleptics, and for patients with psychosis and Parkinson's disease. In addition, since there is some evidence of a beneficial effect of clozapine on symptoms of tardive dyskinesia, patients with psychosis and tardive dyskinesia may represent a patient population in which clozapine is a logical choice as an antipsychotic. Lieberman et al. (1991) reviewed this issue recently in addition to examining the outcome in their own series of 30 patients with tardive dyskinesia treated with clozapine in an uncontrolled study. These authors found that 43% of patients with preexisting tardive dyskinesia had a 50% or greater reduction in their dyskinetic symptoms on an average daily dose of 486 mg of clozapine by the end of the study.

In summary, clozapine is a novel antipsychotic that has distinctive characteristics, including a favorable EPS profile, no reported cases of drug-induced tardive dyskinesia, and apparent increased efficacy in refractory schizophrenia when compared with traditional neuroleptics. However, its ability to induce agranulocytosis limits its indication to subgroups of psychotic patients who are intolerant of EPS or are refractory to currently available antipsychotics. Moreover, in medically ill or geriatric patients, its epileptogenic potential, its highly anticholinergic profile with the attendant risk of anticholinergic delirium, and its tendency to cause orthostatic hypotension must be considered carefully in estimating the drug's risk-benefit ratio. Since weekly white blood cell counts are required in the United States before the drug can be dispensed, their cost must be considered as well. Common drug interactions with neuroleptics are listed in Table 9-3 (Stoudemire, Moran, & Fogel, 1991).

ALTERNATIVES TO NEUROLEPTICS IN MANAGING AGITATION

Although an appropriate choice of neuroleptic can avoid or minimize drug-specific side effects on blood, skin, liver, or blood pressure, the neurotoxicity of neuroleptics is shared by all drugs in the class. Dystonia, parkinsonism, tardive dyskinesia, hyperthermia, and NMS are potential side effects of all standard neuroleptics. The problem of the neurotoxicity of neuroleptics has led to a search for neuroleptic alternatives for the management of psychosis and agitated states. A number of such alternatives have been documented, although none has been as thoroughly studied in large, well-controlled trials as the neuroleptics themselves (Schneider & Sobin, 1991). For most patients, low-dose, high-potency neuroleptics are the standard treatment for acute agitation and psychosis. However, in patients who are intolerant of neuroleptics, or who are at especially high risk for neuroleptic neurotoxicity, alternatives may be preferable.

TABLE 9–3. *Reported Drug Interactions with Neuroleptics*

Medication	Interactive Effect
Type 1A Antiarrhythmics	Chlorpromazine/Thioridazine may prolong cardiac conduction
Tricyclics Beta-blockers Chloramphenicol Disulfiram Fluoxetine Valproate MAO-I[a] Acetaminophen Buspirone	May increase neuroleptic levels
Barbiturates Hypnotics Rifampin Griseofulvin Phenylbutazone Carbamazepine	Lower neuroleptic levels through induction of hepatic enzymes
Gel-Type antacids	May interfere with neuroleptic absorption
Narcotics Epinephine Enflurane Isoflurane	Potentiate hypotensive effects of neuroleptics
Prazosin ACE inhibitors (captopril, enalapril)[a]	Increase hypotensive effect
Narcotics Tricyclics Barbiturates	May increase sedative effects of neuroleptics
Iproniazid	May cause encephalopathy and hepatotoxicity when used with neuroleptics
Guanethidine Clonidine	Neuroleptics may decrease blood pressure control

[a]MAO-I, monoamine oxidase inhibitor; ACE, angiotensin-converting enzyme.

Reprinted by permission from Stoudemire A, Moran MG, & Fogel BS (1991). Psychotropic drug use in the medically ill, Part II. Psychosomatics 32:36.

Even when there is no reasonable alternative to neuroleptics, it is prudent to minimize the dose and, if necessary, augment the therapeutic effect of neuroleptics with other less toxic agents, such as benzodiazepines. Even in schizophrenia, high doses of neuroleptics do not necessarily give better results than low doses, and they definitely cause more side effects (McEvoy, 1986). Therefore, while further studies of alternatives to neuroleptics are in progress, we advise that alternative drug therapies be seriously considered in patients with a history of severe neuroleptic toxicity or with tardive dyskinesia. Other patients who do tolerate neuroleptics should have dosages minimized, and low-dose neuroleptics supplemented with nonneuroleptic adjuncts should be regarded as preferable to high doses of neuroleptics for long-term use.

Neuroleptics are used in medical settings not only for the treatment of schizophrenia but also for the treatment of agitated states, including agitated delirium, dementia complicated by agitation, agitated depression, mania, and overwhelming anxiety. They are also used to treat impulsive violence in retarded or brain-damaged patients. This section suggests alternative treatments for each of these patient groups.

Delirium with Agitation

In agitated delirium, the goal of drug therapy is to diminish agitation so that the patient can be more easily managed while the underlying cause of the delirium is found and treated. Acutely, this can almost always be accomplished with the benzodiazepine lorazepam, 1 to 2 mg orally, intramuscularly or intravenously every hour until the patient is calm and slightly drowsy. Lorazepam, however, can cause anterograde amnesia when given in this manner. Elderly patients are at highest risk for developing confusion or amnesia from benzodiazepines. Benzodiazepines nevertheless can permit the performance of laboratory tests, diagnostic imaging, and physical examinations so that a diagnosis can be reached. Benzodiazepines are relatively contraindicated in patients with pulmonary disease who are at risk for retaining carbon dioxide, in patients with recent head injury in whom the level of consciousness must be monitored as accurately as possible, and in some patients with severe liver disease. In the latter patients, benzodiazepines can precipitate an encephalopathic episode; the risk is considerably greater for agents metabolized by the liver. In virtually all other patients benzodiazepines are safer than neuroleptics. Specific risks avoided by using benzodiazepines include hyperthermia, exacerbation of seizures, and acute dystonia. Benzodiazepines are particularly useful in treating acute delirious agitation in patients with suspected NMS.

If it appears that the delirium will take several days to resolve, and the patient's agitation cannot be easily managed environmentally, a switch to a low-dose, high-potency neuroleptic is reasonable. In these subacute situations, high-potency neuroleptics offer behavior control equal to or better than that of benzodiazepines, with less sedation and amnesia. (See also Chapter 12.)

Depression with Agitation

For severe agitated depression, electroconvulsive therapy (ECT) often is the best therapeutic choice. If ECT is refused or is not available, benzodiazepines may suffice to temporarily manage all but the most severe agitation. When neuroleptics are needed, we recommend low doses (of less than 300 mg chlorpromazine equivalents) and early use of antiparkinsonian medication if any EPS signs or symptoms develop. If high neuroleptic doses appear necessary, ECT is the safer and wiser choice.

Dementia with Agitation

For dementia complicated by agitation, neuroleptics are frequently used and frequently recommended, although the experimental support for their efficacy is limited (Helms, 1985; Salzman, 1987; Schneider, Pollock & Lyness, 1990) and it is well known that the elderly are at high risk for tardive dyskinesia (Kane & Smith, 1982) and for drug-induced parkinsonism (Friedman, 1992). It is well worth the effort to find effective nonneuroleptic therapy for an agitated demented patient, because some form of neurotoxicity is highly likely to occur with long-term use of neuroleptics.

Alternatives include trazodone (Greenwald, Marin, & Silverman, 1986; Pinner & Rich, 1988; Simpson & Foster, 1986), beta-blockers (Greendyke et al., 1989; Weiler, Mungas, & Bernick, 1988), and carbamazepine (Gleason & Schneider, 1990; Leibovici & Tariot, 1988; Patterson, 1988). Although each of these drugs has its own risks and contraindications, it behooves the clinician to check the current state of the literature and plan a carefully monitored trial of a nonneuroleptic agent that is not contraindicated (Schneider & Sobin [1991] offer a thorough review through 1990). If successful, the trial benefits the patient; if unsuccessful, it provides further clinical and ethical justification for the use of neuroleptics. Fogel (1991) offers recommendations for planning and monitoring such therapeutic trials.

Further alternatives exist if the demented person's agitation is due to a superimposed depression. In this situation antidepressant therapies such as tricyclics, monoamine oxidase inhibitors (MAOIs), or ECT may be helpful despite the presence of an underlying dementia. Since demented patients may not describe the cognitive elements of a major depression (Ott & Fogel, 1992), prior psychiatric history, vegetative signs, and observed affect must guide the decision to try antidepressant therapy.

Mania

For mania, the generally recognized treatment of first choice is lithium. Not all patients respond to lithium, however, and lithium response may take as long as 2 weeks—too long if the mania is disrupting needed treatment for a concurrent surgical or medical condition. In these situations, standard practice is to combine lithium with a neuroleptic. When neuroleptics are contraindicated, ineffective, not tolerated, or refused, we find it helpful to combine lithium with an anticonvulsant antimanic drug (carbamazepine, valproate, or clonazepam). All of these drugs have antimanic efficacy, and may work faster to calm agitated behavior in some cases (Chouinard, 1988; McElroy et al., 1988; Post, 1988).

Clonazepam in particular has been reported to reduce manic agitation in *hours*. Clonazepam can be given at the rate of 2 mg po every 2 hours until the patient is calm; the total loading dose is then given each 24 hours on a q8h schedule. The dosage is adjusted downward if the patient becomes excessively drowsy, and is tapered and discontinued once lithium or other antimanic therapy begins to take effect. The major side effects are sedation and ataxia; the only major contraindication is pulmonary disease with the risk of carbon dioxide retention. Controlled studies of its efficacy, however, are lacking.

Carbamazepine is begun at 200 mg every 12 hours in an acutely manic patient; it is increased every 2 to 3 days to obtain a blood level of 8 to 12 μg/ml. Valproate (Depakote) can be started at 250 mg tid or qid in the acutely manic patient; dosages are adjusted upward as tolerated to attain blood levels of 50 to 120 μg/ml. The major acute side effects are gastrointestinal complaints and sedation. Medical considerations concerning carbamazepine and valproate are discussed below in the section "anticonvulsants as psychotropics."

Overwhelming Anxiety

Overwhelming anxiety is sometimes treated with neuroleptics when it fails to respond to benzodiazepines. One alternative is to add a beta-blocker to the benzodiazepine (Kathol et al., 1980; Ouslander, 1981). We have found this to be particularly useful if tachycardia and tremor are among the principal signs of the anxiety state. A second alternative is to substitute clonazepam for more conventional benzodiazepines. Finally, if the anxiety state is subacute or chronic, antidepressant therapy should be considered. An antidepressant trial is an obvious move if panic attacks accompany the generalized anxiety; some evidence suggests that antidepressants may help chronically anxious individuals even without panic attacks (Kahn, McNair, & Lipman, 1986).

Agitation Following Head Injury

Alternatives to neuroleptics are available for patients with agitation following head injury. These patients have been treated successfully with beta-blockers (Yudofsky, 1981; Yudofsky, et al., 1984). One may begin with 40 mg/day of propranolol or nadolol and gradually increase until the resting pulse is approximately 60. Hypotension, bradycardia, and oversedation are the three limiting side effects. Once the patient is established on an adequate dose of a beta-blocker, 4–8 months may elapse before maximum effects on behavior, so the drug trial should not be discontinued prematurely. Gualtieri (1991) and Chandler, Barnhill, and Gaultieri (1988) have reported that amantadine, in doses of 100 to 400 mg/day, has dramatically helped selected patients with agitation during the recovery phase of traumatic brain injury. Gaultieri described amantadine as a "yes/no" drug—either it helps rapidly and dramatically, or it does not help at all.

Agitated or Self-Injurious Behavior

Agitated or self-injurious behavior in mentally retarded or autistic adults has been treated successfully with beta-blockers (Ratey et al., 1986) and with buspirone (Ratey et al., 1989). Dosage of buspirone should begin relatively low (e.g., 5 mg bid), with the drug increased incrementally over several days until symptoms are relieved, side effects develop, or a dose of 20 mg tid is reached. The most common and troublesome side effects in the mentally retarded population are nonspecific dizziness or lightheadedness, and a paradoxical increase in nervousness or agitation.

In these patients with brain damage, an EEG should always be obtained, and anticonvulsants should be tried if there are paroxysmal or epileptiform features on the EEG. Carbamazepine or valproate are the anticonvulsants of first choice because they have calming effects and are much less likely than barbiturates to disinhibit impulsive behavior. In these situations, dialogue with a neurologist may be useful in arriving at an optimal drug choice.

MONOAMINE OXIDASE INHIBITORS

Despite the introduction of many new psychotropics, MAOIs have remained an important alternative for several conditions, including atypical or tricyclic-refractory depression, panic disorder/agoraphobia, and social phobia. Some patients have strikingly positive therapeutic responses to MAOIs, with poor response to all other available medications. When such patients have a significant concurrent medical illness, situations arise in which risks and benefits must be balanced, and clinical strategies must be employed to contain risks that are not completely avoidable. This section deals with several such issues, including: (1) general anesthesia for patients on MAOIs; (2) drug interactions; (3) dietary precautions; and (4) management of medically important side effects. Before discussing these issues, we review some important differences between the MAOIs most commonly prescribed in the United States: phenelzine (Nardil), tranylcypromine (Parnate), and selegiline (Eldepryl). The first two of these drugs are FDA approved for psychiatric indications. The third is approved for the treatment of Parkinson's disease, but is an effective antidepressant when used at doses sufficient to produce nonselective MAO inhibition (i.e., greater than 10 mg/day) (Mann et al., 1989).

The first main difference among the three drugs is that phenelzine and tranylcypromine are nonselective MAOIs, (inhibiting both MAO-A and MAO-B) whereas selegiline is a selective inhibitor of MAO-B at doses up to 10 mg/day. When selegiline is used at this lower, antiparkinson dose, no dietary restrictions or drug interaction precautions are necessary. At higher doses, although precautions may be necessary, tyramine sensitivity may be less than with tranylcypromine at a comparable dose (Bieck & Antonin, 1989).

A second difference concerns reversibility of MAO

inhibition. While all three MAOIs produce enzyme inhibition that is irreversible to some extent, phenelzine-induced inhibition is the most irreversible. In a study by Bieck and Antonin, some subjects receiving 60 mg/day of phenelzine had abnormal sensitivity to oral tyramine 3 months after discontinuation of the drug! However, within 2 weeks, there was sufficient MAO activity to make a food-drug interaction unlikely, with more than 200 mg of oral tyramine being necessary to raise systolic blood pressure by 30 mm Hg. By contrast, subjects receiving selegiline 20 mg/day had attained a similar level of MAO activity within 2 days of discontinuation, and those receiving tranylcypromine had regained that level of activity within a week. Practically, all patients receiving MAO inhibitors should be warned to observe dietary and drug interaction precautions for at least 2 weeks following discontinuation of the drug. However, food-drug or drug-drug interactions following the recommended 2-week period could occur with phenelzine under unusual circumstances; the possibility should be considered if a patient on phenelzine develops symptoms suggesting such an interaction.

A third difference is that tranylcypromine has an amphetamine-like stimulant effect. In fact, amphetamine is actually an active metabolite of tranylcypromine. For these reasons, it is less likely to cause drowsiness than phenelzine, and less likely to cause weight gain (Cantu & Korec, 1988; Tuomisto & Smith, 1986).

Phenelzine, and the less frequently prescribed isocarboxazid (Marplan), are hydrazides, and have potentially greater hepatotoxicity than the other agents. Although the incidence of liver toxicity from MAOIs is low, a nonhydrazide agent might be preferable in a patient with known liver disease.

Common Side Effects

The most common medically relevant side effect of MAOIs is orthostatic hypotension. Symptomatic orthostatic hypotension occurs in 11% to 14% of patients receiving MAOIs (Rabkin et al., 1984); measurable orthostatic hypotension may occur in 50% of patients (Lazarus et al., 1986). Despite this high rate, some patients who cannot tolerate tricyclics because of orthostatic hypotension may be able to tolerate tranylcypromine (Jenike, 1984). Orthostatic hypotension may be particularly problematic in patients with cerebral vascular disease and those at high risk for falls and fractures. If MAOIs are psychiatrically necessary, volume expansion (Rabkin et al., 1985) or coadministration of stimulants (Fawcett & Kravitz, 1985) may permit successful treatment. The latter strategy, however, runs the risk of hypertension, and should be done in the hospital with both medical and psychopharmacologic consultation if the treating psychiatrist is unfamiliar with MAOIs. An initial, simpler step in minimizing MAOI-induced hypotension is to reduce or eliminate other drugs the patient is taking that might have hypotensive effects. For example, reducing or eliminating diuretic or vasodilator therapy may be sufficient to mitigate the problem in a medically ill patient. Unlike tricyclics, MAOIs do not have quinidine-like effects and are relatively safe with patients with heart block. In fact, they may shorten the Q-Tc interval (Robinson, 1982), but this is usually not clinically relevant.

The MAOIs have relatively little anticholinergic effect, although occasional patients treated with phenelzine have developed urinary retention. Other patients report dry mouth, constipation, or impotence (Sheehan et al., 1980).

Hypertensive crisis, the most dreaded complication of MAOI therapy, is relatively rare. Kline et al. (1980) estimated the frequency of severe hypertensive episodes to be 0.3% of treated patients, with the risk of death less than 0.001%. Although most episodes of hypertension have been associated with food-drug or drug-drug interactions, spontaneous hypertensive reactions can occur and patients should be warned about them. The risk may be somewhat greater with tranylcypromine. This may be related to the observation that tranylcypromine was associated with the highest level of tyramine sensitivity of all MAOIs tested by Cooper (1989).

The MAOIs, particularly tranylcypromine, would be relatively contraindicated in patients at particularly high risk for medical complications from a hypertensive crisis, such as patients on oral anticoagulants. If a MAOI were to be prescribed to such a patient, reasonable precautions would include meticulous informed consent, avoiding tranylcypromine, and supplying the patient with a hypotensive drug (such as nifedipine) to have on hand for immediate treatment of a suspected hypertensive crisis.

Monoamine Oxidase Inhibitors and General Anesthesia

Although MAOI therapy necessitates adaptation of anesthetic technique, it is at worst a mild and relative contraindication to general anesthesia. A 2-week period without MAOIs is ideal prior to elective surgery or ECT, but numerous situations arise where such a washout period is not feasible. Either the need for surgery or ECT is too urgent, or the deterioration of the patient's mental status off MAOIs would pose a major problem. Several case series and reviews have suggested that general anesthesia can safely be given to patients on MAOIs if certain basic precautions are taken (El-Ganzouri et al., 1985; Michaels et al., 1984; Stack et al., 1988; Wells & Bjorksten, 1989; Wong, 1986). The reviews by Stack et al. and by Wells and Bjorksten, as well as an earlier article by Janowsky and Janowsky (1985), spell out the precautions to be taken. These are summarized as follows:

1. Meperidine (Demerol) is absolutely contraindicated for postoperative analgesia in patients on MAOIs. Its use in such patients can produce hypertension, hyperthermia, and death (Brown & Cass, 1979; Denton, Borell, & Edwards, 1962, Palmer, 1960; Stack et al., 1988; Taylor, 1962). The problem is relatively specific to meperidine. Morphine, fentanyl, and codeine all have been given without complications to patients on MAOIs (Davidson, Zung, & Walker, 1984; Michaels, 1984; Yousef & Wilkinson 1988). Nonetheless, because of potential interactions, initial dosage of narcotic analgesics should be conservative.

2. Because it has an indirect sympathomimetic effect, curare should be avoided in MAOI-treated patients (Stack et al., 1988). If succinylcholine is given to a patient on phenelzine, monitoring of neuromuscular function and adequacy of respiration must be particularly careful, because phenelzine can reduce psuedocholinesterase levels in some patients (Bodley, Halwax, & Potts, 1969).

3. If hypotension develops intraoperatively, volume expansion or direct sympathomimetics such as norepinephrine are the preferred therapy. Indirect-acting sympathomimetics such as metaraminol should be avoided, because they can produce hypertensive crises (Janowsky & Janowsky, 1985; Sheehan, 1980).

4. Because of concerns about blood pressure stability, patients receiving prolonged anesthesia for major surgery should be monitored with an indwelling arterial catheter (Sides, 1987; Wong, 1986).

5. If significant hypertension develops during surgery, it can be managed with intravenous phentolamine or nitroprusside (Janowsky & Janowsky, 1985).

6. Droperidol should be avoided as an anesthetic adjunct because it can lead to prolonged cardiac and respiratory depression in patients on MAOIs (Janowsky & Janowsky, 1985).

7. Because of the potential for various drug interactions, particular care must be taken to assure that all staff caring for surgical patients are aware of the patient's MAOI therapy and its implications. Prescribing or drug administration errors sometimes take place in the course of a patient's journey from operating room to recovery room to intensive care to surgical floor. At each point along the way, MAOI-related safety issues should be reinforced. As a practical precaution, a sticker stating that the patient is "allergic to Demerol" can be placed on the patient's chart.

Drug Interactions

The MAOIs are associated with a wide range of drug interactions, many of which are displayed in Table 9-4. These can be grouped into three general classes: interactions with indirect-acting sympathomimetics; interactions with serotonin drugs; and pharmacodynamic potentiation of other drugs' CNS side effects.

Monoamine oxidase inhibitors increase intracellular

TABLE 9–4. *Reported Drug Interactions with Monoamine Oxidase Inhibitors*

Medication	Interactive Effect
Meperidine	Fatal reaction
L-Dopa Methyldopa; Dopamine Buspirone; Guanethidine Cyclic antidepressants Carbamazepine; Cyclobenzaprine	Elevation of blood pressure
Direct-Acting Sympathomimetics Epinephrine; Norepinephine Isoproterenol; Methoxamine	Elevation of blood pressure
Indirect-Acting Sympathomimetics Cocaine; Amphetamines Tyramine; Methylphenidate Phenethylamine; Metaraminol Ephedrine; Phenylpropanolamine	Severe hypertension
Direct- and Indirect-Acting Sympathomimetics Pseudoephedrine Metaraminol Phenylephrine	Severe hypertension
Serotonergic Agents Fluoxetine Tryptophan	"Serotonin Syndrome" (Ataxia, nystagmus, confusion, fever, tremor)
Caffeine Theophylline Aminophylline	Mild increase in blood pressure
Hypoglycemic Agents	Lower blood glucose
Anticoagulants	Prolonged PT
Succinylcholine	Phenelzine prolongs action
Diuretics Propranolol Prazosin Calcium channel blockers	Increased hypotensive effect

Reprinted by permission from Stoudemire A, Moran MG, & Fogel BS (1990). Psychotropic drug use in the medically ill, Part I. Psychosomatics, 31:384.

catecholamine stores. When an *indirect*-acting sympathomimetic agent is given, excessive amounts of catecholamines can be released, leading to a hypertensive crisis. Indirect-acting sympathomimetics include cocaine, amphetamines, methylphenidate, ephedrine, pseudoephedrine, and phenylpropanolamine. As a practical point, the most frequent occasion for such reactions is patients' use of over-the-counter (OTC) cold remedies that include a sympathomimetic decongestant (Harrison et al., 1989). The problem is complicated by the fact that many OTC drugs have multiple formulations with similar sounding names, some of which are safe and others of which are not. Table 9-5 lists common OTC preparations that contain ingredients contraindicated for patients on MAOI therapy.

Although not strictly sympathomimetics, L-dopa and carbidopa-levodopa (Sinemet) can also produce hypertensive reactions in combination with MAOIs. Direct-acting antiparkinson drugs, such as bromocriptine, would presumably be safer if a patient with Parkinson's disease required MAOI therapy at antidepressant doses. The use of selegiline in antiparkinson doses of 10 mg/day is of course compatible with L-dopa or carbidopa-levodopa.

Direct-acting sympathomimetics can be used if necessary—for example, as bronchodilators for the treatment of asthma. Inhalers are safer than systemic drugs. Furthermore, a challenge with the inhaled drug, with blood pressure determination before and after, provides even greater security that a particular drug is safe in the face of MAOI therapy.

The second major type of drug interaction with MAOIs concerns their combination with serotonin drugs such as fluoxetine or buspirone. The interaction with meperidine may have a similar mechanism since meperidine blocks serotonin reuptake (Brown & Cass, 1979; Stack et al., 1988).

Clinical symptoms of the "serotonin syndrome" include myoclonus, ataxia, fever, and confusion. Neuromuscular symptoms range from tremor to lead-pipe rigidity. Neurologic examination shows hyperreflexia, sometimes with clonus; nystagmus; altered mental status; and, frequently, rigidity. Blood pressure may be either normal or elevated, but hypertension, when it occurs usually is not extreme. Fever usually is present, and at times there is hyperthermia. However, profuse sweating, so common in NMS, is not part of the syndrome. The syndrome is best prevented by avoiding the addition of serotonin drugs when patients are on MAO inhibitors. Furthermore, MAO inhibitors should be avoided for at least 5 weeks following discontinuation of fluoxetine because of the very long half-life of the latter drug and its active metabolite norfluoxetine. If the safety of an MAOI is in question in a patient who has recently taken fluoxetine, a fluoxetine blood level should obtained. Detectable levels of fluoxetine or norfluoxetine are a contraindication to starting MAO inhibitors.

Treatment begins with instituting supportive measures and discontinuing the drugs that promote the syndrome. Specific therapy has been reported anecdotally with a number of drugs with serotonin-blocking effects, including chlorpromazine and cyproheptadine (Kahn, 1989; Stoudemire, Fogel, & Gulley, 1991).

The third type of drug interaction concerns pharmacodynamic potentiation of other drugs' CNS effects. For example, MAOIs potentiate the anticholinergic effects of atropine (Janowsky & Janowsky, 1985). Central nervous system depressant effects of narcotics, ben-

TABLE 9–5. *Common over-the-counter (OTC) preparations that contain ingredients contraindicated for patients treated with monoamine oxidase inhibitors*

OTC products containing pseudoephedrine, phenylephrine, or phenylpropanolamine		
	Pseudoephedrine	
Actifed	Robitussin-PE	
Contact	Sine-Aid	
CoTylenol	Sinutab	
Vicks Formula 44M	Sudafed	
Vicks Formula 44D	Tylenol Maximum Strength Sinus Medication	
Vicks NyQuil		
	Phenylephrine	
Dimetane Decongestant	Nostril	
Dristan Advanced Formula Tablets & Coated Caplets	Vicks Sinex	
Neo-Synephrine	Robitussin Night Relief	
	Phenylpropanolamine	
Alka-Seltzer Plus	Cheracol Plus	Sine-Off
Acutrim	Coricidin	Triaminic
Allerest	Dexatrim	
OTC products for which there are prohibited combination formulations		
Acceptable product	*Prohibited product*	
Dimetane	Dimetane Decongestant	
Robitussin	Robitussin-PE, -DM, -CF, Night Relief	
Alka-Seltzer	Alka-Seltzer Plus Cold Medicine	
Cheracol D Cough Formula	Cheracol Plus Head Cold/Cough Formula	
Chlor-Trimeton Allergy Tablets	Chlor-Trimeton Decongestant Tablets	
Sucrets Lozenges	Sucrets-Cold Decongestant Formula Lozenges	
Tylenol	CoTylenol	
Coricidin Tablets	Coricidin 'D' Decongestant Tablets	

Reprinted by permission from Stoudemire A, Fogel BS, & Gulley LR (1991). Psychopharmacology in the medically ill. In A Stoudemire & BS Fogel (Eds.), Medical psychiatric practice (Vol. I; Ch. 2, p. 69). Washington, DC: American Psychiatric Press, Inc.

zodiazepines, and barbiturates can be greater in patients taking MAOIs, as can sedative side effects of nonsteroidal anti-inflammatory drugs. The mechanism of these pharmacodynamic interactions is not well understood, but the prevalence of these interactions argues for particular conservatism in dosage of anticholinergics and CNS depressant drugs in MAOI-treated patients.

Monoamine Oxidase Inhibitors and Food-Drug Interactions

The interaction of MAO inhibitors with tyramine-containing foods remains the primary deterrent to their wider use by psychiatrists, and may be a major reason for their nonuse by nonpsychiatric physicians. Moreover, burdensome dietary restrictions often are a reason that patients are reluctant to take MAOIs even when they are clinically helpful. Patients with concurrent medical diseases, such as diabetes or renal failure, that impose additional dietary restrictions may be particularly reluctant to restrict their diets further.

However, many patients' and physicians' problems with MAOI diets are both unnecessary and avoidable. Sullivan and Shulman (1984) surveyed MAOI diets in 22 psychiatric hospitals in several countries and noted that only aged cheese, nonfresh fish, and wine were restricted on all reported diets. They further observed that total proscription could only be justified for cheese, nonfresh fish, wine, yeast extract (e.g., Marmite), and broad beans—precisely those foods that were prohibited on 85% or more of the surveyed diets. Cooper (1989), reviewing this issue, pointed out that the tyramine content of many other foods mentioned on lists is so low that adverse reactions are unlikely unless an unusually large quantity of the food is eaten. Cooper emphasized that cheese, the food most often implicated in hypertensive reactions, varies enormously in its tyramine content. Cream cheese and cottage cheese are free from tyramine. The tyramine content of cheddar cheese in a large sample varied from 0 to 1.4 mg/g of cheese—an astonishing number when one considers that the average threshold for 30 mm Hg blood pressure elevation with tranylcypromine is only 8 mg of tryamine (Bieck & Antonin 1989). Patients therefore should be told that the lack of a hypertensive reaction on one occasion does not imply that a particular aged cheese is safe.

Psychiatrists frequently confront hospital diets or published recommendations that advocate stringent food restrictions for MAO inhibitors that are not supported by a critical review of the literature. Since spontaneous hypertensive reactions do occur with MAOIs, multiple reports of a particular food-drug interaction along with chemical analyses and/or challenge studies are needed to establish conclusively that a particular food is risky. We recommend explaining these issues to patients, encouraging them to be extremely scrupulous about high-risk foods and less concerned about foods less definitely associated with hypertensive reactions.

Management of Side Effects

Side effects of MAOIs that are particularly problematic in medically ill patients include orthostatic hypotension, hypertensive episodes, weight gain, and insomnia. An additional but infrequent side effect is potentiation of insulin-induced hypoglycemia by MAOIs in diabetic patients (Cooper & Ashcroft, 1966). Active management of these side effects can enable patients who benefit from MAOI therapy to continue the treatment.

Orthostatic hypotension is approached initially by reducing or eliminating diuretics, vasodilators, and other medications that may lower blood pressure. Adequate hydration is assured, and salt intake is liberalized if it has been restricted. If these measures are insufficient, two maneuvers have been reported anecdotally to be helpful, at least for the short term. These are the administration of fludrocortisone (Florinef), 0.1 mg/day (Simonson, 1964), and concurrent treatment with metoclopramide (Patterson, 1987).

Episodes of hypertension, whether spontaneous or attributable to food-drug or drug-drug interactions, pose particular risks in patients with cerebral vascular disease and cardiovascular disease and patients on anticoagulants. In these patients especially, we recommend training patients to recognize the early signs of a hypertensive reaction, such as headache and flushing, and to carry with them an oral hypotensive medication. Patients are asked to seek medical attention should they develop an episode, but immediate use of an oral hypotensive agent relieves some of the risk and discomfort they would otherwise face before getting medical attention. Options include nifedipine 10 mg (Clary & Schweizer, 1987; Schenck & Remick, 1989; Ward & Davidson, 1991); verapamil 80 mg (Merikangas & Merikangas, 1988), chlorpromazine 25 to 50 mg, and thioridazine 25 to 50 mg (Kahn, 1989). If nifedipine is used, absorption is promoted by biting the capsule and swallowing its contents (Ward & Davidson, 1991). If symptoms do not resolve within 15 minutes for nifedipine or 30 minutes for the other drugs, a second dose can be taken.

Insomnia is a common side effect of MAOIs that can be particularly problematic for patients with other medical conditions, such as arthritic pain, that keep them awake at night. In such patients, the first step in managing insomnia is to identify and optimally treat medical problems that can interfere with sleep (see Chapter 21). If this is insufficient, options include a short-acting benzodiazepine at bedtime, or trazodone, 25 to 50 mg qhs (Jacobsen, 1990).

Weight gain in patients on MAOIs is particularly a problem for those with diabetes or hyperlipidemia. In such patients, tranylcypromine usually is the MAOI of choice, because it is less likely to cause weight gain then phenelzine. If a patient must be switched to tranylcypromine from another MAOI, there must be a 2-week drug-free interval to avoid a hypertensive reaction. If weight gain persists on tranylcypromine, caloric restriction and exercise can be advised, but these prescriptions often are not successful.

Autonomic side effects such as urinary retention and impotence may be more likely in patients with preexisting autonomic problems, such as those with diabetes. It is rational to attempt management of these side effects with bethanecol, a cholinergic agonist drug that does not cross the blood-brain barrier. Initial dosage should be conservative—5 mg tid or qid; the dose can be increased if tolerated and helpful. The main side effect is nausea.

Finally, the possibility that MAOIs may influence patients' response to insulin leads us to recommend that, when insulin-dependent diabetics are started on an MAOI, blood sugar should be checked with a glucometer four times a day over the first 72 hours of treatment, and for a similar period after each dosage increase.

LITHIUM CARBONATE

Effect on Renal Function

The primary metabolic consideration in the use of lithium in medically ill patients is renal function. Lithium is excreted by the kidney, and rates of excretion are affected by age and creatinine clearance. Before starting lithium, all patients require a routine assessment of renal function via measurement of serum electrolytes, blood urea nitrogen (BUN) and creatinine, and a standard urinalysis. In patients with known or suspected kidney disease, a 24-hour urine collection to determine baseline creatinine clearance should also be obtained. Lithium excretion is primarily determined by glomerular filtration rate (GFR) and proximal reabsorption. Lithium is filtered freely at the glomerulus; then, approximately 55% of filtered lithium is reabsorbed in the proximal tubule (DePaulo, 1984) and a further 15% is reabsorbed in the descending loop. Sodium depletion increases the reabsorbed fraction of both sodium and lithium ions up to 95%, therefore decreasing clearance of lithium. Thiazide diuretics that act primarily at the distal tubule enhance proximal reabsorption of lithium because they deplete sodium, leading to enhanced proximal reabsorption of sodium and lithium. Loop diuretics such as furosemide appear to have less effect on lithium clearance, although they can deplete sodium as well. Potassium-sparing diuretics such as spironolactone and triamterene also may reduce lithium clearance, although they have been less well studied than other diuretics. Reports of lithium toxicity have been reported with nonthiazide diuretics, however, such as indapamide (Hanna, Lobao, & Stewart, 1990).

Medications such as acetazolamide, theophylline, and aminophylline, which act as diuretics by inhibiting proximal tubular reabsorption, increase lithium *excretion* moderately and may therefore *decrease serum levels*. Nonsteroidal anti-inflammatory drugs, including indomethacin, ibuprofen, phenylbutazone, and piroxicam, decrease renal lithium clearance and *increase* lithium levels (Ragheb, 1990; Rogers, 1985). Aspirin and sulindac, however, apparently have no effect on lithium clearance.

Patients on thiazide diuretics usually need approximately 50% less lithium to attain therapeutic levels, but there is considerable interindividual variation. Patients on diuretics should be dosed slowly, and lithium levels should be monitored at least twice a week during the initiation of therapy. Frequent monitoring of levels should be resumed for a few weeks after any change in diuretic dosage or in diet. Patients on diuretics, and their families, deserve especially detailed warnings about the early signs of lithium intoxication; commercially available bracelets with imprinted medical warnings are appropriate for some patients. A recent report showed that furosemide (a loop diuretic) does not increase serum lithium as much as the distal tubular diuretic hydrochlorothiazide (Crabtree et al., 1989).

Lithium is dialyzable. Therefore, lithium should be given to patients after dialysis, with the usual dose being 300 to 600 mg by mouth. The dose need not be repeated until after the next dialysis. Serum levels of lithium should be taken several hours after dialysis, since plasma levels may actually rise in the postdialysis period, when reequilibration with tissue stores occurs (Bennett, Muther, & Parker, 1980). The dialyzability of lithium can be exploited to rapidly reduce lithium levels in life-threatening cases of lithium toxicity.

In virtually all patients on lithium, there is some loss of the kidney's ability to concentrate the urine. Occasionally, this leads to symptomatic polyuria and a diagnosis of nephrogenic diabetes insipidus. Even when it does not, a careful history reveals more frequent urination and larger urine volumes in most patients on lithium. The mechanism of these changes is a direct toxic effect of lithium on the loop of Henle and the distal tubule. The effect is dose-related. In some cases, losses of concentrating ability are irreversible. Polyuria may disrupt work or sleep routines, and can aggravate incontinence in patients with impaired bladder control. Recent research indicates this problem can be partially mitigated by once-daily dosing of lithium.

When lithium-induced polyuria threatens to limit lithium treatment, there are several options open. First, thiazide diuretics can be employed, with suitable precautions, to enhance lithium reabsorption at the proximal tubule, thereby protecting the more distal nephron from high lithium concentrations (Forrest, Cohen, & Torretti, 1974; Lippman, Wagemaker, & Tuker, 1981; MacNeil, Hanson-Nortey, & Paschalis, 1975). The total lithium dose is reduced by as much as 50% if this strategy is used.

A second option is the potassium-sparing diuretic amiloride, which has been reported to be helpful in ameliorating this side effect in a dose of 10 to 20 mg/day (Kosten & Forrest, 1986). In refractory cases, 50 mg/day of hydrochlorothiazide could be added. With or without adjunctive thiazide therapy, amiloride can increase lithium levels, potentially leading to lithium toxicity if levels are not monitored and dosage adjusted. If amiloride is used alone, hyperkalemia is a risk; electrolytes should be rechecked a few times after starting the drug. If amiloride is used with hydrochlorothiazide, lithium dosage must be reduced.

Side Effects of Lithium

Angiotensin converting enzyme (ACE) inhibitors used in the treatment of hypertension have been reported to have a pharmacokinetic interaction with lithium. This may be due to ACE inhibitor–induced lithium retention (Baldwin & Safferman, 1990). If the two classes of drugs are used together, lithium levels must be monitored closely (Douste-Blazy et al., 1986). Similar problems have been reported with most ACE inhibitors, including enalapril, captopril and lisinopril.

Lithium-induced ECG changes include inversion and flattening of T waves. Sinus node dysfunction and sinoatrial node block have been described, as well as rare episodes of ventricular irritability even at therapeutic levels (Mitchell & MacKenzie, 1982). The most common cardiac abnormality associated with lithium in therapeutic doses is sinoatrial node dysfunction; aggravation of ventricular arrhythmias and heart block have been very rarely reported (Horgan et al., 1973; Jaffe, 1977; Tangedahl & Gau, 1972; Tilkian et al., 1976). In contrast, antiarrhythmic effects of lithium have been described (Levenson et al., 1986; Polumbo et al., 1973). Clinically significant cardiovascular side effects of lithium, however, are sufficiently rare that they are seldom relevant to drug choice, even in patients with cardiovascular disease. Electrocardiograms made before and after initiation of lithium therapy are an appropriate precaution for patients with asymptomatic abnormalities of cardiac conduction or repolarization. In patients with symptomatic arrhythmias, appropriate monitoring during initiation of lithium therapy should be worked out in collaboration with a cardiologist.

Both elderly patients and patients with brain disease are particularly susceptible to developing confusion and tremor even at therapeutic lithium levels. Dosage aiming for the lower end of the therapeutic range is recommended for elderly individuals and others susceptible to the neurologic side effects of lithium (DePaulo, 1984). Maintenance doses of lithium for elderly patients usually are about 50% of the maintenance doses required in younger individuals. When lithium is used to potentiate tricyclics in elderly patients with unipolar depression, doses may be even smaller. At times, doses as small as 150 mg/day may yield therapeutic effects, with much less likelihood of CNS toxicity than with full antimanic dosage (Kushnir, 1986).

Lithium can prolong neuromuscular blockade induced by succinylcholine or pancuronium (Blackwell & Schmidt, 1984). Therefore, lithium therapy should be discontinued during ECT treatment and prior to elective surgery. The aggravation of post-ECT confusion and amnesia by lithium in some patients is another reason to discontinue lithium during ECT.

Because lithium can induce hypothyroidism, patients should have baseline thyroid function tests, including a thyroid-stimulating hormone (TSH) level. The TSH level, measured at quarterly intervals, is an excellent screen for lithium-induced hypothyroidism in its early, asymptomatic stages. In patients with known hypothyroidism, lithium is safe provided adequate thyroid replacement is given. Patients with Hashimoto's disease may have fluctuating thyroid levels; when they are treated with lithium, fluctuations may increase because lithium can cause increases or decreases in thyroid antibody levels. Obtaining TSH and free thyroxine levels is advisable whenever there is an unexplained change in physical or mental status in a patient with Hashimoto's disease who is taking lithium (Lazarus, 1986).

Lithium in Chronic Renal Failure

Although lithium is not recommended in the presence of acute renal failure, there is no evidence that it is necessarily contraindicated in chronic renal failure

(DasGupta & Jefferson, 1990). Lithium doses should be adjusted downward to accommodate decreases in creatinine clearance (Csernansky & Hollister, 1985). Since it is not known if lithium accelerates the progression of certain types of renal disease, close monitoring of renal disease is indicated in patients with chronic renal disease, and alternative mood-stabilizing agents such as carbamazepine or valproate should be considered if not contraindicated for other reasons. The use of lithium in hemodialysis appears to be both safe and effective if lithium doses are reduced to the range of 300 to 600 mg and if lithium doses are given after hemodialysis (Lippman, Manshadi, & Gultekin, 1984; Port, Kroll, & Rosenzweig, 1979; Stoudemire & Fogel, 1987).

Since renal transplantation has become a relatively common procedure, clinicians increasingly will encounter renal transplant patients requiring lithium therapy for bipolar disorder or for mood disorders secondary to corticosteroids. Experience with the use of lithium in transplant recipients remains quite limited, but there are a few reports of its successful implementation (Blazer, Petrie, & Wilson, 1976; Koecheler et al., 1986). Koecheler et al. recommended more conservative use of lithium in patients receiving cadaveric transplants because their renal function is more unstable than in those with living related donor transplants. Cyclosporine can elevate lithium levels by decreasing lithium excretion, so lithium doses may need to be adjusted downward in patients receiving this drug (DasGupta & Jefferson, 1990; Dieperink et al., 1987; Vincent, Weimar & Schalekamp, 1987). Close monitoring of lithium levels is essential given that the transplant's function may be unstable for weeks following transplantation.

TABLE 9–6. *Reported Drug Interactions with Lithium* (Li^+)

Medication	Interactive Effect
Thiazide Diuretics Spironolactone Triamterene Enalapril Non-steroidal Anti-Inflammatory Drugs (e.g., indomethacin, ibuprofen phenylbutazone, piroxicam)	Raise Li^+ levels
Acetazolamide Theophylline Aminophylline	Lower Li^+ levels
Calcium channel blockers	May either raise or lower Li^+ levels, effects not clear; verapamil may cause bradycardia when used with Li^+
Metronidazole	May increase lithium levels; may increase chances of nephrotoxicity
Tetracycline	Minor elevation of Li^+ levels

Reprinted by permission from Stoudemire A, Moran MG, & Fogel BS (1991). Psychotropic drug use in the medically ill, Part II. Psychosomatics 32:38.

Drug Interactions

Table 9-6 summarizes clinically significant drug interactions with lithium (Stoudemire, Moran, & Fogel, 1991). Concurrent use of tetracycline for lithium-induced acne may slightly lower lithium levels, but the effect is of no real clinical significance (Frankhauser et al., 1988).

Choreoathetosis has been observed in one case when lithium was combined with the calcium channel blocker verapamil (Helmuth et al., 1989). Other reports have also documented choreoathetosis as a sign of lithium toxicity (Reed, Wise, & Timmerman, 1989), and other involuntary movements have been reported with the combined use of verapamil and lithium (Price & Giannini, 1986).

Lithium and Diabetes

There is evidence to suggest that, at least in some patients, lithium can decrease glucose tolerance (DasGupta & Jefferson, 1990). However, *increased* glucose tolerance also has been reported as a lithium effect (Vendsborg, 1979). Although glucose tolerance curves may be affected by lithium in some patients, there is no strong evidence that this is a clinically significant effect in the vast majority of patients, nor is there evidence to suggest the need for routine glucose monitoring in patients treated with lithium other than diabetic patients with unstable control of blood glucose, who would undoubtedly be monitored closely anyway (DasGupta & Jefferson, 1990).

Lithium and Epilepsy

Although lithium administration has been associated with EEG changes, a recently report of a series of bipolar patients treated with lithium suggests that lithium may be benign in patients with epilepsy (Shukla, Mukherjee, & Decina, 1988). In an open study of eight patients, lithium levels were maintained in the 0.6 to 1.1 mEq/liter range and concurrent anticonvulsants were limited to phenytoin and phenobarbital (no patients were on carbamazepine, sodium valproate, or clonazepam). Lithium prevented the recurrence of affective episodes without increasing seizure frequency in patients with incompletely controlled seizures and did not induce any seizures in well-controlled patients. Although lithium should not be dismissed because of an undue concern about aggravation of seizures, the mood-stabilizing anticonvulsants carbamazepine and sodium valproate are natural alternatives for patients with concurrent bipolar disorder and epilepsy. When lithium and anticonvulsants are given simultaneously, pharmacodynamic interactions may lead to CNS side effects at lithium levels within the therapeutic range (Fogel, 1988), so the upper end of the therapeutic range of lithium levels usually should be avoided.

ANTICONVULSANTS AS PSYCHOTROPICS

In the last several years, anticonvulsants have entered the mainstream of psychopharmacologic practice. Carbamazepine (Tegretol), valproate (Depakote, Depakene), and clonazepam (Klonopin) all have been widely employed for psychiatric indications, although they are not specifically FDA-approved as psychotropics. They have found a particularly important place as alternatives or adjuncts to lithium in the treatment of bipolar disorder (Goodwin & Jamison, 1990; Pope et al., 1991; Prien & Gelenberg 1989). Other experimental applications have included treatment-refractory anxiety and depressive disorders, and behavior disturbances in the mentally retarded (McElroy & Pope, 1988).

Carbamazepine

Issues to be considered when prescribing carbamazepine to medically ill patients include hematologic toxicity, hepatic toxicity, quinidine-like effects on cardiac conduction, antidiuretic actions, enzyme induction leading to drug-drug interactions, clinical interpretation of carbamazepine blood levels, and management of carbamazepine overdose. Drug interactions are summarized in Table 9-7. A particularly common and relevant interaction for medical-psychiatric practice is the interaction of carbamazepine with the calcium channel blockers diltiazem and verapamil. These two drugs, but not nifedipine, raise carbamazepine levels substantially, frequently producing toxicity if carbamazepine dosage is not lowered (Bahls, Ozuna, & Ritchie, 1991).

TABLE 9–7. *Reported Drug Interactions with Benzodiazepines, Psychostimulants, and Carbamazepine*

Medication	Interactive Effect
	Benzodiazepines
Cimetidine Disulfiram Ethanol Isoniazid	May elevate serum levels of benzodiazepines metabolized predominantly by oxidation
Estrogens Cigarettes Methylxanthine derivatives Rifampin	Tend to lower benzodiazepine levels
	Psychostimulants
Guanethidine	Decreased anti-hypertensive effect
Vasopressors	Increased pressor effect
Oral Anticoagulants	Increased prothrombin time
Anticonvulsants	Increased levels of phenobarbital, primidone, phenytoin
Tricyclics	Increased blood levels of CyAD
MAO-Is	Hypertension
	Carbamazepine
Erythromycin	May raise carbamazepine levels and precipitate heart block
Anti-arrhythmics	May have additive effects on cardiac conduction time
Diltiazem Verapamil Danazol	May raise carbamazepine levels to toxic levels
Valproate	Lower levels of valproate when used with carbamazepine; valproate increases metabolites of carbamazepine

Reprinted by permission from Stoudemire A, Moran MG, & Fogel BS (1991). Psychotropic drug use in the medically ill, Part II. Psychosomatics, 32:38.

Hematologic Toxicity

When carbamazepine was first introduced in the United States, the manufacturer recommended frequent blood counts because of concerns about the development of agranulocytosis. However, as evidence accumulated that these potentially fatal side effects were rare and idiosyncratic, this recommendation was dropped. Present practice regarding monitoring for hematologic interactions is based on the idea that there are two different hematologic reactions to carbamazepine. One is a predictable and often transient drop in both red and white blood cell counts; the other is a rare and idiosyncratic failure of the bone marrow that can occur at an unpredictable time after initiation of therapy. Leukopenia occurs in 7% to 12% of treated patients (Seetharam & Pellock, 1991; Sobotka, Alexander, & Cook, 1990). Leukopenia apparently is unrelated to aplastic anemia, which occurs in approximately 1 in 575,000 treated patients per year (Seetharam & Pellock, 1991).

Recent editorials in *Neurology* (Pellock & Willmore, 1991) and in the *Canadian Medical Association Journal* (Camfield et al., 1989) have advised *against* any *routine* blood monitoring, with the latter publication arguing that the cost is excessive and, at worst, patients are denied needed therapy because of transient and benign laboratory abnormalities. Baseline laboratory panels should be used to identify patients at high risk for side effects; they and only they would receive regular monitoring of blood tests on an individualized basis (Sobotka, Alexander, & Cook, 1990). Longer term monitoring focuses on educating patients to report for a complete blood count if symptoms of fever or sore throat develop, suggesting the onset of neutropenia.

When patients have preexisting anemia or neutropenia, the predictable drop in red and white blood cell counts induced by carbamazepine occurs at a lower baseline. However, there is no evidence that, in general, patients with preexisting blood disorders are at greater risk for the life-threatening complications of aplastic anemia and agranulocytosis. Therefore, preexisting cytopenias are relative but not absolute contraindications to carbamazepine. The authors recommend that hematologic consultation be obtained prior to initiating carbamazepine in any patient with a baseline hemoglobin below 12 g/dl or a white blood cell count below 4000/mm^3. The consultant should be asked both for an individualized assessment of risk and for specific guidelines for monitoring and drug discontinuation if carbamazepine is begun.

Patients receiving combined therapy with lithium and carbamazepine may be at somewhat less risk for a lowering of the white cell count because of lithium's stimulatory effects on white cell production (Brewerton, 1986; Vieweg et al., 1986–87). However, leukopenia has been reported with combined therapy (Sheehan & Shelley, 1990).

Hepatic Toxicity

As in the case of hematologic toxicity, hepatic toxicity from carbamazepine comes in two kinds: frequent, predictable, and benign; and rare, idiosyncratic, and life threatening (Dreifuss & Langer 1987). The relatively benign form of toxicity, seen in no more than 5% of patients (Jeavons, 1983), consists of mild, asymptomatic elevations of AST and ALT, usually to less than twice the upper limit of their normal values. The life-threatening toxicity is acute hepatic necrosis with liver failure, occurring in less than 1 in 10,000 treated patients. Only 21 cases of this severe hepatic toxicity were reported in the first 20 years of carbamazepine's clinical use (Jeavons, 1983). Severe hepatic toxicity occurs unpredictably, usually within the first month of therapy but occasionally after several months of uneventful treatment. As with hematologic toxicity, the above-cited editorials recommend against routine blood tests, suggesting that regular blood monitoring be reserved for patients with risk factors for liver disease or with abnormal baseline liver function. All patients should be advised to report for an examination and tests of liver enzymes should they develop anorexia, nausea, vomiting or upper abdominal pain. Elevations of AST and ALT to less than twice the upper limit of normal would not necessitate discontinuation of the drug. Greater elevations would trigger either drug discontinuation or consultation with a gastroenterologist or a specialist in liver diseases.

In regard to monitoring liver enzymes, it should be noted that the gamma-glutamyl transpeptidase (GGTP) level can be markedly elevated by carbamazepine, as well as by other anticonvulsants, in the absence of clinical symptoms of liver disease (Jeavons, 1983). An isolated elevation of GGTP, even to high levels, would indicate consultation with a gastroenterologist but not necessarily discontinuation of the drug. A full panel of liver function tests, including a prothrombin time, should be taken into account when evaluating the significance of an elevated GGTP level.

Prescription of carbamazepine to patients with preexisting liver disease has two risks. The first is that any hepatic reaction to carbamazepine will occur on a lower baseline of liver function, so that a mild reaction could become symptomatic. The second is that carbamazepine will be metabolized more slowly, since its primary route of metabolism is hepatic. For this reason, significant liver disease is a relative contraindication to carbamazepine. Consultation with an internist should be obtained before prescribing carbamazepine to a patient with significant liver disease. In patients such as alcoholics who are at risk for liver disease that is not necessarily apparent on routine screening liver function tests, carbamazepine should be started more slowly than usual, with frequent determinations of liver enzymes, prothrombin time, and carbamazepine levels during the initiation of therapy.

Quinidine-Like Effects on Cardiac Conduction

Carbamazepine, which is similar in chemical structure to the TCAs, also has similar quinidine-like effects on the heart, with the potential for slowing conduction through the AV node and suppressing ventricular automaticity (Benassi et al., 1987). It does not, however, apparently affect the QRS complex or Q-T interval at normal heart rates (Kenneback et al., 1991). Symptomatic heart block has been reported when carbamazepine has been given to patients with known or suspected preexisting cardiac disease (Beerman & Edhag, 1978), and cardiac rhythm disturbances are a feature of severe carbamazepine overdose. Therefore, pretreatment ECGs are warranted prior to carbamazepine therapy. If the ECG shows more severe block than first-degree AV block, right bundle-branch block, left anterior hemiblock, or asymptomatic Mobitz type I

(Wenckebach-type) block, carbamazepine should not be prescribed on an outpatient basis unless the patient has been cleared first by a cardiologist. In patients with benign, asymptomatic forms of heart block, a posttreatment ECG should be obtained to rule out aggravation of the heart block by carbamazepine.

Antidiuretic Actions

Carbamazepine has an antidiuretic action, and is associated both with clinically significant hyponatremia and with mild, asymptomatic reductions in serum sodium (Ashton et al., 1977; Flegel & Cole, 1977; Kalff et al., 1984; Perucca et al., 1978; Stephens et al., 1977; Vieweg & Godleski, 1988; Yassa et al., 1988). The effect is thought to be via a direct action on the renal tubules. Patients with other factors predisposing to hyponatremia, including advanced age, diuretic use, and congestive heart failure, are especially at risk. They should have electrolyte determinations weekly during the first month of carbamazepine therapy, with additional determinations done if there is any change in mental or physical status, or if there are significant changes in carbamazepine dosage or in their other medications. As with neutropenia, the antidiuretic effect of carbamazepine is attenuated when the drug is given together with lithium (Vieweg et al., 1987), because lithium makes the renal tubules less sensitive to antidiuretic hormone (White & Fetner, 1975).

As suggestions for practical management of carbamazepine-induced hyponatremia, the authors offer the following:

1. If the sodium level drops below 125 mEq/liter on carbamazepine, the drug should be discontinued.
2. If the sodium level drops to between 125 and 130 mEq/liter, and carbamazepine appears clinically useful, other drugs that may aggravate hyponatremia, such as thiazide diuretics, should be discontinued if possible. If the sodium level still remains below 130 mEq/liter, carbamazepine should be discontinued.
3. If the sodium level is between 130 and 135 mEq/liter, discontinuation of carbamazepine is not necessary but electrolytes should be followed weekly for 1 month to assure stability of the level. Serum sodium should be rechecked immediately if mental status changes.
4. A work up for SIADH should be carried out if the sodium level remains below 130 mEq/liter after carbamazepine discontinuation and is not otherwise explained (e.g., by congestive heart failure).
5. Discontinuation of long-term carbamazepine, when indicated by a low sodium level, should be *gradual*, to avoid withdrawal phenomena such as cholinergic rebound or seizures.
6. When carbamazepine is being used for the indication of seizures, an alternate anticonvulsant less likely to cause hyponatremia should be initiated prior to tapering carbamazepine (e.g., phenytoin or valproate).

Enzyme Induction Leading to Drug-Drug Interactions

Carbamazepine is known to be a potent inducer of the cytochrome P-450 system. As such, it influences the metabolism of all drugs that rely on this system for their metabolism. One well-known consequence of enzyme induction is that it induces its own metabolism and thereby the need to gradually build up carbamazepine dosage over the first few weeks of treatment to maintain a steady blood level. Two other consequences are clinically significant. The first is that the blood levels of some drugs *may drop* if carbamazepine is added to the patient's medication regimen. This has been reported for alprazolam, with the clinically significant consequence of alprazolam withdrawal when carbamazepine was added to a steady dosage of alprazolam (Arana et al., 1988), for clonazepam (Lai et al., 1978), and for haloperidol (Arana et al., 1986). Valproate levels for a given dose are lower when carbamazepine is given concurrently (Ieiri et al., 1990); a similar effect has been reported on imipramine levels in children (Brown et al., 1990). The second effect is that drug metabolites not ordinarily clinically significant might be present in larger quantities as a result of carbamazepine-induced induction of metabolic enzymes. Hydroxymetabolites of desipramine have been reported to cause ECG changes in a patient concurrently treated with carbamazepine and desipramine, despite a desipramine level in the therapeutic range (Baldessarini et al., 1988).

A practical implication of these observations with carbamazepine is that blood levels of drugs metabolized by the liver should be determined promptly if unexpected toxicity or lack of therapeutic effect occurs in the context of concurrent carbamazepine therapy. Furthermore, toxicity in the presence of apparently therapeutic blood levels is possible on the basis of unusually great concentrations of unmeasured metabolites. Specifically, medically ill patients on combined carbamazepine and tricyclic therapy should have posttreatment ECGs even if tricyclic blood levels appear normal.

Clinical Interpretation of Carbamazepine Blood Levels

Carbamazepine blood levels are published for the use of carbamazepine in the treatment of epilepsy. Typical normal ranges are 8 to 12 μg/ml for single-drug therapy and 4 to 8 μg/ml for combined therapy with other anticonvulsants. However, when carbamazepine is used as a psychotropic, or when it is used together with other medications in medically ill patients, the interpretation of levels is subject to several caveats. First, since carbamazepine is heavily protein-bound, free carbamazepine levels, on which both therapeutic and toxic effects depend, can vary if other drugs displace carbamazepine from its protein binding sites. This has been reported for agents as ubiquitous as aspirin. Second, *pharmacodynamic* interactions can induce neurotoxicity of carbamazepine at therapeutic blood levels when the drug is given in conjunction with other psychotropics (Fogel, 1988). This has been specifically reported for coadministration with haloperidol and with lithium. Recently, the neurotoxicity of combined carbamazepine-lithium therapy has been shown to be additive, rather than synergistic (McGinness, Kishimoto, & Hollister, 1990). Third, the level of carbamazepine needed for maximum psychotropic effect may be greater than the level optimal for seizure control.

Finally, usually unmeasured metabolites, such as carbamazepine 10,11-epoxide, can contribute to both therapeutic and toxic effects. While the ratio of the parent compound to the epoxide is fairly predictable among medically well persons, it may vary in the medically ill, particularly in the setting of liver disease or polypharmacy. In particular, the combination of carbamazepine with valproate was reported to produce toxic levels of carbamazepine 10,11-epoxide in some children and adolescents (Rambeck et al., 1990). For all these reasons, frequent clinical reassessments must supplement blood levels in evaluating carbamazepine effect. Neither free carbamazepine levels nor carbamazepine 10,11-epoxide levels are generally available, and neither can be recommended at this time for routine use as a substitute for scrupulous and frequent clinical monitoring.

Management of Carbamazepine Overdose

Because carbamazepine has been increasingly prescribed both as an anticonvulsant and a psychotropic, the incidence of carbamazepine overdose has been increasing. The problems of carbamazepine overdose include coma, seizures, hypotension, and cardiac arrhythmia. The general approach to supportive management is similar to that used for TCA overdose. Although hemodialysis is of little use in treating carbamazepine overdose, vigorous use of activated charcoal and laxatives is helpful because the absorption of carbamazepine is quite slow, and much may remain in the intestine at the time the patient presents for emergency treatment (Morrow & Routledge, 1989). The phenomenon of prolonged absorption can lead to a recurrence of coma following apparent recovery as a result of the eventual absorption of drug remaining in the intestine (Fisher & Cysyk, 1988; Sethna et al., 1989). For this reason, patients with serious carbamazepine overdose should be observed in the hospital for a full 24 hours following the return of consciousness.

Valproate (Depakote, Depakene)

Issues to be considered when prescribing valproate to medically ill patients include: (1) gastrointestinal side effects, (2) hepatic toxicity, (3) effects on coagulation, (4) drug-drug interactions, (5) clinical interpretation of blood levels, and (6) management of overdose.

Gastrointestinal Toxicity

In comparative studies of anticonvulsant side effects, the most prominent and troublesome side effect of valproate has been nausea, often accompanied by vomiting. Medically ill patients, particularly those with diseases predisposing to nausea, may be at increased risk. Depakote (divalproex sodium) is much less likely to cause gastrointestinal upset than Depakene (valproic acid), and more frequent dosing, preferably after meals, sometimes is better tolerated than larger doses taken fewer times per day. Occasional patients will do better on valproic acid syrup or Depakote "sprinkles," a pediatric preparation designed to be sprinkled on food. However, regardless of the preparation used, a substantial proportion of patients simply will be unable to tolerate the drug.

Hepatic Toxicity

Shortly after valproate was first introduced, there were several deaths from acute hepatic necrosis, and, as of 1988, approximately 100 fatalities had been reported from valproate-induced liver failure (Scheffner et al., 1988). This has led to considerable caution in the use of the drug for fear of this reaction. As experiences have accumulated, however, it appears that hepatic necrosis is a major risk only for children under 2 years, particularly those given multiple-drug therapy for epilepsy. The incidence of hepatic necrosis in adults receiving valproate is well under 1 in 10,000 (Eadie, Hooper, & Dickinson, 1988) and may be as low as 1 in 50,000, with 95% of reported cases developed symptoms within the first 6 months of therapy (Scheffner et al., 1988). Given this very infrequent occurrence of life-threatening hepatic toxicity, routine long-term monitoring of liver function tests does not seem necessary. Periodic liver function tests for the first 6 months are a reasonable precaution but, as noted above, are not recommended by contemporary neurologists as routine for all patients. All patients *should* be warned of the early signs of liver disease, and be told to report immediately for repeat testing of liver function should those signs develop during valproate therapy.

A much more common, although benign, hepatic effect of valproate is an increase in serum ammonia level resulting from valproate's inhibition of urea synthesis (Cotariu & Zaidman, 1988; Hjelm et al., 1986; Kugoh et al., 1986). This elevation in serum ammonia usually is asymptomatic; the effect, however, can be of major concern in individuals with preexisting liver disease, and especially those in whom there is a history of hepatic encephalopathy. Significant liver disease is a relative contraindication to valproate therapy, and close monitoring of both liver enzymes and serum ammonia would be an appropriate precaution with alcoholic patients suspected to have subclinical cirrhosis of the liver. Consultation with a gastroenterologist would be advisable before starting valproate in a patient with known liver disease.

Effects on Coagulation

Valproate therapy can increase the prothrombin time, decrease fibrinogen levels, and reduce the platelet count. These findings, one or more of which may occur in as many as one third of patients receiving valproate (Rochel & Ehrenthal, 1983), rarely lead to clinically significant bleeding. However, valproate-treated patients definitely should have a full coagulation panel, including a platelet count, bleeding time, prothrombin time, and partial thromboplastin time, prior to undergoing surgery or dental work, and patients with preexisting anticoagulant therapy or bleeding diatheses require especially close monitoring during initiation of valproate therapy.

Drug Interactions

In contrast to carbamazepine, which is an enzyme inducer, valproate inhibits liver enzymes that metabolize drugs. *Therefore, it can prolong the half-life of other drugs with mainly hepatic metabolism.* This effect has been documented for diazepam, which has a prolonged half-life in the presence of valproate. In general, the coadministration of long-acting benzodiazepines with valproate may be problematic, both because of the prolongation of benzodiazepine metabolism and because of additive sedation and ataxia. If a benzodiazepine must be given to a patient on valproate, lorazepam is a good choice, since it is not likely to interact significantly. The full range of drugs that might have altered metabolism in the presence of valproate is not known, so the possibility of drug accumulation should be considered with other drugs that rely primarily on hepatic metabolism.

Interactions of valproate with other anticonvulsants have been studied extensively (Bourgeois, 1988). Carbamazepine, phenytoin, and phenobarbital all can lower valproate levels (May & Rambeck, 1985). In contrast, valproate increases levels of phenobarbital by inhibiting its metabolism (Redenbaugh et al., 1980), and raises the free fraction of phenytoin by displacing the drug from protein binding sites. This phenomenon can lead to phenytoin toxicity at apparently 'therapeutic' phenytoin levels (Bruno et al., 1980). Of even greater interest to psychiatrists who might be using valproate together with carbamazepine for treatment-refractory mania is the observation that valproate raises the concentration of the carbamazepine 10,11-epoxide metabolite (Pisani et al., 1986). This metabolite, not usually measured, has additive toxicity with carbamazepine (Bourgeois & Wad, 1984). Thus, when valproate is given concurrently, carbamazepine can produce toxicity at apparently therapeutic levels because of an increased level of its 10,11-epoxide metabolite.

Aspirin in usual antipyretic doses can raise both total and free valproate levels, because of both metabolic enzyme inhibition and displacement of valproate from protein binding sites. Significantly toxicity can result (Goulden et al., 1987). Therefore, alternate agents such as acetaminophen would be preferable for treating fever or minor pain in patients on valproate.

Pharmacodynamic interactions have been reported between valproate and neuroleptics, with the development of an encephalopathic syndrome with diffuse EEG slowing (Van Sweden & Van Moffaert, 1985) or increased parkinsonism (Puzynski & Klosiewicz, 1984). Valproate toxicity also has been reported to develop when erythromycin was given to a patient on valproate, leading to abrupt elevations in valproate levels (Redington, 1992).

Clinical Interpretation of Blood Levels

Therapeutic blood levels for valproate in the treatment of epilepsy usually are reported as 50 to 100 μg/ml. The work of McElroy and colleagues (1989) suggests that effective blood levels for bipolar disorder are similar, and that little clinical effect is seen with blood levels less than 50 μg/ml. Toxic effects are frequently seen with levels greater than 100 μg/ml, so the therapeutic index is quite low. Because of individual variations in metabolism of valproate, blood levels should be obtained routinely during upward titration of valproate dosage, to assure that the blood level is indeed adequate and to avoid toxicity. Both the McLean Hospital (Boston) experience and our own suggest that, even within the therapeutic range, there may be a specific threshold at which therapeutic effects begin. For example, one of the authors (BSF) treated a patient with bipolar disorder and bulimia who had complete relief of bulimia from valproate at a level of 80 μg/ml but no relief at 70 μg/ml.

Considering the relationship between blood levels and toxicity, it should be noted that toxicity may develop at apparently therapeutic levels when the patient is on multiple drugs, whereas levels above the usual therapeutic range may be tolerated in the context of single-drug therapy.

Management of Valproate Overdose

Because valproate is not toxic to the heart, patients receiving aggressive support have tolerated massive valproate overdoses, including ingestions of greater than 50 grams. Current recommendations for managing valproate overdose focus on supportive therapy; although gastric lavage might be considered early in an overdose, it is unlikely to be of much help later in a overdose because absorption of valproate is fairly rapid. One intriguing case report (Alberto et al., 1989) suggests that coma caused by valproate overdose might be reversible by naloxone, because of the latter's effects on gamma-aminobutyric acid (GABA) receptors. Given the low toxicity of naloxone and its empirical use in overdoses of unknown agents, it would be an appropriate consideration as adjunctive therapy even when a patient was known to have taken valproate rather than a narcotic.

Clonazepam

Clonazepam will be discussed in this section as well as in the following section on benzodiazepines. Issues to be considered when prescribing clonazepam to medically ill patients include: (1) consequences of its relatively long half-life and hepatic metabolism, (2) drug-drug interactions, and (3) its effects on respiratory drive.

Effects of Long Half-Life and Hepatic Metabolism

Clonazepam is a long-acting benzodiazepine with a half-life in healthy adults of 20 to 58 hours. The primary route of metabolism is the liver. In elderly patients, or individuals with impaired hepatic function resulting from primary liver disease, congestive heart failure, or metabolic inhibition by other drugs, it may have an even longer half-life. Therefore, it may take well over a week for steady-state levels to be reached. The prescription of a standing dose of clonazepam might be well tolerated in a patient on the first day or two of therapy but could later result in ataxia, sedation, falling, confusion, or stupor as the drug further accumulated.

For this reason, patients receiving clonazepam in the setting of advanced age, primary or secondary liver disease, or metabolic-inhibiting drugs such as valproate (although note that the clonazepam-valproate compound is contraindicated in epileptics), should receive very low initial doses, with up to 2 weeks between upward increments in dosage. Rapid loading with clonazepam to treat acute mania, as suggested by Chouinard (1987), could be problematic in these patients.

Another implication of clonazepam's relatively long half-life is the potential for withdrawal symptoms occurring despite what would seem to be a gradual schedule of drug discontinuation. Wong and Tissen (1989) reported a patient without a prior seizure history who had a withdrawal seizure when clonazepam was withdrawn by 0.5 mg every 4 days. In a series of 40 epileptic children withdrawn from long-term clonazepam, Specht et al., (1989) reported withdrawal symptoms in 47.5% of patients when clonazepam was withdrawn at a rate of 0.003 to 0.16 mg/kg/day (equivalent to 0.2 mg/day or faster in a 70-kg adult). These reports suggest that withdrawal from long-term clonazepam therapy should be even slower; when circumstances permit, the authors taper clonazepam as slowly as 0.25 mg every 2 weeks in patients who have had months or years of clonazepam therapy.

Drug-Drug Interactions

Clonazepam, when added to therapeutic levels of lithium, has been reported to produce a reversible neurotoxic syndrome with ataxia and dysarthria (Koczeroinski et al., 1989); the mechanism was presumably pharmacodynamic rather than via increased lithium levels. The report of additive toxicity should be taken in the context of additive therapeutic effect, as reported by Chouinard (1987, 1988), who did not find an unacceptable rate of adverse interaction and who argues that lithium-neuroleptic combinations may be more dangerous. In general, the literature suggests that all reported combinations of multiple antimanic agents have both greater toxicity, and greater effectiveness for selected treatment-refractory patients, than single agents.

Effects on Respiratory Drive

Like other benzodiazepines, clonazepam can decrease hypoxic respiratory drive. Therefore, it is relatively contraindicated in patients with chronic obstructive pulmonary disease (COPD) who are at risk for carbon dioxide retention. When clonazepam is considered for a patient with COPD, baseline blood gas levels should be obtained; the drug should not be given if there is carbon dioxide retention at baseline. Even if baseline blood levels are normal, follow-up blood gas levels after initiating clonazepam therapy would be a reasonable precaution in patients with significant impairment in pulmonary function.

BENZODIAZEPINES

Benzodiazepines are among the most widely prescribed drugs: more than 80 million benzodiazepine prescriptions are filled each year in this country (Bergman & Wynn, 1987) and, of every 20 prescriptions written in the United States, one is for a benzodiazepine, with 11% of the population per year being treated with one of these medications (Dubovsky, 1990). Benzodiazepines are commonly utilized in medically ill patients as anxiolytics, muscle relaxants, anticonvulsants, and hypnotics, and for sedation for procedures. The medically ill and elderly, however, are at increased risk for adverse side effects of these medications. There is also evidence that, in addition to pharmacokinetic changes related to aging, the elderly may be intrinsically more sensitive to both positive and negative clinical effects of the benzodiazepines (Greenblatt, Harmatz, & Shader, 1991).

As a class, benzodiazepines share more similarities than differences in their intrinsic pharmacologic activity, which is medicated via the benzodiazepine receptor (increasing GABA-ergic tone), and side effect profile. There are, however, considerable differences in the pharmacokinetics of the various benzodiazepines. Thus, the knowledge of an individual benzodiazepine's pharmacokinetic profile may be more important, especially in the medically ill and the elderly, than the somewhat arbitrary distinctions between anxiolytic and hypnotic indications. Pharmacokinetic distinctions are especially significant because there is no convincing evidence that one benzodiazepine is more effective than any others, and no evidence that benzodiazepines marketed as hypnotics for the treatment of insomnia are more effective for sleep than those marketed as anxiolytics, and vice versa.

Metabolic Route and Elimination Half-Life

The most important pharmacokinetic considerations in the use of benzodiazepines involve route of metabolism and biotransformation, elimination half-life, and relative lipophilicity (Table 9-8). Most benzodiazepines, including diazepam, chlordiazepoxide, alprazolam, flurazepam, triazolam, and midazolam, undergo primary biotransformation in the liver via microsomal oxidative pathways. Thus, for these medications, drugs that inhibit hepatic microsomal enzymes, such as alcohol, isoniazid, and cimetidine, will increase plasma concentrations and elimination half-lives. Conversely, drugs that induce microsomal enzymes, such as estrogen, methylxanthines, cigarette smoking, and rifampin, will decrease the concentrations of these benzodiazepines (Stoudemire & Fogel, 1987). An example of the clinical significance of this later observation is the fact that theophylline preparations and benzodiazepines are frequently used together in intensive care settings. If the theophylline were discontinued in a patient who was concurrently receiving a benzodiazepine, the patient would theoretically be at risk for a rise in the plasma concentration of the benzodiazepine that could result in a diminished level of consciousness and respiratory depression (Bonfiglio & Dasta, 1991). Since hepatic biotransformation is impaired by advanced age and hepatic disease such as hepatitis and cirrhosis, benzodiazepines that are eliminated by hepatic oxidation should either be avoided or be given at reduced dosages.

TABLE 9–8. *Commonly Used Benzodiazepines*

	Primary Route of Biotransformation	Elimination Half-Life (hr)
Diazepam (Valium)	oxidation	36–200
Flurazepam (Dalmane)	oxidation	50–120
Halazepam (Praxipam)	oxidation	36–200
Chlordiazepoxide (Librium)	oxidation	30–90
Alprazolam (Xanax)	oxidation	12–15
Triazolam (Halcion)	oxidation	3–5
Clorazepate (Tranxene)	oxidation	36–200
Prazepam (Centrax, Vestram)	oxidation	36–200
Midazolam (Versed)[a]	oxidation	2–4
Quazepam (Doral)	oxidation	20–40
Estazolam (ProSom)	oxidation	8–24
Lorazepam (Ativan)	conjugation	10–20
Temazepam (Restoril)	conjugation	8–12
Oxazepam (Serax)	conjugation	8–12

[a]IM or IV route only.

Reprinted by permission from Stoudemire A, & Fogel BS (1987). Psychopharmacology in the medically ill. In A Stoudemire & BS Fogel (Eds.), Principles of medical psychiatry (Ch. 4, p. 104). Orlando, FL: Grune & Stratton.

The parent compounds of benzodiazepines that are eliminated via oxidative metabolism typically are transformed into active metabolites that may be long lived. For example, diazepam, chlordiazepoxide, clorazepate, and prazepam are metabolized to their common active metabolite, desmethyldiazepam, which has a half life in excess of 50 hours in young healthy patients and up to 175 hours in elderly or medically ill patients (Barbee & McLaulin, 1990). Similarly, the benzodiazepine hypnotics flurazepam and quazepam share the active metabolite desalkylflurazepam, which has an elimination half-life of 48 to 120 hours (Greenblatt, 1991). Two benzodiazepines that are metabolized by oxidative pathways but do not have significantly active metabolites are triazolam and alprazolam, with elimination half-lives of 3 to 5 hours and 12 to 15 hours, respectively.

Three available benzodiazepines are metabolized primarily by conjugation with glucuronic acid: lorazepam, oxazepam, and temazepam. These drugs therefore are less likely to accumulate in the elderly and in patients who have hepatic disease. For example, for a patient who presented with symptoms of alcohol withdrawal and had an unknown liver function status, lorazepam or oxazepam would be preferable to chlordiazepoxide. Lorazepam is also one of the only two benzodiazepines (along with midazolam) that is reliably absorbed intramuscularly, should the patient be unable to take oral medications.

Relative Lipophilicity

In addition to metabolic fate and elimination half-life, the relative lipid solubility of a benzodiazepine is also important to consider. More highly lipophilic drugs, such as midazolam, cross the blood-brain barrier quickly but also may be distributed more widely in the peripheral tissues. Diazepam is highly lipophilic and has a rapid onset of action when given intravenously, whereas lorazepam is relatively less lipophilic with less rapid onset of clinical activity when given intravenously (Greenblatt, 1991). When single doses of highly lipophilic drugs are given, redistribution, rather than elimination, determines the duration of clinical effect.

Elimination half-lives vary widely among individual benzodiazepines. Short half-life benzodiazepines are less likely to accumulate with repetitive dosing since they reach steady state more rapidly (steady state is more than 90% achieved after approximately four times the elimination half-life) (Greenblatt, 1991). The ultra-short-acting benzodiazepines carry the risk of interdose anxiety and rebound anxiety and insomnia.

One aspect of the clinical importance of the distinction between long and short half-life benzodiazepines is the propensity of long half-life medications to accumulate in the elderly and increase the risk of falling. In a large retrospective epidemiologic study, Ray, Griffin, and Downey (1989) examined the relative risk for hip fracture in patients age 65 and older who had suffered a hip fracture and had filled a prescription for benzodiazepines within the past 30 days, utilizing computerized pharmacy records available via the Sasketatchewan, Canada, universal health care plan. These investigators found that current users of long half-life benzodiazepines (diazepam, chlordiazepoxide, flurazepam, clorazepate) had a 70% greater risk of hip fracture than did patients on no psychotropic medications. There was no increased risk in current users of short half-life benzodiazepines (alprazolam, lorazepam, oxazepam, triazolam, bromazepam) (Ray, Griffin, & Downey, 1989). In light of the significant morbidity and mortality associated with hip fracture in the elderly (Kelsey & Hoffman, 1987), these findings emphasize the need to minimize or avoid the use of long half-life benzodiazepines in these patients because of their prolonged elimination, leading to accumulation with daytime drowsiness and/or ataxia.

Triazolam

Triazolam, an ultra-short-acting triazalobenzodiazepine marketed as a hypnotic, has been associated with reports of confusion, delirium, and anterograde amnesia (Patterson, 1987). Bixler et al. (1991) examined 18 poor sleepers placed in three parallel groups treated with 0.5 mg of triazolam, 30 mg of temazepam, or placebo. Immediate recall was similar in all three groups; however, delayed recall was more impaired in the triazolam-treated patients. In addition, five of the six triazolam-treated subjects reported episodes of next-day memory impairment (Bixler et al., 1991). Greenblatt et al. (1991) evaluated the sensitivity to triazolam in the elderly via a double-blind crossover study of 26 healthy young (mean age 30 years) and 21 healthy elderly (mean age 69 years) patients. Impaired psychomotor performance, memory, and increased psychomotor sedation were found in the elderly subjects. The authors recommend reducing the average dose of triazolam in the elderly by 50%, suggesting that clinical efficacy of 0.125 mg in the elderly is equivalent to 0.25 mg in healthy young patients.

It has also been suggested that triazolam may have a lower therapeutic index than that of other benzodiazepines, which as a class have a relatively high therapeutic index. A case report of coma with an overdose of only 0.5 mg has been described, whereas other benzodiazepines require approximately 100 times the clinical dose for serious overdose (Kales, 1990).

Clonazepam

Discussed earlier in this chapter as an anticonvulsant clonazepam is a long-acting benzodiazepine with a half-life of 20 to 58 hours in healthy adults. Since it may take greater than 1 week to achieve steady state, patients may be at increased risk for progressive oversedation, ataxia, and psychomotor slowing (Stoudemire, Fogel, & Gulley, 1991). Although it is possible that withdrawal from long half-life benzodiazepines such as clonazepam may result in delayed withdrawal symptoms, the pharmacodynamic profile of clonazepam may provide a gradual physiologic "self-taper." For example, Patterson (1988) described successful detoxification in ten alprazolam-dependent patients, substituting clonazepam for alprazolam on a milligram-for-milligram equivalent basis and then discontinuing the clonazepam.

Quazepam

Quazepam is a highly lipophilic benzodiazepine approved by the FDA for use as a sedative-hypnotic agent. It is rapidly absorbed and reaches peak plasma concentration 1.5 hours after ingestion of an oral dose. It is extensively metabolized in the liver, with an elimination half-life of 41 hours, although, as noted above, it is metabolized to the same long half-life metabolite as flurazepam, desalkylflurazepam (Kales, 1990). Recommended dosage is 15 mg PO for adults and 7.5 mg PO for elderly patients. Its long elimination half-life diminishes the risk of rebound insomnia and withdrawal symptoms, at the expense of increased risk of accumulation and daytime sedation. Since it is more slowly absorbed and may cost slightly more than flurazepam, it seems to offer no particular advantages over already established hypnotics for regular use (*Medical Letter*, 1990). For occasional use, the relative selectivity of the parent compound and its 2-oxo metabolite for BZ1 benzodiazepine receptors may be relevant. Specifically, it may cause relatively less cognitive and motor side effects than flurazepam after a single dose. However, even this potential advantage is not well-established (Fogel & Stoudemire, in press).

Estazolam

Estazolam, like triazolam, is a triazolobenzodiazepine derivative with sedative-hypnotic efficacy. It is rapidly absorbed, with a mean time to maximum plasma concentration of less than 2 hours, and its half-life is approximately 14 hours (Gustavson & Carrigan, 1990), which is similar to that of temazepam. Estazolam is metabolized via microsomal oxidation and has no significant active metabolites. The usual daily dose is 1 to 2 mg (clinical trials have found 2 mg of estazolam comparable to 30 mg of flurazepam) (Cohn et al., 1991; Scharf et al., 1990). Although it is clinically efficacious, there have been concerns that estazolam may share the increased toxicity and increased risk of withdrawal/rebound syndromes with the other triazolobenzodiazepines, triazolam and alprazolam.

Midazolam

Midazolam is primarily used as a parenteral preanesthetic in induction of general anesthesia, and for sedation before short diagnostic procedures such as endoscopy. However, its sedating and anxiolytic action, its remarkably short onset and duration of action, and the option of parenteral use suggest potential use in psychiatric patients. This notion has been bolstered by case reports in the psychiatric literature of its efficacy in the management of acutely agitated and psychotic patients (Bond, Mandos, & Kurtz, 1989; Mendoza et al., 1987).

Midazolam is highly lipophilic at physiologic pH and, unlike most benzodiazepines, is rapidly and well absorbed intramuscularly (Matson & Thurlow, 1988). It has a very rapid onset of CNS effects, with sedation occurring within 5 to 15 minutes after intramuscular injection (3 to 5 minutes after intravenous administration) and reaching its peak within 30 to 60 minutes. Midazolam is rapidly displaced from benzodiazepine receptors and has a short duration of action of approximately 2 hours, with a range of 1 to 6 hours (Bond, Mandos, & Kurtz, 1989). Its biologic half-life is only 1.3 to 2.2 hours (Beck, Salom, & Holzer, 1983). Midazolam undergoes extensive biotransformation to its major pharmacologically active metabolite, 1-hydroxymethylmidazolam, by way of microsomal oxidation. There is evidence that there are significant age-related differences in the clearance of midazolam, with decreased clearance in elderly patients, especially elderly men (Holazo, Winkler, & Patel, 1988). Lower initial doses are therefore recommended in patients older than 60 years. Although the drug is generally well-tolerated, like other benzodiazepines, respiratory depression (including apnea) has been associated with its use. In addition, there have been case reports of hypotension (Matson & Thurlow, 1988), disinhibition of aggressive behavior (Bobo & Miwa, 1988), transient paranoia and agitation (Burnakis & Berman, 1989), and delirium, especially in the elderly (Patterson, 1987).

Like triazolam, midazolam may induce amnestic episodes. Two reports suggest that it causes amnesia in more than 70% of patients receiving it (Dundee & Wilson, 1980; White et al., 1988).

Midazolam is approximately three to four times as potent per milligram as diazepam, and initial intramuscular dosage is 0.07 to 0.08 mg/kg, with the average dosage in a healthy adult being 5 mg and lower dosages recommended in elderly or debilitated patients. Although dosage guidelines in the psychiatric setting are not yet clear, Bond, Mandos, and Kurtz (1989) reported three cases of use of midazolam in the treatment of mentally retarded patients with acute and refractory aggressivity and violence using 5 to 10 mg of midazolam administered intramuscularly. The patients (a 14-year-old girl, a 17-year-old boy, and a 26-year-old man) showed rapid improvement in aggressive behavior. Mendoza et al. (1987) reported three patients with acute psychotic states with hyperarousal who responded favorably when treated with lower dosages of midazolam in a psychiatric emergency room setting. The patients included a 17-year-old boy, a 38-year-old man, and a 34-year-old woman. The authors noted the onset of sedation in these patients to occur within 6 to 8 minutes, with sedation lasting approximately 90 minutes afer a dose of 2.5 to 3 mg IM (Mendoza et al., 1987).

These above clinical case reports support earlier animal work with midazolam that has shown that administration of the drug can have beneficial effects on experimentally induced aggressive rage. For example, electrical stimulation of the hypothalamus in cats can produce aggressive behavior, a response that is diminished by midazolam (Pieri, 1983).Wyant et al. (1990) reported a randomized, single-blind comparison of 5 mg of midazolam, 250 mg of amobarbital sodium, and 10 mg of halperidol in agitated schizophrenic patients. There were five patients in each group. Midazolam and amobarbital sodium were both more effective than haloperidol in controlling agitation over a 2-hour period.

None of the above studies were double blind and controlled, and none involved medically debilitated or elderly patients. Moreover, midazolam is not specifically FDA approved for psychiatric indications. However, the pharmacologic properties of the drug suggest potential usage for psychiatric patients with acute psychomotor agitation or hyperarousal. A specific use may be in the emergency room, to facilitate medical assessment of patients with acute psychosis, who would otherwise be too agitated to permit examination and testing. It remains for controlled, randomized studies to provide clearer guidelines for its indications and usage in primary psychiatric conditions.

Considerations in Benzodiazepine Usage

All benzodiazepines are Schedule IV controlled substances that can lead to tolerance and psychologic and physiologic dependence. Risk factors for addiction include increased duration of treatment, utilization of higher doses, and a previous history of alcoholism. Physical dependence can occur within as short a time as 2 to 3 weeks, and often does occur within 4 months at two to five times the therapeutic dose. In addition, since benzodiazepines do not "cure" anxiety and insomnia, discontinuation syndromes can include reemergence of symptoms that were targeted by treatment. Withdrawal symptoms in tolerant patients include rebound anxiety or insomnia, and major symptoms such as psychosis and seizures (Salzman, 1991).

Despite their high therapeutic index in healthy patients, usage of benzodiazepines should be considered carefully in medically ill patients. In addition to pharmacodynamic alterations and increased drug interactions, the medically ill may be particularly at risk for sedation, especially if benzodiazepines are used concurrently with other CNS depressants such as barbiturate anticonvulsants or narcotics. In addition, it is well established that benzodiazepines reduce the ventilatory response to hypoxia. In chronically hypercapnic patients who have lost their hypercapnic respiratory drive, the suppression of hypoxic drive may result in respiratory suppression. Therefore, baseline arterial blood gases should be obtained in patients with pulmonary disease, and the use of benzodiazepines avoided if pCO_2 is elevated (Stoudemire & Fogel, 1987). In addition, benzodiazepines are contraindicated in sleep apnea, and may exacerbate symptoms in patients with undiagnosed sleep apnea syndromes (Mendelson, 1987).

A final caveat regarding the use of benzodiazepines in the elderly and the medically ill is that the clinician should give a high priority to the exclusion of underlying treatable pathophysiology before symptomatic treatment of anxiety and insomnia is undertaken (see Chapters 6 and 21). Insomnia, the most common reason for benzodiazepine prescriptions, can be due to underlying medical problems such as restless legs syndrome, the disturbed sleep-wake cycle of delirium or dementia, nocturnal myoclonus, and paroxysmal nocturnal dyspnea, or due to primary psychiatric syndromes such as major depression (Moran & Stoudemire, 1992). Drug interactions with benzodiazepines are listed in Table 9-7 (Stoudemire, Moran, & Fogel, 1991).

BUSPIRONE

Buspirone is a relatively new nonbenzodiazepine anxiolytic free from the side effect of respiratory depression, and which also lacks withdrawal symptoms on discontinuation. Buspirone's mean elimination half-life is 2.1 hours, although its metabolites may have elimination half-lives of 6 to 8 hours (Jann, 1988). While buspirone has significant anxiolytic activity, it is devoid of anticonvulsant, muscle relaxant, and sedative-hypnotic activity. There are no synergistic or additive effects between buspirone and alcohol or other sedative-hypnotics.

The pharmacokinetics of the drug have been studied in patients with impaired liver and kidney function. In patients with hepatic cirrhosis, after a single 20-mg dose, the elimination half-life was 6.21 ± 1.79 hours in patients with cirrhosis compared to 4.19 ± 0.53 hours in healthy subjects (Dalhoff et al., 1987; Goa & Ward, 1986). In patients with impaired renal function, including some who were completely anuric, buspirone clearance decreased between 33% and 50%, with no correlation between the severity of renal impairment and buspirone clearance (Gammans, Mayol, & Labudde, 1986). Hence, small reductions in buspirone dosage are likely to be needed in patients with hepatic and renal disease, although the reported prolongation of elimination half-lives by liver or kidney failure is likely to represent an effect of smaller magnitude than normal interindividual differences in optimal dosage. No clinically significant prolongation of busipirone pharmacokinetics in elderly patients has been observed (Gammans et al., 1989). However, the pharmacokinetics are highly variable in the general population, implying a need to individualize dosage.

Drug Interactions

Buspirone does not induce or inhibit hepatic mixed oxidase enzymatic functions (Molitor et al., 1985). The ability of buspirone to displace phenytoin, warfarin, propranolol, and digoxin from protein binding has been studied in vitro (Gammans et al., 1985), and interaction with these drugs does not appear to be clinically significant. Buspirone also appears free from interaction with antihistamines, bronchodilators, H_2 histamine receptor blockers, oral contraceptives, nonsteroidal antiinflammatory drugs, benzodiazepine hypnotics, digitalis preparations, common antihypertensive drugs, and oral hypoglycemics (Domantay & Napoliello, 1989; Levine & Napoliello, 1988). Pharmacokinetic interactions between buspirone and TCAs have not been found (Gammans, Mayol, & Labudde, 1986). However, buspirone has been observed to increase serum haloperidol concentrations (Sussman, 1987). Elevations in blood pressure have been observed in patients taking buspirone and MAOIs (Knapp, 1987), leading to the manufacturer's recommendation against this combination. There may be some slight prolongation in the metabolism of diazepam when it is used concurrently with buspirone, although this is of doubtful clinical significance (Meltzer & Fleming, 1982). Some slight increase in sedation may occur if buspirone is used concurrently with diazepam (Gershon, 1982).

Buspirone and Pulmonary Disease

Of primary interest is the use of buspirone in patients with pulmonary disease. Although controlled systematic studies utilizing buspirone in the treatment of anxiety in patients with COPD have yet to be published, animal studies suggest that buspirone may serve as a partial respiratory stimulant (Garner et al., 1989). In an open study of 82 patients, buspirone was used safely for reduction of anxiety in patients with chronic lung disease. No problems developed from using buspirone together with bronchodilators such as theophylline and terbutaline (Kiev & Domantay, 1988).

Buspirone and Other Conditions

A number of reports have anecdotally suggested clinical efficacy of buspirone in a variety of conditions other than anxiety disorders. Buspirone has been used to treat autism (Realmuto, August, & Garfinkel, 1989), for the suppression of neuroleptic-induced akathisia (D'Mello, McNeil, & Harris, 1989), and as an antidepressant (Robinson et al., 1989). A number of anecdotal reports have suggested that buspirone might also have a role in suppressing anxiety and agitation in brain-injured patients (Levine, 1988) and in patients with dementia (Colenda, 1988; Tiller, Dakis, & Shaw, 1988). Typically, relatively low doses are needed for this antiagitation effect (5 to 10 mg tid). Buspirone also may be helpful in smoking cessation (Gawin, Compton, & Byck, 1989), and in luteal phase dysphoric disorder (Rickels, Freeman, & Sondheimer, 1989; Yatham, Barry, & Dinan, 1989).

Stimulant Effect of Buspirone

Although buspirone has an anxiolytic effect that is maximal in 4 to 6 weeks, some patients experience early stimulation or agitation from the drug. The authors have seen several of these reactions in patients with diagnoses ranging from panic disorder to depression to mental retardation. The reaction is dose-dependent and can be treated by discontinuing buspirone until the agitation or insomnia resolves, then restarting the drug at a lower dose. A reasonable approach to starting buspirone in panicky or neurologically impaired patients is to start no higher than 2.5 mg tid., and warn the patient or caretaker about possible early and transient agitation. Finally, since buspirone has an antidepressant effect at higher doses (30–90 mg/day), it may precipitate hypomania or mania; a series of reasonably well-documented cases was recently reported (Price & Bielefeld, 1989).

PSYCHOSTIMULANTS

Although use of psychostimulants in the treatment of primary depression in physically healthy patients remains controversial, their use in medically ill patients—at least for short periods of time—is relatively well accepted. It is important to remember, however, that there is no formal FDA approval for utilizing these medicines as antidepressants. The two most commonly used psychostimulants are methylphenidate and dextroamphetamine.

Dextroamphetamine is roughly twice as potent as methylphenidate, with typical therapeutic dosages of 10 to 20 mg/day for dextroamphetamine versus 20 to 40 mg/day for methylphenidate (Jenike, 1985). Dextroamphetamine excretion is influenced by urinary pH, and, when urine is acidic, dextroamphetamine is excreted more rapidly and largely unchanged in the urine. (Thus, ascorbic acid or cranberry juice could be used in the ambulatory treatment of mild dextroamphetamine toxicity.) Its plasma half-life is approximately 12 hours (Chiarello & Cole, 1987), as opposed to the 2-hour average half-life of methylphenidate, which is quickly metabolized by the liver to ritalinic acid, a metabolite with little CNS activity (Goff, 1986).

In addition to methylphenidate and dextroamphetamine, pemoline and diethylpropion hydrochloride are sympathomimetic agents that have been utilized less frequently as psychostimulants (Janowsky, 1988). Pemoline is structurally dissimilar to methylphenidate and the amphetamines; it carries the risk of an associated idiosyncratic hepatoxicity in approximately 2% of cases (Nehra et al., 1990).

Two recent literature reviews have critically examined studies on the efficacy of psychostimulants in the treatment of depression (Chiarello & Cole, 1987; Satel & Nelson, 1989). Satel and Nelson (1989) reviewed 16 controlled and 30 noncontrolled studies, concluding that the majority of placebo-controlled studies showed no significant advantage of stimulants over placebo in primary major depression. High response rates of placebo-treated patients have been noted in many of these studies, suggesting that uncontrolled studies should be interpreted with some skepticism. However, both reviews acknowledge the potential usefulness of stimulants in certain depressive subgroups: apathetic, withdrawn geriatric patients; medically ill depressed patients who are unable to tolerate conventional antidepressants; and possibly as adjuvant therapy in patients with refractory depression.

Masand, Pickett, and Murray (1991) performed a retrospective, uncontrolled review of 4,740 consecutive hospital consultations in patients with secondary depression (i.e., the onset of the depression coincided with or came after their medical illness). Of the 198 patients treated with psychostimulants (primarily dextroamphetamine), 82% experienced some improvement and 70% were retrospectively rated as markedly improved. Side effects encountered included confusion and agitation (four patients), hypomania (three patients), paranoid delusions (three patients), elevated blood pressure and sinus tachycardia (three patients each), and one case of atrial fibrillation. In another retrospective chart review study of 29 patients by Rosenberg, Ahmed, and Hurwitz (1991), although moderate or marked improvement was seen in 55% of patients, significant side effects were noted in 28%, including agitation, tachycardia, and visual hallucinations. The presence of delirium predicted nonresponse (Rosenberg, Ahmed, & Hurwitz, 1991).

Masand, Murray, and Pickett (1991) noted improvement in 82% of 17 poststroke patients studied retrospectively, with equal efficacy with dextroamphetamine and methylphenidate. Improvement typically was rapid, usually within the first 2 days of treatment. Lingam and associates (1988) retrospectively studied the records of 25 patients with poststroke depression treated with methylphenidate at dosages of at least 20 mg/day for 5 consecutive days. These authors found "complete" recovery from depression in 52% of the patients. In a prospective study of another medically ill population, depressed cancer patients, Fernandez et al. (1987) treated 30 patients (7 men and 23 women) at an initial dose of 10 mg of methylphenidate tid, with 23 showing marked or moderate improvement. This study included 11 patients who were treated for 1 year at low doses of 5 to 10 mg/day without evidence of tolerance or abuse. One patients with dementia became more agitated and confused on the psychostimulant (Fernandez et al., 1987).

Two recent reports have targeted patients with acquired immunodeficiency syndrome (AIDS)–related depression as a medically ill depressed subgroup that may respond to psychostimulant therapy (Fernandez,

Levy, & Galizzi, 1988; Holmes, Fernandez, & Levy, 1989). Holmes, Fernandez, and Levy (1989) prospectively studied 17 AIDS-related complex patients treated with dextroamphetamine (initial doses 5 to 15 mg/day) or methylphenidate (initial doses 5 mg tid). These patients had diagnoses of organic mental disorder, adjustment disorder, or major depression and were treated for an average of 8 months. Treatment with either stimulant was effective in achieving a marked to moderate improvement in 79% of the patients; 89% showed some improvement in Clinical Global Impression ratings.

In spite of continued debate over their efficacy, psychostimulants may offer in some patients distinct advantages over traditional antidepressants. Unlike CyADs, psychostimulants produce an improvement in mood that, if it occurs, is rapid, usually within the first 24 to 48 hours of treatment (Woods et al., 1986). In addition, the psychostimulants are relatively well tolerated, even in older medically ill patients who may be unable to tolerate the anticholinergic side effects of TCAs. However, the possibilities of rebound depression after cessation of the drugs, habituation, abuse, and precipitation of paranoid reactions argue against *routine* usage of psychostimulants to treat major depression except in patients whose medical illness contraindicates use of standard antidepressants. There are no data to support their long-term use and no controlled studies for either their safety or their efficacy with long-term use for depression in the medically ill. Drug interactions with psychostimulants are listed in Table 9-7.

REFERENCES

Addonizio G (1991). NMS in the elderly—an under-recognized problem. Int J Geriatr Psychiatry 6:547–548.

Addonizio G, Susman VL, & Roth SD (1987). Neuroleptic maliganant syndrome: Review and analysis of 115 cases. Biol Psychiatry 22:1004–1020.

Adler LA, Angrist B, Reiter S, et al. (1989). Neuroleptic-induced akathisia: A review. Psychopharmacology 97:1–11.

Alberto G, Erickson T, Popiel R, et al. (1989). Central nervous system manifestations of a valproic acid overdose responsive to naloxone. Ann Emerg Med 18:889–891.

American Psychiatric Association (1980). Diagnostic and statistical manual of mental disorders, (3rd ed.). Washington, DC: American Psychiatric Association.

American Psychiatric Association (1985). Task force report on antidepressant drug levels. Am J Psychiatry 142:155–162.

Anderson EA, & Powers PS (1991). Neuroleptic malignant syndrome associated with clozapine use. J Clin Psychiatry 52:102–104.

Antal EJ, Lawson IR, Alderson LM, et al. (1982). Estimating steady-state desipramine levels in noninstitutionalized elderly patients using single dose disposition parameters. J Clin Psychopharmacol 2:193–198.

Arana GW, Epstein S, Molloy M, et al. (1988). Carbamazepine-induced reduction of plasma alprazolam concentrations: A clinical case report. J Clin Psychiatry 49:448–449.

Arana GW, Goff DC, Friedman H, et al. (1986). Does carbamazepine-induced reduction of plasma haloperidol levels worsen psychotic symptoms? Am J Psychiatry 143:650–651.

Ashton MG, Ball SG, Thomas TH, et al. (1977). Water intoxication associated with carbamazepine treatment. Br Med J 1:1134–1135.

Bahls FH, Ozuna J, & Ritchie DE (1991). Interactions between calcium channel blockers and the anticonvulsants carbamazepine and phenytoin. Neurology 41:740–742.

Bajwa WK, & Asnis GM (1991). Desipramine-induced urticaria: A clinical problem. J Nerv Ment Dis 179:108–109.

Baldessarini RJ, & Frankenburg FR (1991). Clozapine: A novel antipsychotic agent. N Engl J Med 324:746–754.

Baldessarini RJ, Teicher MH, Cassidy JW, et al. (1988). Anticonvulsant cotreatment may increase toxic metabolites of antidepressants and other psychotropic drugs [letter]. J Clin Psychopharmacol 8:381–382.

Baldwin C, & Safferman A (1990). A case of lisinopril-induced lithium toxicity. DICP, Annals of Pharmacotherapy 24:946–947.

Balon R, & Berchou R (1986). Hematologic side effects of psychotropic drugs. Psychosomatics 27:119–127.

Barbee JG, & McLaulin JB (1990). Anxiety disorders: Diagnosis and pharmacotherapy in the elderly. Psychiatr Ann 20:439–445.

Beck H, Salom M, & Holzer J (1983). Midazolam dosage studies in institutionalized geriatric patients. Br J Pharmacol 16:133S–137S.

Beerman B, & Edhag O (1978). Depressive effects of carbamazepine on idioventricular rhythm in man. Br Med J 2:171–172.

Benassi E, Bo GP, Cocito L, et al. (1987). Carbamazepine and cardiac conduction disturbances. Ann Neurol 22:280–281.

Bennett WM, Muther RS, & Parker RA (1980). Drug therapy in renal failure: Dosing guidelines for adults: Part II. Sedatives, hypnotics, and tranquilizers; cardiovascular, antihypertensive, and diuretic agents; miscellaneous agents. Ann Intern Med 93:286–325.

Bergman SA, & Wynn RL (1987). A review of benzodiazepines. Compend Contin Educ Dent 8:520–526.

Bernardi F, & DelZampa M (1990). Clozapine in idiopathic Parkinson's disease. Neurology 40:1151.

Bieck PR, & Antonin KH (1989). Tyramine potentiation during treatment with MAO inhibitors: Brofaromine and moclobemide vs. irreversible inhibitors. J Neural Transm (suppl) 28:21–31.

Bigger JT, Giardina EGV, Perel JM, et al. (1977). Cardiac antiarrhythmic effect of imipramine hydrochloride. N Engl J Med 296:206–208.

Bixler EO, Kales A, Manfredi RL, et al. (1991). Next-day memory impairment with triazolam use. Lancet 337:827–831.

Blackwell B, & Schmidt GL (1984). Drug interactions in psychopharmacology. Psychiatr Clin North Am 7:625–637.

Blazer DG, Petrie WM, & Wilson WP (1976). Affective psychoses following renal transplant. Dis Nerv System 37:663–667.

Bobo BL, & Miwa LJ (1988). Midazolam disinhibition reaction. Drug Intell Clin Pharm 22:725.

Bodley RP, Halwax K, & Potts L (1969). Low serum cholinesterase levels complicating treatment with phenelzine. Br Med J 3:510–512.

Bond WS, Mandos LA, & Kurtz MB (1989). Midazolam for aggressivity and violence in three mentally retarded patients. Am J Psychiatry 146:925–926.

Bonfiglio MF, & Dasta JF (1991). Clinical significance of the benzodiazepine-theophylline interaction. Pharmacotherapy 11:85–87.

Boston Collaborative Drug Surveillance Program (1972). Adverse reactions to the tricyclic-antidepressant drugs. Lancet 1:529–531.

Bourgeois BFD (1988). Pharmacologic interactions between valproate and other drugs. Am J Med 84(suppl 1A):29–33.

BOURGEOIS BFD, & WAD N (1984). Individual and combined antiepileptic and neurotoxic activity of carbamazepine and carbamazepine-10,11-epoxide in mice. J Pharmacol Exp Ther 231:411–415.

BOWDEN CL, NEMEROFF CB, & POTTER WZ (1991). Practical clinical guidelines for the management of bipolar disorder. Monograph on Treatment. N. Chicago: Abbott Laboratories.

BRANCONNIER RJ, COLE JO, GHAZVINIAN S, ET AL. (1983a). Clinical pharmacology of bupropion and imipramine in elderly depressives. J Clin Psychiatry 44:130–133.

BRANCONNIER RJ, COLE JO, OXENKRUG S, ET AL. (1983b). Cardiovascular effects of imipramine and bupropion in aged depressed patients. Psychopharmacol Bull 19:658–662.

BREWERTON TD (1986). Lithium counteracts carbamazepine-induced leukopenia while increasing its therapeutic effect. Biol Psychiatry 21:677–685.

BROWN CS, WELLS BG, COLD JA, ET AL. (1990). Possible influence of carbamazepine on plasma imipramine concentrations in children with attention deficit hyperactivity disorder. J Clin Psychopharmacol 10:359–362.

BROWN TCK, & CASS NM (1979). Beware—the use of MAO inhibitors is increasing again. Anaesth Intensive Care 7:65–68.

BRUNO J, GALLO JM, LEE CS, ET AL. (1980). Interactions of valproic acid with phenytoin. Neurology 30:1233–1236.

BUFF DD, BRENNER R, KIRTANE SS, ET AL. (1991). Dysrhythmia associated with fluoxetine treatment in an elderly patient with cardiac disease. J Clin Psychiatry 52:174–176.

BURKE RE (1992). Neuroleptic-induced tardive dyskinesia variants. In AE LANG & WJ WEINER (Eds.), Drug-induced movement disorders (Ch. 6, pp. 167–198). Mt. Kisco, NY: Futura Publishing Co., Inc.

BURNAKIS TG, & BERMAN DE (1989). Hostility and hallucinations as a consequence of midazolam administration. Drug Intell Clin Pharm 23:671–672.

Burroughs Wellcome Company (1989). Wellbutrin (bupropion hydrochloride) tablets insert. Research Triangle Park, NC: Burroughs Wellcome.

BURROWS GD, VOHRA J, & DUMOVIC P (1977). TCA drugs in cardiac conduction. Prog Neuropsychopharmacol 1:329–334.

CAMFIELD P, CAMFIELD C, DOOLEY J, ET AL. (1989). Routine screening of blood and urine for severe reactions to anticonvulsant drugs in asymptomatic patients is of doubtful value. Can Med Assoc J 140:1303–1305.

CANTU TG, & KOREK JS (1988). Monoamine oxidase inhibitors and weight gain. Drug Intell Clin Pharm 22:755–759.

CAPS (Cardiac Arrhythmic Pilot Study Investigators) (1988). Effects of encainide, flecainide, imipramine and moricizine on ventricular arrhythmias during the year after acute myocardial infarction. Am J Cardiol 61:501–509.

CAROFF SN, MANN SC, LAZARUS A, ET AL. (1991). Neuroleptic malignant syndrome: Diagnostic issues. Psychiatr Ann 21:130–147.

CAVANAUGH S VON A (1990). Drug-drug interactions of fluoxetine with tricyclics. Psychosomatics 31:273–276.

CHANDLER MC, BARNHILL JB, & GUALTIERI CT (1988). Amantadine for the agitated head-injury patient. Brain Injury 2:309–311.

CHENG YF, LUNDBERG T, BONDESSON U, ET AL. (1988). Clinical pharmacokinetics of clozapine in chronic schizophrenic patients. Eur J Clin Pharmacol 34:445–449.

CHIARELLO RJ, & COLE JO (1987). The use of psychostimulants in general psychiatry. Arch Gen Psychiatry 44:286–295.

CHOUINARD G (1987). Clonazepam in acute and maintenance treatment of bipolar affective disorder. J Clin Psychiatry 48(suppl):29–37.

CHOUINARD G (1988). The use of benzodiazepines in the treatment of manic-depressive illness. J Clin Psychiatry 49(suppl):15–20.

CHOUINARD G, ANNABLE L, & LANGLOIS R (1981). Absence of orthostatic hypotension in depressed patients treated with bupropion. Prog Neuropsychopharmacol 5:483–490.

CIRAULO DA, & SHADER RI (1990a). Fluoxetine drug-drug interactions, I. Antidepressants and antipsychotics. J Clin Psychopharmacol 10:48–50.

CIRAULO DA, & SHADER RI (1990b). Fluoxetine drug-drug interactions, II. J Clin Psychopharmacol 10:213–217.

CLARY C, & SCHWEIZER E (1987). Treatment of MAOI hypertensive crisis with sublingual nifedipine. J Clin Psychiatry 48:249–250.

CLIFFORD DB, RUTHERFORD JL, & HICKS FG (1985). Acute effects of antidepressants on hippocampal seizures. Ann Neurol 18:692–697.

COHEN BJ, MAHELSKY M, & ADLER L (1990). More cases of SIADH with fluoxetine [letter]. Am J Psychiatry 147:948–949.

COHEN-COLE SA, & STOUDEMIRE A (1987). Major depression and physical illness: Special considerations in diagnosis and biological treatment. Psychiatr Clin North Am 10:1–17.

COHN CK, SHRIVASTAVA R, MENDELS J, ET AL. (1990). Double-blind, multicenter comparison of sertraline and amitriptyline in elderly depressed patients. J Clin Psychiatry 51(12, suppl B):28–33.

COHN JB, WILCOX CS, BREMNER J, ET AL. (1991). Hypnotic efficacy of estazolam compared with flurazepam in outpatients with insomnia. J Clin Pharmacol 31:747–750.

COLENDA CC (1988). Buspirone in treatment of agitated demented patient. Lancet 1:1169.

COOPER AJ (1989). Tyramine and irreversible monoamine oxidase inhibitors in clinical practice. Br J Psychiatry 155(suppl 6):38–45.

COOPER AJ, & ASHCROFT G (1966). Potentiation of insulin hypoglycemia by MAOI antidepressant drugs. Lancet 1:407–409.

COOPER GL (1988). The safety of fluoxetine: An update. Br J Psychiatry 153(suppl 3):77–86.

COOPER T, & SIMPSON GM (1978). Prediction of individual dosage of nortriptyline. Am J Psychiatry 135:333–335.

COTARIU D, & ZAIDMAN JL (1988). Valproic acid and the liver. Clin Chem 34:890–897.

CRABTREE BL, MACK JE, JOHNSON CD, ET AL. (1989). Effects of HCTZ versus furosemide on serum lithium. APA Annual Meeting Syllabus, Abstract NR266 (p. 150). Washington, DC: American Psychiatric Association.

CSERNANSKY JG, & HOLLISTER LE (1985). Using lithium in patients with cardiac and renal disease. Hosp Formulary 20:726–735.

CUTLER NR, & NARANG PK (1984). Implications of dosing trycyclic antidepressants and benzodiazepines in geriatrics. Psychiatr Clin North Am 7:845–861.

CUTLER NR, ZAVADIL AP III, & EISDORFER C (1981). Concentrations of desipramine in elderly women are not elevated. Am J Psychiatry 138:1235–1237.

DALHOFF K, POULSEN HE, GARRED P, ET AL. (1987). Buspirone pharmacokinetics in patients with cirrhosis. Br J Clin Pharmacol 24:547–550.

DASGUPTA K, & JEFFERSON JW (1990). The use of lithium in the medically ill. Gen Hosp Psychiatry 12:83–97.

DAVIDSON J (1989). Seizures and bupropion: A review. J Clin Psychiatry 50:256–261.

DAVIDSON J, ZUNG WW, & WALKER JI (1984). Practical aspects of MAO inhibitor therapy. J Clin Psychiatry 45:81–84.

DAWLING S, CROME P, & BRAITHWAITE RA (1980). Pharmacokinetics of single oral dose of nortriptyline in depressed elderly hospital patients and young healthy volunteers. Clin Pharmacokinet 5:394–401.

DAWLING S, CROME P, HEYER EJ, ET AL. (1981). Nortriptyline therapy in elderly patients: Dosage prediction from plasma concentration at 24 hours after a single 50 mg dose. Br J Psychiatry 139:413–416.

Dawling S, Lynn K, Rosser R, et al. (1982). Nortriptyline metabolism in chronic renal failure: Metabolite elimination. Clin Pharmacol Ther 32:322–329.

Dechant KL, & Clissold SP (1991). Paroxetine: A review of its pharmacodynamic and pharmacokinetic properties, and therapeutic potential in depressive illness. Drugs 41:225–253.

Denton PH, Borell VM, & Edwards NV (1962). Dangers of monoamine oxidase inhibitors. Br Med J 2:1752–1753.

DePaulo JR (1984). Lithium. Psychiatr Clin North Am 7:587–599.

DeVane CL, Laizure SC, Stewart JT, et al. (1990). Disposition of bupropion in healthy volunteers and subjects with alcoholic liver disease. J Clin Psychopharmacol 10:328–332.

Devinsky O, Honigfeld G, & Patin J (1991). Clozapine-related seizures. Neurology 41:369–371.

Dieperink H, Leyssac PP, Kemp E, et al. (1987). Nephrotoxicity of cyclosporin A in humans: Effects on glomerular filtration and tubular reabsorption rates. Eur J Clin Invest 17:493–496.

D'Mello DA, McNeil JA, & Harris W (1989). Buspirone suppression of neuroleptic-induced akathisia: Multiple case reports. J Clin Psychopharmacol 9:151–152.

Domantay AG, & Napoliello MJ (1989). Buspirone for elderly anxious patients: A review of clinical studies. Fam Pract Recertification 11(9, suppl):17–23.

Douste-Blazy PH, Rostin M, Livarek B, et al. (1986). Angiotensin converting enzyme inhibitors and lithium treatment. Lancet 1:1448.

Dreifuss FE, & Langer DH (1987). Hepatic considerations in the use of antiepileptic drugs. Epilepsia 28(suppl 2):S23–S29.

Dubovsky SL (1990). Generalized anxiety disorder: New concepts and psychopharmacologic therapies. J Clin Psychiatry 51(1, suppl):3–10.

Dundee JW, & Wilson DB (1980). Amnesic action of midazolam. Anaesthesia 35:459–461.

Eadie MJ, Hooper WD, & Dickinson RG (1988). Valproate-associated hepatotoxicity and its biochemical mechanisms. Med Toxicol Adverse Drug Exp 3:85–106.

Edwards JG, Long SK, & Sedgwick EM (1986). Antidepressants and convulsive seizures: Clinical, electroencephalographic, and pharmacologic aspects. Clin Neuropharmacol 9:329–360.

El-Ganzouri AR, Ivankovich AD, Braverman B, et al. (1985). Monoamine oxidase inhibitors: Should they be discontinued preoperatively? Anesth Analg 64:592–596.

Ellison JM, Milofsky JE, & Ely E (1990). Fluoxetine-induced bradycardia and syncope in two patients. J Clin Psychiatry 51:385–386.

Fabre LF, Scharf MB, & Itil TM (1991). Comparative efficacy and safety of nortriptyline and fluoxetine in the treatment of major depression: A clinical study. J Clin Psychiatry 52(suppl):62–67.

Fallon BA, & Liebowitz MR (1991). Fluoxetine and extrapyramidal symptoms in CNS lupus. J Clin Psychopharmacol 11:147–148.

Farid FF, Wenger TL, Tsai SY, et al. (1983). Use of bupropion in patients who exhibit orthostatic hypotension on tricyclic antidepressants. J Clin Psychiatry 44:170–173.

Fawcett J, & Kravitz HM (1985). Treatment-refractory depression. In AF Schatzberg (Ed.), Common treatment problems in depression (Ch. 1, pp. 2–27). Washington DC: American Psychiatric Press, Inc.

Feder R (1991). Bradycardia and syncope induced by fluoxetine [letter]. J Clin Psychiatry 52:139.

Feighner JP, & Cohn JB (1985). Double-blind comparative trials of fluoxetine and doxepin in geriatric patients with major depressive disorder. J Clin Psychiatry 46:20–25.

Feighner JP, Boyer WF, Meredith CH, et al. (1988). An overview of fluoxetine in geriatric depression. Br J Psychiatry 153(suppl 3):105–108.

Fernandez F, Adams F, Holmes VF, et al. (1987). Methylphenidate for depressive disorders in cancer patients. Psychosomatics 28:455–461.

Fernandez F, Levy JK, & Galizzi H (1988). Response of HIV-related depression to psychostimulants: Case reports. Hosp Community Psychiatry 39:628–631.

Fisch C (1985). Effect of fluoxetine on the electrocardiogram. J Clin Psychiatry 46:42–44.

Fisch C (1991). Effects of sertraline on the ECG in non-elderly and elderly patients with major depression [abstract]. Biol Psychiatry 29:353S–354S.

Fisher RS, & Cysyk B (1988). A fatal overdose of carbamazepine: Case report and review of literature. J Toxicol Clin Toxicol 26:477–486.

Flegel KM, & Cole CH (1977). Inappropriate antidiuresis during carbamazepine treatment. Ann Intern Med 87:722–723.

Flugelman MY, Tal A, & Pollack S (1985). Psychotropic drugs and long QT syndromes: Case reports. J Clin Psychiatry 46:290–291.

Fogel BS (1983). Caution in the use of drugs in the elderly [letter]. N Engl J Med 308:1600.

Fogel BS (1988). Combining anticonvulsants with conventional psychopharmacologic agents. In SL McElroy & HG Pope (Eds.), Use of anticonvulsants in psychiatry: Recent advances (pp. 77–94). Clifton, NJ: Oxford Health Care.

Fogel B (1991). Treatment of agitation. In E Light & B Lebowitz (Eds.), The elderly with chronic mental illness. New York: Springer.

Fogel BS, Stoudemire A (in press). New psychotropics in the medically ill. In A Stoudemire & BS Fogel (Eds.), Medical psychiatric practice (Vol. 2; Ch. 3). Washington, DC: American Psychiatric Press, Inc.

Forrest JN, Cohen AD, & Torretti K (1974). On the mechanism of lithium-induced diabetes insipidus in man and rat. J Clin Invest 53:1115–1123.

Forster PL, Dewland PM, Muirhead D, et al. (1991). The effects of sertraline on plasma concentration and renal clearance of digoxin [abstract]. Biol Psychiatry 29:355S.

Fowler NO, McCall D, & Chou TC (1976). Electrocardiographic changes and cardiac arrhythmias in patients receiving psychotropic drugs. Am J Cardiol 37:223–230.

Frankhauser MP, Lindon JL, Connolly B, et al. (1988). Evaluation of lithium-tetracycline interaction. Clin Pharm 7:314–317.

Fricchione GL, & Vlay SC (1986). Psychiatric aspects of patients with malignant ventricular arrhythmias. Am J Psychiatry 143:1518–1526.

Friedman JH (1992). Induced parkinsonism. In AE Lang & WJ Weiner (Eds.), Drug-induced movement disorders (Ch. 3, pp. 41–84). Mt. Kisco, NY: Futura Publishing Co.

Friedman JH, & Lannon MC (1989). Clozapine in the treatment of psychosis in Parkinson's disease. Neurology 39:1219–1221.

Gammans RE, Bullen WW, Briner L, et al. (1985). The effects of buspirone binding of digoxin, dilanton, propranolol and warfarin to human plasma. Fed Proc 44:1123.

Gammans RE, Mayol RF, & Labudde JA (1986). Metabolism and disposition of buspirone. Am J Med 80:41–51.

Gammans RE, Westrick ML, Shea JP, et al. (1989). Pharmacokinetics of buspirone in elderly subjects. J Clin Pharmacol 29:72–78.

Ganzini L, Heintz R, Hoffman WF, et al. (1991). The prevalence of tardive dyskinesia in neuroleptic-treated diabetics: A controlled study. Arch Gen Psychiatry 48:259–263.

GARNER SJ, ELDRIDGE FL, WAGNER PG, ET AL. (1989). Buspirone, an anxiolytic drug that stimulates respiration. Am Rev Respir Dis 139:946–950.

GAWIN F, COMPTON M, BYCK R (1989). Buspirone reduces smoking. Arch Gen Psychiatry 46:288–289.

GEORGOTAS A, MCCUE RE, FRIEDMAN E, ET AL. (1987). Electrocardiographic effects of nortriptyline, phenelzine, and placebo under optimal treatment conditions. Am J Psychiatry 144:798–801.

GERSHON S (1982). Drug interactions in controlled clinical trials. J Clin Psychiatry 43:95–98.

GLASSMAN AH, BIGGER JT, & GIARDINA EV (1979). Clinical characteristics of imipramine-induced orthostatic hypotension. Lancet 1:468–472.

GLASSMAN AH, JOHNSON LL, & GIARDINA EV (1983). The use of imipramine in depressed patients with congestive heart failure. JAMA 250:1977–2001.

GLEASON R, & SCHNEIDER LS (1990). Carbamazepine treatment of agitation in Alzheimer's outpatients refractory to neuroleptics. J Clin Psychiatry 51:115–118.

GOA KL, & WARD A (1986). Buspirone: A preliminary review of its pharmacologic properties and therapeutic efficacy as an anxiolytic. Drugs 32:114–129.

GOFF DC (1986). The stimulant challenge test in depression. J Clin Psychiatry 47:538–543.

GOLDEN RN, JAMES SP, SHERER MA, ET AL. (1985). Psychoses associated with bupropion treatment. Am J Psychiatry 142:1459–1462.

GOODWIN FK, & JAMISON KR (1990). Manic-depressive illness. New York: Oxford University Press.

GOULDEN KJ, DOOLEY JM, CAMFIELD PR, ET AL. (1987). Clinical valproate toxicity induced by acetylsalicylic acid. Neurology 37:1392–1394.

GREENBLATT DJ (1991). Benzodiazepine hypnotics: Sorting out the pharmacokinetic facts. J Clin Psychiatry 52(9, suppl):4–10.

GREENBLATT DJ, HARMATZ JS, & SHADER RI (1991). Clinical pharmacokinetics of anxiolytics and hypnotics in the elderly. Part I. Clin Pharmacokinet 21:165–177.

GREENBLATT DJ, HARMATZ JS, SHAPIRO L, ET AL. (1991). Sensitivity to triazolam in the elderly. N Engl J Med 324:1691–1698.

GREENDYKE RM, BERKNER JP, WEBSTER JC, ET AL. (1989). Treatment of behavioral problems with pindolol. Psychosomatics 30:161–165.

GREENWALD BS, MARIN DB, & SILVERMAN SM (1986). Serotoninergic treatment of screaming and banging in dementia. Lancet 2:1464–1465.

GRIMSLEY SR, JANN MW, CARTER JG, ET AL. (1991). Increased carbamazepine plasma concentrations after fluoxetine coadministration. Clin Pharmacol Ther 50:10–15.

GUALTIERI CT (1991). Neuropsychiatry and behavioral pharmacology. New York: Springer-Verlag.

GUSTAVSON LE, & CARRIGAN PJ (1990). The clinical pharmacokinetics of single doses of estazolam. Am J Med 88 (suppl 3A):2S–5S.

HANNA ME, LOBAO CB, & STEWART JT (1990). Severe lithium toxicity associated with indapamide therapy [letter]. J Clin Psychopharmacol 10:379.

HARRISON WM, MCGRATH PJ, STEWART JW, ET AL. (1989). MAOIs and hypertensive crisis: The role of OTC drugs. J Clin Psychiatry 50:64–65.

HELMS PM (1985). Efficacy of antipsychotics in the treatment of the behavioral complications of dementia: A review of the literature. J Am Geriatr Soc 33:206–209.

HELMUTH D, LJALJEVIC Z, RAMIREZ L, ET AL. (1989). Choreoathetosis induced by verapamil and lithium treatment [letter]. J Clin Psychopharmacol 9:454–455.

HIMMELHOCH JM (1981). Cardiovascular effects of trazodone in humans. J Clin Psychopharmacol 1(suppl):76–81.

HJELM M, OBERHOLZER V, SEAKINS J, ET AL. (1986). Valproate-induced inhibition of urea synthesis and hyperammonaemia in healthy subjects. Lancet 2:859.

HOFF GS, RUUD TE, & TORNDER M (1981). Doxepin in the treatment of duodenal ulcer: An open clinical and endoscopic study comparing doxepin and cimetidine. Scand J Gastroenterol 16:1041–1042.

HOLAZO AA, WINKLER MB, & PATEL IH (1988). Effects of age, gender and oral contraceptives on intramuscular midazolam pharmacokinetics. J Clin Pharmacol 28:104–105.

HOLMES VF, FERNANDEZ F, & LEVY JK (1989). Psychostimulant response in AIDS-related complex patients. J Clin Psychiatry 50: 5–8.

HORGAN JH, PROCTOR JD, VELANDIA J, ET AL. (1973). Antiarrhythmic effect of lithium. Arch Int Pharmacodyn Ther 206:105–112.

HWANG AS, & MAGRAW RM (1989). Syndrome of inappropriate secretion of antidiuretic hormone due to fluoxetine [letter]. Am J Psychiatry 146:399.

IEIRI I, HIGUCHI S, HIRATA K, ET AL. (1990). Analysis of the factors influencing anti-epileptic drug concentrations—valproic acid. J Clin Pharmacol Ther 15:351–363.

IRWIN M, & SPAR JE (1983). Reversible cardiac conduction abnormality associated with trazodone administration [letter]. Am J Psychiatry 140:945–946.

JACOBSEN FM (1990). Low-dose trazodone as a hypnotic in patients treated with MAOIs and other psychotropics: A pilot study. J Clin Psychiatry 51:298–302.

JAFFE CM (1977). First-degree atrioventricular block during lithium carbonate treatment. Am J Psychiatry 134:88–89.

JANN MW (1988). Buspirone: An update on a unique anxiolytic agent. Pharmacotherapy 8:100–116.

JANOWSKY D, CURTIS G, & ZISOOK S (1983). Ventricular arrhythmias possibly aggravated by trazodone. Am J Psychiatry 140:796–797.

JANOWSKY DS (1988). (Reply). J Clin Psychopharmacol 8:450.

JANOWSKY EC, & JANOWSKY DS (1985). What precautions should be taken if a patient on an MAOI is scheduled to undergo anesthesia? J Clin Psychopharmacol 5:128–129.

JEAVONS PM (1983). Hepatotoxicity in antiepileptic drugs. In J OXLEY, D JANZ, & H MEINARDI (Eds.), Chronic toxicity of antiepileptic drugs (Ch. 1, pp. 1–46). New York: Raven Press.

JENIKE MA (1984). Monoamine oxidase inhibitors in elderly depressed patients. J Am Geriatr Soc 32:571–575.

JENIKE MA (1985). Handbook of geriatric psychopharmacology (pp. 73–87). Littleton, MA: PSG Publishing.

JENIKE MA (1988). Psychoactive drugs in the elderly: Antipsychotics and anxiolytics. Geriatrics 43:53–65.

JOHNSTON JA, LINEBERRY CG, ASCHER JA, ET AL. (1991). A 102-center prospective study of seizure in association with bupropion. J Clin Psychiatry 52:450–456.

KAHN DA (1989). The transition to and from monoamine oxidase inhibitors in clinical practice: Warnings and recommendations. Curr Affective Illness 8:5–14.

KAHN N, FREEMAN A, JUNCOS JL, ET AL. (1991). Clozapine is beneficial for psychosis in Parkinson's disease. Neurology 41: 1699–1700.

KAHN RJ, MCNAIR DM, & LIPMAN RS (1986). Imipramine and chlordiazepoxide in depressive and anxiety disorders, II. Arch Gen Psychiatry 43:79–85.

KALES A (1990). Quazepam: Hypnotic efficacy and side effects. Pharmacotherapy 10:1–12.

KALFF R, HOUTKOOPER MA, MEYER JWA, ET AL. (1984). Carbamazepine and serum sodium levels. Epilepsia 25:390–397.

KANE J, HONIGFELD G, SINGER J, ET AL. (1988). Clozapine for the treatment resistant schizophrenic. Arch Gen Psychiatry 45:789–796.

KANE JM (1989). The current status of neuroleptic therapy. J Clin Psychiatry 50:322–328.

KANE JM (1990). Psychopharmacologic treatment issues. Psychiatr Med 8:111–112.

KANE JM, & SMITH JM (1982). Tardive dyskinesia. Arch Gen Psychiatry 39:473–481.

KANTOR S, GLASSMAN A, BIGGER J, ET AL. (1978). The cardiac effects of therapeutic plasma concentration of imipramine. Am J Psychiatry 15:534–548.

KATHOL RG, RUSSELL N, & SLYMEN DJ (1980). Propranolol in chronic anxiety disorders. Arch Gen Psychiatry 37:1361–1365.

KATZ IR, SIMPSON GM, JETHANANDANI V, ET AL. (1989). Steady state pharmacokinetics of nortriptyline in the frail elderly. Neuropsychopharmacology 2:229–236.

KECK PE, MCELROY SL, & POPE HG (1991). Epidemiology of neuroleptic malignant syndrome. Psychiatr Ann 21:148–151.

KELSEY JL, & HOFFMAN S (1987). Risk factors for hip fracture. N Engl J Med 316:404–406.

KENNEBACK G, BERGFELDT L, VALLIN H, ET AL. (1991). Electrophysiologic effects and clinical hazards of carbamazepine treatment for neurologic disorders in patients with abnormalities of the cardiac conduction system. Am Heart J 121:1421–1429.

KIEV A, & DOMANTAY AG (1988). A study of buspirone coprescribed with bronchodilators in 82 anxious ambulatory patients. J Asthma 25:281–284.

KLEIN DF, GITTLEMAN R, QUITKIN F, ET AL. (1980). Diagnosis and drug treatment of psychiatric disorders: Adults and children, (2nd ed.). Baltimore: Williams & Wilkins.

KNAPP JE (1987). Monoamine oxidase inhibitor interaction information. Medical update. Evansville, IN: Mead Johnson.

KOCZEROINSKI D, KENNEY SH, & SWINSON RP (1989). Clonazepam and lithium—a toxic combination in the treatment of mania? Int Clin Psychopharmacol 4:195–199.

KOECHELER JA, CANAFAX DM, SIMMONS RL, ET AL. (1986). Lithium dosing in renal allograft recipients with changing renal function. Drug Intell Clin Pharm 20:623–624.

KOSTEN TR, & FORREST JN (1986). Treatment of severe lithium-induced polyuria with amiloride. Am J Psychiatry 143:1563–1568.

KRIKLER DM, & CURRY POL (1976). Torsades de pointes, an atypical ventricular tachycardia. Br Heart J 38:117–120.

KUGOH T, YAMAMOTO M, & HOSOKAWA K (1986). Blood ammonia level during valproic acid therapy. Jpn J Psychiatry Neurol 40:663–668.

KUSHNIR SL (1986). Lithium-antidepressant combinations in treatment of depressed, physically ill geriatric patients. Am J Psychiatry 143:378–379.

KUTCHER SP, REID K, DUBBIN JD, ET AL. (1986). Electrocardiogram changes and therapeutic desipramine and 2-hydroxy-desipramine concentrations in elderly depressives. Br J Psychiatry 148:676–679.

LADER M (1989). Clinical pharmacology of antipsychotic drugs. J Int Med Res 17:1–16.

LAI AA, LEVY RH, & CUTLER RE (1978). Time course of interaction between carbamazepine and clonazepam in normal man. Clin Pharmacol Ther 24:316–323.

LAZARUS JH (1986). Endocrine and metabolic effects of lithium (pp. 99–117). New York: Plenum.

LAZARUS LW, GROVES L, GIERL B, ET AL. (1986). Efficacy of phenelzine in geriatric depression. Biol Psychiatry 21:699–701.

LEIBOVICI A, & TARIOT PN (1988). Carbamazepine treatment of agitation associated with dementia. J Geriatr Psychiatr Neurol 1:110–112.

LEMBERGER L, BERGSTROM RF, WOLEN RL, ET AL. (1985). Fluoxetine: Clinical pharmacology and physiologic disposition. J Clin Psychiatry 46:14–19.

LEMBERGER L, ROWE H, BOSOMWORTH JC, ET AL. (1988). The effect of fluoxetine on the pharmacokinetics and psychomotor responses of diazepam. Clin Pharmacol Ther 43:412–419.

LEVENSON JL (1985). Neuroleptic malignant syndrome. Am J Psychiatry 142:1137–1145.

LEVENSON JL, MISHRA A, BAVERNFEIND RA, ET AL. (1986). Lithium treatment of mania in a patient with recurrent ventricular tachycardia. Psychosomatics 27:594–596.

LEVINE A (1988). Buspirone and agitation in head injury. Brain Injury 2:165–167.

LEVINE S, & NAPOLIELLO MJ (1988). A study of buspirone coprescribed with histamine H2-receptor antagonists in anxious outpatients. Int Clin Psychopharmacol 3:83–86.

LEVY NB (1987). Chronic renal disease, dialysis, and transplantation. In A STOUDEMIRE & BS FOGEL (Eds.), Principles of medical psychiatry (Ch. 27, pp. 583–594). Orlando, FL: Grune & Stratton.

LIEBERMAN E, & STOUDEMIRE A (1987). The use of tricyclic antidepressants in patients with glaucoma. Psychosomatics 28:145–148.

LIEBERMAN JA, COOPER TB, SUCKOW RF, ET AL. (1985). Tricyclic psychopharmacology antidepressant and metabolite levels in chronic renal failure. Clin Pharmacol Ther 37:301–307.

LIEBERMAN JA, KANE JM, & JOHNS CA (1989). Clozapine: Guidelines for clinical management. J Clin Psychiatry 50:329–338.

LIEBERMAN JA, SALTZ BL, JOHNS CA, ET AL. (1991). The effects of clozapine on tardive dyskinesia. Br J Psychiatry 158:503–510.

LINGHAM VR, LAZARUS LW, GROVES L, ET AL. (1988). Methylphenidate in treating post-stroke depression. J Clin Psychiatry 49:151–153.

LIPINSKI JF, MALLYA G, ZIMMERMAN P, ET AL. (1989). Fluoxetine-induced akathisia: Clinical and theoretical implications. J Clin Psychiatry 50:339–342.

LIPMAN BS, DUNN M, & MASSIE E (1984). Clinical electrocardiography (7th ed.). Chicago: Year Book Medical Publishers.

LIPPMAN S, WAGEMAKER H, & TUKER D (1981). A practical approach to management of lithium concurrent with hyponatremia, diuretic therapy, and/or chronic renal failure. J Clin Psychiatry 42:304–306.

LIPPMAN SB, MANSHADI MS, & GULTEKIN A (1984). Lithium in a patient with renal failure on hemodialysis. J Clin Psychiatry 45:444.

LIPSEY JR, ROBINSON RG, & PEARLSON GD (1984). Nortriptyline treatment of post-stroke depression: A double blind study. Lancet 1(8372):297–300.

LOWRY MR, & DUNNER FJ (1980). Seizures during tricyclic therapy. Am J Psychiatry 137:1461–1462.

LUCHINS DJ (1983). Review of clinical and animal studies comparing the cardiovascular effects of doxepin and other tricyclic antidepressants. Am J Psychiatry 140:1006–1009.

LUCHINS DJ, OLIVER AP, & WYATT RJ (1984). Seizures with antidepressants: An in vitro technique to assess relative risk. Epilepsia 25:25–32.

MACNEIL S, HANSON-NORTEY E, & PASCHALIS C (1975). Diuretics during lithium therapy. Lancet 1:1925–1926.

MANN JJ, AARON SF, WILNER PJ, ET AL. (1989). A controlled study of the antidepressant efficacy and side effects of (-)deprenyl: A selective monoamine oxidase inhibitor. Arch Gen Psychiatry 46:45–50.

MANOACH M, NETZ H, VARON D, ET AL. (1986). The effect of tricyclic antidepressants on ventricular fibrillation and collateral blood supply following acute coronary occlusion. Heart Vessels 2:36–40.

MANOACH M, VARON D, NEUMAN M, ET AL. (1989). The cardioprotective features of tricyclic antidepressants. Gen Pharmacol 20:269–275.

MANOACH M, VARON D, NEUMAN M, ET AL. (1987). Reduction of infarct size following acute coronary occlusion by augmenting collateral blood supply induced by infusion of tricyclic antidepressants. Heart Vessels 3:80–83.

MASAND P, MURRAY GB, & PICKETT P (1991). Psychostimulants in post-stroke depression. J Neuropsychiatry 3:23–27.

MASAND P, PICKETT P, & MURRAY GB (1991). Psychostimulants for secondary depression in medical illness. Psychosomatics 32:203–208.

MATSON AM, & THURLOW AC (1988). Hypotension and neurological sequelae following intramuscular midazolam [letter]. Anaesthesia 43:896.

MATTES JA (1989). Clozapine for refractory schizophrenia: An open study of 14 patients treated up to 2 years. J Clin Psychiatry 50:389–391.

MAY T, & RAMBECK R (1985). Serum concentrations of valproic acid: Influence of dose and comedication. Ther Drug Monit 7:387–390.

MAYEUX R, STERN Y, & WILLIAMS JBW (1986). Clinical and biochemical features of depression in Parkinson's disease. Am J Psychiatry 143:756–759.

MCCUE RE, GEORGOTAS A, NAGACHANDRAN N, ET AL. (1989a). Plasma levels of nortriptyline and 10-hydroxynortriptyline and treatment-related electrocardiographic changes in the elderly depressed. J Psychiatr Res 23:73–79.

MCCUE RE, GEORGOTAS A, SUCKOW RF, ET AL. (1989b). 10-Hydroxynortriptyline and treatment effects in elderly depressed patients. J Neuropsychiatry 1:176–1989.

MCELROY SL, KECK PE, POPE HG, ET AL. (1989). Valproate in psychiatric disorders: Literature review and clinical guidelines. J Clin Psychiatry 50 (suppl):23–29.

MCELROY SL, & POPE HG (1988). Use of anticonvulsants in psychiatry. Clifton, NJ: Oxford Health Care, Inc.

MCEVOY JP (1986). The neuroleptic threshold as a marker of minimum effective neuroleptic dose. Compr Psychiatry 27:327–335.

MCGINNESS J, KISHIMOTO A, & HOLLISTER LE (1990). Avoiding neurotoxicity with lithium-carbamazepine combinations. Psychopharmacol Bull 26:181–184.

(1989). Drugs that cause psychiatric symptoms. [Med Lett] 31(808):113–116.

(1990). Quazepam: A new hypnotic. [Med Lett] 32(816):39–40.

MELTZER HY, & FLEMING R (1982). Effect of buspirone on prolactin and growth hormone secretion in laboratory rodents and man. J Clin Psychiatry 43:76–79.

MENDELSON WB (1987). Pharmacotherapy of insomnia. Psychiatr Clin North Am 10:555–563.

MENDOZA R, DJENDEREDJIAN HA, ADAMS J, ET AL. (1987). Midazolam in acute psychotic patients with hyperarousal. J Clin Psychiatry 48:291–292.

MERIKANGAS JR, & MERIKANGAS KR (1988). Calcium channel blockers in MAOI-induced hypertensive crisis [abstract]. Psychopharmacology 96 (suppl):229.

MICHAELS I, SERRINGS M, SHIER N, ET AL. (1984). Anesthesia for cardiac surgery in patients receiving monoamine oxidase inhibitors. Anesth Analg 63:1041–1044.

MICHAELS I, SERRINGS M, SHIER N, ET AL. (1984). Anesthesia for cardiac surgery in patients receiving monoamine oxidase inhibitors. Anesth Analg 63:1041–1044.

MILLER DD (1991). Effect of phenytoin on plasma clozapine concentrations in two patients. J Clin Psychiatry 52:23–25.

MILLER DD, SHARAFUDDIN JA, & KATHOL RG (1991). A case of clozapine-induced neuroleptic malignant syndrome. J Clin Psychiatry 52:99–101.

MITCHELL JE, & MACKENZIE TB (1982). Cardiac effects of lithium therapy in man: A review. J Clin Psychiatry 43:47–51.

MOLITOR JA, GAMMANS RE, CARROLL CM, ET AL. (1985). Effect of buspirone on mixed function oxidase in rats. Fed Proc 44:1257.

MORAN MG, & STOUDEMIRE A (1992). Sleep disorders in the medically ill patient. J Clin Psychiatry 53(6, suppl):29–36.

MORROW JI, & ROUTLEDGE PA (1989). Poisoning by anticonvulsants. Adverse Drug React Acute Poisoning Rev 8:97–109.

MOSHAL MG, & KAHN F (1981). Trimipramine in the treatment of active duodenal ulceration. Scand J Gastroenterol 16:295–298.

MUIR WW, STRAUCH SM, & SCHAAL SF (1982). Effects of tricyclic antidepressant drugs on the electrophysiological properties of dog Purkinje fibers. J Cardiovasc Pharmacol 4:82–90.

NEHRA A, MULLICK F, ISHAK KG, ET AL. (1990). Pemoline-associated hepatic injury. Gastroenterology 99:1517–1519.

NELSON JC, ATILLASOY E, & MAZURE C (1988). Hydroxydesipramine in the elderly. J Clin Psychopharmacol 8:428–433.

NELSON JC, JATLOW PI, BOCK J, ET AL. (1982). Major adverse reactions during desipramine treatment: Relationship to drug plasma concentrations, concomitant antipsychotic treatment and patient characteristics. Arch Gen Psychiatry 39:1055–1061.

NELSON JC, JATLOW PI, & QUENLAN DM (1984). Subjective complaints during desipramine treatment. Arch Gen Psychiatry 41:55–59.

NELSON JC, MAZURE C, & JATLOW PI (1988). Antidepressant activity of 2-hydroxydesipramine. Clin Pharmacol Ther 44:283–288.

NORDIN C, BERTILSSON L, & SIWERS B (1987). Clinical and biochemical effects during treatment of depression with nortriptyline: The role of 10-hydroxynortriptyline. Clin Pharmacol Ther 42:10–19.

NOVESKE FG, HAHN KR, & FLYNN RJ (1989). Possible toxicity of combined fluoxetine and lithium. Am J Psychiatry 146:1515.

OSTERGAARD K, & DUPONT E (1988). Clozapine treatment of drug-induced psychotic symptoms in late stages of Parkinson's disease [letter]. Acta Neurol Scand 78:349–350.

OTT B, & FOGEL B (in press). Measurement of depression in dementia: Self-ratings vs. clinician ratings. Int J Ger Psychiatry.

OUSLANDER JG (1981). Drug therapy in the elderly. Ann Intern Med 94:711–722.

PALMER H (1960). Potentiation of pethidine [letter]. Br Med J 2:944.

PATO MT, & MURPHY DL (1991). Sustained plasma concentrations of fluoxetine and/or norfluoxetine four and eight weeks after fluoxetine discontinuation. J Clin Psychopharmacol 11:224–225.

PATTERSON JF (1987). Triazolam syndrome in the elderly. South Med J 80:1425–1426.

PATTERSON JF (1988). Alprazolam dependency: Use of clonazepam for withdrawal. South Med J 81:830–836.

PELLOCK JM, & WILLMORE LJ (1991). A rational guide to routine blood monitoring in patients receiving antiepileptic drugs. Neurology 41:961–964.

PERUCCA E, GARRATT A, HEBDIGE S, ET AL. (1978). Water intoxication in epileptic patients receiving carbamazepine. J Neurol Neurosurg Psychiatry 41:713–718.

PFEIFFER RF, KANG J, GRABER B, ET AL. (1990). Clozapine for psychosis in parkinsonian patients with dopaminomimetic psychosis. Mov Disord 5:239–242.

PHILLIPSON M, MORANVILLE JT, JESTE DV, ET AL. (1990). Antipsychotics. Clin Pharmacol 6:411–422.

PIERI L (1983). Preclinical pharmacology of midazolam. Br J Clin Pharmacol 16(suppl 1):17S–27S.

Pinner E, & Rich C (1988). Effects of trazodone on aggressive behavior in seven patients with organic mental disorders. Am J Psychiatry 145:1295–1296.

Pisani F, Fazio A, Oteri G, et al. (1986). Sodium valproate and valpromide: Differential interactions with carbamazepine in epileptic patients. Epilepsia 27:548–552.

Pohl R, Balon R, Berchou R, et al. (1987). Allergy to tartrazine in antidepressants. Am J Psychiatry 144:237–238.

Polumbo RA, Branzi A, Schroeder JS, et al. (1973). The antiarrhythmic effect of lithium chloride for experimental ouabain-induced arrhythmias. Proc Soc Exp Biol Med 142:1200–1204.

Pope HG, McElroy SL, Keck PE, et al. (1991). Valproate in the treatment of acute mania. A placebo-controlled study. Arch Gen Psychiatry 48:62–68.

Port FK, Kroll PD, & Rosenzweig J (1979). Lithium therapy during maintenance hemodialysis. Psychosomatics 20:130–131.

Post RM (1988). Effectiveness of carbamazepine in the treatment of bipolar affective disorder. In SL McElroy, HG Pope (Eds.), Use of anticonvulsants in psychiatry: Recent advances (Ch. 1, pp. 1–25). Clifton, NJ: Oxford Health Care, Inc.

Potter WZ, Rudorfer MV, & Lane EA (1984). Active metabolites of antidepressants: Pharmacodynamics and relevant pharmacokinetics. In E Usdin, M Asberg, L Bertilsson, & F Sjogvist (Eds.), Frontiers in biochemical and pharmacological research in depression (Pp. 373–390). New York: Raven Press.

Preskorn S, Weller E, Weller R, et al. (1983). Plasma levels of imipramine and adverse effects in children. Am J Psychiatry 140:1332–1335.

Preskorn SH, & Fast GA (1991). Therapeutic drug monitoring for antidepressants: Efficacy, safety, and cost effectiveness. J Clin Psychiatry 52(6, suppl):23–33.

Preskorn SH, & Othmer SC (1984). Evaluation of bupropion hydrochloride: The first of a new class of atypical antidepressants. Pharmacotherapy 4:20–34.

Preskorn SH (1989). Tricyclic antidepressants: The whys and hows of therapeutic drug monitoring. J Clin Psychiatry 50(7, suppl): 34–42.

Price WA, & Bielefeld M (1989). Buspirone-induced mania. J Clin Psychopharmacol 9:150–151.

Price W, & Giannini RJ (1986). Neurotoxicity caused by lithium-verapamil synergism. J Clin Pharmacol 26:717–719.

Prien RF, & Gelenberg AJ (1989). Alternatives to lithium for preventive treatment of bipolar disorder. Am J Psychiatry 146:840–848.

Puzynski S, & Klosiewicz L (1984). Valproic acid amide in the treatment of affective and schizoaffective disorders. J Affective Disord 6:115–121.

Rabkin J, Quitking F, & Harrison W (1984). Adverse reactions to monoamine oxidase inhibitors: Part I. A comparative study. J Clin Psychopharmacol 4:270–278.

Rabkin JG, Quitkin F, & McGrath P (1985). Adverse reactions to monoamine oxidase inhibitors: Part II. Treatment correlates and clinical management. J Clin Psychopharmacol 5:2–9.

Ragheb M (1990). The clinical significance of lithium-nonsteroidal anti-inflammatory drug interactions. J Clin Psychopharmacol 10:350–354.

Rambeck B, Salke-Treumann A, May T, et al. (1990). Valproic acid-induced carbamazepine-10,11-epoxide toxicity in children and adolescents. Eur Neurol 30:79–83.

Ratey JJ, Nikkelsen EJ, Smith GB, et al. (1986). Beta-blockers in the severely and profoundly retarded. J Clin Psychopharmacol 6:103–107.

Ratey JJ, Sovner R, Nikkelsen EJ et al. (1989). Buspirone therapy for maladaptive behavior and anxiety in developmentally disabled persons. J Clin Psychiatry 50:382–384.

Rausch JL, Pavlinac DM, & Newman PE (1984). Complete heart block following a single dose of trazodone. Am J Psychiatry 141: 1472–1473.

Ray WA, Griffin MR, & Downey M (1989). Benzodiazepines of long and short life elimination and the risk of hip fracture. JAMA 262:3303–3307.

Ray WA, Griffin MR, & Malcolm E (1991). Cyclic antidepressants and the risk of hip fracture. Arch Intern Med 151:754–756.

Realmuto GM, August GJ, & Garfinkel BD (1989). Clinical effect of buspirone in autistic children. J Clin Psychopharmacol 9:122–125.

Redenbaugh JE, Sato S, Penry JK, et al. (1980). Sodium valproate: Pharmacokinetics and effectiveness in treating intractable seizures. Neurology 30:1–6.

Redington K, Wells C, & Petito F (1992). Erythromycin and valproate interaction [letter]. Ann Intern Med 116:877–878.

Reed SM, Wise MG, & Timmerman I (1989). Choreoathetosis: A sign of lithium toxicity. J Neuropsychiatry 1:57–60.

Rees RK, Gilbert DA, & Katon W (1984). Tricyclic antidepressant therapy for peptic ulcer disease. Arch Intern Med 144:566–569.

Regan WM, Margolin RA, & Mathew RJ (1989). Cardiac arrhythmia following rapid imipramine withdrawal. Biol Psychiatry 25:482–484.

Richelson E (1982). Pharmacology of antidepressants in use in the United States. J Clin Psychiatry 43:4–11.

Richelson E (1984). The newer antidepressants: Structures, pharmacokinetics, pharmacodynamics, and proposed mechanism of action. Psychopharmacol Bull 20:213–223.

Rickels K, Freeman E, & Sondheimer S (1989). Buspirone in treatment of premenstrual syndrome. Lancet 1:777.

Roberts HE, Dean RC, & Stoudemire A (1989). Clozapine treatment of psychosis in Parkinson's disease. J Neuropsychiatry 1:190–192.

Robinson DS, Alms DR, Shrotriya RC, et al. (1989). Serotonergic anxiolytics and treatment of depression. Psychopathology 22(suppl 1):27–36.

Robinson DS, Nies A, & Corcelia J (1982). Cardiovascular effects of phenelzine and amitriptyline in depressed outpatients. J Clin Psychiatry 43:8–15.

Rochel M, & Ehrenthal W (1983). Haematological side effects of valproic acid. In J Oxley, D Janz, H Meinardi (Eds.), Chronic toxicity of antiepileptic drugs (Ch. 8, pp. 101–104). New York: Raven Press.

Rogers MP (1985). Rheumatoid arthritis: Psychiatric aspects and use of psychotropics. Psychosomatics 26:915–925.

Roose SP, Dalack GW, Glassman AH, et al. (1991a). Is doxepin a safer tricyclic for the heart? J Clin Psychiatry 52:338–341.

Roose SP, Dalack GW, Glassman AH, et al. (1991b). Cardiovascular effects of bupropion in depressed patients with heart disease. Am J Psychiatry 148:512–516.

Roose SP, Dalack GW, & Woodring S (1991). Death, depression, and heart disease. J Clin Psychiatry 52(6, suppl):34–39.

Roose SP, Glassman AH, Giardina EGV, et al. (1987a). Tricyclic antidepressants in depressed patients with cardiac conduction disease. Arch Gen Psychiatry 44:273–275.

Roose SP, Glassman AH, Giardina EGV, et al. (1987b). Cardiovascular effects of imipramine and bupropion in depressed patients with congestive heart failure. J Clin Psychopharmacol 7:247–251.

Roose SP, Glassman AH, & Giardina EV (1986). Nortriptyline in depressed patients with left ventricular impairment. JAMA 256: 3253–3257.

Rosebush PI, Stewart TD, & Gelenberg AJ (1989). Twenty neuroleptic rechallenges after neuroleptic malignant syndrome in 15 patients. J Clin Psychiatry 50:295–298.

ROSENBERG PB, AHMED I, & HURWITZ S (1991). Methylphenidate in depressed medically ill patients. J Clin Psychiatry 52:263–267.

ROSENSTEIN DL, TAKESHITA J, & NELSON JC (1991). Fluoxetine-induced elevation and prolongation of tricyclic levels in overdose [letter to the editor]. Am J Psychiatry 148:807.

RUDORFER MV, & YOUNG RC (1980). Desipramine: Cardiovascular effects and plasma levels. Am J Psychiatry 137:984–986.

SAKAUYE K (1990). Psychotic disorders: Guidelines and problems with antipsychotic medications in the elderly. Psychiatr Ann 20:456–465.

SALAMA A, & SHAFEY M (1989). A case of severe lithium toxicity induced by combined fluoxetine and lithium carbonate. Am J Psychiatry 146:278.

SALZMAN C (1987). Treatment of agitation in the elderly. In HY MELTZER (Ed.), Psychopharmacology, a generation of progress (Ch. 120, pp. 1167–1176). New York: Raven Press.

SALZMAN C (1991). The APA task force report on benzodiazepine dependence, toxicity and abuse. Am J Psychiatry 148:151–152.

SATEL SL, & NELSON JC (1989). Stimulants in the treatment of depression: A critical overview. J Clin Psychiatry 50:241–249.

SCHARF MB, ROTH PB, DOMINGUEZ RA, ET AL. (1990). Estazolam and flurazepam: A multicenter, placebo-controlled comparative study in outpatients with insomnia. J Clin Pharmacol 30:461–467.

SCHEFFNER D, KONIG S, RAUTERBERG-RUTLAND I, ET AL. (1988). Fatal liver failure in 16 children with valproate therapy. Epilepsia 29:520–542.

SCHENCK CH, & REMICK RA (1989). Sublingual nifedipine in the treatment of hypertensive crisis associated with monoamine oxidase inhibitors. Ann Emerg Med 18:114–115.

SCHENKER S, BERGSTROM RF, WOLEN RL, ET AL. (1988). Fluoxetine disposition and elimination in cirrhosis. Clin Pharmacol Ther 44:353–359.

SCHNEIDER LS, COOPER TB, SEVERSON JA, ET AL. (1988). Electrocardiographic changes with nortriptyline and 10-hydroxynortriptyline in elderly depressed outpatients. J Clin Psychopharmacol 8:402–408.

SCHNEIDER LS, POLLOCK VE, & LYNESS SA (1990). A metaanalysis of controlled trials of neuroleptic treatment in dementia. J Am Geriatr 38:553–563.

SCHNEIDER LS, & SOBIN PB (1991). Non-neuroleptic medications in the management of agitation in Alzheimer's disease and other dementia: A selective review. Int J Geriatr Psychiatry 6:691–708.

SCHWARTZ P, & WOLF S (1978). QT interval prolongation as predictor of sudden death in patients with myocardial infarction. Circulation 57:1074–1077.

SECOR JW, & SCHENKER S (1987). Drug metabolism in patients with liver disease. Adv Intern Med 32:379–406.

SEETHARAM MN, & PELLOCK JM (1991). Risk-benefit assessment of carbamazepine in children. Drug Safety 6:148–158.

SELLERS TD JR, CAMPBELL RWF, BASHORE TM, ET AL. (1977). Effects of procainamide and quinidine sulfate in the Wolf-Parkinson-White syndrome. Circulation 55:15–22.

SETHNA M, SOLOMON G, CEDARBAUM J, ET AL. (1989). Successful treatment of massive carbamazepine overdose. Epilepsia 30:71–73.

SEWELL DD, & JESTE DV (1991). Neuroleptic malignant syndrome: Clinical presentation, pathophysiology, and treatment. In A STOUDEMIRE, BS FOGEL (Eds.), Medical psychiatric practice (Vol. 1; Ch. 12, 425–452). Washington, DC: American Psychiatric Press.

SHEEHAN DV, CLAYCOMB JB, & KOURETAS N (1980–81). Monoamine oxidase inhibitors: Prescription and patient management. Int J Psychiatry Med 10:99–121.

SHEEHAN JD, & SHELLEY RK (1990). Leucopenia secondary to carbamazepine despite concurrent lithium treatment. Br J Psychiatry 157:911–912.

SHUKLA S, MUKHERJEE S, & DECINA P (1988). Lithium in the treatment of bipolar disorders associated with epilepsy: An open study. J Clin Psychopharmacol 8:201–204.

SIDES CA (1987). Hypertension during anaesthesia with monoamine oxidase inhibitors. Anaesthesia 42:633–635.

SILVER PA, & SIMPSON GM: Antipsychotic use in the medically ill. Psychother Psychosom 49:120–136.

SIMONSON M (1964). Controlling MAO inhibitor hypotension. Am J Psychiatry 120:1118–1119.

SIMPSON DM, & FOSTER D (1986). Improvement in organically disturbed behavior with trazodone treatment. J Clin Psychiatry 47:191–193.

SMITH JM, & BALDESSARINI RJ (1980). Changes in prevalence, severity and recovery in tardive dyskinesia with age. Arch Gen Psychiatry 37:1368–1373.

SMITH RC, CHOJNACKI M, & HU R (1980). Cardiovascular effects of therapeutic doses of tricyclic antidepressants: Importance of blood level monitoring. J Clin Psychiatry 41:57–63.

SOBOTKA JL, ALEXANDER B, & COOK BL (1990). A review of carbamazepine's hematologic reactions and monitoring recommendations. DICP 24:1214–1217.

SPECHT U, BOENIGK HE, & WOLF P (1989). Discontinuation of clonazepam after long-term treatment. Epilepsia 30:458–463.

STACK CG, ROGERS P, & LINTER SPK (1988). Monoamine oxidase inhibitors and anaesthesia. Br J Anaesth 60:222–227.

STARK P, & HARDISON CD (1985). A review of multicenter controlled studies of fluoxetine vs. imipramine and placebo in outpatients with major depressive disorder. J Clin Psychiatry 46:53–58.

STARK P, FULLER RW, & WONG DT (1985). The pharmacologic profile of fluoxetine. J Clin Psychiatry 46:7–13.

STEINER W, & FONTAINE R (1986). Toxic reaction following the combined administration of fluoxetine and L-tryptophan: Five case reports. Biol Psychiatry 21:1067–1071.

STEPHENS WP, ESPIR MLE, TATTERSALL RB, ET AL. (1977). Water intoxication due to carbamazepine. Br Med J 1:754–755.

STERN SL, RIBNER HS, COOPER TB, ET AL. (1991). 2-hydroxydesipramine and desipramine plasma levels and electrocardiographic effects in depressed younger adults. J Clin Psychopharmacol 11:93–98.

STOUDEMIRE A, & ATKINSON P (1988). Use of cyclic antidepressants in patients with cardiac conduction disturbances. Gen Hosp Psychiatry 10:389–397.

STOUDEMIRE A, & CLAYTON L (1989). Successful use of clozapine in a patient with a history of neuroleptic malignant syndrome. J Neuropsychiatry 1:303–305.

STOUDEMIRE A, & FOGEL BS (1987). Psychopharmacology in the medically ill. In A STOUDEMIRE & BS FOGEL (Eds.), Principles of medical psychiatry (Ch. 4, pp. 79–112). Orlando, FL: Grune & Stratton.

STOUDEMIRE A, FOGEL BS, & GULLEY LR (1991). Psychopharmacology in the medically ill: An update. In A STOUDEMIRE & BS FOGEL (Eds.), Medical psychiatric practice (Vol. I; Ch. 2, pp. 29–97). Washington, DC: American Psychiatric Press.

STOUDEMIRE A, MORAN MG, & FOGEL BS (1990). Psychotropic drug use in the medically ill: Part I. Psychosomatics 31:377–391.

STOUDEMIRE A, MORAN MG, & FOGEL BS (1991). Psychotropic drug use in the medically ill: Part II. Psychosomatics 32:34–46.

SULLIVAN EA & SHULMAN KI (1984). Diet and monoamine oxidase inhibitors: A re-examination. Can J Psychiatry 29:707–711.

SUSSMAN N (1987). Treatment of anxiety with buspirone. Psychiatr Ann 17:114–117.

SZYMANSKI S, LIEBERMAN JA, PICOU D, ET AL. (1991). A case report of cimetidine induced clozapine toxicity. J Clin Psychiatry 52:21–22.

TANGEDAHL TN, & GAU GT (1972). Myocardial irritability associated with lithium carbonate therapy. N Engl J Med 287:867–869.

TAYLOR DC (1962). Alarming reaction to pethidine in patients on phenelzine. Lancet 2:410–412.

TILKIAN JG, SCHROEDER JS, CAO J, ET AL. (1976). Effect of lithium on cardiovascular performance. A report on extended ambulatory monitoring and exercise testing before and during lithium. Am J Cardiol 38:701–798.

TILLER JWG, DAKIS JA, & SHAW JM (1988). Short-term buspirone treatment in disinhibition with dementia. Lancet 2:510.

TUOMISTO J, & SMITH DF (1986). Effects of tranylcypromine enantiomers on monoamine uptake and release and imipramine binding. J Neural Transm 65:135–145.

VAN SWEDEN B, & VAN MOFFAERT M (1985). Valproate as psychotropic agent. Acta Psychiatr Scand 72:315–317.

VEITH RC, FRIEDEL RO, BLOOM V, ET AL. (1980). Electrocardiogram changes and plasma desipramine levels during treatment of depression. Clin Pharmacol Ther 27:796–802.

VEITH RC, RASKIND MA, & CALDWELL JH (1982). Cardiovascular effects of tricyclic antidepressants in depressed patients with chronic heart disease. N Engl J Med 306:954–959.

VENDSBORG PB (1979). Lithium and glucose tolerance in manic-melancholic patients. Acta Psychiatr Scand 59:306–316.

VERNIER VG (1961). The pharmacology of antidepressant agents. Dis Nerv Syst 22:507–513.

VIEWEG WV, HILLARD JR, HOFFMAN MA, ET AL. (1988). Depression and the Wolff-Parkinson-White syndrome. Psychosomatics 29:113–116.

VIEWEG WVR, GLICK JL, HERRING S, ET AL. (1987). Absence of carbamazepine-induced hyponatremia among patients also given lithium. Am J Psychiatry 144:943–947.

VIEWEG WVR, & GODLESKI LS (1988). Carbamazepine and hyponatremia. Am J Psychiatry 145:1323–1324.

VIEWEG WVR, YANK GR, ROW WT, ET AL. (1986–87). Increase in white blood cell count and serum sodium level following the addition of lithium to carbamazepine treatment among three chronically psychotic male patients with disturbed affective states. Psychiatr Q 58:213–217.

VINCENT HH, WEIMAR W, & SCHALEKAMP MADH (1987). Effect of cyclosporine in fractional excretion of lithium and potassium in kidney transplant recipients. Kidney Int 31:1048.

VISHWANATH BM, NAVALGUND AA, CUSANO W, ET AL. (1991). Fluoxetine as a cause of SIADH [letter]. Am J Psychiatry 148:542–543.

WARD NG, & DAVIDSON RC (1991). Gastric absorption of nifedipine [letter]. J Clin Psychiatry 52:188.

WEBER JJ (1989). Seizure activity associated with fluxoetine therapy. Clin Pharm 8:296–298.

WEILER PG, MUNGAS D, & BERNICK C (1988). Propranolol for the control of disruptive behavior in senile dementia. J Geriatr Psychiatr Neurol 1:226–230.

WELLENS HJ, & DURRER D (1974). Wolf-Parkinson-White syndrome and atrial fibrillation: Relation between refractory period of accessory pathway and ventricular rate during atrial fibrillation. Am J Cardiol 34:777–782.

WELLENS HJJ, BRAAT S, BRUGADA P, ET AL. (1982). Use of procainamide in patients wtih the Wolff-Parkinson-White syndrome to disclose a short refractory period of the accessory pathway. Am J Cardiol 50:1087–1089.

WELLS DG, & BJORKSTEN AR (1989). Monoamine oxidase inhibitors revisited. Can J Anaesth 35:64–74.

WENGER TL, & STERN WC (1983). The cardiovascular profile of bupropion. J Clin Psychiatry 44:176–182.

WENGER TL, BUSTRACK JA, & COHN JB (1982). Comparison of the electrocardiographic effects of amitriptyline and bupropion. Clin Pharmacol Ther 31:280.

WENGER TL, COHN JB, & BUSTRACK J (1983). Comparison of the effects of bupropion and amitriptyline on cardiac conduction in depressed patients. J Clin Psychiatry 44:174–175.

WHITE MG, & FETNER CD (1975). Treatment of the syndrome of inappropriate secretion of antidiuretic hormone with lithium carbonate. N Engl J Med 292:390–392.

WHITE PF, VASCONEZ LO, MATHES SA, ET AL. (1988). Comparison of midazolam and diazepam for sedation during plastic surgery. Plast Reconstr Surg 81:703–710.

WILKERSON RD (1978). Antiarrhythmic effects of tricyclic antidepressant drugs in ouabain-induced arrhythmias in the dog. J Pharmacol Exp Ther 205:666–674.

WILLIAMS RL, & MAMELOK RD (1980). Hepatic disease and drug pharmacokinetics. Clin Pharmacokinet 5:528–547.

WILNER KD, LAZAR JD, APSELOFF G, ET AL. (1991). The effects of sertraline on the pharmacodynamics of warfarin in healthy volunteers [abstract]. Biol Psychiatry 29:354S.

WONG KC (1986). Preoperative discontinuation of monoamine oxidase inhibitor therapy: An old wives tale? Semin Anesth 5:145–148.

WONG T, & TIESSEN E (1989). Seizure in gradual clonazepam withdrawal. Psychiatr J Univ Ottowa 14:484.

WOODS SW, TESAR GE, MURRAY GB, ET AL. (1986). Psychostimulant treatment of depressive disorders secondary to medical illness. J Clin Psychiatry 47:12–15.

WROBLEWSKI BA, MCCOLGAN K, SMITH K, ET AL. (1990). The incidence of seizures during tricyclic antidepressant drug treatment in a brain-injured population. J Clin Psychopharmacol 10:124–128.

WYANT M, DIAMOND BI, O'NEAL E, ET AL. (1990). The use of midazolam in acutely agitated psychiatric patients. Psychopharmacol Bull 26:126–129.

YASSA R, ISKANDAR H, NASTASE C, ET AL. (1988). Carbamazepine and hyponatremia in patients with affective disorder. Am J Psychiatry 145:339–342.

YATHAM LN, BARR S, & DINAN TG (1989). Serotonin receptors, buspirone, and premenstrual syndrome. Lancet 1:1447–1448.

YOUNG RC, ALEXOPOULOS GS, SHAMOIAN CA, ET AL. (1984). Heart failure associated with high plasma 10-hydroxynortriptyline levels. Am J Psychiatry 141:432–433.

YOUNG RC, ALEXOPOULOS GS, SHAMOIAN CA, ET AL. (1985). Plasma 10-hydroxy-nortriptyline and ECG changes in elderly depressed patients. Am J Psychiatry 142:866–868.

YOUSSEF MS, & WILKINSON PA (1988). Epidural fentanyl and monoamine oxidase inhibitors. Anaesthesia 43:210–212.

YUDOFSKY S (1981). Propranolol in the treatment of rage and violent behavior in patients with chronic brain syndrome. Am J Psychiatry 138:218–230.

YUDOFSKY S, STEVENS L, & SILVER J (1984). Propranolol in the treatment of rage and violent behavior associated with Korsakoff's psychosis. Am J Psychiatry 141:114–115.

ZUNG WWK (1983). Review of placebo-controlled trials with bupropion. J Clin Psychiatry 44:104–114.

APPENDIX

TABLE A–1 *Some Drugs That Cause Psychiatric Symptoms*[a]

Drug	Reactions	Comments
Acyclovir (Zoviraxi)	Hallucinations, fearfulness confusion, insomnia, hyperacusis, paranoia, depression	At high doses, particularly in patients with chronic renal failure
Albuterol (Proventil; Ventolin)	Hallucinations, paranoia	Several reports
Alprazolam (Xanax)	See Benzodiazepines	
Amantadine (Symmetrel[b])	Visual hallucinations, paranoid delusions, nightmares, mania, exacerbation of schizophrenia	Several reports; more frequent in elderly
Aminocaproic acid (Amicar[b])	Acute delirium, hallucinations	Following bolus injection in one patient
Amiodarone (Cordarone)	Delirium, hallucinations	In one patient
Amphetamine-like drugs	Bizarre behavior, hallucinations, paranoia, agitation, anxiety, manic symptoms	Usually with overdose or abuse; can occur with inhaler abuse
	Depression	On withdrawal
Amphotericin B (Fungizone)	Delirium	With IV and intrathecal use
Anabolic steroids	Aggression, mania, depression, psychosis	Several reports
Anticonvulsants	Agitation, confusion, delirium, depression, psychosis, aggression, mania, toxic encephalopathy	Usually with high doses or high plasma concentrations
Antidepressants, tricyclic	Mania or hypomania; delirium, hallucinations, paranoia	Mania or hypomania in about 10% of patients; also after withdrawal
Antihistamines	Anxiety, hallucinations, delirium	Especially with overdosage
Asparaginase (Elspar)	Confusion, depression, paranoia	May occur frequently
Atenolol (Tenormin)	See Beta-adrenergic blockers	
Atropine and anti-cholinergics	Confusion, memory loss, disorientation, depersonalization, delirium, auditory and visual hallucinations, fear, paranoia, agitation, bizarre behavior	More frequent in elderly and children with high doses; has occurred with transdermal scopolamine
	Sudden incoherent speech, delirium with high fever, flushed dry skin, hallucinations	From eye drops, particularly when mistaken for nose drops
Baclofen (Lioresal[b])	Hallucinations, paranoia, nightmares, mania, depression, anxiety, confusion	Sometimes with treatment, but usually after sudden withdrawal
Barbiturates	Excitement, hyperactivity, visual hallucinations, depression, delirium-tremens-like syndrome	Especially in children and the elderly, or on withdrawal
Belladonna alkaloids	See Atropine and anticholinergics	
Benzodiazepines	Rage, hostility, paranoia, hallucinations, depression, insomnia, nightmares, anterograde amnesia	During treatment or on withdrawal; may be more common in elderly
Beta-adrenergic blockers	Depression, confusion, nightmares, hallucinations, paranoia, delusions, mania, hyperactivity	With usual doses, including ophthalmic use
Betaxolol (Kerlone)	See Beta-adrenergic blockers	
Bromocriptine (Pariodel)	Mania, delusions, hallucinations, paranoia, aggressive behavior, schizophrenia relapse, depressions, anxiety	Not dose-related; may persist weeks after stopping drug
Buprenorphine (Buprenex)	See Narcotics	
Bupropion (Wellbutrin)	Psychosis, hallucinations, agitation, paranoia	In depressed patients; aggravation of symptoms in schizophrenics
Caffeine	Anxiety, confusion, psychotic symptoms	With excessive doses
Captopril (Capoten)	Severe anxiety, hallucinations, insomnia, mania	Especially in depressed patients
Carbamazepine (Tegretol[b])	See Anticonvulsants	
Cephalosporins	Confusion, disorientation, paranoia, hallucinations	Several reports
Chlorambucil (Leukeran)	Hallucinations, lethargy, seizures, stupor, coma	In five of six patients at high dosage
Chloroprocaine (Nesacane)	See Procaine derivatives	
Chloroquine (Aralen[b])	Confusion, delusions, hallucinations	Several reports
Ciprofloxacin (Cipro)	Delirium	In one patient
Cimetidine (Tagamet)	See Histamine H_2-receptor antagonists	
Clomiphene citrate (Clomid[b])	Schizophrenia-like symptoms, paranoia	Two reports
Clonazepam (Klonopin)	See Benzodiazepines	
Clonidine (Catapres[b])	Delirium, hallucinations, depression	May resolve with continued use
Clorazepate (Tranxene[b])	See Benzodiazepines	
Cocaine	Anxiety, agitation, psychosis	Can occur with topical use

TABLE A–1. (*Continued*)

Drug	Reactions	Comments
Codeine	See Narcotics	
Contraceptives, oral	Depression	In 15% in one study
Corticosteroids (prednisone, cortisone, ACTH, others)	Mania, depression, confusion, paranoia, hallucinations, catatonia	Especially with high doses; can occur on withdrawal or with inhalation
Cyclobenzaprine (Flexeril[b])	Mania, hyperactivity, psychosis	In three patients
Cyclopentolate (Cyclogyl)	See Atropine and anticholinergics	
Cycloserine (Seromycin[b])	Anxiety, depression, confusion, psychosis	Common
Cyclosporine (Sandimmune)	Hallucinations; mania	Each in one patient
Dapsone	Insomnia, agitation, hallucinations, mania, depression	Several reports; may occur even with low doses
Deet (Off[b])	Toxic encephalopathy, mania, hallucinations	With excessive or prolonged use, particularly in infants and children
Diazepam (Valium[b])	See Benzodiazepines	
Diethylpropion (Tenuate)	See Amphetamine-like drugs	
Digitalis glycosides	Nightmares, euphoria, confusion, amnesia, aggression, psychosis, depression	Especially high doses or high plasma levels and in elderly
Diltiazem (Cardizem)	Depression, suicidal thoughts	Reported in eight patients
Disopyramide (Norpace[b])	Agitation, panic, depression, psychosis	Within 24–48 hours after starting
Disulfiram (Antabuse[b])	Delirium, depression, psychosis	Not related to alcohol reactions
Dronabinol (Marinol)	Anxiety, disorientation, psychosis	Most disturbing in the elderly
Ephedrine	Hallucinations, paranoia	Excessive dosage
Ethchlorvynol (Placidyl)	Agitation, hallucinations, paranoia	Related to abuse and withdrawal
Ethionamide (Trecator-SC)	Depression, hallucinations	Depression frequent
Ethosuximide (Zarontin[b])	See Anticonvulsants	
Etretinate (Tegison)	Severe depression	May be potentiated by alcohol
Famotidine (Pepcid)	See Histamine H_2-receptor antagonists	
Fenfluramine (Pondimin)	See Amphetamine-like drugs	
Flecainide (Tambocor)	Visual hallucinations	In one man
Fluoxetine (Prozac)	Mania, hypomania	Several reports
Flurbiprofen (Ansaid)	See Nonsteroidal anti-inflammatory drugs	
Gentamicin (Garamycin[b])	Confusion, disorientation, hallucinations	Reported in three patients with IM use
Halothane (Fluothane)	Depression	Postoperative period
Histamine H_2-receptor antagonists	Hallucinations, paranoia, bizarre behavior, delirium, disorientation, depression, mania	Usually with high dosage, more often in elderly, renal dysfunction
Hydroxychloroquine (Plaquenil)	Irritability, difficulty concentrating, psychosis	One report with high dosage
Ibuprofen (Motrin[b])	See Nonsteroidal anti-inflammatory drugs	
Indomethacin (Indocin[b])	See Nonsteroidal anti-inflammatory drugs	
Interferon alfa (Roferon-A; Intron A[b])	Delirium, paranoia, depression, suicidal thoughts, anxiety	In 10 of 58 patients with viral hepatitis
Iohexol (Omnipaque)	Confusion, disorientation	Infrequent
Iopamidol (Isovue)	Confusion, disorientation	Infrequent
Isocarboxazid (Marplan)	Mania, insomnia, anxiety, paranoid delusions	In two patients on stopping the drug
Isoniazid (INH[b])	Depression, agitation, hallucinations, paranoia	Several reports
Isosorbide dinitrate (Isordil[b])	Hallucinations, depression, suicidal thoughts	In one elderly woman on two occasions
Isotretinoin (Accutane)	Depression	In 6 of 110 patients
Keramine (Ketalar[b])	Nightmares, hallucinations, crying, delirium	Acute; frequent with usual doses
Ketoconazole (Nizoral)	Hallucinations	In one patient
Levodopa (Dopar[b])	Delirium, depression, agitation, hypomania, nightmares, night terrors, hallucinations, paranoia	More frequent in elderly or with prolonged use
L-glutamine	Grandiosity, hyperactivity, hypersexuality	In two men
Lidocaine (Xylocaine[b])	See Procaine derivatives	
Lorazepam (Ativan[b])	See Benzodiazepines	
Maprotiline (Ludiomil[b])	Hypnopompic hallucinations	In three young patients
Mefloquine (Lariam)	Psychosis, acute brain syndrome, depression	Many recent reports
Methandrostenolone	See Anabolic steroids	
Meperidine (Demerol[b])	See Narcotics	
Methadone (Dolophine[b])	See Narcotics	
Methyldopa (Aldomet[b])	Depression, amnesia, nightmares, psychosis	Several reports
Methylphenidate (Ritalin[b])	Hallucinations, paranoia	Several reports in children
Methyltestosterone	See Anabolic steroids	
Methysergide (Sansert)	Depersonalization, hallucination, agitation	Several reports
Metoclopramide (Reglan[b])	Mania, severe depression, crying, delirium	Several reports
Metrizamide (Amipaque)	Confusion, hallucinations, depression, anxiety	May be prolonged
Metronidazole (Flagyl[b])	Depression, agitation, uncontrollable crying, disorientation; hallucinations	Two cases with oral use; hallucinations with high IV doses in one man

TABLE A–1. (*Continued*)

Drug	Reactions	Comments
Midazolam (Versed)	See Benzodiazepines	
Morphine	See Narcotics	
Nabilone (Cesamet)	Anxiety, disorientation, psychosis	Most disturbing in the elderly
Nalidixic acid (NegGram[b])	Confusion, depression, hallucinations	Rare
Nalorphine	See Narcotics	
Naproxen (Anaprox; Naprosyn)	See Nonsteroidal anti-inflammatory drugs	
Narcotics	Nightmares, anxiety, agitation, euphoria, dysphoria, depression, paranoia, hallucinations	Usually with high doses
Nifedipine (Procardia; Adalat)	Irritability, agitation, panic, belligerence, depression	Several reports
Niridazole (Ambilhar)	Confusion, hallucinations, mania, suicide	More likely with higher doses
Nonsteroidal anti-inflammatory drugs	Paranoia, depression, inability to concentrate, anxiety, confusion, hallucinations, hostility	Not reported with all drugs in this class
Norfloxacin (Noroxin)	Depression; anxiety	Anxiety in one patient
Oxandrolone	See Anabolic steroids	
Oxymetazoline (Afrin[b])	Hallucinations, anxiety, insomnia	With nasal decongestants in children
Oxymetholone (Anadrol)	See Anabolic steroids	
Pargyline (Eutonyl)	Manic psychosis	In one patient
Penicillin G Procaine	See Procaine derivatives	
Pentazocine (Talwin)	See Narcotics	
Pergolide (Permax)	Hallucinations, paranoia, confusion, anxiety; depression	On withdrawal
Phenelzine (Nardil)	Paranoia, delusions, fear, mania, rage	Mania or hypomania in about 10% of depressed patients
Phenmetrazine (Preludin)	See Amphetamine-like drugs	
Phentermine (Fastin[b])	See Amphetamine-like drugs	
Phenylephrine (Neo-Synephrine[b])	Depression, hallucinations, paranoia	Overuse of nasal spray
Phenylpropanolamine (Dexatrim[b])	See Amphetamine-like drugs	
Phenytoin (Dilantin[b])	See Anticonvulsants	
Podophyllin	Delirium, paranoia, bizarre behavior	Oral use in child, topical in two adults
Polythiazide (Renese)	Depression	In two patients after 2 weeks use
Prazosin (Minipress[b])	Hallucinations, depression, paranoia	In four patients; two had renal failure
Primidone (Mysoline[b])	See Anticonvulsants	
Procainamide (Pronestyl[b])	See Procaine derivatives	
Procaine derivatives	Terror, confusion, psychosis, agitation, bizarre behavior, depression, panic	Many reports, especially with Penicillin G Procaine
Procarbazine (Matulane)	Mania	In one patient
Promethazine (Phenergan[b])	Hallucinations, terror	In two children
Propoxyphene (Darvon[b])	See Narcotics	
Propranolol (Inderal[b])	See Beta-adrenergic blockers	
Pseudoephedrine (in Actifed)	Hallucinations, paranoia	Reported with usual dosage in children and with overuse in one adult
Quinacrine (Atabrine)	Mania, paranoia, anxiety, hallucinations, delirium	More common with high doses
Quinidine	Confusion, agitation, psychosis	Usually dose-related
Ranitidine (Zantac)	See Histamine H_2-receptor antagonists	
Reserpine (Serpasil[b])	Depression, nightmares	Common with > 0.5 mg/day
Salicylates	Agitation, confusion, hallucinations, paranoia	Chronic intoxication
Scopolamine (Hyoscine[b])	See Atropine and anticholinergics	
Sulindac (Clinoril)	See Nonsteroidal anti-inflammatory drugs	
Theophylline	Withdrawal, mutism, hyperactivity; anxiety; mania	Usually with high serum concentrations
Thiabendazole (Mintezol)	Psychic disturbances	Occasional
Thyroid hormones	Mania, depression, hallucinations, paranoia	Initial doses in susceptible patients
Timolol (Timoptic[b])	See Beta-adrenergic blockers	
Tobramycin (Nebcin)	Delirium, hallucinations, agitation	In one 66-year-old patient
Tocainide (Tonocard)	See Procaine derivatives	
Tranylcypromine (Parnate)	Mania or hypomania	In about 10% of depressed patients
Trazodone (Desyrel[b])	Delirium, hallucinations, paranoia, mania	Several reports
Triazolam (Halcion)	See Benzodiazepines	
Trichlormethiazide (Naqua[b])	Depression; suicidal ideation	In two patients after 2–3 months use
Trihexyphenidyl (Artane[b])	See Atropine and anticholinergics	
Trimethoprim-sulfamethoxazole (Bactrim[b])	Psychosis; depression, disorientation	Few reports
Valproic acid (Depakene[b])	See Anticonvulsants	

TABLE A–1. (*Continued*)

Drug	Reactions	Comments
Verapamil (Isoptin; Calan[b])	Auditory, visual, and tactile hallucinations	In one woman
Vincristine (Oncovin[b])	Hallucinations	Less than 5% of patients; high doses
Zidovudine (Retrovir)	Mania with paranoia, hallucinations	Reported in two patients

[a]For specific detailed citations, refer to the original *Medical Letter* source.
[b]Also available with other brand names or generically.
Reprinted with permission from the [Medical Letter] 31(808):113–116, 1989.

10 Electroconvulsive therapy in the medical and neurologic patient

RICHARD D. WEINER, M.D., PH.D., AND
C. EDWARD COFFEY, M.D.

Electroconvulsive therapy (ECT) involves the use of a series of electrically induced seizures to produce a clinical remission for severe episodes of major depressive disorder, schizophrenia, or mania (American Psychiatric Association [APA], 1987). At present in the United States, ECT is administered each year to 30,000 to 80,000 such individuals (APA, 1978; Thompson & Blaine, 1987).

The fact that ECT remains a clinically viable treatment modality after more than 50 years, and the development of literally dozens of psychopharmacologic alternatives, is a testimonial to its therapeutic potency and to its capacity to evolve over the years into a safer, if not a more acceptable treatment modality. At numerous points over this period, panels of scholars have carefully considered the question of whether a role for ECT still exists in contemporary psychiatric practice. In virtually all of such instances, the answer has been in the affirmative. One of the more recent of these evaluations was undertaken in June 1985 by a joint National Institute of Mental Health (NIMH) and National Institutes of Health (NIH) Consensus Development Panel (Consensus Conference, 1985). Once more, this panel concluded that ECT is, when properly administered, a safe and effective procedure for which there continues to be an established clinical need.

In 1990 a comprehensive set of practice guidelines for ECT was provided by the Task Force or ECT of the American Psychiatric Association (APA, 1990). The recommendations contained within the APA report go a long way toward a rational standardization of the clinical practice of ECT and represent a first step in what will likely be a move toward the development of practice guidelines in other areas of psychiatry, as well as medicine in general. Because of the importance of the APA report, we have endeavored, as much as possible, to keep the present chapter in conformance with its recommendations.

At present, ECT is generally reserved for those who either have not responded to psychopharmacologic trials or cannot tolerate such an attempt, whether for reasons of adverse effects or because of an urgent need for immediate relief (Abrams, 1992; APA, 1990). In practice one frequently finds that such individuals are more severely ill not only from a psychiatric perspective but from a medical or neurologic standpoint as well. More and more often it is the elderly, frail, chronically physically ill patient who is considered for referral for ECT. In such a situation, the referring physician must be able to carefully evaluate the delicate balance of risks and benefits that represents the logical determinant of clinical choice.

The presence of significant concurrent medical illness raises a number of salient questions that must be addressed as part of this process: Is ECT truly indicated for the patient? Is it safe? Should the ECT procedure be modified to minimize risk? It is to these issues, each becoming more commonplace as our population ages, that both this chapter and Chapter 11, which focuses on these questions from an anesthetic perspective, are directed. It is our hope that the reader will (1) arrive at a better understanding of whether and under what circumstances a medically or neurologically ill patient should be referred for ECT; and (2) be better able to administer the safest and most effective ECT possible under such circumstances.

CLINICAL INDICATIONS

Electroconvulsive therapy is mainly used for the management of severe major depressive episodes (APA, 1978;

Thompson & Blaine, 1987). Most remaining ECT referrals are for acute schizophrenic episodes, with mania running a distant third. Regardless of diagnosis, the psychopharmacologic management route is generally followed initially, reserving the use of ECT for times when an adequate therapeutic response is not forthcoming or when such a trial must be terminated owing to either unacceptable levels of toxicity or a substantial deterioration in the patient's condition (APA, 1990). Still, it is important to recognize that situations do exist where a referral to ECT is indicated on a "primary" basis. Such reasons include cases where a trial of an alternative modality is associated with greater risk than ECT, where the urgency of the patient's condition does not allow time for a drug trial (which may take 4 to 6 weeks), and where the patient has a strong history of preferential response to ECT (APA, 1990). In some cases, ECT can be considered truly lifesaving (Roy-Byrne & Gerner, 1981).

As a final general point concerning the clinical utility of this treatment modality, it must be understood that the result of a successful course of ECT is reflected in a remission of the index episode, rather than a "cure" of the underlying disorder, which in most cases is recurrent in nature. Clinical remission of an episode of illness, whether produced by ECT or a trial of psychopharmacologic agents, must be followed by some type of continuation therapy in order to minimize the risk of recurrence. This issue is dealt with later in this chapter, in the context of post-ECT continuation/maintenance therapy.

Depressive Disorders

Electroconvulsive therapy is effective in major depressive episodes of all types (Abrams, 1992; APA, 1990; Brandon et al., 1984; Gregory, Shawcross, & Gill, 1985). It is generally recognized that ECT is more effective in inducing a remission than antidepressant drugs (Janicak et al., 1985; Scovern & Killman, 1980; Siris, Glassman, & Stetner, 1982), and that it can exert a powerful therapeutic effect in drug nonresponders (Avery & Lubrano, 1979; Paul et al., 1981). It was formerly believed that ECT was more effective in certain types of major depression, e.g., melancholia (APA, 1987), but there is now compelling evidence that it has a broader spectrum of action (Pande et al., 1988; Prudic et al., 1989; Zimmerman, Coryell, & Pfohl, 1985).

Although there have been many attempts to arrive at a successful means of predicting a positive ECT response in terms of specific historical, demographic, and symptomatologic factors, these efforts have not met with success when applied across groups of patients from different settings (Abrams, 1992; Fink, 1979). One possible exception to this rule is the presence of delusions, which may represent a marker for severity and has been linked to a favorable response to ECT in some studies (e.g., Clinical Research Centre, 1984). Depressive illness other than a major depressive episode (e.g., dysthymia) does not respond to ECT any better than to alternative approaches. Still, it is important to recognize that patients with dysthymic disorders can develop a superimposed major depressive episode and may thereby become appropriate candidates for ECT. In such cases, one sometimes observes improvement in both conditions.

There has so far been little effort to study the effects of concurrent medical and neurological illness on the efficacy of ECT for treatment of major depressive episodes, or, for that matter, any specific diagnostic entity. Although there is a sense that "secondary" major depressive episodes, i.e., occurring in the context of chronic illnesses of other types (be they mental or physical), may not be as responsive to somatic therapy as are "primary" episodes (Zorumski et al., 1986), it is difficult to incorporate such impressions into clinical practice. It is also unclear whether such a relation would be dependent on an etiologic linkage between the conditions or would exist even when the occurrence of the two is purely incidental.

Schizophrenia

At present ECT in schizophrenia is reserved for patients suffering from an acute psychotic episode or exacerbation of illness who either have not responded to antipsychotic drugs or cannot tolerate such agents (APA, 1990). Even though only a small fraction of patients with schizophrenia are referred for ECT, this condition may account for approximately 10% of the total ECT population (APA, 1978; Thompson & Blaine, 1987). Controlled investigations continue to show that ECT is effective in such cases (Abraham & Kulhara, 1987; Brandon et al., 1985), with the remission-inducing efficacy of ECT similar to that of antipsychotic drug treatment. Still other data point toward the combination of ECT plus antipsychotic drugs as being more effective for the treatment of schizophrenic episodes than either ECT or drugs alone (Small et al., 1982; Smith et al., 1967).

With respect to prediction of response to ECT in individuals with schizophrenic episodes, it is now clear that the presence of affective or catatonic symptoms is associated with an excellent chance of remission (Salzman, 1980; Small, 1985). Chronic schizophrenia, however, especially in conjunction with predominantly residual, or "negative," symptomatology, is unfortunately associated with a poor response to ECT.

Mania

The discovery of lithium's antimanic effects rapidly led to a major decrease in referrals of manic patients for ECT. At present, the use of ECT for mania is largely reserved for those individuals who are refractory to or intolerant of psychopharmacologic alternatives (Abrams, 1992; APA, 1990). Although substantial efficacy for ECT in the treatment of manic episodes has been claimed for many years, it has only been recently that both prospective and retrospective data supporting this contention have become available (Black, Hulbert, & Nasrallah, 1989; Black, Winokur, & Nasrallah, 1987; Mukerjee, Sackeim, & Lee, 1988; Small et al., 1988). These investigations have included the first controlled prospective comparison of ECT and lithium, the standard first-line treatment of mania (Small et al., 1988). This latter study clearly demonstrates that ECT is at least as effective as lithium in its ability to induce a therapeutic remission in manic patients, and it may have a more rapid onset of action.

Interestingly, a switch to hypomania or mania has been reported to occur in bipolar depressed patients during a course of ECT, much as with antidepressant drugs, although the incidence of this phenomenon is unknown. In such a case, the ECT course can be continued on the basis of ECT's own antimanic properties.

Other Psychiatric Conditions

The use of ECT for primary psychiatric disorders other than major depression, schizophrenia, and mania is supported only on an anecdotal basis and represents a minuscule fraction of ECT referrals during the present era. There is no convincing evidence that ECT should be considered in the absence of the primary indications discussed earlier. When making this determination, however, the practitioner must also keep in mind that a major depressive episode may coexist with, and at times even be masked by, other intercurrent psychiatric disorders.

ECT as a Treatment of Medical and Neurologic Conditions

Although the use of ECT in patients with medical and neurologic disorders is covered extensively below, those discussions are for the most part based on the assumption that the ECT is being administered for the treatment of a mental disorder, rather than the coexisting medical or neurologic condition. In fact, the neurobiologic effects of ECT have on occasion been utilized for the treatment of a number of such illnesses. In most of these situations, e.g., catatonia, hypopituitarism, and intractable seizure disorder, the evidence supporting the use of ECT is anecdotal, and ECT should be considered only as a last resort (APA, 1990). Two exceptions to this rule are neuroleptic malignant syndrome (NMS) (Addonizio & Susman, 1987; Greenberg, 1986; Hughes, 1986; Pearlman, 1986) and Parkinson's disease, especially in the presence of the "on-off" syndrome (Andersen et al., 1987). In both cases, it is believed that the powerful dopaminergic potentiation known to be associated with ECT is responsible for the substantial benefits, albeit transient, that are often seen. Even here, though, much more work is necessary before a distinct role for ECT can be firmly established.

CONTRAINDICATIONS AND RISKS

One of the major points of the APA guidelines on ECT is that, rather than "absolute" contraindications to ECT, there exist a variety of situations where the presence of significantly increased risk requires especially careful justification of its use, as contrasted with available treatment alternatives or no treatment at all (APA, 1990). In addition, efforts must be made, where possible, in such cases to modify the ECT procedure to minimize the likelihood of adverse sequelae. *Relative contraindications* to ECT include the following conditions: space-occupying intracerebral lesion, recent (within 3 months) myocardial infarction, leaky or otherwise unstable vascular aneurysm or malformation, recent intracerebral hemorrhage, retinal detachment, pheochromocytoma, or other situations associated with high anesthetic risk (APA, 1990).

The mortality associated with ECT is low, often cited as 1 in 10,000 patients (Abrams, 1992). At the same time, however, it must also be pointed out that the presence of significant intercurrent medical disease may substantially increase the risk of both death and serious morbidity. Major medical sequelae that can be produced by a course of ECT are uncommon (Frederiksen & d'Elia, 1979; Heshe & Roeder, 1976). The most frequent such events are severe delirium, malignant cardiac arrhythmia, and prolonged apnea. Less serious, but also more prevalent, adverse sequelae include transient mild cardiac arrhythmia, oral trauma, injury to dental structures, and mild to moderate levels of delirium. As is described herein, special modifications of ECT technique may prevent or at least ameliorate many of these events.

Adverse effects produced by ECT can be explained in terms of the medical physiology of induced seizures. Profound autonomic surges during and following the ictal discharge lead to systemic hypertension and abrupt transitions in cardiac rate and output (Perrin, 1961). The hypertension, in combination with an ictal loss in cerebrovascular autoregulation, results in expansion of intracerebral blood volume, as well as a transient increase in the permeability of the blood-brain barrier, both of which are associated with an accompanying rise in intracerebral pressure (Bolwig, Hertz, & Westergaard, 1977). In addition, the sudden increase in cardiac output produced by the autonomic activation associated with a generalized seizure may precipitate significant myocardial ischemia, particularly in the presence of a preexisting impairment of myocardial blood supply (Deliyiannis, Eliakim, & Bellet, 1962).

In addition to vascular effects, the ictal discharge transiently disrupts neuronal metabolism, eliciting generalized slowing of a nonspecific nature on the interictal electroencephalogram (EEG) (Weiner, 1983), which builds up over the treatment course and is correlated with behavioral aspects of delirium, such as disorientation. Along with these global cerebral effects, certain areas of the brain, such as the hippocampus, are particularly sensitive to pathophysiologic changes and may account for the more specific form of cerebral impairment (i.e., amnesia) that is often present during and immediately after a course of ECT (Squire, 1986).

Alterations of cerebral function increase the possibility of pathologic changes in cerebral anatomy, a consideration that is of particular concern given the complaints of severe and permanently altered cognitive function elicited from a small number of ECT recipients (Breggin, 1979). A careful evaluation of the literature, however, leads one to the conclusion that ECT is not associated with a significant risk of "brain damage" per se (Coffey et al., 1988, 1991; Weiner, 1984). Still, it is established that a subtle, persistent retrograde amnesia, at least for weeks and months just prior to the ECT, is present in some circumstances. As with other adverse effects already described, this amnesia can be considerably attenuated by modifications in ECT technique.

ECT TECHNIQUE

Pre-ECT Evaluation

A thorough medical evaluation is a necessary prerequisite to a determination of potential risk with ECT (Abramczuk & Rose, 1979; Elliot, Linz, & Kane, 1982). The characteristics of this pre-ECT workup include a complete medical history and physical examination, with special attention to systems affected most by ECT: cardiovascular, respiratory, and neurologic. Pertinent laboratory tests include serum electrolytes, chest roentgenogram, and electrocardiogram (ECG). Additional procedures are used selectively (APA, 1990), for example, EEG, neuroradiologic, and neuropsychologic evaluation of patients in whom cerebral dysfunction is known or suspected. This category includes many elderly depressives as well as patients with depression and cognitive impairment.

Given the use of anesthesia with ECT, a preoperative anesthesia consultation is indicated, unless the treating psychiatrist is adequately trained in the relevant anesthetic and resuscitative skills required with ECT and risk factors are minimal. Because the medically ill patient is at greater risk for complications with ECT, it is important to ascertain in advance the ability of available medical resources to manage the occurrence of potential untoward events. Such resources include not only the individual providing anesthesia and the emergency medical equipment and supplies present in the ECT treatment area but also the institution's ability to provide immediate and appropriate emergency assistance if required. Patients with significant anesthetic risk, for example, should have a well-trained anesthesiologist in attendance at all treatments. For these reasons, seriously medically ill patients are best given ECT in a general hospital rather than a psychiatric hospital setting. Provision of ECT in a surgical or recovery room suite is sometimes necessary.

A final component of the pre-ECT workup, and one that is of great importance, is the informed consent (APA, 1990). At present, ECT is nearly always a voluntary procedure, with the attainment of consent guided by a variety of clinical, ethical, and legal considerations (Parry, 1986; Rothman, 1986; Taub, 1987). Adequate informed consent must include the provision, in terms understandable to the patient, of data on the nature of the treatment, its likely benefits and risks, and whatever treatment alternatives exist in lieu of ECT. This information must be given in sufficient depth to allow a reasonable person to arrive at a decision about whether to agree to the treatment; and the delivery of this information must be documented. Standardized patient information sheets are available for assistance in eliciting informed consent for ECT (APA, 1990). Not inconsequentially, such tools also serve as documentation that the material was provided to the patient.

Establishment of patient competence to provide consent for treatment varies from state to state. In general, some degree of understanding of the existence of the psychiatric illness, the nature of the treatment offered, and the likely consequences of treatment versus no treatment is required. In most jurisdictions a formal judicial guardianship procedure is required in the case of the incompetent patient, with separate regulations available for emergency provision of treatment.

Before the ECT treatments themselves are initiated, a decision must be made concerning the handling of the patient's medications during the treatment course. In addition to concerns about potential drug interactions with anesthetic agents, the evidence that psychotropic medications augment response when given concurrently with ECT is unproved, except possibly for the use of neuroleptics in schizophrenics or other agitated psychotic individuals (Barkai, 1985; Siris et al., 1982; Small et al., 1982). Theoretically, neuroleptics with minimal autonomic effects, such as haloperidol, should be used in such cases, though no investigation of this issue has ever been carried out. Antidepressant agents are generally discontinued prior to onset of ECT, though a prolonged drug-free period is probably not necessary either for cyclic antidepressants or for monoamine oxidase inhibitors (Azar & Lear, 1984; El-Ganzouri et al., 1985; Freese, 1985). When stopping antidepressant drugs, one must keep in mind that an abrupt termination may be associated with withdrawal effects, including cardiac irritability (Raskin, 1984).

Lithium should be discontinued prior to ECT because it may produce neurotoxic effects (Small, Kellams, & Milstein, 1980; Weiner et al., 1980), prolong the effects of muscle relaxants, and lessen the degree of therapeutic response. Sedative hypnotic agents, including barbiturates and particularly benzodiazepines, are also problematic in combination with ECT, as they either decrease seizure duration or increase seizure threshold or both (Price & Zimmer, 1985; Standish-Barry, Deacon, & Snaith, 1985). Should the patient experience extreme anxiety during the ECT course, modest doses of agents with low anticonvulsant properties (e.g., hydroxyzine) or short half-life benzodiazepines (e.g., oxazepam) can be administered.

All of the patient's nonpsychotropic medications should be reviewed prior to ECT as well, and those not deemed indicated should be discontinued. A number of such agents may be associated with specific adverse effects during ECT. Anticonvulsant drugs can create obvious problems with inducing seizures, though if clinically necessary ECT can still be carried out with certain modifications (see below). Lidocaine and, to a lesser degree, some beta-blockers also have a negative effect on seizure duration (Hood & Mecca, 1983; Khanna et al., 1989; Ottosson, 1960), whereas agents with stimulant properties (e.g., theophylline, may increase the risk of prolonged seizures, particularly with high blood levels (Peters, Wochos, & Peterson, 1984). The use of a beta-blocking agent without adequate anticholinergic premedication can produce cardiac asystole (Decina, et al., 1984). Reserpine may be associated with cardiovascular collapse with ECT.

Although most daily medications should be delayed until after the procedure on the day of each treatment, drugs that exert a protective effect during ECT—for example, cardiac and antihypertensive medications, as well as oral corticosteroid preparations—should be administered 2 to 3 hours prior to each treatment with minimal amounts of accompanying fluid. Also, insulin doses in diabetics receiving ECT should be structured to accommodate timing of meals on the ECT days. If ECT is not administered early in the morning, such individuals may also need infusion of an intravenous 5% dextrose solution prior to each treatment.

Number and Frequency of Treatments

The ECT treatments typically are administered at a frequency of three a week on alternate days in the United States, although this pattern is occasionally altered because of ECT-induced cerebral impairment (treatments decreased in frequency) or because of a particularly urgent need for a rapid response (treatments increased in frequency). One seizure is produced per ECT treatment, except in the case of an ECT modification called multiple-monitored ECT, in which several serial seizure inductions are elicited during each treatment session. This practice has not been adequately investigated and remains controversial (Abrams, 1985; APA, 1990; Maletzky, 1981). Multiple-monitored ECT potentially offers the opportunity of a more rapid response with fewer periods of anesthesia, but at the cost of a greater number of induced seizures, a more intense autonomic activation, and greater post-ECT confusion. Fewer treatment sessions makes multiple-monitored ECT theoretically attractive for cases requiring a significant amount of additional supportive services at the time of each treatment, for example, medical specialists, specialized monitoring capabilities, endotracheal intubation, or provision of ECT in the operating room area.

The number or ECT treatments administered is determined by clinical response. Once a plateau in the therapeutic effect has occurred (typically after 6 to 12 treatments for depression and mania), the ECT course may be stopped (Snaith, 1981). The number of treatments that should be given in the absence of a major therapeutic response is open to debate, though some clinicians use 8 to 12, with the lower numbers applicable in cases of absolutely no response. Despite its initial appeal as a potential biological marker for establishing the presence of a treatment response (Papakostas et al., 1981), the dexamethasone suppression test does not appear to be useful for this purpose (Coryell, 1986), probably because of the complex effect of the induced seizures on the pituitary-adrenal axis.

Anesthesia

As indicated earlier, ECT is a procedure that requires general anesthesia. The anesthetic induction and its associated use of oxygenation, muscular relaxation, and (if indicated) anticholinergic premedication is covered in Chapter 11.

Stimulus Electrode Placement

At present, there is a wide spectrum of practices concerning stimulus electrode location with ECT (APA, 1990; Weiner, 1986). The most frequently used type of electrode placement—bilateral—involves the application of stimulus electrodes over both frontotemporal regions. An alternative form is unilateral nondominant (Lancaster, Steinert, & Frost, 1958), where both stimulus electrodes are placed over the hemisphere presumed to be nondominant for speech. This type of placement is associated with considerably less severe and less persistent confusion and memory impairment (Squire, 1986; Weiner et al., 1986b). Unfortunately, there remains considerable controversy about the relative efficacy of these two types of electrode placement. Although many studies claim equal benefits (d'Elia & Raotma, 1975; Pettinati et al., 1986; Weiner et al., 1986b), a significant minority report an advantage for bilateral ECT (Abrams, 1986; Sackeim et al., 1987). Recent data have suggested that "high dose" stimulation may enhance the therapeutic potency of nondominant unilateral technique (Abrams, Swartz, & Vedak, 1991). In addition, for patients with manic symptomatology there is some evidence of a specific therapeutic advantage for bilateral ECT (Milstein et al., 1987), though this point has been disputed (Mukherjee et al., 1988).

It is still unclear whether the therapeutic differences in depressed patients reported by some investigators are based on an intrinsically lower potency for unilateral nondominant ECT or the discrepancies are the result of nonoptimal ECT technique (Weiner & Coffey, 1986). In an attempt to provide at least some temporary guidance on this matter, it has been proposed that, unless a particularly urgent response is needed, patients start on unilateral nondominant ECT and then switch to bilateral ECT if an adequate response is not forthcoming after about six treatments (Abrams & Fink, 1984). The absence of compelling evidence supporting such a switch (Stromgren, 1984) led the APA to equivocate on this issue in its practice guidelines (APA, 1990). In any event, it is clear that for unilateral ECT to be maximally effective it must be delivered in at least a moderately suprathreshold fashion across stimulus electrodes that are placed so as to allow maximal opportunity for seizure generalization (Weiner & Coffey, 1986; Sackeim & Mukherjee, 1986). This technique can be achieved in practice by a frontotemporal-to-high-centroparietal electrode configuration (d'Elia, 1970).

Electrical Stimulus

The nature and intensity of the electrical stimulus itself has been a further source of contention in the ECT literature. Present evidence seems to suggest that a low-energy, interrupted stimulus waveform, such as the brief pulse, is as effective as a standard higher-energy stimulus such as the sine wave, while also being significantly less cerebrotoxic (Weiner et al., 1986a,b). The situation is not, however, completely resolved; and some clinicians report that pulse stimuli, at least when combined with unilateral nondominant electrode placement, may not be as effective for eliciting therapeutic change (Price & McAllister, 1986). It is also recognized that the energy content of the stimulus waveform can be lowered to a point at which equivalent efficacy is clearly not present (Robin & De Tissera, 1982).

A final technique-related factor is the stimulus dosing strategy. The amount of electricity [measured either in charge (millicoulombs) or energy (watt-seconds or joules)] required to produce a seizure is the *seizure threshold*. This parameter varies enormously from patient to patient, being higher in women and increasing with age (Sackeim, Devanand & Prudic, 1991). In addition to interpatient variability, the seizure threshold typically rises over a course of treatments, typically by more than 100%. The choice of stimulus intensity parameters used in clinical practice has traditionally been somewhat arbitrary, with few scientific data available for assistance. Dosing strategies have ranged from barely suprathreshold stimuli to maximum machine settings (Weiner & Coffey, 1986). It is likely that a "moderate" dosage strategy is indicated at the present time, as barely suprathreshold stimuli may not be as therapeutically effective, at least for unilateral electrode placement (Sackeim et al., 1987), and maximally suprathreshold stimuli appear to be more cerebrotoxic (Fink, 1979).

Seizure Monitoring and Potentiation

The electrical stimulus used with ECT is only a means to an end: the triggering of a generalized seizure within the brain. In recent years practitioners have become more cognizant of the advantages of monitoring the electrophysiologic response to electrical stimulation directly, in the form of EEG monitoring, rather than only indirectly, via the motor activity produced during the seizure (Weiner, 1983; Weiner, Coffee & Krystal, 1991). Ictal EEG monitoring, now widespread in the United States following the incorporation of EEG recording capability into most commercial ECT devices, has allowed more precise determination of seizure duration, which is known to be important for ensuring an adequate therapeutic response (Ottosson, 1960). In addition, such monitoring provides a means to reliably detect the occurrence of prolonged seizures, something that has proved difficult using convulsive motor activity alone (Scott, Shering, & Dykes, 1989). For these reasons, the APA has strongly encouraged the routine use of this technology (APA, 1990). In the future, continued focus on ictal EEG monitoring may also allow us to understand which electrophysiologic components of the induced seizure are necessary for a therapeutic response to occur. Such data could be used to optimize electrical dosing so as to maximize benefits and minimize risks.

Brief seizures, usually defined as lasting less than 25 seconds, appear to be associated with diminished efficacy (Ottosson, 1960). Unfortunately, because of the tendency for seizure threshold to increase during ECT, such events not infrequent occur late in an ECT course, even with maximal stimulus intensities. In this regard it is worth reiterating that a number of pharmacologic agents significantly *shorten* seizure duration, e.g., anticonvulsants, sedative-hypnotics, beta-blockers, and lidocaine. There has been a resurgence of interest in means to prolong seizure duration as a way to obviate this problem (Sackeim, Devanand & Prudic, 1991; Weiner, Coffee & Krystal, 1991). Such techniques now include hyperventilation (Bergsholm, Gran, & Bleie, 1984), intravenous caffeine (Coffey et al., 1990; Lurie & Coffey, 1990), and use of ketamine anesthesia (Lunn et al., 1981). Theophylline may be associated with pathologically prolonged seizures and should therefore be avoided or have doses decreased during ECT. Caffeine is typically available as a solution of caffeine sodium benzoate 500 mg per vial (242 mg of which is caffeine). A typical starting dose is caffeine 242 mg IV over 20 seconds, given 2 minutes prior to anesthesia induction. It is generally well tolerated, although it can produce anxiety in patients prone to such symptoms and may accentuate sympathetically mediated cardiovascular changes. Ketamine is typically administered at a dose of 2.2 mg/kg. It is occasionally associated with a transient emergence psychosis.

CLINICAL MANAGEMENT OF PATIENTS AFTER ECT

As already noted, when successful, ECT produces a therapeutic remission in the present episode of the patient's psychiatric illness; it does not cure the underlying disorder. Without continuation therapy, the likelihood of relapse is high, running to 30 to 65% during the first 12 months for major depressive disorder (Imlah, Ryan, & Harrington, 1965). Particularly because many relapses occur during the first few months, patients are typically begun on psychopharmacological agents soon after ECT course has been completed; this treatment is continued, depending on the patient's clinical history, for at least 6 months. There is now convincing evidence that such therapy is effective in decreasing the incidence of relapse (Coppen et al., 1981; Imlah, Ryan, & Harrington, 1965), and more recent data suggest that it is likely that a prophylactic effect will be present even when pharmacotherapy was not beneficial during the index episode (Sackeim et al., 1990). For patients who have a history of early relapse and are either intolerant of, or do not respond to, psychopharmacologic prophylaxis, maintenance ECT may be considered (Clarke et al., 1989; Stevenson & Geoghegan, 1951). However, despite decades of clinical use, there have never been any controlled studies of its efficacy in preventing relapse.

ECT AND SPECIFIC MEDICAL AND NEUROLOGIC CONDITIONS

Electroconvulsive therapy is now being used widely in medically and neurologically ill patients (Abrams, 1989, 1992; Hay, 1989). Although the principles of ECT practice already outlined are relevant to patients with concurrent medical and neurologic illness, there are additional risks, precautions, and indicated modifications of ECT technique that should be kept in mind. Also, as noted earlier, the presence of a specific medical condition itself rarely can be an indication for ECT. To elucidate the use of ECT in the medically and neurologically ill, the following sections focus on those specific medical conditions for which there are data concerning ECT utilization. Further discussion of these issues is available in Chapter 11. For purposes of consistency a physiologic systems orientation is used.

Central Nervous System Disorders

Despite the absence of controlled studies, a growing clinical literature suggests that ECT is an effective treatment for depression and mania in patients with preexisting *dementia* (Coffey et al., 1989; Dubovsky, 1986; Hsiao, Messenheimer, & Evans, 1987; Price & McAllister, 1989). Although there exists a theoretical concern that ECT may worsen the dementia, it appears that most such patients tolerate the treatments without especially severe or long-lasting cognitive side effects, and in fact some patients actually show improvement in their cognitive and memory functions after a successful course of ECT (Coffey et al., 1989; Price & McAllister, 1989). Should unacceptable cognitive side effects occur, a common practice is to reduce the frequency of treatments or switch to brief-pulse unilateral nondominant ECT (Coffey & Weiner, 1990) if this was not the initial modality.

A number of other organic mental syndromes, including *catatonia* and *delirium* due to many different causes, may also improve with ECT (Hsiao et al., 1987). Of course, a careful neuropsychiatric evaluation is required of all such patients in order to clarify the etiology, including those conditions that might be associated with increase risk from ECT. Interestingly, ECT may exert a specific antidelirium effect even in the absence of improvement in the conditions that originally caused the delirium (Kramp & Bolwig, 1981).

The existence of a *space-occupying intracerebral lesion*, such as a brain tumor (Zwill et al., 1990) or a subdural hematoma (Malek-Ahmadi et al., 1990), is cause for particular concern. Probably because of the transient increases in cerebral blood flow and intracerebral pressure produced by the induced seizure, a significant fraction of patients with such lesions experience neurologic sequelae, occasionally of an irreversible nature. Given the increased risks, use of ECT in this situation must be preceded by well-documented clinical justification. Along with a careful reconsideration of clinical indications, the practitioner should also endeavor to implement certain modifications of the ECT procedure, which theoretically might serve to decrease morbidity. These modifications include the addition of antihypertensive agents, steroids, and diuretics and the use of hyperventilation, all of which should attenuate the increase in intracranial pressure occurring during and immediately after the seizure.

Electroconvulsive therapy has often been used in patients with a history of *cerebrovascular disease*, including transient ischemic episodes, infarction, and cerebrovascular anomalies such as aneurysms and angiomas. This use may be in cases in which the cerebrovascular and functional disorders are merely coexisting (Coffey et al., 1989) or in those where psychiatric disturbance appears secondary to the neurological insult (Murray, Shea, & Conn, 1986). With regard to stroke, a variety of review sources suggest that it is prudent to wait a period of weeks to months following a stroke to allow healing to take place before proceeding with ECT (Dubovsky, 1986; Hsiao et al., 1987). This concern is based on the reasonable theoretical assumption that increased cerebrovascular fragility during the acute poststroke period might lead to adverse effects from the increased cerebral blood flow and increased intracranial pressure that generally accompany the induced seizure. The absence of reports of adverse events following the use of ECT in such a setting suggest that if ECT is strongly indicated otherwise and if a significant delay is believed to be dangerous, the procedure may be carried out earlier during the recuperative period. In patients with ischemic cerebrovascular disease, the acute use of antihypertensive agents in conjunction with ECT may be harmful (Webb et al., 1990), whereas in those at risk for cerebral embolism or hemorrhage, such drugs may be of benefit (Drop, Bouckoms, & Welch, 1988).

Because of the typical rise in seizure threshold during a course of ECT is consistent with anticonvulsive properties, it is not surprising that ECT, in its early days, was evaluated as a treatment for *epilepsy* (Kalinowsky & Kennedy, 1943). Although benefits in the form of decreasing seizure frequency and the ability to abort status epilepticus were claimed, neurologists today prefer to consider only pharmacologic management of these disorders. As with cerebrovascular disease, depressive or psychotic episodes in epileptics may or may not be related to the underlying seizure disorder. In either case, ECT appears to be of benefit in inducing a therapeutic response in such episodes (Schnur et al., 1989). In most cases it can be accomplished without a subsequent increase in seizure frequency, although a small degree of risk for this eventuality, along with the theoretical possibility of triggering status epilepticus, may be present.

The use of anticonvulsant medication in patients receiving ECT is problematic in that seizure threshold is raised and seizure duration shortened. The present need for anticonvulsant medication should be reestablished and doses adjusted as low as is clinically justifiable. One major factor in this determination is the baseline frequency of seizures and the potential, by history, for status epilepticus. If difficulties producing adequate seizures ensue, careful attempts can be made to decrease anticonvulsant drug levels. This decrease may be easier to accomplish with carbamazepine because of its relatively short half-life. Alternatively, consideration can be given to prolonging the induced seizures via mechanisms such as hyperventilation (Bergsholm, Gran, & Bleie, 1984), intravenous caffeine (Coffey et al., 1990), or a switch to ketamine anesthesia (Brewer, Davidson, & Hereward, 1972). In general, the morning dose of anticonvulsant agent should be held on the day of ECT, except for patients at high risk for status epilepticus or with a history of frequent major motor seizures.

Patients with a history of *brain injury*, whether traumatic of secondary to surgery, are sometimes referred for ECT treatment. The few anecdotal reports dealing with such occurrences have not indicated the presence of adverse sequelae (Hsiao & Evans, 1984; Ruedrich, Chu, & Moore, 1983). Clinicians have generally been careful to avoid stimulation directly over or adjacent to a skull defect in order to prevent high intracerebral current densities (Coffey et al., 1987a; Hartman & Saldivia, 1990). In patients with indwelling ventricular shunts, it is important to establish the shunt patency prior to ECT (Coffey et al., 1987a) because of the increased intracranial pressure associated with a shunt blockage.

A high incidence of depressive illness occurs in *Parkinson's disease*; and, as mentioned earlier, a number of reports have established a significant degree of efficacy for ECT in such conditions (Douyon et al., 1989). It is of particular interest that the mental improvement may also be associated with improvement in the neurologic symptomatology. Such improvement, which may relate to an increased dopaminergic function, is particularly prominent in cases of the "on-off" syndrome of Parkinson's disease (Andersen et al., 1987). Interestingly, tardive dyskinesia may either worsen or improve with ECT (Holcomb, Sternberg & Heninger, 1983; Yassa, Hoffman, & Canakis, 1990), suggesting that the action of this treatment modality on dopaminergic systems may be more complex than initially believed (Lerer, 1984).

Among other neurological disorders, ECT has a variable effect on psychiatric symptoms and neurological status in patients with *multiple sclerosis* (Coffey et al., 1987b). Whether it is related to the degree to which the psychiatric illness was preexisting is unclear. Because patients with *myasthenia gravis* are sensitive to muscle relaxant drugs, the use of ECT in such cases is potentially hazardous unless appropriate modifications are made in the anesthetic technique, e.g., decrease succinylcholine dose or switch to nondepolarizing relaxant (Pande & Grunhaus, 1990). Patients with incompletely treated *cerebral or meningeal infections* represent yet another increased risk-group for ECT (Paulson, 1967), as disruption of the blood-brain barrier and elevated intracranial pressure may be present. Theoretically, the further transient increased permeability in the blood-brain barrier at the time of ECT might increase the risk of systemic spread of such an infection, though it has not been reported. *Chronic pain syndrome* not infrequently occur in conjunction with severe depressive symptomatology (see Chapter 17); and some (Mandel, 1975) but not all (Salmon et al., 1988) reports suggest that ECT is of benefit for both conditions. Finally, there is some evidence that depressive illness itself can lead to focal neurologic findings in the apparent absence of organic disease. Successful treatment of such a depressive episode with ECT is accompanied by resolution of the neurologic signs (Coffey, 1987).

Cardiovascular Disease

No somatic treatments for depressive disorders are without significant cardiovascular morbidity. As already noted, electrically induced seizures are associated with marked, though transient, increases in cardiac output, pulse, and blood pressure, along with a variety of generally brief disturbances in cardiac rhythm (Prudic et al., 1987; Webb et al., 1990; Welch & Drop, 1989). Although no increase in cardiac enzymes typically occurs with ECT, even in the presence of a history of cardiac disease (Dec, Stern, & Welch, 1985), patients with such a history have been reported to have a higher cardiac morbidity (Gerring & Shields, 1982; Knos et al., 1990). With *cardiac ischemic disease*, some measure of protection from further anoxia during and shortly after the induced seizures can be gained through the use of preoxygenation (McKenna et al., 1970), sufficient muscular relaxation, and medications that reduce cardiac workload (Kovac et al., 1990; Stoudemire et al., 1990).

A history of *myocardial infarction* should not be considered a contraindication for ECT, though if the result is recent or incompletely healed, the damaged myocardium is at risk for further injury as a result of the increased cardiac workload occurring with each seizure. The ideal waiting period for ECT following myocardial infarction is unknown and seems to vary with the nature and extent of myocardial injury, the presence or absence of postinfarction angina, the stability of postinfarction arrhythmias, and the status of blood supply to the surviving myocardium, as determined by nuclear medicine procedures or angiography. In a patient with life-threatening or profoundly disabling depression, a discussion among the psychiatrist, the cardiologist, and the anesthesiologist would be useful to assess the risk-benefit ratio of ECT in any given situation and to plan premedication and anesthetic technique to minimize risk, if ECT is elected. Techniques such as using beta-blockers or calcium-channel blockers may decrease the level of associated morbidity (see Chapter 11). Ongoing cardiac medications should always be given on days of ECT unless otherwise contraindicated (see the discussion that follows).

Patients with preexisting *ventricular ectopy* or *brady-*

cardia should be considered particularly prone to asystole or bradyarrhythmic-related ventricular contractions during both the onset of the seizure and the immediate postictal period, when parasympathetic tonus is greatest. The ongoing use of medications that may lower cardiac rate, such as beta-blockers, makes the need for adequate parasympathetic blockade even more important (Decina et al., 1984). In other cases, the greatest risk period for cardiac arrhythmias is during the seizure and the postictal period of arousal from anesthesia, when the sympathetic nervous system predominates and when tachycardia-related cardiac ischemia is most likely to occur. Beta-blocking agents can be useful in these instances (Weiner et al., 1979), though, again, consideration should be given to adequate anticholinergic premedication (London & Glass, 1985). Though the effect of an induced seizure on a preexisting cardiac arrhythmia is not entirely predictable, adverse changes will most likely be mild and transient if adequate precautions are taken as outlined above. There have even been reports of arrhythmias being "cured" with ECT (O'Leary, 1981), with such beneficial effects possible related to a restabilization in autonomic tone following the seizure-related sympathetic and parasympathetic surges.

Cardiac medications can sometimes interfere with the ECT procedure. Both lidocaine (Hood & Mecca, 1983; Ottosson, 1960) and some beta-blocking agents (Lathers et al., 1989) may shorten the ECT seizure and thereby lessen therapeutic efficacy. The greatest problems occur with lidocaine, which significantly attenuates seizure duration (Hood & Mecca, 1983; Ottosson, 1960). Theoretically, quinidine and, to a lesser degree, digitalis may prolong metabolism of succinylcholine (Packman, Myer, & Verdun, 1978), though use of these agents with ECT has not been reported to produce adverse clinical sequelae.

The presence of a *cardiac pacemaker* in a patient referred for ECT should be a source of relief rather than concern (Abiuso, Dunkelman, & Prober, 1978; Alexopoulos & Frances, 1980), as the regulation of cardiac rhythm during the ECT procedure exerts a protective effect. This effect also obviates the need for anticholinergic premedication. Because of the theoretical possibility of erroneous pacemaker triggering on the basis of either muscle fasciculations or muscle contractions, some sources have recommended that demand pacemakers be switched to a fixed mode prior to ECT, though this procedure may not be necessary for more recently implanted devices. The only pacemaker situation that could be risky from an electrical safety standpoint is the rather rare use of an internal pacemaker unit with transcutaneous pacing electrodes. (See Chapter 11.)

Hypertension has frequently been stated at a risk factor for ECT, although an increased rate of complications has not been documented. It is thus not clear when additional antihypertensive agents should be utilized with ECT (Anton, Uy, & Redderson, 1977). In practice, it is probably best to think of antihypertensive agents as a means of decreasing morbidity due to other medical conditions—for example, aneurysms, intracerebral masses, and hypercoagulable states (see below)—rather than from the direct effects of the hypertensive disease itself. A wide spectrum of specific pharmacologic agents have been advocated for attenuating the rise in blood pressure with ECT, including nitroprusside (Regestein & Lind, 1980), diazoxide (Kraus & Remick, 1982), ganglionic blocking agents (Egbert et al., 1959), beta-blockers (Jones & Knight, 1981; Kovac et al., 1990; Stoudemire et al., 1990), and nitroglycerine (Lee et al., 1985), the last two types being indicated when a mild degree of attenuation is desired. When using these drugs, one must realize that some degree of hypertension may be necessary during the induced seizure in order to provide the increased flow of oxygen and glucose required by the brain at this time. It is also likely that the systolic pressure required to deliver adequate cerebral perfusion in patients with ongoing hypertensive disease is higher than in those without this condition. Because of this situation it is difficult to recommend optimum pressure levels to use as targets. (See also Chapter 11.)

Endocrine Disorders

There has been much controversy over the use of ECT in patients with *diabetes mellitus*. There is a report that ECT lowers blood glucose levels in mild diabetes (Fakhri, Fadhli, & El Rawi, 1980), but the initial high glucose levels may have been secondary to a depression-related stress effect. In individuals with more severe forms of diabetes, particularly those who are insulin-dependent, ECT may transiently elevate blood glucose, requiring increased doses of antidiabetic agents (Finestone & Weiner, 1984). For this reason, blood glucose levels should be followed on a more frequent basis in patients at risk for such a response, with daily or more frequent tests in particularly brittle cases. As noted earlier, pre-ECT loading with glucose may be indicated on the days of treatment, and insulin doses should be adjusted accordingly. (See Chapter 11.)

Because of its ictally mediated hypothalamic activation and associated potentiating effects on pituitary function, ECT has been advocated as a treatment for certain neuroendocrine hypofunctional states, such as *hypopituitarism* (Ries & Bokan, 1979) and *inappropriate antidiuretic hormone secretion* (Brent & Chodroff, 1982).

Because *hypothyroidism* is often associated with depressive symptomatology, patients with this condition occasionally find themselves referred for ECT, particularly if depressive symptoms do not improve with normalization of hormonal levels with replacement therapy or if an urgent clinical response is indicated. ECT can be carried out in such cases without difficultly (Garrett, 1985). Theoretically, *hyperthyroidism* is a cause for concern with ECT because of a potential for provoking a "thyroid storm" via the transient massive sympathetic activation that occurs. In practice, however, this effect can be ameliorated by the use of beta-blocking agents. Asymptomatic mild hyperthyroid states, which sometimes result from stress-related effects associated with the depressive illness, can actually show improvement with ECT (Diaz-Cabal, Pearlman, & Kawecki, 1986).

Pseudohypoparathyroidism does not present a problem with ECT, so long as patients are taking vitamin D_2 to ensure adequate calcium levels (McCall, Coffey, & Maltbie, 1989). In this regard, ECT is known to diminish plasma calcium levels (Carman et al., 1977).

The presence of a *pheochromocytoma* is a source of potential with ECT, and beta-blocking agents should be used in sufficient dosages to obviate the occurrence of a hypertensive crisis (Carr & Woods, 1985). The use of alpha-adrenergic and tyrosine hydroxylase blocking agents should also be considered.

As described earlier, cortisol supplementation should be continued during ECT in patients with steroid-dependent conditions, including *Addison's disease* (Cumming & Kord, 1956). In fact, because induced seizures are associated with a transient increase in corticosteroid requirements, additional supplementation prior to each treatment is indicated. (See also Chapter 29.)

Metabolic Disorders

Patients with metabolic imbalances sometimes present problems during ECT. *Dehydration* has been reported to be associated with elevated seizure threshold. In addition, there are anesthetic implications regarding the volume depletion that is, by definition, present in dehydrated individuals. Electrolyte imbalance tends not to be a great concern, except in the case of potassium imbalance, as both *hypokalemia* and *hyperkalemia* raise the risk of cardiac dysfunction with ECT. Cardiac risk is particularly worrisome with hyperkalemia, because a sudden rise in serum potassium is produced by both succinylcholine and ictal motor contractions. Patients with widespread burns, chronic immobilization, spastic paralysis, or amyotrophic lateral sclerosis may experience a particularly high serum potassium level during ECT (Bali, 1975). The transient rise in serum potassium can be greatly diminished by the use of atracurium instead of succinylcholine. *Hypo*kalemia, on the other hand, creates a risk of prolonged paralysis and apnea after succinylcholine is administered.

The dynamic fluctuations in serum electrolytes in patients undergoing *renal dialysis* necessitates careful monitoring of these parameters during an ECT course. When possible, it is preferable to schedule ECT on the day after dialysis (Pearlman, Carson, & Metz, 1988). Patients with *fever and leukocytosis* should be evaluated prior to ECT for the presence of a treatable primary medical or neurologic disorder. However, idiopathic conditions of this type have been reported in association with bipolar disorder and may be responsive to ECT (Kronfol, Hamden-Allen, & Black, 1988). Finally, a diagnosis of *porphyria* should result in a switch to nonbarbiturate anesthetic agents.

Hematologic Disorders

Because of the risk of dislodging emboli, *thrombophlebitis* was formerly considered a relative contraindication for ECT. With the innovation of anticoagulants, however, it is no longer the case. Until recently, the preferred agent for patients with hypercoagulable states was heparin, based on the facts that (1) its action can be readily reversed by protamine, (2) its metabolism is not affected by anesthetic agents used with ECT, and (3) it has a short half-life (Alexopoulos et al., 1982; Loo, Cuche, & Benkelfat, 1985). Because of these properties it was recommended that patients on warfarin be switched to heparin. By withholding heparin for 6 to 8 hours prior to the treatment, sufficient time would be available to allow coagulability to stabilize. This effect can be monitored by the partial thromboplastin time, which should be around 1.0 to 1.5 times control values by an hour before each ECT treatment (Alexopoulos et al., 1982). This practice is not necessary for patients given small doses of subcutaneous heparin, where virtually normal partial thromboplastin times are found.

However, more recent series of patients on anticoagulants have found that maintaining such individuals on warfarin may be safe after all, given frequent monitoring with prothrombin times and adjustment of warfarin doses to achieve prothrombin times less than twice control values (Hay, 1987; Tancer & Evans, 1989). Regardless of the agent utilized, because of the elevated risk of either embolization or bleeding in such patients, augmented muscle relaxant dosages should be administered and the adjunctive use of antihypertensive drugs considered. (See also Chapter 11.)

With regard to platelet count, ECT has been administered without difficulty in both *thrombocytopenic* (Kardener, 1968) and *thrombocythemic* (Hamilton & Walter-Ryan, 1986) states. With *sickle cell disease* (La Grone, 1990), one should avoid the "cuff" technique for monitoring convulsion duration, as it may lead to sickling distal to the inflated blood pressure cuff. (See also Chapter 31.)

Pulmonary Disorders

Little has been written regarding the use of ECT in patients with pulmonary disorders. *Chronic obstructive pulmonary disease* (COPD), which is prevalent in the age range of patients typically referred for ECT, is reflected clinically in decreased ventilatory capacity during the ECT procedure. Preoxygenation prior to each treatment may minimize potential for hypoxia in such cases. Patients with COPD or asthma are at relatively higher risk for bronchial spasm following ECT. For this reason, it is recommended that bronchodilators be continued, although theophylline may increase the likelihood of prolonged seizures. Theophylline levels should be kept at 15 μg/ml or below, and other bronchodilators should be used if this level is insufficient. The anesthesiologist should also be alerted to the possibility of prolonged seizures so anticonvulsant agents are on hand to deal with such an occurrence. (See Chapter 11.)

Gastrointestinal Disorders

Patients with symptomatic gastric *reflux*, usually associated with hiatal hernia, should receive the histamine-2 antagonist ranitidine 150 mg (or cimetidine 300 mg) the night before and the morning of each ECT treatment to diminish gastric acid secretion. Metoclopramide, an agent that promotes gastric emptying, has been used by some practitioners on an adjunctive basis but has been avoided by others because of its neuroleptic properties. The use of endotracheal intubation to diminish risk of aspiration during ECT anesthesia should be reserved for the most high-risk cases, e.g., patients with a history of aspiration of gastric contents or with a near-term pregnancy. An alternative to intubation in moderate risk situations is the use of supracricoid pressure during the apneic period.

Gastroparesis, where gastric emptying is impaired, may occur in patients on anticholinergic agents and in those whose gastric innervation is compromised, e.g., diabetes. Evaluation of the presence and severity of the condition using an upper GI series should be carried out prior to ECT, and metoclopramide should be used the night before and the morning of each treatment (Zibrak, Jensen, & Bloomingdale, 1988). *Intestinal pseudoobstruction*, a rare complication of severe depressive illness, can be expected to disappear with a successful course of ECT, possibly secondary to ECT's effect on somatostatin (Stoudemire, Hill, & Markon, 1987). Initiation of ECT should, of course, be delayed until a structural cause has been ruled out. A final gastrointestinal condition of pertinence to ECT is *constipation*, which is common in severely depressed patients. In this case, it is important to avoid the presence of fecal impaction during the ECT course, as a sudden increase in intraabdominal pressure during the stimulus or seizure theoretically could lead to intestinal rupture. It must also be understood that the same risk holds for *urinary retention*, where bladder rupture can be avoided by assiduous attention to pre-ECT voiding, along with the use of postvoid catheterization, or even indwelling Foley catheterization, when necessary (Irving & Drayson, 1984).

Musculoskeletal Disorders

Before the days of muscular relaxation, patients with musculoskeletal disease or abnormalities were referred for ECT with great trepidation. Now, potential complications can be prevented by providing a state of complete muscular relaxation in such cases (Coffey et al., 1986). This effect can be achieved by increased dosage of muscle relaxant and can be monitored with the use of a peripheral nerve stimulator (Baker, 1986). The brief contraction of the jaw muscles that often occurs during the direct passage of the stimulus current varies in intensity and may be only partially modified by the muscle relaxant agent. Particularly with patients with poor dentition, *dental fractures* can occur unless a soft rubber mouthpiece is properly inserted and the mandible and maxilla are held in close apposition during passage of the stimulus current (Faber, 1983). Patients with loose or damaged teeth may require more specific dental protection or removal of involved teeth prior to beginning ECT.

Ophthalmologic Disorders

Although ECT produces a transient increase in intraocular pressure of around 20 mm Hg (Epstein et al., 1975), this change is generally of no clinical consequence, even in cases of chronic open-angle *glaucoma* (Edwards et al., 1990; Nathan et al., 1986). It is only with acute closed-angle glaucoma, a much rarer condition, that a significant ophthalmologic risk with ECT is present (Sibony, 1985). Cholinesterase inhibitors typically used for treating glaucoma, such as physostigmine and neostigmine, may prolong the effects of succinylcholine, but generally only to a mild degree; thus they do not usually constitute a major problem with ECT. On the other hand, organophosphorus anticholinesterase agents, such as echothiophate, have a much more profound and long-lasting effect on succinylcholine metabolism (Messer, Stoudemire & Johnson, 1992; Packman, Meyer, & Verdun, 1978), and should be avoided.

Other Conditions

Malignant hyperthermia, a rare hypermetabolic state, is pharmacologically triggered on an idiosyncratic basis and presents with tachycardia, muscle spasms, acidosis, and anoxia. The most common offending agent is succinylcholine, though other anesthetic and even psychotropic drugs have been reported (Yacoub & Morrow, 1986). A personal and, given the inherited predisposition to this condition, family history of similar past presentations can alert the practitioner to a potential high-risk situation (Bidder, 1981). With such occurrences, a muscle biopsy allows a definitive assessment. Because of the similarity in some aspects of the clinical presentation of malignant hyperthermia with that of neuroleptic malignant syndrome (NMS), some anesthesiologists have assumed that there is an increased risk of the former condition in patients with a history of NMS and have insisted that relaxation with ECT be performed with a nondepolarizing muscle relaxant such as atracurium. However, it is now clear that the two syndromes are etiologically distinct (Hermesh et al., 1988), and that succinylcholine can be used in patients with a history of NMS.

ECT AND PREGNANCY

Electroconvulsive therapy has been performed in many pregnant patients during all three trimesters of gestation without difficulty (Dorn, 1985; Remick & Maurice, 1978; Wisner & Perel, 1988). In fact, some practitioners prefer the use of ECT over psychotropic management during the first trimester of pregnancy because of its presumably lower teratogenic risk. The APA has recommended that pre-ECT evaluation with pregnant women include obstetrical consultation to assess the level of potential risk and to plan for treatment modifications and monitoring, if indicated (APA, 1990). In addition, even though premature induction of labor by ECT has not been reported, any facility administering ECT to a woman with a viable fetus should have ready access to an obstetrician. Adequate oxygenation and muscular relaxation should always be ensured when treating the pregnant woman with ECT. Fetal heart rate should be monitored if the gestational age is greater than 10 weeks (APA, 1990). The presence of an obstetrician or additional forms of fetal monitoring, e.g., real-time ultrasonography or uterine muscle dynamometry, should be considered for high-risk or near-term cases (Wise et al., 1984). As noted earlier, near-term pregnancy may also be an indication for endotracheal intubation with ECT because of the markedly elevated risk of aspiration.

ECT IN THE ELDERLY

Advanced age is not a contraindication to ECT (APA, 1990; Fogel 1988). In many ECT centers, the mean age of patients is in the fifties or even sixties. Although it is accepted that elderly patients are at greater risk for adverse sequelae with *either* psychopharmacologic treatment or ECT, the question of whether the increased risk is greater for one type of modality than the other remains unsettled (Alexopoulos et al., 1984; Burke et al., 1987; Cattan et al., 1990; Hay, 1989). Many sources report a subjective belief that ECT is in fact "safer" than drugs in this age group (Weiner, 1982). Medical morbidity with ECT in the elderly is largely due to the presence of concurrent medical disease, particularly chronic cardiac dysfunction.

One potential problem when providing ECT to the elderly is the ability to induce seizures of sufficient duration. At times even a maximum stimulus intensity may prove unsuccessful, although this failure is less likely with the high maximum output provided by most contemporary ECT devices. As already mentioned, a variety of means exist to deal with this situation should it occur. First, one should be certain that the dose of anesthetic agent is not larger than necessary. Second, the stimulus should not be given until at least 2 minutes have elapsed following administration of the barbiturate anesthetic. Third, drugs with anticonvulsant effects should be discontinued whenever possible. Fourth, a stimulant agent such as caffeine can be used. Fifth, an anesthetic with minimal anticonvulsant properties (e.g., ketamine) should be considered.

SUMMARY

Electroconvulsive therapy is a highly effective and generally safe procedure that is being used more and more in patients with preexisting medical and neurological illness. Our present knowledge about both the physiologic effects associated with ECT and the means to alter these effects pharmacologically allows us to substantially minimize the morbidity and mortality of ECT in this population, thereby ensuring the availability of a greater range of treatment options for psychiatrically ill patients with serious medical and neurologic conditions.

REFERENCES

ABIUSO P, DUNKLEMAN R, & PROBER M (1978). Electroconvulsive therapy in patients with pacemakers. JAMA 240:2459–2460.

ABRAHAM KR, & KULHARA P (1987). The efficacy of electroconvulsive therapy in the treatment of schizophrenia: A comparative study. Br J Psychiatry 151:152–155.

ABRAMCZUK JA, & ROSE NM (1979). Preanesthetic assessment and prevention of post ECT morbidity. Br J Psychiatry 134:582–587.

ABRAMS R (1985). Multiple-monitored ECT. Convulsive Ther 1:285–286.

ABRAMS R (1986). Is unilateral electroconvulsive therapy really the treatment of choice in endogenous depression? Ann NY Acad Sci 462:50–55.

ABRAMS R (1989). ECT in the high-risk patient [Editorial]. Convulsive Ther 5:1–2.

ABRAMS R (1992). Electroconvulsive therapy (2nd ed.). New York: Oxford University Press.

ABRAMS R, & FINK M (1984). The present status of unilateral ECT: Some recommendations. J Affective Disord 7:245–247.

ABRAMS R, SWARTZ CM, & VEDAK C (1991). Antidepressant effects of high-dose right unilateral electroconvulsive therapy. Arch Gen Psychiatry 48:746–748.

ADDONIZIO G, & SUSMAN VL (1987). ECT as a treatment alternative for patients with symptoms of neuroleptic malignant syndrome. J Clin Psychiatry 48:102–105.

ALEXOPOULOS GS, & FRANCES RJ (1980). ECT in cardiac patients with pacemakers. Am J Psychiatry 137:1111–1112.

ALEXOPOULOS GS, NASR H, YOUNG RC, ET AL. (1982). Electroconvulsive therapy in patients on anticoagulants. Can J Psychiatry 27:46–47.

ALEXOPOULOS GS, SHAMOLAN CJ, LUCAS J, ET AL. (1984). Medical problems of geriatric psychiatric patients and younger controls during electroconvulsive therapy. J Am Geriatr Soc 32:651–654.

American Psychiatric Association (1978). Electroconvulsive therapy (Task Force Report No. 14). Washington, DC: American Psychiatric Association.

American Psychiatric Association (1987). Diagnostic and statistical manual of mental disorders (3rd ed., rev.). Washington, DC: American Psychiatric Association.

American Psychiatric Association (1990). The practice of ECT: Recommendations for treatment, training, and privileging. Washington, DC: American Psychiatric Press.

ANDERSEN K, BALLDIN J, GOTTFRIES CG, ET AL. (1987). A double-blind evaluation of electroconvulsive therapy in Parkinson's disease with "on-off" phenomena. Acta Neurol Scand 76:191–199.

ANTON AH, UY DS, & REDDERSON CL (1977). Autonomic blockade and the cardiovascular catecholamine response to electroshock. Anesth Analg 56:46–54.

AVERY D, & LUBRANO A (1979). Depression treated with imipramine and ECT: The DeCarolis study reconsidered. Am J Psychiatry 136:549–562.

AZAR I, & LEAR E (1984). Cardiovascular effects of electroconvulsive therapy in patients taking tricyclic antidepressants. Anesth Analg 63:1140.

BAKER NJ (1986). Electroconvulsive therapy and severe osteoporosis: Use of a nerve stimulator to assess paralysis. Convulsive Ther 2:285–288.

BALI IM (1975). The effect of modified electroconvulsive therapy on plasma potassium concentration. Br J Anaesth 47:398–401.

BARKAI AI (1985). Combined electroconvulsive and drug therapy. Compr Ther 11:48–53.

BERGSHOLM P, GRAN L, & BLEIE H (1984). Seizure duration in unilateral electroconvulsive therapy: The effect of hypocapnia induced by hyperventilation and the effect of ventilation with oxygen. Acta Psychiatr Scand 69:121–128.

BIDDER TG (1981). Electroconvulsive therapy in the medically ill patient. Psychiatr Clin North Am 4:391–405.

BLACK DW, HULBERT J, & NASRALLAH A (1989). The effect of somatic treatment and comorbidity on immediate outcome in manic patients. Compr Psychiatry 30:74–79.

BLACK DW, WINOKUR G, & NASRALLAH A (1987). Treatment of mania: A naturalistic study of electroconvulsive therapy versus lithium in 438 patients. J Clin Psychiatry 48:132–139.

BOLWIG TG, HERTZ MM, & WESTERGAARD E (1977). Acute hypertension causing blood brain barrier breakdown in epileptic seizures. Acta Neurol Scand 56:335–342.

BRANDON S, COWLEY P, McDONALD C, ET AL. (1984). Electroconvulsive therapy: Results in depressive illness from the Leicestershire trial. Br Med J 288:22–25.

BRANDON S, COWLEY P, McDONALD C, ET AL. (1985). Leicester ECT trial: Results in schizophrenia. Br J Psychiatry 146:177–183.

BREGGIN PR (1979). Electroshock: Its brain disabling effects. New York: Springer-Verlag.

BRENT RH, & CHODROFF C (1982). ECT as a possible treatment of SIADH: Case report. J Clin Psychiatry 43:73–74.

BREWER CL, DAVIDSON JRT, & HEREWARD S (1972). Ketamine ("ketalar"): A safer anaesthetic for ECT. Br J Psychiatry 120:679–680.

BURKE WJ, RUBIN EH, ZORUMSKI CF, ET AL. (1987). The safety of ECT is geriatric psychiatry. J Am Geriatr Soc 35:516–521.

CARMAN JS, POST RM, GOODWIN FK, ET AL. (1977). Calcium and electroconvulsive therapy of severe depressive illness. Biol Psychiatry 12:5–17.

CATTAN RA, BARRY PP, MEAD G, ET AL. (1990). Electroconvulsive therapy in octogenarians. J Am Geriatr Soc 38:753–758.

CARR ME JR, & WOODS JW (1985). Electroconvulsive therapy in a patient with unsuspected pheochromocytoma. South Med J 78:613–615.

CLARKE TB, COFFEY CE, HOFFMAN GW, ET AL. (1989). Continuation therapy for depression using outpatient electroconvulsive therapy. Convulsive Ther 5:330–337.

Clinical Research Centre (1984). The Northwick Park ECT trial: Predictors of response to real and simulated ECT. Br J Psychiatry 144:227–237.

COFFEY CE (1987). Cerebral laterality and emotion—the neurology of depression. Compr Psychiatry 28:197–219.

COFFEY CE, FIGIEL GS, DJANG WT, ET AL. (1988). Effects of ECT on brain structure: a pilot prospective magnetic resonance imaging study. Am J Psychiatry 145:701–706.

COFFEY CE, FIGIEL GS, DJANG WT, ET AL. (1989). Subcortical white matter hyperintensity on magnetic resonance imaging: Clinical and neuroanatomic correlates in the depressed elderly. J Neuropsychiatr Clin Neurosci 1:135–144.

COFFEY CE, FIGIEL GS, WEINER RD, ET AL. (1990). Caffeine augmentation of ECT. Am J Psychiatry 147:579–585.

COFFEY CE, HOFFMAN G, WEINER RD, ET AL. (1987a). Electroconvulsive therapy in a depressed patient with a functioning ventriculo-atrial shunt. Convulsive Ther 4:302–306.

Coffey CE, & Weiner RD (1990). Electroconvulsive therapy: An update. Hosp Community Psychiatry 41:515–521.

Coffey CE, Weiner RD, Djang WT, et al. (1991). Brain anatomic effects of ECT: A prospective magnetic resonance imaging study. Arch Gen Psychiatry 48:1013–1021.

Coffey CE, Weiner RD, Kalayjian R, et al. (1986). Electroconvulsive therapy in osteogenesis imperfecta: Issues of muscular relaxation. Convulsive Ther 2:207–211.

Coffey CE, Weiner RD, McCall WV, et al. (1987b). Electroconvulsive therapy in multiple sclerosis: A brain magnetic resonance imaging study. Convulsive Ther 3:137–144.

Consensus Conference (1985). Electroconvulsive therapy. JAMA 254:2103–2108.

Coppen A, Abou-Saleh MT, Milln P, et al. (1981). Lithium continuation therapy following electroconvulsive therapy. Br J Psychiatry 139:284–287.

Coryell W (1986). Are serial dexamethasone suppression tests useful in electroconvulsive therapy? J Affective Disord 10:59–66.

Cumming J, & Kord K (1956). Apparent reversal by cortisone of an electroconvulsive refractory state in a psychotic patient with Addison's disease. Can Med Assoc J 74:291–292.

Dec GW Jr, Stern TA, & Welch C (1985). The effects of electroconvulsive therapy on serial electrocardiograms and serum cardiac enzyme values: A prospective study of depressed hospitalized inpatients. JAMA 253:2525–2529.

Decina P, Malitz S, Sackeim HA, et al. (1984). Cardiac arrest during ECT modified by beta-adrenergic blockage. Am J Psychiatry 141:298–300.

d'Elia G (1970). Unilateral ECT. Acta Psychiatr Scand (suppl 215):1–98.

d'Elia G, & Raotma H (1975). Is unilateral ECT less effective than bilateral ECT? Br J Psychiatry 126:83–89.

Deliyiannis S, Eliakim M, & Bellet S (1962). The electrocardiogram during electroconvulsive therapy as studied by radioelectrocardiography. Am J Cardiol 10:187–192.

Diaz-Cabal R, Pearlman C, & Kawecki A (1986). Hyperthyroidism in a patient with agitated depression: Resolution after electroconvulsive therapy. J Clin Psychiatry 47:322–323.

Dorn JB (1985). Electroconvulsive therapy with fetal monitoring in a bipolar pregnant patient. Convulsive Ther 1:217–221.

Douyon R, Serby M, Klutchko B, et al. (1989). ECT and Parkinson's disease revisited: A "naturalistic" study. Am J Psychiatry 146:1451–1455.

Drop LJ, Bouckoms AJ, & Welch CA (1988). Arterial hypertension and multiple cerebral aneurysms in a patient treated with electroconvulsive therapy. J Clin Psychiatry 49:280–282.

Dubovsky SL (1986). Using electroconvulsive therapy for patients with neurological disease. Hosp Community Psychiatry 37:819–824.

Edwards RM, Stoudemire A, Vela MA, et al. (1990): Intraocular pressure changes in nonglaucomatous patients undergoing electroconvulsive therapy. Convulsive Ther 6:209–213.

Egbert D, Wolfe S, Melmed RM, et al. (1959). Reduction of cardiovascular stress during electroshock therapy by trimethaphan. J Clin Exp Psychopathol Q Rev Psychiatr Neurol 20:315–319.

El-Ganzouri AR, Ivankovich AD, Braverman B, et al. (1985). Monoamine oxidase inhibitors: Should they be discontinued preoperatively? Anesth Analg 64:592–596.

Elliot JL, Linz DH, & Kane JA (1982). Electroconvulsive therapy: Pretreatment medical evaluation. Ann Intern Med 142:979–981.

Epstein HM, Fagman W, Bruce DL, et al. (1975). Intraocular pressure changes during anesthesia for electroshock therapy. Anesth Analg 54:479–481.

Faber R (1983). Dental fracture during ECT. Am J Psychiatry 140:1255–1256.

Fakhri O, Fadhli AA, & El Rawi RM (1980). Effect of electroconvulsive therapy on diabetes mellitus. Lancet 2:775–777.

Finestone DH, & Weiner RD (1984). Effects of ECT on diabetes mellitus: An attempt to account for conflicting data. Acta Psychiatr Scand 70:321–326.

Fink M (1979). Convulsive therapy: Theory and practice. New York: Raven Press.

Fogel BS (1988). Electroconvulsive therapy in the elderly: A clinial research agenda. Int J Geriatr Psychiatry 3:181–190.

Frederiksen SO, & d'Elia G (1979). Electroconvulsive therapy in Sweden. Br J Psychiatry 134:283–287.

Freese KJ (1985). Can patients safely undergo electroconvulsive therapy while receiving monoamine oxidase inhibitors? Convulsive Ther 1:190–194.

Garrett MD (1985). Use of ECT in a depressed hypothyroid patient. J Clin Psychiatry 46:64–66.

Gerring JP, & Shields HM (1982). The identification and management of patients with a high risk for cardiac arrhythmias during modified ECT. J Clin Psychiatry 43:140–143.

Greenberg LB (1986). Electroconvulsive therapy in treating neuroleptic malignant syndrome. Convulsive Ther 2:61–62.

Gregory S, Shawcross CR, & Gill D (1985). The Nottingham ECT study: A double-blind comparison of bilateral, unilateral and simulated ECT in depressive illness. Br J Psychiatry 146:520–524.

Hamilton RW, & Walter-Ryan WG (1986). ECT and thrombocythemia. Am J Psychiatry 143:258.

Hartman SJ, & Saldivia A (1990). ECT in an elderly patient with skull deficits and shrapnel. Convulsive Ther 6:165–171.

Hay DP (1987). Anticoagulants and ECT [Letter]. Convulsive Ther 3:236–237.

Hay DP (1989). Electroconvulsive therapy in the medically ill elderly. Convulsive Ther 5:9–16.

Hermesh H, Aizenberg D, Lapidot M, et al. (1988). Risk of malignant hyperthermia among patients with neuroleptic malignant syndrome and their families. Am J Psychiatry 145:1431–1434.

Heshe AJ, & Roeder E (1976). Electroconvulsive therapy in Denmark. Br J Psychiatry 128:241–245.

Holcomb HH, Sternberg DE, & Heninger GR (1983). Effects of electroconvulsive therapy on mood, parkinsonism, and tardive dyskinesia in a depressed patient: ECT and dopamine systems. Biol Psychiatry 18:865–873.

Hood DA, & Mecca RS (1983). Failure to initiate electroconvulsive seizures in a patient pretreated with lidocaine. Anesthesiology 58:379–381.

Hsiao JK, & Evans DL (1984). ECT in a depressed patient after craniotomy. Am J Psychiatry 141:442–444.

Hsiao JK, Messenheimer JA, & Evans DL (1987). ECT and neurological disorders. Convulsive Ther 33:121–136.

Hughes JR (1986). ECT during and after the neuroleptic malignant syndrome: Case report. J Clin Psychiatry 47:42–43.

Imlah NW, Ryan E, & Harrington JA (1965). The influence of antidepressant drugs on the response to electroconvulsive therapy and on subsequent relapse rates. Neuropsychopharmacology 4:438–442.

Irving AD, & Drayson AM (1984). Bladder rupture during ECT. Br J Psychiatry 144:670.

Janicak PG, Davis JM, Gibbons RD, et al. (1985). Efficacy of ECT: A meta-analysis. Am J Psychiatry 142:297–302.

Jones RM, & Knight PR (1981). Cardiovascular and hormonal responses to electroconvulsive therapy: Modification of an exaggerated response in a hypertensive patient by beta receptor blockade. Anesthesia 36:795–799.

Kalinowsky LB, & Kennedy F (1943). Observation in electroshock therapy applied to problems of epilepsy. J Nerv Ment Dis 98:56–67.

KARDENER SH (1968). EST in a patient with idiopathic thrombocytopenic purpura. Dis Nerv Syst 29:465–466.

KHANNA N, RAY A, ALKONDON M, ET AL. (1989). Effect of beta-adrenoceptor antagonists and some related drugs on maximal electroshock seizures in mice. Indian J Exp Biol 27:128–130.

KNOS GB, SUNG YF, COOPER RC, ET AL. (1990). Electroconvulsive-therapy induced hemodynamic changes unmask unsuspected coronary artery disease. J Clin Anesth 2:37–41.

KOVAC AL, GOTO H, GRAKAWA K, ET AL. (1990). Esmolol bolus and infusion attenuates increase in blood pressure and heart rate during electroconvulsive therapy. Can J Anesth 37:58–62.

KRAMP P, & BOLWIG TG (1981). Electroconvulsive therapy in acute delirious states. Compr Psychiatry 22:368–371.

KRAUS RP, & REMICK RA (1982). Diazoxide in the management of severe hypertension after electroconvulsive therapy. Am J Psychiatry 139:504–505.

KRONFOL Z, HAMDAN-ALLEN G, & BLACK DW (1988). Fever and leukocytosis: physical manifestations of bipolar affective disorder. Prog Neuropsychopharmacol Biol Psychiatry 12:887–891.

LAGRONE D (1990). ECT in secondary mania, pregnancy, and sickle cell anemia. Convulsive Ther 6:176–180.

LANCASTER NP, STEINERT RR, & FROST I (1958). Unilateral electroconvulsive therapy. J Ment Sci 104:221–227.

LATHERS CM, STANFFER AZ, THMER N, ET AL. (1989). Anticonvulsant and antiarrhythmic actins of the beta blocking agent timolol. Epilepsy Res 4:42–54.

LEE JT, ERBGUTH PH, STEVENS WC, ET AL. (1985). Modification of electroconvulsive therapy induced hypertension with nitroglycerin ointment. Anesthesiology 62:793–796.

LERER B (1984). ECT and tardive dyskinesia. J Clin Psychiatry 45:188.

LONDON SW, & GLASS DD (1985). Prevention of electroconvulsive therapy-induced dysrhythmias with atropine and propranolol. Anesthesiology 62:819–822.

LOO H, CUCHE H, & BENKELFAT C (1985). Electroconvulsive therapy during anticoagulant therapy. Convulsive Ther 1:258–262.

LUNN RJ, SAVAGEAU MM, BEATTY WW, ET AL. (1981). Anesthetics and electroconvulsive therapy seizure duration: Implications for therapy from a rat model. Biol Psychiatry 16:1163–1175.

LURIE SN, & COFFEY CE (1990). Caffeine-modified electroconvulsive therapy in depressed patients with medical illness. J Clin Psychiatry 51:154–157.

MALEK-AHMADI P, BECEIRO JR, MCNEIL BW, ET AL. (1990). Electroconvulsive therapy and chronic subdural hematoma. Convulsive Ther 6:38–41.

MALETZKY BM (1981). Multiple-monitored electroconvulsive therapy. Boca Raton, FL: CRC Press.

MANDEL MR (1975). Electroconvulsive therapy for chronic pain associated with depression. Am J Psychiatry 132:632–636.

MCCALL WV, COFFEY CE, & MALTBIE AA (1989). Successful electroconvulsive therapy in a depressed patient with pseudohypoparathyroidism. Convulsive Ther 5:114–117.

MCKENNA G, ENGLE RP, BROOKS H, ET AL. (1970). Cardiac arrhythmias during electroshock therapy: Significance, prevention and treatment. Am J Psychiatry 127:530–533.

MESSER GJ, STOUDEMIRE A, & JOHNSON GC (1992). Electroconvulsive therapy and the chronic use of pseudocholinesterase inhibitor (echothiophate iodide) eye drops for glaucoma: A case report. Gen Hosp Psychiatry 14:56–60.

MILSTEIN V, SMALL JG, KLAPPER MH, ET AL. (1987). Uni- versus bilateral ECT in the treatment of mania. Convulsive Ther 3:1–9.

MUKHERJEE S, SACKEIM HA, & LEE C (1988). Unilateral ECT in the treatment of manic episodes. Convulsive Ther 4:74–80.

MURRAY GB, SHEA V, & CONN DK (1986). Electroconvulsive therapy for the poststroke depression. J Clin Psychiatry 47:258–260.

NATHAN RS, DOWLING R, PETERS JL, ET AL. (1986). ECT and glaucoma. Convulsive Ther 2:132–133.

O'LEARY J (1981). Cardiac effects of electroconvulsive therapy. Anaesth Intens Care 9:400–401.

OTTOSSON JO (1960). Experimental studies on the mode of action of electroconvulsive therapy. Acta Psychiatr Neurol Scand 35(suppl 145):1–141.

PACKMAN M, MYER MA, & VERDUN RM (1978). Hazards of succinylcholine administration during electrotherapy. Arch Gen Psychiatry 35:1137–1141.

PANDE AC, & GRUNHAUS LJ (1990). ECT for depression in the presence of myasthenia gravis. Convulsive Ther 6:172–175.

PANDE AC, KRUGLER T, HASKETT RF, ET AL. (1988). Predictors of response to electroconvulsive therapy in major depressive disorder. Biol Psychiatry 24:91–93.

PAPAKOSTAS Y, FINK M, LEE J, ET AL. (1981). Neuroendocrine measures in psychiatric patients: Course and outcome with ECT. Psychiatr Res 4:55–64.

PARRY J (1986). Legal parameters on informed consent for ECT administered to mentally disabled persons. Psychopharmacol Bull 22:490–494.

PAUL SM, EXTEIN I, CALIL HM, ET AL. (1981). Use of ECT with treatment-resistant depressed patients at the National Institute of Mental Health. Am J Psychiatry 138:486–489.

PAULSON GW (1967). Exacerbation of organic brain disease by electroconvulsive treatment. NC Med J 28:328–331.

PEARLMAN CA (1986). Neuroleptic malignant syndrome: A review of the literature. J Clin Psychopharmacol 6:257–273.

PEARLMAN C, CARSON W, & METZ A (1988). Hemodialysis, chronic renal failure, and ECT [letter]. Convulsive Ther 4:332–333.

PERRIN GM (1961). Cardio-vascular aspects of electroshock therapy. Acta Psychiatr Scand 36(suppl 152):1–44.

PETERS SG, WOCHOS DN, & PETERSON GC (1984). Status epilepticus complicating electroconvulsive therapy in the presence of theophylline. Mayo Clin Proc 59:568–570.

PETTINATI HM, MATHISEN KS, ROSENBERG J, ET AL. (1986). Meta-analytical approach to reconciling discrepancies in efficacy between bilateral and unilateral electroconvulsive therapy. Convulsive Ther 2:7–17.

PRICE TRP, & MCALLISTER TW (1986). Response of depressed patients to sequential unilateral nondominant brief-pulse and bilateral sinusoidal ECT. J Clin Psychiatry 47:182–186.

PRICE TRP, & MCALLISTER TW (1989). Safety and efficacy of ECT in depressed patients with dementia: A review of clinical experience. Convulsive Ther 5:61–74.

PRICE WA, & ZIMMER B (1985). Effect of L-tryptophan on electroconvulsive therapy seizure time. J Nerv Ment Dis 175:636–638.

PRUDIC J, DEVANAND DP, SACKEIM HA, ET AL. (1989). Relative response of endogenous and non-endogenous symptoms to electroconvulsive therapy. J Affective Disord 16:59–64.

PRUDIC J, SACKEIM HA, DECINA P, ET AL. (1987). Acute effects of ECT on cardiovascular functioning: Relations to patients and treatment variables. Acta Psychiatr Scand 75:344–351.

RASKIN DE (1984). Dangers of monoamine oxidase inhibitors. J Clin Psychopharmacol 4:238.

REGESTEIN QR, & LIND JF (1980). Management of electroconvulsive treatment in an elderly woman with severe hypertension and cardiac arrhythmias. Compr Psychiatry 21:288–291.

REGESTEIN QR, & REICH P (1985). Electroconvulsive therapy in patients at high risk for physical complications. Convulsive Ther 1:101–114.

REMICK RA, & MAURICE WI (1978). ECT in pregnancy. Am J Psychiatry 135:761–762.

RIES R, & BOKAN J (1979). Electroconvulsive therapy following pituitary surgery. J Nerv Ment Dis 167:767–768.

ROBIN A, & DE TISSERA S (1982). A double-blind controlled comparison of the therapeutic effects of low and high energy electroconvulsive therapies. Br J Psychiatry 141:357–366.

ROTHMAN DJ (1986). ECT: The historical, social, and professional sources of the controversy. Psychopharmacol Bull 22:459–463.

ROY-BYRNE P, & GERNER RH (1981). Legal restrictions on the use of ECT in California: Clinical impact on the incompetent patient. J Clin Psychiatry 42:300–303.

RUEDRICH SL, CHU CC, & MOORE SL (1983). ECT for major depression in a patient with acute brain trauma. Am J Psychiatry 140:928–929.

SACKEIM HA, DECINA P, KANZLER M, ET AL. (1987). Effects of electrode placement on the efficacy of titrated, low-dose ECT. Am J Psychiatry 144:1449–1455.

SACKEIM HA, DEVANAND DP, & PRUDIC J (1991). Stimulus intensity, seizure threshold, and seizure duration: Impact on the efficacy and safety of electroconvulsive therapy. Psychiatric Clin N Am 14:803–843.

SACKEIM HA, & MUKHERJEE S (1986). Neurophysiological variability in the effects of the ECT stimulus. Convulsive Ther 2:267–276.

SACKEIM HA, PRUDIC J, DEVANAND DP, ET AL. (1990). The impact of medication resistance and continuation pharmacotherapy on relapse following response to electroconvulsive therapy in major depression. J Clin Psychopharmacol 10:96–104.

SALMON JB, HANNA MH, WILLIAMS M, ET AL. (1988). Thalamic pain–the effect of ECT. Pain 33:67–71.

SALZMAN C (1980). The use of ECT in the treatment of schizophrenia. Am J Psychiatry 137:1032–1041.

SCHNUR D, MUKHERGEE S, SILVER J, ET AL. (1989). Electroconvulsive therapy in the treatment of episodic aggressive dyscontrol in psychotic patients. Convulsive Ther 5:353–361.

SCOTT AIF, SHERING PA, & DYKES S (1989). Would monitoring by electroencephalogram improve the practice of electroconvulsive therapy? Br J Psychiatry 154:853–857.

SCOVERN AW, & KILMAN PR (1980). Status of ECT: A review of the outcome literature. Psychol Bull 87:260–303.

SIBONY PA (1985). ECT risks and glaucoma. Convulsive Ther 1:283–287.

SIRIS SG, GLASSMAN AH, & STETNER F (1982). ECT and psychotropic medication in the treatment of depression and schizophrenia. In R ABRAMS & WB ESSMAN (Eds.), Electroconvulsive therapy: Biological foundations and clinical applications (Ch. 5, pp. 91–111). New York: Spectrum Publications.

SMALL JG (1985). Efficacy of electroconvulsive therapy in schizophrenia, mania, and other disorders: 1. Schizophrenia. Convulsive Ther 1:263–270.

SMALL JG, KELLAMS JJ, & MILSTEIN V (1980). Complications with electroconvulsive therapy combined with lithium. Biol Psychiatry 15:103–112.

SMALL JG, KLAPPER MH, KELLAMS JJ, ET AL. (1988). Electroconvulsive treatment compared with lithium in the management of manic states. Arch Gen Psychiatry 45:727–732.

SMALL JG, MILSTEIN V, KLAPPER M, ET AL. (1982). ECT combined with neuroleptics in the treatment of schizophrenia. Psychopharmacol Bull 18:34–35.

SMITH K, SURPHLIS WRP, GYNTHER MD, ET AL. (1967). ECT-chlorpromazine compared in the treatment of schizophrenia. J Nerv Ment Dis 144:284–290.

SNAITH RP (1981). How much ECT does the depressed patient need? In RL PALMER (Ed.), Electroconvulsive therapy: An appraisal (Ch. 8, pp. 61–64). Oxford: Oxford University Press.

SQUIRE LR (1986). Memory functions as affected by electroconvulsive therapy. Ann NY Acad Sci 462:307–314.

STANDISH-BARRY HMAS, DEACON V, & SNAITH RP (1985). The relationship of concurrent benzodiazepine administration to seizure duration in ECT. Acta Psychiatr Scand 71:269–271.

STEVENSON GH, & GEOGHEGAN JJ (1951). Prophylactic electroshock–A five-year study. Am J Psychiatry 107:743–748.

STOUDEMIRE A, HILL D, & MARKON C (1987). Intestinal pseudo-obstruction associated with major depression responsive to ECT. Psychosomatics 28:386–387.

STOUDEMIRE A, KNOS G, GLADSON M, ET AL. (1990). Labetolol in the control of cardiovascular responses to electroconvulsive therapy in high-risk depressed medical patients. J Clin Psychiatry 51:508–512.

STROMGREN LS (1984). Is bilateral ECT ever indicated? Acta Psychiatr Scand 69:484–490.

TANCER ME, & EVANS DL (1989). Electroconvulsive therapy in geriatric patients undergoing anticoagulation therapy. Convulsive Ther 5:102–109.

TAUB S (1987). Electroconvulsive therapy, malpractice, and informed consent. J Psychiatry Law 15:7–54.

THOMPSON JW, & BLAINE JD (1987). Use of ECT in the United States in 1975 and 1980. Am J Psychiatry 144:557–562.

WEBB MC, COFFEY CE, CRESS MM, ET AL. (1990). Cardiovascular response to unilateral electroconvulsive therapy. Biol Psychiatry 28:758–766.

WEINER RD (1982). The role of ECT in the treatment of depression in the elderly. J Am Geriatr Soc 30:710–712.

WEINER RD (1983). EEG related to electroconvulsive therapy. In JR HUGHES & WP WILSON (Eds.), EEG and evoked potentials in psychiatry and behavioral neurology (Ch. 6, pp. 101–126). Boston: Butterworth.

WEINER RD (1984). Does ECT cause brain damage? Behav Brain Sci 7:1–53.

WEINER RD (1986). Electrical dosage, stimulus parameters, and electrode placement. Psychopharmacol Bull 22:499–502.

WEINER RD, & COFFEY CE (1986). Minimizing therapeutic differences between bilateral and unilateral nondominant ECT. Convulsive Ther 2:261–265.

WEINER RD, COFFEE CE, & KRYSTAL AD (1991). The monitoring and management of electrically induced seizures. Psychiatric Clin N Am 14:845–869.

WEINER RD, HENSCHEN GM, DELLASEGA M, ET AL. (1979). Propranolol treatment of an ECT-related ventricular arrhythmia. Am J Psychiatry 136:1594–1595.

WEINER RD, ROGERS HJ, DAVIDSON JRT, ET AL. (1986a). Effects of electroconvulsive therapy upon brain electrical activity. Ann NY Acad Sci 462:270–281.

WEINER RD, ROGERS HJ, DAVIDSON JRT, ET AL. (1986b). Effects of stimulus parameters on cognitive side effects. Ann NY Acad Sci 462:315–325.

WEINER RD, WHANGER AD, ERWIN CW, ET AL. (1980). Prolonged confusional state and EEG seizure activity following concurrent ECT and lithium use. Am J Psychiatry 137:1452–1453.

WELCH CA, & DROP LJ (1989). Cardiovascular effects of ECT. Convulsive Ther 5:35–43.

WISE MG, WARD SC, TOWNSEND-PARCHMAN W, ET AL. (1984). Case report of ECT during high-risk pregnancy. Am J Psychiatry 141:99–101.

WISNER KL, & PEREL JM (1988). Psychopharmacologic agents and electroconvulsive therapy during pregnancy and the puerperium. In

RL Cohen (ed.), Psychiatric Consultation in Childbirth Settings: Parent- and Child-Oriented Approaches (pp. 165–206). New York: Plenum.

Yacoub OF, & Morrow DH (1986). Malignant hyperthermia and ECT. Am J Psychiatry 143:1027–1029.

Yassa R, Hoffman H, & Canakis M (1990). The effect of electroconvulsive therapy on tardive dyskinesia: A prospective study. Convulsive Ther 6:194–198.

Zibrak JD, Jensen WA, & Bloomingdale K (1988). Aspiration pneumonitis following electroconvulsive therapy in patients with gastroparesis. Biol Psychiatry 24:812–814.

Zimmerman M, Coryell W, & Pfohl B (1985). The treatment validity of DSM-III melancholic subtyping. Psychiatr Res 16:37–43.

Zorumski CF, Rutherford JL, Burke WJ, et al. (1986). ECT in primary and secondary depression. J Clin Psychiatry 47:298–300.

Zwill AS, Bowring MA, Price TRP, et al. (1990). Prospective ECT in the presence of intracranial tumor. Convulsive Ther 6:299–307.

11 ECT Anesthesia strategies in the high risk medical patient

GUNDY B. KNOS, M.D., AND
YUNG-FONG SUNG, M.D.

Although ECT is a safe procedure for most healthy patients, multiple review articles (Knos & Sung, 1991; Scalafani, 1988; Selvin, 1987; Weiner & Coffey, 1987) have shown that the cardiovascular and autonomic changes that may occur during the procedure can cause morbidity and mortality in medically compromised patients. This chapter is meant to guide the reader in the preparation and anesthetic management of ECT in the medically frail patient with specific conditions that have been associated with morbidity and mortality during ECT. Excellent discussions regarding the anesthetic agents used for ECT and the monitoring techniques used during ECT may be found elsewhere (Gaines & Rees, 1986; Knos & Sung, 1991; Selvin, 1987; Weiner & Coffey, 1987). A treatment guideline for the application of ECT has also been published by the American Psychiatric Association (1990) and it is strongly recommended that the reader be familiar with the contents of this document prior to application of ECT to patients. The psychiatrist should actively collaborate with the anesthetist prior to the ECT procedure in medically fragile patients so preparations to manage risks associated with the procedure can be made in advance.

PHYSIOLOGIC EFFECTS OF ECT

The physiology of electrically induced convulsions influences the risks of ECT in medically ill patients. Figure 11-1, from a review article by Selvin (1987), illustrates the sequence of physiologic changes that occur during and after ECT. Note especially the changes in blood catecholamine levels and the hemodynamic variation that occurs during treatment.

Endocrine Responses to ECT

Electroconvulsive therapy has profound effects on the hypothalamus, pituitary, thyroid glands, and (most importantly from an anesthetic point of view) the adrenal cortex and medulla (Pitts, 1982). Adrenocorticotropic hormone (ACTH) levels have been shown to increase immediately, followed by a later increase in serum cortisol. These elevated corticosteroid concentrations have little clinical consequence except in diabetics, who may respond with increased blood glucose levels (Selvin, 1987).

Cardiovascular Response to ECT

The immediate cardiovascular response to ECT stimulus is due to a dramatic increase in levels of circulating catecholamines. The catecholamines are largely released from the adrenal medulla, with a smaller contribution from the sympathetic nerve endings themselves (Selvin, 1987). An increase in parasympathetic tone during the first 10 to 15 seconds (Selvin, 1987) may produce bradycardia, asystole, and various cardiac dysrhythmias leading to marked hypotension (Gaines & Rees, 1986). The parasympathetic manifestations are soon replaced by the signs of sympathetic stimulation. Plasma epinephrine can increase to 15 times normal and plasma norepinephrine can reach three times normal (Selvin, 1987). Heart rate and blood pressure may soar, and cardiac dysrhythmias may occur in response to the arrhythmogenic effects of the catecholamine surge (Gaines & Rees, 1986; Selvin, 1987). These changes can produce an increase in myocardial oxygen consumption, which may cause myocardial ischemia and even infarction in susceptible patients (Gaines & Rees, 1986). During general surgical procedures it has been shown that atten-

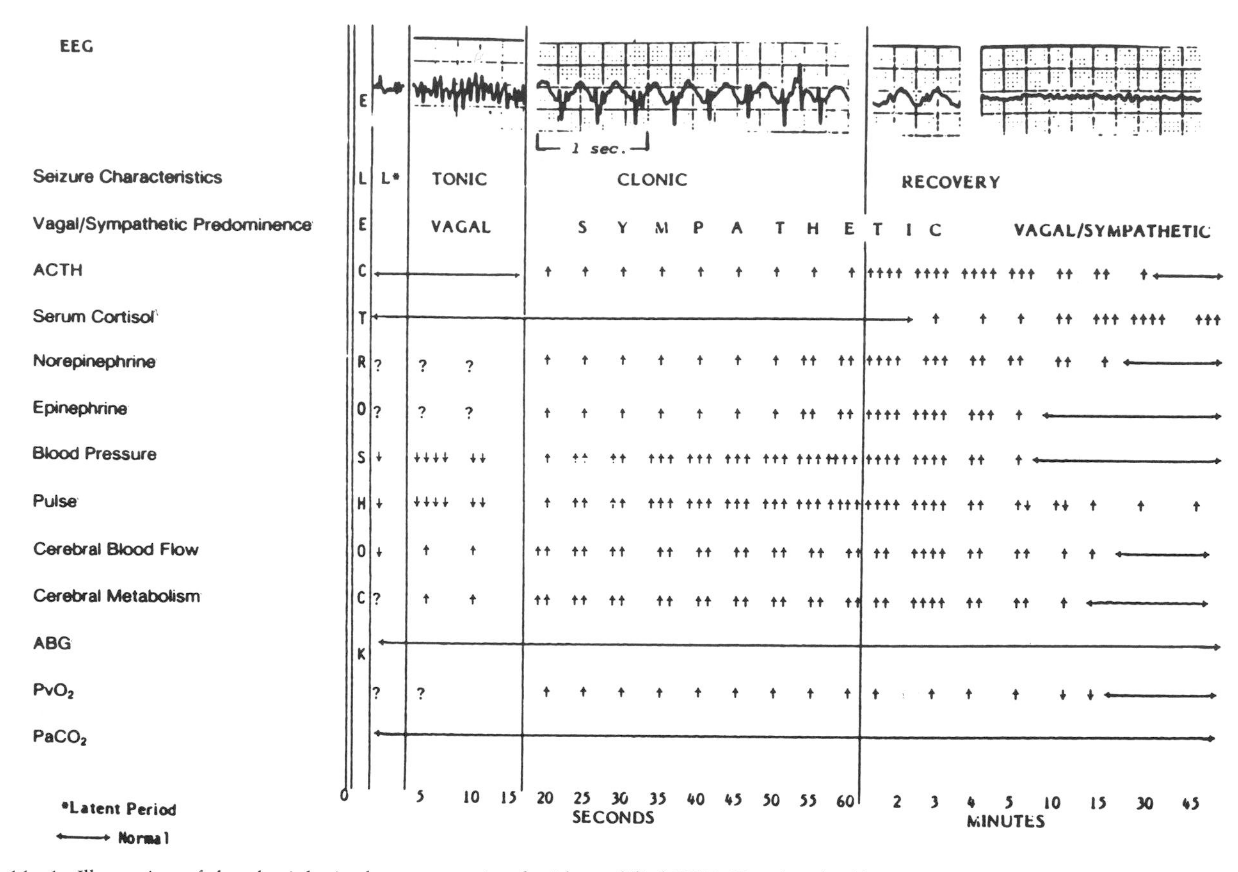

FIG. 11–1. Illustration of the physiologic changes associated with modified ECT. (Reprinted with permission from Selvin BL (1987). Electroconvulsive therapy—1987. Anesthesiology 67:367–385).

uation of these autonomic changes and their secondary effects may decrease anesthetic morbidity and mortality (Rao, Jacobs, & El-Etr, 1983). It is a reasonable assumption that the medically fragile ECT patient may benefit from similar handling.

Cerebrovascular Response to ECT

The cerebrovascular system is also affected by ECT. With the electrical stimulus, there is a brief period of cerebrovascular constriction, followed by a 1.5-to 7.0-fold increase in cerebral blood flow from baseline due to increased systemic arterial pressure and increased cerebral oxygen consumption. The resulting dramatic but transient increase in intracranial pressure that occurs may have profound detrimental effects in ECT patients with cranial arteriovenous malformations, intracranial mass lesions, or preexisting increased intracranial pressure (Gaines & Rees, 1986).

Intraocular and Intragastric Pressure Responses To ECT

Intraocular pressure increases occurring with seizure onset (Epstein et al., 1975) may have serious consequences in a patient with poorly controlled glaucoma. Likewise, seizure onset causes increased intragastric pressure, which may lead to problems with regurgitation and pulmonary aspiration of gastric contents. Patients with symptomatic hiatal hernias who are ventilated with a mask and ambu bag are at particular risk for this complication. Aspiration pneumonitis is a dreaded complication that carries a mortality rate of 28% (Bynum & Pierce, 1976). Patients with symptomatic hiatal hernias with gastric reflux should receive oral H_2 receptor antagonists. These patients should be preoxygenated prior to induction of anesthesia. In more severe cases, anesthetic induction should be followed by rapid endotracheal intubation with cricoid pressure and endotracheal cuff inflation before the electrical stimulus is applied. By sealing the trachea, the endotrachael tube cuff is thought to prevent tracheal trespass of highly acidic gastric contents. A suction apparatus should be ready if regurgitation should unexpectedly occur.

MEDICAL CONDITIONS ASSOCIATED WITH INCREASED RISK WITH ECT

The American Psychiatric Association (APA) Task Force on ECT identified several situations in which ECT may be associated with considerable likelihood of increased morbidity and mortality (APA, 1990). Extensive discussion of the reasons for concern regarding ECT in certain conditions such as recent myocardial infarct and space-occupying intracranial mass lesions has been provided elsewhere (Knos & Sung, 1991; Selvin, 1987; Weiner & Coffey, 1987). Each patient that has a condition thought to be an "absolute" contraindication for ECT must be considered on an individual risk-benefit basis. For example, it is entirely possible that the patient could be dying (wasting away with dehydration and lack of nourishment) from catatonic depression. Patients may also be actively psychotic or suicidal. The relative risks associated with ECT may be significantly less than the morbidity and mortality associated with an intractable depression, hence pushing the risk-benefit ratio in favor of proceeding with ECT. Each case must be individually considered in light of this risk-benefit assessment, with joint involvement of the attending psychiatrist, anesthesiologist, surgeon or cardiologist, and, most especially, the patient and family. All parties involved must be aware that there may be high risk involved in proceeding with ECT under certain circumstances, but that the only other alternative to ECT may be death due to the complications of intractable depression.

Documentation regarding the decision process, alternative possibilities with their probable outcomes, the seriousness of the patient's condition, and the basic elements of the risk-benefit assessment, must be scrupulous. The patient's physical condition must then be optimized and stringent care applied during the ECT procedure and the recovery period.

For patients who have a medical condition that places them at risk for morbidity and mortality during their ECT treatment, serious consideration should be given to performing the procedure within the confines of a general hospital, rather than a free-standing psychiatric facility so the medical personnel, equipment, and facilities for treatment of serious complications can be performed quickly and efficiently. The management of specific medical problems under such circumstances are dealt with later in this chapter.

RELATIVE CONTRAINDICATIONS TO ECT

A diversity of opinion exists as to what constitutes a relative "contraindication" to ECT. Medical conditions frequently mentioned include angina pectoris, congestive heart failure, cardiac pacemakers, glaucoma, retinal detachment, severe osteoporosis, major bone fractures, and severe acute and chronic pulmonary disease (Gaines & Rees, 1986; Hurwitz, 1974). ECT has been successfully performed in all these circumstances, and a guide to the specific management of patients with some of these problems is addressed later in the section on Management of Specific Medical Problems During ECT. (See also Chapter 10.)

PRE-ECT ASSESSMENT

History

Elderly depressed patients referred for ECT often are frail and suffer from cardiovascular or pulmonary diseases. These patients may be malnourished and dehydrated, and chronic diseases may be in poor control because of noncompliance with prescribed medications or diets. Patients often have poor dentition, osteoporosis, and arthritis. If psychotic or catatonic features are present, it may be impossible to obtain a good medical history from the patient. Review of a patient's medication regimen can provide additional information as to the types and severity of a patient's medical problems. A call to the patient's primary physician may be necessary to obtain a clear picture of the patient's health status. Antihypertensive, cardiac, and pulmonary medications should be continued during the patient's hospitalization for ECT. Possible exceptions could include diuretics and oral hypoglycemic agents. This topic is covered in the section on Pre-ECT Orders. The family may need to be questioned about the patient's cardiovascular status, exercise tolerance, allergies, and so on; and any family history suggesting malignant hyperthermia or pseudocholinesterase deficiency should be elicited.

The ECT patient may be on tricyclic antidepressants (TCAs), which block the uptake of norepinephrine, serotonin, and dopamine into presynaptic nerve terminals. These medications have the effect of increasing central and peripheral adrenergic tone (Janowsky, Risch, & Janowsky, 1981). TCAs may augment the effects of barbiturates used prior to ECT stimulus causing increased sleep time and sedation (Gaines & Rees, 1986). They may also have anticholinergic effects. Therapeutic doses of TCAs have been rarely associated with dysrhythmias and tachycardia, and there may be an increased response to direct sympathomimetics used to treat hypotension. The arrhythmogenic effects of epinephrine released during ECT could be potentiated by the presence of TCA. Systemic adaptations in response to chronic TCA therapy are believed to occur and may explain the lack of adverse responses of some patients during ECT, but patients on TCAs for less than 1 month may not be adapted and theoretically may be at higher risk (Selvin, 1987). A prolonged drug-free period from TCAs is probably not necessary and might even be ill-advised in some patients (Azar & Lear, 1984; Gaines & Rees, 1986; Weiner & Coffey, 1987). However, if there is no psychiatric reason for TCA therapy during a course of ECT, they probably should be gradually discontinued (Drop & Welch, 1989), or a lower dose of a mildly sedating TCA, such as nortriptyline, can be used instead for anxiolysis or sleep facilitation.

Lithium therapy has been reported to prolong recovery time in combination with barbiturates when serum lithium levels are above the therapeutic range. The action of muscle relaxants has been shown to be prolonged in patients on lithium therapy, which may lead to prolonged inability of the patient to ventilate adequately after receiving muscle-relaxing drugs for ECT (Gaines & Rees, 1986). Lithium can also decrease the therapeutic response to ECT (Gaines & Rees, 1986; Weiner & Coffey, 1987) and in combination with ECT may induce neurotoxicity and delirium (Drop & Welch, 1989). Therefore lithium should be discontinued prior to ECT (Selvin, 1987).

The attending physician, whether psychiatrist or anesthesiologist, may be the first physician to critically examine the general medical condition of these patients, and patient evaluation should be approached with the same care as that for any other patient prior to surgery. Some patients may not have had consistent medical care due to apathy and disinterest secondary to their depression. Furthermore, medical "clearance" of psychiatric patients often leaves gaps, so neither the psychiatrist nor the anesthesiologist should assume that all medical problems were addressed during the routine examination the patient received on admission to the hospital.

Physical Examination

The patient's airway should be carefully examined, especially with regard to dentition. Due to long-standing depression, some patients may have disregarded their dental hygiene with subsequent loose and broken teeth. Neck range-of-motion should be established, especially in arthritic patients, as neck and jaw mobility may be restricted or there may be instability of the first cervical vertebra on the second cervical vertebra, leading to difficult airway management (Kallos & Smith, 1989). The neck should be examined for thyroid goiter. Signs of hyperthyroidism such as exophthalmos, fine tremor, palpitations, and muscle weakness should be noted, as hyperthyroidism can present with mood disorder and requires special precaution. (Tasch, 1983).

The heart and lungs should be carefully auscultated. Cardiac murmurs and gallops, inspiratory and expiratory wheezes, jugular venous distention, and dependent edema should be noted. Signs of orthopnea can be elicited in patients suspected of having congestive heart failure by placing them in a supine position for 30 minutes.

Patients with a heavy smoking history should be suspected of having significant pulmonary disease. Some patients may develop chronic lung disease and fibrosis from occupational or environmental exposure to various vegetable, animal, or fungal spore dusts. Examples of such diseases are byssinosis (exposure to cotton dust), anthrax (exposure to wool and animal hair), and psittacosis (exposure to infected birds). Signs of significant pulmonary disease are frequent inspirations during normal conversation, pursed lips, barrel chest, and use of accessory muscles during resting ventilation. If pulmonary disease is suspected, an impromptu test of a patient's pulmonary reserve could consist of a brisk walk down the hospital hall or the patient's relative ability to ascend a flight of hospital stairs. Abnormalities detected by clinical examination examination or by the simple clinical tests mentioned above should be followed up by consultation with the appropriate specialist with objective tests as indicated (see below).

The patient's hydration status should be carefully examined, especially in catatonic patients who are taking little food or liquids by mouth. Skin turgor, neck vein distention, lying and standing (if possible) pulses and blood pressures, and appearance of the mucus membranes should be assessed. A dehydrated patient should be adequately rehydrated orally or with intravenous fluids prior to undergoing ECT.

Objective Tests

Laboratory values such as the hematocrit and sodium levels may give additional clues to hydration status. A high hematocrit may indicate severe dehydration. Weekly electrolyte levels are important in patients on diuretics for hypertension or on digitalis for arrhythmias. There is also a correlation between low potassium levels and the incidence of premature ventricular contractions (PVCs) (Rao et al., 1974). Low potassium levels in patients on digitalis have been associated with emergence of digitalis-induced toxic arrhythmias. The possibility of digitalis toxicity should be entertained in any patient on cardiac glycosides exhibiting premature ventricular beats, atrial tachycardia with variable atrioventricular (AV) block, or a heart rate lower than 50 beats/minute due to AV block (Braunwald, 1987). Patients on digitalis or any other cardiac glycosides should have a potassium level of at least 3.5 mEq/L. Hypokalemia and digitalis toxicity should be corrected prior to initiating ECT treatment.

The ECT patient should have a recent ECG, chest roentgenogram, and, if indicated, cervical spine films (particularly in arthritic patients). Studies such as CT scans can be ordered to rule out intracranial neoplasms or aneurysms, as indicated; and in patients with unstable blood pressure, appropriate blood tests to exclude thyroid dysfunction and pheochromocytoma should be considered. Patients to undergo ECT should be medically prepared with the same care as any patient who is to have anesthesia for a surgical procedure. Appropriate specialists should be consulted to maximize the patient's medical condition prior to treatment.

A cardiologist should be consulted to evaluate cardiovascular status and to optimize medication regimens in patients with moderate to severe angina, unstable angina, history of multiple or recent myocardial infarction, poorly controlled hypertension, valvular heart disease, or chronic congestive heart failure. In patients with more serious symptoms than mild, stable angina which is well controlled on medication, a pretreatment evaluation with echocardiography or coronary angiography may be indicated (McPherson & Lipsey, 1988). Coronary artery disease and ability to handle cardiac stress can be evaluated with an ECG stress test. For catatonic patients or those unable to perform the exercise stress test, a dipyridamole-thallium myocardial perfusion scan can disclose areas of ischemia as well as areas of old infarction. For patients with congestive heart failure, a gated blood pool study, chest radiograph, and arterial blood gas assays may be indicated to assess myocardial function (Drop & Welch, 1989). Echocardiography or cardiac catheterization may be necessary to assess the severity of certain critical heart valve lesions, such as aortic stenosis. The frequency and severity of cardiac dysrhythmias noted on pretreatment ECG or suggested by history (e.g., Stokes-Adams attacks) may require evaluation with Holter monitoring for 24 to 48 hours (Adams & Martin, 1987).

For the patient suspected of pulmonary disease, room air arterial blood gas analysis and clinical spirometry are the optimal tests for evaluation of pulmonary reserve prior to ECT (Gal, 1986). A single spirometric study can provide critical information such as the forced vital capacity (FVC), the forced expiratory volume at 1 second (FEV_1), the FEV_1/FVC ratio, maximal voluntary ventilation (MVV), and forced expiratory flow at 25% to 75% of the vital capacity (FEF_{25-75}). Values that indicate an increased risk for morbidity and mortality due to pulmonary disease are shown below.

FVC $<$ 50% predicted
FEV_1 $<$ 50% predicted or $<$ 2.0 L
FEV_1/FVC $<$ 50%
MVV $<$ 50% predicted or $<$ 50 L/min
FEF_{25-75} $<$ 50% predicted

Any patient who does not surpass these minimal criteria should be evaluated and optimized by an internist (Boysen, 1988). Patients with more severe disease usually should be evaluated by a specialist in pulmonary medicine.

PRE-ECT ORDERS

Patients should have nothing by mouth for 6 to 8 hours prior to ECT. Cimetidine 300 mg or ranitidine 150 mg PO may be given at bedtime and on the morning of ECT to increase gastric fluid pH and decrease the risk of aspiration pneumonitis in patients who are to undergo positive pressure ventilation by mask. However, patients with bronchial asthma may develop bronchoconstriction on administration of these H_2-receptor blocking agents (Stoelting, 1986). Therefore the decision to give these medications must be made on an individual basis.

Glycopyrrolate 0.2 mg IM or atropine 0.4 mg IM 30 to 45 minutes prior to ECT is often ordered by the psychiatrist to attenuate the initial bradycardia seen after the ECT stimulus and to decrease airway secretions (Selvin, 1987). Whether pretreatment with anticholinergic agents is absolutely necessary prior to ECT treatment is currently a controversial issue (Gaines & Rees, 1986).

All usual medications, such as antihypertensive agents, cardiac medications, bronchodilators, and thyroid medications, should be continued as usual on the morning of ECT with small sips of water. Beta-blocking agents should be continued, as their abrupt withdrawal may precipitate ischemic myocardial episodes. Exceptions to the rule include the rauwolfia alkaloids, which should be discontinued for 2 weeks prior to ECT to avoid problems of prolonged apnea, hypotension, cardiac dysrhythmias, and death (Goodloe, 1983; Scalafani, 1988). Other medications to control hypertension should be substituted. Diuretics such as furosemide or hydrochlorthiazide should be withheld on the morning of ECT. The insensible fluid loss associated with 8 hours of no oral intake ("NPO past midnight") produces a deficit of more than 1 L in a 70-kg patient at 2 ml/kg/hr (Giesecke & Egbert, 1986) and further diuresis of the patient is detrimentally additive.

Insulin-dependent diabetics, particularly brittle diabetics, should be evaluated by consultation with an internist or endocrinologist. Pretreatment management may require glucose and insulin drips throughout the NPO period. Evening doses of NPH insulin, with peak effects 11 hours later, may need to be withheld prior to ECT. Also, morning doses of regular insulin are withheld until after ECT when the patient is able to eat and drink with no nausea or vomiting. For the care of patients on oral antihypoglycemic agents, such as the sulfonylureas (e.g., acetohexamide, chlorpropamide), one should recall that often these agents have long durations of action (12 to 60 hours) and that their hypoglycemic effects can be severe and profound (Foster, 1987). The decision on how to administer these medications during the treatment period may require expert consultation. When any diabetic patient presents for ECT, a pretreatment glucose level must be checked. These patients should have their ECT performed early in the day so there is minimal disturbance of insulin or medication regimens (Lichtiger, Wetchler, & Philip, 1985). Since glucose is the primary energy substrate of the brain, hypoglycemia in the short run is far more dangerous than hyperglycemia (Foster & Rubenstein, 1986).

In patients on chronic steroid therapy, persistent hypotension has been reported during perioperative stress in the absence of steroid supplements (Knudsen, Christiansen, & Lorentzen, 1981). This problem has led to the practice of giving supplementary steroids (e.g., 100 to 300 mg hydrocortisone phosphate/70 kg body weight) to steroid-dependent patients or any patient who has received steroids within 1 year. The rationale is that steroid supplementation is a low risk procedure (Roizen, 1986) and ECT presents a significant physiologic stress. In the absence of definitive data our practice is to give steroid supplements (as described above) to such patients prior to treatment.

For patients who, for clinical psychiatric reasons, *must* be maintained on monoamine oxidase inhibitors (MAOIs) for their depression, the anesthetist should be aware that an intensified sympathetic response to ECT with hypertension, hypotension, hyperreflexia with convulsions, and hyperpyrexia (Scalafani, 1988; Selvin, 1987) has been reported in the past. More recent articles (El-Ganzouri et al., 1985; Stack, Rogers, & Linter, 1988; Wells & Bjorksten, 1989), however, suggest that patients can be safely anesthetized while taking these drugs and that a prolonged period of withdrawal prior to a procedure is not necessary. However, the potentiation of inducing agents such as barbiturates or of succinylcholine may occur (Wells & Bjorksten, 1989). Also, fatal reactions have been reported in patients on MAOIs who have been given meperidine, and adverse reactions are known to occur with MAOIs and indirect-acting sympathomimetic amines (Stack, Rogers, & Linter, 1988; Wells & Bjorksten, 1989). The issue is discussed in detail in Chapter 9.

MONITORING DURING ECT

The ECT patient should be monitored with the same care as any other patient undergoing a surgical procedure with regional or general anesthesia—perhaps even more closely because of the potentially dramatic hemodynamic changes that can take place over a brief period of time. An ECG and blood pressure cuff are considered necessities along with supplemental oxygen and a means to provide positive pressure ventilation (APA, 1990). A functioning suction system to clear the airway, a peripheral nerve stimulator to monitor muscle relaxation, and a precordial stethoscope to monitor ventilation are highly recommended to increase patient safety. It should also be noted that on January 1, 1990, a means of monitoring blood oxygenation (e.g., pulse oximetry) became standard for basic monitoring during *all* anesthesia (ASA Newsletter, 1989).

The room where ECT is performed should be equipped with intubation equipment, defibrillator, and cardiac resuscitative medications (Scalafani, 1988). The ECT machine itself should be inspected, calibrated, and maintained on a regular basis by a qualified biomedical technician (Kendell, 1981).

Patients with a tendency to rapid oxygen desaturation (e.g., obese patients or those with severe pulmonary disease) or patients with cardiac disease should be given 100% oxygen by mask for a few minutes prior to induction with intravenous anesthetics. If the patient objects excessively to the mask, ventilation with 100% oxygen should commence immediately as loss of consciousness occurs (Selvin, 1987).

One must pay particular attention to the ECG before, during, and after the treatment; and it may be advisable to run a pre-ECT ECG rhythm strip for purposes of comparison in any patient who has risk factors for coronary artery disease. Clinicians who regularly perform anesthesia for ECT are amazed at the high incidence of dysrhythmias, marked tachycardia, and ST segment changes that can be noted with a little close observation of the ECG. Due to the wide range of dysrhythmias and treatment courses available, in-depth discussion of this topic is beyond the scope of this chapter. Generally, dysrhythmias that require treatment are those that produce significant hemodynamic instability (hypotension, for example) or dysrhythmias (such as multifocal PVCs) that have a potential for deteriorating into malignant cardiac rhythms, such as ventricular tachycardia. ECG changes suggestive of myocardial ischemia should also be managed aggressively. The decision of how and when to treat such ECG changes is at the discretion of the attending anesthesiologist. Intravenous lidocaine pretreatment has been shown to be effective in preventing arrhythmias during ECT; however, it has been shown to suppress or shorten the seizure activity in a dose-related manner (Selvin, 1987). Patients who have shown a tendency to develop arrhythmias in the course of their treatment can be given prophylactic lidocaine 1 mg/kg IV as soon as a therapeutic seizure time as been achieved. This dose can be repeated in 20 minutes if arrhythmias appear subsequently (Kofke & Firestone, 1988).

INTRAVENOUS ANESTHETIC AGENTS IN ECT

Intravenous anesthetic agents are usually given prior to induction of ECT to provide amnesia during stimulus administration in case a "missed" or "incomplete" seizure occurs. Regardless of the presence or absence of an anesthetic agent, retrograde amnesia occurs after a successful electrically induced seizure (Selvin, 1987). Ideally, the anesthetic agents used during ECT should have rapid onset with short duration of action, provide rapid return to consciousness after the ECT treatment, and should not shorten the seizure time (as treatment efficacy is partly related to total seizure time) (Selvin, 1987).

Short-acting barbiturates, (methohexital and thiopental) have been the intravenous agents most commonly used. Large doses of intravenous anesthetic agents, such as the barbiturates, have been shown to have no beneficial effect in controlling hypertension and tachycardia during ECT (Gravenstein et al., 1965) and actually shorten seizure time (Lunn et al., 1981). This latter observation should come as no surprise, as these agents are in the same general pharmaceutical class as medications used to treat seizure disorders. With lighter anesthesia, it has been shown that seizure time may be longer due to a lower seizure threshold, possibly making fewer treatments necessary (Maletzky, 1978). We have discussed elsewhere the anesthetic induction agents such as barbiturates, propofol, benzodiazepines and alfentanyl as well as the various muscle relaxing agents used in ECT (Knos & Sung, 1991).

MANAGEMENT OF SPECIFIC MEDICAL PROBLEMS DURING ECT

Intracranial Mass Lesions

In the presence of an intracranial mass lesion or in the situation where the flow of cerebrospinal fluid is obstructed causing increased intracranial pressure (ICP), ECT is associated with a high incidence of morbidity and mortality. These complications are due to rapid increases in ICP, which may occur when patients have marginal cerebral compensatory mechanisms for ICP control (Knos & Sung, 1991). If excision of a symptomatic brain tumor or relief of ICP by surgical means is not possible and ECT is the only hope for life-threatening depression, all persons involved must be fully aware of the high incidence of death and neurologic deterioration in this clinical setting (Maltbie et al., 1980). Stringent efforts to bring the ICP as close as possible to normal is necessary before, during, and after ECT treatment. Pharmacologic maneuvers to decrease intracranial hypertension prior to each treatment may require the intervention of a neurologist. Steroids, such as dexamethasone (1.5 mg four times a day prior to and following ECT), have been used in ECT patients in the past to reduce perifocal edema around malignant brain tumors (Dressler & Folk, 1975; Gutin, 1977; Maltbie et al., 1980). More recently, larger doses of steroids have been recommended: dexamethasone 10 mg IV bolus followed by 4 mg qid (Kastrup, 1991). Mannitol 0.25 to 0.50 mg/kg can be infused 30 minutes prior to the ECT treatment. Mannitol decreases brain volume by increasing plasma osmolality, moving water from the brain into the intravascular space. Its action begins within 10 to 15 minutes, and its effects last for about 2 hours (Newfield, 1987). Furosemide 0.15 to 1.00 mg/kg) can also decrease brain volume by effecting a systemic diuresis and decreasing brain cerebrospinal fluid production (Cottrell et al., 1977).

Hyperventilation prior to ECT induction can decrease brain volume owing to decreased cerebral blood flow from cerebrovascular vasoconstriction. The effects of hyperventilation are almost immediate. This hypnocapnic vasoconstriction may counteract the 1.5- to 7.0-fold increase in cerebral blood flow that occurs during ECT (Gaines & Rees, 1986). However, it should be recalled that the effectiveness of this measure may be compromised by impaired responses to carbon dioxide by diseased vessels around the brain tumor. Also, an arterial carbon dioxide level of less than 30 mm Hg due to excessive hyperventilation may also decrease brain oxygenation because of severe restriction of cerebral blood flow (Newfield, 1987). The patient should be placed in a 15- to 20-degree head-up tilt with the head in the midline position to keep jugular venous drainage open. Coughing and straining causing venous hypertension should be avoided.

The small barbiturate doses used for induction of anesthesia during ECT may help decrease ICP (Bruce, 1983). Glucose-free crystalloid solutions should be used, as there is evidence that should brain ischemia occur it may be exacerbated in the presence of increased blood glucose (De Salles, Muizelaar, & Young, 1987; Pulsinelli et al., 1983). Blood pressure should be controlled during the ECT as blood vessels in and around the tumor may be fragile and prone to rupture. Sublingual nifedipine has been shown to be effective in preventing severe blood pressure increases during and after ECT (Kalayam & Alexopoulos, 1989). Nitroprusside infusion, esmolol, or labetalol may be necessary to maintain tight blood pressure control. Use of these medications is described in the section below on recent myocardial infarction and unstable angina. After ECT, these patients should be carefully observed to ensure that their mental status is stable. Any sudden deterioration in mental status may indicate hemorrhage in the area of the tumor, which may require emergency surgical decompression.

Patients who have had surgical excision of their brain tumor probably can safely undergo ECT 3 months later. Consecutive computed tomography (CT) scans have shown that the postoperative changes of brain surgery are stabilized at approximately 3 months (Jefferies et al., 1980). Prior to that time, postsurgical intracranial bleeding due to the increased cerebral blood flow and pressure changes during ECT can occur (Husum et al., 1983). ECT has been successfully administered to a patient 4 months after surgical excision of an intracranial meningioma (Hsiao & Evans, 1984).

CEREBROVASCULAR INFARCT

Clinical evidence indicates that patients can safely undergo ECT 6 weeks after a cerebral infarction (Alexopoulos, Young, & Abrams, 1989; Anderson & Kissane, 1977). In patients with a recent cerebrovascular accident (CVA), the vascular integrity may be compromised, cerebrovascular fragility may be present, and there is a potential for further adverse neuropathologic changes (Adams & Victor, 1977; Anderson & Kissane, 1977); however, the risk is minimized once healing has occurred (Hsiao, Messenheimer, & Evans, 1987). Also, ECT could theoretically dislodge a second embolus in patients with a history of cerebrovascular embolism due to blood pressure surges (Alexopoulos, Young, & Abrams, 1989). The methods outlined below for stabilizing blood pressure changes during ECT in patients with a recent myocardial infarction should be followed to avoid further insult. Also, sublingual nifedipine may help prevent severe blood pressure increases after ECT (Kalayam & Alexopoulos, 1989).

INTRACRANIAL ANEURYSMS AND CEREBROVASCULAR MALFORMATIONS

The primary consideration for safely performing ECT in the patient with an unclipped cerebral arterial malformation seems to be tight blood pressure control during ECT (Drop & Welch, 1989). The main anesthetic goal during the ECT procedure should be avoiding increased stress on the aneurysmal wall. The principles of management are similar to those for patients with unstable aortic aneurysms outlined below. An increase in the mean arterial pressure or fall in the intracranial pressure (ICP) (as with hyperventilation) increases the transmural pressure, the wall stress, and the risk of aneurysm rupture (Peerless, 1983). Consequently, if there is no increased ICP, maneuvers to decrease ICP below normal in these patients (in contrast to patients with intracranial mass lesions) should *not* be pursued. Intravenous propranolol and hydralazine, oral timolol, and nitroprusside infusions have been used to moderate blood pressure increases in these patients (Drop, Bouckoms, & Welch, 1988; Husum et al., 1983).

ANTICOAGULATED PATIENTS

Occasionally, a patient who is on oral anticoagulation agents for various medical conditions presents for ECT. Because bleeding is the most serious complication from anticoagulant therapy, the concern in these patients has been that there may be potential increased risk of intracerebral hemorrhage with ECT-induced hypertension (Alexopoulous et al., 1982). In patients on long-term anticoagulation therapy, the general incidence of intracerebral hemorrhage (unrelated to ECT) is about 1% (Tancer & Evans, 1989). Intracranial bleeding is also associated with hypertension, recent cerebrovascular accidents, and prolonged coagulation indices (Lieberman et al., 1978; Silverstein, 1979).

Prior to beginning ECT treatments, some authors have

advocated switching from warfarin to heparin. Just prior to treatment, the heparin is then rapidly reversed with protamine sulfate. In this manner, the partial thromboplastin time is at or near control levels when the ECT is performed. This method was reported to be successful over six courses of ECT in five patients (Alexopoulos et al., 1982; Loo, Cuche, & Benkelfat, 1985). However, Hay (1987) reported one incident of recurrence of deep vein thrombosis in an ECT patient who was switched from warfarin to heparin. In 11 reported cases, warfarin was continued without incident during ECT (Tancer & Evans, 1989).

Because there are so few patients who present for ECT while on anticoagulants, it is unlikely that controlled studies assessing the ideal management of these patients will occur. Brain imaging studies may be needed to determine the risk of intracerebral hemorrhage in the anticoagulated ECT patient and to help determine whether intermittent heparin or continuous warfarin is preferable. It has been suggested that anticoagulation be continued during ECT with frequent monitoring of the prothrombin time (PT). The PT should be maintained at less than 2 times control, with the optimal PT range determined on a case-by-case basis (Tancer & Evans, 1989). Hypertension should be well controlled prior to ECT, and pharmacological means (e.g., nitroglycerine drip, labetalol) should be available to attenuate the cardiovascular response to ECT in these patients.

Recent Myocardial Infarction And Unstable Angina

Candidates for ECT with recent (less than 6 months old) myocardial infarction (MI) should be handled with considerable caution. These patients may be at high risk for reinfarction with hemodynamic stress (Goldman et al., 1977; Rao, Jacobs, & El-Etr, 1983; Steen, Tinker, & Tarhan, 1978). No studies specifically examining reinfarction in ECT patients with a recent MI exist in the literature. However, an increased risk of reinfarction in recent MI patients undergoing general anesthesia for other surgical procedures has been associated with intraoperative hypertension and tachycardia (Rao, Jacobs, & El-Etr, 1983). Because tachycardia and transient hypertension are seen on a regular basis in ECT patients, most authors urge caution when proceeding with ECT under these circumstances (Abrams, 1988; Gaines & Rees, 1986; Knos & Sung, 1991; Selvin, 1987). A few reports of successful outcomes of ECT performed 2 to 3 weeks after an MI should be weighed against reports of anesthetic complications following MI for other surgical procedures (Dec, Stern, & Welch, 1985; Kerr et al., 1982). For example, it has been shown that patients undergoing general surgery after a recent MI have a 50% mortality rate should reinfarction occur (Rao et al., 1983).

Preparation prior to ECT in the patient with a recent MI should be thorough and complete. A crash cart with a charged functional defibrillator should be ready. Lidocaine 200 mg should be available in a syringe at the patient's bedside. An arterial line should be strongly considered to aid in meticulous blood pressure control for the first one or two treatments until the patient's response to ECT is determined. [Note that aggressive management of hemodynamic aberrations may be associated with a decreased morbidity and mortality in post-MI patients (Rao, Jacobs, & El-Etr, 1983).] ECG monitoring with a modified V_5 lead for myocardial ischemia is essential. A pre-ECT recording of the patient's ECG is advisable in order to compare any ECG changes noted after ECT.

A nitroglycerine (NTG) infusion is highly advised in this setting. NTG is an organic nitrate that causes relaxation of smooth muscles in venous (primarily) and arterial walls, including coronary vessels, leading to favorable redistribution of coronary blood flow to ischemic areas. NTG can also be used to treat brief hypertensive episodes during ECT. It has the advantage of a rapid onset of action and short half-life with easy titratability in infusion form. NTG is prepared by diluting 50 mg in 250 ml of 5% dextrose in water (D_5W) to yield 200 μg/ml. The infusion should be started at about 10 μg/min and should be titrated to the patient's blood pressure response. It may be initiated before or after application of the electrical stimulus (Abrams & Roberts, 1983). Rapid adjustments in the infusion rate may be necessary and up to 400 μg/min may be required to maintain blood pressure control (Drop & Welch, 1989). The rate of the infusion is tapered off as the seizure and hypertension subside. NTG may produce a reflex tachycardia in response to the decrease in blood pressure (Waller, Kaplan, & Jones, 1979), which may require treatment with a beta-blocking agent. Nitroglycerine 2% ointment applied 45 minutes prior to treatment has also been reported to be effective in moderating ECT-related hypertension (Lee et al., 1985).

Nitroprusside, a nonselective vasodilator, can dramatically control blood pressure during ECT. A nitroprusside infusion is made by mixing 50 mg in 250 or 500 ml of D_5W. The resulting concentrations are 200 or 100 μg/ml, respectively. Nitroprusside has a rapid effect on blood pressure and should be titrated cautiously starting with an initial dose of 0.5 μg/kg/min (Drop & Welch, 1989). Nitroprusside can produce considerable reflex tachycardia, which, as discussed previously, may have considerable detrimental effect on myocardial oxygen demand. It should also be used with caution in patients with potential myocardial ischemia, as it has been reported to produce "intracoronary steal," or shunting of blood away from ischemic myocardium by dilation of coronary vessels to nonischemic myocardium (Waller et al., 1979). Also, sodium nitroprusside has been associated with persistent hypotension after ECT (Ciraulo et al., 1978).

Because of the cardiovascular changes that occur during ECT, beta-blocking agents (e.g., propranolol 0.5 to 4.0 mg IV) have been used to decrease hypertension and incidence of ventricular arrhythmias (Husum et al., 1983; Selvin, 1987). However, one case of cardiac arrest was reported after propranolol 0.5 to 1.0 mg was given prior to ECT (Decina et al., 1984). Propranolol may have a duration of action that lasts too long for the brief hemodynamic disturbance caused by ECT. Two short-acting beta-blocking agents, esmolol and labetalol, have now been shown to effectively control the cardiovascular responses during ECT (Hay, 1989). Esmolol, in a bolus of 1 mg/kg followed by an infusion of 500 μg/kg/min or 24 mg/min has been shown to provide greater hemodynamic stability during ECT (Black, et al., 1989; Klein et al., 1988; Kovac et al., 1988). Labetalol 5 mg IV prior to ECT induction has also been effective in controlling heart rate and blood pressure during ECT (Knos et al., 1990; Stoudemire et al., 1990). Use of these drugs should be considered for any patient with coronary artery disease, poorly controlled hypertension, or tachycardia prior to starting ECT, unless the patient is young and otherwise healthy.

Lidocaine pretreatment is effective in attenuating or preventing ventricular arrhythmias, but it has a dose-related suppression of seizure activity (Selvin, 1987). After seizure induction, lidocaine 2 mg/kg or procainamide may be necessary for treatment of ventricular arrhythmias (Hood & Mecca, 1983).

It should be remembered that preoxygenation and ventilation are paramount for preventing ventricular arrhythmias. Ventricular arrhythmias may occur in as many as 75% of nonoxygenated cardiac patients (McKenna et al., 1970). Finally, one should also be aware of a report of ECT-induced changes in the ECG mimicking an MI without any changes in cardiac isoenzymes or thallium 201 scan indicating that an MI had occurred (Gould et al., 1983).

Hypertension

The patient with uncontrolled hypertension is more likely to have marked fluctuations in blood pressure during anesthesia, and often these blood pressure extremes are associated with ECG signs of myocardial ischemia (Prys-Roberts, Meloche, & Foex, 1971). For this reason, antihypertensive drugs, with the exception of the rauwolfia alkaloids and the possible exception of the diuretics, should be continued throughout the ECT treatment period (see section on pre-ECT orders). For patients with poorly controlled hypertension, an internist should be consulted to optimize blood pressure control before ECT treatments are started. Patients with essential hypertension are at increased risk for coronary artery disease until proved otherwise (Kannel, Schwartz & McNamara, 1969). Those with comparable symptoms or other risk factors should be evaluated for this possibility.

After the ECT treatment, control of hypertension may require hydralazine 2.5 to 20.0 mg IV, sublingual nifedipine 10 mg, esmolol 50 to 100 mg IV, or labetalol 5 to 10 mg IV. It may be necessary to continue infusions of the short-acting vasodilators (nitroglycerine or nitroprusside) with frequent blood pressure monitoring until the longer-acting agents have taken effect. Patients who have required intervention with longer-acting antihypertensive agents should have frequent blood pressure measurements until their pressure has stabilized.

Chronic Congestive Heart Failure

Patients with symptoms (dyspnea on exertion, paroxysmal nocturnal dyspnea, pitting ankle or sacral edema) or a history of congestive heart failure (CHF) should be evaluated by a cardiologist prior to undergoing ECT. The cardiologist should strive for optimal control of the patient's heart failure by properly adjusting the patient's medications, which can include cardiac glycosides, diuretics, calcium channel blockers, ACE inhibitors, or vasodilators. The hemodynamic changes associated with ECT could theoretically cause decompensation and pulmonary edema due to the increase in myocardial work. Control of the hypertensive response via the pharmacologic mechanisms listed in the preceding section can help reduce such risks in patients with CHF (Drop & Welch, 1989). Beta-blocking agents can precipitate cardiac failure in CHF patients owing to their negative inotropic effects (Stoelting, 1987a), so vasodilators and calcium channel blockers are preferable for blood pressure control when CHF is present. Intravenous fluids should be given sparingly, and myocardial depressant drugs such as methohexital should be titrated to the minimal doses that provide anesthesia (Drop & Welch, 1989).

Pacemakers

Electroconvulsive therapy has been successfully performed in patients with cardiac pacemakers (Abiuso,

Dunkelman, & Proper, 1978; Gibson et al., 1973; Pitts, 1982). Animal studies have shown that a cerebral electrical stimulus does not influence pacemaker function probably because of the high electrical resistance of body tissues (Youmans et al., 1969). However, if the pacemaker wire insulation is broken or if a low-resistance pathway is created to the heart by improperly grounded electrical devices, an ECT stimulus could cause pacemaker dysfunction (Drop & Welch, 1989). A chest roentgenogram should be obtained prior to ECT to check for electrode placement and electrode fracture. The ECG should be examined for signs of appropriate pacemaker function evidenced by pacing impulses that can be seen if the patient's heart rate is slower than the pacing rate. Unexpected pauses on the ECG are an indication of pacemaker malfunction (Vandam & Desai, 1989). Pacing impulses should be associated with a myocardial contraction. Palpation of a peripheral pulse while the heart is paced is a simple test of pacemaker function (Zaidan, 1989). Prior to performing ECT on a patient with a pacemaker, the type of pacemaker should be determined. The patient's cardiologist should be contacted for the necessary information, and pre-ECT assessment of pacemaker function by a cardiologist should be considered. Atropine and isoproterenol should be immediately available should artificial cardiac pacemaker function cease during treatment (Stoelting & Miller, 1984).

The *demand pacemaker* is the most common type in use today. Certain considerations apply in the treatment of patients with this type of pacemaker. There should be proper grounding of the ECT device and the ECG monitor. Adequate doses of succinylcholine should be used to prevent seizure-induced muscle potentials of sufficient magnitude to inhibit the demand pacemaker. Three minutes prior to using succinylcholine, a dose of *d*-tubocurarine should be given 3 mg/70 kg IV to minimize succinylcholine-induced muscle fasiculations, which may themselves shut off a demand pacemaker. Patients with pacemakers should not be excessively hyperventilated as it would cause potassium influx into the myocardial cells, producing a further decrease in the resting membrane potential. This situation could produce noncapture of the pacemaker beat, since more current would be required to depolarize the myocardium. A nonreprogrammable demand pacemaker can be converted to a fixed (asynchronous) mode by placing a ring magnet over the pulse generator. This measure eliminates possible interference of pacemaker function by the ECT stimulus, and the fixed-rate mode may avoid ECT-induced severe bradycardia (Drop & Welch, 1989; Selvin, 1987).

More and more patients are now receiving *programmable pacemakers*. A magnet should *not* be used with this type as it may reprogram the pacemaker to an unknown rhythm or no rhythm (Stoelting, Dierdorf, & McCammon, 1988). A cardiologist should be consulted to advise management of patients with this type of pacemaker. An external transthoracic pacer may be needed and so should be available during treatment of these patients.

Automatic Defibrillators

An increasing number of patients have received implantable automatic defibrillators. The now rare circumstance of ECT in these patients is likely to become more common with time. The single short ECT electrical stimulus is unlikely to cause the device to trigger and stimulate the myocardium. However, the tachycardia associated with the ECT stimulus may be interpreted as a ventricular arrhythmia by the device, causing it to fire if the heart rate exceeds 155 beats/minute. For this reason, it may be necessary to deactivate the device just prior to ECT once the patient has been placed on an ECG monitor. Automatic defibrillator models 1510 and 1520 must be deactivated and reactivated with a magnet. Automatic defibrillator models 1550 and 1600 must be interrogated and deactivated with a programmer. A cardiologist familiar with the defibrillator and correct model number must be available for appropriate management of the situation prior to the first ECT treatment. Short-acting beta-blocking drugs may be indicated to prevent excessive tachycardia (Cardiac Pacemakers Inc. Technical Specialist, personal communication, 1991). Suggested beta-blocking agents and appropriate dosages were discussed above in the section on recent myocardial infarction and unstable angina.

Unstable Aortic Aneurysm

Successful ECT in a patient with an unstable aortic aneurysm depends on careful control of blood pressure and heart rate to prevent increased intramural pressure on the weakened aortic wall (Pomeranze et al., 1968). Also, one should be aware that even asymptomatic patients with aortic aneurysms have a high incidence (30% to 60%) of severe coronary atherosclerotic heart disease (Jeffrey et al., 1983). The means to achieve cardiovascular stability in these patients are similar to those outlined for the patient with unstable cardiac disease, with special emphasis on control of blood pressure. In a patient at risk for aneurysmal rupture, ECT should not be performed without a large-bore intravenous line (16 gauge or larger) in place. The first two treatments should be performed in the vicinity of a prepared operating room with several units of blood readily available. Serious consideration should be given to an arterial line and nitroprusside infusion for tight blood pressure control. Because individual patients tend to respond similarly to each ECT in a treatment series, once the patient's response pattern and treatment mode have been determined these intensive precautions, with exception of the large-bore intravenous line, may no longer be necessary.

Pheochromocytoma

If ECT must be performed in the presence of a known pheochromocytoma, an endocrinologist must be consulted to stabilize the patient's condition as much as possible prior to starting ECT therapy. Patients with untreated pheochromocytomas are at risk for cerebrovascular hemorrhage, heart failure, dysrhythmias, or MI due to severe hypertensive crises. Evaluation for pheochromocytoma is discussed in Chapter 29.

Alpha-methyltyrosine may be instituted to reduce catecholamine synthesis by inhibition of tyrosine hydroxylase, which controls the rate-limiting step in catecholamine synthesis. Patients may also be started on a regimen of phenoxybenzamine, a long-acting alpha-blocking agent, started in 10-mg doses tid and increased incrementally until the blood pressure is controlled. *After* alpha-blockade is established, beta-blocking agents are added. Dehydration and plasma volume reduction may occur owing to severe hypertension and renal compensation. Patients must be hydrated carefully, as life-threatening hypotension can occur in response to anesthetic induction agents (Graf & Rosenbaum, 1989). For the first treatments, an arterial line monitor should be established with a nitroprusside infusion available for tight blood pressure control. Phentolamine, a short-acting alpha-antagonist, may also be used in 2- to 5-mg boluses to aid in blood pressure control. Esmolol infusions (discussed above) can be used for control of heart rate (Graf & Rosenbaum, 1989).

Pulmonary Disease

Patients with severe pulmonary disease should have their respiratory function maximized prior to ECT. Patients taking medications for pulmonary disease must have these medications continued during their hospitalization for psychiatric treatment. Patients with asthma and bronchospasm who have had steroid treatment in the past may require renewal of their steroid medications. Aerosol treatments and oral bronchodilators should be continued, possibly requiring evaluation and regulation of medications by a consulting pulmonologist. Several studies have shown that preoperative optimization of patients with pulmonary disease undergoing general anesthesia decreases perioperative respiratory complications (Brown, 1986). Currently, spirometry and arterial blood gas analysis seem to be the most reliable and cost-effective means for evaluating pulmonary risk (Boysen, 1988).

Ongoing reversible airway obstruction and wheezing may require treatment prior to ECT with corticosteroids such as hydrocortisone in a 1 to 3 mg/kg dose. Selective beta$_2$-agonist agents such as albuterol and terbutaline may be used to treat bronchospasm with fewer undesirable side effects (e.g., tachyarrhythmias) seen with agents such as isoproterenol with more beta$_1$ effects. Albuterol can be given orally, but the treatment of choice is two or three puffs of aerosolized albuterol prior to start of the ECT treatment to minimize airway reactivity. Patients with severe bronchospasm may require treatment with aminophylline infusions, a water-soluble salt of theophylline. Prior to starting ECT, aminophylline therapy should be adjusted to obtain serum blood levels in the therapeutic range of 10 to 20 μg/ml. A loading dose of 5.6 mg/kg is recommended followed by an infusion of 0.5 to 0.9 mg/kg/hr to maintain therapeutic levels. Anticholinergic agents such as atropine and glycopyrrolate may be useful in some patients, particularly those with chronic obstructive pulmonary disease with a reactive component. Administration of these drugs via aerosols may avoid some of the undesirable cardiovascular and central nervous system effects. However, these drugs are less effective than the beta$_2$-adrenergic agonists, and they actually plan a minor role in the treatment of pulmonary disease (Hirshman, 1988).

H_2-receptor antagonists, such as cimetidine and ranitidine, may precipitate bronchospasm owing to their inhibition of H_2-receptor-mediated bronchodilation (Stoelting, 1987b). The benefits of using these drugs to increase gastric pH to avoid aspiration pneumonia should be weighed against the severity of the patient's reactive airway disease (Hirshman, 1988). Ranitidine may have a milder effect on airway reactivity than cimetidine (Kastrup, 1991).

Short-acting narcotic agents should be used sparingly in the patient with severe pulmonary disease because of their depressant effect on respiration. Thiopental has been shown to have no increased incidence of bronchospasm in nonintubated patients over patients receiving regional anesthesia (Shnider & Papper, 1961), and its use during ECT induction probably is safe in these patients in the absence of intubation. If beta-antagonist drugs are needed to control heart rate in a patient with mild bronchospastic disease during ECT,

$beta_1$-selective drugs such as metoprolol, atenolol, or esmolol should be used to avoid bronchospasm induced by beta-blockade. In patients with severe bronchospastic disease, any beta-blocking agents are probably contraindicated, and their use should be weighed against the severity of the patient's cardiac disease (Merin, 1986).

COMPLICATIONS OF ECT

Morbidity

The complication rate in ECT has been cited to be 1 in 1700 treatments (Frederiksen & D'Elia, 1978). Before the introduction of modified ECT with short-acting muscle relaxants, the most common complication was fracture of thoracic vertebrae or long bones (Scalafani, 1988). Today, the most common complaint by patients undergoing ECT relates to memory disturbances after treatment (Hood & Mecca, 1983), which seems to be exacerbated by bilateral electrode placement (Squire & Slater, 1978). Skin burns, lacerations of the tongue and oral mucosa, and tooth and eye damage have been described in situations where vigilance by attending personnel has been lax (Scalafani, 1988; Selvin, 1987). Headaches, muscle aches, and anxiety with impending treatment are common complaints. In our institution, where small doses (300 to 600 μg) of alfentanil are used with decreased barbiturate doses for anesthesia induction, nausea has been a problem unless small preinduction doses of droperidol 0.625 mg IV are administered.

Mortality

Mortality due to ECT is considered rare (Scalafani, 1988; Selvin, 1987). Four large studies since the mid-1970s reported no deaths directly attributable to ECT (Rich & Smith, 1981). Current mortality is quoted as 0.03%, or 3 per 10,000 of treated patients, or 0.0045% of individual treatments (Fink, 1979). Four deaths among 2594 patients were reported in a 1984 study from the United Kingdom (Marks, 1984). One study of ECT in the United States cited an incidence of 0.2 deaths per 10,000 treatments (Kramer, 1985). Certainly, these statistics compare favorably to the most recent data citing anesthesia mortality among hospitalized surgical patients as 0.9 deaths per 10,000 cases (Kennan & Boyan, 1985). It should be remembered, however, that these statistics include patients undergoing major surgery and high risk procedures. To compare patients undergoing ECT, a relatively "minor" procedure to a group containing patients undergoing abdominal aortic aneurysm repair or coronary artery bypass surgery may not be appropriate. Rather, we should demand that ECT mortality approach that of ambulatory patients undergoing minor procedures, which was recently quoted as 0.016 deaths per 10,000 cases (Green & Taylor, 1984).

The most frequent case of death during ECT is due to the cardiovascular complications in the form of arrhythmias, myocardial infarction, congestive heart failure, and cardiac arrest. These complications often occur during the recovery period of ECT (Alexopoulos et al., 1984; Gerring & Shields, 1982; Weiner, 1979). Cardiac arrhythmias and myocardial ischemia are not uncommon complications during ECT, with an incidence varying between 8% and 80%, the higher incidence being seen among patients with preexisting cardiac disease (Alexopoulos et al., 1984; Gerring & Shields, 1982; Kitamura & Page, 1984; Pitts, 1982). During the active convulsion, sinus and ventricular tachycardias and PVCs (often multifocal, bigeminy or trigeminy) may occur (Selvin, 1987), possibly related to the arrhythmogenic effects of the elevated circulating catecholamines. The incidence and severity of complications appear to be increased in patients who are hypoxic or hypercapnic, or who have respiratory acidosis. Elderly, hypertensive patients with cardiovascular disease or who are on digitalis or antiarrhythmic or diuretic therapy have a higher incidence of complications (Gerring & Shields, 1982; Kraus & Remick, 1982; Weiner, 1979). Patients also tend to exhibit the same arrhythmia after each treatment (Selvin, 1987).

The best course of action to follow when treating medically frail patients with ECT is to monitor heart rhythm, blood pressure, and oxygenation throughout the treatment until the patient is well awake and stable. (Return to awake status usually requires 30 to 45 minutes of observation after the ECT treatment is finished.) A crash cart with full resuscitative supplies and defibrillator should be readily available. Lidocaine 2% should be ready in a syringe at the patient's bedside in a dose of 2 mg/kg and an individual trained in Advanced Cardiac Life Support (ACLS) should be available within seconds of the ECT area.

CONCLUSION

Electroconvulsive therapy is experiencing a resurgence and is reemerging as a viable treatment modality for refractory depression. As the population ages, with the accompanying increased incidence of serious disease and depression, the clinician performing ECT will be required to treat these severely depressed patients regardless of their increased risk. Meticulous preparation and care of the patient before, during, and after ECT minimizes these risks.

REFERENCES

Abiuso P, Dunkelman R, & Proper M (1978). Electroconvulsive therapy in patients with pacemakers. JAMA 240:2459–2460.

Abrams JJ, & Roberts J (1983). First American conference on nitroglycerine therapy. Am J Med 74:1–66.

Abrams R (Ed) (1988). Electroconvulsive therapy. New York: Oxford University Press.

Adams RD, & Martin JB (1987). Faintness, syncope, and seizures. In E Braunwald, KJ Isselbacher, RG Petersdorf, et al. (Eds.), Harrison's principles of internal medicine (pp. 64–70). New York: McGraw-Hill.

Adams RD, & Victor M (Eds.) (1977). Principles of neurology. New York: McGraw-Hill.

Alexopoulos GS, Shamoian CJ, Lucas J, et al. (1984). Medical problems of geriatric patients and younger controls during electroconvulsive therapy. J Am Geriatr Soc 32:651–654.

Alexopoulos GS, Nasr H, Young RC, et al. (1982). Electroconvulsive therapy in patients on anticoagulants. Can J Psychiatry 27:46–47.

Alexopoulos GS, Young RC, & Abrams RC (1989). ECT in the high-risk geriatric patient. Convulsive Ther 5:75–87.

American Psychiatric Association, Task Force on ECT (1990). The practice of ECT: Recommendations for treatment, training and privileging. Convulsive Ther 6:85–120.

Anderson WD, & Kissane JM (1977). Pathology. St Louis: Mosby.

ASA Newsletter (1989). Standards for gasic intra-operative monitoring (approved by House of Delegates on October 21, 1986) 53:6.

Azar I, & Lear E (1984). Cardiovascular effects of electroconvulsive therapy in patients taking tricyclic antidepressants. Anesth Analg 63:1139.

Black HA, Howie MB, Martin DJ, et al. (1989). Attenuation of electroconvulsive (ECT) autonomic hyperactivity by esmolol. Anesth Analg 68:S30.

Boysen PG (1988). Pulmonary function testing. In MC Rogers (Ed.), Current practice in anesthesiology (pp. 9–11). Toronto: BC Decker.

Braunwald E. (1987). Heart failure. In E Braunwald, KJ Isselbacher, RG Petersdorf, et al. (Eds.), Harrison's principles of internal medicine, (pp. 905–916). New York: McGraw-Hill.

Brown M (1986). Assessment and treatment of the patient with hypoxemia (lecture 132). In American Society for Anesthesiologists (Eds.), 1986 Annual Refresher Course Lectures, American Society of Anesthesiologists (pp. 1–6).

Bruce DA (1983). Head trauma: management. In P Newfield, JE Cottrell (Eds.), Handbook of neuroanesthesia: Clinical and physiological essentials (Ch. 13, pp. 283–301). Boston: Little, Brown.

Bynum LJ, & Pierce AK (1976). Pulmonary aspiration of gastric contents. Am Rev Respir Dis 114:1129–1136.

Ciraulo D, Lind L, Salzman C, et al. (1978). Sodium nitroprusside treatment of ECT-induced blood pressure elevations. Am J Psychiatry 135:1105–1106.

Cottrell JE, Robustelli A, Post K, et al. (1977). Furosemide- and mannitol-induced changed in intracranial pressure and serum osmolarity and electrolytes. Anesthesiology 47:28–30.

Dec DW, Stern TA, & Welch C (1985). The effect of electroconvulsive therapy on serial electrocardiograms and serum cardiac enzyme values. JAMA 253:2525–2529.

Decina P, Malitz S, Sackeim A, et al. (1984). Cardiac arrest during ECT modified by beta-adrenergic blockade. Am J Psychiatry 141:298–300.

DeSalles AA, Muizelaar JP, & Young HF (1987). Hyperglycemia, cerebrospinal fluid lactic acidosis, and cerebral blood flow in severely head-injured patients. Neurosurgery 21:45–50.

Dressler DM, & Folk J (1975). The treatment of depression with ECT in the presence of brain tumor. Am J Psychiatry 132:1320–1321.

Drop LJ, & Welch CA (1989). Anesthesia for electroconvulsive therapy in patients with major cardiovascular risk factors. Convulsive Ther 5:88–101.

Drop LJ, Bouckoms AJ, & Welch CA (1988). Arterial hypertension and multiple cerebral aneurysms in a patient treated with electroconvulsive therapy. J Clin Psychiatry 49:280–282.

El-Ganzouri AR, Ivankovich AD, Braverman B, et al. (1985). Monoamine oxidase inhibitors: should they be discontinued preoperatively? Anesth Analg 64:592–596.

Epstein HM, Fagman W, Bruce DL, et al. (1975). Intraocular pressure changes during anesthesia for electroshock therapy. Anesth Analg 54:479–481.

Fink M (1979). Convulsive therapy: Theory and practice. New York: Raven Press.

Foster DW (1987). Diabetes mellitus. In E Braunwald, KJ Isselbacher, RG Petersdorf, et al. (Eds.), Harrison's principles of internal medicine (pp. 1778–1797). New York: McGraw-Hill.

Foster DW, & Rubenstein AH (1987). Hypoglycemia, insulinoma, and other hormone-secreting tumors of the pancreas. In E Braunwald, KJ Isselbacher, RG Petersdorf, et al. (Eds.), Harrison's principles of internal medicine (pp. 1800–1807). New York: McGraw-Hill.

Frederiksen S, & d'Elia G (1978). Electronconvulsive therapy in Sweden. Br J Psychiatry 134:283–287.

Gaines GY, & Rees DI (1986). Electroconvulsive therapy and anesthetic considerations. Anesth Analg 65:1345–1356.

Gal TJ (1986). Pulmonary function testing. In RD Miller (Ed.), Anesthesia, (2nd Ed., Vol. 3; Ch. 59, pp. 2053–2075). New York: Churchill Livingstone.

Gerring JP, & Shields HM (1982). The identification and management of patients with a high risk for cardiac arrhythmias during modified ECT. J Clin Psychiatry 43:140–143.

Gibson TC, Leaman DM, Devors J, et al. (1973). Pacemaker function in relation to electroconvulsive therapy. Chest 63:1025–1027.

Giesecke AH Jr, & Egbert LD (1986). Perioperative fluid therapy—crystalloids. In RD Miller (Ed.), Anesthesia (2nd Ed., Vol. 2; Ch. 38, pp. 1313–1328). New York: Churchill Livingstone.

Goldman L, Caldera DL, Nussbaum SR, et al. (1977). Multifactorial index of cardiac risk in non-cardiac surgical procedures, N Engl J Med 297:845–850.

Goodloe SL (1983). Essential hypertension. In RK Stoelting & SF Dierdorf (Eds.), Anesthesia and Coexisting disease (Ch. 6, pp. 99–117). New York: Churchill Livingstone.

Gould L, Gopalaswamy C, Chandy F, et al. (1983). Electroconvulsive therapy-induced ECG changes simulating a myocardial infarction. Arch Intern Med 143:1786–1787.

Graf G, & Rosenbaum S (1989). Anesthesia and the endocrine system. In P Barash, BF Cullen, & RK Stoelting (Eds.), Clinical anesthesia (Ch. 44, pp. 1185–1214). Philadelphia: JB Lippincott.

Gravenstein JS, Anton AH, Wiener SM, et al. (1965). Catecholamine and cardiovascular response to electro-convulsive therapy in man. Br J Anaesth 37:833–839.

Green RA, & Taylor TH (1984). An analysis of anesthesia medical liability claims in the United Kingdom 1977–1982. Int Anesthesiol Clin 22:73–90.

GUTIN PH (1977). Corticosteroid therapy in patients with brain tumors. Natl Cancer Inst Monogr 46:151–156.

HAY DP (1987). Anticoagulants and ECT. Convulsive Ther 3:236–237.

HAY DP (1989). Electroconvulsive therapy in the medically ill elderly. Convulsive Ther 5:8–16.

HIRSHMAN CA (1988). Perioperative management of the patient with asthma (lecture 264). In American Society of Anesthesiologists (Eds.), 1988 Annual Refresher Course Lectures, American Society of Anesthesiologists (pp. 1–7).

HOOD DD, & MECCA RS (1983). Failure to initiate electroconvulsive seizures in a patient pretreated with lidocaine. Anesthesiology 58:379–381.

HSIAO JK, & EVANS DL (1984). ECT in a depressed patient after craniotomy. Am J Psychiatry 141:442–444.

HSIAO JK, MESSENHEIMER JA, & EVANS DL (1987). ECT and neurological disorders. Convulsive Ther 3:121–136.

HURWITZ TD (1974). Electroconvulsive therapy; a review. Compr Psychiatry 15:303–314.

HUSUM B, VESTER-ANDERSEN T, BUCHMANN G, ET AL. (1983). Electroconvulsive therapy and intracranial aneurysm. Anaesthesia 38:1205–1207.

JANOWSKY EC, RISCH SC, & JANOWSKY DS (1981). Psychotropic agents. In NT SMITH, RD MILLER, & AN CORBASCIO (Eds.). Drug interactions in anesthesia (pp. 177–195). Philadelphia: Lea & Febiger.

JEFFERIES BF, KISHORE PRS, SINGH KS, ET AL. (1980). Postoperative computed tomographic changes in the brain. Radiology 135:751–753.

JEFFREY CC, KUNSMAN J, CULLEN DJ, ET AL. (1983). Routine coronary angiography prior to elective aortic reconstruction. Anesthesiology 58:462.

KALAYAM B, & ALEXOPOULOS GS (1989). Nifedipine in the treatment of blood pressure rise after ECT. Convulsive Ther 5:110–113.

KALLOS T, & SMITH TC (1989). In P BARASH, BF CULLEN, & RK STOELTING (Eds.). Clinical anesthesia (pp. 1163–1184). Philadelphia: JB Lippincott.

KANNEL WB, SCHWARTZ MJ & MCNAMARA PM (1969). Blood pressure and risk of coronary heart disease. The Framingham Study. Dis Chest 56:43–52.

KASTRUP EK (Ed.) (1991). Drug facts and comparisons. St. Louis: Facts and Comparisons, Inc.

KENDELL RE (1981). The present status of electroconvulsive therapy. Br J Psychiatry 139:265–283.

KENNAN RL, & BOYAN CP (1985). Cardiac arrest due to anesthesia; a study of incidence and causes. JAMA 253:2373–2377.

KERR RA, MCGRAITH JJ, O'KEARNEY RT, ET AL. (1982). ECT: misconceptions and attitudes. Aust NZ J Psychiatry 16:43–49.

KITAMURA T, & PAGE AJF (1984). Electrocardiographic changes following electroconvulsive therapy. Eur Arch Psychiatry Neurol Sci 234:147–148.

KLEIN M, MARTIN D, SOLOFF P, ET AL. (1988). Comparison of the effect of nitroglycerin and esmolol on the cardiovascular response to ECT. Anesthesiology 69:A41.

KNOS GB, SUNG YF, STOUDEMIRE A, ET AL. (1990). Use of labetalol to control cardiovascular responses to electroconvulsive therapy. Anesth Analg 70:S210.

KNOS GB, & SUNG YF (1991). Anesthetic management of the high-risk medical patient receiving electroconvulsive therapy. In A STOUDEMIRE & BS FOGEL (Eds.), Medical psychiatric practice (Vol 1; Ch. 3, pp. 99–144). Washington, DC: American Psychiatric Press.

KNUDSEN L, CHRISTIANSEN LA, & LORENTZEN JE (1981). Hypotension during and after operation in glucocorticoid-treated patients. Br J Anaesth 53:295–301.

KOFKE WA, & FIRESTONE LL (1988). Commonly used drugs—Table A-1. In LL FIRESTONE & CE COOK (Eds.). Clinical anesthesia procedures of the Massachusetts General Hospital (pp. 590–650). Boston: Little, Brown.

KOVAC AL, UNRUH GK, GOTO H, ET AL. (1988). Evaluation of esmolol infusion in controlling increases of heart rate and blood pressure during electroconvulsive therapy. Anesthesiology 69:A895.

KRAMER BA (1985). Use of ECT in California, 1977–1983. Am J Psychiatry 142:1190–1192.

KRAUS R, & REMICK R (1982). Diazoxide in the management of severe hypertension after electroconvulsive therapy. Am J Psychiatry 139:504–505.

LEE JT, ERBGUTH PH, STEVENS WC, ET AL. (1985). Modification of electroconvulsive therapy induced hypertension with nitroglycerin ointment. Anesthesiology 62:793–796.

LICHTIGER M, WETCHLER BV, & PHILIP BK (1985). The adult and geriatric patient. In BV WETCHLER (Ed.), Anesthesia for ambulatory surgery (Ch. 5, pp. 175–224). Philadelphia: JB Lippincott.

LIEBERMAN A, HASS WK, PINT R, ET AL. (1978). Intracranial hemorrhage and infarction in anticoagulated patients with prosthetic heart valves. Stroke 9:18–28.

LOO H, CUCHE H, & BENKELFAT C (1985). Electroconvulsive therapy during anticoagulant therapy. Convulsive Ther 1:258–262.

LUNN RJ, SAVAGEAU MM, BEATTY WW, ET AL. (1981). Anesthetics and electroconvulsive therapy seizure duration; implications for therapy from a rat model. Biol Psychiatry 16:1163–1175.

MALETZKY BM (1978). Seizure duration and clinical effect in electroconvulsive therapy. Compr Psychiatry 19:541–550.

MALTBIE AA, WINGFIELD MS, VOLOW MR, ET AL. (1980). Electroconvulsive therapy in the presence of brain tumor: case reports and evaluation of risks. J Nerv Ment Dis 168:400–405.

MARKS RJ (1984). Electroconvulsive therapy; physiological and anaesthetic considerations. Can Anaesth Soc J 31:541–548.

MCKENNA G, ENGLE RP, BROOKS H, ET AL. (1970). Cardiac arrhythmias during electroshock therapy; significance, prevention and treatment. Am J Psychiatry 127:530–533.

MCPHERSON RW, & LIPSEY JR (1988). Electroconvulsive therapy. In MC ROGERS (Ed.), Current practice in anesthesiology (pp. 212–217). Toronto: BC Decker.

MERIN RG (1986). Pharmacology of the autonomic nervous system. In RD MILLER (Ed.), Anesthesia (2nd ed., Vol 2; Ch. 28, pp. 945–982). New York: Churchill Livingstone.

NEWFIELD P (1987). Anesthetic considerations in patients with increased intracranial pressure (lecture 176). In American Society of Anesthesiologists (Eds.), 1987 Annual Refresher Course Lectures, American Society of Anesthesiologists (pp. 1–7).

PEERLESS SJ (1983). Intracranial aneurysms: neurosurgery. In P NEWFIELD & JE COTTRELL (Eds.), Handbook of neuroanesthesia: Clinical and physiologic essentials (Ch. 8, pp. 173–183). Boston: Little, Brown.

PITTS FN (1982). Medical physiology of ECT. In R ABRAMS & WB ESSMAN (Eds.), Electroconvulsive therapy: Biological foundations and clinical applications (pp. 57–89). New York: Spectrum Publications.

POMERANZE J, KARLINER W, TRIEBEL WA, ET AL. (1968). Electroshock in the presence of serious organic disease: Depression and aortic aneurysm. Geriatrics 23:122–124.

PRYS-ROBERTS C, MELOCHE R, & FOEX P (1971). Studies of anesthesia in relation to hypertension: I. Cardiovascular responses to treated and untreated patients. Br J Anaesth 43:122–137.

PULSINELLI WA, LEVY DE, SIGSBEE B, ET AL. (1983). Increased damage after ischemic stroke in patients with hyperglycemia with or without established diabetes mellitus. Am J Med 74:540–544.

RAO G, FORD WB, ZIKRIA EA, ET AL. (1974). Prevention of arrhythmias after direct myocardial revascularization surgery. Vasc Surg 8:82–89.

RAO TLK, JACOBS KH, & EL-ETR AA (1983). Reinfarction following anesthesia in patients with myocardial infarction. Anesthesiology 59:499–505.

RICH CL, & SMITH NT (1981). Anaesthesia for electroconvulsive therapy: a psychiatric viewpoint. Can Anaesth Soc J 28:153–157.

ROIZEN MF (1986). Anesthetic implications of concurrent diseases. In RD MILLER (Ed.), Anesthesia (2nd Ed., Vol. 1; Ch. 9, pp. 255–357. New York: Churchill Livingstone.

SCALAFANI SG (1988). The patient for electroconvulsive therapy (pp. 9–18). Anesthesiol News.

SELVIN BL (1987). Electroconvulsive therapy—1987. Anesthesiology 67:367–385.

SHNIDER SM, & PAPPER EM (1961). Anesthesia for the asthmatic patient. Anesthesiology 22:886–892.

SILVERSTEIN A (1979). Neurological complications of anticoagulation therapy: a neurologist's review. Arch Intern Med 139:217–220.

SQUIRE LR, & SLATER PC (1978). Bilateral and unilateral ECT: effects on verbal and nonverbal memory. Am J Psychiatry 135:1316–1360.

STACK CG, ROGERS P, & LINTER SPK (1988). Monoamine oxidase inhibitors and anesthesia. Br J Anaesth 60:222–227.

STEEN PA, TINKER JH, & TARHAN S (1978). Myocardial reinfarction after anesthesia and surgery. JAMA 239:2566–2570.

STOELTING RK (1986). Psychological preparation and preoperative medication. In RD MILLER (Ed.), Anesthesia (2nd Ed., Vol. 1; Ch. 11, pp. 381–397. New York: Churchill Livingstone.

STOELTING RK (Ed.) (1987a). Alpha and beta-adrenergic receptor antagonists. In Pharmacology and physiology in anesthetic practice (Ch. 14, pp. 380–293). Philadelphia: JB Lippincott.

STOELTING RK (Ed.) (1987b). Histamine and histamine receptor antagonists. Pharmacology and physiology in anesthetic practice (Ch. 21, pp. 373–384). Philadelphia: JB Lippincott.

STOELTING RK, DIERDORF SF & MCCAMMON RL (1988). Anesthesia and co-existing disease (2nd ed.). New York: Churchill Livingstone.

STOELTING RK, & MILLER RD (1984). Basics of anesthesia. New York: Churchill Livingstone.

STOUDEMIRE A, Knos G, GLADSON M, ET AL. (1990). Labetalol in the control of cardiovascular responses to electroconvulsive therapy in high risk depressed medical patients. J Clin Psychiatry 51:503–512.

TANCER ME, & EVANS DL (1989). Electroconvulsive therapy in geriatric patients undergoing anticoagulation therapy. Convulsive Ther 5:102–109.

TASCH M (1983). Endocrine diseases. In R STOELTING & SF DIERDORF (Eds.), Anesthesia and co-existing disease (Ch. 23, pp. 437–483). New York: Churchill Livingstone.

VANDAM LD, & DESAI SP (1989). Evaluation of the patient and preoperative preparation. In PG BARASH, BF CULLEN, & RK STOELTING (Eds.), Clinical anesthesia. (Ch. 14, pp. 407–438. Philadelphia: JB Lippincott.

WALLER JL, KAPLAN JA, & JONES EL (1979). Anesthesia for coronary revascularization. In JA KAPLAN (Ed.), Cardiac anesthesia. (Ch. 7, pp. 241–280). Orlando, FL: Grune & Stratton.

WEINER RD (1979). The psychiatric use of electrically-induced seizures. Am J Psychiatry 136:1507–1517.

WEINER RD, & COFFEY CE (1987). Electroconvulsive therapy in the medically ill. In A STOUDEMIRE & BS FOGEL (Eds.), Principles of medical psychiatry (pp. 113–134). Orlando, FL: Grune & Stratton.

WELLS DG, & BJORKSTEN AR (1989). Monoamine oxidase inhibitors revisited. Can J Anaesth 36:64–74.

YOUMANS CR, BOURIANOFF G., ALLENSWORTH DC, ET AL. (1969). Electroshock therapy and cardiac pacemakers. Am J Surg 118:931–1937.

ZAIDAN JR (1989). Electrocardiography. In PG BARASH, BF CULLEN, & RK STOELTING (Eds.), Clinical anesthesia (Ch. 22, pp. 587–623. Philadelphia: JB Lippincott.

12 Intensive care

MICHAEL G. GOLDSTEIN, M.D., AND
SCOTT D. HALTZMAN, M.D.

Medical patients with life-threatening illnesses and surgical patients recovering from cardiac, neurosurgical, or other major surgical procedures are likely to experience emotional distress and develop disturbances of thinking and behavior. The high-technology setting of intensive care, with its noise, bright lighting, lack of privacy, impersonality, and absence of environmental cues may exacerbate the patients' difficulties. The intensive care unit staff, too, are stressed by the high degree of responsibility, fast pace, and often grave nature of their patients' conditions.

Most tertiary care hospitals divide critical care facilities into two, often physically separated units: the cardiac care unit (CCU), which includes patients admitted for evaluation of myocardial disease or treatment of acute cardiac injury, and the intensive care units (ICUs), which treat various medical and surgical conditions.

The first section of this chapter describes psychiatric aspects of patient management in the medical intensive care setting, focusing on the CCU. Next, the psychiatric aspects of the management of patients undergoing cardiac surgery are discussed as an example of typical problems in the care of critically ill surgical patients. Subsequent sections describe the general assessment and management of delirium, brief reactive psychosis, stress response syndromes and other common psychiatric disorders in the intensive care setting, the management of patients receiving mechanical ventilation, the management of patients undergoing organ transplantation, and the evaluation and management of ICU staff issues.

CARDIAC CARE UNIT

It has been estimated that 1.5 million people have a myocardial infarction (MI) each year in the United States, and about 1 million survive the acute event (American Heart Association, 1991). A broad range of psychological and psychiatric reactions may occur following an MI. The CCU is the setting for many of these reactions. Such reactions and possible therapeutic interventions are discussed in this chapter. The other phases of care of these patients are discussed in Chapter 23.

ANXIETY

Anxiety is ubiquitous during the first 24 hours of CCU treatment and is most intense just after admission. Many factors contribute to the experience of anxiety, including fears of death or disability, misconceptions about the meaning of an MI and its course, misunderstanding of information provided by staff, misinterpretation of the displays and alarms from cardiac monitors in their rooms, and restriction on ability to perform usual activities, which may be associated with feelings of helplessness and loss of control (Cassem & Hackett, 1971; Krantz, 1980; Stern, 1985). Patients are also concerned about the life problems they were confronting prior to admission and the effects that the illness and hospitalization will have on their ability to handle these problems (Cay et al., 1972; Thomas et al., 1983). Patients who witness a cardiac arrest in a fellow patient may be especially likely to develop anxiety (Stern, 1985).

Anxiety usually diminishes as patients feel more secure with the knowledge that they are being closely monitored. Simply surviving the first hours reduces panic. Some degree of denial, which can be hazardous during the prehospital phase of unstable angina or a myocardial infarction, may actually protect the patient during the CCU phase as it may serve to reduce anxiety and associated cardiovascular stimulation (see below for further discussion of denial). However, transfer from the CCU to a general medical floor or an intermediate care unit is often accompanied by a marked increase in anxiety, attributed by many to the loss of constant supervision and intensive monitoring (Klein et al., 1968; Phillip et al., 1979). Research has demonstrated that the use of a simple algorithm can be used to decide when it is safe to transfer patients from the CCU to nonmonitored beds (Weingarten et al., 1990). This information may be reassuring to CCU patients who are about to undergo such a transfer.

Biological factors also influence the experience of anxiety in the CCU. Patients who are treated with sympathomimetic drugs, such as theophylline or terbutaline for concomitant pulmonary disease, or with isoproterenol for cardiac rhythm disturbances, may develop anxiety as an adverse reaction to these drugs (Thompson et al., 1987). Anxiety may also be a manifestation of alcohol, sedative, opiate, nicotine, or antidepressant withdrawal. The onset of withdrawal symptoms depends on the half-life of the particular drug (see Chapter 8).

Anxiety reactions may be more pronounced in the presence of the Type A behavior pattern (Howard et al., 1990). A low educational level (Guiry et al., 1987) or an anxious family member also appears to correlate with the patient's anxiety state (Frederickson, 1989). Young patients and female patients appear to be at greater risk of developing anxiety in the face of newly discovered cardiac disease (Schocken et al., 1987).

Management

Management of anxiety in the CCU includes both pharmacologic and nonpharmacologic approaches. Nonpharmacologic approaches include providing accurate medical information, explaining the role and meaning of the monitoring equipment, providing emotional support and reassurance, and reinforcing the appropriate use of denial (see the following subsection) (Cohen-Cole, 1985; Cohen-Cole & Bird, 1986a; Mumford, Schlessinger, & Glass, 1982; Thomas et al., 1983). Teaching patients relaxation techniques that they can use as needed may also be an effective strategy to reduce anxiety and enhance self-control and self-efficacy. Patients randomly assigned to brief cognitively oriented psychotherapy in the CCU had fewer manifestations of depression and anxiety as well as shorter hospital stays, fewer medical complications, and less functional disability at follow-up when compared with no-treatment (usual care) controls (Gruen, 1975; Thompson, 1989). Cromwell et al. (1977) found that patients receiving information about their illness had a shorter CCU stay, but only if it was coupled either with an opportunity for participation in their care or for diversion. These observations suggest that if patients are given information about their condition they should also be given something active to do about it.

Studies also suggest that a family member's anxiety may be "transmitted" to the patient (Frederickson, 1989). Staff intervention with family members may help to reduce their fears and misconceptions, improving their capacity to provide support to the CCU patient (Cray, 1989). Attempts at reducing the family's fears may result in a decrease in the patient's anxiety (Thompson, 1989).

Pharmacologic Approaches

Pharmacologic approaches to managing anxiety first entail minimizing the contributions of medical conditions and pharmacologic agents. Chapter 6 describes medical conditions that can mimic anxiety reactions; an evolving MI, hypoglycemia, hypoxia, and pulmonary embolism are common specific causes of anxiety in the CCU. Sympathomimetics, if used, should be maintained at the low range of therapeutic serum levels, as central nervous system toxicity increases with increasing dose. When sympathomimetic agents must be continued, informing the patient about the relation between medication and anxiety symptoms may help the patient to tolerate anxiety. If significant anxiety symptoms persist, anxiolytic medication (e.g., benzodiazepines) may help counter the effects of sympathomimetic agents.

Several authors recommend that anxiolytic medications be used routinely in the CCU (Cassem & Hackett, 1971; Stern, 1985; Stern, Caplan, & Cassem, 1987), but there are no controlled studies assessing their efficacy as being preventive in this setting. Benzodiazepines have long been thought to be effective in reducing or eliminating autonomic reactivity associated with stress (Williams, 1990). Williams suggested that benzodiazepines may reduce cardiac ischemia resulting from the neurochemical change (i.e., increases in plasma norepinephrine, epinephrine, and cortisol) associated with increased arousal. Despite this evidence, we do not recommend *routine* use of benzodiazepines in the CCU because of their potential toxicity (discussed below) and the lack of proved efficacy in the CCU setting.

Although anxiolytics are usually well tolerated in the CCU setting, they may produce excessive sedation, impair cognition, produce rebound effects, or exacerbate delirium (Rickels et al., 1988). Although alprazolam has been found to attenuate the catecholamine response to stress (Stratton & Halter, 1985; Vogel et al., 1984; Williams, 1990), its use is limited by its short duration of action, oral route of administration, and its propensity to produce increased anxiety between dosing intervals (rebound effect). We recommend the use of a benzodiazepine as an adjunct to nonpharmacologic management of anxiety only when there is no evidence of delirium, cognitive impairment, or a history of adverse responses to these drugs. Our own clinical experience supports the choice of lorazepam because of its intermediate half-life, the lack of active metabolites, and the option of parenteral administration. Once stabilized, the patient can easily be switched to multiple daily doses of an oral preparation at an equipotent dose.

Morphine sulfate may additionally be considered for its anxiolytic properties and beneficial hemodynamic properties, especially for the management of patients with acute left heart failure (Guntupalli, 1984). Morphine has minimal long-term applications in the CCU, given the rapid development of tolerance, its tendency to suppress respiration, its unpredictable hypotensive actions, and the low bioavailability of oral preparations. Further discussion of the use of anxiolytics can be found in Chapter 9, on psychopharmacology in the medically ill. (See also Chapter 16.)

Withdrawal States

Withdrawal states and other psychiatric conditions that produce anxiety should also be considered when anxiety develops in the intensive care setting; and when diagnosed, appropriate treatment should ensue. It is especially important to obtain a history of alcohol, sedative, opiate, or nicotine use from the patient or family and to look for physiologic signs of withdrawal if there is any possibility of abuse. The clinician's suspicion of a withdrawal state should increase if an initially calm patient on admission subsequently develops increasing anxiety in association with physiologic signs of alcohol withdrawal (e.g., tremor, increased blood pressure, increased heart rate) or opiate withdrawal (e.g., piloerection, increased gastrointestinal motility, mydriasis). Patients undergoing acute withdrawal from nicotine may also become anxious and irritable and experience insomnia and decreased concentration (Hughes, Higgins, & Hatsukami, 1990). Because nicotine dependence is much more prevalent than other psychoactive substance abuse disorders, especially among a population of patients with cardiovascular disease, nicotine withdrawal probably is the most common withdrawal syndrome encountered in the CCU. Transdermal nicotine patches, recently released in the United States, are likely to improve the management of nicotine withdrawal in the ICU setting (see Chapter 42).

DENIAL

Denial—broadly defined as conscious or unconscious suppression of part or all of the meaning of an event to diminish painful or frightening feelings (Weisman & Hackett, 1961)—is common in patients admitted to the CCU after an MI. The reader should note that this broad definition of denial, used in the consultation-liaison psychiatry literature, differs from the psychoanalytic definition of denial as an unconscious defense mechanism to suppress reality. Denial, however defined, can be an effective mechanism to reduce anxiety and fear (Doehrman, 1977; Krantz, 1980; O'Malley & Menke, 1988). Investigators have found that denial independently predicted rapid medical stabilization of angina (Levenson et al., 1984), and it has been associated with decreased days in the ICU (Levine et al., 1987). Though one study found that denial may protect the patient from death during the immediate postinfarction period (Hackett, Cassem, & Wishnie, 1968), other studies have not confirmed an effect of denial on overall mortality due to MI (Levenson et al., 1984) (see Chapter 23).

During the acute phase of recovery from an MI, denial should not be confronted or challenged. However, information should not be withheld from a patient to facilitate denial, as we believe this practice is unethical. Moreover, information may reduce anxiety associated with uncertainty or worry. Denial may become a problem in the intensive care setting if its leads to impatience and intolerance for restrictions on activity, resistance to medical therapy and recommended procedures, or a desire to sign out of the CCU against medical advice. Struggles between patient and staff over these issues may lead to psychiatric consultation. A combination of education and brief medical psychotherapy usually leads to adequate adherence to the treatment plan (Cohen-Cole, 1985; Cohen-Cole & Bird, 1986a, 1986b). If these efforts are ineffective, a family meeting often is useful, as family members may be able to convince the patient that compliance with treatment is important. Some patients respond to a request to comply "for their family's sake". Chapter 14 discusses such intervention in its section on "personal leverage."

DEPRESSION

Depressive symptoms in the intensive care setting may result from the effects of an acute illness or its treatment, reflect an emotional response to acute stress, or indicate the presence of an adjustment disorder or major depressive disorder (Goldman & Kimball, 1987). Depressed mood in the CCU typically becomes more prominent after the third hospital day, when denial begins to diminish and patients become more aware of the implications of their illness (Cassem & Hackett, 1971). For most patients, dysphoria in the setting of an acute cardiac event is self-limited and does not require pharmacologic intervention. Patients with prolonged stays in the CCU or ICU may develop an adjustment disorder with depressed mood. This disorder usually responds to psychotherapy or to initiation of formal rehabilitation efforts (Razin, 1985; Stern, 1985). Some patients, however, go on to develop a major depressive episode.

As Cohen-Cole has noted in Chapter 4, it may be difficult to make the diagnosis of a major depressive episode in the setting of an acute illness. In such situations, an *inclusive* approach to diagnosis is recommended, counting physical symptoms toward the diagnosis of a major depressive episode. Female patients, patients with low levels of education and those with unskilled occupations appear to be more prone to depressive reactions after an MI (Guiry et al., 1987). One must also consider the possibility that a major depressive episode (or, for that matter, an atypical depression) predated the MI (Carney et al., 1988; Carney, Freedland, & Jaffe, 1990).

Others have reviewed the assessment and treatment of depression in the medical setting (Fogel & Fretwell, 1985; Kathol et al., 1990; Klerman, 1981). Some general guidelines for the management of depression in the intensive care setting follow.

Before initiating pharmacologic treatment for depression, it is important to consider whether any medications or concurrent medical conditions may be contributing to depressive symptoms. Beta-blockers may contribute to the development of depressive symptoms, especially if lipophilic drugs, such as propranolol, are used (Paykel, Fleminger, & Watson, 1982; Petrie, Maffucci, & Woosley, 1982). Less lipophilic beta-blockers, such as atenolol or nadolol, calcium channel blockers, or angiotensin-converting enzyme (ACE) inhibitors are less likely to produce depressive symptoms and should be substituted for lipophilic beta-blockers when possible. If significant symptoms have persisted for more than 2 weeks and if there is no evidence of an organic cause for depression, major depressive episodes should be treated. Because most patients do not stay in the CCU or ICU as long as 2 weeks, antidepressant therapy usually is not initiated in this setting. Though a detailed discussion of pharmacologic treatment of depression is beyond the scope of this chapter, we briefly review issues that relate to choice of pharmacologic agents in the setting of cardiovascular disease.

The type, dose, and timing of psychopharmacological treatment of a major depressive episode, or electroconvulsive therapy (ECT) if needed, depend on the physiological status of the patient (Goldstein & Guttmacher, 1988; Levenson & Friedel, 1985; Stern, 1985) and are discussed in detail in Chapters 9 and 10.

Patients who have been receiving antidepressants when they are admitted to the CCU or ICU present another set of issues. For patients without cardiac conduction disturbances, tachycardia, and orthostatic hypotension in the ICU or CCU, cyclic antidepressants (CyAD) usually can be safely continued, even if congestive heart failure is present (Goldstein & Guttmacher, 1988; Shapiro, 1991; Stoudemire, Fogel, & Gulley, 1991). If a CyAD must be discontinued, it is best to do it gradually because the development of cardiac arrhythmias upon abrupt discontinuation of CyAD has been reported (Babb, Dunlop, & Hoffman, 1990; Regan, Margolin, & Matthew, 1989; Van Sweeden, 1988). Interactions with cardiovascular medications must also be considered. Direct-acting sympathomimetics (e.g., norepinephrine, epinephrine) can be potentiated while patients are taking CyAD, whereas indirect-acting sympathomimetics (e.g., ephedrine, metaraminol) are antagonized by such drugs (Goldstein & Guttmacher, 1988). Diuretics can potentiate the postural hypotensive effects of CyAD, and the anticholinergic and quinidine-like effects of some cardiac medications are potentiated by CyAD that share these properties.

Monoamine oxidase inhibitors (MAOIs) frequently cause orthostitic hypotension and pose a risk for hypertensive crisis if the patient is exposed to tyramine, sympathomimetics, or other pressor agents (Goldstein & Guttmacher, 1988). Therefore these agents are not recommended for the treatment of major depressive episodes in the CCU or ICU.

Although psychostimulants have been advocated as potentially useful psychopharmacologic agents for the treatment of depression in the medically ill (Ayd, 1985; Fernandez et al., 1987; Lingam et al., 1988; Woods et al., 1986), their efficacy as antidepressants remains uncertain (Chiarello & Cole, 1987; Satel & Nelson, 1989; Stoudemire, Fogel, & Gulley, 1991). Masand, Pickett, and Murray (1991) reported that dextroamphetamine and methylphenidate were effective for treating "secondary" depressive symptoms associated with medical illness in a series of 198 acute medical and surgical patients. However, adverse events led to termination of 10% of the trials (Masand, Pickett, & Murray, 1991). We do not recommend the use of psychostimulants to treat major depressive episodes in the CCU or ICU, though they may be used to treat patients in non-intensive-care settings who develop adjustment disorders with depressed mood. If these agents prescribed in the intensive care setting, the treatment team should watch for adverse effects on heart rate and blood pressure. Though alprazolam has been advocated by some authors as a treatment for depression in the medically ill (Freeman et al., 1986; Levy, Davis, & Bidder, 1984), there are not adequate data to assess its clinical utility for treating major depressive episodes. Moreover, potential dependency and withdrawal problems using this drug have been noted earlier.

Electroconvulsive therapy is relatively safe as a somatic treatment of depression in the medically ill (Goldstein & Guttmacher, 1988) (see Chapters 10 and 11) but recent MI, unstable angina, chronic congestive heart failure, pacemakers, unstable aortic aneurysm, severe pulmonary disease, recent cerebrovascular infarction, intracranial aneurysms, and cerebrovascular malformations present special problems (Knos & Sung, 1991). The reader is referred to Knos and Sung's (1991) excellent review of the management of high-risk medical patients undergoing ECT, as well as their chapter in this volume (see Chapter 11), for details regarding the use of this modality in patients requiring care in a CCU or ICU.

DELIRIUM IN THE CCU

Delirium may occur in CCU patients, especially the elderly or those with preexisting organic brain disease (Cassem & Hackett, 1971; Neshkes & Jarvik, 1982). The general assessment and management of delirium is described in Chapter 19. Factors particularly likely to contribute to the development of delirium in the CCU setting are discussed here.

Lidocaine is frequently used in the CCU as treatment for or prophylaxis against ventricular arrhythmias in patients with proved or suspected MI. Systemic side effects have been reported to occur in up to 50% of patients receiving lidocaine in this setting, and central nervous system (CNS) side effects account for most of these (Rademaker et al., 1986; Saravay et al., 1987). Most often, lidocaine produces only minor adverse events, such as dizziness, slurred speech, somnolence, and numbness (Rademaker et al., 1986). Lidocaine toxicity can have a characteristic presentation of "doom anxiety," a morbid sense of impending doom that occurs in as many as 73% of patients with psychiatric reactions to lidocaine (Saravay et al., 1987). A well-controlled study (Rademaker et al., 1986) found the incidence of confusion in CCU patients receiving lidocaine to be 11%, even though patients over age 75 and those with severe heart failure were excluded. More than half of the occurrences of confusion were severe enough to warrant alteration of therapy. Symptoms were much more common during the first 12 hours of administration than during the subsequent 36 hours.

Tocainide (Bikadoroff, 1987) and mexiletine (Schrader & Bauman, 1986), two oral cogeners of lidocaine, may also precipitate delirium (Medical Letter, 1989a). Other antiarrhythmics have also been associated with confusional states, including commonly used agents such as procainamide (Kim & Benowitz, 1990), quinidine (Deleu & Schmedding, 1987), and digoxin (Eisendrath & Sweeney, 1987), even at therapeutic levels. Procainamide's toxicity is related to its anticholinergic effects and the toxic effects of its metabolite *N*-acetylprocainamide, which may accumulate in patients with renal failure or in the elderly (Kim & Benowitz, 1990). Amiodarone also has been reported to be associated with delirium (Trohman et al., 1988), some other antiarrhythmics (e.g., propafenone) have not been associated with this event as yet (Medical Letter, 1989a) (see Table 12-1).

Although studies fail to show a consistent relation between the development of CNS toxicity and levels of antiarrhythmic agents, lidocaine-induced delirium is definitely dose-related (Rademaker et al., 1986). This dose relation is likely to be true for other antiarrhythmics as well. Thus reduction in antiarrhythmic dose may be helpful if delirium occurs (Rademaker et al., 1986). (See also Chapter 23.)

Other Medical Causes of Delirium

Delirium may also result from medical complications of acute cardiac illness, such as hypoxia, hypotension, congestive heart failure, and embolic stroke. There appears to be a relation between the severity of congestive heart failure and the degree of mental status impairment (Vrobel, 1989). Therefore, a complete medical and neurological reexamination is indicated when delirium develops suddenly in a post-MI patient. Withdrawal from alcohol, sedatives, and hypnotics may cause delirium as well. The presence of autonomic arousal or seizures should alert the physician to this possibility. Often withdrawal states are overlooked, as complete alcohol and drug histories frequently are omitted when acutely ill cardiac patients are hospitalized.

Treatment of Agitation Associated with Delirium

While the search for an underlying pathological process to explain the delirium is under way, treatment of agitation is essential. The approach to the agitated critically ill patients is described later in this chapter. From an empirical standpoint, haloperidol is the neuroleptic agent of choice for delirious agitation in patients with cardiac disease, unless the patient has a condition that would be exacerbated by its use (Adams, 1988; Tesar & Stern, 1986) (See section below on management of delirium.)

Sequelae of Cardiac Arrest and Cerebrovascular Events

Special mention should be made of the neurologic and psychiatric sequelae of cardiac arrest, which often are

TABLE 12–1. *Pharmacologic Agents Commonly Contributing to Delirium in the Intensive Care Setting*

Class of Agent	Psychiatric Reactions
Anesthetics	
General: halothane	Depression
Local: all	Terror, confusion, psychosis, agitation, bizarre behavior, depression, panic
Anticholinergics	
Atropine and similar agents: belladonna alkaloids	Confusion, memory loss, disorientation, depersonalization, delirium, auditory and visual hallucinations, fear, paranoia, agitation, bizarre behavior
Anticonvulsants	
Barbiturates	Excitement, hyperactivity, visual hallucinations, depression, delirium-tremens-like syndrome
All others	Agitation, confusion, delirium, depression, psychosis, aggression, mania, toxic encephalopathy
Antihistamines	
Nonselective	
Promethazine Diphenhydramine	Anxiety, hallucinations, delirium
H_2blockers	
Cimetidine Ranitidine Famotidine	Hallucinations, paranoia, bizarre behavior, delirium, disorientation, depression, mania
Cardiac agents	
Antiarrhythmics	
Procaine derivatives: lidocaine, procainamide, mexiletine, tocainide	Terror, confusion, psychosis, agitation, bizarre behavior, depression, panic
Amiodarone	Delirium, hallucinations
Quinidine	Confusion, agitation, psychosis
β-adrenergic blockers (especially lipophilic agents): oral and ophthalmic preparations	Depression, confusion, nightmares, hallucinations, paranoia, delusions, mania, hyperactivity
Digitalis glycosides	Nightmares, euphoria, confusion, amnesia, aggression, psychosis, depression
Antihypertensives	
Diltiazem	Depression, suicidal thoughts
Clonidine	Delirium, hallucinations, depression
Methyldopa	Depression, amnesia, nightmares, psychosis
Nifedipine	Irritability, agitation, panic, belligerence, depression
Prazosin	Hallucinations, depression, paranoia
Verapamil	Auditory, visual, and tactile hallucinations
Corticosteroids	Mania, depression, confusion, paranoia, hallucinations, catatonia
Metoclopramide	Mania, severe depression, crying, delirium
Narcotics: all, especially meperidine	Nightmares, anxiety, agitation, euphoria, dysphoria, depression, paranoia, hallucinations
Respiratory agents	
Albuterol	Hallucinations, paranoia
Theophylline	Withdrawal, mutism, hyperactivity, anxiety, mania
Sedative-hypnotics	
Barbiturates	See Anticonvulsants
Benzodiazepines	Rage, hostility, paranoia, hallucinations, depression, insomnia, nightmares, anterograde amnesia

unrecognized (Reich et al., 1983). Though some studies have found little evidence of neurologic sequelae after cardiac arrest with prompt resuscitation (Bedell et al., 1983; Longstreth et al., 1983), careful assessments of cognitive and psychologic status have shown that these patients may develop subtle signs of persistent cognitive impairment or changes in mood and behavior (Reich et al., 1983; Volpe, Holtzman, & Hirst, 1986). Symptoms and signs include fatigue, distractibility, inability to learn new skills, impaired recall, apathy, irritability, petulance, emotional disinhibition, and disturbances of impulse control, insight, empathy, judgment, and social perceptiveness (Reich, et al., 1983; Volpe, Holtzman, & Hirst, 1986). Because of the relative preservation of recognition memory, the absence of motor findings, and the prominent psychologic and behavioral features, patients with subtle organic sequelae from cardiac arrest may have their symptoms attributed to depression or the effects of medication (Reich et al., 1983; Volpe, Holtzman, & Hirst, 1986).

Cerebrovascular events complicating MI may also produce cognitive impairment or alteration in mental status. A study of 740 consecutive admissions to a CCU found a 2.4% incidence of stroke (Komrad et al. 1984). Ninety percent of strokes appear during the first 2 weeks post-MI (Komrad et al., 1984), and the median time to stroke after MI ranges from 4 to 8 days (Levine, 1989).

The increasing utilization of thrombolytic therapy for treatment of MI has made an impact on the course of post-MI complications. Pilot and clinical trials suggest that adverse neurological events occur in up to 2.5% of patients treated with recombinant tissue-type plasminogen (tPA) in combination with aspirin and heparin (Gore et al., 1991). Higher doses of tPA (150 mg versus 100 mg) were associated with a significant increase in the frequency of intracerebral bleeding (1.5% versus 0.4%), but the higher dose of tPA was not associated with a significant increase in the rate of cerebral infarction (Gore et al., 1991). Studies of 20,768 patients randomized to either streptokinase or tPA showed a significantly higher frequency of strokes in tPA-treated patients (1.3% versus 0.9%) (International Study Group, 1990). Patients with a history of cerebrovascular disease were ineligible for most of these studies, but when such patients were enrolled in a large trial of tPA they experienced an increased frequency of cerebral hemorrhage compared with patients without such histories (3.4% versus 0.5%) (Gore et al., 1991). Therefore thrombolytic therapy should be used with caution in patients who have a history of cerebrovascular disease. Increased vigilance for neurologic impairment is warranted in these patients as well as in all elderly patients because they are more likely to have occult cerebrovascular disease. Investigators of thrombolytic therapy have suggested that these therapies are responsible for adverse neurological events early in the course of an MI but also reduce the overall rate of neurological complications due to severe MI by improving the cardiac prognosis (Gore et al., 1991).

Careful cognitive assessment of patients who have had a cardiac arrest or cerebrovascular event is warranted to improve diagnosis and subsequent management (see Chapter 23). The high sensitivity screening examination of Faust & Fogel (1989) may be helpful for assessing subtle cognitive changes that routine screening examinations [such as the Mini-Mental Status Examination (Folstein, Folstein, & McHugh, 1975)] may overlook.

CARDIAC SURGERY

Cardiac surgery has become a common procedure in the United States. Approximately 353,000 patients underwent coronary artery bypass grafting (CABG) in 1988, and tens of thousands more undergo other cardiac operations (American Heart Association, 1991). The experience of these patients in the intensive care setting is described in the following section.

Delirium and Other Neuropsychiatric Sequelae

The incidence of delirium after cardiac surgery is much higher than after general surgery (Goldman & Kimball, 1985). The reported incidence of delirium after cardiac surgery ranges widely from 1.4% to 70% (Breuer et al., 1983; Calabrese et al., 1987; Dubin, Field, & Gastfriend, 1979; Goldman & Kimball, 1985; Koenfeld, Heller, & Frank, 1978; Smith & Dimsdale, 1989). The wide range in the reported incidence of delirium is probably due to methodologic differences among these studies (e.g., different procedures for identifying delirium, variations in the populations studied). One meta-analysis of 44 studies conducted over the last quarter century revealed a relatively constant prevalence of postcardiotomy delirium (32%) over this period of time (Smith & Dimsdale, 1989). Several factors appear to influence the risk of delirium after cardiac surgery. In their meta-analysis, Smith and Dimsdale (1989) found that only 3 of the 28 factors they reviewed were moderately or highly correlated with postcardiotomy delirium. *Noncongenital heart disease and postoperative electroencephalographic (EEG) abnormalities were positively correlated, whereas preoperative psychiatric intervention was negatively correlated, with delirium.* The higher rates of delirium found in patients undergoing surgery for noncongenital versus congenital heart disease may be explained by the noncongenital patients' older age and more severe cardiac dysfunction (Smith & Dimsdale, 1989). This hypothesis is supported by the slightly positive correlation between age and delirium that was also found in the meta-analysis (Smith & Dimsdale, 1989).

The highly significant negative correlation ($r = -.60$) found between preoperative psychiatric intervention and delirium is an important finding that underscores the benefits of routine psychiatric assessment for these patients. Schindler, Shook, and Schwartz (1989) published the results of a randomized controlled trial of the effects of psychiatric intervention in coronary artery bypass graft surgery patients. They found that medical complications were higher and length of hospital stay was longer in the control group compared to the intervention group, through these differences did not reach statistical significance (Schindler, Shook, & Schwartz, 1989).

Also of note in the results of the meta-analysis reported by Smith and Dimsdale (1989) was the lack of correlation between intraoperative variables and delirium. Such variables as length of time on cardiopulmonary bypass, length of time anesthesia was delivered, brain perfusion flow rate/pressure, and hypothermia were not associated with postoperative delirium. Complexity of the procedure was associated with postoperative delirium in two of three published studies, but the estimated correlation between this variable and delirium was small (Smith & Dimsdale, 1989). Several postoperative variables reviewed in the Smith and Dimsdale meta-analysis were only weakly correlated with delirium. They included severity of illness in the recovery room, sleep deprivation, and low cardiac output (Smith & Dimsdale, 1989). A study of the relation between sleep loss and confusion after cardiac surgery found that confusion was correlated with sleep loss on the day after confusion developed, but was not correlated with sleep loss on the day preceding confusion (Harrell & Othmer, 1987). This study suggests that sleep loss is a consequence of delirium in cardiac surgery patients rather than the other way around.

A number of studies have utilized neuropsychologic tests in an attempt to uncover subtle abnormalities in cognitive function after cardiac surgery. These tests can identify cognitive impairment in patients who may have not developed severe enough dysfunction to produce the syndrome of delirium. Significant abnormalities in cognitive functioning develop shortly after cardiac surgery in 30% to 79% of patients (Nevin et al., 1987; Raymond et al., 1984; Savageau et al., 1982a; Shaw et al., 1986a,b; Smith et al., 1986; Sotaniemi, 1983; Townes et al., 1989). Preoperative correlates of significantly reduced test performance include age greater than 67 years (Savageau et al., 1982a; Townes et al., 1989), the presence of congestive heart failure (Shaw et al., 1986a,b), elevated left ventricular pressure (Savageau et al., 1982a; Shaw et al., 1986a), enlarged heart, and the use of propranolol and chlordiazepoxide (Savageau et al., 1982a).

A case-control study found that a higher preoperative level of depression, measured with the Center for Epidemiologic Studies Depression Scale, was significantly associated with the development of postoperative cognitive dysfunction after coronary artery bypass surgery (Folks et al., 1988).

Perioperative correlates of cognitive dysfunction include total time of operation longer than 7 hours, duration of cardiopulmonary bypass pump time, blood loss, hypotension, difficult intubation, insertion of an intra-aortic balloon pump, hypocapna, and low cerebral perfusion pressure (Nevin et al., 1987; Savageau et al., 1982a). Postoperative factors included electrolyte abnormalities, longer stays in the ICU, bizarre behavior or disorientation, and elevated depression scores on the profile of mood states (Savageau et al., 1982a).

Available evidence suggests, however, that neuropsychologic abnormalities are transient in most patients, with fewer than 5% exhibiting persistent deficits months after surgery (Folks et al., 1988; Raymond et al., 1984; Savageau et al., 1982b; Shaw, Bates, & Cartlidge, 1987; Smith et al., 1986). Two follow-up studies have shown that some individuals actually show improvement in cognitive function from preoperative levels after cardiac surgery (Townes et al., 1989; Willner & Rabiner, 1979). Transient neurologic complications occur in most patients after cardiac surgery (Shaw et al., 1985; Treasure et al., 1989), but persistent disability related to impaired neurologic function occurs in fewer than 5% (Shaw et al., 1986b; Treasure et al., 1989). Controlled studies confirm that cardiac surgery produces more neurologic and neuropsychologic abnormalities immediately after surgery than general, thoracic, or major peripheral vascular surgery, but group differences disappear or decrease to small magnitude after several weeks (Raymond et al., 1984; Smith et al., 1986; Treasure et al., 1989). In one prospective study of 100 consecutive patients, the development of psychopathology or persistent cognitive dysfunction after cardiac surgery was a predictor of mortality during the 5 years after surgery (Willner & Rabiner, 1979). However, further studies are needed to confirm this finding.

Several mechanisms have been proposed to explain the increased incidence of neuropsychiatric sequelae after cardiac surgery when compared to general surgery. They include mobilization of atheroma secondary to the altered flow of cardiopulmonary bypass; microbubbles from cardiopulmonary bypass; air embolism; inadequate cerebral perfusion associated with low-flow, prolonged cardiopulmonary bypass in the presence of extracranial and intracranial arterial occlusive disease; disruption of sleep-wake cycles; effects of anesthetics or other pharmacologic agents used intraoperatively or postoperatively; sensory deprivation; and preoperative psychologic state (Bojar et al., 1983; Goldman & Kimball, 1985; Henriksen, 1984; Larson, 1984; Milano & Kornfeld, 1984; Shaw et al., 1986a; Slogoff, Girgis, & Keats, 1982; Treasure et al., 1989).

When assessing neuropsychiatric problems that develop after cardiac surgery, one should be aware of the contributing factors noted above, while keeping in mind that sleep abnormalities may be a result, rather than a cause, of postcardiotomy confusion (Harrell & Othmer, 1987).

Because of the high incidence of neuropsychological complications after cardiac surgery, the psychiatric evaluation of these patients should always include a bedside screen for cognitive dysfunction. The choice of a specific screening test should be based on the patients' level of impairment and their ability to communicate. Patients who are delirious, somnolent, intubated, or in significant pain cannot fully participate in bedside assessments. (See section on assessment of delirium for our recommendations for cognitive screening in patients with this disorder.) For alert patients who are able to communicate through speech and writing, we recommend that the Mini-Mental Status Examination (Folstein, Folstein, & McHugh, 1975) be used as an initial screen. However, the high sensitivity screen developed by Faust and Fogel (1989) or the Neurobehavioral Cognitive Status Examination (Kiernan et al., 1987; Schwamm et al., 1987) are more likely to identify mild or moderate deficits and should be considered when the clinical history suggests subtle neuropsychological impairment. [See Nelson, Fogel, and Faust (1986) for a detailed review of bedside cognitive screening tests.] While the patient is still in the intensive care unit, comprehensive neuropsychological testing is not recommended.

Developments that may minimize confusional states postoperatively include modification in intensive care settings to decrease noise and sensory monotony; recognition of premorbid factors that increase risk of complications; and preoperative preparation of the patient (Goldman & Kimball, 1985; Lazarus & Hagens, 1968; Milano & Kornfeld, 1984; Smith & Dimsdale, 1989).

ASSESSMENT OF DELIRIUM

Several points regarding the assessment of delirium in the intensive care setting need to be emphasized. As noted earlier, *withdrawal states are sometimes overlooked in the intensive care setting* for several reasons (Tesar & Stern, 1986). Obtaining a history of substance abuse may be difficult with a critically ill or unconscious patient. The autonomic signs that accompany withdrawal may be masked by the patient's acute illness or by antihypertensive medication such as beta-blockers. Finally, toxic screens that establish drug or alcohol use frequently are not done on admission; and if they are requested, it may take several days for the results to return. Moreover, there are no laboratory results that confirm the diagnosis of drug withdrawal.

Patients in the ICU are likely to have impairments of several organ systems, any one of which might produce delirium. *Hypoxia*, *hypercapnia*, and *hypotension*, common sequelae in the critically ill, may produce transient or permanent brain injury. The psychiatrist should personally review vital sign records and blood gas reports to evaluate this possibility. It may be necessary for the psychiatric consultant to recommend or initiate further assessment of physiologic or metabolic status if the treating staff has not pursued all potential physiological derangements that may be contributing to delirium.

Factors that would not produce an organic brain syndrome independently may interact additively or synergistically to produce delirium. In particular, the CNS side effects of pharmacologic agents may be additive or synergistic. Many of the drugs frequently used in the ICU can induce delirium or other organic mental disorders. These drugs are listed in Table 12-1. Special attention should be paid to anticholinergic medications, as studies have highlighted the correlation of anticholinergic activity with cognitive impairment (Golinger, Peet, & Tune, 1987; Tollefson, Montague-Clouse, & Lancaster, 1991). Antiarrhythmics (discussed in the section on delirium in the CCU), corticosteroids, narcotics, and sedatives are other likely offending agents.

Screening for delirium by nursing staff can be done by the Confusion Assessment Method (CAM), a screening tool for detecting delirium which can be easily taught to ICU staff members (Inouye et al., 1990). The CAM can be completed in less than 5 minutes and includes items to assess nine clinical features of delirium derived from *DSM-III-R* criteria: acute onset; inattention; disorganized thinking; altered level of consciousness; disorientation; memory impairment; perceptual disturbances; psychomotor agitation or retardation; and altered sleep-wake cycle (Inouye et al., 1990). The first four of these features are incorporated into an algorithm to guide the diagnostic process. The developers of the CAM report high interobserver reliability, convergent validity, sensitivity, and specificity when used with general medicine inpatients and outpatients at a geriatric assessment center (Inouye et al., 1990).

MANAGEMENT OF DELIRIUM

As in any setting, the first step in managing delirium is to attempt to ameliorate or reverse the medical conditions contributing to the alteration in mental state. Simultaneously, abnormal behavior is addressed with nonpharmacologic and pharmacologic interventions.

Nonpharmacologic Interventions

Nonpharmacologic interventions that are particularly useful in the prevention and management of delirium in the intensive care setting are listed in Table 12-2. They are directed at enhancing patients' cognitive function; enhancing communication among patient, family, and staff; preventing self-harm or harm to staff; min-

TABLE 12–2. *Nonpharmacologic Management of Delirium in the Intensive Care Setting*

Goal of Intervention	Methods
Enhance cognitive function	Reorient frequently Clock, calendar, radio, television in room Provide explanations, education
Enhance communication between family and staff	Encourage writing if unable to speak; use letter board, communication devices, hand or blink signals if unable to speak or write Encourage family visitation
Prevent self-harm or harm to staff	Use mittens and restrain, using the least restraints necessary
Minimize environmental stresses	Provide sensory stimulation but limit noise from alarms, equipment Maintain semblance of day–night cycle (i.e., dim lights at night) Transfer to general medical floor as soon as feasible Engage in nonessential care (baths, dressing changes) during the day Preoperative orientation and visit, when possible
Maximize patients' comfort	Control pain adequately Mobilize (i.e., bed to chair) Permit rest, limit unnecessary awakenings Invite family to stay with the patient to reduce suspiciousness and paranoia
Provide support and reassurance	Empathy, opportunity to ventilate, support, reassurance

imizing environmental stresses; maximizing patients' comfort; and providing support and reassurance.

Explanation and education by staff about the illness, procedures, and technical equipment may help reduce patients' confusion. Family visitation, in general, should be encouraged. Communication with patients in an intensive care setting may be hampered by endotracheal intubation or a neurologic condition that impairs language function. Clinical experience suggests that, for the cognitively intact patient, this inability to express thoughts is often the most stressful aspect of the ICU experience. The staff should actively encourage writing notes to staff and family. Such patients should have a pencil, paper, and writing board at their bedside at all times. If patients are unable to write, they may be able to point to letters or words on a chart that has been specially designed for this purpose. Stovsky, Rudy, and Dragonette (1988) noted that patient satisfaction increased significantly when patients were introduced to the communication board preoperatively. If a communication board is not available, a system of tapping or blinking may permit useful communication. When using this technique, care should be taken to establish that patients are able to maintain their level of attention and respond consistently to questions. Aphasic patients can use picture communication boards.

Technical communication aids have been developed that permit vocalization despite intubation (Venus, 1980; Walsh & Rho, 1985). The Cyber Set Speaking Endotracheal tube (Dacomed Speech Systems, Minneapolis, MN) can be used in intubated patients who have teeth. It is a double-lumen tube that allows air for breathing to flow through an inner cannula. The second lumen is attached to a tone generator that conducts tone to the posterior oral pharynx. The patient can produce speech by articulating with the tongue and teeth (Walsh & Rho, 1985). *Communitrach I* (Implant Technologies, Minneapolis, MN) is a tracheostomy tube with a double lumen that allows air for speaking, supplied from a separate air source, to flow through an outer cannula and through vents toward the vocal cords, enabling the patient to speak. The *Passy-Muir Ventilator Speaking Valve)* (Passy & Passy, Irvine, CA) is an example of a speaking device for ventilator-dependent patients with tracheostomies. It has a one-way valve that fits within the ventilator tubing. The valve closes at the end of the inspiratory cycle, allowing air to be diverted to the larynx. *Olympic Trach-Talk* (Olympic Medical, Seattle, WA) is a one-way-valve speaking device that attaches to a tracheostomy tube externally. However, it can be used only when the patient no longer requires continuous mechanical ventilation. Once extubated, a patient with a tracheostomy may also speak with the assistance of a fenestrated tracheostomy tube or Kistner tube (G. P. Pilling & Son Co., Philadelphia, PA) (Venus, 1980). The Kistner tube is a short tube with a one-way valve that occupies only the distance between the skin to the inside of the tracheal wall (Venus, 1980).

Sensory Stimulation and Sleep in Delirium. Both sensory deprivation and sensory overload, in the form of excessive noise, may lead to behavioral alterations and potentiation of delirium in the ICU (Hansell, 1984; Zegers, 1988). Excessive noise levels, which have been documented to exist in ICUs, may lead to sleep deprivation, irritability, and impaired cognitive performance (Bentley, Murphy, & Dudley, 1977; Hansell, 1984).

Sleep deprivation and disturbance of the normal sleep pattern—suppression of stages 3 and 4 and rapid eye movement (REM) sleep—are ubiquitous in the intensive care and coronary care setting (Aurell & Elmquist, 1985). Though noise, pain, and medications play a major role in contributing to sleep problems in the ICU, some authors suggest that the sleep pattern abnormality is due to the effects of anesthesia or the systemic effects of surgery or illness on the brain's sleep-wake regulating mechanism (Aurell & Elmquist, 1985). Wilson (1972) demonstrated a twofold increase in delirium in patients without windows in their ICU room. To diminish sleep disruption, it is useful to limit noise and maintain some semblance of a day-night cycle (e.g., dim lights at night), provide all routine care during the day, and administer oral medications on a schedule that minimizes nighttime awakenings.

One study attempted to discern if listening to music via head sets would reduce anxiety in patients in the CCU. Patients listening to a tape of "self-selected" music were compared with a group of patients listening to "white noise." There were reductions in anxiety levels in both groups, suggesting that uninterrupted rest may improve both anxiety levels and physiologic profiles (Zimmerman, Pierson, & Marker, 1988).

Though nonpharmacologic management of insomnia is preferred, persistent insomnia in the ICU warrants consideration of pharmacological intervention to help prevent the development of delirium. There is no "perfect" pharmacologic agent for the treatment of insomnia. Anticholinergic agents should be avoided because of possible contributions to delirium. Ultra-short-acting benzodiazepines, such as triazolam, may precipitate amnestic and rebound effects. Our clinical experience suggests lorazepam 0.5 to 2.0 mg, an intermediate-acting benzodiazepine with no active metabolites, is an effective hypnotic. Its availability in a parenteral form is also a benefit. The patient with low levels of anxiety contributing to insomnia may also benefit from low doses of trazodone—25 to 100 mg (Jacobsen, 1990). This sedating antidepressant has a relatively safe cardiac profile, is not anticholinergic, and can be discontinued from this dosage without withdrawal effects. (See Chapter 9.)

Pain, Agitation, and Delirium. It is especially difficult in the ICU to distinguish between agitation associated with pain and that due to delirium. Pain is often undertreated in the hospital setting (Marks & Sachar, 1973). Though narcotic analgesics may contribute to the development or persistence of delirium, analgesia should not be withheld when pain is known to be present or likely. Critically ill patients who have cancer and bone metastases, severe burns, or recent major surgery require narcotics if they are conscious. It is best to continue to treat such patients with a moderate dose of a narcotic such as morphine sulfate or hydromorphone on a regular fixed schedule. This regimen avoids the peaks and troughs of blood levels associated with intermittent, as-needed (PRN) dosing. Meperidine should not be used in repeated doses because of the cumulative CNS toxicity of a major metabolite, normeperidine (Kaiko et al., 1983). Further discussion of acute pain management can be found in Chapter 16.

Pharmacologic Treatment

Controlled studies comparing the effectiveness of pharmacologic agents to placebo for treating delirium in the ICU have not been done, in part because the ICU setting demands that some action be taken acutely to control the agitation associated with delirium. Critical care clinicians have become increasingly comfortable with the use of neuroleptics, benzodiazepines, and opiates for the management of delirium. Much of the treatment currently used stems from clinical case reports.

Neuroleptics. With few exceptions, neuroleptics remain the drugs of choice for management of the agitated intensive care patient with delirium (Cummings, 1985; Eisendrath & Link, 1983; Tesar & Stern, 1986; Wise & Cassem, 1990). Although benzodiazepines are a reasonable alternative to neuroleptics in many medical settings and are increasingly being used concomitantly with neuroleptics, neuroleptics are preferable to benzodiazepines, barbiturates, or opiates alone in the intensive care setting for several reasons. Neuroleptics do not produce significant respiratory depression, and they are less likely than benzodiazepines, barbiturates, and opiates to further impair cognitive function. Although low-potency (e.g., thioridazine and chlorpromazine) phenothiazines have cardiovascular effects, and their use has led to the development of cardiac arrhythmias in some instances (Fowler et al., 1976), haloperidol has been shown to have a relatively safe cardiovascular profile (Menza et al., 1988; Tesar, Murray, & Cassem, 1985; Tesar & Stern, 1986, 1988). The use of haloperidol in agitated ICU patients is discussed in detail below.

Though they are safe in most ICU situations, neuroleptics should be avoided in patients with acute CNS injuries because of the increased risk of respiratory paralysis in these patients (Hershey & Hales, 1984). Alternative agents, such as benzodiazepines, opiates, or paralytic agents, should also be considered for patients who have evidence of preexisting chronic neurologic diseases, such as Parkinson's disease or epilepsy, and those patients who have developed severe toxicity from neuroleptics in the past.

The ideal neuroleptic for ICU use would have minimal cardiovascular, hepatic, or respiratory toxicity and would be available in parenteral forms. Intravenous neuroleptics are most advantageous for two reasons: (1) They permit more rapid relief of signs and symptoms; and (2) they produce more reliable blood levels in critically ill patients, who may have poor absorption from intramuscular injections because of poor muscle perfusion. Haloperidol is the neuroleptic of choice in the intensive care setting because of its minimal effects on cardiac function, blood pressure, and respiratory function and its low incidence of hepatic or renal toxicity (Donlon et al., 1979; Settle & Ayd, 1983; Shulman, 1984). Administration of intravenous haloperidol has also been found to be safe in the ICU (Sanders, Murray, & Cassem, 1991; Tesar & Stern, 1986, 1988; Tesar, Murray, & Cassem, 1985; Thompson & Thompson, 1983). Although there have been three case reports that have suggested an association between haloperidol and the development of atypical ventricular arrhythmias (torsades de pointes) (Fayer, 1986; Henderson, Lane, & Henry, 1991; Kriwisky et al., 1990), this event appears to be rare and idiosyncratic. Although intravenous use of haloperidol has not yet been specifically approved by the U.S. Food and Drug Administration, this lack does not preclude its use when indications for intravenous use are documented. The intravenous line must be flushed with saline before haloperidol is administered if heparin is being used, as haloperidol may precipitate with heparin.

Low-potency phenothiazines are inferior to haloperidol in the intensive care setting for several reasons. They have effects on cardiac conduction (Fowler et al., 1976), anticholinergic effects, which can contribute to delirium (Golinger, Peet, & Tune, 1987), and alpha-adrenergic receptor blocking effects, which can produce hypotension and drug interactions with antihypertensive medications (Richelson, 1984).

Adjuncts to use of intravenous haloperidol may include benzodiazepines or opiate agents. Specific recommendations regarding the use of haloperidol, benzodiazepines, opiates, and anesthetic and paralytic agents are discussed below. Figure 12-1 provides an algorithm for treating severe agitation in the intensive care setting.

Tesar and Stern (1986, 1988) have described a protocol for the use of intravenous *haloperidol* in the intensive care setting that is based on the experience of the psychiatric consultation service at Massachusetts General Hospital. Figure 12-1 includes a modified version of their protocol.

The initial dose of haloperidol is determined by several factors, including the severity of agitation, the risks of agitation to the patient, the patient's age, and the patient's previous response to haloperidol, if known. For mild agitation that is not presenting an immediate danger to the patient, a dose between 0.5 and 2.0 mg IV is chosen, based on the patient's age and history of previous response. In general, elderly patients should be started on a low dose because of the possibility of increased sensitivity to the drug's effects (Salzman, 1984). Evidence suggests that patients with acquired immunodeficiency syndrome (AIDS) encephalopathy may have increased sensitivity to the extrapyramidal side effects of neuroleptics (Hriso et al., 1991). Lower initial doses of haloperidol should be chosen for these patients as well. Clinical experience suggests that patients with chronic exposure to antipsychotics (e.g., chronic schizophrenics) in the past may require doses at the upper end of the range. For moderate or severe agitation or agitation that is presenting an immediate danger to the patient (e.g., an evolving MI, the presence of an intraaortic balloon pump), higher initial doses (2.0 to 10 mg) are usually needed. Some authors advocate the use of even higher initial doses (Tesar & Stern, 1986), but starting doses of more than 10 mg have not been found to be any more effective (Adams, 1988).

After the response to the initial dose has been observed, subsequent doses of haloperidol are based on the level of sedation or calm achieved. Because the peak effect of intravenous haloperidol occurs rapidly, doses can be repeated as often as every 20 to 30 minutes. If there is no change in the level of agitation 20 to 30 minutes after the first dose, the dose is doubled. If some calming is noted, the previous dose is repeated. Once adequate calming has been obtained, the interval between doses is increased. Frequent adjustments usually are needed over the first 24 to 48 hours. Once the patient has been stable for 24 hours, haloperidol should be continued on a regular schedule. A useful formula for estimating the daily haloperidol requirement after rapid sedation is to provide half of what was required over the first 24 hours, in two or three divided doses (Eisendrath & Link, 1983). For example, if 40 mg of haloperidol was necessary to control agitation during the first 24 hours, a total of 20 mg is ordered for the next 24 hours in divided doses (e.g., 10 mg q12h).

The total amount of haloperidol required depends, to a great extent, on the clinical condition of the patient. Single doses as high as 75 mg (Tesar, Murray, & Cassem, 1985) have been reported. There have been reports (Adams, 1988) of the regular use of 480 mg of haloperidol per day, and Fernandez et al. (1988) reported use of a haloperidol drip at rates of 600 mg/day for 5 days. Most patients require less than 100 mg of haloperidol per day (Adams, 1988). To our knowledge, the highest reported daily dose of haloperidol is 1200 mg/day (Sanders, Murray, & Cassem, 1991). In all of the case reports mentioned above, no apparent neurologic or cardiovascular complications emerged related to use of haloperidol alone.

As critical care units become increasingly comfortable with the use of intravenous haloperidol, clinicians have attempted to refine its use (see Figure 12-1). Intravenous haloperidol alone may control agitation; and even in doses of 600 mg daily it causes fewer extrapyramidal symptoms than comparable oral doses (Goldney, Spence, & Bowes, 1986; Menza et al., 1987; Tesar, Murray, & Cassem, 1985). However, the addition of lorazepam potentiates the sedative effects of haloperidol and acts synergistically to calm the agitated ICU patient (Adams, 1988). Controlled studies comparing intravenous haloperidol alone to its use in combination with intravenous lorazepam (in a 4:1 ratio) revealed that combined treatment significantly reduced the incidence of extrapyramidal side effects (Salzman, 1988). Lorazepam may be problematic, however, because of its tendency to produce anterograde amnesia.

In situations where pain complicates the clinical presentation, further control may be obtained by using intravenous opiates. Hydromorphone 0.5 to 4.0 mg q4h, a potent, intermediate acting opiate, is ideal for this purpose (Adams, 1988; Fernandez et al., 1988) (see Figure 12-1).

Research has yet to address the problem of tardive dyskinesia (TD) in patients who have had exposure to high dose intravenous neuroleptics. Risk factors for TD are reviewed elsewhere (Kane & Smith, 1982) and suggest that development of TD may be a dose-related phenomenon. One might, rightfully, have concerns about the risk of the development of a movement disorder in the future. As in all such cases, this risk must be weighed

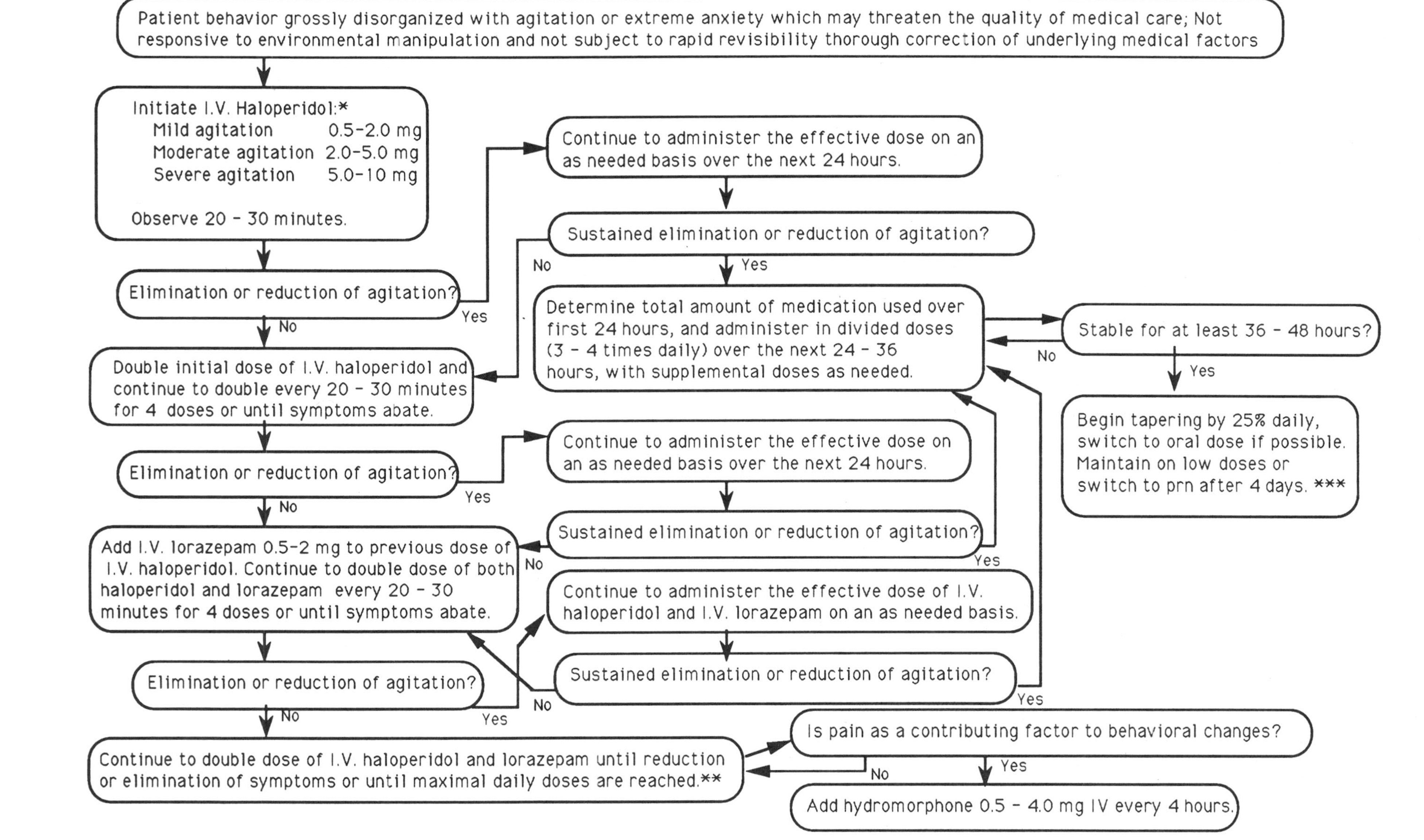

FIG. 12–1. Algorithm for the management of agitation in the intensive care setting.

against the immediate need to bring the patient through an ICU delirium as safely as possible.

Benzodiazepines. Benzodiazepines may be used as alternatives to neuroleptics for managing agitation associated with delirium in the intensive care setting. Some authors actually prefer benzodiazepines to haloperidol in this setting because benzodiazepines are less likely to exacerbate seizures or produce acute dystonic reactions, akathisia, and other extrapyramidal syndromes (Sebastian, 1985; Wells, 1985) (see also Chapter 9). When used in combination with neuroleptics, benzodiazepines may enable the patient to receive a lower total neuroleptic dose while achieving the same degree of behavior control (Adams, 1988; Salzman, 1988). They are the agents of choice for treating alcohol or benzodiazepine withdrawal and are preferred in the presence of status epilepticus, neuroleptic malignant syndrome, and severe Parkinson's disease. They may worsen the cognitive dysfunction of delirium, however, particularly when underlying dementia is present or high doses are used (Salzman, 1984). This is especially true of agents with long half-lives, such as diazepam and chlordiazepoxide, that tend to accumulate, particularly in the elderly (Salzman, 1984). Their respiratory depressant effects, especially in patients with carbon dioxide-retaining obstructive lung disease, and muscle relaxant effects increase the risks of respiratory failure and may make weaning from a ventilator more difficult. Risk is increased when benzodiazepines are combined with other respiratory depressants, such as opiates (Alexander & Gross, 1988; Bailey et al., 1990; Forster et al., 1980; Gross, Weller, & Conrad, 1991).

Lorazepam is our benzodiazepine of choice for treating agitation and delirium in the intensive care setting (see Figure 12-1). Advantages include a relatively short half-life (10 to 12 hours), absence of active metabolites, and elimination mechanisms that are not significantly affected by age or liver disease (Adams, 1988). Lorazepam is also available in parenteral form and, other than midazolam, is the only benzodiazepine that has reliable intramuscular absorption. The initial dose is 0.5 to 2.0 mg. Doses can be repeated every hour intramuscularly and every 30 minutes intravenously, and the dose can be increased to 5 mg hourly if necessary. *Daily doses as high as 240 mg have been used safely* (Adams, 1988). As noted in the section on neuroleptics, above, lorazepam may be used in combination with haloperidol to effectively manage agitation and delirium (Adams, 1988) (see Figure 12-1).

Midazolam is a parenteral benzodiazepine that has been promoted for intravenous sedation for short diagnostic or endoscopic procedures, for sedation before general anesthesia, and as an adjunct to anesthesia regimens (Medical Letter, 1986). Because of its rapidity of action, high degree of lipophilicity, short half-life (elimination half-life of 1 to 4 hours in healthy individuals), and decreased local irritation, it has several advantages for the treatment of delirium in the ICU. Several case reports have shown the efficacy of midazolam for treatment of aggressivity, violence, and hyperarousal (Bond, Mandos, & Kurtz, 1989; Mendoza et al., 1987). Midazolam infusions of doses of 0.1 to 2.0 μg/kg/min have been suggested for treatment of ICU agitation (Shapiro et al., 1986), and it has been used to control delirium tremens in one case with an infusion over 5 days at rates of 20 to 55 mg/hr (Lineaweaver, Anderson, & Hing, 1988). However, like other benzodiazepines, midazolam may also produce respiratory depression, apnea, and hypotension (Alexander & Gross, 1988; Bailey et al., 1990; FDA Drug Bulletin, 1988; Gross, Weller, & Conrad, 1991). Moreover, midazolam's half-life is prolonged in the elderly, in patients with hepatic disease, and in critically ill patients (Byatt et al., 1984; Medical Letter, 1986; Shelly, Mendel, & Park, 1987; Vree et al., 1989). Because midazolam's metabolism is greatly affected by acute liver dysfunction and systemic illness, the potential for iatrogenic overdose and toxicity is greater for midazolam than for lorazepam. It is the primary reason that we and others (Adams, 1988) prefer lorazepam over midazolam for management of delirium and agitation in the ICU, especially in patients who are not intubated.

Benzodiazepine Antagonists. Benzodiazepine antagonists may eventually prove useful for promoting rapid recovery after infusion of benzodiazepines in the ICU. The drug flumazenil has been shown to antagonize the effects of benzodiazepines, such as respiratory depression, and enhance weaning from mechanical ventilation (Editorial, 1984; Gross, Weller, & Conrad, 1991; Kirkwood, 1991). Flumazenil is a competitive antagonist and binds directly to the benzodiazepine receptor. It has a short half-life (30 to 90 minutes). While it can antagonize the effects of benzodiazepines for up to 6 hours (Kirkwood, 1991); other work suggests that the use of a 1-mg IV dose of flumazenil produces only a short-lived (i.e., 60 minutes) reversal of midazolam-induced sedation (Gross, Weller, & Conrad, 1991). In their study, resedation occurred 2 hours after the dose, underlining the importance of repeated dosing when using this drug to treat benzodiazepine toxicity. Side effects include nausea and vomiting, agitation, and seizures, with the latter seen mainly in benzodiazepine dependent patients. Reverse of respiratory effects is not always complete. For these reasons, the availability of flumazene reduces but does not eliminate the role of benzodiazepine sedation in the ICU.

Other Pharmacologic Agents. Delirium or coma associated with anticholinergic toxicity may be reversed with physostigmine, a cholinesterase inhibitor (Duvoisin & Katz, 1968). However, because its effects are short-lived and its use is associated with toxicity (i.e., bradycardia, hypotension, hypersalivation, seizures, vomiting and asystole) (Newton, 1975; Pentel & Peterson, 1980), we do not recommend its use.

Other pharmacologic approaches have also been used to treat agitated patients with delirium in the intensive care setting. When patients have not responded to other approaches, and agitation is interfering with necessary treatments, such as mechanical ventilation or intraaortic balloon counterpulsation, it may be necessary to resort to the use of anesthetic agents or paralytic agents, such as pancuronium bromide, metocurine, or *d*-tubocurarine (Rie & Wilson, 1983; Tesar & Stern, 1986). Risks of paralytic agents include muscle atrophy, axonal degeneration, and myopathic changes after just days (Bachenberg, 1988). These effects may complicate subsequent weaning from mechanical ventilation. Also, because the complete paralysis induced by these agents is frightening, patients receiving these agents must also receive other pharmacologic agents to ensure adequate sedation. Therefore we recommend that paralytic agents be used only when all other measures to treat agitation and delirium have failed; and, if used, they should be discontinued as soon as possible.

Etomidate, a short-acting anesthetic agent, was used as a sedative in some intensive care settings (especially in Europe) until it was found to increase the risk of infection due to its suppression of adrenocortical function (Aitkenhead et al., 1989). Propofol is a new short-acting anesthetic that has been studied as a sedative in the intensive care setting (Aitkenhead et al., 1989; Snellen et al., 1990). Aitkenhead et al. (1989) compared the effects of propofol with midazolam in a randomized controlled trial among 101 patients in ICUs in Britain. All of these patients were receiving mechanical ventilation and morphine for pain management. Propofol was just as effective as midazolam for producing adequate sedation (Aitkenhead et al., 1989). However, patients receiving propofol recovered more quickly, and weaning from mechanical ventilation was achieved significantly earlier in the propofol group than in the midazolam group (Aitkenhead et al., 1989). Two controlled trials comparing propofol and midazolam in coronary artery bypass graft patients reported similar results (McMurray et al., 1990; Snellen et al., 1990). Propofol may become a useful agent for treating severe agitation in the intensive care setting, but we cannot recommend its use as a first-line agent until more data regarding its efficacy and toxicity become available.

Isoflurane is a fluorinated inhalational general anesthetic that has many of the properties of an ideal sedative agent for use in the ICU (Kong, Willats, & Prys, 1989). A controlled trial comparing isoflurane with midazolam conducted by Kong et al. found that isoflurane provided satisfactory sedation in patients requiring mechanical ventilation. Patients treated with isoflurane were more tranquil and cooperative than those treated with midazolam. Moreover, those treated with isoflurane were weaned from ventilator support significantly faster than those treated with midazolam (Kong, Willatts, & Prys, 1989). Because isoflurane must be delivered via inhalation, an anesthesiologist or nurse anesthetist should administer the drug. When considering the use of anesthetic agents, collaboration with an anesthesiologist is also advised.

RESPONSE TO LIFE-THREATENING ILLNESS; DIAGNOSTIC CONSIDERATIONS

Virtually all patients in the intensive care setting are experiencing the psychologic and physical stresses of having a life-threatening serious illness. Such stress is then compounded by the environmental stresses of the intensive care setting. Weakness, fatigue, impairment of cognitive function, restrictions on mobility, and barriers to communication are ubiquitous in the ICU. Serious burns and multiple trauma are especially difficult to cope with because they are often accompanied by severe pain, disfigurement, loss of body parts, multiple surgeries and procedures, and a prolonged hospital and intensive care stay (Avni, 1980; Mendelsohn, 1983; Patterson et al., 1990) (see Chapter 34).

In addition to the experience of psychological stress, many signs and symptoms of psychiatric disorders, such as sleep disturbance, psychomotor changes, and anergia, frequently result from life-threatening illness or medications used to treat them. Therefore psychiatric assessment of the ICU patient involves consideration of those features of the presentation that may be attributed to the medical problem alone, which results in the need to approach psychiatric diagnosis from a broader scope than the *DSM* system permits. Clear-cut diagnoses outside the "organic disorders" rarely can be made in the intensive care setting.

Depression, anxiety disorders, and delirium have been discussed elsewhere in this chapter. Other chapters in this text discuss the assessment of psychiatric disorders in the medically ill (see Chapters 4–6, 8, 13, 14, 17). Table 12-3 describes the more common psychiatric syndromes confronted in the ICU and suggests approaches to the differential diagnosis. In the following sections, we review brief reactive psychosis and stress response syndromes.

Brief Reactive Psychosis

Patients experiencing a combination of the acute stress of a life-threatening illness and the environmental stressors of the intensive care environment may develop symptoms or signs of psychosis. In the absence of delirium or another organic mental disorder, the diagnosis of brief reactive psychosis (American Psychiatric Association, 1987) may be made. Some have used the term "ICU syndrome" or "ICU psychosis" when describing the reactions of some patients to the stress of the intensive care environment. This term is misleading, however (Cassem, 1984; Tesar & Stern, 1986), because even in the absence of a known etiologic agent psychosis occurring in the intensive care setting is almost always a manifestation of a delirium. *Environmental and psychologic stressors and organic factors interact to produce both organic mental disorders and stress response disorders.* Neuroleptics should be used to treat psychosis in the intensive care setting if environmental and psychological interventions do not effectively reduce the signs and symptoms of psychosis.

Stress Response Syndromes

Patients in the ICU may also develop severe stress response syndromes. Horowitz provided a thorough description of stress syndromes and a guide to their treatment (Horowitz, 1986). The diagnostic criteria for posttraumatic stress disorder (PTSD), published by the American Psychiatric Association in the *DSM* system, are based largely on his work.

The diagnosis of posttraumatic stress disorder is made when intrusive thoughts and feelings, maladaptive denial, and associated signs and symptoms are present. When the full syndrome is not present, a diagnosis of an adjustment disorder may be appropriate, though the diagnostic criteria are not easily applied in the medical setting.

TABLE 12–3. *Diagnostic Consideration: Psychiatric Manifestations in the ICU*

Diagnosis	Features Common in ICU Patients	Features Suggestive of Diagnosis	Other Clinical Clues to Diagnosis
Adjustment disorder with anxiety	Presence of stress-inducing event, preoccupation with disease, anxiety, mild autonomic arousal, hypervigilance	Feelings of dread, preoccupation with inappropriate concerns, sleep interrupted by anxiety, autonomic arousal in excess of what would be expected by the medical condition	Increased startle response, anxiety, "worried" facies
Brief reactive psychosis	Presence of stress-inducing event, hypervigilance, impaired communication ability, "emotional turmoil," sleep disturbance	Delusions, hallucinations, disorganized thought process	Oriented, no attentional abnormality, well organized psychotic thought content
Delirium	Disturbed sleep wake cycle, reduced level of consciousness, altered psychomotor activity	Attention deficits, disorganized thinking, perceptual disturbance, incoherence	Confusional Assessment Method; attentional tasks: alternate finger tapping, signaling to odd but not even numbers EEG rhythm changes
Depression	Dysphoria; anxiety; reduced sleep, appetite, concentration; anergia	Hopelessness, anhedonia, guilt, illogical negativism, tearful	Ask patient about family members or pets, ask about plans and future, show pictures of children and assess for reactivity
Posttraumatic stress disorder	Presence of stress-inducing event, sleep problems, difficulty concentrating, psychic distress with recurrence of symptoms, amnesia for traumatic events, sense of foreshortened future	Recurrent dreams, hypervigilance, feelings of detachment from others, restricted affect, irritability, exaggerated startle response	Assess for autonomic arousal as you ask patient to recount events leading to the present admission. Burns, trauma, or unexpected and violent illness onset
Psychoactive substance withdrawal syndrome	Fluctuations in vital signs, anxiety, psychomotor changes	Piloerection, mydriasis, hyperperistalsis; miosis, tremulousness, delirium, myoclonous, seizures	Assess history with family, look for needle tracks, review admission toxicology screen

Horowitz (1986) has described a number of psychotherapeutic, behavioral, and pharmacological interventions to treat a patient who is having intrusive thoughts, recollections, or dreams. Such interventions include reducing external demands on stimulus levels, promoting rest, permitting temporary idealization and dependency, providing information and education, helping the patient to differentiate fantasy from reality, suppressing strong negative emotion with anxiolytics or sedatives, and using desensitization procedures and relaxation. Denial states usually are not challenged in the intensive care setting unless they interfere with effective care (e.g., wanting to sign out of the hospital against advice) or decision making (e.g., to have necessary surgery).

PSYCHOLOGIC CARE OF PATIENTS REQUIRING MECHANICAL VENTILATION AND WEANING FROM VENTILATION

Patients requiring mechanical ventilation because of respiratory failure or other respiratory disorders often experience emotional distress (Cassem, 1984; Gale & O'Shanick, 1985; Mendel & Khan, 1980). Dependence on the machine, loss of control, communication difficulty, fears of death and disability, sensory alteration, and sleep deprivation are some of the psychologic stressors faced by these patients (Cassem, 1984; Gale & O'Shanick, 1985). Many of these patients experience considerable anxiety as they are weaned from the ventilator (Bowden, 1983; Cassem, 1984; Gale & O'Shanick, 1985; Hall & Wood, 1987; Mendel & Khan, 1980). At times, it leads to much difficulty in weaning, as hyperventilation, tightening of chest muscles, and panic may lead to early fatigue and inadequate gas exchange. Anxiety can be exacerbated when patients are not properly informed about the weaning process, if weaning begins before the patient is physically or mentally ready to be weaned, or when trials on a T-tube or a low rate of intermittent mandatory ventilation (IMV) end only when the patient tires. The latter method of weaning leads to the repeated experience of breathlessness and anxiety, which may become associated with the weaning *process*, producing a classically conditioned learned response. Subsequent T-tube or low-rate IMV trials may trigger conditioned anxiety, which leads to early fatigue and more anxiety and breathlessness. The experience of "failure" with this method can frustrate and demoralize a patient if the weaning process is prolonged.

Criteria to assess whether a patient is physically ready to be successfully weaned are readily available and are not discussed in detail here (Francis, 1983; Gale & O'Shanick, 1985; Hall & Wood, 1987; Marini, 1988, 1991; Petty, 1984; Rie & Wilson, 1983; Yang & Tobin, 1991). The patient must be alert and well rested. Delirium or other cognitive problems are likely to make weaning difficult if respiratory function is close to the minimum requirements for weaning. Any agitation should be well controlled before attempts are made to begin the weaning process. Gale and O'Shanick (1985) discussed psychologic parameters to determine weaning readiness and included absence of delirium, understanding by the patient of the weaning process, and a positive and hopeful attitude in the patient regarding weaning.

Patients should be informed well in advance of the procedures to be followed during the weaning process. They should be assured that they will be watched closely and that adjustments will be made if they have any difficulty (Hall & Wood, 1987). To avoid the development of conditioned anxiety and the repeated experience of failure, trials on a T-tube or low-rate IMV should be arranged so they terminate after a predetermined period of time.

The first trial should be no longer than 30 minutes (Hall & Wood, 1987; Petty, 1984). During each trial, especially the first, pulmonary function is monitored and patients are observed closely for signs of fatigue and distress. Irregular breathing, paradoxical motion of the abdomen or rib cage, vigorous use of accessory ventilatory muscles, cardiac tachyarrhythmia, and diaphoresis are clues to distress (Marini, 1991). A disproportionate increase in the respiratory rate relative to the tidal volume during the first minute after the machine is disconnected appears to be a simple, yet reliable indicator that the patient is not yet ready to be weaned (Marini, 1991; Yang & Tobin, 1991).

If patients tolerate the first trial well, the duration of subsequent trials is gradually lengthened until extubation seems likely to succeed. Patients are given an opportunity to rest for at least an hour between trials, and weaning should be suspended during the late evenings and overnight. Encouragement is helpful, and patients should be praised for their efforts and success. If they express discouragement or impatience with the weaning process, it is important to provide concrete feedback about their progress. Sedatives should be avoided, as they may contribute to fatigue and depress respiratory function. Anxiety and fears frequently can be managed with psychotherapy and by reassurance from ICU staff (Gale & O'Shanick, 1985; Mendel & Kahn, 1980). Relaxation training also may help. Biofeedback has been used successfully to promote weaning (Acosta, 1988; Holliday & Hyers, 1990). Acosta (1988) described one case of successful weaning that provided progressive muscle relaxation techniques (contracting and relaxing specific muscle groups) in conjunction with ear oximetry feedback. Holliday & Hyers (1990) reported that 20 patients assigned to a biofeedback intervention had a significant reduction in the mean number of days on the ventilator (21 days versus 33 days) as well as a significant improvement in physiological measures when compared to 20 patients assigned to a control group. The biofeedback intervention consisted of muscle relaxation training with frontalis muscle electromyographic (EMG) feedback as well as tidal volume and diaphragm EMG feedback. Biofeedback was provided daily until extubation or until the patient was placed on no-resuscitation status. These promising results must be confirmed by subsequent studies before we can recommend the routine use of biofeedback to facilitate weaning from mechanical ventilation.

For at least a quarter century (Winnie & Collins, 1966), clinicians have searched for the "pharmacological ventilator"—a medication that would facilitate weaning from the ventilator. However, there have been no definitive studies to show that any agent is superior for this purpose. Because high levels of anxiety and agitation may interfere with weaning, pharmacologic agents that cause minimal CNS depression while reducing anxiety are most advantageous. Benzodiazepines should be avoided, owing to their respiratory depressant effects (Alexander & Gross, 1988). Low dose haloperidol may be one option, as it has minimal sedative properties. Tricyclic antidepressants, such as nortriptyline, may also be tried in low doses (10 mg tid) for their mild anxiolytic effects; the anticholinergic effects may also help by drying upper respiratory secretions. Psychostimulants may be helpful for weaning patients with apathy and inactivity or when patients have a secondary depression (Masand, Pickett, & Murray, 1991). Buspirone, a nonsedating anxiolytic, is of little use when there is an acute need to wean. However, buspirone may be useful in the anxious patient for whom weaning may take several weeks. Buspirone may take weeks to have an effect, and our experience suggests that 4 to 6 weeks at doses of 30 to 60 mg/day are needed to obtain maximal benefit.

ICU CARE OF THE ORGAN TRANSPLANT PATIENT

Rapid advances in transplantation technology have had a dramatic impact on the number of organ transplantations. Psychiatry is playing an increasingly important role in the evaluation and management of these patients. The psychiatric aspects of transplantation are reviewed in Chapters 27 and 36. The ICU plays a role in the postsurgical course of all transplantation patients. For the most part, the problems confronting the transplantation patient do not differ from those of other ICU patients. Nevertheless, certain issues have emerged as unique to the transplantation patient: The patient may be treated both pre- and postoperatively in the ICU; there may be multiple patients involved in live-donor transplantation; and, in contrast to the typical surgical patient, the psychologic adjustment is to a new body part, rather than the modification or removal of a defective organ (House, Trzepacz, & Thompson, 1990).

Despite attempts to understand how transplantation surgery may induce mental changes, the nature of psychiatric illness during the immediate postoperative phase is largely unknown. Common reasons for psychiatric consultation include thought disorder and confusion, depression and thoughts of suicide, anxiety and nervousness, competence issues, and treatment noncompliance (House, Trzepacz, & Thompson, 1990). Two pharmacologic agents used frequently in transplantation patients, cyclosporine and the corticosteroids, have considerable CNS activity. The highly lipophilic properties of cyclosporine increases its tendency to induce delirium (Craven, 1991). Preexisting CNS vulnerability or hypomagnesemia can predispose to CNS toxicity. Despite some reports of competition for binding sites between haloperidol and cyclosporine, haloperidol is still considered the agent of choice for management of cyclosporine-induced delirium. Other CNS symptoms, including hallucinations, are associated with cyclosporine (Berden et al., 1985). CNS cyclosporine toxicity is discussed in more detail in Chapter 36.

Corticosteroid therapy can induce a variety of psychiatric disturbances, especially organic mood disorders (Ling, Perry, & Tsuang, 1981). Even when lower doses of corticosteroids are used, subclinical cognitive deficits may occur that impair patients' understanding of their disease or its treatment (Wolkowitz et al., 1990). In addition, total body irradiation in allogenic marrow graft patients may predispose to cognitive deficits during the posttreatment phase (Jenkins, Linington, & Whittaker, 1991). As with other organically induced psychiatric syndromes, efforts should be made to correct underlying causes whenever possible before providing symptomatic treatment.

STAFF ISSUES

The staff working in intensive care settings are exposed to multiple stresses. Such stresses include the high volume and rapid pace of the work, the high level of clinical responsibility, the severity and high mortality of patients' illnesses, understaffing, and such environmental stresses as noise and overcrowding (Caldwell & Weiner,

1981; Soupios & Lawry, 1987). Interpersonal conflicts with hospital bureaucracy and within hierarchies (e.g., nurse–doctor conflicts) can also add to staff stress (Biley, 1989). Nursing staff members have few opportunities to rest or escape from the ICU for relief. In modern ICUs, nurses are called on to attend to multiple and diverse tasks, which include attending to patient comfort; assessing clinical status; administering medications, intravenous solutions, and blood products; providing enteral and parenteral nutrition; attending to multiple lines and tubes; and monitoring multiple physiologic parameters. They must become familiar with an ever-expanding array of technologic aids that have been developed for the treatment of critically ill patients. Strong feelings about patients and grief over their deaths provide constant stress for these nurses. Dealing with prolonged care of poor-prognosis patients and having to help families deal with their grief are particularly stressful, especially for pediatric intensive care nurses (Caldwell & Weiner, 1981; Woolston, 1984). Intrastaff conflict in the face of great responsibility for clinical care is also a significant stressor (Caldwell & Weiner, 1981; Bailey, Walker, & Madsen, 1980).

Despite impressions that ICU nurses are highly stressed, studies comparing stress among intensive care nurses to stress among nurses working in other settings have yielded inconsistent results (Cronin-Stubbs, & Rooks, 1985; Eisendrath, Link, & Matthay, 1985; Gentry, Foster, & Froehling, 1972). Studies have not found differences in levels of stress between intensive care nurses and medical-surgical nurses (Cronin-Stubbs & Rooks, 1985; Eisendrath et al., 1985). In the neonatal ICU, studies suggest that young nurses and those with less education were more likely to experience job dissatisfaction (Rosenthal, Schmid, & Black, 1989).

The level of stress experienced by physicians in the intensive care setting has not been well studied. One systematic study of house staff stress in the intensive care setting found that ICU rotations were not rated more stressful than other clinical rotations (Eisendrath, et al., 1985). It is of note that the ranking of stress by house officers in this study was fairly consistent with the results of similar studies of intensive care nurses. Prolonged care of poor-prognosis patients was ranked as most stressful. Dealing with death, feeling insecure about clinical responsibility, and communication problems with staff were also rated as rather stressful (Eisendrath et al., 1985). Stressors unique to the critical care unit physician include the dual role of a "manager" of an entire health care team and a "personal physician" in life and death situations (including the assumption of ultimate responsibility for medical malpractice). Professional roles may be confused even more so in intensive care settings, as ICU nursing staff assume responsibilities generally reserved for physicians (Soupios & Lawry, 1987).

Studies that have surveyed strategies for coping with the stress of working in ICUs have also focused primarily on the experience of nurses. Nurse support groups are described as helpful, though controlled studies of their effectiveness are lacking (Weiner & Caldwell, 1981, 1983–1984). Support groups seem to work best if they are initiated in response to a felt need, if the group is highly structured and does not allow early discharge of intense negative feelings, and if the group's problems are primarily interpersonal rather than environmental or administrative (Weiner, Caldwell, & Tyson, 1983). Open-ended, unstructured "gripe groups" are not helpful and may actually increase dissatisfaction.

Comprehensive stress management programs for ICU nurses have been developed that appear to be effective (Bailey, Walker, & Madsen, 1980). These behaviorally oriented programs utilize structured training modules that focus on the development of specific skills. Modules address patient care issues, interpersonal and communication issues, personal stress reduction, and administrative issues. A workshop format is utilized, led by skilled trainers on off-duty time. Written materials are distributed, and skills are demonstrated and practiced during sessions. Evaluation of the stress program is an important component, as it allows participants and trainers to obtain feedback about the impact of the program (Rosenthal, Schmid, & Black, 1989). One study suggested improvement in staff tenure in the unit in response to supportive psychodynamic group therapy (Oehler, Peter, & Seyler, 1989).

Interventions to decrease the stress of ICU physicians have not been studied. However, the study assessing stress among house officers rotating through the ICU found that humor, opportunities to communicate with other staff, the opportunity to recover from sleep deprivation, and regular activities outside the ICU helped house staff to cope with the stresses of the rotation (Eisendrath et al., 1985). The psychiatric consultant can also play a valuable role by providing house staff and attending physicians with emotional support while performing consultations on patients in the intensive care setting.

THE OPTIMAL ICU

In the preceding sections of this chapter, we have described common psychiatric syndromes that occur in the intensive care setting, as well as interventions to help prevent and manage these syndromes. In Table 12-4 we have created a list of interventions, environmental

TABLE 12–4. *Components of an Ideal, Optimal ICU for Preventing and Managing Psychiatric Syndromes*

A. Environmental components
   1. Separate rooms for each patient to enhance privacy and to keep noise and interruptions to a minimum
   2. Lighting synchronized with the day-night cycle (e.g., dim lights at night)
   3. Limited noise from alarms and machines
   4. Clock, calendar, and radio in room to help with orientation
   5. Communication aids available (e.g., pads and writing instruments, communication boards)
   6. Comfortable waiting area for family, with private space for conferences
   7. Preadmission unit visits by patients and family, when possible

B. Assessments
   1. Preadmission or admission screening for cognitive and neurological impairment and substance abuse
   2. Monitoring of mental status and cognitive functioning during ICU stay
   3. Family assessment to determine need for family intervention

C. Staffing
   1. Support group for staff
   2. Adequate break time at a comfortable area away from the unit
   3. Availability of consultants/staff: consultation-liaison psychiatry, behavioral medicine, neuropsychology, neuropsychiatry, anesthesiology, and pharmacy (to help with management of agitation and pain); pulmonary and respiratory therapy (to help with weaning from mechanical ventilation); social work (to help with assessment and management of family issues)
   4. Educational conferences to address psychiatric management issues
   5. Regular administrative meetings of staff to enhance communication and to identify and manage staffing and personnel problems

D. Interventions
   1. Nonpharmacologic
      a. Education: face-to-face, use of written materials
      b. Relaxation training, music to alleviate anxiety
      c. Behavioral interventions and biofeedback for problems with weaning from mechanical ventilation
      d. Supportive psychotherapy for all patients with overt distress
   2. Pharmacologic
      a. Protocols for agitation, delirium, alcohol and drug withdrawal
      b. Appropriate use of psychopharmacologic agents to manage anxiety, depression, psychosis, and adjustment disorders

components, and resources that would be present or readily available in an optimal, ideal ICU. Such a unit is likely to enhance the care and management of critically ill patients while providing a supportive and comfortable working environment.

Acknowledgments

The authors would like to express their appreciation to Barbara Doll for her superb secretarial assistance.

This chapter was supported in part by the National Institute of Drug Abuse grant DA05623 to Michael G. Goldstein and the National Cancer Institute Cancer Prevention Research Consortium grant PO1 CA50087 to James O. Prochaska, Wayne Velicer, David B. Abrams, and Michael G. Goldstein.

REFERENCES

Acosta F (1988). Biofeedback and progressive relaxation in weaning the anxious patient from the ventilator: A brief report. Heart Lung 17:299–301.

Adams F (1988). Emergency intravenous sedation of the delirious, medically ill patient. J Clin Psychiatry 49:22–26.

Aitkenhead AR, Pepperman ML, Willatts SM, et al. (1989). Comparison of propofol and midazolam for sedation in critically ill patients. Lancet 2:704–709.

Alexander CM, & Gross JB (1988). Sedative doses of midazolam depress hypoxic ventilatory response in man. Anesth Analg 67:377–382.

American Heart Association (1991). 1992 Heart and stroke facts. Dallas: American Heart Association National Center.

American Psychiatric Association (1987). Diagnostic and Statistical Manual of Mental Disorders (3rd ed., rev.). Washington, DC: APA

Aurell J, & Elmquist D (1985). Sleep in the surgical intensive care unit: Continuous polygraphic recording of sleep in nine patients receiving postoperative care. Br Med J 290:1029–1032.

Anvi J (1980). The severe burns. Adv Psychosom Med 10:57–77.

Ayd FJ (1985). Psychopharmacology update: Psychostimulant therapy for depressed medically ill patients. Psychiatr Ann 15:462–465.

Babb SV, Dunlop SR, & Hoffman MA (1990). Protracted ventricular arrhythmias occurring after abrupt tricyclic antidepressant withdrawal. Psychosomatics 31:452–454.

Bachenberg KL (1988). Disuse atrophy. Crit Care Med 16:649–650.

Bailey PL, Pace NL, Ashburn MA, et al. (1990). Frequent hypoxemia and apnea after sedation with midazolam and fentanyl. Anesthesiology 73:826–830.

Bailey JT, Walker D, & Madsen N (1980). The design of a stress management program for Stanford intensive care nurses. J Nurs Educ 19:26–29.

Bedell SE, Delbanco TL, Cook EF, et al. (1983). Survival after cardiopulmonary resuscitation in the hospital. N Engl J Med 309:569–576.

Bentley S, Murphy F, & Dudley H (1977). Perceived noise in surgical wards and an intensive care area: An objective analysis. Br Med J 2:1503–1506.

Berden JH, Hoitsma AJ, Merx JC, et al. (1985). Severe central nervous system toxicity associated with cyclosporine. Lancet 1:219–220.

Bikadoroff S (1987). Mental changes associated with tocainide, a new antiarrhythmic. Can J Psychiatry 32:219–221.

Biley FC (1989). Stress in high dependency units. Intens Care Nurs 5:134–141.

Bond WS, Mandos LA, & Kurtz MB (1989). Midazolam for aggressivity and violence in three mentally retarded patients. Am J Psychiatry 156:925–926.

Bojar RM, Najafi H, DeLaria GA, et al. (1983). Neurological complications of coronary revasculatization. Ann Thorac Surg 36:427–432.

Bowden P (1983). Psychiatric aspects of intensive care. In J Tinker

& M Rapin (Eds.), Care of the Critically Ill Patient (pp. 787–797). Berlin: Springer-Verlag.

Breuer AC, Furlan AJ, Hanson MR, et al. (1983). Central nervous system complications of coronary artery bypass graft surgery: Prospective analysis of 421 patients. Stroke 14:682–687.

Byatt CM, Lewis LD, Dawling S, et al. (1984). Accumulation of midazolam after repeated dosage in patients receiving mechanical ventilation in an intensive care unit. Br Med J 289:799–800.

Calabrese JR, Skwerer RG, Gulledge AD, et al. (1987). Incidence of postoperative delirium following myocardial revascularization. Cleve Clin J Med 54:29–32.

Caldwell T, & Weiner MF (1981). Stresses and coping in ICU nursing: I. A review. Gen Hosp Psychiatry 3:119–127.

Carney RM, Rich M, teVelde A, et al. (1988). The relationship between heart rate, heart rate variability and depression in patients with coronary artery disease. J Psychosom Res 32:159–164.

Carney RM, Freedland KE, & Jaffe AS (1990). Insomnia and depression prior to myocardial infarction. Psychosom Med 52:603–609.

Cassem NH (1984). Critical care psychiatry. In WC Shoemaker, WL Thompson, & PR Holbrook (Eds.), Textbook of critical care (pp. 981–989). Philadelphia: WB Saunders.

Cassem NH, & Hackett TP (1971). Psychiatric consultation in a cardiac care unit. Ann Intern Med 75:9–14.

Cay EL, Vetter N, Philip AE, et al. (1972). Psychological status during recovery from an acute heart attack. J Psychosom Res 16:425–435.

Chiarello RJ, & Cole JO (1987). The use of psychostimulants in general psychiatry: A reconsideration. Arch Gen Psychiatry 44:286–295.

Cohen-Cole SA (1985). Interviewing the cardiac patient: I. A practical guide for assessing quality of life. Quality Life Cardiovasc Care 2:7–12.

Cohen-Cole SA, & Bird J (1986a). Interviewing the cardiac patient: II. A practical guide for helping patients cope with their emotions. Quality Life Cardiovasc Care 2:53–65.

Cohen-Cole SA, & Bird J (1986b). Interviewing the cardiac patient: III. A practical guide to educate and motivate patients to cooperate with treatment. Quality Life Cardiovasc Care 2:101–112.

Craven JL (1991). Cyclosporine-associated organic mental disorders in liver transplant recipients. Psychosomatics 32:94.

Cray L (1989). A collaborative project: Initiating a family intervention program in a medical intensive care unit. Focus Crit Care 16:212–218.

Cromwell RL, Butterfield ED, Brayfield FM, et al. (1977). Acute myocardial infarction: Reaction and recovery. St. Louis: Mosby.

Cronin-Stubbs D, & Rooks CA (1985). The stress, social support, and burnout of critical care nurses: The results of research. Heart Lung 14:31–39.

Cummings JL (1985). Acute confusional states. Clinical neuropsychiatry (Ch. 7, pp. 68–74). Orlando, FL: Grune & Stratton.

Deleu D, & Schmedding E (1987). Acute psychosis as idiosyncratic reaction to quinidine: Report of 2 cases. Br Med J 294:1001–1002.

Doehrman SR (1977). Psycho-social aspects of recovery from coronary heart disease: A review. Soc Sci Med 2:199–218.

Donlon PT, Hopkin J, Schaffer CB, et al. (1979). Cardiovascular safety of rapid treatment with intramuscular haloperidol. Am J Psychiatry 136:233–234.

Dubin WR, Field HL, & Gastfriend DR (1979). Postcardiotomy delirium. A critical review. J Thorac Cardiovasc Surg 77:586–594.

Duvoison RC, & Katz R (1968). Reversal of central anticholinergic syndrome in man by physostigmine. JAMA 206:1963–1965.

Editorial (1984). Sedation in the intensive care unit. Lancet 1:1388–1389.

Eisendrath SJ, & Link N (1983). Mental changes in the ICU, detection and management. Drug Ther 13:18–26.

Eisendrath SJ, Link N, & Matthay M (1985). Intensive care unit: How stressful for physicians. Crit Care Med 14:95–98.

Eisendrath SJ, & Sweeney MA (1987). Toxic neuropsychiatric effects of digoxin at therapeutic serum concentrations. Am J Psychiatry 144:506–507.

Faust D, & Fogel BS (1989). The development and initial validation of a sensitive bedside cognitive screening test. J Nerv Ment Dis 177:25–31.

Fayer SA (1986). Torsades de pointes ventricular tachyarrhythmia associated with haloperidol. J Clin Psychopharmacol 6:375–376.

FDA Drug Bulletin (1989). Boxed warning added to midazolam labeling. FDA Drug Bull 18:15–16.

Fernandez F, Adams F, Holmes VF, et al. (1987). Methylphenidate for depressive disorders in cancer patients. Psychosomatics 28:455–461.

Fernandez F, Holmes VF, Adams F, et al. (1988). Treatment of severe, refractory agitation with a haloperidol drip. J Clin Psychiatry 49:239–241.

Fogel BS, & Fretwell M (1985). Reclassification of depression in the medically ill elderly. J Am Geriatr Soc 33:446–448.

Folks DG, Freeman AM, Sokol RS, et al. (1988). Cognitive dysfunction after coronary artery bypass surgery: A case-controlled study. South Med J 81:202–206.

Follick MJ, Gottleib BS, & Fowler JL (1981). Behavior therapy and coronary heart disease: Lifestyle modification. In HJ Sobel (Ed.), Behavior therapy in terminal care: A humanistic approach (pp. 253–298). New York: Harper & Row.

Folstein MF, Folstein SE, & McHugh (1975). Mini-mental state. J Psychiatr Res 12:189–198.

Forster A, Gardaz JP, Suter PM, et al. (1980). Respiratory depression by midazolam and diazepam. Anesthesiology 53:494–497.

Fowler NO, McCall D, Chou T, et al. (1976). Elecrocardiographic changes and cardiac arrhythmias in patients receiving psychotropic drugs. Am J Cardiol 37:223–230.

Francis PB (1983). Acute respiratory failure in obstructive lung disease. Med Clin North Am 67:657–668.

Frederickson K (1989). Anxiety transmission in the patient with myocardial infarction. Heart Lung 18:617–622.

Freeman AM, Fleece L, Folks, DG, et al. (1986). Alprazolam treatment of postcoronary bypass anxiety and depression. J Clin Psychopharmacol 6:39–41.

Gale J, & O'Shanick GJ (1985). Psychiatric aspects of respirator treatment and pulmonary intensive care. Adv Psychosom Med 14:93–108.

Gentry WD, Foster SB, & Froehling S (1972). Psychological response to situational stress in intensive and non-intensive care settings. Heart Lung 1:793.

Goldman LS, & Kimball CP (1985). Cardiac surgery: Enhancing postoperative outcomes. In AM Razin (Ed.), Helping cardiac patients: Behavioral and psychotherapeutic approaches (pp. 113–156). San Francisco: Jossey-Bass.

Goldman LS, & Kimball CP (1987). Depression in intensive care units. Int J Psychiatry Med 17:201–212.

Goldney RD, Spence ND, & Bowes JA (1986). The safe use of high dose neuroleptics in a psychiatric intensive care unit. Aust NZ J Psychiatry 20:370–375.

Goldstein MG, & Guttmacher LB (1988). Treatment of the cardiac impaired depressed patient: Part I. General considerations, heterocyclic antidepressants, and monoamine oxidase inhibitors. Psychiatr Med 6:1–33.

GOLINGER RC, PEET T, & TUNE LE (1987). Association of elevated plasma anticholinergic activity with delirium in surgical patients. Am J Psychiatry 144:1218–1220.

GORE JM, SLOAN M, PRICE TR, ET AL. (1991). Intracerebral hemorrhage, cerebral infarction, and subdural hematoma after acute myocardial infarction and thrombolytic therapy in the Thrombolysis in Myocardial Infarction Study: Thrombolysis in myocardial infarction, phase II, pilot and clinical trial. Circulation 83:448–459.

GROSS JB, WELLER RS, & CONRAD P (1991). Flumazenil antagonism of midazolam-induced ventilatory depression. Anesthesiology 75:179–185.

GRUEN W (1975). Effects of brief psychotherapy during the hospitalization period on the recovery process in heart attacks. J Clin Consult Psychol 43:232–233.

GUIRY E, CONROY RM, HICKEY N, ET AL. (1987). Psychological response to an acute coronary event and its effect on subsequent rehabilitation and lifestyle change. Clin Cardiol 10:256–260.

GUNTUPALLI KK (1984). Acute pulmonary edema. Cardiovasc Clin 2:183–200.

HACKETT TP, CASSEM NH, & WISHNIE HA (1968). The cardiac care unit: An appraisal of its psychologic hazards. N Engl J Med 279:1365–1370.

HALL JB, & WOOD LDH (1987). Liberation of the patient from mechanical ventilation. JAMA 257:1621–1628.

HANSELL HN (1984). The behavioral effects of noise in man: The patient with "intensive care unit psychosis." Heart Lung 13:59–65.

HARRELL RG, & OTHMER E (1987). Postcardiotomy confusion and sleep loss. J Clin Psychiatry 48:445–446.

HENDERSON RA, LANE S, & HENRY JA (1991). Life-threatening ventricular arrhythmia (torsades de pointes) after haloperidol overdose. Hum Esp Toxicol 10:59–62.

HENRIKSEN L (1984). Evidence suggestive of diffuse brain damage following cardiac operations. Lancet 1:816–820.

HERSHEY SC, & HALES RE (1984). Psychopharmacologic approach to the medically ill patient. Psychiatr Clin North Am 7:803–816.

HOLLIDAY JE, & HYERS TM (1990). The reduction of weaning time from mechanical ventilation using tidal volume and relaxation biofeedback. Am Rev Respir Dis 141:1214–1220.

HOROWITZ MJ (1986). Stress response syndromes (2nd Ed.). Northvale, NJ: Jason Aronson.

HOUSE RM, TRZEPACZ PT, THOMPSON TL (1990). Psychiatric consultation to organ transplant services. In A TASMAN, SM GOLDFINGER, & CA KAUFMANN (Eds.), American Psychiatric Press Review of Psychiatry (Vol. 9; Ch. 26). Washington, DC: American Psychiatric Press.

HOWARD JH, RECHNITZER PA, CUNNINGHAM DA, ET AL. (1990). Type A behavior, personality, and sympathetic response. Behav Med 16:149–160.

HRISO E, KUHN T, MASDEU JC, ET AL. (1991). Extrapyramidal symptoms due to dopamine-blocking agents in patients with AIDS encephalopathy. Am J Psychiatry 148:1558–1561.

HUGHES JR, HIGGINS ST, & HATSUKAMI DK (1990). Effects of abstinence from tobacco. In LT KOZLOWSKI, H ANNIS, HD CAPPELL (Eds.), Recent advances in alcohol and drug problems (pp. 317–398). New York: Plenum.

INOUYE SK, VAN DYCK CH, ALESSI CA, ET AL. (1990). Clarifying confusion: The confusion assessment method. Ann Intern Med 113:941–948.

International Study Group (1990). In-hospital mortality and clinical course of 20,891 patients with suspected acute myocardial infarction randomized between aletplase and streptokinase with or without heparin. Lancet 336:71–75.

JACOBSEN F (1990). Low-dose trazodone as a hypnotic in patients treated with MAOIs and other psychotropics: A pilot study. J Clin Psychiatry 51:298–302.

JENKINS PL, LININGTON A, & WHITTAKER JA (1991). A retrospective study of psychosocial morbidity in bone marrow transplant recipients. Psychomatics 32:65.

KAIKO RF, FOLEY KM, GRABINSKI PY, ET AL. (1983). Central nervous system excitatory effects of meperidine in cancer patients. Ann Neurol 13:180–185.

KANE JM, & SMITH JM (1982). Tardive dyskinesia: Prevalence and risk factors, 1959–1979. Arch Gen Psychiatry 39:473–481.

KATHOL RG, NOYES R, WILLIAMS J, ET AL. (1990). Diagnosing depression in patients with medical illness. Psychosomatics 31:434–440.

KIERNAN RJ, MUELLER J, LANGSTON JW, ET AL. (1987). The Neurobehavioral Cognitive Status Examination: A brief but quantitative approach to cognitive assessment. Ann Intern Med 107:481–485.

KIM SY, & BENOWITZ NL (1990). Poisoning due to class IA antiarrhythmic drugs: Quinidine, procainamide and disopyramide. Drug Safety 5:393–420.

KIRKWOOD CF (1991). Flumazenil—a benzodiazepine receptor antagonist. Pharmacol Ther March 243:252.

KLEIN RF, LINER VA, ZIPES DP, ET AL. (1968). Transfer from a cardiac care unit. Arch Intern Med 122:104–108.

KLERMAN GL (1981). Depression in the medically ill. Psychiatr Clin North Am 4:301–317.

KNOS GB, & SUNG Y-F (1991). Anesthetic management of the high-risk medical patient receiving electroconvulsive therapy. In A STOUDEMIRE & BS FOGEL (Eds.), Medical Psychiatric Practice (Vol. 1; Ch. 3, pp. 99–144). Washington, DC: American Psychiatric Press.

KOMRAD MS, COFFEY E, COFFEE KS, ET AL. (1984). Myocardial infarction and stroke. Neurology 34:1403–1409.

KONG KL, WILLATTS SM, & PRYS RC (1989). Isoflurane compared with midazolam for sedation in the intensive care unit. Br Med J 298:1277–1280.

KRANTZ DS (1980). Cognitive process and recovery from heart attack: A review and theoretical analysis. J Hum Stress 6:27–38.

KRIWISKY M, PERRY GY, TARCHITSKY D, ET AL. (1990). Haloperidol-induced torsades de pointes. Chest 98:482–484.

KUGLER J, DIMSDALE JE, HARTLEY LH, ET AL. (1990). Hospital supervised vs home exercise in cardiac rehabilitation: Effects on aerobic fitness, anxiety, and depression. Arch Phys Med Rehabil 71:322–325.

LARSON PB (1984). Coranary heart disease: Etiology, diagnosis and surgical treatment. In JB PIMM & JR FEIST (Eds.), Psychological risks of coronary bypass surgery (pp. 9–20). New York: Plenum.

LAZARUS HJ, & HAGENS JH (1968). Prevention of psychosis following open-heart surgery. Am J Psychiatry 124:1190–1195.

LEVENSON JL, & FRIEDEL RO (1985). Major depression in patients with cardiac disease: Diagnosis and somatic treatment. Psychosomatics 26:91–102.

LEVENSON JL, KAY R, MONTEFERRANTE J, ET AL. (1984). Denial predicts favorable outcome in unstable angina pectoris. Psychosom Med 46:25–32.

LEVINE J, WARRENBURG S, KERNS R, ET AL. (1987). The role of denial in recovery from coronary heart disease. Psychosom Med 49:109–17.

LEVINE SR (1989). Acute cerebral ischemia in a critical care unit: A review of diagnosis and management. Arch Intern Med 149:90–98.

LEVY AB, DAVIS J, & BIDDER TG (1984). Successful treatment of endogenous depression with alprazolam in a patient with recent cardiac disease: Case report. J Clin Psychiatry 45:480–481.

LINEAWEAVER WC, ANDERSON K, & HING DN (1988). Massive doses

of midazolam infusion for delirium tremens without respiratory depression. Crit Care Med 16:294–295.

LING MHM, PERRY PJ, & TSUANG MT (1981). Side effects of corticosteroid therapy. Arch Gen Psychiatry 38:471–477.

LINGAM VR, LAZURUS LW, GROVES L, ET AL. (1988). Methylphenidate in treating post-stroke depression. J Clin Psychiatry 49:151–153.

LONGSTRETH WT, INUI TS, COBB LA, ET AL. (1983). Neurologic recovery after out-of-hospital cardiac arrest. Ann Intern Med 78:173–181.

MARINI JJ (1988). Mechanical ventilation. Curr Pulmonol 9:164–208.

MARINI JJ (1991). Weaning from mechanical ventilation. N Engl J Med 324:1496–1498.

MARKS RM, & SACHAR EJ (1973). Undertreatment of medical inpatients with narcotic analgesics. Ann Intern Med 78:173–181.

MESAND P, PICKETT P, & MURRAY GB (1991). Psychostimulants for secondary depression in medical illness. Psychosomatics 32:203–208.

MCMURRAY TJ, COLLIER PS, CARSON IW, ET AL. (1990). Propofol sedation after open heart surgery: A clinical and pharmacokinetic study. Anaesthesia 45:22–26.

Medical Letter on Drugs and Therapeutics (1989a). Drugs for cardiac arrhythmias. 31:35–40.

Medical Letter on Drugs and Therapeutics (1989b). Drugs that cause psychiatric symptoms. 31:113–118.

Medical Letter on Drugs and Therapeutics (1986). Midazolam. 28:73–76.

MENDEL JG, & KHAN FA (1980). Psychological aspects of weaning from mechanical ventilation. Psychosomatics 21:465–471.

MENDELSOHN IE (1983). Liaison psychiatry and the burn center. Psychosomatics 24:235–243.

MENDOZA R, DJENDEREDJIAN AH, ADAMS J, ET AL. (1987). Midazolam in acute psychotic patients with hyperarousal. J Clin Psychiatry 48:291–292.

MENZA MA, MURRAY, GB, HOLMES VF, ET AL. (1988). Controlled study of extrapyramidal reactions in the management of delirious, medically ill patients; Intravenous haloperidol versus intravenous haloperidol plus benzodiazepines. Heart Lung 17:238–241.

MENZA MA, MURRAY GB, HOLMES VF, ET AL. (1987). Decreased extrapyramidal symptoms with intravenous haloperidol. J Clin Psychiatry 48:278–280.

MILANO MR, & KORNFELD DS (1984). Psychiatry and surgery. In L GRINSPOON (Ed.), Psychiatry update: The American Psychiatric Association annual review (Vol 3; pp. 256–277). Washington, DC: American Psychiatric Press.

MUMFORD E, SCHLESSINGER HJ, & GLASS GV (1982). The effects of psychological intervention on recovery from surgery and heart attacks: An analysis of the literature. Am J Public Health 72:141–152.

NELSON A, FOGEL B, & FAUST D (1986). Bedside cognitive screening instruments: A critical assessment. J Nerv Ment Dis 174:73–83.

NESHKES RE, & JARVIK LF (1982). Clinical psychiatry and cardiovascular disease in the aged. Psychiatr Clin North Am 5:171–179.

NEVIN M, COLCHESTER ACF, ADAMS S, ET AL. (1987). Evidence for involvement of hypocapnia and hypoperfusion in aetiology of neurological deficit after cardiopulmonary bypass. Lancet 2:1493–1495.

NEWTON RW (1975). Physostigmine salicylate in the treatment of tricyclic antidepressant overdosage. JAMA 231:941–943.

OEHLER JM, PETER MA, & SEYLER S (1989). Support groups: Are they really helpful in dealing with NICU stress? Neonat Network 8:21–25.

O'MALLEY PA, & MENKE E (1988). Relationship of hope and stress after myocardial infarction. Heart Lung 17:184–190.

PATTERSON DR, CARRIGAN L, QUESTAD KA, ET AL. (1990). Post-traumatic stress disorder in hospitalized patients with burn injuries. J Burn Care Rehabil 11:181–184.

PAYKEL ES, FLEMINGER R, & WATSON JP (1982). Psychiatric side effects of antihypertensive drugs other than reserpine. J Clin Psychopharmacol 2:14–39.

PENTEL P, & PETERSON CD (1980). Asystole complicating physostigmine treatment of tricyclic antidepressant overdose. Ann Emerg Med 9:588–590.

PETRIE WM, MAFFUCCI RJ, & WOOSLEY RL (1982). Propranolol and depression. Am J Psychiatry 139:92–94.

PETTY TL (1984). Acute respiratory failure in chronic obstructive pulmonary disease. In WC SHOEMAKER, WL THOMPSON & PR HOLBROOK (Eds.), Textbook of critical care (pp. 264–272). Philadelphia: WB Saunders.

RADEMAKER AW, KELLEN J, TAM YK, ET AL. (1986). Character of adverse effects of prophylactic lidocaine in the coronary care unit. Clin Pharmacol Ther 40:71–80.

RAYMOND M, CONKLIN C, SCHAEFFER J, ET AL. (1984). Coping with transient intellectual dysfunction after coronary bypass surgery. Heart Lung 13:531–539.

RAZIN AM (1985). Coronary artery disease: Reducing risk of illness and aiding recovery. In AM RAZIN (Ed.), Helping cardiac patients: Behavioral and psychotherapeutic approaches (Ch. 4, pp. 157–193). San Francisco: Jossey-Bass.

REGAN WM, MARGOLIN RA, & MATHEW RJ (1989). Cardiac arrhythmia following rapid imipramine withdrawal. Biol Psychiatry 25:482–484.

REICH P, REGESTEIN QR, MURAWSKI BJ, ET AL. (1983). Unrecognized organic mental disorders in survivors of cardiac arrest. Am J Psychiatry 140:1194–1197.

RICHELSON E (1984). Neuroleptic affinities for human brain receptors and their use in predicting adverse effects. J Clin Psychiatry 45:331–3336.

RICKELS K, FOX IC, GREENBLATT DJ, ET AL. (1988). Clorazepate and lorazepam: Clinical improvement and rebound anxiety. Am J Psychiatry 145:312–317.

RIE MA, & WILSON RS (1983). Acute respiratory failure. In J TINKER & M RAPID (Eds.), Care of the critically ill patient (pp. 311–340). Berlin: Springer-Verlag.

ROSENTHAL SL, SCHMID KD, & BLACK MM (1989). Stress and coping in a NICU. Res Nurs Health 12:257–265.

SALZMAN C (1984). Clinical geriatric psychopharmacology. New York: McGraw-Hill.

SALZMAN C (1988). Use of benzodiazapines to control disruptive behavior in inpatients. J Clin Psychiatry 49:13–15.

SANDERS KM, MURRAY GB, & CASSEM NH (1991). High-dose intravenous haloperidol for agitated delirium in a cardiac patient on intra-aortic balloon pump. J Clin Psychopharmacol 11:145–146.

SARAVAY SM, MARKE J, STEINBERG MD, ET AL. (1987). Doom anxiety and delirium in lidocaine toxicity. Am J Psychiatry 144:159–163.

SATEL SL, & NELSON JCN (1989). Stimulants in the treatment of depression: A critical overview. J Clin Psychiatry 50:241–249.

SAVAGEAU JA, STANTON BA, JENKIN CD, ET AL. (1982a). Neuropsychological dysfunction following elective cardiac operation: I. Early assessment. J Thorac Cardiovasc Surg 84:585–594.

SAVAGEAU JA, STANTON BA, JENKIN CD, ET AL. (1982b). Neuropsychological dysfunction following elective cardiac operation: II. A six-month reassessment. J Thorac Cardiovasc Surg 84:595–600.

SCHINDLER BA, SHOOK J, & SCHWARTZ GM (1989). Beneficial effects of psychiatric intervention on recovery after coronary artery bypass graft surgery. Gen Hosp Psychiatry 11:358–364.

Schocken DD, Greene AF, Worden TJ, et al. (1987). Effects of age and gender on the relationship between anxiety and coronary artery disease. Psychosom Med 49:118–126.

Schrader BJ, & Bauman JL (1986). Mexilitine: A new type I antiarrhythmic agent. Drug Intell Clin Pharm 20:255–260.

Schwamm LH, Van Dyke C, Kiernan RJ, et al. (1987). The Neurobehavioral Cognitive Status Examination: Comparison with the Cognitive Capacity Screening Examination and the Mini-Mental State Examination in a neurosurgical population. Ann Intern Med 107:486–491.

Settle EC, & Ayd FJ (1983). Haloperidol: A quarter century of experience. J Clin Psychiatry 44:440–448.

Shapiro PA (1991). Nortriptyline treatment of depressed cardiac transplant recipients. Am J Psychiatry 148:371–373.

Shapiro JM, Westphal LM, White PF, et al. (1986). Midazolam infusion for sedation in the intensive care unit: Effect on adrenal function. Anesthesiology 64:394–398.

Shaw PJ, Bates D, Cartlidge NEF, et al. (1985). Early neurological complications of coronary bypass surgery. Br Med J 291:1384–1387.

Shaw PJ, Bates D, Cartlidge NEF, et al. (1986a). Early intellectual dysfunction following coronary bypass surgery. Q J Med 58:59–68.

Shaw PJ, Bates D, Cartlidge NEF, et al. (1986b). Neurological complications of coronary artery bypass graft surgery: Six month follow-up study. Br Med J 293:165–167.

Shaw PJ, Bates D, Cartlidge NEF (1987). Long-term intellectual dysfunction following coronary artery bypass graft surgery: A six-month follow-up study. Q J Med 62:259–268.

Shelly MP, Mendel L, & Park GR (1987). Failure of critically ill patients to metabolize midazolam. Anaesthesia 42:619–626.

Shulman R (1984). Haloperidol therapy for agitated cardiac patients. Int Drug Ther Newlett 19:8.

Slogoff S, Girgis KZ, & Keats AS (1982). Etiologic factors in neuropsychiatric complications associated with cardiopulmonary bypass. Anesth Analg 61:903–911.

Smith LW, & Dimsdale JE (1989). Postcardiotomy delirium: Conclusions after 25 years? Am J Psychiatry 146:452–458.

Smith PLC, Treasure T, Newman SP, et al. (1986). Cerebral consequences of cardiopulmonary bypass. Lancet 1:823–825.

Snellen F, Lauwers P, Demeyere R, et al. (1990). The use of midazolam versus propofol for short-term sedation following coronary artery bypass grafting. Intens Care Med 16:312–316.

Sotaniemi K (1983). Cerebral outcome after extracorporeal circulation: Comparison between prospective and retrospective evaluation. Arch Neurol 40:75–77.

Soupios MA, & Lawry K (1987). Stress on personnel working in a critical care unit. Psychiatr Med 5:187–195.

Stern TA (1985). The management of depression and anxiety following myocardial infarction. Mt Sinai J Med 52:623–633.

Stern TA, Caplan RB, & Cassem NH (1987). Use of benzodiazepines in a coronary care unit. Psychosomatics 28:19–23.

Stoudemire A, Fogel BS, & Gulley LR (1991). Psychopharmacology in the medically ill: An update. In A Stoudemire & BS Fogel (Eds.), Medical psychiatric practice (Vol. 1; Ch. 2, pp. 29–97). Washington, DC: American Psychiatric Press.

Stovsky B, Rudy E, & Dragonnette P (1988). Comparison of two types of communication methods used after cardiac surgery with patients with endotracheal tubes. Heart Lung 17:281–289.

Stratton JR, & Halter JB (1985). Effect of a benzodiazepine (alprazolam) on plasma epinephrine and norepinephrine levels during exercise stress. Am J Cardiol 56:133–139.

Tesar GE, Murray GB, & Cassem NH (1985). Use of high-dose intravenous haloperidol in the treatment of agitated cardiac patients. J Clin Psychopharmacol 5:344–347.

Tesar GE, & Stern TA (1986). Evaluation and treatment of agitation in the intensive care unit. J Intens Care Med 1:137–148.

Tesar GE, & Stern TA (1988). Rapid tranquilization of the agitated intensive care unit patient. J Intens Care Med 3:195–201.

Thomas SA, Sappington E, Gross HS, et al. (1983). Denial in coronary care patients—an objective reassessment. Heart Lung 12:74–80.

Thompson DR (1989). A randomized controlled trial of in-hospital nursing support for first time myocardial infarction patients and their partners: Effects on anxiety and depression. J Adv Nurs 14:291–297.

Thompson DR, Webster RA, Cordle CJ, et al. (1987). Specific sources and patterns of anxiety in male patients with first myocardial infarction. Br J Med Psychol 60:343–348.

Thompson TL, & Thompson WL (1983). Treating postoperative delirium. Drug Ther 13:30–40.

Tollefson GD, Montague-Clouse J, & Lancaster SP (1991). The relationship of serum anticholinergic activity to mental status performance in an elderly nursing home population. J Neuropsychiatry Clin Neurosci 3:314–319.

Townes BD, Bashein G, Hornbein TF, et al. (1989). Neurobehavioral outcomes in cardiac operations: A prospective controlled study. J Thorac Cardiovasc Surg 98:744–782.

Treasure T, Smith PL, Newman S, et al. (1989). Impairment of cerebral function following cardiac and other major surgery. Eur J Cardiothorac Surg 3:216–221.

Trohman RG, Castellanos D, Castellanos A, et al. (1988). Amiodarone-induced delirium. Ann Intern Med 108:68–69.

Van Sweden B (1988). Rebound antidepressant cardiac arrhythmia. Biol Psychiatry 24:363–364.

Venus B (1980). Five year experience with Kistner tracheostomy tube. Crit Care Med 8:106–109.

Vogel WH, Miller J, DeTurck KH, et al. (1984). Effects of psychoactive drugs on plasma catecholamines during stress in rats. Neuropharmacology 23:1105–1108.

Volpe BT, Holtzman JD, & Hirst W (1986). Further characterization of patients with amnesia after cardiac arrest: Preserved recognition memory. Neurology 36:408–411.

Vree TB, Shimoda M, Driessen JJ, et al. (1989). Decreased plasma albumin concentration results in increased volume of distribution and decreased elimination of midazolam in intensive care patients. Clin Pharmacol Ther 46:537–544.

Vrobel TR (1989). Psychiatric aspects of congestive heart failure: Implications for consulting psychiatrists. Int J Psychiatry Med 19:211–225.

Walsh JJ, & Rho DS (1985). A speaking endotracheal tube. Anesthesiology 63:703–705.

Weiner MF, & Caldwell TA (1981). Stresses and coping in ICU nursing: 2. Nurse support groups on intensive care units. Gen Hosp Psychiatry 3:129–134.

Weiner MF, & Caldwell T (1983–1984). The process and impact of an ICU nurse support group. Int Psychiatry Med 13:47–55.

Weiner MF, Caldwell T, & Tyson J (1983). Stresses and coping in ICU nursing: Why support groups fail. Gen Hosp Psychiatry 5:179–83.

Weingarten S, Ermann B, Bolus R, et al. (1990). Early "stepdown" transfer of low-risk patients with chest pain. A controlled intervention trial. Ann Intern Med 113:283–289.

Weisman AD, & Hackett TP (1961). Death and dying as a psychiatric problem. Psychosom Med 23:232–257.

Williams RB Jr (1990). Do benzodiazepines have a role in the prevention or treatment of coronary heart disease and other major medical disorders? J Psychiatr Res 24(suppl 2):51–56.

WILLNER AE, & RABINER CJ (1979). Psychopathology and cognitive dysfunction five years after open-heart surgery. Compr Psychiatry 20:408–418.

WILSON L (1972). Intensive care delirium. Arch Intern Med 130:225–226.

WINNIE AP, & COLLINS VJ (1966). The search for a pharmacologic ventilator. Acta Anaesthesiol Scand (suppl) 23:63–71.

WISE MG, & CASSEM ND (1990). Psychiatric consultation to critical-care units. In A TASMAN, SM GOLDFINGER, & CA KAUFMANN (Eds.), American Psychiatric Press review of psychiatry (Ch. 22). Washington, DC: American Psychiatric Press.

WOLKOWITZ OM, REUS VI, WEINGARTNER H, ET AL. (1990). Cognitive effects of corticosteroids. Am J Psychiatry 147:1297–1303.

WOODS SW, TESAR GE, MURRAY GB, ET AL. (1986). Psychostimulant treatment of depressive disorders secondary to medical illness. J Clin Psychiatry 47:12–15.

WOOLSTON JL (1984). Psychiatric aspects of a pediatric intensive care unit. Yale J Biol Med 57:97–110.

YANG KL, & TOBIN MJ (1991). A prospective study of indexes predicting the outcome of trials of weaning from mechanical ventilation. N Engl J Med 324:1445–1450.

ZEGERS OD (1988). Espacio y tiempo en la unidad de cuidado intensivo. Actas Luso Esp Neurol Psiquiatr 16:246–254.

ZIMMERMAN LM, PIERSON MA, & MARKER J (1988). Effects of music on patient anxiety in coronary care units. Heart Lung 17:560–566.

13 Somatoform disorders, factitious disorders, and malingering

DAVID G. FOLKS, M.D., AND CARL A. HOUCK, M.D.

Somatoform disorders, factitious disorders, and malingering can be conceptualized as a continuum of abnormal illness behavior, distinguished from one another by whether signs and symptoms are consciously or unconsciously produced, and whether conscious or unconscious motivations account for the production of signs and symptoms (Pilowsky, 1990). These distinct diagnostic categories represent various degrees of illness behavior characterized by the process of somatization. *Somatization* may be defined as an individual's conscious or unconscious use of the body or bodily symptoms for psychological purposes or personal gain. The observed prevalence of somatization varies greatly according to the clinical setting and the medical specialty, with reported figures ranging between 5% and 40% of patient visits (Ford, 1983). A survey of more than 260,000 patients in 327 hospitals suggested that 5.2% were somatizers (Wallen et al., 1987). Somatization undoubtedly is more prevalent among clinical populations presenting to general practice, primary care, or internal medicine clinicians. Bain and Spaulding (1967) reported that the most common presenting symptoms in 4000 consecutive new patients in a general medical setting were abdominal and chest pain, dyspnea, headache, fatigue, cough, back pain, dizziness, and nervousness; 30% of these symptoms were attributed to "psychiatric disturbance." Specialty clinics, perhaps pain clinics, approach the higher rate of 40% (Coen & Sarno, 1989). Somatizing disorders result in an increased usage of both inpatient and outpatient medical services, significantly affecting the cost of medical care (Lipowski, 1988). Conservative estimates indicate that at least 10% of all medical services are provided for patients who have no evidence of organic disease; these figures do not include services provided for patients with identified psychiatric syndromes (Smith, Monson, & Ray, 1986).

Somatization is prominent in cultures that accept physical disease as an excuse for disability but reject psychological symptoms as acceptable for entry into the sick role. Similarly, third-party payers readily provide financial restitution for medical expenses or approve disability payments for physical problems, but they deny benefits for disturbances that are psychiatric. Several other secondary gain issues contribute significantly to the development and process of somatization and illness behavior. Ford (1986) has elucidated the specific motivations for somatization as follows: (1) manipulation of interpersonal relationships; (2) privileges of the sick role, including sanctioned dependency; (3) financial gain; (4) communication of ideas or feelings that are somehow blocked from verbal expression; and (5) the influence of intrapsychic defense mechanisms. Any consideration of etiologic factors and somatization must also include an appreciation of the distinction between the concepts of illness and disease (Eisenberg, 1977). *Disease* is defined as objectively measurable anatomic deformations and pathophysiologic states. *Illness* refers to those experiences associated with disease. Therefore illness takes into account the personal nature of suffering, alienation from one's usual gratifying activities, and a decreased capacity to participate in society, all of which significantly affect quality of life.

Irrespective of the underlying motivation(s), all of the possible explanations for somatization encourage a thorough diagnostic investigation and therapeutic approach that focuses on the psychosocial history while formulating the extent of the patient's disease, the magnitude of the illness, and the degree to which the patient is suffering and unable to engage in his or her usual activities (Folks, Ford, & Houck, 1990). Brodsky (1984) outlined family factors that predispose to somatization as depicted in Table 13-1. Steinhausen et al. (1989) examined these family factors in a controlled study of conversion disorder; these investigators found that in

TABLE 13–1. *Developmental Factors Within the Family Constellation Predisposing to Somatization*

1. Somatizers are present in the family of origin.
2. Parental figure(s) are demanding and unrewarding except during illness.
3. Parental figure(s) suffered a significant illness.
4. Coping mechanisms, except for illness behavior, are absent or unacceptable in response to a psychosocial crisis.
5. Disengagement from usual activities or the expression of anger or aggression is expressed through a repertoire of somatic reactions.
6. Illness is consciously feigned in order to achieve an identifiable goal, i.e., obtain secondary goals or avoid punishment or responsibilities.

TABLE 13–2. *Somatization: Principles of Clinical Management*

1. Presentation is considered in the context of psychosocial factors, both current and past.
2. Diagnostic procedures and therapeutic interventions are based on objective findings.
3. Therapeutic alliance is fostered and maintained involving the primary care and psychiatric physician.
4. Social support system and relevant life quality domains[a] are carefully reviewed during each patient contact.
5. Regular appointment schedule is maintained for outpatients, irrespective of clinical course.
6. Patient dialogue and examination and the assessment of new symptoms or signs are engaged judiciously; they usually primarily address somatic rather than psychological concerns.
7. Need for psychiatric referral is recognized early, especially for cases involving chronic symptoms, severe psychosocial consequences, or morbid types of illness behavior.
8. Any associated, coexisting, or underlying psychiatric disturbance is assiduously evaluated and steadfastly treated.
9. Significance of personality features, addictive potential, and self-destructive risk is determined and addressed.
10. Patient's case is redefined in such a way that management rather than cure is the goal of treatment.

[a]Quality of life is an elusive concept but includes the psychosocial domains of occupation, leisure, family, marriage, and health, as well as sexual and psychologic functioning.

From Folks et al. (1990). With permission.

comparison both to other psychiatric diagnostic groups and to normal controls, a greater proportion of psychiatric and medical diseases were present among conversion patients and their family members. Thus the family constellation and developmental history are important with respect to the development of somatization.

Several recommendations evolved during the 1980s regarding the clinical approach to cases involving somatization; these therapeutic insights have enabled primary care physicians and psychiatric consultants to collaborate more effectively. The ten guiding principles outlined in Table 13-2 have been summarized in order to achieve a consistent and successful therapeutic outcome (Folks, Ford, & Houck, 1990). Also, the clinical assessment of quality of life is an elusive exercise but generally includes an appreciation of seven psychosocial domains: occupational, leisure, family, marital, health, sexual and psychological functioning (Derogatis & Lopez, 1983). Of course, somatoform and factitious disorders, or malingering that involves somatization, may be associated with medical comorbidity, an underlying psychiatric syndrome, a coexisting personality disorder, or a psychosocial stressor of diagnostic or therapeutic significance, all of which have additional implications for clinical management (Lipowski, 1990; Stewart, 1990).

SOMATOFORM DISORDERS

Somatoform disorders are characterized by physical complaints lacking known organic basis or demonstrable physical findings in the presence of psychologic factors judged to be etiologic or important in the initiation, exacerbation, or maintenance of the disturbance. A comparison of the individual subtypes of somatoform disorder is presented in Table 13-4. The clinical features outlined in this table serve to facilitate the following discussion. Somatoform disorders emerged as a new category in the third edition of the American Psychiatric Association's (APA) *Diagnostic and Statistical Manual* (APA, 1980) and were expanded in the revised third edition (APA, 1987). Martin (1991) has conceptualized these disorders, emphasizing the evolution toward *DSM-IV* as shown in Table 13-3. Both *DSM-IV* and *ICD-10* (1989) are scheduled for publication in 1993 (Frances, Widiger, & Pincus, 1989; Sartorius, 1989) (The *ICD-9* was published in 1977.) Interestingly, an expanding international cooperation has now resulted in a common language with respect to somatoform and related disorders (Dauncey & Cooper, 1990).

Somatization Disorder

Formerly called "hysteria," or Briquet's syndrome, somatization disorder is a polysymptomatic disorder beginning in early life, affecting mostly women, and characterized by recurrent, multiple somatic complaints reflected in a diffusely positive review of systems (Guze, 1975). Estimates from the National Institutes of Mental Health collaborative Epidemiologic Catchment Area (ECA) studies show an estimated lifetime prevalence of approximately 0.4% for somatization disorder (Swartz, Hughes, & George, 1986). Although accurate point prevalence figures are not readily available, a 1% to 2% prevalence rate is suggested for women, with the female-male predominance approaching 20:1. Sampling difficulties and differential effects of diagnostic

TABLE 13–3. *Somatoform Disorders in ICD and the DSM Diagnostic System*

DSM-III—R Somatoform Disorders (1987)	DSM-I (1952)	DSM-II (1968)	ICD-9 (1977)	DSM-III (1980)	ICD-10 (1989, draft)	DSM-IV (predraft)
Body dysmorphic disorder				Atypical somatoform disorder: Dysmorphophobia	Included under: Hypochondriacal disorder	Body dysmorphic disorder
Conversion disorder	Conversion reaction	Hysterical neurosis: conversion type	Conversion disorder	Conversion disorder	Conversion disorder	Conversion disorder
Hypochondriasis	Psychoneurotic reaction: Other	Hypochondriacal neurosis	Hypochondriasis	Hypochondriasis	Hypochondriacal disorder	Hypochondriasis
Somatization disorder			Other neurotic disorders: somatization disorder, Briquet's disorder	Somatization disorder	Somatization disorder	Somatization disorder
Somatoform pain disorder		Hysterical neurosis: conversion type	Psychalgia, psychogenic pain	Psychogenic pain disorder	Persistent pain disorder	Pain disorder associated with psychological factors

Source: Reprinted with permission from Martin, RI (1991). Somatoform disorders in the general hospital setting. In Judd & Burrows (eds.), *Handbook of Studies on General Hospital Psychiatry*. New York: Elsevier (Biomedical Division).

instruments, i.e., *DSM-III-R* criteria versus Feighner and Research Diagnosis criteria continue to confound research efforts within this diagnostic category (Brown & Smith, 1991).

Somatization disorder is more commonly observed in lower socio-economic groups. Between 5% and 10% of primary care ambulatory populations have been reported to meet *DSM-III-R* diagnostic criteria, suggesting that somatization is the fourth most common diagnostic group seen in an ambulatory medical setting (Brown & Smith, 1991). A familial pattern is observed to affect 10% to 20% of female first-degree biologic relatives of females with somatization disorder; male relatives of females with this disorder show an increased risk of antisocial personality disorder and substance use disorders (Bohman, Cloninger, & Von Knorring, 1984). Adoption studies have indicated that both genetic and environmental factors contribute significantly to the risk for the disorder. Several other investigators have reported the tendency for somatization disorder to be associated with sociopathy, alcoholism, and drug addiction (Folks, Ford, & Houck, 1990). Although no specific data exist to establish the true economic impact of somatization disorder, its prevalence and its association with surgery and consumption of advanced technology for diagnosis or treatment must represent a great cost to society.

The most important diagnostic feature of somatization disorder is recurrent, multiple somatic complaints of several years' duration for which medical attention has been sought (Smith & Brown, 1990). Liskow et al. (1986) observed that this disorder is amazingly heterogeneous; that is, other psychiatric illnesses are likely to coexist. Brown, Golding, and Smith (1990) suggested that risk ratios are highest for panic disorder, major depression, schizophrenia, and obsessive compulsive disorder. Of course, the association of major depression with most if not all of the somatizing disorders is well established in studies of primary care populations in ambulatory clinics (Katon & Russo, 1989). Cases diagnosed as multiple personality also are known to be associated with somatization and other related disorders (Ross et al., 1989). Histrionic and antisocial personality disorder are the most frequently associated personality disorders and may appear in conjunction with symptoms of anxiety or depressed mood, as well as substance abuse (Folks, Ford, & Houck, 1990; Morrison, 1989a). Conversion symptoms also may be a prominent clinical feature with somatization disorder (Folks, Ford, & Regan, 1984). Antisocial behavior and occupational, interpersonal, or marital difficulties also are frequently observed (Ford, 1986). Patients with somatization disorder also may present in the context of acute illness, psychophysiologic symptoms, or other chronic medical conditions. (Manu, Lane, & Matthews, 1989). Thus a mix of primarily psychogenic and organic symptoms is the rule rather than the exception, making diagnosis extraordinarily challenging. The use of psychological testing in patients has not been studied extensively enough to be of diagnostic value. Personality testing may reveal traits that impinge negatively on the course of somatization (e.g., antisocial traits, passive-

TABLE 13–4. *Somatoform or Factitious Disorder of Malingering: Comparison of Clinical Features*

Diagnostic Subtype	Clinical Presentation	Demographic/ Epidemiologic Features	Diagnostic Features	Management Strategy
Somatoform disorders				
Somatization disorder	Polysymptomatic Recurrent/chronic "Sickly" by history	Younger age Female predominance (20:1) Familial pattern; 5–10% incidence in primary care populations	ROS profusely positive Multiple clinical contacts Polysurgery	Therapeutic alliance Regular appointments Crisis intervention
Conversion disorder	Monosymptomatic Mostly acute Simulates disease	Highly prevalent Female predominance Younger age Rural/lower social class Less educated/ psychologically unsophisticated	Simulation incompatible with known physiologic mechanisms or anatomy	Suggestion and persuasion Multiple techniques
Pain disorder associated with psychological factors	Pain syndrome simulated	Female predominance (2:1) Older: 4th or 5th decade Familial pattern Up to 40% of pain populations	Simulation or intensity incompatible with known physiologic mechanisms or anatomy	Therapeutic alliance Redefine goals of treatment Antidepressant medications
Hypochondriasis	Disease concern or preoccupation	Previous physical disease Middle or older age Male/female ratio equal	Disease conviction amplifies symptoms Obsessional	Document symptoms Psychosocial review Psychotherapy
Body dysmorphic disorder	Subjective feelings of ugliness or concern with body defect	Adolescence or young adult ?Female predominance Largely unknown	Pervasive bodily concerns	Therapeutic alliance Stress management Psychotherapy Antidepressant medications
Factitious disorders				
Factitious with physical symptoms	Feigned or simulated physical symptoms or signs or disease	Female, younger, socially conforming Employed in medical field Social supports often available	Feigned illness No external goal of simulation is obvious Organ mode of presentation varies but is physical	Confront as appropriate Redefine illness as psychiatric Psychiatric referral
Factitious with psychological symptoms	Multiple hospitalizations	Female, younger, socially conforming Employed in medical field Social supports often available	Feigned illness No external goal of simulation is obvious Mode of presentation varies but is psychiatric	Confront as appropriate Redefine illness as psychiatric Psychiatric referral
Münchausen syndrome	Multiple hospitalizations	Male, younger, socially nonconforming Social supports often unavailable	Feigned illness Pathologic liar Geographic wandering Antisocial features Frequently leaves against medical advice	Recognize Confront Avoid invasive or iatrogenic procedures or treatments Social work referral
Malingering	Feigned or simulated with physical or psychological symptoms	?Male predominance Psychosocial stress or failure present	Feigned illness External incentives for disease present	Confront Consider psychiatric or psychosocial problems

Source: Reprinted with permission from Folks DG, Ford CV, & Houch CA (1990). Somatoform Disorders, Factitious Disorders, and Malingering. In Stoudemire A (ed.), *Clinical Psychiatry for Medical Students*. Philadelphia: JB Lippincott.

TABLE 13–4. *(Continued)*

Prognostic Outlook	Associated Disturbances	Primary Differential Diagnosis	Psychologic Processes Contributing To Symptoms	Motivation for Production
Poor to fair	Histrionic personality Sociopathy Substance/alcohol use Many life problems Conversion	Physical disease Depression	Unconscious Cultural/developmental	Unconscious psychologic factors
Excellent except for chronic conversion	Drug/alcohol dependence Sociopathy Somatization disorder Histrionic personality	Depression Schizophrenia Neurologic disease	Unconscious Psychologic stress or conflict may be present	Unconscious psychologic factors
Guarded, variable	Depression Substance/alcohol use Dependent/histrionic personality	Depression Psychophysiologic disorder Physical disease Malingering/disability syndrome	Unconscious Acute stressor/ developmental Physical trauma may predispose	Unconscious psychologic factors
Fair to good Waxes and wanes	Obsessional "neurosis" Depression-anxiety	Depression Physical disease Personality disorder Delusional disorder	Unconscious Stress—bereavement Developmental factors	Unconscious psychologic factors
Unknown	Anorexia nervosa Psychosocial distress Avoidant/compulsive personality disorder	Delusional psychosis Depression Somatization disorder	Unconscious Self-esteem factors	Unconscious psychologic factors
Fair to good except Münchausen subtype	Depression Borderline or other personality disorder	Malingering Conversion disorder Hypochondriasis Depression Schizophrenia	Unconscious Developmental—family factors Masochism, dependency, and mastery are utilized	Conscious effort to assume patient status
Fair to good except Münchausen subtype	Schizophrenia Borderline or other personality disorder	Malingering Conversion disorder Hypochondriasis Depression Schizophrenia	Unconscious Developmental—family factors Masochism, dependency, and mastery are utilized	Conscious effort to assume patient status
Poor	Antisocial, histrionic, or borderline personality	Malingering Conversion disorder Hypochondriasis Depression Schizophrenia	Unconscious Developmental—family factors Masochism, dependency, and mastery are utilized	Conscious effort to assume patient status
N/A	Antisocial personality Substance abuse/ dependence	Factitious disorder Personality disorder Ganser syndrome Münchausen syndrome Major psychosis Disability syndrome	Conscious but may display other psychopathology	Conscious response to external incentives

aggressive traits, paranoid traits) and alter the classic presentation of somatization disorder and the therapeutic alliance.

The etiologic foundations of somatization disorder are not readily discernible, although the forementioned familial incidences and association with antisocial and histrionic personality disorder, as well as substance and alcohol use disorders, suggest a biologic predisposition (Ruegg, 1990). The general use of somatizing behavior in the family of origin or culture also may predispose to the syndrome (Brodsky, 1984). Somatization disorder develops in the context of a lifelong pattern in which somatic symptoms are readily utilized to communicate distress and cope with ongoing psychosocial stressors and interpersonal problems. This concept is supported by a controlled study in which significantly more women with somatization disorder had been molested as children (Morrison, 1989b). The ability to use verbal language to communicate moods and emotions is gained through a complex process theorized to be influenced by a number of factors: sociocultural, psychodynamic, developmental, cognitive, and behavioral. It has been postulated that, for a variety of reasons, some patients never develop the ability to describe feelings with affective language but, rather, describe feelings with somatic language (Lesser, Ford, & Friedmann, 1979). This tendency to speak in a somatic vocabulary to represent emotional distress has been discussed in more detail (Stoudemire, 1991) and has been termed "somatothymia."

Attention has been postulated to be abnormal in somatization disorder, and a neurophysiologic study of attention in somatization disorder patients was conducted presenting stimuli to subjects whose cortical potentials were measured by electroencephalography (EEG). The study found a difference in mismatch negativity, defined as the difference in cortical potentials evoked by background stimuli and those evoked by target stimuli, with somatizers showing smaller mismatch negativity. The authors concluded that one interpretation of this difference is that somatizers distinguish relevant and irrelevant stimuli less well than normals, and they questioned whether this finding suggests a subtle neuropsychological disturbance in somatizers warranting further study (James et al., 1989). Perhaps newer concepts and refinements in diagnosis, methodology, and technology will enable similar lines of inquiry to continue regarding the etiology and pathogenesis of this disorder.

Smith, Monson, & Ray (1986) and Quill (1985) have suggested similar management principles in the clinical approach to somatization disorder, as shown in Table 13-5. As depicted, psychiatric consultation or referral

TABLE 13–5. *Somatization Disorder: Principles of Management*

1. Diagnose patient correctly.
2. Identify associated diagnoses (comorbidity).
3. Accept patient's somatizing uncritically.
4. Ensure adequate medical evaluation.
5. Reassure patient of continued physician involvement.
6. Offer patient face-saving, benign somatic explanations (muscle strain, chest tension).
7. Use analogies to introduce concept of stress and its effects on the body, e.g., blushing.
8. Inform patient of realistic goals, i.e., improved work, decreased medical visits.
9. Avoid illusion of pursuing a cure.
10. Use physical examination regularly. Base tests and procedures, especially invasive ones, on objective physical abnormalities.
11. Praise adaptive, healthy behavior while ignoring or minimizing illness behavior.
12. Use benign treatments,when possible, on a time-limited basis, e.g., physical therapy, exercise.
13. Encourage regular medical visits to monitor symptoms and psychosocial variables.
14. Discourage use of alcohol, drugs, and over-the-counter medicines for self-treatment.
15. Discourage doctor-shopping and polypharmacy; encourage utilization of one primary physician.
16. Refer to psychiatrist if agreeable or for major or treatment-resistant psychiatric problems.
17. Consider the possibility of "true" physical disease concurrent with somatized symptoms.

Source: Adapted from Quill (1985) and Smith (1985).

is recommended only for severe cases, or for patients with significant psychosocial complications or coexisting psychiatric disturbances. Unfortunately, many patients are highly resistant to psychiatric referral, so treatment is often by the primary care physician using the strategy outlined in the introductory section. A cure is seldom achieved, but recurrent debilitating symptoms can be relinquished and perhaps exchanged for controlled dependence on a clinic or a specified clinician or therapist. The psychiatric consultant's role usually involves crisis intervention or attention to associated disturbances, with the key therapeutic interventions resulting from the therapeutic alliance with the primary clinician.

Conversion Disorder

Conversion symptoms have been described since antiquity and represent a type of somatoform disorder in which there is a loss or alteration in physical functioning suggesting a physical disorder that cannot be explained on the basis of known physiologic mechanisms. Conversion disorder usually is seen in ambulatory settings or emergency departments, and it frequently runs a rather short-lived course, responding to nearly any therapeutic modality that offers a suggestion of cure. Conversion symptoms are common in medical practice; prevalence rates of 20% to 25% have been estimated for patients admitted to a general medical setting (Ford, 1983). Hospitalized patients have consistently been observed to show conversion symptoms ranging between 5 and 14% of all psychiatric consultations (Folks, Ford, & Regan, 1984). Conversion symptoms also are ubiquitous among randomly selected psychiatric clinic patients; and they are particularly prevalent among patients with drug abuse, sociopathy, alcoholism, and somatization disorder or "hysteria" (Ford & Folks, 1985). Conversion disorder occurs more typically in females; when present in males it tends to be associated with a history of industrial accidents, or in the context of military duty or occupational distress. Conversion disorder reportedly encompasses ages ranging from early childhood into the ninth decade. The disorder appears more frequently in lower socioeconomic groups and in rural or less psychologically sophisticated populations. The more primitive and grossly nonphysiologic conversion symptoms are observed in patients of rural background; by contrast, conversion symptoms observed in better-educated populations more closely simulate known disease.

Diagnostic descriptions and terminology applied to conversion have changed markedly since the 1960s. The diagnosis is unique, implying that specific psychodynamic mechanisms account for the disturbance. In contrast to somatization disorder, which is chronic and polysymptomatic with many organ systems involved, conversion disorder generally is sporadic and monosymptomatic, possessing a symbolic relation between the underlying psychological conflict and the disturbance in physical functioning. A number of traditional clinical features previously associated with conversion—for example, secondary gain, histrionic personality, and "la belle indifference"—appear to have no diagnostic significance; these features are now regarded only as supportive of the diagnosis, having no firm diagnostic validity (Folks, Ford, & Regan, 1984). Moreover, Folks, Ford, and Regan (1984) noted that 20% of their study population of 50 cases with conversion symptoms had coexisting neurologic disorders, including seizures, delirium, head trauma, vascular disease, and cerebral palsy in the rank order of their existence. In an earlier study, Raskin, Talbott, and Meyerson (1966) also discouraged reliance on the traditional features, noting that 22% of their study group "of hysterics" were later found with organic illness judged responsible for the conversion symptoms.

The diagnosis of conversion must rest ultimately on positive clinical findings clearly indicating that the symptom does not derive from organic disease, for example, the demonstration of normal motor function in patients with "paralysis" (Weintraub, 1983). Common examples of conversion symptoms include paralysis, abnormal movements, aphonia, blindness, deafness, or pseudoseizures. Pseudoseizures or psychogenic seizures represent one of the more difficult diagnostic entities within this nosologic category (Desai, Porter, & Penry, 1982). Lempert and Schmidt (1990) reported on 50 patients definitively diagnosed with psychogenic seizures, focusing their study on the natural history and clinical outcome. After following these cases for a mean of two years, concomitant epilepsy was "definite" in 8% and "possible" in 14% while 50% were being prescribed antiepileptic drugs. During the period of follow-up, coexisting psychiatric disturbances emerged in most cases (66%). Depression was the most common psychiatric disorder (24%), whereas "hysterical personality" was found less frequently (8%). The low rate of hysterical or histrionic personality disorder among patients with conversion disorder is consistent with prior reports (Ford & Folks, 1985). Also consistent with previous findings was the observation that individuals with acute onset of psychogenic seizures and readily identifiable stressors were significantly more likely to become seizure-free, in contrast to cases that involved a combination of long-standing physical, psychic, and social problems—56% of the cases studied.

Conversion symptoms almost always conform to a *patient's* concept of disease rather than to pathophysiologic mechanisms or anatomical patterns. For example, one report noted the unique and complex manifestations of tremor in 24 patients, all of whom showed unusual temporal profiles, absence of other neurologic signs, and inconsistent and incongruent symptomatology (Koller et al., 1989). When conversion symptoms occur in isolation, it is appropriate to assign a diagnosis of conversion disorder; however, conversion symptoms also may occur as a part of other major psychiatric syndromes (e.g., somatization disorder, schizophrenia, depression, or cognitive impairment disorders) or as a component of medical or neurologic disease. Evidence of current or prior neurologic disorder also is frequently identified, and a number of neurologic cases have been reported in the literature noting that the cases were initially misdiagnosed as conversion disorder (Merskey & Buhrich, 1975).

Clinical descriptions of conversion phenomena date back to at least 1900 BC, at which time the Egyptian papyri attributed symptoms to "wandering of the uterus." Conversion symptoms have been conceptualized to arise from stressful environmental events acting on the *affective*, or right, hemisphere of the brain in predisposed individuals (Flor-Henry et al., 1981; Ludwig, 1972; Whitlock, 1967). Some patients' symptoms conform to

Freud's concept of "conversion" in reference to the concept that conversion results from the substitution of a somatic symptom for a repressed idea or psychologic conflict (Ford & Folks, 1985). Conversion also may be a means to express forbidden feelings or ideas, as a kind of communication via pantomime or mimicry when direct verbal communication is blocked (Ford & Folks, 1985), or it may simply serve as an acceptable means of enacting the sick role or as an acute entry into illness behavior (Kimball & Blindt, 1982). Individuals with conversion symptoms achieve secondary gains by avoiding certain responsibilities or noxious situations and frequently are able to control or manipulate the behavior of others. Thus classical conditioning paradigms provide possible explanations for some conversion phenomena. Learned symptoms of illness are used later as a means of coping with particularly stressful situations. Cases with newly learned illness behavior resulting in conversion symptoms also may occur, where secondary gain exerts primary control over symptom selection. This concept was illustrated in a report of five patients who manifested psychogenic respiratory distress that was superimposed on other preexisting psychogenic or conversion symptoms on a neurological ward. The emergence of the newer and more severe conversion symptoms resulted in longer hospital stays (Walker et al., 1989). Manifestations of conversion symptoms also may be culture-bound, e.g., the Jewish variant of dybbuk possession (Bilu & Beit-Hallahmi, 1989) or beliefs about sorcery described in southern India, where the cultural influence affects the onset and course of conversion disorder (Keshavan, Narayanan, & Gangadhar, 1989). These reports of culture-bound syndromes in which conversion is manifested are congruent with earlier diagnostic presumptions (Ford & Folks, 1985). More recent theories about the etiology and pathogenesis of conversion have proposed social, communication, and sophisticated neurophysiological mechanisms in a detailed discussion (see Folks, Ford, & Regan, 1984). Refinements in diagnostic tools used for imaging and neurophysiologic exploration have undoubtedly improved our ability to test of these theories. However, their discriminating abilities in differentiated diagnosis and specificity in identifying underlying diagnosis remains a topic of further study.

A wide variety of treatment techniques have been used successfully in cases of conversion disorder. Brief psychotherapy focusing on stress and coping style, and suggestive therapy—sometimes using hypnosis or amobarbital interviews that focus on symptom removal—are commonly employed with *amazing efficacy*. More recently, pharmacologic interventions have been introduced for the treatment of target symptoms, i.e., anxiety, mood, or sleep disturbances accompanying conversion disorder (Lazarus, 1990). A short hospital admission also can be helpful, particularly when symptoms are significantly disabling or alarming. Hospitalization may serve to remove the patient from the stressful situation, demonstrate to the family that the matter is important, and facilitate resolution of the psychologic trauma. Many patients experience spontaneous remission of symptoms or demonstrate marked or complete recovery after a brief therapeutic intervention. In fact, prompt recovery is the rule, and few patients require long-term management. For example, one case involved a healthy 29-year-old farmer who experienced acute blindness when confronted with a mortgage foreclosure notice; this individual's conversion blindness was ultimately "cured" with saline eyedrops. The psychiatric consultant merely told the patient that the prescribed "special" eyedrops had "cured several others with a similar presentation in a matter of days."

Hypnosis often has been utilized as a diagnostic probe with respect to etiologic considerations and confirms the importance of psychologic contributors to the syndromes (Ford & Folks, 1985). Moreover, hypnosis can be used successfully as a suggestive technique for both diagnosis and treatment during psychotherapy or for symptom removal. The amobarbital interview can be used similarly and has been shown to be a useful suggestive technique for diagnosis, therapy, and symptom removal (Dickes, 1977; Folks, Ford, & Regan, 1984; Stevens, 1968). However, such therapeutic techniques that achieve only symptom removal may carry significant clinical consequences if pertinent biopsychosocial issues or problems are not adequately addressed. Menza (1989) illustrated this concept describing a case where a suicide attempt followed the removal of conversion paralysis with the use of an amytal interview.

In contrast to acute episodes of conversion, chronic conversion disorder often carries a poor prognosis and is notoriously difficult to treat, more prognostically similar to somatization disorder and somatoform pain disorder. Behavior modification and psychotherapeutic techniques focusing on the family constellation may be particularly beneficial with recalcitrant conversion symptoms, even when maintained with significant secondary gain (Ford & Folks, 1985). Again, suggestion and persuasion, irrespective of technique, are useful in these cases (Ford & Folks, 1985).

Pain Associated with Psychological Factors

Estimates by the National Institute of Neurological Communication Disorders and Stroke suggest that as many as 75 million Americans are afflicted with chronic pain, at a cost of some $40 billion per annum (Bonica, 1976). Multidisciplinary pain clinics have developed in which patients can be effectively evaluated and treated by consultants from a variety of disciplines: anesthesia, neurology, neurosurgery, orthopedics, psychiatry, psychology, social work. A variety of diagnostic categories for pain are encountered in clinical practice, including previously termed psychogenic or idiopathic (somatoform) pain, muscle tension syndromes, masked depression, and varying degrees of disability syndromes or malingering to be discussed in the concluding section of this chapter. Of course, pain arising from neuropathy or sequelae medical-surgical illness (i.e., posttherapeutic, phantom limb) often require and respond to a specific neurologic intervention.

No substantial information exists on the prevalence of pain in association with accentuating psychiatric factors, although it is estimated that a significant proportion of pain patients—perhaps as many as 40%—have pain that is aggravated by psychologic factors (Stoudemire & Sandhu, 1987). Pain syndromes more frequently present during the fourth or fifth decade of life, usually appearing acutely with increasing severity and pervasiveness (Chaturvedi, 1989). These disturbances are diagnosed in females twice as frequently as in males. Evidence now exists to suggest a familial pattern, with first-degree biologic relatives being at higher risk for developing a pain syndrome with psychologic factors predominating. A known familial pattern that includes a history of depression or alcohol dependence occurs with greater frequency than might be expected within the general population (Folks et al., 1990).

The prominent diagnostic feature of a pain associated with psychologic factors is preoccupation with pain in the absence of significant physical findings that might account for the pain or its intensity. As with conversion symptoms, the pain is manifested inconsistently with respect to known anatomic distribution; or if it mimics a known disease entity, such as angina or sciatica, the pain cannot be adequately explained on the basis of any existing organic pathology. Furthermore, pathophysiologic mechanisms do not fully account for the pain; and as with conversion, the etiologic impact of any psychologic factors must be appreciated, whether they (1) are clearly identified to precipitate the pain; (2) permit the individual to avoid some unacceptable or noxious activity; or (3) result in significant psychosocial support that might not otherwise be forthcoming.

The differential diagnosis for pain must include consideration of other psychiatric syndromes, e.g., somatization disorder, mood and anxiety disorder, schizophrenia, or other syndromes in which complaints of pain are common. A significant minority of patients presenting with pain ultimately are found to be malingering. In these cases the symptoms are manifested for the sole purpose of obtaining an obviously explainable or recognizable goal or are feigned in an attempt to secure narcotic analgesics or other addictive substances. Other important differential diagnoses are psychophysiologic disorders or other conditions prominently affected by psychologic factors, i.e., muscle contraction headache, muscular spasm back pain, irritable bowel syndrome, proctalgia fugax, or other recognizable syndromes that may involve a specific pathophysiologic mechanism that reasonably accounts for the pain.

Major depression or depressive symptoms often are present as an underlying or coexisting disorder (Benjamin et al., 1988). The changes in mood and in personality that accompany pain frequently mirror the symptoms of major depression (Reuler, Girard, & Nardone, 1980). Indeed, pain patients in general seem to report symptoms or changes in their physiologic response, with the emergence of vegetative symptoms, e.g., sleep, appetite, energy, and libido disturbances. Of course, personality disorders also may coexist with pain; histrionic and dependent personality traits have been identified as more frequent in this clinical population (Ford, 1983; Reich, Tupin, & Abramowitz, 1983).

Pain may be conceptualized on the basis of psychosocial features associated with the case (Simpson & Gjerdingen, 1989); evidence of past somatization, the presence of a symptom model, prominent guilt, and a history of physical or psychological abuse by either a parent or a spouse are thought to carry some diagnostic weight. However, these historical findings together with negative physical or laboratory or radiographic findings do not necessarily imply that the pain symptoms are attributable to psychosocial factors. Moreover, as with all forms of somatization, a complete separation of etiology into organic or psychogenic causes may be difficult and in some ways unnecessary. Full appreciation of the etiologic factors involved in any form of pain is complicated, because the clinician must account for the economy of secondary gain or reinforcement, understand abnormal illness behavior, and evaluate the role of unconscious motivations and primary gain. Individuals receiving or likely to procure compensation are prone to confound attempts at clinical management; individuals with compensation neurosis, adjustment disorder with work inhibition, or other, more pervasive disability syndromes also are more difficult to diagnose.

Pain is a subjective experience and cannot be measured easily; pain patients typically perceive or describe their symptoms such that the presentation is compatible with previous illnesses or the symptoms are congruent with a conceptualization of what other known persons may have experienced. The marked variation observed in the perception of pain is possible owing to the neuroanatomy of the brain, which has several associated pathways linking sensory cortex, the limbic system, and the spinal cord.

The bridging of clinical observation and laboratory research is responsible for much of the progress during the 1980s in understanding pain syndromes (Fields, 1988). Of course Melzack and Wall (1965) had provided earlier their gate control hypothesis. However, the modulation of pain and the concept of stimulation-produced analgesia provided a breakthrough in demonstrating that an animal's response to a noxious stimulus could be blocked (Mayer, Akil, & Liebeskind, 1974; Reynolds, 1969). Subsequently, a similar response was noted to be possible in chronic pain patients (Hosobuchi, Adams, & Linchitz, 1977; Richardson & Akil, 1977). Fields (1988) expanded this concept, noting that nociceptive transmission at the level of the spinal cord is presumably modulated in a bidirectional fashion by two cell types that enhance or decrease pain perception, respectively. Interestingly, this model supports a neural mechanism to amplify pain; that is, a patient might actually increase the intensity of pain or perpetuate pain through this proposed mechanism.

Generally, etiologic theories are not mutually exclusive and may be deemed congruous, representing our current understanding of neuroanatomic and physiologic mechanisms, psychodynamics, and learning theories about pain. An eloquent illustration from learning theory uses transactional analysis concepts in which pain is thought to be used for interpersonal manipulation and control (Sternbach, 1974). Psychological tests, most popularly the Minnesota Multiphasic Personality Inventory (MMPI), are routinely used in pain clinics to identify pertinent psychologic factors. Elevations of scores on the three scales labeled hypochondriasis (Hy), depression (D), and hysteria (Hs) often are seen in patients with pain. The scores often show a great elevation of the Hy and Hs scales, with a lesser elevation of the interposed D scale, leading to a configuration termed the "conversion V" (Leavitt, 1985). It has been postulated that these patients have significant emotional distress, as shown in the high D scale, being expressed in the somatic form of hypochondriasis or hysteria, the scores for which are elevated even higher than the depression scale.

However, psychologic tests cannot be carried out or interpreted in a vacuum, and they are unlikely to be useful diagnostically without other supporting data. In short, correlations do *not* distinguish organic from nonorganic patients for whom "conversion" or psychogenic forces are assumed to account for the pain. For example, in a study comparing MMPI profiles of 102 patients with organic back pain with 93 patients with nonorganic back pain, the conversion V was of no use for discriminating between them (Leavitt, 1990).

The use of electromyography, nerve conduction studies, imaging studies (i.e. magnetic resonance imaging or thermography) have proved useful in selected cases of somatoform pain disorder (Bonica & Ventafridda, 1979; Goldberg, Sokol & Cullen, 1987). Numerous treatments for pain have been suggested in the literature (Goldberg, 1990; Reich, Tupin, & Abramowitz, 1983; Stoudemire & Sandhu, 1987).

General principles of treatment include a psychotherapeutic approach focusing on either or both of the following psychologic factors: (1) identified precipitants, activities, or responsibilities that are avoided because of the pain; and (2) the psychosocial support system including family constellation, marital and sexual problems, depression, and other associated syndromes that are identified and vigorously treated. Thus, treatment should not reinforce maladaptive pain syndromes with psychologic determinants. (See Chapter 17.)

A number of specific nonpsychopharmacological treatments are useful in pain patients: transcutaneous nerve stimulation, nerve blocks, biofeedback, and other forms of behavioral or psychotherapy. Irrespective of the selected intervention, attention must be given to the patient's psychosocial, marital, and family situation and to the meaning and significance of the pain itself. Regarding pharmacologic modalities, narcotics or other addicting substances are rarely indicated, but the psychotropics may serve as adjuvants. The cyclic antidepressants, particularly those that act preferentially via serotonin systems, often afford pain relief at a dosage below that generally believed to be effective for depression, i.e., doxepin or desipramide 50 to 75 mg/day. The use of antidepressant therapy continues to gain support in the literature as a primary intervention (Pilowsky & Barrow, 1990; Valdes et al., 1989). Studies such as the one conducted by Mellerup et al. (1990) may lend further support to a pharmacologic approach; they showed that depressed pain patients demonstrated serotonergic abnormalities (in this case lowered imipramine binding) when in pain compared to patients with only depression. Monoamine oxidase (MAO) inhibitors and other agents with serotonergic selectivity, e.g., buspirone, may be reasonable alternatives or may be combined with some of the aforementioned nonpharmacologic therapies (Folks, Ford, & Houck, 1990). The management of chronic pain is also discussed at length in Chapters 16 and 17.

Hypochondriasis

Many inconsistencies and contradictions are found in the literature on hypochondriasis. The actual prevalence of hypochondriasis as a disorder distinct from other somatoform disorders is unknown, with estimates varying with culture and diagnostic criteria. One study suggested a 6-month prevalence rate between 4.2% and 6.3% in a general medicine clinic, a rate somewhat lower than that for anxiety and similar to that for alcohol abuse (Barsky, Wyshak, & Klerman, 1990b).

Hypochondriasis typically begins during middle or older age and is equally common in men and women—features that serve to distinguish it from somatization and conversion disorder. The phenomenon of "transient hypochondriasis" complicates our understanding of this nosologic category, as does the potent interaction between this diagnostic disorder and other Axis I and Axis II disorders (Barsky, Wyshak, & Klerman, 1990a). No significant relation between personality disorder or any Axis I disorder including depression is clearly established (House, 1989). Only a few twin studies have been reported, and inadequate evidence exists for conclusions about the importance of genetic factors in hypochondriacal populations. Developmental and other predisposing factors consistently reveal the importance of parental attitudes toward disease, previous experience with physical disease, lower social class, and culturally acquired attitudes relevant to the epidemiology and etiology of the disorder per se (Ball & Clare, 1990).

Pilowsky (1970) defined hypochondriasis as "a concern with health or disease in one's self that is present for the major part of the time." The disturbance may prove to be adaptive or to be consistent with strong motivations to assume the sick role as outlined in the introductory section of this chapter. The somatic or bodily preoccupation in hypochondriasis must be unjustified with respect to the organic pathology present, and these patients do not respond more than temporarily to clear reassurance concerning their health status. The core features of hypochondriasis appear to consist of a complex of attitudes: disease fear, disease conviction, and bodily preoccupation or absorption associated with multiple somatic complaints. The medical history often includes magnificent and elaborate details of doctor-shopping, deteriorating doctor-patient relationships, and associated feelings of frustration and anger. Anxiety, depressed mood, and obsessive personality features often are observed during the clinical encounter. The clinical course of hypochondriasis is chronic, with waxing and waning of symptoms (Kellner et al., 1989). Complications may arise secondary to numerous exposures to medical care and the dangers of repeated diagnostic or therapeutic procedures.

The most important step in the differential diagnosis of hypochondriasis is the exclusion of physical disease. A number of organic diagnoses can be difficult to identify early in their course, including myasthenia gravis, multiple sclerosis, slowly deteriorating degenerative diseases of the neurologic system, endocrinopathies, systemic diseases (e.g., systemic lupus erythematosus) and occult neoplastic disorders. However, the diagnosis should not be one of exclusion; a positive diagnosis can be achieved through a careful history in the absence of objective physical findings and with the recognition that an emotional component positively and definitively contributes to the disorder. Among psychiatric diagnoses, the most important differential diagnosis is major depression; the entity of masked depression, or "secondary hypochondriasis," is repeatedly cited in the literature (Pilowsky & Barrow, 1990). Panic disorder also is frequently diagnosed in individuals with prominent cardiorespiratory complaints or other symptoms of anxiety (Katon, 1990). Syndromes are well established when hypochondriasis qualifies as a delusional state, presumably with prominent psychobiologic contributors (King, 1990) or psychotic syndromes of hypochondriasis that mimic body dysmorphic disorder (De Leon, Bott, & Simpson, 1989).

Patients with hypochondriasis seem to perceive their bodily functions more acutely than others. Barsky's notion of the "amplifying somatic style" emphasizes that these individuals selectively perceive bodily functions and attribute their symptoms to physical disease (Barsky, 1979). Preoccupation or conviction that a disease process exists and absorption in health concerns represent powerful motives for attending selectively to bodily sensations. Anxiety escalates and further serves as a motive for selective perception. Related ongoing fears include concerns about aging, fear of death, and vulnerability to disease (Barsky & Wyshak, 1989).

Depression or dysphoria also may emerge as the hypochondriacal patient suffers and feels helpless or hopeless as the illness continues to evolve. Anger also may arise in the course of hypochondriasis as a result of unmitigated distress, conflicting diagnoses, ineffective treatments, and experiences of encountering impatient, rejecting, or hostile physicians. These emotions are compounded by doctor-shopping, medication problems, conflicting opinions from physicians, and iatrogenic phenomena, as well as the specific personality features that are a focus of clinical concern (Kellner, 1987).

A critical therapeutic technique when caring for the hypochondriacal patient is the inclusion of a legible psychosocial history in a prominent place in the patient's record (Brown & Vaillant, 1981). The general therapeutic principles outlined for somatization apply to most individuals diagnosed with hypochondriasis. Kellner (1987) also has provided an excellent review of the important strategies for the nonpsychiatric physician when working with hypochondriacal patients (see Table 13-6). Moreover, the effectiveness of any therapeutic strategy is likely improved if the patient's obsessional features are appreciated with respect to bodily complaints.

The treatment recommendations above, together with an understanding of the fascinating displacement of psychodynamics involved in the symptom formation and an appreciation of the psychosocial history, also are important to clinical management. Thus an effective treatment strategy often takes place in the context of the collaboration of a consulting psychiatrist and a primary care physician, who may continue to offer regular appointments to the patient. The possibility of concurrent organic disease or intercurrent illness exists; indeed, such disorders do occur eventually as life progresses. Therefore adequate physical examination on a regular and reasonable basis is helpful, and the judicious and coordinated use of other nonpsychiatric consultants is appropriate in order to evaluate new or justifiable physical complaints. Again, hypochondriacal patients seem to be managed best in a primary care or medical setting. Psychiatric consultation should be considered, especially when the patient requires adjunctive psychiatric treatment—usually for anxiety, depression, or psychosocial distress—or when the primary physician becomes concerned about patient suicide or overt symptoms of depression or is unable to manage his or her own emotional response to the hypochondriacal patient (Folks, Ford, & Houck, 1990).

The prognosis for hypochondriasis is favorable in a substantial portion of patients, perhaps as high as 80% of those who accept treatment (House, 1989). As noted for conversion and pain, young patients, those with chronic psychiatric disturbance, or individuals receiving disability compensation are not likely to do as well as those with hypochondriasis found also to have significant mood or anxiety disturbances potentially more amenable to treatment (Folks, Ford, & Regan, 1984; House, 1989). House (1989) suggested that particular emphasis be placed on the discussion of attitudes toward previous medical contacts, the value of physical investigation, and the psychiatric referral itself. Cognitive behavioral approaches have proved effective, particularly for maintaining an initial good response to psychiatric intervention (Warwick, 1989; Warwick & Salkovskis, 1990). Cognitive-behavioral strategies also are useful for the treatment of clinically significant depression (House, 1989; Kenyon, 1965; Pilowsky, 1970). The therapist must focus on social function, personal distress, and preoccupation with physical symptoms rather than on the physical symptoms themselves.

TABLE 13–6. *Hypochondriasis: Management*

1. Search for underlying psychiatric disorder potentially amenable to treatment.
2. Consider the emergence/coexistence of "true" physical disease.
3. Provide additional time over and above that allowed the average patient.
4. Allow discussion about somatic concerns.
5. Adopt empathic, accepting attitude.
6. Attempt to challenge/change patient's disease conviction.
7. Ensure regularly scheduled appointment times.
8. Physically examine patient briefly on each visit.
9. Judiciously utilize diagnostic tests and therapeutic procedures based only on objective evidence.
10. Provide reassurance, even if rejected, and education about patient's symptoms and sensations and their origins and meanings.
11. Involve both a primary physician and psychiatrist if patient is willing or if an identifiable psychodynamic issues or a comorbid disorder exists.
12. Refer to another colleague if a therapeutic impasse develops.

Source: Adapted from Kellner (1987) and Folks et al. (1990).

Body Dysmorphic Disorder

Body dysmorphic disorder, or dysmorphophobia, has in the past revised version of psychiatric nomenclature been regarded as a hypochondriacal subtype or unspecified somatoform syndrome. The disorder has been reviewed extensively by De Leon, Bott, and Simpson (1989), but systematic data regarding this disorder are lacking. Because most of the references involve single case reports, descriptive accounts of body dysmorphic disorder offer little substantial information regarding epidemiologic or etiologic factors (Folks, Ford, & Houck, 1990; Thomas, 1984). Onset typically occurs during adolescence, but the initial presentation may occur as late as the third decade of life. The condition can persist for years, significantly affecting social or occupational functioning. Polysurgery or unnecessary surgical procedures complicate cases. Currently no information is available on the predisposing factors, sex ratio, or familial pattern. Andreasen and Bardach (1977) estimated that about 2% of patients seeking cosmetic surgery are afflicted with this disorder.

The fundamental diagnostic feature of body dysmorphic disorder is primarily a pervasive subjective feeling of ugliness or physical defect. The patient genuinely believes that these "changes" are readily apparent to others (Thomas, 1984). The nomenclature proposed to divide this diagnostic population into two separate disorders: delusional disorder, somatic subtype (described in the section on hypochondriasis as "nonneurotic") and dysmorphic disorder per se, a true somatoform disorder. De Leon and colleagues (1989) have discussed the difficulty with this nosologic distinction. Nevertheless, criteria consistently applied to the diagnosis of body dysmorphic disorder are as follows: (1) preoccupation with some imagined defect in appearance in a normal-appearing person (if a slight physical anomaly is present, the person's concern is grossly excessive); (2) the belief in the defect is not of delusional intensity, as in delusional disorder, somatic type (i.e., the person can acknowledge the possibility that he or she may be exaggerating the extent of the defect or that there may be no defect at all); and (3) the occurrence of the disorder is not exclusively observed during the course of anorexia nervosa or transsexualism. Commonly, the symptoms of body dysmorphic disorder involve facial flaws, such as wrinkles, spots on the skin, excessive facial hair, shape of the nose, mouth, jaw, or eyebrows, and swelling of the face. Rarely, body dysmorphic disorder includes complaints involving the feet, hands, breasts, back, or some other body part. Some clinicians opine that body dysmorphics are simply manifesting overvalued ideas (De Leon, Bott, & Simpson, 1989). A slight physical defect actually may be present, but the concern expressed is grossly in excess of what might be considered appropriate. Similar to somatization disorder, this diagnostic category of somatoform disorder is characterized by much diagnostic and symptomatic heterogeneity (Block & Glue, 1988). This disorder also presents with much emotional distress displayed in the clinical interview, but this mental status finding should not be confused with transient feelings commonly experienced and related by adolescents (Fitts et al., 1989).

The differential diagnoses entertained with hypochondriasis apply equally to body dysmorphic disorder. However, anxiety syndromes (particularly phobias), personality disorders (especially avoidant and compulsive types), major depression, monosymptomatic delusional hypochondriasis, and other nonpsychotic somatizing disorders frequently are observed (De Leon, Bott, & Simpson, 1989). Obsessive-compulsive disorder shares some of the symptomatic and psychodynamic features of this disorder (Block & Glue, 1988). Moreover, this resemblance parallels the observed preferential response of some body dysmorphic cases to the serotonin reuptake blockers, e.g. clomipramine and fluoxetine (Hollander et al., 1989) or to the antipsychotic agent pimozide (King, 1990). Body dysmorphic disorder also may accompany anorexia nervosa and transsexualism; in this case the individual displays unfounded beliefs about body weight or gender-related physical characteristics, or both (Folks, Ford, & Houck, 1990).

Individuals with body dysmorphic disorder who indeed appear normal somehow develop a low sense of aesthetic perception, whereas those who are somewhat abnormal regard their appearance in the context of a heightened sense of aesthetic perception. Avoidance of social or occupational situations due to phobic or anticipatory anxiety, apprehension, or embarrassment about the defect is the rule. Thus a "neurotic" syndrome operates in which secondary features of anxiety and depression often emerge. Some authors have regarded this syndrome as ominous, indicating that it may represent a prodrome of schizophrenia (Folks, Ford, & Houck, 1990). Furthermore, the belief in the physical defect sometimes approaches a delusional intensity. As argued in the previous section, it is unclear whether different disorders can be clearly distinguished on the basis of whether the belief is quasidelusional or a clear delusion (De Leon, Bott, & Simpson, 1989). Symptoms bordering on psychosis are presumed to develop more often in individuals who possess schizoid, narcissistic, or obsessional personality traits, but they also must be distinguished from the features of personality disorder.

Individuals diagnosed with body dysmorphic disorder repeatedly visit their primary care physician or specialists (e.g., dermatologists or plastic surgeons) in an effort to correct the defect. Depressive and obsessive personality traits and psychosocial distress frequently coexist with the disorder and constitute a focus of treatment. Psychiatric consultation may be most useful for identifying and treating depression, anxiety, and other primary or secondary disturbances that require pharmacologic or psychotherapeutic intervention. Surgery actually has been recommended as a treatment for this disorder, in addition to pharmacotherapy and psychotherapy (De Leon, Bott, & Simpson, 1989). Brotman and Jenike (1984) suggested that patients with persistent anxiety or depressive symptoms be started on a trial of an antidepressant with serotonergic-active properties, including the possible use of an MAO inhibitor (MAOI). In addition to the MAOIs, the antipsychotic drug pimozide, as mentioned, has produced startling and sustained improvement in some cases, particularly in those cases where neurotic preoccupation has become quasidelusional (De Leon, Bott, & Simpson, 1989). However, no other reported antipsychotic or other pharmacologic agent has emerged as effective for treatment of this disorder. Psychiatric intervention with individual or group therapy may be useful, again focusing on psychosocial functions and body image complaints while supporting individuals' efforts to understand their "use" of the defect to cope and obtain secondary gain. Family conferences or therapy also are required in cases where the individual is not yet emancipated from the family of origin, particularly if an eating disorder, personality disorder, or other coexisting disturbance is identified.

FACTITIOUS DISORDERS

Most clinicians at some point in their career encounter a case of factitious disorder. These somatizing states essentially are characterized by the voluntary production of signs, symptoms, or disease for no apparent goal other than to assume the patient role. By contrast, the somatoform disorders are viewed collectively as having symptoms that are manifested unconsciously (see Table 13-3). Münchausen syndrome, the more severe subtype of factitious disorder, is characterized by a triad of features involving simulation of disease, pathological lying, and wandering. These cases frequently involve men of lower socioeconomic class who have had a lifelong pattern of social maladjustment. However, authorities generally concur that most factitious disorders involve socially conforming young women of a higher socioeconomic class who are intelligent, educated, and frequently employed in a medically related field. Thus one rarely encounters the socially nonconforming "wanderers" who satisfy the Münchausen syndrome criteria depicted in Table 13-7.

The epidemiologic data on factitious disorders are inadequate but strongly suggest a preponderance of young adults, most of whom are female and likely to be employed in the health professions. The available literature provides only a few indications of the incidence of factitious illness. These disorders appear to be far more common than was once generally believed, perhaps because of the progress in medical technology and the popular medical journalism that is readily available to the lay public. The paucity of systematic studies and disproportionate number of case reports on the Münchausen syndrome have resulted in contradictory data on the age and sex ratio. Age ranges for this disorder average approximately 30 years, from adolescence to old age.

Factitious illness is not real, genuine, or natural. Thus physical or psychological symptoms are controlled voluntarily, and clinical presentations are simulated to deceive the physician. The characteristic presenting modes and organ system subtypes of factitious disorder are listed in Table 13-8. Factitious disorder with physical symptoms is the most commonly diagnosed subtype. A dramatic and detailed history of present symptoms and the primary goal of assuming the patient role are the pertinent diagnostic features. Eisendrath (1984) suggested that factitious presentations may manifest at one of three levels of enactment: (1) a fictitious history; (2) a simulated disease; or (3) the presence of certifiable pathophysiology.

In pediatric settings Münchausen by proxy also may be encountered; in these cases a parent or caregiver presents a child with a factitious illness (Folks & Freeman, 1985; Meadow, 1982). Descriptions of Münchausen by proxy have tended to focus exclusively on medical and psychiatric assessments of the involved child

TABLE 13–7. *Münchausen Syndrome: Diagnostic Features*[a]

Essential features
Pathologic lying (pseudologia fantastica)
Peregrination (traveling or wandering)
Recurrent, feigned, or simulated illness
Supporting features[b]
Borderline and/or antisocial personality traits
Deprivation during childhood
Equanimity for diagnostic procedures
Equanimity for treatments or operations
Evidence of self-induced physical signs
Knowledge of or experience in a medical field
Most likely to be male
Multiple hospitalizations
Multiple scars (usually abdominal)
Police record
Unusual or dramatic presentation

[a]Patient meets *DSM* criteria for a chronic factitious disorder or an atypical factitious disorder.
[b]May also support the diagnosis of other factitious disorders.
Source: Reprinted with permission from Folks DG, Ford CV, & Houck CA (1990). Somatoform disorders, factitious disorders, and malinger. In Stoudemire A (ed.), *Clinical Psychiatry for Medical Students*. Philadelphia: JB Lippincott.

TABLE 13–8. *Commonly Presenting Features of Chronic Factitious Illnesses*

Organ System Subtypes	Demeanor or Behavior
Abdominal[a]	Bizarre
Cardiac	Demanding
Dermatological[b]	Dramatic
Genitourinary	Evasive
Hematological[a,b]	Medically sophisticated
Infectious	Self-mutilating
Neurological[a]	Unruly
Psychiatric	
Self-medication[a,c]	

[a]Original subtypes identified.
[b]Currently reported to be more common.
[c]Especially insulin, thyroid, vitamins, diuretics, and laxatives.
Source: Reprinted with permission from Folks DG, Ford CV, & Houck CA (1990). Somatoform disorders, factitious disorders, and malingering. In A Stoudemire (ed.), *Clinical Psychiatry for Medical Students*. Philadelphia: JB Lippincott.

and perpetrating parent. However, family assessments may suggest that a possible systemic constellation syndrome is inherent in some cases (Griffith, 1988). This approach essentially is based on the notion that a mother or other caretaker already possessing the potential for illness behavior and somatization joins an enmeshed, authoritarian family system having a systemic history of exploitation of children. Consequently, the paradigm is set for factitious illness: Münchausen by proxy. Of course, measures to protect the afflicted child in these cases must carefully consider the family constellation issues as well as the psychopathology of the individual.

Poorly defined distinctions between factitious disorder, somatization disorder, and malingering have likely contributed to the diversity of diagnoses included in the differential diagnosis of factitious disorder. A diagnosis of inclusion, not exclusion, of a factitious disorder with physical symptoms requires an appropriate index of suspicion, recognition of the clinical features, and clinical perseverance once the diagnosis is established. In addition to the possibility that a true disease exists and accounts for the presentation, other possible differential diagnoses include malingering, pseudomalingering, conversion disorder, and hypochondriasis. Factitious cases seen in psychiatric consultation may have been already misdiagnosed as a range of psychiatric disorders, including conversion disorder, somatization disorder, malingering, schizophrenia, or another major psychosis. Histrionic, schizotypal, borderline, antisocial, or masochistic personality disorders are often, if not always, present; borderline personality disorder is the most common type of coexisting personality disorder (Folks & Freeman, 1985). Psychiatric or psychologic symptoms may replace physical ones and dominate the clinical presentation of factitious disorder. Many cases involving psychologic presentations of factitious illness have been reported (Folks & Freeman, 1985), but these reports often lack clearly delineated inclusion, exclusion, and outcome criteria and seriously compromise the diagnostic specificity of this disorder (Rogers, Bagby, & Rector, 1989). Nevertheless, factitious grief and psychosis commonly are encountered by primary care and psychiatric physicians. Posttraumatic stress disorder also has emerged as a newer, commonly seen subtype, most frequently in Veterans Administration hospitals (Pankratz, 1990).

An attempt to understand the etiology of factitious disorder requires careful consideration of any developmental disturbance, personal history, and current life stressors, as well as an appreciation of primary psychodynamic mechanisms: masochism, dependency, and mastery. The desire to be the center of interest and attention, a grudge against physicians and hospitals that is somehow satisfied by frustrating and deceiving the staff, a desire for drugs, a desire to escape the police, and a desire to obtain free room and board while tolerating the consequences of various therapeutic investigations and treatment are some of the more frequently listed reasons that might motivate the self-destructive behaviors of patients with a factitious disorder. Although patients often possess borderline or other personality traits, one often obtains a childhood history of emotional insecurity, excluding or rejecting parents, and broken homes leading to foster home placement or adoption and subsequent delinquency, antisocial behavior, or failure in psychosexual development. The possible enactment of past or present developmental disturbances within the medical setting also should be considered.

Essentially, factitious patients, through their illness, primarily seek to compensate for developmental traumas and secondarily escape from and make up for stressful life situations. Psychodynamic explanations suggest that the factitiously disordered patient experiences satisfaction from manipulating as many aspects of his or her own medical and surgical care as possible (mastery), receives strong sexual gratification from diagnostic and therapeutic procedures (masochism), and enjoys the warm and personal but ambivalent care inherent in the doctor-patient relationship (dependency), culminating in the excitement of the hospital experience. Case reports involving infections, self-medication, and dermatological lesions continue to reflect these common features (Bialer & Wallack, 1990; Castor et al., 1990; Frumkin & Victoroff, 1990; Harrington, Folks, & Ford, 1988; Koblenzer, 1990; Warrens, Ron, & Dawling, 1990). Newer cases, however, include other more novel presentations that reflect a focus on current medical problems, e.g., acquired immunodeficiency syndrome (Chiarello, 1989; Frumkin & Victoroff, 1990; Nickoloff, Neppe, & Ries, 1989; Silva et al., 1989) or posttraumatic stress syndrome (Folks, Ford, & Houck, 1990; Pankratz, 1990).

The general therapeutic approach to individuals with somatization, as outlined in the introductory section of this chapter, is applicable to cases of factitious disorder; a comparison of clinical features of somatoform disorders is presented in Table 13-3. The clinical approach usually requires an index of diagnostic suspicion that a factitious disorder is indeed present (Gattaz, Dressing, & Hewer, 1990). Essentially, a high index of suspicion and prudent techniques of differential diagnosis are the only keys to detecting factitious illness in emergency or hospital settings. The ability to protect the patient from iatrogenic risk is often the only acceptable or viable treatment, especially for the Münchausen subtype. Psychiatric consultation should be encouraged with all cases; and if confrontation is advisable, the primary physician (not the consultant) should confront the patient in a nonpunitive manner (Hollender & Hersh, 1970). Patients usually are less difficult to confront than might be expected and do not show the intense anger, impulsivity, or instability of interpersonal relationships that is commonly reported with the more extreme cases representing Münchausen syndrome (Reich & Gottfried, 1983). If confronted with the factitious nature of the illness, the patient may deny it, refuse psychiatric intervention, and resume the same behavior; may admit that the factitious illness is present but refuse psychiatric intervention; or may acknowledge the factitious nature of the illness and cooperate with psychiatric intervention (Ford, 1983; Hollender & Hersh, 1970).

Confrontation is not necessarily appropriate for all cases; the psychiatrist simply may attempt to build rapport with the patient while the primary physician continues any necessary noninvasive medical treatment (Earle & Folks, 1986). This technique may include the use of inexact interpretations of psychological defenses, use of double-blind therapy, or the use of techniques that allow the patient to give up the factitious symptoms without losing face (Eisendrath, 1989). Use of the amytal interview has been suggested as a possible adjunct to the use of "gradual and gentle" confrontation (Marriage et al., 1988). Selective confrontation is more often favored in the hospitalized patient who has the intelligence, psychosocial supports, and personal attributes necessary for a more mature adaptation (Van Moffaert, 1989). Integration of psychotherapy with psychotropic drug treatment in self-mutilating patients is a necessity (Earle & Folks, 1986). Antidepressant medicines may be helpful; and in view of the frequent correlation of factitious behavior with an underlying borderline personality structure, significant disorganization, and possible micropsychotic episodes, one may rationally consider the use of a trial of an antipsychotic medicine in the hopes of containing the penchant for self-destruction (Earle & Folks, 1986). The psychiatric approach includes assisting the medical staff with its negative reactions to the factitiously disordered patient. In turn, the medical staff can protect the patient from himself or herself by avoiding potentially dangerous diagnostic or operative procedures. Family members can be particularly helpful in providing pertinent history or assisting the patient in the maintenance of acceptable limits on illness behavior (Harrington, Folks, & Ford, 1988). Treatable psychopathology, e.g., anxiety disorders, depressive syndromes, conversion symptoms, and major psychoses, should be assiduously evaluated and steadfastly treated whenever possible (Folks & Freeman, 1985).

The prognosis for factitious illness generally has been considered poor. However, careful exclusion of malingerers, severe borderline personalities, wandering patients with Münchausen syndrome, the chronic medically ill, and the more evasive miscreant individuals leave a prognostically favorable subgroup of potentially treatable patients, as outlined in Table 13-9 (Folks & Freeman, 1985). Although many of these patients have borderline and other primitive personality types falling within the narcissistic/histrionic/antisocial cluster, many cases potentially are responsive to confrontation or intervention. Reich and associates (1983) observed that once the diagnosis was established, even some of the more severe and chronic cases responded well to a combined medical and psychiatric approach. Finally, as noted with other forms of somatization, the possibility of coexisting physical disease or intercurrent illness must be considered.

MALINGERING

The essential clinical feature of malingering is the intentional production of illness or grossly exaggerated physical or psychological symptoms that is motivated by external incentives such as avoiding military duty, obtaining financial compensation through litigation or disability, evading criminal prosecution, obtaining drugs, or simply securing better living conditions (Gorman, 1982). Malingering should be strongly suspected in the

TABLE 13–9. *Aspects of Factitious Illness Potentially Amenable to Treatment*

1. Presence of treatable psychiatric syndromes, including:
   a. Mood disorders
   b. Anxiety disorders
   c. Psychotic disorders
   d. Conversion disorders
   e. Substance abuse disorders
   f. Organic mental disorders
2. Personality organization closer to compulsive, depressive, or histrionic rather than borderline, narcissistic, or antisocial
3. Stability in psychosocial support system as manifested by marriage, stable occupation, and family ties, in contrast to the single, unemployed wanderer
4. Ability to cope with confrontation or some redefinition of the illness behavior
5. Capability of establishing and maintaining rapport with the treating clinicians

Source: Reprinted with permission from Folks DG, Ford CV, & Houck CA (1990). Somatoform disorders, factitious disorders, and malingering. In Stoudemire A (ed.), *Clinical Psychiatry for Medical Students*, Philadelphia: JB Lippincott.

following circumstances: (1) a medicolegal context overshadows the presentation; (2) a marked discrepancy exists between the clinical presentation and the objective findings; (3) a lack of cooperation is experienced with diagnostic efforts or in compliance with medical regimen; or (4) the psychosocial history suggests the presence of an antisocial personality disorder. Thus malingering can be viewed fundamentally as a sociopathic feigning of illness—the fraudulent simulation or exaggeration of physical or mental disease or defect consciously produced to achieve a specific goal. The reasons for the illness behavior in individual circumstances can be readily understood by an objective observer. Unlike the patient with factitious disorder, who merely wishes to assume the patient role, a malingerer has a more clearly external motivation, and the illness behavior is intentional and consciously produced to achieve a consciously desired goal. A comparison of several clinical features of malingering and somatoform and factitious disorders is shown in Table 13-3.

Malingering is sometimes adaptive and may be observed, to a lesser degree, in apparently normal children, students, test subjects, and employees; thus malingering behavior does not always represent a maladaptive or malignant form. A few clinicians have promulgated the theory that pure malingering is a mental disease worthy of a therapeutic response. Efforts to develop psychometric scales to distinguish between malingering individuals and patients with credibility (Lees-Haley, 1990) or to utilize discrepant findings on the MMPI to distinguish malingerers (Lees-Haley & Fox, 1990) are of questionable diagnostic benefit (Folks, Ford, & Houck, 1990). The validity of findings may rest solely on the level of cooperativeness of the individual being tested (Hawk & Cornell, 1989). Concerted efforts to refine the diagnostic process in a variety of treatment settings have met with limited success; and future systematic research of diagnostic techniques with malingering is certainly necessary, particularly if a diagnostic format is ever to be skillfully applied to forensic cases (Bigler, 1990; Kahn, Fox, & Rhode, 1988; Perry & Kinder, 1990).

The role of psychologic and personality factors in malingering has resulted in a conceptual model that views malingering as a continuum with conversion and other somatoform disorders (Ford, 1986). Briefly, the conscious effort to falsify symptoms, in some cases, includes rather complex motivations, originating in part from the unconscious. Malingering also is likely to arise in individuals with antisocial personality disorder or other various forms of feigned illness (e.g., Ganser syndrome) or as a component of cognitive impairment disorders (Wasyliw & Cavanaugh, 1989). These coexisting psychiatric disturbances may become the focus of evaluation and treatment for these "mentally ill" malingerers. Also, some basic legal principles should be considered in the clinical approach to malingering. In particular, before reporting malingering one should steadfastly follow the basic rules of confidentiality and privilege. A professional and therapeutic posture when approaching the malingering patient initially must include an examination of one's own feelings of anger, disgust, or humiliation, recognizing that the malingering behavior often threatens the very foundations of the doctor-patient relationship. Once the physician is convinced of the disingenuous nature of the symptoms, intervention is indicated (Lande, 1989). Frank, nonjudgmental communication between the physician and the malingerer and awareness that the behavior may be an ongoing reaction to stress or due to a psychiatric disorder may lead to an open discussion of the patient's needs—which may, in turn, provide a basis for an adequate therapeutic alliance or a solution to the problem (Mark et al., 1987).

Other important aspects of malingering concern the way in which physicians perceive such behavior and then their moral judgments. Situations certainly can be considered in which such an extreme form of somatization is regarded as acceptable, constructive, or even praiseworthy, as is the case of the prisoner of war who malingers to protect his country's interests.

Finally, a number of authors have described simulation among persons seeking compensation for work-related injuries or disease (Kreger-Wexhler & Foreman, 1988; Weighill, 1983). However, the psychological difficulties in these cases vary greatly, and the physician must be able to appreciate and assess a number of background factors, i.e., severity of injury, preexisting personality traits, developmental characteristics, social class, attitudinal response, and the pertinent family, social, and employment factors, as well as the actual progress of the physical condition or legal process of settlement (Hicks, 1988). Work-related injuries invariably are related to protracted disability; however, the psychologic factors contributing to this phenomenon are poorly understood. Leavitt (1990) has reported that psychologic disturbances may not differ significantly among patients injured while away from work compared to those whose injury took place at the work site. Fabricated posttraumatic stress syndrome in veterans is yet another illustration of the complexities of such cases where disability and compensation are relevant factors.

Interestingly, the use of MMPI or other psychometrics has not proved to be uniformly sensitive for discriminating malingers from those who have bona fide psychiatric disturbance (Perconte & Goreczny, 1990). Faust, Hart, and Guilmette (1988), in a study of neuropsychologists' capacity to detect malingering in children who had been instructed to "fake bad," found that 93% of the neuropsychologists diagnosed abnormality, and none detected malingering. Thus the clinicians' capacity to detect malingering remains highly questionable and subject to debate (Faust & Guilmette, 1990; Faust, Hart, Guilmette, 1988). Only one of the studies utilizing a neuropsychologic battery to detect malingering has yielded positive results (Goebel, 1983). Weintraub (1983) and Bigler (1990) have discussed the merits of further efforts to develop and validate reliable methods to assess malingering, particularly neuropsychological techniques in relation to other neurodiagnostic measures, i.e., magnetic resonance imaging, computerized electroencephalography, and clinical assessment.

REFERENCES

American Psychiatric Association (1952). Diagnostic and statistical manual of mental disorders (1st ed.). Washington, DC: American Psychiatric Association.

American Psychiatric Association (1968). Diagnostic and statistical manual of mental disorders (2nd ed., rev.). Washington, DC: American Psychiatric Association.

American Psychiatric Association (1980). Diagnostic and statistical manual of mental disorders (3rd ed.). Washington, DC: American Psychiatric Association.

American Psychiatric Association (1987). Diagnostic and statistical manual of mental disorders (3rd ed., rev.). Washington, DC: American Psychiatric Association.

American Psychiatric Association Diagnostic and statistical manual of mental disorders (4th ed., rev.). Washington, DC: American Psychiatric Association.

Andreasen NC, & Bardach J (1977). Dysmorphophobia: symptom or disease? Am J Psychiatry 134:673–676.

Bain ST, & Spaulding WB (1967). The importance of coding presenting symptoms. Can Med Assoc J 97:953–959.

Ball RA, & Clare AW (1990). Symptoms and social adjustment in Jewish depressives. Br J Psychiatry 156:379–383.

Barsky AJ (1979). Patients who amplify bodily sensations. Ann Intern Med 91:63–70.

Barsky AJ, Wyshak G, & Klerman GL (1990a). Transient hypochondriasis. Arch Gen Psychiatry 47:746–752.

Barsky AJ, Wyshak G, & Klerman, et al. (1990b). The prevalence of hypochondriasis in medical outpatients. Soc Psychiatry Psychiatr Epidemiol 25:89–94.

Barsky AJ, & Wyshak G (1989). Hypochondriasis and related health attitudes. Psychosomatics 30:412–420.

Benjamin S, Barnes D, Berger S, et al. (1988). The relationship of chronic pain, mental illness and organic disorders. Pain 32:185–195.

Bialer PA, & Wallack JJ (1990). Mixed factitious disorder presenting as AIDS. Hosp Community Psychiatry 41:552–553.

Bigler ED (1990). Neuropsychology and malingering: Comment on Faust, Hart, and Guilmette (1988). J Consult Clin Psychol 58:244–247.

Bilu Y, & Beit-Hallahmi B (1989). Dybbuk-possession as a hysterical symptom: Psychodynamic and socio-cultural factors. Isr J Psychiatry Relat Sci 26:138–149.

Block S, & Glue P (1988). Psychotherapy and dysmorphophobia: A case report. Br J Psychiatry 152:271–274.

Bohman M, Cloninger R, Von Knorring A-L, et al. (1984). An adoption study of somatoform disorders: III. Cross-fostering analysis and genetic relationship to alcoholism and criminality. Arch Gen Psychiatry 41:872–878.

Bonica JJ (1976). Organization and structure of a multidisciplinary pain clinic. In M Weisenberg & B Tursky (Eds.): Pain: New perspectives in therapy and research. New York: Plenum.

Bonica JJ, Ventafridda V (Eds.) (1979). Advances in Pain Research and Therapy (Vols. 2–5). New York: Raven Press.

Brodsky CM (1984). Sociocultural and interactional influences on somatization. Psychosomatics 24:673–680.

Brotman AW, & Jenike MA (1984). Monosymptomatic hypochondriasis treated with tricyclic antidepressants. Am J Psychiatry 141:1608–1609.

Brown FW, Golding JM, & Smith GR (1990). Psychiatric comorbidity in primary care somatization disorder. Psychosom Med 52:445–451.

Brown FW, & Smith CR (1991). Diagnostic concordance in primary care somatization disorder. Psychosomatics 32:191–195.

Brown HN, & Vaillant GE (1981). Hypochondriasis. Arch Intern Med 141:723–736.

Castor B, Ursing J, Aberg M, et al. (1990). Infected wounds and repeated septicemia in a case of factitious illness. Scand J Infect Dis 22:227–232.

Chaturvedi SK (1989). Psychologic depressive disorder: A descriptive and comparative study. Acta Psychiatr Scand 9:98–102.

Chiarello RJ (1989). Malingering doubt about factitious AIDS [letter; comment]. N Eng J Med 320:1423.

Coen SJ, & Sarno JE (1989). Psychosomatic avoidance of conflict in back pain. J Am Acad Psychoanalyt 17:359–376.

Dauncey MK, & Cooper JE (1990). Diagnostic problems in liaison psychiatry and the ICD-10. J Psychosom Res 34:287–294.

De Leon J, Bott A, & Simpson GM (1989). Dysmorphophobia: Body dysmorphic disorder, somatic subtype. Comp Psychiatry 30:457–472.

Derogatis LR, & Lopez MC (1983). The psychosocial adjustment to illness scale (PAIS & PAIS-SR). Baltimore: Johns Hopkins School of Medicine.

Desai BT, Porter RJ, & Penry K (1982). Psychogenic seizures: A study of 42 attacks in six patients, with intensive monitoring. Arch Neurol 39:202–209.

Dickes RA (1977). Brief therapy of conversion reactions: An inhospital technique. Am J Psychiatry 13:584–586.

Earle JR, & Folks DG (1986). Factitious disorder and coexisting depression: A report of successful psychiatric consultation and case management. Gen Hosp Psychiatry 8:448–450.

Eisenberg L (1977). Disease and illness: Distinctions between professional and popular ideas of sickness. Cult Med Psychiatry 1:9–23.

Eisendrath SJ (1984). Factitious illness: A clarification. Psychosomatics 25:110–116.

Eisendrath SJ (1989). Factitious physical disorders: Treatment without confrontation. Psychosomatics 30:383–387.

Faust D, & Guilmette TJ (1990). To say it's not so doesn't prove that it isn't: Research on the detection of malingering; reply to Biler. J Consult Clin Psychol 58:248–250.

FAUST D, HART K, & GUILMETTE TJ (1988). Pediatric malingering: The capacity of children to fake believable deficits on neuropsychological testing. J Consult Clin Psychol 56:578–582.

FIELDS HL (1988). Sources of variability in the sensation of pain. Pain 33:195–200.

FITTS SN, GISON P, REDDING CA, ET AL. (1989). Body dysmorphic disorder: Implications for its validity as a DSM-III-R clinical syndrome. Psychol Rep 64:655–658.

FLOR-HENRY P, FROWN-AUGH D, TEPPER M, ET AL. (1981). A neuropsychological study of the stable syndrome of hysteria. Biol Psychiatry 16:601–626.

FOLKS DG, FORD CV, & HOUCK CA (1990). Somatoform disorders, factitious disorders and malingering. In A STOUDEMIRE (Ed.). Clinical psychiatry for medical students (Ch. 8, pp. 237–268). Philadelphia: JB Lippincott.

FOLKS DG, FORD CV, & REGAN WM (1984). Conversion symptoms in a general hospital. Psychosomatics 25:285–295.

FOLKS DG, FREEMAN AM (1985). Münchausen syndrome and other factitious illness. Psychiatr Clin North Am 8:263–278.

FORD CV (1983). The somatizing disorders: Illness as a way of life. New York: Elsevier.

FORD CV (1986). The somatizing disorders. Psychosomatics 27:327–337.

FORD CV, & FOLKS DG (1985). Conversion disorder: An overview. Psychosomatics 26:371–383.

FORD CV (1983). The somatizing disorders: Illness as a way of life. New York: Elsevier.

FRANCES AJ, WIDIGER TA, & PINCUS HA (1989). The development of DSM-III. Arch Gen Psychiatry 46:373–375.

FRUMKIN LR, & VICTOROFF JI (1990). Chronic factitious disorder with symptoms of AIDS. Am J Med 88:694–696.

GATTAZ WF, DRESSING H, & HEWER W (1990). Münchausen syndrome: Psychopathology and management. Psychopathology 23:33–39.

GOEBEL RA (1983). Detection of faking on the Halstead-Reitan Neuropsychological Test Battery. J Clin Psychol 39:731–742.

GOLDBERG RJ (1990). Basic principles of pain management. In A STOUDEMIRE (Ed.), Clinical psychiatry for medical students (Ch. 22, pp. 595–608). Philadelphia: JB Lippincott.

GOLDBERG RJ, SOKOL MS, & CULLEN LO (1987). Acute pain management. In A STOUDEMIRE & BS FOGEL (Eds.), Principles of medical psychiatry (pp. 365–388). Orlando, FL: Grune & Stratton.

GORMAN WF (1982). Defining malingering. J Forensic Sci 27:401–407.

GRIFFITH JL (1988). The family systems of Münchausen syndrome by proxy. Fam Process 27:423–437.

GUZE SB (1975). The validity and significance of the clinical diagnosis of hysteria (Briquet's syndrome). Am J Psychiatry 132:138–141.

HARRINGTON TM, FOLKS DG, & FORD CV (1988). Holiday factitial disorder: Management of factitious gastrointestinal bleeding. Psychosomatics 29:438–441.

HAWK GL, & CORNELL DG (1989). MMPI profiles of malingerers diagnosed in pretrial forensic evaluations. J Clin Psychol 45:673–678.

HICKS A (1988). Problems in writing medico-legal reports: I. Clinical exaggeration and malingering. East Afr Med J 65:51–56.

HOLLANDER E, LIEBOWITZ MR, WINCHEL R, ET AL. (1989). Treatment of body-dysmorphic disorder with serotonin reuptake blockers. Am J Psychiatry 146:768–770.

HOLLENDER MH, & HERSH SP (1970). Impossible consultation made possible. Arch Gen Psychiatry 23:343–345.

HOSOBUCHI Y, ADAMS JE, & LINCHITZ R (1977). Pain relief by electrical stimulation of the central gray matter in humans and its reversal by naloxone. Science 197:183–186.

HOUSE A (1989). Hypochondriasis and related disorders: Assessment and management of patients referred for a psychiatric opinion. Gen Hosp Psychiatry 11:156–165.

International classification of disease, 9th revision: Clinical modification (ICD-9-CM) (1977). Washington, DC: U. S. Department of Health and Human Services.

International classification of disease, 10th revision: Clinical modification (ICD-10-CM) (1989, draft). Washington, DC: U. S. Department of Health and Human Services.

JAMES L, GORDEN E, KRAIUHIN J, ET AL. (1989). Selective attention and auditory event-related potentials in somatization disorder. Compr Psychiatry 30:84–89.

KAHN MW, FOX H, & RHODE R (1988). Detecting faking on the Rorschach: Computer versus expert clinical judgment. J Pers Assess 52:516–523.

KATON W, & RUSSO J (1989). Somatic symptoms and depression. J Fam Pract 29:65–69.

KATON WJ (1990). Chest pain, cardiac disease, and panic disorder. J Clin Psychiatry 51:27–30.

KELLNER R, ABBOTT P, WINSLOW WW, ET AL. (1989). Anxiety, depression, and somatization in DSM-III hypochondriasis. Psychosomatics 30:57–64.

KELLNER R (1987). Hypochondriasis and somatization. JAMA 258:2718–2722.

KENYON FE (1965). Hypochondriasis: A survey of some historical, clinical and social aspects. Br J Med Psychol 38:117–133.

KESHAVAN MS, NARAYANAN HS, & GANGADHAR BN (1989). 'Bhanamati' sorcery and psychopathology in south India: A clinical study. Br J Psychiatry 154:218–220.

KIMBALL CP, & BLINDT K (1982). Some thoughts on conversion. Psychosomatics 23:647–649.

KING BH (1990). Hypothesis: Involvement of the serotonergic system in the clinical expression of monosymptomatic hypochondriasis. Pharmacopsychiatry 23:85–89.

KOBLENZER CS (1990). Factitial leg ulcers associated with an unusual sleep disorder [letter]. Arch Dermatol 126:396–397.

KOLLER W, LANG A, VETERE-OVERFIELD B, ET AL. (1989). Psychogenic tremors. Neurology 39:1094–1099.

KREGER-WEXHLER LM, & FOREMAN SM (1988). Malingering vs. the factitious personality in chiropractic practice. J Manipulative Physiol Ther 11:416–421.

LANDE RJ (1989). Malingering. J Am Osteopath Assoc 89:483–488.

LAZARUS A (1990). Somatic therapy for conversion disorder [letter]. Psychosomatics 31:357–358.

LEAVITT F (1985). The value of the MMPI conversion "V" in the assessment of psychogenic pain. J Psychosom Res 29(2):125–131.

LEAVITT F (1990). The role of psychological disturbance in extending disability time among compensable back injured industrial workers. J Psychosom Res 34:447–453.

LEES-HALEY PR (1990). Provisional normative data for a credibility scale for assessing personal injury claimants. Psychol Rep 66:1355–1360.

LEES-HALEY PR, & FOX DD (1990). MMPI subtle-obvious scales and malingering: Clinical versus simulated scores. Psychol Rep 66:907–911.

LEMPERT T, & SCHMIDT D (1990). Natural history and outcome of psychogenic seizures: A clinical study in 50 patients. J Neurol 237:35–38.

LESSER LM, FORD CV, & FRIEDMANN CTH (1979). Alexithymia in somatizing patients. Gen Hosp Psychiatry 1:256–261.

LIPOWSKI ZJ (1988). Somatization: The concept and its clinical application. Am J Psychiatry 145:1358–1368.

LIPOWSKI ZJ (1990). Somatization and depression. Psychosomatics 31:13–21.

LISKOW B, OTHMER E, PENICK EC, ET AL. (1986). Is Briquet's syndrome a heterogeneous disorder? Am J Psychiatry 143:626–269.

LUDWIG AM (1972). Hysteria: A neurobiologic theory. Arch Gen Psychiatry 27:771–777.

MANU P, LANE TJ, & MATTHEWS DA (1989). Somatization disorder in patients with chronic fatigue. Psychosomatics 30:388–395.

MARK M, RABINOWITZ S, ZIMRAN A, ET AL. (1987). Malingering in the military: Understanding and treatment of the behavior. Milit Med 152:260–262.

MARRIAGE K, GOVORCHIN M, GEORGE P, ET AL. (1988). Use of an amytal interview in the management of factitious deaf mutism. Aust NZ J Psychiatry 22:454–456.

MARTIN RL (1991). Somatoform disorders in the general hospital setting. In JUDD & BURROWS (Eds.), Handbook of studies on general hospital psychiatry (Ch. 19, pp. 251–266). New York: Elsevier.

MAYER DJ, AKIL H, & LIEBESKIND JC (1974). Pain reduction by focal electrical stimulation of the brain: An anatomical and behavioral analysis. Brain Res 68:73–93.

MEADOW R (1982). Münchausen syndrome by proxy. Arch Dis Child 57:92–98.

MELLERUP ET, DAM H, KIM MY, ET AL. (1990). Imipramine binding in depressed patients with psychogenic pain. Psychiatry Res 32:29–34.

MELZACK R, & WALL PD (1965). Pain mechanisms: A new theory. Science 150:971–979.

MENZA MA (1989). A suicide attempt following removal of conversion paralysis with amobarbital [letter]. Gen Hosp Psychiatry 11:137–138.

MERSKEY H, & BUHRICH NA (1975). Hysteria and organic brain disease. Br J Med Psychol 48:359–366.

MORRISON J (1989a). Histrionic personality disorder in women with somatization disorder. Psychosomatics 30:433–437.

MORRISON J (1989b). Childhood sexual histories of women with somatization disorder. Am J Psychiatry 146:239–241.

NICKOLOFF SE, NEPPE, & RIES RK (1989). Factitious AIDS. Psychosomatics 30:342–345.

PANKRATZ L (1990). Continued appearance of factitious posttraumatic stress disorder [letter]. Am J Psychiatry 147:811–812.

PERCONTE ST, & GORECZNY AJ (1990). Failure to detect fabricated posttraumatic stress disorder with the use of the MMPI in a clinical population. Am J Psychiatry 147:1057–1060.

PERRY GG, & KINDER BN (1990). The susceptibility of the Rorschach to malingering: A critical review. J Pers Assess 54:47–57.

PILOWSKY I, & BARROW N (1990). A controlled study of psychotherapy and amitriptyline used individually and in combination in the treatment of chronic intractable psychogenic pain. Pain 40:3–19.

PILOWSKY I (1970). Primary and secondary hypochondriasis. Acta Psychiatr Scand 46:273–285.

PILOWSKY I (1990). The concept of abnormal illness behavior. Psychosomatics 31:207–213.

QUILL TE (1985). Somatization disorder: One of medicine's blind spots. JAMA 254:3075.

RASKIN M, TALBOTT JA, & MEYERSON D (1966). Diagnosed conversion reactions: Predictive value of psychiatric criteria. JAMA 197:530–534.

REICH P, & GOTTFRIED LA (1983). Factitious disorders in a training hospital. Ann Intern Med 99:250–247.

REICH J, TUPIN JP, & ABRAMOWITZ SI (1983). Psychiatric diagnosis of chronic pain patients. Am J Psychiatry 140:1495–1498.

REULER JB, GIRARD DE, & NARDONE DA (1980). The chronic pain syndrome: Misconceptions and management. Ann Intern Med 93:588–596.

REYNOLDS DV (1969). Surgery in the rat during electrical analgesia induced by focal brain stimulation. Science 164:444–445.

ROGERS R, BAGBY RM, & RECTOR N (1989). Diagnostic legitimacy of factitious disorder with psychological symptoms. Am J Psychiatry 146:1312–1314.

RICHARDSON DE, & AKIL H (1977). Pain reduction by electrical brain stimulation in man: Part 2. Chronic self-administration in the periventricular gray matter. J Neurosurg 47:184–194.

ROSS CA, HEBER S, NORTON GR, ET AL. (1989). Somatic symptoms in multiple personality disorder. Psychosomatics 30:154–160.

RUEGG RJ (1990). Etiology of somatization disorder [letter]. Am J Psychiatry 147:126.

SARTORIUS N (1989). Making of a common language for psychiatry: Development of the classification of mental, behavioral and developmental disorders in the 10th revision of the I.C.D. WPA Bull 1:3–6.

SILVA JA, LEONG GB, WEINSTOCK R, ET AL. (1989). Factitious AIDS in a psychiatric inpatient. Can J Psychiatry 34:320–322.

SIMPSON DE, & GJERDINGEN DK (1989). Family physicians' and internists' consideration of psychosocial hypotheses during the diagnostic process. Fam Pract Res J 8:55–61.

SMITH GR, & BROWN FW (1990). Screening indexes in DSM-III-R somatization disorder. Gen Hosp Psychiatry 12:148–152.

SMITH GR JR, MONSON RA, & RAY DC (1986). Psychiatric consultation in somatization disorder: A randomized controlled study. N Engl J Med 314:1407–1413.

SMITH RC (1985). A clinical approach to the somatizing patient. J Fam Pract 21:294–301.

STEINHAUSEN HC, VONASTER M, PFEIFFER E, ET AL. (1989). Comparative studies of conversion disorders in childhood and adolescence. J Child Psychol Psychiatry 30:615–621.

STERNBACH RA (1974). Varieties of pain games. Adv Neurol 4:423–430.

STEVENS H (1968). Conversion hysteria: A neurologic emergency. Mayo Clin Proc 43:54–64.

STEWART DE (1990). The changing faces of somatization. Psychosomatics 31:153–158.

STOUDEMIRE A (1991). Somatothymia: Parts I and II. Psychosomatics 32:365–381.

STOUDEMIRE A, & SANDHU J (1987). Psychogenic/idiopathic pain syndromes. Gen Hosp Psychiatry 9:79–86.

SWARTZ M, HUGHES D, GEORGE L, ET AL. (1986). Developing a screening index for community studies of somatization disorder. J Psychiatr Res 20:335–343.

THOMAS CS (1984). Dysmorphophobia: A question of definition. Br J Psychiatry 144:513–516.

VALDES M, GARCIA L, TRESERRA J, ET AL. (1989). Psychogenic pain and depressive disorders: An empirical study. J Affective Disord 16:21–25.

VAN MOFFAERT M (1989). Management of self-mutilation: Confrontation and integration of psychotherapy and psychotropic drug treatment. Psychother Psychosom 51:180–186.

WALKER FD, ALESSI AG, DIGRE KB, ET AL. (1989). Psychogenic respiratory distress. Arch Neurol 46:196–200.

WALLEN J, INCUS HA, GOLDMAN HH, ET AL. (1987). Psychiatric consultations in short-term general hospitals. Arch Gen Psychiatry 44:163–168.

WARRENS AN, RON MA, & DAWLING S (1990). Positive diagnosis of self-medication with homatropine eye drops. Br J Psychiatry 156:124–125.

Warwick HM (1989). A cognitive-behavioural approach to hypochondriasis and health anxiety. J Psychosom Res 33:705–711.

Warwick HM, & Salkovskis PM (1990). Hypochondriasis. Behav Res Ther 28:105–117.

Wasyliw OE, & Cavanaugh JL (1989). Simulation of brain damage: Assessment and decision rules. Bull Am Acad Psychiatry Law 17:373–386.

Weighill VE (1983). "Compensation neurosis." A review of the literature. J Psychosom Res 27:97–104.

Weintraub MI (1983). Hysterical conversion reactions: A clinical guide to diagnosis and treatment (p. 186). New York: SP Medical and Scientific Books.

Whitlock F (1967). The aetiology of hysteria. Acta Psychiatr Scand 43:144–162.

14 Personality disorders in the medical setting

BARRY S. FOGEL, M.D.

General hospital psychiatrists have long been interested in the impact of personality traits on adherence to medical treatment recommendations. The classic article of Kahana and Bibring (1964), known to most psychiatric consultants, describes prototypes of different personality styles (e.g., hysterical, compulsive) and indicates how treatment instructions can be adapted to the patient's personality to promote treatment adherence. Since that article appeared, the diagnosis of personality disorders has been operationalized in the *DSM*, a hierarchy of defense mechanisms has been empirically validated (Vaillant, 1986), the epidemiology of personality disorders has been studied in a variety of medical settings, and the overlap of Axis I and Axis II has become better appreciated. Furthermore, techniques for circumventing resistance, such as reframing and paradoxical intention, introduced by family therapists, have entered the mainstream of psychiatric practice.

This chapter discusses the core ideas of managing patients with personality disorders in the medical setting, emphasizing modern concepts of diagnosis, and presents guidelines for individualized management strategies. The first part of the chapter deals with epidemiology and principles of descriptive and psychodynamic personality diagnosis. The second part presents specific management strategies and suggests how they can be selected based on personality diagnosis. The third part of the chapter deals with the use of medications in patients with personality disorders. The chapter concludes with the discussion of liaison issues, countertransference, special treatment issues with the elderly, and legal considerations. The chapter presupposes knowledge of the *DSM* diagnostic system, as well as of the hierarchy of defense mechanisms originally outlined by Anna Freud (1946) and more recently elaborated and validated by Vaillant (1986).

PRESENTATION OF PERSONALITY DISORDERS

In medical settings, patients with personality disorders are brought to the attention of the psychiatrist because (1) they display angry, manipulative, or self-destructive behavior; (2) they adhere poorly to treatment recommendations; (3) they evoke frustration and anger in their caretakers; (4) they develop severe anxiety, depression, or intractable physical complaints; or (5) they present with concurrent alcohol and drug abuse problems.

PREVALENCE

About 1 in 10 Americans has a personality disorder (Merikangas & Weissman, 1986). Because individuals with personality disorders are more likely to seek medical and psychiatric treatment than those without them, these disorders are highly prevalent among patients seen by psychiatrists in general hospitals. In one series of patients presenting to a general hospital psychiatry service, 36% had personality disorders according to *DSM* Axis II criteria (Koenigsberg et al., 1985). In this same series, patients with alcohol abuse disorders had a 46% rate of concurrent personality disorder, and patients with nonalcohol substance abuse had a 61% rate.

COMORBIDITY WITH AXIS I

Major depression and panic disorder, probably the most common primary psychiatric disorders in general medical practice, are associated with comorbid personality disorder at estimated rates ranging from 23% to 65% for depression and 35% to 58% for panic (Klein et al., 1988; Mavissakalian, Hamann, & Jones, 1990; Noyes et al., 1990; Pilkonis & Frank, 1988; Wetzler et al., 1990). Somatizing patients and those with behavior likely to lead to illness or injury are particularly likely to have personality disorders. Included in this group are drug abusers, recurrent suicide attempters, self-mutilators, and eating disorder patients. Table 14-1 displays estimated rates of personality disorders in several of these populations. In addition, numerous individuals *almost*

TABLE 14–1. *Personality Disorder Rates in Various Clinical Populations*

Population	Diagnostic Tool	Rate of Axis II Dx (%)	Most frequent disorders	Source
Somatizers referred for psychiatric consultation (n = 75)	Clinical	32	Compulsive; histrionic	de Leon et al. (1987)
Cocaine abusers entering outpatient treatment (n = 76)	SCID-II	58	Antisocial; passive-aggressive; borderline	Kleinman et al. (1990)
Anabolic steroid users (n = 20)	PDQ	85	Antisocial; paranoid; histrionic; borderline	Yates et al. (1990)
Normal weight bulimia (n = 52)	DIPD	48	Borderline; avoidant; compulsive	Zanarini et al. (1990)
Chronic pain (n = 43)	Semistructured interview	37	Dramatic and anxious clusters	Reich & Thompson (1987)
Inpatients with eating disorders (n = 35)	PDE	57	Borderline; self-defeating; avoidant	Gartner et al. (1989)
Normal weight bulimia outpatients (n = 300)	PDQ	75	Schizotypal; borderline; dependent	Yager et al. (1989)
Inpatient cocaine and opiate abusers (VA) (n = 117)	SCID II	31 (cocaine) 79 (opioid)	Borderline; antisocial; paranoid	Malow et al. (1989)
Repetitive self-injurious behavior (n = 83)	Clinical (review)	88	Borderline	Konicki & Schulz (1989)
Posttraumatic stress disorder (VA) (n = 536)	Clinical	31	Mixed; borderline; histrionic	Faustman & White (1989)
Inpatient trauma surgery (n = 112)	Clinical	21		Malt et al. (1987)
Suicide attempters (n = 60)	PAS	65		Casey (1989)
Male alcoholic VA inpatients (n = 74)	DIS	34	Antisocial personality	Herz et al. (1990)

meet criteria for *DSM* Personality Disorders (subthreshold conditions). Zimmerman and Coryell (1990) studied 797 relatives of psychiatric patients and never-ill control subjects, using the Structured Interview for *DSM-III* Personality Disorders (SIDP). They found that for schizotypal, borderline, and compulsive personalities, relaxing the *DSM* criterion level by just one item would more than double the number of cases diagnosed.

Patients with subthreshold maladaptive personality traits (falling short of Axis II diagnoses) also are commonly encountered in medical-psychiatric populations. For example, among 609 patients at a general-hospital-based psychiatry outpatient clinic, 51% had personality disorders, an additional 13% had "almost met" *DSM-III* criteria, and another 24% had maladaptive traits listed among *DSM-III* Axis II criteria (Kass et al., 1985). Patients with maladaptive traits might reveal these traits under the stress of medical illness or hospitalization, or in the context of the physician-patient relationship.

DETECTION IN MEDICAL SETTINGS

Personality disorders are particularly likely to go undiagnosed or misdiagnosed by primary care physicians. In a British study (Casey & Tyrer, 1990), general practitioners asked to diagnose all of their patients who had *any* conspicuous psychiatric morbidity identified 5.3% of them as suffering from a personality disorder. When the patients with psychiatric morbidity were interviewed by psychiatrists using questions probing for personality traits, 28% were found to meet diagnostic criteria for a personality disorder. Andersen and Harthorn (1989) studied the psychiatric diagnostic practice of 59 primary care physicians and 38 mental health professionals. Basing their diagnoses on written vignettes of cases meeting *DSM-III-R* criteria, only 13% of the primary care physicians correctly classified the personality disorder vignettes, compared with 50% of the mental health professionals.

Personality disorders have been associated with how people use medical care. Not only are people with personality disorders more likely to be hospitalized, but women with personality disorders of the dramatic cluster (narcissistic, histrionic, borderline, antisocial) are particularly likely to consult primary care physicians for emotional complaints (Reich et al., 1989). Because the personality pathology of these patients may be unappreciated or poorly understood by primary care physicians, the prescribed treatment may fail to take personality into account. The most common situation is one in which either depression or anxiety is treated without awareness of self-defeating personality traits; the self-defeating behavior undermines the treatment through noncompliance or medication misuse. This point may account for many of the disappointments faced by primary care physicians in their office treatment of mental health problems.

Problems of Diagnostic Assessment

The diagnosis of personality disorders in the medical setting is beset by theoretical and practical problems. Theoretical issues include distinguishing normality from pathology; choosing a categorical, dimensional, or prototypical diagnostic system (Frances, 1982; Livesley, 1985, 1986); and the question of whether cultural and sex biases may unduly influence diagnostic schemes (Kaplan, 1983; Presly & Walton, 1973). Empirically, dimensional judgments of personality traits are substantially more reliable than categorical diagnoses of personality disorders (Heumann & Morey, 1990; O'Boyle & Self, 1990). Unfortunately, dimensional measurements are not as easily incorporated into clinicians' thinking as categorical labels. Regarding bias, there is empirical evidence that the histrionic label is differentially applied to women and the narcissistic label to men (Adler, Drake, & Teague, 1990). Furthermore, some *DSM-III-R* categories, such as borderline and schizotypal, or dependent and avoidant, overlap significantly (George & Soloff, 1986; Reich, 1990a).

Practical issues concerning accurate diagnosis include difficulties obtaining an accurate history from severely ill patients, the confounding effects of stress, and effects of the hospital environment, which can aggravate regressive behavior. Acute and chronic pain can exacerbate maladaptive personality traits (Bellissimo & Tunks, 1984). For these reasons, firm diagnoses of personality disorder should not be recorded in patients' charts unless there is adequate historical evidence of persistent maladaptive traits antedating the current episode of medical illness. Such information often can be obtained from past psychiatric records, from discussions with significant others, and from other physicians.

Diagnostic reliability for personality disorders varies widely and depends on the adequacy of the database as well as how typical the patient is for a particular personality syndrome. For example, investigations using structured diagnostic interviews of patients and informants obtained kappa coefficients for interrater agreement of .70 or higher for histrionic, borderline, and dependent personalities (Stangl et al., 1985). In contrast, when three psychiatrists compared personality diagnoses obtained in everyday clinical settings, the maximum kappa coefficient was .49 for antisocial personality; and all other diagnoses were even less reliable (Mellsop et al., 1982).

When gathering data regarding personality diagnoses, information from informants other than the patient is essential. It is easily gathered in hospital settings, as nurses offer a ready source of information about behavior, and visiting family and friends can be approached (with the patient's consent) for additional historical data. In one series of patients evaluated with a structured interview for personality disorders, diagnoses were changed in 20% of cases after further information was obtained from the structured interview of an informant. Interrater reliability, however, was not adversely affected by the additional information (Zimmerman et al., 1986). Moreover, patients with severe personality disorders are more likely to lie than other patients (Ford, King, & Hollender, 1981).

Self-report questionnaires, such as the Personality Disorder Questionnaire (PDQ) (Hyler & Reider, 1984) and the Millon Clinical Multiaxial Inventory (MCMI) (Millon, 1983) appear to be sensitive to personality pathology but are relatively non-specific with regard to distinguishing normality from pathology and distinguishing among closely related personality disorders (Hyler et al., 1989, 1990a; 1990b; Wetzler & Dubro, 1990). In view of their high sensitivity but relatively low specificity, these questionnaires are most appropriate for screening individuals in high risk populations for personality disorder. Patients with questionnaire evidence of personality disorder would be followed up with clinical interviews of the patient and an informant. Self-report questionnaires for Axis II disorders increasingly are being used in general psychiatric research to obtain a measure of personality to use for more precise modeling of treatment outcome. In view of the relevance of Axis II problems to treatment outcome (see below), personality measures should be employed more widely in treatment outcome studies in psychiatry of the medically ill.

Descriptive Versus Psychodynamic Perspectives

Although a categorical diagnostic system such as the *DSM* is helpful for epidemiological study and clinical research, it has drawbacks that may be particularly evident in medical settings. Under the stress of severe medical illness, trauma, or surgery, individuals may transiently display regressive, primitive, "lower level" defenses typical of borderline personalities, although they lack the long-standing history of impaired function necessary for a formal *DSM* diagnosis of borderline personality disorder. In this situation recognition and management of primitive ego defenses are essential, whereas a firm decision about categorical diagnosis is not as important and should be deferred. Severe stress and the regressive situation of the hospital may bring out paranoid behavior in ordinarily schizoid or obsessional individuals or induce typically borderline behavior in previously high-functioning histrionic or narcissistic characters. Under other, less stressful circumstances, these same individuals would not display traits that would be sufficient to warrant a formal Axis II diagnosis.

When a patient's history permits an Axis II diagnosis, the assignment of a diagnostic category may be helpful for anticipating management difficulties and choosing therapeutic strategies. In the following sections, strategies are presented for managing patients with personality disorders, matched to specific *DSM* disorders. When applying these strategies, however, a descriptive diagnosis should be combined with an assessment of defense mechanisms and object relations, in an effort to place the patient on a continuum of adaptive personality functioning. If a patient whose history suggests compulsive personality currently is utilizing paranoid defenses, management strategies for paranoid personalities would be appropriate. Likewise, if an ordinarily histrionic individual is able to mobilize more mature defenses in response to the challenge of a physical illness, management strategies should take the apparent improvement into account.

These intervention strategies are presented contrasting patients with predominantly "more mature" and "less mature" defense mechanisms and object relations. The therapeutic strategies apply to patients with maladaptive personality traits of syndromal proportions and to patients with no formal psychiatric diagnosis who use more maladaptive defenses under the acute stress of medical illness, surgery, or trauma. The hierarchical classification of defense mechanisms as more mature (healthy or neurotic) or less mature (borderline or psychotic) is described in detail by Vaillant (1986; pp. 111–117) (see Table 14-2).

TABLE 14–2. *Vaillant's Glossary of Defenses*

"Psychotic" defenses

Delusional projection: frank delusions about external reality, usually of a persecutory type

Denial (psychotic): denial of external reality

Distortion: grossly reshaping external reality to suit inner needs

"Immature" Defenses

Projection: attributing one's own unacknowledged feelings to others

Schizoid fantasy: tendency to use fantasy, autistic retreat, and imaginary relationships for the purpose of conflict resolution and gratification

Hypochondriasis: the transformation of reproach toward others arising from bereavement, loneliness, or unacceptable aggressive impulses into first self-reproach and then complaints of pain, somatic illness, and neurasthenia

Passive-aggressive behavior: aggression toward others expressed indirectly and ineffectively through passivity

Acting out: direct expression of an unconscious wish or impulse in order to avoid being conscious of the affect or the ideation that accompanies it

Dissociation: temporary but drastic modification of one's character or of one's sense of personal identity to avoid emotional distress

"Neurotic" defenses

Repression: seemingly inexplicable naivete, memory lapse, or failure to acknowledge input from a selected sense organ

Displacement: the redirection of feelings toward a relatively less cared-for (less cathected) object than the person or situation arousing the feelings

Reaction formation: conscious affect and/or behavior that is diametrically opposed to an unacceptable instinctual (id) impulse

Intellectualization: thinking about instinctual wishes in formal, bland terms that leave the associated affect unconscious

"Mature" defenses

Altruism: vicarious but constructive and instinctually gratifying service to others

Humor: overt expression of feelings without individual discomfort or immobilization and without unpleasant effect on others

Suppression: the capacity to hold all components of a conflict in mind and then to postpone action, affective response, or ideational worrying

Anticipation: realistic anticipation of or planning for future inner discomfort

Sublimation: indirect or attenuated expression of instincts without adverse consequences or marked loss of pleasure

Source: Adapted from Vaillant GE (Ed.) (1986). *Empirical Studies of Ego Mechanisms of Defense* (Appendix III; pp. 11–120). Washington, DC: American Psychiatric Association Press.

Bedside Assessment

When evaluating problematic patients with suspected personality disorders several issues deserve emphasis in the initial interviews.

1. What is the patient's view of the problem? Is there a legitimate, practically resolvable issue? Is there a misunderstanding between the patient and the medical staff that needs clarification?

2. Are there practical, negotiable issues compounded by emotional conflicts? Could the practical problem be resolved if emotional conflicts were identified and separated from the practical issues?

3. Is there evidence of intercurrent delirium, dementia, psychosis, or major mood disorder? Is the patient under the influence of prescribed medications, alcohol, or illicit drugs—or withdrawing from such substances?

4. What is the patient's perception of the physician-patient relationship and of the patient's role in the treatment?

5. Who has the more relevant personality problem—the patient or a member of the medical or nursing staff?

When information from the interview, collateral sources, professional caretakers, and the medical record is synthesized, a tentative personality diagnosis may be possible, as well as a concise statement of the behavioral problem needing resolution. Specific strategies can be employed to address behavioral problems according to the patient's personality type while concurrent Axis I problems receive appropriate pharmacologic or psychotherapeutic attention, or both.

CONCURRENT AXIS I AND AXIS II DIAGNOSES

Reciprocal relations of Axis I and Axis II have received increasing attention in the psychiatric literature. For example, problems on Axis I may make a definitive Axis II diagnosis difficult, and the presence of personality disorder may alter the prognosis and treatment strategy for an Axis I disorder (Hyler & Frances, 1985). There is also a frequent concurrence of alcohol and drug abuse with personality disorders. Brief reactive psychosis is associated with borderline personality, as are episodes of major depression (Gunderson & Elliott, 1985; Perry, 1985). Patients with all types of personality disorders are especially likely to develop adjustment disorders in reaction to stress, including the stress of acute medical illness.

Identification of superimposed major depression is of particular importance because the mood disorder may itself exacerbate maladaptive personality traits (Hirschfeld et al., 1983; Libb et al., 1990). A patient who appears dependent or histrionic before antidepressant treatment might be seen as within normal limits after recovery. In addition, a significant subgroup of borderline personalities may represent variants of mood disorders ("subaffective disorders") that respond to antidepressants. High premorbid functioning prior to stress-induced onset of an apparent borderline personality may predict a good response to somatic antidepressant treatment (Novac, 1986).

The presence of a concurrent personality disorder has been associated repeatedly with worse results in the treatment of Axis I disorders, especially depression (Marin et al., 1989; Shea et al., 1990) and panic disorder (Reich, 1988; Reich & Green, 1991). Furthermore, panic disorder patients with personality disorders have less placebo response than those without personality disorders (Reich, 1990b). Among treatment-responsive patients with major depression, those with comorbid personality disorder are more likely to have a slow or incomplete response (Pilkonis & Frank, 1988).

Organic Factors

Secondary ("organic") mental disorders may exacerbate personality traits to syndromal proportions in the context of medical illness. For example, pernicious anemia and lupus have been observed to exacerbate paranoid traits, and hypothyroidism to increase dependent behavior. Exacerbations of personality disorder traits in diabetes mellitus by nocturnal hypoglycemia also have been reported (Krahn & Mackenzie, 1984). Untreated temporal lobe epilepsy can produce idiosyncratic experiences suggesting schizotypal personality or inappropriate rages mimicking those of borderline personality. Passive-aggressive phenomena can be produced by frontal or right-hemisphere lesions causing denial and inattention. Secondary mental disorders also can cause disinhibited behavior, magnifying latent problematic personality traits.

Adolescents and adults with attention deficit hyperactivity disorder (ADHD) may display angry, impulsive, uncooperative behavior in the hospital environment because of the combination of overstimulation and enforced passivity. Patients with borderline personality are particularly likely to suffer from concurrent ADHD, with one study showing a prevalence rate of 25.5% (Andrulonis et al., 1982). Although, in general, stimulant therapy of ADHD improves behavior, patients with borderline personality and ADHD represent a special problem because such patients can display markedly increased sensitivity to the psychotomimetic effects of stimulants. Dosages of as little as 30 mg of dextroamphetamine a day or 0.3 mg/kg of methylphenidate have induced formal thought disorder in experimental studies of borderline patients. Patients with schizotypal features are particularly vulnerable (Lucas et al., 1987; Schulz et al., 1988).

A practical approach to behaviorally significant ADHD in the medical setting is to use stimulants for patients without borderline or schizotypal personalities and without medical contraindications to stimulants. When stimulants are contraindicated and the need for behavioral control is acute, low-dose neuroleptics are a reasonable choice. In the author's experience, thioridazine is effective and is better tolerated than high potency agents, which are likely to produce restlessness and akathisia in ADHD patients.

Diagnostic Pointers

It is assumed that the reader is familiar with the *DSM* classification of personality disorders and with the ways in which patients with particular personality disorders are likely to show problematic behavior in medical settings. When the presence of a personality disorder or maladaptive personality traits impedes medical treatment, the specific personality diagnosis may assist in selecting management strategies but must be supplemented by a current assessment of defensive style and the quality of interpersonal relationships. This assessment is most important for individuals with narcissistic, histrionic, dependent, and passive-aggressive traits, who may have varying levels of personality organization. For example, histrionic or "hysterical" patients with relatively good impulse control and a more mature defensive structure require different management from less mature, impulsive histrionic patients with borderline features (Kernberg, 1986). Diagnostic points to be remembered include the following.

1. Antisocial personalities, diagnosed on the basis of history, always should be presumed to have "psychotic" or "immature" defenses, even if they appear to be functioning at a higher level on a single cross-sectional assessment.

2. The formal diagnosis of borderline personality disorder by *DSM* criteria is less important for medical-psychiatric management than identification of the patient as one with a poor quality of interpersonal relationships and a tendency to use the defenses of splitting and projective identification. Patients with a history of childhood sexual abuse, multiple personality, factitious illness, or multiple suicide attempts are highly likely to have a borderline personality organization (Barnard & Hirsch, 1985; Benner & Joscelyne, 1984; Briere & Zaidi, 1989; Task Force of American Psychiatric Association, 1980).

3. Passive-aggressive and dependent personalities may have either more mature or less mature defensive structures and object relations. More primitive passive-aggressive patients may passively attempt to sabotage medical treatment as an expression of rage against the physician, who serves as a transference figure. By contrast, the passive-aggressive patient with more "neurotic" defenses primarily may have conflicts over control and autonomy, or may be identifying with a passive parent. Management should be in accord with the patient's defensive style and specific emotional conflicts.

4. Schizoid and avoidant personalities may be difficult to distinguish on a single cross-sectional assessment. Both, however, fear closeness and may fear being controlled by others. The enforced intimacy of medical settings and the attendant loss of control lead to anxiety, which is expressed according to the patient's defensive style.

5. Narcissistic personalities may have either a low or a high level of function. More severely impaired individuals may display demanding behavior, entitlement, exploitiveness, and dishonesty that disrupt medical treatment and engender angry countertransference. Better-functioning narcissistic personalities may be less dramatically "entitled" but may reveal an excessive vulnerability to criticism and disappointment, and be prone to feelings of shame (Svrakic, 1987). They may overvalue particular physical or mental attributes and have an unusually difficult time adjusting to the limitations of a chronic illness. The physician, if not devalued, may be idealized to the point of discomfort.

SPECIAL DIAGNOSTIC ISSUES IN THE ELDERLY

As problematic as personality disorder diagnosis may be in general medical patients, it can be even more difficult in elderly medical patients because neither diagnostic criteria nor their application are age-neutral. The criminal behavior that is required for a diagnosis of antisocial personality is less prevalent during old age, as are the identity disturbances typical of borderline personality in young adults (Arboleda-Florez, 1990). Yet personality traits are relatively stable with age (Costa & McCrae, 1980; McCrae, Costa & Arenberg, 1980), suggesting that personality disorders do not disappear altogether but change their form according to the patient's life stage. Antisocial personalities may give up crime and turn to alcoholism and hypochondriasis (Maddocks, 1970), or borderline personalities may develop stable social and occupational functioning but fail to form close personal relationships (McGlashan, 1986). The focus of patients' concerns and conflicts shifts with life stage: Narcissistic individuals may inflate past accomplishments or the talent of their grandchildren; borderline personalities may apply splitting and projective identification to their caregivers and to their children instead of to their parents or lovers.

Age-specific events may invoke symptoms of personality disorder in the elderly. Such situations include bereavement, institutionalization, forced intimacy with caretakers, and cognitive changes, particularly those of early dementia. Personality disorders thus may emerge during old age, especially in individuals who had subsyndromal traits earlier in life.

Because personality disorder diagnoses may assist in planning and in making prognoses, it is worth the effort to attempt personality diagnosis in elderly patients. This exercise can be aided by: (1) obtaining information on the patient's earlier adaptation, especially from reliable informants or old records; (2) considering traits and defensive style and deemphasizing whether formal criteria are met; and (3) considering the possibility that disordered personality can emerge during old age as a function of changed life circumstances or alterations in brain function (Sadavoy & Fogel, 1992).

MANAGEMENT STRATEGIES

Based on data from the patient interview and from collateral sources, the psychiatrist can develop a tentative personality diagnosis that comprises an assignment of an Axis II diagnosis and, if possible, an assessment of current defensive operations and the quality of the patient's interpersonal relations. Axis I disorders and organic aggravating factors also are diagnosed, and their treatment is initiated as appropriate. The psychiatrist then must develop a short-term management strategy that permits medical treatment with a minimum of interference from the patient's maladaptive personality traits.

An individualized strategy is developed by combining strategies, most of which have a long history in the literature of psychiatric consultation (Groves, 1978). One group of strategies is particularly appropriate for patients with "psychotic" or "immature" defenses and poor quality interpersonal relationships; it is especially useful for patients with borderline personality organization or antisocial personality. A second group of strategies is useful for patients with more mature defenses and better quality interpersonal relationships. These basic strategies are described in detail in the following subsections. The choice of strategy is summarized in Table 14-3.

TABLE 14–3. *Matching Diagnosis and Intervention Strategies*

Level of Defenses and Object Relations	Diagnosis	Interventions
Least Mature	Borderline Antisocial Paranoid	Giving a sense of control Being consistent Taking the "one-down" position Limit setting Antisplitting maneuvers
Less Mature	Schizoid Avoidant	Straightforward, matter-of-fact relationship; avoiding excessive closeness
Variable	Narcissistic Passive-aggressive	If less mature: borderline strategies If more mature: strategic reframing
	Histrionic Dependent "Hysterical"	If less mature: borderline strategies If more mature: personal leverage or strategic reframing
More Mature	Compulsive Maladaptive traits of subsyndromal proportions	Strategic reframing

Strategies for Managing Patients with Less Mature Defenses

When managing a patient with less mature defenses, the physician must not assume, a priori, a trusting relationship or the patient's capacity to be consistent. Consistency and appropriateness must be provided by the physician, who must not be unduly moved by pleas, demands, manipulations, or threats, or be personally affronted by the patient's mistrust (Dawson, 1988).

Because patients with less mature defenses may react dramatically to disappointment, it is all-important to clarify treatment expectations. The patient's understanding of the implicit therapeutic contract between physician and patient should be explored and the goals and limits of the patient's medical treatment made clear at the outset. Conflict, both intrapsychic and interpersonal, can be maximized by frequently reorienting the patient, other physicians, nursing staff, and family toward the basic, circumscribed treatment contract (Selzer, Koenigsberg, & Kernberg, 1987). Whenever possible, this implicit contract should be stated in the patient's own words. Management strategies useful for patients with less mature defenses involve giving patients a sense of control, being consistent, taking the "one-down" position, limit-setting, avoiding hospitalization, and preventing and managing splitting of staff (Moss, 1989).

Giving a Sense of Control to the Patient

Giving a sense of control is maximizing the appropriate influence the patient has over the treatment situation. For outpatients, the physician may let the patient decide within reasonable limits a schedule of visits, or let the patient choose a specific medication if there is more than one acceptable alternative (Dawson, 1988). For inpatients, the physician may let the patient decide when to take certain medication, when to go to physical therapy, or the date of discharge (Kahana & Bibring, 1964; Staleniam & Youngs, 1974). Allowing a sense of control is aided by recalling that the physician is responsible for good professional judgment and advice but is not responsible for the patient's decision to follow the advice (Strain, 1978).

Being Consistent

Being consistent means minimizing changes in plan and personnel, thus reducing opportunities for confusion or misunderstanding. Consistency is facilitated by using the minimum number of drugs and orders. Being consistent is made easier for everyone concerned by having written instructions and plans and by allowing patients to restate the plans in their own words. When several caretakers are involved, this strategy means carefully coordinating the efforts of all caretakers, using written communications whenever possible, and sharing the agreements with the patient. The use of written instructions and plans is of documented value for improving compliance of all patients (Hayes, Taylor, & Sackett, 1979) but is especially important for patients with less mature defenses (Hall, 1977). Numerous reports have described "treatment contracts" and the principles involved in writing them (McEnany & Tescher, 1985; O'Brien, Caldwell, & Transea, 1985).

Taking the "One-Down" Position

Taking the "one-down" position means realistically recognizing one's limitations in affecting another person's behavior, and approaching the patient with genuine and appropriate humility. It includes "allowing" some degree of pathologic behavior to go unchallenged if it is not immediately or severely harmful to the patient. This approach—more an attitude than a technique—forestalls many potential control struggles with personality-disordered patients (David, 1979; Ellard, 1977; Schwartz, 1979; Strain, 1978; Watzlawick, Weakland, & Fisch, 1974).

For example, an elderly man with a ventricular arrhythmia was noncompliant with hospitalization because of mistrust of his doctors' motives. The working diagnosis was paranoid personality. The strategies selected were taking the one-down position and giving control. The patient was told that the physician could not and would not force him to stay in the hospital, and that the physician did not expect him to believe his medical diagnosis. The physician wished to give him the opportunity to learn for himself whether medication could prevent him from having further fainting spells. The patient could undertake this regimen on an outpatient basis, just as soon as a few tests were done. Furthermore, if the patient liked, he could have his wife check all medications before giving them to him to take. The patient accepted this arrangement and over the next 48 hours was stabilized on an antiarrhythmic drug. He was then discharged from the hospital and at follow-up was continuing to take his antiarrhythmic drugs.

Taking the one-down position also includes setting appropriately modest expectations regarding all treatments, both psychiatric and medical. Informed consent procedures should be scrupulous, giving weight to side effects and the possibility that treatment will not be effective. All treatment is seen as a "treatment trial" (Sweeney, 1987).

Limit Setting

Limit setting is the trump card—to be played if a patient is seriously impeding treatment or is endangering someone (Strain, 1978). For outpatients, such problematic behavior often involves abuse of prescribed drugs (Sternbach, 1974). For inpatients, limit setting is an appropriate response to such behavior as deliberately pulling out intravenous lines, smoking in a room with oxygen, or screaming all night in a room full of sick patients.

In the inpatient case, all involved caretakers are told of the problem and decide together what limits will be set. The patient is told without anger which behaviors are unacceptable "in this hospital". If limits are set angrily, the patient responds more to the anger than to the limit. Kahana and Bibring (1964) warned, "Great care should be exercised not to introduce limits as if they were an expression of impatience or punitiveness." Reference to "this hospital" restricts the control struggle to the hospital rather than to the whole world. In the outpatient case, the patient is told that the behavior is unacceptable within that physician's practice. The patient is told that the physician will not continue to treat the patient or prescribe medication if the behavior continues. (Appropriate alternative sources of care would, of course, be discussed with the patient to forestall charges of abandonment.)

In the variation called sympathetic limit setting, the physician adds that there must be good reasons for the patient's behavior, *but* the behavior is unacceptable (Lipp, 1977). With either type of limit setting, the inpatient is told that discharge (termination of the treatment contract) is inevitable if the unacceptable behavior continues. If involuntary psychiatric treatment is indicated, a request for temporary certification is completed, and the patient is referred to an appropriate psychiatric facility. If the patient is too medically unstable for such a referral, control of destructive behavior is maintained with neuroleptics, special observation, or restraints until the patient's physical condition permits either discharge or referral to a secure mental health facility.

The main obstacle to effective limit setting is countertransference (Selzer, Koenigsberg, & Kernberg, 1987). The two most common problems are excessive expression of the clinician's anger or a failure to set limits because of either: (1) the clinician's fear of the patient's anger; or (2) the clinician's oversympathetic defense against his or her *unconscious* anger. One authority on the treatment of personality disorders has stated, "Consistency and limit setting are equally as important as benignness, availability, and good intentions. When each demand of the patient was met with yielding or bending of the rules, as with demanding, angry, overtly hostile patients in an effort to keep them from 'exploding,' the result was greater loss of control, (and) even more demanding behavior" (Hall, 1977).

Limit setting, though often essential for treating patients with less mature defenses, such as borderline personalities, can have negative effects on patients with more mature defenses and should be used with them only when other methods have failed.

Avoiding or Limiting Hospitalization

Dawson (1988) has pointed out that hospitalization itself can aggravate primitive behavior in borderline patients. He observed that it is difficult to get borderline patients out of a helpless, dependent position once they are in it. The environment of the hospital often evokes less mature defenses, either through increasing anxiety or through providing powerful transference figures (physicians and nurses) or situations that evoke imagery of past traumatic events. For this reason, both hospitalizations and their length should be limited, even when it means ignoring patients' threats or demands aimed at initiating or prolonging a stay in hospital. Strict criteria of medical necessity should be applied when making the decision to hospitalize a borderline patient. The goals of medical hospitalization usually should be limited to carrying out those procedures and treatments that can be done only in hospital; the goals of psychiatric hospitalizations usually should be limited to confirming a diagnosis and initiating treatment of those Axis I conditions that are most disabling or dangerous.

Preventing and Managing Staff Splitting

Patients with borderline personality organization frequently employ splitting as a defense (Stoudemire & Thompson, 1982). The important people in their world are divided into "good" and "bad" objects: the former loved and idealized, the latter hated and devalued. In a medical setting, the patient with borderline personality organization may induce among physicians and staff intense differences in attitude and feeling toward the patient, so the staff are split into two factions with strongly held and opposing views regarding the patient's diagnosis or treatment. Issues frequently disputed include pain management, the need for psychotropic medication, the validity of the patient's physical complaints, and the limits that should be set on the patient's behavior.

Splitting of physicians and other staff may arise in three ways. First, the patient may manipulate some caretakers into a particular viewpoint, whereas others, unaffected by the manipulation, disagree and may view the patient negatively as "manipulative." Second, caretakers may have differing countertransference reactions to the patient's behavior. Some may respond with sympathy and protectiveness, whereas others respond with anger and annoyance. Third, the patient may actually show different affect or behavior with different caretakers because of different transferences, thereby inducing conflicting opinions in the absence of any deliberate manipulation.

However splitting arises, it can impede optimal care of the patient. Consequences of splitting can include inconsistent care, inappropriate medication, and staff inefficiency due to intrastaff conflict. Therefore when splitting of caretakers occurs, "antisplitting maneuvers" are required.

The basic antisplitting maneuver is the network meeting. All physicians and staff involved in the patient's care are brought together in one room. The psychiatrist identifies the problem of splitting and initially encourages ventilation of feelings and disagreements. The psychiatrist then points out that these feelings and disagreements are induced by the patient's behavior and that they provide valuable information about the patient. The psychiatrist then explains that the patient must be treated with the utmost consistency. A rational treatment plan must be agreed on and carried out, regardless of the patient's varying reactions to different caretakers and despite any attempts made at manipulation. The psychiatrist works with the staff to develop a consensus on an appropriate care plan, and this plan is written down. The psychiatrist or the primary medical physician then informs the patient of the meeting and reviews the key points of the care plan with the patient. Depending on the patient's level of function, it is sometimes reasonable to invite the patient into the staff meeting to hear the consensus directly.

Meetings of involved physicians and staff may be required periodically if the patient's length of stay in the hospital is relatively long. Splitting that arises concerning one issue may be resolved, but another issue may arise a few days later.

The most common error during network meetings held to combat splitting is the failure to include an important caretaker. When this error occurs, those who have attended the meeting and those who have not may wind up on opposite sides of a split. If several physicians are involved regularly in the patient's care, all must be present. If nurses on day and evening shifts are split, both shifts must be represented at the meeting.

Splitting may occur in outpatient settings when patients with borderline personality organization see multiple physicians for their various medical problems. They seek opinions from each physician, attending especially to any differences or discrepancies among physicians' opinions. These perceptions may lead to noncompliance or to the patient's devaluing one physician while idealizing another.

The psychiatrist can counteract this splitting by bringing together the physicians most involved in the patient's care, writing or telephoning those more peripherally involved, and developing a consensus among all physicians regarding the patient's diagnosis and appropriate treatment. If consensus is not possible, the basis of differences of opinion is clarified. Then the psychiatrist and the primary medical physician meet with the patient and review the information together. Family members may be included in the conference if they are higher functioning than the patient.

When confronted with antisplitting measures, some patients become anxious or enraged. Some defect from treatment. Others remain in treatment but may require short-term neuroleptic therapy to contain the anxiety that emerges when the splitting defense is effectively confronted. Despite these problems, antisplitting maneuvers probably are safer for the patient than ignoring the problem, because significant caretaker splitting may lead to errors in physician or nursing judgment that increase the risk of a poor medical outcome.

Constructing a Management Strategy for Patients with Less Mature Defenses

For patients with less mature defenses, begin by clarifying expectations, being consistent, giving a sense of control, and taking the "one-down" position. Even if there is a past history of inappropriate behavior in the medical setting, there usually is no need to set limits until the patient acts or threatens to act destructively. Then limit setting can be done, sympathetically in the case of borderline and antisocial patients and matter-of-factly with paranoid and schizotypal patients. Sympathetic limit setting should not be used with paranoid, avoidant, and schizotypal patients because sympathy may arouse further anxiety by implying closeness (McGrath, 1978). If sympathetic limit setting is ineffective with a borderline or antisocial patient, one should proceed with matter-of-fact limit setting. Limits framed so they save face for the patient often are embraced gratefully by patients with more primitive narcissistic personalities. When staff splitting occurs, antisplitting maneuvers should be instituted.

Strategies for Patients with More Mature Defenses

Strategies for patients with more mature defenses and better-quality interpersonal relationships presuppose a reasonable degree of trust and consistency; they attempt to use these patients' personality styles and personal issues as leverage to help them comply with prescribed therapeutic interventions. Strategies for these personalities include use of strategic reframing statements and personal leverage.

Strategic Reframing

Strategic reframing means presenting medical instructions in language that is most consistent with the patient's characteristic coping style, or cleverly framing medical instructions in a manner designed to discourage subversion of the treatment plan by the patient's maladaptive defenses. The former is "positive reframing," and the latter is "paradoxical reframing".

Positive Reframing. After giving a narcissistic patient medical instructions, one might say, "Many patients would have difficulty with these instructions. Only a special person could really follow this regimen as it was intended" (Lipp, 1977). The medical need for compliance is expressed in language appealing to the patient's sense of entitlement to special treatment. A compulsive patient might be asked to carry out an instruction because "it will save you time and money." In this case, compliance is linked to the patient's concern with efficiency.

Positive reframing statements may emphasize illness as an occasion for personal growth or the mastery of illness as a route to greater independence. These types of statements are particularly appealing to individuals with more mature defenses who have sympathy for the ideas of humanistic psychology and the concept of the physician and patient as peers. Strategic reframing statements preferably use the patient's own language and focus on the patient's individual concerns. The process of creating reframing statements is similar to that used when developing hypnotic suggestions and self-hypnotic routines and may be combined with formal trance procedures by clinicians with sufficient expertise in hypnosis. Applications of self-induced trance in medical settings are described by Spiegel and Spiegel (1978).

Paradoxical Reframing. Disabled patients with dependent personalities might be given a detailed plan for a gradual return to normal activity and be told that it would be difficult for them to carry it out and that they could probably do it only by having weekly visits with the psychiatrist for support. Even then, if they did succeed, it would be largely due to their own perseverance in the face of a frightening situation. The psychiatrist would see them, however, only if they complied with the rehabilitative activities. Dependent patients with a lack of self-confidence are thus offered a situation in which they will get all the credit if they succeed but no blame if they fail. Also, the personal need to depend on others is linked with rehabilitation, through the psychiatrist's offer of weekly support. Furthermore, noncompliance is linked with the threatened loss of a dependent relationship.

Paradoxical Intention. Also known as "prescribing the symptom," paradoxical intention is a special type of strategic reframing. This controversial technique is anecdotally effective in patients with compulsive and passive-aggressive traits but may be risky in borderline patients, whose humorless reception of paradoxical requests may lead to regression or acting out (Greenberg & Pies, 1983).

Personal Leverage

Using personal leverage means motivating the patient with poor self-esteem by relating medically therapeutic behavior to an important personal relationship. The source of leverage may be the primary physician, a nurse, a consultant, a house officer, or a family member. Personal leverage may involve suggestions, direct orders, recommendations (Balint, 1957), reassurance, or handwritten instructions (Hayes, Taylor, & Sackett, 1979). Typical statements are as follows.

1. "As your personal physician, I recommend X" (Kahana & Bibring, 1964).
2. "I really believe that someone like you could handle this procedure just fine."
3. "Dr. Z., our eminent consultant, has reviewed your case and recommends Y."
4. "Do it for your children."

In some cases, having a personally influential person talk to the patient about the medical treatment or personal problem is most helpful. For instance, a spouse might be invited by the doctor to help explain a procedure to the patient, assuming that the spouse has previously been shown to be positively influential toward the patient's overall health.

Personal leverage may be useful with dependent, histrionic, and better-functioning narcissistic patients, especially if the source of the personal leverage is someone who is currently in a stable positive relationship with the patient (Hall, 1977). Personal leverage usually is not effective with compulsive patients, who tend to mistrust personal appeals and are much more comfortable with a rational, factually oriented approach.

Appeals to religious or philosophical beliefs, expectations of one's family or community, and other transpersonal concerns are variants of personal leverage that can be extraordinarily effective if the clinician has sufficient understanding of the patient's cultural milieu and belief system.

Considerations in Choice of Strategy

For patients with more mature defenses, the psychiatric interview determines the personal issue underlying the patient's noncompliance or emotional difficulty. The physician then attempts to describe the proposed treatment in a way that addresses the personal issue. Frequently occurring personal issues are mentioned in personality disorder descriptions; however, it is always best if a specific personal issue can be identified during the initial consultation.

For patients who avoid intimacy, whether they be schizoid, avoidant, schizotypal, or paranoid, it is useful for the physician to adopt a neutral, predictable, matter-of-fact style. In inpatient settings, where other caretakers are involved with patients, the psychiatrist might recommend that they do the same. This practice avoids the anxiety these patients experience when confronted with excessive intimacy (McGrath, 1978). By reducing their anxiety, it may diminish the intensity of other troublesome traits.

Usage of Psychotropic Medications

The use of psychotropic medications to treat concurrent Axis I disorders in patients with personality disorders is well established and is recommended but with a few qualifications. First, particular care must be given to the psychodynamics of prescribing. Second, people with poor impulse control usually should not be given refillable prescriptions or prescriptions for quantities of medications that would be dangerous if taken all at once, because of the risks of overdose and medication misuse. Third, benzodiazepines usually should be avoided in borderline personalities because of the potential for abuse (Tyrer, 1989), as well as the risk of disinhibition of impulsive behavior (Cowdry & Gardner, 1988; Gardner & Cowdry, 1985). Fourth, depressed borderline patients treated with antidepressants should be monitored carefully for irritability or impulsiveness, which can be aggravated, particularly in treatment nonresponders (Mann & Kapur, 1991; Soloff, 1987). It is not established that any particular class of antidepressant provokes suicidal behavior more than any other (Mann & Kapur, 1991). However, stimulating antidepressants such as bupropion and fluoxetine (Lippinski et al., 1989) may be more likely than sedating agents to cause agitation early in treatment, and borderline patients treated with such drugs should be warned about the possibility of agitation and reassessed for suicide potential and evidence of psychosis if agitation develops. Fifth, short-acting benzodiazepines, because of their tendency to cause rebound anxiety, may exacerbate drug-seeking behavior in patients with personality disorders. A preferred benzodiazepine may therefore be a longer-acting agent such as clorazepate, given twice a day in a relatively low dose, without as-needed (PRN) doses. In patients with severe autonomic accompaniments of anxiety, beta-blockers may be used alone or in combination with low-dose benzodiazepines. Of course, medical contraindications to beta-blockers, such as asthma, should be excluded.

Evidence favoring pharmacologic treatment of Axis II disorders themselves is beginning to accumulate (Liebowitz, Stone, & Turkat, 1986). At present it is strongest for the use of low-dose neuroleptics in patients with borderline personality disorder, particularly those with severe hostility, behavioral dyscontrol, psychotic features, or a history of micropsychotic episodes (Cowdry & Gardner, 1988; Goldberg et al., 1986; Gunderson, 1986; Soloff, 1987; Soloff et al., 1986).

Although there is less evidence for its efficacy, the work of Cowdry and Gardner (1988) suggests that carbamazepine could be helpful in the longer-term control of impulsive behavior in some patients with borderline personality disorder. The use of carbamazepine in this situation is not based on a diagnosis of epilepsy or bipolar disorder; EEG data may be of interest but have not been demonstrated to predict outcome. Of course, if a patient has clinical or EEG evidence of limbic epilepsy and a borderline personality, carbamazepine clearly would be the drug of choice.

Monoamine oxidase inhibitors can be useful for patients with histrionic, avoidant, dependent, passive-aggressive, or narcissistic personalities if multiple "neurotic" symptoms such as fears, compulsions, anxiety, and depression are present (Klar & Siever, 1984). The treatment of these personalities with MAO inhibitors is likely to become more attractive when less toxic, reversible, selective MAO inhibitors such as moclobemide become available in the United States.

In patients with chronic anxiety or depression aggravating their maladaptive behavior, the serotonergic drugs buspirone and fluoxetine or sertraline may be worth a trial, especially because of their lower toxicity than tricyclics and their lower abuse potential than benzodiazepines. Evidence for reduced central serotonergic function in patients with personality disorder and impulsive aggression (Coccaro et al., 1989) has heightened interest in these drugs. Beta-blockers also are useful in occasional patients with personality disorders aggravated by anxiety (Liebowitz et al., 1986). Benzodiazepines, though undoubtedly useful for some patients, are likely to become a long-term habit when used by dependent or avoidant personalities (Tyrer, 1989). Avoidant personalities, many of whom may suffer from a form of social phobia, may respond to MAO inhibitors or to fluoxetine (Deltito & Stam, 1989).

Pragmatic psychiatrists often employ neuroleptics for treating borderline, schizotypal, and paranoid individuals who decompensate in medical settings. When disruptive behavior or distorted thinking preclude the patient's cooperation with medically necessary treatment, the benefit-risk ratio for low-dose neuroleptics is high. Relatively low doses minimize the risk of severe extrapyramidal symptoms and seem to be effective in patients without concurrent schizophrenia.

Use of Analgesics

Although many physicians do not regard analgesics as psychotropics, analgesics can profoundly affect behavior and mood, and their use may be especially problematic in patients with personality disorders. The association of pain with regressive behavior has already been noted. Although analgesic management is covered in Chapters 16 and 17, a few special points are worth emphasizing here.

Acute pain should always be treated adequately though preferably not with PRN schedules. Concerns about dependency can be addressed later, after the acute medical situation is resolved. Inadequate analgesia can provoke severe acting-out in patients with personality disorders of the dramatic cluster.

Only one physician should be responsible for prescribing analgesia, especially for patients with less mature defenses. That physician should reevaluate the dosage schedule frequently. For most patients, medication should be given at regularly scheduled intervals and not PRN. Regular dose schedules permit patients to have autonomy without reinforcing pain complaints if they offer medications at specific times rather than on demand, but permit patients the option of refusing or delaying doses, or of choosing a smaller dose. Such schedules may also permit nursing staff to withhold or delay medication if a patient is oversedated.

Pentazocine and meperidine can produce excitement and hallucinations; patients with severe personality disorders may be at high risk for overt display of psychotoxicity. If these agents must be used in patients with severe personality disorders, monitoring for mental and behavioral side effects should be especially close.

Psychodynamics of Prescribing

In patients with personality disorders, it is especially worthwhile to determine the psychologic meaning of medication to the patient. Medication compliance and medication abuse are frequent and often can be understood in terms of difficulties with impulse control, idiosyncratic ideas about medication, personal issues related to medication taking, or problems in the physician-patient relationship. A compulsive personality might focus excessively on the expense of medication while subconsciously resenting the loss of control implied in "depending" on a drug. A borderline personality might oppose discontinuation of a medication, despite medically unacceptable side effects, because the pill is a consistent anchoring point in a fragmented inner world (Adelman, 1985). Many problems can be avoided by prospective discussion of feelings about medication before initiating treatment and by prediction of the individual patient's confounding issues. The psychologic management strategies applied above can then be applied when making the prescription.

LIAISON ISSUES AND PSYCHIATRIC PRIMARY CARE

Patients with less mature defenses and poor interpersonal relations are particularly likely to produce anger in caretakers; the consequence of their splitting defenses is the polarization of staff. For these patients, psychiatric consultation rarely helps unless the reactions of attending physicians and nursing staff are addressed explicitly. Attention to the consultee, liaison work with nursing staff, and ongoing availability of the consultant are needed to forestall crises. Weekend and holiday coverage for the consultant must be arranged to promote maximum consistency of advice given to the caretakers.

When dealing with caretakers, the consultant should permit ventilation of anger and disagreement and then emphasize that these reactions provide diagnostic information about the patient (Gallop, 1985; Groves, 1975). The consultant should help caretakers "cool down," so the basic principles of being consistent, giving a sense of control, and setting limits without anger can be emphasized. Further details are offered in the earlier section on Preventing and Managing Staff Splitting.

Often suggestion of a pharmacologic intervention, such as recommending low-dose neuroleptics for the patient, permits a face-saving reappraisal of the patient by the staff. If the patient can be viewed as "stabilized" by an external agent, anger over the patient's past behavior can be rationalized. These pharmacologic interventions must not be carried out solely for their effect on staff, but should be indicated by the patient's diagnosis and behavior.

Sometimes consultation, even with liaison and follow-up, is insufficient to contain the self-destructive, disruptive, or noncompliant behavior of the medically ill patient with a severe personality disorder. At other times, as with some pain patients, personality problems can prevent proper diagnosis and management in a purely medical setting. In these situations primary care of the patient by the psychiatrist can facilitate resolution of the problem. When this arrangement is chosen, all orders are written by the psychiatrist, who sees the patient daily and uses medical specialists as consultants. At hospitals where there is a medical-psychiatric unit, the patient usually can be transferred there. Psychiatric primary care can greatly improve consistency of management because it puts the psychiatrist in the front-line position of deciding whether particular physical complaints need medical intervention or should be dealt with as somatic expressions of a psychiatric disorder (Fogel & Goldberg, 1983–1984).

Psychiatric primary care is unsuitable if the medical problem is unstable; in such a case concurrent care is better. Medical-psychiatric units may do better at containing the behavior of personality-disordered patients than they do at changing their maladaptive defenses; longer-stay, milieu-oriented units and outpatient therapy are more suited to promoting enduring behavioral change. Therefore, patients with severe personality disorders and active medical illness usually should be stabilized on a medical ward or medical-psychiatric unit and then transferred to a conventional psychiatric unit if milieu-oriented inpatient treatment is still needed after medical stabilization.

Psychiatric primary care at times is suitable for long-term outpatient treatment. The patient's coping with the chronic illness and health care system can be used as a focus for psychotherapeutic exploration and behavioral change. (See Chapter 1 for details of medical psychotherapy.)

Countertransference Reactions

Countertransference reactions to patients with personality disorders comprise a wide range of possibilities, including hate, anger, desire to avoid or abandon the patient, personal attraction, overprotectiveness, and taking sides with the patient against family members or other professional caretakers. The detection of countertransference reactions may be more difficult when they are defended against with reaction formation. For example, compulsive overconcern or excessive inclusiveness during a medical workup may be a defense against an unacceptable desire to avoid the patient.

Gallop and Wynn (1987), observing the reactions of nurses and psychiatry residents to patients with severe personality disorders, noted that nurses were likely to be explicitly angry, rejecting, and devaluing of the patients. The psychiatry residents were likely to feel incompetent and powerless, and to develop conflicts with their supervisors.

Even the *label* of personality disorder has effects on countertransference. Including the term "personality disorder" in a case vignette of a patient with a major depression led to a lower rate of psychiatrists' recommendations for antidepressant drug treatment and to inferences that the patient was not motivated, or even seriously ill (Lewis & Appleby, 1988).

In consultation-liaison work, the psychiatric consultant must tactfully assist the patient's professional caretakers to understand their countertransference, learning to see it as information about the patient rather than an implication of their inadequacy.

In medical-psychiatric settings, the focus shifts to maintaining a clear awareness of countertransference issues despite the intense distractions offered by the patient's medical care. On medical-psychiatric units, psychiatrists and nurses may lose sight of countertransference issues that would not escape their notice in a strictly psychiatric setting. In fact, experience with psychiatric primary care in medical-psychiatric settings is an excellent way for a psychiatrist to build empathy for medical consultees.

Special Treatment Issues in the Elderly

As discussed above, personality disorders endure into old age, although they may be less commonly diagnosed because the Axis I diagnoses of depression, dementia, and delirium occupy the foreground in geriatric psychiatry. Elderly medical patients' personality problems cannot be ignored, because they can disrupt medical treatment just as they do in younger patients. Therapists of the elderly (Kernberg, 1986) have suggested that elderly patients with some forms of personality disorder may be as responsive to psychotherapeutic treatment as younger patients.

Once a personality disorder is recognized, the same management techniques may be applied to elderly patients as to younger patients. There are, however, some differences in emphasis. Because of the high prevalence of secondary mental disorders in the elderly, the search for aggravating organic factors must be aggressive and should always include adequate testing of cognition and laboratory evaluation for toxic and metabolic factors.

Special considerations apply to treating elderly patients with personality disorders who reside in nursing homes or other chronic care institutions (Sadavoy, 1987; Sadavoy & Dorian, 1983). Special problems include the forced intimacy of the institution, the unattractiveness of the patient, and the frequent impossibility of transfer. In these situations, a consistent, regular supportive psychotherapeutic relationship can have a remarkable stabilizing effect on the patient's behavior. Indeed, many of these patients had supportive relationships with family members or primary care physicians prior to being institutionalized. Liaison interventions should focus on reducing the caregivers' level of affect and intimacy with the patient when it is clear that the patient is overwhelmed and the situation is becoming chaotic. Neuroleptic therapy, when considered because of acting-out behavior, should be kept to the low doses appropriate for treating personality disorders; full antipsychotic doses rarely are necessary and are associated with a high rate of disabling side effects.

Age prejudice may lead to insensitivity to elderly patients' personality problems, whereas oversolicitousness may lead to an underutilization of confrontation and limit-setting. Psychiatrists must carefully examine their own attitudes toward the elderly as well as the attitudes of the caregivers seeking consultation. The presence of positive and negative biases toward elderly patients must be sought and gently explored with consultees. For example, many clinicians believe that elderly patients are more hypochondriacal than younger patients, but systematic studies of the issue do not support this view (Costa & McCrae, 1985). Elderly neurotics, like young neurotics, exaggerate physical problems. The greater number of complaints by the elderly is related to a higher prevalence of true somatic disease.

LEGAL ISSUES

Patients with less mature defenses and personality disorders who have unstable interpersonal relations may rapidly turn against their physicians because of disappointment or disagreement. If there has been an error of judgment or an unexpectedly poor medical outcome, the patient may sue the physician or may threaten litigation with manipulative intent. Lawyers, not understanding the dynamics of borderline personality, may with good intentions pursue patient claims that are based more on near-psychotic transference than on reality (Gutheil, 1985).

In the consultation-liaison setting, the psychiatrist usually is called after the physician-patient relationship has begun to deteriorate and threats have been made. While interventions with patient and caretakers are made to "cool down" the situation, the attending physician should be asked to obtain a second opinion on the medical management and to document thoroughly the reasons for all medical judgments made to date (Schwarcz & Halaris, 1984). The psychiatrist reviews with the attending physician the alternatives of discharging the patient or transferring the patient's care to another doctor. The psychiatrist then explores the patient's feelings toward the attending physician and explicitly discusses alternative treatment options. This discussion is documented thoroughly. The patient's responsibility for subsequent treatment choices thus is established. Limits then can be set if necessary, because alternatives already have been worked out that would not constitute abandonment of the patient.

Occasionally, the psychiatrist has an opportunity for primary prevention. If a patient with a borderline or antisocial personality is idealizing a doctor, the psychiatrist can advise the doctor to review the treatment contract with the patient, gently disillusion the patient about the idealized traits ascribed, scrupulously document consent discussions, obtain second opinions, and use the language of humble collaboration rather than the language of rescue. With repeated consultations, consultees learn to recognize and forestall dangerous levels of idealization.

When the psychiatrist is the primary caretaker of a patient with combined medical illness and personality disorder, a special issue arises around termination of treatment and discharge planning because physical symptoms, rather than suicide threats, may be the currency of manipulation. Psychiatrists are at risk for medical malpractice if they minimize the patient's physical complaints and an untoward medical outcome is the result. In this situation, the author recommends that the psychiatrist call in a medical specialist to examine the patient and share responsibility for the decision that discharge is reasonable. Then, responsible limit-setting can be employed despite threats of litigation.

REFERENCES

Adelman SA (1985). Pills as transitional objects: A dynamic understanding of the use of medication in psychotherapy. Psychiatry 48:246–253.

Adler DA, Drake RE, & Teague GB (1990). Clinicians' practices in personality assessment: Does gender influence the use of DSM-III axis II? Compr Psychiatry 31:125–133.

Andersen SM, & Harthorn BH (1989). The recognition, diagnosis, and treatment of mental disorders by primary care physicians. Med Care 27:869–886.

Andrulonis PA, Glueck BC, Stroebel CF, et al. (1982). Borderline personality subcategories. J Nerv Ment Dis 170:670–679.

Arboleda-Florez J (1990). Anti-social burnout: An exploratory study. Presented at the Annual Meeting of the American Psychiatric Association, New York.

Balint M (1957). The doctor, his patient and illness. New York: International Universities Press.

Barnard C, & Hirsch C (1985). Borderline personality and victims of incest. Psychol Rep 57:715–718.

Bellissimo A, & Tunks E (1984). Chronic pain: The psychotherapeutic spectrum (p. 118). New York: Praeger.

Benner DG, & Joscelyne B (1984). Multiple personality as a borderline disorder. J Nerv Ment Dis 172:98–104.

Briere J, & Zaidi LY (1989). Sexual abuse histories and sequelae in female psychiatric emergency room patients. Am J Psychiatry 146:1602–1606.

Casey PR (1989). Personality disorder and suicide intent. Acta Psychiatr Scand 79:290–295.

Casey PR, & Tyrer P (1990). Personality disorder and psychiatric illness in general practice. Br J Psychiatry 156:261–265.

Coccaro EF, Siever LJ, Kiar HM, et al. (1989). Serotonergic studies in patients with affective and personality disorders. Arch Gen Psychiatry 49:587–599.

Costa PT, & McCrae RR (1980). Still stable after all these years: personality as a key to some issues in adulthood and old age. In PB Baites & OC Brim (Eds.), Life-span development and behavior (Vol. 3). Orlando, FL: Academic Press.

Costa PT, & McCrae RR (1985). Hypochondriasis, neuroticism, and aging: When are somatic complaints unfounded? Am Psychol 40:1928.

Cowdry RW, & Gardner DL (1988). Pharmacotherapy of borderline personality disorder. Arch Gen Psychiatry 45:111–119.

David DS (1979). Humility and the physician. J Chronic Dis 32:541–542.

Dawson DF (1988). Treatment of the borderline patient, relationship management. Can J Psychiatry 33:370–374.

de Leon J, Saiz-Ruiz J, Chinchilla A, et al. (1987). Why do some psychiatric patients somatize? Acta Psychiatr Scand 76:203–209.

Deltito JA, & Stam M (1989). Psychopharmacological treatment of avoidant personality disorder. Compr Psychiatry 30:498–504.

Ellard J (1977). How to deal with personality disorders. Mod Med Asia 13:2833.

Faustman WO, & White PA (1989). Diagnostic and psychopharmacological treatment characteristics of 536 inpatients with posttraumatic stress disorder. J Nerv Ment Dis 117:154–159.

Fogel BS, & Goldberg RJ (1983–1984). Beyond liaison: A future role for psychiatry in medicine. Int J Psychiatry Med 13:185–192.

Ford CV, King BH, & Hollender MH (1988). Lies and liars: Psychiatric aspects of prevarication. Am J Psychiatry 145:554–562.

Frances A (1982). Categorical and dimensional systems of personality diagnosis: A comparison. Compr Psychiatry 23:516–527.

Freud A (1946). The ego and the mechanisms of defense. New York: International Universities Press.

Gallop R (1985). The patient is splitting: Everyone knows and nothing changes. J Psychosoc Nurs Ment Health Serv 23:6–10.

Gallop R, & Wynn F (1987). The difficult inpatient: Identification and response by staff. Can J Psychiatry 32:211–215.

Gardner DL, & Cowdry RW (1985). Alprazolam-induced dyscontrol in borderline personality disorder. Am J Psychiatry 142:98–100.

Gartner AF, Marcus RN, Halmi K, et al. (1989). DSM-III personality disorders in patients with eating disorders. Am J Psychiatry 146:1585–1591.

George A, & Soloff PH (1986). Schizotypal symptoms in patients with borderline personality disorders. Am J Psychiatry 143:212–215.

Goldberg SC, Schulz SC, Schulz PM, et al. (1986). Borderline and schizotypal personality disorders treated with lowdose thiothixene vs. placebo. Arch Gen Psychiatry 43:680–686.

Greenberg RP, & Pies R (1983). Is paradoxical intention risk-free? A review and case report. J Clin Psychiatry, 44:66–69.

Groves JE (1975). Management of the borderline patient on a medical or surgical ward: The psychiatric consultant's rule. Int J Psychiatry Med 32:178–183.

Groves JE (1978). Taking care of the hateful patient. N Engl J Med 298:883–887.

Gunderson JG (1986). Pharmacotherapy for patients with borderline personality disorder. Arch Gen Psychiatry 43:698–699.

Gunderson JG, & Elliott GR (1985). The interface between borderline personality disorder and affective disorder. Am J Psychiatry 142:277–288.

Gutheil TG (1985). Medicolegal pitfalls in the treatment of borderline patients. Am J Psychiatry 142:914.

Hall A (1977). The psychotherapy of character disorder. Aust NZ J Psychiatry 11:175–178.

Hayes B, Taylor D, & Sackett D (Eds.) (1979). Compliance in health care. Baltimore: Johns Hopkins University Press.

Herz LR, Volicer L, D'Angelo N, et al. (1990). Additional psychiatric illness by diagnostic interview schedule in male alcoholics. Compr Psychiatry 30:72–79.

Heumann KA, & Morey LC (1990). Reliability of categorical and dimensional judgments of personality disorder. Am J Psychiatry 147:498–500.

Hirschfeld RMA, Klerman GL, Clayton PJ, et al. (1983). Assessing personality: Effects of the depressive state on trait measurement. Am J Psychiatry 140:695–699.

Hyler SE, & Frances A (1985). Clinical implications of Axis I–Axis II interactions. Compr Psychiatry 26:345-351.

Hyler SE, & Reider RC (1984). Personality diagnostic questionnaire revised (PDQR). New York State Psychiatric Institute, 722 W. 168th St., New York, NY 10032.

Hyler SE, Lyons M, Rieder RO, et al. (1990a). The factor structure of self-report DSM-III axis II symptoms and their relationship to clinicians' ratings. Am J Psychiatry 147:751–757.

Hyler SE, Rieder RO, Williams JBW, et al. (1989). A comparison of clinical and self-report diagnoses of DSM-III personality disorders in 552 patients. Compr Psychiatry 30:170–176.

Hyler SE, Skodol AE, Kellman HD, et al. (1990b). Validity of the personality diagnostic questionnaire—revised: Comparison with two structured interviews. Am J Psychiatry 147:1043–1048.

Kahana RJ, & Bibring GL (1964). Personality types in medical management. In NE Zinberg (Ed.), Psychiatry and medical practice in a general hospital (pp. 108–123). New York: International Universities Press.

Kaplan M (1983, July). A woman's view of DSM III. Am Psychol 38:786–792.

Kass F, Skodol AE, Charles E, et al. (1985). Scaled ratings of DSMIII personality disorders. Am J Psychiatry 142:627–630.

Kernberg OF (1986). Severe personality disorders: Psychotherapeutic strategies. New Haven: Yale University Press.

Klar H, & Siever LJ (1984). The psychopharmacologic treatment of personality disorders. Psychiatr Clin North Am 7:791–801.

Klein DN, Taylor EB, Harding K, et al. (1988). Double depression and episodic major depression: Demographic, clinical, familial, personality, and socioenvironmental characteristics and short-term outcome. Am J Psychiatry 145:1226–1231.

Kleinman PH, Miller AB, Millman RB, et al. (1990). Psychopathology among cocaine abusers entering treatment. J Nerv Ment Dis 178:442–447.

Koenigsberg, Kaplan RD, Gilmore MM, et al. (1985). The relationship between syndrome and personality disorder in DSM-III: Experience with 2,462 patients. Am J Psychiatry 142:207–212.

Konicki PE, & Schulz SC (1989). Rationale for clinical trials of opiate antagonists in treating patients with personality disorders and self-injurious behavior. Psychopharmacol Bull 25:556–563.

Krahn DD, & Mackenzie TB (1984). Organic personality syndrome caused by insulin related nocturnal hypoglycemia. Psychosomatics 25:711–712.

Lewis G, & Appleby L (1988). Personality disorder: The patients psychiatrists dislike. Br J Psychiatry 153:44–49.

Libb JW, Stankovic S, Freeman A, et al. (1990). Personality disorders among depressed outpatients as identified by the MCMI. J Clin Psychiatry 46:277–284.

Liebowitz MR, Stone MH, & Turkat ID (1986). Treatment of personality disorders. In AL Frances & RE Hales (Eds.), American Psychiatric Association annual review (Vol. 5; pp. 356–393). Washington, DC: APA Press.

Lippinski JF, Mallya G, Zimmerman P, et al. (1989). Fluoxetine-induced akathisia: clinical and theoretical implications. J Clin Psychiatry 50:339–342.

Lipp MR (1977). Respectful treatment: The human side of medical care. Hagerstown, MD: Harper & Row.

Livesley WJ (1985). The classification of personality disorder: 1. The choice of category concept. Can J Psychiatry 30:353–358.

Livesley WF (1986). Trait and behavioral prototypes of personality disorder. Am J Psychiatry 143:728–732.

Lucas PB, Gardner DL, Wolkowitz OM, et al. (1987). Dysphoria associated with methylphenidate infusion in borderline personality disorder. Am J Psychiatry 144:1577–1579.

Maddocks PD (1970). A five-year follow-up of untreated psychopaths. Br J Psychiatry 116:511–575.

Malow RM, West JA, Williams JL, et al. (1989). Personality disorders classification and symptoms in cocaine and opioid addicts. J Consult Clin Psychol 57:765–767.

Malt U, Myhrer T, Blikra G, et al. (1987). Psychopathology and accidental injuries. Acta Psychiatr Scand 76:261–271.

Mann JJ, & Kapur S (1991). The emergence of suicidal ideation and behavior during antidepressant pharmacotherapy. Arch Gen Psychiatry 48:1027–1033.

Marin DB, Widiger TA, Frances AJ, et al. (1989). Personality disorders: issues in assessment. Psychopharmacol Bull 25:508–514.

Mavissakalian M, Hamann MS, & Jones B (1990). A comparison of DSM-III personality disorders in panic/agoraphobia and obsessive-compulsive disorder. Compr Psychiatry 31:238–244.

McCrae RR, Costa PT, & Arenberg D (1980). Constancy of adult personality structure in males: Longitudinal cross-sectional and times-of-measurement analyses. J Gerontol 35:877–883.

McEnany GW, & Tescher BE (1985). Contracting for care: One nursing approach to the hospitalized borderline patient. J Psychosoc Nurs Ment Health Serv 23:11–18.

McGrath WB (1978). The paranoid personality. Ariz Med 35:604.

McGlashan TH (1986). Borderline personality disorder and unipolar affective disorder: Long-term effects of comorbidity. J Nerv Ment Dis 175:467–473.

Mellsop G, Varghese F, Joshua S, et al. (1982). The reliability of Axis II of DSM-III. Am J Psychiatry 139:1360–1361.

MERIKANGAS KR, & WEISSMAN MM (1986). Epidemiology of DSM-III Axis II personality disorders. In AJ FRANCES & RE HALES (Eds.), American Psychiatric Association annual review (Vol. 5; pp. 258–278). Washington, DC: APA Press.

MILLON T (1983). Millon clinical multiaxial inventory. Minneapolis, MN: National Computer Systems.

MOSS JH (1989). Borderline personality disorder: When medical care is complicated by mental illness. Postgrad Med 85:151–158.

NOVAC A (1986). The pseudoborderline syndrome: A proposal based on case studies. J Nerv Ment Dis 174:84–91.

NOYES R, REICH J, CHRISTIANSEN J, ET AL. (1990). Outcome of panic disorder: Relationship to diagnostic subtypes and comorbidity. Arch Gen Psychiatry 47:809–818.

O'BOYLE M, & SELF D (1990). A comparison of two interviews for DSM-III-R personality disorders. Psychiatry Res 32:85–92.

O'BRIEN P, CALDWELL C, & TRANSEA G (1985). Destroyers: Written treatment contracts can help cure self-destructive behaviors of the borderline patient. J Psychosoc Nurs Ment Health Serv 23:19–23.

PERRY JC (1985). Depression in borderline personality disorder: Lifetime prevalence at interview and longitudinal course of symptoms. Am J Psychiatry 142:15–21.

PILKONIS PA, & FRANK E (1988). Personality pathology in recurrent depression: Nature, prevalence, and relationship to treatment response. Am J Psychiatry 145:435–441.

PRESLY AS, & WALTON JH (1973). Dimensions of abnormal personality. Br J Psychiatry 122:269–276.

REICH J (1990a). Relationship between DSM-III avoidant and dependent personality disorders. Psychiatry Res 34:281–292.

REICH J (1990b). The effect of personality on placebo response in panic patients. J Nerv Ment Dis 178:699–702.

REICH J, BOERSTLER H, YATES W, ET AL. (1989). Utilization of medical resources in persons with DSM-III personality disorders in a community sample. Int J Psychiatry 19:1–9.

REICH JH (1988). DSM-III personality disorders and the outcome of treated panic disorder. Am J Psychiatry 145:9114–9115.

REICH JH, & GREEN AI (1991). Effect of personality disorders on outcome of treatment. J Nerv Ment Dis 179(2):74–82.

REICH JH, & THOMPSON WA (1987). Differential assortment of DSM-III personality disorder clusters in three populations. Br J Psychiatry 150:471–475.

SADAVOY J (1987). Character disorders in the elderly: An overview. In J SADAVOY & M LESZCZ (Eds.), Treating the elderly with psychotherapy: The scope for change in later life. Madison, CT: International Universities Press.

SADAVOY J, & DORIAN B (1983). Treatment of the elderly characterologically disturbed patient in the chronic care institution. J Geriatr Psychiatry 16:233–240.

SADAVOY J, & FOGEL B (1992). Personality disorders in old age. In JE BIRREN, RB SLOAN, & G COHEN (Eds.), Handbook of mental health and aging (2nd ed.). Orlando, FL: Academic Press.

SCHULZ SC, CORNELIUS J, SCHULZ PM, ET AL. (1988). The amphetamine challenge test in patients with borderline disorder. Am J Psychiatry 145:809–814.

SCHWARCZ G, & HALARIS A (1984). Identifying and managing borderline personality patients. Am Fam Physician 29:203–208.

SCHWARTZ DA (1979). The suicidal character. Psychiatry Q, 51:64–70.

SELZER MA, KOENIGSBERG HW, & KERNBERG OF (1987). The initial contract in the treatment of borderline patients. Am J Psychiatry 144:927–930.

SHEA MT, PILKONIS PA, BECKHAM E, ET AL. (1990). Personality disorders and treatment outcome in the NIMH treatment of depression collaborative research program. Am J Psychiatry 147:711–718.

SOLOFF PH (1987). Neuroleptic treatment in the borderline patient: Advantages and techniques. J Clin Psychiatry 48:26–30.

SOLOFF PH, GEORGE A, NATHAN RS, ET AL. (1986). Progress in pharmacotherapy of borderline disorders. Arch Gen Psychiatry 43:691–697.

SPIEGEL H, & SPIEGEL D (1978). Trance and treatment. New York: Basic Books.

STALENIAM MN, & YOUNGS DD (1974). Psychiatric consultation with patients who refuse medical care. Int J Psychiatry Med 5:115–123.

STANGL D, PFOHL B, ZIMMERMAN M, ET AL. (1985). A structured interview for the DSM-III personality disorders: A preliminary report. Arch Gen Psychiatry 42:591–596.

STERNBACH RA (1974). Pain patients: Traits and treatment. Orlando, FL: Academic Press.

STOUDEMIRE A, & THOMPSON T (1982). The borderline personality in the medical setting. Ann Intern Med 96:76–79.

STRAIN JJ (1978). Psychological interventions in medical practice. Norwalk, CT: Appleton-Century-Crofts.

SVRAKIC DM (1987). Clinical approach to the grandiose self. Am J Psychoanal 47(2):167–179.

SWEENEY DR (1987). Treatment of outpatients with borderline personality disorder. J Clin Psychiatry 48:32–35.

Task Force on Nomenclature and Statistics of the American Psychiatric Association (1980). Diagnostic and statistical manual of mental disorders (3rd ed.). Washington, DC: American Psychiatric Association.

TYRER P (1989). Risks of dependence on benzodiazepine drugs: The importance of patient selection. Br Med J 298:102–105.

VAILLANT GE (Ed.) (1986). Empirical studies of ego mechanisms of defense. Washington, DC: American Psychiatric Press.

WATZLAWICK P, WEAKLAND JH, & FISCH R (1974). Change: Principles of problem formation and problem resolution. New York: WW Norton.

WETZLER S, & DUBRO A (1990). Diagnosis of personality disorders by the Million Clinical Multiaxial Inventory. J Nerv Ment Dis 178:361–363.

WETZLER S, KAHN R, CAHN W, ET AL. (1990). Psychological test characteristics of depressed and panic patients. Psychiatry Res 31:179–192.

YAGER J, LANDSVERK J, EDELSTEIN CK, ET AL. (1989). Screening for Axis II personality disorders in women with bulimic eating disorders. Psychosomatics 30:255–262.

YATES WR, PERRY PJ, ANDERSEN KH, ET AL. (1990). Illicit anabolic steriod use: a controlled personality study. Acta Psychiatr Scand 81: 548–550.

ZANARINI MC, FRANKENBURG FR, POPE HG, ET AL. (1990). Axis II comorbidity of normal-weight bulimia. Compr Psychiatry 30:20–24.

ZIMMERMAN M, & CORYELL WH (1990). DSK-III personality disorder dimensions. J Nerv Ment Dis 178:686–692.

ZIMMERMAN M, PFOHL B, STANGL D, ET AL. (1986). Assessment of DSM-III personality disorders: The importance of interviewing an informant. J Clin Psychiatry 47:261–263.

15 Sexual dysfunction in the medically ill

PETER J. FAGAN, Ph.D., AND
CHESTER W. SCHMIDT, Jr., M.D.

Sexual functioning of the medical patient is now recognized as an integral part of the patient's psychosocial adjustment to illness. Among community-based samples, disorders of sexual arousal in men and inhibited orgasm in both men and women range from 5% to 10% and premature ejaculation from 36% to 38%; no data are available for female arousal, vaginismus, or dyspareunia (Spector & Carey, 1990). Not surprisingly, in medical populations the prevalence of sexual dysfunction is significantly greater (see Table 15-1).

Reports on the effects on sexual functioning of specific diseases and conditions, such as diabetes (Jensen, 1985), multiple sclerosis (Barrett, 1982), coronary artery bypass surgery (Papadopoulos et al., 1986), and diseases of the reproductive organs (Stoudemire, Techman, & Graham, 1985; Wise, 1985), reflect a growing concern about the sexual lives of patients. In addition to the impact of the medical conditions on sexual function, many medications have sexual side effects that complicate the clinical picture. Psychiatrists are receiving more requests for consultations about sexual problems in patients being treated by their medical colleagues (Sullivan & Lukoff, 1990). Therefore a complete psychiatric evaluation of a medical or surgical patient should include an assessment of sexual functioning.

This chapter provides a methodologic guide for the psychiatrist evaluating the sexual functioning of medical or surgical patients. Diagnostic procedures, tests, and issues are also discussed. For more extensive descriptions of sexual problems associated with specific medical illnesses and surgical procedures, the reader is referred to texts and reviews on sexual medicine (Kolodny, Masters, & Johnson, 1979; Schover & Jensen, 1988; Wise & Schmidt, 1985). This chapter also suggests treatment approaches for sexual dysfunction in the medical patient. Although we repeat what is available elsewhere in terms of general sexual therapy (Hawton, 1985; Leiblum & Pervin, 1980; Meyer, Schmidt, & Wise, 1983), an attempt is made to adopt some of the therapeutic techniques and approaches generally utilized to the special needs and limitations of the medical patient.

A conscious attempt was made to discuss issues of sexuality for male and female medical patients equally. Nevertheless, the chapter reflects the bias of currently available research (and therefore knowledge), which focuses on men with physical illness. The source of this bias in the literature is unknown, but it may relate to the ready availability of measurement for penile tumescence or to men's greater tendency to complain about sexual dysfunction in the context of physical and neurologic illness.

Elaboration of the Sexual Problem

The first task in the assessment of sexual functioning is to define the problem clearly, which is not always easy. Patients and physicians often collude in settling for a vague description. "I can't stay hard long enough" and "Sex with him is unenjoyable for me" are not definitive statements of a sexual dysfunction.

The Disorder and the Sexual Response Pattern

One way of obtaining a behavioral description of the sexual problem is to ask the patient to describe the most recent attempt at sexual intercourse. The data are then compared to the four-stage human sexual response cycle described by Masters and Johnson (1966) and amplified by Kaplan (1979). The resultant five-stage sexual cycle of (1) desire, (2) arousal, (3) plateau, (4) orgasm, and (5) resolution provides a framework for the diagnosis of psychosexual dysfunctions. Figure 15-1 displays the commonly observed physiologic arousal patterns of the female and male sexual response cycles (Kolodny, Masters, & Johnson, 1979). A major difference between male and female sexual response is that females have

TABLE 15–1. *Some Medical Conditions and Sample Prevalence of Sexual Dysfunctions*

Dysfunction	Epilepsy[a] (%)	Diabetes[b] (%)	Multiple Sclerosis[c] (%)	Ostomy[d] (%)
Men	$N = 38$	$N = 80$	$N = 68$	
Erectile dysfunction	3	34	68	
Inhibited sexual desire	3	31	48	
Premature ejaculation	0	4	—	
Retarded ejaculation	3	3	44	
Dyspareunia	—	—	—	
Women	$N = 48$	$N = 80$	$N = 149$	
Inhibited sexual desire	19	24	41	
Orgasmic dysfunction	19	11	37	
Inhibited arousal	—	—	35	
Vaginismus	0	0		
Sexual aversion	2	8		
Dyspareunia	—	—	—	
Men and women: 131 ($M = 59$; $F = 72$)				
Sex related problems				42
Less enjoyment				71
Decreased frequency				78
Marital discord				14

[a]Jensen et al. (1990)
[b]Jensen (1981).
[c]Valleroy, & Kraft (1984).
[d]Follick, Smith, & Turk (1984).

the capacity for multiple orgasms within a single response cycle.

Sexual Dysfunctions

Disorders of Desire

Disorders of desire (libido) are expressed as a loss of interest in sexual activity. The disorder is usually manifested by a global decrease in cognitive, emotional, and physiologic readiness to initiate or respond to sexual experience. Sexual fantasy life is minimal. The patient usually is able to describe a previous time when sexual interest and activity were greater. The distress shown about the loss of desire ranges from slight to great, but most often the patient expresses a lack of determination to alter the level of sexual activity. In the setting of illness, malaise, pain, fatigue, and depression often lower sexual desire. In relationships where the loss of desire is long-standing, the consultation is usually initiated by a dissatisfied partner.

Loss of sexual desire does not mean the patient cannot function sexually. Neurologic and endocrine disorders, as well as temporal lobe epilepsy or a pituitary adenoma, can selectively affect libido while leaving physical functioning intact. There are indications that temporal lobe epileptiform discharges can interfere with the hypothalamic regulation of pituitary secretion in men (Herzog et al., 1990), and can produce polycystic ovarian syndrome in women (Herzog et al., 1984). Patients on antihypertensive drugs report decreased libido and frequency despite apparent erectile competence or ability to be aroused.

Given an adequate amount of stimulation of the genitals, the resultant vasocongestion may cause arousal with erection or vaginal lubrication sufficient for satisfying coitus. Genital stimulation produces reflexogenic (sacral) arousal that is processed cortically and subcortically, further augmenting the arousal and intensifying erection or vaginal lubrication.

Sexual desire disorders can also be caused by interpersonal and psychosocial factors. Reaction to the effects of the illness on the relationship (e.g., spouse as patient and partner) may contribute to decreased sexual desire. A study of diabetes and female sexuality (Schreiner-Engel et al., 1985) reported that diabetic women had relatively little impairment in sexual responses but were significantly lower than matched controls in measures of sexual desire, psychosexual functioning, and satisfaction with their primary relationships. In contrast, epileptic patients have been described as generally hyposexual presumably because of its previously mentioned effects on hypothalamic function (Herzog et al., 1984; 1990). However, in a carefully designed study

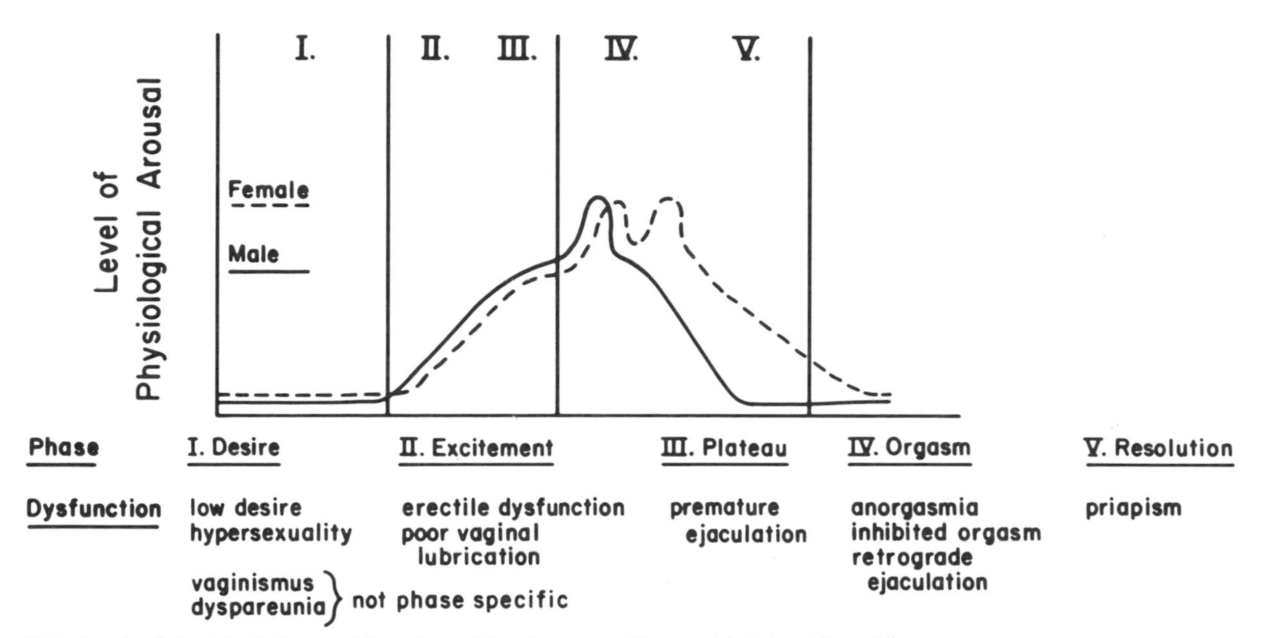

FIG. 15-1. Levels of physiological arousal in males and females engaged in sexual activity. Adapted from Kolodny RC, Masters WH, & Johnson VE (1979). *Textbook of Sexual Medicine*. Boston: Little, Brown.

involving biopsychosocial variables, Jensen et al. (1990) concluded that epilepsy does not necessarily increase the risk of sexual dysfunction in men and women.

Disorders of Arousal

Partial or complete failure to attain erection or the vaginal lubrication-swelling response following adequate sexual stimulation defines this dysfunction. For men there is insufficient vasocongestive reaction to permit engorgement of the two cylindrical corpora cavernosa in the shaft of the penis and the corpus spongiosum, which enlarges the glans penis. In women, similar insufficient vasocongestion is manifested by limited (or no) vaginal lubrication and by an absence of swelling in the labia majora.

The self-report of firm, lasting morning erections (corroborated by spouse if possible) and the presence of turgid noncoital erections are the two most important items in the sexual history indicative of adequate physiologic erectile capability (Segraves, Schoenberg, & Ivanoff, 1983). There is not always sufficient information for determining the etiology of erectile dysfunction among certain groups of medically ill men. For example, men with Alzheimer's disease may be at higher risk for erectile dysfunction that is not associated with degree of cognitive impairment, age, depression, medications, or other physical problems (Zeiss et al., 1990). Sexual activity may occur so infrequently in some subgroups, such as masturbation attempts by older diabetic men, that the history alone is not sufficient for diagnosis. In conditions such as diabetes, hyperprolactinemia, or decreased serum testosterone, men may be impotent with their spouses; but with the strong stimulus of an extramarital partner they may be able to function adequately (Abel et al., 1982; Schwartz, Bauman, & Kolodny, 1981; Segraves, 1982). Women with arousal disorders may also respond to novel sexual experiences.

Sexual Pain Disorders

Dyspareunia (pericoital genital pain) can occur in both sexes. In men pain is usually associated with Peyronie's disease (angulation of the penis caused by localized fibrosis of the corpora cavernosa). The fibrosis can occur following trauma, or it may be idiopathic. Men who complain of pain during orgasm may be suffering from occult chronic prostatitis in which sex-related pain or non-sex-related penile discharge is the only symptom. The contraction of the infected prostate during the emission phase of the sexual response is the likely source of the dyspareunia. Referral to a urologist is appropriate. Coital pain in women is often associated with lack of lubrication (a disorder of arousal), ulcerative erosions of episiotomy scars, vestibular adenitis, and a variety of uterine disorders. In both sexes every effort should be made to identify a physical cause of the pain before a diagnosis of psychogenic pain is made.

Vaginismus is the involuntary spasmodic contraction of the musculature around the vaginal outlet and the outer third of the vaginal barrel. It is a reflex phenomenon comparable to the blink of an eye at an approaching object. Vaginismus renders penile penetration impossible, whereas dyspareunia does not. Prolonged contraction of the musculature often results in cramping pain, and certainly pain results if penetration is attempted during the spasms.

Disorders of Orgasm

Disorders of orgasm include male premature ejaculation and inhibited orgasm in both sexes. Premature ejaculation occurs before the individual wishes because of recurrent absence of voluntary control in sexual activity. Viewed in this way, premature ejaculation may also be considered a disorder of the plateau phase. There is insufficient control of the plateau arousal prior to the stage of ejaculatory inevitability reached during the late plateau phase. As premature ejaculation becomes more problematic to the partner, the orgasmic pleasure accompanying ejaculation often correspondingly decreases to minimal levels.

Inhibited orgasm is marked by a delay or absence of orgasm following a normal sexual arousal phase that is adequate in focus, intensity, and duration. Primary (lifelong) anorgasmia is found in both sexes, although it is rare in men. In men, orgasm with normal pleasurable sensations may occur without ejaculation, e.g., in the presence of spinal cord lesions, retrograde ejaculation caused by sphincter muscle neuropathy, or thioridazine treatment. In women, orgasm may occur with noncoital clitoral stimulation alone. If this response pattern has been lifelong, such women should not be judged a priori to be abnormal or dysfunctional in their orgasmic response (Alzate, 1985; Derogatis et al., 1986).

Anorgasmia has been reported following the use of serotonergic drugs such as clomipramine and fluoxetine. Segraves (1989) has suggested that serotonin has a modulating effect on sexual response, especially orgasm, but that the relation between dosage and sexual function may not be linear. A clear understanding of the underlying mechanism is not yet available.

The history of morning or noncoital erections does not imply normal orgasmic function. Patients on beta-blockers or monoamine oxidase inhibitors may have drug-induced orgasmic dysfunction despite normal morning erections.

Disorders of Resolution

Disorders of resolution are those in which the anatomic and physiologic processes of the arousal and plateau phases are not reversed within a normal time period. A refractory period follows ejaculation, during which further ejaculation is impossible, although partial or full erection may be maintained. This refractory period lengthens with age, and some men become distressed because they cannot have coitus as frequently during a night as when they were younger. The condition of priapism, a prolonged and painful erection that lasts more than several hours, is the main dysfunction of the male resolution phase. Although, properly speaking, not a "dysfunction," the longer and more gradual resolution phase of women (see Figure 15-1) may become a source of distress if a postorgasmic male partner simply rolls over and goes to sleep. Interpersonal tension produced by this inattentiveness often leads to sexual withdrawal.

Other Sexual Problems

The medical patient with acquired cortical or subcortical dysfunction may present with a problem such as gender dysphoria, hypersexuality, transvestism, or other paraphilias of acute onset apparently coinciding with the medical illness. If the primary illness, e.g., temporal lobe epilepsy with hypoactive sexual desire, is treated medically without resolution of the gender/sexual disorder, the patient should be referred to a specialist in sexual disorders.

Elaboration of the presenting sexual problem has three goals: (1) a detailed description of the behavior; (2) determination of its onset relative to the illness, surgery, or medication regimen; and (3) information about any prior episodes of this problem during the life of the patient. It is for the last two goals that a complete psychosexual and psychosocial history is obtained.

Psychiatric Interview

The preferred sequence for the interview is first to meet with the patient and sexual partner together for an initial exploration of the sexual problem or concern; next to meet separately with the patient and the partner; and finally to discuss issues of treatment with both.

During the initial meeting the psychiatrist establishes rapport with the couple and seeks to answer the following questions: Who initiated seeking professional help for the sexual problem? What are the investments each has in their sexual life together? Who leads the description of the problem? How is responsibility for it apportioned? What level of communication exists between them? How have they responded as a couple to any limitations the medical illness may impose presently or in the future?

The individual interview with the identified patient includes a complete psychosexual and psychosocial history as well as a mental status examination to document the presence of signs and symptoms of a coexisting psychiatric disorder. The history also provides a developmental framework within which it is helpful to understand the patient's reaction to both the medical illness and the sexual dysfunction (Fagan, Meyer, & Schmidt, 1986). It includes an assessment of object relations and defensive style. Of particular importance is, of course, the sexual history per se. Table 15-2 outlines one possible format for obtaining a complete sexual history.

The individual interview with the partner follows the same format as that of the identified patient: psychosexual and psychosocial histories (with sexual data stressed) as well as mental status examination. Although some adjustments are made, dictated by clinical sensitivity (e.g., decrease of emphasis with first-rank schizophrenic symptomatology), it is better to err on the side of overinclusiveness in the interview with the partner. The partner, even of a medically ill patient, may be implicated in the etiology of the dysfunction and most certainly should be called on to collaborate in its remedy. Even if the treatment is a change in medications, the partner must be willing to participate in a renewed sexual life.

The final phase of the interview sequence is a conjoint session in which the treatment recommendations are made to the couple. Although the recommendations may seem relatively straightforward to the psychiatrist (e.g., change medications, have nocturnal penile tumescence studies performed), they are often complex to the patients. One must be prepared to spend time responding to anxious questions, especially from obsessive patients. Time and patience expended at this point yield greater compliance with treatment recommendations, especially when the clinician proposes marital or sexual therapy. When a medically ill patient with a sexual dysfunction does not have a sexual partner, there must be a gentle investigation of why the patient is solitary. Issues of shame, isolation, or fear of spreading disease may socially isolate patients. Evaluation of their social situation is necessary especially if they are in a residential institution.

TABLE 15–2. *Data Covered for Sexual History*

Childhood
- First sex play with peers
- Family sleeping arrangements
- Sex play with siblings
- Sexual abuse, incest, or molestation
- Parental attitudes toward sex
- Sources of sexual knowledge

Puberty
- Menarche or first ejaculation: subjective reaction
- Secondary sex characteristics: subjective reaction
- Body image
- Masturbation fantasies and frequency
- Homoerotic fantasies and behavior
- Dating experiences (physical intimacies)
- Intercourse
  - Age at first occurrence
  - Reaction to first intercourse

Young adulthood
- Lengthy or live-in relationships
- Pattern of sexual activities with others
- Paraphilic behaviors
- Previous marriages
- Courtship
- Parental attitudes toward spouse
- Sexual activity (dysfunction?)
- Reasons for termination of marriage
- Venereal disease

Adulthood
- Present primary sexual relationship
- Development of relationship
  - Significant nonsexual problems (e.g., money, alcohol, in-laws)
  - Infertility; contraceptive practices
  - Children problems
- Sexual behaviors
  - Extramarital affairs
  - Intercourse frequency during relationship
  - Variety of sexual behaviors (e.g., orogenital, masturbation)
  - Homosexual activity
  - Possible exposure to HIV (no. of partners, IVDU, anal receptive)
- Elaboration of onset and history of present problem (without partner present)
  - Previous dysfunction (in either partner)
  - Perception of partner's reaction to problem

Other Diagnostic Procedures

The psychiatrist must either perform a complete physical examination (see Table 15-3) or verify that one has been done by another physician. Likewise, the psychiatrist should make sure that an adequate laboratory information is available to permit a comprehensive diagnosis that integrates psychologic, social, or biomedical factors.

Medical History

The medical history of the patient should center on the symptomatology or history of endocrine, vascular, or neurologic diseases that may impair sexual functioning as well as the medications that have been used to treat those diseases or symptoms. A detailed substance ingestion history should be obtained, including nicotine use.

A family history of chronic diseases should be obtained. Medical conditions that commonly affect sexual response in both sexes are outlined in Table 15-4. Drugs that commonly affect sexual function are listed in Table

TABLE 15–3. *Components of Physical Examination of Sexually Dysfunctional Patients*

Organ System	Physical Examination
Endocrine system	Hair distribution Gynecomastia Testes Thyroid gland
Vascular system	Peripheral pulses
Gastrointestinal system	Hepatomegaly; or atrophic liver with peripheral neuropathy due to alcoholism
Genitourinary system	Prostate (in male) Pelvic examination (in female)
Nervous system	
Sacral innervation	
S1–S2	Mobility of small muscles of the foot
S2–S4	Internal, external anal sphincter tone; bulbocavernosus reflex (in male)
S2–S5	Perianal sensation
Peripheral sensation	
Deep tendon reflexes	
Long tract signs	

Source: Wise TN, & Schmidt CW (1985). Diagnostic approaches to sexuality in the medically ill. In RC Hall & TP Beresford (Eds.), *Handbook of psychiatric diagnostic procedures* (Vol. 2). New York: Spectrum.

TABLE 15–4. *Some Organic Factors Affecting Sexual Response*

Men	Women	Both
Peyronie's disease	Atrophic vaginitis	Chronic systemic disease
Urethral infections	Infections of the vagina	Chronic pain
Testicular disease	Cystitis, urethritis	Diabetes mellitus
Hypogonadal androgen-deficient states	Endometriosis	Angina pectoris
Hydrocele	Episiotomy scars, tears	Hypertension
Lumbar sympathectomy	Uterine prolapse	Multiple sclerosis
Radical perineal prostatectomy	Infections of external genitalia	Hyperprolactinemia Spinal cord lesions Alcoholism Substance abuse

Source: Schmidt CW (1988). Sexual disorders. In A Harvey, A Owens, & VA McKusick (Eds.), *Principles and Practices of Medicine* (22nd ed.). Norwalk, CT: Appleton-Century-Crofts.

15-5. Substance abuse, beta-adrenergic blockers, centrally acting antihypertensives, and antiandrogens are among the best established exogenous causes of sexual dysfunction. The use of tobacco or the excessive consumption of alcohol (Fagan et al., 1988; Fahrner, 1987) may contribute to disorders of sexual arousal because of their cardiovascular and neurologic effects; patterns of their use should be thoroughly documented during the history.

Endocrine Studies

Endocrine studies of medical patients should include measurement of fasting blood glucose and assays of follicle-stimulating hormone (FSH), luteinizing hormone (LH), testosterone, and prolactin, as well as a general survey, such as the SMA-12 or SMA-18, for liver and renal disease. The free or bioavailable testosterone level may be a more significant indicator of hormonal deficiency than the serum testosterone level (Carani et al., 1990) in older men. For menstruating women, the interpretation of LH and FSH values depends on the time in the cycle when the blood sample is obtained. It is suggested that LH and FSH values taken 5 to 7 days into the menstrual cycle usually avoids the LH surge of ovulation and other premenstrual fluctuations as well as providing an indication of polycystic ovary syndrome if one obtains a markedly elevated LH/FSH ratio (A. Stoudemire & B. S. Fogel, personal communication, 1991). Formal endocrine consultation may be helpful when planning the hormonal evaluation.

Vascular Studies

Standard vascular studies are the Doppler studies of penile blood flow and a measurement of the penile-brachial index (PBI: the ratio of penile systolic pressure to brachial systolic pressure). Gerwertz and Zarins (1985) suggested that a vascular etiology for erectile impotence is possible when the PBI is less than 0.75. In the presence of normal values, other invasive procedure, such as pelvic angiography, should be deferred unless abnormal results are obtained from nocturnal penile tumescence studies.

Neurologic Assessment

Neurologic assessment includes a review of the motor, sensory, and autonomic nervous system. As shown in Table 15-3, the lumbosacral spinal pathways merit special attention as essential to sexual functioning. Cystometrography or urinary flow studies can grossly define autonomic function in this area. Advances in determining the location of autonomic nerves from the pelvic plexus to the corpora cavernosa (Lepor et al., 1985) have made possible a nerve-sparing technique for retropubic prostatectomy (Walsh, Lepor, & Eggleston, 1983). Such patients should therefore not necessarily be considered a priori surgically impotent if their surgery involved this technique. Neurologic patients with bladder and spasticity problems are more likely to have sexual dysfunction (Valleroy & Kraft, 1984). Reduced vibrotactile penile thresholds have been reported in both

TABLE 15–5. *Some Drugs That May Affect Sexual Response*

Drug	Sexual Effect
Antihypertensive and cardiovascular agents	
Acetazolamide	Decreased desire
Atenolol	Inhibited arousal
Bethanidine	Inhibited arousal
Chlorthalidone	Decreased desire
Clofibrate	Decreased desire
Clonidine	Inhibited arousal
Digoxin	Decreased desire
Disopyramide	Inhibited arousal
Guanethidine	Inhibited arousal, no ejaculation
Hydralazine	Inhibited arousal
Methyldopa	Decreased desire, inhibited arousal, inability to ejaculate
Pentolinium	Inhibited arousal, inability to ejaculate
Phenoxybenzamine	No ejaculation
Prazosin	Inhibited arousal
Propranolol	Decreased desire, inhibited arousal
Reserpine	Decreased desire, inhibited arousal, decreased or no ejaculation, breast enlargement
Spironolactone	Decreased desire, inhibited arousal
Thiazide diuretics	Inhibited arousal
Timolol	Decreased desire, inhibited arousal, no ejaculation
Drugs often abused	
Alcohol	Decreased desire, inhibited arousal
Heroin	Inhibited arousal
Methadone	Decreased desire, inhibited arousal
Sedative-hypnotics	Inhibited arousal
Psychoactive drugs	
Antidepressants	
Amitriptyline	Changes in libido, inhibited arousal, inability to ejaculate
Bupropion	May facilitate sexual response
Clomipramine	Decreased libido, inhibited arousal, delayed ejaculation, anorgasmia
Cyclobenzaprine	Changes in libido
Doxepin	Changes in libido, inhibited arousal
Fluoxetine	Delayed or no orgasm
Imipramine	Changes in libido, inhibited arousal, inability to ejaculate
Phenelzine	Inhibited arousal, anorgasmia
Trazodone	Spontaneous erections or priaprism in men; increased libido in women
Anxiolytic agents: benzodiazepines	Decreased libido, inhibited arousal, delayed ejaculation, anorgasmia
Antipsychotic agents: The following antipsychotic agents have been associated with decreased libido, inhibited arousal, and inability to ejaculate: chlorpromazine, fluphenazine, haloperidol, mesoridazine, perphenazine, pimozide, prochlorperazine, thioridazine, trifluoperazine, thiothixine. In addition:	
Thioridazine	Inhibition of emission, painful ejaculation, priapism
Lithium	Decreased libido, inhibited arousal
Gastrointestinal drugs	
Chlordiazepoxide	Decreased desire
Cimetidine	Decreased desire, inhibited arousal
Dicyclomine hydrochloride	Inhibited arousal
Methantheline bromide	Inhibited arousal
Propantheline bromide	Inhibited arousal
Hormonal drugs	
Estrogens	Decreased desire in men
Hydroxyprogesterone caproate	Inhibited arousal
Methandrostenolone	Decreased desire
Norethandrolone	Decreased desire, inhibited arousal
Norethindrone	Decreased desire, inhibited arousal
Progesterone	Decreased desire, inhibited arousal
Others	
Aminocaproic acid	No ejaculation
Fenfluramine	Decreased desire, inhibited arousal
Homatropine methylbromide	Inhibited arousal
Metronidazole	Decreased desire
Naproxen	Inhibited arousal, no ejaculation
Phenytoin	Decreased sexual activity

Source: DxRx, 1, 2, (1985); Wise (1984). How drugs can help or hinder sexual function. Drug Therapy (pp. 137–149). See also Drugs that cause sexual dysfunction (1992). Medical Letter on Drugs and Therapeutics. 2, 73–78, for a more complete list with references.

aging and diabetic men (Rowland et al., 1989). (See also Chapter 40 on spinal cord injuries and sexual dysfunction.)

Nocturnal Penile Tumescence Studies

Nocturnal penile tumescence (NPT) studies (Karacan, 1982) are helpful for ruling out predominantly organic etiology in medical patients. Presented with an identified pathologic condition, the physician may be inclined to attribute the erectile dysfunction to the medical illness or postsurgical status. With the two exceptions of hyperprolactinemia and vascular steal syndrome, a normal NPT can be interpreted to indicate adequate neurologic and vascular competence for coital erections. NPT studies also incidentally provide polysomnographic data that may assist in the diagnosis and treatment of a sleep apnea or major depression, as suggested by decreased rapid eye movement (REM) latency and increased REM density. In addition to data on the frequency, duration, and rigidity of tumescence episodes, NPT studies can establish the relation between penile blood flow and bulbocavernosus and ischiocavernosus muscle activity (Karacan, Aslan, & Hirshkowitz, 1983). This information is especially important for the assessment of neurologically impaired men.

Interpretation of NPT studies is complex. Concerns about technical difficulties and basic assumptions of NPT have been raised that may impinge on reliability and validity. Segraves, Schoenberg, and Segraves (1985) cited previous studies in which, although NPT studies showed only partial sleep tumescence, the subjects later reported full erections with masturbation and intercourse. They described one insulin-dependent diabetic with documented peripheral neuropathy who had a virtually flat nocturnal tumescence record. Subsequently, however, he was able to have intercourse in an extramarital situation. As the authors (1985, p. 177) noted, "These cases are a potent reminder that we have minimal information concerning the general physiological activation responsible for nocturnal erections and have no evidence that this arousal is equivalent in intensity to environmental stimuli." Depression has been shown to depress NPT responses, suggesting an "organic" erectile disorder. Depression in medically ill patients may thus confound NPT data (Thase et al., 1987).

A review of the literature (Schiavi, 1988) supported the validity of the polysomnographic assessment of NPT for the differential diagnosis of erectile disorder but also suggested that there is much more to be learned about the central neural and endocrine processes. Schiavi (1990) added to his cautions about interpretations when he reported that a high proportion of men over age 65 failed to obtain full NPT even though they and their partners independently reported having intercourse regularly.

If NPT studies are abnormal and there are no additional signs or symptoms of organic etiology or major depression, further vascular and neurologic studies are indicated. Doppler blood flow and the penile-brachial index are usually obtained at the physical examination or as part of the NPT procedure. At the present time there is no noninvasive method for determining the arterial or venous etiologies of impotence. The combination of a pulsed Doppler and an intracavernosal injection of an vasoactive agent offers the most promise for the highest reliability with the least amount of invasion (Buvat et al., 1990). Other studies include arteriography of the aortic bifurcation and the internal iliac and internal pudendal arteries. When venous drainage problems are suspected, as when the glans penis does not engorge, cavernosography should be performed (DePalma, 1984). Other neurologic studies include cystometrography and the testing of pudendal nerve conduction velocity (sacral latency testing).

A home monitor for continuous measurement of NPT and penile rigidity is available (Bradley et al., 1985; Kaneko & Bradley, 1986). Although the assumption of this device—that radial stiffness is related to axial rigidity—awaits further validation, the unit may provide reliable and valid NPT studies to those patients who did not previously have access to sleep laboratories because of geographic or financial restrictions.

Other home devices, such as the "stamp test" or tension gauges to assess nocturnal tumescence, are frequently used. There is little consensus about their reliability and validity. The working hypothesis is that circumferential tumescence equates with tip-to-base rigidity, which may not be the case, especially in some medical conditions. In diabetic men, for example, an engorged but still flaccid penis might break the devices and be considered a positive stamp test yet be inadequate for coital penetration. The results of the stamp test for impotence should be interpreted with care and may be suggestive of organic impotence only when no stamps are broken over three nights (Carroll, Baltish, & Bagley, 1990). The use of NPT is also discussed in Chapter 39.

Intracavernosal Injection

Intracavernosal injection of vasodilators such as phenoxybenzamine (alpha-adrenergic blocker) (Brindley, 1983) and papaverine hydrochloride with phentolamine mesylate (Zorgniotti & LeFleur, 1985) have been employed both diagnostically and therapeutically for presumed vascular and neurogenic impotence. At the present stage of development of pharmacologic erection challenge, caution should be used when interpreting negative (no erection) results. In one study, 50% of those with normal NPT did not have any response to intracavernosal papaverine injection (Buvat et al., 1986). (See Chapter 39.)

Technology and Sexual Diagnostics

The evaluation of sexual disorders by refined diagnostic technologies has exponentially increased the possibilities of identifying an organic cause of impotence. Table 15-6 summarizes the procedures that are available to urologists and other specialities. In addition to what has been described above, further discussion of these methods is to be found in Chapter 39 of this text. As Buvat et al. (1990, p. 289) concluded in their review of assessment methods of erectile disorder, "All of these pharmaco-Doppler and ultrasonographic investigations are very recent, and their values remain as yet very uncertain. The possibility of false positive results due to stress should be kept in mind."

The remarkable development in technology that provides diagnostic information about the organic etiology of sexual dysfunction carries with it a special challenge to the mental health professional involved in the evaluation and treatment of impotence. In an article entitled "In Pursuit of the Perfect Penis," Leonore Tiefer has warned that the increasing medicalization of male sexuality may reinforce the equation of masculine self-adequacy with penile adequacy (Tiefer, 1986). This warning is especially applicable to the man with a se-

TABLE 15–6. *Technological Methods of Assessing Erectile Disorders*

Method of Assessment	Description
Arteriography	Locates presumed arterial obstruction for vascular surgery. Invasive.
Penile-brachial index (PBI)	Ratio of penis to arm blood pressure. PBI < 0.75 suggestive of arterial obstruction.
Pulse volume recordings	Records tracings of arterial pulse measured by small cuff around penis.
Biothesiometry	Electromagnetic test probe place on penis to detect threshold at which skin vibrations are perceived.
Doppler sonography	Measures penile arterial pressure. Requires skill to detect deep penile arteries.
Duplex ultrasonography	Measures arterial diameters by more accurate pulsed Doppler analysis. Invasive when erection attained through intracavernosal injection.
Dynamic infusion cavernosography	Assessment of venous drainage by measuring pressure of saline in the cavernous bodies during induced erections.

rious medical illness. The illness is likely to have initially inflicted a certain narcissistic injury, and a subsequent erectile dysfunction may be related to lack of self-worth. The clinician treating men (and women) with a medical illness and sexual dysfunction should labor to help the patient understand, first, the meaning of sexual activity in his or her life and, second, what the necessary limitation of this activity means for self and partner. Cognitive restructuring of the "meanings" attributed to sex and self-worth may prove helpful to the patient as medical/surgical treatments are sought.

Lastly, the extent of effort, time, and financial resources put into diagnostic studies of erectile and other sexual dysfunction should be limited by treatment options. What use the physician and patient can make of the knowledge gained is a question that should be asked prior to each diagnostic procedure.

Psychologic Assessment

Psychologic self-report instruments that assess either psychosocial distress or sexual functioning (Derogatis & Melisaratos, 1979) provide a valuable component in the full assessment of the medical patient with a sexual disorder. In some instances a patient admits to a sexual behavior in a self-report format that concern for social desirability prevents in a face-to-face interview (DeLeo & Magni, 1983). Conte (1983) has written an excellent review and evaluation of self-report measurements of sexual functioning.

Multiple attempts to divide the etiologies of erectile dysfunction into strictly organic and psychogenic categories have not proved replicable. This failure may reflect a fault in the theoretic dichotomy of etiologies. Organic and psychologic factors in sexual functioning are not dichotomous, but orthogonal, factors. Each contributes to the sexual response—even in those patients whose sexual dysfunction appears directly caused by a medical condition.

The premorbid personality of the patient determines greatly the response to treatment interventions and compliance. A particularly helpful self-report personality inventory we have found useful for treatment of sexual dysfunction in men is the NEO Personality Inventory (NEO-PI) (Costa et al., 1992; Costa & McCrae, 1985). The NEO-PI measures five factors or normal personality: Neuroticism, Extraversion, Openness, Agreeableness, and Conscientiousness. An "ideal" profile of the candidate for the treatment of sexual dysfunction would be one with enough Neuroticism to provide the motivation (distress) to work in therapy; be of at least average Extraversion, so the exposure of therapy is not too threatening; have high Openness to new experiences (change in therapy); and, lastly, be Conscientious enough to follow a treatment program. Deviations from this "ideal" profile alert the therapist and patient to probable areas of resistance in therapy. Treatment goals are more easily realizable when a realistic assessment of the patient's stable (and premorbid) personality is taken into account. Rater (e.g., spouse) forms of the NEO-PI are available to enhance the validity of the findings, an especially helpful advantage of the NEO-PI in the case of a medical patient who may be experiencing severe emotional state reactions to the illness, and a clear picture of the premorbid personality is sought.

Of particular utility with medical patients is the Psychological Adjustment to Illness Scale (PAIS). The PAIS (Derogatis, 1986) can be administered in a 20 to 30-minute semistructured interview or given in a self-report (PAIS-SR) format. It contains seven subscales (including a sexual relationship scale) that measure different domains of psychosocial adjustment following illness. The PAIS provides normative values for the following clinical groups: lung cancer patients, renal dialysis patients, acute burn patients, and essential hypertensives.

TREATMENT ISSUES

The initial treatment issue for the sexual problem of a medical patient is to ensure that the pathologic processes of the illness are ameliorated as much as possible. The second issue is to convey to the patient that the goal of sexual therapy (regardless of the mode of intervention) is to maximize whatever sexual abilities remain that can be developed within the patient's aesthetic and ethical parameters. With illness taking control from some areas of the patient's life, he or she should experience sexual therapy as a means of regaining initiative in a relationship and securing pleasure for self and partner.

The premorbid baseline of the patient's sexual function should be noted. It is patently more difficult to recover and improve a premorbid low baseline than assist a patient back to a relatively sound baseline that may have existed before the illness.

Both men and women can be affected by psychotropic medications such that sexual dysfunction may occur (see Table 15-5). Anorgasmia in women and erection and ejaculation in men may be associated with the use of cyclic antidepressants, MAO inhibitors, benzodiazepines, and neuroleptics (Segraves, 1989; Segraves, 1988).

Recent Psychotropic Drugs and Sexual Dysfunction

Fluoxetine 20 to 80 mg/day has been associated with anorgasmia in both men and women (Herman et al., 1990; Kline, 1989; Lydiard & George, 1989; Musher, 1990). Anecdotal reports have claimed success in treating fluoxetine anorgasmia with cyproheptadine as well as with amantadine.

Bupropion 300 to 600 mg/day does not appear to cause sexual dysfunction, and some patients who have had sexual dysfunction on other antidepressants had a return of sexual function when bupropion replaced the original antidepressant (Gardner & Johnston, 1985). Bupropion may also improve the sexual functioning of men and women who present with subclinical depression (Crenshaw, Goldberg, & Stern, 1987).

Buspirone in the treatment of generalized anxiety has not been associated with sexual dysfunction (Othmer & Othmer, 1987). Trazodone has been associated with anorgasmia in a depressed woman (Jani & Wise, 1988) and increased libido (Gartrell, 1986). It is also associated with priapism in men (Sacks, et al., 1985). It is discussed below.

If one suspects that the sexual dysfunction is pharmacologically caused (see Table 15-5), a drug holiday, if possible, may be employed as a diagnostic-therapeutic measure. When a drug is implicated, efforts should be made to achieve the same effects with a different drug or the same drug at a lower dosage. For example, although the beta-blockers atenolol, metoprolol, pindolol, and propranolol all decrease testosterone in normal volunteers, Rosen, Kostis, and Jekelis (1988) suggested that, in addition, some subjects are vulnerable to sexual dysfunction in association with propranolol. The sexual inhibiting effects of some drugs (e.g., antihypertensives) can cause patients to unilaterally discontinue the regimen and thereby put themselves at serious medical risk. Concern about the effects of drugs on sexual functioning therefore is not "merely" a concern about patient's sexual lives. If the etiology of the sexual disorder is alcohol, nicotine, or other drug abuse, treatment should be aimed at controlling the abuse.

Illness, injury, and surgery, especially that involving breasts and genitalia, may have severe effects on body image and on the patient's self-concept as spouse or lover (Stoudemire, Techman, & Graham, 1985). Sexual desire often is decreased even though the capacity for sexual functioning may remain intact. For the premorbidly psychologically healthy, a brief schedule of individual therapy (five or six sessions) followed by conjoint therapy (three or four sessions) usually is sufficient to assist the patient and partner through the adjustment crisis. Those with characterologic or severe neurotic traits require more extensive psychotherapy when faced with such trauma.

Another group for whom problems of body image exist are the head-injured. Problems of body image and sexual identity in the head-injured have been treated successfully with cognitive restructuring, social skills, training, assertiveness training, and behavioral assignments (Valentich & Gripton, 1986). Disturbances of body image also are discussed by Riether & McDaniel in Chapter 34.

Medical-Surgical Treatments

Disorders of Desire

Inhibited sexual desire due to illness-related fatigue and malaise should be acknowledged. With chronic debilitating illnesses such as multiple sclerosis, a realistic candor about the inadequacy of waiting for spontaneous sex is in the patient's interest. The clinician may encourage the patient and spouse to plan for sex by conserving energy and allotting time for sexual togetherness.

Estrogen replacement therapy in women who have undergone hysterectomy and oophorectomy does not appear to improve libido and sexual function (Dennerstein, Wood, & Burrows, 1977; Utian, 1975), but topical or systemic estrogen reduces the occurrence of dyspareunia due to lack of lubrication and vaginal atrophy in posthysterectomy and postmenopausal women. Systemic estrogen replacement therapy may have a positive effect on a woman's mood and sense of well-being and secondarily, therefore, on her baseline of interest in and desire for sexual activity. A review of studies (Walling, Andersen, & Johnson, 1990) suggests that androgen increases the sexual responsiveness of women, especially among younger surgically postmenopausal women (Sherwin, Gelfand, & Brender, 1985). The samples studied in the review varied in both characteristics and treatment such that definite conclusions regarding the hormone replacement regimen is not yet available. It does seem likely, however, that further research in androgen treatment with postmenopausal women will provide guidelines for the treatment of this group as well as further the understanding the relation of the "normal" range of testosterone in premenopausal women.

A man with abnormally low plasma testosterone levels not due to pituitary or hypothalamic disease may be administered testosterone cypionate (200 mg IM every other week). This treatment is for decreased desire in the presence of abnormally low testosterone levels. Testosterone replacement is not a treatment for erectile dysfunction.

Patients with pituitary disease may require postsurgical replacement therapy. To restore fertility in cases of pituitary destruction, Jones (1985) described a regimen of 1500 to 2000 IU of human chorionic gonadotropin intramuscularly twice weekly until normal plasma testosterone levels are restored. Then one ampule of human menopausal gonadotropin is given every other day for at least 6 months.

Bromocriptine (1.25 mg daily up to 2.5 mg twice daily) may be helpful for restoring both libido and erectile function in patients with hyperprolactinemia and low plasma testosterone levels (Carter et al., 1978). Patients with high prolactin levels and demonstrable tumors may require both surgical resection and bromocriptine treatment (Prescott et al., 1982).

The use of centrally acting agents to increase sexual desire and performance is yet inconclusive, though it appears that most interest has been generated in the aphrodisiac effects of serotonin antagonists and dopamine agonists (Segraves, 1989; Segraves et al., 1985). For example, in experimental conditions, subcutaneous apomorphine, a dopamine receptor agonist, was shown to elicit penile erections (Lal et al., 1984) and cyproheptadine, a serotonin antagonist, has been employed successfully (4 mg four times a day) for antidepressant-induced anorgasmia (Sovner, 1984). In his careful review, Segraves (1989, pp. 277–278) concluded, however, that research evidence is "loosely compatible with the hypothesis that increased dopaminergic activity is associated with the erectile response in humans . . . [but there is as yet] . . . no convincing evidence to date that serotonergic activity is related to sexual activity in humans."

Many urologists prescribe yohimbine, an alpha$_2$-andrenergic blocker (2 mg PO three times daily), as the first-line pharmacologic treatment for impotence, (Condra et al., 1986). Some candidly attribute a placebo effect to the treatment. Although it is clear that yohimbine enhances the sexual motivation in male rats, it is not so evident that it is a robust agent for the treatment of erectile dysfunction in men. Among men with psychogenic impotence it restored complete sexual functioning (intercourse) in 25% (Reid et al., 1987). Although 40% of organically impotent men had increased breakage of snap gauges, only 5% of 215 consecutive medical patients had complete subjective improvement on 16.2 mg/day (Sonda, Mazo, & Chancellor, 1990). When these authors increased the yohimbine to 21.6 mg/day, complete improvement was obtained by 13% of those who had achieved no or partial improvement on the lower dosage. After 11 months only mild and reversible side effects were recorded in 3% of the patients (Sonda, Mazo, & Chancellor, 1990). There is one report of the successful yohimbine treatment of a depressed man with clomipramine-induced anorgasmia in which the yohimbine, acting as an enhancer of norepinephrine, facilitated the serotonergic–noradrenergic interaction in the patient's sexual functioning (Price & Grunhaus, (1990). (See also Chapter 39.)

Disorder of Arousal

Little is known about the response to treatment of women with disease processes that impair sexual arousal. As in men, side effects of drugs can be minimized by decreased dosage, substitution, or discontinuance of the drug. Dyspareunia due to vulvovaginal conditions and atrophic vaginitis can usually be eliminated with topical lubricants or estrogens. A nonpetroleum vaginal lubricant, rather than jellies, which may provide a nidus for infection, should be prescribed for women whose diseases have been implicated in inhibited sexual arousal. If recurrent urinary tract infection is associated with the disease, as in diabetes and multiple sclerosis, cunnilingus should be avoided since it is a possible vehicle of infection transmission.

The development of pharmacologically induced erections by intracavernosal autoinjections of vasodilating drugs (Zorgniotti & LeFleur, 1985) such as papaverine hydrochloride and phentolamine mesylate appears to a safe and effective treatment of organic impotence (Cole, 1990). Although some vascular compromise is compatible with a good response to vasodilator injection, a minimum level of vascular competence is required. A PBI-pulsation ratio of less than 0.5 coupled with a maximum penile rigidity of less than 350 grams successfully discriminated those who responded unsuccessfully to papaverine in one report (Brendler et al., 1986). (See Chapter 39 for more details.)

Pharmacologic erection treatment is effective for men with neurogenic impotence due to multiple sclerosis, pelvic nerve dysfunction, spinal cord injury, transverse myelitis, and diabetes mellitus. Goldstein et al. (1985) reported on the pharmacologic erection treatment of 22 men with neurologic impotence over a 15-month period. Using test doses of 0.42 to 0.83 mg of phentolamine mesylate and 12.5 to 25.0 mg of papaverine hydrochloride in a total volume of 0.5 to 1.0 ml, erections were maintained for a period of 45 minutes to 6 hours, and ejaculation was achieved by most. Priapism due to the initial dosage was managed with corporal epinephrine (10 μg) irrigation. Goldstein et al. (1985) reported no systemic complications or major local penile problems despite multiple injections. Other (Brendler et al., 1986) have reported a high sensitivity of neurologic patients to the papaverine and reduced the dosage according to the response of the patient.

Papaverine, phentolamine, and more recently prostaglandin E_1 are the drugs most commonly employed in the intracavernosal injections. Men with idiopathic impotence who are not responsive to sex therapy may also be helped by intracavernosal injection therapy.

The two side effects for which one should be alert are priapism and penile fibrosis (at needle-stick sites), which in one study was palpable in 57% of the patients after a 12-month period (Levine et al., 1989). The dropout rate from autoinjection therapy in a 2-year study was reported to be 46%, half of the attrition occurring within the first four administrations (Althof et al., 1989). (See Chapter 39.)

Penile Prostheses

The surgical implantation of a penile prosthesis enables a man to have penetration and, to the extent possible prior to surgery, ejaculation and orgasm (Coleman et al., 1985). Subsequent to surgery, however, the man cannot have an erection without the prosthesis. Similarly, prospective protheses recipients should be counseled that their masculine self-esteem may be shored up with the prosthesis (Tiefer, Pedersen, & Melman, 1988), but the resultant sexual function may be less than they recall regarding premorbid erectile fullness and ejaculatory sensation (Steege, Stout, & Carson, 1986). Patients with manual dexterity limitations (or at risk for them because of disease such as multiple sclerosis) should be advised against a prosthesis that requires fine digital movements for inflation or deflation.

Reports of patient satisfaction with penile prostheses have been consistently high, although most studies have serious methodologic weaknesses (Collins & Kinder, 1984). One exception was a preoperative and postoperative study done by Berg and colleagues (1984) to identify predictors of successful adjustment following implantation of a penile prosthesis. Favorable predictors were (1) patient ambivalence about the operation; (2) a mature relationship between patient and partner; (3) some sexual activity before and after the onset of impotence; and (4) good mental health. Indicators of unfavorable prognosis were (1) a psychogenic component to the erectile dysfunction; and (2) unrealistic expectations of what the prosthesis would accomplish for sexual and relational satisfaction.

When consulting urologists about penile implantations and pharmacological erection treatment, psychiatrists should urge the involvement of spouses in the preoperative discussion of implantation and should assess carefully the quality of expectations for the treatment held by the patient and the wife.

External vacuum therapy consists of a hand-operated pump that produces a vacuum in a plastic cylinder placed over a flaccid penis. The resulting engorgement (erection) is preserved and utilized during coitus by a constriction band slipped onto the base of the penis after the erection is achieved. The vacuum pump has proved to be an efficient noninvasive treatment of erectile dysfunction. It is especially helpful for men whose impotence is vasculogenic and who are assisted by the constriction of the rubber band at the base of the engorged penis. Further information can be obtained from Nu-Potent, (Augusta, Georgia). (See also Chapters 39 and 40 for a critical review of penile prostheses.)

Disorder of Orgasm

Apart from managing local vaginal conditions and discontinuing possible causal drugs, there are no organic therapies for anorgasmia in women secondary to a disease process. In those women with identified neuronal damage, including diabetic neuropathy, the goal of therapy should therefore be to assist the patient (and spouse) to adjust to the permanent limitations caused by the disease.

If a woman who can have orgasm only clitorally seeks to expand her sexual responsiveness to include vaginal orgasm, she should be referred to a skilled sex therapist. The goal of such therapy typically is to assist the woman either to obtain vaginal orgasm or to enjoy more fully the (nonorgasmic) contribution of vaginal stimulation in her sexual response.

In men, lack of ejaculatory reflex should be distinguished from retrograde ejaculation, and the condition should be explained to the patient. If retrograde ejaculation is anticipated as a result of impending surgery (e.g., retroperitoneal lymphadenectomy), adequate fertility counseling and sperm bank information should be given to the patient. Imipramine 25 to 50 mg/day has been used to restore anterograde ejaculation in some of these patients (Nijman et al., 1982). Failure of emission in a patient with multiple sclerosis was corrected with ephedrine sulfate, a medication that stimulates the sympathetic nerves controlling emission (Schover et al., 1988).

Premature ejaculation is treated behaviorally with sensate focus techniques (Masters & Johnson, 1970), which are sometimes complemented by anxiolytic medication. Basing his intervention on reports that clomipramine was associated with retarded ejaculation, Assalian (1988) successfully treated five men with premature ejaculation.

In men, primary or long-standing anorgasmia (without an identified organic cause) requires intensive psychotherapy to effect improvement.

Disorders of Resolution

Priapism, unless treated within 4 hours, may lead to penile ischemia, fibrosis, and permanent impotence. It is treated with an injection of epinephrine directly into the penis. Trazodone can cause priapism as a side effect during the treatment of depression. Because of the risk of permanent erectile impairment, this side effect must be regarded as a medical emergency. Male patients being treated with trazodone should be informed of this potential side effect and advised to seek immediate treatment if it occurs. Two patients on trazodone reported 30- to 60-minute erections with no subjective sexual arousal (Sacks et al., 1985). They were switched to nortriptyline without further episodes of abnormal erections.

Psychologic Treatment

Offering psychologic treatment to medical patients with sexual dysfunction does not imply that there is a clear choice between medical-surgical interventions and those of a psychologic nature. Treatment is determined by etiology. For the medical patient (as for the healthy individual), sexual dysfunction is a biopsychosocial disorder. Treatment of the sexual dysfunction is concerned with the patient's physical condition and psychologic state, as well as the social and relational adjustment the patient is making to the illness.

As is the case with so many prognoses, those who develop a sexual problem subsequent to an illness have a better chance of positive resolution of the dysfunction if they have a good baseline of sexual behavior. In a review of the sexual rehabilitation of 284 consultations in a cancer center, patients who were young, were not clinically depressed, and who had less-conflicted marriages were those who had more positive outcome (Schover, Evans, & von Eschenbach, 1987).

A useful model for treatment of the sexual problems of the medical patient is the *PLISSIT* model developed by Annon and Robinson (1978). The physician gives permission (*P*) to the patient to discuss his or her sexual concerns and apprehensions by initiating the discussion with an appropriate and solicitous question. In response to the questions the patient and spouse may have about the effects of the illness or surgery on sexual functioning, the clinician provides limited information (*LI*), which is sufficient factual information to enable informed choice without overloading the patient with accurate but excessive detail. As rapport is established between physician and patient, specific suggestions (*SS*) are given by the clinician regarding maximizing sexual relations given the limitations imposed by the medical condition. In some cases, the patient must be referred to someone who can provide intensive therapy (*IT*) in an individual or couple modality. Each of these interventions can be demonstrated using cardiovascular disease as an example.

1. *Permission.* Without the clinician's inquiry into their sexual concerns and problems, most patient remain silent about them. DeLeo and Magni (1983) reported that only 10% of patients on antihypertensive medication spontaneously reported impotence. Systematic questioning increased the incidence to 26%. When the patients were given a questionnaire to complete in privacy, 47% admitted impotence.

Spouses also need to be given the opportunity to air their concerns about the patient's illness and sexuality. This situation is especially true for partners of patients who have had a myocardial infarction (MI) or coronary artery bypass surgery (Papadopoulos et al., 1986).

2. *Limited information.* MI patients and their spouses have less anxiety and less fear of "coital death" when a health professional makes concerted efforts to address their sexuality as part of the rehabilitation program. Hellerstein and Friedman (1970) have described the cardiovascular demands of sexual intercourse. During coitus, blood pressure generally rises to approximately 160 mm Hg systolic with a concurrent heart rate of 150 beats/minute. It has been compared to a brisk walk around a city block or climbing two flights of stairs. Exercise testing establishing sufficient aerobic exercise capacity can thus be used to establish the probable cardiac safety of resuming intercourse. Prior to discharge from hospital, such patients and their spouses should

be given some concrete means of appreciating the cardiovascular demands of intercourse. Full sexual activity can be resumed 8 to 12 weeks after an acute infarct. In the interim, couples should continue to engage in caressing and other expressions of sexual tenderness to which they have been accustomed.

For the first few months of recovery, apprehension and reluctance are typical feelings about sexual activity. Beyond 6 months, male MI patients have been reported to have no significant difference in incidence in sexual dysfunction when matched with controls for age, hypertension, diabetes, and smoking habits (Dhabuwala, Kumar, & Pierce, 1986). Men who were sexually active before their MI who persist in avoiding sexual activity beyond 6 months after MI should thus be considered to have sexual dysfunction. They should be educated as to the limited risk that coitus entails for them. Problems persisting after education require psychotherapy or sexual therapy.

3. *Specific suggestions.* When intercourse is attempted by a post-MI patient, the couple may find that a side-by-side or female-superior position places less isometric tension and cardiovascular demand on the male patient and is thus preferable for them. Prophylactic nitroglycerin can be suggested to patients who experience mild angina during coital activity.

In addition to cardiovascular patients, those with chronic illnesses or conditions, such as multiple sclerosis or ostomy, can be encouraged to participate in one of the numerous patient self-help organizations. Such organizations regularly feature speakers or workshops on sexual problems commonly experienced by persons with the illness particular to the group. In addition to these specific medical illness groups, Impotence Anonymous (IA, 119 South Ruth Street, Maryville, TN 37801-5746) is oriented toward surgical implants and may be particularly helpful for patients who are considering a penile prosthesis.

4. *Intensive therapy.* The information, suggestions, and reassurance that can be obtained during two or three medical office visits suffices for many medical patients to remedy the sexual dysfunction. For others, the psychologic component of the sexual problem requires referral to someone who can provide a specialized mode of treatment. Formal psychiatric consultation should be obtained at this point. Psychiatrists and other sexual therapists with specific, detailed knowledge of the patient's medical problems obtain the best results.

The initial psychiatric differential diagnosis is whether the sexual dysfunction is a psychosexual dysfunction or is secondary to another Axis I disorder, such as major mood disorder, which must be treated first. Medical patients referred for evaluation of a sexual dysfunction may be depressed and the sexual problem only one aspect of the dysphoria. A careful assessment should be made to determine whether a major depression is present or the low mood is an understandable reaction to the illness.

If, however, there is no other Axis I disorder that requires prior treatment and the etiology of the sexual dysfunction is presumed intrapsychic, individual psychotherapy is indicated, subsequently supplemented by group therapy if necessary.

If interpersonal factors are implicated in the etiology of the sexual dysfunction, a conjoint modality is indicated. If marital or relational issues predominate, they must be attended to first by employing the techniques of family therapy. If sexual dysfunction is fairly discrete, more behavioral therapy (e.g., sensate focus) may suffice. In practice, conjoint therapy initiated because of a sexual problem almost always involves both marital and specifically sexual therapy.

Acknowledgments

The authors wish to thank Dr. Thomas N. Wise for his helpful comments and suggestions concerning this chapter.

REFERENCES

ABEL GG, BECKER JV, CUNNINGHAM-RATHER J, ET AL. (1982). Differential diagnosis of impotence in diabetics: The validity of sexual symptomatology. Neurol Urodynam 1:57–69.

ALTHOF SE, TURNER LA, LEVINE SB, ET AL. (1989). Why do so many people drop out from auto-injection therapy for impotence? J Sex Marital Ther 15:121–129.

ALZATE H (1985). Vaginal eroticism and female orgasm: A current appraisal. J Sex Marital Ther 11:271–284.

ANNON JS, & ROBINSON CH (1978). The use of vicarious learning in the treatment of sexual concerns. In J LOPICCOLO & L LOPICCOLO (Eds.), Handbook of sex therapy (Ch. 3, pp. 35–56). New York: Plenum.

ASSALIAN P (1988). Clomipramine in the treatment of premature ejaculation. J Sex Res 24:213–215.

BARRETT M (1982). Sexuality and multiple sclerosis. New York: National Multiple Sclerosis Society.

BERG R, MINDUS P, BERG G, ET AL. (1984). Penile implants in erectile impotence: Outcome and prognostic indicators. Scan J Urol Nephrol 18:277–282.

BRADLEY WE, TIMM GW, GALLAGHER JM, ET AL. (1985). New method for continuous measurement of nocturnal penile tumescence and rigidity. Urology 26:4–9.

BRENDLER CB, ALLEN RP, ENGEL RM, ET AL. (1986). NPT predicts response to pharmacological erection. Paper read at the Mid-Atlantic American Urological Association annual meeting, Bermuda.

BRINDLEY GS (1983). Cavernosal alpha-blockade: A new technique for investigating and treating erectile impotence. Br J Psychiatry 143:332–337.

BUVAT J, BUVAT-HERBAUT M, DEHAENE JL, ET AL. (1986). Intracavernous injection of papaverine a reliable screening test for vascular impotence? J Urol 135:476–478.

BUVAT J, BUVAT-HERBAUT A, LEMAIRE A, ET AL. (1990). Recent developments in the clinical assessment and diagnosis of erectile dysfunction. Annu Rev Sex Res 1:265–308.

CARANI C, ZINI D, BALDINI A, ET AL. (1990). Effects of androgen treatment in impotent men with normal and low levels of free testosterone. Arch Sex Behav 19:223–233.

CARROLL JL, BALTISH MH, & BAGLEY DH (1990). The use of PotenTest in the multidisciplinary evaluation of impotence: is it a reliable measure? J Sex Marital Ther 16:181–187.

CARTER JN, TYSON JE, TOLIS G, ET AL. (1978). Prolactin-secreting tumors and hypogonadism in 22 men. N Engl J Med 299:847–852.

COLE HM (Ed.) (1990). Vasoactive intracavernous pharmacotherapy for impotence: Papaverine and phentolamine. JAMA 264:752–754.

COLEMAN E, LISTIAK A, BRAATZ G, ET AL. (1985). Effects of penile implant surgery on ejaculation and orgasm. J Sex Marital Ther 11:199–205.

COLLINS GF, & KINDER BN (1984). Adjustment following surgical implantation of a penile prosthesis: A critical overview. J Sex Marital Ther 10:255–271.

CONDRA M, MORALES A, SURRIDGE DH, ET AL. (1986). The unreliability of nocturnal penile tumescence recording as an outcome of measurement in the treatment of organic impotence. J Urol 135:280–282.

CONTE HR (1983). Development and use of self-report techniques for assessing sexual functioning: A review and critique. Arch Sex Behav 12:555–576.

COSTA PT, FAGAN PJ, & PIEDMONT RL, ET AL. (1992). The five-factor model of personality and sexual functioning in outpatient men and woman. Psychiatr Med 10:199–215.

COSTA PT, MCCRAE RR (1985). The NEO Personality Inventory Manual. Odessa, FL: Psychological Resources.

CRENSHAW TL, GOLDBERG JP, STERN WC (1987). Pharmacologic modification of psychosexual dysfunction. J Sex Marital Ther 13:239–250.

DELEO D, & MAGNI G (1983). Sexual side effects of antidepressant drugs. Psychosomatics 140:1076–1082.

DENNERSTEIN L, WOOD C, & BURROWS GD (1977). Sexual response following hysterectomy and oophorectomy. Obstet Gynecol 49:92–96.

DEPALMA RG (1984). Vascular assessment of sexual dysfunction. Paper read at Society for Sex Therapy and Research annual meeting, New York.

DEROGATIS LR (1986). The psychosocial adjustment to illness scale (PAIS). J Psychosom Res 30:77–91.

DEROGATIS LR, FAGAN PJ, SCHMIDT CW, ET AL. (1986). Psychological subtypes of anorgasmia: A marker variable approach. J Sex Marital Ther 12:197–210.

DEROGATIS LR, & MELISARATOS N (1979). The DSFI: A multidimensional measure of sexual functioning. J Sex Marital Ther 5:244–280.

DHABUWALA CB, KUMAR A, & PIERCE JM (1986). Myocardial infarction and its influence on male sexual functioning. Arch Sex Behav 15:499–504.

FAGAN PJ, MEYER JK, & SCHMIDT CW JR (1986). Sexual dysfunction in an adult developmental perspective. J Sex Marital Ther 12:1–12.

FAGAN PJ, SCHMIDT CW, WISE TN, ET AL. (1988). Alcoholism in patients with sexual disorders. J Sex Marital Ther 14:245–252.

FAHRNER EM (1987). Sexual dysfunction in male alcohol addicts: prevalence and treatment. Arch Sex Behav 16:247–257.

FOLLICK MJ, SMITH TW, & TURK DC (1984). Psychosocial adjustment following ostomy. Health Psychol 3:505–517.

GARDNER E, & JOHNSTON J (1985). Buproprion—an antidepressant without sexual pathophysiological action. J Clin Psychopharmacol 5:25–29.

GARTRELL N (1986). Increased libido in women receiving Trazodone. Am J Psychiatry 143:781–782.

GERWERTZ BL, & ZARINS CK (1985). Vasculogenic impotence. In RT & HW SCHOENBERG (Eds.), Diagnosis and treatment of erectile disturbances (Ch. 5, pp. 105–114). New York: Plenum.

GOLDSTEIN I, PAYTON T, SAENEZ DE TEJADA I, ET AL. (1985). Neurologic impotence: An advance in treatment. Paper read at the annual meeting of the Society for Sex Research and Therapy, Minneapolis.

HAWTON K (1985). Sex therapy: A practical guide. New York: Oxford University Press.

HELLERSTEIN HK, & FRIEDMAN EH (1970). Sexual activity and the postcoronary patient. Arch Intern Med 125:987–999.

HERMAN JB, BROTMAN AW, POLLACK MH, ET AL. (1990). Fluoxetine-induced sexual dysfunction. J Clin Psychiatry 51:25–27.

HERZOG AG, DRISLANE FW, SCHOMER DL, ET AL. (1990). Abnormal pulsatile secretion of luteinizing hormone in men with epilepsy: relationship laterality and nature of paroxysmal discharges. Neurology 40:1557–1561.

HERZOG AG, SEIBEL MM, SCHOMER D, ET AL. (1984). Temporal lobe epilepsy: an extrahypothalamic pathogenesis for polycystic ovarian syndrome? Neurology 34:1389–1393.

HESLINGA K, SCHELLEN AM, & VERKUYL A (1974). Not made of stone. Springfield, IL: Charles C Thomas.

JANI NN, & WISE TN (1988). Antidepressants and inhibited female orgasm: A literature review. J Sex Marital Ther 14:279–283.

JENSEN SB (1981). Diabetic sexual function: A comparative study of 160 insulin-treated diabetic men and women and an age-matched control group. Arch Sex Behav 10:493–504.

JENSEN SB (1985). Sexual relationships in couples with a diabetic partner. J Sex Marital Ther 11:259–270.

JENSEN P, JENSEN SB, SORENSEN PS, ET AL. (1990). Sexual dysfunction in male and female patients with epilepsy: a study of 86 outpatients. Arch Sex Behav 19:1–14.

JONES TM (1985). Hormonal factors in erectile dysfunction. In RT SEGRAVES & HW SCHOENBERG (Eds.), Diagnosis and treatment of erectile disturbances (Ch. 6, pp. 115–158). New York: Plenum.

KANEKO S, & BRADLEY WE (1986). Evaluation of erectile dysfunction with continuous monitoring of penile rigidity. J Urol 136:1026–1029.

KAPLAN HS (1979). Disorders of sexual desire. New York: Brunner/Mazel.

KARACAN I (1982). Nocturnal penile tumescence as a biological marker in assessing erectile dysfunction. Psychosomatics 23:349–360.

KARACAN I, ASLAN C, & HIRSHKOWITZ M (1983). Erectile mechanisms in man. Science 220:1080–1082.

KLINE M (1989). Fluoxetine and anorgasmia. Am J Psychiatry 146:804–805.

KOLODNY RC, MASTERS WH, & JOHNSON VE (1979). Textbook of sexual medicine. Boston: Little, Brown.

LAL S, ACKMAN D, THAVUNDAYIL JX, ET AL. (1984). Effect of apomorphine, a dopamine agonist, on penile tumescence in normal subjects. Prog Neuropsychopharmacol Biol Psychiatry 8:695–699.

LEIBLUM SR, & PERVIN LA (1980). Principles and practice of sex therapy. New York: Guilford Press.

LEPOR H, GREGERMAN M, CROSBY R, ET AL. (1985). Precise localization of the autonomic nerves from the pelvic plexus to the corpora cavernosa: A detailed anatomical study of the adult male pelvis. J Urol 4:207–212.

Levine SB, Althof SE, Turner LA, et al. (1989). Side effects of self-administration of intracavernous papaverine and phentolamine for the treatment of impotence. J Urol 141:54–57.

Lydiard R, & George M (1989). Fluoxetine-related anorgasmia South Med J 82:933–934.

Masters WH, & Johnson VE (1966). Human sexual response. Boston: Little, Brown.

Masters WH, & Johnson VE (1970). Human sexual inadequacy. Boston: Little, Brown.

Meyer JK, Schmidt CW, & Wise TN (Eds.) (1983). Clinical management of sexual disorders (2nd ed.). Baltimore: Williams & Wilkins.

Musher J (1990). Anorgasmia with the use of fluoxetine. Am J Psychiatry 147:948.

Nijman JM, Jager S, Boer, PW, et al. (1982). The treatment of ejaculation disorders after retroperitoneal lymph node dissection. Cancer 50:2967–2971.

Othmer E, & Othmer S (1987). Effect of buspirone on sexual dysfunction inpatients with generalized anxiety disorder. J Clin Psychiatry 48:201–203.

Papadopoulos C, Shelly SI, Piccolo M, et al. (1986). Sexual activity after coronary bypass surgery. Chest 90:681–685.

Prescott RW, Kendall-Taylor P, Hall K, et al. (1982). Hyperprolactinemia in men: Response to bromocriptine therapy. Lancet 1:245–249.

Price J, & Grunhaus LJ (1990). Treatment of clomipramine-induced anorgasmia with yohimbine: a case report. J Clin Psychiatry 51:32–33.

Reid K, Surridge DH, Morales A, et al. (1987). Double-blind trial of yohimbine in treatment of psychological impotence (Vol. 2). Lancet 8556:421–423.

Rosen RC, Kostis JB, & Jekelis AW (1988). Beta-blocker effects on sexual function in normal males. Arch Sex Behav 617:241–255.

Rowland DL, Greenleaf W, Mas M, et al. (1989). Penile and finger sensory thresholds in young, aging and diabetic males. Arch Sex Behav 18:1–12.

Sacks M, Miller F, Gunn J, et al. (1985). Unusual erectile activity as a side effect of trazodone. Hosp Community Psychiatry 36:298.

Schiavi RC (1988). Nocturnal penile tumescence in the evaluation of erectile disorders: A critical review. J Sex Marital Ther 14:83–97.

Schiavi RC (1990). Sexuality and aging in men. Annu Rev Sex Res 1:227–249.

Schover LR, Evans RB, & von Eschenbach AC (1987). Sexual rehabilitation in a cancer center: Diagnosis and outcome in 384 consultations. Arch Sex Behav 16:445–461.

Schover LR, & Jensen SB (1988) Sexuality and chronic illness: A comprehensive approach. New York: Guilford Press.

Schover LR, Thomas AJ, Lakin MM, et al. (1988). Orgasm phase dysfunctions in multiple sclerosis. J Sex Res 25:548–554.

Schreiner-Engel P, Schiavi RC, Vietorisz D, et al. (1985). Diabetes and female sexuality: A comparative study of women in relationships. J Sex Marital Ther 11:165–175.

Schwartz MF, Bauman J, & Kolodny RC (1981). Prolactin level in men presenting with sexual dysfunction. Paper read at annual meeting of Society for Sex Therapy and Research, New York.

Segraves RT (1982). Male sexual dysfunction. Paper read at annual meeting of Society for Sex Therapy and Research, Charleston, SC.

Segraves RT (1988). Psychiatric drugs and inhibited female orgasm. J Sex Marital Ther 14:202–207.

Segraves RT (1989). Effects of psychotropic drugs on human erection and ejaculation. Arch Gen Psychiatry 46:275–284.

Segraves RT, Madsen R, Carter CS, et al. (1985). Erectile dysfunction associated with pharmacological agents. In RT Segraves & HW Shoenberg (Eds.), Diagnosis and treatment of erectile disturbances (Ch. 2, pp. 23–63). New York: Plenum.

Segraves RT, Schoenberg HW, & Ivanoff J (1983). Serum testosterone and prolactin levels in erectile dysfunction. J Sex Marital Ther 9:19–26.

Segraves RT, Schoenberg HW, & Segraves KA (1985). Evaluation of the etiology of erectile failure. In RT Segraves & HW Schoenberg (Eds.), Diagnosis and treatment of erectile disturbances (Ch. 8, pp. 165–196). New York: Plenum.

Sherwin BB, Gelfand MM, & Brender W (1985). Androgen enhances sexual motivation in females: A prospective, crossover study of sex steroid administration in the surgical menopause. Psychosom Med 47:339–351.

Sonda LP, Mazo R, & Chancellor MB (1990). The role of yohimbine for the treatment of erectile impotence. J Sex Marital Ther 16:15–21.

Sovner R (1984). Treatment of tricyclic-induced orgasmic inhibition with cyproheptadine. J Clin Psychiatry 4:169.

Spector HP, & Carey MP (1990). Incidence and prevalence of the sexual dysfunctions: A critical review of the empirical literature. Arch Sex Behav 19:389–408.

Steege JF, Stout AL, & Carson CC (1986). Patient satisfaction in Scott and Small-Carrion penile implant recipients: A study of 52 patients. Arch Sex Behav 15:393–399.

Stoudemire A, Techman T, & Graham SD (1985). Sexual assessment of the urologic oncology patient. Psychosomatics 26:405–410.

Sullivan G, & Lukoff D (1990). Sexual side effects of antipsychotic medication: Evaluation and interventions. Hosp Community Psychiatry 41:1238–1241.

Thase ME, Reynolds CF, Glanz LM, et al. (1987). Nocturnal penile tumescence in depressed men. Am J Psychiatry 144:89–96.

Tiefer L (1986). In pursuit of the perfect penis. Am Behav Sci 29:579–599.

Tiefer L, Pedersen B, & Melman A (1988). Psychosocial follow-up of penile prosthesis implant patients and partners. J Sex Marital Ther 14:184–201.

Utian WH (1975). Effect of hysterectomy, oophorectomy and estrogen therapy on libido. Int J Obstet Gynecol 13:97–100.

Valentich M, & Gripton J (1986). Facilitating the sexual integration of the head-injured person in the community. Sexuality Disability 7:28–42.

Valleroy ML, & Kraft GH (1984). Sexual dysfunction in multiple sclerosis. Arch Phys Med Rehabil 65:125–128.

Walling M, Andersen BL, & Johnson SR (1990). Hormonal replacement therapy for postmenopausal women: A review of sexual outcomes and related gynecologic effects. Arch Sex Behav 19:119–137.

Walsh PC, Lepor H, & Eggleston JC (1983). Radical prostatectomy with preservation of sexual function: anatomical and pathological considerations. Prostate 4:473–485.

Wise TN (1985). Sexual dysfunction following diseases of the reproductive organs. Adv Psychosom Med 12:136–139.

Wise TN, & Schmidt CW (1985). Diagnostic approaches to sexuality in the medically ill. In RC Hall & TP Beresford (Eds.), Handbook of psychiatric diagnostic procedures (Vol. 2; Ch. 13, pp. 237–272). New York: Spectrum Publications.

Zeiss AM, Davies HD, Wood M, et al. (1990). The incidence and correlates of erectile problems in patients with Alzheimer's disease. Arch Sex Behav 19:325–331.

Zorgniotti AW, & LeFleur RS (1985). Auto-injection of the corpus cavernosum with a vasoactive drug combination for vasculogenic impotence. J Urol 133:39–41.

16 Acute pain management

RICHARD J. GOLDBERG, M.D.

The psychiatrist in the medical-surgical setting encounters a variety of issues involving difficulties with the management of acute pain. These issues include recognition and alleviation of psychosocial stresses and depression; advocacy of proper medical diagnoses and management; systems problems that result from lack of coordination among providers; the need for the provision of adjunctive nonpharmacologic approaches to pain, such as hypnosis and relaxation; and the education of physicians and nursing staff regarding the proper use of analgesics. This chapter addresses the most common problems that involve analgesic pharmacology, providing the basic information needed for their solutions. Most of these problems involve recognizing inadequate prescribing practices of narcotics and managing the complications associated with their use.

DOSE AND FREQUENCY

The most common cause of inadequate acute pain management is the underuse of narcotic analgesics. In fact, many doctors prescribe less than half the effective analgesic dose for their patients with pain (Marks & Sachar, 1973; Sriwatanakul et al., 1983). In addition to doses that are inadequate, narcotics often are prescribed at time intervals that extend beyond the effective half-life of the drug. The use of suboptimal doses and excessive time between dose intervals may result from physician orders, nursing administration, or patient bias. Some patients attempt to endure pain and earn the respect of the nurses and doctors as a "good" patient. It is not unusual to find patients on a q4–6h schedule of meperidine, when this drug usually requires administration every 3 to 4 hours. When patients start to complain of severe pain 3.5 hours after their last dose, they often are thought of as "addicted" or excessively preoccupied with medication. In fact, such patients may only be expressing the fact that their pain has reemerged because their dose time has gone beyond the effective duration of the narcotic. Table 16-1 lists usual dose ranges for commonly prescribed analgesics, and Table 16-2 lists the average duration of action.

The consultant dealing with a pain patient who continues to complain despite "usually adequate treatment" should first look at the dose and frequency of analgesic delivered. Ask patients what level of relief they get within the first hour after their dose. If relief during that time is inadequate, the dose is too low. If pain relief is complete but reemerges before the next dose, the duration between doses is too long. It is indispensable to look directly at the drug administration record, not at orders or progress notes, to confirm what has actually been given to the patient and on what schedule. Narcotics should be given on a schedule corresponding to their analgesic half-life (see Table 16-2). Regarding dose, there is no rule that 75 mg of meperidine, for example, is the right dose for everyone. Some patients require 75 mg and others 150 mg for the same effect.

One concern about giving high narcotic doses is respiratory depression, especially if the patient is also on benzodiazepines. If the patient is in severe pain, however, significant respiratory depression is unlikely. Respiratory depression can be treated by naloxone, the drug of choice for reversal of respiratory depression secondary to narcotics. Because rapid reversal can precipitate acute withdrawal, however, it is suggested that a dilute solution of naloxone (0.4 mg in 10 ml of saline) be given slowly, titrated against the patient's respiratory rate until just enough of the solution has been given to reverse the respiratory depression (Foley, 1985a). Patients should then be followed closely with frequent respiratory rate checks, as naloxone has a short half-life; the patients may have to be redosed if taking longer-acting agents, such as methadone or levorphanol (Foley, 1985a; Martin et al., 1973).

FIXED VERSUS AS-NEEDED DOSE SCHEDULES

When acute pain has been diagnosed, physicians must consider whether analgesics should be prescribed on an as-needed (PRN) or a fixed schedule. Severe, continuous pain is better treated on a fixed schedule for the following reasons (Goldberg & Tull, 1983).

TABLE 16–1. *Usual Adult Dose Ranges of Commonly Prescribed Narcotics*

Agent	Usual Dose Range (mg)	Comment
Codeine	30–60 PO q4–6h	Usually combined with an NSAID
Oxycodone	5–10 PO q4–6h	Comes as tablets with 5 mg oxycodone plus aspirin or acetaminophen
Pentazocine	25–100 PO q3–4h	Mixed agonist-antagonist; can cause psychotic symptoms[a]
Butorphanol	1–4 IM q3–4h 1–2 intranasally q3–4h	Mixed agonist-antagonist; less respiratory depression
Meperidine	50–150 SQ or IM q3–4h 50–150 PO q3–4h	Rapid onset of action; has a psychotoxic metabolite
Morphine	2–20 SQ or IM q3–4h 5–75 PO q4h	Sedating; may lower blood pressure
Hydromorphone	1–2 SQ or IM q3–4h 1–2 IV q3–4h 2–4 PO q3–4h	
Methadone	3 rectally q6–8h 5–20 PO q6–8h	Long half-life; lower dose in renal failure

Note: Commonly used dose ranges are *not* equianalgesic (see Table 16–4). Higher or lower doses may be appropriate for particular patients, depending on severity of pain, duration of treatment, tolerance, body weight, pharmacokinetics, drug interactions, and use of adjuncts.
[a]Mixed agonist-antagonists must not be given together with other narcotics.

TABLE 16–2. *Analgesic Duration (with Oral Dosing)*

Analgesic	Duration of Action (hr)
Morphine	2.5–7
Meperidine	2–4
Methadone	4–7 (longer with chronic use)
Hydromorphone	4–6
Pentazocine	3–4
Codeine	4–7
Propoxyphene	4–7
Oxycodone	3–5

1. A schedule based on the half-life of the narcotic prevents reemergence of pain before the next dose.

2. The dose required to treat reemergent pain resulting from an as-needed schedule often is larger than would be required to prevent it using a fixed schedule (Reuler, Girard, & Nardone, 1980; Shimm et al., 1979).

3. Patients on an as-needed schedule are in a dependent position, which requires them to ask for medication in a way that can be interpreted as excessive preoccupation with medication, and can be experienced by the patient as humiliating.

4. Elderly patients and those who are cognitively impaired and in pain may lose track of time or have difficulty calling the nursing staff to ask for medication.

Exceptions to the usual fixed-dosage schedule are patients with impaired hepatic metabolism (extensive liver disease or patients older than 80 years), in whom the drug's half-life may be prolonged, and patients with nocturnal pain only (Twycross & Lack, 1984).

Narcotic analgesics should be given during the night for patients whose sleep is being interrupted by pain or who awaken in the morning with excessive reemergent pain. Lack of sleep creates a vicious cycle of increased pain, which causes more lack of sleep (Moldofsky et al., 1975). Conversely, if a patient is sleeping well and is not awakening with increased perception of pain and its consequent anxiety, there is no need to schedule analgesic doses during the patient's sleep time.

Some clinicians double the nighttime narcotic dose to maintain analgesic control without unnecessary awakening of patients. For the elderly or frail patient, the bedtime dose should be increased by no more than 50% (Twycross & Lack, 1984). With the newer "slow-release" morphine preparations (Roxanol SR, MS Contin) q8–12h dosing allows for continuous analgesia without nighttime interruption.

It is also important to ascertain if the pain problem is the result of some intermittent procedure such as débridement, dressing changes, or physical therapy. In such instances the most important intervention would be to provide a narcotic dose prior to the intervention. Chapter 35 describes analgesia for burn débridement and dressing changes in detail.

An alternative narcotic scheduling method is known as "reverse PRN." This method involves offering the medication to the patient on a scheduled basis, allowing the patient to accept, refuse or delay the dose. This reverse PRN method ensures that the medication is made available but also helps the patient maintain some autonomy over titrating the dose (Goldberg & Tull, 1983; Sriwatanakul et al., 1983).

A major concern among physicians and nurses is the fear of creating an addicted patient, a fear that is largely unwarranted. For example, in a large review of 11,882 nonaddicted Medicare inpatients who received narcotics, in only four cases was iatrogenic narcotic addiction documented (Porter & Jick, 1980). Fear of addiction and lack of knowledge of pharmacokinetic profiles are common among nursing staff, who frequently choose lower doses when dose range choices are ordered (Sriwatanakul et al., 1983).

EFFECTIVE USE OF NARCOTICS

The effective use of analgesics is based on an assessment of whether the pain is acute or chronic, a definition of the specific pain syndrome, and an understanding of the clinical pharmacology of the drug prescribed (Foley, 1982). For patients with severe, acute pain, morphine represents the standard narcotic against which all others are compared. Unlike the mixed agonist-antagonist analgesics, morphine has no ceiling effect. In addition, it is neither more nor less likely to cause physical dependence than any other equianalgesic narcotic. Analgesic narcotics that are more potent, with shorter half-lives, such as hydromorphone, are much more difficult to titrate. Narcotics with longer half-lives, such as methadone, tend to accumulate in tissues and precipitate more central nervous system (CNS) and respiratory side effects (Gourlay, Cherry, & Cousins, 1986).

Sustained-release morphine has become available in the form of 30-mg controlled-release tablets. This medication is recommended to be given every 12 hours, or every 8 hours if there is a pattern of breakthrough pain. A sustained-release preparation usually is not the drug of choice for acute, severe pain, but it is a good option for the patient whose acute pain has been stabilized on oral morphine or other oral narcotics and is likely to remain severe and continuous for several days or longer. The obvious advantage of such a medication is the convenience in dosing time for both patient and staff. Moreover, the slow-release mechanism may result in fewer peak-and-trough fluctuations and therefore may cause fewer side effects, such as nausea and vomiting (Walsh, 1984). Finally, due to the fact that the tablets are slow-release and the morphine sulfate is contained in a matrix, the controlled-release morphine is less likely to become an illicit street drug and may be more palatable and socially acceptable to the patient than the current morphine sulfate aqueous solution or morphine sulfate USP tablets. The problem with sustained-release morphine for acute pain is that it does not permit precise titration of dosage to meet the fluctuating needs for analgesia that characterize acute illness and its hospital treatment.

MIXED OPIOID AGONIST-ANTAGONIST DRUGS

The mixed opioid agonist-antagonist drugs have moderate-to-strong analgesic activity (Houde, 1979). The group consists of a number of drugs, including pentazocine, buprenorphine, nalbuphine, and butorphanol.

1. *Pentazocine* was the first drug of this group to be available. A single dose of 30 to 60 mg IM is equivalent to 10 mg of morphine but with a slightly shorter duration of activity. Its drawbacks include the following: It can increase intrabiliary pressure; it can increase left ventricular end-diastolic pressure; and it is associated with a high incidence of psychotomimetic side effects (up to 20% of patients) (Kane & Pokorny, 1975; Taylor et al., 1978).
2. *Buprenorphine* (Buprenex) is a parenteral opioid analgesic, with 0.3 mg equivalent to 10 mg morphine IM. Effects usually are seen within 15 minutes, peak at about 1 hour, and are gone by 6 hours. In addition to its opiate agonist activity, buprenorphine demonstrates narcotic antagonist activity, similar to naloxone. Because of its slow rate of dissociation from narcotic receptors, buprenorphine-induced respiratory depression may not be reliably reversed with naloxone. Psychotomimetic reactions also are a risk.
3. *Nalbuphine* (Nubain) is another parenteral mixed agonist-antagonist narcotic. It is essentially equivalent to morphine in dosage. Its onset of action (intramuscularly) is within 15 minutes and its duration of action between 3 and 6 hours. Nalbuphine injections contain sulfites, which are allergens to some patients.
4. *Butorphanol* has an onset of analgesia of about 10 minutes with a duration of about 3 hours. It is equipotent with intramuscular morphine (Vandam, 1980). It can be given either as an injection or as a nasal spray.

All of the members of this group cause respiratory depression and can lead to physical dependence with prolonged use (Nagashima et al., 1976; Popio et al., 1978). Abrupt discontinuation in chronic users can lead to a narcotic withdrawal syndrome (Heel et al., 1978); therefore these drugs should be tapered gradually after chronic use.

Overall, the mixed agonist-antagonist group is likely to be used as an analgesic adjunct to anesthesia for relatively short-term pain problems. The major clinical issue is to be sure not to add a member of this group to narcotic agonists, as abrupt withdrawal can be precipitated.

ORAL VERSUS PARENTERAL USE

Oral analgesic administration offers simplicity and economy, avoids the discomforts and potential complications of repeated injections, makes the patient less dependent on others for care, and even obviates the need for hospitalization in some cases (Beaver, 1980; Twycross & Lack, 1986). There are, however, two specific instances when the use of parenteral narcotics is warranted. The first is in the patient with acute, severe pain requiring immediate relief, as in the trauma, burn, myocardial infarction, or postoperative patient. The second instance is the patient with severe pain in which oral medications fail to provide adequate relief, when tolerance has developed, or when the patient is no longer able to swallow (Moertel, 1980).

When one has determined that parenteral narcotics are needed to control pain, several variables should be considered. Parenterally administered narcotics have a more rapid onset of action and a shorter duration of effect than do orally administered narcotics. Peak concentrations may vary three- to fivefold between patients, and the time required to reach peak concentration varies as well. Moreover, for a given patient, differences in meperidine concentration as small as 0.05 μg/ml can represent the differences between no relief and complete analgesia (White, 1985). One study has documented that meperidine concentrations, when given intramuscularly every 3 to 4 hours, equals or exceeds the minimal analgesic concentration for only 35% of the dosing period because of the highly variable absorption and a narrow therapeutic window (Austin, Stapleton, & Mather, 1980).

TRANSDERMAL FENTANYL

The first transdermal narcotic is now available: the fentanyl citrate patch. Transdermal fentanyl systems delivery fentanyl at a rate of 75 μg/hour. It has been demonstrated, using double-blind, placebo-controlled methods, to be comparable to conventional intramuscular morphine for the control of postoperative pain (Caplan & Southam, 1990). A dose of 75 μg of fentanyl transdermal per hour is equivalent to about 50 mg of intramuscular morphine given over 24 hours (Foley, 1985b). The fentanyl should be applied several hours before surgery to ensure adequate postoperative levels. Its slowly declining plasma level after discontinuation can be an advantage in the changeover to oral medication. If prompt discontinuation is needed, an antagonist infusion is necessary because of the prolonged release. It takes 17 hours or more for the fentanyl serum concentration to fall by 50% after removal of the patch.

INTRAVENOUS NARCOTICS

The intravenous route is best reserved for the following situations (Portenoy, 1986).

1. Acute trauma, burn, or myocardial infarction patients
2. Severely cachectic patients by whom intramuscular injections are not tolerated
3. Patients with severe bleeding diatheses or low platelet counts, for whom intramuscular injections or suppositories are contraindicated.
4. Cases in which the intramuscular route is not desirable (as with pediatric patients) and the oral route is not possible.

Unfortunately, the intravenous bolus route causes a duration of analgesia that is short-lived because of rapid drug uptake by tissue and rapid elimination. As a result of this diminished duration of action, larger and more frequent doses are needed for breakthrough pain, thereby precipitating a higher incidence of respiratory depression, sedation, nausea, and vomiting. Finally, the demand for repeated intravenous narcotic boluses may become a time-consuming function for nursing personnel.

As a result of these disadvantages of intravenous bolus narcotic administration, analgesic researchers have devised several models of continuous opioid infusion. Intravenous drips may be rate-controlled by medical and nursing staff or in some cases by the patients themselves. According to Graves et al. (1983), minimal sedation and fewer side effects can be achieved with patient-controlled analgesia. Moreover, the potential for overdose can be minimized if small bolus doses (morphine 1 to 2 mg/hour) are used, with a mandatory dosing-free interval between successive doses (White, 1985).

Patient-controlled analgesia (PCA) has gained widespread use in several areas of acute pain management. Computerized programmable pumps allow bolus doses of narcotic to be administered by the patient. This dosing may be coupled with a background basal infusion of narcotic. Bolus dose limits, "lock-out" intervals, and a maximum total hourly dose all are programmable by medical staff. Obviously, patients who are confused are not good candidates for this method of drug administration. In addition, patients with a drug addiction history or tendency may be unable to limit their utilization of the PCA system to analgesic dosage only. If patients with a drug abuse history are given PCA, they should be observed cautiously for their resistance to timely tapering of PCA use.

Table 16-3 presents a method for initialing an intravenous morphine drip. For a complete review of the guidelines used in the management of continuous opioid infusions, refer to White (1985) and Portenoy (1986). For the most part, it is preferable to maintain patients on oral analgesics for as long as possible, because administration by the intravenous (or intrathecal) routes has been associated with rapid development of tolerance (Foley, 1985b).

An area of expanding use for narcotic analgesics is via the spinal route (Cullen et al., 1985). Opiates may be administered intrathecally but are most often delivered through a catheter placed percutaneously into the lumbar epidural space. Analgesia is achieved via diffusion of drug into cerebrospinal fluid (CSF) and subsequently into spinal cord opiate receptors located in the substantial gelatinosa. Bolus or continuous administration of narcotic is possible. This technique has found widespread use in the management of postoperative and trauma patients. Profound analgesia with extended duration of action (up to 24 hours from a single bolus dose of morphine) using low narcotic doses is the principal benefit from this technique. Side effects include respiratory depression, pruritus, and urinary retention. The concern regarding delayed respiratory depression when hydrophilic agents such as morphine are employed has limited the use of this technique to intensive care unit (ICU) settings in many hospitals. Special training for nursing personnel has, however, expanded its area of use to all medical and surgical wards in some institutions. Its use should be reversed for patients who have failed good trials of narcotic medication by other routes.

TABLE 16–3. *Intravenous Morphine Drip*

1. Discontinue previous narcotics and sedatives.
2. Obtain baseline vital signs and level of alertness.
3. Loading dose: 2–5 mg IV q15min until pain is relieved.
4. Maintenance dose: Total loading dose/hour; check vital signs every 15 minutes for first hour, then every hour for the next 6 hours, then every 6 hours for the next 24 hours, then every shift.

Calculation of morphine infusion rates to be given in D_5W (may also be given in D_5NS)

Morphine (mg)	*Solution (ml)*	*Rate (ml/hr or microdrop/min)*
250	1,000	Hourly morphine × 4
500	1,000	Hourly morphine × 2

5. Dosage should not be increased by more than one-third of previous hourly dosage. If one increases hourly dosage rate, vital signs must be rechecked every 15 minutes for 1 hour and every hour for the next 6 hours.
6. As the mixing of such solutions requires time and care by hospital pharmacy, one must give adequate advance notice to the hospital pharmacy for IV morphine solution preparation.
7. Upper range of infusion generally is 300 mg/hr, although most persons are controlled on much lower doses (and some infusion rates may be higher in tolerant individuals).

Source: Modified from Citron ML (1984, December). How would you treat severe cancer pain refractory to oral analgesics? *Drug Therapy* 11, 15, 19.

CONVERSIONS AMONG NARCOTICS

There are several situations where improper narcotic conversion leads to psychiatric consultation. Typically, these conversions involve a change from the parenteral to the oral route, encountered frequently in postoperative situations. When a patient is changed precipitously from parenteral meperidine to oral acetaminophen and codeine, the acute decrease in narcotic dose can be dramatic, leading to iatrogenic withdrawal, a dramatic increase in pain complaints, and an agitated patient. In the case of a switch from the parenteral to the oral route, one must recognize that increases in total milligrams per dose are necessary to maintain the same level of analgesia. Differences in oral-to-parenteral analgesic efficacy are due to limited gastrointestinal absorption and to first-pass hepatic metabolism. Original data by Houde, Wallenstein, and Beaver (1965) indicated a 6:1 oral/parenteral ratio for morphine, but the results of this single-dose study do not hold for regular administration. Sixfold increased oral doses cause oversedation and respiratory depression. For patients on regular schedules of morphine, the oral/parenteral ratio can be considered 3:1; in the case of meperidine, this ratio is 4:1.

When calculating conversion from a parenteral to an oral narcotic, or to sort out the total narcotic dose in situations where many analgesics have been used, it is helpful to convert all analgesics to a standard reference (Goldberg et al., 1986). This conversion among narcotics is easily facilitated by converting all drugs to oral morphine equivalents (see Table 16-4). From a methodologic viewpoint, the information for this table comes primarily from pain relief studies in cancer patients, whose acute pain needs and analgesia metabolism may be different from those of the noncancer patient. In addition, the time–effect curves for studying various doses of analgesia generally result from single-dose experiments. Single-dose studies neglect the kinetic changes that take place following repeated dosing, especially with narcotics that have long half-lives, such as meth-

TABLE 16–4. *Conversion of Narcotics to Oral Morphine Equivalents and Relative Potencies*

To convert oral dose of any narcotic to an equivalent oral dose of morphine, multiply by:

0.15	for propoxyphene
0.2	for meperidine
1/3	for codeine
1/2	for pentazocine
2	for oxycodone
3	for methadone
8	for hydromorphone

To convert intramuscular dose of any narcotic to an equivalent oral dose of morphine, multiply by:

1.5	for pentazocine
6.0	for methadone
40.0	for hydromorphone
3.0	for morphine
0.8	for meperidine
30.0	for butorphanol

Divide by 3 to get equivalent intramuscular dose of morphine. Equivalent doses are approximate; higher oral doses may be needed when initiating therapy.

adone. Finally, the conversion table presented does not acknowledge the probable differences in pain relief associated with age, sex, race, or quality of pain (Kaiko et al., 1982). For all these reasons, when switching from parenteral to oral narcotics, it is advisable to begin at about two-thirds of the predicted equivalent dose and then titrate upward if needed.

DEVELOPMENT OF TOLERANCE

Another variable that accounts for pain relief problems is the development of tolerance. The tolerant patient is the one who notices a shortened duration of analgesic effect and an eventual decrease in pain relief. The rate of development of tolerance varies; however, "since tolerance to most of the adverse dose-limiting effects of narcotics, including respiratory and general CNS depressant effects, develops concomitantly with tolerance to analgesic effect, even substantial tolerance can usually be surmounted and adequate analgesia restored by increasing the narcotic dose" (Beaver, 1980). Tolerance must be differentiated, however, from loss of pain control due to new or advancing disease. Cross-tolerance is incomplete with most narcotics. Switching parenteral narcotics therefore may be helpful. Because cross-tolerance is incomplete, Foley (1985b) recommended calculating the dosage of the new drug via morphine equivalents equal to the new preparation and reducing the dose by one-half to start. The dosage of the new narcotic may then be titrated upward, based on the patient's response. When a patient is not responding to one narcotic, however, the problem usually is inadequate dose or duration rather than tolerance. Yet any patient who is exposed to continuous doses of narcotics may develop tolerance, often in as soon as 5 to 7 days. It is often necessary therefore to increase the dose in order to obtain the same pain relief. Unfortunately, staff often become anxious about continuing narcotics for a long period and attempt to decrease the dose just at the time when the patient has developed tolerance and an increase is called for. The consultant must explain this phenomenon and should anticipate some resistance to the recommendation to increase narcotics for a "problem" patient just at the time when the staff expected a solution that included tapering narcotics. If tolerance has developed in a patient, a greater-than-expected oral dose may be required when switching from parenteral administration. Alternatively, cross-tolerance may not be complete in some patients, which may account for an overly sedated or "narcotic toxic" patient following a switch in medications.

PREVENTION OF NARCOTIC-INDUCED CONSTIPATION

Constipation is a problem that faces all patients on narcotic analgesics, and the effect seems to be dose-dependent. When narcotics bind to opiate receptors in the gut (Foley, 1982), the result is increased tone with markedly diminished propulsive contractions in both the small and large intestines. The resulting delay in the passage of the intestinal contents causes considerable desiccation of the feces, which in turn retards their advance through the colon. Moreover, greatly enhanced anal sphincter tone, combined with inattention to the normal sensory stimuli for the defecation reflex due to narcotic-induced CNS effects, further contribute to narcotic-induced constipation (Jaffe & Martin, 1980). The constipating effect of narcotic analgesics is not due entirely to local effects on intestinal action. In rats, naloxone injected into the cerebral ventricles acts to abolish the constipating effects of morphine injected in small quantities into the cerebral ventricles (Parolaro, Sala, & Gori, 1977).

Constipation can be uncomfortable and disturbing to patients and can lead to a need for manual disimpaction. Severe cramping pain also can result, or narcotics may lead to a functional ileus. Therefore severe constipation should be avoided and treated preventively. There is one circumstance in which narcotic-induced changes in colonic motility may be life-threatening. Patients with chronic ulcerative colitis, when treated with opioids during acute attacks, can develop a life-threatening toxic dilation of the colon, termed *toxic megacolon* (Garrett, Sauer, & Moertel, 1967).

Stimulant cathartics, by virtue of their direct and local effects on longitudinal peristalsis, are useful for treating narcotic-induced constipation. All anthraquinone derivatives, which are stimulant cathartics, effectively prevent constipation; but senna compounds (e.g., Perdiem granules or Senokot tablets) are better accepted than cascara or aloe. Some clinicians have proposed that there is a fixed dose relation: one-half of a Senokot (concentrated senna) tablet reverses the constipating effect of morphine 4 mg IM or its equivalent (Maguire, Yon, & Miller, 1981).

The most practical and useful regimen to prevent and treat narcotic-induced constipation is the following, adapted from the description by Twycross and Lack (1984, pp. 56–63).

1. Ask about patient's usual bowel habits and use of laxatives.
2. Perform a rectal examination to check for fecal impaction.
3. Be aware that fluid stool, especially with incontinence, may mean high fecal impaction.
4. Record bowel motions each day in the appropriate log.
5. Encourage fluids (i.e., water, prune juice) and foods that have a laxative effect if bowels do not move spontaneously.
6. For patients on morphine or alternative strong narcotic, prescribe:
    a. Casanthranol 30 mg with docusate 100 mg (Peri-Colace), one capsule 1 to 4 times a day. Dose is adjusted by the nurse according to result.
    b. If casanthranol with docusate causes abdominal cramps, change to docusate alone (available as a generic agent) 100 mg 1 to 4 times daily.
    c. If casanthranol with docusate (two capsules qid) is ineffective, give milk of magnesia and cascara suspension (25 ml milk of magnesia; 5 ml cascara aromatic fluid extract) 30 ml qhs in addition.
7. If no movement by the next morning, manually access location of stool in rectum. If good contact with rectal mucosa seems possible, use a bisacodyl (Dulcolax) 10 mg suppository.
8. If a suppository is not feasible or is ineffective, administer an oil retention enema or Fleets enema (or both) followed by soap-suds enema (SSE) if no result.
9. Manually disimpact if necessary.
10. Whenever possible, strong patient preferences for bowel care should be honored and a requisite physician order sought.

The compositions of available laxatives are listed in Table 16-5. This regimen may be followed on a chronic basis as long as narcotics are needed for analgesia.

USE OF PLACEBOS AND HYPNOSIS

Placebo (Latin: "I shall please,") is a term that evokes emotion in both health care givers and recipients. A placebo is an inert substance, given in any form and without inherent pharmacologic property, that, by virtue of the environment in which it is taken, the psychologic state of the recipient, and the specific neurophysiologic processes of the recipient, may have a desired effect on a given disorder. Unfortunately, there remains significant misunderstanding within the medical community regarding placebos and their purpose (Goldberg, Leigh, & Quinlan, 1979). Physicians and nurses generally underestimate the percentage of patients who experience pain relief when given a placebo. Placebos typically are given to disliked patients suspected of exaggerating their pain, who fail to respond to usual medical regimens, or both. Placebo use often is rationalized by the staff as a means of avoiding "dangerous" or "possibly addicting" treatment. Positive responses to

TABLE 16–5. *Composition of Laxatives*

Laxative	Composition
Peri-Colace	
Capsule	Casanthranol 30 mg Docusate 100 mg
Solution (per 5 ml)	Casanthranol 10 mg Docusate 20 mg
Colace (docusate sodium)	
Capsule	Docusate 100 mg
Solution (per 5 ml)	Docusate 50 mg
Dialose (docusate potassium)	
Capsule	Docusate 100 mg
Senokot	
Tablet	Sennoside B 7.5 mg
Dulcolax	
Tablet	Bisacodyl 5 mg
Suppository	Bisacodyl 10 mg
Milk of magnesia	
Suspension (per 5 ml)	Magnesium hydroxide 350 mg
Concentrate (per 5ml)	Magnesium hydroxide 1050 mg
Magnesium citrate	
Solution (per 10 oz)	Magnesium citrate 17.45 g
Cascara sagrada and milk of magnesia	
Suspension (per 5 ml)	Magnesium hydroxide 350 mg in cascara sagrada
Metamucil	
Powder packets	Psyllium mucilloid 5 g/10 g Dextrose 5 g/10 g

placebos often are misinterpreted by physicians as evidence that the pain has no physiologic basis. Several studies have shown that overdemanding and complaining patients are, if anything, less likely to respond to placebo than patients well liked by the hospital staff (Goodwin, Goodwin, & Vogel, 1979).

Thirty to forty percent of patients are placebo responders. In addition, not only does pain respond to placebo, but in the studies comparing morphine 10 mg IM to placebo, nausea and anxiety respond as well (Beecker, 1955). In an enlightening study of pain relief after oral surgery in 51 subjects, placebo analgesia was shown to be reversible by naloxone in a dose-dependent fashion. It was therefore concluded that placebos activate endogenous endorphins and have a specific "true" physiologic basis (Levine, Gordon, & Fields, 1978). In addition, placebos, like opioids, gradually become less effective at the same dose, thus displaying a tolerance effect.

When placebo use is encountered or contemplated, the physician should ask, "What is happening in this treatment system?" Typical answers include the following: Inadequate narcotic medication is being used; the personality style of the patient has promoted the staff to become unduly suspicious; the patient and staff are caught in ongoing interpersonal conflict (Goldberg & Tull, 1983). The only appropriate clinical role for placebos is in research protocols, conducted with proper informed consent.

Patients in acute pain generally are unable to learn hypnotic techniques that would allow them to reduce their pain. However, instruction in these techniques prior to predictably pain free procedures may be useful in decreasing and need for medication.

ALLERGIC AND BEHAVIORAL SIDE EFFECTS OR NARCOTICS

Psychiatric consultation may be requested when the patient develops some behavioral disturbance and the issue is raised whether it represents a reaction to narcotics. Idiosyncratic reactions, such as nausea, vomiting, and dizziness, to morphine and related opiods often are described by patients as an "allergy" to narcotics. Actually, true allergic phenomena to narcotics are uncommon and usually are manifested by urticaria, skin rashes, and dermatitis. More severe anaphylactoid reactions are rare. In a review of the medical literature from entries in *Index Medicus* from 1966 to 1982, there were no documented cases of anaphylaxis following oral or intramuscular administration of specific narcotics (Levy & Rockoff, 1982). True anaphylaxis has been described for both intravenous codeine and intravenous meperidine (Levy & Rockoff, 1982; Shanahan, Marshall, & Garrett, 1983). It is thought that those persons who are know to have the "aspirin-intolerance syndrome" have a higher likelihood of having more severe allergic reactions to morphine, morphine derivatives, and codeine (Samter & Beers, 1968).

The mechanism by which opiates causes allergic reactions is histamine release. Comparatively, codeine appears to have a more potent histamine-releasing action than morphine in equianalgesic doses (Shanahan, Marshall, and Garrett, 1983). In vitro and in vivo studies have shown that increased doses of morphine cause increased histamine release from human skin that is not antagonized by naloxone. This finding implies that the mechanism of histamine release is not via specific receptors and that naloxone is unlikely to be of value in blocking an allergic phenomenon (McLelland, 1986). Recognizing that allergic reactions to morphine are due to histamine release, potent H_1-antihistamines, such as diphenhydramine or doxepin, are effective inhibitors of the common allergic responses (Sullivan, 1982). More recently, oral ketotifen, a histamine release inhibitor used commonly outside the United States, has been demonstrated to suppress the wheal-and-flare reactions induced by intradermal codeine (Wang, Wang, & Chang, 1985).

When severe pain requires narcotics in a narcotic-allergic patient, there are several options available. Obviously, a nonnarcotic analgesic should be considered, recognizing the hematologic and gastric risks associated with high-dose aspirin and nonsteroidal agents. Another option when narcotics cannot be used is the parenteral phenothiazine analgesic methotrimeprazine (Rogers, 1981). This drug appears to have some analgesic effect (possibly by raising the pain threshold), but published reports are equivocal and do not show the analgesic effect to be independent of the sedative or tranquilizing effects. This drug is described further in the following section.

No research data are available describing the degree of cross-reactivity of allergic responses within the family of opioid narcotics. If one needs to use a narcotic analgesic, an allergic reaction to morphine does not necessarily imply an allergy to a chemically related congener, such as meperidine. As there are other families of narcotic analgesics, an alternative would be to use a member of the phenylpiperidine family (e.g., meperidine) or a member of the agonist-antagonist family (e.g., pentazocine).

SPECIFIC PSYCHOTOXICITIES OF NARCOTICS

From time to time, patients appear to develop problems associated with specific narcotics. Not specifically allergic responses, these problems may be seen as some poorly understood form of "toxicity," such as confusion, restlessness, or some generalized dysphoria. In such cases the patient may respond better by switching to another narcotic at an equianalgesic dose.

All narcotics have some psychotoxic potential, with more specific problems associated with meperidine, pentazocine, nalbuphine, and butorphanol tartrate. Narcotics produce analgesia, drowsiness, changes in mood, and mental clouding, though analgesia should occur without loss of consciousness. As the dose is increased, the subjective effects, including pain relief, become more pronounced. Moreover, in some individuals the euphoric effects become greater. For a given degree of analgesia, the mental clouding produced by therapeutic doses of morphine is considerably less pronounced and of a different character than that produced by alcohol or barbiturates. Morphine and its related analgesics rarely produce the garrulous, jocular, and emotionally labile behavior seen with alcohol or barbiturate intoxication. Finally, high doses of morphine (and of most related congeners) produce convulsions at dose levels in excess of those required for analgesia (Jaffe & Martin, 1980).

Meperidine is a phenylpiperidine analgesic with CNS effects similar to those of morphine, with some differences. The corneal reflex during systemic meperidine analgesia tends to be obtunded or abolished, and some patients develop dysphoria. More specifically, the accumulation of normeperidine, the *N*-demethylated metabolite, especially in patients with renal failure or cancer, can cause CNS excitation with seizures, myoclonus, or toxic psychosis. These CNS excitatory signs generally occur with immeperidine doses of more than 1200 mg/day. The CNS toxicity of meperidine is much more likely to occur when the drug is given IV. With IV administration, it may occur at doses as low as 350 mg/day (Szeto et al., 1977).

Life-threatening reactions may follow the administration of meperidine to patients being treated with monoamine oxidase inhibitors. These reactions consist of hypertension, excitation, delirium, hyperpyrexia, convulsions, respiratory depression, and death (Jaffe & Martin, 1980). Because meperidine blocks serotonin reuptake, there is a theoretical possibility of a serotonin syndrome when combining it with drugs such as fluoxetine or clomipramine. If such cases developed, it would be reasonable to treat them by stopping the drugs and using a serotonin antagonist such as cyproheptadine.

Although there are a few scattered reports of adverse interactions between other narcotics and MAOIs, there is no consistent documentation of a problem. Therefore narcotics other than meperidine may be used concurrently with MAOIs, though it would be judicious to observe the patient carefully for any potential adverse effects.

METHODS TO AUGMENT NARCOTIC ANALGESIA

Although potentiators of narcotic analgesics have been used for more than two decades, there have been relatively few controlled studies of their effectiveness. In most cases, analgesic adjuvants for acute pain are best reserved for situations where it is important for the patient to have a lower narcotic dose because of respiratory suppression, nausea, constipation, sedation, or other side effects.

Hydroxyzine

Hydroxyzine (Atarax; Vistaril) is an antihistamine often used as a coanalgesic. The mechanism of analgesic effects of antihistamines is unknown and remains speculative. These drugs may act as analgesics through interaction with an unknown histaminergic system (i.e., H_3-receptors), through an interaction with biogenic amines or the sympathetic nervous system, or through an interaction with one or more chemical mediators, such as substance P, bradykinin, or opioids (Rumore & Schlichting, 1986). Although it is common clinical practice to prescribe hydroxyzine in 25 mg IM doses in combination with intramuscular narcotics, this practice appears to be based mostly on anecdotal support. Some clinical studies have compared hydroxyzine and morphine to morphine alone, or hydroxyzine and meperidine to meperidine alone, using a dose of 100 mg of hydroxyzine. These studies support an additive analgesic effect for hydroxyzine at that dose (Hupert, Yacoub, & Turgeon, 1980; Stambaugh, 1979). The analgesic effect of 100 mg of hydroxyzine has been demonstrated to be equivalent to about 8 mg of IM morphine (Belville et al., 1979). Finally, when using hydroxyzine with a narcotic analgesic, the dose should be given on the same schedule as the narcotic. The primary side effect is sedation.

Stimulants

Stimulants represent another class of analgesic adjuvants most useful in patients with coexisting sedation, loss of alertness, or depressive symptoms. In a large double-blind study involving 450 patients, Forrest et al. (1977) demonstrated the augmentation of morphine analgesia with either 5 or 10 mg of oral dextroamphetamine. At these doses the analgesia was judged to be, respectively, 1.5 and 2.0 times the analgesia of the given morphine dose alone. Moreover, dextroamphetamine offset the sedating effects of the morphine without significantly changing vital signs (Forrest et al., 1977). Amphetamine coprescribing is not common practice because of several concerns. Tolerance of its alerting effects has been recognized to occur within weeks, leading to the need for higher doses; and at doses of more than 40 mg paranoia and hallucinations have been known to occur (Ellinwood, 1969). Furthermore, amphetamines can suppress sleep and appetite. To avoid the carryover of alerting effects into the evening, dextroamphetamine or methylphenidate should be prescribed starting with 5 mg PO in the morning and at noontime, increasing the dose slowly if necessary. Frail, elderly patients with coexisting heart disease can begin with 2.5 mg morning and noon.

Some of the concerns about the side effects of stimulants may be exaggerated. A retrospective review of 66 medical and surgical patients treated with dextroamphetamine or methylphenidate for depression demonstrated minimal side effects. Of those patients who did develop possibly drug-related side effects, the effects were mild, were reversible with drug discontinuation, and showed no correlation with the dosage prescribed or the age of the patient. These side effects included maculopapular rash, nausea, sinus tachycardia, and confusion in two patients with concurrent dementia (Woods et al., 1986). The use of psychostimulants also is discussed in Chapter 9.

Cocaine

Cocaine is the most powerful euphoriant known and has been touted as an analgesic augmenter for years. In fact, Coca-Cola originally contained cocaine and was marketed as both a stimulant and an analgesic. Unfortunately, despite its intermittent use in Brompton's solution, no study has demonstrated any general analgesic properties of cocaine. Morphine alone has been demonstrated to be as effective as Brompton's mixture in advanced cancer patients (Melzack, Mount, & Gordon, 1979). Cocaine may, however, induce mood elevation that may be interpreted by some people as pain relief (Warfield, 1985).

Steroids

Steroids are important analgesic adjuncts for acute pain syndromes associated with metastatic bone disease, epidural cord compression, headache due to increased intracranial pressure, and tumor infiltration of brachial and lumbar plexuses or other peripheral nerves. Steroids are thought to provide analgesia via their peripheral antiinflammatory effects and their central effects on neurotransmitters (Foley, 1985a). For patients with epidural cord compression, 100 mg of dexamethasone IV given with irradiation provides significant pain relief in up to 85% of patients (Foley, 1985a). After initial treatment, this dose can be tapered and maintained at 16 mg of dexamethasone while irradiation is completed. For nerve infiltration, dosages of 4 to 16 mg of dexamethasone during radiation therapy may significantly reduce pain. Steroids may have the additional benefits of producing increased appetite, weight gain, and improved mood. However, care must be taken with the coadministration of nonsteroidal antiinflammatory drugs because of the increased risk of gastrointestinal (GI) side effects (ulcers and GI bleeding).

Brompton's Mixture

Perhaps the most widely publicized "analgesic cocktail" is the "Brompton cocktail." This combination was first advocated in 1896 by Snow to treat pain associated with advanced cancer and was reintroduced during the twentieth century at the Brompton Chest Hospital in Great Britain. Brompton's original formula consisted of an elixir containing morphine or heroin combined with cocaine in a vehicle of syrup, alcohol, and chloroform water. This elixir often is combined with a phenothiazine, providing agents that control pain and counteract side effects at multiple levels within the nervous system. Some people have thought that heroin may provide analgesia superior to that of the morphine in the cocktail, but this advantage has never been documented (Kaiko et al., 1981). Moreover, there appears to be no advantage of the Brompton cocktail over a simple aqueous morphine solution for treating intractable pain in patients with advanced malignant disease (Melzack, Mount, & Gordon, 1979; Schad, 1980). Finally, Twycross and Lack (1984) stated, "It is more nauseating (due to the syrup content) and may cause a burning sensation in the throat (due to the alcohol)." They concluded, "The traditional mixture of morphine sulfate and cocaine in a vehicle of syrup, alcohol, and chloroform water offers no advantages over a simple solution of morphine sulfate in chloroform or tap water."

Neuroleptics

Neuroleptics represent another class of narcotic potentiators. Despite their widespread use, Moore and Dundee (1961) drew attention to the fact that phenothiazines do not have significant analgesic properties. In fact, no well-controlled study exists to document the coanalgesic effects of phenothiazines in pain patients. The primary use of phenothiazines in terminal cancer patients remains antiemetic or anxiolytic in conjunction with morphine (Goldberg & Cullen, 1986).

A unique member of the phenothiazine group is methotrimeprazine (Levoprome). It has been reported to have two-thirds the analgesic potency equivalent of morphine in single-dose studies using 15 mg (Foley, 1985a). It may be especially useful in patients who have pain and nausea, narcotic tolerance, constipation, respiratory depression, or narcotic allergy. Moreover, its mechanism of action appears to be outside the realm of direct opiate-receptor analgesia (Foley, 1985a). Treatment should begin with a test dose of 5 mg IM to see if significant orthostatic hypotension or sedation results. If not, doses of 10 to 20 mg IM q6h may be used (Foley, 1985a). This drug is available for intramuscular use only.

Fluphenazine, at doses of 1 mg q6–8h in combination with amitriptyline, has been reported to be useful for the treatment of certain neuropathic conditions, such as diabetic neuropathy and postherpetic neuralgia (Davis et al., 1977; Taub, 1973). Again, no controlled studies have documented this effect.

The basis of the use of haloperidol as a coanalgesic largely rests on anecdotal reports. In one controlled study using 34 patients, haloperidol was compared to placebo as a postoperative analgesic in doses of 5 and 10 mg. The results of this study demonstrated no difference in postoperative analgesic, although good sedation and antiemesis were obtained (Judkins & Karmer, 1982). In a retrospective review of 424 cancer patients receiving morphine alone or morphine and haloperidol in doses up to 30 mg, reduction in the need for morphine was seen at both low and high doses of haloperidol (Hanks et al., 1983). Although haloperidol has been reported to be useful as a coanalgesic in doses of 0.5 to 1.0 mg two or three times a day (Foley, 1985a), its use seems most appropriate for the management of agitated delirium in the patient requiring pain management.

Antidepressants

Antidepressants also are used as analgesic adjuvants (Lee & Spencer, 1977). The analgesic effects of antidepressants possibly are mediated centrally by enhancing the transmission in serotonin pathways (Basbaum & Fields, 1978; Carasso, Yehuda, & Streifler, 1979; Watson & Evans, 1985). Moreover, the inability of naloxone to influence their analgesic actions in a rat nociceptive assay supports a nonopiate mechanism (Spiegel, Kolb, & Pasternak, 1983). Despite the first report (Paoli, Darcourt, & Corsa, 1960) describing the use of imipramine for chronic pain, there have been no other studies in humans demonstrating enhancement of morphine analgesia. Naturally, depression itself may lead to an exaggerated perception of pain; and chronic pain may, conversely, lead to depressive symptoms. The demonstration of independent analgesic effects thus is difficult. Amitriptyline and imipramine, however, in doses of 50 to 100 mg, have been shown in various uncontrolled studies to be effective in the treatment of postherpetic neuralgia, headaches, and diabetic neuropathy. In one double-blind study, desipramine, but not amitriptyline, was shown in augment morphine analgesia (Levine et al., 1986).

The analgesic effects of antidepressants have been reviewed by France, Houpt, and Ellinwood (1984). In patients for whom there are no contraindications amitriptyline, imipramine or desipramine may be used as an analgesia potentiator at a starting dose of 25 to 50 mg, increasing every 4 to 7 days by 25 mg, provided anticholinergic and sedative side effects are tolerated. Although there are no clinical studies using trazodone, it also may be a useful coanalgesic in light of its strong serotonergic effects and low anticholinergic profile. Doxepin is primarily active via serotonin mechanisms and tends to have less anticholinergic side effects than amitriptyline; it also might be a rational alternative. (See Chapter 17 for more details.)

Other Agents

Two other drugs may find use as narcotic alternatives or coanalgesics in the future. One controlled study described the successful use of clonazepam to treat neuralgic pain with allodynia (Bouckoms & Litman, 1985). Finally, an experimental cholceystokinin antagonist, proglumide, was studied in 80 human subjects in doses of 50 to 100 mg IV and was found to significantly potentiate the analgesic effects of small doses of morphine (Price et al., 1958). More work along these lines may elucidate new mechanisms for controlling pain. The use of adjuvant analgesics in the management of chronic pain is further discussed in Chapter 17.

Nonnarcotic Analgesics

The adjunctive use of nonnarcotic analgesics should not be overlooked, even in situations of severe acute pain, though the nonnarcotic analgesics generally are used for mild to moderate pain.

Nonsteroidal antiinflammatory drugs (NSAIDs), of which aspirin in the salicylate family is best known, control pain via the inhibition of prostaglandin synthetase function, thus preventing the formation of prostaglandin E_2 (PGE_2). This compound is known to sensitize certain tissues to the pain-producing effects of substances such as bradykinin (Inturrisi, 1985). They also have central effects, which are well understood. Most painful conditions are associated with some inflammation, which is remediated by these drugs. Although marketed for antiarthritic conditions, these drugs have a role in treating a variety of painful conditions, including postoperative pain, dysmenorrhea, headaches, and metastatic bone pain. Moreover, in contrast to narcotic analgesics, NSAIDs do not cause dependence or physical tolerance. These agents, however, can cause upper gastrointestinal bleeding, diminished platelet function, renal dysfunction, hepatitis, and salt and water retention.

The NSAIDs include the following drug families with the following representative members: salicylates (aspirin), paraaminophenol (acetaminophen), pyrroles (indomethocin and sulindac), propionic acids (ibuprofen and naproxen), and oxicams (piroxicam). Of all these drugs, acetaminophen is the least effective antiinflammatory agent, with the pyroles, propionic acids, and oxicams differing from aspirin primarily with respect to their higher analgesic potential and better patient tolerance (Inturrisi, 1985). (See Table 17–2.)

The nonaspirin NSAIDs differ primarily in their pharmacokinetics and duration of analgesic action, with their side effects and toxicities being essentially the same (Amadio, 1984). They can cause CNS side effects, including nervousness, anxiety, insomnia, drowsiness, and visual disturbance (Kruis & Barger, 1980; Sotsky & Tossell, 1984; Steele & Morton, 1986; Thornton, 1980). The most severe CNS side effects are associated with indomethacin and include severe frontal headaches, vertigo, mental confusion, severe depression, psychosis, and hallucinations (Flower, Moncada, & Vane, 1980). The psychotomimetic effect of indomethacin may be related to its serotonin-like chemical configuration (Kantor, 1982).

Although there is a body of literature describing increased analgesic efficacy when combining NSAIDs with narcotic analgesics for alleviating cancer-related pain (Ferrer-Brechner & Ganz, 1984; Weingart, Sorkness, & Earhart, 1985), there are no significant data describing increased analgesic efficacy when combining NSAIDs and antidepressants. Such combinations, however, may be useful in patients with moderate pain and depressive symptomatology. If pain is due to a significant component of inflammation or to bony metastases, pain relief is more likely. By avoiding narcotics if possible for such pain, one avoids the side effects, such as drowsiness and constipation, as well as the possibility of physical dependence and tolerance. (NSAID's are also discussed in detail in Chapter 17.)

Ketorolac Tromethamine

Ketorolac tromethamine (Toradol) is the first member of the group of NSAIDs available for intramuscular injection. Available concentrations are 15% and 30% (15 and 30 mg/ml). Although pain relief may be noted as soon as 10 minutes after injection, peak analgesic effect is at 45 to 90 minutes. The plasma elimination half-life is 4 to 6 hours in young adults and 5 to 9 hours in the elderly. Steady-state plasma levels are reached after 1 day of dosing every 6 hours. To minimize time delay in initial analgesia, a loading dose is recommended, giving twice the maintenance dose as the first dose (i.e., 30 to 60 mg as a loading dose). Because the primary route of excretion is through the kidney, patients with renal impairment may require lower doses.

In acute settings, 30 mg IM of Toradol is about equivalent to 100 mg IM of meperidine or 12 mg IM of morphine. With continued use, 30 mg of Toradol is equivalent to about 9 mg IM of morphine. Side effects are similar to those of other NSAIDs, including drowsiness, and nausea/dyspepsia. Ketorolac may be a useful alternative to narcotics in postoperative patients for whom narcotics are contraindicated because of allergy, respiratory distress, or addiction concerns. (See also Table 17–2.)

NARCOTIC WITHDRAWAL

If narcotics are taken on a regular basis, physical dependence develops. It generally occurs in patients taking oral narcotics for more than 3 to 4 weeks and, along with tolerance, is a characteristic feature of the opioids. Physical dependence on opioids is best defined as a condition that is associated with withdrawal symptoms after an abrupt discontinuation or significant decrease in dosage. The emergence of withdrawal symptoms occurs more rapidly and more intensely with the use of drugs with shorter half-lives. Withdrawal symptoms are best described as a noradrenergic and cholinergic hyperactivity state and include abdominal pain, diarrhea, muscle aching, yawning, rhinorrhea, and lacrimation.

Narcotic withdrawal (or abstinence) symptoms may create an acute management problem and occur on a predictable and observable basis. They usually appear in order of severity. The consultant often is faced with a decision about "covering" the habit of an alleged addict admitted to the hospital. Many addicts exaggerate their habit in order to obtain a free "high" in the hospital. The naive clinician who conscientiously converts what the addict claims to a maintenance dose may inadvertently overdose the patient, as the actual amount may be much less than the patient claims, either because of lying or because the actual amount of narcotic in a "bag" of heroin is less than advertised. It is prudent, therefore, to watch for objective signs of withdrawal before initiating zealous coverage. Objective withdrawal symptoms can be rated using a checklist such as that in Table 16-6. Significant objective symptoms—but not subjective anxiety or drug craving alone—warrant attention with increased narcotic coverage.

There are several options available for discontinuing narcotics in patients who have become physically dependent on them. One possibility is slow weaning from the specific narcotic to minimize the development of withdrawal symptoms. Bedtime doses should be decreased last in order to minimize sleep disruption. For morphine, a 10% to 25% decrease each day is reasonable. The main drawbacks are the need to tolerate some withdrawal symptoms and the extension of drug use over time, especially when the process is initiated on the day the patient is otherwise medically ready to leave the hospital.

Another option is a methadone withdrawal. Methadone was first synthesized by German chemists and used clinically by the late 1940s. The pharmacologic actions of single doses of methadone are qualitatively identical to those of morphine. The outstanding properties of methadone are its effective analgesic activity, its efficacy by the oral route, its extended duration of action in suppressing withdrawal symptoms in physically dependent individuals, and its tendency to show persistent effects with repeated administration (Jaffe & Martin, 1980). A narcotic-dependent patient may be put on an equivalent dose of oral methadone (see Table 16-4). Because of an approximately 15-hour half-life after a single dosage and a 22-hour half-life after chronic administration (Inturrisi & Verebey 1972; Verebey, et al., 1975), the withdrawal symptoms associated with methadone are less intense but more prolonged. The withdrawal process frequently takes up to 30 days, and many drug-dependent individuals require months before they are free of abstinence symptoms.

TABLE 16–6. *Narcotic Withdrawal Checklist*

Name:

Date and time:

Vital signs
- Blood pressure
- Temperature
- Respirations
- Pulse rate

Symptoms
- Opiate craving
- Anxiety
- Yawning
- Perspiration
- Lacrimation
- Rhinorrhea
- Interrupted sleep
- Mydriasis
- Goose flesh
- Tremors
- Hot/cold flashes
- Aching bones/muscles
- Anorexia
- Restlessness
- Nausea/vomiting
- Diarrhea
- Spontaneous orgasm

Dose & Comments:

Use of Clonidine for Narcotic Withdrawal

The attractiveness of using clonidine is that opiates can be withdrawn rapidly. Clonidine is an alpha-2-adrenergic agonist that substantially diminishes withdrawal symptomatology by replacing opiate-mediated inhibition with alpha-2-adrenergic-mediated inhibition of brain noradrenergic activity, primarily in the locus ceruleus (Gold et al., 1980). Clonidine has no affinity for opiate receptors and thus supports a parallel function that opiate receptors and nonadrenergic receptors have in the brain. The disadvantages of using clonidine are the development of sedation and orthostatic hypotension, a slight risk of inducing hallucinations (which stop on drug discontinuation) and some increase in the more subjective signs of withdrawal, including insomnia, restlessness, anorexia, muscle aching, and craving sensations (Brown, Salmon, Rendell, 1980; Jasinski, Johnson, & Kocher, 1985). Additionally, it is possible that the concurrent use of tricyclic antidepressants may antagonize clonidine's opiate withdrawal-blocking effects via the same mechanism by which they are known to antagonize clonidine's antihypertensive action: direct inhibition of alpha-adrenergic receptors (Gold, Redmond, & Kleber, 1978; Haggerty & Slatkoff, 1981).

The technique for clonidine detoxification (see Table 16-7) involves stopping all opiates and beginning clonidine at a q4–6h dosage of 6 μg/kg/24 h (doses of clonidine are rounded to the nearest tenth of a milligram). This dosage may be increased gradually to 17 μg/kg/24 hr (and the schedule may be spread up to every 8 hours) over the next week, depending on the individual's objective and subjective signs of opiate withdrawal. If the signs of opiate withdrawal seem to be absent within a week, clonidine must be tapered over

TABLE 16–7. *Sample Inpatient Clonidine Withdrawal Regimens for 70-kg Man*

Day 1:	0.4 mg PO test dose in morning, with a repeat 0.4 mg PO at bedtime if no significant side effects from clonidine
Day 2–10:	0.5 mg every morning (8 a.m.) 0.2 mg every afternoon (4 p.m.) 0.5 g every evening (11 p.m.)
Day 11–14:	Reduce dose on a daily basis by 50% of the previous day's dose until stopping on day 14.
	For moderate to severe opioid dependence
Day 1:	0.3 mg PO q3h while awake
Day 2:	0.2 mg PO q6h while awake
Day 3:	0.2 mg PO q8h while awake
Day 4:	0.1 mg PO q12h
Day 5:	Discontinue medication
	For habit of less than 7 bags of heroin/day
Day 1:	0.2 mg PO q3h while awake
Day 2:	0.2 mg PO q6h while awake
Day 3:	0.2 mg PO q8h while awake
Day 4:	0.1 mg PO q12h
Day 5:	Discontinue medication

Note: Hypotension, sedation, and dry mouth are the major side effects.
Source: Gelenberg AJ (Ed.) (1985). Clonidine (Catapres) in detoxification. *Biological Therapies in Psychiatry,* 8, 13, 16. With permission.

3 to 4 days to prevent the mild hyperadrenergic state caused by abrupt clonidine discontinuation. During the first 24 to 48 hours of clonidine therapy, close attention must be paid to the patient's blood pressure in order to recognize clinically significant hypotension. The day-by-day withdrawal procedure is well described in more detail in the literature (Charney & Kleber, 1980; Gelenberg, 1985; Traub, 1985). Chapter 8 offers further details.

Two new opiate withdrawal regimens have recently been described in the literature. The first regimen combines clonidine and naltrexone at the onset of the withdrawal regimen (Charney, Heninger, & Kleber, 1986). This regimen is generally too complex for most general hospital settings, especially in patients with concurrent medical illness. In any case, elective narcotic withdrawal should not be undertaken in patients who are acutely ill, as it adds physiologic stress and may complicate the interpretation of signs and symptoms of other disorders.

Another regimen for opiate withdrawal is the use of a clonidine analogue, lofexidine. Lofexidine is still an investigational drug and has been administered on an experimental basis in dosages of 0.4 to 2.0 mg/day in divided doses q2–6h. According to one open trial using 30 opiate-dependent outpatient volunteers, 21 successfully completed the detoxification, with only two of the nine noncompleters attributing their treatment failure to unacceptable withdrawal discomfort. Lofexidine showed significant antiwithdrawal efficacy but no dizziness, profound sedation, or lowering blood pressure, even at does as high as 20 mg/day. Although the mechanisms that differentiate it from clonidine has not been worked out, lofexidine might prove to be a safer, more clinically useful agent than clonidine (Washton, Resnick, & Geyser, 1983). Finally, although clonidine is widely accepted as a treatment for opiate withdrawal, it is not currently recognized by the Food and Drug Administration as a pharmacotherapeutic agent for that purpose.

Treatment of Acute Pain in Narcotic Addicts

The need to treat acute pain in a methadone maintenance patient or in an illicit narcotic addict raises both pharmacologic and systems problems. It is not common for such patients to be over- or undertreated because of a lack of understanding of addiction management.

Treating pain in patients receiving methadone or naltrexone for drug addiction requires knowledge of the pharmacology of these drugs. Methadone is a synthetic opiate agonist that has strong analgesic properties but produces much less euphoria and sedation than morphine. It is used, therefore, in the detoxification and maintenance treatment of addicts as a substitute for other narcotics, to suppress the opiate-agonist abstinence syndrome. Naltrexone is a pure opiate antagonist that attenuates or produces a complete but reversible block of the pharmacologic effects of narcotics, including euphoria and analgesia. It may precipitate mild to severe withdrawn in individuals physically dependent on opiates. It is thought that naltrexone displaces opiates from their CNS receptors in a competitive fashion. Thus it is used in formerly addicted patients to block the euphoric effects of narcotics, thereby eliminating the reinforcement for continued use.

In patients on methadone maintenance, it is useful to keep the management of addiction separate from the management of pain. The methadone can be continued for opiate maintenance therapy and a different narcotic chosen to treat the pain. When methadone maintenance is established, the pain can be treated as if the patient were not an addict, except that a 50% greater than normal analgesic dose must be used for each dose of analgesic (because of tolerance). Generally, the physician or staff are reluctant to be seen as supporting the patient's "habit"; therefore addicted patients often are given too little analgesia, whereas the pharmacology of the addiction actually necessitates a greater analgesic dosage (Goldberg, 1980). Once the underlying cause of pain is resolved, a daily 20% reduction in the methadone dose may be attempted during hospitalization if one wishes to withdraw methadone maintenance. Ideally, the patient can be discharged to a methadone program.

For patients on naltrexone, the use of narcotic analgesics may be ineffective. Unfortunately, no literature has been found to describe the specifics of treatment of severe pain in patients on naltrexone. It is possible to overcome naltrexone blockade of opiate receptors by using large doses of narcotic. The alternatives that are available are limited. Obviously, if one can use a non-narcotic analgesic (e.g., aspirin or an NSAID) with a sedating coanalgesic (e.g., 100 mg IM hydroxyzine), adequate analgesia may be possible. Another possibility is to use methotrimeprazine (Levoprome) in doses up to 20 mg IM q6h, as discussed elsewhere in this chapter.

REFERENCES

AMADIO P JR (1984). Peripherally acting analgesics: In appropriate management of pain in primary care practice (symposium). Am J Med 77:38–53.

AUSTIN KL, STAPLETON JV, & MATHER LE (1980). Multiple intramuscular injections—a major source of variability in analgesic response to meperidine. Pain 8:47–62.

BASBAUM AI, & FIELDS HL (1978). Endogenous pain control mechanisms: Review and hypothesis. Ann Neurol 4:451–462.

BEAVER WT (1980). Management of cancer pain with parenteral medication. JAMA 244:2653–2657.

BEECKER HK (1955). The powerful placebo. JAMA 159:1602–1606.

BELVILLE JW, DOREY F, CAPPARELL D, ET AL. (1979). Analgesic effects of hydroxyzine compared to morphine in man. J Clin Pharmacol 19:290–296.

BOUCKOMS AJ, & LITMAN RE (1985). Clonazepam in the treatment of neuralgic pain syndrome. Psychosomatics 26:933–936.

BROWN MJ, SALMON D, & RENDELL M (1980). Clonidine hallucinations. Ann Intern Med 93:456–457.

CAPLAN RA, & SOUTHAM M (1990). Transdermal drug delivery and its application to pain control. In C BENEDETTI, ET AL. (Eds.), Advances in pain research and therapy (Vol. 14; pp. 223–240). New York: Raven Press.

CARASSO RL, YEHUDA S, & STREIFLER M (1979). Clomipramine and amitriptyline in the treatment of severe pain. Int J Neurosci 9:191–194.

CHARNEY DS, HENINGER GR, & KLEBER HD (1986). The combined use of clonidine and naltrexone as a rapid, safe, and effective treatment of abrupt withdrawal from methadone. Am J Psychol 143:831–837.

CHARNEY DS, & KLEBER HD (1980). Iatrogenic opiate addiction: Successful detoxification with clonidine. Am J Psychol 137:989–990.

CULLEN ML, STAREN ED, EL-GANZOURI, ET AL. (1985). Continuous epidural infusion for analgesia after major abdominal operations: A randomized, prospective, double-blind study. Surgery 98:718–726.

DAVIS JL, LEWIS SB, GERICH JE, ET AL. (1977). Peripheral diabetic neuropathy treated with amitriptyline and fluphenazine. JAMA 238:2291–2292.

ELLINWOOD EH JR (1969). Amphetamine psychosis: A multidimensional process. Semin Psychiatry 1:208–266.

FERRER-BRECHNER T, & GANZ P (1984). Combination therapy with ibuprofen and methadone for chronic cancer pain. Am J Med 77:78–83.

FLOWER RJ, MONCADA S, & VANE JR (1980). Analgesic-antipyretics and anti-inflammatory agents: Drugs employed in the treatment of gout. In AG GILMAN, LS GOODMAN, & A GILMAN (Eds.), The pharmacological basis of therapeutics (pp. 682–728). New York: Macmillan.

FOLEY KM (1982). The practical use of narcotic analgesics. Med Clin North Am 66:1091–1104.

FOLEY KM (1985a). Non-narcotic and narcotic analgesics: Applications. In KM FOLEY (Ed.), Management of cancer pain (pp. 135–148) (syllabus of postgraduate course). New York: Memorial Sloan Kettering Cancer Center.

FOLEY KM (1985b). The treatment of cancer pain. N Engl J Med 313:84–95.

FORREST WH JR, BROWN BW JR, BROWN CR, ET AL. (1977). Dextroamphetamine with morphine for the treatment of postoperative pain. N Engl J Med 296:712–715.

FRANCE RD, HOUPT JL, & ELLINWOOD EH (1984). Therapeutic effects of antidepressants in chronic pain. Gen Hosp Psychiatry 6:55–63.

GARRETT JM, SAUER WG, & MOERTEL CG (1967). Colonic motility in ulcerative colitis after opiate administration. Gastroenterology 53:93–100.

GELENBERG AJ (Ed.) (1985). Clonidine (Catapres) in detoxification: Caveats and issues. Biol Ther Psychiatry 8:15–16.

GOLD MS, POTTASH AC, SWEENEY DR, ET AL. (1980). Opiate withdrawal using clonidine. JAMA 243:343–346.

GOLD MS, REDMOND DE JR, & KLEBER HD (1978). Clonidine blocks acute opiate withdrawal symptoms. Lancet 2:599–602.

GOLDBERG RJ (1980). Strategies in psychiatry for the primary physician. Darien, CT: Patient Care Publications.

GOLDBERG RJ, & CULLEN LO (1986). Use of psychotropics in cancer patients (No. 8 in a series). Psychosomatics 27:687–700.

GOLDBERG RJ, LEIGH H, & QUINLAN D (1979). The current status of placebo in hospital practice. Gen Hosp Psychiatry 1:196–201.

GOLDBERG RJ, MOR V, WIEMANN M, ET AL. (1986). Analgesic use in terminal cancer patients: Report from the National Hospice Study. J Chronic Dis 39:37–45.

GOLDBERG RJ, & TULL RM (1983). The psychosocial dimensions of cancer. New York: Free Press.

GOODWIN JS, GOODWIN JM, & VOGEL AV (1979). Knowledge and use of placebos by house officers and nurses. Ann Intern Med 91:106–110.

GOURLAY GK, CHERRY DA, & COUSINS MJ (1986). A comparative study of the efficacy and pharmacokinetics of oral methadone and morphine in the treatment of severe pain in patients with cancer. Pain 25:297–312.

GRAVES DA, FOSTER TS, BATENHORST RL, ET AL. (1983). Patient-controlled analgesia. Ann Intern Med 99:360–366.

HAGGERTY JL, & SLATKOFF S (1981). Clonidine therapy and meperidine withdrawal. Am J Psychiatry 138:698.

HANKS GW, THOMAS PJ, TRUEMAN T, ET AL. (1983). The myth of haloperidol potentiation. Lancet 2:523–524.

HEEL RC, BROGDEN RN, SPEIGHT TM, ET AL. (1978). Butorphanol: A review of its pharmacological properties and therapeutic efficacy. Drugs 16:473–505.

HOUDE RW (1979). Analgesic effectiveness of the narcotic agonist-antagonists. Br J Clin Pharm 7:2975–3085.

HOUDE RW, WALLENSTEIN SL, & BEAVER WT (1965). Clinical measurements of pain. In G de Stevens (Ed.), Analgetics. Orlando, FL: Academic Press.

Hupert C, Yacoub M, & Turgeon LR (1980). Effect of hydroxyzine on morphine analgesia for the treatment of postoperative pain. Anaesth Analg 59:690–696.

Inturrisi CE (1982). Narcotic drugs. Med Clin North Am 66:1061–1071.

Inturrisi CE (1985). Non-narcotic and narcotic analgesics. In KM Foley (Ed.), Management of cancer pain (pp. 135–148). New York: Memorial Sloan Kettering Cancer Center.

Inturrisi CE, & Verebey K (1972). Disposition of methadone in man after a single oral dose. Clin Pharmacol Ther 13:923–930.

Jaffe JH, & Martin WR (1980). Opioid analgesics and antagonists. In AB Gilman, LS Goodman, & A Gilman (Eds.), The pharmacological basis of therapeutics. New York: Macmillan.

Jasinski DR, Johnson RE, & Kocher TR (1985). Clonidine in morphine withdrawal. Arch Gen Psychiatry 42:1063–1066.

Judkins KC, & Karmer M (1982). Haloperidol as an adjunct in the management of postoperative pain. Anaesthesia 37:1118–1120.

Kaiko RF, Rogers AG, Wallenstein SL, et al. (1981). Analgesic and mood effects of heroin and morphine in cancer patients with postoperative pain. N Engl J Med 304:1501–1504.

Kaiko RF, Wallenstein SL, Rogers AG, et al. (1982). Narcotics in the elderly. Med Clin North Am 66:1079–1089.

Kane FJ, & Pokorny A (1975). Mental and emotional disturbance with pentazocine (Talwin) use. South Med J 68:808–811.

Kantor TG (1982). Control of pain by nonsteroidal anti-inflammatory drugs. Med Clin North Am 66:1053–1059.

Kruis R, & Barger R (1980). Paranoid psychosis with sulindac [letter to the editor]. JAMA 243:1420.

Lee R, & Spencer PSJ (1977). Antidepressants and pain. J Intern Med Res 5(suppl 1):146–156.

Levine JD, Gordon NC, & Fields HL (1978). The mechanism of placebo analgesia. Lancet 2:654–657.

Levine JD, Gordon NC, Smith R, et al. (1986). Desipramine enhances opiate postoperative analgesia. Pain 27:45–49.

Levy JH, & Rockoff MA (1982). Anaphylaxis to meperidine. Anaesth Analg 61:301–303.

Maguire LC, Yon JL, & Miller E (1981). Prevention of narcotic-induced constipation [letter]. N Engl J Med 305:1651.

Marks RM, & Sachar EJ (1973). Undertreatment of medical inpatients with narcotic analgesics. Ann Intern Med 78:173–181.

Martin WR, Jasinski DR, Haertzen CA, et al. (1973). Methadone—a re-evaluation. Arch Gen Psychiatry 28:286–295.

McLelland J (1986). The mechanism of morphine-induced urticaria. Arch Dermatol 122:138–139.

Melzack R, Mount BM, & Gordon JM (1979). The Brompton mixture versus morphine solution given orally: Effects of pain. Can Med Assoc J 115:435–438.

Moertel CG (1980). Treatment of cancer pain with orally administered medications. JAMA 244:2448–2450.

Moldofsky H, Scarisbrick P, England R, et al. (1975). Musculoskeletal symptoms of non-REM sleep disturbance in patients with "fibrositis syndrome" and healthy subjects. Psychosom Med 38:341–351.

Moore J, & Dundee JW (1961). Alterations in response to somatic pain associated with anaesthesia: 7. The effects of nine phenothiazine derivatives. Br J Anaesth 33:422–431.

Nagashima H, Karamanian A, Malovany R, et al. (1976). Respiratory and circulatory effects of intravenous butorphanol and morphine. Clin Pharmacol Ther 19:738–745.

Paoli F, Darcourt G, & Corsa P (1960). Note préliminaire sur l'action de l'imipramine dans les états douloureux. Rev Neurol (Paris) 102:503–504.

Parolaro D, Sala M, & Gori E (1977). Effect of intracerebroventricular administration of morphine upon intestinal motility in rats and its antagonism with naloxone. Eur J Pharmacol 46:329–338.

Popio KA, Jackson DH, Ross AM, et al. (1978). Hemodynamic and respiratory effects of morphine and butorphanol. Clin Pharmacol Ther 23:281–287.

Porter J, & Jick H (1980). Addiction rare in patients treated with narcotics. N Engl J Med 302:123.

Portenoy RK (1986). Continuous infusion of opioid drugs in the treatment of cancer pain: Guidelines for use. J Pain Symptom Management 1:223–228.

Price DD, von der Gruen A, Miller J, et al. (1985). Potentiation of systemic morphine analgesia in humans by proglumide, a cholecystokinin antagonist. Anaesth Analg 64:801–806.

Reuler JB, Girard DE, & Nardone DA (1980). The chronic pain syndrome: Misconceptions and management. Ann Intern Med 93:588–596.

Rogers A (1981). Twenty-one problems in pain control and ways to solve them (pp. 65–75). Your Patient and Cancer. September.

Rumore MM, & Schlichting DA (1986). Clinical efficacy of antihistaminics as analgesics. Pain 25:7–22.

Samter M, & Beers RF (1968). Intolerance to aspirin: Clinical studies and consideration of its pathogenesis. Ann Intern Med 68:975–983.

Schad R (1980). Brompton's mixture for chronic pain. South Med J 73:1420–1421.

Shanahan EC, Marshall AG, & Garrett COP (1983). Adverse reactions to intravenous codeine phosphate in children. Anaesthesia 38:40–43.

Shimm DS, Logue GL, Maltbie AA, et al. (1979). Medical management of chronic cancer pain. JAMA 241:2408–2412.

Snow H (1896). Opium and cocaine in the treatment of cancerous disease. Br Med J 11:718.

Sotsky SM, & Tossell JW (1984). Tolmetin induction of mania. Psychosomatics 25:626–628.

Spiegel K, Kolb R, & Pasternak GW (1983). Analgesic activity of tricyclic antidepressants. Ann Neurol 13:462–465.

Sriwatanakul K, Weis OF, Alloza JL, et al. (1983). Analysis of narcotic analgesic usage in the treatment of postoperative pain. JAMA 250:926–929.

Stambaugh JE (1979). Pharmacologic and pharmacokinetic consideration in analgesic regimens. Hosp Pract 14:35–39.

Steele TE, & Morton WA Jr (1986). Salicylate-induced delirium. Psychosomatics 27:455–456.

Sullivan TJ (1982). Pharmacologic modulation of the whealing response to histamine in human skin: Identification of doxepin as a potent in vivo inhibitor. J Allergy Clin Immunol 69:260–264.

Szeto HH, Inturrisi CE, Houde R, et al. (1977). Accumulation of normeperidine, an active metabolite of meperidine, in patients with renal failure or cancer. Ann Intern Med 86:738–741.

Taub A (1973). Relief of postherpetic neuralgia with psychotropic drugs. J Neurosurg 39:235–239.

Taylor M, Galloway DB, Petrie JC, et al. (1978). Psychotomimetic effects of pentazocine and dihydrocodeine tartrate. Br Med J 2:1198.

Thornton TL (1980). Delirium associated with sulindac [letter to the editor]. JAMA 243:1630–1631.

Traub SL (1985). Clonidine for opiate withdrawal. Hosp Formulary 1:77–80.

Twycross RG, & Lack SA (1984). Therapeutics in terminal cancer (Ch. 9). London: Pitman Publishing.

Twycross RG, & Lack SA (1986). Oral morphine in advanced cancer. Beaconsfield, Bucks, England: Beaconsfield Publishers.

Vandam LD (1980). Butorphanol. N Engl J Med 302:381–384.

Verebey K, Volavka J, Mule S, et al. (1975). Methadone in man: Pharmacokinetic and excretion studies in acute and chronic treatment. Clin Pharmacol Ther 18:180–190.

Walsh TD (1984). Oral morphine in chronic cancer pain. Pain 18:1–11.

Wang SSM, Wang SR, & Chang BN (1985). Suppressive effects of oral ketotifin on skin responses to histamine, codeine, and allergen skin tests. Ann Allergy 55:57–61.

Warfield CA (1985). Psychotropic agents for pain control: Clinical guidelines. Hosp Pract 20:141–143.

Washton AM, Resnick RB, & Geyer G (1983). Opiate withdrawal using lofexidine, a clonidine analogue with fewer side effects. J Clin Psychiatry 44:335–337.

Watson CPN, & Evans RJ (1985). A comparative trial of amitriptyline and zimelidine in post-herpetic neuralgia. Pain 23:387–394.

Weingart WA, Sorkness CA, & Earhart RH (1985). Analgesia with oral narcotics and added ibuprofen in cancer patients. Clin Pharm 4:53–58.

White PF (1985). Patient controlled analgesia: A new approach to the management of postoperative pain. Semin Anaesth 4:255–266.

Woods SW, Tesar GE, Murray GB, et al. (1986). Psychostimulant treatment of depressive disorders secondary to medical illness. J Clin Psychiatry 47:12–15.

17 Chronic pain management

RUSSELL K. PORTENOY, M.D.

Chronic pain is a highly prevalent clinical problem experienced by an extraordinarily heterogeneous patient population. Clinicians of every discipline encounter patients who describe frequent or persistent pain as either the primary condition for which they seek treatment (e.g., chronic headache or low back pain) or a distressing complication of an underlying medical disorder (e.g., rheumatoid arthritis or cancer). In many disciplines, pain management has evolved as a subspecialty to meet the complex needs of these patients.

The management of chronic pain requires specific skills, including the ability to perform a comprehensive pain assessment and implement a multimodal treatment approach designed to address the pain and associated physical, psychosocial, and behavioral problems of the patient. These elements of pain management are discussed in this chapter with particular reference to the population with chronic nonmalignant pain. The comprehensive management of cancer pain has been reviewed elsewhere (Foley, 1985; Portenoy & Foley, 1989; World Health Organization, 1986) and is also discussed in Chapter 24.

DEFINITION OF PAIN

Pain has been defined by the International Association for the Study of Pain as "an unpleasant sensory and emotional experience associated with actual or potential tissue damage, or described in terms of such damage" (Merskey, 1986). Implicit in this definition is the recognition that the relation between tissue damage and pain is not uniform or predictable.

Nociception, Pain, and Suffering

Following a careful assessment, the pain reported by the patient may be considered to be appropriate, less than expected, or excessive for the degree of tissue damage demonstrable. This observation can be clarified by the distinctions among nociception, pain, and suffering (Loeser & Egan, 1989).

Nociception is the activity produced in the afferent nervous system by potentially tissue-damaging stimuli (Besson & Chaouch, 1987; Willis, 1985). These stimuli may be mechanical, thermal, or chemical; when nociception is persistent, the chemical mediators of inflammation probably are involved.

Pain is the perception of nociception. Similar to other perceptions, pain is determined by more than the activity induced in the sensorineural apparatus by external or internal stimuli. Just as extensive tissue damage can exist in the absence of pain (e.g., immediately following acute trauma), pain can be perceived by the clinician to be disproportionate to the nociception apparent during clinical evaluation. As discussed more fully below, this nonnociceptive pain may be considered to have an unequivocal organic cause (e.g., neuropathic pains such as postherpetic neuralgia), a likely organic cause that remains to be identified, a likely psychological cause, or some combination of these processes. In many cases, however, a comprehensive assessment fails to yield enough evidence of underlying pathophysiology to permit a diagnosis; these pains are best termed "idiopathic" (Arner & Myerson, 1988).

It is important to recognize that most nonnociceptive pains, including those believed to have a predominantly psychological cause and nearly all of those that remain idiopathic, are not factitious or reported by malingerers. These pains are "real" and involuntary experiences for the patient. The challenge of the pain assessment is to clarify the contribution of the nociceptive and nonnociceptive factors that may be sustaining the pain, thereby targeting treatment appropriately. Interventions that attempt to lessen nociception but neglect nonnociceptive contributions to the pain can lead to a poor outcome, in which tissue damage is ameliorated but pain continues.

Recognition of the diverse nociceptive and nonnociceptive contributions to pain has led to several models that attempt to clarify various dimensions of pain. Melzack and Casey (1968) postulated that there are three broad dimensions to the pain experience—termed sensory-discriminative, motivational-affective, and cognitive-evaluative—and that each may be subserved by distinct somatosensory pathways in the central nervous system (CNS). The sensory-discriminative dimension reflects that adaptive, informational component that may characterize some types of pain (particularly acute pain due to tissue damage), in which the qualities of the pain clarify the spatial, temporal, and quantitative aspects of a noxious stimulus. This information probably is transmitted in rapidly conducting afferent pathways, which have been well characterized in animal models of nociception (Besson & Chaouch, 1987; Willis, 1985). The motivational-affective dimension, which reflects the reactive component of pain, may be mediated by polysynaptic afferent pathways that are known to interconnect with brainstem reticular neurons and the limbic system. Inputs into the cortex presumably mediate the cognitive-evaluative dimension, but little is known of the neuroanatomical substrate of this aspect of pain.

Suffering is a more global experience that can usefully be likened to impaired quality of life (Portenoy, 1990b). The degree of suffering may be determined by any of numerous problems in addition to pain (e.g., other symptoms, loss of physical function, familial dissolution, or financial concerns) or by the experience of a concurrent psychological disorder (e.g., depression or anxiety). Evaluation of these factors is another component of pain assessment. Just as an exclusive focus on nociception, rather than pain, may suggest treatments that ultimately reduce tissue damage without enhancing comfort, an emphasis on pain management alone in patients with profound suffering determined by other factors may fail to improve the overall quality of life even if pain is relieved.

Chronic Pain

There have been numerous attempts to define chronic pain; and, to some extent, inconsistencies in the nomenclature have confused matters in the clinical setting. A definition that encompasses many earlier ones designates a pain as chronic if it persists for 1 month beyond the usual course of an acute illness or a reasonable duration for an injury to heal, if it is associated with a chronic pathologic process, or if it recurs at intervals for months or years (Bonica, 1990). The variable characteristics of patients with chronic pain are discussed in following sections.

CATEGORIES OF PATIENTS WITH PAIN

A useful perspective on chronic pain is gained from a brief description of the various categories of patients who seek treatment for unrelieved pain. Although the broad distinctions among these categories are clinically relevant, they should not obscure the enormous variability of patients within any category.

Acute Monophasic Pain

Acute pains are short-lived or are anticipated to be short-lived given the natural history of the underlying pathological process. Acute pains that require clinical intervention are typically associated with surgery, major trauma, and burns. Notwithstanding the physiologic and psychologic complexities of these pains (Cousins, 1989), the clinical assessment usually focuses on the obvious physical pathology, and treatment emphasizes the provision of comfort through the administration of opioid drugs. The potential efficacy of pain treatment is high, but undertreatment is common (Edwards, 1990; Perry & Heidrich, 1982).

Recurrent Acute Pains

Pain syndromes characterized by recurrent acute pains are both prevalent and diverse. Such pains include headache, dysmenorrhea, and pain associated with sickle cell anemia, inflammatory bowel disease, and some arthritides or musculoskeletal disorders (e.g., hemophilic arthropathy). As noted previously, some experts label these disorders as another form of chronic pain. Although it is clearly appropriate in some cases, many patients benefit from an approach to assessment and therapy that is more appropriate to repeated discrete episodes of acute pain than chronic pain of any other type. The overall approach to the patient, and hence the most informative label (recurrent acute pain versus chronic pain), should be determined by the degree to which the individual manifests the physical, psychosocial, and behavioral characteristics that parallel the disturbances observed in chronic pain syndromes.

Chronic Pain Associated with Cancer

A comprehensive assessment is essential in patients with chronic cancer pain. This assessment must elucidate the underlying organic lesion that is presumed to exist in almost all of these patients (Gonzales et al., 1991) and clarify the relation between pain and suffering. Through this process, an opportunity for effective antineoplastic therapy may be discovered, and primary analgesic therapies can be combined with other appropriate treatments. Like acute monophasic pain, opioid therapy is the major analgesic approach in patients with cancer pain (Foley, 1985; Health and Public Policy Committee, 1983; McGivney & Grooks, 1984; Portenoy & Foley,

1986; Swerdlow & Stjernsward, 1982; World Health Organization, 1986).

Chronic Pain Due to Progressive Nonmalignant Medical Diseases

Chronic pain can occur in the setting of progressive medical diseases other than cancer. Examples are the chronic pains that may develop in patients with sickle cell anemia, hemophilia, and some connective tissue diseases. In these patients, like those with cancer, the extent and course of the underlying disease and its psychologic implications must be addressed as elements of pain therapy. There should be an ongoing assessment of these processes, and treatments targeted at the underlying pathology should be considered repeatedly. At the same time, however, the clinician must guard against the tendency to minimize the role of psychosocial and behavioral disturbances in the expression of the pain. Although organic factors are important in these patients and may be the only issue of concern in some, the potential relevance of psychosocial and rehabilitative concerns must be recognized. Similar to chronic nonmalignant pain of other types, the therapy should have dual goals, namely comfort and rehabilitation.

Chronic Pain Associated with a Nonprogressive Organic Lesion

Many patients experience pain associated with an organic lesion that is neither rapidly progressive nor life-threatening. Included in this category are numerous musculoskeletal pain syndromes (e.g., treatment-refractory osteoporosis and spondylolisthesis) and neuropathic pain syndromes (e.g., postherpetic neuralgia, painful polyneuropathy, central pain, and reflex sympathetic dystrophy). In these patients an intensive initial assessment that characterizes the underlying organic contribution to the pain is usually best followed by a shift of the therapeutic focus away from ongoing assessment of the organic disease and onto symptomatic improvement and rehabilitation.

Chronic Nonmalignant Pain Syndrome

As discussed previously, a large group of patients experience pain that is both perceived to be excessive for the degree of evident organic pathology and is associated with psychologic and behavioral disturbances that are presumed to contribute to the pain and disability. Occasional patients experience chronic pain in the complete absence of an organic lesion. Specific appellations have been applied to these syndromes based on the site of the pain, including atypical facial pain, failed low back syndrome, chronic tension headache, and chronic pelvic pain of unknown etiology. Some of these patients fulfill criteria for discrete psychiatric disorders, such as hypochondriasis or somatization disorder (American Psychiatric Association, 1980). Only a small proportion, however, have no identifiable organic lesion, and the existence of a concurrent organic process greatly augments the diagnostic challenge for the clinician. (Pain associated with somatoform disorders is addressed in Chapter 13.) To a large extent, the multidisciplinary approach to pain management evolved to address the complex physical and psychologic needs of these patients.

PRINCIPLES OF PAIN ASSESSMENT

Given the complexity of chronic pain, a comprehensive assessment should be considered an essential element of management. This assessment clarifies the organic and psychologic contributions to the pain and characterizes the range of problems—physical, psychological, sexual, familial, social, financial, and so on—that also may require treatment. In some patients with multiple interacting problems, a useful way of conceptualizing the goal of this assessment is the development of a pain-oriented problem list (see Figure 17-1).

The information required to develop the pain-oriented problem list derives from the history, physical examination, and selected laboratory and radiographic procedures. The history of the pain must include reference to temporal features (onset, course and daily pattern), location, severity, quality, and factors that provoke or relieve it. The history of the present illness should be supplemented by queries intended to elicit a past history of persistent pain, prior pain treatments, and previous use of licit drugs (including alcohol, tobacco, and over-the-counter and prescription medicines) and illicit drugs. A psychosocial assessment should characterize premorbid mental illness or personality disorder, coping styles demonstrated during earlier episodes of physical disease or psychological stress, the current mental state with particular reference to anxiety and depression, current resources (social, familial, and financial), and present functional status. The patient's activities during the day should be enumerated to help clarify the degree of physical inactivity and social isolation.

The physical examination of patients with chronic pain should aim to clarify the underlying organic contributions to the pain. If further evaluation is necessary, the information obtained from the history and this examination can guide the selection of appropriate laboratory and imaging procedures to confirm and char-

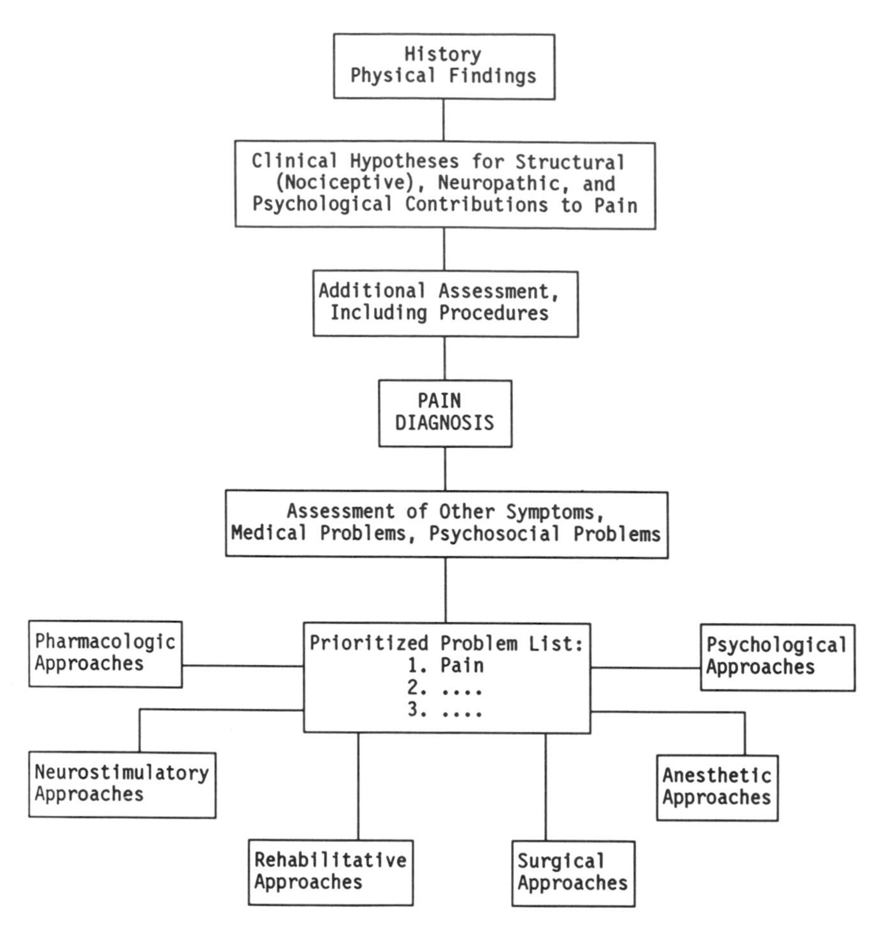

FIG. 17–1. Pain-oriented problem list.

acterize these problems. Likewise, the initial assessment may suggest the need for referral to other specialists, who may be able to better characterize the underlying organic processes or provide guidance to the management of specific types of pain. For example, evaluation of a neuropathic pain syndrome requires a detailed neurologic assessment, which may necessitate referral to a specialist who can clarify the nature of the underlying neurologic lesion and may offer suggestions for primary therapy that may have analgesic consequences.

It must be emphasized that the discovery of a lesion during this comprehensive assessment does not indicate that the predominating pathogenesis for the pain is organic. A competent evaluation identifies potentially treatable organic conditions and clarifies the degree to which pain and disability can be ascribed to these factors or to other identifiable pathology, including psychologic disturbances.

The information obtained from this assessment may be further illuminated by consideration of several specific clinically relevant aspects of the pain. The most important are the following.

Temporal Features

The previous discussion of pain definition and patient categories emphasized that the distinction between acute and chronic pain. To some extent, these temporal features determine the phenomenology of the pain and the options for therapy. Acute pain usually has a well- defined onset, a readily identifiable cause (e.g., surgical incision), and a duration anticipated to be no more than the time required for the injury to heal, usually less than several weeks. There is an association between acute pain and specific pain behaviors (e.g., moaning, grimacing, and splinting of the painful part), signs of sympathetic hyperactivity (including tachycardia, hypertension, and diaphoresis), and the affect of anxiety.

In contrast, chronic pain is usually characterized by an ill-defined onset and a duration that has either already exceeded the healing period of the inciting lesion or is anticipated to continue indefinitely into the future. Pain behaviors associated with acute pain are absent, and typically there are no signs of sympathetic hyperactivity. Sleep disturbance is common, and some patients develop other vegetative signs, such as lassitude or anorexia. A clinical depression evolves in some patients, but this phenomenon is not uniform (Romano & Turner, 1985).

Among patients with chronic pain, different temporal profiles also occur. Most patients have fluctuating pain with some pain-free intervals. Even those with continuous pain experience large fluctuations in the intensity of pain or have acute episodes of severe exacerbation that the patient perceives as distinct from the baseline pain. In a survey of patients with cancer, for example, almost two-thirds experienced transitory flares of pain, which varied greatly in duration, frequency, and quality (Portenoy & Hagen, 1990). Thus it is common for patients with chronic pain to have periods that present as superimposed acute pain. These temporal profiles may be important in the development of a therapeutic approach to the patient.

Topographic Features

Some topographic features of the pain have great clinical relevance. For example, the distinctions among focal pains, multifocal pains, and generalized pains may determine the potential utility of some treatments, such as nerve blocks and cordotomy, which depend on the specific location and extent of the pain.

The distinction between focal and referred pain is similarly important. Focal pains are experienced superficial to the underlying organic lesion, whereas referred pains are experienced at a distant site. Numerous subtypes of referred pain can be distinguished, including pain referred anywhere along the course of an injured peripheral nerve (such as pain in the foot from a sciatic nerve lesion); pain referred along the course of a nerve supplied by a damaged nerve root (known as radicular pain); pain referred to the lower part of the body, usually the feet and legs, from a lesion involving the spinal cord; and pain referred to a site remote from the lesion and outside the dermatome affected by the lesion. In the latter case, the causative lesion may involve any of diverse tissues (Kellgren, 1939; Torebjork, Ochoa & Schady, 1984). For example, shoulder pain may arise owing to compression of the median nerve at the wrist, irritation of the ipsilateral diaphragm, or trigger points in the sternocleidomastoid muscle. Knowledge of pain referral patterns is needed to target appropriate assessment procedures.

Etiological Features

Whenever possible, the organic pathology believed to contribute to the pain, or to explain it fully, should be identified. The importance of this consideration is particularly evident in patients with cancer pain and those with pain related to other progressive medical diseases. In these patients, identification of a lesion may present options for primary therapy that could have analgesic consequences. A survey of patients with cancer pain noted that the pain evaluation identified previously unsuspected lesions in 63% of patients, approximately 20% of whom received a new or additional primary therapy as a result (Gonzales et al., 1991).

Syndromic Features

Syndrome identification during pain assessment may provide information about underlying organic processes, suggest an efficient evaluation, guide the selection of treatments, and indicate prognosis. The development of criteria for syndrome identification has been one of the major goals of the taxonomy for pain that has been developed by the International Association for the Study of Pain (Merskey, 1986).

Pathophysiologic Features

Increasing attention has focused on the classification of pain on the basis of putative mechanisms. Although it is likely that most chronic pain is multiply determined by a complex interaction between organic and psychologic disturbances, it is useful to label a syndrome according to the predominating mechanism imputed to be involved.

Pain syndromes perceived to have a predominating organic contribution have been termed nociceptive or neuropathic. Nociceptive pain is believed to be commensurate with the degree of ongoing tissue damage from an identifiable peripheral lesion that involves either somatic or visceral structures. Nociceptive pain originating from somatic structures (somatic pain) is typically described as aching, stabbing, throbbing, or pressure-like. The quality of visceral pain is determined by the structures involved: obstruction of a hollow viscus usually causes a gnawing or crampy pain, whereas distention or torsion of organ capsules or the mesentery produces pain typically described as aching. Designation of pain as nociceptive has clinical implications, as pain of this type can often be reduced through interventions that improve the peripheral nociceptive lesion. For example, joint replacement can usually relieve refractory arthritic pain.

Neuropathic pain is related to aberrant somatosensory processes induced by an injury to the peripheral or central nervous system (Portenoy, 1991). Several major subtypes can be identified, including those in which the abnormal focus of neural activity responsible for perpetuating the pain is in the peripheral nervous system ("peripheral neuropathic pain"), those in which the focus is in the CNS ("deafferentation pain"), and those in which the pain is dependent on efferent activity in the sympathetic nervous system ("sympathetically maintained pain"). The pains are often dysesthetic (abnormal pain, unfamiliar to the patient) and disproportionate to any evident nociceptive lesion. Identification of a neuropathic mechanism is important in clinical management, as a number of specific therapies may be particularly useful in this situation (see below).

As discussed previously, most patients who lack the criteria to fulfill a pathophysiologic diagnosis of nociceptive or neuropathic pain can be said to have an idiopathic pain syndrome. Some of these patients have pain that is believed to have a predominantly psychologic pathogenesis, including a subgroup who meet criteria for specific psychiatric diagnoses.

An effort to extend this concept of pathophysiologic classification of chronic pain patients has attempted to integrate medical-physical, psychosocial, and behavioral dimensions of the pain. Specifically, a polydiagnostic approach to classification has been proposed (Turk & Rudy, 1990). In this approach, a system termed the multiaxial assessment of pain (Turk & Rudy, 1988), which is an empirically derived psychosocial and behavioral classification that categorizes patients into three groups ("dysfunctional," "interpersonally distressed," and "adaptive copers"), is combined with a medical-physical classification, such as the aforementioned taxonomy of pain developed by the International Association for the Study of Pain (Merskey, 1986). The objective is to develop a pathophysiologic classification that could be useful for targeting a multimodal therapeutic approach based on the character of these various disturbances. This approach has yet to be validated in clinical practice, but it affirms the complexity of chronic pain patients and the need to assess the various processes contributing to both the expression of pain and the patient's response to it.

Management of Chronic Pain

Six broad therapeutic approaches, each comprising numerous treatments, may be explored in the management of chronic pain (see Table 17-1). The comprehensive assessment guides the selection of a multimodal treatment approach. In some situations, specifically those in which chronic pain is experienced without severe disability or psychologic disturbance, the optimal approach to the pain may involve administration of one or more treatments by a single knowledgeable clinician. For example, most patients with cancer pain and many of those with recurrent vascular headache can be well managed by the individual practitioner using drugs alone. If such a patient might benefit from the use of an analgesic modality outside the purview of the clinician, referral to another individual clinician for this modality would be appropriate.

TABLE 17–1. *Approaches Used for the Management of Chronic Pain*

Primary therapies directed against the underlying etiology
Primary analgesic therapies
Pharmacologic approaches
Anesthetic approaches
Surgical approaches
Physiatric approaches
Neurostimulatory approaches
Psychologic approaches

Should the assessment indicate that the pain patient has a chronic pain syndrome, that is, pain associated with affective and behavioral disturbances, with a high degree of disability, optimal care may require the development of a multidisciplinary, multimodal approach that emphasizes both patient comfort and the rehabilitation of physical, psychologic, and social functioning. Importantly, many such patients believe that the only goal of therapy is relief of pain, and that comfort alone can immediately and dramatically improve function. This outcome is unlikely to occur in the patient who has been disabled by pain, and an effort must be made from the start of treatment to impose a clinical agenda that includes physical and psychosocial rehabilitation as one of the major goals of therapy.

Thus the traditional multidisciplinary pain management approach, which is now promulgated through a large number of pain clinics throughout the United States and other countries, should be considered potentially optimal care for a subpopulation of patients with chronic nonmalignant pain associated with disturbances in diverse areas of function. Many of these patients present such complex interactions among symptoms, physical impairments, and functional disturbances that the skills of experienced clinicians in several disciplines are needed to adequately assess clinical needs and elaborate a multimodal therapeutic approach. It is the responsibility of the primary clinician to perform a comprehensive pain assessment, and thereby determine the need for this approach.

Pharmacological Approaches

Expertise in several therapeutic approaches provides the foundation for the clinician to actively participate in the management of the patient. For the psychiatrist, these approaches usually comprise pharmacologic therapies and psychological interventions. There have been many advances in the pharmacotherapy of pain, and this approach is emphasized here.

The drugs used to treat pain can be divided into three broad categories: (1) NSAIDs; (2) adjuvant analgesics; and (3) opioids. Adjuvant analgesics can be defined as drugs that have primary indications other than pain but are analgesic in selected circumstances. Along with NSAIDs, the adjuvant analgesics are emphasized in the management of chronic nonmalignant pain.

NSAIDs

The NSAIDs are believed to provide analgesia through a peripheral mechanism mediated by inhibition of the enzyme cyclooxygenase; this activity reduces inflammatory mediators known to sensitize or activate peripheral nociceptors (Higgs & Moncada, 1983; Vane, 1971). There is substantial evidence, however, that these drugs also operate through a central mechanism (Willer et al., 1989). These central effects may provide an explanation for the marked disproportion between the antiinflammatory effects and the analgesia observed with some of these drugs, such as acetaminophen.

Analgesia with NSAIDs is characterized by a ceiling dose, beyond which additional increments fail to yield greater pain relief, and by a lack of demonstrable physical dependence or tolerance. It is important to recognize that *the ceiling dose, like the minimal effective dose and the dose associated with adverse effects, is unknown in any individual patient and may be higher or lower than the standard recommended dose.* As discussed below, this observation has important implications for the clinical use of these drugs.

Numerous NSAIDs are currently available (Sunshine & Olson, 1989) (see Table 17-2). Specific guidelines for the selection and administration of these drugs can be summarized as follows.

Drug Selection. The NSAIDs are useful independently for mild to moderate pain and provide additive analgesia when combined with opioid drugs for the treatment of severe pain (Ferrer-Brechner & Ganz, 1984). Relatively few NSAIDs have been specifically approved by the U.S. Food and Drug Administration (FDA) as analgesics, but clinical experience suggests that any can potentially be useful for this indication.

All NSAIDs should be used cautiously in patients with renal insufficiency; if indicated in this setting, acetaminophen is the preferred drug, notwithstanding data establishing the potential for renal toxicity from this drug (Sandler, et al., 1989). Acetaminophen is also preferred in patients with a bleeding or ulcer diathesis. If the risk of bleeding or ulcer formation is relatively small and an NSAID with potent antiinflammatory effects would be preferred over acetaminophen, the safest choice is one of two nonacetylated salicylates: choline magnesium trisalicylate or salsalate. These two agents have far less ulcer-producing potential than other NSAIDs, and at usual clinical doses they do not impair platelet aggregation (Danesh et al., 1987). All NSAIDs, except acetaminophen, may cause or exacerbate encephalopathy and may be problematic in patients at risk from volume overload; they must therefore be administered cautiously to patients with preexisting encephalopathy and those with congestive heart failure, peripheral edema, or ascites. Finally, it should be noted that significant hepatic disease mandates prudence regarding administration of acetaminophen.

The most common toxicity associated with the NSAIDs is gastrointestinal (GI). NSAID-induced gastropathy ranges from dyspepsia to life-threatening hemorrhage. Despite the high prevalence and serious nature of these disorders, the use of prophylactic agents remains controversial. Misoprostol (Cytotec) a prostaglandin E_1 analogue, prevents NSAID-induced gastric ulceration (Graham, Agrawal, & Roth, 1988), but one study suggested that its cost may be excessive in patients with no prior episode of GI hemorrhage (Edelson, Tosteson, & Sax, 1990); this drug also produces distressing GI side effects (nausea, diarrhea) in a high proportion of patients. A study in healthy volunteers suggests that cimetidine, an H_2-blocker, can reduce the risk of NSAID-induced gastropathy (Frank, et al., 1989), but controlled trials have failed to demonstrate benefit during long- term NSAID therapy (Robinson et al., 1989; Roth et al., 1987). Sucralfate reduces symptoms but not the incidence of ulceration (Caldwell et al., 1987). Omeprazole, a new hydrogen-potassium ATPase pump inhibitor, blocks gastric acid production and may be protective, but it has not been adequately tested.

Given the limited data, it is reasonable to consider prophylactic treatment with misoprostol in those patients receiving long-term NSAID therapy who are at relatively high risk for gastric ulceration. Based on epidemiologic data, this high risk group includes the elderly (over age 60), smokers, those with rheumatoid arthritis, those concurrently receiving a corticosteroid, those with recent upper abdominal pain, and those with distressing GI side effects from NSAIDs (Freis et al., 1989); patients with a history of ulcer disease or gastritis might also be considered in this category. On theoretical grounds, patients at high risk who cannot tolerate misoprostol might be considered for treatment

TABLE 17–2. *Nonsteroidal Antiinflammatory Drugs*

Generic Name	Approximate Half-Life (hr)	Dosing Schedule	Recommended Starting Dose (mg/day)[a]	Maximum Recommended Dose (mg/day)	Comment
p-Aminophenol derivatives					
Acetaminophen[b]	2–4	q4–6h	1400	6000	Overdosage produces hepatic toxicity. Not antiinflammatory and therefore not preferred as first-line analgesic or co-analgesic in pts with bone pain. Lack of GI or platelet toxicity, however, may be important in some cancer pts. At high doses, platelet counts and liver function tests should be done monthly.
Salicylates					
Aspirin[b]	3–12[c]	q4–6h	1400	6000	Standard for comparison. May not be tolerated as well as some of the newer NSAIDs.[d]
Diflunisal[b]	8–12	q12h	1000 once, then 500 q12h	1500	Less GI toxicity than aspirin.[d]
Choline magnesium trisalicylate[b]	8–12	q12h	1500 once, then 1000 q12h	4000	Unlike other NSAIDs, choline Mg trisalicylate and salsalate have minimal GI toxicity and no effect on platelet aggregation, despite potent antiinflammatory effects. They are therefore particularly useful in some cancer pts.[d]
Salsalate	8–12	q12h	1500 once, then 1000 q12h	4000	
Propionic acids					
Ibuprofen[b]	3–4	q4–8h	1200	4200	Available over-the-counter.[d]
Naproxen[b]	13	q12h	500	1000	Available as a suspension. Some studies show greater efficacy of higher doses, specifically 1500 mg/day, with little to no increase in adverse effects; long-term efficacy of this dose and safety in a medically ill population is unknown, however, and it should be used cautiously in selected pts.[d]
Naproxen sodium[b]	13	q12h	550	1100	Some studies show greater efficacy of higher doses, specifically 1650 mg/day, with little to no increase in adverse effects; long-term efficacy of this dose and safety in a medically ill population is unknown, however and it should be used cautiously in selected pts.[d]
Fenoprofen	2–3	q6h	800	3200	[d]
Ketoprofen	2–3	q6–8h	150	300	[d]
Flurbiprofen[b]	5–6	q8–12h	100	300	Experience too limited to evaluate higher doses, though it is likely that some patients would benefit.[d]
Acetic acids					
Indomethacin	4–5	q8–12h	75	200	Available in sustained-release and rectal formulations. Higher incidence of side effects, particularly GI and CNS, then propionic acids.[d]
Tolmetin	1	q6–8h	600	2000	[d]
Sulindac	14	q12h	300	400	Less renal toxicity than other NSAIDs.[d]
Suprofen[b]	2–4	q6h	600	800	Experience too limited to evaluate higher doses, though it is likely that some pts would benefit.[d]
Diclofenac	2	q6h	75	200	[d]

TABLE 17–2. *(Continued)*

Generic Name	Approximate Half-Life (hr)	Dosing Schedule	Recommended Starting Dose (mg/day)[a]	Maximum Recommended Dose (mg/day)	Comment
Ketorolac	4–7	q4–6h	150	120	Parenteral formulation available. Experience limited to the treatment of acute pain, for which the maximum first-day dose has been 150 mg and the maximum daily dose thereafter has been 120 mg. There is yet no experience with the chronic administration of this drug, and efficacy and safety of long-term administration remain to be determined. Experience is also too limited to evaluate higher doses.
Etodolac	3–11	q6–8h	600	1200	Experience too limited to evaluate higher doses.
Oxicams					
Piroxicam	45	q24h	20	40	Administration of 40 mg for > 3 weeks is associated with a high incidence of peptic ulcer, particularly in the elderly.[d]
Fenamates					
Mefenamic acid[b]	2	q6h	500 once, then 250 q6h	1000	Not recommended for use longer than 1 week and therefore not indicated as cancer pain therapy.[d]
Meclofenamic acid	2–4	q6–8h	150	400	[d]
Pyrazoles					
Phenylbutazone	50–100	q6–8h	300	400	Not a first-line drug due to risk of serious bone marrow toxicity. Not preferred for cancer pain therapy.

[a]Starting dose should be one-half to two-thirds recommended dose in the elderly, those on multiple drugs, and those with renal insufficiency. Doses must be individualized. Low initial doses should be titrated upward if tolerated and clinical effect is inadequate. Doses can be incremented weekly. Studies of NSAID in the cancer population are meager; dosing guidelines are thus empiric.

[b]Pain is approved indication.

[c]Half-life of aspirin increases with dose.

[d]At high doses, stool guaiac should be done bimonthly and liver function tests, BUN, creatinine, and urinalysis checked every 1–2 months.

with a combination of sucralfate and an H_2-blocker (Roth, 1988). As noted, the use of a nonacetylated salicylate also is appropriate in this group.

Several other factors should be considered when selecting an NSAID. First, it should be recognized that the pyrazole subclass, specifically phenylbutazone, is associated with a greater risk of adverse effects than the other NSAIDs and has been supplanted by newer drugs. Second, the great variability in the response to different drugs suggests that a favorable prior experience with a particular agent is an indication to use the same drug again. Third, concern about compliance with a prescribed regimen may be lessened by the use of an NSAID with once-daily (e.g., piroxicam) or twice-daily dosing (e.g., choline magnesium trisalicylate, diflunisal, naproxen, and others). Finally, the cost of the different NSAIDs varies widely, and when cost is an important issue, pharmacies can be screened for a schedule of charges so this factor may be taken into consideration.

An injectable formulation of an NSAID, ketorolac, has become available in the United States. Studies have demonstrated that this drug can have efficacy comparable to that of morphine in the postoperative setting (O'Hara et al., 1987). At this time, experience with ketorolac is limited to short-term use for acute pain, where it is particular valuable for patients predisposed to adverse effects from opioids. (See Chapter 16.)

Explore the Dose-Response Relationship. Given the observation that the standard recommended dose of an NSAID is empirically derived from dose-ranging studies and may or may not be appropriate for the individual patient, it is prudent to consider trials of escalating doses, particularly in patients who may be predisposed to averse effects (e.g., the elderly or those with mild renal insufficiency). By initiating treatment with a dose lower than the standard recommended dose and then increasing the dose at intervals while monitoring effects, the minimal effective dose, ceiling dose, and toxic dose can be identified. The risk of serious toxicity is reduced by the use of the lowest dose capable of providing meaningful therapeutic effects.

Clinical experience suggests that doses should be increased on a weekly basis or less frequently, particularly with long-acting drugs such as piroxicam. Should there be no additional analgesia following an increase in dose, the ceiling has presumably been reached and the dose should be lowered to the previous level and continued, or the drug discontinued. The existence of dose-related toxicities, combined with the paucity of information available about the long-term safety of high doses, suggests that this process of dose exploration should be limited by an empirical maximum dose. A reasonable guideline limits upward dose titration to 1.5 to 2 times the standard dose. Long-term therapy above the standard dose should be monitored especially carefully with tests for occult fecal blood, urinalysis, and serum tests of renal and hepatic function on a bimonthly basis.

Duration of Trials. Several weeks are needed to judge the efficacy of a dose when NSAIDs are used for the treatment of grossly inflammatory lesions, such as arthritis. Clinical experience suggests that a briefer period, such as a week, may be adequate for the same purpose in pain patients without a grossly inflammatory lesion.

Switching Drugs. Given the interindividual variability in drug response, it is reasonable to follow failure with one of the NSAIDs by a trial with another. This guideline suggests that the clinician should become familiar with a small group of drugs, which can then be administered sequentially in patients with refractory pain.

Adjuvant Analgesics

The adjuvant analgesics comprise many drug classes, within which specific agents have been discovered to have analgesic effects (see Table 17-3).

TABLE 17–3. *Adjuvant Analgesics*

Drug Class	Examples
Antidepressants	
Tricyclic antidepressants	Amitriptyline Doxepin Imipramine Nortriptyline Desipramine
"Newer" antidepressants	Trazodone Fluoxetine Maprotiline
MAO inhibitors	Phenelzine
Anticonvulsants	Carbamazepine Phenytoin Valproate Clonazepam
Oral local anesthetics	Mexiletine Tocainide
Neuroleptics	Fluphenazine Haloperidol Methotrimeprazine Pimozide
Muscle relaxants	Orphenadrine Carisoprodol Methocarbamol Chlorzoxazone Cyclobenzaprine
Antihistamines	Hydroxyzine
Psychostimulants	Caffeine Methylphenidate Dextroamphetamine
Corticosteroids	Dexamethasone Prednisone Methylprednisolone
Sympatholytic drugs	Prazosin Phenoxybenzamine
Calcium channel blockers	Nifedipine
Miscellaneous drugs	Baclofen Clonidine Capsaicin

Antidepressants. Tricyclic antidepressants (TCAs) have been demonstrated to be analgesic in a large number of chronic pain states (Butler, 1984; Getto, Sorkness, & Howell, 1987; Max et al., 1991; Portenoy, 1990c). Analgesic effects were first believed to be related to the reversal of depression (Evans et al., 1973), but this notion has been refuted by several observations, including the occurrence of analgesia earlier than mood change in chronic pain patients, the appearance of analgesia in the absence of mood changes in depressed patients with pain, and the development of analgesia in chronic pain patients without depression (Couch, Ziegler, & Hassanein, 1976; Kishore-Kumar et al., 1990; Max et al., 1987; Watson et al., 1982). Analgesic effects have also been demonstrated in animal models (Spiegel, Kalb, & Pasternak, 1983).

The analgesic action of TCA drugs may relate to their effects on central monoamines in endogenous pain-modulating pathways. The best characterized of these monoamine-dependent pain-modulating systems are serotonergic and noradrenergic tracts that originate in the rostroventral medulla, descend to the dorsal horn of the spinal cord, and inhibit transmission of nociceptive information at the first central synapse (Besson & Chaouch, 1987; Hammond, 1985). TCA drugs block the reuptake of these transmitters with varying selectivity. It is interesting that the specific tricyclics most widely accepted as analgesics are the tertiary amine compounds, such as amitriptyline (see Table 17-3), whose activity is greatest at serotonin synapses. However, secondary amine TCAs, which have relatively selective effects at norepinephrine synapses such as desipramines, are also analgesic (Kishore-Kumar et al., 1990; Max et al., 1991), and the relative importance of serotonin and norepinephrine remains uncertain. In addition, TCAs bind to a large variety of other receptors (Charney, Menkes, & Heninger, 1981; Richelson, 1979), some of which employ neurotransmitters and neuromodulators that have been implicated in the process of pain modulation (Besson & Chaouch, 1987; Sosnowski & Yaksh, 1990). The potential interactions of the TCAs with endogenous pain-modulating pathways are thus complex and have thus far prevented elucidation of the analgesic mechanism of these drugs.

The analgesic efficacy of the TCAs has been established through numerous controlled trials. Favorable studies of amitriptyline have been performed in patients with diabetic neuropathy (Max et al., 1987), postherpetic neuralgia (Watson et al., 1982), tension and migraine headache (Couch et al., 1976; Diamond & Baltes, 1971), myofascial pain (Carette et al., 1986), and psychogenic pain (Pilowsky et al., 1982). Imipramine has been found to be effective in patients with arthritic pain (Gingras, 1976) and painful diabetic neuropathy (Kvinsdahl et al., 1984); and doxepin has been found to be analgesic in patients with psychogenic headache (Okasha, Ghaleb, & Sadek, 1973) or coexistent chronic pain and depression (Hameroff et al., 1982). Desipramine was analgesic in a studies of patients with postherpetic neuralgia (Kishore-Kumar et al., 1990) and diabetic neuropathy (Max et al., 1991); and clomipramine has been noted to be analgesic in several pain syndromes (Langohr, Stohr, & Petruch, 1982).

The "newer" antidepressants have also been studied, with somewhat more equivocal results. The analgesic efficacy of maprotiline was favorably compared with clomipramine in idiopathic pain (Eberhard et al., 1988) and amitriptyline in postherpetic neuralgia (Watson et al., 1990). A controlled clinical trial has indicated that amitriptyline and trazodone can both be effective in patients with cancer pain (Ventafridda et al., 1987), but another trial of trazodone for dysesthetic pains in patients with traumatic myelopathy did not demonstrate a favorable effect (Davidoff et al., 1987). Zimelidine, which is similar to trazodone in its relatively selective action at the serotonin synapse, was found to be analgesic in a controlled trial of patients with mixed organic and psychogenic pain syndromes (Johansson & Van Knorring, 1979) but was ineffective in an open-label comparison against amitriptyline in patients with postherpetic neuralgia (Watson & Evans, 1985). Convincing analgesic effects were observed in a controlled trial of another selective serotonin reuptake inhibitor, paroxetine (Sindrup et al., 1990). Zimelidine is not currently available in the United States. There has been limited assessment of the analgesic potential of fluoxetine, but clinical experience with this drug suggests that it is not effective as a primary therapy for pain.

Monoamine oxidase inhibitors (MAOIs) also increase the availability of central monoamines, and there are some data to suggest analgesic effects from these compounds as well. They have achieved some measure of acceptance in refractory migraine (Anthony & Lance, 1969), and their analgesic potential has been supported by a controlled trial in patients with atypical facial pain (Lascelles, 1966). Given their potential for toxicity, however—specifically hypertensive crisis after ingestion of tyramine-containing foodstuffs—their use cannot be recommended except for severe refractory vascular headache or atypical facial pain. If an MAOI is considered, dietary restrictions must be explained to the patient and a list provided of those prescription and over-the-counter medications that may be interact adversely with these drugs. For patients with pain, the potential interaction of an MAOI with meperidine should be particularly emphasized; this combination can produce a serious hyperpyrexic syndrome. (See Chapter 9.)

A TCA trial should be considered for virtually all chronic pain syndromes unless contraindicated by specific medical conditions, including significant cardiac arrhythmias, symptomatic prostatic hypertrophy, or narrow angle glaucoma. Clinical experience suggests that the indication is strongest in patients with chronic neuropathic pain characterized by continuous dysesthesias and in individuals with pain of any type complicated by depression.

Given the abundant data from controlled studies, a tertiary amine TCA, specifically amitriptyline, doxepin, imipramine, or clomipramine, is preferred for the initial trial; experience is greatest with amitriptyline. Patients unable to tolerate these agents due to their anticholinergic, sedative, or hypotensive effects should undergo a trial with a secondary amine TCA, specifically nortriptyline or desipramine.

Dosing guidelines are empirical. Initial doses of tertiary amine TCAs should be low, such as 10 to 25 mg at night. Doses should be increased gradually over 1 to 2 weeks to 50 to 75 mg at night, then held steady for a week or so before resuming upward titration. Analgesic effects usually occur within 4 to 7 days of reaching an effective dose. For all the aforementioned drugs except nortriptyline, this typically requires a dose of 50 to 150 mg per night (Kishore-Kumar et al., 1990; Max et al., 1987; Watson et al., 1982). The usually effective dose for nortriptyline is not known but may be somewhat less. Studies have suggested that analgesia is a dose-dependent effect (Max et al., 1987), and an effort should be made to continue upward titration of the dose if neither analgesia nor intolerable side effects occur in the usual therapeutic range. Higher doses may also be needed if a coexistent depression is prominent. The existence of a therapeutic window for analgesia during treatment with nortriptyline has been suggested anecdotally, but confirmatory data are lacking and this effect has not been observed with other drugs. Most patients can be adequately treated with a single nighttime dose, but some demonstrate waning of analgesic effects toward evening and are better managed with twice-daily dosing. Although the range of plasma levels associated with analgesic effects is unknown for all agents, a plasma level can be determined to ensure that noncompliance, poor absorption, or unusually rapid catabolism is not compromising efforts to achieve adequate plasma levels. Finally, clinical experience suggests that patients may demonstrate widely varying responses to different TCA drugs; failure to respond to adequate doses of one should therefore be followed by a trial with another.

Anticonvulsants. Anticonvulsant medications (see Table 17-3) have become widely accepted for the management of chronic neuropathic pain, particularly those characterized as lancinating (shooting, stabbing) pains (Swerdlow, 1984). Their mode of analgesic action in these syndromes is not known but presumably relates to the same mechanisms that yield anticonvulsant effects, such as the capacity to suppress paroxysmal discharges, neuronal hyperexcitability, or spread of abnormal discharges (Weinberger, Nicklas, & Berl, 1976). The possibility that the latter processes are involved in neuropathic pain has been suggested by recordings of spontaneous electrical activity induced by nerve injury in both experimental and clinical studies (Albe-Fessard & Lombard, 1982; Loeser, Ward, & White, 1968; Nystrom & Hagbarth, 1981; Wall & Gutnick, 1974).

Phenytoin was the first anticonvulsant used to treat neuropathic pain. Case series suggested efficacy in trigeminal neuralgia (Blom, 1963; Braham & Saia, 1960), and favorable effects were reported in isolated cases or small groups of patients with glossopharyngeal neuralgia, tabetic lightning pains, paroxysmal pain in postherpetic neuralgia, thalamic pain, postsympathectomy pain, and posttraumatic neuralgia (Cantor, 1972; Green, 1961; Hatangdi, Boas, & Richards, 1976; Raskin et al., 1974; Swerdlow & Cundill, 1981; Taylor et al., 1977). Controlled trials in patients with painful neuropathy from Fabry's disease (Lockman et al., 1973) and painful diabetic neuropathy (Chadda & Mathur, 1978) had similar outcomes. As noted, most of the neuropathic pains in these syndromes were characterized by a prominent lancinating component.

Controlled trials have established the efficacy of carbamazepine in trigeminal neuralgia (Campbell, Graham, & Zilkha, 1966; Killian & Fromm, 1968; Rockcliff & Davis, 1966), in the lancinating (but not the continuous) pains of postherpetic neuralgia (Killian & Fromm, 1968), and in painful diabetic neuropathy (Rull et al., 1969). Published case series and anecdotal reports have also suggested benefit in glossopharyngeal neuralgia, tabetic lightning pains, paroxysmal pain in multiple sclerosis, postsympathectomy pain, and lancinating pains due to cancer and posttraumatic neuralgia (Ekbom, 1972; Elliot, Little, & Milbrandt, 1976; Espir & Millac, 1970; Mullan, 1973; Raskin et al., 1974; Swerdlow & Cundill, 1981; Taylor et al., 1977).

Uncontrolled clinical trials and anecdotal reports have similarly suggested that clonazepam and valproate may be effective in neuropathic pains characterized by lancinating dysesthesias. Clonazepam has been reported to be useful for treatment of trigeminal neuralgia, paroxysmal postlaminectomy pain, and posttraumatic neuralgia (Caccia, 1975; Martin, 1981; Swerdlow & Cundill, 1981). Valproate has been beneficial in the management of trigeminal neuralgia, postherpetic neuralgia, and other lancinating neuropathic pains (Peiris et al., 1980; Raftery, 1979; Swerdlow & Cundill, 1981).

Three other drugs that are not classified as anticonvulsants have been demonstrated to have analgesic effects in trigeminal neuralgia. They or their congeners are therefore considered with the aforementioned agents as potential treatments for patients with lancinating neuropathic pains. Baclofen is a gamma-aminobutyric acid (GABA) agonist primarily indicated for the treatment of spasticity. It is widely considered to be a second-line agent in trigeminal neuralgia (Fromm, Terence, & Chatta, 1984) and is used for lancinating pains of other types. Tocainide is an oral local anesthetic chemically related to lidocaine. The efficacy of this drug in trigeminal neuralgia (Lindstrom & Lindblom, 1987) is often used to justify a trial of mexiletine, another oral local anesthetic with a more favorable therapeutic index (see below). Finally, pimozide is a neuroleptic demonstrated to be analgesic in trigeminal neuralgia (Lechin et al., 1989). As discussed below, the potential adverse effects of the neuroleptics during long-term administration suggests that the use of pimozide for lancinating neuropathic pains should be reserved for patients who fail on other agents.

Clinical guidelines for the selection and administration of these agents in patients with neuropathic pain are empiric. As noted, the clearest indication for a trial of any of these drugs is a lancinating neuropathic pain, the prototype for which is trigeminal neuralgia. The use of these drugs in nonneuropathic pains is not supported by available data, and there are only meager data suggesting potential efficacy against neuropathic pains characterized by continuous dysesthesias. Nonetheless, clinical experience suggests that one or more of these drugs can occasionally be useful for the latter type of pain, and a trial is reasonable in refractory cases.

Surveys (Swerdlow & Cundill, 1981) and clinical experience indicate that patients may have markedly different analgesic responses to the various drugs. Should one fail, therefore, a trial of another is in order.

In all cases, drugs administered for these neuropathic pains are administered as they are for their primary indication; for example, the anticonvulsants are prescribed as if seizures, rather than pain, were the clinical indication. A conservative oral loading dose of phenytoin is reasonable, but the other drugs should be initiated at low doses and then gradually titrated upward until limiting side effects occur or the plasma concentration exceeds the therapeutic range. Although it is possible that isolated patients could respond to higher doses, clinical experience suggests that patients who fail to benefit at these doses are unlikely to obtain relief.

Oral Local Anesthetics. The introduction of oral local anesthetic drugs for the management of cardiac arrhythmias has revived interest in the analgesic potential of this class. This potential has been suggested by a large experience in the use of parenteral local anesthetics to treat pains of diverse etiology (Glazer & Portenoy, 1991). Two oral analogues of lidocaine have been tested in well-controlled clinical trials. As noted previously, the efficacy of tocainide for trigeminal neuralgia has been established (Lindstrom & Lindblom, 1987). More recently, a carefully controlled trial demonstrated the efficacy of mexiletine in patients with painful diabetic polyneuropathy (Dejgard, Petersen, & Kastrup, 1988). Interestingly, these data suggest that oral local anesthetics may be effective in both the lancinating and continuous dysesthesias of neuropathic pain states.

Although additional controlled investigations are needed to clarify the analgesic potential of these and other local anesthetics, it is presently reasonable to view them as second-line agents for the treatment of refractory neuropathic pain syndromes of any type. Mexiletine is the safest compound (Kreeger & Hammill, 1987) and should be administered first. Dosing is empiric at the present time and should mimic that applied in the treatment of cardiac arrhythmias (Kreeger & Hammill, 1987).

Neuroleptics. Numerous anecdotal reports and case series have suggested the value of haloperidol, fluphenazine, perphenazine, thioridazine, chlorprothixene, and chlorpromazine for the management of patients with a variety of painful disorders, most of which were neuropathic (Kocher, 1976; Margolis & Gianascol, 1956; Nathan, 1978; Weis, Sriwatanakul, & Weintraub, 1982). Few controlled studies, however, have confirmed the potential for analgesic effects from these or other neuroleptic drugs. Controlled studies have established the analgesic efficacy of single doses of methotrimeprazine for both acute and chronic pain (Beaver et al., 1966; Bloomfield et al., 1964; Lasagna & DeKornfeld, 1961). This drug, which is available only in a parenteral formulation and has prominent sedative and hypotensive effects, has been used in patients with pain due to far-advanced cancer but cannot be recommended for treatment of nonmalignant pain syndromes. A controlled single-dose trial of another phenothiazine, chlorpromazine, failed to demonstrate analgesic effects against cancer pain (Houde & Wallenstein, 1966). As discussed previously, pimozide has been suggested to be effective for trigeminal neuralgia (Lechin et al., 1989) and can be considered a second-line agent for the management of lancinating neuropathic pains.

Given the limited data supporting the analgesic efficacy of neuroleptic drugs in chronic nonmalignant pain and the risks associated with these agents (most importantly, the potential for refractory movement disorders), it is appropriate to restrict their use to trials in patients with intractable neuropathic pain. Clinical experience is greatest with fluphenazine and haloperidol. Doses reported to be effective are generally low, and in contrast to the drugs described previously, there are no clinical data to support the concept of upward dose titration. (See also Chapter 16.)

Muscle Relaxants. Muscle relaxant drugs commonly used for treatment of musculoskeletal pains must be distinguished from drugs useful for treatment of spasticity. Spasticity, which results from a lesion in the central nervous system, is clearly different from the muscle "spasms" that occur in the setting of peripheral nerve or myofascial injury. Muscle spasms represent focal areas of increased muscle activity that are associated with tenderness and splinting of the painful part. Discrete trigger points may be related to these areas. Drugs useful for treatment of true spasticity, such as baclofen and dantrolene, are not indicated for common myofascial pains associated with spasm, and muscle relaxant drugs are not indicated for treatment of true spasticity. The exception is diazepam, which is commonly used in both settings.

The muscle relaxant drugs used in the United States include orphenadrine, carisoprodol, chlorzoxazone, methocarbamol, and cyclobenzaprine (see Table 17-3). The mechanism of the analgesic effects produced by these drugs is conjectural. They suppress polysynaptic reflexes in experimental preparations (Smith, 1965), but the clinical relevance of this phenomenon is not known. There is no evidence that they actually reduce contraction of striated muscle. Nonetheless, controlled studies have demonstrated efficacy greater than placebo for each of these agents in the treatment of musculoskeletal pain of one type or another (Bercel, 1977; Gold, 1978). Some studies have demonstrated analgesia above that provided by aspirin or acetaminophen, or analgesia from the combination of a muscle relaxant and either aspirin or acetaminophen above that provided by the analgesic alone (Birkeland & Clawson, 1968). Thus these drugs are clearly analgesic, but the available data are inadequate to determine whether these favorable responses are due to a nonspecific sedative action, a primary analgesic effect, a partial lessening of muscle tension, or some combination of these effects. Moreover, there are no studies comparing the efficacy or side effect liability of the drugs within this class and none comparing any of these drugs to an NSAID (other than aspirin), an opioid, or a sedative/hypnotic drug. The long-term consequences of chronic administration of any of these compounds to patients with persistent pain similarly is unknown.

The available data suggest that patients with acute musculoskeletal pain may benefit from the administration of a muscle relaxant, but there is no support at the present time for long-term administration. Chronic treatment therefore should be considered only in patients with refractory pain who demonstrate clear-cut long-term benefits. Regardless of the duration of therapy, the potential sedative and anticholinergic effects of these drugs must be considered prior to administration and monitored after treatment begins; these effects can be additive to other centrally active drugs and may further impair function or distress the patient sufficiently to undermine any benefit from these drugs. The selection of a specific drug and the dosing are empirical, and there is no evidence that upward dose titration beyond usually recommended doses produces any other effect than progressive sedation, with an unknown degree of accruing risk.

Antihistamines. Numerous studies, some controlled, have demonstrated the analgesic efficacy of antihistamine drugs (Rumore & Schlichting, 1986). In addition to orphenadrine, which is classified as both a muscle relaxant and an antihistamine, controlled trials of diphenhydramine and hydroxyzine have established the analgesic potential of these drugs (Birkeland & Clawson, 1968; Gold, 1978; Stambaugh & Lance, 1983). Although the mechanisms by which these drugs provide analgesia are not known, it is interesting to speculate that this effect may be mediated by specific histamine receptors in the endogenous pathways for pain modulation (Rumore & Schlichting, 1986).

Notwithstanding the experimental data demonstrating analgesic effects from antihistaminic drugs, clinical experience with these agents in patients with chronic pain has been disappointing. Some antihistamines are used in combination products marketed as minor analgesics, usually for headache, and orphenadrine is often administered as a muscle relaxant. With these two exceptions, antihistamines usually are not applied to the treatment of chronic pain. (See also Chapter 16.)

Miscellaneous Drugs. Other drugs are analgesic in specific clinical circumstances. The quality of the evidence supporting such an effect varies with the agent involved. Clonidine, for example, is an alpha-2-adrenergic agonist that is unequivocally analgesic (Max et al., 1988) and has been used to treat chronic and recurrent pains (Shafar, Tallett, & Knowlson, 1972; Tan & Croese, 1986). Long-term trials of oral or transdermal clonidine in patients with chronic pain have not yet appeared, but clinical experience suggests that some patients, including some with refractory neuropathic pains, respond favorably to this drug. In patients with intractable neuropathic pain and no medical contraindications, a therapeutic trial of oral or transdermal clonidine is reasonable.

Patients with sympathetically maintained pain (reflex sympathetic dystrophy or causalgia) who are unable to tolerate nerve blocks or who have failed to benefit from these procedures have been empirically treated with sympatholytic drugs. Case reports and clinical series have suggested that phenoxybenzamine (Ghostine et al., 1984), prazosin (Abram & Lightfoot, 1981), oral guanethidine (Tabira, Shibasaki & Kuroiwa, 1983), and propranolol (Simson, 1974) may be useful for these pains. Other, nonsympatholytic drugs—specifically corticosteroids (Kozin et al., 1981) and nifedipine (Prough et al., 1985)—have also been successfully administered according to anecdotal reports.

Studies in the postoperative setting have suggested analgesic potential for some of the benzodiazepine drugs, including diazepam and midazolam (Miller et al., 1986; Singh et al., 1981). One survey observed that alprazolam provided analgesia to patients with neuropathic cancer-related pain (Fernandez, Adams, & Holmes, 1987). In contrast, a small controlled study of repeated doses of chlordiazepoxide in chronic pain failed to demonstrate any benefit (Yosselson-Superstine, Lipman, & Sanders, 1985). In an experimental pain paradigm that used signal detection methodology to distinguish a change in sensorineural function from psychologic effects (response bias), the analgesic effects produced by diazepam could be ascribed to psychologic influences alone, whereas morphine analgesia could be explained by both shifts in sensory discriminability and response bias (Yang et al., 1979). Thus benzodiazepine drugs may have analgesic effects that may be produced through a mechanism different than that of routine analgesics, specifically one linked to the psychotropic effects of these agents. Although pain relief may be reported by the patient administered one of these drugs, their use has not become accepted for the long-term management of pain, largely due to concerns about adverse cognitive effects and abuse liability. Clinical experience supports the use of diazepam for short-term management of acute musculoskeletal pain.

Drugs with sympathomimetic effects, specifically dextroamphetamine, methylphenidate, and caffeine, are also analgesic (Bruera et al., 1987; Forrest et al., 1977; Laska et al., 1984). Caffeine is commonly added to combination products used for the treatment of headache, and both dextroamphetamine and methylphenidate are used to reverse opioid-induced sedation and to provide coanalgesic effects in patients with cancer pain.

Finally, analgesic effects have been established for both cannabinoid drugs (Noyes et al., 1975) and L-tryptophan (Seltzer et al., 1983), but neither are in clinical use for the indication of pain. Cannabinoids, which are used as antiemetics in the cancer population, often produce psychotomimetic side effects, which have prevented additional trials in patients with pain. L-tryptophan has been associated with the development of a rare systemic disease, eosinophilia-myalgia syndrome, and has been removed from the market in the United States.

Opioid Analgesics

Opioid analgesics are accepted therapy for acute pain and chronic cancer pain. Guidelines for chronic administration have achieved a consensus among cancer pain experts (Foley, 1985; Portenoy & Foley, 1989; World Health Organization, 1986). Favorable results have been observed following optimal administration of these drugs in this population.

Treatment of chronic nonmalignant pain with opioid drugs has traditionally been rejected because of persistent concerns about the potential for loss of efficacy over time (i.e., the development of tolerance), adverse pharmacologic effects, and, most importantly, the development of addiction. Case series, however, have been reported in which patients with diverse painful disorders refractory to usual therapies were maintained for long periods on opioid drugs (see Table 17-4) (France, Urban & Keefe, 1984; Portenoy & Foley, 1986; Taub, 1982; Tennant & Uelman, 1983; Tennant et al., 1988; Urban et al., 1986; Wan Lu, Urban, & France, 1988). These patients obtained partial analgesia, which in some cases seemed to facilitate better functioning. Opioid toxicity was never viewed as a problem, and management difficulties compatible with the development of addictive behaviors occurred in only a small number of patients.

This experience has been supplemented by a critical review of the medical literature (Portenoy, 1990a) that finds limited support for the concerns that have been used heretofore to justify the absolute rejection of this approach. For example, the notion that analgesic tolerance inevitably undermines long-term opioid treatment owing to the gradual loss of efficacy is contraindicated by numerous surveys performed in the cancer population (Brescia et al., 1992; Kanner & Foley, 1981; Twycross, 1974), as well as the aforementioned surveys of patients with nonmalignant pain. For reasons that are not understood, analgesic tolerance is rarely the cause of dose escalation in the clinical setting; in the absence of a progressive lesion capable of increasing pain, opioid doses usually remain stable. Concern about a need for escalating opioid doses therefore does not justify the decision to withhold chronic opioid therapy.

Perhaps more important, there is strong evidence that true addiction—which may be defined as a psychologic and behavioral syndrome characterized by psychological dependence, compulsive drug use, and other aberrant drug-related behaviors—is an extraordinarily rare outcome among patients with no history of substance abuse who are administered opioids for painful medical disease (Chapman, 1989; Perry & Heidrich, 1982; Portenoy, 1990a; Porter & Jick, 1980). The concern that long-term exposure to these drugs, by itself, can lead to addiction is also negated by the enormous, highly favorable experience in the management of cancer pain. It is now generally accepted that addiction represents the final outcome of a complex set of determinants, some related to properties inherent in a drug and others related to genetic, psychological, social, and situational characteristics of the patient. Clinical experience suggests that a history of substance abuse may signal an increased risk of problem behaviors during long-term

TABLE 17–4. *Published Surveys of Chronic Opioid Therapy for the Treatment of Nonmalignant Pain*

Author	No. of Pts.	Diagnosis	Opioids	Daily Dose Equivalent	Duration	Analgesic Efficacy	Adverse Effects	Comments
Taub (1982)	313	Mixed	Mixed	Mean 10–20 mg (max. 40 mg) oral methadone	Up to 6 yr	Few details; all said to benefit	No toxicity; abuse in 13	Prior drug abuse in 8 of 13 problem cases
Tennant & Uelman (1983)	22	Not stated	Not stated	Not stated	Not stated	Not stated	No abuse	All had failed pain clinics; 15 returned to work
France et al. (1984)	16	Most back pain	Mixed	Mean 8 mg (range 3–20) oral methadone	Mean 13 mo (range 6–22)	Pain relief >50% in all, >75% in 13; sustained in 12	No toxicity; no abuse	Improved function in 12 of 16; higher doses needed in 5 over time
Portenoy & Foley (1986)	38	Mixed; some neuropathic	Mixed	Median 10–20 mg (range <10–60) parenteral morphine	Median 3–4 yr; range 6 mo to 10 yr	Adequate 11; partial 13; inadequate 14	No toxicity; abuse in 2	Both problem cases had prior drug abuse; marked gains in function uncommon but few details
Urban et al. (1986)	5	Phantom pain	Methadone	10–20 mg	Mean 22 mos (range 12–26)	At follow-up, all >50% relief	No toxicity; no abuse	Over time, relief waned in 1
Portenoy (1990a)	20	Mixed; some neuropathic	Mixed	Median 10–20 mg (range <10–50) parenteral morphine	Median <2 yr; range 6 mo to 10 yr	Adequate 9; partial 10; inadequate 14	Personality change in 1; myoclonus in 1; abuse in 2	Marked gains in function uncommon but few details; with warning, both problem cases stopped abuse
Tennant et al. (1988)	52	Mixed	Mixed	Highly variable; range 10–240 mg oral methadone	Duration of program not noted; opioid use averaged >12 yr	Adequate 88%; partial 12%	Constipation 20; edema 12; adrenal insufficiency 1; abuse 9	Many medically ill, accounting for side effects; no dose increases needed
Wan Lu et al. (1988)	76	Mixed; most low back pain	Mixed; most methadone	Oral methadone <20 mg	Mean 29 mo (range 6–76)	Pain relief 25–99% in 87%; relief sustained at follow-up	Side effects in 36%; abuse behavior by 7	Activity said to increase but few details; no dose escalation needed

opioid therapy for nonmalignant pain (Taub, 1982), but that most of those who comprise the remainder of this population incur an extremely small risk of this outcome.

It is thus conceivable that chronic opioid therapy is an alternative approach capable of providing enhanced comfort to highly selected patients with refractory nonmalignant pain. Based on clinical experience, guidelines for management have been suggested (see Table 17-5), the practical application of which is evolving. Although clinical experience has suggested that long-acting opioids, such as methadone or controlled-release oral morphine, may be most salutary for this indication (allowing less frequent dosing and potentially subjecting the patient to smaller fluctuations in pain), there is no evidence that the therapy is more successful with one type of drug than another. The use of agonist-antagonist drugs such as pentazocine is acceptable but has no strong justification; the lower likelihood of physical dependence and abuse with these drugs, which has been identified in studies of addicts and is used for the marketing of these agents, is not a meaningful advantage in most patients with no history of substance abuse. Similarly, the use of a "weak" opioid, such as propoxyphene, may increase the comfort of the clinician during the administration of this therapy but has no funda-

TABLE 17–5. *Proposed Guidelines for Management of Opioid Maintenance Therapy for Nonmalignant Pain*

1. Should be considered only after all other reasonable attempts at analgesia have failed.
2. A history of substance abuse should be viewed as a relative contraindication.
3. A single practitioner should take primary responsibility for treatment.
4. Patients should give informed consent before the start of therapy; points to be covered include recognition of the low risk of psychological dependence as an outcome, the potential for cognitive impairment with the drug alone and in combination with sedative/hypnotics, and understanding by female patients that children born while the mother is on opioid maintenance therapy will likely be physically dependent at birth.
5. After drug selection, doses should be given on an around-the-clock basis. Several weeks should be agreed on as the period of initial dose titration; and although improvement in function should be continually stressed, all should agree to at least partial analgesia as the appropriate goal of therapy.
6. Failure to achieve at least partial analgesia at relatively low initial doses in the nontolerant patient raises questions about the potential treatability of the pain syndrome with opioids.
7. Emphasis should be placed on attempts to capitalize on improved analgesia by gains in physical and social function.
8. In addition to the daily dose determined initially, patients should be permitted to escalate dose transiently on days of increased pain; two methods are acceptable: (a) Prescription of an additional four to six "rescue doses" to be taken as needed during the month. (b) Instruction that one or two extra doses may be taken on any day but must be followed by an equal reduction of dose on subsequent days.
9. Patients must be seen and drugs prescribed at least monthly. Patients should be assessed for the efficacy of treatment, adverse drug effects, and the appearance of either misuse or abuse of the drugs during each visit. The results of the assessment should be clearly documented in the medical record.
10. Exacerbations of pain not effectively treated by transient, small increases in dose are best managed in the hospital, where dose escalation, if appropriate, can be observed closely and the return to baseline doses can be accomplished in a controlled environment.
11. Evidence of drug hoarding, acquisition of drugs from other physicians, uncontrolled dose escalation, or other aberrant behaviors should be followed by tapering and discontinuation of opioid maintenance therapy.

mental advantage over the use of the "strong" pure agonist agents.

There is a consensus that the oral route of administration is preferred for the long-term management of nonmalignant pain, as it is for malignant pain. Repetitive parenteral administration is typically eschewed because of the logistics involved and the likelihood of prominent fluctuations in plasma drug concentration. Continuous subcutaneous infusion of an opioid and spinal opioid administration are rarely used for chronic treatment of nonmalignant pain and should be implemented only by experienced clinicians.

Controlled prospective studies are needed to evaluate the role of chronic opioid therapy for nonmalignant pain. Given the enormous concerns about drug abuse in the United States, there is a compelling need for professionals engaged in the treatment of these challenging patients to undertake an ongoing, dispassionate, and scientifically grounded discussion about the medical role of opioid drugs.

Anesthetic Approaches

Analgesic techniques traditionally considered to be within the purview of anesthesiologists play an important role in the management of chronic pain. These techniques comprise neural blockade and intraspinal modalities, including the long-term administration of opioids or local anesthetics.

Neural blockade describes a diverse group of procedures that transiently or more permanently block sympathetic or somatic nerves (or both) (Cousins, Dwyer, & Gibb, 1988; Raj, 1988). Temporary somatic nerve blocks with local anesthetic may be diagnostic, prognostic, or therapeutic. Diagnostic blocks elucidate the afferent pathways involved in the experience of pain. Prognostic blocks are implemented prior to a neurolytic procedure. Although extensive clinical experience indicates that a favorable response to a temporary block does not predict permanent relief following neurolysis, the failure to achieve pain relief with local anesthetic is commonly taken to contraindicate a subsequent destructive procedure. Repeated local anesthetic blocks are used therapeutically in patients who obtain substantial and fairly prolonged relief after each procedure. Repeated sympathetic blocks with local anesthetic are considered a mainstay approach for sympathetically maintained neuropathic pains (i.e., reflex sympathetic dystrophy or causalgia). More permanent nerve blocks produced by neurolytic solutions (e.g., phenol or alcohol) are an important modality in cancer pain management but are only rarely considered in the population with nonmalignant pain.

Local anesthetics have been used to provide more prolonged neural blockade through techniques of perineural or epidural infusion. Epidural local anesthetic infusion, either alone or in combination with opioid drugs, has become an accepted modality in cancer pain management. As noted previously, rare patients with chronic nonmalignant pain have been treated with these approaches.

Although trigger-point injections are typically classified as an anesthetic approach, these simple techniques may be considered within the scope of all practitioners (Travell & Simons, 1983). Trigger points in muscle are a common cause of pain (the myofascial pain syndrome), and the use of local anesthetic injections into painful trigger points can be a useful approach in patients with these pains.

Neurostimulatory Approaches

It has been appreciated for some time that stimulation of afferent neural pathways may eventuate in analgesia. The best known application of this principle is transcutaneous electrical nerve stimulation (TENS). Other approaches include counterirritation (systematic rubbing of the painful part), percutaneous electrical nerve stimulation, dorsal column stimulation, deep brain stimulation, and acupuncture. Surveys of the techniques that have been used most extensively for nonmalignant pain—TENS, acupuncture, and dorsal column stimulation—have suggested that most patients achieve analgesia soon after the approach is implemented, but only a few can obtain prolonged relief. A controlled trial that compared TENS to a program of stretching exercises in patients with chronic low back pain failed to identify any positive effect from electrical stimulation (Deyo et al., 1990).

Thus the data supporting the efficacy of stimulation procedures are ambiguous. Clinical experience suggests that a small number of patients can benefit for a prolonged period from any of these approaches. Given the safety of TENS, it is reasonable to recommend a therapeutic trial of this technique in most patients with chronic pain, notwithstanding the likelihood of an inadequate result. Occasional patients opt to continue stimulation following a brief exposure. It is important to recognize that an adequate trial typically involves several weeks, during which time the patient should experiment with different electrode placements and stimulation parameters. There is no one correct approach, and it is possible that many patients who could potentially benefit do not because the trial has been too limited. The application of other neurostimulatory approaches should be implemented by experienced practitioners knowledgeable about the management of chronic pain.

Physiatric Approaches

The potential for analgesic effects from physiatric therapies, including the use of orthoses or prostheses, occupational therapy, and physical therapy, often is not recognized. The rehabilitative consequences of these approaches are evident, however, and they are often used as part of a multimodal treatment approach. The potential for enhanced comfort, particularly in those with myofascial pains, provides another justification for continued emphasis on these therapeutic modalities. When chronic pain patients are encountered in psychiatric practice who have not had recent treatment with physiatric approaches, referral for assessment (or reassessment) is warranted.

Neurosurgical Approaches

Procedures designed to surgically denervate the painful part have been developed for every level of the nervous system, from peripheral nerve to cortex (Gybels & Sweet, 1989). These procedures have been extensively applied to the management of cancer pain; in this setting, cordotomy is the most useful. These modalities are rarely appropriate for patients with nonmalignant pain, as permanent relief is uncommon and the risks associated with the procedure, including the risk of inducing a new persistent pain, are too great given the long life expectancies of these patients. The potential exceptions to this view include the use of peripheral neurectomy in patients with painful peripheral mononeuropathies, such as that caused by neuroma, and the use of the dorsal root entry zone lesion in patients with avulsion of neural plexus (Young, 1990).

It is also useful to note the existence of other neurosurgical procedures, such as lobotomy or cingulotomy, which are not antinociceptive but, rather, reduce the affective concomitants of the pain. These approaches are now performed rarely.

Psychologic Approaches

An essential aspect of the comprehensive assessment is to clarify the psychologic needs of the patient and appropriately target psychologic interventions. Specific cognitive and behavioral approaches have become widely accepted in the management of pain (Turk, Meichenbaum, & Genest, 1983). Cognitive approaches comprise numerous procedures, including relaxation training, distraction techniques, hypnosis, and biofeedback, all of which may enhance a patient's sense of personal control and potentially reduce pain. Behavioral therapy may be effective in improving the functional capabilities of the patient with chronic pain (Fordyce et al., 1986). These cognitive and behavioral approaches, combined with intensive physiatric therapies, are the foundation of the multidisciplinary pain management clinic. As discussed previously, the multidisciplinary pain clinic is appropriately considered to offer optimal treatment to a subgroup of patients with chronic nonmalignant pain associated with affective disturbances and profound disability.

The use of cognitive or behavioral psychologic interventions may also be appropriate outside the usual multidisciplinary pain management setting. The purpose of the comprehensive pain assessment described previously is to characterize the functional and psychosocial concomitants of the pain and the resources presented by the patient, and from this information determine the most efficient therapeutic program that targets specific problems, including pain. Those patients whose level of disability is not sufficient to warrant referral to a multidisciplinary pain program, or who are not amenable to treatment in such a setting, may still benefit from psychologic interventions designed to lessen discomfort, improve coping, or enhance function. For example, many patients could probably benefit from simple cognitive techniques, such as relaxation training or distraction, which may provide periods of increased comfort and bolster a sense of personal control over the pain. Other types of cognitive therapy, such as biofeedback training or hypnosis, clearly help some patients as well. There is no strong evidence that any specific cognitive technique is better than any other, and patients usually receive training in whatever approaches are available to the clinician.

In a similar way, patients who could benefit from behavioral therapy can be offered such an approach as part of multimodal pain treatment organized by the practitioner. For example, a graduated exercise program can be implemented through the use of an activity diary maintained by the patient. The patient is provided with detailed instructions about the timing and nature of the exercise (e.g., swimming, walking, stretching) and uses the diary to document adherence to the program. By referring to the diary, the clinician can identify areas in need of additional assistance and can chart the patient's progress, the demonstration of which may be strong encouragement to the patient.

Finally, comprehensive pain assessment may indicate the need for other types of psychological treatments, including individual insight-oriented therapy or family therapy. Again, these approaches may be viewed as part of a multimodal pain-oriented therapy, the goals of which encompass both enhanced comfort and improved function—physical and psychosocial.

CONCLUSION

Management of the patient with chronic pain requires a comprehensive assessment and the ability to implement a multimodal treatment plan that is targeted to the specific disturbances presented by the patient. Cooperation among individuals in different disciplines is often needed to provide optimal care to these patients. Although a subgroup of patients require the type of intensive management and expertise available at a multidisciplinary pain clinic, many patients can be greatly helped through the efforts of a single concerned clinician who can organize a suitable pain management program.

REFERENCES

ABRAM SE, & LIGHTFOOT RW (1981). Treatment of longstanding causalgia with prazosin. Regional Anesth 6:79–81.

ALBE-FESSARD D, & LOMBARD MC (1982). Use of an animal model to evaluate the origin of deafferentation pain and protection against it. In JJ BONICA, U LINDBLOM, & A IGGO, (Eds.), Advances in pain research and therapy: Proceedings of the third world congress on pain (Vol. 5; pp. 691–700). New York: Raven Press.

American Psychiatric Association (1980). Diagnostic and statistical manual of mental disorder (3rd ed.). Washington, DC: American Psychiatric Association.

ANTHONY M, & LANCE JW (1969). MAO inhibition in the treatment of migraine. Arch Neurol 21:263–268.

ARNER S, & MYERSON BA (1988). Lack of analgesic effect of opioids on neuropathic and idiopathic forms of pain. Pain 33:11–23.

BEAVER WT, WALLENSTEIN SM, HOUDE RW, ET AL. (1966). A comparison of the analgesic effects of methotrimeprazine and morphine in patients with cancer. Clin Pharmacol Ther 7:436–446.

BERCEL NA (1977). Cyclobenzaprine in the treatment of skeletal muscle spasm in osteoarthritis of the cervical and lumbar spine. Curr Ther Res 22:462–468.

BESSON J-M, & CHAOUCH A (1987). Peripheral and spinal mechanisms of nociception. Physiol Rev 67:67–185.

BIRKELAND IW, & CLAWSON DK (1968). Drug combinations with orphenadrine for pain relief associated with muscle spasm. Clin Pharmacol Ther 9:639–646.

BLOM S (1963). Tic douloureux treated with new anticonvulsant. Arch Neurol 9:285–290.

BLOOMFIELD S, SIMARD-SAVOIE S, BERNIER J, ET AL. (1964). Comparative analgesic activity of levomepromazine and morphine in patients with chronic pain. Can Med Assoc J 90:1156–1159.

BONICA JJ (1990). Definitions and taxonomy of pain. In JJ BONICA (Ed.), The management of pain (Ch. 2, pp. 18–27). Philadelphia: Lea & Febiger.

BRAHAM J, & SAIA A (1960). Phenytoin in the treatment of trigeminal and other neuralgias. Lancet 2:892–893.

BRESCIA FJ, PORTENOY RK, RYAN M, ET AL. (1992). Pain, opioid use and survival in hospitalized patients with advanced cancer. J Clin Oncol 10:149–155.

BRUERA E, CHADWICK S, BRENNEIS C, ET AL. (1987). Methylphenidate associated with narcotics for the treatment of cancer pain. Cancer Treat Rep 71:67–70.

BUTLER S (1984). Present status of tricyclic antidepressants in chronic pain. In CR BENEDETTI, CR CHAPMAN, & G MORICCA (Eds.), Advances in pain research and therapy: Recent advances in the management of pain (Vol. 7; pp. 173–198). New York: Raven Press.

CACCIA MR (1975). Clonazepam in facial neuralgia and cluster headache: Clinical and electrophysiological study. Eur Neurol 13:560–563.

CALDWELL JR, ROTH SH, WU WC, ET AL. (1987). Sucralfate treatment of nonsteroidal anti-inflammatory drug-induced gastrointestinal symptoms and mucosal damage. Am J Med 83 (suppl 3B):74–82.

CAMPBELL FG, GRAHAM JG, & ZILKHA KJ (1966). Clinical trial of carbamazepine (Tegretol) in trigeminal neuralgia. J Neurol Neurosurg Psychiatry 29:265–267.

Cantor FK (1972). Phenytoin treatment of thalamic pain. Br Med J 2:590.

Carette S, McCain GA, Bell DA, et al. (1986). Evaluation of amitriptyline in primary fibrositis. Arthritis Rheum 29:655–659.

Chadda VS, & Mathur MS (1978). Double blind study of the effects of diphenylhydantoin sodium in diabetic neuropathy. J Assoc Physicians India 26:403–406.

Chapman CR (1989). Giving the patient control of opioid analgesic administration. In CS Hill & WS Fields (Eds.), Advances in pain research and therapy: Drug treatment of cancer pain in a drug-oriented society (Vol. 11; pp. 339–352). New York: Raven Press.

Charney DS, Menkes DB, & Heninger FR (1981). Receptor sensitivity and the mechanism of action of antidepressant treatment. Arch Gen Psychiatry 38:1160–1180.

Couch JR, Ziegler DK, & Hassanein R (1976). Amitriptyline in the prophylaxis of migraine effectiveness and relationship of antimigraine and antidepressant effects. Neurology 26:121–127.

Cousins M (1989). Acute and postoperative pain. In PD Wall & R Melzack (Eds.), Textbook of pain (Ch. 18, pp. 284–305). Edinburgh: Churchill Livingstone.

Cousins MJ, Dwyer B, & Gibb D (1988). Chronic pain and neurolytic neural blockade. In MJ Cousins & PO Bridenbaugh (Eds.), Neural blockade in clinical anesthesia and management of pain (2nd ed., Ch. 29.2, pp. 1053–1084). Philadelphia: JB Lippincott.

Danesh BJZ, Saniabadi AR, Russell RI, et al. (1987). Therapeutic potential of choline magnesium trisalicylate as an alternative to aspirin for patients with bleeding tendencies. Scott Med J 32:167–168.

Davidoff G, Guarracini M, Roth E, et al. (1987). Trazodone hydrochloride in the treatment of dysesthetic pain in traumatic myelopathy: A randomized, double-blind, placebo-controlled study. Pain 29:151–161.

Dejgard A, Petersen P, & Kastrup J (1988). Mexiletine for treatment of chronic painful diabetic neuropathy. Lancet 1:9–11.

Deyo RA, Walsh NE, Martin DC, et al. (1990). A controlled trial of transcutaneous electrical nerve stimulation (TENS) and exercise for chronic low back pain. N Engl J Med 332:1627–1634.

Diamond S, & Baltes BJ (1971). Chronic tension headache—treatment with amitriptyline—a double blind study. Headache 11:110–116.

Eberhard G, et al. (1988). A double-blind randomized study of clomipramine versus maprotiline in patients with idiopathic pain syndromes. Neuropsychobiology 19:25–34.

Edelson JT, Tosteson ANA, & Sax P (1990). Cost-effectiveness of misoprostol for prophylaxis agains nonsteroidal anti-inflammatory drug-induced gastrointestinal tract bleeding. JAMA 264:41–47.

Edwards WT (1990). Optimizing opioid treatment of postoperative pain. J Pain Symptom Manage 5:S24–S36.

Ekbom K (1972). Carbamazepine in the treatment of tabetic lightning pains. Arch Neurol 26:374–378.

Elliot F, Little A, & Milbrandt W (1976). Carbamazepine for phantom limb phenomena. N Engl J Med 295:678.

Espir MLE, & Millac P (1970). Treatment of paroxysmal disorders in multiple sclerosis with carbamazepine (Tegretol). J Neurol Neurosurg Psychiatry 33:528–531.

Evans W, Gensler F, Blackwell B, et al. (1973). The effects of antidepressants drugs on pain relief and mood in the chronically ill. Psychosomatics 14:214–219.

Fernandez F, Adams F, & Holmes VF (1987). Analgesic effect alprazolam in patients chronic, organic pain of malignant origin. J Clin Psychopharmacol 7:167–169.

Ferrer-Brechner T, & Ganz P (1984). Combination therapy with ibuprofen and methadone for chronic cancer pain. Am J Med 77:78–83.

Foley KM (1985). The treatment of cancer pain. N Engl J Med 313:84–95.

Fordyce WE, Brockway J, Bergman J, et al. (1986). A control group comparison of behavioral versus traditional management methods of acute back pain. J Behav Med 9:127–140.

Forrest WH, Brown B, Brown C, et al. (1977). Dextroamphetamine with morphine for the treatment of postoperative pain. N Engl J Med 296:712–715.

France RD, Urban BJ, & Keefe FJ (1984). Long-term use of narcotic analgesics in chronic pain. Soc Sci Med 19:1379–1382.

Frank WO, Wallin BA, Berkowitz JM, et al. (1989). Reduction of indomethacin induced gastroduodenal mucosal injury and gastrointestinal symptoms with cimetidine in normal subjects. J Rheumatol 16:1249–1252.

Freis JF, Miller SF, Spitz PW, et al. (1989). Toward an epidemiology of gastropathy associated with nonsteroidal anti-inflammatory drug use. Gastroenterology 16:815–823.

Fromm GH, Terence CF, & Chatta AS (1984). Baclofen in the treatment of trigeminal neuralgia. Ann Neurol 15:240–247.

Getto CJ, Sorkness CA, & Howell T (1987). Antidepressants and chronic nonmalignant pain: A review. J Pain Symptom Manage 2:9–18.

Ghostine SY, Comair YG, Turner DM, et al. (1984). Phenoxybenzamine in the treatment of causalgia. J Neurosurg 60:1263–1268.

Gingras MA (1976). A clinical trial of Tofranil in rheumatic pain in general practice. J Int Med Res 4:41–49.

Glazer S, & Portenoy RK (1991). Systemic local anesthetics in pain control. J Pain Symptom Manage 6:30–39.

Gold RH (1978). Treatment of low back syndrome with oral orphenadrine citrate. Curr Ther Res 23:271–276.

Gonzales GR, Elliott KJ, Portenoy RK, et al. (1991). The impact of a comprehensive evaluation in the management of cancer pain. Pain 47:141–144.

Graham DY, Agrawal NM, & Roth SH (1988). Prevention of NSAID- induced gastric ulcer with misoprostol: Multicentre, double-blind, placebo-controlled trial. Lancet 2:1277–1280.

Green JB (1961). Dilantin in the treatment of lightning pains. Neurology (Minneap) 11:257–258.

Gybels JM, & Sweet WH (1989). Neurosurgical treatment of persistent pain. Basel: Karger.

Hameroff SR, Cork RC, Scherer K, et al. (1982). Doxepin effects on chronic pain, depression and plasma opioids. J Clin Psychiatry 43:22–27.

Hammond DL (1985). Pharmacology of central pain-modulating networks (biogenic amines and nonopioid analgesics). In HL Fields, R Dubner, & F Cervero (Eds.), Advances in pain research and therapy: Proceedings on the fourth world congress on pain (Vol. 9; pp. 499–513). New York: Raven Press.

Hatangdi VS, Boas RA, & Richards EG (1976). Postherpetic neuralgia: management with antiepileptic and tricyclic drugs. In JJ Bonica et al. (Eds.), Advances in pain research and therapy (Vol. 1; pp. 583–587). New York: Raven Press.

Health and Public Policy Committee, American College of Physician (1983). Drug therapy for severe chronic pain in terminal illness. Ann Intern Med 99:870–873.

Higgs GA, & Moncada S (1983). Interaction of arachidonate products with other pain mediators. In JJ Bonica, U Lindblom, A Iggo (Eds.), Advances in pain research and therapy (Vol. 5; pp. 617–626). New York: Raven Press.

HOUDE RW, & WALLENSTEIN SL (1966). Analgesic power of chlorpromazine alone and in combination with morphine [abstract] Fed Proc 14:353.

JOHANSSON F, & VON KNORRING L (1979). A double-blind controlled study of serotonin uptake inhibitor (zimelidine) versus placebo in chronic pain patients. Pain 7:69–78.

KANNER RM, & FOLEY KM (1981). Patterns of narcotic drug use in a cancer pain clinic. Ann NY Acad Sci 362:161–172.

KELLGREN JG (1939). On distribution of pain arising from deep somatic structures with charts of segmental pain areas. Clin Sci 4:35–46.

KILLIAN JM, & FROMM GH (1968). Carbamazepine in the treatment of neuralgia: Use and side effects. Arch Neurol 19:129–136.

KISHORE-KUMAR R, MAX MB, SCHAFER SC, ET AL. (1990). Desipramine relieves postherpetic neuralgia. Clin Pharmacol Ther 47:305–312.

KOCHER R (1976). Use of psychotropic drugs for the treatment of chronic severe pain. In JJ BONICA ET AL. (Eds.), Advances in pain research and therapy (Vol. 1; pp. 279–282). New York: Raven Press.

KOZIN F, RYAN LM, CARERRA GF, ET AL. (1981). The reflex sympathetic dystrophy syndrome (RSDS): III. scintigraphic studies, further evidence for the therapeutic efficacy of systemic corticosteroids, and proposed diagnostic criteria. Am J Med 70:23–29.

KREEGER W, & HAMMILL SC (1987). New antiarrhythmic drugs: Tocainide, mexiletine, flecainide, encainide, and amiodarone. Mayo Clin Proc 62:1033–1050.

KVINSDAHL B, MOLIN J, FROLAND A, ET AL. (1984). Imipramine treatment of painful diabetic neuropathy. JAMA 251:1727–1730.

LANGOHR HD, STOHR M, & PETRUCH F (1982). An open and double-blind cross-over study on the efficacy of clomipramine (Anafranil) in patients with painful mono- and polyneuropathies. Eur Neurol 21: 309–317.

LASAGNA L, & DEKORNFELD TJ (1961). Methotrimeprazine—a new phenothiazine derivative with analgesic properties. JAMA 178:119–122.

LASCELLES RG (1966). Atypical facial pain and depression. Br J Psychiatry 122:651–659.

LASKA EM, SUNSHINE A, MUELLER F, ET AL. (1984). Caffeine as an analgesic adjuvant. JAMA 251:1711–1718.

LECHIN F, VAN DER DYS B, LECHIN ME, ET AL. (1989). Pimozide therapy for trigeminal neuralgia. Arch Neurol 9:960–962.

LINDSTROM P, & LINDBLOM U (1987). The analgesic effect of tocainide in trigeminal neuralgia. Pain 28:45–50.

LOCKMAN LA, HUNNINGHAKE DB, DRIVIT W, ET AL. (1973). Relief of pain of Fabry's disease by diphenylhydantoin. Neurology (Minneap) 23:871–875.

LOESER JD, & EGAN KJ (1989). History and organization of the University of Washington Multidisciplinary Pain Center. In JD LOESER & KJ EGAN (Eds.), Managing the chronic pain patient: Theory and practice at the University of Washington Multidisciplinary Pain Center (Ch. 1, pp. 3–20). New York: Raven Press.

LOESER JD, WARD AA, & WHITE LE (1968). Chronic deafferentation of human spinal cord neurons. J Neurosurg 29:48–50.

MARGOLIS LH, & GIANASCOL AJ (1956). Chlorpromazine in thalamic pain syndrome. Neurology (Minneap) 6:302–304.

MARTIN G (1981). The management of pain following laminectomy for lumbar disc lesions. Ann R Coll Surg Engl 63:244–252.

MAX MB, CULNANE M, SCHAFER SC, ET AL. (1987). Amitriptyline relieves diabetic neuropathy pain in patients with normal or depressed mood. Neurology 37:589–594.

MAX MB, KISHORE-KUMAR R, SCHAFER SC, ET AL. (1991). Efficacy of desipramine in painful diabetic neuropathy: A placebo-controlled trial. Pain 45:3–9.

MAX MB, SCHAFER SC, CULNANE M, ET AL. (1988). Association of pain relief with drug side effects in post-herpetic neuralgia: A single-dose study of clonidine, codeine, ibuprofen, and placebo. Clin Pharmacol Ther 43:363–371.

MCGIVNEY WT, & GROOKS GM (1984). The care of patients with severe chronic pain in terminal illness. JAMA 251:1182–1188.

MELZACK R, & CASEY KL (1968). Neurophysiology of pain. In RA STERNBACH (Ed.), The Psychology of Pain (2nd ed., pp. 1–24). New York: Raven Press.

MERSKEY H (1986). Classification of chronic pain: Descriptions of chronic pain syndromes and definitions of pain terms. Pain [suppl] 3:S1–225.

MILLER R, EISENKRAFT JB, COHEN M, ET AL. (1986). Midazolam as an adjunct to meperidine analgesia for postoperative pain. Clin J Pain 2:37–43.

MULLAN S (1973). Surgical management of pain in cancer of the head and neck. Surg Clin North Am 53:203–210.

NATHAN PW (1978). Chlorprothixene (Taractan) in postherpetic neuralgia and other severe pains. Pain 5:367–371.

NOYES R, BRUNK SF, AVERY DH, ET AL. (1975). The analgesic properties of delta-9-tetrahydrocannabinol and codeine. Clin Pharmacol Ther 18:84–89.

NYSTROM B, & HAGBARTH KE (1981). Microelectrode recordings from transected nerves in amputees in phantom limb pain. Neurosci Lett 27:211–216.

O'HARA DA, FRAGEN RJ, KINZER M, ET AL. (1987). Ketorolac tromethamine as compared with morphine sulfate for treatment of postoperative pain. Clin Pharmacol Ther 41:556–561.

OKASHA A, GHALEB AA, & SADEK A (1973). A double-blind trial for the clinical management of psychogenic headache. Br J Psychiatry 122:181–183.

PEIRIS JB, PERERA GLS, DEVENDRA SV, ET AL. (1980). Sodium valproate in trigeminal neuralgia. Med J Aust 2:278.

PERRY S, & HEIDRICH G (1982). Management of pain during debridement: A survey of U.S. burn units. Pain 13:267–280.

PILOWSKY I, HALLET EC, BASSETT KL, ET AL. (1982). A controlled study of amitriptyline in the treatment of chronic pain. Pain 14:169–179.

PORTENOY RK (1990a). Chronic opioid therapy in non-malignant pain. J Pain Symptom Manage 5:S46–S62.

PORTENOY RK (1990b). Pain and quality of life: clinical issues and implications for research. Oncology 4:172–178.

PORTENOY RK (1990c). Pharmacologic management of chronic pain. In HL FIELDS (Ed.), Pain syndromes in neurology (Ch. 11, pp. 257–278). London: Butterworths.

PORTENOY RK (1991). Issues in the management of neuropathic pain. In A BASBAUM & J-M BESSON (Eds.), Towards a new pharmacotherapy of pain (Ch. 12, pp. 393–416). New York: John Wiley & Sons.

PORTENOY RK, & FOLEY KM (1986). Chronic use of opioid analgesics in non-malignant pain: Report of 38 cases. Pain 25:171–186.

PORTENOY RK, & FOLEY KM (1989). The management of cancer pain. In JC HOLLAND & JH ROLAND (Eds.), Handbook of psychooncology: Psychological care of the patient with cancer (Ch. 32, pp. 369–382). New York: Oxford University Press.

PORTENOY RK, & HAGEN NA (1990). Breakthrough pain: Definition, prevalence and characteristics. Pain 41:273–281.

PORTER J, & JICK H (1980). Addiction rare in patients treated with narcotics. N Engl J Med 302:123.

PROUGH DS, MCLESKEY CH, BORSHY GG, ET AL. (1985). Efficacy of oral nifedipine in the treatment of reflex sympathetic dystrophy. Anesthesiology 62:796–799.

RAFTERY H (1979). The management of postherpetic pain using sodium valproate and amitriptyline. J Irish Med Assoc 72:399–401.

RAJ PP (1988). Prognostic and therapeutic local anesthetic block. In MJ COUSINS & PO BRIDENBAUGH (Eds.), Neural blockade in clinical anesthesia and management of pain (2nd ed., Ch. 27.2, pp. 899–934). Philadelphia: JB Lippincott.

RASKIN NH, LEVINSON SA, HOFFMAN PM, ET AL. (1974). Postsympathectomy neuralgia: Amelioration with diphenylhydantoin and carbamazepine. Am J Surg 128:75–78.

RICHELSON E (1979). Tricyclic antidepressants and neurotransmitter receptors. Psychiatry Ann 9:186–194.

ROBINSON MG, GRIFFIN JW, BOWERS J, ET AL. (1989). Effect of ranitidine gastroduodenal mucosal damage induced by nonsteroidal anti-inflammatory drugs. Dig Dis Sci 34:424–428.

ROCKLIFF BW, & DAVIS EH (1966). Controlled sequential trials of carbamazepine in trigeminal neuralgia. Arch Neurol 15:129–136.

ROMANO JM, & TURNER JA (1985). Chronic pain and depression: Does the evidence support a relationship? Psychol Bull 97:18–34.

ROTH SH (1988). NSAID and gastropathy: A rheumatologist's review. J Rheumatol 15:912–919.

ROTH SH, BENNETT RE, MITCHELL CS, ET AL. (1987). Cimetidine therapy in nonsteroidal anti-inflammatory drug gastropathy: Double-blind long-term evaluation. Arch Intern Med 147:1798–1801.

RULL JA, QUIBRERA R, GONZALEZ-MILAN H, ET AL. (1969). Symptomatic treatment of peripheral diabetic neuropathy with carbamazepine (Tegretol): Double blind cross-over trial. Diabetologia 5:215–218.

RUMORE MM, & SCHLICHTING DA (1986). Clinical efficacy of antihistamines as analgesics. Pain 25:7–22.

SANDLER DP, SMITH JC, WEINBERG CR, ET AL. (1989). Analgesic use and chronic renal disease. N Engl J Med 320:1238–1243.

SELTZER J, DEWAR T, POLLACK RL, ET AL. (1983). The effects of dietary tryptophan on chronic maxillofacial pain and experimental pain tolerance. J Psychiatr Res 17:181–186.

SHAFAR J, TALLETT ER, & KNOWLSON PA (1972). Evaluation of clonidine in prophylaxis of migraine. Lancet 1:403–407.

SIMSON G (1974). Propranolol for causalgia and Sudek's atrophy. JAMA 227:327.

SINDRUP SH, GRAM LF, BROSEN K, ET AL. (1990). The selective serotonin reuptake inhibitor paroxetine is effective in the treatment of diabetic neuropathy symptoms. Pain 42:135–144.

SINGH PN, SHARMA P, GUPTA PK, ET AL. (1981). Clinical evaluation of diazepam for relief of postoperative pain. Br J Anaesth 53:831–836.

SMITH CM (1965). Relaxants of skeletal muscle. In WS ROOT & FG HOFFMANN (Eds.), Physiological pharmacology (Vol. 2; Ch. 1, pp. 2–96). Orland, FL: Academic Press.

SOSNOWSKI M, & YAKSH TL (1990). Spinal administration of receptor-selective drugs as analgesics: New horizon. J Pain Symptom Manage 5:204–213.

SPIEGEL K, KALB R, & PASTERNAK GW (1983). Analgesic activity of tricyclic antidepressants. Ann Neurol 13:462–465.

STAMBAUGH JE, & LANCE C (1983). Analgesic efficacy and pharmacokinetic evaluation of meperidine and hydroxyzine, alone and in combination. Cancer Invest 1:111–117.

SUNSHINE A, & OLSON NZ (1989). Non-narcotic analgesics. In PD WALL & R MELZACK (Eds.), Textbook of pain (2nd ed., Ch. 47, pp. 670–685). New York: Churchill Livingstone.

SWERDLOW M (1984). Anticonvulsant drugs and chronic pain. Clin Neuropharmacol 7:51–82.

SWERDLOW M, & CUNDILL JG (1981). Anticonvulsant drugs used in the treatment of lancinating pains: A comparison. Anesthesia 36:1129–1132.

SWERDLOW M, & STJERNSWARD J (1982). Cancer pain relief—an urgent problem. World Health Forum 3:325–330.

TABIRA T, SHIBASAKI H, & KUROIWA Y (1983). Reflex sympathetic dystrophy (causalgia) treatment with guanethidine. Arch Neurol 40:430–432.

TAN Y-M, & CROESE J (1986). Clonidine and diabetic patients with leg pains. Ann Intern Med 105:633.

TAUB A (1982). Opioid analgesics in the treatment of chronic intractable pain of non-neoplastic origin. In LM KITAHATA & D COLLINS (Eds.), Narcotic analgesics in anesthesiology (Ch. 10, pp. 199–208). Baltimore: Williams & Wilkins.

TAYLOR PH, GRAY K, BICKNELL RG, ET AL. (1977). Glossopharyngeal neuralgia with syncope. J Laryngol Otol 91:859–868.

TENNANT FS, ROBINSON D, SAGHERIAN A, ET AL. (1988). Chronic opioid treatment of intractable non-malignant pain. Pain Management Jan/Feb:18–36.

TENNANT FS, & UELMAN GF (1983). Narcotic maintenance for chronic pain: Medical and legal guidelines. Postgrad Med 73:81–94.

TOREBJORK HE, OCHOA JL, & SCHADY W (1984). Referred pain from intraneural stimulation of muscle fascicles in the median nerve. Pain 18:145–156.

TRAVELL JG, & SIMONS DG (1983). Myofascial pain and dysfunction: The trigger point manual. Baltimore: Williams & Wilkins.

TURK DC, & RUDY TE (1990). Toward an empirically derived taxonomy of chronic pain patients: Integration of psychological assessment data. J Consult Clin Psychol 56:233–238.

TURK DC, & RUDY TE (1990). The robustness of an empirically derived taxonomy of chronic pain patients. Pain 43:27–35.

TURK DC, MEICHENBAUM D, & GENEST M (1983). Pain and behavioral medicine: A cognitive-behavioral perspective. New York: Guilford Press.

TWYCROSS RG (1974). Clinical experience with diamorphine in advanced malignant disease. Int J Clin Pharmacol Ther 9:184–198.

URBAN BJ, FRANCE RD, STEINBERGER DL, ET AL. (1986). Long-term use of narcotic/antidepressant medication in the management of phantom limb pain. Pain 24:191–197.

VANE JR (1971). Inhibition of prostaglandin synthesis as a mechanism of action for aspirin-like drugs. Nature [New Biol] 231:232–235.

VENTAFRIDDA V, BONEZZI C, CARACENI A, ET AL. (1987). Antidepressants for cancer pain and other painful syndromes with deafferentation component: Comparison of amitriptyline and trazodone. Ital J Neurol Sci 8:579–587.

WALL PD, & GUTNICK M (1974). Ongoing activity in peripheral nerves: 2. The physiology and pharmacology of impulses originating in a neuroma. Exp Neurol 43:580–593.

WAN LU C, URBAN B, & FRANCE RD (1988). Long-term narcotic therapy in chronic pain. Presented at the Canadian Pain Society and American Pain Society Joint Meeting, Toronto.

WATSON CPN, & EVANS RJ (1985). A comparative trial of amitriptyline and zimelidine in postherpetic neuralgia. Pain 23:387–394.

WATSON CPN, EVANS RJ, REED K, ET AL. (1982). Amitriptyline versus placebo in postherpetic neuralgia. Neurology 32:671–673.

WATSON CPN, EVANS RJ, & WATT VR (1990). Amitriptyline versus maprotiline in postherpetic neuralgia. Presented at the Ninth Annual Scientific Meeting of the American Pain Society, St. Louis, October 25–28.

WEINBERGER J, NICKLAS WJ, & BERL S (1976). Mechanism of action of anticonvulsants. Neurology (Minneap) 26:162–173.

WEIS O, SRIWATANAKUL K, & WEINTRAUB M (1982). Treatment of postherpetic neuralgia and acute herpetic pain with amitriptyline and perphenazine. A Afr Med J 62:274–275.

Willer J-C, De Broucker T, Bussel B, et al. (1989). Central analgesic effect of ketoprofen in humans: Electrophysiological evidence for a supraspinal mechanism in a double-blind and cross-over study. Pain 38:1–8.

Willis WD (1985). The pain system: the neural basis of nociceptive transmission in the mammalian nervous system. Basel: Karger.

World Health Organization (1986). Cancer pain relief. Geneva: WHO.

Yang JC, Clark WC, Ngai SH, et al. (1979). Analgesic action and pharmacokinetics of morphine and diazepam in man: An evaluation by sensory decision theory. Anesthesiology 51:495–502.

Yosselson-Superstine S, Lipman AG, & Sanders SH (1985). Adjunctive antianxiety agents in the management of chronic pain. Isr J Med Sci 21:113–117.

Young RF (1990). Clinical experience with radiofrequency and laser DREZ lesions. J Neurosurg 72:715–720.

III Neuropsychiatry

18 Neurologic assessment, neurodiagnostic tests, and neuropsychology in medical psychiatry

BARRY S. FOGEL, M.D., AND DAVID FAUST, Ph.D.

NEUROPSYCHIATRIC ASSESSMENT IN MEDICAL PSYCHIATRY

Because there are a number of different reasons for undertaking neurologic assessment, no "cookbook" approach to neurologic diagnosis is appropriate. Instead, when embarking on neurologic assessment of the psychiatric patient, the psychiatrist should begin with a set of diagnostic hypotheses generated by the psychiatric history and the mental status examination, and should form a set of specific questions to be answered by the neurologic assessment. For example, if a patient with suspected senile dementia is referred for a computerized tomography (CT) scan with no specific question posed, the radiologist often reports "cortical atrophy, consistent with age." If the psychiatrist poses the question "This patient has severe dementia; is the degree of cortical atrophy compatible with severe Alzheimer's disease?", this focuses the radiologist on a more quantitative and statistical assessment of the CT scan findings. If multiple sclerosis (MS) is suspected in an apparently hysterical patient, specialized history taking and examination may be needed to detect mild problems with eye movements or visual function that probably would not be tested on a "routine" neurologic examination. Thus, neurologic assessment for diagnostic purposes should begin with a differential diagnosis in mind and a set of specific questions on which to focus with the neurologic consultant.

Specific neurologic tests and examination procedures are best selected with reference to a particular population and a set of potential diagnoses that occur frequently in that population. For example, neurologic examination of elderly patients with psychiatric problems always should include several tests for diffuse cerebral dysfunction, a careful cognitive mental status examination to exclude dementia or delirium, and careful testing of gait because of the high prevalence of gait disorders in the elderly. By contrast, a meticulous sensory examination rarely would be warranted or reliable. As another example of tailoring evaluation to a specific population, evaluation of alcoholics with acute mental status changes always should include a therapeutic trial of thiamine, and might include tests of vitamin status. Recorded "routine" neurologic examinations of patients with autopsy-proved Wernicke-Korsakoff syndrome were normal in 27% of 64 patients undiagnosed in life, and an additional 34% of those patients showed mental status changes only, without ataxia or eye signs (Harper, Giles, & Finlay-Jones, 1986). A thiamine trial, although certainly not "routine," is indicated for alcoholics because of their high base rate of thiamine deficiency and the insensitivity of routine examination to its CNS effects.

As a final example of population-specific evaluation, patients who have been treated with neuroleptics always require a special examination for abnormal involuntary movements. Published standard neurologic examinations give some attention to the problem of involuntary movements, but it has been demonstrated that a formal movement disorder assessment, such as the Abnormal Involuntary Movement Scale (AIMS), can greatly enhance the detection of tardive dyskinesia and related disorders (Munetz & Schulz, 1986). Because of the extraordinary medical-legal issues raised by prescribing neuroleptics in the face of tardive dyskinesia, the examination for tardive dyskinesia cannot be omitted or abbreviated. The tardive dyskinesia issue illustrates the point of focusing the neurologic examination, particularly if it is to be done by someone else, such as a neurologist or internist. Comments about tardive dyskinesia or its absence often fail to appear in internists' or neurologists' consultation notes unless the psychiatrist has specifically raised the question as one to be addressed in the consultation.

Role of Formal Consultation with Neurologists and Neuropsychologists

It is becoming more usual for psychiatrists to perform an initial neurologic assessment themselves, and to order the more common ancillary tests, such as brain imaging (CT or magnetic resonance imaging [MRI]), the electroencephalogram (EEG), and even such procedures as brain mapping or single-photon emission CT (SPECT). The neurologist usually is called when a definite or potential neurologic problem has been identified by the psychiatrist's initial assessment and further diagnostic confirmation or management is desired. A potential pitfall exists when psychiatrists "sign off" the "organic" parts of their cases after the neurologist has been consulted. First, the psychiatrist, who meets most regularly with the patient, is in the best position to observe changes in neurologic status and to collect additional history of neurologic relevance from the patient and significant others. Second, the psychiatrist also is in the best position to determine whether the neurologic consultation and therapy have adequately addressed the patient's concerns. Third, many neurologic diseases are progressive or episodic, so the disease actually changes after the neurologic consultation. Patients in psychiatric treatment often are comfortable relating new physical problems and symptoms to their psychiatrist, rather than returning to a neurologic consultant they have seen only once or twice. Finally, the psychiatrist is best positioned to assess the overall impact of neurologic therapies on a patient's function, mental status, and quality of life. Appropriate feedback to the neurologist can lead to an adjustment of neurologic management to patients' individual requirements.

Neurologic patients, regardless of the source of referral, frequently display mixtures of neurologic disease, psychiatric disorders with somatization, and abnormal illness behavior. In a sample of 133 female neurologic inpatients, Creed et al. (1990) found one-third to have either a mixture of neurologic and psychiatric illness, or a diagnosable psychiatric disorder that did not fully explain the neurologic symptoms. In such cases, it is essential to avoid premature closure of neurologic diagnostic issues while proceeding to treat the psychiatric problems that are identified. Creed et al. observed that neurologists frequently assign such patients to a "functional" category, rather than explicitly identifying symptoms that remain undiagnosed. The psychiatrist consulting a neurologist often can further the diagnostic process by identifying in the referral the psychiatric problems that are known, and explicitly asking the neurologist to identify additional neurologic problems, or to determine whether the patient's neurologic symptoms are fully explained by the known psychiatric problem.

When using consultants, psychiatrists should be aware that neurologists vary greatly in their level of interest in behavior. Some neurologists perform a minimal mental status examination and have a very high threshold for diagnosing any neurologic disease in patients with known psychiatric problems. Others are trained behavioral neurologists with a deep interest in brain-behavior relationships; they conduct meticulous cognitive mental status assessments and occasionally may even overemphasize the organic components in a patient's illness. When a choice of consultants is available, the psychiatrist should bear these distinctions in mind. An individual with gross neurologic disease in need of management does not require a behavioral neurologist, particularly if the behavioral aspects of the problem will be managed by the psychiatrist. In contrast, the patient with obscure symptoms on the borderline between the neurologic and the psychiatric can profit greatly from a consultation with a behavioral neurologist, particularly if this consultation leads to critical dialogue between the behavioral neurologist and the psychiatrist.

Neuropsychologists also vary in their interests and special skills, and the variation may be even greater than for neurologists, because there are no professional regulations in psychology determining who may call himself a neuropsychologist. Some neuropsychologists restrict their activities to forming an opinion about whether brain damage is present or absent; others go considerably further and address lateralization, localization, and detailed description of the mechanism of cognitive dysfunction. Some neuropsychologists use fixed batteries for all patients; others adapt testing procedures to patient characteristics and referral questions, and may in some cases use clinically-based rather than standardized, normed testing procedures. Test batteries themselves differ in their coverage of particular areas of cognitive disorder and are more or less sensitive or specific depending on the population to which they are applied.

Neuropsychologists also vary in their involvement in cognitive rehabilitation and in the provision of educational, therapeutic, and rehabilitative services to patients with cognitive disorders and their families. If cognitive rehabilitation or similar therapeutic intervention is contemplated, it is generally best to have the neuropsychologic evaluation done by a neuropsychologist with a significant interest in cognitive rehabilitation. The assessment then can be tailored to answer specific questions particularly relevant to rehabilitation or therapy.

The next several sections of this chapter discuss the neurologic examination of the psychiatric patient, offer specific advice for improving the neurologic examination, and review the advantages and limitations of different common neurodiagnostic tests. Neuropsychologic assessment is treated separately and in greater detail later in this chapter.

NEUROLOGIC EXAMINATION OF THE PSYCHIATRIC PATIENT

Specific Features of the Neurologic Examination

The comprehensive neurologic examination of the psychiatric patient is distinguished by several specific features. The extent to which they are addressed is a measure of the comprehensiveness of the examination. Although more limited neurologic examinations are indicated in many circumstances, a comprehensive examination is necessary before a psychiatric patient can be declared neurologically intact. Its features are: (1) careful attention to the evaluation of cognitive mental status; (2) specific assessment of frontal lobe function; (3) systematic attention to extrapyramidal signs, including tremors, rigidity, involuntary movements, and disturbances of gait and posture; (4) elicitation of "soft signs" of neurologic dysfunction; (5) detailed examination of symptomatic areas (e.g., sensory examination in patients complaining of numbness); and (6) efforts to circumvent uncooperativeness or behavioral problems to gather diagnostically relevant data. Each of these features is discussed in turn.

Cognitive Mental Status

The routine evaluation of cognitive mental status by general neurologists often is similar to the Mini-Mental State Examination, emphasizing verbal tests of attention, memory, language, and calculation. Tests of right hemisphere and frontal lobe functions are done routinely by behavioral neurologists, but frequently are omitted by general neurologists. Since neurologic disorders of psychiatric relevance may have cognitive signs limited to right hemisphere or frontal functions, this omission is acceptable only when the focus of neurologic consultation is limited (e.g., to evaluation of peripheral neuropathy) or when a full cognitive mental status examination will be done by someone else (a psychiatrist or neuropsychologist) and *subsequently integrated* into the diagnostic formulation.

Typical bedside tests of right hemisphere functions include clock drawing and figure copying tasks. Typical bedside tests of frontal lobe cognitive functions include: (1) list generation tasks; (2) alternation tasks such as Trail-Making B, or writing script *m*s and *n*s in an alternating sequence; (3) motor sequence tasks such as fist-palm-side; (4) go–no go tasks such as alternately responding and not responding to repetitive stimuli; and (5) processing conflicting stimuli, as when the patient is asked to tap twice if and only if the examiner taps once, and vice versa, or when the patient is asked to make a movement quickly when given the command in a soft voice and slowly when commanded in a loud voice (Faust & Fogel, 1989; Fogel & Eslinger, 1991; Merriam et al., 1990).

Frontal Lobe Function

Specific evaluation of frontal lobe functions includes assessment of motor as well as cognitive signs of frontal dysfunction. Both neurologic disorders such as MS and psychiatric disorders such as schizophrenia can present with frontal signs as the sole or predominant abnormality. Motor signs of frontal dysfunction include: (1) palmar or plantar grasp reflexes, (2) apraxia of gait with difficulty of initiating walking and abnormal placing of the feet, and (3) impersistence or perseveration in carrying out various commands on the neurologic examination, such as holding up the hands, maintaining lateral gaze to command, or performing rapid alternative movements.

Since many of the "frontal signs" can be produced by diffuse cerebral dysfunction, localization of the dysfunction to the frontal lobe requires demonstration of the intactness of other functions. Thus, impersistence on a motor task is more likely to represent frontal dysfunction if the patient does not show behavioral signs of delirium, and if some attentional tasks, such as counting backward, are done adequately. The grasp reflex, a motor sign of frontal dysfunction, is more specific diagnostically if it is unilateral.

Extrapyramidal Signs

The neurologic examination of the psychiatric patient emphasizes extrapyramidal signs for two major reasons. The first is that disorders of the basal ganglia, which produce such abnormalities, frequently present with psychiatric symptoms. The second is that numerous drugs used to treat psychiatric disorders produce or aggravate involuntary movements, or cause disturbances of muscle tone or gait. Full assessment of involuntary movements requires observation of the patient under various conditions—while walking, while doing a task requiring mental concentration, and while moving an uninvolved body part. Assessment of rigidity must include axial muscles, which usually are tested by examination of the neck.

TABLE 18–1. *Soft Signs*

Dysarthric speech
Right-left confusion
Adventitious overflow
Difficulty with finger-thumb opposition
Finger mirror movements
Awkward foot tapping
Extinction of one of two simultaneous examiner finger touches
Pronation/supination (rapid alternating movement) difficulties
Mixed lateral dominance
Awkward gait, including difficulty with tandem gait
Difficulties hopping on one foot
Dysgraphesthesia

Soft Signs

Soft signs of neurologic dysfunction, in contrast to hard signs such as the Babinski, are less consistently associated with demonstrable brain abnormalities on imaging or at neuropathologic examination. However, soft signs may be produced by "hard" brain damage. For example, soft signs may represent the earliest manifestation of progressive disorder, such as a degenerative dementia or brain tumor, or may be residual following recovery from traumatic brain injury or stroke.

In physically healthy psychiatric patients, soft signs most often represent individual differences in brain function not associated with gross anatomic findings on brain imaging. Epidemiologists take soft signs as suggestive evidence of an "organic" basis for various psychiatric syndromes. In medical psychiatry, soft signs must be interpreted with regard to the population of the patients being treated. Soft signs in patients with acquired immunodeficiency syndrome (AIDS) might herald the onset of AIDS dementia, particularly if they were not present when the patient was first diagnosed with human immunodeficiency virus (HIV) infection. In general, soft signs are more likely to represent specific neurologic disease if they are new, because they are so prevalent and nonspecific in the general population. Medical-psychiatric patients at high risk for central nervous system (CNS) complications may benefit from a baseline assessment of soft signs. Such signs easily elicited on a general neurologic examination include difficulty with rapid alternating movements, impaired fine finger movements such as tapping the thumb and forefinger together or subsequently touching the thumb to each of the other fingers, mild reflex asymmetry, and difficulty with graphesthesia.

A list of soft signs employed by Gardner, Lucas, and Cowdry (1987) in a study of borderline personality disorder is presented in Table 18-1. In this study 55% of normal control subjects had one or more soft signs and 32% had two or more. However, subjects with borderline personality disorder had significantly more soft signs. Excessive soft signs also have been reported in schizophrenia (Rossi et al., 1990), and have been correlated with a positive family history of psychiatric problems (Woods, Yurgelun-Todd, & Kinney, 1987) and with a greater probability of violence (Krakowski et al., 1989).

Related to the observation of soft signs is the recording of handedness. Mixed dominance or left-handedness is associated with an increased risk of tardive dyskinesia in patients treated with neuroleptics (Joseph, 1990). It is conveniently assessed with the Edinburgh Inventory (Oldfield, 1971), shown in Table 18-2.

Whether soft signs should trigger additional neurodiagnostic assessment of the medical-psychiatric patient is a crucial judgment that must include consideration of the epidemiologic setting and of the combination of the soft signs with the history and other findings. Dys-

TABLE 18–2. *Handedness Assessment Questions of the Edinburgh Inventory*

The patient is asked which hand is used for the following tasks. (Question 11 refers to which foot and question 12 to which eye.)

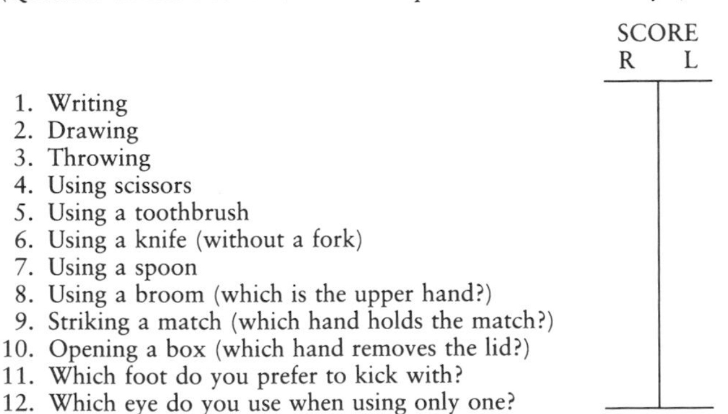

	SCORE R	SCORE L
1. Writing		
2. Drawing		
3. Throwing		
4. Using scissors		
5. Using a toothbrush		
6. Using a knife (without a fork)		
7. Using a spoon		
8. Using a broom (which is the upper hand?)		
9. Striking a match (which hand holds the match?)		
10. Opening a box (which hand removes the lid?)		
11. Which foot do you prefer to kick with?		
12. Which eye do you use when using only one?		
TOTALS		

SCORING:
For each question exclusive use of the right hand is scored as +2, exclusive use of the left hand as −2, and mixed use as +1 and −1. The scores are placed in the appropriate column and added. The laterality score is equal to the addition of the total for each column.

Reprinted by permission from Oldfield RC (1971). The assessment and analysis of handedness: The Edinburgh Inventory. Neuropsychologia 9:97–113.

arthria, for example, would trigger an MRI in a patient suspected of having MS because of a prior episode of unilateral numbness; it might simply be observed if it were a long-standing phenomenon in a chronically mentally ill patient with other soft signs and a history of school failure.

Examination of Symptomatic Areas

Detailed examination of symptomatic areas is essential because a screening neurologic examination cannot possibly test every part of the nervous system in detail. In addition to detailed assessment of cognition in patients with primary mental and behavioral complaints, other commonly occurring situations for detailed examination include: (1) examination of eye signs in patients suspected of having MS; (2) evaluation of various sensory modalities, including temperature sensation and two-point discrimination, in patients suspected of having peripheral neuropathy, such as patients suspected of alcoholism or solvent exposure; (3) detailed gait evaluation in patients with possible parkinsonism, including all those who are on neuroleptic therapy; (4) mapping of pain and sensory change in patients suspected of having nerve entrapments such as the carpal tunnel syndrome (Katz et al., 1990); and (5) olfactory testing in patients with altered smell or taste, or a history of closed head injury and possible olfactory nerve damage. Laboratory tests such as computerized visual fields, nerve conduction tests, and computerized balance testing can be viewed as a further extension of focused examination.

The literature suggests that focused bedside assessment increases the yield of abnormal findings, and that laboratory assessment increases the yield yet more. Navarro and Kennedy (1991) examined patients with type I diabetes and showed that of 46 patients with no abnormalities on standard clinical examination, 26 had abnormalities on laboratory tests of thermal sensitivity and 15 had abnormalities on nerve conduction tests. Caligiuri, Bracha, and Lohr (1989) studied rigidity in medicated schizophrenic patients. By testing muscle tone in one limb while the other was exercised, they found increased tone in 8 of 19 patients under age 50; an additional six were found to show abnormally increased tone when tested with a mechanical drive measuring the amount of force needed to displace the index finger by a fixed angle. Grenman et al. (1988) showed that electronystagmography (ENG) documented abnormalities of eye movements in patients with definite MS who had no demonstrable findings of brainstem or cerebellar abnormality on clinical examination.

Uncooperative Patients

Psychiatrists, particularly those working in emergency, acute inpatient, or public sector settings, frequently encounter patients who are uncooperative with neurologic examinations. Examination in these situations may necessarily be incomplete, but two points deserve comment. First, it is almost always possible to see a patient walk. Careful observations of gait may reveal important data about motor asymmetries, ataxia, and involuntary movements. Second, it is crucial to identify parts of a neurologic examination that were not done, and to explicitly defer a neurologic diagnosis if necessary data are missing. If urgent neurologic problems cannot be excluded from the differential diagnosis with the data available, it may be necessary to conduct further examination with the patient in restraints, or following rapid tranquilization. Formal recording of the neurologic examination should indicate when nonstandard procedures were employed in the examination.

Reliability of Neurologic Assessment

The reliability of neurologic signs depends on the experience and skills of the examiner, as well as the "hardness" of the signs in question. In a study of the accuracy of bedside diagnoses in 100 consecutive neurologic inpatients, no attending neurologists were found to make diagnostic errors based on an incomplete or inaccurate examination (Chimowitz, Logigian, & Caplan, 1990). However, abnormal sensory findings in workers exposed to organic solvents were significantly more likely to be found by a neurologist than by an occupational health physician (Maizlish et al., 1987).

The study by Chimowitz, Logigian, and Caplan (1990) also evaluated the accuracy of neurologists' bedside diagnoses in 40 patients who had unequivocal diagnoses after full neurodiagnostic investigation. Junior residents were correct 65% of the time, senior residents 75% of the time, and attending neurologists 77% of the time. The most common reason for attending neurologists' error was poor diagnostic reasoning rather than errors in examination. This suggests that medical psychiatrists should review their neurologic consultants' data as well as their impressions.

Enhancing the Neurologic Examination

We assume that the medical psychiatrist will personally perform a neurologic examination. In doing so, the psychiatrist is likely to gain little from the rote application of a textbook neurologic examination, but may learn much of value from a hypothesis-testing examination, which tailors tests and observations to the patient's presenting problem and epidemiologic setting. The following paragraphs present several suggestions for more precise and effective neurologic examination of the psychiatric patient.

The sensitivity and specificity of neurologic signs and the detection of occult organic brain disease depend on the clinical population and the diseases suspected. For instance, when MS is suspected, as it ought to be in "hysterics," bedside examination of eye movements may be insensitive to diagnostically relevant abnormalities. Measurement of eye movements in 84 patients with MS and 21 patients with optic neuritis using a simple portable electrophysiologic apparatus revealed a subclinical eye movement disorder in 80% of the definite, 74% of the probable, and 60% of the possible MS patients (Ruelen, Sanders & Hogerhuis, 1983).

Tests infrequently performed may detect confirmatory evidence of an organically-based disorder, whereas commonly used screening tests are often insensitive to psychiatrically relevant diffuse cerebral pathology. Jenkyn et al. (1977) correlated evidence of diffuse cerebral disorder as measured by the Halstead-Reitan Neuropsychological Battery with findings on neurologic examination and found 13 physical signs with a false-negative rate of less than 60%. These were the glabellar blink, the nuchocephalic reflex, the suck reflex, impairment of upward gaze, impairment of downward gaze, a tendency to keep the raised arm up after the signal to drop it, impaired visual tracking, impersistence of lateral gaze, trouble accurately recalling three items after a time delay with intervening distraction, paratonia of both arms, difficulty spelling *world* backward, paratonia of both legs, and difficulty in giving past presidents in reverse order. By contrast, they found an extremely high false-negative rate—over 85%—for such popularly used tests as the Babinski sign, double simultaneous stimulation, orientation to place and data, and recall of current presidents.

Enlarged pupils may be the only neurologic sign besides the abnormal mental state in cases of anticholinergic delirium (Brizer & Manning, 1982). In such cases, vital signs are abnormal but somatic motor and reflex signs are absent. In many other acute confusional states that include cognitive, sensory, memory, and attentional aberrations, there are no focal or lateralizing sensory or motor signs because the neurologic disorder is diffuse or multifocal.

Amnesia after an accident with head injury is sufficient for diagnosis of concussion even if there is no history of loss of consciousness (Rimel & Tyson, 1979). Experience in consulting at a busy regional trauma center suggests that evidence of concussion is not always sought by the trauma team when the patient arrives at the emergency room awake and other physical injuries need immediate care. The psychiatrist, if called afterward for assessment of posttraumatic behavior changes, must make the diagnosis retrospectively from evidence of posttraumatic amnesia. If the patient is examined soon enough after the injury, *anterograde* amnesia may still be present. If it is, neurologic dysfunction is unequivocally demonstrated.

The ability to communicate with gestures in the presence of muteness has no value in distinguishing functional from organic etiologies. A study of the nine mute patients included in 350 consecutive general hospital admissions for head injury revealed that all of the patients with mutism could communicate somewhat with gestures. However, all who were tested showed impaired writing (Levin et al., 1983). All nine recovered some speech, although the four with basal ganglion lesions made a less complete recovery. Unlike the ability to gesture, the ability to write a coherent sentence to dictation is, in general, a good high-sensitivity screening test for organically-based language disorder.

Specific visual examinations are superior to screening eye examinations. In a review (Ruelen, Sanders, & Hogerhuis, 1983) of 18 patients with probable or definite MS and 20/20 vision, 16 had one or more abnormalities on other measures of ocular function, such as visual evoked potentials and psychophysical tests of contrast sensitivity. Fifty-nine percent of these had pallor of the optic nerve head or changes in the nerve fiber layer on funduscopy. Normal visual acuity does not exclude visual system disorder. Careful funduscopic examination is mandatory to properly screen the visual system. It certainly should be done in all patients suspected of having MS—a group that includes all patients suspected of having a conversion disorder with neurologic symptoms.

The determination of pallor of the optic nerve head is easy and is frequently valuable in differential diagnosis. A reasonable set of skills for the psychiatrist would thus include examination of the optic disk, assessment of visual fields by confrontation, and assessment of visual acuity. Theoretical knowledge of other parts of the visual examination is helpful in assessing the completeness or incompleteness of examinations done by other physicians. Even eye examinations by ophthalmologists may not include all of the relevant points for assessment of MS, unless the ophthalmologist is specifically asked by the psychiatrist to look for evidence of MS.

Gait disorder may be the only hard sign of a treatable neurologic illness. Dubin, Weiss, and Zeccardi (1983) studied 1,140 patients evaluated by an emergency psychiatric service after having been cleared medically by an emergency room physician. Thirty-nine of these showed signs of disorientation, abnormal vital signs, or clouding of consciousness, and, of these, 38 were found to have had gait disturbance, weight loss, hypertension, abnormal vital signs, or a significant medical history.

In a neurologic evaluation of 50 patients over the age of 70 with gait disturbance of previously undetermined etiology, 24% had an illness that could either be treated or palliated (Sudarsky & Ronthal, 1983). Specific observation of gait is helpful in the diagnosis of Parkinson's disease and cerebellar atrophy. Robins (1983) contended that 10% to 33% of patients presenting for evaluation of dementia have a potentially reversible cause. A summary of six studies involving 503 patients revealed the leading causes of reversible dementia to be normal-pressure hydrocephalus, depression, subdural hematoma or intracranial mass, other psychiatric disorders, drugs, and thyroid disease. The neurologic conditions in this list may not produce focal motor or sensory signs, but may show only disturbances of mental status and gait. Normal-pressure hydrocephalus virtually always affects gait.

For these reasons, examination of gait is essential in any complete psychiatric evaluation. Observation of the patient walking into a room and taking a seat is an easy, unobtrusive part of any psychiatric interview. If subtle abnormalities are noticed, they can be brought out by asking the patient to perform special tasks. Mild weakness of one side can be brought out by having the patient walk on tiptoes and on heels. The foot may droop on the weak side.

In younger patients with good premorbid coordination, recently acquired problems with strength or coordination can be demonstrated by having them stand on one foot for 10 seconds, or hop on one foot 10 times. These tests are relatively easy for physically fit people, but are easily disrupted by significant neurologic disorders affecting gait or balance.

Patients who appear slightly off balance can be asked to walk on an imaginary line, touching heel to toe, as in the familiar test for intoxication. Patients suspected of parkinsonism should be observed making turns, stopping and starting on command, and speeding up and slowing down on command. Elderly patients with diffuse cerebral dysfunction and a slow, shuffling gait may not have the same difficulties as parkinsonian patients do when turning and changing pace.

Finally, it is worth mentioning that involuntary movements and posturing of the arms sometimes are most evident when the patient is walking. This evidence sometimes can help in the diagnosis of dystonic disorders, Huntington's disease, and tardive dyskinesia with limb involvement.

Neurologic examination of the very elderly patient is best interpreted with knowledge of the norms for neurologic function in advanced age. Cranial nerve examination in normal elderly persons may reveal small pupils, a slowing of pupillary reflexes, diminished upward gaze, sensorineural hearing loss, and diminished sense of smell. Elderly patients also may show a variety of visual problems related to cataracts, macular degeneration, or glaucoma (Wright & Henkind, 1983). Other abnormalities, such as nystagmus, facial weakness, diminished downward gaze, or hemianopic field defects, cannot be explained by advanced age and are presumptive signs of neurologic disease. Sensory deficits, although they may be due to age-associated changes in sensory organs, are important in the overall management of the patient, because they may significantly affect the patient's ability to communicate and to correctly perceive the environment.

Motor examination of elderly persons shows that, as a group, they are not as strong as young adults. There is great individual variability, however. The detection of moderate to severe weakness does not pose any problem for the examiner; the problems lie in distinguishing mild but significant weakness from normal age-associated muscular change. In this area, the most useful tests are tests of function, such as having the patient arise from bed, get up from a chair, or climb stairs. Some loss of muscle bulk is normal for age, particularly in the intrinsic muscles of the hands. Mild impairment in coordination is occasionally encountered in normal elderly persons, although it should not be accepted as age-related without a search for other evidence of motor system disease. Bradykinesia is common in patients over 75 but, in the absence of other parkinsonian signs, does not imply the presence of Parkinson's disease. Action tremors in the elderly usually imply benign essential tremor rather than cerebellar dysfunction. The latter should not be diagnosed without clear-cut ataxia or dysmetria. (Dysmetria is the inability of a patient to target limb movement precisely from one point to another.)

Decreased vibratory sensitivity is common in the elderly and does not necessarily imply posterior column disease, unless the abnormality is severe. Age-related losses are usually prominent only in the lower extremities, however, and a clear-cut loss of vibration sense in the upper extremities warrants careful investigation. More than mild loss of joint position sense should not be attributed to age alone.

Normal elderly persons may have palmomental or snout reflexes and often do not have ankle jerks. Babinski signs, in contrast, are rare in normal elderly persons. It is not uncommon, however, to encounter Babinski signs caused by cervical spondylosis with mild cord compression, particularly in elderly individuals with severe degenerative disease of the cervical spine. In the absence of pain, weakness, bladder dysfunction, or gait disorder, the spinal disease would not require specific treatment. The Babinski sign in this case does not imply cerebral disease.

Normal elderly persons, even active ones, may be unable to stand on one foot with their eyes closed; this may be due to changes in the proprioceptive system with aging. The bradykinesia that accompanies advanced age may result in a slow, cautious gait, with diminished arm swing. Significant instability of gait, asymmetry of gait, and failure of postural reflexes, however, always imply CNS disease. Unfortunately, many elderly people with gait disturbance and a tendency to fall have multifactorial gait disturbance, with sensory, motor, and cognitive factors all contributing. It is nonetheless worthwhile to investigate all elderly persons with gait disturbance for specific etiologies, particularly such treatable ones as vitamin B_{12} deficiency, hypothyroidism, normal-pressure hydrocephalus, Parkinson's disease, and mild toxic-metabolic delirium.

The material just presented is elaborated and abundantly referenced in a review by Wolfson and Katzman (1992).

Many neurologic disorders are diagnosable more on the basis of history than on the basis of the examination. In this area, psychiatrists may have an edge over neurologists because of their fine history-taking skills and their habit of using collateral information. When a group of neurologists independently performed neurologic histories and examinations on a group of acute stroke patients, there was substantial disagreement between examiners on basic historic points, such as the history of past stroke or transient ischemic attack (TIA), and even on whether the patient had a headache at the time of onset. Interobserver variability was less on the physical examination, particularly regarding weakness and global aphasia; sensory examination was more variable. Disagreement between independent examiners increased when patients were more difficult to examine (Shinar, Gross, & Mohr, 1985). When a patient is difficult to examine on initial evaluation, the treating psychiatrist usually has the opportunity to repeat the neurologic examination later in the patient's course when there is better cooperation. An important example of a subtle historic point of considerable diagnostic significance is the history of fatigue in patients suspected of having MS. Fatigue in patients with MS has specific distinguishing features, such as worsening with heat and association with the worsening of other neurologic symptoms (Krupp et al., 1988, 1989).

Patients giving a history of unexpected functional decline superimposed on baseline neurologic abnormality need a full neurologic reassessment. Important clinical situations of this type include the development of hydrocephalus in a patient with a prior head injury, the development of a glioma in a patient with tuberous sclerosis, and the development of cord compression in a patient with cerebral palsy and spinal deformity (Reese et al., 1991).

Orthostatic hypotension is a feature of many medical and neurologic disorders with behavioral manifestations; it should be carefully sought in all medical-psychiatric patients. Patients should have blood pressures measured while lying down, after lying for at least 3 minutes, and while standing; a drop on standing of more than 10 mm Hg diastolic or 20 mm Hg systolic is abnormal. While prescribed medications probably are the most frequent cause of orthostatic hypotension in medical-psychiatric patients, unexpected orthostatic hypotension may be a clue to autonomic neuropathy (diabetes, alcoholism, porphyria, etc.), occult carcinoma, extrapyramidal disorders, or Wernicke's encephalopathy. In patients with dizzy spells or falls, the possibility of delayed orthostatic hypotension should be investigated. This involves having the patient remain standing and rechecking the blood pressure at 5 and 10 minutes.

A systematic examination for tardive dyskinesia should be performed by the psychiatrist on every patient with a history of neuroleptic exposure, and on every elderly patient prior to initiating neuroleptic therapy. A structured examination, such as the AIMS (Baldessarini, Cole, & Davis, 1979), is best. In a study at the Western Psychiatric Institute, it was demonstrated that the detection rate of tardive dyskinesia rose substantially when formal screening for tardive dyskinesia was instituted (Munetz & Schulz, 1986). The study is particularly compelling because a consultation service for movement disorders had been actively promoting its services in the baseline year prior to institution of formal screening; evidently many cases had not been identified or referred to the consultation service. Elderly patients are at increased risk for tardive dyskinesia, and occasional elderly patients have dyskinetic or choreatic movements without a history of neuroleptic exposure (D'Allesandro, Benassi, & Cristina, 1986). Therefore, a baseline examination for involuntary movements should be done before initiating neuroleptic drug therapy in an elderly patient.

SPECIFIC NEURODIAGNOSTIC PROCEDURES

Neurodiagnostic procedures used by medical psychiatrists include anatomic brain imaging (CT and MRI), EEG-based procedures (routine EEG, evoked potentials, and computerized EEG techniques), nuclear medicine procedures (positron emission tomography [PET] and SPECT), and cerebrospinal fluid (CSF) examination. All of these are discussed in this section. Polysomnography is not discussed in this chapter, but is mentioned in Chapters 21 and 25. Magnetoencephalography is not included because it is not yet a clinically available test.

Brain Imaging Procedures: Computed Tomography and Magnetic Resonance Imaging

Computed tomography and MRI are the two major clinical techniques for anatomic brain imaging. Computed tomography, the earlier of the two procedures, is based on roentgenography. It is rapid, widely available, and relatively inexpensive and has virtually no contraindications. It is the brain imaging test of choice in acute trauma and other neurologic emergencies in medically unstable patients. It is also the recommended test for screening patients for gross intracranial disease, as in the assessment for treatable dementia.

Magnetic resonance imaging, which produces images by a different technology using magnetic stimulation of protons rather than ionizing radiation, produces images of superior resolution to those of CT and is the diagnostic test of choice in nonemergency cases with medically stable patients, in whom the need to precisely characterize lesions justifies the increased expense of the procedure. Metal aneurysm clips, other metal foreign bodies, and cardiac pacemakers are contraindications. Moreover, patients with severe claustrophobia may not tolerate the procedure. The patient is inside a large magnet during the scan, and therefore is not easily accessible to emergency medical intervention should a crisis arise.

Magnetic resonance imaging is particularly superior to CT in imaging brain areas adjacent to bone because it is free from the artifacts associated with bone on CT scan. (Bone is transparent to magnetism.) These areas include the structures of the posterior fossa as well as the orbital frontal and anterior temporal regions. Because these areas frequently are relevant to diagnosis in neuropsychiatry, MRI has become the standard imaging procedure in the intensive but nonemergent evaluation of altered mental status. A second reason for the emergence of MRI as the diagnostic procedure of choice is that many pathologic processes, most notably demyelination, do not necessarily alter the radiodensity of the brain and therefore may be invisible on CT scan, yet easily be visualized on MRI.

The next two subsections discuss specific issues related to the application of these two imaging procedures to medical psychiatry.

Computed Tomography

Despite the greater resolution and sensitivity of MRI, the CT scan remains the standard brain imaging procedure in the workup of dementia (Gado & Press, 1986), in the evaluation of subacute mental status changes in patients with possible metastatic cancer, in patients with suspected stroke, and in head trauma patients. In these situations, the speed, availability, and lower cost of the CT scan are positives, and the sensitivity of the test is adequate to answer the clinical questions for which brain imaging tests usually are ordered. When the CT scan is negative in one of these situations but the patient's clinical status continues to raise the concern of a gross brain lesion, MRI would be considered as a follow-up.

When CT is used to evaluate dementia or acute head injury, contrast usually is not necessary. However, contrast is the rule in evaluating subacute mental status changes, particularly when concerns of brain tumor, recent stroke, or subdural hematoma are high in the differential diagnosis. In these situations, relevant lesions may be isodense and visualized only with contrast.

The urgency with which a CT scan is ordered in a patient with delirium or a subacute mental status change depends on three factors: the underlying medical disease, the presence of focal findings on neurologic examination, and the likelihood of a toxic or metabolic explanation for the mental status change. Computerized tomographic scanning is performed earlier in patients who are immunosuppressed or who have primary tumors known to metastasize to the brain. By contrast, CT scanning is appropriately deferred in a delirious patient with known hyponatremia, no focal neurologic signs, and an underlying disease, such as congestive heart failure, that is not directly associated with brain lesions.

Weinberger (1984) suggested six indications for CT scanning in psychiatric patients without known medical or neurologic conditions or risk factors. They were: "1) confusion and/or dementia of unknown cause, 2) first episode of a psychotic disorder of unknown etiology, 3) movement disorder of uncertain etiology, 4) anorexia nervosa, 5) prolonged catatonia, and 6) first episode of a major affective disorder or personality change after age 50." However, there are many nuances in phrases like "of unknown cause" or "of uncertain etiology."

There is substantial evidence to suggest that, in general psychiatric practice, patients with typical psychiatric syndromes and with no historic or neurologic examination findings suggesting neurologic disease should *not* routinely receive brain imaging procedures. McClellan, Eisenberg, and Giyanani (1988) reported results of CT scanning in 261 psychiatric inpatients who comprised consecutive admissions, excluding those patients with prior neurologic diagnoses, focal neurologic examination findings, or a positive history of seizures, increasing headaches, or other suggestive symptoms. Of the 261, 230 (88.1%) had normal scans and 27 (10.4%) showed atrophy only. Four (1.5%) had abnormal findings: two had basal ganglia calcification, one had an old lacunar infarct, and one had an osteoma. None of the four abnormalities were relevant to the patient's psychiatric syndrome.

Skjodt, Torfing, and Teisen (1988) performed CT scans on 181 patients with dementia clinically attributable to toxic exposure, all but one of whom had been demented for at least 1 year. Only five patients showed focal hypodense lesions, three suggesting old trauma and two suggesting old strokes. No treatable dementias were identified. Their findings suggest that, even in the dementia workup, the CT scan might be unnecessary if the dementia is long standing and an etiology seems clear on clinical grounds.

A reasonable conclusion based on presently available knowledge is that the CT scan should not be regarded as a screening test to be conducted on all psychiatric patients. The choice of brain imaging as a diagnostic procedure should follow a psychiatric and neurologic examination, and should be based on a differential diagnosis. In a patient with a classic long-standing psychiatric history and no relevant risk factors or physical signs, brain imaging would not be done. In other cases, one might proceed directly to MRI, as, for example, when demyelinating disease was high in the differential diagnosis.

Although CT scans in patients presenting with apparent primary psychiatric illness rarely reveal treatable neurologic disease, they are not infrequently abnormal. An extensive and at times contradictory literature supports a few general conclusions: (1) some, but not all patients with schizophrenia or with bipolar disorder have enlarged ventricles; (2) ventricular enlargement in these patients is not progressive at a greater rate than general age-related changes; and (3) patients with larger ventricles tend to show (but do not always show) more cognitive impairment, fewer positive and more negative signs of schizophrenia, and more retardation if depressed. Clark, Marbeck, and Li (1990) compared 13 different measurements of ventricular area dimensions in 100 normal subjects, and extracted four factors of ventricular size, all of which were positively correlated with age. They observed that different diseases may show different patterns of ventricular enlargement, so that studies of ventricular enlargement in psychiatric disease will yield different results depending on the specific measures chosen in a given study. The question remains open whether nonspecific CT abnormalities eventually may be helpful in clinical management or in the assessment of prognosis in patients with schizophrenia or bipolar illness (Pandurangi, Dewan, & Boucher, 1986).

In evaluating CT evidence of cerebral atrophy, the influence of alcohol and medication should be considered. Even moderate drinkers may show mild cortical atrophy that reverses on total cessation of drinking (Cala, Jones, & Burnes, 1983). An overall increase in CSF volume was correlated significantly with the length of alcohol addiction in a recent study of 42 male alcoholics (Mann et al., 1989). Corticosteroid therapy may lead to cerebral atrophy as well, which is reversible with discontinuation of steroids (Bentson, Reza, & Winter, 1978; Heinz, Martinez, & Hawnggeli, 1979; Okuno, Masatoski, & Koniski, 1980).

Despite the diagnostic nonspecificity of cerebral atrophy and ventricular enlargement, clinicians frequently do use their presence in support of a diagnosis of degenerative dementia. This is not recommended, unless quantitative measurements show that the atrophy is well beyond age norms or serial scans show significant progressive loss of brain tissue. In the latter case, it is important to ascertain that serial scans were done using the same equipment and technique. It is sobering to realize that, since the first edition of this book, no quantitative assessment of cerebral atrophy on CT has been generally accepted. Recently developed image-processing software should permit the development of standardized, economical quantitative measures over the next few years. If and when they are available, quantitative measurements of cerebral atrophy should enter into diagnostic reasoning, much as other quantitative data (EEG background rhythm, cognitive test scores) do now.

Interpretations of CT scans, like all diagnostic imaging data, are influenced by the skill and bias of the interpreter. Shinar et al. (1987) studied interobserver reliability in the interpretation of CT scans in stroke patients. Although they showed excellent agreement for the detection of infarcts and intracerebral hemorrhage, agreement was lower on more quantitative judgments such as lesion size or the extent of edema. Strikingly, there was not complete agreement among the six neurologists in the study as to whether scans were normal or abnormal—the kappa reliability of the normal/abnormal judgment was .60 and the reliability for side of lesion was .65. Such data strongly support a direct review of the CT scan images when reported imaging findings do not agree well with the history and examination.

Magnetic Resonance Imaging

Magnetic resonance imaging has become the "gold standard" of brain imaging in psychiatry because of its high sensitivity to brain lesions of behavioral relevance. Magnetic resonance imaging can demonstrate lesions not seen on CT scans in a number of psychiatrically important conditions, including MS, encephalitis, meningitis, brain contusion, and focal epilepsy (Baker, Berquist, & Kispert, 1985; Gabarski, Gabrielsen, & Gilman 1985; Jacobs et al., 1986; Laster, Penry, & Moody, 1985; Lesser, Modic, & Weinstein, 1986; Margulis & Amparo, 1985; Sheldon, Siddharthan, & Tobia, 1985; Sperling, Wilson, & Engel, 1986).

However, the conduct of MRI studies and the interpretation of MRI results are substantially more complex than CT evaluation. One reason is that the image itself depends critically on the details of how the brain is stimulated magnetically, and when and how the resonance is recorded. Further, MRI is so sensitive that it generates numerous nonspecific findings whose interpretation depends on clinical and epidemiologic context. The medical psychiatrist regularly reviewing MR images can profit greatly from a review of basic principles of MRI, to enable constructive dialogue with the radiologist performing the test. Two useful references are Daniels, Haughton, and Naidich (1987) and Bradley and Bydder (1990).

In the author's experience, the three most common issues arising concerning the conduct and interpretation of MRI in medical-psychiatric settings are: (1) the interpretation of white matter hyperintensities; (2) the role of special views and imaging sequences; and (3) the use of paramagnetic contrast (gadolium-labeled DTPA).

White Matter Hyperintensities. White matter hyperintensities, best visualized on T_2-weighted images, can be critical diagnostic findings in subcortical stroke, MS, and HIV-related cognitive impairment (Kieburtz et al., 1990; Post et al., 1988). However, it was observed early in the history of MRI that elderly persons with no signs of dementia or cerebral vascular disease could have white matter abnormalities, including "rims" (periventricular hyperintensity on T_2-weighted images), "caps" (increased T_2 signals around the poles of the lateral ventricles), and "UBOs" (hyperintense bright patches in subcortical white matter on T_2-weighted images) (Kertesz et al., 1988). Pathologic studies have shown that UBOs are nonspecific and may represent infarctions, gliosis, demyelination, or no abnormality at all (Braffman et al., 1988). Grafton et al. (1991) performed MRI examinations postmortem on several elderly people, then examined their brains pathologically. Periventricular, subependymal areas of high signal intensity described as rims and caps on MRI were associated pathologically with areas of myelin pallor, gliosis, and dilated perivascular spaces. These findings, although definitely abnormal, do not necessarily imply the presence of neurologic disease, and at times have been associated with aging alone. Areas of hyperintensity in the centrum semiovale had no consistent neuropathologic correlate at all. Some areas abnormal on MRI were normal pathologically, and post-mortem brain examination showed areas of gliosis and even infarcts that were not visualized on MRI.

Whereas some studies have found no difference between normal and demented elderly subjects with regard to the prevalence of white matter hyperintensities (Leys et al., 1990), a number of studies suggest that white matter hyperintensities are more common, and more severe, in people with dementia (Bondareff et al., 1990; Kertesz et al., 1990; Zubenko et al., 1990). In the study by Kertesz et al., patients with periventricular hyperintensities performed more poorly than those without them on tests of comprehension and attention. White matter abnormalities also have been found in patients with major depression referred for electroconvulsive therapy (ECT) (Coffey et al., 1990), and in patients with late-onset psychosis (Miller et al., 1989). It is speculated, but not certain, that clinically normal individuals with severe white matter hyperintensities are at increased risk for subsequent development of dementia or psychiatric illness.

The lack of diagnostic specificity of white matter abnormalities has led to efforts to quantitate the white matter lesions (Bondareff et al., 1990; Coffey et al., 1992). At this time no such scheme has been widely accepted. Early hopes that white matter hyperintensities would distinguish between Alzheimer's disease and vascular dementia have not been fulfilled.

At present, pathologic significance of white matter hyperintensities is likely if the patient is under 50; there is an extensive differential diagnosis. In older patients, attribution of pathologic significance to white matter hyperintensities should be made only if the lesions are extensive, or if they clearly correlate with localized neurologic or cognitive findings. In other cases, they are best regarded as nonspecific signs of brain aging and/or age-associated brain disease.

Special Views and Imaging Sequences. A second major consideration for medical psychiatrists is the use of special views or special imaging techniques for MRI. If special techniques or views are not requested, most MRI centers will supply a standard set of horizontal and sagittal views of the brain, with at least one T_1-weighted sequence and one T_2-weighted sequence. The conventional views and imaging sequences were selected for efficient detection of lesions in patients typically referred for MRI studies. However, special views and techniques have been used with success to make observations not possible with routine imaging. For example, Squire, Amaral, and Press (1990) used special high-resolution views of the temporal lobe and hypothalamic region to compare patients with Korsakoff's syndrome and those with non-Korsakoff amnesia. The Korsakoff patients had shrunken mamillary bodies with normal-size hippocampal formations, whereas the other group had the opposite findings. Fontaine et al. (1990) studied the temporal lobes of patients with panic disorder using multiple axial and coronal sections through the temporal lobes. They found abnormalities in 40% of their patients and only 10% of controls. These results stand in contrast to the usual experience of normal routine MRI findings in patients with anxiety disorders. Another special technique used by Fontaine et al. was the administration of 2 mg of clonazepam to the panic patients before the MRI, to minimize the risk of claustrophobia.

A second major type of special assessment involves quantitative analysis of MR data. In one such type of analysis, T_2 relaxation times are directly calculated from measurement of signal intensity on two scans with different T_2 weighting. This technique was applied to demonstrate an abnormally short left caudate T_2 in patients with tardive dyskinesia (Bartzokis et al., 1990). Regional T_1 measurements were compared with diagnoses and neuropsychologic findings by Besson et al. (1989), who showed increased T_1 values in the parietal and temporal regions in patients with Alzheimer's disease, as well as an association between visual-spatial impairments and right parietal T_1 values. The findings were a function of the quantitative approach; the brain regions would not necessarily look different on the MR images. Bondareff et al. (1988) showed that total mean T_2 value was strongly correlated with the Blessed-Roth Dementia Scale in a group of patients with probable Alzheimer's disease.

Another type of quantitative measurement is computerized volumetric analysis, which was used by Tanna et al. (1991) to demonstrate increased ventricular volume in patients with dementia. The computerized image processing technique requires a human operator identifying regions of interest with a mouse; its labor-intensive character suggests that it probably will not become available as a routine MRI service. However, the technique is likely to become increasingly important in neuropsychiatric research, and may be essential to proper interpretation of PET scan findings (Tanna et al., 1991).

Use of Paramagnetic Contrast. The final technical concern frequently faced by medical psychiatrists is the decision to use paramagnetic contrast. Gadolium-labeled DTPA is a contrast agent that produces an intensive image on T_2-weighted sequences but does not cross the normal blood-brain barrier. In the normal brain, it causes increased brightness of the gray matter on T_2 sequences because of the greater vascularity of brain matter. If blood vessels are occluded, contrast will not reach the tissue supplied by that vessel. If the blood-brain barrier is abnormally permeable, contrast will accumulate in an extravascular location. Thus, gadolium contrast may aid in the visualization and characterization of tumors and recent infarcts in which the blood-brain barrier has broken down (Elster et al., 1989). Inflammatory lesions also may be better visualized with contrast; contrast has been used successfully to distinguish new from old lesions resulting from MS (Miller et al., 1988).

The contrast medium has very low toxicity, with the most consistent findings being a completely asymptomatic and transient rise in serum iron concentration for less than 24 hours following the injection (Bydder, 1990). Because of the low toxicity and improved visualization of some lesions, many centers use contrast routinely. However, MRI is so sensitive that most lesions of clinical relevance can be visualized without contrast. Thus, if a patient objects to a contrast injection, it usually can be omitted. Obtaining both contrast and noncontrast scans becomes particularly relevant when evaluating a progressive or unstable condition, such as a brain tumor or inflammatory lesion.

Virtually all clinically available MRI techniques rely on the magnetic resonance of protons. In the near future, however, MRI based on the resonance of phosphorous-31 will permit the noninvasive in vivo assessment of energy metabolism and phospholipid status. Although this remains a research procedure at the present time, recent work by Pettegrew et al. (1991) establishes the practicability of the technique.

Electroencephalography-Based Procedures

Electroencephalography

The EEG, one of the earliest ancillary physiologic tests in psychiatry, remains one of the most useful in the clinical assessment of the psychiatric patient. Although it lacks the visual impact of anatomic brain imaging, it actually is more useful in demonstrating abnormalities of brain function. In addition, it is an inexpensive test that is readily available in small hospitals without CT scanners, and in psychiatric hospitals with limited medical laboratory facilities.

In cases of reversible diffuse brain dysfunction on a toxic-metabolic basis, the CT is normal while the EEG may be diagnostic. The detection of mild deliria presenting psychiatrically and the differentiation of cognitive dysfunction secondary to depression from degenerative cerebral disease, are two well-established indications for the use of the EEG in psychiatry (Kiloh, McComas, & Osselton, 1981). Mild diffuse brain damage resulting from hypoxia and hypotension may contribute to psychiatric disorders following open heart surgery; the EEG may be the only laboratory evidence of such diffuse damage. For example, in a study of 100 patients evaluated following open heart surgery for cardiac valvular disease, 38 thought to be neurologically asymptomatic received EEGs postoperatively; 14 showed abnormalities. In the same study, 15 of the 49 who received careful postoperative neuropsychologic evaluation showed cognitive deficits. Yet, only 4 of the 100 patients were diagnosed as cognitively disordered by their primary physicians (Sotaniemi, 1983). The "hard" abnormality of an abnormal EEG can be useful diagnostically in linking cognitive disorder to probable hypotensive brain injury.

Abnormal EEGs are found in two-thirds of untreated patients with pernicious anemia and in virtually all with significant cognitive impairment (Evans, Edelsohn, & Golden, 1983). In a patient with a psychiatric disorder and a low vitamin B_{12} level, the EEG might thus provide confirmatory evidence that the low B_{12} level was psychiatrically relevant. Hypothyroidism, when it produces an organic mental disorder, is associated with EEG slowing. The EEG may be particularly valuable in linking thyroid disorder and mental symptoms in cases in which the serum thyroxine (T_4) level is normal but the thyroid-stimulating hormone (TSH) level is elevated (Haggerty, Evans, & Prange, 1986). Herpes simplex encephalitis, which may present with nonspecific symptoms of psychiatric illness, can produce an abnormal EEG in an early stage while the CT is negative and the CSF is not diagnostic. In suspected cases, therefore, all three tests are relevant, and the EEG should not be omitted when the CT and CSF exam are normal (Griffith & Ch'ien, 1983). Drury, Klass, and Westmoreland (1985) described four patients with acute psychosis who showed striking periodic EEG abnormalities and who recovered clinically and electroencephalographically within a week. Patients with apparent "brief reactive psychosis" deserve an EEG to exclude unsuspected encephalopathy.

Although routine EEG is an excellent test for metabolic encephalopathy, it is sometimes normal in mild metabolic disorders. For example, 3 of 12 delirious liver transplant candidates had normal EEGs (Trzepacz, Maue, & Coffman, 1986). In many such cases of mild metabolic encephalopathy with normal EEG, a prior baseline EEG tracing may permit a diagnosis. If it is known that a patient's baseline alpha rhythm is 11 to 12 Hz, a current alpha rhythm of 9 Hz indicates a significant change compatible with a metabolic encephalopathy (Markand, 1984). For this reason, patients with chronic medical diseases known to cause intermittent metabolic encephalopathies (e.g., liver disease, renal disease) should be considered for a baseline EEG, particularly if they appear vulnerable to psychiatric complications.

An increased rate of EEG abnormalities has been encountered in a number of "functional" mental disorders, including obsessive-compulsive disorder (Marks & Kett, 1986) and borderline personality (Cowdry, Pickar, & Davies, 1985–86). These EEG abnormalities do not imply a neurologic diagnosis, however, nor do they necessarily have prognostic or therapeutic implications. The relevance of EEG findings must be established empirically for each disorder; the mere presence of sharp abnormalities on an EEG does not imply a diagnosis of epilepsy or that the psychiatric disorder will improve with anticonvulsants.

The literature on EEG for some psychiatric disorders shows wide variations in the prevalence of abnormality. Whereas some investigators find a high prevalence of EEG abnormalities in borderline personality (Cowdry, Pickar, & Davies, 1985–86), others find no significant difference from controls (Cornelius, Brenner, & Soloff, 1986). These discrepancies probably reflect differences in the specific populations being studied. Treatment-refractory patients with any psychiatric diagnosis might be expected to have a higher incidence of EEG abnormality than a random population sample with that diagnosis.

Proper interpretation of EEGs in elderly patients requires a knowledge of EEG norms for age. The dominant posterior background rhythm in the EEG may slow with age, but does not go below 8 Hz in the absence of disease. Slowing with age is gradual when it occurs, and a decrease of more than 1 Hz in background rhythm over a period of less than 1 year should arouse suspicion of CNS disease or metabolic disturbance. The general organization of the EEG is preserved with normal aging. By contrast, patients with Alzheimer's disease may show slowing of the dominant posterior rhythm below 8 Hz, as well as increased theta and delta activity and poorer organization of the record (Coben, Danziger, & Storandt, 1985; Nolfe & Giaquinto, 1986).

Diagnosis of Epilepsy. Notwithstanding its value in the diagnosis of delirium, the most frequent use of the EEG by medical psychiatrists is the assessment of suspected epileptic seizures. There are four common clinical situations, each with different implications for the effective use of EEG.

The first situation is the evaluation of the patient with apparent generalized tonic-clonic convulsions, who is suspected to have pseudoseizures. The crucial point is that generalized tonic-clonic seizures of epileptic origin always are accompanied by typical ictal EEG changes and postictal diffuse slowing of the EEG. If the emergency EEG obtained immediately after an attack is normal, the diagnosis of pseudoseizures is confirmed. Intensive monitoring, with simultaneous video or direct observation of behavior, is the definitive test. The usual technical problem in the interpretation of the ictal EEGs in such patients is muscle or movement artifact. However, postictal EEG activity should be readily seen because of the postictal lethargy and somnolence that follow generalized seizures.

The second situation concerns the evaluation of the patient with stereotypical attacks of sensory, autonomic, or psychic symptoms without alterations of consciousness—episodes that can represent simple partial seizures but can also be a manifestation of panic disorder or a somatoform disorder. Ictal EEGs in simple partial seizures usually are abnormal when the seizures originate from the somatosensory, auditory, or visual cortex, but usually are normal when they arise from the limbic cortex. In the latter case the diagnosis of epilepsy usually is confirmed when the simple partial seizure spreads and produces alteration of consciousness, since complex partial seizures usually are associated with abnormal ictal EEGs. Ictal findings, when present, usually consist of focal spike-and-wave discharges. However, deep limbic foci either may produce no surface manifestation or may present with nonspecific theta or delta rhythms. These reflect the transmission with distortion of spike-and-wave activity from deeper structures (Engel, 1989, pp. 147–148).

Lieb et al. (1976) reported that 14% of complex partial seizures and 82% of simple partial seizures, confirmed as epileptic by depth electrode, did not manifest any surface EEG abnormalities. Thus, in evaluating a patient with suspected simple partial seizures, ictal EEG findings establish the diagnosis, nonspecific temporal slowing supports but does not confirm the diagnosis, and a normal EEG does not rule out the diagnosis.

The third situation concerns the evaluation of the patient with suspected complex partial seizures—attacks in which generalized disturbances of sensory, motor, autonomic, or psychic function are accompanied or followed by impairment of consciousness. Complex partial seizures usually come from temporal lobe foci, but are not synonymous with temporal lobe epilepsy because they may also arise from extratemporal foci, and because temporal foci may give rise to simple partial seizures only. Ictal findings can range from clear-cut spike-and-wave activity to relatively nonspecific theta or delta slowing. Interictal findings are similar to those seen in simple partial seizures.

A single routine EEG in a patient with complex partial seizures may fail to record any ictal or interictal events, or may record nonspecific abnormalities of unclear significance. To improve diagnostic accuracy, one can: (1) record multiple EEGs or record EEGs for longer periods, assuring that drowsiness and sleep are recorded (to increase the yield of interictal discharges); (2) use provocative maneuvers such as sleep deprivation to increase the likelihood of both interictal and ictal discharges; (3) add electrodes in different locations, to increase the likelihood of demonstrating spikes rather than nonspecific sharp or slow activity; or (4) do simultaneous recording of EEG and behavior, to establish an electroclinical correlation when the electrical findings are nondiagnostic. McNamara (1991) reviewed measures to increase the yield of abnormal EEGs in patients suspected of having epilepsy.

With provocative techniques and/or intensive monitoring, EEG diagnosis is possible in perhaps as many as 90% of patients with genuine complex partial epilepsy (Driver & McGillivray, 1982). However, there are unequivocal cases of partial complex seizures in which surface EEG recordings are persistently normal despite special procedures (Engle, Driver, & Falconer, 1975). A practical approach to diagnosing a strongly suspected case with persistently negative EEG findings is to treat the patient with anticonvulsants and observe the outcome. If the patient obtains a stable remission of seizure activity, a diagnosis can be made. If the response is poor or impersistent, and the attacks continue to cause distress or disability, referral to a specialized epilepsy center is indicated.

The fourth clinical situation arises when evaluating a patient with well-established and treated epilepsy who has an impairment in cognitive mental status. The differential diagnosis of such impairment includes the possibility that subclinical seizure discharges are interfering with cognitive processes; other considerations include anticonvulsant drug toxicity, concurrent pathology, or a progressive brain disease. A common example is the child with known petit mal epilepsy who shows impaired attention and school performance without frank disturbances of consciousness. The diagnostic problem is that interictal EEG discharges are not necessarily relevant to the patient's cognitive complaint. In this situation it is helpful diagnostically to record the EEG while doing formal cognitive tests sensitive to the patient's reported difficulties. If the observed electrical discharges are relevant, test performance will covary along with the electrical changes.

Even when the EEG supports a clinical diagnosis of focal epilepsy, accurate localization of the focus in the brain often cannot be accomplished by surface EEG alone. Spencer, Williamson, & Bridger (1985) reported that even ictal EEGs, read by experienced electroencephalographers, correctly locate the lobe of onset in less than half of cases. Depth electrodes, subdural electrodes, and/or PET scan definitively establish the focus in cases destined for surgery.

Diagnosis of Dementia. Recent guidelines for the evaluation of dementia do not require EEG as a necessary component of the dementia evaluation. This position is based on the observation that the EEG is either normal or nonspecifically abnormal in Alzheimer's disease and vascular dementia, the two most common kinds. Nonetheless, there are several specific situations in which the EEG can aid in the differential diagnosis of dementia. Rather than being done routinely, EEGs in dementia patients should be ordered for specific reasons, several of which are outlined here.

First, a baseline EEG is a reasonable step in a patient with questionable dementia. In this situation, an abnormally slow posterior rhythm, and excessive slow activity elsewhere, supports the suspicion of a brain disease affecting the cortex. More severe diffuse abnormality suggests a metabolic encephalopathy, prompting a search for an underlying cause. Excess fast beta activity anteriorly suggests benzodiazepine or barbiturate use, that could be contributing to the cognitive problems. A completely normal EEG is possible in questionable or mild dementia, but in this case a follow-up study 6 months to 1 year later showing progressive abnormality can help establish the diagnosis of Alzheimer's disease. The demonstration of change over time may permit diagnosis of a progressive process when both tests are within the broad limits of normal or at worst mildly and nonspecifically abnormal.

Second, the EEG is useful diagnostically in patients with dementia accompanied by myoclonus. If the patient has Creutzfeldt-Jacob disease, the EEG is likely to show periodic sharp wave bursts characteristic of that disease. Alzheimer's disease patients with myoclonus do not show this EEG abnormality (Markand, 1990; McNamara, 1991).

Third, the EEG is helpful in assessing patients who show dementia of subacute onset, wherein metabolic, infectious, or epileptic causes are substantially more likely than in a chronic, insidious deterioration. In this situation, the EEG findings occasionally are diagnostically specific.

Fourth, the EEG is helpful in evaluating sudden deterioration in a patient with a known mild to moderate dementia. Marked EEG abnormality with diffuse delta activity is not ordinarily found in the earlier stages of Alzheimer's disease; neither are definite focal abnormalities. A markedly slow EEG suggests a toxic-metabolic problem, and focal findings suggest a second superimposed CNS disease.

Finally, the EEG can be useful in studying patients who are thought to be demented but whose behavior or clinical course suggests that the diagnosis may have been in error. A normal EEG is incompatible with long-standing Alzheimer's disease or Alzheimer's disease of moderate or greater severity. Such a finding should lead to reassessment of the dementia diagnosis. Either the patient suffers from pseudodementia, or the cause of dementia is a subcortical process that leaves the surface EEG intact.

Effects of Psychotropic Drugs. Virtually all psychotropic drugs have been reported to have an effect on the EEG (Glaze, 1990). Because these effects are both common and nonspecific, they complicate the interpretation of EEG findings in medical-psychiatric patients. Electroencephalographic abnormalities attributable to neurologic disease sometimes are inappropriately ignored because of a misattribution of the findings to drug effects; alternatively, drug-induced abnormalities may erroneously be used in support of neurologic diagnoses. A summary of psychotropic drug effects on the EEG is presented in Table 18-3, pp. 384–385.

In approaching the interpretation of the EEG in a patient on psychotropic drugs, the following general principles are helpful:

1. Intoxication with virtually any psychotropic drug can produce slowing of the EEG. When this is accompanied by markedly increased anterior beta activity, an overdose with a benzodiazepine, barbiturate, or tricyclic antidepressant should be considered as a likely cause.

2. At therapeutic doses, benzodiazepines, barbiturates, and stimulants produce increased beta activity. They do not cause focal or paroxysmal changes.

3. At therapeutic doses, neuroleptics cause either no change or mild generalized slowing, but changes are less likely with lower doses and long-term treatment. For this reason, EEG abnormalities should not be dismissed if the patient has been on a modest dose of a neuroleptic for a long time. Tricyclic antidepressants at therapeutic doses either have no effect on the visually analyzed EEG or produce a combination of excessive beta activity and increased theta activity. Thus, slow activity unaccompanied by increased beta activity in a patient on tricyclics would suggest a separate, non–drug-related cause for the EEG slowing.

4. Lithium has been associated with a particularly high frequency of paroxysmal EEG changes, which can include high-voltage bursts of diffuse slow activity and even spikes. Although more severe abnormalities are associated with either toxic lithium levels or clinical evidence of neurotoxicity, patients can have significant EEG abnormalities at therapeutic lithium levels without gross signs of toxicity. The association of mood disturbance with partial complex seizures complicates the issue, with a possibility of diagnostic error in either direction. When epileptiform activity occurs in a patient on lithium and a diagnosis of epilepsy is in question, it is appropriate to discontinue lithium, managing the mood disorder with some other drug, and to repeat the EEG after a few weeks.

5. Paroxysmal activity, including both focal and generalized spikes, spike-and-wave discharges, and sharp waves, can occur as a rare complication of either antidepressant or neuroleptic treatment, even with dosages in the therapeutic range. Such phenomena are most commonly seen with chlorpromazine among the neuroleptics and amitriptyline among the antidepressants. These phenomena are more common in patients with a history of seizures, but can occur in patients without such history. When they do, they may be indicating an increased vulnerability to seizures, but no definitive neurologic diagnosis can be made. A prudent approach is to discontinue the drug or substitute a different psychotropic and repeat the EEG after several weeks.

Suggestions for Improving the Diagnostic Value of Electroencephalography. Brain electrical activity is a dynamic phenomenon that changes with time, circumstances, and the patient's state of arousal. The diagnostic efficacy of the EEG is greatest when the test is timed with this in mind. To take an obvious example, the patient who is said to awaken from sleep with brief episodes of bizarre behavior requires a sleep EEG to be properly investigated for nocturnal epilepsy. Ideally, an all-night sleep study would be done, which could in addition assess the patient for abnormalities of sleep architecture. The following paragraphs elaborate on this theme.

1. *Electroencephalograms are most valuable in the identification of organic brain disease if obtained when patients are symptomatic.* A psychiatrist wishing to evaluate a neurologic etiology for suspected paroxysmal hysterical or anxiety symptoms should obtain not only a routine EEG but also one made under the circumstances a patient believes will elicit an attack. Fariello et al. (1983) reported on 32 cases of patients with paroxysmal symptoms who were studied as they reenacted the situations that provoked their paroxysmal attacks while having simultaneous polygraphic recording of the EEG, electrocardiogram (ECG), and respiration. The investigators found it necessary to change the diagnoses in 19 cases, many of which had been previously diagnosed as hysterical. Provocative circumstances included having patients take a bath, stand up quickly, read, and undergo venipuncture. Although a careful history and awareness of the existence of unusual seizure types is helpful in diagnosing epilepsy, ictal EEG abnormality during the time of the suspected seizure facilitates an unequivocal diagnosis.

2. *An emergency EEG is a definitive procedure when partial complex status is suspected.* Nonconvulsive status epilepticus should be suspected in instances of acute mental status change with diminished responsiveness or automatic behavior, particularly in patients with a past history of seizures. Partial complex status may present as a confusional state, as aphasia, as decreased responsiveness with staring, or as catatonia. Although a history of seizures is often present, there are well-documented cases with no past history of epilepsy. The interictal EEG was abnormal in eight cases reported by Ballenger, King, and Gallagher (1983) and in 9 of the

12 cases reviewed from literature. The neurologic examination was normal in four of the eight cases reported. The ictal EEG, however, was abnormal in every case in which it was performed. Lee (1985) described a series of 11 patients ages 42 to 76 who acutely developed prolonged confusional states with prominent behavioral abnormalities. Their EEGs showed runs of 1- to 2.5-Hz generalized spike-wave and polyspike activity, establishing a diagnosis of nonconvulsive status epilepticus.

3. *Patients with disabling paroxysmal symptoms of questionable etiology and patients with known epilepsy refractory to standard treatment should be referred for intensive neurodiagnostic monitoring* (Gumnit, 1986). Intensive neurodiagnostic monitoring, which comprises EEG telemetry—with or without simultaneous video recording—and ambulatory cassette EEG recording, may resolve a diagnostic or therapeutic impasse when routine EEG procedures fail. Simultaneous video-EEG describes a procedure in which at least 10 channels of EEG are simultaneously recorded with a video picture of the patient, with the information from the two sources synchronized. Telemetered EEG refers to EEG recording from a wireless transmitter worn by the patient while normally active in a supervised medical setting. An ambulatory cassette EEG may be recorded on either an inpatient or an outpatient basis, using a Holter monitor-like device.

These procedures can be used to identify the nature of a paroxysmal event, to identify the type of seizure in patients with epilepsy, to locate the part of the brain in which a seizure begins, and to quantify the number of seizures. Intensive neurodiagnostic monitoring is best performed at a comprehensive epilepsy center, or at least under the direction of an experienced electroencephalographer with a major commitment to the procedure. Although intensive neurodiagnostic monitoring is extraordinarily sensitive, the presence of numerous artifacts and technical difficulties make it unsatisfactory for occasional use by a neurologist without special expertise in clinical neurophysiology.

Selection between inpatient video-EEG telemetry and ambulatory cassette monitoring is best made by an experienced neurologist, considering the specifics of the case. Most patients with a mixture of genuine and psychogenic seizures will best be sorted out by several days of video-EEG monitoring in an inpatient setting with anticonvulsants withdrawn. Relatively mild but frequent paroxysmal events in reliable patients often can be satisfactorily investigated with outpatient ambulatory cassette monitoring.

Ambulatory cassette monitoring and inpatient telemetry are the most sensitive detectors of interictal electrical abnormalities and brief subclinical seizure activity in patients with epilepsy (Ebersole & Bridgers, 1985; Ebersole & Leroy, 1983). They would not ordinarily be employed, however, in a patient with typical epileptic symptoms who responded well to standard anticonvulsant treatment. The expense and inconvenience of long-term monitoring become justified when seizure control is difficult to attain, putting either the diagnosis or the therapy in doubt. One particular example is when a patient with automatisms fails to respond to the usual drugs for TLE. Some of these patients turn out, on long-term monitoring, to have generalized spike-and-wave discharges that indicate a form of epilepsy better managed with valproic acid than with carbamazepine or phenytoin.

Evoked Potentials

Evoked potentials are derived by repeatedly presenting a stimulus to the patient, recording the EEG for a fixed period following each stimulus, and then averaging a large number of responses. The averaging usually is carried out by a microcomputer, which digitizes the EEG signal and calculates the average at a number of closely spaced points, simulating a smooth curve. The averaging procedure cancels out the EEG background activity, leaving a pattern specifically related to the presented stimulus. Stimuli may be simple, such as flashing lights, or complex, such as stimuli for a discrimination task. The auditory, visual, and somatosensory modalities all can be used for the stimulus.

Evoked potentials can be divided into early and late potentials. Early potentials depend on neuronal activity in the primary sensory pathways and are relatively independent of attentional state and even level of consciousness, although pattern-shift visual evoked responses require ocular fixation. By contrast, late evoked potentials depend on level of consciousness, attentional state, and, in certain circumstances, mental set. Evoked responses usually are reported in terms of specific positive or negative waves, described in terms of latency of onset and peak amplitude. Early evoked potentials usually are recorded over the appropriate sensory area for the modality of presentation, while late evoked potentials can be recorded anywhere over the scalp, depending on the purpose of the recording (see the section on "Brain Electrical Activity Mapping").

Early Evoked Potentials. Early evoked potentials will show delayed latency if there is delayed conduction in the primary sensory pathways; they are particularly sensitive to demyelinating lesions. Visual evoked potentials show delayed latency in optic neuritis that will persist

TABLE 18–3. *Drug Effects on the EEG*

Drug Class (Subclass and Individual Drugs)	Therapeutic Doses	Intoxication
I. Neuroleptics ("major tranquilizers") A. Phenothiazines Chlorpromazine Fluphenazine Thioridazine Trifluoperazine B. Butyrophenones Haloperidol C. Thioxanthene derivatives Thiothixene D. Indole derivatives Molindone	Visual analysis: little or no change in most patients; may see slight slowing of alpha rhythm and slight increase in voltage of theta activity. Computer-assisted EEG analysis: increased slow activity in delta and theta bands. Rarely, paroxysmal patterns, including the appearance or increased frequency of focal spikes or sharp waves, bilateral spike discharges, and generalized spike-and-wave complexes. Rarely, thioridazine increased beta activity.	Diffuse slowing, with increased amounts of theta and delta activity and decreased amounts of alpha rhythms.
II. Antidepressants A. Tricyclic antidepressants Amitriptyline Desipramine Imipramine Nortriptyline Protriptyline B. MAO inhibitors Isocarboxazid Tranylcypromine C. Others Amoxapine	Most commonly, increased theta and beta activity; sometimes, decreased alpha rhythms. Very rarely, appearance or increased frequency of paroxysmal activity, including focal spikes or sharp waves, bilateral spikes, or generalized spike-and-wave discharges.	Diffuse slowing (delta) with superimposed beta or irregular 9- to 10-Hz activity.
III. Anxiolytics A. Benzodiazepines Chloradiazepoxide Diazepam Flurazepam HCl B. Glycol derivatives Meprobamate C. Others Glutethimide Methaqualone Ethchlorvynol Methyprylon Chloral hydrate	Increased amounts and voltage of beta activity. Highly variable from patient to patient.	Diffuse slowing, with increased beta and delta activity and decreased alpha activity, with superimposed beta activity and then, rarely, alpha-coma pattern or isoelectric EEG.
IV. Antimanic Lithium	Visual analysis reveals minimal change or decreased amount and frequency of alpha activity.	Diffuse slowing, with increased amounts of theta and delta activity and decreased alpha rhythms. Sharp waves. Rarely, periodic complexes similar to those of Jakob–Creutzfeldt disease.
V. CNS stimulants *d*-Amphetamine Methylphenidate Cocaine Caffeine Nicotine	Increased beta and alpha activity. Visual analysis: little or no change.	With overdose of stimulant drugs, diffuse slowing (theta and delta activity). Cocaine can cause seizures, even in low doses.
VI. Alcohol	Chronic alcoholism: little or no change in the EEG.	During acute intoxication and in association with high blood levels of alcohol, diffuse slowing, with increased theta and delta activity and decreased alpha activity. EEG findings in Wernicke's encephalopathy: varying degrees of background slowing with increased amounts of theta and delta activity.

TABLE 18–3 *(Continued)*

Drug Class (Subclass and Individual Drugs)	Therapeutic Doses	Intoxication
VII. Hallucinogens		
Lysergic acid diethylamide (LSD)	At low and high doses, little or no change in the EEG; sometimes, increased beta activity.	Continuous rhythmic theta activity with periodic slow-wave complexes.
Marijuana	Little or no change.	Increased theta and delta activity.
Phencyclidine	Rhythmic theta and delta activity and increased voltage of beta activity.	
Psilocybin	Increased voltage of beta activity.	
VIII. Analgesics/drugs of abuse Morphine Heroin Methadone Codeine	Variable changes: single dose may cause minimal change, while multiple doses may cause slowing of the alpha rhythm and background activity; with tolerance, alpha rhythm may return to normal.	Diffuse slowing with increased theta and delta activity and decrease in alpha rhythm.
Meperidine	Slowing of the background activity with increased amounts of theta and delta activity.	Alpha-coma pattern may be seen with overdose of morphine.
IX. Anticonvulsants A. Barbiturates Phenobarbital Primidone	Small dose; increased amount and amplitude of beta activity; with high doses, increased beta activity and theta/delta activity with decreased amount and frequency of alpha rhythm.	Increased delta activity with superimposed beta activity. Alpha-coma pattern (rare); burst-suppression and isoelectric patterns have been reported.
B. Benzodiazepines Diazepam Clonazepam	See above, III.	
C. Bromides	Increased voltage of beta activity and amount of theta activity.	Generalized slowing with theta and delta activity.
D. Phenytoin	Visual analysis reveals little effect; may cause slow alpha rhythm and increased theta activity.	Slowing of the alpha rhythm and increased generalized delta activity.
E. Carbamazepine	Slow alpha rhythm and increased theta and delta activity.	Generalized slowing with increased theta and delta activity and slowing of the alpha rhythm.
F. Ethosuximide	No appreciable effects.	Generalized slowing with increased theta and delta activity and decreased alpha rhythm.
G. Valproate	Little or no change; may be associated with increased theta activity.	Generalized slowing with increased theta and delta activity and decreased alpha rhythm.
X. Other drugs A. Antihistamines Diphenhydramine	At therapeutic doses, little change; may be associated with increase in voltage of beta activity. 6-Hz spike-and-wave activity reported.	Generalized slowing with increased theta and delta activity and decreased alpha rhythm.
B. Antibiotics Isoniazid Penicillin	No significant change.	Diffuse slowing with increased theta and delta activity. Paroxysmal abnormalities, including bilateral sharp waves and status epilepticus.
C. Corticosteroids	Acute doses: slight slowing of alpha rhythm.	
D. Bronchodilators Aminophylline Theophylline	No significant change	Diffuse slowing with increased theta and delta activity and paroxysmal abnormalities, including status epilepticus.

Reprinted by permission from Glaze DG (1990). In Daly DD & Pedley TA (eds.), Current Practice of Clinical Electroencephalography (Ch. 15, pp. 489–512). New York: Raven Press.

after symptoms have remitted. Auditory evoked potentials may show delayed latency if there is demyelination of auditory pathways. In the case of auditory pathways, it may even be possible to locate the approximate site of the lesion, depending on which waves are delayed.

The early evoked potentials—pattern-shift visual evoked potentials, brainstem auditory evoked potentials, and short-latency somatosensory evoked potentials—can demonstrate abnormal sensory system function when the history and neurologic examination are equivocal. They may thus serve to document a second, discrete lesion in suspected cases of MS, assisting the diagnosis and at times averting more invasive tests such as myelography. Evoked potential batteries, when used to evaluate suspected MS, have less sensitivity and predictive power overall than MRI (Lee et al., 1991). However, they may identify areas of functional abnormality not seen on MRI, because the lesions are beneath the resolution of the scanner. Evoked potentials also can be used to define the anatomic distribution of the disease process or monitor objective changes in patients' status (Chiappa & Young, 1985).

Evoked potentials have anatomic specificity but not etiologic specificity. Abnormal latencies for brainstem auditory evoked potentials imply impaired function of brainstem pathways but do not distinguish among vascular, demyelinating, metabolic, and neoplastic etiologies. Symmetric delays in visual evoked potentials can be produced by vitamin B_{12} deficiency, the effects of toxins, inflammatory diseases, and even normal aging (Celesia, 1986). Thus, despite the wide use of evoked potentials in the diagnosis of MS, abnormal evoked potentials should not be equated with the presence of demyelinating disease (Stockard & Iragui, 1984).

Early evoked potentials can be applied to investigate sensory losses suspected to be due to hysteria or malingering. The principle is that sensory losses, if they are sufficiently severe, reliably produce delays in latency of evoked potentials, or occasionally an absence of evoked potentials. Normal pattern-shift visual evoked potentials imply visual acuity of at least 20/120. A patient who claimed only the ability to see light and movement but had normal visual evoked potentials would thus have no organic basis for the symptom. The threshold for producing auditory evoked potentials correlates with the degree of hearing loss; the audiometric threshold is within 5 or 10 dB of the threshold for the appearance of the PIV-V complex of the evoked potential. (The PIV-V complex is a particular portion of the brainstem auditory evoked response that appears consistently in the absence of auditory system disease, provided that the auditory stimulus is sufficiently strong.) Hysterical, malingered, or functionally elaborated deafness would thus be detectable as a discrepancy between the audiometric threshold and the threshold for appearance of the PIV-V complex. An exception would exist for patients with cortical deafness or auditory agnosia; however, these patients always show abnormalities on EEG or brain imaging. Somatosensory evoked potentials test the posterior columns and can be expected to be abnormal if there is total anesthesia or significant loss of joint position sense. Sensory deficits not affecting joint position sense are compatible with normal somatosensory evoked potentials. Nonorganic sensory loss would be suspected if normal somatosensory evoked potentials were obtained in a patient who complained of total anesthesia or loss of joint position sense.

Malingering patients may alter their visual evoked potentials by failing to fix their vision on the stimulus; they may alter somatosensory evoked potentials by dislodging the stimulus electrodes. If malingering is suspected, the patient should be continuously observed by a skilled observer throughout the recording process in order to exclude patient-produced artifacts. For further details, consult the superb review by Howard and Dorfman (1986).

In addition to aiding in the diagnosis of MS and the differential diagnosis of hysteria, early evoked potentials may be useful in demonstrating the physiologic effect of toxins of psychiatric interest. For example, some patients with tardive dyskinesia will have abnormal brainstem auditory evoked responses (Zeitlhofer, Brainan, & Reisner, 1984). Chronic alcoholics may show abnormal brainstem auditory evoked responses, and patients with abnormal responses are much more likely to show brainstem or cerebellar atrophy on CT scan (Chu, 1985).

Early evoked responses are reliable when interpreted by an experienced clinical neurophysiologist. The tests are relatively inexpensive and noninvasive and can be administered by a technician. The psychiatrist should consider them when the hypothesis to be tested concerns the intactness of primary sensory pathways.

Late Evoked Potentials. Late evoked potentials, a subject of endless fascination for researchers in clinical neurophysiology, have yet to establish a secure role in medical psychiatry. The main reason is that no incremental diagnostic or predictive value has been established for any late evoked potential measure. One of the best studied late evoked potentials is the P300, a positive wave recorded at the vertex from 265 to 500 msec following a target auditory stimulus. Decreased amplitude, increased latency, or both have been reported for P300 in dementia (Kraiuhin, Gordon, & Meares, 1986), focal brain lesions (Ebner, Haas, &

Lucking, 1986), borderline personality (Blackwood, St. Clair, & Kutcher, 1986), and schizophrenia (Blackwood et al., 1987). Parameters for patient and control groups may overlap, however. In the discrimination of demented patients from normal controls, evoked potentials were less effective than psychometrics (Kraiuhin, Gordon, & Meares, 1986) or even routine EEG (Visser, Van Tilburg, & Hooijer, 1985). The P300 evoked potential, therefore, is of physiologic interest but of limited value as a diagnostic test.

Late evoked potentials recently have been applied to demonstrate neurophysiologic correlates of hypnosis (Spiegel 1991). In a number of studies, subjects given hypnotic suggestions were able to modify the amplitude of P300 responses to stimuli in various sensory modalities. Only highly hypnotizable subjects were able to bring about such changes. These observations further limit the applicability of the P300 to the diagnosis of fixed neurologic disease.

Topographic Electroencephalographic Spectral Analysis

The EEG contains far more information than can be readily analyzed visually. This is particularly true of EEG background rhythms, which comprise activity in several different frequency bands, combining to produce rather complex waveforms. A mathematical procedure called the fast Fourier transform, done by a computer using a digitized EEG signal, can produce an analysis of the EEG in terms of relative power in several different frequency bands. The principle of breaking a complex waveform into frequency bands is familiar to stereo buffs, who use graphic equalizers to vary the volume of music in several different frequency bands to adjust to the acoustics of any given space.

Topographic EEG spectral analysis displays the breakdown of the EEG into several frequency bands at a number of different points on the scalp, corresponding to the different EEG electrodes. Computerized graphic displays interpolate values between actual data points, generating a picture that can be displayed either in black and white or in colors representing different frequencies or levels of intensity. The end result is a graphic display that permits a facile visual analysis of asymmetries and changes in the frequency spectrum of a patient's EEG and also provides a set of quantitative data that can be used for statistical analyses of various kinds.

Topographic EEG methods have been able to discriminate groups of patients with Alzheimer-type dementia from elderly healthy controls (Coben, Danziger, & Storandt, 1985); they also can distinguish dyslexic boys from age-matched controls without reading problems (Duffy, 1985). In both of these situations, visual inspection of the EEG would not have reliably distinguished the two groups. In the latter case, topographic EEG revealed differences from controls in the medial frontal region, an unsuspected finding given the previous belief that dyslexia reflected a temporal lobe problem only. Subsequent study of dyslexics by the same investigator showed that clinically different subtypes of dyslexia had distinguishable topographic EEG patterns.

Topographic EEG spectral analysis is a more sensitive measure of background rhythm disturbances than is a visually interpreted EEG (Jerrett & Corsak, 1988). Since the data come from EEG electrodes on the scalp, however, this method is no more able than a routine EEG to detect deep cerebral events far from the recording electrodes. It also shows considerable test-retest variability resulting from physiologic state, vigilance of the patient at the time of recording, various artifacts, and circadian fluctuations. The dominant posterior rhythm, however, shows considerable stability, even with a 1-year retest interval (Gasser, Bacher, & Steinberg, 1985).

Because of its ease of interpretation and automatic production of quantitative data, topographic EEG is a popular research tool for investigating drug effects on the EEG and for providing neurophysiologic measures that discriminate different groups of patients. In the medical-psychiatric setting, it may find its greatest utility in work with specific, well-defined clinical populations in which a neurophysiologic abnormality may or may not be present. For example, topographic EEG might be useful in a group of patients with chronic lung disease, helping to assess which patients with psychiatric complications have a subtle metabolic encephalopathy and which are suffering mainly from reactive emotional symptoms. Although these applications seem rational and promising, topographic EEG has not been applied often with medically ill psychiatric patients and cannot be advised as a routine clinical test at this time (American Academy of Neurology, 1989, pp. 1100–1101). Clinicians working in hospitals or clinics with experienced clinical neurophysiologists available may wish to adapt topographic EEG to specific clinical problems they encounter in their practice.

Brain Electrical Activity Mapping

Brain electrical activity mapping (BEAM) is a specific modification of topograhic EEG that includes not only EEG spectral analysis but also a topographic display of evoked potentials. Evoked potentials are calculated every 4 msec for a total of 512 msec after the presentation of a stimulus, and the amplitude of the evoked potential is determined at each point in the EEG electrode montage. Amplitudes are then interpolated between the actual data points, and a colored computer graphic is generated based on those interpolations. Thus BEAM may test any form of visual, auditory, or somatosensory evoked potential. The result is presented as an endless-loop animated picture representing the spread of the evoked potential through the brain over the 512-msec period of the overall recording. The pictures are fascinating to watch and lend themselves to instant qualitative interpretation, even by the inexperienced observer. A full BEAM study includes both topographic EEG spectral analysis and BEAM evoked potentials in several sensory modalities.

In patients with brain tumors, BEAM shows delayed arrival of the evoked potential at the tumor site, with late persistence of a potential at that site. BEAM could predict the recurrence of a malignant glioma several weeks before it appeared on CT scan (Duffy, 1985). BEAM studies also can distinguish patients with Alzheimer's disease, dyslexia, and epilepsy from appropriate controls.

Because BEAM evaluates late potentials, it is vulnerable to all of the confounding variables that affect the conventional late evoked potentials, including attentional state, medications, level of consciousness, cooperation, and mental set. It also is vulnerable to artifactual data produced by eye movement, muscle activity, and 60-Hz electrical signals (Morihisa, 1989).

In general, guidelines for BEAM are similar to those for late evoked potentials and topographic EEG. Specifically, BEAM can be used to test physiologic hypotheses and distinguish among patients suspected to have a specific neurophysiologic dysfunction. It is etiologically nonspecific and cannot be regarded as a clinical diagnostic test for any disease entity at this time. Because of the extremely large number of variables measured by the BEAM apparatus, "statistically significant" differences between groups may represent nothing more than the chance effects expected when one has a large number of variables. The ultimate development of BEAM as a true diagnostic test will hinge on the development of large banks of normative data for different patient populations. Even then, artifacts may arise from failure to adequately standardize the conditions of recording.

At the present time, a psychiatrist should not simply order a BEAM study. Rather, specific hypotheses should be discussed with a competent clinical neurophysiologist, who might then help devise an adaptation of BEAM methodology suitable for investigating the patient or group of patients.

Nuclear Medicine Procedures

Positron Emission Tomography

Positron emission tomography (Holcomb et al., 1989) is a nuclear medicine procedure that makes use of a special radionuclide produced in an on-site cyclotron dedicated to the purpose. The radionuclide produces positrons, which collide with nearby electrons and produce diametrically opposed pairs of photons. The photons are detected by an array of detectors, and the fact that they occur in diametrically opposed pairs permits location of their source. Technology similar to that used in CT and MRI permits the computer reconstruction of an image. In contrast with CT and MRI, however, the PET scan image does not represent anatomic structures. It is a representation of the distribution of a radionuclide.

By incorporating suitable radionuclides in physiologically or pharmacologically active substances, the PET scan technique permits the collection of data corresponding to physiologic and pharmacologic phenomena, such as glucose metabolism or receptor binding. Many reported studies have been done with [^{18}F]fluorodeoxyglucose. This compound is incorporated in metabolically active cells in the brain and remains within those cells. Thus, the distribution of fluorodeoxyglucose permits mapping of brain areas according to their cerebral metabolic activity. Because of the efficiency of the detection process and the high activity of the radionuclides used, PET scanning can be carried out with a very small dose of tracer compound.

Positron emission tomography scans also can be conducted with inhaled ^{15}O or ^{15}O-labeled water; these scans can be used to measure regional oxygen metabolism and blood flow (Holcomb et al., 1989). Dopamine receptor occupancy can be estimated by administering a dopamine-receptor blocker labeled with ^{11}C (Farde et al., 1986; Sedvall et al., 1986). Benzodiazepines and opiate-receptor binding ligands, similarly labeled, also have been employed (Holcomb et al., 1989). Weinberger et al. (1991) have mapped the distribution of cerebral muscarinic acetylcholine receptors with PET. The PET scan procedure takes about 30 minutes. At present, it requires an on-site cyclotron to produce the radionuclide, and the overall cost of the procedure is substantially greater than the cost of a CT or MRI study. Experts in the field have predicted that this cost will come down if and when the test becomes more widespread (TerPogossian, 1985).

The fluorodeoxyglucose PET scan, because it measures cerebral metabolic activity, is sensitive to the level of consciousness and the mental and physical activity of the patient being scanned. For example, continual movement of the right hand throughout a PET scan would be expected to produce increased glucose utilization over the left motor strip. Because so many patient variables may affect the PET scan, comparisons between patients and control groups require substantial control over the circumstances of testing. This and the 30-minute scan period limit the utility of the fluorodeoxyglucose PET scan in uncooperative patients.

Although the PET scan is inferior to CT and MRI in resolution (TerPogossian, 1985), its unique ability is that it can measure metabolic changes that antedate anatomic changes or occur in their absence. Patients with Alzheimer's disease may show PET scan changes in the posterior temporal and parietal regions at a time when the CT scan is normal for age or shows only nonspecific diffuse atrophy (Duara, Grady, & Haxby, 1984; Friedland, Bodinger, & Brant-Zawadski, 1984; Haxby, Grady, & Duara, 1986). Positron emission tomography shows diffuse hypometabolism in normal-pressure hydrocephalus; this contrasts with temporoparietal hypometabolism in Alzheimer's disease (Jagust, Friedland, & Bodinger, 1985). It remains to be seen, however, whether PET scans will be predictive of surgical outcome, or add anything to a careful history and examination in selecting patients for shunts.

The PET scan may show decreased metabolism in a temporal lobe seizure focus that is not visualized on MRI or CT. Subsequent removal of foci shown on PET scan may improve the epilepsy, and pathologic examination may show the underlying cause of the epileptic condition (Sperling, Wilson, & Engel, 1986).

Positron emission tomography scans may show abnormal caudate metabolism in patients at risk for Huntington's disease, prior to the onset of clinical symptoms (Clark et al., 1986). If patients with these findings all subsequently develop Huntington's disease, the PET scan may prove to be the most sensitive diagnostic test for this disorder (Kuhl, Phelps, & Markham, 1982; Martin, 1985). In other movement disorders, such as dystonia, parkinsonism, and hereditary chorea, PET studies in a research context may illuminate the pathophysiology, but PET would not be considered as a diagnostic test (Martin, 1985). In rare cases in which it was unclear whether a movement disorder was of organic origin, a positive PET study would document abnormal physiology, but a negative study would not be conclusive, because a PET scan may be normal in some movement disorders of unquestionable organic etiology.

Applied to stroke or trauma patients, the PET scan may show hypometabolism in larger areas than appear clinically involved by brain imaging techniques (Langfitt et al., 1986). Ultimately, PET may aid in the differential diagnosis of the neurobehavioral sequelae of brain injury because it may show persistent functional abnormality in the limbic and the cortical structures at a time when a normal MRI or CT would favor a "functional" diagnosis.

Positron emission tomography scan findings in primary psychiatric illness yield variable results because of the daunting complexities of study design and data interpretation (Morihisa, 1991). In the 1989 review of Holcomb et al., numerous interesting findings were reported but few have been replicated multiple times. Probably the best replicated findings concern hypometabolism of several brain regions in patients with schizophrenia: these include the frontal lobes, basal ganglia, and temporal lobes (Buchsbaum, 1990). Despite tantalizing progress, the PET scan is far from being a clinical test for any primary psychiatric disorder.

By contrast, several areas of definite clinical utility have emerged for PET scanning of patients with primary *neurologic* disease. These clinical applications are based on well-replicated studies, and are summarized in a recent report of the American Academy of Neurology's Subcommittee on Therapeutics and Technology Assessment (American Academy of Neurology, 1991). They regard the following indications for PET studies as "clinically safe and useful in the evaluation of neurologic disease":

1. *The presurgical evaluation of patients with refractory seizure disorders.* In this situation, PET can localize seizure foci, sparing patients expensive and invasive depth electrode studies. In some cases, PET is the only localizing study identifying lesions whose etiologic role is later confirmed by successful surgery.
2. *The differential diagnosis of dementia.* All major dementing illnesses have distinctive patterns of hypometabolism on PET scanning. In cases of questionable dementia, definitive abnormalities can be demonstrated when anatomic brain images are normal.
3. *Differential diagnosis of movement disorders.* Positron emission tomography can physiologically distinguish between Parkinson's disease with dementia and progressive supranuclear palsy. It can also offer confirmatory but nonspecific evidence of caudate hypometabolism in patients with suspected Huntington's disease.
4. *Management of brain tumors.* Positron emission tomography scanning can localize tumor tissue to guide biopsy. It can assist in the estimation of the grade of malignancy of tumors and in differentiating recurrent

tumor from necrosis induced by chemotherapy radiation. This latter use is of major clinical significance since it reduces the risk of giving unnecessary treatment to a patient with ambiguous findings on anatomic brain imaging.

In the first and the last of these indications, PET scanning not only is clinically useful but is evidently cost effective because it obviates invasive procedures. In the other two cases mentioned, the advisability of a PET scan depends on how urgently the incremental information is needed for clinical or scientific reasons. Most cases of early or questionable dementia can simply be observed, once specifically treatable etiologies have been ruled out. Nonetheless, some patients may have compelling emotional or even legal reasons to know as certainly as possible the specific disease from which they suffer. Of course, if and when specific treatments become available to arrest progression of degenerative diseases, early and precise diagnosis will become essential.

In addition to the applications endorsed by the American Academy of Neurology, another promising application of PET scanning is in the diagnosis of neuropsychiatric lupus. In a recent series of 13 patients, 10 of whom had neurologic involvement, Stoppe et al. (1990) showed that all 10 had abnormal findings on PET scans, sometimes when MRI or CT scans were normal. Moreover, one case showed reversal of the PET scan abnormality when her condition improved with immunosuppressant treatment. If this application is replicated and interpretive criteria are standardized, it will be rational to apply PET scanning to the differential diagnosis of neuropsychiatric lupus versus functional or steroid-induced psychosis. The high cost of PET might be justified by the high cost, and high risk, of inappropriately chosen treatment in this situation.

Because of the high cost of PET, and its lack of coverage by many third-party payers, its availability is restricted to a relatively small number of tertiary care medical centers. As long as this situation remains, the cost and inconvenience of transport to a PET center must figure additionally in the clinician's decision to consider applying this technique.

Single-Photon Emission Computed Tomography

Single-photon emission computed tomography is a nuclear medicine procedure based on a photon-emitting radionuclide rather than a positron-emitting radionuclide. This difference has several implications. Since photon-emitting radionuclides are readily available and may be fairly stable, an on-site cyclotron is not necessary to conduct SPECT studies. However, precise quantitation of regional radioactivity is done accurately more with PET than with SPECT. The reason is that with SPECT the intensity of radiation from the single-photo-emitting source falls off as the square of the distance and in proportion to the density of the material through which the photon travels. In a structure like the brain, which is not homogeneous, precise mathematical reconstruction of regional radioactivity is virtually impossible. Positron emission tomography is based on the detection of photon pairs that are diametrically opposed. This permits a more straightforward correction for attenuation of photons by tissue and leads to more accurate quantification.

Although SPECT does not permit as precise a quantitation of receptor counts, blood flow, or metabolic rates as does PET, it is cheaper and more widely available. SPECT is easily performed in a community hospital nuclear medicine department (Devous, 1989). Unfortunately, the quality of equipment for SPECT varies, and many community hospitals have scanners with insufficient resolution for reliable clinical application to neuropsychiatric questions (Morihisa, 1991). Psychiatrists ordering SPECT studies must assure themselves of the experience of the imaging team and the quality of its equipment.

Radionuclides used for SPECT (Devous, 1989) include xenon-133, [^{123}I]iodoamphetamine (^{123}IMP), ^{99m}Tc-HMPAO, and receptor ligands such as ^{123}I-labeled quinuclidinol, which binds to cholinergic receptors. Xenon-133, administered by inhalation, permits true quantitative assessment of regional cerebral blood flow. However, it can only be used five to 10 times in the lifetime of the patient because it exposes the airway to a relatively high radiation dose. The lipophilic intravenous agents ^{123}IMP and ^{99m}Tc-HMPAO are taken up by brain tissue in proportion to cerebral blood flow, and then undergo a conformational change that temporarily traps them within brain cells. Their process of distribution is not well understood, so imaging with these agents permits only semiquantitative assessment of regional cerebral blood flow. Of the two agents, ^{99m}Tc-HMPAO has a longer half-life, and so is more suitable for studies that might be conducted hours after the injection of the radionuclide, or for doing multiple studies from one vial of radionuclide. Receptor ligand radionuclides show differential binding as expected by the regional distribution of the receptors to which they bind. Both dopamine D_2 receptors and cholinergic receptors have been successfully imaged by SPECT, but no clinical tests have as yet been developed using receptor ligand radionuclides.

Research findings with SPECT in primary psychiatric disorders have paralleled those with PET: a variety of abnormalities in regional cerebral blood flow have differentiated groups of patients with mood disorders or schizophrenia from groups of normal subjects. However, no SPECT procedure can be used to diagnose an individual patient as suffering from a specific psychiatric condition (Devous, 1989; Morihisa, 1991). As with PET, applications to neurologic diagnoses are somewhat better established, and at times are appropriate additions to clinical practice.

In clinical situations, SPECT using ^{123}IMP or ^{99m}Tc-HMPAO is used to test hypotheses regarding regional cerebral blood flow that may explain discrepancies between neuropsychiatric findings and anatomic imaging results. The development of such hypotheses requires neurologic evaluation and a neurologic differential diagnois; SPECT results unrelated to such a context are diagnostically nonspecific and at worst misleading. Typical applications include the following:

1. Identification of stroke in patients with a recent onset of focal neurologic findings and negative CT or MRI.
2. Identification of areas of compromised perfusion in patients with a stroke or head trauma with neurologic deficits disproportionate to lesions visualized on CT or MRI.
3. Confirmation of focal or multifocal brain function abnormalities in patients with AIDS and suspected AIDS dementia (Pohl et al., 1988).
4. Demonstration of parietal lobe hypoperfusion in patients with Alzheimer's disease (Battistin et al., 1990; Hellman et al., 1989; Johnson et al., 1988; Perani et al., 1988; Jagust, Budinger & Reed, 1987). Using SPECT, patients with Alzheimer's disease can be differentiated from patients with multi-infarct dementia, who have multifocal perfusion deficits and relatively less severe parietal lobe hypoperfusion.
5. Identification of areas of hypoperfusion in patients with cocaine abuse and cognitive dysfunction and nondiagnostic anatomic brain imaging (O'Connell et al., 1991; Tumeh et al., 1990).

In applying SPECT for these purposes, it should be expected that SPECT evidence may support a diagnostic impression but will not be diagnostic in itself. A patient with a SPECT scan showing patches of hypoperfusion, a negative MRI, and no clear-cut explanation, such as cerebral vascular disease or cocaine abuse, represents a diagnostic conundrum with no solution apart from careful clinical follow-up.

Lumbar Puncture and Cerebrospinal Fluid Examination

Cerebrospinal fluid examination, although less frequently done now than in the past, remains the essential test for the diagnosis of acute or chronic meningitis, for encephalitis, and for neoplastic involvement of the meninges. It is regularly performed by neurologists and internists when infection is suspected; the psychiatrist's role initiating the CSF examination is that of raising a suspicion of infectious or inflammatory CNS disease based on mental status changes. Mild deliria with mainly psychiatric manifestations may be viewed as functional by nonpsychiatrists, leading to delays in performing necessary CSF examinations. In our experience, this is most likely to happen with young adults with acute-onset psychoses or personality changes. These patients may have encephalitis or meningoencephalitis, with a raised CSF cell count supporting the diagnosis. Such elevated cell counts were found in almost one-fifth of a sample of patients admitted to a Finnish hospital with acute psychosis of less than 2 weeks' duration (Ahokas, Koskiniemi, & Vaheri, 1985). The Finnish investigators also found that 41% of their sample had immunoglobulin IgG oligoclonal bands in the CSF, suggesting an inflammatory or immunologic process. Although afebrile psychiatric patients with mild lymphocytic pleocytosis in the CSF, or elevated oligoclonal bands, usually do not have specifically treatable CNS infections, their positive CSF findings identify them as a group requiring comprehensive neuropsychiatric assessment and follow-up. It is likely that such patients may have a different course, prognosis, and response to treatment than those without CNS inflammation underlying their psychopathology.

Cerebrospinal fluid findings in viral meningoencephalitis are nonspecific and do not permit ascertainment of the causal agent (Griffith & Ch'ien, 1984). Viral cultures can be negative in definite cases, and in any case their results are not available quickly enough to guide management. For this reason, CSF examination sometimes is omitted when a typical systemic viral infection is complicated by headache and mild changes in cognitive mental status. This is appropriate, but psychiatric considerations may indicate CSF examination even when a viral infection such as infectious mononucleosis already has been diagnosed. Specifically, when psychotic-level behavior changes take place, CSF evidence of inflammation helps make the attribution of the mental disorder to organic CNS involvement, rather than to a functional psychosis triggered by the stress of physical illness. Prognostic and management implications follow.

Patients with disorders such as malignancies, lupus, and AIDS that are known to frequently involve the CNS may require diagnostic CSF examination if brain imaging, serologic studies, and physical examination cannot establish the presence of CNS involvement in a patient with manifest mental changes. For example, cell count may be elevated or C4 levels decreased from baseline in patients with CNS lupus (Rodnan & Schumacher, 1983). Psychiatrists involved in the care of such patients should initiate CSF examination when the information would be necessary to determine whether mental symptoms were likely to be due to direct CNS involvement. Because psychiatrists must choose treatment strategies based on this determination, their need to know the CSF findings may be greater than that of the primary physician at particular points in the patient's course. When gross, nonmental signs of brain involvement supervene, the neurologist or internist would naturally make the decision regarding the need for CSF examination.

In a patient suspected of having MS, the finding of elevated IgG or oligoclonal bands in the CSF supports the diagnosis (Caroscio, Kochwa, & Sacks, 1986; Whitaker, Benveniste & Zhou, 1990). Although MRI and evoked potentials are sufficient to diagnose MS in most cases, there are occasional patients in whom CSF examination is necessary to establish the diagnosis. Certainly, CSF examination for oligoclonal bands and IgG index should be conducted when there is a difficult differential diagnosis between MS and hysteria.

Chemical analysis of CSF has a limited role in the diagnosis of toxic-metabolic encephalopathy. Cerebrospinal fluid glutamine or ammonia levels are elevated in hepatic encephalopathy and Reye's syndrome, and determination of the CSF glutamine level might make sense in diagnosing a patient with known liver disease and mental status changes for whom the EEG, physical examination, and blood ammonia level were not decisive (Fishman, 1992, pp. 320–322; Tarter et al., 1985).

Measurement of CSF concentrations of neurotransmitters and their metabolites has increasingly been pursued in research laboratories, but as yet no clinical tests have emerged from this research (Fishman, 1992, pp. 240–244; Stahl, 1985). One might anticipate, however, that appropriately selected CSF neurotransmitter measures would help in distinguishing emotional reactions from mental signs of altered CNS chemistry in patients with psychiatric complications of systemic disease. Development of such tests would require systematic study of subgroups of patients at risk, much as should be done for BEAM and other sensitive but nonspecific tests.

The diagnosis of fungal infection of the CSF should be considered when subacute mental changes develop in an immunocompromised patient. The psychiatrist may become involved in such cases because the initial CSF examination was normal or nonspecifically abnormal, raising concerns about psychiatric contributions to the change in mental status. In assessing such patients, it is useful for the psychiatrist to know that the diagnosis of fungal meningitis may require multiple lumbar punctures and CSF cultures, with a relatively high volume (20 ml or more) of CSF taken for each culture (Greenlee, 1991). The cryptococcal antigen, the most popular CSF test for the most common treatable fungal meningitis, is falsely negative in 42% of patients with culture-positive disease. Fungal meningitis thus cannot be ruled out by a single CSF examination, and follow-up examinations should be performed if clinical suspicion is high (McGinnis, 1985). CSF ferritin levels are consistently elevated in patients with bacterial and fungal meningitis but not in patients with viral meningitis (Campbell, Skikne, & Cook, 1986). This work suggests that analysis of CSF proteins might aid in the early diagnosis of CSF infections in immunocompromised patients. Ferritin also is increased in herpes simplex encephalitis (Sindic et al., 1985). If measurement of CSF ferritin level is available, it might be considered in patients with acute psychosis and temporal lobe abnormalities on EEG or brain imaging that suggest herpes encephalitis, particularly in cases where routine CSF findings are equivocal.

Cerebrospinal fluid examination is not useful in diagnosing typical cases of Alzheimer's disease (Becker Feussner, & Mulrow, 1985; Hammerstrom & Zimmer, 1985). In an exhaustive review of studies of CSF examinations in Alzheimer's disease, Van Gool & Bolhuis (1991) concluded that there is no specific diagnostic CSF test for Alzheimer's disease. Alzheimer's disease patients have shown a variety of differences from control groups on CSF levels of various substances, including proteins found in senile plaques and neurofibrillary tangles. Notwithstanding, insufficient sensitivity and specificity have limited the application of these findings to clinical diagnosis. Lumbar puncture in dementia patients should be performed if there are atypical features, such as rapid progression, fever, meningeal signs, or a positive blood serology for syphilis. Lumbar puncture is an essential part of the workup of patients with dementia and hydrocephalus on brain imaging.

The technique and precautions for lumbar puncture are described in standard neurology textbooks and are not repeated here. We will mention, however, that the amount of fluid removed for CSF studies bears no relation to the subsequent development of headache, so that there is no reason not to obtain sufficient fluid for cultures or biochemical studies in patients for whom a lumbar puncture is indicated (Fishman, 1992, pp. 172–177). The subsequent occurrence of headache does depend on the size of needle used. Prolonged recumbency does not affect the likelihood of developing a post-puncture headache, so outpatient lumbar puncture is reasonable, particularly if a small-gauge needle is used (Gibb, 1984).

IMPROVING THE FORMULATION OF THE NEUROLOGIC DIFFERENTIAL DIAGNOSIS

The formulation of psychosocial diagnosis requires integration of information from the developmental history, the family history, and the clinical interview. The development of a useful and plausible neurologic differential diagnosis similarly requires an integration of diverse information from the history of present illness, the family history, the mental status examination, the neurologic examination, and laboratory tests. Schizophrenia is not diagnosed from mental status findings alone, nor should mental status findings or laboratory tests be used in a simplistic way to categorize a case as either organic or functional (Fogel, 1990). It is more useful to generate for each case a plausible list of organic conditions that might be causal or contributory to the patint's mental disorder. Contributory problems are more common than causal problems, and may be more often neglected because the primary psychiatric factors in the case may be of impressive magnitude.

A useful discipline is to think in every psychiatric case of three plausible organic conditions the patient might have, given the epidemiologic setting, the history, the examination, and the laboratory findings. The points that follow have been useful in formulating neurologic differential diagnoses for psychiatric patients.

Hard neurologic signs should never be dismissed as irrelevant. Borson (1982) described a man recurrently treated for various functional psychiatric disorders while receiving a medical diagnosis of inactive Behçet's disease. Despite a deteriorating course, with psychiatric symptoms accompanied at times by abnormal gait and rigid posture, the psychiatric disorder was not related by his physicians to Behçet's disease. This seems puzzling in view of the known involvement of the CNS by that disorder. (Behçet's disease is a form of vasculitis that usually presents with oral and genital ulcers and recurrent ocular inflammation. Central nervous system involvement, which may include meningoencephalitis, occurs in approximately 25% of patients [Petersdorf, Adams & Braunwald, 1983].) In another instance (Shraberg & D'Souza, 1982), a woman with coma induced by opiate overdose was misdiagnosed as hysterical because of apparent intermittent alertness and forcible eye closure when the examiner tried to open her eyes. Hard neurologic signs included hyperreflexia, difficulty swallowing, and coma vigil. The EEG was diffusely abnormal and the CT scan showed cerebral edema. Other cases have been reported of psychotic patients with cerebellar disease on CT scan, dysarthria, ataxia, and unilateral difficulty on tests of coordination who were diagnosed as functional (Hamilton et al., 1983), and of a subdural hematoma ultimately diagnosed by EEG and CT scan that presented with depression, ataxia, and fluctuating mental status with periods of relative alertness erroneously taken as evidence for a functional etiology (Alarcon & Thweatt, 1983).

Soft neurologic signs and nonspecific psychiatric symptoms are most meaningful when occurring in combinations suggestive of specific diagnoses. Neurologic signs and symptoms must be interpreted in a proper context. Repeated examinations and a search for independent confirmatory signs may be needed. Organic illnesses that are most frequently misdiagnosed include myasthenia gravis, hyperthyroidism, MS, normal-pressure hydrocephalus, brain tumor, chronic subdural hematoma, and pancreatic carcinoma (Martin, 1983). Clues often exist but may be subtle, such as tremor with hyperthyroidism, gait disturbance with normal-pressure hydrocephalus, and gait disturbance and headache with subdural hematoma.

While unilateral signs usually are diagnostically meaningful, bilateral signs do not necessarily imply the presence of neurologic disease. Frontal release signs such as the snout reflex are found in a significant proportion of the normal population (Jacobs & Glassman, 1980). Babinski signs are present after strenuous exercise in 7% of normal subjects (Elliot & Walsh, 1925; Farrell, 1941).

In patients with considerable abnormality of behavior, it may be impossible to perform the subtle sensory and visual-field examinations that may be needed to identify parietal lobe lesions. In one reported case of parietal infarctions presenting psychiatrically, the onset of psychiatric illness in midlife coupled with a history of hypertension suggested the neurologic etiology (Tippin & Dunner, 1981).

The clinical diagnosis of MS often is based on a diagnostic history plus signs that, in themselves, would be nonspecific. Kellner et al. (1983) reported two cases of patients hospitalized with rapid cycling bipolar illness who eventually were found to have MS. In the first instance, the patient had right-sided muscle spasms at age 22 following an episode in her teens of visual disturbance. She was originally diagnosed as suffering a conversion reaction. The second patient developed a mood disorder at age 46 and at age 49 developed paresthesia and stiffness in her legs and poor coordination. At age 52, in light of a progressive gait disturbance, she received a lumbar puncture, at which time oligoclonal bands were found in the CSF. Multiple sclerosis may be associated with unipolar or bipolar mood disorder. A typical scenario is presentation with "hysteria" and neurologic symptoms out of proportion to neurologic signs, followed after a variable period by harder neurologic findings. The combination of a vague intermittent neurologic history with a mood disorder should lead a psychiatrist to consider MS. The diagnosis should be pursued with particular vigor if bedside examination shows soft signs such as hyperactive reflexes and mild incoordination. In this situation, ancillary tests such as evoked potentials and special eye examinations are worthwhile.

Patients whose behavior symptoms get worse on neuroleptics may have an undiagnosed organic disorder. Walker (1982) cited three patients with organic disease (one with a left basal ganglion infarct and two with cortical atrophy) who experienced a deterioration of mental status on phenothiazines that was not reversed with antiparkinsonian drugs. Once a neuroleptic is given, significant neurologic abnormalities may be attributed to the drug rather than to the possibility of underlying brain disease. The neurologic assessment of the unmanageable catatonic patient, particularly in the emergency setting, is facilitated by administering a benzodiazepine rather than a neuroleptic. Neuroleptics may further confuse the clinical picture and cause a clinician to miss a treatable neurologic disease. The neuroleptic malignant syndrome can mimic catatonia. In this instance, a benzodiazepine could help, whereas neuroleptics could worsen that condition (McEvoy & Lohr, 1984).

Diagnostic exclusion of an organic basis for patients' symptoms is impossible early in the course of many illnesses. Psychiatric symptoms may be harbingers of organic illness such as lupus, demyelinating disease (e.g., MS), degenerative brain disease (e.g., Huntington's chorea), and other illnesses—months to years before the emergence of hard neurologic signs. Oommen, Johnson, and Ray (1982) reported on a case of herpes type II viral encephalitis that presented as a confusional psychosis without other neurologic signs and with a normal CT scan and EEG. The CSF was positive and the diagnosis was confirmed at autopsy. Four other such cases were found in the literature. Clinicians cannot assume that viral encephalitis is not present because of the absence of hard neurologic signs and abnormal laboratory tests. Yik, Sullivan, and Troster (1982) reported a recurrent psychiatric disorder (including exhibitionism) commencing in a woman at age 42 that was ultimately attributable to hepatic encephalopathy. The covert diagnosis of hepatic portal vein occlusion was made 5 years after the onset of psychiatric symptoms. Neurologic signs were present that waxed and waned, but there were no peripheral stigmata of liver disease early in the course of the illness.

Depression with profound weight loss without a past personal or family history of depression raises a suspicion of brain tumor, pancreatic carcinoma, a hormone-producing tumor, or endocrine cancer. Frontal and temporal cerebral metastases are particularly likely to present with psychiatric symptoms (Peterson & Perl, 1982). In one study, 22% of patients with frontal tumors presented with psychiatric symptoms (Strauss & Keshner, 1935). Spinal cord tumors may be misdiagnosed as conversion disorders (Epstein, Epstein, & Postel, 1971).

Peroutka (1982) reported a case of a 72-year-old woman with the onset of auditory hallucinations and paranoid delusions attributable to a right temporoparietal-occipital lesion found on CT scan. Only an isolated neuropsychologic test finding and a history of a focal seizure and reversible focal motor dysfunction pointed toward a neurologic disorder.

Among lupus patients, 20% to 40% develop a psychiatric disorder such as delirium, dementia, or depression (Bresnahan, 1982). Electroencephalography and CSF examination usually show abnormalities but occasionally are normal in definite cases. Neurologic signs pointing to an organic etiology for psychosis may appear some time after the presentation of schizophreniform psychoses in patients with numerous diseases of the basal ganglia and adjacent regions, such as idiopathic calcification, Wilson's disease, Huntington's chorea, postencephalitic parkinsonism, bilateral subthalamic infarctions, and brainstem encephalitis (Cummings, Gosenfield, & Houlihan, 1983). The absence of definite neurologic abnormality on initial screening does not permit the clinician to relax vigilance regarding the subsequent development of neurologic signs. Particularly in cases of Huntington's disease and Wilson's disease, diagnoses have been missed because the subsequent development of parkinsonian signs or involuntary movements was attributed to neuroleptics rather than considered as possibly indicating further evolution of an organic disease with psychiatric manifestations.

NEUROPSYCHOLOGIC ASSESSMENT IN MEDICAL PSYCHIATRY

The field of neuropsychology is dedicated to the study of brain-behavior relations in illness and health. Of particular interest are higher level cortical or intellectual functions such as memory and reasoning, as opposed to elementary functions such as sensation. This section offers guidelines for the use of neuropsychologic tests in medical psychiatry, and addresses special topics such as mild head injury, toxic exposure, and the detection of malingering that are of particular concern to medical psychiatrists.

An Overview of the Field and the Purposes of Assessment

Intellectual Assessment

Perhaps the most important intent of neuropsychologic assessment is to provide a detailed description of intellectual or cognitive functions. What is often referred to as intellectual assessment, comprising one of the standardized intelligence tests (e.g., the Wechsler Adult Intelligence Scale–Revised [WAIS-R]) and possibly a very broad screening test for "organicity" (e.g., the Bender Gestalt), offers general information only and, by itself, may be of limited practical value. The Wechsler scales provide scores on a series of subtests that sample a range of functions, as well as overall Verbal, Performance, and Full-Scale IQ scores. One can draw some general conclusions, and perhaps more specific hypotheses, based on the results of intelligence testing. For example, one might be able to conclude that overall level of intellectual functioning is high or very high and that the patient has generally stronger verbal than nonverbal skills. The problem, however, is what one cannot say.

Suppose one refers a patient with a past history of moderate or severe head injury, primarily involving frontal areas. The patient presents with considerable variability in mood, impulsiveness, forgetfulness, and an apparent lack of insight into his condition. The psychiatrist asks for intellectual assessment or neuropsychologic screening. What comes back is a normal intelligence test and a Bender test judged to be of marginal quality but probably unremarkable. One can only conclude that the individual appears normal, or near normal, in the areas assessed on the intelligence test and the Bender test. Moderate or even severe deficits could exist, however, attributable to frontal dysfunction, that were missed by intelligence testing. Because of the lack of comprehensive coverage and the highly structured nature of most of the tasks on intelligence tests, they can be insensitive to deficits associated with frontal dysfunction. Their lack of sensitivity is not limited to frontal disorders. For example, fairly gross, but select, language disorders can also be missed.

The foregoing is not meant to criticize the intelligence test, which is an exemplary psychometric instrument, but to point out that it cannot substitute for neuropsychologic assessment. First, as noted, it may miss significant areas of deficit. Second, it does not provide specific enough information. It may be useful to know that a patient performed poorly on measures emphasizing verbal production, but one needs to know what exactly is the problem. Is it a difficulty with articulation, fluency, or naming? Does the patient have a general disorder in the retrieval of previously stored information, language material included, that interferes with performance on a wide variety of tasks? A third, related, point is that coverage is inadequate. Many areas of cognitive functioning simply are not assessed at all or are assessed only superficially. For example, intelligence tests do not sample reading or writing; also, the comprehension and production of grammatic structure are not assessed directly, nor are scores derived in these areas. Many authors have described or reported on the limits of intelligence testing as the sole means of assessing cognitive functions (e.g., Lezak, 1983; Reitan, 1985; Warrington, James, & Maciejewski, 1986).

Assessment of Memory

To give some idea of what might be considered as comprehensive and detailed coverage of cognitive functions, memory and learning are used as an example. These are complex areas of functioning, and disruption can occur in very limited or specific areas. Thus, one or a few memory tasks cannot provide the sampling that is needed. A comprehensive examination certainly would assess immediate, recent, and remote memory. By immediate memory, we mean the simple repetition of material in temporary "storage," or the repetition or reproduction of material that is kept actively in mind. Repeating a phone number given by the information operator is an immediate memory task. Recent memory tasks call for the reproduction of material that has not been kept actively in mind, but rather has been stored and must be retrieved. If one is first asked to count backward from 10 to 1 before repeating the phone number, this becomes a recent memory task. A precise delineation between recent and remote memory is not possible, but remote memory usually refers either to information that has been stored in memory in the past (e.g., the words one knows) or to the recall of newly presented information after some period of delay, usually at least 30 minutes.

There is good reason to assess these three aspects of memory. Not uncommonly, a patient shows intact immediate memory but gross problems in recent and remote memory. Furthermore, a patient who performs very poorly on recent memory tasks may do so because immediate memory is highly problematic, perhaps as a result of gross deficiencies in attention. In fact, the patient's recent memory may be intact but, because the information never "got in" in the first place, there is nothing to store in recent memory. There can thus be marked dissociations among immediate, recent, and remote memory deficits, although their superficial features may be similar (e.g., poor retention on a long-term basis), and unless one examines all three, erroneous classification may result.

A comprehensive memory assessment must also distinguish between the storage and retrieval of information. When asked to reproduce presented material, a patient may perform poorly. When cues are provided, however, such as multiple-choice questions about the presented material, a patient may answer most, or all, of these questions correctly. These correct answers would clearly suggest that the presented material was stored in some manner and that the patient's essential problem involves retrieval. A patient who has not stored information cannot retrieve it, but a patient who cannot retrieve information spontaneously has not necessarily failed to store it. Problems in storage are different from problems in retrieval and have different implications for diagnosis, treatment, management, and perhaps prognosis.

Comprehensive memory testing must include not only verbal but also visual materials. Patients with intact functioning in one of these two areas have more compensatory options open than those patients with both visual and verbal memory deficits. They may be able to learn to rely on the stronger modality or supplement the weaker modality with the stronger one. Furthermore, failure to assess visual memory may result in the misreading or misinterpretation of clinically relevant cognitive and perceptual deficits. For example, a patient's disorientation may be partially due to severe impairments in visual memory, which make adjustment to new surroundings (e.g., the hospital setting) highly problematic.

Additional evaluation would compare memory for rote materials, such as a list of unrelated words, to memory for meaningful materials, such as sentences or narratives. Patients with left temporal abnormalities may have inordinate difficulties memorizing rote verbal materials but may perform at an average or near-average level with conceptually-based materials (Luria, 1980). With the latter types of materials, context, redundancy, or the potential for storing material as conceptual units may facilitate recall. In contrast, patients with mild frontal disorders might perform normally with rote materials but have marked difficulties with conceptually-based materials. The latter may provide a richer field for competing associations or may require a more organized approach, thereby bringing to the fore inadequacies in self-regulation, difficulties resisting the pull of irrelevant associations, or problems maintaining or shifting mental sets.

Comprehensive assessments will also cover incremental learning, or gains in retention following repeated presentations of materials. For example, the examiner may present a list of 15 unrelated words, assess the number of words retained after the first presentation, present the list again, and then again assess retention. The procedure can be repeated until the patient retains all of the words, or for a set number of presentations. Some patients will show poor first-trial learning; that is, they will remember very little information after the first presentation of material. Following subsequent presentations, however, they may show substantial gains in retention, and perhaps by the third or fourth presentation they may retain as much, or nearly as much, as normal individuals. Such patients thus often benefit greatly from an additional repetition or two of information (the physician may wish to keep this fact in mind when explaining a medication schedule). Other patients will show normal or near-normal retention following the first presentation of information but minimal gains over repeated tasks. For example, a patient with frontal disorder and an associated deficit in planning may not utilize even the simplest natural memory strategy for increasing retention following repeated exposure to material (e.g., paying closer attention to the material missed the first time). In such a case, the problem is not with memory per se; instead, there is a failure to approach memory tasks with adequate planning. If intervention is implemented to improve memory, it thus must focus on these strategic shortcomings. Obviously, such an approach would be very different from that used with the patient who simply needs an additional presentation of material.

The above discussion shows the level of detail that might be required to achieve a thorough assessment of memory. It also suggests the contrast between this depth and the relatively superficial coverage provided by mental status examination. General intelligence tests, such as the WAIS-R, offer only indirect coverage of memory, on such items as the digit-symbol task or vocabulary. The same can be said for other areas of functioning, which are described more briefly.

Assessment of Motor and Sensorimotor Functions

In a comprehensive neuropsychological examination, motor and sensorimotor functions may be covered in some detail. In the motor area, one wishes to examine speed and dexterity and, in particular, to compare right- and left-sided functioning. Many neuropsychologists will also examine what Luria (1973, 1980) has referred to as kinetic melody, or the ability to execute a series of shifting and sequential motor movements. For example, one might have the patient extend an arm and, at the same time, extend one finger, then two fingers, and then the whole hand. Sequential and shifting movements are pertinent to a range of everyday tasks, such as writing. Sensorimotor functions may be examined by assessing for stereognosis, graphesthesia, or discrimination of stimuli presented to both sides of the body simultaneously. It is in the examination of motor and sensorimotor functions that neuropsychologic examinations may most closely resemble neurologic examinations, although many of the techniques or procedures the neuropsychologist employs yield quantitative scores.

Assessment of Visual-Motor Functions

Visual-motor functions are typically examined through copying tasks and spontaneous drawing tasks. Both types of task must be used because the patient may perform well on one but not the other. For example, a patient may produce well-practiced or sparse objects that show no pathognomonic signs on a spontaneous task, but the same patient may show significant deficits on copying a complex figure. Abnormal results on visual-motor tasks usually are nonspecific (Lezak, 1983) because functioning in this general area can be disrupted by a wide range of problems. In order to begin sorting out the factors underlying problem performance, the neuropsychologist might administer parallel tasks that place little or no demand on motor functions and instead stress visual analysis (e.g., the reproduction of designs using easily handled sticks) and parallel tasks that place little or no demand on visual analysis and instead stress motor requirements (e.g., tracing the presented figures).

Language Assessment

In the language area, at minimum, one wishes to assess naming, fluency, repetition, spontaneous speech, oral comprehension, the understanding and production of grammatic structure, reading, writing to dictation, and spontaneous writing. Any one of these areas can be assessed in considerable detail. For example, seemingly simple problems in naming may be select and involve only certain categories or classes of words. Various problems can disrupt reading, including impaired visual scanning or spatial analysis, problems forming visual-auditory associations, and aphasic conditions.

Assessment of Attention, Concentration, and Vigilance

Attention, concentration, and vigilance also demand detailed assessment. Attention, for example, is not a unitary function, and selective problems can occur. Problems might involve focusing on any material, focusing on the most pertinent material, avoiding distraction, sustaining attention over time, and so on.

Assessment of Higher-Level Reasoning

The assessment of higher-level reasoning is of major importance. One wishes to determine whether the patient is able to think abstractly, to form concepts, and to generate and test plausible hypotheses. One wishes to assess these functions in both the verbal and nonverbal areas. For example, in the verbal area, one might examine the patient's capacity to form analogies, to interpret proverbs, and to draw inferences from written material. In the verbal and nonverbal areas, one might examine the capacity to execute a series of consecutive reasoning operations, to think inductively and deductively, and to form and maintain a complex set. In this latter area, for example, one might determine whether the patient can perform alternating addition by threes and sixes, that is, 1–4–10–13–19, and so on. Not only is higher-level reasoning critical to many areas of everyday functioning but, because it is one of the most advanced and complex of the cognitive functions, it is often one of the first areas to be affected by disease processes.

Assessment of Executive Functions

Finally, one wishes to examine self-regulation and planning, functions that often are disrupted by frontal disorders, and also by other conditions including attention deficit-hyperactivity disorder. Can the patient inhibit impulsive responses to prominent stimuli and instead deliberate before responding? Does the patient jump to impulsive conclusions in thinking? Does the patient act according to a plan, or does the patient's behavior quickly deteriorate into a disorganized jumble when all but the simplest requirements for action are presented? Assessment of self-regulation and planning shades into some of the areas already mentioned, such as higher-level reasoning, in part because it is a prerequisite for numerous intellectual functions.

Practical Use of Neuropsychological Test Results

The information obtained through neuropsychologic assessment can be put to various uses, although not all of these are equally well served. Frequently, results are used to make a determination, or to form a probability statement, regarding the presence of organically-based brain impairment. Along these same lines, one can attempt to distinguish organic from functional disturbances in cognitive functioning. One can also attempt to localize brain damage.

How well does neuropsychological assessment achieve these aims? More importantly, how well does neuropsychologic assessment compare with other means for detecting and localizing brain damage, such as the CT scan or other neurodiagnostic techniques? The answer depends on the presenting features of the patient, the type of neuropsychologic assessment that is performed, perhaps the skills of the examiner, and of course the accuracy of the competing methods that are available. Different variations of neuropsychologic batteries are discussed later in this chapter. For now, we briefly review some findings on the accuracy of one of the more popular batteries, the Halstead-Reitan Neuropsychological Battery.

The Halstead-Reitan is a standardized battery that has been researched extensively. With certain types of patient populations, the Halstead-Reitan usually detects brain damage and infrequently misidentifies normal individuals as brain damaged. Most typically, positive studies involve patients who are shown to have brain damage through some independent technique (e.g., CT scan) and patients who are normal according to these techniques. Patients with equivocal diagnoses often are dropped or eliminated from the investigation. For example, a study might involve patients with left-hemisphere tumors and with right-hemisphere tumors, who are compared with normal individuals. With such patient populations, "hit rates" at or exceeding 80% or even 90% are not uncommon. (As is discussed further below, however, with subtle or less clear-cut cases, accuracy rates may be considerably lower.) Attempts to localize the damage generally are less successful. In these studies, the brain is usually divided into four quadrants: left anterior, left posterior, right anterior, and right posterior. This is clearly a rather gross and somewhat arbitrary division, and it is thus rather disappointing that accuracy rates in many studies are not impressive. For example, it is not unusual for accuracy rates to be about 50%, although substantially higher and lower rates have been reported.

Attempts to distinguish brain-damaged individuals from individuals with purely psychiatric disturbance using the Halstead-Reitan Battery have met with mixed success but, in general, accuracy rates are lower than those achieved when comparing brain-damaged individuals to normal subjects. Of course, individuals with psychiatric disturbance can show intellectual deficits, and a certain percentage of psychiatric patients have CNS dysfunction. Any "clean" dichotomy between brain-damaged and psychiatric patients is a mistaken one, and one is left to question how these studies *should* come out: Could a clear separation between the two groups on neuropsychologic measures actually mean that these measures are demonstrating an unacceptable rate of false-negative errors in the psychiatric group?

A negative finding—a trustworthy demonstration that cognitive functioning is normal or that brain damage is not present—can be very useful, as well as reassuring to patients. However, a positive finding, by itself, may be of little value because the most pressing questions are often left unanswered: What is the impact on everyday functioning? Are intervention strategies available? What can be done to help the patient? What is the active disease process? If the purpose of neuropsychological assessment is limited to determining the presence or absence of brain damage, then the potential information derived from these tests, in most cases, is not being fully exploited.

Neuropsychologic assessment also can be directed toward delineating and describing cognitive functioning in a comprehensive manner. The phrase "cognitive functioning," rather than "organic dysfunction," is used deliberately. In some cases, based on the cognitive profile, one may be able to draw relatively confident inferences about the presence or absence of brain damage and about localization. In many cases, however, especially if an initial assessment is being performed, one simply may not know the cause of the dysfunction. For example, the patient may present with a history of head injury and prior learning disability, medication side effects, and gross mood disorder. With so many factors in the mix, one often cannot determine what is causing what. However, knowing that dysfunction is present, describing it in detail, creating a baseline for future comparison, and drawing implications for patient management or treatment can be of significant value. It is in the description of cognitive status and dysfunction that neuropsychologic assessment, whatever its limits, has no real competition and adds depth and detail to even the most thorough mental status examination.

The description of cognitive functioning produced by neuropsychologic assessment can be applied in a number of ways. For example, many cases of moderately severe memory disorder are not diagnosed prior to neuropsychologic assessment. Some of these patients have been receiving verbal psychotherapy which is likely to be of limited benefit in patients who cannot remember much from their sessions. There are also many cases in which patients with language disorders that are subtle but impede speech comprehension are given complex verbal instructions by hospital staff or by family members, which result in poor compliance as a result of misunderstanding. In other cases, physicians explain complex medication schedules to patients with memory impairment. The result, of course, is likely to be noncompliance or, even worse, actions that endanger the patient's health. By understanding more about the patient's cognitive status, the physician can make appropriate environmental and therapeutic adjustments. If impaired cognition and/or memory is recognized, inpatient staff might simplify the patient's environment and provide frequent orienting information, the psychotherapist might adjust the complexity of oral communications or decide to forego certain approaches entirely, or the family therapist might help family members better understand the patient's strengths and weaknesses and aid in the facilitation of effective communication.

A detailed assessment of cognitive functions also can help in treatment of the target symptom or symptoms, especially regarding functions, such as attention, that are hard to measure precisely without formal procedures. Suppose a patient has trouble on measures of continuous attention, and that the attentional difficulties interfere with performance on a range of everyday tasks. If pharmacologic intervention is attempted, quantitative measures of attentional functions can be repeated once a therapeutic level of the drug has been established. To a considerable extent, these measures reduce the guesswork involved in evaluating drug effects on the target symptom or symptoms. With a patient in the early stages of a dementing condition, one might try a pharmacologic intervention to improve memory functioning. Determination of baseline neuropsychologic test performance for purposes of comparison of pretreatment and posttreatment functioning permits objective evaluation of the drug trial.

Neuropsychologic assessment can assist in determining the course of an illness affecting cognitive functioning over time, and in this manner can assist in differential diagnosis. If a detailed baseline of cognitive functioning is established, retesting at a later date can provide much more exact information than would be available otherwise. Repeated neuropsychologic assessment might thus be used to follow the course of an individual who presents with equivocal cognitive complaints or findings that might or might not represent the early stages of a dementing condition. A steady decline in scores over time might provide the first hard evidence of a progressive disease process.

Variations in Approaches to Neuropsychological Assessment

When one requests neuropsychologic consultation, the report that one receives varies with the practitioner who performs the assessment and the particular procedures that are utilized. Four distinct approaches are described here.

The quantitative approach to neuropsychology developed with the intent of identifying measures sensitive to the cognitive and behavioral manifestations of brain dysfunction. Empirical procedures were used to uncover tests or test items that achieved maximal separation between brain-damaged and non-brain-damaged individuals, and among individuals with different forms of brain damage.

The Halstead-Reitan Neuropsychological Battery

The prototypical, and arguably the most advanced, battery emerging from the quantitative approach is the Halstead-Reitan. Actually, there is no single Halstead-Reitan. Rather, the term encompasses three different batteries that are applied to individuals at different age levels. Each of these batteries includes a number of tests, plus an additional, rather gross screening test for language disorder and constructional disorder. Many practitioners supplement the Halstead-Reitan with one of the Wechsler intelligence scales, and many practitioners also add further procedures, such as measures to assess memory functions. For the adult version of the battery, one can compute an Impairment Index, based on five tests from which one derives seven scores. The Impairment Index reflects the number of scores that fall within the impaired range. Cutoffs can be used to categorize individuals as most probably normal or brain damaged. As noted, use of the Halstead-Reitin Impairment Index results in relatively high accuracy rates among certain patient groups. General overviews of the battery are provided by Boll (1981) and Reitan (1986); Reitan and Wolfson (1985) provide more exhaustive coverage.

One considerable strength of the Halstead-Reitan is that it provides quantitative scores for all tests. As a result, it is relatively easy to compare the results of one patient with those of another, to compare different sets of results for one patient over time, and to conduct research on the battery. However, because testing procedures may differ in nontrivial ways across examiners (Snow, 1987), one should try to ascertain whether identical methods were used when different examiners have been involved. The Halstead-Reitan is also by far the most thoroughly researched battery in neuropsychology, and across a range of topics (e.g., the effects of alcohol on cognitive functions), there is a much larger background literature than is available on other methods. (This extensive research base also has demonstrated the limitations of the approach, which may be raised by champions of other approaches that need not deal with such negative evidence—or positive scientific evidence for that matter—because their methods are untested.) An additional advantage of the Halstead-Reitan is the availability of standardized decision rules for certain interpretations.

One obvious problem with the Halstead-Reitan is that it does not adequately sample certain critical areas of cognitive functioning. For example, coverage of memory and higher level reasoning is quite restricted. Another potential problem is that the battery is not tailored to particular referral questions or other case features. When the battery is not supplemented with other measures, every adult patient, for example, receives the identical set of tests. A 34-year-old referred to rule out brain damage will thus receive the same measures as a 71-year-old referred to obtain a more detailed understanding of memory dysfunction. Additional problems include a high frequency of false-positive diagnoses among older and less educated groups and limited relations between performance on the battery and everyday functioning (Faust, Ziskin, & Hiers, 1991, Chapter 6).

The Luria Approach

An approach to neuropsychologic assessment first described in detail by A. R. Luria stands in sharp contrast to the Halstead-Reitan and other quantitative approaches. Luria's approach emphasizes theory, observation, flexibility, and the clinical method, as opposed to quantification and standardization. Luria's approach is much too complicated to review in any detail here, and the interested reader should consult the original sources (Luria, 1973, 1980).

Neuropsychological assessment, as conducted from the Luria approach, is founded on a particular theory of brain-behavior relations. In essence, Luria's view is that, although specific brain areas are responsible for specific functions, virtually all meaningful cognitive activities are complex and require the integrated working of multiple brain areas. A breakdown in any of the required components, therefore, can result in a breakdown in the entire activity. For example, even a simple activity such as repeating a set of digits requires adequate comprehension of the task, adequate auditory acuity and auditory discrimination, the temporary storage of information, the transfer of information to centers for motor encoding, and so on. A deficiency in any one of these functions, each of which has a corresponding or underlying brain area for its execution, can disrupt the entire activity. Functional impairments can appear superficially alike, regardless of which particular component is deficient. For example, two patients who fail to repeat digits correctly may make similar errors but do so for different reasons.

Starting with this set of assumptions, one first conducts a broad-based assessment to discover which, if any, complex functions are disrupted. This is done primarily through a series of clinical tasks, although some of the tasks are structured and standardized. For example, to assess various facets of memory, one might use a set of tasks that covers the full range of memory functions. If problems are observed, the examiner proceeds to disentangle, or isolate, the specific underlying components that account for the manifest difficulties. In the example of problems with repeating digits, one would look at the possible contributing factors one at a time. If one hypothesized that the patient's failure might be due to difficulties in motor encoding, one might simply have the patient read off a series of visually presented digits. A failure to articulate these digits properly would point toward motor encoding as one contributing factor. One proceeds in this manner, uncovering any general difficulties in cognitive functioning and then probing to determine underlying contributions or components. A great variety of clinical tasks might be administered. In one work, Luria (1980) described scores, if not hundreds, of such procedures. The ultimate goal is to determine the specific areas of difficulty and their full range of impact on cognitive or intellectual functioning.

One strength of the Luria approach is its extreme flexibility in tailoring assessment to a patient's presentation and initial performance. In fact, it is not unusual for an examiner to literally invent procedures on the spot that allow information to be obtained about specific questions raised by ongoing results. Furthermore, because assessment is based on a theory of brain-behavior relationships, such innovations can be guided and organized by theoretical postulates so that they will not deteriorate into a shotgun approach. Additionally, the derived information can be rich and detailed and often seems directly related to everyday functioning and planning of remediation.

Among the potential shortcomings of the Luria approach is that the usefulness or accuracy of the assessment is highly dependent on the skills of the examiner. Although Luria may have used this approach to great advantage, Luria was an exceptionally gifted individual. Whether others, and how many others, can achieve comparable results is uncertain. A long and intensive period of training with high-quality supervision is probably a minimal requirement for competency with the approach.

Additionally, and the same applies to the other approaches described below, there is a lack of formal scientific research validating the method or delineating the results that might be obtained with different types of patients or conditions. One does not know what such research might show, and the unstructured nature of the method may foster problematic judgment practices or biases that tend to increase as structure decreases (Dawes, Faust, & Meehl, 1989).

The Flexible Battery Approach

Lezak (1983) is a prominent advocate of a popular orientation to neuropsychologic assessment, often referred to as the "flexible battery approach." The assessment method and specific tests are tailored to the patient's presenting characteristics and the nature of the referral question. To the extent possible, however, standardized, and validated tests are used so that quantitative scores can be derived. The examiner begins with a broad set of instruments, which are supplemented in relation to the patient's presentation. For example, additions to the typical group of memory tests are made for a patient with an apparent decline in memory functioning and with subjective complaints about forgetfulness. Furthermore, depending on the initial results, additional testing of problematic areas is carried out. If standard, validated measures are not available, experimental or less-established procedures are used. The reader should consult Lezak (1983) for a full description of the flexible battery approach.

Lezak provides a strong rationale for the flexible battery approach, pointing out that it provides rich descriptive data that incorporate both qualitative and quantitative information and that it specifically addresses the questions leading to the consultation. That many neuropsychologists use the flexible battery approach testifies to the positive perception of the method, but we currently do not know, or have not determined empirically, the extent to which these positive perceptions are warranted. There has been virtually no research on the flexible battery approach, and in fact, it is fairly hard to imagine how many types of critical questions about it could be subjected to empirical test. Recent studies that have failed to support some of Lezak's positions, such as her proposed method for determining prior functioning (Mortensen, Gade, & Reinisch, 1991), reinforce caution against accepting too much on faith.

The Process/Qualitative Approach

A final approach to neuropsychologic assessment is a process-oriented, or qualitative, approach, championed by Edith Kaplan. As described by Milberg, Hebben, and Kaplan (1986), this approach uses a core set of tests but, depending largely on the patient's presentation, additional tests are also used to attain greater clarity about specific problem areas and to cross-validate impressions or hypotheses derived from the core tests. Standard tests are commonly modified for use with cognitively impaired individuals, in order to provide more detailed information about the cognitive processes underlying test performance. For example, after a test has been administered in standard fashion, the same test might be repeated with extra time allowed for the patient to complete it. Encouragement or cues might be provided to examine their effects on performance and thereby achieve a greater understanding of the patient's disabilities and rehabilitative needs.

The process/qualitative approach is founded, in part, on traditions within psychology that have yielded noteworthy insights. For example, Jean Piaget, the famous developmentalist, became interested in the systematic patterns of reasoning underlying children's failures on standard test items. The approach also incorporates Werner's (1937) landmark work on cognitive development. This overall effort to incorporate findings from various areas of cognitive science and experimental neuropsychology predated, and helped create, the current emphasis on the cognitive neurosciences. As is the case with Luria's approach and the flexible battery approach, the process/qualitative approach has been subjected to little formal empiric testing. From this standpoint, the approach remains largely unproven.

Special Issues

Depression and Neuropsychologic Tests

Because medical-psychiatric patients frequently have depression secondary to primary somatic illness, their neuropsychologic assessments must take into account the confounding effect of depression. Whereas profound depression can impair patients' ability to cooperate with testing, less severe depression permits patients to cooperate but impairs their performance. Depression can impair neuropsychologic performance globally, or can produce relatively selective deficits in motor skills, attention and concentration, memory, abstract reasoning, or visual-spatial functions (Cassens, Wolfe, & Zola, 1990). It can also impair verbal fluency, but does not produce true aphasia (Stoudemire et al., 1989).

Distinction between depression-related cognitive dysfunction and deficits produced by gross brain disease is most often possible when the cognitive deficits have a pathognomonic quality, as with frequent paraphasic errors in speech or consistent hemineglect. Neuropsychologic assessment usually can produce estimates of the likelihood that a cognitive deficit is due entirely to depression, or that it more likely represents a deficit of another cause that is aggravated by depression. In all cases, retesting after optimal treatment of depression is necessary to make an accurate assessment of fixed deficits. A more common clinical scenario, however, than the "dementia versus depression" problem is the quite common occurrence of depression coexisting with a preexisting dementing illness (Alzheimer's disease, multi-infarct dementia, Parkinson's disease). If underlying dementia is present, concurrent depression usually will exacerbate the degree of cognitive dysfunction present. If any doubt exists regarding the relative contribution of depression to the patient's condition, the depression—if suspected—should be aggressively treated. Cognitive dysfunction related to the effects of depression usually will remit with response to antidepressant treatment with either cyclic antidepressants or ECT (Stoudemire et al., 1991).

Computerized Assessment

Computerized test administration and interpretation are becoming increasingly popular. Within the context of neuropsychological evaluation, computer assistance is most commonly used to score and administer personality tests, which often are included in neuropsychologic test batteries. For example, many neuropsychologists administer the Minnesota Multiphasic Personality Inventory (MMPI), and various services have developed MMPI interpretive programs. Computerized scoring or interpretation of neuropsychologic tests per se is less common.

Whatever the type of test, when test *administration* is computerized, the computerized format should yield scores equivalent to those obtained with the standard or traditional testing format. Equivalence is not merely an academic concern because some studies show that changes in format can affect testing results (e.g., Baker et al., 1985). For example, some elderly individuals may show a performance decrement when testing is shifted to a computerized format (Hofer & Green, 1985).

Equivalence should not be assumed, but rather established through sufficient scientific testing. For many tests, equivalence has not been evaluated. If equivalence has not been established or evaluated, there is doubt that one can apply the background knowledge and research derived from the standard or traditional testing format. In such cases, use of a computerized format adds uncertainty to methods that are usually already less certain than one might wish.

There is much confusion about computerized test interpretation, and it is sometimes accorded a status it does not deserve. In essence, computerized interpretation is *automated* interpretation. Automated interpretations may or may not be verified or formally validated. For example, a program may simply automate a particular clinician's interpretive strategies or approach. If clinician Jones has the odd and unverified belief that a solicitous smile demonstrates pent up hostility toward communists, he can program the computer to do as he does—to render the interpretation "communist hater"—whenever the code sheet lists this "sign" as present. In contrast, the computer may be programmed to follow validated decision rules.

A computerized interpretive program should be evaluated against the same basic criteria as other interpretive methods. For example, it should demonstrate adequate sensitivity and specificity. The wrapping or testimonials that may accompany a program are no substitute for more formal, scientific support. Some brochures emphasize the program developer's credentials but say little or nothing about scientific validation of the interpretive system. As considerable research shows, one cannot assume that highly experienced or credentialed clinicians are necessarily accurate (Ziskin & Faust, 1988, Chapter 8), including those who go in for designing computerized interpretive systems.

For computerized personality test interpretation, the validation research is typically thin. Some programs lack any formal scientific testing, and in many cases, because of proprietary interests, the basis on which programs derived interpretive statements is not shared with program users. For lack of such information, it is very difficult to appraise services in an informed manner. When research is conducted, it is often limited to "consumer validation" studies, or an examination of clinicians' satisfaction with interpretive reports, without independent verification of diagnostic accuracy.

Validation research is even more sparse for computerized neuropsychologic test interpretation, and what is available, at least thus far, is not particularly impressive (e.g., Adams, Kvale, & Keegan, 1984). There seems to be fairly broad consensus that these programs are not yet sufficiently developed for routine clinical use.

Some practitioners who wish to have easy access to one or another personality test, such as the MMPI or the Millon Clinical Multiaxial Inventory, may be tempted to use a computerized interpretive service. We would suggest at minimum that the clinician check to determine the amount of published research available on the interpretive program and not be distracted by packaging or claims that lack scientific proof. If the program lacks adequate background research, one cannot assume it produces accurate or useful interpretations, no matter how impressive the reports might look or sound. Reviews of programs for interpreting personality tests can be found in the *Mental Measurements Yearbook* (Conoley & Kramer, 1989).

Neuropsychologic Deficits following Mild Head Injury

About 5 to 10 years ago, a number of studies appeared suggesting that mild head injury often leads to serious and persistent neuropsychologic impairment (e.g., Barth et al., 1983). Prior to this, separate research indicated that seemingly minor head injury can produce microscopic brain lesions (e.g., Oppenheimer, 1968). Studies with primates also showed that experimentally induced acceleration/deceleration forces, even in the absence of direct head impact, can lead to similar microscopic injury (Gennarelli et al., 1982). It was natural to draw a cause-and-effect relation between the phenomena seemingly shown in these separate lines of investigation.

However, the functional significance of microscopic injuries of the type these studies demonstrated remains unclear, and the background literature sometimes cited to bolster positions, such as Oppenheimer's work, may be less supportive of claims than might be assumed. For example, in contrast to the argument that even one rather minor blow to the head may well cause serious and lasting deficit, Oppenheimer (1968) stated, "If such injuries are *repeated* (as they may be, for instance, in an *unsuccessful* boxer), one would anticipate that a progressive, cumulative loss of tissue, and of nervous function, would occur" (p. 306). In other cases, to support the argument that a low-speed motor vehicle accident with *g* forces probably numbering in the single digits caused serious, permanent injury, individuals will cite animal studies with *g* forces running into the hundreds.

Additionally, these neuropsychologic studies suffered from a critical methodologic defect. They contained a disproportionate number of individuals who tend to perform poorly on neuropsychologic tests whether or not they have suffered head injury. Neuropsychologic test performance correlates with such variables as education and socioeconomic status (Leckliter & Matarazzo, 1989), with less educated and less affluent individuals overrepresented among the head injured. For this and other reasons, such studies on mild head injury usually require a matched control group.

A number of subsequent studies with the needed control groups have indicated that serious and lasting deficit is uncommon following mild head injury (e.g., Gentilini et al., 1985; Levin et al., 1987). For example, in their multicenter study, Levin et al. found that mildly head injured subjects showed greater difficulties than controls on a series of neuropsychologic tests shortly after their injuries. However, on 1-month follow-up these difficulties had diminished substantially, and by the 3-month follow-up there were few, if any, significant differences in neuropsychologic test performance between the groups. Levin et al. (1987) concluded that "a single uncomplicated minor head injury produces no permanent disabling neurobehavioral impairment in the great majority of patients who are free of preexisting neuropsychiatric disorder and substance abuse" (p. 234).

Ambiguities remain, however. Some individuals with mild head injuries have persistent complaints or complaints that exceed documentable injury. Some individuals may be more vulnerable to the effects of mild injuries, such as older individuals or those who have experienced prior head injuries. Individuals may also experience subtle difficulties that usually impact little on everday functioning but may impede performance under certain circumstances—for example, when the individual is fatigued, must respond rapidly or under pressure, or must execute demanding tasks. The clarification of such issues awaits future research.

Despite these uncertainties, this more recent and better controlled research does not support the notion of frequent, persistent, and major deficits following mild head injury. Rather, if problems persist, they are likely to be mild or subtle, and their etiology is uncertain (see Levin, Eisenberg, & Benton (1989) for an overview of the area). In a number of these studies, the definitional criteria for mild head injury have included length of unconsciousness up to 30 minutes. Therefore, despite what is sometimes claimed, the clinician can feel confident that an event that leads to no loss of consciousness, no retrograde amnesia, and no posttraumatic amnesia is unlikely to produce serious, lasting deficit.

There appears to be considerable risk that patients referred to neuropsychologists for complaints associated with mild head injury will be overdiagnosed, or seen as more organically impaired or functionally disabled than is actually the case (Faust, Ziskin, & Hiers, 1991, Chapter 4). Distinguishing between subtle deficit and normal variation can be difficult given the state of the art (see further below), and interpretive practices often flow from the belief that it is worse to miss disorder than to overdiagnose disorder. Overdiagnosis creates the associated risk of iatrogenic dysfunction: the label of brain damage and the message that one is impaired (or worse off than one thought) can have adverse effects. As Modlin and Sargent (1986) stated, "Overdiagnosis and overtreatment are common and understandable reactions that medical practitioners have to such a complex clinical state, and . . . recovery is sometimes facilitated through the cessation of medical ministrations" (p. 57).

In cases of mild head injury, some basic steps seem to facilitate return to adequate functioning and to decrease the likelihood of psychiatric complications (such as anxiety reactions or dysfunction produced through secondary gain). Around the time of injury, one can explain the nature of postconcussive symptoms to patients, who can be reassured that their difficulties are very likely to be time limited. If indicated, individuals can also be told that their work efficiency and endurance may be temporarily reduced, and that they should return to work or resume a full workday as they are ready. A 1- or 2-week follow-up appointment can be scheduled to monitor progress. In essence, one provides education and support, and keeps patients who push too hard from pushing too hard right away and patients who do not push hard enough pushing hard enough. Research by such individuals as Gronwall (1986) suggests that approaches of this type promote full symptom relief in the great majority of individuals.

Neuropsychology of Subtle Effects of Toxic Exposure

As in cases of mild head injury, individuals exposed to toxic agents who do not display hard signs of disorder may be referred to a neuropsychologist for evaluation for the presence of brain damage. To understand the problems created for the neuropsychologist who attempts this differential diagnosis when other methods are unable to detect brain damage, one must understand something about typical methods for validating neuropsychologic tests. Most often, one starts with a group of subjects for whom the presence of brain damage has been established with certainty or virtual certainty through some other method, such as the CT scan. One then examines agreement between neuropsychologic testing results and patient status.

In contrast, how does one evaluate the accuracy of neuropsychologic tests when there is no independent, definitive criteria for establishing brain damage? Although some approaches are possible (e.g., following patients with suspected progressive conditions over time), they have rarely been used, and most often one simply does not know how accurate neuropsychologic assessment methods are in subtle cases of this type. There are, however, some studies that examine the accuracy of neuropsychologic methods in cases in which brain damage is detected only through the most sensitive means currently available. Studies involving these less gross and obvious cases tend to show considerably lower accuracy rates than those obtained with less subtle or more serious cases, with results sometimes exceeding chance accuracy only slightly (Faust, Ziskin & Hiers, 1991, Chapter 4). Furthermore, errors occur in both directions, with cases of brain damage missed and controls misdiagnosed as brain damaged. Thus, when other methods are unable to determine the presence of brain damage, one generally should not count on neuropsychologic methods to do so reliably, at least not yet.

The neuropsychology of toxic effects is still in a very early stage of development (Hawkins, 1990). There is not yet strong or consistent evidence that neuropsychologic methods can detect toxic effects when other methods cannot, or that such methods can be used to differentiate the actions of one toxic agent versus another or the effects of toxic agents versus alternate etiologies. Clear patterns of test results have not emerged that distinguish one toxic agent from another. Consequently, as applies in a variety of other areas, one might use neuropsychologic tests to establish a baseline and follow a person over time, or in an attempt to describe and delineate dysfunction in cases of known involvement. However, using neuropsychologic assessment results to conclude that brain damage is present in cases of suspected toxic exposure, when other methods yield negative results, is questionable.

Neuropsychologic Screening

Given the time and expense involved in comprehensive neuropsychologic assessment and the limited availability of practitioners, efforts have been made to develop methods that screen for brain damage or cognitive dysfunction, or that help in determining the need for more comprehensive assessment (for a review of screening methods, see Berg, Franzen, & Wedding, 1987). A number of these efforts have taken the form of bedside examinations (Nelson, Fogel, & Faust 1986). Some bedside procedures are best used to identify or quantify delirium or dementia; other methods are designed to detect less severe impairment.

The current authors have developed one such method, the High Sensitivity Cognitive Screen (Faust & Fogel, 1989). The measure requires about 20 minutes to administer and covers such areas as memory, language, attention and concentration, visual-motor and spatial functions, and self-regulation and planning. Classification of screen results as normal or abnormal, which is based on structured decision rules, shows a high rate of agreement with the results of more comprehensive neuropsychologic evaluation. Therefore, a normal result on the screen can eliminate the need for comprehensive neuropsychologic assessment. The screen is less effective in making more precise determinations, such as identifying specific areas of dysfunction. For further details about the screen and other methods of bedside examination, see Faust and Fogel (1989), Fogel (1991) and Nelson, Fogel, and Faust (1986).

Detection of Malingering

Although in most contexts patients stand to benefit by representing their symptoms honestly, there are exceptions. For example, in the context of litigation, the examinee has much to gain or lose depending on the clinician's impression. A diagnosis of brain damage or functional impairment may mean millions of dollars or, in a criminal case, the difference between living and dying. Even outside the litigation context, a patient may feign dysfunction in order to fulfill some simple instrumental aim (e.g., the desire for a warm hospital bed and food) or to pursue a pathologic aim (e.g., to be operated on).

Research suggests that individuals can alter scores on psychologic tests and, in so doing, fool clinicians into diagnosing abnormality or brain damage (Rogers, 1988). For example, in Heaton et al.'s (1978) study, clinicians performed at chance level to 20% above chance level when attempting to discriminate the neuropsychologic test protocols of individuals with brain damage versus those faking brain damage. Some research suggests that even children can succeed in faking brain damage (Faust, Hart, & Guilmette, 1988). Although one might reassure oneself that faking can be detected on interview, studies are not at all encouraging on this score (Rogers, 1988).

In contrast, the MMPI has a number of scales of demonstrated utility in the detection of malingering (Greene, 1988). For example, the F scale consists of items that are rarely endorsed by normal individuals and are endorsed more frequently, but not necessarily often, by disordered individuals. A number of these items refer to phenomena that a lay person might falsely view as typical of aberrant individuals (e.g., Sometimes my voice goes away even when I am not ill), leading to endorsement on the part of fakers. Thus, one indication of malingering is a highly elevated score on the F scale, outside the range usually obtained by seriously disordered individuals. Other scales assess the relative balance between self-disclosure and defensiveness, the similarity of response patterns to those of known fakers, and the tendency to endorse items that obviously refer to pathology versus items that similarly tap pathology but seem to refer to neutral material.

For many years, the MMPI was the only psychologic method for the detection of malingering that had been studied in some detail. The MMPI is a personality test, and an individual faking neuropsychologic disorder will not necessarily fake emotional disorder. Recently, substantial effort has gone into the development of additional detection methods. One approach is to design a task that appears much harder than it is. For example, one might present the patient with an array of 15 numbers, letters, and figures. One explains that exposure to the materials will be brief and that, given the difficulty of the task, one does not expect complete recall. However, the first row might consist of the numbers A, B, and C, the second row the Roman numerals I, II, and III, and so on. The idea is that the malingerer may overplay the role and perform much more poorly than is plausible. A related approach is to examine performance for deviations from the patterns expected when effort is adequate. For example, one can determine whether success rate corresponds with item difficulty when difficulty is disguised or made difficult to discern. Research on such methods is mixed to date, with some showing promising results (Franzen, Iverson, & McCracken, 1990). Approaches are becoming more sophisticated and may well achieve a high level of accuracy before long. One potential problem with these and other methods is that a malingerer who discerns the task design or gains information about the method may well be able to beat it.

Another approach that shows promise is the forced-choice procedure (Pankratz, 1983). One designs a task on which 50% correct performance represents chance performance. For example, with a patient who complains about the loss of sensation in a hand or finger, one can move the hand up or down with the patient's eyes closed and have the patient guess the position. The malingerer may overact the part and perform below chance level to the point that results are statistically improbable. For example, to produce only 30% correct responses over 50 trials, one generally has to know the correct answers to provide the incorrect ones with sufficient frequency. The forced-choice format can be adapted to many types of complaints. For example, if the patient reports memory loss, one can present a list of 15 words orally, and then provide a written list of 30 words and ask the patient to indicate which of the words were presented before. The patient should get half of the items right by chance alone. Although a negative result on forced-choice procedures does not rule out malingering, performance well below chance indicates a high likelihood of faking on that task.

Given the increasing frequency of litigation and other circumstances in which the honesty of examinees is in question, it is often wise to use the MMPI to evaluate for malingering. If use of the MMPI is not feasible, one can use one of these alternative methods. In many instances, one should also obtain outside or collateral information to verify complaints. For example, the individual in litigation who indicates that his work problems started with the accident may have been fired from a number of jobs in the past.

Methods for Evaluating Everyday Functioning

When neuropsychology first evolved, a major, or the major, goal was to develop methods that could identify and localize brain damage. However, as imaging techniques and related methods became more advanced and less dangerous, the neuropsychologist was called on less often to aid in this diagnostic task. For this and other reasons, attention has turned increasingly to the assessment of everyday functioning. Indeed, in many cases, the psychiatrist or neurologist turns to the neuropsychologist for exactly this purpose.

Research suggests that traditional neuropsychologic assessment methods do not do a particularly good job of predicting functioning in life. The majority of this research has involved the Halstead-Reitan Battery, and results have been mixed, at best (Faust, Ziskin & Hiers, 1991, Chapter 4). Some studies have shown some capacity to predict standing on coarse categories (employment versus unemployment), but not to make more refined judgments, such as the particular everyday living tasks the patient is likely to find troubling. It may be that traditional tests are not suited to the purpose because the factors associated with sensitivity in detecting brain impairment, such as the requirement to perform multiple functions simultaneously, oppose the features that produce differential or *specific* predictive power. For most traditional tests or assessment approaches, there is a paucity of research, pro or con, on prediction of everyday functioning.

Given these limits, there has been increasing effort to design tests and evaluative procedures that directly parallel everyday activities. For example, Wilson et al. (1989) have developed a memory test battery that includes such everyday activities as remembering material from a newspaper, recalling a new route, and remembering the placement of a hidden belonging. Larrabee and colleagues (e.g., Crook & Larrabee, 1988) have also developed procedures intended to provide a closer approximation to real-life memory demands. It might be considered a landmark in neuropsychology that, in 1990, a book appeared entitled *The Neuropsychology of Everyday Life* (Tupper & Cicerone).

Methods for evaluating everday functioning are generally still in an early or experimental phase of development. However, especially if used for purposes of hypothesis formation, or to aid in constructing initial rehabilitative plans that can be modified as necessary, these procedures can serve as a useful adjunct to traditional neuropsychologic tests. Again, this is another area in which developments may well be rapid and improved methods may be available before too long.

Determination of Suitable Patients for Neuropsychologic Assessment and Procedures for Referral

Who should receive neuropsychologic assessment? Although virtually every patient seen by a psychiatrist deserves detailed mental status testing, not every patient is appropriate for neuropsychologic services. Referring patients on a routine basis is a disservice both to the patient and to the neuropsychologist. Aside from wasted time and expense, the patient may worry needlessly about the possibility of brain damage. Only patients with suspected or definite cognitive impairment for whom more exact or detailed assessment is clinically relevant need to be referred.

Ideally the psychiatrist will use a sufficiently detailed and sensitive mental status examination or cognitive screening procedure to reliably rule out cognitive dysfunction. Unfortunately, the items typically included in mental status examinations or brief bedside screening tests lack sensitivity, and even moderate deficits may go undetected. For example, discrete right-hemisphere lesions that selectively affect constructional praxis are likely to be missed. A review of bedside cognitive screening tests has been provided by Nelson, Fogel, and Faust (1986).

Since bedside or office screening has many false-negatives, the presenting problem and the history may indicate the need for neuropsychologic assessment even if bedside screening shows no deficit. Patients complaining of a decline in cognitive function frequently require neuropsychologic assessment. For example, neuropsychologic assessment would be indicated for a patient with complaints of decreasing memory functions beyond normal changes with aging. A past history of injuries or diseases with significant CNS effects, combined with a current behavioral problem, is also reason for referral. Patients with a history of moderate head injury, unexplained school failure, or an episode of encephalitis usually should be referred. Chronic drug or alcohol abuse is frequently associated with cognitive dysfunction; substance abusers often need neuropsychologic assessment to help plan therapy and rehabilitation. Patients with significant occupational exposure to toxic agents and current neuropsychiatric complaints also should be studied.

What findings on mental status examination should initiate referral? Detection of cognitive impairment on bedside mental status examination does not necessarily indicate that a patient needs neuropsychologic assessment. If cognitive defects detected at the bedside are consistent with the patient's medical and psychiatric diagnoses and do not require further delineation or quantification for treatment planning, no further neuropsychologic testing is required. Furthermore, patients should not be referred for neuropsychologic assessment when their status is rapidly fluctuating, as in drug withdrawal or delirium.

There is a preferred way to make a referral to a neuropsychologist. It is problematic if one does not provide any background information or specify the questions of interest. As previously discussed, many neuropsychologists tailor assessment procedures to the features of the patient and to the referral question. Failure to specify the referral questions may thus place the neuropsychologist at a disadvantage. Furthermore, if referral questions are stated explicitly, the neuropsychologist can alert the psychiatrist to potential circumstances that preclude satisfactory answers. A good referral might be something like the following:

> The patient is a 68-year-old woman; previously functioned very well in advanced executive position, but has shown dramatic decline in work performance over last year. Described by husband as very forgetful, overlooks important work details. Physical examination reveals several vegetative signs of depression and markedly depressed mood. Depression seems to have predated declining performance. Also scheduled for neurologic examination. Please provide baseline for future comparison, and impressions regarding dementia versus pseudodementia.

Identifying Appropriate Neuropsychologic Services

There are no formal restrictions on who can be called a neuropsychologist, although guidelines and regulations have been suggested. Some psychologists who indicate they offer neuropsychologic services may not specialize in this area, and there is no set formula for finding a well-qualified or competent practitioner. Guilmette et al. (1990), who conducted a comprehensive survey of those practicing neurospychology, stated, "Probably the most important question to ask a clinician is not how long they have been practicing neuropsychology, but rather how much of their practice is devoted to neuropsychology. In general, the more one's professional activities are devoted to neuropsychology, the better one is trained" (p. 391).

CONCLUSION

Neurologic and neuropsychologic investigations work best when used to test specific hypotheses. Furthermore, soft data are most meaningful when they are corroborated by other data obtained independently. Even with adequate corroboration and the use of statistical criteria for abnormality of test results, it may be difficult to establish a firm neuropsychiatric diagnosis on a single evaluation of a patient. The greatest resource of the clinician in this situation is reexamination of the patient at another time. Replicability of results after an interval supports their meaningfulness, and the presence or absence of progression may confirm or disconfirm a diagnostic hypothesis.

Even with the best testing procedures, psychiatric manifestations of organic illness may appear before tests of organic dysfunction turn positive. Reexamination after an interval sometimes reveals specific neurologic disease where nothing definite was revealed at the time of the psychiatric presentation. For this reason, it is important to avoid premature diagnostic closure. The psychiatrist should maintain an open mind about the etiology of the patient's illness and be prepared to reevaluate the patient neurologically should new symptoms develop or the treatment response be less than expected.

REFERENCES

Adams KM, Kvale VI, & Keegan JF 1984). Relative accuracy of three automated systems for neuropsychological interpretation. J Clin Neuropsychology 6:413–431.

AHOKAS A, KOSKINIEMI M-L, & VAHERI A (1985). Altered white cell count, protein concentration, and oligoclonal IgG bands in the cerebrospinal fluid of many patients with acute psychiatric disorders. Neuropsychobiology 14:1–4.

ALARCON RD, & THWEATT RW (1983). A case of subdural hematoma mimicking severe depression with conversion-like symptoms. Am J Psychiatry 140:1360–1361.

American Academy of Neurology (1989). Assessment: EEG brain mapping. Neurology 39:1100–1101.

American Academy of Neurology Subcommittee on Therapeutics & Technology Assessment (1991). Assessment: Positron emission tomography. Neurology 41:163–167.

BAKER EL, LETZ RE, FIDLER AT, ET AL. (1985). A computer-based neurobehavioral evaluation system for occupational and environmental epidemiology: Methodology and validation studies. Neurobehav Toxicol Teratol 7:369–377.

BAKER HL, BERQUIST TH, & KISPERT DB (1985). Magnetic resonance imaging in a routine clinical setting. Mayo Clin Proc 60:75–90.

BALDESSARINI RJ, COLE JO, & DAVIS JM (1979). Tardive dyskinesia (Task Force Report 18). (Pp. 177–199). Washington, DC: American Psychiatric Press, Inc.

BALLENGER CE, KING DW, & GALLAGHER BB (1983). Partial complex status epilepticus. Neurology 33:1545–1552.

BARTH JT, MACCIOCCHI SN, GIORDANI B, ET AL. (1983). Neuropsychological sequelae of minor head injury. Neurosurgery 13:529–532.

BARTZOKIS G, GARBER JH, MARDER SR, ET AL. (1990). MRI in tardive dyskinesia: Shortened left caudate T_2. Biol Psychiatry 28:1027–1036.

BATTISTIN L. PIZZOLATO G, DAM M, ET AL. (1990). Regional cerebral blood flow study with ^{99m}Tc-hexamethyl-propyleneamine oxime single photon emission computed tomography in Alzheimer's and multi-infarct dementia. Eur Neurol 30:296–301.

BECKER PM, FEUSSNER JR, & MULROW CD (1985). The role of lumbar puncture in the evaluation of dementia: The Durham Veterans Administration/Duke University Study. J Geriatr Soc 33:392–396.

BENTSON J, REZA M, & WINTER, J (1978). Steroids and apparent cerebral atrophy on computed tomography scans. J Comput Assist Tomogr 2:16–23.

BERG RA, FRANZEN MD, & WEDDING D (1987). Screening for brain impairment. New York: Springer Publishing.

BESSON JAO, CRAWFORD JR, PARKER DM, ET AL. (1989). Magnetic resonance imaging in Alzheimer's disease, multi-infarct dementia, alcoholic dementia and Korsakoff's psychosis. Acta Psychiatr Scand 80:451–458.

BLACKWOOD DHR, ST CLAIR DM, & KUTCHER SP (1986). P300 event-related potential abnormalities in borderline personality disorder. Biol Psychiatry 21:557–560.

BLACKWOOD DH, WHALLEY LJ, CHRISTIE JE, ET AL. (1987). Changes in auditory P3 event-related potential in schizophrenia and depression. Br J Psychiatry 150:154–160.

BOLL TJ (1981). The Halstead-Reitan Neuropsychological Battery. In SB FILSKOV & TJ BOLL (Eds.), Handbook of clinical neuropsychology (Ch. 18, pp. 577–607). New York: John Wiley & Sons.

BONDAREFF W, RAVAL J, COLLETTI PM, ET AL. (1988). Quantitative magnetic resonance imaging and the severity of dementia in Alzheimer's disease. Am J Psychiatry 145:853–856.

BONDAREFF W, RAVAL J, WOO B, ET AL. (1990), Magnetic resonance imaging and the severity of dementia in order adults. Arch Gen Psychiatry 47:47–51.

BORSON S (1982). Behcet's disease as a psychiatric disorder: A case report. Am J Psychiatry 139:1348–1349.

BRADLEY WG, & BYDDER G (Eds.) (1990). MRI atlas of the brain. New York: Raven Press.

BRAFFMAN BH, ZIMMERMAN RA, TROJANOWSKI JQ, ET AL. (1988). Brain MR: Pathologic correlation with gross histopathology, 2: Hyperintense white matter foci in the elderly. Am J NR 9:629–636.

BRANT-ZAWADSKI M, FEIN G, & VAN DYKE C (1985). MR imaging of the aging brain: Patchy white-matter lesions and dementia. Am J Neuroradiol 6:675–682.

BREITNER JCS, HUSAIN MM, FIGIEL GS, ET AL. (1990). Cerebral white matter disease in late-onset paranoid psychosis. Biol Psychiatry 28:266–274.

BRESNAHAN B (1982). CNS lupus. Clin Rheum Dis 8:183–195.

BRIZER DA, & MANNING DW (1982). Delirium induced by poisoning with anticholinergic agents. Am J Psychiatry 139:1343–1344.

BUCHSBAUM MS (1990). The frontal lobes, basal ganglia, and temporal lobes as sites for schizophrenia. Schizophrenia Bull 16:379–389.

BYDDER G (1990). Use of gadolinium-DTPA. In BRADLEY WG, & BYDDER G (Eds.), MRI Atlas of the brain (Ch. 9, pp. 302–325). New York: Raven Press.

CALA LA, JONES B, & BURNES P (1983). Results of computerized tomography, psychometric testing, and dietary studies in social drinkers with emphasis on reversibility with abstinence. Med J Aust 2:264.

CALIGIURI MP, BRACHA HS, & LOHR JB (1989). Asymmetry of neuroleptic-induced rigidity: Development of quantitative methods and clinical correlates. Psychiatry Res 30:272–284.

CAMPBELL DR, SKIKNE BS, & COOK JD (1986). Cerebrospinal fluid ferritin levels in screening for meningism. Arch Neurol 43:1257–1263.

CAROSCIO JT, KOCHWA S, & SACKS H (1986). Quantitative cerebrospinal fluid IgG measurements as a marker of disease activity in multiple sclerosis. Arch Neurol 43:1129–1137.

CASSENS G, WOLFE L, & ZOLA M (1990). The neuropsychology of depression. J Neuropsychiat Clin Neurosci 2:202–213.

CELESIA GG (1986). EEG and event-related potentials in aging and dementia. J Clin Neurophysiol 3:99–112.

CHIAPPA KH, & YOUNG RR (1985). Evoked responses overused, underused, or misused? Arch Neurol 42:76–77.

CHIMOWITZ MI, LOGIGIAN EL, & CAPLAN LR (1990). The accuracy of bedside neurological diagnoses. Ann Neurol 28:78–85.

CHU NS (1985). Computed tomographic correlates of auditory brain stem responses in alcoholics. J Neurol Neurosur Psychiatry 48:348–353.

CLARK C, MARBECK R, & LI D (1990). An empirical model for analysing and interpreting ventricular measures. J Neurol Neurosurg Psychiatry 53:411–415.

CLARK CM, HAYDEN MR, STOSSL AJ, ET AL. (1986). Regression model for predicting dissociation of regional cerebral glucose metabolism in individuals at risk for Huntington's disease. J Cereb Blood Flow Metab 6:756–762.

COBEN LA, DANZIGER W, & STORANDT M (1985). A longitudinal EEG study of mild senile dementia of the Alzheimer type: Changes at one year and at 2.5 years. Electroencephalogr Clin Neurophysiol 61:101–112.

COFFEY CE, WILKINSON WE, PARASHOS IA, ET AL. (1992). Quantitative cerebral anatomy of the aging brain: A cross-sectional study using magnetic resonance imaging. Neurology 42:527–536.

COFFEY CE, FIGIEL GS, DJANT WT, ET AL. (1990). Subcortical hyperintensity on magnetic resonance imaging: A comparison of normal and depressed elderly subjects. Am J Psychiatry 147:187–189.

CONOLEY JC & KRAMER JJ (Eds.) (1989). The tenth mental measurements yearbook. Lincoln NB: The Bures Institute of Mental Measurements, The University of Nebraska–Lincoln.

CORNELIUS JR, BRENNER RP, & SOLOFF PH (1986). EEG abnormalities in borderline personality disorder: Specific or nonspecific. Biol Psychiatry 21:977–980.

Cowdry Rw, Pickar D, & Davies R (1985-86). Symptoms and EEG findings in the borderline syndrome. Int J Psychiatry Med 15:201–211.

Creed F, Firth D, Timol M, et al. (1990). Somatization and illness behaviour in a neurology ward. J Psychosom Res 34:427–437.

Crook TH, & Larrabee GJ (1988). Interrelationships among everyday memory tests: Stability of factor structure with age. Neuropsychology 2:1–12.

Cummings JL, Gosenfield LF, & Houlihan JP (1983). Neuropsychiatric disturbances associated with the ideopathic calcification of the basal ganglia. Biol Psychiatry 18:591–601.

D'Aessandro R. Benassi G, & Cristina E (1986). The prevalence of lingual-facial-buccal dyskineasias in the elderly. Neurology 36:1350–1351.

Daniels DL, Haughton VM, & Naidich TP (1987). Cranial and spinal magnetic resonance imaging. An atlas and guide. New York: Raven Press.

Dawes RM, Faust D, & Meehl PE (1989). Clinical versus actuarial judgment. Science 243:1668–1674.

Devous MD (1989). Imaging brain function by single-photon emission computer tomography. In NC Andreasen (Ed.), Brain imaging: Applications in psychiatry (Ch. 4, pp. 147–234). Washington, DC: American Psychiatric Press Inc.

DiChiro G (1985). Magnetic resonance imaging: A time for assessment. Mayo Clin Proc 60:135–136.

Driver MV, & McGillivray BB (1982). Electroencephalography. In J Laidlaw & A Richens (Eds.), A textbook of epilepsy (2nd ed., Ch. 5, pp. 155–194). New York: Churchill Livingstone.

Drury I, Klass DW, & Westmoreland BF (1985). An acute syndrome with psychotic symptoms and EEG abnormality. Neurology 35:911–914.

Duara R, Grady C, & Haxby J (1984). Human brain glucose utilization and cognitive function in relation to age. Ann Neurol 16:703–713.

Dubin WR, Weiss KJ, & Zeccardi JA (1983). Organic brain syndrome: A psychiatric imposter. JAMA 249:60–62.

Duffy FH (1985). The BEAM method of neurophysiological diagnosis. Ann NY Acad Sci 457:19–34.

Ebersole JS, & Bridgers SL (1985). Direct comparison of 3- and 8-channel ambulatory cassette EEG with intensive impatient monitoring. Neurology 35:846–854.

Ebersole JS, & Leroy RF (1983). An evaluation of ambulatory, cassette EEG monitoring: 2. Detection of interictal abnormalities. Neurology 33:8–18.

Ebner A, Haas JC, & Lucking CH (1986). Event-related brain potentials (P300) and neuropshychological deficit in patients with focal brain lesions. Neurosci Lett 64:330–334.

Elliott TR, & Walsh Fm (1925). The Babinski or extensor form of plantar response in toxic states apart from organic disease of the pyramidal tract or systems. Lancet 1:65–68.

Elster AD, Moody DM, Ball MR, et al. (1989). Is Gd-DTPA required for routine cranial MR imaging? Radiology 173:231–238.

Engel J (1989). Seizures and epilepsy. Philadelphia: FA Davis.

Engel J, Driver MV, & Falconer MA (1975). Electrophysiological correlates of pathology and surgical results in temporal lobe epilepsy. Brain 98:129.

Epstein BS, Epstein JA, & Postel DM (1971). Tumors of the spinal cord simulating psychiatric disorders. Dis Nerv System 32:742–743.

Evans DL. Edelsohn GA, & Golden RN (1983). Organic psychoses without anemia or spinal cord symptoms in patients with vitamin B_{12} deficiency. Am J Psychiatry 140:218–221.

Farde L, Hall H, Ehrin E, et al. (1986). Quantitative analysis of D2 dopamine receptor binding in the living human brain by PET. Science 231:258–261.

Fariello RG, Booker HE, Chun RWM, et al. (1983). Reenactment of the triggering situations for the diagnosis of epilepsy. Neurology 33:878–884.

Farrell MJ (1941). Influence of locomotion on the plantar reflex in normal and physically and mentally inferior persons. Arch Neurol 46:22–23.

Faust D, & Fogel BS (1989). The development and initial validation of a sensitive bedside cognitive screening test. J Nerv Ment Dis 177:25–31.

Faust D, Hart K, & Guilmette TJ (1988). Pediatric malingering: The capacity of children to fake believable deficits on neuropsychological testing. J Consult Clin Psychol 56:578–582.

Faust D, Ziskin J, & Hiers JB Jr (1991). Brain damage claims: Coping with neuropsychological evidence (Vols. 1 & 2). Los Angeles: Law & Psychology Press.

Fishman RA (1992). Cerebrospinal fluid in diseases of the nervous system. Philadelphia: WB Saunders

Fogel BS (1990). Major depression versus organic mood disorders: A questionable distinction. J Clin Psychiatry, February: 53–56.

Fogel BS (1991). The high sensitivity cognitive. Int'l Psychogeriatr 3:273–288.

Fogel BS, & Eslinger P (1991). Diagnosis and management of patients with frontal lobe syndromes. In A Stoudemire & BS Fogel (Eds.), Medical psychiatric practice, (Vol. 1; Ch. 10, pp. 349–392). Washington, DC: American Psychiatric Press, Inc.

Fontaine R, Breton G, Dery R, et al. (1990). Temporal lobe abnormalities in panic disorder: An MRI study. Biol Psychiatry 27:304–310.

Franzen MD, Iverson GL, & McCracken LM (1990). The detection of malingering in neuropsychological assessment. Neuropsychol Rev 1:247–279.

Friedland RP, Bodinger TF, & Brant-Zawadski M (1984). The diagnosis of Alzheimer-type dementia: Positron emission tomography versus NMR. JAMA 252:2750–2752.

Gabarski SS, Gabrielsen TO, & Gilman S (1985). The initial diagnosis of multiple sclerosis: Clinical impact of magnetic resonance imaging. Ann Neurol 17:469–474.

Gado MH, & Press GA (1986). Computed tomography in the diagnosis of dementia. Geriatr Med Today 5(7):47–73.

Gardner D, Lucus PB, & Cowdry RW (1987). Soft sign neurological abnormalities in borderline personality disorder and normal control subjects. J Nerv Ment Dis 175:177–180.

Gasser T, Bacher P, & Steinberg H (1985). Test-retest reliability of spectral parameters of the EEG. Electroencephalogr Clin Neurophysiol 60:312–319.

Gennarelli TA, Thibault LE, Adams JH, et al. (1982). Diffuse axonal injury and traumatic coma in the primate. Ann Neurol 12:564–574.

Gentilini M, Nichelli P, Schoenhuber R, et al. (1985). Neuropsychological evaluation of mild head injury. J Neurol Neurosurg Psychiatry 48:137–140.

Gibb WRG (1984). Current practice of diagnostic lumbar puncture. Br Med J 289:530.

Glaze DG (1990). Drug effects. In DD Daly & TA Pedley (Eds.), Current practice of clinical electroencephalography (Ch. 15, pp. 489–512). New York: Raven Press.

Grafton ST, Sumi SM, Stimac GK, et al. (1991). Comparison of postmortem magnetic resonance imaging and neuropathologic findings in the cerebral white matter. Arch Neurol 48:293–298.

Greene RL (1988). Assessment of malingering and defensiveness by objective personality inventories. In R Rogers (Ed.), Clinical assessment of malingering and deception (pp. 123–158). New York: Guilford Press.

GREENLEE JE (1991). Cerebrospinal fluid in central nervous system infections. In WM SCHELD, RJ WHITLEY, & DT DURACK (Eds.), Infections of the central nervous system (Ch. 36, pp. 861–886). New York: Raven Press.

GRENMAN R, AANTAA E, KATEVUO K, ET AL. (1988). Otoneurological and ultra low field MRI findings in multiple sclerosis patients. Acta Otolaryngol 449:77–83.

GRIFFITH JF, & CH'IEN LT (1983). Herpes simplex virus encephalitis: Diagnostic and treatment considerations. Med Clin North Am 67:991–1008.

GRIFFITH JF, & CH'IEN LT (1984). Viral infections of the central nervous system. In GJ GALASSO, TC MERIGAN, & RA BUCHANAN (Eds.), Antiviral agents and viral diseases of man (Ch. 10, pp. 399–432). New York: Raven Press.

GRONWALL D (1986). Rehabilitation programs for patients with mild head injury: Components, problems, and evaluation. J Head Trauma Rehabil 1:53–62.

GUILMETTE TJ, FAUST D, HART K, ET AL. (1990). A national survey of psychologists who offer neuropsychological services. Arch Clin Neuropsychol 5:373–392.

GUMNIT RJ (1986). Intensive neurodiagnostic monitoring: Role in the treatment of seizures. Neurology 36:1340–1346.

HAGGERTY JG, EVANS DL, & PRANG EHA (1986). Organic brain syndrome associated with marginal hypothyroidism. Am J Psychiatry 143:785–786.

HAMILTON NG, FRICK RB, TAKAHASHI T, ET AL. (1983). Psychiatric symptoms in cerebellar pathology. Am J Psychiatry 140:1322–1326.

HAMMERSTROM DC, & ZIMMER B (1985). The role of lumbar puncture in the evaluation of dementia: The University of Pittsburgh Study. J Am Geriatr Soc 33:397–400.

HARPER CG, GILES M, & FINLAY-JONES R (1986). Clinical science in the Wernicke-Korsakoff complex role in a retrospective analysis of 131 cases diagnosed in necropsy. J Neurol Neurosurg Psychiatry 49:341–345.

HAWKINS KA (1990). Occupational neurotoxicology: Some neuropsychological issues and challenges. J Clin Exp Neuropsychol 12:664–680.

HAXBY JV, GRADY CL, & DUARA R (1986). Neocortical metabolic abnormalities precede non-memory cognitive defects in early Alzheimer's-type dementia. Arch Neurol 43:882–885.

HEATON RK, SMITH HH JR, LEHMAN RA, ET AL. (1978). Prospects for faking believable deficits on neuropsychological testing. J Consult Clin Psychol 46:892–900.

HEINZ E, MARTINEZ J, & HAWNGGELI A (1977). Reversibility of cerebral atrophy in anorexia nervosa and Cushing's syndrome. J Comput Assist Tomogr 1:415–418.

HELLMAN RS, TIKOFSKY RS, COLLIER BD, ET AL. (1989). Alzheimer disease: Quantitative analysis of I-123-iodoamphetamine SPECT brain imaging. Radiology 172:183–188.

HOFER PJ, & GREEN BF (1985). The challenge of competence and creativity in computerized psychological testing. J Consult Clin Psychol 53:826–838.

HOLCOMB HH, LINKS J, SMITH C, ET AL. (1989). Postron emission tomography: Measuring the metabolic and neurochemical characteristics of the living human nervous system. In NC ANDREASEN (ED.), Brain imaging: Applications in psychiatry (Ch. 5, pp. 235–370). Washington, DC: American Psychiatric Press, Inc.

HOWARD JE, & DORFMAN LJ (1986). Evoked potentials in hysteria and malingering. J Clin Neurophysiol 3:39–50.

JACOBS L, & GLASSMAN MD (1980). Three primitive reflexes in normal adults. Neurology 30:184–188.

JACOBS L, KINKEL WR, POLACHINI I, ET AL. (1986). Correlations of nuclear magnetic resonance imaging, computerized tomography, and clinical profiles in multiple sclerosis. Neurology 36:27–34.

JAGUST WJ, BODINGER TF, & REED BR (1987). The diagnosis of dementia with single photon emission computed tomography. Arch Neurol 44:258–262.

JAGUST WJ, FRIEDLAND RP, & BODINGER TF (1985). PET with [^{18}F]fluorodeoxyglucose differentiates normal pressure hydrocephalus from Alzheimer-type dementia. J Neurol Neurosurg Psychiatry 48:1091–1096.

JENKYN LR, WALSH DB, CULVER CM, ET AL. (1977). Clinical signs of diffuse cerebral dysfunction. J Neurol Neurosurg Psychiatry 40:956–966.

JERRETT SA, CORSAK J(1988). Clinical utility of topographic EEG brain mapping. Clin Electroencepholography 19:134–143.

JOHNSON KE, HOLMAN BL, MUELLER SP, ET AL (1988). Single photon emission computed tomography in Alzheimer's disease. Arch Neurol 45:392–396.

JOSEPH AB (1990). Non-right-handedness and maleness correlate with tardive dyskinesia among patients taking neuroleptics. Acta Psychiatr Scand 81:530–533.

KATZ JN, LARSON MG, SABRA A, ET AL. (1990). The carpal tunnel syndrome: Diagnostic utility of the history and physical examination findings. Ann Intern Med 112:321–327.

KELLNER CHR, DAVENPORT MSW, POST RM, ET AL. (1983). Rapidly cycling bipolar disorder in multiple sclerosis. Am J Psychiatry 141:112–113.

KERTESZ A, BLACK SE, TOKAR G, ET AL. (1988). Periventricular and subcortical hyperintensities on magnetic resonance imaging. Arch Neurol 45:404–408.

KERTESZ A, POLK M, & CARR T (1990). Cognition and white matter changes on magnetic resonance imaging in dementia. Arch Neurol 47:387–391.

KIEBURTZ KD, KETONEN L, ZETTELMAIER AE, ET AL. (1990). Magnetic resonance imaging findings in HIV cognitive impairment. Arch Neurol 47:643–645.

KILOH LG, MCCOMAS AJ, & OSSELTON JW (1981). The values and limitations of electroencephalography. In Clinical Electroencephalography (4th ed.). London: Butterworths & Co.

KRAIUHIN C, GORDON E, & MEARES R (1986). Psychometrics vs ERPs in the diagnosis of dementia. J Gerontol 41:154–162.

KRAKOWSKI MI, CONVIT A, JAEGER J, ET AL. (1989). Neurological impairment in violent schizophrenic inpatients. Am J Psychiatry 146:849–853.

KRUPP LB, ALVAREZ LA, LAROCCA NG, ET AL. (1988). Fatigue in multiple sclerosis. Arch Neurol 45:435–437.

KRUPP LB, LAROCCA NG, MUIR-NASH J, ET AL. (1989). The fatigue severity scale. An application to patients with multiple sclerosis and systemic lupus erythematosus. Arch Neurol 46:1121–1123.

KUHL DE, PHELPS ME, & MARKHAM CH (1982). Cerebral metabolism and atrophy in Huntington's disease determined by "FDG and computed tomographic scan. Ann Neurol 12:425–434.

LANGFITT TW, ORBRIST WD, ALAVI A, ET AL. (1986). CT, MRI, and PET in the study of brain trauma. J Neurosurg 64:760–767.

LASTER DW, PENRY JK, & MOODY DM (1985). Chronic seizure disorders: Contribution of MR imaging when CT is normal. Am J Neurol Res 6:177–180.

LECKLITER IN, & MATARAZZO JD (1989). The influence of age, education, IQ, gender, and alcohol abuse on Halstead-Reitan Neuropsychological Test Battery performance. J Clin Psychol 45:484–512.

LEE SI (1985). Nonconvulsive status epilepticus: Ictal confusion in late life. Arch Neurol 42:778–781.

LEE KH, HASHIMOTO SA, HOOGE JP, ET AL. (1991). Magnetic resonance imaging of the head in the diagnosis of multiple sclerosis: a prospective 2-year follow-up with comparison of clinical evaluation, evoked potentials, oligoclonal banding, and CT. Neurology 41:657–660.

LESSER RP, MODIC MT, & WEINSTEIN MA (1986). Magnetic resonance imaging (1.5 Telsa) in patients with intractable seizures. Arch Neurol 43:367–371.

LEVIN HS, EISENBERG HM, & BENTON AL (Eds.) (1989). Mild head injury. New York: Oxford University Press.

LEVIN HS, MADISON CF, BAILEY CB, ET AL. (1983). Mutism after closed-head injury. Arch Neurol 40:601–606.

LEVIN HS, MATTIS S, RUFF RM, ET AL. (1987). Neurobehavioral outcome following minor head injury: A three-center study. J Neurosurg 66:234–243.

LEVIN R, LEATON EM, & LEE SI (1986). The value of nasophyaryngeal recording in psychiatric patients. Biol Psychiatry 21:1236–1238.

LEYS D, SOETAERT G, PETIT H, ET AL. (1990). Petriventricular and white matter magnetic resonance imaging hyperintensities do not differ between Alzheimer's disease and normal aging. Arch Neurol 47:524–527.

LEZAK MD (1983). Neuropsychological assessment (2nd ed.). New York: Oxford University Press.

LIEB JP, WALSH GO, BABB TL, ET AL. (1976). A comparison of EEG seizure patterns recorded with surface & depth electrodes in patients with temporal lobe epilepsy. Epilepsia 17:137–160.

LURIA AR (1973). The wroking brain (B Haigh, Trans.). New York: Basic Books.

LURIA AR (1980). Higher cortical functions in man (2nd ed.). (B Haigh, Trans.). New York: Basic Books.

MAIZLISH NA, FINE LJ, ALBERT JW, ET AL. (1987). A neurological evaluation of workers exposed to mixtures of organic solvents. Br J Ind Med 44:14–25.

MANN K, OPITZ H, PETERSEN D, ET AL. (1989). Intracranial CSF volumetry in alcoholics: Studies with MRI and CT. Psychiatry Res 29:277–279.

MARGULIS AR, & AMPARO E (1985). Magnetic resonance imaging in clinical practice. Postgrad Med 78:127–132.

MARKAND ON (1984). Electroencephalography in diffuse encephalopathies. J Clin Neurophysiol 1:357–407.

MARKAND ON (1990). Organic brain syndromes and dementias. In DD DALY & TA PEDLEY (Eds.), Current practice of clinical electroencephalography (Ch. 13, pp. 401–424). New York: Raven Press.

MARKS IM, & KETT PA (1986). Neurological factors in obsessive-compulsive disorder. Bri J Psychiatry 149:315–319.

MARTIN MJ (1983). Brief review of organic diseases masquerading as functional illness. Hosp Community Psychiatry 34:328–332.

MARTIN WRW (1985). Positron emission tomography in movement disorders. Can J Neurol Sci 12:6–10.

MCCLELLAN RL, EISENBERG RL, & GIYANANI VL (1988). Routine CT screening of psychiatry inpatients. Radiology 169:99–100.

MCEVOY JP, & LOHR JB (1984). Diazepam for catatonia. Am J Psychiatry 141:284–285.

MCGINNIS MR (1985). Detection of fungi in cerebrospinal fluid. Am J Med 75:129–138.

MCNAMARA ME (1991). Advances in EEG-based diagnostic technologies. In A STOUDEMIRE & BS FOGEL (Eds.), Medical psychiatric practice (Vol. 1; Ch. 5, pp. 163–192). Washington, DC: American Psychiatric Press.

MERRIAM AE, KAY SR, OPLER LA, ET AL. (1990). Neurological signs and the positive-negative dimension in schizophrenia. Biol Psychiatry 28:181–192.

MILBERG WP, HEBBEN N, & KAPLAN E (1986). The Boston process approach to neuropsychological assessment. In I GRANT & KA ADAMS (Eds.), Neuropsychological assessment and neuropsychiatric disorders (Pp. 65–86). New York: Oxford University Press.

MILLER BL, LESSER IM, BOONE K, ET AL. (1989). Brain white-matter lesions and psychosis. Br J. Psychiatry 155:73–78.

MILLER DM, RUDGE P, JOHNSON G, ET AL. (1988). Serial gadolinium enhanced magnetic resonance imaging in multiple sclerosis. Brain 11:927–939.

MODLIN HC, & SARGENT J (1986). Neuropsychological assessment in a head injury case. Bull Menninger Clin 50:50–57.

MORIHISA JM (1989). Computerized EEG and evoked potential mapping. In NC ANDREASEN (Ed.), Brain imaging: applications in psychiatry (Ch. 3, pp. 123–146). Washington, DC: American Psychiatric Press, Inc.

MORIHISA JM(1991). Advances in neuroimaging technologies. In A STOUDEMIRE & BS FOGEL (Eds.), Medical psychiatric practice (Vol. 1; Ch. 1, pp. 3–28). Washington, DC: American Psychiatric Press.

MORTENSEN EL, GADE A, & REINISCH JM (1991). A critical note on Lezak's 'best performance method' in clinical neuropsychology. J Clin Exp Neuropsychol 13:361–371.

MOUTSOPOULOS HM (1987). Reiter's syndrome and Behçet's syndrome. In E BRAUNWALD, KJ ISSELBACHER, RG PETERSDORF, ET AL. (Eds.), Harrison's principles of internal medicine (11 ed., Ch. 268, pp. 1436–1438). New York: McGraw-Hill.

MUNETZ MR, & SCHULZ SC (1986). Screening for tardive dyskinesia. J Clin Psychiatry 47:75–77.

NAVARRO C, & KENNEDY WR (1991). Evaluation of thermal and pain sensitivity in type I diabetic patients. J Neurol Neurosurg Psychiatry 54:60–64.

NELSON A, FOGEL B, & FAUST D (1986). Bedside cognitive screening instruments: A critical assessment. J Nerv Ment Dis 174:73–83.

NOLFE G, & GIAQUINTO S (1986). The EEG and the normal elderly: A contribution to the interpretation of aging and dementia. Electroencephalogr Clin Neurophysiol 63:540–546.

O'CONNELL RA, SIRECI SN, FASTOV ME, ET AL. (1991). The role of SPECT brain imaging in assessing psychopathology in the medically ill. Gen Hosp Psychiatry 13:305–312.

OKUNO T, MASATOSHI I, & KONISHI Y (1980). Cerebral atrophy following ACTH therapy. J Comput Assist Tomogr 4:20–23.

OLDFIELD RC (1971). The assessment and analysis of handedness: The Edinburgh Inventory. Neuropsychologia 9:97–113.

OOMMEN KJ, JOHNSON PC, & RAY CG (1982). Herpes simplex type II virus encephalitis presenting as psychosis. Am J Med 73:445–448.

OPPENHEIMER DR (1968). Microscopic lesions in the brain following head injury. J Neurol Neurosurg Psychiatry 31:299–306.

PANDURANGI AK, DEWAN MJ, & BOUCHER M (1986). A comprehensive study of chronic schizophrenic patients: 2. Biological, neuropsychological, and clinical correlates of CT abnormality. Acta Psychiatr Scand 73:161–171.

PANKRATZ LC (1983). A new technique for the assessment and modification of feigned memory deficit. Percept Mot Skills 57:367–372.

PEROUTKA SJ (1982). Hallucinations and delusions following a right temporoparietal occipital infarction. Johns Hopkins Med J 151:181–185.

PETERSDORF RG, ADAMS RD, & BRAUNWALD E (Eds.) (1983). Harrison's principles of internal medicine (10th ed.). New York: McGraw-Hill.

PETERSON LG, & PERL M (1982). Psychiatric presentations of cancer. Psychosomatics 23:601–604.

PETTEGREW JW, KESHAVAN MS, PANCHILINGAM K, ET AL. (1991). Alterations in brain high-energy phosphate and membrane phospholipid metabolism in first-episode, drug-naive schizophrenics. Arch Gen Psychiatry 48:563–568.

POHL P, VOGL G, FILL H, ET AL. (1988). Single photon emission computed tomography in AIDS dementia complex. J Nucl Med 29:1382–1386.

POST MLD, TATE LG, QUENCER RM, ET AL. (1988). CT, MR, and pathology in HIV encephalitis and meningitis. AJR 151:373–380.

Reese ME, Msall ME, Owen S, et al. (1991). Acquired cervical spine impairment in young adults with cerebral palsy. Dev Med Child Neurol 33:153–166.

Reitan RM (1985). Relationships between measures of brain functions and general intelligence. J Clin Psychol 41:245–253.

Reitan RM (1986). Theoretical and methodological bases of the Halstead-Reitan Neuropsychological Battery. In I Grant & KM Adams (Eds.), Neuropsychological assessment of neuropsychiatric disorders (Ch. 1, pp. 3–30). New York: Oxford University Press.

Reitan RM, & Wolfson D (1985). The Halstead-Reitan Neuropsychological Test Battery: Theory and clinical interpretation. Tucson, AZ: Neuropsychology Press.

Rimel RW, & Tyson GW (1979). The neurologic examination in patients with a central nervous system trauma. J Neurosurg Nurs 11:148–155.

Robins PV (1983). Reversible dementia in the misdiagnosis of dementia: A review. Hosp Community Psychiatry 34:830–835.

Rodnan GP, & Schumacher HR (Eds.) (1983). Primer on the rheumatic diseases (8th ed.). Atlanta: Arthritis Foundation.

Rogers R (Ed.) (1988). Clinical assessment of malingering and deception. New York: Guilford Press.

Rossi A, DeCataldo S, DiMichele V, et al. (1990). Neurological soft signs in schizophrenia. Br J Psychiatry 157:735–739.

Ruelen JPH, Sanders ACM, & Hogerhuis LAH (1983). Eye movement disorders in multiple sclerosis and optic neuritis. Brain 106:121–140.

Sedvall G, Farde L, Persson A, et al. (1986). Imaging of neurotransmitter receptors in the living human brain. Arch Gen Psychiatry 43:995–1005.

Sheldon JJ, Siddharthan R, & Tobia J (1985). Magnetic resonance imaging of multiple sclerosis: A comparison with clinical and CT examinations in 74 patients. Am J Roentgenol 145:957–964.

Shinar D, Gross CR, Hier DB, et al. (1987). Interobserver reliability in the interpretation of computed tomographic scans of stroke patients. Arch Neurol 44:149–155.

Shinar D, Gross CR, & Mohr JP (1985). Inter-observer variability in the assessment of neurological history and examination in the Stroke Data Bank. Arch Neurol 42:557–565.

Shraberg D, & D'Souza T (1982). Coma vigil masquerading as psychiatric diagnosis. J Clin Psychiatry 43:375–376.

Sindic CJ, Kevers L, Chalon MP, et al. (1985). Monitoring and tentative diagnosis of herpetic encephalitis by protein analysis of cerebrospinal fluid. Particular relevance of the assays of ferritin and S-100. J Neurol Sci 67:359–369.

Skjodt T, Torfing KF, & Teisen H (1988). Computed tomography in patients with dementia probably due to toxic encephalopathy. Acta Radiol 29:495–496.

Snow WG (1987). Standardization of test administration and scoring criteria: Some shortcomings of current practice with the Halstead-Reitan Test Battery. Clin Neuropsychol 1:250–262.

Sotaniemi KA (1983). Cerebral outcome after extra-corporeal circulation. Arch Neurol 40:75–77.

Spencer SS, Williamson PD, & Bridger SL (1985). Reliability and accuracy of localization by scalp ictal EEG. Neurology 35:1567–1575.

Sperling MR, Wilson G, & Engel J (1986). Magnetic resonance imaging in intractable partial epilepsy: Correlative studies. Ann Neurol 20:57–62.

Spiegel D (1991). Neurophysiological correlates of hypnosis and dissociation. J Neuropsychiatry Clin Neurosci 3:440–446.

Squire LR, Amaral DG, & Press GA (1990). Magnetic resonance imaging of the hippocampal formation and mammillary nuclei distinguish medial temporal lobe and diencephalic amnesia. J Neurosci 10:3106–3117.

Stahl Sm (1985). Can CSF measures distinguish among schizophrenia, depression, movement disorders, and dementia? Psychopharmacol Bull 21:396–399.

Stockard JJ, & Iragui VJ (1984). Clinically useful application of evoked potentials in adult neurology. J Clin Neurophysiol 1:159–202.

Stoppe G, Wildhagen K, Seidel JW, et al. (1990). Positron emission tomography in neuropsychiatric lupus erythematosus. Neurology 40:304–308.

Stoudemire AS, Hill CD, Morris R, et al. (1991). Cognitive outcome following tricyclic and electroconvulsive treatment of major depression in the elderly. Am J Psychiatry 148:1336–1340.

Stoudemire A, Hill C, Gulley LR, et al. (1989). Neuropsychological and biomedical assessment of depression-dementia syndromes. J Neuropsychiat Clin Neurosci 1:347–361.

Strauss I, & Keshner M (1935). Mental symptoms in cases of tumor of the frontal lobe. Arch Neurol Psychiatry 33:986–1005.

Sudarsky L, & Ronthal M (1983). Gait disorders among elderly patients. Arch Neurol 40:740–743.

Tanna NK, Kohn MI, Horwich DN, et al. (1991). Analysis of brain and cerebrospinal fluid volumes with MR imaging: Impact on PET data correction for atrophy. Part II: Aging and Alzheimer dementia. Radiology 178:123–130.

Tarter, Hegedus AM, Van Thiel DH, et al. (1985). Portal-systemic encephalopathy: Neuropsychiatric manifestations. Int J Psychiatry Med 15:265–275.

TerPogossian MM (1985). PET, SPECT, and NMRI: Competing or complementary disciplines? J Nucl Med 26:1487–1498.

Tippin J, & Dunner FJ (1981). Biparietal infarction in a patient with catatonia. Am J Psychiatry 138:1386–1287.

Trzepacz PT, Maue FR, & Coffman G (1986). Neuropsychiatric assessment of liver transplantation candidates: Delirium and other psychiatric disorders. Int J Psychiatry Med 16:101–111.

Tumeh SS, Nagel JS, English RJ, et al. (1990). Cerebral abnormalities in cocaine abusers: Demonstration by SPECT perfusion brain scintigraphy. Radiology 176:821–824.

Tupper DE, & Cicerone KD (Eds.) (1990). The neuropsychology of everyday life: Assessment and basic competencies. Boston: Klewer Academic Publishers.

van Gool WA, & Bolhuis PA (1991). Cerebrospinal fluid markers of alzheimer's disease. J Am Geriatr Assoc 39:1025–1039.

Visser SL, Van Tilburg W, & Hooijer C (1985). Visual evoked potentials in senile dementia (Alzheimer type) and in nonorganic behavior disorders in the elderly; comparison with EEG parameters. Electroencephalogr Clin Neurophysiol 60:115–121.

Walker WR (1982). Phenothiazine therapy in latent organic brain syndrome. Psychosomatics 23:962–968.

Warrington EK, James M, & Maciejewski C (1986). The WAIS as a lateralizing and localizing diagnotic instrument: A study of 656 patients with unilateral cerebral lesions. Neuropsychologia 24:223–239.

Weinberger DR (1984). Brain disease and psychiatric illness: When should a psychiatrist order a CT scan? Am J Psychiatry 141:1521–1527.

Weinberger DR, Gibson R, Coppola R, et al. (1991). The distribution of cerebral muscarinic acetylcholine receptors in vivo in patients with dementia. Arch Neurol 48:169–176.

Werner H (1937). Process and achievement: A basic problem of education and developmental psychology. Harvard Educ Rev 7:350–368.

Wilson B, Cockburn J, Baddeley A, et al. (1989). The development and validation of a test battery for detecting and monitoring everyday memory problems. J Clin Exp Neuropsychol 11:855–870.

WOLFSON LI, & KATZMAN R (1992). The neurologic consultation at age 80. In R KATZMAN, & JW ROWE (Eds.), Principals of geriatric neurology (pp. 75–88). Philadelphia: FA Davis.

WOODS BT, YURGELUN-TODD D, & KINNEY DK (1987). Relationship of neurological abnormalities in schizophrenics to family psychopathology. Biol Psychiatry 21:325–331.

WRIGHT BE, & HENKIND P (1983). Aging changes and the eye. In R KATZMAN & RD TERRY (Eds.), The neurology of aging. Philadelphia: FA Davis.

YIK KY, SULLIVAN SN, & TROSTER M (1982). Neuropsychiatric disturbance due to occult occlusion of the parietal vein. Can Med Assoc J 126:50–52.

ZEITLHOFER J, BRAINAN M, & REISNER T (1984). Brain stem auditory evoked responses in tardive dyskinesia. J Neurol 231:266–268.

ZISKIN J, & FAUST D (1988). Coping with psychiatric and psychological testimony (Vols. 1–3; 4th ed.), Marina del Rey, CA: Law & Psychology Press.

ZUBENKO GS, SULLIVAN P, NELSON JP, ET AL. (1990). Brain imaging abnormalities in mental disorders of late life. Arch Neurol 47:1107–1111.

Parts of the section on neurologic history and examination were previously published in Fogel BS (1986). Neurological screening of psychiatric patients. In L EXTEIN & M GOLD (Eds.), Medical Mimics of Psychiatric Disorders (Ch. 2, pp. 15–32). Washington, DC: American Psychiatric Press, Inc.

19 Dementia and delirium

ANDREW EDMUND SLABY, M.D., Ph.D., M.P.H., AND
STEVEN ROBERT ERLE, M.D.

DEMENTIA

Dementia is a chronic impairment of cognitive ability attributable to cerebral cortical or subcortical cellular dysfunction (Slaby, 1982, 1986). The prevalence of dementia is not precisely known. Estimates vary greatly depending on whether populations sampled include only individuals over 65, whether subjects are from the general community or from institutional populations, whether attempts have been made to differentiate subtypes of dementing illnesses, and whether standardized diagnostic criteria for dementia have been employed (Henderson, 1986).

Over 50 surveys of dementia prevalence have been reported (Henderson, 1990). Review of 47 surveys conducted between 1945 and 1985 by Jorn's group indicated that the prevalence of dementia:

1. Increased according to an exponential model, doubling every 5.1 years up to age 95.
2. Varied greatly between surveys, partly because of differences in populations studied and differences in methodology employed.
3. Was equal in men and women, but dementia of the Alzheimer type was more frequent in women.
4. Of both multiinfarct and Alzheimer type varied by region, with Soviet and Japanese researchers unique in reporting higher rates of the multiinfarct type than of the Alzheimer type (Henderson, 1990).

The National Institute of Mental Health Multisite Epidemiologic Catchment Area 6-month study indicated the prevalence rate for mild dementia in community dwellers over age 65 to be 11.5% to 18.4%. Five to 6% had severe dementia (Kallman & May, 1989).

The prevalence of severe dementias increases dramatically in the institutionalized elderly: 15% in retirement communities, 30% in nursing homes, and 54% in state hospitals (Cummings & Benson, 1983). Comorbid dementia and delirium are frequent in hospitalized patients with multiple medical problems and medications, and are associated with poor prognosis and higher mortality than uncomplicated dementia (Rabins & Folstein, 1982; Trzepacz, Teague, & Lipowski, 1985). In one study (Erkinjuntti et al., 1986) of 2,000 consecutive medical admissions, 9.1% were demented and, of the demented, 41.4% were delirious on admission. Of the delirious, 24.9% were demented.

The cost of dementia to the United States in 1989 was estimated to be $13.26 billion for direct costs, plus twice that amount for indirect costs of home care. The average annual cost of care for an individual patient in 1989 was estimated to be $17,643.

In addition to delirium, a number of other psychiatric disorders, particularly depression, may occur concomitantly with dementia, obfuscating the clinical picture. In a British study (Mowry & Burvill, 1988) using the Mini-Mental State Examination, Subtest B of the Raven's Matrices, and other measures of cognitive functioning, almost 30% of those found demented were also found to have an additional psychiatric disorder, usually depression. Older depressed people report more memory problems than normal subjects and actually report more mental slowing, impaired concentration, and indecisiveness than demented subjects without depression (O'Connor et al., 1990). Memory complaints and memory performance on structured examination, however, correlate poorly in both normal and depressed subjects, suggesting that older people may be likely to complain of memory problems even when objective signs are not present.

Dementia is a *syndrome* diagnosis that requires an etiologic differential diagnosis. At least 20% of cases diagnosed as dementia are at least partially reversible (McCartney & Palmateer, 1985).

Clinical Description

Personality changes can occur early in the course of a dementia in response to awareness of cognitive changes, as a defense against conscious awareness of the evolving intellectual limitations, or as a direct consequence of brain disease. Depression, when it occurs, usually is insidious in onset (Cummings, 1989; Kretschmar,

Kretschmar, & Stuhlmann, 1989; Reynolds & Hoch, 1987; Tobias, Lippmann, & Pary, 1989), and may or may not be accompanied by neurovegetative signs such as sleep and appetite disturbance (Greenwald et al., 1989; Lazarus et al., 1987; Stoudemire et al., 1989). Diagnosis-specific treatment of the mood disorder usually will produce some clinical improvement even if true dementia is present.

Early signs of dementia include lack of initiative, increasing irritability, loss of interest in life activities, and problems learning new information or carrying out activities requiring original thought (Murray, 1987; Read, Small, & Jarvik, 1985). Early dementing processes are best detected through serial mental status examinations, supplemented by regular inquiry about activities of daily living such as the handling of personal finances, arrangement of transportation and driving, cooking, shopping, housework and chores, and communication with friends and relatives. Such inquiry enables the physician to detect changes that may be largely denied by patients themselves. Interviews with collateral sources often are necessary for diagnosis.

Problems exist in both the underdiagnosis and overdiagnosis of the syndrome of dementia (Homer, Honavar, & Lantos, 1988; Martin & Guthrie, 1988), and a number of treatable causes of cognitive change can mimic dementia. In one study nearly half of patients diagnosed to have dementia in life were found not to have neuropathologic changes at necropsy consistent with a dementing disorder (Homer, Honavar, & Lantos, 1988). However, postmortem confirmation of an Alzheimer disease diagnosis is considerably more accurate when the diagnosis was made at a neurologic referral center, employing standard criteria (Risse et al., 1990).

A common personality change early in the course of dementia with depression is hypochondriasis (Ben-Arie, Swartz, & Dickman, 1987). Other changes include subtle disorganization of thought that may suggest a primary thought disorder (Emery, 1988). Some patients show outbursts of aggression even during an examination (Hamel et al., 1990), and increased psychomotor activity resembling mania (Cummings, 1987). However, most patients with a progressive dementia become more passive and less spontaneous (Horowitz & Charcot, 1988; Petry et al., 1988). Delusions, agitation, diurnal rhythm disturbances (Reisberg et al., 1987), and regression to earlier forms of behavior (Petry et al., 1989) also appear, although psychotic symptoms seldom appear without definite cognitive impairment (Rubin & Kinscherf, 1989).

Differential Diagnosis

The challenge in the etiologic evaluation of a patient with dementia is early identification of potentially treatable or reversible causes. Although a less extensive workup may be appropriate when encountering an end-stage demented person in a nursing home, it is ethically incumbent on the physician to investigate conditions likely to respond to specific treatment with meaningful improvement in the patient's function, cognition, behavior, or quality of life. In one study (Larson et al., 1984), 15 of 83 outpatients designated as having irreversible dementias were found to have potentially reversible or at least treatable dementing processes, including hypothyroidism, subdural hematomas, transient ischemic attacks, rheumatoid vasculitis, bipolar disorder, and drug toxicity. Decompensation in any organ system (cardiovascular, renal, pulmonary, gastrointestinal, etc.) can cause or exacerbate cognitive dysfunction in the elderly. A complete list of the possibilities would include practically all medical illnesses with systemic effects on the central nervous system (CNS). More common problems, however, include vitamin B_{12} and folate deficiency, iron-deficiency anemia, congestive heart failure, chronic obstructive pulmonary disease, Parkinson's disease, urinary tract infections, dehydration, and acquired immunodeficiency syndrome (AIDS).

The original concept of differentiation between treatable dementias and untreatable ones has given way to a view that emphasizes the opportunities for partial remediation that often exist even in patients with irreversible brain disease. Estimates of so-called reversible dementias range from 10% to 30% of all dementing illnesses (Barry & Moskowitz, 1988; Roca et al., 1984; Sloane, 1983). In addition, another 25% to 30% of patients may have partially treatable but not fully reversible dementing disorders (Sloane, 1983). The most common reversible causes in a review of 32 studies involving 2,889 subjects by Clarfield (1988) were drugs (28%), depression (26%), and metabolic disorders (16%). In the 11 studies with follow-up, 11% of dementias reversed partially and 3%, fully. Treatable or reversible cognitive dysfunction is more likely in milder dementias of relatively brief duration (Roca et al., 1984), and when the presence of clouding of consciousness suggests a superimposed delirium (Freeman & Rudd, 1982). The issue of "treatable dementia" is best seen in terms of searching for any drug-induced, toxic, or metabolic disturbance that could be *contributing* to cognitive dysfunction in the patient, regardless of whether or not a primary underlying dementia is present.

The list of potential causes of dementia is lengthy.

Although many psychiatrists refer patients with suspected dementia to a neurologist or internist for further diagnostic workup, even those psychiatrists who do not conduct a full etiologic investigation can be instrumental in finding potential aggravating factors in dementia that other specialists may overlook. For example, mild degrees of hypothyroidism may go undetected by the physician who orders only a thyroxine level and resin triiodothyronine uptake measure in a patient who appears euthyroid by physical examination. The psychiatrist's suspicion, however, may lead to obtaining a thyroid-stimulating hormone (TSH) level. Elevated TSH levels almost always imply hypothyroidism, even when circulating levels of thyroid hormone are normal. The prevalence of this occurrence is 2% to 5%, especially in women over age 60 (Wilson & Jefferson, 1985). Cognitive impairment and other psychiatric symptoms often occur with hypothyroidism, despite the absence of other physical stigmata of thyroid disease.

Pseudodementia

"Pseudodementia" has been defined as intellectual impairment attributable to a primary "nonorganic" psychiatric disorder (Caine, 1981; Ron, 1990). The term probably should be abandoned because these individuals almost never have symptoms that fit diagnostic criteria for true dementia. In addition, many patients appear to have both organic and functional causes of cognitive impairment. In pseudodementia associated with depression the cognitive deficits usually take the form of attention, concentration, and mild memory problems unless there is an underlying primary dementia, such as Alzheimer's disease.

Pragmatically, the diagnosis of *depressive* pseudodementia is made by a response to psychotropic medication or electroconvulsive therapy (ECT). Since dementia and depression frequently occur together (Shraberg, 1978), however, improvement of symptoms with antidepressant treatment does not automatically imply that dementia is not present. Reevaluation following successful antidepressant treatment may reveal persistent cognitive impairment. Reifler (1986) has proposed a classification of mixed cognitive-affective disorders. Patients with "type I" mixed disturbance have depression with associated cognitive disturbance attributable to the depression alone, and patients with "type II" have both depression and dementia. Treatment of "type I" depression results in improvement of both mood and cognitive deficits, whereas treatment of "type II" depression results in full improvement only of mood and vegetative symptoms. Reifler, Larson, and Hanley (1982) estimated the prevalence of coexistent depression and dementia in geriatric psychiatry clinics at 19%.

Depressive and psychotic symptoms have been reported in 30% to 40% of patients with Alzheimer's disease (Wragg & Jeste, 1989). Isolated symptoms are estimated to be two to three times more common than the full syndrome of mood disorder or psychotic disorder. Using a modified version of the Present State Examination, Rovner's group (1989) found a prevalence rate for major depression of 17% in a series of 144 patients with Alzheimer's disease. Depressed Alzheimer's patients are more cognitively impaired and more disabled than those not depressed. With respect to affective symptoms, patients with dementia and depression appear to respond to treatment as well as those with depression alone (about 70%). Their cognitive ratings, although improved, may remain in the demented range on standard tests (Greenwald et al., 1989).

Strategy for Evaluation

Of the 15% of the geriatric population and the smaller proportion of younger patients who demonstrate intellectual decline (Horowitz & Charcot, 1988) a number will have clues indicating that a more intense investigation for one or more possible etiologic factors may be beneficial (Barry & Moskowitz, 1988; Gordon & Freedman, 1990; Kokmen, 1989). For patients without such clues, clinicians might follow the recommendations of a National Institute on Aging Task Force, and carry out the following tests: automated biochemical screening (e.g., SMA-22), urinalysis, stool for occult blood, chest roentgenogram, electrocardiogram (ECG), and computerized tomography (CT) (Clarfield, 1988). Clues for further evaluation come from the history and the results of general diagnostic screens.

The genetic family history frequently is contributory. Down's syndrome patients, for example, have a greater incidence of Alzheimer's disease with aging. This has in fact led some to posit that Down's syndrome is a congenital form of Alzheimer's disease and that both disorders result from a ubiquitous pathogen that affects susceptible individuals (Mozer, Bal, & Howard, 1987). A similar genetic component is believed to play a role in Pick's disease, in which first-degree relatives of index cases are at significantly greater risk for the disease than are second-degree relatives (Heston, White, & Mastri, 1987). A relatively common autosomal dominant gene is considered to contribute to a familial form of Alzheimer's disease wherein morbid risk among first-degree relatives reaches 50% by 90 years of age (Mohs et al., 1987).

The aim of diagnostic evaluation of cognitive impairment in a patient is fourfold: quantification of the cognitive impairment; ascertainment of whether the deficit represents dementia, delirium, or depression; establishment of the diagnostic category; and evaluation of the extent to which the deficit interferes with functioning (Ramsdell et al., 1990). Serial evaluation can lead to therapeutic interventions as well as recognition of new and possibly reversible vascular or affective components (Kallman & May, 1989).

Physical and Neurologic Examination in Dementia

Physical examination provides clues to the underlying disorder. Many authors suggest that the physician looking for neurologic signs of dementia may find such primitive release signs as the snout and palmomental reflexes in adult patients with impaired higher cortical function. However, the palmomental reflex may occur in up to 60%, and the snout reflex in up to 33% of normal elderly volunteers, so the presence of primitive reflexes in elderly patients is not diagnostic (Wolfson & Katzman, 1983). With this in mind, it is important to be aware of the physical changes that occur with natural aging and to be able to differentiate such changes from pathologic processes. Furthermore, classic responses to acute disease may be masked in the elderly patient, so medical disease may present primarily as a confusional state. An elderly patient with an acute infection may not have an elevated white blood count or a fever but may become increasingly cognitively impaired (Wolfson & Katzman, 1983).

Parietal lobe dysfunction is suggested by aphasia, astereognosis, sensory extinction, difficulty in distinguishing weight and texture, and problems with two-point discrimination. Left hemispheric dysfunction is indicated by impairment of language and conceptualization. Right hemispheric lesions may produce neglect of the left side. The combination of agraphia, acalculia, finger agnosia, and right-left confusion (Gerstmann syndrome) indicates a lesion of the superior parietal lobe of the dominant hemisphere. Neglect of the left side seen in lesions of the right parietal lobe may be demonstrated by asking a patient to draw a clock. The left side of the clock may be distorted or absent in the presence of right parietal lobe lesions. The subtlety of impairment early in the course of a dementia cannot be overemphasized. Demented individuals with a greater premorbid repertoire of well-practiced skills will perform better on neuropsychologic assessment as well as bedside cognitive tests.

Changes in gait often accompany aging and include problems with balance, as noted by problems with one-legged standing and tandem walking. Deterioration of gait leading to falls may provide an important clue to the underlying pathology of dementing processes. Gait apraxia is characterized by difficulty in initiating the act of walking, leading to slow, hesitating, and sliding steps in which the patient's feet appear to stick to the floor (Wolfson & Katzman, 1983). This type of gait suggests bilateral frontal lobe dysfunction, as seen with subdural hematomas, multiple infarcts, meningiomas, butterfly gliomas, or late-stage Alzheimer's disease. Gait apraxia with urinary incontinence in the presence of dementia is characteristic of normal-pressure hydrocephalus (Adams et al., 1965). A shuffling, festinating gait accompanied by tremor, cogwheel rigidity, bradykinesia, and masklike facies suggests Parkinson's disease or drug-induced extrapyramidal dysfunction. A waddling gait may be due to proximal muscle weakness, as seen in hypothyroidism (Wolfson & Katzman, 1983). An ataxic gait with a dementing process may be due to multiple infarctions that include the cerebellum or its connections, or to chronic alcoholism. Acute onset of a small-step gait accompanied by slowness, brisk reflexes, extensor plantar responses, dysarthria, dysphagia, and pathologic laughing or crying in a patient with a history of hypertension also suggests multiinfarct dementia (Hachinski, Lassen, & Marshall, 1974).

Other specific neurologic signs that may be of diagnostic significance in the demented patient include (Kaufman, 1981):

1. Myoclonus, which can occur with Creutzfeldt-Jacob disease, postanoxic events, or myoclonic epilepsy, and which complicates about 10% of Alzheimer's disease cases.
2. Asterixis, which can occur with hepatic encephalopathy, uremia, carbon dioxide retention, or hypoxia.
3. Chorea, which may suggest Huntington's chorea, Wilson's disease, or other disorders of the basal ganglia.
4. Peripheral neuropathy with dementia, which suggests Wernicke-Korsakoff syndrome, infectious processes such as tertiary syphilis, systemic vasculitis, heavy metal or organic solvent exposure, or a toxic effect of such medications as isoniazid or phenytoin. If a patient presents with peripheral neuropathy that includes impaired vibratory and position sense in the lower extremities, vitamin B_{12} deficiency should be considered.
5. Extrapyramidal signs that occur independently of neuroleptics, suggesting basal ganglia disease or system degenerations.

Vital signs may assist in the differential diagnosis. Hypertension increases the likelihood of multiinfarct dementia. In the presence of rapid pulse with fever, it suggests sepsis or sedative-hypnotic withdrawal (Goldfrank et al., 1982). Tachycardia in the cognitively impaired patient warrants a rectal or tympanic membrane temperature because many demented patients are unable to keep an oral thermometer under their tongues long enough to give an accurate reading.

Pupillary signs may also provide diagnostic clues: pupillary mydriasis may suggest anticholinergic drug side effects, and pupillary constriction may suggest opiate toxicity. The Argyll-Robertson pupil, which fails to react to light but constricts on accommodation, suggests tertiary syphilis (Victor & Adams, 1974). Further discussion of the neurologic examination is provided in Chapter 18.

Mental Status Examination

Symptoms seen with dementia, regardless of etiology, relate both to the location and extent of impaired brain function and to individual adaptation to it. Premorbid defenses and psychopathology, the sociocultural matrix, and concurrent medical illnesses are other contributing factors. Cognitive changes can be insidious in onset and proceed over months, as in the instance of Creutzfelt-Jakob syndrome, or years, as in cases of Alzheimer's disease. Changes persist, regardless of stress, although environmental factors exaggerate or diminish symptoms.

Consciousness is not impaired, except in advanced cases. Memory disturbance, greater for recent events, is a cardinal feature. This may be disregarded as "absent-mindedness" early in the course of a dementia. A careful mental status examination should be performed to document the extent of changes and to provide a baseline on which to gauge progression of illness. Acquisition of new skills eventually becomes impossible; individuals become more dependent on earlier acquired skills to function.

Disorientation is a late but predictable outcome of progressive dementia. Disorientation to time occurs first, followed by disorientation to place and person. Judgment becomes impaired, and critical perspective is lost. Family members may be upset by sexual indiscretions and violations of norms of modesty. Individuals with major monetary or administrative responsibilities are more likely to be diagnosed earlier than those whose lives have been characterized by routine. Mental status examinations should be tailored to an individual's unique status in life. A cook can be examined on details of recipes for various dishes, and a stockbroker on market trends.

Affect is altered, and premorbid defenses may be exaggerated. Those who project may become paranoid; those who withdraw, schizoid. Those who use denial may take on a manic quality. The catastrophic response, a defense against conscious recognition of the cognitive deficit, is characterized by anxiety, anger, or even rage, despite initial amiability, when a patient is confronted with a task he cannot perform.

Depression can occur alone and present as dementia or can accompany a dementiform process, enhancing its severity. Concrete thinking, paucity of association, negativism, withdrawal, and memory impairment are symptoms of profound depression as well as dementia. The diagnosis of mood disorder is suggested by prior personal history or family history of depression and by vegetative signs of a mood disorder. The definitive diagnosis of depression as a complicating or primary etiologic factor, however, may not be confirmed until a response to antidepressant therapy occurs. Complex delusions are more common in patients with intact intellectual function, whereas in psychosis with cognitive impairment the paranoid beliefs are less complicated (Cummings, 1988; Thal, 1988).

The addition of formal cognitive tests can both enhance the ability to detect an early dementia and quantify the extent of deterioration (Christensen, 1989; Davis, Morris, & Grant, 1990; Galasko et al., 1990; Reisberg, 1985). Some questions have high specificity (e.g., orientation), high sensitivity (e.g., spelling *world* backward), and high sensitivity and specificity (e.g., memory for three objects) (Roca, 1987). Wolf-Klein, Silverstone, and Levy (1989) found that clock face drawing yielded a sensitivity of 86.7% and a specificity of 97.2% for Alzheimer's disease.

Electroencephalography

Bilateral slowing on the electroencephalogram (EEG) is seen with some but not all cases of degenerative brain disease; profound cognitive deficits have been reported with relatively little EEG abnormality (Slaby, 1986; Slaby & Wyatt, 1974). The EEG is nevertheless a useful tool in differentiating "pseudodementia" from dementia (Goodnick, 1985; Kiloh, 1961; Post, 1975; Wells, 1977, 1979), because approximately 70% of patients with "pseudodementia" have normal EEGs.

It should be noted that, although definitely abnormal EEGs are not consistent with normal aging, 30% to 40% of normal persons over age 60 may show brief runs of irregular focal anterior temporal slowing in the delta to theta range, especially on the left, occasionally with some sharp components that appear maximally with drowsiness and that disappear with sleep. When present to a mild degree, these are not believed to be pathologic in this age group (Busse & Orbrist, 1963).

Early EEG abnormalities often seen with senile dementia include a decrease in the frequency of the posterior rhythm and a decrease in the amplitude of background activity. As the dementia progresses, the EEG may show an increase in generalized slowing with a disappearance of fast activity. Such changes may lag behind the clinical picture of deterioration, although on a population basis the extent and presence of such changes is correlated with the degree of cognitive deterioration (Weiner, 1983).

Absence of cerebral dysrhythmia does not exclude true dementia. The EEG is often normal in early cases of such true dementias as Pick's disease (Mahendra, 1984). In Huntington's disease, the EEG will show gradual flattening, which correlates with the degree of cortical atrophy, until only very low-voltage, irregular delta and theta activity is seen (Shista et al., 1974). Triphasic waves can be seen in hepatic encephalopathy, as well as in some heavy-metal poisonings and subacute encephalitis (Cummings & Benson, 1983; Lesse, Hoefer, & Austin, 1958). Finally, the presence of spike or sharp wave activity may indicate an epileptogenic focus suggestive of either an occult seizure disorder or a cerebral lesion attributable to tumor or infarction, which may indicate specific treatment. Patients with Alzheimer's disease occasionally have seizures with focal slow EEG changes (Cummings & Benson, 1983).

A number of studies have evaluated both the sensitivity and validity of the EEG in the differential diagnosis of dementia. In a study of 94 demented patients with different etiologies, Hughes et al. (1989) found abnormalities in the EEG, usually diffuse slow waves, in 83% of all patients. These correlated well with the Modified Hachinski Test and the Mattis Dementia Rating Scale and its subtests, but poorly with the Mini-Mental State Examination. Focal EEG abnormalities in the temporal lobes were characteristic of alcoholic dementia. The 17% of patients with normal tracings scored well on the Modified Hachinski Tests but not on the Mini-Mental State Examination. In a study of the sensitivity of the EEG in the differential diagnosis between Alzheimer's disease and vascular dementia (Erkinjuntti et al., 1988), the mean frequency of background activity decreased in all groups with increasing degrees of dementia and, quantitatively, the percentage of alpha power decreased and that of theta and delta increased in proportion to the degree of dementia. Electroencephalograms of those patients with multiinfarct dementia, when compared to EEGs of Alzheimer's patients, exhibited more slow wave and irritative (i.e., sharp wave and/or spike paroxysms) and focal abnormalities.

Patients with depressive pseudodementia often are distinguished clinically from those with true cognitive deficiency by a sleep pattern showing more early morning awakening (Reynolds et al., 1988). Sleep EEGs demonstrate more rapid eye movement (REM) and greater phasic REM activity intensity in pseudodemented compared to demented patients. Buysse et al. (1988) found a longer first REM period in depressive pseudodemented patients, permitting correct identification in 88.5% of patients. Waking EEGs of depressed, pseudodemented patients usually are normal or only mildly abnormal (Brenner, Reynolds, & Ulrich, 1989).

Computerized Tomography

Computerized tomography scans assist in the diagnosis of focal brain lesions or hydrocephalus. Cortical atrophy, the most frequent finding, is nonspecific. Mild cortical atrophy and ventricular dilation can be found in older patients without a dementing process, and conversely, cerebral atrophy and ventricular dilatation can be minimal in true dementia (Tomlinson, Blessed, & Roth, 1968, 1970).

Brain weight decreases with normal aging, with associated anatomic changes, including widening sulci and ventricular enlargement; such changes are generally greater in patients with senile dementia than in normal persons of the same age. Ventricular enlargement correlates better than sulcal enlargement with the severity of the dementia (McGreer, 1986). Attempts have been made to quantify ventricular volume and shape, to measure age-corrected ventricle-brain ratios, and to measure gray-white discriminability, to permit differentiation of changes associated with normal aging from those attributable to senile dementia. At the time of this writing, such measurements are not routinely available, nor has their diagnostic utility been confirmed (McGeer, 1986). Serial CT scans may permit the tracking of increasing ventricular size associated with progressive atrophy from advancing dementing illness (Brinkman & Largen, 1984). Demonstration of progressive enlargement of ventricles may help establish the presence of a dementing process when this is in doubt.

The CT scan is excellent for the detection of intracranial masses and blood collections (subdural hematomas) within the cranium. Contrast enhancement will increase visibility of focal changes in most conditions, with the exception of static or degenerative processes, such as scarring from past trauma or old strokes (Cummings & Benson, 1983). In the detection of infarcts, the use of the CT scan can be problematic. If the infarct is large, it may be well visualized, but smaller infarcts, such as the lacunar infarcts of multiinfarct dementia, may go undetected. Most lesions must be at least 0.5 cm in diameter before they are visualized (Cummings & Benson, 1983). In addition, after infarction tissues may be isodense for up to 10 days before the infarct is demonstrable. Although contrast enhancement may overcome this problem (Ambrose, 1973), it is not always successful in doing so (Oldendorf, 1980). Computed tomography scans obtained 3 or more weeks after the insult provide greater yield (Cummings & Benson, 1983). Magnetic resonance imaging (MRI) is currently the most useful test for the detection of multiple small infarcts.

In hydrocephalus with blockage at the level of the temporal hiatus, CT scan findings include enlarged third and lateral ventricles, ballooning of the temporal horns, and obliteration of the cerebral sulci over the convexities (Gado & Press, 1986). When cerebrospinal fluid flow is blocked over the higher convexities, the sulci can enlarge, minimizing cerebral atrophy. In differentiating normal-pressure hydrocephalus from the atrophy noted with Alzheimer disease, enlargement of the temporal horns along with a pattern of periventricular lucency suggests the diagnosis of normal-pressure hydrocephalus (Gado & Press, 1986). Radionuclide cisternography is used to distinguish between degenerative conditions and cerebrospinal fluid flow obstruction. Bilateral circumscribed atrophy is found in the frontotemporal regions on CT scan in cases of Pick's disease.

The radiologic diagnosis of Huntington's disease is based on a quantitative measure of caudate atrophy (Gado & Press, 1986). This measure is the ratio of the length of the line connecting the two anterior corners of the frontal horn to the length of the bicaudate line measured between the medial borders of the two caudate nuclei. The mean ratio in normal subjects is 2.48. In persons with Huntington disease, the mean is approximately 1.33. A ratio of less than 1.6 suggests Huntington's disease. Individuals with parkinsonian dementia have a normal ratio.

Normal-pressure hydrocephalus is characterized by gait apraxia, urinary incontinence, and dementia (Adams et al., 1965) and is generally differentiated from high-pressure hydrocephalus by a normal opening pressure on lumbar puncture of less than 180 mm of water, and from Alzheimer's disease by a CT scan presentation of markedly enlarged ventricles with minimal cortical atrophy (Black, 1982). However, if the cerebrospinal fluid flow is blocked at the higher convexities or the superior sagittal sinus, the cerebral sulci can enlarge and mimic cerebral atrophy (Gado & Press, 1986). In such cases, cisternography may be useful. In this technique, a radioactive material (usually indium) is injected into the cerebrospinal fluid via lumbar puncture. Normally, the isotope will rarely be detected in the ventricular system because of the one-way flow away from the ventricles (Cummings & Benson, 1983). The isotope will diffuse over the convexities of the brain, to be absorbed and disappear within 48 hours (Black, 1982). In normal-pressure hydrocephalus, the isotope rapidly refluxes into the ventricles, with failed or delayed visualization over the convexities, even at 48 to 72 hours and as late as 96 hours (Tyler & Tyler, 1984). Large-volume lumbar puncture (the removal of 30 to 50 ml of cerebrospinal fluid), performed every 2 to 3 days, may also aid in the diagnosis by demonstration of an improved clinical picture with the cerebrospinal fluid removal (Tyler & Tyler, 1984). Although shunting of the cerebrospinal fluid has offered marked improvement in many patients, it has failed to improve others. Several studies have been done to determine prior to surgery which candidates will be helped most by shunting, and it appears that patients with the complete triad of gait disturbance, dementia, and incontinence do better than patients with dementia alone (Black, 1980). Positive prognostic factors for shunting include established etiology, relatively acute onset, low cerebrospinal fluid outflow, small sulci, and periventricular hypodensity (Thomsen et al., 1986).

It has been suggested that CT scan diagnosis of normal-pressure hydrocephalus be based on evidence of (1) more than moderately dilated ventricles without evidence of ventricular obstruction, (2) obliteration of the cerebral sulci, (3) areas of periventricular low density, and (4) "rounding" of the frontal horns of the lateral ventricles (Vassilouthis, 1984). Patients who satisfied at least one of the first two clinical criteria in addition to showing CT scan evidence of dilated ventricles underwent ventriculoperitoneal shunting, with a favorable response. Computed tomography findings may thus confirm diagnosis. Regarding surgical therapy, better clinical results have been reported when low-pressure (as opposed to medium-pressure) shunts are employed (McQuarrie, Saint-Louis, & Scherer, 1984). The most frequently encountered complications of shunting are subdural hematomas, shunt malfunction, infection, and postoperative seizures (Black, 1980; Tyler & Tyler, 1984).

Many cases of dementing processes may not be visible on CT scan. Those include chronic meningitis, vasculitis, chronic alcohol abuse, multiple sclerosis, and toxic metabolic encephalopathies. We have observed that, after a CT scan is found to be "negative," the diagnostic evaluation sometimes stops and other causes of treatable dementia are not sought. Knowledge of the limitations of CT scanning should help prevent premature diagnostic closure.

It is clear that CT scanning is not required or sufficiently sensitive to make the diagnosis of dementia. Its use in evaluation is to detect major treatable lesions such as subdural hematomas and meningiomas (Letton,

1988). In an evaluation of the contribution of CT to differential diagnosis Roberts & Caird (1990) studied 280 confused elderly patients, 94% of whom were diagnosed as demented, with the others predominantly suffering from receptive dysphasia. Tumors, hygromas, subdural hematomas, and other space-occupying lesions were found in 11%. In those with other intracranial and extracranial causes, 64% were potentially treatable. Approximately one third of 170 patients with a duration of confusion of less than 1 year had potentially reversible dementias, compared to only 1% of 110 patients with dementia of longer duration.

Dexamethasone Suppression Test

The dexamethasone suppression test has been used to attempt to differentiate cases of depression with secondary cognitive impairment from true dementia (McAllister & Price, 1985; McAllister et al., 1982; Rudorfer & Clayton, 1981). Clinical usefulness, however, is limited, since the DST is abnormal in many demented but nondepressed patients.

Cerebrospinal Fluid Examination

Although many neurology texts suggest the routine use of lumbar puncture in the diagnostic workup of dementia, its clinical utility has been cast into doubt. Recent studies suggest that, although complications from this procedure are infrequent, it is expensive and the yield of useful diagnostic information is low (Hammerstrom & Zimmer, 1985). For this reason, cerebrospinal fluid evaluation is suggested only for cases of dementia with rapid onset or progression, for cases with positive serology for syphilis, for patients under the age of 55, for suspected cases of viral encephalitis, for patients with headache, meningeal signs, elevated white blood cell count, pulmonary infiltrates on chest roentgenogram, or fever suggesting low-grade fungal meningitis (Hammerstrom & Zimmer, 1985), and for immunosuppressed patients.

Sodium Amytal Test

Sodium amytal can be used in the evaluation of suspected pseudodementia. Specifically, it helps in diagnosing a psychiatric basis for cognitive symptoms in younger individuals with suspected conversion or psychosis. A 5% solution (500 mg of amytal dissolved in 10 ml of sterile water) is administered at a rate not faster than 1 ml/min (50 mg/min) (Perry & Jacobs, 1982). Dosage is sufficient when yawning, slurred speech, or lateral nystagmus occurs. Dosage is then stopped and the patient is reevaluated. Pseudodemented patients may show improvement in memory and orientation and increased verbalization (Snow & Wells, 1981; Ward, Rowlett, & Burke, 1978; Wells, 1979). Patients with cognitive impairment disorders show no improvement or transient deterioration. False-positive, false-negative, and inconclusive results limit the test's usefulness (McAllister et al., 1982; Ward, Rowlett, & Burke, 1978). In some instances, the difficulty appears due to faulty technique, such as injecting the drug too slowly or too rapidly. Symptoms that are due to malingering will not be distinguished by the amytal test (Herman, 1938; Lambert & Rees, 1944; Ward, Rowlett, & Burke, 1978).

The amytal test is likely to be most useful in differential diagnosis in young to middle-aged patients with atypical features of cognitive disturbance. In the more typical elderly patient with coexistent depression and dementing or other neurologic illness, the amytal test usually is uninterpretable.

Other Diagnostic Tests

Magnetic resonance imaging and positron emission tomography (PET) are leading to improvements in the detection and differential diagnosis of Alzheimer's disease. Magnetic resonance imaging provides a more detailed visualization of the brain and facilitates discrimination between white and gray matter (Friedland et al., 1984). In addition, MRI offers better definition of discrete brain lesions (McGeer, 1986). The MR image is superior to the CT scan for the detection of infarcts, tumor edema, posterior fossa lesions, and the demyelination of multiple sclerosis (McGeer, 1986). Thus, MRI has clear advantages over CT in the differential diagnosis of dementia, although the MR image is not specifically diagnostic for Alzheimer's disease. If the clinician must choose one brain imaging procedure without regard to cost or inconvenience, it should be MRI. Studies of Alzheimer patients using MRI have shown statistically significant regional reductions in the percentage of gray matter and increases in cerebrospinal fluid volume (Rusinek et al., 1991). They have also shown specific hippocampal and parahippocampal atrophy (Kesslak, Nalcioglu, & Cotman, 1991). Magnetic resonance imaging is expensive, however, and not available at all hospitals, and is not deemed necessary if clinical, laboratory, and CT information already suggest primary degenerative dementia (Anderson, 1988).

Positron emission tomography offers the greatest promise of providing new information regarding in vivo brain metabolism in patients with Alzheimer's disease (McGeer, 1986). The PET image is a map of brain glucose metabolism (through the use of [^{18}F]fluoro-2-deoxy-D-glucose) or of cerebral blood flow and oxygen metabolism (through the use of oxygen-15) (Cutler et al., 1984; McGeer, 1986). Brain glucose metabolism by PET imaging does not appear to change significantly with age (Cutler et al., 1984; McGeer, 1986), offering a clear advantage over the use of the CT scan, which may show atrophy in healthy elderly persons. There is a marked reduction in local cerebral glucose metabolism in patients with dementia, although the patterns of cortical involvement are variable (Friedland et al., 1984; McGeer, 1986). In mild to moderate Alzheimer's dementia, memory deficits often precede reductions in brain glucose metabolism (Cutler et al., 1984). Patients with Alzheimer's or Pick's disease have distinctive patterns of decreased cortical oxygen metabolism, which differentiate them from patients with other dementing illnesses (Cummings & Benson, 1983). Because the PET scanner requires the use of a cyclotron, it is available only at a few research centers. With time, however, the PET scanner may permit the confirmation of early Alzheimer dementia before any CT scan or MRI evidence of atrophy is noted.

Neuropsychologic Testing

Neuropsychologic testing is a valuable adjunct in the evaluation of the demented patient when the bedside mental status examination is equivocal. Documentation of specific cognitive deficits, quantification of the extent of impairment, and measurement of the rate of cognitive deterioration may be used not only to clarify the diagnosis, but also to develop management strategies and to measure response to treatment (Huppert & Tym, 1986). Neuropsychologic testing may also differentiate the cognitive changes associated with depression from those of Alzheimer's disease by identifying such specific cortical signs as apraxia and aphasia. Neuropsychologic testing is also useful in disability evaluations and in gathering the necessary documentation for legal decisions regarding competency or guardianship. Further discussion is provided in Chapter 18.

Management

Rational inpatient and outpatient management of dementia entails development of a biopsychosocial formulation to direct management. Effective treatment is based on: (1) diagnosis of the primary illness and concomitant psychiatric disorders, such as delirium, depression, mania, and psychosis and comorbid medical or surgical illness; (2) assessment of auditory and visual impairment and premorbid personality style; (3) measurement of the nature, extent, and progression of cognitive deficit; (4) evaluation of functions of adult daily living, including ability to care for self, and sphincter control; (5) assessment of gait, stance, balance, and impairment of ambulation and movement; and (6) evaluation of social and family relationships and other significant community supports (Mahendra, 1984; Roth, 1982).

Environmental Strategies

Most demented patients can live at home in the early stages (Mahendra, 1984). Performance of daily activities may be better if memory is assisted by written daily reminders and better organization of daily routine.

Families frequently express a concern that their relatives with dementia may be harmful to themselves, others, or property because of lack of attention to stoves, heaters, and electrical appliances. Signs with large letters and repeated reminders accompanied by demonstrations of how to properly handle appliances facilitate remembering to turn on and off appliances and to lock and unlock windows and doors (Mahendra, 1984). Durable powers of attorney and other transfers of legal authority should be established early in the course of illness to maximize patients' participation in the decision-making process (see Chapter 45).

Caretakers and family must learn to accept patients' slower pace with patience, tact, and understanding, while supervising daily activities such as personal hygiene, dressing, and eating. Special attention must be paid to maintaining adequate nutrition, hydration, and exercise. Instructions and medical procedures should be explained in simple, nontechnical language. Repetition followed by written explanation enhances comprehension. Problems arise in medical and surgical services when time is not taken to explain procedures or tests to patients with dementia. Care must be taken not to infantilize elderly demented patients, despite their increased dependency. For example, calling older people by their first names when there is no close personal relationship is inappropriate.

Adequate management requires social support from the family. Responsible family members should be advised of the diagnosis, the prognosis, and possible problems they may encounter because of patients' cognitive deficits. Frequent but brief visits to homes of patients living alone by caretakers, friends, or neighbors minimizes likelihood of accidents and allows monitoring of patients for falls, malnutrition, dehydration, apathy, alcoholism, and self-neglect arising from loneliness and depression.

Patients may, out of neglect, deteriorate and develop

incontinence, emaciation, dehydration, bedsores, infections (e.g., pneumonia), contractures, and hip fractures, leading to further depression and immobility. Nursing homes and chronic care hospitals become primary care providers in end-stage disease.

Research suggests that the burden of care borne by families, rather than deterioration of patients' mental or physical capacity, is the leading factor in institutionalization (Ross & Kedward, 1977). Relocation of demented patients to institutional care settings can be disastrous, because demented patients have considerable difficulty learning to adapt to new environments (Borup, Gallego, & Hefferman, 1979, 1980). Patients may become more agitated and confused, often requiring psychotropic medication for behavior management (Zarit & Zarit, 1984).

Role of the Psychiatrist

The psychiatrist, along with other mental health professionals, can help develop environmental management strategies in addition to providing adequate diagnosis and proper medication. For patients living at home, the psychiatric treatment team can educate family and other caretakers about environmental manipulations that, when used in conjunction with appropriate medication, will improve the patient's performance and decrease the need for nursing home placement.

In hospital and nursing home settings, the treatment team can apply environmental strategies similar to those used in the home, such as calendars, nightlights, primary nursing to increase familiarity with staff, written instructions to the patient in simple language regarding procedures and medications, and other reminders such as signs. Other techniques that have been employed successfully in the nursing home setting have included token economies, in which tokens to purchase privileges, food, candy, or clothing are given for desired behaviors; milieu therapy, with cash used as a reinforcer; and "reality orientation," with staff reinforcing orientation and current information in daily conversation.

Hospital and nursing home staff must first be made aware of the patient's memory deficits because they are often not obvious without formal testing, and frustrated staff may misconstrue noncompliance attributable to memory problems as passive-aggressive or uncooperative behavior. Staff may be instructed that verbal repetition or written instruction may be necessary to overcome the patient's impairment. Additionally, the treatment team can educate staff on the special psychologic issues of the demented patient, to aid them in developing management plans (Herst & Moulton, 1985). In the nursing home setting, this is particularly important because, although it has been estimated that over three quarters of the patients in skilled nursing homes have diagnosable psychiatric illness (Teeter et al., 1976), nursing home staff are often poorly educated regarding psychiatric diagnosis and treatment and are not comfortable around patients with behavioral disturbances. Optimal environments attend to consistency, structured daily living plans, and activities, with special attention to the patient's need for privacy and personal dignity. Programs for dementia units in nursing homes are described and evaluated in a recent monograph by Sloane and Mathew (1991).

Family Therapy

Dementia is a source of emotional suffering for the caregiver. For this reason, "caregiver burden" (O'Quin & McGraw, 1985) has become a major focus of attention. The most troublesome behaviors for caregivers to manage include aimless wandering (Arie, 1986)—often worse at night, with disruption of the caregiver's sleep (Sanford, 1975)—dependency, confusion, falls, rages, urinary incontinence, apathy, and sullen moods. Family members also find constant repetition of questions, searches for lost items, and the need for frequent reorientation of the demented patient a source of constant frustration (Zarit, Reever, & Bach-Peterson, 1980). Although problems of memory loss and disorientation are the primary reason for families to seek help for demented patients, the major basis for institutionalization is not cognitive, but is families' becoming overwhelmed by caregiving responsibilities. In three independent samples of caregivers of elderly mentally infirm patients attending or about to attend day hospital, prevalence levels of significant emotional distress varied from 57% to 73%. High General Health Questionnaire scores were usually associated with diagnosable psychiatric illness (Gilleard et al., 1984). Caregivers are at risk for depression, chronic fatigue, insomnia, and appetite changes (Fiore, Becker, & Coppel, 1983; Rabins, Mace, & Lucas, 1982). The psychiatrist, therefore, may have a unique role in the alleviation of caregiver burden, particularly through the diagnosis and treatment of major depression, adjustment disorders, and reversible cognitive impairment disorders in the caregiver.

Formal family therapy can aid in reducing caregiver distress, enhance environmental management, and prevent elder abuse and premature institutionalization. Contextual family therapy (Boszormenyi-Nagy & Krasner, 1986), which addresses the issues of rights and obligations of family members to one another within the circumstances of illness, may be helpful. Other useful techniques include educating families about Alzheimer's disease and related disorders, incorporating family participation in the behavioral management of the patient, individual counseling for the primary care provider, and multifamily support groups (Fuller et al., 1979).

Another experimental approach is supporter endurance training (Levine et al., 1984), a cognitive-behavioral training program designed to teach anxiety reduction by using social skills training. Educational materials for the patient and family are also provided by the Alzheimer's Disease and Related Disorders Association (70 East Lake Street, Chicago, Il 60601; 1-800-621-0379 outside Illinois; in Illinois, 1-800-572-6037).

In spite of these interventions, care providers still require respite, which may be provided through brief hospital or nursing home admission for the demented patient, adult day care, hired caregivers who provide care in the home, and hospice care for terminally ill demented patients (Arie, 1986).

Individual Psychotherapy

Individual psychotherapy may be a viable treatment option for the mildly demented patient who still retains sufficient cognitive capacity to participate in this form of therapy. The psychotherapist should maintain a flexible therapeutic approach that makes allowances for (1) communication problems secondary to hearing deficits or mild aphasia, (2) repetition to compensate for memory loss, (3) somatization by the patient as a resistance to further exploration, and (4) transference to the therapist, who may be viewed as a child or grandchild (Charatan, 1985; Goldberg & Cullen, 1986; Kroetsch & Shamoian, 1986).

Common issues in psychotherapy include dependency (Cohen, 1981), injury to self-esteem related to loss of intellectual capacity, mourning of multiple "partial losses"—including loss of spouse, friends, and employment, diminishing physical function, and change in family role (Berezin, 1972; Goldberg & Cullen, 1986)—and realistic concerns regarding the future. Some advocate the use of "life review" therapy, with the therapeutic use of reminiscence to repair wounds in self-esteem through the review of accomplishments, to resolve intrapsychic conflicts, and to help patients to come to terms with the meaning of their lives (Butler, 1963; Lewis & Butler, 1974). The skilled therapist must balance potential benefits of reminiscence against the possibility that review of the past may be met with regret for the life not lived, guilt, and a greater sense of loss (Cohen, 1981). Because of the diminished capacity of the mildly demented patient to reason abstractly, interpretations and clarifications should be straightforward and simple. In supportive psychotherapy, the goal of better adaptation may be enhanced by engaging the family as collaborators or participants in the therapy; by using telephone interventions for reassurance, support, and advice; and by identifying the specific social, psychologic, or biologic problems that are most amenable to change and that, therefore, are most likely to respond to focused intervention (Kroetsch & Shamoian, 1986).

Group therapy other than to enhance reality testing has limited value in the management of demented patients (Krebs-Roubicek, 1989).

Nutrition

Demented patients neglect food, particularly if indigent or living alone. Vitamin deficiencies may be compounded by excess use of alcohol. Alcohol may be taken to attempt to relieve depression and insomnia, or with unconscious or conscious self-destructive intent. Frequent weighings to determine food intake and monitoring of skin turgor and other signs of dehydration may be necessary. Supplemental vitamins are recommended for all patients with dementia. Dysphagic patients require semisoft or soft diets. Tube feeding may be necessary if swallowing is markedly impaired.

Severe protein-calorie malnutrition predisposes the patient with senile dementia of the Alzheimer's type to bedsores, impaired cellular and humoral immunity, and increased mortality rates (Johnson, 1985). Laboratory tests available for the assessment of protein-calorie malnutrition include serum albumin level, total lymphocyte count, creatinine-height index (the ratio of 24-hour creatinine clearance to the patient's height in centimeters), and serum transferrin level. These tests, however, are affected by aging, acute and chronic disease, infection, and anemia (Johnson, 1985). For this reason, although some suggest serum albumin levels and total lymphocyte counts be done to determine those patients with malnutrition at risk for bedsores, others suggest that protein-calorie malnutrition be assessed more simply by serial weights, with 10% loss of usual body weight indicating a malnourished state (Johnson, 1985). Formal nutritional support consultation should be obtained in patients with profound weight loss or severe decubiti.

Sensory Augmentation

Decreased sensory input stresses the demented patient. Eyeglasses of correct prescription and hearing aids facilitate functioning. Quiet, dark nights provoke anxiety and agitation (called "sundowning"). Night lights calm patients and may help them stay oriented. When institutional placement is required, the presence of familiar objects, such as family photographs or bed lamps from home, facilitates functioning. Visits by friends and family members well known to a patient reduce paranoia and help maintain orientation.

Institutional Care

Institutionalization is often required in the final stages of dementia. Friends and families require guidance in identifying when and where to place the patient. Financial concern, guilt, helplessness, and despair plague families as the demented approach death.

Abuse of Elderly Patients

Abuse represents an extreme reaction to the burden of care (Rathbone-McCuan & Goodstein, 1985). The spectrum of abuse includes physical beatings, starvation, verbal abuse, overmedication, tying patients in bed or locking them in a room for excessive periods of time, withholding of basic care, and financial exploitation or theft of money and property (Rathbone-McCuan & Voyles, 1982). Physical, behavioral, social, and emotional clues, assisted by home visits, facilitate discovery of victimization. Clues include (Rathbone-McCuan & Goodstein, 1985): (1) rope, chain, cigarette, and iron burns or marks of the neck or extremities suggestive of being tied to bed, a chair, or the toilet; (2) unusual patterns of bruises and welts or lacerations on the back, legs, arms, or face, suggesting beatings by electric cords, hangers, and other objects; (3) scalp hemorrhage or bald patches on the scalp, suggestive of hair pulling; (4) frequent emergency room visits for fractures and injuries; (5) caregivers' reluctance to promptly seek medical attention for injuries; (6) excess fear of care providers, evinced by hypervigilance or moving away from the caregiver as if dodging a blow; and (7) excessive control or restriction by caregivers of out-of-home contact, such as preventing friends and relatives from seeing the patient, locking doors and phones, and not allowing the patient to go out of doors. Abusers may be under the influence of drugs themselves and obsessed with losing control, thus justifying their overmedicating the patients under their care. Abusers commonly have histories of having been abused themselves. In addition to frustration resulting from the current burden of care, anger may exist toward a debilitated parent because of unsatisfactory past parenting.

Some communities have departments of elderly affairs with teams to make home assessments and hospital-based protocols for abuse detection. Brief hospitalization is sometimes advocated to provide families respite and to de-escalate chaotic home situations (Arie, 1986).

The stress of a demented patient's behavior varies over the course of the illness (Haley & Pardo, 1989). The instrumental self-care deficits that begin early in dementia usually are less physically demanding on the caregiver than the basic self-care deficits that emerge with increasing severity of dementia. Distressing behavioral symptoms, in contrast, may diminish late in the course of the illness. Incontinence is a particularly distressing symptom and plays an important role in the decision to institutionalize (Ouslander et al., 1990). Patients with new-onset incontinence deserve a thorough evaluation for treatable causes (AHCPR, 1992). Nonspecific techniques such as education in use of toilet schedules and diapers reduces the stress burden and enables more patients to be managed at home.

Group classes in stress management techniques are useful to many caregivers. Commonly taught strategies include the regular use of respite care, and the development of collaborative relationships among potential caregivers to reduce the time any one caregiver spends and to minimize redundant efforts (Harris & Rabins, 1989). In one study of the impact of caregiver-focused health care education, assistance with problem solving, education about dementia and caregiving, regularly scheduled in-home respite, and a self-help family caregiver group, Mohide and her group found that there was a clinically important improvement in quality of life, greater satisfaction with nursing care, longer mean time to long-term institutionalization, and fewer problems in the caregiver role (Mohide et al., 1990). Other studies have comparably indicated that caregiver support groups decrease level of emotional distress (Russell, Proctor, & Moniz, 1989), reduce the psychologic morbidity of the caregiver, and delay the placement of patients in institutions without increasing use of health services by either the patient or the caregiver (Brodaty & Gresham, 1989).

Sleep Disturbance in Dementia

Sleep disturbance in the demented patient is a special problem that deserves careful evaluation. While dementing illness itself is associated with sleep difficulties, other factors of normal aging, medication, or concurrent disease may exacerbate the problem. With normal aging, total sleep time decreases, sleep latency increases, sleep is disturbed by more frequent awakenings, and decreased time is spent in sleep stages 3 and 4 (Jenike, 1985; Thompson, Moran, & Nies, 1983). In advanced

Alzheimer's disease, there is commonly pervasive disruption of the circadian sleep/wake rhythms, with increased daytime sleepiness (Vitiello & Prinz, 1989). Sometimes educating the older patient that this is a normal occurrence of aging will improve sleep quality without the addition of medication. Other nonpharmacologic interventions include (Jenike, 1985): (1) the elimination of alcohol, caffeine, and cigarette smoking; (2) discouragement of daytime naps and provision of regularly scheduled daytime activities; (3) regular exercise, which should be encouraged when physically permissible but should be avoided in the evening, when it may lead to problems with falling asleep; (4) regularly set times for going to bed and arising in the morning; (5) correction of urinary problems and restriction of evening fluid intake, which may help to decrease nighttime awakenings caused by nocturia; and (6) correction or maximal control of pain or itching, nocturnal dyspnea of congestive heart failure, respiratory problems, angina, or other chronic diseases associated with disturbed sleep. Further discussion of sleep hygiene is provided in Chapter 21.

The patient should also be evaluated for depression or superimposed delirium or psychosis with behavioral disturbance and nighttime agitation. Intractable insomnia can be treated with benzodiazepine sedative-hypnotic preparations. Barbiturates are not recommended because of problems with paradoxical excitation and suppression of REM sleep, leading to rebound insomnia and nightmares. The elderly patient's increased sensitivity to barbiturates can also lead to ovcrdosing (Kales & Kales, 1974; Thompson, Moran & Nies, 1983). In addition, barbiturates are habituating (Kales & Kales, 1974).

Benzodiazepine pharmacokinetics vary widely. Long-acting agents such as diazepam, flurazepam, clorazepate, and chlordiazepoxide have half-lives that may be as long as 90 hours in the elderly; the half-life increases with aging because of decreased hepatic metabolism and increased volume of distribution (Thompson, Moran, & Nies, 1983). The long-acting benzodiazepines may accumulate and produce unwanted daytime sedation not immediately reversible with cessation of the drug. Diazepam can produce sedation up to 2 weeks after cessation in healthy elderly volunteers (Salzman et al., 1983). For these reasons, shorter acting benzodiazepine sedative-hypnotics such as temazepam may be preferred, especially in patients with liver impairment, because they are not metabolized by the hepatic mixed-function oxidase system (Thompson, Moran & Nies, 1983). Even short-acting benzodiazepines, such as lorazepam, oxazepam, and triazolam, may have problems, however, including retrograde and anterograde amnesia, rebound insomnia, and greater problems with withdrawal symptoms with the abrupt cessation of usage. In addition, both long- and short-acting benzodiazepines may worsen cognitive impairment in mild dementia and occasionally may produce delirium. Benzodiazepines will worsen sleep apnea, which occurs with relatively high prevalence in patients with dementia (Reynolds et al., 1985). Thus, long-term use is discouraged.

Low doses of sedating antidepressants can be useful for elderly patients with insomnia, especially if depression or anxiety appears to be a primary determinant of insomnia. Trazodone, 50 to 100 mg given after a light snack to improve absorption, may be used. Other drugs used to treat insomnia, such as sedating antihistamines and chloral hydrate, deserve mention. Elderly patients are often prescribed 25 to 50 mg of diphenhydramine hydrochloride (Benadryl). Because of anticholinergic effects, however, this drug may precipitate delirium or aggravate memory loss in Alzheimer disease patients, who already have impaired cholinergic systems. Chloral hydrate has been a safe and effective hypnotic with a short half-life (8 hours) and seldom is reported to produce delirium. Problems with chloral hydrate include gastrointestinal discomfort and excessive flatus, and hepatic enzyme induction with increased rate of metabolism drugs, including anticoagulants (Jenike, 1985). Chloral hydrate's metabolites can displace acidic drugs such as diphenylhydantoin or warfarin from plasma proteins, producing transient increases in drug effect and shorter drug half-lives (Thompson, Moran, & Nies, 1983).

Psychopharmacotherapy

Psychopharmacologic agents are used in dementia primarily for behavior and symptom control (Maletta, 1985). Prior to use of medication for behavioral control, psychiatrists must ascertain if signs and symptoms observed are caused or exacerbated by concomitant medical illness or the medications used in treatment. If medication was used for a previous psychiatric disorder, was it effective? Did any adverse reactions occur? Patients tend to be prescribed medication by several physicians. Psychiatrists thus need to ascertain all medications currently taken. The "bathroom cupboard test" (Miller, 1984) entails asking demented patients or their families to bring in all medication bottles in order to evaluate all possible drug interactions and side effects. Patients prescribed psychotropics are evaluated at frequent intervals to assess continued need for medication.

Initial dosages of psychotropic drugs should generally be one third to one half the dose for younger, healthier adults. Dosage is gradually raised until symptoms are relieved or excessive side effects emerge (Mahendra, 1984). On achievement of symptom control, the minimum effective dose is prescribed. After patients are symptom free for a week or two, the dose is tapered downward until the least amount required to prevent symptom exacerbation is defined. This may be as little as one third of the original therapeutic dose (Thompson, Moran, & Nies, 1983).

Psychopharmacologic management of the demented patient has three specific goals: (1) management of agitation; (2) diagnostic-specific treatment of any affective, anxiety, or psychotic component; and (3) amelioration of the cognitive dysfunction where possible.

Management of Agitation in the Demented Patient

Antipsychotic medication is indicated for aggression, hostility, assaultiveness, restlessness, agitation, and hyperactivity, with or without such evidence of psychosis as delusions, hallucinations, and paranoia. Family members should be aware that antipsychotic medication is being prescribed for behavioral symptoms and will not reverse intellectual deterioration (Maletta, 1985). Furthermore, trials of neuroleptics for agitation should be for a maximum of several weeks for three reasons: (1) there is no evidence that they are of long-term benefit for patients with dementia, (2) the risks of neurologic side effects are significant, and (3) agitation in many patients with dementia is often a transient response to medical illness or environmental change. Finally, if a neuroleptic does not work for the problem after several weeks, it is unlikely to become effective with continued administration. An attempt to withdraw neuroleptics, therefore, should be made as soon as the medical, social, or environmental precipitant has been identified and corrected.

Selection of antipsychotic medication depends on disease state, history of previous exposure and response to a particular drug, and side effects of the drug (Maletta, 1985, 1990b). Metaanalysis of neuroleptic treatment in demented patients has failed to identify any single neuroleptic as being superior to others (Schneider, Pollock, & Lyness, 1990). However, some psychiatrists favor thioridazine for agitated elderly patients because of its low incidence of extrapyramidal side effects (Mahendra, 1984). Others contend that since Alzheimer patients have problems with decreased acetylcholine, it is illogical to use highly anticholinergic drugs, which further decrease cholinergic transmission (Maletta, 1985, 1990b). Even mildly demented patients can develop central anticholinergic problems in the absence of peripheral anticholinergic signs. Because anticholinergic side effects for thioridazine increase with advancing dosages, patients who require more than 100 mg of thioridazine to control psychotic processes usually should be switched to high-potency neuroleptics, even though these agents do present a greater likelihood of parkinsonian side effects. Although both haloperidol and thioridazine are equally effective in managing target behaviors, fatigue and extrapyramidal symptoms are greater with haloperidol than with thioridazine (Steele, Lucas, & Tune, 1986).

Patients given high-potency neuroleptics in doses over 100 chlorpromazine equivalents per day probably will need an antiparkinsonian agent for optimal physical function. Amantadine at dosages of 100 mg orally twice a day may be preferable to benztropine (Cogentin) or trihexyphenidyl (Artane) for this purpose, because of amantadine's lesser anticholinergic effects. Haloperidol is given in doses of 0.5 to 1 mg at bedtime or twice a day and thiothixene in doses of 2 to 4 mg at bedtime or twice a day, with a gradual increase in dose over several days until a response occurs or limiting side effects develop. Drug choice is determined by the prescriber's preference as well as the specific side effects of the neuroleptic. Thiothixene is as efficacious as haloperidol and may have fewer extrapyramidal side effects. Although both loxapine and thioridazine are equally effective for management of anxiety, excitement, emotional lability, and uncooperativeness in disturbed, demented patients, a greater incidence of sedation and extrapyramidal symptoms may occur with loxapine, and more problems with orthostatic hypotension occur with thioridazine (Barnes et al., 1982).

Sedating neuroleptics may help agitated patients with difficulty falling asleep. In some patients, however, sedation increases confusion and disorientation, actually aggravating agitation. Coadministration of sedating neuroleptics with hypnotics, analgesics, and antihistaminics further depresses CNS function (Salzman, 1982). Orthostatic hypotension is a serious side effect. It predisposes to falls, resulting in fractures; to stroke; and to myocardial infarction. Vulnerability to falls is increased in patients with cervical osteoarthritis, patients with low cardiac output, and patients concomitantly on cyclic antidepressants (Salzman, 1982; Zuckerman et al., 1987). Risk is reduced by using higher potency neuroleptics with less alpha-adrenergic blocking activity (e.g., haloperidol, thiothixene, or molindone), concurrent use of supportive stockings, avoidance of diuretics, and liberalized salt intake. When feasible, patients should be encouraged to rise slowly from the recumbent position, particularly at night.

Fear exists that certain neuroleptics, especially thioridazine, may aggravate cognitive dysfunction. An open crossover study, however, comparing efficacy and side effects of haloperidol and thioridazine in the management of behavioral symptoms in Alzheimer's disease patients, indicated that thioridazine, up to 75 mg/day for 2 weeks, did not affect intellectual function (Steele, Lukas, & Tune, 1986).

A trend exists in general psychopharmacotherapy to reduce neuroleptic dose by addition or substitution of benzodiazepines, such as lorazepam and clonazepam. These are both used for control of acute psychotic agitation in delirium and manic states. In the elderly, however, there is a significant risk of oversedation—particularly with clonazepam—and of paradoxical aggressive responses (Salzman, 1988; Van Praag, 1977). Well-controlled studies of this approach are lacking, but it could be considered in patients who cannot tolerate neuroleptics. Serotonergic agents may turn out to be safer neuroleptic adjuncts or alternatives for agitation in dementia. Trazodone (Salzman, 1988), buspirone, clonazepam, and fluoxetine have all been reported to be useful in management of agitation, presumably by affecting serotonergic systems (Sobin, Schneider, & McDermott, 1989). There have been few controlled studies, however. Beta-blockers have also been used with success in some patients, as has carbamazepine. This issue is discussed further in Chapter 9.

In assessing psychopharmacologic options, it is crucial to evaluate the patient carefully for depression. Specific therapy for agitated depression is successful in about two thirds of cases, a far higher proportion than the proportion of agitated patients who respond to neuroleptics (Fogel, 1991a).

Two major problems commonly encountered in the management of the demented patient, particularly in a nursing home, are that all drug therapy is abandoned if an initial choice fails to prove effective, or alternately that medication is continued even if not successful.

Management of Depression and Psychosis in Dementia

The management of depression in dementia requires standard therapies with conservative initial dosage. Doses that may be administered without problems in a nondemented patient may cause a severe exacerbation of dementia or agitation with a demented patient.

After medical causes of depression have been ruled out, the depression is vigorously treated in hope of not only elevating mood but also improving cognitive functioning. Depression is frequently aggravated or precipitated by prescription drugs, with over 100 prescription drugs, including digoxin, cimetidine, reserpine, methyldopa and propranolol, associated with causing or aggravating depression (Fogel, 1991b). In these instances evaluation of the role of the drug in the genesis of depression should be evaluated before an antidepressant is commenced. Less anticholinergic antidepressants (e.g., fluoxetine, sertraline, bupropion) are favored because of increased susceptibility of demented patients to delirium. The newer antidepressants may have less negative impact on tracking performance and vigilance and produce less drowsiness than tricyclics such as amitriptyline (Burns, Moskowitz, & Jaffe, 1986). Monoamine oxidase inhibitors may be better tolerated than tricyclics in older patients than younger ones (probably because monoamine oxidase activity increases with increasing age), and may be effective in treating depression accompanying dementia (Fogel, 1991a). Electroconvulsive therapy may be used safely for depression in most older patients, including the cognitively impaired, although documentation of pre-ECT cognitive function is recommended.

Neuroleptics are effective for managing psychoses in demented patients, including those associated with paranoia. Patients with cognitive impairment disorders are more sensitive to neuroleptic side effects than they are to antidepressant side effects, and therefore dosage must be kept low (Fogel, 1991b). This general guideline should also be followed in young patients with the AIDS dementia complex. Because of increased incidence of akinesia, akathisia, and rigidity in this group, dopamine agonist antiparkinsonian medication may be required if high-potency, low-dosage neuroleptics are used. The heightened sensitivity to development of an atropine psychosis dictates use of neuroleptics only in instances of psychotic thinking and not for general sedation if an alternate therapy is available and efficacious. If agitation persists despite use of the neuroleptic for the psychosis, adjuncts such as benzodiazpines or serotonergic agents deserve consideration (Fogel, 1991a).

Management of Cognitive Dysfunction in Dementia

A number of drugs have been advocated for treating the cognitive deterioration associated with senile dementia of the Alzheimer type. Ergoloid mesylates (Hydergine) has been the most widely studied and used over the last 30 years (Battaglia et al., 1989; Dysken, 1987; Jenike, 1985). In 1990, Hydergine was reported to be the 11th most prescribed drug in the world (Walker, 1990). There are anecdotal reports of increased cerebral glucose utilization, as measured by PET, following the administration of Hydergine.

Hydergine is given in dosages averaging 3 to 6 mg/day but has been given in doses as high as 12 mg without serious side effects. In some studies, Hydergine has shown promise, with improvements reported in mood, dizziness, locomotion, and self-care (Arrigo et al., 1989; Kopelman & Lishman, 1986; Yesavage et al., 1979). Some also report improved memory and orientation. Thienhaus and his associates (1987) reported greatest impact in enhancing short-term memory. The most significant improvement, however, appears to be in lifting depression (Jenike, 1985). The studies supporting Hydergine's efficacy are limited, and some patients are reported to improve only if a trial of 3 to 6 months is attempted. Other investigators advocate discontinuing Hydergine after 2 or 3 months if there has been no improvement because there has been at least one report of possible drug toxicity leading to deterioration in function (Walker, 1990). A dose of 2 mg three times a day is recommended if it is tolerated because some patients respond to 6 mg/day but do not respond to 3 mg/day. Mildly demented patients with less-than-major depression may be most likely to benefit. However, even at best benefits are small and transient and the emerging consensus of American geriatric psychiatrists is against using the drug at all.

Clear, simple instructions on how medication is to be taken should be provided. Family, friends, or attendants are required to administer medication in advanced stages of dementia. Failure to monitor drug intake may result in drug automatism, in which a patient in a semitoxic state repeats dosages and becomes confused and sedated. In the extreme, such behavior can lead to death by unintentional overdose.

Referral for Experimental Therapies

A number of experimental therapies for Alzheimer's disease are under investigation. Many are based on efforts to enhance cholinergic transmission; others try to improve cerebral metabolism. Oral tetrahydroaminoacridine (THA; tacrine, Cognex), an anticholinesterase, and other cholinergic agonists have been proposed as means of arresting cognitive, functional, and behavioral decline in Alzheimer's disease. Several large clinical trials suggest that tacrine gives functionally significant cognitive improvement to a minority of patients who can tolerate it (Davis et al., 1992; Farlow et al., 1992). Abnormal liver enzymes are a common and significant side effect. The clinical use of tacrine was recently reviewed by Fogel & Stoudemire (1993). The use of THA continues to cause controversy although it is available in the United States under a treatment IND.

A superb review of experimental treatments is *Treatment Development Strategies for Alzheimer's Disease*, edited by Crook et al. (1986). Patients with mild to moderate Alzheimer's disease may be considered for referral to a research center if they are motivated and their expectations are realistically moderate. Patients with Alzheimer's disease should not be referred to clinical research centers when they have deteriorated severely and all other interventions have failed. The patients who are most likely to benefit from referral are relatively young patients with memory loss who are otherwise in good physical health and do not yet have severe problems with basic activities of daily living. Experimental therapies have the greatest likelihood of benefiting patients early in the course of the disease (Jenike, 1985).

Administration of acetylcholine precursors was a common experimental approach in the 1980s (Whalley, 1989). Treatment with choline and lecithin, precursors of acetylcholine, is hypothesized to alleviate symptoms of Alzheimer's disease by increasing the brain acetylcholine level in much the same way L-dopa reduces symptom intensity in Parkinson's disease. Outcome studies, however, have been either negative or equivocal (Gauthier, 1990; Jenike et al., 1986).

The cholinomimetic physostigmine appears to cause some improvement in those in whom oral physostigmine increased cholinergic activity (Mohs et al., 1985).

3′-Azido-2′,3′-dideoxythymidine (AZT) has been used effectively in both adults and children in the management of the AIDS dementia complex (Culliton, 1989; Yarchoan et al., 1988). Evidence indicates improvement with AZT not only of the cognitive deficit seen with the human immunodeficiency virus but also of nerve conduction in patients with peripheral neuropathy and paraplegia (Yarchoan et al., 1988). This indicates need for continued studies of the impact of antiretroviral chemotherapy.

Multiinfarct Dementia

The second leading cause of dementia, multiinfarct dementia, generally is distinguishable from Alzheimer's disease by its suddenness of onset, stepwise deterioration of intellectual functioning, and greater association with hypertension and cardiac disease, including a history of myocardial infarction, angina, and congestive heart failure (Tresch et al., 1985). However, multiinfarct dementia is not ruled out by a history of gradual and insidious deterioration (Fischer et al., 1990). A history of diabetes is common. Depression and neurologic abnormalities, such as asymmetric reflexes, dysarthria, gait abnormalities, limb spasticity, and focal deficits secondary to stroke, are also seen (Read, Small, & Jarvik, 1985). Personality is often preserved in the early stages (Sloane, 1983), although depression is common. The male-female ratio of incidence of multiinfarct dementia is 2:1. Onset is earlier than for Alzheimer's disease (Sloane, 1983). Magnetic resonance imaging should be considered in patients with suspected multiinfarct dementia because of its superiority to CT in detecting small infarcts (McGeer, 1986).

Treatment focuses on blood pressure control and reduction of stroke risk by management of transient ischemic attacks and cardiac conditions increasing stroke incidence, such as cardiomyopathies, chronic left heart failure, aortic valve disease, atrial myxoma, rheumatic heart disease, and subacute bacterial endocarditis, as well as complications of prosthetic heart valve insertion. Control of other stroke risk factors, such as diabetes, smoking, and the hyperlipidemias, is also important (Mahendra, 1984).

Reduction of blood pressure should be undertaken cautiously. Patients with arteriosclerotic disease may require higher baseline blood pressures to perfuse their brains. Longitudinal studies of hypertensive patients with multiinfarct dementia suggest that systolic blood pressure be maintained in a high-normal range (135 to 150 mm Hg) to improve cognition and to prevent clinical deterioration associated with blood pressure reduction below this range. Aspirin may be used as an antiplatelet agent to prevent stroke. Warfarin is given to patients with prosthetic valve and valvular heart disease that predisposes to emboli. Carotid endarterectomy may be beneficial to patients with symptomatic carotid stenosis; treatable carotid lesions are detected by ultrasound studies and magnetic resonance angiography (MRA).

Often, primary care physicians become less aggressive about the prevention of future strokes and controlling stroke risk factors when dementia and psychiatric complications supervene. The psychiatrist has a crucial role in reiterating to primary care providers that multiinfarct dementia, unlike Alzheimer's dementia, has a better course when proper attention is given to such risk factors (Meyer et al., 1986). In their study of 52 patients with multiinfarct dementia, it was found among the hypertensive patients that improved cognition correlated with control of systolic blood pressure within the upper limits of normal (135 to 150 mm Hg).

Forensic Issues and Competency

Need for power of attorney, guardianship, and conservatorship, and concerns regarding testamentary capacity and competency are legal issues in all instances of dementia. If dementia is diagnosed early and deterioration is spotty, patients may be able to make some decisions regarding the distribution of assets on their death and the selection of a guardian to handle their affairs later in their course. Individuals who live with a fantasy of eternal good health are sometimes struck down with wills absent or so dated that property and money remain tied up in litigation for years after death while care providers attempt to recoup the cost of care. Families should be advised early in the course of illness to seek legal help in protecting the rights and property of patients while providing access to funds for the care required.

The psychiatrist plays a dual role in the legal management of the patient with dementia. First, the psychiatrist can assess the patient's degree of cognitive impairment and the effects that this impairment has on the patient's judgment and decision-making capacity with regard to person and property. Second, the psychiatrist can intervene with the patient and the family to work out a mutually agreeable legal mechanism to protect the patient's rights. This is often done by the psychiatrist working together with an attorney. These issues are discussed at length in Chapter 45.

DELIRIUM

Delirium (also called acute confusional state, toxic psychosis, acute organic brain syndrome with psychosis, toxic-metabolic encephalopathy, acute brain syndrome, and acute cerebral insufficiency) (Ellison, 1984; Engel & Romano, 1959: Lipowski, 1990) represents the behavioral response to widespread disturbances in cerebral metabolism (Strub, 1982). The term "delirium" is derived from the Latin words for "off the track" (Tobias, Lippman, & Pary, 1989). Patients who are delirious are confused and disoriented and have short-term memory defects and fluctuating states of arousal (Lipowski, 1990). Autonomic and other neurologic signs appear, with visual and tactile hallucinations (Crump et al., 1984). Characteristically, symptoms are global, appear abruptly, and are of relatively brief duration (usually less than 1 month). A global disorder of attention is seen in most cases (Lipowski, 1990).

The presence of delirium in critically ill surgical patients is significantly correlated with increased serum anticholinergic activity as indicated by radioreceptor assay (Golinger, Peet, & Tune, 1987). The prognosis for delirium associated with medical illness is much graver than the prognosis of postsurgical delirium, which is usually transient and often attributed to drug intoxication or withdrawal (Table 19-1), or a reaction to stress of the procedure (Table 19-2 below). Delirium can lead to complications, including patient injury, assault, and suicide. Diagnostic signs include fluctuating

TABLE 19–1. *Drugs Associated with Delirium (Partial Listing)*

Adrenocorticotropic hormone	Ethinamate
Alcohol	Glutethimide
Aminophylline	Heroin
Amphetamines (speed, white crosses, black beauties, ice, crystal meth, crank)	Histamine H_2-receptor antagonists
	Ibuprofen
Antiarrhythmic agents	Indomethacin
Antibiotics	Jimson pod
Anticonvulsants	Ketamine
Antifungals	Levodopa
Antihypertensives	Lidocaine
Antimalarials	Lithium carbonate
Antiparkinsonian agents	Local anesthetic agents
Antispasmodics	Lysergic acid diethylamide
Antituberculosis drugs	Methylenedioxyamphetamine (Ecstasy)
Antivirals	Methyprylon
Asthma powders	Meperidine hydrochloride
Atropine	Meprobamate
Baclofen	Mescaline
Barbiturates	Methadone
Benzodiazepines	Methaqualone
Benzquinamide	Methyldopa
Benztropine	Morning glory seeds
Biperiden	Mydriatics
Bismuth salts	Narcotic analgesics
Bromides	Nonnarcotic analgesics
Butyrophenones (e.g., haloperidol)	Nutmeg
Caffeine	Paraldehyde
Camphor	Pentazocine
Cannabis	Percodan
Chloral hydrate	Peyote (cactus buttons)
Cimetidine	Phencyclidine (PCP, angel dust)
Corticosteroids	Phenelzine
Clozapine	Phenothiazines (e.g., chlorpromazine, thioridazine)
Cocaine (crack)	Phenylpropanolamine hydrochloride
Colchicine	Podophyllin
Contrast agents	Procainamide hydrochloride
Cyclopentolate	Procyclidine
Cycloplegics	Promethazine
Cyclosporine	Proprietary hypnotics
Cytotoxic chemotherapeutic agents	Psilocybin (mushrooms)
Digitalis	Quaaludes (ludes)
Dimenhydrinate	Scopolamine
Dimethyltryptamine	Sodium thiopental
Diphenhydramine	Sulfasalazine
Disulfiram	Theophylline
Ephedrine	Trazodone
Ergot alkaloids	Tricyclic antidepressants
Ergotamine hydrochloride	Trihexyphenidyl
Ethchlorvynol	Tripelennamine

List compiled from Lipowski (1990), Murray (1987), and Slaby, Lieb, & Tancredi (1985).

level of awareness, worsening of symptoms at night, disruption of the sleep-wake cycle, and agitation or lethargy (Lipowski, 1987, 1990).

Estimates of the prevalence of delirium vary. In one study 33% of medically ill patients were found to be cognitively impaired (Knights & Folstein, 1977); in another (Massie, Holland, & Glass, 1983) 85% of terminally ill cancer patients met criteria for delirium. Figures cited in a review of incidence studies by Lipowski (1990) vary from 16.6% to 18.9% for two large Swiss studies of medical and surgical patients to 35% for patients 65 and older being admitted to a number of different British hospitals. In another study cited (Lipowski, 1989), about 40% of patients 55 years and older who were demented were delirious on admission to the hospital, whereas 2.5% of those who were delirious were demented. Other recent estimates include one fourth of patients over 65 admitted to two acute care hospitals in Edmonton, Canada (Rockwood, 1989), 13% to 100% of postcardiotomy patients (Theobald, 1987), and 26.3% of a group of postsurgical patients on whom consultation was sought for mood disturbance (Golinger, 1986). Even in settings such as acute care hospitals, where prevalence is high, cognitive def-

TABLE 19–2. *Differential Diagnosis of Delirium*

Syndromal
Atypical psychosis
Brief reactive psychosis
Conversion disorder
Dementia
Depression with cognitive impairment
Dissociative disorders
Factitious disorders
Hypomania with cognitive impairment
Mania
Post-traumatic stress disorder
Psychoactive substance hallucinosis
Schizophreniform disorder or relapsing schizophrenia

Etiologic
Acidosis or alkalosis
Anesthesia
Brain lesions (stroke, subdural, tumor, etc.)
Brain trauma
Cerebrovascular diseases (including subarachnoid hemorrhage)
Collagen-vascular disease (especially SLE)
Dehydration
Drug toxicity
Drug withdrawal
Electrolyte imbalance: Ca, K, Mg, Na, PO_4
Endocrine disease (especially adrenal, pituitary, thyroid)
Hyperosmolar states
Hypoglycemia or Hyperglycemia
Hypoxemia or Hypercarbia
Infections (cerebral or systemic)
Medication effects (anticholinergics, steroids, psychotropics, etc.)
Migraine
Nutrition: protein-calorie malnutrition, vitamin deficiency
Organ failure: heart, liver, pancreas, kidney
Seizures (ictal or post-ictal)
Sensory deprivation
Surgery
Trauma

icits may be missed in as many as 79% of patients (O'Brien, 1989).

Death rates for delirious patients exceed rates for demented, depressed, and cognitively intact controls at index admission and a 1-year follow-up (Rabins & Folstein, 1982). Delirious patients have a greater incidence of diffusely slow EEGs, hyperthermia, low mean blood pressures, and tachycardia than the comparison groups. Fatality rates of the delirious are greater than those of control groups matched for age, race, sex, ward status, and medical diagnosis, indicating that medical diagnosis alone does not explain findings. Delirium suggests a more serious form of medical illness and at times is a harbinger of death. A review (Black, Wanack, & Winokur, 1985) of 543 delirious patients indicated that they, particularly if under 40, are predisposed to early death when compared with controls. Risk is greatest in the first 2 years of follow-up, with higher risk of death from heart disease and cancer in women and from pneumonia and influenza in men. In a classic study, 40% of elderly demented patients died soon after developing delirium (Post, 1965). Twenty-five percent of patients with delirium referred for psychiatric consultation in the study by Trzepacz, Teague, and Lipowski (1985) died within 6 months of consultation, with 66% of those dying during the index admission. Weddington, Muelling, and Moosa (1982) reported a 33% 3-month mortality rate following psychiatric consultation for delirium. These studies suggest that those florid forms of delirium most likely to trigger psychiatric consultation tend to be associated with severe physical illness and high mortality rate.

Delirium (Lipowski, 1967, 1983) mimics other psychiatric disturbances and misleads physicians by depressive, paranoid, schizophrenic, phobic, and hysterical symptoms. Agitation, belligerence, and bizarre behavior or depressed, stuporous, or apathetic behavior may occur in the same patient, depending on the time the patient is evaluated (Dubin, Weiss, & Zeccardi, 1983). Hallucinations predominantly appear in withdrawal delirium; they are less frequent in delirium associated with cardiac or pulmonary disease (Lipowski, 1990; Sirois, 1988).

Psychosocial stress precipitates medical, as well as psychiatric, illness, and the two can occur concurrently, with overt signs of delirium discounted because symptoms are attributed to the psychiatric disorder (Daniel & Rabin, 1985). Physical examinations of patients with lupus cerebritis, hypothyroidism, hypoglycemia, and phenobarbital toxicity may be indistinguishable from those with primary psychiatric disorders. Clues suggesting underlying medical illness in the acutely psychotic patient include onset after age 40 with no prior psychiatric history and the occurrence of cardinal features of delirium (i.e., rapid onset, global cognitive impairment, fluctuating levels of consciousness, disturbance of the sleep/wake cycle, disorientation, worsening of psychosis at night, tremors, pathologic reflexes, and visual, tactile, or olfactory hallucinations). In one survey (Trzepacz, Teague, & Lipowski, 1985) of general hospital psychiatric consultations resulting in the diagnosis of delirium, 67% were referred because of noncognitive psychiatric symptoms. Thirty-five percent of these were referred for affective symptoms: depression, suicidal ideation, and tearfulness. The majority of the patients were perceived as manipulative or "a nuisance" by referring staff.

Delirium is confused with dementia because it is often superimposed on dementia (Lipowski, 1983). Symptoms of dementia may remain after delirium has cleared (Gallant, 1985). Patients presenting with acute, as opposed to insidious, onset of change in mental status and fluctuating levels of consciousness should be evaluated for delirium, with appropriate correction of underlying medical problems. Trzepacz, Bauer, and Greenhouse (1988) have developed a 10-item clinician-rated scale for the diagnosis of delirium. Delirious patients have been found to score significantly higher than demented patients, schizophrenics, or normal controls.

Clinical Description

Fluctuating clouding of consciousness is the hallmark of delirium (Adams, 1984; Fogel, 1991b; Lipowski, 1967; Murray, 1987; Strub, 1982). However, insomnia, nightmares, intermittent nighttime disorientation, and anxiety often appear first. Review of nursing progress notes will often document transient fluctuations of symptoms in suspected cases.

Symptoms, especially agitated behavior and visual hallucinations, are more likely at night. At one moment, patients appear combative and suspicious, shouting obscenities. At another, they are stuporous, mumble incoherently, pick at clothing, and drift off to sleep in midsentence. Confusion may alternate with periods of relative lucidity. Medical care is compromised by the patient's suspicion of staff, medication, and medical procedures, and by accidental or intentional disconnection of catheters, intravenous lines, and tubes (Adams, 1984). Associated features include purposeless movements of arms and legs and multifocal myoclonus. Tremor is often present (Ellison, 1984). Other neurologic signs are relatively uncommon in the absence of a primary CNS cause of delirium. Asterixis, a motor disturbance marked by intermittent lapses of a sustained posture, such as flapping movements of hyperextended hands, accompanies hepatic encephalopathy, hypoxia, uremia, and some other metabolic disturbances.

It should be noted that the mildly delirious patient may present differently from the patient with a full-blown delirium. In the mildly delirious patient, abnormalities of mood and demanding or nuisance behaviors are often prominent. If the clinician does not test for cognitive disturbances in these patients, such disturbances are often overlooked or misdiagnosed as "functional."

The principal clues in clinically distinguishing mild delirium from alternative psychiatric diagnoses are the facts that the cognitive dysfunction is disproportionate to other psychopathology and that attention and orientation fluctuate, often from hour to hour (Fogel, 1991b). Mildly delirious patients usually exhibit the characteristic slowing of the dominant background rhythm of the EEG. Sleep charts and standardized clinician-rated mental status examinations are helpful in establishing the diagnosis.

Differential Diagnosis

Older people are at the greatest risk for delirium (Wells & Duncan, 1980), but the very young are also predisposed. Mortality of medical (opposed to perioperative) delirium ranges from 7% to 12% (Daniel & Rabin, 1985), and is due to the underlying illness. Preexisting brain disease; drug and alcohol addiction; chronic hepatic, renal, cardiac, and pulmonary dysfunction; and organ failure are other risk factors for delirium (McEvoy, 1981; Wells & Duncan, 1980). Elderly patients display more cognitive accompaniments of neurologic and metabolic problems than younger patients. Common physical illnesses leading to delirium are pneumonia, cancer, urinary tract infection, hyponatremia, dehydration, congestive heart failure, uremia, and stroke (Lipowski, 1990).

Before it is possible to establish the exact etiology of a delirium, it is first necessary to establish that the apparent cognitive impairment is due to delirium rather than one of the disorders provided in the syndromal differential diagnosis displayed in Table 19-2, such as dementia, depression, or a psychosis (Fogel, 1991b). Dementia and pseudodementia can resemble quiet delirium (Berrios, 1989). Longitudinal assessment of symptom clusters facilitates this differential diagnosis. Acquired immunodeficiency syndrome can present with a number of disorders both in the syndromal differential diagnosis (e.g., dementia, depression, anxiety disorders, delirium) and in the etiologic differential diagnosis (e.g., pneumonia, encephalitis) (Dickson & Tanseen, 1990). The stepwise progression of multiinfarct dementia may suggest delirium with the emergence of new symptoms, but symptom intensity does not fluctuate over hours as does delirium (Zubenko, 1990). The intermittent irritability seen in many Alzheimer's patients can suggest the behavioral fluctuations of delirium. It may resemble that seen with mild dementia and may be seen regardless of whether a patient exhibited any premorbid bad temper (Burns et al., 1989).

Visceral illness can produce a delirium with no abnormalities in routine blood tests. In these cases, patients are often referred to psychiatrists for the evaluation of "functional illness" because of mental status changes that are not explained by abnormal laboratory data. It is then the task of the psychiatrist to refocus the primary care physician's attention back on the possibility that there may still exist undiagnosed physical disease. Alcohol intoxication and withdrawal are common causes of delirium in young populations, and CNS effects of polypharmacy are common in the elderly.

Drug-Induced Delirium

Drugs contribute to delirium in both young and old (Cassem & Hackett, 1987; Trzepacz, Teague, & Lipowski, 1985). There are a number of points of particular note in understanding and treating drug-induced deliria.

1. *Narcotic analgesics can produce agitated delirium.* Meperidine, in particular, has been noted to produce delirium, including tremor, seizures, and multifocal myoclonus (Foley, 1982). This complication has been reported to be secondary to the accumulation of meperidine's active metabolite, normeperidine, with repeated (particularly intravenous) dosing, and it appears to be more common in patients with renal insufficiency. Often, substitution of a different narcotic analgesic will eliminate the problem.

Another narcotic analgesic that has psychotomimetic properties is pentazocine (Talwin). The psychotomimetic effect of this narcotic is noted with higher doses of the drug and therefore limits its usefulness for patients with severe or chronic pain (Foley, 1982). In addition to this, pentazocine is a narcotic with agonist-antagonist properties and therefore can precipitate an acute withdrawal state in a patient who was previously receiving chronic narcotic agonists of the morphine type (Foley, 1982). For this reason, care must be taken in switching a patient from other morphine-type narcotics to pentazocine; this problem is often overlooked by primary care physicians (Foley, 1985).

2. *Blood levels of medication can suggest toxicity for such agents as anticonvulsants, digoxin, theophylline, lithium, and tricyclics.* However, it is important to note that, particularly in the elderly, delirium may occur with digitalis and quinidine at serum levels well within the normal limits of most clinical laboratories. Although psychotoxicity generally is dose-dependent, the threshold for toxicity varies from individual to individual. Even more importantly, CNS toxicity may precede signs of ECG or gastrointestinal disturbance; digitalis has been noted to be associated with visual disturbances that may precede ECG clues that such intoxication is occurring (Volpe & Soave, 1979).

3. *Confusion and delirium may occur with propranolol, especially in the elderly.* This drug has been noted to produce depression or visual hallucinations, as well as toxic confusional states (Gershon et al., 1979; Paykel, Fleminger, & Watson, 1982). Although neurotoxic reactions to propranolol have been most often reported with large doses, the CNS side effects also can occur in low or therapeutic dose ranges (i.e., less than 160 mg/day) (Remick, O'Kane, & Sparling, 1981). Timolol, another lipid-soluble beta-blocker, can have similar effects, whether given orally for hypertension or as eye drops for glaucoma.

4. *Another class of medication not often considered as a cause of delirium is the antibiotics.* Central nervous system toxicity and delirium have been seen with high-dose intravenous penicillin. In addition to hallucinations, other neurotoxic effects include generalized seizures, myoclonus, and hyperreflexia. These effects are more common in patients with renal insufficiency (Snavely & Hodges, 1984). Psychiatric symptoms have also been noted with chloramphenicol, although the occurrence of the symptoms is infrequent, even with prolonged use. Neurotoxic reactions have also been seen with ofloxacin, norfloxacin, ciprofloxacin, gentamycin, tobramycin, and metronidazole (McCartney, Hatley, & Kessler, 1982; Snavely & Hodges, 1984).

5. *Anticonvulsants can produce delirium; this usually occurs at high serum levels.* Paradoxical excitation has, however, occurred with barbiturates even within the normal therapeutic range (Penry & Newmark, 1979), particularly in individuals with preexisting brain damage or mental retardation. Concurrent administration of drugs, including chloramphenicol, sulfamethoxazole, phenylbutazone, and disulfiram, can inhibit the metabolism of phenytoin, increasing the risk of delirium (Delgado-Escueta, Treimara, & Walsh, 1983). Phenytoin levels have also been reported to be elevated by the concurrent administration of estrogen-containing oral contraceptives, phenothiazines, and warfarin anticoagulants (Rivinus, 1982).

6. *Cimetidine can produce neuropsychiatric side effects, including auditory and visual hallucinations, agitation, paranoia, disorientation, and stupor, as well as depression and anxiety states* (Weddington, Muelling, & Moosa, 1982). Central nervous system problems from cimetidine are seen more commonly in the very young, the elderly, and patients with cirrhosis or uremia (Strum, 1984). Often, when such a reaction occurs, the psychiatrist will switch the patient to ranitidine because there have been some reports of reduction in problems with CNS side effects by switching agents. Ranitidine, however, has been reported to reduce hepatic blood flow by 20% to 40% (Goff, Garber, & Jenike, 1985; McCarthy, 1983), thus leading to a potential for retarded metabolism of other drugs. Ranitidine may also cause depression (Billings & Stien, 1986), and it has been shown to interfere with the metabolism of a num-

ber of drugs, including acetaminophen, benzodiazepines, nifedipine, metoprolol, and warfarin (McCarthy, 1983; Rubin, 1984).

7. *Antihypertensives can precipitate delirium.* Although it is more common for the diuretics to produce delirium via electrolyte disturbances, other antihypertensives, such as reserpine and methyldopa, can cause delirium by direct action on the CNS (Paykel, Fleminger, & Watson, 1982). If there is a sudden alteration of mental status in a patient who is receiving diuretics, electrolyte levels, including calcium, should be checked.

8. *Mental status changes have also been noted with nonsteroidal antiinflammatory agents.* Psychotic reactions have been reported with sulindac, indomethacin, tolmetin, and naproxen ("Drugs That Cause Psychiatric Symptoms," 1981; Kruis & Barger, 1980; Sotsky & Tossell, 1984; Steele & Morton, 1986). With salicylates, blood levels in the toxic range are associated with major psychiatric side effects. Levels are particularly helpful because the pharmacokinetics of salicylates are highly variable among individuals.

9. *Caffeine can produce a delirium.* Central nervous system toxic effects of caffeine have appeared in some individuals with indigestion of doses as small as 50 mg of caffeine a day. One gram of caffeine (8 to 10 cups of coffee or five 200-mg over-the-counter caffeine tablets) may be sufficient to produce frank delirium (Goldfrank et al., 1981; Stillner, Popkin, & Pierce, 1978). The exact incidence of caffeine delirium is, however, unknown at this time.

10. *Theophylline, which is used in patients with asthma and chronic obstructive pulmonary disease, can cause a delirium that is dependent on its serum level and is unusual at levels within the therapeutic range.* For this reason, plasma concentrations should be carefully monitored. Coarse tremor in conjunction with a delirious state suggests theophylline toxicity in patients on this drug (Park, 1986).

11. *Corticosteroids very frequently cause mental status changes ranging from delirious psychosis to mania and depression.* The reaction has been more commonly recorded in females than in males. A prior history of psychiatric illness does not necessarily predispose the patient to the development of toxic delirium on corticosteroids. In addition, although steroid-induced delirium usually has been reported with higher dosages (i.e., 40 mg of prednisone or more), one study suggests that neither the dosage nor the duration of treatment appears to affect the time of onset, the severity, the type of mental disturbance, or the duration of mental disturbance in patients receiving corticosteroids (Ling, Perry, & Tsuang, 1981). Steroid-induced delirium usually responds to dosage reduction or drug cessation. For patients in whom corticosteroids must be used, lithium may prevent steroid-related psychosis (Goggans, Weisberg, & Koran, 1983); others have found low-dose neuroleptics helpful.

12. *Polypharmacy, especially with drugs having anticholinergic side effects, contributes to delirium in the elderly* (Birren & Bernstein, 1979; Blazer et al., 1983; Lipowski, 1983). Coingestion of neuroleptics and tricyclics both therapeutically and in intentional overdose leads not only to increased incidence of drug-induced delirium, but also to increased risk of cardiac consequences (Wilens, Stern, & O'Gara, 1990). Deliria caused by sedative-hypnotic agents, impaired dopaminergic transmission, serotonin excess, lithium toxicity, and anticholinergic toxicity represent the most common psychotropic drug-induced delirious states (Fogel, 1991b). All patients suffering from these states require basic supportive care: cessation of the drug ingestion, protection from injury to self and others, monitoring vital signs, and care to minimize dehydration.

13. *When patients become delirious while taking a neuroleptic, alone or in combination with another agent such as lithium, a tricyclic, or an anticonvulsant, the neuroleptic malignant syndrome (NMS) must be considered.* Muscular rigidity, diaphoresis, unstable vital signs, and increased creatine kinase (CK) support the diagnosis of NMS.

14. *Sweating induced by exercise or summer heat, diuretic use, and placement on a low-sodium diet for hypertension are two of a number of instances that may lead to toxicity in a patient maintained on lithium.* As lithium replaces the sodium lost, serum levels rise and mental status changes. Concurrent use of anticonvulsants, anticholinergic agents, and neuroleptics with lithium may lead to toxicity at "therapeutic" levels. Severe lithium toxicity includes rigidity, tremor, and involuntary movements resembling NMS and, because risks of renal and systemic complications are similar, must be treated as aggressively as NMS. Serum CK and urine myoglobin levels corroborate the rhabdomyolysis (muscle breakdown) seen with NMS. Individuals with preexisting brain injury are at a particular risk for toxicity at therapeutic lithium levels.

15. *Increased use of serotonin agonists (e.g., fluoxetine and L-tryptophan) has led to higher incidence of delirium related to drugs in this class, especially when the drugs are given with monoamine oxidase inhibitors (MAOIs).* The serotonin syndrome is characterized by fever, nystagmus, tremor, myoclonus, dysarthria, and, rarely, rigidity. There is no established specific treatment, although there is anecdotal support for cyproheptadine, a serotonin receptor blocker. Benzodiazepines, rather than neuroleptics, are used for agitation

because of the possibility of pharmacologic interaction between neuroleptics and serotonergic agents. This same syndrome may occur during the washout period following cessation of fluoxetine if a MAOI is given before serum fluoxetine and norfluoxetine levels fall (Fogel, 1991b).

16. *Sedative-hypnotics produce delirium most commonly on withdrawal, although they may occasionally do so with intoxication.* Elderly and brain-damaged patients are at greatest risk and may become confused or amnesic after as little as one dose of a short-acting hypnotic (e.g., triazolam). Withdrawal may be precipitated by the admission to a hospital of someone who did not use more than the prescribed amount of a sedative-hypnotic just as it may in a moderate alcohol user by sudden cessation of a CNS depressant. Clonazepam is particularly useful in alprazolam withdrawal. Its longer duration of action mitigates the withdrawal syndrome. Benzodiazepines all can cause respiratory depression; and arterial blood gases are indicated if the patient is stertorous or has a history of lung disease.

17. *The drugs used to treat AIDS are reported to produce hallucinations, confusion, and frank delirium* (Grant & Atkinson, 1990). Acyclovir, amphotericin B, 5-flucytosine, pilocaylbuzine, trimethoprim, sulfamethoxazole, thiabendazole, vincristine, and a number of antimicrobial agents have been implicated. Zidovudine (AZT) itself has not been found to have any detectable negative impact on cognitive functioning and, in fact, may be beneficial in some cases. The impact of these drugs is sometimes difficult to differentiate from symptoms of AIDS encephalopathy or opportunistic brain infection. Haloperidol and/or lorazepam given parenterally have been used for the delirium in AIDS (Fernandez, Levy, & Mansell, 1989).

18. *Anticholinergic drugs not only place older people at greater risk of confusion but also impair learning of new material even in the healthy elderly* (McEvoy et al., 1987). Therefore, they should be minimized in older patients, particularly if they are demented.

19. *Delirium has been reported to be caused by antiarrhythmic agents, including tocainide* (Trohman, 1988). Psychiatric side effects of antiarrhythmics are discussed further in Chapter 23.

"Pseudodelirium"

The syndromal differentiation of delirium from primary psychiatric disorders presenting with features of delirium is not always easy (Fogel, 1991a; Lipowski, 1990). Cognitive impairment of delirious proportions may be a manifestation of schizophrenia or bipolar illness alone, but patients with a history of recurrent episodes of psychiatric illness may also have a superimposed delirium attributable to a coexistent medical problem. If the current disorder is phenomenologically unlike past episodes, the patient definitely requires a medical evaluation for causes of delirium.

Pseudodelirium accompanies mania (Bond, 1980), schizophrenia, depression, paranoid disorders, atypical and brief reactive psychoses (Lipowski, 1990), and hysteria (Wells & Duncan, 1980). Delirious mania is identified (Bond, 1980) by acute onset, irritability, insomnia, emotional withdrawal, other hypomanic and manic symptoms, personal or family history of affective illness, and responsivity to treatment for mania. Fluctuating levels of consciousness and visual hallucinations are rare with schizophrenia. Schizophrenic hallucinations tend to be auditory (Wells & Duncan, 1980) and, despite the schizophrenic's marked disorganization of thought, attention and awareness of the examiner are usually maintained. The generalized cognitive and memory impairment seen with delirium is not found with schizophrenia. Younger schizophrenics usually are not disoriented. When they age, they may develop disorientation with a bizarre content. Familiar situations and surroundings may appear unfamiliar to the schizophrenic, whereas delirious patients more often mistake the unusual for the familiar. When paranoid, the delirious perceive danger for all; schizophrenics personalize danger. Flat, distant affect is absent in people with delirium, whose affect tends to be labile, intense, and transient. Nocturnal worsening of symptoms characteristic of delirium may be absent in schizophrenics, who often sleep well despite symptomatic exacerbation on arousal. Finally, schizophrenics do not have the multifocal myoclonus and bilateral asterixis considered to be pathognomic of diffuse CNS dysfunction.

Orientation and cognitive functioning usually remain intact in hysteria (Wells & Duncan, 1980), and asterixis, myoclonus, and usually hallucinations are absent. Hallucinations, when they occur, generally carry an air of drama. Impairment of cognition and orientation, if present, is inconsistent with what is known of organic dysfunction. A hysterical person may be disoriented to person but not to place or time. This is rarely true in delirium. Cognitive dysfunction in hysteria is out of keeping with the patient's level of alertness and responsivity to examiner and environment.

Laboratory tests may be abnormal in delirious presentation of primary psychiatric illness. Thyroxine can be transiently elevated in acute delirious mania. Creatine kinase can be mildly elevated in schizophrenics. Major elevation of CK in a treated schizophrenic, however, suggests NMS. (See Chapter 7 for further discussion of laboratory evaluation.)

History

The patient and collateral sources, including the primary physician, primary nurse, family, and significant friends and neighbors, provide the history of delirium. For hospitalized patients, chart review often reveals previous episodes of "confusion" in relation to specific drugs or procedures. It is often more helpful to review medications directly from the nurse's medication record rather than from the order sheet, because this will aid in detecting transcription or administration errors and will document the use of PRN medication. Nursing notes document episodes of transient fluctuations in mental status, nighttime disorientation, agitation, or insomnia, giving clues to etiology. Hospital discharge summaries also provide important information regarding previous medical problems, hospital course, and discharge medications, as well as a record of previous psychiatric consultations. Family and collateral sources provide data on past psychiatric history and information on drug or alcohol abuse or dependence, which the patient may deny.

The incidence of postoperative delirium depends on the surgical procedure and the patient population. Calabrese (1987) reported that only 4 of 59 (6.8%) patients undergoing myocardial revascularization developed confusion. In contrast, Smith and Dimsdale (1989) reported a 32% rate of delirium following cardiotomy.

Vital sign changes, such as fever, hypotension, and elevation of blood pressure and pulse, reported in current hospital records provide leads to identification of withdrawal states, hypertensive encephalopathy, thyroid abnormalities, and sepsis.

Oral temperatures are often misleading, because the delirious patient may not be able to keep the thermometer in the mouth; rectal or tympanic membrane temperatures are therefore preferred. Furthermore, elderly patients may have severe inflammation or serious infection without fever or leukocytosis (Blass & Plum, 1983). Even relatively mild infections, such as lower urinary tract infections, can produce delirium in the elderly (Blass & Plum, 1983).

Patients with recent surgery have records of anesthesia that record anesthetic agents used, blood loss and replacement, hypoxia, cardiac arrhythmias, and blood pressure fluctuations (Murray, 1987). Prolonged hypoxia or hypotension is particularly relevant to prolonged postoperative delirium.

Physical Examination

A complete physical examination, including vital signs, is necessary when seeking an etiology of delirium. Deep tendon reflexes and plantar reflexes are tested, and meningeal signs are sought. Asterixis is elicited by having the patient hyperextend the wrist and spread the fingers, while the examiner observes for flapping or subtle, abnormal, irregular jerking movements. Multifocal myoclonus is most often observed in muscles of the face and shoulders. Gentle massage of the lids over the closed eyes or placement of a blank, white sheet of paper a few inches in front of the open eye sometimes evokes vivid visual images in a delirious patient.

Patients are observed for attention and spontaneous movement. Do patients attend to examiners, or do they stare off into space and startle easily? Are they alert to the calling of their names, only to drift off to sleep? Do patients talk to themselves and stare at the ceiling or windows as if communicating with an unseen stimulus? Do patients mumble incoherently, pick at their clothes, or grasp unseen objects in the air? Are myoclonic movements present?

Does calling patients' names alert them when they appear stuporous, or is painful stimulation required? Is left-right orientation preserved, with respect to both the self and the examiner? Is speech coherent and logical or pressured, soft, or slurred? Are patients dysphasic or circumlocutory? Are thoughts goal-directed, and is conversation relevant to the current situation, or is there a paucity of spontaneous speech? If a patient is respirator dependent or if vocal cords are absent or paralyzed, written responses can be used. Is handwriting legible or does it show signs of tremor? Do patients print or use cursive writing? Are spelling and syntax correct, or is there evidence of perseveration by repetition of letters or words? In metabolically induced cognitive dysfunction, handwriting is disproportionately more impaired than speech (Adams, 1984).

Mental Status Examination

Formal mental status examination should be performed, including evaluation of mood and affect; level of consciousness; motor behavior; rate, pressure, and rhythm of speech; grammar and syntax; perceptual disturbances, including hallucinations and illusions; delusions and psychotic thinking; memory (immediate, recent, and remote); attention and concentration; abstraction; ability to name objects, both on visual confrontation and from description; writing; reading; calculations; visual-spatial orientation; orientation to person, place, and time; ability to perform commands on verbal and visual instruction; grasp of the current situation; thought content; and judgment. Formal examination provides clues to diagnosis and establishes a baseline for repeated testing to enable detection of improvement or deterioration.

A number of formal mental status examinations, although not specific for delirium, are available. These include the Mini-Mental State Examination (Folstein, Folstein, & McHugh, 1975) and the Cognitive Capacity Screening Examination (Jacobs et al., 1977). The major value of these structured instruments is that they can be repeated over time by the primary care physician or by nonphysician staff in order to document either improvement or deterioration of the patient's mental status. Reexamination of the patient at intervals often yields more diagnostic information than even the most detailed examination at one point in time. It should be noted, however, that these structured instruments often are insensitive to the milder forms of delirium. In such cases, mild delirium is better diagnosed by careful bedside testing of attention and concentration, as by having the patient write sentences to dictation or draw a clock, and by observing the performance on multistep activities done to verbal and visual commands.

In delirium, visual-spatial skills can be assessed by having the patient copy a geometric figure or by having the patient draw a clock face with hands set to a specific time, such as 10 minutes after 11:00 o'clock. Some common tests, such as proverb interpretation, naming of presidents, serial sevens, or spelling the word *world* backward, all of which are often part of routine mental status examination, are without proven value if given in isolation (Adams, 1984). A sufficiently extensive examination is necessary for diagnostic confidence.

It is notable that the stress of taking neuropsychologic tests itself has been reported sufficient to induce delirium (Rozzini, Zanetti, & Trabucchi, 1989). Reported success in distinguishing delirium from other psychiatric disorders by formal mental status examination varies by the examination used but tends to be greatest in those circumstances when scores obtained are combined with other data. Johnstone and her group (1988) using the Present State Examination, found few significant differences in scores obtained from patients with organic illness when compared to matched controls of manic, schizophrenic, and depressed patients. Trzepacz, Sclabassi, and Van Thiel (1989) found significantly worse scores on the Mini-Mental State Examination and Trail-Making Tests in delirious patients versus nondelirious controls. In this study the mean peak activity of the delirious patients was lower on the computerized spectral analysis of the EEG when compared to the nondelirious controls, who showed a bimodal distribution of latency values and a greater proportion of abnormal values of the brainstem auditory evoked potentials. Mean brainstem auditory evoked potentials were the same for both groups, but the somatosensory evoked potentials were abnormal for those with delirium but not for controls. They inferred that the basic pathophysiology of delirium in the patients was subcortical.

Clinical Laboratory Examinations

Laboratory tests are ordered with attention to specific suspected etiologic agents. In addition to a complete blood count (CBC) and differential, levels of electrolytes, blood urea nitrogen (BUN), CK, fasting blood glucose, arterial blood gases, and serum ammonia, erythrocyte sedimentation rate, syphilis serology, thyroid and liver function studies, urine and serum toxicology screens, and lumbar puncture may be indicated (Murray, 1987).

Lumbar puncture is not a first-line test in patients with acute changes in mental status. It should be performed only after thorough clinical evaluation and serious consideration of the value and hazards of performing such a procedure (Fishman, 1980). It is an emergency procedure in patients who have an acute mental status change in which fungal or bacterial meningitis is suspected, and it is a serious consideration in patients with fever of unknown origin. Lumbar puncture is occasionally used in patients with suspected acute subarachnoid hemorrhage, although the CT scan detects this problem in most cases. Cerebrospinal fluid examination also helps in the evaluation of encephalitides, neurosyphilis, and unexplained seizures (Fishman, 1980).

Electroencephalography, ECG, and CT scan with and without contrast are helpful in diagnosing the presence of delirium and its cause. If a CT scan without contrast is negative but there is concern regarding the presence of a mass lesion or new stroke that remains initially isodense on CT scan, it may be better visualized with the use of contrast.

It is not necessary to perform a CT scan on every patient with delirium, because toxic and metabolic disorders are the most common cause, and the diagnosis of such disorders may be readily confirmed by laboratory tests or history in most cases. In the presence of any physical sign of increased intracranial pressure, however, such as the presence of papilledema, signs of meningeal irritation, a history of recent head trauma or headache, focal neurologic findings, or coma, a CT scan is indicated early in the workup. Eventually, however, any prolonged, undiagnosed delirium deserves a CT scan or MRI.

Recent evidence suggests CT may reveal some specific finding that may help to discern a delirious patient from the majority of nondelirious patients. Significant differences among those with delirium include ventricular dilation and cortical atrophy. There is a statistically significant correlation between the width of the sylvian fissure and Mini-Mental State Examination scores (Koponen et al., 1989a). These findings corroborate the fact that individuals with Parkinson's disease, multiinfarct dementia, or Alzheimer's disease are at greater risk for delirium. Positron emission tomography scans provide additional etiologic information. Whereas CT picks up cortical and ventricular atrophy, PET scans detect hypometabolic regions. This metabolic dysfunction may be the first clue to the presence of dementias such as Alzheimer's disease (Fazekas et al., 1989; Hoffman et al., 1989; Johnson, 1990; Kertesz, Polkinia, & Carr, 1990; Pohl et al., 1988).

Koponen's group (1989c, 1989d) has investigated use of evaluation of cerebrospinal fluid beta-endorphin–like immunoreactivity in delirium and found it to be significantly lower in delirious patients than controls, and to have a significant positive correlation with cognitive ability as measured by the Mini-Mental State Examination.

Electroencephalography

In the differential diagnosis of delirium, special attention is accorded the EEG. Slowing of background activity with mixed spikes and sharp waves is seen with hyperactive delirious patients. Quiet patients usually show slow background activity. Periodic high-amplitude sharp waves are seen in herpes simplex encephalitis, triphasic waves in hepatic encephalopathy, and paroxysmal slow waves in dialysis encephalopathy. Computer-analyzed EEG may reveal changes correlating with the hallucinatory syndrome and withdrawal seen with acute termination of heavy alcohol use (Spehn & Stemmier, 1985), although the routine EEG may be read as normal.

Abnormal EEGs do not rule out coexistent psychiatric and medical illness; conversely, certain deliria, such as delirium tremens or alcohol withdrawal delirium, do not present with slowing on routine EEG (Allahyari, Deisenhammer, & Weiser, 1976; Lipowski, 1967; Weiner, 1983). Most patients with delirium, however, do have generalized diffuse background slowing, and the degree of slowing is correlated with the degree of cognitive impairment (Lipowski, 1967). Electroencephalography may not be as useful in the differential diagnosis of dementia from delirium because generalized slowing may occur in primary degenerative dementia.

Behavioral abnormalities parallel levels of CNS abnormality in delirium (Weiner, 1983). Slowing of the posterior alpha rhythm is found early in delirium. Subsequent changes, in order of progression, are generalized delta slowing, decrease in EEG level of reactivity, and, finally, loss of fast (alpha and beta) activity concomitant with diffuse, very slow (delta) activity. Moderate behavioral impairment is seen with fluctuating amounts of frontal, intermittent, rhythmic delta activity superimposed on a slow background. Low-voltage, irregular delta activity with suppression-burst activity occurs with coma (Weiner, 1983). Alpha waves (8 to 12 Hz) and faster beta waves (greater than 13 Hz) are predominant in EEGs of normal adults in the awake state. Prominent theta waves (4 to 7 Hz) and delta waves (less than 4 Hz) are abnormal in an awake patient. Specific abnormalities characterize certain disease states. Hepatic encephalopathy and renal and pulmonary failure may produce bilateral synchronous triphasic waves (Weiner, 1983). Occasional triphasic waves and frontal, intermittent, rhythmic delta activity are seen with moderate to severe hypercalcemia. Excess fast beta waves, especially in the frontocentral areas, are seen in states of barbiturate and benzodiazepine intoxication; a milder accentuation of beta activity occurs even at therapeutic levels of these drugs. Anoxic brain injury may be accompanied by suppression-burst activity. State-dependent EEG changes facilitate diagnosis of comatose patients where history is not readily available.

Quantitative bedside EEG of patients with delirium reveals significant reduction of the alpha percentage and increased delta and theta activity and a decrease in mean and peak frequencies (Koponen, et al., 1989b). Cognitive impairment in delirium and spectral EEG changes are positively correlated. These changes contrast with the absence of EEG changes with pseudodementia and pseudodelirium (Brenner, Reynolds, & Ulrich, 1989). Electroencephalographic findings consistent with delirium, as with dementia, do not exclude an affective component that requires antidepressant or psychosocial therapy. In some instances resolutions of the depression may resolve with remission of the delirium, confounding both diagnosis and the assessment impact of treatment (Borchardt & Popkin, 1987).

Toxic Screens

Toxic screens are often helpful in acute mental status changes, particularly in young patients in whom there is a higher probability of drug or alcohol abuse, and especially in the absence of previous psychiatric history. Serum levels of specific drugs, such as digoxin, phenytoin, theophylline, anticonvulsants, barbiturates, lithium, and other psychotropic medications, may be useful in determining their role in the production of a delirious state. There may be major variations in sensitivity and accuracy among laboratories, however. In the elderly, toxicity may take place within the therapeutic range of many drugs, and for patients on multiple drugs toxicities may be additive, although each drug alone may be in the therapeutic range. Multiple drugs with anticholinergic effects may produce postoperative delirium, which is associated with excessive cholinergic receptor binding by cholinergic antagonists in the plasma (Tune et al., 1981).

Management

Management of delirium is contingent on correction of underlying medical problems and on treating agitation by environmental manipulation and medication. Patients are placed in safe, structured settings with limited sensory input. Private rooms, adequate lighting, and removal of monitors or equipment to areas outside patients' rooms or to nursing stations serve this end. One-to-one nursing or the companionship of family members reduces patients' anxiety, reorients them to familiar persons, and prevents wandering away and accidents. Misinterpretations are corrected at the time they are voiced. Orientation is facilitated by well-placed calendars and wall clocks. Treatment plans are simple and repeatedly explained, allowing for memory limitations (Lazarus & Hagens, 1968; Massie, Holland & Glass, 1983). The basis of delirium should be explained to frightened family and friends of patients, who may fear that the patient is "going crazy" or will be chronically psychotic.

It has long been known that cholinergic blocking agents play a role in delirium (Lipowski, 1990). More recent evidence also suggests that activity of the cholinergic system is decreased in dementia (Antuono et al., 1979). This has specific implications in understanding the unique susceptibility of individuals with dementia to delirium and provides further support for the observation that the first sign of a primary degenerative dementia may be delirium. This fact further suggests the need to be particularly sensitive to minimizing use of neuroleptics with strong anticholinergic properties in the management of a delirious patient. Lorazepam (Ativan) may be as effective or more so to calm delirious patients. Monitoring of respiration and/or blood gases is necessary when benzodiazepines are given to patients especially prone to respiratory depression (Adams, 1988a).

There are a number of illnesses, among them AIDS and cancer, in which cognitive deterioration is seen as part of an inevitable progression to permanent brain damage. Krenz et al. (1988) found in a follow-up study of dementia that 8% of cases remitted partially and 3% fully. The most common causes of cognitive changes mislabeled as dementia were drugs (28.2%), metabolic (15.5%), and depression (26.2%). Given a bias toward less aggressive treatment of many illnesses, cancer and AIDS in particular, in the geriatric population (Maletta, 1990a), it is probable that many people labeled and treated as incurably dysfunctional could live more fulfilling lives if the diagnosis of delirium was more aggressively pursued and treated.

Cancer patients in intensive care units who present agitated and delirious, for example, should receive rapid tranquilization with safe drugs that have minimal side effects (Adams, 1988b). Phenothiazines are potent alpha-adrenergic blockers and can enhance confusion via hypotension. Lorazepam is safer unless there are respiratory problems, in which case it may enhance delirium if hypoxia or hypercapnia occur. Meperidine and, less frequently, morphine for pain can produce delirium; meperidine also can cause seizures (Adams, 1988a). Four-point restraints should be avoided if possible because they increase fear and, therefore, agitation. Use of clocks or a television set has not been demonstrated to have measurable therapeutic value in management of delirium (Adams, 1988a). A combination of a benzodiazepine, and a neuroleptic, given intravenously has been reported to be most effective and to minimize need for large doses of either. Use of a combination of haloperidol, lorazepam, and hydromorphone is recommended for management of the critically ill cancer patient who is agitated, delirious, and in pain (Adams, 1988b).

Of all the cognitive impairment disorders associated with AIDS, delirium is the most common and potentially reversible (Fernandez, Levy, & Mansell, 1989). Restlessness, irritability, insomnia, and difficulty thinking are harbingers of frank confusion. Treatment at the prodromal stage can obviate later problems. The most frequent causes are CNS infections such as cryptococcal meningitis, cerebral toxoplasmosis, and herpes meningoencephalitis. Neoplasia, progressive multifocal leukoencephalopathy, and cerebral vascular accidents also can result from decreased immunosurveillance. Neurotoxicities from antiviral agents ("Drugs for Viral Infections," 1990), antifungals, antibiotics, and other chemotherapeutic agents; hypoxemia; electrolyte imbalance; and hyper- or hypoglycemia may be life threatening and, if untreated, may result in permanent brain damage (Fernandez, Levy, & Mansell, 1989). Again, triple therapy (lorazepam, haloperidol, and hydromorphone) or haloperidol alone intravenously or intramuscularly or orally may be particularly helpful as an adjunct to diagnosis-specific treatment (e.g., restoration of electrolyte balance when disturbed). The principal goal is to ameliorate the disturbances of cognition, behavior, and affect that are associated with increased mortality and morbidity with AIDS. Recommended doses (Fernandez, Levy, & Mansell, 1989) include up to 10 mg each of haloperidol and lorazepam hourly if required for severe agitation. Total 24-hour doses of 240 mg each of lorazepam and haloperidol have been reported without major adverse side effects. Symptom control is usually attained within 48 hours and then the dosage is tapered gradually (Fernandez, Levy, & Mansell, 1989). However, many AIDS patients do not tolerate high neuroleptic doses. (See Chapter 32.)

The incidence of postoperative and post–myocardial infarction delirium has been decreased by preoperative psychiatric interviews conducted 2 to 3 days prior to surgery (Kornfeld et al., 1974; Lazarus & Hagens, 1968). Recommendations are then made to nursing staff and attending physicians based on the patients' concepts of their illness, the treatment and prognosis, the patients' personality style and past psychiatric history, and their current life situation and stressors (Lazarus & Hagens, 1968). Maintenance of adequate sleep is encouraged (Lazarus & Hagens, 1968; Parker & Hodge, 1967). Mobility is allowed within reasonable limits. Patients respond best to movement to a side room or general medical unit with family members in attendance for long periods of time to approximate a nearly normal environment (Parker & Hodge, 1967). In some instances diurnal variation in hospital lighting levels, simulating night and day, may be helpful.

Sensitivity to the confused individual as person may reduce the intensity of confusion. A sense of dignity is enhanced when it is explained to both the patient and family and friends that the bizzare behavior is not of any great psychologic meaning (Adams, 1988a). Assuring the patient that uncharacteristic behavior is due to a medical condition and that every effort will be made to reduce the confusion and protect him and others from harm will reduce fear and shame and, therefore, agitation (Adams, 1988a; Richeimer, 1987). Delirium, like dementia, is a stressful life event (Richeimer, 1987) for both patients and their significant others. Patient stress and stress observed among the patients' visitors increase anxiety. Anxiety begets fear, and fear begets agitation and confusion.

Psychopharmacotherapy

There is no established drug of choice for the management of agitation in delirious patients (Adams, 1984; Lipowski, 1990). Some prefer benzodiapezines; others, neuroleptics. Of the benzodiazepines, diazepam is usually avoided because of its long half-life, poor intramuscular (IM) absorption, and problems with tissue depot accumulation (Adams, 1984). Lorazepam given in 1- to 2-mg increments orally or intramuscularly every hour until low-level sedation is achieved is a better alternative, because it is reliably absorbed by the IM route and does not require hepatic metabolism. Lorazepam, however, may cause anterograde amnesia. Benzodiazepines as a group are relatively contraindicated in patients with respiratory compromise who are at risk for carbon dioxide retention (Barbee & McLaughlin, 1990). Benzodiazepines raise seizure threshold, unlike neuroleptics, and do not carry the risk of extrapyramidal side effects.

Of the neuroleptics, low-potency phenothiazines are considered least preferable because of their hypotensive and anticholinergic effects (Adams, 1984). Haloperidol and thiothixene are preferred because of ease of administration and relative lack of cardiac, pulmonary, and hemodynamic side effects. They are employed at low doses, 0.5 to 2 mg orally or IM every 30 to 60 minutes. When given intravenously, haloperidol is administered at 1 mg/min (Massie, Holland & Glass, 1983). No extrapyramidal, cardiac, pulmonary, or hypotensive side effects were noted in a series of over 1,000 patients with advanced malignancy (some of whom were on dopamine to sustain blood pressure) treated with intravenous haloperidol and lorazepam (Adams, 1984). In fact, in a study reported by Moulaert (1989) of surgical intensive care patients, heart rate, respiratory rate, and systolic-diastolic arterial blood pressure all returned to normal values when acute nonspecific delirium was managed with intravenous haloperidol. The average dose to calm a patient was 38 mg. Although much of the literature has focused on use of haloperidol, thiothixene has been reported to be equally effective and without any more significant side effects in the management of delirium (Peterson & Bongar, 1989) (See also Chapter 12).

One procedure for the rapid intravenous treatment of agitation entails initial administration of 3 mg of haloperidol and 0.5 to 1 mg of lorazepam, given in less than 1 minute each (Adams, 1984). If no response is seen in 30 minutes, the dose is increased to 5 mg of haloperidol and 0.5 to 2 mg of lorazepam. If in another 30 minutes there is no response, another 10 mg of haloperidol and 0.5 to 10 mg of lorazepam are given until sedation is obtained. No further medication is given until restlessness recurs. If symptoms reemerge in 4 hours, the patient is provided a dose every 4 hours for the next 12 to 18 hours, after which time lorazepam is stopped and haloperidol dosing intervals are lengthened.

In a review of over 2,000 cases of combined intravenous administration of haloperidol and lorazepam, Adams (1988b) reported successful sedation with lowered required dosages of the neuroleptic in treatment of delirium resulting from cancer and multisystem failure. In one instance, control was obtained with combined therapy after failure with 350 mg of haloperidol alone. Hourly doses of 10 mg of both medications have been given for as long as 15 days without problems. In most instances one or two doses are all that is required. Rarely, greater than 100 mg/day is given. If intractable pain is present, intravenous hydromorphone may be given adjunctively, as mentioned earlier. In a prospective study of delirious medically ill patients given intravenous haloperidol and lorazepam together (Menza et al., 1988), those receiving combination therapy had significantly fewer extrapyramidal side effects without any adverse cardiac or respiratory problems.

In treating the agitated, delirious patient, the psychiatrist should always be suspicious of the possibility of NMS in those patients who deteriorate when treated with neuroleptics. The incidence of NMS has been estimated to be 0.5% to 1% of those patients who receive neuroleptics. It carries a high mortality rate—14% in patients receiving oral neuroleptics and 38% in patients receiving intramuscular neuroleptics (Mueller, 1985). (Treated mortality probably is lower.) It is more common with depot and high-potency neuroleptics and can occur with drug combinations, especially lithium carbonate and haloperidol (Smego & Durack, 1982). The most typical laboratory abnormalities are an elevated serum CK level, an elevated white blood cell count, and an elevated urine myoglobin level. Creatine kinase levels appear to correlate with intensity of the NMS; therefore, it has been suggested that serial CK measurements be used to follow the course of the syndrome (Harpe & Stoudemire, 1987). Some authors suggest that there are patients who appear to have an atypical form of NMS, without the classic signs of fever or muscular rigidity; others suggest that NMS is actually a misnomer and probably represents a spectrum that includes: (1) patients with medical problems that may cause fever, accompanied by severe extrapyramidal symptoms; (2) patients with medical problems unlikely to cause fever, with severe extrapyramidal symptoms; and (3) patients who present with the neuroleptic-induced syndrome in the absence of any other medical disorder (Levinson & Simpson, 1986). In other cases, authors report incomplete episodes that may be a *forme fruste* of a NMS, heralding the possibilities of future episodes (Levinson & Simpson, 1986). This distinction is important because NMS must be differentiated from catatonic functional disorders, as well as other infections, vascular, neoplastic, or toxic-metabolic disease (Smego & Durack, 1982).

Neuroleptic malignant syndrome must be differentiated from viral encephalitis, tetanus, bacterial and fungal meningitis, heat stroke (secondary to neuroleptics), tetany resulting from hypocalcemia, hyperthyroidism, severe forms of Parkinson's disease, malignant hyperthermia, and allergic drug reactions (Caroff, 1980; Levinson, 1985). Fever in the presence of a neuroleptic should not be automatically ascribed to the use of the neuroleptic; concomitant medical illness should be sought. Any extrapyramidal disorder should be treated promptly, as should hyperthermia and dehydration. Patients who develop severe extrapyramidal symptoms with immobility, rigidity, and impaired cognition, whether fever is present or not, should be treated vigorously for their symptoms. It has been recommended that, in mild cases, the neuroleptic dosage be reduced and regular dosages of anticholinergic drugs or amantadine be instituted, or increased if the patient is already on them. If the patient appears to have akinesia, inability to speak or eat, or respiratory problems, and if these symptoms do not respond immediately to the administration of antiparkinsonian agents, then neuroleptics should be discontinued and more vigorous therapies instituted, including dantrolene sodium or bromocriptine or both. These drugs should be given for at least 8 to 12 days following clinical improvement because cessation of the drug may cause a relapse (Levinson & Simpson, 1986). With oral neuroleptic cessation alone, NMS may last 5 to 10 days after cessation of therapy; with depot neuroleptics, NMS may persist for as long as a month (Caroff, 1980). Sewell & Jeste (1991) reviewed the diagnosis and treatment of NMS in detail.

Forensic Issues

There are far fewer legal issues involved in care of individuals with delirium than in those with dementia, in which problems regarding testamentary capacity, right to treatment, right to refuse treatment, commitment, and guardianship regularly arise. Delirium usually is a medical emergency, and the patient therefore can usually be treated without informed consent, under the common law doctrine of implied consent (Fogel, Mills, & Landen, 1986). This doctrine states that, in a true medical emergency, a temporarily incompetent person can be treated "as a reasonable person would choose to be treated." In urgent situations, the doctrine of implied consent often can be extended if measures are taken to safeguard patients' interests, including involving family members in treatment decisions, obtaining second opinions, and seeking administrative consultation. A temporary guardian is necessary if time permits and if the medical risks and alternatives are nontrivial. In any case, careful documentation is required. Ethical and existential questions arising regarding quality of life and concern for limitation of supportive and resuscitative efforts in the late stages of dementia are not of concern in instances of delirium, unless the patient is suffering from a terminal medical illness. (See also Chapter 45.)

REFERENCES

ADAMS F (1984). Neuropsychiatric evaluation and treatment of delirium in the critically ill cancer patient. Cancer Bull 36:156–160.

ADAMS F (1988a). Emergency intravenous sedation of the delirious, mentally ill patient. J Clin Psychiatry 49:22–26.

ADAMS F (1988b). Neuropsychiatric evaluation and treatment of delirium in cancer patients. Adv Psychosom Med 18:26–36.

ADAMS RD, FISHER CM, HAKIM S, ET AL. (1965). Symptomatic occult hydrocephalus with "normal" cerebrospinal fluid pressure: A treatable syndrome. N Engl J Med 273:117–126.

Agency for Health Care Policy & Research (AHCPR) (1992). Clinical practice guideline: Urinary incontinence in adults. Rockville, MD: U.S. Dept. of Health & Human Services.

ALEXOPOULOS GS, ABRAMS RC, YOUNG RC, ET AL. (1988). Cornell scale for depression in dementia. Biol Psychiatry 23:271–284.

ALLAHYARI H, DEISENHAMMER E, & WEISER G (1976). EEG examination during delirium tremens. Psychiatr Clin 9:21–31.

AMBROSE J (1973). Computerized transverse axial scanning (tomography): 2. Clinical application. Br J Radiol 46:1023–1047.

AMCHIN J, & POLAN HJ (1986). A longitudinal account of staff adaption to AIDS patients on a psychiatric unit. Hosp Community Psychiatry 37:1235–1238.

ANDERSON B (1988). MRI and dementia. Neurology 38:166–167.

ANTUONO P, AMADUCCI L, PAZZAGLI A, ET AL. (1979). Psychopharmacological prospectives in the treatment of dementia. Prog Neuropsychopharmacol 3:75–80.

ARIE T (1986). Management of dementia: A review. Br Med Bull 42:91–96.

ARRIGO A, CASALE R, GIORGI I, ET AL. (1989). Effects of intravenous high dose co-dergocrine mesylate ('Hydergine') in elderly patients with severe multi-infarct dementia: A double-blind, placebo controlled trial. Curr Med Res Opin 11:491–500.

ASHBY D, WEST CR, & AMES D (1989). The ordered logistic regression model in psychiatry: Rising prevalence of dementia in old people's homes. Stat Med 8:1317–1326.

ASKE D (1990). The correlation between Mini-Mental Status Examination Scores and Katz ADL Status among dementia patients. Rehabil Nurs 15:140–142, 146.

BARBEE JG, & MCLAUGHLIN JB (1990). Anxiety disorders: Diagnosis and pharmacotherapy in the elderly. Psychiatr Ann 20:439–445.

BARNES RE, VEITH R, OKIMOTO J, ET AL. (1982). Efficacy of antipsychotic medications in behaviorally disturbed demented patients. Am J Psychiatry 139:1170–1174.

BARRY PP, & MOSKOWITZ MA (1988). The diagnosis of reversible dementia in the elderly: A critical review. Arch Intern Med 148:1914–1918.

BATT LJ (1989). Managing delirium. Implications for geropsychiatric nurses. J Psychosoc Nurs Ment Health Serv 27:22–25.

BATTAGLIA A, BRUNI G, ARDIA A, ET AL. (1989). Nicergoline in mild to moderate dementia. A multicenter, double-blind, placebo-controlled study. J Am Geriatr Soc 37:295–302.

BAUMGARTEN M (1989). The health of persons giving care to the demented elderly: A critical review of the literature. J Clin Epidemiol 42:1137–1148.

BAUMGARTEN M, BECKER R, & GAUTHIER S (1990). Validity and reliability of the dementia behavior disturbance scale. J Am Geriatr Soc 38:221–226.

BEN-ARIE O, SWARTZ L, & DICKMAN BJ (1987). Depression in the elderly living in the community—its presentation and features. Br J Psychiatry 150:169–174.

BEREZIN MA (1972). Psychodynamic considerations of aging and the aged: An overview. Am J Psychiatry 128:33–41.

BERNSTEIN JG (1973). Antipsychotic drugs in the general hospital: Uses and cautions. Psychosomatics 20(5):335–347.

BERRIOS G (1989). Non-cognitive symptoms and the diagnosis of dementia—historical and clinical aspects. Br J Psychiatry 154:11–16.

BILLINGS RF, & STIEN MD (1986). Depression associated with ranitidine. Am J Psychiatry 143:915–916.

BIRREN JE, & BERNSTEIN L (1979). Health and aging in our society: Perspectives on mortality and the emergence of geriatrics. Trans Assoc Life Insur Med Dir Am 62:135–153.

BLACK DW, WANACK G, & WINOKUR G (1985). The Iowa record-linkage study II: Excess mortality among patients with organic mental disorders. Arch Gen Psychiatry 42:78–81.

BLACK PM (1980). Idiopathic normal pressure hydrocephalus: Results of shunting in 62 patients. J Neurol Neurosurg Psychiatry 52:371–377.

BLACK PM (1982). Idiopathic normal pressure hydrocephalus: Current understanding of diagnostic tests and shunting. Postgrad Med 7:57–67.

BLASS JP, & PLUM KC (1983). Metabolic encephalopathies in older adults. In R KATZMAN & R TERRY (Eds.), The neurology of aging (Ch. 9, pp. 189–220). Philadelphia: FA Davis.

BLAZER DG, FREDERSPIEL CF, RAY WA, ET AL. (1983). The risk of anticholinergic toxicity in the elderly: A study of prescribing practices in two populations. J Gerontol 38:31–35.

BOND TC (1980). Recognition of acute delirious mania. Arch Gen Psychiatry 37:553–554.

BORCHARDT CM, & POPKIN MK (1987). Delirium and the resolution of depression. J Clin Psychiatry 48:373–375.

BORUP JH, GALLEGO D, & HEFFERMAN P (1979). Relocation and its effect on mortality. Gerontologist 19:135–140.

BORUP JH, GALLEGO D, & HEFFERMAN P (1980). Relocation and its effect on health functionings and mortality. Gerontologist 20:468–479.

BOSZORMENYI-NAGY I, & KRASNER B (1986). Between give and take: A clinical guide to contextual therapy. New York: Brunner/Mazel.

BRENNER R, REYNOLDS C, & ULRICH R (1989). EEG findings in depressive pseudomentia and dementia with secondary depression. Electroencephalogr Clin Neurophysiol 72:298–304.

BRINKMAN SD, & LARGEN JW (1984). Changes in brain ventricular size with repeated CT scans in suspected Alzheimer's disease. Am J Psychiatry 141:81–83.

BRODATY H, & GRESHAM M (1989). Effect of a training program to reduce stress in carers of patients with dementia. Br Med J 299:1375–1379.

BROWN RG, & MARSDEN CD (1988). Subcortical dementia: The neuropsychological evidence. Neuroscience 25:363–387.

BRUST JC (1988). Vascular dementia is overdiagnosed. Arch Neurol 45:799–801.

BUCHNER DM, & LARSON EB (1987). Falls and fractures in patients with Alzheimer-type dementia. JAMA 257:1492–1495.

BURNS A, FOLSTEIN S, BRANDT J, ET AL. (1989). Clinical assessment of irritability, aggression, and apathy in Huntington and Alzheimer disease. J Nerv Ment Dis 178:20–26.

BURNS M, MOSKOWITZ H, & JAFFE J (1986). A comparison of the effects of trazadone and amitriptyline on skills performance by geriatric subjects. J Clin Psychiatry 47:252–254.

BUSSE EW, & OBRIST WD (1963). Significance of focal electroencaphalographic changes in the elderly. Postgrad Med 34:179–182.

BUTLER RN (1963). The life review: An interpretation of reminiscence in the aged. Psychiatry 26:65–76.

BUYSSE DJ, REYNOLDS CF, KUPFER DJ, ET AL. (1988). Electroencephalographic sleep in depressive pseudodementia. Arch Gen Psychiatry 45:568–575.

CAINE ED (1981). Pseudodementia: Current concepts and future directions. Arch Gen Psychiatry 38:1359–1364.

CALABRESE J (1987). Incidence of postoperative delirium follow myocardiotomy, and revascularization. Cleve Clin J Med 54:29–32.

CAROFF SN (1980). The neuroleptic malignant syndrome. J Clin Psychiatry 41:79–83.

CASEY DA, & FITZGERALD BA (1988). Mania and pseudodementia. J Clin Psychiatry 49:73–74.

CASSEM NH, & HACKETT TP (1987). The setting of intensive care. In NH Cassem & TP Hackett (Eds.), Massachusetts General Hospital handbook of general psychiatry (Ch. 18, pp. 353–379). Littleton, MA: PSG Publishing.

CAVENAR, JO, MALTBIE AA, & AUSTIN L (1979): Depression simulating organic brain disease. Am J Psychiatry 136:895–900.

CHAPIN JK, & WOODWARD DJ (1989). Ethanol withdrawal increases sensory responsiveness of single somatosensory cortical neurons in the awake, behaving rat. Alcoholism 13:8–14.

CHARATAN FB (1985). Depression and the elderly: Diagnosis and treatment. Psychiatr Ann 5(5):313–316.

CHRISTENSEN K (1989). A new approach to the measurement of cognitive deficits in dementia. Clin Geriatr Med 5:519–530.

CLARFIELD A (1988). The reversible dementias: Do they reverse? Ann Intern Med 99:476–486.

CLARFIELD AM (1989). Diagnosing and treating dementia [letter]. Br Med J 298:600.

COHEN GD (1981). Perspectives on psychotherapy with the elderly. Am J Psychiatry 138:347–350.

CROOK T, BARTUS R, FERRIS S, ET AL. (1986). Treatment development strategies for Alzheimer's disease. Madison, CT: Mark Powley Associates.

CRUMP GL, PELLEGRINI AJ, LIPPMANN S, ET AL. (1984). Diagnosing delirium in acute mental disturbance. J Ky Med Assoc 22:168–169.

CULLITON BJ (1989). AZT reverses AIDS dementia in children [news]. Science 246:21–23.

CUMMINGS JL (1987). Dementia syndromes: Neurobehavioral and neuropsychiatric features. J Clin Psychiatry 48:3–8.

CUMMINGS JL (1988). Organic psychosis. Psychosomatics 29:16–26.

CUMMINGS JL (1989). Dementia and depression. J Neuropsychiatry 1:236–242.

CUMMINGS JL, & BENSON DF (1983). Dementia: A clinical approach. Boston: Butterworth.

CUTLER NR, DUARA R, CREASY H, ET AL. (1984). Brain imaging: Aging and dementia. Ann Intern Med 101:355–369.

DANIEL DG, & RABIN PL (1985). Disguises of delirium. South Med J 78:666–671.

DAVIS P, MORRIS J, & GRANT E (1990). Brief screening tests versus clinical staging in senile dementia of the Alzheimer type. J Am Geriatric Soc 38:129–135.

DAVIS KL, THAL LJ, GAMZEE ER, ET AL. (1992). A double-blind placebo-controlled multicenter study of tacrine for Alzheimer's disease. NEJM 327:1253–1259.

DELGADO-ESCUETA AV, TREIMARA DM, & WALSH GO (1983). The treatable epilepsies (2nd of 2 parts). N Engl J Med 308:1576–1579.

DICKSON L, & RANSEEN J (1990). An update on selected organic mental syndromes. Hosp Community Psychiatry 41:290–300.

Drugs for viral infections. (1990). Med Lett 32:73–78.

Drugs that cause psychiatric symptoms. (1981). Med Lett Drugs Ther 23(3):9–12.

DUBIN WR, WEISS KJ, & ZECCARDI JA (1983). Organic brain syndrome—the psychiatric imposter. JAMA 249:60–62.

DYSKEN M (1987). A review of recent clinical trials in the treatment of Alzheimer's dementia. Psychiatr Ann 17:178–191.

ELLISON JM (1984). DSM-III and the diagnosis of organic mental disorders. Ann Emerg Med 13:521–528.

EMERY OB (1988). The deficit of thought in senile dementia Alzheimer's type. Psychiatr J Univ Ottawa 13:3–8.

ENGEL GL, & ROMANO J (1959). Delirium: A syndrome of cerebral insufficiency. J Chronic Dis 9:260–277.

ERKINJUNTTI T, LARSEN T, SULKAVA R, ET AL. (1988). EEG in the differential diagnosis between Alzheimer's disease and vascular dementia. Acta Neurol Scand 77:36–43.

ERKINJUNTTI T, WIKSTROM J, PALO J, ET AL. (1986). Dementia among medical inpatients. Evaluations of 2000 consecutive admissions. Arch Intern Med 146:1923–1926.

FARLOW M, GRACON SI, HERSHEY LA, ET AL. (1992). A controlled trial of tacrine in Alzheimer's disease. JAMA 268:2523–2529.

FAZEKAS F, ALAVI A, CHAWLUK JB, ET AL. (1989). Comparison of CT, MR, and PET in Alzheimer's dementia and normal aging. J Nucl Med 30:1607–1615.

FERNANDEZ F, LEVY J, & MANSELL P (1989). Management of delirium in terminally ill patients. Int J Psychiatry Med 19:165–172.

FIORE J, BECKER J, & COPPEL DB (1983). Social network interactions: A buffer or a stressor. Am J Community Psychol 11:423–439.

FISCHER P, GATTERER G, MARTERER A, ET AL. (1990). Course characteristics in the differentiation of dementia of the Alzheimer type and multi-infarct dementia. Acta Psychiatr Scand 81:551–553.

FISHMAN RA (1980). Clinical examination of cerebrospinal fluid. In Cerebrospinal fluid in diseases of the nervous system (Ch. 5, pp 141–167). Philadelphia: WB Saunders.

FOGEL BS, & STOUDEMIRE A (1993). New psychotropics in the medically ill. In A Stoudemire & BS Fogel (Eds.), Medical Psychiatric Practice (Vol. 2; Ch. 3). Washington, DC: American Psychiatric Press, Inc.

FOGEL BS (1991a). Beyond neuroleptics: The treatment of agitation. In E Light & BD Lebowitz (Eds.), The elderly with chronic mental illness. Chap. 10, pp. 167–190.

FOGEL B (1991b). Organic mental disorders. In L Sederer (Ed.), Inpatient psychiatry (Chap. 8, pp. 121–252). Baltimore: Williams & Wilkins.

FOGEL BS, MILLS MJ, & LANDEN JE (1986). Legal aspects of the treatment of delirium. Hosp Community Psychiatry 37:154–158.

FOLEY KM (1982). The practical use of narcotic analgesics. Med Clin North Am 66(5):1091–1104.

FOLEY KM (1985). The treatment of cancer pain. N Engl J Med 313:84–95.

FOLSTEIN MF, FOLSTEIN SE, & MCHUGH PR (1975). Mini-mental state: A practical method for grading the cognitive state of patients for the clinician. J Psychiatr Res 12:189–198.

Food and Drug Administration (1991). Panel rejects approval of tacrine. P&T 16(5):397.

FREEMAN FR, & RUDD SM (1982). Clinical features that predict potentially reversible progressive intellectual deterioration. J Am Geriatr Soc 30:449–451.

FRIEDLAND RP, BUDINGER TR, BRAT-ZAWADSKI M, ET AL. (1984). The diagnosis of Alzheimer's-type dementia: A preliminary comparison of positron emission tomography and proton magnetic resonance. JAMA 252:2750–2752.

FULLER J, WARD E, EVANS A, ET AL. (1979). Dementia: Supportive groups for relatives. Br Med J 1:1684–1685.

GADO MH, & PRESS GA (1986). Computed tomography in the diagnosis of dementia. Geriatr Med Today 5(7):47–73.

GALASKO D, KLAUBER MR, HOFSTETTER CR, ET AL. (1990). Mini-Mental Status Examination in the early diagnosis of Alzheimer's disease. Arch Neurol 47:49–52.

GALLANT DM (1985). Differential diagnosis of confusion in the geriatric patient. Geriatr Med Today 4(4):72–81.

GAUTHIER S (1990). THA-lecithin combination treatment in patients with intermediate stage Alzheimer's disease. N Engl J Med 322:1272–1276.

GERSHON ES, GOLDSTEIN RE, MOSS AJ, ET AL. (1979). Psychosis with ordinary doses of propranolol. Ann Intern Med 90:938–939.

GILLEARD CJ, BELFORD H, GILLEARD E, ET AL. (1984). Emotional distress amongst the supporters of the elderly mentally ill. Br J Psychiatry 145:172–177.

GOFF DC, GARBER HJ, & JENIKE MA (1985). Partial resolution of ranitidine-associated delirium with physostigmine: Case report. J Clin Psychiatry 46:400–401.

GOGGANS FC, WEISBERG LJ, & KORAN LM (1983). Lithium prophylaxis of prednisone psychosis: A case report. J Clin Psychiatry 44:111–112.

GOLDBERG RJ, & CULLEN LO (1986). Depression in geriatric cancer patients: Guide to assessment and treatment. Hospice J 2(2):79–98.

GOLDFRANK L, FLOMENBAUM N, LEWIN N, ET AL. (1982). Withdrawal? Hosp Physician 18:12–34.

GOLDFRANK L, LEWIN N, MELINEK M, ET AL. (1981). Caffeine. Hosp Physician 17:42–59.

GOLINGER R (1986). Delirium in surgical patients seen at psychiatric consultation. Surg Gynecol Obstet 163:104–106.

GOLINGER R, PEET T, & TUNE L (1987). Association of elevated plasma anticholinergic activity with delirium in surgical patients. Am J Psychiatry 144:1218–1220.

GOODNICK PJ (1985). Pseudodementia. Geriatr Med Today 4(10):31–40.

GORDON M, & FREEDMAN M (1990). Evaluating dementia: What price testing? Can Med Assoc J 142:1367–1369.

GRANT I, & ATKINSON J (1990). Neurogenic and psychogenic behavior correlates of HIV infection. Immunol Mech Neurol Psychiatr Dis 00:291–304.

GREENWALD BS, KRAMER-GINSBERG E, MARIN DB, ET AL. (1989). Dementia with coexistent major depression. Am J Psychiatry 146:1472–1478.

HACHINSKI VC, LASSEN NA, & MARSHALL J (1974). Multi-infarct dementia, a cause of mental deterioration in the elderly. Lancet 2:207–210.

HALEY WE, & PARDO KM (1989). Relationship of severity of dementia in caregiving stressors. Psychol Aging 4:389–392.

HAMEL M, GOLD DP, ANDRES D, ET AL. (1990). Predictors and consequences of aggressive behavior by community-based dementia patients. Gerontologist 30:206–211.

HAMMERSTROM DC, & ZIMMER B (1985). The role of lumbar puncture in the evaluation of dementia: The University of Pittsburgh study. J Am Geriatr Soc 33:397–400.

HARPE C, & STOUDEMIRE A (1987). Aetiology and the treatment of neuroleptic malignant syndrome. Med Toxicol 2:166–176.

HARRIS KA, & RABINS PV (1989). Dementia: Helping family caregivers cope. J Psychosoc Nurs Ment Health Serv 27:7–12.

HENDERSON AS (1986). The epidemiology of Alzheimer's disease. Br Med Bull 42(1):3–10.

HENDERSON AS (1990). Epidemiology of dementia disorders. Adv Neurol 51:15–25.

HERMAN M (1938). The use of intravenous sodium amytal in psychogenic amnesia states. Psychiatr Q 12:738–742.

HERST L, & MOULTON P (1985). Psychiatry in the nursing home. Psychiatr Clin North Am 8(3):551–561.

HESTON LL, WHITE JA, & MASTRI AR (1987). Pick's disease. Arch Gen Psychiatry 44:409–411.

HOFFMAN JM, GUZE BH, BAXTER LR, ET AL. (1989). [18 F]-flurodeoxy glucose (FDG) and positron emission tomography (PET) in aging and dementia. A decade of studies. Eur Neurol 29(suppl 3):16–24.

HOMER AC, HONAVAR M, & LANTOS PL (1988). Diagnosing dementia: Do we get it right? Br Med J 297:894–896.

HOROWITZ GR, & CHARCOT JM (1988). What is a complete workup for dementia? Clin Geriatr Med 4:163–180.

HUGHES JR, SHANMUGHAM S, WETZEL LC, ET AL. (1989). The relationship between EEG changes and cognitive functions in dementia: A study in a VA population. Clin Electroencephalogr 20:77–85.

HUPPERT FA, & TYM E (1986). Clinical and neuropsychological assessment of dementia. Br M Bull 42(1):11–18.

IBE O, & KITCHEN AD (1983). Differentiation of delirium from dementia [Letter to the editor]. JAMA 250:1393–1394.

IKEDA T, FURUKAWA Y, MASHIMOTO S, ET AL. (1990). Vitamin B_{12} levels in serum and cerebral spinal fluid of people with Alzheimer's disease. Acta Psychiatr Scand 82:327–329.

IVERSEN LL (1988). Test models and new directions in dementia research. Psychopharmacol Ser 5:196–203.

JACOBS JW, BERNHARD MR, DELGADO A, ET AL. (1977). Screening for organic mental syndromes in the medically ill. Ann Intern Med 86:40–46.

JENIKE MA (1985). Handbook of geriatric psychopharmacology. Littleton, MA: PSG Publishing.

JENIKE MA, ALBERT MS, HELLER H, ET AL. (1986). Combination therapy with lecithin and ergoloid mesylates for Alzheimer's disease. J Clin Psychiatry 47:249–251.

JOHNSON J (1985). Nutrition as a factor of mortality in senile dementia of the Alzheimer's type. Psychiatr Ann 15(5):323–330.

JOHNSON JC (1990). Delirium in the elderly. Emerg Med Clin North Am 8:255–265.

JOHNSTONE EC, COOLING NJ, FRITH CD, ET AL. (1988). Phenomenology of organic and functional psychoses and the overlap between them. Br J Psychiatry 153:770–776.

KALES A, & KALES JS (1974). Sleep disorders: Recent findings in the diagnosis and treatment of disturbed sleep. N Engl J Med 280:487–499.

KALLMAN H, & MAY H (1989). Mental status assessment in the elderly. Prim Care 16:329–347.

KAUFMAN DM (1981). Clinical neurology for psychiatrists. Orlando, FL: Grune & Stratton.

KERTESZ A, POLKINIA M, & CARR T (1990). Cognition and white matter changes on magnetic resonance imaging in dementia. Arch Neurol 47:387–391.

KESSLAK JP, NALCIOGLU O, & COTMAN CW (1991). Quantification of magnetic resonance scans for hippocampal and parahippocampal atrophy in Alzheimer's disease. Neurology 41:51–54.

KILOH LG (1961). Pseudodementia. Acta Psychiatr Scand 37:336–361.

KNIGHTS EB, & FOLSTEIN MF (1977). Unsuspected emotional and cognitive disturbance in medical patients. Ann Intern Med 87:723–724.

KOKMEN E (1989). Etiology, diagnosis, and management of dementia. Compr Ther 15(9):59–69.

KOPELMAN MD, & LISHMAN WA (1986). Pharmacological treatments of dementia (noncholinergic). Br Med Bull 42(1):101–105.

KOPONEN H, HURRI L, STENBACK U, ET AL. (1989a). Computed tomography findings in delirium. J Nerv Ment Dis 17:226–230.

Koponen H, Partanen J, Paakkonen A, et al. (1989b). EEG spectral analysis in delirium. J Neurol Neurosurg Psychiatry 52:980–985.

Koponen H, Stenback U, Mattila E, et al. (1989c). CSF beta-endorphin-like immunoreactivity in delirium. Biol Psychiatry 25:938–944.

Koponen H, Stenback U, Mattila E, et al. (1989d). Delirium among elderly persons admitted to a psychiatric hospital: Clinical course during the acute stage and one-year follow-up. Acta Psychiatr Scand 79:579–585.

Kornfeld DS, Heller SS, Frank KA, et al. (1974). Personality and psychological factors in postcardiotomy delirium. Arch Gen Psychiatry 31:249–253.

Krebs-Roubicek EM (1989). Group therapy with demented elderly. Prog Clin Biol Res 317:1261–1272.

Krenz C, Larson EB, Buchner DM, et al. (1988). Characterizing patient dysfunction in Alzheimer's-type dementia. Med Care 26:453–461.

Kretschmar C, Kretschmar JH, & Stuhlmann W (1989). Diagnostic instruments differentiating dementia-depression. Prog Clin Biol Res 317:71–78.

Kroetsch P, & Shamoian CA (1986). Psychotherapy for the elderly. Med Aspects Hum Sex 20(3):62–65.

Kruis R, & Barger R (1980). Paranoid psychosis with sulindac [Letter to the Editor]. JAMA 243:1420.

Lambert C, & Rees WL (1944). Intravenous barbiturates in the treatment of hysteria. Br Med J 2:70–73.

Larson EB, Reifler BV, Featherstone HJ, et al. (1984). Dementia in elderly outpatients: A prospective study. Ann Intern Med 100:417–423.

La Rue A, D'Elia LF, Clark EO, et al. (1986). Clinical tests of memory in dementia, depression, and healthy aging. J Psychol Aging 1:69–77.

Lazarus HR, & Hagens JH (1968). Prevention of psychosis following open heart surgery. Am J Psychiatry 124:76–81.

Lazarus LW, Newton N, Cohler B, et al. (1987). Frequency and presentation of depressive symptoms in patients with primary degenerative dementia. Am J Psychiatry 144:41–45.

Lesse S, Hoefer PFA, & Austin JH (1958). The electroencephalogram in diffuse encephalopathies. Arch Neurol Psychiatry 79:359–375.

Letton P (1988). CT scan policy in assessment of dementia. Lancet 1:991–992.

Levine NB Gendron CE, Dastoor DP, et al. (1984). Existential issues in the management of the demented elderly patient. Am J Psychiatry 38:215–223.

Levinson DF, & Simpson GM (1986). Neuroleptic-induced extrapyramidal symptoms with fever: Heterogeneity of the "neuroleptic malignant syndrome." Arch Gen Psychiatry 43:839–848.

Levinson JL (1985). Neuroleptic malignant syndrome. Am J Psychiatry 142:1137–1145.

Lewis Mi, & Butler RN (1974). Life review therapy. Geriatrics 29:165–173.

Ling MHM, Perry PJ, & Tsuang MT (1981). Side-effects of corticosteroid therapy: Psychiatric aspects. Arch Gen Psychiatry 38:471–477.

Lipowski ZJ (1967). Delirium, clouding of consciousness and confusion. J Nerv Ment Dis 145:227–255.

Lipowski ZJ (1983). Transient cognitive disorders (delirium, acute confusional states) in the elderly. Am J Psychiatry 140:1426–1436.

Lipowski ZJ (1987). Delirium (acute confusional states). JAMA 258:1789–1792.

Lipowski ZJ (1989). Delirium in the elderly patient. N Engl J Med 320:578–582.

Lipowski ZJ (1990). Delirium: Acute confusional states. New York: Oxford University Press.

Mahendra B (1984). Dementia. A survey of the syndrome of dementia. Lancaster, England: MTP Press.

Maletta GJ (1985). Medication to modify at-home behavior of Alzheimer's patients. Geriatrics 40(12):31–42.

Maletta G (1990a). The concept of "reversible" dementia. J Am Geriatric Soc 38:136–140.

Maletta GJ (1990b). Pharmacologic treatment and management of the aggressive demented patient. Psychiatr Ann 20:446–455.

Martin RA, & Guthrie R (1988). Office evaluation of dementia. How to arrive at a clear diagnosis and choose appropriate therapy. Postgrad Med 84:176–180, 183–187.

Massie M, Holland J, & Glass E (1983). Delirium in terminally ill cancer patients. Am J Psychiatry 140:1048–1050.

McAllister TW, Ferrell RB, Price TRP, et al. (1982). The dexamethasone suppression test in two patients with severe pseudodementia. Am J Psychiatry 139:479–481.

McAllister TW, & Price TRP (1985). Severe depressive pseudodementia with and without dementia. Am J Psychiatry 139:626–629.

McCarthy DM (1983). Ranitidine or cimetidine. Ann Intern Med 99:551–553.

McCartney CF, Hatley LH, & Kessler JM (1982). Possible tobramycin delirium. JAMA 247:1319.

McCartney JR, & Palmateer LM (1985). Assessment of cognitive deficit in geriatric patients: A study of physician behavior. J Am Geriatr Soc 33:467–471.

McEvoy JP (1981). Organic brain syndromes. Ann Intern Med 95:212–220.

McEvoy JP, McCue M, Spring B, et al. (1987). Effects of amantadine and trihexyphenidyl on memory in elderly normal volunteers. Am J Psychiatry 144:573–577.

McGeer PL (1986). Brain imaging in Alzheimer's disease. Br Med Bull 42(1):24–28.

McQuarrie IG, Saint-Louis L, & Scherer PB (1984). Treatment of normal pressure hydrocephalus with low versus medium pressure cerebrospinal fluid shunts. Neurosurgery 15:484–488.

Menza MA, Murray GB, Holmes VF, et al. (1988). Controlled study of extrapyramidal reactions in the intravenous haloperidol versus intravenous haloperidol plus benzodiazepines. Heart Lung 17:238–241.

Meyer JS, Judd BW, Tawakha T, et al. (1986). Improved cognition after control of risk factors for multi-infarct dementia. JAMA 256:2203–2209.

Miller E (1984). Psychological aspects of dementia. In JHS Pearce (Ed.), Dementia: A clinical approach. Boston, MA: Blackwell Scientific Publications. Chap. 10, pp. 1135–1531.

Mohide EA, Pringle DM, Streiner DL, et al. (1990). A randomized trial of family caregiver support in the home management of dementia. J Am Geriatr Soc 38:446–454.

Mohs RC, Breitner JCS, Silverman JM, et al. (1987). Alzheimer's disease. Arch Gen Psychiatry 44:405–408.

Mohs RC, Davis BM, Johns CA, et al. (1985). Oral physostigmine treatment of patients with Alzheimer's disease. Am J Psychiatry 142:28–33.

Moulaert P (1989). Treatment of acute nonspecific delirium with i.v. haloperidol in surgical intensive care patients. Acta Anaesthesiol Belg 40:183–186.

Mowry BJ, & Burvill PW (1988). A study of mild dementia in the community using a wide range of diagnostic criteria. Br J Psychiatry 153:328–334.

Mozer HN, Bal DG, & Howard JT (1987). Perspectives on the etiology of Alzheimer's disease. JAMA 257:1503–1507.

MUELLER PS (1985). Neuroleptic malignant syndrome. Psychosomatics 26:654–662.

MURRAY GB (1987). Confusion, delirium, and dementia. In TP Hackett & NH Cassem (Eds.), Massachusetts General Hospital handbook of general hospital psychiatry (2nd ed.). Littleton, MA: PSG Publishing. Ch. 6, pp. 84–115.

O'BRIEN JG (1989). Evaluation of acute confusion (delirium). Prim Care 16:349–460.

O'CONNOR DW, POLLITT PA, ROTH M, ET AL. (1990). Memory complaints and impairment in normal, depressed, and demented elderly persons identified in a community survey. Arch Gen Psychiatry 47:224–227.

OLDENDORF WH (1980). The quest for an image of the brain. New York: Raven Press.

O'QUIN J, & MCGRAW KO (1985). The burdened caregiver: An overview. In TJ Hutton & AD Kenny (Eds.), Senile dementia of the Alzheimer type: Proceedings of the Fifth Tarbox Symposium. New York: Alan R Liss. 00:65–75.

OUSLANDER JG, ZARIT SH, ORR NK, ET AL. (1990). Incontinence among elderly community-dwelling dementia patients—characteristics, management, and impact on caregivers. J Am Geriatr Soc 38:440–445.

PARK GD (1986). Rheumatic diseases. In R Spector (Ed.), The scientific basis of clinical pharmacology: Principles and examples. Boston: Little, Brown & Co. 20:360–378.

PARKER DL, & HODGE JR (1967). Delirium in a coronary care unit. JAMA 20:132–133.

PAYKEL ES, FLEMINGER R, & WATSON JP (1982). Psychiatric side-effects of antihypertensive drugs other than reserpine. J Clin Psychopharmacol 2:14–39.

PENRY JK, & NEWMARK ME (1979). The use of anti-epileptic drugs. Ann Intern Med 90:207–218.

PERRY JC, & JACOBS D (1982). Overview: Clinical applications of the amytal interview in psychiatric emergency settings. Am J Psychiatry 139:552–559.

PETERSON LG, & BONGAR B (1989). Navane versus haldol. Treatment of acute organic mental syndromes in the general hospital. Gen Hosp Psychiatry 11:412–417.

PETRY S, CUMMINGS JL, HILL MA, ET AL. (1988). Personality alterations in dementia of the Alzheimer type. Arch Neurol 45:1187–1190.

PETRY S, CUMMINGS JL, HILL MA, ET AL. (1989). Personality alterations in dementia of the Alzheimer type: A three-year follow-up study. J Geriatr Psychiatry Neurol 2:203–207.

POEWE W, BESWLTE T, KARAMAT E, ET AL. (1990). CSF somatostatin-like immunoreactivity in dementia of Parkinson's disease. J Neurol Neurosurg Psychiatr 53:1105–1106.

POHL P, VOGL G, FILL H, ET AL. (1988). Single photon emission computed tomography in AIDS dementia complex. J Nucl Med 29:1382–1386.

POST F (1965). Clinical psychiatry in late life. Oxford, England: Pergamon Press.

POST F (1975). Dementia, depression and pseudodementia. In DF Benson & D Blumer (Eds.), Psychiatric aspects of neurologic diseases. New York: Grune & Stratton. 6:99–120.

RABINS P, MACE N, & LUCAS M (1982). The impact of dementia on the family. JAMA 248:333–335.

RABINS PV, & FOLSTEIN MF (1982). Delirium and dementia: Diagnostic criteria and fatality rates. Br Psychiatry 140:149–153.

RAMSDELL JW, ROTHROCK JF, WARD HW, ET AL. (1990). Evaluation of cognitive impairment in the elderly. J Gen Intern Med 5:55–64.

RATHBONE-MCCUAN E, & GOODSTEIN RK (1985). Elder abuse: Clinical considerations. Psychiatr Ann 15(5):331–339.

RATHBONE-MCCUAN E, & VOYLES B (1982). Case detection of abused parents. Am J Psychiatry 139:189–192.

READ SL, SMALL GW, & JARVIK LF (1985). Dementia syndrome. In RE Hales & AJ Frances (Eds.), Psychiatry update: The American Psychiatric Association annual review (Vol. 4; Ch. 11, pp. 211–226). Washington, DC: American Psychiatric Press.

REIFLER BV (1986). Mixed cognitive-affective disturbances in the elderly: A new classification. J Clin Psychiatry 47:354–356.

REIFLER B, LARSON E, & HANLEY R (1982). Coexistence of cognitive impairment and depression in geriatric outpatients. Am J Psychiatry 139:623–626.

REISBERG B (1985). Alzheimer's disease update. Psychiatr Ann 15:319–322.

REISBERG B, BORENSTEIN J, SALOB SP, ET AL. (1987). Behavioral symptoms in Alzheimer's disease: Phenomenology and treatment. J Clin Psychiatry 48:9–15.

REMICK RA, O'KANE J, & SPARLING TG (1981). A case report of toxic psychosis with low-dose propranolol therapy. Am J Psychiatry 138:850–851.

REYNOLDS CF, & HOCH CC (1987). Differential diagnosis of depressive pseudodementia and primary degenerative dementia. Psychiatr Ann 17:743–749.

REYNOLDS CF, HOCH CC, KUPFER DL, ET AL. (1988). Bedside differentiation of depressive pseudodementia from dementia. Am J Psychiatry 145:1099–1103.

REYNOLDS CF III, KUPFER DJ, TASKA LS, ET AL. (1985). Sleep apnea in Alzheimer's dementia: Correlation with mental deterioration. J Clin Psychiatry 46:257–261.

RICHEIMER S (1987). Psychological intervention in delirium. Postgrad Med 81:173–180.

RISSE SC, RASKIND MA, NOCHLIN D, ET AL. (1990). Neuropathological findings in patients with clinical diagnoses of probable Alzheimer's disease. Am J Psychiatry 147:168–172.

RIVINUS TM (1982). Psychiatric effects of the anticonvulsants regimens. J Clin Psychopharmacol 2:165–192.

ROBERTS MA, & CAIRD (1990). The contribution of computerized tomography to the differential diagnosis of confusion in elderly patients. Age Ageing 19:50–56.

ROCA R (1987). Bedside cognitive examination. Psychosomatics 28:71–76.

ROCA RP, KLEIN LE, KIRBY SM, ET AL. (1984). Recognition of dementia among medical patients. Arch Intern Med 144:73–75.

ROCKWOOD K (1989). Acute confusion in elderly medical patients. J Am Geriatr Soc 37:150–154.

RON MA (1990). Suspected dementia: Psychiatric differential diagnosis. J Neuropsychiatry 2:214–220.

ROSS HE, & KEDWARD HB (1977). Psychogeriatric hospital admissions for the community and institutions. J Gerontol 32:420–427.

ROTH M (1982). Perspectives in the diagnosis of senile and presenile dementia of Alzheimer's type. Adv Med 18:268–287.

ROVNER BW, BROADHEAD J, SPENCER M, ET AL. (1989). Depression and Alzheimer's disease. Am J Psychiatry 146:350–353.

ROZZINI R, ZANETTI O, & TRABUCCHI M (1989). Delirium induced by neuropsychological tests [letter]. J Am Geriatr Soc 37:666.

RUBIN CE (1984). Cimetidine and ranitidine [letter to the Editor]. JAMA 251:2211–2212.

RUBIN EH, & KINSCHERF DA (1989). Psychopathology of very mild dementia of the Alzheimer type. Am J Psychiatry 146:1017–1021.

RUDORFER MV, & CLAYTON PJ (1981). Depression, dementia and dexamethasone suppression [letter to the Editor]. Am J Psychiatry 138:701.

RUSINEK H, DELEON M, GEORGE A, ET AL. (1991). Alzheimer disease: Measuring loss of cerebral gray matter with MR imaging. Radiology 178:109–114.

Russell V, Proctor L, & Moniz E (1989). The influence of a relative support group on carers' emotional distress. J Adv Nurs 14:863–867.

Salzman C (1982). A primer on geriatric psychopharmacology. Am J Psychiatry 139:67–74.

Salzman C (1988). Treatment of agitation, anxiety, and depression in dementia. Public 24:39–41.

Salzman C, Shader RI, Greenblatt DJ, et al. (1983). Long versus short half-life benzodiazepines in the elderly: Kinetics and clinical effects of diazepam and oxazepam. Arch Gen Psychiatry 40:293–297.

Sanford JRA (1975). Tolerance of debility in elderly dependents by supporters at home. Br Med J 3:471–473.

Schneider L, Pollock V, & Lyness S (1990). A meta analysis of controlled trials of neuroleptic treatment in dementia. J Am Geriatr Soc 38:553–563.

Sewell DD, & Jeste DV (1991). Neuroleptic malignant syndrome: Clinical presentation, pathophysiology, and treatment. In A Stoudemire & BS Fogel (Eds.), Medical psychiatric practice (Vol. 1; Ch. 12, pp. 425–454). Washington DC: American Psychiatric Press, Inc.

Shista SK, Troupe A, Marszalek KS, et al. (1974). Huntington's chorea: An electroencephalographic and psychometric study. Electroencephalogr Clin Neurophysiol 36:387–393.

Shraberg D (1978). The myth of pseudodementia: Depression and the aging brain. Am J Psychiatry 135:601–603.

Sirois F (1988). Delirium: 100 cases. Can J Psychiatry 33:375–378.

Slaby AE (1982). Dementia. In AJ Giannini (Ed.), Neurologic, neurogenic, and neuropsychiatric disorders. Garden City, NY: Medical Examination Publishing.

Slaby AE (1986). Dementia. In LI Sederer (Ed.), Inpatient, psychiatry: Diagnosis and treatment (2nd ed.; Ch. 7, pp. 150–170). Baltimore: Williams & Wilkins.

Slaby AE, Lieb J, Tancredi AE (1985). The handbook of psychiatric emergencies (3rd Ed.). New York: Medical Examination Publishing Co.

Slaby AE, & Wyatt RJ (1974). Dementia in the presenium. Springfield, IL: Charles C Thomas.

Sloane PD, Mathew LJ (Eds.). (1991). Dementia units in long-term care. Baltimore: The Johns Hopkins University Press.

Sloane RB (1983). Organic mental disorders. In L Grinspoon (Ed.), Psychiatric update: The American Psychiatric Association annual review (Vol. 2; Ch. 8, pp. 106–118). Washington, DC: American Psychiatric Press, Inc.

Smego RA, & Durack DT (1982). The neuroleptic malignant syndrome. Ann Intern Med 142:1183–1185.

Smith LW, & Dimsdale JE (1989). Postcardiotomy delirium: Conclusions after 25 years? Am J Psychiatry 146:452–458.

Snavely SR, & Hodges GR (1984). The neurotoxicity of antibacterial agents. Ann Intern Med 101:92–104.

Snow SS, & Wells CE (1981). Case studies in neuropsychiatry: Diagnosis and treatment of coexistent dementia and depression. J Clin Psychiatry 42:439–441.

Sobin P, Schneider L, & McDermott H (1989). Fluoxetine in the treatment of agitated dementia [letter]. Am J Psychiatry 146:1636.

Sommer BR, Satlin A, Friedman L, et al. (1989). Glycopyrrolate versus atropine in post-ECT amnesia in the elderly. J Geriatr Psychiatry Neurol 2:18–21.

Sotsky SM, & Tossell JW (1984). Tometin induction of mania. Psychosomatics 25:626–628.

Spehn W, & Stemmier G (1985). Post-alcoholic diseases: Diagnostic relevance of computerized EEG. Electroencephalogr Clin Neurophysiol 60:106–114.

Steele C, Lucas M, & Tune L (1986). Haloperidol vs thiordazine in the treatment of behavioral symptoms in senile dementia of the Alzheimer's type: Preliminary findings. J Clin Psychiatry 47:310–312.

Steele TE, & Morton WA (1986). Salicylate-induced delirium. Psychosomatics 27:455–456.

Stillner V, Popkin MK, & Pierce CM (1978). Caffeine-induced delirium during prolonged competitive stress. Am J Psychiatry 137:855–856.

Stoudemire A, Hill C, Gulley LR, et al. (1989). Neuropsychological and biomedical assessment of depression-dementia syndromes. J Neuropsychiatry 1:347–361.

Strub RL (1982). Acute confusional state. In FD Benson & D Blumen (Eds.), Psychiatric aspects of neurologic disease (Vol. 2; Ch. 1, pp. 1–23). New York: Grune & Stratton.

Strum WB (1984). Cimetidine and ranitidine. JAMA 251:2212.

Teeter RB, Garetz FK, Miller WR, et al. (1976). Psychiatric disturbances of aged patients in skilled nursing homes. Am J Psychiatry 133:1430–1434.

Thal LJ (1988). Dementia update: Diagnosis and neuropsychiatric aspects. J Clin Psychiatry 49:5–7.

Theobald D (1987). Delirium: Definition, evaluation, and management in the critically ill patient. Indiana Med 80:526–528.

Thienhaus OJ, Wheeler BG, Simon S, et al. (1987). A controlled double-blind study of high-dose dihydroergotoxine mesylate (Hydergine) in mild dementia. J Am Geriatr Soc 35:219–223.

Thompson TL II, Moran MG, & Nies AS (1983): Drug therapy: Psychotropic drug use in the elderly. N Engl J Med 308:134–138, 194–199.

Thomsen A, Borgesen S, Bruhn P, et al. (1986). Prognosis of dementia in normal-pressure hydrocephalus after a shunt operation. Ann Neurol 20:304–310.

Tobias CR, Turns DM, & Lippmann S (1988). Psychiatric disorders in the elderly. Psychopharmacologic management. Postgrad Med 83:313–319.

Tomlinson BE, Blessed G, & Roth M (1968). Observations on the brain of nondemented old people. J Neurol Sci 7:331–356.

Tomlinson BE, Blessed G, & Roth M (1970). Observations on the brains of demented old people. J Neurol Sci 11:205–242.

Tresch DD, Folstein MF, Rabins PV, et al. (1985). Prevalence and significance of cardiovascular disease and hypertension in elderly patients with dementia and depression. J Am Geriatric Soc 33:530–537.

Trohman R (1988). Amiodarene-induced delirium. Ann Intern Med 108:68–69.

Trzepacz P, Bauer R, & Greenhouse J (1988). A symptom rating scale for delirium. Psychiatry Res 23:89–97.

Trzepacz PT, Sclabassi RJ, & Van Thiel DH (1989). Delirium: A subcortical phenomenon. J Neuropsychiatry 1:283–290.

Trzepacz PT, Teague GB, & Lipowski ZJ (1985). Delirium and other organic mental disorders in a general hospital. Gen Hosp Psychiatry 7:101–106.

Tune LE, Holland A, Folstein MF, et al. (1981). Association of postoperative delirium with raised serum levels of anticholinergic drugs. Lancet 2:651–653.

Tyler KL, & Tyler HR (1984). Differentiating organic dementia. Geriatrics 39:38–52.

Van Praag HM (1977). Psychotropic drugs in the aged. Compr Psychiatry 18:429–442.

Vassilouthis J (1984). The syndrome of normal pressure hydrocephalus. J Neurosurg 61:501–509.

Victor M, & Adams RD (1974). Common disturbances of vision, ocular movement, and hearing. In MM Wintrobe, GW Thron, RD

ADAMS, ET AL. (Eds.), Harrison's principles of internal medicine (7th ed., Ch. 20, pp. 100–110). New York: McGraw-Hill.

VITIELLO M, & PRINZ P (1989). Alzheimer's disease: Sleep and sleep/wake patterns. Clin Geriatr Med 592:289–299.

VOLPE BT, & SOAVE R (1979). Formed visual hallucinations and digitalis toxicity. Ann Intern Med 91:865–866.

WALKER C (1990). Ergoidmesylates vs Alzheimer's: The latest round. Geriatrics 45:22–24.

WARD NG, ROWLETT DB, & BURKE P (1978). Sodium amylobarbitone in the differential diagnosis of confusion. Am J Psychiatry 135:75–83.

WEDDINGTON WW, MUELLING AE, & MOOSA HH (1982). Adverse neuropsychiatric reactions to cimetidine. Psychosomatics 23:49–53.

WEINER RD (1983). EEG in organic brain syndrome. In JR HUGHES & WP WILSON (Eds.), EEG and evoked potential in psychiatry and behavioral neurology. (Ch. 1, pp. 1–24). Boston: Butterworth.

WELLS CE (1977). Dementia. Definition and description. In CE WELLS (Ed.), Contemporary neurology series: Dementia (2nd ed., Ch. 1, pp. 1–14). Philadelphia: FA Davis.

WELLS CE (1979). Pseudodementia. Am J Psychiatry 136:895–900.

WELLS CE, & DUNCAN GW (1980). Delirium. In Neurology for psychiatrists (Ch. 3, pp. 45–64). Philadelphia: FA Davis.

WHALLEY LJ (1989). Drug treatments of dementia. Br J Psychiatry 155:595–611.

WILENS TE, STERN TA, & O'GARA PT (1990). Adverse cardiac effects of combined neuroleptic ingestion and tricyclic antidepressant overdose. J Clin Psychopharmacol 10:51–54.

WILSON WH, & JEFFERSON JW (1985). Thyroid disease, behavior, and psychopharmacology. Psychosomatics 26:481–492.

WOLF-KLEIN G, SILVERSTONE F, & LEVY A (1989). Screening for Alzheimer's disease by clock drawing. J Am Geriatr Soc 37:730–734.

WOLFSON LI, & KATZMAN R (1983). The neurological consultation at age 80. In R KATZMAN & RD TERRY (Eds.), The neurology of aging (Ch. 10, pp. 221–244). Philadelphia: FA Davis.

WRAGG RE, & JESTE DV (1989). Overview of depression and psychosis in Alzheimer's disease. Am J Psychiatry 146:577–587.

YARCHOAN R, THOMAS RV, GRAFMAN J, ET AL. (1988). Long-term administration of 3′-azido-2′,3′-dideoxythymidine to patients with AIDS-related neurological disease. Ann Neurol 23:582–587.

YESAVAGE JA, TINKLENBERG JR, HOLLISTER LE, ET AL. (1979). Vasodilators in senile dementias: A review of the literature. Arch Gen Psychiatry 36:220–223.

ZARIT S, REEVER K, & BACH-PETERSON J (1980). Relative of the impaired elderly: Correlates of feelings of burden. Gerontologist 20:649–655.

ZARIT SH, & ZARIT JM (1984). Psychological approaches to families of the elderly. In MG EISENBERG, LC SUTKIN, & MA JANSEN (Eds.), Chronic illness and disability through the life span: Effects on self and family (Ch. 13, pp. 269–288). New York: Springer Publishing.

ZUBENKO G (1990). Progression of illness in the differential diagnosis of primary dementia. Am J Psychiatry 147:435–438.

ZUCKERMAN JD, FABIAN DR, SAKALES SR, ET AL. (1987). Comprehensive care of geriatric hip fracture patients. Hospimedica 5:37.

SUGGESTED READINGS

ABERNETHY D, & TODD E (1986). Doxepin cimetidine interaction: Increased doxepin bioavailability during cimetidine treatment. J Clin Pharmacol 6:8–12.

ANTES J (1989). Ciprofloxacin and delirium. Ann Intern Med 110:170–171.

ANTON RF, WAID LR, FOSSEY M, ET AL. (1986). Case report of carbamazepine treatment of organic brain syndrome with psychotic features. J Clin Psych 6:232–234.

BASAVARAJU N, & PHILLIPS S (1989). Cortisol deficient state—a cause of reversible cognitive impairment and delirium in the elderly. J Am Geriatr Soc 37:49–51.

BERESFORD TP, HOLT RE, HALL RC, ET AL. (1985). Cognitive screening at the bedside: Usefulness of a structured examination. Psychosomatics 26:319–324.

BERRIOS GE, & QUEMADA JI (1990). Depressive illness in multiple sclerosis: Clinical and theoretical aspects of the association. Br J Psychiatry 156:10–16.

BLASS JP, GLEASON P, BRUSH D, ET AL. (1988). Thiamine and Alzheimer's disease: A pilot study. Arch Neurol 45:833–835.

BONDAREFF W (1990). Magnetic resonance imaging and the severity of dementia in older adults. Arch Gen Psychiatry 47:47–51.

BROWN C, KAMINSKY M, FEROLI E, ET AL. (1983). Delirium with phenytoin and disulfiram administration. Ann Emerg Med 12:310–313.

CAMERON DJ, THOMAS RI, MULVIHILL M, ET AL. (1987). Delirium: A test of the Diagnostic and Statistical Manual III criteria on medical inpatients. J Am Geriatric Soc 35:1007–1010.

COHEN CI, & CASIMIR GJ (1989). Factors associated with increased hospital stay by elderly psychiatric patients. Hosp Community Psychiatry 40:741–751.

COHEN GD (1990). Prevalence of psychiatric problems in older adults. Psychiatr Ann 28:433–438.

CRIPPEN DW (1990). The role of sedation in the ICU patient with pain and agitation. Crit Care Clin 6:369–392.

D'ALESSANDRO R, GALLASSI R, BENASSI G, ET AL. (1988). Dementia in subjects over 65 years of age in the Republic of San Marino. Br J Psychiatry 153:182–186.

DEC GW, & STERN TA (1990). Tricyclic antidepressants in the intensive care unit. J Intensive Care Med 5:69–81.

DUBOVSKY SL, FRANKS RD, & ALLEN S (1987). Verapamil: A new antimanic drug with potential interactions with lithium. J Clin Psychiatry 48:371–372.

DUFFY FH (1989). Clinical value of topographic mapping and quantified neurophysiology. Arch Neurol 46:1133–1134.

ELMER M (1989). Management of the behavioral symptoms associated with dementia. Prim Care 16:431–449.

ERKINJUNTTI T (1987). Short formable mental status. J Am Geriatr Soc 35:412–416.

ERKINJUNTTI T, AUTIO L, & WIKSTROM J (1988). Dementia in medical wards. J Clin Epidemiol 41:123–126.

ESSA M (1986). Carbamazepine in dementia. J Clin Psychopharmacol 6:234–236.

ESSER SR, & VITALIANO PP (1988). Depression, dementia, and social supports. Int J Aging Hum Dev 26:289–301.

EVANS DL, & NEMEROFF CB (1984). The dexamethasone suppression test in organic affective syndrome. Am J Psychiatry 141:1465–1467.

FINDER RL, & BENNETT CR (1984). Use of scopolamine for dental anesthesia and analgesia techniques. J Oral Maxillofac Surg 42:802–804.

FISMAN M, GORDON B, FELEKI V, ET AL. (1985). Hyperammonemia in Alzheimer's disease. Am J Psychiatry 142:71–73.

FOGEL BS, & SATEL SL (1985). Age, medical illness and the DST in depressed general hospital inpatients. J Clin Psychiatry 46:95–97.

FOGEL BS, SATEL SL, & LEVY S (1985). Occurrence of high concentrations of post-dexamethasone cortisol in elderly psychiatric inpatients. Psychiatr Res 15:85–90.

FRACKOWIAK R (1989): PET: Studies in dementia. Psychiatr Res 29:353–355.

Francis J, & Kapoor WN (1990). Delirium in hospitalized elderly. J Gen Intern Med 5:65–79.

Francis J, Martin D, & Kapoor W (1990). A prospective study of delirium in hospitalized elderly. JAMA 263:1097–1101.

Freeman AM, Folks DG, Sokol RS, et al. (1985). Cognitive function after coronary bypass surgery: Effect of decreased cerebral blood flow. Am J Psychiatry 142:110–112.

Freinhar JP, & Alvarez WA (1986). Clonazepam treatment of organic brain syndromes in three elderly patients. J Clin Psychiatry 47:525–526.

Gabuzda DH, Levy SR, & Chiappa KH (1988). Electroencephalography in AIDS and AIDS-related complex. Clin Electroencephalogr 19:1–6.

Georgotas A, McCue RE, Kim OM, et al. (1986). Dexamethasone suppression in dementia, depression, and normal aging. Am J Psychiatry 143:452–456.

Gibson GE, Sheu KFR, Blass JP, et al. (1988). Reduced activities of thiamine-dependent enzymes in the brain and peripheral tissues of patients with Alzheimer's disease. Arch Neurol 45:836–840.

Gilhooly M, & Whittich J (1989). Expressed emotion in caregivers of the demented elderly. Br J Med Psychol 62:265–272.

Gilleard CJ, Gilleard E, & Whittich JE (1984). Impact of psychogeriatric day hospital care on the patient's family. Br J Psychiatry 145:487–492.

Goldman LS, & Luchins MD (1984). Depression in the spouses of demented patients. Am J Psychiatry 141:1467–1468.

Goldney R, & Lander H (1979). Pseudodelirium. Med J Aust 1:630.

Goto K, Umegaki H, & Suetsugu M (1976). Electroencephalographic and clinicopathological studies in Creutzffeldt-Jakob syndrome. J Neurol Neurosurg Psychiatry 39:931–940.

Gottfries CG (1989). Pharmacological treatment strategies in dementia disorders. Pharmacopsychiatry 22:129–134.

Graves AB, White E, Koepsell TD, et al. (1990). The association between head trauma and Alzheimer's disease. Am J Epidemiol 131:491–501.

Grubb BP (1987). Digitalis delirium in an elderly woman. Postgrad Med 81:329–330.

Guze SB, & Cantwell DP (1984). The prognosis in "organic brain" syndromes. Am J Psychiatry 120:878–881.

Guze SB, & Daengsurisri S (1986). Organic brain syndromes prognostic significance in general medical patients. Arch Gen Psychiatry 17:365–366.

Hachinski VC (1990). The decline and resurgence of vascular dementia. Can Med Assoc J 142:107–111.

Hales RE, Polly S, & Orman D (1988). An evaluation of patients who received an organic mental disorder diagnosis on a psychiatric consultation-liaison service. Gen Hosp Psychiatry 11:88–94.

Hein MD, & Jackson IMD (1990). Review: Thyroid function in psychiatric illness. Gen Hosp Psychiatry 12:232–244.

Ho D, Bredesen D, Vinters H, et al. (1989). The acquired immuno deficiency syndrome (AIDS) dementia complex. Ann Intern Med 111:400–410.

Hodkinson E, McCafferty FG, Scott JN, et al. (1988). Disability and dependency in elderly people in residential and hospital care. Age Aging 17:147–154.

Jacobs L, Conti D, Kinkel WR, et al. (1976). "Normal pressure" hydrocephalus relationship of clinical and radiographic findings to improvement following shunt surgery. JAMA 235:510–512.

Jain KK, & Fischer B (1989). Diagnostic steps of dementia in the commentary by CG Gottfies. Alzheimer Dis Associated Disord 3:228–230.

Jayalumar P, et al. (1989). Multi-infarct dementia: A computed tomographic study. Acta Neurol Scand 73:292–295.

Jayaram G, Coyle J, & Tune L (1986). Relapse in chronic schizophrenics treated with fluphenazine decanoate is associated with low serum neuroleptic levels. J Clin Psychiatry 47:247–248.

Jeff O, & Roth M (1969). The new psychogeriatric unit at Newcastle General Hospital. Occup Ther 32:21–26.

Johnson JC, Gottlieb GL, Sullivan E, et al. (1990). Using DSM-III criteria to diagnose delirium in elderly general medical patients. J Gerontol 45:M113–119.

Johnson KA, Holman BL, Rosen TJ, et al. (1990). Iofetamine I123 single photon emission compound tomography is accurate in the diagnosis of Alzheimer's disease. Arch Intern Med 150:752–756.

Jorm AF, & Jacomb PA (1989). The informant questionnaire on cognitive decline in the elderly (IQCODE): socio-demographic correlates, reliability, validity and some norms. Psychol Med 19:1015–1022.

Karlsson I, Brane G, Melin E, et al. (1988). Effects of environmental stimulation on biochemical and psychological variables in dementia. Acta Psychiatr Scand 77:207–213.

Karnaze D, & Caroner R (1987). Low serum cobalamin levels in primary degenerative dementia. Arch Intern Med 147:429–431.

Kashani JH, Beck NC, Hoeper EW, et al. (1987). Psychiatric disorders in a community sample of adolescents. Am J Psychiatry 144:584–589.

Katon W, & Raskind M (1980). Treatment of depression in the medically ill elderly with methylphenidate. Am J Psychiatry 137:963–965.

Khanna S (1988). Hypopituitarism presenting as delirium. Int J Psychiatry Med 18:89–92.

Kirn TF (1989). Dementias appear to have individual profiles in single photon emission computed tomography. JAMA 261:965, 968.

Knopman DS, & Sawyer-DeMaris S (1989). Practical approach to managing behavioral problems in dementia patients. Geriatrics 45:27–30, 35.

Kokmen E, Beard CM, Offord KP, et al. (1989). Prevalence of medically diagnosed dementia in a defined United States population: Rochester, Minnesota, January 1, 1975. Neurology 39:773–776.

Kokmen E, Chandra V, & Schoenberg BS (1988). Trends in incidence of dementing illness in Rochester, Minnesota, in three quinquennial periods, 1960–1974. Neurology 38:975–980.

Kral VA (1986). Therapeutic modalities in modern geriatric psychiatry. Psychiatr J Univ Ottawa 11:86–89.

Kurita H (1988). The concept and nosology of Heller's syndrome: Review of articles and report of two cases. Jpn J Psychiatry 42:785–793.

Lacey JH (1982). Anorexia nervosa and a bearded female saint. Br Med J 285:1816–1817.

Lawton MP, Moss M, Fulcomer M, et al. (1982). A research and service oriented multilevel assessment instrument. J Gerontol 37:91–99.

Levine PM, Siberfarb PM, & Lipowski ZJ (1978). Mental disorders in cancer patients: A study of 100 psychiatric referrals. Cancer 42:1385–1391.

Levkoff SE, Safran C, Cleary PD, et al. (1988). Identification of factors associated with the diagnosis of delirium in elderly hospitalized patients. J Am Geriatr Soc 36:1099–1104.

Levy RM, Bredese DE, & Rosenblum ML (1990). Neurologic complications of HIV infection. Am Fam Physician 41:517–536.

Lopez OL, Boller F, Becker JT, et al. (1990). Alzheimer's disease and depression: Neuropsychological impairment and progression of the illness. Am J Psychiatry 147:855–860.

Markstein R (1989). Pharmacological approaches in the treatment of senile dementia. Eur Neurol 29:33–41.

Mayeux R (1990). Therapeutic strategies in Alzheimer's disease. Neurology 40:175–180.

MAZUR A (1986). U.S. trends in feminine beauty and overadaptation. J Sex Res 22:281–303.

MCKEITH J (1984). Clinical use of the DST in a psychogeriatric population. Br J Psychiatry 145:389–393.

MIGNOGIA MJ (1986). Integrity versus despair: The treatment of depression in the elderly. Clin Ther 8:248–260.

MORRIS L, MORRIS R, & BRITTON P (1989). Cognitive and perceived control in spouse caregivers of dementia sufferers. Br J Med Psychol 62:173–179.

MORRIS RG, MORRIS LW, & BRITTON PG (1988). Factors affecting the emotional wellbeing of the caregivers of dementia sufferers. Br J Psychiatry 153:147–156.

MORTIMER JA (1990). Epidemiology of dementia: Cross-cultural comparisons. Adv Neurol 51:27–33.

MURRAY GB, SHEA V, & CONN DK (1986). Electroconvulsive therapy for poststroke depression. J Clin Psychiatry 47:258–260.

NAGASAWA H, KOGURE K, KAWASHIMA K, ET AL. (1990). Effects of co-dergocrine mesylate chyosrginer in multi-infarct dementia as evaluated by positron emission tomography. Tohoku J Exp Med 162:223–225.

NEARY D (1990). Dementia of frontal lobe type. J Am Geriatr Soc 38:71–72.

NICHOLSON CD (1990). Pharmacology of nootropics and metabolically active compounds in relation to their use in dementia. Psychopharmacology 101:147–159.

O'CONNOR DW, POLLITT PA, HYDE JB, ET AL. (1989a). The prevalence of dementia as measured by the Cambridge Mental Disorders of the Elderly Examination. Acta Psychiatr Scand 79:190–198.

O'CONNOR DW, POLLITT PA, HYDE JB, ET AL. (1989b). The reliability and validity of the Mini-Mental Health Status Exam in a British community survey. J Psychiatr Res 23:87–96.

OLAFSSON K, KORNER A, BILLE A, ET AL. (1989). The GBS Scale in multi-infarct dementia and senile dementia of Alzheimer type. Acta Psychiatr Scand 79:94–97.

OWENS JF, & HUTELMYER CM (1982). The effect of preoperative intervention on delirium in cardiac surgical patients. Nurs Res 31:60–62.

PACHNER AR, DURAY P, & STEERE AC (1989). Central nervous system manifestations of Lyme disease. Arch Neurol 46:790–795.

PALLETT P (1990). A conceptual framework for studying family caregiver burden in Alzheimer's type disease. J Nurs Scholarship 22:52–58.

PANEGYRES PK, GOLDSWAIN P, & KAKULAS BA (1983). Adult-onset adrenoleukodystrophy manifesting as dementia. Am J Med 87:481–483.

PARKER PE, WALTER-RYAN WG, PITTMAN CS, ET AL. (1986). Lithium treatment of hyperthyroidism and mania. J Clin Psychiatry 47:264–266.

PATTERSON C, & LE CLAIR JK (1989). Acute decompensation in dementia: Recognition and management. Geriatrics 44:20–26, 31–32.

PEARCE JMS (1984). Management. In Dementia: A clinical approach. Boston: Blackwell Scientific Publications. 11:154–165.

PETTINATI HM, & BONNER KM (1984). Cognitive functioning in depressed geriatric patients with a history of ECT. Am J Psychiatry 141:49–52.

PHILLIPSON M, MORANVILLE JT, & JESTE DV (1990). Antipsychotics. Clin Geriatr Med 6:411–422.

PICCININ G, FINARI G, & PICCIRILLI M (1990). Neuropsychological effects of L-deprenyl in Alzheimer's type dementia. Clin Neuropharmacol 13:147–163.

PRICE R, & BREW B (1988). The AIDS dementia complex. J Infect Dis 158:1079–1083.

Psychogeriatrics. (1972). World Health Organization Technical Report 507. Geneva: World Health Organization.

RAMSAY R, & KATONA CLE (1989). Diagnosing and treating dementia. Br Med J 298:51–52.

RASKIND MA, & RISSE SC (1986). Antipsychotic drugs and the elderly. J Clin Psychiatry 47:17–22.

REGLAND B, GOTTFRIES C, ORELAND L, ET AL. (1988). Low B_{12} levels related to high activity of platelet MAO in patients with dementia disorders. Acta Psychiatr Scand 78:451–457.

REICH P, REGESTEIN QR, MURAWSKI BJ, ET AL. (1983). Unrecognized organic mental disorders in survivors of cardiac arrest. Am J Psychiatry 140:1194–1197.

ROCCA WA, BONAIUTO S, LIPPI A, ET AL. (1990). Prevalence of clinically diagnosed Alzheimer's disease and other dementing disorders: A door-to-door survey in Appignano, Macerata Province, Italy. Neurology 40:626–631.

ROSEN WG, MOHS RC, & DAVIS KL (1984). A new rating scale for Alzheimer's disease. Am J Psychiatry 141:1356–1364.

ROSERS M, & REICH P (1986). Psychological intervention with surgical patients: Evaluation outcome. Adv Psychosom Med 15:23–50.

ROZZINI R, INZOLI M, & TRABUCCHI M (1988). Delirium from transdermal scopolamine in an elderly woman. JAMA 260:478.

SAKAUYE K (1990). Psychotic disorders: Guidelines and problems with antipsychotic medications in the elderly. Psychiatr Ann 20:456–465.

SALZMAN C (1987). Treatment of the elderly agitated patient. J Clin Psychiatry 148:19–22.

SALZMAN C (1990). Antidepressants. Clin Geriatr Med 6:399–410.

SARA G, KRAIUHIN C, GORDON E, ET AL. (1988). The P300 event related potential component in the diagnosis of dementia. Aust NZ J Med 18:657–660.

SCHAERF FW, MILLER RR, LIPSEY JR, ET AL. (1989). ECT for major depression in four patients infected with human immunodeficiency virus. Am J Psychiatry 146:782–784.

SCICUTELLA A, & DAVIES B (1988). Characterization of monoclonal antibodies to galactolipids and uses in studies of dementia. J Neuropathol Exp Neurol 47:406–419.

SHELINE Y (1990). Quantifying undiagnosed organic mental disorder in geriatric inpatients. Hosp Community Psychiatry 41:1004–1008.

SHEU KFR, CLARKE DD, KIM YT, ET AL. (1988). Studies of transketolase abnormality in Alzheimer's disease. Arch Neurol 45:841–845.

SIER H (1988). Primary hyperparathyroidism and delirium in the elderly. J Am Geriatr Soc 36:157–170.

SIMPSON CJ, & KELLETT JM (1987). The relationship between preoperative anxiety and post-operative delirium. J Psychosom Res 31:491–497.

SLABY AE, & CULLEN LO (1987). Dementia and delirium. In A STOUDEMIRE & BS FOGEL (Eds.), Principles of medical psychiatry. Orlando, FL: Grune & Stratton. Ch. 6, pp. 135–175.

SMALL GW (1988). Psychopharmacological treatment of elderly demented patients. J Clin Psychiatry 49:8–12.

SOININEN H, PARTANEN J, LAULUMAA V, ET AL. (1989). Longitudinal EEG spectral analysis in early stage of Alzheimer's disease. Electroencephalogr Clin Neurophysiol 72:290–297.

SPAR JE, & GERNER R (1982). Does the dexamethasone test distinguish dementia from depression? Am J Psychiatry 139:283–290.

STARKSTEIN SE, BERTHIER ML, PREZIOSI TJ, ET AL. (1989). Depression in patients with early versus late onset of Parkinson's disease. Neurology 39:1441–1445.

Starkstein SE, Bolduc PL, Preziosi TJ, et al. (1989). Cognitive impairments in different stages of Parkinson's disease. J Neuropsychiatry Clin Neurosci 1:143–248.

Starkstein SE, Rabins PV, Berthier ML, et al. (1989). Dementia of depression among patients with neurological disorders and functional depression. J Neuropsychiatry 1:263–268.

Stroletz LJ, Reyes PF, & Zalewska M (1990). Computer analysis of EEG activity in dementia of the Alzheimer's type and Huntington's disease. Neurobiol Aging 11:15–20.

Sunderland T, Alterman IS, Yount D, et al. (1988). A new scale for the assessment of depressed mood in demented patients. Am J Psychiatry 145:955–959.

Sunderland T, Molchan SE, Martinez RA, et al. (1990). Treatment approaches to atypical depression in the elderly. Psychiatr Ann 20:474–481.

Swedlow DB, & Schreiner MS (1985). Management of Reye's syndrome. Crit Care Clin 1:285–311.

Tariot P, Cohen RM, Sunderland T, et al. (1987). L-Deprenyl in Alzheimer's disease. Arch Gen Psychiatry 44:427–433.

Tatemichi TK, Foulkes MA, Mohr JP, et al. (1990). Dementia in stroke survivors in the Stroke Data Bank cohort. Stroke 21:858–866.

Thal L, Salmon DP, Lasker B, et al. (1989). The safety and lack of efficacy of vinpocetine in Alzheimer's disease. J Am Geriatr Soc 37:515–520.

Tobias CR, Lippmann S, & Pary R (1989). Dementia in the elderly. Postgrad Med 86:97–98, 101–108.

Trappler B, Viswanathan R, & Sher J (1986). Alzheimer's disease in a patient on long-term hemodialysis: A case report. Gen Hosp Psychiatry 8:57–60.

Trick G, Barris M, & Blarer-Bluth M (1989). Abnormal pattern electroencephalograms in patients with senile dementia of the Alzheimer type. Ann Neurol 26:226–231.

Tross S, Price TW, Navia B, et al. (1988). Neuropsychological characterization of the AIDS dementia complex: A primary report. Gower Academic J 2:81–88.

Turner DA, & McGeachie (1988). Normal pressure hydrocephalus and dementia-evaluation and treatment. Clin Geriatr Med 4:815–830.

Volicer L, Rheaume Y, Brown J, et al. (1986). Hospice approach to the treatment of patients with advanced dementia of the Alzheimer type. JAMA 256:2210–2213.

Weinstein H, Hijora A, Van Royen E, et al. (1989). Determination of cerebral blood flow by SPECT. Clin Neurol Neurosurg 91:13–18.

Wolff HG, & Curran D (1935). Nature of delirium and allied states: The dysergastic reaction. Arch Neurol Psychiatry 33:1175–1215.

Wooten V (1990). Evaluation and management of sleep disorders in the elderly. Psychiatr Ann 20:466–473.

Wragg RE, & Jeste DV (1988). Neuroleptics and alternative treatments. Psychiatr Clin North Am 11:195–213.

Zorumski CF, Rubin EH, Burke WJ (1988). Electroconvulsive therapy for the elderly: A review. Hosp Community Psychiatry 39:643–647.

20 Clinical neurology

MARY EILEEN McNAMARA, M.D.

The interface between neurology and psychiatry has received much scholarly attention in the last two decades because of the growth of behavioral neurology and neuropsychology, and advances in basic neurosciences that permit a rational exploration of pathophysiology. This chapter discusses some common neurologic disorders for which psychiatric complications are prevalent: Parkinson's disease, multiple sclerosis, stroke, myasthenia gravis, and epilepsy.

PARKINSON'S DISEASE

> A more melancholy object I never beheld. The patient, naturally a handsome, middle-sized, sanguine man, of cheerful disposition, and an active mind, appeared much emaciated, stooping, and dejected. (Parkinson, 1817)

Parkinson's disease has been a puzzle to medical science since its description in 1817 by James Parkinson, who reported a clinical syndrome he termed "paralysis agitans." The apparent contradiction inherent in the title—paralysis with agitation—conveys the sense of the seemingly paradoxical symptoms of the illness. On the one hand, victims of the illness have profoundly slowed voluntary movement, with akinesia, rigidity, and abulia, coupled with a masked face and unblinking stare. On the other hand, the patients may have tremor at rest and with movement, explosive pseudobulbar weeping and laughing, pressured pallilalic speech, festination, and akathisia, all suggesting an agitated state. Sacks, in his philosophically profound and emotionally compelling book *Awakenings* (1973), wrote, "The Parkinsonian has indeed lost, and quite fundamentally, his inner sense of scale and pace. Hence the incontinent accelerations and retardations, the magnifications and minifications, to which he is prone." As Sacks indicates, it is both simplistic and incorrect to view Parkinson's disease as purely a disorder of movement, because the observed muscular activity and the mental state of the patient are inextricably linked. Although this point is often overlooked by physicians, it is well known to patients. Physical slowing is reflected in slowing of thought processes, and bursts of intense emotion are coupled with increased psychomotor activity. This is of course true for many other states: the abulia of depression, the restless movements of mania, the nervous fidgeting of worry, and the increase in tardive dyskinesia with stress.

Stevens (1973) has discussed the striking anatomic and functional similarities of the two of the brain's major dopaminergic systems, the nigrostriatal and the mesolimbic. This anatomic homology may underlie some of the linkages of motor activity and emotion that are seen in Parkinson's disease, but the details are unknown.

There has been an intense reemergence of interest in Parkinson's disease in recent years because of the discovery of the toxic effects of MPTP (*N*-methyl-4-phenyl-1,2,3,6-tetrahydropyridine). Beginning in the late 1970s, sporadic outbreaks of severe Parkinson's syndrome were noted in young drug abusers in California, Maryland, and Vancouver, British Columbia, Among these was a 23-year-old college student who in 1977 developed severe bradykinesia, rigidity, and mutism and was initially thought to have catatonic schizophrenia until a response to L-dopa revealed the surprising diagnosis of Parkinson's syndrome. The patient was referred to the National Institute of Mental Health (NIMH), where the patient's psychiatrist, G. Davis, in a skilled piece of detective work, obtained a history that the patient had been synthesizing illicit drugs. Investigating this clue, Davis visited the patient's home and collected the glassware used for the production of synthetic meperidine. Contaminating the glassware were found several pyridines, including MPTP. Eventually, MPTP was shown by Langston and others to be the culprit of the patient's illness, causing degeneration of the substantia nigra in humans and producing Parkinson's syndrome (Ballard, Tetrud, & Langston, 1985; Langston, et al., 1983). Prior research on Parkinson's disease had always been hampered by a lack of animal models for the disease. Now, with the availability of MPTP, there has been a veritable explosion in parkinsonian research. MPTP appears to cause harm to neurons after it is converted by monoamine oxidase B (MAO-B) in nerve terminals and glial cells, where it becomes "MPP+" (1-methyl-4-phenlypyridine) (Snyder & D'Amato, 1986). MPP+ binds to neuromelanin; the pathologic changes of Parkinson's disease are highest in pigmented brainstem nuclei. Neuromelanin, which results from the oxidative catabolism of catecholamines, is found in the dopaminergic cell bodies of the substantia nigra and in the noradrenergic locus ceruleus. Moreover, the degradation of catecholamines, particularly dopamine, is accompanied by the formation of free radicals and cytotoxic quinones, which might also account for the particular vulnerability of these cells.

The experience with MPTP led to speculation that idiopathic Parkinson's disease may be due to an environmental toxin. MPTP and MPP+ are simple chemicals closely related to many substances occurring in nature and industry (Golbe, 1990). Genetic differences in metabolism might make some patients less able to detoxify certain exogenous or endogenous agents and, hence, more vulnerable to Parkinson's disease. In support of this, Steventon et al. (1989) reported abnormalities in defects in the detoxification pathways involving cysteine, and in sulfur metabolism in Parkinson's patients. Of note, a defect of cysteine degradation also exists in Hallervorden-Spatz syndrome, which also shows nigral damage. Parker, Boyson, and Parks (1989) reported abnormalities of the electron transport chain in Parkinson's disease. Complex I, the proximal portion of the mitochondrial electron transport chain, is also inhibited by MPTP. The genome for this chain is mitochondrial, rather than nuclear, which might explain why Parkinson's disease, which has a low-level familial incidence, does not follow mendelian inheritance.

Classically, Parkinson's disease has been regarded as idiopathic degeneration of dopaminergic neurons in the substantia nigra of the midbrain, resulting in loss of "motor impulse" to the corpus striatum. Actually, there is evidence of widespread catecholamine dysfunction. Degeneration and loss of dopamine are also seen in the ventral tegmental area, which projects to the cingulate, entorhinal, and frontal cortex and to the nucleus accumbens, olfactory tubercle, and central amygdaloid nucleus (Javoy-Agid & Agid, 1980; Synder & D'Amato, 1986). Furthermore, there is evidence of degeneration of the locus ceruleus, which has extensive noradrenergic projections (Riederer et al., 1977). Cell loss is seen in the serotonergic raphe nucleus, the substance P–containing neurons in the lateral reticular formation of the medulla, and multiple other sites (Halliday et al., 1990). Additionally, Parkinson's disease patients have further evidence of serotonergic alteration, with decreased binding sites for serotonin in the basal ganglia and some regions of the cerebral cortex (Raisman, Cash, & Agid, 1986); the cerebrospinal fluid shows evidence of decreased serotonin turnover (Fornu, 1986; Mayeux et al., 1984). Cholinergic neurons in the pedunculopontine nucleus and in the basal forebrain are also decreased (Zweig et al., 1989). High rates of phagocytosing microglia in parkinsonian brains suggest that Parkinson's disease is an active process and not just the remote effect of a prior insult, as was previously thought (McGreer et al., 1988). With such evidence of involvement of several neurotransmitters known to be important in behavior, the association of Parkinson's disease with neuropsychiatric disorders is not surprising.

Clinically, Parkinson's disease usually occurs in mid- to late life and gradually progresses over 10 to 20 years. Males are affected somewhat more frequently (Diamond et al., 1990), with a prevalence of about 1% of the population over 50 and 2% of the population over 70. Approximately 5% of patients with Parkinson's disease in the United States, and 10% in Japan, have an onset of disease before age 40. Young-onset patients tend to have a more gradual progression of parkinsonian signs and symptoms, and a lesser frequency of dementia, than patients with older onset Parkinson's disease. However, they more often have dystonia as an early sign, and tend to have earlier problems with dyskinesias and motor fluctuations with levodopa therapy (Golbe, 1991). Occasionally, patients with typical signs of parkinsonism presenting at an early age are misdiagnosed because the physician is unaware of the existence of young-onset cases. Such cases can present to psychiatrists, who are asked to see them for mental manifestations of apathy and depression or because of tremors thought to be of psychogenic origin.

The clinical manifestations of Parkinson's disease begin slowly and subtly, and it can be several years before the diagnosis is evident. The illness often begins unilaterally, and may present as a slowness of gait, a change in handwriting, a dystonic posture of a foot, a tendency to fall, a blunting of facial expression, and a vague sense of fatigue. Commonly, these can be misdiagnosed as evidence of a functional illness, particularly when the symptoms are noted to exacerbate with stress. Over time, Parkinson's disease becomes more frankly evident with the development of cogwheel rigidity, "pill-rolling" hand tremor, dysarthria, a stooped posture, a shuffling, "festinating" gait, and retropulsion, coupled with a blank, nonblinking facial stare and marked diminution of movement. Associated autonomic symptoms can include oiliness of the skin, hyperhidrosis, excessive production of saliva with drooling, orthostatic hypotension, constipation, and bladder dysfunction with difficulty initiating stream.

The early response to medication is usually excellent,

and the parkinsonian patient can typically become symptom free for a while. The illness continues to progress, however, and with time the halcyon of relief vanishes. Patients become less responsive to treatment, and bizarre complications supervene, including dyskinesias, dystonias, and the "on-off" phenomenon wherein the patient alternates between periods of movement and frozen rigidity. The relative contribution of medication to both symptoms and progression of this neurodegenerative illness is a subject of active debate, and beyond the scope of this chapter. The interested reader is referred to Lesser (1979), Juncos et al. (1989), Nutt, Gancher, and Woodward (1988), Olanow (1990), and Caraceni, Scigliano, and Musicco (1991).

Dementia in Parkinson's Disease

Dementia has consistently been reported as a concomitant of Parkinson's disease, with estimates of prevalence varying but correlated with the length of illness. Sweet et al. (1976) found dementia in one third of 100 patients followed for 6 years, with deficits increasing over time, similar to the results of Ebmeier et al. (1990), who found defined dementia in 23% of parkinsonian patients followed for 3 years. Mortimer et al. (1982) studied over 60 patients with idiopathic Parkinson's disease who were receiving dopaminergic medications and found that they had marked cognitive problems with memory, abstract reasoning, and concept formation, consistent with a subcortical dementia. Cognitive impairment can be found early in the course of the illness (Levin, Llabre, & Weiner, 1989). As noted above, it is somewhat less common in young-onset Parkinson's disease (Golbe, 1991).

The dementia of Parkinson's disease may have diverse etiologies, and its classification and pathogenesis are the subject of intense investigation. Depletion of neurons and formation of Lewy bodies in the nucleus basalis of Meynert occur in idiopathic Parkinson's disease (Nakano & Hirona, 1984), and cell death is most severe in demented patients (Whitehouse et al., 1983). Rinne et al. (1989), in contrast, believe the dementia of Parkinson's disease is due to cell loss in the medial substantia nigra, which projects to the caudate. Some patients have evidence of both Parkinson's disease and Alzheimer's disease (de la Monte et al., 1989). Hornykiewicz and Kish (1984) reported that demented Parkinson's disease patients, irrespective of the presence or absence of Alzheimer neuropathology, have a disturbance of central cholinergic function. Moreover, parkinsonian patients show significant memory impairments in response to small doses of scopolamine that cause no effects in matched controls, and anticholinergic medications cause severe impairment on tests believed to assess frontal lobe functions (Dubois et al., 1987, 1990).

Regardless of etiology, the development of dementia in the later stages of Parkinson's disease represents a difficult problem for patient and physician. The dementia is inexorable and does not respond to antiparkinsonian medication. Rather, such patients may develop delirium, agitation, and psychosis on small doses of the drugs required to ameliorate motor symptoms, thus presenting a therapeutic impasse.

Differential Diagnosis of Parkinsonism Plus Dementia

In assessing dementia in Parkinson's disease, the clinician should first consider whether the patient indeed has true Parkinson's disease or instead has any of a variety of illnesses that produce extrapyramidal signs accompanied by dementia. Chief among these is *progressive supranuclear palsy* (PSP), also known as Steele-Richardson syndrome, which presents with bradykinesia, retropulsion, and rigidity, and may be accompanied by a fulminant dementia (Jackson, Jankovic, & Ford, 1983). Early loss of vertical eye movements and absence of tremor help to differentiate this illness. Progressive supranuclear palsy responds poorly to dopaminergic medication.

Another mimic of Parkinson's disease is *multiinfarct dementia*, particularly with lacunar infarcts of the basal ganglia. Such patients also have stooped posture, shuffling gait, rigidity, and dementia (Murrow et al., 1990; Tolosa & SantaMaria, 1984). Patients usually, but not always, have a history of hypertension (Fisher, 1982). The classic stepwise deterioration in function is a history often difficult to obtain in clinical practice. A computerized tomographic (CT) head scan may be useful in detecting small lacunar strokes, but results may be normal. Magnetic resonance imaging (MRI) is a superior test, detecting up to eight times the number of lacunae visualized on CT (DeWitt et al., 1984). Therapy for multiinfarct dementia is directed toward prevention of further lacunar infarcts. Postsynaptic dopamine agonists, such as bromocriptine, may be helpful in improving bradykinesia, and drug treatment of associated depression can improve mood and function. (See Chapter 19.)

Several other illnesses may also mimic Parkinson's disease, including olivopontocerebral degeneration (Pascual et al., 1991), the rigid forms of Huntington's disease and Wilson's disease, carbon monoxide intoxication (Klawans et al., 1982), and tumors (Leenders, Findley, & Cleeves, 1986). Extrapyramidal syndromes can be induced by neuroleptics, which include not only antipsychotics but also antiemetic agents such as metoclopramide (Patel, Seth, & Meador, 1986).

Association with Psychopathology

The concept of the relationship of psychiatric and motor symptoms in Parkinson's disease has evolved considerably over the past century, as has psychiatry itself. (For a review of this subject, see McNamara, 1991a). Despite differing diagnostic standards over time, a consistently high rate of depression has been reported, varying between 29% and 90% (Gotham, Brown, & Marsden, 1986). As noted by Taylor and Saint-Cyr (1990), the usual methods of evaluating depression are not always adequate because the cognitive and motor slowness intrinsic to Parkinson's disease mimic affective disorder. For example, parkinsonian patients have problems initiating movement and so appear apathetic. Patients attempting to walk might be "frozen" until another person gently touches them, at which point they can begin to move but then might have trouble stopping. Likewise, parkinsonian patients can have trouble initiating plans and endeavors and thus also be considered apathetic, although they might well be able to enjoy activities initiated by someone else. A review of thought content is a more sensitive way of determining depression, with feelings of hopelessness, despair, suicidal ideation, and guilt being evidence of depression. A life review will often reveal evidence of depression well before the onset of Parkinson's disease Mayeux et al. (1986) have reported that a quarter of depressed Parkinson's disease patients had depression antedating the onset of motor symptoms. Clearly, for these patients, depression could not be reactive to a motor disability that did not yet exist.

The evidence of mood disturbance antedating the motor symptoms of Parkinson's disease is reminiscent of studies earlier in the century of "the parkinsonian personality." Such patients were noted to be somewhat rigid and inflexible in character, as if their motor symptoms became an outward reflection of their inward personality. This continues to be noted (Paulson & Dadmehr, 1991; Taylor & Saint-Cyr, 1990), and calls into question how the words "premorbid" and "antedate" can be used in an illness for which we do not know the cause or onset. Could the "parkinsonian personality" actually be the first manifestation of a catecholaminergic deficit that later also produces motor effects? About 80% of nigral dopamine must be lost before motor symptoms appear. Backward extrapolation of the subsequent rate of progression suggests that the onset of the illness might actually be in adolescence or the 20s (Golbe, 1990).

Schiffer et al. (1988) also noted that the manifestations of depression in parkinsonian patients are frequently atypical, with high rates of panic and anxiety. Their study was replicated by Kurlan et al. (1990). This is of particular interest since MAO inhibitors have been reported to be particularly effective for atypical depression, and MAO-B activity has been implicated in the progression of Parkinson's disease.

Depressed parkinsonian patients also show reduced metabolism in the inferior frontal lobes on positon emission tomography, as do patients with primary mood disorders (Mayberg et al., 1990). Mayeux and colleagues have shown that depression in parkinsonian patients is correlated with low cerebrospinal fluid (CSF) levels of 5-hydroxyindoleacetic acid (5-HIAA), suggesting a reduction in central serotonin (Mayeux et al., 1981). Treatment with a serotonin precursor (L-5-hydroxytryptophan) resulted in an improvement of depressive symptoms and an increase in CSF 5-HIAA levels (Mayeux et al., 1986). Not all patients with low CSF 5-HIAA levels develop depression, however, suggesting that some other factor must come into play.

There are also suggestions that depression may exacerbate Parkinson's disease. Rats depleted of dopamine may appear and behave normally until stressed, at which time parkinsonian signs emerge. Neurologic impairments were related both to the extent of dopamine depletion and to the intensity of stress in an additive fashion in this "biopsychosocial" rat model (Snyder, Stricker, & Zigmond, 1985). Clinicians may note an analogous phenomenon in patients who appear to have increased symptoms when stressed or depressed. Furthermore, Starkstein et al. (1989) have found that patients with an early onset of Parkinson's disease have a higher rate of major and minor depressive symptoms, as has also been reported by SantaMaria, Tolosa, and Valles (1986), who suggested "aminergic dysfunction" can make the patient with Parkinson's disease, already having a loss of nigral cells, develop parkinsonian symptoms earlier.

Psychosis in Parkinson's Disease

Psychosis is a frequent concomitant of late Parkinson's disease but is not well characterized in the literature. A common presentation is of "Lilliputian" visual hallucinations with impaired reality testing in a clear sensorium. The patient will report little people under the furniture, at the window, and so on. The hallucinations usually are entertaining, but occasionally paranoid delusions and agitation may accompany the presentation. The symptoms are usually related to dopamine agonist treatment but at times are seen with anticholinergics alone.

Depressive psychosis is also seen, with mood-congruent auditory hallucinations and paranoia. Dementia and antiparkinsonian pharmacotherapy appear to be frequent, if not universal, risk factors for psychosis.

The treatment of psychosis with Parkinson's disease may be quite difficult. Neuroleptics and other dopamine antagonists that ameliorate psychosis will exacerbate motor symptoms. Moreover, such patients are generally in an advanced stage of their illness and thus are exquisitely sensitive to medications, with side effects of delirium, orthostatic hypotension, and urinary retention. Clozapine, a novel neuroleptic that is relatively selective for D_1 dopamine receptors with less effect on D_2 and sigma receptors than conventional neuroleptics, can at times be helpful (Wolters et al., 1989). (See Chapter 9.)

Neuropsychiatric Complications of Antiparkinsonian Drugs

The pharmacotherapy of Parkinson's disease may itself be the cause of considerable psychiatric morbidity. Anticholinergics, helpful for parkinsonian tremor and rigidity, can produce a toxic confusional state with agitation and impairment of memory. In demented parkinsonian patients, this occurs very frequently, perhaps because the patients already have loss of cholinergic neurons in the nucleus basalis of Meynert (Nakano & Hirano, 1984). As noted earlier, anticholinergics can also cause significant cognitive symptoms in parkinsonian patients even in the absence of dementia or delirium. Beta-blockers such as propranolol, which are prescribed to control action tremors of Parkinson's disease, can also be associated with confusion, depression, and other mental status changes (Avorn, Everitt, & Weiss, 1986).

Dopaminergic agents, such as amantadine, bromocriptine, and L-dopa, also can produce hallucinations, psychosis, and paranoia. Again, demented patients appear more sensitive to these effects (Goodwin, 1971). Some patients seem differentially sensitive to psychiatric side effects of different dopaminergic agents, and a switch from bromocriptine to L-dopa, or the converse, could be considered when psychosis develops. Sometimes, a combination of low doses of bromocriptine and L-dopa is better tolerated than a higher dose of either one alone (Rinne, 1987). Levodopa modifies the plasma levels of bromocriptine, resulting in smoother blood levels with less oscillation (Rabey et al., 1989).

There are currently a number of new medications recently introduced for the treatment of Parkinson's disease. These include controlled-release Sinemet, or carbidopa-levodopa (Bush et al., 1989), and pergolide (Permax) (Goetz, 1990). There is little psychiatric literature on these two newer drugs, but their side effects appear to be similar to those of other dopaminergic agents (Stern et al., 1984). Selegiline is a relatively selective MAO-B inhibitor at low doses, and is used now in Parkinson's disease in an attempt to halt the progression of the disease. Results from ongoing trials have been mixed (Elizan et al., 1989; Langston, 1990). Selegiline has also been reported to produce euphoria in parkinsonian patients, perhaps by degradation to amphetamine (Karoum et al., 1982).

Also on the horizon are several other possible neuropharmacologic interventions for Parkinson's disease. Enhancement of gamma-aminobutyric acid (GABA) transmission also improves parkinsonian symptoms in animal models, and may in humans (Bennett, 1986; Bennett, Ferrara, & Cruz, 1987). Glutamate blockade by *n*-methyl-*d*-aspartate (NMDA) antagonists also may be beneficial (Klockgether & Turski, 1990). Also, cholecystokinin analogues have been reported to be useful for L-dopa–induced dyskinesias in parkinsonian animal models (Boyce et al., 1990).

Also extremely interesting is the possibility of "brain transplant" for the treatment of Parkinson's disease. More properly expressed, the technique involves placement of catecholaminergic neurons derived from adrenal medulla or other sources in the patient's caudate nucleus. Results have been mixed (Ostrosky-Solis et al., 1988).

Psychiatric side effects of all parkinsonian drugs are to a certain extent dose-related, and it is prudent to use as low a dose of medication as possible to achieve acceptable function. In end-stage Parkinson's disease, the clinician may be faced with the unfortunate choice of producing a patient who can move but is psychotic, or a patient with normal affect who is frozen in immobility. Not infrequently, the difference in medication to produce either of these two alternatives is very small. At these times a drug holiday may be rational (Weiner et al., 1980), although the long-term value of such holidays is debatable. Drug holidays should be conducted in a hospital because the patient may experience pronounced bradykinesia while drug free and be at risk for pulmonary embolus, aspiration, atelectasis, decubiti, and ileus. The purpose of a drug holiday is to increase sensitivity of striatal dopaminergic receptors that may have altered their receptivity after chronic stimulation by medications.

Treatment of Depression in Parkinson's Disease

The treatment of psychiatric illness in parkinsonian patients should have as its goal the improvement of mood and function while avoiding exacerbation of motor symptoms and delirium. Tricyclic antidepressants have been most frequently reported as helpful for the treatment of depression in Parkinson's disease. Tricyclics, which have monoamine agonist and cholinergic antagonist properties, can also have a direct beneficial effect on motor functions. Imipramine, desipramine, and nortriptyline all have been reported as effective antidepressants (Anderson et al., 1980; Laitinen, 1969; Strang, 1965).

Triyclics should be started at low dosages and increased gradually. Parkinson's disease is frequently complicated by orthostatic hypotension, which can be aggravated by tricyclics, resulting in syncope. Other autonomic problems of Parkinson's disease can also be aggravated by anticholinergic agents. These include urinary retention, erectile failure, and constipation.

Since serotonergic metabolism appears altered in Parkinson's disease, some of the newer, serotonergic antidepressants might be a better choice. These include trazodone, fluoxetine, and sertraline. There is little literature on this subject, however.

Electroconvulsive therapy (ECT) has also been reported to be useful in treatment of depression in patients with Parkinson's disease. Interestingly, motor symptoms may also improve with ECT, with the motor improvement occasionally antedating the improvement in mood (Asnis, 1977; Lebensohn & Jenkins, 1975). An increase in dopamine turnover as a result of ECT has been postulated as the mechanism of action. Patients may be more sensitive to the development of post-ECT confusion, as are other patients with central nervous system (CNS) disorders. Post-ECT confusion and agitated delirium can, at times, be quite pronounced and are a significant limiting factor to this procedure. (See also Chapter 10.)

MULTIPLE SCLEROSIS

Epidemiology and Etiology

Young adults are the chief victims of multiple sclerosis (MS). Epidemiologic studies suggest that the disease is probably acquired in childhood, and that there follows a clinically quiescent period of several years before the onset of symptoms. Onset of clinical MS is unusual before the age of 10 and in senescence. The risk of acquiring MS is highly correlated with the geographic location of the patient's childhood, and to a certain extent the risk of MS is a function of distance from the equator. Patients who live in New Orleans until the age of 15 have a 1:10,000 risk of developing MS, whereas the risk for patients who were reared in Canada is nearly 13 times higher. Patients who migrate in adulthood from a high-risk to a low-risk area retain an increased risk for MS, supporting the hypothesis of an infectious etiology with a long latency period. Detailed epidemiologic investigations of discrete epidemics of MS in the Faroe Islands (associated with the influx of the British Army) and in Newfoundland (linked with outbreaks of canine distemper) also provide indirect but supportive evidence of an infectious etiology (Kurtzke & Hyllested, 1986; Pryse-Phillips, 1986).

Other lines of evidence suggest an autoimmune factor in MS. Cerebrospinal fluid studies of patients frequently indicate elevated gamma-globulin, composed of heterogeneous molecules of polyclonal origin. Subsets of gamma-globulin with restricted heterogeneity (oligoclonal bands) are found in more than 90% of patients with MS. The presence of such bands, however, is only modestly related to the activity of the disease and is not specific to MS. The bands are found in many active neurologic diseases, including acute stroke, infection, and neuropathies, where their presence is thought to represent an immune response to neurologic injury (Chu, Sever, & Madden, 1983; Farrell et al., 1985). Furthermore, when gamma-interferon was administered intravenously (IV) to 19 MS patients, six (32%) experienced an acute flare of their disease, suggesting a partial role of immunologic function in pathogenesis (Panitch et al., 1987) Conversely, immunosuppression with adrenocorticotropin (ACTH) or cyclophosphamide has been somewhat effective in arresting acute exacerbations of the illness.

There is substantial evidence that genetic factors also contribute to the development of MS. The prevalence of MS is significantly higher among relatives of MS patients than in the general population (Doolittle et al., 1990), with first-degree relatives having risks of 30 to 50 times that of the general population (Sadovnick & Baird, 1988). Magnetic resonance imaging studies of asymptomatic family members showing unsuspected demyelination suggest this rate may be even higher (Lynch et al., 1990).

Furthermore, the concordance rate of MS is 11 times higher in monozygotic than dizygotic twins, 26% versus 2.3% (Ebers et al., 1986). Since the concordance rate for monozygotic twins is not 100%, however, environmental factors must play a role in the pathogenesis of MS.

Studies of experimental allergic encephalomyelitis, a research model that resembles MS, have led some researchers to propose that viral proteins bearing structural similarity to myelin components initiate an autoimmune response resulting in demyelination (Waksman, 1985). Possibly, subsets of the population have human lymphocyte antigen (HLA), or other endogenous markers, resembling the putative viral proteins, which would engender an autoimmune response. Population studies have revealed an association of MS with HLA antigens A3, B7, and DR2, and to polymorphism of the myelin basic protein gene (Boylan et al., 1990).

Pathology

The characteristic lesions of MS are plaques of scattered, circumscribed demyelination in the CNS. Such plaques occur predominantly in white matter, with a predilection for periventricular areas, optic nerves, and spinal cord. They can occur anywhere in the CNS, however, including cranial and spinal nerve roots. Microscopically, there is destruction of myelin sheaths and, to a lesser extent, of neuronal fibers. Lymphocytes, macrophages, immunoglobins, and complement are found in active disease. Later, acute inflammation may be replaced by reactive gliosis. Signs and symptoms are chiefly due to interference with neuronal conduction by the demyelinating plaques.

The great variations in locations, numbers, and recurrences of plaques among individual patients account for the very broad spectrum of the clinical presentation. Multiple sclerosis has justifiably been proposed to replace syphilis as "the Great Imitator" (Paulson, 1980). Initial symptoms may include motor or sensor impairment, visual loss, diplopia, ataxia, incontinence, vertigo, and changes in mental status. Later, as the disease progresses, initial symptoms may be partially or completely resolved, only to be replaced or joined by a panoply of new symptoms as fresh plaques occur elsewhere in the CNS.

Multiple sclerosis may be more common than previously realized. Postmortem studies and newer diagnostic techniques indicate that the characteristic demyelinating plaques of MS are found more frequently than expected from clinical history (Gebarski et al., 1985). Recent studies are also calling into question the basic course of the disease, once regarded as a relapsing and remitting illness. Poser (1985), in fact, suggested that the "clinical manifestations of MS are the externally observable parts of an underlying, essentially continuous process."

Consistent with this, the old distinction between "chronic progressive" and "relapsing-remitting" MS is becoming blurred as newer neuroimaging techniques demonstrate active lesion formation in asymptomatic patients who are clinically considered "in remission" (Willoughby et al., 1989).

Diagnosis

To improve diagnosis, numerous criteria for the classification of MS have been proposed. The most widely accepted systems divide patients into clinically definite, probable, and possible categories (Rose et al., 1976; Schumacher et al., 1965), which reflect persistent difficulties in confidently diagnosing MS. Older criteria were based on clinical findings and history alone, whereas the most recent Poser criteria incorporate findings on neurodiagnositic tests such as evoked potentials, neuroimaging, and immunologic studies with clinical findings (Poser et al., 1983).

Neurodiagnostic techniques may assist in accurate differential diagnosis, not only to discriminate MS from other neurologic illness but also to confirm the presence of disease. This is particularly important in patients whose symptoms are vague and fleeting. Not infrequently, such patients are initially thought to be hysterical.

Examination of the CSF is mandatory for investigating a diagnosis of MS when clinical evaluation and neuroimaging are inconclusive. In addition to the usual routine studies, specific immunologic studies can be helpful in confirmation of MS (Whitaker, 1985), and form part of the Poser criteria. The CSF of patients with MS contains qualitatively altered and quantitatively elevated immunoglobulin G (IgG). Up to 90% of patients with active MS show discrete bands on electrophoresis, termed "oligoclonal bands" (Farrell et al., 1985). Immunoglobulin A bands are found both in active disease and in remission (Grimaldi et al., 1985). Oligoclonal banding is also seen in other diseases in which there is chronic antigenic stimulation, such as neurosyphilis, subacute sclerosis, panencephalitis, varicella, mumps, and cryptococcal meningitis, so the presence of oligoclonal bands in itself is not sufficient for a diagnosis of MS (Chu, Sever, & Madden, 1983).

Myelin basic protein (MBP) can also be detected frequently in the CSF of patients with active MS (Martin-Mandiere et al., 1987). It is a component of brain tissue released into the CSF when there is white matter tissue destruction. Elevated MBP levels are also seen in other destructive processes such as encephalitis, hypoxia, or stroke. When tissue destruction is marked, MBP can even be detected in the blood (Jacque et al., 1982).

Evoked potentials (EPs) are noninvasive means of detecting MS. Multiple sclerosis is defined by the existence of two or more lesions separated by space and time. For the patient with only one clinical lesion, the definition of a second, clinically silent lesion helps confirm the diagnosis. Evoked potentials (EPs) are modality specific because they represent the electrical potential evoked by sensory input. They can be elicited by visual, tactile, or auditory input, and conduction delays in expected potentials are taken as indications of disease. (For further discussion of EPs, see McNamara, 1991b). Chiappa (1980) showed 81% of patients with definite MS had abnormal visual EPs, 47% had abnormal brainstem auditory EPs, and 68% of patients had abnormal somatosensory EPs. Again, abnormal EPs are not specific for MS, but are seen in a wide variety of CNS and other illnesses.

Neuroradiologic studies also contribute to the diagnosis of cerebral MS. On nonenhanced CT scans, hypodensities in the white matter corresponding to established MS plaques may be visualized. Far more sensitive than CT scans for the detection of MS plaques is MRI. The better resolution of MR images and their ability to detect lesions in the brainstem and spinal cord usually obscured by artifacts on CT scans make this procedure a clearly superior technique (Kirshner et al., 1985; Lee et al., 1991).

However, normal MR image of the head does not rule out MS. Approximately 70% to 95% of clinically definite cases of MS show multiple white matter lesions on MRI, and only 50% to 60% of clinically probable cases of MS show these lesions. Furthermore, patients can show lesions of the spinal cord but have a normal MR image of the brain. The detection of lesions can be increased by the use of contrast enhancement. Plaques that enhance with CT contrast, or with gadolinium on MRI, are associated with impairment of the blood-brain barrier, and are more acute lesions (Bastianello et al., 1990; Koopsmans et al., 1989).

Bursts of enthusiasm for new technologies should not obscure the maxim that laboratory investigations are never a replacement for clinical acumen. None of the above tests are specific for MS, and several disorders may imitate MS. A full discussion of these illnesses is beyond the scope of this chapter, and readers interested in the extensive differential diagnosis of MS are referred to standard neurologic texts and the following references: Gilbert and Sadler (1983), Rudick et al. (1986), Poser, Roman, and Vernant (1990), Powell et al. (1990), and Berger et al. (1992).

Psychiatric Aspects of Multiple Sclerosis

The neuropsychiatric symptoms of MS are as wide and varied as the motor and sensory symptoms. Euphoria was once regarded as the classic mental symptom of MS since its description by Cottrell and Wilson in 1926. "Euphoria" may be the presentation of parietal lobe anosognosia or of frontal lobe disinhibition, but detailed neuropsychiatric definitions are lacking. The presence of euphoria has been found to correlate with intellectual deterioration (Surridge, 1969)

Of interest, in recent years depression, rather than euphoria, has come to be regarded as the cardinal symptom of MS. This has resulted from changes in the definitions of both MS and euphoria (McNamara, 1991a).

Prevalence

In a recent report of large series of patients attending an MS clinic, Sadovnick et al. (1991) reported that 47% of patients died of a complication of their disease. Of the remaining other deaths, 29% were due to suicide. The proportion of suicides among MS patients was 75 times that of the age-matched general population. Clearly, the psychiatric complications of MS are important (Sadovnick et al., 1991).

Using strict Research Diagnostic Criteria, the Schedule for Affective Disorders and Schizophrenia-Lifetime interview, and multiple other ratings scales, Joffe et al. (1987) reported, in a study of 100 MS clinic patients, that 72% of the patients had some psychiatric diagnosis. Forty-two percent of the patients had a lifetime history of depression, and 13% met criteria for bipolar disorder. Dysthymia, panic, and anxiety were also commonly noted.

Reischies et al. (1988) and Honer et al. (1987) have linked the degree of depression to the density of plaques seen on MRI. Dalos and colleagues showed that depression was linked to disease activity when they reported that emotional disturbance was present in 90% of patients with progressive illness, compared to 39% of stable patients (Dalos et al., 1983). Foley et al. (1990) showed in a prospective study that psychologic distress is correlated with immune status in MS, reporting increases in CD4+/CD8+ lymphocyte ratios as depression increased.

Confounding the psychiatric presentation of MS patients is a high rate of dementia. Intellectual impairment has long been described in MS but, despite abundant research, has been poorly defined until recently (Ron, 1986). The degree of cognitive impairment does not correlate very closely with the degree of other neurologic impairment (Jacobs et al., 1986; Peyser et al.,

1980; Rao et al., 1989), with lesion number on MRI (Huber et al., 1987), or with degree of depression (Lyon-Caen et al., 1986). IQ scores and intelligence tests, when used alone, are also relatively insensitive (Ron, 1986). The Mini-Mental State Examination, which commonly is used to define dementia in a number of reports, has a sensitivity rate of only 18% (Rao et al., 1990). When patients with MS are carefully examined with detailed neuropsychologic testing, however, a high rate of cognitive dysfunction is seen. Stringently defining impairment as performance at or below the fifth percentile of the control group, Rao et al. (1991) reported that 43% of MS patients were impaired on four or more of 31 cognitive indices. Memory, attention, and conceptual reasoning were most commonly impaired.

Therapy of Multiple Sclerosis

Although more scientific investigation is needed to delineate the incidence and clinical features of psychiatric disturbance in MS, several features of the illness suggest plausible hypotheses both for etiology and for treatment. At a psychologic level, MS is a fearsome illness. It strikes young adults at a crucial time in their psychosocial development. Furthermore, the prognosis for an MS patient is uncertain, so planning for the future is difficult. The uncertainty of when one will have the next symptom, or become blind, demented, or crippled, can be terrifying. The symptoms of MS—gait problems, incontinence, blindness, and so on—can be demoralizing to patients and their families and increase their feelings of helplessness.

Physical and occupational therapy may improve mood as well as maximize body functioning. The National Multiple Sclerosis Society, with its many local branches, affords patients and their families support and education as well.

The medications used to treat MS are potentially another common source of neuropsychiatric symptoms. Corticosteroids, a frequently-used palliative therapy for acute exacerbations of MS, are well-known causes of depression, mania, psychosis, and cognitive impairment (Wolkowitz et al., 1990). Previous treatments of ACTH to stimulate endogenous corticosteroid production, and oral prednisone, often administered over weeks, are associated with significant morbidity. In 1980, Dowling, Bosch, & Cook introduced high-dose pulse intravenous methylprednisolone treatment, given over a shorter period of time (Thompson et al., 1989), which may have fewer complications.

Baclofen, used to treat spasticity, is associated with encephalopathy, auditory and visual hallucinations, paranoia, psychosis, and dyskinesias, possibly as a result of alterations in central dopamine metabolism. Neuropsychiatric side effects of baclofen occur both in administration and in withdrawal (Abarbanel, Herishanu & Frisher, 1985; Kirubakaran, Mayfield, & Rengachary, 1984).

More recently, attempts have been made to suppress the progress of MS with immunosuppressants such as plasmapheresis (Khatri et al., 1991; Weiner et al., 1983), total lymphoid irradiation (Cook et al., 1986), cyclophosphamide (MS Study Group, 1990), and peptides immunologically similar to fragments of MBP (Bornstein et al., 1991), each with some modest success, although the role of immunosuppression in the treatment of MS remains unclear (Goodin, 1991). The risks of profound immunosuppression include opportunistic infection, renal and hepatic toxicity, bone marrow suppression and the possible long-term development of neoplasia. The psychiatric treatment of the MS patient involved in experimental protocols mandates careful consultation with the investigators, not only to avoid invalidating the protocol but also to prevent the coadministration of drugs that are potentially harmful.

Treatment of Psychiatric Complications

There is little specific literature on the treatment of psychiatric illness in MS. Schiffer (1987) has pointed out the necessity of identifying the neurologic, psychologic, and social components of each patient's presentation. A careful history of the development of psychiatric symptoms relative to the time frame of the development of MS is crucial, as is a review of medications and physical examination. Neuropsychologic testing and an interview with family members also are often indicated.

Psychotherapy must be appropriate to the patient's ability to understand and remember the process. Depression may have a more profound effect on the patient's social and vocational functioning than the symptoms of MS itself. Schiffer (1987) has pointed out that, while many patients fear loss of independence, particularly those with active disease, other patients in whom the disease is quiescent may become depressed, fearing loss of dependent gratifications as their neurologic function improves.

Issues in family therapy frequently revolve around grief, anger, and frustration. Moreover, if discrimination between organically-determined and "voluntary" behavior is difficult for the clinician, it is even more so for family members of the patients. Considerable frustration, hurt feelings, anger, and retribution are not uncommon when families do not understand the patient's behavior. Proper diagnosis assists not only the clinician but also the family. For example, pseudobulbar lability is common in patients with MS. In this symptom, there is a relative or an absolute uncoupling of the felt emotion, mood, and expressed affect. Patients may burst into disinhibited laughter or tears at grossly inappropriate moments, often significantly alienating family members in the process. In families in which these symptoms are not understood as involuntary and beyond the conscious control of the patient, anger and isolation result. This adds to the considerable stress that can be expected whenever a chronic disabling illness occurs in a family member.

Psychopharmacologic Issues

Although there are few systematic studies of the psychopharmacology of MS, several clinical maxims are evident from an understanding of the pathophysiology of MS. First, autonomic problems are quite common in MS. These include orthostatic hypotension, sexual dysfunction, and neurogenic bladder (Blaivas, 1980). Bladder infections, exacerbated by emptying problems and catheterization, are among the chief causes of death in MS patients, mandating particular attention to these problems. Most psychotropics have autonomic effects. Patients with MS can be exquisitely sensitive to anticholinergic and hypotensive side effects, resulting in orthostasis, syncope, urinary retention, constipation, etc. Monitoring of blood pressure, heart rate, and urologic functioning is crucial, and heavily anticholinergic agents such as amitriptyline and chlorpromazine generally should be avoided.

Because up to 25% of plaques undercut gray matter, MS patients may be vulnerable to seizures. Up to 10% of MS patients have seizures at some point in their illness (Kinnunen & Wikstrom, 1986). The physician should consider an electroencephalogram (EEG) prior to the institution of psychotropics, and consider the use of antiepileptic drugs when an ictal focus is found.

Since many of the more severely disabled MS patients may also be malnourished and/or have renal impairment from prior urinary tract infections, the physician should also give some consideration to albumin levels and binding fractions when adjusting psychotropic dosage. Moreover, probably because of multiple cerebral lesions, MS patients can be quite sensitive to the cognitive side effects of medications, with excessive sedation, confusion, or delirium occurring.

Trazodone, which has been reported as helpful in elderly patients with dementia and behavior disturbance (Simpson & Foster, 1986), may also be helpful in MS, although this has not been specifically addressed in the literature. Excessive sedation and ataxia are limiting factors in its use with MS patients.

Use of MAO inhibitors can be problematic in MS because of orthostatic hypotension. Moreover, the MS patient with frontal lobe dishibition and poor judgment might have difficulty complying with a tyramine-free diet.

Fluoxetine, which is not usually sedating and has not anticholinergic side effects, has been a useful antidepressant in the author's experience, although again there is little research literature support for its use in MS. As with tricyclics, fluoxetine can precipitate seizures in susceptible patients, mandating an EEG before its use, and an review of an MR image for lesions that involve or undercut the cortex. The long half-life and high protein binding suggest slow dosage titration, beginning either with less than daily dosage of the 20-mg capsule or dilution with liquid to give a dose of 5 to 10 mg/day. Sertraline (Zoloft), a recently released serotonin uptake inhibitor, has similar advantages but has a significantly shorter half-life that allows easier and quicker dosage titration.

Bupropion (Wellbutrin), a stimulating antidepressant with virtually no sedative or anticholinergic effects, would be another rational option for the MS patient with depression, particularly if the depression were accompanied by apathy and fatigue. Because of the risk of seizures with bupropion, concurrent antiepileptic prophylaxis would be advisable if there were any history of seizures, an abnormal EEG, or lesions in or near the cortex.

Carbamazepine is frequently used to treat some of the paroxysmal symptoms of MS, such as trigeminal neuralgia; a mood-stabilizing effect has been observed when this is done. Lithium tends to produce ataxia, tremor, and confusion in MS patients and often is poorly tolerated. The polyuria that frequently results from lithium use can aggravate problems with incontinence. For MS patients with bipolar symptoms, carbamazepine might thus be preferable to lithium. Carbamazepine, which lowers the white blood cell count, should probably be avoided if the patient is on immunosuppressants, however. If lithium is used, a relatively low blood level is advisable, and amiloride can be used if polyuria aggravates incontinence.

Valproate, another antiepileptic drug with mood-stabilizing properties, is also helpful for some patients with MS. The most troublesome side effects are weight gain and tremor. However, the drug's relative lack of cognitive and hematologic side effects make it an attractive drug for some patients.

Multiple sclerosis patients also appear to be particularly sensitive to the extrapyramidal side effects of neuroleptics, so relatively low doses should be used, and prophylactic antiparkinsonian medication should be strongly considered, particularly with high-potency neuroleptics. Amantadine would be the first choice because it has relatively few anticholinergic effects and independently may reduce fatigue in MS patients, possibly through dopamine agonism (Cohen & Fisher, 1989).

Despite these concerns, depression in MS patients can and should be treated because the depression can be more a source of anguish, and can be more disabling, than the MS itself. As a general principle, after thorough medical and neurologic evaluation, the dosage should be increased slowly under careful monitoring. In the only systematic study of antidepressant therapy in MS, Schiffer and Wineman (1990) reported successful treatment with desipramine, although side effects limited dosage for half of the treated patients. Elsewhere, Schiffer et al. reported that low-dose amitriptyline can improve pseudobulbar lability (Schiffer, Hernedon, & Rudick, 1985).

Finally, in addition to relieving the despair, depression, and confusion that commonly accompany MS, there remains the frankly hypothetical but reasonable hope that treating the psychiatric problems of MS might affect the course of MS itself. There is substantial evidence that immunologic factors, which clearly play a role in MS, are affected by psychiatric status. As noted above, researchers have found that disease activity in MS is associated with a worsened emotional status (Dalos et al., 1983), and, as a correlate, that depression in MS patients is associated with lower CD8+ lymphocytes, and higher CD4+/CD8+ ratios (Foley et al., 1990).

Given the uncertain nature of MS, perhaps the most useful thing the clinician can do is to provide stable and consistent support, advice, and advocacy for the patient. Families may also benefit from an opportunity to work through the stress, anxiety, and guilt that may be engendered. Divorce, loss of employment, poverty, and social isolation also cripple patients with MS and add to their despair. Knowing that there is an informed and caring person who will not abandon them may make their losses more bearable.

STROKE

Etiology of Stroke

Cerebral infarction has many etiologies. Most commonly, carotid artherosclerotic plaques ulcerate, dislodging "white emboli" that travel distally and lodge in cerebral blood vessels, causing occlusion and ischemia. Alternatively, atheromata may cause stenosis and occlusion of carotid arteries with compromise of cerebral blood flow. As the incidence of atherosclerotic vascular disease increases with age, so does the incidence of stroke (Barnett, 1980; Beal et al., 1981; Bogousslavsky & Regli, 1986).

Cardiac disease is also a significant cause of cerebral embolism, often resulting in large and functionally devastating strokes. Atrial fibrillation with hemostasis and development of clot within a dilated left atrium accounts for approximately half of cardiogenic emboli. Other etiologies include rheumatic valvular disease, cardiomyopathy, endocarditis, and acute myocardial infarction.

Another common cause of stroke, with particular prevalence in the elderly, is hyaline degeneration of the small midline perforating arteries of the brain, resulting in lacunar infarctions. At times, stenosis of the os of the small perforators, or emboli from atheromata or cardiac sources, also can cause lacunar infarction (Chimowitz et al., 1991). These small infarcts have both a pathophysiology and a presentation distinct from large-vessel infarctions. Lacunae most commonly occur in the periventricular white matter of the frontal lobe and at the head of the caudate nucleus. Because of their anatomic locations, which interrupt frontal neuronal pathways, lacunar infarcts most often result in frontal lobe symptoms such as emotional lability, gait apraxia, abulia, and incontinence. At times, small infarcts may be so numerous that diffuse softening of the subcortical white matter occurs, a condition known as Binswanger disease.

Hypertension is the most common underlying cause of lacunar infarction. Other etiologies include high blood viscosity, vasculitis, and amyloid angiopathy (Caplan, 1980; Cosgrove et al., 1985; Estes et al., 1991; Ghika, Bogousslavsky, & Regli, 1990; Hendricks, Franke, & Theunissen, 1990; Ishii, Nishihara, & Imamuia, 1986; Kinkel et al., 1985; Vonsattel et al., 1991).

Stroke is not just a disease of the elderly. Three percent of all strokes occur in young adults, and the incidence is probably increasing as a result of the increase in cocaine abuse. Migraine, alcohol, mitral valve prolapse, trauma, autoimmune disease, antiphospholipid antibodies, cigarette use, and the use of oral contraceptives have been identified as other contributing factors (Adams et al., 1986; Briley, Coull, & Goodnight, 1989; Dorfman, Marshall, & Enzmann, 1979; Hillbom & Kaste, 1981; Jackson, Boughner, & Barnett, 1984; Levine et al., 1991a, 1991b; Sloan et al., 1991).

Psychiatric Presentations of Stroke

The abrupt onset of hemiparesis in middle cerebral artery infarction is dramatic and immediately obvious, but the presentation of cerebrovascular ischemia can at times be far more subtle. Posterior cerebral artery infarctions may produce no motor symptoms, and a left posterior infarct may have as its role overt manifestation the acute onset of a fluent aphasia. Patients suffering such infarcts, whose speech is confused and nonsensical, are occasionally misdiagnosed as psychotic and admitted to psychiatric units. Since examination of the patient's visual fields usually reveals a right homonymous hemianopsia, it is lamentable that clinicians often omit this simple test (Benson, 1973).

Right hemisphere infarctions may also be mistaken for psychiatric illness. Delirium, delusions, and bizarre behavior may be the presentation of right-sided stroke, particularly if there is preexisting cortical atrophy (Hier, Mondlock, & Caplan, 1983; Levine & Grek, 1984).

Rostral basilar artery infarction may also result in disturbances of behavior and vision with a paucity of motor symptoms. With high brainstem infarction, patients may have vivid visual and auditory hallucinations with intact reality testing. Other patients present with agitated delirium, confusion, and disorientation (Caplan, 1980; Cascino & Adams, 1986; Geller & Bellur, 1987; Mehler, 1989). Bilateral paramedian thalamic infarction from occlusion of a Y-shaped branch of the posterior circulation, the thalamosubthalamic perforating artery, can produce profound subcortical dementia from very small lesions (Guberman & Stuss, 1983).

The prompt diagnosis of stroke has more than nosologic significance. Regardless of etiology, whether carotid, cardiac, lacunar, aneurysmal, or systemic disease, strokes often recur. Further neurologic damage can be prevented by timely medical intervention.

Cerebral edema develops usually between 2 and 5 days after a large stroke, and it may be so severe as to cause brain herniation and death. Treatment of this edema can be lifesaving (Ropper & Shafran, 1984). Moreover, evidence is accumulating that part of the neuronal injury that occurs in stroke may not take place immediately but may occur over several hours to days, and may be reversible if treated rapidly. A cascade of biochemical events if initiated at the onset of stroke, and there is solid animal evidence that it is this subsequent chain of events, not the ischemia itself, that causes much of the lesion (Scheinberg, 1991). One critical mechanism involves excessive influx of calcium into compromised neurons through channels gated by excitatory neurotransmitters, particularly glutamate and aspartate (Benveniste et al., 1988; Stys, Waxman, & Ransom, 1991). Moreover, restoration of blood flow after ischemia, as Hallenbeck and Dutka (1990) noted, "has a dark side." Further cell death can occur from the interaction of blood and damaged tissue. These potential mechanisms, involving free radical generation, leukocyte-mediated injury, and altered endothelial function, are complex but suggest further potential interventions. Potentially, pharmacologic intervention acutely poststroke might greatly lessen the resultant deficit. For example, Simon and Shiraishi (1990) reported that blockading excitatory amino acid receptors 5 minutes before middle cerebral artery occlusion in rats decreased the size of stroke by 50% to 82%. If and when clinically effective neuroprotective treatments are developed early diagnosis of stroke will be crucial.

Because it is impractical to refer all patients with behavioral abnormalities to a neurologist, psychiatrists should be familiar with methods to confirm or exclude a diagnosis of stroke. A screening neurologic examination with visual fields should be performed on every patient. It is an error to believe that CT scan will exclude the possibility of cerebrovascular infarctions, because often new strokes will not be visible for several days. Moreover, lacunar strokes are often below the resolution of CT scans, although most are detected with MRI. An EEG may show focal slowing before changes are demonstrated on radiographic studies, but lacunar infarcts usually do not produce any EEG changes (Schaul et al., 1986). A high index of suspicion is warranted when patients are elderly, hypertensive, or diabetic; are substance abusers; or have a history of cardiac disease.

Some mention should be made of the evolving understanding of the significance of perivascular hyperintensities on MR images. Initially, it was thought that all such findings indicated small strokes. More recent work has questioned this, although such hyperintense lesions are more common in dementia (Steingart et al., 1987). It appears that some of the white matter hyperintensities seen on an MR image may represent infarcts, but some reflect changes in myelin or localized edema. It is the number, location, and etiology of these hyperintensities, not their presence alone, that indicates pathology. This issue is discussed at length in Chapter 18.

Poststroke Depression

The most prevalent and most studied of poststroke psychiatric syndromes is depression. Folstein, Maiberger, and McHugh (1977) examined 20 consecutive stroke patients admitted to a rehabilitation hospital, and found a 45% prevalence of depression indicated by the Present State Examination and Hamilton Rating Scale. Orthopedic patients with a similar degree of physical disability had only a 10% prevalence of depression. Finklestein et al. (1982), in a similar investigation, found a 48% prevalence of depression.

In an effort to better define the clinical characteristics

of poststroke depression, Robinson and associates evaluated patients hospitalized for stroke and followed them longitudinally, defining depression by DSM-III criteria, but ignoring the exclusion for cases of organic etiology. At the time of initial examination, 26% of 103 patients met criteria for major depression and a further 20% met criteria for dysthymia. A 6-month follow-up, the prevalence of major depression had increased to 34% and another 26% of patients had dysthymia. Only 40% of all patients remained nondepressed. A 12-month study of participants in a stroke clinic for ambulatory patients suggests that major depression that develops after a stroke has a natural course of about 8 months (Robinson, Lipsey, & Price, 1985; Robinson & Price, 1982).

Although depressed mood in stroke patients has often been attributed to a reactive response to physical disability, available data do not support this conclusion. Degree of depression has not correlated with degree of disability. Patients with physical disability from other causes have not demonstrated such a high incidence of depression.

The etiology of depression after cerebrovascular accident is unknown, although several hypotheses present themselves. The abrupt onset of a stroke, often without premonitory warnings, coupled with the resultant devastating and obvious impairments, can be a traumatic and frightening experience for patients and families. Whereas other illnesses, such as myocardial infarction, may threaten life, brain injury threatens the sense of self. Patients will often say that they would rather die than "become a vegetable," and the threat of another stroke, despite therapy, is always a possibility.

Some authors have noted an increase in depression associated with specific locations of stroke, and they postulate a neuroanatomic etiology of mood disorder in stroke. Robinson et al. (1984) reported that the highest frequency of depression was seen with left anterior brain infarctions and that, the closer the left hemisphere lesion was to the frontal pole, the more severe was the depression. When infarcts occurred in the right hemisphere, depressive symptoms correlated with proximity to the occipital pole. This study, however, excluded patients with severe comprehension deficits—presumably patients with left posterior infarctions. Also, patients with right anterior frontal-parietal infarctions may appear apathetic and euphoric, so their depressions may be difficult to detect with standard, unmodified interviews and rating scales.

Starkstein et al. (1990) reported two patients who had both an anosognosia and clinical depression, noting that awareness of physical disability did not necessarily correlate with awareness of emotional disability, but did not address the sensitivity and specificity of this approach. It is therefore unclear if there is truly an anatomic left-right or anterior-posterior gradient for depression, or whether the observed association is an artifact of patient selection and diagnostic procedure. This ambiguity has recently been noted by others as well (Stern & Bachman, 1991).

Animal models, however, suggest that there is a true asymmetry in the brain in response to stroke that may underlie the development of depression. In rats with experimentally induced right middle cerebral artery ischemic lesions, there was widespread subsequent depletion of norepinephrine and dopamine, in both injured and uninjured cortex, both ipsilateral and contralateral to the lesion site (Robinson, 1979). Cortical norepinephrine was decreased within 12 hours by 75%, with similar depletions in the locus ceruleus, where the noradrenergic cell bodies are located. A similar widespread depletion of dopamine, to below 50% of control levels, also occurred in the substantia nigra and ventral tegmental area. Interestingly, identical lesions made in the left middle cerebral artery did not produce significant changes in norepinephrine and dopamine. Furthermore, this hemispheric asymmetry in response to stroke was observed only in male rats (Lipsey et al., 1986). Female rats had depletions with both right- and left-side lesions.

Studies have examined poststroke patients by using radiolabeled ligands to examine receptor bindings on positron emission tomography scans. Preliminary investigations suggest the poststroke patients have decreased serotonin receptor bindings with left hemisphere infarcts, with the opposite finding—increased serotonin binding—in right hemisphere infarcts. Such as asymmetry might underlie the observed increase in depression following left hemisphere stroke (Riesenberg, 1986).

Other Poststroke Psychopathology

Other psychiatric disturbances have also been reported to follow stroke, but these are less well studied. Mania has been reported with right hemisphere lesions, but the discrimination between true mania and its mimic, frontal lobe disinhibition, has been problematic. In other reports it is unclear whether the excitable behavior observed is due to a central lesion per se or is more related to secondary epileptogenesis (Cummings & Mendez, 1984; Jampala & Abrams, 1983).

Paranoia has also been observed as a consequence of stroke, particularly in patients with left posterior lesions and fluent aphasia. These patients are unable to comprehend language and, as Benson (1973) noted, "in the most severe cases the patient literally does not understand that he does not understand." Patients may believe that others are plotting against them and, unaware that the problem lies within themselves, become delusional and agitated. Similar phenomena have been observed in deafness and may underlie some instances of late-onset schizophreniform psychoses (Benson, 1973; Miller et al., 1986).

Psychiatric diagnosis in the stroke patient may be made difficult by the presence of neurologic impairments that complicate the presentation. Infarcts of the left hemisphere, producing aphasia, may render patients incapable of communicating their moods verbally. Lesions of the right hemisphere—which appears to be dominant for communication of mood by gestures, facial expression, and prosody—may cause a dissociation between subjective feelings of mood and overt expressions of affect. Flat, monotonous, and uninflected speech may mask the presence of a mood disorder. Similarly, bland denial of illness (anosognosia) produced by right parietal infarctions may also obscure the presentation of depression (Ross, 1981; Ross & Rush, 1981). Multiple lacunar infarcts often produce organic emotional lability or abulia, which can confound diagnosis (Ishii, Nishihara, & Imamuia, 1986).

Disturbances of sleep and appetite and other "vegetative" symptoms, usually helpful in the diagnosis of depression, may also be misleading in the poststroke patient. These functions may be affected by neurologic lesions themselves, by medications, and by the environment. Disinterest in self-care, poor motivation, withdrawal, negativism, irritability, anhedonia, anxiety, and sluggishness appear to be more consistent indicators of depression in the poststroke patient. Interviews with the family and the caretakers of the patients often provide valuable information about mood and behavior that the patient is unable to give (Finklestein et al., 1982; Lipsey et al., 1986; Reding et al., 1985, 1986; Ross & Rush, 1981).

Treatment of Poststroke Depression

The treatment of poststroke depression is similar to that of other depressive disorders, but with some special considerations. The leading cause of death in poststroke patients is cardiac disease. Vascular disease that occurs in the carotids also occurs in the coronary arteries, and myocardial infarction, angina, congestive heart failure, and conduction abnormalities are common associated diagnoses. The stroke patient should have a cardiologic evaluation, and psychotropic drug therapy should take cardiac risk factors into account (Heyman et al., 1984; Rokey et al., 1984). (See Chapter 9.)

Hypertension is another common associated finding in stroke. Either hypertension or hypotension may be hazardous to the patient with coronary and cerebrovascular disease. Moreover, stroke patients may be more sensitive to the hypotensive side effects of psychiatric medications than other patients. Those with hemodynamically significant stenosis of the carotid artery may be at particular risk of stroke with an episode of hypotension.

Seizures develop in more than 10% of all stroke patients. Strokes that involve or undercut cortex increase the risk for seizure to 26% (Faught et al., 1989; Olsen, Hogenhauen, & Thage, 1987). Since tricyclics and neuroleptics affect seizure threshold, the psychiatrist should assess the risk of seizures, usually in consultation with the patient's neurologist, prior to the institution of these drugs. Lesion location, and possibly an EEG, are the most useful guides to decide whether antiepileptic drugs should first be administered. Of course, short-term neuroleptics may occasionally be necessary to control agitation before an EEG can be obtained, and low dosages of high-potency neuroleptics are preferable in this situation (Cocito, Favale, & Reni, 1982).

Another significant risk to stroke patients is the development of dysphagia and recurrent aspiration. Patients with bilateral strokes are at particular risk because a full two thirds of such patients aspirate (Horner, Massey, & Brazer, 1990). Multiple psychotropics may further impair the swallowing reflex, potentially placing the patient at risk of inanition, pneumonia, or asphyxia. Examination of the gag reflex, voluntary cough, or the presence of dysphonia assist in identifying patients at risk (Moss & Green, 1982).

Stroke patients frequently are on multiple medications at the time of evaluation. Propranolol, reserpine, clonidine, and methyldopa, used in the treatment of hypertension, can produce depression and other psychiatric symptoms. Digitalis, tocainide, calcium channel blockers, and other cardiac medications also are associated with significant psychotropic effects (DeMuthy & Ackerman, 1983; Erman & Guggeheim, 1981; Snyder & Reynolds, 1985; Vincent, 1985).

Medications can interact to affect protein binding and clearance rates. Oral anticoagulants (e.g., warfarin) may interact with a number of psychiatric medications, with subsequent changes in coagulation times. Prothrombin times therefore should be checked more frequently in patients on anticoagulants when psychiatric medication is being initiated or changed. Propranolol can elevate levels of chlorpromazine and other neuroleptics. Because many drug-drug interactions remain to be explored, relevant serum drug levels should be measured whenever new medications are introduced or dosages are changed (Wood & Feely, 1983).

There is little evidence in the literature of successful treatment of poststroke depression. Careful readings of reports of such treatment reveal a high rate of drug side effects and treatment failure. Lipsey et al. (1984) administered nortriptyline to poststroke patients, beginning with 20 mg at bedtime and increasing slowly over 4 weeks to 100 mg and/or until therapeutic blood levels were achieved. Patients who were able to tolerate the medication evidenced a significant decrease in depression as measured by the Hamilton Rating Scale and the Zung Depression Scale. Seven of 17 patients were unable to complete treatment because of side effects, chiefly delirium. Trazodone also has been reported to be efficacious in poststroke depression. Reding and colleagues (1986) employed 50 mg, increased every 3 days to a maximum of 200 mg, but only a small minority of patients were able to tolerate this dose. A significant proportion of patients developed side effects, chiefly oversedation. The causes of such failures are unclear but do give some pause.

The ideal timing for initiation of antidepressant therapy after stroke is unknown. It is advisable to wait at least until the acute effects of stroke, such as autonomic destabilization and cerebral edema, have resolved and the pathophysiology of the stroke has been determined.

Electroconvulsive therapy has also been given to patients with poststroke depression, but the safety of this procedure for patients with cerebrovascular disorders is not fully established because long-term follow-up studies have not been reported. Murray, Shea, and Conn (1986) retrospectively reported the successful use of ECT in 14 patients, but there was a lengthy average time between stroke and ECT of up to 12 years. Given the reduced life expectancy for most poststroke patients, one must question whether these patients who survived so long were representative of typical stroke patients. Only one patient was reported to have suffered a complication of ECT—cardiac arrhythmia—but the criteria for patient exclusion were not documented. The incidence of cardiac disease, carotid stenosis, vascular malformations, and other lesions commonly associated with cerebral infarction were not reported. Given that ECT is commonly associated with transient marked elevations in blood pressure, as well as transient cardiac arrhythmia, the presence of such lesions may be a significant risk factor. Weisberg, Elliot, and Mielke (1991) have reported stroke with lobar hemorrhage associated with ECT. In our experience, both the age and the etiology of the stroke are important determinants of the advisability of ECT. Chapters 10 and 11 discusses relevant precautions in anesthetic management of ECT in high-risk patients.

In conclusion, although clinical considerations warrant the empirical drug treatment of poststroke depression, more rational treatment will await further clarification of pathophysiology and further controlled studies of treatment outcome.

MYASTHENIA GRAVIS

Myasthenia gravis (MG) is a disease that begins quietly and insidiously, and can terminate in catastrophe. The illness starts at the microscopic level, with immune blockade at the neuromuscular junction. The movement of large muscles depends on biochemical events at millions of receptors on muscle fibers, which implement signals received from terminal nerve fibers. Normally, nerve terminals release acetylcholine, which floats across a tiny synaptic space and is received by receptors on muscle end plates and initiates muscle contraction. In MG the ability of acetylcholine to transmit this impulse from nerve to muscle is impaired because the receptors for acetylcholine are damaged.

Initially, when relatively few receptors are damaged, the symptoms of MG are few, nonspecific, and evanescent. The patient, often a young woman, notes fatigue at the end of the day, after exercise, or after emotional upset. Patients may also note occasional visual blurring or diplopia.

The neurologic examination is entirely normal when the patient is asymptomatic, and even when symptoms are marked there are normal reflexes, normal sensory examination, and little or no muscle wasting. Even individual muscle testing can be normal unless the examination is done repetitively. Therefore, misdiagnosis of MG is common in the early stages.

With progression of the illness, signs become more overt and muscle weakness more constant. Involvement of extraocular muscles is common, with ptosis, dysconjugate gaze, and diplopia in a variety of patterns. Ocular symptoms are present in 84% of patients at presentation and are the *only* presenting symptoms in 54% (Bever et al., 1983; Kaminski et al., 1990). Weakness of other muscles controlled by cranial nerves is also seen, with facial weakness and palatal and pharyngeal dysfunction resulting in difficulties in mastication and swallowing. The illness is life threatening when weakness involves the muscles of respiration.

Myasthenia can begin at any age, but most commonly begins in young adulthood, when it affects women three times more frequently than men. When MG begins in later life, the sex ratio is reversed and male patients predominate. Weakness, which can begin focally, usually becomes generalized within 13 months of onset and maximal within 3 years (Grob, Brumner, & Namba,

1981). Prior to the advent of modern critical care methods, MG was frequently fatal, often from "myasthenic crisis" with profound weakness and respiratory failure.

In MG, acetylcholine receptors are injured by autoantibodies. Why the patient begins producing these autoantibodies is unknown, but features of MG can be induced by immunizing animals with acetylcholine receptors, producing antibodies. Once damaged, the ability of receptors to respond to acetylcholine is decreased and altered. One way to demonstrate this altered function, and a key diagnostic test, is electromyography (EMG). Normally, electrical stimulation of a motor nerve causes a release of acetylcholine and contraction of the muscle fiber, which can be recorded by EMG as a muscle action potential. In the same way that a patient with MG fatigues with exertion, in EMG, repetitive nerve stimulation causes an abnormal decrement of the expected potentials. However, routine EMG is somewhat insensitive for mild generalized or ocular MG and will miss about 30% to 40% of such cases (Kelly et al., 1982). Single-fiber EMG, a recent refinement, can be helpful in equivocal cases, detecting 88% or more of patients (Sanders, Howard, & Johns, 1979).

Another way to diagnose MG is to detect directly the circulating autoantibodies. Acetylcholine receptor (AChR) autoantibodies are many (polyclonal) but include IgG antibodies directed against a particular portion of the receptor, the extracellular portion of the alpha subunit. The presence of AChR antibodies correlates roughly with the severity of the illness. Using conventional methods, antibodies are found in 24% of cases in remission, 50% of cases of active ocular myasthenia, 80% of cases of mild generalized MG, and 89% of severe cases (Tindall, 1981). Enzyme-linked immunosorbent assay (ELISA) (Engel, 1984) is much more sensitive, detecting IgG AChR antibodies in the serum of 85% to 90% of myasthenic patients (Steck, 1990). A proportion of the remainder may have IgM antibodies (Yamamoto et al., 1991) not detected by the usual methods.

Antibodies to striated muscles also can be present in MG. Although their role in pathogenesis is uncertain, their association with thymoma is a useful clinical feature.

An otherwise rare tumor, thymoma, is present in 10% to 15% of myasthenic patients. Most are benign tumors but about a third are malignant (Keesey et al., 1980). Another 50 % of patients show hyperplasia of the thymic medulla. Although the role of the thymus in MG remains somewhat unclear, it seems likely that sensitization to AChRs occurs in the thymus (Sommer et al., 1990). Removal of the thymus, particularly in the early phase of the illness in young adults, often ameliorates MG. Detection of a thymoma by radiologic studies is an absolute indication for thymectomy at any stage of the illness or age of the patient because of the possibility of malignancy (Miller et al., 1991; Olanow, Lane, & Roses, 1982).

Psychiatric Implications

A major concern for psychiatrists is that "hysterical" patients referred for "neurasthenia," with a normal neurologic examination, might actually have myasthenia. As discussed, the diagnosis of mild cases of myasthenia can be exceedingly difficult. This was also noted by Sneddon in 1980 when she reported that 30% of myasthenic patients had initially been misdiagnosed as psychiatrically ill. Nicholson, Wilby, and Tennant (1986) interviewed 32 newly diagnosed myasthenics and found that 46% of patients who presented with generalized symptoms were initially misdiagnosed, and that they had seen at least five doctors and waited over 3 years for a definitive diagnosis. They noted that misdiagnosed patients had higher evidence of anxiety and anger. They properly noted that the failure of diagnosis itself may have contributed to these characteristics.

Accurate diagnosis is crucial, however, because of many of the psychotropics and other medications used by psychiatrists can acutely worsen MG, sometimes with catastrophic results. The imperiled AChR receptor can be further compromised by drugs that add to neuromuscular blockade. These include lithium (Neil, Himelhoch, Licata, 1976), chlorpromazine (McQuillen, Gross, & Johns, 1962), propranolol, phenelzine, phenytoin, and others (Argov & Mastaglia, 1979). (The use of contrast in CT can also aggravate MG). Santy (1983) reported two cases of women who presented with weakness and anxiety, worsened by stress. Fear and anxiety are symptoms of panic attacks, but are also common symptoms of respiratory failure, and both these patients were found to have MG. Both women became entirely asymptomatic when treated with antimyasthenic drugs. The use of respiratory depressants in such cases, such as benzodiazepines, could potentially result in respiratory failure.

Proper care of the myasthenic patient requires close collaboration with the neurologist. Since the illness is subject to fluctuations and occasional abrupt, precipitous worsening, any treatment plan needs continuous reassessment.

Since ventilatory failure is the most serious risk of MG, agents that depress respiratory drive, such as benzodiazapines, lithium (Weiner et al., 1983), and sedatives, generally should be avoided. Patients with moderate to severe bulbar or generalized myasthenia are most likely to develop "myasthenic crisis," which can occur with intercurrent infection, thymectomy, or steroid initiation but may also suddenly happen for no apparent reason. Serial measurements of vital capacity and blood gases can assist in discriminating respiratory symptoms of hypoxia and hypercarbia from psychogenic symptoms of anxiety, fatigue, or confusion. Furthermore, while daytime respiratory function can be adequate, breathing in sleep can be impaired and the patient should be carefully monitored for hypersomnia, nonrestorative sleep, apneas, and daytime somnolence (Quera-Salva et al., 1992). Further insults to respiratory function, particularly cigarette smoking and obesity, can be treated with appropriate psychiatric intervention.

Since MG has a predilection for facial and bulbar muscles, swallowing is often impaired, putting the patient at risk for aspiration. Neuroleptics impair swallowing and the gag reflex, and aggravate myastenia, and should be avoided.

Several other reports suggest that worsening of myasthenia is precipitated by emotional stress. Mackenzie, Martin, and Howard (1969) reported that statements such as, "I don't dare show my temper because it makes me too weak," were typical. Tennant, Wilby, and Nicholson (1986) reported similar statements in a small controlled study using the Coultard rating scale, concluding that myasthenic patients were more likely to suppress anger and depression. The implication is that the patients felt they needed to suppress their emotions to minimize their motor signs.

Although more work remains to be done on whether emotional stress causes myasthenia, it is intuitively obvious that myasthenia can cause emotional stress. Martin and Flegenheimer (1971), who were psychiatric consultants to the myasthenia gravis service at Mount Sinai Hospital, have nicely discussed the complexity of emotional response to myasthenia. This included initial relief when the diagnosis is finally confirmed, which can be followed by feelings of helplessness, dependency, or frustration. They noted that sexual and family functioning may be impaired.

The majority of young adults with MG are women, who, in addition to having the usual problems of dealing with chronic illness, may be stressed by the knowledge that a small proportion of infants born to myasthenic mothers also develop transient myasthenia, probably as a result of transfer of maternal autoantibodies. Although the neonatal myasthenic symptoms spontaneously resolve, their occurrence can be very distressing to families.

There has been some suggestion that myasthenia might involve the CNS in addition to the peripheral nervous system. Nicotinic AChR are present in the brain, particularly the hippocampus, hypothalamus, midbrain, and cortex, but are genetically distinct from peripheral AChR (Lindstrom, Schoepfer, & Whiting, 1987). Acetylcholine receptor antibodies are found in the CSF as well as the serum (Lefvert & Pirskanen, 1977), although their significance is disputed (Keesey et al., 1978). In a small series, Tucker et al. (1988) reported impaired memory in 12 myasthenics and suggested central cholinergic dysfunction. Lewis, Ron, and Newsom-Davis (1989) tested auditory vigilance in five myasthenics and found no evidence of impairment, which did not support a hypothesis of central cholinergic deficits.

The medications used to treat MG are a potential source of neuropsychiatric impairment. The anticholinesterase inhibitor pyridostigmine bromide (Mestinon) is the mainstay of therapy in MG. It has negligible lipid solubility and does not cross the blood-brain barrier (Riker, 1989), and therefore lacks CNS side effects. One caveat should be mentioned, however. Sometimes patients appear paradoxically to deteriorate with increasing doses of pyridostigmine. This should never be assumed to be psychogenic without further investigation. There is a therapeutic window for proper anticholinesterase dosage, and when levels are too high the patient can develop "cholinergic crisis," with further muscle weakness. Symptoms from too much versus too little drug are usually discriminated by the use of edrephonium (Tensilon). A small injection of this short-acting anticholinesterase will improve symptoms if oral therapy has been too little, and increase weakness if the patient has been overtreated.

Prednisone, ACTH, methylprednisolone, and other steroids used as immune suppressants have well-described cognitive and affective side effects. Moreover, there is usually an initial worsening of muscle weakness at the initiation of corticosteroid therapy, prior to the development of immunosuppression and improvement. This early deterioration, which is probably the result of a direct effect on excitation-contraction coupling and ionic channels in the AChRs (Miller, Milner-Brown, & Mirka, 1986), can be terrifying for the patient even if forewarned. As with thymectomy, patients may become transiently respirator-dependent. Anxiety over an inability to move, coupled with feelings of suffocation, cannot be underestimated, and myasthenic patients may go through this experience several times. Supportive psychiatric intervention prior to, during, and after a period of ventilator dependency can reduce patients' psychic difficulty. (Chapter 12 discusses weaning from ventilators.)

Steroid therapy frequently becomes chronic (Sghirlanzoni, Peluchetti, & Mantegazza, 1984), and further attempts at immunosuppression may be employed, such as plasma exchange (Rodnitzky & Goeken, 1982) and azathioprine (Hohlfeld et al, 1985). Although such therapeutic measures are usually well tolerated, these patients, as any immunosuppressed patients, must be carefully monitored for CNS or opportunistic infections that initially can masquerade as psychiatric illness, such as fungal meningitis.

There are few reports on the effective pharmacologic treatment of psychiatric illness in myasthenic patients. Some principles of treatment can be derived from an understanding of the pathophysiology of the illness. Medications that further impair neuromuscular transmission, as discussed above, are contraindicated. Respiratory depressants such as benzodiazepines, lithium, and narcotics should also be avoided or used with care. As noted, MG is an illness that can fluctuate in severity, and a medication that can be used safely in one phase of the illness might be hazardous at other times.

The newer, serotonergic antidepressants may be safer drugs than tricyclics. Fluoxetine may cause bradycardia (Ellison, Milofsky, & Ely, 1990) or diarrhea, effects that potentially could be additive with side effects of pyridostigmine or other anticholinesterases.

EPILEPSY

Epidemiology and Pathophysiology

The population of treated epileptics in the United States, by conservative estimates, numbers over 2 million, yielding a lifetime prevalence of approximately 1%, which is roughly equal to the prevalence of schizophrenia and bipolar disorder. This may be a low estimate of the true prevalence of the disorder. A European epidemiologic study suggested that two thirds of the total population of patients with seizures are not under treatment at any given time, one third because of noncompliance and a remarkable one third because the epilepsy has never been diagnosed at all (Mesulam, 1985).

Complex partial seizures (CPS), previously referred to by the less accurate term "temporal lobe epilepsy," are the most common form of epilepsy in adults. Although most CPS arise from the structures of the temporal lobe, seizures may also arise from the frontal, occipital, or parietal areas or may secondarily involve those areas. The manifestations of CPS can hence be protean. Moreover, the complex partial epilepsies can be the most difficult to control. While the primary generalized epilepsies usually respond well to anticonvulsants, seizures with a focal onset can be difficult to suppress completely.

The temporal lobe is not uniform, but is composed of several evolutionarily and functionally distinct regions. Some of the temporal cortex is involved in auditory and vestibular processing, and to a certain extent in speech and memory; much of the rest of the temporal cortex is densely interconnected by association fibers to other lobes of the brain. Deep within the temporal lobe are major structures of the limbic system, constituting the anatomic substrate of emotion (Papez, 1937).

High fevers in infancy appear to cause mesial temporal sclerosis, which underlies an estimated 50% of cases of CPS. Closed head trauma, often minor, may be another common cause of CPS. With trauma, the inferomedial and polar surfaces of the brain are injured by impact against the rough surfaces of the skull. Some viruses, chiefly herpes simplex, have a predilection for the temporal lobe. The middle cerebral artery, which supplies most of the temporal lobe, is the usual route for emboli from cardiac and carotid disease. The vasculature of the temporal lobe is vulnerable to compression, and neuronal structures in this region are vulnerable to hypoxia (Glaser, 1964; Ounstead et al., 1985).

The temporal lobe, both cortical and limbic areas, is heavily involved in higher cortical functions such as speech, memory, and affect; damage or seizures in these areas give rise to a number of neuropsychiatric conditions. A seizure is a rapid and synchronous discharge of neurons, and the manifestation of a seizure will correlate with the function of the portion of the brain that is discharging. A discharge in the part of the motor cortex that represents the arm will result in movements of the arm; a discharge in the visual cortex will result in changes in vision. Since the temporal lobe subserves affective, associative, and mnestic functions, a seizure in this region will result in changes in mood, cognition, memory, and behavior (Gloor et al., 1982). The amygdala is particularly epileptogenic, often rapidly involved in a seizure discharge that spreads from other regions of the brain. Experiments on animals indicate that the amygdala is involved in coordination of rage and fear responses, which may underlie certain manifestations of CPS. According to Stevens et al. (1969), the normal temporal cortex appears to "suppress and modulate primitive automatic emotionally charged patterns of fight, flight, feeling good and sexuality." Most seizures are short lived; the average complex seizure lasts less than 2 minutes. It would appear, then, that seizure activity could not account for more long-lasting behavioral changes. There are, however, both clinical and theoretical exceptions to this point.

Status epilepticus can occur not only for major motor (grand mal) seizures but for CPS as well. Once believed rare, complex partial status is now thought to be of higher prevalence than recognized previously (Van Rossum, Ockhuysen, & Arts, 1985). Partial status may last for a week, or even for 2 months by one report.

Psychomotor status onset is usually acute and presents with an alteration of consciousness, confusion, and an absence of focal neurologic findings. The patient is often initially thought to be suffering from psychosis or some other psychiatric disturbance. Behavior can be bizarre and can be accompanied by subjective feelings of anxiety, paranoia, or euphoria. Prolonged periods of bizarre behavior followed by amnesia for the event can suggest the diagnosis of hysteria (Drury et al., 1985; Lee, 1985; Mayeux et al., 1979). The EEG documents the epileptic origin of the behavior, and episodes may be terminated with anticonvulsants. New-onset changes in mental status symptoms in patients with known epilepsy always constitute an indication for an EEG.

Association of Seizures and Psychopathology

A seizure discharge may also produce prolonged alterations of the brain and behavior after the actual discharge has ceased. The effects of ECT bear witness to this point. That this is also true of partial seizures was demonstrated by Stevens and colleagues in 1969. Depth electrodes were placed in the amygdalas; brief stimulation produced changes in mood and behavior that persisted for several weeks. Attendant EEG changes were not seen, confirming that the persisting behavioral changes, although the result of seizure, were not themselves seizures. Among the several changes were severe depression and manic excitability with pressure of speech (Stevens et al., 1969).

Along another line of evidence of persistent behavioral effects of seizures is the concept of "cyclic psychosis," first noted over 60 years ago (Glaus, 1931), when it was observed that some patients experienced periods of psychosis and no seizures, alternating with periods of sanity accompanied by convulsions, which led Glaus and others to conclude that epilepsy and psychosis were "antagonistic." This observation has been reconfirmed many times (Flor-Henry, 1969a, 1969b; Kristensen & Sindrup, 1978), and formed the basis of the origin of ECT.

Multiple mechanisms may explain this phenomenon. It is possible that, in the periods of psychosis without overt seizures, epileptic activity may persist in deep structures of the temporal lobe, altering function, while the surface EEG remains unchanged or demonstrates only slow waves (Laitinen & Toivakka, 1980). Weiser (1983), using depth records, demonstrated ictal rage, hallucinations, and laughter that were not reflected by the surface EEG. Without the use of depth electrodes, such behavior would probably have been regarded as nonepileptic. Persistent epileptic activity cannot, however, explain all interictal behavior changes.

There is evidence that the brain has homeostatic mechanisms to prevent further seizures. Otherwise, an epileptic focus would seize continuously; in reality this appears to happen only occasionally. Although some of these inhibitory mechanisms may be local, some appear to be mediated by the corpus callosum, as if one side of the brain were attempting to suppress seizures in the other hemisphere. Perhaps these local and contralateral mechanisms contribute to psychiatric disturbance as well (Spencer et al., 1984b; Stevens, 1983).

That epilepsy is associated with psychiatric disturbance is an ancient notion. Herodotus, writing about an epileptic king, said, "It would not be likely that if the body suffered from a great disease the mind was not sound either." Aretaeia of Cappadocia wrote that "they become languid, spiritless, stupid, inhuman, asociable, not disposed to hold intercourse, not to be sociable at any period of life, sleepless, subject to many horrid dreams, without appetite and with bad digestion, pale, of leaden color, slow to learn from torpidity of the understanding and of the senses."

Since then, the association of psychiatric problems with seizures has been the subject of persistent and sometimes vociferous debate. Epilepsy has carried a severe social stigma, which no one wishes to perpetuate, and researchers who have described psychiatric problems as a result of epilepsy have been criticized in this regard. If psychosis, depression, and other psychiatric syndromes, however, are in fact as much a manifestation of an epileptic focus as tonic-clonic movements and staring spells, it would be a disservice to patients to fail to diagnose and treat them.

Trimble (1985) has reviewed the literature on psychiatric disturbance in epilepsy and noted the difficulties in determining the incidence and prevalence of possible associations. One difficulty is the precise delineation and diagnosis of psychiatric syndromes. Such difficulty is of course not unique to the epileptic population. Other problems include separating ictal and interictal symptoms from effects of medication. Despite such difficulties in definition, and in patient selection, a large number of papers attest to various ways in which epilepsy can be complicated by psychiatric disturbance.

Psychosis

It has been estimated that psychosis occurs in approximately 7% of patients with epilepsy (McKenna, Kane, & Parrish, 1985). Most of these patients appear to have the origin of their ictus in the temporal lobe (Flor-Henry, 1969b; Stoudemire, Nelson, & Houpt, 1983). There has been argument as to whether this is true, but given the limitations of localization by scalp EEG (Spencer, Williamson, & Bridgers, 1985), and given that the amygdala is easily involved by a seizure disorder regardless of the original ictal focus, and that it is known that other, nonconvulsive disorders of the temporal lobe can also result in psychosis, it appears likely that the temporal lobe is etiologic in the production of psychotic disorders. Psychosis appears to develop more frequently, but not exclusively, when seizures originate in the left temporal lobe (Flor-Henry, 1969a, 1969b; Levine & Finklestein, 1982). An excess of left-handedness in psychotic epileptics has also been reported and may suggest a shift in cerebral dominance (Sherwin, Magnan, & Bancaud, 1982).

Chronic interictal psychosis (as opposed to the brief psychotic and confusional states seen with seizures and in the postictal period) has been reported to develop usually some years after the onset of seizures (Slater, Beard, & Glitheroe, 1963), but has also been reported to develop within much shorter intervals—a month or less (Levine & Finklestein, 1982).

The psychotic disorder associated with epilepsy appears schizophreniform, but with qualifying differences that can assist in differential diagnosis. Many patients have paranoia, occurring in a clear or clouded sensorium, without hallucinations, loose associations, and the other accompaniments of true schizophrenia. Other patients may have the positive symptoms of schizophrenia, such as auditory hallucinations and delusions, but usually do not have the negative symptoms, such as withdrawal and affective blunting. The psychosis may be associated with normal affect and the ability to form and maintain social relationships, or may be accompanied by such disturbances of mood as depression, rage, anxiety, or euphoria (Glaser, 1964; Pond, 1957). Premorbid personality is more likely to have been normal than in "functional" schizophrenia (Toone, Garralda, & Ron, 1982).

Treatment of Psychosis

The chronic interictal psychosis of epilepsy does not, as a rule, respond to anticonvulsants (Levine & Finkelstein, 1982; Mendez, Cummings, & Benson, 1984; Stevens, 1983) but may respond to neuroleptics. Highly anticholinergic drugs such as chlorpromazine can promote seizures by lowering the seizure threshold and should be avoided (Mendez, Cummings, & Benson, 1984; Remick & Fine, 1979). Neuroleptics are highly protein bound, and their possible displacement of anticonvulsants from binding sites and consequent effect on free drug levels is largely unknown (Simpson & Yadalam, 1985). Moreover, neuroleptics and anticonvulsants can compete for metabolism by the liver, resulting in either increases or decreases in blood levels. Phenothiazines, for example, will elevate phenytoin levels by this mechanism (Vincent, 1980). The opposite effect of anticonvulsants on neuroleptic levels has also been reported. Carbamazepine will cause a decrease in plasma haloperidol levels by an average of 60%, which may result in a reemergence of psychosis (Arana et al., 1986). Phenytoin and phenobarbital have also been reported to reduce haloperidol levels by 50% or more and to reduce levels of the thioridazine metabolite mesoridazine (Linnoila et al., 1980). Relatively nonanticholinergic agents, such as fluphenazine and molindone, are preferred for the treatment of epileptic psychosis. Anticonvulsant levels should be rechecked following dosage changes, in view of possibilities of drug-drug interaction.

Clonazepam has been reported to be helpful in the psychotic disorders of epilepsy (Frykholm, 1985). A long-acting benzodiazepine, clonazepam has a wide spectrum of effects; it has been reported to be effective for seizures, myoclonus, mania (Chouinard, 1985; Dreifuss, 1985), and chronic pain (Beaudry et al., 1986), as well as psychosis. It may act both by elevating serotonin levels and by facilitation of GABA transmission (Browne, 1978). Clonazepam has the advantages of once-a-day dosage and meaningful therapeutic blood levels and does not subject the patient to the risk of tardive dyskinesia.

Resection of the epileptic temporal lobe has not uniformly ameliorated an associated psychosis (Falconer, 1973). More recently, the presence of psychosis has excluded patients otherwise considered for temporal lobectomy (Delgado-Escueta & Walsh, 1985), because the long-term functional outcome of lobectomy appears worse in psychotic patients. However, some centers do perform epilepsy surgery on otherwise-suitable patients whose psychoses are stabilized on neuroleptics.

Finally, when the epileptic patient presents with psychosis or any other new psychiatric disturbance, the physician should consider alternative diagnoses. Seizure patients are prone to developing brain tumors, often years after the initial presentation of epilepsy (Spencer et al., 1984a). Psychoses have also been linked to anticonvulsant toxicity, even with "therapeutic" blood levels (Franks & Richter, 1979). Thus, a comprehensive neurologic reevaluation is therefore indicated whenever an epileptic patient presents with a new major psychiatric syndrome.

Depression

Several studies, reviewed by Robertson and Trimble (1983), indicate that depression is the most common psychiatric disturbance in epilepsy. Although estimates of prevalence vary between approximately 20% and 70%, most investigators agree that depression occurs in epileptics far more commonly than in the general population (Kogeorgos, Fonagy, & Scott, 1982).

Despite this concurrence of investigators, the phenomenology of depression in seizure patients remains poorly defined. Most studies have rated depression with standardized scales such as the Minnesota Multiphasic Personality Inventory (MMPI) or the Beck Depression Inventory (Perini & Mendius, 1984), or by retrospective tally of assigned diagnoses in hospital records (Schiffer & Babigian, 1984). It remains unclear whether the depression of epilepsy is a mood state alone or is usually accompanied by the changes in sleep, appetite, cognition, libido, and other variables that characterize major mood disorders. Himmelhoch (1984) has done some of the clearest and most detailed descriptions of affective changes that can be seen with epilepsy, and reports that "most patients" have enough atypical or organiform symptoms to vitiate a simple diagnosis of major depressive disorder. In a review of 748 patients referred to a mood disorders clinic, 10% were found to have epilepsy. Of these patients, 60% had been previously diagnosed as bipolar, 30% as dysthymic, and 10% as unipolar depressed. Lability of affect, paradoxical response to antidepressants, confusional states, and other associated symptoms helped to distinguish these patients from nonepileptics.

Mendez, Cummings, and Benson (1986) reported that patients with epilepsy also have a risk of suicide that is five times higher than that in the general population and that 30% of depressed epileptic patients have attempted suicide.

The etiology of the association of mood disorders with seizures is unknown, but several lines of evidence suggest possible links. It is known from analysis of cerebral cortex resected at temporal lobectomy for intractable seizures that there is a reduction of adrenergic receptors not only in epileptogenic neocortex but also in surrounding, otherwise normal brain (Briere et al., 1986). Such a reduction in adrenoceptors results in a focal diminution of responsiveness of cortical neurons to norepinephrine. Repeated experimental seizures in rats also led to reduction in adrenoceptors (Bergstrom & Kellar, 1979). It is tempting to speculate that such a focal change is manifested behaviorally as depression.

Depression may exacerbate seizures by disruption of the sleep architecture. Unipolar depression is associated with a decrease in total sleep time and with sleep continuity disturbance, and both unipolar and bipolar depression have changes in the distribution of sleep stages that resemble those of sleep deprivation, with shortened rapid eye movement (REM) latency and altered REM distribution. Since sleep deprivation and fatigue are well-known precipitants of convulsions in epileptic patients, it is interesting to speculate whether such changes in sleep underlie an increase in seizure frequency that is often observed when epileptics become depressed.

Phenytoin, phenobarbital, and carbamazepine have all been reported to cause depression, as well as difficulties with memory, concentration, and alertness (Committee on Drugs, 1985). Depression is most common with phenobarbital, less common with phenytoin, and infrequent with carbamazepine, which actually can improve mood in epileptic patients. As a general rule, the depressed epileptic on phenobarbital should be switched to a less psychotoxic anticonvulsant.

In children and adolescents and in retarded or brain-injured adults, barbiturates can aggravate hyperactivity, impulsiveness, or aggression. When these behaviors are problematic, an alternate anticonvulsant should be substituted.

Phenytoin less consistently causes depression, but some individuals may experience rather severe depression or cognitive impairment. In some cases, these symptoms are due to phenytoin-induced folate deficiency. Phenytoin may also interfere with thiamine effects, so supplementation of both B vitamins and folate is rational for patients on phenytoin. Folate levels should be obtained when psychiatric status changes in a patient on phenytoin without vitamin supplementation.

Phenytoin can produce a number of disfiguring changes in patients' physical appearance, including coarseness of facial features, hirsutism, and gingival hyperplasia. These can produce reactive depression, particularly in adolescents. In general, carbamazepine is preferable to phenytoin in children and adolescents for this reason.

Of all the common anticonvulsants, valproic acid (Depakene or Depakote) may be the most benign in terms of cognitive and somatic side effects. Weight gain and tremor are relatively common, but the incidence of hyponatremia, hepatitis, and bone marrow suppression may be significantly less common than with carbamazepine and phenytoin. Originally approved for use only in primary generalized seizures, valproic acid may be effective in some cases of CPS. Although a discussion of the proper use, indications, and precautions in the use of antiepileptic drugs is beyond the scope of this chapter, valproic acid appears a promising drug for multiple neuropsychiatric conditions.

One disconcerting report suggests that the neuropsychiatric side effects of anticonvulsants may not always be reversible. Bourgeois et al. (1983) prospectively studied 72 newly diagnosed epileptic children over 4 years with psychologic interviews and neuropsychologic tests. Eight of these children suffered a decline in IQ of 10 points or more over the time of the study. Intellectual deterioration did not correlate with seizure frequency per se, as confirmed by others (Ellenberg, Hirtz & Nelson, 1986), but rather loss of IQ correlated with repeated episodes of drug toxicity.

Such reports would indicate that the physician should attempt to use the minimal number of dosage of anticonvulsants required for seizure control. Recent reports have shown that single-drug therapy is often as effective as the use of two or more anticonvulsants and that it is less toxic (Albright & Bruni, 1985). The interested reader is referred to the multicenter cooperative study chaired by Mattson for a comprehensive, detailed, and elegant comparative review of anticonvulsant efficacy and toxicity (Mattson et al., 1985).

Choice of Antidepressants

Antidepressants, when employed for the treatment of depression in epilepsy, must be chosen with care. As a general rule, all antidepressants have the potential to lower the seizure threshold and to affect antiepileptic drug levels. Furthermore, there are few or no large studies on the effects of psychotropics on epileptics, and the results of in vitro and in vivo studies on seizure threshold appear to conflict.

As with neuroleptics, anticholinergic antidepressants such as imipramine and amitriptyline tend to lower the seizure threshold and may result in convulsions (Edwards, 1979). Conflicting reports exist about the safety of desipramine. Although in vitro desipramine appears relatively safe, in overdosage desipramine is reported to cause five times more seizures than other tricyclics (Wedin et al., 1986). Maprotiline also appears to be significantly epileptogenic, although the mechanism of action is unclear (Dessain et al., 1986; Hoffman & Wachsmuth, 1982; Trimble, 1980). If tricyclics are to be used, nortriptyline would be an attractive choice because it is less anticholinergic than the tertiary tricyclics and has meaningful blood levels. The latter is particularly helpful when pharmacokinetic interactions with anticonvulsants are suspected.

Trazodone, which is highly serotonergic, has no incidence of convulsions in overdosage (Wedin et al., 1986), but its efficacy in depressions of epilepsy has not been specifically reported (Mendez, Cummings, & Benson, 1984). Initial dosages should be small (50 mg), and dosages should be increased slowly because epileptic patients on multiple medications can be exquisitely sensitive to sedation. Even with very cautious increments, patients may be unable to tolerate sufficient medication to produce an antidepressant effect, and in such cases the drug must be discontinued. This also can occur with tricyclics.

Monoamine oxidase inhibitors are also often well tolerated and effective in depressed epileptics. Monoamine oxidase inhibitors are reported to have anticonvulsant properties by some authors (Prockop, Shore, & Brodie, 1959), and proconvulsant properties by others (Tagashira et al., 1982). In the author's experience, some patients with CPS have a short-lived flurry of sensory auras and increased somnolence, which may clear over days and be replaced by an antidepressant effect. In some, exacerbation of epileptic symptoms persists and the drug must be discontinued. Patients who can tolerate MAO inhibitors often prefer these drugs because of the relative lack of sedation and anticholinergic side effects. Hypothetical concerns of a possible interaction between MAO inhibitors and carbamazepine, based on the latter's chemical similarity to tricyclics, have not been documented.

Lithium can be a problematic drug in epilepsy because it further lowers the seizure threshold. Sedation, tremor, and weight gain can be additive to the effects of antiepileptic drugs. When possible, use of a mood-stabilizing antiepileptic drug allows lithium to be omitted entirely. However, it would be prudent not to start carbamazepine in a fully MAO inhibited patient.

The newer nontricyclic antidepressants, fluoxetine, sertraline, and especially bupropion, appear to lower the seizure threshold as well, and may precipitate seizures in susceptible individuals. Their lack of sedation and weight gain, however, make them otherwise attractive drugs. Because of their propensity to lower the seizure threshold, however, they are seen by many as contraindicated in patients with seizure disorders. However, in the author's opinion, they may be utilized, if necessary, in a patient with well-controlled epilepsy, if careful attention is given to maintaining adequate antiepileptic drug blood levels, and the medication is built up slowly and cautiously. Patients should be informed of the risk of aggravation of seizures. (See also Chapter 9.)

Electroconvulsive therapy also has been used to treat depression in epileptics. Acutely, the induced convulsion of ECT can both suppress further seizures and reverse depression (Sackeim, 1983), perhaps by evoking brain inhibitory mechanisms to protect against further seizures. Epileptic patients appear to have a particularly high seizure threshold during the administration of ECT when compared with nonepileptic patients. Electroconvulsive therapy probably should be reserved for severely ill patients who are refractory to medical treatment and who are at imminent risk for suicide, cachexia, or other major problems (Robertson & Trimble, 1985). Concerns have been raised that ECT, like other seizures, may kindle an epileptic focus leading to the development of a new seizure disorder in nonepileptic patients and, by extension, that it may ultimately exacerbate preexisting seizures (Davinsky & Duchowny, 1983). Electroconvulsive therapy should not be withheld because of these theoretical concerns, however, if it is strongly indicated psychiatrically. Careful decisions must be made, after consultation with a neurologist, about how to withdraw and reinstitute antiepileptic drugs prior to ECT. Effective ECT is virtually impossible with therapeutic antiepileptic drug levels, but the patient can be at risk of seizures or status epilepticus in their absence. (See Chapter 10.)

Despite all the precautions, depression can and should be treated in people with epilepsy. Depression can be far more disabling and painful than seizures, and patients' quality of life can be significantly improved by proper psychiatric care.

Personality Disorder and Epilepsy

There is a vast and contentious literature on personality changes associated with temporal lobe epilepsy. Impressions of a characteristic epileptic personality are shared by some clinicians working with epileptic patients, but are inconsistently confirmed by rigorous studies using structured diagnostic instruments (Benson, 1991). The personality traits associated with temporal lobe epilepsy include altered sexuality (usually hyposexuality), hypergraphia, interpersonal "stickiness," religious and philosophical preoccupations, and intense emotionality. Although these features have been attributed by some to hyperconnection of the limbic system with association areas, there are other explanations, including coincidental psychiatric disorder, emotional reactions to illness, effects of antiepileptic drugs, and complications of brain damage separate from effects of seizures.

REFERENCES

ABARBANEL J, HERISHANU Y, & FRISHER S (1985). Encephalopathy associated with baclofen. Ann Neurol 17:617–618.

ADAMS HP, BUTLER MJ, BILLER J, ET AL. (1986). Nonhemorrhagic cerebral infarction in young adults. Arch Neurol 43:793–796.

ALBRIGHT P, & BRUNI J (1985). Reduction of polypharymacy in epileptic patients. Arch Neurol 42:797–799.

ANDERSON J, AABRO E, GULMANN N, ET AL. (1980). Antidepressant treatment of Parkinson's disease. Acta Neurol Scand 62:210–219.

ARANA G, GOFF DC, FRIEDMAN H, ET AL. (1986). Does carbamazepine-induced reduction of plasma haloperidol levels worsen psychotic symptoms? Am J Psychiatry 143:650–651.

ARGOV Z, & MASTAGLIA FL (1979). Disorders of neuromuscular transmission caused by drugs. N Engl J Med 301:409–413.

ASNIS G (1977). Parkinson's disease, depression and ECT: A review and case study. Am J Psychiatry 134:191–195.

AVORN J, EVERITT DE, & WEISS S (1986). Increased antidepressant use in patients prescribed betablockers. JAMA 255:357–360.

BALLARD PA, TETRUD JW, & LANGSTON JW (1985). Permanent human Parkinsonism due to 1-methyl-4-phenyl-1,2,3,6-tetrahydropyridine (MPTP): Seven cases. Neurology 35:949–956.

BARNETT HJM (1980). Progress towards stroke prevention: Robert Wartenberg Lecture. Neurology 30:1212–1225.

BASTIANELLO S, POZZILLI C, BERNARDI S, ET AL. (1990). Serial study of gadolinium-DTPA MRI enhancement in multiple sclerosis. Neurology 40:591–595.

BEAL MF, WILLIAMS RS, RICHARDSON EP, ET AL. (1981). Cholesterol embolism as a cause of transient ischemic attacks and cerebral infarction. Neurology 31:860–865.

BEAUDRY P, FONTAINE R, CHOUINARD G, ET AL. (1986). Clonazepam in the treatment of patients with recurrent panic attacks. J Clin Psychiatry 47:83–85.

BENNETT JP (1986). Striatal dopamine depletion, dopamine receptor stimulation, and GABA metabolism: Implications for the therapy of Parkinson's disease. Ann Neurol 19:194–197.

BENNETT JP, FERRARA MB, & CRUZ CJ (1987). GABA-mimetic drugs enhance sponophine-induced contralateral turning in rats with unilateral nigrostriatal dopamine denervation: Implications for the therapy of Parkinson's disease. Ann Neurol 21:41–45.

BENSON DF (1983). Psychiatric aspects of aphasia. Br J Psychiatry 123:555–556.

BENSON F (1991). The Geschwind syndrome. Adv Neurol 55:411–422.

BENVENISTE H, JORGENSEN MB, DIEMER NH, ET AL. (1988). Calcium accumulation by glutamate receptor activation is involved in hippocampal cell damage after ischemia. Acta Neurol Scand 78:529–536.

BERGER JR, TORNATORE C, MAJOR E, ET AL. (1992). Relapsing and remitting human immunodeficiency virus-associated leukoencephalomyelopathy. Ann Neurol 31:34–38.

BERGSTROM DA, & KELLAR KJ (1979). Effect of electroconvulsive shock on monaminergic receptor binding sties in rat brain. Nature 278:464–466.

BEVER CT, SQUINO AV, PENN AS, ET AL. (1983). Prognosis of ocular myasthania. Ann Neurol 14:516–519.

BLAIVAS JG (1980). Management of bladder dysfunction in multiple sclerosis. Neurology 30:12–18.

BOGOUSSLAVSKY J, & REGLI F (1986). Borderzone infarctions distal to internal carotid artery occlusion: Prognostic implications. Ann Neurol 20:346–350.

BORNSTEIN MB, MILLER A, SLAGLE S, ET AL. (1991). A placebo-controlled, double-blind, randomized two-center, pilot trial of Cop 1 in chronic progressive multiple sclerosis. Neurology 41:533–539.

BOURGEOIS BFD, PRENSKY AL, PALKES HS, ET AL. (1983). Intelligence in epilepsy: A prospective study in children. Ann Neurol 14:438–444.

BOYCE S, RUPNIAK NMJ, STEVENTON M, ET AL (1990). CCK-85 inhibits L-dopa-induced dyskinesias in Parkinsonian squirrel monkeys. Neurology 40:717–718.

BOYLAN KB, TAKAHASHI N, PATY DW, ET AL. (1990). DNA length polymorphism 5′ to the myelin basic protein gene is associated with multiple sclerosis. Ann Neurol 27:291–297.

BRIERE R, SHERWIN AL, ROBITAILLE Y, ET AL. (1986). Alpha-I adrenoceptors are decreased in human epileptic foci. Ann Neurol 19:26–30.

BRILEY DP, COULL BM, & GOODNIGHT SH (1989). Neurological disease associated with antiphospholipid antibodies. Ann Neurol 25:221–227.

BROWNE TR (1978). Drug therapy: Clonazepam. N. Engl J Med, 299:812–816.

BUSH DF, LISS CL, MORTON A, ET AL. (1989). An open multicenter long-term treatment evaluation of Sinemet CR. Neurology 39 (suppl 2):101–104.

CAPLAN LR (1980). "Top of the basilar" syndrome. Neurology 30:72–79.

CARACENI T, SCIGLIANO G, & MUSICCO M (1991). The occurrence of motor fluctuations in parkinsonian patients treated long term with levodopa: Role of early treatment and disease progression. Neurology 41:380–384.

CASCINO GD, & ADAMS RD (1986). Brainstem auditory hallucinosis. Neurology 36:1042–1047.

CHEN H, CHOPP M, & WELCH KMA (1991). Effect of mild hyperthermia on the ischemic infarct volume after middle cerebral artery occlusion in the brain. Neurology 41:1133–1135.

CHIAPPA KH (1980). Pattern shift visual, brainstem auditory and short-latency somatosensory evoked potentials in multiple sclerosis. Neurology 30:110–123.

CHIMOWITZ MI, FURLAN AJ, SILA CA, ET AL. (1991). Etiology of motor or sensory stroke: A prospective study of the predictive value of clinical and radiological features. Ann Neurol 30:519–525.

CHOUINARD G (1985). Antimanic effects of clonazepam. Psychosomatics 26 (suppl):7–12.

CHU AB, SEVER JL, & MADDEN DL (1983). Ogliclonal IgG bands in cerebrospinal fluid in various neurological diseases. Ann Neurol 13:434–439.

COCITO L, FAVALE E, & RENI L (1982). Epileptic seizures in cerebral arterial occlusive disease. Stroke 13:189–195.

COHEN RA, & FISHER M (1989). Amantadine treatment of fatigue associated with multiple sclerosis. Arch Neurol 46:676–680.

Committee on Drugs (1985). Behavioral and cognitive effects of anticonvulsant therapy. Pediatrics 76:644–646.

COOK SD, TROIANO R, ZITO G, ET AL. (1986). Effect of total lymphoid irradiation in chronic progressive multiple sclerosis. Lancet 1:1405–1409.

COSGROVE GR, LEBLANC R, MEAGHER-VILLEMURE K, ET AL. (1985). Cerebral amyloid angiopathy. Neurology 35:625–631.

COTTRELL SS, & WILSON SAK (1926). The affective symptomatology of disseminated sclerosis. J Neurol Psychopathol 7:1–30.

CUMMINGS JL, & MENDEZ MF (1984). Secondary mania with focal cerebrovascular lesions. Am J Psychiatry 141:1084–1087.

DALOS NP, RABINS PV, BROOKS BR, ET AL. (1983). Disease activity and emotional state in multiple sclerosis. Ann Neurol 13:573–577.

DE LA MONTE SM, WELLS SE, WHYTE ETH, ET AL. (1989). Neuropathological distinction between Parkinson's dementia and Parkinson's plus Alzheimer's disease. Ann Neurol 26:309–320.

DELGADO-ESCUETA AV, & WALSH GO (1985). Type I complex partial seizures of hippocampal origin: Excellent results of anterior temporal lobectomy. Neurology 35:143–154.

DEMUTH GW, & ACKERMAN SH (1983). Alphamethyldopa and depression. Am J Psychiatry 140:534–538.

DESSAIN EC, SCHATZBERG AF, WOODS BT, ET AL. (1986). Maprotiline treatment of depression. Arch Gen Psychiatry 43:86–90.

DEVINSKY O, & DUCHOWNY MS (1983). Seizures after convulsive therapy, a retrospective case survey. Neurology 33:921–925.

DEWITT LD, BUONANNO FS, KISTLER JP, ET AL. (1984). Nuclear magnetic resonance imaging in evaluation of clinical stroke syndromes. Ann Neurol 16:535–545.

DIAMOND SG, MARKHAM CH, HOEHN MM, ET AL. (1990). An examination of male-female differences in progression and mortality of Parkinson's disease. Neurology 40:763–766.

DOOLITTLE TH, MYERS RH, LEHRICH JR, ET AL. (1990). Multiple sclerosis sibling pairs: Clustered onset and familial predisposition. Neurology 40:1546–1552.

DORFMAN LJ, MARSHALL WH, & ENZMANN DR (1979). Cerebral infarction and migrane: Clinical and radiologic correlations. Neurology 29:317–322.

DOWLING PC, BOSCH VV, & COOK SK (1980). Possible beneficial effect of high dose intravenous steroid therapy in acute demyelinating disease and transverse myelitis. Neurology 30:33–36.

DREIFUSS FE (1985). Treatment of seizure disorders and myoclonus. Psychosomatics 26 (suppl):30–35.

DRURY I, KLASS DW, WESTMORELAND BF, ET AL. (1985). An acute syndrome with psychiatric symptoms and EEG abnormalities. Neurology 35:911–914.

DUBOIS B, DANZE F, PILLON B, ET AL. (1987). Cholinergic-dependent cognitive deficits in Parkinson's disease. Ann Neurol 22:26–30.

DUBOIS B, PILLON B, LHERMITTE F, ET AL. (1990). Cholinergic deficiency and frontal dysfunction in Parkinson's disease. Ann Neurol 29:117–121.

EBERS GC, BULLMAN DE, SADOVNICK AD, ET AL. (1986). A population based study of multiple sclerosis in twins. N Engl J Med 315:1638–1642.

EBMEIER KP, CALDER SA, CRAWFORD JR, ET AL. (1990). Clinical features predicting dementia in idiopathic Parkinson's disease: A follow-up study. Neurology 40:1222–1224.

EDWARDS JG (1979). Antidepressants and convulsions. Lancet 22:1368–1369.

ELIZAN TS, MELVIN MD, MOROS DA, ET AL. (1989). Selegiline use to prevent progression of Parkinson's disease. Arch Neurol 46:1275–1279.

ELLENBERG JH, HIRTZ DG, & NELSON KB (1986). Do seizures in children cause intellectual deterioration? N Engl J Med 314:1085–1088.

ELLISON JM, MILOFSKY JE, & ELY E (1990). Fluoxetine-induced bradycardia and syncope in two patients. J Clin Psychiatry 51:385–386.

ENGEL AG (1984). Myasthenia gravis and myasthenic syndromes. Ann Neurol 16:519–534.

ERMAN MK, & GUGGENHEIM FG (1981). Psychiatric side effects of commonly used drugs (pp. 55–64). Drug Therapy, November.

ESTES ML, CHIMOWITZ MI, AWAD IA, ET AL. (1991). Sclerosing vasculopathy of the central nervous system in nonelderly demented patients. Arch Neurol 48:631–636.

FALCONER MA (1973). Reversibility by temporal-lobe resection of the behavioral abnormalities of temporal lobe epilepsy. N Engl J Med 289:451–455.

FARRELL MA, KAUFMANN JCE, GILBERT JJ, ET AL. (1985). Oligoclonal bands in multiple sclerosis: Clinical-pathological correlation. Neurology 35:212–218.

FAUGHT E, PETERS D, BARTOLUCCI, ET AL. (1989). Seizures after primary intracerebral hemorrhage. Neurology 39:1059–1093.

FINKLESTEIN S, BENOWITZ L, BALDESSARINI RJ, ET AL. (1982). Mood, vegetative disturbance, and dexamethasone suppression test after stroke. Ann Neurol 12:463–468.

FISHER CM (1982). Lacunar strokes and infarcts: A review. Neurology 32:871–876.

FLOR-HENRY P (1969a). Psychosis and temporal lobe epilepsy. Epilepsia 10:363–395.

FLOR-HENRY P (1969b). Schizophrenic-like reactions and affective psychoses associated with temporal lobe epilepsy: Etiological factors. Am J Psychiatry 126:148–152.

FOLEY FW, LAROCCA NJ, & SMITH CR, ET AL. (1992). A prospective study of depression and immune dysregulation in multiple sclerosis. Arch Neurology 49:238–244.

FOLEY FW, TRAUGOTT V, LAROCCA NG, ET AL. (1990). Depression

and immune status in multiple sclerosis: A prospective study [abstract]. Neurology 40 (suppl 1):140

FOLSTEIN MF, MAIBERGER R, & MCHUGH PR (1977). Mood disorder as a specific complication of stroke. J Neurol Neurosurg Psychiatry 40:1018–1020.

FORNU LS, LANGSTON JW, DELANNEY LE, ET AL. (1986). Locus ceruleus lesions and eosinophilic inclusions in MPTP-treated monkeys. Ann Neurol 20:449–455.

FRANKS RD, & RICHTER AJ (1979). Schizophrenia-like psychosis associated with anticonvulsant toxicity. Am J Psychiatry 136:973–974.

FRYKHOLM B (1985). Clonazepam-antipsychotic effect in a case of schizophrenia-like psychosis with epilepsy and in three cases of atypical psychosis. Acta Psychiatr Scand 71:539–542.

GEBARSKI SS, GABRIELSEN TO, GILMAN S, ET AL. (1985). The initial diagnosis of multiple sclerosis: Clinical impact of magnetic resonance imaging. Ann Neurol 17:469–474.

GELLER TJ, & BELLUR SN (1987). Peduncular hallucinosis: Magnetic resonance imaging confirmation of mesencephalic infarction during life. Ann Neurol 21:602–604.

GHIKA JA, BOGOUSSLAVSKY J, & REGLI F (1990). Deep perforators from the carotid system. Arch Neurol 47:1097–1100.

GILBERT JJ, & SADLER M (1983). Unsuspected multiple sclerosis. Arch Neurol 40:533–536.

GLASER GH (1964). The problem of psychosis in psychomotor temporal lobe epileptics. Epilepsia 5:271–278.

GLAUS A (1931). Uber Kombinationen von Schizophrenia und Epilepsie. Z Ges Neurol Psychiatrie 135:450–500.

GLOOR P, OLIVIER A, QUESNEY LF, ET AL. (1982). The role of the limbic system in experiential phenomena of temporal lobe epilepsy. Ann Neurol 12:129–144.

GOETZ CG (1990) Dopaminergic agonists in the treatment of Parkinson's disease. Neurology 40(suppl 3):50–54.

GOLBE LI (1990). The genetics of Parkinson's disease: A reconsideration. Neurology 40(suppl 3):7–14.

GOLBE LI (1991). Young-onset Parkinson's disease: A clinical review. Neurology 41;168–173.

GOODIN DS (1991). The use of immunosuppressive agents in the treatment of multiple sclerosis: A critical review. Neurology 41:980–985.

GOODWIN FK (1971). Psychiatric side effects of levodopa in man. JAMA 218:1915–1919.

GOTHAM AM, BROWN RG, & MARSDEN CD (1986). Depression in Parkinsons's disease: A quantitative and qualitative analysis. J Neurol Neurosurg Psychiatry 49:381–389.

GRIMALDI LME, ROOS RP, NALEFSKI EA, ET AL. (1985). Oligoclonal IgA bands in multiple sclerosis and subacute sclerosing panencephalitis. Neurology 35:813–817.

GROB D, BRUNNER NG, & NAMBA T (1981). The natural course of myasthenia gravis and effect of therapeutic measures. Ann NY Acad Sci 377:652–669.

GUBERMAN A, & STUSS D (1983). The syndrome of bilateral paramedian thalamic infarction. Neurology 33:540–546.

HALLENBECK JM, & DUTKA AJ (1990). Background review and current concepts of reperfusion injury. Arch Neurol 47:1245–1254.

HALLIDAY GM, LI YW, BLUMSBERG PC, ET AL. (1990). Neuropathology of immunohistochemically identified brainstem neurons in Parkinson's disease. Ann Neurol 27:373–385.

HARRIS JO, FRANK JA, PATRONAS N, ET AL. (1991). Serial gadolinium-enhanced magnetic resonance imaging scans in patients with early, relapsing-remitting multiple sclerosis: Implications for clinical trials and natural history. Ann Neurol 29:548–555.

HENDRICKS HT, FRANKE CL, & THEUNISSEN PHMH (1990). Cerebral amyloid angiopathy: Diagnosis by MRI and brain biopsy. Neurology 40:1308–1310.

HEYMAN A, WILKINSON WE, HURWITZ BJ, ET AL. (1984). Risk of ischemic heart disease in patients with TIA. Neurology 4:626–630.

HIER DB, MONDLOCK J, & CAPLAN LR (1983). Behavioral abnormalities after right hemisphere stroke. Neurology 33:337–344.

HILLBOM M, & KASTE M (1981). Ethanol intoxication: A risk factor for ischemic brain infarction in adolescents and young adults. Stroke 12(4):422–425.

HIMMELHOCH JM (1984). Major mood disorders related to epileptic changes. In D BLUMER (Ed.), Psychiatric aspects of epilepsy (pp. 271–294). Washington, DC: American Psychiatric Press, Inc.

HOFFMAN BF, & WACHSMUTH R (1982). Maprotiline and seizures. J Clin Psychiatry 43:117–118.

HOHLFELD R, TOYKA KV, BESINGER VA, ET AL. (1985). Myasthenia gravis: Reactivation of clinical disease and of autoimmune factors after discontinuation of long-term azathioprine. Ann Neurol 17:238–242.

HONER WG, HURWITZ T, LI DKB, ET AL. (1987). Temporal lobe involvement in multiple sclerosis patients with psychiatric disorders. Arch Neurol 44:187–190.

HORNER J, MASSEY EW, & BRAZER SR (1990). Aspiration in bilateral stroke patients. Neurology 40:1686–1688.

HORNYKIEWICZ O, & KISH SJ (1984). Neurochemical basis of dementia in Parkinson's disease. Can J Neurol Sci II (suppl 1): 185–190.

HUBER SJ, PAULSON GW, SHUTTLEWORTH EG, ET AL. (1987). Magnetic resonance imaging correlates of dementia in multiple sclerosis. Arch Neurol 44:732–736.

ISHII N, NISHIHARA Y, & IMAMUIA T (1986). Why do frontal lobe symptoms predominate in vascular dementia with lacunes? Neurology 36:340–345.

JACKSON AC, BOUGHNER DR, & BARNETT HJM (1984). Mitral valve prolapse and cerebral ischemic events in young patients. Neurology 34:784–787.

JACKSON JA, JANKOVIC J, & FORD J (1983). Progressive supranuclear palsy: Clinical features and response to treatment in 16 patients. Ann Neurol 13:273–278.

JACOBS L, KINKEL WR, POLACHINI J, ET AL. (1986). Correlations of nuclear magnetic resonance imaging, computerized tomography, and clinical profiles in multiple sclerosis. Neurology 36:27–36.

JACQUE C, DELASSALLE A, RANCUREL G, ET AL. (1982). Myelin basic protein in CSF and blood. Arch Neurol 30:557–560.

JAMPALA VC, & ABRAMS R (1983). Mania secondary to left and right hemisphere damage. Am J Psychiatry 140:1197–1199.

JAVOY-AGID F, & AGID Y (1980). Is the mesocortical dopaminergic system involved in Parkinson's disease? Neurology 30:1326–1330.

JOFFE RT, LIPPERT GP, GRAY TA, ET AL. (1987). Mood disorder and multiple sclerosis. Arch Neurol 44:376–378.

JUNCOS JL, ENGBER TM, RAISMAN R, ET AL. (1989). Continuous and intermittent levodopa differentially affect basal ganglia function. Ann Neurol 25:473–478.

KAMINSKY HJ, MAAS E, SPIEGEL P, ET AL. (1990). Why are eye muscles frequently involved in myasthenia gravis? Neurology 40:1663–1669.

KAROUM F, CHUANG LW, EISLER T, ET AL. (1982). Metabolism of (−)deprenyl to amphetamine and methamphetamine may be responsible for deprenyl's therapeutic benefit: A biochemical assessment. Neurology 32:503–509.

KEESEY JC, BEIN M, MINK J, ET AL. (1980). Detection of thymoa in myasthenia gravis. Neurology 30:233–239.

KEESEY JC, TOURELLOTTE WW, HERMANN C, ET AL. (1978). Acetylcholine receptor antibody in cerebrospinal fluid. Lancet 1:777.

Kelly JJ, Daube JR, Lennon VA, et al. (1982). The laboratory diagnosis of mild myasthenia gravis. Ann Neurol 12:238–242.

Khatri BO, McQuillen MP, Hoffan RG, et al. (1991). Plasma exchange in chronic progressive multiple sclerosis: A long term study. Neurology 41:409–414.

Kinkel WR, Jacobs L, Polachini I, et al. (1985). Subcortical arteriosclerotic encephalopathy (Binswanger's disease). Arch Neurol 42:951–959.

Kinnunen E, & Wikstrom J (1986). Prevalence and prognosis of epilepsy in patients with multiple sclerosis. Epilepsia 27:729–733.

Kirshner HS, Tsai SI, Runge VM, et al. (1985). Magnetic resonance imaging and other techniques in the diagnosis of multiple sclerosis. Arch Neurol 42:859–863.

Kirubakaran V, Mayfield D, & Rengachary S (1984). Dyskinesia and psychosis in a patient following baclofen withdrawal. Am J Psychiatry 141:692–693.

Klawans HL, Stein RW, Tanner C, et al. (1982). A pure parkinsonian syndrome following acute carbon monoxide intoxication. Arch Neurol 39:302–304.

Klockgether T, & Turski L (1990). NMDA antagonists potentiate anti-Parkinsonian actions of L-dopa in monoamine-depleted rats. Ann Neurol 28:539–546.

Kogeorgos J, Fonagy P, & Scott DF (1982). Psychiatric symptom patterns of chronic epileptics attending a neurological clinic: A controlled investigation. Br J Psychiatry 140:236–243.

Koopsmans RA, Oger JJF, Mayo J, et al. (1989). The lesion of multiple sclerosis: Imaging of acute and chronic stages. Neurology 39:959–963.

Kristensen O, & Sindrup EH (1978). Psychomotor epilepsy and psychosis. I. Physical aspects. Acta Neurol Scand 57:361–369.

Kurlan R, Kersun J, Henderson R, et al. (1990). Anxiety and depression commonly co-exist in Parkinson's disease. Neurology 40:423–424.

Kurtzke JF, & Hyllested K (1986). Multiple sclerosis in the Faroe islands. Neurology 36:307–328.

Laitinen L (1969). Desipramine in treatment of Parkinson's disease. Acta Neurol Scand 45:109–113.

Laitinen L, & Toivakka E (1980). Slowing of scalp EEG after electrical stimulation of amygdala in man. Acta Neurochir 80 (suppl 30):177–181.

Langston JW (1990). Selegiline as neuro protective therapy in Parkinson's disease. Neurology 40(suppl 3):61–66.

Langston JW, Ballard P, Tetrud JW, et al. (1983). Chronic parkinsonism in humans due to a product of meperidine-analog synthesis. Science 25:979–980.

Lebensohn Z, & Jenkins RB (1975). Improvement of parkinsonism in depressed patients treated with ECT. Am J Psychiatry 132:283–285.

Lee KH, Hashimoto SA, Hooge JP, et al. (1991). Magnetic resonance imaging of the head in the diagnosis of multiple sclerosis: A prospective 2 year follow-up with comparison of clinical evaluation, evoked potentials, oligoclonal banding, and CT. Neurology 41:657–660.

Lee SI (1985). Non convulsive status epilepticus. Ictal confusion in later life. Arch Neurol 42:778–781.

Leenders KL, Findley LJ, & Cleeves L (1986). PET before and after surgery for tumor-induced parkinsonism. Neurology 36:1074–1078.

Lesser RP, Fahn S, Snider ST, et al. (1979). Analysis of the clinical problems in parkinsonism and the complications of long-term levodopa therapy. Neurology 29:1253–1260.

Lefvert AK, & Pirskanen R (1977). Acetylcholine-receptor antibodies in cerebrospinal fluid of patients with myasthenia gravis. Lancet 2:351–352.

Levin BE, Llabre MM, & Weiner WJ (1989). Cognitive impairments associated with early Parkinson's disease. Neurology 39:557–561.

Levine DN, & Finklestein S (1982). Delayed psychosis after right temporoparietal stroke or trauma: Relation to epilepsy. Neurology 32:267–273.

Levine DN, & Grek A (1984). The anatomic basis of delusions after right cerebral infarction. Neurology 34:577–582.

Levine SR, Brust JCM, Futrell N, et al. (1991a). A comparative study of the cerebrovascular complications of cocaine: Alkaloidal versus hydrochloride—a review. Neurology 41:1173–1177.

Levine SR, Fagan SC, Pessin MS, et al. (1991b). Accelerated intracranial occlusive disease, oral contraceptives, and cigarette use. Neurology 41:1893–1901.

Lewis SW, Ron MA, & Newsom-Davis J (1989). Absence of central functional cholinergic deficits in myasthenia gravis. J Neurol Neurosurg Psychiatry 52:258–261.

Lindstrom J, Schoepfer R, & Whiting P (1987). Molecular studies of the neuronal nicotinic acetylcholine receptor family. Mol Neurobiol 1:281–337.

Linnoila M, Vivkari M, Vaisanen K, et al. (1980). Effect of anticonvulsants on plasma heloperidol and thioridazine levels. Am J Psychiatry 137:819–821.

Lipsey JR, Robinson RG, Pearlson GD, et al. (1984). Nortriptyline treatment of post-stroke depression: A double blind study. Lancet 1:297–300.

Lipsey JR, Spencer WC, Rabins PV, et al. (1986). Phenomenological comparison of poststroke depression and functional depression. Am J Psychiatry 143:527–529.

Lynch SG, Rose JW, Smoker W, et al. (1990). MRI in familial multiple sclerosis. Neorology 40:900–903.

Lyon-Caen O, Jouvent R, Hauser S, et al. (1986). Cognitive function in recent onset demyelinating diseases. Arch Neurol 43:1138–1141.

Mackenzie KR, Martin MJ, & Howard FM (1969). Myasthenia gravis: Psychiatric concomitants. Can Med Assoc J 100:988–991.

Martin RD, & Flegenheimer WV (1971). Psychiatric aspects of the management of the myasthenic patient. Mount Sinai J Med 38:594–601.

Martin-Mandiere C, Jacque C, Delassalle A, et al. (1987). Cerebrospinal myelin basic protein in multiple sclerosis. Arch Neurol 44:276–278.

Mattson RH, Cramer JA, Collins JF, et al. (1985). Comparison of carbamazepine, phenobarbital, phenytoin, and primidone in partial and secondarily generalized tonic-clonic seizures. N Engl J Med 313:145–151.

Mayberg HS, Starkstein SE, Sadzot B, et al. (1990). Selective hypometabolism in the inferior frontal lobe in depressed patients with Parkinson's disease. Ann Neurol 28:57–64.

Mayeux R, Alexander MP, Benson DF, et al. (1979). Poromania. Neurology 29:1616–1619.

Mayeux R, Stern Y, Cote L, et al. (1984). Altered serotonin metabolism in depressed patients with Parkinson's disease. Neurology 34:642–646.

Mayeux R, Stern Y, Rosen J, et al. (1981). Depression, intellectual impairment and Parkinson disease. Neurology 31:645–650.

Mayeux R, Stern Y, Williams JB, et al. (1986). Clinical and biochemical features of depression in Parkinson's disease. Am J Psychiatry 143:758–759.

McGeer PL, Itagaki S, Akiyama H, et al. (1988). Rate of cell death in parkinsonism indicates active neuropathological process. Ann Neurol 24:574–576.

McKenna PJ, Kane JM, & Parrish K (1985). Psychotic syndromes in epilepsy. Am J Psychiatry 142:895–904.

McNamara ME (1991a). Psychological factors affecting neurological conditions: Depression and stroke, multiple sclerosis, Parkinson's disease, and epilepsy. Psychosomatics 32:255–267.

McNamara ME (1991b). Advances in EEG-based diagnostic technologies. In A Stoudemire & BS Fogel (Eds.), (Vol. 1) Medical Psychiatric Practice (pp. 163–189). Washington, DC: American Psychiatric Press, Inc.

McQuillen MP, Gross M, & Johns RJ (1962). Chlorpromazine induced weakness in myasthenia gravis. Arch Neurol 8:286–290.

Mehler MF (1989). The rostral basilar artery syndrome: Diagnosis, etiology, prognosis. Neurology 39:9–16.

Mendez MF, Cummings JL, & Benson DF (1984). Epilepsy: Psychiatric aspects and use of psychotropics. Psychosomatics 25:883–894.

Mendez MF, Cummings JL, & Benson DF (1986). Depression in epilepsy, significance and phenomenology. Arch Neurol 43:766–770.

Mesulam MM (1985). Principles of behavioral neurology (p. 291). Philadelphia, FA Davis Co.

Miller BL, Benson DF, Cummings JL, et al. (1986). Late-life paraphrenia: An organic delusional syndrome. J Clin Psychiatry 47:204–207.

Miller RG, Filler-Katz A, Kiprov D, et al. (1991). Repeat thymectomy in chronic refractory myasthenia gravis. Neurology 41:923–924.

Miller RG, Milner-Brown HS, & Mirka A (1986). Prednisone-induced worsening of neuromuscular function in myasthenia gravis. Neurology 36:729–732.

Mortimer JA, Pirozzolo FJ, Hansch E, et al. (1982). Relationship of motor symptoms to intellectual deficits in Parkinson disease. Neurology 32:133–137.

Moss HB, & Green A (1982). Neuroleptic-associated dysphagia confirmed by esophageal menometry. Am J Psychiatry 139:515–516.

MS Study Group (1990). Efficacy and toxicity of cyclosporine in chronic progressive multiple sclerosis: A randomized, double-blinded, placebo-controlled clinical trial. Ann Neurol 27:591–605.

Murray GB, Shea V, & Conn DK (1986). Electroconvulsive therapy for poststroke depression. J Clin Psychiatry 47:258–260.

Murrow RW, Schweiger GD, Kepes JJ, et al. (1990). Parkinsonism due to a basal ganglia lacunar state. Neurology 40:897–900.

Nakano I, & Hirano A (1984). Parkinson's disease: Neuron loss in the nucleus basalis without concomitant Alzheimer's disease. Ann Neurol 15:415–418.

Neil JF, Himmelhoch JM, & Licata SM (1976). Emergence of myasthenia gravis during treatment with lithium carbonate. Arch Gen Psychiatry 33:1090–1092.

Nicholson GA, Wilby J, & Tennant C (1986). Myasthenia gravis: The problem of a "psychiatric" misdiagnosis. Med J Aust 144:632–638.

Nutt JG, Gancher ST, & Woodward WR (1988). Does an inhibitory action of levodopa contribute to motor fluctuations? Neurology 38:1553–1557.

Olanow CW (1990). Oxidation reactions in Parkinson's disease. Neurology 40(suppl 3):32–37.

Olanow CW, Lane RS, & Roses AD (1982). Thymectoy in late onset myasthenia gravis. Arch Neurol 39:82–83.

Olsen TS, Hogenhauen H, & Thage O (1987). Epilepsy after stroke. Neurology 37:1209–1211.

Ostrosky-Solis F, Quintanar L, Madrazo I, et al. (1988). Neuropsychological effects of brain autograft of adrenal medullary tissue for the treatment of Parkinson's disease. Neurology 38:1442–1450.

Ounstead C, Glaser G, Lindsay J, et al. (1985). Focal epilepsy with mesial temporal sclerosis after acute meningitis. Arch Neurol 42:1058–1060.

Panitch HS, Hirsch RL, Schindler J, et al. (1987). Treatment of multiple sclerosis with gamma interferon: Exacerbations associated with activation of the immune system. Neurology 37:1097–1102.

Papez J (1937). A proposed mechanism of emotion. Arch Neurol Psychiatry 38:725–743.

Parker WD, Boyson SJ, & Parks JK (1989). Abnormalities of the electron transport chain in idiopathic Parkinson's disease. Ann Neurol 26:719–723.

Parkinson J (1817). An essay on the shaking palsy. London: Sherwood.

Pascual J, Pazos A, Olmo ED, et al. (1991). Presynaptic parkinsonism in olivoportocerebellar atrophy: Clinical, pathological, and neurochemical evidence. Ann Neurol 30:425–428.

Patel BR, Seth KD, & Meador KJ (1986). Metoclopramide-induced parkinsonism. Neurology 36(suppl 1):75.

Paulson GW (1980). Multiple sclerosis: The great imitator. Hosp Med, August, pp. 48–59.

Paulson GW, & Dadmehr N (1991). Is there a premorbid personality typical for Parkinson's disease? Neurology 41(suppl 2):73–76.

Perini G, & Mendius R (1984). Depression and anxiety in complex partial seizures. J Nerv Ment Dis 172:287–290.

Peyser JM, Edwards ER, Poser CM, et al. (1980). Cognitive function in patients with multiple sclerosis. Arch Neurol 37:577–579.

Pond DA (1957). Psychiatric aspects of epilepsy. J Indian Med Profession 3:1441.

Poser CM (1985). The course of multiple sclerosis [letter]. Arch Neurol 42:1035.

Poser CM, Paty D, Scheinberg L, et al. (1983). New diagnostic criteria for multiple sclerosis: Guidelines for research protocols. Ann Neurol 13:27–31.

Poser CM, Roman GC, & Vernant JC (1990). Multiple sclerosis or HTLV-1 myelitis? Neurology 40:1020–1022.

Powell BR, Kennaway NG, Rhead WJ, et al. (1990). Juvenile multiple sclerosis-like episodes associated with a defect of mitochondrial beta oxidation. Neurology 40:487–491.

Prockop DJ, Shore DA, & Brodie BB (1959). Anticonvulsant properties of monoamine oxidase inhibitors. Ann NY Acad Sci 80:643–651.

Pryse-Phillips WEM (1986). The incidence and prevalence of multiple sclerosis in Newfoundland and Labrador, 1960–1984. Ann Neurol 20:323–328.

Quera-Salva MA, Guilleminault C, Chevret S, et al. (1992). Breathing disorders during sleep in myasthenia gravis. Ann Neurol 31:86–92.

Rabey JM, Oberman Z, Scharf M, et al. (1989). The influence of levodopa in the pharmacokinetics of bromocriptine in Parkinson's disease. Clin Neuropharmacol 12:440–447.

Raisman R, Cash R, & Agid Y (1986). Parkinson's disease, decreased density of ^{3}H-imipramine and ^{3}H-paroxetine binding sites in putamen. Neurology 36:556–560.

Rao SM, Leo GJ, Bernardin L, et al. (1990). Prevalence of cognitive dysfunction in multiple sclerosis [abstract]. Neurology 40(suppl 1):140.

Rao SM, Leo GJ, Bernardin L, et al. (1991). Cognitive dysfunction in multiple sclerosis. I: Frequency, patterns, and prediction. Neurology 41:685–691.

Rao SM, Leo GJ, Haughton VM, et al. (1989). Correlation of magnetic resonance imaging with neuropsychological testing in multiple sclerosis. Neurology 39:161–166.

Reding M, Orto L, Willensky P, et al. (1985). The dexamethasone suppression test, an indicator of depression in stroke but not a predictor of rehabilitation outcome. Arch Neurol 42:209–212.

Reding MJ, Orto LA, Winter SW, et al. (1986). Antidepressant therapy after stroke. Arch Neurol 43:763–765.

Reischies FM, Baum K, Brau H, et al. (1988). Cerebral magnetic resonance imaging findings in multiple sclerosis: Relation to disturbance of affect, drive, and cognition. Arch Neurol 45:1114–1116.

Remick RA, & Fine SH (1979). Antipsychotic drugs and seizures. J Clin Psychiatry 40:78–80.

Riederer P, Birkmayer W, Seemann D, et al. (1977). Brain noradrenaline and 3-methoxy-hydrophenylglycol in Parkinson's syndrome. J Neural Transm 41:241–251.

Riesenberg D (1986). Radiolabeled ligands expand PET exploration of numerous normal, abnormal brain functions. JAMA 256:969–970.

Riker WF (1989). Memory impairment in myasthenia gravis [letter]. Neurology 39:611.

Rinne JO, Rommukainen J, Paljarui L, et al. (1989). Dementia in Parkinson's disease is related to neuronal loss in the medial substantia nigra. 26:47–50.

Rinne UK (1987). Early combination of bromocriptine and levodopa in the treatment of Parkinson's disease: A 5 year follow-up. Neurology 37:826–828.

Robertson MM, & Trimble MR (1985). Depressive illness in patients with epilepsy: A review. Epilepsia 24(suppl):S109–S116.

Robinson RG (1979). Differential behavioral and biochemical effects of right and left hemispheric cerebral infarction in the rat. Science 205:707–710.

Robinson RG, Kubos KL, Starr LB, et al. (1984). Mood disorders in stroke patients: Importance of location of lesion. Brain 107:81–93.

Robinson RG, Lipsey JR, & Price TR (1985). Diagnosis and clinical management of post-stroke depression. Psychosomatics 26:769–778.

Robinson RG, & Price TR (1982). Post-stroke depressive disorders: A follow-up study of 103 patients. Stroke 13:635–641.

Rodnitzky RL, & Goeken JA (1982). Complications of plasma exchange in neurological patients. Arch Neurol 39:350–354.

Rokey R, Rolak LA, Harati Y, et al. (1984). Coronary artery disease in patients with cerebrovascular disease: A prospective study. Ann Neurol 16:50–53.

Ron MA (1986). Multiple sclerosis: Psychiatric and psychometric abnormalities. J Psychosom Res 30:3–11.

Ropper AH, & Shafran B (1984). Brain edema after stroke. Arch Neurol 41:26–29.

Rose AS, Ellison GW, Myers LW, et al. (1976). Criteria for the clinical diagnosis of multiple sclerosis. Neurology 2:20–22.

Ross ED (1981). The aprosodias. Arch Neurol 38:561–569.

Ross ED, & Rush JA (1981). Diagnosis and neuroanatomical correlates of depression in brain-damaged patients. Arch Gen Psychiatry 38:1344–1354.

Rudick RA, Schiffer RB, Schwetz KM, et al. (1986). Multiple sclerosis. The problem of incorrect diagnosis. Arch Neurol 43:578–583.

Sackeim HA, Decina P, Prohovnik I, et al. (1983). Anticonvulsant and antidepressant properties of electroconvulsive therapy: A proposed mechanism of action. Biol Psychiatry 18:1301–1310.

Sacks O (1973). Awakenings (p. 250). New York: EP Dutton.

Sadovnick AD, & Baird PA (1988). The familial nature of multiple sclerosis: Age-corrected empiric recurrences risks for children and siblings of patients. Neurology 38:990–991.

Sadovnick AD, Eisen K, Ebers GC, et al. (1991). Causes of death in patients attending multiple sclerosis clinics. Neurology 41:1193–1196.

Sanders DB, Howard JF, & Johns TR (1979). Single-fiber electromyography in myasthenia gravis. Neurology 29:68–76.

Santamaria J, Tolosa E, & Valles A (1986). Parkinson's disease with depression, a possible subgroup of idiopathic parkinsonism. Neurology 36:1130–1133.

Santy PA (1983). Undiagnosed myasthenia gravis in emergency psychiatric referrals. Ann Emerg Med 12:397–398.

Schaul N, Green L, Peyster R, et al. (1986). Structural determinants of electroencephalographic findings in acute hemispheric lesions. Ann Neurol 20:703–711.

Scheinberg P (1991). The biologic basis for the treatment of acute stroke. Neurology 41:1867–1873.

Schiffer RB (1987). The spectrum of depression in multiple sclerosis. An approach for clinical management. Arch Neurol 44:596–599.

Schiffer RB, & Babigian HM (1984). Behavioral disorders in multiple sclerosis, temporal lobe epilepsy and amyotrophic lateral sclerosis: An epidemiologic study. Arch Neurol 41:1067–1069.

Schiffer RB, Herndon RM, & Rudick RA (1985). Treatment of pathologic laughing and weeping with amitriptyline. N Engl J Med 312:1480–1482.

Schiffer RB, Kurlan R, Rubin A, et al. (1988). Evidence for atypical depression in Parkinson's disease. Am J Psychiatry 145:1020–1022.

Schiffer RB, & Wineman NM (1990). Antidepressant pharmacotherapy of depression associated with multiple sclerosis. Am J Psychiatry 147:1493–1497.

Schumacher GA, Beebe G, Kibler RF, et al. (1965). Problems of experimental trials of therapy in multiple sclerosis. Ann NY Acad Sci 12:552–568.

Sghirlanzoni A, Peluchetti D, Mantegazza R, et al. (1984). Myasthenia gravis: Prolonged treatment with steroids. Neurology 34:170–174.

Sherwin I, Magnan PP, & Bancaud J (1982). Prevalence of psychosis in epilepsy as a function of the laterality of the epileptogenic lesion. Arch Neurol 39:621–625.

Simon R, & Shiraishi K (1990). *N*-methyl-*O*-Aspartate antagonist reduces stroke size and regional glucose metabolism. Ann Neurol 27:606–611.

Simpson DM, & Foster D (1986). Improvement in organically disturbed behavior with trazadone treatment. J Clin Psychiatry 47:191–193.

Simpson GM, & Yadalam K (1985). Blood levels of neuroleptics: State of the art. J Clin Psychiatry 46(5, Sec 2):22–28.

Slater E, Beard AW, & Glitheroe E (1963). The schizophrenia-like psychoses of epilepsy. Br J Psychiatry 109:95–150.

Sloan MA, Kittner SJ, Rigamonti D, et al. (1991). Occurrence of stroke associated with use/abuse of drugs. Neurology 41:1358–1364.

Sneddon J (1980). Myasthenia gravis: A study of social, medical, and emotional problems in 26 patients. Lancet 1:526–528.

Snyder AM, Stricker EM, & Zigmond MJ (1985). Stress induced neurological impairments in an animal model of parkinsonism. Ann Neurol 18:544–551.

Snyder SH, & D'Amato RJ (1986). MPTP, a neurotoxin relevant to the pathophysiology of Parkinson's disease. Neurology 36:250–258.

Snyder SH, & Reynolds IJ (1985). Calcium antagonist drugs. N Engl J Med 313:995–1002.

Sommer N, Willcox N, Harcourt GC, et al. (1990). Myasthenic thymus and thymoma are selectively enriched in acetylcholine receptor-reactive T cells. Ann Neurol 28:312–319.

Spencer DD, Spencer SS, Mattson RH, et al. (1984a). Intracerebral masses in patients with intractable partial epilepsy. Neurology 34:432–436.

Spencer SS, Spencer DD, Glaser GH, et al. (1984b). More intense focal seizure type after callasal section: The role of inhibition. Ann Neurol 16:686–693.

Spencer SS, Williamson PD, & Bridgers SL (1985). Reliability and accuracy of localization by scalp ictal EEG. Neurology 35:1567–1575.

Starkstein SE, Berthier ML, Bolduc PL, et al. (1989). Depression in patients with early versus late onset of Parkinson's disease. Neurology 39:1441–1445.

Starkstein SE, Berthier ML, Fedoroff P, et al. (1990). Anosognosia and major depression in two patients with cerebrovascular lesions. Neurology 40:1380–1382.

Steck AJ (1990). Antibodies in the neurology clinic. Neurology 40:1489–1492.

Steingart A, Hachinski VC, Lau C, et al. (1987). Cognitive and neurologic findings in subjects with diffuse white matter lucencics on tomographic scan (leuko-araiosis). Arch Neurol 44:32–35.

Stern RA, & Bachman DL (1991). Depressive symptoms following stroke. Am J Psychiatry 148:351–356.

Stern Y, Mayeux R, Ilson J, et al. (1984). Pergolide therapy for Parkinson's disease: Neurobehavioral changes. Neurology 34:201–204.

Stevens J (1973). The anatomy of schizophrenia. Arch Gen Psychiatry 29:177–189.

Stevens JR (1983). Psychosis and epilepsy. Ann Neurol 14:347–348.

Stevens JR, Mark VH, Erwin F, et al. (1969). Bitemporal stimulation in man. Arch Neurol 21:157–169.

Steventon GB, Heafield MT, Waring RH, et al. (1989). Xenobiotic metabolism in Parkinson's disease. Neurology 39:883–887.

Stoudemire A, Nelson A, & Houpt JL (1983). Interictal schizophrenia-like psychoses in temporal lobe epilepsy. Psychosomatics 24:331–335.

Strang RR (1965). Imipramine in treatment of Parkinsonism: A double-blind placebo study. Br Med J 2:33–34.

Stys PK, Waxman SG, & Ransom BR (1991). $Na^+ - Ca^{2+}$ exchanger mediates Ca^{2+} influx during anoxia in mammalian central nervous system white matter. Ann Neurol 30:375–380.

Surridge D (1969). An investigation into some psychiatric aspects of multiple sclerosis. Br J Psychiatry 115:749–764.

Sweet RD, McDowell FH, Feigenson JS, et al. (1976). Mental symptoms in Parkinson's disease during chronic treatment with levodopa. Neurology 26:305–310.

Tagashira E, Hiramori T, Urano T, et al. (1982). Specific action of tranylcypromine to phenobarbital withdrawal convulsions. Psychopharmacology (Berlin) 77:101–104.

Taylor AE, & Saint-Cyr JA (1990). Depression in Parkinson's disease: Reconciling physiological and psychological perspectives. J Neuropsychiatry Clin Neurosci 2:92–98.

Tennant C, Wilby J, & Nicholson GA (1986). Psychological correlates of myasthenia gravis: A brief report. J Psychosom Res 30:575–580.

Thompson AT, Kennard C, Swash M, et al. (1989). Relative efficacy of intravenous methylprednisolone and ACTH in the treatment of acute relapse of MS. Neurology 39:969–971.

Tindall RSA (1981). Humoral immunity in myasthenia gravis: Biochemical characterization of acquired antireceptor antibodies and clinical correlations. Ann Neurol 10:437–447.

Tolosa ES, & SantaMaria J (1984). Parkinsonism and basal ganglia infarcts. Neurology 34:1516–1518.

Toone BK, Garralda ME, & Ron MA (1982). Psychoses of epilepsy and the functional: A clinical and phenomenological comparison. Br J Psychiatry 141:256–261.

Trimble MR (1980). New antidepressant drugs and the seizure threshold. Neuropharmacology 19:1227–1228.

Trimble MR (1985). The psychoses of epilepsy and their treatment. Clin Neuropharmacol 8:211–220.

Tucker DM, Roeltgen DP, Wann D, et al. (1988). Memory dysfunction in myasthenia gravis: Evidence for central cholinergic effects. Neurology 38:1173–1177.

Van Rossum J, Ockhuysen AA, & Arts RJ (1985). Psychomotor status. Arch Neurol 00:989–993.

van Swieten JC, Geyskes GG, Derix MA, et al. (1991). Hypertension in the elderly is associated with white matter lesions and cognitive decline. Ann Neurol 30:825–830.

Vincent FM (1980). Phenothiazine-induced phenytoin intoxication. Ann Intern Med 93:56–57.

Vincent FM (1985). Tocainide encephalopathy. [Letter]. Neurology 35:1804–1805.

Vonsattel JPG, Myers RH, Whyte ETH, et al. (1991). Cerebral amyloid angiopathy without and with cerebral hemorrhages: A comparative histological study. Ann Neurol 30:637–649.

Waksman B (1985). Mechanisms in multiple sclerosis. Nature 318:104–105.

Wedin GP, Oderda GM, Schwartz WK, et al. (1986). Relative toxicity of cyclic antidepressants. J Emerg Med 15:797–804.

Weiner HL, Dau P, Birnbaum G, et al. (1983). Plasma exchange in acute multiple sclerosis. Design of a cooperative study. Arch Neurol 40:691–692.

Weiner M, Chauson A, Wolpert E, et al. (1983). Effect of lithium on the responses to added respiratory resistances. N Engl J M 308:319–322.

Weiner WJ, Koller WC, Perlik S, et al. (1980). Drug holiday and management of Parkinson's disease. Neurology 30:1257–1261.

Weisberg LA, Elliot D, & Mielke D (1991). Intracerebral hemorrhage following electroconvulsive therapy [letter]. Neurology 4:1849.

Weiser HG (1983). Depth recorded limbic psychopathology. Neurosci Biobehav Rev 7:427–440.

Whitaker JN (1985). Quantitation of the synthesis of immunoglobin G within the central nervous system. Ann Neurol 17:11–12.

Whitehouse PJ, Hedreen JC, White CL, et al. (1983). Basal forebrain neurons in the dementia of Parkinson disease. Ann Neurol 13:243–248.

Willoughby EW, Grochowski E, Li DKB, et al. (1989). Serial magnetic resonance scanning in multiple sclerosis: A second prospective study in relapsing patients. Ann Neurol 25:43–49.

Wolkowitz OM, Reus VI, Wingartner H, et al. (1990). Cognitive effects of corticosteroids. Amer J Psychiatry 147:1297–1303.

Wolters EC, Hurwitz TA, Mak E, et al. (1989). Clozapine in the treatment of Parkinsonian patients with dopaminomiretic psychosis. Neurology 40:832–834.

Wood AJ, & Feely J (1983). Pharmacokinetic drug interaction with propranolol. Clin Pharmacokinet 8:253–262.

Yamamoto T, Vincent A, Ciulla T, et al. (1991). Seronegative myasthenia gravis: A plasma factor inhibiting agonist-induced acetylcholine receptor function copurifies with IgM. Ann Neurol 30:550–557.

Yip PK, He YY, Hsu CY, et al. (1991). Effect of plasma glucose on infarct size in focal cerebral ischemia-reperfusion. Neurology 41:899–905.

Zweig RM, Jankel WR, Hedreen JC, et al. (1989). The pedunculoportine nucleus in Parkinson's disease. Ann Neurol 26:41–46.

21 Sleep disorders in the medically ill

QUENTIN R. REGESTEIN, M.D.

EPIDEMIOLOGY

In medical patients, the worse the symptoms the worse the sleep. Those with cardiovascular, neurologic, and musculoskeletal problems may be particularly affected (Johns et al., 1970). The elderly have increased chances of being ill, and have an increased prevalence of insomnia and sleep apnea (Ancoli-Israel et al., 1981; Bixler et al., 1979, 1984; Carskadon et al., 1980; Johns et al., 1970; Karacan et al., 1976; Lugaresi et al., 1980; Williams, Karacan, & Hursch, 1974). Medical inpatients suffer further disruption of sleep because of acute illness and hospital routines. For example, 38% of hospital inpatients in one teaching hospital took sleeping pills on one randomly chosen weekday, in addition to the minor tranquilizers taken by 20% (Salzman, 1981). This compares with 3% of adults in epidemiologic surveys reporting use of prescription hypnotics within the previous year (Mellinger, Balter, & Uhlenhuth, 1985).

PHYSIOLOGY

Sleep is part of a 24-hour sleep/wake rhythm possibly governed by two bodily clocks or by multiple clocks coordinated by a master clock (Moore-Ede, Sulzman, & Fuller, 1982). These "internal clocks" coordinate daily oscillations of all physiologic functions in a way that is controlled, economical, and adjusted to the individual's ecologic niche (Aschoff, 1964; Wever, 1975a, 1975b, 1975c, 1979; Halberg, 1960). The clocks obtain their timing information primarily from the light/dark cycle, although social schedule may have some influence in man, except in less socially oriented individuals, such as those with chronic schizophrenia (Gordon et al., 1986; Morgan, Mirors, & Waterhouse, 1980; Wever, 1970). Mealtimes, if regular, also provide timing information. When timing signals or sensitivity to them are weak or inconstant, endogenous, longer-than-24-hour cycles emerge, causing slower physiologic rhythms. Conditions such as winter light in northern latitudes, social isolation, or lack of timed work obligations will promote this departure from 24-hour cycles, and thus progressively later bedtimes and arising times. The temporal coordination among physiologic systems becomes less secure under these circumstances, and "internal desynchrony," the breaking apart of the usual synchronous relationships (e.g., among sleep, body temperature, hormonal, and electrophysiologic parameters) may occur. One subject under isolation had a 33.2-hour sleep/wake cycle but a 24.9-hour body temperature cycle (Hauri, 1982). Desynchrony may also occur in normal circumstances (Czeisler et al., 1986). Such desynchrony is more likely in elderly and neurotic individuals, and may be associated with work inefficiency (Czeisler, Moore-Ede, & Coleman, 1982) and depression (Lund, 1974; Wehr et al., 1979; Wever 1975c). The increased prevalence of depression and suicide at more polar latitudes, known since the 19th century, may be due in part to the shorter periods of daylight there. Various drugs delay or advance circadian phases, including steroids, caffeine, alcohol, and benzodiazepines (Moore-Ede, Sulzman, & Fuller, 1982; Seidel et al., 1984).

Of the two basic types of sleep, non-rapid eye movement (NREM) sleep is a continuous spectrum from light to deep sleep. By definition, it is composed of polygraphic sleep stages 1 through 4 which are characterized by increasingly slow electroencephalographic (EEG) waves (Rechtschaffen & Kales, 1968). Slow wave sleep occurs early in the night, increases disproportionately after sleep deprivation, preempts rapid eye movement (REM) sleep during restricted sleep regimens, and is increased with increased metabolism as associated with exercise, increased body temperature (Horne & Shackell, 1987), high-normal thyroid indices, and hyperthyroidism (Baekeland & Lasky, 1966; Bunnell & Horvath, 1985; Dunleavy et al., 1974; Hobson, 1968; Horne, 1980; Horne & Reid, 1984; Johns et al., 1975; Oswald, 1980). Growth hormone is secreted mostly during stage 4 sleep (Parker, et al., 1969). Some have postulated that slow wave (deep) sleep is necessary for body restitution (Hartmann, 1973) or to enforce the idleness needed for restitution (Horne, 1980).

In contrast to NREM sleep, REM sleep manifests more signs of arousal, such as EEG fast waves, continual eye movements, dreams, increased oxygen consumption, and increased blood pressure (Brebbia & Altschuler, 1965; Hartmann, 1973; Khatri & Freis, 1969; Littler et al., 1975; Snyder, 1971). Variability in heart rate and blood pressure increase 50% in REM compared with NREM sleep (Snyder et al., 1964). This increased variability may relate to the increase in both cardiac arrhythmias (Regestein et al., 1981) and angina attacks (Murao et al., 1972) during REM sleep (Armstrong et al., 1965).

Originally, sleep may have evolved to conserve energy or to keep individuals within their own ecological time niche (Allison & Twyver, 1970). It is regulated at basic hindbrain levels. Serotonergic neurons in the raphe nuclei and in the preoptic area of the hypothalamus are necessary for normal sleep. Stimulation of the nucleus of the tractus solitarius, which receives vagal afferents, synchronizes forebrain EEG rhythms and causes decreased arousal, which suggests one possible mechanism for postprandial drowsiness. The suprachiasmatic nuclei of the hypothalamus receive information about light directly from the retina and regulate circadian rhythms (Moore-Ede, Sulzman, & Fuller, 1982). An oscillating, reciprocally inhibiting pair of pontine cell groups, the cholinergic gigantotegmental field cells of the mesencephalon and noradrenergic locus ceruleus, may time the cycle between NREM and REM sleep (Hobson, 1983). Disease processes that involve centers in the brainstem may increase specific sleep stages (Hobson, 1975) or decrease (Aldrich et al., 1989b) or abolish (Fischer-Perroudon, Mouret, & Jouvet, 1974) sleep. Recently, however, sleep has been conceived as deriving from multiple cortical and subcortical mechanisms (Kelly, 1985; McGinty et al., 1985) that exert an influence on brainstem structures and are in turn affected by them.

Neurochemically, serotonergic neurons seem responsible for the initiation of NREM sleep and the priming of REM sleep, and some adrenergic neurons play a role in cortical arousal and REM sleep (Jouvet, 1974). Clinically, this translates into increased sleep when the dietary serotonin precursor, L-tryptophan, is administered (Hartmann, 1974) and into insomnia as a frequent side effect of methysergide, an antiserotonergic drug. There is an inverse relationship between the amount of REM sleep present and levels of brain catecholamine-affecting drugs. Raising central nervous system (CNS) levels of acetylcholine by physostigmine injections increases arousal; that is, during slow wave sleep such injections cause REM, and during REM they cause wakefulness (Sitaram et al., 1976).

COMMON SLEEP DISORDERS

Some sleep disorders, such as narcolepsy or sleep apnea, warrant medical or surgical intervention, but more commonly a sleep disorder—for example, one that causes insomnia—will complicate the clinical status of a superimposed medical condition. Problems such as steroid treatments, rheumatoid arthritis, and renal dialysis will lead to worsened sleep. Factors predisposing to disordered sleep are discussed in the following subsections as a background for understanding the relationships between them and common medical problems.

The classification for sleep disorders (American Psychiatric Association, 1987; Association of Sleep Disorders Clinics, 1990) involves dyssomnias (i.e., poor sleep quality) and parasomnias (i.e., conditions provoked by sleep). The uncovering of common mechanisms—for example, between obstructive and central sleep apnea (e.g., Sanders, 1984; Sanders, Rogers, & Pennock, 1985)—has suggested that sleep problems should be thought of in terms of both symptoms and mechanisms; similar pathologic processes may underlie insomnia in one patient but hypersomnia in another. The following discussion will thus enumerate sleep disturbances that present as complaints of insufficient night sleep, daytime sleepiness, or behavioral problems during sleep, or as complaints from others about the patient's behavior during sleep (Table 21-1).

Dyssomnias

Insomnia is a complaint of too little sleep. It usually results from a combination of predispositions and precipitating causes. The most common predisposing factors are the use of CNS-acting drugs, advancing age, schedule irregularities, and hyperarousal. More direct causes include various psychiatric, neurologic, and endocrine conditions, irritating symptoms of any disease, and rare primary sleep disorders.

Hypersomnia means abnormally large amounts of sleep, often associated with excessive daytime sleepiness. Excessively sleepy patients develop low standards for wakefulness and often fail to seek help for sleepiness until others complain about their continual dozing, or an accident caused by sleepiness occurs.

Drug Effects (Table 21-2)

Among the drugs that often diminish sleep quality, the most common offenders are caffeine, alcohol, and nic-

TABLE 21–1. *Major Sleep Disorders*

Insomnia	Subjective complaint of disordered sleep May have minimal objective findings May be associated with any cause of poor sleep quality
Excessive sleepiness	Pathologically augmented tendency to fall asleep Often recognized as abnormal by patient and others Also associated with diverse medical causes of poor sleep quality, and with some idiopathic conditions
Sleep/wake schedule disorders	Adequate sleep, but at the wrong time Due to circadian rhythm impairments or artificial displacements of sleep schedule (e.g., by hospital routines, treatment schedules, work shifts)
Parasomnias	Difficulties that occur during sleep: motor automatisms or fixed action patterns such as bruxism, enuresis, walking, or emotional arousal (e.g., night terrors)

TABLE 21–2. *Drugs that Commonly Diminish Sleep Quality*

Household drugs
Caffeine
Alcohol
Nicotine
Over-the-counter drugs
Nasal decongestants
Appetite suppresants
Hypnotics
Prescription drugs
(any with CNS effects)

otine. Caffeine is consumed by over 80% of the adult population (Dews, 1982). It has a 12- to 20-hour duration of action (Hollingsworth, 1912), lessens sleep quality (Brezinova, 1974), correlates by dose with insomnia prevalence (Shirlow & Mathers, 1985), and increases next-morning drowsiness (Goldstein, Kaizer, & Whitby, 1969). It increases insomnia symptoms in hospital patients (Victor, Lubetsky, & Greden, 1981) and may worsen their psychologic symptoms (Charney, Heninger, & Jaflow, 1985; Neil et al., 1978). Coffee, tea, chocolate, and caffeine-containing medications should be discontinued in insomnia patients. The caffeine withdrawal syndrome can last months and entail a significant loss of energy and initiative; overt depression can occur. Paradoxically, caffeine apparently induces sleepiness in some patients (Regestein, 1989).

Alcohol ingestion lessens sleep quality and provokes sleep apnea (Scrima, Hartman, & Hiller, 1989). After only one drink in the evening, some patients will awaken in the middle of the night, sometimes with a feeling of excess warmth or palpitations. This may be due to increased adrenergic activity, as suggested by increased cerebrospinal fluid catecholamine metabolites 6 to 14 hours after alcohol intake (Borg, Krande, & Sedval, 1981). Alcoholism causes decreased sleep quality and at the extreme of severe alcoholism there is "fragmented" sleep, with frequent awakenings and shifts among sleep stages (Mello & Mendelson, 1970; Pokorny, 1978). Years after alcohol withdrawal, alcoholics continue to suffer diminished sleep quality (Adamson & Burdick, 1973; Vitiello et al., 1990).

Nicotine is a stimulant (Henningfield, 1984), which partly accounts for its addicting and appetite-suppressing properties. Daily sleep length diminishes with increased use of tobacco (Palmer & Harrison, 1983). Smokers take longer to fall asleep at night than nonsmokers, feel sleepy during withdrawal (Cummings et al., 1985), and may improve their sleep within a week after quitting (Soldatos et al., 1980). Daily sleep length diminishes with increased use of tobacco (Palmer & Harrison, 1983).

Of these common drugs, caffeine and alcohol usually are discontinued by the motivated insomnia patient, but nicotine retains a tenacious hold, having an addiction profile and a recidivism rate paralleling that of heroin (Henningfield, 1984). Nicotine chewing gum reportedly increased the abstinence rate of patients in a smoking cessation clinic (Hjalmarson, 1984), and clonidine may suppress cigarette craving during the withdrawal period (Glassman et al., 1984). Further discussion of smoking cessation can be found in Chapter 42.

Some over-the-counter drugs aggravate insomnia, especially nasal decongestants and appetite suppressants. Over-the-counter sleeping pills are basically placebo medications (Mendelson, 1980), but their anticholinergic effects may diminish sleep quality in the elderly.

Of the prescription medications, catecholamine agonists and blockers, antibronchospastic agents, stimulating antidepressants, and antiarrhythmia drugs all commonly disrupt sleep, as can antidepressant withdrawal (Dilsaver, 1989). Methyldopa (Smirk, 1963), propranolol (Petrie, Maffucci, & Woosley, 1982), xanthine derivatives, and beta-agonists are frequent offenders, and thyroid hormone, contraceptives, and methysergide deserve suspicion in individual cases. Use of diuretics commonly underlies restless legs syndrome, a motor disorder that may disrupt sleep. Antibiotics that interfere with protein synthesis and that enter the nervous system (e.g., tetracyclines) may reduce deep sleep (Nonaka, Nakazawa, & Kotorii, 1983) or induce nightmares (Williams, 1988).

The shorter acting benzodiazepines, especially triazolam, can cause nocturnal agitation (Regestein & Reich, 1985), early morning insomnia (Kales et al., 1983), and next-day anxiety (Morgan & Oswald, 1982), presumably by provoking a rapidly appearing withdrawal syndrome. "Next-night" rebound insomnia can occur after triazolam discontinuation. About 60% or more of elderly triazolam users ingest it daily (Baker & Oleen, 1988), with paranoid, amnestic episodes and other severe reactions being possibly more common with such regimens than previously thought, particularly at doses above 0.125 mg. (Oswald & Adam, 1989).

Given the panoply of drug effects on sleep, any use of CNS-acting drugs should be considered a possible factor in aggravating insomnia. The clinical dilemma between therapeutic and unwanted effects is sometimes resolved by switching to peripherally-acting drugs. For example, in the insomniac asthmatic patient, careful titration of antibronchospastic drug doses to minimal effective levels and the use of shorter acting agents toward evening may lessen insomnia. Inhalers interfere less with sleep than systemic agents. Theophylline should be avoided when possible.

Age Effects (Table 21-3)

Sleep normally becomes lighter and more disrupted with aging. From the neonatal period through senescence, it takes progressively longer to fall asleep and there is increasing nocturnal wakening and decreasing deep sleep (Williams, Karacan, & Hursch, 1974). The 30-year-old, for instance, obtains less than one half the stage 4 sleep obtained by the 20-year-old (Gaillard, 1978), although the clinical significance of lessened deep sleep in the elderly is unclear (Spiegel, Koeberle, & Allen, 1986). The decline of sleep with age parallels the decline of cerebral functioning as measured by cognitive functioning (Prinz, 1977) and cerebral blood flow (Prinz, Obrist, & Wang, 1975). Since behavioral and neurologic intactness covary in old people (Tomlinson, Blessed, & Roth, 1970), and since dementia amplifies the sleep deficits of aging (Reynolds et al., 1985), it would seem that the usual diminution of sleep with age derives from age-associated neuronal and dendritic loss (Brady & Vijayashankar, 1977). This structural decline with age, combined with waning biologic clock functioning (Wever, 1975c), predisposes older people to sleep disorders. It is not age per se, however, but state of health, that determines sleep quality. Almost any physiologic measurement shows wider variance with age (Comfort, 1968). For instance, sleep quality correlates with intellectual intactness much more than with age among elderly people (Prinz, 1977), and sleep apnea is found much more in random samples than in healthy samples of elderly (Ancoli-Israel et al., 1981; Reynolds et al., 1985) A reduction in ventilation rate, however, and an increase in disturbed breathing during sleep have been noted even in normal aging subjects (Bliwise et al., 1984; Krieger, Mangin, & Kurtz, 1980).

TABLE 21–3. *Mechanisms of Decreased Sleep Quality with Age*

Normal neuronal loss
Lessened competence of bodily clocks
Less sensitivity to light or social cues
Less coordination among different physiologic timing rhythms
Increased disease
Increased use of CNS-acting drugs
Increased sensitivity to CNS-acting drugs
Psychosocial changes
Fewer timed obligations
Less activity/prolonged in-bed times
Regression (e.g., passive acceptance of poor sleep, reliance on sedatives, avoidance of behavioral improvements)

The psychologic regression that may accompany old age, the absence of timed work obligations, and a passive attitude toward controlling one's own habits all promote insomnia and lessen adherence to insomnia treatments (Regestein, 1980). Studies indicate that many aged individuals sleep 9.5 hours during a 12-hour inbed time (Stramba-Badiale, Ceretti, & Forni, 1979; Webb & Swinburne, 1971), which likely lessens the refreshment of sleep, since prolonged in-bed times may be detrimental to overall sleep quality. For instance, healthy experimental subjects tested in the afternoon on vigilance and performance tasks did worse after they were allowed to sleep late than when they arose after a more usual length of sleep (Taub et al., 1971).

Schedule Effects

The entrainment of a 24-hour sleep/wake rhythm requires arising at the same clock time daily (Webb & Agnew, 1974). Over a quarter of U.S. workers work outside the usual daytime work schedules (Gordon et al., 1986). They are out of synchrony with the daylight/night cycle, many of them are on rotating shifts (Czeisler, Moore-Ede, & Coleman, 1982). Night workers sleep less than others and worse than others (Tepas et al., 1982), possibly because their off-work hours are less organized (Kripke, Cook, & Lewis, 1971). Some individuals without timed work obligations keep highly erratic schedules (Regestein, 1982) or else sleep progressively later until they come into conflict with daytime obligations. Such progressively later sleeping reflects the underlying longer-than-24-hour circadian rhythm found in individuals deprived of physiologic timing signals (e.g., periodic bright light and regular mealtimes). Among the elderly, an opposite tendency, toward increasingly earlier sleep schedules, is found (Tune, 1968), probably reflecting the shortening of the underlying endogenous circadian rhythm that occurs with age (Weitzman et al., 1982). Habitual late sleeping in the morning also predisposes to depression (Globus, 1969; Wehr et al., 1979). This may contribute to the depression of people unable to work regularly because of physical illness. When possible, patients should arise before 8 a.m. even if they do not have scheduled work obligations.

Hyperarousal

Arousal is that state in which focused thought and sustained productive efforts occur. It is more than simple wakefulness. The arousal profile of an individual is related to neurophysiologic properties, such as the percentage of EEG alpha rhythm manifested under standard conditions (Gray, 1967) or the electrocortical responsiveness to stimuli (Tecce, 1971), and is expressed in particular behavioral tendencies (Eysenck, 1972). For instance, high-arousal individuals ordinarily have lower sensory thresholds and more prolonged duration of responses (Mangan, 1972), qualities likely to keep them up at night and to turn stress into insomnia (Healy et al., 1981). It is the intensity rather than the distress of stimuli that causes insomnia in the hyperaroused individual. Pleasurable but intense events in an evening will thus lengthen the time needed to fall asleep. For instance, watching exciting television dramas in the evening reportedly worsens sleep (Saletu, Gruenberger, & Anderer, 1983). Autonomic indices of arousal during the night also differ in poor sleepers, who show higher skin resistances (Monroe, 1967). The rapidity with which the higher-arousal individual forms conditioned connections may worsen the insomnia, once its starts, through the formation of a learned association between going to bed and not falling asleep (Hauri & Fisher, 1986).

Psychopathology

Although some attribute most insomnia to psychopathology (Sweetwood et al., 1980; Tan et al., 1984) and others do not (Carskadon et al., 1976; Zorick et al., 1981), it may be that both insomnia and psychopathology result from some common factor (Hauri, 1979). In one series of hospitalized patients requiring psychiatric consultation, 80% had disturbed sleep (Berlin et al., 1984). Many insomnia patients have a common profile of psychologic traits. They describe themselves as thoughtful, conscientious, worried anticipators (Regestein et al., 1985) who are more nervous, strained, brooding, and hopeless and who often feel anxious rather than sleepy at bedtime (Kales et al., 1984b). They have difficulty falling asleep for daytime naps as well as at night (Stepanski et al., 1988). The sleep polygraphic findings of primary insomniacs—persons with insomnia of no obvious cause—resemble those of anxiety patients (Reynolds et al., 1984). Unlike affective disorders in dementia, this pattern may be present when the cause of insomnia is relatively covert and the daytime functioning of the patient is relatively intact. Psychopathology per se does not cause insomnia, although depression, anxiety, and dementia induce sleep problems (Reynolds, 1987). Some severe psychopathology, such as that found in personality disorder or schizophrenia, often does not.

Hospitalization

Patients hospitalized with acute medical problems are likely to suffer noise, frequent disturbances, somatic symptoms, psychologic distress, enforced sleeping positions, casts or drainage tubes, polypharmacy, and a lessened distinction between night and day, any of which may lessen sleep. Even patients who sleep well in the hospital may crave more sleep (Cumming, 1984). Sleep probably promotes healing (Adam & Oswald, 1984), whereas partial sleep deprivation, in addition to its psychologic consequences, also worsens somatic symptoms (Moldofsky & Scarisbrick, 1976). Despite the salutary effects of sleep, however, medical care may necessarily interfere with the sleep of the most seriously ill patients. Intensive care unit patients average about five interruptions per hour during the calmest night hours (Dlin et al., 1971) and get less total sleep than patients on other wards (Broughton & Baron, 1978). Elective surgery patients average less than 6 hours of sleep before as well as after surgery (Ellis & Dudley, 1976; Murphy et al., 1977). After open-heart surgery, sleep may become light, fragmented and short, as little as 1 to 4 hours on the first postoperative day (Johns et al., 1974). Since such subtleties as the view from the patient's window influence surgical convalescence (Ulrich, 1984), gross disruption of sleep probably does as well, although systematic confirmation of this effect has yet to be made.

Special Medical Problems

Various medical problems predictably worsen sleep via fever, discomfort, drug effects, worry, and the like. The sleep consequences of medical problems vary widely. The subjective nature of insomnia (Regestein & Reich,

1978), the small amount of sleep that suffices for some people (Jones & Oswald, 1968), and the admixture of psychologic regression during illness all complicate complaints of sleeplessness. Many people with medical reasons for insomnia actually sleep well, and one must therefore consider preexisting tendencies before attributing insomnia entirely to the medical problem at hand.

Breathing Disorders (Table 21-4)

The existence of sleep breathing disorders has long been recognized (e.g., Caton, 1889). Sleep decreases the ventilatory response (Cherniack, 1981) and thus predisposes to disordered breathing. The recumbent position itself increases the work of breathing (Anch, Remmers, & Bunce, 1982) and thus may aggravate any breathing disorder. Many other impediments may amplify breathing disorder during sleep, thus damaging sleep quality (Roehrs et al., 1985). Nasal blockage forces mouth breathing, which impairs sleep quality (Olsen, Kern, & Westbrook, 1981) and impedes breathing during sleep, occasionally to the extent of apnea and death (Wynne, Block, & Boysen, 1980). Sleep apnea sometimes may be reversed simply by repair of nasal septal deviation (Heimer et al., 1983). The normal dilation of the upper airway during inspiration may be impaired by topical nasal anesthesia (Oomen, Abu-osba, & Thach, 1982) or by sedatives, especially alcohol (Taasan et al., 1981).

Structural or functional impairment anywhere along the respiratory tract makes disturbed breathing during sleep more likely. The narrowing of the airway by obesity, large adenoids or tonsils, an enlarged soft palate or uvula, patulous pharyngeal mucosa, micrognathia (Pollack et al., 1987; Spier et al., 1986), cosmetic jaw surgery (Guilleminault, Riley, & Powell, 1985), peritonsillar abscesses (Lau, 1987), mandibular cyst (Alving, 1985), macroglossia, or the thickened tissues of acromegaly thus all increase the risk of apnea (Fairbanks, 1984; Orr, 1983), as may the narrower glottis of women (Haponik et al., 1981). Restrictive lung disease (e.g., from interstitial fibrosis, kyphoscoliosis, neuromuscular disease, obesity, or pregnancy) diminishes the volume of functioning lung, thus predisposing to hypoxemia during sleep and sleep apnea (George & Kryger, 1987; Sawicka & Branthwaite, 1987). The impairment of ventilatory control mechanisms in states of hypoxia, hypercapnia, or slowed circulation time also increase the risk of sleep apnea (Chada, Birch, & Sackner, 1985; Cherniack, 1981; Strohl, Cherniack, & Gothe, 1986).

TABLE 21–4. *Major Risk Factors for Sleep Breathing Disorders*

Obesity
Upper airway structural impediment to air flow (e.g., large uvula, enlarged tonsils, redundant pharyngeal mucosa, nasal blockage, retrognathia)
Male sex
Older age
Use of alcohol/sedatives
Lung disease (e.g., with hypercapnia or hypoxia)
Heart failure
CNS pathology (central apnea)

Combinations of structural abnormalities and functional dysregulation often occur. For instance, brisk inspiratory efforts, possibly driven by hypoxemia, may be made against a compromised upper airway (e.g., in snorers) (Skatbud & Dempsey, 1985), thus generating a high negative interthoracic pressure that sucks the hypopharnyx closed, thereby obstructing breathing (Remmers et al., 1978). This negative pressure also diminishes cardiac output by reducing left ventricular filling and stroke volume (Tolle et al., 1983). In the presence of any further heart failure, this may delay feedback information about carbon dioxide levels to central chemoreceptors, setting up ventilatory drive overshoot and thus risking further periodic breathing. Conversely, ventilatory drive may be diminished by the effects of obesity (Lopata & Onal, 1982), alcohol (Chan et al., 1989), or apnea-induced hypoventilation itself (Berthon-Jones & Sullivan, 1987; Guilleminault et al., 1989; Jones et al., 1985), possibly by increases in central opioids found in this condition (Gislason et al., 1989).

Since airway compromise and impaired ventilatory control mechanisms damage sleep quality, patients with chronic lung disease sleep poorly (Klink & Quan, 1987; Perez-Padilla et al., 1985), especially when there is significant hypoxia during wakefulness (Wynne, Block & Boysen, 1980). Prolonged oxygen desaturation during sleep may lead to gasping and prompt awakening, although patients who breathe poorly during sleep and complain of insomnia reportedly have less desaturation than those who complain of excessive sleepiness (Roehrs et al., 1985). Nocturia may derive from atrial natriuretic peptide release presumably from right atrial dilation in sleep apnea (Krieger et al., 1989), although sleep apnea is also associated with nephrotic syndrome (Chaudhary et al., 1988). Coma, EEG abnormalities, and anoxic spasms have been reported from apnea-caused oxygen desaturation (Cirignotta et al., 1989).

Because of the multiplicity of the above-mentioned medical problems that induce sleep-related breathing disorders, they rise in frequency from about 1.5% in the general population (Jeong & Dimsdale, 1989; Lavie, 1983) to as much as 39% in elderly subjects (Ancoli-Israel et al., 1981), in whom they may be stable (Mason, Ancoli-Israel, & Kripke, 1989) or may cause earlier death (Ancoli-Israel et al., 1989; Bliwise et al., 1988).

Although sleep apnea has been specifically connected with insomnia in some patients (Guilleminault et al., 1973), it may also present as hypersomnia (see below). The presence of a breathing disorder clearly worsens sleep quality, but mechanisms by which it causes complaints of insomnia or hypersomnia in given patients are not entirely clear (Orr, 1983). There are asymptomatic normal and obese subjects with sleep apnea (Block, 1979; Cook et al., 1982), which suggests that breathing disorder does not explain complaints related to poor sleep quality. Presumably, nervous and psychologic factors are interposed between poor sleep quality of any cause and complaints about it. For instance, frequent, brief, unremembered nocturnal arousals may induce next-day sleepiness unaccompanied by complaints of insufficient sleep.

Denial of symptoms is frustratingly common in sleep apnea patients. Their apnea possibly is caused by the many cognitive and affective disorders documented among them (Guilleminault & Dement, 1978; Kales et al., 1985). Depression associated with sleep apnea may be reversed on relief of the disorder by continuous positive airway pressure (Millman et al., 1989).

Snoring

Snoring results from a high-frequency opening and closing of the oropharynx (Sauerland & Harper, 1976). It is augmented by compromise of the upper airway from any structural narrowing (e.g., at the nostril) (Petrusen, 1990) or flaccidity and is associated with obesity, hypertension, and heart and lung disease (Norton & Dunn, 1985). It increases with age (Lugaresi et al., 1980), genetic predisposition (Deray, Duchowny, & Resnick, 1987; Kaprio et al., 1987); and dementia (Erkinjuntti et al., 1987). It is worsened by states of severe exhaustion, the use of alcohol (Issa & Sullivan, 1984; Scrima, Broudy, & Cohn, 1982), and other sedatives. Sleep apnea becomes likely with loud snoring, especially in those with flabbier pharyngeal tissues (Brown et al., 1985; Lugaresi et al., 1979), although a third of loud snorers had no sleep apnea in one series (Miles & Simmons, 1984). Previously considered a benign inconvenience, snoring is associated with breathing disturbance during sleep (Berry et al., 1989) and augmented risk of hypertension and coronary artery disease (Waller & Bhopal, 1989), but remains relatively little investigated even among inpatients, in whom it is easily detected. A nostril dilating device worn at night may reduce snoring (Petrusen, 1990). Sleep-related breathing disorder, particularly obstructive sleep apnea, is discussed in Chapter 25.

Obesity

Possibly from increased bogginess of mucosa in the supine position and fat deposition at the tongue base, obesity increases upper airway resistance and thus the risk of breathing disorder during sleep. The inefficiency of a broad, flat diaphragm and the resistance of abdominal fat may underlie the diminished ventilatory response found with obesity (Lopata & Onal, 1982). The presence of sleep apnea, in turn, contributes to the diurnal obesity-hypoventilation syndrome (Jones et al., 1985). A "night eating syndrome," composed of much eating after the evening meal, insomnia, and morning anorexia, is reportedly frequent in the obese (Stunkard, Grace, & Wolff, 1955).

Cardiovascular Disorders

Coronary artery disease is associated with failure to nap during the afternoon (Trichopoulos et al., 1987) and with breathing impairment during sleep, even in the absence of symptomatic lung disease (Olazabal et al., 1982; Peter et al., 1986). The mechanism of this impairment is unclear. However, treatment of coronary artery disease with bypass grafting has reversed sleep apnea (Gugger & Hess, 1989). Lying down diminishes lung functional residual capacity, potentially increasing hypoxia and thus sleep apnea (George & Kryger, 1987). Patients with angina may have painful episodes that awaken them, especially from dreaming or REM sleep (Murao et al., 1972; Nowlin et al., 1965), possibly as a consequence of episodically increased heart rate (Quuyumi et al., 1984), but such patients sleep worse than their pain episodes alone would explain (Karacan, Williams, & Taylor, 1969). When oxygen desaturation occurs during sleep in coronary artery disease, the myocardial blood flow requirement may be similar to those of maximal exercise (Shepard et al., 1984), predisposing toward anginal pain. Chronic hypoxia during sleep may also induce pulmonary hypertension (Coccagna & Lugaresi, 1978), leading to right-sided heart failure (Hall, 1986; Sullivan & Issa, 1980). Heart failure itself, possibly from delayed feedback from lung to central chemoreceptors, induces an undershoot/overshoot cyclicity in breathing and thus increases the incidence of sleep apnea (Peter et al., 1986).

In studies of the hypertensive male population, about 30% have manifested undiagnosed sleep apnea (Fletcher et al., 1985; Jeong & Dimsdale, 1989; Kales et al., 1984b), whereas 50% of those with sleep apnea have reportedly had moderate hypertension (Guilleminault & Dement, 1978). This likely explains the associations found between snoring and hypertension (Lugaresi et al., 1980). Normal sleep generally lessens the frequency and grade of cardiac arrhythmias (Lown et al., 1973; Regestein, DeSilva, & Lown, 1981). However, in cases in which chronic hypoxia shifts individuals to the sharply falling slope of the oxygen-hemoglobin dissociation curve, lessened ventilation during sleep can cause profound oxygen desaturation (Fleetham & Kryger, 1981), at which time ventricular ectopy may emerge (Shepard et al., 1985). A bradycardia-tachycardia pattern is frequently found during episodes of sleep-impaired breathing, although any type of cardiac arrhythmia may emerge (Tilkian et al., 1977).

Because of the association of sleep apnea with cardiac arrhythmia, readers of sensational articles in the lay press occasionally insist on immediate consultation for snoring spouses. Sleep breathing problems are common and chronic and are associated with high-grade ventricular arrhythmia only in a small minority of cases, often those with severe apneas (Otsuka, Sadakane, & Ozawa, 1987) and other risk factors such as massive obesity and hypertension.

Endocrine-metabolic Factors

Exercise increases slow-wave (deep) sleep, but more in physically fit than in less fit individuals (Horne & Moore, 1985; Trinder et al., 1982; Walker et al., 1978). The mechanism may be metabolic, resulting from raised body temperature (Bunnell & Horvath, 1985; Horne & Staff, 1983) and consequently increased metabolism (Oswald, 1980). Increased thyroid indices and hyperthyroidism also are associated with more (Dunleavy et al., 1974; Johns et al., 1975) and hypothyroidism with less (Kales et al., 1967) deep sleep. Clinically, however, hyperthyroidism may cause severe insomnia (Regestein, 1982) and goiter has been associated with sleep apnea (Young et al., 1986). Fever induces broken sleep in some patients and prolonged sleep in others. Sleep itself decreases cerebral glucose metabolic rate, according to preliminary positron emission tomography results, except increases of cerebral metabolism were noted in conjunction with nightmares (Heiss et al., 1985).

The autonomic neuropathy of diabetes mellitus increases the risk of sleep apnea (Rees et al., 1981), presumably by interference with muscular responses of the pharyngeal airway (Guilleminault et al., 1981). Two of the author's insulin-dependent sleep apnea patients submitted to surgical revision of the upper airway atypically had no subjective postoperative improvement, presumably because such treatment left the basic neuropathic impairment of the pharyngeal inspiratory response untouched.

Androgens presumably cause the increase in sleepiness after puberty (Carskadon et al., 1980), as well as the arousal differences found between the sexes in adulthood (Broverman, Klaiber, & Kobayashi, 1968). The large preponderance of men (e.g., 15:1 in a typical series) (Guilleminault & Dement, 1978) suggests that androgenic hormones also play an etiologic role in obstructive sleep apnea. Exogenous androgen administration has reportedly induced obstructive apnea in a woman (Johnson, Anch, & Remmers, 1984), although this is not typical (Millman, 1985).

Exogenous growth hormone (GH) administration during sleep in normal subjects as well as inpatients with acromegaly has resulted in much-decreased deep sleep, possibly because of a GH–deep sleep feedback mechanism (Carlson et al., 1972; Mendelson et al., 1980). Acromegaly further increases the risk of obstructive sleep apnea, possibly because of thickened pharyngeal tissues (Cadieux et al., 1982). However, sleep complaints improve little after hypophysectomy, possibly because upper airway changes remain (Pekkarinen et al., 1987).

Menopause carries a high risk of sleep disturbance, even among questionnaire respondents who have been screened for psychopathology (Ballinger, 1976; Lauritzen, 1976). This is worse in women with hot flashes (Shaver et al., 1988) and may be due to estrogen deficiency in many patients, since it may be remedied by replacement estrogen (Campbell, 1976; Thompson & Oswald, 1977). In most menstruating women, however, cyclical fluctuations in estrogen level have little effect on sleep (Billiard & Passouant, 1974).

Other endocrinologic conditions may affect nervous system functioning to provoke a variety of arousal-related syndromes, including sleep disturbance (e.g., the premenstrual state [Abraham, 1983; Rubinow & Roy-Byrne, 1984], which may cause insomnia or hypersomnia [Billiard, Guilleminault, & Dement, 1975; Endicott et al., 1981]), and Cushing's syndrome (Regestein, Rose, & Williams, 1972; Starkman, Schteingart, & Schork, 1981). A covarying time course of the endocrine condition and the sleep disturbance supports an etiologic relationship.

Neurologic Conditions (Table 21-5)

Although sleep is most directly regulated by the nervous system, neurologic disorders disturb sleep surprisingly less than might be expected compared with breathing or endocrine disorders, except when sleep regulating centers are directly affected. For instance, extensive pontine infarctions may occur without much polygraphic sleep abnormality (Markand & Dyken, 1976; Marquardsen & Harvald, 1964), although bilateral lesions of the midpontine tegmentum can produce severe

TABLE 21–5. *Some Neurologic Causes of Impaired Sleep Quality*

Direct involvement of major sleep regulating mechanisms (rare) (e.g., with somnolence of secondary narcolepsy)
Secondary effects of motor dysfunction (muscle disease, late effects of polio, parkinsonism)
Enforced inactivity
Difficulties changing sleep positions or arising for nocturia
Sleep breathing impairment
Peripheral neuropathy
Sleep movement disorders
Diminished upper airway function
Aggravation of gastric emptying problems by recumbency
Generalized cortical impairment
Insensitivity to timing cues such as social schedules and meal times, with disruption of circadian patterns
Increased daytime naps
Increased likelihood of taking CNS-acting drugs
Diminished diurnal activity, increased tendency to retreat to bed
Seizure disorders
Emergence of excitatory phenomena during sleep (e.g., awakenings, automatisms)
Effects of anticonvulsant medications

sleep abnormalities. Lesions more rostral in the core arousal system induce more stimulus-modulated impairments, (e.g., medial thalamic strokes induce disruptable drowsiness rather than sleep) (Masson et al., 1987; Prendes & Rosenberg, 1986). Vertebral basilar embolus may induce drowsiness (Gulcman et al., 1987) as well as coma, although states of somnolence/akinetic mutism may be distinguished from amnestic confusional states according to which branches of the distal basilar arterial tree are affected (Prendes & Rosenberg, 1986). Diencephalic lesions may provoke narcolepsy (Aldrich & Naylor, 1989). Often neurologic conditions impair sleep through indirect effects, such as breathing problems associated with parkinsonian rigidity, refusal to follow good sleep habits in the mildly demented, or enforced inactivity with irregular napping in the motorically disabled.

The polygraphic abnormalities of sleep in various neurologic diseases may not be highly relevant to the clinical status of the individual patient, and may be highly variable. For instance, Mouret (1975), who found slower sleep onset, less deep sleep, and increased blepharospasms during REM sleep in Parkinson disease patients, noted that there is an "absence of clear and constant alterations of sleep patterns in this disease," as subsequently confirmed (Apps et al., 1985). Others have reported frequent sleep abnormalities in parkinsonism, including longer sleep onset, more wakefulness during sleep, and less sleep spindle activity (Kales et al., 1971; Puca, Bricolo, & Turella, 1973). The sleep disturbances correlate with clinical symptoms (Schneider et al., 1974); the rest tremor disappears during sleep (Stern, Roffwarg, & Duvoisir, 1968). The relief of restless legs syndrome by bromocriptine (Walters et al., 1988) or L-dopa (Akpinar, 1987) suggests it may be a parkinsonian equivalent.

A marked heterogeneity of effects on sleep is also found in seizure disorders. Surveyed epilepsy patients report poor sleep quality and excessive daytime sleepiness (Ruiz-Primo, Coria, & Torres, 1985), possibly because of their lighter, less stable sleep (Declerck, 1986; Touchon et al., 1987). Sleepwalking and night terrors, normally rare in adults, are reported with some frequency (Hoeppner, Garron, & Cartwright, 1984). Patients with partial rather than generalized seizure disorders have more disturbed sleep. Epilepsy patients in general have lighter-than-normal polygraphically recorded sleep patterns (Declerck et al., 1982), and those with temporal lobe epilepsy in particular show frequent stage shifts and numerous awakenings (Baldy-Moulinier, 1982). Frequent brief awakenings may be related to spike-and-wave paroxysms (Burr, Stefan, & Renin, 1986). Insomnia and nightmare problems have been observed in association with temporal lobe epilepsy, sometimes with frequent brief awakenings throughout laboratory sleep recordings. Such problems are probably not caused by anticonvulsant medications, which have indeed occasionally relieved such problems, although these agents exert various minor direct effects on sleep (Harding, Alfon, & Powell, 1985; Johnson, 1982).

Sleep deprivation activates epilepsy, although the mechanism is uncertain (Dahl & Dam, 1985). It may act by inducing subsequent sleep, or by some independent mechanism such as increased susceptibility to kindling (Shouse, 1988). Sleep itself leads to more epileptic discharges, with increased spiking activity and generalization during NREM sleep. However, the desynchronized EEG activity of REM sleep limits and focalizes epileptic discharges and thus inhibits seizures (DeClerck, 1986; Vieth, 1986).

Posttraumatic coma may ablate normal sleep polygraphic patterns; this is a poor prognostic sign (Bergamasco et al., 1968). Even lesser degrees of head trauma without coma may lead to excessive daytime sleepiness (Guilleminault et al., 1983). Trauma can elicit a vast range of sleep abnormalities including sleep apnea (Rosomoff, 1986); the Kleine-Levin syndrome, which involves relapsing, several-day episodes of somnolence (Will, Young, & Thomas, 1988); and secondary narcolepsy (Maccario, Ruggles, & Meriwether, 1987). Whiplash has been associated with subsequent minor sleep problems (Coren & Searleman, 1985). Brain tumors may disrupt normal sleep EEG patterns with delta activity. Amplitude and regularity of rhythm during sleep may differ between cerebral and posterior fossa lesions (Daly, 1968; Ohgami, 1973). Thalamotomy lesions may cause ipsilateral alterations in sleep recordings (Jurko, Orlando, & Webster, 1971).

Impairments of cortical functioning consistently alter sleep patterns (Lipowski, 1980), but in a variety of ways. In mental retardation, for instance, total sleep time reportedly increases in Down's syndrome but decreases in aminoacidurias and idiopathic mental retardation (Petre-Quadens & Jouvet, 1967). Rapid eye movement sleep in particular is diminished by the last two conditions, but is increased in Prader-Willi syndrome, which involves hypothalamic dysfunction (Vela-Bueno et al., 1984).

Sleep quality is much impaired by dementia in general (Feinberg, Koresko, & Heller, 1967; Prinz et al., 1982; Reynolds et al., 1985) and by Jakob-Creutzfeldt disease in particular (Vitrey, Huquet, & Dollfus, 1970). Severe dementia may be associated with either abbreviated or prolonged sleep as well as multiple naps and nocturnal waking periods (Regestein, 1987a). Specific dementing illnesses that have been associated with polygraphically disordered sleep include Parkinson's disease (Apps et al., 1985), Huntington's disease (Hansotia, Wall, & Berendes, 1985), Shy-Drager syndrome (Coccagna et al., 1985), progressive supranuclear palsy (Aldrich et al., 1989b), spinocerebellar degeneration (Katayama et al., 1986), olivopontocerebellar degeneration (Salazar-Gruesco, Rosenberg, & Roos, 1988), and cerebral Whipple's disease (Daiss, Wiethoelten, & Schuma, 1986).

Diminished arousal in dementia patients presumably impairs their integration of ventilatory function and their reflex response to airway obstruction. Although apneic episodes are longer and more frequent in some Alzheimer's disease patients compared with controls (Hoch et al., 1989), they nevertheless tend to be clinically inconsequential (Bliwise et al., 1989). Of 80 mostly middle-aged sleep apnea patients, however, 24% showed psychometric evidence of "mild to severe impairment of brain functioning" (Kales et al., 1984a). The severe oxygen desaturation associated with some sleep apnea (Cirignotta et al., 1989) presumably aggravates preexisting cerebral impairments.

Sleep apnea has been associated with other neurologic diseases, such as Tourette's syndrome (Silvestri et al., 1987) and stroke (Kapen, Maas & Nichols, 1990). It also has been reported as a late complication of bulbar polio (Guilleminault & Murtta, 1978). In sum, it seems likely that brain disease of many kinds can cause or aggravate sleep apnea, and that sleep apnea in turn impairs arousal functions and cognitive performance.

A disconnection between dreaming and verbalizing occurs in lobotomy patients, who either fail to recall dreams on being awakened from REM sleep or else remember the feeling of the dream but not the dream content (Jus et al., 1972).

Acute encephalitis that involves the basilar sleep-regulating systems can induce somnolence and sleep/wake cycle disturbance associated with cranial nerve signs, as distinguished from hyperkinetic reactions (Howard & Lees, 1987). This may involve a paucity of motor responsiveness and EEG alpha patterns superimposed on sleep-like states (Al-Mateen et al., 1988). Late encephalitic complications include sleep apnea (Guilleminault & Motta, 1978) and Kleine-Levin syndrome (Merriam, 1986). Disturbances of consciousness are found more in caudal than rostral basilar abscesses, as part of a polysymptomatic devastation (Arseni & Ciurea, 1988). The late effects of poliomyelitis entail disturbance of sleep directly or through breathing disturbance (Cosgrove, Alexander, & Kitts, 1987).

Despite the range of sleep effects produced by various neurologic conditions, a basic degenerative sleep pattern remains common to a host of disorders. This involves alterations of the sleep/wake cycle, increased time in bed with diminished sleep time, prolonged sleep onset, increased awakening during sleep, decreased deep sleep, decreased REM sleep, fewer sleep spindles, and disordered breathing. Some or all of these features have been observed in association with stroke (Koerner et al., 1986).

Some connection between multiple sclerosis (MS) and narcolepsy has been postulated since the majority of MS patients have not only sleep attacks but cataplexy (Poirier et al., 1987). However, MS patients reportedly do not have the genetic haplotype or sleep onset characteristics of narcolepsy (Rumbach et al., 1989). Less devastating neurologic syndromes may also alter sleep quality. Migraine headache awakens the patient from REM sleep and may lead to excessive sleeping in the recovery phase, which distinguishes it in some cases from tension headache, which may occur after a period of wakefulness (Blau, 1982; Dexter & Weitzman, 1970). The aberrant catecholamine hypersensitivity observed in association with prolapsed mitral valve syndrome may induce lighter sleep and middle-of-the-night waking episodes with anxiety, sweating, and tachycardia (Clark et al., 1980). These nocturnal panic attacks are often confused with psychogenic insomnia.

Pregnancy

First-trimester pregnancy often increases sleep and sleepiness, sometimes enough to reverse an insomnia problem, presumably as a result of augmented levels of progesterone, which has mild sedating properties. In the third trimester, however, urinary frequency, movements of the unborn, and interference with accustomed sleeping positions combine to lessen sleep quality. Rapid eye movement and stage 4 sleep decrease, and wakefulness and sleep disturbances increase (Berlin, 1988). These abnormalities may persist postpartum and possibly exacerbate postpartum emotional disturbances (Karacan & Williams, 1970). Further sleep disturbance may arise from night sweating, which has reportedly been observed in a majority of normal obstetric patients (Lea & Aber, 1985).

Gastrointestinal Problems

Recumbency provokes the pain of hiatus hernia, presumably from diminished esophageal drainage. Thus esophagitis patients double their chances of recovery in 6 weeks if the head of the bed is raised up on blocks (Harvey et al., 1987). Remarkably, experimental infusions of weak hydrochloric acid via a nasal-esophageal catheter, which are indistinguishable from water infusions by the awake subject, provoke arousals during sleep with greater frequency in the esophagitis patient (Orr et al., 1988). This suggests a mechanism for awakening that is unassociated with pain so that the cause is unsuspected by the patient.

Overall, secretion of gastric acid falls during sleep (Stacher, Presslich, & Starker, 1975). Increased secretion of gastric acid during REM sleep, mentioned previously, may characterize some patients (Armstrong et al., 1965) but not others (Orr et al., 1976). Nevertheless, gastric acid pain may disturb sleep, sometimes as a rebound hypersecretion effect from use of antacids, or possibly from the inhibition of esophageal clearing by sedatives (Orr, Robinson, & Randall, 1985).

Functional conditions of the large bowel worsen sleep (Johnsen, Jacobson, & Forde, 1986). Frequent urges to defecate can also aggravate other sleep problems. Patients with difficulty falling asleep are particularly likely to remain awake when awakened by the urge to defecate. The sleep of hepatic failure patients is often short, light, and disrupted (Bergonzi et al., 1978; Kurtz et al., 1972) and worsens with worsening hepatic status. False neurotransmitters resulting from enzymes produced by gut bacteria and normally destroyed by the liver may be shunted into the circulation and thence to the brain, where they presumably interfere with neurochemical regulation of sleep (Fisher & Baldessarini, 1971). This mechanism may also explain the occasional neurally intact patient whose sleep lessens after portocaval shunt, or the great improvement in sleep found in the patient with hepatic failure placed on a minimum protein diet.

Miscellaneous Medical Problems

In general, symptoms of medical problems, such as pain, itch, or dyspnea, or the effects of treatments, such as use of CNS-acting drugs, may worsen sleep. In the acutely ill patient who previously obtained restful sleep, a covarying time course between the illness and the sleep problem supports a causal connection. In the chronically ill patient, the relationship between a medical problem and disturbed sleep may be less clear. Responses even to chronic confinement and immobility may vary from healthy sleep to alteration of specific sleep stages (Rotenberg & Kobrin, 1985).

Unrefreshing sleep and fatigue is common in rheumatic diseases (Moldofsky, 1989). Fibrositis (or fibromyalgia) syndrome patients complain of sleep disturbances in 60% to 90% of cases (Goldenberg, 1987) and continue to so in long-term follow-up (Felson & Goldenberg, 1986). They also have worse pain and mood symptoms when they obtain less stage 4 sleep (Moldofsky & Lue, 1980); and experimental stage 4 sleep deprivation of normal subjects can produce fibrositis-like symptoms (Moldofsky & Scarisbrick, 1976). Accidental trauma may provoke a fibrositis-like syndrome, including the sleep disturbance (Saskin, Moldofsky, & Lue, 1986). (See also Chapter 33.)

Poor sleep quality likely diminishes tolerance of most other disease symptoms. There is some evidence, for instance, that sleep quality is worse in osteoarthritis patients who complain about morning joint pain and stiffness compared with those who do not (Moldofsky, Lue, & Saskin, 1987). Osteoarthritis patients also tend to move more during their sleep (Leigh et al., 1988). Abnormal sleep EEG patterns in which alpha rhythm appears during slow (delta) wave sleep ("alpha-delta" sleep) are seen more in arthritis patients, perhaps underlying the diminished refreshment of their sleep.

Chronic uremic patients show less deep sleep, less REM sleep, and many myoclonic jerks; these features are all diminished but not abolished by hemodialysis) Passouant et al., 1970). Sleep is worse the night after a hemodialysis session, possibly because of direct effects of the hemodialysis but partly because of schedule shifts and other encumbrances of the method (Daly & Hassall, 1970).

Irritative symptoms of prostatism compel nocturia, but an associated problem with sleep results not from the nocturia per se but from an inability to return to sleep within a tolerable interval. Occasionally, bladder training can be used successfully to increase the urinary volume at which the urinary urge is experienced, lessening disruptions of sleep by nocturia.

Sensory impressions of the external world direct attention away from the internal states (Hernandez-Peon, 1966).

After the lights go out at bedtime, internal sensations may thus intensify, worsening disease symptoms. This especially augments itch and pain. Chronic itch delays sleep onset and disrupts sleep. Patients with pruritus, when visually and polygraphically monitored, show scratching during all stages of sleep, and diminished deep sleep (Brown & Kalucy, 1975; Savin, Peterson, & Oswald, 1973).

Table 21–6 summarizes the most common causes of disturbed sleep in medical-surgical patients.

Idiopathic Disorders of Impaired Wakefulness (Table 21-7)

Medical problems tend to provoke lethargy rather than sleepiness. The ill patient dozes by default, because of lassitude, rather than because of an active supervening of sleep. In narcolepsy and idiopathic hypersomnia, however, sleep preempts all other activity. In narcolepsy, REM sleep supervenes, involving part or all of the REM state, including sleep, hallucinations or dreams, and bilateral atonia. In idiopathic hypersomnia, polygraphically normal sleep persistently occurs both in multiple naps during the day and prolonged sleep at night (more than 10 hours).

Individual presentations of narcolepsy vary from cases with one sleep attack daily to those with frequent 5- to 15-minute sleep attacks, cataplexy (i.e., brief bilateral atonia precipitated by sudden emotion or surprise), visual hallucinations at sleep onset, and sleep paralysis. Other features of narcolepsy include disrupted nocturnal sleep, nocturnal automatic behavior, and diminution of social and occupational competence (Broughton et al., 1981).

Idiopathic hypersomnia sleep attacks last hours rather than minutes. Nocturnal sleep is not disrupted and neither cataplexy nor hallucinations are present (Roth, 1980). About half the patients have "sleep drunkenness," a severe inability to function on first awakening mornings. This contrasts sharply with narcoleptic patients, who tend to function best on first arising.

Both conditions are associated with depression. Narcolepsy also may be associated with a higher incidence of MS (Poirier et al., 1986) and diabetes (Honda et al., 1986), and can be acquired following head trauma or other CNS damage (Aldrich & Naylor, 1989; Kowatch, Parmelee, & Morin, 1989; Mitler et al., 1986).

Narcolepsy and idiopathic hypersomnia most often begin before age 35 and remain present daily thereafter. This history distinguishes them from effects of acute medical illness.

Restless legs syndrome, compulsive leg movements prior to sleep onset, may provoke sleep deprivation (Ekbom, 1960).

TABLE 21–6. *Common Causes of Sleep Disorders in Medical and Surgical Patients*

INSOMNIA
- External influences
  - nursing care
  - commotion on intensive care units
  - enforced sleeping positions, attached equipment
- Nervous system causes
  - hyperarousal
  - arousal-inducing psychopathology
  - cortical impairments
  - lesions of sleep control mechanisms
  - acute trauma
  - catecholamine hypersensitivity (e.g., prolapsed mitral valve syndrome)
  - seizure disorders
  - CNS-acting drugs
- Timing system disorder
  - irregular sleep schedule
  - systematic and progressive delay or advance of sleep phase timing
  - insufficient sensitivity to timing information
- Breathing disorder
  - airway obstruction, upper or lower respiratory tract
  - pulmonary disease with hypercapnia or hypoxia
  - heart failure with slowed circulation time
  - constricted mechanics of breathing
    - obesity
    - kyphoscoliosis
    - muscle disease
    - parkinsonism
    - diminished ventilatory response
- Metabolic factors
  - increased cerebral metabolism
  - thyroid abnormalities
  - corticosteroid abnormalities
  - diabetic neuropathy with airway function disturbance
  - premenstrual and menopausal factors
- Gastrointestinal problems
  - effects of large meal
  - hiatus hernia and esophageal acid reflux
  - peptic ulcer pain
  - nocturnal symptoms of bowel disease
- Other problems
  - nocturnal disturbance from any cause (pain, itch, renal dialysis, dyspnea, nocturia)
  - liver disease

HYPERSOMNIA (complaint of excessive daytime sleepiness)
- Idiopathic sleep disorder (narcolepsy, idiopathic hypersomnia)
- Postencephalitic syndromes
- Fever or increased cerebral metabolic rate (some patients)
- Drugs
- States of severe lethargy and debilitation
- Severe CNS disorder (toxic, vascular, traumatic, neurodegenerative)
- Breathing disorders
- Metabolic disorders
- Poor nocturnal sleep quality from any medical or surgical cause
- CNS inflammation (encephalitis lethargica, multiple sclerosis)

EVALUATION

The Sleep History

The patient's sleep pattern should be described, including regularity of bedtimes, untoward behavior during

TABLE 21–7. *Distinguishing Characteristics of Narcolepsy and Idiopathic Hypersomnia*

	Narcolepsy	Idiopathic Hypersomnia
Night sleep	Average length Disrupted by awakenings	Prolonged Uninterrupted
Day naps	15–20 minutes	20–120 minutes
Typical naps/day	2–5	1–2
State on awakening mornings	Refreshed	Sleep drunk in half of cases
Sleep latency in day-recorded naps	Often <5 minutes	>5 minutes
Polygraph features	REM at sleep onset	Normal pattern
Haplotypes	Virtually all DR2 or DQW1 positive	None specific
Motor symptoms	Cataplexy Sleep paralysis	None

sleep (e.g., twitching, loud snoring, gasping, struggling, walking, night terrors, bed-wetting), bad dreams or nightmares, timing of any night wakings, and associated phenomena such as palpitations, emotional arousal associated with anxiety, hot flashes, sweating, pain, discomfort, or other symptoms. The regularity of rising times and sleep patterns should be documented with a month's sleep chart of bed and arising times in cases of chronic insomnia. This can serve as a baseline for comparison after treatment efforts are made. Patients with obstructive sleep apnea may have symptoms on first awakening that include headache, extremely dry mouth, and grogginess, often accompanied by nasal blockage or a sense of sinus fullness.

Dysphoria in the morning occurs in some depressive states. Grogginess on awakening is worsened by heavy use of caffeine, delayed sleep phase syndrome, or generally poor sleep quality.

Excessive daytime sleepiness is provoked especially by monotonous situations such as turnpike driving, boring meetings, and evening television watching. Hyperarousal is more likely in careful, organized, thoughtful, introspective people. Psychopathology and drug use are evaluated in the diagnostic interview. Childhood onset of insomnia suggests neurologic dysfunction (Hauri & Olmstead, 1980; Regestein & Reich, 1983).

The patient's story may neglect sleep difficulties found prior to the onset of any medical condition. Although some insomnia patients with psychopathology deny psychologic causes for their insomnia (Kales & Kales, 1984), others may diagnose themselves as distress-prone neurotics bound to suffer irremediable sleep problems. Many excessively sleepy patients develop low standards for wakefulness, deny any impairment, and are pushed into medical consultation by relatives.

The patient's bed partner may provide information about snoring, twitching, and other behavioral signs of sleep disorder and may provide objective information on the sleep problem and its effect on the patient's functioning.

Narcolepsy may be diagnosed by history alone or in conjunction with treatment response (Regestein, Reich, & Mufson, 1983). Cataplexy is virtually pathognomonic, but an unrelenting pattern of brief sleep attacks, frequent sleepiness, poor school or work performance, and self-management with preventive naps and careful use of caffeine or over-the-counter stimulants strongly suggests narcolepsy. Some families have several diagnosed narcoleptic or unusually sleep-prone individuals.

The diagnosis of idiopathic hypersomnia rests primarily on a history of prolonged nocturnal sleep and obligatory naps, with a sleepiness tendency from youth and sleep drunkenness.

In nocturnal myoclonus a 1 to 3-second muscle twitch, usually in the legs, appears every 20 to 30 seconds during sleep. This is sometimes an incidental polygraphic finding that increases in incidence with age (Roehrs et al., 1983). The history usually will include the patient's frequently kicking the covers off the bed. Alterations of central neurotransmitter levels may underlie this condition (Vardi et al., 1978).

Objective Evaluation of Sleep Disorder (Table 21-8)

In cases of insomnia, measuring the time it takes the patient to fall asleep (the sleep latency) and the proportion of recording time in which the patient slept (the sleep efficiency) provides most of the therapeutically useful information. One night's sleep is one data point in the continuing sleep/wake cycle and thus yields less information, often, than a month's sleep chart. The reason is that the exact polygraphic stages of sleep are poorly correlated across nights (Moses et al., 1972) and rarely provide much therapeutic information (Jacobs et al., 1988; Regestein, 1988). Insomnia is a subjective complaint, poorly understood through laboratory monitoring (Regestein & Reich, 1978). In one series of 150 insomnia patients, for instance, laboratory sleep monitoring neither specifically nor sensitively identified insomnia patients (Bixler et al., 1986).

Considering the difficulties with laboratory monitoring, a month's chart of sleep is the best measure of insomnia. Bedtimes, waking times, and estimated sleep during the night may be charted with a horizontal line on a date-hour matrix such as that used in Figure 21-1. Relevant information, such as episodes of awakening with gasping or palpitations, drug use, and medical symptom ratings, can be recorded for each date.

TABLE 21–8. *Common Methods of Investigation for Sleep Disorders*

Method	Rationale
Sleep chart	Insomnia, excessive sleepiness, timing of parasomnias
Every 1/2–1 hour brief observation of patient	Quantify sleep, discern temporal pattern of sleep
Actigraphy	Quantify week-long sleep pattern
All-night polygraphy	Describe sleep pattern, quantify breathing or movement disorder, detect cerebral or cardiac dysrhythmias
24-hour ambulatory six-channel polygraphy	Describe sleep and nap pattern, detect cerebral or cardiac dysrhythmias
Nocturnal ambulatory monitoring of airflow, O_2 saturation, and EKG	Detect sleep breathing disorders
Multiple sleep latency test	Quantify sleepiness, detect REM at sleep onset
Haplotyping	Helps rule out narcolepsy
Continuous performance tests	Quantify arousal impairment

Figures 21-2 and 21-3 provide two sample charted sleeping diary logs. Such records correlate with laboratory monitoring results (Lewis, 1969) and provide a baseline against which to judge treatment outcome.

Home tape recording of a patient's loud snoring serves to convince him of the problem as well as provide an indication that specific investigation for sleep breathing disorder is warranted.

Portable wrist-motion recorders permit long-term, inexpensive, and accurate ambulatory monitoring (Kripke et al., 1978) and may be combined with continuous ambulatory body temperature data to demonstrate circadian desynchrony. Normally, temperature will peak in late afternoon and be lowest in the early morning, an hour or two prior to normal arising times.

Hospital and nursing home inpatients may be monitored by an hourly 5-second nurse observation of whether the patient is asleep or awake (e.g., Regestein & Morris, 1987; Webb & Swinburne, 1971). Patterns such as early night insomnia, fitful sleep, and daytime napping may be noted, and snoring and breathing disorders may be detected for further investigation.

Breathing disorder during sleep may be investigated with ambulatory devices that record wrist motion, breathing, blood oxygen saturation, and electrocardiogram. The ambulatory cassette EEG is technically more difficult because of its relatively low signal voltage and many artifacts. Continual light densitometry of the earlobe or finger measures blood oxygen saturation, and can provide a key estimate of sleep breathing. In cases of mild obstructive apnea in the absence of cardiovascular complications and severe sleepiness, ambulatory monitoring may be sufficient and convenient, but in more severe cases of sleep breathing disorder, quantitative sleep laboratory assessment of apnea episodes, oxygen desaturation, and grade of cardiac arrhythmias may be necessary to make a sound judgment regarding the need for surgery and other treatments.

The level of daytime alertness provides a test of nocturnal sleep quality. Sustained attention tasks are most sensitive to impaired sleep (Wilkinson, 1965). A continuous performance test in which the subject decides each second whether to respond to a visual stimulus (Rosvold et al., 1956) sensitively detects poor sleep quality. Patients with chronic sleep problems may follow their daytime functions by regularly recording results of convenient arousal tests, such as performance in computer games, exercise performance, or grip strength, measured against the patient's own baseline.

The multiple sleep latency test (MSLT) consists of four or five daytime trials to estimate polygraphically recorded time to sleep onset (sleep latency) (Richardson et al., 1978). It measures sleepiness rather than arousal. Of the primary sleep disorders, narcolepsy yields an average sleep latency of 5 minutes or less, and idiopathic hypersomnia shows an average sleep latency of 8 or 9 minutes. Sleep-onset REM periods may be elicited by this test and are found in states of central catecholamine depletion (e.g., with use of certain antihypertensive drugs, stimulant withdrawal, narcolepsy, or severe mood disorder). The MSLT should be used to confirm narcolepsy where the diagnosis is unclear (e.g., in the absence of cataplexy), but false-negative results (Regestein, Reich, & Mufson, 1983; Rhodes et al., 1989) prevent the use of this test to rule out narcolepsy.

A leukocyte antigen, DR-2, is reportedly present in almost all patients with narcolepsy and may help distinguish this condition from other hypersomnias (Mitler et al., 1986; Montplaisir et al., 1986; Rubin, Hajdukovich, & Mitler, 1988).

IMPROVING SLEEP QUALITY

Sleep disorders in the medical patient are relieved by neutralizing predispositions to poor sleep, treating med-

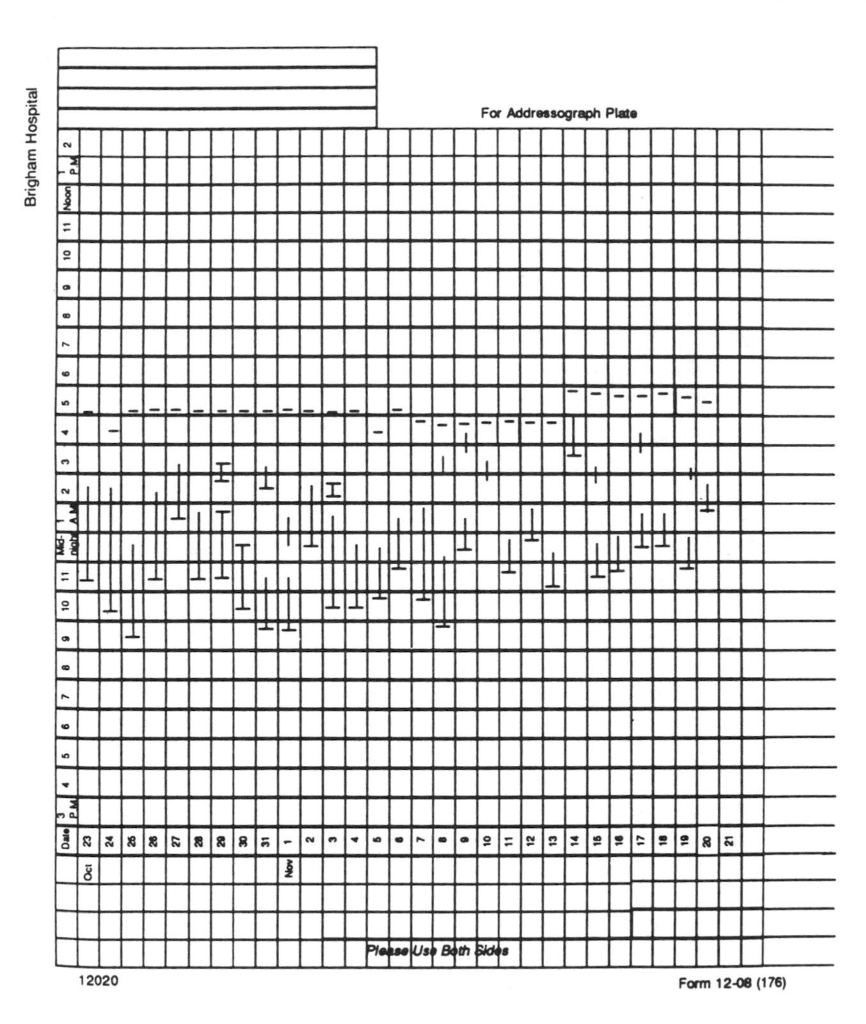

FIG. 21-1. The sleep chart of a 57-year-old man with excessive daytime sleepiness after posttraumatic coma. The chart reveals awakenings, low sleep efficiency, and absence of deep sleep. (Note: Patients enter a vertical mark each time they get in bed or arise. Horizontal line indicates estimated time of sleep. Mistakes are consistent with mild cognitive impairment.)

ical illness, and ameliorating nocturnal symptoms if total cure is not possible.

Drugs

Withdrawal from caffeine and alcohol is easily done in the motivated patient. Nicotine withdrawal requires distinction between the low-dose smoker with fluctuating blood nicotine levels, who may quit abruptly, and the high-dose smoker who almost always requires gradual withdrawal and a sophisticated multimodality treatment approach. However, most who quit do so without formal treatment, and 85% of quitters quit abruptly (Fiore et al., 1990) (see Chapter 42). When recreational drugs worsen sleep disorder, appropriate psychiatric intervention fosters abstinence. Therapeutic dilemmas may arise when prescription regimens cause insomnia. Revised drug regimens are usually possible, however, as previously noted.

Age

Among the aged, a helpless passivity can be encountered as part of a psychologically regressed picture that involves demands for hypnotic drugs coupled with refusal to keep sleep charts, discontinue naps and sleep-disrupting drugs, maintain regular rising times, and adhere to medical treatment regimens. The sleep disorder often represents a confluence of disease manifestations and a lack of *raison d'être* (Reynolds et al., 1985). Since insomnia in the elderly is often attributed to "a mix-

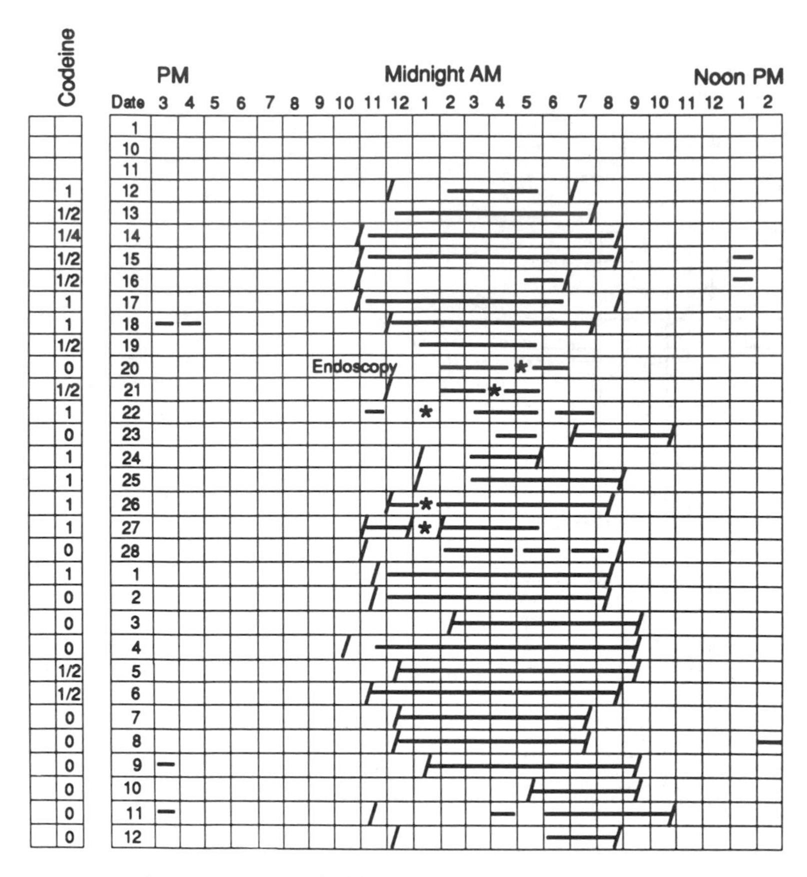

FIG. 21-2. The patient went to bed between 10:30 p.m. and 1:30 a.m. but did not fall asleep for hours. He arose around 9:00 a.m. but as early as 7:30 a.m. and as late as 10:30 a.m.

The columns to the left of the date are to note drug use. The numbers are the number of codeine pills taken at bedtime.

ture of medical, psychological, social, and developmental factors" (Institute of Medicine, 1979), it is often deemed poorly treatable. In fact, physicians overestimate the social problems of the elderly (Branch, 1977), and good sleep habits are essentially mechanical, simple, cheap, and thus easy to institute for the motivated elderly.

The elderly are prescribed roughly double their proportionately necessary amount of hypnotic drugs (Mellinger, Balter, & Uhlenhuth, 1985). This reflects, in part, their more continuous rather than intermittent use of such agents (Baker & Oleen, 1988). Hypnotics, however, have disadvantages in elderly patients. The gut absorption of such agents is slower in elderly persons because of decreased gastric acid, absorptive area, gut blood flow, and enzyme transport system function (Salzman, 1979). For instance, chlordiazepoxide has an absorption half-life of 20 minutes in the elderly, compared within 5 minutes in others (Shader, Greenblatt, Harmatz, et al., 1977). Rapidity of action is necessary for the relief of bedtime insomnia, however, and patients prefer rapid-acting drugs (Hollister, 1978), so slow absorption can lead to demands for a "better" pill. Delayed excretion in the elderly may lead to drug accumulation great enough to cause severe psychologic or neurologic disturbance (Learoyd, 1972). One cannot generalize, however. Although longer acting hypnotics are sometimes incriminated as causing next-day sedation in the elderly (Fillingim, 1982), the observations are often based on short-term studies that insufficiently reflect actual use. Next-day functioning after long-term use of such agents may actually show improved function on some measures, such as mental arithmetic tests (Bliwise et al., 1984).

The total dose of the drug may be more important than pharmacokinetics in determining next-day hypnotic drug effects (Johnson & Chernik, 1982). After long-term use of hypnotics in a geriatric nursing home population, next-day differences among hypnotics of different classes were small, with barbiturates inducing slightly better ratings than benzodiazepines in "motility" and "continence" (Berg et al., 1977). The hypnotic

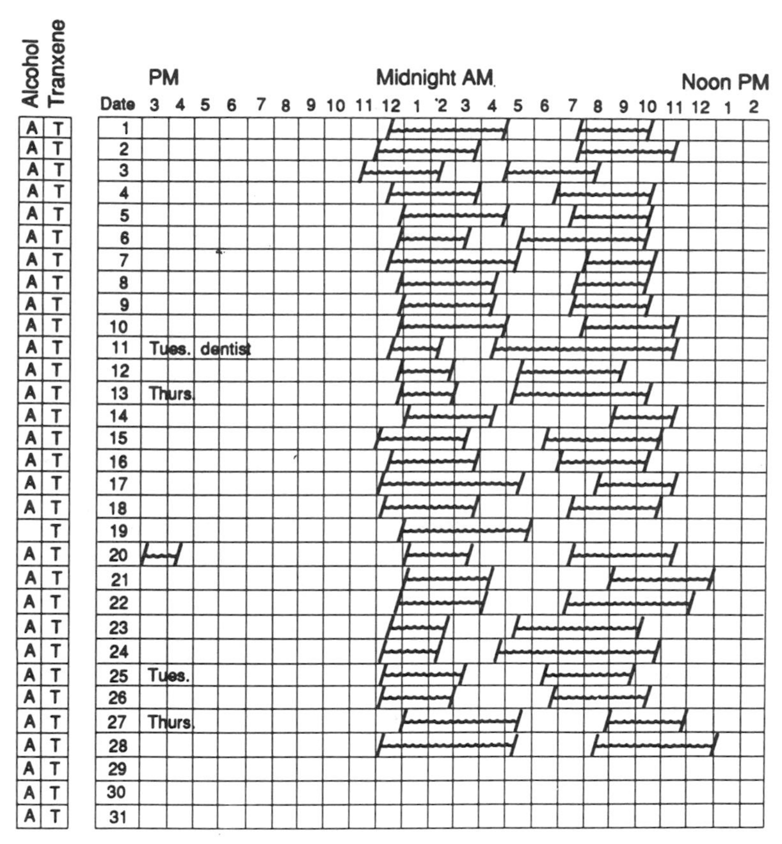

FIG. 21-3. The patient went to bed between midnight and 1:00 a.m. and got out of bed in the middle of the night, returned to bed and finally arose between 9:30 a.m. and 1:00 p.m.

drug literature consists mostly of younger/normal volunteers (see Johnson & Chernik, 1982; Williams & Karacan, 1976), and thus provides an inadequate guide for use of hypnotic agents in the elderly. Into this dearth of information sometimes steps confident opinion—for example, the decrying of hypnotics (Orr, Altschuler, & Stahl, 1982), particularly their chronic use (Kripke, 1985; Williams, 1980). The disadvantages of drug treatment must be balanced, however, against those of chronic sleep deprivation (Adam, 1979).

Hypnotics in the elderly, in lower initial doses (Greenblatt et al., 1991), should be prescribed only after obtaining regular charted sleep schedules, remedying nocturnal disease symptoms and specific sleep problems, and optimizing prescribed drug regimens. In general, drug tolerance may be prevented by restricting hypnotics to intermittent use. Inability to stop relying on hypnotics after several months warrants further illumination of the unremedied sleep problem.

For the working or retired elderly person without nervous system impairment, diazepam is rapidly absorbed and thus rapid in onset. Oxazepam and temazepam are intermediate-acting benzodiazepines whose slowness of onset can be mitigated by using the powder taken out of the gel capsule. Chloral hydrate is much less expensive and comes in syrup form for those who have difficulty swallowing pills. For dependable patients, barbiturates may be useful and pose fewer amnestic effects (Bixler et al., 1975). At this writing, zolpidem is a promising new rapid-onset, short-acting hypnotic.

For the elderly with mild cognitive impairment, lorazepam should be avoided because of its relatively greater amnestic effects (Roth, 1980; Scharf et al., 1984). Flurazepam may cause next-day effects on memory (Bixler et al., 1979) and motility (Crowley & Hydinger-Macdonald, 1979), although its longer action has the benefit of shortening sleep latency on withdrawal nights (Bliwise et al., 1984). It is therefore more likely to foster intermittent drug use by the patient than are shorter acting agents, which usually become a nightly habit.

For the geriatric outpatient with obvious cognitive deficits, hypnotic drugs become more perilous, but a careful empirical approach, using a family member to dispense the medication and rating prescription benefit by sleep chart outcome, remains possible. Oxazepam, temazepam or zolpidem are perhaps the safest agents to use in such cases.

Almost half of institutionalized elderly are prescribed hypnotics (Marttila et al., 1977). Some of this might presumably be avoided at the cost of analyzing each individual sleep problem. In one hospital, staff education and delayed evening drug rounds reduced hypnotic doses per patient by almost half (Sheerin, 1972). The mildly sedative actions of antihistamines are occasionally prescribed for insomnia in the elderly, but this induces more drowsiness than sleep (Roehrs et al., 1984) and may increase sleep movements (Borbely & Youmbi-Balderer, 1988). For chronically institutionalized and supervised elderly persons, in whom abuse and overdose are unlikely, the list of possible hypnotics is long, but hypnotic treatment should be coupled with good sleep habits and the avoidance of prolonged in-bed times, naps, and sleep-disrupting drugs. Almost any hypnotic agent may be chosen to titrate against quantified observations of sleep quality and daytime functioning. Individual variation may be more salient than the research averages described above; even wine served with supper, despite the previously mentioned lessening of sleep by alcohol, has reportedly improved sleep of geriatric patients (Mishara & Kastenbaum, 1974). The complexities in cases of the medically ill, insomniac elderly must be considered individually (Regestein, 1984).

Scheduling Problems

Scheduling problems involve a departure from the ideal of arising at a regular clock time daily after an in-bed time of fixed length. Irregular rising times usually can be made regular. A systematic delay of bedtimes and wake times often found in "night people." The "delayed-sleep-phase syndrome" is often amplified in patients taking caffeine, xanthine drugs, nicotine, beta-agonists, or other long-acting stimulants. It may be remedied by early morning bright light exposure (Rosenthal et al., 1990) or a gradual delay of wake times around the clock (Czeisler, Richardson, & Coleman, 1981), 2 or 3 hours later every day, in keeping with the delaying tendency of the bodily clocks. Advances to earlier bedtimes, in extremely small steps (e.g., 5 minutes a day) can also accomplish the resetting of sleep time to earlier hours (Regestein, 1976). This obviates diurnal sleeping periods but takes several weeks to accomplish. In cases of evening or night work, patients must sleep the same clock hours on vacation days to maintain regular circadian rhythms. Rotating shifts should be changed as infrequently as possible and toward later time periods—that is, evenings to night rather than to days—again in keeping with the delaying tendency of the bodily clocks.

Hyperarousal

Hyperarousal is the likely diagnosis when sleep is delayed after bedtime but subsequently uninterrupted. It may be relieved by arousal-lowering routines instituted 2 or 3 hours prior to bedtime. Television watching may decrease sleep quality and impair next-day performance (Saletu, Gruenberger, & Anderer, 1983) and should be avoided evenings, along with stimulating activities. Meditation and muscle relaxation procedures (Kales & Kales, 1984), hot water soaks (Bunnel & Horvath, 1985), and a predictable, relaxing evening routine are all useful remedies. Evening exercise has reportedly induced brief wakeful periods early in the night (Baekeland & Lasky, 1966). Practically, this does not occur in all hyperarousal insomniac patients, but for those so affected, exercise earlier in the day improves subsequent sleep. Hypnotic drugs may be used intermittently if arousing evening activity is unavoidable.

Psychopathology

When anxious patients need sleeping pills, the longer acting agents, such as flurazepam, diazepam, clorazepate, or phenobarbital, may provide next-day therapeutic sedation. L-Tryptophan is an essential amino acid with measurable hypnotic effects (Hartmann & Spinweber, 1979), especially useful as a first hypnotic medication for patients who dislike "drugs." It exerts near-maximal effect after 1 to 2 grams taken at bedtime (Hartmann, Cravens, & List, 1974). The eosinophilia-myalgia syndrome likely caused by L-tryptophan impurity (Swygert et al., 1990) will presumably be prevented in the future. Patients requiring antidepressants may take one that is sedating at bedtime. More stimulating agents, such as buproprion, fluoxetine, sertraline, tranylcypromine, protriptyline, and desipramine, should be taken earlier in the day to avoid aggravating insomnia. Alprazolam may provide both hypnotic and antidepressant actions in some patients (Feighner, 1984). In potentially suicidal patients prescribed a sedative, benzodazepines have proved relatively safe in acute overdose (Greenblatt et al., 1977), provided they are not taken concurrently with alcohol or other sedatives.

Triazolam is better avoided in patients with significant psychopathology, since its rapid elimination time provokes a mini-withdrawal syndrome manifested by early morning insomnia (Kales et al., 1983), next-day anxiety (Morgan & Oswald, 1982) or paranoia (Adam & Oswald, 1989), nocturnal agitated and amnestic episodes, and restlessness (Bayer et al., 1986). It may cause insidious types of agitation mimicking common presentations of anxiety (Poitras, 1980). Side-effects are frequent among inpatients (Blatter et al., 1988; Mauro & Sperlogano, 1987; Patterson, 1987).

The increased likelihood of amnestic toxicity with more powerful, low-dose, short-acting benzodiazepines (Morris & Estes, 1987; Poitras, 1980; Roth et al., 1980; Scharf et al., 1984; Schneider-Helmert, 1985) and the problem of next-day effects from the longer acting benzodiazepines (Carskadon et al., 1982; Martilla et al., 1977) make both of these agents less than ideal as hypnotics, especially in the elderly. Healthy elderly persons sustain higher plasma concentrations and therefore greater effects from triazolam than do younger individuals (Greenblatt et al., 1991). In elderly inpatients, triazolam may induce more agitation compared with other drugs (Bayer et al., 1986), leading to descriptions of "triazolam syndrome in the elderly" (Patterson, 1987). A single night of triazolam given routinely before surgery resulted in more agitation the next morning compared with an intermediate-acting benzodiazepine (Mauro & Sperlogano, 1987). For an initial choice of hypnotic, therefore, an intermediate-acting benzodiazepine such as temazepam or estazolam might avoid the potential disadvantages of both short- and long-acting drugs (Table 21-9).

Occasionally, antidepressants remedy insomnia in patients who are not apparently depressed. Of course, the neurotransmitter systems affected by these agents regulate many functions besides mood, including pain modulation and the functioning of biologic clocks. Patients helped by antidepressants, however, often do have occult depressions (Ware, 1983). In such cases, a sedating antidepressant (e.g., nortriptyline, doxepin, or trazodone), taken at bedtime and gradually increased to adequate doses, serves to relieve insomnia, among other depressive symptoms. To permit proper assessment of emerging side effects, hypnotics should not be simultaneously begun. Where insomnia is relieved specifically by antidepressants, hypnotic drugs preexisting in the drug regimen can be discontinued when insomnia remits.

TABLE 21–9. *Benzodiazepine Agonists (by Elimination Half-Life)*

Short (1.5–6 hours)
adinazolam (Deracyn)*
brotizalam (Lendormin)*
midazolam (Dormicam)
triazolam (Halcion)
zolpidem (Stilnox)*
zopiclone (Immovane)*
Intermediate (5–40 hours)
alprazolam (Xanax)
chlordiazepoxide (Librium)
clonazepam (Klonopin)
estazolam (ProSom)
flunitrazepam (Rohypnol)*
loprazolam (Dormonoct)*
lorazepam (Ativan)
nitrazepam (Mogadon)*
oxazepam (Serax)
temazepam (Restoril)
Long (20–150 hours)
clorazepate (Tranxene)
diazepam (Valium)
flurazepam (Dalmane)
halazepam (Paxipam)
ketazolam (Loftran)*
prazepam (Centrax)
quazepam (Doral)

*Not available in the United States as of December, 1992.

Hospitalization

Hypnotic drugs are routinely prescribed "as needed" for hospital inpatients, and constitute the major treatment for disordered sleep. In one study of medical-surgical inpatients in a university referral hospital, 38% were given a hypnotic medication, and of these, over half took diazepam or another sleep medication (Salzman, 1981). Discharged on hypnotic drugs, many patients continue to take them, thus accounting for about 20% of hypnotic drug dependency (Clift, 1975).

The environments of tertiary care hospitals interfere with optimal management of sleep disturbance. However, staff can try to promote dark, quiet nights and brightly lit days; withhold sedative-hypnotics from daytime nappers; discontinue caffeine and nicotine; substitute reading for evening television; promote evening hot tub baths, relaxation exercises, or back rubs; relieve nocturnal disease symptoms such as itch, pain, or dyspnea; delay evening sedation rounds; and wait until the patient requests a hypnotic rather than leaving "as needed" hypnotics at the patient's bedside.

Medical Problems

Sleep Apnea

Specific diagnostic investigation for sleep apnea is warranted in any loud-snoring individual who has problems possibly remedied by treatment of sleep apnea (e.g., excessive sleepiness, hypertension, cardiac arrhythmias, and personality changes toward apathy or irritability). Patients with clusters of possibly such apnea-related features, especially in the presence of apnea risk factors (see Table 21-4), deserve diagnostic investigation. Candidates for investigation are readily found among inpatients in whom clinical suspicion of sleep apnea remains basically nonexistent (Chikkalingaiah & Regestein, 1991). Central sleep apnea may be suspected in poorly functioning individuals who have any problems such as severe lung disease or heart failure likely to lead to Cheyne-Stokes breathing.

Even very mild breathing disturbance (e.g., 5 to 10 apneic episodes per hour) when symptomatic, merits treatment, usually by weight loss, sedative discontinuation, or surgery (Orr & Moran, 1985). Surgery is indicated by typical structural impairment of the upper airway (Fujita et al., 1985) or pharyngeal closure (visualized on fiberoptic endoscopy) during inspiration against a closed glottis in the recumbent position. Treatment is indicated until symptoms remit and apneic episodes fall below 20 per hour; this can almost always be accomplished without permanent tracheostomy.

The lung and cardiovascular diseases that induce disturbances in breathing and sleep quality should be controlled. Continuous positive airway pressure (CPAP) devices, which raise the inspired air pressure 5 to 10 cm of water (Sanders, 1984; Sullivan & Issa, 1980), splint open the pharynx (Abbey, Cooper, & Kwentus, 1989), quickly reverse snoring and sleep breathing disorder (Aldrich et al., 1989a), and maintain the improvement (Conway et al., 1985). CPAP is indicated in the absence of surgical indications, after surgery fails, or during weight-loss treatment. Air pressure through a tight-fitting nasal mask is empirically raised in the sleep laboratory until snoring and apneas are abolished (Sullivan & Grunstein, 1989). Higher pressures induce more mask leaks and discomfort; they are necessitated by hyperobesity and stuffed nose and are obviated by upper airway revision surgery. Frequent follow-up during the start-up period and at 1 year fosters compliance and treatment refinement. Low-flow nasal oxygen may help both the apnea and the hypoxia in some cases in which airway obstruction is not major (Martin et al., 1982).

Surgical correction of upper airway obstruction—for instance, deviation of the nasal septum (Heimer et al., 1983) or jaw-tongue retrusion (Spire, Kuo, & Campbell, 1983) as well as thickened or redundant tissues (Fujita et al., 1985; Simmons, Guilleminault, & Miles, 1984)—relieves snoring and sleep apnea. Subjective improvement is more frequent than changes in number of apneic episodes suggest (Simmons, Guilleminault, & Miles, 1984; Thawley, 1985); even apnea nonresponders may show objectively better daytime functioning (Regestein et al., 1988). Therefore, multiple measures of breathing and sleep disturbance better assess treatment (Wetmore et al., 1986). A higher mortality rate in patients with more than 20 apneic episodes per hour mandates their postoperative reassessment (He et al., 1988). Operative complications in the author's experience usually are minor and transient, although serious, occasionally fatal complications have been reported (Esclamado et al., 1989).

Where sleep-disordered breathing causes oxygen desaturation, with severe hypertension, high-grade arrhythmias, or unsafe degrees of daytime sleepiness, tracheostomy may induce rapid relief (Guilleminault & Cummiskey, 1982). However, there are psychologic difficulties that must be addressed in almost all cases of ambulatory tracheostomy, and a careful home tracheostomy care routine must be developed (Dye, 1983).

Weight reduction by diet (Aubert-Tulkens et al., 1989) or surgery has also been used to relieve obstructive sleep apnea (Peiser et al., 1984). Although medroxyprogesterone, which stimulates respiration, or protriptyline, which limits REM sleep, drys the airway, and provides mild stimulation, may alleviate sleep apnea (Brownell et al., 1982; Hensley, Sanders, & Strohl, 1980), such treatments rarely provide uncomplicated long-term relief (Cook, Benich, & Wootensa, 1989; Millman, 1989).

Menopause

When sleep is disordered at the menopause, it may be ameliorated by replacement estrogen (Campbell, 1976). Sleep quality is improved (Thompson & Oswald, 1977), and insomnia relief is associated with shorter sleep latencies (Regestein et al., 1981).

Narcolepsy

The treatment of narcolepsy requires continual record keeping, empirical sleep and nap scheduling, physician support, and various drugs. A regimen of stimulants and antidepressants must be empirically fashioned (Regestein, Reich, & Mufson, 1983). Stimulants more effectively maintain than restore wakefulness, and therefore should be given early on a fixed rather than an as-needed schedule. Stimulant drug abusers feigning narcolepsy are rare but easy to spot by their relative lack of concern or sincerity, inconsistent history, lack of family confirmation, and history of difficulties associated with acts of commission rather than of omission. Perhaps because of specific arousal abnormalities, narcolepsy patients themselves do not sustain much euphoric effect from stimulants and do not tend to abuse such drugs. Idiopathic hypersomnia is treated similarly to narcolepsy.

Nocturnal Myoclonus

Nocturnal myoclonus should be treated only if there is a clearly demonstrable connection between it and diminished sleep quality. A number of muscle-relaxing drugs, such as clorazepate, diazepam, and baclofen, have been tried, often with impressive results. L-Dopa (Boivin, Montplaisir, & Poirier, 1989), opioids (Kavey et al.,

1988), phenoxybenzamine (Ware, Blumoff, & Pittard, 1988), and nitrazepam (Moldofsky et al., 1986) also show promise.

SLEEP HYGIENE

A number of nonspecific treatments, summarized as sleep hygiene, are frequently offered to patients with insomnia problems of all varieties (Hauri, 1985). They are as follows:

1. Rising at a regular clock time daily.
2. Limitation of total daily in-bed time to the usual amount of sleep prior to the onset of the problem.
3. Discontinuation of CNS-acting drugs, such as caffeine, alcohol, nicotine, nasal decongestants, beta-blockers, and beta-antagonists.
4. Regular mealtimes, with a relatively small meal at night and avoidance of after-supper eating. (Small bedtime snacks help some but worsen sleep in others.)
5. A regular, predictable evening routine. Overstimulating television should be avoided in favor of music or relaxed reading.
6. Regular daytime exercise as appropriate for age and physical condition. (Evening exercise has variable effects.)
7. Daytimes of purposeful activity or arousing involvement.
8. Evening relaxation routines, such as meditation techniques, progressive muscle relaxation, and hot tub soaks.

Many patients will claim regular arising times but forget weekend irregularities. Unemployment or night work is a large factor in irregular wake times. These approximate guidelines are inapplicable to some patients. Some, for instance, are made rather uncomfortable by body-temperature-raising tub soaks; others, also made uncomfortable, nevertheless continue them because they sleep better.

Where the recumbent position of sleep worsens hiatus hernia pain or elicits the reflex wakening induced by mild acid in the esophagus, raising the head of the bed on blocks and using extra pillows can give relief. Nocturnal peptic ulcer pain may be suppressed by bedtime administration of a histamine blocker such as ranitidine, although doxepin has reportedly resolved ulcers previously unresponsive to histamine blockers alone (Mangla & Pereira, 1982), presumably because of its powerful anticholinergic, antihistaminic, and REM-suppressing effects. Trimipramine has been used similarly.

The nocturia associated with prostatism may be treated with bladder training. Progressively increasing the interval between the urge to urinate and urination—for example, with 20-minute increments each week to perhaps a 100-minute interval—plus subsequent morning fluid loading and evening fluid restriction will raise the threshold against the urinary reflex and in many cases relieve nocturia.

The properties of specific sedative-hypnotic drugs render them less likely to cause side effects in some disease states, and more likely in others. Diazepam (Klotz, 1975)—but neither oxazepam (Schuil, Wilkinson, & Johnson, 1971) nor temazepam—excretion is slowed in liver disease. For those with renal impairments, hexobarbital is rapidly excreted. Chloral hydrate can aggravate symptoms of stomach irritation and affect the metabolism of anticoagulants (Mendelson, 1980). Patients using antacids may have delayed absorption of benzodiazepines (Shader et al., 1978), as will those taking pills after meals. For patients taking major tranquilizers or antidepressants, a sedating agent taken near bedtime might substitute for hypnotic medication. Seizure disorder patients may take their phenobarbital, clonazepam, or other sedating anticonvulsant near bedtime.

For brief, intermittent, nonspecific suppression of insomnia by drugs there are "no hard and fast guidelines for their use" (Mendelson, 1987). The physician's experience and comfort will likely determine the optimal drug (Meyer, 1982; Rickels, 1986). Thus he may familiarize himself with a long or intermediate-acting benzodiazepine (e.g. clonazepam, nitrazepam, diazepam or zopiclone) (Lader & Lawson, 1987; Van der Kleijn, 1989) and an alternative such as trazodone (Mouret et al., 1988), thioridazine (Reynolds et al., 1988), buspirone (DeRoeck et al., 1989), chlormetmazole (Liljenberg et al., 1986), or aspirin (Hauri & Silberfarb, 1980). Further refinement of drug treatment relates to medical (Regestein, 1984), genetic (Harris & Allan, 1989), or idiosyncratic (Regestein, 1987b) factors. The 0.9% of the U.S. population who use hypnotics more than 29 days per year (Mellinger, Balter, & Uhlenhuth, 1985) have exceptional problems or habituation unlikely to be remedied by drugs alone. Thus benzodiazepines or barbituates may be used intermittently and rarely after sleep hygiene is in place, in cases of CNS electrical abnormality connected with insomnia, in sleep movement disorders where next-day vigilance test scores and functioning are improved, and in the hospital, accompanied by those sleep hygiene measures possible and combined with a follow-up program to prevent subsequent dependence. Hypnotics disrupt assessment of drug effects when simultaneously started with antidepressants; a sedating antidepressant is preferable to a combination of an antidepressant and a hypnotic. Where drugs cause insomnia, it is better to refine the medical regimen than to add sedatives.

Despite ubiquitous ritual advice that hypnotics should be used intermittently, 62% of 1,726 elderly outpatients on triazolam or oxazepam used them daily, as did 42% of 534 of those on flurazepam (Baker & Oleen, 1988). The difference may be due to lessened disruption of sleep after withdrawal of flurazepam (Schneider-Helmert, 1988) or lessened anxiety on withdrawal of long-acting in comparison to short-acting drugs (Rickels et al., 1990). Reduced sleep is a cardinal symptom of the benzodiazepine withdrawal syndrome (Petrusson & Lader, 1981; Tyrer, Owen, & Dowling, 1983). It may be attenuated by gradual withdrawal (Greenblatt et al., 1987) but not eliminated in every case.

The tendency of the patient toward passivity and regression will much determine his use of hypnotics (Regestein, 1983) and tolerance of drug withdrawal. Withdrawal reactions may be much more severe and frequent after lengthy courses than after shorter term usage (Rickels et al., 1983). Detailed aspects of benzodiazepine selection based on pharmacokinetic factors relevant to medically ill patients are discussed in chapter 9.

The regressed chronic complainer will more likely be prescribed the most hypnotics with the least benefit. The psychologically mature patient may use them the least but benefit the most. Hypnotic-dependent patients should probably be sent to a sleep disorders clinic for detection of occult causes of insomnia, the most common of which are undetected affective disorders (Ford & Kamerow, 1989) and poorly appreciated drug side-effects.

REFERENCES

Abbey NC, Cooper KR, & Kwentus JA (1989). Benefit of nasal CPAP in obstructive sleep apnea is due to positive pharyngeal pressure. Sleep 12:420–422.

Abraham GE (1983). Nutritional factors in the etiology of the premenstrual tension syndromes. J Reprod Med 28:446–464.

Adam K (1979). Do drugs alter the restorative value of sleep? In P Passouant & I Oswald (Eds.), Pharmacology of states of alertness (Ch. 2, pp. 105–111). New York: Pergamon Press.

Adam K, & Oswald I (1984). Sleep promotes healing. Br Med J 2:1400–1401.

Adam K, & Oswald I (1989). Can a rapidly-eliminated hypnotic cause daytime anxiety? Pharmacopsychiatry 22:115–119.

Adamson J, & Burdick JA (1973). Sleep of dry alcoholics. Arch Gen Psychiatry 28:146–149.

Akpinar S (1987). Restless legs syndrome treatment with dopaminergic drugs. Clin Neuropharmacol 10:69–79.

Aldrich M, Eisler A, Lee M, et al. (1989a). Effects of continuous, positive airway pressure on phasic events of REM sleep in patients with obstructive sleep apnea. Sleep 12:413–419.

Aldrich M, Foster NL, White RF, et al. (1989b). Sleep abnormalities in progressive supranuclear palsy. Ann Neurol 25:577–581.

Aldrich MS, & Naylor MW (1989). Narcolepsy associated with lesions of the diencephalon. Neurology 18:1505–1508.

Allison T, & Twyver H (1970). The evolution of sleep. Natural History 79:56–65.

Al-Mateen M, Gibbs M, Dietrich R, et al. (1988). Encephalitis lethargica-like illness in a girl with mycoplasma infection. Neurology 38:1155–1158.

Alving J (1985). Obstructive sleep apnea syndrome caused by oral cyst. Acta Neurol Scand 71:408–410.

American Psychiatric Association (1987). *Diagnostic and statistical manual of mental disorders* (3rd ed., rev.). Washington, DC: American Psychiatric Association.

Anbert-Tulkens G, Calec C, & Rodenstein DO (1989). Cure of sleep apnea syndrome after long-term nasal continuous positive airway pressure therapy and weight loss. Sleep 12:216–222.

Anch AM, Rammers JE, & Bunce H (1982). Supraglottic airway resistance in normal subjects and patients with obstructive sleep apnea. J Appl Physiol 53:1158–1163.

Ancoli-Israel S, Klamber MR, Kripke DI, et al. (1989). Sleep apnea in female patients in a nursing home. Chest 96:1054–1058.

Ancoli-Israel S, Kripke DF, Mason W, et al. (1981). Sleep apnea and nocturnal myoclonus in a senior population. Sleep 4:349–58.

Apps MC, Sheaff PC, Ingram DA, et al. (1985). Respiration and sleep in Parkinson's disease. J Neurol Neurosurg Psychiatry 48:1240–1245.

Armstrong RH, Burnap DB, Jacobson A, et al. (1965). Dreams and gastric secretions in duodenal ulcer patients. New Physician, 14:241–243.

Arseni C, & Ciurea AV (1988). Cerebral absesses secondary to otorhinological infections Z61. Neurochirurgia 49:22–36.

Aschoff J (1964). Survival value of diurnal rhythms. In OG Edholm (Ed.), The biology of survival (Ch. 3, pp. 79–98). London: Academic Press.

Association of Sleep Disorders Clinics Steering Committee (1990). International Classification of Sleep Disorders. Rochester, MN: American Sleep Disorders Association.

Baekeland F, & Lasky R (1966). Exercise and sleep patterns in college athletes. Percept Mot Skills 23:1203–1207.

Baker MI, & Oleen MA (1988). The use of benzodiazepine hypnotics in the elderly. Pharmacotherapy 8:241–247.

Baldy-Moulinier M (1982). Temporal lobe epilepsy and sleep organization. In MB Sterman & MN Shouse (Eds.), Sleep and epilepsy (Ch. 41, pp. 329–337). New York: Academic Press.

Ballinger CB (1976). Subjective sleep disturbance at the menopause. J Psychosom Res 20:509–513.

Bayer AJ, Bayer EM, Pathy MSJ, et al. (1986). A double-blind controlled study of chlormethiazole and triazolam as hypnotics in the elderly. Acta Psychiatr Scand 73(suppl 329):104–111.

Bergamasco B, Bergamini L, Doriguzzi T, et al. (1968). EEG sleep patterns as a prognostic criterion in post-traumatic coma. Electroencephalogr Clin Neurophysiol 24:374–377.

Bergonzi P, Bianci A, Mazza S, et al. (1978). Night sleep organization in patients with severe hepatic failure. Eur Neurol 17:271–275.

Berlin RM (1988). Sleepwalking disorder during pregnancy: A case report. Sleep 11:298–300.

Berlin RM, Litovitz GL, Diaz M, et al. (1984). Sleep disorders on a psychiatric consultation service. Am J Psychiatry 141:582–584.

Berry DTR, Phillips BA, Cook VR, et al. (1989). Sleep-disordered breathing in healthy aged persons: One year follow-up of daytime sequelae. Sleep 12:211–215.

Berthon Jones M, & Sullivan CE (1987). Time course of change in ventilatory response to CO_2 with long term CPAP therapy for obstructive sleep apnea. Am Rev Respir Dis 135:144–147.

Billiard M, Guilleminault C, & Dement WL (1975). A menstruation-linked periodic hypersomnia. Neurology 25:436–443.

Billiard M, & Passouant P (1974). Sexual hormones and sleep in women. Rev Electroencephalogr Neurolophysiol Clin 4:89–106.

Bixler EO, Kales A, Jacoby JA, et al. (1984). Nocturnal sleep and wakefulness: Effects of age and sex in normal sleepers. Int J Neurosci 23:33–42.

Bixler EO, Kales A, Kales J, et al. (1986). Insomnia: Validation of sleep laboratory criteria. In American Psychiatric Association, New Research Paper Abstracts (p. 71). Washington, DC.

Bixler EO, Scharf MF, Leo L, et al. (1975). Hypnotic drugs and performance. In F Kagan, T Harwood, K Rickels, et al. (Eds.), Hypnotics (Ch. 11, pp. 175–196). New York: Spectrum.

Bixler EO, Scharf MB, Soldatos C, et al. (1979). Effects of hypnotic drugs on memory. Life Sci 25:1379–1388.

Blatter M, Hoigne R, Hess T, et al. (1988). Side effects of frequently administered hypnotics, sedatives and anxiolytics. Schweiz Med Wochensch 188:1859–1864.

Blau JN (1982). Resolution of migraine attacks: Sleep and the recovery phase. J Neurol Neurosurg Psychiatry 43:223–226.

Bliwise DL, Bliwise NG, Partinen M, et al. (1988). Sleep apnea and mortality in an aged cohort. Am J Public Health 78:544–547.

Bliwise D, Seidel W, Greenblatt DJ, et al. (1984). Nighttime and daytime efficacy of flurazepam and oxazepam in chronic insomnia. Am J Psychiatry 141:191–195.

Bliwise DL, Yesavage JA, Tinklenberg JR, et al. (1989). Sleep apnea in Alzheimer's disease. Neurobiol Aging 10:343–346.

Block AJ (1979). Sleep apnea, hypopnea and oxygen desaturation in normal subjects. N Engl J Med 300:513–517.

Boivin DG, Montplaisir J, & Poirier G (1989). The effects of L-dopa on periodic leg movements and sleep organization in narcolepsy. Clin Neuropharmacol 12:339–345.

Borbely AA, & Youmbi-Balderer G (1988). Effect of diphenhydramine on subjective sleep parameters and on motor activity during bedtime. Int J Clin Pharmacol Ther Toxicol 26:392–396.

Borg S, Krande H, & Sedval G (1981). Central norepinephrine metabolism during alcohol intoxification in addicts and healthy volunteers. Science 213:1135–1137.

Brady H, & Vijayashankar N (1977). Anatomical changes in the nervous system. In CE Finch & L Hayflick (Eds.), Handbook of the biology of aging (Ch. 13, pp. 241–261). New York: Van Nostrand Reinhold.

Branch LG (1977). Updating the needs of Massachusetts elderly. N Engl J Med 297:838–840.

Brebbia DR, & Altschuler KZ (1965). Oxygen consumption rate and electroencephalographic stage of sleep. Science 150:1621–1623.

Brezinova V (1974). Effect of caffeine on sleep: EEG study in late middle age. Br J Clin Pharmacol 1:203–208.

Broughton R, & Baron R (1978). Sleep patterns in the intensive care unit and the ward after acute myocardial infarction. Electroencephalogr Clin Neurophysiol 45:348–360.

Broughton R, Ghanem Q, Hishikawa Y, et al. (1981). The socioeconomic and related life effects in 180 patients with narcolepsy from North America, Asia, and Europe compared to matched controls. In I Karacan (Ed.), Psychophysiological aspects of sleep (Ch. 12, pp. 96–105). Park Ridge, NJ: Nays.

Broverman DM, Klaiber EL, & Kobayashi V (1968). Roles of activation and inhibition in sex differences in cognitive abilities. Psychol Rev 75:23–50.

Brown DG, & Kalucy RS (1975). Correlation of neurophysiological and personality data in sleep scratching. Proc R Soc Med 68:530–532.

Brown IG, Bradley TD, Phillipson EA, et al. (1985). Pharyngeal compliance in snoring subjects with and without obstructive sleep apnea. Am Rev Respir Dis 132:211–215.

Brownell LG, West P, Sweatman P, et al. (1982). Protriptyline in obstructive sleep apnea. N Engl J Med 307:1037–1042.

Bunnell DE, & Horvath SM (1985). Effects of body heating during sleep interruption. Sleep 8:274–282.

Burr W, Stefan H, & Renin H (1986). Epileptic activity during sleep and wakefulness. Eur Neurol 25(suppl 2):141–145.

Cadieux RJ, Kales A, Santen RJ, et al. (1982). Endoscopic findings in sleep apnea associated with acromegaly. J Clin Endocrinol Metab 55:18–22.

Campbell S (1976). Double blind psychometric studies on the effect of natural estrogens on post-menopausal women. In S Campbell (Ed.), The management of the menopause and post-menopausal years (Ch. 13, pp. 149–158). Baltimore: University Park Press.

Carlson HE, Gillin JC, Gordon P, et al. (1972). Absence of sleep-related growth hormone peaks in aged normal subjects and in acromegaly. J Clin Endocrinol Metab 34:1102–1105.

Carskadon MA, Dement W, Mifler M, et al. (1976). Self-reports versus sleep laboratory findings in 122 drug-free subjects with complaints of chronic insomnia. Am J Psychiatry 133:1382–1388.

Carskadon MA, Harvey K, Duke P, et al. (1980). Pubertal changes in daytime sleepiness. Sleep 2:453–460.

Carskadon MA, Seidel WF, Greenblatt DJ, et al. (1982). Daytime carryover of triazolam and flurazolam in elderly insomniacs. Sleep 5:361–371.

Caton R (1889). Narcolepsy. Br Med J 1:358.

Chada TS, Birch S, & Sackner MD (1985). Periodic breathing triggered by hypoxia in normal aware adults. Chest 88:16–23.

Chan CS, Grunstein RR, Bye PTP, et al. (1989). Obstructive sleep apnea with severe chronic airflow limitation. Am Rev Respir Dis 140:1274–1278.

Charney DS, Heninger GR, & Jaflow PI (1985). Increased anxiogenic effects of caffeine in panic disorders. Arch Gen Psychiatry 42:233–243.

Chaudhary BA, Sklan AH, Chaudhary TK, et al. (1988). Sleep apnea, proteinuria and nephrotic syndrome. Sleep 11:69–74.

Cherniack NA (1981). Respiratory dysrhythmias during sleep. N Engl J Med 305:325–330.

Chikkalingaiah R, & Regestein QR (1991). Incidence of suspected sleep apnea among inpatients. Sleep Res 20:225.

Cirignota F, Zucconi M, Mondin S, et al. (1989). Cerebral anoxia attacks in sleep apnea syndrome. Sleep 12:400–404.

Cirignotta F, Zucconi M, Mondini S, et al. (1989). Cerebral amoxic attacks in sleep apnea syndrome. Sleep 12:400–404.

Clark RW, Boudoulas H, Schaal SF, et al. (1980). Adrenergic hyperactivity and cardiac abnormality in primary disorders of sleep. Neurology 30:113–119.

Clift AD (1975). Dependence on hypnotic drugs in general practice. In AD Clift (Ed.), Sleep disturbance and hypnotic drug dependence (Ch. 4, pp. 71–95). Amsterdam: Excerpta Medica.

Coccagna G, & Lugaresi E (1978). Arterial blood gases and pulmonary and systematic arterial pressure during sleep in chronic obstructive pulmonary disease. Sleep 1:117–124.

Coccagna G, Mantonelli P, Zucconi M, et al (1985). Sleep-related respiratory and hemodynamic changes in Shy-Drager syndrome: A case report. J Neurol 232:310–313.

Comfort A (1968). Physiology, homeostasis and aging. Gerontologia 14:224–234.

Conway W, Fujita S, Sorick F, et al. (1985). Uvulopalatopharyngoplasty, one year follow-up. Chest 88:385–387.

Cook DB, Huse J, Roudebush C, et al. (1982). Asymptomatic sleep hypoventilation in morbid obesity. Chest 82:236.

Cook WR, Benich JJ, & Wootensa H (1989). Indices of severity of obstructive sleep apnea syndrome do not change during medroxyprogesterone therapy. Chest 96:262–266.

Coren S, & Searleman A (1985). Birth stress and self-reported sleep difficulty. Sleep 8:222–226.

Cosgrove JL, Alexander MA, & Kitts EL (1987). Late effects of poliomyelitis. Arch Phys Med Rehabil 68:4–7.

Crowley TJ, & Hydinger-Macdonald M (1979). Bedtime flurazepam and the human circadian rhythm of spontaneous motility. Psychopharmacology (Berlin) 62:157–161.

Cumming G (1984). Sleep promotion, hospital practice, and recovery from illness. Med Hypotheses 15:31–37.

Cummings KM, Giovino G, Jaen CR, et al. (1985). Reports of smoking withdrawal symptoms over a 21 day period of abstinence. Addict Behav 10:373–381.

Czeisler CA, Allan JR, Strogatz SH, et al. (1986). Bright light resets the human circadian pace-maker independent of the timing of the sleep-wake cycle. Science 233:667–671.

Czeisler CA, Moore-Ede MC, & Coleman RM (1982). Rotating shift work schedules that disrupt sleep are improved by applying circadian principles. Science 217:460–463.

Czeisler CA, Richardson GS, & Coleman RM (1981). Chronotherapy: Resetting the circadian clocks of patients with delayed sleep phase insomnia. Sleep 4:1–21.

Dahl M, & Dam M (1985). Sleep and epilepsy. Ann Clin Res 17:235–242.

Daiss W, Wiethoelten H, & Schuma F (1986). Cerebral Whipple's disease. Nervenarzt 57:476–479.

Daly DD (1968). The effect of sleep on the electroencephalogram in patients with brain tumors. Electroencephalogr Clin Neurophysiol 25:521–529.

Daly RJ, & Hassall C (1970). Reported sleep in maintenance hemodialysis. Br Med J 2:508–509.

Declerck AL (1986). Interaction sleep and epilepsy. Eur Neurol 25(suppl 2):117–127.

Declerk AC, Wauguier A, Sijben-kiggen R, et al. (1982). A normative study of sleep in different forms of epilepsy. In MB Sterman & MN Shouse (Eds.), Sleep and epilepsy (Ch. 42, pp. 329–337). New York: Academic Press.

Deray MJ, Duchowny MS, & Resnick T (1987). Snoring in fathers of infants with SIDS and apnea. Sleep Res 16:176.

DeRoeck J, Cluydts R, Schotte C, et al. (1989). Explorative single-blind study on the sedative and hypnotic effects of buspirone in anxiety patients. Acta Psychiatr Scand 79:129–135.

Dews PB (1982). Caffeine. Annu Rev Nutr 2:323–341.

Dexter JD, & Weitzman ED (1970). The relationship of nocturnal headaches to sleep stage patterns. Neurology 20:513–518.

Dilsaver SL (1989). Anti-depressant withdrawal syndromes. Acta Psychiatr Scand 79:113–117.

Dlin BM, Rosen H, Dickstein K, et al. (1971). The problems of sleep and rest in the intensive care unit. Psychosomatics 12:155–163.

Dunleavy DLF, Oswald I, Brom P, et al. (1974). Hyperthyroidism, sleep and growth hormone. Electroencephalogr Clin Neurophysiol 36:259–263.

Dye JP (1983). Living with a tracheostomy for sleep apnea. N Engl J Med 308:1167–1168.

Ekbom KA (1960). Restless legs syndrome. Neurology 10:868–873.

Ellis BW, & Dudley HAF (1976). Some aspects of sleep research in surgical stress. Psychosom Res 20:303–308.

Endicott J, Halbreigh U, Schacht S, et al. (1981). Premenstrual changes and affective disorders. Psychosom Med 43:519–529.

Erkinjuntti T, Pantinen M, Sulkava R, et al. (1987). Snoring and dementia. Age Ageing 16:305–310.

Esclamado RM, Glenn MG, McCulloch TM, et al. (1989). Perioperative complications and risk factors in the surgical treatment of obstructive sleep apnea syndrome. Laryngoscope 99:1125–1129.

Eysenck HJ (1972). Human typology, higher nervous activity, and factor analysis. In VD Nebylitsyn & JA Gray (Eds.), Biological bases of individual behavior (Ch. 11, pp. 165–181). New York: Academic Press.

Fairbanks DN (1984). Snoring: Surgical vs nonsurgical management. Laryngoscope 94:1188–1192.

Feighner JP (1984). Open label study of alprazolam in severely depressed inpatients. J Clin Psychiatry 44:332–334.

Feinberg I, Koresko R, & Heller N (1967). EEG sleep patterns as a function of normal and pathological aging in man. J Psychiatr Res 5:107–144.

Felson DT, & Goldenberg DL (1986). Natural history of fibromyalgia. Arthritis Rheum 29:1522–1526.

Fillingim JM (1982). Double-blind evaluation of temazepam, flurazepam and placebo in geriatric insomniacs. Clin Ther 4:369–380.

Fiore MC, Novotny TE, Pierce JP, et al. (1990). Methods used to quit smoking in the United States. JAMA 263:2760–2765.

Fischer-Perroudon C, Mouret J, & Jouvet M (1974). On a case of agrypnia (4 months without sleep) in the course of Morvan's disease. Electroencephalogr Clin Neurophysiol 36:1–18.

Fisher JE, & Baldessarini RJ (1971). False neurotransmitters and hepatic failure. Lancet 2:75–79.

Fleetham JA, & Kryger MH (1981). Sleep disorders in chronic airflow obstruction. Med Clin North Am 65:549–561.

Fletcher E, DeBehnke R, Lovoi S, et al. (1985). Undiagnosed sleep apnea in patients with essential hypertension. Ann Intern Med 103:190–195.

Ford DE, & Kamerow PB (1989). Epidemiologic study of sleep disturbances and psychiatric disorders. JAMA 262:1479–1484.

Fujita S, Conway W, Sicklested J, et al. (1985). Evaluation of the effectiveness of uvulopalatopharyngoplasty. Laryngoscope 95:70–74.

Gaillard JM (1978). Chronic primary insomnia: Possible physiopathological involvement of slow wave sleep efficiency. Sleep 1:133–147.

George CF, & Kryger MH (1987). Sleep in restrictive lung disease. Sleep 10:409–418.

Gislason T, Anquist M, Boman G, et al. (1989). Increased CSF opioid activity in sleep apnea syndrome. Chest 96:250–254.

Glassman A, Jackson WK, Walsh BT, et al. (1984). Cigarette craving, smoking withdrawal and clonidine. Science 126:864–866.

Globus GG (1969). A syndrome associated with sleeping late. Psychosom Med 31:528–535.

Goldenberg DL (1987). Fibromyalgia syndrome. JAMA 257:2782–2787.

Goldstein A, Kaizer S, & Whitby O (1969). Psychotropic effects of caffeine in man: IV. Quantitative and qualitative differences associated with habituation to coffee. Clin Pharmacol Ther 10:489–497.

Gordon NP, Cleary PD, Parlan CF, et al. (1986). The prevalence and health impact of shift work. Am J Public Health 76:1225–1228.

Gray JA (1967). Strength of the nervous system, introversion-extroversion, conditionability and arousal. Behav Res Ther 5:151–169.

Greenblatt DI, Allen MD, Noel BJ, et al. (1977). Acute overdose with benzodiazepine derivatives. Clin Pharmacol Ther 21:497–514.

Greenblatt DJ, Harmatz JS, Engelhardt N, et al. (1991). Sensitivity to triazolam in the elderly. N Engl J Med 324:1691–1698.

Greenblatt DJ, Harmatz JS, Zinny MA, et al. (1987). Effect of gradual withdrawal on the rebound sleep disorder after discontinuation of triazolam. N Engl J Med 317:722–728.

Gugger M, & Hess CW (1989). Regression of sleep apnea symptoms after coronary artery bypass surgery. Schweiz Med Wochenschr 119:1124–1127.

Guilleminault C, Briskin JG, Greenfield M, et al. (1981). The impact of autonomic nervous system dysfunction on breathing during sleep. Sleep 4:263–278.

Guilleminault C, & Cummiskey J (1982). Progressive improvement of apnea index and ventilatory response to CO_2 after tracheostomy in obstructive sleep apnea syndrome. Am Rev Respir Dis 126:14–20.

Guilleminault C, & Dement WC (1978). Sleep apnea syndromes and related disorders. In RC William & I Karacan (Eds.), Sleep disorders, diagnosis and treatment (Ch. 1, pp. 9–28). New York: John Wiley & Sons.

Guilleminault HC, Eldridge FL, Dement WC, et al. (1973). Insomnia with sleep apnea: A new syndrome. Science 181:856.

Guilleminault C, Faull KF, Miles L, et al. (1983). Posttraumatic excessive daytime sleepiness: A review of 20 patients. Neurology 33:1584–1589.

Guilleminault C, & Motta J (1978). Sleep apnea syndrome as a long term sequela of poliomyelitis. In C. Guilleminault (Ed.), Sleep apnea syndromes (Ch. 34, pp. 309–315). New York: Alan R. Liss.

Guilleminault C, Riley R, & Powell N (1985). Sleep apnea in normal subjects following mandibular osteotomy with retrusion. Chest 88:777–778.

Guilleminault C, Stoohs R, Schneider H, et al. (1989). Central alveolar hypoventilation and sleep. Chest 96:1210–1212.

Gulcman L, Beringer VM, Szendro G, et al. (1987). Prominent somnolence and a cerebellar syndrome in subclavian artery thrombosis—a case report. Angiology 38:912–915.

Halberg F (1960). The 24 hour scale: A time dimension of adaptive functional organization. Perspect Biol Med 3:491–527.

Hall JB (1986). The cardiopulmonary failure of sleep-disordered breathing. JAMA 255:930–933.

Hansotia P, Wall R, & Berendes J (1985). Sleep disturbance and severity of Huntington's disease. Neurology 35:1672–1674.

Haponik EF, Bleecker ER, Allen RP, et al. (1981). Abnormal inspiratory flow-volume curves in patients with sleep-disordered breathing. Am Rev Respir Dis 124:571–574.

Harding GFA, Alfon CA, & Powell G (1985). The effect of sodium valproate on sleep, reaction times and visual evoked potential in normal subjects. Epilepsia 26:597–601.

Harris RA, & Allan AM (1989). Genetic differences in coupling of benzodiazepine receptors to chloride channels. Brain Res 490:26–32.

Hartmann EL (1973). The functions of sleep (p. 94). New Haven, CT: Yale University Press.

Hartmann EL (1974). Hypnotic effects of L-tryptophan. Arch Gen Psychiatry 31:394–397.

Hartmann EL, Cravens J, & List S (1974). Hypnotic effects of L-tryptophan. Arch Gen Psychiat 31:394–397.

Hartmann E, & Spinweber CL (1979). Sleep induction by L-tryptophan. J Nerv Ment Dis 167:497–499.

Harvey RF, Hadey N, Gill TR, et al. (1987). Effects of sleeping with the bed-head raised and of rontididine in patients with severe peptic oesophogitis. Lancet 2:1200–1202.

Hauri P (1979). What can insomniacs teach us about the function of sleep? In M Drucker-Colin, MF Shkurowich, & MB Sterman (Eds.), The functions of sleep (Ch. 11, pp. 151–227). New York: Academic Press.

Hauri P (1982). The sleep disorders (p. 11). Upjohn, Kalamazoo.

Hauri P (1985). Primary sleep disorders and insomnia. In TL Riley (Ed.), Clinical aspects of sleep and sleep disturbance (Ch. 5, pp. 81–112). Boston: Butterworth.

Hauri P, & Fisher J (1986). Persistent psychophysiologic (learned) insomnia. Sleep 9:38–53.

Hauri P, & Olmstead F (1980). Childhood-onset insomnia. Sleep 3:59–65.

Hauri PJ, & Silberfarb PM (1980). Effects of aspirin on the sleep of insomniacs. Curr Ther Res 28:867–874.

He J, Kryger MH, Zorick FJ, et al. (1988). Mortality and apnea index in obstructive sleep apnea. Chest 94:9–14.

Healey ES, Kales A, Monroe LJ, et al. (1981). Onset of insomnia: Role of life-stress. Psychosom Med 43:439–451.

Heimer D, Scharf SM, Lieberman A, et al. (1983). Sleep apnea syndrome treated by repair of deviated nasal septum. Chest 84:184–185.

Heiss WD, Dawlik G, Herholz K, et al. (1985). Regional cerebral glucose metabolism in man during wakefulness, sleep and dreaming. Brain Res 327:362–366.

Henningfield JE (1984). Pharmacologic basis and treatment of cigarette smoking. J Clin Psychiatry 45(12, Sec 2):24–34.

Hensley MJ, Sanders MH, & Strohl TP (1980). Methoxyprogesterone treatment of obstructive sleep apnea. Sleep 3:441–446.

Hernandez-Peon R (1966). Physiological mechanisms of attention. In R Russell (Ed.), Frontiers of physiological psychology (Ch. 5, pp. 121–144). New York: Academic Press.

Hjalmarson AI (1984). Effect of nicotine chewing gum in smoking cessation. JAMA 252:2835–2838.

Hobson JA (1968). Sleep after exercise. Science 162:1503–1505.

Hobson JA (1975). Dreaming sleep attack, and desynchronized sleep enhancement: Report of a case of brain stem signs. Arch Gen Psychiatry 32:1421–1424.

Hobson JA (1983). Sleep: Order and disorder. Behav Biol Med 1:1–36.

Hoch CC, Reynolds CF, Nebes RD, et al. (1989). Clinical significance of sleep-disordered breathing in Alzheimer's disease. J Am Geriatr Soc 37:138–144.

Hoeppner JB, Garron DC, & Cartwright RD (1984). Self-reported sleep disorder symptoms in epilepsy. Epilepsia 25:434–437.

Hollingsworth HL (1912). The influence of caffeine on mental and motor efficiency. Arch Psychol 20:1–166.

Hollister LE (1978). Clinical pharmacology of psychotherapeutic drugs (Ch. 2, pp. 25–28). New York: Churchill-Livingstone.

Honda Y, Doi Y, Ninomiya R, et al. (1986). Increased frequency of non-insulin-dependent diabetes mellitus among narcoleptic patients. Sleep 9:254–259.

Horne JA (1980). Sleep and body restitution. Experientia 36:11–13.

Horne JA, & Moore J (1985). Sleep EEG effects of exercise with an additional body cooling. Electroencephalogr Clin Neurophysiol 60:33–38.

Horne JA, & Reid AS (1984). Night-time sleep EEG changes following body heating in a warm bath. Electroencephalogr Clin Neurophysiol 60:154–157.

Horne JA, & Shackell BS (1987). Slow wave sleep elevations after body heating: Proximity to sleep and effects of aspirin. Sleep 10:383–392.

Horne JA, & Staff LHE (1983). Exercise and sleep: Body-heating effects. Sleep 6:36–46.

Howard RS, & Lees AJ (1987). Encephalitis lethargica. Brain 110:19–33.

Institute of Medicine (1979). Sleeping pills, insomnia and medical practice (p. 121). Washington, DC: National Academy of Sciences.

Issa FG, & Sullivan CF (1984). Upper airway closing pressures in snorers. J Appl Physiol Respir Environ Exerc Physiol 57:528–535.

Jacobs EA, Reynolds CF, Kupler DD, et al. (1988). The role of polysomnography in a differential diagnosis of chronic insomnia. Am J Psychiatry 145:346–349.

Jeong DU, & Dimsdale JE (1989). Sleep apnea and essential hypertension. Clin Exp Hypertens [A] 11:1301–1323.

Johns MW, Egan P, Gay TJA, et al. (1970). Sleep habits and symptoms in male medical and surgical patients. Br Med J 2:509–512.

Johns MW, Large AA, Masterton JP, et al. (1974). Sleep and delirium after open heart surgery. Br J Surg 61:377–381.

Johns MW, Masterson JP, Paddle-Ledinek JE, et al. (1975). Variations in thyroid function and sleep in healthy young men. Clin Sci Molec Med 49:629–632.

Johnsen R, Jacobsen BK, & Forde OH (1986). Associations between symptoms of irritable colon and psychological and social conditions and lifestyle. Br Med J 292:1633–1635.

Johnson LC (1982). Effects of anti-convulsant medication on sleep: In MB Sterman & M Shouse (Eds.), Sleep and epilepsy (Ch. 28, pp. 381–394). New York: Academic Press.

Johnson LC, & Chernik DA (1982). Sedative-hypnotics and human performance. Psychopharmacology 76:101–113.

Johnson MW, Anch AM, & Remmers JE (1984). Induction of the obstructive sleep apnea syndrome in a woman by exogenous androgen administration. Am Rev Respir Dis 128:1023–1025.

Jones HS, & Oswald I (1968). Two cases of healthy insomnia. Electroencephalogr Clin Neurophysiol 24:378–380.

Jones JB, Wilhoit SC, Findley LJ, et al. (1985). Oxyhemoglobin saturation during sleep in subjects with and without the obesity-hypoventilation syndrome. Chest 88:9–15.

Jouvet M (1974). The role of monoaminergic neurons in the regulation and function of sleep. In O Petre-Quadens & J Schlag (Eds.), Basic sleep mechanisms (Ch. 9, pp. 207–236). New York: Academic Press.

Jurko MF, Orlando J, & Webster CL (1971). Disordered sleep patterns following thalamotomy. Clin Electroencephalogr 2:213–217.

Jus A, Jus K, Villeneuve P, et al. (1972). Absence of dream recall in lobotomized patients. Lancet 1:955–956.

Kales A, Ansel RD, Markham CH, et al. (1971). Sleep in patients with Parkinson's disease and normal subjects prior to and following levo-dopa administration. Clin Pharmacol Ther 12:397–406.

Kales A, Cadieux R, Shaw L, et al. (1984a). Sleep apnea in a hypertensive population. Lancet 2:1005–1008.

Kales A, Caldwell AB, Cadieux RJ, et al. (1985). Severe obstructive sleep apnea II: associated psychopathology and psychosocial consequences. J Chronic Dis 38:427–434.

Kales A, Henser G, Jacobson A, et al. (1967). All night sleep studies in hyperthyroid patients before and after treatment. J Clin Endocrinol 27:1593–1599.

Kales A, & Kales JD (1984). Evaluation and treatment of insomnia. New York: Oxford University Press.

Kales A, Soldatos CR, Bixler EO, et al. (1983). Early morning insomnia with rapidly eliminated benzodiazepines. Science 220:95–97.

Kales JD, Kales A, Bixler EO, et al. (1984b). Biopsychobehavioral correlates of insomnia: V. Clinical characteristics and behavioral correlates. Am J Psychiatry 141:1371–1375.

Kapen S, Maas C, & Nichols C (1990). Obstructive sleep apnea is a major risk factor for stroke. Neurology 40(suppl 1):136.

Kaprio J, Koskenvuo KJ, Partinen M et al. (1987). A twin study of snoring. Sleep Res 16:365.

Karacan I, Thornby JI, Anch M, et al. (1976). Prevalence of sleep disturbance in a primary urban Florida county. Soc Sci Med 10:239–244.

Karacan I, & Williams RL (1970). Current advances in theory and practice relating to post partum syndromes. Psychiatr Med 1:307–328.

Karacan I, Williams RL, & Taylor WJ (1969). Sleep characteristics of patients with angina pectoris. Psychosomatics 10:280–284.

Katayama S, Yokoyana S, Hiramo Y, et al. (1986). TRH and sleep abnormalities in spinocerebellar degeneration (pp. 227–230). Amsterdam: Elsevier.

Kavey N, Walters AS, Hening W, et al. (1988). Opioid treatment of periodic movements in sleep in patients without restless legs. Neuropeptides 11:181–184.

Kelly DD (1985). Sleep and dreaming. In ER Kandel & JH Schwartz (Eds.), Principles of neural science (Ch. 49, pp. 648–657). New York: Elsevier.

Khatri IM, & Freis ED (1969). Hemodynamic changes during sleep in hypertensive patients. Circulation 39:785–790.

Klink M, & Quan SF (1987). Prevalence of reported sleep disturbances in a general adult population and their relationship to obstructive airway diseases. Chest 91:540–546.

Klotz U (1975). Effects of age and liver disease on disposition and elimination of diazepam in adult man. J Clin Invest 55:347–359.

Koerner E, Flooh E, Reinhart B, et al. (1986). Sleep alterations in ischaemic stroke. Eur Neurol 25(suppl 2):104–110.

Kowatch R, Parmelee DX, & Morin CM (1989). Narcolepsy in a child following removal of a craniopharyngioma. Sleep Res 18:250.

Krieger J, Laks L, Wilcox J, et al. (1989). Atrial natriuretic peptide release during sleep in patients with obstructive sleep apnea before and during treatments with nasal continuous positive airway pressure. Clin Sci 77:407–411.

Krieger J, Mangin P, & Kurtz D (1980). Respiratory changes in the course of sleep in the normal aged subject. Rev Electroencephalogr Neurophysiol 10:177–185.

Kripke DF (1985). Chronic hypnotic use: The neglected problem. In WP Koella, E Ruether, & H Schultz (Eds.), Sleep '84 (pp. 338–340). Stuttgart: Gustave Fischer.

Kripke DF, Cook B, & Lewis OF (1971). Sleep of night workers: EEG recordings. Psychophysiology 7:377–384.

Kripke DF, Mullaney DJ, Messin S, et al. (1978). Wrist actigraphic measures of sleep and rhythm. Electroencephalogr Clin Neurophysiol 44:674–676.

Kurtz D, Zenglein JP, Imler M, et al. (1972). Study of night sleep in the course of porto-caval encephalopathy. Electroencephalogr Clin Neurophysiol 33:167–178.

Lader M, & Lawson C (1987). Sleep studies and rebound insomnia: Methodological problems, laboratory findings and clinical implications. Clin Neuropharmacol 10:291–312.

Lau SK (1987). Sleep apnea due to peritonsillar abscess. J Laryngol Otol 101:617–18.

Lauritzen CR (1976). Female climacteric syndrome: Significance, problems and treatment. Acta Obstet Gynecol Scand [suppl] 51:48–61.

Lavie P (1983). Incidence of sleep apnea in a presumably healthy working population: A significant relationship with excessive daytime sleepiness. Sleep 6:312–318.

Lea MJ, & Aber RC (1985). Descriptive epidemiology of night sweats upon admission to a university hospital. South Med J 78:1065–1067.

Learoyd BM (1972). Psychotropic drugs and the elderly patient. Med J Aust 1:1131–1133.

LEIGH TJ, BIRD HA, HINDMARCH I, ET AL. (1988). Measurement of nocturnal body motility: Behavior of osteoarthritic patients and healthy controls. Rheumatol Int 8:67–70.

LEWIS SA (1969). Subjective estimates of sleep: An EEG evaluation. Br J Psychology 60:203–208.

LILJENBERG B, ALMQUIST M, BROMAN JE, ET AL. (1986). The effect of chlormethiazole in EEG recorded sleep in normal elderly volunteers. Acta Psychiatr Scand [suppl] 329:34–39.

LIPOWSKI ZJ (1980). Delirium (Ch. 6, pp. 152–197). Springfield, IL: Charles Thomas.

LITTLER WA, HONOUR AJ, CARTER RD, ET AL. (1975). Sleep and blood pressure. Br Med J 3:346–348.

LOPATA M, & ONAL E (1982). Mass loading, sleep apnea and the pathogenesis of obesity hypoventilation. Am Rev Respir Dis 126:640–645.

LOWN B, TYKOCINSKI M, GARFEIN A, ET AL. (1973). Sleep and ventricular premature beats. Circulation 48:691–701.

LUGARESI E, CIRIGNOTTA F, COCCAGNA C, ET AL. (1980). Some epidemiological data on snoring and cardiocirculatory disturbances. Sleep 3:221–224.

LUGARESI E, COCCAGNA G, FARNETI P, ET AL. (1979). Snoring, Electroencephalogr Clin Neurophysiol 39:59–64.

LUND R (1974). Personality factors and desynchronization of circadian rhythms. Psychosom Med 36:224–228.

MACCARIO M, RUGGLES KH, & MERIWETHER JA (1987). Post-traumatic narcolepsy. Milit Med 152:370–371.

MANGAN GL (1972). The relationship of strength-sensitivity of the visual system to extraversion. In VD NEBYLITSYN & JA GRAY (Eds.), Biological bases of individual behavior (Ch. 17, pp. 254–261). New York: Academic Press.

MANGLA JC, & PEREIRA M (1982). Tricyclic anti-depressants in the treatment of peptic ulcer disease. Arch Intern Med 142:273–275.

MARKAND ON, & DYKEN ML (1976). Sleep abnormalities in patients with brain stem lesions. Neurology 26:769–776.

MARQUARDSEN J, & HARVALD B (1964). The electroencephalogram in actue vascular lesions of the brain stem and the cerebellum. Acta Neurol Scand 40:58–68.

MARTIN RJ, SANDERS MH, GRAY BA, ET AL. (1982). Acute and long-term ventilatory effects of hyperoxia in the adult sleep apnea syndrome. Am Rev Respir Dis 125:175–180.

MARTTILA JK, HAMMEL RJ, ALEXANDER B, ET AL. (1977). Potential untoward effects of long-term use of flurazepam in geriatric patients. J Am Pharmacol Assoc 17:692–695.

MASON WJ, ANCOLI-ISREAL S, & KRIPKE D (1989). Apnea revisited: A longitudinal follow-up. Sleep 12:423–429.

MASSON C, MEAR JY, MASSON M, ET AL. (1987). Symptomatic hypersomniac ictus of a posterior thalamic hemorrhage. Presse Med 16:79–80.

MAURO C, & SPERLOGANO P (1987). Controlled clinical evaluation of two triazole hypnotic benzodiazepines, estazolam and triazolam, used the night before surgery. Minn Med 78:1381–1384.

MCGINTY DJ, DRUCKER-COLIN R, MORRISON A, ET AL. (EDS.) (1985). Brain mechanisms of sleep. New York: Raven Press.

MELLINGER GD, BALTER MB, & UHLENHUTH EH (1985). Insomnia and its treatment. Arch Gen Psychiatry 42:225–232.

MELLO NK, & MENDELSON JH (1970). Behavioral studies of sleep pattern in alcoholics during intoxication and withdrawal. J Pharmacol Exp Ther 175:94–112.

MENDELSON WB (1980). The use and misuse of sleeping pills. New York: Plenum.

MENDELSON WB (1987). Pharmacotherapy of insomnia. Psychiatr Clin North Am 10:555–563.

MENDELSON WB, SLATER S, GOLD P, ET AL. (1980). The effect of growth hormone administration on human sleep: A dose response study. Biol Psychiatry 15:613–618.

MERRIAM AE (1986). Kleine-Levin syndrome following acute viral encephalitis. Biol Psychiatry 21:1301–1304.

MEYER BR (1982). Benzodiazepines in the elderly. Med Clin North Am 66:1017–1035.

MILES LE, & SIMMONS FB (1984). Evaluation of 198 patients with loud disruptive snoring. Sleep Res 13:154.

MILLMAN RP (1985). Sleep apnea in hemodialysis patients: The lack of testosterone effect. Nephron 40:401–410.

MILLMAN RP (1989). Medroxyprogesterone and obstructive sleep apnea. Chest 96:225.

MILLMAN RP, FOGEL BS, MCNAMARA ME, ET AL. (1989). Depression as a manifestation of obstructive sleep apnea: Reversal with nasal continuous positive airway pressure. J Clin Psychiatry 50:348–351.

MISHARA BL, & KASTENBAUM R (1974). Wine in treatment of long-term geriatric patients in mental institutions. J Am Geriatr Soc 22:88–94.

MITLER MM, SHAFOR R, SOBERA M, ET AL. (1986). Human leucocyte antigen (HLA) studies in excessive somnolence: Narcolepsy versus sleep apnea. Sleep Res 15:148.

MOLDOFSKY H (1989). Sleep and fibrositis syndrome. Rheum Dis Clin N Am. 15:91–103.

MOLDOFSKY H, & LUE FA (1980). The relationship of alpha and delta EEG frequencies to pain and mood in fibrositis patients treated with chlorpromazine and L-tryptophan. Electroencephalogr Clin Neurophysiol 50:71–80.

MOLDOFSKY H, LUE FA, & SASKIN P (1987). Sleep and morning pain in primary osteoarthritis. J Rheumatol 14:124–128.

MOLDOFSKY H, & SCARISBRICK P (1976). Induction of neurasthenic musculoskeletal pain syndrome by selective sleep stage deprivation. Psychosom Med 38:35–44.

MOLDOFSKY H, TULLIS C, QUANCE G, ET AL. (1986). Nitrazepam for periodic movements in sleep. Can J Neurol Sci 13:52–54.

MONROE LJ (1967). Psychological and physiological differences between good and poor sleepers. J Abnorm Psychol 72:255–264.

MONTPLAISIR J, POIRER G, DECARY F, ET AL. (1986). Association between HLA antigens and different types of hypersomnia. JAMA 255:2295–2296.

MOORE-EDE M, SULZMAN FM, & FULLER CA (1982). The clocks that time us. Cambridge, MA: Harvard University Press.

MORGAN K, & OSWALD I (1982). Anxiety caused by a short-life hypnotic. Br Med J 284:942.

MORGAN R, MIRORS DS, & WATERHOUSE JM (1980). Does light rather than social factors synchronize temperature rhythm of psychiatric patients? Chronobiologia 7:331–335.

MORRIS HH, & ESTES ML (1987). Traveler's amnesia. Transient global amnesia secondary to triazolam. JAMA 258:945–946.

MOSES J, LUBIN A, NAITOH P, ET AL. (1972). Reliability of sleep measures. Psychophysiology 9:78–82.

MOSKO SS, DICKEL MJ, & ASHURST J (1988). Night-to-night variability in sleep apnea and sleep-related periodic leg movements in the elderly. Sleep 11:340–348.

MOURET J (1975). Differences in sleep in patients with Parkinson's disease. Electroencephalogr Clin Neurophysiol 38:653–657.

MOURET J, LEMOIRE P, MINUIT MP, ET AL. (1988). Effects of trazadone on the sleep of depressed subjects. Psychopharmacology 95:S37–S43.

MURPHY F, BENTLEY S, ELLIS BW, ET AL. (1977). Sleep deprivation in patients undergoing operation: A factor in the stress of surgery. Br Med J 2:1521–1527.

NEIL JF, HIMMELHOCH JM, MALLINGER PG, ET AL. (1978). Caffeinism complicating hypersomnia depressive episodes. Compr Psychiatry 19:377–385.

NONAKA K, NAKAZAWA Y, & KOTORII T (1983). Effects of antibiotics, minocycline and ampicillin, on human sleep. Brain Res 288:253–259.

NORTON PG, & DUNN V (1985). Snoring as a risk factor for disease: An epidemiological survey. Br Med J 291:630–632.

NOWLIN JB, TROYER WG, COLLINS WS, ET AL. (1965). The association of nocturnal angina pectoris with dreaming. Ann Intern Med 63:1040–1046.

OHGAMI S (1973). Change of delta activity during rapid eye movement sleep in patients with brain tumor. Electroencephalogr Clin Neurophysiol 34:153–162.

OLAZABAL JR, MILLER MJ, COOK WR, ET AL. (1982). Disordered breathing and hypoxia during sleep in coronary artery disease. Chest 82:548–551.

OLSEN KD, KERN EB, & WESTBROOK PR (1981). Sleep and breathing disturbance secondary to nasal obstruction. Otolaryngol Head Neck Surg 89:804–810.

OOMEN MP, ABU-OSBA YK, & THACH BT (1982). Genioglossus muscle responses to upper airway pressure changes: Afferent pathways. J Appl Physiol Respir Environ Exerc Physiol 52:445–450.

ORR WC (1983). Sleep related breathing disorders. Chest 84:475–479.

ORR W, ALSCHULER K, & STAHL M (1982). Managing sleep complaints (p. 67). Chicago: Year Book Publishers.

ORR WC, HALL WH, STAHL ML, ET AL. (1976). Sleep patterns and gastric acid secretions in ucler disease. Arch Intern Med 136:655–660.

ORR WC, LACKEY C, ROBINSON MG, ET AL. (1988). Esophogeal acid clearance during sleep in patients with Barrett's esophogus. Dig Dis Sci 33:654–659.

ORR WC, & MORAN WB (1985). Diagnosis and management of obstructive sleep apnea. Acta Otolaryngol 111:583–588.

ORR WC, ROBINSON MG, & RANDALL OH (1985). Arousal responses to endogenous stimulation during sleep. Sleep Res 14:51.

OSWALD I (1980). Sleep as a restorative process. Prog Brain Res 53:279–288.

OSWALD I, & ADAM K (1989). Can a rapidly-eliminated hypnotic cause daytime anxiety. Pharmacopsychiatry 22:115–119.

OTSUKA K, SADAKANE N, & OZAWA T (1987). Arrhythmogenic properties of disorder breathing during sleep in patients with cardiovascular disorders. Clin Cardiol 10:771–782.

PALMER CD, & HARRISON GA (1983). Sleep latency and lifestyle in Oxfordshire villages. Ann Hum Biol 10:417–428.

PARKER DC, SASSIN JF, MACE JW, ET AL. (1969). Human growth hormone release during sleep in chronic uremics undergoing an extrarenal dialysis. Electroencephalogr Clin Neurophysiol 29:441–449.

PASSOUANT P, CADILHAC J, BALDY-MOULINIER M, ET AL. (1970). Study of nocturnal sleep in chronic uremics undergoing an external dialysis. Electroenceph Clin Neurophysiol 29:441–449.

PATTERSON JF (1987). Triazolam syndrome in the elderly. South Med J 80:1425–1426.

PEISER J, LAVIE P, OVNAT A, ET AL. (1984). Sleep apnea syndrome in the morbidly obsese as an indication for wight reduction surgery. Ann Surg 199:112–115.

PEKKARINEN T, PARTINEN M, PELKINEN N, ET AL. (1987). Sleep apnea and daytime sleepiness in acromegaly: Relationship to endocrinological factors. Clin Endocrinol 27:649–654.

PEREZ-PADILLA R, WEST P, LERTZMAN M, ET AL. (1985). Breathing during sleep in patients with interstitial lung disease. Am Rev Respir Dis 132:224–229.

PETER JH, FUCHS E, KOEHLER U, ET AL. (1986). Studies in the prevalence of sleep apnea activity (SAA): Evaluation of ambulatory screening results. Eur J Respir Dis 69(suppl 146):451–456.

PETRE-QUADENS O, & JOUVET M (1967). Sleep in the mentally retarded. J Neurol Sci 4:354–357.

PETRIE WM, MAFFUCCI RJ, & WOOSLEY RC (1982). Propranolol and depression. Am J Psychiatry 139:92–93.

PETRUSEN B (1990). Snoring can be reduced when the nasal airflow is increased by the nasal dilator nozovent. Arch Otolaryngol Head Neck Surg 116:462–464.

PETURSSON H, & LADER MH (1981). Withdrawal from long-term benzodiazepine treatment. Br Med J 283:643–645.

POIRIER G, MONTPLAISIR J, DUMONT M, ET AL. (1987). Clinical and sleep laboratory study of narcoleptic symptoms in multiple sclerosis. Neurology 37:693–695.

POIRIER G, MONTPLAISIR J, DUQUETTE P, ET AL. (1986). Narcoleptic symptoms, HLA and MSLT data in multiple sclerosis. Sleep Res 15:156.

POITRAS R (1980). On episodes of anterograde amnesia associated with use of triazolam. Union Med Can 109:427–429.

POKORNY AD (1978). Sleep disturbances, alcohol, and alcoholism: A review. In RL WILLIAM & I KARACAN (Eds.), Sleep disorders, diagnosis and treatment (Ch. 9, pp. 233–260). New York: John Wiley & Sons.

POLLAK PT, VINCKEN W, MUNRO IR, ET AL. (1987). Obstructive sleep apnea by hemarthrosis-induced micrognathia. Eur J Respir Dis 70:117–121.

PRENDES JL, & ROSENBERG SJ (1986). Rip van Winkle syndrome: Confusion & irresistable somnolence after stroke. South Med J 79:1162–1164.

PRINZ PN (1977). Sleep patterns in the healthy aged: Relationship with intellectual functioning. J Gerontol 32:179–186.

PRINZ PN, OBRIST WL, & WANG H (1975). Sleep patterns in healthy elderly subjects. Individual differences as related to other neurological variables. Sleep Res 4:132.

PRINZ P, PESKIND E, VITALIANO P, ET AL. (1982). Changes in the sleep and waking EEG's of nondemented and demented elderly subjects. J Am Geriatr Soc 30:86–93.

PUCA FM, BRICOLO A, & TURELLA G (1973). Effect of L-dopa or amantadine therapy on sleep spindles in parkinsonism. Electroencephalogr Clin Neurophysiol 35:327–330.

QUUYUMI A, MOCKUS LJ, WRIGHT CA, ET AL. (1984). Mechanisms of nocturnal angina pectoris. Lancet 1:1207–1209.

RECHTSCHAFFEN A, & KALES A (Eds.) (1968). A manual of standardized terminology, techniques and scoring system for sleep stages of human subject. Los Angeles: Brain Information Service, Brain Research Institute, University of California at Los Angeles.

REES PJ, COCHRANE GM, PRIOR JG, ET AL. (1981). Sleep apnea in diabetic patients with autonomic neuropathy. J Roy Soc Med 74:192–195.

REGESTEIN QR (1976). Treating insomnia: A practical guide for managing sleeplessness, circa 1975. Compr Psychiatry 17:517–526.

REGESTEIN QR (1980). Sleep and insomnia in the elderly. J Geriatr Psychiatry 13:153–171.

REGESTEIN QR (1981). Sleep and insomnia in the elderly. J Geriatr Psychiat 23:153–171.

REGESTEIN QR (1982). Diagnosis and treatment of chronic insomnia. In R GALLON (Ed.), Psychosomatic approach to illness (Ch. 6, p. 132). New York: Elsevier-North Holland.

REGESTEIN QR (1984). Treatment of insomnia in the elderly. In C SALZMAN (Ed.), Clinical geriatric pharmacology (Ch. 8, pp. 149–170). New York: McGraw-Hill.

REGESTEIN QR (1987a). Relationship between psychological factors and cardiac rhythm and electrical disturbances. Compr Psychiatry 16:137–148.

REGESTEIN QR (1987b). Specific effects of sedative/hypnotic drugs in the treatment of incapacitating chronic insomnia. Am J Med 83:909–913.

Regestein QR (1988). Polysomnography in the diagnosis of chronic insomnia. Am J Psychiatry 145:483.

Regestein QR (1989). Pathological sleepiness induced by caffeine. Am J Med 87:586–588.

Regestein Q, DeSilva R, & Lown B (1981). Cardiac ventricular ectopic activity increases during REM sleep. Sleep Res 10:58.

Regestein QR, Ferber R, Johnson TS, et al. (1988). Relief of sleep apnea by revision of the adult upper airway. Arch Otolaryngol Head Neck Surg 114:1109–1113.

Regestein Q, Hallett M, Mufson M, et al. (1985). A hyperarousal scale. Sleep Res 14:135.

Regestein QR, & Morris J (1987). Sleep patterns observed in institutionalized . J Am Geriatr Soc 35:767–772.

Regestein QR, & Reich P (1978). Current problems in the diagnosis and treatment of chronic insomnia. Perspect Biol Med 21:232–239.

Regestein QR, & Reich P (1983). Incapacitating childhood-onset insomnia. Compr Psychiatry 24:244–248.

Regestein QR, & Reich P (1985). Agitation observed during treatment with newer hypnotic drugs. J Clin Psychiatry 46:280–283.

Regestein QR, Reich P, & Mufson MJ (1983). Narcolepsy: An initial clinical response. J Clin Psychiatry 44:166–172.

Regestein QR, Rose LI, & Williams GH (1972). Psychopathology in Cushing's syndrome. Arch Intern Med 130:114–117.

Regestein QR, Schiff I, Tulchinski D, et al. (1981). Relationships among estrogen-induced psychophysiological changes in hypogonadal women. Psychosom Med 43:147–155.

Remmers JE, deGroot WJ, Sauerland EK, et al. (1978). Pathogenesis of upper airway occlusion during sleep. J Appl Physiol Respir Environ Exerc Physiol 44:931–938.

Reynolds CF (1987). Sleep and affective disorders. Psychiatr Clin North Am 10:583–591.

Reynolds CF, Kupfer DJ, Hoch CC, & Sewitch DE (1985). Sleeping pills for the elderly: Are they ever justified? J Clin Psychiat 46(2 pt 2):9–12.

Reynolds C, Kupfer DJ, Taska LS, et al. (1982). Pills for the elderly: Are they ever justified? Psychiatr Clin North Am 10:583–561.

Reynolds CF, Kupfer DJ, Taska LS, et al. (1985). Sleep of healthy seniors: A revisit. Sleep 8:20–29.

Reynolds CF, Hoch CC, Stack J, et al. (1988). The nature and management of sleep/wake disturbance in Alzheimer's dementia. Psychopharmacol Bull 24:43–48.

Reynolds CF, Taska LS, Sewitch DE, et al. (1984). Persistent psychophysiologic insomnia: Preliminary research diagnostic criteria and EEG sleep data. Am J Psychiatry 141:804–805.

Rhodes NP, Hayes B, Guilleminault C, et al. (1989). The narcoleptic tetrad: Objective polysomnographic tests. Sleep Res 18:294.

Richardson GS, Carskadon MA, Flagg W, et al. (1978). Excessive daytime sleepiness in man: Multiple sleep latency measurement of narcoleptic and control subjects. Electroencephalogr Clin Neurophysiol 45:621–627.

Rickels K (1986). Clinical use of hypnotics: Indication for use and the need for a variety of hypnotics. Acta Psychiatr Scand [suppl] 332:132–141.

Rickels K, Case WG, Downing RW, et al. (1983). Long-term diazepam therapy and clinical outcome. JAMA 250:767–771.

Rickels K, Schmeizer E, Case WL, et al. (1990). Long-term therapeutic use of benzodiazepines. Arch Gen Psychiatry 47:899–907.

Roehrs T, Conway W, Wittig R, et al. (1985). Sleep-wake complaints in patients with sleep-related respiratory disturbances. Am Rev Resp Dis 132:520–523.

Roehrs T, Tietz EI, Zorick FJ, et al. (1984). Daytime sleepiness and anti-histamines. Sleep 7:137–141.

Roehrs T, Zorick MD, Sicklesteel J, et al. (1983). Age-related sleep-wake disorders at a sleep disorder center. J Am Geriatr Soc 31:364–370.

Rosenthal NE, Joseph-Vanderpool JR, Levendosky AA et al. (1990). Phase-shifting effects of bright morning light as treatment for delayed sleep phase syndrome. Sleep 13:354–361.

Rosomoff HL (1986). Occult respiratory and autonomic dysfunction in craniovertebral anomalies and upper cervical spinal disease. Spine 11:345–347.

Rosvold HD, Mirsky AF, Sarason I, et al. (1956). A continuous performance test of brain damage. J Consult Psychology 20:343–350.

Rotenberg VS, & Kobrin VI (1985). The structure of night sleep and cardiac rhythm in patients suffering for many years from hypo- and akinesia. Activ Nerv Sup 27:179–85.

Roth B (1980). Narcolepsy and hypersomnia. Activitas Nervosa Superiora. Basel, Switzerland: S Karger.

Roth T, Hartse KM, Saab PG, et al. (1980). The effects of flurazepam, lorazepam, and triazolam on sleep and memory. Psychopharmacology 70:231–237.

Rubin RL, Hajdukovich RM, & Mitler MM (1988). HLA-DR2 association with excessive somnolence in narcolepsy does not generalize to sleep apnea. Clin Immunol Immunopathol 49:149–158.

Rubinow DR, & Roy-Byrne P (1984). Premenstrual syndromes: Overview from a methodologic perspective. Am J Psychiatry 141:163–172.

Ruiz-Primo ME, Coria S, & Torres O (1985). Prevalence of subjective sleep disorders in poorly controlled chronic epileptics. Sleep Res 14:243.

Rumbach L, Tongio MM, Warter JM, et al. (1989). Multiple sclerosis, sleep latencies and HLA antigens. J Neurol 236:309–310.

Salazar-Gruesco EF, Rosenberg RS, & Roos RP (1988). Sleep apnea in olivopontocerebellar degeneration: Treatment with trazodone. Am Neurol 23:399–401.

Saletu B, Gruenberger J, & Anderer P (1983). Evening television and sleep. Med Welt 34:866–870.

Salzman C (1979). Update on geriatric psychopharmacology. Geriatrics 34:87–90.

Salzman C (1981). Psychotropic drug use and polypharmacy in a general hospital. Gen Hosp Psychiatry 3:1–9.

Sanders MH (1984). Nasal CPAP effect on patterns of sleep apnea. Chest 86:839–844.

Sanders MH, Rogers RM, & Pennock BE (1985). Prolonged expiratory phase in sleep apnea. Am Rev Respir Dis 131:401–408.

Saskin P, Moldofsky H, & Lue FA (1986). Sleep and posttraumatic pain modulation disorder (fibrositis syndrome). Psychosom Med 48:319–323.

Sauerland EK, & Harper RM (1976). The human tongue during sleep: Electromyographic activity of the genioglossus muscle. Exp Neurol 51:160–170.

Savin JA, Peterson WD, & Oswald I (1973). Scratching during sleep. Lancet 2:296–297.

Sawicka EH, & Branthwaite MA (1987). Respiration during sleep in kyphoscoliosis. Thorax 42:801–808.

Scharf MB, Khosla N, Brocker N, et al. (1984). Differential amnestic properties of short and long-acting benzodiazepines. J Clin Psychiatry 45:51–53.

Schneider E, Maxion H, Ziegler B, et al. (1974). The relationship between sleep and Parkinson's disease and its alteration by L-dopa. J Neurol 207:95–108.

Schneider-Helmert D (1988). Why low-dose benzodiazepine-dependent insomniacs can't escape their sleeping pills. Acta Psychiatr Scand 78:706–711.

Schuil HJ, Wilkinson GR, & Johnson R (1971). Normal disposition of oxazepam in acute hepatitis and cirrhosis. Ann Intern Med 84:420–425.

Scrima L, Broudy KN, & Cohn MA (1982). Increased severity of obstructive sleep apnea after bedtime alcohol ingestion: Diagnostic potential and proposed mechanism of action. Sleep 5:318–328.

Scrima L, Hartman PG, & Hiller FC (1989). Effect of three alcohol doses on breathing during sleep in 30–49 year old non-obese snorers and non-snorers. Alcoholism 13:420–427.

Seidel WF, Roth T, Roehrs T, et al. (1984). Treatment of a 12-hour shift of sleep schedule with benzodiazepines. Science 224:1262–1264.

Shader RI, Georgotus A, Greenblatt DL, et al. (1978). Impaired desmethyldiazepam from chlorazepate by magnesium aluminum hydroxide. Clin Pharmacol Ther 24:308–315.

Shader RI, Greenblatt DJ, Harmatz JS, et al. (1977). Absorption and disposition of chlordiazepoxide in young and elderly male volunteers. J Clin Pharmacol 17:709–718.

Shaver J, Giblin E, Lentz M, et al. (1988). Sleep patterns and stability in perimenopausal women. Sleep 11:556–561.

Sheerin E (1972). A programme which led to a reduction in night sedation at a major hospital. Med J Aust 2:678–681.

Shepard JW, Garrison MN, Grither DA, et al. (1985). Relationship of ventricular ectopy to oxyhemoglobin saturation in patients with obstructive sleep apnea. Chest 88:335–340.

Shepard JW, Schweitzer PK, Keller CA, et al. (1984). Myocardial stress: Exercise versus sleep in patients with chronic obstructive lung disease. Chest 86:366–374.

Shirlow MJ, & Mathers CD (1985). A study of caffeine consumption and symptoms: Indigestion, palpitations, tremor, headache and insomnia. Int J Epidemiol 14:239–247.

Shouse MN (1988). Sleep deprivation increases susceptibility to kindling and penicillin seizure events during all waking and sleep states in cats. Sleep 11:162–171.

Silvestri R, DeDomenico P, Giugliotta MA, et al. (1987). Gilles da la Tourette's syndrome: Arousal and sleep polygraphic findings. Acta Neurol 9:263–272.

Simmons FB, Guilleminault C, & Miles CE (1984). A surgical treatment for snoring and obstructive sleep apnea. West J Med 140:43–46.

Sitaram N, Wyatt RJ, Dawson S, et al. (1976). REM sleep induction by physostigmine infusion during sleep. Science 191:1281–1283.

Skatbud JB, & Dempsey JA (1985). Airway resistance and respiratory muscle function in snores during NREM sleep. J Appl Physiol 59:328–335.

Smirk H (1963). Hypotensive action of methyldopa. Br Med J 1:146–151.

Snyder F (1971). The physiology of dreaming. Behav Sci 16:31–43.

Snyder F, Hobson JA, Morrison DF, et al. (1964). Changes in respiration, heart rate, and systolic blood pressure in human sleep. J Appl Physiol 19:417–422.

Soldatos CR, Kales JD, Scharf MB, et al. (1980). Cigarette smoking associated with sleep difficulty. Science 207:551–553.

Spiegel R, Koeberle S, & Allen SR (1986). Significance of slow wave sleep: Considerations from a clinical viewpoint. Sleep 9:66–79.

Spier S, Rivlin J, Rowe RD, et al. (1986). Sleep in Pierre Robin syndrome. Chest 90:711–715.

Spire JP, Kuo N, & Campbell N (1983). Maxillo-facial surgical approach: An introduction and review of mandibular advancement. Bull Eur Physiopathol Respir 19:604–606.

Stacher G, Presslich B, & Starker H (1975). Gastric acid secretion and sleep stages during natural night sleep. Gastroenterology 68:1449–1455.

Starkman MN, Schteingart DE, & Schork MA (1981). Depressed mood and other psychiatric manifestations of Cushing's syndrome: Relationship to hormone levels. Psychosom Med 43:3–18.

Stepanski E, Zorick F, Roehrs T, et al. (1988). Daytime alertness in patients with chronic insomnia compared with asymptomatic control subjects. Sleep 11:54–60.

Stern M, Roffwarg H, & Duvoisir R (1968). The parkinsonian tremor in sleep. J Nerv Ment Dis 147:202–210.

Stramba-Badiale M, Ceretti A, & Forni G (1979). Aspects of sleep in the aged and very aged (long-lived) subject. Minerva Med 70:2551–2554.

Strohl KP, Cherniack NS, & Gothe B (1986). Physiologic basis of therapy for sleep apnea. Am Rev Respir Dis 134:791–802.

Stunkard A, Grace W, & Wolff H (1955). The night eating syndrome. Am J Med 19:78–86.

Sullivan CE, & Grunstein RR (1989). Continuous positive airway pressure in sleep disordered breathing. In MH Kryger, T Roth, & WC Dement (Eds.), Principles and practice of sleep medicine (Ch. 59, pp. 559–570). Philadelphia: WB Saunders.

Sullivan C, & Issa F (1980). Pathophysiological mechanisms in obstructive sleep apnea. Sleep 3:235–246.

Sweetwood H, Grant I, Gerst MS, et al. (1980). Sleep disorder over time: Psychiatric correlates among males. Br J Psychiatry 136:456–462.

Swygert LA, Maes EF, Sewell LL, et al. (1990). Eosinophilia-myalgia syndrome. JAMA 264:1698–1703.

Taasan VC, Block AJ, Bayser PG, et al. (1981). Alcohol increases sleep apnea and oxygen desaturation in asymptomatic men. Am J Med 71:240–245.

Tan T, Kales J, Kales A, et al. (1984). Biopsychobehavioral correlates of insomnia: IV. Diagnosis based on DMS-III. Am J Psychiatry 141:357–362.

Taub JM, Globus GG, Phoebus E, et al. (1971). Extended sleep and performance. Nature 233:142–143.

Tecce JJ (1971). Contingent negative variation and individual differences. Arch Gen Psychiatry 24:1–16.

Tepas DL, Stock CG, Maltese JW, et al. (1982). Reported sleep of shift workers: A preliminary report. Sleep Res 7:313.

Thawley SE (1985). Surgical treatment of obstructive sleep apnea. Med Clin North Am 69:1337–1358.

Thompson J, & Oswald I (1977). Effect of oestrogen on the sleep, mood, and anxiety of menopausal women. Br Med J 2:1317–1319.

Tilkian A, Guilleminault C, Schroeder JS, et al. (1977). Sleep induced apnea syndrome. Am J Med 63:343–358.

Tolle FA, Judy WV, Yu PL, et al. (1983). Reduced stroke volume related to pleural pressure in obstructive sleep apnea. J Appl Physiol Respir Environ Exerc Physiol 55:1718–1724.

Tomlinson BE, Blessed G, & Roth M (1970). Observations of the brains of demented old people. J Neurol Sci 11:205–242.

Touchon J, Baldy-Moulinier M, Billiaed M, et al. (1987). Organization of sleep in recent epilepsy pre- and posttreatment by carbamazepine. Rev Neurol 143:462–467.

Trichopoulos D, Tzonu A, Christopoulos C, et al. (1987). Does a siesta protect from coronary heart disease? Lancet 2:269–270.

Trinder J, Stevenson J, Paxton SJ, et al. (1982). Physical fitness, exercise and REM sleep cycle length. Psychophysiology 19:89–93.

Tune GS (1968). Sleep and wakefulness in normal human adults. Br Med J 2:269–271.

Tyrer P, Owen R, & Dowling S (1983). Gradual withdrawal of diazepam after long-term therapy. Lancet 1:1402–1406.

Ulrich RS (1984). View through a window may influence recovery from surgery. Science 224:420–421.

Van der Kleijn, E (1989). Effects of zopiclone and temazepam on sleep, behavior and mood during the day. Eur J Clin Pharmacol 36:247–251.

Vardi J, Glaubman H, Rabey JM, et al. (1978). Myoclonic attacks induced by L-dopa and bromocryptine in Parkinson patients. J Neurol 218:35–42.

Vela-Bueno A, Kales A, Soldatos C, et al. (1984). Sleep in the Prader-Willi syndrome. Arch Neurol 41:294–296.

Victor BS, Lubetsky M, & Greden JF (1981). Somatic manifestations of caffeinism. J Clin Psychiatry 42:185–188.

Vieth J (1986). Vigilance, sleep and epilepsy. Eur Neurol 25(suppl 2):128–133.

Vitiello MR, Prinz P, Personius JP, et al. (1990). Nightime hypoxia is increased in abstaining chronic alcoholic men. Alcoholism 14:38–41.

Vitrey JM, Huquet P, & Dollfus D (1970). Recording of two sleep records in the course of Jacob Creutzfelt's disease. Rev Neurol 122:528–529.

Walker JM, Floyd TC, Fein G, et al. (1978). Effects of exercise on sleep. J Appl Physiol Respir Environ Exerc Physiol 44:945–951.

Waller PC, & Bhopal RS (1989). Is snoring a cause of vascular disease? Lancet 1:143–146.

Walters AS, Hening WA, Kavey N, et al. (1988). A double-blind randomized crossover trial of bromocriptive and placebo in restless legs syndrome. Arch Neurol 24:455–458.

Ware JC (1983). Tricyclic anti-depressants in the treatment of insomnia. J Clin Psychiatry 44(9, Sec 2):25–28.

Ware JC, Blumoff R, & Pittard JT (1988). Peripheral vasconstriction in patients with sleep-related periodic leg movement. Sleep 11:182–187.

Webb WB, & Agnew HW (1974). Regularity in the control of the free-running sleep-wakefulness rhythm. Aerospace Med 45:701–704.

Webb WB, & Swinburne H (1971). An observational study of the sleep aged. Percept Mot Skills 32:895–898.

Wehr TA, Wirz-Justice A, Goodwin FK, et al. (1979). Phase advance of the circadian sleep-wake cycle as an anti-depressant. Science 206:710–713.

Weitzman ED, Moline ML, Czeisler CA, et al. (1982). Chronobiology of aging: Temperature sleep-wake rhythms and entertainment. Neurobiol Aging 3:299–309.

Wetmore SJ, Sorima L, Snyderman NL, et al. (1986). Postoperative evaluation of sleep apnea after uvulopalatopharyngoplasty. Laryngosope 96:738–741.

Wever R (1970). On the Zeitgeber strength of a light/dark cycle for circadian periodicity in man. Pfluegers Arch 321:133–142.

Wever RA (1975a). Autonomous circadian rhythms in man. Naturwissenschaften 62:443–444.

Wever R (1975b). The circadian multi-oscillator system of man. Eur J Chronobiol 3:19–55.

Wever R (1975c). The meaning of circadian periodicity for old people. Verh Dtsche Ges Pathol 59:160–180.

Wever RA (1979). The circadian system of man: Results of experiment under temporal isolation. New York: Springer.

Wilkinson RT (1965). Sleep deprivation. In OG Edholm & AL Bacharach (Eds.), The physiology of human survival (Ch. 14, pp. 399–429). New York: Academic Press.

Will RG, Young JPR, & Thomas DJ (1988). Kleine-Levin syndrome. Report of two cases with onset of symptoms precipitated by head trauma. Br J Psychiatry 152:410–412.

Williams NR (1988). Erythromycin: A case of nightmares. Br Med J 296:214.

Williams RL (1980). Sleeping pill insomnia. J Clin Psychiat 41:153–154.

Williams RL, & Karacan I (1976). Pharmacology of sleep (p. 142). New York: John Wiley & Sons.

Williams RL, Karacan I, & Hursch CJ (1974). Electroencephalography (EEG) of human sleep: Clinical applications. New York: John Wiley & Sons.

Wynne JW, Block AJ, & Boysen PG (1980). Oxygen desaturation in sleep: Sleep apnea and COPD. Hosp Pract 15(10):77–85.

Young R, Waldron J, Baer S, et al. (1986). Obstructive sleep apnea in association with retrosternal goitre and acromegaly. J Laryngol Otol 100:861–863.

Zorick F, Roth T, Hartze K, et al. (1981). Evaluation and diagnosis of resistant insomnia. Am J Psychiatry 138:769–773.

22 Traumatic brain injury

C. THOMAS GUALTIERI, M.D.

There are more than 2 million cases of traumatic brain injury (TBI) in the United States each year. Five hundred thousand cases are so severe that admission to the hospital is required (Goldstein, 1990). Each year, 70,000 to 90,000 people with TBI will suffer lifelong debilitating loss of function (Goldstein, 1990). Even relatively mild head injuries may lead to permanent neuropathic changes and prolonged disability (Marshall & Marshall, 1985). Neurobehavioral deficits contribute disproportionately to chronic disability caused by TBI (63% of cases), accounting for more disability overall than deficits like aphasia and hemiparesis (Jennet et al., 1981; Levin 1987).

The statistics make a strong case for brain injury as one of the most important neuropsychiatric conditions: the large (and growing) number of patients, their tendency to be injured when they are young, and the lifetime of disability they have to endure. However, the neurobehavioral sequelae of TBI, such as anterograde amnesia and inattention, depression, psychosis, and personality change, are amenable to treatment. An appropriate rehabilitation program or a timely, well-directed psychiatric intervention may spell the difference between a productive life and one of chronic disability (Gualtieri, 1990c).

The focus of this chapter is closed head injury (CHI), the most prevalent form of brain injury (Kraus, 1987). The more general term, "traumatic brain injury," refers to CHI and also to penetrating brain injuries (PBIs) and open head injuries (OHIs). The term "traumatic brain injury" may sometimes be used to refer to diffuse encephalopathic injury caused by anoxia (diffuse anoxic injury), thermal, electrical, or radiation injury. The terminology may not be very precise, but the important issues in rehabilitation and psychiatry are very similar.

EPIDEMIOLOGY

Traumatic brain injury patients tend to be young: the peak incidence is from age 15 to 24 (Kraus, 1987). There are secondary peaks for infants and children and for the elderly (Kraus, 1987). At every age, males are twice as likely as females to be victims of TBI (Bourke, 1988).

The most common causes of TBI are transport-related accidents (motor vehicles, bicycles), falls, interpersonal assaults (especially gunshot wounds), and sporting injuries (Bourke, 1988; Kraus, 1987). Falls are a common cause of TBI in children and the single most common cause of TBI in the elderly.

Factors that predispose to the occurrence of TBI may also complicate the process of diagnosis and treatment. Alcoholism, drug abuse, and risk-taking behavior are obvious elements in TBI related to vehicular accidents (Kraus, 1987). Elderly people who are most prone to injury from falling frequently have premorbid evidence of degenerative brain disease (Masdeu et al., 1989). The likelihood of TBI is higher in people who have had a previous TBI, an association that may be mediated by behavioral characteristics (e.g., risk taking, substance abuse) or by deficits in attention, perception, and response time that are attributable to the initial TBI (Annegers et al., 1980). This is important, because the neurobehavioral effects of more than one brain injury may be more than simply additive (Carlsson, 1987). It also speaks to an important issue in treatment: to warn a TBI patient of the increased risk of a second injury.

PATHOLOGY

The majority of TBI cases in developed countries are caused by motor vehicle accidents or by falls. They are, therefore, acceleration-deceleration injuries, most of which are CHIs. Closed head injury is characterized by neuropathologic changes that differ substantially from those described in PBIs. (Studies of PBI have usually been done in combat victims [e.g., Lishman, 1987].)

In CHI the brain is injured by contusion, the result of direct and contrecoup impact. The cortical lesions that derive from such an impact are more extensive, as a rule, than the lesions of a PBI, but they are qualitatively similar. In addition, CHI also produces diffuse, microscopic damage to the brain, especially to the subcortical white matter (Bourke, 1988). Cortical impact injury is transmitted with particular intensity to axial structures in the brainstem, generating shear forces that cause scattered multifocal subcortical and brainstem axonal damage ("diffuse axonal injury") (Blumbergs, Jones, & North, 1989; Bourke, 1988). The combination of cortical, subcortical, and brainstem injury is characteristic of CHI.

The axonal tracts that are disrupted in CHI comprise monoaminergic projections from the brainstem to cortical and subcortical structures. Thus, experimental studies in laboratory animals, as well as clinical experiments, have been consistent in demonstrating persistent deficits in monoamine neurotransmission after CHI (Gualtieri, 1990c; Silver, Yudofsky, & Hales, 1991).

Cortical regions that are particularly vulnerable to contusive damage in an acceleration-deceleration injury are the rostral poles and the inferior surfaces of the frontal and temporal lobes. They are damaged by impact against the frontal, sphenoid, or temporal bones or by movement along the irregular surface of the cribriform plate. Thus, the behavioral characteristics of frontal and temporal lobe lesions are common among CHI victims.

The anatomic description of a TBI patient comprises the cortical and subcortical lesions that may be demonstrable in the neurologic examination, by brain imaging, and by neuropsychologic testing. Diffuse microscopic damage may be difficult to demonstrate in the living patient. However, if the cortical lesions are extensive, or if there has been a prolonged period of unconsciousness, the occurrence of diffuse axonal injury may be inferred. Sometimes, magnetic resonance imaging (MRI) will demonstrate multiple small areas of increased T_2 signal in subcortical or brainstem white matter; computed tomography (CT) scans rarely show these white matter abnormalities (Levin et al., 1987a).

In patients who have had relatively mild CHI, with no demonstrable signs of cortical deficit, there may still be microscopic damage to axial neurons (Adams et al., 1982). Such patients might present with relatively nonspecific complaints of inattention, fatigue, irritability, and depression. It is easy to attribute symptoms like these to "reactive depression," "compensation neurosis," or "posttraumatic stress," especially when the neurologic exam is normal and the CT scan and the MRI are normal. It may be possible to demonstrate areas of statistically significant abnormality with techniques such as positron emission tomography (PET), single-photon emission CT (SPECT), or computerized electroencephalographic (EEG) brain mapping. However, the anatomic and clinical significance of such abnormalities are insufficiently established by research to warrant the routine use of these imaging methods in the clinic evaluation of mild CHI (see Chapter 18).

The irony of CHI, and a dilemma to a great many patients, is that clinically significant neuropathic injury may occur, even with a mild injury, with only a few moments' loss of consciousness (LOC), or none at all (Barth et al., 1983; Marshall & Marshall, 1985). It is even possible for brain trauma to occur in acceleration-deceleration injuries (e.g., whiplash), in which the head itself never strikes a rigid surface (Yarnall & Rossie, 1988). Psychiatrists are especially likely to see patients in whom brain injury has occurred in relatively minor accidents. Notwithstanding such clinically important exceptions, the duration of LOC and the depth of coma, defined by the Glasgow Coma Scale (Jennett et al., 1981), are the most important elements in predicting the extent of neuropathic damage (Eisenberg, 1985).

RECOVERY AND OUTCOME

The outcome of any given TBI is determined by four elements: (1) the extent and location of gross brain damage, (2) the extent of diffuse microscopic brain damage, (3) the patient's premorbid history and level of functioning, and (4) the nature and extent of therapeutic and rehabilitative efforts.

The topography of the specific cortical and subcortical lesions is determined directly by anatomic brain imaging techniques such as CT and MRI. Magnetic resonance imaging is more sensitive than CT, and it is the test of choice if it is feasible and not contraindicated. Since TBI-related cortical lesions most frequently occur adjacent to bone, the transparency of bone to magnetism gives MRI a substantial advantage over CT. Indirect evidence of brain lesions can be obtained by PET, SPECT, computerized EEG brain mapping, or neuropsychologic testing. However, anatomic correlation with these tests is imprecise.

The extent of diffuse microscopic brain damage is difficult to assess by imaging procedures. It is most commonly inferred from the duration of LOC, posttraumatic amnesia (PTA), or posttraumatic confusion (PTC), or from the severity of the patient's postconcussive symptoms.

The most relevant dimensions of premorbid state are age, IQ, and preexisting neuropsychiatric conditions. The effects of TBI are less devastating to a young person, for example. Children in the first decade of life can be expected to recover completely from a lesion to the language-dominant hemisphere, presumably because the cortical substrates of language are capable of "redistribution" or "vicariation" (LeVere, Gray-Silva, &

LeVere, 1988). Traumatic brain injury in an elderly person may unmask or accelerate the development of dementia. In contrast, children may be more prone to posttraumatic epilepsy than adults (Young et al., 1983). High IQ is predictive of good recovery (Alves & Jane, 1985; Kraus, 1987; Levin, 1987). Preexisting psychiatric conditions are almost invariably aggravated by a TBI (Alves & Jane, 1985), although there are occasional reports of a psychiatric cure in patients after TBI (e.g., Lewis et al., 1990). Posttraumatic stress symptoms in TBI patients are more likely to occur in those who have a preexisting psychiatric condition (Keshavan, Channabasavanna, & Reddy, 1981).

The final determinant of outcome following TBI are the therapeutic and rehabilitative efforts. Although there has been continuing scepticism in the medical community about the value of some aspects of TBI rehabilitation (Berrol, 1990; Volpe & McDowell, 1990), an attitude of therapeutic nihilism is hardly appropriate. The proper treatment of TBI patients, by professionals who understand the unique requirements of this class of patients, may result in some degree of success. The inappropriate treatment of TBI may have unfortunate consequences. Scepticism about the success of TBI rehabilitation may arise simply because the course of TBI recovery is prolonged and complicated.

Proper treatment for the TBI patient requires an appreciation of the time required for recovery to occur. Several months may be required for a patient to recover from a mild postconcussion syndrome. A severely injured patient may be capable of functional gains even 5 years after the event (Gualtieri & Nygard, 1988).

The "trajectory of recovery" from TBI (Gualtieri, 1990c) is a term that describes the dynamic thrust of the patient's condition over the months and years. It is not a linear process. Rather, it is like the process of development and maturation, given to fits and starts. Periods of rapid recovery may alternate with "plateaus" during which little change may occur. There are times when one should respect a plateau, as if it were a necessary period of stabilization and retrenchment. During such a period, the thrust of rehabilitation should be oriented toward maintenance. If a patient is stuck on a prolonged plateau, however, intensive rehabilitation or even pharmacologic treatment might be directed to move the patient off dead center, and to veer the trajectory of recovery back onto the incline.

During the recovery process, it may become clear that some functions are lost completely or severely impaired, and that the loss is irreversible. The orientation of treatment then is directed toward acceptance, compensation, and maintenance. The goal is to maintain the status quo, to make the best of it, and to prevent secondary deterioration.

The trajectory of recovery from TBI may also take a downward turn, and it is clear that some patients experience secondary deterioration, months or years after the injury itself. The mechanisms for secondary deterioration are not well understood, and are discussed elsewhere (Gualtieri & Cox, 1991).

NEUROBEHAVIORAL SEQUELAE OF TRAUMATIC BRAIN INJURY

The neurobehavioral consequences of TBI fall into three groups: the transient neurobehavioral syndromes, the static encephalopathies, and the delayed neurobehavioral sequelae. The transient syndromes include the postconcussion syndrome (PCS) and the agitated states that occur during recovery from coma. The personality changes that follow injuries to the frontal and temporal lobes are examples of the permanent, fixed sequelae, or static encephalopathies. The delayed neurobehavioral sequelae of TBI are secondary complications. Examples are depression, delayed amnesia, psychosis, epilepsy and dementia.

Transient Neurobehavioral Syndromes

Postconcussion Syndrome

Postconcussion syndrome is a transient condition that follows a relatively mild blow to the head or a severe whiplash injury. There may be only a few seconds of LOC or a few minutes of posttraumatic confusion. The symptoms of PCS include headache, dizziness, fatigue, hypacusis, photosensitivity, irritability, emotional lability, insomnia, and alcohol intolerance. Although the symptoms are temporary, they may persist for 12, 18, or 24 months. Although the symptoms may be mild, they are by no means trivial because they can compromise the patient's family life and ability to work.

The vast majority (75% to 90%) of TBIs are, in fact, mild head injuries (LOC not exceeding 20 minutes), and the majority of mild TBI cases reporting to emergency rooms for care display symptoms of PCS: 59% at the time of discharge from the emergency room, 66% at 3 months later, 50% at 6 months, and 46% at 12 months (Alves et al., 1986). (The increase in symptoms from time zero to 3 months later is dealt with below.)

Postconcussion syndrome, then, is the most common TBI syndrome, and the most frequent TBI-related condition in primary care practice. Patients with PCS are often referred to psychiatrists because the symptoms are persistent and do not respond to simple reassurance. The conventional neurologic examination and the CT scan are normal, so the condition is believed to be "functional" or "psychogenic." In fact, most PCS patients are discovered to have subtle neurologic deficits if one performs a sufficiently sensitive clinical examination for cortical dysfunction (Gualtieri, 1990c) or administers a series of neuropsychologic tests of memory, attention, reaction time, and speed of mental processing (Jakobsen et al., 1987; Levin, 1985).

The basic approach to PCS is symptomatic and supportive. Counseling is the central element: educating the patient and the family about the course of PCS; helping them deal with legal and compensation issues; and discussing problems such as irritability, social withdrawal, hyposexuality, job failure, and dismay over the peculiar neurologic symptoms that occur from time to time.

Psychotropic drugs should be avoided, as a rule, early in the course of PCS, because problems like insomnia and depression frequently remit spontaneously and because psychotropic drugs may obscure the clinical picture at a time when careful, serial neuropsychologic evaluation is central to treatment and prognosis. In particularly difficult cases, or when symptoms persist for 6 months or longer, one may wish to consider trazodone for insomnia (Gualtieri, 1990c), stimulants for memory and attention problems and for chronic fatigue and anergia (Gualtieri & Evans, 1988), buspirone for akathisia (Gualtieri, 1991a), and fluoxetine for depression, anxiety, or posttraumatic stress disorder (Cassidy, 1989). There was a time when the prescription of benzodiazepines was discouraged in this class of patients because of sedation, depression, disinhibition, or dependence (Gualtieri 1988), but the use of low-dose lorazepam or clonazepam, for the usual indications, occasionally is beneficial. The typical doses of various psychotropics for PCS patients are shown in Table 22-1. They are, in general, within the usual ranges for treatment of primary psychiatric disorders.

The late development of major depression is probably the most important psychiatric consequence of PCS (see below).

TABLE 22–1. *Psychotropic Drug Doses for PCS Symptoms*

Methylphenidate	10–30 mg/day bid or tid
Dextroamphetamine	5–15 mg/day bid or tid
Trazodone	50–150 mg qhs
Fluoxetine	20 mg/day
Buspirone	5–10 mg tid
Lorazepam	1 mg bid prn
Clonazepam	0.5–1 mg bid prn

Agitation during Coma Recovery

When the victim of a severe TBI emerges from coma, there is an extremely unstable period when the patient is confused and disoriented; this is usually accompanied by behavioral and emotional instability. The patient's level of arousal is low, although he may appear to be hypervigilant, intense, and hyperreactive to environmental stimuli. The attention span is short and distractibility is high. There may be a dense anterograde amnesia. Stereotyped motor behaviors are common, including elements of the Klüver-Bucy syndrome (Gerstenbrand, Aichner, & Saltuari, 1983; Gualtieri, 1990c). Emotional lability is the rule. There may be intense reactions to people and to therapeutic interventions, with hostility, assaultiveness, and violent epithets. If the patient is ambulatory, he may wander with no regard to personal safety or to the norms of decency. The behavior is disorganized, and the most common element of affective expression is agitation (Gualtieri, 1990c).

This condition is a phase of the recovery process. It does not call for a heavy therapeutic hand—for example, neuroleptic drugs. The patient ideally is treated in a setting with environmental or behavioral intervention, rather than controlled or suppressed with sedating drugs or restraints. Principles of specialized treatment include maintaining a small, familiar staff, limiting stimulation, restricting therapy time to what the patient can tolerate, and waiting for the condition to pass.

Drug treatment should be guided by the presumption that the state of agitation represents a general failure of arousal and mental processing. Therefore, drugs that sedate a patient or reduce his clarity of thought are relatively contraindicated. These include the sedating anticonvulsants, antispasticity drugs, benzodiazepines, metoclopramide, and the neuroleptics.

If the problem of postcoma agitation is severe and prolonged, and if the problem cannot be mitigated by behavior management or withdrawing unnecessary medications, pharmacologic intervention may be necessary. The neuropsychiatrist has a number of alternatives to choose among: the psychotropic anticonvulsants (carbamazepine and valproate), lithium, and beta-blockers all have been successful in some cases (Gualtieri, 1988). None has achieved priority, however, since patient response is unpredictable, and it may not be possible to explore each of the possible alternatives. There has been recent research, however, to support an additional alternative: the dopamine agonists, especially amantadine. The rate of success for amantadine in one case series was greater than 50% (Gualtieri et al., 1989). The rationale of treatment with direct agonists such as amantadine and bromocriptine is to restore the tonic monoaminergic stimulation from brainstem structures that are disrupted in CHI by the process of diffuse axonal anjury (Gualtieri et al., 1989). Typical doses of psychotropics for postcoma agitation are listed in Table 22-2.

Acute agitation in TBI patients in general hospital intensive care unit settings often requires urgent treatment to prevent injury. In this situation, serial trials of dopamine agonists and other nonsedating drugs may not be feasible. In such cases, neuroleptics and benzodiazepines are the main options, and are used as in the treatment of any agitated delirium (see Chapters 12 and 19). However, since neuroleptics and benzodiazepines have negative effects on cognition and motor ability, they are not appropriate for long-term treatment. Furthermore, there is one preclinical study that suggests that neuroleptics may retard the course of recovery from brain injury (Feeney, Gonzales, & Law, 1982). Therefore, strong consideration should be given to tapering and discontinuing neuroleptics or benzodiazepines once the crisis is over, and to exploring the nonsedating alternatives if agitation recurs and interferes with rehabilitation.

Static Neurobehavioral Syndromes

There is an extensive literature that describes the personality changes associated with frontal and temporal lobe lesions (e.g., Lezak, 1983; Lishman, 1987). Closed head injury patients frequently display secondary personality changes because of the frequency with which the frontal and temporal cortex are contused by acceleration-deceleration injuries. Obviously, patients with established frontal or temporal lobe injury are referred for psychiatric care. More often, however, patients with subtle elements of the frontal and temporal lobe syndromes are referred to psychiatrists with little suspicion of organic etiology because the neurologic examination is "nonfocal" and the CT scan is normal.

Lesions of the frontal convexity typically are associated with abulia, lack of motivation and initiative, anergia, failure to plan, and inability to execute complex behavioral programs such as getting up in the morning, getting dressed, having breakfast, and going to work. This clinical presentation may be misinterpreted as an anergic depression or as an "inadequate personality."

TABLE 22–2. *Typical Psychotropic Drug Doses for Postcoma Agitation*

Drug	Dose
Amantadine	100–400 mg/day bid or tid
Bromocriptine	5–10 mg tid
Haloperidol	1–2 mg bid
Carbamazepine	Blood level 4–12 mEq/liter
Valproic acid	Blood level 50–150 mEq/liter
Lithium	Blood level 0.6–1.2 mEq/liter

Something that a psychiatrist may refer to as "denial" or "lack of insight" is the phoenomenon of anosognosia, usually associated with lesions of the right frontal lobe or the right parietal lobe (Prigatano & Schacter, 1991). Anosognosia is the inability to appreciate the nature of one's impairment, or even comprehend that one is impaired. The typical presentation is an unconcerned patient with an intensely concerned family.

Lesions of the orbitomedial surface of the frontal lobes typically are associated with inattention and distractibility, excitability, impulsiveness, disinhibition, and socially inappropriate behavior. It is not difficult for a psychiatrist to interpret this syndrome as sociopathic behavior or as attention-deficit hyperactivity disorder.

The personality changes associated with temporal lobe epilepsy have been dealt with extensively in the psychiatric literature (Bear & Fedio, 1977); the same attributes may also be seen with temporal lobe lesions that do not give rise to overt seizures (Barnhill & Gualtieri, 1989). In addition to the typical "interictal personality," there are other personality variants associated with temporal lobe syndromes. Somatic preoccupations, even frank hypochondriasis, are well-established symptoms of temporal lobe disease (Lezak, 1983).

Emotional instability is a central element of some temporal lobe syndromes. Temporal lobe patients are prone to the full range of anxiety disorders; to extreme emotionalism, or pseudobulbar affect (which may resemble an agitated depression); to explosive rage attacks; and to affective and schizophreniform psychoses (Delgado-Escueta, Mattson, & King, 1981; Flor-Henry, 1979; House et al., 1989; Krauthammer & Klerman, 1978).

Akathisia may be caused by trauma to the basal ganglia. Psychiatrists are familiar with the problem of missing the diagnosis of akathisia in patients treated with neuroleptic drugs. They may not be aware that it also can be a syndrome associated with TBI (Gualtieri, 1991).

The diffuse effects of CHI may exaggerate preexisting personality traits. In this sense, it is similar to early dementia, which also may exaggerate premorbid traits. A temperamental person may develop severe moodswings; a shy person might become increasingly withdrawn; a lazy person may now be simply impossible to motivate.

The reader is referred to the literature in this area, where treatment issues are dealt with extensively (Gualtieri, 1990c; Levin, Benton & Grossman, 1982; Levin, Eisenberg & Benton, 1989; Prigatano & Schachter, 1991). Assessment and management of frontal lobe syndromes was recently reviewed by Fogel and Eslinger (1991).

Delayed Neurobehavioral Sequelae

There is variation in the onset of symptoms following mild TBI (Alves et al., 1986) and severe TBI (Gualtieri & Cox, 1991). Not all of the behavioral syndromes of TBI begin immediately with the recovery of consciousness. This is an important point, because delayed-onset syndromes are often misdiagnosed.

Delayed Symptoms of Postconcussion Syndrome

Most of the time, symptoms of PCS develop soon after the injury, within the first day or so. In some patients, however, PCS symptoms such as headache, memory loss, inattention, or hyposexuality may not arise until several days or weeks have passed. Such an event commonly is taken as evidence for a "psychogenic" etiology, but it is equally possible that some cases of delayed PCS are the result of delayed reporting or delayed appreciation. Patients who are seen in emergency rooms with mild TBI are sent home with advice to look for signs of acute neurologic deterioration, but very few are told about the subtle symptoms of PCS. It may take the patient a while to realize that he does not retain the information he reads, or that he forgets things, or that his sexual drive has diminished.

There is also a neuropathic basis for delayed onset of behavioral sequelae. The clinical effects of diffuse anoxal injury may not be manifest for days or even weeks following TBI (Alves et al., 1986). The theory of diaschisis or "late functional depression" following brain injury is based on experiments in animals with cortical injuries, with the demonstration of late-onset hypometabolism in cerebellum, red nucleus, and locus ceruleus (Feeney & Sutton, 1988). The hypothesis of "late functional depression" is consistent with the descriptions of deficits in monoamine metabolism following TBI. Altered levels of catecholamines, for example, may result from damage to catecholaminergic neurons, with a shift from neurotransmitter production to protein synthesis for repair (Feeney & Sutton, 1988; Feeney et al., 1985). The gradual evolution of functional depression occurs over a span of weeks following TBI.

"Functional depression" of monoamine systems may account for reduced arousal, inattention, difficulty concentrating, memory lapses, fatigue, anergia, comprehension difficulties, lack of flexibility, extreme sensitivity to environmental stressors like bright light and loud noises, and irritability. Whatever the pathophysiology, the symptoms of PCS tend to be reported with greater frequency after weeks have passed than immediately after the injury (Alves et al., 1986).

Posttraumatic Depression

Depression, in the psychiatric sense, is another delayed-onset syndrome. After mild TBI, depression develops as the persistent symptoms of PCS exhaust the patient's emotional reserve, or, possibly, his monoaminergic reserve. After moderate and severe TBI, the psychologic and neuropathic processes are more complex. Mood disorders arise more frequently, and they are, as a rule, more severe.

In one study, the prevalence of depressive symptoms in patients who sustained mild head injury was 34% to 39% (Levin et al., 1987b). In another study, more than half of the patients with TBI complained of depressive symptoms at 3, 6, and 12 months following the injury (McKinlay et al., 1981). The rate is higher for patients with moderate or severe head injuries, and the severity of the injury is directly related to the severity and the prevalence of residual mood disorder (Levin & Grossman, 1978). Posttraumatic depression can be severe and long lasting. In a study of wartime veterans with TBI followed over 18 years, the cumulative rate of suicide was no less than 14% (Vauhkonen, 1959).

These are very high rates, especially when they are compared to the lifetime prevalence rates for the general population, about 5% for major depressive disorder and about 3% for dysthymic disorder (Regier et al., 1988). Traumatic brain injury, like all of the major CNS insults of adult life, carries a significant increase in the probability of occurrence of depression. Depression rates among TBI patients are similar to those for stroke (Robinson, Starr & Price, 1984; Robinson & Szetala, 1981), Parkinson disease (Mayeux et al., 1981), multiple sclerosis (Joffe et al., 1987), and dementia (Zubenko, Moossy, & Kopp, 1990).

The comparison of posttraumatic depression to depression after stroke is particularly interesting, since both represent sudden and catastrophic events occurring in an otherwise healthy brain, and the course of depression is not aggravated by the ongoing effects of a neurodegenerative disease. As with TBI, there is an increase in the prevalence of depression even in the second year after stroke (Robinson, Starr, & Price, 1984). The paradigm that may be relevant to both conditions is Feeney et al.'s (1985) hypothesis of late functional depression.

There may be a difference, however, in terms of treatment response. Mood disorders associated with stroke tend to be prolonged and difficult to treat (Robinson et al., 1983). As a general rule, patients with depression and neurologic disease are less likely to respond favorably to antidepressants than depressives who do not have neurologic disease (Berrios & Samuel, 1987). Posttraumatic depression, however, may have a better prognosis. There are no controlled studies, but psychiatrists who specialize in the treatment of TBI patients tend to be confidant about the likelihood of a favorable response to conventional antidepressants (Cassidy, 1989; Silver, Yodofsky, & Hales, 1991).

Deficits in monoaminergic neurotransmission that occur in brainstem nuclei following TBI are known to be amenable to treatment with simple monoaminergic drugs, such as the psychostimulants (Feeney & Sutton, 1988). The stimulants are known to enhance recovery from experimentally-induced brain injury (Feeney, Gonzales, & Law, 1982), and double-blind studies have shown favorable effects of stimulants on depressive symptoms and cognitive function in TBI patients (Gualtieri & Evans, 1988). Drugs that enhance monoaminergic neurotransmission, then, such as the psychostimulants and the antidepressants, might represent rational pharmacotherapy for posttraumatic depression (Feeney, Gonzales, & Law, 1982; Gualtieri, 1988). There are few controlled studies, however.

Delayed Amnesia

The other delayed-onset syndrome known to arise in the first 2 years after TBI is delayed amnesia (Gualtieri & Cox, 1991). Memory disturbance is one of the most prominent residual effects after CHI, and the frequency ranges from 23% to 79% (Levin, 1990). There is, however, remarkable variation in individual rates of recovery of memory functions following TBI. In one study based on patients enrolled in the National Coma Data Bank, memory performance recovered to a normal level of performance in about 50% of patients; 18% showed a flat or a minimal trajectory of recovery; and the remaining patients showed performance that improved initially but then deteriorated in the second year after TBI (Ruff et al., 1990a). Clinical observations of anxiety and tension were notable in the "peak-drop" group, but it is not likely that the decrement in performance was entirely explicable in terms of emotional distress or mood disorder.

It usually is assumed that memory deficits will recover, at least to a degree, after TBI, and various psychologic treatments and medications have been brought to bear to effect a more efficient return. The TBI victim, therefore, may hope to achieve at least some degree of improvement in memory function or, if an impairment is resistant to therapy, the adoption of specific compensatory strategies.

What the TBI victim does not usually consider, and his doctors do not either, is the possibility of delayed deterioration in memory. This new concept is developing as an entirely novel idea, that some TBI patients actually lose cognitive function, especially memory, in the second year following injury. This appears to be a specific effect, not a dementing condition, and it is not the function of some other delayed sequela like depression, psychosis, or epilepsy. It would seem that serial memory assessment over the first 2 or 3 years is a necessary measure for patients who have sustained a severe TBI. The cause of delayed amnesia is not known.

Posttraumatic Epilepsy

Posttraumatic epilepsy is another delayed sequela of TBI that is pertinent to psychiatric practice. Sudden changes in mood, explosive rage attacks, attentional lapses, and personality disorganization may, in fact, be signs of a seizure disorder.

The lifetime prevalence of epilepsy after CHI is estimated to be 2% (Annegers et al., 1980) to 5% (Jennett, 1979). Severe CHIs may have a higher rate: 7.1% at 1 year and 11.5% after 5 years in one study (Annegers et al., 1980); and 7% at 1 year and 10% after 2 years in another (McQueen et al., 1983). Although most cases arise in the first year following TBI, the risk of posttraumatic epilepsy continues for many years, especially in patients who have had penetrating injuries. For example, the onset of posttraumatic epilepsy in Korean war veterans was 57% in the first year, 18% between 1 and 5 years, and 7% between 5 and 10 years (Salazar et al., 1985).

The relatively high rates of posttraumatic epilepsy among wartime victims who had penetrating head injuries and among patients who have had neurosurgical procedures gave rise to the venerable custom of "anticonvulsant prophylaxis," an idea that has lost its value in the present era of CHI treatment. Although the prescribing habits of many practitioners appear to ignore the results of clinical studies, it does not appear that anticonvulsant prophylaxis is likely to diminish the likelihood of posttraumatic epilepsy (McQueen at al., 1983; Salazar et al., 1985; Young et al., 1983). This prescribing practice may cause depression or neuropsychologic deficits in convalescent TBI patients as a result of the side effects of "prophylactic" anticonvulsants. When a TBI patient on "prophylactic" phenytoin or phenobarbital presents with depression, it is usually a good idea to reconsider the necessity of anticonvulsant treatment (Temkin et al., 1990).

Late-Onset Psychosis

Not all of the psychotic disorders associated with TBI are delayed in onset. A mild CHI can precipitate acute psychosis, manic or schizophreniform, absent any focal neurologic findings. The patient usually is confused and disoriented. This acute psychosis, therefore, is similar to the agitated state that often accompanies coma recovery in patients with severe TBI. Typical features include hallucinations, poorly systematized delusions, misidentification of family members and medical personnel, agitation, combativeness, and a fluctuating course. Most of these patients meet diagnostic criteria for delirium, but others simply have a brief reactive psychosis. Patients usually require antipsychotic drugs acutely for control, but efforts should be made to taper and discontinue the medication after a few weeks, since most patients recovery rapidly (Cummings, 1985).

It is well known that virtually all of the variants of schizophrenia and the affective psychoses may arise following TBI, and that a period of years may elapse between the time of injury and the first development of psychotic symptoms (Krauthammer & Klerman, 1978; Lishman, 1987). Although TBI patients with preinjury histories of psychosis are probably at greater risk, there is no evidence that preexisting schizoid traits are a risk factor for posttraumatic psychosis (Thomsen 1984, 1987).

Some cases of posttraumatic psychosis resemble the schizophreniform psychosis of temporal lobe epilepsy (Slater, Beard, & Glithero, 1983). Psychosis may occur in patients with trauma to the temporal lobes who have never had a seizure (Barnhill & Gualtieri, 1989). The phenomenon of limbic kindling may be relevant to the pathophysiology of both conditions.

There is not much new information about the proper treatment of posttraumatic psychosis, although most clinicians rely on the psychotropic anticonvulsants, carbamazepine and valproate, as "maintenance drugs," augmented by neuroleptics during periods of extreme disorganization or disabling positive symptoms. There is a recent report that buspirone may augment the psychotropic effects of carbamazepine and valproate in patients with temporal lobe disease (Gualtieri, 1991b).

Dementia

There are only three established risk factors to contribute to the development of Alzheimer's disease: a family history of Alzheimer disease or of Down's syndrome, or a history of severe TBI earlier in life (Amaducci & Lippi, 1988). The odds ratio—that is, the degree to which TBI will increase one's risk of developing a dementing condition—has been estimated at 4.5 (French et al., 1985) or 5.3 (Heyman et al., 1984). Henderson (1986) reported that 15% of Alzheimer's disease patients, but only 3.8% of controls, had a previous history of TBI. The TBI patients who go on to develop dementia are more likely to have had a severe injury, or more than one head injury, or a concurrent illness such as alcoholism or atherosclerosis (Violon & Demol, 1987). A careful history for early severe TBI should be part of the dementia workup, especially in cases of presenile dementia. The mechanism of the association between TBI and subsequent dementia is unknown.

Older TBI patients also are more likely to develop a posttraumatic dementia. The confusion, amnesia, and disorientation that accompany the acute recovery period may be more severe and last longer than in younger patients. The trajectory of recovery is flat, or even downward, and the clinical picture grows increasingly to resemble that of Alzheimer disease.

ASSESSMENT

The delineation of a lesion with a standard neurologic examination or with direct imaging techniques such as CT scanning and MRI is a straightforward and simple way to establish an organic basis for the complaints of a CHI patient. Nevertheless, there are many patients whose symptoms are severe and debilitating but whose examination and scans are normal or equivocal. It is increasingly appreciated that functional imaging methods such as PET, SPECT, and computerized EEG brain mapping may define areas of abnormality not seen on anatomic studies. Further development of functional imaging techniques may eventually resolve a number of common diagnostic problems. Until these techniques are better validated, however, the physician must rely on the clinical history, the mental status examination, and neuropsychologic testing to infer the locus and extent of brain injury, particularly in cases of mild CHI.

The key element of the clinical history is the description of a traumatic event; whether the patient lost consciousness, and if so, for how long; the depth of coma, defined by the Glasgow Coma Scale; the occurrence and the duration of posttraumatic amnesia and posttraumatic confusion; and the temporal course of symptom onset after the injury.

Acquiring the history of neurobehavioral sequelae requires familiarity with the specific syndromes associated with CHI. The CHI patient will seldom report all symptoms and deficits directly; they have to be drawn out by concise and pertinent questioning. Questionnaires and rating scales, such as the Neurobehavioral Rating Scale developed by Levin, Benton, and Grossman (1982) and the Rebound Neurobehavioral Rating Scale (Gualtieri, 1990c, 1991) (Table 22-3), are available to complement the clinical history.

It also is necessary to get information about the patient from family members, from people who have known the patient before the injury, and, sometimes, from school transcripts, standardized tests (e.g., California Achievement Test or the Scholastic Aptitude Test), employment records, or previous medical records. Anosognosia and amnesia are TBI sequelae that will inevitably compromise the patient's accuracy as a historian. Current disability should not be attributed to a specific traumatic event unless one is quite sure that there was no preexisting morbid condition.

A structured mental status examination, such as the Mini-Mental State Examination (Folstein, Folstein, & McHugh, 1975), the Neurobehavioral Cognitive Assessment, and the mental status examination (MSE) developed by Strub and Black (1985), are suitable for patients with moderate to severe deficits, but they are not, as a rule, sufficiently sensitive to pick up deficits in a mildly impaired individual. The psychiatrist examining a CHI patient should complement the MSE with tests that capture subtle injury to frontal and temporal lobe structures. These include tests of auditory and visual memory; measures of verbal, semantic, and figural fluency; and tasks requiring imitation and performance in complex motor tasks (what is referred to, in the Luria examination, as "kinetic melody"). One examines the patient for reflex activity suggestive of "frontal release": the palmomental, Hoffman's, grasp, snout, suck, and corneal-mandibular reflexes and the glabellar tap. One also looks for signs of ataxia or nystagmus, problems with balance, and indications of subtle motor weakness. As a rule, one emphasizes the mental and motor examinations; since the lesions of mild CHI are most likely to be frontal and temporal, an extensive sensory examination probably is not going to be revealing. (See Chapter 18 for further discussion of the neurologic examination.)

The office or bedside MSE, even when augmented by additional tests of frontal and temporal lobe function, still lacks tests for reaction time and processing speed impairments, two keys areas in CHI-induced impairment. Personal computer–based assessment systems show promise in addressing this deficit in assessment (Baker, Letz, & Fidler, 1985). An advantage of a computerized office-based cognitive assessment is that it can be administered serially, to measure recovery after TBI, and also to measure the effects of drug administration. However, the computerized battery is no substitute for an experienced neuropsychologist. It is an assessment tool, not a diagnostic instrument.

A complete neuropsychologic battery given by a neuropsychologist who has experience with TBI patients is probably the most sensitive diagnostic instrument, but it is arduous and expensive. There are several different approaches to the problem of neuropsychologic assessment, and many good neuropsychologists have adopted an eclectic style that includes individual tests from several different batteries. (See Chapter 18.) The principal roles of formal neuropsychologic testing in TBI patients is to delineate specific and sometimes subtle areas of functional weakness, to help in planning specific treatment strategies, and to assess patients' courses and response to treatment.

The diagnosis of clinically significant TBI is not based on any one of the elements described above. It is based on a consistent clinical picture drawn from all of the sources, with the necessary skepticism concerning the importance of a single, isolated finding. The clinical utility, for example, of frontal release signs is limited in an elderly patient; neurologic "soft signs" such as dysdiadochokinesia are of limited utility in young patients or patients with developmental disabilities. The significance of many MRI findings may be ambiguous. Even the description of the patient's present problems may be confused, or contradictory, since people may minimize or exaggerate a symptom, or they may make attributions about a problem that color its actual relation to a traumatic event. The Minnesota Multiphasic Personality Inventory (MMPI) may indicate "hypochondriasis," but the patient with chronic pain and disability often exaggerates symptoms, especially if physicians are skeptical or uncaring.

In general, the clinical evaluation of mild TBI does not require expensive neurodiagnostic procedures or comprehensive neuropsychologic testing. However, if the issue is litigation, one may be urged to spare no expense to identify an "objective" neuropathic finding. Professionals can be pulled unwittingly into the adversarial process, and there are inevitable pressures to put "spin" on equivocal findings that will benefit the plaintiff or the defense.

If the purpose is disability determination, the process may be equally daunting, since there are no objective criteria to measure the degree to which mild symptoms such as fatigue, hyperacusis, and motor incoordination will affect the performance of a construction worker; or the degree to which frontal lobe symptoms such as disorganization, inattention, difficulty planning, and in-

TABLE 22–3. *The Rebound Neurobehavioral Rating Scale*

Name: ____________________ Date: ____________________

Rater: ____________________ Position: ____________________

Current Medications: ____________________

Consider the individual's behavior in settings you are familiar with, over the past week. Circle the appropriate number for each item. Answer all items.

0-N/A —not a functional deficit or a problem
1-Mild —behavior inappropriate, does not interrupt treatment or prevent community/family activities, can be redirected
2-Moderate—behavior interrupts activities, requires intervention, not potentially harmful, requires close supervision
3-Severe —behavior interrupts all activities and is potentially harmful to self, others, or property, requires close supervision

	N/A	Mild	Moderate	Severe	
COGNITIVE					
1. Disoriented—confused about person, place, or time	0	1	2	3	
2. Short attention span, distractable	0	1	2	3	
3. Low arousal, hard to arouse	0	1	2	3	
4. Memory deficit	0	1	2	3	
5. Disorganized thinking—confused, disconnected, loose, tangential	0	1	2	3	
					______
PERFORMANCE					
6. Speech articulation deficit	0	1	2	3	
7. Expressive language deficit	0	1	2	3	
8. Comprehension deficit	0	1	2	3	
9. Motor coordination deficit	0	1	2	3	
10. Low energy level, easily fatigued	0	1	2	3	
					______
SOCIAL INTERACTIONS					
11. Withdrawn, socially isolated	0	1	2	3	
12. Noncompliant, uncooperative	0	1	2	3	
13. Hostile	0	1	2	3	
14. Disinhibited—inappropriate comments or actions	0	1	2	3	
EMOTIONAL STATE					
15. Decreased initiative/motivation	0	1	2	3	
16. Depressed mood	0	1	2	3	
17. Anxiety/tension	0	1	2	3	
18. Excitable	0	1	2	3	
19. Lability of mood	0	1	2	3	
20. Emotional incontinence	0	1	2	3	
21. Irritability	0	1	2	3	
22. Low frustration tolerance	0	1	2	3	
23. Agitation	0	1	2	3	
					______
BEHAVIOR					
24. Restless, overly active, fidgety	0	1	2	3	
25. Impulsive	0	1	2	3	
26. Temper outbursts, explosive behavior	0	1	2	3	
27. Destructive	0	1	2	3	
28. Aggressive	0	1	2	3	

Total Score ______

Comments:

One of several rating scales useful for serial patient evaluation, for example in drug studies (Gualtieri, 1990c, 1991).

decision will compromise the function of an executive or a professional.

Assessment of disability following TBI is best made over time, with serial assessment by the treating physician. The patient's difficulties should be assessed in terms of real-world performance at home, in school, at work, or at play. Disability cannot be established by neuropsychologic tests. The validity of tests in predicting social and occupational performance has not been very well established. Consulting professionals should give due weight to the observations of the people who live and work with the patient. The sequelae of frontal lobe lesions, in particular, may be more evident in real life than in the artificial setting of the professional's office.

TREATMENT

Patients with mild TBI should be expected to make a complete recovery. Patients who have had moderate or severe injuries can usually be returned to a useful, independent life, sometimes in competitive employment. With the proper development of group homes and sheltered workshops, even patients who had profound injuries and have major residual handicaps can be happy and productive citizens. Treatment for TBI patients is a gloomy prospect only in communities where appropriate services have not yet been developed. Treatment of TBI is, in fact, one of the most encouraging fields in neuropsychiatry.

Rehabilitation

Rehabilitation is one of the most active fields in modern medicine, and physiatry (physical medicine and rehabilitation) is one of the fastest growing specialties. The emphasis, though, is on physical medicine, so the problems of rehabilitation for the TBI patient often fall to the neuropsychologist, the rehabilitation neurologist, or the psychiatrist.

Rehabilitation in specialized centers is carried out by a team of master's-level professionals in physical therapy, occupation therapy, speech and language therapy, therapeutic recreation, and vocational rehabilitation. The team's efforts are coordinated by a comprehensive individualized rehabilitation plan, or IRP. Physicians and psychologists may lead the interdisciplinary team, may consult to it, or they may be team members. Usually, there is a nonphysician case manager who oversees the management of the plan, monitors outcome, and endeavors to maintain funding.

The advantage of specialized residential rehabilitation is that various services can be delivered in integrated fashion. At best, therapeutic results are synergistic. However, if residential services are delivered far from the patient's home, there may be a problem with community reintegration. If there are no suitable follow-up therapies, functional gains may be lost. Also, since TBI recovery is not linear, a heavy investment in residential rehabilitation may be expended while the patient is on a "plateau," that is, a period of limited recovery. When the patient enters a more active recovery phase, months later, an important opportunity for improvement may be wasted for lack of appropriate therapies.

The solution to the problem will be the development of community-based rehabilitation and outpatient treatment opportunities for TBI patients. This is a natural evolution, from inpatient to community-based service, and it will lead to substantial cost savings and better services for a much larger population of patients. The need for specialized residential rehabilitation will continue, of course, for extremely difficult patients with multiple problems, for patients from areas of the country where there are no outpatient services, and especially for TBI patients who have severe neuropsychiatric problems, such as psychosis, aggression, disinhibition, and socially inappropriate behavior.

Virtually all of the residential rehabilitation programs will present data that demonstrate good outcomes for their patients. Virtually none of these data, however, have been published in peer-reviewed journals. There are no controlled studies to demonstrate a clear advantage to residential treatment compared to comprehensive outpatient treatment, using even the simplest elements of a research design, such as random assignment of subjects, control for level of severity, or independent ratings of outcome. In the absence of validated data, the decision to recommend residential rehabilitation must be made on the basis of an individualized estimate of the costs versus the anticipated benefits.

Psychologic Therapies

For the PCS patient, the most important aspect of counseling is to educate the patient about the recovery process, to alleviate the transient adjustment problems that arise, and to advocate for an appropriate resolution of issues surrounding disability and compensation.

For the patients with persistent neurologic deficits, the goal of counseling is to help the patient adjust to a life of diminished capacity, to acquire compensatory strategies, and to develop a positive, optimistic world view. The patient with posttraumatic amnesia needs strategies to cope with this deficit. The patient with persistent behavioral or emotional instability must learn new strategies for self-control and redirection. For the patients and for their families, there is all of the necessary grief work, grappling with outrageous fortune, picking up the pieces, and carrying on.

For the coma recovery patient who is agitated, confused, and sometimes psychotic, the family needs an extraordinary level of support through a truly monstrous problem. In the face of a family member whose behavior is incomprehensible, and an outcome that looks dire, families may react sharply and behave poorly. Family therapy frequently is needed. The fact that some of the risk factors for TBI (e.g., alcohol abuse) may also be associated with dysfunctional families increases the need for this treatment modality. Nurses and other staff benefit from training in dealing with the informational needs and the emotional reactions of family members who come to visit the patient.

Counseling the TBI patient, at any stage, requires an appreciation of the trajectory of recovery, because that is what the therapist must interpret to the patient and family. The therapist must define a safe therapeutic environment and an appropriate rehabilitation program.

"Cognitive remediation" is a relatively new and unproven therapy that trains TBI patients in repetitive but incremental tasks in areas of their neuropsychologic deficit, such as memory. It is usually administered by computer. Patients who are treated several times a week tend to improve on the tasks the computer gives them. The improvements are not necessarily sustained, however, after the treatment is ended, and the benefits derived are not necessarily generalizable to other, more meaningful areas of the patient's life.

The emphasis in rehabilitation for TBI patients is increasingly oriented toward the development of real-world skills that will allow the patient to live independently, to get along with people, and to work at a meaningful job. The author's preference is always to get the patient home and back to work as soon as possible, and then deal with problems as they arise.

The families of TBI patients are in a unique position, because they must contend with a medical catastrophe that may be as devastating as cancer, dementia, or acquired immunodeficiency syndrome. The difference is, the condition is not terminal. Sooner or later, they must confront the reality of living with someone who is different from before, or even a mere shadow of his former self, and someone with whom it may be extremely hard to get along. The only experience that is even remotely similar is the birth of a retarded child. However, the fact of sudden change from good health to severe disability, the months of medical treatment, hopes for recovery that wither as the years pass—all lend a unique air of tragedy to the TBI family. The passage of time tends to diminish a family's sympathy for a disabled patient, or at least its energy to care for the patient's extraordinary needs.

There have been good articles published on the professional care of families of TBI victims (Brooks, 1991). There is also a growing movement across the country to develop family support groups, advocacy groups, and self-help groups. To a very large degree, this is the result of the work of the National Head Injury Foundation (NHIF). The requirements of TBI victims, and their families, are not dissimilar to the requirements of developmentally disabled people and their families, and the NHIF has borrowed some of the methods that were pioneered by the Association for Retarded Citizens. The NHIF, however, is wrestling with a double loyalty. It is a consumer-based group at heart, but it also functions as a trade association for the TBI industry, and there may be a conflict therein. The constituent Head Injury Foundations of the individual states seem to be more inclined toward social advocacy, group homes, sheltered employment, and community-based treatment facilities.

NEUROPSYCHOPHARMACOLOGY

Effects of Neuroactive Drugs on Recovery

In TBI, the guiding principle in drug treatment is concern over the effect of neuroactive drugs on the course or trajectory of recovery. Traumatic brain injury patients commonly are treated with psychoactive drugs, anticonvulsants, and antispasticity drugs. In one survey, the average TBI patient admitted to a rehabilitation center was taking no fewer than 2.5 neuroactive drugs (Gualtieri, 1990a). The first responsibility of the psychiatric consultant, then, is to evaluate the medications a TBI patient is taking, and how they may be affecting the recovery process. Is the drug necessary? Does it have behavioral toxicity or negative effects on cognition and motor performance? Is there an alternative that is potentially less toxic? It is a prevailing belief that TBI patients, like elderly patients and very young children, have a low threshold to the encephalopathic effects of neuroactive drugs, and especially drug combinations (Gualtieri, 1988). Like developmentally handicapped people, their rehabilitation can be compromised by ill-chosen psychopharmacology.

Psychopharmacology for TBI patients is guided also by the principle that cerebral recovery is a very complicated process, and the effect of a centrally active drug on the process may be hard to predict. Drug interventions, therefore, are like small experiments; they should be guided by a trial-and-error approach, with frequent follow-up visits and careful measurement of the results of drug treatment. A psychotropic drug may control a particular symptom, such as aggression or agitation, but if the patient's progress in rehabilitation therapies slows down, the psychiatrist ought to consider an alternative treatment.

There is a belief in the field that certain drugs may actually enhance the recovery process, whereas others may retard the process or derail it entirely (Gualtieri, 1988). In the first instance, the focus of attention is on the monoaminergic drugs, especially the dopamine agonists. The origin of this belief is a seminal paper by Feeney, Gonzales, and Law (1982), who demonstrated that amphetamine accelerated the recovery rate in experimental animals with brain injury, that haloperidol retarded recovery, and that simultaneous treatment with amphetamine and haloperidol tended to undo the negative effects of the latter drug. The idea was developed in a series of controlled studies of the psychostimulant methylphenidate by the author (Evans, Gualtieri, & Amara, 1986; Evans, Gualtieri, & Patterson, 1987; Gualtieri & Evans, 1988) and in two clinical series concerned with the dopamine agonist amantadine (Chandler, Barnhill, & Gualtieri, 1988; Gualtieri et al., 1989). Additional evidence has accrued to support the belief that dopamine agonists (the stimulants bromocriptine, amantadine, L-dopa, and apomorphine) enhance recovery from brain injury in animals (Maeda & Maki, 1986, 1987) and in stroke victims (Crisostomo et al., 1988), and that they may accelerate recovery from coma (Horiguchi, Inami, & Shoda, 1990), hepatic encephalopathy (Morgan et al., 1977), prolonged states of akinetic mutism (Ross & Stewart, 1981), hemispatial neglect (Fleet et al., 1987), and apathy-abulia (Catsman-Berrevoets & Van Harskamp, 1988; Gualtieri et al., 1989).

It has been proposed that dopamine agonists represent a rational (as opposed to symptomatic) treatment for some TBI patients, based on this argument: brainstem injury is usually associated with damage to nuclei that are responsible for monoamine neurotransmission, and deficits in monoamine metabolism have been found to characterize the post-TBI state in patients and in experimental animals (Gualtieri, 1988; Van Woerkum, Teelken, & Minderhoud, 1977; Vecht et al., 1975). The stimulants and the dopamine agonists are a rational treatment for patients who are in a state of persistent hypoarousal, because hypoarousal usually is related to brainstem injury with a persistent deficit in catecholaminergic neurotransmission.

Stimulants are known to improve memory, mood, and motor performance in patients with sequelae of mild TBI (Gualtieri & Evans, 1988). As a general rule, the psychostimulants methylphenidate and dextroamphetamine are helpful for patients whose injuries were comparatively mild; direct agonists such as amantadine and bromocriptine seems to be more effective in severely injured patients (Gualtieri et al., 1989).

Whether the dopamine agonists actually accelerate the recovery process remains an open question. It may be taken as well established, however, that they are effective treatments for many of the clinical problems associated with TBI. In mildly injured patients the stimulants are prescribed for problems such as fatigue, hypersomnolence, lack of endurance, inattention, inability to concentrate, disinhibition, poor memory, and anergic depression. In more severely injured patients, drugs such as amantadine may be prescribed for persistent hypoarousal, abulia, mutism, disinhibition, and agitation during coma recovery. Even if it has not been proven that dopaminergic drugs actually enhance the recovery process, there is no reason not to consider a medication trial for patients who have stalled on a plateau of partial recovery (Gualtieri, 1990c).

The other result of Feeney, Gonzalez, and Law's (1982) seminal article has been the persistent belief that there are centrally acting compounds that may retard the recovery process. Naturally, if dopamine agonists are a rational treatment for TBI recovery, then dopamine blockers might be predicted to exert the opposite effect. Although there is preclinical evidence that they do, there is no direct evidence from clinical studies in TBI patients. Nevertheless, neuroleptics, including metoclopramide, tend to be used very conservatively, if at all, in head injury centers (Gualtieri, 1990b).

Two other commonly prescribed drugs also are considered to be of dubious value for TBI patients. The anticonvulsants phenobarbital and phenytoin are known to exercise negative effects on memory, mood, and motor performance (Temkin et al., 1990), and chronic use of phenytoin may be associated with cerebellar degeneration (Masur et al., 1989), normal-pressure hydrocephalus (Kalanie et al., 1986), or dementia (Reynolds, 1975). If an antiepileptic drug is required, carbamazepine or valproic acid are preferred (Gualtieri, 1990a; Massagli, 1991). However, carbamazepine and valproate are not entirely free of negative behavioral or cognitive effects, and they are not always superior to phenytoin as anticonvulsants (Dodrill & Troupin, 1991).

Problem-Specific Drug Treatment

Psychoactive drug treatment is necessarily oriented toward the alleviation of a specific problem. Problem definition is often sufficient, because the psychiatric categories of the DSM are not always appropriate for the TBI patient. Diagnoses such as "organic (secondary) personality disorder" or "atypical psychosis" do little even to describe the clinical situation, let alone to convey its specific neuropathic origins.

The problems for which a psychopharmaceutical may be appropriate may be defined in several different contexts. Indication for psychoactive drug therapy include:

1. To advance the trajectory of recovery, moving the patient off a plateau.
2. To improve a specific neuropsychologic deficit, such as inattention, anterograde amnesia, or motor incoordination.
3. To alleviate a neurologic symptom, such as central pain or akathisia.
4. To alleviate a functional symptom, such as insomnia, irritability, fatigue, emotional lability, or emotional incontinence.
5. To treat a specific neurobehavioral syndrome, such as the apathy and abulia that arises from lesions to the frontal convexity, the disinhibition and socially inappropriate behavior that arise from an orbitomedial frontal lesion, or the explosive behavior that arises from a temporal lobe lesion.
6. To treat a specific psychiatric disorder, such as major depression (Silver, Yudofsky, & Hales, 1991), secondary mania (Bamrah & Johnson, 1991; Krauthammer & Klerman, 1978), or late-onset psychosis (Barnhill & Gualtieri, 1989).

To treat a TBI patient properly, a wide range of psychopharmacologic interventions are directed against an array of clinical conditions that are not included in the psychiatric nosology. Two recent monographs address the issue in much greater detail (Gualtieri, 1988, 1990c).

Although there are virtually no controlled studies of psychoactive drugs in TBI patients, aside from the stimulant studies we have referred to above, there is a developing consensus among psychiatrists in the field regarding certain drugs. Amitriptyline is an extremely effective treatment for poststroke emotionalism and for pseudobulbar affect associated with TBI (Schiffer, Herndon, & Rudick, 1985), but it has strong anticholinergic properties, and anticholinergic drugs can impair mood, memory, and motor performance (Katz et al., 1985). It is not at all clear why amitriptyline is so effective, or even how it compares to its sister drug, nortriptyline, but it has always been a favorite among neurologists, and the doses they often use (20 to 50 mg/day) are well below those that usually are necessary to treat major depression. The alternative serotonergic antidepressants (e.g., fluoxetine) seem to be equally effective for emotionalism but probably are less likely to produce cognitive toxicity (Allen & Lader, 1989). Whether fluoxetine will compete successfully with the venerable amitriptyline is a matter for practice to decide.

Trazodone is an excellent hypnotic for TBI cases, especially for mild CHI patients with PCS (Gualtieri, 1990b). Traumatic brain injury patients, especially patients with PCS, have a typical kind of insomnia. They almost invariably describe racing thoughts when they lie down and try to sleep. Trazodone is the drug of choice, in doses ranging from 50 to 300 mg or higher. Lithium is said to be an alternative, in low doses (300 or 600 mg/day), with no requirement for determination of therapeutic serum levels unless the patient has renal impairment.

Buspirone can be effective for akathisia associated with PCS, and as an adjunct to carbamazepine or valproate in patients with temporal lobe disease (Gualtieri, 1991). The effective dose usually is relatively low (15 to 30 mg/day) and the latency to respond is short. Headache and light-headedness are two side effects that may cause patients to reject the drug.

Fluoxetine is an excellent antidepressant, with little risk of cognitive toxicity or fatal overdose (Allen & Lader, 1989; Cassidy, 1989). Since it is a serotonergic, it is sometimes useful for aggression and for self-injurious behavior. It requires very close monitoring, though, because of side effects such as akathisia, insomnia, anorexia, suicidality, and hypomania. Patients have heard a lot of "bad things" about fluoxetine and they may not be willing to try it unless their physician is committed to very close supervision.

Naltrexone has been proposed for a variety of indications, especially "organic bulimia" (actually, hyperphagia), but one's clinical success rarely matches that of reported cases in the literature (Childs, 1986).

The gap the neuroleptics have left behind for the treatment of psychosis, agitation, aggression, disorganization, and paranoia is usually met by the psychotropic anticonvulsants (Gualtieri, 1990a), the beta-blockers (Mattes, 1986), clonidine (Bakchine et al., 1989), lithium (Hale & Donaldson, 1982), and the benzodiazepines (Gualtieri, 1990b). The psychotropic anticonvulsants, when they are prescribed for psychiatric disorders, should be in the usual therapeutic blood level range (4 to 12 μg/ml for carbamazepine and 50 to 100 μg/ml for valproate).

The reluctance to prescribe neuroleptics in some quarters amounts to an obsession, just as in other quarters neuroleptic prescription is virtually automatic. There obviously is a role for the neuroleptics in the TBI patient, in judicious doses and usually for only a short period of time. The indications are: emergency treatment of acute agitation, psychotic disorders that do not respond to the alternatives listed above, posttraumatic Tourette's syndrome, and the state of hyperkinesia (resembling Huntington's disease) that occasionally occurs during coma recovery.

Benzodiazepines were, at one time, thought to be relatively contraindicated in TBI cases; because they are sedating, they may have negative effects on neuropsychologic performance, and they are apt to produce paradoxical or disinhibiting effects (Gualtieri, 1988). However, this probably is an overstatement of the downside. There is an appropriate role for benzodiazepines, for the usual psychiatric indications, for at least some TBI patients. Lorazepam is a current favorite, because of its favorable pharmacologic profile, because the intramuscular form is rapidly absorbed, and because it has no active metabolites (see Chapters 6 and 9). Clonazepam is another, probably because it is an anticonvulsant and because it has been recommended for the treatment of bipolar disorder.

Pharmacotherapy of Posttraumatic Epilepsy

The TBI patient with posttraumatic epilepsy, or with a likelihood of developing posttraumatic seizures, represents an important problem for the psychopharmacologist. There are three rules to guide practice in this class of patients:

1. Any behavior disorder that arises in a seizure patient should first be attributed to the seizure disorder itself, or at least to the irritative focus that has given rise to the seizure disorder. The first treatment is to optimize anticonvulsant therapy, either by adjusting the serum level or by taking advantage of the psychotropic effects of carbamazepine and/or valproate.
2. If adjunctive psychopharmacology is required, the clinician should try to use a psychotropic that will enhance seizure control, and try to avoid psychotropics that may lower the seizure threshold.
3. If a psychotropic drug is added to the anticonvulsant regimen, blood levels of the latter must be checked to rule out drug interaction (Fogel, 1988).

The interested reader is referred to a monograph on this subject (Gualtieri, 1990a). The likeliest adjuncts to the antiepileptic drugs are the stimulants and dopamine agonists, the beta-blockers, buspirone, lithium, and, of course, the benzodiazepines, especially clonazepam. Most of the neuroleptics tend to lower the seizure threshold, but fluphenazine, molindone, and pimozide might be exceptions to this rule (Oliver, Luchins & Wyatt, 1982). Data on the antidepressants are not very clear to the author, although, as with the stimulants, low to moderate doses may improve seizure control whereas high doses may lower the threshold. Bupropion is relatively contraindicated in this population.

TRAUMATIC BRAIN INJURY AND SUBSTANCE ABUSE

Substance abuse, particularly of alcohol, is strongly implicated as the cause of most head injuries (Sparadeo, Strauss, & Barth, 1990). Between 29% and 58% of TBI patients with positive blood alcohol levels taken on admission to the emergency department are legally intoxicated, and there is evidence of alcohol addiction in 25% to 68% of these cases (Brismar, Engstrom, & Rydberg, 1982; Field, 1976; Sparadeo & Gill, 1989; Sparadeo, Strauss, & Barth, 1990). Rimel et al. (1982) reported that no fewer than 34% of patients with moderate head injury were problem drinkers before injury.

There also is evidence that alcohol can have a negative effect on the acute stage of the recovery process from TBI. Victims of TBI with positive blood alcohol levels at the time of injury have longer duration of coma and a lower LOC during the acute stage of recovery (Edna, 1982), and they have a longer length of stay in the hospital (Sparadeo & Gill, 1989). There is even evidence to suggest that a high blood alcohol level at the time of injury may have a detrimental effect on the patient's ultimate recovery of neuropsychologic function—for example, of memory (Brooks et al., 1989). Excessive alcohol users have a higher mortality rate after TBI than nonusers or moderate users; they have a higher rate of mass lesions on CT scan; and they have a lower rate of good outcomes after TBI (Ruff et al., 1990b). It is not entirely clear whether these data indicate a direct neurotoxic effect of alcohol, occurring at the time of acute injury, or whether they indicate a premorbid "weakness" in the alcoholic patient that renders him more vulnerable to the direct effects of the injury. Studies of patients who are alcoholics, but who have negative blood levels at the time of injury, are contradictory (Sparadeo, Strauss, & Barth, 1990).

It also appears that resumption of alcohol abuse will retard cognitive recovery (Parsons, 1987). This is important, because about 50% of TBI patients will return to preinjury levels of alcohol consumption within a year of their injury (Sparadeo & Gill, 1989). Patients with physical disabilities as the consequence of TBI are likely to turn to alcohol as a way of alleviating emotional distress (Greer, 1986: Hackler & Tobis, 1983). Furthermore, TBI patients seem to be more sensitive to the intoxicating effects of alcohol than they were prior to the injury (Oddy, Coughlin, & Tyerman, 1985). Alcohol also can interact with other medications that are prescribed to TBI patients (Gualtieri, 1990b).

The treatment of the alcoholic TBI patient certainly is problematic, since there is consensus among rehabilitation professionals that traditional approaches to alcoholism treatment usually are inappropriate for the TBI patient with cognitive impairment, particularly those with lack of insight (anosognosia), poor judgment, and poor safety awareness. Many rehabilitation centers have developed special programs for the patient with a "dual diagnosis," but there is little in the way of systematic research to allow the physician to recommend any particular approach (or any approach at all) (Sparadeo, Strauss, & Barth, 1990). Clearly, if the patient's premorbid drinking habits were symptomatic of a psychiatric disorder, such as major depression, bipolar disorder, panic disorder, or attention-deficit hyperactivity, the appropriate treatment would be obvious. It is not at all clear, however, that this comprises a significant number of the alcoholic patients one will see in rehabilitation facilities. In the author's experience, it is not.

Other drugs of abuse play a role of uncertain magnitude in the epidemiology of TBI, although drugs such as cannabis and cocaine are known to have a negative influence on the progress of recovery (Gualtieri, 1990b). Even the "innocent" psychoactive drugs, such as nicotine (Gualtieri, 1990b) and caffeine (Gualtieri, 1990b), may interact with the recovery process and the complex of cognitive and behavioral difficulties that arise during long-term rehabilitation.

LEGAL ISSUES

In compensation hearings and tort proceedings, physicians are expected to answer the following questions: (1) What, exactly, is the patient's deficit? (2) What is the prognosis? (3) Is the patient's problem the result of a specific TBI? and (4) Are there other elements to explain the patient's deficits? "Reasonable medical certainty" that a specific deficit is attributable to a specific traumatic event is all that a court requires, but the notion of "certainty" is problematic to physicians, who are trained, after all, in the scientific method. In science, *post hoc ergo propter hoc* is a fallacy, not proof.

Physicians also are limited by their professional point of view. To a neurologist, a normal examination means that the patient's problem is not "neurologic." Therefore, it is functional. If it is "functional," then it is psychiatric. To the psychiatrist, if a problem is traumatic in origin, it is "posttraumatic stress disorder," especially if the neurologist has found nothing and has referred the patient for therapy. This simple algorithm is unfair to the increasing numbers of neurologists who are behaviorally oriented, and to the increasing numbers of neuropsychiatrists, but it remains a problem that fuels contention and litigation.

It never seems difficult for counsel on one side or the other to find an "expert" who will hold to one view, or its diametric opposite: (1) whether the patient has even had a brain injury; (2) if so, whether it is responsible for the functional deficits that have ensued; (3) if so, whether it is responsible for all of the patient's deficits; and (4) how severe those deficits are. Not all of the disagreements between professionals involved in litigation can be attributed to intellectual loyalty to the interests of the person who is paying one's fee. At least some of the problem may be the tendency of professionals to argue from their own (limited) training and experience.

However, even if the process of TBI assessment were removed from the adversarial arena, there still would be important legal issues: whether a TBI patient can drive safely or not; whether the patient can operate heavy machinery; whether the patient is competent to handle personal finances or the proceeds of a generous award. The patient with anosognosia may refuse treatment because he really does not believe there is anything wrong. When should such a patient be compelled to undergo treatment, and how likely is the treatment to succeed if the patient refuses to acknowledge that nothing is wrong? Another issue is whether the TBI patient has lost the capacity to function as a parent, and whether custody arrangements should be made to protect children from a parent who has no safety awareness or who is volatile and aggressive.

REFERENCES

Adams JH, Graham DL, Murray LS, et al. (1982). Diffuse axonal injury due to non-missile head injury: An analysis of 45 cases. Ann Neurol 12:557–563.

Allen D, & Lader M (1989). Interactions of alcohol with amitryptiline, fluoxetine and placebo in normal subjects. Int Clin Psychopharmacol 4 (suppl 1):7–14.

Alves WA, Colohan ART, O'Leary TJ, et al. (1986). Understanding posttraumatic symptoms after minor head injury. J. head Trauma Rehabil 1:1–12.

Alves WM, & Jane JA (1985). Mild brain injury: Damage and outcome. In D Becker & J Povlishock (Eds.), CNS trauma status report, 1985 (pp. 255–269). Bethesda, MD: National Institute of Neurological and Communicative Disorders and Stroke.

Amaducci L, & Lippi A (1988). Risk factors and genetic background for Alzheimer's disease. Acta Neurol Scand (suppl) 116:13–18.

Annegers JF, Grabow JD, Kurland LT et al. (1980). The incidence, causes and secular trends of head trauma in Olmsted County, Minnesota. Neurology 30:912–919.

Bakchine S, Lacomblez L, Benoit N, et al. (1989). Manic-like state after bilateral orbitofrontal and right temporoparietal injury: Efficacy of clonidine. Neurology 39:777–781.

BAKER EL, LETZ, R, & FIDLER AT (1985). A computer-administered neurobehavioral evaluation system for occupational and environmental epidemiology. J Occup Med 27:206–212.

BAMRAH JS, & JOHNSON J (1991). Bipolar affective disorder following head injury. Br J Psychiatry 158:117–119.

BARNHILL LJ, & GUALTIERI CT (1989). Late-onset psychosis after closed head injury. Neuropsychiatry Neuropsychol Behav Neurol 2:211–218.

BARTH JT, MACCIOCHI SN, GIORDANI B, ET AL. (1983). Neuropsychological sequelae of minor head injury. Neurosurgery 13:529–533.

BEAR DM, & FEDIO P (1977). Quantitative analysis of interictal behavior in temporal lobe epilepsy. Arch Neurol 34:454–467.

BERRIOS GI, & SAMUEL C (1987). Affective disorder in the neurological patient. J Nerv Ment Dis 175:173–176.

BERROL S (1990). Issues in cognitive rehabilitation. Arch Neurol 47:219–220.

BLUMBERGS PC, JONES NR, & NORTH JB (1989). Diffuse axonal injury in head trauma. J Neurol Neurosurg Psychiatry 52:838–841.

BOURKE RS (1988). Head injury. Bethesda, MD: National Institute of Neurological and Communicative Disorders and Stroke.

BRISMAR B, ENGSTROM A, & RYDBERG U (1982). Head injury and intoxication: A diagnostic and therapeutic dilemma. Acta Chir Scand 149:11–14.

BROOKS N, SYMINGTON C, BEATTIE A, ET AL. (1989). Alcohol and other predictors of cognitive recovery after severe head injury. Brain Injury 3:235–246.

BROOKS NJ (1991). The head-injured family. J Clin Exp Neuropsychol 13:155–188.

CARLSSON GS (1987). Long-term effects of head injuries sustained during life in three male populations. J Neurosurg 67:197–205.

CASSIDY JW (1989). Fluoxetine: A new serotonergically active antidepressant. J Head Trauma Rehabil 4:67–70.

CATSMAN-BERREVOETS CE, & VAN HARSKAMP F (1988). Compulsive pre-sleep behavior and apathy due to bilateral thalamic stroke: Response to bromocriptine. Neurology 38:647–649.

CHANDLER MC, BARNHILL JB, & GUALTIERI CT (1988). Amantadine for the agitated head-injury patient. Brain Injury 2:309–311.

CHILDS A (1986). Naltrexone in organic bulimia. Unpublished manuscript.

CRISOSTOMO EA, DUNCAN PW, PROPST M, ET AL. (1988). Evidence that amphetamine with physical therapy promotes recovery of motor function in stroke patients. Ann Neurol 23:94–97.

CUMMINGS JL (1985). Organic delusions. Br J Psychiatry 146:184–197.

DELGADO-ESCUETA AV, MATTSON RH, & KING L (1981). The nature of aggression during epileptic seizures. N Engl J Med 305:711–716.

DODRILL CB, & TROUPIN AS (1991). Neuropsychological effects of carbamazepine and phenytoin: A reanalysis. Neurology 41:141–143.

EDNA T (1982). Alcohol influence and head injury. Acta Chir Scand 148:209–212.

EISENBERG HM (1985). Outcome after head injury: General considerations and neurobehavioral recovery. In D BECKER & J POLISHOCK (Eds.), CNS trauma staus report, 1985 (pp. 271–280). Bethesda, MD: National Institute of Neurological and Communicative Disorders and Stroke.

EVANS RW, GUALTIERI CT, & AMARA I (1986). Methylphenidate and memory: Dissociated effects in hyperactive children. Psychopharmacology 90:211–216.

EVANS RW, GUALTIERI CT, & PATTERSON DR (1987). Treatment of chronic closed head injury with psychostimulant drugs: A controlled case study and an appropriate evaluation procedure. J Nerv Ment Dis 175:106–110.

FEENEY DM, GONZALEZ A, & LAW WA (1982). Amphetamine, haloperidol and experience interact to affect rate of recovery after motor cortex surgery. Science 217:855–857.

FEENEY DM, & SUTTON RL (1988). Catecholamines and recovery of function after brain damage. In DG STEIN & BA SABEL (Eds.), Pharmacological approaches to the treatment of brain and spinal cord injuries (pp. 121–142). New York: Plenum.

FEENEY DM, SUTTON RL, BOYESON MG, ET AL. (1985). The locus coeruleus and cerebral metabolism: Recovery of function after cortical injury. Physiol Psychol 13:197–203.

FIELD J (1976). Epidemiology of head injury in England and Wales: With particular application to rehabilitation. Leicester, England: Willsons.

FLEET WS, VALENSTEIN E, WATSON RT, ET AL. (1987). Dopamine agonist therapy for neglect in humans. Neurology 37:1765–1770.

FLOR-HENRY P (1979). On certain aspects of the localization of the cerebral systems regulating and determining emotions. Biol Psychiatry 14:677–698.

FOGEL (1988). Combining anticonvulsants with conventional psychopharmacologic agents. In McElroy SL & Pope HG (Eds.), Anticonvulsants in Psychiatry. Clifton, NJ: Oxford Health Care, Inc.

FOGEL & ESLINGER (1991). Diagnosis and management of patients with frontal lobe syndromes. In Stoudemire A & Fogel BS (Eds.), Medical Psychiatric Practice, (Vol. 1). Washington, DC: American Psychiatric Press.

FOLSTEIN MF, FOLSTEIN SE, & MCHUGH PR (1975). Mini mental state—a practical method of grading the cognitive state of patients for the clinician. J Psychiatr Res 12:189–198.

FRENCH LR, SCHUMAN LM, MORTIMER JA, ET AL. (1985). A case-control study of dementia of the Alzheimer type. Am J Epidemiol 121:414–421.

GERSTENBRAND F, AICHNER PF, & SALTUARI L (1983). Kluver-Bucy syndrome in man: Experiences with posttraumatic cases. Neurosci Biobehav Rev 7:413–417.

GOLDSTEIN M (1990). Traumatic brain injury: A silent epidemic. Ann Neurol 27:327.

GREER BG (1986). Substance abuse among people with disabilities: A problem of too much accessibility. J Disability 14:34–38.

GUALTIERI CT (1988). Pharmacotherapy and the neurobehavioral sequelae of closed head injury. Brain Injury 2:101–129.

GUALTIERI CT (1990a). The behavioral pharmacology of traumatic brain injury and the specific role of carbamazepine. American Psychiatric Association National Meeting, May 1990.

GUALTIERI CT (1990b). The neuropharmacology of inadvertent drug effects in patients with traumatic brain injuries. J Head Trauma Rehabil 5:32–40.

GUALTIERI CT (1990c). Neuropsychiatry and behavioral pharmacology. Berlin: Springer-Verlag.

GUALTIERI CT (1991a). Buspirone for the behavior problems of patients with organic brain disorders. J Clin Psychopharmacol 11:280–281.

GUALTIERI CT (1991b). Buspirone for the neurobehavioral sequelae of temporal lobe injuries. Unpublished manuscript.

GUALTIERI CT, CHANDLER M, COONS T, ET AL. (1989). Amantadine: A new clinical profile for traumatic brain injury. Clin Neuropharmacol 12:258–270.

GUALTIERI CT, & COX DR (1991). The delayed neurobehavioral sequelae of traumatic brain injury. Brain Injury 5:219–232.

GUALTIERI CT, & EVANS RW (1988). Stimulant treatment for the neurobehavioral sequelae of traumatic brain injury. Brain Injury 2:273–290.

GUALTIERI CT, & NYGARD NK (1988). Rehabilitation for head injury, five years after the event. Rebound Q Res Rep 1:16–26.

HACKLER E, & TOBIS JS (1983). Re-integration into the community. In M ROSENTHAL, E GRIFFITH, M BOND, ET AL. (Eds.), Rehabilitation of the head injured adult. Philadelphia: FA Davis.

HALE MS, & DONALDSON JO (1982). Lithium carbonate in the treatment of organic brain syndrome. J Nerv Ment Dis 170:362–365.

HENDERSON AS (1986). The epidemiology of alzheimer's disease. Br Med Bull 42:3–10.

HEYMAN A, WILKINSON WE, STAFFORD JA, ET AL. (1984). Alzheimer's disease: A study of epidemiologic aspects. Ann Neurol 15:335–341.

HORIGUCHI J, INAMI Y, & SHODA T (1990). Effects of long-term amantadine treatment on clinical symptoms and EEG of a patient in a vegetative state. Clin Neuropharmacol 13:84–88.

HOUSE A, DENNIS M, MOLYNEUX A, ET AL. (1989). Emotionalism after stroke. Br Med J 298:991–994.

JAKOBSEN J, BAADSGAARD SE, THOMSEN S, ET AL. (1987). Prediction of post-concussional sequelae by reaction-time test. Acta Neurol Scand 75:341–345.

JENNETT B (1979). Posttraumatic epilepsy. Adv Neurol 22:137–147.

JENNETT B, SNOEK J, BOND MR, ET AL. (1981). Disability after severe head injury: Observations on the use of the Glasgow Outcome Scale. J Neurol Neurosurg Psychiatry 44:285–293.

JOFFE RT, LIPPERT GP, GREY TA, ET AL. (1987). Mood disorder and multiple sclerosis. Arch Neurol 44:376–378.

KALANIE H, NIAKAN E, HARATI Y, ET AL. (1986). Phenytoin-induced benign intracranial hypertension. Neurology 36:443.

KATZ IR, GREENBERG WH, BARR GA, ET AL. (1985). Screening for cognitive toxicity of anticholinergic drugs. J Clin Psychiatry 46:323–326.

KESHAVAN MS, CHANNABASAVANNA SM, & REDDY GNN (1981). Post-traumatic psychiatric disturbances: Patterns and predictors of outcome. Br J Psychiatry 138:157–160.

KRAUS JF (1987). Epidemiology of head injury. In PR COOPER (Ed.), Head injury (Ch. 1, pp. 1–19). Baltimore: Williams & Wilkins.

KRAUTHAMMER C, & KLERMAN GL (1978). Secondary mania: Manic syndromes associated with antecedent physical illness or drugs. Arch Gen Psychiatry 35:1333–1338.

LEVERE ND, GRAY-SILVA S, & LEVERE TE (1988). Infant brain recovery: The benefit of relocation and the cost of crowding. In S FINGER, TE LEVERE, CR ALMLI, ET AL. (Eds.), Brain injury and recovery (pp. 133–150). New York: Plenum.

LEVIN HS (1985). Neurobehavioral recovery. In D BECKER & J POLISHOCK (Eds.), CNS trauma status report, 1985 (pp. 281–299). Bethesda, MD: National Institute of Neurological and Communicative Disorders and Stroke.

LEVIN HS (1987). Neurobehavioral sequelae of head injury. In PR COOPER (Ed.), Head injury (pp. 442–463). Baltimore: Williams & Wilkins.

LEVIN HS (1990). Memory deficit after closed-head injury. J Clin Exp Neuropsychol 12:129–153.

LEVIN HS, AMPARO E, EISENBERG HM, ET AL. (1987a). Magnetic resonance imaging and computerized tomography in relation to the neurobehavioral sequelae of mild and moderate head injury. J Neurosurg 66:706–713.

LEVIN HS, BENTON AL, & GROSSMAN RG (1982). Neurobehavioral consequences of closed head injury. New York: Oxford University Press.

LEVIN HS, EISENBERG HM, & BENTON AL (1989). Mild Head Injury. New York: Oxford University Press.

LEVIN HS, GARY HE, HIGH WM, ET AL. (1987b). Minor head injury and the postconcussional syndrome: Methodological issues in outcome studies. In HS LEVIN, J GRAFMAN, & HM EISENBERG (Eds.), Neurobehavioral recovery from head injury (pp. 262–275). New York: Oxford University Press.

LEVIN HS, & GROSSMAN RG (1978). Behavioral sequelae of closed head injury: A quantitative study. Arch Neurol 35:720–727.

LEWIS SW, HARVEY I, RON M, ET AL. (1990). Can brain damage protect against schizophrenia? A case report of twins. Br J Psychiatry 157:600–603.

LEZAK MD (1983). Neuropsychological assessment. New York: Oxford University Press.

LISHMAN WA (1987). Organic psychiatry, The psychological consequences of cerebral disorder. Oxford, England: Blackwell Scientific Publications.

MAEDA H, & MAKI S (1986). Dopaminergic facilitation of recovery from amygdaloid lesions which affect hypothalamic defensive attack in cats. Brain Res 363:135–140.

MAEDA H, & MAKI S (1987). Dopamine agonists produce functional recovery from septal lesions which affect hypothalamic defensive attack in cats. Brain Res 407:381–385.

MARSHALL LF, & MARSHALL SB (1985). Current clinical head injury research in the United States. In D BECKER & J POLISHOCK (Eds.), CNS trauma status report, 1985 (pp. 45–51). Bethesda, MD: National Institute of Neurological and Communicative Disorders and Stroke.

MASDEU JC, WOLFSON L, LANTOS G, ET AL. (1989). Brain white-matter changes in the elderly prone to falling. Arch Neurol 46:1292–1296.

MASSAGLI T (1991). Neurobehavioral effects of phenytoin, carbamazepine, and valproic acid: Implications for use in traumatic brain injury. Arch Phys Med Rehabil 72:219–226.

MASUR H, ELGER CE, LUDOLPH AC, ET AL. (1989). Cerebellar atrophy following acute intoxication with phenytoin. Neurology 39:432–433.

MATTES JA (1986). Propanalol for adults with temper outbursts and residual attention deficit disorder. J Clin Psychopharmacol 6:299–302.

MAYEUX R, STERN Y, ROSEN J, ET AL. (1981). Depression, intellectual impairment and Parkinson disease. Neurology 31:645–650.

MCKINLAY WW, BROOKS DN, MARTINAGE DP, ET AL. (1981). The short-term outcome of severe blunt head injury as reported by relatives of the injured person. J Neurol Neurosurg Psychiatry 44:527–533.

MCQUEEN JK, BLACKWOOD DHR, HARRIS P, ET AL. (1983). Low risk of late posttraumatic seizures following severe head injury: Implications for clinical trials of prophylaxis. J Neurol Neurosurg Psychiatry 46:899–904.

MORGAN MY, JAKOBOVITS A, ELITHORN A, ET AL. (1977). Successful use of bromocriptine in the treatment of a patient with chronic portasysyemic encephalopathy. N Engl J Med 296:793–794.

ODDY M, COUGHLAN T, & TYERMAN A (1985). Social adjustment after closed head injury: A further follow-up seven years after injury. J Neurol Neurosurg Psychiatry 48:564–568.

OLIVER AP, LUCHINS DJ, & WYATT RJ (1982). Neuroleptic-induced seizures: An in vitro technique for assessing relative risk. Arch Gen Psychiatry 39:206–209.

PARSONS O (1987). Do neuropsychological deficits predict alcoholics' treatment course and posttreatment recovery? In O PARSONS, N BUTTERS, & P NATHAN (Eds.), Neuropsychology of alcoholism: Implications for diagnosis and treatment. New York: Guilford.

PRIGATANO GP, & SCHACTER DL (1991). Awareness of deficit after brain injury. New York: Oxford University Press.

REGIER DA, BOYD JH, BURKE JD, ET AL. (1988). One-month prevalence of mental disorders in the United States based on five epidemiologic catchment area sites. Arch Gen Psychiatry 45:977–986.

Reynolds EH (1975). Chronic anti-epileptic toxicity: A review. Epilepsia 16:319–352.

Rimel RW, Giordani B, Barth JT, et al. (1982). Moderate head injury: Completing the clinical spectrum of brain trauma. Neurosurgery 11:344–351.

Robinson RG, Starr LB, Kubos KL, et al. (1983). A two-year longitudinal study of post-stroke mood disorders: Findings during the initial evaluation. Stroke 14:736–741.

Robinson RG, Starr LB, & Price TR (1984). A two year longitudinal study of mood disorders following stroke: Prevalence and duration at six months follow-up. Br J Psychiatry 144:256–262.

Robinson RG, & Szetela B (1981). Mood change following left hemispheric brain injury. Ann Neurol 9:447–453.

Ross ED, & Stewart RM (1981). Akinetic mutism from hypothalmic damage: Successful treatment with dopamine agonists. Neurology 31:1435–1439.

Ruff RM, Marshall LF, Gautille T, et al. (1990a). Verbal memory deficits following severe head injury: What recovery occurs after one year? Unpublished manuscript.

Ruff RM, Marshall LF, Klauber MR, et al. (1990b). Alcohol abuse and neurological outcome of the severely head injured. J Head Trauma Rehabil 5:21–31.

Salazar AM, Jabbari B, Vance SC, et al. (1985). Epilepsy after penetrating head injury. I. Clinical correlates: A report of the Vietnam Head Injury Study. Neurology 35:1406–1414.

Schiffer RB, Herndon RM, & Rudick RA (1985). Treatment of pathologic laughing and weeping with amitryptiline. N Engl J Med 312:1480–1482.

Silver JM, Yudofsky SC, & Hales RE (1991). Depression in traumatic brain injury. Neuropsychiatry Neuropsychol Behav Neurol 4:12–23.

Slater E, Beard AW, & Glithero E (1983). The schizophrenia-like psychoses in epilepsy. Br J Psychiatry 10:93–97.

Sparadeo F, & Gill D (1989). Effects of prior alcohol use on head injury recovery. J Head Trauma Rehabil 4:75–82.

Sparadeo FR, Strauss D, & Barth JT (1990). The incidence, impact and treatment of substance abuse in head trauma rehabilitation. J Head Trauma Rehabil 5:1–8.

Strub RL, & Black FW (1985). The mental status examination in neurology. Philadelphia: FA Davis Company.

Temkin NR, Dikmen SS, Wilensky AJ, et al. (1990). A randomized, double-blind study of phenytoin for the prevention of post-traumatic seizures. N Engl J Med 323:497–502.

Thomsen IV (1984). Late outcome of very severe blunt head trauma: A 10–15 year second follow-up. J Neurol Neurosurg Psychiatry 47:260–268.

Thomsen IV (1987). Late psychosocial outcome in severe blunt head trauma. Brain Injury 1:131–143.

van Woerkom TCAM, Teelken AW, & Minderhoud JM (1977). Difference in neurotransmitter metabolism in frontotemporal-lobe contusion and diffuse cerebral contusion. Lancet 1:812–813.

Vauhkonen K (1959). Suicide among the male disabled with war injuries to the brain. Acta Psychiatr Neurol Scand 137(suppl):90–91.

Vecht CJ, van Woerkom TCAM, Teelken AW, et al. (1975). Homovanillic acid and 5-hydroxyindoleacetic acid cerebrospinal fluid levels. Arch Neurol 32:792–797.

Violon A, & Demol J (1987). Psychological sequelae after head trauma in adults. Acta Neurochir 85:96–102.

Volpe BT, & McDowell FH (1990). The efficacy of cognitive rehabilitation in patients with traumatic brain injury. Arch Neurol 47:220–222.

Yarnell PR, & Rossie GV (1988). Minor whiplash head injury with major debilitation. Brain Injury 2:255–258.

Young B, Rapp RP, Norton JA, et al. (1983). Failure of prophylactically administered phenytoin to prevent post-traumatic seizures in children. Childs Brain 10:185–192.

Zubenko GS, Moossy J, & Kopp U (1990). Neurochemical correlates of major depression in primary dementia. Arch Neurol 47:209–214.

IV Specific Diseases and Medical Subspecialties

or crippling pessimism, the psychologic tasks of changing life-style and modifying risk factors still remain difficult. Smoking, overeating, and compulsive working often are fueled by anxiety or depression, and are especially hard to give up when the new stress of cardiac disease has been added. Bad habits create a maladaptive, hard-to-change behavioral homeostasis. Cessation of smoking may exacerbate overeating and depression. Attempting to exercise while still overweight and smoking can be self-defeating and demoralizing, as can trying to lose weight without exercising.

Psychologic Factors Affecting Cardiac Disease

A vast body of research has focused on Type A behavior as a risk factor for coronary disease. In this author's opinion, recent epidemiologic studies have not strongly supported Type A behavior as a coronary risk factor, and most angiographic studies have failed to find an association between Type A behavior and the extent of coronary artery disease (CAD) (Dimsdale, 1988). This issue is discussed in detail elsewhere (see Chapter 12). Less attention has focused on depression and anxiety, but they have been shown to have important influences over the onset and course of coronary heart disease (Appels, 1990; Booth-Kewley & Friedman, 1987). Depression predicts major cardiac events (Carney et al., 1988), and work disability after MI (Schleifer et al., 1989).

One specific pathway by which psychologic factors can affect cardiac disease has been demonstrated experimentally. Acute mental stress can precipitate silent myocardial ischemia (i.e., ischemic changes on the electrocardiogram [ECG] without symptoms of angina) (Freeman et al., 1987; Rozanski et al., 1988). Silent ischemia has important implications for the management and outcome of coronary disease, and may be partly a consequence of cognitive or defensive traits such as denial, visceral hyposensitivity, or systematic misperception of angina (Barsky et al., 1990).

Denial is common in patients with cardiac disease, varying in its timing, strength, and adaptive value. Some "silent" or "atypical" MIs and sudden deaths may occur when denial prevents the individual from acknowledging the symptoms and promptly seeking medical care. The length of delay between the onset of symptoms of an MI and hospitalization is a powerful predictor of morbidity and mortality (Doehrman, 1977). The patient's spouse or others close by may be able to overcome the patient's denial and convince the patient to seek care. Unfortunately, physicians seldom have an opportunity to affect such denial, except through education of the general public. Denial *during* hospitalization has adaptive value, perhaps even reducing morbidity and mortality (Levenson et al., 1984, 1989; Levine et al., 1987). Patients with such denial minimize their symptoms, displace them to less threatening organ systems (angina becomes dyspepsia), and pay little heed to cardiac monitors and alarms.

Psychologic stressors can play an important role in precipitating life-threatening ventricular arrhythmias (Follick et al., 1988; Lown et al., 1980). Sudden cardiac death following psychologic distress has long been anecdotally reported, but scientific study is difficult. A systematic review of published cases of ventricular fibrillation in patients without known cardiac disease could identify preceding psychologic distress in 22% (Viskin & Belhassen 1990). Whether Type A personality traits predict sudden cardiac death after MI remains controversial (Ahern et al., 1990; Brackett, 1988). In patients with known significant arrhythmias, clinical depression is correlated with subsequent mortality (Kennedy et al., 1987). In the National Heart, Lung and Blood Institute's Cardiac Arrhythmia Pilot Study (CAPS), depression was an independent risk factor for death or cardiac arrest (Ahern et al., 1990) but did not appear to have any effect on ventricular premature contraction rates or response to antiarrhythmic therapy (Follick et al., 1990). Although cardiac variables remain the most significant factors leading to ventricular arrhythmias and sudden death, psychiatric symptoms and disorders do appear important in predisposing some individuals.

Physician's Countertransference

The psychiatrist evaluating a cardiac patient with psychologic symptoms should consider the possibility of relevant countertransference reactions in the internist or cardiologist responsible for the patient's primary care. These reactions at times can aggravate the patient's behavioral difficulties or interfere with appropriate medical treatment. Physicians old enough to be at risk for coronary disease will tend to identify with some of their cardiac patients. Although this could potentially enhance their ability to empathize with patients, it may instead lead them to distance themselves. If patients are very frightened by their heart disease, physicians may withdraw to avoid their own resonant anxiety. If patients are strong deniers, physicians unconsciously worried about their own mortality may collude in the denial, distancing themselves not only from the patients but from the disease as well.

Younger physicians are more likely to err in the other direction. Enthusiastically launching an attack on risk factors, they may become almost messianic in their approach to the patient. Concerned over some patients' denial of illness, they may try to directly overcome their defenses. Benevolently, even feeling morally obligated to do so, the physician may attempt to "reason with" (i.e., scare) the patient by reciting a litany of disastrous consequences if the patient will not stop smoking, lose weight, and so on. This usually increases the patient's anxiety, in turn increasing the need to deny illness. Frustrated, the physician may then become angry, communicating (sometimes nonverbally) to the patient that the disease is self-induced, the result of an indulgent, undisciplined, self-destructive life-style. Even a subtle, unconscious tendency to blame patients for their illness may affect them adversely, because many patients are already feeling guilty, hopeless, and blameworthy. (See Chapters 42 and 43.)

BIOLOGIC ASPECTS

Heart failure causes symptoms that may be misinterpreted as representing a primary psychiatric disorder. Mild to moderate heart failure produces insomnia (orthopnea and paroxysmal nocturnal dyspnea), anorexia, fatigue, weakness, and constipation, symptoms that may be mistakenly attributed to depression. In more severe heart failure, subtle or overt encephalopathy may occur, including confusion, cognitive dysfunction, drowsiness, apprehension, poor judgment, and delirium with psychosis. These symptoms are the consequence of ischemia—decreased perfusion of the brain and other organs—which includes hypoxia, decreased delivery of nutrients (particularly glucose), and increased accumulation of waste products (leading to acidosis). Hepatic congestion in heart failure can impair the ability to metabolize and excrete many drugs, further aggravating encephalopathy. With left heart failure, pulmonary congestion further exacerbates hypoxia.

Heart failure may be acute or chronic. In chronic, end-stage heart failure, the ensuing cardiac cachexia may be very difficult to distinguish from severe depression, with extreme weakness and lethargy, anorexia and weight loss, social withdrawal, and loss of the will to live. These symptoms may reflect not only brain ischemia, but poor perfusion of kidneys, liver, and other organs, leading to multiple organ dysfunction. The diagnostic distinction between cardiac cachexia and depression is an important one, because antidepressants will not relieve symptoms attributable to low cardiac output, and risk causing cardiac side effects and further delirium. The presence of cardiac cachexia is suggested on physical examination by palpating extremities that are cool, pale, or cyanotic, with weak or absent pulses, reflecting poor perfusion. Elevated circulating tumor necrosis factor recently has been shown to be a marker for cardiac cachexia, but a routine test is not yet available (Levine et al., 1990).

Objective estimation of cardiac index and ejection fraction (by arteriography or radionuclide studies) can help in deciding whether a patient's symptoms are "proportionate" to the cardiac disease. Cachexia cannot be explained solely on a cardiac basis if the ejection fraction is 40% but might well be with an ejection fraction in the range of 10% to 20%. A cardiac index greater than 2 liters/min/m^2 is inconsistent with cardiac cachexia. (These are general guidelines; exact cutoffs cannot be given because the interpretation of ejection fraction depends on heart size and other factors.)

Besides heart failure, neuropsychologic deficits may result from congenital heart disease (Newberger et al., 1984), cardiac arrest (Bass, 1985), and endocarditis (Bademosi et al., 1979). Older patients with coronary artery disease often also have other vascular disease, especially cerebrovascular disease, and even minor reductions in blood flow can sometimes result in cognitive dysfunction.

Neuropsychiatric complications are common after cardiac surgery (Smith & Dimsdale, 1989). Although the incidence of serious neuropsychiatric sequelae probably has declined over the past 25 years, 25% to 35% of patients have some cognitive deficits, "organic" psychosis, and/or delirium after open heart surgery (Willner & Rodewald, 1991). Microemboli, increased time on cardiopulmonary bypass, and preoperative cerebrovascular disease are all risk factors for neuropsychiatric sequelae. Cognitive deficits in patients undergoing coronary artery bypass graft surgery are common but usually subtle, and generally resolve by 6 months (Hammeke & Hastings 1988).

A variety of acute cardiovascular events, including arrhythmias, coronary ischemia (angina), acute valve dysfunction, and systemic and pulmonary embolism, may lead to acute sympathetic discharge, with a surge in circulating catecholamines, which may be experienced by the patient as generalized anxiety or a panic attack. In the elderly and in patients with known heart disease, the new abrupt onset of acute anxiety should lead the physician to consider the possibility of an acute cardiac event, and not to just suppress the anxiety with an antianxiety drug.

Substance abuse plays a major role in the etiology and exacerbation of heart disease, with smoking as the best known example. Alcoholism is associated with an increased risk of heart failure, arrhythmias, sudden death, labile hypertension, and stroke (Regan 1990). Cocaine abuse has caused MI, coronary artery spasm, ventricular arrhythmias, cardiomyopathy, and cardiac arrest

(Gradman, 1988). Although there has been some uncertainty whether high caffeine consumption poses any cardiovascular risks, moderate use of caffeine does not adversely affect the heart, even in patients with ventricular arrhythmias (Chelsky et al., 1990).

PRACTICAL DIAGNOSIS AND MANAGEMENT

Depression

Cardiac disease, as already noted, can mimic psychiatric disorders. Congestive heart failure frequently includes symptoms that can be mistaken for major depression, particularly in the elderly, although, of course, patients commonly have depression *and* cardiac disease. The presence of a "true" depression is suggested by pronounced feelings of guilt or worthlessness, suicidal ideation, anhedonia, and functional disability and affective disturbance that are out of proportion to the degree of heart failure (see Chapter 4). The dexamethasone suppression test, thyrotropin-releasing hormone stimulation test, and sleep studies are not helpful in determining whether a cardiac patient has a major depression. Besides their limited sensitivity even in psychiatric populations, these tests are less specific in the medically ill. False-positives can result from many medical illnesses and medications (APA Task Force on Laboratory Tests in Psychiatry, 1987; Levenson & Friedel, 1985).

Cognitive Dysfunction in Cardiac Disease

Reversible cognitive impairments associated with low cardiac output must also be distinguished from dementia in the patient with heart failure. The elderly are especially at risk for being considered primarily demented. The simplest method for determining whether the individual is temporarily encephalopathic (delirious) rather than suffering from dementia is to perform serial mental status examinations as the degree of heart failure varies. If improvement in cardiac output corrects cognitive deficits, the patient is not likely to have a primary dementia. Although an electroencephalogram (EEG) would be likely to show diffuse, nonfocal slowing in a reversible, low-output encephalopathy, this finding is not specific and does not rule out dementia. If the cognitive diagnosis remains in doubt, formal neuropsychologic assessment should be obtained. Physicians should also keep in mind that many cardiac drugs can worsen cognitive impairment either directly or by further contributing to systemic hypotension.

Panic Attacks/Arrhythmias

Panic attacks and cardiac arrhythmias share many of the same symptoms (shortness of breath, palpitations, light-headedness, and autonomic arousal) and can be confused with each other. Psychologic stressors can play an important role in the precipitation of either. Most often, it is panic disorder that is misdiagnosed as an arrhythmia, usually as paroxysmal atrial tachycardia (PAT). Both panic disorder and PAT occur mostly in a young, otherwise healthy, predominantly female population. Diagnostic confusion can be compounded by treatment, since some drugs for PAT, such as beta-blockers, may partially ameliorate panic symptoms. A careful history will usually lead to a correct diagnosis. Syncope, with actual loss of consciousness, is unusual in panic disorder but common with serious arrhythmias. Attacks of arrhythmia tend to be less stereotypical than panic attacks. Panic attacks last typically 5 to 10 minutes, whereas arrhythmias vary from seconds to days. The development of agoraphobia points to a diagnosis of panic disorder.

When in doubt, ambulatory ECG (Holter) monitoring can be used to document the presence or absence of arrhythmias. Hyperventilation can be used to try to provoke typical symptoms during the period of monitoring. A hyperventilation test without some form of monitoring is inconclusive, since both panic attacks and arrhythmias may sometimes be induced by hyperventilation. The relationship between panic attacks and mitral valve prolapse is discussed later in this chapter.

Sleep Disturbance

Insomnia is a frequent symptom in both psychiatric and cardiac illness, and may mislead both diagnosis and treatment. The cardiac patient's sleep difficulties usually have some specific characteristics. Heart failure produces orthopnea, that is, shortness of breath and inability to sleep in the recumbent position. The patient characteristically uses several pillows in order to sleep. With more severe failure, paroxysmal nocturnal dyspnea (PND) develops, with severe shortness of breath and wheezing, not always relieved immediately on sitting up. Nocturnal angina awakens the patient with typical chest pain with or without associated features such as left arm numbness. If the clinician fails to discern the cardiac cause of insomnia and treats the patient with hypnotics, at best there will be little or no benefit and the correct intervention will be missed. At worst, the hypnotic will further compromise an already impaired respiratory drive or exacerbate sleep apnea.

Somatoform Disorders

Somatoform disorders (hypochondriasis, somatization disorder, conversion, pain disorder-psychological type, and factitious disorder) and cardiac disease are sometimes mistaken for each other but also frequently coexist. A somatizing patient can learn how easy it is, by complaining of the right symptoms to a physician, to precipitate the "rule-out-MI" cascade, with a guaranteed stay in the coronary care unit. Although the physical examination and ECG can "rule in" cardiac disease, they cannot exclude it acutely. There may also be errors in the other direction, with cardiac symptoms misdiagnosed as psychogenic. For example, patients with variant angina or coronary vasospasm may have "clear coronaries" on arteriography, and ECG changes are usually manifested only during an attack. The condition is diagnosed either by recording ECG changes during a spontaneous attack, or during an attack provoked by a vasoconstrictor challenge. Early cardiomyopathy can be difficult to discern on physical examination, and the patient's weakness may be misattributed to depression or hypochondriasis. If the history, examination, or ECG raises the suspicion of decreased cardiac output, an echocardiogram or radionuclide angiography is the next step in documenting impaired left ventricular performance.

Atypical Chest Pain

Psychiatrists often are asked to see individuals with "atypical" chest pain, to aid in correct diagnosis. A substantial minority (10% to 30%) of patients with chest pain undergoing diagnostic cardiac catheterization have normal or near-normal coronary arteries. Patients with negative cardiac diagnostic studies and no prior history of organic heart disease have high prevalence rates (40% to 50%) for panic disorder, major depression, and/or multiple phobias (Cormier et al., 1988). Cardiologists may employ a standard exercise ECG treadmill test, or more recent tests that combine radionuclide imaging with exercise stress—for example, thallium imaging to assess myocardial perfusion and radionuclide ventriculography to measure changes in left ventricular ejection fraction and wall motion during exercise (Froelicher, 1983; Patterson et al., 1986).

The sensitivity and specificity of these tests for detecting CAD vary with the specific criteria utilized, the degree of effort made by the exercising patient, and the prevalence of CAD in the population studied. Radionuclide imaging tests have fewer false negatives than the ECG treadmill test in determining whether CAD is present and avoid the problem of submaximal exercise during the treadmill test (Patterson et al., 1986). For a symptomatic patient who is at low risk for CAD (e.g., a young woman with nonanginal or atypical chest pain and few or no risk factors), a radionuclide study would be very helpful in ruling out CAD. However, for a patient with a high probability of disease (e.g., a 45-year-old man with a history typical of angina pectoris), noninvasive exercise tests are not useful for establishing a cardiac basis for chest pain because the high likelihood of having CAD reduces the tests' negative predictive value. Coronary catheterization would be more appropriate.

Special Considerations in Psychotherapy

Psychologic interventions with cardiac patients are made by nonpsychiatric physicians as well as by mental health professionals, and are aimed at alleviating anxiety or depression, addressing denial or non-compliance, and/or modifying risk factors. The balance between support and confrontation should depend on the phase of illness and whether defenses are currently maladaptive. For acutely or recently ill patients overwhelmed by anxiety or depression, the first therapeutic task may be to strengthen the patient's defenses, whereas for those with excessive denial interfering with rehabilitation or risk factor reduction, a confrontation of defenses may be desirable (see Chapter 1).

Anxiety can result from a multitude of different specific fears. If physicians presume without asking that they know why their patients are anxious, their patients are likely to feel misunderstood. For one patient the fear of sudden death may be most frightening, but another may be preoccupied with fears of physical and/or sexual limitations; the two should not be approached in the same way. Reassurance given in a facile manner, without investigating the patient's fears, can undermine the physician-patient relationship, because the patient may regard it as empty reassurance offered by a physician not really interested in what the patient is actually experiencing. Appropriate psychotherapeutic interventions are matched to specific fears. Unrealistic fears often can be reduced by straightforward cognitive interventions—for example, educating the patient about the safety of sex after a heart attack. Medical psychotherapy is indicated if fears persist despite reasonable efforts at education and reassurance. (See Chapter 6.)

Effective therapeutic intervention in depression requires an understanding of its roots for the individual cardiac patient, just as in anxiety. For some, therapy is aimed at improving damaged self-esteem and self-denigrating pessimism after an MI. For others, unresolved grief over previous losses, sometimes with survivor guilt, must be worked through. The narcissistic injury of the MI may result in an especially hostile, irritable depression in those with preexisting narcissistic personality traits.

If not excessive, denial reduces anxiety but does not prevent the patient from accepting and cooperating with medical treatment. For many of these patients, defensive denial is firm and requires from physicians only that they not interfere with it. For others, it is fragile, and the physician must judge whether denial should be supported and strengthened, or whether the patients would do better giving up the defense to be able to share their fears with someone directly. When denial is extreme, patients may refuse vital treatment or threaten to leave against medical advice. Here the physician must try to help reduce denial, but not through a direct assault on the patient's defenses. Since such desperate denial of reality usually reflects intense underlying anxiety, trying to scare the patient into cooperating will intensify the denial. A better strategy is to avoid directly challenging the patient's claims while simultaneously reinforcing the staff's concern for the patient, setting limits where necessary, and maximizing the patient's sense of control over nonessential details of care.

Denial after hospital discharge also can be either helpful or harmful. Too little denial with unstable maladaptive defenses may leave the patient flooded with fears of disability and death, and result in a "cardiac neurosis." Excessive denial may result in a counterphobic rush back to full-time work, disregard for the cardiac rehabilitation plan, ignoring of modifiable risk factors, and a cavalier attitude toward cardiac drugs. The physician's task is to help patients balance the need for adaptive denial with the need to express their fears and losses, promoting realistic assessment of their health while maintaining hope through rehabilitation.

Denial is only one of several factors that can interfere with treatment adherence and risk-factor reduction. Depression, anxiety, substance abuse, character style, family dynamics, unrecognized drug side effects, and sociocultural factors all can impede a change in lifestyle. Simply cajoling patients to change without regard to the source of resistance often results in guilt, demoralization, and alienation from the physician. Smoking, for example, is an addictive disorder and should be treated as such, not just as an indication of self-destructiveness or weak willpower. (See Chapter 42.)

Psychosocial and Behavioral Interventions

Psychosocial interventions can be aimed at psychosocial as well as medical risk factors for coronary diseases. Social isolation and life stresses are associated with increased mortality in coronary disease (Ruberman et al., 1984), and their reduction may improve outcome (Frasure-Smith & Prince, 1985). Although psychotherapeutic (especially behavioral) treatments have shown promise in reducing these psychosocial risk factors, the small number of studies, with their methodologic limitations, have not provided conclusive evidence of the effectiveness of such treatments (Blanchard & Miller, 1977; Razin, 1984, 1985a, 1985b). More recent long-term studies have shown positive benefits of altering Type A behavior in preventing recurrences in post-MI patients (Friedman, Thoresen, & Gill, 1986). Recently it has been recognized that depression is an important complicating factor in smoking, accounting for a significant proportion of cessation failures (Anda et al., 1990; Glassman et al., 1990). Its treatment may improve the chances of success in smoking cessation efforts. (See Chapter 42.)

A variety of nonpharmacologic treatments involving behavior change, safe but underutilized, can help reduce hypertension (Kaplan, 1985), particularly behavioral/stress management therapies such as biofeedback and relaxation techniques (Engel, Glasgow, & Gaardner, 1983; Patel, Marmot, & Terry, 1981). The more successful studies have shown 10- to 26-mm Hg drops in systolic pressure and 5- to 15-mm Hg drops in diastolic pressure at up to 4 years follow-up (Weiss 1988).

Behavioral treatments are not as effective as drug therapy (De-Ping Lee et al., 1988). A major consideration is compliance. Behavior therapies for hypertension are ineffective for patients who are unmotivated to continue practicing the techniques on their own; they should receive drug treatment. Biofeedback and relaxation therapies should be reserved for highly motivated patients with mild hypertension who did not wish to take antihypertensive medication because of side effects or because they put a high value on a drug-free approach. Behavioral treatment may also be helpful for moderate to severe hypertension as adjunctive treatment in addition to antihypertensive drugs, especially in patients with anxiety-induced blood pressure lability.

Cardiac Rehabilitation

Cardiac rehabilitation, an essential part of overall treatment, involves early ambulation during hospitalization, outpatient prescriptive exercise training, and education of patient and family (Wenger & Hellerstein, 1984), often including sexual counseling (McLane, Krop, & Mehta, 1980). Rehabilitation programs that reduce risk factors and improve physical fitness have been shown to reduce mortality (Oldridge et al., 1988; O'Connor et al., 1989), but psychosocial outcome has not been well studied (Greenland & Chu, 1988). Although exercise training is often said to reduce mild anxiety and depression in cardiac patients (Cassem & Hackett, 1977; Wenger & Hellerstein, 1984), the demonstrable benefits are mainly physical (Langosch 1988).

Major psychopathology can prevent some patients from benefiting from traditional rehabilitation programs (Cay, 1982). For the psychiatrically ill cardiac patient, optimal treatment combines specific psychiatric treatment with a cardiac rehabilitation program designed for the individual. For those with major psychopathology, psychiatric intervention should precede entry into rehabilitation, to reduce symptoms that could interfere with participation. For patients with severe personality disorders, it is usually best to individualize rehabilitation, with realistic tailoring of goals. A schizoid or paranoid patient, for example, may have serious difficulty in participating in group meetings.

DRUG TREATMENT

Antidepressants

Tricyclic Antidepressants

Among the antidepressants, the tricyclics (TCAs) have been available the longest and are therefore the best known. The well-established cardiac effects of tricyclics include changes in conduction (quinidine-like effects), increases in heart rate, and orthostatic hypotension (Glassman & Bigger, 1981). (See Chapter 9.)

Antidepressants do not have clinically significant effects on stable congestive heart failure (Dalack & Roose, 1990; Veith et al., 1982). Many studies of the effects of TCAs on the ECG have shown statistically significant but generally clinically insignificant increases in the P-R, QRS, and Q-T intervals. However, most reported studies were done in patients without cardiac disease. Studies of antidepressant treatment in patients with chronic heart disease generally have *not* found clinically significant increases in the P-R, QRS, or Q-T intervals (Hayes et al., 1983; Veith et al., 1982). Increases in heart rate were similarly found to occur less frequently than expected. Sertraline, a new serotonin uptake inhibitor, does not appear to cause any significant ECG abnormalities (Guy & Silke 1990), and does not cause hypotension. The majority of studies of antidepressants in cardiac disease have examined patients with stable New York Heart Association class I or II disease, so one must be cautious in extrapolating results to unstable or more severe (class III or IV) disease, especially with newer agents. For example, tricyclics theoretically would be expected to have a negative inotropic effect, and, although this is generally clinically insignificant, it may become a problem in patients with very severe unstable heart failure (ejection fraction below 20%). The effects of TCAs and newer agents such as fluoxetine and bupropion on the ECG are discussed in detail in Chapter 9.

Orthostatic hypotension is the most common serious cardiovascular effect of TCAs encountered in cardiac patients, especially in the elderly. Depressed cardiac patients are much more likely than other patients with depression to have hypotension as a side effect, often related to their concurrent use of cardiac or antihypertensive drugs or to impaired left ventricular function (Glassman et al., 1982). In the elderly, orthostatic hypotension can contribute to insufficient cerebral perfusion, which in turn can lead to syncope and falls. The patient may become bedridden or fracture a hip. These risks can be reduced by addressing other causes of hypotension in the elderly cardiac patient, such as overtreatment with diuretics, nitrates, or antihypertensives, and by giving specific attention to adequate hydration. In some hypertensive patients, the hypotensive effects of tricyclics actually can be beneficial (Zachariah & Rosenbaum, 1982).

Orthostatic blood pressure changes occur rarely from fluoxetine or bupropion in patients without cardiac disease. Bupropion has been reported to be safe in patients with some impairment of left ventricular function (Roose et al., 1987a). However, elevations in blood pressure can occur with bupropion.

While at one time tricyclics were withheld in patients with premature ventricular contractions, most tricyclics possess type 1A antiarrhythmic activity and tend to reduce rather than aggravate the arrhythmia. Imipramine has been the most studied in this regard (Connolly et al., 1984), and although it is an effective antiarrhythmic, side effects render it less well tolerated than conventional antiarrhythmic drugs (Cardiac Arrhythmia Pilot Study Investigators, 1988). If a patient already taking a tricyclic acutely develops cardiac disease, it should be noted that abrupt withdrawal of TCAs can precipitate serious ventricular arrhythmias (Babb, Dunlop, & Hoffman 1990). Discontinuation of tricyclics is not automatically necessary on the occurrence of an arrhythmia. When discontinuation is advised, a gradual taper usually is preferable.

A small number of cases of sudden death have been reported in patients treated with tricyclics, recently including cases in children ("Sudden death," 1990). The most plausible theoretical explanation is that tricyclics, like quinidine, may greatly prolong the Q-T interval in a few particular patients, resulting in the "long Q-T syndrome." However the evidence for a causal relationship between tricyclics and sudden death is far from convincing, and thus it would be reasonable to regard this as a possible but extraordinarily rare event (Warrington, Padgham, & Lader 1989). Because of the rarity of sudden death, its discussion is not a routine part of informed consent to antidepressant drug therapy.

Before starting a patient on an antidepressant, congestive heart failure should be stabilized and potassium, digitalis, and antiarrhythmic levels should be within normal limits. Patients with serious conduction delays are at risk for increased degrees of heart block if treated with TCAs (Roose et al., 1987b; Vieweg, Yazel, & Ballenger, 1984). Fluoxetine, sertraline, bupropion, and trazodone are clearly preferred in patients with clinically significant cardiac conduction abnormalities. Heart block occurred in one study in 9% of TCA-treated patients with preexisting bundle-branch block versus 0.7% with normal ECGs (Roose et al., 1987b). Tricyclics have been used safely with close cardiac monitoring in patients with stable, uncomplicated, isolated left or right bundle-branch block (without any evidence of second- or third-degree block). In more advanced or unstable degrees of heart block, a pacemaker should be inserted first, or an alternative psychiatric treatment pursued first (Stoudemire & Atkinson, 1988). If cardiac function is stable, antidepressant treatment can often be initiated on an outpatient basis with careful monitoring of the ECG until therapeutic blood levels are achieved. With new, changing, or unstable cardiac disease, the patient should be monitored in the hospital during the introduction of antidepressants.

A low starting dose of a TCA (generally 10 to 25 mg at bedtime) should be used. For most medically stable patients, the dose can be raised by 25 mg every few days until a therapeutic dose is reached or limiting side effects are encountered. With serious unstable cardiac disease, such as a new MI, it is advisable to delay the introduction of antidepressants. How long to delay after MI depends on the stability of cardiac function. After an uncomplicated MI with rapid recovery, antidepressants usually can be started safely after approximately 4 weeks. Complications of the MI such as threatened extension of the infarct, hypotension, uncontrolled arrhythmias, severe heart failure, and major new conduction disturbances would be contraindications to beginning tricyclics. Whether fluoxetine, trazodone, or bupropion can be used safely earlier or with complicated MIs remains to be demonstrated. The use of TCAs in the post-MI period is also discussed in detail in Chapter 9.

There is little evidence to support the widespread beliefs that one or another of the TCAs is much safer in the cardiac patient (Luchins, 1983). The nature of the cardiac disease and the character of the depression should influence the particular choice of antidepressant. For example, for a patient who has premature ventricular beats and depression-related insomnia, imipramine would be a good choice because of its antiarrhythmic (Cardiac Arrhythmia Pilot Study Investigators, 1988) and mildly sedating properties, although this drug has a high degree of associated orthostatic hypotension. Trazodone, bupropion, sertraline, or fluoxetine would be useful for patients when the anticholinergic or quinidine-like effects of TCAs were unwanted. Although there were early case reports of trazodone worsening preexisting ventricular ectopy (Janowsky et al., 1983) subsequent research has not supported such a risk (Himmelhoch, Schectman, & Auchenbach, 1984). Bupropion appears to be free of cardiac side effects. Symptomatic bradycardia has recently been reported anecdotally in two patients receiving fluoxetine (Ellison, Milofsky, & Ely, 1990) and as well as atrial fibrillation (Buff et al. 1991). Side effect profiles of the TCAs with respect to their use in medically compromised cardiac patients are discussed in detail in Chapter 9.

Lithium

At therapeutic and toxic levels lithium may cause changes in the ECG and occasionally may cause arrhythmias. Lithium has been used safely in patients with severe cardiac disease by beginning with a low dose, increasing slowly, and monitoring carefully for side effects (Levenson et al., 1986). Because of age-related decline in the renal glomerular filtration rate, elderly patients generally require about two thirds the maintenance dose of lithium used for younger patients. If cardiac disease further decreases renal perfusion, the maintenance dose should be even lower. Conservative use of lithium in the elderly cardiac patient also is warranted because of their frequent use of low-salt diets and diuretics, and less stable fluid and electrolyte status, all of which can affect lithium levels. Lithium is particularly difficult to use in patients being treated for acute heart failure, and alternatives generally should be substituted. (See also Chapter 9.)

Monoamine Oxidase Inhibitors

There is much less clinical experience or study of the use of monoamine oxidase (MAO) inhibitors to treat depression in the presence of cardiac disease, particularly in the elderly. Monoamine oxidase inhibitors include several different drugs, each with a different side effect profile. All, however, can cause significant resting and orthostatic hypotension (Goldman, Alexander, & Luchins, 1986), raising the same concerns discussed with regard to the TCAs. In some hypertensive patients, this may be a beneficial rather than an adverse effect. Monoamine oxidase inhibitors appear to have little effect on heart rate or cardiac conduction (Goldman, Alexander, & Luchins, 1986). An additional consideration is that the patient must be willing and able to follow the dietary and drug restrictions imposed by the use of a MAO inhibitor.

Antipsychotics

Although antipsychotics occasionally can affect cardiac conduction or rhythm, hypotension is the major adverse cardiovascular effect. This occurs much less frequently with higher potency, less anticholinergic drugs (e.g., haloperidol), which can be used safely even with severe acute cardiac disease or soon after cardiac surgery (Risch, Groom, & Janowsky, 1982). In patients without cardiac disease, clozapine has caused tachycardia in about 25% (about 10- to 15-beats/min increase), hypotension in 9%, hypertension in 4%, and minor ECG changes in only 1%. Pimozide and thioridazine appear more likely to prolong the Q-T interval than other neuroleptics.

Benzodiazepines and Buspirone

Given orally, benzodiazepines are essentially free of cardiovascular side effects, and numerous studies have documented their safety even in the immediate post-MI period (Risch, Groom, & Janowsky, 1982). Except for lorazepam and midazolam, benzodiazepines should not be given intramuscularly, because they are erratically absorbed, particularly if cardiac output is diminished. Intravenous administration, if too rapid, can cause acute hypotension. Buspirone has no recognized cardiovascular side effects and may have a place in treatment of cardiac patients with anxiety disorders. (See Chapter 6.)

Stimulants

Stimulants such as dextroamphetamine and methylphenidate have been advocated by some for the short-term treatment of depression in the seriously medically ill (Woods et al., 1986). Because careful longitudinal studies have not clearly established their antidepressant properties, and because of the potential for drug abuse and dependence, their use has been controversial. Many clinicians have been especially reluctant to try stimulants in elderly patients with cardiac disease. Their safe use has been described, however, in patients with arrhythmias, congestive heart failure, coronary artery disease, or hypertension (Woods et al., 1986). Stimulants seem to be most appropriate when one cannot afford to wait for a response to TCAs. Most patients who respond favorably do so within 24 to 48 hours. Stimulant treatment may be especially helpful when depressive immobilization impedes medical recovery. The use of psychostimulants is reviewed in Chapter 9.

Disulfiram

Disulfiram itself has no significant cardiovascular side effects. The concurrent ingestion of even small amounts of alcohol, however, may produce heart failure, shock, arrhythmias, MI, respiratory depression, and sudden death. Disulfiram is relatively contraindicated in cardiac patients, and should be reserved for those alcoholic patients who are highly unlikely to drink while on disulfiram but are unable to abstain from alcohol without it.

Other Drugs

Carbamazepine, a tricyclic compound with quinidine-like properties, may exacerbate congestive heart failure or CAD and may produce hypotension or hypertension, arrhythmias, and atrioventricular block. Precautions similar to those for TCAs should therefore be taken with this drug.

Valproic acid does not have adverse cardiac effects. Beta-blockers such as propranolol should not be used as antianxiety agents in patients with congestive heart failure, because they may exacerbate it.

ELECTROCONVULSIVE THERAPY

Electroconvulsive therapy (ECT) may be safer than TCAs for some patients with cardiac disease. It produces predictable but brief increases in heart rate and blood pressure, and can cause arrhythmias or conduction changes during the seizure and immediately thereafter. However, the risks are limited to a specific time during which the patient is carefully monitored, in contrast to the continuous risk associated with drug therapy.

The risks associated with ECT are greater in elderly cardiac patients than in younger, healthier patients (Burke et al., 1985). Small case series continue to show that ECT can be used safely in patients with stable cardiac disease (Dec, Stern, & Welch, 1985; Guttmacher & Greenland, 1990), but the answer to the question of which is safer, ECT or antidepressants, is not settled, especially in acute, unstable cardiac disease (Dec, Stern, & Welch 1985; Levenson & Friedel, 1985). If more extensive clinical experience confirms the relative lack of cardiac side effects of fluoxetine and bupropion, they will represent the safest initial choices. Electroconvulsive therapy will be reserved for more severely ill or treatment-unresponsive patients.

In medication nonresponders for whom ECT proves necessary, the post-ECT cardiovascular response can be attenuated with labetalol before seizure induction (Stoudemire, Knos, & Gladson, 1990). Caffeine pretreatment to augment seizure response in ECT can be used safely in cardiac patients (Lurie & Coffey, 1990), which is not surprising since moderate amounts of caffeine have no effect on ventricular arrhythmias (Chelsky et al., 1990). Decisions about the use of ECT in cardiac patients should be made through close collaboration between psychiatrist and cardiologist (see also Chapters 10 and 11).

PSYCHIATRIC SIDE EFFECTS OF CARDIOVASCULAR DRUGS[1]

Psychiatric side effects are common with cardiovascular drugs (Table 23-1). Essentially all of the drugs shown also can cause dizziness. Among the antiarrhythmic agents, lidocaine is the most common cause of central nervous system (CNS) excitement and psychosis. It is well known that digitalis toxicity may include the visual illusion of yellow haloes around lights, but it can also produce anorexia, depression, and delirium. The visual symptoms may sometimes be present but not be volunteered by the patient unless the physician specifically asks about them. It is often unappreciated that mental symptoms may be the earliest symptoms of digitalis intoxication and may serve as an early warning of potentially life-threatening toxicity. Psychiatric symptoms can occur even at "normal" serum drug levels (Eisendrath & Sweeney, 1987).

CNS symptoms are the most common side effects of antihypertensives, occurring to varying degrees with almost all antihypertensive drugs (Gengo & Gabos, 1988). Depressive symptoms are often seen in patients receiving antihypertensive drugs, especially reserpine, beta-blockers, methyldopa, and clonidine (Avorn, Everitt, & Weiss, 1986); DeMuth & Ackerman, 1983; Goodwin & Bunney, 1971). Most commonly, such "depressions" are characterized by vegetative symptoms (insomnia, weakness, poor concentration, lethargy, decreased libido), with less frequent or absent psychologic symptoms (self-deprecation, hopelessness, guilt, suicidal ideation). Hydrophilic beta-blockers that enter the CNS less readily (e.g., atenolol, nadolol) may cause less depression and less of other psychiatric side effects than lipophilic beta-blockers (e.g., propranolol, metoprolol), but not all agree that there is a difference (Dimsdale, Newton, & Joist, 1989; Gengo & Gabos 1988; "Choice of a beta-blocker." [1986]).

Among the antihypertensive agents, angiotensin-converting enzyme (ACE) inhibitors have fewer psychiatric side effects (including sexual dysfunction) than methyldopa or propranolol (Croog et al., 1986), but they occasionally have been reported to elevate or depress mood (Williams, 1988). A calcium-channel blocker or ACE inhibitor usually would be the best first choice for the treatment of hypertension in a patient with a major psychiatric disorder. The abrupt discontinuation of antihypertensives, especially beta-blockers, can produce withdrawal symptoms, including severe anxiety, restlessness, palpitations, vivid dreams, and tremor in addition to hypertension, angina, nausea, vomiting, and headache (Houston & Hodge, 1988).

When diuretics cause hypokalemia or hypovolemia, a similar secondary mood disorder may occur. Diuretic-induced hypokalemia is frequently overlooked, especially in the elderly. Anorexia, lethargy, weakness, constipation, and depressive affect are common. Less often, irritability, anxiety, or delirium occur. When the motor weakness associated with (unrecognized) hypokalemia is severe, sometimes extending to paralysis, a misdiagnosis of conversion disorder may even be made (Lishman, 1987). Hyponatremia, a potential side effect of thiazides, frequently presents with abnormal mental status.

Some patients taking antihypertensives or diuretics who are thought to be "reactively depressed" over their cardiovascular disease are actually suffering iatrogenic symptoms. Other cardiac drugs (including calcium channel blockers, ACE inhibitors, tocainide, flecainide, and propafenone) may also produce anorexia and weight loss as a result of drug-induced abnormalities of taste and smell (Funck-Bretano et al., 1990; Levenson, 1985). The appendix to this chapter contains a list of antihypertensives and antiarrhythmics with their indications, doses and side effects.

Drug Interactions

The most common adverse effect of combining psychotropic and cardiac drugs is hypotension, since many in each class can themselves produce hypotension (antidepressants, antipsychotics, MAO inhibitors; antihypertensives, diuretics, antiarrhythmics, vasodilators). Since TCAs, and to a lesser extent antipsychotics, may prolong the P-R, QRS, and Q-T intervals, clinicians

[1]As a convenience for the reader, tables of medications used in the treatment of cardiovascular disease (antihypertensives, antiarrhythmics, etc.) are included as an appendix to this chapter.

TABLE 23–1. *Psychiatric Side Effects of Cardiac Drugs*

Drugs	Psychosis	Anxiety or Agitation	Depression	Cognitive Deficits or Confusion	Insomnia
Antiarrhythmics					
Lidocaine	×	×		o	
Tocainide	×	×		×	
Flecainide	o	×			
Encainide					
Phenytoin	o	o		×	o
Mexiletine	o	×			×
Quinidine	o	×		×	
Procainamide	o	o	o		
Disopyramide		o	o		o
Amiodarone			o	o	
Bretylium	o	o		o	
Digitalis	o	o	o	o	o
Moricizine	o	o	o	o	o
Propafenone				o	
Antihypertensives					
Methyldopa	o		×	×	×
Beta-blockers	o		×	o	×
Reserpine	×	o	×	o	o
Clonidine	o	o	×	o	×
Hydralazine	o	o	o	o	
Prazosin	o	o	×		
Minoxidil					
Guanethidine			o		
Sodium nitroprusside		o			
ACE inhibitors	o	o	o	o	o
Captopril					
Enalapril					
Lisinopril					
Calcium channel blockers	o	o	o	o	o
Diltiazem					
Nifedidipine					
Nicardipine					
Verapamil					
Diuretics					
Furosemide					
Hydrochlorothiazide			×[a]		
Bumetanide	×[b]			×[b]	
Inotropes					
Amrinone, milrinone, dopamine, dobutamine					
L-Dopa	×	×	×	×	×

× = occasional, o = rare to uncommon.
[a]Via electrolyte/fluid imbalance.
[b]In patients with impaired hepatic function.
See also tables in the appendix of this chapter.

should be aware of their synergistic effects on conduction time with other antiarrhythmics (e.g., quinidine, procainamide, and amiodarone) to avoid producing atrioventricular block or a long Q-T syndrome. If a psychotropic is being initiated in a cardiac patient who is already on one of these antiarrhythmics, it would be prudent to initiate treatment in the hospital. The new drug propafenone competes with weak bases for binding to alpha$_1$-acid glycoprotein (Funck-Brentano et al., 1990) and therefore might increase free tricyclic levels. Tricyclics that cause dry mouth may reduce the effectiveness of sublingual nitrates because of failure to dissolve medication.

Fluoxetine, which is tightly bound to plasma protein, may theoretically affect digitoxin and warfarin activity (both of which are highly protein bound) by shifts in plasma concentrations. The protein binding of digoxin is approximately 27%; hence the activity of digoxin would not be expected to be affected by fluoxetine in a major way. Digitoxin, which is 90% protein bound, might be displaced by fluoxetine, leading to toxicity. Reports of clinically significant reactions between fluoxetine and warfarin appear to be extremely rare, as are significant interactions with digoxin.

Antidepressants and some phenothiazines block the antihypertensive effects of guanethidine and bethanidine by blocking the uptake of the latter drugs into neurons. Methyldopa has precipitated symptoms of lithium toxicity in patients receiving both drugs, even at nontoxic lithium levels. Methyldopa may also cause haloperidol toxicity on a previously well-tolerated dose. In general, methyldopa should be avoided in patients with psychiatric disorders.

Thiazide diuretics and salt-restricted diets may both raise lithium levels into the toxic range, but neither constitutes an absolute contraindication to lithium treatment, because lithium dosing can be adjusted downward. Nonthiazide diuretics (e.g., furosemide) have less pronounced effects on lithium levels. During an acute diuresis, such as that produced in congestive heart failure when cardiac drugs rapidly improve cardiac output and renal perfusion, adjustment of lithium dosage can be very difficult because of rapid and unpredictable changes in sodium and fluid balance. In such a case, a neuroleptic is safer than lithium until the period of acute diuresis is over. In patients who chronically receive diuretics, lithium treatment can be safely continued as long as the dose is adjusted for any further change in diuretics, salt intake, or cardiac output (see Chapter 9 for further treatment of this topic). Angiotensin-converting enzyme inhibitors (e.g., captopril, enalapril) occasionally can increase lithium levels. Calcium channel blockers have been reported sometimes to decrease lithium levels, but in other cases to amplify lithium neurotoxicity. Verapamil inhibits the metabolism of carbamazepine, raising the carbamazepine level (MacPhee et al., 1986).

It is well known that the administration of sympathomimetic agents to patients on MAO inhibitors can cause a hypertensive crisis. The combination of MAO inhibitors with beta-blockers, methyldopa, or reserpine has also been warned against, but few data are available. Phenothiazines and beta-blockers together may produce mutual enhanced drug effects through mutual inhibition of their hepatic metabolism, and some beta-blockers reduce the metabolism of benzodiazepines. Barbiturates will tend to decrease the effects of digoxin and quinidine via the induction of hepatic enzymes. Trazodone may increase blood levels of phenytoin. Drug interactions with summary tables can be found in Chapter 9.

SPECIAL TOPICS

Pacemakers and Implantable Defibrillators

Permanent pacemakers and implantable cardioverter-defibrillators involve the implantation of a foreign body and produce different stresses than other types of cardiac surgery (Phibbs & Marriott, 1985). In other types of cardiac surgery, the patient can feel "cured," but a pacemaker is a constant reminder of illness and possible sudden death. Anxiety is common early after implantation. Depression may appear after hospital discharge, especially in patients with strong needs to be independent and in control (Blacher & Basch, 1970). Those with implanted defibrillators primarily experience fear of shocks or battery failure (Morris et al., 1991; Tchou et al., 1989), but panic attacks, machine dependency, and counterphobic acting out have all been described (Fricchione, Olson, & Vlay, 1989).

Mitral Valve Prolapse and Panic Disorder

Mitral valve prolapse (MVP) and panic disorder share many symptoms, including tachycardia, palpitations, light-headedness, atypical chest pain, dyspnea, and fatigue. Patients with either may appear highly anxious. The relationship between MVP and panic is not entirely clear. Some, but not all, studies have found a significantly increased prevalence of MVP in panic disorder, with a range of zero to 50% (Dager, Comess, & Dunner, 1986; Liberthson et al., 1986) and an average of 18% (Margraf, Ehlers, & Roth, 1980). In contrast, studies of patients with MVP have not found an increased prevalence of panic disorder (Devereux, Kramer-Fox, & Kligfield, 1989; Margraf, Ehlers, & Roth, 1988). Although there do appear to be some differences between patients with panic disorder and MVP versus patients with panic disorder without MVP, these are of doubtful clinical significance (Gorman et al., 1988; Matuzas et al., 1989). The association may not be specific since MVP has been found to be more common in other psychiatric disorders, including bipolar disorder, generalized anxiety disorder, and anorexia nervosa (Margraf, Ehlers, & Roth 1988).

The conflicting and confusing nature of these results arises from several factors. The criteria used for the diagnosis of MVP vary considerably. Echocardiographic diagnosis of MVP, which is the usual procedure, is unreliable (Dager et al., 1989). Many studies have suffered from small and biased samples, poor controls, and nonblinded ratings. Although the association between panic disorder and MVP remains controversial, most would agree that it involves mild, nonstructural (i.e., nonpathologic) forms of MVP. It is not clear whether one predisposes toward the other, whether they share a common underlying etiologic factor, or whether they are just frequent but unrelated comorbidities. In a patient with panic disorder with mild MVP seen on echocardiogram and no history of heart disease, antibiotic prophylaxis for dental and other procedures is not necessary. More significant MVP may indicate antibiotic prophylaxis; a cardiologist should be consulted in doubtful cases.

The appropriate workup of the patient with panic disorder and a systolic murmur depends both on the murmur and on the clinical context. Patients with personal or family risk factors for valvular heart disease; those with symptoms of palpitations, syncope, or chest pain; and those with loud or atypical murmurs all should have echocardiograms.

REFERENCES

ABRAMOV LA (1976). Sexual life and sexual frigidity among women developing acute myocardial infarction. Psychosom Med 38:418–425.

AHERN DK, GORKIN L, ANDERSON JL, ET AL. (1990). Biobehavioral variables and mortality on cardiac arrest in the cardiac arrhythmia pilot study (CAPS). Am J Cardiol 66:59–62.

ANDA RF, WILLIAMSON DF, ESCOBEDO LG, ET AL. (1990). Depression and the dynamics of smoking. JAMA 264:1541–1545.

APA Task Force on Laboratory Test in Psychiatry (1987). The dexamethasone suppression test: An overview of its current status in psychiatry. Am J Psychiatry 144:1253–1262.

APPELS A (1990). Mental precursors of myocardial infarction. Br J Psychiatry 156:465–471.

AVORN J, EVERITT DE, & WEISS S (1986). Increased antidepressant use in patients prescribed β-blockers. JAMA 255:357–360.

BABB SV, DUNLOP SR, & HOFFMAN MA (1990). Protracted ventricular arrhythmias occurring after abrupt tricyclic antidepressant withdrawal. Psychosomatics 31:452–454.

BADEMOSI O, FALASE AO, JAIYESIMI F, ET AL. (1979). Neuropsychiatric manifestations of infective endocarditis: A study of 95 patients at Ibadan, Nigeria. J Neurol Neurosurg Psychiatry 39:325–329.

BARSKY AJ, HOCHSTRASSER B, COLES NA, ET AL. (1990). Silent myocardial ischemia. Is the person or the event silent? JAMA 264:1132–1135.

BASS C, WADE C, & GARDNER WN (1983). Unexplained breathlessness and psychiatric morbidity in patients with normal and abnormal coronary arteries. Lancet 1:605–608.

BASS E (1985). Cardiopulmonary arrest: Pathophysiology and neurological complications. Ann Intern Med 103:902–927.

BENSON H, & MCCALLIE DP (1979). Angina pectoris and the placebo effect. N Engl J Med 300:1424–1429.

BLACHER RS (1978). Paradoxical depression after heart surgery: A form of survivor syndrome. Psychoanal Q 47:267–283.

BLACHER RS, & BASCH SH (1970). Psychological aspects of pacemaker implantation. Arch Gen Psychiatry 22:319–323.

BLANCHARD EB, & MILLER ST (1977). Psychological treatment of cardiovascular disease. Arch Gen Psychiatry 34:1402–1413.

BOOTH-KEWLEY S, & FRIEDMAN HS (1987). Psychological predictors of heart disease: A quantitative review. Psychol Bull 101:343–362.

BRACKETT CD (1988). Psychosocial and psychological predictors of sudden cardiac death after healing of acute myocardial infarction. Am J Cardiol 61:979–983.

BUFF DD, BRENNER R, KIRTANE SS, ET AL. (1991). Dysthythmia associated with fluoxetine treatment in an elderly patients with cardiac disease. J Clin Psychiatry 52:174–176.

BURKE WJ, RUTHERFORD JL, ZORUMSKI CF, ET AL. (1985). Electroconvulsive therapy in the elderly. Compr Psychiatry 26:480–486.

Cardiac Arrhythmia Pilot Study Investigators (1988). Effects of encainide, flecainide, imipramine and moricizine on ventricular arrhythmias during the year after acute myocardial infarction: The CAPS. Am J Cardiol 61:501–509.

CARNEY RM, RICH MW, FREEDLAND KE, ET AL. (1988). Major depressive disorder predicts cardiac events in patients with coronary artery disease. Psychosom med 50:627–633.

CARNEY RM, RICH MW, TEVELDE A, ET AL. (1987). Major depressive disorder in coronary artery disease. Am J Cardiol 60:1273–1275.

CASSEM NH, & HACKETT TP (1977). Psychological aspects of myocardial infarction. Med Clin North Am 61:711–721.

CAY EL (1982). Psychological aspects of cardiac rehabilitation: Unsolved problems. Adv Cardiol 31:237–241.

CHELSKY LB, CUTLER JE, GRIFFITH MN, ET AL. (1990). Caffeine and ventricular arrhythmias. An electrophysiological approach. JAMA 264:2236–2240.

Choice of a beta-blocker. (1986). Med Lett 28:20–22.

CONNOLLY SJ, MITCHELL LB, SWERDLOW CD, ET AL. (1984). Clinical efficacy and electrophysiology of imipramine for ventricular tachycardia. Am J Cardiol 53:516–521.

CORMIER LE, KATON W, RUSSO J, ET AL. (1988). Chest pain with negative cardiac diagnostic studies: Relationship to psychiatric illness. J Nerv Ment Dis 176:358–361.

CROOG SH, LEVINE S, TESTA MA, ET AL. (1986). The effects of antihypertensive therapy on the quality of life. N Engl J Med 314:1657–1664.

DAGER SR, COMESS KA, & DUNNER DL (1986). Differentiation of anxious patients by two-dimensional echocardiographic evaluation of the mitral valve. Am J Psychiatry 143:533–535.

DAGER SR, COMESS KA, SAAL AK, ET AL. (1989). Diagnostic reliability of M-mode echocardiography for detecting mitral valve prolapse in 50 consecutive panic patients. Compr Psychiatry 30:369–375.

DALACK GW, & ROOSE SP (1990). Perspectives on the relationship between cardiovascular disease and affective disorder. J Clin Psychiatry 51(7, suppl):4–9.

DEC GW JR, STERN TA, & WELCH C (1985). The effects of electroconvulsive therapy on serial electrocardiograms and serum cardiac enzyme values. A prospective study of depressed hospitalized inpatients. JAMA 253:2525–2529.

DEMUTH GW, & ACKERMAN SH (1983). Alphamethyldopa and depression: A clinical study and review of the literature. Am J Psychiatry 140:534–538.

DE-PING LEE D, DE QUATTRO V, ALLEN J, ET AL. (1988). Behavioral vs. β-blocker therapy in patients with primary hypertension: Effects on blood pressure, left ventricular function and mass, and the pressor surge of social stress anger. Am Heart J 116:637–644.

DEVEREUX RB, KRAMER-FOX R, & KLIGFIELD P (1989). Mitral valve prolapse: Causes, clinical manifestations, and management. Ann Intern med 111:305–317.

DIMSDALE JE (1988). A perspective on type A behavior and coronary disease. N Engl J Med 318:110–112.

DIMSDALE JE, NEWTON A, & JOIST B (1989). Neuropsychological side effects of beta-blockers. Arch Intern Med 149:514–525.

DOEHRMAN SR (1977). Psychosocial aspects of recovery from coronary heart disease: A review. Soc Sci Med 11:199–218.

DUBIN WR, FIELD HL, & GASTFRIEND DR (1979). Postcardiotomy delirium: A critical review. J Thorac Cardiovasc Surg 77:586–594.

EISENDRATH SJ, & SWEENEY MA (1987). Toxic neuropsychiatric effects of digoxin at therapeutic serum concentrations. Am J Psychiatry 144:506–507.

ELLISON JM, MILOFSKY JE, & ELY E (1990). Fluoxetine-induced bradycardia and syncope in two patients. J Clin Psychiatry 5:385–386.

ENGEL BT, GLASGOW MS, & GAARDNER KR (1983). Behavioral treatment of blood pressure: 3. Follow-up results and treatment recommendations. Psychosom Med 45:23–29.

FOLLICK MJ, AHERN DK, GORKIN L, ET AL. (1990). Relation of psychosocial and stress reactivity variables to ventricular arrhythmias in the Cardiac Arrhythmia Pilot Study (CAPS). Am J Cardiol 66:63–67.

FOLLICK MJ, GORKIN L, CAPONE RJ, ET AL. (1988). Psychological distress as a predictor of ventricular arrhythmias in a post-myocardial infarction population. Am Heart J 116:32–36.

FRASURE-SMITH N, & PRINCE R (1985). The ischemic heart disease life stress monitoring program: Impact on mortality. Psychosom Med 47:431–435.

FREEMAN LJ, NIXON PG, SALLABARK P, ET AL. (1987). Psychological stress and silent myocardial ischemia. Am Heart J 114:477–482.

FREIDMAN M, THORESEN CE, & GILL JJ (1986). Alteration of type A behavior and its effect in cardiac recurrences in post myocardial infarction patients: Summary results of the Recurrent Coronary Prevention Project. Am Heart J 112:653–665.

FREYHAN FA, GIANELLI S JR, O'CONNELL RA, ET AL. (1979). Psychiatric complications following open heart surgery. Compr Psychiatry 12:181–195.

FRICCHIONE GL, OLSON LC, & VLAY SC (1989). Psychiatric syndromes in patients with the automatic internal cardioverter defibrillator: Anxiety, psychological dependence, abuse, and withdrawal. Am Heart J 117:1411–1214.

FROELICHER VF (1983). Exercise testing and training. Chicago: Year Book Medical Publishers.

FUNCKE-BRENTANO C, KROEMER HK, LEE JT, ET AL. (1990). Propafenone. N Engl J Med 322:518–525.

GENGO FM, & GABOS C (1988). Central nervous system considerations in the use of β-blockers, antiotensin-converting enzyme inhibitors, and thiazide diuretics in managing essential hypertension. Am Heart J 116:305–310.

GIARDINA EG, BIGGER JT JR, GLASSMAN AH, ET AL. (1979). The electrocardiogram and antiarrhythmic effects of imipramine hydrochloride at therapeutic plasma concentrations. Circulation 60:1045–1052.

GLASSMAN AH, & BIGGER JT (1981). Cardiovascular effects of therapeutic doses of tricyclic antidepressants: A review. Arch Gen Psychiatry 38:815–820.

GLASSMAN AH, HELZER JE, COVEY LS, ET AL. (1990). Smoking, smoking cessation, and major depression. JAMA 264:1546–1549.

GLASSMAN AH, WALSH BJ, ROOSE SP, ET AL. (1982). Factors related to orthostatic hypotension associated with tricyclic antidepressnats. J Clin Psychiatry 43(2):35–38.

GOLDMAN LS, ALEXANDER RC, & LUCHINS DJ (1986). Monoamine oxidase inhibitors and tricyclic antidepressants: Comparison of their cardiovascular effects. J Clin Psychiatry 47:225–229.

GOODWIN FK, & BUNNEY WE (1971). Depressions following reserpine: A re-evaluation. Semin Psychiatry 3:435–448.

GORDON T, & KANNEL WB (1971). Premature mortality from coronary heart disease: The Framingham study. JAMA 215:1617–1625.

GORMAN JM, GOETZ RR, FYER M, ET AL. (1988). The mitral valve prolapse—panic disorder connection. Psychosom Med 50:114–122.

GRADMAN AH (1988). Cardiac effects of cocaine: A review. Yale J Biol Med 61:137–147.

GREENLAND P, & CHU JS (1988). Efficacy of cardiac rehabilitation services, with emphasis on patients after myocardial infarction. Ann Intern Med 109:650–663.

GUTTMACHER LB, & GREENLAND P (1990). Effects of electroconvulsive therapy on the electrocardiogram in geriatric patients with stable cardiovascular diseases. Convulsive Ther 6:5–12.

GUY S, & SILKE B (1990). The electrocardiogram as a tool for therapeutic monitoring: A critical analysis. J Clin Psychiatry 51(12 suppl B):37–39.

HACKETT TP, CASSEM NH, & WISHNIE HA (1968). The coronary-care unit: An appraisal of its psychological hazards. N Engl J Med 279–1365–1370.

HAMMEKE TA, & HASTINGS JE (1988). Neuropsychological alterations after cardiac operation. J Thorac Cardiovasc Surg 96:326–331.

HAYES RL, GERNER RH, FAIRBANKS L, ET AL. (1983). ECG findings in geriatric depressives gives trazodone, placebo, or imipramine. J Clin Psychiatry 44:180–183.

HIMMELHOCH JM, SCHECHTMAN K, & AUCHENBACK R (1984). The role of trazodone in the treatment of depressed cardiac patients. Psychopathology 17(suppl 2):51–63.

HOUSTON MC, & HODGE R (1988). Beta-adrenergic blocker withdrawal syndromes in hypertension and other cardiovascular diseases. Am Heart J 116:515–522.

JANOWSKY D, CURTIS G, ZISOOK S, ET AL. (1983). Ventricular arrhythmias possibly aggravated by traxodone. Am J Psychiatry 148:796–797.

Joint National Committee on Detection, Evaluation, and Treatment of High Blood Pressure (1984). Hypertension prevalence and the status of awareness, treatment and control in the United States.

KANNEL WB, & THOM TJ (1990). Incidence, prevalence, and mortality of cardiovascular diseases. In JW HURST, RC SCHLANT, CE RACKLEY, ET AL. (Eds.), The heart, arteries, and veins (7th ed., Ch. 37, pp. 627–638). New York: McGraw-Hill.

KAPLAN NM (1985). Non-drug treatment of hypertension. Ann Intern Med 102:359–373.

KASHANI JH, LABABIDI Z, & JONES RS (1982). Depression in children and adolescents with cardiovascular symptomatology: The significance of chest pain. J Am Acad Child Psychiatry 21:187–189.

KENNEDY GJ, HOFER MA, COHEN D, ET AL. (1987). Significance of depression and cognitive impairment in patients undergoing programmed stimulation of cardiac arrhythmias. Psychosom Med 49:410–421.

LANGOSCH W (1988). Psychological effects of training in coronary patients: A critical review of the literature. Eur Heart J 9(suppl M):37–42.

LEVENSON JL (1985). Dysosmia and dysgeusia presenting as depression. Gen Hosp Psychiatry 7:171–173.

LEVENSON JL, & FRIEDEL RO (1985). Treating major depression in cardiac patients. Psychosomatics 26:91–102.

LEVENSON JL, KAY R, MONTEFERRANTE J, ET AL. (1984). Denial predicts favorable outcome in unstable angina pectoris. Psychosom Med 46:25–32.

LEVENSON JL, MISHRA A, BAUERNFEIND RA, ET AL. (1986). Lithium treatment of mania in a patient with recurrent ventricular tachycardia. Psychosomatics 27:594–596.

LEVENSON JL, MISHRA A, HAMER RM, ET AL. (1989). Denial and medical outcome in unstable angina. Psychosom Med 51:27–35.

LEVINE B, KALMAN J, MAYER L, ET AL. (1990). Elevated circulating levels of tumor necrosis factor in severe chronic heart failure. N Engl J Med 323:236–241.

LEVINE J, WARRENBURG S, KERNS R, ET AL. (1987). The role of denial in recovery from coronary heart disease. Psychosom Med 49:109–117.

LIBERTHSON R, SHEEHAN DV, KING ME, ET AL. (1986). The prevalence of mitral valve prolapse in patients with panic disorders. Am J Psychiatry 143:511–515.

LISHMAN WA (1987). Organic psychiatry: The psychological consequences of cerebral disorder (2nd ed., Ch. 11, pp. 476–477). Oxford, England: Blackwell Scientific Publications.

LLOYD GG, & CAWLY RH (1978). Psychiatric morbidity in men one week after first acute myocardial infarction. Br Med J 2:1453–1454.

LOWN B, DESILVA RA, REICH P, ET AL. (1980). Psychophysiologic factors in sudden cardiac death. Am J Psychiatry 137:1325–1335.

LUCHINS DJ (1983). Review of clinical and animal studies comparing the cardiovascular effects of doxepin and other tricyclic antidepressants. Am J Psychiatry 140:1006–1009.

LURIE SN, & COFFEY CE (1990). Caffeine-modified electroconvulsive therapy in depressed patients with medical illness. J Clin Psychiatry 51:154–157.

MACPHEE GJA, THOMPSON GG, MCINNES GT, ET AL. (1986). Verapamil potentiates carbamazepine toxicity: A clinically important inhibitory interaction. Lancet 1:700–703.

MARGRAF J, EHLERS A, & ROTH WT (1988). Mitral valve prolapse and panic disorder: A review of their relationship. Psychosom Med 50:93–113.

MATUZAS W, AL-SADIR J, UHLENHUTH EH, ET AL. (1989). Correlates of mitral valve prolapse among patients with panic disorder. Psychiatry Res 28:161–170.

MCLANE M, KROP H, & MEHTA J (1980). Psychosexual adjustment and counseling after myocardial infarction. Ann Intern Med 92:514–519.

MEHTA J, & KROP H (1979). The effect of myocardial infarction on sexual functioning. Sexual Disability 2:115–121.

MILLER SP (1978). Amenorrhea and anniversary reaction in a woman presenting with chest pain. Am J Psychiatry 135:120–121.

MORRIS P, BADGER J, CHMIELEWSKI C, ET AL. (1991). The automatic implantable cardioverter defibrillator. Psychosomatics 32:58–64.

National Heart, Lung, and Blood Institute (1982). Tenth report of the director, National Heart, Lung, and Blood Institute. Vol 2: Heart and vascular disease. Bethesda, MD: National Institutes of Health.

NEWBERGER JW, SILBERT AR, BUCKLEY LP, ET AL. (1984). Cognitive function and age at repair of transposition of the great arteries in children. N Engl J Med 310:1495–1499.

O'CONNOR GT, BURING JE, YUSUF S, ET AL. (1989). An overview of randomized trials of rehabilitation with exercise after myocardial infarction. Circulation 80:234–244.

OLDRIDGE NB, GUYATT GH, FISCHER ME, ET AL. (1988). Cardiac rehabilitation after myocardial infarction: Combined experience of randomized clinical trials. JAMA 260:945–950.

PATEL C, MARMOT MC, & TERRY DJ (1981). Controlled trial of biofeedback-aided behavioral methods in reducing mild hypertension. Br Med J Clin Res 282:2005–2008.

PATTERSON RE, LIBERMAN HA, MORRIS DC, ET AL. (1986). Noninvasive assessment of coronary artery disease. In AR LEFF (Ed.) Cardiopulmonary exercise testing (Ch. 8, pp. 139–181). Orlando, FL: Grune & Stratton.

PHIBBS B, & MARRIOTT HJL (1985). Complications of permanent transvenous pacing. N Engl J Med 312:1428–1432.

RAZIN AM (1984). Psychotherapeutic intervention in angina: 1. A critical review. Gen Hosp Psychiatry 6:250–257.

RAZIN AM (1985a). Coronary disease: Reducing risk of illness and aiding recovery. In AM RAZIN (Ed.), Helping cardiac patients (Ch. 4, pp. 157–197). San Francisco: Jossey-Bass.

RAZIN AM (1985b). Psychotherapeutic intervention in angina. 2: Implications for research and practice. Gen Hosp Psychiatry 7:9–14.

REGAN TJ (1990). Alcohol and the cardiovascular system. JAMA 264:377–381.

RISCH SC, GROOM GP, & JANOWSKY DS (1982). The effects of psychotropic drugs on the cardiovascular system. J Clin Psychiatry 43:16–31.

ROOSE SP, GLASSMAN AH, GIARDINA EGV, ET AL. (1987a). Cardiovascular effects of imipramine and bupropion and depressed patients with congestive heart failure. J Clin Psychopharmacol 7:247–251.

ROOSE SP, GLASSMAN AH, GIARDINA EGV, ET AL. (1987b). Tricyclic antidepressants in depressed patients with cardiac conduction disease. Arch Gen Psychiatry 44:273–275.

ROZANSKI A, BAIREY CN, KRANTZ DS, ET AL. (1988). Mental stress and the induction of silent myocardial ischemia in patients with coronary artery disease. N Engl J Med 318:1005–1012.

RUBERMAN W, WEINBLATT E, GOLDBERG JD, ET AL. (1984). Psychological influences on mortality after myocardial infarction. N Engl J Med 311:552–559.

SCHLEIFER SJ, MACARI-HINSON MM, COYLE DA, ET AL. (1989). The nature and course of depression following myocardial infarction. Arch Intern Med 149:1785–1789.

SCHOVER LR, & JENSEN SB (1988). Sexuality and chronic illness. A comprehensive approach (Ch. 9, pp. 203–216). New York: Guilford Press.

SMITH LW, & DIMSDALE JE (1989). Postcardiotomy delirium: Conclusions after 25 years? Am J Psychiatry 146:452–458.

STERN MJ, PASCALE L, & ACKERMAN A (1977). Life-adjustment postmyocardial infarction. Arch Intern Med 137:1680–1685.

STOUDEMIRE A, & ATKINSON P (1988). Use of cyclic antidepressants in patients with cardiac conduction disturbances. Gen Hosp Psychiatry 10:389–397.

STOUDEMIRE A, KNOS G, & GLADSON M (1990). Labetalol in the control of cardiovascular responses to electroconvulsive therapy in high-risk depressed medical patients. J Clin Psychiatry 51:508–512.

Sudden death in children treated with a tricyclic antidepressant. (1990). Med Lett 32:53.

TCHOU PJ, PIASECKI E, GUTMANN M, ET AL. (1989). Psychological support and psychiatric management of patients with automatic implantable cardioverter defibrillators. Int J Psychiatry Med 19:393–407.

U.S. Department of Health and Human Services, Social Security Administration (1983). Characteristics of Social Security disability beneficiaries (SSA Publ. No. 13-11947). Washington, DC: U.S. Government Printing Office.

VAZQUEZ-BARQUERO JL, ACERO JAP, OCHOTECO A, ET AL. (1985). Mental illness and ischemic heart disease: Analysis of psychiatric morbidity. Gen Hosp Psychiatry 7:15–20.

VEITH RC, RASKIND M, CALDWELL J, ET AL. (1982). Cardiovascular effects of the tricyclic antidepressants in depressed patients with chronic heart disease. N Engl J Med 306:954–959.

VIEWEG WVR, YAZEL JJ, & BALLENGER JC (1984). Tricyclic antidepressant use in a patient with bundle branch blocks and ventricular ectopy. J Clin Psychiatry 45:353–355.

VISKIN S, & BELHASSEN B (1990). Idiopathic ventricular fibrillation. Am Heart J 120:661–671.

WARRINGTON SJ, PADGHAM C, & LADER M (1989). The cardiovascular effects of antidepressants. Psychol Med 19(suppl 16):1–40.

WEISS SM (1988). Stress management in the treatment of hypertension. Am Heart J 116:645–649.

WENGER NK, & HELLERSTEIN HK (1984). Rehabilitation of the coronary patient (2nd ed.). New York: John Wiley & Sons.

WILLIAMS GH (1988). Converting-enzyme inhibitors in the treatment of hypertension. N Engl J Med 319:1517–1525.

WILLNER A, & RODEWALD G (EDS.) (1991). The impact of cardiac surgery on the quality of life: Neurological and psychological aspects. New York: Plenum.

WOODS SW, TESAR GE, MURRAY GB, ET AL. (1986). Psychostimulant treatment of depressive disorders secondary to medical illness. J Clin Psychiatry 47:12–15.

YUSUF S, WITTES J, & FRIEDMAN L (1988). Overview of results of randomized clinical trials in heart disease. JAMA 260:2088–2093.

ZACHARIAH PK, & ROSENBAUM AH (1982). Stabilization of high blood pressure with tricyclic antidepressants and lithium combinations in hypertensive patients. Mayo Clin Proc 57:625–628.

APPENDIX

Some Oral Antihypertensive Drugs

Drug	Daily Adult Dosage	Frequent or Severe Adverse Effects[a]
ANGIOTENSIN-CONVERTING ENZYME (ACE) INHIBITORS		
Captopril (*Capoten*)	Initial: 25–50 mg in 2 doses Maintenance: 50–300 mg in 2 doses	Cough; hypotension, particularly with a diuretic or volume depletion; loss of taste with anorexia; skin rash; bronchospasm; acute renal failure with bilateral renal artery stenosis or stenosis of the artery to a solitary kidney; cholestatic jaundice; angioedema; hyperkalemia if also on potassium supplements or potassium-retaining diuretics; blood dyscrasias and renal damage are rare except in patients with renal dysfunction, and particularly in patients with collagen-vascular disease; may increase fetal mortality and should not be used during pregnancy
Enalapril (*Vasotec*)	Initial: 2.5–5 mg in one dose Maintenance: 5–40 mg in 1 or 2 doses	Similar to captopril (manufacturer of enalapril claims lower incidence of some effects, but comparative data are limited); pancreatitis
Lisinopril (*Prinivil, Zestril*)	5–40 mg in 1 dose	Similar to captopril
Ramipril (*Altace*)	Initial: 2.5 mg in 1 dose Maintenance: 2.5–20 mg in 1 or 2 doses	Similar to captopril
CALCIUM-CHANNEL BLOCKERS		
Diltiazem (*Cardizem SR*)	60–180 mg twice daily	Similar to verapamil, but less likely to cause constipation
Isradipine (*DynaCirc*)	Initial: 5 mg in 2 doses Maintenance: 5–20 mg in 2 doses	Similar to nifedipine
Nicardipine (*Cardene*)	20–40 mg three times daily	Similar to nifedipine
Nifedipine (*Procardia XL*)	30–120 mg once daily	Similar to verapamil, but more likely to cause edema, and less likely to cause constipation or AV block; may cause tachycardia; arthralgias
Verapamil (*Calan, Isoptin, Verelan,* others)	120–480 mg in 2 or 3 doses (*Isoptin SR, Calan SR,* and *Verelan* can be given 240 mg once or twice daily)	Heart failure; hypotension; AV block; constipation; dizziness; edema; headache, bradycardia
DIURETICS		
THIAZIDE-TYPE		Hyperuricemia; hypokalemia; hypomagnesemia; hyperglycemia; hyponatremia; hypercalcemia; pancreatitis; rashes and other allergic reactions; increased serum low density lipoprotein cholesterol and triglyceride concentrations, may be transient; depression; impotence
Bendroflumethiazide (*Naturetin*)	2.5–5 mg	
Benzthiazide (*Exna*, others)	12.5–50 mg	
Chlorothiazide (*Diuril*, others)	125–500 mg	
Cyclothiazide (*Anhydron*)	1–2 mg	
Hydrochlorothiazide (*Esidrix*, others)	12.5–50 mg	
Hydroflumethiazide (*Saluron*, others)	12.5–50 mg	
Methyclothiazide (*Enduron*, others)	2.5–5 mg	
Polythiazide (*Renese*)	2–4 mg	
Trichlormethiazide (*Naqua*, others)	1–4 mg	
Chlorthalidone (*Hygroton*, others)	12.5–50 mg	
Indapamide (*Lozol, Lozide*)	2.5 mg	
Metolazone (*Zaroxolyn, Diulo*) (*Mykrox*)	1.25–10 mg 0.5–1 mg	
Quinethazone (*Hydromox*)	25–200 mg	
LOOP DIURETICS		Dehydration; circulatory collapse; thromboembolism; hypokalemia; hypomagnesemia; hypocalcemia; hyperglycemia; metabolic alkalosis; hyperuricemia; blood dyscrasias; rashes; lipid changes as with thiazide-type diuretics
Bumetanide (*Bumex*)	0.5–10 mg in 2 to 3 doses	
Furosemide (*Lasix*, others)	40–320 mg in 2 to 3 doses	

Some Oral Antihypertensive Drugs (continued)

Drug	Daily Adult Dosage	Frequent or Severe Adverse Effects[a]
POTASSIUM-RETAINING		
Amiloride (*Midamor*, others)	5–10 mg in 1 dose	Hyperkalemia; GI disturbances; rash; headache
Spironolactone (*Aldactone*, others)	25–100 mg in 1 or 2 doses	Hyperkalemia; hyponatremia; gynecomastia; agranulocytosis; menstrual abnormalities; GI disturbances; rash; tumorigenic in rats
Triamterene (*Dyrenium*)	50–200 mg in 1 or 2 doses	Hyperkalemia; GI disturbances; increased blood urea nitrogen; metabolic acidosis; nephrolithiasis
SOME COMBINATIONS		Similar to individual components
Aldactazide (hydrochlorothiazide 25 or 50 mg and spironolactone 25 or 50 mg)	1–2 tablets daily	
Dyazide (hydrochlorothiazide 25 mg and triamterene 50 mg)	1–2 tablets daily	
Maxzide (hydrochlorothiazide 25 or 50 mg, triamterene 37.5 or 75 mg)	1 tablet daily	
Moduretic (hydrochlorothiazide 50 mg and amiloride 5 mg)	½–1 tablet daily	
DRUGS WITH PERIPHERAL SYMPATHOLYTIC ACTION		
BETA-ADRENERGIC BLOCKING DRUGS		
Acebutolol (*Sectral*)	200–1200 mg in 1 or 2 doses	Similar to propranolol, but has intrinsic sympathomimetic activity and is relatively cardioselective, with less lipid changes and resting bradycardia and more antinuclear antibodies; occasional drug-induced lupus
Atenolol (*Tenormin*)	25–100 mg in 1 dose	Similar to propranolol; relatively cardioselective at ≤100 mg/day
Betaxolol (*Kerlone*)	10–40 mg in 1 dose	Similar to propranolol; relatively cardioselective
Carteolol (*Cartrol*)	2.5–10 mg in 1 dose	Similar to propranolol, but has intrinsic sympathomimetic activity and less resting bradycardia and lipid changes; asthenia and muscle cramps
Labetalol (*Trandate, Normodyne*)	200–1200 mg in 2 doses	Similar to propranolol, but has intrinsic sympathomimetic activity and more orthostatic hypotension; fever; hepatotoxicity
Metoprolol (*Lopressor*)	50–200 mg in 1 or 2 doses	Similar to propranolol; at doses <200 mg is relatively cardioselective
Nadolol (*Corgard*)	20–160 mg in 1 dose	Similar to propranolol
Penbutolol (*Levatol*)	20–80 mg in 1 dose	Similar to propranolol, but has intrinsic sympathomimetic activity and less resting bradycardia and lipid changes
Pindolol (*Visken*)	10–40 mg in 2 doses	Similar to propranolol, but has intrinsic sympathomimetic activity and less resting bradycardia and lipid changes
Propranolol (*Inderal*, others)	40–320 mg in 2 doses (*Inderal LA* can be given in one daily dose)	Fatigue; depression; bradycardia; decreased exercise tolerance; congestive heart failure; aggravates peripheral vascular disease; GI disturbances; increased airway resistance; masks symptoms of hypoglycemia; Raynaud's phenomenon; vivid dreams or hallucinations; organic brain syndrome; rare blood dyscrasias and other allergic disorders; increased serum triglycerides, decreased HDL cholesterol; generalized pustular psoriasis; transient hearing loss; sudden withdrawal can lead to exacerbation of angina and myocardial infarction
Timolol (*Blocadren*, others)	10–40 mg in 2 doses	Similar to propranolol
PERIPHERAL ADRENERGIC NEURON ANTAGONISTS		
Guanadrel (*Hylorel*)	10–50 mg in 2 doses	Similar to guanethidine, but less diarrhea, aggravation of asthma has not been reported
Guanethidine (*Ismelin*)	10–300 mg in 1 dose	Orthostatic hypotension; exercise hypotension; diarrhea; may aggravate bronchial asthma; bradycardia; sodium and water retention; retrograde ejaculation
Reserpine (*Serpasil*, others)	0.05–0.25 mg in 1 dose	Psychic depression; nightmares; nasal stuffiness, drowsiness; GI disturbances; bradycardia

Some Oral Antihypertensive Drugs (continued)

Drug	Daily Adult Dosage	Frequent or Severe Adverse Effects[a]
ALPHA-ADRENERGIC BLOCKERS		
Prazosin (*Minipress*, others)	First day: 1 mg at bedtime Maintenance: 1–20 mg in 2 doses	Syncope with first dose; dizziness and vertigo; palpitations; fluid retention; headache, drowsiness; weakness; anticholinergic effects; priapism; urinary incontinence
Terazosin (*Hytrin*)	First day: 1 mg at bedtime Maintenance: 1–20 mg in 1 dose	Similar to prazosin
Doxazosin (*Cardura*)	First day: 1 mg at bedtime Maintenance: 1–16 mg in 1 dose	Similar to prazosin, but with less hypotension after first dose
DRUGS WITH CENTRAL SYMPATHOLYTIC ACTION		
Clonidine (*Catapres*, others) (*Catapres TTS*, transdermal)	0.1–0.6 mg in 2 doses One patch weekly (0.1–0.3 mg/day)	Severe insomnia and rebound hypertension; headache; cardiac arrhythmias after sudden withdrawal; CNS reactions similar to methyldopa, but more sedation and dry mouth; bradycardia; contact dermatitis from patches
Guanabenz (*Wytensin*)	8–32 mg in 2 doses	Similar to clonidine
Guanfacine (*Tenex*)	1–3 mg in 1 or 2 doses	Similar to clonidine, but milder
Methyldopa (*Aldomet*, others)	500 mg–2 grams in 2 doses	Sedation and other CNS symptoms; fever; orthostatic hypotension; bradycardia; GI disorders, including colitis; hepatitis; cirrhosis; hepatic necrosis; Coombs' positive hemolytic anemia; lupus-like syndrome; immune thrombocytopenia; red cell aplasia
DIRECT VASODILATORS		
Hydralazine (*Apresoline*, others)	40–200 mg in 4 doses	GI disturbances; tachycardia; aggravation of angina; headache and dizziness, fluid retention; nasal congestion; rashes and other allergic reactions; lupus-like syndrome; hepatitis glomerulonephritis
Minoxidil (*Loniten*, others)	First day: 5 mg Maintenance: up to 100 mg/day; usually 10–40 mg is adequate	Tachycardia; aggravation of angina; marked fluid retention; possible pericardial effusion; hair growth on face and body; coarsening of facial features; thrombocytopenia; leukopenia

[a]Most antihypertensive drugs can probably cause sexual dysfunction (*Medical Letter* 29:65, 1987).
Reprinted by permission. Med Lett 33:35–36 (1991).

Calcium-Channel Blockers for Hypertension

Drug	Initial Dosage	Cost[a]
Diltiazem–*Cardizem SR* (Marion Merrell Dow)	60–120 mg bid	$36.75
Isradipine—*DynaCirc* (Sandoz)	2.5 mg bid	27.60
Nicardipine—*Cardene* (Syntex)	20 mg tid	31.96
Nifedipine—*Procardia XL* (Pfizer)	30 mg once	31.59
Verapamil—average generic price (range: $6.07 to $22.59)	80 mg tid	18.11
Calan (Searle)		35.59
Isoptin (Knoll)		35.59
sustained release—*Calan SR* (Searle)	180 mg once	28.66
Isoptin SR (Knoll)		28.66
Verelan (Lederle)	240 mg once	29.73

[a]Cost to the pharmacist for 30 days' treatment of hypertension at the manufacturer's recommended initial dosage, according to Average Wholesale Price listings in *First DataBank Price Alert*, April 15, 1991.
Reprinted by permission from Isradipine for hypertension (1991). Med Lett 33:51.

Beta Blockers Available in the United States

Drug	Activity	Half-life (hours)	Maintenance Dosage[1]	Cost of 30 Days' Treatment[1]
Acebutolol (*Sectral*—Wyeth-Ayerst)[2]	$Beta_1$-selective, intrinsic sympathomimetic	3–4	400 mg once/day	$22.17
Atenolol (*Tenormin*—ICI)	$Beta_1$-selective	6–7	50 mg once/day	17.71
Carteolol (*Cartrol*—Abbott)	Nonselective, intrinsic sympathomimetic	5–6	2.5 mg once/day	17.82

(*Beta Blockers Available in the United States continued*)

Drug	Activity	Half-life (hours)	Maintenance Dosage[1]	Cost of 30 Days' Treatment[1]
Labetalol (*Normodyne*—Key) (*Trandate*—Allen Hanburys)	Nonselective beta- and selective $alpha_1$-blocking activity	6–8	200 mg bid	23.98 22.40
Metoprolol (*Lopressor*—Geigy)[3,4]	$Beta_1$-selective	3–7	100 mg once/day	16.71
Nadolol (*Corgard*—Princeton)[4]	Nonselective	20–24	40 mg once/day	19.14
Penbutolol (*Levatol*—Reed & Carnrick)	Nonselective, intrinsic sympathomimetic	5	20 mg once/day	16.51
Pindolol (*Visken*—Sandoz)	Nonselective, intrinsic sympathomimetic	3–4	10 mg bid[5]	34.56
Propranolol[6]—average generic price (range: $1.22 to $20.84)	Nonselective	4	60 mg bid	5.61
(*Inderal*—Wyeth-Ayerst)				23.37
extended release generic		10	120 mg once/day	16.78
(*Inderal LA*—Wyeth-Ayerst)				20.64
Timolol[3]—average generic price (range: $13.16 to $22.08)	Nonselective	4–5	10 mg bid	17.65
(*Blocadren*—Merck)				23.26

[1]Lowest usual maintenance dosage recommended by the manufacturer for treatment of hypertension is used to calculate the cost to the pharmacist, according to Average Wholesale Prices listed in *Drug Topics Red Book* 1989 and July *Update* or *Red Book Data Base Services.* The cost to the patient will be higher.
[2]Also marketed for treatment of cardiac arrhythmias.
[3]Also marketed for use following myocardial infarction.
[4]Also marketed for treatment of angina pectoris.
[5]Manufacturer does not specify maintenance dosage.
[6]Also marketed for use in management of arrhythmias, angina pectoris, migraine, hypertrophic subaortic stenosis, and following myocardial infarction.
Reprinted by permission from Carteolol and penbutolol for hypertension (1989). Med Lett 31:71.

ACE Inhibitors for Hypertension

Drug	Daily Dosage	Dosage for Renal Impairment	Cost[a]
Benazepril—*Lotensin* (Ciba)	Initial: 10 mg Usual: 20–40 mg O.D. or bid Maximum: 80 mg	Serum creatinine >3 mg/dl: 5–40 mg	$19.09
Captopril—*Capoten* (Squibb)	Initial: 25 mg bid or tid Usual: 25–150 mg bid or tid Maximum: 450 mg	Decreased	34.05
Enalapril—*Vasotec* (Merck)	Initial: 5 mg Usual: 10–40 mg O.D. or divided Maximum: 40 mg	Serum creatinine >3 mg/dl: 2.5–40 mg	25.09
Fosinopril—*Monopril* (Mead Johnson)	Initial: 10 mg Usual: 20–40 mg O. D. or bid Maximum: 80 mg	No change	22.75
Lisinopril—*Prinivil* (Merck) *Zestril* (Stuart)	Initial: 10 mg Usual: 20–40 mg O. D. Maximum: 80 mg	Serum creatinine >3 mg/dl: 5–40 mg	22.66 22.66
Ramipril—*Altace* (Hoeschst)	Initial: 2.5 mg Usual: 5–10 mg O. D. or bid Maximum: 20 mg	Serum creatinine >2.5 mg/dl: 1.25–5 mg	22.25

[a]Cost to the pharmacist for 30 days' treatment with the lowest usual dosage, according to Average Wholesale Prices listed in *Drug Topics MicREData*, August 1991.
Reprinted by permission from Three new ACE inhibitors for hypertension (1991). Med Lett 33:84.

Drugs of Choice for Common Arrhythmias

Arrhythmia	Drug of Choice	Alternatives	Remarks
Atrial fibrillation or atrial flutter[1]	Digoxin to slow ventricular response	Verapamil or a beta-blocker to slow ventricular response[2]	Digoxin, verapamil, and possibly beta-blockers may be dangerous for patients with Wolff-Parkinson-White syndrome.
		Quinidine, procainamide, or disopyramide for long-term suppression[3]	Amiodarone, flecainide, encainide, and propafenone are also effective for suppression.[3]

Drugs of Choice for Common Arrhythmias (continued)

Arrhythmia	Drug of Choice	Alternatives	Remarks
Supraventricular tachycardia[4]	Adenosine or verapamil[5] IV for termination	Esmolol, another beta-blocker, or digoxin for termination	Cardioversion or atrial pacing may be required for some patients. Quinidine, procainamide, disopyramide, diltiazem, propranolol, acebutolol, verapamil, flecainide, encainide, propafenone, or digoxin may be effective for long-term suppression.
Ventricular premature complexes (VPCs) or nonsustained ventricular tachycardia	No drug therapy indicated for asymptomatic patients[6]	For symptomatic patients, a beta-blocker	There is no evidence that prolonged suppression with drugs prevents sudden cardiac death. For post-MI patients, treatment with a beta-blocker has decreased mortality, and treatment with encainide or flecainide has increased it.
Sustained ventricular tachycardia[7,8]	Lidocaine for acute treatment	Procainamide, bretylium	Cardioversion is the safest and most effective therapy. A beta-blocker, procainamide, quinidine, amiodarone, disopyramide, or mexiletine can be used for long-term suppression.
Ventricular fibrillation[9]	Lidocaine	Procainamide, amiodarone, bretylium	Cardiopulmonary resuscitation with rapid defibrillation is essential.
Cardiac glycoside-induced ventricular tachyarrhythmias[8,10]	Lidocaine	Phenytoin, procainamide, a beta-blocker	Self-limited if short-acting digitalis stopped. Digoxin-immune Fab (digoxin antibody fragments—*Digibind*) should be used to treat life-threatening digoxin or digitoxin intoxication (*Medical Letter*, 28:87, 1986). Avoid cardioversion and bretylium, except for ventricular tachycardia. A beta-blocker or procainamide can make heart block worse.
Torsades de pointes	Magnesium[11]	Cardiac pacing, isoproterenol[12]	Magnesium may be effective even in absence of hypomagnesemia. Potassium may also be needed and causative agents (eg, quinidine) should be discontinued.

[1]Cardioversion is the treatment of choice for recent onset with compromised circulation due to an excessively rapid ventricular rate. For patients with atrial flutter, atrial pacing can also be effective.

[2]May be preferred for some patients (RH Falk and JI Leavitt, *Ann Intern Med*, 114:573, 1991).

[3]A recent analysis of several older studies suggests that use of quinidine for this indication may increase mortality. Whether other drugs used for this condition might also increase mortality remains to be determined.

[4]Vagotonic maneuvers (such as carotid sinus massage, gagging, the Valsalva maneuver, or increasing venous return by straight leg raising) may be tried first. Two recent reports indicate that radio frequency current delivered through a catheter electrode may cure some patients with supraventricular tachycardia, particularly those with Wolff-Parkinson-White syndrome (WM Jackman et al. *N Engl J Med*, 324:1605, June 6, 1991; H Calkins et al. *N Engl J Med*, 324:1612, June 6, 1991).

[5]Verapamil is contraindicated for patients receiving intravenous beta-blockers or those with congestive heart failure and should be used with caution in patients taking oral quinidine.

[6]Some authorities give lidocaine early during acute myocardial infarction to prevent ventricular fibrillation.

[7]Cardioversion is preferred by most cardiologists for sustained ventricular tachycardia causing hemodynamic compromise, but some first try a chest thump, IV lidocaine, or both.

[8]Some ventricular tachycardias can be caused or exacerbated by bradycardia or heart block. In the presence of high-grade heart-block, antiarrhythmic drugs can cause cardiac standstill. When high-grade heart block is present, therefore, a temporary pacemaker should be inserted before using antiarrhythmic drugs; pacing may abolish the arrhythmia. When a drug must be used in the presence of heart block, lidocaine or phenytoin is least likely to increase the block.

[9]Defibrillation is the treatment of choice; drugs are for prevention of recurrence.

[10]KCl can be given carefully, usually 20 mEq/hr IV, to patients with low or normal serum potassium concentrations. Extreme care must be taken to keep serum potassium below 5.5 mEq/L. In the presence of heart block not associated with paroxysmal atrial tachycardia, potassium should be withheld if the serum concentration is greater than 4.0 to 4.5 mEq/L, since high serum potassium may increase atrioventricular block.

[11]D Tzivoni et al, *Circulation*, 77:392, 1988.

[12]A Keren et al, *Circulation*, 64:1167, 1981.

Reprinted by permission from Drugs for cardiac arrhythmias (1991). Med Lett 33:55–56.

Dosage and Adverse Effects of Antiarrhythmic Drugs

Drug	Usual Dosage[a] and Interval	Effect on ECG	Adverse Effects	Serum Concentrations[b]
IA. Quinidine (many manufacturers)	PO: 200–400 mg q4–6h (sulfate) 324–648 mg q8–12h (gluconate)	Prolongs QRS, QT and (±) PR	Diarrhea and other GI symptoms, cinchonism, hepatic granulomas and necrosis, thrombocytopenia, rashes, hypotention, heart block tachyarrhythmias, torsades de pointes, fever, lupus-like syndrome	2–5 μg/ml
Procainamide (*Pronestyl*, and others)	PO: 50–100 mg/kg/day in divided doses, q3–4H or q6h (sustained-release) IV Loading: no more than 100 mg q5min to 1 gram (~ 12 mg/kg) IV Maintenance: 2–4 mg/min	Prolongs QRS, QT and (±) PR	Lupus-like syndrome, confusion, disorientation, GI symptoms, rash, hypotension, arrhythmias, torsades de pointes, blood dyscrasias, fever	4–10 μg/ml (NAPA: active metabolite 10 to 20 μg/ml)
Disopyramide (*Norpace*, and others)	PO: 100–200 mg q6–8h (long-acting formulatin available)	Prolongs QRS, QT and (±) PR	Anticholinergic effects, hypotension, heart failure, tachyarrhythmias, torsades de pointes, heart block, nausea, vomiting, diarrhea, cholestatic jaundice, agranulocytosis, constipation, hypoglycemia, nervousness	2–5 μg/ml
1B. Lidocaine (*Xylocaine*, and others)	IV Loading: 1 mg/kg given over 2 min, then 2 mg/kg over 20 min *or* 50 mg given over 1 min and repeated every 5 min × 3 *or* 20 mg/min infused over 10 min IV Maintenance: 1–4 mg/min	No significant change	Drowsiness or agitation, slurred speech, tinnitus, disorientation, coma, seizures, paresthesias, cardiac depression, especially with excessive accumulation in heart failure or liver failure or infusions for more than 24 hours, bradycardia/asystole	1.5–5 μg/ml
Phenytoin (*Dilantin*, and others)	PO Loading: 14 mg/kg PO Maintenance: 200–400 mg/day IV Loading: 50 mg q5min to total dose of 1000 mg (~ 12 mg/kg) IV Maintenance: 200–400 mg/day	No significant change	Ataxia, nystagmus, drowsiness, coma, blood dyscrasias, cardiac toxicity with rapid IV injection, fever, rash, hepatic granulomas and necrosis	5–20 μg/ml
Mexiletine (*Mexitil*)	PO Initial dose: 100–200 mg q8h taken with food PO Maintenance: 100–300 mg q6–12 h, maximum 1200 mg/day	No significant change	GI upset, fatigue, nervousness dizziness, tremor, sleep upset, seizures, visual disturbances, psychosis, fever, hepatic toxicity, blood dyscrasias	0.5–2 μg/ml
Tocainide (*Tonocard*)	PO Initial dose: 200–400 mg q8h PO Maintenance: 200–600 mg q8h, maximum 2400 mg/day	No significant change	GI upset, paresthesias, dizziness, tremor, confusion, nightmares, psychotic reactions, coma, seizures, rash, fever, arthralgia, agranulocytosis, aplastic anemia, thrombocytopenia, hepatic granulomas, interstitial pneumonitis	3–10 μg/ml

Dosage and Adverse Effects of Antiarrhythmic Drugs (continued)

Drug	Usual Dosage[a] and Interval	Effect on ECG	Adverse Effects	Serum Concentrations[b]
1C. Flecainide (*Tambocor*)	PO Initial dose: 100 mg q 12h, increase q4–6 days if required, by 50 mg q12h PO Maintenance: ≤ 400 mg/day	Prolongs PR and QRS	Bradycardia, heart block, new ventricular fibrillation, sustained ventricular tachycardia, heart failure, dizziness, blurred vision, nervousness, headache, GI upset, neutropenia	0.2–1 μg/ml
Encainide (*Enkaid*)	PO Initial dose: 25 mg q8h, increase q4–6 days if required to 35 mg q8h, and then to 50 mg q8h PO Maintenance: ≤ 200 mg/day	Prolongs PR and QRS	Bradycardia, heart block, new ventricular fibrillation, sustained ventricular tachycardia, heart failure, dizziness, headache, visual disturbances, diarrhea, GI upset, glucose intolerance	Active metabolites preclude establishment
Propafenone (*Rythmol*)	PO Initial dose: 150 mg q8h, increase q3–4 days if required PO Maintenance: 150–300 mg q8h	Prolongs PR and QRS	Bradycardia, heart block, new ventricular fibrillation, sustained ventricular tachycardia, heart failure, dizziness, lightheadedness, metallic taste dysgeusia, GI upset, bronchospasm	Active metabolites preclude establishment
OTHER I.				
Moricizine (*Ethmozine*)	PO Initial dose: 200 mg q8h, increase q3–4 days if required PO Maintenance: 200–300 mg q8h	Prolongs PR and QRS	Bradycardia, heart failure, new ventricular fibrillation, sustained ventricular tachycardia, nausea, dizziness, headache	Not established
II. BETA-ADRENERGIC BLOCKERS				
Propranolol (*Inderal*, and others	PO: 10–80 mg q6h (long-acting formulation available) IV: 1–5 mg total (1 mg/min)	Prolongs PR (±) No Change QRS Shortens QT Bradycardia	Heart block, hypotension, heart failure, bronchospasm	Not established
Acebutolol (*Sectral*)	PO: 200 mg bid, increase gradually to 600–1200 mg/day	Bradycardia	Hypotension, bradycardia, bronchospasm, ANA, arthritis, myalgia, arthralgia, lupus-like syndrome, pulmonary complications	Not established
Esmolol (*Brevibloc*)	IV Loading: 500 μg/kg over one min followed by 25 μg/kg/min; increase by 25 to 50 μg/kg/min q4 min to desired effect Usual maintenance: 100 μg/kg/min; maximum300 μg/kg/min	Bradycardia	Hypotension, heart block, heart failure, bronchospasm, pain at infusion site	0.15–2 μg/ml
III. Amiodarone (*Cordarone*)	PO Loading: 800–1600 mg/day for 1–3 weeks then 600–800 mg/day for 4 weeks PO Maintenance: 100–400 mg/day	Prolongs PR, QRS, and QT Sinus bradycardia	Acute pulmonary toxicity, pulmonary fibrosis, bradycardia, heart block, new ventricular fibrillation, sustained ventricular tachycardia, torsades de pointes (unusual), hyper- or hypothyroidism, GI upset, alcoholic-like hepatitis, phospholipidosis, peripheral neuropathy, ataxia, tremor, dizziness, photosensitivity, blue-gray skin, corneal microdeposits	Not established

Dosage and Adverse Effects of Antiarrhythmic Drugs (continued)

Drug	Usual Dosage[a] and Interval	Effect on ECG	Adverse Effects	Serum Concentrations[b]
Bretylium (*Bretylol*, and others; *Bretylate* in Canada)	IV Loading: 5 mg/kg with additional doses of 10 mg/kg to maximum of 30 mg/kg (effect may be delayed) IV Maintenance: 5–10 mg/kg q6h or continuous infusion 1–2 mg/min	No change Sinus bradycardia	Orthostatis hypotension, nausea and vomiting, increased sensitivity to catecholamines, initial increase in arrhythmias	Not established
IV. CALCIUM-CHANNEL BLOCKER				
Verapamil HCl (*Isoptin*, *Calan*, and others)	IV Initial dose: 5–10 mg over 2–3 min; repeat in 30 min, if necesssary IV infusion: Initial dose is followed by 0.375mg/min for 30 min IV Maintenance: 0.125 mg/min PO: 40–120 mg tid or qid (long-acting formulation available)	Prolongs PR	Heartblock, hypotension, asytole, bradycardia, dizziness, headache, fatigue, peripheral edema, nausea, constipation, Stevens-Johnson syndrome	Probably 100–300 ng/ml
V. OTHERS				
Adenosine (*Adenocard*)	IV: 3–6 mg; may repeat 6–12 mg rapid bolus injection	Prolongs PR	Transient dyspnea, chest discomfort (non-myocardial), hypotension	Not established
Digoxin (*Lanoxin*, and others)	PO (tablets): 1–1.5 mg over 24 hr in 3–4 divided doses Maintenance: 0.125–0.5 mg/day	Prolongs PR Depresses ST segment Flattens T wave	Anorexia, nausea, vomiting, diarrhea, abdominal pain, headache, confusion, abnormal vision, bradycardia, AV block, arrhythmias	1–2ng/ml

[a]Patients with decreased hepatic or renal function may require lower dosage.
[b]Range of usually effective and tolerated concentrations.
Reprinted by permission from Drugs for cardiac arrhythmias (1991). Med Lett 33:59–60.

24 Oncology

LYNNA M. LESKO, M.D., Ph.D,
MARY JANE MASSIE, M.D., AND
JIMMIE HOLLAND, M.D.

Thirty percent of Americans now living will develop some form of malignancy, and three out of four families will be affected. Four out of 10 patients who get cancer will be surviving the illness 5 years after diagnosis. Several cancers that had a very poor prognosis 20 years ago, such as acute lymphocytic leukemia in children, Hodgkin's disease, Ewing's and osteogenic sarcoma, Wilms' tumor, and testicular cancer, are being cured today.

Cancer has remained, however, a disease equated with hopelessness, pain, fear, and death. Its diagnosis and treatment often produce psychologic stresses resulting from the actual symptoms of the disease, as well as the patient's and family's perceptions of the disease and its stigma. The universal patient fears have been termed "the six D's" (Holland et al., 1979): *death*; *dependency* on family, spouse, and physician; *disfigurement* and change in body appearance and self-image, sometimes resulting in loss of changes in sexual functioning; *disability* interfering with achievement of age-appropriate tasks in work, school, or leisure roles; *disruption* of interpersonal relationships: and, finally *discomfort* or pain in later stages of illness. Currently, cancer treatment includes multimodal treatment regimens (surgery, chemotherapy, radiation), drugs used in high doses or administered by various routes (intrathecal, intracarotid, or intrahepatic), and innovative procedures such as bone marrow transplantation, all adding to a large number of cancer survivors. Consequently, as the study of survivorship in cancer patients is begun, a seventh issue can be added—*disengagement* from the cancer diagnosis, treatment, hospital, and the patient role and reentry into a near-normal life-style.

The patient's ability to manage these stresses depends on medical, psychologic, and social issues. These include: (1) the disease itself (i.e., site, symptoms, clinical course, type of treatment required); (2) prior level of adjustment, especially to medical illness; (3) the threat that cancer poses in attaining age-appropriate developmental tasks and goals (i.e., adolescence, career, family); (4) cultural and religious attitudes; (5) the presence of emotionally supportive persons in the patient's environment; (6) the patient's potential for physical and psychologic rehabilitation; and (7) the patient's own personality and coping style (Holland, 1982a).

Why are physicians beginning to focus on the psychologic concerns of patients with cancer? Figure 24-1 summarizes data from the National Cancer Institute on cancer mortality trends in the United States from 1954 through 1976. Cancer deaths in persons under the age of 30 years have dropped sharply since 1966, largely because of dramatic improvements in survival from certain cancers. Patients under 45 years have had a clear but less impressive decline in cancer mortality. Those under 60 show a plateau in mortality, which began in 1968. Optimism about improved survival has increased, especially concerning pediatric and elderly patients. Treatments are now more rigorous, involving multimodal approaches and multidrug regimens. Procedures such as bone marrow transplantation, once experimental, are becoming standard treatments employed earlier in the course of a patient's disease.

More rigorous treatment, however, requires a higher level of patient participation and responsibility. In keeping with the patients' increased responsibility is a more frank disclosure of diagnosis, prognosis, and treatment options by physicians. Patients and families appear more interested in treatment issues and quality of life, both during and after treatment, than in the past. Patients no longer accept global statements such as "You should be grateful just to be alive." Since patients survive longer, there are more delayed effects of treatment. Consequently, professional attention has turned from death and dying issues to the emotional, physical, and behavioral consequences of rigorous treatment regimens. Quality of life and delayed treatment effects have become important clinical and research issues.

In this chapter are reviewed the normal responses to cancer; the most frequently encountered psychiatric disorders; psychotherapeutic, pharmacologic, and behavioral management of these disorders; and several special

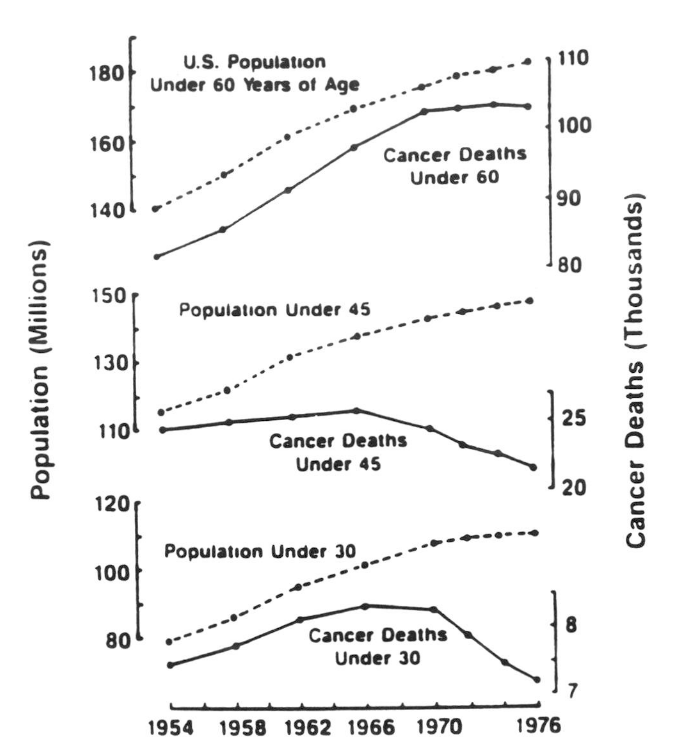

FIG. 24-1. Cancer mortality and U.S. population trends, 1954–1976. (*Source:* U.S. Public Health Statistics.)

topics, such as central nervous system (CNS) side effects of cancer treatment, nausea and vomiting, anorexia, drug interactions, unconventional cancer treatments, terminal illness, and survivor issues.

PSYCHIATRIC DISORDERS IN CANCER PATIENTS

Prevalence

There have been many myths about the psychologic state of cancer patients. Assumptions have varied from "All patients are depressed and need psychiatric intervention" to "Patients manage well and few need help." Prevalence studies counter these attitudes. The Psychosocial Collaborative Oncology Group (PSYCOG) reported a study of 215 randomly selected hospitalized and ambulatory patients at three major cancer centers (Derogatis et al., 1983). Using DSM-III diagnostic criteria, 47% of patients met criteria for a psychiatric disorder and 53% did not. Of the 47% who had a recognizable psychiatric disorder, 68% had an adjustment disorder with depressed, anxious, or mixed mood; 13% had major depression; 8% had an organic mental disorder; 7% had a personality disorder; and 4% had an anxiety disorder. The spectrum of depressive disorders, encompassing adjustment disorder with depressed mood and major depression, accounted for the majority of diagnoses. Nearly 90% of the psychiatric disorders observed were either reactions to or manifestations of disease or treatment. Only 11% represented prior psychiatric problems, such as personality and anxiety disorders (Figure 24-2). Patients with cancer are thus largely psychologically healthy individuals who have emotional distress related to illness.

Bukberg, Penman, and Holland (1984) examined the prevalence of depression in hospitalized cancer patients, using modified DSM-III criteria that eliminated physical symptoms. These authors found that 24% of these patients were severely depressed, 18% were moderately depressed, and 14% had depressive symptoms of "sadness." The remaining 44% showed no symptoms of depression despite their cancer. The factor most significantly related to severe depression was physical function. Seventy-seven percent of those who were most depressed were also the most physically impaired, although distinguishing vegetative signs of depression from physical symptoms is difficult at greater levels of illness, making interpretation of findings difficult. These findings are similar to those of a study by Plumb and Holland (1977), who found that 23% of cancer patients were significantly depressed. Two studies of patients on general medical floors found a similar prevalence of depression, supporting the belief that patients with malignancies are no more or less depressed than patients with equally physically debilitating illness. In summary, approximately 25% of all cancer patients, irrespective of their hospital and physical status, may be experiencing significant depression (Moffic & Paykel, 1975; Schwab et al., 1967).

To date it is impossible to stratify the prevalence of psychiatric disorders with certain cancer types because of the paucity of comparison studies. However, there has been much debate about the increased prevalence of depression in pancreatic cancer patients. Holland and colleagues (1986) conducted a controlled study to compare the prevalence of depression in a large cohort of advanced pancreatic cancer patients ($N = 107$) compared to a cohort of advanced gastric cancer patients ($N = 111$). Medical information, sociodemographic variables, and data from psychosocial inventories indicated that patients with pancreatic cancer had significantly greater general psychologic distress (depression and anxiety). Possible mechanisms for this increased prevalence of depression or altered affective state include a tumor-mediated paraneoplastic syndrome, antibody production induced against a protein released by tumor that crosses the blood-brain barrier to bind

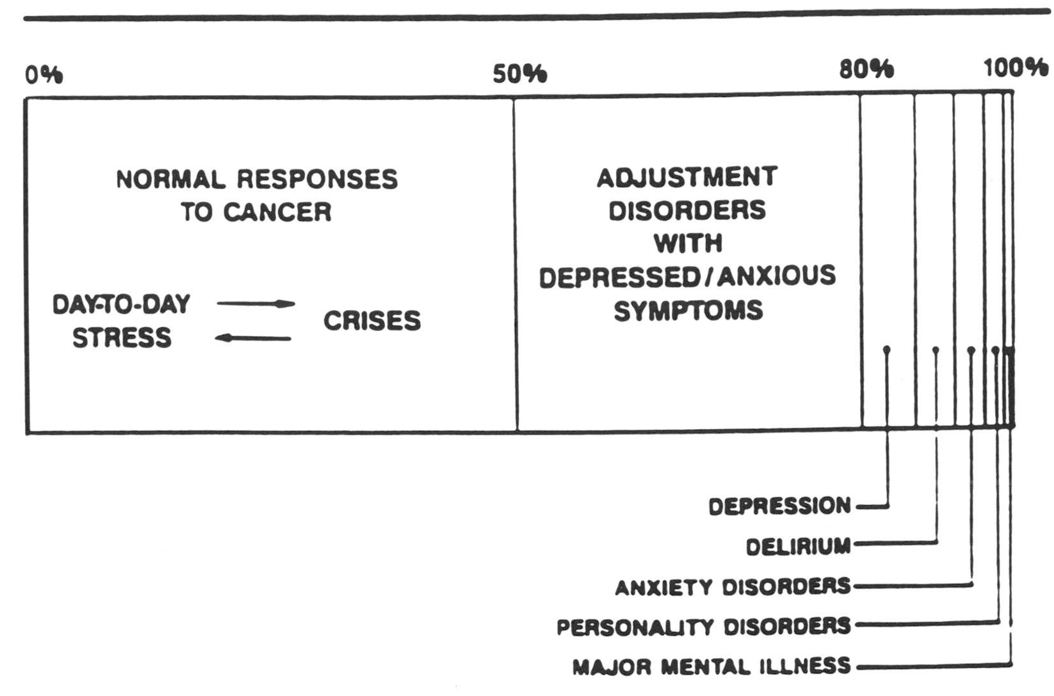

FIG. 24-2. Spectrum of psychiatric disorders in cancer. (Derived from PSYCOG prevalence data [Derogatis et al., 1983].)

with serotonin receptors, or, finally, autoantibody production that stimulates antiidiotypic antibodies that act as an alternative receptor for serotonin and reduce its synaptic availability (Holland, 1989). Diagnosis of depression in physically ill patients is also discussed in Chapter 4.

Cancer-Related Suicide

In the United States, suicide is the ninth leading cause of death (12.5 deaths for every 100,000 people) (Resnik, 1980), but accurate data are difficult to obtain about suicide in patients with cancer. There is a general consensus among caregivers in the cancer setting that relatively few patients actually commit suicide. Three studies, in Finland (Louhivouri & Hakama, 1979), Connecticut (Fox et al., 1982), and Sweden (Bolund, 1985), indicate that, although few cancer patients commit suicide, they may be at a somewhat greater risk than the general population. Death certificates often do not indicate suicide in patients with advanced illness, however, and the actual rate may be underreported.

Factors associated with an increased risk of suicide in cancer patients include an advanced stage of disease, a poor prognosis, mild delirium with poor impulse control, poorly controlled pain, depression, preexisting psychiatric or personality disorders (prior alcohol abuse or suicide attempts), physical and emotional exhaustion, and social isolation. Patients will risk factors should be monitored more closely as being at higher risk. Most of the patients who have suicidal ideation do not carry out attempts, and the thought of suicide provides a sense of ultimate control: "If it gets too bad, I can always commit suicide" (Holland, 1982a). Suicidal risk is slightly increased in patients with head and neck cancer. Since tumors in the mouth and pharynx are associated with alcohol and tobacco abuse, often with preexisting personality disorder, these patients may be at greater risk for that reason. Breitbart (1989) found that patients with acquired immunodeficiency syndrome (AIDS) constituted a large proportion of patients seen in consultation for evaluation of suicidal risk.

Clinical evaluation and management principles, in the presence of advanced cancer or AIDS, must take into account the patient's short expected survival. Although the psychiatrist must assess the patient for major depression or other predisposing conditions and treat them, suicidal risk must be evaluated in light of short expected survival and the desirability of keeping the patient at home with family or significant others. At times, the psychiatrist must simply recognize the risk of rational suicide and discuss it with the family and the oncologist, rather than recommend psychiatric hospitalization. Evaluation and management of suicidal patients in the medical setting are discussed in Chapter 5.

DIAGNOSIS AND MANAGEMENT

The most frequent disorders in cancer patients are anxiety, depression, delirium, nausea and vomiting, anorexia, and pain. Each is discussed in the following subsections in relation to their diagnosis and management.

Anxiety

Anxiety in the oncologic setting is either acute, related to disease symptoms or treatment, or chronic, which antedates the cancer and is exacerbated by the illness. Acute anxiety occurs at several points: (1) while awaiting the diagnosis of cancer, (2) while awaiting procedures and tests (bone marrow aspiration, wound débridement, lumbar puncture), (3) prior to major treatment (surgery, chemotherapy, radiation), (4) while awaiting test results, (5) with change of treatment, (6) after learning of relapse, and (7) on the anniversary of illness-related events. Pain, hypoxia, endocrine abnormalities, drug withdrawal, and medications may also produce symptoms of anxiety.

The chronic anxiety states that may cause problems during cancer include generalized anxiety disorders, simple phobias (e.g., claustrophobia during diagnostic imaging procedures), fear of needles, and panic states provoked by stressful events. All may require use of anxiolytic drugs, supplemented in many cases by behavioral techniques (relaxation or distraction) or psychotherapy. Hypnosis (self-hypnosis) may be useful in selected patients (see Chapter 6).

Acute and chronic anxiety states in cancer usually are treated with benzodiazepines. Antipsychotic medications, however, in low doses (e.g., thioridazine 10 mg tid) can be used for severe anxiety when a therapeutic dose of a benzodiazepine is not effective. Antihistamines can be used for patients who have serious respiratory impairment and for those in whom physicians have other reasons to be concerned about suppression of central respiratory mechanisms by the benzodiazepines. Anxiety accompanied by hyperventilation frequently occurs in lung cancer patients who are short of breath. Concerns about compromising respiratory function should not preclude a trial of an anxiolytic, since the patient's shortness of breath may be markedly improved when the secondary anxiety is improved. Low-dose neuroleptics are the preferable alternative in patients with carbon dioxide retention. Tricyclic antidepressants (e.g., imipramine) have anxiolytic, antipanic, and antiphobic effects in patients for whom they are not medically contraindicated (see Chapter 9). Buspirone (a nonbenzediazepine anxiolytic) may be a useful adjunct to psychotherapy in patients with chronic anxiety; however, patients who have been treated previously with benzodiazepines may find buspirone a less immediately effective treatment.

In patients with hepatic disease related to cancer or its treatment, the benzodiazepines of choice are the short-acting compounds that are metabolized primarily by conjugation and are excreted by the kidney (e.g., oxazepam and lorazepam). In control of anxiety before chemotherapy or painful procedures, the short-acting benzodiazepine lorazepam is useful because it produces anterograde amnesia after both oral and intravenous administration (Healey et al., 1983). It can be given intravenously as part of the antiemetic regimen with some chemotherapeutic agents that produce severe emesis (e.g., cisplatin). Lorazepam is also rapidly absorbed via the sublingual route. Lorazepam does not decrease the number or frequency of emetic events; however, patients remember little of their vomiting episodes (Laszlo et al., 1985). Lorazepam reduces the motor restlessness caused by metoclopramide, commonly given to reduce emesis. Intravenous anesthetics such as fentanyl or ketamine have been used effectively in children undergoing bone marrow aspiration or biopsy. They have an anesthetic and tranquilizing effect within minutes of intravenous administration. Ketamine, however, can produce dream-like states with visual hallucinations, which are distressing to most adults.

Anxiety during terminal illness may be due to poorly controlled pain or hypoxia. It responds to intravenous morphine sulfate and a benzodiazepine or neuroleptic as needed. Recommended starting doses of morphine sulfate are highly variable (0.5–100 mg/hr); maximum doses vary widely (4 to 480 mg/hr) (see Portenoy et al., 1986). Benzodiazepines should usually be avoided in terminally ill patients with delirium because of their tendency to aggravate confusion; low-dose neuroleptics are preferred (see Chapter 16).

Depression

Depression in cancer patients results from: (1) stress related to the cancer diagnosis and treatment, (2) medications (steroids, interferon, or other chemotherapeutic agents), (3) a biologically determined depression (endogenous depression), which is not related to a precipitating event, or (4) recurrence of a bipolar mood disorder. The first two are the most common, but it can be very difficult to determine with certainty whether a depression appearing in cancer is related to a preexisting mood disorder. Whereas the diagnosis of depression in physically healthy patients depends heavily on the somatic symptoms of anorexia, fatigue, and weight loss, these indicators are of little value in the assessment of a cancer patient, since they are common to both cancer and depression. Diagnosis must rest on psychologic, not somatic symptoms: anhedonia, dysphoric mood, and feelings of helplessness or hopelessness, loss of self-esteem, worthlessness, suicide, or guilt. Psychotic depression is rare in the cancer patient, except when associated with steroids. Cancer patients who are at higher risk for depression are those with poor physical condition, inadequately controlled pain, advanced stages of illness, and preexisting mood disorders.

The antidepressant agents that are used in the oncology setting are (1) non–monoamine oxidase inhibitor (MAOI) antidepressants, (2) lithium, (3) MAOIs, and (4) sympathomimetic stimulants. The cyclic antidepressants (e.g., nortriptyline or desipramine) are used most commonly. For reasons that are unclear, many patients with neoplasms and depression respond to doses far below those effective in physically healthy depressed patients. The starting dose is therefore low, beginning with 10 to 25 mg at bedtime, increased by 25 mg every 2 to 3 days until a beneficial effect is seen. Usually this is achieved by 75 to 125 mg/day, although wide variations in dose response exist and serum levels should be monitored, particularly in patients who are either sensitive or refractory to treatment. Elderly patients and those prone to constipation (e.g., patients on opiate analgesics) will usually benefit from concurrent use of 100 to 300 mg/day of docusate. Patients are maintained on their antidepressant for 4 to 6 months after their symptoms have improved; then consideration should be given to gradually tapering and discontinuing the medication. Patients should be followed carefully for signs of relapse, and patients prone to recurrent depression should be maintained on antidepressants indefinitely.

The choice of antidepressant depends on the nature of the depressive symptoms, concurrent medical problems, the side effects of the medication, and the route of access. As do other medically ill patients, the depressed patient who is agitated or has insomnia will benefit from a drug with sedating effects (e.g., nortriptyline, doxepin or trazodone). Patients with psychomotor retardation will benefit from compounds with less sedating effects, such as desipramine. The patient who has stomatitis secondary to radiotherapy or chemotherapy, or urinary retention or slow intestinal motility, should be given a tricyclic with low anticholinergic potency, such as nortriptyline or desipramine. Patients who are unable to take medication in pill form because of oral, pharyngeal, or esophageal surgery, restricted oral intake, or severe esophagitis may be able to take an elixir (nortriptyline, doxepin) or an intramuscular (IM) form (amitriptyline, imipramine). Because of the discomfort caused by the volume of the injected vehicle, 50 mg is usually the maximum dosage that can be delivered per IM injection.

Although intravenous use of tricyclic antidepressants has not been approved in the United States, several studies from Europe indicate their efficacy and safety via this route (Santos, Beliles, & Arana, in press). Because cancer patients have a number of reasons to have altered pharmacokinetics, tricyclic blood levels should be obtained in patients who fail to respond to usual dosages or who show unusually severe side effects at low doses.

Bupropion is a relatively new drug in the United States and its use in cancer patients is as yet limited. At present, it is not the first drug of choice for depressed patients with cancer; however, we would consider prescribing bupropion if patients have a poor response to a reasonable trial of other antidepressants. Bupropion may be somewhat activating in medically ill patients. It should be avoided in patients with seizure disorders and brain tumors and in those who are malnourished since it is a drug that is known to suppress appetite.

Fluoxetine, a selective inhibitor of neuronal serotonin uptake, has fewer sedative and autonomic effects than the tricyclics. The most common side effects are mild nausea and a brief period of increased anxiety. Fluoxetine can cause appetite suppression usually lasting for a period of several weeks. Some of our patients have experienced transient weight loss, but weight usually returns to baseline level. The anorexic properties of fluoxetine have not been a limiting factor in our use of this drug in cancer patients. In general, the side effect profile of fluoxetine may make it a more favorable treatment for depressed medically ill patients (see Chapter 9).

Patients who have been receiving lithium prior to their cancer diagnosis may be maintained on it. Close monitoring is especially important during pre- and postoperative periods, when fluid intake is low; during chemotherapy, when hydration is mandatory; and when patients are receiving a nephrotoxic drug such as cisplatin. The use of lithium to stimulate granulocyte production in neutropenic patients or to prevent leukopenia during chemotherapy has been attempted. The stimulatory effects, however, appear to be minimal and transient; mood changes have not been noted in these patients. Lithium for this use has been supplanted by granulocyte colony stimulating factor. Corticosteroids (prednisone and dexamethasone), commonly used in cancer treatments can cause a range of psychiatric disorders, including mania and depression. There are several reports in the literature regarding the potential usefulness of lithium in prevention of steroid-induced mood changes (Goggans, Wishberg, & Koran, 1983). We have not found lithium helpful in preventing steroid-related affective symptoms in more than a few patients, and prefer to use neuroleptics should psychotic symptoms appear.

If a cancer patient has responded to a MAOI prior to the medical illness, its continued use is appropriate. Most psychiatrists in the oncology setting, however, are reluctant to prescribe a MAOI in patients who often already have dietary restrictions or nutritional deficiencies. If MAOIs are used, meperidine should not be prescribed for pain, because of the possibility of a fatal hypertensive crisis. Furthermore, dosages of other narcotics may have to be reduced because MAOIs may potentiate the effects and retard the metabolism of narcotics.

The psychostimulants dextroamphetamine and methylphenidate are sometimes used with cancer patients who are depressed. Used in the terminal phase of illness, amphetamines often improve appetite, promote an increased sense of well-being, and counter the sedative effects of opiates. The starting dose is 2.5 mg of dextroamphetamine or methylphenidate given twice daily (in the morning and early afternoon); dosages are increased each day until the desired effect is obtained, or until side effects such as tachycardia or insomnia limit further increases (see also Chapter 9). Pemoline, a less potent psychostimulant, has little abuse potential (Chiarello & Cole, 1987). One potential advantage of pemoline is that it is not a controlled substance. It comes in a chewable tablet so patients who have difficulty swallowing can absorb the drug through the buccal mucosa. We have begun to use pemoline frequently in a population of cancer patients with depressive symptoms and it appears to be as effective as methylphenidate and dextroamphetamine. Pemoline should be used with caution in patients with renal impairment, and liver function tests should be monitored periodically with longer term treatment. Psychostimulants can potentiate the analgesic effects of narcotic analgesics while counteracting unwanted daytime sedation (Bruera, 1989). Occasionally, however, they can produce nightmares, insomnia, and even psychosis.

Electroconvulsive therapy (ECT) can be used in depressed cancer patients in whom treatment with antidepressants poses unacceptable side effects or for patients with psychotic or dangerously suicidal features. (see Chapters 10 and 11).

Hematologic Side Effects of Psychotropic Drugs

The incidence of hematologic problems associated with psychotropic drugs is very small (e.g., the incidence of agranulocytosis varies from 1:10,000 to 1:3,000) (Balon & Berchou, 1986). The exact rates are unknown among medically ill patients, and it is presumed that cancer does not contribute to a patient's susceptibility to these side effects. Danielson et al. (1984) reported on drug-induced blood disorders (pancytopenia, hemolytic anemia, thrombocytopenia, and granulocytopenia) at a health maintenance organization in the northwestern United States from 1972 through 1982. During this 10-year period of surveillance of 200,000 members, only 26 persons were hospitalized for hematologic side effects secondary to medications. The rate of side effects was 1:100,000 persons for each year. Most cases reported were of thrombocytopenia caused by sulfa and quinidine. Only 1 of the 26 cases of hematologic side effects was related to a psychotropic drug (amitriptyline).

According to Balon and Berchou (1986), the most frequently reported hematologic side effects of psychotropic medications are agranulocytosis, leukopenia, eosinophilia, thrombocytopenia, purpura, and anemia; less frequently reported side effects are thrombocytosis, leukocytosis, altered platelet function, and immunologic alterations. White middle-aged women may be at a greater risk for development of agranulocytosis. Hematologic toxicity usually develops after several weeks of drug therapy and involves various mechanisms: bone marrow suppression, direct toxicity to bone marrow elements, liver function impairment, immune-related destruction, or direct peripheral blood toxicity. The most frequently reported hematologic side effects are associated with neuroleptics and antidepressants. Agranulocytosis secondary to antidepressants is more rare than that caused by neuroleptics.

In summary, hematologic side effects of psychiatric drugs are very rare and usually disappear after discontinuation of the medication, without adverse sequelae. Psychotropics are often prescribed for life-threatening periods of delirium or depression and should not be withheld because of concern for rare hematologic side effects. The use of psychotropic agents in patients undergoing bone marrow transplants is discussed by House in Chapter 36.

Delirium

Delirium is common in cancer as a result of both the direct involvement of the CNS by tumor and the indirect effects on the CNS of toxic-metabolic consequences of the disease or its treatment. The prevalence of delirium in cancer patients has ranged from about 5% to 25% in various studies (Table 24-1). These differences are based on variations in populations sampled. Some studies report prevalence rates based on screening all hospital admissions sequentially for diagnosable psychiatric disorders (Derogatis, et al., 1983; Folstein, et al., 1984), whereas others report the frequency of cognitive impairment disorders among patients referred for a suspected psychiatric problem (Hinton, 1972; Levine, Silberfarb, & Lipowski, 1978; Massie & Holland, 1984). Older age, preexisting dementia, and levels of physical disability are confounding factors that prohibit interpretation of data to arrive at a uniform prevalence rate. At Memorial Sloan-Kettering Hospital, 15% of 546 inpatients seen by a psychiatric consultation service met diagnostic criteria for delirium (Massie & Holland, 1984).

As in other medical settings, early symptoms of delirium are often unrecognized or misdiagnosed by medical or nursing staff as depression. Yet, early recognition of delirium is important, since the underlying etiology may be a treatable complication of cancer. When faced with an abrupt behavioral change in a cancer patient, the physician must investigate a number of potential causes of delirium. Those causes can be classified as shown in Table 24-2. Particularly frequent metabolic problems include hyponatremia, hypercalcemia, malnutrition, and liver failure. Thyroid or adrenal status may be altered. Patients with hematologic malignancies and AIDS are at especially high risk for opportunistic infection. Other cancers, such as those of the lung and the breast, frequently metastasize to the brain. The design of the "delirium workup" is tailored to the patient's specific cancer.

Chemotherapeutic agents can cause delirium; this effect is dealt with in detail later in this chapter (Table 24-3).

As they are for other medically ill patients, neuroleptics are the mainstay in the management of delirious agitation. Patients who are debilitated require a low starting dose. Haloperidol is the most commonly pre-

TABLE 24–1. *Prevalence of Delirium in Hospitalized Cancer Patients*

Author	Number of Patients	State or Type of Disease	Cognitive Impairment Disorders (CID) (%)	Comments
Hospitalized patients (referred for psychiatric consultation)				
Massie, Holland, and Glass (1983)	334	All stages	25	
Levine, Silberfarb, and Lipowski (1978)	100	All stages	40	All "chronic" and "acute" organic brain syndromes (DSM-II)
Shevits, Silberfarb, and Lipowski (1976)	1,000	Cancer and noncancer (gen. hosp.)	16	All "chronic" and "acute" organic brain syndromes (DSM-II)
Massie and Holland (1987)	546	All stages	20	
Hinton (1972)	50	Terminal illness	10	
Massie, Holland, and Glass (1983)	13	Terminal illness	85	
Hospitalized patients (not only those referred for psychiatric consultation)				
Davis, et al. (1973)	46	Advanced cancer	27	Patients screened for CID & depression
Folstein et al. (1984)	83	Consecutive admissions	26	
Derogatis et al. (1983)	215		8	Included delirium, dementia with depression, organic mood syndrome, atypical organic brain syndrome, organic personality disorder
Posner (1979)		CNS complications of cancer	15	
Hospitalized patients on general medical and surgical services (cancer and noncancer patients)				
Adams (1984)		Gen. surgical patients ≥ 65 years	10–15	
Lipowski (1983)		Gen. medical patients "elderly"	16	

TABLE 24–2. *Causes of Delirium in Cancer Patients*

- Direct effects
  - Primary tumor
  - Metastatic lesions
- Indirect effects
  - Infections (CNS and non-CNS)
  - Vascular complications
  - Metabolic problems
    - Organ failure
    - Electrolyte imbalances
  - Treatment effects
    - Chemotherapeutic agents
    - Steroids
    - Radiation
    - Medications: analgesics, anticholinergic drugs, antibiotics
  - Nutrition
    - Malnutrition of chronic illness
  - Remote effects
    - Paraneoplastic syndromes
    - Tumors that secrete psychoactive substances

scribed antipsychotic drug in this setting because of its low incidence of cardiovascular and anticholinergic effects and because of the availability of several forms of administration (elixir, pill, parenteral). Although not specifically approved by the Food and Drug Administration (FDA) for intravenous (IV) administration, it is often administered by the IV route in cancer centers when patients are thrombocytopenic and unable to take medications IM, when stomatitis or extensive oropharyngeal surgery makes them unable to take any drug orally (PO), and when rapid sedation is mandatory, as with severely agitated patients with thrombocytopenia. The use of IV haloperidol is discussed in detail in Chapter 12.

Nausea and Vomiting

Nausea and vomiting are frequent and distressing complications of some cancers, and are also common side effects of chemotherapy and radiation. If untreated, vomiting leads to dehydration, disturbances in serum electrolytes, malnutrition, metabolic imbalance, aspiration pneumonia, and occasionally esophageal tears. Sometimes oncologists are forced to give less than an optimal dose of a chemotherapeutic regimen because of nausea or vomiting. If nausea or vomiting is severe and untreated, some patients will abandon traditional chemotherapeutic treatment (because of an intolerable

TABLE 24–3. *CNS Toxicity of Major Chemotherapeutic Agents Used in Oncology*

	Delirium	Lethargy	Hallucinations	Dementia	Cognitive Impairment	Depression	Extrapyramidal Symptoms	Cerebellar Dysfunction
Methotrexate[b]	X	X		X				
Fluorouracil[a]	X						X	X
Vincristine[a]	X	X	X					
Vinblastine[a]	X	X	X			X		
Bleomycin	X							
Carmustine (BCNU)	X			X				
Cisplatin	X							
Dacarbazine				X				
Hydroxyurea			X					
Asparaginase[a]	X	X	X		X			
Procarbazine	X	X	X			X		
Prednisone[a]	X	X	X			X		
Cytosine arabinoside (Ara-C)[a]	X	X			X			X
Fludarabine	X			X	X			
Isophosphamide[b]	X	X	X					
5-Azacytidine					X			
Aminoglutethimide					X			
Interferon	X	X	X					
Interleukin	X	X						
Cyclophosphamide	X							

[a]*Common* agents associated with psychiatric side effects.
[b]*Most common* agents associated with psychiatric side effects.

quality of life) and seek out unproven remedies (which are a potential risk for shortened survival). Considerable effort has gone into finding safe and effective antiemetics to ensure patients' ability to tolerate curative treatment regimens.

The etiology of nausea and vomiting in the oncologic patient encompasses (1) physiologic and metabolic, (2) treatment-related, and (3) psychologic causes. The most common causes are listed in Table 24-4.

Metabolic and physiologic causes of vomiting are easily identified, but may be difficult to treat. They include bowel obstruction by tumor, metastases, or mechanical or drug-related ileus, fluid and electrolyte imbalance, metabolic abnormalities of ketoacidosis and uremia or hepatic dysfunction, hypercalcemia, high fever, endocrine dysfunction (i.e., adrenocortical insufficiency), and CNS dysfunction (primary brain tumor, metastatic disease, increased intracranial pressure). Since early satiety is a feature of advanced abdominal cancer, large meals can contribute to anorexia and nausea. Frequent small and visually appealing meals may reduce nausea and result in greater caloric intake. Causes of vomiting unrelated to cancer diagnosis (gastritis, ulcer, pancreatitis, renal or biliary colic, myocardial infarction) should never be overlooked.

Treatment-related nausea or vomiting is one of the most common problems faced by the cancer patient. Radiation, chemotherapy agents, narcotic drugs, and some IV antibiotics all may produce nausea. Figure 24-3 lists chemotherapy agents commonly used in cancer patients and their potential for producing emesis (Gralla, 1983). Their route of administration and the individual variation among patients and treatment cycles affect the incidence and severity of emesis. Chemotherapeutic agents also vary in their mechanism of emetic action. For example, cisplastin, a highly emetic chemotherapeutic agent, acts primarily on the gastrointestinal tract, producing vomiting by stimulation of the vagus nerve. In contrast, 5-fluorouracil causes vomiting by directly stimulating the chemoreceptor trigger zone (CTZ).

TABLE 24–4. *Common Causes of Nausea and Vomiting in Cancer Patients*

Physiologic and metabolic
Bowel obstruction
Uremia
Hepatic dysfunction
CNS disorders (tumor, metastases, increased intracranial pressure)
Fluid and electrolyte imbalance
High fever
Endocrine abnormalities
Treatment related
Chemotherapeutic agents
Radiation
Analgesics
Antibiotics
Psychologic (psychophysiologic)
Anticipatory nausea and vomiting
Anxiety
Anorexia nervosa, bulimia

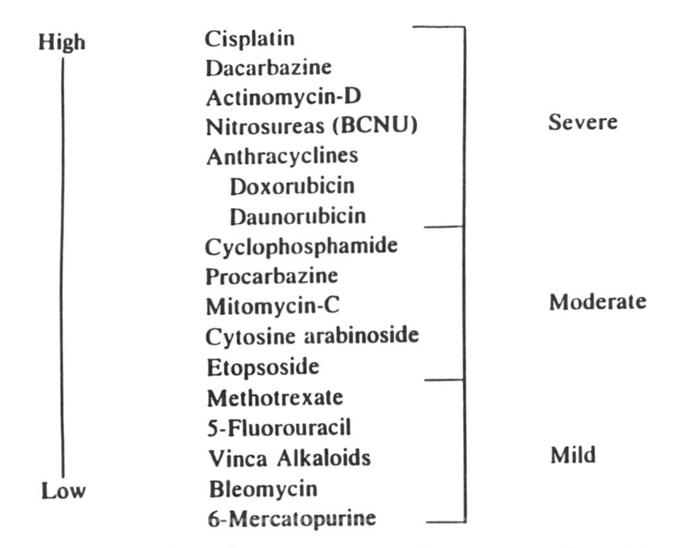

FIG. 24-3. Chemotherapy agents causing nausea and vomiting. (Modified from Gralla J (1985). The control of nausea and vomiting. In Bosel [Ed.], Current concepts in medical oncology (pp. 47–54). New York: Gold Publishers.

Behavioral and psychologic causes of nausea and vomiting are also common. Nausea and vomiting in anticipation of chemotherapy or radiation follows a classical behavioral conditioning paradigm (see the following subsection, "Anorexia," and a later subsection, "Behavioral Interventions"). Present treatment approaches use antiemetics and anxiolytics combined with relaxation or other psychologic and behavioral techniques. Anxiety related to unfamiliar treatments (e.g., radiation equipment or IV infusions and new procedures) can be extreme enough to produce nausea. It is well managed by anxiolytics such as alprazolam, lorazepam, or diazepam. Finally, there is the rare patient who develops cancer in the context of a preexisting eating disorder in which one of the symptoms is vomiting (anorexia nervosa or bulimia).

There are several classes of pharmacologic agents with proven efficacy in the management of chemotherapy-related nausea and vomiting; phenothiazines, butyrophenones, cannabinoids, or combinations of steroids, antihistamines, anxiolytics, and phenothiazines (Table 24-5). Each requires a specific dosage and schedule to be effective. In addition, patients vary in their responses, so no one drug or dose can be assumed to be the best

TABLE 24–5. *Antiemetics Useful for Chemotherapy-Related Nausea and Vomiting*

Dopamine antagonists
- Chlorpromazine
- Prochlorperazine
- Metoclopramide
- Haloperidol

Adjuvants
- Anticholinergics and antihistamines
  - Hydroxyzine pamoate
  - Scopolamine
  - Diphenhydramine
- Adrenergic stimulants
  - Dextroamphetamine
- Cannabinoids
  - Delta-9-tetrahydrocannabinol (THC)
- Benzodiazepines
  - Lorazepam
- Steroids
  - Dexamethasone
  - Methylprednisolone
- 5-Hydroxytryptamine receptor antagonist (ondansetron)

for all patients. The choice of antiemetic agent depends on (1) the emetogenic potential of the chemotherapy agent and its route of action, (2) the etiology of the symptoms, (3) the mode of administration and the side effects of the antiemetic drug, and (4) concurrent medical problems.

Phenothiazines produce their antiemetic effects by partial inhibition of the CTZ via their antidopaminergic properties. They are effective for patients receiving nitrosoureas and methotrexate, but are usually ineffective when used alone in patients receiving the most highly emetogenic chemotherapeutic agent, cisplatin. Since antiemetic effects of this class of neuroleptics require a dosage similar to that used in antipsychotic treatment, extrapyramidal side effects of acute dystonias, akathisia, muscle rigidity, or akinesia are common. They are less common with the low-potency phenothiazines. Prochlorperazine and chlorpromazine can be administered parenterally, rectally, and orally (tablet or elixir).

Metoclopramide, a procainamide derivative and CNS dopamine antagonist with peripheral gastrointestinal cholinergic effects, was used for years in diabetic patients to increase the tone of the lower esophageal sphincter, promote gastric emptying, and increase the motility of the upper gastrointestinal tract. It was found more recently to be highly effective in controlling chemotherapy-related emesis when given in higher doses (Gralla et al., 1981). Given at 2 mg/kg IV 30 minutes prior to chemotherapy, and repeated every 2 hours for two to five doses, it is more effective for emesis resulting from cisplatin than is chlorpromazine (3.0 mg/kg) or haloperidol (1.0 mg/kg) (Gralla et al., 1981). As with other dopamine antagonists, side effects of long-term use of metoclopramide may include tardive dyskinesia. With chronic oral use, metoclopramide may cause a drug-induced Parkinson's syndrome (particularly in the elderly) and akathisia. Butyrophenones, especially haloperidol, are potent inhibitors of the CTZ. Haloperidol at 1 to 3 mg IV or PO q3–6h has shown significant antiemetic activity. Its effect against the highly emetogenic drugs, such as doxorubicin and cisplatin, however, has not been substantiated in controlled studies.

Cannabinoids, such as delta-9-tetrahydrocannabinol (THC) and synthetic cannabinols, have proved to have antiemetic efficacy for several chemotherapeutic agents (e.g., high-dose methotrexate) when compared with placebo. It is believed that the cannabinoids act centrally to raise the nausea and vomiting threshold of the emetic center. The average oral THC dose is 5 to 10 mg/m^2 (7.5 to 15 mg), given 2 hours before chemotherapy and repeated every 3 to 4 hours for 24 to 48 hours. Side effects are considerable for many patients, especially the elderly, who complain of dysphoria and confusion. Some authors believe that smoking marijuana is more effective than taking oral tablets because of incomplete intestinal absorption of the pill.

Antihistamines (diphenhydramine or hydroxyzine) commonly used in labyrinthine-induced vomiting, or motion sickness, have a minor role in the management of oncologic patients with treatment-induced emesis. As sole agents they are inferior to the more commonly used dopamine antagonists.

Other antiemetic agents, such as steroids and anxiolytics, have been effective adjunctive antiemetics when used in combinations. Dexamethasone (3 to 20 mg PO or IV) and methylprednisolone (200 to 600 mg IV) have some antiemetic activity. Parenteral lorazepam (2 to 5 mg IV) given prior to or concurrently with the infusion of chemotherapy may reduce vomiting and produces a mild amnesia for vomiting episodes, thus appearing promising as a way to inhibit the development of anticipatory nausea and vomiting. The most effective regimen for emesis control currently combine several agents: metoclopramide, steroids, and lorazepam (Gralla, 1983). These medications are given only a few hours before chemotherapy, and they are continued by IV infusion during treatment and up to 24 to 36 hours after completion of the chemotherapy infusion. Most combined antiemetic regimens designed for use with highly emetogenic chemotherapy protocols (Strum et al., 1984) are very effective, but many produce side effects in themselves. It is impossible to predict which patient may develop any one of the numerous side effects and in response to what doses of these agents. Clinical practice indicates that young men and elderly patients develop the most extrapyramidal side effects to the dopaminergic agents and the elderly are most sensitive to the amnestic side effects of the benzodiazapines. Ondansetron (see next paragraph) appears to be the most advantageous of these antiemetic agents because of its efficacy and side effect profile.

Recently, researchers have been exploring newer antiemetic agents. Ondansetron (GR 38032F), a selective 5-hydroxytryptamine receptor antagonist, appears to be more effective than metoclopramide for prophylaxis of acute nausea and emesis secondary to cisplatin (Cubeddu et al., 1990; Marty et al., 1990). These studies indicate the role of peripheral serotonin in mediating chemotherapy-induced nausea and vomiting. Besides its promise in efficacy, ondansetron does not produce side effects of sedation, diarrhea, akathisia, and other extrapyramidal symptoms. This side effect profile is particularly attractive in treating the young adult (under 30) cancer population, where 25% of patients receiving metoclopramide develop extrapyramidal symptoms. The current cost of this drug often governs and limits its administration to *only* the day that chemotherapy is given.

Current state-of-the-art antinausea regimens at Memorial Sloan-Kettering Cancer Center (Kris, 1991, personal communication) are shown in Table 24-6.

Movement Disorders Secondary to Antiemetic Agents

All the antiemetics that block dopamine receptors can produce movement disorders, classically described in the psychiatric literature on neuroleptic side effects. As mentioned before, chlorpromazine, haloperidol, and metoclopramide are all antiemetics with dopamine antagonist properties, thereby potentially producing dystonic reactions, akathisias, and tardive dyskinesia.

Dystonic reactions and akathisia have been reported as common side effects with such agents; however, there have been few reports of long-term side effects, such as tardive dyskinesia. Short-term side effects have responded to anticholinergic medications. Of note, metoclopramide is capable of producing tardive dyskinesia not only after long-term oral use (Lazzara et al., 1986; Wilholm et al., 1984), but also after short-term, high-dose parenteral use (Breitbart, 1986).

Anorexia

Anorexia is one of the most common symptoms of cancer. Anorexia and related weight loss can be caused by the disease, by its treatment, or by psychologic disorders. Medical causes are related to tumor, bowel obstruction, fever, reduced food intake, metabolic abnormalities (hepatic and renal dysfunction), and ectopic hormone production by tumors. Loss of appetite develops because of surgery, radiation, and chemotherapy. Head and neck surgery may change facial architecture and limit food intake. Gastrectomy, pancreatectomy, and bowel resection may produce malabsorption and anorexia. Radiation produces acute side effects of glossitis, stomatitis, esophagitis, and altered sense of taste, all making it difficult to eat. Many of the chemotherapeutic agents, besides causing ulcerations of the gastrointestinal tract, produce nausea, vomiting, and anorexia. Antibiotics, antifungal agents, and pain medications produce anorexia. Cancer treatment alters the taste of food, the pleasure of eating, and the normal anatomic and metabolic processes of digestion.

Psychiatric syndromes and behavioral dynamics are often overlooked as the cause of appetite loss. Sometimes anorexia occurs in the context of depression or anxiety in the cancer patient. In such cases it may be difficult to determine cause and effect. Anorexia, in rare instances, can emerge as a symptoms of a previous psychiatric illness. Adolescents and young adults with eating disorders, such as anorexia nervosa and bulimia, bring to the cancer setting a complex set of preexisting psychologic and behavioral dynamics.

Learned food aversions also may play an important role in cancer-related anorexia. "Learned food aversions are acquired aversions to specific foods or tastes which develop as a result of the association of those foods with unpleasant internal responses (nausea and vomiting)" (Bernstein, 1986). This behavioral phenomenon is similar to classical conditioning paradigms, in which animals learn to associate a conditioned stimulus (taste) with an unconditioned stimulus (symptoms of the illness). Bernstein and colleagues have examined learned food aversions in pediatric cancer patients receiving chemotherapy, constituting the first demonstration of such a phenomenon in humans. Children were randomly assigned to a control group or an experimental group. Those in the experimental group were offered a novel flavor of ice cream shortly before their scheduled chemotherapy treatment. All children were tested for food aversion at 1 to 4 weeks. Those exposed to the novel food were three times as likely to have developed a food aversion.

Management of anorexia can involve educational, behavioral, and pharmacologic approaches. Consultation from a nutritionist provides very valuable ideas for patients—for example, small frequent meals with high caloric content, appetizing recipes, visually appealing presentations, and avoidance of strongly flavored foods. Often a nutritional consultant is mandatory when a patient has undergone head and neck surgery and re-

TABLE 24–6. *Recommended Combination Antiemetics for Patients Receiving Chemotherapy Causing Severe and Moderate Emesis**

GENERAL APPROACH: NEUROTRANSMITTER BLOCKING AGENT + CORTICOSTEROID + BENZODIAZEPINE OR ANTIHISTAMINE	
I. SINGLE-DAY ANTIEMETIC PROGRAMS	
REGIMEN A:	
Metoclopramide:	3 mg/kg IVPB, 30 min prior to chemotherapy, and repeat dose 1½ hr after the start of chemotherapy (2 doses total)
Dexamethasone:	20 mg IVPB, 20 min prior to chemotherapy × 1 dose
Lorazepam:	1.5 mg/m² (max. 3 mg) IVPB, prior to chemotherapy × 1 dose only
Diphenhydramine:	50 mg PO, IV, or IM, q4h *prn* for restlessness or acute dystonic reaction
OR	
REGIMEN B: Recommended for patients <40 years of age and individuals of any age with a history of extrapyramidal symptoms after receiving antiemetics	
Ondansetron:	0.15 mg/kg IVPB, 30 mins prior to chemotherapy, and repeat doses 1½ and 3½ hr after the start of chemotherapy (3 doses total)
Dexamethasone:	20 mg IVPB, 20 min prior to chemotherapy × 1 dose
II. MULTIPLE-DAY ANTIEMETIC PROGRAM (for *each* day of chemotherapy)	
Ondansetron:	0.15 mg/kg IVPB, 30 min prior to chemotherapy, and repeat doses 1½ and 3½ hr after the start of chemotherapy (3 doses total)
Dexamethasone:	20 mg IVPB, 20 min prior to chemotherapy × 1 dose
III. DELAYED EMESIS REGIMEN FOLLOWING CISPLATIN (to start 24 hours after cisplatin)	
Metoclopramide:	0.5 mg/kg PO, qid × 2 days, then q4h *prn*
Dexamethasone:	8 mg PO, bid × 2 days, then 4 mg PO, bid × 2 days
Diphenhydramine:	50 mg PO, q4h *prn* for restlessness or acute dystonic reaction

*Adapted from the Formulary Committee, MSKCC (1992).

quires special pureed foods or prosthetic devices. Quite often, simple behavioral techniques are overlooked by the apathetic patient or distraught family member. The ambience of the mealtime can be very important in encouraging eating. Having a family member share a meal, serving a favorite wine or beer, or using special table settings can add social aspects that individuals associate with a pleasant meal. Operant conditioning, in which rewards such as family visits or watching TV or movies are contingent on weight gain or caloric intake, have been successful in patients with anorexia nervosa, but there has been no effort to apply these behavioral methods to patients with cancer.

Anorexia may also result from anxiety or worry about poor intake, anticipatory anxiety, or nausea before a meal. Fear, anxiety, worry, and anorexia may become coupled to one another. In these situations, behavioral techniques such as relaxation and self-hypnosis can lower anxiety and the anticipatory phenomena around eating.

Several pharmacologic agents are useful in promoting weight gain: antihistamines, steroids, cannabinoids (THC), and tricyclic antidepressants. Murphy (1990) in an American Cancer Society newsletter reports that dronabinol, an antiemetic, has also been shown to improve weight gain and reduce weight loss (by 1 pound per month). Appetite is increased by 50% to 70%. Dronabinol, an active substance of marijuana, is marketed in the United States as Marinol.

Dobin and Kleinman (1991) have reported on a random sample of oncologists measuring their attitudes and experiences concerning marijuana as an effective antiemetic in cancer chemotherapy. This survey of over 1,000 oncologists indicated that 44% recommended the use of marijuana to control emesis and 48% would prescribe it if legal. As a group they believed that smoked marijuana is more effect than the legally available oral synthetic dronabinol (THC) and is as safe. Smoked marijuana may be more efficacious than oral THC because of (1) the active agents in crude marijuana, which are absent in synthetic THC; (2) greater bioavailability of

THC absorbed via the lung versus the gastrointestinal tract; and (3) the patients' ability to control or titrate the amount or rate of dosage.

Preliminary data have suggested that megestrol acetate, a progestational drug used in women with breast cancer, promotes increased appetite and subsequent nonfluid weight gain in patients with various oncologic illnesses and AIDS (Loprinzi et al., 1990; Tchekmedyian et al., 1986; Von Roenn et al., 1988). Megestrol acetate, in doses of 480 to 1,600 mg/day for 6 weeks or more, was shown to result in an average weight gain of 11 pounds in a small group of women with breast cancer. Megestrol acetate, at a dose of 160 mg/day, resulted in a 5-pound weight gain in a significant proportion of cancer patients (none with breast cancer) (Tchekmedyian et al., 1986). In a large randomized, double-blind, placebo-controlled trial of megestrol acetate in patients with cancer-associated anorexia and cachexia, patients receiving 800 mg/day of megestrol experienced significant appetite stimulation and weight gain (10% gained 15 pounds or more) with little associated toxicity (Loprinzi et al., 1990). A dose of 160 to 320 mg/day of megestrol acetate is more commonly used. Patients' loss of anorexia and reversal of some cachexia frequently improved their overall sense of well-being; their impact on cancer survival is unclear. Pharmacologic management of anorexia is more often used in terminally ill patients, however, whereas educational or behavioral techniques are used earlier in the course of illness and during active treatment.

Patients with significant malnutrition who remain anorectic despite pharmacologic and environmental manipulations require formal nutritional support, such as gastrostomy feedings or total parenteral nutrition (TPN). Untreated malnutrition can produce or exacerbate both mood disorders and delirium.

Pain

In the PSYCOG study (Derogatis et al., 1983), each of the 215 patients rated the severity of their pain. Of patients who received a psychiatric diagnosis, 39% indicated the severity of their pain as greater than 50 mm on a 100-mm visual analog scale. The psychiatric diagnosis of those patients was depressed or mixed mood (69%) or major depression (15%). In contrast, only 19% of patients who did *not* receive a psychiatric diagnosis had significant pain. This finding of increased frequency of psychiatric disturbance in patients with pain has been reported by others. Ahles, Blanchard, and Ruckdeschel (1983) compared cancer patients with pain and without pain and found that patients with pain obtained higher scores on measures of depression, anxiety, hostility, and somatization. Sternbach (1974) noted that anxiety symptoms often accompany acute pain, whereas depression is found in patients with chronic pain. Psychologic symptoms of anxiety or depression may be either a consequence of or a contributor to pain, and treating both usually has the effect of reducing pain (Peteet et al., 1986).

The data just cited confirm clinical observations that psychiatric symptoms (e.g., anxiety, depression) of patients who are in pain must *initially* be considered as a *consequence* of uncontrolled pain. Acute anxiety, depression, despair (when the patient believes the pain means disease progression), irritability, agitation, uncomfortableness, anger, and change in sleep patterns are common emotional and behavioral symptoms of pain. *Psychiatrists should first assist in the management of pain and then reassess the patient's mental state after pain is adequately controlled to determine whether the patient's symptoms are due to psychiatric illness.* In summary, the cancer patient with pain has an enhanced risk of developing psychiatric disorders commonly seen in cancer. Clearly, depression, anxiety, and mixed symptoms of depression and anxiety are the most common problems.

Pharmacologic therapy remains the mainstay of treatment of acute and chronic pain. The analgesic medications most commonly used are divided into these major categories (Moulin & Foley, 1984): (1) aspirin, acetaminophen, and nonsteroidal antiinflammatory drugs, which produce analgesia primarily by peripheral mechanisms, via inhibition of the enzyme prostaglandin synthetase; (2) narcotic agonists and antagonists, which act centrally and peripherally by binding to opiate receptors and activating pain modulatory systems; and (3) adjuvant drugs (antidepressants, antipsychotics) that act centrally to control pain by poorly understood mechanisms (Foley, 1984).

Foley (1986) advocated an extensive assessment of the pain in a patient prior to attempting to treat it. Pain can be considered in five categories: (1) acute cancer-related pain secondary to disease or treatment (surgery, chemotherapy, radiation); (2) chronic pain related to tumor progression or treatment; (3) cancer-related pain and preexisting chronic pain; (4) cancer-related pain with history of drug addiction; and (5) pain in terminally ill patients.

Excellent guidelines for pharmacologic management of pain are available (Foley, 1984, 1986) (see also Chapters 16 and 17). These principles include:

1. Treating the psychologic as well as the physical symptoms of the pain.

2. Using a specific drug for the specific pain (aspirin or nonsteroidal antiinflammatory drugs for mild to moderate pain, and oral or parenteral narcotics for moderate to severe pain).

3. Learning the pharmacology of specific pain medications—analgesic doses for each route of administration, peak time and duration of analgesia, pharmacokinetics, and toxic side effects (e.g., sedation, respiratory depression, nausea, vomiting, constipation).

4. Administering analgesics on a regular schedule.

5. Being flexible in changing medications but allowing each analgesic an adequate trial.

6. Watching for development of analgesic tolerance, physical dependence, and withdrawal.

7. *Not using placebos.*

8. Considering use of combinations of drugs to enhance analgesia (e.g., narcotic analgesic with nonnarcotic analgesic, dextroamphetamine, hydroxyzine, or amitriptyline; [Walsh, 1986]).

9. Adequately managing pain in terminal illness by morphine infusions (Portnoy et al., 1986).

NEUROPSYCHIATRIC AND PSYCHOLOGIC SIDE EFFECTS OF CHEMOTHERAPEUTIC AGENTS

Many chemotherapeutic agents do not cross the blood-brain barrier to any significant degree, and therefore produce few direct side effects on the CNS. However, delirium is associated with the use of methotrexate (especially with intrathecal or high-dose intravenous administration). The mechanism of this syndrome is not known. However, high-dose methotrexate often is given in combination with radiation. The pairing of these two treatments may alter the blood-brain barrier to allow more drug to cross into the CNS. Methotrexate can produce acute and chronic toxicity to the nervous system. Acute side effects of methotrexate include transverse myelopathy and meningeal irritation. The most devastating chronic complication is leukoencephopathy, a syndrome that may develop during the first year after therapy and that usually is accompanied by focal neurologic findings, dysarthria, ataxia, and, frequently dementia. The cause of this usually irreversible syndrome is unknown; most researchers speculate that prior radiation to the CNS alters the blood-brain barrier permeability, allowing higher and more toxic doses of methotrexate to enter the brain (Patchel & Posner, 1989).

Other agents causing delirium are 5-fluorouracil, the vinca alkaloids (vincristine, vinblastin), bleomycin, carmustine (BCNU) cisplatin, L-asparaginase, procarbazine, cytosine arabinoside (Ara-C), ifosfamide, and prednisone (Fleishman & Lesko, 1989; Kaplan & Wiernik, 1983; Moore, Fowler, & Crumpler, 1990; Young & Posner, 1980; Zalupski & Baker, 1988). Cerebellar ataxia occurs infrequently from Ara-C, 5-fluorouracil, BCNU, and procarbazine. The vinca alkaloids, chemotherapeutic agents derived from the periwinkle plant used extensively in the treatment of leukemia and lymphoma, produce a peripheral neuropathy that can be severe and extremely painful for years after a treatment is discontinued (see Table 24-3).

Corticosteroids are used in cancer treatment for several purposes: (1) to reduce cerebral edema associated with metastatic or primary brain tumor, (2) as chemotherapeutic agents for leukemia and lymphoma, (3) combined with radiation therapy to the spine to treat spinal cord compression, and (4) increasingly, as an adjunct antiemetic agent in combination with metoclopramide, haloperidol, diphenhydramine, or lorazepam for chemotherapy protocols that utilize highly emetogenic drugs.

The initial psychologic response to steroids, euphoria and irritability, occurs independently of the dose. Some effects are beneficial, providing a sense of well-being, increased appetite, and weight gain. Many effects, however, are uncomfortable: insomnia, restlessness, hyperactivity, muscle weakness, fatigue, and depression. Severe effects are usually seen with higher doses but can occur at small doses. Cessation of steroids can produce depression. Exaggerated responses to steroids—for example, profound mood disturbance, such as mania, severe depression, or delirium—are less common. These drug-induced changes may be accompanied by hallucinations, paranoia, and delusions and can be clinically indistinguishable from primary mood disorder. These responses may appear when the dose is abruptly increased, tapered, or discontinued. Patients who require steroids as part of their chemotherapy, or who develop a spinal cord compression that demands emergency treatment, should be given steroids even if they have a history of affective lability. Patients' moods should be carefully monitored, and psychopharmacologic treatment should be prescribed when necessary. If steroid reactions occur during a rapid taper, the steroid dose should be increased and then lowered more gradually. The management of steroid withdrawal syndrome is discussed in Chapter 29.

IMMUNOTHERAPY

Biologic response modifiers and adoptive immunotherapy represent promising new developments in cancer treatment. Neuropsychiatric disturbances have been reported with several immunologic agents, including interferon (Adams, Quesada, & Gutterman, 1984; Quesada et al., 1986) and interleukin-2 (Denicoff et al., 1987b). To date interferons have been used in clinical studies with hairy cell leukemia, chronic myelocytic leukemia, and Kaposi sarcoma. Intramuscular administration of interferon in dosages of 2 to 50 million units/day produces flu-like symptoms of lethargy, anorexia, headaches, nausea, or depression (Rohatiner et al., 1983; Smedley et al., 1983). A study of tumor necrosis factor in hepatitis B carriers in whom social class variables were controlled, conducted by McDonald, Mann and Thomas (1987) using the General Health Questionnaire, found that interferon did increase psychiatric morbidity. Nonpsychotic symptoms of fatigue, poor concentration, anxiety, and depression were the major symptoms. Symptoms of interferon appear to be dose related and disappear on discontinuation of the drug.

Adoptive immunotherapy with interleukin-2 and lymphokine-activated killer (LAK) cells can be complicated by cognitive impairment, disorientation, confusion, and mental slowing as well visual hallucinations, particularly when administered in high doses. These neuropsychiatric toxicities can be treatment limiting and often appear several days after treatment has begun. This latency period is characterized by more subtle preliminary changes that include irritability and decreased concentration and attention (Denicoff et al., 1987a). For a more comprehensive review of the literature, the reader is referred to Holland and Lesko (1989).

OTHER MEDICATIONS CAUSING CENTRAL NERVOUS SYSTEM SIDE EFFECTS

Amphotericin B is used regularly for the treatment of fungal infections in immunologically compromised cancer patients. Because it is poorly absorbed via the gastrointestinal tract, it is administered intravenously. It can cause anaphylaxis, fever, rigors, anorexia, and impaired renal function. Various neurologic side effects, including delirium, have been reported with intrathecal administration. Patients treated with the methyl ester of amphotericin B may be particularly susceptible to CNS side effects (Ellis, Sobel, & Nielsen, 1982). Ellis and colleagues reported 14 patients who developed progressive severe neurologic dysfunction, including dementia, akinesia, mutism, hyperreflexia, tremor, and white matter deterioration. Symptoms were dose dependent and were most severe with doses greater than 9.0 g. Often, it is difficult to distinguish CNS effects of antifungal medication from those of fever, CNS infection, or metabolic abnormalities.

Acyclovir is a relatively new antiviral drug that has proved efficacious for the prophylaxis and treatment of herpes simplex and varicella zoster virus. It is used widely in patients who are immunologically compromised by AIDS, bone marrow transplantation, or leukemia. The use of parenteral acylovir (750 to 3,000 mg/m^2/day) has been associated with minimal drug toxicity. Wade and Myers (1983) reported reversible neurotoxicity in 6 of 143 bone marrow transplant recipients studied. Symptoms, which developed a median of 8 days after the initiation of treatment, included lethargy (n = 5), agitation (n = 5), tremor (n = 5), disorientation (n = 1), and transient hemiparosthesias (n = 1). The authors noted that these patients may have been predisposed to neurologic side effects by previous intrathecal methyltrexate therapy, total body irradiation, preexisting CNS leukemia, herpes virus infections, of the concurrent use of interferon. Improvement in all patients with discontinuation of acyclovir, however, suggested that the transient CNS toxicity was secondary to the antiviral agent rather than to other causative agents.

NEUROPSYCHIATRIC SIDE EFFECTS OF RADIATION TREATMENT

Whole-brain radiation, used to treat primary as well as metastatic lesions of the brain, can be complicated by a radiation-induced encephalopathy. Three types of encephalopathy syndromes have been described: (1) acute encephalopathy (seen immediately after first radiation treatment); (2) early-delayed encephalopathy (beginning 6 to 16 weeks after treatment); and (3) late-delayed encephalopathy (seen 6 months to several years later) (Patchell & Posner, 1989; Sheline, 1980).

Acute encephalopathy can occur during the immediate course of high-dose radiation therapy (RT). Patients can become lethargic and complain of headache, nausea and vomiting, and fever. It is thought that this type of acute reaction is due to increased intracranial pressure secondary to radiation-induced changes in the blood-brain barrier. Left untreated, it can lead to worsening of neurologic deficits and even brain herniation. Corticosteroids, particularly dexamethasone, are the treatment of choice for this syndrome.

An early-delayed encephalopathy can begin 1 to 4 months after radiation treatment but has been reported earlier or later. Symptoms consist of lethargy, headache, nausea, and vomiting. In children who receive whole-brain RT prophylactically for leukemia, the picture is usually one of generalized somnolence and headache. Patients who receive more focal RT to the brain can present with symptoms of focal neurologic disease suggestive of recurrence of tumor. The cause of early-delayed radiation encephalopathy is unknown but may be related to radiation-induced edema or demyelination. Improvement in symptoms usually occurs spontaneously in 1 to 6 weeks. Steroids may be helpful in treating symptoms as well as for prophylaxis prior to or during RT.

A late-delayed encephalopathy (usually severe and permanent) may develop 6 months to 3 years (average 12 months) after radiation therapy. Late-delayed reactions may be related to total radiation dose, fractionated dose of radiation (amount given per day), time frame in which the radiation was given, presence of concurrent intracranial disease, concurrent chemotherapy, and variations in sensitivity of the brain as a result of tumor effects. This syndrome is characterized by symptoms that suggest a focal neurologic lesion, accompanied by personality change and headache. Seizures can also complicate the picture. Differential diagnosis includes recurrent tumor, infarct, or abscess. Computed tomography (CT) scan of the brain usually reveals a hypodense lesion in the white matter. Biopsy of brain has shown necrosis. Clinically, radiation necrosis of the brain may present not only with symptoms of a cognitive impairment disorder but also with personality changes and depression. The depressive component of the syndrome has been reported to respond to antidepressant treatment despite the presence of underlying brain damage (McMahon & Vahora, 1986). Steroids can help symptomatically; however, surgical resection of the necrotic mass is often necessary.

DeAngelis, Delattre, and Posner (1989) reported 12 cases of delayed CNS complications in 360 disease-free patients treated for cerebral metastases with whole-brain radiation. All 12 patients developed progressive ataxia, dementia, and urinary incontinence within 5 to 36 months. These symptoms were severe enough to cause death in seven of the patients. Cortical atrophy and hypodense white matter were again the major CT findings. The authors described a clinical syndrome similar to subcortical dementia, as in Parkinson's disease and AIDS. Symptoms of dizziness, fatigue, mild headaches, and mild recent memory changes progressed to dissolving ataxia and incontinence. The authors stressed the role of corticosteroids in decreasing radionecrosis and capillary permeability.

Rowland et al. (1984) retrospectively studied children with acute lymphocytic leukemia who were treated with intrathecal methotrexate and cranial radiation as prophylaxis against CNS recurrence. They found that children who received both radiation and intrathecal methotrexate had a mean IQ 10 points lower than that of children who received only intrathecal methotrexate. Soft neurologic signs and abnormalities in growth hormones were also evident. Radiation causes a transient syndrome in children, known as radiation somnolence syndrome. This occurs several weeks after radiation and is caused by demyelination.

SPECIAL CONSIDERATIONS IN PSYCHOTHERAPEUTIC TREATMENT

The therapeutic approaches used in cancer are a combination of psychotherapeutic, pharmacologic, and behavioral techniques. Psychopharmacologic management is effective for anxiety syndromes, depression, delirium, schizophrenia, bipolar disorder, pain syndromes, nausea, vomiting, and insomnia. Behavioral interventions, including relaxation, biofeedback, systematic desensitization, hypnosis, and guided imagery, are helpful for pain and anxiety during procedures, nausea and vomiting, and cancer-related eating disorders. Psychotherapeutic approaches include professionally led groups, individual therapy, and self-help individual treatment (patient-to-patient volunteers) and self-help groups (Make Today Count, Cansurmount, Candlelighters).

Psychotherapeutic Interventions with Cancer Patient

Psychotherapy

Psychotherapy with cancer patients includes utilizing educational techniques, answering questions, correcting facts, and giving reassurance. Interpretation and explanation of psychologic dynamics often leads to exploration of more effective coping mechanisms. The psychiatrist often explains laboratory results, treatment options, and predicted side effects of treatment, provides information about hospital procedures and community support services, helps patients and families learn to negotiate the medical system, and defines normal symptoms of psychologic stress.

Such treatment begins by focusing on current issues; however, exploration of reactions to cancer often includes exploration of situations unrelated to illness. Some patients with cancer may become interested in more exploratory or analytically-oriented psychotherapy. The young cancer patient frequently chooses to continue psychotherapy to explore life issues of growing up, separating from family, and starting career, family, and marriage while under active cancer treatment or during survivorship.

In addition, the therapist must be knowledgeable of the medical aspects of the patient's disease—its prognosis, treatment, and common side effects. The therapist must be flexible in approach; the focus of treatment shifts as illness changes. The therapist often becomes the patient's advocate in the medical system, with family members, or with employers. Patient cancellations or no-shows are quickly followed up by phone rather than left to interpretation, since physical status can change quickly. Sometimes psychotherapy is continued by telephone for those patients too ill to come to the office regularly. (See also Chapter 1.)

Group Psychotherapy

Group therapy may be advantageous for the cancer patient, allowing the patient to receive support from others (patients or nonpatients) who have experienced and have conquered similar problems of medical illness. The cancer patient in a group setting can easily learn that there are a range of normal reactions to illness and a range of healthy adaptive coping styles and strategies that others have employed to make adjustment to illness easier. Group participation helps decrease the sense of isolation and alienation because the cancer patient and the family can see that they are not alone in adjusting to illness. Groups for cancer patients and families are often formed for patients with the same disease (e.g., Hodgkin's disease, breast cancer, or leukemia) or for patients at the same stage of different diseases or separately for family members. Initially, therapists had some concern about having the dying patient in a group with patients who were fairly recently diagnosed or whose long-term prognosis was good. Spiegel, Bloom, and Yalom (1981) have reported on their success in leading groups for patients with breast cancer at all stages of disease, ranging from recently diagnosed patients to the terminally ill. These types of groups, when directed by skilled leaders, can be highly rewarding for many patients. The principles of group therapy for the non–medically ill person apply for group therapy with cancer patients. (See also Chapter 3.)

Self-Help and Mutual Support Programs

Self-help and mutual support programs provide alternative support for patients and families. Life crises, such as bereavement, separation, divorce, drug addiction, or life-threatening illness, often provide the impetus for individuals to seek emotional support from others experiencing the same trauma. Self-help and mutual support programs for cancer patients started after World War II, when the American Cancer Society began a visitors program offering practical help for patients at home. The International Laryngectomy Association was founded in 1942, and Reach to Recover was officially sponsored by the American Cancer Society in 1952 to meet the needs of women undergoing mastectomies. As colostomies became a common practice during gastrointestinal cancer, ostomy clubs were started.

Most self-help support networks for cancer patients work closely with professional medical services, thereby offering social support as an adjunct to medical care. Self-help groups include those run by parents of pediatric center patients (Candlelighters), cancer patient visitors' programs (Reach to Recovery, International Laryngectomy Society, ostomy clubs), and others (Make Today Count, Cansurmount, Compassionate Friends). At the Memorial Sloan-Kettering Cancer Center, the Patient-to-Patient Volunteer Program has been implemented; it is a program in which volunteers visit every newly admitted cancer patient. Volunteers see patients who have the same cancer diagnosis as they once had, or one similar to it. Such volunteers help decrease the sense of alienation and isolation of patients because of their unique knowledge and sensitivity, which comes with having had the same experience. Veteran patient volunteers facilitate coping in the newly diagnosed patient by: (1) being a source of credible information based on their own past patient experiences, (2) demonstrating constructive ways of managing and living despite illness, (3) providing the motivation for rehabilitation and enhancement of self-worth, (4) encouraging patients to participate in their own treatment, and (5) serving as a "surrogate patient," with whom spouses and family members can ask questions and express feelings. The veteran patient also provides education about the needs of cancer patients (Mastrovito & Moynihan, 1989).

Sexual Counseling and Rehabilitation

Gynecologic cancers make up at least 17% of all malignancies of women and include (in decreasing order of incidence) cancers of the uterus, the ovary, the cervix, the vulva, the vagina, and the fallopian tube. Treatment is usually surgical, followed by radiation or chemotherapy or both. Many gynecologic cancers have a good prognosis. Increasingly, there is a recognition of the need for psychosexual rehabilitation after gynecologic cancer treatment, irrespective of patient age. A psychiatrist trained as a sex therapist brings special expertise to surgical teams caring for women undergoing radical gynecologic procedures and men undergoing prostate surgery. All health professionals, however, can learn how to identify and refer a patient whose concerns about sexuality and possible sexual dysfunction require additional professional help (Auchincloss, 1984).

Auchincloss (1984) suggested it is the job of the oncologic professional to initiate the subject of sexuality with all patients. Patients are often embarrassed or may even assume that after cancer one's sex life should no longer exist. Married patients may have had precancer sexual problems, which are compounded by cancer treatment. Single patients face dating and potential marriage (partner) issues. Elderly patients and patients who have undergone extensive gynecologic reconstructive surgery or those who are sterile secondary to radiation are likely to be ignored by staff. Evaluation includes obtaining a sexual history (including that of the partner), noting existing sexual problems, and addressing the patient's current treatment-specific questions and fears. The therapist must be informed about side effects of sequelae of surgery, radiation, or chemotherapy in order to supply information and practical suggestions. Patients with persistent concerns about body image, sexual desire, and sexual function should be referred to professionals who are trained in sexual rehabilitation. Treatment techniques include supportive psychotherapy, behavioral techniques such as relaxation, and gradual reeducation in sexual functioning. Sexual aspects of surgery are also discussed in Chapters 34 and 40.

Behavioral Interventions

Behavioral techniques include passive relaxation with visual imagery, progressive muscle relaxation, electromyographic (EMG) feedback, systematic desensitization, and cognitive distraction (Burish & Lyles, 1981; Morrow & Morrell, 1982; Redd, Andresen, & Minagawa, 1982). These methods are useful as adjuvant treatments, combined with pharmacologic agents for pain and during chemotherapy infusions, and as a primary intervention for children undergoing painful procedures such as bone marrow biopsies or venipuncture.

Several studies under the direction of Redd and Jacobsen at Memorial Sloan-Kettering Cancer Center utilize behavioral distraction methods: (1) using party blowers in small children undergoing painful procedures such as venipuncture, chemotherapy infusions, and lumbar punctures; (2) video games for adolescents undergoing chemotherapy to relieve anticipatory anxiety and nausea, and (3) the use of novel fragrances for adults undergoing diagnostic procedures in confining (and often anxiety-provoking) scanning equipment (CT and magnetic resonance imaging scanners).

During the course of chemotherapy, anywhere between 25% and 65% of the patients became significantly sensitized to their treatment and develop anticipatory symptoms of anxiety, nausea, or vomiting (Redd, 1989). Research efforts have been directed toward the usefulness of passive relaxation and cognitive distraction during chemotherapy infusions and for anticipatory nausea and vomiting, which may occur before hospital visits. Anticipatory nausea and vomiting before chemotherapy or radiation follows a classic behavioral conditioning paradigm and is a far more common symptom than has been previously recognized (Pratt et al., 1984). It can occur hours of even days before chemotherapy administration and often occurs in patients who receive *repeated* cycles of a chemotherapy regimen or *highly* emetogenic drugs, in those who experience *severe* posttreatment nausea and vomiting, and in those who have high levels of anxiety or alterations of taste and smell associated with the chemotherapy infusion. Studies at Memorial Sloan-Kettering Cancer Center involving long-term survivors of Hodgkin's disease reveal that symptoms of nausea and vomiting can be elicited as long as 10 years after treatment has been completed (Cella, Pratt, & Holland, 1986). Present treatment approaches for this phenomenon include use of combinations of behavioral methods with antiemetic and antianxiety agents (Stoudemire, Cotanch, & Laszlo, 1984). The reader is encouraged to review an excellent chapter by Redd (1989) that summarizes the array of anticipatory phenomena patients can experience, the prevalence of anticipatory side effects, the mechanisms involved in their development, and the various behavioral methods used in their treatment—hypnosis with imagery (passive relaxation), progressive muscle relaxation training with imagery, and cognitive or attentional distraction.

Systematic desensitization, a counterconditioning procedure often used in treating phobias, has been used in treating cancer-related phenomena of needle phobias and anticipation of procedures. The technique, somewhat time consuming and labor intensive, involves training the patient in a relaxation technique such as progressive muscle relaxation, constructing with them a hierarchy of stimuli related to the feared event, and finally having the patient practice relaxation training while visualizing the "least to most" aversive stimuli within the hierarchy (Redd, 1989).

Unconventional Cancer Treatments

Any disease that has a largely unknown cause, high likelihood of fatality, and uncertain cure causes great fear in the public's mind. Cancer and mental illness have historically been most feared; AIDS, for the same reasons, has become the most feared in the 1990s. Consequently, such diseases, for which traditional medicine cannot provide a cure, have always elicited an array of nontraditional or unconventional treatments (Holland,

Geary, & Furman, 1989). The history of cancer quackery is of great psychiatric interest since these therapies have flourished over centuries. They have only changed in their types and nature. In general, the popular unconventional treatments of a particular period reflect the public's perception of cancer and medicine at that time (Cassileth, 1986; Wharton, 1987). Thus, balms, tonics, and electrical waves, popular early in this century, gave way to krebiozen in the 1960s, laetrile in 1970s, and, most recently, "natural" approaches that enhance the body's defenses. The alternative therapy community exists separately (often secretively) from mainstream medicine, with strong adversarial polarized positions, often leaving patients confused as to how to view the two worlds.

There has been a recent effort to try to bridge this gap by preparation of a thoughtful and comprehensive report (which should be read by anyone interested in the area) of therapies in the United States, undertaken by the Office of Technological Assessment at the request of the U.S. Congress (1990). The report elucidates the present status of these therapies, outlining the four major areas of alternative therapies: psychological and behavioral, nutritional, herbal, and pharmacologic/biologic. Many therapies, such as those of Gerson and Kelly, combine several of these approaches and include a spiritual or religious context as well. The best known of the psychologic approaches are Simonton's visualization methods to enhance immune function; Siegel's Exceptional Cancer Patients, to improve survival by positive emotional expression; and Commonweal, where Lerner combines group, yoga, touch, relaxation, and visualization into a comprehensive week-long therapeutic experience.

It is important to those working in the psychosocial and behavioral aspects of cancer to have a clear understanding of the issues, since some of the psychologic and behavioral methods that are used in mainstream medicine (psychological support, visual imagery, relaxation) for enhancing quality of life and symptom control are also included in unconventional therapies with the promise of curing cancer or extending survival through their use. This "gray zone" grows even broader in view of psychoimmune studies that show the impact of stress in healthy subjects on immune function and Spiegel and colleagues' (1989) study showing greater survival among women with advanced breast cancer who attended weekly psychotherapeutic group meetings. It is hard for frightened individuals to see these findings as interesting research data that warrant further study. The stress they feel, plus the promise of benefit from practitioners of alternative cancer treatments, make it difficult to resist "as long as it doesn't hurt."

Within this confusing picture, the conscientious mental health practitioner has a difficult job advising patients about the use of unconventional therapies, wishing not to take away hope and faith in these methods but also not to condone an unproven treatment based on anecdotal data that places a burden on the patient via the notion that their personality and behavior are the vehicle for their recovery or tumor progression. Gray and Doan (1990) have offered a helpful perspective, suggesting that cancer is a frightening disease that elicits in some individuals the feeling of becoming the "heroic warrior" fighting the "dragon" cancer. These patients gain by the warrior stance and should be encouraged to maintain it, emphasizing to them that, indeed, there are many things we do not know and, if they find the cancer of self-help books, visualization, mental attitude, and diet regimens helpful, they should use them. However, there are other patients for whom the "warrior" myth is distressing. They can't "love enough" or be "strong enough," and feel guilty that they cannot "fight" hard enough, sometimes resulting in anger from family. These individuals need to be encouraged to understand that there are many ways of coping with cancer, not one, and that data today do not support evidence of longer survival with the unconventional therapies (Bagenal et al., 1990, Morgenstern et al., 1984).

There are several key points a mental health professional should know about unconventional therapies in order to teach and care for patients:

1. The patients who use alternative therapies are not poorly educated individuals with advanced disease who are grasping at straws. Today, they are educated, seeking all available information, and usually are using conventional therapies at the same time (Cassileth, 1986). In western countries where it has been examined, about 10% to 20% of patients with cancer use these treatments. Among them, a small percentage appear to stop conventional therapy for an unconventional one.
2. Practitioners vary widely, and some charge large amounts of money (e.g., Burton's Immunoaugmentative Therapy) for treatment that some believe may be deleterious. However, many approaches aim to improve quality of life and are offered by individuals who feel a mission to use some special approach.
3. Many patients turn to these treatments because they find too little psychologic support given by mainstream medicine; they usually find it in the alternative setting.

In summary, the psychiatrist has a key role in:

1. Educating staff about patients' psychologic needs and outlining the support they need in the context of their care.

2. Teaching staff to discuss the pros and cons of alternative therapies with the patient, never condemning or judging angrily.

3. Maintaining a list for staff to use of local unconventional practitioners and therapists (since they change often and are different in each part of the country) so that they become knowledgeable and can give advice. A therapy that is harmful should be condemned; a therapy that is aimed at quality of life, and is to be used as an adjunct to conventional therapy, can be condoned. However, patients should be warned that if the approach makes them upset (guilty, depressed that they are failing), they should discontinue it. The negative side of these approaches is blaming the victim: "You caused the cancer and now you can cure it."

4. Understanding and explaining to staff the complex emotions of patients who seek alternatives, the need for physicians to react with sensitivity and provide guidance to patients in this area, and the need to be certain that patients are not leaving mainstream care because they do not feel psychologically supported in their medical care.

Elderly Patients

The psychologic and social impact of cancer in the elderly is significant, since the elderly cancer patient must bear two stigmas in our society: having cancer and being old. One common myth about cancer in the elderly is that it is a uniformly fatal disease. Data suggest that cancer is now a far more treatable disease in older individuals than it was in the past.

Cancer treatment in the elderly is complicated by: (1) concurrent chronic diseases such as hypertension, heart disease, and diabetes; (2) greater toxicity of chemotherapeutic agents excreted by the kidneys (e.g., cisplatin, methotrexate); (3) physical debilitation, which prevents vigorous rehabilitation after surgery or during convalescent care; and (4) lack of family involvement, sometimes because of geography. Early diagnosis and treatment of cancers in the elderly produce the best prognosis, but, unfortunately, older individuals more often neglect early symptoms (e.g., pain, blood in stool, coughing of blood, weakness and fatigue) of cancer than do younger individuals. Procrastination in care and treatment also results from fear of the financial burden of treatment, less education, or pessimism and fatalistic attitudes. Lack of awareness of community health care services and absence of an existing relationship with a physician also contribute to delay in the elderly.

Terminal Illness: Issues of Death and Dying

In the care of the terminally ill patient, the art and science of medicine and psychiatry are blended. Often the art is strongly influenced by the caregiver's own personality and by the history that family and patient bring with them, as well as their psychosocial values and issues (Holland, 1982b). Conflicts that arise regarding decisions about care may lead to requests for psychiatric consultation and intervention.

Terminally ill patients are those who have not responded to known curative measures, and treatment is aimed at providing maximal comfort during their limited life span. Physicians, patients, and family attitudes variably define the life span remaining for a terminal patient as a few hours to a few months. As more effective resuscitative and life support measures have become available in critical care medicine, *terminal illness* has become more difficult to define. Potential death and recovery exist simultaneously, especially in critical care and dialysis units in cancer hospitals. When the label "dying" is assigned to a patient, attitudes and behaviors of staff, family, and friends often assume a different character. These attitudes may tend to isolate or alienate the patient from those whom the patient most needs at a crucial time. We often forget that the person who is dying has not changed, only the life expectancy has changed; emotional needs only intensify.

To Tell or Not To Tell

Should patients know of their critical state? What should they be told, and how and when? There is a trend toward greater candor in discussions of medical facts with both the patient and the family. Novack et al. (1979) conducted a study in Rochester, New York, and reported that 97% of physicians favor telling cancer patients their diagnosis, in contrast to 90% who did not tell in 1961. Most patients prefer to know about their illness. Those who need to deny the truth will continue to do so.

The "conspiracy of silence" still remains in much of Europe and the East. Recent articles in the lay press have again emphasized the patient's right to be told the truth. The necessity of each physician's knowing each patient and that patient's capacity to "hear the truth" cannot be overemphasized. Most patients, even children, want to be told the seriousness of their illness, but they also want to hear the ways in which family and professionals will help. By discussing with patients and families the availability of such supportive measures as pain control and alternative forms of nutrition, by offering flexibility in providing care outside the hospital, and by having an understanding of the patient's perceptions of their stages of illness, physicians communicate their pledge to provide continued care for dying patients. Currently, there is a trend in the United

States to allow terminally ill children to die at home; this requires a compassionate and sophisticated team approach to caring for the patient's medical symptoms and the psychologic needs of the patient's family and friends.

Perceptions of Terminal Illness

A patient's *perception* of terminal illness depends on several issues: medical or physical factors in the course of illness (pain, relapse) and patient-related concerns (cultural attitudes about death, coping mechanism, developmental life stage). The patient's or family's perception of a "sense" of terminal disease results from an increase in the severity of medical problems, delirium, development of pain, organ failure, a rapid course irrespective of adequate medical management, or the level of care given by the health care team. A patient's and family's *acceptance* of terminal illness depends on many cultural and religious factors.

The psychologic stages of denial, depression, anger, bargaining, acceptance, and resolution, as defined by Kubler-Ross (1969), do not occur in any patterned progression. In a study of 90 patients with advanced cancer, Plumb and Holland (1977) were unable to relate nearness to death to level of depressive symptoms, acceptance, or decathexis. More typically, patients show daily changes in optimism and pessimism, related to medical events. Intellectual problem solving and information gathering are mechanisms to provide control over unknown prognosis, side effects, drugs, and general feelings of hopelessness. Patients who "cope well" during terminal illness have the ability to blend problem-solving methods with the right amount of denial and hope.

The stage of life cycle in which fatal illness occurs has a strong impact on how death will be perceived, since both attitudes and views toward death change with age (Holland et al., 1979). The young adult, like the adolescent, recognizes the finality of death. Death is apt to be viewed as the adversary that must be fought, and anger, hostility, and bitterness may be strong in terminal stages. The mature adult, often at the peak of intellectual output and work, often has established social and family roles. Here, death strikes at the time of feeling one has "made it" in terms of work, family, and children. The older adult, sensing diminishing responsibilities and thinking toward the "rewards of older age," feels that death must be faced, but "not yet". There are also an increased sense of vulnerability to disease and an altered sense of "time left" instead of "time since birth." Since old age is normally a time for taking stock and increased introspection, death is approached more realistically and can be an openly accepted outcome. The elderly adult occasionally has a realistic view of life expectancy and may be more concerned about the stages of terminal illness and the manner of dying than about death itself. Of great concern are thoughts of not being a burden, particularly to one's children. The frequency of losses of significant others—especially spouse, friends, and siblings—may contribute to a pervasive sadness, and sometimes the meaning of living is substantially reduced. Life review in a few supportive sessions may permit the patient to put memories in order and rework past events—functions that become more urgent as terminal illness ensues (Viederman, 1982).

Palliative Care

Care of the terminally ill has traditionally been designated to and accepted as a responsibility by various religious groups. Calvary Hospital in New York, started by Dominican nuns in the late 1800s, is a specialized facility for care of terminally ill patients, primarily those with cancer. Twenty years ago, the issues of dying patients were made publicly visible by Cicely Saunders, who founded the St. Christopher's Hospice in London, England. This was the beginning of the hospice movement, which has since spread to the United States. Unfortunately, the decision as to whether a terminally ill patient is cared for at home or in a hospice setting can be made hastily, with only a few family members present and without involving the patient. The decision can be governed by financial restraints or job requirements of supporting family or by the lack of such family. Ideally, a mental health professional should be consulted at this stage to best fit the needs of the patients, family needs, and doctor and community resources.

Irrespective of whether a terminally ill patient is cared for at home or in the hospice setting, psychiatric consultation can be beneficial in the care of patients who are terminally ill and their families. Issues that should be addressed by such consultants in the care of such patients are:

1. What is the extent of the patient's and family's understanding of terminal illness, their coping mechanisms, and their cultural or religious mind-set?
2. Is there a preexisting psychiatric syndrome (of the patient or a family member)—a problem with pain control, drug abuse, or delirium—that makes management of such a patient complex for the physician?
3. Has the pattern of family function been one that suggests the ability to maintain cohesion and retain ad-

equate social support during this period and previous periods of stressful illness and death?

4. Are there internal or external family resources to indicate that the family's self-image can be maintained throughout the illness and during bereavement?

5. Do the patient and the family have an open communication with professional staff and their primary physicians so that a sense of trust in care and decision making can be sustained throughout the illness?

6. Do they have the physical and psychologic stamina, the financial support, and the educational ability to juggle work schedules and changes in work and family responsibilities and to implement changes in medication and physical care, besides assuming day-to-day responsibilities?

The psychiatrist has an important consultative role in the palliative care of the patient with cancer. There are two distinct areas for which intervention is requested: for help in decisions about treatment and care, and for assistance in control of distressing symptoms associated with advanced and terminal stages of cancer (Lederberg, Massie, & Holland, 1990). The first issue for which a consultation is requested is often to confront the reality that treatment must now be aimed at comfort rather than cure. The patient, family, and physician have been aimed in a course of treatment with hope for cure or, at the least, control of disease. The transition to comfort care is a painful emotional realization of altered goals for all concerned. It is accompanied by confrontation with death and a sense of hopelessness and helplessness, which may have been denied to this point.

A psychiatrist is helpful to the patient and staff in several areas: to help the patient and family decide *where* care is to be given, how *aggressive* they wish life-sustaining efforts to be, and, how actively the family will be able to *participate* in care of the patient at home. The psychiatrist also has a unique opportunity to get to know the patient's family, which serves as a useful bridge later for acceptance of continued support during bereavement. Ideally, both patients and families are better off emotionally if terminal illness can be managed by keeping the patients at home. Home care often requires active professional support to see the family through it (Coyle, Loscalzo, & Bailey, 1989; Farkas, 1989). Preexisting psychiatric disorders in the patient or family will require special, and often intensive, management during this time.

The second role for the psychiatrist is symptom control. This requires evaluation of mental status and mood to recognize changes. Delirium, anxiety, depression, and pain require control. Physical symptoms of nausea, vomiting, pain, hiccoughs, bowel alterations, anorexia, and insomnia require attention, often using behavioral or pharmacologic approaches. Specific interventions are well described for control of these symptoms (Holland & Rowland, 1989).

Survivor Issues

Given the advancement of innovative procedures such as bone marrow transplantation and the progress in combination chemotherapy, some patients with cancer can expect a lengthened or near-normal life span. This improvement in long-term survival has altered the perception of cancer from a uniformally fatal illness in all to a chronic illness for several age groups. This alteration in perception of cancer has, in turn, changed the role that mental health professionals serve in an oncology setting. Rather than helping family members deal with the inevitable death of the patient, a greater focus is being placed on the long-term psychosocial and mental health needs of this new and growing population of survivors.

However, increased survival may be associated with an increased risk for delayed medical complications, including organ failure, CNS dysfunction, sterility, secondary malignancies, and decreased physical stamina (Meadows & Hobbie, 1986; Meadows & Silber, 1985). In addition, as a result of the stressors inherent in illness and treatment, survivors can experience prolonged psychologic, interpersonal, and vocational dysfunction. Clinical reports and empirical studies have emphasized the following psychosocial sequelae: fears of disease recurrence, diminished self-esteem, preoccupation with death, heightened psychologic distress, job-related difficulties, social isolation, and difficulties reentering school, family, and friendship networks (Cella & Tross, 1986; Lesko et al., 1992) (Table 24-7). In short, cancer and its treatment have been associated with psychosocial disruptions in school, peer, vocational, familial, and mental health domains. Paradoxically, having been suc-

TABLE 24–7. *Psychological Issues of Survivorship*

Concern with termination of treatment
Fear of relapse
Preoccupation with somatic symptomatology and the sense of physical damage
Reentry into developmental life tasks
Transition from patient to healthy status
Sense of being on one's own
Hidden concerns: financial security, job discrimination, social withdrawal, life or health insurance difficulties
Tenuous sense of longevity
Lingering affinity with death
Survivor's syndrome (guilt)

cessfully treated for cancer may also represent a stressful period in the course of the cancer experience.

Authors Cella, Tross, and Lesko have carefully studied young adult survivors across several stages of lymphoma, leukemia, and testicular cancer. In following patients within single diseases, differential adjustment relationships between stage of disease or treatment options were studied. One surprising finding is that most adults are psychologically resilient after treatment but do report a chronic level of psychologic distress. This distress usually is not of pathologic proportions nor does it interfere with reentry to family, social, leisure, school, or employment networks.

The above studies represent the first comprehensive attempt to directly examine the psychosocial functioning of young adults with cancer. When compared with normative samples of nonpatients, these survivors (taken as a whole group) reported heightened levels on several indicators of psychologic distress. Although not entirely consistent across different psychologic measures, these young people were generally one standard deviation above the mean for psychologic distress, but, when compared to normative samples of psychiatric outpatients, these survivors reported significantly less psychologic distress. For instance, survivors reported less intrusive and avoidant cognitions associated with the stressor of being diagnosed and treated for cancer than did patients experiencing post-traumatic stress disorders. In aggregate, these findings again suggest that the psychosocial adjustment of survivors is quite variable. Finally, while group comparisons shed light on the psychosocial functioning of survivors, in general, wide variability in psychosocial adjustment may mask identification of a cohort of cancer survivors most at risk for psychosocial dysfunction. Sociodemographic, disease/treatment, and psychologic distress variables only partially explain this variability.

These psychologic findings, though, are primarily from studies that have looked at middle-aged adult survivors or adult survivors of childhood cancers. Special attention to adolescent and young adult survivors is needed for a number of crucial reasons. First, healthy adolescents and young adults are normally at increased risk for psychosocial difficulties by virtue of the rapid changes associated with their developmental state. Although most negotiate developmental milestones quite adequately, research suggests that, if too many stressors occur simultaneously, psychosocial adjustment may be threatened. Thus, when a cancer diagnosis precedes or occurs during this time, the potential for psychosocial dysfunction is increased. Adolescents who have been treated for cancer not only have the substantial physical, cognitive, emotional, and interpersonal tasks faced by all adolescents, but have the added burden of integrating a life-threatening disease into their experiences. Persistent body image concerns, somatic preoccupation, disruptions in heterosexual relationships, and deficits in social competence have all been documented in this age group (Fritz & Williams, 1988; Mulhern et al., 1989; Rait et al., 1992). Second, assessment during adolescence or young adulthood, rather than later adulthood, allows for attention to potential problems that may surface closer to the original stressor of the diagnosis and treatment.

Fritz and Williams (1988) assessed 52 survivors of childhood cancer and their families who were more than 2 years past treatment. Two thirds of their patients had excellent psychosocial functioning without serious social issues. Many expressed a positive effect of their illness and reported little depression. Illness and related variables were not predictive of psychosocial outcome, unlike some psychosocial variables such as communication patterns and peer support. Lesko and colleagues have underscored many of the same issues (Rait et al., 1992). Their studies examined family functioning, mental health, self-esteem, social competence, and problem behaviors of 61 adolescents. This group perceived their families as less adaptable and flexible than families of medically well adolescents. Family functioning was related to posttreatment psychological adjustment. Finally, these teenagers thought their families were less cohesive than was the case with their counterparts.

At best in the adolescent population surviving cancer, the present findings have highlighted the variability in patient response/adaptation to illness and the need to accept individual response patterns. An adolescent's developmental trajectory may be difficult to predict until more longitudinal studies are done. In summary, there are probably no universal or cookbook intervention programs developed for adolescents. A few programs have been developed for adult posttreatment patients; one example is the Post Treatment Resource Program at Memorial Sloan-Kettering Cancer Center (Zampini, 1991, personal communication).

Acknowledgments

The authors wish to acknowledge and show appreciation of other Memorial Sloan-Kettering Cancer Center colleagues whose works and clinical experiences are mentioned in this chapter: Sarah Auchincloss, M.D., William Breitbart, M.D., David Cella, Ph.D., Stewart Fleishman, M.D., Rene Mastrovito, M.D., Paul Jacobsen, Ph.D., William Redd, Ph.D., and Julia Rowland, Ph.D. Ms. Linda Maxwell prepared the final editing and typing; her technical assistance is again greatly appreciated.

REFERENCES

Adams F (1984). Neuropsychiatric evaluation and treatment of delirium in the critically ill cancer patient. Cancer Bulletin of the

University of Texas, M D Anderson Hospital and Tumor Institute 36:156–160.

Adams F, Quesada Jr, & Gutterman JU (1984). Neuropsychiatric manifestations of human leukocyte interferon therapy in patients with cancer. JAMA 252:938–941.

Ahles TA, Blanchard EB, & Ruckdeschel JC (1983). Multidimensional nature of cancer-related pain. Pain 17:227–288.

Auchincloss S (1984). Gynecological cancer: Psychological and sexual sequelae and management (pp. 25–26). In MJ Massie & LM Lesko (Eds.), Current concepts in psychooncology. New York: Gold Publishers.

Bagenal FS, Easton DF, Harris E, et al. (1990). Survival of patients with breast cancer attending Bristol Cancer Help Centre. Lancet 336:606–610.

Balon R, & Berchou R (1986). Hematologic side effects of psychotropic drugs. Psychosomatics 27:119–127.

Bernstein I (1986). Etiology of anorexia in cancer. Cancer 58:1881–1886.

Bolund C (1985). Suicide and cancer: 1. Demographic and social characteristics of cancer patients who committed suicide in Sweden 1973–1976. J Psychosoc Oncol 3:17–30.

Breitbart W (1986). Tardive dyskinesia associated with high dose intravenous metoclopramide. N Engl J Med 315:518.

Breitbart W (1989). Suicide in patients with cancer. In JC Holland & JH Rowland (Eds.), Handbook of Psychooncology (Ch. 24, pp. 291–299). New York: Oxford University Press.

Bruera E (1989). Use of methylphenidate as an adjuvant to narcotic analgesics in patients with advanced cancer. J Pain Symptom Management 4:3–6.

Bukberg JB, Penman DT, & Holland JC (1984). Depression in hospitalized cancer patients. Psychosom Med 46:199–212.

Burish TG, & Lyles JN (1981). Effectiveness of relaxation training in reducing adverse reactions to cancer chemotherapy. J Behav Med 4:65–78.

Cassileth B (1986). Unorthodox cancer medicine. Cancer Invest 4:591–598.

Cella D, Pratt A, & Holland JC (1986). Persistent anticipatory nausea, vomiting and anxiety in cured Hodgkins disease patients after completion of chemotherapy. Am J Psychiatry 143:641–643.

Cella D, & Tross S (1986) Psychological adjustment to survival from Hodgkin's disease. J Consult Clin Psychol 54:616–622.

Chiarello RJ, & Cole JO (1987). The use of psychostimulants in general psychiatry: A reconsideration. Arch Gen Psychiatry 44:286–295.

Coyle N, Loscalzo M, & Bailey L (1989). Supportive home care for the advanced cancer patient and family. In JC Holland & JH Rowland (Eds.), Handbook of psychooncology (Ch. 48, pp. 598–606). New York: Oxford University Press.

Cubeddu LX, Hoffman IS, Fuenmayor NT, et al. (1990). Efficacy of ondansetron (GR 38032F) and the role of serotonin in cisplatin-induced nausea and vomiting. N Engl J Med 322:810–816.

Danielson DA, Douglas SW, Herzog P, et al. (1984). Drug induced blood disorders. JAMA 252:3257–3260.

Davies RK, Quinlan DM, McKegney FD, et al. (1973). Organic factors and psychological adjustment in advanced cancer patients. Psychosom Med 35:464–471.

DeAngelis LM, Delattre J, & Posner JB (1989). Radiation-induced dementia in patients cured of brain metastases. Neurology 39:789–796.

Denicoff KD, Rubinow DR, Pappa M, et al. (1987a). Neuropsychiatric toxicity of interleukin-2/Lak. CME Syllabus and Proceedings Summary, Abstract No. 31A. Presented at the 140th Annual Meeting of the American Psychiatric Association, Chicago, IL.

Denicoff KD, Rubinow DR, Pappa MZ, et al. (1987b). The neuropsychiatric effects of treatment with interleukenin-2 and lymphokine-activated killer cells. Ann Intern Med 107:293–300.

Derogatis LR, Morrow GR, Fetting J, et al. (1983). The prevalence of psychiatric disorders among cancer patients. JAMA 249:751–757.

Dobin RE, & Kleinman MAR (1991). Marijuana as an antiemetic medicine: A survey of oncologists' experiences and attitudes. J Clin Oncol 9:1314–1319.

Ellis WG, Sobel RA, & Nielsen SL (1982). Leukoencephalopathy in patients treated with amphotericin B methyl ester. J Infect Dis 136:125–137.

Farkas C (1989). Management of special psychiatric problems in terminal care: Role for a psychiatric nurse-clinician. In JC Holland & JH Rowland (Eds.), Handbook of psychooncology (Ch. 49, pp. 607–611). New York: Oxford University Press.

Fleishman SB, & Lesko LM (1989). Delirium and dementia. In J Holland & J Rowland (Eds.), Handbook of psychooncology (Ch. 30, pp. 342–355). New York: Oxford University Press.

Foley K (1984). Pharmacologic management of pain. In MJ Massie & LM Lesko (Eds.), Current concepts in psychooncology (pp. 29–32). New York: Gold Publishers.

Foley K (1986). Non-narcotic and narcotic analgesics: Applications. In K Foley (Ed.), Management of cancer pain (pp. 135–152). New York: Gold Publishers.

Folstein MF, Fetting JH, Lobo A, et al. (1984). Cognitive assessment of cancer patients. Cancer 53:2250–2257.

Fox BH, Stanek EJ, Boldy SC, et al. (1982). Suicide rates among cancer patients in Connecticut. Chronic Dis 35:85–100.

Fritz B, & Williams J (1988). Issues of adolescent development for survivors of childhood cancer. J Am Acad Child Adolesc Psychiatry 27:712–715.

Goggans FC, Wishberg LJ, & Koran LM (1983). Lithium prophylaxis of prednisone psychosis: A case report. J Clin Psychiatry 44:111–112.

Gralla RJ (1983). Metoclopramide as an antiemetic agent. In RJ Gralla (Ed.), Supportive care of the cancer patient. New York: Biomedical Information.

Gralla RJ (1985). The control of nausea and vomiting in patients. In GR Bosel (Ed.), Current concepts in medical oncology (pp. 47–54). New York: Gold Publishers.

Gralla RJ, Itri LM, Pisko SE, et al. (1981). Antiemetic efficacy of high dose metoclopramide: Randomized trials with placebo and prochloroperazine in patients with chemotherapy-induced nausea and vomiting. N Engl J Med 305:905–909.

Gray RE, & Doan BD (1990). Heroic self-healing and cancer: Clinical issues for the health professions. J Palliat Care 6:32–41.

Healey M, Pickens R, Meisch R, et al. (1983). Effects of clorazepate, diazepam, lorazepam and placebo on human memory. J Clin Psychiatry 44:4436–4439.

Hinton J (1972). Psychiatric consultation of fatal illness. Proc R Soc Med 65:29–32.

Holland JC (1982a). Psychological aspects of cancer. In JF Holland & E Frei (Eds.), Cancer medicine (2nd ed., pp. 1175–1203). Philadelphia: Lea & Febiger.

Holland JC (1982b). Psychological issues in the care of the terminally ill (Lesson 25). Directions in Psychiatry (Available from Editorial Offices, 420 E. 51st St., New York, NY 10022).

Holland JC (1989). Gastrointestinal cancer. In JC Holland & JH Rowland (Eds.), Handbook of psychooncology (Ch. 15, pp. 208–217). New York: Oxford University Press.

Holland JC, Geary N, & Furman A (1989). Alternative cancer therapies. In JC Holland & JH Rowland (Eds.), Handbook of psychooncology (Ch. 42, pp. 508–515). New York: Oxford University Press.

HOLLAND JC, KORZUN AH, TROSS S, ET AL. (1986). Comparative psychological disturbance in patients with pancreatic and gastric cancer. Am J Psychiatry 143:982–986.

HOLLAND JC, & LESKO LM (1989). Chemotherapy, endocrine therapy and immunotherapy. In JC HOLLAND & J ROWLAND (Eds.), Handbook of psychooncology (Ch. 10, pp. 146–162). New York: Oxford University Press.

HOLLAND JC, & ROWLAND JH (EDS.) (1989). Handbook of psychooncology. New York: Oxford University Press.

HOLLAND JC, ROWLAND J, LEBOVITS A, ET AL. (1979). Reactions to cancer treatment: Assessment of emotional response to adjunct radiotherapy. Psychiatr Clin North Am 2:347–358.

KAPLAN RS, & WIERNIK PH (1983). Neurotoxicity of antineoplastic drugs. Semin Oncol 9:103–130.

KUBLER-ROSS E (1969). On death and dying. New York: Macmillan.

LASZLO J, CLARK RA, HANSON DC, ET AL. (1985). Lorazepam in cancer patients treated with cisplatin: A drug having antiemetic, amnesic and anxiolytic side effects. J Clin Oncol 3:864–869.

LAZZARA RR, STOUDEMIRE A, MANNING D, ET AL. (1986). Metoclopramide induced tardive dyskinesia: A case report. Gen Hosp Psychiatry 8:107–109.

LEDERBERG MS, MASSIE MJ, & HOLLAND JC (1990). Psychiatric consultation to oncology. In A TASMAN, SM GOLDFINGER, & CA KAUFMANN (Eds.), American Psychiatric Press Review of Psychiatry (Vol. 9, Ch. 25, pp. 491–514). Washington, DC: American Psychiatric Press.

LESKO LM, OSTROFF JS, MUMMA GH, ET AL. (1992). Long-term psychological adjustment of acute leukemia survivors: Impact of bone marrow transplantation vs. chemotherapy. Psychosom Med 54:30–47.

LEVINE PM, SILBERFARB PM, & LIPOWSKI ZJ (1978). Mental disorders in cancer patients: A study of 100 psychiatric referrals. Cancer 42:1385–1391.

LIPOWSKI ZJ (1983). Transient cognitive disorders (delirium, acute confusional states) in the elderly. Am J Psychiatry 140:1426–1436.

LOPRINZI CL, ELLISON NM, SCHIAD DJ, ET AL. (1990). Controlled trial of megestrol acetate for treatment of cancer anorexia and cachexia. J Natl Cancer Inst 82:1127–1132.

LOUHIVOURI KA, & HAKAMA M (1979). Risk of suicide among cancer patients. Am J Epidemiol 109:59–65.

MARTY M, POUILLART P, SCHOLL S, ET AL. (1990). Comparison of the 5-hydroxytryptamine (serotonin) antagonist ondansetron (GR 383032F) with high-dose metoclopramide in the control of cisplatin-induced emesis. N Engl J Med 322:816–821.

MASSIE MJ, & HOLLAND JC (1984). Current concepts in psychiatric oncology. In L GREENSPAN (Ed.), Psychiatry update III (Ch. 16, pp. 239–256). Washington, DC: American Psychiatry Press.

MASSIE MJ, & HOLLAND JC (1987). Consultation and liaison issues in cancer care. Psychiatr Med 5:343–359.

MASSIE MJ, HOLLAND J, & GLASS E (1983). Delirium in terminally ill cancer patients. Am J Psychiatry 140:1048–1050.

MASTROVITO R, & MOYNIHAN R (1989). Self-help and mutual support programs in cancer. In JC HOLLAND & JH ROWLAND (Eds.), Handbook of psychooncology (Ch. 41, pp. 502–507). New York: Oxford University Press.

MCDONALD EM, MANN AH, & THOMAS HC (1987). Interferons as mediators of psychiatric morbidity: An investigation in the trial of recombinant alpha interferon in hepatitis-B carriers. Lancet 11:1175–1178.

MCMAHON T, & VAHORA S (1986). Radiation damage to the brain: Neuropsychiatric aspects. Gen Hosp Psychiatry 8:437–441.

MEADOWS A, & HOBBIE N (1986). The medical consequences of cure. Cancer 58:524–528.

MEADOWS A, & SIBLER J (1985). Delayed consequences of therapy for childhood cancer. Cancer 58:524–528.

MOFFIC H, & PAYKEL ES (1975). Depression in medical inpatients. Br J Psychiatry 126:346–353.

MOORE DH, FOWLER WC, & CRUMPLER LS (1990). Fluorouracil neurotoxicity. Gynecol Oncol 36:152–154.

MORGENSTERN H, GELLERT GA, WALTER SD, ET AL. (1984). The impact of a psychosocial support program on survival with breast cancer: The importance of selection bias in program evaluation. J Chronic Dis 37:273–282.

MORROW GR, & MORRELL BS (1982). Behavioral treatment for the anticipatory nausea and vomiting induced by cancer chemotherapy. N Engl J Med 306:1476–1480.

MOULIN D, & FOLEY K (1984). Management of pain in patients with cancer. Psychiatr Ann 14:815–822.

MULHERN R, WASSERMAN A, FRIEDMAN A, ET AL. (1989). Social competence and behavioral adjustment of children who are long-term survivors of cancer. Pediatrics 83:18–25.

MURPHY G (1990). Medical Affairs Newsletter II (ii), B.

NOVACK DH, PLUMER R, SMITH RL, ET AL. (1979). Changes in the physicians' attitudes toward telling the cancer patient. JAMA 241:897–900.

PATCHEL RA, & POSNER JB (1989). Cancer and the nervous system. In J HOLLAND & J ROWLAND (Eds.), Handbook of psychooncology (Ch. 29, pp. 327–341). New York: Oxford University Press.

PETEET J, TAY V, COHEN G, ET AL. (1986). Pain characteristics and treatment in an outpatient cancer population. Cancer 57:1259–1265.

PLUMB MM, & HOLLAND JC (1977). Comparative studies of psychological function in patients with advanced cancer. 1. Self-reported depression symptoms. Psychosom Med 39:264–276.

PORTENOY RK, MOULIN DF, ROGER A, ET AL. (1986). Intraveneous infusion of opioids in cancer pain: Clinical review and guidelines for use. Cancer Treat Rep 70:575–581.

POSNER JB (1979). Delirium and exogenous metabolic brain disease. In PB BEESON, W MCDERMOTT, & JB WYNGAARDEN (Eds.), Cecil textbook of medicine (Ch. 220, pp. 644–651). Philadelphia: WB Saunders.

PRATT A, LAZAR R, PENMAN D, ET AL. (1984). Psychological parameters of chemotherapy-induced conditioned nausea and vomiting: A review. Cancer Nurs 1:483–490.

QUESADA JR, TALPAZ M, RIOS A, ET AL. (1986). Clinical toxicity of interferons in cancer patients: A review. J Clin Oncol 4:234–243.

RAIT D, OSTROFF J, SMITH K, ET AL. (1992). Lives in a balance: Family functioning and the psychosocial adjustment of adolescent cancer survivors. In press.

REDD WH (1989). Management of anticipatory nausea and vomiting. In J HOLLAND & J ROWLAND (Eds.), Handbook of psychooncology (Ch. 35, pp. 423–433). New York: Oxford University Press.

REDD WH, ANDRESEN GV, & MINAGAWA Y (1982). Hypnotic control of anticipatory emesis in patients receiving cancer chemotherapy. J Consult Clin Psychol 50:14–19.

RESNIK HLP (1980). Psychiatric emergencies—suicide. In I KAPLAN, A FREEDMAN, & BJ SADDOCH (Eds.), Comprehensive textbook of psychiatry (Vol. 3, Ch. 29, pp. 2085–2097). Baltimore: Williams & Wilkins.

ROHATINER AZS, PRIOR PF, BURTON AC, ET AL. (1983). Central nervous system toxicity of interferon. Br J Cancer 47:419–442.

ROWLAND JH, GLIDEWELL OJ, SIBLEY RF, ET AL. (1984). Effects of different forms of central nervous system prophylaxis and neuropsychologic function in childhood leukemia. J Clin Oncol 2:1327–1335.

SANTOS AB, BELILES KE, & ARANA GW (in press). Parenteral use of psychotropic agents. In A Stoudemire, & BS Fogel (Eds.), Medical psychiatric practice (Vol. 2; Ch. 4). Washington, DC: American Psychiatric Press, Inc.

SCHWAB JJ, BIALOW M, BROWN J, ET AL. (1967). Diagnosing depression in medical inpatients. Ann Intern Med 67:695–707.

SHELINE GE (1980). Irradiation injury of the human brain: A review of clinical experience. In J GILBERT & AR KAGAN (Eds.), Radiation damage to the nervous system (pp. 39–58). New York: Raven Press.

SHEVITZ SA, SILBERFARB PM, & LIPOWSKI ZJ (1976). Psychiatric consultants in a general hospital: A report of 1000 referrals. Dis Nerv System 37:295–300.

SMEDLEY H, KATRAK M, SIKORA K, ET AL. (1983). Neurological effects of recombinant human interferon. Br Med J 286:262–264.

SPIEGEL D, BLOOM J, & YALOM I (1981). Group support for patients with metastatic cancer. Arch Gen Psychiatry 45:333–339.

SPIEGEL D, KRAEMER H, BLOOM JR, ET AL. (1989). Effect of psychosocial treatment on survival of patients with metastatic breast cancer. Lancet 2:888–891.

STERNBACH RA (1974). Pain patients: Traits and treatment. New York: Academic Press.

STOUDEMIRE A, COTANCH P, & LASZLO J (1984). Recent advances in the pharmacologic and behavioral management of chemotherapy induced emesis. Arch Intern Med 144:1029–1033.

STRUM SF, MCDERMED JE, STENG BR, ET AL. (1984). Combination metoclopramide and dexamethasone: An effective antiemetic regimen in outpatients receiving cisplatin chemotherapy. J Clin Oncol 2:1057–1063.

TCHEKMEDYIAN NS, TAIT N, MOODY M, ET AL. (1986). Appetite stimulation with megestrol acetate in cachetic cancer patients. Semin Oncol 13:37–43.

U.S. Congress, Office of Technology Assessment (1990). Unconventional cancer treatments (OTA-H-405). Washington, DC: U.S. Government Printing Office.

VIEDERMAN M (1982). Psychotherapeutic management of depression in the mentally ill. In JC HOLLAND (Ed.), Current concepts in psychooncology (pp. 29–39). New York: Gold Publishers.

VON ROENN JH, MURPHY RL, WEBER KM, ET AL. (1988). Megestrol acetate for treatment of cachexia associated with human immunodeficiency virus (HIV) infection. Ann Intern Med 109:840–849.

WADE JC, & MYERS JD (1983). Neurological symptoms associated with parenteral acyclovir treatment after marrow transplantation. Ann Intern Med 98:921–925.

WALSH TD (1986). Controlled study of imipramine and morphine in chronic pain due to advanced cancer. In Proceedings of the American Society of Clinical Oncology (Vol. 5, Abstract No. 929, p. 237). Chicago: American Society of Clinical Oncology.

WHARTON JC (1987). Traditions of folk medicine in America. JAMA 257:1632–1635.

WILHOLM B-E, MORTIMER O, BOETHIUS G, ET AL. (1984). Tardive dyskinesia associated with metoclopramide. Br Med J 288:545–547.

YOUNG DF, & POSNER JB (1980). Nervous system toxicity of the chemotherapeutic agents. In PJ VINKEN & GW BRUYN (Eds.), Handbook of clinical neurology. Vol. 39: Neurological manifestations of systemic diseases, part II (pp. 91–129). New York: Elsevier Biomedical Press.

ZALUPSKI M, & BAKER LH (1988). Ifosfamide. J Natl Cancer Inst 80L:556–566.

25 Pulmonary disease

WENDY L. THOMPSON, M.D., AND
TROY L. THOMPSON II, M.D.

Difficulty in breathing has many psychiatric implications. Patients react emotionally to the discomfort of dyspnea, the loss of functional capacity, and the threat of suffocation and death, while hypoxia, hypercarbia, hyperventilation, respiratory failure, and many pulmonary medications all have direct effects on the brain. Depression is frequently associated with pulmonary disease (Klerman, 1981; Lindegard, 1982). Associated depression can range from a mild dysthymia or an adjustment disorder with depressed mood to a major depressive episode (Thompson & Thompson, 1984).

Another psychiatric symptom that frequently accompanies pulmonary disease is anxiety. Most patients with respiratory disease have episodes accompanied by anxiety when they feel unable to breathe adequately. Dyspnea, like pain, is subjective, clearly influenced by emotional and psychiatric factors, and not necessarily related to measurements of actual pulmonary function or blood gases. If severe, dyspnea may be felt as a sensation of suffocating, strangling, or drowning, and is overwhelmingly frightening, sometimes leading to panic reactions (Thompson & Thompson, 1985a). Another frequently associated problem in those with respiratory disease is sexual dysfunction, including inhibited sexual excitement, inhibited orgasm, and premature ejaculation. Cognitive impairment disorders often accompany respiratory disease as well, especially in debilitated patients or geriatric patients, who may also have other significant physical disorders.

PSYCHOLOGIC ASPECTS OF PULMONARY DISEASE

The psychologic ramifications of pulmonary disease depend on the specific disease, the age of onset, the etiology of the illness, and the severity of the illness. The psychiatrist must do a thorough assessment of both the individual and the family to determine whether there are specific psychodynamic conflicts, behavioral triggers, or environmental issues that contribute to exacerbations of the respiratory illness or its symptoms (Stoudemire, 1985). It is important to become aware of common familial conflicts or characterologic pathology that may play a role in preventing optimal medical management of the patient (Thompson & Thompson, 1985a).

Some pulmonary disease, such as reversible obstructive airways disease (ROAD, or asthma), may have a significant "psychosomatic" component (Alexander, French, & Pollock, 1968). Earlier studies hypothesized a specific central psychologic conflict leading to later development of asthma (French & Alexander, 1939–1941). This conflict was thought to be strong, unconscious dependency wishes toward the mother, coupled with a fear of separation from her. Although such a "specificity hypothesis," associated with asthma and other illnesses with psychosomatic components, is generally considered to be obsolete, the psychodynamic issues set forth by French and Alexander (1939–1941) may still be relevant to some patients.

Another important consideration in psychiatric assessment of patients with pulmonary disease is the developmental or psychosocial life stage during which the patient develops asthma. For example, an infant or child with severe respiratory disease, whether asthma or cystic fibrosis, is likely to be perceived and treated differently by family, friends, and relatives. This may lead to significant alterations in the relationship between the mother and the child and adversely affect advancement through the early stages of intrapsychic development, leading to later susceptibility to the trauma of separation or other psychologic impairments.

Asthma may also develop in middle age or old age, and the vicissitudes of the particular psychosocial stage must be considered when evaluating the impact of the asthma. It may disrupt the patient's identity as primary wage earner or, when it occurs around the time of retirement, it can disrupt long-standing plans, leading to acrimony in the family and sometimes contributing to depression in the patient. It often forces alterations of ingrained family roles and expectations. Grandparents may fear being around grandchildren for fear of contracting an upper respiratory infection, leading to a severe asthma episode. As with childhood asthma, the illness can vary from mild, intermittent episodes to severe refractory asthma requiring oral steroids. In older patients especially, possible interactions with other drugs that they are taking should be carefully evaluated.

The experience or fear of suffering respiratory distress with a concomitant sensation of strangling or drowning, along with frequent urgent trips to the emergency room and multiple hospitalizations, can lead to pervasive anxiety in the child and the child's parents. The sedating side effects of some medications, as well as sleep disturbances resulting from respiratory distress, can impair learning, even when a child is able to attend school. Asthmatic children often are prone to becoming scapegoats among their friends because of restrictions on activities at school or at home. Problems may also arise within the family because of the alterations in life-style required by family members dealing with an asthmatic child (Fritz, 1983). For this reason, the families of some asthmatic children may benefit from a formal psychiatric assessment and educational discussions. Respiratory symptoms in a child may relate to the child's conflicts or fantasies, perhaps around the death of another child by sudden infant death syndrome, or may reflect parental conflicts as well as inherent biologic susceptibility (Wilson, 1980–1981). Fritz, Rubinstein, and Lewiston (1987) found, in three cases of death resulting from childhood asthma, that unsupportive families, depression, emotional precipitation of attacks, and a tendency to deny symptoms of asthma played a significant part.

In adulthood, studies have focused on patients who have suffered near-death from asthma. An initial study (Yellowlees et al., 1988) indicated no increased psychiatric morbidity among this group, but a follow-up study (Yellowlees & Ruffin, 1989) found a 40% incidence of psychiatric disorders. All these patients had a high level of denial and, following the episode, had either increased anxiety or increased denial. Sibbald et al. (1988) found that negative attitudes toward asthma were correlated with higher morbidity, which may contribute to further exacerbations of the asthma. Significant asthma morbidity also increased negative attitudes toward asthma, fostering a vicious cycle.

How these psychodynamic factors can contribute to both the onset and later exacerbations of asthma is currently being studied. The asthmatic reaction is thought to result from a complex interplay between the autonomic nervous system and such substances as epinephrine, norepinephrine, histamine, bradykinin, leukotrienes, sensory neuropeptides, and prostaglandins (Knapp, 1985; Stein, 1982; Weiner, 1987). The central nervous system is directly involved in stimulating production or regulating metabolism of these substances. Brain stimulation, particularly in the hypothalamus and limbic system formation, affects the level of many of these neurotransmitters, which in turn influence the autonomic nervous system (Martin, Reichlin, & Brown, 1977). This may be the pathway whereby conflicts and emotions affect the development and course of the asthma by affecting immunologic or airway reactivity. Emotional states may also affect the balance between functional sympathetic and functional parasympathetic activity (Moran, 1991).

A syndrome of vocal cord dysfunction also exists, which has usually been misdiagnosed over a period of years as intractable asthma. Psychiatric disorders are present in most of these patients, and the patients tend to respond best to psychotherapy and to speech therapy (Appelblatt & Baker, 1981; Christopher et al., 1983). The possibility of vocal cord dysfunction should be suspected in patients who complain that their throat closes and who, on auscultation, have upper airway wheezing that is secondarily transmitted to the lungs. They often present with a history of "asthma" that has been refractory to usual medical management. This syndrome is diagnosed through direct visualization of the vocal cords during inspiration and expiration, which shows inappropriate opposition of the cords during breathing. This can have important implications for both psychiatric and medical management of the patient.

Psychodynamic and Personality Factors

Most psychodynamic explanations in pulmonary patients have focused on those with asthma. Little has been done to explore psychodynamic aspects of chronic obstructive pulmonary disease (COPD) or other chronic pulmonary illnesses. While COPD patients often have been reported to have increased anxiety, depression, fatigue, difficulty coping, and somatic preoccupations, they do not appear to differ significantly in these areas from other chronically ill patients (Sandhu, 1986). In those COPD patients with chronic dyspnea, it has been hypothesized by Dudley, Sitzman, and Rugg (1985) that the patient restricts both activating (anger/anxiety) and nonactivating (depression/withdrawal) affects in order to avoid the experience of dyspnea, suggesting that a "personality trait" may result from reactions to the illness, rather than be a cause of the illness.

Personality traits have been found to influence the perception by patients of added resistive loads in breathing. Those who are more dependent and anxious, whether or not they have asthma, tend to have greater thresholds both for inspiration and for expiration than do those who have more adaptive personality styles or who are rigidly independent (Hudgel, Cooperson, & Kinsman, 1982). These results are not always consistent over time and do not always distinguish between those who have asthma and those who do not. Nonetheless, personality traits can influence the effort exerted on pulmonary function tests, which produce less valid results if a full effort is not made.

Regardless of the physiologic severity of the disease, there tends to be a correlation between the patient's rating of the severity of the illness and the degree of emotional disturbance found (Plutchik et al., 1978). There also appears to be an increasing incidence and severity of neurotic and psychosomatic symptoms associated with an increased amount of asthma medication required to control the disease (Teiramaa, 1978). The latter finding makes one wonder whether the psychiatric symptoms cause the severity of the asthma to increase and, therefore, increase the amount of medication required, whether the psychiatric symptoms interfere with medical evaluation and patient compliance, and/or whether the asthma medication is aggravating the psychiatric symptoms. The psychiatric assessment must address these issues. Some of these questions relate to the panic-fear studies of Kinsman and associates (Dirks, Jones & Kinsman, 1977; Kinsman et al., 1973, 1980a). These studies have divided panic-fear symptomatology into that which is characterologic and that which is specifically illness related. Panic and fear appear to be associated with a basic lack of ego resources and with dependency conflicts, emotional lability, and a tendency toward pervasive anxiety. Patients with "characterologic" panic relate to a variety of situations in their lives with anxiety, panic, and a sense of helplessness or dependency. In contrast, illness-specific anxiety seems to function as "signal anxiety" and is specific to the situation of breathing difficulties.

Anxiety: Common Factors in Respiratory Illness

Common to almost all patients with respiratory illness is anxiety related to the sensation of dyspnea, fear of being placed on a respirator, and, ultimately, a fear of death. Paradoxically, epinephrine and corticosteroids, the substances produced during states of anxiety or fear, are also used to treat asthma, yet these states can induce an asthma attack. It is possible that some individuals respond to emotional stress with a parasympathetic response, involving increased release of acetycholine (Vingerhoets, 1985), or that some asthmatics do not increase production of epinephrine during stress (Mathe & Knapp, 1969).

Anxiety may cause agoraphobia or an overreporting of symptoms to the physician, leading to increased and often unnecessary use of medications, with the attendant side effects (Thompson & Thompson, 1985a). Many COPD patients demonstrate symptoms of anxiety sufficient to interfere with their daily lives (Agle & Baum, 1977).

Among asthmatics at a pulmonary referral hospital, 42% reported anxiety focused on breathing, compared to 20% of COPD patients (Kinsman et al., 1983). Some studies have found no increase in the prevalence of anxiety *disorders* in COPD patients compared with the general population (Karajgi et al., 1990), whereas others report an increased incidence (34%) of anxiety disorders (Yellowlees et al., 1987). There seems to be more agreement that there is an increased prevalence of panic disorders among COPD patients, although the reported prevalence varies from 8% (Karajgi et al., 1990) to 24% (Yellowlees et al., 1987). In some normal individuals, increased carbon dioxide concentration through inhalation of carbon dioxide may increase discharge from the locus ceruleus, a part of the noradrenergic system that may be the generator of panic attacks (Gorman et al., 1984). Hypercapnia associated with COPD might lead to increased activity of the locus ceruleus, causing panic attacks in susceptible individuals (see Table 25-1).

Patients with end-stage COPD often have severe anxiety, and it may be difficult to separate this from the chronic sensation of dyspnea they experience. In order to make such patients comfortable, it is often necessary to use medication to reduce their anxiety. Buspirone may be helpful in these patients since it does not depress respiratory drive and has anxiolytic effects. Benzodiazipines are risky to use in these patients, whose res-

TABLE 25–1. *Psychiatric Symptoms Commonly Associated with Hypoxia and Hypercapnia*

Partial Pressure of O_2 (% of Sea Level)	Mental Symptomatology Attributable to Resultant Hypoxia
90	Altered visual dark adaptation
85	Loss of judgment, inappropriate behavior
75	Decreased ability to carry out complex tasks
65	Impaired short-term memory
50	Severe loss of judgment
30–40	Unconsciousness
Partial Pressure of CO_2	**Mental Symptomatology Attributable to Resultant Hypercapnia**
Normal	No change
Moderately elevated	Headache, drowsiness, indifference or inattention, perceptual changes, forgetfulness
Severely elevated	Stupor or coma

Modified from Gale and O'Shanick (1985).

piratory functioning is already severely compromised, unless they are chronically respirator dependent. In some patients, low-dose antidepressant therapy may be beneficial, and low-dose neuroleptics can be helpful in psychotic or extremely agitated patients.

Although there is virtually no research on the impact of anxiety on tuberculosis (TB) reported in the English literature, Pecyna (1989) in the Polish literature, suggested that patients with neurotic anxiety reactions demonstrate a "diminution of psychic resistance" that may decrease the likelihood of achieving sputum negativity during treatment.

Anxiety and Asthma

Those patients who have both high illness-specific anxiety and high characterologic anxiety tend to have more hospitalizations and lengthier hospital stays and to take higher doses of medications, regardless of the severity of their illness; these factors may lead to increased morbidity and mortality. Baron et al. (1986) adapted these studies to children with asthma and found those with high panic-fear personality features also seemed to be prescribed higher doses of medication, especially corticosteroids. In contrast, patients with excessively low panic-fear symptomatology tend to ignore their symptoms and not respond appropriately, or at an appropriate time, to the warning symptoms of their illness. They have a tendency, therefore, to underuse medications, not only those prescribed as needed but also those prescribed on a regular schedule. They have a tendency to get discharged from hospitals prematurely but also to have a higher rate of rehospitalization; this also leads to higher rates of morbidity and mortality. The patients with the best medical outcome tend to be those with high illness-specific panic-fear but average levels of characterologic panic-fear. Such patients are able to attend to their respiratory symptoms in a timely and appropriate fashion and to elicit appropriate responses from their medical caregivers as well.

Physicians usually do not assess the patient's panic-fear level in determining when to alter medications. Often, when there is an exacerbation, the patient will speak to the physician over the telephone to adjust medications. If the psychiatrist has evaluated the degree of panic-fear present, the psychiatrist can enhance the physician's management of the patient. For example, for high panic-fear patients, a mini-peak flow meter might be used to establish an objective measure of dyspnea. For low panic-fear patients, assistance of a family member might be enlisted. With the high panic-fear patient, anxiolytics or relaxation techniques may be employed along with psychotherapy to reduce the anxiety. In low panic-fear patients, therapy focusing on their need for denial and their counterdependent personality style may be helpful. Primary care by the psychiatrist, with ongoing pulmonary consultation, may be a useful management alternative for some asthmatics who have serious characterologic impediments to treatment.

Chronic Obstructive Pulmonary Disease and Anxiety

Kinsman et al. (1983) developed a Bronchitis-Emphysema Symptom Checklist, similar to the Asthma Symptom Checklist, in an attempt to describe how patients experience and cope with chronic bronchitis and emphysema. They found that hopelessness and helplessness generally accompanied a failure to cope with the illness as a result of increasing pulmonary insufficiency, decreased personal and social resources, or both. In contrast to Dudley, Sitzman, and Rugg (1985), they believed that withdrawal and isolation were not defensive, but since they usually occurred in conjunction with increased feelings of hopelessness, helplessness, resignation, and anxiety, they often indicated instead a breakdown of psychologic defenses and a consequent poor prognosis.

In COPD patients, anxiety does not appear to have the same "signal anxiety" function as in asthma patients, because the illness may take more of a downward course, and any deviation from a neutral affect may exacerbate the dyspnea (Sandhu, 1986). Williams (1989) and Williams and Bury (1989) found a low correlation between lung function and degree of disability, but a high correlation between dyspnea and disability. Patients with chronic airflow obstruction tended to have fewer hypochondriacal concerns than a matched group of family practice patients but more severe somatic symptoms and more anxiety and depression (Kellner, Samet, & Pathak, 1987). COPD patients who are coping poorly seem to have a number of factors in common with asthmatic patients. Among these factors are a tendency toward chronic anxiety, which may be related to high panic-fear as well. In particular, such patients often have a failure to accept their illness and to mourn what they have lost, manifested by an inability to shift their expectations and goals. This may lead to chronic anxiety and externalization of responsibility for their feelings and behaviors—all of which may lead to poor compliance with the medical regimen (Post & Collins, 1981–1982).

Clearly, such variables as psychiatric disturbance, personality structure, reactions to medical caregivers, and expectations about treatment have some prognostic significance for medical outcome (Rutter, 1979), and timely psychiatric intervention can have an important impact on outcome. In some COPD patients, anxiety and fear of shortness of breath (often associated with a fear of and signal of imminent death) may result in avoidance of even minimal physical activity (Agle & Baum, 1977). When a patient becomes so fearful about being short of breath, this in and of itself can interfere with the ability to breathe, either by leading to hyperventilation or by increasing the sensation of shortness of breath and decreasing the patient's efficiency in breathing. Any anxiety resulting from the shortness of breath makes the perception of the dyspnea even more acute.

Reactions to Illness

It has been hypothesized that asthma may be a reaction to a loss or disappointment in the patient's life, and, although these observations are a matter of conjecture, clinicians have often observed that many patients with asthma seem to have suffered some kind of loss or disappointment prior to the onset of their asthma, and these should be assessed in the psychiatric evaluation. Disappointments in a close personal relationship have been found to be associated with acute onset of asthma, whereas arrival of a new family member or duration of up to 3 years in marriage has been associated with a more insidious onset of asthma (Teiramaa, 1981).

Asthma can be an episodic illness in which, between episodes, the patient appears to be a relatively normal individual, and yet must continue to make concessions to the illness in terms of ongoing treatment. COPD and cystic fibrosis, in contrast, are chronic and eventually terminal illnesses, wherein the patient can expect little improvement in baseline functioning but may experience periodic exacerbations. Other chronic respiratory diseases, such as TB, atypical TB, or other infectious diseases, require coping with a subacute or relatively long-standing illness, but one in which there is both the promise of recovery, with the possibility of return to full function, as well as the possibility of ongoing impaired function for the rest of the individual's life.

At times, the treatment for pulmonary illness may also provoke a significant psychiatric reaction. Often, patients with severe COPD may require ongoing oxygen therapy. This is a difficult adjustment for many patients since the oxygen tubing and tanks are a visible reminder of their illness. They may react with embarrassment, anxiety, depression, or narcissistic injury, and react maladaptively by not utilizing the oxygen. Even those requiring only nocturnal oxygen may have difficulty, especially around sexual activities. Often, those patients who have not accepted their illness or aging, have many unresolved narcissistic issues, and/or have an unaccepting spouse tend to have the most difficulty with home oxygen. For many other patients, oxygen therapy provides a means to increase their activity level, thereby increasing their sense of well-being, both physical and psychologic, and ability to enjoy life.

Sexual Dysfunction

Anxiety associated with dyspnea may extend to sexual situations as well, making many patients fearful about sexual activities or performance, but anxiety or other reactive psychiatric symptoms are not the only cause of sexual dysfunction in patients with chronic respiratory disease. Especially in asthmatics, some of the sexual dysfunction may be caused by exercise-induced bronchospasm. Beta-adrenergic agents often can prevent this bronchospasm. Some COPD patients experiencing sexual inadequacy would have had the same problem regardless of their pulmonary status (Kass, Updegraff, & Muffly, 1972). For other patients, a decrease in pulmonary function is reflected in worsening of sexual function in the absence of identifiable psychiatric factors (Agle & Baum, 1977; Fletcher & Martin, 1982). COPD affects as many as 15% of older men, and many of these older patients have other illnesses, such as diabetes, which may affect their sexual performance as well as interfere with the medical management of their respiratory disease. Older patients also tend to take more types of medication than younger patients and thus have an increased risk of drug interactions and side effects, including drug-induced sexual dysfunction (Fritz, 1983).

Other problems with sexual functioning may reflect overall concerns of the patient regarding self-image and self-esteem. As with any chronic illness, there may be a dramatic disruption of the established family roles and interactions, and patients with COPD often feel that their lives are out of their control (Dudley, Sitzman, & Rugg, 1985; Thompson, 1986).

Depression

In addition to difficulty with self-esteem, some respiratory patients also may experience significant depression. This may range from an adjustment disorder with depressed mood to a severe and disabling major depression (Thompson & Thompson, 1984). Although depression may significantly impair sexual functioning, some depressed individuals may actually desire more close physical contact and feel an increased need to be held and caressed (Hollender & Mercer, 1976). Depression may also predate the illness itself and may thereby complicate the medical management of the illness.

Mood instability is a frequent side effect of many pulmonary medications, especially corticosteroids. Corticosteroids may decrease the patient's immune response, including the immune response to upper respiratory tract infections, and most patients with chronic respiratory disease will have an exacerbation of their underlying illness whenever they have a superimposed upper respiratory infection. In both monkeys (Reite, Harbeck, & Hoffman, 1981) and humans (Kiecolt-Glaser & Glaser, 1986), depression also has been found to suppress immune response, and thus it may cause more frequent upper respiratory infections (Reite, Harbeck, & Hoffman, 1981), which may then produce further exacerbation of underlying illness.

The depression of pulmonary patients may be manifested in a variety of ways (Covino et al., 1982), and many physicians have a tendency to view depression that is related to receiving an emotionally traumatic new diagnosis as understandable and, thus, not to treat it appropriately, let alone vigorously. Depression may interfere with appropriate medical management, regardless of whether the depression is "warranted" or not (see Chapter 4). A chart review study of asthmatic patients found that many had at least fleeting suicidal ideation and that there was a significantly higher incidence of suicide and suicide attempts than in a group of hypertensive patients (Levitan, 1983); this study probably underestimated the actual figures. Suicidal ideation may not be expressed directly but may be expressed in a more passive fashion by poor adherence to medical treatment. Among Southwestern Native Americans, isoniazid (INH) is involved in 7% of suicide attempts and 19% of successful suicides (Holdiness, 1987). This may be an expression of depression related in part to the stress of having TB or, in some patients, to depressive side effects of INH.

The diagnosis of major depression may be difficult in patients with symptoms of chronic respiratory disease, since many depressive and pulmonary symptoms overlap, including fatigue, lassitude, weight loss, anorexia, and loss of interest in usual activities. Sleep disturbance may be caused by episodes of sleep apnea or nocturnal coughing, as well as by nocturnal asthma attacks. Depression may be closely associated with the multiple significant losses sustained by an individual with chronic respiratory disease. As mentioned previously, these may include loss of occupation and earning capacity as well as loss of physical strength and changes in physical appearance. Some of these losses are due to the physical wasting caused by chronic illness or to impairment of pulmonary function, and others are attributable to the side effects of medications, especially corticosteroids, taken to treat chronic respiratory disease. The development of moon facies, hirsutism, acne, buffalo hump, and striae as a result of steroid use may be devastating to some individuals, particularly adolescents.

Other Psychiatric Conditions

Mild or moderate respiratory disease per se does not cause psychotic episodes, although the stress of impending hospitalization, and the resultant breakdown of normal psychologic defenses, may make patients more prone to an episode of mania, depression, or schizophrenia if they are otherwise susceptible to these disorders. Severe pulmonary disease may produce cognitive impairment syndrome with psychotic symptoms. Of course, psychotic symptoms often significantly impair medical management. At least one study (Agle & Baum, 1977) found paranoid thinking and paranoid psychosis to be prominent in a sample of patients with COPD; perhaps the combination of dyspnea, other organic factors, and functional impairment exacerbated paranoid responses.

Alcoholism is also a fairly common problem in patients with COPD. It is difficult to ascertain how much of this is cause and how much of this is effect. Some patients claim that their drinking problem was a response to developing chronic respiratory disease. Most alcoholics also have been smokers, which may have produced their COPD. Chronic alcoholics are more susceptible to aspiration pneumonia, which may lead to a chronic respiratory disease, including bronchiectasis.

Other Biologic Aspects of Respiratory Disease that May Affect Physical or Mental Functioning

Normal alterations in breathing in individuals without respiratory disease do not produce any corresponding subjective sensation of mental change, except during hyperventilation, when an individual may experience dizziness or dissociative feelings, but many psychiatric symptoms are associated with various levels of hypoxia and hypercapnia (see Table 25-1). Chronic hypoxia often leads to changed and reduced neuropsychologic functioning (Fix et al., 1982). Many patients with chronic respiratory disease, however, may have had multiple episodes of respiratory arrest or severe hypoxia, which are not manifested by their present arterial blood gas values. These patients frequently present with some cognitive impairment, particularly memory loss and a tendency to "sundown" (get worse in the evening or at night), especially when hospitalized or in some other unfamiliar environment. A study (Grant et al., 1982) of patients with hypoxemic COPD found that those studied performed significantly worse than controls on virtually all neuropsychologic tests. The higher cognitive functions, such as abstracting ability and complex perceptual-motor integration, were most severely affected, and this was found to be correlated with low p_aO_2; therefore, it seems that development of cognitive symptoms is most closely related to hypoxemia.

Patients with chronic respiratory disease and dementia are particularly susceptible to cerebral derangement caused by superimposed problems. This is especially true of elderly patients, who, as mentioned earlier, tend to have more additional medical problems and to take a larger number of medications than younger patients (Thompson, Moran, & Nies, 1983). In addition to the disease itself, the medications usually used to treat respiratory illness can affect both cognitive and affective functioning. However, at times it may be difficult to distinguish between a drug effect and the effect of the respiratory illness itself.

Those with nocturnal asthma often benefit from use of longer acting beta-adrenergic agents (see Table 25-2) before bedtime, and nocturnal oxygen may help those with nocturnal desaturation.

Especially in the elderly and those with acquired immunodeficiency syndrome (AIDS), TB may present with vegetative signs suggestive of depression. The initial depressive symptoms are often weight loss, lethargy, lack of interest in usual activities, and, at times, confusion. Many of these patients complain of sleep disturbance, which may be a related to night sweats and a low-grade fever. In elderly patients with systemic symptoms and possible past exposure to TB, a chest roentgenogram is essential, along with sputum cultures for both *Mycobacterium tuberculosis* and atypical forms of TB, such as *M. avium intracellulare* and *M. kansasii*. Factors that favor reactivation of dormant TB include immunosuppressive therapy, silicosis, diabetes mellitus, leukemia, lymphomas, gastric resection, poor nutritional status, severe alcoholism, emotional stress, and AIDS (Moran, 1985).

TABLE 25–2. *Drugs Commonly Used to Treat Asthma and COPD*

Generic Name	Proprietary Names
$Beta_2$-agonists	
Relatively short acting	
albuterol (salbutamol)	Proventil, Ventolin
isoetharine	Bronkometer, Bronkosol
isoproterenol	Isuprel, Norisodrine
metaproterenol	Alupent, Metaprel
pirbuterol	Maxair
terbutaline	Brethaire, Brethine, Bricanyl
Longer acting	
bitolterol	Tornalate
fenoterol[a]	Berotec
formoterol[a]	
Anticholinergic	
atropine	
ipratropium bromide	Atrovent
Theophylline preparations (bronchodilator)	
aminophylline	Somophyllin
dyphylline	Dilor
oxtriphylline	Choledyl
theophylline	Slo-Phyllin, Elixophyllin, Theo-Dur, Uniphyl
Mast cell degranulation inhibitors	
cromolyn	Intal
nedocromil	Tilade
Steroids (inhaled)	
beclomethasone	Becotide, Beclovent, Vanceril
flunisolide	Aerobid
triamcinolone	Azmacort
Steroids (oral)	
dexamethasone	Decadron
hydrocortisone	Hydrocortone
methylprednisolone	Medrol
prednisone	Deltasone, Sterapred
Adjunct to oral steroids	
troleandomycin	TAO

[a]Not approved for use in the United States.

Sleep Apnea

Sleep apnea and nocturnal desaturation are often present in individuals with pulmonary disease. Sleep apnea may be either obstructive (peripheral) or central in origin. Sleep apnea should be suspected in an individual with excessive daytime sleepiness, sleep disturbances, a history of snoring and sudden awakening, depression, irritability, and obesity. Often, the patient is unaware of the sleep disturbance but will complain of excessive daytime drowsiness and may be at increased risk for automobile and other accidents and for difficulty at work. Such patients may have significant cognitive impairments, including difficulty with concentration, attention, and recall (Doghramji, 1989). These individuals may have normal daytime oxygen saturation, but the degree of impairment seems related to the severity of the nocturnal hypoxia and sleep disturbance (Findley et al., 1986). The same may be true for asthmatics with nocturnal asthma attacks or for those with nocturnal oxygen desaturation without frank sleep apnea. Generally, these deficits tend to improve with successful treatment of the sleep apnea.

The use of continuous positive airway pressure (CPAP) is the most significant recent discovery in the treatment of obstructive sleep apnea (Sullivan et al., 1981). This provides room air at increased pressure and is applied through a nasal mask. The pressure usually required to eliminate sleep apnea is between 2.5 and 15 cm of water

(Doghramji, 1989) and is determined in a sleep lab through polysomnography with a variable pressure device. Once the appropriate pressure is determined, the patient then uses a portable device at home. The mechanism of action of CPAP is still uncertain but is thought to be due to a number of factors, including creation of positive transmural pressure across the pharynx, stimulation of mechanoreceptors, or an increase in lung volume with a theoretical increase in pharyngeal lumen size (Hoffstein, Zamel, & Phillipson, 1984; Rapoport, Garay, & Goldring, 1983; Sullivan et al., 1981). CPAP has rapidly become the preferred treatment for obstructive sleep apnea. The major drawbacks of CPAP have to do with local irritation and discomfort from the nasal mask. Some patients may not follow through with the sleep study after experiencing CPAP briefly; this may be due to anxiety generated by the sensation of CPAP. In those who follow through with a home program, CPAP has been found to significantly reduce the number of apneic episodes, increase the percentage of delta sleep time, and reduce snoring and daytime somnolence (Doghramji, 1989).

Central sleep apnea generally presents with apneic episodes without an attempt to inspire, and snoring may be absent or mild. Awakening may be accompanied by gasps for air, and daytime somnolence is usually absent (White, 1985). Central sleep apneas usually occur along with obstructive apneas and, in these patients, CPAP may also be an effective treatment (Sanders, 1984). Otherwise, treatment efficacy has been mixed but has generally consisted of use of nortriptyline, clomipramine, medroxyprogesterone, acetazolamide, and/or supplemental oxygen. At least one case report (Hoffstein & Slutsky, 1987) suggests that CPAP may be useful for central apnea alone. For a more complete discussion of sleep disorders, see Chapter 21.

DIAGNOSTIC AND MANAGEMENT CONSIDERATIONS

Since respiratory dysfunction may significantly affect oxygen saturation, blood pH, and acid-base balance, as well as the oxygen–carbon dioxide ratio, it is imperative to consider an organic etiology for central nervous system dysfunction in the face of any psychiatric symptomatology in a respiratory patient, especially if the symptomatology is of recent onset. In older patients, but in most younger patients as well, evaluation should include a thorough screening for other common causes of cognitive impairment disorders, including determination of electrolytes, routine blood chemistries, serum vitamin B_{12} and folate levels, liver and renal function tests, thyroid function tests, venereal disease serology, antibodies to human immunodeficiency virus (HIV), complete blood count (CBC), urinalysis, and erythrocyte sedimentation rate (ESR) (Thompson & Thompson, 1986). The patient should have arterial blood gas determinations early in the course of an evaluation, along with pulmonary function testing. A detailed medication and other drug use history must be taken to completely evaluate possible drug interactions and the effects that drugs may have on respiratory and mental function. It is important to remember that symptoms associated with chronic respiratory disease, such as fatigue, lethargy, and loss of interest in usual activities, are also frequently signs or symptoms of depression, and that sleep apnea or nocturnal asthma episodes can mimic a sleep disorder caused by depression, or vice versa. A history of steroid use and a history of reactions either to high-dose steroids or to steroid withdrawal are also important information to obtain in evaluating chronic respiratory disease patients.

During history taking, the patient's panic-fear level should be assessed. This may be ascertained by discussing the individual's reactions to asthma attacks and by assessing the patient's overall personality style. At times, however, it may be helpful to give the patient the Asthma Symptom Checklist (Kinsman et al., 1973) or the Bronchitis-Emphysema Symptom Checklist (Kinsman et al., 1983) to more fully and objectively evaluate the patient's reaction to an asthma attack. These tests can sometimes help clarify whether or not a patient reacts to respiratory distress with withdrawal and anger, thereby distancing those who may attempt to aid him, and may help assess panic-fear reactions. This is most useful as an adjunct to the clinical interview.

In general, the dexamethasone suppression test (DST) is not useful in patients with chronic respiratory disease. In those with severe illness whose symptoms most closely mimic those of depression, the patient is generally either on corticosteroids or severely cachectic, and both of these factors tend to invalidate DST results. Urinary methoxyhydroxyphenylglycol (MHPG) levels also have not been standardized for diagnostic purposes and are not routinely recommended (Thompson & Thompson, 1985a).

Special Considerations in Psychotherapeutic Treatment

Often, patients with severe COPD have a narrow range within which they can maintain their optimal pulmonary status, and they may have difficulty tolerating the emotional stress caused by psychotherapy (Dudley, Sitzman, & Rugg, 1985). The chronic respiratory disease patient's fear that a dyspneic episode may be precipitated by stress may interfere with the ability and willingness to participate in psychotherapy; these concerns may become an important focus of the therapy itself if the patient can be engaged. In treating such patients psychotherapeutically, it is important to attempt to disengage the dyspnea-anxiety cycle and to encourage optimal self-care and avoidance of pessimism (Agle & Baum, 1977). It is important to have the patient understand that dyspnea does not necessarily mean imminent danger or death but can be merely a signal to slow down and pace activities. Dyspnea, like pain, is subjective and may not be directly correlated with objective measurement of pulmonary function. Physical therapy (PT) may need to be combined with psychotherapy in order for this message to reach the patient, and PT may function as a form of systematic desensitization. Physical therapy usually consists of a slowly graduated program of exercise, including such things as walking progressively longer distances, swimming, or riding a stationary bicycle, depending on the severity of the illness. The physical therapist can help the patient learn when to rest, when to slow down, and when to stop, often with the aid of objective measurements such as ear oximetry.

In COPD patients, the goals of psychotherapy may be somewhat different than in patients with other medical disorders. For patients who associate expressions of affect with dyspnea, an important goal of treatment is for the patient to learn that expressing emotions does not necessarily lead to shortness of breath (Dudley, Sitzman, & Rugg, 1985). Many COPD patients avoid interpersonal contact in order to decrease the amount of stress they experience and, as a result, may become socially isolated. They may be perceived by others as rude or socially inappropriate, which may elicit negative or frankly hostile responses, further reinforcing the patient's withdrawal. Therefore, desensitization to social experiences may be useful in treating some patients, and those who have become more socially isolated may benefit from group therapy. Groups of other pulmonary patients may be particularly useful because withdrawn patients learn that they are not the only person with their disorder and that others have developed effective means of coping with situations that stress them. Group or family therapy is the psychotherapeutic treatment of choice for many COPD patients (Post & Collins, 1981–1982). Group therapy also is helpful for those who cannot tolerate the intensity of individual psychotherapy. Family therapy is especially recommended when dysfunctional family relationships interfere with medical management or exacerbate the patient's psychopathology, although some recommend involvement of the spouse in most instances (Sandhu, 1986).

Psychotherapy

It has been found that the leading factors in precipitating asthmatic attacks are infection (40%), allergy (30%), and emotion (30%) (Weiner, 1977), so there is clearly a role for psychotherapy in many patients with chronic respiratory disease to address the emotional component. In general, psychotherapy of these patients may follow the regular patterns and styles of individual, group, family, or behavioral therapy, with a few exceptions.

In asthmatic patients, anxiety caused by dealing with conflicts in psychotherapy may sometimes lead to hyperventilation and may even provoke an asthma attack. This is not an absolute contraindication to psychotherapy, but the patient should come to a psychotherapy session prepared for this eventuality and bring as-needed (PRN) medications, such as bronchodilator inhalers, to abort an episode. At times, the therapist may have to be more cautious and go more slowly in exploring significant conflicts or defenses to avoid precipitating an attack. These patients are often very sensitive to emotional stimuli, so it is imperative to take a more active approach than might occur with more standard psychoanalytically-oriented psychotherapy patients.

Although psychoanalysis has been useful to some patients with asthma, standard psychotherapies are generally as effective in symptom management when combined with pharmacologic therapy and other approaches. In asthmatics, individual psychotherapy is recommended when psychodynamic conflicts play a significant role in preventing the patient from coping realistically with the illness. Family therapy may be essential in some cases, especially with asthmatic children (Liebman, Minuchin, & Baker, 1974) or when a pathologically enmeshed family is involved. (See Chapter 2.)

Education

Often times, an educational approach is important and useful in helping respiratory patients regain a sense of control over their lives and in lessening their sense of anxiety or panic. However, educational interventions should be part of a more comprehensive program. A comprehensive pulmonary rehabilitation program for COPD patients including education, physical and respiratory therapy instruction, psychosocial support and exercise training has been found to increase exercise endurance more than education alone in a control group (Toshima, Kaplan, & Ries, 1990). Education about the proper use of medications, the breathing apparatus, and basic pulmonary physiology requires multiple sessions rather than a one-time intervention. Education and an emphasis on self-care have been demonstrated to result in a decrease in the number of emergency room visits and hospital admissions (Agle et al., 1973).

Behavioral Therapies

Behavioral approaches have also been utilized in COPD patients, with mixed results. Techniques aimed specifically at targeting musculature, such as inspiratory resistance, with a target feedback device have been effective in increasing ventilatory muscle strength (Belman & Shadmehr, 1988). However, this is probably not any more effective than using devices to build up respiratory muscle strength without the use of an electronic feedback device. Progressive muscle relaxation was found by Renfroe (1988) to have equivocal results in a controlled study. It helped decrease dyspnea, anxiety, heart rate, and respiratory rate at the end of each training session, but only the decreases in respiratory rate persisted after several weeks. Acosta (1988) and Malatesta, West, and Malcolm (1989) found biofeedback and progressive relaxation to be helpful in improving respiratory parameters, but, in each, this was in a single case during ventilator weaning.

The panic—fear studies discussed earlier suggest that behavioral therapy might be very helpful in many pulmonary patients. However, even something as seemingly simple as anxiety reduction via relaxation training may be potentially dangerous in asthmatics, because decreasing signal anxiety around the illness, thereby decreasing vigilance toward early symptoms, may actually make the patient experience more and worse episodes (Kinsman et al., 1980b).

However, many forms of behavioral therapy have been tried with asthmatics, with variable success. The more effective treatments seem to be those targeting symptom discrimination of asthma signals and self-management of asthma-related behaviors (Dahl, Gustafsson, & Melin, 1990), and possibly biofeedback to increase correlation between peak expiratory flow rate and medication usage (Kotses et al., 1988) and hypnotherapy (Morrison, 1988). Feedback of respiratory resistance was ineffective in a group of asthmatics (Erskine-Milliss & Cleary, 1987), probably because endurance of respiratory musculature is not a significant factor in asthmatics as compared to COPD patients. Cluss (1986) reviewed the behavioral literature on treatment of asthma and found biofeedback, operant interventions, and systemic desensitization to be successful in altering pulmonary function, asthma symptoms, and/or asthma-related behaviors. Relaxation training was found to be of limited effectiveness. Many studies combine relaxation training with biofeedback; a controlled study found this to be effective in reducing the frequency of emotional precursors to asthma attacks (Lehrer et al., 1986), with improvement on methacholine challenge in those with large airway obstruction. It is unclear whether this has any long-term significance on decreasing the severity or number of asthma attacks, improving daily functioning, or decreasing rates of hospitalization.

Most of these studies lack methodologic sophistication, lack comparable control groups, or are case reports, so it is difficult to definitively assert that behavioral treatment is effective for asthmatics. Certainly, it may be helpful in many instances, and in pulmonary patients it is often most helpful to have a multifaceted and integrated treatment approach, incorporating dynamic resolution of conflicts, pharmacotherapy, and behavioral techniques (Meany et al., 1988).

Selection and Integration of Therapies

The choice of therapy is illustrated by the following examples. An asthmatic patient who is suffering from depression with lowered self-esteem as a result of an inability to work, who uses a significant amount of denial and a counterdependent stance, and who tends to underutilize medication would probably be an appropriate candidate for individual psychotherapy and possibly antidepressant pharmacotherapy.

Chronically anxious asthmatic patients with little or no insight into the origins of their anxiety would probably do best with behavioral techniques (or antianxiety medication) as an adjunct to supportive group or individual therapy. Efficacy is measured in terms of improvement in medical status, fewer hospitalizations or emergency room visits, improved relationship with the primary physician, satisfaction of the patient and the family, and decrease in psychiatric symptomatology.

In some chronic respiratory disease patients, treatment to aid smoking cessation may be particularly important. No particular treatment method has been proved to be consistently more effective than others. Any form of smoking cessation treatment, however, is usually more effective than no treatment (Raw, 1978), and most successful smoking cessation programs have several principles in common. These include an opportunity at the beginning of treatment for the smokers to examine their motives for smoking and for stopping, an explanation of the plan and expectations of the patient, and a brief period of preparation for the day of stopping smoking. Nicotine-containing gum has been used successfully as an adjunct to group and individual approaches, as have hypnosis and behavioral techniques for some patients (Levine & Johnson, 1985). Smoking cessation is discussed in more detail in Chapter 42.

CONSIDERATIONS IN PHARMACOTHERAPEUTICS

Psychiatric Side Effects of Pulmonary Medications

Psychiatric side effects caused by respiratory drugs may range from minimal to severe and incapacitating. Table 25-2 lists some commonly used drugs to treat asthma and COPD. Table 25-3 lists some psychiatric symptoms caused by a number of these pulmonary medications. Current treatment recommendations ("Drugs for ambulatory asthma," 1991) suggest using inhaled beta$_2$-selective agents, such as albuterol, terbutaline, pirbuterol, and bitolterol, as the initial drug of choice for occasional acute symptoms of asthma and for prophylaxis of exercise induced asthma. Other longer acting beta$_2$-agonists such as formoterol are currently being studied (McAlpine & Thomson, 1990) and may shortly appear on the market. These drugs have fewer cardiac stimulant effects. For chronic asthma, these medications can be supplemental with inhaled steroids, inhaled cromolyn, or oral theophylline. Generally, psychiatric side effects tend to diminish and then disappear within a few days to a few weeks after the drug is stopped or the dosage is adjusted downward.

TABLE 25–3. *Some Psychiatric Symptoms Caused by Selected Pulmonary Drugs*

Drug	Psychiatric Symptoms Produced
Albuterol	Paranoia, hallucinations
Antihistamines	Anxiety, hallucinations, delirium
Atropine	Confusion, memory loss, delirium, paranoia, and tactile, visual, and auditory hallucinations
Beta$_2$-agonists	Anxiety, insomnia
Cephalosporins	Paranoia, confusion, disorientation
Chloramphenicol	Memory impairment, confusion, depersonalization, hallucinations
Corticosteroids	Depression, mania, emotional lability, hallucinations, paranoia, catatonia
Cycloserine	Depression, anxiety, confusion, hallucinations, paranoia, agoraphobia
Ephedrine	Hallucinations, paranoia
Ethionamide	Depression, psychosis
Gentamicin	Confusion, hallucinations
Isoniazid (INH)	Depression, anxiety, paranoia, hallucinations, confusion
Penicillin G procaine	Hallucinations, disorientation, agitation, confusion, bizarre behavior
Phenylephrine	Depression, hallucinations, paranoia
Pseudoephedrine	Hallucinations, paranoia (more in children)
Theophylline	Anxiety, withdrawal, hyperactivity, psychosis, delirium

Reprinted by permission from Thompson WL, & Thompson TL II (1985). Use of medications in patients with chronic pulmonary disease. Adv Psychosom Med 14:136–148.

The pulmonary drugs that most frequently produce psychiatric side effects are the corticosteroids. Steroids are used to treat many pulmonary illnesses, ranging from asthma to idiopathic pulmonary fibrosis or sarcoidosis. The type of psychiatric symptoms presented by the patient taking steroids tend to be quite varied and may vary within the same patient at different times. In patients receiving 40 mg or less of prednisone a day, the incidence of psychosis has been found to be less than 1%. The incidence rises steadily, however, to about 28% in patients receiving 80 mg or more of prednisone daily (Boston Collaborative Drug Surveillance Program, 1972). The incidence of psychosis, as well as other side effects, may be less if an alternate-day regimen is used, which is the rationale for adding troleandomycin (TAO) to methylprednisolone.

More recent data tend to disagree with the classic study by Rome and Braceland (1952) that defined four discrete grades of steroid psychosis. Recent research confirms the clinical impression that there are not discrete grades or stages of steroid psychosis. Psychiatric symptoms attributable to steroids, including anxiety, agitation, emotional lability, auditory and visual hallucinations, delusions, hypomania, apathy, memory loss, and depression and suicidal ideation, tend to fluctuate widely over relatively short periods of time. The onset of steroid psychosis tends to be approximately 5.9 days after the institution of steroid therapy, with twice as many cases occurring within 5 days or less as after 6 days or more (Hall et al., 1979). These steroid psychoses are generally treated by reducing and stopping the steroids and giving small doses of neuroleptics. Cyclic antidepressants may worsen these steroid-induced psychiatric symptoms. The psychiatric symptoms tend to remit rapidly when the steroids are stopped, and such a toxic reaction is not predictive of any future psychiatric disorder. It does suggest a diathesis for, though does not reliably predict, future similar reactions to steroids.

Steroid psychosis also can occur with withdrawal or reduction in dosage of steroids, and during withdrawal reactions patients may exhibit a wide variety of mental symptoms (Sharfstein, Sack, & Fauci, 1982). Steroid withdrawal symptoms can be treated either by increasing the steroids or by reducing the rate of dosage reduction; psychotropic drugs in small doses may be effective, if necessary, to alleviate mental symptoms (see also Chapter 29).

Theophylline preparations and sympathomimetic bronchodilators may cause jitteriness and a sensation of anxiety, restlessness, and irritability and may interfere with sleep. It is important to distinguish drug-induced "anxiety" from somatic anxiety. Drug-induced symptoms tend to be dose related and temporally related to taking the offending medication. In general, most individuals can tolerate mild jitteriness and tremor associated with theophylline and inhaled bronchodilators, especially if the side effects are explained to them. Management of these side effects consists of attempting to lower the dosage to the minimum required to manage the respiratory illness and, if necessary, to alter dosage timing. The specific type of theophylline preparation given also may make a difference, and some patients feel less jittery with one inhaled bronchodilator than another. If the anxiety is too great to be tolerated, the patient can often be managed without oral theophylline, if inhaled cromolyn or ipratropium, or inhaled steroids are added.

Theophylline toxicity, which generally occurs with blood levels greater than 20μg/ml, is characterized by marked anxiety and severe nausea. It is managed by stopping theophylline until the symptoms abate and the blood level returns to the therapeutic range. Theophylline toxicity may also cause a delirium with severe agitation and psychotic symptoms.

Some patients who became exceedingly tremulous on theophylline preparations have essential tremor, which is aggravated by theophylline. Benzodiazepine or barbiturate treatment of the tremor may improve the patient's tolerance of theophylline. Of course, many beta-blockers, the other major treatment for essential tremor, are relatively contraindicated in the chronic pulmonary population due to their potential constricting effects on the airways.

It has long been known that antituberculosis drugs, such as INH and *para*-aminosalicylate (PAS), may cause a toxic psychosis (Duncan & Kerr, 1962; Pugh et al., 1952). With the increased incidence of AIDS, more patients are requiring antituberculous drugs. The drugs usually used to treat *M. tuberculosis*, as well as atypical TB, are listed in Table 25-4. Ball and Rosser (1989) have found that patients from an Afro-Caribbean background seem especially susceptible to psychosis resulting from INH and PAS. This toxic psychosis has no definitive pattern but has symptoms ranging from paranoid delusions, mania, and hallucinations to confusion; addition of pyridoxine did not alter this psychosis.

As mentioned previously, INH is known to cause depression in some patients and has been associated with a significant proportion of suicide attempts and deaths among Southwestern Native Americans. Some psychiatric patients have had a worsening of their psychiatric disorder on rifampicin, INH, pyrazinamide, or ethionamide. However, INH is a monoamine oxidase inhibitor (MAOI), and some patients whose antidepressant was stopped when they began treatment for TB had good antidepressant results when given INH (Anonymous, 1989), presumably as a result of a MAOI antidepressant effect of INH. Cycloserine has been found to cause neurologic and psychiatric syndromes in about 50% of patients, in a dose-related fashion (Holdiness, 1987).

TABLE 25–4. *Drugs Commonly Used to Treat TB and Atypical TB*

Isoniazid (INH)	Amikacin
Rifampicin	Clofanzimine
Pyrazinamide (PZA)	*Para*-aminosalicylate (PAS)
Streptomycin	
Ethambutol (ETA)	
Thiacetazone	
Cycloserine	
New and experimental drugs	
Azithromycin	Clarithromycin
Rifapentine	Ciprofloxacin
Rifabutin (Ansamycin)	Sparfloxacin
Oxflacin	
Augmentin (amoxacillin + beta-lactamase inhibitor)	

Adapted from data in Mitchison et al. (1988) and Young and Inderlied (1990).

Other Adverse Drug Interactions

A number of additional adverse drug interactions may occur between psychotropic drugs and drugs prescribed for chronic respiratory disease. The most frequent agents involved in these adverse interactions are the MAOIs. In patients taking a MAOI, the pressor effects of indirectly acting sympathomimetic amines, such as tyramine, are well known. What is less clear is whether MAOIs enhance the pulmonary action of epinephrine, since epinephrine is metabolized by monoamine oxidase (MAO), as well as catechol-O-methyltransferase (COMT) (Baldessarini, 1985). Some of the active metabolites of epinephrine also are degraded by MAO, so MAOIs may prolong the action of these active metabolites as well (Sharman, 1973). MAOIs also can intensify and prolong the central effects of antihistamines and anticholinergic agents (Baldessarini, 1985).

Isoniazid, the prototype of MAOIs, is potentially problematic when used with tricyclic antidepressants. In addition, tricyclic antidepressants can potentiate the anticholinergic effects of atropine (Baldessarini, 1985), which is often used as an inhaled bronchodilator, and possibly newer atropine derivatives, such as ipratropium. Tricyclics also may potentiate the pressor effects of epinephrine to a slight degree (Boakes et al., 1973). Although epinephrine is no longer the drug of choice in treating an acute asthmatic exacerbation, care should still be exercised if a tricyclic antidepressant is given to a patient with severe cardiac disease who is receiving concomitant epinephrine injections (Rossing, Fanta, & Goldstein, 1980). Progestational agents are sometimes given to patients with sleep apnea and may potentiate the action of tricyclic antidepressants by interfering with metabolism of tricyclics by the liver (Baldessarini, 1985). It is safe to use antidepressants, particularly tricyclics, and serotonergic agents, such as fluoxetine and sertraline, with more selective $beta_2$–agonists (e.g., terbutaline, metaproterenol, albuterol, and isoetharine).

MAOIs should be avoided if possible when sympathomimetic preparations, such as phenylephedrine, pseudoephedrine, or ephedrine, are used, and should be utilized cautiously with metaproterenol. Neuroleptic drugs (especially those with more pronounced anticholinergic effects, such as thioridazine and chlorpromazine) can have additive anticholinergic effects with atropine, ipratropium, or other anticholinergic compounds and may also potentiate the effects of antihistamines (Thompson & Thompson, 1985b). Also, although not reported in the literature, there have been several cases in which therapy for atypical TB, utilizing a six-drug regimen, caused the blood level of imipramine to increase to a toxic level, probably because of inhibition of the microsomal enzyme system of the liver (W. L. Thompson, personal communication). Psychotropic drug treatment is discussed further in Chapter 9.

Effects of Psychotropic Drugs on Chronic Pulmonary Disease

The anxiety associated with pulmonary disease must be carefully evaluated to try to distinguish psychologically based anxiety from biomedically based anxiety symptoms caused by such factors as hypoxia or medications (Greenblatt, Shader, & Abernethy, 1983). Some degree of anxiety may not always be detrimental to the management of chronic respiratory disease; however, in patients who are on a respirator and in many patients with COPD, judicious use of small amounts of benzodiazepines may greatly increase their comfort and decrease their sensation of dyspnea. Many of these patients are elderly or severely debilitated and, therefore, may require lower-than-usual doses of benzodiazepines, since sensitivity to benzodiazepine effects at a given serum level increases with aging.

Buspirone, an azapirone antianxiety drug, should be the first-line drug in respiratory patients with chronic anxiety. Several studies have shown buspirone to be as effective as benzodiazepines, such as diazepam, in relieving anxiety (Gammans et al., 1989; Pecknold et al., 1989; Robinson, Napoliello, & Schenk, 1988). In addition, buspirone has been found to be well tolerated in the elderly and to not have consistently different pharmacokinetics in elderly subjects; therefore, there is generally no need to decrease the initial dose because of increased age alone (Gammans et al., 1989; Robinson, Napoliello, & Schenk, 1988). In animal models, buspirone has been found to stimulate respiration by increasing tidal activity and frequency (Garner et al., 1989; Mendelson, Martin, & Rapoport, 1990) and a shift of the apneic threshold to a lower level of pCO_2. In contrast, diazepam depresses respiration and shifts the apneic threshold to a higher pCO_2 (Garner et al., 1989). This benzodiazepine effect occurs when subjects are awake as well as asleep; the latter may prolong sleep apneic episodes and be dangerous. In contrast, buspirone improved the respiratory status of patients with sleep apnea in a pilot study (Mendelson, Maczaj, & Holt, 1991). Another study found buspirone to be effective and well tolerated in combination with bronchodilators, such as theophylline and terbutaline (Kiev & Domantay, 1988). Buspirone should not be used with MAOIs.

Other anxiolytics may suppress respiratory activity in patients with respiratory disease. Promethazine may reduce breathlessness and improve exercise tolerance without altering lung function in certain patients with nonreversible airways obstruction, whereas diazepam may have no effect on breathlessness and may reduce exercise tolerance (Woodcock, Gross, & Geddes, 1981). In general, in respiratory patients with acute anxiety, it may be better to use a benzodiazepine with a shorter half-life, such as oxazepam, lorazepam, or temazepam, if buspirone is not the drug of choice. Respiratory depression may be less likely to result with these agents, and, if it does, the adverse effects can be reversed in a shorter interval. Propranolol shouldbe avoided in patients with ROAD because it causes bronchoconstriction.

Antidepressants are also frequently used to treat patients with chronic respiratory disease. The particular antidepressant to be used should be chosen on the basis of its side effect profile. Patients with compromised pulmonary status, especially COPD or sleep apnea, should generally receive the less sedating antidepressants. Close attention should be paid in such patients to possible synergistic effects with other sedating drugs, such as antianxiety medications and sedative-hypnotics, which may interact to reduce the respiratory drive. Most antidepressants, however, have little or no effect on respiratory status, and some, such as doxepin, may act as mild bronchodilators as a result of their anticholinergic properties. Fluoxetine has been found to be a generally safe and effective antidepressant, and there have been no reports of adverse interactions with it and pulmonary medications (Cooper, 1988; Feighner et al., 1988).

In fact, fluoxetine has been found to increase arterial oxygen concentration (Ciraulo & Shader, 1990). Protriptyline and tricyclic antidepressants are often used in patients with sleep apnea (Thompson & Thompson, 1984, 1985b) because of their effect in suppressing rapid eye movement (REM) sleep. However, there are no double-blind, controlled studies demonstrating that protriptyline is superior to other tricyclics for sleep apnea.

Since inhaled atropine has been shown to be a potent bronchodilator, either in combination with beta-adrenergic agents or by itself, there is some theoretical benefit to using an antidepressant with more potent anticholinergic effects in patients who have a degree of reversibility to their symptoms, especially in patients who have nocturnal symptoms. As with the benzodiazepines, the dose of an antidepressant may need to be smaller than in the medically healthy individual. Serum levels should be monitored if severe side effects persist at a relatively low dosage, if there is no response to a usually therapeutic dosage (and compliance is believed to be good), if an increase in dosage above usual recommended dosage levels is contemplated, or if the patient is on TB medications (see also Chapter 9).

Most of the neuroleptics in common usage can be employed with a few cautionary notes in patients with chronic respiratory disease. The main concern with neuroleptic use is the provocation of tardive dyskinesia, especially because antipsychotic medications may cause a tardive dyskinesia that affects the respiratory musculature (Jann & Bitar, 1982). Although this is rare and generally occurs with long-term neuroleptic use, it can be devastating to patients with reduced respiratory capacity. Also, some of the higher potency neuroleptics, such as haloperidol, may cause bronchoconstriction (Steen, 1976), but otherwise do not tend to have a significant respiratory depressive effect.

Laryngeal dystonia is an extremely rare form of acute dystonic reaction caused by neuroleptics that presents as acute dyspnea. It generally occurs, like other dystonic reactions, within 24 to 48 hours after neuroleptic therapy is initiated or, in a small number of cases, when dosage is increased. Higher neuroleptic doses are more likely to produce such effects. This can create a life-threatening situation, but the condition usually responds dramatically to intramuscular injection of antihistaminic or antiparkinsonian agents.

Tartrazine

Tartrazine (FD&C Yellow #5) is a dye used to color many foods, beverages, and drugs, including some psychotropic drugs. It was reported to provoke severe bronchospasm in susceptible individuals (Pohl et al., 1987; Settipane, 1983, 1987). However, critical literature reviews, and more recent double-blind studies (Simon, 1986; Virchow et al., 1988) suggest that only a very small percentage at best of asthmatic individuals respond to tartrazine with flares of asthma or urticaria. Another literature review by Robinson (1988) found little evidence to suggest that tartrazine sensitivity was a significant problem, even in those with confirmed aspirin sensitivity. Three of 19 children in a double-blind study demonstrated a reaction to tartrazine: urticaria in one, abnormal behavior in another, and asthma and abdominal pain in the third (Wilson & Scott, 1989). Pollock and Warner (1990) found subtle behavioral changes in all 19 children in their double-blind study of the effects of artificial food coloring agents. In addition, pharmaceutical companies have been phasing out the use of tartrazine, so this coloring agent does not seem to pose a significant risk for asthmatics, as was formerly believed.

ENVIRONMENTAL ALLERGIES

Environmental allergies, also called "total allergy syndrome" or "20th-century disease," are conditions that were initially described in the 1950s by Randolph and Rollins (1950) and Randolph (1952). While originally described as an allergy to food or petroleum products, a wide range of chemical and environmental agents have been implicated by "clinical ecologists." These agents are as diverse as synthetic fabrics, smoke, disinfectants, cosmetics, paints, dust, organic solvents, multiple foods and additives, formaldehyde, and many more substances. Likewise, the symptoms are diverse, ranging from mental and behavioral to multiple somatic symptoms (Table 25-5).

A new specialty of "clinical ecology" has been developed to treat these patients. The symptoms have been

TABLE 25–5. *Physical and Mental Symptoms Associated with Environmental Allergies*

Anxiety	Memory loss
Concentration difficulties	Nausea/vomiting
Confusion	Nose/throat irritation
Depression	Pain
Diarrhea	Rage/anger
Dyspnea	Recurrent respiratory tract infections
Fatigue	Swelling/bloating
Headache	Tinnitus
Insomnia	Weight change
Irritability	

Adapted from information in Black, Rathe, and Goldstein (1990) and Simon, Katon and Sparks (1990).

attributed to damage to the immune system caused by environmental toxins (Terr, 1986), possibly resulting in non-immunoglobulin E hypersensitivity to multiple substances. However, the methods used for diagnosing and treating these disorders have been questioned. Diagnosis has been made by hair analysis, blood tests, and sublingual or injection provocation tests (Ferguson, 1990). The procedure of symptom provocation and neutralization has been described by several investigators (Ferguson, 1990; Jewett, Fein, & Greenberg, 1990; Terr, 1986) and consists of administering a test dose of a suspected allergen either sublingually or by intracutaneous or subcutaneous injection. The patient then reports any symptoms that occur. Other doses of the same substance, either higher or lower, are then administered until the proper dose is found to neutralize the reaction. Results are based on subjective reports of symptoms by the patient, and most clinicians do not utilize placebo.

However, a recent double-blind study of this method (Jewett, Fein, & Greenberg, 1990) indicated that the responses obtained were placebo effects and were the result of suggestion and chance. Earlier studies, such as that by Kailin and Collier (1971), also found equal relief from active and placebo injections. In addition, Terr (1986) studied the immune status of a series of patients with the diagnosis of environmental allergies and found that their immune status did not differ from that of normal individuals, except for a few individuals with a history of infections, who had elevated immunoglobulin A and lymphocyte levels. In addition, these patients did not improve on intensive treatment regimens, including neutralization, elimination diets, and environmental manipulation.

Another approach taken to these patients has been to evaluate their psychiatric status along with assessing their supposed allergies. Several such studies have found an extremely high incidence (approaching 100%) of psychiatric diagnoses among those whose food or other allergies could not be attributed to a specific syndrome, such as atopic food allergy or salicylate intolerance (Rix, Pearson, & Bentley, 1984; Simon, Katon, & Sparks, 1990; Stewart & Raskin, 1985). Such diagnoses include somatoform disorders, mood disorders, anxiety disorders, and psychosis. Black, Rathe, and Goldstein (1990) found mood, anxiety, and somatoform disorders to be the most prevalent.

The etiology of pervasive environmental "allergies" is still somewhat confusing. It may be seen as the modern equivalent of "neurasthenia" or "hysteria," and may be related to "chronic fatigue syndrome." It can be viewed as a form of expression of a somatoform disorder or of a masked depression. "Compensation neurosis" may also be involved in those patients with pending litigation. Simon, Katon, and Sparks (1990) suggested that the development of this syndrome is more related to an underlying propensity for symptom amplification and a history of prior psychologic distress than to a current psychiatric disorder. This might cause increased sensitivity to a perceived environmental toxin and may provide an explanation for already existing, chronic symptoms. Such individuals may also respond to life stresses with higher levels of both physical and psychologic symptoms.

It is particularly important to determine psychiatric disorders or disability in this patient population because the treatment itself is extremely restrictive and may heighten already existing anxieties, as well as socially disable the patient. Exclusion or rotation diets often interfere with family meals, limit the ability to eat in restaurants, and focus the individual and family on diet to the exclusion of much else. Likewise, development of environmentally safe havens at home or recommendations of a move may disrupt the family, isolate the patient, and sever other social and emotional ties as well as foster the inability to work. Therefore, misconceptions about food and chemical allergies and their treatment may cause significant harm (Pearson, 1988). However, like patients with other types of somatoform disorders, these patients are often extremely reluctant to follow through with psychiatric evaluation or treatment. Ferguson (1990) has found many patients with environmental allergies to be relieved when told authoritatively that they do not have a food allergy; however, many other patients have organized their lives around this diagnosis and are extremely resistant to giving it up. Many patients were extremely satisfied with their clinical ecologist and believed this individual had more to offer than traditional medical practitioners (Black, Rathe, & Goldstein, 1990). Now that it seems apparent that environmental allergy is, for the most part, a psychiatric, rather than medical syndrome, the focus needs to be on further assessment of psychiatric treatments beyond the methods currently used to treat patients with somatoform disorders.

HIGH ALTITUDE SICKNESS

Altitude sickness is generally thought of as a syndrome that occurs above altitudes of 12,000 to 14,000 feet above sea level. The incidence of acute mountain (or high altitude) sickness has been found to be 53% in unacclimatized hikers ascending to about 14,000 feet (Hackett & Rennie, 1976). However, there appears to be a continuum of symptoms, attributable to cerebral hypoxia, preceding development of the acute syndrome (see Table 25-1). These symptoms include altered problem-solving abilities, impaired memory and judgment, altered mood, and affective disturbance (Flynn & Thompson, 1990), and these alterations may begin at altitudes of 5,000 to 8,000 feet in healthy, young individuals (Shukitt & Banderet, 1988). This is the altitude to which many commercial airlines pressurize their planes and is the altitude of many cities, such as Santa Fe, New Mexico; Denver, Colorado; Flagstaff, Arizona; Mexico City, Mexico; and Johannesburg, South Africa.

In contrast to the sudden development of acute severe high altitude sickness, the onset of symptoms in milder forms may be insidious. It may be heralded by more subtle alterations in decision making and amplification of usual personality traits. Loss of impulse control and deficits in judgment may lead to dangerous or erratic behavior, even at lower altitudes (Flynn & Thompson, 1990; Hecht, 1971). The central nervous system is most sensitive to altitude-induced hypoxia, and neurologic impairment is manifested by the above mental alterations as well as a progression of physical symptoms, including malaise, fatigue, loss of appetite and insomnia (Singh et al., 1969) or sleepiness (Shukitt & Banderet, 1988).

In normal, healthy individuals at sea level, the arterial oxygen level (p_aO_2) is usually around 95 mm Hg. This can drop to 70 mm Hg at an elevation of about 1 mile. At 8,000 feet, the p_aO_2 further drops to 50 mm Hg (Hecht, 1971). In individuals with respiratory disease, such as acute asthma, bronchiectasis, COPD, and cystic fibrosis, the p_aO_2 may drop to 30 to 40 mm Hg at even lower altitudes (Gong, 1984). Delirium often develops when the p_aO_2 reaches these levels. If such individuals use medications or substances that depress the central nervous system, such as alcohol, cognitive changes will occur more readily.

Even in normal, healthy women who ascended to altitudes of 20,500 feet, complex cognitive tasks and psychosocial functioning were found to be significantly affected (Petiet et al., 1988). One study (Pigman & Karakla, 1990) found blacks who had been raised at a low altitude were more susceptible to developing mountain sickness.

When altitude sickness progresses past the mild form, it is manifested by physical symptoms including headache, malaise, fatigue, anorexia, oliguria, and nausea and vomiting (McDonnell, 1990). If untreated, symptoms may progress to pulmonary edema (severe dyspnea with cough) and/or cerebral edema. Severe, sustained cerebral hypoxia may result in a progression of symptoms through increasing anxiety, impaired recent memory, confusion and disorientation, obtundation, seizure, and coma (Flynn & Thompson, 1990).

Physicians practicing at lower altitudes (who may not usually think about high altitude–related disorders) may be called on to educate and treat their patients who are planning airline travel or trips to higher altitudes, especially those patients who are elderly; have compromised pulmonary function, cardiovascular or other diseases, such as anemia, cardiac valvular disease or cardiac anomalies, or seizures; or have a history of stroke or transient ischemic attacks. Such education should include a discussion of potential symptoms and should include, if possible, the patient's traveling companion. A slower ascent is recommended for at risk patients, as well as restriction of activities until the patient has begun to acclimatize to the new altitude. Alcohol and tobacco should be avoided, and doses of any central nervous system depressants (such as benzodiazepines or barbiturates) should be minimized or eliminated (Flynn & Thompson, 1990).

A program of physical conditioning prior to traveling to high altitudes may be helpful, along with maintaining adequate hydration and nutrition. Prophylaxis with acetazolamide at doses of 500 to 1,000 mg daily in divided doses also may be helpful in susceptible or at risk individuals (Hackett & Rennie, 1976; Pigman & Karakla, 1990). Acetazolamide, a carbonic anhydrase inhibitor, causes an alkaline diuresis, thereby creating metabolic acidosis with subsequent increased respiratory drive, which in turn raises the arterial oxygen pressure (Birmingham Medical Research Expeditionary Society Mountain Sickness Study Group, 1981) and seems to offer some protection against the symptoms of high altitude sickness. The main side effects of acetazolamide are tingling of the fingers and circumorally and an altered taste of carbonated beverages. Patients can be informed of these side effects, which will help foster compliance and allay anxiety (Flynn & Thompson, 1990).

Acetazolamide can also be given to treat symptoms of acute altitude sickness, which usually leads to cessation of symptoms within 24 hours (Pigman & Karakla, 1990). Other than immediate descent, the usual treatment for acute, more severe, mountain sickness is dexamethasone, although its efficacy is not conclusively proven (Levine et al., 1989; Rock et al., 1987). Dexamethasone (4 mg every 6 hours) has been found to reduce the symptoms of acute mountain sickness by 63% but did not alter oxygenation, sleep apnea, urinary catecholamine levels, hematologic profiles, the appearance of chest roentgenograms, or the results of psychometric tests (Levine et al., 1989).

Although prophylactic dexamethasone may reduce the symptoms of acute mountain sickness, if it is abruptly discontinued, the symptoms usually recur (Rock et al., 1987). Therefore, dexamethasone is recommended for acute mountain sickness only when immediate descent is impossible or to facilitate cooperation with evacuation (Levine et al., 1989). Supplemental oxygen may be helpful in some individuals but should be used with caution in patients with COPD or other chronic pulmonary disease because their respiratory drive may have become dependent on the degree of hypoxia or hypercarbia.

The definitive treatment for high altitude sickness is prompt descent to a lower altitude, which will generally cause cessation of symptoms. Brief exposure to high altitude rarely causes permanent cerebral damage or cognitive deficits, if promptly treated (Clark, Heaton & Wiens, 1983). For those with an agitated delirium, administration of haloperidol (or other neuroleptic) at doses of 1 to 2 mg every 30 minutes to hourly, and titrated as needed, may calm the patient and facilitate more effective treatment, including cooperation with transport to a lower altitude (Flynn & Thompson, 1990).

REFERENCES

Acosta F (1988). Biofeedback and progressive relaxation in weaning the anxious patient from the ventilator: A brief report. Heart Lung 17:299–301.

Agle DP, & Baum GL (1977). Psychological aspects of chronic obstructive pulmonary disease. Med Clin North Am 61:749–758.

Agle DP, Baum GL, Chester EH, et al. (1973). Multidiscipline treatment of chronic pulmonary insufficiency 1. Psychologic aspects of rehabilitation. Psychosom Med 35:41–49.

Alexander F, French TM, & Pollock G (1968). Psychosomatic specificity: Experimental study and results (Vol. 1). Chicago: University of Chicago Press.

Anonymous (1989). Psychosis and antituberculous therapy. Lancet 2:623.

Appelblatt NH, & Baker SR (1981). Functional upper airway obstruction: A new syndrome. Arch Otolaryngol 107:305–306.

Baldessarini R (1985). Drugs and the treatment of psychiatric disorders. In AG Gilman, LS Goodman, TW Rall, et al. (Eds.), The pharmacological basis of therapeutics (7th ed., Ch. 19, pp. 387–445). New York: Macmillan.

Ball R, & Rosser R (1989). Psychosis and antituberculous therapy. Lancet 2:105.

Baron C, Lamarre A, Veilleux P et al. (1986). Psychomaintenance of childhood asthma: A study of 34 children. J Asthma 23:69–79.

Belman MJ, & Shadmehr R (1988). Targeted resistive ventilatory muscle training in chronic obstructive pulmonary disease. J Appl Physiol 65:2726–2735.

Birmingham Medical Research Expeditionary Society Mountain Sickness Study Group (1981). Acetazolamide in control of acute mountain sickness. Lancet 1:180–183.

Black DW, Rathe A, & Goldstein RB (1990). Environmental illness. A controlled study of 26 subjects with "20th Century disease." JAMA 264:3166–3170.

Boakes AJ, Laurence DR, Teoh PC, et al. (1973). Interactions between sympathomimetic amines and antidepressant agents in man. Br Med J 1:311–315.

Boston Collaborative Drug Surveillance Program (1972). Acute adverse reaction to prednisone in relation to dosage. Clin Pharmacol Ther 13:694–697.

Christopher KL, Wood RP II, Eckert RC, et al. (1983). Vocal cord dysfunction presenting as asthma. N Engl J Med 308:1566–1570.

Ciraulo DA, & Shader RI (1990). Fluoxetine drug-drug interactions II. J Clin Psychopharmacol 10:213–217.

Clark CF, Heaton RK, & Wiens AN (1983). Neuropsychological functioning after prolonged high-altitude exposure in mountaineering. Aviat Space Environ Med 54:202–207.

Cluss PA (1986). Behavioral interventions as adjunctive treatments for chronic asthma. Prog Behav Modif 20:12–16.

Cooper GL (1988). The safety of fluoxetine—an update. Br J Psychiatry Suppl 3:77–86.

Covino NA, Dirks JF, Kinsman RA, et al. (1982). Patterns of depression in chronic illness. Psychother Psychosom 37:144–153.

Dahl J, Gustafsson D, & Melin L (1990). Effects of a behavioral treatment program on children with asthma. J Asthma 27:41–46.

Dirks JF, Jones NF, & Kinsman RA (1977). Panic-fear: A personality dimension related to intractability in asthma. Psychosom Med 39:120–126.

Doghramji K (1989). Sleep disorders: A selective update. Hosp Community Psychiatry 40:29–40.

Drugs for ambulatory asthma. (1991). Med Lett 33:9–12.

Dudley DL, Sitzman J, & Rugg M (1985). Psychiatric aspects of patients with chronic obstructive pulmonary disease. Adv Psychosom Med 14:64–77.

Duncan H, & Kerr D (1962). Toxic psychosis due to isoniazid. Br J Dis Chest 56:131–138.

Erskine-Milliss JM, & Cleary PJ (1987). Respiratory resistance feedback in the treatment of bronchial asthma in adults. J Psychosom Res 31:765–775.

Feighner JP, Boyer WF, & Meredith CH, et al. (1988). An overview of fluoxetine in geriatric depression. Br J Psychiatry Suppl 3:150–158.

Ferguson A (1990). Food sensitivity or self-deception? N Engl J Med 323:476–478.

Findley LJ, Barth JT, Powers DC, et al. (1986). Cognitive impairments in patients with obstructive sleep apnea and associated hypoxemia. Chest 90:686–690.

Fix AJ, Golden CJ, Daughton D, et al. (1982). Neuropsychological deficits among patients with chronic obstructive pulmonary disease. Int J Neurosci 16:99–105.

Fletcher EC, & Martin RJ (1982). Sexual dysfunction and erectile impotence in chronic obstructive pulmonary disease. Chest 81:413–421.

Flynn CF, & Thompson TL II (1990). Effects of acute increases in altitude on mental status: Prevention and treatment. Psychosomatics 31:146–152.

French TM, & Alexander F (1939–1941). Psychogenic factors in bronchial asthma. Psychosom Med Monogr 4:1–96.

Fritz GK (1983). Childhood asthma. Psychosomatics 24:959–967.

Fritz GK, Rubinstein S, & Lewiston NJ (1987). Psychological factors in fatal childhood asthma. Am J Orthopsychiatry 57:253–257.

Gale J, & O'Shanick GJ (1985). Psychiatric aspects of respirator treatment and pulmonary intensive care. Adv Psychosom Med 14:93–108.

Gammans RE, Westrick ML, Shea JP, et al. (1989). Pharmacokinetics of buspirone in elderly subjects. J Clin Pharmacol 29:72–78.

Garner SJ, Eldrige FL, Wagner PG, et al. (1989). Buspirone,

an anxiolytic drug that stimulates respiration. Am Rev Respir Dis 139:946–950.

Gong H Jr (1984). Air travel and patients with chronic obstructive pulmonary disease (editorial). Ann Intern Med 100:595–596.

Gorman JM, Askanazi J, & Leibowitz MR, et al. (1984). Response to hyperventilation in a group of patients with panic disorder. Am J Psychiatry 141:857–861.

Grant I, Heaton RK, McSweeny AJ, et al. (1982). Neuropsychologic findings in hypoxemic chronic obstructive pulmonary disease. Arch Intern Med 142:1470–1476.

Greenblatt DJ, Shader RI, & Abernethy DR (1983). Current status of benzodiazepines. N Engl J Med 309:354–358, 410–416.

Hackett PH, & Rennie D (1976). The incidence, importance, and prophylaxis of acute mountain sickness. Lancet 2:1149–1154.

Hall RCW, Popkin MK, Stickney SK, et al. (1979). Presentation of the steroid psychoses. J Nerv Ment Dis 167:229–236.

Hecht HH (1971). A sea level view of altitude problems. Am J Med 50:703–708.

Hoffstein V, & Slutsky AS (1987). Central sleep apnea reversed by continuous positive airway pressure. Am Rev Respir Dis 135:1210–1212.

Hoffstein V, Zamel N, & Phillipson EA (1984). Lung volume dependence of pharyngeal cross-sectional area in patients with obstructive sleep apnea. Am Rev Respir Dis 130:175–178.

Holdiness MR (1987). Neurological manifestations and toxicities of the antituberculous drugs. A review. Med Toxicol 2:33–51.

Hollender MH, & Mercer AJ (1976). The wish to be held and the wish to hold in men and women. Arch Gen Psychiatry 33:49–51.

Hudgel DW, Cooperson DM, & Kinsman RA (1982). Recognition of added resistive loads in asthma: The importance of behavioral styles. Am Rev Respir Dis 126:121–125.

Jann MW, & Bitar AH (1982). Respiratory dyskinesia. Psychosomatics 23:764–765.

Jewett DL, Fein G, & Greenberg MH (1990). A double-blind study of symptom provocation to determine food sensitivity. N Engl J Med 323:429–433.

Kailin EW, & Collier R (1971). "Relieving" therapy for antigen exposure. JAMA 217:78.

Karajgi B, Rifkin A, Doddi S, et al. (1990). The prevalence of anxiety disorders in patients with chronic obstructive pulmonary disease. Am J Psychiatry 147:200–201.

Kass I, Updegraff K, & Muffly RB (1972). Sex in chronic obstructive pulmonary disease. Med Aspects Hum Sex 7:33–38.

Kellner R, Samet JM, & Pathak D (1987). Hypochrondriacal concerns and somatic symptoms in patients with chronic airflow obstruction. J Psychosom Res 31:575–582.

Kiecolt-Glaser JK, & Glaser R (1986). Psychological influences on immunity. Psychosomatics 27:621–624.

Kiev A, & Domantay AG (1988). A study of buspirone coprescribed with bronchodilators in 82 anxious ambulatory patients. J Asthma 25:281–284.

Kinsman RA, Dirks JF, Dahlem NW, et al. (1980a). Anxiety in asthma: Panic-fear symptomatology and personality in relation to manifest anxiety. Psychol Rep 46:196–198.

Kinsman RA, Dirks JF, Jones NF, et al. (1980b). Anxiety reduction in asthma: Four catches to general application. Psychosom Med 42:397–405.

Kinsman RA, Fernandez E, Schocket M, et al. (1983). Multidimensional analysis of the symptoms of chronic bronchitis and emphysema. J Behav Med 6:339–357.

Kinsman RA, Luparello T, O'Banion K, et al. (1973). Multidimensional analysis of the subjective symptomatology of asthma. Psychosom Med 35:250–267.

Klerman GL (1981). Depression in the medically ill. Psychiatr Clin North Am 4:301–317.

Knapp PH (1985). Psychophysiologic aspects of bronchial asthma: A review. In EB Weiss, MS Segal, & M Stein (Eds.), Bronchial asthma: Mechanisms and therapeutics (2nd ed., Ch. 78, pp. 914–931). Boston: Little, Brown.

Kotses H, Harver A, Creer TL, et al. (1988). Measures of asthma severity recorded by patients. J Asthma 25:373–376.

Lehrer PM, Hochron SM, McCann B, et al. (1986). Relaxation decreases large-airway but not small-airway asthma. J Psychosom Res 30:13–25.

Levine BD, Yoshimura K, Kobayashi T, et al. (1989). Dexamethasone in the treatment of acute mountain sickness. N Engl J Med 321:1707–1713.

Levine DJ, & Johnson RW (1985). Psychiatric aspects of cigarette smoking. Adv Psychosom Med 14:48–63.

Levitan H (1983). Suicidal trends in patients with asthma and hypertension: A chart study. Psychother Psychosom 39:165–170.

Liebman R, Minuchin S, & Baker L (1974). The use of structural family therapy in the treatment of intractable asthma. Am J Psychiatry 131:535–540.

Lindegard B (1982). Physical illness in severe depressives and psychiatric alcoholics in Gothenburg, Sweden. J Affect Disord 4:383–393.

Malatesta VJ, West BL, & Malcolm RJ (1989). Application of behavioral principles in the management of suffocation phobia: A case study of chronic respiratory failure. Int J Psychiatry Med 19:281–289.

Martin JB, Reichlin S, & Brown GM (1977). Hypothalamic control of anterior pituitary secretion. Clin Neuroendocrinol 14:13–44.

Mathe AA, & Knapp PH (1969). Decreased plasma free fatty acids and urinary epinephrine in bronchial asthma. N Engl J Med 281:234–238.

McAlpine LG, & Thomson NC (1990). Prophylaxis of exercise-induced asthma with inhaled formoterol, a long acting β_2 adrenergic agonist. Respir Med 84:293–295.

McDonnell L (1990). Altitude sickness. Aust Fam Physician 19:205, 208–210.

Meany J, McNamara M, Burks V, et al. (1988). Psychological treatment of an asthmatic patient in crisis. Dreams, biofeedback, and pain behavior modification. J Asthma 25:141–151.

Mendelson WB, Maczaj M, & Holt J (1991). Buspirone administration to sleep apnea patients. J Clin Psychopharmacol 11:71–72.

Mendelson WB, Martin JV, & Rapoport DM (1990). Effects of buspirone on sleep and respiration. Am Rev Respir Dis 141:1527–1530.

Mitchison DA, Ellard GA, & Grosset J (1988). New antibacterial drugs for the treatment of mycobacterial disease in man. Br Med Bull 44:757–774.

Moran MG (1985). Psychiatric aspects of tuberculosis. Adv Psychosom Med 14:109–118.

Moran MG (1991). Psychological factors affecting pulmonary and rheumatologic diseases. Psychosomatics 32:14–23.

Morrison JB (1988). Chronic asthma and improvement with relaxation induced by hypnotherapy. J R Soc Med 81:701–704.

Pearson DJ (1988). Psychologic and somatic interrelationships in allergy and pseudoallergy. J Allergy Clin Immunol 81:351–360.

Pecknold JC, Matas M, Howarth BG, et al. (1989). Evaluation of buspirone as an antianxiety agent: Buspirone and diazepam versus placebo. Can J Psychiatry 34:766–771.

Pecyna MB (1989). Dynamics of somatic and psychic symptoms in patients hospitalized for pulmonary tuberculosis complicated with neurotic anxiety reactions during antituberculous treatment. Wiad Lek 42:284–291.

PETIET CA, TOWNES BD, BROOKS RJ, ET AL. (1988). Neurobehavioral and psychosocial functioning of women exposed to high altitude in mountaineering. Percept Mot Skills 67:443–452.

PIGMAN EC, & KARAKLA DW (1990). Acute mountain sickness at intermediate altitude: Military mountainous training. Am J Emerg Med 8:7–10.

PLUTCHIK R, WILLIAMS MH JR, JERRETT I, ET AL. (1978). Emotions, personality and life stresses in asthma. J Psychosom Res 22:425–431.

POHL R, BALON R, BERCHOU R, ET AL. (1987). Allergy to tartrazine in antidepressants. Am J Psychiatry 144:237–238.

POLLOCK I, & WARNER JO (1990). Effect of artificial food colours on childhood behaviour. Arch Dis Child 65:74–77.

POST L, & COLLINS C (1981–1982). The poorly coping COPD patient: A psychotherapeutic perspective. Int J Psychiatry Med 11:173–182.

PUGH DL, EDWARDS GF, MCLAREN RG, ET AL. (1952). Toxic psychiatric manifestations in the treatment of tuberculosis with sodium para-aminosalycylate. Tubercle 33:369–376.

RANDOLPH TG (1952). Sensitivity to petroleum including its derivatives and antecedents, abstracted. J Lab Clin Med 40:931–932.

RANDOLPH TG, & ROLLINS JP (1950). Beet sensitivity: Allergic reactions from the ingestion of beet sugar (sucrose) and monosodium glutamate of beet origin. J Lab Clin Med 35:407–415.

RAPOPORT DM, GARAY SM, & GOLDRING RM (1983). Nasal CPAP in obstructive sleep apnea: Mechanisms of action. Bull Eur Physiopathol Respir 19:616–620.

RAW M (1978). The treatment of cigarette dependence. In Y ISRAEL, FB GLASER, H KALANT, ET AL. (Eds.), Research Advances in alcohol and drug problems (Vol. 4; Ch. 11, pp. 441–478). New York: Plenum Press.

REITE M, HARBECK R, & HOFFMAN A (1981). Altered cellular immune response following peer separation. Life Sci 29:1133–1136.

RENFROE KL (1988). Effect of progressive relaxation on dyspnea and state anxiety in patients with chronic obstructive pulmonary disease. Heart Lung 17:408–413.

RIX KJ, PEARSON DJ, & BENTLEY SJ (1984). A psychiatric study of patients with supposed food allergy. Br J Psychiatry 145:121–126.

ROBINSON D, NAPOLIELLO MJ, & SCHENK J (1988). The safety and usefulness of buspirone as an anxiolytic drug in elderly versus young patients. Clin Ther 10:740–746.

ROBINSON G (1988). Tartrazine—the story so far. Food Chem Toxicol 26:73–78.

ROCK PB, JOHNSON TS, CYMERMAN A, ET AL. (1987). Effect of dexamethasone on symptoms of acute mountain sickness at Pikes Peak, Colorado (4,300 m). Aviat Space Environ Med 58:668–672.

ROME HP, & BRACELAND FJ (1952). The psychological response to ACTH, cortisone, hydrocortisone and related steroid substances. Am J Psychiatry 108:641–651.

ROSSING TH, FANTA CH, & GOLDSTEIN DH (1980). Emergency therapy of asthma: Comparison of the acute effects of parenteral and inhaled sympathomimetics and infused aminophylline. Am Rev Respir Dis 122:365–371.

RUTTER BM (1979). The prognostic significance of psychological factors in the management of chronic bronchitis. Psychol Med 9:69–70.

SANDERS MH (1984). Nasal CPAP effect on patterns of sleep apnea. Chest 86:839–844.

SANDHU HS (1986). Psychosocial issues in chronic obstructive pulmonary disease. Clin Chest Med 7:629–642.

SETTIPANE GA (1983). Aspirin and allergic diseases: A review. Am J Med 74:102–109.

SETTIPANE GA (1987). The restaurant syndromes. N Engl Regional Allergy Proc 8:39–46.

SHARFSTEIN SS, SACK DS, & FAUCI AS (1982). Relationship between alternate-day corticosteroid therapy and behavioral abnormalities. JAMA 248:2987–2989.

SHARMAN DF (1973). The catabolism of catecholamines: Recent studies. Br Med Bull 29:110–115.

SHUKITT BL, & BANDERET LE (1988). Mood states at 1,600 and 4,300 meters terrestrial altitude. Aviat Space Environ Med 59:530–532.

SIBBALD B, WHITE P, PHAROAH C, ET AL. (1988). Relationship between psychosocial factors and asthma morbidity. Fam Pract 5:12–17.

SIMON GE, KATON WJ, & SPARKS PJ (1990). Allergic to life: Psychological factors in environmental illness. Am J Psychiatry 147:901–906.

SIMON RA (1986). Adverse reactions to food additives. N Engl Regional Allergy Proc 7:533–542.

SINGH I, KHANNA PK, SRIVASTAVA MC, ET AL. (1969). Acute mountain sickness. N Engl J Med 280:175–184.

STEEN SN (1976). The effects of psychotropic drugs on respiration. Pharmacol Ther 2:717–741.

STEIN M (1982). Biopsychosocial factors in asthma. In LJ WEST & M STEIN (Eds.), Critical issues in behavioral medicine (Ch. 12, pp. 159–182). Philadelphia: Lippincott.

STEWART DE, & RASKIN J (1985). Psychiatric assessment of patients with "20th-century disease" ("total allergy syndrome"). Can Med Assoc J 133:1001–1006.

STOUDEMIRE A (1985). Psychosomatic theory and pulmonary disease: Asthma as a paradigm for the biopsychosocial approach. Adv Psychosom Med 14:1–15.

SULLIVAN CE, ISSA FG, BERTHON-JONES M, ET AL. (1981). Reversal of obstructive sleep apnea by continuous positive airway pressure applied through the nares. Lancet 1:862–865.

TEIRAMAA E (1978). Psychic disturbances and severity of asthma. J Psychosom Res 22:401–408.

TEIRAMAA E (1981). Psychosocial factors, personality and acute-insidious asthma. J Psychosom Res 25:43–49.

TERR AI (1986). Environmental illness. A clinical review of 50 cases. Arch Intern Med 146:145–149.

THOMPSON TL II, MORAN MG, & NIES AS (1983). Psychotropic drug use in the elderly. N Engl J Med 308:134–138, 194–199.

THOMPSON TL II, & THOMPSON WL (1986). Treating dementia in the elderly. Female Patient 11:62–77.

THOMPSON WL (1986). Sexual problems in chronic respiratory disease: Achieving and maintaining intimacy. Postgrad Med 79:41–52.

THOMPSON WL, & THOMPSON TL II (1984). Treating depression in asthmatic patients. Psychosomatics 25:809–812.

THOMPSON WL, & THOMPSON TL II (1985a). Psychiatric aspects of asthma in adults. Adv Psychosom Med 14:33–47.

THOMPSON WL, & THOMPSON TL II (1985b). Use of medications in patients with chronic respiratory disease. Adv Psychosom Med 14:136–148.

TOSHIMA MT, KAPLAN RM, & RIES AL (1990). Experimental evaluation of rehabilitation in chronic obstructive pulmonary disease: Short-term effects on exercise endurance and health status. Health Psychol 9:237–252.

VINGERHOETS AJJM (1985). The role of the parasympathetic division of the autonomic nervous system in stress and the emotions. Int J Psychosom 32:28–34.

VIRCHOW C, SZCZEKLIK A, BIANCO S, ET AL. (1988). Intolerance to tartrazine in aspirin-induced asthma: Results of a multicenter study. Respiration 53:20–23.

WEINER H (1977). Psychobiology and human disease. New York: Elsevier.

Weiner HM (1987). Stress, relaxation and asthma. Int J Psychosom 34:21–24.

White DP (1985). Central sleep apnea. Med Clin North Am 69:1205–1219.

Williams SJ (1989). Chronic respiratory illness and disability: A critical review of the psychosocial literature. Soc Sci Med 28:791–803.

Williams SJ, & Bury MR (1989). "Breathtaking": The consequences of chronic respiratory disorder. Int Disabil Stud 11:114–120.

Wilson CP (1980–1981). Parental overstimulation in asthma. Int J Psychoanal Psychother 8:602–621.

Wilson N, & Scott A (1980). A double-blind assessment of additive intolerance in children using a 12 day challenge period at home. Clin Exp Allergy 19:267–272.

Woodcock AA, Gross ER, & Geddes DM (1981). Drug treatment of breathlessness: Contrasting effects of diazepam and promethazine in pink puffers. Br Med J 283:343–346.

Yellowlees PM, Alpers JH, Bowden JJ, et al. (1987). Psychiatric morbidity in subjects with chronic obstructive airflow. Med J Aust 146:305–307.

Yellowlees PM, Haynes S, Potts N, et al. (1988). Psychiatric morbidity in patients with life-threatening asthma: Initial report of a controlled study. Med J Aust 149:246–249.

Yellowlees PM, & Ruffin RE (1989). Psychological defenses and coping styles in patients following a life-threatening attack of asthma. Chest 95:1298–1303.

Young LS, & Inderlied CB (1980). *Mycobacterium avium* complex infections: A comprehensive overview of diagnosis and treatment. AIDS Patient Care 4:10–18.

26 Gastroenterology

STEVEN A. EPSTEIN, M.D.,
THOMAS N. WISE, M.D.,
AND RICHARD L. GOLDBERG, M.D.

Humans have always been fascinated by what they eat, what they excrete, and what disorders upset their gastrointestinal systems. The interplay between psychologic factors and the gastrointestinal system has been accepted for centuries, and current culture emphasizes the relationship between emotions and gastrointestinal disease. Recent research has corroborated links between emotional phenomena and gastrointestinal events. For example, experimentally induced emotional stress has been shown to affect gastrointestinal motility (Clouse, 1988). In addition, an expanding literature has been establishing comorbidity between gastrointestinal symptoms and psychiatric phenomena. For example, panic disorder often presents with gastrointestinal symptoms (Katon, 1984; van Valkenburg et al., 1984). Psychologic etiologies have also been studied as predictors of functional gastrointestinal symptoms. A recent study found that 44% of patients in a gastrointestinal clinic had been physically or sexually abused. In this population, patients with functional bowel disorders had a greater prevalence of physical abuse than those with organic disorders (Drossman et al., 1990). Furthermore, behavioral problems such as alcohol ingestion and tobacco use clearly foster gastric, liver, and pancreatic disorders. Thus, it is not surprising that up to 60% of the gastroenterologist's clinical activities are devoted to complaints that are primarily of psychologic origin (Switz, 1976).

ESOPHAGEAL DISORDERS

Dysphagia, the sensation of difficulty swallowing, is a symptom of a number of physical disorders (Table 26-1). Dysphagia for solids alone suggests obstruction, especially when it is progressive. If solids and liquids are not tolerated, the problem is probably motor dysfunction. A barium swallow or esophagogram is usually sufficient to diagnose mechanical blockage. Esophageal manometry must be performed to rule out a motility disorder such as achalasia (Klinger & Strang, 1987).

The esophageal disorders of greatest psychiatric interest are the disorders of esophageal motility. Chest pain, difficulty swallowing solids, difficulty ingesting liquids, heartburn, and regurgitation are regularly found in individuals with esophageal motility abnormalities (Reidel & Clouse, 1985). "Nutcracker esophagus," a manometric pattern characterized by high-amplitude peristaltic contractions in the distal esophagus, is the most common motility abnormality (Browning & Members of the Patient Care Committee, 1990). Acute stressors (intermittent bursts of white noise and difficult cognitive problems) were shown to increase esophageal contraction amplitude in controls and to a greater extent in nutcracker esophagus patients (Anderson et al., 1989). Heightened emotional arousal increases respiration and swallowing rates, which may augment esophageal distress. Anxiety that increases swallowing rates may thus exacerbate esophageal disorders (Fonagu & Calloway, 1986). Personality profiles of nutcracker esophagus patients reveal susceptibility to gastrointestinal symptoms during emotional stress (Richter et al., 1986). There is no clearly effective specific treatment for nutcracker esophagus. Treating associated anxiety disorders, if present, is rational.

A related condition is diffuse esophageal spasm. This is characterized by simultaneous repetitive contractions and other manometric abnormalities, with peristalsis only intermittently normal. In contrast, patients with nutcracker esophagus have normal peristaltic waves of excessively high amplitude (Bradley, McDonald, & Richter, 1990).

Reflux esophagitis, which is distal esophageal inflammation that results from repeated contact with refluxed gastric and duodenal contents, presents with heartburn, regurgitation, and retrosternal pain. Approximately 10% of the population have heartburn daily (Wesdorp 1986). Contributory factors include increased reflux (e.g., as a result of decreased lower esophageal sphincter pres-

TABLE 26–1. *Medical Differential Diagnosis of Dysphagia*

	Oropharyngeal	Esophageal
Neuromuscular	Multiple sclerosis Parkinson disease Postpolio syndrome (Sonies & Dalakas, 1991) Muscular dystrophy Motor neuron disease (amyotrophic lateral sclerosis) Brainstem stroke	Achalasia Diffuse esophageal spasm Nutcracker esophagus
Obstructive		Carcinoma Foreign body Esophageal webs Cervical osteoarthritis
Systemic	Thyrotoxicosis Sarcoidosis Trichinosis	Chagas disease
Collagen vascular	Polymyositis-dermatomyositis	Scleroderma

Adapted from Klinger and Strang (1987).

sure), gastroduodenal factors (e.g., increased gastric acid), and diminished esophageal clearance (e.g., resulting from medications with anticholinergic effects).

Treatment of reflux is best approached in a stepwise manner (Lieberman, 1990). One should begin with conservative measures—for example, avoiding late night snacks and lying down after meals; elevating the head of the bed (Harvey et al., 1987); reducing fatty foods, chocolate, alcohol, and cigarettes; and antacids as needed. Anticholinergic agents such as tertiary amine tricyclic antidepressants and low-potency neuroleptics should be avoided if possible. If the above measures fail, histamine-2 (H_2) blocker therapy (e.g., ranitidine) may be tried. Omeprazole (see section "Peptic Ulcer Disease") may be used for severe or refractory disease. If these measures are unsuccessful, one could add a promotility drug such as the widely used metoclopramide. Although its mechanism of action is not clear, metoclopramide appears to sensitize tissues to acetylcholine, thereby increasing lower esophageal sphincter pressure and increasing gastric emptying. Its antiemetic properties are due to its antagonism of dopamine-2 receptors. Pharmacologic effects generally last for 1 to 2 hours. Approximately 10% of patients who are regularly prescribed metoclopramide experience restlessness, drowsiness, and fatigue. According to the Reglan package insert (A.H. Robins), extrapyramidal symptoms may occur in 1:500 patients and are usually dystonic reactions. However, one might speculate that the common side effect of restlessness might be akathisia. A recent review of 16 patients who had metoclopramide-induced movement disorders included 10 cases of tardive dyskinesia and five with parkinsonism. These side effects often go unrecognized by the prescribing physician (Miller & Jankovic, 1989). There have been reports of depression and other psychiatric disturbances as well. If metoclopramide or another promotility agent (e.g., bethanechol) is unsuccessful, antireflux surgery to increase lower esophageal sphincter competence may be tried.

Psychiatric disorders are widely recognized in a variety of esophageal disorders (Bradley, McDonald, & Richter, 1990). Globus (the sensation of a lump in one's throat) may be associated with panic disorder and depression (Greenberg, Stern, & Weilburg, 1988). Mood disorders and generalized anxiety disorder are commonly seen in individuals with esophageal motility disorders (Clouse & Lustman, 1982). Individuals with symptomatic hiatal hernias often have generalized anxiety and depressive disorders (Nielzen et al., 1986). Psychopharmacologic and psychotherapeutic treatment of associated anxiety or depression will often substantially alleviate esophageal symptoms. Brown et al. (1986) reported on three patients with the globus hystericus syndrome who responded to antidepressants. Trazodone has been found to be effective in the esophageal spasm syndrome (Clouse et al., 1987).

PEPTIC ULCER DISEASE

The contemporary association between emotional stress and onset of abdominal pain has its roots in the observations of William Beaumont, a Canadian surgeon who documented the correlation between anger and increased gastric fluid output by observing the gastric fistula of an injured Canadian lumberjack, Alexis St. Martin (Beaumont, 1833). General etiologic factors in peptic ulcer disease include cigarette smoking, aspirin, and corticosteroids. Alcohol itself has not been proven to cause ulcers, but binge drinking causes gastritis and people with alcoholic cirrhosis have an increased prevalence of ulcers. Both caffeinated and decaffeinated coffee stimulate acid secretion. However, most research has not shown a relationship between coffee drinking and ulcers (Schubert, 1984). Diet is no longer considered to be a factor in ulcer production. *Helicobacter pylori*, formerly called *Campylobacter pylori*, is a gram-negative bacillus that causes gastritis and may be a factor in the development of peptic ulcer disease. Most who are infected are asymptomatic, but some have non-ulcer dyspepsia or peptic ulceration. There is no clearly superior treatment regimen, but therapy may include bismuth, metronidazole, and amoxicillin or tetracycline (Peterson, 1991).

Gastric and duodenal ulcers differ. Duodenal ulcers may be associated with excess secretion of hydrochloric acid by the stomach; delayed gastric emptying promotes gastric ulcers (Isenberg, 1981). Duodenal ulcers are more frequent in males, but there is no sex bias in gastric ulcers; duodenal ulcers also occur more commonly in people with type O blood. Animal studies have provided a tentative link between brain chemistry and ulcer production; for example, increased brain thyrotropin-releasing hormone synthesis may correlate with gastric ulcer formation (Hernandez, 1989).

Although there are a variety of other gastrointestinal ulcers such as stress ulcers and Cushing's ulcers, peptic ulcer disease has been the primary focus of psychosomatic investigation. Early psychosomatic studies were hampered by imprecise definitions and lacked documentation of ulcer type and location. Franz Alexander (1950) is best known for his psychosomatic hypotheses. He hypothesized that the patient with a duodenal ulcer has frustrated wishes to be loved and cared for, resulting in persistent oral dependent needs. The onset of the ulcer disease occurred when such unsatisfied cravings were augmented by an increase in responsibility or frustrated dependency needs. Weiner et al. (1957) investigated 2,073 Army inductees utilizing psychologic testing and serum pepsinogen levels, which correlated with basal rates of gastric acid secretion. Supporting Alexander's theory, they found that the Army inductees who later developed duodenal ulceration were those with high baseline pepsinogen levels, as well as those with the most intense dependency needs and conflicts with authority. These investigators thus integrated a somatic vulnerability with a vulnerable personality characteristic and a stressful life event, namely induction into the Army.

Recent investigators have also examined the relationship between stress and psychosocial factors in the development of peptic ulcer disease. Discriminant analysis of patients with peptic ulcer disease versus controls showed depression to be the best discriminator; other major factors were increased perception of the negative impact of life events (although frequency of life events themselves did not differ), number of relatives with peptic ulcer disease, and serum pepsinogen I concentration. Behavioral risk factors such as smoking, alcohol use, and aspirin use were also significantly higher in peptic ulcer disease patients. Thus, peptic ulcer disease appears to be mediated by psychologic, behavioral, and physiologic factors (Feldman et al., 1986; Walker et al., 1988). Experimentally, emotional stress in the form of dichotic listening has been shown to increase gastric acid secretion in patients with duodenal ulcers (Rask-Madsen et al., 1990).

Psychologic Treatment

Psychologically and behaviorally oriented treatments for peptic ulcer disease are difficult to assess because the majority of duodenal and gastric ulcers heal with no medication within 6 weeks. Psychologic treatments may nevertheless offer some help in the management of peptic ulcer disease. Chapell et al. (1936) utilized dietary management and group therapy with an educational component similar to Buchanan's (1978) two-step methodology for the group treatment of patients with organic illnesses. This method allowed group members to learn about the physiologic aspects of the disease in a formal didactic fashion. The approach promoted discussions of life stresses and difficulties in a group setting. The treated group improved more than untreated controls. More recently, others have documented that stress management training utilizing relaxation therapy and assertiveness training is effective in diminishing ulcer pain and in reducing antacid ingestion (Brooks & Richardson, 1980).

Although psychoanalytic psychotherapy offers a method to delineate unconscious conflicts presumably associated with peptic ulcer disease, the actual effects of psychoanalytic treatment on the course of the disease are unclear. Sjodin et al. (1986) investigated the use of pharmacotherapy alone versus pharmacotherapy plus short-term dynamic and cognitive/educational psychotherapeutic approaches. Both groups improved, but the individuals who underwent psychotherapy had less abdominal pain and, after 1 year of treatment, appeared to have higher global ratings of both physical health and freedom from emotional complaints.

Direct behavioral interventions for peptic ulcer disease are in an experimental stage. Investigators have shown that biofeedback techniques may reduce gastric acid secretion as well as gastric motility (Whitehead, Renault, & Goldiamond, 1975). Relaxation training may also diminish gastric acid secretion and thus provide a therapeutic rationale for these behavioral interventions. The technical difficulties and cost of such strategies, however, prevent their widespread practical application (Stracher et al., 1975).

Medical Treatment

Histamine-2 Blockers

Medical management of ulcer disease is the presently accepted approach. The H_2 blockers cimetidine and ranitidine are primary treatments for unperforated ulcer disease. These drugs control gastric acid secretion and can help heal the acute lesion as well as prevent ulcer relapse. There is some evidence, albeit controversial, that ranitidine is more effective in preventing duodenal ulcer relapse (Gough et al., 1984). Side effects of the drugs include confusional states and possibly depression, although depression is poorly substantiated. Initially, it was thought that only cimetidine produced delirium, since it crossed the blood-brain barrier. Recently, however, ranitidine has also been documented to cause confusion (Silverstone, 1984), but apparently much less frequently than cimetidine (Lipsy, Fennerty, & Fagan, 1990). The risk of delirium is greater in elderly patients and in individuals with liver disease. Both drugs also can cause erectile dysfunction, which may limit compliance with the prescribed drug regimen (Assael et al., 1974). Symptoms remit when the drug is stopped.

Cimetidine inhibits oxidative hepatic metabolism of other drugs. Therefore, the levels of those benzodiazepines metabolized primarily by oxidation, and levels of tricyclic antidepressants, may increase (Greenblatt et al., 1984; Henaver & Hollister, 1984) (Table 26-2). The general clinical significance of such slight elevations is doubtful, but the interaction deserves consideration when side effects develop or when patients are otherwise at high risk for toxicity. Lorazepam utilizes an alternative metabolic pathway (glucuronide conjugation), as do oxazepam and temazepam; they are probably preferred if a benzodiazepine is needed in conjunction with this H_2 blocker (Klotz & Reimann, 1980). Chlorpromazine may be malabsorbed if taken with cimetidine, and lowered blood levels will thus result (Howes et al., 1983). Transient increases in carbamezepine levels may occur when cimetidine is added (Macphee et al., 1984). Ranitidine, however, does not impair the clearance of diazepam, lorazepam (Abernethy et al., 1984), or carbamazepine (Webster et al., 1984). It may be given on a twice-daily schedule, in contrast to cimetidine, which is given four times a day. The twice-a-day schedule enhances compliance, so ranitidine is preferable in patients having difficulty with treatment adherence.

The two new H_2 antagonists, famotidine and nizatidine, appear to be as efficacious as cimetidine and ranitidine in the treatment of ulcer disease (Brazer et al., 1989; Lipsy, Fennerty, & Fagan, 1990). There is a recent report of reversible confusion in two elderly patients with mild renal insufficiency who received intravenous famotidine (Hennan, Carpenter, & Janda, 1988). Similarly to ranitidine, famotidine does not affect the hepatic mixed-function oxidase systems (Langtry, Grant, & Goa, 1989). Nizatidine interactions have not been studied, but there is no evidence to suggest they will be different from those of ranitidine (Lipsy, Fennerty, & Fagan, 1990). Dosage of H-2 blockers should be decreased in the context of renal failure. Otherwise, an increased incidence of delirium due to these agents will result.

TABLE 26–2. *Drug Interactions*

Gastrointestinal Drug	Psychiatric Drug	Effect on Psychiatric Drug	Mechanism	Reference
antacids	chlordiazepoxide	decreased peak effect of single dose	decreased absorption	Greenblatt et al. (1976)
	diazepam	decreased peak effect of single dose	decreased absorption	Greenblatt et al. (1978)
antacids	chlorpromazine	decreased level	decreased absorption	Forrest, Forrest, and Serra (1970)
cimetidine	imipramine	increased level	inhibition of hepatic oxidative metabolism	Henaver and Hollister (1984)
	diazepam	increased level	same as imipramine	Greenblatt et al. (1984)
	chlordiazepoxide	increased level	same as imipramine	Desmond et al. (1980)
	carbamazepine	transient increased level	same as imipramine	Macphee et al. (1984)
	chlorpromazine	decreased level	?impaired absorption	Howes et al. (1983)
omeprazole	diazepam	increased level	inhibition of hepatic oxidative metabolism	Maton (1991)

Other Medical Therapies

Omeprazole inhibits gastric acid secretion by altering activity of H^+/K^+-ATPase in gastric parietal cells. While it has been shown to be effective in the treatment of duodenal ulcer (Graham et al., 1990; McFarland et al., 1990), it is approved in the United States only for severe or refractory gastroesophageal reflux and for prolonged use in pathologic hypersecretory states such as Zollinger-Ellison syndrome. Side effects are minimal, but it inhibits the metabolism of diazepam (Maton, 1991).

Misoprostol, a synthetic E prostaglandin, prevents gastric ulcers induced by nonsteroidal antiinflammatory drugs (NSAIDs) (Graham, Agrawal, & Roth 1988). No significant central nervous system effects have as yet been reported (Monk & Clissold, 1987).

The antiulcer (cytoprotective) drug sucralfate is a safe and effective alternative in susceptible patients (Garnett, 1982). The agent acts locally to coat and protect ulcer sites from gastric acid and has little systemic absorption. The drug is given in a dose of 1 g before meals and at bedtime.

Tricyclic antidepressants such as doxepin also possess H_2 receptor-blocking capabilities and have been used experimentally as antiulcer agents (Shrivastava, Shah, & Siegal, 1985). The efficacy of these drugs as potent H_2 blockers is controversial (Ries, Gilbert, & Katon, 1984), and they should not be considered primary treatment, although they may augment and work synergistically with formal medical regimens, particularly in anxious or depressed patients with high psychophysiologic gastrointestinal reactivity.

Complications from now outmoded medical interventions for peptic ulcer disease include the milk-alkali syndrome, wherein persistent antacid ingestion and milk drinking create hypercalcemia and its resultant mental symptoms. Anticholinergic agents such as propantheline used to treat peptic ulcer disease can precipitate the well-known syndrome of anticholinergic toxicity.

Finally, gastrectomy is used to treat ulcer patients with intractable bleeding, perforation, or nonresponse to medical therapy. Patients with gastrectomies are vulnerable to vitamin B_{12} deficiency, which can present with a secondary depression or cognitive impairment.

ABDOMINAL PAIN, DYSPEPSIA OF UNKNOWN ETIOLOGY, AND DELAYED GASTRIC EMPTYING

Chronic abdominal pain is often seen in medical practice. Patients for whom no clear etiology can be found present a perplexing management problem. Drossman (1982) has reviewed his experience with 24 individuals considered to have psychogenic chronic abdominal pain. His patients were often noted to have evidence of incompletely resolved grief as a result of prior losses. Most of these patients did not consider their complaints to be of psychologic origin and avoided psychiatric care.

Eisendrath et al. (1986) noted that patients with psychogenic abdominal pain more often had a past history of somatization of psychologic distress as well as a family history of alcoholism. Patients with medically unexplained abdominal pain deserve psychiatric consultation. Certainly, exploratory surgery should not be performed on such patients without prior psychiatric evaluation. Somatoform and idiopathic abdominal pain syndromes have been recently reviewed elsewhere (Stoudemire & Sandhu, 1987).

The couvade syndrome presents with acute gastrointestinal symptoms and may be mistaken for organic disease. The syndrome, found in men whose wives are pregnant, is characterized by nausea, vomiting, abdominal bloating, and fatigue (Enoch, Trethowan, & Barker, 1967). Lipkin and Lamb (1982) demonstrated that physicians often miss the relationship of the symptoms to the patient's expectant fatherhood. Couvade is best managed by very conservative measures and reassurance that the gastrointestinal difficulties will remit following the birth of the child (Lipkin & Lamb, 1982). Patients refractory to explanation and reassurance require psychiatric assessment and medical psychotherapy.

Before attributing abdominal pain to psychologic origins, however, great care should be taken to rule out organic or anatomic etiologies that may have been overlooked in the medical evaluation. Acute intermittent porphyria, for example, presents with a history of chronic abdominal pain (often with a record of multiple negative exploratory laparotomies), peripheral neuropathy, and delirium. Occult intestinal adhesions should also be considered, as well as occult endometriosis in women where the boundaries between abdominal and pelvic pain syndromes are blurred.

Dyspepsia of unknown etiology (DUO) accounts for some 30% to 40% of cases presenting to gastroenterologists when there is a complaint of chronic upper abdominal pain (Soll & Isenberg, 1983). It appears to be caused by a host of psychologic factors that are likely to influence gut functioning. It has been suggested that DUO is secondary to psychiatric illness, particularly depression (Drossman, 1982). Other authors believe that DUO is the result of excessive autonomic arousal, anxiety, neurotic personality traits or problems with unexpressed anger (Gomez & Dally, 1977; Stockton, Weinman, & McColl, 1985; Talley et al., 1986). Most recently, Langeluddecke, Goulston, and Tennant (1990) carefully examined the association of such psychologic factors in a group of patients with DUO and a matched group with peptic ulcer disease. The DUO patients had significantly more symptoms of anxiety and tension and higher scores for trait tension and hostility than the peptic ulcer group. However, the groups did not differ significantly in depressive symptoms, neuroticism, psychoticism, or suppression of negative affects.

Patients with DUO usually respond poorly to psychopharmacologic agents other than antidepressants. It is particularly important to minimize the use of opiate medications since DUO patients can easily become drug dependent. Supportive psychotherapy can be generally helpful to the patient and family, but it usually does not lead to persistent pain relief.

Delayed gastric emptying may present with nausea, vomiting, bloating, postprandial fullness, early satiety, and anorexia. Etiologies include diabetic gastroparesis, gastroesophageal reflux, postvagotomy sequelae, and gastric ulcer. Anorexia nervosa patients have been shown to have delayed gastric emptying, possibly as a result of gastric neuromuscular involvement. Depression may cause delayed gastric emptying, probably on a neurogenic basis. Medications with significant anticholinergic activity, such as imipramine, may delay emptying. Treatment includes eliminating anticholinergic drugs and using promotility medications such as metoclopramide. Surgical procedures are occasionally used in refractory cases (Ricci & McCallum, 1988). Recently, renutrition alone has been shown to improve gastric emptying in anorexia nervosa (Rigaud et al., 1988).

FOOD ALLERGIES AND PSEUDO–FOOD ALLERGY

Many patients complain of gastrointestinal problems such as indigestion, nausea, and flatulence, and attribute such symptoms to food allergies or malabsorptive syndromes resulting from enzyme deficiencies (Parker, Sussman, & Krondl, 1988). Food allergies, which are more commonly found in infants and children, are most often due to eggs, cow's milk, legumes, and nuts. In adults, shellfish are reported to be common allergens. However, well-controlled laboratory investigations of putative food allergies could confirm that only 25% to 40% of such patients had the attributed allergy. The evaluation for food allergy is subtle and requires carefully controlled challenges. Skin tests correlate poorly with the severity of symptomatology produced by foods.

Lactose intolerance is a common condition wherein a variety of abdominal symptoms occur after ingestions of mammalian milk (Buller & Grand, 1990). Decreased lactase-phlorizin hydrolase (an enzyme that splits lactose into glucose and galactose) is the cause of this syndrome. The diagnosis can be made by assessing the capacity for lactose absorption using the lactose absorption test. The lactose breath hydrogen test also elucidates lactose nonabsorption and is easier to carry out in children. Treatment of food allergies is elimination of dietary allergens while maintaining a balanced diet. For lactose intolerance, use of enzyme substitutes such as yeast beta-galactosidase can reduce symptoms when milk products are ingested. A number of effective over-the-counter enzyme preparations are widely available.

In addition to the well-documented food allergies and lactose malabsorption syndromes, the syndrome of pseudo–food allergy has been described. Rix, Pearson, and Bentley (1984) examined a series of patients complaining of food allergies and found that most did not have a definable allergy. This false disease conviction can lead patients to seek out a variety of treatments that do not focus on the actual problem of abnormal illness beliefs, irritable bowel syndrome, or a somatoform disorder. Chronic hyperventilation, depression, or fatigue resulting from malnutrition often accompany such presentations. A pernicious pseudo–food allergy by proxy has been described wherein a parent falsely believes a child is allergic to certain foods and only allows a diet that causes malnutrition. Most commonly adults with pseudo–food allergy are women between 35 and 50 who have a mild mood disorder. Management includes careful attention to illness beliefs, treatment of the depression, and ensuring a balanced diet, since such patients often restrict their food intake to dangerously low nutritional levels.

HEPATIC DISEASE

Although there are a variety of hepatic diseases that result from circulatory abnormalities, biliary obstruction, and hepatotoxic agents, parenchymal liver disease is the most common. The various causes of parenchymal liver disorders include viral hepatitis, neoplastic disease, pyogenic abscesses, genetic errors of metabolism, toxic hepatitis, and alcoholic cirrhosis. In recent immigrants to the United States from less developed countries, parasitic diseases must always be considered.

Alcoholic Liver Disease

Alcoholic liver disease, including fatty liver, alcoholic hepatitis, and alcoholic cirrhosis, is the most common form of liver disease and ranks among the five leading causes of death in the United States (Galambos, 1979). The major mental syndrome seen with alcoholic cirrhosis is portal-systemic encephalopathy (Tarter et al.,

1985–1986). The psychologic symptoms of hepatic encephalopathy are varied and may be mistaken for primary psychiatric disorders. The early stages of portal-systemic encephalopathy are characterized by cognitive and attentional impairment, with irritability and affective lability; later stages are stupor and coma. Sleep disturbances, tremor, asterixis, and motor incoordination are also common in the early stages. Verbal coherence is often preserved in early encephalopathic states, so a complete cognitive assessment is essential for diagnosis. Read et al. (1967) noted that individuals with hepatic encephalopathy were often misdiagnosed either as schizophrenic or bipolar.

Mean blood ammonia level is significantly greater in encephalopathic patients than in nonencephalopathic patients with comparably severe liver disease (Walker & Schenker, 1970). While it is difficult to establish a close relationship between the actual blood ammonia level and the severity of encephalopathy, this correlation is improved by obtaining arterial blood samples and by avoiding any delays in analysis (Seligson & Hirahare, 1957). The concentration of glutamine in the cerebrospinal fluid shows an even better correlation with severity of encephalopathy (Gilon et al., 1959). The electroencephalogram (EEG) may be helpful in doubtful cases if it shows characteristic diffuse slowing and/or triphasic waves of metabolic delirium.

Treatment interventions involve both direct and supportive measures. Pharmacologic approaches, such as oral neomycin and lactulose, are helpful in modifying the encephalopathy. When agitation is severe, low-dose haloperidol can be considered. Benzodiazepines are relatively contraindicated because they may worsen the encephalopathy. The risk is greater with those benzodiazepines metabolized by oxidation, such as diazepam and chlordiazepoxide. In a single-dose study, peak plasma concentrations of buspirone were 16 times higher in patients with cirrhosis than in healthy controls. This effect was likely due to portal-systemic shunting and decreased hepatic elimination. Although subjects were twice as old as controls, this study constitutes some evidence that buspirone should be used at reduced doses in patients with liver disease (Dalhoff, Poulsen, & Garred, 1987). It has been reported that flumazenil, a gamma-aminobutyric acid (GABA) antagonist, can abate the signs of encephalopathy even when diet is unrestricted (Bansky et al., 1989; Ferenci et al., 1989). These findings favor a predominant role of increased GABAergic tone in hepatic encephalopathy.

Supportive measures include environmental interventions appropriate to delirium (see Chapter 19) and efforts to educate the family. When the etiology of the liver disease is alcoholic, significant others should be educated regarding the crucial role of abstinence, and a formal family intervention should be considered. Because of the encephalopathic patient's cognitive defects, direct confrontation of the patient regarding the alcohol problem is usually unproductive.

A particularly difficult issue is whether or when to provide liver transplantation for alcoholic patients with end-stage hepatic disease. The Consensus Development Panel of the National Institutes of Health in 1983 concluded that alcoholic hepatitis and cirrhosis should be viewed as indications for liver transplant in only rare instances (Members of the Consensus Development Panel, 1983). Two criteria were selected as indicating which patients might be considered appropriate: (1) those with established indicators of fatal outcome, and (2) those likely to abstain from alcohol. Beresford et al. (1990) use the following criteria to measure the likelihood of the alcoholic transplant candidate's future abstinence: patient and family recognitition of alcoholism; social stability; and changes in style of living. Most of the applicants for hepatic transplant who suffer from alcohol dependence have impressed these authors with their potential for recovery. (See also Chapter 36.)

Viral Hepatic Disease

Viral infections, most commonly hepatitis B virus, can cause serious hepatic disease. In the acute syndrome, transient neuropsychologic changes may occur, which may be manifested by irritability, poor concentration, and lethargy (Apstein, Koff, & Koff, 1979). Chronic viral hepatitis may produce fatigue and symptoms suggestive of depression. Cyclic antidepressants may help the mood symptoms but can aggravate the hepatitis itself. For this reason, patients with chronic hepatitis receiving cyclic antidepressants should have weekly liver function tests during the first month of treatment, and then as needed thereafter if hepatic symptoms relapse. Tricyclic blood levels should also be considered because the hepatitis can impair drug metabolism.

Hepatitis B virus, which can be transmitted by parenteral sources, such as intravenous drug abuse, transfusions, or sexual activity, may affect an individual's social status, leading to social isolation. Psychologic difficulties, such as fatigue, depression, and demoralization, combine with difficulties in sexual functioning because of the fear of transmitting infection. All patients with hepatitis B warrant testing for acquired immunodeficiency syndrome (AIDS) because the two diseases are transmitted by the same routes and are prevalent in the same populations.

Infectious mononucleosis is frequently associated with hepatomegaly. The disorder is due to a primary infection with the Epstein-Barr virus (EBV). Mental symptoms in infectious mononucleosis include depression, which may mimic primary major depression, as well as acute psychosis, which includes cognitive features such as confusion and marked memory loss (Allen & Tilkian, 1986). Pharmacologic treatment of these psychiatric syndromes may be needed; if it is, blood counts and liver enzymes should be monitored during the early weeks of treatment.

Chronic fatigue, a syndrome that has recently been the object of much controversy, is the so-called chronic active EBV infection, which is characterized by severe cyclic fatigue, sore throat, myalgias, headaches, paresthesias, arthralgias, depression, insomnia, confusion, and gastrointestinal complaints (Buchwald, Sullivan, & Komaroff, 1987). Despite the presence of antibody titers specific to EBV-specific antigens in many of these patients, there is no conclusive evidence that EBV is causally related to the syndrome. Authorities have therefore recommended interpreting antibody titers to EBV with caution until more definitive data are available. Nevertheless, systemic viral syndromes of this nature should be considered in the differential diagnosis of cyclical mood disturbances associated with physical complaints of chronic fatigue and lassitude.

Gold et al. (1990) found that patients who did demonstrate significant titers of antibodies to EBV and chronic fatigue manifested higher in vitro natural killer cell activity and lower in vitro interleukin-2 production than controls. Also, these patients had a high frequency of depressive illness meeting DSM criteria. Over 50% of patients followed with chronic fatigue in this study improved during follow-up despite negligible changes in titers of EBV proteins. The authors concluded that, clinically, most of these patients improve over time (see Chapter 32).

INFLAMMATORY BOWEL DISEASE

Inflammatory bowel disease refers to two disorders, ulcerative colitis and regional enteritis, or Crohn's disease (Farmer, 1981). Ulcerative colitis is generally limited to the mucosal lining of the large intestine. Its dominant symptoms are diarrhea and hematochezia. In severe disease states, the individual will have multiple watery stools, which can lead to dehydration and anemia. Crohn's disease is a transmural process that may affect any part of the alimentary canal but most often involves the distal ileum and proximal colon. Intestinal obstruction, fistulas, and abscesses may occur, causing such symptoms as abdominal pain, fever, and severe weight loss. Ulcerative colitis increases the risk of colonic cancer to a greater extent than Crohn's disease. Crohn's disease can recur relentlessly despite long periods of apparent remission, whereas ulcerative colitis may actually be cured by colectomy.

Psychologic Factors in Inflammatory Bowel Disease

Despite the early literature identifying ulcerative colitis as a psychosomatic disease, this conceptualization must be strongly questioned. North et al. (1990) recently reviewed the English language literature for associations between psychiatric factors and ulcerative colitis. They criticized virtually all studies for methodologic flaws. None of the seven well-controlled studies (e.g., Andrews, Barczak, & Allan, 1987; Tarter et al., 1987) found an association between ulcerative colitis and psychopathology.

Contemporary investigators have tried to partition psychologic factors in Crohn's disease from those in ulcerative colitis. Whybrow, Kane, and Lipton (1968) found that a variety of noxious life changes exacerbated Crohn's disease episodes. Those patients with more depressive symptomatology had a greater length of illness. The essential problem with most of these studies, however, is that it is difficult to separate reactive phenomena from etiologic factors. McKegney, Gordon, and Levine (1970) compared ulcerative colitis patients with those with Crohn's disease and found no significant differences from psychosocial, psychiatric, or behavioral perspectives. They documented that those with more severe physical disabilities had greater depressive symptomatology. Andrews, Barczak, and Allan (1987), using a self-report measure and the Structured Clinical Interview for DSM-III-R, found that physical illness in Crohn's disease was associated with psychiatric morbidity, but the association was not significant for ulcerative colitis patients. As with most other diseases, therefore, psychologic factors in the etiology of the conditions remain speculative.

Two recent, well-controlled studies examined psychiatric diagnoses in consecutive cases of ulcerative colitis and Crohn's disease, and compared rates with those obtained in control patients with other chronic medical illnesses (Helzer et al., 1982, 1984). The authors used structured interviews and operational diagnostic criteria. Whereas ulcerative colitis patients were no more likely than medically ill controls to have psychiatric diagnoses, Crohn's disease patients had significantly higher prevalence of depression than controls. The severity of gastrointestinal symptoms and psychiatric symptoms were independent, and there was no consistent temporal sequence of involvement. The findings suggest that depression and Crohn's disease are associated, but that attribution of causality is not justified (Helzer et al., 1982, 1984).

A recent study used the Diagnostic Interview Schedule in patients with Crohn's disease, patients with ulcerative colitis, and normal subjects (Tarter et al., 1987). Crohn's disease patients had a significantly greater prevalence of anxiety and depression than normal subjects, whereas ulcerative colitis patients did not have excess prevalence of any psychiatric disorder. Bipolar illness and Crohn's disease have been reported to co-exist in a few patients; the first manic episode often was associated with steroid use (Holroyd & DePaulo, 1990). Nutritional problems, treatment of chronic pain, and steroid withdrawal may cause or exacerbate psychiatric symptoms in these patients. Depression and fatigue often occur after the withdrawal of long-term steroid therapy (e.g., after successful colectomy). In such a case, the patient should first be evaluated for adrenal insufficiency. If this evaluation is negative, antidepressants should be considered (B. Fogel, personal communication).

Psychologic Treatment

The essential differences between regional enteritis—Crohn's disease—and ulcerative colitis must be understood in providing psychologic support for these patients. Specifically, the individual with regional enteritis will have to cope with an uncertain and relentless disease that has exacerbations and remissions. The relatively favorable prognosis of the ulcerative colitis patient after colectomy stands in contrast to that of the Crohn's disease sufferer, whose illness may persist. In one study, the three most prominent concerns of inflammatory bowel disease patients were having an ostomy bag, a low energy level, and having surgery (Drossman et al., 1989). Those with Crohn's disease were significantly more concerned with pain; ulcerative colitis patients were significantly more concerned with loss of bowel control and developing cancer. Thus, psychologic treatment must be directed toward coping with specific life stresses, problems with the actual treatment itself, and the existential nature of the chronic disease (Zisook & DeVaul, 1977). Individuals who become depressed in the setting of chronic bowel pain may benefit from antidepressant medication. Careful behavioral focus on the activities of daily living may help in restoring proper dietary habits and may minimize the exhaustion and fatigue that occur in the setting of anemia and dehydration. In both ulcerative colitis and Crohn's disease, steroids are often used, leading to their well-known mental side effects (Lewis & Smith, 1983).

Utilization of intravenous hyperalimentation is sometimes necessary in fulminant cases of regional enteritis or for individuals with a short bowel syndrome. The psychologic effects of this treatment have been catalogued by Perl et al. (1980). Depression, fear, distortion of self-concept, loss of appetite, and marital stress are common in individuals who undergo total parenteral nutrition for periods of a few months or longer (Hall & Beresford, 1987). Patients on chronic hyperalimentation also have been found to have mild cognitive impairment. Both mood changes and cognitive dysfunction thus can arise during chronic hyperalimentation. In almost all cases, these are due to metabolic problems, infection, or the disease that led to the need for total parenteral nutrition. Depression should be treated with antidepressants and psychotherapy. The presence of cognitive dysfunction should lead to a comprehensive assessment of fluid, electrolyte, and vitamin status, as well as screening for infection related to the central line.

Psychotropic agents may need to be used parenterally in patients receiving total parenteral nutrition. (For an overview of parenteral use of psychotropics, see Santos, Beliles, & Arana, in press (1993)). Lorazepam, chlordiazepoxide, and diazepam may be given intravenously. Lorazepam is the only benzodiazepine that is reliably absorbed intramuscularly. Alprazolam and lorazepam may be given sublingually as well. Antipsychotic medications that are available intramuscularly include haloperidol, fluphenazine, trifluoperazine, thiothixene, perphenazine, loxitane, and chlorpromazine. Haloperidol is widely used intravenously, particularly in intensive care settings. Dosages begin at 0.5 to 2.0 mg and should be doubled every 20 minutes until psychosis or agitation from delirium is treated. Efficacious intravenous doses are often much higher than oral or intramuscular doses. Frequency of administration must be adjusted to the particular situation (Tesar & Stern, 1986). (See Chapter 12.)

Antidepressants usually do not need to be given parenterally. Psychotherapy can help relieve depression as one waits to begin or resume an antidepressant. If necessary, amitriptyline or imipramine can be used intramuscularly. Electroconvulsive therapy should be considered for severe depression in a patient who cannot take medications orally. Carbamazepine and valproate are not available in parenteral preparations. Electroconvulsive therapy or the combination of a neuroleptic and a benzodiazepine could be used for acute mania.

In both ulcerative colitis and Crohn's disease, institution of an ostomy demands significant adaptation by the patient. Patients may react with shame, depression, and anger at the alteration of their excretory system. Prior to surgery, it is most helpful to have the ostomy therapist talk to both patient and spouse. Utilization of patient organizations, such as the United Ostomy Association or the National Ileitis and Colitis Foundation, can be an invaluable adjunct to medical treatment. Such groups provide patient advocates who have had personal experience with their own ostomies and can aid the patient with concerns such as leakage, skin breakdown, and odors. Inclusion of the spouse is important in managing the patient with an ostomy. Studies have noted that men are more comfortable having their wives view their ostomy and appliance, whereas women seem to experience more shame and do not like their spouses to see their ostomy (Dlin, Perlman, & Ringold, 1969; Druss et al., 1968). Psychologic complications of ostomy surgery are discussed at length in Chapter 34.

THE IRRITABLE BOWEL SYNDROME AND CONSTIPATION

Irritable Bowel Syndrome

The irritable bowel syndrome (IBS) is a heterogenous condition of uncertain etiology that accounts for 50% of ambulatory cases seen by gastroenterologists (Thompson, 1979). The syndrome is composed of a cluster of bowel symptoms that reflect alternating diarrhea, constipation, and abdominal pain. It may be partitioned into diarrhea-predominant, constipation-predominant, and mixed varieties. Classification of this disorder is difficult because a variety of investigations have shown that normative populations frequently report bowel habits similar to those of patients with operationally defined IBS (Thompson & Heaton, 1980). Two recent studies indicate that, whereas patients with IBS have psychologic problems, nonpatients who meet IBS criteria do not (Drossman et al., 1988; Whitehead et al., 1988). Barsky (1987) believes that IBS is best seen as dimensional rather than categorical.

It thus may be that patients with IBS overestimate their excretory problems and seek medical care because of abnormal illness behavior rather than the disease itself. Other data, however, suggest that individuals with IBS may have abnormal myoelectric activity of the colon, abnormalities of gastric hormones, or food allergies (Wise, 1986). Noncolonic gastrointestinal symptoms are often present in IBS, perhaps because motility disturbances may be seen throughout the gastrointestinal tract; extragastrointestinal symptoms are common as well (Whorwell, 1989). One study found that 28% of patients met criteria for somatization disorder (Young, Alpers, & Norland, 1976). Most investigators would agree that IBS involves a combination of psychologic and physiologic factors.

Patients with IBS have an increased incidence of psychologic disturbances when compared with those with inflammatory bowel disease (Walker et al., 1990; Young, Alpers, & Norland, 1976). It appears that psychiatric illness often precedes the onset of gastrointestinal symptoms (Walker et al., 1990). Studies such as this one suggest that mental symptoms are associated with the disorder rather than merely a reaction to the discomfort of symptoms. (For an excellent review, see Walker, Roy-Byrne, & Katon, 1990.) Patients with IBS are more anxious and depressed than normal controls. There is a marked tendency toward hypochrondriacal concerns in patients with IBS, since they view minor illnesses more seriously than other individuals and consult physicians more often for a variety of illnesses. There is no clear evidence that stress causes a different colonic, small bowel, or stomach motor response in IBS than in controls (Camilleri & Neri, 1989).

Various treatments have been utilized for IBS (Wise, 1986). These include dietary management with high-fiber diets that avoid gas-producing legumes, anticholinergic medication, bulking agents, and utilization of minor tranquilizers or tricyclic antidepressants. Gastrointestinal symptoms in panic disorder patients, many of which are consistent with IBS, have been shown to respond to antipanic pharmacotherapy (Lydiard et al., 1986; Noyes et al., 1990). Pharmacotherapy studies in IBS are generally poorly designed, so it is difficult to draw definitive conclusions about any particular psychotropic agent (Klein, 1988). However, there is evidence that irritable bowel complaints may respond to antidepressants whether or not the patient is clinically depressed. Medications with anticholinergic effects, such as doxepin, are preferred for diarrhea-predominant irritable bowel. Those with low anticholinergic effects, such as desipramine or fluoxetine, are preferred in patients with a constipation-predominant condition. Many patients who do not have concurrent major depression will respond symptomatically to relatively low doses of a cyclic antidepressant. Low-dose antidepressants are also excellent antianxiety agents for this population, often obviating the need for a benzodiazepine. Buspirone, which has been shown to have antidepressant and anxiolytic effects with minimal side effects (Arana & Hyman, 1991), may prove to be helpful in this population.

Focal problem-solving psychotherapy techniques, similar to those prescribed for patients with peptic ulcer disease, have been shown to be an effective strategy in modifying IBS symptoms (Hislop, 1980). Combined cognitive/behavioral therapy may be helpful in some patients (Schwarz et al., 1990). A combined approach utilizing group therapy, relaxation therapy, and focal problem solving may diminish somatic concerns and anxiety in persons with irritable bowel despite the persistence of symptoms (Wise, Cooper, & Ahmed, 1982). Hypnotherapy has also been successful for some patients with refractory symptoms (Harvey et al., 1989). In patients with marked hypochondriacal concerns, prevention of doctor-shopping and small, but frequent, doses of physician interaction may be useful.

One prospective study found that, of patients managed with high-fiber diet, bulking agents, antispasmodics, and education, 68% were virtually symptom free at 5-year follow-up. Good prognostic signs were male sex, predominant constipation, short history, and history of acute illness at the onset of the bowel disturbance (Harvey, Mauad, & Brown, 1987b).

Constipation

Constipation, defined as infrequent or difficult-to-pass stools, is an extremely common problem in medical and psychiatric settings. There are numerous pathophysiologic bases of constipation, the most common of which is a typical U.S. diet: high in refined food and low in fiber (Lennard-Jones, 1985). Psychotropic medications that are highly anticholinergic often cause constipation. For the patient with chronic idiopathic constipation, highly anticholinergic medications (such as amitriptyline and clozapine) should be avoided. Constipation may be associated with psychopathology such as depression (Fisher et al., 1989) or anxiety. One recent study found constipated patients' Minnesota Multiphasic Personality Inventory profiles to be high on Hypochondriasis and Hysteria and low on Depression, the so-called Psychosomatic V (Derroede et al., 1989).

To minimize constipation, it is important to exercise regularly, drink plenty of fluids, and defecate at about the same time each day (preferably after breakfast, since colonic motility is highest then) (Spiller, 1990). Treatment is primarily with a high-fiber diet (Taylor, 1990). However, side effects such as flatulence and bloating may be intolerable, and high fiber intake is simply ineffective for some people. Some who have difficulty relaxing the anal sphincter may benefit from biofeedback. Psychotherapy and/or psychotropic medication may help those for whom constipation is associated with psychiatric morbidity. Stool softeners, bowel retraining programs, bisacodyl (a contact stimulant), and occasional use of other laxatives may be needed in difficult cases (Spiller, 1990).

CARCINOMA OF THE PANCREAS

Carcinoma of the pancreas is a particularly lethal disease since spread beyond the pancreas has occurred in 85% to 90% of patients at the time of clinical presentation (Malagelada, 1979). Whether a clinical depression often predates the development of any physical signs and symptoms of pancreatic cancer has generated a lively debate (Brown & Paraskevas, 1982; Holland, 1982; Jacobson & Ottoson, 1971; Joffe et al., 1986; Sachar, 1975). Fras, Litin and Pearson (1967) utilized a controlled prospective design to demonstrate that depression, anxiety, and loss of ambition were present in 76% of pancreatic cancer patients prior to surgery as compared with 20% of patients with other abdominal neoplasms. More recently, utilizing the Profile of Mood States, Holland et al. (1986) compared 107 patients with advanced pancreatic cancer with 111 patients with advanced gastric cancer and found that those with pancreatic cancer experienced significantly greater general psychologic disturbance. Coffman and Starkman (1990), however, using similar methodology, found pancreatic cancer patients to be no more depressed or anxious than age-matched controls with other abdominal cancers.

Mechanisms underlying psychologic disturbance in pancreatic cancer patients remain to be elucidated. Some authors have suggested that a paraneoplastic syndrome might account for altered mood states (Petty & Noyes, 1981). Others have suggested that the underlying mechanism could be autoimmune. According to this hypothesis, antibodies produced against a protein released by pancreatic cancer cells possess cross-reactivity with central nervous system tissues such as serotonin receptors, thereby altering mood states (Brown & Paraskevas, 1982). Further research in this area may help shed light on basic mechanisms of depression and other psychiatric illnesses. (See also Chapter 24.)

COLONIC CARCINOMA

Issues of colonic cancer and of its surgical treatment are discussed in Chapters 24 and 34, respectively.

GENERAL GUIDELINES FOR TREATMENT

Gastrointestinal phenomena such as abdominal pain, dyspepsia, or diarrhea often occur concurrently with subjective emotional symptoms. To delineate such phenomena, it is best that the patient keep a behavioral diary, which is simply a self-report inventory that can be tailored to each patient's specific needs (Kanfer & Saslow, 1965). Elements that need to be included in such a diary are: (1) the specific behavior to be monitored (such as abdominal pain, anxiety, depression, flatulence, or diarrhea); (2) the antecedents of the behavior; and (3) the consequences of the behavior. Such an inventory elicits specific information and allows the patient to make linkages between emotional phenomena and physiologic data. Such a diary can be developed from a small note pad, which can be written in once or twice each day. In addition, dietary patterns and other information that the patient may initially overlook can be entered.

The medical psychotherapy of patients with gastrointestinal disorders often combines didactic, insight-oriented, supportive, and behavioral strategies. The therapist must be flexible and understand the physiology of each disease in order to fully comprehend the patient's own experience. Illness beliefs and behaviors must be fully understood as well. Strategies for medical psychotherapy are discussed in Chapter 1, and selection of patients for behavior therapy is discussed in Chapter 41. In addition to these individual interventions, referral of patients with similar illnesses to group therapy may be appropriate, and is especially suitable for socially isolated patients. As noted earlier, gastrointestinal symptoms in many of these patients will "mask" primary Axis I anxiety and depressive disorders of syndromal and subsyndromal proportions. Sorting out this particular group of patients is essential, since many will respond to formal psychiatric interventions despite the presence of multiple psychophysiologic complaints and a relative lack of psychologic insight.

REFERENCES

Abernethy DR, Greenblatt DJ, Eshelman FN, et al., (1984). Rantidine does not impair oxidative or conjugative metabolism: Noninteraction with antipyrine, diazepam, and lorazepam. Clin Pharmacol Ther 35:188–192.

Alexander F (1950). Psychosomatic medicine. New York: Norton.

Allen AD, & Tilkian SM (1986). Depression correlated with cellular immunity in systemic immunodeficient Epstein-Barr virus syndrome. J Clin Psychiatry 47:133–135.

Anderson KO, Dalton CB, Bradley LA, et al. (1989). Stress induces alteration of esophageal pressures in healthy volunteers and non-cardiac chest pain patients. Dig Dis Sci 34:83–91.

Andrews H, Barczak P, & Allan RN (1987). Psychiatric illness in patients with inflammatory bowel disease. Gut 28:1600–1604.

Apstein MD, Koff E, & Koff RS (1979). Neuropsychological dysfunction in acute viral hepatitis. Digestion 19:349–358.

Arana GW, & Hyman SE (1991). Handbook of psychiatric drug therapy (2nd ed., p. 141). Boston: Little, Brown.

Assael M, Bass D, Fischel RE, et al. (1974). Impotence and peptic ulcer. Int J Psychiatry Med 5:377–387.

Baldessarini RJ (1985). Chemotherapy in psychiatry. Cambridge, MA: Harvard University Press.

Bansky G, Meier PJ, Riederer E, et al. (1989). Effects of the benzodiazepine receptor antagonist flumazenil in hepatic encephalopathy in humans. Gastroenterology 97:44–50.

Barksy AJ (1987). Investigating the psychological aspects of the irritable bowel syndrome [editorial]. Gastroenterology 93:902–904.

Beaumont W (1833). Experiments and observations on the gastric juice and physiology of digestion. Plattsburgh, NY: FP Allen.

Beresford TP, Turcotte RM, Merion R, et al. (1990). A rational approach to liver transplantation for the alcoholic patient. Psychosomatics 31:241–254.

Bradley LA, McDonald JE, & Richter JE (1990). Psychophysiological interactions in the esophageal diseases: Implications for assessment and treatment. Semin Gastrointest Dis 1:5–22.

Brazer SR, Tyor MP, Pancotto FS, et al. (1989). Randomized, double-blind comparison of famotidine with ranitidine in treatment of acute, benign gastric ulcer disease. Dig Dis Sci 34:1047–1052.

Brooks GR, & Richardson FC (1980). Emotional skills training: A treatment program for duodenal ulcer. Behav Ther 11:198–207.

Brown JH, & Paraskevas F (1982). Cancer and depression: Cancer presenting with depressive illness: An autoimmune disease? Br J Psychiatry 141:227–232.

Brown SR, Schwartz JM, Summergrad P, et al. (1986). Globus hystericus syndrome responsive to antidepressants. Am J Psychiatry 143:917–918.

Browning TH, & Members of the Patient Care Committee of the American Gastroenterological Association (1990). Diagnosis of chest pain of esophageal origin. Dig Dis Sci 35:289–293.

Buchanan DC (1978). Group therapy for chronically physically ill patients. Psychosomatics 19:425–431.

Buchwald D, Sullivan JL, & Komaroff AL (1987). Frequency of "chronic active Epstein-Barr virus infection" in a general medical practice. JAMA 257:2302–2307.

Buller HA, & Grand RJ (1990). Lactose intolerance. Annu Rev Med 41:141–148.

Camilleri M, & Neri M (1989). Motility disorders and stress. Dig Dis Sci 34:1777–1786.

Chappell NM, Stefano JJ, Rogerson JS, et al. (1936). The value of group psychological procedures in the treatment of peptic ulcer. Am J Dig Dis 3:813–817.

Clouse RE (1988). Anxiety and gastrointestinal illness. Psychiatr Clin North Am 11:399–417.

Clouse RE, & Lustman PJ (1982). Psychiatric illnesses and contraction abnormalities of the esophagus. N Engl J Med 309:1337–1342.

Clouse RE, Lustman PJ, Eckert TC, et al. (1987). Low-dose trazodone for symptomatic patients with esophageal contraction abnormalities. Gastroenterology 92:1027–1036.

Coffman KL, & Starkman NN (1990). Pancreatic cancer and depression [abstract]. Psychosom Med 52:236.

Dalhoff K, Poulsen ME, & Garred P (1987). Buspirone pharmacokinetics in patients with cirrhosis. Br J Clin Pharmacol 24:547–550.

Derroede G, Girard G, Bonchoncha M, et al. (1989). Idiopathic constipation by colonic dysfunction. Dig Dis Sci 34:1428–1433.

Desmond PV, Patwardhan RV, Schenker S, et al. (1980). Cimetidine impairs elimination of chlordiazepoxide in man. Ann Intern Med 93:266–268.

Dlin BM, Perlman A, & Ringold E (1969). Psychosexual response to ileostomy and colostomy. Am J Psychiatry 126:3–9.

Drossman DA (1982). Patients with psychogenic abdominal pain: Six years observation in the medical setting. Am J Psychiatry 139:1549–1557.

Drossman DA, Leserman J, Nachman G, et al. (1990). Sexual and physical abuse in women with functional or organic gastrointestinal disorders. Ann Intern Med 113:828–833.

DROSSMAN DA, MCKEE DC, SANDLER RS, ET AL. (1988). Psychosocial factors in the irritable bowel syndrome. Gastroenterology 95:701–708.

DROSSMAN DA, PATRICK DL, MITCHELL CM, ET AL. (1989). Health-related quality of life in inflammatory bowel disease. Dig Dis Sci 34:1379–1386.

DRUSS RG, O'CONNOR J, PRUDDEN JF, ET AL. (1968). Psychologic response to colectomy. Arch Gen Psychiatry 18:53–59.

EISENDRATH SJ, WAY LW, OSTROFF JW, ET AL. (1986). Identification of psychogenic abdominal pain. Psychosomatics 27:705–712.

ENOCH MD, TRETHOWAN WH, & BARKER JC (1967). Some uncommon psychiatric syndromes. Bristol, England: John Wright.

FARMER RG (1981). Factors in long term prognosis of patients with inflammatory bowel disease. Am J Gastroenterol 75:97–109.

FELDMAN M, WALKER P, GREEN JL, ET AL. (1986). Life events stress and psychosocial factors in men with peptic ulcer disease. Gastroenterology 91:1370–1379.

FERENCI P, GRIMM G, MERYN S, ET AL. (1989). Successful long-term treatment of portal-systemic encephalopathy by the benzodiazepine antagonist flumazenil. Gastroenterology 96:240–243.

FISHER SE, BRECKON K, ANDREWS MA, ET AL. (1989). Psychiatric screening for patients with faecal incontinence or chronic constipation referred for surgical treatment. Br J Surg 76:352–355.

FONAGU P, & CALLOWAY SP (1986). The effect of emotional arousal on spontaneous swallowing rates. J Psychosom Res 30:183–188.

FORREST FM, FORREST IS, & SERRA MT (1970). Modification of chlorpromazine metabolism by some other drugs frequently administered to psychiatric patients. Biol Psychiatry 2:53.

FRAS I, LITIN EM, & PEARSON JS (1967). Comparison of psychiatric symptoms in carcinoma of the pancreas with those in some other intraabdominal neoplasms. Am J Psychiatry 123:1553–1562.

GALAMBOS J (1979). Cirrhosis: Epidemiology. In L SMITH (Ed.), Major problems in internal medicine (Ch. 6, pp. 91–127). Philadelphia: WB Saunders.

GARNETT WR (1982). Sucralfate—alternative therapy for peptic ulcer disease. Clin Pharm 1:307–314.

GILON E, SZEINBERG A, TAUMAN G, ET AL. (1959). Glutamine estimation in cerebrospinal fluid in cases of liver cirrhosis and hepatic coma. J Lab Clin Med 53:714.

GOLD D, BOWDEN R, SIXBEY J, ET AL. (1990). Chronic fatigue: A prospective clinical and virologic study. JAMA 264:48–53.

GOMEZ J, & DALLY P (1977). Psychologically mediated abdominal pain in surgical and medical outpatient clinics. Br Med J 1:1451–1453.

GOUGH KR, BARDHAN KD, CROWE JP, ET AL. (1984). Ranitidine and cimetidine in prevention of duodenal ulcer relapse. Lancet 2:659–662.

GRAHAM DY, AGRAWAL NM, & ROTH SH (1988). Prevention of NSAID-induced gastric ulcer with misoprostol: Multicentre, double-blind, placebo-controlled trial. Lancet 2:1277–1280.

GRAHAM DY, MCCULLOUGH A, SKLAR M, ET AL. (1990). Omeprazole versus placebo in duodenal ulcer healing. Dig Dis Sci 35:66–72.

GREENBERG DB, STERN TA, & WEILBURG JB (1988). The fear of choking: Three successfully treated cases. Psychosomatics 29:126–129.

GREENBLATT DJ, ABERNATHY DR, MORSE DS, ET AL. (1984). Clinical importance of the interaction of diazepam and cimetidine. N Engl J Med 310:1639–1643.

GREENBLATT DJ, ALLEN MD, MACLAUGHLIN DS, ET AL. (1978). Diazepam absorption: Effect of antacids and food. Clin Pharmacol Ther 24:600–609.

GREENBLATT DJ, SHADER RI, HARMATZ JS, ET AL. (1976). Influence of magnesium and aluminum hydroxide mixture on chlordiazepoxide absorption. Clin Pharmacol Ther 19:234–239.

HALL RCW, & BERESFORD TP (1987). Psychiatric factors in the management of long-term hyperalimentation patients. Psychiatry Med 5:211–217.

HARVEY RF, GORDON PC, HADLEY N, ET AL. (1987a). Effects of sleeping with the bed-head raised and of ranitidine in patients with severe peptic oesophagitis. Lancet 2:1200–1203.

HARVEY RF, HINTON RA, GUNARY RM, ET AL. (1989). Individual and group hypotherapy in treatment of refractory irritable bowel syndrome. Lancet 1:424–425.

HARVEY RF, MAUAD EC, & BROWN AM (1987b). Prognosis in the irritable bowel syndrome: A 5-year prospective study. Lancet 1:963–965.

HELZER JE, CHANNAS S, NORLAND CC, ET AL. (1984). A study of the association between Crohn's disease and psychiatric illness. Gastroenterology 86:324–330.

HELZER JE, STILLINGS WA, CHAMMAS S, ET AL. (1982). A controlled study of the association between ulcerative colitis and psychiatric diagnoses. Dig Dis Sci 27:513–518.

HENAVER SA, & HOLLISTER LE (1984). Cimetidine interaction with imipramine and nortriptyline. Clin Pharmacol Ther 35:183–186.

HENNAN NE, CARPENTER DU, & JANDA SM (1988). Famotidine-associated mental confusion in elderly patients. Drug Intell Clin Pharm 22:976–978.

HERNANDEZ DE (1989). Neurobiology of brain-gut interactions. Dig Dis Sci 34:1809–1816.

HISLOP IG (1980). Effect of very brief psychotherapy on the irritable bowel syndrome. Med J Aust 2:620–623.

HOLLAND JCB (1982). Psychologic aspects of cancer. In JF HOLLAND & E FREI (Eds.), Cancer medicine (2nd ed., pp. 1175–1203). Philadelphia: Lea & Febiger.

HOLLAND JCB, KORZUN AH, TROSS S, ET AL. (1986). Comparative psychological disturbance in patients with pancreatic and gastric cancer. Am J Psychiatry 143:982–986.

HOLROYD S, & DEPAULO JR (1990). Bipolar disorder and Crohn's disease. J Clin Psychiatry 51:407–409.

HOWES CA, PULLAR T, SOURINDHRIU I, ET AL. (1983). Reduced steady state plasma concentrations of chlorpromazine and indomethacin in patients receiving cimetidine. Eur J Clin Pharmacol 24:99–102.

ISENBERG JI (1981). Peptic ulcer. Dis Mon 28:1–58.

JACOBSON L, & OTTOSON JO (1971). Initial mental disorder in carcinoma of the pancreas and stomach. Acta Psychiatr Scand 220:120–127.

JOFFEE RT, RUBINOW DR, DENICOFF KD, ET AL. (1986). Depression and carcinoma of the pancreas. Gen Hosp Psychiatry 8:241–245.

KANFER FH, & SASLOW G (1965). Behavioral analysis. Arch Gen Psychiatry 12:529–538.

KATON W (1984). Panic disorder and somatization. Am J Med 77:101–106.

KLEIN KB (1988). Controlled treatment trials in the irritable bowel syndrome: A critique. Gastroenterology 95:232–241.

KLINGER RL, & STRANG JP (1987). Psychiatric aspects of swallowing disorders. Psychosomatics 28:572–576.

KLOTZ U, & REIMANN I (1980). Delayed clearance of diazepam due to cimetidine. N Engl J Med 302:1012–1014.

LANGELUDDECKE P, GOULSTON K, & TENNANT C (1990). Psychological factors in dyspepsia of unknown cause: A comparison with peptic ulcer disease. J Psychosom Res 34:215–222.

LANGTRY HD, GRANT SM, & GOA KL (1989). Famotidine: An updated review of its pharmacokinetic properties, and therapeutic use in peptic ulcer disease and other allied diseases. Drugs 38:551–590.

LENNARD-JONES JE (1985). Pathophysiology of constipation. Br J Surg 72(suppl):S7–S8.

LEWIS DA, & SMITH RE (1983). Steroid induced psychiatric syndromes. J Affect Disord 5:319–332.

LIEBERMAN D (1990). Treatment approaches to reflux esophagitis. Drugs 39:674–680.

LIPKIN M, & LAMB GS (1982). The couvade syndrome: An epidemiologic study. Ann Intern Med 96:509–511.

LIPSY RJ, FENNERTY B, & FAGAN TC (1990). Clinical review of histamine-2 receptor antagonists. Arch Intern Med 150:745–751.

LYDIARD RB, LARAIA MT, HOWELL EF, ET AL. (1986). Can panic disorder present as irritable bowel syndrome? J Clin Psychiatry 47:470–473.

MACPHEE GJ, THOMPSON GG, SCOBIE G, ET AL. (1984). Effects of cimetidine on carbamazepine auto and heteroinduction in man. Br J Clin Pharmacol 18:411–419.

MALAGELADA J (1979). Pancreatic cancer: An overview of epidemiology, clinical presentation, and diagnosis. Mayo Clin Proc 54:459–467.

MATON PN (1991). Omeprazole. N Engl J Med 324:965–975.

MCFARLAND RJ, BATESON MC, GREEN JRB, ET AL. (1990). Omeprazole provides quicker symptom relief and duodenal ulcer healing than ranitidine. Gastroenterology 98:278–283.

MCKEGNEY FP, GORDON RO, & LEVINE SM (1970). A psychosomatic comparison of patients with ulcerative colitis and Crohn's disease. Psychosom Med 32:153–166.

Members of the Consensus Development Panel (1983). Liver transplanation consensus conference. JAMA 250:2961–2964.

MILLER LG, & JANKOVIC J (1989). Metoclopramide-induced movement disorders: Clinical findings with a review of the literature. Arch Intern Med 149:2486–2492.

MONK JP, & CLISSOLD SP (1987). Misoprostol. Drugs 33:1–30.

NIELZEN S, PETTERSSON KI, REGNELL G, ET AL. (1986). The role of psychiatric factors in symptoms of hiatus hernia or gastric reflux. Acta Psychiatr Scand 73:214–220.

NORTH CS, CLOUSE RE, SPITZNAGEL EL, ET AL. (1990) The relation of ulcerative colitis to psychiatric factors: A review of findings and methods. Am J Psychiatry 147:974–981.

NOYES R, COOK B, GARVEY M, ET AL. (1990). Reduction of gastrointestinal symptoms following treatment for panic disorder. Psychosomatics 31:75–79.

PARKER SL, SUSSMAN GL, & KRONDL M (1988). Dietary aspects of adverse reactions to food in adults. Can Med Assoc J 139:711–718.

PERL M, HALL RCW, DUDRICK SJ, ET AL. (1980). Psychological aspects of long-term home hyperalimentation. J Parenter Enteral Nutr 4:554–560.

PETERSON WG (1991). *Helicobacter pulori* and peptic ulcer disease. N Engl J Med 324:1043–1048.

PETTY F, & NOYES R JR (1981). Depression secondary to cancer. Biol Psychiatry 16:1203–1220.

RASK-MADSEN C, BRESNICK WH, LOSS MA, ET AL. (1990). The effect of emotional stress on gastric acid secretion in normal subjects and duodenal ulcer patients [AGA abstracts] Gastroenterology 98:A382.

READ A, SHERLOCK S, LAIDLAW J, ET AL. (1967). The neuropsychiatric syndromes associated with chronic liver disease and an extensive portal-systemic collateral circulation. Q J Med 36:135–150.

REIDEL WL, & CLOUSE RE (1985). Variations in clinical presentation of patients with esophageal contraction abnormalities. Dig Dis Sci 30:1065–1071.

RICCI DA, & MCCALLUM RW (1988). Diagnostic and treatment of delayed gastric emptying. Adv Intern Med 33:357–384.

RICHTER JE, OBRECHT WF, BRADLEY LA, ET AL. (1986). Psychological similarities between patients with the nutcracker esophagus and irritable bowel syndrome. Dig Dis Sci 31:131–138.

RIES RK, GILBERT DA, & KATON W (1984). Tricyclic antidepressant therapy for peptic ulcer disease. Arch Intern Med 144:566–569.

RIGAUD D, BEDIG G, MERROUCHE M, ET AL. (1988). Delayed gastric emptying in anorexia nervosa is improved by completion of a renutrition program. Dig Dis Sci 33:919–925.

RIX KJB, PEARSON DF, & BENTLEY SJ (1984). A psychiatric study of patients with supposed food allergy. Br J Psychiatry 145:121–126.

SACHAR E (1975). Evaluating depression in the medical patient. In J STRAIN & S GROSSMAN (Eds.), Psychological care of the medically ill (Ch 6, pp. 64–75). New York: Appleton-Century-Crofts.

SANTOS AB, BELILES KE, & ARANA GW (in press). Parenteral use of psychotropic agents. In A Stoudemire, & BS Fogel (Eds.), Medical psychiatric practice (Vol. 2; Ch. 4). Washington, DC: American Psychiatric Press, Inc.

SCHUBERT TT (1984). Update in treatment of peptic ulcer disease. Mo Med November:723–727.

SCHWARZ SP, TAYLOR AL, SCHARFF L, ET AL. (1990). Behaviorally treated irritable bowel syndrome patients: A four-year follow-up. Behav Res Ther 28:331–335.

SELIGSON D, & HIRAHARE K (1957). The measurement of ammonia in whole blood, erythrocytes and plasma. J Lab Clin Med 49:962.

SHRIVASTAVA RK, SHAH BK, & SIEGAL H (1985). Doxepin and cimetidine in the treatment of duodenal ulcer: A double-blind comparative study. Clin Ther 7:181–189.

SILVERSTONE PH (1984). Ranitidine and confusion. Lancet 1:1071.

SJODIN L, SVENDLUND J, OTTOSSON JO, ET AL. (1986). Controlled study of psychotherapy in chronic peptic ulcer disease. Psychosomatics 27:187–200.

SOLL AH, & ISENBERG JI (1983). Duodenal ulcer diseases. In M SLEISINGER & J FORDRAN (Eds.), Gastrointestinal diseases (Ch. 40, pp. 625–672). Philadelphia: WB Saunders.

SONIES BC, & DALAKAS MC (1991). Dysphagia in patients with the post-polio syndrome. N Engl J Med 324:1162–1167.

SPILLER R (1990). Management of constipation: 2. When fibre fails. Br Med J 300:1064–1065.

STOCKTON M, WEINMAN J, & MCCOLL I (1985). An investigation of psychosocial factors in patients with upper abdominal pain: A comparison with other groups of surgical outpatients. J Psychosom Res 29:191–198.

STOUDEMIRE A, & SANDHU J (1987). Psychogenic/idiopathic pain syndromes. Gen Hosp Psychiatry 9:79–86.

STRACHER G, BERNER P, NASKE R, ET AL. (1975). Effect of hypnotic suggestion of relaxation on basal and betazole-stimulated gastric acid secretion. Gastroenterology 68:656–661.

SWITZ (1976). What the gastroenterologist does all day. Gastroenterology 70:1048–1050.

TALLEY NJ, FUNG LH, GILLIGAN IJ, ET AL. (1986). Association of anxiety, neuroticism, and depression with dyspepsia of unknown cause. Gastroenterology 90:886–892.

TARTER RE, HEGEDOS PM, VAN THIEL DH, ET AL. (1985–1986). Portal-systemic encephalopathy neuropsychiatric manifestations. Int J Psychiatry Med 15:265–275.

TARTER RE, SWITALA J, CARRA J, ET AL. (1987). Inflammatory bowel disease: Psychiatric status of patients before and after disease onset. Int J Psychiatry Med 17:173–181.

TAYLOR R (1990). Management of constipation: 1. High fibre diets work. Br Med J 300:1063–1064.

TESAR GE, & STERN TA (1986). Evaluation and treatment of agitation in the intensive care unit. J Intensive Care Med 1:137–148.

THOMPSON WG (1979). The irritable gut. Baltimore: University Park Press.

THOMPSON WG, & HEATON KW (1980). Functional bowel disorders in apparently healthy people. Gastroenterology 79:283–288.

VAN VALKENBURG C, WINOKUR G, BEHAR D, ET AL. (1984). Depressed women with panic atacks. J Clin Psychiatry 45:367–369.

WALKER CO, & SCHENKER S (1970). Pathogenesis of hepatic encephalopathy with special reference to the role of ammonia. Am J Clin Nutr 23:619.

WALKER EA, ROY-BYRNE PP, & KATON WJ (1990). Irritable bowel syndrome and psychiatric illness. Am J Psychiatry 147:565–572.

WALKER EA, ROY-BYRNE PP, KATON WJ, ET AL. (1990). Psychiatric illnes and irritable bowel syndrome: A comparison with inflammatory bowel bowel disease. Am J Psychiatry 147:1656–1661.

WALKER P, LUTHER J, SAMLOF IM, ET AL. (1988). Life events stress and psychosocial factors in men with peptic ulcer disease. II. Relationships with serum pepsinogen concentrations and behavioral risk factors. Gastroenterology 94:323–330.

WEBSTER LK, MIHALY GW, JONES DB, ET AL. (1984). Effect of cimetidine and ranitidine on carbamazepine and sodium valproate pharmacokinetics. Eur J Clin Pharmacol 27:341–343.

WEINER H, THALER M, REISER MF, ET AL. (1957). Etiology of duodenal ulcer I. Relation of specific psychological characteristics to rate of gastric secretion (serum pepsinogen). Psychosom Med 19:1.

WESDORP ICE (1986). Reflux esophagitis: A review. Postgrad Med J 62(suppl 2):43–55.

WHITEHEAD WE, BOSMAJIAN L, ZONDERMAN AB, ET AL. (1988). Symptoms of psychologic distress associated with irritable bowel syndrome. Gastroenterology 95:709–714.

WHITEHEAD WE, RENAULT PF, & GOLDIAMOND I (1975). Modification of human gastric acid secretion with operant conditioning procedures. J Appl Behav Anal 8:147–152.

WHORWELL PJ (1989). Diagnosis and management of irritable bowel syndrome: Discussion paper. J R Soc Med 82:613–614.

WHYBROW PC, KANE TJ, & LIPTON MA (1968). Regional ileitis and psychiatric disorders. Psychosom Med 30:209.

WISE TN (1986). Psychological management of IBS. Pract Gastroenterol 10:40–50.

WISE TN, COOPER JN, & AHMED S (1982). The efficacy of group therapy for patients with irritable bowel syndrome. Psychosomatics 23:465–469.

YOUNG SJ, ALPERS DH, & NORLAND CC (1976). Psychiatric illness and the irritable bowel syndrome. Gastroenterology 70:162–166.

ZISOOK S, & DEVAUL RA (1977). Emotional factors in inflammatory bowel disease. South Med J 70:716–719.

27 Chronic renal failure and its treatment: Dialysis and transplantation

NORMAN B. LEVY, M.D.

Chronic renal failure strikes about 200 of every 1 million people each year, or about 50,000 Americans and 220,000 Chinese. These figures include renal failure caused by systemic illnesses, such as generalized arteriosclerosis and metastatic cancer. The causes of kidney failure include diabetic nephrosclerosis, chronic glomerulonephritis, chronic pyelonephritis, polycystic kidney disease, hypertensive nephropathy, and lupus nephritis. In recent years diabetes has emerged as the most common cause of renal failure in the United States. The two primary technologic treatments for chronic renal failure are kidney transplantation and dialysis. Renal transplantation usually is the treatment of choice for chronic renal failure. If a transplant is successful, the patient's quality of life is always better than it was on dialysis.

Transplanted kidneys may come from a living donor, who is almost always closely related, or from a cadaver. Kidney survival is greater with living related donors, but greater availability makes cadavers the more common source of kidneys (see Table 27-1). Most cadaveric kidneys are "harvested" from the highways as a result of motor vehicle accidents. Nephrectomies are performed on recently brain-dead individuals. There is some debate in the transplant community concerning living unrelated donors, such as the spouse of the patient. Some authorities believe that such donors need not subject themselves to surgery, since cadaveric kidneys survive almost as well. The issue of commercially available organs from individuals who were willing to sell their kidneys brought an outcry, and these sales have been banned by federal statute in the United States and in many other countries.

Immunosuppressants are used postoperatively to inhibit a foreign body reaction by the recipient, which would otherwise result in rejection of the organ. The principal medications used for this are prednisone and azathioprine (Imuran). Cyclosporine has also been shown to offer substantial additional suppression of rejection. A consideration in its use is the cost, about $5,000 per year. However, the first year of cyclosporine therapy is paid for by Medicare, which covers nearly all patients with renal failure in the United States. Subsequently, the cost of cyclosporine is paid by private insurance, Medicaid, or the patients themselves. Newer immunosuppressants include the monoclonal antibody OKT3, antilymphocyte globulin, and FK-506 (Starzl et al., 1990) (see also Chapter 36).

DIALYSIS

Peritoneal dialysis and hemodialysis are the two forms of dialysis treatment. In peritoneal dialysis, dialysate fluid is introduced into and removed from the peritoneal space via an indwelling catheter. The peritoneum serves as a semipermeable membrane through which fluid and wastes pass, which are removed together with dialysate. Peritoneal dialysis may be performed intermittently, by a machine in a hospital, or continuously at home. In continuous ambulatory peritoneal dialysis (CAPD), patients introduce 2 liters of dialysate fluid three to four times a day into their peritoneal space and retrieve the fluid by gravity. Hemodialysis may be performed at the patient's home or at dialysis units in which there may be varying degrees of self-care. Home dialysis requires the participation of a spouse, a parent, another relative, a friend, or a stable paid caregiver, and requires 4 to 5 hours 3 days each week. Newer "high-flux" and "high-efficiency" hemodialysis can now be performed in as little as 2.5 to 3 hours for each treatment. However, these machines are not generally available for home use.

TABLE 27–1. *Renal Graft Survival—United Network for Organ Sharing (UNOS)*[a]

	One Year	Two Years
Cadaveric (first one received)	78%	70%
Living donor	90%	85%[b]
All transplants to diabetics	77%	
All transplants to people with glomerulonephritis	77%	

[a]There was no difference in graft survival in men and women. However, women had better graft outcomes when retransplanted, but the difference was significant only during the first year. The age of the recipient was significant in graft outcome only at its extremes: pediatric versus those over 60.

[b]Estimated.

Adapted from Cecka JM, and Terasaki PI (1990). The UNOS scientific renal transplant registry—1990. In PI Terasaki (Ed.), Clinical transplants (Ch. 1, pp. 1–10). Los Angeles: UCLA Tissue Typing Laboratory.

Stresses of Dialysis

In peritoneal dialysis the patient's abdomen is filled with fluid, which is then permitted to drain. In hemodialysis a "lifeline" connects the patient's blood supply to a dialysis machine, and blood flows freely between the living person and the mechanical device. Both procedures place patients in unusually dependent positions, both on the procedure itself and on the individuals responsible for the procedure (Reichsman & Levy, 1972). Changes in the form of treatment may be made, such as switching from a dialysis center to home dialysis, from peritoneal dialysis to hemodialysis (or vice versa), or from either dialysis procedure to undergoing transplantation.

When a highly independent patient is placed on a form of treatment that involves total dependence, emotional difficulties can ensue (DeNour, Shaltiel & Czaczkes, 1968; Levy, 1976). By contrast, more passive or dependent patients can have emotional problems if pushed into self-care. Only the possibility of transplantation gives renal patients an opportunity to pursue the independent life they had prior to renal failure, and it is the ideal procedure for most individuals since most adults experience a great need for independence. Astute nephrologists assess the character structure of their patients to determine the best form of treatment for each individual (Czaczkes & DeNour, 1978). This aspect of care is discussed in the section "Psychologic Treatment of Dialysis and Transplant Patients".

Patients on dialysis are placed on low-sodium, low-potassium, low-phosphate, low-protein, and low-fluid diets. For those on hemodialysis, the diet means the virtual elimination of all fruits and vegetables; eating of only small amounts of meats, fish, and dairy products; and the cessation of drinking directly from a glass in favor of sucking on cracked ice. Patients on peritoneal dialysis usually are given a more lenient diet, since that treatment is more successful than hemodialysis in removing water and the wastes resulting from eating and drinking. In both treatments, most patients find that their diets place them in situations of deprivation. Some patients may use denial in coping with their dietary regimen by pushing out of mind the necessity of diet and thereby eating freely. Such noncompliant behavior usually is readily evident to dialysis staff, who can easily monitor and measure dietary indiscretion by serum potassium levels and weight changes between dialysis runs.

Although more frequent dialysis runs could permit liberalization of diet, Medicare will only fund three runs weekly. More frequent runs could be considered as an intervention in rare cases in which the patient was unable to adhere to a diet but could pay for additional dialysis sessions.

Another stress of dialysis results from the lack of respite from treatment and illness (Levy, 1976). Patients are constantly reminded of their illness by receiving treatment either continually or three times a week, being on a restrictive diet, and having to take medications. Many other chronic illnesses afford their victims some break from continued awareness of their illness. For example, patients with metastatic cancer may have periods of remission during which they do not require any radiotherapy or chemotherapy. Patients with cardiac conditions often have long periods of being asymptomatic at rest, even if their activity is limited.

Patients on dialysis almost universally experience a diminution in their ability or interest in performing sexually, which is discussed later in this chapter. This sexual dysfunction can be devastating for some patients and is undesirable for most.

About two-thirds of hemodialysis patients who were employed prior to illness no longer work full time (Gutman, Stead & Robinson, 1981). Most choose to have themselves declared to be permanently disabled, even though they may engage in some part-time activity, which may be on or "off the books". Since disability payments always are less than earnings from employment, the advent of chronic disability has negative economic effects. In the face of less income, patients also have expenses for which they may not be compensated by insurance. These include medications, the use of taxis, telephone toll calls, and special diets. While some of these expenses may be paid by Medicaid for those at or near the poverty line, most dialysis patients face some economic hardship.

Peritoneal dialysis has special stresses of its own. This treatment distorts the patient's body by infusing fluid into the abdominal space. Although clinical experience indicates that, in general, patients tolerate such changes in their bodies reasonably well, patients who are greatly invested in their appearance can have difficulty accepting a change in abdominal girth. Patients on peritoneal dialysis also face the pain and inconvenience of recurrent episodes of peritonitis, which require hospitalization at a mean frequency of once every 11 months.

Psychologic Complications of Dialysis

Dialysis patients, like virtually all other chronically ill people, experience anxiety and depression, which are the most common psychologic symptoms seen in the physically ill (Lefebvre, Nobert, & Crombez, 1972). Anxiety may be present during dialysis runs, particularly in hemodialysis. This anxiety occurs because hemodialysis involves the circulation of blood continuously through a machine in a procedure that can produce major medical complications. Although medical problems also can occur in peritoneal dialysis, such as peritonitis, more dramatic complications can occur during hemodialysis, such as stroke and cardiac emergencies.

Early in the history of hemodialysis, patients were dialyzed over periods of 10 hours each run. Patients tended to sleep, or to attempt to sleep, in dialysis units, which were open throughout the night in order to enable patients to pass their time by sleeping. Insomnia resulting from anxiety was an exceedingly common symptom in those days. Anxiety associated with the procedure can also be expressed by masturbatory behavior, more commonly occurring in men but not restricted to them. Such activity is a method of handling anxiety by a sexual outlet, much like the masturbation seen in men about to face combat situations during war (Reichsman & Levy, 1972). Anxiety may also be seen in connection with the uncertainty patients have about the future, fear about their sexual performance, and fear about their ability to cope with the stress of dialysis and the expectations of dialysis staff and family (Levy, 1983).

Depressed mood and the depressive syndromes occur frequently (Sacks, Peterson, & Kimmel, 1990). Since depression often occurs in response to loss, it is understandable that depression is frequent, since dialysis patients have had loss of strength, energy, sexual ability, work ability, physical freedom, and life expectancy. The evaluation of dialysis patients for depression is complicated by the fact that the signs and symptoms of renal failure and its treatment may be identical to the vegetative signs and symptoms of depression, such as diminished appetite, dryness of the mouth, constipation, and diminished sexual interest and ability. The evaluation of depression in medically ill patients is discussed extensively in Chapter 4.

Suicide

Suicide has been reported to be more common among dialysis patients than in the general population and probably is more common than in most other chronic diseases (Abram, Moore, & Westervelt, 1971; Haenel, Brunner, & Battegay, 1980). Unfortunately, there have been no recent studies of suicide in dialysis or transplant patients; improvements in outcomes in the treatment of chronic renal failure may have improved the situation.

The suicide rate in U.S. dialysis patients has been attributed to the "democratization" of dialysis in this country. With the expansion of medical insurance via Medicare since 1973, essentially all patients who need it now receive treatment. These include two groups at relatively higher suicide risk: those with addictive disorders and those with serious systemic medical illnesses.

Optimistically, the shortening of dialysis runs and the use of erythropoietin have improved the quality of life for many dialysis patients. However, dialysis patients have the means for their demise readily at hand. Death may ensue if they sever their fistulas and exsanguinate or go on potassium "binges" and fail to show up for the next two dialysis runs.

Sexual Dysfunction

Observations have been made that most men on dialysis experience impotence. The frequency of impotence is in the range of 70% of adult male patients (Abram et al., 1975; Levy, 1973). There are problems in both sexes concerning sexual desire, and in women there is a marked diminution in orgasm during sexual intercourse. Men and women both have a marked decrease in frequency of sexual intercourse after initiation of dialysis treatment, in comparison with their frequency before becoming uremic.

The cause of these sexual problems is not well understood, but they are probably due to physical illness as well as its psychologic complications. Tests of nocturnal penile tumescence suggest a high prevalence of impaired erectile mechanisms (Procci et al., 1983). The cause of the organic sexual impairment is not fully understood. There is a decrease in testosterone and an increase in other hormones, but not enough to explain the extent of these patients' sexual difficulties (Antoniou & Shalhoub, 1978; Lim, Auletta, & Kathpolia, 1978; Massry et al., 1977). In particular, testosterone replacement usually does not help. A frequent iatrogenic cause of sexual dysfunction relates to the use of antihypertensives that can diminish libido in both men and women and cause impotence in men. Psychologic factors that cause impotence include depression, reversal of family roles, and feelings related to the cessation of urination in male patients (Levy, 1983).

Most patients on dialysis have total cessation of urination. Since the organ of urination in men is the same as that of sexuality, there may be major psychologic ramifications to this most-often-used function of this organ completely ceasing (DeNour, 1969). Here, too, in the more tenuously masculine individual, such a change may be seen as a major blow to the male identity and may affect sexual function (see also Chapter 15).

The frequent work disability of male patients often forces the spouse either to return to work or to increase her outside work activity. This places the male patient in a situation of having to participate to a greater extent in household activities such as shopping, cleaning, cooking, and caring for the children. For many patients this is quite acceptable. For others, however, especially those who are insecure about their masculinity, such a reversal of family roles may lead to feelings of inadequacy and emasculation (Reiss, 1990).

Cognitive Impairment Disorders

Cognitive impairment disorders occur in a variety of situations. Patients involved in intellectual work activity will tend to note progressive impairment as their day for dialysis approaches. After dialysis they often experience a short period of delirium termed "dysequilibrium syndrome," caused by the rapid change of fluid and electrolytes that has taken place during the dialysis run. This syndrome may last from minutes to hours. The presence of a transient metabolic encephalopathy can be confirmed by an electroencephalogram, especially if a premorbid tracing is available for comparison; the typical finding is diffuse slowing. Computed tomography and magnetic resonance imaging do not help confirm a dialysis-related delirium but may be helpful in screening for opportunistic central nervous system infection or cerebrovascular events in patients with *persistent* changes in neurologic function. The dysequilibrium syndrome is not seen in CAPD because this procedure is slow and continuous, without sudden fluxes of fluid and electrolytes.

Among the most serious cognitive disorders seen in dialysis patients is dialysis encephalopathy. This is an often fatal neurologic syndrome that may occur in patients who have been on hemodialysis for at least 2 years. The early signs are dysarthria, stuttering speech, memory impairment, depression, and psychosis. These patients often develop bizarre limb movements, asterixis, and generalized tremulousness. The disease is progressive and leads to a neurologic death if not successfully treated. Its cause is not fully understood. Aluminum, from trace amounts in dialysate water and from phosphate binding gels, is the most likely culprit (Alfrey, 1986). There are conflicting data, however, concerning aluminum toxicity, since aluminum tends to be present in higher-than-normal quantities in the brain in other types of dementia, including Alzheimer disease. Trace amounts of tin and zinc have also been blamed as a cause of this syndrome. In dialysis "dementia" the EEG shows typical slow-wave bursts in the early stages (Alfrey, 1986).

Treatment of dialysis encephalopathy may include chelation therapy with deferoxamine, reduction in dose of aluminum-containing phosphate binding gels, and use of deionized water (free of aluminum traces) to prepare dialysis solutions. Early cases may be reversible with this treatment (Alfrey, 1986; O'Hare, Callaghan, & Murnaghan, 1983). The psychiatric treatment should address the symptomatic manifestations. Antidepressant and antipsychotic medications should therefore be used when depressive or psychotic syndromes appear. Dialysis dementia has markedly decreased in incidence since aluminum-free dialysates have been in general use.

Treatment of Dialysis-Induced Anemia

In the past 2 years recombinant human erythropoietin (EPO) has become available and is now widely used to ameliorate the severe anemia seen in patients undergoing dialysis. With relief of their anemia, patients have experienced improvement in a wide variety of life functions and in the quality of their lives (Evans et al., 1990; Wolcott et al., 1990). Treatment with EPO is indicated for all dialysis patients except those with polycythemia and those with uncontrolled hypertension, in whom the lower blood viscosity caused by anemia may protect against stroke.

RENAL TRANSPLANTATION

Renal transplantation is the preferred ultimate treatment for many if not most patients. In addition, many nephrologists believe that diabetics are better treated with transplantation than with dialysis, although the transplant procedure itself is associated with a greater morbidity and mortality in diabetic than nondiabetic patients. Children, who compose less than 1% of the total number of people with renal failure, tend to be better off developmentally with transplantation. Patients who have an identical twin who is in good health are most assured of success in receiving the donated kidney from their twin sibling, without the use of either steroids of cyclosporine. The factor that interferes with the delivery of renal transplantation is the short supply of available kidneys, both live and cadaveric (Fox & Swazey, 1974).

The Donor

The selection of the donor is often a very difficult matter for the patient with renal failure to handle. The patient expects that people will come forth by themselves, certainly without any personal coaxing. Fortunately, this is usually the case, and it is uncommon for patients with renal failure to confront relatives with their need for the other's kidney. Such a request is usually made by an intermediary person, such as a family member or a physician. Mothers often want to donate their kidneys to their children, to enable the child to have an opportunity for a new life. At times the "black sheep," usually a sibling of the patient, is selected by the family to "undo" past grievances by making the sacrifice of undergoing surgery and kidney donation (Fellner & Marshall, 1968).

The decision to donate a kidney is often a spontaneous one, seemingly without much thought or deliberation as to its consequences, inconvenience, pain, or complications. There has been some debate among renal specialists about the long-term complications of a kidney donation (Simmons, 1981; Simmons, Klein, & Simmons, 1977). There is no evidence, however, that having only one kidney places the donor in greater danger of either hypertension or kidney failure.

Preoperatively, the donor is in the situation of being near center stage, as a person whose generosity may make it possible for another to lead a new life. Postoperatively, the donor leaves center stage and is replaced by the recipient. Some donors experience a period of "blues" as they perceive what they think is ingratitude in the recipients' preoccupation with their own lives and with the survival of the new kidney. In a follow-up of 130 related donors, it was reported that, in addition to rapidly making the decision to donate, the great majority had no difficulty in coming to that decision and expressed gratification in being donors (Simmons, 1981). Twenty percent of recipients, however, experience unhappiness at never being able to repay the donor in an equivalent manner. Simmons stated that male donors, especially if married, tend to be somewhat dissatisfied about having made a donation. The author has not personally encountered this reaction. If the kidney is viable after a year, there tends to be less dissatisfaction with donation. Remarkably, only 18% of donors of rejected kidneys expressed dissatisfaction at having given their kidney for transplantation (Simmons, 1981).

In the case of a cadaveric kidney, matters are a good deal different. As previously mentioned, cadaver kidneys are harvested largely on the highways from young, brain dead adults. The recipient generally is not told much about the donor except the age and sex. Some recipients become very curious about the identity of the donor and may pursue the matter by reading old newspaper articles about automobile accidents and interviewing nurses and other medical people involved in their receiving the kidney. At times recipients experience survivor guilt at having reaped a reward out of the misfortune of another. Psychologic reactions to transplantation are also discussed in Chapter 36.

Organ Rejection or Acceptance

A number of investigators have examined the idea that psychologic factors may play a role in either acceptance or rejection of donated organs (Basch, 1973; Muslin, 1971). It has been reported (Viederman, 1974) that one patient rejected an organ probably as a result of having "given up" psychologically. In another study (Eisendrath, 1967), 8 of 11 patients who died following transplant surgery experienced a sense of pessimism and panic about their transplant to a degree not observed in the surviving recipients. Steinberg, Levy, and Radvila (1981) attempted to test the notion that there is a connection between psychologic acceptance or rejection of the organ. The absence of ambivalence toward the impending surgery and about the donor, realistic expectations concerning the future, reasonable optimism concerning the upcoming operative procedure, and the relative lack of serious mental symptoms were hypothesized to favor acceptance of the donated organ. Based on this system, statistical significance was *not* shown between the predicted outcome and actual outcome at 1-year follow-up.

Postoperative Psychologic Problems

Recipients of kidney transplants have a "sword of Damocles" hanging over them in that they live in the shadow of possible organ rejection. The natural forces of the body work to reject foreign substances. The body responds with great vigor to "protect" itself against the transplanted kidney. As time goes on and rejection is not experienced, the recipient feels relatively relieved. The recipient is never, however, completely safe from the possibility of rejection. Also, many diseases causing renal failure can recur in the transplanted kidney, causing another risk to the success of the transplant.

The immunosuppressant medications may also cause difficulty. This is particularly true of prednisone, which must be taken for the rest of the patient's life, initially and at times of rejections in large doses and later in relatively small ones. The numerous medical and psychiatric complications of prednisone therapy are well known (see Chapter 29). Patients with a personal history of depression or with dysthymic symptoms are more vulnerable to developing a major depressive disorder while receiving prednisone (Levy, 1986). Patients with a history of psychosis or with borderline personalities are more vulnerable to psychosis with prednisone. These psychiatric complications of steroids should be treated symptomatically. Antidepressant, antipsychotic, and antimanic medications should be given if symptoms warrant them (Levy, 1985).

Patients who develop severe psychiatric side effects from steroids are candidates for immunosuppression with cyclosporine, because cyclosporine-treated patients require less prednisone and azathioprine (Canadian Multicenter Transplant Study Group, 1986). In general, cyclosporine deserves consideration in transplant recipients with severe steroid side effects of any kind (Amend, Suthanthiram, & Gambertoglio, 1986). However, cyclosporine has been associated with delirium, seizures, and tremors (Canadian Multicenter Transplant Study Group, 1986). Anxiety and depression can also rarely occur with cyclosporine. The clinical manifestations and management of cyclosporine toxicity are discussed further in Chapter 36.

Patients on prednisone and other immunosuppressants after transplantation are vulnerable to opportunistic infections and may have bacterial infections with fewer systemic signs than usual. New psychiatric symptoms in transplant recipients, if accompanied by cognitive deficits, indicate the need for careful screening for infectious diseases. The psychiatrist can play a critical role in initiating the delirium workup by identifying the patient's mental symptoms as being of probable organic origin. When the mental status examination leaves the issue of organic etiology unclear, an EEG may help (see Chapters 18 and 19).

Sexual Functioning

The sexual function of patients undergoing renal transplantation is impaired in comparison with function prior to kidney failure, but is usually much better than it was while the patients were on dialysis (Levy, 1973; Salvatierra, Fortmann, & Belzer, 1975). Since posttransplant kidney function may vary, especially in cases of rejection, sexual function may be dependent on the degree of renal function of the individual. In one study, 46% of 56 male transplant patients were either partially or totally impotent (Levy, 1973). The cause of sexual dysfunction in these individuals is not clearly understood, but physiologic factors seem to be greatly implicated because of hormonal and other changes in renal failure (Procci et al., 1983). The more successful the transplant is in normalizing renal function, the less likely it is that there will be sexual dysfunction.

PSYCHOLOGIC TREATMENT OF DIALYSIS AND TRANSPLANT PATIENTS

An evaluation of the personality and family structure of patients can result in appropriately tailoring the medical treatment to psychologic needs. The highly independent patient should be treated by a self-care form of dialysis such as home hemodialysis or CAPD or should receive transplantation (Levy, 1980). Profoundly narcissistic patients who are preoccupied with body shape and appearance probably should not receive peritoneal dialysis. The patient who is phobic of blood may have major difficulties tolerating hemodialysis and may require behavioral therapy. An assistant, usually from the family, plays an essential role in home hemodialysis. In order for such treatment to take place, there needs to be sufficient physical space for the machine at the patient's home and a consistent supportive "significant other" who is relatively unambivalent about helping with the treatment.

Patients with renal failure should be told that they may develop such problems as depression or sexual dysfunction. If prepared, the patient, in the face of the presentation of these problems, will tend to see them as complications of treatment and will more readily call them to professional attention (Levy, 1984).

A factor of importance in any therapy of these patients is that they feel very much "overdoctored," spending many hours a week in the dialysis procedure. Any form of psychotherapy will have greater success if it takes place in conjunction with clinic visits or dialysis runs (Freyberger, 1983). Psychotherapists may therefore find themselves conducting sessions in a unit in which there is no sound barrier between the patient and others.

Treatment of Sexual Dysfunction

In the case of sexual dysfunction, careful consideration should be given to the use of behavioral sexual techniques, as described in Chapter 15. Such techniques are based on the fact that patients who have sexual dysfunctions tend to withdraw altogether from sexual situations and even avoid nonsexual intimacy with their partners (McKevitt, 1976). This therapy encourages a closeness of partners and attempts to restore sexuality (Berkman, 1978). Restoration is accomplished by assuring the partners that the goal of sexuality is not necessarily intercourse but physical and emotional intimacy.

Consideration should also be given to the surgical treatment of intractable impotence by a penile prosthesis. The two systems used are either an implantation of a Silastic rod in the body of the penis or a hydraulic system in which the patient, by squeezing a mechanism implanted at the base of his testicles, pumps fluid from an abdominal reservoir into two collapsed sacs implanted into both sides of the cavernous portion of the penis, thereby creating an erection. Since both techniques involve the destruction of erectile tissue, surgeons performing these procedures need to establish the absence of nocturnal penile tumescence prior to engaging in surgery (see Chapters 15 and 39).

There has been no systematic study of dialysis patients who have received these procedures. The clinical experience of this author has been quite favorable. In general, dialysis patients tend not to be adequately informed by their medical professionals about this option. Nevertheless, by self-selection those who would profit greatly by such a procedure seem to be the people who have had it performed.

Psychotropics in Renal Failure

Pharmacokinetics refers to the absorption, distribution, degradation, and excretion of drugs and their metabolites (Bennett et al., 1980). Several of the factors constituting the pharmacokinetics of psychotropic medications can be altered in renal failure (Levy, 1990–1991). Although not all psychotropics have been adequately evaluated in patients with kidney failure, it is known that many medications have diminished absorption from the small bowel because of excess gastric alkalinization seen in kidney failure. Drug distribution can be affected by ascites and edema, which can increase the apparent volume of distribution, requiring higher doses of medication. In the case of dehydration and muscle wasting, a lower dosage is needed because of a decreased volume of distribution. Most medications bind to albumin and other plasma proteins, leaving the unbound drug as the active substance. In renal failure, perhaps because of diminished albumin concentrations, there is decreased protein binding of medications. This is particularly significant because almost all psychotropics have high protein binding affinity. This means that, in renal failure, a higher proportion of a given dose of medicine will be unbound and available for either efficacy or toxicity.

Fortunately, the metabolism and excretion of most psychotropic medications is not seriously affected by renal failure. Most psychotropic medications are fat soluble, pass the blood-brain barrier, are highly protein bound, and are detoxified by the liver and eventually eliminated in the bile fluid. There are notable exceptions (see Table 27-2). These include lithium, which is entirely excreted by the kidney and not changed by the liver. However, because of its small molecular size it is completely dialyzed. In part because of the protein binding diminution in kidney failure, the general rule is to give a renal failure patient not more than two thirds of the maximum dose one would give an individual with normal renal function (Brater, 1983).

Antidepressants

Virtually all the antidepressants may be used in renal failure. The greatest body of experience is with the tricyclics. Because of the many complicating factors affecting the blood level of these medications, nortrip-

TABLE 27–2. *Pharmacokinetics of Commonly Prescribed Psychotropics*

Drug	Plasma Protein Binding (Percent)	Excretion	Half-Life in Hours (Normal/Renal Failure)
Anti-anxiety			
Alprazolam	70–80	Hepatic	10–19/?
Buspirone	?	Hepatic	2–5/?
Chlordiazepoxide	47	Hepatic[a]	5–30/?
Diazepam	94–98	Hepatic[a]	20–90/?
Lorazepam	90	Hepatic	9–16/32–70
Meprobamate	0–20	Hepatic[b]	6–17/16–17
Oxazepam	97	Hepatic	6–25/25–90
Antidepressants			
Amitriptyline	96	Hepatic	32–40/?
Desipramine	90	Hepatic	12–54/?
Doxepin	93–95	Hepatic	8–25/10–30
Fluoxetine	94.5	Hepatic	1–10/?
Imipramine	96	Hepatic	6–20/?
Maprotiline	?	Hepatic	48/?
Nortriptyline	95	Hepatic	18–93/15–66
Antipsychotics			
Chlorpromazine	91–99	Hepatic	11–42/11–42
Haloperidol	90–92	Hepatic[b]	10–36/?
Lithium			
Lithium Carbonate	0	Renal	14–28/?

[a]Converted to active metabolite *N*-desmethydiazepam, excreted by the kidney.
[b]A minor amount is excreted by the kidneys.
Adapted from Bennett et al., (1987). Drug prescribing in renal failure: Dosing guidelines for adults. Philadelphia: American College of Physicians.

tyline is particularly useful because of its meaningful blood levels. The "therapeutic window" for nortriptyline is the same in renal failure patients as it is in medically well patients. Fluoxetine has been shown to be both nontoxic and efficacious in renal patients (Bergstrom et al., 1991). It also has the added attractions of not affecting heart conduction, having a very low anticholinergic activity, and having a very high therapeutic-to-toxic dose ratio, making it most difficult for fluoxetine be used as a means of suicide in a population with a relatively high suicide risk. Sertraline would be predicted to be equally benign in this population.

Recently, attention has been focused on the hydroxylated metabolites of the tricyclics. In patients with normal renal function, it is believed that they contribute to both the therapeutic and toxic effects of these medications (Nelson, Atillasoy, & Mazure, 1988; Nordin, Bertilsson, & Siwers, 1987). Renal failure patients may be more sensitive to tricyclics than those with normal kidney function because of these metabolites. The use of psychotropics in patients with renal failure is discussed in more detail in Chapter 9.

Lithium

Because it is a small molecule that is readily dialyzed, lithium is eliminated in the process of dialysis (Port, Kroll, & Rosenzweig, 1979). It is entirely excreted by the kidney and therefore retained in anuric patients after a single dose (Procci, 1977). Lithium may be given in a single dose of generally 600 mg after each dialysis run. In the process of dialysis, it can be entirely eliminated from the body. Lithium levels before and after dialysis are used to establish the proper dose (Das Gupta & Jefferson, 1990; Lippmann, Manshadi, & Gultekin, 1984; Stoudemire, Fogel, & Gulley, 1991).

REFERENCES

ABRAM HS, HESTER LR, EPSTEIN GM, ET AL. (1975). Sexual functioning in patients with chronic renal failure. J Nerv Ment Dis 166:220–226.

ABRAM HS, MOORE GL, & WESTERVELT FB JR. (1971). Suicidal behavior in chronic dialysis patients. Am J Psychiatry 127:1199–1204.

ALFREY A (1986). Dialysis encephalopathy. Kid Int 29(suppl):S53–S57.

AMEND WJC, SUTHANTHIRAM M, & GAMBERTOGLIO JG (1986). Immunosuppression following renal transplantation. In MR GOROVY & RD GUTTMAN (Eds.), Renal transplantation (Ch. 3, pp. 73–92). New York: Churchill Livingstone.

ANTONIOU LD, & SHALHOUB RJ (1978). Zinc in the treatment of impotence in chronic renal failure. Dial Transplant 7:912–915.

BASCH SH (1973). The intrapsychic integration of a new organ: A clinical study of kidney transplantation. Psychoanal Q 42:364–384.

BENNETT WM, MUTHER RS, PARKER RA, ET AL. (1980). Drug therapy in renal failure: Dosing guidelines for adults: 2. Sedatives, hypnotics and tranquilizers; cardiovascular, antihypertensive and diuretic agents; miscellaneous agents. Ann Intern Med 93:286–325.

BERGSTROM RF, BEASLEY CM JR, LEVY NB, ET AL. (1991). Fluoxetine pharmacokinetics after daily doses of 20 mg fluoxetine in patients with severely impaired renal function. Pharm Res. In press.

BERKMAN A (1978). Sex counseling with hemodialysis patients. Dial Transplant 7:924.

BRATER DC (1983). Drug use in renal disease. Sydney, Australia: Aidis Health Science Press.

CANADIAN MULTICENTER TRANSPLANT STUDY GROUP (1986). A randomized clinical trial of cyclosporine in cadaveric renal transplantation. N Engl J Med 314:1219–1225.

CZACZKES JW, & DENOUR AK (1978). Chronic hemodialysis as a way of life. New York: Brunner/Mazel.

DAS GUPTA K, & JEFFERSON JW (1990). The use of lithium in the medically ill. Gen Hosp Psychiatry 12:83–97.

DENOUR AK (1969). Some notes on the psychological significance of urination. J Nerv Ment Dis 148:615–623.

DENOUR AK, SHALTIEL J, & CZACZKEKS JW (1968). Emotional reactions of patients on chronic hemodialysis. Psychosom Med 30:521–533.

EISENDRATH RM (1967). The role of grief and fear in the death of transplant patients. Am J Psychiatry 126:381–387.

EVANS RW, RADER B, MANNINEN DL, & THE COOPERATIVE MULTICENTER EPO CLINICAL TRIAL GROUP (1990). The quality of life of hemodialysis recipients treated with recombinant human erythropoietin. JAMA 263:825–830.

FELLNER CH, & MARSHALL JR (1968). Twelve kidney donors. JAMA 206:2703–2707.

FOX RC, & SWAZEY JP (1974). The courage to fail: A social view of organ transplants and dialysis. Chicago: University of Chicago Press.

FREYBERGER H (1983). The renal transplant patients: Three-stage model and psychotherapeutic strategies. In NB LEVY (Ed.), Psychonephrology 2: Psychological problems in kidney failure and their treatment (Ch. 22, pp. 259–265). New York: Plenum.

GUTMAN RA, STEAD W, & ROBINSON RR (1981). Physical activity and employment status of patients on maintenance dialysis. N Engl J Med 304:309–313.

HAENEL T, BRUNNER F, & BATTEGAY R (1980). Renal dialysis and suicide: Occurrence in Switzerland and Europe. Compr Psychiatry 21:140–145.

LEFEBVRE P, NOBERT A, & CROMBEZ JC (1972). Psychological and psychopathological reactions in relation to chronic hemodialysis. Can Psychiatr Assoc J 17:9–13.

LEVY NB (1973). Sexual adjustment to maintenance hemodialysis and transplantation: National survey by questionnaire. Trans ASAIO 19:138–142.

LEVY NB (1976). Coping with maintenance hemodialysis—psychological considerations in the care of patients. In SG MASSRY & AL SELLERS (Eds.), Clinical aspects of uremia and dialysis (Ch. 3, pp. 53–68). Springfield, IL: Charles C Thomas.

LEVY NB (1980). The "uncooperative" patient with ESRD, causes and treatment. Proc Eur Dial Transplant Assoc 17:523–530.

LEVY NB (1983). Sexual dysfunctions of hemodialysis. Clin Exp Dial Apheresis 7:275–288.

LEVY NB (1984). Psychological complications of dialysis: Psychonephrology to the rescue. Bull Menninger Clin 48:237–250.

LEVY NB (1985). Use of psychotropics in patients with kidney failure. Psychosomatics 26:669–709.

LEVY NB (1986). Renal transplantation and the new medical era. Adv Psychosom Med 15:167–179.

LEVY NB (1990–1991). Psychopharmacology in patients with renal failure. Int J Psychiatry Med 20:303–312.

LIM VS, AULETTA F, & KATHPOLIA S (1978). Gonadal dysfunction in chronic renal failure: An endocrinological review. Dial Transplant 7:896–907.

LIPPMANN SB, MANSHADI MS, & GULTEKIN A (1984). Lithium in a patient with renal failure on hemodialysis [letter]. J Clin Psychiatry 45:444.

MASSRY SG, GOLDSTEIN DA, PROCCI WR, ET AL. (1977). Impotence in patients with uremia: A possible role for parathyroid hormone. Nephron 19:305–310.

MCKEVITT PM (1976). Treating sexual dysfunction in dialysis and transplant patients. Health Soc Work 1:133–157.

MUSLIN HL (1971). On acquiring a kidney. Am J Psychiatry 127:1185–1188.

NELSON JC, ATILLASOY E, & MAZURE C (1988). Hydroxydesipramine in the elderly. J Clin Psychopharmacol 8:428–433.

NORDIN C, BERTILSSON L, & SIWERS L (1987). Clinical and biochemical effects during treatment of depression with nortriptyline: The role of 10-hydroxynortriptyline. Clin Pharmacol Ther 42:10–19.

O'HARE JA, CALLAGHAN NM, & MURNAGHAN DJ (1983). Dialysis encephalopathy. Medicine 62:129–141.

PORT FK, KROLL PD, & ROSENZWEIG J (1979). Lithium therapy during maintenance hemodialysis. Psychosomatics 20:130–132.

PROCCI WR (1977). Mania during maintenance hemodialysis successfully treated with oral lithium carbonate. J Nerv Ment Dis 164:355–358.

PROCCI WR, GOLDSTEIN DA, KLETZKY OA, ET AL. (1983). Impotence in uremia: Preliminary results of a combined medical and psychiatric investigation. In NB LEVY (Ed.), Psychonephrology 2: Psychological problems in kidney failure and their treatment (Ch. 20, pp. 235–246). New York: Plenum.

REICHSMAN F, & LEVY NB (1972). Problems in adaptation to maintenance hemodialysis: A four-year study of 25 patients. Arch Intern Med 130:850–865.

REISS D (1990). Patient, family and staff responses to end-stage renal disease. Am J Kidney Dis 15:194–200.

SACKS C, PETERSON RA, & KIMMEL PL (1990). Perception of illness and depression in chronic renal disease. Am J Kidney Dis 15:31–39.

SALVATIERRA O, FORTMANN JL, & BELZER FO (1975). Sexual function in males before and after transplantation. Urology 5:64–66.

SANTOS AB, BELILES KE, & ARANA GW (1993). Parenteral use of psychotropic agents: Clinical applications of basic pharmacokinetic principles. In A STOUDEMIRE & B FOGEL (Eds.), Medical Psychiatric Practice (Vol. 2; Ch. 5). Washington DC, American Psychiatric Press, Inc. (in press).

SIMMONS RG (1981). Psychological reactions to giving a kidney. In NB LEVY (Ed.), Psychonephrology 1: Psychological factors in hemodialysis and transplantation (Ch. 18, pp. 227–245). New York: Plenum.

SIMMONS RG, KLEIN SD, & SIMMONS RL (1977). Gift of life: The social and psychological impact of organ transplantation. New York: Wiley Interscience.

STARZL TE, FUNG J, JORDON M, ET AL. (1990). Kidney transplantation under FK506. JAMA 264:63–67.

STEINBERG J, LEVY NB, & RADVILA A (1981). Psychological factors affecting acceptance or rejection of kidney transplants. In NB LEVY (Ed.), Psychonephrology 1: Psychological factors in hemodialysis and transplantation (Ch. 18, pp. 185–193). New York: Plenum.

STOUDEMIRE A, FOGEL BS, & GULLEY L (1991). Psychopharmacology in the medically ill: An update. In A STOUDEMIRE & B FOGEL (Eds.), Medical psychiatric practice (Vol. 1; Ch. 2, pp. 29–97). Washington, DC: American Psychiatric Press, Inc.

VIEDERMAN M (1974). The search for meaning in renal transplantation. Psychiatry 37:283–290.

WOLCOTT DL, MARSH JT, LA RUE A, ET AL. (1990). Recombinant human erythropoietin treatment may improve quality of life and cognitive function in chronic hemodialysis patients. Am J Kidney Dis 14:478–485.

28 Obstetrics and gynecology

JANICE PETERSEN, M.D.

Women turn to their physicians for care of a variety of obstetric and gynecologic problems. The reproductive system, so intimately related to a woman's sense of identity, is highly invested emotionally. The concerned physician, aware of the psychologic implications of obstetric and gynecologic conditions, seeks to treat these problems as well as the physical symptoms of the patient. With more distressed or problematic patients, the physician or patient may pursue psychiatric consultation for diagnostic and treatment recommendations. This chapter describes psychiatric aspects of common obstetric and gynecologic disorders, as well as psychiatric treatment approaches.

INFERTILITY

Infertility can be a devastating condition for the 15% of American couples it affects. Strictly defined, infertility is usually considered a lack of conception after 1 year of unprotected intercourse. Eighty percent of couples achieve pregnancy after 1 year, and an additional 5 to 10% after 2 years (Behrman & Kistner, 1975). Of those who seek treatment, only 40% can be helped by current treatment methods (Collins et al., 1983).

The ability to bear children is an unquestioned expectation of most couples. When this highly valued experience is disrupted by infertility, a major narcissistic injury and sense of loss occur. Reactions are complex and diverse; they include shock, denial, grief, guilt, anger, anxiety, depression, and emotional hypersensitivity. The individual's sense of identity may be shaken. For some, a sense of defectiveness and guilt about past experiences that may have contributed to infertility may be overwhelming. Others may project the blame onto the partner. The lack of control over such a basic life function is highly distressing (Stotland & Smith, 1990).

As the couple realize that they are having difficulty becoming pregnant, they may feel increasingly pressured to perform sexually at the time of ovulation. What formerly was an enjoyable activity becomes a necessary chore; sexual dysfunction may eventually develop and compound the narcissistic injury. If the couple seeks treatment, they must tolerate the physician's inquiry and intervention into ordinarily private experiences. Treatment for the woman may involve daily basal body temperatures, blood or urine samples for luteinizing hormone (LH) determinations, oral or intramuscular medications, frequent pelvic examinations, ultrasound examinations, and laparoscopy or laparotomy with attendant anesthesia; the man's role in treatment may involve urologic evaluation and masturbation on demand for a sperm sample for artificial insemination or in vitro fertilization. Couples with male infertility must deal with the issues surrounding artificial insemination by a donor. Infertility treatment is expensive, physically taxing, time-consuming, and often depersonalizing. When the woman's menstrual period comes at the end of the month despite these efforts, the couple experiences a sense of loss and failure.

Women commonly are more distressed by infertility than their spouses, and this difference contributes to difficulties for each partner in understanding the other's reaction. If one member of the pair has been diagnosed as having the problem, that partner may feel particularly injured and guilty, and the other may feel deprived and angry. Their families and friends may have little understanding of what they are experiencing and may make well-meaning but unempathic suggestions. The couple may have difficulty associating with others who are pregnant or who have children. The infertile couple often feel isolated in its distress. It is not surprising that many such couples experience significant discord and some divorce as a result of infertility.

Some contributions to the literature have addressed the issue whether personality factors contribute to infertility. These studies are flawed by lack of adequate control groups and failure to determine whether personality findings were present previously or arose after the infertility emerged. Psychologic causes of infertility, such as psychogenic amenorrhea, nonconsummation, vaginismus, psychogenic impotence, and avoidance of intercourse, account for fewer than 5% of infertility cases (Seibel & Taymor, 1982). In gynecologic practice, the most common psychological management problem is the couple's grief reaction as they face the potential loss of a life goal and try to cope with a stressful treatment plan.

Evaluation of infertility requires a detailed history and examination of both members of the couple. For the psychologic evaluation, the psychiatrist should determine why the couple has presented for treatment now, who is more concerned, what life changes have occurred in anticipation of pregnancy, how the problem has affected the marital relationship, how they have coped with it, and what they will do if they do not become pregnant (Downing, 1987). Major psychiatric disturbances must be identified and addressed, particularly depression, which affects approximately 6% of women of child-bearing age (Myers et al., 1984) and which may be more prevalent in women with infertility.

When dealing with infertility, the psychiatrist should encourage both members of the couple to be involved in the consultation, as infertility is a problem for both of them. The gynecologic and urologic evaluation processes, diagnoses, treatments, and options should be reviewed. If the couple is willing to continue treatment but is having difficulty coping with it, they may need assistance negotiating with the gynecologist for a treatment plan that gives them more control, or they may simply need permission to "take a break." If the couple's main difficulty is conflict between the spouses, couples therapy may be needed to clarify the issues, each member's contributions to them, and their options for responding to them. If the couple's main difficulty is disagreement over treatment options (e.g., whether to pursue artificial insemination by donor), these issues can be explored and worked through, and a plan can be negotiated. For the couple who is considering termination of treatment for infertility, basic grief work may be needed to assist them in coming to terms with their loss and considering their other options: adoption or child-free living.

Another approach that can be helpful is participation in groups such as Resolve, a nation-wide infertility peer support group. Such groups supply information and a chance to talk with other people who have similar experiences. For selected individuals this form of assistance may be the most appropriate.

PSYCHOLOGIC REACTIONS TO STERILIZATION

Although sterilization by tubal ligation is a generally well accepted contraceptive choice, many women report feelings of sadness about the loss of their reproductive potential. It is usually of mild degree, however, and subsides with time (Wilcox et al., 1991). There are 2 to 7% of women with tubal ligations who go on to experience more severe regret that can lead them to seek reversal of the procedure. The main risk factor for such regret is age less than 30 years at the time of the procedure. Other factors associated with increased rate of regret are marital discord and combination of the sterilization procedure with other surgery, such as cesarean section. The most common reason for request for reversal of sterilization is the desire to bear children in a new relationship (Thomson & Templeton, 1978). For couples who unsuccessfully seek reversal of sterilization, the disappointment can be significant, and they face many of the issues described above in the infertility section.

REPRODUCTIVE ENDOCRINE DISORDERS

Polycystic ovary disease (PCOD) is a common disorder characterized by polycystic ovaries, hirsutism, acne, obesity, menstrual irregularity, and infertility. Some authors (Orenstein & Raskind, 1983; Orenstein et al., 1986) have associated this disorder with Briquet's or somatization disorder. Thus patients with multiple unexplained somatic complaints who demonstrate features of PCOD should be evaluated further. Hirsutism must be specifically asked about or observed on physical examination. Useful laboratory indicators include elevated LH, decreased follicle-stimulating hormone (FSH), and increased plasma and free testosterone or urinary 17-ketosteroids (Gold & Josimovich, 1987). Pelvic ultrasonography demonstrates ovarian cysts. Treatment for PCOD can involve oral contraceptives, glucocorticoids or antiandrogen agents, all of which suppress the excessive androgen activity associated with this disease.

Hyperprolactinemia is a condition wherein excess prolactin is secreted by the pituitary gland. Whereas this condition may be asymptomatic, it can also cause galactorrhea, menstrual irregularity, amenorrhea, and infertility. Some authors have associated hyperprolactemia with hostility, anxiety, and depression (Kellner et al., 1984; Thienhaus & Hartford, 1986). Patients with symptomatic anxiety or depression plus any of the typical symptoms of hyperprolactemia should be further evaluated. Galactorrhea is a symptom that must be specifically asked about, as the patient may not spontaneously report it. An elevated prolactin level is diagnostic. Of special concern is the possibility of pituitary tumors, which should be ruled out by imaging in cases of prolactin levels above 50 ng/ml. Bromocriptine, a dopamine antagonist, has been shown to be effective in lowering prolactin levels (Gold & Josimovich, 1987). (For further details see Chapter 29.)

PREMENSTRUAL SYNDROME

Premenstrual syndrome (PMS) is comprised of affective, behavioral, and physical symptoms that occur prior to menses and are significantly reduced with the onset of menses or shortly thereafter. Myriad symptoms have been attributed to PMS. Psychological symptoms include depression, anxiety, fatigue, irritability, and food cravings; somatic symptoms are abdominal pain, breast tenderness, bloating, weight gain, and edema (Hamilton et al., 1984). In order to qualify for a diagnosis of PMS, the symptoms must be charted prospectively for 2 to 3 months. The constellation of symptoms varies from patient to patient but generally occurs with consistent timing from cycle to cycle. Approximately 90% of women experience some premenstrual symptoms, but fewer than 4% have symptoms that are rated severe (Johnson et al., 1988). The incidence of PMS is thought to increase with age, having its peak in the 35- to 45-year-old age group (Andersch et al., 1986).

The American Psychiatric Association in 1987 proposed a more narrow diagnostic category—late luteal phase dysphoric disorder (LLPDD)—which was described in the appendix of the *DSM-III-R* (American Psychiatric Association, 1987). This diagnosis involved similar timing of the symptoms before and resolving after menstruation. LLPDD requires five of the following symptoms to be present: lability, irritability, anxiety, depressed mood, decreased interest in usual activities, decreased energy, difficulty concentrating, change in appetite, insomnia or hypersomnia, or physical symptoms such as breast tenderness, headaches, joint pain, bloating, or weight gain. In addition, the symptoms must interfere with functioning, not be merely an exacerbation of another psychiatric disorder, and be measured prospectively by daily ratings. Although the diagnosis of LLPDD is useful for research on the affective symptoms of PMS, most patients and primary care clinicians continue to use the more general term, PMS, when working with patients with cyclic menstrual symptoms. The term LLPDD will undergo significant revision in DSM-IV.

Because the timing of the symptoms is essential for making the diagnosis of PMS, daily prospective charting is necessary. It can be accomplished by having the patient daily note the severity of her symptoms (none, mild, moderate, or severe) on a calendar that the clinician can then examine at subsequent visits. Charting in itself can increase the patient's self-awareness and thus be therapeutic.

When evaluating the complaint of PMS, the clinician must distinguish between patients who simply believe they have this problem from those who truly meet the criteria. Several university-affiliated PMS clinic studies indicate that at least half of women presenting with the complaint of "PMS" are suffering from a *continuous* mental disorder (Dejong et al., 1985; Severino & Moline, 1989, p. 85; Steege, Stout, & Rupp, 1988). These women most commonly suffer from premenstrually exacerbated depression. Other psychiatric disorders that can present with a complaint of PMS include dysthymia, bipolar disorder, cyclothymia, personality disorder, and adjustment disorder. These patients may be mistakenly associating their mood symptoms with menstrual cycle or they may be focusing on premenstrual exacerbations of their ongoing disorder. Identification of the underlying psychiatric disorder allows appropriate treatment for that disorder.

The etiology of premenstrual syndrome has been the subject of intense interest. Previous theories of hormonal imbalance have not held up to scrutiny; in particular, there is no difference between estrogen or progesterone levels in rigorously defined PMS patients and control groups (Rubinow et al., 1988). The possibility of a predisposing biologic vulnerability is supported by an increased rate of previous and lifetime affective disorders in PMS patients compared to control groups (Severino & Moline, 1989, pp. 84–85). Why do some women get PMS and not others? A biopsychological model suggests that a predisposing biologic vulnerability is aggravated by psychologic and social stresses that may make these women more sensitive to the psychotropic effects of the normal hormonal fluctuations of the menstrual cycle. Steroid hormones in general have profound effects on behavior, and estrogen and progesterone do in particular. In animal studies both hormones are shown to significantly affect neurotransmitter levels, receptors, and degradative enzymes (Severino & Moline, 1989, p. 92). In a sensitive nervous system, the five- to tenfold rise and fall of these hormones during the luteal phase triggers behavioral symptoms (Severino & Moline, 1989, p. 6).

Special attention should be given to identifying major psychiatric disorders, such as depression, cyclothymic disorder, bipolar disorder (especially the rapid cycling subgroup), anxiety disorders, and personality disorders, as these entities can be exacerbated premenstrually and may present with a chief complaint of PMS. Similarly, one must rule out physical conditions, such as dysmenorrhea, endometriosis, hypothyroidism, or mastodynia (breast tenderness), that also may be exacerbated premenstrually. A physical examination should be performed. Endocrine tests for estrogen, progesterone, prolactin, and FSH levels have been recommended in the past, but they currently are thought to be of little value unless there are specific symptoms or signs of an endocrine disorder, such as hirsutism, markedly irregular periods, galactorrhea, or ovarian cysts. The only hormone levels that should be routinely determined are thyroid-stimulating hormone (TSH), triiodothyronine (T3), and thyroxine (T4) as there is evidence that thyroid dysfunction, a common endocrine disorder of women, can present with a complaint of PMS (Schmidt et al., 1990).

Treatment of rigorously defined PMS first requires an alliance with the patient that she does indeed have a problem deserving assistance. Many women have been frustrated by being told that their symptoms are psychosomatic or by being given multiple unsuccessful treatments. They may appreciate being educated about our current understanding of etiology, in particular the possible role of biologic vulnerability, stressors, and effects of normal hormonal fluctuation. During the period of observation and charting, the patient can be instructed about nonspecific measures, such as multiple small, balanced meals during the day and avoidance of concentrated sweets, caffeine, alcohol, and excessive salt. Although controlled studies are lacking to prove the efficacy of these interventions, many patients nevertheless report improvement on this regimen, and there are few side effects (Severino & Moline, 1989, pp. 165–167). Another nonspecific intervention that can improve symptoms is regular aerobic exercise: at least 20 minutes three times weekly. Although conclusive studies are not available, existing outcome studies suggest that this regimen can lead to reduction in PMS symptoms (Severino & Moline, 1989, pp. 164–165).

For patients with clear stressors or conflicts that contribute to their premenstrual symptoms, psychotherapy can be considered. There are no data supporting psychotherapy in general for PMS, but women may wish to explore psychotherapy on the issues for which it is indicated.

In the past various pharmacologic approaches have been touted. Most widely known is the use of progesterone suppositories for 7 to 10 days prior to menstruation. Well designed studies, however, consistently show that progesterone is no better than placebo in alleviating PMS symptoms (Severino & Moline, 1989, pp. 168–178). Other agents that have been used include diuretics, nonsteroidal antiinflammatory drugs (NSAIDs), pyrodoxine, and magnesium. Although diuretics may improve symptoms of weight gain, edema, and bloating, study reports are mixed about the effect on mood symptoms (Severino & Moline, 1989, pp. 196–203). Similarly, NSAIDs may reduce abdominal, back, and joint pain associated with menstruation, but studies are unclear about the effect on psychological symptoms (Severino & Moline, 1989, pp. 203–205). Pyridoxine and magnesium supplements have been popular for the treatment of PMS, but well designed studies show no evidence of deficiency of either of these compounds in PMS patients or of efficacy compared to placebo (Severino & Moline, 1989, pp. 213–222). Many other agents that have been used for PMS have similarly not been shown to be better than placebo (see the excellent review by Severino & Moline, 1989).

Ovulation suppression is one approach that has yielded positive results. The use of ovariectomy, estradiol implants, gonadotropin-releasing factor (GRF), and danazol treatment are supported by well designed studies showing their efficacy. Unfortunately, use of these modalities is generally impractical: Ovariectomy is limited by invasiveness, infertility, and irreversibility of the procedure; estradiol implants are not available in the United States; GRF treatment is associated with menopause-like symptoms of hot flashes; and danazol is associated with androgen-like side effects such as hirsutism, acne, and voice deepening (Severino & Moline, 1989, pp. 181–190). The most commonly used form of ovulation suppression, oral contraceptive pills, has not been well studied for the treatment of PMS. Preliminary studies suggest that some patients benefit from them (Severino & Moline, 1989, pp. 181–184). An empiric trial can be considered for the patient who is interested in this approach, but the clinician should be aware that in some patients the symptoms worsen and frank depression may develop with modality.

Well designed studies suggest that alprazolam is effective for the treatment of some symptoms associated with PMS (Severino & Moline, 1989, pp. 226–228). Alprazolam in low doses (0.25 two or three times a day) can be titrated to the patient's needs during the symptomatic part of the cycle. It can effectively decrease symptoms of anxiety, irritability, sleeplessness, and dysphoria, although it does not significantly diminish physical symptoms such as breast tenderness, edema, and joint pain. Caution must be used for patients with a history of substance abuse because of alprazolam's dependency potential and the severe withdrawal syndromes that have been described with this high potency agent. The other benzodiazepines have not been studied for the treatment of PMS, so the generalizability of these results to related compounds is unknown, but it is unlikely that alprazolam has any selective efficacy in this regard over other benzodiazepines. Furthermore, the reduction in symptoms reported above should be considered symptomatic only—and hardly "curative."

Perhaps the most promising direction for the treatment of PMS is suggested by three studies of antidepressants (Harrison, Endicott, & Nee, 1989; Richels et al., 1990; Stone, Pearlstein, & Brown, 1991). These studies examined the use of nortriptyline and fluoxetine when given consistently throughout the menstrual cycle in doses similar to those used for treatment of depression. The results show a significant decrease in depression, irritability, tension, and sleep disturbance that is more marked over several months of use. Again, physical symptoms are not as dramatically improved. Clinical experience suggests that these agents are well tolerated by most patients. Two other available studies regarding psychotropic medications and PMS suggest that the use of buspirone (Richels, Freeman, & Sondheimer, 1989) and clomipramine (Eriksson et al., 1990) is also promising.

These last four medications act on the serotonergic system. Whether serotonergic activity is essential for improvement in PMS symptoms is unclear, as nonserotonergic agents, such as desipramine and bupropion, have not been studied. Nevertheless, one can speculate that menstrual hormonal fluctuations destabilize the serotonergic system, resulting in PMS symptoms and associated affective perturbations, and that serotonergic antidepressants restore the equilibrium.

Is rigorously defined PMS a mood disorder? In that its major symptoms are mood symptoms common to other affective disorders suggests that it is. The fact that these symptoms respond to antidepressants is also suggestive. More research is needed, though, to determine the precise relation between PMS and affective disorders.

PELVIC PAIN

Pelvic pain is a common complaint in the gynecology clinic. Renaer (1981) estimated that approximately 25% of noncontraceptive gynecologic visits involve some kind of pain complaint, and that 50% of laparoscopies are done to evaluate pain. Some complaints, such as dysmenorrhea or acute pelvic pain due to ectopic pregnancy, respond to current pharmacologic and surgical treatments. This section addresses chronic pelvic pain, defined as pain of more than 6 months' duration, which has been described as "one of the most perplexing problems facing the gynecologist" (Lundberg, Well, & Mathers, 1973).

The role of personality in pelvic pain has long intrigued clinicians. Many anecdotal reports in the literature lack matched controls or comparisons. Several studies utilizing control groups (Castelnuovo-Tedesco & Krout, 1970; Renaer et al., 1979) show higher levels of psychopathology (hypochondriasis, hysteria, paranoia, psychasthenia, and schizoid features) in patients with pelvic pain than those without. These characteristics, however, do not discriminate pain patients with clear pelvic pathology from those with none. Furthermore, elevated Minnesota Multiphasic Personality Inventory (MMPI) profiles of both of these groups resembles those of patients with chronic pain in other organ systems (Pasnau, Soldinger, & Andersen, 1985). Whether the psychopathology preceded the pain or is a result of the pain has not been addressed.

Patients with no discernible pelvic pathology comprise 17 to 28% of all pelvic pain patients (Kresch et al., 1984; Renaer, 1980; Rosenthal et al., 1984). Although many theories have been proposed regarding the biologic, psychologic, or "psychosomatic" etiology of this syndrome, current research remains inconclusive. Clinical experience suggests that this group is diverse, representing a spectrum of biologic and psychological contributions (Stoudemire & Sandhu, 1987). Beware the assumption that all of these patients have "psychogenic pain." Although a small number of the patients exhibit a conversion-like syndrome, most do not. The label "psychogenic" often has a perjorative connotation, unfair to the patient and inaccurately suggesting a known etiology. Finally, it should be noted that a history of childhood incest is common in patients with chronic idiopathic pelvic pain, so the early sexual history should be thoroughly investigated as part of the basic evaluation (Walker et al., 1988).

Psychiatric evaluation of the pelvic pain patient should focus on the possibility of depression, somatization disorder, substance abuse, personality disorder, and history of sexual abuse. One group has suggested that an initially multidisciplinary approach to evaluation and treatment is more effective than standard gynecologic evaluation with subsequent referral for psychologic treatment (Peters, VanDoist, & Jellis, 1991).

For chronic pelvic pain of unknown etiology, treatment should focus on improving the patient's functional level. One promising avenue is the use of antidepressants (Walker, Roy-Burne, & Katon, 1990). Reliance on narcotic analgesics should be avoided because it is not an effective chronic treatment strategy and may lead to narcotic dependence and escalating doses. Activity should be encouraged. Many patients benefit from grief work about their pain problem and its effects on energy level, mood, daily activities, and intimate relationships. Couples may need counseling to help in their adjustment to the effects of the pain. Life stresses that exacerbate the pain may benefit from stress management or supportive psychotherapy. Clinical experience suggests that patients with a history of sexual abuse may benefit from specialty group therapy as a primary treatment modality or adjunct to individual therapy (C. Henderson, personal communication, 1991; M. Keeling, personal communication, 1991).

Treatment of the patient with a clear etiology for her pelvic pain should focus on acknowledgment of her difficulties, education about her treatment options, and encouragement to maintain maximum activity level. For some, treatment involves hysterectomy or exogenous hormones that impair fertility, and hence may present a difficult choice. Others with impaired sexual dysfunction may require medical psychotherapy or couples therapy to adjust to their situation. Treatment of chronic pain is further discussed in Chapter 17.

Endometriosis is one of the most common causes of chronic pelvic pain, and *laparoscopic evaluation should be an essential part of the evaluation of any patient suffering from chronic pelvic pain.* Danazol and nafarelin, medications used to treat this disorder, have been associated with side effects of insomnia and mood swings that are thought to be secondary to their antiestrogenic effects. When these symptoms are significant, the patient may wish to pursue other treatment modalities, such as depot progestational agents or oral contraceptives.

ABORTION

More than 1.5 million surgical abortions are performed in the United States each year. This procedure accounts for 30% of pregnancy outcomes for all females and 45% of the outcomes for females age 11 to 19 (U.S. Bureau of the Census, 1985). It is more commonly utilized by unmarried adolescents and young adults and by multiparous women in their late thirties and forties.

Why do such unwanted pregnancies occur? Adolescent sexuality has become increasingly common, and most adolescents are not mature enough to take responsibility for contraception. If current trends continue, it is estimated that 40% of adolescent girls will become pregnant by age 19 (Guttmacher Institute, 1981). Young adult women may also fail to use contraception in an appropriate way or may ambivalently wish for a pregnancy that they later regret. Often these women are unmarried, with no prospect of marriage to the father of the fetus, and are unable to support a child.

Older women who become pregnant may have completed their families and be unwilling to start again; for many of them, contraceptive failures have occurred. Termination of pregnancy may also be sought for reasons of danger to maternal health or knowledge of an abnormal fetus.

Although much has been written about the psychologic sequelae of abortion, the procedure is associated with surprisingly low psychiatric morbidity, as measured by the incidence of postabortion major depression or psychosis (Brewer, 1977; McCance, Olley, & Edward, 1973). Although the woman seeking abortion is often highly distressed before the procedure, this distress subsides rapidly afterward for most (Adler, 1980; Simon, 1972). It is not, however, a psychologically painless procedure: some experience guilt and grief in the aftermath (Ashton, 1980). The most frequent complaint 1 year later is a sense of regret, but most of the women say they would do it again if faced with similar circumstances. Some may experience an anniversary grief reaction, which may occur at the time of the year abortion occurred or at the time the child would have been born. The idea that abortion causes de novo major psychiatric disorders has not held up under critical evaluation; severe postabortion reactions are associated with previous psychiatric difficulties (Ashton, 1980; Ekblad, 1955; Kravitz et al., 1976).

Psychiatric evaluation of abortion patients should include the history of the circumstances of the pregnancy and an exploration of the patient's relationship with her family and the father of the child. Other issues to be addressed are what the patient thinks about the pregnancy and her options and kind of support she has. An evaluation for major psychiatric disorders and for suicidal ideation must be performed and appropriate treatment initiated if the evaluation reveals significant psychopathology.

Counseling for the patient seeking abortion should be routinely provided by the agency or physician performing the procedure. Psychiatric consultation is thus reserved for patients with unusual circumstances, including the following.

1. Major psychiatric disorders such as schizophrenia, bipolar disorder, severe personality disorder, and mental retardation
2. Marked adjustment reactions prior to the procedure
3. Unresolvable conflict between the patient and her family or the father of the fetus
4. Unresolvable ambivalence about whether to have the procedure
5. Persistent contraceptive noncompliance and repeated abortions
6. Marked postabortion reactions
7. Pregnancies that have resulted from rape or incest

In patients with major psychiatric disorders the concerns involve the following questions: Is this patient competent to make the decision? (Competence evaluation is discussed in Chapters 44 and 45.) Will she be able to tolerate the procedure? Is there adequate psychiatric follow-up care? Most psychiatric patients *can* make informed decisions; such patients are at no additional risk for hospitalization following abortion if they receive adequate treatment and support. Consultation in this instance serves the function of assuring the gynecologic staff that psychiatric patients are receiving adequate care.

Patients with severe adjustment reactions (including disorganized behavior, suicidal ideation, or homicidal ideation toward the father of the fetus) require evaluation, crisis intervention, and mobilization of support systems. Hospitalization may be necessary.

Severe conflict between the patient and the family or the father of the fetus can be accompanied by angry, threatening, disorganized, or suicidal behavior. An individual evaluation of each party is then necessary. Clarification of legal rights regarding the abortion decision is indicated. The woman has the right to decide about her own body unless she is an unemancipated minor, in which case the parents have the right to make the decision. The dissenting party must then accept the decision. Limits and consequences of inappropriate behavior should be made clear.

Clinical experience suggests that the patient with unresolvable ambivalence wishes to bear the child but is uncertain how she will be able to care for it. Because the time needed to make such a decision may interfere with the feasibility of the abortion, it is useful to empathize with the patient's mixed feelings and suggest that she consider bearing the child to give herself the time to think about it. If she decides that she is unable to care for the child, she can relinquish it after delivery. Psychotherapy then centers on understanding the meaning of the child to the patient and assisting her to determine if it is realistic for her to keep it. If the patient decides to keep the child under questionable circumstances, the physician should arrange a social service evaluation to monitor the placement and to pursue court-ordered foster care if needed. Clinicians should recall that ambivalently regarded infants, especially those for whom abortion was considered, are at greater risk for child abuse and neglect.

Clinical experience suggests that patients with contraceptive noncompliance and repeated therapeutic abortions fall into two categories: a chronically disorganized, low-functioning group and a higher-functioning group with neurotic conflicts. For women in the former group, contraceptive noncompliance is just one of many characterologic problems. Although focused counseling about the consequences of repeated abortions and the appropriate use of contraceptives may be helpful, these women should also be confronted about their generalized self-destructive behavior and be encouraged to seek psychotherapy. Women in the latter group need to understand the meaning of the neurotic symptom so they can begin to resolve the conflict. Although the identification and interpretation of the conflict may occur during an initial evaluation session, formal individual psychotherapy is advisable.

Rarely, a patient demonstrates major psychiatric symptoms after an abortion, such as depression or psychosis. Treatment of these conditions should follow standard approaches for the specific diagnosis, with particular emphasis on the meaning of the loss, mourning it, and dealing with guilt.

For those women whose pregnancies result from rape and incest, additional supportive counseling is indicated. The families of incest victims should be evaluated by a social services organization so appropriate social and legal intervention can be applied.

PSYCHOLOGIC SIDE EFFECTS OF ORAL CONTRACEPTIVES

Although numerous studies of the psychologic side effects of oral contraceptives (OCPs) have been performed, they are marred by many methodologic flaws (see reviews by Glick & Bennet, 1982; Slap, 1981; Warnes & Fitzpatrick, 1979; Weissman & Slaby, 1973). They nevertheless suggest that most women taking OCPs do not experience major psychologic side effects. Many but not all of the studies, however, report an increased incidence of depression in women taking OCPs. One study reported that OCPs with strong progestins are associated with an increased incidence of depression and decreased libido (Grant & Pryse-Davies, 1968). Another study suggested that women with premenstrual irritability before starting OCPs have a greater incidence of depression on more strongly estrogenic OCPs and do better on more progestational agents. A third study concluded that women without premenstrual irritability felt better on more estrogenic pills (Cullberg, 1972). Pyridoxine, a cofactor in catecholamine synthesis, has been reported to be deficient in OCP users (Parry & Rush, 1979; Rose et al., 1972). One double-blind placebo-controlled study showed the pyridoxine supplementation alleviated depression associated with OCP use (Adams et al., 1973).

Although the etiology of these OCP-related complaints cannot be discriminated with certainty, it is possible that some are related to pharmacologic effects of these agents. Fewer complaints of OCP-related depression may be noted now compared to the 1960s and 1970s owing to the introduction of lower-dosage OCPs (Mishell, 1989). The new long-term contraceptive agent

Norplant has been associated with few reports of psychological side effects (Shoupe & Mishell, 1989).

The physician can adopt an empiric approach to the management of OCP-related mood complaints. If another birth control method is appropriate and acceptable, the patient can discontinue OCPs and shift to it. If she continues to desire the use of OCPs, she can be given a trial of other OCP preparations with a lower estrogen or progestin component (or both). Occasionally a higher estrogen component is needed (Hatcher et al., 1986). Another option would be to supplement the OCP with pyridoxine 20 mg twice a day (Adams et al., 1973). Perhaps most important is developing an alliance with the patient to help her find what works best for her.

POSTHYSTERECTOMY REACTIONS

Hysterectomy is the most common surgical procedure of women and accounts for more than 650,000 operations a year (National Center for Health Statistics, 1985). Although some are performed for emergencies and malignancies, most are elective procedures for such indications as endometriosis, menorrhagia, fibroids, dysfunctional uterine bleeding, and chronic pelvic pain. Postoperative psychiatric reactions, particularly depression, have historically been associated with hysterectomy. Studies have reported that psychiatric morbidity after hysterectomy was twice as common as after nongynecologic surgery (Baker, 1968; Richard, 1974).

The incidence of psychiatric disorders following simple hysterectomy, however, is similar to that following combined hysterectomy and bilateral salpingo-oophorectomy, regardless of whether replacement hormones are provided (Baker, 1968; Gath et al., 1982a,b; Martin, Roberts, & Clayton, 1980; Richard, 1974). Cessation of ovarian hormones thus does not appear to contribute significantly to the development of posthysterectomy major psychiatric disorders. The increased rate of previously reported depression for women after hysterectomy compared to control groups may be related to a higher rate of depression preexisting in that group prior to hysterectomy (Lalinec-Michaud & Englesmann, 1985).

Research has compared the preoperative psychiatric functioning of women undergoing elective hysterectomies to their postoperative functioning. Multiple studies show a surprising level of psychopathology in the preoperative state (25 to 57%), which drops significantly 6 months to 1 year later (10 to 35%) (Gath, Cooper, & Day, 1982b; Lalinec-Michaud & Englesmann, 1984; Martin, Roberts, & Clayton, 1980; Moore & Tolley, 1976). In these studies hysterectomy appeared to improve rather than impair emotional functioning for most women. This improvement may be related to relief of pelvic symptomatology. Nevertheless, a small subgroup did experience a postoperative psychiatric reaction. Age, marital status, number of children, social class, and type of gynecologic pathology did not predict these reactions; the strongest predictor was previous psychiatric history and preoperative mental state (Gath & Rose, 1985). The hysterectomy may play a precipitating role in bringing these difficulties to the physician's attention.

For the vulnerable woman, hysterectomy may be decompensating for a number of reasons. A grief reaction may be triggered by the loss of the genital organ and the ability to bear children. For the woman whose major source of self-esteem is childbearing, loss of the uterus means loss of her feminine identity. For others, damage to the genitalia means impaired sexuality and loss of sexual value to the spouse. Particularly if her spouse or family has a negative attitude about the procedure and views her as "damaged goods," she will feel injured and devalued (Stotland & Smith, 1990).

Psychiatric evaluation may be requested before or after hysterectomy. In addition to a standard history, the evaluation should explore why the procedure is being considered now, what the indications are, and what the patient expects from the procedure, both benefits and detriments. What is the meaning of the hysterectomy for the patient? Did she want more children? Is she afraid of altered relations with her husband? Is she concerned about an altered sense of femininity? Is she worried about a change in her sexual responsiveness? What do her husband and family think about the procedure? What previous reactions to stress has she had, and what was helpful for coping with them? What support systems are available to her? Are there other situational stresses affecting the patient to indicate that surgery should be postponed? The differential diagnosis should include somatization disorder; a disproportionate number of women with this disorder undergo hysterectomy, and they are thought to be at risk for unnecessary procedures (Martin, Roberts, & Clayton, 1980).

Psychiatric treatment prior to surgery should include standard pharmacologic and psychotherapeutic approaches to the specific psychiatric diagnosis. If the surgery is elective and the patient would benefit from psychotherapy or psychotropic drugs, the hysterectomy should be postponed. Ventilation, problem solving, and grief work may be needed for life crises occurring coincidentally with the gynecologic problems (e.g., marital discord, conflicts with adolescent children, concerns about aging parents, or deaths of significant others). Regarding the procedure itself, education and reality testing about its effects can be helpful. In particular, the patient should know that although many patients are concerned about femininity and sexuality there is no physical reason why these areas should be affected. If the patient continues to be distressed about these issues, psychotherapeutic exploration of her concerns is indicated.

For postoperative reactions a similar approach can be taken. Postoperative cognitive disorders must be ruled out. Then medical psychotherapy, which often includes grief work, is offered, supplemented with pharmacologic treatment if a major psychiatric disorder is diagnosed. Psychiatric aspects of surgery in women, particularly for cancer, is discussed in Chapter 34, as is sexual dysfunction associated with gynecologic procedures.

PSYCHOTROPIC MEDICATIONS DURING PREGNANCY

The pregnant woman who requires psychotropic medications for treatment of a serious mental illness raises a dilemma for the physician. On one hand, management of her mental condition without medication may be problematic and potentially restrictive (i.e., requiring inpatient hospitalization and seclusion). On the other hand, the potential risks for the fetus include physical and behavioral teratogenicity, as well as side effects in the newborn. Data on these topics are difficult to obtain in a systematic fashion and often consist of case series or case reports. Although information may be available on malformations and neonatal toxicity, little is known about the long-term behavioral effects of fetal exposure to these agents. Some estimate of risk to the fetus can be made, however. The balance of risks and benefits for each patient must be assessed on an individual basis (Cohen, Heller, & Rosenbaum, 1991).

Neuroleptics are generally not thought to increase the rate of physical malformation (Calabrese & Gulledge, 1985; Hauser, 1985). Although there are a few case reports of malformation associated with neuroleptic use, several reviewers conclude that neuroleptic medication is relatively safe during pregnancy (Briggs, Freedman, & Yaffe, 1990; Calabrese & Gulledge, 1985; Cohen, Heller, & Rosenbaum, 1991). There is no information about effects of human fetal neuroleptic exposure on subsequent behavior. Neonatal side effects that have been noted include transient jaundice, sedation, motor excitement, agitation, hypotonia, and extrapyramidal symptoms (Calabrese & Gulledge, 1985; Cohen, Heller, & Rosenbaum, 1991).

Similarly, although there are case reports of anomalies associated with *tricyclic antidepressants*, these agents have not been documented to increase the rate of congenital malformations (Calabrese & Gulledge, 1985; Cohen, Heller, & Rosenbaum, 1991). Again, there are no studies available on the effects of prenatal human exposure to these agents and subsequent behavior. Side effects in the newborn may include cardiac failure, tachycardia, myoclonus, respiratory distress, and urinary retention (Briggs, Freedman, & Yaffe, 1990; Calabrese & Gulledge, 1985). A fetal withdrawal syndrome consisting of colic, cyanosis, rapid breathing, and irritability has also been described (Briggs, Freedman, Yaffe, 1990). These problems relatively contraindicate tricyclics during the days immediately before delivery.

Little experience with *monoamine oxidase inhibitors* (MAOIs) has been described. One small study reported an increased risk of malformation associated with these agents (Heinonen, Slone, & Shapiro, 1977). MAOIs have been considered contraindicated during pregnancy because of the risk of hypertensive episodes in fetus and mother and the availability of other treatment modalities (Hauser, 1985).

Lithium has been associated with increased rate of malformations in the past. Of 217 babies in the Register of Lithium Babies, 7 were stillborn, 2 had Down's syndrome, and 18 had cardiovascular malformations, 6 of which were Ebstein's anomaly (Calabrese & Gulledge, 1985). Registry data are thought to be enriched for adverse outcomes. Reports now view the risk of Ebstein's anomaly to be considerably less than the Registry suggested (Zalstein et al., 1990). One study of children who had been exposed to lithium during the first trimester showed no developmental difficulties compared to controls (Schou, 1976). Side effects of lithium in the newborn include transient cyanosis, lethargy, hypotonia, jaundice, hypothermia, low Apgar scores, and altered thyroid and cardiac function (Briggs, Freedman, & Yaffe, 1990). Lithium levels should be carefully monitored immediately after delivery, as the rapid shifts in blood volume can lead to lithium toxicity in the mother.

Use of *benzodiazepines* during the first trimester has been associated with conflicting reports about increased incidence of malformations. *Diazepam* has been associated with an increased rate of cleft palate (Briggs, Freedman, & Yaffe, 1990; Calabrese & Gulledge, 1985). In addition, a syndrome that resembles the fetal alcohol syndrome and results from in utero exposure to benzodiazepines has been described, consisting of growth retardation, dysmorphism, malformations of facial regions, and central nervous system (CNS) dysfunction (Laegreid et al., 1987). Because of the possible association with such malformations, and because their use is rarely urgent, benzodiazepines should be avoided during the first trimester (Calabrese & Gulledge, 1985). Few data are available on the effects of *alprazolam* (for discussion of the management of panic attack patients see Cohen, Heller, & Rosenbaum, 1991). Side effects of benzodiazepines in the newborn include CNS and respiratory depression (Briggs, Freedman, & Yaffe, 1990; Calabrese & Gulledge, 1985). A withdrawal syndrome from diazepam has been described; it consists in irritability, jitteriness, tremor, diarrhea, and vomiting (Rementeria & Bhatt, 1977).

Phenobarbital has been commonly used in the past as a sedating agent during pregnancy. Although its use has been associated with an increased rate of malformations in epileptic patients, it is not thought to pose increased risk in nonepileptic patients (Briggs, Freedman, & Yaffe, 1990). It is associated, however, with severe, even fatal, hemorrhagic disease in the newborn, which can be treated with vitamin K during pregnancy or labor. Phenobarbital has also been associated with neonatal withdrawal symptoms (Briggs, Freedman, & Yaffe, 1990).

With the increasing use of *anticonvulsants* for the treatment of bipolar disorder, there are increasing questions about the use of these agents during pregnancy. Although *carbamazepine* was once considered to be a relatively safe anticonvulsant, data have indicated a risk of craniofacial abnormalities, postnatal growth deficiency, and developmental delay (Jones et al., 1989).

Valproate is another agent that is increasingly being used. It is associated with intrauterine growth retardation and increased rate of a variety of malformations. Most prominent are neural tube defects, in particular a 2% incidence of spinal bifida (Briggs, Freedman, & Yaffe, 1990). Amniocentesis to determine levels of amniotic fluid alpha-fetoprotein can be used to screen for neural tube defects (including spinal bifida) at a time when termination of pregnancy would be feasible if desired. High-resolution ultrasound also aids in prenatal diagnosis.

Phenytoin, an anticonvulsant, is associated with a two- to threefold increased rate of malformation after first trimester fetal exposure in epileptic patients. It is not known if this risk extends to nonepileptic patients. Fetal hydantoin syndrome is a recognizable pattern of malformations noted in infants prenatally exposed to phenytoin. It consists in facial anomalies and hypoplasia of the distal phalanges. It also can cause severe or fatal hemorrhagic disease of the newborn due to suppression of vitamin K-dependent coagulation. This disorder can be treated by administering vitamin K to the mother toward the end of pregnancy and during labor (Briggs, Freedman, & Yaffe, 1990).

Clonazepam and *lorazepam* are benzodiazepines that are used increasingly for management of psychotic patients. There are few data available on the rate of malformations in babies exposed to these agents during the first trimester. However, because they resemble diazepam, it is possible that facial anomalies, as noted above, may also be a risk factor.

Pharmacologic management of psychiatric illness during pregnancy involves several basic principles (see Table 28-1). Although the overall risk of psychotropic medication use may be low, it nevertheless entails some uncertainty. The known risk of individual agents should be reviewed with the patient and her spouse. Such information was well summarized by Briggs, Freedman, & Yaffe (1990) and by Cohen, Heller, and Rosenbaum (1991). The prudent physician and couple consider whether the psychiatric illness can be managed without medication. Whatever the treatment plan agreed upon, informed consent should be obtained from the patient and spouse. For some patients with depression or bipolar disorder, electroconvulsive therapy may be an appropriate alternative (see following section). If medication is deemed necessary, postponement until the end of the first trimester is advisable if possible. Most fetal organ differentiation is completed during the first trimester, and the fetus now is not vulnerable to physical teratogenic effects. The brain does continue to develop throughout fetal life, however, so use of the medication with the lowest toxicity, finding the lowest effective dosage, and considering divided doses to lessen the peak levels of fetal exposure are essential. For patients who use medications throughout their pregnancy, consider tapering the dosage near the expected time of delivery to decrease the level and possibilities of severe side effects in the neonate. The pediatrician should be alerted to the maternal medication use so that appropriate neonatal care can be rendered (see excellent discussion by Cohen, Heller, & Rosenbaum, [1991]).

ELECTROCONVULSIVE THERAPY IN PREGNANCY

In general, electroconvulsive therapy (ECT) is considered safe during pregnancy (Fink, 1981; National Institute of Mental Health Consensus Conference, 1985; Repke & Berger, 1984; Varan et al., 1985; Wise et al., 1984). This conclusion was drawn from a series of case reports in the literature rather than from controlled studies. There is nevertheless a striking absence of reported untoward effects. Because of the teratogenic effects of pharmacotherapy, ECT may actually be safer for the fetus during the first trimester, but again controlled studies are lacking (Remick & Maurice, 1978). Developmental follow-up on children who were in utero during ECT has shown no abnormalities (Forrssman,

TABLE 28–1. *Guidelines for Psychotropic Drug Use during Pregnancy*

General guidelines
1. Obtain informed consent of patient and spouse. Advise about risk of major malformation, as well as uncertainty about effects of medication on developing fetal brain.
2. Avoid medication during the first trimester, which is when fetal organogenesis occurs.
3. Maximize supportive interventions, which may include psychotherapy, social intervention, and hospitalization.
4. Consider medication particularly when there is homicidal or suicidal ideation, an inability to care for self, or an inability to comply with prenatal care.
5. If medication is required:
   a. Use lowest effective dosage.
   b. Divide dosage to minimize peak levels of fetal exposure.
   c. Use those with lowest fetal and maternal toxicity.
   d. Prefer older drugs with a longer track record.
   e. Minimize number of medications.
   f. Taper before delivery to minimize neonatal burden.
   g. Arrange pediatric support at delivery.
6. Avoid medications during lactation; they are excreted in breast milk and have unknown effects on the developing neonatal brain.

Psychosis
1. Although antipsychotic medications show no increased risk of major malformation, their effect on the developing fetal brain and subsequent behavior is not well studied.
2. Maintenance low-dose antipsychotic therapy may offset the risk of relapse and the need for higher doses and possible noncompliance with prenatal care.

Mania
1. Antimanic agents, e.g., lithium, carbamazepine, and valproate, all have been associated with physical malformation and should be avoided when feasible. The literature, however, now suggests that the risk of lithium may have been overestimated.
2. Prior to conception, evaluate the need for prophylaxis.
3. First trimester
   a. Avoid lithium and carbamazepine. If there is fetal exposure to lithium before week 12, consider cardiac ultrasonography at week 20.
   b. If medication is needed, consider antipsychotic medication or lorazepam.
   c. Consider electroconvulsive therapy (ECT).
   d. If the illness is not controlled with these interventions, the benefit of lithium may outweigh the risk.
   e. Patients who take carbamazepine or valproate during the first trimester should be screened with a serum alpha fetoprotein for increased rise of fetal neural tube defects.
4. Second and third trimesters
   a. If treatment is needed after week 12, consider lithium, carbamazepine, or valproate.
   b. Monitor blood lithium levels, as the dosage increases as the pregnancy progresses.
   c. Decrease lithium during and after delivery to prevent a toxic level in the mother.

Depression
1. Although antidepressant medications show no increased risk of major malformation, their effect on the developing brain and subsequent behavior is not well studied.
2. If medication is required, use secondary, rather than tertiary, amine tricyclics.
3. Prefer tricyclics over new compounds, which have a shorter track record.
4. Avoid monoamine oxidase inhibitors.
5. Consider ECT as a safe alternative.

Anxiety disorders
1. Older anxiolytic medications are controversial regarding their risk of major malformation. Their effect on the developing brain and subsequent behavior is not well studied. Limited information is available for alprazolam and clonazepam.
2. Taper the anxiolytic drug prior to conception, which may require a shift to a tricyclic or clonazepam.
3. If the patient is on an anxiolytic when pregnancy is confirmed, attempt to taper drug usage.

Adapted from Cohen, Heller, and Rosenbaum (1991). With permission.

1955; Impastato, Gabriel, & Landaro, 1964; Sobel, 1960).

Recommendations for safe ECT in the pregnant patient include a complete physical and pelvic examination, an obstetrician on the treatment team, and fetal monitoring for third-trimester patients (Remick & Maurice, 1978). ECT is relatively contraindicated in patients at high risk of premature labor. Depending on maternal and fetal risk factors, patients may require endotracheal intubation, electrocardiograph (ECG) monitoring, arterial blood gas measurements, Doppler ultrasonography of fetal heart rate, tocodynamometer recording of uterine tone, and tests of fetal lung maturity. Glycopyrrolate may be safer than atropine if an anticholinergic premedication is needed (Wise et al., 1984). (See also Chapters 10 and 11.)

PSYCHOTROPIC DRUGS DURING LACTATION

Few case reports of breast-feeding mothers taking psychotropic drugs have shown adverse effects on the infants. The American Academy of Pediatrics (1989) has concluded that the risk of ingestion of psychotropic drugs by a lactating mother "is unknown but may be of concern." The only exception is lithium, which is considered contraindicated. Single cases have been reported, however, of infant lethargy and weight loss due to accumulation of diazepam and of infant lethargy due to chlorpromazine (American Academy of Pediatrics, 1989). Long-term effects of infant exposure to psychotropic agents have not been studied.

The most conservative approach is that psychotropic drugs should be avoided during lactation because of these unknown risks. If these agents are necessary for maternal health, the patient should be informed of the uncertainty regarding long-term effects on the infant and given the option of discontinuing breast feeding. If the patient insists on breast feeding, the lowest possible dosage of medication should be utilized, the drug dose should follow the infant's time at the breast, and the infant should be monitored for untoward effects.

One report studied desipramine and its metabolite 2-hydroxydesipramine in human breast milk and in the nursing infant's serum during administration of 300 mg/day to the mother (Stancer & Reed, 1986) Even though it was estimated that the infant in this study was ingesting 1/100 of the dose of the mother each day (6 mg/kg), derived from the milk, no detectable levels of either metabolite could be detected in the infant's serum. Other studies focusing on tricyclic antidepressants have failed to find detectable serum levels in nursing infants of mothers taking these medications, indicating levels of less than 10 ng/ml. (Bader & Newman, 1980). One clinical report, however, described an 8-week-old nursing infant with respiratory depression who appeared to correlate with accumulation of the desmethyl metabolite of doxepin after the dose in the mother had been increased from 10 mg/day to 75 mg/day; doxepin itself was undetectable in the child's serum (Matheson, Pande, & Alertsen, 1985). The consensus, however, is that imipramine, desipramine, amitriptyline, and the monoamine oxidase inhibitor tranylcypromine can be, in general, safely given to nursing mothers (Ananth, 1978). More detailed reviews of the literature may be found in articles by Ananth (1978), Stancer and Reed (1986), and Briggs, Freedman, and Yaffe, 1990.

PREVENTION OF CHILD ABUSE AND NEGLECT

There has been a revolution in our awareness of child abuse and neglect. Although efforts to detect child abuse and neglect are usually centered in the pediatric clinic, high-risk families can often be identified during pregnancy in the prenatal clinic as well. The obstetrician may turn to the psychiatric consultant for assistance evaluating and coordinating the treatment of such patients. Early identification allows more effective intervention. One large, controlled study (Gray et al., 1979) showed that for high-risk families such interventions as regular pediatric appointments every 2 weeks, encouragement to call when problems arose, referrals for appropriate medical and mental health treatment, and lay health visitors significantly reduced the incidence of serious injuries, from 10% in the control group to 0% in the experimental group. Psychosocial risk factors, as noted in Table 28-2, should be explored during the prenatal history. Another important source of information about potential abuse is the observation of parental reactions to the infant at the time of delivery and during the immediate postpartum period. A hostile, negative, or disappointed reaction is significantly correlated with later abuse (Gray et al., 1979).

Evaluation of the high risk patient should involve a thorough history to assess strengths and vulnerabilities, especially impulse control. Particular attention to current stresses and living situation can clarify the social interventions that may be needed. Determine the supports available to the patient, including the father of the child, the patient's family, her friends, and social agencies, such as a community mental health center. The patient's wishes and fantasies about the baby should be explored, as should her plans for providing appropriate food, clothing, and shelter for the infant. Major mental illnesses, such as schizophrenia, mood disorders, substance abuse, and severe personality disorders, should be identified and treated. The possibility of mental retardation should also be considered.

TABLE 28–2. *Risk Factors for Perinatal Child Abuse*

1. Abuse and neglect in the parent's background
2. Abuse, failure to thrive, relinquishment, or foster care for previous children
3. Previously expressed desire to abort or relinquish the current pregnancy
4. History of mental illness
5. Mental retardation
6. History of institutionalization
7. History of spouse abuse
8. Alcohol or drug abuse
9. Chaotic living situation
10. Criminal record
11. Social isolation
12. Lack of financial resources
13. Age less than 17 years
14. Lack of anticipative behavior
15. Abnormal expectations for the child
16. Single parenthood

Data from Aycub and Pfeifer (1977); Gray, Cutler, Dean, et al. (1979); and Soumenkoff, Marnefe, Gerard, et al., 1981.

Management of these patients requires a comprehensive psychosocial approach. Major categories of interventions include (1) developing an alliance; (2) improving social problems; (3) improving parenting behavior; (4) treating mental disorders; (5) coordinating postpartum follow-up; and (6) obtaining court-ordered measures such as involuntary mental health treatment or protective custody for the child. A number of health professionals may become involved in such a case, so coordination of the various efforts by the psychiatrist is invaluable.

Developing an alliance is the single most important factor when working with women or couples at risk for child abuse and neglect. Many of them have been abused and neglected themselves and may react with suspicion and hostility at the physician's concern. Many are pleased, however, that someone is taking an interest in them. An essential aspect of the therapeutic alliance is continual positive reinforcement of the patient's self-esteem and acknowledgement of her efforts to be a good parent. It is also important to encourage the patient's connection to the clinic staff by calling her to reschedule missed appointments and by encouraging her to call if problems arise. In many ways the patient needs the clinic staff members to serve as models for her own parenting.

Assisting the patient to solve problems in a structured manner is another important intervention. Clarify any existing problems with finances, living situation, or support systems. Many women need information about the services that are available and the steps they must take to get them. They may need encouragement for making an appointment with social services to pursue Medicaid, food stamps, and Aid to Families with Dependent Children (AFDC). They may need suggestions about budgeting and places to stay. Sometimes an arrangement can be made with the family, the father of the child, or friends to provide resources, financial and otherwise. Other potential supports can be explored and engaged.

A third area for therapeutic attention is the development of the parenting identity and behavioral repertoire. The health professional's mirroring of the patient's wish to be a good mother and promoting the patient's interest in anticipative behaviors, such as crib and baby clothes, helps build that aspect of the self. The patient's fantasies about what the child will be like should be explored; and reality testing or problem solving is provided when needed. If the patient is amenable, parenting classes at community agencies can be considered an additional way of educating her and providing support.

For prenatal patients with a diagnosed mental disorder, specific psychiatric treatment should be arranged. If the consulting psychiatrist chooses to provide such care, appointments can be scheduled to coincide with the prenatal clinic visits. The issues of termination of therapy after delivery and arrangements for subsequent treatment must also be addressed.

Coordination of postpartum follow-up is needed in many other areas as well. The inpatient obstetric nursing staff must be informed about high-risk patients so they can observe interactions and provide maximum education and support. The pediatric clinic staff also needs to be advised of the patient's situation so an appropriate alliance can be formed. Many patients benefit from postpartum outreach to their homes by a visiting nurse. In some areas volunteer workers may be available to provide a supportive, nonthreatening relationship during the postpartum period.

For patients who are seriously disturbed and either unable to take care of themselves or in imminent danger of harming their infants, a court-ordered intervention must be obtained. A certification for involuntary inpatient psychiatric treatment should be considered if the patient is gravely disabled or dangerous to self or others, including the fetus. If the patient does not require hospitalization, yet there is grave concern about her competence to care for the child, a protective custody order can be sought that requires foster home placement or child protective services monitoring. A protocol for assessing competence to parent was suggested by Gabinet (1986). Clinical experience suggests that the courts are interested in the information and opinions provided by the psychiatrist. Because the court generally wishes to preserve the mother-infant relationship, it may use compliance with a proposed treatment plan as a condition of regaining custody. Such conditions of custody can have a powerful structuring and motivating effect on the mother.

Although helping families at high risk for child abuse and neglect is often complicated and time-consuming, there is significant potential for reducing serious injury and even death. Perhaps our present efforts will decrease the incidence of child abuse and neglect for the next generation.

SUBSTANCE ABUSE AND PREGNANCY

Substance abuse is of special concern in pregnant women because of the increased incidence of fetal loss, malformation, and growth retardation and because of neonatal abstinence syndromes and altered neonatal development (James, 1991; Zuspan & Rayburn, 1986). Substance abuse is also associated with poor general health and nutrition for the mother and an increased frequency of complications of labor and delivery including preterm labor (Zuspan & Rayburn, 1986). Female substance abusers may demonstrate low self-esteem, impulsiveness, and provocative behavior that impedes adherence to medical treatment. The obstetrician may request psychiatric consultation for assistance in managing these patients and arranging appropriate treatment plans.

Perhaps the most common error when dealing with substance abuse is failure to identify it. A substance abuse history should be routinely obtained in the prenatal history as well as in every psychiatric consultation. Other tip-offs to substance abuse include requests for drugs, evidence of intoxication, and suggestive physical findings (e.g., "tracks," a perforated nasal septum, or stigmata of liver disease). The substance abuse history should explore the history of drug use, the current pattern of use, the presence of tolerance and withdrawal symptoms, the perceived beneficial effect of substance use, and social factors that reinforce substance use. A thorough psychosocial evaluation should inquire about the patient's current financial status and living situation, as well as the attitudes of significant others and the availability of potential supports. Because substance abuse may serve the function of self-medicating an underlying psychiatric disorder, the differential diagnosis should include mood disorders, personality disorders, anxiety disorders, and psychoses. See Chapter 8 for further discussion of the evaluation and treatment of alcohol and drug abuse.

Establishing an alliance is a crucial step when treating substance abuse. Whatever clinicians' personal reactions, they should avoid being judgmental. The patient should be informed about confidentiality of medical evaluation and treatment. Responsiveness to the patient's concerns can be conveyed by open-ended interview techniques for at least part of the interview. If she indicates that she is concerned about the risk to the baby, this concern can be reinforced and used to help her maintain motivation to seek treatment. Once an empathic interaction has been established, the diagnosis of substance abuse can be clarified. For some patients, recognition of this problem is a genuine surprise. Others are relieved that someone is taking an interest. Some, however, deny the problem and require a more confrontational approach: "I realize that you aren't concerned about drinking every day, but it is definitely harmful to you and the baby, and you need help with it. Let's try to understand what we can do about it." (See excellent discussion by Senay, 1983.)

The patient can benefit from education about the nature of substance abuse and the type of treatment and life changes that are needed. The fact that dealing with a substance abuse problem is a long-term process should be explained. Inpatient detoxification may be indicated. Residential treatment programs should be recommended if they are available. Education about the referrals to long-term support programs, such as Alcoholics Anonymous or Narcotics Anonymous, should be included. The patient may need considerable support to deal with her reactions to such treatment programs and to pursue treatment in these settings.

Treatment for some substance abuse disorders must be modified during pregnancy. For patients with narcotic abuse, rapid withdrawal should be avoided, as it is associated with an increased incidence of fetal loss. Methadone maintenance titrated to a low dosage, 20 to 30 mg, has been recommended in this group to block craving for street drugs as well as the withdrawal syndrome (Zuspan & Rayburn, 1986). For alcohol-dependent patients, gradual detoxification on long-acting benzodiazepines or phenobarbital is indicated despite some reports of an increased rate of malformations associated with these agents. The use of disulfiram, however, is contraindicated during pregnancy because of its toxic effects on the fetus (Briggs, Freedman, & Yaffe, 1990).

Cocaine abusers may benefit from dopamine agonists to help decrease the cocaine craving. Little information is available about the effects of levodopa and amantadine on the fetus. Bromocriptine, however, has been more widely used during pregnancy and is not associated with an increased rate of physical or behavioral problems in the child. Other agents that have been used for management of cocaine abstinence include carbamazepine, which has been associated with fetal malformation, and desipramine, which is not associated with increased rate of malformation (James, 1991). (See previous section, Psychotropic Medications During Pregnancy.)

Another important point is understanding the patient's social situation so changes that support drug-free living can be made. Some patients may need assistance in obtaining Medicaid, food stamps, or AFDC. Others may need assistance in finding a place to stay apart from substance-abusing acquaintances. Potential non-substance-using supports, such as family and friends, should be explored. The patient's "significant other" should be included in the evaluation process. If the expectant father is also a substance abuser, an attempt to engage him in treatment should be made as well.

Pregnant women who continue to use drugs present a dilemma for health professionals. On one hand, this behavior is abusive and dangerous to the fetus and raises the issue whether to hold the woman for involuntary mental health treatment or incarcerate her for child abuse. On the other hand, if all women who continue to use substances during pregnancy were detained, many of these women would avoid prenatal care altogether. Although a few such cases receive media attention, legal decisions and clinical practice have favored the woman's right to autonomy and thus have supported her access to medical care. Clinicians working with this population have focused their efforts on improving alliances in order to maximize the chance that the patient will enter substance abuse treatment.

Substance-abusing women who are disorganized, self-destructive, and noncompliant also raise the question of potential child abuse and neglect after delivery. In this instance the child protective service agency should be advised of the high-risk situation and a court-ordered evaluation and protective custody order considered (see previous section, Prevention of Child Abuse and Neglect).

Although dealing with substance-abusing patients can be emotionally taxing and time-consuming, the physician can render an important service by being willing to be concerned. Not only may the child be spared the untoward effects of maternal substance abuse, but the alliance achieved during prenatal care may create a crucial turning point in the mother's life. (Further discussion of substance abuse can be found in Chapter 8.)

TOBACCO USE DURING PREGNANCY

As cigarette smoking has become increasingly prevalent among women of childbearing age, primarily due to the malignant marketing of tobacco companies, birth defects similar to the fetal alcohol syndrome have emerged as a potential complication of smoking during pregnancy. Smoking during pregnancy is also associated with prematurity and low birth weight (Nieburg et al., 1985). Smokers who become pregnant should thus not only be counseled to stop smoking but should be referred for smoking cessation interventions if they are unable to stop on the physician's advice alone. The use of nicotine-containing gum is contraindicated during pregnancy because of toxic effects on the fetus. Smoking cessation interventions are discussed further in Chapter 42.

PREGNANCY LOSS

More than one million women a year experience an unsuccessful pregnancy. A surprisingly high percentage of pregnancies result in loss: 15 to 20% in miscarriage, 1% in stillbirth, and 1% in perinatal death (Borg & Lasker, 1982; Friedman & Gradstein, 1982). Furthermore, with modern, highly sensitive pregnancy tests, women are increasingly aware of early pregnancy loss.

Even an early pregnancy is a child for a woman who desires one. Thus loss of the pregnancy can be like loss of a beloved person. The normal grieving reaction may include a sense of shock and disbelief, intense sadness and crying, anger toward the physician and others, irritability, somatic symptoms of weight loss, sleep disturbance, decreased sexual interest, and guilt and a sense of inadequacy at having been unable to sustain the fetus ("What did I do wrong?"). The woman usually has a much more pronounced reaction to miscarriage than her spouse. Her bodily changes make the pregnancy a more tangible reality for her. The husband's subdued reaction may cause the woman to feel isolated and abnormal in her grief. Friends and family members, unless they have suffered miscarriage themselves, may have little sympathy for the patient and may minimize the patient's distress with the comment, "You can get pregnant again." The grief reaction is most intense during the first few weeks after the loss and continues for several months, gradually subsiding. After 6 months, severe symptoms that continue to disrupt the patient's previous level of functioning should be considered a pathologic grief reaction and deserve a full psychiatric evaluation.

Pregnancy loss during the second or third trimester, after fetal movement has been felt and an ultrasound image seen, may be associated with even more intense grieving because of the potentially increased attachment to the physically perceived fetus. Stillbirth and perinatal death are even more intense losses. After carrying the child for 9 months and preparing her life for the new arrival, the woman has invested a great deal emotionally in the child. Instead of the expected happy event, the woman must face loss, the reactions of the family and friends, the making of funeral arrangements, and the return home to an empty crib. She may be unable to tolerate being on an obstetric unit, where she can see other women and their newborns. Such a loss can exacerbate preexisting marital conflict, especially when grief handling styles are not complementary, and may lead to marital discord, sexual dysfunction, and in some cases divorce.

Late fetal loss, stillbirth, and perinatal death are also difficult for physicians and nurses, who may be overwhelmed by the patient's feelings and their own personal reactions. They also may feel a sense of guilt, wondering if there was something they could have done to prevent the loss. The lack of control that occurs with fetal demise is especially disturbing for medical professionals, who are accustomed to being in control. They sometimes react by blaming the patient or minimizing the patient's loss. It is crucial that the staff members handle their own reactions in a constructive way, because the patient and her spouse are vulnerable to their comments and depend on them for emotional support.

Although psychological factors are not thought to be contributory to fetal loss in general, there is some indication that women with multiple spontaneous abortions can improve their chances of carrying a pregnancy to term with psychotherapy, particularly psychotherapy that facilitates grieving of the previous losses (Tupper & Weil, 1962). Unfortunately, there are no recent replications of this work.

Psychiatric treatment of pregnancy loss involves a number of interventions. The events of the pregnancy, the fetal loss, and subsequent circumstances should be reviewed. The explanation for the loss should be discussed, and any uncertainties about the events clarified, especially regarding the patient's being at fault. An important factor for facilitating a grief reaction is allowing and empathizing with the patient's feelings. At times it may seem overwhelming for the facilitator, but it is useful and deeply appreciated by the patient. She should be educated that she is not mentally ill, but that she is experiencing a normal grief reaction and that it will be intense at first and then gradually subside over several months. The fact that she and her husband may have different grieving styles can be discussed. If the spouse and the family can be involved in the evaluation and similarly informed, they will be better able to work through their own grief and support the patient as well.

If the patient has had a stillbirth or perinatal death,

the couple should be helped to decide whether they wish to see and hold the body, take a picture of it, and name the child. Although the couple may want to "forget" the unfortunate event, they should be educated that many couples later found it helpful to have tangible memories of their child (Macey & Harmon, 1987). They may need assistance with practical issues, such as making funeral arrangements or memorial services and what to tell family and friends. The psychiatrist can help them anticipate the potential reactions of others, so the couple are less vulnerable to unempathic comments. It is also useful to anticipate with the couple how they may feel when they return home and what they will do with any preparations they have made for the child.

The psychiatrist might also talk with obstetric staff members to assess their reactions and allow ventilation so they are better able to deal with the patient. Some hospitals have found it useful to arrange case conferences to review these cases and provide support for the staff.

After discharge, the couple should be encouraged to call if they desire. Follow-up obstetric appointments 1 week and 1 month after discharge can be arranged; more frequent appointments should be arranged if the couple or the woman wishes them. The couple can be educated about the importance of supportive or group therapy in decreasing symptomatology (Forrest, Standizer, & Baum, 1982). If the couple desires further mental health treatment, it can be arranged as well.

The question of when to seek a new pregnancy may be raised. Although the wish to replace the lost child and urgency related to other conditions, such as infertility, may push the couple to see a new pregnancy immediately, these concerns must be balanced against the importance of grief reaction resolution for facilitating attachment to a new child. Particularly for third trimester losses, stillbirths, and perinatal losses, the couple should be advised of this point and encouraged to give themselves some time, perhaps 6 months, before seeking a new pregnancy (Macey & Harmon, 1987).

POSTPARTUM PSYCHIATRIC DISORDERS

Postpartum psychiatric disorders span a spectrum of conditions that occur from a few days to several months after delivery. There are three main categories.

1. "*Maternity blues*," a transient condition, affects 50 to 80% of new mothers within the first week after delivery and usually resolves within 2 to 3 weeks. It is characterized by emotional lability (particularly tearfulness and irritability), sleep disturbance, fatigue, and mild confusion.

2. *Postpartum depression*, a condition that is more common than generally realized, affects 10 to 15% of new mothers. Its initial presentation may resemble "the blues," but is distinguished by its persistence beyond the first postpartum month and by the severity of the symptoms. This condition is consistent with *DSM* criteria for major depression or bipolar disorder, depressed.

3. *Postpartum psychosis*, a much less common condition, affects only 0.1% of new mothers. It is defined as a psychosis occurring within 6 months of delivery and is consistent with *DSM* diagnoses of brief psychotic reaction, schizophreniform disorder, schizophrenia, and bipolar disorder with manic episode (Inwood, 1985).

The role of biologic contributions to these disorders has long intrigued clinicians. The biologic changes associated with childbirth, particularly the rapid drop of estrogen and progesterone levels, have been hypothesized to affect CNS neurotransmitter levels and to play a "predisposing, if not causal role" in the etiology of these disorders (Campbell & Winokur, 1985). Efforts to prevent and treat these disorders with hormone replacement have, however, yielded equivocal results. Genetic vulnerability does seem to play a role, as there is an increased lifetime incidence of depression and a higher rate of positive family histories of mental disorders in those who develop postpartum disorders (Campbell & Winokur, 1985). Psychosocial factors also affect the development of these disorders. Risk factors include an unwanted pregnancy, an unstable or absent marital relationship, and a lack of social supports. Women who have had a complicated delivery or a premature, abnormal, sick, or "difficult" child are more at risk for postpartum disorders than are those who have lost the child through stillbirth or perinatal death (Pitt, 1985). Women who have a poor relationship and identification with their mothers are also thought to be more at risk for psychologic difficulties (Melges, 1968).

Psychiatric evaluation of postpartum disorders should include a detailed history of the pregnancy, the delivery, and postpartum adjustment, as well as information about past psychiatric illness, family history, and social history, especially the availability of such supports as the father of the child, family members, and friends. A complete mental status examination should be performed. Although cognitive impairment disorders are now uncommon sequelae of childbirth, they must be ruled out. The possibility of hypothyroidism due to thyroiditis or pituitary insufficiency should be evaluated with a thyrotropin (TSH) level. Endocrine consultation is relevant if abnormal vital signs or laboratory values suggest pituitary disease. The presence of psychotic symptoms, such as delusions and hallucinations, should be addressed. The woman's ability to care for the child must also be evaluated. Reports of difficulty feeding and of infant weight loss may be early indicators of failure to thrive. Special attention must be given to evaluating suicidal or infanticidal ideation, as they constitute psychiatric emergencies.

The mainstays of treatment for the "maternity blues" are education and reassurance that the feelings will subside in a matter of weeks. If symptoms last longer than 3 to 4 weeks, however, full psychiatric evaluation is indicated.

Management of a postpartum depression involves the standard approaches to major depression, including psychotherapy and medication. If the patient can be treated as an outpatient, support systems must be mobilized to help provide appropriate care for the child. If psychiatric hospitalization is needed, consider admitting the child concurrently in order to facilitate maternal-infant attachment and reinforce mothering behavior in the patient. Involve the father as much as possible in the treatment plan. If the child is not admitted, arrangement of the child's care must be made with the father, family, friends, or social services. In patients who have had documented episodes of postpartum depression, consideration of instituting antidepressants immediately after delivery of subsequent children should be considered. The safety of breast feeding, however, must be evaluated as discussed earlier.

The treatment of postpartum psychosis also requires standard treatment approaches for the specific diagnosis. Because of the unpredictable nature of psychosis, most of these patients require hospitalization. As already mentioned, care for the child must be arranged. If the mother's mental condition appears to be chronic, questions about her ability to care for the child are likely to be raised during discharge planning. For some patients, social services must be involved at this time, to evaluate the need for foster care or supervised placement of the child with the mother. For others, regular pediatric follow-up is needed to monitor the situation. If visiting nurse service is available, it can also provide support and monitoring.

MENOPAUSE

Menopause is the natural cessation of menstruation that occurs during the fifth and sixth decades of life, the average age being 51 years. Historically, the stereotype of this time in a woman's life emphasized mood instability, "involutional melancholia," and impaired social functioning. In contrast, more modern data indicate that not only is there no increase in the rate of depression there is actually a decrease in the rate of depression compared to that in younger cohorts (Klerman & Weismann, 1989; Weissman, 1979). Furthermore, general population surveys indicate that there is *no* increase in somatic or mood symptoms at this transition (Matthews et al., 1991; McKinlay, McKinlay, & Brambilla, 1987). Menopause is not the unpleasant state, or "disease," it was once thought to be.

For many women the menopause is a relief because of the cessation of menstrual bleeding, dysmenorrhea, and need for contraception. Some miss the sign of their childbearing potential, and others wonder about their femininity and sexual attractiveness; but most women make this transition without difficulty.

Nevertheless, some patients do seek their physician's care. Whereas primary care physicians are comfortable answering questions about hormone replacement, hot flashes, osteoporosis, and cardiovascular risks related to menopause, they may call for psychiatric consultation when complaints of mood instability, irritability, depression, fatigue, poor concentration, and anxiety fail to respond to hormone replacement strategies. What is the significance of these complaints in light of the studies cited above?

Although most women experience no difficulty with menopause, those seeking assistance are truly distressed. In one study (Anderson et al., 1987) more than 30% of the women who attended a menopause clinic met criteria for depression. Clinical experience suggests that such women have a preexisting depression that is either exacerbated by the menopause or comes to the physician's attention because of menopause. Indeed, risk factors for menopausal depression include previous depression and negative beliefs about menopause. Cultural expectation of difficulty at menopause gives legitimacy to a woman's complaints at that time. Some women encounter stressful life events during that period, such as caretaking or loss of elderly parents, marital difficulties, financial difficulties, or health problems; and they may attribute normal emotional reactions to menopause. For these women, the experience of irregular uterine bleeding and sleep disturbance due to night sweats may make coping all the more difficult.

Evaluation of the menopausal woman should include a complete investigation of all physical and psychologic complaints. The FSH level is elevated in perimenopausal women, so a FSH assay can help establish that the woman is in fact menopausal. A thorough psychosocial history helps evaluate contributions from situational issues. The differential diagnosis should include mood disorders, personality disorders, anxiety disorders, and substance abuse. Cognitive disorders must also be ruled out, particularly those due to thyroid dysfunction. The menopausal woman with marked paroxysmal symptoms, such as dizzy spells and memory lapses, should undergo electroencephalography (EEG) as a screen for temporal lobe epilepsy, a disorder that can be exacerbated at menopause.

Treatment involves education about the range of symptoms associated with menopause and their relation to other factors that contribute to the patient's distress. Hormone replacement therapy (HRT) (with estrogen and progestin) can be considered especially for those with hot flashes, night sweats, and insomnia. HRT is now recommended for all menopausal women because of its benefits for osteoporosis and reducing cardiovascular risks. Exceptions are women who have suffered breast cancer or uterine cancer, liver disease, or thrombophlebitis. Women who are menopausal secondary to hysterectomy require estrogen replacement only, as the progestin component is not needed to prevent endometrial cancer. Anecdotally, some women are reported to do better with hormone replacement therapy that includes a low-dose androgen to replace the ovarian androgens lost at menopause (see below).

If a mental disorder is present, the clinician must clarify that the problem is distinct from menopause. The patient should then receive the treatment appropriate for that disorder (e.g., antidepressant medication and psychotherapy for depression). This step should occur independently of the decision to pursue HRT, as there is no evidence that HRT is an effective treatment for any mental disorder (Ballinger, 1990). For women whose difficulties are related to situational factors, psychotherapy addressing those issues is indicated.

Sexual dysfunction is another complaint associated with menopause. If this symptom is related to thinning and drying of the vaginal epithelium due to reduced levels of estrogen, it can be treated with topical estrogen cream or HRT if desired. Another potential complaint is decreased sexual desire. This symptom can be related to lowered androgen levels when the ovaries cease production of gonadal steroids. Although the differential diagnosis of this symptom should include the standard evaluation of decreased sexual desire applied to any age group, when other problems such as marital discord or medical illness have been ruled out there is evidence that low doses of replacement testosterone with estrogen can be beneficial (Ballinger, 1990, p. 783).

As women are living longer and enjoying better health, we will see continued efforts to "depathologize" menopause and view it as a continuation of normal growth and development—the transition to a productive middle age and later life.

REFERENCES

Adams P, Rose D, Folkard J, et al. (1973). Effect of pyridoxine on depression associated with oral contraceptives. Lancet 1:897–904.

Adler N (1980). Psychosocial issues of therapeutic abortion. In DD Young Ehrhardt (Eds.), Psychosomatic obstetrics and gynecology (Ch. 11, pp. 159–180). New York: Appleton-Century-Crofts.

American Academy of Pediatrics, Committee on Drugs (1989). Transfer of drugs and other chemicals into human milk. Pediatrics 84:924–936.

American Psychiatric Association (1987). Diagnostic and statistical manual (3rd ed., revised) (pp. 367–369). Washington, DC: American Psychiatric Association.

Ananth J (1978). Side effects in the neonate from psychotropic agents excreted through breast feeding. Am J Psychiatry 135:801–805.

Andersch B, Wenderstam C, Hahn L, et al. (1986). Premenstrual complaints—prevalence of premenstrual symptoms in a Swedish urban population. J Psychosom Obstet Gynecol 5:39–40.

Anderson E, Hamburger, H, Liu JH, et al. (1987). Characteristics of menopausal women seeking assistance. Am J Obstet Gynecol 156:428–433.

Ashton J (1980). The psychological outcome of induced abortion. Br J Obstet Gynecol 87:1115–1122.

Aycub C, & Pfeifer D (1977). The prophylaxis of child abuse and neglect. Child Abuse Negl 1:71–75.

Bader TF, & Newman K (1980). Amitriptyline in human breast milk and the nursing infant's serum. Am J Psychiatry 137:855–856.

Baker MG (1968). Psychiatric illness after hysterectomy. BMJ 2:91–95.

Ballinger CB (1990). Psychiatric aspects of the menopause. Br J Psychiatry 156:773–787.

Behrman S, & Kistner R (1975). Progress in infertility. Boston: Little, Brown.

Borg S, & Lasker J (1982). When pregnancy fails. Boston: Beacon Press.

Brewer C (1977). Incidence of post-abortion psychosis: a prospective study. BMJ 1:476–477.

Briggs G, Freedman R, & Yaffe S (Eds) (1990). Drugs in pregnancy and lactation (2nd ed.). Baltimore: Williams & Wilkins.

Calabrese J, & Gulledge A (1985). Psychotropics during pregnancy and lactation: a review. Psychosomatics 26:413–426.

Campbell J, & Winokur G (1985). Postpartum affective disorders: Selected biological aspects. In D Inwood (Ed.), Postpartum psychiatric disorders (Ch. 2, pp. 19–39). Washington, DC: American Psychiatric Press.

Castlenuovo-Tedesco P, & Krout B (1970). Psychosomatic aspects of chronic pelvic pain. Psychiatr Med 1:109–126.

Cohen L, Heller V, & Rosenbaum J (1991). Psychotropic drug use in pregnancy: an update. In A Stoudemire & B Fogel (Eds.), Medical psychiatric practice (Ch. 18, pp. 615–634). Washington, DC: American Psychiatric Press.

Collins J, Wrixon W, James L, et al. (1983). Treatment independent pregnancy among infertile couples. N Engl J Med 309:1201–1205.

Cullberg J (1972). Mood changes and menstrual symptoms with different gestagen/estrogen combinations: a double blind comparison with placebo. Acta Psychiatr Scand 236:1–86.

DeJong R, Rubinow DR, Roy-Burne P, et al. (1985). Premenstrual mood disorder and psychiatric illness. Am J Psychiatry 142:1359–1361.

Downing J (1987). Impaired fertility. Unpublished manuscript.

Ekblad M (1955). Induced abortion on psychiatric grounds: follow-up study of 479 women. Acta Psychiatr Scand (suppl 99):1–238.

Eriksson E, Lisjo P, Sundblad C, et al. (1990). Effect of clomipramine on premenstrual syndrome. Acta Psychiatr Scand 81:85–87.

Fink M (1981). Convulsive and drug therapies of depression. Annu Rev Med 32:405–412.

Forest G, Standizer E, & Baum J (1982). Support after perinatal death: a study of support and counseling after perinatal bereavement. BMJ 2:1475–1479.

Forrssman H (1955). Follow-up of sixteen children whose mothers were given electric convulsive therapy during gestation. Acta Psychiatr Neurol Scand 30:437–441.

Friedman R, & Gradstein B (1982). Surviving pregnancy loss. Boston: Little, Brown.

Gabinet L (1986). A protocol for assessing competence to parent a newborn. Gen Hosp Psychiatry 8:263–272.

Gath D, Cooper P, & Bond A, et al. (1982a). Hysterectomy and psychiatric disorder. II. Demographic, psychiatric and physical factors in relation to psychiatric outcome. Br J Psychiatry 140:343–350.

Gath D, Cooper P, & Day A (1982b). Hysterectomy and psychiatric disorder. I. Levels of psychiatric morbidity before and after hysterectomy. Br J Psychiatry 140:335–342.

Gath D, & Rose N (1985). Psychological problems and gynecological surgery. In R Priest (Ed.), Psychological disorders in obstetrics and gynecology (Ch. 2, pp. 31–48). London: Butterworth.

Glick ID, & Bennett S (1982). Psychiatric complications of progesterone and oral contraceptives. J Clin Psychopharmacol 1:350–367.

Gold J, & Josimovich J (1987). Gynecologic endocrinology. New York: Plenum.

Grant E, & Pryse-Davies J (1968). Effects of oral contraceptive on endometrial monoamine oxidase and phosphates. BMJ 3:777–780.

Gray JD, Cutler C, Dean J, et al. (1979). Prediction and prevention of child abuse and neglect. J Soc Issues 35:127–139.

Guttmacher Institute (1981). Teenage Pregnancy: The problem that hasn't gone away. New York: Alan Guttmacher Institute.

Hamilton J, Parry B, Alagna S, et al. (1984). Premenstrual mood changes: a guide to evaluation and treatment. Psychosomatics 14:426–435.

Harrison W, Endicott J, & Nee J (1989). Treatment of premenstrual depression with nortriptyline: a pilot study. J Clin Psychiatry 50:136–139.

Hatcher R, Guest F, Stewart F, et al. (1986). Contraceptive technology 1986–87 (13th ed). New York: Irvington Publishers.

Hauser L (1985). Pregnancy and psychotropic drugs. Hosp Community Psychiatry 36:817–818.

Heinonen O, Slone D, & Shapiro S (1977). Birth defects and drugs in pregnancy (Ch. 24, pp. 336–337). Littleton, MA: Publishing Sciences Group.

Impastato D, Gabriel A, & Landaro H (1964). Electric and insulin shock therapy during pregnancy. J Nerv Ment Dis 25:542–546.

Inwood D (1985). The spectrum of postpartum psychiatric disorders. In D Inwood (Ed.). Postpartum psychiatric disorders (Ch. 1, pp. 1–18). Washington, DC: American Psychiatric Press.

James, ME (1991). Cocaine abuse during pregnancy: psychiatric considerations. Gen Hosp Psychiatry 13:399–409.

Johnson SR, McChesney C, & Dean JA (1988). Epidemiology of premenstrual symptom in a non-clinical sample. I. Prevalence, natural history, and help-seeking behavior. J Reprod Med 33:340–346.

Jones K, Lacro R, Johnson R, et al. (1989). Pattern of malformations in children of women treated with carbamazepine during pregnancy. N Engl J Med 320:1661–1666.

Kellner R, Buchman M, Fava G, et al. (1984). Hyperprolactinemia, distress and hostility. Am J Psychiatry 141:759–763.

Klerman G, & Weissman MW (1989). Increasing rates of depression. JAMA 261:2229–2235.

Kravitz A, Notman M, Anderson J, et al. (1976). Outcome following therapeutic abortion. Arch Gen Psychiatry 33:725–733.

Kresch A, Steifer D, Sach L, et al. (1984). Laparoscopy in 100 women with chronic pelvic pain. Obstet Gynecol 64:672–674.

Laegreid L, Olegard R, Wahlstrom J, et al. (1987). Abnormalities in children exposed to benzodiazepines in utero. Lancet 1:108–109.

Lalinec-Michaud M, & Engelsmann F (1984). Depression and hysterectomy. Psychosomatics 25:550–558.

Lalinec-Michaud M, & Engelsmann F (1985). Anxiety, fears, and depression related to hysterectomy. Can J Psychiatry 30:44–47.

Lundberg W, Well J, & Mathers J (1973). Laparoscopy in evaluation of pelvic pain. Am J Obstet Gynecol 42:872–876.

Macey T, & Harmon R (1987). Perinatal loss. Unpublished manuscript.

Martin R, Roberts W, & Clayton PJ (1980). Psychiatric status after hysterectomy: one year prospective follow-up. JAMA 244:350–353.

Matheson I, Pande H, & Alertsen AR (1985). Respiratory depression caused by N-desmethyldoxepin in breast milk [letter]. Lancet 2:1124.

Matthews KA, Wing RR, Kuller LH, et al. (1990). Influences of natural menopause on psychological characteristics and symptoms of middle aged healthy women. J Consult Clin Psychol 58:345–351.

McCance C, Olley P, & Edward V (1973). Long-term psychiatric follow-up. In Y Horobin (Ed.), Experience with abortion (Ch. 6, pp. 245–300). London: Cambridge University Press.

McKinlay JB, McKinlay SM, & Brambilla DJ (1987). Health states and utilization behavior associated with menopause. Am J Epidemiol 125:110–121.

Melges F (1968). Postpartum psychiatric syndromes. Psychosom Med 30:95–108.

Mishell D (1989). The pharmacologic and metabolic effects of oral contraceptive. Int J Fertil 34(suppl):21–26.

Moore J, & Tolley D (1976). Depression following hysterectomy. Psychosomatics 17:86–89.

Myers J, Weissman M, Tischer G, et al. (1984). Six month prevalence of psychiatry disorders in three communities. Arch Gen Psychiatry 41:959–967.

National Center for Health Statistics (1985). Detailed diagnosis and surgical procedures for patients discharged from short stay hospitals: United States, 1983. Vital and Health Statistics Series 13, No. 82. Washington, DC: US Government Printing Office.

National Institute of Mental Health Consensus Conference on Electroconvulsive Therapy (1985). JAMA 254:2103–2108.

Nieburg P, Marks JS, McLaren NM, et al. (1985). The fetal tobacco syndrome. JAMA 253:2998–2999.

Orenstein H, & Raskind M (1983). Polycystic ovary disease in two patients with Briquet's disorder. Am J Psychiatry 140:1201–1204.

Orenstein H, Raskind M, Wylie D, et al. (1986). Polysymptomatic complaints and Briquet's syndrome in polycystic ovary disease. Am J Psychiatry 143:768–771.

Parry B, & Rush A (1979). Oral contraceptives and depressive symptomatology: biologic mechanisms. Comp Psychiatry 20:347–358.

Pasnau R, Soldinger S, & Andersen B (1985). Pelvic pain. In R Priest (Ed.), Psychological disorders in obstetrics and gynecology (Ch. 3, pp. 49–69). London: Butterworth.

Peters A, VanDoist E, & Jellis B (1991). A randomized clinical trial to compare two different approaches to women with chronic pelvic pain. Obstet Gynecol 77:740–744.

PITT B (1985). The puerperium. In R PRIEST (Ed.), Psychological disorders in obstetrics and gynecology (Ch. 6, pp. 147–172). London: Butterworth.

REMENTERIA J, BHATT K (1977). Withdrawal symptoms in neonates from intrauterine exposure to diazepam. J Pediatr 90:123–126.

REMICK R, & MAURICE W (1978). ECT in pregnancy. Am J Psychiatry 125:761–762.

RENAER M (1980). Chronic pelvic pain without obvious pathology in women. Eur J Obstet Gynecol Reprod Biol 10:415–463.

RENAER M (1981). Chronic pelvic pain in women. Berlin: Springer-Verlag.

RENAER M, VERTOMMEN H, NIJS P, ET AL. (1979). Psychological aspects of chronic pelvic pain in women. Am J Obstet Gynecol 134:75–80.

REPKE J, & BERGER N (1984). Electroconvulsive therapy in pregnancy. Obstet Gynecol 63(suppl 3):39S–41S.

RICHARD DH (1974). A post-hysterectomy syndrome. Lancet 2:983–985.

RICHELS K, FREEMAN E, & SONDHEIMER S (1989). Buspirone in treatment of premenstrual syndrome [Letter]. Lancet 1:777.

RICHELS K, FREEMAN EW, SONDHEIMER S, ET AL. (1990). Fluoxetine in the treatment of premenstrual syndrome. Curr Ther Res 48(1):161–166.

ROSE D, STRONG R, ADAMS P, ET AL. (1972). Experimental vitamin B_6 deficiency and the effect of estrogen containing oral contraceptives on tryptophan metabolism and vitamin B_6 requirements. Clin Sci 42:465–477.

ROSENTHAL R, LING F, ROSENTHAL T, ET AL. (1984). Chronic pelvic pain: psychological features and laparoscopic findings. Psychosomatics 25:833–841.

RUBINOW DR, HOBAN MC, GROVER GN, ET AL.. (1988). Changes in plasma hormones across the menstrual cycle in patients with menstrually-related mood disorder and in control subjects. Am J Obstet Gynecol 158:5–11.

SCHMIDT PJ, ROSENFELD D, MULLER KL, ET AL. (1990). A case of autoimmune thyroiditis presenting as menstrual related mood disorder. J Clin Psychiatry 51:434–436.

SCHOU M (1976). What happened later to lithium babies? A follow-up study of children born without malformations. Acta Psychiatr Scand 54:193–197.

SEIBEL MM, & TAYMOR ML (1982). Emotional aspects of infertility. Fertil Steril 37:137–145.

SENAY E (1983). Substance abuse disorders in clinical practice. Littleton, MA: John Wright, PSG.

SEVERINO S, & MOLINE M (1989). Premenstrual syndrome: a clinician's guide. New York: Guilford Press.

SHOUPE D, & MISHELL D (1989). Norplant: Subdermal implant system for long term contraception. Am J Obstet Gynecol 160:1286–1292.

SIMON E (1972). A follow-up study of women who request abortion. Am J Orthopsychiatry 43:574–585.

SLAP G (1981). Oral contraceptives and depression. J Adolesc Health Care 2:53–64.

SOBEL D (1960). Fetal damage due to ECT, insulin, coma, chlopromazine, and reserpine. Arch Gen Psychiatry 2:606–610.

SOUMENKOFF G, MARNEFE C, GERARD M, ET AL. (1981). A coordinated attempt for prevention of child abuse at the antenatal care level. Child Abuse Negl 6:87–94.

STANCER HC, & REED KL (1986). Desipramine and 2-hydroxydesipramine in human breast milk and the nursing infant's serum. Am J Psychiatry 143:1594–1600.

STEEGE JF, STOUT AC, & RUPP SL (1988). Clinical features. In WR JR KEYE (Ed.), The premenstrual syndrome (Ch. 7, pp. 113–127). Philadelphia: WB Saunders.

STONE AB, PEARLSTEIN TB, & BROWN WA (1991). Fluoxetine in the treatment of late luteal phase dysphoric disorder. J Clin Psychiatry 52:290–293.

STOTLAND NL, & SMITH TE (1990). Psychiatric consultation to obstetrics and gynecology: systems and syndromes. In A TASMAN, SM GOLDFINGER, & CA KAUFMANN (Eds.), Review of psychiatry (Vol 9; Ch. 27, pp. 537–566). Washington, DC: American Psychiatric Press.

STOUDEMIRE A, & SANDHU J (1987). Psychogenic/idiopathic pain syndromes. Gen Hosp Psychiatry 9:79–86.

THIENHAUS O, & HARTFORD J (1986). Depression in hyperprolactinemia. Psychosomatics 27:663–664.

THOMSON P, & TEMPLETON A (1978). Characteristics of patients requesting reversed sterilization. Br J Obstet Gynecol 85:161–164.

TUPPER C, & WEIL RJ (1962). The treatment of habitual aborters by psychotherapy. Am J Obstet Gynecol 83:421–424.

U.S. Bureau of the Census (1985). Statistical abstract of the United States, 1986. Washington, DC: US Government Printing Office.

VARAN L, GILLIESON M, SKENE D, ET AL. (1985). ECT in an acutely psychotic pregnant woman with actively aggressive (homicidal) impulses. Can J Psychiatry 30:363–367.

WALKER E, KATON W, HARROP GRIFFITHS J, ET AL. (1988). Relationship of chronic pelvic pain to psychiatric diagnosis and childhood sexual abuse. Am J Psychiatry 145:75–80.

WALKER E, ROY-BURNE P, & KATON W (1990). An open trial of nortriptyline and fluoxetine in women with chronic pelvic pain. Abstracts Annual Meeting of American Society for Psychosomatic Obstetrics and Gynecology, New York City.

WARNES H, & FITZPATRICK C (1979). Oral contraceptives and depression. Psychosomatics 20:187–194.

WEISSMAN MM (1979). The myth of involutional melancholia. JAMA 242:742–744.

WEISSMAN M, & SLABY A (1973). Oral contraceptive and psychiatric disturbance: evidence from research. Br J Psychiatry 123:513–518.

WILCOX L, CHU S, EAHER E, ET AL. (1991). Risk factors for regret after tubal sterilization: 5 years of followup in a prospective study. Fertil Steril 55:927–933.

WISE M, WARD S, TOWNSEND-PARCHMAN W, ET AL. (1984). Case report of ECT during high risk pregnancy. Am J Psychiatry 141:99–101.

ZALSTEIN E, KOREN Z, EINARSON T, ET AL. (1990). A case controlled study on the association between first trimester exposure to lithium and Ebstein's anomaly. Am J Cardiol 65:817–818.

ZUSPAN F, & RAYBURN W (1986). Drug abuse during pregnancy. In W RAYBURN & F ZUSPAN (Eds.), Drug therapy in obstetrics and gynecology (Ch. 4, pp. 37–52). Norwalk, CT: Appleton-Century-Crofts.

29 Endocrine disorders

SUSAN G. KORNSTEIN, M.D., AND
DAVID F. GARDNER, M.D.

Psychiatric symptomatology in endocrine disorders has long been recognized. Cognitive, affective, and behavioral changes were noted in the earliest accounts of Addison (1868), Cushing (1932), Sheehan (1939), and others. These symptoms often precede or present simultaneously with the more definitive signs and symptoms of endocrine disease and may confuse even the most astute diagnostician. Symptoms may be mislabeled during early stages as "neurotic" in nature and during more advanced stages as dementia or psychotic disorders. Particularly in the elderly, in whom the physical manifestations of an endocrinopathy may not be evident and the mental changes mistaken for senility, the diagnosis of an endocrine disorder may easily be missed.

In this chapter, we focus on the endocrine disorders in which psychiatric syndromes are most often encountered, specifically, those involving the thyroid and adrenal glands, and disorders of calcium homeostasis. We also examine complications of abnormal pituitary function, diabetes mellitus, hypoglycemia, pheochromocytoma, and male hypogonadism. In addition, anabolic steroids and hirsutism are briefly discussed. Readers should also note that endocrine tests are also discussed in some detail in Chapter 7.

The emphasis of the chapter is on the differential diagnosis of the psychiatric syndromes and on screening diagnostic studies. Our contention is that, whenever possible, treatment should first be directed at the underlying endocrine disorder, as resolution of the psychiatric disturbance usually follows. However, some clinicians hold that when disabling mental symptoms exist, psychotropics should be used until endocrine homeostasis is achieved. When solid scientific evidence exists for either efficacy or significant complications of such use, these studies are highlighted.

THYROID DISEASE

Hyperthyroidism

Hyperthyroidism (or thyrotoxicosis) is a clinical syndrome of diverse etiologies that have in common an increase in circulating thyroid hormone concentrations. The clinical signs and symptoms of this disorder reflect the effects of these excess thyroid hormone levels on a variety of organ systems. Various surveys have estimated the annual incidence of hyperthyroidism to be approximately 0.2 per 1000 patients per year, with women affected five to seven times more frequently than men (McKenzie & Zakarija, 1989). The peak incidence in women occurs between 20 and 40 years of age.

The most common cause of hyperthyroidism is Graves' disease, a systemic autoimmune disorder characterized by diffuse thyroid enlargement and extrathyroidal manifestations involving primarily the eyes (ophthalmopathy) and skin (dermopathy). Graves' disease, toxic nodular goiter, subacute thyroiditis, and excessive thyroid hormone replacement account for more than 95% of all cases of hyperthyroidism (Spaulding & Lippes, 1985). Other rare causes include thyrotropin-secreting pituitary tumors, trophoblastic tumors, struma ovarii, and iodine-induced thyrotoxicosis. The differential diagnosis should also include consideration of factitious thyroid hormone ingestion (Mariotti et al., 1982).

Symptoms of hyperthyroidism vary from patient to patient, but the following are most often reported: nervousness, increased sweating, heat intolerance, fatigue, dyspnea, palpitations, weight loss despite increased appetite, eye symptoms, muscle weakness, hair loss, and hyperdefecation. Menstrual cycle abnormalities, including anovulation, amenorrhea, oligomenorrhea, and menometrorrhagia, may also be observed. The term "masked," or "apathetic," hyperthyroidism describes the atypical hyperthyroid patient who lacks many of the classic symptoms of the disorder (Thomas, Mazzaferri, & Skillman, 1970). These patients are usually elderly and present with unexplained weight loss, muscle weakness, atrial fibrillation, and heart failure. The diagnosis is often overlooked because the patients lack the more typical symptoms of adrenergic overactivity and may in fact appear hypothyroid.

The patient's general appearance may be the first clue that thyroid hyperfunction is present. The typical patient is hyperactive and appears to have lost weight. Speech may be rapid and rambling, and facial features may reflect anxiety and apprehension. Other important physical findings include a resting tachycardia (occasionally associated with atrial fibrillation), warm, smooth, moist skin, often with a velvety texture, fine hair, thyroid enlargement, fine tremor of the hands, and hyperactive deep tendon reflexes. The nature of the thyroid enlargement depends on the etiology of the hyperthyroidism (e.g., diffuse in Graves' disease, nodular in toxic multinodular goiter, tender in subacute thyroiditis). Lid lag and lid retraction, resulting in a characteristic stare, may be seen in all hyperthyroid patients. Infiltrative eye signs (exophthalmos, conjunctival inflammation, and extraocular muscle palsies) are specific for Graves' disease. Patients with long-standing, untreated hyperthyroidism may manifest a proximal myopathy associated with frank muscle wasting.

Mental changes in hyperthyroidism are common and may be the initial complaint (Hall, 1983). The most frequent psychiatric presentation is anxiety with irritability, emotional lability, a feeling of apprehension, and inability to concentrate. Cognitive changes include short attention span, distractibility, and decreased recent memory, but the impairment is generally less severe than in hypothyroidism. Psychosis is also less common and less severe; there may be rapid speech and psychomotor agitation or frank psychosis with paranoid and grandiose delusions. With thyroid storm, visual hallucinations and delirium are not uncommon.

Symptoms of hyperthyroidism are most frequently misdiagnosed as anxiety disorders. Features that help distinguish hyperthyroidism from an anxiety disorder include the findings of cognitive impairment, tachycardia, fatigue accompanied by a desire to be active, constant rather than intermittent anxiety, and palms that are warm and dry, rather than cool and clammy (Popkin & Mackenzie, 1980). Apathetic hyperthyroidism may be mistaken for a primary depressive disorder or dementia. Other psychiatric misdiagnoses of hyperthyroidism include bipolar disorder, anorexia nervosa, schizophrenia, substance abuse, psychotic depression, and cognitive impairment disorders of other etiology. In the case of a manic-like presentation of hyperthyroidism, treatment with lithium has been reported to produce dramatic short-term improvement; however, rapid relapse into the hyperthyroid state occurs (Hall, 1983).

The diagnosis of hyperthyroidism rests on the demonstration of elevated concentrations of circulating thyroid hormones. Two different, though complementary, diagnostic strategies are now available for the initial screening of patients with suspected hyperthyroidism. The first involves the demonstration of elevated concentrations of circulating thyroid hormones. An elevated serum *free* thyroxine (T_4) in a patient with appropriate signs and symptoms is sufficient for the diagnosis of hyperthyroidism in most clinical settings. Elevation of the serum *total* T_4 concentration is not diagnostic, as alterations in serum thyroid hormone binding proteins, particularly thyroxine-binding globulin (TBG), may increase the total T_4 concentration without affecting the free hormone concentration. Because it is the free hormone that is biologically active, initial studies should measure the free T_4 in serum. In most clinical laboratories, the free T_4 concentration is estimated using the free thyroxine index (FTI), which is calculated from the total T_4 and triiodothyronine (T_3) resin uptake (T_3RU) test. In addition, there are now several methods for *direct* measurement of the free T_4 concentration, although most hospital laboratories continue to use the FTI to estimate free T_4 concentration.

The availability of "sensitive" thyroid-stimulating hormone (TSH) assays has provided an alternative approach to the initial diagnosis of hyperthyroidism (Toft, 1988). In the past, the clinical utility of serum TSH measurements in the diagnosis of hyperthyroidism was limited by the inadequate sensitivity of the assay. Newer assays, with sensitivities below 0.1 μU/ml, can now readily discriminate between the low values occasionally seen in normals and the markedly depressed values seen in patients with hyperthyroidism. A low serum TSH must be followed by measurement of the free T_4 concentration, as not all patients with suppressed serum TSH concentrations are hyperthyroid (Ehrmann & Sarne, 1989; Sawin, Geller, & Kaplan, 1991). A normal serum TSH concentration, however, is strong evidence against the diagnosis of hyperthyroidism.

An occasional hyperthyroid patient with a low TSH level has a normal serum free T_4 concentration. This patient should be screened with a serum T_3 measurement. The syndrome of hyperthyroidism with a normal free T_4 concentration and an elevated T_3 concentration has been called T_3-hyperthyroidism, or T_3-thyrotoxicosis, and probably accounts for about 5% of all patients presenting with hyperthyroidism (Medeiros-Neto, 1986). Patients with normal serum free T_4 and T_3 concentrations but suppressed TSH levels have been categorized as having "subclinical hyperthyroidism."

To a great extent, the sensitive TSH assay has replaced the thyrotropin-releasing hormone (TRH) test for clinical evaluation of patients with suspected thyroid dysfunction. In normals, the TSH concentration rises rapidly after intravenous TRH, usually peaking between 15 and 30 minutes. In hyperthyroid patients, the TSH response is blunted or absent. A normal TSH response effectively excludes the diagnosis of hyperthyroidism. Several studies have now documented that a low basal serum TSH level is highly predictive of a blunted TSH response to TRH, so the more cumbersome and costly TRH test can now be avoided in most patients with suspected hyperthyroidism. It should be noted that approximately 25% of euthyroid patients with depression may also demonstrate blunted TSH responses to TRH (Hein & Jackson, 1990). Blunted responses occur in other clinical settings, including anorexia nervosa, alcohol withdrawal, hypopituitarism, starvation, and glucocorticoid excess states, and in normal men over age 60 (Hein & Jackson, 1990).

Transient hyperthyroxinemia has been reported in several series of patients with acute psychiatric disorders (Spratt, Pont, & Miller, 1982). The elevated free thyroxine index is usually associated with a normal serum T_3 concentration, although elevated T_3 levels have also been reported. In most of these patients, serum thyroid hormone concentrations return to normal within 1 to 2 weeks without specific therapy, suggesting that these patients are not truly hyperthyroid. A study of 87 newly hospitalized psychiatric patients has shown that TSH levels, using a sensitive assay, are usually not suppressed at the time of hospital admission (Chopra, Solomon, & Tien-Shang, 1990). Fourteen patients (16%) had elevated TSH levels, and only one had an undetectable level. As in previous studies, these abnormalities returned to normal after several weeks of observation. These data suggest that the phenomenon of acute hyperthyroxinemia is the result of a central abnormality in hypothalamic-pituitary-thyroid regulation, rather than a primary thyroid abnormality.

A detailed discussion of therapy for hyperthyroidism is beyond the scope of this review. Although the ultimate goal of all therapeutic strategies is normalization of circulating thyroid hormone levels, initial therapy is often symptomatic in nature. Beta-adrenergic blocking agents have no direct effects on thyroid hormone secretion, but they do effectively reverse many of the signs and symptoms of hyperthyroidism. Patients report rapid alleviation of nervousness, palpitations, sweating, and tremor, as well as an improvement in their overall sense of well-being. Definitive therapy for most patients is either antithyroid drugs (propylthiouracil or methimazole) or radioactive iodine. The role of surgery is limited, reserved primarily for children and pregnant women not responding to or having an allergic reaction to antithyroid drugs. It should be noted that patients receiving radioactive iodine treatment are at lifelong risk for developing hypothyroidism. Hypothyroidism usually presents during the first 1 to 2 years after therapy, although its appearance may be delayed for more than 5 to 10 years.

As thyroid function returns to normal, anxiety, affective symptoms, and cognitive deficits usually improve (Kathol, Turner, & Delahunt, 1986). Mental impairment may persist in patients with severe or prolonged illness (Ettigi & Brown, 1978). Use of psychotropic medications should be tempered by the risk of potentially lethal complications. Tricyclic antidepressants may intensify tachycardia and evoke cardiac arrhythmias in the context of hyperthyroidism (Loosen, 1988). Although no definitive studies currently exist, potential sensitization of tissues to catecholamines by coincident exposure to a high concentration of thyroid hormone may be the cause of this phenomenon (Whybrow & Prange, 1981). There are few data on the use of monoamine oxidase inhibitors, but caution against their use is warranted because of possible effects on blood pressure (Wilson & Jefferson, 1985). Neuroleptics may exacerbate tachycardia and cardiac arrhythmias (Wilson & Jefferson, 1985). Increased risk of severe dystonic reactions with hyperthyroidism has been reported with haloperidol and fluphenazine (Witschy & Redmond, 1981).

When psychotropic medications are indicated for management of severe psychiatric symptoms, conservative regimens should be employed, with frequent monitoring of blood levels, vital signs, and the electrocardiogram (ECG). Adrenergic blockers are helpful for symptomatic relief, although the risk of further psychiatric complications must be considered (Utiger, 1984). Benzodiazepines appear to be safe and may be useful adjuncts to manage anxiety and agitation.

Hypothyroidism

Hypothyroidism is the clinical syndrome resulting from a decrease in circulating thyroid hormone concentrations. It is a common disorder, occurring most frequently in women between the ages of 40 and 60. The overall incidence is about 1.0% in women and 0.1% in men. The prevalence of the disorder increases significantly with age and may be as high as 6.0% in women over age 60 (Sawin, Castelli, & Hershman, 1985).

The most common cause of primary hypothyroidism in adults in the United States is autoimmune thyroiditis (Hashimoto's disease, or chronic lymphocytic thyroiditis). Hypothyroidism is also common following radioactive iodine therapy or surgery for hyperthyroidism. From a global perspective, iodine deficiency is an important cause of hypothyroidism, but it is rare in industrialized societies. Less common causes of hypothyroidism include external neck irradiation for malignancies, medications (e.g., lithium, iodine-containing drugs), and infiltrative disorders of the thyroid. Finally, hypothyroidism may be secondary to pituitary or hypothalamic disorders and is then referred to as secondary or tertiary hypothyroidism, respectively.

Typical symptoms of hypothyroidism include weakness, fatigue, sleepiness, cold intolerance, weight gain, constipation, hair loss, hoarseness, and nonspecific muscle aches, stiffness, and cramps. Menstrual disturbances are also common and are characterized by anovulation, oligomenorrhea, and menorrhagia. Physical findings may include bradycardia, dry, cool skin, facial puffiness, thickening of the tongue, slow speech, nonpitting edema, and delayed relaxation of the deep tendon reflexes (pseudomyotonia). Onset of signs and symptoms of hypothyroidism is often subtle and insidious; thus it may go unnoticed by the patient and relatives. This point is particularly true in the elderly, in whom a clinical diagnosis of hypothyroidism may be difficult. Because of the high prevalence of the disorder in the elderly, its subtle clinical manifestations, and ease of treatment, some authors have advocated routine screening for hypothyroidism in patients over age 60.

Mental disturbance is a prominent feature of hypothyroidism and may be the presenting complaint (Hall, 1983). Progressive cognitive dysfunction characterized by slowness in comprehension and impairment of attention, recent memory, and abstract thinking is often found. Affective disturbance is perhaps the most common clinical presentation, with dysphoric mood, psychomotor retardation or agitation, sleep disturbance, crying spells, anhedonia, decreased libido, and suicidal thoughts. A manic-like state, with agitation, paranoid ideation, and hypersexuality, is occasionally observed. Insidious personality changes may be noted, such as irritability, suspiciousness, anxiety, and social withdrawal. Severe and long-standing or rapidly progressive disease may present with generalized agitation, paranoid delusions, auditory or visual hallucinations, and clouded sensorium.

Given the wide range of psychiatric presentations and the usual subtlety of physical findings, the diagnosis of hypothyroidism can be easily missed. Personality changes may be mistaken for neurosis, cognitive changes for dementia, affective disturbance for major mood disorder (depression or mania), and psychotic features for schizophrenia or an acute paranoid state. Studies have shown that a significant number of patients referred to psychiatrists for treatment of a "primary" depression have overt, mild, or subclinical hypothyroidism (Gold & Pearsall, 1983): Patients with rapid cycling bipolar illness have also been shown to have a high prevalence of hypothyroidism (Bauer, Whybrow, & Winokur, 1990).

Definitive diagnosis of hypothyroidism in the patient with suggestive signs and symptoms requires laboratory confirmation. As most patients with hypothyroidism have primary hypothyroidism (i.e., hypothyroidism due to intrinsic thyroid disease), the test with the highest sensitivity and specificity for accurate diagnosis is the serum TSH concentration. An elevated TSH level should be confirmed by measurement of the free thyroxine index (FTI, see above). In most patients, an elevated serum TSH is associated with a low FTI. Patients with mild degrees of hypothyroidism may have an elevated serum TSH with free T_4 concentrations within the normal range. This combination of laboratory findings has been referred to as "subclinical hypothyroidism" (Cooper et al., 1984); but this term is a misnomer, as many of these patients have significant clinical manifestations of thyroid hormone deficiency. Even mild degrees of hypothyroidism may be associated with affective disturbance. Therefore in the absence of any medical contraindication, psychiatric patients with even mildly elevated serum TSH levels should undergo a therapeutic trial of thyroid hormone supplementation. Because the basal serum TSH concentration is elevated in patients with primary thyroid failure, TRH testing is usually unnecessary.

There is a subset of patients in whom measurements of serum T_4, T_3, FTI, and TSH are entirely normal but who demonstrate an exaggerated TSH response to TRH, suggesting a mild degree of primary hypothyroidism. Among a series of 100 consecutive patients admitted to a psychiatric service for depression or anergia, 15% were found to have exaggerated TSH responses to TRH, with normal baseline thyroid function studies (Gold, Pottash, & Extein, 1982). Nine of the fifteen (60%) had positive antithyroid microsomal antibody studies, suggesting an underlying autoimmune thyroiditis. Subsequent studies have also confirmed a higher than normal prevalence of antimicrosomal thyroid antibodies in patients admitted to psychiatric services with depression (Nemeroff et al., 1985). Whether depressed patients with autoimmune thyroiditis and exaggerated TSH responses to TRH should all be treated with thyroid hormone remains unclear, although such a recommendation was supported by one study (Targum, Greenberg, and Harmon, 1984). Prospective, controlled, blinded studies are needed to definitively answer this question. Several older studies have confirmed the benefits of T_3 or T_4 administration for augmenting the response to tricyclics (particularly in women) even in the absence of documented thyroid function abnormalities (Goodwin, Prange, & Post, 1982). A preliminary study has also shown improvement with high-dose levothyroxine in patients with rapid cycling bipolar disorder who are refractory to lithium and carbamazepine (Bauer & Whybrow, 1990). Until replicated, this study should be considered as mainly heuristic.

The one situation in which screening for hypothyroidism with a serum TSH level can be misleading is the rare patient with secondary or tertiary hypothyroidism. In these patients, the serum TSH concentration may be normal or low despite clinical hypothyroidism and a low FTI. Thus patients with convincing clinical evidence for hypothyroidism who *do not* have an elevated serum TSH should be further screened with measurement of the FTI. The combination of a low FTI and a normal or low TSH suggests underlying hypothalamic or pituitary disease, and these patients should be referred for more detailed endocrinologic and radiologic investigation.

The diagnosis of autoimmune thyroiditis can be confirmed by the presence of circulating antithyroid antibodies. The single most sensitive test is the antithyroid microsomal antibody titer. High titers of antithyroid antibodies in a patient with primary hypothyroidism or a goiter are virtually diagnostic of autoimmune thyroiditis (Hay, 1985). Patients with this disorder are at increased risk for other autoimmune endocrine disorders, including Addison's disease, hypoparathyroidism, type I diabetes mellitus, and premature ovarian failure (Trence, Morley, & Handwerger, 1984). For reasons noted above, some authors have advocated *routine* screening of depressed patients for the presence of antithyroid antibodies in addition to routine thyroid function studies (Hein & Jackson, 1990). Patients with antibodies or an elevated serum TSH concentration would then be considered for a trial of thyroid hormone therapy. Detailed prospective studies are needed before this recommendation should be generally applied.

In general, psychotropic agents have few direct effects on circulating thyroid hormone concentrations. The major exception is lithium, which has important inhibitory effects on thyroid hormone synthesis and secretion. Although the incidence of hypothyroidism in lithium-treated patients varies among series, approximately 5 to 10% of patients receiving long-term lithium therapy develop hypothyroidism (Yassa, Saunders, & Nastase, 1988). The risk of developing hypothyroidism appears to be increased in women (Reus, 1989), in patients with rapid cycling bipolar disease (Cowdry et al., 1983), and in patients positive for circulating antithyroid antibodies (Emerson, Dyson, & Utiger, 1973). *All* patients on long-term lithium therapy should be periodically screened for thyroid dysfunction, even in the absence of overt clinical hypothyroidism. The development of hypothyroidism is not a contraindication for continuing lithium therapy. If a significant therapeutic response to lithium has been observed, thyroid hormone supplementation may be instituted to correct the hypothyroidism and should have no adverse impact on the treatment.

Replacement therapy in hypothyroidism should be in the form of L-thyroxine. Titration of the replacement dose is based on serial serum TSH determinations, with the goal of therapy being restoration of the serum TSH to the normal range (Watts, 1989). A suppressed TSH concentration in a patient receiving thyroid hormone replacement, even if associated with normal serum thyroid hormone levels, indicates overtreatment and a need to decrease the dose of thyroid supplement. Elderly patients and those with underlying cardiac disease should be started at a low initial daily dose of L-thyroxine, 0.025 to 0.050 mg per day, because of the risk of precipitating an ischemic cardiac event with full replacement doses. The dose of thyroid hormone can then be gradually titrated upward until an adequate clinical and biochemical response has been achieved. Most hypothyroid patients can be adequately replaced with a total daily dose in the range of 0.10 to 0.15 mg.

With thyroid hormone replacement, both physical and mental symptoms improve, although emotional and cognitive disturbances may persist for weeks to months. When disease has been long-standing, residual cognitive deficits may exist (Wilson & Jefferson, 1985). Psychotic symptoms usually clear within 2 to 3 weeks, especially if an antipsychotic medication is added. Exacerbation of the psychosis may occur early in the course of replacement therapy. If thyroid replacement is given in too high a dose or too rapidly, an acute delirium with psychosis may be precipitated (Josephson & Mackenzie, 1979).

If affective symptoms persist after a euthyroid state has been achieved, psychiatric intervention may be indicated. Use of tricyclic antidepressants or lithium while the patient is still hypothyroid does not augment resolution of symptoms and may induce rapid cycling (Cowdry et al., 1983; O'Shanick & Ellinwood, 1982). An additional concern is the sensitivity of hypothyroid patients to the hypotensive, sedative, and anticholinergic side effects of psychotropic medications. Chlorpromazine was reported to induce hypothermic coma in myxedematous patients in one study (Jones & Meade, 1964). Use of benzodiazepines is not typically associated with untoward effects in this group, although hypothyroid patients can have an increased sensitivity to all sedative drugs. In addition, it has been reported that hypothyroidism may precipitate or exacerbate obstructive sleep apnea (McNamara, Southwick, & Fogel, 1987); in this situation, use of benzodiazepines may worsen the sleep apnea and hence the secondary mood symptoms. If psychotropic medications are used in patients with hypothyroidism, low starting doses and less sedating drugs are recommended.

Effects of Nonthyroidal Illness and Medication on Thyroid Function

A variety of nonthyroidal illnesses may have a profound effect on circulating thyroid hormone concentrations. Even transient, mild medical illnesses may be associated with a fall in the serum T_3 concentration owing to decreased conversion of T_4 to T_3 in peripheral tissues. In more severely ill patients, decreases in the total T_4 and FTI may accompany the decline in serum T_3 levels. Most of these patients have no evidence of underlying pituitary or thyroid disease and on physical examination appear euthyroid.

This constellation of findings in critically ill patients has been termed the *euthyroid sick syndrome* (Cavalieri, 1991) and is a common source of confusion for both the internist and psychiatrist trying to find treatable pathology in these patients. Studies have documented that the serum T_4 concentration has a strong inverse correlation with mortality (Slag, Morley, & Elson, 1981). Serum TSH levels are usually normal or minimally elevated, but some reports using a sensitive TSH assay have also described low or undetectable levels in some patients (Wehrmann, Gregerman, & Burns, 1985).

In most patients with severe nonthyroidal illness, the diagnosis of euthyroid sick syndrome is suggested by the combination of a low total T_4 and FTI with a normal or minimally elevated serum TSH concentration, in the absence of clinical evidence of hypothyroidism. Laboratory studies in these patients are thus suggestive of either secondary or tertiary hypothyroidism, and it is the clinical setting that determines whether investigation of the hypothalamic-pituitary axis is indicated. Additional thyroid studies might include direct measurement of the free T_4 concentration by the equilibrium dialysis technique, which is usually normal, or measurement of the reverse T_3 concentration, which is usually elevated in these patients (Chopra et al., 1979).

Although the exact pathophysiology of this syndrome remains to be defined, it appears to be an adaptive mechanism to preserve body nitrogen during states of severe stress (Utiger, 1980). Endocrine consultation should be obtained if thyroid hormone therapy is contemplated, as there is currently no evidence to support the benefits of thyroid supplementation in these patients (Brent & Hershman, 1986).

The association of acute psychiatric illness with transient hyperthyroxinemia and the blunting of the TSH response to TRH in depressed patients has been discussed above. The pathophysiology of these alterations in thyroid function has yet to be explained.

The effects of drugs on thyroid function tests has been reviewed in detail (MacAdams, 1988). Glucocorticoids and dopamine have significant inhibitory effects on TSH secretion in normal subjects and may be the etiology of suppressed TSH concentrations in euthyroid hospitalized patients (Ehrmann & Sarne, 1989). Lithium, as previously discussed, has inhibitory effects on thyroid hormone synthesis and secretion. Finally, the anticonvulsants phenytoin and carbamazepine have important effects on circulating thyroid hormone concentrations (Curran & DeGroot, 1991). Specifically, patients receiving long-term therapy with these agents often have decreased serum T_4 and free T_4 concentrations with normal serum TSH levels, changes consistent with a "secondary" form of hypothyroidism (Hein & Jackson, 1990). Despite these changes in serum thyroid hormone levels, patients usually appear clinically euthyroid. A variety of effects of phenytoin on thyroid hormone binding and metabolism have been described, but to date a totally satisfactory explanation of the observed changes in serum thyroid hormone and TSH concentrations is not available.

Whereas lithium exerts its effects directly on the thyroid, carbamazepine appears to affect thyroid function by accelerating thyroxine degradation (Curran & DeGroot, 1991) and by inhibiting TSH secretion at the level of the pituitary (Joffee et al., 1984). In patients being treated concurrently with lithium and carbamazepine, carbamazepine could theoretically prevent the compensatory rise in TSH in the presence of a low T_4 caused by lithium. A decision to treat patients on anticonvulsants with thyroid hormone should therefore be based on clinical assessment in addition to laboratory values. In borderline cases, a therapeutic trial of L-thyroxine should be considered.

ADRENAL DISORDERS

Cushing's Syndrome

Cushing's syndrome is the clinical disorder resulting from prolonged exposure of body tissues to inappropriately elevated plasma glucocorticoid levels. It may occur spontaneously (endogenous hypercortisolism) or secondary to the chronic administration of glucocorticoids for the treatment of a variety of clinical disorders. The differential diagnosis of spontaneous Cushing' syndrome includes (1) excessive pituitary ACTH secretion, usually caused by a pituitary adenoma (Cushing's disease); (2) primary neoplasms of the adrenal cortex; and

(3) ectopic production of ACTH by malignancies (most commonly, small-cell carcinoma of the lung). In adults, Cushing's disease accounts for 60 to 70% of all cases, adrenal tumors for approximately 20%, and the ectopic ACTH syndrome for the remaining 10 to 20% (Loriaux, 1990a).

Typical clinical manifestations include truncal obesity, moon facies, hypertension, purple striae, muscle weakness (primarily proximal), excessive bruising, facial plethora, acne, hirsutism, menstrual disturbances (usually amenorrhea), and peripheral edema. Significant hyperpigmentation suggests the ectopic ACTH syndrome but is occasionally seen with Cushing's disease. Metabolic consequences of Cushing's syndrome include carbohydrate intolerance, hypokalemic alkalosis, and osteoporosis, and it is one of these sequelae that often brings the patient to medical attention. It should be emphasized that the clinical presentation of Cushing's syndrome is variable, and there is no specific feature that appears in all cases.

Psychiatric disturbances occur in more than 50% of patients (Starkman & Schteingart, 1981). Although in the past, psychiatric findings have been attributed to the disruption of physical and social function caused by disfigurement, mental symptoms may in fact precede any physical signs or symptoms. A wide range of psychiatric symptoms have been reported, and variation in symptomatology may occur during the course of a single patient's illness (Haskett, 1985). Depression is the most frequent presentation, accompanied by irritability, insomnia, crying spells, low energy, poor concentration and memory, decreased libido, and suicidal ideation. Psychomotor retardation may alternate with periods of agitation. Rapid mood fluctuations, with episodes of acute anxiety, may be seen. Euphoria and manic excitement are occasionally observed, occurring typically prior to the onset of depression, sometimes with an intervening period of remission. Psychoses are usually depressive, but paranoid states and schizophreniform syndromes rarely occur.

Cognitive impairment disorders with confusion and disorientation are also seen. Patients may show evidence of varying degrees of diffuse bilateral cerebral dysfunction on neuropsychological testing, with the most striking impairment being in nonverbal visual-ideational and visual memory functions (Whelan et al., 1980). Pneumoencephalographic studies have demonstrated cerebral and cerebellar atrophy in some patients (Momose, Kjellberg, & Kliman, 1971). Computed tomography (CT) scanning has shown reversible atrophic changes that may be caused by fluid and electrolyte changes or protein loss (Heinz, Martinez, & Haenggeli, 1977). Internal hydrocephalus and cortical atrophy have been reported at autopsy (Soffer, Iannaccone & Gabrilove, 1961). Medical complications of the illness, such as electrolyte disturbances, diabetic ketosis, and hypertensive encephalopathy may cause additional mental impairment.

Cushing's syndrome can present a difficult diagnostic challenge to even the most astute clinician because of the marked diversity of its clinical manifestations. With the frequency of psychiatric symptoms as initial complaints, patients may be misdiagnosed as having major depression, mania, bipolar depression, schizophrenia, or a variety of secondary, toxic, or metabolic conditions. Definitive diagnosis requires the demonstration of cortisol overproduction and an abnormality in the suppression of cortisol secretion.

Two convenient screening tests are available for the assessment of patients with suspected Cushing's syndrome (Carpenter, 1988). The overnight dexamethasone suppression test involves the administration of 1 mg of dexamethasone at 11:00 p.m. and measurement of the serum cortisol concentration at 8:00 a.m. the following morning. Normal individuals demonstrate suppression of the serum cortisol concentration below 5 μg/dl; a normal result virtually excludes the diagnosis of Cushing's syndrome. False-positive results (i.e., serum cortisol levels higher than 5 μg/dl in the absence of Cushing's syndrome) may occur with obesity, depression, acute stress (e.g., hospitalization), or alcohol withdrawal, and in patients taking estrogens, phenytoin, or a number of other drugs. In the past, use of the dexamethasone suppression test (DST) was advocated for the diagnosis of melancholia (Carroll et al., 1981), but more recent evidence finds the DST to be of relatively little use in clinical settings (APA Task Force, 1987). The 24-hour urinary free cortisol assay is another excellent screening test for Cushing's syndrome, although both false-positive and false-negative results have been reported. All patients with positive results on either of the above screening studies should be referred for further endocrinologic investigation.

Pseudo-Cushing's syndrome is a term that has been applied to the hypercortisolism seen in patients with major affective disorders and alcohol abuse (Loriaux & Niemann, 1990). Elevated plasma and urinary cortisol concentrations and lack of cortisol suppression following dexamethasone have been well documented in both of these clinical settings. Obese, hirsute patients with depression who demonstrate increased urinary free cortisol levels can be virtually impossible to distinguish from patients with mild Cushing's syndrome. Studies suggest that ACTH responses to corticotropin-releasing hormone (CRH) can be used to discriminate between patients with depression and those with Cushing's disease (Gold, Loriaux, & Roy, 1986). Depressed patients tend to have blunted ACTH responses to CRH, whereas patients with Cushing's disease generally have exaggerated responses compared to normal controls. There is, however, significant overlap between the groups, with 20% of depressed patients having peak ACTH responses in the range of those seen with Cushing's disease, and 27% of the Cushing's patients having peak ACTH responses in the range of those of the depressed patients. It should be emphasized that the abnormal endocrine profiles in depressed patients and those with alcohol-induced pseudo-Cushing's syndrome revert to normal with resolution of the depression or with alcohol abstinence.

Successful therapy of Cushing's syndrome requires an accurate determination of the cause of the hypercortisolism. In patients with primary adrenal tumors or the ectopic ACTH syndrome, therapeutic intervention is directed at removing the underlying neoplasm. In patients in whom it cannot be done, hypercortisolism can be treated "medically" with inhibitors of cortisol biosynthesis, specifically aminoglutethimide, metyrapone, and more recently ketoconazole. The latter agent has been used successfully in patients with ectopic ACTH syndrome, with relatively few side effects (Sonino, 1987). Transsphenoidal pituitary surgery has become the treatment of choice for most patients with Cushing's disease (Mampalam, Tyrrell, & Wilson, 1988). Preoperative confirmation and localization of the pituitary adenoma can be successfully performed via inferior petrosal sinus sampling for ACTH (Oldfield et al., 1985). This radiographic technique is particularly helpful in cases in which preoperative testing does not clearly discriminate between Cushing's disease and an ectopic source of ACTH.

As cortisol levels are reduced to normal, psychiatric symptoms usually remit. Low doses of neuroleptics, or electroconvulsive therapy may be helpful for symptomatic relief. Use of tricyclic antidepressants may precipitate a manic episode, and these drugs should be employed only if depression is severe or persists after cortisol levels are normalized.

Exogenous or iatrogenic Cushing's syndrome results from the prolonged administration of either ACTH or supraphysiologic doses of glucocorticoids. The duration of therapy and the dose administered determine the likelihood of developing clinical features of Cushing's syndrome. Table 29-1 summarizes equivalent doses of available glucocorticoid preparations. Hydrocortisone doses greater than 30 mg per day (or prednisone 7.5 mg per day) should be considered supraphysiologic; they have the potential not only for inducing Cushing's syndrome but also for causing suppression of the hypothalamic-pituitary-adrenal axis. Physical manifestations of exogenous Cushing's syndrome are similar to those of the endogenous disorder, except that myopathy, glucose intolerance, and osteoporosis tend to be more prominent clinical features. Cataracts and aseptic necrosis of the femoral head may also be seen with exogenous Cushing's syndrome, whereas both of these problems are rare with spontaneous cortisol excess. Finally, hypertension is less common, and signs of masculinization and hyperpigmentation do not occur. The diagnosis of exogenous Cushing syndrome is based on the history of prolonged administration of supraphysiologic doses of a glucocorticoid preparation. As with other metabolic disorders, cases of factitious hypercortisolism have been reported.

TABLE 29–1. *Equivalent Glucocorticoid Doses*

Glucocorticoid	Dose (mg)
Cortisone	25
Hydrocortisone	20
Prednisone	5
Methylprednisolone	4
Triamcinolone	4
Dexamethasone	0.75

As many as 5 to 14% of patients receiving high doses of glucocorticoids develop significant psychiatric symptoms at some time during their treatment (Falk, Mahnke, & Poskanzer, 1979). Occurrence of these symptoms is clearly related to the dose of drug administered (Boston Collaborative Drug Surveillance Program, 1972). Patients receiving prednisone or its equivalent (Table 29-1) at a dosage of more than 40 mg per day are at greatest risk. Premorbid personality, a history of previous steroid psychosis, or a history of previous psychiatric disorder did not affect risk in one series (Hall et al., 1979). There is some evidence that females are more prone to these disturbances than males (Ling, Perry, & Tsuang, 1981).

Mental changes can be seen at any point during the course of glucocorticoid therapy but appear to be most common within the first 5 days of steroid treatment (Hall et al., 1979). Alterations in mood are the most frequent finding, with euphoria occurring in most cases, often accompanied by irritability, increased appetite, increased libido, and insomnia. A prodromal state of "cerebral excitability" has been reported (Goolker & Schein, 1953). Depression is less common but may be severe with significant suicidality; it may follow a period of euphoria. Cognitive impairment may also be found (Wolkowitz et al., 1990). Some patients present with a "spectrum psychosis" marked by rapid shifts of symptoms from affective to schizophreniform to organic. These symptoms may include agitation, emotional lability, distractibility, memory impairment, pressured speech, mutism, auditory and visual hallucinations, delusions, and body image distortions (Hall et al., 1979).

After cessation of glucocorticoid therapy, spontaneous remission of psychiatric and behavioral symptoms occurs over several weeks to months. Reduction of dosage alone may cause symptoms to remit. Treatment with low to moderate doses of neuroleptics results in significant improvement with or without discontinuation of steroids and has been reported to shorten the duration of steroid-induced psychotic states (Hall et al., 1979).

Electroconvulsive therapy has been reported to ameliorate steroid psychosis, whereas tricyclic antidepressants have been reported to cause exacerbation of symptoms (Hall, Popkin, & Kirkpatrick, 1978). Prophylactic treatment with lithium at psychiatric doses reduced the psychiatric complications in one study of patients receiving ACTH for multiple sclerosis (Falk, Mahnke, & Poskanzer, 1979). No conclusive selection criteria exist. Although the study of Hall et al. (1979) suggests that there is no correlation of previous psychosis due to steroids to subsequent events, lithium prophylaxis of patients with previous steroid psychoses should be considered in some cases.

Addison's Disease (Adrenocortical Insufficiency)

Inadequate production of adrenal corticosteroids may result from a variety of primary adrenal disorders (Addison's disease) or be secondary to inadequate ACTH secretion (secondary adrenal insufficiency). Whereas in the past tuberculosis was the major cause of Addison's disease, 70% of new cases now fall into the idiopathic category (Loriaux, 1990b). Most of these cases are believed to be secondary to autoimmune destruction of the adrenal gland. Many of the patients have other autoimmune endocrine deficiencies (e.g., hypothyroidism, hypoparathyroidism, premature ovarian failure) as part of an autoimmune polyendocrine deficiency syndrome (Trence, Morley, & Handwerger, 1984). Other causes of adrenal insufficiency include fungal diseases (especially histoplasmosis), metastatic tumor, adrenal hemorrhage associated with anticoagulant therapy, sepsis (meningococcemia), infiltrative disorders (amyloidosis, sarcoidosis, hemochromatosis), and previous adrenal surgery. There have been several reports documenting primary adrenal insufficiency in patients with acquired immunodeficiency syndrome (AIDS), perhaps related to cytomegalovirus infection involving the adrenal glands (Greene, Cole, & Greene, 1984).

Typical clinical manifestations include generalized weakness and fatigue, weight loss, anorexia, vomiting, hyperpigmentation, abdominal pain, hypotension with orthostatic dizziness, nonspecific joint and muscle aches, and perceptual abnormalities. Common laboratory abnormalities include hyperkalemia and hyponatremia; an occasional patient also has hypercalcemia. In patients with secondary adrenal insufficiency associated with primary hypothalamic-pituitary abnormalities, hyperkalemia and hyperpigmentation are absent.

Psychiatric symptoms are observed in 60 to 90% of patients with Addison's disease (Popkin & Mackenzie, 1980). Development of symptoms is insidious and often precedes the more classic somatic features. Early manifestations include apathy, social withdrawal, fatigue, anxiety, irritability, negativism, and poverty of thought. Moderate to severe depression is present in 30 to 50% of patients. Often there is cognitive impairment, particularly of memory. Mental changes tend to be episodic and fluctuating in severity. During Addisonian crisis, delirium may occur with confusion, disorientation, and frank psychosis. Because of the slow onset of illness and the frequent initial appearance of psychiatric symptoms, the diagnosis of Addison's disease may be overlooked (Varadaraj & Cooper, 1986). Common misdiagnoses include depression, hypochondriasis, and conversion disorders. Autoimmune thyroiditis may coexist with autoimmune Addison's disease. As potentially lethal complications (i.e., Addisonian crisis) can occur with thyroid replacement alone in this situation, a high index of suspicion should be maintained for their coexistence.

The diagnosis of adrenal insufficiency may be suspected on the basis of a low 8:00 a.m. serum cortisol concentration or decreased 24-hour urinary free cortisol or 17-hydroxysteroid excretion. Definitive diagnosis requires an ACTH challenge. The rapid ACTH test with synthetic ACTH (Cortrosyn) is an excellent screening study, with a normal result virtually excluding the diagnosis of primary or secondary adrenal insufficiency. An abnormal result, however, requires confirmation, and endocrinologic consultation should be obtained for definitive studies with prolonged ACTH infusions. The availability of reliable ACTH radioimmunoassays has simplified the workup to some extent. Just as an elevated serum TSH concentration is diagnostic of primary hypothyroidism, a clearly elevated serum ACTH level in a patient with an abnormal Cortrosyn test is virtually diagnostic of primary adrenal insufficiency, and prolonged ACTH infusions are generally not necessary for diagnosis.

Treatment of Addison's disease usually requires both glucocorticoid and mineralocorticoid replacement. A typical glucocorticoid replacement regimen would be hydrocortisone 20 mg in the morning and 10 mg in the late afternoon. A single morning dose of a more potent, longer-acting glucocorticoid such as prednisone (5.0 to 7.5 mg) is also effective. The only available oral mineralocorticoid preparation is fludrocortisone (Florinef); the dose of this agent is adjusted for the individual patient based on blood pressure and serum potassium determinations. Patients with secondary adrenal insufficiency require only glucocorticoid replacement.

Treatment of Addison's disease with both glucocorticoid and mineralocorticoid replacement produces rapid resolution of mood and cognitive disturbances (Ettigi & Brown, 1978). Secondary psychoses may persist for several weeks. Only rarely do irreversible mental changes occur. Psychotropic medications should not be employed except with careful monitoring and low initial dosages, as most are likely to exacerbate hypotension (Thompson, 1973).

Steroid Withdrawal Syndrome

Prolonged exposure to supraphysiologic doses of glucocorticoids may result in suppression of the hypothalamic-pituitary-adrenal (HPA) axis. While exposure to the equivalent of 20 to 30 mg of prednisone per day for more than 5 days may cause abnormalities in tests of pituitary-adrenal reserve (Streck & Lockwood, 1979), *clinically significant* HPA suppression is unlikely to occur with less than 2 weeks of treatment. Any patient exposed to supraphysiologic doses of glucocorticoids for more than 2 weeks should be considered at risk for adrenal suppression. Although no specific quantitative guidelines can be offered, the longer the duration of therapy, the higher the dose, and the longer-acting the glucocorticoid preparation, the more likely it is that HPA suppression will occur.

The consequences of steroid withdrawal fall into three categories: (1) precipitation of adrenocortical insufficiency; (2) corticosteroid withdrawal syndrome; and (3) flare-up of the underlying disease for which steroids were administered. Although there are few well-documented cases of acute adrenal insufficiency after prolonged steroid therapy, patients who have been on supraphysiologic doses of glucocorticoids who are subject to acute stress (e.g., surgery, serious infection) should be "covered" with pharmacologic doses of hydrocortisone. For the stress of general anesthesia, for example, 100 mg of hydrocortisone intravenously every 8 hours is sufficient to prevent acute adrenal insufficiency. Rapid tapering should be encouraged as the patient recovers from the acute stress.

The corticosteroid withdrawal syndrome refers to a symptom complex that occurs during the tapering or cessation of exogenous glucocorticoids; it includes anorexia, nausea, arthralgias, myalgias, headache, fever, weakness, weight loss, and postural hypotension (Axelrod, 1990). Mental changes, including depression, agitation, anxiety, and psychotic reactions, may also be seen (Ling, Perry, & Tsuang, 1981; Wolkowitz & Rapaport, 1989). According to Dixon and Christy (1980), the diagnosis is made "when HPA responsiveness is shown to be normal, there is no recurrence of underlying disease, and symptoms are relieved only by ingesting amounts of corticosteroid that are above the physiologic level." The syndrome is believed to be one of true physical or psychological drug dependence (Flavin et al., 1983). Clinically, it may be difficult to differentiate this dependency syndrome from true adrenocortical insufficiency.

Recovery of the HPA axis after prolonged therapy with high doses of glucocorticoids may take as long as 12 months (Axelrod, 1990). The time course of recovery varies with the total duration of previous steroid therapy and the total previous steroid dose, but it is impossible to predict the pattern of recovery for any given patient. Therefore one should assume suppression of the HPA axis for 12 months in any patient who has received corticosteroids for more than 2 weeks; appropriate steroid coverage should then be provided for any stressful medical illness or surgery.

There is no proven method for hastening the recovery of the HPA axis. The most effective way to manage patients during the recovery phase of steroid therapy is to utilize a small morning dose of hydrocortisone (e.g., 10 to 20 mg) until recovery can be documented using a Cortrosyn test (see above). A 3-month period between tests is a convenient interval, and recovery can be assumed when the post-Cortrosyn serum cortisol concentration exceeds 20 μg/dl. At this time, exogenous glucocorticoid supplementation can be safely discontinued.

DISORDERS OF CALCIUM METABOLISM

Hyperparathyroidism

Primary hyperparathyroidism is a clinical disorder characterized by hypercalcemia due to excessive parathyroid hormone (PTH) secretion. It is the most common cause of hypercalcemia in the ambulatory population, with an overall prevalence of approximately 0.1% (Fitzpatrick & Bilezikian, 1990). Women are affected more frequently than men, and the prevalence of the disorder in women over age 60 may be as high as 1 in 600 (Heath, Hodgson, & Kennedy, 1980). In approximately 85% of cases, excessive PTH secretion is due to a single parathyroid adenoma. Fifteen percent of cases are due to parathyroid hyperplasia; and, rarely, hyperparathyroidism is associated with multiple adenomas or a parathyroid carcinoma. The epidemiology of hypercalcemia in hospitalized patients is different, with malignancy being responsible for a far greater proportion of cases than in the ambulatory care setting. The malignancies most commonly associated with hypercalcemia are lung, breast, prostate, renal cell, cervix, multiple myeloma, and head and neck cancers. Studies have documented that hypercalcemia in many patients with cancer is the result of the production of a PTH-related protein (PTHrP) by the underlying malignancy (Insogna, 1989). This peptide has structural homology with native PTH and can now be readily assayed by many commercial laboratories.

A detailed discussion of the differential diagnosis of hypercalcemia is beyond the scope of this chapter and has been reviewed elsewhere (Klee, Kao, & Heath, 1988). Causes other than hyperparathyroidism and malignancy include hyperthyroidism, sarcoidosis (and other granulomatous diseases), vitamin A and vitamin D intoxication, immobilization, drugs (thiazides, lithium), Addison's disease, familial hypocalciuric hypercalcemia (FHH), acute renal failure with rhabdomyolysis, and the now-rare "milk-alkali" syndrome.

There has been a significant change in the pattern of clinical presentation of patients with primary hyperparathyroidism over the last 25 to 30 years. Prior to 1970, more than 90% of patients presented with either renal disease (kidney stones and nephrocalcinosis) or bone disease. With the introduction of "routine" biochemical screening of asymptomatic patients during the 1960s, the clinical presentation of this disorder changed dramatically (Heath, 1989). More than 50% of patients currently diagnosed with primary hyperparathyroidism are entirely asymptomatic; 20 to 30% have vague, nonspecific symptoms, and fewer than 20% now present with renal or bone disease. In many patients the symptoms are subtle, and their onset is insidious. Patients may complain of nonspecific muscle weakness, fatigue, lethargy, anorexia, nausea, constipation, increased thirst, and vague abdominal and musculoskeletal pains. More acute symptoms may be associated with a kidney stone, peptic ulcer disease, pancreatitis, or a pathologic fracture.

Reports of the occurrence of psychiatric symptoms in patients with hyperparathyroidism vary from 5% to 65%. Generally, changes in mental status parallel the degree of elevation of serum calcium (Brown, Preisman, & Kleerekoper, 1987). With mild hypercalcemia, personality changes, loss of spontaneity, and lack of initiative are typical complaints. With moderate hypercalcemia (12 to 16 mg/dl), symptoms are principally depressive, with dysphoria, anhedonia, apathy, anxiety, irritability, impaired concentration and recent memory, and sometimes suicidal ideation. With severe hypercalcemia (16 to 19 mg/dl) or following a precipitous rise in serum calcium, psychotic and cognitive symptoms predominate, including confusion, disorientation, catatonia, agitation, paranoid ideation, delusions, and auditory and visual hallucinations. At levels above 19 mg/dl, stupor and coma are common. In some patients, significant elevations in serum calcium may be seen without observable mental changes. One study suggested that the psychiatric symptomatology of hyperparathyroidism may be related to changes in CNS turnover of monoamines, as evidenced by low cerebrospinal fluid (CSF) concentrations of monoamine metabolites—5-hydroxyindoleacetic acid, homovanillic acid, and 3-methoxy-4-hydroxyphenylglycol (MHPG)—which increase after parathyroid surgery (Joborn et al., 1988b).

In view of the insidious onset and frequent lack of specific symptomatology, the diagnosis of hyperparathyroidism is often overlooked (Gatewood, Organ, & Mead, 1975). For this reason and because of its relatively low cost, a screening serum calcium determination should be obtained for most psychiatric patients during their initial evaluation. Common misdiagnoses include neuroses, hypochondriasis, mood disorders, schizophrenia, and cognitive impairment disorders of other etiology. Correction of hypercalcemia results in rapid reversal of many of the psychiatric manifestations (Borer & Bhanot, 1985; Peterson, 1968). A trial of oral phosphate therapy that induced a detectable reduction in serum calcium with concurrent reversal of psychiatric symptoms might be useful for determining if parathyroid surgery would be helpful. No published data appear to exist on this issue.

A serum calcium determination remains the single best screening test for primary hyperparathyroidism. *An "ionized" calcium determination may be helpful in patients with abnormal serum albumin concentrations and acid-base disorders, but should not be ordered routinely.* The definitive diagnosis of hyperparathyroidism requires the exclusion of other causes of hypercalcemia and demonstration of elevated serum PTH concentration. The combination of hypercalcemia and an elevated serum PTH level by either the "mid-molecule" or "intact" PTH assays is diagnostic in most patients (Endres, Villanueva, & Sharp, 1989). Older, indirect tests of parathyroid function, such as the 24-hour urinary cyclic AMP excretion or tubular reabsorption of phosphate (TRP), should be avoided as they lack adequate diagnostic sensitivity and specificity. Patients on chronic lithium therapy may present a clinical picture indistinguishable from that of primary hyperparathyroidism (Mallette & Eichhorn, 1986). Discontinuation of lithium in these patients should result in return of serum calcium levels to the normal range.

Definitive therapy for hyperparathyroidism is surgical, with a cure rate of 90 to 95% in the hands of an experienced surgeon (Heath, 1989). The indications for surgery remain controversial, particularly in patients with mild hypercalcemia and in the elderly. The evidence regarding the reversibility of psychiatric symptoms following parathyroidectomy is conflicting (Heath, 1989). Two small series that examined elderly patients with dementia, however, suggested improvement in functional status in most of the patients who underwent successful surgery (Heath, Wright, & Barnes, 1980; Joborn, Hetta, & Johansson, 1988a). Studies to localize parathyroid adenomas should be limited to patients for whom the decision has already been made to proceed with surgery. Noninvasive localization techniques include neck imaging by computed tomography (CT) and magnetic resonance imaging (MRI), as well as radioisotopic imaging by thallium-technetium scanning (Eisenberg et al., 1989).

Hypoparathyroidism

Hypoparathyroidism is characterized by a decrease in the serum calcium concentration in association with hyperphosphatemia (Downs & Levine, 1990). The most common cause of hypoparathyroidism is accidental damage to or removal of the parathyroid glands during thyroid surgery or radical neck surgery for cancer. Other etiologies include magnesium deficiency (frequently associated with alcoholism), parathyroid aplasia (DiGeorge syndrome), neck irradiation, infiltrative disorders (e.g., amyloid, sarcoidosis, hemosiderosis), and autoimmune destruction, usually as part of the autoimmune polyglandular deficiency syndrome (Trence, Morley, & Handwerger, 1984). Pseudohypoparathyroidism is a rare inherited disorder in which there is impaired end-organ responsiveness to PTH. Other conditions causing hypocalcemia include malabsorption syndromes, vitamin D deficiency, acute pancreatitis, multiple transfusions, and chronic renal failure.

Clinical manifestations of hypoparathyroidism primarily reflect underlying hypocalcemia, which typically results in increased neuromuscular irritability. Symptoms range in severity from mild paresthesias to tetany, muscle cramps, and carpopedal spasms. More severe hypocalcemia may result in laryngeal stridor and generalized seizures. Findings on physical examination may include a positive Chvostek or Trousseau sign (Parfitt, 1989), cataracts, and a parkinsonian-like syndrome. The latter is most often seen in patients with basal ganglia calcification associated with long-standing hypocalcemia.

Psychiatric findings occur in 30 to 50% of hypoparathyroid patients (Popkin & Mackenzie, 1980). The rapidity of change in serum calcium and other electrolytes seems to be the most important factor in determining the severity of complaints. Symptoms may include anxiety, depression, irritability, emotional lability, social withdrawal, phobias, and obsessions. In severe cases delirium with confusion, disorientation, agitation, paranoia, and auditory and visual hallucinations may be present. Intellectual deterioration is found in one-third of patients with primary hypoparathyroidism, the result of a long duration of illness prior to diagnosis and treatment. With surgically induced hypoparathyroidism, cognitive changes are rare, as the condition is usually treated promptly. Symptoms of hypoparathyroidism may be mistaken for neuroses, hypochondriasis, conversion disorders, anxiety disorders, depression, schizophrenia, dementia, and cognitive impairment disorders of other etiology.

The diagnosis of primary hypoparathyroidism is confirmed by the findings of a low serum calcium concentration in association with a low serum PTH level. In patients with a normal serum albumin and normal acid-base status, ionized calcium determinations are usually unnecessary.

Treatment of hypoparathyroidism consists of oral vitamin D and calcium supplements, with the goal being return of the serum calcium concentration to the low-normal range. Normalization of serum calcium levels typically results in resolution of symptoms. When intellectual impairment is present, residual cognitive deficits often persist. With regard to the use of psychotropic medications, patients with hypoparathyroidism have shown an increased sensitivity to acute dystonias secondary to phenothiazines (Schaaf & Payne, 1966).

DISORDERS OF GLUCOSE METABOLISM

Diabetes Mellitus

Diabetes mellitus is a heterogeneous group of disorders characterized by hyperglycemia and a variety of potentially disabling complications. These complications include specific microvascular disorders (diabetic retinopathy and nephropathy), macrovascular diseases (accelerated atherosclerosis causing coronary artery, cerebrovascular, and peripheral vascular disease), and various other complications including neuropathy, increased susceptibility to infections, and high-risk pregnancies. An excellent overview of the medical management of diabetes mellitus and its complications can be found in a published symposium (Rizza & Greene, 1988).

Hyperglycemia results in symptoms often misdiagnosed in their early stages as hypochondriasis. Polyuria, polyphagia, and polydipsia are often combined with anorexia, fatigue, blurred vision, and paresthesias. Ketoacidosis rarely (in fewer than 5% of cases) presents with delirium, whereas hyperosmolar nonketotic states more often present with altered mental status. Schizophreniform and depressive syndromes have also been observed. A significant diagnostic dilemma exists in the case of a patient whose sole presenting symptom is impotence (Martin, 1981). Diabetic autonomic neuropathy and vascular disease may coexist with psychogenic etiologies, particularly depression. The presence of morning erections, situation-specific impotence, and abrupt onset suggest psychogenic components. Clinical polysomnography with assessment of nocturnal penile tumescence and penile arterial Doppler studies often define organic impotence in the diabetic patient (Karacan et al., 1978). (See also Chapters 15 and 39.)

The role of psychosocial factors in diabetes has been a controversial issue. Although the evidence suggesting a role for psychosocial factors in the onset of diabetes is inconclusive, many studies have shown a relation between psychosocial variables (e.g., psychiatric illness, stress) and metabolic control (Helz & Templeton, 1990). The use of glycosylated hemoglobin levels (Larsen, Horder, & Mogensen, 1990) as an objective measure of diabetic control has greatly enhanced research in this area. Whether psychosocial problems play a causal role or are simply effects of the illness has been difficult to differentiate, but studies suggest that psychosocial factors do affect metabolic control both directly via neuroendocrine effects and indirectly by influencing patient compliance (Helz & Templeton, 1990). Psychosocial interventions, including psychotropic medications, electroconvulsive therapy, and various forms of psychotherapy (individual, group, family, stress-reduction techniques) have been shown to have a beneficial effect on diabetic control in some patients (Helz & Templeton, 1990). The possible impact of family dynamics on the neuroendocrine control of diabetes, as described by Minuchin (Minuchin, Rosman, & Baker, 1978), continues to be debated (Coyne & Anderson, 1988; Rosman & Baker, 1988).

The high prevalence of psychiatric illness in diabetic patients has been recognized only in recent years. In a study of 114 adult diabetic patients, 81 (71%) had a lifetime prevalence of psychiatric disorder, with the most common diagnoses being depressive and anxiety disorders (Lustman et al., 1986). The presence of psychiatric illness was associated with poorer diabetic control, as measured by glycosylated hemoglobin levels, and decreased accuracy in the reporting of diabetic symptoms. Other studies have found a high prevalence of eating disorders in young female diabetic patients (Lloyd, Steel, & Young, 1987; Rosmark et al., 1986). The coexistence of eating disorders and diabetes can be particularly dangerous, especially as many patients induce glycosuria by omitting or reducing their insulin as a method of purging. Clues that an eating disorder may be present include erratic or poor glycemic control, unexplained elevated glycosylated hemoglobin, weight fluctuations, and overconcern with body shape and weight. An excellent review of eating disorders in diabetics has been published (Marcus & Wing, 1990).

The use of psychotropic agents in diabetic patients must take into account the medical complications of the illness. For example, anticholinergic activity of the antipsychotics and antidepressants may have adverse effects in patients with diabetic gastroparesis or neurogenic bladders, exacerbating vomiting and urinary retention, respectively. Use of metoclopramide may also result in extrapyramidal side effects. Alpha-adrenergic blocking activity of some psychotropic agents may worsen orthostatic hypotension. Selection of medications with both low anticholinergic and alpha-adrenergic blocking profiles in selected patients may be critical to successful overall management (see Chapter 9). Tricylic antidepressants should be used with appropriate precautions in diabetic patients with heart disease and a predisposition to cardiac arrhythmias. Caution should also govern the use of beta-blockers in insulin-dependent patients, as they may mask autonomic symptoms due to hypoglycemia.

Diabetes is associated with a variety of painful neuropathic syndromes. The most common diabetic neuropathy is a bilateral, symmetric, predominantly sensory syndrome characterized by paresthesias and loss of sensation. Superimposed on this relatively asymptomatic condition, however, may be numerous painful neuropathies, typically associated with burning and shooting pains in the lower extremities. Anticonvulsants (phenytoin, carbamazepine) have shown some success in the treatment of these pain states. The role of antidepressant and antipsychotic therapy for these pain syndromes remains controversial (Maciewitz, Bouckoms, & Martin, 1985). Studies favor the use of tricyclic antidepressants whose net effect is to increase the availability of serotonin at the synapse (e.g., amitriptyline, doxepin, trazodone, imipramine); fluoxetine and sertraline have not been specifically studied as yet in this population. Inconsistencies abound, however, with regard to the need for "antidepressant" dosages in these situations. Careful monitoring of nondepressed patients indicates that therapeutic benefit may be derived from "subtherapeutic" doses of these agents (Kvinesdal et al., 1984). Consideration of anticholinergic activity is essential for optimizing clinical care. Use of neuroleptics such as haloperidol and fluphenazine for chronic neuropathic pain has been less encouraging. Concern related to the development of tardive dyskinesia in these patients has relegated their use to only the most severe cases in which other therapeutic modalities have failed. Refractory pain has been relieved in some patients with lidocaine or oral tocainide or mexilitine. Capsaicin is an approved topical agent for the management of painful diabetic neuropathies (Chad et al., 1990). The lack of good controlled trials documenting its efficacy, the need for application three or four times per day, and its expense warrant a conservative approach to its use at this time. However, its total lack of CNS side effects make it attractive for those patients whose pain responds to it.

Hypoglycemia

Hypoglycemia is a syndrome best defined by the following triad of clinical findings: (1) appropriate symptoms (see below); (2) plasma or serum glucose concentration less than 50 mg/dl; and (3) reversal of symptoms by glucose administration. Reactive (postabsorptive, postprandial) hypoglycemia refers to hypoglycemia that occurs within 5 hours of food ingestion. Fasting (spontaneous) hypoglycemia occurs more than 6 hours after food ingestion.

Hypoglycemic symptoms are generally divided into two categories: adrenergic and neuroglycopenic (Field, 1989). Reactive hypoglycemia syndromes are most often characterized by adrenergic symptoms, whereas the fasting hypoglycemias are typically associated with neuroglycopenic symptoms. Although the symptoms of hypoglycemia in adults are variable, in any given patient the symptom complex tends to be consistent. Adrenergic or catecholamine-mediated symptoms may include tachycardia, anxiety, diaphoresis, tremor, weakness, hunger, irritability, and palpitations. Symptoms may be suggestive of a panic disorder or hyperventilation. An inadequate supply of glucose to the CNS, or neuroglycopenia, may result in faintness, headache, blurred vision, lethargy, confusion, dizziness, weakness, incoordination, bizarre behavior, reversible focal neurologic findings, seizures, and coma.

Fasting hypoglycemia may be associated with a variety of disorders, including insulin-secreting islet cell tumors of the pancreas, large non-islet-cell tumors, severe liver or renal disease, endocrine deficiency states (e.g., adrenal insufficiency, hypopituitarism), severe inanition, autoimmune disorders associated with antibodies to insulin or to the insulin receptor, and drugs or toxins (Field, 1989). In this last category, the most common offending agents are those used for routine management of diabetes mellitus—insulin and sulfonylureas. Alcohol, salicylates, and propranolol are also implicated in a significant number of cases (Seltzer, 1989). Finally, the differential diagnosis of fasting hypoglycemia must include surreptitious administration of either insulin or an oral hypoglycemic agent. Factitious hypoglycemia secondary to one of these agents must be considered prior to pancreatic exploration for an islet cell tumor in any patient with hyperinsulinism. The presence of antiinsulin antibodies or low C-peptide levels at the time of hypoglycemia strongly suggests a factitious etiology (Horwitz, 1989). Screening of urine or blood for sulfonylureas is available for patients suspected of surreptitious oral hypoglycemic agent ingestion.

Reactive hypoglycemia is generally divided into three major categories: (1) alimentary hypoglycemia, usually associated with previous gastric surgery; (2) mild diabetes, characterized by a delayed, but excessive, insulin response to glucose; and (3) idiopathic, or "functional," hypoglycemia (Hofeldt, 1989). Rarely, reactive hypoglycemia has been described in patients with hypothyroidism, cortisol deficiency, hyperinsulinism, or alcohol ingestion.

Perhaps the major clinical issue regarding hypoglycemia is its overdiagnosis, which has been the subject of numerous reviews (Cahill & Soeldner, 1974; Gastineau, 1983; Nelson, 1985). This problem has been particularly true for reactive hypoglycemia, which became a fashionable diagnosis to account for a variety of poorly defined physical and psychological ills, including depression, anxiety, fatigue, sexual dysfunction, and overall loss of vitality (Yager & Young, 1974). The Minnesota Multiphasic Personality Inventory (MMPI) profiles of patients being evaluated for reactive hypoglycemia have been reported to differ significantly from those of general medical patients and to show the "conversion-V" triad (scales Hs, D, Hy) suggestive of underlying emotional disturbance as the basis for their somatic complaints (Johnson et al., 1980).

The diagnosis of hypoglycemia should *not* be made on the basis of the results of an oral glucose tolerance test. As summarized in an editorial, "the test is inadequate for detecting postprandial hypoglycemia: healthy persons may have low nadir blood glucose concentrations, there is a poor correlation between the nadir glucose concentration and the occurrence of symptoms, and symptoms may occur after the ingestion of a placebo or mixed meals as well as after the ingestion of glucose in the absence of low blood glucose concentrations" (Service, 1989). Ideally, patients should be evaluated with a blood glucose determination at the time of symptoms after a normal mixed meal (Hogan, Service, & Sharbrough, 1983). Alternatively, several random glucose determinations should be obtained 2 to 4 hours after meals, in addition to those obtained at the time of maximal symptoms. Almost invariably, these measurements fail to show any correlation between symptoms and low blood glucose concentrations.

Some authors have suggested the term "postprandial syndrome" (Service, 1989) to describe the symptom complex that occurs in some patients after ingestion of food. A possible role for enteric hormones in the pathogenesis of postprandial symptoms requires further investigation. It should always be remembered that there are a few patients with symptomatic hypoglycemia (occurring after meals) who require some specific therapy. Many of these patients respond to a low carbohydrate diet, with frequent small meals.

An important issue in the psychopharmacologic management of patients with hypoglycemia is the risk of beta-blocker therapy. Early misdiagnosis of an anxiety disorder and treatment with agents whose action blocks the normal response to hypoglycemia may prevent subjective experience of potentially lethal hypoglycemia.

DISORDERS OF PITUITARY FUNCTION

Pituitary tumors account for most of the clinically significant disorders of pituitary function. Patients with pituitary tumors typically present with evidence of endocrine dysfunction or neurologic manifestations, or both. Endocrine presentations are divided into those associated with pituitary hormone hypersecretion (e.g., acromegaly, Cushing's disease, hyperprolactinemia) and those associated with inadequate pituitary hormone secretion (hypopituitarism). The most common neurologic manifestations of pituitary tumors are headaches and visual abnormalities, the latter related to the proximity of the optic nerves and optic chiasm to the sella turcica. Less common neurologic findings, usually found in patients with large invasive pituitary tumors, include cranial nerve abnormalities (associated with tumor extension into the cavernous sinus), hypothalamic abnormalities, and, rarely, hydrocephalus associated with obstruction of the foramen of Munro. A published symposium provides an excellent summary of important issues concerning the diagnosis and management of pituitary tumors (Molitch, 1987).

Hyperprolactinemia is the most common hypothalamic-pituitary abnormality encountered in clinical practice, and prolactin-secreting pituitary adenomas (prolactinomas) are the most common functioning pituitary tumors. Clinical manifestations associated with hyperprolactinemia are primarily related to gonadal dysfunction. Women typically present with menstrual dysfunction (amenorrhea or oligomenorrhea), galactorrhea, or infertility. Other symptoms may include sexual dysfunction (decreased libido, dyspareunia) and hirsutism (Evans, Carlsen, & Ho, 1990). Psychological disturbances such as hostility, depression, and anxiety have been reported to occur more often in women with hyperprolactinemia and amenorrhea than in women with normal prolactin levels and amenorrhea (Fava et al., 1981). Men with hyperprolactinemia usually have evidence of gradually progressive gonadal dysfunction, manifested by decreased libido and potency. Symptoms may be mistaken for a primary sexual dysfunction. An elevated prolactin level may be discovered as part of a work up for infertility in men, and a small percentage of men with hyperprolactinemia have galactorrhea. In patients with an underlying pituitary tumor as the etiology of the hyperprolactinemia, headaches and visual symptoms may accompany these endocrine manifestations. The most prominent neuropsychiatric feature of pituitary tumors in men is apathy, typically without dysphoric mood (Cohen, Greenberg, & Murray, 1984). Irritability and impulsiveness may also be seen.

The differential diagnosis of hyperprolactinemia is extensive (Vance & Thorner, 1987) but can be divided into five major categories: (1) hypothalamic disorders interfering with normal dopaminergic inhibition of prolactin secretion; (2) prolactin-secreting pituitary tumors (prolactinomas); (3) "neurogenic" factors, such as chest wall lesions and nipple stimulation; (4) endocrine causes, such as hypothyroidism and normal pregnancy; and (5) drugs. Categories of drugs associated with hyperprolactinemia include neuroleptics (especially phenothiazines and butyrophenones), antidepressants (amitriptyline, imipramine, amoxapine), antihypertensive agents (methyldopa, reserpine), and dopamine receptor antagonists (metoclopramide, domperidone, sulpiride). Miscellaneous drugs associated with increased prolactin levels include verapamil, cimetidine, estrogens, and opiates. Most of the nonpituitary disorders listed above, including patients with drug-induced hyperprolactinemia, are associated with serum prolactin concentrations of less than 150 ng/ml. A serum prolactin level higher than 200 ng/ml is virtually diagnostic of a prolactin-secreting pituitary adenoma. However, patients with small pituitary tumors may have prolactin levels of 30 to 200 ng/ml, overlapping the values seen with other causes of hyperprolactinemia.

A common clinical dilemma is the patient on a phenothiazine or butyrophenone who develops menstrual dysfunction or galactorrhea associated with an elevated serum prolactin concentration. The simplest way to distinguish drug-induced hyperprolactinemia from a prolactinoma is simply to discontinue the drug and determine whether the serum prolactin level returns to normal, virtually excluding the possibility of a pituitary tumor. In patients in whom the discontinuation of antipsychotic therapy is deemed unsafe, a TRH challenge test may be useful for distinguishing patients with drug-induced hyperprolactinemia from those with pituitary tumors (Lankford et al., 1981).

The major therapeutic options for management of a patient with a prolactin-secreting pituitary adenoma are transsphenoidal pituitary surgery and medical therapy with bromocriptine. The success of surgery is primarily related to the size of the tumor, with reported cure rates of 70 to 80% for microadenomas and 0 to 40% for macroadenomas. There has been considerable concern about recurrences of hyperprolactinemia after apparently successful surgery (Vance & Thorner, 1987); and in many centers it has resulted in the use of bromocriptine as the primary therapeutic modality for these tumors. Bromocriptine is an orally administered ergot alkaloid with potent dopamine agonist properties. Numerous series have documented its efficacy in lowering serum prolactin levels in patients with prolactin-secreting microadenomas and macroadenomas. The drug is well tolerated after oral administration, with the most common side effects being nausea and orthostastic hypotension. These side effects rarely require cessation of therapy. In addition to lowering serum prolactin levels, it is now well established that bromocriptine therapy can also effectively decrease the size of prolactinomas, often reversing visual field abnormalities, headaches, and other neurologic symptoms (Vance & Thorner, 1987).

The decrease in serum prolactin levels with treatment in women with prolactinomas is usually associated with a return of normal ovulatory menstrual cycles and cessation of galactorrhea. Fertility is generally restored, and many of these women experience perfectly normal term pregnancies. Careful monitoring for symptoms of pituitary expansion during pregnancy is essential, although complication rates in women with microadenomas are low. Normalization of serum prolactin concentrations in men similarly results in restoration of normal gonadal function, assuming there is no impairment of gonadotropin secretion from the primary pituitary tumor.

Controversy remains as to the optimum form of initial therapy in prolactinoma patients, with some centers advocating dopamine agonist therapy and other centers favoring transsphenoidal surgery. Patients not cured by surgery remain good candidates for bromocriptine.

Acromegaly is the clinical syndrome that results from sustained hypersecretion of growth hormone by a pituitary adenoma (Baumann, 1987). Rarely, the syndrome results from ectopic secretion of growth hormone-releasing hormone. Important clinical features include acral enlargement, soft-tissue overgrowth, hyperhidrosis, prognathism, and arthralgias. Symptoms related to an expanding pituitary mass may also be present (e.g., headaches, visual disturbances). Glucose intolerance is common, and there is an increased incidence of hypertension and cardiovascular disease. There are few reports of psychiatric disturbances associated with acromegaly. Personality changes, including apathy, lack of initiative and spontaneity, and mood lability, have been reported (Bleuler, 1951). Patients may also experience psychological reactions to their altered physical appearance, particularly their facial distortion. The diagnosis of acromegaly is based on (1) failure of serum growth hormone levels to be suppressed to less than 1 ng/ml after an oral glucose load; and (2) an elevated serum somatomedin-C level.

For most acromegalic patients, transsphenoidal pituitary surgery is the treatment of choice. External pituitary irradiation is considered an adjunctive form of therapy in patients not cured by pituitary surgery. There has been increased interest in medical therapy for acromegalic patients. A potent long-acting, somatostatin analogue, octreotide (Sandostatin), is now available; it has potent inhibitory effects on growth hormone secretion in acromegalic patients (O'Dorisio & Redfern, 1990). Primarily because of expense and the need for parenteral administration, octreotide is generally reserved for patients failing maximal conventional therapy (i.e., surgery and irradiation). The overall side effect profile is low, with nausea and diarrhea reported most frequently. Psychiatric side effects are rare, occurring in fewer than 1% of patients.

Hypopituitarism may result from a variety of destructive processes affecting the pituitary and hypothalamus. Common etiologies include large pituitary tumors, pituitary surgery, pituitary irradiation, and serious head injury. Less common causes include hypothalamic tumors, infiltrative disorders such as sarcoidosis and hemochromatosis, lymphocytic hypophysitis, metastatic tumors to the pituitary (most often breast or lung), and postpartum pituitary infarction (Sheehan syndrome). Cavernous sinus thrombosis, temporal arteritis, and carotid artery aneurysms have also been associated with hypopituitarism. Clinical manifestations of hypopituitarism in adults primarily reflect target gland deficiencies (i.e., hypothyroidism, hypoadrenalism, and hypogonadism). Diabetes insipidus may result from involvement of the posterior pituitary and hypothalamus. The diagnosis of hypopituitarism is frequently delayed because of the nonspecific nature of the symptoms and their insidious onset. Psychological symptoms may include amotivation, dysphoria, and cognitive impairment. Rapidly developing cases may present with psychoses or acute confusional states (Khanna et al., 1988).

The diagnostic approach to the patient with hypopituitarism is two-pronged. Imaging studies (CT scans or MRI) are essential for defining pituitary and hypothalamic anatomy. Endocrine function can be assessed by establishing the adequacy of target gland function, as well as by provocative testing of pituitary hormone reserve. Screening studies should include thyroid function tests, 8:00 a.m. serum cortisol, and serum testosterone (in males). Use of psychotropic agents is suggested only after correction of the endocrine abnormality.

MISCELLANEOUS ENDOCRINE DISORDERS

Pheochromocytoma

Pheochromocytomas are rare catecholamine-secreting tumors that arise most often from chromaffin tissue of the adrenal medulla (Benowitz, 1990). The estimated incidence of these tumors in the hypertensive population is considerably less than 1%, and routine screening of hypertensive patients for pheochromocytomas, in the absence of suspicious symptoms, should be discouraged. They occur equally in both sexes, with most patients presenting between the ages of 30 and 50.

The hallmark of a pheochromocytoma is hypertension, which may be sustained or paroxysmal; it is typically labile. The most common symptoms are headache, palpitations, and sweating. Other symptoms may include nausea, vomiting, tremor, anxiety, and abdominal pain. Metabolic consequences may include hyperglycemia and weight loss. Cardiac arrhythmias and hypertensive crises account for the significant incidence of sudden death in undiagnosed patients. The paroxysms that characterize the disorder may be confused with seizures, anxiety attacks, or hyperventilation. Occasionally, the clinical picture suggests another "hypermetabolic" disorder, hyperthyroidism. Psychological symptoms occur, although one study found minimal true overlap with criteria-defined anxiety disorders (Starkman et al., 1985). Affective lability is most commonly described in association with signs of sympathetic arousal.

Pheochromocytoma should be seriously considered as a diagnosis in patients with: (1) hypertension refractory to therapy; (2) hypertension accompanied by hypermetabolism; (3) hypertension accompanied by paroxysmal headaches and sweats; and (4) patients with other disorders associated with pheochromocytoma, such as neurofibromatosis and medullary carcinoma of the thyroid. Patients with anxiety attacks not accompanied by hypertension generally should not be evaluated for pheochromocytoma unless attacks include the physical symptoms of headache, sweating, or vomiting, or they are unresponsive to psychiatric treatment.

Diagnosis of a pheochromocytoma depends on the demonstration of increased levels of catecholamines or their metabolites in urine or blood (Sheps, Jiang, & Klee, 1988). The best screening test for the disorder is a 24-hour urine assay for catecholamines, metanephrines, and vanillylmandelic acid (VMA). Normal results in a patient with suspicious symptoms necessitate additional investigation, such as plasma catecholamine determinations or a "clonidine suppression" test (Sheps et al., 1990). Other pharmacologic tests using phentolamine (Regitine), histamine, and glucagon are inherently dangerous and generally should be avoided.

Finally, it should be remembered that pheochromocytomas may occur as part of a variety of familial disorders, including the multiple endocrine neoplasia syndromes (types 2A and 2B), neurofibromatosis (von Recklinghausen's disease), von Hippel-Landau disease, tuberous sclerosis, and Sturge-Weber syndrome. The most common of these familial disorders, type 2 multiple endocrine neoplasia, also includes hyperparathyroidism and medullary carcinoma of the thyroid (Sheps, Jiang, & Klee, 1988). (Chapter 7 provides further discussion of the evolution of pheochromocytoma.)

Male Hypogonadism

An understanding of male hypogonadism requires a review of the two major functions of the testes: secretion of testosterone and production of sperm. Hypogonadism refers to a deficiency of either or both of these essential functions. Inadequate sperm production alone usually presents as infertility, whereas the consequences of testosterone deficiency depend on the stage of sexual development of the patient. An excellent review of this complex area has been published (Plymate & Paulsen, 1990), and this section focuses primarily on the initial approach to the adult male patient with hypogonadism manifested by a decreased serum testosterone concentration.

Hypogonadal disorders in men can generally be grouped into those associated with diseases involving the testes directly (primary hypogonadism) and those associated with disorders of the hypothalamic-pituitary axis (secondary hypogonadism). As noted above, the clinical manifestations of testosterone deficiency vary with the age of the patient. In teenage boys, testosterone deficiency is characterized by failure of normal secondary sexual characteristics to develop (i.e., delayed puberty). In postpubertal males, testosterone deficiency is accompanied by symptoms including decreased libido and energy, sexual dysfunction, and decreased body hair and muscle mass.

Primary gonadal failure is typically associated with elevated serum gonadotropin concentrations and is therefore referred to as hypergonadotropic hypogonadism. The most common cause of this disorder in young men is Klinefelter's syndrome, associated with an abnormal chromosomal constitution of 47,XXY. Other causes of primary gonadal failure include myotonic dystrophy, mumps orchitis, testicular trauma or irradiation, cancer chemotherapy, and autoimmune testicular failure. Uremia and chronic alcohol abuse have features suggestive of both primary and secondary hypogonadism. Hypogonadotropic hypogonadism may be due to either primary pituitary or hypothalamic pathology. Pituitary causes include destructive pituitary tumors, prolactinomas (see above), hemochromatosis, sarcoidosis, therapeutic irradiation or surgery, and severe systemic illness. The classic hypothalamic cause of hypogonadotropic hypogonadism is Kallmann's syndrome. This syndrome is characterized by a deficiency of the hypothalamic peptide gonadotropin-releasing hormone (GnRH). Patients with Kallman's syndrome present as prepubertal eunuchs with minimal development of secondary sexual characteristics. Associated features may include anosmia or hyposmia and cleft palate and lip.

Male hypogonadism of any etiology may cause significant psychological distress and impaired social adjustment. Low self-esteem and self-confidence and feelings of inadequacy, isolation and alienation are common. Klinefelter's syndrome has been reported in association with a wide variety of psychiatric disorders, including mental retardation, personality disorders, depression, bipolar disorder, sexual deviance, neuroses, alcoholism, paranoid states, and schizophrenia (Caroff, 1978; Swanson & Stipes, 1969). The most consistent association is a form of personality disorder characterized by passivity, dependency, and low social drive; it is thought that such traits may be related to androgen deficiency, although the exact mechanism remains unclear (Swanson & Stipes, 1969). Male patients with recently acquired loss of libido and drive not explained by major depression or drugs deserve a screening testosterone level and attention paid to gonadal size on physical examination. If these screening measures suggest hypogonadism, further workup is indicated.

An appropriate *initial* diagnostic evaluation of a patient with suspected hypogonadism should include a careful history and physical examination, focusing on the disorders outlined above. The degree of sexual development, testicular size, and evidence of pituitary dysfunction should be the focus of the physical examination. Laboratory studies should include a serum testosterone, gonadotropin levels (LH and FSH), and a serum prolactin level. The results of these studies should define the level of the primary pathologic disturbance (i.e., a primary testicular disorder versus a central abnormality in hypothalamic-pituitary function). Patients falling into the latter category usually require radiographic imaging of the pituitary either by CT scanning or MRI. All patients with definite hypogonadism should be referred for further endocrinologic investigation.

The appropriate therapy for testosterone deficiency is a long-acting injectable testosterone preparation such as testosterone cypionate (Snyder, 1990). Oral androgens are not the agents of choice for hypogonadal men, as they are less effective in restoring normal virilization and have considerably greater toxicity (see below). Hypogonadotropic hypogonadal men desiring fertility should be treated with clomiphene, gonadotropins or pulsatile GnRH rather than with androgens.

Anabolic Steroids

The widespread use of anabolic steroids to improve athletic performance has been the focus of considerable controversy (Hallagan, Hallagan, & Snyder, 1989). An excellent review of this subject has been published (Wilson, 1988). Surveys of male high school students suggest that as many as 7% have experimented with these agents (Buckley et al., 1988), and usage may be as high as 15 to 20% in athletes engaging in intercollegiate sports (Windsor & Dumitru, 1989). The pattern of abuse varies among the various sports and geographic areas. Most athletes obtain their drugs (many of which are veterinary drugs) illicitly and engage in a practice called "stacking," in which oral and parenteral forms of several drugs are combined, enabling the athletes to achieve levels that are 10 to 100 times the ordinary pharmacologic range. Despite the common belief that androgen use enhances athletic performance via an increase in lean body weight and muscle strength, this fact has yet to be demonstrated consistently in controlled double-blind studies (Wilson, 1988). More importantly, these drugs have been shown to have significant toxicity, which far outweighs any possible benefit. This toxicity is primarily hepatic and includes cholestatic jaundice, peliosis (blood-filled hepatic cysts), and hepatoma. Other side effects may include lipid alterations (decreased HDL and increased LDL cholesterol), acne, gynecomastia, impotence, oligospermia, sleep apnea, and decreased glucose tolerance.

Psychiatric disturbances, including secondary mood disorders (depression or mania), psychoses, irritability, and aggression (sometimes to the point of violence), have been reported (Pope & Katz, 1988). In addition, studies suggest that the use of anabolic steroids may result in physical and psychological drug dependence; withdrawal symptoms include fatigue, decreased sex drive, depressed mood, insomnia, anorexia, dissatisfaction with body image, and a desire for more steroids (Brower et al., 1990). The possibility of steroid use and dependence should be considered in athletes and bodybuilders who present with psychiatric symptoms.

Hirsutism

Hirsutism is a clinical syndrome of diverse etiologies characterized by the excessive growth of hair in women in androgen-dependent areas. These areas typically include the upper lip, chin, neck, chest, lower abdomen, and perineum. Hirsutism should be distinguished from hypertrichosis, which is a generalized increase in vellous, but not terminal, hair. It is an increase in the number and thickness of coarse, dark terminal hairs that is responsible for the cosmetic disturbance in women with hirsutism. The psychosocial effects of hirsutism in terms of body image and social interaction are significant and frequently cause patients to seek consultation. It is essential to take into account racial, ethnic, and cultural norms when evaluating these patients.

It is important to distinguish hirsutism from virilization. Although both of these disorders are associated with excessive androgen production, the magnitude of the androgen excess is greatly increased in women with virilization. In addition to hirsutism, the virilized woman has temporal hair recession ("male-pattern" baldness), deepening of the voice, increased muscle mass, clitoral enlargement, and signs of defeminization (e.g., loss of female body contour, breast atrophy).

Causes of hirsutism generally fall into four major categories: (1) ovarian; (2) adrenal; (3) drugs; and (4) idiopathic. A detailed discussion of this differential diagnosis is beyond the scope of this chapter and has been discussed elsewhere in several excellent reviews (Mechanick & Dunaif, 1990; Rittmaster & Loriaux, 1987). Polycystic ovary syndrome, probably the single most common cause of hirsutism, is discussed elsewhere in this volume (see Chapter 28). The syndrome of hyperandrogenism + insulin resistance + acanthosis nigricans (HAIR-AN) is a rare cause of hirsutism that may be associated with a secondary mood disorder (Levin, Terrell & Stoudemire, 1992).

Drugs causing hirsutism are divided into those that are androgen-independent and those associated with intrinsic androgenic properties. In the former category are minoxidil, diazoxide, phenytoin, glucocorticoids, and cyclosporine. Androgenic agents causing hirsutism include anabolic steroids taken to enhance athletic performance, and norgestrel and levonorgestrel, progestins found in some commonly prescribed birth control pills (e.g., Ovral and Triphasil, respectively).

The decision to pursue diagnostic studies in a given patient is based on the severity and duration of the hirsutism, the presence of associated menstrual disturbances, family history, and ethnic background. A reasonable *screening* diagnostic approach includes the following laboratory studies: serum testosterone, dehydroepiandrosterone sulfate, LH, FSH, and prolactin. Patients with stigmata of Cushing's syndrome should undergo an overnight dexamethasone suppression test. More detailed hormonal profiling should be performed in consultation with an endocrinologist or gynecologist. Women with evidence of virilization and those with rapidly progressive hirsutism should be referred immediately because of the significant risk of an underlying ovarian or adrenal neoplasm.

CONSULTATION BETWEEN ENDOCRINOLOGY AND PSYCHIATRY

As noted above, evaluation of endocrine disorders has traditionally occurred *after* the demonstration of systemic signs and symptoms. Skin changes, electrolyte and metabolic disturbances, and other evidence of hormonal abnormalities have become the "activation threshold" for the diagnostic workup. Expanding awareness of cognitive, affective, and behavioral changes associated with endocrinopathies should facilitate early recognition of these treatable disorders.

Personal experience of the psychiatrist determines his or her level of comfort and confidence when administering and interpreting endocrine tests. If a psychiatrist chooses not to pursue an endocrine evaluation when clinically indicated, endocrine consultation should be obtained. With collaboration, both the psychiatrist and the endocrinologist can define limits of confidence with regard to the evaluation and treatment of secondary mental disorders due to endocrine dysfunction.

SUMMARY

Endocrine disturbances frequently mimic psychiatric disturbances. "High risk" groups, which require careful endocrine evaluation, include the following:

1. Patients with affective symptoms and coexistent cognitive dysfunction, especially geriatric patients
2. Patient with inconsistent or atypical presentation of psychiatric disorders
3. Patients with mental symptoms that are refractory to standard psychiatric treatments
4. Patients with symptoms of dementia
5. Patients with known preexisting endocrine abnormalities
6. Patients with affective symptoms after closed head injury

REFERENCES

ADDISON T (1868). On the constitutional and local effects of disease of the suprarenal capsules. In S WILKES & E DALDEY (Eds.), A collection of the unpublished writings of Thomas Addison (Vol. 36). London: New Sydenham Society.

APA Task Force on Laboratory Tests in Psychiatry (1987). The dexamethasone suppression test: an overview of its current status in psychiatry. Am J Psychiatry 144:1253–1262.

AXLEROD L (1990). Corticosteroid therapy. In KL BECKER (Ed.), Principles and practice of endocrinology and metabolism (Ch. 79, pp. 613–623). Philadelphia: Lippincott.

BAUER MS, & WHYBROW PC (1990). Rapid cycling bipolar affective disorder. II. Treatment of refractory rapid cycling with high-dose levothyroxine: a preliminary study. Arch Gen Psychiatry 47:435–440.

BAUER MS, WHYBROW PC, & WINOKUR A (1990). Rapid cycling bipolar affective disorder. I. Association with grade I hypothyroidism. Arch Gen Psychiatry 47:427–432.

BAUMANN G (1987). Acromegaly. Endocrinol Metab Clin North Am 16:685–703.

BENOWITZ NL (1990). Pheochromocytoma. Adv Intern Med 35:195–226.

BLEULER M (1951). The psychopathology of acromegaly. J Nerv Ment Dis 113:497–511.

BORER MS, & BHANOT VK (1985). Hyperparathyroidism: neuropsychiatric manifestations. Psychosomatics 26:597–601.

Boston Collaborative Drug Surveillance Program (1972). Acute adverse reaction to prednisone in relation to dosage. Clin Pharmacol 13:694–697.

BRENT GA, & HERSHMAN JM (1986). Thyroxine therapy in patients with severe nonthyroidal illness and low serum thyroxine concentration. J Clin Endocrinol Metab 63:1–8.

BROWER KJ, ELIOPULOS GA, BLOW FC, ET AL. (1990). Evidence for physical and psychological dependence on anabolic androgenic steroids in eight weight lifters. Am J Psychiatry 147:510–512.

BROWN GG, PREISMAN RC, & KLEEREKOPER M (1987). Neurobehavioral symptoms in mild primary hyperparathyroidism: related to hypercalcemia but not improved by parathyroidectomy. Henry Ford Hosp Med J 35:211–215.

BUCKLEY WE, YESALIS CE III, FRIEDL KE, ET AL. (1988). Estimated prevalence of anabolic steroid use among high school seniors. JAMA 260:3441–3445.

CAHILL GF JR, & SOELDNER JS (1974). A non-editorial on non-hypoglycemia. N Engl J Med 291:904–905.

CAROFF SN (1978). Klinefelter's syndrome and bipolar affective illness: a case report. Am J Psychiatry 135:748–749.

CARPENTER PC (1988). Diagnostic evaluation of Cushing's syndrome. Endocrinol Metab Clin North Am 17:445–472.

CARROLL BJ, FEINBERG M, GREDEN JF, ET AL. (1981). A specific test for the diagnosis of melancholia: standardization, validation, and clinical utility. Arch Gen Psychiatry 38:15–22.

CAVALIERI RR (1991). The effects of nonthyroid disease and drugs on thyroid function tests. Med Clin North Am 75:27–39.

CHAD DA, ARONIN N, LUNDSTROM R, ET AL. (1990). Does capsaicin relieve the pain of diabetic neuropathy? Pain 42:387–388.

CHOPRA IJ, SOLOMON DH, HEPNER GW, ET AL. (1979). Misleadingly low free thyroxine index and usefulness of reverse triiodothyronine measurement in nonthyroidal illness. Ann Intern Med 90:905–912.

CHOPRA IJ, SOLOMON DH, & TIEN-SHANG H (1990). Serum thyrotropin in hospitalized psychiatric patients: evidence for hyperthyrotropinemia as measured by an ultrasensitive thyrotropin assay. Metabolism 39:538–543.

COHEN LM, GREENBERG DB, & MURRAY GB (1984). Neuropsychiatric presentation of men with pituitary tumors (the "four A's"). Psychosomatics 25:925–928.

COOPER DS, HALPERN R, WOOD LC, ET AL. (1984). L-Thyroxine therapy in subclinical hypothyroidism: a double-blind, placebo controlled study. Ann Intern Med 101:18–24.

COWDRY RW, WEHR TA, ZIS AP, ET AL. (1983). Thyroid abnormalities associated with rapid-cycling bipolar illness. Arch Gen Psychiatry 40:414–420.

COYNE JC, & ANDERSON BJ (1988). The "psychosomatic family" reconsidered: diabetes in context. J Marital Fam Ther 14:113–123.

CURRAN PG, & DEGROOT LJ (1991). The effect of hepatic enzyme-inducing drugs on thyroid hormones and the thyroid gland. Endocr Rev 12:135–150.

CUSHING H (1932). The basophil adenomas of the pituitary body and their clinical manifestations. Johns Hopkins Med J 50:137–195.

DIXON RB, & CHRISTY NP (1980). On the various forms of corticosteroid withdrawal syndrome. Am J Med 68:224–230.

DOWNS RW JR, & LEVINE MA (1990). Hypoparathyroidism and other causes of hypocalcemia. In KL BECKER (Ed.), Principles and practice of endocrinology and metabolism (Ch. 58, pp. 447–457). Philadelphia: Lippincott.

EHRMANN DA, & SARNE DH (1989). Serum thyrotropin and the assessment of thyroid status. Ann Intern Med 11:179–181.

EISENBERG H, PALLOTTA J, SACKES B, ET AL. (1989). Parathyroid localization, three-dimensional modeling, and percutaneous ablation techniques. Endocrinol Metab Clin North Am 18:659–700.

EMERSON CH, DYSON WL, & UTIGER RD (1973). Serum thyrotropin and thyroxine concentrations in patients receiving lithium carbonate. J Clin Endocrinol Metab 36:338–346.

ENDRES DB, VILLANUEVA R, & SHARP CF (1989). Measurement of parathyroid hormone. Endocrinol Metab Clin North Am 18:611–629.

ETTIGI PG, & BROWN GM (1978). Brain disorders associated with endocrine dysfunction. Psychiatr Clin North Am 1:117–136.

EVANS WS, CARLSEN E, & HO KY (1990). Prolactin and its disorders. In KL BECKER (Ed.), Principles and practice of endocrinology and metabolism (Ch. 18, pp. 134–139). Philadelphia: Lippincott.

FALK WE, MAHNKE MW, & POSKANZER DC (1979). Lithium prophylaxis of corticotropin-induced psychosis. JAMA 241:1011–1012.

FAVA GA, FAVA M, KELLNER R, ET AL. (1981). Depression, hostility and anxiety in hyperprolactinemic amenorrhea. Psychother Psychosom 36:122–128.

FIELD JB (1989). Hypoglycemia: definition, clinical presentations, classification, and laboratory tests. Endocrinol Metab Clin North Am 18:27–41.

FITZPATRICK LA, & BILEZIKIAN JP (1990). Primary hyperparathyroidism. In KL BECKER (Ed.), Principles and practice of endocrinology and metabolism (Ch. 56, pp. 430–437). Philadelphia: Lippincott.

FLAVIN DK, FREDRICKSON PA, RICHARDSON JW, ET AL. (1983). Corticosteroid abuse—an unusual manifestation of drug dependence. Mayo Clin Proc 58:764–766.

GASTINEAU CF (1983). Is reactive hypoglycemia a clinical entity? Mayo Clin Proc 58:545–549.

GATEWOOD JW, ORGAN CH, & MEAD BT (1975). Mental changes associated with hyperparathyroidism. Am J Psychiatry 132:129–132.

GOLD MS, & PEARSALL HR (1983). Hypothyroidism—or is it depression? Psychosomatics 24:646–656.

GOLD MS, POTTASH ALC, & EXTEIN I (1982). "Symptomless" autoimmune thyroiditis in depression. Psychiatry Res 6:261–269.

GOLD PW, LORIAUX DL, & ROY A (1986). Responses to corticotropin releasing hormone in the hypercortisolism of depression and Cushing's disease. N Engl J Med 314:1329–1335.

GOODWIN FK, PRANGE AJ, & POST RM (1982). Potentiation of antidepressant effects by L-triiodothyronine in tricyclic nonresponders. Am J Psychiatry 139:34–38.

GOOLKER P, & SCHEIN J (1953). Psychic effects of ACTH and cortisone. Psychosom Med 15:589–613.

GREENE LW, COLE W, & GREENE JB (1984). Adrenal insufficiency as a complication of the acquired immunodeficiency syndrome. Ann Intern Med 101:497–498.

HALL RCW (1983). Psychiatric effects of thyroid hormone disturbance. Psychosomatics 24:7–18.

HALL RCW, POPKIN MK, & KIRKPATRICK B (1978). Tricyclic exacerbation of steroid psychosis. J Nerv Ment Dis 166:738–742.

HALL RCW, POPKIN MK, STICKNEY SK, ET AL. (1979). Presentation of the "steroid psychoses." J Nerv Ment Dis 17:229–236.

HALLAGAN JB, HALLAGAN LF, & SNYDER MB (1989). Anabolic-androgenic steroid use by athletes. N Engl J Med 321:1042–1045.

HASKETT RF (1985). Diagnostic categorization of psychiatric disturbance in Cushing's syndrome. Am J Psychiatry 142:911–916.

HAY ID (1985). Thyroiditis: a clinical update. Mayo Clin Proc 60:836–843.

HEATH DA (1989). Primary hyperparathyroidism: clinical presentation and factors influencing clinical management. Endocrinol Metab Clin North Am 18:631–646.

HEATH DA, WRIGHT AD, & BARNES AD (1980). Surgical treatment of primary hyperparathyroidism in the elderly. BMJ 280:1406–1408.

HEATH H III, HODGSON SF, & KENNEDY MA (1980). Primary hyperparathyroidism: incidence, morbidity, and potential economic impact in a community. N Engl J Med 302:189–193.

HEIN MD, & JACKSON I (1990). Review: thyroid function in psychiatric illness. Gen Hosp Psychiatry 12:232–244.

HEINZ ER, MARTINEZ J, & HAENGGELI A (1977). Reversibility of cerebral atrophy in anorexia nervosa and Cushing's syndrome. J Comput Assist Tomogr 1:415–418.

HELZ JW, & TEMPLETON B (1990). Evidence for the role of psychosocial factors in diabetes mellitus: a review. Am J Psychiatry 147:1275–1282.

HOFELDT FD (1989). Reactive hypoglycemia. Endocrinol Metab Clin North Am 18:185–201.

HOGAN MJ, SERVICE FJ, & SHARBROUGH F (1983). Oral glucose tolerance test compared with a mixed meal in the diagnosis of reactive hypoglycemia: a caveat on stimulation. Mayo Clin Proc 58:491–496.

HORWITZ DL (1989). Factitious and artifactual hypoglycemia. Endocrinol Metab Clin North Am 18:203–210.

INSOGNA KL (1989). Humoral hypercalcemia of malignancy: the role of parathyroid hormone-related peptide. Endocrinol Metab Clin North Am 18:779–794.

JOBORN C, HETTA J, & JOHANSSON H (1988a). Psychiatric morbidity in primary hyperparathyroidism. World J Surg 12:476–481.

JOBORN C, HETTA J, RASTAD J, ET AL. (1988b). Psychiatric symptoms and cerebrospinal fluid monoamine metabolites in primary hyperparathyroidism. Biol Psychiatry 23:149–158.

JOFFEE RT, GOLD PW, UHDE TW, ET AL. (1984). The effect of carbamazepine on the thyrotropin response to thyrotropin-releasing hormone. Psychiatry Res 12:161–166.

JOHNSON DD, DORR KE, SWENSON WM, ET AL. (1980). Reactive hypoglycemia. JAMA 243:1151–1155.

JONES JH, & MEADE TW (1964). Hypothermia following cholorpromazine therapy in myxedematous patients. Gerontol Clin 6:252–256.

JOSEPHSON AM, & MACKENZIE TB (1979). Appearance of manic psychosis following rapid normalization of thyroid status. Am J Psychiatry 136:846–847.

KARACAN I, SALIS PJ, WARE JW, ET AL. (1978). Nocturnal penile tumescence and diagnosis in diabetic impotence. Am J Psychiatry 135:191–197.

KATHOL RG, TURNER R, & DELAHUNT J (1986). Depression and anxiety associated with hyperthyroidism: response to antithyroid therapy. Psychosomatics 27:501–505.

KHANNA S, AMMINI A, SAXENA S, ET AL. (1988). Hypopituitarism presenting as delirium. Int J Psychiatry Med 18:89–92.

KLEE GG, KAO PC, & HEATH H III (1988). Hypercalcemia. Endocrinol Metab Clin North Am 17:573–600.

KVINESDAL B, MOLIN J, FRELAND A, ET AL. (1984). Imipramine treatment of painful diabetic neuropathy. JAMA 251:1727–1739.

LANKFORD HV, BLACKARD WG, GARDNER DF, ET AL. (1981). Effects of thyrotropin-releasing hormone and metoclopramide in patients with phenothiazine-induced hyperprolactinemia. J Clin Endocrinol Metab 53:109–112.

LARSEN ML, HORDER M, & MOGENSEN EF (1990). Effect of long-term monitoring of glycosylated hemoglobin levels in insulin-dependent diabetes mellitus. N Engl J Med 323:1021–1025.

LEVIN TR, TERRELL TR, & STOUDEMIRE A (1992). Organic mood disorder associated with the HAIR-AN syndrome. J Neuropsychiatry Clin Neurosci 4:51–54.

LING MH, PERRY PJ, & TSUANG MT (1981). Side effects of corticosteroid therapy: psychiatric aspects. Arch Gen Psychiatry 38:471–477.

LLOYD GG, STEEL JM, & YOUNG RJ (1987). Eating disorders and psychiatric morbidity in patients with diabetes mellitus. Psychother Psychosom 48:189–195.

LOOSEN PT (1988). Thyroid function in affective disorders and alcoholism. Endocrinol Metab Clin North Am 17:55–82.

LORIAUX DL (1990a). Cushing's syndrome. In KL BECKER (Ed.), Principles and practice of endocrinology and metabolism (Ch. 76, pp. 595–600). Philadelphia: Lippincott.

LORIAUX DL (1990b). Adrenocortical insufficiency. In KL BECKER (Ed.), Principles and practice of endocrinology and metabolism (Ch. 77, pp. 600–604). Philadelphia: Lippincott.

LORIAUX DL, & NIEMANN L (1990). Cushing's syndrome: advances in diagnosis and treatment. Adv Endocrinol Metab 1:23–33.

LUSTMAN PJ, GRIFFITH LS, CLOUSE RE, ET AL. (1986). Psychiatric illness in diabetes mellitus: relationship to symptoms and glucose control. J Nerv Ment Dis 174:736–742.

MacAdams MR (1988). Effects of medications and nonthyroidal illness on thyroid function testing. Med Grand Rounds 1:31–42.

Maciewicz R, Bouckoms A, & Martin JB (1985). Review: drug therapy of neuropathic pain. Clin J Pain 1:39–49.

Mallette LE, & Eichhorn E (1986). Effects of lithium carbonate on human calcium metabolism. Arch Intern Med 146:770–776.

Mampalam TJ, Tyrrell JB, & Wilson CB (1988). Transsphenoidal microsurgery for Cushing's disease: a report of 216 cases. Ann Intern Med 109:487–493.

Marcus MD, & Wing RR (Eds.) (1990). Eating disorders and diabetes: diagnosis and management. Diabetes Spect 3:361–400.

Mariotti S, Martino E, Cupini C, et al. (1982). Low serum thyroglobulin as a clue to the diagnosis of thyrotoxicosis factitia. N Engl J Med 307:410–412.

Martin LM (1981). Impotence in diabetes: an overview. Psychosomatics 22:318–329.

McKenzie JM, & Zakarija M (1989). Hyperthyroidism. In LJ DeGroot (Ed.), Endocrinology (Ch. 43, pp. 646–682). Philadelphia: WB Saunders.

McNamara E, Southwick SM, & Fogel BS (1987). Sleep apnea and hyperthyroidism presenting as depression in two patients. J Clin Psychiatry 48:164–165.

Mechanick JI, & Dunaif A (1990). Masculinization: a clinical approach to the diagnosis and treatment of hyperandrogenic women. Adv Endocrinol Metab 1:129–173.

Medeiros-Neto GA (1986). Triiodothyronine thyrotoxicosis. In SH Ingbar & LE Braverman (Eds.), The thyroid (Ch. 65, pp. 1429–1438). Philadelphia: Lippincott.

Minuchim S, Rosman BL, & Baker L (1978). Psychosomatic families: anorexia nervosa in context. Cambridge: Harvard University Press.

Molitch ME (1987). Pituitary tumors: diagnosis and management [symposium]. Endocrinol Metab Clin North Am 16:475–828.

Momose KJ, Kjellberg RN, & Kliman B (1971). High incidence of cortical atrophy of the cerebral and cerebellar hemispheres in Cushing's disease. Radiology 99:314–348.

Nelson RL (1985). Hypoglycemia: fact or fiction? Mayo Clin Proc 60:844–850.

Nemeroff CB, Simon JS, Haggerty JJ, et al. (1985). Antithyroid antibodies in depressed patients. Am J Psychiatry 142:840–843.

O'Dorisio TM, & Redfern JS (1990). Somatostatin and somatostatin-like peptides: clinical research and clinical applications. Adv Endocrinol Metab 1:175–230.

Oldfield EH, Chrousos GP, Schulte HM, et al. (1985). Preoperative lateralization of ACTH-secreting pituitary microadenomas by bilateral and simultaneous inferior petrosal venous sinus sampling. N Engl J Med 312:100–103.

O'Shanick GJ, & Ellinwood EH (1982). Persistent elevation of thyroid stimulating hormone in women with bipolar affective disorder. Am J Psychiatry 139:513–514.

Parfitt AM (1989). Surgical, idiopathic, and other varieties of parathyroid hormone-deficient hypoparathyroidism. In LJ DeGroot (Ed.), Endocrinology (Ch. 64, pp. 1049–1064). Philadelphia: WB Saunders.

Peterson P (1968). Psychiatric disorders in primary hyperparathyroidism. J Clin Endocrinol 28:1491–1495.

Plymate SR, & Paulsen CA (1990). Male hypogonadism. In KL Becker (Ed.), Principles and practice of endocrinology and metabolism (Ch. 119, pp. 948–970). Philadelphia: Lippincott.

Pope HG Jr, & Katz DL (1988). Affective and psychotic symptoms associated with anabolic steroid use. Am J Psychiatry 145:487–490.

Popkin MK, & Mackenzie TB (1980). Psychiatric presentations of endocrine dysfunction. In RCW Hall (Ed.), Psychiatric presentations of medical illness (Ch. 9, pp. 139–156). New York: Spectrum Publications.

Reus VI (1989). Behavioral aspects of thyroid disease in women. Psychiatr Clin North Am 12:153–165.

Rittmaster RS, & Loriaux DL (1987). Hirsutism. Ann Intern Med 106:95–107.

Rizza RA, & Greene DA (1988). Diabetes mellitus [symposium]. Med Clin North Am 72:1271–1607.

Rosman BL, & Baker L (1988). The "psychosomatic family" reconsidered: diabetes in context—a reply. J Marital Fam Ther 14:125–132.

Rosmark B, Berne C, Holmgren S, et al. (1986). Eating disorders in patients with insulin-dependent diabetes mellitus. J Clin Psychiatry 47:547–550.

Sawin CT, Castelli WP, & Hershman JM (1985). The aging thyroid: thyroid deficiency in the Framingham study. Arch Intern Med 145:1386–1388.

Sawin CT, Geller A, & Kaplan MM (1991). Low serum thyrotropin (TSH) in older persons without hyperthyroidism. Arch Intern Med 151:165–170.

Schaaf M, & Payne C (1966). Dystonic reactions to prochlorperazine in hyperparathyroidism. N Engl J Med 275:991–994.

Seltzer HS (1989). Drug-induced hypoglycemia: a review of 1418 cases. Endocrinol Metab Clin North Am 18:163–183.

Service FJ (1989). Hypoglycemia and the postprandial syndrome. N Engl J Med 321:1472–1474.

Sheehan HL (1939). Simmonds disease due to post partum necrosis of the anterior pituitary. Q J Med 8:277–307.

Sheps SG, Jiang NS, & Klee GG (1988). Diagnostic evaluation of pheochromocytoma. Endocrinol Metab Clin North Am 17:397–414.

Sheps SG, Jiang NS, Klee GG, et al. (1990). Recent developments in the diagnosis and treatment of pheochromocytoma. Mayo Clin Proc 65:88–95.

Slag MF, Morely JE, & Elson MK (1981). Hypothyroxinemia in critically ill patients as a predictor of high mortality. JAMA 245:43–45.

Snyder PJ (1990). Treatment of male hypogonadism. Adv Endocrinol Metab 1:245–260.

Soffer LJ, Iannaccone A, & Gabrilove JL (1961). Cushing's syndrome: a study of fifty patients. Am J Med 30:129–146.

Sonino N (1987). The use of ketoconazole as an inhibitor of steroid production. N Engl J Med 317:128–818.

Spaulding SW, & Lippes H (1985). Hyperthyroidism: causes, clinical features, and diagnosis. Med Clin North Am 69:937–951.

Spratt DI, Pont A, & Miller MB (1982). Hyperthyroxinemia in acute psychiatric disorders. Am J Med 73:41–48.

Starkman MN, & Schteingart DE (1981). Neuropsychiatric manifestations of patients with Cushing's syndrome. Arch Intern Med 141:215–219.

Starkman MN, Zelnik TC, Nesse RM, et al. (1985). Anxiety in patients with pheochromocytomas. Arch Intern Med 145:248–252.

Streck WF, & Lockwood DH (1979). Pituitary adrenal recovery following short-term suppression with corticosteroids. Am J Med 66:910–914.

Swanson DW, & Stipes AH (1969). Psychiatric aspects of Klinefelter's syndrome. Am J Psychiatry 126:82–90.

Targum SD, Greenberg RD & Harmon RL (1984). Thyroid hormone and the TRH stimulation test in refractory depression. J Clin Psychiatry 45:345–346.

Thomas FB, Mazzaferri EL, & Skillman TG (1970). Apathetic thyrotoxicosis: a distinctive clinical and laboratory entity. Ann Intern Med 72:679–685.

THOMPSON WF (1973). Psychiatric aspects of Addison's disease: report of a case. Med Ann DC 43:62–64.

TOFT AD (1988). Use of sensitive immunoradiometric assay for thyrotropin in clinical practice. Mayo Clin Proc 63:1035–1042.

TRENCE DL, MORLEY JE, & HANDWERGER BS (1985). Polyglandular autoimmune syndromes. Am J Med 7:107–116.

UTIGER RD (1980). Decreased extrathyroidal triiodothyronine production in nonthryoidal illness: benefit or harm? Am J Med 69:807–810.

UTIGER RD (1984). Editorial retrospective: beta-adrenergic antagonist therapy for hyperthyroid Graves' disease. N Engl J Med 310: 1597–1598.

VANCE ML, & THORNER MO (1987). Prolactinomas. Endocrinol Metab Clin North Am 16:731–753.

VARADARAJ R, & COOPER AJ (1986). Addison's disease presenting with psychiatric symptoms [letter]. Am J Psychiatry 143:553–554.

WATTS NB (1989). Using a sensitive thyrotropin assay for monitoring treatment with levothyroxine. Arch Intern Med 149:309–312.

WEHRMANN RE, GREGERMAN RI, & BURNS WH (1985). Suppression of thyrotropin in the low-thyroxine state of severe nonthyroidal illness. N Engl J Med 312:546–552.

WHELAN TB, SCHTEINGART DE, STARKMAN MN, ET AL. (1980). Neuropsychological deficits in Cushing's syndrome. J Nerv Ment Dis 168:753–757.

WHYBROW PC, & PRANGE AJ (1981). A hypothesis of thyroid-catecholamine receptor interaction. Arch Gen Psychiatry 38:106–113.

WILSON JD (1988). Androgen abuse by athletes. Endocr Rev 9:181–199.

WILSON WH, & JEFFERSON JW (1985). Thyroid disease, behavior and psychopathology. Psychosomatics 26:481–492.

WINDSOR R, & DUMITRU D (1989). Prevalence of anabolic steroid use by male and female adolescents. Med Sci Sports Exerc 21:494–497.

WITSCHY JK, & REDMOND FC (1981). Extrapyramidal reaction to fluphenazine potentiated by thyrotoxicosis. Am J Psychiatry 138:246–247.

WOLKOWITZ OM, & RAPAPORT M (1989). Long-lasting behavioral changes following prednisone withdrawal [letter]. JAMA 261:1731–1732.

WOLKOWITZ OM, REUS VI, WEINGARTEN H, ET AL. (1990). Cognitive effects of corticosteroids. Am J Psychiatry 147:1297–1303.

YAGER J, & YOUNG RT (1974). Non-hypoglycemia is an epidemic condition. N Engl J Med 291:907–908.

YASSA R, SAUNDERS A, & NASTASE C (1988). Lithium-induced thyroid disorders: a prevalence study. J Clin Psychiatry 49:14–16.

30 Dermatology

MADHULIKA A. GUPTA, M. D.

The skin occupies an important and powerful position as an organ of communication, and it plays an important role in socialization throughout the life cycle. The skin may be used to communicate psychological distress. Alternately, development of a disfiguring cutaneous condition can result in social disapproval and increased self-consciousness, which in turn can result in academic underachievement, social withdrawal, and serious psychological problems especially when the skin disorder occurs during a developmentally critical period such as adolescence. Such factors are obviously important with disorders such as acne and psoriasis.

This chapter examines several aspects of psychodermatology, including psychiatric aspects of primary dermatologic disorders (urticaria, atopic dermatitis, pruritus, psoriasis, acne), dermatologic manifestations of psychiatric disorders (body dysmorphic disorder, eating disorders), and dermatologic signs of medical illnesses that have psychiatric symptoms (AIDS, Lyme disease, syphilis). It also discusses dermatologic side effects of psychotropic drugs. Readers are referred to Tables 30-1 through 30-5 for summary information and basic descriptions of common psychodermatologic problems discussed in the text.

SKIN AND PSYCHE DURING EARLY DEVELOPMENT

The skin is composed of epidermis and dermis. The epidermis and epidermal appendages (e.g., the pilosebaccous units, eccrine and apocrine glands), and nervous system develop from the embryonic ectoderm. The dermis is derived from the mesoderm, except for the cutaneous nerves, which are derived from the ectoderm. Neuropeptides such as substance P, vasoactive intestinal peptide (VIP), and enkephalins play an important role in pruritus, flushing, and sweating as a result of their central nervous system (CNS) and peripheral actions. The importance of psychoneuroimmunologic factors in dermatology is supported by the well-documented observation that psychosocial stress can exacerbate certain dermatologic conditions that have an immunologic component, for example, atopic dermatitis, chronic urticaria, and psoriasis (Gupta & Voorhees, 1990; Medansky & Handler, 1981).

Cutaneous stimulation during infancy is an important factor for cell growth and CNS maturation (Pauk et al., 1986). In preweanling rat pups, temporary interruption of active tactile stimulation, or "maternal deprivation," is associated with reduced ornithine decarboxylase activity, which is a sensitive index of cell growth and CNS development (Pauk et al., 1986). Heavy stroking reduces or prevents these effects (Evoniuk, Kuhn, & Schanberg, 1979). One study has shown that preterm human neonates who received tactile and kinesthetic stimulation gained weight more rapidly, were more alert, and exhibited more mature neurologic reflexes than controls (Field et al., 1986).

PSYCHIATRIC ASPECTS OF PRIMARY DERMATOLOGIC DISORDERS

Urticaria

Urticaria and angioedema (see Table 30-1) have a lifetime prevalence of about 15 to 20% (Monroe & Jones, 1977) in the general population. Psychological factors have been observed to play the most important role in chronic versus acute urticaria, cholinergic urticaria, and possibly "adrenergic" urticaria (see Table 30-2). There have been a few case reports of a close association between angioedema (Table 30-1) and neurologic symptoms, suggesting both focal and generalized CNS involvement (Sunder, Balsam, & Vengrow, 1982).

Psychiatric Aspects

Severe emotional stress is known to exacerbate urticarial reactions, regardless of the primary cause for the urticaria (see Table 30-3) (Arnold, Odom, & James, 1990).

Psychiatric Aspects of Therapy

"Antidepressant" drugs such as doxepin, usually at doses around 10 mg tid for 2 weeks (Gupta, Gupta, & Ellis,

TABLE 30–1. *Some Dermatologic Disorders Influenced by Psychiatric Factors*

Urticaria (or hives)[a]: Wheals, which are generally associated with itching, stinging, or a pricking sensation. Wheals are caused by localized edema of the dermis, secondary to increased capillary permeability and extravasation of proteins and fluids into the dermis. Nonimmunologic and immunologic factors, including type I hypersensitivity mediated by IgE and activation of the complement cascade, are important in pathogenesis.

Angioedema (previously "angioneurotic edema")[a]: Represents edema of the subcutaneous tissue; it may occur alone or in conjunction with urticaria. Usually acute, affecting easily distensible tissues such as the lips, mucous membranes of the mouth, and external genitalia. Edema of the tongue or larynx may lead to respiratory compromise.

Atopic dermatitis (also called atopic eczema, infantile eczema, disseminated neurodermatitis)[a]: A genetically determined disorder of unknown cause associated with an increased tendency to form IgE and an increased susceptibility to asthma and hay fever. It has three stages.

1. *Infantile atopic dermatitis* (ages 2 months to 2 years). Symptoms usually disappear after 2 years. Usually begins as an itchy erythema of the cheeks. Rash eventually manifests as moist, crusted areas and may spread to other parts of the body.
2. *Childhood atopic dermatitis* (ages 2 to 10 years). Pruritis is a prominent feature. Rash is less exudative and often lichenified and scaly. Itching is often paroxysmal, and the scratching impulse reaches compulsive proportions. A vicious itch-scratch cycle is often established.
3. *Adolescence and adult-onset atopic dermatitis.* Rash is typically dry and lichenified, and pruritus is a prominent feature. Condition typically improves with age.

Psoriasis[a]: A chronic recurrent inflammatory disorder characterized by circumscribed, erythematous, scaly plaques. Lesions may affect any body region but have a predilection for the scalp, extensor surfaces of the limbs, and the sacral region. Itching and burning are frequent complaints. The rate of epidermal cell replication is markedly accelerated in active lesions.

Acne[b]: A very common disorder associated with an increase in the rate of sebum secretion by the sebaceous glands. It generally affects sebaceous-gland rich areas such as the face and back. The characteristic lesions include papules, open and closed comedones, pustules, nodules, cysts and scars that often are cosmetically disfiguring.

[a]Arnold, Odom, & James (1990).
[b]Cunliffe & Cotterill (1975).

TABLE 30–2. *Urticarial Reactions Associated with Psychiatric Factors*

Chronic idiopathic urticaria[a]: Urticaria of unidentifiable cause recurring daily for 6 weeks or more. A definite cause is not identifiable in up to 70% of patients with urticaria. Emotional factors can precipitate or perpetuate this disorder.

Cholinergic urticaria: Tiny distinctive wheal within a large flare of erythema, associated with sympathetic stimulation. It represents the response to acetylcholine released by autonomic nerves. The lesions occur with heat, exercise, or emotional stress and represent 5% of the cases of urticaria.

"Adrenergic" urticaria[b]: A rare entity that presents as widespread, pruritic papules surrounded by striking white haloes. The lesions, associated with elevated levels of plasma norepinephrine and epinephrine, occur at times of emotional stress.

[a]Arnold, Odom, & James (1990).
[b]Shelley & Shelley (1985).

1987), may be effective treatment for chronic idiopathic urticaria and cold urticaria. Anxiolytic and antihistaminic drugs are often used individually or in combination with other therapies, often with limited success. The antihistaminic and anticholinergic properties of the tricyclic antidepressants are the likely basis for their efficacy in the urticarias. For example, dermal blood vessels possess both histamine H1 and H2 receptors, and it has been observed that combined H1 plus H2 antihistamine therapy is significantly more effective than H1 antihistamines alone for chronic urticaria (Gupta, Gupta, & Ellis, 1987). Doxepin, trimipramine, and amitriptyline are some of the most potent histamine H1 receptor antagonists among the tricyclics and are also H2 receptor antagonists (Gupta, Gupta, & Ellis, 1987). Doxepin, for example, is 800 times more potent an H1 receptor antagonist than the commonly used antihistamine diphenhydramine, 67 more times more potent than the H1 receptor antagonist hydroxyzine, and six times more potent than the H2 receptor antagonist cimetidine. The anticholinergic effect of doxepin is also significantly greater than that of the antihistamines hydroxyzine or diphenhydramine.

Atopic Dermatitis

Atopic dermatitis afflicts 7 to 24 individuals per 1000 and accounts for 20% of all patients treated in dermatology clinics (Faulstich & Williamson, 1985). Immunologic problems and abnormalities of the sympathetic nervous system are among the factors that have been implicated in the pathogenesis of atopic dermatitis. The role of immunologic factors in atopic dermatitis is not clear. The most consistent finding has been an impairment of T cell function, with a reduced or normal number of T cells (Solomon, 1985).

Some patients with atopic dermatitis have high psychophysiologic reactivity, as measured by electromyography, heart rate, and anxiety scores (Faulstich et al., 1985); and they react abnormally to certain pharmacologic agents such as acetylcholine (Solomon, 1985).

Psychiatric Aspects

The importance of psychological factors (Table 30-3) in atopic dermatitis has been cited on many occasions (Faulstich & Williamson, 1985). Factors such as the absence of adequate tactile stimulation and a rejecting

TABLE 30–3. *Psychiatric Aspects of Dermatologic Disorders*

Urticaria and angioedema

Psychosocial stress: Stress was associated with onset in 51% of patients versus 8% of controls (Rees, 1957); linkage with exacerbation variously reported among 7% (Juhlin, 1981) and 67% (Rees, 1957) of patients.

Psychopathology: "Severe psychiatric problems" including depression among 16% of 330 patients studied (Juhlin, 1981). Possible association with "premenstrual tension" (Rees, 1957).

Atopic dermatitis

Psychosocial stress: Stressful life events preceded the onset of atopic dermatitis in 70% (Faulstich & Williamson, 1985) of cases. Disease-related stress and stress resulting from the family environment are important predictors of symptom severity among children (Gil et al., 1987).

Psychopathology: Neurosis or psychosis (DSM-II) among 60% of 10 patients studied versus 20% of controls, with 30% prevalence of schizophrenia (Ullman, Moore, & Reidy, 1977). No correlation between the severity of psychopathology and the severity or chronocity of the skin disorder (Ullman, Moore, & Reidy, 1977). Atopic patients have higher state and trait anxiety (Faulstich & Williamson, 1985).

Psoriasis

Psychosocial stress: Stress was implicated in the onset and exacerbation of symptoms in 39% (Seville, 1977) of patients versus 10% of controls. In comparison to low-stress reactors, high-stress reactors reported greater psoriasis-related daily stress, greater interpersonal dependence, and less overt anger (Gupta et al., 1989a). High-stress reactors have psoriasis in more easily visible and emotionally charged body regions, e.g., scalp, face, neck, forearms, hands, and genital regions (Gupta et al., 1989a). Psoriatics experienced "significantly higher strain levels" compared to healthy controls when exposed to a stress-provoking situation and had increased urinary epinephrine levels (Arnetz et al., 1985).

Psychopathology: Pruritis of psoriasis is directly correlated with the severity of depression in the patient (Gupta et al., 1988). There is an 18% prevalence of alcoholism among psoriatics versus 2% prevalence among other dermatologic controls (Morse, Perry, & Hurt, 1985).

Acne

Psychosocial stress: Acne can cause significant psychosocial stress as a result of its negative impact on the body image and self image of the patient (Rubinow et al., 1987) in both severe cystic acne and mild to moderate acne vulgaris (Gupta et al., 1991a).

Psychopathology: Acne may affect overall body image unrelated to the skin and precipitate an eating disorder among psychologically vulnerable subjects (Gupta et al., 1992b). Acne excoriée may serve as a "protective mechanism" among patients having difficulty coping with the social and sexual demands of adulthood (Sneddon & Sneddon, 1983) and may be used as an excuse to avoid social contact.

mother figure (Ullman, Moore, & Reidy, 1977) have been reported to contribute to childhood atopic dermatitis. Atopic patients have also been found to scratch themselves more readily in response to an itch stimulus than nonatopic controls, suggesting that they develop a conditioned scratch response sooner than nonatopic controls (Jordan & Whitlock, 1974). Studies that have evaluated the role of psychiatric factors in atopic dermatitis have been uncontrolled (Table 30-3) and in most cases have not defined the nosologic criteria used to diagnose the psychopathologic syndromes in question (Table 30-3). Although psychological factors may perpetuate the disorder in some cases, as suggested by some well designed studies (Gil et al., 1987), based on the current literature it is not possible to make a strong causal connection between psychological factors and atopic dermatitis (Solomon, 1985).

Psychiatric Aspects of Therapy

Childhood atopic dermatitis may be associated with dysfunctional family dynamics. One study reported that 45% of atopic patients whose mothers received counseling had clear skin versus 10% of the group receiving standard therapies (Williams, 1951). Details of the counseling techniques used were not provided. In another case series (Koblenzer & Koblenzer, 1988) of infants and children with intractable atopic dermatitis, the authors noted that psychotherapy aimed at "increasing parental insight into their conflicting feelings," especially in relation to the child's demand for attention and tendency to want to share the parental bed at night, led to improvement of the skin condition. Among adults, the importance of identifying specific life stresses and allowing the patient to express emotional symptoms related to the stressor has been emphasized (Brown & Bettley, 1971).

Behavior modification has been used to reduce scratching behaviors that tend to perpetuate the itch-scratch cycle and exacerbate atopic dermatitis (Bar & Kuypers, 1973; Faulstich & Williamson, 1985). Among adults, standard medical treatments combined with progressive relaxation and hypnosis (Brown & Bettley, 1971) were associated with greater improvement of the skin at 14 months follow-up than standard dermatologic treatments alone. There exist no definite reported guidelines for the initiation of behavior therapies in atopic dermatitis. When scratching is assessed as being a significant perpetuating factor or when the patient reports a strong behavioral overlay, the addition of behavior therapies to the standard medical treatments may be of benefit to the patient.

Psychotropic drugs are used in atopic dermatitis to obtain general sedation, for example, anxiolytics and major tranquilizers (Gupta, Gupta, & Haberman, 1986). In a 4-week double-blind placebo-controlled study (Medansky, 1971) there was no significant difference between perphenazine 2 mg tid, amitryptyline 10 mg tid, and chlordiazepoxide 10 mg tid in comparison to placebo for management of skin symptoms due to atopic dermatitis and a range of other skin disorders.

Pruritus

Pruritus, or itching, is reported to be the most common symptom of dermatologic disease and second only to disfigurement as a source of distress for the patient (Gilchrest, 1982).

Psychiatric Aspects

Pruritus that has a psychogenic component is described as being paroxysmal, severe, and stopping completely as soon as pain is induced by scratching (Arnold, Odom, & James, 1990). In cases of psychogenic pruritus such as lichen simplex chronicus (also called neurodermatitis circumscripta) and prurigo nodularis, psychological stress, anxiety, and depression are believed to precipitate scratching of a localized area of skin, which initiates the scratch-itch-scratch cycle and eventually results in lichenification of the skin or multiple itchy nodules, respectively (Arnold, Odom, & James, 1990). Certain body regions such as the ear canals, eyelids, nostrils, and perianal and genital regions are especially susceptible to pruritus. In the older literature, pruritus ani, scroti, and vulvae were often considered to be psychogenic; however, other causes such as candidiasis must be ruled out. Pruritus with an accompanying rash has been reported to be the symptom of an epidemic hysteria in an elementary school setting (Robinson et al., 1984).

Pruritus associated with awakening from sleep is typically considered to have a primary dermatologic or medical basis rather than a psychogenic cause. One study, however, refuted this generally accepted criterion for the "organicity" of pruritus and showed that with psoriasis, where pruritus is reported to have a "psychosomatic" component, complaints of a sleep disturbance correlated with severity of depression and other sleep physiologic parameters rather than primary clinical dermatologic parameters (Gupta et al., 1989b).

Opiate ingestion can provoke generalized pruritus through a CNS mechanism (Rothman, 1941) or by degranulation of mast cells peripherally. Pruritus can be a presenting complaint among heroin addicts (Rathod, Alarcon, & Thomson, 1978). The effect of opiates on pruritus can be blocked by the opiate antagonist naloxone (Bernstein et al., 1982).

Psychiatric Aspects of Therapy

Psychogenic pruritus is not a diagnosis of exclusion; it is a symptom that may be present with or without cutaneous lesions. Various primary dermatologic disorders (e.g., psoriasis and atopic dermatitis) can be associated with pruritus that is strongly influenced by psychosomatic factors. In many instances, such as lichen simplex chronicus and prurigo nodularis, the primary consideration is the control of pruritus with standard dermatologic agents such as intralesional steroids and emollients. Tranquilizers and antihistamines may be used, but they generally are not effective (Gilchrest, 1982; Gupta et al., 1986a). Antidepressant drugs, especially those that are strongly antihistaminic such as doxepin, may be effective in certain pruritic states (Gupta, Gupta, & Ellis, 1987).

Psoriasis

Psoriasis is a chronic recurrent skin disorder (Table 30-1) with a 1 to 2% prevalence among the general population. The cause of psoriasis is not known, and both genetic and environmental factors are believed to play important roles (Arnold, Odom, & James 1990). Psoriasis has been associated with decreased responsiveness of the beta-adrenergic receptors in the epidermal cells (Voorhees & Duell, 1971). The beta-adrenergic receptors are linked with the adenylate cyclase-cyclic AMP system; this may be one reason that lithium, which has an inhibitory effect on adenylate cyclase, has been shown to exacerbate psoriasis. Other abnormalities of psoriasis include those involving the arachidonic acid transformation cascade (Voorhees, 1983). Lithium affects the phosphoinositide pathway by inhibiting the enzyme inositol monophosphatase, thereby slowing the rate of resynthesis of phosphatidylinositol. This action can affect the arachidonic acid transformation cascade and lead to an exacerbation of psoriasis.

Substance P, a neuropeptide involved in itch and pain perception and the modulation of inflammation, has been implicated in psoriasis, especially in cases when the lesions follow a symmetric dermatomal distribution (Farber et al., 1986). The role of substance P in psoriasis requires further confirmation.

Psychiatric Aspects

Psychosocial factors have been implicated as being important in the onset and exacerbation of psoriasis in 40 to 80% of patients (Gupta, Gupta, & Haberman, 1987a) (Table 30-3). It must be emphasized that the cosmetic disfigurement caused by psoriasis can profoundly affect the quality of life of the patient. Psoriasis has been reported to affect most major areas of patients' lives, for example, the patients' ability to find jobs they liked in 64% of cases, work performance in 43%, overall socialization in 37%, relations with spouse or partner in 26%, and sexual activity in 41% (Gupta et al., 1990a).

Among associated psychopathologic factors, pruritus severity correlates most strongly with the degree of depression in the psoriasis patient (Gupta et al., 1988) (Table 30-3).

Psychiatric Aspects of Therapy

The patient with psoriasis must be evaluated within a developmental context. The social, occupational, and close interpersonal functioning of the patient must be assessed, as must the mental state and suicide risk (Gupta et al., 1992d). Assessment of the day-to-day difficulties faced by the patient is especially important because, in the group of "high-stress reactors," having to cope with the daily stress of psoriasis may in turn exacerbate the skin condition (Gupta et al., 1989a). Various psychological interventions, including hypnosis-induced relaxation, psychotherapy, and temperature biofeedback, have been reported to help psoriasis (Gupta, Gupta, & Haberman, 1987a). Antidepressant medications may prove to be helpful for treatment of the pruritus and sleep disturbances associated with psoriasis (Gupta et al., 1990a).

Topical corticosteroids comprise the standard form of therapy for psoriasis and other dermatologic conditions. Systemic absorption sufficient to suppress the hypothalamic-pituitary-adrenal axis can occur in varying degrees with both low- and high-potency topical steroids (Garden & Freinkel, 1986). Some of the factors that enhance absorption (Garden & Freinkel, 1986; Yohn & Weston, 1990) and produce iatrogenic Cushing syndrome include (1) the use of high-potency steroids; (2) application over skin where the epidermal barrier has been disrupted by disease; (3) application over inflamed skin or normal skin regions where absorption is high (e.g., intertriginous areas) or where the body surface area is large in comparison with the body mass (e.g., in infants and children); (4) use of occlusive vehicles such as ointments rather than creams or lotions; and (5) the use of occlusive plastic coverings and (6) instances where the metabolism of the steroid is impaired (e.g., hepatic failure). The iatrogenic excess of systemic glucocorticoids can result in neuropsychiatric symptoms (e.g., depression, agitation, and psychosis), and such effects are generally dose-related. Some of the most potent topical glucocorticoids are clobetasol proprionate and betamethasone diproprionate; one of the least potent is hydrocortisone. According to the manufacturer, clobetasol can cause suppression of the hypothalamic-pituitary-adrenal (HPA) axis with doses as low as 2 g per day.

Methotrexate is used for severe psoriasis. Given orally or parenterally, neurotoxicity related to methotrexate has not been observed (Weiss, Walker, & Wiernik, 1974). Neurotoxicity is reported with intrathecal administration in high doses, a situation that never arises during treatment of psoriasis (see Chapter 24).

Psoriasis precipitated or exacerbated by *lithium* is typically resistant to conventional antipsoriatic treatment (Selmanowitz, 1986). Psoriasis can appear within the first few months of lithium treatment and usually within the first few years (Skoven & Thormann, 1979). Usually there is no family history of psoriasis (Selmanowitz, 1986). If lithium is discontinued, the psoriatiform rash usually remits within a few months (Selmanowitz, 1986) or reverts back to its premorbid state. For the bipolar patient discontinuing lithium, alternative prophylactic treatments may become necessary (e.g., the use of other mood stabilizers, such as carbamazepine and sodium valproate) as these drugs do not exacerbate psoriasis (Abel et al., 1986).

Phenothiazine antipsychotics (Kahn & Davis, 1970; Pavlidakey et al., 1985) and some antidepressants (Kochevar, 1980) can cause a photosensitive skin rash when the patient is exposed to ultraviolet (UVA) light. Because UVA with psoralens (PUVA) is frequently used as a treatment for psoriasis, this side effect may also interfere with the management of psoriasis.

Acne

Acne has a peak incidence during midadolescence, with a prevalence of about 40% among adolescents (Cunliffe & Cotterill, 1975). Acne excoriée is a syndrome where self-excoriation and squeezing of the acne lesions results in permanent scars.

Psychiatric Aspects

Acne can result in severe problems with self-image and in some psychopathologic reactions (Table 30-3).

Psychiatric Aspects of Therapy

In the adolescent with acne, it is important to evaluate the effect of even mild acne on the patient's body image, mental state, and social and vocational functioning. Acne can have a profound effect on the self-image of the patient, which in turn may lead to a depressive reaction (Gupta et al., 1991a; Rubinow et al., 1987). In some cases this situation leads to concerns about aspects of the body image unrelated to the skin, such as body weight and shape, which in turn may lead to an eating disorder (Gupta et al., 1991a). A preliminary report has observed that bulimic behaviors can exacerbate acne

(Gupta et al., 1991a), possibly owing to the endocrine changes associated with severe dietary restriction and binge eating. As noted earlier, acne excoriée is a syndrome where scars result from self-excoriation of acne pimples, and these scars can often be disfiguring. Acne excoriée may be used as an excuse by patients to avoid social contact (Sneddon & Sneddon, 1983), especially when the adolescent acne patient has difficulty coping with the emerging social and other demands of adulthood.

Isotretinoin is used to treat severe cystic acne and recalcitrant acne. There have been reports of a 10% prevalence of "insomnia and minor depression" (Bruno, Beacham, & Burnett, 1984) and the development of a major depressive syndrome (Scheinman et al., 1990) in association with oral isotretinoin therapy. The onset of depression was not related to dosage or duration of treatment; and depressive symptoms resolved within 1 week of cessation of the medication irrespective of dosage (Scheinman et al., 1990). The development of a major depressive disorder represents an idiosyncratic rather than a predictable side effect of isotretinoin (Scheinman et al., 1990).

Pregnancy must be avoided by any female taking isotretinoin. It is recommended that effective methods of birth control be used for at least 1 month before starting isotretinoin, during isotretinoin use, and for at least 2 months after stopping isotretinoin. It is recommended that the patient either abstain from sexual intercourse or use two reliable methods of birth control at the same time.

Lithium has been associated with a range of dermatologic reactions (see Table 30-4). Acneiform eruptions are often dose-dependent and should alert the clinician to a possible underlying state of lithium toxicity (Heng, 1982). Such reactions have been observed in several patients with serum lithium levels in the range of 1.5 to 2.5 mEq/liter. Lithium-induced acne tends to be pustular with an erythematous base. Serious lithium toxicity has been reported as a result of combining lithium with tetracycline, a standard acne therapy that sometimes has a nephrotoxic effect (McGennis, 1978).

Erythromycin, one of the standard treatments for acne, can interact with *carbamazepine* and result in markedly elevated blood levels of carbmazepine in some patients (Wroblewski, Singer, & Whyte, 1986).

DERMATOLOGIC MANIFESTATIONS OF PSYCHIATRIC DISORDERS

A wide range of psychiatric disorders (see Table 30-5) (American Psychiatric Association, 1987) can be associated with cutaneous signs and symptoms. The cutaneous syndromes may be (1) a primary psychiatric symptom such as a delusion or hallucination; (2) secondary to pathologic behaviors resulting from a psychiatric disorder (e.g., dermatitis artefacta); or (3) a clinically associated disorder with no established causal relation (e.g., cutis verticis gyrata in schizophrenia). Some of the syndromes where a definitive psychiatric intervention can be made are summarized briefly in Table 30-5.

Body Dysmorphic Disorder

Body dysmorphic disorder, or dysmorphophobia, presents as a preoccupation with some imagined defect in the appearance that often appears in the skin, for example, wrinkles, spots on the skin, and excessive facial hair (American Psychiatric Association, 1987; see also Chapter 13). Aging skin can lead to interpersonal sensitivity and adversely affect body image (Gupta, 1990; Gupta et al., 1991b). (See also Chapter 13.)

Anorexia Nervosa and Bulimia Nervosa

Dermatologic changes in anorexia nervosa and bulimia nervosa may be the first signs that give the clinician a clue that an eating disorder is present, as many of these patients either minimize or deny their symptoms and often refuse to comply with treatment (Gupta, Gupta, & Haberman, 1987c). The dermatologic signs are a result of (1) starvation or malnutrition e.g., lanugo-like body hair, asteatotic skin, brittle hair and nails, carotenodermia and pruritus (Gupta, Gupta, & Voorhees, 1992c); (2) self-induced vomiting e.g., hand calluses, dental enamel erosion, gingivitis, and a Sjögren-like syndrome; (3) use of laxatives such as phenolphthalein (which can result in a fixed drug reaction), diuretics such as the thiazide diuretics (which can result in a photosensitivity reaction), and the emetic ipecac (which can be associated with a dermatomyositis-like syndrome); and (4) other concomitant psychiatric illness e.g., hand dermatitis due to compulsive hand washing with anorexia nervosa and trichotillomania with bulimia nervosa. Complaints of skin conditions such as acne associated with multiple food "allergies" have been present in atypical forms of anorexia nervosa (Terr, 1986). Some of these patients complain of other skin-related symptoms from "environmental allergies" and may have "twentieth century disease" (Black, Rathe, & Goldstein, 1990).

Cutaneous Manifestations of AIDS

The prevalence of dermatologic symptoms in acquired immunodeficiency syndrome (AIDS) patients may be as

TABLE 30–4. *Dermatologic Side Effects of Psychotropic Drugs*

Antidepressants

1. Papular rashes, urticaria, petechiae, and pruritus that may respond to antihistamines or substitution with a structurally dissimilar antidepressant. Urticaria and angioedema with or without bronchospasm are reported with fluoxetine; 2.7% on fluoxetine reported pruritus.
2. Cutaneous vasculitis, especially with maprotiline.
3. Toxic epidermal necrolysis with amoxapine, mianserin, carbamazepine.
4. Exfoliative dermatitis, several reports with carbamazepine, rarely associated with imipramine and desipramine.
5. Leukocytoclastic vasculitis with trazodone and fluoxetine.
6. Combination of amitriptyline and minocycline (used for acne therapy) may lead to hyperpigmentation of skin.
7. Cutaneous photosensitivity to ultraviolet A in a few cases, especially with protriptyline and after exposure to sunlight with imipramine.
8. Leukonychia and erythema multiforme with trazodone. Erythema multiforme and Stevens-Johnson syndrome with fluoxetine.
9. Acne with maprotiline.
10. Sweat gland necrosis and blisters during coma induced by amitriptyline and clorazepate.
11. Pathologic sweating in up to 25% of patients; 8.4% of patients on fluoxetine versus 3.8% on placebo reported excessive sweating.

Antipsychotics

1. Blue-gray discoloration of skin associated with long-term, high-dose phenothiazines, especially chlorpromazine.
2. Photosensitive dermatitis with phenothiazines, especially chlorpromazine, usually in the ultraviolet A range. May necessitate switching to a nonphenothiazine antipsychotic.
3. Lupus-like syndrome reported with perphenazine, chlorprothixene, and chlorpromazine. Antinuclear antibody titer positive in 63% patients receiving long-term antipsychotic drugs.
4. Allergic rash, usually maculopapular, erythematous, and pruritic 2–10 weeks after initiation of antipsychotic drugs. Pruritus may result from hepatic cholestasis.
5. Contact dermatitis with phenothiazines, especially liquid preparations.
6. Seborrheic dermatitis with drug-induced parkinsonism.
7. Miscellaneous reports of erythema multiforme and Stevens-Johnson syndrome, nonthrombocytopenic purpura, urticaria, angioedema, palmar erythema, and telangiectasia.

Lithium

1. Precipitation or exacerbation of acneiform eruptions and psoriasis.
2. Other maculopapular eruptions and generalized pruritus.
3. Diffuse or localized hair loss.
4. Increased prevalence of positive antinuclear antibody titer in lithium-treated patients; some case reports of lupus-like syndrome.
5. Increased growth of warts, inflammatory verrucous hyperplasia, exfoliative dermatitis.

Anxiolytic and hypnotic drugs

Benzodiazepines

1. Exacerbation of porphyria with chlordiazepoxide, fixed drug eruption, maculopapular rashes, photosensitivity, urticaria, erythema multiforme, erythema nodosum.
2. Hyperpigmentation in previously dermabraded scars with diazepam.
3. Bullous lesions with diazepam overdose.

Barbiturates

1. Precipitation or exacerbation of porphyria.
2. Bullous lesions in comatose patients.
3. Fixed drug eruption.
4. Miscellaneous reactions including erythema multiforme, toxic epidermal necrolysis, lupus-like syndrome, purpura, photosensitivity, maculopapular eruptions, acne.

Other drugs

1. Bullous lesions with glutethimide and ethchlorvynol overdose.
2. Methaqualone associated with erythema multiforme.
3. Bromides associated with increased pigmentation, acneiform rash, pustular folliculitis, vaculitis, urticaria, erythema nodosum.

Adapted from Gupta, Gupta, & Haberman (1986); Gupta, Gupta, & Ellis (1987); and Dista Products Company (1990).

high as 98% (Kerdel & Penneys, 1989). It is important for the clinician to be aware of some of the cutaneous signs, as they may aid in the differential diagnosis of-neuropsychiatric symptoms that may be AIDS-related. The dermatologic symptoms can be infectious, neoplastic, allergic, or nonspecific (Kerdel & Penneys, 1989). An existing dermatologic disorder may worsen in association with AIDS. Some infective syndromes include oral hairy leukoplakia, which usually presents as an asymptomatic plaque on the side of the tongue, herpes simplex infectious, herpes zoster or severe extensive primary varicella, molluscum contagiosum, extensive warts, candidiasis, syphilis, and scabetic lesions in an atypical distribution. Neoplasms can be present in as many as 40% of cases, the most common one being the Kaposi sarcoma. Some of the nonspecific skin rashes include seborrheic dermatitis, generalized pruritus with a multitude of prurigo-like lesions, and xerosis of the skin. AIDS is also associated with an exacerbation of previously existing psoriasis. The cutaneous manifestations of AIDS are extensive, and a complete discussion is not within the scope of this chapter (see Chapter 32).

Cutaneous Manifestations of Lyme Disease

Lyme disease is a spirochetal infection that is associated with neuropsychiatric manifestations and skin lesions

TABLE 30–5. *Dermatologic Manifestations of Psychiatric Disorders*

Primary psychiatric symptom

Delusions of parasitosis (Bishop, 1983): Patient has a delusion that he or she is suffering from a parasitic infestation of the skin. One or more people living in close proximity to the patient may develop a shared delusional disorder. Underlying problems include a paranoid disorder, major depressive disorder with psychotic features, schizophrenia, multi-infarct dementia, sensory problems, e.g., visual or auditory impairment and xerosis of the skin. *Management*: Treat underlying disorder. Rule out medical causes (Freinhar, 1984), e.g., lymphoma, diabetes mellitus, renal or liver disease, brain tumor, neurosyphilis, pellagra, vitamin B_{12} deficiency. Antipsychotic drugs (e.g., pimozide) generally are only moderately effective.

Tactile hallucinations: Typically a feature of schizophrenia or an acute cognitive disorder resulting from the abuse of sympathomimetic drugs such as amphetamines, cocaine, and phencyclidine.

Lesions secondary to psychiatric disorders (Gupta, Gupta, & Haberman, 1987b)

Dermatitis artefacta: Cutaneous lesions that are wholly self-inflicted, but the self-infliction is typically denied by the patient. Lesions have wide ranging morphologic features (e.g., erythema, blisters, sinuses, nodules) depending on the means used to create them. Some lesions are bizarre, with sharp geometric borders surrounded by normal-looking skin. A consistent psychiatric feature is an immature personality. Other associated disorders include mental retardation, depressive disorder, psychosis, Munchausen syndrome, malingering, and child abuse. *Management*: A supportive empathic approach, avoiding direct discussion regarding the self-inflicted nature of the lesions, along with treatment of the underlying psychiatric disorder.

Neurotic excoriations: Lesions produced as a result of repetitive self-excoriation that may have been initiated by an itch or an urge to excoriate a benign irregularity on the skin. Patients acknowledge the self-inflicted nature of the lesions. Lesions are not bizarre and may range in number from a few to several hundred. They are most commonly associated with a compulsive personality and a depressive disorder. *Management*: An empathic, supportive approach is more effective than an insight-oriented approach. Benzodiazepines, amitriptyline, clomipramine, fluoxetine (Gupta & Gupta, 1992a) and pimozide have been used to treat these excoriations.

Trichotillomania (traumatic alopecia): Nonscarring alopecia that results from compulsion to pluck out one's own hair. Extracted hair may be chewed or swallowed. Hair of the scalp, eyebrows, eyelashes, beard, or pubic area may be affected. Patients typically deny that the alopecia is self-induced. The disorder has previously been classified as an impulse control disorder not elsewhere classified (DSM-III-R, 1987). Among children, the syndrome may represent a disturbed parent-child relationship. *Management*: Family therapy, individual psychotherapy, and behavior modification have all been effective. Clomipramine (Swedo et al., 1989) and fluoxetine may be effective because of their antiobsessive properties.

in addition to many other systemic symptoms. The following is a brief overview of some of the dermatologic manifestations that may aid the clinician in the differential diagnosis of some neuropsychiatric symptoms.

Lyme disease presents with the classic lesions of erythema chronicum migrans (ECM). ECM is the unique clinical marker for Lyme disease, occurring approximately 7 days after a tick bite. It starts as a red maculopapule at the site of the tick bite, and the area of the erythema gradually expands. The red outer border of the lesion may be flat or raised with partial clearing, resulting in a ringed lesion. Sites of predilection include the groin, thigh, and axilla. Several days after the primary lesions of ECM, approximately 50% develop multiple annular secondary lesions. ECM usually appears first, followed by other systemic symptoms including neurologic symptoms. (See also Chapter 32.)

Autoimmune Disorders

Autoimmune disorders, such as systemic lupus erythematosis (SLE), can present with a range of dermatologic reactions in addition to psychiatric symptoms. If an autoimmune disorder is suspected, the psychiatrist should order screening tests, such as a complete blood count, erythrocyte sedimentation rate, and rheumatoid factor and antinuclear antibody titers, before referring the patient to an internist or dermatologist. When drug-induced SLE is suspected, an antihistone antibody level should also be determined. A skin biopsy from a representative lesion helps confirm the presence of SLE. (See also Chapter 33.)

DERMATOLOGIC SIDE EFFECTS OF PSYCHOTROPIC DRUGS

A survey of 347 patients with 358 cutaneous reactions revealed that 94% of the skin reactions were a generalized morbilliform rash, 5% were urticarial, and 1% were generalized pruritus (Bigby et al., 1986). The most commonly cited drugs implicated in the cutaneous reactions were not in the psychotropic category. Cutaneous drug reactions can arise as a result of immunologic (or allergic) and nonimmunologic mechanisms, for example, by cumulative toxicity, drug interactions, or metabolic alterations. Nonallergic mechanisms are more common than allergic mechanisms (Wintroub & Stern, 1985). Because the skin reacts in a limited manner to a wide variety of stimuli, it is often not possible to specify the responsible drug or pathogenic mechanism on the basis of the clinical appearance of the skin lesions alone. Table 30-4 provides a summary of dermatologic reactions to psychotropic drugs.

Morbilliform eruptions, the most common drug-induced reaction, do not always recur upon reinstitution of the offending drug (Wintroub & Stern, 1985). The rash typically occurs within 1 week of starting therapy and sometimes may even decrease as therapy is continued with the responsible agent (Wintroub & Stern, 1985). On other occasions the drug may have to be discontinued if the rash persists. Urticarial drug reactions, when associated with angiodema, may be part of a life-threatening anaphylactic reaction. Other, less frequent drug-induced skin reactions include erythema multiforme, vasculitis, toxical epidermal necrolysis, and bullous eruptions (Wintroub & Stern, 1985). The agent causing these cutaneous reactions should be stopped and a medical consultation sought.

Drug-induced photosensitivity reactions are adverse cutaneous responses to the combined actions of the drug and light. The cutaneous reactions are usually most marked in sun-exposed areas but may extend to sun-protected areas (Wintroub & Stern, 1985). In some cases, cutaneous photosensitivity accompanies other drug-induced syndromes, for example, drug-induced porphyria and drug-induced SLE. The presence of underlying systemic complications must therefore be ruled out in cases of drug-induced photosensitivity. Sunscreens against both ultraviolet B and ultraviolet A should be considered, with a sun-protection factor of 15 or higher (Harber & Bickers, 1989). Sunscreens block absorption of the sunlight spectrum responsible for cutaneous synthesis of vitamin D. A preliminary study suggests that long-term use of sunscreens can lower body stores of vitamin D (Matsuoka et al., 1988).

REFERENCES

Abel EA, DiCicco LM, Orenberg EK, et al. (1986). Drugs in exacerbation of psoriasis. J Am Acad Dermatol 15:1007–1022.

American Psychiatric Association (1987). Diagnostic and statistical manual of mental disorders (3rd ed, revised). Washington, DC: American Psychiatric Association.

Arnetz BB, Fjellner F, Eneroth P, et al. (1985). Stress and psoriasis: psychoendocrine and metabolic reactions in psoriatic patients during standardized stressor exposure. Psychosom Med 47:528–541.

Arnold HL, Odom RB, & James WD (1990). Diseases of the skin (8th ed). Philadelphia: WB Saunders.

Bar LHJ, & Kuypers BRM (1973). Behavior therapy in dermatological practice. Br J Dermatol 88:591–598.

Bernstein JE, Swift RM, Soltani K, et al. (1982). Antipruritic effect of an opiate antagonist, naloxone hydrochloride. J Invest Dermatol 78:82–83.

Bigby M, Jick S, Jick H, et al. (1986). Drug-induced cutaneous reactions. JAMA 256:3358–3363.

Bishop ER Jr (1983). Monosymptomatic hypochondriacal syndromes in dermatology. J Am Acad Dermatol 9:152–158.

Black DW, Rathe A, & Goldstein RB (1990). Environmental illness: a controlled study of 26 subjects with "20th century disease." JAMA 264:3166–3170.

Brown D, & Bettley F (1971). Psychiatric treatment of eczema: a controlled trial. BMJ 2:729–734.

Bruno NP, Beacham BE, & Burnett JW (1984). Adverse effects of isotretinoin therapy. Cutis 33:484–486.

Cunliffe WJ, & Cotterill JA (1975). The acnes, clinical features, pathogenesis and treatment. Major Probl Dermatol 6:151–172.

Evoniuk GE, Kuhn CM, & Schanberg SM (1979). The effect of tactile stimulation on serum growth hormone and tissue ornithine decarboxylase activity during maternal deprivation in rat pups. Commun Psychopharmacol 3:363–370.

Farber EM, Nickoloff BJ, Recht B, et al. (1986). Stress, symmetry and psoriasis: possible role of neuropeptides. J Am Acad Dermatol 14:305–311.

Faulstich ME, & Williamson DA (1985). An overview of atopic dermatitis: toward a bio-behavioral integration. J Psychosom Res 29:647–654.

Faulstich ME, Williamson DA, Duchmann EG, et al. (1985). Psychophysiological analysis of atopic dermatitis. J Psychosom Res 29:415–417.

Field TM, Schanberg SM, Scafid F, et al. (1986). Tactile kinesthetic stimulation effects on preterm neonates. Pediatrics 77:654–658.

Freinhar JP (1984). Delusions of parasitosis. Psychosomatics 25:47–53.

Garden JM, & Freinkel RK (1986). Systemic absorption of topical steroids. Arch Dermatol 122:1007–1010.

Gil KM, Keefe FJ, Sampson HA, et al. (1987). The relation of stress and family environment to atopic dermatitis symptoms in children. J Psychosom Res 31:673–684.

Gilchrest BA (1982). Pruritus: pathogenesis, therapy, and significance in systemic disease states. Arch Intern Med 142:101–105.

Gupta MA (1990). Fear of aging: a precipitating factor in late onset anorexia nervosa. Int J Eating Disord 9:221–224.

Gupta MA & Gupta AK (1992a). Fluoxetine is an effective treatment for neurotic excoriations: A case report. Curtis. In press.

Gupta MA, Gupta AK, & Ellis CN (1987). Antidepressant drugs in dermatology. Arch Dermatol 123:647–652.

Gupta MA, Gupta AK, & Haberman HF (1986). Psychotropic drugs in dermatology: a review and guidelines for use. J Am Acad Dermatol 14:633–645.

Gupta MA, Gupta AK, & Haberman HF (1987a). Psoriasis and psychiatry: an update. Gen Hosp Psychiatry 9:157–166.

Gupta MA, Gupta AK, & Haberman JH (1987b). The self-inflicted dermatoses: a critical review. Gen Hosp Psychiatry 9:45–52.

Gupta MA, Gupta AK, & Haberman HF (1987c). Dermatologic signs in anorexia nervosa and bulimia nervosa. Arch Dermatol 123:1386–1390.

Gupta MA, Gupta AK, Kirkby S, et al. (1988). Pruritus in psoriasis: a prospective study of some psychiatric and dermatologic correlates. Arch Dermatol 124:1052–1057.

Gupta MA, Gupta AK, Kirkby S, et al. (1989a). A psychocutaneous profile of psoriasis patients who are stress reactors: a study of 127 patients. Gen Hosp Psychiatry 11:166–173.

Gupta MA, Gupta AK, Kirkby S, et al. (1989b). Pruritus associated with nocturnal wakenings: organic or psychogenic? J Am Acad Dermatol 21:479–484.

Gupta MA, Gupta AK, Ellis CN, et al. (1990a). Some psychosomatic aspects of psoriasis. Adv Dermatol 5:21–32.

Gupta MA, & Voorhees JJ (1990). Psychosomatic dermatology: is it relevant? Arch Dermatol 126:90–93.

Gupta MA, Gupta AK, Ellis CN, et al. (1992b). Acne and bulimia nervosa may be related: a case report. Can J Psychiatry 37:58–61.

Gupta MA, Gupta AK, Schork NJ, et al. (1991a). Psychiatric aspects of the treatment of mild to moderate facial acne: some preliminary observations. Int J Dermatol 29:719–721.

Gupta MA, Goldfarb MT, Schork NJ, et al. (1991b). Treatment of mild to moderately photoaged skin with topical trentinoin has favourable psychosocial effect: a prospective study. J Am Acad Dermatol 24:780–781.

Gupta MA, Gupta AK & Voorhees JJ (1992c). Starvation-associated pruritus: A clinical feature of eating disorders. J Am Acad Dermatol 27:118–120.

Gupta MA, Schork NJ, Gupta AK, & et al. (1992d). Suicidal ideation in psoriasis. Int J Dermatol. In press.

Harber LC, & Bickers DR (1989). Photosensitivity Diseases (2nd ed). Philadelphia: BC Decker.

Heng MCY (1982). Cutaneous manifestations of lithium toxicity. Br J Dermatol 106:107–109.

Jordan JM, & Whitlock JA (1974). Atopic dermatitis, anxiety and conditioned scratch responses. J Psychosom Res 18:297–299.

Juhlin L (1981). Recurrent urticaria: clinical investigations of 330 patients. Br J Dermatol 104:369–381.

Kahn G, & Davis BP (1970). In vitro studies on long-wave ultraviolet light-dependent reactions of the skin photosensitizer chlorpromazine with nucleic acids, purines and pyrimidines. J Invest Dermatol 55:47–52.

Kerdel FA, & Penneys NS (1989). Cutaneous manifestations of AIDS in adults and children. Curr Probl Dermatol 1:101–119.

Koblenzer CS, & Koblenzer PJ (1988). Chronic intractable atopic eczema. Arch Dermatol 124:1673–1677.

Kochevar IE (1980). Possible mechanisms of toxicity due to photochemical products of protriptyline. Toxicol Appl Pharmacol 54:258–264.

Matsuoka KY, Wortsman J, Hanifan N, et al. (1988). Chronic sunscreen use decreases circulation concentration of 25-hydroxyvitamin D: a preliminary study. Arch Dermatol 124:1802–1804.

McGennis AJ (1978). Lithium carbonate and tetracycline interaction. BMJ 1;1183.

Medansky RS (1971). Emotions and skin—a double-blind evaluation of psychotropic agents. Psychosomatics 12:326–329.

Medansky RS, & Handler RM (1981). Dermatopsychosomatics: classification, physiology and therapeutic approaches. J Am Acad Dermatol 5:125–136.

Monroe EW, & Jones HE (1977). Urticaria, an updated review. Arch Dermatol 113:80–90.

Morse RM, Perry HO, & Hurt RD (1985). Alcoholism and psoriasis. Alcoholism 9:396–399.

Pauk J, Kuhn CM, Field TM, et al. (1986). Positive effect of tactile versus kinesthetic or vestibular stimulation on neuroendocrine and ornithine decarboxylase activity in maternally-deprived rat pups. Life Sci 39:2081–2087.

Pavlidakey GP, Hashimoto K, Heller GL, et al. (1985). Chlorpromazine-induced lupus-like disease: case report and review of the literature. J Am Acad Dermatol 13:109–115.

Rathod NH, Alarcon RDE, & Thomson IG (1978). Signs of heroin usage detected by drug users and their parents. Lancet 2:1411.

Rees L (1957). An aetiological study of chronic urticaria and angioneurotic oedema. J Psychosom Res 2:172–189.

Robinson P, Szewczyk M, Haddy L, et al. (1984). Outbreak of itching and rash: epidemic hysteria in an elementary school. Arch Intern Med 144:1959–1962.

Rothman S (1941). Physiology of itching. Physiol Rev 21:357–381.

Rubinow DR, Peck GL, Squillace KM, et al. (1987). Reduced anxiety and depression in cystic acne patients after successful treatment with oral isotretinoin. J Am Acad Dermatol 17:25–32.

Scheinman PL, Peck GL, Rubinow DR, et al. (1990). Acute depression from isotretinoin. J Am Acad Dermatol 22:1112–1113.

Selmanowitz VJ (1986). Lithium, leukocytes and lesions. Clin Dermatol 4:170–175.

Seville RH (1977). Psoriasis and stress. Br J Dermatol 97:297–302.

Shelley WB, & Shelley ED (1985). Adrenergic urticaria: a new form of stress-induced hives. Lancet 2:1031–1033.

Skoven I, & Thormann J (1979). Lithium compound treatment and psoriasis. Arch Dermatol 115:1185–1187.

Sneddon J, & Sneddon I (1983). Acne excoriée: a protective device. Clin Exp Dermatol 8:65–68.

Solomon LM (1985). Atopic dermatitis. In SL Moschella & YHJ Hurley (Eds.), Dermatology (Sect. III, pp. 334–353). Philadelphia: WB Saunders.

Sunder TR, Balsam MJ, & Vengrow MI (1982). Neurological manifestations of angiodema. JAMA 247:2005–2007.

Swedo SE, Leonard HL, Rapoport JL, et al. (1989). A double-blind comparison of clomipramine and desipramine in the treatment of trichotillomania (hair pulling). N Engl J Med 321:497–501.

Terr AI (1986). Food allergy: a manifestation of eating disorder? Int J Eating Disord 5:575–579.

Ullman KC, Moore RW, & Reidy M (1977). Atopic eczema: a clinical psychiatric study. J Asthma Res 14:91–99.

Voorhees JJ (1983). Leukotrienes and other lipoxygenase products in the pathogenesis and therapy of psoriasis and other dermatoses. Arch Dermatol 119:541–547.

Voorhees JJ, & Duell EA (1971). Psoriasis as a possible defect of the adenyl cyclase-cyclic AMP cascade. Arch Dermatol 104:352–358.

Weiss HD, Walker MD, & Wiernik PH (1974). Neurotoxicity of commonly used antineoplastic agents. N Engl J Med 291:75–81.

Williams D (1951). Management of atopic dermatitis in children: control of the maternal rejection factor. Arch Dermatol Syphilol 63:545–547.

Wintroub BU, & Stern R (1985). Cutaneous drug reactions: pathogenesis and clinical classification. J Am Acad Dermatol 13:167–179.

Wroblewski B, Singer W, & Whyte J (1986). Carbamazepine-rythromycin interaction: case studies and clinical significance. JAMA 255:1165–1167.

Yohn JJ, & Weston WL (1990). Topical glucocorticoids. Curr Probl Dermatol 2:31–63.

31 Hematologic disorders

ELISABETH J. SHAKIN, M.D., AND
TROY L. THOMPSON II, M.D.

Blood is realistically and symbolically linked to life itself. It permeates every other organ system, and so it may be more difficult for a patient to deny or split off a blood disorder from conscious concerns. Blood and, therefore, blood disorders are especially frightening to many individuals. The sight of blood following an injury usually is a child's first concrete indication of the body's vulnerability, and even the sight of blood evokes strong emotional responses, including fainting, in many individuals throughout life.

This chapter addresses major aspects of the psychiatric diagnosis and treatment of several hematologic disorders that are associated with psychiatric manifestations. The psychiatric consultant may be asked to evaluate patients with hemophilia and thalassemia in hemophilia treatment centers. Patients with sickle cell disease, porphyria, vitamin B_{12} deficiency, folate deficiency, and hyperviscosity syndromes sometimes are seen by consulting psychiatrists on general medical units and in outpatient medical clinics. Psychogenic purpura (autoerythrocyte sensitization) and anticoagulant malingering are two infrequent disorders for which psychiatrists also may be consulted. Hematologic malignancies have been discussed at length elsewhere and are not discussed here (Lesko, Massie, & Holland, 1987).

HEMOPHILIA

Hemophilia A and B constitute about 65% of the inherited coagulation disorders. The two disorders are clinically indistinguishable. Hemophilia A (classic hemophilia) and hemophilia B (Christmas disease) result from deficient clotting activity of factors VIII and IX, respectively. Hemophilia is inherited recessively on the X chromosome and therefore affects 50% of the sons of asymptomatic carrier mothers (Hilgartner, Aledort, & Giardina, 1985; Mattsson & Kim, 1982).

Comprehensive hemophilia treatment centers have included psychosocial services since at least the late 1970s. Interventions aimed at preventing psychopathology by promoting normal development and providing early intervention when psychiatric distress develops have been key factors in improved emotional functioning of many young patients with this chronic, life-threatening disorder (Agle, 1984; Agle & Heine, unpublished; Hernandez, Gray, & Lineberger, 1989).

Although the exact role of stress in the exacerbation of bleeding episodes in hemophiliacs has not been vigorously studied, anecdotal reports have described "spontaneous" bleeding in response to both negative and positive emotional stressors (e.g., anxiety, school vacations) (Agle, 1964; Agle & Heine, unpublished; Browne, Mally, & Kane, 1960; Mattsson & Gross, 1966a). Use of antihemophilic factor does not correlate with disease severity (Handford, Charney, & Ackerman, 1980). For example, blood product use decreases in some patients who have been treated with hypnosis and psychotherapy for anxiety reduction (Handford, Charney, & Ackerman, 1980; LaBaw, 1975; LeBaron & Zeltzer, 1984; Lichstein & Eakin, 1985; Lucas, 1965, 1975; Swirsky-Sacchetti & Margolis, 1986).

Disease-Related Considerations

Major physical trauma, surgery, and dental extractions often produce increased bleeding in patients with mild factor deficiencies. In patients with severe factor deficiencies (less than 1% of factor VIII or IX), hemorrhaging can occur spontaneously or with minor trauma. Bleeding into large joints and muscles is accompanied by acute pain; repeated episodes often lead to chronic pain and arthritis. Whereas acute pain usually lessens after stasis (due to factor replacement), chronic pain requires multimodal treatments, which may include behavioral approaches, transcutaneous electrical nerve stimulators (TENS), analgesics, and orthopedic rehabilitation, sometimes including joint reconstruction

This chapter contains revised materials that originally appeared in Shakin EJ, & Thompson TL II (1991). Psychiatric aspects of hematologic disorders. In A Stoudemire & BS Fogel (Eds.), *Medical-Psychiatric Practice* (Vol. 1; pp. 193–242). Washington, DC: American Psychiatric Press.

(Choiniere & Melzack, 1987; Holdredge & Cotta, 1989; Roche, Gijsbers, & Belch, 1985; Varni, 1981).

Bleeding into the pharyngeal soft tissue can produce respiratory compromise, and bleeding into the central nervous system (CNS) may result in increased intracranial pressure. Early deaths related to CNS hemorrhage and disabling arthritis were common sequelae of hemophilia before the advent of factor replacement therapy (see Figure 31-1). Currently, a normal life-span may be expected in human immunodeficiency virus (HIV)-seronegative hemophiliacs (S. Shapiro, personal communication, 1990).

Factor VIII replacement therapy currently is thought to be relatively safe; however, prior to 1985 exposure to pooled blood products increased the patient's risk of exposure to hepatitis or HIV. A complicating aspect of factor replacement therapy is that the development of an antibody or inhibitor to transfused factors may lead to rapid consumption of factor concentrate; treatment, therefore, is problematic in these patients (Hilgartner, 1980).

TABLE 31–1. *Common Psychosocial Issues in Hemophilia*

Children
Fears of immobilization with orthotic devices
Medical education
Reporting trauma earlier to allow prompt factor replacement
Participation in home treatment
Pain: disease-related, treatment-related
School: minimize absences due to illness
Separation from family with hospitalizations
Teenagers
Acting out: risk-taking behavior
Normal developmental tasks versus vulnerability from disease: jobs, sexual performance, self-esteem, masculinity
All Ages
CNS complications of AIDS
Psychosocial complications of AIDS
CNS hemorrhage
Drug dependence and abuse
Family coping
Pain management
Psychiatric disorders
Sexual dysfunction
Staff–patient issues
Utilization of factor transfusions: underutilization or excessive amount or frequency

Psychosocial Considerations

Patients with hemophilia must cope with different disease-related issues at different ages (see Table 31-1). The family can be understandably stressed, particularly at the time of diagnosis (Reis, Linhart, & Lazerson, 1982). As children with hemophilia are educated about the disease and its management, they must integrate the normal tasks of development with the impact of having a chronic and potentially serious disease (Agle, 1964; Handford & Mayes, 1989; Holdredge & Cotta, 1989; Mattsson, 1984; Mattsson & Gross, 1966a, 1966b; Mattsson, Gross, & Hall, 1971).

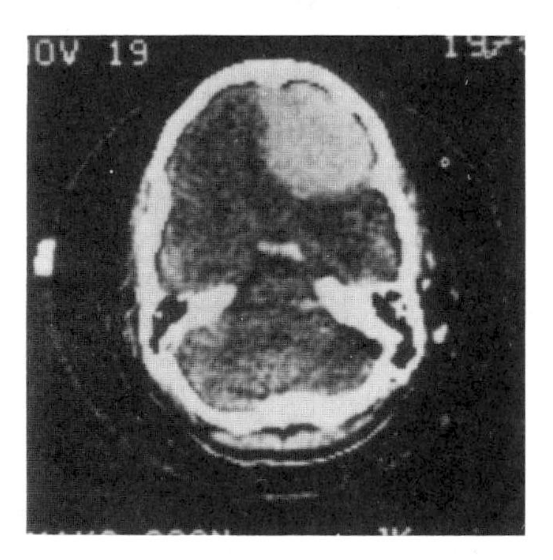

FIG. 31-1. CNS bleed of a hemophiliac. An epidural hematoma of the right frontal lobe, secondary to a right frontal skull fracture, is shown on this computed tomogram (CT scan) of the head of a 19-year-old male patient with factor VIII deficiency. (From Gilchrist & Piepgras, 1976. With permission.)

The actual incidence and prevalence of psychiatric disorders in hemophiliacs are unknown at this time. Most children and adolescents cope fairly well with the social stigmatization and the psychological and physical vulnerability due to the disease (Agle & Mattsson, 1970; Handford, Mayes, & Bixler, 1986; Hilgartner, Aledort, & Giardina, 1985; Magrab, 1985; Mattsson, 1984; Mattsson & Gross, 1966a; Mattsson & Kim, 1982; Shakin & Thompson, 1991; Steinhausen, 1976). Psychosocial support and education aim to modify the natural inclination toward overprotection of these vulnerable children and to provide a sense of mastery of practical skills that can be used for home care. All family members are encouraged to participate in these efforts, as the impact of the child's disability and sometimes early death can generate fear, anxiety, anticipatory grief, marital discord, and financial hardship due to the costs of treatment and the need to make accommodations in the family's work and leisure activities in order to take care of a sick child (Agle, 1964; Agle & Mattsson, 1968; Handford & Mayes, 1989; Hilgartner, Aledort, & Giardina, 1985; Mattsson, 1984; Mattsson & Gross, 1966a; Mattsson, Gross, & Hall, 1971; Mattsson & Kim, 1982; Meijer, 1980–81; Salk, Hilgartner, & Granich, 1972). Other literature sources provide more detailed discussion of these issues (Jacobs, 1991; Shakin & Thompson, 1991). (See also Chapter 2.)

Deviations in usual factor replacement product usage, new medical or psychiatric symptoms, and problems with drugs, alcohol or analgesic abuse may warrant formal psychiatric assessment and intervention (Simon, 1984). As patients with equally severe hemophilia generally are expected to require comparable amounts of factor products, if patients repeatedly treat the same bleeding episode with an excessive amount of replacement factor or otherwise changes their product consumption over time, it may indicate that increased anxiety is influencing product usage. Underuse of appropriate factor replacement likewise may suggest a depressive disorder.

Treatment

Although most routine psychiatric interventions can be useful with hemophiliacs, group therapy with parents may be especially useful, not only as a forum for ventilating emotional concerns but also as part of a psychoeducational model of management (Mattsson & Agle, 1972; Mattsson & Kim, 1982).

The dosages of antidepressants and antipsychotic agents used in these patients may need to be adjusted according to their renal and hepatic functioning. Drug interactions with analgesics may include oversedation with a number of psychotropic agents, possibly resulting in falls. Hypertensive crises are a concern when analgesics, especially meperidine, are given with monoamine oxidase inhibitors (MAOIs), possibly resulting in an increased risk of intracranial hemorrhages (Handford & Mayes, 1989; Jonas, 1977, 1989).

Clinicians often are fearful of creating or contributing to drug dependence and addiction in these patients. However, in practice, analgesics and benzodiazepines generally may be used safely in this population (Brunner, Schapera, & Gruppa, 1982). Increased risk for dependence and abuse occurs in older patients who have a history of psychiatric hospitalizations and drug overdoses. Obtaining a complete drug history is important for preventing drug withdrawal, as these patients tend to be polysubstance users and sometimes abusers (Jonas, 1989).

Any change in mental status should lead to both a diagnostic evaluation and simultaneous initiation of treatment. Clinical neuropsychiatric assessments may help to distinguish primary from secondary psychiatric problems, but a computed tomography (CT) or magnetic resonance imaging (MRI) scan of the head is necessary for definitive diagnosis of a CNS hemorrhage, or for confirmation of a neurologic complication of acquired immunodeficiency syndrome (AIDS).

Linear fractures may require a routine skull roentgenogram for detection. Seizures, aphasia, impaired cognition, hydrocephalus, and even death are possible sequelae of CNS hemorrhages in these patients. However, half of hemophiliac patients with intracranial hemorrhage have no antecedent history of head trauma (Gilchrist, Piepgras, & Roskos, 1989; Handford & Mayes, 1989; Mattsson & Kim, 1982).

In addition to the CNS complications and opportunistic infections associated with HIV disease, there are many psychosocial consequences of being HIV-seropositive. About 70 to 90% of hemophilia A patients and 30 to 50% of hemophilia B patients with moderate to severe disease who regularly used blood products before 1985 are thought to be HIV-seropositive (S. Shapiro, personal communication, 1990); most probably will develop AIDS. About 6% of patients with severe factor VIII deficiency already have AIDS (Agle, 1989; Jason et al., 1989; P. H. Levine, 1985).

Hemophiliac patients and their families may be faced with social ostracism, an increased sense of vulnerability, and job and insurance discrimination. Although they may react with increased anxiety, depression, acting out, hypochondriasis, or noncompliance, most families are able to cope fairly well with this added burden (Agle, 1989; Agle, Gluck, & Pierce, 1987; Mayes, Handford, & Kowalski, 1988). Occasionally, anger and concern lead to a breakdown of trust in the health care workers, who were reassuring in the past about the safety of blood products; issues of distrust must be confronted early to avoid later difficulties with treatment. Alternate care may not be available as a solution for a breakdown in the doctor-patient relationship, as treatment centers are usually the only providers of care for these patients.

In addition to facing their own fears of exposure to HIV, health care workers are dealing with the guilt of having unknowingly administered HIV-infected products and with their own discomfort about issues of sexual conduct. The full spectrum of sexual behaviors, from abstinence to engaging in "unsafe" sex, may be seen in hemophiliac patients, and some authors have recommended that staff should consider breaching confidentiality, if necessary, in order to inform sexual partners of their potential risk of HIV exposure (S. B. Levine, 1984; Simon, 1989). Legislation regarding the liability of physicians, either for disclosure or nondisclosure of HIV risk status, varies considerably among states. In summary, as a result of the increasing numbers of hemophiliacs with AIDS, psychiatrists and other health care workers must now provide safe sex counseling, address the neuropsychiatric complications of AIDS, manage increasingly complex staff-patient relationships, and deal with issues of death and dying in young adults (Agle, 1989; Faulstick, 1987).

SICKLE CELL DISEASE

Sickle hemoglobin is present in a variety of conditions referred to as the sickle cell syndromes. At least 80,000 children worldwide die from sickle cell anemia each year (Huntsman, 1985). Sickle cell disease (SCD) includes sickle cell anemia, sickle cell–beta thalassemia, sickle cell–hemoglobin C, and other variants. Sickle cell anemia (SCA) is discussed here as a model for addressing and managing the other sickle cell syndromes, as its high prevalence, significant morbidity, and early mortality result in a variety of psychosocial complications. Genetic transmission of SCA is recessive; therefore the gene for sickle hemoglobin must be inherited from both parents for SCA to result. The heterozygous child is a carrier of the disease. Some sickle cell carriers make decisions about life style, marriage, and having children based on the mistaken assumption that they have the disease. As a result of this misconception, this group frequently experiences unnecessary anxiety and impaired self-esteem (Huntsman, 1985; Whitten & Nishiura, 1985). Public education is important for preventing this source of distress in carriers and for ensuring availability of genetic counseling to all individuals.

Anemia, painful vasoocclusive crises, and organ damage are responsible for the major manifestations of SCA (see Table 31-2). Reduced erythropoietic activity in the bone marrow leads to potentially life-threatening aplastic crises. There is both increased red blood cell fragility and shortened survival. Although patients with SCA have both normal and sickled cells at any given time, under certain conditions microvascular occlusion results in painful *vasoocclusive crises*. More recently, the hemolytic and clinical severity of SCA patients has been attributed to intracellular polymerization of hemoglobin S, rather than cell sickling. Deoxygenation may lead to polymerization, which results in irreversible membrane damage and dehydration. It is the polymerized cells that are thought to be responsible for microvascular occlusion, and so current treatment strategies (e.g., hydroxyurea) are aimed at reducing polymerization (Rodgers, 1991).

TABLE 31–2. *Frequent Medical Complications of Sickle Cell Anemia*

Birth to 1 year old
Aplastic crises: anemia
Vasoocclusive crises
Organ damage: lungs, kidneys, bones, CNS, penis (priapism)
Dactylitis: painful, swollen hands and feet
Increased infections
Older children
Isosthenuria: inability to concentrate urine
Bedwetting
Cerebrovascular accidents (CVAs)
Intellectual impairment
Retinal damage: progressive loss of sight
Motor impairment
Sensory impairment
Growth retardation or no change
Delayed puberty

Painful crises can be precipitated by infection, dehydration, low temperatures, other physical stressors, and possibly anxiety; but often an exact precipitant cannot be identified. Ischemia and necrosis lead to pain and *organ damage* in areas normally supplied by the now occluded vasculature. Supportive management of such patients includes analgesics and behavioral techniques to reduce pain, broad-spectrum antibiotics to treat infections, transfusions for severe anemia secondary to hemolysis or aplastic crises, and hydration (Powars, 1975; Whitten & Nishiura, 1985).

Enuresis occurs in up to 45% of 8- to 11-year-olds with SCA. It probably relates to drinking large amounts of fluid in order to compensate for isosthenuria (Hurtig & White, 1986). CNS damage has been associated not only with intellectual impairment in about 5% of SCA patients, but also with progressive loss of vision secondary to retinal damage (Ballas et al., 1982; Mattsson & Kim, 1982; Powars, 1975; Whitten & Nishiura, 1985). Growth retardation and delayed puberty also have been reported, but at least one prospective study has not shown differences in physical growth (Kramer, Rooks, & Pearson, 1978). The neuropsychiatric and hematologic complications of folate deficiency, resulting from increased folate turnover, can be prevented with folate supplementation in most cases (Rodgers, 1991).

Cognitive impairment in patients with SCA may result from cerebrovascular accidents (CVAs) or chronic microvascular insults without CVAs. Standard tests of intellectual functioning (e.g., Wechsler Intelligence Scale for Children, or WISC) may not reveal such deficits (Chodorkoff & Whitten, 1963). Children with SCD have more crossed-dominance, visuomotor deficits, and decreased attention spans compared to controls. Even in children matched for sex, race, age, and socioeconomic status, spelling and reading skills may be more frequently impaired in children with SCA (Fowler, Whitt, & Lallinger, 1988; Leavell & Ford, 1983). The role of psychosocial factors, including family dynamics, socioeconomic status, and increased fatigue and absenteeism from disease should not be underestimated when evaluating performance in school and on cognitive testing. Neuropsychological testing and MRI scanning may provide more specific clues to understanding the nature and cause of such deficits.

Psychosocial Issues

Unlike hemophilia patients, patients with SCA usually lack access to comprehensive treatment centers. Private physicians often provide the entire management for the psychosocial problems of SCA patients and their families (see Table 31-3). Generally, children with SCA adjust well socially and personally (Kumar, Powars, & Allen, 1976; Lemanek, Moore, & Gresham, 1986; Whitten & Nishiura, 1985). However, increased levels of distress, particularly depression, have been reported in both healthy siblings and parents of these children (Treiber, Mabe, & Wilson, 1987). Hence, any significant psychiatric complaint should lead to a screening evaluation of other close family members.

Self-esteem may be impaired in children with SCA owing to the lack of control over disease exacerbations, fear of abandonment with hospitalizations, and teasing by peers, often related to delayed puberty and growth retardation (Whitten & Fischhoff, 1974; Whitten & Nishiura, 1985). Adaptive coping can be encouraged by allowing expression of emotional concerns and by supporting the child's independent functioning between crises.

In adolescents with SCA there may be an increase in symptoms of depression, social withdrawal, and body image concerns (Morgan & Jackson, 1986). Whereas some patients develop a compensatory pseudomaturity, others act out and develop problems with drug dependence, anxiety, and hypochondriasis (Hurtig & White, 1986; Williams, Earles, & Pack, 1983).

Adults with SCA often are viewed as being more difficult to manage than other medical inpatients. They frequently are inappropriately labeled as addicts, not only because of the stigma of the disease, but also because of the need for narcotics to treat the disease. The latter probably relates to the small fraction of patients with severe disease who often are hospitalized as often as every 2 to 4 weeks. These patients may make up about 6 to 7% of the total sickle cell population. Problems with pain control and drug dependence may contribute to manipulative or demanding behaviors in these patients. Guidelines for managing these patients are discussed at length under pain management in SCA. Further elaboration of these problems is discussed elsewhere (Ballas, 1990).

Accurate estimates of the prevalence of specific psychiatric disorders in SCA are not yet available. One methodologic problem when doing research on patients with chronic diseases is that one must attempt to distinguish premorbid psychopathy from psychosocial and other medical complications of the disease. Despite this dilemma, however, there is consensus that increased environmental stressors lead to disease exacerbations and to increased anxiety, depression, and preoccupation with death and dying (Barrett et al., 1988; Whitten & Fischhoff, 1974). The full range of intellectual functioning and a small percentage of patients with maladaptive personality traits were described in one series (Leavell & Ford, 1983).

TABLE 31–3. *Common Psychosocial Issues in Children with SCA and Their Families*

Children
Fear of abandonment with hospitalization
Threatened sense of self-esteem
Adolescents with SCA
Acting out
Body image concerns
Drug dependence and abuse
Hypochondriasis
Pseudomaturity
Social withdrawal
Adults with SCA
Dealing with death and dying
Drug dependence and abuse
Intellectual functioning
Family members and patients of all ages
Anxiety
Dealing with chronic illness
Depression
Pain control

Pain Management

It is difficult to predict the clinical course and severity of painful episodes in SCA (Benjamin, 1989). Many SCA crises do not have identifiable precipitants. In children the signs and symptoms of a crisis may include fever, loss of bowel sounds, abdominal rebound tenderness, and joint involvement. Adjunctive measures may be sufficient for controlling such episodes and include whirlpool treatments, increased hydration, heat, and emotional support. In adults, localized findings are seen in fewer than 50% of patients in SCA crisis. Whereas medications may not be required for mild episodes, hospitalization with parenteral narcotics may be needed during severe crises (Powars, 1975; Whitten & Fischhoff, 1974; Whitten & Nishiura, 1985).

Clinicians often are poorly skilled in managing pain because they fear they will contribute to or cause iatrogenic addiction and, as a result, are suspicious of the patients' motives behind requests for narcotics. Unfortunately, there is no definitive means for distinguishing a true crisis from malingering. Stoic patients and patients who continually watch television as a distraction technique, for example, may be perceived as not being in severe enough pain to warrant the use of narcotics; overly dramatic patients may be seen as manipulative or malingering. When patients fall asleep—sometimes the first sign of relief from pain—medications should not be withheld or they may awaken with severe recurrent pain (Ballas, 1990; Benjamin, 1989).

Despite these pitfalls, routine dosages of narcotics are reported to reduce pain significantly in 80% of these patients. General guidelines for pain management are outlined in Table 31-4. The patient may report that only certain narcotics at specific doses have been effective for them in prior episodes. This is often misinterpreted as drug-seeking behavior; instead, it may reflect the patient's knowledge of side effects, effectiveness in pain reduction, and desire to avoid medications associated with drug abuse (e.g., methadone). Conveying a supportive attitude toward pain management encourages early treatment of painful crises. Patients who perceive in their caretakers mistrust or other negative attitudes often avoid seeking treatment out of a sense of anger and helplessness; such patients may act out in attempts to procure needed medication.

Despite the reluctance of most physicians to prescribe narcotics, substance abuse and dependence occasionally do occur in SCA patients (Benjamin, 1989). Combining nonnarcotic analgesics—e.g., acetaminophen, nonsteroidal antiinflammatory agents (NSAIDs)—which act primarily on the peripheral nervous system, with narcotic analgesics, which act on the CNS, can produce additive or synergistic analgesic effects and may allow lower dosages of both medications. Patients should be discharged on equianalgesic oral dosages of their parenteral medications. If this is not done, they may return to the same hospital or to other emergency rooms seeking additional analgesics, as oral dosages must be higher than parenteral dosages to provide equal analgesia. Although it is appropriate occasionally to provide patients with "as needed" narcotic prescriptions, this approach generally is best avoided. Exceptions can be made for patients with frequent crises who do not have daily access to outpatient clinics. Ballas (1990), Benjamin (1989), and Goldberg, Sokol, & Cullen (1987) have discussed these topics further. (See also Chapters 16 and 17.)

TABLE 31–4. *General Principles of Pain Management for Sickle Cell Anemia*

1. Obtain a complete history of the pain: quality, intensity, location, precipitants and alleviating factors (if any), duration, prior episodes and their treatment, psychological factors.
2. Early treatment reduces psychological and physical debilitation.
3. Starting dose depends on prior analgesic requirements.
4. Dosage interval should be frequent at first, around the clock to maintain adequate plasma levels.
5. Consider patient-controlled analgesia or the subcutaneous use of hydromorphone to avoid abscess formation, tissue induration, and myositis-myonecrosis due to frequent narcotic injections.
6. Start with aspirin or an NSAID; then switch to mild narcotics (e.g., oxycodone with aspirin or acetaminophen, codeine). If pain remains uncontrolled, then switch to stronger opioids (e.g., hydromorphine, meperidine, morphine).
7. Increase efficacy and reduce side effects with drug combinations.
   a. Hydroxyzine may reduce nausea and vomiting from the narcotics and potentiate their analgesic effects.
   b. Bowel regimens may reduce constipation from narcotics.
   c. Anticonvulsants may reduce the seizure risk in patients with CVAs or history of seizures in patients on narcotics.
   d. Phenothiazines and narcotics both may lower the seizure threshold, so avoid such combinations if possible.
8. Consider antianxiety measures to reduce anxiety and pain, e.g., thermal biofeedback, relaxation, self-hypnosis (Cozzi, Tyron, & Sedlacek, 1987; Thomas, Koshy, & Patterson, 1984).
9. Consider adjunctive analgesics, e.g., tricyclic antidepressants, trazodone, hydroxyzine, fluoxetine, antihistamines.
10. Taper medications slowly as the crisis resolves.
11. Keep dosage intervals and routes of administration constant if possible; use equianalgesic dosages when converting from parenteral to oral routes and vice versa.
12. Avoid dispensing narcotics to outpatients to treat pain that *might* occur, but also do not allow pain to escalate to a severe degree before implementing analgesics.
13. Educate patients and staff about general management strategies for the pain of SCA.
14. Consider "prophylactic" transfusions to reduce the frequency of painful crises for patients with frequent painful episodes.
15. Consider referral to drug rehabilitation or individual counseling for patients who are willing and who have clinically significant substance abuse problems.
16. Consider the use of individualized, plasticized, wallet-sized cards to expedite effective treatment by improving communication and information to emergency room and house staff physicians (Ballas, 1990).

Adapted from Ballas (1990); Benjamin (1989); Goldberg, Sokol, & Cullen (1987); Mattsson & Kim (1982); Shakin & Thompson (1991).

THALASSEMIA

Like hemophilia, thalassemia is a chronic illness that requires ongoing dependence on blood products and the medical community. Although it is relatively uncommon in the United States, increased numbers of thalassemics have appeared in the San Francisco, Chicago, and areas of the northeastern United States. (It affects about 1 in 40,000 live births in the United States.) Traditionally, the disease has affected Greeks and Italians in the United States; however, increasing numbers of U.S. patients are expected owing to increasing immigration from Southeast Asia and the Far East (Hilgartner, Aledort, & Giardina, 1985).

Disease-Related Issues

Thalassemia is transmitted via autosomal recessive inheritance by two parents who are carriers of the disease. Homozygotes are severely affected males or females who cannot synthesize adequate amounts of the globin chains of hemoglobin. Alpha, beta, gamma, and delta are the designations used for subtypes of thalassemia, corresponding to the type of globin chain that is abnormal. For example, beta-thalassemia major refers to patients who are homozygotes for the beta gene for thalassemia and who cannot synthesize sufficient amounts of beta globin chains. Beta-thalassemia is the most common variant of the disease and is known also as Cooley's anemia or Mediterranean anemia. The anemia results from ineffective erythropoiesis and intramedullary hemolysis. Progressive anemia can be accompanied by irritability, poor appetite, and failure to thrive during the first year of life. Increased red blood cell (RBC) production in the spleen, liver, bone marrow, and lymph nodes is not adequate to compensate for the anemia. Hypersplenism further contributes to RBC destruction, thereby adding to the decreased tissue oxygenation from anemia. The increased erythropoiesis in the bone marrow can lead to fractures of long bones and distort the cranium and the child's facial features.

The goal of treatment with repeated transfusions and sometimes splenectomy is to maintain a normal hemoglobin level. Premature death can occur as a result of repeated infections, splenectomy, or cardiac failure. Although maintaining a near-normal hemoglobin level minimizes facial distortion, iron overload with multiple transfusions often is complicated by cirrhosis, discolored skin, cardiac morbidity, and endocrine dysfunction. Associated cardiac problems include pericarditis, heart failure, and a variety of arrhythmias. Hypothyroidism, diabetes, hypoparathyroidism with resultant hypocalcemia, and delayed growth and sexual development are endocrine abnormalities associated with thalassemia. Psychiatric sequelae of endocrine dysfunction are described elsewhere in this text. (See Chapter 29).

Desferoxamine mesylate (Desferal), which is available for intravenous administration or by a subcutaneous infusion pump, retards iron overload by chelating iron. Although it does not prevent the increased risk of exposure to HIV and hepatitis due to multiple blood transfusions, these risks have been minimized by current blood screening practices. Specific blood components also can result in transfusion reactions (Pochedly, 1986). Bone marrow transplantation is another treatment approach; gene therapy is still at the experimental level.

As with many of the other hematologic disorders, the prevalence of psychiatric disorders in thalassemia has not been well studied. Cerebral iron deposition and the CNS effects of anemia are speculated to play a role in the reduced verbal intelligence quotients (IQs) seen in some thalassemic children (Sherman, Koch, & Giardina, 1985). More specific organic mental syndromes (or cognitive impairment disorders) also might be expected to occur, but because of the small number of patients, the few clinicians focusing on these disorders, or the early mortality of patients, especially prior to the 1970s, such syndromes have not yet been reported.

Psychosocial Issues

Many of the issues faced by patients with thalassemia and their families are similar to those issues already discussed in this chapter concerning other disorders (see Table 31-5). Financial stress and day-to-day worries about the patient's condition are the biggest concerns raised by families of thalassemics (Giordano, 1985; Hilgartner, Aledort, & Giardina, 1985). Parents usually are willing to participate in genetic counseling to prevent another child being born with the disease. Many families with an affected child choose not to have additional offspring. Amniocentesis and chorionic villus biopsy are the most reliable methods of prenatal diagnosis using direct analysis of fetal DNA (Hilgartner, Aledort, & Giardina, 1985; Rowley, Loader, & Walden, 1985; Schwartz, & Benz, 1991).

Studies to date have described anxiety, depression, impaired body image and self-esteem, and acting out or extreme passivity in adolescents with thalassemia. Uncertainties about new treatments and the shortened life-span of affected children are common causes of such reactions. Some degree of guilt about having transmitted the disease is normal in the parents, but a greatly exaggerated sense of guilt may be seen as part of a depressive reaction which may occur more frequently in the mothers. The emotional reactions of fathers of such children have not been reported in the literature. There is one report of thalassemia minor occurring in

TABLE 31–5. *Frequent Psychosocial Issues Associated with Thalassemia*

Treatment-related issues
Adolescents: passivity; denial and acting out
Body image concerns
Emotional stresses: anxiety; day-to-day concerns; depression
Impaired self-esteem
Patient-related issues
Concerns about treatments: genetic screening; prenatal diagnosis
Financial burdens
Limited access to medical care

a patient who also had bipolar disorder (Becker, Cividalli, & Cividalli, 1980; Hilgartner, Aledort, & Giardina, 1985; Joffe, Horvath, & Tarvydas, 1986; Mattsson & Kim, 1982).

Generic psychiatric interventions are aimed primarily at improving communication and interpersonal support within thalassemic families. Outside referrals for group, individual, or family therapy should be considered when appropriate.

PORPHYRIA

Porphyria is transmitted via autosomal dominant inheritance and results from the synthesis of abnormal quantities of porphyrins in the liver and bone marrow. The porphyrins are synthesized in almost all cells and act as intermediate metabolites in the synthesis of the heme portion of hemoglobin. Partial blocks of this pathway or specific enzyme deficiencies result in accumulation of protoporphyrin and its precursors (see Figure 31-2). Only the type III isomers have a functional role in heme synthesis. Patients who have biochemical signs of the disease but no clinical symptoms are described as having "latent" porphyria (Goldberg et al., 1983; Lishman, 1987a; Meyer, 1980; Robinson, 1977; Woo & Cannon, 1991).

Disease-Related Issues

The terms erythropoietic, hepatic, and erythrohepatic refer to the sites of abnormal synthesis of the various porphyrins (see Table 31-6). Three of the four hepatic porphyrias have associated neuropsychiatric findings and therefore are discussed in some detail here. With acute intermittent porphyria (AIP), hereditary coproporphyria (HCP), and variegate porphyria (VP), neurologic symptoms (see Table 31-7) have variable progression but often are preceded by acute, life-threatening attacks of abdominal pain. Abdominal pain can be accompanied by persistent constipation, vomiting, diaphoresis, headaches, fever, tachycardia, postural hypotension, hypertension, and urinary retention (Lishman,

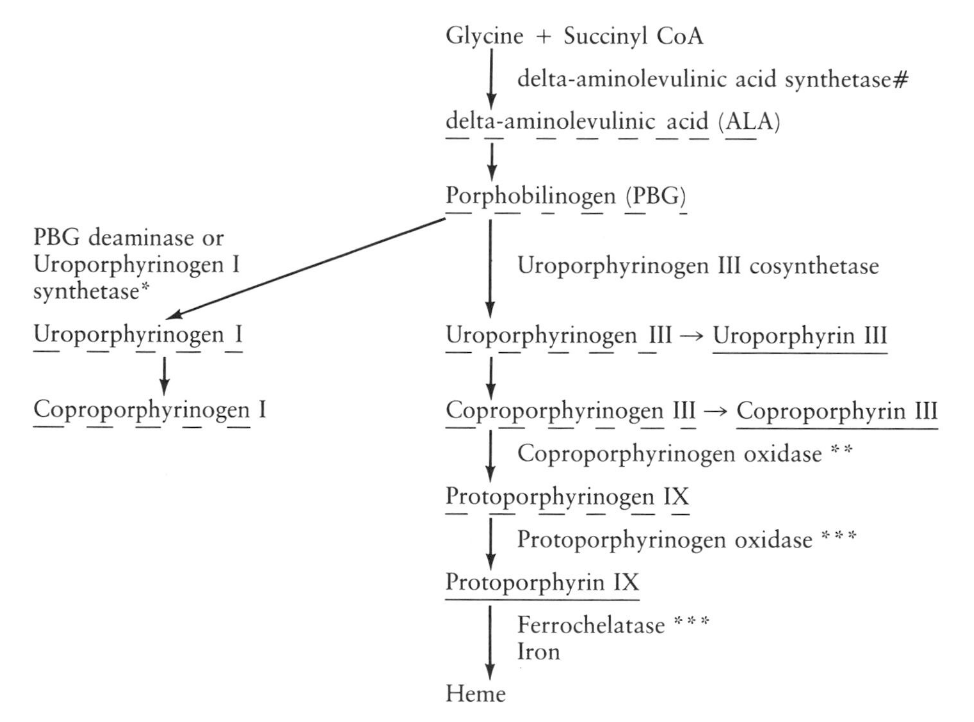

FIG. 31-2. Synthesis of porphyrins in acute intermittent porphyria (AIP), hereditary coproporphyria (HCP), and variegate porphyria (VP). (Adapted from Labbe & Lamon, 1986, p. 1591; Meyer, 1980, p. 496; Robinson, 1977, p. 154; and Woo & Cannon, 1991, p. 150. From Shakin & Thompson, 1991. With permission.) #, site of action for hematin and glucose; *, site of deficiency in AIP. Uroporphyrinogen I synthetase catalyzes the deamination of the PBG side chain and also leads to condensation and cyclization of four molecules. Uroporphyrinogen I synthetase is sometimes referred to as PBG deaminase; **, site of deficiency in HCP; ***, possible sites of deficiency in VP. Colorless compounds are indicated by broken underline (_ _ _ _); colored compounds are indicated by solid underline (________).

TABLE 31–6. *Porphyrias*

Erythropoietic porphyria
- Congenital erythropoietic porphyria (CEP)

Hepatic porphyrias
- Acute intermittent porphyria (AIP)*
- Hereditary coproporphyria (HCP)*
- Variegate porphyria (VP)*
- Porphyria cutanea tarda (PCT)

Erythrohepatic porphyria
- Protoporphyria (PP)

*Associated with neuropsychiatric findings.

TABLE 31–7. *Frequent Neuropsychiatric Symptoms Due to Porphyria*

Cranial nerve impairment
- Ophthalmoplegia
- Dysphagia
- Optic nerve atrophy

Coma
Delirium
Dementia
Other Secondary Mental Disorders
Paralysis
- Bulbar compromise
- Respiratory paralysis

Peripheral neuropathy (usually motor)
Seizures
Sensory signs and symptoms
- Hypesthesia
- Pain in the extremities
- Parethesias

Data from Jefferson & Marshall (1981a) and Meyer (1980).

1987a; Meyer, 1980). Mistaken diagnoses are common and have included appendicitis, renal colic, pancreatitis, intestinal obstruction, Guillain-Barré syndrome, poliomyelitis, and hypertensive encephalopathy (Jefferson & Marshall, 1981a; Lishman, 1987a; Sack, 1990). Unfortunately, abdominal pain and the neuropsychiatric findings are similar in these three disorders, so the presence of dermatologic findings in conjunction with laboratory data are necessary to differentiate AIP, HCP, and VP.

Dermatologic findings of porphyria include hyperpigmentation, as well as blisters and abrasions, which often develop secondary infections following minimal trauma. Hypertrichosis may occur in women with porphyria. Dermatologic problems are common in VP, often presenting simultaneously with neuropsychiatric symptoms. Photosensitivity occurs in about one-third of patients with HCP, and photodermatitis does not develop in AIP.

The assessment of acute attacks of AIP, HCP, and VP should include a thorough search for possible precipitants, such as infection, pregnancy, menstruation, dieting or fasting, alcohol ingestion, physical stress (including surgery), and a review of recent medications to determine if any have been ingested that can incite attacks (see Table 31-8). Nonpsychiatric medications that have been associated with acute attacks are numerous: chloramphenicol, chlorpropamide, cimetidine, ergot preparations, methyldopa, metronidazole, oral contraceptives, progesterone, pyrazinamide, rifampicin, sulfonamides, and theophylline. Unlike with other chronic hematologic disorders, emotional distress has not been cited as a possible precipitant of porphyria exacerbations (Ackner, Cooper, & Gray, 1962; Sack, 1990). Attacks may resolve completely, lead to long-term neuropsychiatric impairment, or sometimes cause death (Goldberg et al., 1983; Jefferson & Marshall, 1981a; Lishman, 1987a; Massey, 1980; Meyer, 1980; Robinson, 1977).

Although abdominal pain usually is the presenting complaint in AIP, psychiatric disturbances may present

TABLE 31–8. *Psychoactive Drugs that Should and Should Not Be Used in Porphyria Patients*[a]

Safe (do not precipitate attacks of porphyria)	Unsafe (may precipitate attacks)
Analgesics	Alcohol
Dihydrocodeine	Anticonvulsants
Meperidine	Carbamazepine
Morphine	Phenytoin
Other opiates	Hydantoins: phenytoin, other
Antihistamines	Primidone
Diphenhydramine	Antidepressants
Antihypertensives	Amitriptyline
Propranolol	Imipramine
Reserpine	Barbiturates[b]
Antipsychotics	Benzodiazepines
Chlorpromazine	Chlordiazepoxide
Promazine	Nitrazepam
Trifluoperazine	Oxazepam
Miscellaneous	Hormones
Atropine	Contraceptives
Diazepam (for seizures)	Estrogens
Nitrous oxide	Other steroids
	Progesterone
	Miscellaneous
	Amphetamines
	Cocaine
	Dichloralphenazone ("Welldorm")
	Ergot derivatives
	Ethchlorvynol
	Meprobamate
	Methyldopa
	Pentazocine
	Theophylline

[a]A more complete table of other unsafe drugs can be found in Goldberg et al. (1983).
[b]Often still used with amytal interviews, electroconvulsive therapy (ECT), and laparotomy.
Adapted from Eilenberg & Scobie (1960); Goldberg et al. (1983); Lishman (1987a); Meyer (1980); Robinson (1977); Sack (1990); Wells & Duncan (1980). Modified with permission (Shakin & Thompson, 1991).

alone or simultaneously. About 30 to 70% of patients with AIP have concomitant psychiatric complaints, and AIP has a higher prevalence in psychiatric inpatients compared to the general population. Although paranoia, auditory hallucinations, and catatonia usually occur late in the disease, AIP has been mistaken for schizophrenia, schizoaffective disorder and atypical psychosis. Other mistaken diagnoses have included conversion disorder, anxiety, and depression (Ackner, Cooper, & Gray, 1962; Jefferson & Marshall, 1981a; Lishman, 1987a; Meyer, 1980; Tishler, Woodward, & O'Connor, 1985). Psychiatric symptoms may represent disease exacerbations (Ackner, Cooper, & Gray, 1962), may be secondary to the stress of a chronic illness, or may be related to iatrogenic complications (e.g., unnecessary surgery) (Cashman, 1961; Whittaker & Whitehead, 1956). If psychiatric symptoms are thought to be related to organic factors, the clinician should do an appropriate medical and neurologic evaluation. Finally, there also has been long-standing and unresolved debate as to whether porphyria patients are at increased risk for developing personality disorders and neuroses (APA, 1991; Copeman, 1891; Jefferson & Marshall, 1981a; Roth, 1945; Wetterberg & Osterberg, 1969).

Laboratory Findings in AIP, HCP, and VP

A variety of laboratory abnormalities have been reported with AIP (see Table 31-9). Although the electroencephalogram (EEG) may be normal during acute attacks, repeated studies are warranted because abnormal EEGs are common (Ackner, Cooper, & Gray, 1962; Lishman, 1987a). Neuropsychological testing is useful for defining cognitive impairments and for confirming dementia (Tishler, Woodward, & O'Connor, 1985).

TABLE 31–9. *Laboratory Findings Associated with AIP*

Anemia: normochromic, normocytic
Catecholamines: increased excretion, with hypertension, and tachycardia
Glucose tolerance tests: abnormal
Hypercholesterolemia
Hypomagnesemia
Hyponatremia: secondary to vomiting or syndrome of inappropriate antidiuretic hormone secretion (SIADH)
Leukocytosis
Liver function tests: normal or elevated bilirubin, transaminases, alkaline phosphatase
Thyroxine: elevated without hyperthyroidism

Data from Goldberg et al., 1983; Meyer, 1980; Woo & Cannon, 1991).

More specific laboratory tests on urine, feces, and blood are used to distinguish AIP, HCP, and VP. Elevated porphobilinogens are colorless in fresh urine. With AIP, urine that is acidified, heated, left standing, or exposed to sunlight turns pink, red, or even black with the formation of uro- and coproporphyrins (see Figure 31-2). This observation led to the development of *qualitative tests*, including the Watson-Schwartz and Hoesch tests, which are used to screen urine for porphobilinogens during acute attacks. The qualitative test detects elevated aminolevulinic acid (ALA) and porphobilinogen (PBG) in an acute attack of AIP, HCP, or VP (see Table 31-10). Occasionally, ALA, PBG, or both are elevated in asymptomatic patients with AIP, but generally these tests are not useful between attacks. One also should consider that many diseases or conditions increase the level of circulatory porphyrins by interference with heme synthesis or excretion. These disorders include lead poisoning, iron deficiency anemia, chronic and acute alcohol use, sideroblastic anemias, megaloblastic anemias, hemolytic anemias, leukemia, sickle cell anemia, polycythemia, hereditary tyrosinemia, liver disease, Dubin-Johnson syndrome, Rotor syndrome, Gilbert's disease, and malabsorption. Phenothiazines also can result in false-positive porphyria tests. Finally, some psychiatric patients (e.g., schizophrenics) have increased urinary excretion of porphyrin-like substances (e.g., haemopyrrole lactam), but do not have porphyria (Goldberg et al., 1983; Reio & Wetterberg, 1969; Rosse et al., 1989a).

If qualitative tests are negative, the clinician should request *quantitative measurement* of ALA and PBG using urine chromatography. A 24-hour urine sample is collected in a bottle designed to prevent interaction of PBG with light. False-negative qualitative tests are found with latent porphyria and during remissions of AIP where urinary excretion of PBG and ALA may be normal or only mildly elevated. Unfortunately, quantitative measurements of ALA and PBG do not correlate with either the presence or the severity of neuropsychiatric symptoms (Jefferson & Marshall, 1981a; Lishman, 1987a; Meyer, 1980). If both qualitative and quantitative tests are negative, erythrocytes, lymphocytes, or skin fibroblasts may be used to measure the decreased activity of uroporphyrinogen I synthetase (sometimes referred to as PBG deaminase). PBG deaminase can be measured in RBCs (see Figure 31-2), thereby confirming the diagnosis of AIP (Meyer, 1980; Rosse et al., 1989a; Sack, 1990; Tishler, Woodward, & O'Connor, 1985). Increased levels of ALA synthetase and decreased activity of delta-5-alpha-reductase also have been suggested to play a role in AIP but are not used for laboratory diagnosis (Woo & Cannon, 1991).

TABLE 31–10. *Some Characteristics of Porphyrias Causing Neuropsychiatric Symptoms*

Parameter	AIP	HCP	VP
Enzyme deficiency	PBG deaminase or UROgen I synthetase	COPROgen oxidase	PROTOgen oxidase or ferrochelatase
Photodermatitis	No	Infrequent	Yes
Laboratory tests			
Erythrocyte levels			
Protoporphyrins	Normal	Normal	Normal
Coproporphyrins	Normal	Sometimes increased during attacks	Normal
Urine levels			
ALA	Very high during acute attacks	Raised only during acute attacks	Raised during acute attacks
PBG	Very high during acute attacks	Raised only during acute attacks	Raised during acute attacks
Uroporphyrin	Usually raised	Sometimes during acute attacks	Sometimes raised
Coproporphyrin	Sometimes raised	Usually raised; always raised during acute attacks	Sometimes raised
Feces			
Porphyrin	Normal	Sometimes raised, especially if photosensitivity present	Very high during acute attacks
Protoporphyrin	Sometimes raised	Raised or normal	Very high
Coproporphyrin	Sometimes raised	Very high	Raised

PBG = porphobilinogen; UROgen = uroporphyrinogen; COPROgen = coproporphyrinogen; PROTOgen = protoporphyrinogen; ALA = aminolevulinic acid.

Adapted from Meyer (1980, p. 495) and Goldberg et al. (1983, p. 9.84). Reproduced with permission (Shakin & Thompson, 1991).

If *HCP* is suspected based on the above results, high coproporphyrin III levels in feces and urine may distinguish it from AIP and VP (see Table 31-10). ALA, PBG, and porphyrins are detected intermittently in the urine of HCP patients. Qualitative tests are positive during acute attacks and negative during remissions. Partial deficiency of coproporphyrinogen oxidase can be measured in leukocytes and skin fibroblasts, thereby confirming the diagnosis. Induction of ALA synthetase also has been suggested to play a role in HCP, but this assay is not used for laboratory diagnosis of the condition (Goldberg et al., 1983; Meyer, 1980; Woo & Cannon, 1991).

Varigate porphyria should be suspected if qualitative tests remain positive between attacks. Levels of urinary ALA, PBG, and porphyrins are normal or increased during remissions of VP and in VP patients with only dermatologic complaints. In VP patients with only a few symptoms, significant elevation of proto- and coproporphyrin also should suggest the diagnosis. VP is distinguished from HCP by the relatively greater elevation of fecal protoporphyrins in VP. Measurement of specific enzyme activity usually is not necessary to diagnose VP, although increased ALA synthetase has been measured in such patients (Goldberg et al., 1983; Meyer, 1980; Woo & Cannon, 1991).

Treatment

Preventive measures in the treatment of porphyria include genetic counseling, education about known precipitants, wearing protective clothing and other measures to avoid direct sunlight in dermatologic variants, and careful diagnosis to prevent unnecessary surgery or psychiatric hospitalization (Meyer, 1980). Patients can die from acute exacerbations of porphyria. In the psychiatric setting, porphyria should be considered as a possible diagnosis before the administration of barbiturates with electroconvulsive therapy or with sodium amytal interviews (Hirsch & Dunsworth, 1955; Mann, 1961). Propofol (Diprivan) is a nonbarbiturate, sedative-hypnotic agent whose effectiveness seems comparable to that of the benzodiazepines and barbiturates in the induction of general anesthesia, but whose mechanism of action is unknown (Diprivan, 1989; Langley & Heel, 1988; Larajani et al., 1989; Raeder & Misvaer, 1988; Rolly & Versichelen, 1985). The safety of propofol for patients with suspected porphyria might be tested when barbiturates are contraindicated.

Acute porphyria attacks are treated in the hospital with intravenous glucose (300 to 500 g/day). Hematin is used if neuropsychiatric symptoms do not abate. In addition to halting the acute attack, these agents reduce the excess production of porphyrins and porphyrin precursors. In animals, they also suppress induction of ALA synthetase (see Figure 31-2). Supportive measures include narcotic analgesics for abdominal pain, phenothiazines and chloral hydrate for nausea and vomiting, maintenance of fluid and electrolyte balance, beta-blockers for tachycardia and hypertension, and cardiorespiratory support if needed for severe complications (Bissell, 1988; Sack, 1990).

VITAMIN B_{12} DEFICIENCY

Metabolism of Vitamin B_{12} and Folate

Vitamin B_{12} and folate metabolism may be better understood in conjunction with one another, so they are discussed together. Vitamin B_{12} is present in fish, meat, leguminal nodules, and dairy products. Body stores of vitamin B_{12} (2 to 5 mg in the typical adult male liver) far exceed the body's requirements for it (approximately 2.5 μg per day), so deficiency states usually take years to develop in previously healthy individuals (see Table 31-11). In contrast, body stores of folate (5 to 10 mg in the liver) do not far exceed requirements for it (50 μg per day). As folate is available only in liver, yeast, and leafy green vegetables, deficiency states may develop over several months, instead of over years as is the case for vitamin B_{12} (see Table 31-12) (Cooper & Rosenblatt, 1987; Hillman, 1980; Jefferson & Marshall, 1981b). Alcohol and anticonvulsants are common causes of folate deficiency (Hunter, Jones, & Jones, 1967).

Ingested vitamin B_{12} is isolated from other proteins and bound to intrinsic factor in the stomach. After ileal absorption, various transcobalamins bind vitamin B_{12} for subsequent distribution. Vitamin B_{12} has predominantly two active forms in the body, methylcobalamin and adenosylcobalamin. Folate has both an inactive form, methyltetrahydrofolate (MTHF), and an active form, tetrahydrofolate (THF). The formation of THF from MTHF generates a methyl group that is used in the formation of methylcobalamin. Methylcobalamin is the donor of the methyl group that is used as a cofactor to convert homocysteine to methionine. In vitamin B_{12} deficiency, methylcobalamin is deficient, so homocysteine may become elevated and folate remains in its inactive form, leading to functional folate deficiency as well. Furthermore, as adenosylcobalamin is needed for converting methylmalonyl-CoA to succinyl-CoA, elevated methylmalonic acid can be used as a measure of vitamin B_{12} deficiency (Carmel, 1983; Chanarin, 1987; Hillman, 1980). Vitamin B_{12} metabolism is discussed further elsewhere (Hillman, 1980; Shakin & Thompson, 1991).

Clinical symptoms of vitamin B_{12} and folate deficiency can be understood on the basis of their metabolism. Megaloblastic anemia probably results from impaired deoxyribonucleic acid (DNA) synthesis, as THF is used to make purines and pyrimidines. THF also is important in amino acid metabolism (e.g., methionine) and functions as a coenzyme in neurotransmitter synthesis (e.g., norepinephrine, serotonin), which may account for the psychiatric manifestations seen in these deficiency states. Neurologic deficits also can result from impaired myelin synthesis, as adenosylocobalamin is needed for lipid and carbohydrate metabolism (Carmel, 1983; Chanarin, 1987; Gross, 1987; Hillman, 1980; Thornton & Thornton, 1978).

TABLE 31–11. *Major Causes of Vitamin B_{12} Deficiency*

Deficiency state: < 80 pg/ml serum
Normal: 175–725 pg/ml

1. Decreased ingestion
   a. Decreased food intake (e.g., starvation, prolonged dieting)
   b. Strict vegetarians
2. Decreased absorption
   a. Decreased intrinsic factor
      (1) Pernicious anemia (most common cause)
      (2) Gastrectomy
      (3) Congenital absence of intrinsic factor
   b. Ileal defects
      (1) Sprue
      (2) Regional enteritis (Crohn's disease)
      (3) After surgery involving the ileum
      (4) Malignancy involving the ileum
   c. Increased competition for vitamin B_{12}
      (1) Overgrowth of intestinal flora blind loops
      (2) Fish tapeworm
   d. Drugs that interfere with absorption, e.g., *p*-aminosalicylic acid, neomycin, colchicine
3. Impaired utilization
   Abnormal or deficient binding proteins

Adapted from Babior and Bunn (1980) and Gross (1987). From Shakin and Thompson (1991). With permission.

TABLE 31–12. *Major Causes of Folate Deficiency*

Deficiency state
Serum <3 mg/ml (normal >6 mg/ml)
RBCs <140 mg/ml (normal >160 mg/ml)

1. Decreased ingestion of folate (most common cause)
   a. Alcoholics
   b. Poverty
   c. Elderly
   d. Overcooking food
2. Decreased absorption
   a. Malabsorption syndromes: sprue, Crohn's disease
   b. Blind loops
   c. Drugs: phenytoin, ethanol, barbiturates
3. Impaired utilization
   a. Folate antagonists: pyrimethamine, methotrexate, triamterene, pentamidine, trimethoprim
   b. Enzyme deficiency
   c. Vitamin B_{12} deficiency
   d. Scurvy
   e. Alcohol
4. Increased requirements
   a. Pregnancy
   b. Infancy
   c. Malignancy
   d. Increased hematopoiesis (hemolytic anemia)

Adapted from Babior & Bunn (1980); Gross (1987); Jefferson & Marshall (1981b). Reprinted with permission (Shakin & Thompson, 1991).

Clinical Manifestations

Although megaloblastic anemia is the classic hematologic abnormality associated with vitamin B_{12} deficiency, neutrophil hypersegmentation, leukopenia, thrombocytopenia, and ineffective erythropoesis with hemolysis also have been reported (Gross, 1987). Megaloblastosis can be masked by folate supplementation or iron deficiency, and normal blood smears and bone marrow specimens may occur when true vitamin B_{12} deficiency is present (Jefferson & Marshall, 1981b).

Further details of the possible neurologic manifestations of vitamin B_{12} deficiency are discussed elsewhere (Adams & Victor, 1985). In general, the associated neurologic symptoms of vitamin B_{12} deficiency include myelopathy (subacute combined degeneration), encephalopathy, and peripheral neuropathy. Intellectual impairment may be moderate to severe. Visual impairment can be the initial presenting complaint, and visual evoked potentials may be abnormal, even in asymptomatic patients. A variety of gastrointestinal, cardiac, and respiratory symptoms also can be seen in deficiency states (Adams & Victor, 1985; Jefferson & Marshall, 1981b; Roos, 1978; Shakin & Thompson, 1991).

Despite commonplace descriptions of associated neurologic and hematologic abnormalities, psychiatric symptoms can be the sole presenting complaint of vitamin B_{12} deficiency (Lindenbaum et al., 1988; Reynolds, 1976a; Strachan & Henderson, 1965; Zucker, Livingston & Nakra, 1981). However, some reports include not only psychiatric patients whose psychiatric symptoms did not respond to vitamin B_{12} supplementation, but also patients whose deficiency was thought to be secondary to inadequate nutrition, commonly seen in chronic psychiatric populations and sometimes secondary to their psychiatric disorder (Carney & Sheffield, 1970; Edwin, Holten, & Norum, 1965; Elsborg, Hansen, & Rafaelsen, 1979; James, Golden, & Sack, 1986; Lishman, 1987b; Shulman, 1967a; Zucker, Livingston, & Nakra, 1981). Shulman (1967b) reported improvements in the depression of some vitamin B_{12}-deficient patients with reassurance and education alone, even before the vitamin was given.

Psychiatric presentations of vitamin B_{12} deficiency have included dementia, delirium, personality changes, paranoid psychosis, mania, and depression (Addison, 1849; Roos, 1978; Rosse et al., 1989b; Zucker, Livingston, & Nakra, 1981). The latter conditions probably should be diagnosed as organic delusional syndrome or organic mood disorder, manic or depressive subtype (in DSM-IV, secondary psychotic disorder due to a nonpsychiatric medical condition and secondary mood disorder due to a nonpsychiatric medical condition). As many as 75% of patients with vitamin B_{12} deficiency have nonpsychotic symptoms, and up to 16% have frank psychosis. Suicide attempts have been reported in one-fifth of some vitamin B_{12}-deficient populations. Psychiatric symptoms of this deficiency often are nonspecific and have been attributed mistakenly to a primary psychiatric illness (Jefferson, 1977; Jefferson & Marshall, 1981b; Roos & Willanger, 1977). Given the relatively high incidence of neuropsychiatric symptoms, early recognition of vitamin B_{12} deficiency states can prevent long-term morbidity by means of appropriate treatment (Hart & McCurdy, 1971).

Routine screening of medical and psychiatric patients for vitamin B_{12} deficiency is sometimes advocated, given the potential benefits of treatment. Most evidence suggests that screening is reasonable and indicated for patients at highest risk, including patients who are elderly, postgastrectomy, cognitively impaired, or who have megaloblastic anemia or neurologic impairments (Geagea & Ananth, 1975; Jefferson & Marshall, 1981b; Phillips & Kahaner, 1988; Shulman, 1967c).

Laboratory Diagnosis

Laboratory evaluation of patients with suspected vitamin B_{12} deficiency may include measurement of the vitamin in various body fluids, measurement of absorption (Schilling test), and nonspecific tests (e.g., electroencephalography). Radioisotope-dilution methods measure both active and inactive cobalamin in the serum; therefore, up to 20% of patients with true vitamin B_{12} deficiency have normal serum vitamin B_{12} levels. Hence the serum measurement is a false-negative test for vitamin B_{12} deficiency in this situation (Kolhouse, Kondo, & Allen, 1978). Also, administration of parenteral vitamin B_{12} can result in normal serum levels before body stores are replenished (Jefferson & Marshall, 1981b).

A low normal serum vitamin B_{12} level should lead to further investigation if there are clinical grounds for suspecting B_{12} deficiency (see Figure 31-3). Low serum levels have been noted with both senile and presenile dementia (Abalan & Delile, 1985); when a low level is encountered, a serum methylmalonic acid can help establish the causal relation to mental changes. If pure intrinsic factor were used as the binding protein, it might eliminate the problem of false-normal serum vitamin B_{12} levels (Cooper & Whitehead, 1978). False measurement of these serum levels in deficiency states are particularly a diagnostic problem in patients without anemia. Elevated serum methylmalonic acid and total homocysteine are more sensitive measures of vitamin B_{12} deficiency and can be used when the vitamin B_{12}

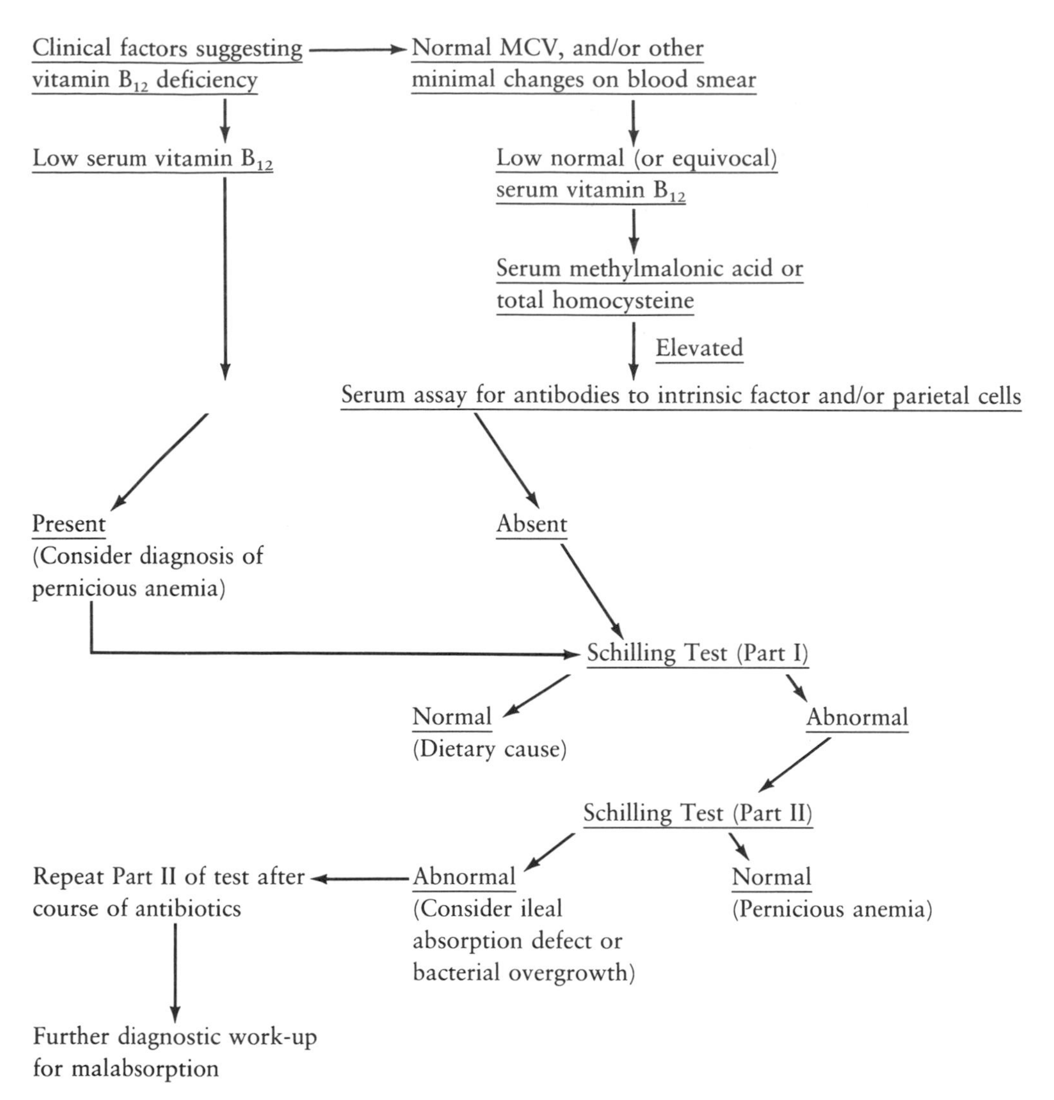

FIG. 31-3. Laboratory evaluation of vitamin B_{12} deficiency. Data from Shakin & Thompson (1991) and Antony (1991).

levels are in the low normal range. Furthermore, they can detect systemic vitamin B_{12} deficiency for up to 24 hours after parenteral vitamin B_{12} treatment has been initiated. These tests are available in about five national laboratories at this time, and cost about the same as each part of the Schilling test (J. Lindenbaum, personal communication, 1991). Whether antibody measurement and the Schilling test will continue to be part of the traditional workup if these tests become more widely available remains to be seen. Measurement of urinary methylmalonic acid and homocysteine is considered too cumbersome and is no longer used to test for vitamin B_{12} deficiency (Carmel, 1983; Lindenbaum et al., 1988).

Pernicious anemia (due to decreased intrinsic factor) can be differentiated from dietary vitamin B_{12} deficiency and ileal disease with a Schilling test (a quantitative measure of vitamin B_{12} absorption). In the first part of the test, radiolabeled vitamin B_{12} is given orally, followed 2 hours later by flushing parenterally with the unlabeled vitamin. Increased urinary excretion of the radiolabeled vitamin is assayed as a measure of gastrointestinal absorption. Details are provided in Chapter 7.

Electroencephalographic (EEG) abnormalities are seen in 48 to 64% of patients with vitamin B_{12} deficiency. These abnormalities tend to normalize with vitamin B_{12} administration, but their clinical and diagnostic relevance are unclear (Evans, Edelson, & Golden, 1983; Roos & Willanger, 1977; Walton, Kiloh, & Osselton, 1954).

Treatment

Treatment is initiated with a commercial preparation of vitamin B_{12} (1000 μg IM), which is given daily for 1 to 2 weeks, then weekly until the anemia is corrected, then monthly, usually for the rest of the patient's life. Since the cost of vitamin B_{12} treatment is relatively low and the risks or side effects are negligible; it seems prudent to treat the demented, geriatric patient who has a low serum vitamin B_{12} level, even when other metabolic parameters are normal. However, the likelihood of a response is small. An elevated serum methylmalonic acid or total homocysteine level in such patients might be used as an indication for more aggressive treatment (J. Lindenbaum, personal communication, 1991). Neurologic symptoms may require biweekly injections for the first 6 months (Gross, 1987).

Vitamin B_{12}-deficient patients frequently describe a sense of well-being during the first day of treatment. Within 2 weeks the megaloblastosis disappears, reticulocytosis starts, and the hematocrit normalizes. Folate may cause partial reversal of hematologic problems; but it also may worsen, precipitate, or not alter the neurologic picture in vitamin B_{12} deficiency. Neurologic and psychiatric problems may take months to reverse and sometimes are irreversible (Gross, 1987; Hillman, 1980; Katz, 1985).

FOLATE DEFICIENCY

Clinical Manifestations

Vitamin B_{12} and folate deficiency may occur in the absence of anemia or bone marrow changes; and although hematologic findings are similar for the two disorders, the spectrum of neuropsychiatric illness is somewhat different. Subacute combined degeneration and peripheral neuropathy are less commonly associated with folate deficiency (Botez, Fontaine, & Botez, 1977; Manzoor & Runcie, 1976; Shorvon, Carney, & Chanarin, 1980; Strachan & Henderson, 1967). Depression and cognitive disorders are the most frequent psychiatric diagnoses in psychiatric inpatients with folate deficiency; conversely, folate deficiency has been noted in up to 30% of psychiatric inpatients (Carney, 1967; Reynolds, 1976b; Reynolds, Preece, & Bailey, 1970; Thornton & Thornton, 1978). Folate deficiency is present in 20% of geriatric admissions and in an even greater percentage of geriatric patients with psychiatric symptoms (Reynolds, 1976b).

Depression also is the most common finding associated with folate deficiency on inpatient medical units; however, the relation between depression and folate deficiency is complex. Depressed patients may have impaired appetites, which may contribute to dietary folate deficiency. Folate deficiency may lead to fatigue, either worsening a preexisting depression or causing a secondary mood disorder. Depression also may result from disturbances of CNS serotonin and catecholamine metabolism, resulting from folate deficiency. Finally, folate deficiency and depression may occur independently of one another in the same patient (Botez, Fontaine, & Botez, 1977; Reynolds, 1976b; Reynolds, Preece, & Bailey, 1970; Shorvon, Carney, & Chanarin, 1980).

The secondary mental disorders that have been associated with folate deficiency include dementia, organic delusional syndromes that may mimic schizophrenia (secondary psychotic disorders), and nonspecific cognitive deficits, including impaired concentration and memory. Mental retardation and minor CT abnormalities also have been reported. It is unclear whether such deficits are reversible with folate administration (Botez, Fontaine, & Botez, 1977; Feigenbaum, Lee, & Ho, 1988; Freeman, Finkelstein, & Mudd, 1975; Lishman, 1987b; Reynolds, 1967a, 1976b; Strachan & Henderson, 1967).

Laboratory Diagnosis

Microbiologic techniques are available to determine serum folate levels; however, RBC levels may more accurately detect persistent folate deficiency (see Chapter 7). RBC levels remain low for a longer time than serum levels and may be useful for making a diagnosis of deficiency if folate already has been administered. Approximately 95% of folate is stored in RBCs, and ingestion of folate can falsely elevate serum levels before stores are replenished. Unfortunately, both RBC folate levels and serum homocysteine levels are not routinely available. Cerebrospinal fluid (CSF) levels may reflect CNS stores more accurately, but their clinical relevance is unknown. Measurement of folate and vitamin B_{12} levels should proceed simultaneously, as there is significant overlap between the two deficiency states (Carmel, 1983; Freeman, Finkelstein, & Mudd, 1975; Gross, 1987; Lindenbaum et al., 1988; Reynolds et al., 1972; Weckman & Lehtovaara, 1969; Wells & Casey, 1967).

Treatment

Standard treatment of folate deficiency is 1 to 5 mg of folate daily by mouth for 4 to 5 weeks. Some patients have been started on 30 mg per day, and others have received lifetime treatment (Botez, Fontaine, & Botez, 1977; Gross, 1987).

Treatment improves the hematologic problems, but it remains unclear how folate affects some neurologic and mental symptoms. Some studies have shown an increased frequency of seizures in patients who are treated with folate who had a history of seizures. The psychiatric consultant may wish to confer with a neurologist or a hematologist when choosing the appropriate treatment for patients with low folate, anemia and a history of seizures (Lishman, 1987b; Reynolds, 1967b).

The clinical course of hematologic improvement is similar following treatment in patients with vitamin B_{12} and folate deficiency (Gross, 1987; Hillman, 1980). Psychiatric symptoms usually are alleviated during the initial weeks, whereas it may take months for neurologic symptoms to abate. Psychiatric inpatients who were treated with folate and carried diagnoses of depression, schizophrenia, and secondary delusional disorder were in better condition and were discharged earlier than patients with the same diagnoses but who did not receive folate (Carney & Sheffield, 1970). Improved mood, concentration, self-confidence, drive, and sociability have been noted after folate administration; however, increased aggressive behavior, severe mood swings, confusion, and even psychosis also have developed after folate treatment. The causal links between folate and neuropsychiatric symptoms remain unclear (Gross, 1987; Manzoor & Runcie, 1976; Reynolds, 1967b).

HYPERVISCOSITY SYNDROME

The hyperviscosity syndrome (HVS) should be suspected when a patient presents with a bleeding diathesis, retinopathy, hypervolemia with cardiac failure, a variety of neurologic symptoms, and possibly renal failure (Bergsagel, 1983a; Crawford, Cox, & Cohen, 1985; Euler, Schmitz, & Loeffler, 1985; Stern, Purcell, & Murray, 1985). HVS can be caused by elevated serum immunoglobulins (e.g., Waldenström's macroglobulinemia, multiple myeloma), elevated blood cell counts (e.g., leukemia, polycythemia), or decreased blood cell deformability (e.g., sickle cell disease) (see Table 31-13) (Fahey, Barth, & Solomon, 1965; Forconi, Pieragalli, & Guerrini, 1987; Mueller, Hotson, & Langston, 1983).

HVS-induced neuropsychiatric symptoms can include fatigue, slowed thinking, headaches, visual disturbances, and partial complex seizures. These symptoms can precede or accompany delirium or dementia; and serial assessments of the patient's cognitive functioning should be documented in this setting (Bergsagel, 1983b; Silberfarb & Bates, 1983). The EEG may be abnormal in HVS patients who have dementia or complex partial seizures.

As only mild elevation of serum viscosity may be present with severe cognitive impairments, it is useful to measure whole blood viscosity as well (Mueller, Hotson, & Langston, 1983). Serum viscosity usually is elevated; that is, greater than 4.0 relative to distilled water

TABLE 31–13. *Causes of Hyperviscosity Syndrome in Adults*

Causes
Elevated serum proteins or lipoproteins
Cryoglobulinemia
Diabetes
Hyperlipidemia?
Lupus erythematosus
Macroglobulinemia
Multiple myeloma
Polymyositis-polysynovitis
Rheumatoid arthritis
Sjögren's syndrome
Elevated blood cell count
Leukemia
Polycythemia, primary and secondary
Decreased blood cell deformability
Diabetes
Sickle cell disease

From Mueller, Hotson, & Langston (1983). With permission.

(normal is less than 1.8) with hemostatic and circulatory disturbances (Bergsagel, 1983c). Rouleaux formation may be present on the blood smear. It is probably prudent to measure serum and whole blood viscosity in any patient who presents with dementia and a disease known to cause HVS (Crawford, Cox, & Cohen, 1985; Euler, Schmitz, & Loeffler, 1985; Mueller, Hotson, & Langston, 1983; Stern, Purcell, & Murray, 1985).

Although the incidence of dementia and other cognitive disorders associated with HVS is still unknown, cognitive impairment and other neurologic phenomena may be reversed by treatment of HVS and concurrent reduction of serum viscosity. In patients with Waldenström's macroglobulinemia, for example, fatigue or neurologic symptoms in association with serum viscosity that is greater than five times normal, would be an indication for plasmapheresis (Alexanian, 1980). Phlebotomy (with plasma reinfusion) can be used to reduce RBC volume in severely symptomatic patients with polycythemia, who have extremely high hematocrits. It is done extremely slowly to avoid oxygen desaturation (e.g., in patients with severe pulmonary outflow obstruction or tetralogy of Fallot) (Braunwald, 1980). Other treatments for HVS include hydration, chemotherapy, and plasmapheresis, depending on the underlying condition.

PSYCHOGENIC PURPURA (AUTOERYTHROCYTE SENSITIZATION)

Psychogenic purpura originally was thought to result from autoerythrocyte sensitization, as similar bruising may occur with the intracutaneous injection of autologous blood (Gardner & Diamond, 1955). To date, however, evidence for an immunologic, infectious, or renal mechanism that would account for this phenomenon has not been documented. Furthermore, no antierythrocyte antibodies or evidence of hemostatic deficiency have been found.

Typically, repeated crops of painful bruises occur in women who complain of many accompanying physical symptoms. Although a history of trauma often can be elicited, the induction of bruises with hypnotic suggestion or emotional distress suggests that psychophysiologic factors may have a role in the pathogenesis of these lesions. Although there have been case reports that document the voluntary production of such lesions, in most patients factitious production of lesions or malingering cannot be proved. Patients may have histrionic and masochistic character traits with secondary depression and anxiety. A history of sexual trauma, marked intrapsychic conflicts, and difficulty controlling hostile impulses sometimes are present in such patients; and, in general, a history of severe emotional stressors is the rule (Agle, Ratnoff, & Wasman, 1967; Mattsson & Agle, 1979; Ratnoff & Agle, 1968; Sorenson, Newman, & Gordon, 1985).

Treatments for patients with psychogenic purpura have included combinations of medications (e.g., antidepressants, hormones, steroids, antihistamines), biofeedback, and psychotherapy or, at the other extreme, reassurance alone. No one mode of treatment is consistently more effective, despite the fact that case reports have included patients treated with a variety of approaches who have had no recurrence of bruising after 2 years of follow-up (Mattsson & Agle, 1979; Sorenson, Newman, & Gordon, 1985). This finding may indicate that the condition is a final common pathway, and that various approaches can be effective depending on the individual's underlying psychopathology. Perhaps, the attention received during a course of treatment, regardless of the specific treatment, is truly the therapeutic agent.

ANTICOAGULANT MALINGERING

When excessive bleeding cannot be accounted for by usual medical conditions, a prolonged prothrombin time (PT) should alert the clinician to the possibility of self-administration of warfarin (Coumadin). This problem was first described in 1951 (Stafne & Moe, 1951) and subsequently labeled as anticoagulant malingering, a misnomer. These patients more correctly should be diagnosed as having a factitious disorder. They usually are not consciously seeking to obtain secondary gain by ingesting warfarin and therefore technically should not be diagnosed as malingerers. Although they voluntarily control their behavior, the underlying motivation usually is an unconscious need to be in the patient role (i.e., primary gain).

Such patients often have masochistic and histrionic personality traits. Some fantasize about causing bleeding in others, and further exploration may reveal violent accounts of past bleeding episodes in family members or themselves. Typically, they have access to warfarin derivatives from their jobs or from prescriptions dispensed to them or their families. This possibility should be considered, especially in health care workers (Agle, Ratnoff, & Spring, 1970).

The diagnosis may be confirmed by detecting the anticoagulant in the patient's plasma (e.g., a significant warfarin level in a patient not receiving the medication). The level may not be detectable, however, despite the fact that the PT can be elevated for up to 3 days after ingestion. Parenteral vitamin K_1 easily corrects deficiencies of factors II, VII, VIII, and IX by displacing the warfarin compounds (Agle, Ratnoff, & Spring, 1970; Roschlau & Sellers, 1980).

Once bleeding is corrected, the primary physician should gently discuss the situation with the patient. Initially, one should anticipate denial, rage, paranoia, or a wide range of other relatively primitive psychological defenses. It should be remembered that these patients have "selected" a maladaptive and dangerous means for coping with stress and for seeking attention, so a logical rational response should not be expected immediately. The primary physician may wish to discuss the patient's responses with a psychiatrist; and if the patient can be encouraged to see a psychiatrist (with the primary physician present, if desired), a psychiatric consultant should do a comprehensive evaluation, including assessing whether the patient poses a danger to self or others, and whether psychotropic medication or hospitalization is warranted. Although conjoint medical and psychiatric evaluation and management sometimes is not successful owing to the complexity of these patients, it continues to be the recommended course of management (Agle, Ratnoff, & Spring, 1970).

CONCLUDING REMARKS

The major issue at the interface of psychiatry and hematology in recent years has been the unintentional iatrogenic infection of thousands of hemophiliacs with HIV. These patients almost certainly will die of AIDS, whereas they could have expected a normal or almost normal life-span if they had continued to have only hemophilia. This situation is probably the largest and most tragic example of iatrogenic patient deaths in modern medicine. The transmission of this deadly condition to these patients has caused major strains in many doctor-patient relationships and with other staff at hemophilia treatment centers. The health care professionals often feel guilty and anxious about what else they may still be transmitting that has yet to be identified by medical science. Many patients and families who previously felt hopeful about their hemophilia and grateful to their medical community now feel betrayed, angry, and frightened.

A few other issues also should be briefly mentioned here. Sickle cell anemia is a condition in which the degree of trust present in the doctor-patient relationship frequently is tested; this situation often is played out around how much pain the patient "should be" experiencing in a given clinical situation and what analgesic and how much of it is needed for adequate pain management.

Thalassemia is a model condition that can be used to study the degree of parental guilt experienced by producing a child with an inheritable genetic disorder. If this type of guilt or other emotional responses are exaggerated or prolonged, it should spur the physician to look for other associated emotional factors that might explain symptom amplification, and to consider psychiatric consultation if appropriate.

Porphyria often causes psychiatric symptoms, but psychiatric stress does not seem to precipitate episodes of porphyria. Because a combination of abdominal, dermatologic, neurologic, and psychiatric symptoms may appear, porphyria is a model condition that illustrates the importance of psychiatrists closely collaborating with other specialists to ensure optimal patient diagnosis and treatment.

Vitamin B_{12} and folate deficiencies also are conditions that underline the importance of consultation psychiatrists being active with other medical specialists. Many people in the United States have become increasingly health and diet conscious. Because vitamin B_{12} and folate deficiencies can cause neuropsychiatric and multiple somatic complaints, it is important that psychiatrists and other medical specialists be alert for these conditions. Vitamin B_{12} and folate deficiencies also provide an opportunity for psychiatrists to interact effectively with and more broadly educate the health and diet conscious public.

Acknowledgments

The authors acknowledge the assistance of Dr. Susan Eshleman and the technical assistance of Marianne Bermender in the preparation of this chapter.

REFERENCES

Abalan F, & Delile JM (1985). B_{12} deficiency in presenile dementia. Biol Psychiatry 20:1251.

Ackner B, Cooper JE, & Gray CH (1962). Acute porphyria: a neuropsychiatric and biochemical study. J Psychosom Res 6:1–24.

Adams RD, & Victor M (1985). Diseases of the nervous system due to nutritional deficiency. In Principles of Neurology (Ch. 38, pp. 760–786). New York: McGraw-Hill.

Addison T (1849). Anemia: disease of the supraenal capsules. Lond Med Gaz 8:517–518.

Agle DP (1964). Psychiatric studies of patients with hemophilia and related states. Arch Intern Med 114:76–82.

Agle DP (1984). Hemophilia—psychological factors and comprehensive management. Scand J Haematol [suppl 40] 33:55–63.

Agle DP (March 1989). The hemophiliac population and HIV infection. Presentation at the American Psychosomatic Society, San Francisco.

Agle D, Gluck H, & Pierce GF (1987). The risk of AIDS: psychological impact on the hemophilic population. Gen Hosp Psychiatry 9:11–17.

Agle DP, & Heine P (unpublished). The psychosocial dimensions of the hemophilia treatment center project [brochure]. The hemophilia treatment center history project, co-sponsored by the office of Maternal and Child Health, U.S. Public Health Department and the National Hemophilia Foundation.

Agle DP, & Mattsson A (1968). Psychiatric and social care of patients with hereditary hemorrhagic disease. Mod Treat 5:111–124.

Agle DP, & Mattsson A (1970). Psychiatric factors in hemophilia: methods of parental adaptation. Bibl Haematol 34:89–94.

Agle DP, Ratnoff OD, & Spring GK (1970). The anticoagulant malingerer: psychiatric studies of three patients. Ann Intern Med 73:67–72.

Agle DP, Ratnoff OD, & Wasman M (1967). Studies in autoerythrocyte sensitization: the induction of purpuric lesions by hypnotic suggestion. Psychosom Med 29:491–503.

Alexanian R (1980). Plasma cell neoplasms and related disorders. In KJ Isselbacher, RD Adams, & E Braunwald (Eds.), Harrison's principles of internal medicine (Ch. 63, pp. 333–338). New York: McGraw-Hill.

American Psychiatric Association. Task Force on DSM-IV (1991). DSM IV options book. Washington, DC: American Psychiatric Association.

Antony AC (1991). Megaloblastic anemias. In R Hoffman, E Benz, S Shattil, et al. (Eds.), Hematology, basic principles and practice (Ch. 32, pp. 392–417). New York: Churchill Livingstone.

Babior BM, & Bunn HF (1980). Megaloblastic anemias. In KJ Isselbacher, RD Adams, & E Braunwald (Eds.), Harrison's principles of internal medicine (Ch. 311, pp. 1518–1525). New York: McGraw-Hill.

Ballas SK (1990). Treatment of pain in adults with sickle cell disease. Am J Hematol 34:49–54.

Ballas SK, Lewis CN, Noone AM, et al. (1982). Clinical, hematological and biochemical features of Hb SC disease. Am J Hematol 13:37–51.

Barrett DH, Wisotzek IE, Abel GG, et al. (1988). Patients with sickle cell disease. South Med J 81:745–750.

Becker RD, Cividalli G, & Cividalli NJ (1980). Psychological considerations in the management of patients with thalassemia major. 1. Reactions to chronic illness in childhood and adolescence: a report of a case. Arts Psychother 7:165–195.

Benjamin LJ (1989). Pain in sickle cell disease. In KM Foley & RM Payne (Eds.), Current therapy of pain (pp. 90–104). Toronto: B.C. Decker.

BERGSAGEL D (1983a). Macroglobulinemia. In WJ WILLIAMS, E BEUTLER, & AJ ERSLER (Eds.), Hematology (3rd ed., Ch. 123, pp. 1104–1109). New York: McGraw-Hill.

BERGSAGEL D (1983b). Plasma cell myeloma. In WJ WILLIAMS, E BEUTLER, & AJ ERSLER (Eds.). Hematology (3rd ed., Ch. 122, pp. 1078–1104). New York: McGraw-Hill.

BERGSAGEL D (1983c). Plasma cell neoplasms—general considerations. In WJ WILLIAMS, E BEUTLER, & AJ ERSLER (Eds.), Hematology (3rd ed., Ch. 121, pp. 1067–1078). New York: McGraw-Hill.

BISSELL DM (1988). Treatment of acute hepatic porphyria with hematin. J Hepatol 6:1–7.

BOTEZ MI, FONTAINE F, & BOTEZ T (1977). Folate-responsive neurological and mental disorders: report of 16 cases. Eur Neurol 16:240–246.

BRAUNWALD E (1980). Cyanosis, hypoxia, and polycythemia, In KJ ISSELBACHER, RD ADAMS, & E BRAUNWALD (Eds.), Harrison's principles in internal medicine (Ch. 28, pp. 166–171). New York: McGraw-Hill.

BROWNE WJ, MALLY MA, & KANE RP (1960). Psychosocial aspects of hemophilia: study of 28 children and their families. Am J Orthopsychiatry 30:730–740.

BRUNNER RL, SCHAPERA NE, & GRUPPA RA (1982). Pain, psychosocial adjustment and drug use: their interactions in hemophilia. Psychosom Med 44:120.

CARMEL R (1983). Clinical and laboratory features of the diagnosis of megaloblastic anemia. In J LINDENBAUM (Ed.), Nutrition in hematology (Ch. 1, pp. 1–31). New York: Churchill Livingstone.

CARNEY MWP (1967). Serum folate values in 423 psychiatric patients. BMJ 4:512–516.

CARNEY MWP, & SHEFFIELD BF (1970). Associations of subnormal serum folate and vitamin B_{12} values and effects of replacement therapy. J Nerv Ment Dis 150:404–412.

CASHMAN MD (1961). Psychiatric aspects of acute porphyria. Lancet 1:115–116.

CHANARIN I (1987). Megaloblastic anemia, cobalamin, and folate. J Clin Pathol 40:978–984.

CHODORKOFF J, & WHITTEN CF (1963). Intellectual status of children with sickle cell anemia. J Pediatr 63:29–35.

CHOINIERE M, & MELZACK R (1987). Acute and chronic pain in hemophilia. Pain 31:317–331.

COOPER BA, & ROSENBLATT DS (1987). Inherited defects of vitamin B_{12} metabolism. Annu Rev Nutr 7:291–320.

COOPER BA, & WHITEHEAD VM (1978). Evidence that some patients with pernicious anemia are not recognized by radiodilution assay for cobalamin in serum. N Engl J Med 299:816–818.

COPEMAN SM (1891). In proceedings of the pathological society of London. Lancet 1:196–197.

COZZI L, TYRON WW, & SEDLACEK K (1987). The effectiveness of biofeedback-assisted relaxation in modifying sickle cell crises. Biofeedback Self-Regul 12:51–61.

CRAWFORD J, COX EB, & COHEN HS (1985). Evaluation of hyperviscosity in monoclonal gammopathies. Am J Med 79:13–22.

DIPRIVAN R (PROPOFOL) (1989). Package insert. Wilmington, DE: Stuart Pharmaceuticals.

EDWIN E, HOLTEN K, & NORUM KR (1965). Vitamin B_{12} hypovitaminosis in mental diseases. Acta Med Scand 177:689–699.

EILENBERG MD, & SCOBIE BA (1960). Prolonged neuropsychiatric disability and cardiomyopathy in acute intermittent porphyria. BMJ 1:858–860.

ELSBORG L, HANSEN T, & RAFAELSEN OJ (1979). Vitamin B_{12} concentrations in psychiatric patients. Acta Psychiatr Scand 59:145–152.

EULER HH, SCHMITZ N, & LOEFFLER H (1985). Plasmapharesis in paraproteinemia. Blut 50:321–330.

EVANS DL, EDELSON GA, & GOLDEN RN (1983). Organic psychosis without anemia or spinal cord symptoms in patients with vitamin B_{12} deficiency. Am J Psychiatry 140:218–221.

FAHEY JL, BARTH WF, & SOLOMON A (1965). Serum hyperviscosity syndrome. JAMA 192:464–467.

FAULSTICK ME (1987). Psychiatric aspects of AIDS. Am J Psychiatry 144:551–556.

FEIGENBAUM LZ, LEE D, & HO J (1988). Routine assessment of folate levels in geriatric assessment for dementia. J Am Geriatr Soc 36:755.

FORCONI S, PIERAGALLI D, & GUERRINI M (1987). Primary and secondary blood hyperviscosity syndromes, and syndromes associated with blood hyperviscosity. Drugs (suppl 2) 33:19–26.

FOWLER MG, WHITT JK, & LALLINGER RR (1988). Neuropsychologic and academic functioning of children with sickle cell anemia. J Dev Behav Pediatr 9:213–220.

FREEMAN JM, FINKELSTEIN JD, & MUDD SH (1975). Folate-responsive homocystinuria and "schizophrenia." N Engl J Med 292:491–496.

GARDNER FH, & DIAMOND LK (1955). Autoerythrocyte sensitization: a form of purpura producing painful bruising following autosensitization to red cells in certain women. Blood 10:675–690.

GEAGEA K, & ANANTH J (1975). Response of a psychiatric patient to vitamin B_{12} therapy. Dis Nerv Syst 36:343–344.

GILCHRIST GS, & PIEPGRAS DG (1976). Neurologic complications in hemophilia. In M HILGARTNER & C POCHEDLY (Eds.), Progress in pediatric hematology and oncology (Vol. 1; Ch. 6, pp. 79–97). Hemophilia in children. Littleton, MA: Publishing Sciences Group.

GILCHRIST GS, PIEPGRAS DG, & ROSKOS RR (1989). Neurologic complications in hemophilia. In M HILGARTNER & C POCHEDLY (Eds.), Hemophilia in the child and adult (Ch. 3, pp. 45–68). New York: Raven Press.

GIORDANO V (1985). Psychosocial impacts on a thalassemic patient's life. Ann NY Acad Sci 445:324–326.

GOLDBERG A, MOORE MR, MCCOLL KEL, ET AL. (1983). Porphyrin metabolism and the porphyrias. In DJ WEATHERALL, JG LEDINGHAM, & DA WARRELL (Eds.), Oxford textbook of medicine (Vol. 1; Ch. 9.81, pp. 9.81–9.89). New York: Oxford University Press.

GOLDBERG RJ, SOKOL MS, & CULLEN LO (1987). Acute pain management. In A STOUDEMIRE & BS FOGEL (Eds.), Principles of medical psychiatry (Ch. 16, pp. 365–388). Orlando, FL: Grune & Stratton.

GROSS LS (1987). Neuropsychiatric aspects of vitamin deficiency states. In RE HALES & SC YUDOFSKY (Eds.), Textbook of neuropsychiatry (Ch. 18, pp. 327–338). Washington, DC: American Psychiatric Press.

HANDFORD HA, CHARNEY D, & ACKERMAN L (1980). Effect of psychiatric intervention on use of antihemophilic factor concentrate. Am J Psychiatry 137:1254–1256.

HANDFORD HA, & MAYES SD (1989). The basis of psychosocial programs in hemophilia. In M HILGARTNER & C POCHEDLY (Eds.), Hemophilia in the child and adult (Ch. 11, pp. 195–212). New York: Raven Press.

HANDFORD HA, MAYES SD, & BIXLER EO (1986). Personality traits of hemophilic boys. J Dev Behav Pediatr 4:224–229.

HART RJ, & MCCURDY PR (1971). Psychosis in vitamin B_{12} deficiency. Arch Intern Med 128:596–597.

HERNANDEZ J, GRAY D, & LINEBERGER HP (1989). Social and economic indicators of well-being among hemophiliacs over a 5 year period. Gen Hosp Psychiatry 11:241–247.

HILGARTNER MW (1980). Comprehensive care for hemophilia. U.S. Department of Health, Education and Welfare Publ. No. 79-5129. Washington, DC: U.S. Government Printing Office.

HILGARTNER MW, ALEDORT L, & GIARDINA PJV (1985). Thalassemia and hemophilia. In N HOBBS & JM PERRIN (Eds.), Issues in the care of children with chronic illness (Ch. 15, pp. 299–323). San Francisco: Jossey-Bass.

Hillman RS (1980). Vitamin B_{12}, folic acid, and the treatment of megaloblastic anemias. In AG Gilman, LS Goodman, & A Gilman (Eds.), The pharmacologic basis of therapeutics (Ch. 57, pp. 1331–1346). New York: Macmillan.

Hirsch S, & Dunsworth FA (1955). An interesting case of porphyria. Am J Psychiatry 111:703.

Holdredge SA, & Cotta S (1989). Physical therapy and rehabilitation in the care of the adult and child with hemophilia. In M Hilgartner & C Pochedly (Eds.), Hemophilia in the child and adult (Ch. 14, pp. 235–262). New York: Raven Press.

Hunter R, Jones M, & Jones TG (1967). Serum B_{12} and folate concentrations in mental patients. Br J Psychiatry 113:1291–1295.

Huntsman RG (1985). Hemoglobinopathies. Med North Am 29:3998–4008.

Hurtig AL, & White LS (1986). Psychosocial adjustment in children and adolescents with sickle cell disease. J Pediatr Psychol 11:411–427.

Jacobs J (1991). Family therapy in the context of childhood medical illness. In A Stoudemire & B Fogel (Eds.), Medical-Psychiatric Practice (Vol. 1; Ch. 14, pp. 483–506). Washington, DC: American Psychiatric Press.

James SP, Golden RN, & Sack DA (1986). Single case study: vitamin B_{12} deficiency and the dexamethasone suppression test. J Nerv Ment Dis 174:560–561.

Jason L, Lui K-J, Ragni MV, et al. (1989). Risk of developing AIDS in HIV-infected cohorts of hemophilic and homosexual men. JAMA 261:725–727.

Jefferson JW (1977). The case of the numb testicles. Dis Nerv Syst 38:749–751.

Jefferson JW, & Marshall JR (1981a). Metabolic disorders. In Neuropsychiatric features of medical disorders (Ch. 8, pp. 201–206). New York: Plenum.

Jefferson JW, & Marshall JR (1981b). Vitamin disorders. In Neuropsychiatric features of medical disorders (Ch. 11, pp. 231–257). New York: Plenum.

Joffe RT, Horvath Z, & Tarvydas I (1986). Bipolar affective disorder and thalassemia minor. Am J Psychiatry 143:7.

Jonas DL (1977). Psychiatric aspects of hemophilia. Mt Sinai J Med 44:457–463.

Jonas DL (1989). Drug abuse in hemophilia. In M Hilgartner & C Pochedly (Eds.), Hemophilia in the child and adult (Ch. 13, pp. 227–233). New York: Raven Press.

Katz M (1985). Megaloblastic anemia. Med North Am 29:3966–3979.

Kolhouse JF, Kondo H, & Allen NC (1978). Cobalamin analogues are present in human plasma and can mask cobalamin deficiency because current radioisotope dilution assays are not specific for true cobalamin. N Engl J Med 299: 785–792.

Kramer MS, Rooks Y, & Pearson HA (1978). Growth and development in children with sickle cell trait. N Engl J Med 299:686–689.

Kumar S, Powars D, & Allen J (1976). Anxiety, self-concept, and personal and social adjustments in children with sickle cell anemia. J Pediatr 88:859–863.

LaBaw WL (1975). Autohypnosis in hemophilia. Hematologica 9:103–110.

Labbe RF, & Lamon JM (1986). Porphyrins and disorders of porphyrin metabolism. In N Tietz (Ed.), Textbook of clinical chemistry (Ch. 16, pp. 1589–1614). Philadelphia: WB Saunders.

Langley MS, & Heel RC (1988). Propofol: a review of its pharmacodynamic and pharmacokinetic properties and use as an intravenous anaesthetic. Drugs 35:334–72.

Larajani GE, Gratz I, Afshar M, et al. (1989). Clinical pharmacology of propofol: an intravenous anesthetic agent. DICP Ann Pharmacother 23:743–49.

Leavell SR, & Ford CV (1983). Psychopathology in patients with sickle cell disease. Psychosomatics 24:24–37.

LeBaron S, & Zeltzer LK (1984). Research on hypnosis in hemophilia—preliminary success and problems: a brief communication. Int J Clin Exp Hypn 32:290–295.

Lemanek KL, Moore SL, & Gresham FM (1986). Psychological adjustment of children with sickle cell anemia. J Pediatr Psychol 11:397–409.

Lesko LM, Massie MJ, & Holland JC (1987). Oncology. In A Stoudemire & BS Fogel (Eds.), Principles of medical psychiatry (Ch. 23, pp. 495–520). Orlando, FL: Grune & Stratton.

Levine PH (1985). The acquired immunodeficiency syndrome in persons with hemophilia. Ann Intern Med 103:723–726.

Levine SB (1984). Introduction to the sexual consequences of hemophilia. Scand J Hematol [suppl 40] 33:75–82.

Lichstein KL, & Eakin TL (1985). Progressive versus self-control relaxation to reduce spontaneous bleeding in hemophiliacs. J Behav Med 8:149–162.

Lindenbaum J, Healton EB, Savage DG, et al. (1988). Neuropsychiatric disorders caused by cobalamin deficiency in the absence of anemia or macrocytosis. N Engl J Med 318:1720–1728.

Lishman WA (1987a). Endocrine diseases and metabolic disorders. In Organic psychiatry: the psychological consequences of cerebral disorders (Ch. 11, pp. 482–485). Oxford: Blackwell Scientific Publications.

Lishman WA (1987b). Vitamin deficiencies. In Organic psychiatry: the psychological consequences of cerebral disorder (Ch. 12, pp. 501–507). Oxford: Blackwell Scientific Publications.

Lucas ON (1965). Dental extractions in the hemophiliac: control of emotional factors by hypnosis. Am J Clin Hypn 7:301–307.

Lucas ON (1975). The use of hypnosis in hemophilia dental care. Ann NY Acad Sci 240:263–266.

Magrab PR (1985). Psychosocial development of chronically ill children. In N Hobbs & JM Perrin (Eds.), Issues in the care of children with chronic illness (Ch. 33, pp. 698–716). San Francisco: Jossey-Bass.

Mann J (1961). Acute porphyria provoked by barbiturates given with electroshock therapy. Am J Psychiatry 118:509–511.

Manzoor M, & Runcie J (1976). Folate-responsive neuropathy: report of 10 cases. BMJ 1:1176–1178.

Massey EW (1980). Neuropsychiatric manifestations of porphyria. J Clin Psychiatry 41:208–213.

Mattsson A (1984). Hemophilia and the family: life-long challenges and adaption. Scand J Hematol [suppl 40] 33:65–74.

Mattsson A, & Agle D (1972). Group therapy with parents of hemophiliacs: therapeutic process and observations on parental adaption to chronic illness in children. J Am Acad Child Adolesc Psychiatry 11:558–571.

Mattsson A, & Agle DP (1979). Psychophysiologic aspects of adolescence: hemic disorders. Adolesc Psychiatry 7:269–280.

Mattsson A, & Gross S (1966a). Adaptional and defensive behavior in young hemophiliacs and their parents. Am J Psychiatry 122:1349–1356.

Mattsson A, & Gross S (1966b). Social and behavioral studies on hemophilic children and their families. J Pediatr 68:952–964.

Mattsson A, Gross S, & Hall TW (1971). Psychoendocrine study of adaptation in young hemophiliacs. Psychosom Med 33:215–225.

Mattsson A, & Kim SP (1982). Blood disorders. Psychiatr Clin North Am 5:345–56.

Mayes SD, Handford HA, & Kowalski C (1988). Parent attitudes and child personality traits in hemophilia: a six-year longitudinal study. Int J Psychiatry Med 18:339–355.

MEIJER A (1980–81). Psychiatric problems of hemophilic boys and their families. Int J Psychiatry Med 10:163–172.

MEYER UA (1980). Porphyrias. In KJ ISSELBACHER, RD ADAMS, & E BRAUNWALD (Eds.), Harrison's principles of internal medicine (Ch. 96, pp. 494–500). New York: McGraw-Hill.

MORGAN SA, & JACKSON J (1986). Psychological and social concomitants of sickle cell anemia in adolescents. J Pediatr Psychol 11:429–440.

MUELLER J, HOTSON JR, & LANGSTON JW (1983). Hyperviscosity-induced dementia. Neurology 33:101–103.

O'SHANICK GJ, GARDNER DF, & KORNSTEIN SG (1987). Endocrine disorders. In A STOUDEMIRE & BS FOGEL (Eds.), Principles of medical psychiatry (Ch. 30, pp. 641–658). Orlando, FL: Grune & Stratton.

PHILLIPS SL, & KAHANER KP (1988). An unusual presentation of vitamin B_{12} deficiency. Am J Psychiatry 145:529.

POCHEDLY C (July 1986). Key questions on chronic blood transfusions in children. Resident Staff Physician 32:65–68.

POWARS DR (1975). Natural history of sickle cell disease—the first ten years. Semin Hematol 12:267–285.

RAEDER JC, & MISVAER G (1988). Comparison of propofol induction with thiopentone or methohexitone in short outpatient general anaesthesia. Acta Anaesthesiol Scand 32:607–13.

RATNOFF OD, & AGLE DP (1968). Psychogenic purpura: a re-evaluation of the syndrome of autoerythrocyte sensitization. Medicine 47:475–500.

REIO L, & WETTERBERG L (1969). False porphobilinogen reactions in the urine of mental patients. JAMA 207:148–150.

REIS E, LINHART R, & LAZERSON J (1982). Using a standard form to collect psychosocial data about hemophilia patients. Health Soc Work 7:206–214.

REYNOLDS EH (1967a). Schizophrenia-like psychoses of epilepsy and disturbances of folate and vitamin B_{12} metabolism induced by anticonvulsant drugs. Br J Psychiatry 113:911–919.

REYNOLDS EH (1967b). Effects of folic acid on the mental state and fit-frequency of drug-treated epileptic patients. Lancet 1:1086–1088.

REYNOLDS EH (1976a). The neurology of vitamin B_{12} deficiency. Lancet 2:832–833.

REYNOLDS EH (1976b). Neurologic aspects of folate and vitamin B_{12} metabolism. Clin Hematol 5:661–696.

REYNOLDS EH, GALLAGHER BB, MATTSON RH, ET AL. (1972). Relationship between serum and cerebrospinal fluid folate. Nature 240:155–157.

REYNOLDS EH, PREECE JM & BAILEY J (1970). Folate deficiency in depressive illness. Br J Psychiatry 117:287–292.

ROBINSON SH (1977). Hypochromic anemias, II. Heme metabolism and the porphyrias. In WS BECK (Ed.), Hematology. Cambridge, MA: MIT Press. Lecture 7, pp. 153–164.

ROCHE PA, GIJSBERS K, & BELCH JJF (1985). Modification of haemophiliac haemorrhage pain by transcutaneous electrical nerve stimulation. Pain 21:43–48.

RODGERS GP (1991). Recent approaches to the treatment of sickle cell anemia. JAMA 265:2097–2101.

ROLLY G, & VERSICHELEN L (1985). Comparison of propofol and etomidate for induction of anesthesia in premedicated patients. Anaesthesia 40:945–48.

ROOS D (1978). Neurological complications in patients with impaired vitamin B_{12} absorption following partial gastrectomy. Acta Neurol Scand [suppl 69] 59:4–77.

ROOS D, & WILLANGER R (1977). Various degrees of dementia in a selected group of gastrectomized patients with low serum B_{12}. Acta Neurol Scand 55:363–376.

ROSCHLAU WHE, & SELLERS EM (1980). Anticoagulants. In P SEEMAN, EM SELLERS, & WHE ROSCHLAU (Eds.), Principles of medical pharmacology (Ch. 39, pp. 351–362). Toronto: University of Toronto Press.

ROSSE RB, GIESE AA, DEUTSCH SI, ET AL. (1989a). Biochemical evaluations in psychiatry. In Laboratory diagnostic testing in psychiatry (Ch. 4, pp. 36–49). Washington, DC: American Psychiatric Press.

ROSSE RB, GIESE AA, DEUTSCH SI, ET AL. (1989b). Hematologic measures of potential relevance to psychiatrists. In Laboratory diagnostic testing in psychiatry (Ch. 3, pp. 31–35). Washington, DC: American Psychiatric Press.

ROTH N (1945). The neuropsychiatric aspects of porphyria. Psychosom Med 7:291–301.

ROWLEY PT, LOADER S, & WALDEN ME (1985). Toward providing parents the option of avoiding the birth of the first child with Cooley's anemia: response to hemoglobinopathy screening and counseling during pregnancy. Ann NY Acad Sci 445:408–416.

SACK GH (1990). Acute intermittent porphyria. JAMA 264:1290–1293.

SALK L, HILGARTNER MW, & GRANICH B (1972). The psychosocial impact of hemophilia on the patient and his family. Soc Sci Med 6:491–505.

SCHWARTZ E & BENZ EJ (1991). The thalassemia syndromes. In R HOFFMAN, EJ BENZ, SJ SHATTIL, ET AL. (Eds.). Hematology, basic principles and practise (Ch. 31, pp. 368–392). New York: Churchill Livingstone.

SHAKIN EJ, & THOMPSON TL (1991). Psychiatric aspects of hematologic disorders. In A STOUDEMIRE & B FOGEL (Eds.), Medical-psychiatric practice (Vol. 1; Ch. 6, pp. 193–242). Washington DC: American Psychiatric Press.

SHERMAN M, KOCH D, & GIARDINA P (1985). Thalassemic children's understanding of illness: a study of cognitive and emotional factors. Ann NY Acad Sci 445:327–336.

SHORVON SD, CARNEY MWP, & CHANARIN I (1980). The neuropsychiatry of megaloblastic anemia. BMJ 281:1036–1038.

SHULMAN R (1967a). Vitamin B_{12} deficiency and psychiatric illness. Br J Psychiatry 113:252–256.

SHULMAN R (1967b). Psychiatric aspects of pernicious anemia: a prospective controlled investigation. BMJ 3:266–270.

SHULMAN R (1967c). A survey of vitamin B_{12} deficiency in an elderly psychiatric population. Br J Psychiatry 113:241–251.

SILBERFARB PM, & BATES GM (1983). Psychiatric complications of multiple myeloma. Am J Psychiatry 140:788–789.

SIMON R (1984). Hemophilia and the family system. Psychosomatics 25:845–849.

SIMON RM (1989). The family and hemophilia. In M HILGARTNER & C POCHEDLY (Eds.), Hemophilia in the child and adult (Ch. 12, pp. 213–226). New York: Raven Press.

SORENSEN RU, NEWMAN AJ, & GORDON EM (1985). Psychogenic purpura in adolescent patients. Clin Pediatr 24:700–704.

STAFNE WA, & MOE AE (1951). Hypoprothrombinemia due to dicumarol in a malingerer: a case report. Ann Intern Med 35:910–911.

STEINHAUSEN H (1976). Hemophilia: a psychological study in chronic disease in juveniles. J Psychosom Res 20:461–467.

STERN TA, PURCELL JJ, & MURRAY GB (1985). Complex partial seizures associated with Waldenström's macroglobulinemia. Psychosomatics 26:890–892.

STRACHAN RW, & HENDERSON JG (1965). Psychiatric syndromes due to avitaminosis B_{12} with normal blood and marrow. Q J Med 34:303–317.

STRACHAN RW, & HENDERSON JG (1967). Dementia and folate deficiency. Q J Med 36:189–204.

Swirsky-Sacchetti T, & Margolis CG (1986). The effects of a comprehensive self-hypnosis training program on the use of factor VIII in severe hemophilia. Int J Clin Exp Hypn 34:71–83.

Thomas JE, Koshy M, & Patterson L (1984). Management of pain in sickle cell disease using biofeedback therapy: a preliminary study. Biofeedback Self Regul 9:413–420.

Thornton WE, & Thornton BP (1978). Folic acid, mental function and dietary habits. J Clin Psychiatry 39:315–322.

Tishler PV, Woodward B, & O'Connor J (1985). High prevalence of intermittent acute porphyria in a psychiatric patient population. Am J Psychiatry 142:1430–1436.

Treiber F, Mabe PA, & Wilson G (1987). Psychological adjustment of sickle cell children and their siblings. Children's Health Care 16:82–88.

Varni JW (1981). Self-regulation techniques in the management of chronic arthritic pain in hemophilia. Behav Ther 12:185–194.

Walton JN, Kiloh LG, & Osselton JW (1954). The electroencephalogram in pernicious anemia and subacute combined degeneration of the cord. Electroencephalogr Clin Neurophysiol 6:45–64.

Weckman N, & Lehtovaara R (1969). Folic acid and anticonvulsants. Lancet 1:207–208.

Wells CE, & Duncan GW (1980). Other neurologic disorders important for the psychiatrist. In Neurology for psychiatrists (Ch. 14, pp. 203–222). Philadelphia: Davis.

Wells DG, & Casey HJ (1967). Lactobacillus casei CSF folate activity. BMJ 3:834–836.

Wetterberg L, & Osterberg E (1969). Acute intermittent porphyria: a psychometric study of twenty-five patients. J Psychosom Res 13:91–93.

Whittaker SRF, & Whitehead TP (1956). Acute and latent porphyria. Lancet 1:547–551.

Whitten CF, & Fischhoff J (1974). Psychosocial effects of sickle cell disease. Arch Intern Med 133:681–689.

Whitten CF, & Nishiura EN (1985). Sickle cell anemia. In N Hobbs & JM Perrin (Eds.), Issues in the care of children with chronic illness (Ch. 12, pp. 236–260). San Francisco: Jossy-Bass.

Williams I, Earles AN, & Pack B (1983). Psychological considerations in sickle cell disease. Nurs Clin North Am 18:215–229.

Woo J, & Cannon DC (1991). Metabolic intermediates and inorganic ions. In JB Henry (Ed.), Clinical diagnosis and management by laboratory methods (Ch. 8, pp. 140–172). Philadelphia: WB Saunders.

Zucker DK, Livingston RL, & Nakra R (1981). B_{12} deficiency and psychiatric disorders: case report and literature review. Biol Psychiatry 16:197–205.

32 Human immunodeficiency virus and other infectious disorders affecting the central nervous system

PAUL SUMMERGRAD, M.D., SCOTT L. RAUCH, M.D.,
AND REBECCA R. NEAL, M.D., M.S.W.

Infectious diseases represent an important cause of neuropsychiatric disturbance. A wide variety of viral, bacterial, and fungal, and parasitic organisms can afflict the central nervous system (CNS), contributing greatly to human suffering. For many patients, agitation, confusion, or lethargy may accompany the early stages of CNS infection and may bring the patient to psychiatric attention. Elderly patients with mild, stable dementias may become agitated or delirious after developing a urinary tract infection, pneumonia, or dehydration associated with fever. Such patients are commonly seen in general hospitals. Any comprehensive discussion on the neuropsychiatric impact of infectious diseases must include these disorders and therefore would be lengthy.

Important as such conditions are, we limit ourselves, in this chapter, to four disorders, three of which are of unquestioned infectious etiology: human immunodeficiency virus infection; syphilis; Lyme disease; and one of possible infectious etiology, the chronic fatigue syndrome. These disorders are likely to bring patients to medical-psychiatric attention, or they may be areas of intense concern for patients seen by psychiatrists. We review information on their epidemiology, pathogenesis, clinical description, diagnosis, and treatment, and on clinical dilemmas in the care of patients with these disorders.

HUMAN IMMUNODEFICIENCY VIRUS INFECTION

Infection with the human immunodeficiency virus (HIV), the causative agent of acquired immunodeficiency syndrome (AIDS), represents one of the major public health problems of this era. The expanding burden and changing demography of HIV infection, its association with a wide variety of other infectious disorders, its pleomorphic presentation, and the high frequency of neuropsychiatric illness associated with it make detailed knowledge of HIV infection essential for the psychiatrist.

The acquired immunodeficiency syndrome was first described in 1981, after outbreaks of a heretofore rare infectious disorder, *Pneumocystis carinii* pneumonia, and a previously rare vascular tumor, Kaposi's sarcoma, in young homosexual men in New York and California (Friedman-Kien et al., 1981; Gottlieb et al., 1981; Siegal et al., 1981). Although initial reports focused on these and other symptoms of immune compromise, it rapidly became clear that patients with AIDS developed a high frequency of neuropsychiatric illness (Nurnberg et al., 1984; Snider et al., 1983). This illness burden was distributed in four areas: (1) psychiatric disturbances, particularly depression, but also affective and schizophreniform psychoses presumably directly related to effects of the virus on the CNS; (2) a chronic dementing disorder, variously called AIDS or HIV encephalopathy, or the AIDS-dementia complex, also related to the direct CNS effects of the virus; (3) a large number of other opportunistic infections and malignancies of the CNS associated with significant neuropsychiatric symptoms; and (4) psychological reactions to the risk or actuality of HIV infection and its consequences (Faulstich, 1987; Forstein, 1984; Navia, Jordan, & Price, 1986).

In this section, we focus on the neuropsychiatric consequences of HIV infection, especially the AIDS-dementia complex, and other psychiatric syndromes directly related to HIV infection. We review other opportunistic CNS processes, especially as they can confound the evaluation and management of patients with HIV infection. Psychological reactions to HIV infection and their management are also discussed (Dilley & Forstein, 1990). Finally, we review issues related to serologic testing, psychopharmacologic management of HIV infection, and emergent medical-psychiatric management dilemmas involving patients with HIV-related illness.

Epidemiology

Over 2 million Americans are believed to be infected with the HIV (Rubin, 1988).

At first, HIV infected individuals were diagnosed as having AIDS on the basis of infections or malignancies indicative of significant immune compromise. More recently, neuropsychiatric illness, that is, HIV encephalopathy, also known as the AIDS-dementia complex, has also been included in Centers for Disease Control (CDC) case surveillance definition for AIDS (CDC, 1989e; Rubin, 1988). Estimates of the number of Americans likely to meet diagnostic criteria for AIDS have varied widely. In any event, large numbers of HIV-infected patients are likely to present for medical-psychiatric evaluation over the years.

Initially, the epidemic appeared concentrated among white homosexual men, but in recent years the rate of new cases has leveled off in this population. Increasingly, HIV infection has become a disorder of poor, minority group men and women who often are burdened with other health problems, and it is strongly associated with intravenous and other drug abuse. The routes of transmission have remained constant, however. The routes remain sexual contact, including vaginal intercourse and anal intercourse, congenital acquisition, and transfusions or blood product exchange related either to medical treatment or to intravenous drug abuse. The role of health care workers in transmitting the virus to uninfected patients is unclear, although the cases of patients who were possibly infected by their dentists have been reported (CDC; JAMA, MMWR, 1991). There is no evidence for casual transmission of the virus outside these routes (CDC, 1989e; Rubin, 1988).

Neuropsychiatric Manifestations

A wide variety of neuropsychiatric manifestations of the infection have been reported. Of them, what has been referred to as the AIDS-dementia complex (ADC) is probably the most important, in terms of clinical burden and frequency (Gabuzda & Hirsch, 1987; Levy et al., 1985; McArthur 1987; Navia, Jordan, & Price, 1986; Snider et al., 1983). As Perry observed, there are differences between CDC criteria for ADC and DSM-III-R criteria for Organic Mental Disorder, with criteria for the latter having been drawn more closely around familiar psychiatric syndromes (S. W. Perry, 1990).

The AIDS-dementia complex is characterized by evidence of an insidious onset dementia, difficulties of motor function, and affective and behavioral disturbances. As described by Navia et al. in their review of 46 patients at Memorial Sloan Kettering in New York, patients with ADC present initially with memory impairment and poor concentration. They develop increasing difficulties with attention and a feeling of slowed mental functioning. Many patients have early motor difficulties, including leg weakness, loss of balance, and impaired coordination. Patients frequently have evidence of social withdrawal, apathy, or depressed mood. Anecdotally, patients have been reported as having irritability, lability, or frank affective psychoses early in their course. Navia et al. and others have reported these symptoms as the first manifestation of HIV infection in up to one-third of reported cases (Navia, Jordan, & Price, 1986). Occasionally, no other evidence of immune compromise is found at presentation (Navia & Price, 1987).

In most cases the disorder progresses along with evidence of increasing immune compromise in an indolent but steady fashion. In a few cases, the disorder may have an abrupt onset or accelerate after a previously slow course (Morgan, Clark, & Hartman, 1988; Navia & Price, 1987; Navia, Jordan, & Price, 1986). As the disorder progresses, many patients become profoundly ill with global dementia, marked slowing of mental processes, incontinence, hypertonia, hyperreflexia, and weakness. Terminally, patients may become paraplegic and remain in a bedridden, mute state. Some patients, however, continue to have prominent psychotic symptoms including visual hallucinations and delusions. Patients may have periods of agitation and may develop myoclonus or seizures (McArthur, 1987; Navia et al., 1986).

It is estimated that 40 to 70% of patients with AIDS develop ADC during the course of their illness (Levy, Bredesen, & Rosenblum, 1988; McArthur, 1987; Navia, Jordan, & Price, 1986; Petito et al., 1986). However, exact percentages of those who do develop ADC are not known nor are predictors of who will progress to a severe debilitated state. Symptoms of ADC do not necessarily parallel the systemic symptoms of AIDS.

It has been suggested that ADC represents a form of

subcortical dementia. Subcortical dementias, first described by Martin Albert, are characterized by memory disturbance and slowed mental function, often with significant affective symptoms bearing some similarity to depressive disorders (Albert, Feldman, & Willis, 1974). Subcortical dementias have been reported in a variety of neuropsychiatric conditions, including Parkinson's disease, Huntington's disease, and Binswanger's disease (Cummings & Benson, 1984, 1988; Summergrad & Peterson, 1989).

One area of ongoing intense research interest has been the question of whether patients who are HIV positive, but otherwise asymptomatic, show evidence of cognitive disorder. Attempts to resolve this issue have been complicated by studies that have suffered from poorly selected patient and control populations, the confounding variables including drug and alcohol abuse and head trauma, inconsistent controls for the level of intercurrent and related medical illness, and failure to use similar batteries of neuropsychological tests.

A number of reports with improved methodologic design suggest that there is little evidence of cognitive disorder in medically asymptomatic patients who are HIV-seropositive. (McArthur et al., 1989). McArthur found no evidence of statistically significant differences between seronegative and asymptomatic seropositive subjects in regard to neurologic symptoms, neurologic examinations, or magnetic resonance imaging (MRI) of the brain. Cerebrospinal fluid (CSF) examination of seropositive patients with and without neurologic symptoms showed mild pleocytosis, elevated protein, elevated immunoglobulin (Ig), and CNS synthesis of IgG directed against HIV-1; thus the CSF findings did not correlate with neurologic symptoms.

Koralnick et al. (1990), examining *symptomatic* HIV-positive patients, found frequent abnormalities on electroencephalography (EEG): slowing of fundamental activity, poor anterior spread of alpha rhythms, unusual anterior theta activities, and diffuse slow theta dysrhythmias of low amplitude. Additionally, statistically significant abnormalities in otoneurologic electrophysiologic parameters suggested subclinical lesions in the auditory, vestibuloocular, and ocular pathways.

Evidence from several studies supports the presence of neuropsychological abnormalities in patients with ARC and AIDS. Whether medically asymptomatic seropositive patients have evidence of abnormalities on neuropsychological evaluation remains a subject of active investigation. Several studies have found no evidence of neuropsychological abnormalities or MRI abnormalities comparing asymptomatic seropositive patients and seronegative controls (Miller, Selves, & McArthur, 1990; Selves, Miller, & McArthur, 1990; Sidths & Price, 1990). However, Stern et al. (1991) and Lunn et al. (1991) have reported abnormalities in asymptomatic seropositive patients on neuropsychologic testing.

Other Direct CNS Manifestations

A variety of other conditions are associated with the direct effects of HIV infection.

Aseptic Meningitis

Aseptic meningitis is a flu-like illness that, at the time of seroconversion, is associated with rash, lymphadenopathy, headache, and stiff neck. The illness is generally brief; CSF reveals a pleocytosis, generally mononuclear, with mild elevations of protein and normal glucose (Gabuzda & Hirsch, 1987; McArthur, 1987).

Vacuolar Myelopathy

Patients with vacuolar myelopathy often present with progressive gait ataxia, leg weakness, hyperreflexia, and signs referable to the posterior spinal columns. This picture is usually associated with dementia; pathologically there is evidence of vacuolation of the white matter of the posterior and lateral columns (Gabuzda & Hirsch, 1987; McArthur, 1987).

Peripheral Neuropathy

A wide variety of peripheral neuropathies related to HIV infection have been described, including distal polyneuropathy, often causing paresthesias in the hands and feet. Some patients have also developed demyelinating polyneuropathies or mononeuritis multiplex. Additionally, patients may develop a Guillain-Barré type of ascending polyneuritis but without the albuminocytologic dissociation classically seen in typical Guillain-Barré patients (Gabuzda & Hirsch, 1987; McArthur, 1987).

Pathogenesis

A range of studies suggests that HIV itself is the cause of ADC. Studies of affected areas of human brain have shown evidence of HIV by immunofluorescent in situ hybridization, Southern blotting, and antigen expression studies. Elevated intrathecal synthesis of HIV-specific IgG has been reported, as have isolation and growth of the virus from CSF-affected patients (Bukasa et al., 1988; Gabuzda et al., 1986; Ho et al., 1985; Koenig et al., 1986; Levy et al., 1985; Shaw et al., 1985). Historically, there was controversy about the contribution of cytomegalovirus (CMV) infection to the development of ADC. However, most authorities currently agree that the pattern and type of neuropathologic involvement, as well as the evidence from antigen and molecular studies, do not support CMV as the causative agent of ADC (de la Monte et al., 1987).

Although a number of cell types are affected in the CNS, monocyte/macrophage lines are of particular importance (de la Monte et al., 1987; Koenig et al., 1986). It is unclear whether there is direct viral spread to the CNS (as during the period of aseptic meningitis) or whether the primary route of entry to the CNS is via monocytes and macrophages. The mechanism by which HIV causes neurologic impairment remains unclear. A variety of etiologies have been advanced, including cell death via calcium-channel-mediated neurotoxicity (Dreyer et al., 1990), vascular insufficiency in the CNS microcirculation (Smith et al., 1990), and elevation of a neurotoxic tryptophan metabolite, quinolinic acid (AIDS dementia, 1990). Other mechanisms may include direct cytoxicity or T cell-modulated cytotoxicity, neurotoxic monokine production, alterations in the blood-brain barrier, and other CNS (CMV and JC) virus infections (Ho et al., 1989). Histologic studies have revealed a subacute encephalitis in 30 to 90% of patients dying with AIDS. The range of frequencies is most likely a function of sampling differences in studies, as well as variable pathologic criteria. Most studies report evidence of white matter pallor, demyelination, microglial nodules, and multinucleated cells, primarily in subcortical gray and white matter. Not all patients with pathologic abnormalities have had clinical evidence of neuropsychiatric disorders during life (Anders et al., 1986; de la Monte et al., 1987; Navia et al., 1986; Petito et al., 1986).

Neuroradiologic and Neurophysiologic Features

Many patients with ADC have evidence of cortical and central atrophy on computed tomography (CT) scanning. MRI reveals evidence of periventricular white matter abnormalities in some cases, although these abnormalities did not distinguish individuals who were HIV-seropositive from seronegative controls (McArthur et al., 1989). As noted above, Koralnick found evidence of increased EEG abnormality and abnormalities on electrophysiologic otoneurologic examinations in HIV-seropositive patients compared with controls (Koralnick et al., 1990). Several reports have noted defects in cortical and subcortical regions using nuclear magnetic resonance (NMR) spectroscopy, single photon emission CT (SPECT), and positron emission tomography (PET). CNS glucose metabolism, assessed by PET scan, may improve following zidovudine therapy (Bottomly, Hardy, & Cousins, 1990; Brunetti et al., 1989; Editorial, 1989a; Pohl et al., 1988).

CSF

Many patients who are HIV-seropositive show CSF findings of pleocytosis, increased protein, isolatable HIV, and intrathecal synthesis of HIV-specific antibodies. However, these abnormalities have not distinguished asymptomatic seropositive patients from seropositive patients with neuropsychological or neurologic impairment (McArthur et al., 1989). Other markers of active, symptomatic HIV CNS infection are being sought at present.

Psychiatric Disturbances

A wide variety of psychiatric disorders have been reported in association with HIV infection, including affective and schizophreniform psychoses, paranoia, mania, and irritable-anxious states. Most of these reports have been case studies and case series (Beckett et al., 1987; Buhrich, Cooper, & Freed, 1988; Cummings et al., 1987; Faulstich, 1987; Holland & Tross, 1985; Joffe et al., 1986; Navia, Jordan, & Price 1986; Perry and Jacobsen, 1986; Rundell, Wise & Ursano, 1986). In a review of the published literature on HIV-infected patients with psychotic illness, Harris et al., found 31% of all such patients had psychotic symptoms as their presenting manifestation of HIV infection (Harris, Gleghorn, & Jeste, 1989).

In addition to these more severe conditions, a number of other disorders have also been noted. A high frequency of depression and dysthymic symptoms have been reported in patients with HIV infection (Dilley et al., 1985; Holland & Tross, 1985; S.W. Perry & Tross, 1984). Others have noted frequent anxiety disorders and obsessional concerns regarding the risk of AIDS in traditional risk groups and in others at significantly lower risk (Brotman & Forstein, 1988; Jenike & Pato, 1986).

Studies attempting to control for the premorbid level psychiatric disturbance have been performed with some groups of homosexual men. Atkinson et al. (1988) found high rates of anxiety and depressive disorder in seropositive homosexual men at various stages of illness, with especially high levels of anxiety in men with AIDS-related complex (ARC). Interestingly, they also found high lifetime prevalence rates for anxiety and depression in homosexual men who were seronegative for HIV. A similar finding has been reported by Williams et al. (1991). It is unclear whether these findings reflect the general impact of the AIDS epidemic on homosexual men, differential rates of anxiety and depressive disorder among homosexual men compared to the general population, or other methodologic factors specific to these studies. Ostrow et al. (1989), in a multicenter study, reported an association between psychological distress and self-reports of nonspecific but possibly HIV-related physical symptoms, such as fatigue, new skin rash, night sweats, or a new, dry cough. In this study, neither self-report of physical symptoms nor measures of psychological distress differentiated seropositive from seronegative subjects. Dew et al. found more depression, anxiety, anger, and hostility in HIV-positive hemophiliac men than in hemophiliacs who were HIV-negative. Past personal or family history of psychiatric illness, low social support, low educational level, and loss of or diminished social and family support were associated with increased symptomatology and accounted for much of the variance between the HIV-positive and HIV-negative subjects (Dew, Ragni, & Nimorwicz, 1990).

Psychological Aspects of HIV Testing

Serologic testing is a highly stressful event for individuals concerned enough about their risk for HIV infection to undergo it. Many patients being tested for HIV, however, have preexisting psychiatric disorders, and many actually feel better once their HIV status is ascertained. Perry et al. found high lifetime rates of mood disorder and drug abuse among individuals in a longitudinal study of response to HIV antibody testing (Perry et al., 1990a). They also found *decreased* rates of anxiety and depression after serologic testing even among individuals found to be seropositive. (Perry et al., 1990b). In another report on the follow-up of patients who had undergone serologic testing, Perry et al. (1990c) found *decreasing* rates of suicidal ideation among seropositive and seronegative individuals during the 2 months after testing. Continued suicidal ideation was associated with evidence of a depressive disorder.

These studies suggest caution in the attribution of all psychiatric symptoms to direct CNS effects of HIV infection (Atkinson et al., 1988; Ostrow et al., 1989). Notwithstanding, men with AIDS do represent a group at increased risk for suicide, even when compared with men who suffer from other chronic diseases (Glass, 1988; Marzuk et al., 1988).

Other CNS Processes in HIV-Related Illness

Not all neuropsychiatric syndromes in patients with AIDS are direct results of HIV infection or manifestations of primary psychiatric illness. In approximately 30% of patients other opportunistic infections or malignancies complicate the course of the HIV infection. Among them are cerebral toxoplasmosis, primary CNS lymphoma, cryptococcal meningitis, herpes simplex encephalitis, CMV infection, progressive multifocal leukoencephalopathy, tuberculous meningitis, and infections with atypical mycobacterial species (Dalakas, Wichman, & Sever, 1989; Gabuzda & Hirsch, 1987; Levy et al., 1985; McArthur, 1987; Snider et al., 1983). Other viral, fungal, and parasitic disorders have been reported to afflict patients with AIDS. Additionally, patients with Kaposi's sarcoma metastatic to brain have been reported.

Less well recognized is that patients with AIDS may also suffer from cerebrovascular complications, including intracerebral hemorrhage, thromboembolic disease, hemorrhagic infarction, and infarction secondary to nonbacterial thrombotic endocarditis (Dalakas, Wichman, & Sever, 1989; Levy et al., 1985; Snider et al., 1983). It is beyond the scope of this chapter to review these disorders in detail, but the psychiatrist must be alert to the possible development of these other AIDS-related CNS illnesses. Changes in mental status, particularly increased confusion, should not be presumed to be a consequence of direct HIV CNS effects or metabolic phenomena alone. In patients with many of the above disorders, levels of alertness may diminish and patients may develop focal neurologic signs including hemisyndromes. Patients who develop acute changes in mental status require a neurologic examination and brain imaging, followed in most cases by examination of the CSF, unless the presence of an intracranial mass or severe thrombocytopenia contraindicate a lumbar puncture. Many of these disorders appear as focal lesions on CT scanning (e.g., single or multiple ring-enhancing lesions in cerebral toxoplasmosis, and areas of white matter hypodensity in progressive multifocal leukoencephalopathy). Patients with CSF lymphoma generally show more uniform enhancement on contrast CT studies. Treatment of opportunistic infections, once identified, usually requires collaboration between a neurologist and an infectious disease specialist.

Serologic Studies

More than 90% of patients with AIDS or HIV infection have evidence of antibodies to HIV, usually within 6 months of infection. Tests generally available for the detection of AIDS antibodies are the enzyme-linked immunosorbent assay (ELISA), which is used for initial screening, and the Western blot, which is used to confirm positive ELISAs.

The ELISA has a sensitivity of 99.7% and a specificity of 98.5% (CDC, 1989a). False-positive ELISAs are most common in patients with a history of multiple prior transfusions, organ transplants, and previous pregnancies. Western blotting has a sensitivity of 99.3% and a specificity of 97.8% but can generally screen out weakly reactive ELISAs (CDC, 1989a). The Centers for Disease Control (CDC) recommends that no results of serologic testing be reported to patients unless a screening test such as the ELISA has been reactive on two or more tests and those initial results have been confirmed by the Western blot. Occasionally, the Western blot test may be reported as indeterminate, reflecting a partial or atypical banding pattern. In such cases, some patients may be in the process of seroconverting. However, if the Western blot remains indeterminate on repeated testing for 6 months, that person is "almost certainly *not* infected with HIV-1" unless other clinical or historical factors suggest otherwise (CDC, 1989a). As with all diagnostic tests, the predictive value of ELISAs and Western blotting are diminished in a population of low disease prevalence. It is still unclear what percentage of patients actually infected with HIV are seronegative on repeated antibody testing. The possibility should be considered in patients at high risk who have symptoms consistent with HIV infection (CDC, 1988a, 1989a; Rubin, 1988).

Other molecular, antigen capture, and antibody detection methods are available or under development but are not in routine clinical use. Increases in the P24 antigen level appear related to disease activity, but it is not a particularly useful screening measure (Hjelle & Busch, 1989; Jackson, 1990). Serial T cell measures are recommended for initiation of prophylaxis against *Pneumocystis carinii* pneumonia, but no such recommendations regarding T cell measures and neuropsychiatric illness are currently available (CDC, 1989b). Issues related to the diagnosis and treatment of syphilis in HIV-infected individuals are reviewed in the section under syphilis below.

Approach to the Patient with Possible HIV Infection

When should a psychiatrist consider initiating an evaluation for possible HIV neuropsychiatric illness, and what tests should be performed? Answers to these questions depend in part on the desire of patients to know their HIV status, as well as the clinical urgency of having accurate information.

Although the epidemiology of AIDS and HIV infection may be of some benefit in these discussions, any patient with a history of frequent multiple partner heterosexual or homosexual contact, drug abuse, or other sexually transmitted disease can be considered to be at increased risk for HIV infection. The diagnosis also should be considered in patients who present without a personal or family history of psychiatric illness, who have new-onset psychosis, mania, and major depression accompanied by cognitive impairment or focal neurologic deficits.

All such patients should be specifically asked about symptoms of forgetfulness, mental slowing, difficulty concentrating, and minor neurologic complaints such as clumsiness, diminished fine motor capacity, or gait disturbance. Patients who show evidence of such abnormalities by either history or examination should be considered for further evaluation. A careful history of changes in physical health should also be obtained with special reference to symptoms of fatigue, fever, night sweats, diarrhea, skin changes, swollen lymph glands, new dry cough, weight loss, and shortness of breath. Careful physical, neurologic, and bedside cognitive examinations are clearly important. Patients with established histories of psychiatric illness should be assessed clinically for HIV-related illness if they undergo new cognitive or neurologic changes.

If patients have symptoms that arouse concern, they should be asked to have HIV serologic testing; consent should be obtained and appropriate pre- and posttest counseling provided. Perry et al. (1991) have found benefit from stress prevention training (formal training in coping strategies) as part of the counseling procedure for patients later found to be seropositive (Perry et al., 1991). In addition, a CBC and platelet count should be obtained to screen for lymphopenia, anemia, and thrombocytopenia. Serologic studies for hepatitis and syphilis should be done, as these disorders are associated with significant neuropsychiatric abnormalities, as well as being associated with increased risk for HIV infection. Likewise, other general laboratory measures such as liver function tests, thyroid function tests, and screening chemistries should be obtained because of the high rate of occult medical illness among psychiatric patients. (See Chapter 7.)

If a patient with new or increased psychiatric symptoms is HIV-positive, neuroradiologic testing should begin with an MRI. The MRI is more sensitive than CT to the periventricular white matter abnormality seen in the ADC, and it can also screen for mass lesions associated with CNS lymphoma, toxoplasmosis, or brain abscess. If mass lesions are seen on MRI, further evaluation by contrast-enhanced CT or gadolinium-enhanced MRI may be of benefit in defining their etiology. There should be prompt infectious disease and neurologic consultation.

If patients do not have a mass lesion that precludes

CSF examination, they should undergo lumbar puncture. In addition to the usual CSF cell count, protein, and glucose, a CSF VDRL is performed, as are tuberculosis, bacteriologic, and fungal cultures. CSF cytology should be performed if malignancy is suspected. CSF HIV cultures, oligoclonal bands, and IgG levels have all revealed abnormalities in patients with CNS HIV involvement. However, in one study, McArthur et al. (1989) found that increased rates of pleocytosis, elevated IgG, oligoclonal bands, and positive CSF HIV culture did not distinguish men with clinical neuropsychiatric abnormalities from men who were seropositive but asymptomatic. Thus the CSF examination is for identification of specifically treatable infectious and neoplastic complications, *not* for confirmation of the presence of CNS HIV involvement.

Neuropsychological testing should be performed to provide evidence of current cognitive status, as well as to provide a baseline for later studies. Because of the variability in the current literature, no specific tests have been uniformly used to assess cognitive function. However, timed tests of subcortical or frontal lobe function such as Trail Making A and B, the cancellation test, and the grooved pegboard test have been reported as abnormal in patients with HIV infection. Also of some value may be the WAIS-R, particularly the Verbal and Digit Span subtest (Marotta & Perry, 1989; Miller, Selves, & McArthur, 1990). The NIMH Work Group on Neuropsychological Issues in HIV infection and AIDS has recommended extended and abbreviated versions of the NIMH neuropsychological battery in an attempt to standardize the evaluation of patients with HIV neuropsychiatric abnormalities (see Table 32-1).

These tests are designed to pick up early and subtle cognitive changes, especially in the subcortical functions of attention and processing speed. Attempts are under way to evaluate the results in a clinical (and not just group norm) fashion. It is being done to reduce the chance of missing abnormalities in patients with stage II or III disease. Attempts to control for intercurrent psychiatric illness and other confounding variables are included in the assessment instruments. The extended version, which takes 7 to 9 hours to complete, is clearly designed for specialized research settings. The abbreviated battery, which takes 1 to 2 hours to perform, hopefully can be integrated into clinical services and a national database (Adams & Heaton, 1990; Butters et al., 1990).

When obtaining baseline measures of neuropsychological function, it is important to gather data on daily functioning from patients, relatives, significant others, and close friends. Such data can be particularly important for discharge planning. In a study of psychiatric inpatients with HIV-related illness at San Francisco General Hospital, none of the 18 patients with dementia could be discharged to outpatient care (Baer, 1989). Follow-up studies for patients with documented HIV neuropsychiatric impairment should be obtained if there is deterioration in clinical status as expressed by impaired activities of daily living, new cognitive difficulties, new or recurrent psychiatric difficulties, or focal neurologic events.

TABLE 32–1. *Initial Evaluation of the Psychiatric Patient with Possible HIV-Related Illness*

Careful medical and neurologic history and examination
HIV serology; ELISA—if reactive two times, confirm with Western blot
VDRL; hepatitis serology
CBC, platelets, BUN, creatinine, electrolytes, liver function tests
Thyroid tests
MRI scan of the head—if evidence of mass lesions, patients should have contrast-enhanced CT or MRI
Lumbar puncture with studies, including cell count, protein, glucose, CSF, VDRL, IgG, oligoclonal bands, HIV, CSF culture, and other bacteriologic, cytologic, and fungal studies as indicated
Neuropsychological testing
WAIS-R; Verbal and Digit Span subsets
Trail Making A and B
Cancellation Test
Grooved Pegboard Test
Assessment of capacity for activities of daily living
Follow-up examination, including repeat neuropsychologic and neuroimaging testing, if the history suggests cognitive or functional decline

Psychotherapeutic Issues

When HIV-positive individuals seek psychiatric treatment, there are several important questions to be included in the initial interview. The clinician must ask how long the patient has known he or she was infected, when the infection might have occurred, by what means, and if high risk behavior is continuing. Additional pertinent data include a careful assessment of the patient's support network, which includes identification of individuals who know about the infection, if any, and the patient's perception of their response to the news. It is also important to determine if there have been any symptoms and if the patient has a primary care physician who is well informed about HIV. It is not uncommon to encounter patients who can offer extensive detail about their medical conditions including their most recent CD4 counts. Given the prevalence of HIV infection and clustering of cases in high risk groups, the clinician should be sensitive to the likelihood that the patient has friends who have died or are currently ill because of the virus.

Much of the routine content of the initial psychiatric interview takes on a specific meaning when the patient has HIV infection. The symptoms and history of present illness must be assessed in light of what is known about the natural history of the infection as well as the patient's premorbid ego maturation and his or her response to the catastrophe of the infection.

Although we lack specific epidemiologic research on possible correlations between stage of HIV infection and psychiatric diagnosis, clinicians and researchers have made several important observations. The initial emotional response to diagnosis of the infection depends on previously existing coping skills, the psychiatric history, and the nature of the existing social network. Predictably, there is an initial period of emotional upheaval, but its persistence beyond a few months seems determined by the patient's premorbid functioning and the presence or absence of effective supports. There are some data to suggest a greater incidence of psychological morbidity among patients with CDC stage II and III infection. Possible explanations include the cumulative effects of psychosocial stressors in concert with the development of physical manifestations of disease. It is not known if the psychiatric symptoms reflect specific subcortical organic pathology.

As noted above, a good history of cognitive function is an indispensable part of the evaluation. The psychiatrist needs to inquire specifically about slowing of thought processes as well as of motor functioning. Forgetfulness and inattention are common complaints; changes in speech and writing are less common but nevertheless can occur. In one case, stuttering was the first indication of an HIV-related CNS lesion (C. Frank, personal communication, 1991).

The risk of suicide in the HIV-infected patient merits special attention. Marzuk and colleagues have documented unexpectedly high rates of suicide among CDC stage IV individuals (Marzuk et al., 1988). Dilley and Forstein (1990) have suggested that suicidal ideation may represent the patients' search for a means of control in their lives as well as an attempt to cope with grief and loss. The clinician who treats HIV-infected people is frequently confronted with the dilemma of how to respond to the patient's thoughts of "rational suicide" as the illness progresses. From the onset, the clinician must attend to signals of hopelessness or detachment and inquire specifically about suicidal thoughts and intent, regardless of the course of other psychiatric symptoms. It is also prudent to attend to evidence of noncompliance with appropriate medical care (i.e., decisions to avoid proved treatments such as AZT). Noncompliance may include ongoing high risk behaviors including indiscriminate sexual activity and continued intravenous drug abuse. At worst, such behavior represents explicit suicidal intent.

To date, there has been no systematic study of the efficacy of psychotherapy for treatment of HIV infection and coexisting psychopathology. Clinical experience (R. R. Neal, unpublished data, 1991) suggests that the patients' request for psychiatric assistance provides a unique opportunity to help with unresolved conflicts in their lives and to affect their ability to cope with the evolution of the illness in a positive manner. Attention to the issues discussed above may reassure patients that the clinician is knowledgeable and capable of understanding their unique situation. Comfort with medical and psychosocial aspects of the illness equips the clinician with several possibilities for active interventions, which are often appreciated by the individual who may have inadequate or inaccurate information about HIV itself or about the available medical and psychosocial resources.

Consciously or unconsciously, the HIV diagnosis raises the idea of a finite life-span. It not only mobilizes anxiety about dependency and loss of control, it provides motivation to tend to one's unfinished business in life. In such a situation, the clinician has the mandate to offer a treatment plan that considers the need to contain unwanted impulses and to support growth and maturation. Depending on the patient's request and the psychiatric assessment of need, the clinician must maintain a flexible stance that utilizes referrals to medical, inpatient psychiatric, financial, housing, and family services as well as the offer of short- or long-term psychotherapy and psychopharmacologic treatment.

Pharmacologic Treatment

Treatment for ADC is limited at present. There is some evidence that AZT can reduce progression or induce partial remission of the cognitive and neurologic symptoms. There is, at present, no clear evidence to suggest that AZT therapy can prevent asymptomatic patients or patients with mild neuropsychiatric symptoms from developing ADC. AZT has been reported to cause confusion, anxiety, tremulousness, seizures, and a Wernicke's encephalopathy-like picture among patients receiving it; hence it may be limited in its utility because of these and hematologic side effects. Many of the other medications that are used to treat symptoms of HIV infection or AIDS have been associated with encephalopathic changes, including acyclovir and ganciclovir. Amantadine has been associated with anxiety; and DDI (dideoxyinosine) and DDC (dideoxycytidine) both have been associated with painful peripheral neuropathy. Interferon has been associated with fatigue and depression (Bridge et al., 1989; Drugs for viral infection, 1990; Schmitt et al., 1988; Yarchoan et al., 1987, 1988, 1989).

Treatment of neuropsychiatric disorders in HIV-positive patients is at present empiric. Some patients have been reported to show positive response to psychostimulants, including improved cognitive function (Fernandez, Levy, & Galizzi, 1988; Fernandez, Levy, & Mansell, 1989; V. F. Holmes, Fernandez, & Levy, 1989). The latter finding may have been related in part to increased speed of performance, findings also seen with other subcortical dementias treated with antidepressant or stimulant agents. Dextroamphetamine appeared to be of less utility than methylphenidate because of the higher frequency of dyskinesias associated with its use (Fernandez, Levy, & Galizzi, 1988). The benefits of stimulant use must be weighed carefully against the possibility that these drugs may be abused, especially by patients with a drug abuse history. Improved mood and speed of cognitive performance appear to be useful endpoints when monitoring continued prescription of stimulant agents. Anecdotal reports have suggested that AIDS patients may have an increased susceptibility to tricyclic-related side effects including confusion. This state was confirmed in a study by Fernandez, Levy, & Mansell (1985), who noted the high rates of confusion developing among patients with AIDS, especially those taking the more potent anticholinergic agents, such as amitriptyline. Notwithstanding, Rabkin and Harrison (1990) reported successful use of imipramine in depressed men with HIV infection. Studies using agents with less anticholinergic activity are currently under way. In the Fernandez study cited above, depressed patients with ARC had the greatest treatment response. Schaerf et al. (1989) have reported successful ECT treatment of four HIV-positive depressed inpatients, one of whom had AIDS.

Treatment of anxiety disorders and psychosis in patients with HIV infection should be guided at present by the cautious use of standard agents for these syndromes. Anecdotal evidence suggests that patients may be sensitive to the side effects of pharmacologic agents: hence doses should initially be low and titrated upward cautiously, unless an emergent state requires more intensive pharmacotherapy. There have been some reports of neuroleptic malignant syndrome and increased extrapyramidal symptoms when neuroleptics are used in patients with HIV infection. The likelihood of these reactions may be greater in patients with AIDS, especially if they are treated with high potency neuroleptics (Baer, 1989; Bernstein & Scherokman, 1986).

Special Considerations with HIV-Related Illness

A number of ethical, legal, and clinical dilemmas arise during the treatment of HIV-infected patients. Adler and Beckett (1989) have reviewed many of these issues in outpatient psychotherapy settings. Binder (1987) and others have reviewed difficulties that arise on inpatient units. Of particular concern in the latter setting have been issues of countertransference, antibody testing, confidentiality, and transfer of patients. Sacks et al. (1990) have noted a high rate of HIV risk factors among acute psychiatric inpatients. In particular, bipolar disorder and, to a lesser extent, borderline personality disorder were associated with history of high risk behavior.

Psychiatrists in hospital settings are likely to be concerned with the ability of inpatients to control bodily secretions and with the management of AIDS-related medical illness. Particularly problematic are HIV-infected patients or patients in high risk groups who attempt to harm themselves or others by exposure to blood. Also of concern are patients with impaired judgment secondary to neuropsychiatric illness who engage in high risk behavior such as sexual contact with HIV-positive patients or patients in high risk groups.

Although these problems can occur on open units, they are more likely to occur on locked units, especially those in state hospitals (Cornos et al., 1989; Horworth et al., 1989). Sophisticated medical care often is unavailable in such settings, making the medical management of HIV-positive patients problematic. Given the demographics of state hospitals with their high rate of homeless and drug-abusing patients, and the association of certain diagnostic groups (borderline, bipolar) with impulsive high risk behavior, these settings may harbor a disproportionate number of high risk patients. Studies suggesting both a high level of severe mental illness, substance abuse, and medical illness among homeless patients raise the possibility that HIV-positive patients may be likely to be overrepresented in state hospitals (Breakey et al., 1989; Susser, Struening, & Conover, 1989). Psychiatrists working in such facilities must be alert to those issues. Related public health issues, such as the increased rate of tuberculosis among HIV-positive patients, are of concern as well (Ruder et al., 1989; Theuer et al., 1990). The recent increase in the national prevalence of tuberculosis (TB) may be accounted for in large measure by the increased rate of TB in patients with HIV infection. In contrast to TB in the non-HIV-infected patient, extrapulmonary TB (largely occurring in association with pulmonary TB) is more common, especially in patients who are immunocompromised. Sites of extrapulmonary involvement include lymph nodes, blood, bone marrow, genitourinary system and the CNS. Tuberculous abscesses, or tuberculomas, can be found in the CNS and may be demonstrated on CT as ring enhancing lesions or hypodense areas. Tuberculous meningitis also can occur.

Not all patients with extrapulmonary involvement have abnormal chest roentgenograms or positive tuberculin skin reactions. It is especially important in inpatient psychiatric settings that patients with suspected TB receive expert diagnostic assistance for their own well-being as well as for that of other patients who may be exposed to them. Such patients often require transfer to a medical service of a general hospital for adequate isolation and sputum collection for diagnosis and treatment. The CDC is currently recommending tuberculosis screening for patients at high risk of HIV infection and for residents of mental health facilities. (Barnes et al., 1991; Dooley et al., 1990).

SYPHILIS

Syphilis occupies a unique position in psychiatry. During the late nineteenth century, neurosyphilis was responsible for a significant number of admissions to psychiatric hospitals. The early twentieth century saw the identification of the causative agent, *Treponema pallidum*, and the introduction of ameliorative treatments. Syphilis was among the first common neuropsychiatric illnesses for which a specific organic etiology was demonstrated.

An understanding of the stages of syphilis is critical not only for clinical diagnosis but for determining the value of serologic and other diagnostic tests. The stages of syphilitic infection are reviewed below.

Epidemiology

In 1989 syphilis affected 18.4 per 100,000 persons in the United States (Rolfs & Nahashima, 1990). This rate has increased since a nadir was reached during the late 1950s (CDC, 1988b; Relman, Schoolnik, & Swartz, 1988a). During the early 1970s, the rate of syphilis increased primarily because of increased infection among homosexual white men. Since the development of the AIDS epidemic, rates of syphilis infection have decreased among this group. Rates, however, have been increasing significantly among minority group individuals, especially in inner city areas (CDC, 1988b; Relman, Schoolnik, & Swartz, 1988a). Rates among blacks have increased since 1986. In 1989 rates for blacks were 121.8 cases per 100,000 (Rolfs & Nahashima, 1990).

Syphilis is spread primarily via sexual contact with an infected individual, during the primary or secondary stages of infection. Blood transmission occurs rarely.

Pathogenesis

Syphilis is caused by the spirochete *Treponema pallidum*. The organism enters the skin after direct contact with an active syphilitic lesion, such as a chancre. In approximately 25 to 40% of cases, syphilis invades the CNS during the primary, secondary, or early latent stages (Holmes & Lukehart, 1987; Lukehart et al., 1988; Relman, Schoolnik, & Swartz, 1988a). It may occur as part of the widespread dissemination of the bacteria during the secondary stage.

Clinical Manifestations

Primary Syphilis

Primary syphilis occurs 3 to 6 weeks after contact with the organism. The usual symptom is a chancre at the point of entry and local lymphadenopathy. During this stage, *T. pallidum* can be directly demonstrated in lesions by darkfield examination, and patients are infectious (Holmes & Lukehart, 1987; Larsen & Beck-Sague, 1988; Relman, Schoolnik, & Swartz, 1988a).

Secondary Syphilis

Secondary syphilis occurs 6 to 8 weeks after the end of the primary stage, and its hallmark is a widely disseminated maculopapular rash over the trunk, extremities, and later the palms and soles. Other mucocutaneous lesions may also be present. Patients are often systemically ill during this period with fever, generalized lymphadenopathy, fatigue, sore throat, headache, and meningismus. Secondary syphilis occasionally is asymptomatic, especially if its rash is not noted. The frequency of asymptomatic secondary infection is unknown. The secondary stage persists for 2 to 6 weeks and is followed by the stage of latent syphilis (Holmes & Lukehart, 1987; Larsen & Beck-Sague, 1988; Musher, 1978; Relman, Schoolnik, & Swartz, 1988a).

Latent Syphilis

Latent syphilis is defined by a normal physical examination, absence of cardiac, neurologic, or CSF evidence of syphilis, a history of primary or secondary syphilis, and a positive specific (treponemal) serology. This period is divided into early latent syphilis, which is the first year after infection, and late latent syphilis thereafter. Patients whose CSF is normal (i.e., without pleocytosis, elevated protein, or positive VDRL) at 2 years after infection are at low risk of progression to neurosyphilis.

Late Syphilis

Late syphilis is caused by ongoing inflammatory disease, most usually in the aorta (syphilitic aortitis) or the nervous system (neurosyphilis), although other organs can also be involved. Neurosyphilis, which occurs in approximately 10% of patients, can take one of several forms.

Asymptomatic Neurosyphilis. Neurosyphilis is characterized by a normal neurologic examination in the face of evidence of ongoing meningeal inflammation: pleocytosis, elevated protein, and reactive VDRL. The use of lumbar puncture for early or possible asymptomatic neurosyphilis is reviewed in the clinical dilemma section, below. Some patients have clinical evidence of syphilitic meningitis during the early or latent stage. In this case the symptoms are those of acute meningitis, with headache and confusion possibly accompanied by cranial nerve abnormalities.

Meningovascular Syphilis. Meningovascular syphilis generally occurs 6 to 7 years after infection, although in the series of Adams and Victor (1989) from the Boston City Hospital, it occurred as early as 6 months and as late as 10 to 12 years after infection. Symptoms include stroke syndromes of subacute onset with a preceding encephalitic picture, including psychiatric disturbances such as lability or personality changes (Lishman, 1988; Simon, 1985; Swartz, 1984). There is significant inflammation of the meninges, as well as vasculitis, which can lead to vascular occlusion. Meningovascular syphilis should always be considered in a young person with new onset of stroke, especially in the face of positive syphilis serology.

Parenchymal Neurosyphilis. Parenchymal neurosyphilis has two major forms: general paresis (also known as the general paralysis of the insane, or dementia paralytica) and tabes dorsalis.

General Paresis. General paresis usually begins 20 years after infection. Its initial symptoms are those of a dementia with memory disturbance, dysarthria, myoclonus, and hyperreflexia. Personality changes and irritability frequently are noted. These symptoms progress to frank dementia, abnormal motor function, and psychosis. During the prepenicillin era, it was estimated that two-thirds of patients would become psychotic during this phase. These psychoses typically took either an expansive, grandiose, manic form, or resembled depressive psychosis with somatic delusions. The endstage of the illness left untreated patients in a bedridden, helpless state (Adams & Victor, 1989; Hooshmand, Escobar, & Kopf, 1972; Lishman, 1988; Simon, 1985; Swartz, 1984).

Tabes Dorsalis. Tabes dorsalis usually develops 25 to 30 years after the initial infection. The cardinal findings in tabetic neurosyphilis include loss of position and vibration sense, absent lower extremity reflexes, ataxia, incontinence, and sharp, rapidly occurring pains in many areas of the body, often called lancinating or lightning pains.

Perhaps because of prior treatment with antibiotics, patients may present with combinations of the above presentations. Some authorities believe that atypical presentations are the rule in modern neurosyphilis. These atypicalities are discussed further under clinical dilemmas below (Adams & Victor, 1989; Hooshmand, Escobar, & Kopf, 1972; Lishman, 1988; Simon, 1985; Swartz, 1984).

Serology and Diagnosis

Serologic tests for syphilis are divided into two general groups: nonspecific reaginic tests and specific treponemal tests. Both types have important functions to play in the diagnosis of syphilis, although they provide different types of information.

Nonspecific reaginic tests

The nonspecific reaginic tests in most frequent use are the rapid plasma reagin test (RPR), and the Venereal Disease Research Laboratory test (VDRL). These tests measure IgG and IgM antibodies against a lipoidal antigen produced by an organism–host interaction.

Of these tests, the VDRL has been the most widely studied. The VDRL generally becomes positive within several weeks of primary infection and is positive in virtually all patients with secondary syphilis. It continues to remain positive in latent and late syphilis, although it may be negative in serum in as many as 20 to 30% of patients with late disease (Larsen & Beck-Sague, 1988; Relman, Schoolnik, & Swartz, 1988a; Sparling, 1971). The VDRL is particularly useful as an index of disease activity and treatment, and as an indicator of CNS infection. VDRL results are reported as reactive or nonreactive; and if reactive, the dilution titer is reported as well (e.g., 1:8, 1:32). In general, patients with primary or active syphilis show a fourfold or greater increase in titer during the 4- to 8-week period between initial conversion and the height of the secondary stage. Patients with secondary syphilis usually have a titer of 1:32 or higher. The VDRL titer can be used as a guide to treatment efficacy. Falling titers of reaginic antibodies provide evidence of successful treatment, although some patients may remain reactive at a low titer despite adequate treatment (Larsen & Beck-Sague, 1988). In CSF the presence of a positive VDRL is considered diagnostic of neurosyphilis.

The reaginic tests are, unfortunately, liable to false-positive reactions. In general, false-positive reactions are rare in patients whose titers are higher than 1:8 (Holmes & Lukehart, 1987; Larsen & Beck-Sague, 1988; Relman, Schoolnik, & Swartz, 1988a). The incidence of false-positive tests is variable but may be as high as 40%. Patients with connective tissue disorders, especially systemic lupus erythematosus, are likely to show false-positive serologies; they also can occur with other recent infections and other treponemal disorders. To our knowledge, there are no reports of Lyme disease causing a false-positive syphilis serology. Concern over false-positive VDRL tests has led to increased use of specific treponemal tests.

Specific Treponemal Tests

A variety of specific treponemal tests are available, including the fluorescent treonemal antibody–absorbed test (FTA-ABS), hemagglutination assays, and *Treponema pallidum* immobilization (TPI). Of these tests, the FTA-ABS is the most widely used and studied. These tests measure the presence of antibodies specific for *T. pallidum*. The FTA-ABS has a far lower level of false positives than the VDRL, although it can be abnormal in patients with connective tissue disorders, particularly lupus (Holmes & Lukehart, 1987; Larsen & Beck-Sague, 1988; Relman, Schoolnik, & Swartz, 1988a). The FTA-ABS is reported as nonreactive, borderline, or reactive, but reactive sera are not reported as titers so the FTA-ABS is not helpful for determining disease activity or response to treatment.

The FTA-ABS generally becomes positive during the primary stage and has a sensitivity of 98% (Larsen & Beck-Sague, 1988). It then remains positive for life despite treatment. Classically, the FTA-ABS has been thought to be nearly always reactive with secondary syphilis, although this point has been questioned in patients with concurrent HIV infection (see below). The number of individuals with false-negative FTA-ABS during late syphilis is low, certainly much lower than the 20 to 30% false-negative VDRL rate during late syphilis.

The microhemagglutination *T. pallidum* test (MHA-TP) is similar in sensitivity and specificity to the FTA-ABS for the diagnosis of syphilis. The MHA-TP can be used reliably only on serum and not on CSF (S. Larsen, personal communication, 1991; Larsen et al., 1981; Rein et al., 1980).

Thus the VDRL or other reaginic tests are best used as initial screening tests (because of their lesser expense and ease of performance) and as indicators of disease activity and of response to treatment. The VDRL has a special role in CSF evaluation, which is reviewed further below. The major drawback of the VDRL is its high rate of false-positive results and its tendency to become negative during late neurosyphilis.

The FTA-ABS is more specific and sensitive than the VDRL, but it does not provide information about the degree of disease activity or response to treatment. Its role in CSF evaluation is controversial. The stages of syphilis and serology results seen during those stages are outlined in Table 32-2. Data regarding false-positive and false-negative serologic results are shown in Table 32-3.

CSF Evaluation

Evaluation of the CSF is critical to the treatment of syphilis of prolonged duration (Adams & Victor, 1989). Studies have confirmed that upward of 40% of patients with primary or secondary syphilis have CSF abnormalities, and in approximately 30% the causative organism can be directly demonstrated in CSF (Lukehart et al., 1988). Studies of untreated patients found asymptomatic neurosyphilis in up to 30% of those with disease of 2 years or more duration (Holmes & Lukehart, 1987). Untreated, a significant fraction of these cases progress to symptomatic neurosyphilis. Therefore the evaluation of patients with asymptomatic neurosyphilis is critical to prevent further disease progression (Adams & Victor, 1989).

Patients with active neurosyphilis usually have a pleocytosis of 10 to 200 cells/cu mm, with the cells virtually all lymphocytes or mononuclear cells, elevated protein of up to 200 mg/dl, and a reactive VDRL. It is rare for the glucose to be low. Not all patients have all these CSF findings simultaneously, and the degree of pleocytosis and the protein elevation generally closely parallel the degree of disease activity (Adams & Victor, 1989; Relman, Schoolnik, & Swartz, 1988a; Simon, 1985).

The VDRL is the gold standard for the diagnosis of CSF neurosyphilis. If the CSF VDRL is reactive after a well performed, nontraumatic lumbar puncture, the patient has neurosyphilis. In most patients with late syphilis who have CNS involvement, the CSF VDRL is

TABLE 32–2. *Syphilis: Stages and Serology*

Syphilis Stage (Time from infection)	Duration	Positive VDRL (serum)	FTA-ABS (serum)
Primary (3–6 weeks)	1–5 Weeks	80–100%	90+%
Secondary (8–12 weeks)	2–10 Weeks	Nearly 100% positive titer >1:32	100%
Latent (>1 year)	Variable	Decreasing titer 75–90%	95–100%
Meningovascular (6 months to 10 years)	Variable—months	70–75+%	95–100%
Parenchymal neurosyphilis (20–30 years)	Years, especially if untreated	70–75+%	95–100%

TABLE 32–3. *False-Positive and False-Negative Syphilis Serology*

VDRL	FTA-ABS
False positive generally <1:8 Viral illnesses (20–40%): mycoplasma, hepatitis, systemic lupus erythmatosus, other connective tissue disorders, aging, pregnancy, other autoimmune disorders, infectious mononucleosis, malaria	False positive Usually borderline reactive 15% of patients SLE: are positive often with atypical banding patterns
False negative Late syphilis (20–30%) Latent syphilis (10–25%) Concurrent HIV infection	False negative Rare

reactive. However, in a significant minority, especially those with very late syphilis, the VDRL may be nonreactive (Davis & Schmitt, 1989; Simon, 1985). The FTA-ABS, unfortunately, cannot replace the VDRL for evaluation of neurosyphilis. The presence of a positive FTA-ABS may simply reflect passive transfer of antibody from serum to CSF (Jaffe et al., 1978).

Treatment with antibiotics generally returns CSF parameters to normal, although a few patients who otherwise appear fully treated still have a weakly reactive CSF VDRL. Whether this group requires further treatment remains controversial (Adams & Victor, 1989).

Clinical Dilemmas in the Evaluation of Neurosyphilis

The psychiatrist dealing with a patient with possible syphilitic infection or a history of syphilis must focus concern on the question of direct CNS involvement and on adequacy of prior treatment. The fact of prior treatment for primary or secondary syphilis, even with currently recommended standard regiments, is not proof that a patient has not developed neurosyphilis. Low serum VDRL titers do not rule out neurosyphilis. Current recommendations for treatment of primary or secondary syphilis have come under sharp attack (Bayne, Schmidley, & Goodin, 1986; Guinan, 1987; Lukehart, et al., 1988; Markovitz et al., 1986; Moskovitz et al., 1982; Musher, 1988). The demonstration of *T. pallidum* in the CSF of patients previously treated for primary or secondary syphilis has increased concern that standard treatment regimens may be inadequate. This doubt is further supported by pharmacologic studies suggesting that benzathine penicillin achieves low penetration into the CSF (Jaffe & Kabins, 1982; Lukehart et al., 1988). As such, patients must still be considered at some risk for neurosyphilis if previous low-dose benzathine penicillin or oral antibiotic regimens were used. Additionally, patients whose VDRL titers fell after treatment for primary or secondary syphilis by less than four- to eightfold, probably are at increased risk for treatment failure (Guinan, 1987).

In general, patients with syphilis of unknown or greater than 1 year's duration who present for psychiatric treatment should have a CSF evaluation performed to detect asymptomatic neurosyphilis. A negative CSF examination—acellular with normal protein and serology—makes active CNS disease unlikely, with progression to later symptomatic neurosyphilis a rare event (Adams & Victor, 1989; Relman, Schoolnik, & Swartz, 1988a).

Older patients presenting with some combination of neurologic and psychiatric disturbances require a CSF examination if they have evidence of prior syphilitic infection as documented by a reactive serum VDRL, FTA-ABS, or a reliable clinical history. Clinical dilemmas may arise when patients have a nonreactive CSF VDRL. The mere presence of a positive CSF FTA-ABS is not enough to warrant treatment in a patient who has no other CSF abnormalities and a neuropsychiatric examination not suggestive of neurosyphilis. Conversely, patients with a negative CSF VDRL, especially with other CSF abnormalities on examination consistent with neurosyphilis, deserve presumptive treatment, especially if they have evidence of prior syphilitic infection by serum FTA-ABS (Adams & Victor, 1989;

Davis & Schmitt, 1989; Simon, 1985). Because of the significant false-negative VDRL rate in late syphilis, a serum FTA-ABS should be obtained when clinical suspicion of neurosyphilis is high (Larsen & Beck-Sague, 1988; Sparling, 1971).

Special Populations

Primary or Secondary Syphilis

Examination of the CSF in patients with primary or secondary syphilis is not routine, although patients with meningitis due to secondary syphilis would undoubtedly undergo a CSF examination. Some authorities are now recommending routine use of high-dose initial therapy to decrease the risk of later progression to neurosyphilis (Musher, 1988).

HIV Infection

Of particular concern to psychiatrists are reports of the unusual behavior of syphilitic infections in patients co-infected with HIV. In 1987 Johns et al. reported four cases of neurosyphilis in patients with concomitant HIV infection who had fallen ill despite standard treatment for neurosyphilis, as described by Berry et al. (1987) and Johns, Tierney, & Felsenstein (1987). Subsequent reports have amplified on these early reports, describing other abnormalities in the presentation of syphilis in patients with HIV infection. It is not known if these reports represent a true change in the behavior of neurosyphilis in the face of HIV infection (Hook, 1989). Patients with secondary syphilis and HIV infection may be seronegative for syphilis antibodies (Hicks, Benson, & Lipton, 1987). Haas et al. (1990) reported that 7% of HIV-seropositive men lost evidence of prior syphilis reactivity, as did 38% of symptomatic HIV-seropositive men. Although they were cautious about drawing a conclusion, individuals caring for persons at risk for HIV infection must be aware that the diagnosis of syphilis and perhaps its clinical course and response to treatment may be altered in this population. When HIV-seropositive patients are treated for syphilis, the CDC recommends clinical and serologic follow-up with a quantitative nontreponemal reaginic test (VDRL, RPR) at 1, 2, 3, 6, 9, and 12 months after treatment. Patients with early syphilis whose titers increase or fail to decrease fourfold within 6 months should undergo CSF examination and be retreated. (CDC, 1989e, pp. 11–12). It remains unclear if all HIV-infected patients with early syphilis should have a CSF examination (CDC, 1989e). However, given the substantial co-morbidity between HIV and syphilis and the evidence that genital ulcers (as in syphilis) promote the acquisition of HIV infection, psychiatrists must be particularly alert to the issue of syphilis-HIV co-infection (CDC, 1988b).

Treatments

As noted above, the recommendations for treatment of primary and secondary syphilis have been undergoing some modifications (Musher, 1988). In general, for neurosyphilis the most effective regimen is one that uses aqueous penicillin G in doses of 12 million to 24 million units in divided intravenous doses for at least 10 to 14 days, followed by weekly benzathine penicillin G 2.4 million units intramuscularly per week for 3 weeks (Adams & Victor, 1989; Holmes & Lukehart, 1987). Other regimens, especially those using oral agents or non-aqueous penicillin, are likely to be less effective (Lukehart et al., 1988; Musher, 1988). Patients treated for syphilis prior to 1980 with low-dose therapy (parenteral or oral) should be regarded as inadequately treated. The Jarisch-Herxheimer reaction, consisting of fever, chills, headache, and rash, occurs briefly after initial penicillin treatment. It is less likely to occur with late syphilis infections (Relman, Schoolnik, & Swartz, 1988b). Treatment efficacy for neurosyphilis is documented by observing the return of CSF parameters to normal within 1 year (Adams & Victor, 1989; Relman, Schoolnik, & Swartz, 1988b). As noted above, the VDRL may remain weakly reactive, and it is unclear if it is an indication for further pharmacotherapy. The return of pleocytosis, increase in protein, or VDRL titer is indicative of failed treatment. Pharmacotherapy for the psychiatric disturbances associated with syphilis is empiric and should follow principles outlined for HIV infection, above, especially in regard to symptom constellation, dosage, and close observation for neuropsychiatric side effects.

CHRONIC FATIGUE SYNDROME

In 1985 reports of focal epidemics of a syndrome characterized by prolonged fatigue, low-grade fever, lymphadenopathy, impaired concentration, and confusion appeared in the medical literature (Jones et al., 1985; Straus et al., 1985; Tobi et al., 1982). Review of the medical literature from the past century reveals descriptions of strikingly similar syndromes known by different names (Greenberg, 1990). In 1869 George Beard introduced the term neurasthenia, a concept modified by Janet and renamed psychasthenia in 1906. During World War I and World War II, shell shock and operational fatigue referred to homologous constellations of symptoms. Between the world wars, the term neurocirculatory asthenia was coined in New England and studied by White and Cohen at Massachusetts General Hospital. Over the last 50 years, indistinguishable syndromes have variably been called myalgic encephalomyelitis, benign epidemic neuromyasthenia, Icelandic disease and Royal Free Disease (Hickie et al., 1990; Holmes et al., 1987).

The reports of the 1980s were unique in that a number of these cases were associated with elevated Ebstein-Barr virus (EBV) titers or occurred after acute infectious mononucleosis. Consequently, the syndrome rapidly became known as the chronic EBV syndrome, or chronic mononucleosis syndrome (Dubois et al., 1984; Jones et al., 1985). Popularization of this syndrome in the media led to a large number of patients presenting with complaints of chronic fatigue and requesting scrologic evaluation of antibodies to EBV.

A high frequency of these patients reported depressed mood and other psychiatric symptoms. Therefore an association between the syndrome, mood disorder, and reactivation of EBV was considered. Psychiatrists were being asked to see these patients and to obtain EBV serologies on them. Hence the nosologic status of this disorder as a medical illness and its relation to psychiatric illness, became the subject of much study and controversy.

Studies have since cast doubt on the association of this syndrome with reactivation of prior EBV infection and led a working group at the CDC to rename the condition chronic fatigue syndrome (CFS). They have also defined strict criteria for the syndrome as an aid to clinical diagnosis and research (see Table 32-4) (Holmes et al., 1988).

TABLE 32–4. *CDC Working Case Definition for Chronic Fatigue Syndrome*

A case must meet both major criteria and 6 of the 11 symptom criteria plus 2 or more of the 3 physical criteria or 8 or more of the 11 symptom criteria.

Major criteria

1. New persistent fatigue lasting 6 months or more and reducing patient's activity to below 50% of premorbid functioning.
2. Full evaluation, including laboratory studies to rule out other causes of fatigue.
   a. Malignancy
   b. Localized or systemic infection
   c. HIV-related illness
   d. Chronic psychiatric disorders
   e. Inflammatory disorders
   f. Autoimmune disorders
   g. Endocrine disease
   h. Substance abuse
   i. Other chronic medical illness

Minor criteria

Symptom criteria

1. Mild fever: oral temperature 37.5°–38.6°C
2. Sore throat
3. Painful lymph nodes in anterior or posterior cervical or axillary areas
4. Generalized muscle weakness
5. Myalgia
6. Prolonged fatigue (>24 hours) after moderate exercise
7. Generalized headaches
8. Migratory arthralgias
9. Neuropsychologic complaints
10. Sleep disturbance
11. Rapid development of the main symptom complex (hours to a few days)

Physical criteria (documented by physician on two occasions, 1 month apart or more)

1. Low grade fever: temperature 37.6°–38.6°C
2. Nonexudative pharyngitis
3. Palpable or tender anterior or posterior cervical or axillary nodes (< 2 cm in diameter)

Clinical Diagnosis

As noted, CFS bears close similarity to various reported syndromes characterized by fatigue (Henderson & Shelokov, 1959). The hallmark of CFS is chronic disabling fatigue that can impair normal household or occupational activities. Patients may describe a feeling of feverishness, have tender or enlarged lymph nodes and recurrent sore throats, and complain of depressed mood or other neuropsychiatric symptoms. Many patients state that the onset of fatigue followed a severe flu-like syndrome from which they never fully recovered (CDC, 1989c). Given the subjective nature of the presenting complaints, attempts to document the severity of the syndrome objectively have proved difficult. In one study regarding the effect of acyclovir on symptoms of CFS, investigators could find no evidence of elevated temperature despite patients' reports of feverishness (Straus et al., 1988). Likewise, a study of muscle strength in CFS found no objective evidence of diminished muscle strength or easy fatigability despite patients' subjective reports of weakness (Lloyd, Phales, & Gandevia, 1988).

The CDC diagnostic criteria for CFS require that patients fulfill the two major criteria and eight minor criteria. Exclusion criteria require that other medical disorders, including malignancy, infection, autoimmune disease, endocrinopathies, sleep apnea, and others, be ruled out. Special studies, such as overnight polysomnography, may be necessary in certain cases to definitively rule out alternative diagnoses. Patients with evidence of a primary psychiatric disorder, including major depression, an anxiety disorder, histrionic personality disorder, or schizophrenia, are to be excluded according to the CDC criteria for CFS. Furthermore, use of psychotropic medicines (e.g., lithium, antipsychotics, or antidepressants) constitutes grounds for exclusion as well (Holmes et al., 1988).

In a study of 135 patients evaluated at a university clinic for chronic fatigue, only 6 patients met these new and strict criteria for CFS. In 7 patients, fatigue was secondary to an identifiable medical disorder, such as sleep apnea, seizures, polymyalgia rheumatica, or panhypopituitarism. Ninety-one patients were excluded because of psychiatric diagnoses including major depression, panic disorder, somatization disorder, and dysthymia (Manu, Lane, & Matthews, 1988). Therefore when CDC exclusion criteria for CFS pertaining to psychiatric diagnoses are disregarded (as many investigators have done), the composition of the CFS study population is drastically altered. When psychiatric comorbidity is not used as a basis for exclusion, most of the CFS population meets diagnostic criteria for a concurrent psychiatric illness (Gold et al., 1990).

Of further nosologic concern are findings of Goldenberg et al. (1990), who pointed out the resemblance between CFS and fibromyalgia. They reported that up to 70% of patients presenting with chronic fatigue also have signs and symptoms of fibromyalgia and suggested that considerable overlap between these syndromes may exist.

Serologic Studies

Early case series reported unusual profiles of EBV antibodies in patients with chronic EBV syndrome (Jones, et al., 1985; Tobi et al., 1982). EBV infects almost all adults (CDC, 1989c; Merlin, 1986). When individuals are infected during adolescence or early adulthood, 30 to 45% develop infectious mononucleosis, a severe flu-like syndrome characterized by intense sore throat, fatigue, lymphadenopathy, and occasionally splenomegaly. During the period of acute infection, serologic detection of Paul-Bunnell (heterophil) antibodies can confirm the diagnosis (Schooley, 1987).

A variety of antibodies are characteristically produced during the different phases of EBV infection. During the early replicative phase of infection, patients initially show high titers of IgM antibody to EBV capsid antigen (VCA), which disappears by 1 to 2 months after infection. The development of IgG to VCA also occurs early and then persists at lower titers. Shortly after infection, patients develop antibodies to early antigens (EA), which may persist at low levels for many years. Several months after infection, antibodies to EBV nuclear antigens (EBNA 1 & 2) are produced and persist for life. Therefore the typical adult with prior exposure to EBV but who does not have active infection has absent IgM to VCA, low IgG to VCA, low titers of antibody to EA, and lifetime antibodies to EBNA 1 & 2 (Sumaya, 1986).

In contrast, patients thought to have chronic EBV infection were found to have elevated levels of replicative enzyme antibodies (VCA and EA) and in some cases absent EBNA antibodies (Jones et al., 1985; Straus et al., 1985). This pattern, which was similar to several reported cases of a chronic severe form of infectious mononucleosis with pneumonitis and hematologic abnormalities, suggested that chronic EBV might represent an ongoing replication of the virus leading to clinical symptoms (Henle et al., 1987; Miller et al., 1987; Schooley et al., 1986). More recent studies have cast doubt on this hypothesis. Although levels of replicative antibodies are higher in subjects than controls, it has only rarely reached the level of statistical significance, and there is significant overlap in antibody levels between patients and asymptomatic controls (Buchwald, Sullivan, & Komaroff, 1987; Gold et al., 1990).

Related studies have demonstrated elevated antibody levels to a variety of viruses in patients purported to have chronic EBV infection. They include elevated antibody titers to CMV, herpes simplex, and measles. Such findings suggest a general change in immune function potentially causing reactivation of these viruses (Holmes et al., 1987). There have been some reports of an association with human herpes virus 6, though these findings were not substantiated in a large scale prospective study. A variety of other subtle changes in immune function have been noted in patients with CFS. Changes in T cell helper/suppressor ratios have been reported (in both directions). Investigators have found mild IgA and IgG deficiencies, altered natural killer cell activity (in both directions), low in vitro interleukin 2 production, and other anomalies. Such findings have been inconsistent and are of uncertain significance (CDC 1989c; Komaroff, Geiger, & Wormsely, 1988; Straus, 1988).

Psychiatric Diagnoses

Patients presenting for treatment of CFS have a high frequency of psychiatric symptoms, particularly those associated with depression (Gold et al., 1990; Manu, Lane, & Matthews, 1985). Kruesi et al. found 75% of subjects with CFS had DSM-III diagnoses when evaluated using the Diagnostic Interview Schedule. Of these patients, 35% had psychiatric diagnoses that preceded the onset of their fatigue. Lifetime rates of depression exceeded those for patients with diabetes. Major depression, dysthymia, simple phobia, and somatization disorder were the most frequent diagnoses (Kreusi, Dale, & Straus, 1989). In contrast, Hickie et al. (1990) studied 48 patients with CFS and 48 controls suffering from major depression. They used structured interviews of subjects and subjects' relatives, as well as psychometric tests. In the context of a retrospective paradigm, they found premorbid prevalence rates of major depression (12.5%) and total psychiatric disorders (24.5%) in CFS subjects that were no higher than those observed in a parallel medically ill population and significantly different from a psychiatric control population.

Neuropsychiatric complaints in CFS include impaired attention and difficulty concentrating and abstracting. These parameters were studied in 21 CFS subjects using a battery of neuropsychological instruments. Despite subjective complaints to the contrary, results indicated that subjects performed significantly better than age-matched groups in terms of the normative data. Whereas superior performances may have reflected above average levels of education in the study population, the findings showed no evidence of gross cognitive impairment and again highlighted the discrepancy between subjective reporting and objective deficits in this population (Altay et al., 1990). There have been reports of brain white matter hyperintensity on MRI in patients with CFS. Still, in such cases, structural abnormalities have not been correlated with neuropsychiatric symptoms (Straus, 1988).

In treatment studies using acyclovir, Straus et al. (1988) found a correlation between changes in the profile of mood states and symptoms of CFS. The relation between affective symptoms and chronic fatigue is complex and the direction of causation unclear.

Preexisting literature suggests that psychiatric factors can adversely affect immune function (Imboden, Canter, & Cluff, 1961; Krontol et al., 1983; Schleifer et al., 1984). Imboden and colleagues found increased rates of depression, measured by the MMPI, in patients who later had delayed recovery from influenza (Imboden, Canter, & Cluff, 1961). Thus psychiatric factors may affect the immunologic profile of some patients with complaints of chronic fatigue. At present, data do not support the idea that EBV is a necessary or causative agent in affective illness (Amsterdam, Henle, & Winoner, 1986; Cooke, Langlet, & McLaughlin, 1988).

Treatment

Specific treatments for CFS are not available. Straus et al. (1988) studied acyclovir infusions in a double-blind placebo-controlled treatment study in patients with CFS. There was no significant difference in outcome or clinical well-being between subjects in the treatment or placebo groups. Intravenous infusion of immunoglobulin or repletion of magnesium in patients with CFS have yielded contradictory or equivocal results (Cox, Campbell, & Dawson, 1991; Editorial, 1991; Lloyd et al., 1990; Peterson et al., 1990). Anecdotally, low-dose antidepressants have been reported to alleviate CFS (Goodnick, 1990). Goldenberg found benefit from low-dose amitriptyline in patients with fibromyalgia, a condition that may have some overlap with CFS (Goldenberg, Felson, & Dinnerman, 1986; Hudson, Pope, & Goldenberg, 1991).

Treatment aimed at reducing symptomatology is warranted whether for psychiatric or other symptoms. Use of antidepressants or other psychopharmacologic agents may be helpful for specific psychiatric syndromes, although controlled studies of antidepressants in CFS have not been reported. CFS patient responses to the Illness Behavior Questionnaire suggest that CFS patients are particularly invested in the notion that they suffer from a physical disease. The results predict that CFS subjects tend to be especially resistant to reassurance or psychological explanations as means of intervention (Hickie et al., 1990). Indeed, patients are often highly reluctant to consider a psychiatric diagnosis but may accept psychopharmacologic assistance for specific symptoms if it is presented in a matter-of-fact, nonthreatening manner.

Clinical Dilemmas

Clinicians must be wary of foreclosing judgment on patients with CFS. Clearly, the complaint of chronic fatigue is common among patients seen in general medical practice, and psychiatric disorders are common in unselected samples of patients with complaints of fatigue (Buchwald, Sullivan, & Komaroff, 1987; Holmes et al., 1987). The new criteria for CFS are drawn in a purposefully narrow fashion in hopes of selecting patients with a relatively homogeneous clinical syndrome, thereby enhancing the likelihood of delineating a discrete etiology and pathophysiology through research.

Psychiatrists must be cautious when ascribing all psychiatric symptoms to primary psychiatric illness. The diagnostic criteria for major depression in the setting of medical illness are controversial (see Chapter 4). Furthermore, the evidence of abnormal immune activation in patients with CFS suggests that, at least for some patients, symptoms may be secondary to a somatic pathophysiologic process, whatever its ultimate etiology proves to be. Likewise the high frequency of mood and anxiety disorders in nonselected patients presenting with the complaint of fatigue must be considered when evaluating these patients. Despite the reluctance of these patients to consider a psychiatric diagnosis, the high morbidity and mortality of untreated psychiatric illness, and the availability of effective treatment for these disorders, it behooves the clinician to diagnose and vigorously treat psychiatric disorders in patients with fatigue (Greenberg, 1990). Finally, psychiatrists must be alert to other medical disorders presenting as chronic fatigue and be certain that patients have an adequate evaluation so as to exclude alternative etiologies. The current balance of evidence speaks against the routine use of serologic tests for EBV in the evaluation of CFS, other than for research purposes.

LYME DISEASE

In 1975 an unusual cluster of arthritis among children and adults in Lyme, Connecticut led to the characterization of a new form of arthritis called Lyme arthritis (Steere et al., 1977). Since its initial discovery, the clinical description of the illness, including neuropsychiatric symptoms, epidemiology, pathogenesis, serology, and treatment, have been well described (Steere 1987, 1989; Steere et al., 1977).

Epidemiology

Lyme disease has been found in 40 states in the United States and has been reported in Europe, Australia, and other countries. Since 1982 there have been 13,825 cases of Lyme disease reported in the United States. Most of the United States cases have clustered in the Northeast, Northwest, and Upper Midwest (CDC, 1989d; Relman, Schoolnik, & Swartz, 1988b; Schmid et al., 1985; Steere, 1987, 1989). Most cases are seen in May, June, July, or August, with the peak incidence during July. Most victims live in, or have traveled through, heavily wooded areas during these months, increasing the likelihood of encountering the vector, although infected ticks have also been reported on lawns in endemic areas (Falco & Fish, 1988).

Pathogenesis

Lyme disease is caused by the spirochete *Borellia burgdorferi. B. burgdorferi* is carried by the mature deer ticks *Ioxides dammini, I. pacificus,* and *I. ricinus.* The deer tick transmits the spirochete via a bite. Clinical symptoms are associated with replication of the spirochete, as well as its dissemination to other organ systems. Immune complexes may play a role in the pathogenesis of some specific symptoms (Harden, Steere, & Malawiste, 1979; Steere, 1987).

Clinical Description

The first manifestation of Lyme disease, for 60 to 80% of patients, is a distinctive rash, erythema chronicum migrans (ECM). Other patients may not recall a rash or may not have had one. The rash appears within 3 to 32 days after a tick bite and appears first at the site of the bite (Berardi, Weeks, & Steere, 1988; Steere, 1987, 1989; Steere et al., 1983a). The rash is annular and red, and may be up to 56 cm in diameter. Over time, there is central clearing of the rash, leaving an erythematous ring in a so-called bulls-eye pattern. Many patients then develop secondary lesions, primarily on the thighs and trunk (Steere, Broderick, & Malawista, 1978). Associated symptoms include fever, fatigue, myalgia, and headache. In patients who develop more severe constitutional symptoms, CSF examination shows lymphocytic pleocytosis, elevated protein, and normal glucose (Steere et al., 1983b). Three months after infection, approximately 15% of patients progress to a second phase of illness, which may include significant neurologic symptoms including meningitis, encephalitis, radiculitis, central or peripheral neuropathy, and myelitis; cardiac symptoms occur in 8% (Relman, Schoolnik, & Swartz, 1988b; Steere, 1987). Symptoms of meningoencephalitis, including headaches and stiff neck (when they occur) are accompanied by irritability, depression, confusion, and emotional lability. Patients may also have an associated facial palsy or another peripheral manifestation of illness. If this phase of illness is not treated with high-dose antibiotics, symptoms can persist for up to a year (Steere et al., 1983b). CSF findings are similar to those described above.

Cardiac symptoms most commonly include conduction system abnormalities, including first degree AV block, second degree block (both type 1 and type 2), and complete (third degree) AV block. Patients may also experience some left ventricular dysfunction. Valvular disease has not been reported (McAlister et al., 1989; Relman, Schoolnik, & Swartz, 1988b; Steere, 1987; Steere et al., 1980).

A third phase of illness occurs approximately 6 months after the initial infection. It can consist in an oligoarthritis (which may begin earlier), affecting primarily the knees but also the small joints in a symmetric pattern. In some cases, patients experience a migratory arthritis or migratory myalgic symptoms. An encephalopathic state can also occur during this stage (see below) (Steere, 1987; Steere et al., 1977, 1979).

The classic stages described above may not occur in a typical fashion, as some patients may not have any symptoms until stage 2 or 3. Also, patients may not have neurologic symptoms until stage 3 but may have arthritic symptoms as early as stage 2. In a modified staging system, stage 1 occurs early and is associated with ECM. Stage 2 occurs within days to weeks after stage 1, with symptoms not developing at times until months later. Stage 3 is defined as late infection, beginning 1 year after infection (Steere, 1987).

Neuropsychiatric Manifestations

Mild encephalopathic states characterized by subjective and objective evidence of difficulty in memory, orientation, calculation, and construction are among the most frequent presentations at Lyme disease clinics (Halperin, Luft, & Anand, 1989). This state may occur late in the course of illness after other manifestations of Lyme disease have cleared, making diagnosis more difficult. Ten of thirteen patients seen by Halperin et al. with encephalopathy who had *B. burgdorferi*-specific IgG measured in CSF also had evidence of increased specific IgG production in the CNS compared to that in the periphery. Seven of seventeen encephalopathic patients examined by MRI had white matter abnormalities, which in three cases cleared on repeat MRI after appropriate antibiotic therapy.

Pachner et al. also found variable CNS manifestations of Lyme disease often occurring months or years after initial infection. Such manifestations have included psychiatric syndromes, including violent behavior, emotional lability with inappropriate laughter, depression, and compulsive behavior. High-dose intravenous antibiotics were associated with clearing of psychiatric symptoms in at least two cases (Pachner, Duray, & Steere, 1989). Cases of Lyme CNS illness also have been reported to end in a chronic dementing syndrome (Steere, 1989).

Halperin et al. (1990) studied nonselected patients referred for evaluation of possible CNS Lyme disease. Of those who were seropositive, only a small fraction had evidence of CNS Lyme disease as documented by intrathecal antibody production.

Logigian et al. found that 89% of patients with known Lyme disease and chronic neurologic symptoms, had evidence of an encephalopathy with mood, sleep, and memory disturbances. Most of the patients had evidence of increased intrathecal antibody production to *B. burgdorferii* and of radiculitis or distal paresthesias (Logigian, Kaplan, & Steere, 1990).

Thus, encephalopathy can be a major consequence of Lyme infection. However, its frequency in a nonselected population is likely to be low (Halperin et al., 1990).

Serodiagnosis

Serologic studies may provide some aid in the diagnosis of Lyme disease. However, because most patients with early Lyme disease lack evidence of antibody to *B. burgdorferi*, clinical diagnosis of early Lyme disease remains critical.

Serologic studies commonly performed for Lyme disease are the ELISA and the indirect immunofluorescence assay (IFA) for Lyme-related IgG. Standard ELISA can detect antibody in approximately 40 to 50% of early cases (Barbour, 1989; Grodzicki & Steere, 1988; Relman, Schoolnik, & Swartz, 1988b; Shrestha, Grodzicki, & Steere, 1985; Steere, 1987). In patients with later manifestations of Lyme disease, serologic testing is more useful, with a sensitivity of 60%. Major sources of false-positive findings include patients infected with other spirochetal organisms, including other *Borellia* species, *T. pallidum*, and *Leptospira* species (Magnenelli, Anderson, & Johnson, 1987). Studies of an antibody capture enzyme immunoassay test suggests that this test will improve the serologic yield for early Lyme disease to 67%, and to 93% by convalescence; unfortunately this test is not yet clinically available (Berardi, Weeks, & Steere, 1988). Other sources of potential error in serologic testing are variable reference standards and the lack of inter- and intra-laboratory reliability. False-negative results, even with IgM capture assays, are common especially during early illness. One group of investigators reported that patients who had previously been treated with oral antibiotics for Lyme disease lacked antibodies but had a demonstrable T cell response to *B. burgdorferi* (Datwyler et al., 1988; Editorial, 1989b). False-positive results can occur in the presence of autoimmune disorders, neurologic disorders, and other spirochetal illnesses including syphilis. The Western blot procedure may serve as a confirmatory test to eliminate false-positive findings. The polymerase chain reaction for the detection of spirochetal DNA eventually may allow more accurate and rapid diagnosis of Lyme disease but is not currently clinically available (Sigal, 1991).

Issues in Neuropsychiatric Diagnosis

In patients with classic symptoms of Lyme disease, such as ECM, diagnosis is a somewhat simple matter. A patient with a history of ECM, meningoencephalitis with facial nerve palsy, and oligoarthritis is likely to have Lyme disease (Barbour, 1989). However, given the nonspecific nature of many early complaints of patients with Lyme disease, patients who lack a history of ECM (as do up to 40% of patients) are more difficult to diagnose. Clues to the diagnosis include travel in wooded, endemic areas and onset of illness during the summer months.

Patients with irritability, confusion, or lability as a sole or secondary manifestation of Lyme disease are more likely to come to psychiatric attention (Pachner

& Steere, 1985; Reik, Burgdorfer, & Donaldson, 1986; Reik et al., 1979). The presence of symptoms of meningoencephalitis in association with cranial nerve palsy (including bilateral facial nerve palsy) or radiculoneuritis is suggestive of Lyme disease. Close attention to the course of symptoms, history of travel or residence in endemic areas, and history of rash may be helpful. In patients without rash, it is easy to confuse the diagnosis with that of aseptic meningitis. In the latter, however, symptoms are usually acute. With Lyme disease the acute phase of illness often is followed by persistent symptoms of chronic headache, stiff neck, and constitutional symptoms. It is also of assistance to the psychiatrist that serologic studies done during this phase of illness are significantly more likely to be abnormal (Pachner & Steere, 1985; Reik, Burgdorfer, & Donaldson, 1986; Reik et al., 1979).

Particular diagnostic uncertainty may occur in cases where patients have late-onset encephalopathic or psychiatric symptoms (Halperin, Luft, & Anand; 1989; Pachner, Duray, & Steere, 1989). Given the limited knowledge about the cognitive changes in Lyme disease, the absence of pathognomonic neuropsychiatric findings, and the paucity of confirmed psychiatric cases, patients with a serious question of neuropsychiatric Lyme disease require a systematic hypothesis-testing investigation.

In addition to gathering a detailed history of prior travel, rashes, and other symptoms, evidence of active infection may be inferred from typical CSF abnormalities. Patients with stage 2 focal neurologic Lyme disease should undergo lumbar puncture. In stage 3 encephalopathic or psychiatric cases where patients are seropositive, or seronegative with a clear history of Lyme disease, especially as documented by ECM, the CSF should be examined. Evidence of pleocytosis and elevated protein represent nonspecific evidence of CNS infection (as does MRI evidence of white matter hyperintensity). CSF that shows evidence of increased *B. burgdorferi*-specific IgG in CSF is probably the best evidence for ongoing CNS infection. Further confirmatory evidence is suggested by clearing of white matter lesions on MRI after appropriate antibiotic therapy (Halperin, Luft, & Anand, 1989; Pachner, Duray, & Steere, 1989; Steere, 1989).

Treatment

The treatment of Lyme disease is based on adequate antibiotics directed against the causative organism. Patients with early Lyme disease generally respond well to tetracycline 250 mg PO qid for 10 to 20 days. Alternative regimens with erythromycin or penicillin are also useful but are associated with a higher rate of late illness. Patients who have evidence of active CNS disease should be treated with 20 million units per day of aqueous penicillin G in divided doses for 14 days or with ceftriaxone 2 g IV for 14 days, and should receive expert infectious disease and neurologic consultation. Ceftriaxone, though more expensive than penicillin, produces higher CSF antibiotic levels and may be preferable (Pfister et al., 1989). This regimen is associated with a rapid resolution of meningeal signs and symptoms, although focal neurologic symptoms may persist (Steere, 1987; Steere, Pachner, & Malawista, 1983). Follow-up CSF examination should be carried out to confirm the adequacy of treatment. Halperin (1989) found evidence of recurrence of neuropsychiatric and encephalopathic symptoms in patients with CNS Lyme disease treated with antimicrobial agents.

Lyme disease is rarely fatal, but when fatalities do occur they are most often due to cardiac involvement, such as complete heart block, acute myopericarditis, and cardiomegaly. Some patients require temporary pacing wires until conduction deficits normalize (McAlister et al., 1989; Steere, 1989). Thus all patients who present for psychiatric attention with a possible diagnosis of Lyme disease should undergo a cardiac evaluation including an electrocardiogram. The possibility of heart block is particularly important to consider if one plans to treat depression with tricyclic antidepressants in a patient with recent or concurrent Lyme disease.

REFERENCES

Adams KM, & Heaton RK (1990). Statement concerning the NIMH neuropsychological battery. Clin Exp Neuropsychol 12:960–962.

Adams R, & Victor M (1989). Principles of neurology (4th ed., pp. 573–580). New York: McGraw-Hill.

Adler C, & Beckett A (1989). Psychotherapy of the patient with an HIV infection: some ethical and therapeutic dilemmas. Psychosomatics 30:203–208.

AIDS dementia may be linked to metabolite of tryptophan (1990). JAMA 264:305–306.

Albert ML, Feldman RG, & Willis A (1974). The 'subcortical dementia' of progressive supranuclear palsy. J Neurol Neurosurg Psychiatry 37:121–130.

Altay HT, Toner BB, Booker H, et al. (1990). The neuropsychological dimensions of postinfectious neuromyasthenia (chronic fatigue syndrome): a preliminary report. Int J Psychiatry Med 20:141–149.

Amsterdam JD, Henle W, Winoner A (1986). Serum antibodies to Epstein-Barr virus in patients with major depressive disorder. Am J Psychiatry. 143:1593–1596.

Anders KH, Guerra WF, Tomiyasu U, et al. (1986). The neuropathology of AIDS: UCLA experience and review. Am J Pathol 124:537–558.

Atkinson JH, Grant I, Kennedy CJ, et al. (1988). Prevalence of psychiatric disorders among men infected with human immunodeficiency virus: a controlled study. Arch Gen Psychiatry 45:859–864.

Baer JW (1989). Study of 60 patients with AIDS or AIDS related complex requiring psychiatric hospitalization. Am J Psychiatry 146:1285–1288.

Barbour AG (1989). The diagnosis of Lyme disease: rewards and perils. Ann Intern Med 110:501–502.

Bayne LL, Schmidley JW, & Goodin DS (1986). Acute syphilitic meningitis: its occurrence after clinical and serologic cure of secondary syphilis with penicillin G. Arch Neurol 43:137–138.

Beckett A, Summergrad P, Manschreck T, et al. (1987). Symptomatic HIV infection of the central nervous system in a patient without evidence of immune deficiency. Am J Psychiatry 144:1342–1344.

Berardi VP, Weeks KE, & Steere AC (1988). Serodiagnosis of early Lyme disease: analysis of IgM and IgG antibody responses by using an antibody-capture enzyme immunoassay. J Infect Dis 158:754–760.

Bernstein WB, & Scherokman B (1986). Neuroleptic malignant syndrome in a patient with acquired immunodeficiency syndrome. Acta Neurol Scand 73:636–637.

Berry CD, Hooton TM, Collier AC, et al. (1987). Neurologic relapse after benzathine penicillin therapy for secondary syphilis in a patient with HIV infection. N Engl J Med 316:1587–1589.

Binder RL (1987). AIDS antibody tests on inpatient psychiatric units. Am J Psychiatry 144:176–181.

Bottomley PA, Hardy CJ, & Cousins JP (1990). AIDS dementia complex: brain high-energy phosphate metabolite deficit. Radiology 176:407–411.

Breakey WR, Fischer PJ, Kramer M, et al. (1989). Health and mental health problems of homeless men and women in Baltimore. JAMA 262:1352–1357.

Bridge TP, Heseltine PNR, Parker ES, et al. (1989). Improvement in AIDS patients on peptide T. Lancet 2:226–227.

Brotman AW, & Forstein M (1988). AIDS obsessions in depressed heterosexuals. *Psychosomatics* 29:428–431.

Brunetti A, Bey G, Dichiw G, et al. (1989). Reversal of brain metabolic abnormalities following treatment of AIDS dementia complex with 3′-azido-2:3′ dideoxythymidine (AZT, Zidovudine): a PET-FDG Study. J Nucl Med 30:581–590.

Buchwald D, Sullivan JL, Komaroff AL (1987). Frequency of chronic active Epstein-Barr virus infection in a general medical practice. JAMA 257:2302–2307.

Buhrich N, Cooper DA, & Freed E (1988). HIV infection associated with symptoms indistinguishable from functional psychosis. Br J Psychiatry 152:649–653.

Bukasa KS, Sindic CJM, Bodeus M, et al. (1988). Anti-HIV antibodies in the CSF of AIDS patients: a serologic and immunoblotting study. J Neurol Neurosurg Psychiatry 51:1063–1068.

Butters N, Grant I, Haxby J, et al. (1990). Assessment of AIDS related cognitive changes: recommendations of the NIMH workshop on neuropsychological assessment approaches. J Clin Exp Neuropsychol 12:963–978.

CDC (1986). Classification system for human T-lymphotropic virus type III/lymphadenopathy-associated virus infections. MMWR 35:334–339.

CDC (1988a). Update: Serologic testing for antibody to human immunodeficiency virus. MMWR 36:833–840, 845.4.

CDC (1988b). Continuing increase in infectious syphilis—United States. MMWR 37:35–38.

CDC (1989a). Interpretation and use of the Western blot assay for serodiagnosis of human immunodeficiency virus type 1 infections. MMWR 38(S-7):1–7.

CDC (1989b). Guidelines for prophylaxis against Pneumocystis carinii pneumonia for persons infected with human immunodeficiency virus. MMWR 38(S-5).

CDC (1989c). Chronic fatigue syndrome (pp. 1–10). Atlanta: CDC.

CDC (1989d). Lyme disease—United States, 1987 and 1988. MMWR 38:668–672.

CDC (1989e). MMWR. 38(S-8):1–38.

CDC (1991). Update: Transmission of HIV infection during an invasive dental procedure–Florida: leads from the MMWR. JAMA 265:563–568.

Cooke RG, Langlet F, & McLaughlin BJM (1988). Age-specific prevalence of Epstein-Barr virus antibodies in adult patients with affective disorders. J Clin Psychiatry 49:361–363.

Cornos F, Enfield M, Horworth E, et al. (1989). The management of HIV-1 infection in state psychiatric hospitals. Hosp Commun Psychiatry 40:153–157.

Cox IM, Campbell MJ, & Dowson D (1991). Red blood cell magnesium and chronic fatigue syndrome. Lancet 337:757–760.

Cummings JL, & Benson DR (1984). Subcortical dementia: review of an emerging concept. Arch Neurol 41:874–879.

Cummings JL, & Benson DF (1988). Psychological dysfunction accompanying subcortical dementias. Annu Rev Med 39:53–61.

Cummings MA, Cummings KL, Rapaport MH, et al. (1987). Acquired immunodeficiency syndrome presenting as schizophrenia. West J Med 146:615–618.

Dalakas M, Wichman A, & Sever J (1989). AIDS and the nervous system. JAMA 261:2396–2399.

Datwyler RJ, Volkman DJ, Luft BJ, et al. (1988). Dissociation of specific T and B lymphocyte responses to Borrelia burgdorferi. N Engl J Med 319:1441–1446.

Davis LE, & Schmitt JW (1989). Clinical significance of cerebrospinal fluid tests for neurosyphilis. Ann Neurol 25:50–55.

De la Monte SM, Ho DD, Schooley RT, et al. (1987). Subacute encephalomyelitis of AIDS and its relation to HTLV-III infection. Neurology 37:562–569.

Dew MA, Ragni MV, & Nimorwicz P (1990). Infection in the human immunodeficiency virus and vulnerability to psychiatric distress: a study of men with hemophilia. Arch Gen Psychiatry 47:737–744.

Dilley JW, & Forstein M (1990). Psychosocial aspects of the human immunodeficiency virus (HIV) epidemic. In A Tasman, SM Goldfinger, & CA Kaufman (Eds.), Annual review of psychiatry (Vol 9; pp. 631–685). Washington, DC: American Psychiatric Press.

Dilley JW, Ochitill HN, Perl M, et al. (1985). Findings in psychiatric consultations with patients with acquired immune deficiency syndrome. Am J Psychiatry 142:82–86.

Dooley SW Jr, Carter UL, Hutton MD, et al. (1990). Guidelines for preventing the transmission of tuberculosis in health care setting with specific focus on HIV-related issues. MMWR 17:1–29.

Dreyer EB, Kaiser PK, Offerman JT, et al. (1990). HIV-1 coat protein neurotoxicity prevented by calcium channel antagnostics. Science 248:364–367.

Drugs for viral infection. Med Lett Drugs Ther 32:75–78.

Dubois RE, Seeley JK, Brus I, et al. (1984). Chronic mononucleosis syndrome. South Med J 77:1376–1382.

Editorial (1989a). Clinical trials of zidovudine in HIV infection. Lancet 2:483–484.

Editorial (1989b). Diagnosis of Lyme disease. Lancet 2:198–199.

Editorial (1991). Chronic fatigue syndrome—false avenues and dead leads. Lancet 337:331–332.

Falco RC, & Fish D (1988). Prevalence of Ixodes daminni near the homes of Lyme disease patients in Westchester County, New York. Am J Epidemiol 127:826–830.

Faulstich ME (1987). Psychiatric aspects of AIDS. Am J Psychiatry 144:551–556.

Fernandez F, Levy JK, & Galizzi H (1988). Response of HIV-related depression to psychostimulants: case reports. Hosp Comm Psychiatry 39:628–631.

Fernandez F, Levy JK, & Mansell PWA (1989). Response to antidepressant therapy in depressed persons with HIV infection. Abstract presented to the 5th International Aids Conference, Montreal, Canada, June, 1985.

Forstein M (1984). The psychosocial impact of the acquired immunodeficiency syndrome. Semin Oncol 11:77–82.

Friedman-Kien A, Laubastein L, Rubinstein P, et al. (1981). Disseminated Kaposi's sarcoma in homosexual men. Ann Intern Med 96:693–700.

Gabuzda DH, & Hirsch MS (1987). Neurologic manifestations of infections with human immunodeficiency virus: clinical features and pathogenesis. Ann Intern Med 107:383–391.

Gabuzda DH, Ho DD, de la Monte SM, et al. (1986). Immunohistochemical identification of HTLV-III antigen in brains of patients with AIDS. Ann Neurol 20:289–295.

Glass RM (1988). AIDS and suicide. JAMA 259:1369–1370.

Gold D, Bowden R, Sixbey J, et al. (1990). Chronic fatigue: a prospective clinical and virologic study. JAMA 264:48–53.

Goldenberg DL, Felson DT, Dinnerman H (1986). A randomized controlled trial of amitriptyline and naproxen in the treatment of patients with fibromyalgia. Arthritis Rheum 29:1371–1377.

Goldenberg DL, Simms RW, Geiger A, et al. (1990). High frequency of fibromyalgia in patients with chronic fatigue seen in a primary care practice. Arthritis Rheum 33:381–387.

Goodnick PJ (1990). Bupropion in chronic fatigue syndrome. Am J Psychiatry 147:1091.

Gottlieb MS, Schroff R, Shanker HM, et al. (1981). Pneumocystis carinii pneumonia and mucosal candidiasis in previously healthy homosexual men: evidence of a new acquired cellular immunodeficiency. N Engl J Med 305:1425–1431.

Greenberg DB (1990). Neurasthenia in the 1980's: chronic mononucleosis, chronic fatigue syndrome and anxiety and depressive disorders. Psychosomatics 31:129–137.

Grodzicki RI, & Steere AC (1988). Comparison immunoblotting and indirect enzyme-linked immunosorbant assay using different antigen preparations for diagnosing early Lyme disease. J Infect Dis 157:521–525.

Guinan ME (1987). Treatment of primary and secondary syphilis: defining failure at three and six month follow up. JAMA 257:359–360.

Haas JS, Bolan G, Larsen SA, et al. (1990). Sensitivity of treponemal test for detecting prior treated syphilis during human immunodeficiency virus infection. J Infect Dis 162:862–866.

Halperin JJ (1989). Abnormalities of the nervous system in Lyme disease: response to antimicrobial therapy. Rev Infect Dis 11(S6):S1499–S1504.

Halperin JJ, Krupp LB, Golightly MG, et al. (1990). Lyme borreliosis-associated encephalopathy. Neurology 42:1340–1343.

Halperin JJ, Luft BJ, & Anand AK (1989). Lyme neuroborelliosis: central nervous system manifestations. Neurology 39:753–759.

Harden JA, Steere AC, & Malawista SE (1979). Immune complexes in the evolution of Lyme arthritis dissemination and localization of abnormal C1q binding activity. N Engl J Med 301:1358–1363.

Harris MJ, Gleghorn A, Jeste DV (1989). HIV-related psychosis. Presented at the 142nd meeting of the American Psychiatric Association, San Francisco.

Henderson DA, & Shelokov A (1959). Epidemic neuromyasthenia: a clinical syndrome? N Engl J Med 260:757–764.

Henle W, Henle G, Andersson J, et al. (1987). Antibody responses to Epstein-Barr virus determined nuclear antigen (EBNA)-1 and EBNA-2 in acute and chronic Epstein-Barr virus infection. Proc Natl Acad Sci USA 84:570–574.

Hickie I, Lloyd A, Wakefield D, et al. (1990). The psychiatric status of patients with the chronic fatigue syndrome. Br J Psychiatry 156:534–540.

Hicks CB, Benson AM, & Lipton GP (1987). Seronegative secondary syphilis in a patient infected with the human immunodeficiency virus (HIV) with Kaposi sarcoma: a diagnostic dilemma. Ann Intern Med 107:492–495.

Hjelle B, Busch M (1989). Direct methods for detection of HIV-1 infection. Arch Pathol Med 113:975–980.

Ho DD, Bredeson DE, Vinters HV, et al. (1989). The acquired immunodeficiency syndrome (AIDS) dementia complex. Ann Intern Med 111:400–410.

Ho DD, Roth TR, Schooley RT, et al. (1985). Isolation of HTLV-III from cerebrospinal fluid and neurol tissues of patients with neurologic syndromes related to the acquired immunodeficiency syndrome. N Engl J Med 313:1493–1497.

Holland JC, & Tross S (1985). The psychosocial and neuropsychiatric sequence of the acquired immunodeficiency syndrome and related disorders. Ann Intern Med 103:760–764.

Holmes GP, Kaplan JE, Gantz NM, et al. (1988). Chronic fatigue syndrome: a working case definition. Ann Inter Med 108:387–389.

Holmes GP, Kaplan JE, Stewart JA, et al. (1987). A cluster of patients with a chronic mononucleosis-like syndrome. JAMA 57:2297–2302.

Holmes KK, & Lukehart SA (1987). Syphilis. In E Braunwald, KJ Isselbacher, & RG Petersdorf et al. (Eds.), Harrison's principles of internal medicine (11th ed., pp. 639–649). New York: McGraw-Hill.

Holmes VF, Fernandez F, & Levy JK (1989). Psychostimulant response in AIDS-related complex patients. J Clin Psychiatry 50:5–8.

Hook EW (1989). Treatment of syphilis: current recommendations, alternatives and continuing problems. Rev Infect Dis 11(S6):S1511–S1517.

Hooshmand H, Escobar MR, & Kopf SW (1972). Neurosyphilis: a study of 241 patients. JAMA 219:726–729.

Horworth E, Kramer M, Cornos F, et al. (1989). Clinical presentation of AIDS and HIV-infection in state psychiatric facilities. Hosp Commun Psychiatry 40:502–506.

Hudson JI, Pope HG Jr, & Goldenberg DL (1991). Chronic fatigue syndrome. JAMA 265:357–358.

Imboden JB, Canter A, & Cluff LE (1961). Convalescence from influenza: a study of the psychological and clinical determinants. Arch Intern Med 108:393–399.

Jackson JB (1990). Human immunodeficiency virus type-1 antigen and culture essays. Arch Pathol Lab Med 114:249–253.

Jaffe HW, & Kabins SA (1982). Examination of cerebrospinal fluid in patients with syphilis. Rev Infect Dis 4:S842–847.

Jaffe HW, Larsen SA, Peters M, et al. (1978). Tests for treponemal antibody in CSF. Arch Intern Med 138:252–255.

Jenike MA, & Pato C (1986). Disabling fear of AIDS responsive to imipramine. Psychosomatics 27:143–144.

Joffe RT, Rubinow DR, Squillace K, et al. (1986). Neuropsychiatric aspects of AIDS. Psychopharmacol Bull 22:684–688.

Johns DR, Tierney M, & Felsenstein D (1987). Alteration in the natural history of neurosyphilis by concurrent infection with the human immunodeficiency virus. N Engl J Med 316:1569–1572.

Jones JF, Ray G, Minnich LL, et al. (1985). Evidence for active Epstein-Barr virus infection in patients with persistent, unexplained illnesses: elevated anti-early antigen antibodies. Ann Intern Med 102:1–6.

Komaroff AL, Geiger AM, & Wormsely S (1988). IgG subclass deficiencies in chronic fatigue syndrome. Lancet 1:1288–1289.

Koralnick IJ, Beaumanoir A, Hausler R, et al. (1990). A controlled study of early neurologic abnormalities in men with symptomatic human immunodeficiency virus infection. J Engl J Med 323:864–870.

Krontol Z, Silva J, Greden J, et al. (1983). Impaired lymphocyte function in depressive illness. Life Sci 33:241–247.

Kruesi MJP, Dale J, & Straus SE (1989). Psychiatric diagnosis in patients with chronic fatigue syndrome. J Clin Psychiatry 50:53–56.

Larsen SA, & Beck-Sague CM (1988). Syphilis. In A Balows, WT Hausler, M Okaski, et al. (Eds.), Laboratory diagnosis of infectious diseases; principles and practice (Vol I; pp. 490–503). New York: Springer-Verlag.

Larsen SA, Hambie EA, Pettit DE, et al. (1981). Specificity, sensitivity and reproducibility among the fluorescent treponemal antibody absorbtion test, the microhemagglutination assay for Treponema pallidum antibodies and the hemagglutination treponemal test for syphilis. J Clin Microbiol 14:441–445.

Levy RM, Bredesen DM, Rosenblum ML (1988). Opportunistic central nervous system pathology in patients with AIDS. Ann Neurol 23(suppl):S7–S12.

Levy RM, Redesen DE, & Rosenblum ML (1985). Neurologic manifestations of the acquired immunodeficiency syndrome (AIDS): experience at UCSF and review of the literature. J Neurosurg 62:475–495.

Lishman WA (1988). Organic psychiatry (2nd ed). London: Blackwell.

Lloyd A, Hickie I, Wakefield D, et al. (1990). A double-blind placebo controlled trial of intravenous immunoglobulin therapy in patients with chronic fatigue syndrome. Am J Med 89:561–568.

Lloyd AR, Phales J, & Gandevia SC (1988). Muscle strength, endurance, and recovery in the post infection fatigue syndrome. J Neurol Neurosurg Psychiatry 51:1316–1322.

Logigian EL, Kaplan RF, & Steere AC (1990). Chronic neurologic manifestations of Lyme disease. N Engl J Med 323:1438–1444.

Lukehart SA, Hook EW, Baker-Zander SA, et al. (1988). Invasion of the central nervous system by Treponema pallidum: implications for diagnosis and treatment. Ann Intern Med 109:855–862.

Lunn S, Skydsbjerg M, Schulsinger H, et al. (1991). A preliminary report on the neuropsychologic sequelae of human immunodeficiency virus. Arch Gen Psychiatry 48:135–142.

Magnanelli LA, Anderson JF, & Johnson RC (1987). Cross reactivity in serological tests for Lyme disease and other spirochetal infections. J Infect Dis 156:183–188.

Manu P, Lane TJ, & Matthews DA (1988). The frequency of chronic fatigue syndrome in patients with symptoms of persistent fatigue. Ann Intern Med 109:554–556.

Markovitz DM, Beutnar KR, Maggio RP, et al. (1986). Failure of recommended treatment for secondary syphilis. JAMA 255:1767–1768.

Marotta R, & Perry S (1989). Early neuropsychological dysfunction caused by human immunodeficiency virus. J Neuropsychiatry 1:225–235.

Marzuk PM, Tierney H, Tardiff K, et al. (1988). Increased risk of suicide in persons with AIDS. JAMA 259:1333–1337.

McAlister HF, Klementowitz PT, Andrews C, et al. (1989). Lyme carditis: an important cause of reversible heart block. Ann Intern Med 110:339–345.

McArthur JC (1987). Neurologic manifestations of AIDS. Medicine 66:407–437.

McArthur JC, Cohen BA, Selves OA, et al. (1989). Low prevalence of neurologic and neuropsychological abnormalities in otherwise healthy HIV-1 infected individuals: results from the multicenter AIDS cohort study. Ann Neurol 26:601–611.

Merlin TL (1986). Chronic mononucleosis: pitfalls in the laboratory diagnosis. Hum Pathol 17:2–8.

Miller EN, Selves OA, & McArthur JC (1990). Neuropsychological performance in HIV-1 infected homosexual men: the multicenter AIDS cohort study (MACS). Neurology 40:197–203.

Miller G, Grogan E, Rowe D, et al. (1987). Selective lack of antibody to a component of the EB nuclear antigen in patients with chronic active Epstein-Barr virus infection. J Infect Dis 156:26–35.

Morgan MK, Clark ME, & Hartman WL (1988). AIDS related dementia: a case report of rapid cognitive decline. J Clin Psychol 44:1024–1028.

Moskovitz BL, Klinch JJ, Goldman RL, et al. (1982). Meningovascular syphilis after "appropriate" treatment of primary syphilis. Arch Intern Med 142:139–140.

Musher DM (1978). Evaluation and management of an asymptomatic patient with a positive VDRL reaction. Curr Clin Top Infect Dis 9:147–157.

Musher DM (1988). How much penicillin cures early syphilis? Ann Intern Med 109:849–850.

Navia BA, Cho ES, Petito CU, et al. (1986). The AIDS dementia complex. II. Neuropathology. Ann Neurol 19:525–535.

Navia BA, Jordan BD, & Price RN (1986). The AIDS dementia complex. I. Clinical features. Ann Neurol 19:517–524.

Navia BA, & Price RW (1987). The acquired immunodeficiency syndrome dementia complex as the presenting or sole manifestation of human immunodeficiency virus infection. Arch Neurol 44:65–69.

Nurnberg HG, Prudic J, Fiori M, et al. (1984). Psychopathology complicating acquired immune deficiency syndrome. Am J Psychiatry 141:95–96.

Ostrow DG, Monjan A, Joseph J, et al. (1989). HIV-related symptoms and psychological functioning in a cohort of homosexual men. Am J Psychiatry 146:737–742.

Pachner AR, Duray P, & Steere AC (1989). Central nervous system manifestations of Lyme disease. Arch Neurol 46:790–795.

Pachner AR, & Steere AC (1985). The triad of neurologic manifestations of Lyme disease, meningitis, cranial neuritis, and radiculoneuritis. Neurology 35:47–53.

Perry S, Fishman B, Jacobsberg L, et al. (1991). Effectiveness of psychoeducational interventions in reducing emotional distress after human immunodeficiency virus antibody testing. Arch Gen Psychiatry 48:143–147.

Perry S, Jacobsburg LB, Fishman B, et al. (1990a). Psychiatric diagnosis before serologic testing for the human immunodeficiency virus. Am J Psychiatry 147:89–93.

Perry S, Jacobsberg LB, Fishman B, et al. (1990b). Psychological response to serological testing for HIV. AIDS 4:145–152.

Perry S, Jacobsberg LB, Fishman B, et al. (1990c). Suicidal ideation and HIV testing. JAMA 263:679–682.

Perry S, & Jacobsen P (1986). Neuropsychiatric manifestations of AIDS-spectrum disorders. Hosp Commun Psychiatry 37:135–141.

Perry SW (1990). Organic mental disorders caused by HIV: update on early diagnosis and treatment. Am J Psychiatry 147:696–710.

Perry SW, & Tross S (1984). Psychiatric problems of AIDS inpatients at the New York Hospital: a preliminary report. Public Health Rep 99:200–205.

Peterson PK, Shepard J, Macres M, et al. (1990). A controlled trial of intravenous immunoglobulin G in chronic fatigue syndrome. Am J Med 89:554–560.

Petito CK, Cho ES, Lemann W, et al. (1986). Neuropathology

of acquired immunodeficiency syndrome (AIDS): an autopsy review. J Neuropathol Exp Neurol 45:635–646.

Pfister HW, Preac-Mursic V, Wilshe B, et al. (1989). Cefataxime versus penicillin G for acute neurologic manifestations in Lyme borreliosis: a perspective randomized study. Arch Neurol 46:1190–1194.

Pohl P, Vogl G, Fill H, et al. (1988). Single photon emission computed tomography in AIDS dementia complex. J Nucl Med 29:1382–1386.

Rabkin JG, & Harrison WM (1990). Effect of imipramine on depression and immune status in a sample of men with HIV infection. Am J Psychiatry 147:495–497.

Reik L, Burgdorfer W, & Donaldson JO (1986). Neurologic abnormalities in Lyme disease without erythema chronicum migrans. Am J Med 81:73–78.

Reik L, Steere AC, Bartenhagen NH, et al. (1979). Neurologic abnormalities of Lyme disease. Medicine 58:281–294.

Rein MF, Banks GW, Logan LC, et al. (1980). Failure of the Treponema pallidum immobilization test to provide additional diagnostic information about contemporary problem sera. Sex Transm Dis 7:101–105.

Relman DA, Schoolnik GK, & Swartz MN (1988a). Lyme disease. Sci Am Med 7(VII):6–11.

Relman DA, Schoolnik GK, & Swartz MN (1988b). Syphilis in nonvenereal treponematoses. Sci Am Med 7(VI):1–10.

Rolfs RT, & Nahashima AK (1990). Epidemiology of primary and secondary syphilis in the United States, 1981–1989. JAMA 264:1432–1437.

Rubin RH (1988). Acquired immune deficiency syndrome. Sci Am Med 7(XI):1–19.

Ruder HL, Cauthen GM, Kelly GD, et al. (1989). Tuberculosis in the United States. JAMA 262:385–389.

Rundell JR, Wise MG, & Ursano RJ (1986). Three cases of AIDS-related psychiatric disorders. Am J Psychiatry 143:777–778.

Sacks MH, Perry S, Grover R, et al. (1990). Self-reported HIV-related risk behaviors in acute psychiatric inpatients: a pilot study. Hosp Commun Psychiatry 41:1253–1255.

Schaerf FW, Miller RR, Lipsey JR, et al. (1989). ECT for major depression in four patients infected with human immunodeficiency virus. Am J Psychiatry 146:782–784.

Schleifer SJ, Keller SE, Myerson AT, et al. (1984). Lymphocyte function in major depressive disorder. Arch Gen Psychiatry 41:484–486.

Schmid GP, Horsley R, Steere AC, et al. (1985). Surveillance of Lyme disease in the United States, 1982. J Infect Dis 151:1144–1149.

Schmitt FA, Bigley JW, McInnis R, et al. (1988). Neuropsychological outcome of zidovudine (AZT) treatment of patients with AIDS and AIDS related complex. N Engl J Med 319:1573–1578.

Schooley RT (1987). Epstein-Barr virus infection, including infectious mononucleosis. In E Braunwald, RJ Isselbacher, & RG Petersdorf (Eds.). Harrison's principles of internal medicine (11th ed., pp. 699–703). New York: McGraw-Hill.

Schooley RT, Carey RW, Miller G, et al. (1986). Chronic Epstein-Barr virus infection associated with fever and interstitial pneumonitis: clinical and serologic features and response to antiviral chemotherapy. Ann Inter Med 104:636–643.

Screening for tuberculosis and tuberculosis infection in high-risk populations and the use of preventive therapy for tuberculosis infection in the United States. (1990). MMWR 39:1–12(1-3).

Selves OA, Miller E, & McArthur J (1990). HIV-1 infection: no evidence of cognitive decline during the asymptomatic stages. Neurology 40:204–208.

Shaw GM, Harper ME, Hahn BH, et al. (1985). HTLV-III infection in brains of children and adults with AIDS encephalopathy. Science 227:177–182.

Shrestha M, Grodzicki RL, & Steere AC (1985). Diagnosing early Lyme disease. Am J Med 78:235–240.

Sidths JJ, & Price RW (1990). Early HIV-1 infections in the AIDS dementia complex. Neurology 40:323–326.

Siegal FP, Lopez C, Hammer GS, et al. (1981). Severe acquired immunodeficiency in male homosexuals, manifested by chronic perianal ulcerative herpes simplex lesion. N Engl J Med 305:1439–1444.

Sigal LH (1991). Summary of the fourth international symposium on Lyme borreliosis. Arthritis Rheum 34:367–370.

Simon RP (1985). Neurosyphilis. Arch Neurol 42:606–613.

Smith TW, DeGinolami U, Henin D, et al. (1990). Human immunodeficiency virus (HIV) leukoencephalopathy and the microcirculation. J Neuropathol Exp Neurol 49:357–370.

Snider WD, Simpson DM, Nielsen S, et al. (1983). Neurologic complications of acquired immune deficiency syndrome: analysis of 50 patients. Ann Neurol 14:403–418.

Sparling PF (1971). Diagnosis and treatment of syphilis. N Engl J Med 284:642–653.

Steere AC (1987). Lyme disease. In E Braunwald, RJ Isselbacher, RG Petersdorf, et al. (Eds.), Harrison's principles of internal medicine (11th ed., pp. 657–659). New York: McGraw-Hill.

Steere AC (1989). Lyme disease. N Engl J Med 321:586–596.

Steere AC, Bartenhagen NH, Craft JE, et al. (1983a). The early clinical manifestations of Lyme disease. Ann Intern Med 99:76–82.

Steere AC, Batsford WP, Weinberg M, et al. (1980). Lyme carditis: cardiac abnormalities of Lyme disease. Ann Intern Med 93(part I):8–16.

Steere AC, Broderick TF, & Malawista SE (1978). Erythema chronicum migrans and Lyme arthritis: epidemiologic evidence for a tick victor. Am J Epidemiol 108:312–321.

Steere AC, Hardin JA, Ruddy S, et al. (1979). Lyme arthritis: correlating serum and cryoglobulin IgM with activity and serum IgG with remission. Arthritis Rheum 22:471–483.

Steere AC, Hutchinson GJ, Rahn DW, et al. (1983b). Treatment of the early manifestation of Lyme disease. Ann Intern Med 99:22–26.

Steere AC, Malawista SE, Snydman DR, et al. (1977). Lyme arthritis an epidemic of oligoarticular arthritis in children and adults in three Connecticut communities. Arthritis Rheum 20:7–17.

Steere AC, Pachner AR, & Malawista SE (1983). Neurologic abnormalities of Lyme disease: successful treatments with high dose intravenous penicillin. Ann Intern Med 99:767–772.

Stern Y, Marder K, Bell K, et al. (1991). Multidisciplinary baseline assessment of homosexual men with and without human immunodeficiency virus infection. III. Neurologic and Neuropsychological Findings. Arch Gen Psychiatry 48:131–138.

Straus SE (1988). The chronic mononucleosis syndrome. J Infect Dis 157:405–412.

Straus SE, Dale JK, Tobi M, et al. (1988). Acyclovir treatment of the chronic fatigue syndrome: lack of efficacy in a controlled trial. N Engl J Med 319:1697–1698.

Straus SE, Trosato G, Armstron G, et al. (1985). Persisting illness and fatigue in adults with evidence of Epstein-Barr infection. Ann Intern Med 1:7–16.

Sumaya CV (1986a). Epstein-Barr virus serologic testing: diagnostic indications and interpretations. Pediatric Infect Dis 5:337–341.

Summergrad P, & Peterson B (1989). Binswanger's disease. I. The clinical recognition of subcortical arteriosclerotic encephalopathy in elderly neuropsychiatric patients. J Geriatr Psychiatry Neurol 2:123–133.

SUSSER E, STRUENING EL, & CONOVER S (1989). Psychiatric problems in homeless men lifetime psychosis, substance use, and current distress in new arrivals at New York City shelters. Arch Gen Psychiatry 46:845–850.

SWARTZ M (1984). Neurosyphilis. In KK HOLMES (Ed.), Sexually transmitted diseases (pp. 313–314). New York: McGraw-Hill.

THEUER CP, HOPEWELL PC, ELIAS D, ET AL. (1990). Human immunodeficiency virus infection in tuberculosis patients. J Infect Dis 162:8–12.

TOBI M, MORAG A, RAVID Z, ET AL. (1982). Prolonged atypical illness associated with serologic evidence of persistent Epstein-Barr virus infection. Lancet 1:61–64.

WILLIAMS JBW, RABKIN JG, REMIEN RH, ET AL. (1991). Multidisciplinary baseline assessment of homosexual men with and without human immunodeficiency virus infection. II. Standardized clinical assessment of current and lifetime psychopathology. Arch Gen Psychiatry 48:124–130.

YARCHOAN R, BERG G, PROWERS P, ET AL. (1987). Response of human immunodeficiency virus associated neurologic disease to 3′ azido-2′,3′ dideoxythymidine. Lancet 1:132–134.

YARCHOAN R, MITSUYA H, MYERS CE, ET AL. (1989). Clinical pharmacology of 3′ azido-2′-3′ dideoxythymidine (Zidovudine) and related dideoxynucleosides. N Engl J Med 321:726–738.

YARCHOAN R, THOMAS RV, GRAFTMAN J, ET AL. (1988). Longterm administration of 3′ azido, 2′,3′ dideoxythymidine to patient with AIDS related neurological disease. Ann Neurol 23(suppl):S82–S87.

SUGGESTED READINGS

BARNES PF, BLOCH AB, DAVIDSON PT, ET AL. (1991). Tuberculosis in patients with human immunodeficiency virus infection. N Engl J Med 324:1644–1650.

GRANT I, ATKINSON JH, HESSELINK JR, ET AL. (1987). Evidence for early central nervous system involvement in the acquired immunodeficiency syndrome (AIDS) and other human immunodeficiency virus (HIV) infections: studies with neuropsychological testing and magnetic resonance imaging. Ann Intern Med 107:828–836.

IRWIN MR, SMITH TL, & GILLIN C (1987). Reduced natural killer cytotoxicity in depressed patients. Life Sci 41:2127–2133.

JANSSEN RS, SAYKIN AJ, & CANNON L (1989). Neurological and neuropsychological manifestation of HIV-1 infections: associations in the AIDS-related complex but not asymptomatic HIV-1 infection. Ann Neurol 26:592–600.

KOENIG S, GENDELMAN HE, ORENSTEIN JM, ET AL. (1986). Detection of AIDS virus in macrophages in brain tissue from AIDS patients with encephalopathy. Science 233:1089–1093.

LEVY JA, HOLLENDER H, SHIMABUKURO J, ET AL. (1985). Isolation of AIDS associated retroviruses from cerebrospinal fluid and brain of patients with neurologic symptoms. Lancet 2:586–588.

MASUR H, MICHELIS MA, GREENE JB, ET AL. (1981). An outbreak of community acquired Pneumocystis carinii pneumonia: initial manifestation of cellular immune dysfunction. N Engl J Med 305:1431–1438.

MCARTHUR JC, COHEN BA, FARZEDEGAN H, ET AL. (1988). Cerebrospinal fluid abnormalities in homosexual men with and without neuropsychiatric findings. Ann Neurol 23(suppl):S34–S37.

SACKS MH, SILBERSTEIN C, WEILER P, ET AL. (1990). HIV-related risk factors in acute psychiatric inpatients. Hosp Commun Psychiatry 41:449–451.

TODD J (1989). AIDS as a current psychopathological theme: a report on five heterosexual patients. Br J Psychiatry 154:253–255.

Update (1989). Acquired immunodeficiency syndrome—United States, 1981–1988. MMWR 38:229–236.

33 Connective tissue diseases

MICHAEL G. MORAN, M.D., AND SHERRY N. DUBESTER, M.D.

Formerly referred to as the collagen vascular diseases, the connective tissue diseases comprise rheumatoid arthritis, systemic lupus erythematosus, scleroderma, Sjögren's syndrome, other forms of arthritis, polymyositis, polymyalgia rheumatica, and various vasculitides. Their common clinical features include connective tissue inflammation, small vessel damage, serosal surface inflammation, and, frequently, involvement of the heart, lung, and kidneys. The etiology of many of these illnesses is unknown; but autoimmune processes have been implicated in almost all, and the laboratory evaluation usually demonstrates immunologic abnormalities. Although the inciting immunologic events are largely unknown, subsequent steps in immunologic pathogenic mechanisms include the formation of antigen-antibody complexes that fix complement, followed by an influx of cellular components of the inflammatory response. Vascular damage and tissue destruction result. The symptoms of the various diseases coincide with the involvement of specific organs or organ systems and can be varied; the diagnosis can be elusive until findings characteristic of the particular disease appear (Rodnan, 1973).

Neuropsychiatric manifestations of these diseases are complex. There are three categoric presentations: direct involvement of the central nervous system (CNS), with affective disturbances, delirium, or other cognitive impairment disorders; psychological sequelae of the patient's awareness of the illness or its impact; and psychiatric side effects of the drugs used to treat the illness. The prevalence and nature of psychiatric comorbidity in connective tissue diseases is unclear. Studies that have used tools such as the Minnesota Multiphasic Personality Inventory (MMPI) have been criticized for not discriminating adequately between symptoms related to the underlying medical illness and those due to primary psychiatric disturbances (Pincus, 1986). With systemic lupus erythematosus, distinguishing between psychological reactions to illness and neuropsychiatric lupus is often difficult even when sophisticated testing is used. Specific diagnostic dilemmas and guidelines for their evaluation are discussed in this chapter.

LABORATORY DIAGNOSIS

Although the laboratory findings specific to each disease are discussed in the sections on the individual illnesses, some general introduction to these tests may be helpful. For the psychiatrist presented with a patient who has psychiatric symptoms, and in whom a connective tissue disease is suspected, history and physical findings, as in any other clinical setting, should guide the selection of laboratory tests ordered (see Table 33-1).

Changes in serum proteins termed acute phase reactants are associated with elevation of the *erythrocyte sedimentation rate* (ESR). An elevated ESR is a nonspecific finding that can serve as a rough guide to the activity of several connective tissue diseases: rheumatoid arthritis, systemic lupus erythematosus (SLE), giant cell arteritis, polymyalgia rheumatica, and the vasculitides.

As mentioned before, many of these diseases are thought to be mediated by antibodies to various cell components, such as intranuclear structures. When the antibodies react with appropriate antigen, immune complexes are formed, with activation of the *complement* sequence. Measured levels of complement reflect an equilibrium between rates of synthesis and catabolism. Levels can serve as a rough guide to disease activity, most reliably for SLE. All three commonly assayed levels (CH50, C4, and C3) are low in SLE. With drug-induced lupus the complement levels are normal (Volanakis, 1986).

Antinuclear antibodies (ANAs) are antibodies directed against different nuclear components. The lupus erythematosus (LE) assay, or "LE prep" was the first test for antinuclear antibodies; however, because it lacks both sensitivity and specificity it is no longer used as a screening test. ANAs now are detected by immunofluorescence. Fluorescent antinuclear antibody assays (FANAs) yield a visual picture of the autoantibody specificity; the nuclei of the patient's test cells fluoresce with patterns categorized as homogeneous, peripheral, speckled, centromere, and nucleolar (see Table 33-2). The *homogeneous* pattern is associated with antibodies to

TABLE 33–1. *Overview of the Patient with Suspected Connective Tissue Disease*

1. History
   a. Musculoskeletal pain
   b. Morning stiffness
   c. Swelling
   d. Weakness
   e. Fatigue
   f. Also: pulmonary, gastrointestinal, and neurologic symptoms; fever, weight loss, anorexia; mucocutaneous lesions, photosensitivity, dry eyes or mouth; family history
2. Physical examination
   a. General examination including thorough neurologic evaluation.
   b. Examination of specific joints for deformity, swelling, tenderness, warmth, range of motion, and crepitus
3. Laboratory evaluation
   a. Chemistry profile
   b. CBC
   c. Specific tests for the following disorders (Schumacher, 1988)
   (1) RA: RF, ANA, ESR
   (2) SLE: ANA (if ANA positive, then antibodies to double-stranded DNA) (Schur, 1990)
   (3) Fibromyalgia: ESR, TFTs, RF, ANA (Yunus, 1988)
   (4) Polymyositis: aldolase, CPK with MM fraction
   (5) Temporal arteritis: ESR
   (6) Scleroderma: ANA, RF
4. Consultation with a rheumatologist (when the history suggests possible connective tissue disease)

TABLE 33–2. *Patterns of Antinuclear Antibodies with Certain Diseases*

ANA	Disease
Homogeneous	SLE Drug-induced lupus
Peripheral	SLE
Speckled	Scleroderma Mixed connective tissue disease SLE Sjögren's syndrome
Centromere	CREST variant of scleroderma
Nucleolar	Raynaud's phenomenon Scleroderma Sjögren's syndrome

Adapted from Schur (1990).

the DNA-histone complex, which is most often seen in SLE, drug-induced lupus, and rheumatoid arthritis (Schur, 1990); the homogeneous pattern is also seen in the fewer than 5% of normal individuals who are ANA-positive (Rodnan & Schumacher, 1983). The *peripheral*, or rim, pattern reflects antibodies to DNA. The *speckled* pattern is associated with antibodies to several nuclear components; it is most often seen with scleroderma, mixed connective tissue disease, SLE, and Sjögren's syndrome. The *centromere* pattern, reflecting antibody to chromosomal centromeres, is seen in the CREST (calcinosis, Raynaud's phenomenon, esophageal dysmotility, sclerodactyly, telangiectasias) variant of scleroderma. The *nucleolar* pattern, due to antibodies to nuclear RNA, may be seen with scleroderma, Raynaud's phenomenon, and Sjögren syndrome (Schur, 1990). One can also order specific determinations of serum antibodies to double or single-stranded DNA or, when drug-induced lupus is suspected, to histones (Pincus, 1986). Note that approximately 75% of patients with SLE have antibodies to double-stranded DNA, a finding that is almost unique to SLE, whereas antibodies to single-stranded DNA are much less specific (Schur, 1990).

Rheumatoid factors designate autoantibodies of several classes, most often immunoglobulin M (IgM), that react with the Fc portion of IgG. The autoantibodies form in response to conformationally-altered immunoglobulins. Though in some conditions the autoantibodies are thought to contribute to tissue damage, rheumatoid factors in some cases are protective by decreasing the quantity and pathogenicity of circulating immune complexes. Rheumatoid factors are increased in most chronic inflammatory disorders and any illness with continuous immune complex formation (Schur, 1990). Examples of conditions in which rheumatoid factors are present include infective endocarditis, tuberculosis, syphilis, and sarcoidosis. Rheumatoid factors are usually absent in enteropathic arthritis, Reiter syndrome, and ankylosing spondylitis (Schumacher, 1988). The test is positive in 84% of patients with rheumatoid arthritis, with high titers suggesting RA as the diagnosis. Elderly patients often have increased levels of rheumatoid factors, making this test less helpful in the geriatric population with arthritis symptoms (Schur, 1990).

Muscle destruction and inflammation play an important role in some connective tissue diseases, such as polymyositis. Evidence of such inflammation can be found in the serum: intracellular muscle enzymes are liberated by cellular destruction. Aldolase and creatine phosphokinase (CPK) are the tests most commonly ordered. One should ask for "fractionated" CPK: MM patterns reflect CPK of skeletal muscle origin.

SYSTEMIC LUPUS ERYTHEMATOSUS

When the American Rheumatism Association offered criteria for the classification of systemic lupus erythematosus (SLE), it tried to bring order to the understanding of a protean illness. The clinical presentations of the disease are varied and include skin rash (especially the classic malar erythematous rash), polyarthralgias and arthritis, serositis (pericarditis and pleurisy), renal, cardiac, and neurologic abnormalities, and anemia and thrombocytopenia (Tan, 1985) (see Table 33-1).

Women are at much greater risk for contracting SLE than are men, with the prevalence ratio about 8:1. Female sex hormones may modulate the immune response in such a way as to increase the risk of SLE.

In general, noncaucasian races have a higher prevalence of SLE than Caucasians. Asians are at three times the risk of Caucasians. Of all groups, black women are at highest risk, being about three to four times more likely to contract SLE than age-matched white female controls.

Familial aggregation seen in SLE probably has a genetic basis. In one survey of 225 SLE patients, 24% had relatives with the disease or another autoimmune illness. Relatives without overt symptoms often have abnormal laboratory immunologic profiles (Ball & Koopman, 1986). Some HLA tissue types (HLA-DR2, HLA-DR3, and HLA-DQ1) have a frequent association with SLE and with certain specific complement component deficiencies (Ahearn et al., 1982). Environmental factors do not appear to be of prime importance in accounting for the frequent patterns of SLE occurrence among close relatives.

Pathogenesis

Although incompletely understood, SLE seems to result from an immunoregulatory disturbance associated with a polyclonal activation of B cells, which in turn results in an increased production of autoantibodies. Genetic, environmental, and hormonal factors probably play a role in causing the immunoregulatory disturbance. The autoantibodies and corresponding antigens form immune complexes, with subsequent deposition in joints, small vessels, and glomeruli (Schumacher, 1988). Deposition is followed by activation of the complement system, which produces a specific type of cellular inflammatory response that disrupts vascular structure. Tissues supplied by the vessels and subject to perivascular inflammation suffer nutrient and oxygen deficits leading to infarction, causing organ malfunction. Organs involved in this manner in SLE include skin, joints, serous membranes, kidneys, and the nervous system. However, organ malfunction is not entirely on a vascular basis.

Another mechanism involves the direct interaction between autoantibodies and antigen at target sites; the antigen may be a cell constituent to which there are autoantibodies, or circulating antigen may be deposited in basal membranes (Schumacher, 1988). Activation of the complement system follows with the appearance of an inflammatory response. The association of deposited immune complexes with tissue damage would explain the lack of association between levels of *circulating* immune complexes and the degree of disease activity during exacerbations of SLE (Ball & Koopman, 1986). The presence of antibodies specific to components of the cell nucleus, or ANAs, is considered a hallmark of SLE and forms the basis of several important screening tests for the disease.

Laboratory Diagnosis

Autoantibodies have been demonstrated to most of the major known components of the cell nucleus. The LE cell phenomenon is the result of the interaction between antibodies to deoxyribonucleoprotein and the exposed contents of cell nuclei; in the LE prep this reactive mixture is then phagocytized by a polymorphonuclear leukocyte. The distended leukocyte is the LE cell (Hargraves, Richmond, & Morton, 1948). Although the test is relatively easy to perform, it is neither sensitive nor specific for the diagnosis of SLE. About 50 to 75% of SLE patients demonstrate the LE cell phenomenon. This test is no longer recommended for screening (Tan, 1985).

The FANA is much more sensitive as a screen, being positive in almost all patients with SLE. Titers above 1:160, when associated with clinical signs of disease, are significant and suggest the need for obtaining the anti-DNA antibody test to confirm the diagnosis of SLE. The homogeneous pattern is common but has no diagnostic specificity. When FANA is positive in rheumatoid arthritis patients, the pattern is usually speckled and the titer low (1:160 to 1:320). Rim patterns, at *any* titer, are suggestive of the diagnosis of SLE and should prompt the anti-DNA antibodies assay. Among asymptomatic and clinically normal persons, the FANA test is positive at a rate of 2 to 5%; false positives are more common in the elderly. In clinically normal individuals with positive ANA, titers are usually low (< 1:80) (Pincus, 1986). If the ANA is positive but at low titers and there are no symptoms to suggest a connective tissue disease, no further workup is indicated.

A test more specific for SLE detects the presence of antibodies to double-stranded DNA (Ceppelini, Polli, & Celada, 1957). Although only 50 to 80% of SLE patients have the antibodies to double-stranded DNA, these antibodies are seen so seldom in other conditions as to make their presence diagnostic of SLE.

Antibodies to histones (intranuclear polypeptides) are present in about half of all patients with lupus, and in about 75% of patients with drug-induced lupus (Pincus, 1986). Rheumatoid factor is present in a small number of lupus patients (Stage & Mannik, 1973). Almost all patients demonstrate an elevated ESR, which varies somewhat with the activity of the illness. About 20% of SLE patients have a chronic biologic false-positive test for syphilis; it may be positive years before other manifestations of the disease.

The clinician faced with the possible diagnosis of SLE should start with the FANA, as it is the best screening tool. Antibodies to double-stranded DNA confirm the diagnosis, although they are not always present.

Psychiatric Aspects of SLE

Systemic lupus erythematosus is of unknown etiology and is complex, poorly predictable, and often inexorable in its progression. It results in debility, economic hardship, suffering, pain, and death. Although it affects up to one million Americans, most people have never heard of it and have no experience with it. One would be hard put to construct a disease that would more severely test the psychological defenses of its victims.

The literature on the psychiatric aspects of SLE is confusing; usage varies regarding the terms psychosis, organic brain syndrome, delirium, and "toxic reactions" (Hall, Stickney, & Gardner, 1981; O'Connor, 1959; Stern & Robbins, 1960). Few studies are prospective, and many are simple chart reviews. In addition, figures for the prevalence of neuropsychiatric disease often omit affective disturbances, such as depressive responses (adjustment disorder with depressed mood, organic mood disorder, major depression). Studies are often small, consisting of 20 or fewer patients. Most surveys report the incidence of neuropsychiatric symptoms in SLE as being between 33% and 60% (Ball, 1986; O'Connor, 1959) and higher when affective disorders are included. The psychiatrist who wants a firm grasp on the neuropsychiatric aspects of SLE must often make extrapolations from studies of other rheumatic diseases, especially rheumatoid arthritis, about which much has been written. SLE makes cameo appearances in tables and charts outlining medical problems that present with psychiatric symptoms, but rarely gets full attention (Lipkin, 1985).

Neuropsychiatric abnormalities occur in approximately 50% of lupus patients and can reflect direct effects of the SLE, side effect of medications used to treat SLE, secondary metabolic disturbances, or psychological reactions to the illness. The neuropsychiatric disorder may be transient or may be associated with permanent neurologic deficits. Neuropsychiatric symptoms are unpredictable and may occur at any time during the illness, or even precede diagnosis. The most common autopsy findings in the brains of patients with cerebral lupus are microscopic infarcts and hemorrhages. Neuropsychiatric symptoms and brain pathology are loosely associated. A normal psychiatric presentation may coexist with an abnormal electroencephalogram (EEG), or positron emission tomography (PET) scan, or later positive findings at autopsy (Harris & Hughes, 1985).

From a descriptive standpoint, the clinical presentations of the neuropsychiatric aspects of lupus fall into four categories: secondary mental disorders, adjustment to illness issues, seizures, and peripheral nervous system disorders.

Secondary ("Organic") Mental Disorders

Secondary mental disorders in SLE range from subtle cognitive impairment to delirium to mood disorders and psychosis. No significant correlation has been found between cognitive impairment and psychological disturbance (Denburg, Carbotte, & Denburg, 1987). The cognitive impairment, as measured by neuropsychological testing, is not necessarily reflected in clinical evidence of the neurologic disturbance. However, cognitive impairment has been associated with the presence of lymphocytotoxic antibodies (Long et al., 1990).

The severity of delirium seen in SLE patients can vary widely, from a mildly disturbed level of attention and concentration to profound disorganization with agitation and hallucinations. The most common differential diagnostic problem with delirium in the SLE patient is determining whether the delirium is secondary to an exacerbation of the primary disease ("lupus cerebritis") or to CNS infection, metabolic derangement, or corticosteroid side effects. A thorough history, physical examination, and mental status examination early in the course provide a baseline with which subsequent examinations can be compared. Workup for the delirium usually should include vital signs, a metabolic profile such as an SMA-20, brain imaging, and a lumbar puncture with cerebrospinal fluid (CSF) cell count, glucose, protein, and IgG, as well as cultures for bacterial and nonbacterial infections.

Diagnosis of CNS lupus as the cause of the delirium must generally follow a thorough search for other causes. Clinically, presentations include meningitis with sterile CSF and cerebral dysfunction secondary to vasculitis, hemorrhage, or infarct. The appearance of new systemic symptoms of lupus or new vasculitic lesions, coincident with the appearance of the delirium, is diagnostically helpful. Elevation of CSF IgG favors the diagnosis, as does thrombocytopenia (Zvaifler, 1986). The EEG is rarely specific; delirium of almost any cause produces diffuse slowing. Moreover, the CSF of CNS lupus is usually normal, and there are no pathognomonic findings. The most common abnormalities are a mild pleocytosis (usually fewer than 50 mononuclear cells) and a slight elevation of protein (usually to less than 80 mg/dl).

Vascular and hemorrhagic sequelae of SLE involvement of the brain include subarachnoid hemorrhage, lupus meningitis (noninfectious meningeal inflammation that can present with a typical meningitic picture with increased intracranial pressure), or infarcts with stroke symptomatology. All can be associated with delirium; focal lesions from infarction produce the behavioral, cognitive, or sensorimotor symptoms expected from the location of the lesion. Subarachnoid hemorrhage and lupus meningitis are diagnosed by lumbar puncture. Subarachnoid hemorrhage in patients with SLE usually is associated with low CSF glucose (Ball & Koopman, 1986).

Computed tomography (CT) scans and magnetic resonance imaging (MRI) scans are routinely used for the differential diagnosis of delirium in these complex clinical situations (Reinitz, Hubbard, & Zimmerman, 1984); it is important to emphasize that radiologic studies must be interpreted in the context of the patient's history. In one study using CT scans, lupus cerebritis was associated with marked atrophy in more than 75% of patients. None of the cerebritis patients had normal scans. The control group—SLE patients without CNS symptoms—had normal scans in almost 75% of cases (Gaylis et al., 1982). Another CT study, however, suggested that there was a stronger association between steroid use and atrophy than between cerebritis and atrophy (Carette et al., 1982). Another group demonstrated atrophy only in cerebritis patients who had symptoms of gradual onset (Weisberg, 1986).

One MRI study suggested three patterns of cerebritis lesions: large areas of increased T2 signal, suggestive of infarcts; multiple small areas of increased T2 signal, suggestive of microinfarcts; and focal gray matter areas of increased T2 signal, which appeared to resolve on subsequent scans. No autopsy findings were reported in this study. Comparison with CT scans in the same group showed greater sensitivity of MRI for the detection of lesions. Of the seven patients who had lesions on MRI, only two had a positive CT scan (Aisen, Gabrielsen, & McCune, 1985). Another comparative study showed that the MRI procedure was more sensitive for detecting lesions in SLE patients than was the CT scan. Of the nine patients studied, MRI found lesions in eight patients, and CT found lesions in six. Both techniques showed one patient to be free of lesions, and lesions were of similar degree in three additional patients. Three patients showed more areas of involvement on MRI than on CT. In addition, MRI showed all lesions with greater clarity than did CT (Vermess et al., 1983). Another study concluded that the combination of PET and MRI provided an even more sensitive workup for neuropsychiatric SLE (Stoppe et al., 1990).

Metabolic abnormalities are a potential cause of delirium, especially in those lupus patients with severe renal involvement. Renal failure, with its consequent electrolyte imbalances and potential for fluid overload, can produce delirium in the SLE patient. Laboratory examinations reveal elevated blood urea nitrogen (BUN) and creatinine; vital signs show hypertension or postural changes in blood pressure. Physical examination can demonstrate edema in cases of fluid overload or decreased skin turgor with volume depletion. Dialysis can be an effective treatment for many of the neuropsychiatric symptoms that result from renal failure. However, if dialysis is accomplished too rapidly, with marked shifts in extracellular volume and electrolyte composition, transient affective and cognitive disturbances may result (see Chapter 27, regarding dialysis and associated mental changes). Liver failure due to hepatic involvement by lupus can produce delirium that is especially difficult to treat. Restriction of dietary protein, judicious use of steroids, and control of intestinal flora with oral antibiotics can be helpful for managing delirium or coma due to hepatic failure.

Other sources of delirium in the lupus patient are systemic and CNS infections. Because of their immunocompromised state, lupus patients are susceptible to infection by a variety of pathogens—bacterial, parasitic, and fungal. The use of corticosteroids and immunosuppressive agents such as cyclophosphamide to treat the disease may further inhibit the immune system. In many reported series, infection was the most common cause of death (Ball, 1986). Sites of infection include the gut, lung, skin, and urinary tract. Hypoxemia in cases of pulmonary infection or fibrosis can worsen the delirium; frequent blood gas determinations are necessary in such instances.

The treatment of delirium caused by CNS lupus is to give corticosteroids or to increase the dosage if the patient is already taking them. If steroid dosage has already been increased at the time of the consultation but the symptoms of delirium preceded the dosage increase, the patient should be observed for a few days at the increased dose, as most instances of delirium so treated remit. Acute management of the agitation and psychotic symptoms can be achieved with low doses of neuroleptics. Failure to respond should prompt further diagnostic evaluation, as the workup in this clinical setting is chiefly a process of elimination of likely causes. (See Chapter 19 for a comprehensive discussion of the general topic of delirium and its management. The focus here is on those features of the differential diagnosis of delirium that have special relevance for the SLE patient.)

Delirium due to cortiscosteroid side effects usually follows a change in dosage, usually an increase. Confusion, alterations in the personality, and occasionally mood changes are seen. The history of recent initiation or increase in dosage of steroids is crucial to the diagnosis. Psychotic disorder from steroids is more frequent at high doses. Treatment includes lowering the dosage of steroids and adding a small dose of a neuroleptic, for example, haloperidol 2 to 4 mg per day (Hall, Stickney, & Gardner, 1981). The occurrence of psychotic delirium in these cases is not reliably predictive of its recurrence should the steroids again be increased (Hall et al., 1979). A mild delirium accompanied by paranoid ideation and derealization is seen occasionally in patients rapidly tapered down from a lengthy high-dose course (e.g., a drop to 20 mg of prednisone a day from 80 mg a day over a period of fewer than 4 days) (W. Thompson, personal communication, May 1987).

When evaluating the SLE patient with acute onset of psychosis or depression, testing for antibodies to ribosomal P proteins has been demonstrated to be clinically useful (Schneebaum et al., 1991). Elevated levels of these autoantibodies are associated with severe depression and psychosis in SLE. If present, treatment should be directed at the SLE; improvement in mental status and a decrease in the level of autoantibodies should occur with treatment of the SLE. If autoantibodies to ribosomal P proteins are not present in the clinical setting of severe depression or psychosis, other causes for the mental status changes should be actively sought and treatment appropriately directed.

Observed or self-reported depressive affect is a common finding among SLE patients. The prevalence of mood disorders is difficult to determine, however, as most studies of prevalence have not been based on standardized diagnostic criteria. Reports suggest that 50 to 80% of SLE patients complain of depressed mood at some point during their illness. One large study of ambulatory outpatients revealed self-reported "lowered spirits" in 70% of patients. In that same group, MMPI results showed abnormal depression scores in 41% (Liang et al., 1984). In order of decreasing frequency, the patient's most frequently reported fears were of worsening disease, death, and disability. Complaints about limitations imposed by the disease, such as easy fatigability and the need to avoid the sun, were also prominent. Loss of physical functioning, the lack of freedom to plan a family (especially when pregnancy resulted in a relapse), and reduced social functioning were highly correlated with elevated depression scores. Sexual activity decreased significantly after the onset of the illness. However, one study showed that SLE patients can maintain healthy social functioning if the difficulties inherent in the illness are recognized and addressed (Stein et al., 1986).

Some patients believe that the illness has a positive effect on their lives, having made them aware of their mortality in a new and useful way. This revelatory experience has been described with other illnesses (Liang et al., 1984).

Dysthymic disorder and adjustment disorders with depressed mood are best treated initially with medical psychotherapy alone; antidepressants should be prescribed for patients with vegetative symptoms that persist despite psychotherapy. Major depression usually requires use of antidepressants in addition to measures that address the primary causes: addition of steroids (for cases of CNS involvement), and correction of metabolic problems or drug toxicity.

Seizures

Convulsions are a common complication with lupus patients, occurring in about 15% of cases. Presentations can be focal, akinetic, or grand mal in type. Generalized motor seizures, the most common type, may result from the underlying disease, uremia, sepsis, or hypertensive encephalopathy. Convulsions may occur before the diagnosis of SLE and tend to occur early in the illness. Seizures are rare or absent in some patients but a recurrent, relapsing problem for others (Harris & Hughes, 1985).

Intracerebral hemorrhage (made more likely because of thrombocytopenia), meningoencephalitis, and cerebral infarctions can produce seizures as complications. Seizures should thus be seen as ominous signs of possible intracranial pathology and should be evaluated with extensive physical and laboratory examinations, including CT or MRI scans. Perhaps the most common cause of seizures is a metabolic derangement: uremia (Johnson & Richardson, 1968; Rodnan, 1973). Treatment consists chiefly of hemodialysis. End-stage hepatic failure can also result in convulsions. Use of anticonvulsant medication should not be substituted for a thorough evaluation of the origin of the seizures. If anticonvulsants are given, it is important to recognize the potential interaction of steroids with some anticonvulsants: Phenytoin and phenobarbital, for example, alter the metabolism of oral corticosteroids and decrease their efficacy. Therefore higher doses of steroids may be required if these anticonvulsants are used. There is no known interaction between carbamazepine and corticosteroids. When a patient with SLE who is taking anticonvulsants fails to responds to treatment, drug levels should be checked and a clinical pharmacologist should be considered.

Peripheral Nervous System Disorders

Neurologic disturbances outside the CNS are fairly common, especially in patients with severe disease (Rodnan, 1973). Thrombocytopenia may be a marker for those at greatest risk for these complications and perhaps for CNS disease as well (Sergent et al., 1975). Peripheral neuropathies, myopathy, and myasthenia gravis are among the clinical presentations. Transverse myelitis is an especially ominous occurrence and can be a cause of death. Arteritis and thrombosis of spinal vessels, spinal subarachnoid hemorrhages, and myelomalacia can also occur, producing varying syndromes of hemiplegia and sensory disturbances. Brachial plexus neuropathy presents with decreased strength and sensation in the arms and hands; CT scans may have a place in its diagnosis (Pillemer et al., 1984). Management for most of these complications is supportive, with special attention to the development of infection in those patients who are immobilized or who require surgical measures as part of the treatment. Some of these disorders, such as the uncomplicated peripheral neuropathies, may respond to corticosteroid treatment.

Other neurologic disturbances that can occur with SLE are brainstem strokes, coma, aseptic meningitis, chorea, and encephalomyelitis (Ball, 1986).

Steroid Withdrawal Syndrome

Rapid tapering of steroids can produce a withdrawal syndrome consisting of flu-like symptoms (myalgias, low grade fever, anorexia) and even joint effusions and postural hypotension. The withdrawal symptoms are occasionally difficult to differentiate from a recrudescence of the underlying illness. However, by holding the steroid dose constant, one can generally distinguish between the withdrawal symptoms, which go away in a few days to a week or two, and the underlying illness, which persists or worsens. Nonsteroidal antiinflammatory drugs (NSAIDs) may help with the withdrawal symptoms. (See also Chapter 29.)

As a rule, alternate-day steroid administration has no place in the treatment of SLE because alternate-day administration does little to suppress disease activity. However, in those patients already on such regimens, the most common neuropsychiatric disturbance is mild insomnia occurring on the day of the steroid dose.

Drug-Induced Lupus-like Syndromes

Drug-induced lupus syndromes were initially described during the 1940s and 1950s in relation to the administration of penicillin and the sulfonamides. Other drugs have subsequently been implicated, including procainamide, isonaizid, hydralazine, methyldopa, quinidine, chlorpromazine, beta-blockers, anticonvulsants (especially the hydantoins) and the antithyroid drugs propylthiouracil and methimazole (Hahn et al., 1972). Procainamide is the most frequent offender, with 50% of patients developing antinuclear antibodies and approximately one-half of those developing clinical symptoms.

The syndrome usually consists of polyarthralgias and myalgias initially, followed by fever, pleuritic pain, and migraines (Schumacher, 1988). Renal disease is uncommon, though deaths from severe glomerulonephritis and renal failure have occurred. CNS involvement is rare in drug-induced lupus. Positive LE cell reactions and antibodies to double-stranded DNA may be present. Some patients develop biologic false-positive tests for syphilis. Antibodies to histones are the most sensitive test for the drug-induced syndrome, being positive in about 75% of patients.

RHEUMATOID ARTHRITIS

A disease of unknown etiology, rheumatoid arthritis (RA) causes chronic, symmetric nonsuppurative inflammation of the synovium of the diarthrodial joints, as well as many extraarticular manifestations. Inflammation of the synovium is the primary pathologic event, resulting in pain and swelling with accumulation of fluid in the joint space. The chronically inflamed synovium is called a pannus. Recurrences cause proliferation of the pannus, with invasion and erosion of periarticular bone, gradual destruction of joint cartilage, and weakening of tendons and ligaments. This process results in poorly functioning or nonfunctional joints, immobility, and crippling. In some patients the disease pursues a malignant course, with rapidly destructive arthritis and diffuse vasculitis. Other variants of the disease are milder, with frequent remissions and only mild relapses easily controlled by antiinflammatory agents.

Rheumatoid arthritis is a common disorder, with a population prevalence between 0.3% and 1.5%. Two to three times as many women as men are affected. The disease can present insidiously or rapidly; in most patients symptoms begin gradually, with malaise, fatigue, and diffuse musculoskeletal pain often reported before the onset of specific joint symptoms. The course of RA is variable, though typically it is at first intermittent, becoming more constant over time. Most patients have periods of at least partial remissions; after continuous symptoms for 2 years, however, a complete remission is improbable (Schumacher, 1988).

The prognosis in patients who have persistent RA seems poor, although studies vary as to observed outcomes. Many see RA as producing occupational disability in more than half of the patients by 10 years, with most therapeutic endeavors having little effect on disability (Hardin, 1986).

Adult patients often have their first arthritic symptoms in the hands, with pain, swelling, stiffness, and warmth as prominent features. Symptoms are usually worse in the morning. Knees, ankles, wrists, and elbows are involved in the disease process. Bilateral symmetry is the rule, with sparing of the distal interphalangeal (DIP) joints. Subcutaneous nodules can be found along the olecranon processes and the ulna in about one-fourth of patients. Joint deformity is common and is visible in the classic ulnar deviation of the fingers (Rodnan, 1973).

Rheumatoid arthritis is a systemic disease; extraarticular complications develop in more than three-fourths of patients and are probably more common in those with high titers of rheumatoid factor (see the subsection on laboratory diagnosis) or with subcutaneous nodules. Neuropsychiatric complications, more common in those without rheumatoid factor, are exceptional. Organ involvement can include the heart, lung, eye, and nervous system, as well as certain systemic complications. Heart involvement includes pericarditis, cardiomyopathy, and valvular lesions (especially aortic). Pulmonary disease may manifest as pleurisy, rheumatoid nodules in the parenchyma, progressive fibrosis, and pneumoconiosis with nodules (Caplan syndrome) (Caplan, Payne, & Withey, 1962). In the eye, scleritis and iridocyclitis occur.

Neurologic involvement occurs when neurologic structures or their vascular supply are affected by vasculitis or rheumatoid nodule formation. In addition, joint, ligament, and tendon failure in the area of major nerves can cause compressive neuropathy or myelopathy, the former seen most commonly in the median nerve (with thenar wasting), ulnar nerve, and peroneal nerve. Myelopathy may occur if cervical spine involvement leads to cord compression. CNS involvement is rare, but vasculitis and rheumatoid nodule-like granulomas in the meninges have been reported (Schumacher, 1988). A mild, chronic sensory polyneuropathy can occur, as well as an acute severe sensorimotor neuropathy (Chamberlain & Bruckner, 1970).

Pathogenesis

Although the primary etiology of RA remains unknown, many factors point to its being an immunologically based disease. A microvascular injury is probably the first event in the pathogenetic pathway. Synovial membrane proliferation occurs because of repeated inflammation. Local antibody production follows, with immune complex formation between rheumatoid factors and IgGs. The complement sequence is activated, and vascular fluid and cellular components invade the joint space. Proteolytic enzymes are released, with protein destruction and further inflammation. At extraarticular sites, circulating immune complexes may contribute to the pathologic features (Hardin, 1986; Panush, Bianco, & Schur, 1971).

The strong association of RA with human lymphocyte antigen HLA-DR4 suggests a hereditary immunologic predisposition to the disease. This association is seen in most ethnic groups; African-Americans with RA are an exception. Though HLA-DR4 individuals have a greater chance of developing RA, only few such individuals are affected (Schumacher, 1988).

Systemic complications include the anemia of chronic disease, keratoconjunctivitis sicca, amyloidosis, and Felty syndrome (leukopenia with splenomegaly). RA patients may have fatigue and myalgias as part of the symptom complex. As the illness progresses, muscle wasting, weight loss, and decreased appetite can occur.

Functional capacity in patients with RA was classified by Steinbrocker, Traeger, and Batterman (1949):

Class I	No impairment, all activities are performed.
Class II	Normal duties are performed, but there is slight limitation or discomfort in one or more joints, no assistive devices are needed.
Class III	Only a few duties of job or self-care are possible; assistive devices may be needed.
Class IV	Incapacitated, bedridden, or wheelchair-bound; no self-care is possible.

Laboratory Diagnosis

The most characteristic laboratory finding in RA is the presence of the rheumatoid factor (RF). Depending on which of the various available tests are used, 70 to 90% of cases that are strictly selected on clinical data are seropositive. RF is composed of autoantibodies (chiefly IgM) that react with the Fc protein of IgC. Although it is not certain that the ability of RF to form immune complexes is of specific etiologic importance in the disease, evidence is mounting that supports this assertion. Results of the RF assay are reported as the titer, and a higher titer is correlated with greater likelihood of correct diagnosis of RA, tissue type HLA-DR4, aggressive joint disease, nonneuropsychiatric extraarticular disease, subcutaneous nodules, neuropathy, and Felty syndrome (Spalding & Koopman, 1986). In the section on psychiatric aspects of the disease, these subjects are discussed relative to seropositivity.

Rheumatoid factor is not specific for RA, being found in other inflammatory diseases as well: infective endocarditis, leprosy, syphilis, sarcoidosis, and tuberculosis. Other rheumatic diseases may have RF as well, including SLE, scleroderma, and polymyositis (Cohen, 1975). Among the general population without rheumatic disease, seropositivity occurs at a rate of about 3%. With aging, the prevalence of RF increases, to about 14% in men and 9% in women over 70 years of age. Titers among the asymptomatic seropositive elderly are low (1:40 to 1:80).

The ESR is a nonspecific index that may vary proportionally with the activity of the disease. ANA is present in about 30% of RA patients, usually in those positive for RF. Speckled patterns are most typical, at titers from 1:160 to 1:320.

Psychiatric Aspects of RA

Psychiatrists may be asked to deal with the psychiatric aspects of RA from at least three perspectives: as emotional contributors to the illness or its exacerbations; as the emotional consequences of the disease and its implications; and as psychiatric complications of the treatment, especially side effects of medications (Anderson et al., 1985). Physicians may believe that certain psychological events or stressors bring on exacerbations of the disease. Patients commonly report the occurrence of severe emotional distress just prior to the first clinical evidence of the illness. In addition, the psychiatrist may be consulted regarding the way in which the patient's emotional life affects rehabilitation and treatment efforts. Knowledge of the patient's defensive and adaptive styles is crucial to long-term management of the case.

Decades of interest in RA as one of the classic "psychosomatic" illnesses has resulted in numerous studies of RA patients' psychological characteristics (Halliday, 1942; Thomas, 1936). Enthusiasm reigned in the search for a "rheumatic personality" and thus generated many data. RA patients were characterized as self-sacrificing, masochistic, inhibited, perfectionistic, retiring, and interested in outdoor sports (Moos, 1964; Moos & Solomon, 1964). They were thought to be people whose emotions were "unavailable" to them (Silverman, 1985). Women with RA actively suppressed anger and sexuality; men were depressive. Gardiner (1980) found RA patients to be "more neurotic, more likely to give socially undesirable responses, and more prone to psychiatric disturbances" than the general population. Other work seems to contradict these findings, showing that RA patients' personalities hardly differ from those of a nonpatient population (Cassileth et al., 1984; Crown, Crown, & Fleming, 1975). The concept that specific personality types or traits are associated with this disease, as well as other so-called psychosomatic illnesses, is now considered anachronistic.

It is tempting, nevertheless, when speaking of a disease that seems to be immunologically mediated, to look for ways in which the immune system may be affected by stressors in RA patients. Indeed, many have looked for such a causal relation (Adler, 1981; Solomon, 1969), and effects on the immune system have been found in studies not restricted to RA patients (Rogers, Dubey, & Reich, 1979). Lymphocyte proliferation in response to mitogen stimulation, immunoglobulin concentrations, and the course of adjuvant arthritis in rats have been shown to respond to environmental stressors (Amkraut, Solomon, & Kraemer, 1971; Silverman, 1985). Natural experiments in humans have implied autonomic or other neurologic determinants in the course of RA. Arthritic limbs may improve after denervation; even in disease of long duration hemiparetic limbs are unaffected (Rodnan, 1973; Silverman, 1985).

A variety of psychological reactions to the stresses of RA have been described (Kahana & Bibring, 1964; Lipowski, 1975; Meenan, 1981). About 40 to 50% of RA patients report significant depressive affect on psychological testing, making depression the most common emotional complaint (Silverman, 1985). Psychotic disturbances are seen only rarely. Some researchers found elevated MMPI scores on the hypochondriasis, depression, and hysteria scales in RA patients; however, the MMPI responses that were typically responsible for the abnormal scores may be more consistent with the presence of RA than of psychopathology (Pincus, Leight, & Bradley, 1986). One prospective study found that symptoms of depression and anxiety actually predicted a good outcome, whereas externalized hostility predicted a worse outcome (McFarlane, Kalucy, & Brooks, 1987). Other work contradicts previous findings that the presence of RA correlates with psychopathology (Pow, 1987).

One study suggested that measures of social stress and RA severity offer the most accurate prediction of psychiatric disturbance (Creed, Murphy, & Jayson, 1990). Researchers have demonstrated the importance of social support as both a predictor of psychiatric difficulties and as an important focus of treatment. Murphy and coworkers found that psychiatric disorders in RA patients correlated significantly with social stress and lack of support; psychiatric symptoms included abnormal illness behavior (Murphy, Creed, & Jayson, 1988). Patient adjustment to illness was found in one study to be significantly related to the attitude of the spouse: RA patients who perceived their spouse as being supportive, regardless of actual spousal support, coped more effectively with their disease (Beckham, 1987).

Almost all patients believe that their mobility is undermined, and many rate impaired mobility as their "biggest problem" (Liang et al., 1984). This loss of potential loss of mobility obviously directly affects patients' sense of independence. Such a loss may be especially difficult for the elderly (Weiner, 1975). "Total disability" was rated as the biggest fear by one group of patients (Liang et al., 1984). The psychological implications of this dreaded consequence are many: Ultimate dependence on others, fear of burdening others, fear of precipitating a divorce, and loss of control with resulting helplessness are but a few (Cobb, Miller, & Wieland, 1959; Medsger & Robinson, 1972). The repeated losses (of function, ability, mobility, relationships, and sense of body integrity) can produce a state virtually identical to the grief seen with object loss, and it should be treated as such. Medical psychotherapy focused on working through the losses associated with the illness is helpful. (See Chapter 1.)

Assessment and Treatment Strategies

If symptoms of a major depressive disorder appear, antidepressant medication should be considered (Rimon & Laakson, 1984; Robinson, 1977). Sleep is often disturbed by joint pain and stiffness. Treatment should begin with antiinflammatory agents, but occasionally patients need sedative-hypnotics (Rogers, 1985).

The importance of taking a sexual history in these patients may be missed by those physicians who do not see the person with RA as still having a sexual life. Even those who are wheelchair-bound may remain sexually active (Yoshino & Uchida, 1981). Fear of imposing on or repulsing a spouse, of becoming too fatigued during sex, or of failing in sexual performance may make sex a conflicted arena for these patients, but many retain their interest (Baum, 1982; Wolpaw, 1960). Sexual partners of RA patients, too, may develop fears about sex: of being too demanding with a sick person or hurting or humiliating the partner. The internist or rheumatologist can play a facilitative role in getting the patient and partner to discuss their fears or in referring the couple for psychotherapy. (See Chapter 15.)

Analgesics, used for other than antiinflammatory effect, may have a place in the treatment of RA. The general experience is that persons with RA rarely become addicted "chronic pain" patients, although we may wonder why (Rogers, 1985; Rogers, Liang, & Partridge, 1982). The alterations in physical appearance in RA are sometimes apparent, even to the lay observer. Prominent rheumatoid nodules and the use of a cane or wheelchair are difficult for many patients to tolerate. Patients may react with shame and attempt to conceal their bodies or restrict their social contacts. Some degree of denial is obviously helpful in any adaptive effort, and some studies of RA patients suggest that those who see themselves as persons other than "arthritic patients" and do not adopt permanent images of themselves as sick (or deformed) may actually have better outcomes (Rogers, Liang, & Partridge, 1982). Patients who restructure their long-term goals do better functionally and psychologically than those who blame themselves or hope for unrealistic outcomes (Parker et al., 1988).

Some arthritic patients suffer disturbed sleep, often due to pain or stiffness, and may request a medication. We recommend low doses of sedating antidepressants, rather than benzodiazepines, if the medication is going to be used more than a few weeks. In these situations, one should caution the patient about anticholinergic side effects.

Some specific psychological interventions and approaches may be helpful. When the diagnosis is first made, the physician should be aware of the impact of so significant an event; the patient is being told he or she has a chronic, potentially debilitating disease. "Dosing" of the news and allowing the patient to react are essential. Many patients feel guilty and wonder what they might have done to bring this calamity on themselves. An understanding attitude and a clear explanation of the disease are needed. The physician should describe the course of treatment and make it clear that the patient's effort and participation are necessary to achieve a good response. Goals of disease control, not cure, are usually discussed at this point. The aim is the prevention of disability. Most internists (one should also read here *rheumatologists*) can deal with the expectable emotional distress, but they should refer to a psychiatrist when anxiety, depression, or denial severely interfere with functioning or participation in the treatment program (Rogers, Liang, & Partridge, 1982).

The ongoing and chronic financial costs of the illness impose severe burdens on many patients. Loss of insurance and the ability to pursue their regular occupations may aggravate this problem (Baum, 1982; Meenan, 1981). Effects of financial strain may be insidious. Finances are frequently offered as a reason for not buying needed medications, for not making office visits, and for missing physical therapy and other rehabilitation appointments. Patients with advanced RA incur medical expenses many times the national average, but they are rarely referred to social service organizations (Meenan et al., 1978). Early awareness of how financial difficulties and social problems are interfering with continuing care can enable the physician to intervene and make timely referrals (e.g., to social services) that can help restore the patient's full participation in the treatment program (Moran, 1987).

Rehabilitation outcome has been studied extensively in RA patients (Moldofsky & Chester, 1970; Rosillo & Vogel, 1971; Vogel & Rosillo, 1969, 1971). Not surprisingly, patients who are highly motivated tend to have better outcomes. Those who are rated as "intelligent" also fare better in their rehabilitation. Patients who have dysphoric moods unassociated with pain have significantly worse outcomes than those with dysphoric moods only at times of pain. Cognitive factors in chronic pain syndromes have been found to be useful predictors of pain and disability (Flor & Turk, 1988). Passive coping strategies in the presence of significant pain contributes to the highest degree of depression (Brown, Nicassio, & Wallston, 1989).

High ego strength on psychological testing predicts a good rehabilitation outcome. Poor impulse control is associated with a poor prognosis (Silverman, 1985). Those patients who see their treatment goals as adjustable and flexible, rather than rigid, achieve more in rehabilitation programs. A positive attitude toward the rehabilitation personnel is also conducive to good outcome (Rosillo & Vogel, 1971; Vogel & Rosillo, 1969, 1971).

Some centers incorporate as part of their medical treatment teams a psychiatrist, a social worker, and a psychiatric nurse. The social worker can help assess the patient's personal, family, and financial resources, and the psychiatric nurse can help the patient identify stress-inducing situations and design coping strategies. Occupational and physical therapists also have a crucial role in the overall plan. Staff meetings, with the psychiatrist in attendance, allow early detection of affective disturbances and psychiatric side effects of medications, as well as subtle manifestations of emotional distress that, if not recognized, may mark the beginnings of "noncompliance." This comprehensive approach to the RA patient is especially useful in complex cases (Silverman, 1985).

Psychiatric Side Effects of Medications

In a review, Rogers (1985) discussed in detail the psychiatric side effects of drugs used to treat RA, as well as the use of psychotropics in RA patients. Other excellent reviews have appeared (Cuthbert, 1974; Drew, 1962; Lewis & Smith, 1983). The salient points are discussed here. Three categories of drugs are typically employed to treat RA: simple analgesics, antiinflammatory drugs, and so-called remittive drugs.

Analgesics are often necessary adjuncts to the commonly used antiinflammatory medications. Acetaminophen and codeine are typical examples of analgesics. The antiinflammatory drugs form the foundation of drug treatment and serve to interdict the pathologic process, although they are generally held to be incapable of inducing a remission. Salicylates were once the standard of this category; they are used in doses higher than those for common analgesia. NSAIDs are now the most commonly used agents of this type; ibuprofen and indomethacin are examples. Corticosteroids also have a place in the treatment of RA, both systemically and locally administered.

Those patients with severe disease may be considered candidates for "remittive drugs." Examples of these drugs include gold sodium thiomalate, D-penicillamine, the antimalarials, azathioprine, sulfasalazine, and methotrexate (Hardin, 1986; Medical Letter, 1989). Usually not employed for mild disease because of side effects and expense, these medications work by mechanisms still poorly understood and often require months to evoke an effect. Sometimes an induced remission is difficult to differentiate from spontaneous remission.

Nonpsychiatric side effects of the salicylates are commonly known: gastric irritation, with mucosal erosion and bleeding, and disturbed platelet functioning. The local irritation can be somewhat reduced by the use of enteric coated preparations. Blood levels of 15–30 mg/dl are considered therapeutic; levels are measured 2–3 hours after the last dose. NSAIDs can cause bone marrow suppression, hepatic and renal dysfunctions or gastrointestinal distress. Gold toxicity manifests in renal, skin, and hematologic side effects. D-Penicillamine can cause autoimmune diseases such as SLE. The antimalarials can cause retinal damage.

As for psychiatric side effects, salicylates can cause delirium that can reach psychotic proportions (Greer, Ward, & Corbin, 1965). The timing of the occurrence of toxic symptoms correlates poorly with the blood salicylate level, but such levels are helpful for determining the overall severity of the intoxication. Virtually all patients will experience some symptoms of toxicity at blood salicylate levels higher than 50 mg/dl. The NSAIDs cause dizziness, vertigo, and headache in up to 50% of patients. Sedation and delirium, with associated sleep disruption, have also been reported. Longer-acting agents, such as piroxicam, have half-lives of up to 86 hours and may take two weeks or more to reach steady state. As a result, side effects may not appear for many days after beginning the medication. The gastrointestinal distress seen in RA patients is sometimes due to anxiety or depression, which would require separate treatment. Paranoia, depression, decreased concentration, anxiety, confusion, hallucinations, and hostility have been reported with many NSAIDs (Medical Letter, 1989).

Gold has been reported to cause confusion and hallucinations. Behavioral abnormalities may occur with high-dose regimen of methotrexate. Poorly described psychiatric disturbances have been reported with penicillamine. Among the antimalarials, psychiatric side effects seem to be more common with chloroquine than with its hydroxylated form, hydroxychloroquine (Good & Shader, 1977). Affective disturbances, psychosis, nightmares, delirium, and suicide have been reported (Medical Letter, 1989). Azathioprine has no known psychiatric side effects. Steroid side effects have been covered in the section in SLE and in Chapter 19, on delirium.

NSAIDs can increase steady-state plasma lithium levels by decreasing the renal clearance of lithium. Patients on lithium and NSAIDs should be monitored especially closely during any change of NSAID dose. Other antiarthritis drugs are not known to interact with commonly used psychotropic medications.

FIBROMYALGIA SYNDROME

Fibromyalgia is a common syndrome of musculoskeletal pain, falling in the category of nonarticular rheumatism. Once considered controversial, fibromyalgia is now seen as a specific entity with characteristic historic features and physical findings. The main clinical features are musculoskeletal pain, typically in an axial distribution; stiffness, often worse upon awakening; and fatigue, related to nonrestorative sleep. Other symptoms include subjective soft tissue swelling, tender spots, headaches, irritable bowel symptoms, dysmenorrhea, paresthesias, anxiety, and depression. Physical examination reveals few findings except for discrete spots of exquisite sensitivity called "tender points." The prevalence of fibromyalgia in the general population may be as high as 5%. Approximately 85% of reported fibromyalgia cases are in women. One worker described most cases as beginning during the childbearing years, whereas others have found "middle-aged" women to be at risk (Hug & Gerberg, 1990; Waxman & Zatzkis, 1986). Fibromyalgia is underdiagnosed in the elderly (Wolfe, 1988; Yunus et al., 1988).

Diagnostic criteria for fibromyalgia were proposed in 1990 by the American College of Rheumatology (ACR) based on the results of a multicenter study examining 558 patients. The diagnosis is made only when a patient presents with widespread pain of at least 3 months' duration and pain in 11 of 18 tender points on palpation; these criteria yielded a sensitivity of 88% and specificity of 81%. Widespread pain is defined as left- and right-sided pain, above and below the waist, as well as axial skeletal pain. The tender points are bilateral and at the following sites: occiput, low cervical, trapezius, supraspinatus, second rib, lateral epicondyle, gluteal, greater trochanter, and knee (Wolfe et al., 1990). To distinguish the tender points of FMS from the trigger points of myofascial pain syndrome, the following test can be used: palpation of tender points elicits local pain; trigger points cause referred pain.

The ACR 1990 criteria also abolished the distinction between primary, secondary, and concomitant fibromyalgia. Previously, fibromyalgia with no underlying causative condition was distinguished from fibromyalgia produced by an underlying condition or concurrent with but not secondary to another illness. The ACR criteria allows the diagnosis of fibromyalgia regardless of other diagnoses and without prior exclusion of other disorders (Wolfe et al., 1990).

It is important to realize that FMS is often found in conjunction with other rheumatic diseases. The approach to the fibromyalgia patient should include a thorough history, physical examination, and laboratory assessment consisting of CBC, chemistry profile, ESR, ANA, RF, and thyroid function tests. Laboratory findings in fibromyalgia are absent, though ANA is occasionally present in low titer (Yunus, 1988). Electromyography and muscle biopsy are indicated only when there are objective signs of muscle weakness or a metabolic myopathy (Schumacher, 1988).

The etiology of fibromyalgia is unknown. Patients who meet strict criteria for FMS usually complain of nonrestorative sleep, which has led investigators to examine the disrupted sleep itself for a possible etiologic role. EEG sleep studies in these patients often have revealed intrusion of alpha waves into the usual slow wave pattern of stage 4 sleep. This pattern, termed alpha-delta sleep, is distinctly different from that of patients with major depression, whose sleep EEGs demonstrate decreased rapid eye movement (REM) latency in the majority of cases. Experimental selective deprivation of stage 4 sleep in healthy volunteers produces a syndrome similar to FMS, consisting of aches, fatigue, and tender points. In the normal volunteers, physical fitness seems to protect against development of the FMS-like syndrome. Successful pharmacologic treatment of FMS often involves agents that increase stage 4 sleep (Schumacher, 1988).

Neurotransmitters have been postulated to play a role

in FMS. Serotonin may be important, as it is involved in pain perception, sleep, and the regulation of affective states (Yunus, 1988). The CSF level of substance P was found to be significantly higher in fibromyalgia patients than in normal controls (Vaery et al., 1988). Other neurobiochemical and hormonal disturbances may be involved. There is evidence of sympathetic dysregulation over the cutaneous microcirculation as manifested by an increase in Raynaud's phenomenon (Vaery et al., 1988), and a less than expected vasoconstriction response following an auditory stimulation test and left hand cold pressor test (Vaery et al., 1989). One group found that norepinephrine excretion was significantly higher in fibromyalgia patients than in normal controls (Russell et al., 1986). An estrogen deficit was implicated in another study as a promoting factor after most of the patients reviewed had undergone menopause prior the diagnosis of FMS; the investigators proposed that an estrogen deficit could disrupt sleep and affective stability, with subsequent somatization. Estrogen therapy was recommended for a subset of FMS patients (Waxman & Zatzkis, 1986).

Other studies have focused on predictive factors in FMS patients. Pain, psychological disturbances, and functional disability predicted the severity of illness in one study (Hawley, Wolfe, & Cathey, 1988). Functional disability in FMS may be partially caused by the anxiety and vulnerability associated with having a vague, poorly understood illness (Robbins, Kirmayer, & Kapusta, 1990).

Serum concentrations of procollagen type III aminoterminal peptide in FMS patients were examined because of earlier reports of low levels in some patients with FMS. The study found that fibromyalgia patients with low serum concentration of procollagen type III amino-terminal peptide had more disturbed sleep, more symptoms, and an increased number of tender points compared to fibromyalgia patients with normal serum concentrations (Jacobsen et al., 1990).

The association between FMS and psychpathology is also incompletely understood. The frequency of what seem to be psychiatric complaints or personality disturbances suggested to some authors a psychological origin to the illness. Other authors have seen the same psychiatric symptoms and signs as the expected results of life with a chronic illness that is difficult to diagnose and treat. The prevalence of depressive affect, dependent behavior, obsessive-compulsive traits, and intense response to loss (Alfici, Sigal, & Landau, 1989), and immature defenses and problems with aggression (Egle et al., 1989), have been interpreted as both causes and results of FMS. The controversy extends even to the use of standard psychiatric assessment tools, such as the MMPI, for sorting out this dilemma. Some workers suggest that these familiar tests are helpful for detecting antecedent psychopathology in FMS patients (Leavitt & Katz, 1989), whereas others suggest they are not (Goldenberg, 1989).

Psychopathology is not seen as a requisite feature of FMS. With regard to the prevalence of depression, for example, two recent studies found no significant difference between patients with FMS and those with rheumatoid arthritis (Kirmayer, Robbins, & Kapusta, 1988; Yunus, 1988).

Goldenberg found that depressive and somatic symptoms were not more common in FMS than in other chronic medical illnesses. He reported a high incidence of depression in the fibromyalgia patients with only a few depressed at the time of illness; prior depressive episodes typically preceded the onset of fibromyalgia symptoms by several years. Most of the FMS patients did not meet criteria for a psychiatric diagnosis. However, Goldenberg did find the incidence of depression in the primary relatives of FMS patients and patients hospitalized for depression to be comparable, raising the possibility of a common neurochemical disturbance in depression and FMS (Goldenberg, 1989).

Thus a common clinical problem for the clinician treating the patient *in whom FMS has already been diagnosed* is that of when to treat for depression. Because of this significant overlap between physical and vegetative symptoms in FMS and depressed patients, the psychiatrist should use mental status disorders for depression as the guide. The presence of feelings of helplessness, hopelessness, and worthlessness are useful indicators for the existence of depression in the chronically ill patient. These symptoms suggest the need for psychotherapy and pharmacotherapy for depression.

Patients with FMS often seek medical care for months or years before accurate diagnosis. The physician must convey confidence in the diagnosis and the possibility for significant improvement. The patient should assume a central role in the treatment plan. Environmental interventions minimize exacerbating factors, such as prolonged inactivity, exposure to cold weather, poor sleep, overwork, and other forms of stress (Yunus, 1988). Physical therapies (including the use of heat massage and liniments), stretching exercises, and massage help the patient gradually become more physically active, an important goal for those who avoid exercise and associate activity with pain. Biofeedback may be useful for treatment of fibromyalgia, particularly in the absence of significant psychopathology (Ferraccioli et al., 1987). Systemic corticosteroid treatment is not indicated for FMS. Aspirin and NSAIDs may be helpful and should be tried in order to break the cycle of chronic pain. One study found that the combination of carisoprodol, paracetamol (acetaminophen), and caffeine was effective in the treatment of FMS, producing improved sleep quality, an increased pain threshold at tender points, and a general decrease in malaise (Vaery et al., 1989). There is no role for opiates in the management of FMS (Schumacher, 1988).

Double-blind, placebo-controlled studies have demonstrated that low-dose amitriptyline is associated with improvement in sleep and decreased pain in FMS (Carette, 1986; Goldenberg, Felson, & Dinerman, 1986). Amitriptyline was effective at doses of only 25 or 50 mg. The exact mechanism of action of tricyclics in FMS is unclear, though treatment of the underlying sleep disorder may be the essential mechanism. Other antidepressants that are serotonergically active may also be effective, although there have been no controlled trials. If there is a significant component of depression, full antidepressant doses are likely to be needed. Cyclobenzaprine, a tricyclic compound, has been shown to be effective in the treatment of FMS; typical doses are 10 to 20 mg at bedtime (Campbell et al., 1984).

When a major psychiatric disorder such as depression or severe anxiety is suspected by the primary physician or rheumatologist, the psychiatrist can help with diagnosis and appropriate treatment. Even in the absence of such conditions, the psychiatrist can facilitate treatment by promoting the patient's maximal effort and participation.

OTHER CONNECTIVE TISSUE DISEASES

Neuropsychiatric manifestations of other connective tissue diseases generally derive from the patient's psychological response to the nonneurologic manifestations of the disease (see, however, Sjögren's syndrome and the vasculitides, below). Drug side effects can also present psychiatric concerns, which are similar to those already discussed. In most cases, the CNS is not directly involved. Early in the course of most of these diseases, nonspecific symptoms may dominate the clinical picture. Specific diagnostic evidence may be absent. In this setting, the internist or rheumatologist may believe that the picture is best explained as a psychiatric illness and may refer the patient. It is then incumbent on the psychiatrist to be aware of the differential diagnostic problem at hand and to be alert to the appropriate diagnostic measures. Several diseases are discussed here; and if specific aspects of an illness are notable, representative features are discussed.

Progressive systemic sclerosis, or *scleroderma*, affects connective tissue in a widespread manner, causing esophageal dysfunction, pursed and retracted facies, sclerodactyly, and renal and heart disease. Trigeminal neuropathy and peripheral mononeuropathies have been reported. CNS involvement is rare and most commonly appears with end-stage disease, with delirium as a consequence of renal or hepatic failure (Ochtill & Amberson, 1978). ANAs are present in more than 50%, with a nucleolar pattern; one-third are positive for RF.

Polymyositis is a systemic disease of unknown etiology that progressively affects the proximal musculature in an inflammatory process. The symptoms—weakness and pain in various muscle groups—typically present as difficulty climbing stairs or rising from a chair. Serum aldolase and CPK are elevated. The electromyogram (EMG) is abnormal, and muscle biopsy is usually diagnostic. Muscles of the heart and gastrointestinal tract can be involved. Nervous system disease is absent. Early presentations of polymyositis, with malaise, weakness, and weight loss, may be misdiagnosed as a major depressive disorder.

Dermatomyositis is closely related to polymyositis in terms of its pathologic findings and clinical presentation. The dusky red, raised rash seen over the elbows and hand and knee joints distinguishes this entity clinically from polymyositis proper. In addition, those dermatomyositis patients over 40 years of age are at a 50% risk of having carcinoma. The myositis can present as long as 2 years before the malignancy can be detected. Lung cancer and breast cancer are the most commonly associated malignancies.

Primary Sjögren's syndrome results from the inflammatory cell infiltration of the lacrimal and salivary glands, causing xerophthalmia and xerostomia. Secondary Sjögren's syndrome appears in association with another connective tissue disease, especially RA. The diagnosis is made by demonstrating abnormally low tear and saliva production, and by lacrimal or salivary gland biopsy. Neuropsychiatric symptoms can be associated with this disease; they include affective disturbances, cognitive dysfunction, and focal neurologic deficits. MRI studies demonstrate subcortical and periventricular white matter lesions and seem to be more sensitive in patients with focal deficits than is either the CT scan or cerebral angiography (Alexander et al., 1988).

The *vasculitides* are illnesses of widespread systemic distribution, characterized by vascular inflammation, tissue necrosis, and organ dysfunction. Giant cell arteritis is an important example; symptoms include recurrent headaches, tenderness along the temporal artery, and elevated ESR (readings over 100 are common). Associated symptoms are transient blindness, cerebral ischemic attacks, delirium, and stroke (Cochran, Fox, & Kelly, 1978). Giant cell arteritis can be effectively treated with corticosteroids; high doses (prednisone equivalent 1 mg/kg/day) are considered necessary to prevent blindness and stroke. The patient may need to take the steroids for life. Temporal artery biopsy should be performed on any patient in whom the diagnosis is considered. If the patient has visual symptoms or evidence of transient ischemic attacks, steroids should be begun even before biopsy is performed. Lengthy specimens of the artery (4 to 6 cm) may be necessary, as may be a biopsy of the occipital artery should both temporal arteries be negative.

There is considerable clinical overlap between giant cell arteritis (GCA) and *polymyalgia rheumatica* (PMR) (Allen & Studenski, 1986). Some refer to the two as one disease entity (polymyalgia rheumatica/temporal arteritis). PMR is characterized by fatigue and aching stiffness of the neck muscles and limb girdles, especially severe in the morning. Apart from the highly elevated ESR, other laboratory findings include anemia, decreased serum albumin, mild liver function abnormalities, and increased levels of serum α_2-globulin and plasma von Willebrand factor (Schumacher, 1988). Muscle enzymes, muscle histology, and EMG are normal; physical examination reveals normal muscle strength. These procedures distinguish PMR from polymyositis. Approximately 50 to 60% of temporal arteritis patients have PMR, and about 25 to 75% of PMR patients have evidence of arteritis on temporal artery biopsy. Even among those PMR patients without symptoms of arteritis, biopsy is positive in 10 to 20% (Guyton & Ball, 1986).

The ESR is high (100 mm/hour or greater) in almost 90% of PMR/GCA patients. The few who have a normal ESR and are asymptomatic have probably either been treated recently with corticosteroids or have another illness. If the physician suspects PMR/GCA on the basis of clinical findings and history, and the ESR is not high, causes of a falsely decreased ESR (e.g., liver disease with poor fibrinogen production) should be sought. Fibromyalgia can mimic PMR/GCA but can be excluded on the basis of tender points. Depression, also in the differential diagnosis, usually presents with intrapsychic symptoms of worthlessness and hopelessness.

Even though NSAIDs improve the symptoms of PMR/GCA, only high-dose steroids can prevent stroke and blindness. Patients presumed to have the illness should be told to contact the physician upon the occurrence of any new symptoms; and they should be started on steroids for any new neurologic symptoms, headache, or visual disturbances. For PMR *without* GCA, patients should not be treated with corticosteroids unless there is no response to a 2- to 4-week trial of NSAIDs. If steroids are used in these instances, prednisone doses usually range from 5 to 15 mg per day.

Essential mixed cryoglobulinemia (EMC) is a syndrome comprising the triad of palpable purpura, arthritis or arthralgia, and Raynaud's phenomenon. Cryoglobulins are present and are assumed to be the pathogenic entities; they are serum proteins, often immunoglobulins, that reversibly precipitate with exposure to cool temperatures, even room temperature. The purpura and joint symptoms are precipitated clinically by physical exertion or cold. Renal involvement with vasculitis can cause significant impairment. The psychiatrist may see the patient because of referral for symptoms without an obvious organic basis. Laboratory features include elevated ESR, increased immunologlobulins on serum protein electrophoresis, and mild anemia. If there is renal involvement, renal function tests may be abnormal. Rheumatoid factor is often present in a high titer. The mere detection of cryoglobulins is not diagnostic of EMC, as their presence often is secondary to another disease, such as multiple myeloma, lymphoma, or chronic infection. Elevated cryoglobulins in a patient with palpable purpura and arthritis or arthralgias virtually makes the diagnosis. Care must be taken when obtaining of the blood sample, however, so as to avoid false-negative results: Upon obtaining the sample, it should immediately be stored at 37°C to avoid precipitating the cryoglobulins.

Many patients with EMC require no treatment, or only NSAIDs. Those with renal, or the less common peripheral neurologic involvement (peripheral neuropathy) may require trials of corticosteroids or cytotoxic drugs. Occasionally, plasmapheresis is helpful.

REFERENCES

ADLER R (1981). Psychoneuroimmunology. Orlando, FL: Academic Press.

AHEARN JM, PROVOST JT, DORSCH SA, ET AL. (1982). Interrelationships of HLA-DR, MB, and MT phenotypes, autoantibody expression, and clinical features in systemic lupus erythematosus. Arthritis Rheum 25:1031.

AISEN AM, GABRIELSEN TO, MCCUNE WJ (1985). MR imaging of systemic lupus erythematosus involving the brain. AJR 144:1027–1031.

ALEXANDER EL, BEALL SS, GORDON B, ET AL. (1988). Magnetic resonance imaging for cerebral lesions in patients with Sjögren syndrome. Ann Intern Med 108:815–823.

ALFICI S, SIGAL M, & LANDAU M (1989). Primary fibromyalgia syndrome—a variant of depressive disorder? Psychother Psychosom 51(3):156–161.

ALLEN NB, & STUDENSKI SA (1986). Polymyalgia rheumatica and temporal arteritis. Med Cli North Am 70:369–384.

AMKRAUT AA, SOLOMON GF, & KRAEMER HC (1971). Stress, early experience and adjuvant-induced arthritis in the rat. Psychosom Med 33:203–214.

ANDERSON KO, BRADLEY LA, YOUNG LD, ET AL. (1985). Rheumatoid arthritis: review of psychological factors related to etiology, effects, and treatment. Psychol Bull 98:358–387.

BALL GV (1986). Systemic lupus erythematosus (SLE). In GV BALL & WJ KOOPMAN (Eds.), Clinical rheumatology. Philadelphia: WB Saunders.

BALL GV, & KOOPMAN WJ (1986). Clinical rheumatology. Philadelphia: WB Saunders.

BAUM J (1982). A review of the psychological aspects of rheumatic diseases. Semin Arthritis Rheum 11:352–361.

BECKHAM JC (1987). Stress and rheumatoid arthritis: can a cognitive coping model help explain a link? Semin Arthritis Rheum 17(2):81–89.

BROWN GK, NICASSIO PM, & WALLSTON KA (1989). Pain coping strategies and depression in rheumatoid arthritis. J Consult Clin Psychol 57:652–657.

CAMPBELL SM, GATTER RA, CLARK S, ET AL. (1984). A double blind study of cyclobenzaprine versus placebo in patients with fibrositis. Arthritis Rheum (suppl)27:576.

CAPLAN A, PAYNE RB, & WITHEY JL (1962). A broader concept of Caplan's syndrome related to rheumatoid factors. Thorax 17:205–212.

CARETTE S (1986). Evaluation of amitriptyline in primary fibrositis: a double blind placebo-controlled study. Arthritis Rheum 29:655–659.

CARETTE S, UROWITZ MB, GROSMAN H, ET AL. (1982). Cranial computerized tomography in systemic lupus erythematosus. J Rheumatol 9:855–859.

CASSILETH BR, LUSK EJ, STROUSE TB, ET AL. (1984). Psychological status in chronic illness: a comparative analysis of six diagnostic groups. N Engl J Med 311:506–511.

CEPPELINI R, POLLI E, & CELADA F (1957). DNA reacting factor in serum of a patient with lupus erythematosus diffusus. Proc Soc Exp Biol Med 96:572.

CHAMBERLAIN MA, & BRUCKNER FE (1970). Rheumatoid neuropathy: clinical and electrophysiological features. Ann Rheum Dis 29:609–616.

COBB S, MILLER M, & WIELAND M (1959). On the relationship between divorce and rheumatoid arthritis. Arthritis Rheum 2:214–218.

COCHRAN JW, FOX JH, & KELLY MP (1978). Reversible mental symptoms in temporal arteritis. J Nerv Ment Dis 166:446–447.

COHEN AS (1975). Laboratory diagnostic procedures in the rheumatic diseases. Boston: Little, Brown.

CREED F, MURPHY S, & JAYSON MV (1990). Measurement of psychiatric disorder in rheumatoid arthritis. J Psychosom Res 34(1):79–87.

CROWN S, CROWN JM, & FLEMING A (1975). Aspects of the psychology and epidemiology of rheumatoid disease. Psychol Med 5:291–299.

CUTHBERT MF (1974). Adverse reactions to nonsteroidal anti-rheumatic drugs. Curr Med Res Opin 2:600–610.

DENBURG SD, CARBOTTE RM, & DENBURG JA (1987). Cognitive impairment in systemic lupus erythematosus: a neuropsychological study of individual and group deficits. J Clin Exp Neuropsychol 9:323–339.

DREW JF (1962). Concerning the side effects of antimalarial drugs in the extended treatment of rheumatic disease. Med J Aust 2:618–620.

EGLE UT, RUDOLF ML, HOFFMANN SO, ET AL. (1989). Personality markers, defense behavior and illness concept in patients with primary fibromyalgia. Z Rheumatol 48(2):73–78.

FERRACCIOLI G, GHIRELLI L, SCITA F, ET AL. (1987). EMG biofeedback in training in fibromyalgia syndrome. J Rheumatol 14:820–825.

FLOR H, TURK DC (1988). Chronic back pain and rheumatoid arthritis: predicting pain and disability from cognitive variables. J Behav Med 11:251–265.

GARDINER BM (1980). Psychological aspects of rheumatoid arthritis. Psychol Med 10:159–163.

GAYLIS NB, ALTMAN RD, OSTROV S, ET AL. (1982). The selective value of tomography of the brain in cerebritis due to systemic lupus erythematosus. J Rheumatol 9:850–854.

GOLDENBERG DL (1989). Psychological symptoms and psychiatric diagnosis in patients with fibromyalgia. J Rheumatol (suppl)16:127–130.

GOLDENBERG DL, FELSON DT, & DINERMAN H (1986). A randomized controlled trial of amitriptylline and naproxen in the treatment of patients with primary fibromyalgia. Arthritis Rheumatol 29:1371–1377.

GOOD MI, & SHADER RI (1977). Behavioral toxicity and equivocal suicide associated with chloroquine and its related derivatives. Am J Psychiatry 134:798–801.

GREER HD, WARD HP, & CORBIN KB (1965). Chronic salicylate intoxication in adults. JAMA 193:555–558.

GUYTON JM, & BALL GV (1986). Vasculitis. In GV BALL & WJ KOOPMAN (Eds.), Clinical rheumatology. Philadelphia: WB Saunders.

HAHN BH, SHARP GC, IRWIN WS, ET AL. (1972). Immune responses to hydralazine and nuclear antigens in hydralazine-induced lupus erythematosus. Ann Intern Med 76:365–374.

HALL RCW, POPKIN MK, STICKNEY SK, ET AL. (1979). Presentation of the steroid psychoses. J Nerv Ment Dis 167:229–236.

HALL RCW, STICKNEY SK, & GARDNER ER (1981). Psychiatric symptoms in patients with systemic lupus erythematosus. Psychosomatics 22:15–24.

HALLIDAY JL (1942). Psychological aspects of rheumatoid arthritis. Proc R Soc Med 35:455–457.

HARDIN JG (1986). Rheumatoid arthritis. In GV BALL & WJ KOOPMAN (Eds.), Clinical rheumatology. Philadelphia: WB Saunders.

HARGRAVES MM, RICHMOND H, & MORTON R (1948). Presentation of two bone marrow elements: the "tart" cell and the "LE" cell. Proc Mayo Clinic 23:23–28.

HARRIS EN, & HUGHES GBV (1985). Cerebral diseases in systemic lupus erythematosus. In Springer seminars in immunopathology. Springer-Verlag 8:251–266.

HAWLEY DJ, WOLFE F, & CATHEY MA (1988). Pain, functional disability, and psychological status: a 12 month study of severity in fibromyalgia. J Rheumatol 15:1551–1556.

HUG C, & GERBERG NJ (1990). Fibromyalgia syndrome, a frequently misdiagnosed entity. Schweitz Med Wochenschr 120:395–401.

JACOBSEN S, JENSEN LT, FOLDAGER M, ET AL. (1990). Primary fibromyalgia clinical parameters in relation to serum procollagen type III aminoterminal peptide. Br J Rheumatol 29:174–177.

JOHNSON RT, & RICHARDSON EP (1968). The neurological manifestations of systemic lupus erythematosus. Medicine 47:337–369.

KAHANA RJ, & BIBRING GL (1964). Personality types in medical management. In RGN ZINBI (Ed.), Psychiatry and medical practice in a general hospital (pp. 108–123). New York: International University Press.

KIRMAYER LJ, ROBBINS JM, & KAPUSTA MA (1988). Somatization and depression in fibromyalgia. Am J Psychiatry 145:950–954.

LEAVITT F, & KATZ RS (1989). Is the MMPI invalid for assessing psychological disturbance in pain related organic conditions? J Rheumatol 16:521–526.

LEONHARDT T (1966). Long-term prognosis of systemic lupus erythematosus. Acta Med Scand 445:440–443.

LEWIS DA, & SMITH RE (1983). Steroid-induced psychiatric syndromes—a report of 14 cases and a review of the literature. J Affect Disord 5:319–332.

LIANG MH, ROGERS M, LARSON M, ET AL. (1984). The psychosocial impact of systemic lupus erythematosus and rheumatoid arthritis. Arthritis Rheum 27:13–19.

LIPKIN M (1985). Psychiatry in medicine. In HI KAPLAN & BJ SADOCK (Eds.), Comprehensive textbook of psychiatry (pp. 1263–1277). Baltimore: Williams & Wilkins.

LIPOWSKI ZJ (1975). Physical illness, the patient and his environment: psychosocial foundations of medicine. In S ARIETI (Ed.), American handbook of psychiatry (pp. 3–42). New York: Basic Books.

LONG AA, DENBURG SD, CARBOTTE RM, ET AL. (1990). Serum lymphocytotoxic antibodies and neurocognitive function in systemic lupus erythematosus. Ann Rheumatol Dis 49:249–253.

MCFARLANE AC, KALUCY RS, & BROOKS PM (1987). Psychological predictors of disease course in rheumatoid arthritis. J Psychosom Res 31:757–764.

Medical Letter (1989). Drugs for rheumatoid arthritis. Med Lett Drugs Ther 31:61–64.

MEDSGER AR, & ROBINSON H (1972). A comparative study of divorce in rheumatoid arthritis and other rheumatic diseases. J Chronic Dis 25:269–275.

MEENAN RF (1981). The impact of chronic disease: a sociomedical profile of rheumatoid arthritis. Arthritis Rheum 24:544–549.

MEENAN RF, YELIN EH, HENKE CJ, ET AL. (1978). The costs of rheumatoid arthritis: a patient-oriented study of chronic disease costs. Arthritis Rheum 21:827–833.

MILLER ML, & GLASS DN (1981). The major histocompatibility complex antigens in rheumatoid arthritis and juvenile arthritis. Bull Rheum Dis 31:21–25.

MOLDOFSKY H, & CHESTER WJ (1970). Pain and mood patterns in patients with rheumatoid arthritis. Psychosom Med 32:309–317.

MOOS RH (1964). Personality factors associated with rheumatoid arthritis: a review. J Chronic Dis 17:41–55.

MOOS RH, & SOLOMON GF (1964). Personality correlates of the rapidity of progression of rheumatoid arthritis. Ann Rheum Dis 23:145–151.

MORAN MG (1987). Treatment noncompliance in asthmatic patients: an examination of the concept and a review of the literature. Semin Respir Med 8:271–277.

MURPHY S, CREED F, & JAYSON MI (1988). Psychiatric disorder and illness behaviour in rheumatoid arthritis. Br J Rheumatol 27:357–363.

OCHTILL HN, & AMBERSON J (1978). Acute cerebral symptomatology, a rare presentation of scleromyxedema. J Clin Psychiatry 39:471–475.

O'CONNOR JF (1959). Psychoses associated with systemic lupus erythematosus. Ann Intern Med 51:526–536.

PANUSH RS, BIANCO NE, & SCHUR PH (1971). Serum and synovial fluid IgG, IgA, and IgM antigammaglobulins in rheumatoid arthritis. Arthritis Rheum 14:737–747.

PARKER J, MCRAE C, SMARR K, ET AL. (1988). Coping strategies in rheumatoid arthritis. J Rheumatol 15:1376–1383.

PILLEMER SR, ASHBY P, GORDON DA, ET AL. (1984). Brachial plexus radiculopathy and computed tomography [Letter]. Ann Intern Med 100:619.

PINCUS T (1986). Antinuclear antibodies. In GV BALL & WJ KOOPMAN (Eds.), Clinical rheumatology. Philadelphia: WB Saunders.

PINCUS T, LEIGHT FC, & BRADLEY AL (1986). Elevated MMPI scores for hypochondriasis, depression and hysteria in a patient with RA reflect disease rather than psychological status. Arthritis Rheumatol 29:1456–1466.

POW JM (1987). The role of psychological influences in rheumatoid arthritis. J Psychosom Res 31:223–229.

RAGAN C, & FARRINGTON E (1962). The clinical features of rheumatoid arthritis. JAMA 181:663–667.

REINITZ E, HUBBARD D, & ZIMMERMAN RD (1984). Central nervous system diseases in systemic lupus erythematosus: axial tomographic scan as an aid to differential diagnosis [Letter]. J Rheumatol 11:252–253.

RIMON R, & LAAKSON RL (1984). Overt psychopathology in rheumatoid arthritis: a fifteen-year follow-up study. Scand J Rheumatol 13:324–328.

ROBBINS JM, KIRMAYER LJ, & KAPUSTA MA (1990). Illness worry and disability in fibromyalgia syndrome. Int J Psychiatry Med 20:49–63.

ROBINSON ET (1977). Depression in rheumatoid arthritis. J R Coll Gen Practitioners 27:423–427.

RODNAN GP (1973). Primer on rheumatic diseases. JAMA [Special issue] 224:662–812.

RODNAN GP, & SCHUMACHER HR (Eds.) (1983). Primer on the rheumatic diseases. Atlanta: Arthritis Foundation.

ROGERS MP (1985). Rheumatoid arthritis: psychiatric aspects and use of psychotropics. Psychosomatics 26:915–925.

ROGERS MP, DUBEY D, & REICH P (1979). The influence on the psyche and the brain on immunity and disease susceptibility: a critical review. Psychosom Med 41:147–164.

ROGERS MP, LIANG MH, & PARTRIDGE AJ (1982). Psychological care of adults with rheumatoid arthritis. Ann Intern Med 96:344–348.

ROSILLO RH, VOGEL ML (1971). Correlation of psychological variables and progress in physical rehabilitation. 4. The relation of body image to success of physical rehabilitation. Arch Phys Med Rehabil 52:182–186.

RUSSELL IJ, VIPRAIO GA, MORGAN WW, ET AL. (1986). Is there a metabolic basis for fibrositis syndrome? Am J Med 15:163–173.

SCHNEEBAUM AB, SINGLETON JD, WEST SG, ET AL. (1991). Association of psychiatric manifestations with antibodies to ribosomal P proteins in systemic lupus erythematosus. Am J Med 90:54–62.

SCHUMACHER HR (Ed.) (1988). Primer on the rheumatic diseases (9th ed.). Atlanta: Arthritis Foundation.

SCHUR PH (1990). Serologic tests in the evaluation of rheumatic diseases. Brigham Womens Hosp Med Update II 9:1–6.

SERGENT JS, LOCKSHIN MD, KLEMPNER MW, ET AL. (1975). Central nervous system disease in systemic lupus erythematosus. Am J Med 58:644–645.

SHORT CL, BAUER W, & REYNOLDS WE (1957). Rheumatoid arthritis. Cambridge: Harvard University Press.

SILVERMAN AJ (1985). Rheumatoid arthritis. In HI KAPLAN & BJ SADOCK (Eds.), Comprehensive textbook of psychiatry. Baltimore: Williams & Wilkins.

SOLOMON GF (1969). Stress and antibody response in rats. Int Arch Allergy Appl Immunol 35:97–104.

SPALDING DM, & KOOPMAN WJ (1986). Rheumatoid factor, cryoglobulinemia, and erythrocyte sedimentation rate. In GV BALL & WJ KOOPMAN (Eds.), Clinical rheumatology. Philadelphia: WB Saunders.

STAGE DE, & MANNIK M (1973). Rheumatoid factors in rheumatoid arthritis. Bull Rheum Dis 23:720–725.

STEIN H, WALTERS K, DILLON A, ET AL. (1986). Systemic lupus erythematosus—a medical and social profile. J Rheumatol 13:570–576.

STEINBROCKER O, TRAEGER CH, & BATTERMAN RC (1949). Therapeutic criteria in rheumatoid arthritis. JAMA 140:659–662.

STERN M, & ROBBINS ES (1960). Psychoses in systemic lupus erythematosus. Arch Gen Psychiatry 3:205–212.

Stoppe G, Wildagen K, Seidel JW, et al. (1990). Positron emission tomography in neuropsychiatric lupus erythematosus. Neurology 40:304–308.

Tan EM (1985). Systemic lupus erythematosus: immunologic aspects. In DJ McCarty (Ed.), Arthritis and allied conditions. Philadelphia: Lea & Febiger.

Thomas GW (1936). Psychic factors in rheumatoid arthritis. Am J Psychiatry 93:693–710.

Vaery H, Abrahamsen A, Free O, et al. (1989). Treatment of fibromyalgia (fibrositis syndrome): a parallel double blind trial with carisoprodol, paracetamol and caffeine (Somadril comp) versus placebo. Clin Rheumatol 8:245–250.

Vaery H, Helle R, Free O, et al. (1988). Elevated CSF levels of substance P and high incidence of Raynaud phenomenon in patients with fibromyalgia: new features for diagnosis. Pain 32:21–26.

Vaery H, Qiao ZG, Mrkrid L, et al. (1989). Altered sympathetic nervous system response in patients with fibromyalgia (fibrositis syndrome). J Rheumatol 16:1460–1465.

Vermess M, Bernstein RM, Bydder GM, et al. (1983). Nuclear magnetic resonance (NMR) imaging of the brain in systemic lupus erythematosus. J Comput Assist Tomogr 7:461–467.

Vogel ML, & Rosillo RH (1969). Correlation of psychological variables and progress in physical rehabilitation. 2. Motivation, attitude and flexibility of goals. Dis Nerv Syst 30:593–601.

Vogel ML, & Rosillo RH (1971). Correlation of psychological variables and progress in physical rehabilitation. 3. Ego functions and defensive and adaptive mechanism. Arch Phys Med Rehabil 52:15–21.

Volanakis JE (1986). The complement system. In GV Ball & WJ Koopman (Eds.), Clinical rheumatology. Philadelphia: WB Saunders.

Waxman J, & Zatzkis SM (1986). Fibromyalgia and menopause: examination of the relationship. Postgrad Med 80:170–171.

Weiner CL (1975). The burden of rheumatoid arthritis: tolerating the uncertainty. Soc Sci Med 9:97–104.

Weisberg LA (1986). The cranial computed tomographic findings in patients with neurologic manifestations of systemic. Comput Radiol 10:63–68.

Wolfe F (1988). Fibromyalgia in the elderly: differential diagnosis and treatment. Geriatrics 43(6):57–60.

Wolfe F, Smythe HA, Yunus MB, et al. (1990). The American College of Rheumatology 1990 criteria for the classification of fibromyalgia: report of the Multicenter Criteria Committee. Arthritis Rheum 33:160–172.

Wolpaw RL (1960). The arthritic personality. Psychosomatics 1:195–197.

Yoshino S, & Uchida S (1981). Sexual problems of women with rheumatoid arthritis. Arch Phys Med Rehabil 62:122–123.

Yunus MB (1988). Diagnosis, etiology, and management of fibromyalgia syndrome: an update [Review]. Compr Ther 14(4):8–20.

Yunus MB, Holt GS, Masi AT, et al. (1988). Fibromyalgia syndrome among the elderly: comparison with younger patients. J Am Geriatr Soc 36:987–995.

Zvaifler NJ (1986). Neuropsychiatric manifestations of SLE, common, hard to identify. Clin Psychiatry News May:26.

V Surgical Subspecialties and Trauma

34 Surgery and trauma: General principles

ANNE MARIE RIETHER, M.D., AND
J. STEPHEN McDANIEL, M.D.

Basic issues in the psychiatric care of surgical patients are explored in this chapter. The preoperative psychological preparation and psychiatric considerations of anesthesia are discussed, with the areas of focus including surgery specific to females (see also Chapter 28), surgically related sexual dysfunction (see also Chapter 15), amputation, aesthetic surgery, ostomy patients, cancer surgery (see also Chapter 24), burn surgery (see also Chapter 35), and the spinal cord injured patient (see also Chapter 40).

The following postoperative complications are addressed as well: postoperative delirium and intensive care unit (ICU) syndromes, including etiologies and preventive strategies (see also Chapters 12 and 19) and management of postoperative pain. Acute pain management is further discussed in Chapter 16. Psychiatric aspects of transplant surgery are discussed separately in Chapters 27 and 36. A review of psychiatric aspects of anesthesia can be found in a companion volume (Rogers, 1993).

PSYCHIATRIC ASPECTS OF ANESTHESIA

Although patients facing surgery expect that anesthesia will prevent pain and render them amnestic of the procedure, ironically, anesthesia is also feared. A part of this fear is concern over the loss of control involved in putting one's self in the hands of another person who has the power of life and death—the anesthesiologist. Blacher (1987) emphasized that with the loss of control during anesthesia many patients are afraid they will reveal personal secrets. Although it does not occur, it frequently is a source of worry to patients and should be addressed if the physician suspects a concern.

Another contributing factor to preoperative anxiety is the patient's lack of familiarity with the anesthesiologist. Most patients have the opportunity to establish a relationship with their surgeons during the initial physical examination and surgical planning, thereby facilitating comfort and even some degree of idealization of the surgeon. Therefore, because the anesthesiologist is actually the person "putting the patient to sleep," a preoperative anesthesiology visit can have a calming effect. One study found a preoperative visit to be a much more powerful "tranquilizing agent" than pentobarbital (2 mg/kg IM) given 1 hour before the operation (Egbert, Battit, Turndorf, et al., 1963). Patients given medication in lieu of a visit by the anesthesiologist may come to the operating room anxious and sedated, requiring more anesthetic agent than their visited counterparts. Egbert, Battit, and Turndorf, et al. (1963) have written extensively about the technique of the anesthesiologist's preoperative interview addressing patients' expectations and fears, preoperative events, and anesthesia risks in an effort to alleviate anxiety. For those patients in whom a preoperative anxiolytic is needed, benzodiazepines without active metabolites, i.e., lorazepam 1.0 mg PO/IM or oxazepam 15 mg PO, generally are associated with fewer complications than longer-acting benzodiazepines with active metabolites.

Value of Preoperative Teaching and Preparation

A large body of research has demonstrated the beneficial effect of preoperative education on postoperative outcome (Hathaway, 1986). Surgical patients who receive preoperative preparation have been shown to have less anxiety, stress, and pain; to be more cooperative with postoperative activities such as positioning, deep breathing, coughing, leg exercises, and ambulation; to have fewer postoperative complications; to experience better and quicker recoveries; and to be more satisfied with their care (Haines & Viellion, 1990; Rothrock, 1989; Swindale, 1989). Research prior to 1980 that found a relationship between preoperative psychoeducational interventions and reduced hospital stays has not been supported by more recent studies, however, perhaps because length of stay increasingly is being mandated by health insurers (Devine & Cook, 1983; Rothrock, 1989).

One of the goals of preoperative teaching is to provide patients with information, which may be either procedural or sensory in nature. Procedural information consists of concrete facts about what is going to occur. Instruction in specific behaviors, such as incisional splinting and pulmonary exercises, enhance the patient's ability to participate in the postoperative regime. "Sensory information" focuses on what the patient experiences—on what might be seen, felt, tasted, or smelled. Research suggests that patients who experience a low level of anxiety benefit most from procedural information, whereas patients who experience a high level of anxiety benefit most from sensory information and psychological interventions (Hathaway, 1986).

Psychological interventions focus on an exploration of the patient's feelings and attitudes regarding the surgical experience. Psychological strategies such as teaching relaxation techniques may help increase the patient's sense of control over a stressful situation. Mogan, Wells, and Robertson (1985) found that patients experienced less distress from painful sensations postoperatively when they were instructed preoperatively in a simple relaxation technique utilizing imagery, rhythmic breathing, and distraction. Each preoperative teaching plan should be tailored to meet the needs of the individual patient. Whether the emphasis is on procedural information, sensory information, or psychological interventions, teaching strategies help ease anxiety in the surgical patient by reducing the gap between the anticipated experience of surgery and the actual event (Rothrock, 1989).

In these days of increased patient medical acuity and staff shortages, preoperative teaching, although important, sometimes is given a relatively lower priority than other patient care activities. One of the goals of current research has been to develop efficient and effective strategies for continuing to meet the psychoeducational needs of surgical patients. For example, Good-Reis and Pieper (1990) compared an individualized structured teaching program to less formalized teaching methods for vascular surgery patients. They found that structured instruction tended to be fatiguing for patients, was costly in terms of time, and had no clear advantage over unstructured teaching when ICU stay, hospital stay, and analgesic use were the outcome measures. Group instruction was found by Linderman (1972) to be as effective and more efficient than individual teaching. An advantage of group teaching is that patients undergoing similar surgical procedures are able to share their feelings with and gain reassurance from one another. Filmstrips, videotapes, pamphlets, and booklets have been used to teach patients about the surgical experience. Patients who are anxious tend to remember little of what they are told verbally and are able to recall more if they are given written informational material in conjunction with oral presentations (Ley, Bradshaw, Eaves, et al., 1973).

Many patients today are admitted to the hospital 24 hours or less before their scheduled surgeries, reducing the time available for preoperative preparation. It can be a period of intense anxiety for patients, making them less receptive to psychoeducational interventions. Several researchers have investigated different strategies for meeting the educational needs of patients in this difficult situation. Haines and Viellion (1990) successfully offered preoperative teaching at the time of preadmission screening 10 to 14 days before admission for elective surgery. Most of the patients who participated in this program believed that it helped prepare them for the surgical experience. Williams (1986) provided preoperative education for orthopedic patients in the home setting. She found no significant differences between the postoperative recoveries of home-educated patients and those of similar patients who first received preoperative teaching in the hospital.

The use of anatomic models is becoming more routine in many hospitals and helps to clarify in the patient's mind the nature of the proposed surgery. At our institution, we use preoperative videos to prepare patients for cardiothoracic and heart transplant surgery; because of the success of these teaching tapes, other departments are contracting with our health education service to produce their own.

As patients become more informed consumers, they are becoming more health conscious, and increased education may facilitate a better working relationship between staff and patients. Pritchett and Hull have been publishing patient teaching materials since 1973. Their materials are available to individuals, hospital staff, and educators. The current toll free number is 1-800-241-4925.

Intraoperative Awareness and Recall

An interesting complication that has developed as a result of "balanced anesthesia" (the combined use of inhalation agents, intravenous narcotics, and muscle relaxants, i.e., a combination of halothane, fentanyl, and succinylcholine or curare) is the phenomenon of patients regaining some level of consciousness during surgery and being unable to move because of the relaxant (Guerra, 1986). This experience of intraoperative awareness with muscle paralysis has been well studied. Patients with histories of alcoholism, chronic benzodiazepine use, and chronic narcotic use are at increased risk of experiencing this phenomenon because of a heightened tolerance to anesthetic agents. Blacher (1975, 1984) coined the term "postoperative traumatic neurosis" for this intraoperative awareness. The condition is characterized by (1) repetitive nightmares usually of disguised versions of an operative situation; (2) anxiety and irritability; (3) a preoccupation with death; and (4) a concern with sanity that leaves patients reluctant to seek help. Blacher suggested that open discussion of awareness with the patient when it occurs results in symptom resolution. Treatment consists of direct confirmation to patients that indeed they had been awake and paralyzed during part of their surgery (Blacher, 1984). In each of his cases (1975), symptoms abated after one session. However, one should be alert to the possibility of this postoperative "posttraumatic stress disorder" (PTSD) exacerbating preexisting psychiatric conditions.

Numerous studies have examined the ability of patients to experience recall during surgery. Pearson (1961) demonstrated the efficacy of intraoperative suggestion for reducing the duration of hospitalization in general surgical patients. With the use of audio tapes played through earphones, one group of anesthetized patients heard positive suggestions regarding rapid recovery and early discharge. These suggestions pointed out to the patients that the amount of comfort they had during the postoperative period depended to a great extent on them; and the more they could relax, the more comfortable they would be. Specific items discussed were the ability to empty the bladder and bowel, ambulation, injections, pain, and dressing changes. A placebo group heard either music or blank tape. Patients in the experimental group who had received suggestions were discharged an average of 2.4 days sooner than the control group.

The implications of the possibility that the patient might hear what is said during surgery is clear. Anesthesiologists and surgeons should insist on decorum in the operating room so that patients' unconscious perceptions are positive.

Neuropsychiatric Complications of Anesthetic Agents

A number of anesthetic agents are known to produce neuropsychiatric side effects. (See Table 34-1 for a list of common perioperative medications used for patients undergoing anesthesia). Barbiturates extensively used as hypnotics may also act as *convulsants* depending on their chemical structure. Methohexital, commonly used during electroconvulsive therapy (ECT), for example, increases central nervous system (CNS) excitability (Steen & Michenfelder, 1979). Among the inhalation agents, halothane can cause headaches, ataxia, lethargy, and diffuse electroencephalographic (EEG) changes. Agents such as atropine must be closely monitored for central anticholinergic toxicity. A relatively new intravenous agent, propofol, has been associated with acute postoperative opisthotonos (Saunders & Harris, 1990).

Ketamine is an intravenous anesthetic agent with the most well-known psychiatric side effects. As many as 20% of patients receiving this drug experience symptoms ranging from hyperexcitability and vivid, unpleasant dreams to acute psychosis and delirium upon emergence from anesthesia. The syndrome is more likely in children under 12 years of age and adults over 70 years.

As these symptoms are limited to the emergence phase, these patients should be allowed to emerge from anesthesia quietly. The use of intravenous lorazepam or midazolam given at the end of the surgical procedure has been shown to avert these unpleasant CNS effects (Houlton, Downing, & Brock-Utne, 1978). If psychotic symptoms persist, low-dose neuroleptics are indicated. There are no known chronic neuropsychiatric sequelae from the appropriate medical use of ketamine. However, the use of ketamine as a recreational drug has increased in recent years, particularly in large urban areas. There are no current data on neuropsychiatric sequelae of recreational ketamine use, but the possible consequences of psychosis should be considered when treating this group of patients.

Another frequently used agent, droperidol, a butyrophenone, has a side effect profile similar to that of other neuroleptics. It was at one time frequently used for preoperative sedation; however, it has been associated with dysphoria and akathisia, which at times leads to refusal of surgery (Lee & Yeakel, 1975). These effects are largely dose-related and have been diminished by using very low doses, i.e., 0.625 mg q4h as needed. Droperidol is now administered for its antiemetic effect rather than its sedative properties. Benzodiazepines have largely replaced droperidol as preoperative sedatives.

A host of other drugs can contribute to perioperative psychiatric presentations. Although not considered primary anticholinergics, phenothiazines, antihistamines, and analgesics can block muscarinic receptors. Corticosteroids often are used perioperatively and have well known psychiatric side effects, including mood lability, depression, irritability, insomnia, poor concentration, and psychosis (Lam & Parkin, 1981, Ling, Perry, & Tsuang, 1981). The incidence of these reactions has been shown to correlate with the dosage administered (Kostic & Levic, 1989). Steroid delirium is also

TABLE 34–1. *Perioperative Medications of Patients Undergoing Anesthesia*

Medication	Comment
Anticholinergic drugs	
Atropine Scopolamine Glycopyrrolate	Restlessness, irritability, delirium are possible side effects. (Glycopyrrolate does not cross the blood-brain barrier, so it produces fewer CNS side effects.)
Benzodiazepines	
Midazolam Diazepam Lorazepam	Safe CNS side effect profile. These agents can be used for preoperative sedation, induction, or maintenance of anesthesia. They produce amnesia, cause minimal depression of circulation, and act at specific sites as anticonvulsants. Midazolam has $t_{1/2}$ of 1.2 hours with rapid onset (only available IV).
Opiates	
Preoperative agents Morphine Meperidine	Opioids generally act as cerebral vasoconstrictors, producing decreased cerebral blood flow and decreased intracranial pressure. These agents may stimulate the chemoreceptor trigger zone, causing nausea and vomiting. Normeperidine, the active toxic metabolite of meperidine, can cause CNS excitation and associated confusional states.
Induction/maintenance agents	
Fentanyl Sufentanyl Alfentanyl	
Agonist-antagonist opiates	
Pentazocine Butorphanol Nalbuphine Buprenorphine	Because of their ceiling effect, these agents have limited analgesic properties so are rarely used for induction or maintenance of anesthesia. They are used for postoperative analgesia with minimal risk of ventilatory depression. Induction of opiate withdrawal is a potential side effect in addicted patients.
Barbiturates	
Thiopental Methohexital Thiamylal	Barbiturates are used for induction of anesthesia. As potent cerebral vasoconstrictors, they reduce cerebral blood flow and volume as well as intracranial pressure. Methohexital can activate epileptic foci.
Other Intravenous Agents	
Ketamine	Phencyclidine derivative, this agent produces dissociative anesthesia. Patients appear to be in cataleptic states with open eyes and slow nystagmic gaze. There is a significant incidence of unpleasant visual, auditory, and proprioceptive illusions that can progress to delirium upon emergence.
Etomidate	This agent produces rapid induction and prompt awakening and is therefore useful for outpatient anesthesia. It suppresses adrenocortical function for up to 8 hours after the initial dose.
Propofol	Generally awakening occurs in 4–8 minutes. There are minimal residual sedative effects; however, propofol has been associated with acute postoperative opisthotonos.
Inhalation agents	
Nitrous oxide Halothane Enflurane Isoflurane	Nitrous oxide must be supplemented with other agents for surgical anesthesia; it is often used as the sole anesthetic agent for brief procedures. Halothane and enflurane are potent cerebral vasodilators that increase cerebral blood flow. They both produce CNS stimulation manifested by increased electrical activity on EEG. Mild impairment in cognitive functioning and reaction times for up to 2 days postoperatively has been associated with the latter agents.
Muscle relaxants	
Succinylcholine	Muscle relaxants are used to provide skeletal muscle relaxation or paralysis to facilitate intubation or optimal operative conditions. They produce no anesthetic or analgesic effects.
Neuroleptics	
Droperidol Promethazine Prochlorperazine	These agents are used for their antiemetic effects. Acute extrapyramidal symptoms are potential side effects.

associated with withdrawal or dosage reduction in patients who have been administered maintenance corticosteroids.

Cimetidine has been implicated in a spectrum of reversible confusional states, especially in elderly patients and in those with renal or hepatic failure (Ducluzeau, Vial, Pobel, et al., 1990). The newer H2 antagonists, ranitidine and famotidine, demonstrate little or no evidence of interaction with the hepatic microsomal enzyme system (Humphries, 1987). However, like cimetidine, ranitidine has been implicated in causing acute reversible confusional states (Mandal, 1986; Sonnenblick & Yinnou, 1986). Metoclopramide, used perioperatively for gastroesophageal reflux and nausea, can produce extrapyramidal symptoms including acute dystonic reactions through central dopamine receptor antagonism. The incidence and severity of metoclopramide's side effects increase with the dose and the time course over which the patient is treated (Scheller & Sears, 1987). Because the active metabolites of metoclopramide are primarily excreted through the kidneys and are therefore dependent on creatinine clearance, dosage reduction should be considered in elderly patients. However, even small doses of oral or intravenous metoclopramide given preoperatively have been associated with acute neurologic signs and symptoms (Barnes & Braude, 1982; Fouilladieu, Hosanki, McGee, et al., 1985). (See also Rogers 1993, for a more extended discussion.)

Postoperative Delirium

Postoperative delirium is not rare, nor is it limited to specific surgical procedures or a specific type of patient. One study found a 78% incidence of delirium in a random sample of 200 general surgical patients (Titchaer, Zwerling, Gottschalk, et al., 1956); and another study found a 10 to 15% incidence of general surgical patients age 65 or older to be delirious after their operations (Millar, 1981). A metaanalytic review by Cryns, Gorey & Goldstein (1990) examined the effects of surgery on the mental status of older persons. After analyzing data from 18 empiric studies, these authors concluded that surgery has a significant decompensating impact on the mental status of geriatric patients. Interestingly, some of the investigations to clarify the incidence and mechanisms of postoperative delirium have inadvertently provided interventions that have themselves altered the responses being investigated. For example, Lazarus and Hagens (1968) found that one psychiatric interview conducted 2 to 3 days prior to surgery reduced the incidence of postoperative psychosis by 50%. Similarly, Layne and Yudofsky (1971) found that a single 1-hour therapeutic interview before surgery resulted in a 50% reduction in postoperative psychosis among cardiotomy patients.

The type of operation to be performed must in many instances be considered a risk factor for delirium. The most widely recognized is "postcardiotomy" delirium. In a review of the literature examining 44 studies that investigated postcardiotomy delirium, the prevalence of this postoperative complication was shown to have remained fairly constant over time at 32% (Smith & Dimsdale, 1989). This number is not insubstantial given that 170,000 patients per year undergo coronary artery bypass grafts in the United States alone (Kinchla & Weiss, 1985). Other high risk surgeries include "black patch delirium" following ophthalmologic surgery and the increased incidence of perioperative cognitive dysfunction in cancer patients as described by Holland and Mastrovito (1980) and Silberfarb (1983). Other high risk groups include burn victims, women undergoing abortion or hysterectomy, transplant patients, and those with chronic pain (Siebert, 1986).

Postoperative delirium may not appear until the third or fourth postoperative day and may be related to the cumulative effect of medications, metabolic disorders, decreased sensory input, and decreased sleep. Usually delirium comes on abruptly and is short-lived. Periods lasting longer than a week should alert the physician to search for additional specific causes prolonging the delirium.

When assessing the etiology of postoperative delirium, an accurate record of all the patient's medicines with their total daily doses should be recorded, with special attention being paid to narcotic analgesics, cardiac drugs, and any medications with anticholinergic activity. This record should be based on nurses' records rather than the doctor's order sheet, as errors in transcription or administration are occasionally made. The usual spectrum of toxic, metabolic, and infectious causes should be considered, as well as embolic phenomena in patients at risk. Table 34-2 (Tune & Folstein, 1986) provides a list of etiologic factors for the preoperative and postoperative periods associated with delirium. The assessment and management of delirium is further discussed in Chapters 12 and 19. Table 34-3 lists preventive strategies for postoperative delirium.

Coexisting Psychiatric Illness

It has now been well established that psychiatric comorbidity in medical and surgical patients is associated with increases in length of hospitalization (Fulop, Strain, Fahs, et al., 1989). In one study, patients with concurrent psychiatric illnesses not only were found to have longer hospital stays but were also significantly more likely to die or to be discharged to a long-term care facility (Fulop, Strain, Vita, et al., 1987). Therefore it is important that psychiatric intervention be an integral part of perioperative care.

Psychiatrists are often faced with the dilemma of perioperative psychopharmacologic management of patients with preexisting psychiatric illness. Generally, treatment with cyclic antidepressants, lithium, or neuroleptics need not be discontinued before administra-

TABLE 34–2. *Factors Associated with Postoperative Delirium*

Factors originating during the preoperative period	Factors originating intraoperatively or during the postoperative period
*Increasing age (>50) Psychiatric illness (psychosis, personality disorder)	*Medications and drug combinations, especially anticholinergics
*Preoperative factors causing special vulnerabilities	Sleep deprivation, sensory deprivation, and immobilization, low cardiac output postoperatively
Preoperative dementia Susceptibility to toxins and metabolic abnormalities	Cerebrovascular accidents Length of time under anesthesia (8 + hours, 4 + hours bypass time)
Acute intermittent porphyria Drug sensitivities	Operative hypotension, hypercapnia, hypoxia, hypovolemia
Chronic renal disease Electrolyte abnormality Uremia Anemia Dialysis dysequilibrium	Choice of anesthetic
Neoplasm Metastasis	Type of surgery (higher incidence with eye operations/complex surgeries) ? SICU environment
Disseminated intravascular coagulopathy Nonbacterial endocarditis	Postoperative electrolyte abnormality, acid-base disturbances, anemia, hypoxia, abnormal PCO_2 hypoglycemia, uremia, hyperammonemia, endocrinopathies
Neurosyphilis Jarisch-Herxheimer reaction	*Fever and infection
*Addiction Withdrawal Thiamine deficiency	Cardiac, pulmonary, or abdominal complications
Traumatic injury Concussion Contusion Subdural hematoma Fat embolism	
Seizure disorders Postictal confusion	

From Tune & Folstein (1986). With permission.
SICU = surgical intensive care unit.
*Most common factors.

TABLE 34–3. *Preventive Strategies for Postoperative Delirium*

- Preoperative psychiatric evaluation and intervention
- Perioperative treatment of any drug addiction
- Preoperative education
- Correction of metabolic abnormalities
  - Electrolyte disturbance
  - Acid-base problems
  - Anemia
  - Hypoxia
  - Abnormal PCO_2
  - Hypoglycemia
  - Uremia
  - Hyperammonemia
  - Endocrinopathies
- Treatment of any infections or toxicities
- Careful monitoring of potentially complicating drugs
  - Anticholinergics
  - Corticosteroids
  - Histamine H2 antagonists
  - Opiates
  - Benzodiazepines
- Control of fever
- Avoidance of sleep deprivation
- Attention to appropriate sensory input

tion of anesthesia for elective operations (Stoelting, Dierdorf, & McCammon, 1988). On the other hand, it has been traditionally recommended that monoamine oxidase inhibitors (MAOIs) be discontinued at least 14 to 21 days before elective surgery to allow regeneration of new monoamine oxidase (Michaels, Serrins, Shier, et al., 1984). However, this practice poses a dilemma in severely depressed patients in whom medication discontinuation is contraindicated, and for emergency surgeries, when planned discontinuation is impossible. In cases where MAOIs are continued, it is important to adjust the choice and doses of drugs used during anesthesia to prevent adverse effects. Meperidine should be avoided during the preoperative and intraoperative periods. Induction of anesthesia with intravenous barbiturates is acceptable, but CNS and respiratory depression are likely to be accentuated. Succinylcholine must be used cautiously in view of the possibility of decreased plasma cholinesterase activity. Generally, nitrous oxide combined with volatile anesthetics is the accepted means of maintenance anesthesia in these cases. For postoperative analgesia, morphine is the preferred drug (Stoelting, Dierdorf, & McCammon, 1988). *The use of meperidine with MAOIs is absolutely contraindicated* owing to a frequently severe and often fatal excitatory interaction (Stack, Rogers, & Linter, 1988; Sedgwick, Lewis, & Linter, 1990).

POSTOPERATIVE ANALGESIA

The role of the psychiatrist in postoperative pain management can be a complex one depending on the pain management services available at particular institutions. Ideally, there should be psychiatric representation on the pain management service, a team usually coordinated within the department of anesthesiology. By combining the disciplinary interests of psychiatry and anesthesiology in a team approach, potential interprofessional conflicts can be avoided. The team can best determine and implement treatment options. However, if such a multidisciplinary team is not available, the psychiatrist can best manage pain by closely focusing on maintaining analgesia while evaluating possible psychiatric comorbidity including addiction.

Despite many advances in our knowledge of pharmacology, pain during the early postoperative period remains a serious problem. The physiologic consequences of postoperative pain are well substantiated. The sympathetic stimulation that can accompany severe pain can compromise patients' cardiac status, respiratory function, gastrointestinal motility, and peripheral vascular permeability (Lutz & Lamer, 1990). Although parentally administered opioids are the most frequently used means of postoperative pain control, conventional doses are often ineffective (Marks & Sacher, 1973). In the classic study by Marks and Sacher (1973), data were gathered from 37 medical inpatients and from questionnaires sent to a representative group of 102 house physicians in New York teaching hospitals. The doctors fully subscribed to the belief that patients should be pain-free. Despite the fact that their doctors prescribed analgesic regimens, 32% of the patients studied experienced severe discomfort, and 41% were in moderate discomfort. The authors attributed these findings to the doctors' underestimation of effective dose ranges of analgesics, overestimation of duration of action, and exaggeration of the danger of addiction. Furthermore, studies have shown that adequate analgesia to alleviate postoperative pain contributes to earlier ambulation and earlier dismissal from the hospital (Broadman, Hannallah, Norden, et al., 1987).

Systemic Analgesics

There are currently a number of techniques used to manage postoperative pain. Although parenteral opioids can be highly effective, complications may arise from the use of inadequate analgesic doses and dosage intervals, as described in Chapter 16. One way to eliminate gaps in analgesia is to give opioids on a time-contingent basis. If doses are given by the clock, timed to peak as the previous dose falls below the therapeutic level, constant relief can be achieved. Still less fluctuation in blood levels can be achieved by using long-acting agents such as levorphanol and methadone, rather than the more commonly used morphine or meperidine. A relatively large parenteral dose of methadone given intraoperatively can produce prolonged postoperative analgesia, which can facilitate subsequent fixed dosing (Gourlay, Wilson, & Glyn, 1982).

Because of the common fears of addiction and ventilatory depression associated with opioid administration, drugs with mixed agonist-antagonist properties, such as pentazocine and butorphanol, have been introduced in an effort to reduce these risks. However, the price for the added safety is a ceiling on the efficacy of these preparations. Increasing the dose beyond the usual amounts results in only a slight increase in analgesic efficacy, a definite drawback with patients with severe pain (Boas, Holford, & Villiger, 1985). Care must be taken not to use these mixed agonist-antagonist agents in opioid-addicted patients, as withdrawal symptoms may ensue. Similarly, systematic tapering to prevent withdrawal is often indicated if these agents are used for prolonged periods. A detoxification regimen of progressive dosage tapering is particularly indicated for prolonged use of butorphanol, a significantly more potent analgesic than nalbuphine or pentazocine. Generally, these mixed agonist-antagonists can be abruptly discontinued if used for short-term management of acute pain.

Once postoperative patients are able to take medications orally, they can be shifted to oral opioids, even when pain is fairly severe (Abram, 1989). It is even more important to use oral analgesics on a time-contingent basis, as their onset of action is even more prolonged, and significant periods of inadequate analgesia occur if the patient waits for the pain to return before requesting the next dose. As pain becomes less severe, some of the less efficacious opioids such as codeine and propoxyphene can provide adequate analgesia. These drugs are frequently combined with nonopioid analgesics such as aspirin, acetaminophen, or nonsteroidal antiinflammatory drugs (NSAIDs). Table 34-4 (Abram, 1989) lists prescribing information for typical narcotic agonists as well as most agonist-antagonist drugs.

Ketorolac, a new NSAID, is a promising alternative to oral nonsteroidal agents and a nonnarcotic alternative to opioid analgesics. Its analgesic effectiveness has been shown to be similar or superior to that of morphine, meperidine, or pentazocine in single dose studies of patients with postoperative pain or renal colic (Resman-Targoff, 1990). Other advantages of ketorolac over opiates are the absence of respiratory depression and lack of drug abuse potential. Ketorolac is indicated for short-term pain management. It is given intramuscularly with an initial loading dose of 30 to 60 mg followed by a dosage of one-half the loading dose given every 6 hours. The maximum daily dosage is 120 mg with the exception of the first day when a total of 150 mg may be administered. It is also available in an oral form.

Patient-Controlled Analgesia

Because of the aforementioned problems in providing sustained pain relief with systemic analgesia, researchers during the 1960s began testing systems that allowed patients to self-administer small intravenous doses of narcotics (Sechzer, 1971). Today the patient-controlled analgesia (PCA) device is a widely used and efficient

TABLE 34–4. *Oral and Parenteral Narcotic Analgesics for Severe Pain*

Drug	Route*	Acute Equianalgesic Dose (mg)	Relief Duration (hr)	Elimination Half-Life (hr)	Comments
Narcotic agonists					
Morphine	IM PO	10 60	2.5–7 2.5–7	2.9 ± 0.5	Standard for comparison; also available in slow-release tablets; IM use not recommended
Codeine	IM PO	130 200	4–6 4–7	3–3.5	Biotransformed to morphine, useful as an initial narcotic analgesic
Oxycodone	IM PO	15 30	 3–5	—	Short-acting; available alone or as 5-mg dose in combination with aspirin and acetaminophen
Heroin	IM PO	5 60	4–5 4–5	0.5	Illegal in U.S.; high solubility for parenteral administration
Levorphanol (Levo-Dromoran)	IM PO	2 4	4–6 4–7	11.4	Good oral potency, requires careful titration during initial dosing because elimination half-life exceeds duration of relief accumulation
Hydromorphine (Dilaudid)	IM PO	1.2 7.5	4–5 4–6	2–3 2–4	Available in high-potency injectable form (10 mg/ml) for cachectic patients and as rectal suppositories; more soluble than morphine
Oxymorphone (Numorphan)	IM PR	1 10	4–6 4–6	2–3	Available in parenteral and rectal suppository forms only
Meperidine (Demerol) Normeperidine	IM PO	75 300	2–4 3–4	3–4 12–16	Contraindicated in patients with renal disease; accumulation of active toxic metabolite, normeperidine, produces CNS excitation
Methadone (Dolophine)	IM PO	10 20	6–8 6–8	15–30	Good oral potency; requires careful titration of the intial dose to avoid drug accumulation
Mixed agonist-antagonist drugs					
Pentazocine (Talwin)	IM PO	60 180	2–4 3–4	2–4	Limited use for cancer pain; psychotomimetic effects with dose escalation; can precipitate withdrawal in narcotic dependent patients
Nalbuphine (Nubain)	IM PO	10 —	4–6	5	Not available orally; less severe psychotomimetic effects than pentazocine; can precipitate withdrawal in narcotic dependent patients
Butorphanol (Stadol)	IM Intranasal	2 2	3–4	3–9	Not available orally; produces psychotomimetic effects; can precipitate withdrawal in narcotic-dependent patients; Nasal spray formulation approved by FDA in 1992. Dosage intervals must be longer in elderly
Partial agonists					
Buprenorphine (Buprenex)	IM SL	0.3 0.6	4–6 5–6	5	No psychotomimetic effects; can precipitate withdrawal in narcotic-dependent patients; large doses of naloxone needed to reverse its respiratory depression

Modified from Abram (1989). Some data from 1992 AMA Drug Evaluations (AMA, 1992).
*IM = intramuscular; PO = oral; PR = rectal; and SL = sublingual.
Note: Pharmacokinetics are variable and dosage must be individualized. Lower doses and/or longer intervals should be considered with advanced age or impaired hepatic or renal function.

method of postoperative pain management (Gleason, Rodwell, Shaw, et al., 1990). This device consists of an electronically controlled infusion pump with a timing device. When patients experience pain, they trigger the device by pushing a button on a cord extended from the machine. A preset amount of analgesic agent is then delivered intravenously. A regulatory timer is programmed to prevent administration of additional doses until a predetermined interval has elapsed, thus preventing administration of a second dose until after the first dose has exerted its maximal analgesic effect. Morphine and meperidine are widely used in PCA devices because of their rapid onset of action and minimal ceiling effect. The risk of clinically significant postoperative respiratory depression has been shown to be low (White, 1987). However, the use of meperidine in these devices must be closely monitored because the accumulation of the active toxic metabolite normeperidine can cause CNS excitation and associated acute confusional states.

Generally, PCA devices are contraindicated for those individuals who cannot understand the concept of how to use the device. Most other patients are suitable candidates for this method of analgesia. Those patients who do poorly with PCA devices are those who are excessively dependent on others to meet their needs. In our experience, PCA devices can be safely used in patients with opioid addiction. The device provides a good estimation of tolerance and allows self-titration in this patient population.

Other Approaches

Other effective methods of postoperative pain management usually administered by anesthesiologists, include epidural or intrathecal administration of spinal opioids and selective analgesia achieved with the use of various types of neural blockade, i.e., brachial plexus blockade, interscalene blockade, axillary plexus blockade, and intercostal nerve blocks (Lutz & Lamer, 1990).

Although generally reserved for acute non-surgical and chronic pain management, cognitive behavioral techniques can be useful for the management of postoperative pain for those patients who are able to participate. Imagery and relaxation techniques when taught preoperatively, can be helpful adjuvants to overall pain management (Spencer, 1989).

BREAST SURGERY

Women who undergo breast surgery are similar in some respects to those who choose elective aesthetic surgery on their breasts. Both of these groups of women experience surgical trauma and have the task of incorporating their reshaped breasts into their new body image.

Breast cancer is the most common cancer of women, and it is the leading cause of death in American women between the ages of 40 and 44 (Walbroehl, 1985). From the initial finding of a "suspicious lump" through diagnosis and treatment, breast cancer can have a profound psychological impact on a woman. It may trigger issues of earlier losses, as well as fears about future physical attractiveness and sexual functioning.

The symbolic importance of a woman's breasts can be understood in terms of conscious and unconscious conflicts that arise when breast surgery is recommended. These conflicts depend on the woman's existing feelings about her body and in particular her breasts (Stevens, 1987). A woman's breasts may be symbolic of early childhood nurturance, femininity, attractiveness, and sexuality (Nadelson & Notman, 1979). Diseases of the breast may therefore have emotional or psychological effects that go beyond the confines of the diseased tissue regardless of the woman's age (Stevens, 1987). Most breast lumps are discovered by the patients themselves (Stevens, 1987) and this conscious realization is often preceded by unconscious knowledge. Review of dream material may be helpful in bringing to consciousness a woman's experience of the cancer. Sometimes dreams, fantasies, or sudden fears that something is wrong with her body may foretell conscious knowledge of the cancer (Stevens, 1987).

The degree and duration of psychological and physical disabilities resulting from gynecologic cancer also depends on the stage and grade of the tumor, the surgical procedure, and postoperative medical treatment (Cain, Kohorn, Quinlan, et al., 1983). Pharmacologic or radiologic interventions after mastectomy can lead to loss of hair, appetite disturbance, and more bodily changes. Steroids can cause weight gain as well as changes in mood. Further loss of the feminine appearance can result from androgen therapy.

For patients with advanced disease and metastatic spread, psychiatric intervention is helpful for dealing with anxiety and depression. In addition, patients with terminal disease frequently need treatment for bone pain; hypercalcemia and metastatic spread sometimes lead to cognitive impairment disorders.

Today, many women diagnosed with breast cancer have the option of breast-conserving surgical procedures. Frequently, success rates comparable to those associated with the more radical mastectomies previously recommended are obtained (Veronesi, Saccozzi, Del Vecchio, et al., 1981). These newer treatment approaches—including segmental resections—often are recommended with adjuvant irradiation or chemotherapy. Unfortunately, women choosing these less radical surgical alternatives sometimes report less psychosocial support from peers and partners, who equate the loss of less tissue as being less traumatic (Rowland & Holland, 1989). When choosing a less extensive resection there is also the fear of leaving cancer cells. In addition, women may have concerns about choosing a less extensive surgical procedure, especially as no clear treatment consensus exists among the physician experts undertaking breast cancer treatments. To deal with these conflicts some women may wish their physicians to make a surgical choice, or they rely heavily on their psychosocial support persons for advice and guidance.

Each woman diagnosed with breast cancer responds with psychological reactions dependent on her prior associations with breast cancer (Rowland & Holland,

1989), previous methods of coping (Meyerowitz, 1980), and the point in her life cycle at which the cancer is diagnosed (Rowland & Holland, 1989). Her style is also dependent on real or fantasied losses, including death. There is some literature to support the hypothesis that a woman with a more "confrontational or tackling stance" adjusts better (Penman, 1979), at least during the initial rehabilitative and adjustment stages.

Traditionally, mastectomy was the treatment of choice for nearly all breast cancers and continues to be frequently recommended for many patients. Studies of postmastectomy patients describe alternations in mood, body image, sexuality, femininity, and social and occupational functioning (Jamison, Wellisch, and Pasnau, 1978; Meyerowitz, 1980; Rowland & Holland, 1989; Stevens, 1987). Interestingly, these psychological changes are similar to those of patients experiencing other types of amputation (Findlay, Lippman, Danforth, et al., 1985; Meyerowitz, 1980; Wolberg, Romsaas, Tanner, et al., 1989).

As might be expected, women with previous psychiatric morbidity seem to be at greater risk for psychological morbidity that requires intervention (Bloom, Cook, Fotopoulis, et al., 1987). Women with additional concurrent illness, with less expected family and peer support, and with feelings of being out of control seem to have more psychological difficulty after mastectomy. Conversely, those women with stable self-esteem and sexual identity and a history of coping under stress (and with no history of a psychiatric disorder) show few serious long-term psychopathologic sequelae (Bloom, Cook, Fotopoulis, et al., 1987).

Simultaneous reconstruction has become more available and better accepted for women with early breast carcinoma, and for many it has reduced the psychological morbidity postoperatively (Stevens, 1987).

The psychological process for women with mastectomy who are to have simultaneous breast reconstruction is different than the process for the women undergoing mastectomy without reconstruction. This alternative may give the women a sense of "restoration" and not merely "replacement" of the lost breast (Stevens, 1987). For the most part this option is available to women with early breast cancer and no evidence of metastatic spread. Generally, discussion of reconstruction procedures for suitable patients should be discussed with the patient at the time the cancer surgery is planned. Currently fewer than 10% of all women undergoing mastectomy have reconstruction, although data collected by the American Society of Plastic and Reconstructive Surgery suggests that this situation may be changing (Rowland & Holland, 1989).

Immediate reconstruction has been shown to reduce some of the negative emotional consequences of this cancer surgery (Snyderman & Guthrie, 1971), and it is becoming a more frequent option at the time of surgery if there is no evidence on frozen section of microscopic spread to the axillary lymph nodes. For women with immediate reconstruction there may be fewer depressive symptoms and fewer concerns about the deformity left by the mastectomy (Stevens, McGrath, Druss, et al., 1984). In one study 100% of the patients with delayed reconstruction thought that the external prosthesis was a burden and a discomfort and did not experience it as a restoration for breast tissue (Stevens, McGrath, Druss, et al., 1984).

Although every woman deals with the loss of a breast differently, in some cases patients with delayed reconstruction have more difficulty integrating this change into a stable body image. Interestingly, neither immediate nor delayed reconstruction appears to alter the grief or mourning process seen with the loss of tissue, and in both groups the mourning for the loss of a breast begins before surgery, as the loss is anticipated (Stevens, McGrath, Druss, 1984).

As more women are offered and choose breast reconstruction, two psychological patterns have begun to emerge. Interestingly, women who sought reconstructive consultation but refused for nonmedical reasons tend to be older, single, and of lower socioeconomic class (Holland & Rowland, 1987; Jacobs, Holland, Rolland, et al., 1983).

Patients with breast carcinoma seem to have a period of at least 16 months of emotional, social, and vocational adjustment; educational and counseling groups may be helpful during this period of psychological recovery (Wolberg, Romsaas, Tanner, et al., 1989). Several self-help groups, such as Reach to Recovery, encourage more participation by women in their treatment and advocate peer "counseling" and support. These groups provide additional psychological support from other women with first-hand knowledge who can aid in the psychological recovery. (See also Chapter 3.)

Several support groups specifically concerned with reconstruction after mastectomy have developed in many cities. AWEAR (Ask a Woman to Explain About Reconstruction) and RENU (Reconstruction Education for National Understanding) are two that provide information and emotional support for women contemplating immediate or delayed breast reconstruction.

GYNECOLOGIC SURGERY

Gynecologic cancer surgery can have a profound psychological impact on a woman, and gynecologic cancer is the fourth most common form of cancer diagnosed in women (Auchincloss, 1989). Treatment and outcome depend on the site and stage of the tumor, although life expectancy has improved for most cancers.

The degree of psychological distress experienced by a woman varies, but the circumstances by which a woman is informed of her cancer may influence her reaction. Notification by telephone of an abnormal Papanicolaou smear, a carcinoma, or a cancer recurrence is frightening, and most women would rather be told in person by their primary physician (Cain, Kohorn, Quinlan, et al., 1983).

The psychological impact of hysterectomy is well documented in the literature. Surgery within the pelvis is rarely a neutral emotional event, even in women who are postmenopausal and undergoing treatment for benign disease (Notman, 1987). Women with carcinoma may harbor unconscious guilt about possibly causing the gynecologic cancer by some past romantic affair, abortion, or promiscuity.

Although the primary concern of the surgical treatment for gynecologic malignancy is survival, addressing psychological concerns and future sexual functioning is important. The issues of surgically induced menopause, loss of childbearing capacity, and the impact of a life-threatening illness are important to psychological recuperation.

The posthysterectomy syndrome (Richards, 1973), which includes psychological and somatic complaints after this surgery, has been debated, and there are some reports (Lalinec-Michaud, Engelsmann, & Marino, 1988) observing that women undergoing hysterectomy may not differ from women undergoing other surgery in terms of general psychiatric morbidity. Paradoxically, women with nonmalignant pathology seem to be at higher risk for depression after hysterectomy (Nadelson & Notman, 1979), having undergone a major loss for a "benign" condition. Most of these depressive symptoms decrease during the first year.

Factors that may place the women at high risk for posthysterectomy depression include age under 40 (Nadleson & Notman, 1990; Polivy, 1974), high levels of anxiety (Lalinec-Michaud & Engelsmann, 1985), and a psychiatric history. Furthermore, the length of time interval between being informed that a hysterectomy is recommended and the actual surgery may also be significant. Women who undergo emergency hysterectomy seem to have more psychic distress (Tang, 1985), and it may be that a woman needs to work through her anticipatory grief prior to the surgical loss.

The site, stage, grade, and recommended treatment are also important predictors of emotional distress (Cain, Kohorn, Quinlan, et al., 1983). If an oophorectomy is done, surgical menopause ensues; and in some patients hormone replacement leads to significant improvement in feelings of well-being. Even when estrogen is being replaced, some women do better psychologically on more or less estrogen, or with cyclic progestin therapy added. Bilateral oophorectomy, however, does not imply estrogen deficiency, as fragments of ovarian tissue as well as the adrenal glands continue to produce estrogen (Martin, Roberts, & Clayton, 1980). Loss of adrenal or ovarian function can and does occur after some gynecologic surgery or as a result of chemotherapy or irradiation to the pelvis. This loss can lead to vaginal dryness; and, with time, estrogen insufficiency leads to atrophy of the vaginal mucosa (Auchincloss, 1989). Local or systemic estrogen replacement may be helpful, but may be contraindicated for the breast or gynecologic cancer survivor (Auchincloss, 1989).

Some other studies have found linear trends between grade of tumor (or histologic differentiation) and depression and psychosocial impairment. The higher the tumor grade, the greater is the psychological impairment (Cain, Kohorn, Quinlan, et al., 1983). Although the women in this study were not informed directly about the histology, they may have reacted to the surgeon's unconscious communication of a less optimistic attitude for poorly differentiated tumors (Cain, Kohorn, Quinlan, et al., 1983).

Regardless of the patient's current sexual activity or sexual preference, every women has sexual desires, daydreams, and fantasies. After gynecologic surgery women may temporarily lose interest in sex, this may resolve spontaneously or may require specific treatment. Male patients with cancer more frequently have a loss of their excitement phase, whereas women are more likely to have a decreased desire phase (Auchincloss, 1989). Sexual dysfunction is a possible complication after hysterectomy, as the surgery is often done in combination with radiation treatment. Postoperative irradiation can lead to sexual dysfunction because of fibrosis, vaginal stenosis, and decreased lubrication (Seibel, Freeman, & Graves, 1980). Unfortunately, women may not bring up sexual issues or concerns. In a study of cervical cancer patients, 80% of women reported desiring more information about sexual functioning; however, only 5% reported that they would bring it up themselves (Vincent, Vincent, Greiss, et al., 1975).

For the woman who has undergone extensive pelvic exenteration surgery, the physical assault may be extensive, including the removal of the rectum, distal colon, bladder, ileac vessels, uterus, tubes, and ovaries in addition to the entire pelvic floor. Ostomies are created for urinary and fecal excretion. Such radical surgery requires psychological support and education (Silberfarb, 1984). Vaginal reconstructive surgery should be discussed with appropriate patients, and it may lessen the feelings of mutilation and help the patient reconstruct her body image. In one small study of vaginal reconstruction and rehabilitation after pelvic exenteration, six of eight women became sexually active and orgasmic within 6 months (Lamont, De Petrillo, & Sargeant, 1978). Alternative sources of sexual stimulation, such as intramammary, intrathigh, or anal intercourse, as well as oral and manual stimulation, are all possible sources of sexual satisfaction for a couple after radical gynecologic surgery and should be discussed with patients and their partners.

Frank, open discussions of sexual concerns and issues should begin with the treating physician or psychiatric consultant. Questions about sexual well-being, the couple's previous sexual relations, and intimacy could be introduced with the simple question: How has this surgery affected your sexual life? (Auchincloss, 1989). Unfortunately, the patient may feel that her first priority is to worry about the cancer and that sexual issues are a personal matter not to be discussed with the doctor. Scanty information may lead to misconceptions by the patient as well as her partner.

For some patients who have loss of sexual desire, a serum prolactin level and thyroid profile are diagnostically useful. For women with loss of desire or excitement after treatment-related ovarian loss, testosterone therapy may prove helpful (Auchincloss, 1989). Masculinizing side effects, however, are severe in some women. Some patients may avoid sex because of dyspareunia due to decreased lubrication after irradiation, and fears of "catching" the radiation or cancer or anxiety about hurting her may lead to sexual withdrawal by the partner.

Psychological and sexual recovery can be enhanced with psychotherapy and education. Frequently lubricants, regular vaginal dilation, and the female superior (woman-on-top) position may be helpful. Sexual therapy over an extended period may be necessary in some patients, depending on the sexual diagnosis. Although decreased sexual desire is the most frequently seen in women cancer patients, there may be also dysfunction in the excitement or orgasm stages. Sometimes problems in one of these areas can lead to difficulties in all three. For example, if because of dyspareunia a woman has an excitement phase problem, she may lose all desire and consequently have no orgasm (Auchincloss, 1989).

Finally, psychiatric consultants to gynecologic surgical patients must also take into consideration coexisting medical problems. Diabetes or renal or hepatic involvement by cancer can further impair sexual functioning. Also, thyroid hormone levels may be low, and psychotropics, antihistamines, and anticholinergic drugs can decrease vaginal lubrication. (Sexual dysfunction in the medically ill patient is discussed in Chapter 15.) Table 34-5 applies the triphasic model to the assessment of sexual dysfunction in men and women (Auchincloss, 1989).

AMPUTATION

Amputation is one of the first surgical procedures recorded in ancient history. Throughout history amputations have been performed as a punishment or torture (Shukla, Sahu, Tripothi, et al., 1982); therefore we tend to view this loss consciously or unconsciously as a punishment. Loss of a body part may not only interfere with function but also can lead to the grief reaction that occurs with any acute loss (Wilson & Krebs, 1983). As with other types of surgery, if possible the patient undergoing an amputation should be evaluated for preexisting psychiatric difficulty and ego strengths. As much as possible, the meaning of the amputation with regard to the patient's self-concept and body image should be determined. Amputation of a lower extremity of a dancer, for instance, produces more grief than amputation of a nonfunctional extremity in a paraplegic. Similarly, a traumatic amputation more greatly affects an active child than a retired sedentary individual. Irrespective of age, however, there are issues of incorporating a new body image; and regardless of the precipitating causes leading to the amputation, the patient will be dealing with the loss and with grief for the amputated limb (Riether & Stoudemire, 1987).

A significant prevalence of major depressive reactions was found by Kashani, Frank, & Kashani et al. (1983) in a prospective study of 65 amputees. The most common cause for the amputations in their series was vascular disease, and 34% of the patients were found to be depressed on clinical interview, according to *DSM-III* criteria for major depression. There were also higher frequencies of depression among the female amputees (50%); only 30% of the males had depressive symptoms. In general, it appears that geriatric amputees are more likely to have significant psychiatric morbidity; and, when found, vegetative symptoms or "depressive equivalents" should be treated with antidepressant drugs and psychotherapy.

There are two broad groups of amputation patients: those with disease-related amputations and those with traumatic amputations. There are differences between these two populations. For elderly patients with vascular disease and nonhealing ulcers, an amputation may be a welcome relief from pain; for individuals with

TABLE 34–5. *Application of the Triphasic Model to Assessment of Sexual Dysfunction in Cancer*

Phase	Sexual dysfunction	Relevance to cancer patients and survivors
Desire	Inhibited sexual desire	Not unusual when patient is in active treatment
Sexual thoughts, fantasies, daydreaming; finding potential partner attractive	Loss of interest in sex Few or no thoughts about sex Negative ("antisexual") attitudes about sex Anxious, panicky feeling about sex Avoidance of sexual situations	After treatment, loss of desire may be related to the cancer itself, treatment side effects, psychological factors (depression, anxiety), and partner issues Often requires longer treatment of couple by sex therapist because of prominent psychological component
Excitement	Inhibited sexual excitement	
Penile erection in men	Erectile impotence in men: difficulty attaining or maintaining an erection	Requires thorough medical evaluation, including review of medications Common after bladder, prostate, colorectal surgery, and radiation therapy May have psychological component even when physical cause is present Supportive partner is essential Treatment depends on cause counseling to decrease anxiety, decrease focus on performance For complete organic impotence, consider penile prosthesis
Vaginal lubrication and engorgement in women	Impaired vaginal lubrication and engorgement in women	Common after surgery, irradiation to pelvis, or any treatment that causes ovarian loss or failure Patient may complain of dry, sore vagina or painful intercourse Treatment: estrogens (local or systemic), lubricant, taking more time for foreplay, dealing with communication issues
Orgasm		
Reflex muscle contractions, associated with pleasure, ejaculation, and emission in men, pleasurable sensation in women	Inhibited female orgasm: anorgasmia	May be related to fatigue, depression, stress, medication, anxiety Need for longer or more direct stimulation of clitoris Address the need for time and relaxation, communication issues with partner
	Inhibited male orgasm: retarded ejaculation	May be related to fatigue, depression, stress, medication, anxiety
	Premature ejaculation: inability to control timing of orgasm	Rare complaint in cancer patients, unless preexisting Easily treated by sex therapist—good prognosis with brief therapy Retrograde ejaculation and anejaculatory orgasm occur after some abdominal and pelvic cancer surgeries
Other	Dyspareunia: pain with intercourse	Leads to sexual avoidance unless treated promptly Requires thorough gynecologic evaluation and treatment of cause (surgical change in vagina, irradiation changes, estrogen lack) Practice "no painful sex" rule (i.e., no intercourse unless medical cause is adequately treated)
	Vaginismus: vaginal muscle spasm, making penetration painful or impossible	Response to pain or fear of pain with penetration Good prognosis with combined relaxation and sequenced penetration treatment, done by patient herself, then with partner

From Auchincloss (1989). With permission.

cancer, it may be a sacrifice in exchange for the hope of a surgical cure.

Despite the limb being nonfunctional or painful, there may be ambivalence about this surgery in which a body part "disappears." There may be fantasies about what was done with the amputated limb, and optimism before the surgery may be replaced by anger or despair when the limb is gone (Lundberg & Guggenheim, 1986). These patients may express helplessness and hopelessness, and they may have resigned themselves to disability and nonfunction. Furthermore, the chronically ill patient may have assimilated the extremity with vascular disease into a stable sense of self, and they may have difficulty adjusting to a stump. In these cases it may be difficult for them to become motivated for rehabilitation (Lundberg & Guggenheim, 1986).

A study of aged lower limb amputees found that depression is common and, if left untreated, interfered with rehabilitation (Caplan & Hackett, 1963). Pathologic reactions that may signal the need for additional therapy include no reaction at all to the loss, a state of euphoria or unfounded optimism, prolonged depression or anxiety, severe overreaction to the loss, aggressive or asocial behavior, or masochism.

For the patient with a traumatic amputation, surgery is often a sudden life-saving measure to prevent sepsis in the case of gangrene, or the end result of a failed revascularization procedure in a trauma patient. In most cases of traumatic injury, anxiety is the predominant symptom (Schweitzer & Rosenbaum, 1982). These patients, who most often are younger than the patient with vascular disease, face unique stresses. They are often admitted emergently with no preoperative teaching, and suffer sudden disruption to their body integrity. Although grateful for being alive, these patients may become overwhelmed by the consequences of the loss. Thus with little or no time to prepare for the amputation, patients with a traumatic injury may have difficulty coping (Wilson & Krebs, 1983).

In a prospective study of 72 amputees (50 of whom had had limbs amputated during or following accidents), nearly all had phantom limb phenomena; and nearly two-thirds had psychiatric symptoms of depression. A longer time may be necessary for these patients to grieve their loss as they struggle with anger, frustration, and blame. For some patients, nightmares and other symptoms of posttraumatic stress disorder may occur. Themes of guilt, punishment, anger, and blame predominate on projective testing for some traumatic amputation patients (Lundberg & Guggenheim, 1986).

One of the most difficult challenges for the psychiatric consultant is phantom phenomena (Wilson & Schirger-Krebs, 1987). The persistence of pain or sensation in a phantom limb often occurs immediately after amputation, and it may decrease over time or become a persistent problem. Phantom pain occurs in 2 to 48% of cases (Lundberg & Guggenheim, 1986; Parkes, 1973) and there is some research to suggest that premorbid psychological factors may influence it. Patients with a high degree of neuroticism, with a rigid, compulsive personality style who have difficulty asking for help, often have a less successful recovery (Morgenstern, 1970; Parkes, 1976). Review of the assessment and treatment of phantom limb syndromes may be found elsewhere (Harwood, Hanumanthu, & Stoudemire, 1992).

Patients with a long history of illness and surgical operations on the body part that is subsequently amputated are most likely to have phantom pain, according to some researchers (Melzack, 1971). The persistent phantom sensation also may be secondary to an unresolved grief process in which the loss is denied, and the persisting sensations can be thought of as a wish fulfillment hallucination (Kolb, 1952; Lundberg & Guggenheim, 1986). Causalgia, a posttraumatic pain syndrome also can contribute to phantom pain. Causalgia patients may respond to the combination of an antidepressant and a phenothiazine. Analgesics, nerve blocks, and behavioral treatments are also helpful. When phantom pain becomes chronic, patients may benefit from regularly scheduled pain clinic appointments, where a multidisciplinary approach to pain control is undertaken.

Rehabilitation is the goal following any amputation; not only does the level of the amputation relate to the rehabilitation potential, but so does cognitive ability. Psychological testing may be valuable in the assessment of elderly amputees with vascular disease to assess cognitive impairment (Pinzur, Graham, & Osterman, 1988). In one study of 60 adult patients who received below- or above-the-knee amputations for peripheral vascular insufficiency, 28% were found to have cognitive deficits hampering their rehabilitation potential, sometimes prohibiting them from attaining prosthetic competence and gait training (Pinzur, Graham, & Osterman, 1988).

In a retrospective study of the rehabilitation of patients with lower extremity amputations, patients with the most successful rehabilitation were those who were able to grieve for their lost limb (Parkes, 1975). Others have found that the greater the support systems available to the patient and the less psychological impairment prior to amputation, the better is the postsurgical coping (Wilson & Krebs, 1983).

Almost all amputees, regardless of age or the reason for the amputation, want a prosthesis (Lundberg &

Guggenheim, 1986; Mazet, 1967), but the value of the prosthesis depends on its functional value. It may help to psychologically restore "the missing part", restoring a completed body image both physically and psychologically, or it may be a burden. Patients with narcissistic personality organization may find the loss so devastating that they have difficulty moving on to successful rehabilitation with a prosthesis, continuing to see themselves as defective.

In some cases of traumatic injury resulting in limb amputation, replantation or the surgical restoration of a severed appendage is possible (Schweitzer & Rosenbaum, 1982). Physically, these patients have their body part disconnected and then reattached. There is no sensation in the reattached limb, and often function does not return for several months if at all.

Unfortunately, these patients for whom limb salvage is possible do not escape psychiatric morbidity. Not surprisingly, the presence of previous difficulty coping, marital dysfunction, previous psychiatric difficulty, and a stressful life event within the year antedating the accident are associated with adverse postoperative outcomes (Schweitzer & Rosenbaum, 1982).

Unlike the patient with an amputation for a chronic illness, patients with traumatic amputation followed by replantation have longer hospital stays and often must undergo several operations to restore function. The patient may have difficulty reintegrating the retransplanted limb back into their body schema, especially if the integrity of the replanted limb is in question.

When a traumatic amputation occurs secondary to a life-threatening event, there are often fears of death. Later these anxieties are replaced by concerns over the viability of the replanted limb and the effects on function, occupation, appearance, and sexual attractiveness (Schweitzer & Rosenbaum, 1982). Data from semistructured psychiatric interviews of 30 patients with replanted limbs showed that most were assessed as needing psychiatric intervention. In this series there was a high incidence of stressful life events as well as psychopathology that predated the trauma, and those patients with a poorer postoperative emotional adjustment had psychopathology predating the accident.

The psychological treatment for a patient who has undergone an amputation needs the coordinated effort of the surgical team, rehabilitation staff, and the psychiatric liaison service. Many amputees overcome the difficulties that this surgery presents, but psychiatric and psychosocial intervention should begin early in the rehabilitative process as a form of tertiary prevention (Lundberg & Guggenheim, 1986).

AESTHETIC (COSMETIC) SURGERY

Elective aesthetic surgery patients differ from patients who have had traumatic surgery or other general surgery because most of the time these patients are not medically ill and are not undergoing a procedure to relieve pain or prolong life (Riether & Stoudemire, 1987). This surgery is not limited to a particular organ system, and the purpose of aesthetic surgery is to change how patients and others view their bodies. All aesthetic surgery is done for psychological reasons, at least in part (Pruzinsky & Edgerton, 1990) and its purpose is to "improve upon the normal" (Goin & Goin, 1986). It is a surgery that alters body image, so it can be expected to have impact on an individual's self-concept and identity. The most common elective cosmetic surgical procedures are liposuction, breast augmentation, blepharoplasty, rhinoplasty, and rhytidectomy (ASPRS, 1989; Pruzinsky & Edgerton, 1990).

When the aesthetic plastic surgeon produces a change in the outward appearance, or one's mirror image, there is not always a corresponding change in body image. The concept of body image is a concept of psychological variables, which includes cognitive understanding, sensory input, and comparisons made by the individual between self and others. A person's body image is shaped by the emotional responses he or she receives from peers (Belfer, Harrison, & Murray, 1979; Murray, Mulliken, Kaban, et al., 1979; Riether & Stoudemire, 1987; Schilder, 1950). These reactions cannot be underestimated, as research has shown that attractive individuals are more likely to be seen as intelligent, friendly, successful, sensitive, interested, and outgoing (Arndt, Travis, Lefebvre, et al., 1986; Dion, Berscheid, & Walster, 1972).

There may be temporary alterations in body image, such as pregnancy, acne, or broken bones; but generally these changes are not incorporated into the "psychological picture of oneself" (Henker, 1979). This body image is not static, developing as one matures; there is some literature to suggest that the younger the patients are when they undergo bodily changes, the more easily this change takes place. As Pruzinsky and Edgerton (1990) reported: "Younger patients have not had as long to develop the bond between their appearance and identity. That is, early in life, the body image is more malleable."

Murray, Mulliken, Kaban, et al. (1979), in their review of the literature, reasoned that a patient must work through several stages before the mental concept of body image undergoes change. There is the initial decision to have an operation, followed by the surgical experience. Next, there is the immediate postoperative period, followed by the reintegration phase, during which the patient adjusts to or incorporates a new body image. When the individual has a congenital or traumatic deformity, early correction may be important to prevent distortion of body image, isolation from peers, and psychological difficulties (Murray, Mulliken, Kaban, et al., 1979).

Most of us would like to change some aspect of our appearance. As adults many of us spend money on clothing, hair styles, bleaches, cosmetics, tans, and health spas to change the way we look to ourselves and others (Goin & Goin, 1989). It is no wonder then that children are fascinated with toys called "transformers" whose main appeal is the transformations that take place when the arms or legs of these figures are moved. Likewise, the literature is replete with examples of characters undergoing bodily changes: Pinocchio, Alice in Wonderland, and the Toad-Prince (Goin & Goin, 1989).

Women request cosmetic surgery more than men; in 1988 men accounted for only 16% of cosmetic surgery (ASPRS, 1989; Pruzinsky & Edgerton, 1990). In addition to gender-specific differences in the frequency of these requests, men and women seem to have different motivations. Women frequently express the wish to feel more attractive to themselves; and although more critical of their appearance, they are more psychologically comfortable with bodily changes (Fisher, 1986; Pruzinsky & Edgerton, 1990). Men, on the other hand, are more threatened by body changes (Fisher, 1986) and seem to be more often concerned with how others view them (Gorney, 1989). There is some literature to suggest that men are also more likely to be more psychologically disturbed both before and after aesthetic surgery than women (Edgerton & Langman, 1982; Pruzinsky & Edgerton, 1990; Wright, 1987).

When all patients requesting cosmetic surgery are combined, there appear to be certain concerns leading to requests for this surgery. In a survey of more than 3000 men and women, the three primary reasons for both groups requesting aesthetic plastic surgery were self-consciousness in interpersonal relationships, social isolation (often from members of ethnic groups and the elderly), and employment reasons (including wanting to look younger to avoid the appearance of aging and its implied relation to efficiency (Reich, 1991).

Unfortunately, many people continue to believe cosmetic improvement is the right of only the rich and famous or a product of pathologic narcissism. Patients requesting this "elective" surgery may face familial disapproval when requesting to "change the family nose" or other characteristic feature (Goin & Goin, 1989). Cosmetic "improvement" is by definition a judgment that the current feature is not attractive or is "defective" in some way. Families may fear that changing the outward physical appearance will somehow alter familial, cultural, or racial ties (Goin & Goin, 1989). There may also be concerns that the patient is putting himself or herself at risk for a "nonessential" procedure (Reich, 1991). As much as possible, these conflicting feelings require resolution prior to surgery, so the patient is not burdened with postoperative guilt and rejection by his or her support system.

Although it is true that beauty is in the eye of the beholder, there are some patients who have a disproportionate concern about a minimal deformity. Sometimes this distorted body image may have delusional proportions and cause the patient to greatly exaggerate a minimal defect. In the DSM this disorder is designated body dysmorphic disorder and is characterized by the belief that the body is grossly deformed. Because as many as 2% of cosmetic surgical consultations may be related to this disorder, a consultation for psychiatric assessment may be helpful for the surgeon to better understand the patient's motivation (Andreasen & Bardach, 1977; Connolly & Gipson, 1978). The role of the consulting psychiatrist frequently is to sort out the relative contributions of reality and fantasy in producing the patient's dissatisfaction (Belfer, 1987). In some instances this preoperative psychiatric referral clarifies unrealistic or magical expectations (Pruzinsky, 1988). It may also bring to consciousness any psychological conflict for which surgery was requested. Body dysmorphic disorder is also discussed in Chapter 13.

Many times an individual unhappy about some physical attribute shows anxiety and tends to focus on that deformity (Reich, 1991). After awhile they may even reject or try to conceal that body part (Reich, 1991)—which some have nicknamed the "camouflage technique" (Harris, 1982). Some patients become socially avoidant and forbid photographs, and others shy away from situations that would prohibit them from covering up their deformity, such as wearing a bathing suit at the beach. After a time, this behavior may lead to denial, or depersonalization, of the deformity. These individuals may refuse to believe that they really look like the image in a photograph. They may have an unrealistic mental image of what they currently look like, and their expectations of a postsurgical result may be impossible to achieve with surgery. It is essential to clarify a patient's preoperative expectations regarding the surgical change. In some cases operative procedures can be staged so there is not such an abrupt physical change (Pruzinsky & Edgerton, 1990).

There are two broad types of cosmetic surgical procedures: those that restore what once was, and those that create an appearance that the person has not previously had (Pruzinsky & Edgerton, 1990). Generally, restorative procedures are thought to be associated with less psychological morbidity because the "restored image" exists in the memory of the patient's experience (Pruzinsky & Edgerton, 1990). Type-changing surgery, such as cleft-palate repair or changing the shape of one's nose, may cause patients difficulty adjusting to the change, especially if their expectation of the "ideal" was different from the surgical outcome.

As plastic surgical techniques advance, more type-changing procedures are possible, including massive bony-structure reconstructive techniques for aesthetic purposes (Pruzinsky & Edgerton, 1990). Some surgeons are utilizing computer programs to help the patient visualize these changes before surgery, to help them assimilate their new "body" well before they enter the operating room.

The timing of the surgical request is usually significant, and useful information can be gained by asking why the patient presents for surgery at this time (Belfer, 1987). Sometimes stressful life situations that involve a loss or change in a relationship, such as a divorce, may lead to requests for cosmetic improvement. Webb, Slaughter, Meyer, et al. (1965) found that 90% of face-lift patients over 50 years of age had experienced the death of a significant other within the preceding 5 years. In these patients there were often psychological concerns about aging and death (Goin, Burgoyne, Goin, et al., 1980; Goin & Goin, 1989; Webb, Slaughter, Meyer, et al., 1965).

There are some who believe that physical "rehabilitation" can aid psychological rehabilitation (Reich, 1991); and although it is not a good idea to operate on patients who are severely depressed, some researchers have found marked improvement in patients' self-esteem, confidence, and social relationships after surgery (Goin, Burgoyne, Goin, et al., 1980; Robin, Copas, Jack, et al., 1988).

Guidelines for patient selection are found in many textbooks of plastic surgery, but it is generally accepted that those patients judged to be poor candidates on psychological grounds are those who are excessively fearful, have unrealistic or magical expectations, have minimal deformities, or are currently psychotic or severely depressed (Goin, Burgoyne, Goin, et al., 1980). The patient with a history of many surgical procedures is also traditionally thought of as a high-risk patient who may have a somatization or personality disorder. These patients have been nicknamed "insatiable" cosmetic surgery patients (Edgerton, Langman, & Pruzinsky, 1990a).

As psychiatric consultation becomes more available to plastic surgeons, some patients who traditionally have been considered high risk from a psychological point of view are being accepted for surgery. Edgerton, Langman, and Pruzinsky (1990b) chose to perform aesthetic surgery on 87 patients who traditionally would be considered "psychologically disturbed," including patients with neurotic, psychotic, and severe personality disorders. These patients were followed an average of 6.2 years, and 82% of them had a positive psychological outcome, with only 12% reporting minimal improvement, and only 3% reporting being negatively affected by the surgery. The surgeons found that combined surgical-psychological treatment of these patients was helpful, and that there was little correlation between the magnitude of the anatomic deformity and either the degree of preoperative emotional disturbance or post-operative psychological benefit (Edgerton, Langman, & Pruzinsky, 1990b). The five major patient groups excluded from surgery, according to their methodology, were noncompliant or noncommunicating patients, the unrealistic patients, individuals who requested a technically "too difficult" operation, and those who had economic or geographic restrictions.

Although more psychologically disturbed patients may be accepted for surgery, those with personality disorders may require more psychological support perioperatively, including patients with borderline, obsessive, dependent, narcissistic, hysterical, or paranoid personality traits. Some of these patients may even request surgery on account of their personality disorder (Reich, 1991). They may become regressed, self-destructive, or dissatisfied with the surgery depending on their life-long characterologic style. Pruzinsky and Edgerton (1990) well described the confusion that some borderline patients experience regarding their identity and their body image when they reported, "to attempt to match the surgical results to the borderline desires is tantamount to hitting a moving target" (Pruzinsky & Edgerton, 1990).

Although alterations of any part of the body may have psychological impact, the symbolic importance of the organ varies. Structures such as the face, breast, and genitalia are more cathectic. The face is the organ most often used for identification, and it is intimately tied with recognition. Our profiles, facial movements, and characteristic features are the ways other people see us. Similarly, we recognize emotions in others by "reading" their faces.

The face also has the sensory structures that enable us to see, touch, smell, hear, and taste. Facelifts are done to recreate a previously existing condition rather than to change a body part into a shape or relation that has never existed (Goin & Goin, 1989). One's body image may not "age" as quickly as the physical body; a wrinkled and sagging face may conflict with the individual's self-concept leading to the request for a facelift to "look like I did 10 years ago."

Postoperatively, facelift patients experience a transient hypesthesia of the skin of the face and neck (Goin & Goin, 1986). They may have difficulty when sleeping because of the decreased sensation of their face on the pillow. The issue of body image or bodily boundaries may become blurred, as it is unclear where skin stops and the pillow begins. These physical sensations cannot be emphasized enough because, in addition to "seeing" the changes, the patient also "feels" different (Pruzinsky & Edgerton, 1990).

The link between self-esteem and body image was assessed by Gillies (1984) when she asked patients to identify the part of their bodies they thought was their "self core," or the seat of their identity. Most responded that it was behind the bridge of their nose or in their midchest. Surgery in this area may therefore be influencing not only how patients look but also how they feel about themselves.

The body part that is being altered by surgery may symbolize an unconscious conflict for the individual. Some psychoanalytic authors believe that, because the nose contains erectile tissue similar to the genitals, sexual feeling may be displaced that influence the subjective evaluation of their appearance (MacGregor & Shaffner, 1980).

Specific surgeries of the breast and genitalia are reviewed in other sections of this chapter, but the surgeon must be mindful of the symbolic meanings of these structures in the patient's mind. The patient's fantasies and expectations regarding surgery should always be assessed, and the individual hoping that the surgeon's knife may mend a broken marriage or improve business success should be considered at high risk for postoperative psychological problems. On the other hand, for the patient with a marked deformity, plastic surgery may "reconstruct" the body image and improve self-image.

Although most authors report few long-term psychiatric complications in appropriately screened cosmetic surgical patients (Reich, 1969), there may be a period of reactive anxiety and depression immediately after surgery (Thomson, Knorr, & Edgerton, 1978). In a study of 599 patients, Reich (1969) reported reactive depression and anxiety in 31% of patients. These reactions, however, may be a reflection of preoperative difficulties. In a prospective psychological study of 50 female facelift patients, Goin, Burgoyne, Goin, et al. (1980) found that there was a high frequency of postoperative depressive reactions in patients who showed depressive features on preoperative psychological testing. They placed greater emphasis on the premorbid function and characterologic styles of the patients than on the symbolic meaning of the operation.

OSTOMY SURGERY

Patients undergoing surgery in which an ostomy is planned not only experience a change in body image but also must struggle with a variety of complex emotions and feelings regarding the ostomy. Unlike aesthetic surgery, no one "wants" an ostomy; and there may be concerns about sexual attractiveness and social acceptability, as well as a fear of recurrence of the disease for which the surgery was performed (Watson, 1985). Often patients have only a vague understanding, prior to surgery, of how an ostomy functions; and they may appear more concerned with the technical aspects of surgery, anesthesia, and the surgical "cure." During the immediate postoperative period, the patient may focus mainly on incisional pain, advancement of diet, and regaining strength. Sometimes it is not until patients are being taught how to care for their stoma that they begin to deal with the loss of the diseased body part and a changed body image. Postoperatively, patients may be not only dealing with a break in their body integrity but also attempting to incorporate into their body image a new surgically made opening in the anterior abdominal wall. The patient must also grieve the loss of control over one of the most basic of human bodily functions: elimination. Because the stoma has no sphincter mechanism, the colostomy patient may worry about passing flatus, having an odor, or leaking. The patient may feel overwhelmed by learning new ostomy management skills and dealing with this previously hidden internal organ. Some patients who refer to the stoma as "it" or the "bag" may be reacting to the ostomy in a detached way, as if it belonged to the surgeons who created it (Gloeckner, 1984).

This difficulty in redefining one's body image can have a profound impact on the patient's acceptance of responsibility for the care of the ostomy. An enterostomal therapist may provide invaluable assistance in helping the patient deal with these technical and emotional issues. In addition, self-help groups affiliated with the Ileostomy Association, the Urinary Conduit Association, and the International Ostomy Association are found in many communities. Members of these groups share first-hand experiences, providing emotional and psychological support for the new ostomy patient. In addition, they may provide hints regarding irrigation techniques, dietary practices to prevent constipation, diarrhea, or excessive flatus, and overall stomal care.

Issues of altered body image, worries about sexual

attractiveness, and fear of odor or leakage are similar for patients with ileal conduits, colostomies, and ileostomies. In a retrospective study of 40 ostomy patients, the most negative statements regarding body image were made during the first year after surgery (Dlin & Perlman, 1971), followed by perceived improvement in attractiveness during the following years (Gloeckner, 1984). Those with an ileostomy had the highest self-rating scores on sexual attractiveness. Patients who were not symptomatic preoperatively and who had the most complications and problems with their ostomy after surgery had the lowest postoperative attractiveness scores (Gloeckner, 1984).

Although the impact of mutilating surgery may be traumatic for the emotionally unprepared patient, it appears that most patients who are given emotional support and psychiatric intervention when needed can resume full, well-adjusted lives. In a long-term follow up of 344 ileostomy patients at the Mayo Clinic, 84% thought that the management of their ileostomy was not a major problem, and 92% were satisfied with their way of living. Most patients said they were not restricted by their ostomy (82%) and 95.6% returned to their previous employment. Sexual habits did not change in 87.2%, and one-third of the married women became pregnant and had normal vaginal deliveries (Roy, Sauer, Beahrs, et al., 1970).

Showing patients the ostomy appliance they will most likely be wearing after surgery may raise some important questions preoperatively, including sexual concerns. In a survey that assessed the psychosexual response to ileostomies and colostomies among 500 ostomy association members, it was found that there was little or no discussion with patients before or after surgery about sexual issues (Dlin, Perlman, & Ringold, 1969).

Discussion of sexual issues is best handled when the psychiatrist is familiar with the sexual problems encountered by ostomy patients. To postpone these discussions until the patient is ready to go home or to wait until the patient brings up specific questions may lead to misconceptions and undue anxiety in the patient and spouse. A complete psychosocial history should be taken from these patients; it may be the best introduction to detailed discussion of sexual functioning after surgery. Whenever possible, the patient's sexual partner should be included in discussions of sexual function to alleviate miscommunication and fears about "breaking the news." Information should be given regarding possible problems with impotence, ejaculation, changes in libido, dyspareunia, fertility, and childbearing. In a study of 40 patients with a permanent ostomy, 30% stated that their partner reacted with "fear of hurting me," which could have been alleviated by including the sexual partner in the preoperative and postoperative discussions (Gloeckner & Starling, 1982).

Although coital practice and frequency among married ileostomates apparently do not differ markedly from those among the general population (Gruner, 1983; Gruner, Naas, Fretheim, et al., 1977), some patients may have psychological problems related to their stoma, especially during the first year after surgery. They may experience depression or anxiety and may worry about sexual attractiveness. It is during this critical time that psychotherapy can help patients return to their previous level of sexual functioning (Dlin & Perlman, 1971).

A survey conducted by the Ileostomy Association of Great Britain and Ireland found that only 40% of the patients with sexual problems after ileostomies felt comfortable discussing these issues with their primary physician (Burnham, Lennard-Jones, & Brooke, 1977). Other studies reported that frequently patients have little or no preoperative discussion of possible sexual problems and little or no sexual guidance (Jones, Breckman, & Hendry, 1980). Dlin and Perlman (1971), in their interview of 146 ostomy patients over age 50, found that only 11% had received any sexual counseling before or after surgery.

Patients about to undergo extensive dissection for carcinoma will be dealing with the emotional crisis of having cancer as well as worries and fears about postoperative dysfunction. Patients are often reluctant to bring up issues of possible sexual dysfunction because of modesty, embarrassment, or the feeling that the doctor is interested in discussing "more important things." The reported incidence of sexual difficulties following abdominal-perineal resection for cancer ranges from 53% to 100%, with an average of 75% (Bernstein, 1972). Although there have been few studies on the sexual dysfunction of women after pelvic exenteration, dyspareunia sometimes occurs, possibly as a result of the perineal scarring. There may also be decreased lubrication because of pain or psychological factors in women after abdominal-perineal resection (Young, 1982). For the woman with dyspareunia, the female-superior position might be more comfortable, and decreased lubrication can be alleviated by saliva or a vaginal lubricant (Young, 1982). For women who are concerned about conception and pregnancy after surgery, most previously fertile women can conceive and give birth to healthy babies following ostomy surgery (Burnham, Lennard-Jones, & Brooke, 1977). Although Daly (1968) thought that fertility might be reduced in women with ileostomies, he found that primary closure of the pelvic peritoneum might improve fertility. Women who do not wish to become pregnant should be informed that birth control pills have, at times, been found unabsorbed in

TABLE 34–6. *Types of Ostomies and Possible Sequelae and Complications*

Type of Ostomy	Disease	Procedure	Excreta	Continence	Physical impairment of sexual function	Other complications
Colostomy	Colorectal cancer	Abdominoperineal resection	± Formed stool	± With irrigation	+	Skin irritation, bowel obstruction, prolapse or retraction of stoma, stenosis of stoma
Ileostomy	Inflammatory bowel diseases (ulcerative colitis, Crohn's disease)	Total colectomy	Fluid stool	– – With conventional care + With Kock pouch	±	Renal stone formation* and the above
Urostomy (ileal conduit)	Bladder cancer	Radical cystectomy	Urine	—	+	Urinary infections and all the above
Colostomy and urostomy (ileal conduit)	Gynecologic cancer (advanced and recurrent)	Pelvic exenteration	± Formed stool Urine	± With irrigation – –	+ +	Combination of all the above

From Holland JC (1989). Gastrointestinal Cancer. In Holland JC & Rowland JH (eds). *Handbook of Psychooncology*. New York: Oxford University Press. Ch. 15, pp. 208–217. With permission. (Originally appeared in Hurny & Holland, 1985).

– – = absent in all patients; ± = present in half; + + = present in all patients; + = present in most.

*Due to chronic asymptomatic water and salt depletion.

the ostomy appliance because of increased intestinal motility and that the failure to be absorbed negates their contraceptive effectiveness (Young, 1982).

For the man with partial postoperative impotence, a change in position while having sex might help. Also, a penile or penoscrotal ring can be helpful by decreasing venous return, permitting erection and penetration. For total impotence in the male patient, a penile prosthesis may be considered (Carson, 1983; Collins & Kinder, 1984; Merril, 1983; Schlamowitz, Beutler, Scott, et al., 1983). For the male homosexual couple, although they may continue with manual and oral sexual activity, the ostomy should not be used for intercourse because of the risk of disrupting it (Young, 1982). The various types of ostomies and possible sequelae and complications are listed in Table 34-6 (Hurny & Holland, 1985), and the applications of the triphasic model to assess sexual dysfunction in cancer patients is found in Table 34-5 (Auchincloss, 1989). The treatment of sexual dysfunction is discussed further in Chapter 15. For further reading, the *Handbook of Psychooncology* (Holland & Rowland, 1989) is an excellent reference. (See also Chapter 39 on impotence.)

REFERENCES

ABRAM SE (1989). Pain-acute and chronic. In PG BARASH, BF CULLEN, & RK STOELTING (Eds.), Clinical Anesthesia (Ch. 53, pp. 1427–1454). Philadelphia: JB Lippincott.

American Medical Association, Division of Drugs and Toxicology (1992). Drug Evaluations Subscription. Chicago: American Medical Association.

American Psychiatric Association (1987). Diagnostic and Statistical Manual of Mental Disorders (3rd ed., revised). Washington, DC: APA.

American Society for Plastic & Reconstructive Surgery (1989). Estimated number of cosmetic surgery procedures performed by ASPRS members. (Report of surgical procedures conducted by Board Certified Plastic and Reconstructive Surgeons.) Chicago: ASPRS, Director of Communications Executive Office.

ANDREASEN NC, & BARDACH J (1977). Dysmorphophobia: symptom or disease? Am J Psychiatry 134:673–676.

ARNDT EM, TRAVIS F, LEFEBVRE A, ET AL. (1986). Beauty and the eye of the beholder: social consequences and personal adjustments for facial patients. Br J Plast Surg 39:81–84.

AUCHINCLOSS SS (1989). Sexual dysfunction in cancer patients; issues in evaluation and treatment. In JC HOLLAND & JH ROWLAND (Eds.), Handbook of Psychooncology (Ch. 33, pp. 383–415). New York: Oxford University Press.

BARNES TRE, & BRAUDE WM (1982). Acute akathisia after oral droperidol and metoclopramide preoperative medication. Lancet 2:48–49.

BELFER ML (1987). Psychological consideration of the plastic surgery patient. In SA SOHN (Ed.), Fundamentals of aesthetic plastic surgery (Ch. 2, pp. 9–13). Baltimore: Williams & Wilkins.

BELFER ML, HARRISON AM, & MURRAY JE (1979). Body image and the process of reconstructive surgery. Am J Dis Child 133:532–535.

BERNSTEIN WC (1972). Sexual dysfunction following radical surgery for cancer of rectum and signoid colon. Med Aspects Hum Sex 6:156–163.

BLACHER RS (1975). On awakening paralyzed during surgery: a syndrome of traumatic neruosis. JAMA 234:67–68.

BLACHER RS (1984). Awareness during surgery [Editorial]. J Anesthesiol 61:1.

BLACHER RS (1987). General surgery and anesthesia: the emotional experience. In RS Blacher (Ed.), The psychological experience of surgery (Ch. 1, pp. 1–26). New York: John Wiley & Sons.

Bloom J, Cook M, Fotopoulis S, et al. (1987). Psychological response to mastectomy: a prospective comparison study. Cancer 59:189–196.

Boas RA, Holford NHG, & Villiger JW (1985). Opiate drug choice and drug use. Clin J Pain 1:117.

Broadman LM, Hannallah RS, Norden JM, et al. (1987). "Kiddie caudals": experience with 1154 consecutive cases without complications [abstract]. Anesth Analg 66(suppl):518.

Burnham WR, Lennard-Jones JE, & Brooke BN (1977). Sexual problems among married ileostomists: survey conducted by the Ileostomy Association of Great Britain and Ireland. Gut 18:673–677.

Cain EN, Kohorn EI, Quinlan DM, et al. (1983). Psychosocial reactions to the diagnosis of gynecologic cancer. Obstet Gynecol 62:635–641.

Caplan LM, & Hackett TP (1963). Emotional effects of lower limb amputation in the aged. N Engl J Med 269:1166–1171.

Carson CCM (1983). Inflatable penile prosthesis: experience with 100 patients. South Med J 76:1139–1141.

Collins GF, & Kinder BN (1984). Adjustment following surgical implantation of a penile prosthesis: a critical overview. J Sex Marital Ther 10:255–271.

Connolly FH, & Gipson M (1978). Dysmorphophobia—a long-term study. Br J Psychiatry 132:568–570.

Cryns AG, Gorey KM, & Goldstein MZ (1990). Effects of surgery on the mental status of older persons: a meta-analytic review. J Geriatr Psychiatry Neurol 3:184–191.

Daly DW (1968). The outcome of surgery for ulcerative colitis. Ann R Coll Surg 42:38–57.

Devine EC, & Cook TD (1983). A meta-analytic analysis of psychoeducational interventions on length of postsurgical hospital stay. Nurs Res 32:267–274.

Dion K, Berscheid E, & Walster E (1972). What is beautiful is good. J Pers Soc Psychol 24:285–290.

Dlin BM, & Perlman A (1971). Emotional response to ileostomy and colostomy in patients over the age of 50. Geriatrics 26:112–118.

Dlin BM, Perlman A, & Ringold E (1969). Psychosexual response to ileostomy and colostomy. Am J Psychiatry 126:374–381.

Ducluzeau R, Vial T, Pobel D, et al. (1990). Neuropsychic effects of cimetidine and ranitidine. J Toxicol Clin Exp 10:37–39.

Edgerton MT Jr, & Langman MW (1982). Psychiatric considerations. In EH Courtiss (Ed.), Male aesthetic surgery (Ch. 4, pp. 17–38). St Louis: Mosby.

Edgerton MT, Langman MW, & Pruzinsky T (1990a). Patients seeking symmetrical recontouring for "perceived" deformities in the width of the face and skull. Aesthet Plast Surg 14:59–73.

Edgerton MT, Langman MW, & Pruzinsky T (1990b). Plastic surgery and psychotherapy in the treatment of 100 psychologically disturbed patients. Paper presented at the American Association of Plastic Surgeons meeting, Hot Springs, Virginia.

Egbert LD, Battit GE, Turndorf H, et al. (1963). Value of preoperative visit by anesthetist: study of doctor-patient rapport. J Am Psychiatr Assoc 185:553–555.

Findlay PA, Lippman ME, Danforth D, et al. (1985). Mastectomy versus radiotherapy as treatment for stage I-II breast cancer: a prospective randomized trial at the National Cancer Institute. World J Surg 9:671–675.

Fisher S (1986). Development and structure of the body image (Vol. 2; pp. 348–350). Hillsdale, NJ: Lawrence Erlbaum.

Fouilladieu JL, Hosanki M, McGee K, et al. (1985). Possible potentiation by hydroxyzine of metoclopramide's undesirable side effects. Anesth Analg 64:1227.

Fulop G, Strain JJ, Fahs M, et al. (1989). Medical disorders associated with psychiatric comorbidity and prolonged hospital stay. Hosp Commun Psychiatry 40:80–82.

Fulop G, Strain JJ, Vita J, et al. (1987). Impact of psychiatric comorbidity on length of hospital stay for medical/surgical patients: a preliminary report. Am J Psychiatry 144:878–882.

Gillies DA (1984). Body image changes following illness and injury. J Enterost Ther 11:186–189.

Gleason RP, Rodwell S, Shaw R, et al. (1990). Post-cesarean analgesia using a subcutaneous pethidine infusion. Int J Gynaecol Obstet 33:13–17.

Gloeckner MR (1984). Perceptions of sexual attractiveness following ostomy surgery. Res Nurs Health 7:87–92.

Gloeckner MR, & Starling JR (1982). Providing sexual information to ostomy patients. Dis Colon Rectum 25:575–579.

Goin MK, Burgoyne RW, Goin JM, et al. (1980). A prospective psychological study of 50 female face-lift patients. Plast Reconstr Surg 65:436–442.

Goin MK, & Goin JM (1986). Psychological effects of aesthetic facial surgery. Adv Psychosom Med 15:84–108.

Goin JM, & Goin MK (1989). Psychological aspects of aesthetic plastic surgery. In JR Lewis (Ed.), The art of aesthetic plastic surgery (Ch. 6, pp. 39–47). Boston: Little, Brown.

Good-Reis DV, & Pieper BA (1990). Structured vs unstructured teaching: a research study. AORN J 51:1334–1339.

Gorney M (1989). Guidelines for preoperative screening of patients. In JR Lewis (Ed.), The art of aesthetic plastic surgery (Ch. 7, pp. 49–53). Boston: Little, Brown.

Gourlay GK, Wilson PR, & Glyn CJ (1982). Methadone produces prolonged postoperative analgesia. BMJ 284:630.

Gruner OP (1983). Sexual problems after rectal surgery and operated colitis. Praxis 72:948–952.

Gruner OP, Naas R, Fretheim B, et al. (1977). Marital status and sexual adjustment after colectomy: results in 178 patients operated on for ulcerative colitis. Scand J Gastroenterol 12:193–197.

Guerra F (1986). Awareness and recall. Int Anesthesiol Clin 24:4.

Haines N, & Viellion G (1990). A successful combination: preadmission testing and preoperative education. Orthop Nurs 9:53–57.

Harris DL (1982). The symptomatology of abnormal appearance: an anecdotal survey. Br J Plast Surg 35:312–323.

Harwood DD, Hanumanthu S, & Stoudemire A (1992). Pathophysiology and management of phantom limb pain. Gen Hosp Psychiatry 14:107–118.

Hathaway D (1986). Effect of preoperative instruction on postoperative outcomes: a meta-analysis. Nurs Res 35:269–275.

Henker FO III (1979). Body-image conflict following trauma and surgery. Psychosomatics 20:812–820.

Holland JC, & Mastrovito R (1980). Psychologic adaptation to breast cancer. Cancer 46:1045.

Holland JC, & Rowland JH (1987). Psychological reactions to breast cancer and its treatment. In JR Harris, S Hellman, IC Henderson, et al. (Eds.), Breast diseases (2nd ed., Ch. 16, pp. 632–647). Philadelphia: JB Lippincott.

Holland JC, & Rowland JH (Eds.) (1989). Handbook of psychooncology. New York: Oxford University Press.

Houlton PJC, Downing JW, & Brock-Utne JG (1978). Intravenous ketamine anesthesia for major abdominal surgery: an assessment of a technique and the influence of ataractic drugs on the psychomimetic effects of ketamine. Anesth Intens Care 6:222.

Humphries TJ (1987). Famotidine: a notable lack of drug interactions. Scand J Gastroenterol 134:55–60.

Hurny C, & Holland JC (1985). Psychosocial sequelae of ostomies in cancer patients. CA 35:170–180.

JACOBS E, HOLLAND J, ROLLAND J, ET AL. (1983). Who seeks breast reconstruction? A controlled psychological study [Abstract]. Psychosom Med 45:80.

JONES MA, BRECKMAN B, & HENDRY WF (1980). Life with an ileal conduit: results of questionnaire survey of patients and urological surgeons. Br J Urol 52:21–25.

KASHANI JH, FRANK RG, KASHANI SR, ET AL. (1983). Depression among amputees. J Clin Psychiatry 44:256–258.

KINCHLA J, & WEISS T (1985). Psychological and social outcomes following coronary artery bypass surgery. J Cardiopul Rehabil 5:274–283.

KOLB LC (1952). The psychology of the amputee: a study of phantom phenomena, body image and pain. Coll Papers Mayo Clin 44:586–590.

KOSTIC VS, & LEVIC Z (1989). Psychiatric disorders associated with corticosteroid therapy. Neurologia 38:161–166.

LALINEC-MICHAUD M, & ENGELSMANN F (1985). Anxiety, fears and depression related to hysterectomy. Can J Psychiatry 30:44–47.

LALINEC-MICHAUD M, ENGELSMANN F, & MARINO J (1988). Depression after hysterectomy. Psychosomatics 29:307–314.

LAM AM, & PARKIN JA (1981). Cimetidine and prolonged postoperative somnolence. Can Anesthesiol Soc J 28:450.

LAYNE OL, & YUDOFSKY SC (1971). Postoperative psychosis in cardiotomy patients. N Engl J Med 284:518–520.

LAZARUS HR, & HAGENS JH (1968). Prevention of psychosis following open heart surgery. Am J Psychiatry 124:1190–1195.

LEE CM, & YEAKEL AE (1975). Patients refusal of surgery following Innovar premedication. Anesth Analg 54:224.

LEY P, BRADSHAW P, EAVES D, ET AL. (1973). A method for increasing patients' recall of information presented by doctors. Psychol Med 3:217–220.

LINDERMAN CA (1972). Nursing intervention with the presurgical patient: effectiveness and efficiency of group and individual preoperative teaching—phase two. Nurs Res 21:196–209.

LING MHM, PERRY PJ, & TSUANG MT (1981). Side effects of corticosteroid therapy. Arch Gen Psychiatry 38:471–477.

LUNDBERG SG, & GUGGENHEIM FG (1986). Sequelae of limb amputation. Adv Psychosom Med 15:199–210.

LUTZ LJ, & LAMER TM (1990). Management of postoperative pain: review of current techniques and methods. May Clin Proc 65:584–596.

MACGREGOR FC, & SHAFFNER B (1980). Screening patients for nasal plastic operations. Psychosom Med 12:277–291.

MANDAL SK (1986). Psychiatric side effects of ranitidine. Br J Clin Pract 40:260.

MARKS RM, & SACHER EJ (1973). Undertreatment of medical inpatients with narcotic analgesics. Ann Intern Med 78:173–181.

MARTIN RL, ROBERTS WV, & CLAYTON PJ (1980). Psychiatric status after hysterectomy: a one-year prospective follow up. JAMA 244:350–353.

MAZET R JR (1967). The geriatric amputee. Artif Limbs 11:33–41.

MELZACK R (1971). Phantom limb pain: implications for treatment of pathologic pain. Anesthesiology 35:409–419.

MERRILL DC (1983). Clinical experience with Scott inflatable penile prosthesis in 150 patients. Urology 4:371–375.

MEYEROWITZ BE (1980). Psychosocial correlates of breast cancer and its treatments. Psychol Bull 87:108–131.

MICHAELS I, SERRINS M, SHIER NQ, ET AL. (1984). Anesthesia for cardiac surgery in patients receiving monoamine oxidase inhibitors. Anesth Analg 63:1014–1024.

MILLAR HR (1981). Psychiatric morbidity in elderly surgical patients. Br J Psychiatry 138:17–20.

MOGAN J, WELLS N, & ROBERTSON E (1985). Effects of preoperative teaching on postoperative pain: a replication and expansion. Int J Nurs Stud 22:267–280.

MORGENSTERN FS (1970). Chronic pain. In OW HILL (Ed.), Modern trends in psychosomatic medicine (Ch. 14, pp. 225–245). London: Butterworth.

MURRAY JE, MULLIKEN JB, KABAN LB, ET AL. (1979). Twenty year experience in maxillocraniofacial surgery: an evaluation of early surgery on growth function and body image. Ann Surg 190:320–330.

NADELSON CC, & NOTMAN MT (1979). Disease and illnesses specific to women. In G USDIN & JM LEWIS (Eds.), Psychiatry in general medical practice (Ch. 21, pp. 475–497). New York: McGraw-Hill.

NADELSON CC, & NOTMAN MT (1990). The psychiatric aspects of obstetrics and gynecology. In R MICHELS, AM COOPER, SB GUZE, ET AL. (Eds.), Psychiatry (Vol. 2, Ch. 120, pp. 1–11). Philadelphia: JB Lippincott.

NOTMAN MT (1987). Emotional impact of gynecological surgery. In RS BLACHER (Ed.), The psychological experience of surgery (Ch. 7, pp. 99–115). New York: John Wiley & Sons.

PARKES CM (1973). Factors determining the persistence of phantom pain in the amputee. J Psychosom Res 17:97–108.

PARKES CM (1975). Psycho-social transitions: comparison between reactions to loss of a limb and loss of a spouse. Br J Psychiatry 127:204–210.

PARKES CM (1976). The psychological reaction to loss of a limb: the first year after amputation. In JG HOWELLS (Ed.), Modern perspectives in the psychiatric aspects of surgery (Ch. 24, pp. 515–532). New York: Brunner/Mazel.

PEARSON RE (1961). Response to suggestions given under general anesthesia. Am J Clin Hypn 4:106.

PENMAN DT (1979). Coping strategies in adaptation to mastectomy. Dissertation Abstr Int 40:5825B.

PINZUR MS, GRAHAM G, & OSTERMAN H (1988). Psychologic testing in amputation rehabilitation. Clin Orthop 229:236–240.

POLIVY J (1974). Psychological reactions to hysterectomy: a critical review. Am J Obstet Gynecol 118:417–426.

PRUZINSKY T (1988). Collaboration of plastic surgeon and medical psychotherapist: elective cosmetic surgery. Med Psychother 1:1–13.

PRUZINSKY T, & EDGERTON MT (1990). Body-image change in cosmetic plastic surgery. In TF CASH & T PRUZINSKY (Eds.), Body images: development, deviance, and change (Ch. 10, pp. 217–236). New York: Guilford Press.

REICH J (1969). The surgery of appearance: psychological and related aspects. Med J Aust 2:5–13.

REICH J (1991). The aesthetic surgical experience. In JW SMITH & SJ ASTON (Eds.), Plastic surgery (Ch. 4, pp. 127–140). Boston: Little, Brown.

RESMAN-TARGOFF BH (1990). Ketorolac: a parenteral nonsteroidal anti-inflammatory drug. DICP 23:1098–1104.

RICHARDS DH (1973). Depression after hysterectomy. Lancet 2:430–433.

RIETHER AM, & STOUDEMIRE A (1987). Surgery and trauma. In A STOUDEMIRE & BS FOGEL (Eds.), Principles of medical psychiatry (Ch. 19, pp. 423–449). Orlando, FL: Grune & Stratton.

ROBIN AA, COPAS JB, JACK AB, ET AL. (1988). Reshaping the psyche: the concurrent improvement in appearance and mental state after rhinoplasty. Br J Psychiatry 152:539–543.

ROGERS C (1993). Psychiatric aspects of anesthesia. In A STOUDEMIRE & BS FOGEL (Eds.), Medical-Psychiatric Practice (Vol. 2). Washington DC: American Psychiatric Press, Inc.

ROTHROCK JC (1989). Perioperative nursing research. Part I: Preoperative psychoeducational interventions. AORN J 49:597–619.

ROWLAND JH, & HOLLAND JC (1989). Breast cancer. In JC HOLLAND

& JH Rowland (Eds.), Handbook of psychooncology (Ch. 14, pp. 188–207). New York: Oxford University Press.

Roy P, Sauer W, Beahrs OH, et al. (1970). Experience with ileostomies: evaluation of long term rehabilitation in 497 patients. Am J Surg 119:77–86.

Saunders PR, & Harris MN (1990). Opisthotonos and other unusual neurological sequelae after outpatient anesthesia. Anesthesia 45:552–557.

Scheller MS, & Sears KL (1987). Postoperative neurologic dysfunction associated with preoperative administration of metaclopramide. Anesth Analg 66:274–276.

Schilder P (1950). The image and appearance of the human body (pp. 283–304). New York: International Universities Press.

Schlamowitz KE, Beutler LE, Scott FB, et al. (1983). Reactions to the implantation of an inflatable penile prosthesis among psychogenetically and organically impotent men. J Urol 129:295–298.

Schweitzer I, & Rosenbaum MB (1982). Psychiatric aspects of replantation surgery. Gen Hosp Psychiatry 4:271–279.

Sechzer PH (1971). Studies in pain with the analgesic-demand method. Anesth Analg 50:1–10.

Sedgwick JV, Lewis IH, & Linter SPK (1990). Anesthesia and mental illness. Int J Psychiatry Med 20:209–225.

Seibel MM, Freeman MG, & Graves WL (1980). Carcinoma of the cervix and sexual function. Obstet Gynecol 55:484–487.

Shukla GD, Sahu SC, Tripathi RP, et al. (1982). A psychiatric study of amputees. Br J Psychiatry 141:50–53.

Siebert CP (1986). Recognition, management, and prevention of neuropsychological dysfunction after operation. Int Anesthesiol Clin 24:4.

Silberfarb PM (1983). Chemotherapy of cognitive deficits in cancer patients. Annu Rev Med 34:35.

Silberfarb PM (1984). Psychosexual impact of gynecologic cancer. Med Aspects Hum Sex 18:212–226.

Smith LW, & Dimsdale JE (1989). Postcardiotomy delirium; conclusions after 25 years? Am J Psychiatry 146:452–458.

Snyderman RK, & Guthrie RH (1971). Reconstruction of the female breast following radical mastectomy. Plast Reconstr Surg 47:565–567.

Sonnenblick M, & Yinnou A (1986). Mental confusion as a side effect of ranitidine. Am J Psychiatry 143:257.

Spencer KE (1989). Postoperative pain: the alternatives to analgesia. J Prof Nurs 4:479–480.

Stack GG, Rogers P, & Linter SPK (1988). Monoamine oxidase inhibitors and anesthesia—a review. Br J Anaesth 60:222–227.

Steen PA, & Michenfelder JD (1979). Neurotoxicity of anesthetics. Anesthesiology 50:437.

Stevens LA (1987). The psychological aspects of breast surgery. In RS Blacher (Ed.), The psychological experience of surgery (Ch. 6, pp. 87–98). New York: John Wiley & Sons.

Stevens LA, McGrath MH, Druss RG, et al. (1984). The psychological impact of immediate breast reconstruction for women with early breast cancer. Plast Reconstr Surg 73:619–626.

Stoelting RK, Dierdorf SF, & McCammon RL (1988). Psychiatric illness. In RK Stoelting, SF Dierdorf, & RL McCammon (Eds.), Anesthesia and co-existing disease (pp. 717–728). New York: Churchill Livingstone.

Swindale JE (1989). The nurse's role in giving pre-operative information to reduce anxiety in patients admitted to the hospital for elective minor surgery. J Adv Nurs 14:899–905.

Tang GWK (1985). Reactions to emergency hysterectomy. Obstet Gynecol 65:206–210.

Thomson JA, Knorr NJ, & Edgerton MT (1978). Cosmetic surgery: the psychiatric perspective. Psychosomatics 19:7–15.

Titchaer JL, Zwerling I, Gottschalk L, et al. (1956). Psychosis in surgical patients. Surg Gynecol Obstet 102:59–65.

Tune L, & Folstein MF (1986). Postoperative delirium. In FG Guggenheim (Ed.), Psychological aspects of surgery (pp. 51–89). New York: Karger.

Veronesi U, Saccozzi R, Del Vecchio M, et al. (1981). Comparing radical mastectomy with quadrantectomy, axillary dissection, and radiotherapy in patients with small cancers of the breast. N Engl J Med 305:6–11.

Vincent CE, Vincent B, Greiss FC, et al. (1975). Some marital-sexual concomitants of carcinoma of the cervix. South Med J 68:552–558.

Walbroehl GS (1985). Sexuality in cancer patients. Am Fam Physician 31:153–158.

Watson PG (1985). Meeting the needs of patients undergoing ostomy surgery. J Enterost Ther 12:121–124.

Webb WL, Slaughter R, Meyer E, et al. (1965). Mechanisms of psychosocial adjustment in patients seeking "face-lift" operation. Psychosom Med 27:183–192.

White PF (1987). Mishaps with patient-controlled analgesia. Anesthesiology 66:81–83.

Williams D (1986). Preoperative patient education: in the home or in the hospital? Orthop Nurs 5:37–41.

Wilson PG, & Krebs MJS (1983). Coping with amputation. Vasc Surg 17:165–175.

Wilson PG, & Schirger-Krebs MJ (1987). Psychological factors in lower limb amputations. In RS Blacher (Ed.), The psychological experience of surgery (Ch. 5, pp. 73–86). New York: John Wiley & Sons.

Wolberg WH, Romsaas EP, Tanner MA, et al. (1989). Psychosexual adaptation to breast cancer surgery. Cancer 63:1645–1655.

Wright MR (1987). The male aesthetic patient. Arch Otolaryngol Head Neck Surg 113:724–727.

Young CH (1982). Sexual implications of stoma surgery, Part 1. Clin Gastroenterol 2:383–395.

35 Burn trauma

LAWSON F. BERNSTEIN, M.D.

Burn trauma is the fourth leading cause of injury death in the United States, with more than 6000 fatalities a year. Another 100,000 patients are hospitalized annually for burn injury, and 50% of them sustain substantial disabilities. Burn trauma results in an expenditure of billions of health care dollars annually (Committee on Trauma Research, 1985). Risk factors for burn injury include drug and alcohol abuse, preexisting physical or neurologic infirmity, poverty, old age, and childhood (Brodzka, 1983; Feck & Baptiste, 1979; MacArthur & Moore, 1975). Men are burned in greater numbers than women; and African-Americans are burned more frequently than whites (Rossignol, Boyle, Locke, et al., 1986; Rossignol, Locke, & Burke, 1986). A substantial minority of burns are due to criminal assault (Cesare, Morgan, Felice, et al., 1990; Purdue & Hunt, 1990).

Approximately 20% of burn patients have a preexisting psychiatric diagnosis, which is a significant risk factor for burn injury and for increased postburn morbidity (Andreasen, Noyes, & Hartford, 1972a; Brezel, Kassenbrock, & Stein, 1988). Substance abuse is the most frequent diagnosis, followed by personality disorder, schizophrenia, and mood disorders (Berry, Patterson, Wochtel, et al., 1984). Suicidal self-immolation is relatively rare but often lethal (Andreasan & Noyes, 1975; Davidson & Brown, 1985). As would be expected, the incidence of psychopathology is high in individuals who set themselves on fire.

PSYCHOLOGICAL ASPECTS OF THE DISEASE

Prolonged survival after massive trauma is a modern phenomenon. Lindenmann and Cobb were the first to describe the neuropsychiatric sequelae of burn injury after the Coconut Grove fire of 1943 (Cobb & Lindemann, 1943). Their description of delirium, acute grief reactions, and posttraumatic stress disorder began the modern era of psychiatric burn care. The overwhelming nature of the trauma results in a predictable series of acute, subacute, and chronic psychological reactions (Hamburg, Artz, Reiss, et al., 1953; Steiner & Clark, 1977) (see Table 35-1).

Resuscitative Phase

The resuscitative phase of burn injury spans the time from acute trauma through fluid and electrolyte repletion and reestablishment of stable vital signs and renal function. During this phase the patient often is critically ill and intubated (see Table 35-2). The patient may appear profoundly withdrawn or may become delirious. Denial of injury and projection predominate as psychological coping styles and should not be challenged unless they compromise adequate medical treatment. Although some authors have advocated questioning patients about wishes for heroic care during lucid intervals (Imbus & Zawacki, 1977), it is often impractical and stands on tenuous legal ground regarding informed consent. Initial psychiatric consultation should include a conference with the patient's family. Explanation of the usual burn course and assessment of the patient's condition, in concert with other members of the burn team, is helpful for orienting the family to a new and often frightening environment. This consultation should not be done in the patient's presence. As the shock of the trauma wanes, the family may begin to mourn the loss of the "intact" patient as well as the loss of other family members who may have died in the conflagration. The psychotherapeutic role here is primarily supportive.

Acute Phase

The acute period begins after the initial trauma, fluid resuscitation, and stabilization of vital signs. The patient may have embarked on an arduous course of surgical debridement, grafting, and physical therapy. Although some degree of physical autonomy returns, the patient still is largely dependent on the burn team for care. Patients with different character styles react accordingly. The previously compliant patient may become irritable and demanding as he strives to regain autonomy, whereas the regressed patient may become even more so. A more realistic assessment of injury and disfigurement begins, and patients may start to mourn lost limbs or facial features. A supportive stance still is appropriate, but the therapist should be attuned to dynamic themes and occasionally perform confrontation.

TABLE 35–1. *Chronology of Burn Trauma Care*

Phase of Injury	Resuscitative	Acute	Convalescent
Setting	Emergency room, ICU, burn unit	ICU, burn unit	Hospital ward, home, rehabilitation facility
Surgical care	Fluid resuscitation, maintenance of vital signs; urine output; débridement of necrotic tissue; antibiotic therapy	Débridement, hydrotherapy, skin grafting and reconstructive surgery, physical therapy	Skin grafting and reconstructive surgery, physical therapy, occupational therapy
Patient response			
Physiologic	Shock, initial decrease in renal function and increase in hepatic metabolism, varied acute serum protein changes with net decrease	Stabilization of vital signs with persistent hypermetabolism, increase in renal clearance, and decrease in hepatic oxidative metabolism; variable presence of bone marrow suppression	Hypermetabolic response present up to 1–2 years
Endocrine	Dramatic increase in endorphins, corticosteroids, and catecholamines	Persistent elevation in corticosteroids, decrease in endorphins	Unknown
Psychological	Denial, projection	Regression, mourning, coping	Coping, adjustment, transition to home
Psychiatric diagnoses	Substance withdrawal, delirium, other cognitive disorders, PTSD	Pain syndromes, PTSD, secondary mood disorders, adjustment disorders, delirium	Adjustment reactions, affective disorders, PTSD, marital/family/ work problems
Family response	Shock, denial, anticipatory mourning	Mourning, coping, anger, emotional support of patient	Adjustment/acceptance, transition to home and work
Psychiatric interventions	Pharmacotherapy, liaison with family and burn team	Psychotherapy, pharmacotherapy, behavioral therapy	Psychotherapy, pharmacotherapy, marital and group therapies

TABLE 35–2. *Burn Injury According to Total Injury**

Degree of Burn	Cause	Surface Appearance
First degree: All are considered minor unless <18 months, >65 years, or with severe fluid loss	Flash flame Ultraviolet light (sunburn)	Dry, no blisters, edema: erythema
Second degree (partial thickness) Minor: <15% of body surface area in adults Moderate: 15–30% in adults or <15% with involvement of face, hands, feet, or perineum Severe: >30% of body surface	Contact with hot liquids or solids, flash flame to clothing, direct flame, chemicals	Moist blebs, blisters, white to pink or cherry red color
Third degree (full thickness) Minor: <2% Moderate: 2–10%; any involvement of face, hands, feet, or perineum Severe: >10% and major chemical or electrical burns	Contact with hot liquids or solids, flame to clothing, chemicals, electricity	Dry with leathery eschar under débridement, charred blood vessels visible under eschar, white (waxy pearly) dark or mixed color

*Combination of first through third degree burns possible, and the degree of burn may change with time.
From Freund and Marvin (1990). With permission.

Convalescent Phase

During the convalescent phase the patient anticipates leaving the hospital setting and returning home or convalescing in a rehabilitation facility. Issues about coping and life style changes predominate. Adjustment reactions with depressed mood may arise, as well as frank major depression, although it apparently is rare (Andreasan, Norris, & Hartford, 1971; Chang & Herzog, 1976). Family therapy involves reintegration of the "changed" family member. Support groups comprised of other burn-injured patients may be helpful during this period.

SPECIAL DIAGNOSTIC CONSIDERATIONS

History-Taking

The problem of assessing the burned patient often begins with his arrival in a comatose state, intubated, with a dearth of clinical information. The critical nature of burn injury and the mental status derangements that accompany it may make traditional history-taking impossible. It is imperative that the consultant use all available informants (including family, police and fire officials, ambulance crews) for gathering clinical data. If there is a history of previous suicide attempt or recent major psychiatric illness, a staff member should stay with the patient until further assessment of suicidality is possible. This includes the intensive care unit (ICU) setting, unless the patient is so moribund that the need is obviated. Even in a well-staffed ICU, suicidal patients left without direct observation can harm themselves.

Causes of Agitation

The cause of agitation in the acutely burned patient can be difficult to determine. Delirium, alcohol or drug withdrawal, PTSD, and pain can present with psychic and motor agitation. Because of the hypermetabolic response to burn injury, tremor, hyperreflexia, and fever are seen early in the burn course and may obscure the diagnostic picture. Frequently, several causes of agitation occur together.

Alcohol and Drug Withdrawal

The consultant should evaluate all acutely agitated burn patients for occult alcohol or drug withdrawal. The prevalence of alcohol and substance abuse in this population has been estimated at anywhere from 20% to 80% (Brezel, Kassenbrock, & Stein, 1988; Howland & Hingson, 1987; Waller, 1972). The prevalence of positive blood alcohol levels in new admissions to the New York Hospital Burn Center is approximately 10%; and most of these patients meet objective criteria for alcohol abuse. Additionally, 10% of alcohol-seronegative patients meet criteria for alcohol abuse (Bernstein et al., 1992): thus nearly 20% of the patients on this clinical service had alcohol-related problems.

Different forms of drug abuse are associated with characteristic patterns of burn injury. Cocaine and methamphetamine abusers frequently sustain their burns as a result of using flammable inhalants in their drug use (free-basing). Therefore flash burns of the face and chest with a history of drug abuse are suggestive of cocaine or stimulant abuse. Puncture wounds around veins are suggestive of intravenous drug use. Flame burns sustained while smoking in bed suggest alcohol abuse. Alcohol and drug abuse, especially when it results in unrecognized or inadequately treated withdrawal, is a potent risk factor for the development of "burn delirium" (Andreasen, Noyes, & Hartford, 1972a; Steiner & Clark, 1977).

Delirium

It has been estimated that up to 70% of burn patients become delirious during the acute phase of their illness (Andreasan, Hartford, & Knott, 1977; Antoon, Volpe, & Crawford, 1972; Blank & Perry, 1984). In my experience, approximately 20% of burn patients develop a florid and severe delirium (APA, 1987a). Causes for burn delirium are legion, but some of the more common are hypoxia, hypovolemia, hyponatremia, sepsis, and acidosis (Antoon, Volpe, & Crawford, 1972; May, Ehleben, DeClemente, 1984). In addition, changes in the sleep-wake cycle engendered by an ICU setting may predispose the burn patient to delirium (Miller, Gardner, & Mlott, 1976).

Pain

Pain is ubiquitous in acute burn care and a powerful stimulus for agitation, anxiety, and anger in the burn population. Surprisingly, chronic pain syndromes are not characteristic of surviving burn patients (Marvin & Heimback, 1985). Pain management in the acutely burned patient is complicated by the respiratory depressant effect of opiate analgesics, which are the treatment of choice for acute burn pain. An equally difficult issue is the management of iatrogenic pain engendered by dressing changes and debridement procedures necessary to burn care. The burn patient agitated by pain or the conditioned response to anticipated pain can be violent and disruptive. The issue of adequacy of analgesia always should be considered in an agitated or delirious burn patient.

Neurologic Diagnoses: Anoxia and Postconcussional Syndromes

Burn trauma often occurs in the setting of a major physical catastrophe, such as an explosion or vehicle crash (Committee on Trauma Research, 1985). During the initial hours and days of burn care, those burn patients suffering concomitant central nervous system (CNS) trauma either as an extension of the burn (electrical burn injury), due to metabolic disturbances (anoxia, acidosis), or due to mechanical trauma (closed head injury) may present with nonspecific agitation, disinhibition, or delirium (Haberal, 1986). Mental changes due to acute CNS injury (see Table 35-3) can mimic functional psychiatric illness. Subtle cognitive changes and personality changes also may be encountered (Martyn, 1986).

PTSD

Posttraumatic stress disorder (PTSD) is a well-characterized sequela of burn trauma and treatment, and a cause of agitation and social withdrawal in burn patients. Approximately 20% of burn patients suffer symptoms of PTSD (Andreasan, Norris, & Hartford, 1971). PTSD can occur days to months after the burn injury. Some patients complain of sleep disturbance and nightmares in which they relive the traumatic event, whereas others experience this phenomenon while awake (APA, 1987b). Burn patients experiencing a "flashback" can appear psychotic, but lack of a premorbid psychiatric history and intact reality testing help differentiate this group from burn patients with psychotic disorders or delirium.

Diagnosing Depression

Profound psychomotor withdrawal, depressed mood, and sleep disturbance are characteristic of patient behavior after burn injury (Adler, 1943; Hamburg, Hamburg, & DeGoza, 1953; Steiner & Clark, 1977). It can be difficult to differentiate major depression from this "adjustment disorder" response to burns. Generally, this state is treated with antidepressants only if it persists past the period of acute burn injury or interferes with the patient's participation in physical therapy and rehabilitation. This depression-like response to burn injury may represent a secondary mood disorder due to stress-mediated neuroendocrine changes.

Children and the Elderly

Child Abuse

A significant portion of burned children are child abuse victims (Caniano, Beaver, & Boles, 1986; Showers & Garrison, 1988; Stuart, Kenney, & Morgan, 1987). They tend to be boys more than girls and are likely to live with a single parent of lower socioeconomic status (Lenofski & Hunter, 1977; Peclet, Newan, Eichelberger, et al., 1990; Purdue, Hunt, & Prescott, 1988). Abused children tend to have characteristic patterns of burns that should alert the consultant (see Table 35-4). The abusing caretaker usually is late in getting the child to medical attention and may give an inconsistent or evasive history or respond hysterically to the interview. The abuser usually has a history of being abused as a child (Stuart, Kenney, & Morgan, 1987). Where abuse is suspected, the appropriate child welfare agency should be notified to investigate the incident. The psychiatric consultant should avoid a punitive stance with the family.

Characteristics of the Elderly Burn Patient

The elderly are likely to be burned in the setting of physical illness and debilitation (Rossignol, Locke, & Burke, 1986). Elderly burn victims tend to suffer greater

TABLE 35–3. *Secondary Mental Disorders Associated with Burn Trauma*

Syndrome	Cardinal Features	Associated Diagnoses
Secondary personality syndrome	Change or accentuation of previous characteristic traits, which may involve affective instability, recurrent outbursts of aggression, marked impaired social judgment, apathy and indifference, suspiciousness or paranoid ideation	Head trauma Hypoxemia
Amnestic syndrome	Impairment of short- and long-term memory, remote memory usually retained, not occurring during course of delirium and not meeting criteria for dementia	Thiamine deficiency Subcortical stroke Head trauma Electrical injury

TABLE 35–4. *Characteristics of Burns due to Child Abuse*

Accident when child alone
Injury attributed to sibling
History incompatible with observed injury or motor development of child
Stressed or unstable family
Presentation for treatment with someone other than parent
Sharply demarcated burns of hands, feet, buttocks, perineum, or nonexploratory part of child

area of burns due to relative heat insensitivity and poor mobility (Anous & Heimbach, 1986), and they are more likely to die from their burns (Hong, Zamboni, Erikkson, et al., 1986; Manktelow, Meyer, Herzog, et al., 1989; Ostrow, Bongard, Sacks, et al., 1987). There may be a delay between injury and treatment due to physical impairment or social isolation. The consultant asked to evaluate the burned elderly patient should suspect premorbid medical or neurologic illness, especially dementia. A particularly careful review of medication use is helpful both initially and upon any change in mental status. Questions of competency and right to refuse treatment often arise with this population.

PATHOPHYSIOLOGY

Stress Response

Burn trauma engenders a hypermetabolic stress response characterized by massive increases in circulating catecholamine and corticosteroid levels (Chance, Nelson, Kim, et al., 1987). Untreated drug or alcohol withdrawal also engenders a hypermetabolic state that may be additive with that produced by the burn. This hypermetabolic state initially is accompanied by behavioral agitation usually followed by psychomotor retardation, social withdrawal, depressed mood, and cognitive and sleep disturbance suggesting major depression (Charlton, Klein, & Gagliard, 1981). These mood changes may be a direct consequence of the endocrine changes and may reflect monoamine depletion. Traditionally, this state of depression and withdrawal has not been treated and appears to wane with time. Long-term follow-up studies of burn patients have not documented an increased incidence of major depression (Andreasan, Norris, & Hartford, 1971).

Drugs given in the acute burn setting to alleviate pain and agitation actually may prevent untoward consequences of the stress response. In a rat model, burn-induced acute respiratory distress syndrome is ameliorated by a variety of psychoactive agents including benzodiazepines, opiates, neuroleptics, and beta-blockers (de Oliveira, Shimans, de Oliviera Antonio, et al., 1986). This point argues for prompt treatment of alcohol and drug withdrawal, severe pain, and agitation in the acutely burned patient.

Chronic Anxiety, Phobic Avoidance, and PTSD

Both burn trauma and its treatment are risk factors for the development of PTSD, phobic avoidance, and other manifestations of anxiety (Andreasen, Norris, & Hartford, 1971; Horowitz, Wilner, & Alvarez, 1986; Stoddard, Norman, Murphy, et al., 1989). Debridement of necrotic burned tissue and dressing changes are the procedures likely to induce a phobic response in burned patients; they also are risk factors for PTSD (Perry & Heidrich, 1982). Extended periods in the ICU, particularly if accompanied by intubation and treatment with paralyzing agents, also are potent risk factors for the development of acute or chronic anxiety states. In addition, the use of high doses of short-acting benzodiazepines in these patients increases their risk of subclinical withdrawal, the presentation of which may be a PTSD-like syndrome preferentially responsive to benzodiazepines or carbamazepine (Klein, Uhde, & Post, 1986).

Developmental Issues in Burned Children and Adolescents

The reaction of children to burn injury depends on their developmental stage. The metabolic response to burn injury often results in physical growth retardation and may delay the attainment of developmental milestones (Stoddard, 1982). It is expected that a burned child will regress psychologically in the setting of physical trauma and separation from home and parents while in the hospital. Physically abused children are at greater risk for regressive behavior. When the consultant is called to assist in the management of an "acting out" child, he or she should bear in mind that screaming and fighting are age-appropriate responses to pain and separation. The issue of autonomy for the burned child also may be acted out through recalcitrance with dressing changes, feeding, and rehabilitation.

Adequate management of a pediatric burn pain is the first consideration, as there is evidence that most burned children receive inadequate analgesia (Perry & Heidrich, 1982). Burned neonates may present with a failure-to-thrive picture due to the catabolic response to burn injury as well as to psychological factors (Caniano, Beaver, & Boles, 1986). Burned children, in contrast to adults, appear to carry an increased risk for anxiety and mood disorders, estimated at 10 to 20% five-year prevalence (Herndon, Lemaster, Beard, et al., 1986; Stoddard, Norman, Murphy, et al., 1989). This difference may be due to the lack of extended follow-up studies of adult burn patients resulting in a spuroiusly low estimation of mood disorders in this group. Risk factors for postburn mood disorders in children and adolescents appear to be the same as those for burn injury, including premorbid psychiatric diagnosis and inadequate analgesia in the setting of prolonged surgical treatment.

The burned adolescent faces additional problems of coping, as body image changes and peer group acceptance are particularly important to this age group. Surprisingly, a few studies have suggested that burned adolescents are relatively hardy in their response to burn injury (Sutherland, 1988). Facial disfigurement is the single greatest factor in predicting poor postburn adolescent adjustment (Sawyer, Minde, & Zuker, 1982).

Psychiatric consultation is desirable for most children with significant burns, as the consequence of untreated psychiatric illness may be a lifetime of dysfunction. It is important to brief parents prior to discharge as to warning signs for outpatient consultation (e.g., nightmares or poor school performance). Scheduled checkups with the psychiatrist coordinated with surgical outpatient follow-up visits are an efficient way to monitor what can be a protracted period of psychosocial adjustment.

Burn Injury in the Elderly

The lesser ability to mount an appropriate physiologic response to burn trauma places the elderly population at high risk for death and long-term disability after burn injury (Manktelow, Meyer, Herzog, et al., 1989). Elderly burn patients also are at high risk for the development of delirium and residual cognitive impairment after the resolution of the delirium. In many elderly patients, burn injury heralds the onset of institutional care away from home, family, and friends. Adjustment and dependency issues, as well as fears of death and dying, predominate in this age group.

SPECIAL CONSIDERATIONS IN PSYCHOTHERAPEUTIC OR PSYCHOPHARMACOLOGIC TREATMENT

General Pharmacologic Considerations and Drug Interactions in Burn Trauma

A number of physiologic changes attendant to burn injury have implications for psychiatric pharmacotherapy (see Table 35-5). These considerations are most critical in the resuscitative and acute settings but persist in a significant way through surgical burn care. During acute burn injury, cardiac output decreases, resulting in reduced blood flow to (and drug extraction by) the liver (Moncrief, 1973). Hepatic microsomal degradation (phase I metabolism) appears to be impaired, whereas conjugation (phase II metabolism) is unchanged (Williams, 1984). Protein binding is affected variably based on the protein subunit involved and the degree of hypoproteinemia (Bloedow, Hansbrough, Hardin, et al., 1986; Martyn, 1986). Renal function initially is impaired and then enhanced if shock-induced renal insufficiency does not supervene (Goodwin, Aulick, Becke, et al., 1980).

After the resuscitative phase, the hypermetabolic response to burn injury increases cardiac output, thereby increasing extraction-dependent drug clearance. Hepatic oxidative dysfunction continues, however, resulting in a net decrease in hepatic drug metabolism (Williams, 1984).

Treatment of Delirium

The treatment of any delirium begins with an aggressive search for its underlying etiology (see Tables 35-6 and 35-7 for diagnostic workup and list of deliriogenic drugs). Burn delirium is more common during the acute phase of treatment but can occur at any time and may have a chronic and relapsing course. The most common behavioral impairment necessitating treatment of the delirious burn patient is psychic or motor agitation that compromises care. What follows is a regimen for treatment of agitated delirium. The quietly delirious patient may require little or no pharmacotherapy.

The delirious patient is by no means uncommunicative or unable to assimilate information during lucid intervals. All delirious patients should be oriented frequently; and, to the extent possible, normal lighting and sleep/wake cycles should be maintained. In those instances where physical restraint is necessary, it should be used sparingly given the poor skin integrity of the burned patient. It is helpful to explain to both the patient and the family that delirium is a common consequence of burn injury and that it will pass.

Benzodiazepines Versus Neuroleptics

The traditional treatment for burn delirium has been high-potency neuroleptics, usually haloperidol. This drug has been used for its rapid tranquilizing properties, its ease of parenteral use, and its favorable hemodynamic, respiratory, and gastrointestinal side effect profile. An-

TABLE 35–5. *Drug Metabolism After Burn Injury*

Agent	Metabolism and Administration	Acute Effect of Burn	Chronic Effect of Burn	Comments
Lorazepam	Glucuronide conjugation PO/IM/IV	None (Martyn & Greenblatt, 1988)	None	No interactions with burn chemotherapeutic agents Consider gradual taper after prolonged use
Diazepam	Hepatic oxidation PO/IM/IV	Increased half-life, decreased sedative effect due to protein binding changes (Martyn, Abernethy, & Greenblatt, 1984)	Increased half-life, prolonged sedative effect with saturation of fat stores	Markedly sclerosing to vein with IV administration, prolonged sedation relatively contraindicated in acute setting, cimetidine impairs metabolism (Martyn, Greenblatt, & Quinby, 1983)
Chlordiazepoxide	Hepatic oxidation PO/IM	Increased half-life	Increased half-life, prolonged sedative effect with saturation of fat stores	Erratically absorbed IM, not available IV, cimetidine further impairs metabolism (Martyn, Greenblatt, & Quinby, 1983); avoid in ICU setting
Midazolam	Hepatic oxidation PO/IM/IV	Increased half-life	Increased half-life	Ranitidine impairs metabolism (Martyn, 1986)
Lithium	Renal PO	Increased half-life	Decreased half-life if renal function intact	Electrolyte imbalance and diuretic use affect excretion. Follow levels daily in ICU setting
Carbamazepine	Hepatic oxidation PO	Variable	Unknown	May increase excretion of other hepatically metabolized drugs; bone marrow suppression may be additive with that of burn injury
Valproate	Hepatic oxidation PO	Variable	Unknown	Risk of SIADH, platelet dysfunction
Neuroleptics	Hepatic oxidation PO/IM/IV	Increased half-life	Increased half-life	H2 blockers may enhance excretion and lower serum levels; may lower seizure threshold after CNS injury; potentiate opiate-induced respiratory depression. Possibly increased risk of dystonia due to burn-associated postsynaptic cholinergic supersensitivity (Martyn, 1986).
Anticholinergics	Hepatic oxidation PO/IM/IV	Unknown	Unknown	May cause or exacerbate hypotension, ileus, delirium.
Tricyclic antidepressants	Hepatic oxidation PO/some IM/IV	Decreased serum levels due to protein binding abnormalities (Martyn, Abernethy, & Greenblatt, 1984)	Increased half-life	Excretion can be enhanced by H2 blockers; additive cardiac conduction effects with quinidine-like antiarrhythmics. Follow levels and ECG carefully; avoid tertiary amine tricyclics in acute setting.
Fluoxetine	Hepatic oxidation & conjugation (norfluoxetine renally excreted) PO	Increased half-life	Increased half-life	Has superior side effect profile, but complicated protein binding interactions need to be studied; may enhance opiate analgesia with concomitant reduction in respiratory depression (Hynes, Lochner, Bernisk, et al., 1985).

other benefit is its relative lack of adverse drug interactions with commonly used burn chemotherapeutic agents. Haloperidol's downside lies in a putative increased risk of dystonia (Huang & Demling, 1987) and perhaps neuroleptic malignant syndrome (NMS) in the burn population, as well as the enhancement of the respiratory depressant effect of opiates (Keats, Telford, & Kurow, 1961). The intravenous route is preferable in emergent situations, as oral absorption may be impaired and intramuscular use traumatizes already injured skin and muscles.

Although some studies suggest that benzodiazepines are deliriogenic in this population (Stanford & Pine, 1988), the contribution of undiagnosed alcohol and sedative-hypnotic withdrawal in burn delirium has provided a rationale for the use of these agents (Lineaweaver, Anderson, & Hing, 1988). Specifically, short-acting parenteral benzodiazepines such as lorazepam

TABLE 35–6. *Workup for Burn Delirium*

Physical and neurologic examinations
Chest roentgenogram
ECG
Metabolic parameters including vitamin B_{12}/folate, magnesium, arterial ammonia, arterial blood gases (including carboxyhemoglobin during resuscitative phase), VDRL/FTA, CBC with differential, electrolytes, liver function tests, blood cultures, toxicology screen if not previously done, therapeutic drug levels, urinalysis
Consider CT/LP/EEG (diffuse slowing with most deliria but can be normal or fast with alcohol withdrawal delirium)

TABLE 35–7. *Burn Chemotherapeutic Agents Associated with Mental Status Changes*

Opiate analgesics: meperidine, morphine, levorphanol, others
Adjuvant pain medications: ketamine, nitrous oxide
Antibiotics: cephalosporins
Histamine blockers: cimetidine, ranitidine, hydroxyzine
Benzodiazepines: lorazepam, diazepam, midazolam, others

and midazolam may be used when delirium is suspected to be due to withdrawal (see Table 35-8).

Both benzodiazepines and neuroleptics can cause or potentiate respiratory depression in burned patients treated with opiates. Although no studies exist comparing the relative risks of one agent versus another in this regard, my clinical experience indicates that high-potency benzodiazepines carry a relatively greater risk of causing clinically significant respiratory depression in patients at risk for pulmonary failure.

In those instances where the intubated patient's agitation is life-threatening and unresponsive to pharmacotherapy, the use of a nondepolarizing neuromuscular relaxant such as *d*-tubocurarine may be indicated. Depolarizing agents such as succinylcholine can cause fatal hyperkalemia and are contraindicated in burn patients (Tolmied, 1967). Burned patients treated for long periods with these agents may be at greater risk for PTSD than other burn patients, so alternative therapy for agitation should be pursued once the crisis is over.

Treatment of Alcohol and Sedative-Hypnotic Withdrawal

Alcohol and sedative-hypnotic abuse are risk factors for burn injury (Brodzka, 1983; Darko, Wachtel, Ward, et al., 1986; Howland & Hingson, 1987). Detoxifying the burned patient may be complicated by concomitant sepsis, hypotension, or respiratory damage from inhalation injury. Benzodiazepines are the treatment of choice for alcohol and sedative-hypnotic withdrawal; they have trivial effects on blood pressure but do have respiratory depressant effects. *The key to detoxifying the burned patient is to carefully titrate dosage until amelioration of target symptoms is achieved with minimal sedation.* Short-acting agents are preferable in critically ill burned patients because of the greater pharmacologic control inherent in their short half-life and their ease of parenteral use. Longer-acting agents may be used in less severely burned patients. In instances of combined illicit opiate/sedative-hypnotic abuse, opiate analgesics can be used to cover opiate withdrawal while benzodiazepines

TABLE 35–8. *Treatment of Delirium**

Agent	Route	Dosage	Comments
Haloperidol	PO/IM/IV	0.5–10.0 mg q30min until calmed; then 2–10 mg q6–12h, titrated to effect	Possible increased risk of dystonia, NMS. May potentiate opiate induced respiratory depression.
Lorazepam	PO/IM/IV	0.5–2.0 mg q30min until calmed; then 0.5–2.0mg q4–6h titrated to effect	Additive CNS and respiratory depression with opiates, possibly deliriogenic, treats underlying sedative-hypnotic withdrawal. Few untoward drug interactions.
Midazolam	PO/IM/IV	Consult anesthesiologist	Use only in ICU setting with ventilatory equipment available. Excretion impaired by histamine blockers; may need up to 25ml/hr in adults.
d-Tubocurarine	IV	Consult anesthesiologist	Neuromuscular blocking agent for use in ICU setting with intubated patients for life-threatening agitation unresponsive to above. May increase risk of PTSD.

NMS = neuroleptic malignant syndrome; PTSD = posttraumatic stress disorder.
*Can consider combination of these protocols with appropriate dose reductions. These doses are suggested for average weight adults, which must be adjusted for patient status and the clinical situation.

TABLE 35–9. *Protocol for Substance Detoxification in the Acute Setting**

Substance	Agent/Route	Dosage	Comment
Alcohol/sedative-hypnotics	Lorazepam PO/IM/IV	0.5–2.0mg PO/IV q1h until cessation of withdrawal symptoms, then q4th around the clock with 0.5–1.0 mg q1h prn objective signs/symptoms or withdrawal for 24 hours; then divide total 24-hr dose by 6 and administer q4h. Detox. over 3–4 days for alcohol, 5–7 days for sedative-hypnotics.	Use conservative dosage in resuscitative and acute setting, given the increased risk of respiratory compromise. Titrate dose to amelioration of symptoms (e.g., tremor, hyperreflexia) *without concomitant CNS depression*; consider holding dose if latter is present. IV more potent than PO/IM preparations.
Opiates	Methadone PO/IM/IV	In IV drug abusers, consider increased dosage standing and PRN narcotics for pain and withdrawal symptom control. If opiate withdrawal symptoms persist, transfer patent to ICU and start methadone 10 mg PO/IM or 5 mg IV q4h (not to exceed 40 mg PO/IM [20 mg IV] on 1st day) in addition to short-acting narcotics. Titrate to cessation of objective withdrawal criteria (rhinnorhea, diarrhea, piloerection).	Use only in ICU with ventilatory equipment available in the resuscitative and acute setting. Titrate to objective criteria for withdrawal to *minimize sedation and respiratory depression*. When stable, total 24-hour dose and give half the dose q12h. Decrease methadone by 5 mg/day. Note PO/IM preparations approximately half as potent as IV. Monitor blood pressure and respiratory status carefully. Give naloxone IV slow push for obtundation; repeat as necessary and titrate to effect.
Cocaine/methamphetamine	None	Consider midazolam infusion and/or neuroleptic for psychotic features.	Acute intoxication requires cardiac monitoring. Treatment is symptomatic.

*Can consider combination of these protocols with appropriate dose reductions. These dosages are suggested and must be titrated to patient status and the clinical situation.

cover sedative-hypnotic withdrawal (see Table 35-9). Opiate-specific withdrawal symptoms such as pain, rhinorrhea, and piloerection can be used as a guide for when more opiates, rather than more benzodiazepines, are needed to stabilize the patient. Substance-abusing burn patients often are nutritionally debilitated and then made NPO or placed on parenteral nutrition because of postburn ileus. It is imperative that these patients receive prophylaxis for Wernicke's encephalopathy (thiamine, 100 mg IV qd) until such time as normal nutritional intake is resumed.

Treatment of Opiate Withdrawal

Opiate abusers and burn patients on methadone maintenance present particular problems in burn care. Methadone maintenance dosages should be confirmed prior to reinstitution, as patients have been unwittingly overdosed. In addition, opiate analgesic needs are greater in these patients. A regimen for titrating methadone dose in intravenous drug users is provided below. However, given the acute respiratory depressant effect of methadone and its long half-life, it is preferable, when possible, to avoid its use with active intravenous drug users in the resuscitative setting and to increase standing and as-needed opiate analgesic doses instead. Once the patient is stable, a portion of the total narcotic dose can be converted to methadone and tapered as appropriate.

Treatment of Stimulant Intoxication and Withdrawal

The advent of free-basing (the use of flammable solvents for volatilizing cocaine or amphetamine) has resulted in a greater proportion of abusers being burned. Stimulant abusers presenting after extended periods of abuse may appear profoundly sedated or even delirious. In addition, many of these patients sustain chemical burns to their eyes that adds to disorientation. It is my experience that these "pseudodelirious" patients often are arousable and oriented, and that no pharmacotherapy is needed other than sleep. The need for the latter can be profound, lasting up to 3 to 4 days. In those patients who present with acute stimulant intoxication, cardiac monitoring is necessary, as there is an increased risk of arrhythmias and sudden death (Isner, Estes, & Thompson, 1985). The stimulant-intoxicated patient can be paranoid, psychotic, and hallucinatory. Symptomatic treatment as noted in the delirium section is appropriate (see Table 35-8). Literature on sedation of acutely psychotic patients with midazolam infusion (Mendoza, Olenderedjian, Adams, et al., 1987) suggests its efficacy for stimulant-induced psychosis (see Table 35-8).

Treatment of Burn Pain

Pain is a ubiquitous and unavoidable consequence of burn injury. The effective management of burn pain is an area in which clinical practice varies, and few well-designed studies exist. There is evidence to suggest that burn patients, especially children, are chronically undermedicated for pain (Perry & Heidrich, 1982), and that physiologic changes that accompany burn convalescence actually may decrease pain tolerance (Andreasen, Noyes, Hartford, et al., 1972b; Coderre & Melzack, 1985). The management of pain is complicated by the difficulty of assessing pain in those patients who are unable to communicate verbally with the consultant.

Types of Pain: Acute Injury, Procedural, and Background Pain

It is helpful to distinguish between the pain of acute burn injury, burn wound healing, and burn wound care, as each has its own particular set of physical and psychological responses. In the setting of trauma, the onset of acute burn pain may be delayed (Marvin & Heimback, 1985). After the resuscitative phase, patients usually report a baseline level of pain that is tolerable at rest. The primary source of excruciating pain becomes the bedside surgical procedures, such as débridement of tissue and dressing change, hydrotherapy, and physical therapy (Beale, 1983; Syzfelbein, Osgood, & Carr, 1985). This distinction is important because the anticipation of this pain can engender a host of conditioned behavioral responses that may lower the patient's pain threshold. This "pain–anxiety–pain" cycle is the crucial issue for clinical pain management in burn patients (Shires & Black, 1984; Wall, 1979). In addition, the physiology of pain perception changes as the patient heals, so cutaneous pain response is enhanced in healing burned tissue. This fact, coupled with progressive opiate tolerance, usually necessitates increasing amounts of analgesia during burn recovery (Syzfelbein, Osgood, & Carr, 1985).

Assessing Pain

The cardinal rule of pain management is to trust the patient's perception of pain severity. Psychiatric symptomatology attributed to mood disorder, character pathology, or even delirium actually may result from undermedicated pain. A diagnosis of "drug-seeking behavior" is one of exclusion in this population, even in the setting of known drug abuse.

For those patients who can communicate with the consultant by speaking or writing, it is helpful to note the time of onset of the pain (chronic versus situational, or both), character of pain (pricking, burning, dull), continuity (intermittent, continuous), and associated pain behaviors (e.g., anxiety, avoidance, sleep disturbance, anger, agitation, tantrums in children). Formal pain rating scales, such as the McGill Pain Questionnaire, may be helpful for following treatment efficacy (Wilkie, Savedra, Holzemer, et al., 1990). The isolation of affect and social withdrawal associated with PTSD can confound pain assessment by verbal interview alone (Perry, Cella, Falkenberg, et al., 1987). Visual pain rating scales such as the Visual Analog Scale (VAS) (see Figure 35-1) may help assess these patients (Revill & Robinson, 1976).

It is important to educate caretakers to query the patient specifically and often about pain control, especially during procedures, and to provide prompt analgesia (Simons, McFadd, Frank, et al., 1978; Syzfelbein, Osgood, & Carr, 1985). For patients unable to communicate, assessment is less straightforward. Pain may be inferred if the blood pressure and pulse are elevated with procedures or chronically elevated from a known baseline. This parameter is especially helpful for neonates and toddlers, as well as intubated patients who are pharmacologically paralyzed. Poorly controlled pain should lead the differential diagnosis for agitation with hypertension or tachycardia in this population. However, lack of vital sign abnormalities does not rule out inadequate analgesia, as many burn patients have other conditions (e.g., sepsis, hypovolemia) that can mask pain-associated tachycardia and hypertension. If an intubated patient can respond with writing or gestures, the VAS can be used to assess the level of pain.

In situations in which the pain is procedurally related, continued observation and frequent questioning as to pain experienced during the procedure, even as frequently as each minute, may help ensure adequate analgesia. Plasma levels of opiate analgesics and serum immunoreactive beta-endorphin have been utilized to assess pain control (Perry & Inturrisi, 1983; Syzfelbein, Osgood, & Carr, 1985). Low levels of the former and elevation of the latter have been associated with inadequate analgesia. This test is not readily available or practical at this time, however.

Pain Pharmacotherapy

Acute burn pain management is based on opiate analgesics, which often are used at doses far exceeding those seen in nontrauma patients. Morphine sulfate is the

(10 cm)

FIG. 35–1. Pain assessment: visual analog scale (VAS). The patient marks the point that best reflects the current level of pain or pain relief from an analgesic dose. The score is measured by the distance in centimeters from the extreme left of the line to the patient's mark. Relative scores over time are used to assess pain treatment efficacy. (From Huskisson, 1974. With permission.)

drug of choice, although other, longer-acting preparations are gaining favor for managing "background" pain. Meperidine, which is frequently used in many burn units, should be avoided except for procedural pain prophylaxis because of CNS toxicity associated with its metabolite, normeperidine (Inturrisi & Umans, 1986). Parenteral administration is preferable because of predictable absorption and greater rapidity of analgesic onset. An around-the-clock schedule of administration should be instituted with as-needed "rescue" doses as indicated. This regimen results in a relatively stable serum level and avoids periods of inadequate analgesia and breakthrough pain. Rote patterns of prescription should be avoided because of the wide variability of patient analgesic response. In pharmacologically paralyzed intubated patients, a constant narcotic infusion may provide added efficacy with a net decrease in total narcotic dose used per 24–hour period (Syzfelbein, Osgood, & Carr, 1985; Tamsen, Hartvig, Fagerlund, et al., 1982). Representative opiate regimens are given in Table 35-10. Adequate prophylaxis for constipation is indicated for all patients receiving opiates.

Long-acting opiate preparations have gained favor in the treatment of the "background pain" associated with burn injury. Most burn patients report a continuous, though usually low, level of discomfort that is exacerbated with movement and burn treatment procedures (Perry & Heidrich, 1982). By using a long-acting preparation such as methadone or MS Contin, a constant level of analgesia is attained without the "breakthrough pain" associated with shorter-acting agents. With this schema, short-acting drugs such as morphine and sufentanyl are still used for acute situational pain (see Figure 35-2). Converting part of the patient's daily dose of opiates to a long-acting opiate drug not only provides better background pain control but makes it easier to taper the patient when opiate analgesia is no longer required.

A number of protocols have been designed to deliver continuous patient-controlled opiate analgesia, either by intravenous infusion or indwelling infusion pump (Filkins, Cosgrave, Marvin, et al., 1981). Epidural instillation of opiate preparations may provide even greater efficacy in pain control (Dyer, Anderson, Michell, et al., 1990). These methods decrease the net amount of opiate used, and patients report that analgesia is enhanced. In general, the method is contraindicated in delirious and drug-abusing patients and is not recommended for the resuscitative and early acute phases of injury. It also requires a setting where vital signs can be meticulously monitored.

Procedural Pain

The pain associated with burn dressing changes (BDC) and débridement is unique and requires special consideration. Many patients report this pain to be the most excruciating pain experienced during the hospitalization (Beale, 1983; Perry & Heidrich, 1982). Ultra-short-acting narcotics and anesthetics are the mainstay of procedurally induced pain management.

It is important to note that even short-acting opiates such as morphine require up to 0.5 hour after intravenous administration to achieve adequate pain prophylaxis. The timing of morphine bolus administration must anticipate the actual time of the procedure and not be "on call" to the procedure room or physical therapy. During the procedure, frequent pain reassessment by the staff and as-needed medication for breakthrough pain are necessary. Pain is a powerful antagonist to the respiratory depressant effect of morphine; nevertheless, naloxone should be available for intravenous administration, during and after the procedure period, to reverse any untoward respiratory depression. Naloxone administration is titrated to effect rather than given in a fixed dosage, thereby preventing iatrogenic precipitation of opiate withdrawal.

A variety of opiate and nonopiate agents are used for procedural pain prophylaxis. Sufentanyl, an ultra-short-acting opiate used for anesthetic induction, has gained favor as an analgesic agent for burn procedures (Perry & Heidrich, 1982). Its ease of parenteral use and short duration of effect make it ideal for pain prophylaxis during BDCs. However, repeated dosing may be necessary. The inhalation agent nitrous oxide has been used as an adjunct to opiates and as a primary analgesic agent for BDCs and procedural pain (Filkins, Cosgrave, Marvin, et al., 1981; Marvin, Engrav, & Heimbach, 1984). Bone marrow suppression and a megaloblastic anemia

TABLE 35–10. *Pharmacologic Treatment of Burn Pain*

Agent/Class	Pain	Suggested Dosage	Comment
Morphine/opiate*	Background	Adults: 5–25 mg IM/SC or 2.5–15.0 mg IV q4–6h Children: 0.1–0.2 mg/kg IM/SC or one-half of dose IV q4–6h	For moderate to severe pain. Standing doses should be given around the clock with prn rescue doses q2h. Onset of action 30 minutes after administration. Titrate dose to effect.
	Procedural	Adults: 5–35 mg IM/2–25 mg IV 30 minutes prior to procedure and as needed Children: 1–10 mg IM/IV 30 minutes prior to procedure and as needed	Sedation, respiratory depression, nausea are common side effects. Plasma levels 50–100 mg/ml associated with adequate pain prophylaxis (Perry & Inturrisi, 1983). Consider continuous IV infusion for severe pain in an intubated patient.
Meperidine/opiate	Procedural	Adults: 50–200mg IM 30 minutes prior to procedure and as needed Children: 10–50 mg IM 30 minutes prior to procedure and as needed	For moderate to severe pain. Poor choice for background pain given that metabolite normeperidine is a CNS irritant associated with agitation and myoclonus. Oral dosage approximately 25% as potent as IM. Absolutely contraindicated with history of recent MAOI use.
Levorphanol tartrate/opiate	Background	Adults: 2–5 mg IM, 4–10 mg PO q4h Children: 0.5–4.0 mg IM, 1–8 mg PO q4h and q2h prn rescue doses	Kinetics similar to those of morphine, but it is 4–8 times as potent IM; relatively more effective PO than morphine. Approximately one-half as potent PO as IM.
	Procedural	Adults: 2–8 mg IM, 4–16 mg PO 30 minutes prior to procedure and as needed Children: 0.5–0.6mg IM, 1–12 mg PO 30 minutes prior to procedure and as needed	
Codeine/opiate	Background	Adults: 30–120 mg PO/IM q4h Children: 5–50 mg PO/IM q4h	For mild background pain relief only, often given with antipyretic (acetominophen/NSAID).
Methadone/opiate	Background	Dosage range not established. Initially, convert one-third daily morphine dose to methadone and divide bid (after 10 mg test dose). Can then titrate methadone dose up as indicated.	Long-acting narcotic used for management of moderate to severe background pain in conjunction with shorter-acting agents for procedural and background pain. As narcotic need decreases, decrease methadone 5 mg/day while monitoring for withdrawal symptoms. Lower doses in renal failure.
Sufentanyl/opiate	Procedural	Adults: 1–6 μg/kg 5–10 minutes prior to procedure, then 0.5–1.0 μg/kg as needed, or begin 0.03–0.05 μg/kg/min infusion 15 minutes prior to procedure.	Ultra-short-acting opiate used for moderate to severe procedural pain. Powerful respiratory depressant. Should be used in an ICU setting with an anesthetist and naloxone available.
Naloxone/opiate antagonist	N/A	Dilute 400-μg ampul in 10 ml saline (40 μg/ml solution). Administer 40 μg IV slow push q30–45 sec to reverse opiate-induced respiratory depression; repeat as necessary.	Opiate antagonist used to reverse opiate-induced respiratory depression. Can precipitate iatrogenic withdrawal if given by bolus. Short duration of effect may necessitate repeated dosing.
Nitrous oxide/inhalant anesthetic	Procedural	Given in inhalant mixture with 50% oxygen during painful procedures. Can be self-administered	Short-acting inhalant anesthetic used for moderate to severe procedural pain. Associated with rare megaloblastic anemia. Use requires anesthetist and ventilatory equipment.
Ketamine/intramuscular anesthetic	Procedural	Adults & children: 4 mg/kg IM; repeat as necesssary	Short-acting IM anesthetic with excellent analgesic and amnestic properties; associated with tachyphylaxis requiring increased dosage over time; rare psychotomimetic anesthesia emergence phenomena reported.
Halothane/inhalant general anesthetic	Procedural	Consult anesthesiologist	Consider general anesthesia for massive or repeated débridement procedures. May have additive effect with tricyclics/neuroleptics in lowering seizure threshold in susceptible patients.

*These doses are *suggested* dose ranges that must be adjusted for patient status and clinical situation.

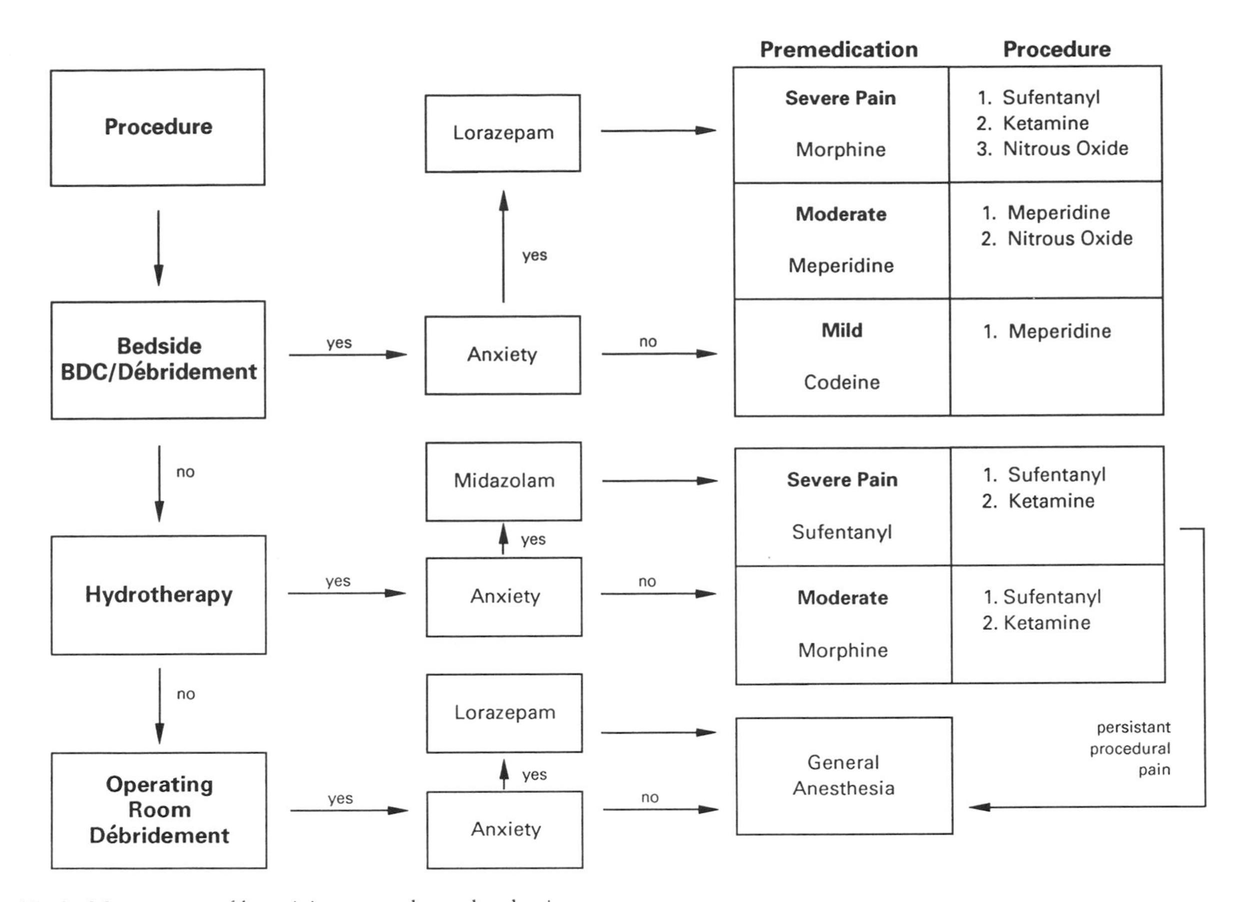

FIG. 35–2. Management of burn injury procedure related pain.

similar to that of vitamin B_{12} deficiency have been associated with chronic nitrous oxide use but usually are not clinical concerns in the burn trauma setting. Ketamine is an intramuscular anesthetic agent used for BDCs and procedure pain management (Demling, Ellerbe, & Jarret, 1978; Ward & Diamond, 1976). Ketamine provides excellent analgesia but is associated with tachyphylaxis and may have psychotomimetic properties. Oral antipyretic/analgesic preparations such as Percocet generally are not adequate agents for procedure pain prophylaxis. In situations such as massive or repeated eschar débridement and grafting, it may be preferable to place the patient under general anesthesia. Consultation with the surgeon and an anesthesiologist is required.

Adjuvant Pain Treatments

A variety of agents have been used to augment narcotic analgesia, although none has been studied systematically. Table 35-11 and Figure 35-3 display the commonly used alternatives.

Along with burn wound healing comes pruritus, which may reach intolerable proportions. Antihistamines such as hydroxyzine are the agents of choice. Neuroleptics, especially chlorpromazine, also have antipruritic properties, but their hypotensive and extrapyramidal effects relatively contraindicate their routine use as antipruritics. Hydroxyzine has both intrinsic analgesic and narcotic-enhancing properties (Bonica & Albe-Fessard, 1976).

Tricyclic antidepressants such as amitriptyline have been used to relieve the "depression" associated with burn pain (Andreasen, Noyes, Hartford, et al., 1972b; Steiner & Clark, 1977), although their direct analgesic properties may underlie their efficacy in burn pain. The anticholinergic, alpha-adrenergic-blocking, and quinidine-like properties of these agents make them a poor choice in the resuscitative and early acute setting.

Neuroleptics have been used as adjuvants to opiate analgesia. They may potentiate opiate-induced respiratory depression, and their use generally is inferior to using a higher opiate dosage.

Findings suggest that serotonergic agonists ameliorate pain and enhance analgesia (Hosobuchi, Pomeranz, & Bascom, 1980; Messing & Lytle, 1977; Sternbach, Janowsky, Huey, et al., 1976). Clonazepam, an ultra-

TABLE 35–11. *Pharmacologic Adjuvants to Burn Pain Treatment*

Agent	Use	Suggested Dosage	Comment
Hydroxyzine/ antihistamine	Background pain and pruritus	Adults: 100–200 mg IV bid/tid Children: 1 mg/kg IV bid/tid Give with narcotic dose	Antihistamine with analgesic properties. Useful as antipruritic and in IV form as adjuvant to opiate analgesia. May cause or exacerbate ileus or delirium. Consider in acute setting.
Benzodiazepines	Background pain associated anxiety	Dosage varies with agent and physiologic state of patient. Consider clonazepam 0.5 mg PO qd/bid.	Variety of short- and long-acting agents used to prevent or treat pain associated anxiety. No evidence of intrinsic analgesic properties.
	Procedural pain associated anxiety	Consider lorazepam 0.5–1.0 mg PO/IM/ IV 15–30 minutes prior to procedure in conjunction with other analgesics.	CNS/respiratory depression additive with opiates. Consider short-acting agents (e.g., lorazepam) for procedure-related anxiety. Administer 15–30 minutes prior to procedure.* Avoid long-acting agents in acute setting.
Tricyclic antidepressants	Background pain	Generally, adjuvant pain dosage lower than that used for antidepressant effect. Consider nortriptyline 25 mg PO bid.	Avoid use in resuscitative and early acute setting. Consider full therapeutic trial of serotonergic tricyclic (nortriptyline) at antidepressant dose for nonresponders.*
Neuroleptics	Procedural and background pain, pruritus	Dosage varies with agent and physiologic state of patient. Generally, low dose of low-potency agent (chlorpromazine 50–100 mg PO bid/ tid) used as adjuvant pain treatment.	Avoid low-potency agent in acute setting. May enhance opiate-induced respiratory depression. Antihistaminic properties ameliorate pruritus.*
Serotonergic agonists	Background pain	Consider fluoxetine 5 mg/day (use liquid), titrating slowly upward according to response; maximum 40 mg/day.*	Experimental evidence that serotonergic agonists (e.g., tryptophan, fluoxetine, clonazepam, trazodone) may enhance opiate analgesia).*

These doses are *suggested* dose ranges that must be titrated according to patient status and the clinical situation.
*See Table 36–5 for metabolism and drug interactions.

long-acting benzodiazepine, has serotonergic activity that might enhance analgesia. Fluoxetine has been shown to augment opiate analgesia while attenuating respiratory depression in burned rats (Hynes, Lochner, Bemis, et al., 1985). Neither agent has been systematically tested in burn patients.

Transcutaneous electrical nerve stimulation (TENS) (Kimball, Drew, Walker, et al., 1987) and nerve blocks have been used to manage chronic burn pain unresponsive to pharmacotherapy.

"Pain–Anxiety–Pain" Cycle

Pain and anxiety are inextricably linked in burn care (Andreasen, Noyes, Hartford, et al., 1972b; Choinere, Melzack, Rondeau, et al., 1989; Shires & Black, 1984). Those situations that are likely to provoke or inflict pain, such as BDC or physical therapy, require both analgesia and anxiolytic management if this cycle is to be broken. A variety of benzodiazepines are used prior to painful procedures for anxiolytic and amnestic purposes and as an adjuvant for sleep in the burn patient. It is preferable to use a short-acting benzodiazepine such as lorazepam 0.5 to 1.0 mg PO or IV for procedure-associated anxiety. It should be given 15 to 30 minutes prior to the event so that adequate prophylaxis of anxiety-augmented pain perception is achieved. In instances such as hydrotherapy, where amnesia for the procedure is desired, premedication with midazolam may be preferable to lorazepam, given the former's powerful amnesic effect. Pharmacotherapy also helps to prevent the anticipatory conditioned anxiety that may be a risk factor for PTSD.

Staff Attitudes and Analgesia Administration

Staff attitudes toward pain assessment and management are the greatest single factor in inadequate analgesia administration (Perry & Heidrich, 1982). Fears ranging from iatrogenic addiction to unresolved conflict and denial around pain-inflicting roles can adversely affect drug administration practice (Porter & Jick, 1980). For any pain consultation, staff attitudes should be assessed in a nonjudgmental way, and instruction in both pain assessment and adequate management (including the patient's anticipated need for increased analgesia with

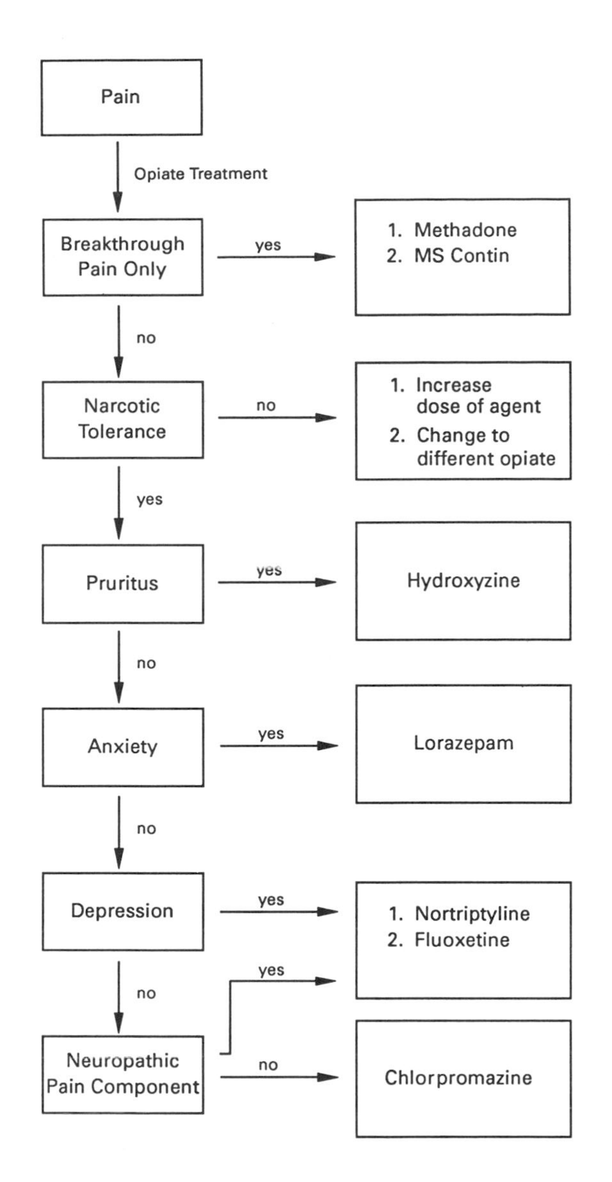

FIG. 35–3. Management of burn injury background pain.

time) should be provided. Training staff to recognize pain-associated behaviors (including avoidance and isolation), to assess pain with rating instruments such as the VAS (see Figure 35-1), and to manage pain pharmacotherapy promote adequate analgesia administration. Exploring staff anger toward difficult and demanding patients and devising behavioral plans to deal with such situations is necessary for patient management and for the staff's sense of control.

Behavioral Strategies of Pain Management

The technique of stress inoculation has met with success when managing procedural pain in some burn patients. This structured series of cognitive and behavioral interventions has resulted in decreased pain behaviors and anxiety, and increased patient compliance. During phase 1 the rationales and goals for burn treatment are presented. Phase 2 involves teaching physical coping skills to be used during painful procedures: deep breathing and muscle relaxation. Phase 3 involves cognitive strategies to be used during painful procedures, such as attention diversion and mental distraction (the patient imagines he is somewhere pleasant and conjures up that scene). Phase 4 involves rehearsal of these skills while imagining undergoing a painful procedure; and phase 5 involves coaching the patient during the procedure. For the latter, a nurse or attendant also trained in the exercise can assume the role of coach (Wernick, Jaremko, & Taylor, 1981).

Hypnosis has been used to manage the chronic and procedural pain of burn injury. This type of pain management should be taught to all patients with low levels of concurrent psychopathology or with substance abuse. I have generally requested a consultation from a colleague skilled in hypnosis when this treatment is indicated. The reader is referred to other monographs for further instruction (Margolis, 1979; Wakeman & Kaplan, 1978). (See also Wain [1993] for an extended discussion of medical hypnosis.)

Neonates and Children

The keystone of pediatric burn care is adequate pharmacologic management of pain. There are, however, a number of behavioral strategies that enhance children's compliance with painful procedures and also may minimize pain perception. Neonates should be gently held, caressed, and cooed to during BDC and other procedures. Studies have shown that having young children help in the BDC procedure (e.g., hold the dressing, apply the bandage) and providing them with rewards for cooperation (e.g., a treat or wheelchair ride) enhances compliance and attenuates pain (Kavanaugh, 1983; Lasoff & McEttrick, 1986; Varni, Bessman, Russo, et al., 1980). In those situations where it is not possible, allowing the child to watch television or listen to a story has been noted to decrease the incidence of recalcitrant behavior during BDC. Behavior contracting, with rewards for compliance, is helpful (Simons, McFadd, Frank, et al., 1978; Wernick, Brantley, & Malcolm, 1980; Zide & Pardoe, 1975). Generally, these measures are attempted only with school-age children. In all instances, an appropriate explanation by a favorite nurse or caretaker prior to the procedure may minimize behavioral acting out. These techniques are predicated on adequate levels of analgesia.

Psychodynamic Issues and Psychotherapy with Burn Trauma Victims

The overwhelming nature of burn trauma threatens the patient's emotional integrity with loss of life, physical attributes, bodily self-image, and self-determination. The psychiatrist must help the patient regain both physical and psychic homeostasis while coping adequately with the enormity of the trauma suffered. Predictable patterns of psychological adjustment provide a framework for this work.

Common to all physical injury is a loss of body integrity and the self-concept and control it implies. Burn trauma leaves the patient visibly and often permanently changed. An attempt to integrate this profound experience with the patient's self-concept assists the process of healing. Often a sense of control is regained if the patient can rationalize the injury as necessary to the rescue of others, or as making the patient stronger or unique for having survived it.

Grief and Mourning

Profound grief and mourning is common in acutely burned patients and their families. In many instances the patient has survived a conflagration in which other family members have died or are now gravely ill. The patient may have attempted (and failed) to rescue trapped family members and may experience guilt over survivorship. In general, the patient should be told if family members or others have died if he specifically asks. However, in some instances the patient may make it clear that he does not wish to know or will deny the fact that others have died. This denial and repression should not be challenged unless others forms of coping are available to the patient.

Disfigurement

As the patient allows himself awareness of his physical limitations or disfigurement due to burns, he will begin to mourn those physical attributes lost to the burn. Disfigurement most troubling to the patient usually involves the face, genitals, and hands. It is rare not to encounter fear and concern regarding current and future physical appearance and limitations.

Facially burned patients may wait weeks or months before asking to see their reflections. Mirrors should not be present in the patient's room until one is requested. The badly disfigured patient usually gains a sense of his injuries by gauging the expressions of shock or revulsion he sees in family members (especially young children, in whom this is a common and predictable occurrence). When the patient does request a mirror, it is best that a supportive nurse or family member be in attendance. Minimizing the patient's fear of disfigurement should be avoided; a frank but hopeful evaluation of current disfigurement and future benefits from reconstructive surgery is helpful. Patients often are concerned as to whether they will be able to return to work or even venture out in public. It is often helpful to have a "graduate" burn patient with similar wounds visit as an example of someone who has survived the transition. Failure to successfully rationalize and come to terms with facial disfigurement often leads to increasing social isolation and withdrawal. Diagnostically, the assessment of mood and affect in the facially disfigured patient can be difficult because of the absence of subtle visual cues. It is often helpful when assessing these patients to use depression rating scales.

Burns of the perineum, female breasts, and buttocks can crystallize a variety of conflicts regarding sexuality. Patients not only fear lessened desirability but may harbor concerns about reproductive capabilities. In those vulnerable patients uncomfortable or in conflict about their sexuality, feelings of guilt and retribution may predominate. In instances where genital appearance or function has been impaired, consultation with plastic and urologic or gynecologic surgeons is indicated to discuss prognosis and the future possibility of reconstruction. (See Chapters 15, 34, 39 and 40.)

Burned hands adversely affect the patient to a surprising degree. A loss of ability to physically manipulate the world engenders a sense of powerlessness. This is of greatest concern in those who work with their hands, but hand burns are distressing to all patients. Psychotherapy stressing issues of passivity and loss of control, in conjunction with physical therapy, can ameliorate continued symptoms.

A significant amount of burn trauma is due to self-immolatory attempts by psychotic or personality disordered patients (Andreasan & Noyes, 1975; Davidson & Brown, 1985). Not all of these attempts are suicidal in nature, but the action often represents a psychotic attempt to "cleanse" or change the patient in some profound way. I remember in particular a handsome young man with a diagnosis of schizophrenia with dysmorphic delusions, who set himself afire to "make myself more attractive to women." Often these patients have unrealistic or even magical conceptions about the benefits of reconstructive surgery. Gentle confrontation, coordinated with preoperative consultation with the surgeon, may help to minimize unrealistic expectations.

It can be overwhelming for both experienced and novice consultants to confront the disfigurement and pain of the patient. The consultant's countertransference may manifest in preoccupation with biologic aspects of patient management, or avoidance of the patient. There is no best way to manage this response except to be cognizant of it and, if necessary, to titrate one's amount of exposure to the patient to a tolerable level. Fear and repugnance usually attenuate with time. It is diminished by spending time with the burn treatment team.

Adjustment Disorders, Depression, and PTSD

The differential diagnosis of major depression versus adjustment disorder with depressed mood is a difficult one in the burn trauma setting. It is rare for any burn patient not to experience some period of sleep disturbance and neurovegetative symptoms during recovery. Anxiety over recovery and family care predominate during the acute phase, but an often overlooked period of worsening mood can accompany the departure from the hospital when the patient leaves an environment where he is is accepted and cared for, returning to one of often unrealistic expectations. If this transition is not anticipated by both patient and family, transient dysphoria or worse can occur. In some instances a series of passes prior to discharge or a gradually decreasing number of visits back to the burn unit may ease this transition.

PTSD patients often suffer profound guilt and depression in addition to their anxiety symptoms. Reliving the trauma can include intrusive thoughts over failed rescue attempts or witnessing the grisly death of a loved one. Avoidance is the usual response to these events, and can be extraordinarily resistant to psychotherapy. Dreams may express conflicts over passivity, such as dreaming of not escaping from the fire. A supportive stance that encourages the patient to express these concerns provides an avenue for further exploration. A brief trial with a short-acting benzodiazepine such as lorazepam can be prescribed for transient sleep disturbance, but treatment with therapeutic doses of a benzodiazepine, monoamine oxidase inhibitor, or tricyclic antidepressant may be needed for persistent flashbacks, chronic anxiety, and mood disturbance. Supportive group therapy for PTSD patients may facilitate psychotherapeutic treatment.

Family Reactions and Treatment

In addition to the initial supportive and informative role the consultant may play with families of burned patients, there are other predictable times of family stress that often require psychiatric intervention. The first occurs when the patient has survived the resuscitative and acute phases of burn care and is beginning physical therapy. The family role changes from that of advocacy and vigilance to a more passive one of observation and encouragement. Families may begin to feel powerless and frightened as they, often for the first time, confront the enormity of the patient's injury and the task of rehabilitation (Cella, Perry, Kulchycky, et al., 1988). They may begin to bear the brunt of the patient's anger as he leaves the passive and dependent role of critical care patient. The next identifiable family stressor often comes, paradoxically, with the patient's departure from the hospital. Unrealistic family expectations coupled with the patient's physical and emotional limitations may precipitate anger or depression in both. Ongoing family conferences with discussion and anticipation of these stressors ease this transition.

Social Services

Much patient and family anxiety stems from the disruption of home life and economic security. Planning for economic maintenance of the family and subsequent posthospitalization rehabilitation for the patient should begin as soon as possible after injury. Early social work assessment and intervention as required reduce a major source of anxiety.

REFERENCES

ADLER A (1943). Neuropsychiatric complications in victims of Boston's Cocoanut Grove Disaster. JAMA. 123:1098–1101.

ANDREASAN N, HARTFORD C, & KNOTT J (1977). EEG changes associated with burn delirium. Dis Nerv Syst 38:27–31.

ANDREASAN N, NORRIS A, & HARTFORD C (1971). Incidence of long-term psychiatric complications in severely burned adults. Ann Surg 174:785–793.

ANDREASAN N, & NOYES R (1975). Suicide attempted by self-immolation. Am J Psychiatry 132:554–556.

ANDREASEN N, NOYES R, & HARTFORD C (1972a). Factors influencing adjustment of burn patients during hospitalization. Psychosom Med 34:517–525.

ANDREASEN N, NOYES R, HARTFORD C, ET AL. (1972b). Management of emotional reactions in seriously burned adults. N Engl J Med 286:65–69.

ANOUS M, & HEIMBACH D (1986). Causes of death and predictors in burned patients more than 60 years of age. J Trauma 26:135–139.

ANTOON A, VOLPE J, & CRAWFORD J (1972). Burn encephalopathy in children. Pediatrics 50:609–616.

APA (1987a). Diagnostic and statistical manual of mental disorders (p. 103). Washington, DC: American Psychiatric Association.

APA (1987b). Diagnostic and statistical manual of mental disorders (pp. 250–251). Washington, DC: American Psychiatric Association.

BEALE J (1983). Factors influencing the expectation of pain among patients in a children's burn unit. Burns. 9:187–192.

BERNSTEIN L, JACOBSBERG L, ASHMAN T, ET AL. (1992). Detection of alcoholism in burn patients. Hosp Comm Psychiatry 43:255–256.

BERRY C, PATTERSON T, WACHTEL T, ET AL. (1984). Behavioral factors in burn mortality and length of stay in hospital. Burns 10:409–414.

BLANK K, & PERRY S (1984). Relationship of psychological process during delirium to outcome. Am J Psychiatry 141:843–847.

BLOEDOW D, HANSBROUGH J, HARDIN T, ET AL. (1986). Post-burn serum drug binding and serum protein concentrations. J Clin Pharmacol 26:147–151.

BONICA J, & ALBE-FESSARD D (1976). Comparison of the analgesic effects of morphine, hydroxyzine, and their combination in patients with post-operative pain. Recent Studies of the Nature and Management of Acute Pain: A Symposium. New York: HP Publishing.

BREZEL BM, KASSENBROCK JM, & STEIN JM (1988). Burns in substance abusers and in neurologically and mentally impaired patients. Burn Care Rehabil 9:169–171.

BRODZKA W (1983). Burn; causes and risk factors in an inner city population. Arch Phys Med Rehabil 64:487–498.

CANIANO D, BEAVER B, & BOLES E (1986). Child abuse. Ann Surg 203:219–224.

CELLA D, PERRY S, KULCHYCKY S, ET AL. (1988). Stress and coping in relatives of burn patients: a longitudinal study. Hosp Commun Psychiatry 39:159–166.

CESARE J, MORGAN A, FELICE S, ET AL. (1990). Characteristics of blunt and personal violent injuries. J Trauma. 30:176–182.

CHANCE W, NELSON J, KIM M, ET AL. (1987). Burn-induced alterations in feeding, energy expenditure, and brain amine neurotransmitters in rats. J Trauma 27:503–509.

CHANG F, & HERZOG B (1976). Burn morbidity: a followup study of physical and psychological disability. Ann Surg 183:34–37.

CHARLTON J, KLEIN R, & GAGLIARD G (1981). Assessment of pain relief in patients with burns: Proceedings of the Third World Congress on Pain. Pain 1:181–187.

CHOINERE M, MELZACK R, RONDEAU J, ET AL. (1989). The pain of burns: characteristics and correlates. J Trauma 29:1531–1539.

COBB S, & LINDEMANN E (1943). Symposium on management of Cocoanut Grove burns at Massachusetts General Hospital: neuropsychiatric observations. Ann Surg 117:814–824.

CODERRE T, & MELZACK R (1985). Increased pain sensitivity following heat injury involves a central mechanism. Behav Brain Res 15:259–262.

COMMITTEE ON TRAUMA RESEARCH, C. O. L. S., NATIONAL RESEARCH COUNCIL AND THE INSTITUTE OF MEDICINE (1985). Injury in America. Washington, DC: National Academy Press.

DARKO D, WACHTEL D, WARD H, ET AL. (1986). Analysis of 585 burn patients hospitalized over a 6 year period. Part III. Psychosocial data. Burns 12:395–401.

DAVIDSON T, & BROWN L (1985). Self-inflicted burns: a 5-year retrospective study. Burns 11:157–160.

DEMLING R, ELLERBE S, & JARRET F (1978). Ketamine anesthesia for tangential excision of burn eschar: a burn unit procedure. J Trauma 18:269–270.

DE OLIVEIRA G, SHIMANO L, & DE OLIVEIRA ANTONIO M (1986). Acute respiratory distress syndrome: the prophylactic effects of neurodepressant agents. J Trauma 26:451–457.

DYER R, ANDERSON B, MICHELL W, ET AL. (1990). Postoperative pain control with a continuous infusion of epidural sufentanil in the intensive care unit: a comparison with epidural morphine. Anesth Analg 71:130–136.

FECK G, & BAPTISTE M (1979). The epidemiology of burn injury in New York. Public Health Rep 94:312–318.

FILKINS S, COSGRAVE P, MARVIN J, ET AL. (1981). Self-administered anesthetic: a method of pain control. Burn Care Rehabil 2:33–34.

FREUND P, & MARVIN J (1990). Postburn pain. In J BONICA (Ed.), The management of pain (2nd ed., Ch. 26, pp. 481–489). Malverne, PA: Lea & Febiger.

GOODWIN C, AULICK L, BECKE R, ET AL. (1980). Increased renal perfusion and kidney size in convalescent burn patients. JAMA 65:67–75.

HABERAL M (1986). Electrical burns: a five year experience—1985 Evans Lecture. J Trauma 26:103–109.

HAMBURG D, ARTZ C, REISS E, ET AL. (1953). Clinical importance of emotional problems in the care of patients with burns. N Engl J Med 248:355–359.

HAMBURG D, HAMBURG B, & DEGOZA S (1953). Adaptive problems and mechanisms in severely burned patients. Psychiatry 16:1–20.

HERNDON D, LEMASTER J, BEARD S, ET AL. (1986). The quality of life after major thermal injury in children: an analysis of 12 survivors with >80% total body, 70% third degree burns. J Trauma 26:609–619.

HONG Q, ZAMBONI W, ERIKKSON E, ET AL. (1986). Burns in patients under 2 and over 70 years of age. Ann Plast Surg 17:39–45.

HOROWITZ M, WILNER N, & ALVAREZ W (1986). Stress-response syndromes: a review of posttraumatic and adjustment disorders. Hosp Commun Psychiatry 37:241–249.

HOSOBUCHI Y, POMERANZ B, & BASCOM D (1980). Tryptophan loading may reverse tolerance to opiate analgesics in humans: a preliminary report. Pain 9:161–169.

HOWLAND J, & HINGSON R (1987). Alcohol as a risk factor for injuries or death due to fires and burns: review of the literature. Public Health Rep 102:475–483.

HUANG V, & DEMLING R (1987). Haloperidol complications in burn patients. J Burn Care Rehabil 8:269–273.

HUSKISSON E (1974). Measurement of pain. Lancet. 2:1127–1131.

HYNES M, LOCHNER M, BEMIS K, ET AL. (1985). Fluoxetine, a selective inhibitor of serotonin reuptake, potentiates morphine analgesia without altering its discriminative stimulus properties or affinity for opioid receptors. Life Sci 36:2317–2323.

IMBUS S, & ZAWACKI B (1977). Autonomy for burn patients when survival is unprecedented. N Engl J Med 297:308–311.

INTURRISI C, & UMANS J (1986). Meperidine biotransformation and central nervous system toxicity in animals and humans. Adv Pain Res Ther 8:143–153.

ISNER J, ESTES N, & THOMPSON P (1985). Cardiac consequences of cocaine: premature myocardial infarction, ventricular tachyarrhythmias, myocarditis and sudden death. Circulation [suppl 3] 72:415.

KAVANAUGH C (1983). A new approach to dressing change in the severely burned child and its effect on burn related psychopathology. Heart Lung. 12:612–619.

KEATS A, TELFORD J, & KUROW Y (1961). Potentiation of meperidine by promethazone. Anesthesiology 22:34–41.

KIMBALL K, DREW J, WALKER S, ET AL. (1987). Use of TENS for burn reduction in burn patients receiving travase. J Burn Care Rehabil 8:28–31.

KLEIN D, UHDE TW, & POST R (1986). Preliminary evidence for the utility of carbamazepine in alprazolam withdrawal. Am J Psychiatry 142:235–236.

LASOFF E, & MCETTRICK M (1986). Participation versus diversion during dressing changes: can nurses' attitudes change? Issues Comprehen Pediatr Nurs 9:391–398.

LENOFSKI E, & HUNTER K (1977). Specific patterns of inflicted burn injuries. J Trauma 17:842–846.

LINEAWEAVER W, ANDERSON K, & HING D (1988). Massive doses of midazolam infusion for delirium tremens without respiratory depression. Crit Care Med 16:294–296.

MACARTHUR J, & MOORE F (1975). Epidemiology of burns. JAMA 231:259–270.

Manktelow M, Meyer A, Herzog S, et al. (1989). Analysis of life expectency and living status of elderly patients surviving a burn injury. J Trauma 29:203–207.

Margolis C (1979). Hypnosis in the treatment of burns. Burns 6:253–254.

Martyn J (1986). Clinical pharmacology and drug therapy in the burned patient. Anesthesiology 65:67–75.

Martyn J, Abernethey D, & Greenblatt D (1984). Plasma protein binding of drugs after severe burn injury. Clin Pharmacol Ther 35:534–536.

Martyn J, & Greenblatt D (1988). Lorazepam conjugation unimpaired in burned patients. Pharmacol Ther 43:250–255.

Martyn J, Greenblatt D, & Quinby W (1983). Diazepam kinetics in patients with severe burns. Anesth Analg 62:293–297.

Marvin J, Engrav L, & Heimbach D (1984). Self administered nitrous oxide analgesia for débridement: a five year experience. Presented at the 16th Annual Meeting of the American Burn Association.

Marvin J, & Heimbach D (1985). Pain control during the intensive care phase of burn care. Crit Care 1:G1–G6.

May S, Ehleben C, & De Clemente F (1984). Delirium in burn patients isolated in a plenum laminar air flow ventilation unit. Burns 10:331–338.

Mendoza R, Dlenderedjian A, Adams J, et al. (1987). Midazolam in acute psychotic patients with hyperarousal. J Clin Psychiatry 48: 291–292.

Messing R, & Lytle L (1977). Serotonin-containing neurons: their possible role in pain and analgesia. Pain 4:1–21.

Miller W, Gardner N, & Mlott S (1976). Psychosocial support in the treatment of severely burned patients. J Trauma 15:722–725.

Moncrief J (1973). Burns. N Engl J Med 288:444–454.

Ostrow L, Bongard F, Sacks S, et al. (1987). Burns in the elderly. AFP 35:149–154.

Peclet M, Newman K, Eichelberger M, et al. (1990). Patterns of injury in children. J Pediatr Surg 25:85–91.

Perry S, Cella D, Falkenberg J, et al. (1987). Pain perception in patients with stress disorders. J Pain Sympt Management. 2:22–23.

Perry S, & Heidrich G (1982). Management of pain during debridement: a survey of US burn units. Pain 13:267–280.

Perry S, & Inturrisi C (1983). Analgesia and morphine disposition in burn patients. J Burn Care Rehabil 4:276–279.

Porter J, & Jick H (1980). Addiction rare in patients treated with narcotics. N Engl J Med 302:123–127.

Purdue G, & Hunt J (1990). Adult assault as a mechanism of burn injury. Arch Surg 125:268–269.

Purdue G, Hunt J, & Prescott P (1988). Child abuse by burning—an index of suspicion. J Trauma 28:221–224.

Revill S, & Robinson J (1976). The reliability of the linear analysis for evaluating pain. Anesthesia 31:1191–1198.

Rossignol A, Boyle C, Locke J, et al. (1986). Hospitalized burn injuries in Massachusetts: an assessment of injury and product involvement. Am J Public Health 76:1341–1343.

Rossignol A, Locke J, & Burke J (1986). A comparison of age-specific burn injury rates in five Massachusetts communities. Public Health Rep 101:1146–1150.

Sawyer M, Minde K, & Zuker R (1982). The burned child—scared for life? A study of the psychosocial impact of a burn injury at different developmental stages. Burns 9:205–213.

Shires G, & Black E (1984). Pain and anxiety in the burn injured. J Trauma 24:5168–5197.

Showers J, & Garrison K (1988). Burn abuse, a four year study. J Trauma 28:351–355.

Simons R, McFadd A, Frank H, et al. (1978). Behavioral contracting in a burn care facility: a strategy for patient participation. J Trauma 18:257–260.

Stanford G, & Pine P (1988). Post burn delirium associated with intravenous lorazepam. J Burn Care Rehabil 9:160–161.

Steiner H, & Clark W (1977). Psychiatric complications of burned adults: a classification. J Trauma 17:134–143.

Sternbach J, Janowsky D, Huey L, et al. (1976). Effects of altering brain serotonin activity on human chronic pain. Adv Pain Res Ther 1:601–606.

Stoddard F (1982). Coping with pain: a developmental approach to the treatment of burned children. Am J Psychiatry 139:736–740.

Stoddard F, Norman D, Murphy J, et al. (1989). Psychiatric outcome of burned children and adolescents. J Am Acad Child Adolesc Psychiatry 28:589–595.

Stuart J, Kenney J, & Morgan R (1987). Pediatric burns. AFP 36:139–149.

Sutherland S (1988). Burned adolescents' description of their coping strategies. Heart Lung 17:150–157.

Syzfelbein S, Osgood P, & Carr D (1985). The assessment of pain and plasma β-endorphin immunoreactivity in burned children. Pain 22:173–182.

Tamsen A, Hartvig P, Fagerlund C, et al. (1982). Pain controlled analgesic therapy. Part 1. Pharmacokinetics of pethidine in the pre- and postoperative periods. Clin Pharmacokinet 7:149–163.

Tolmied J (1967). Succinyl choline danger in the burned patient. Anesthesiology 28:467–470.

Varni J, Bessman C, Russo D, et al. (1980). Behavioral management of chronic pain in children: case study. Arch Phys Med Rehabil 61:375–379.

Wain HJ (1993). Hypnosis in medical disorders. In A Stoudemire & BS Fogel (Eds.), Medical-Psychiatric Practice (Vol. 3). Washington, DC: American Psychiatric Press.

Wakeman R, & Kaplan J (1978). An experimental study of hypnosis in painful burns. Am J Clin Hypn 21:21–25.

Wall P (1979). On the relationship of anxiety to pain. Pain 6:253–264.

Waller J (1972). Nonhighway injury-fatalities. I. The roles of alcohol and problem drinking drugs and medical impairment. J Chronic Dis 26:33–45.

Ward C, & Diamond A (1976). An appraisal of ketamine in the dressing of burns. Postgrad Med J 52:222–223.

Wernick R, Brantley P, & Malcolm R (1980). Behavioral techniques in the psychological management of burn patients. Int J Psychiatry Med 10:145–150.

Wernick R, Jaremko M, & Taylor P (1981). Pain management in severely burned adults: a test of stress inoculation. J Behav Med 4:103–109.

Wilkie D, Savedra M, Holzemer W, et al. (1990). Use of the Mcgill Pain Questionnaire to measure pain: a meta-analysis. Nurs Res 39:36–41.

Williams R (1984). Drug administration in hepatic disease. N Engl J Med 390:1616–1622.

Zide B, & Pardoe R (1975). The use of behavior modification therapy in a recalcitrant child. Plastic Reconstr Surg 57:378–382.

36 Transplantation surgery

ROBERT M. HOUSE, M.D.

Few procedures in medicine evoke as much awe as organ transplantation, and few confront a patient with as much stress. Rapid advances in transplantation have led to an increase in patient survival and improvement in quality of life, which has led to an increased demand for transplants (see Table 36-1).

Basic issues of organ transplantation are discussed in this chapter, including epidemiology, patient evaluation and selection, the waiting period, and psychological reactions of donors. Specific consideration is then given to heart, liver, bone marrow, and pancreatic transplantation. Renal transplantation is covered in Chapter 27.

THE RECIPIENT

Transplant surgeons are more willing to transplant patients who in the past would have not been considered, such as those with cancer or in whom alcohol or drug abuse may have caused organ failure. For the younger patient, organ failure is often a result of a congenital defect. Many have experienced a lifelong struggle with illness, and transplantation is simply one more in a series of medical interventions. In contrast, organ failure in the adult may be the result of an acute disease process or a complication arising from a long-standing illness.

Selection

One of the most difficult tasks of evaluating patients for transplant is deciding on the selection criteria. Because of the scarcity of organs and the high cost of transplantation it is important to select patients who will best be able to handle the various stresses and compliance requirements inherent in transplantation. There are no generally accepted criteria indicating how psychiatric history and mental status data should be used in the evaluation of a prospective recipient. Added to this is the sobering realization that recommending against transplantation can mean condemning the transplant candidate to a premature death as a result of organ failure. The decision to transplant must be individualized, and care must be taken to protect patients against discrimination because of the existence, past or present, of a psychiatric disorder (House & Thompson, 1988). Patients with various psychiatric disorders, such as bipolar disorder, mental retardation, and schizophrenia, have been successfully transplanted (Merrikin & Overcast, 1985; Surman, 1987).

Factors that must be considered in the selection process include the patient's wish for the transplant; availability of supports such as family and friends; realistic expectations about the proposed procedure; understanding of the risks and benefits in order to provide valid informed consent; ability to comply with treatment including the treatment of existing drug or alcohol problems; absence of behavior that would be incompatible with maintaining the transplant, such as the refusal of an alcoholic patient to alter his use of alcohol; and willingness of the patient to make life style changes that transplantation may require, such as a possible change in occupation. Patients who present with these factors tend to have relatively little difficulty after surgery in incorporating the transplanted organ and viewing themselves as transplant survivors (Christopherson & Lunde, 1971; House, Dubovsky, & Penn, 1983). On the other hand, the absence of any of these factors should not by itself be considered a reason for exclusion. Although each is important, the complexity of the individual case requires a careful clinical evaluation rather than reliance on a rigid set of exclusion criteria.

Exclusion

From a psychiatric standpoint, there are few reasons for excluding a transplant candidate; and they center on a patient's ability to comply with medication, rehabilitation, and avoidance of unnecessary risks to the transplant. Commonly mentioned reasons include chronic psychosis, significant mental retardation, irreversible cognitive disorders such as Alzheimer's disease, and severe personality disorders such as borderline personality disorder and antisocial personality disorder in which there may be difficulty with compliance, impulse control, or abuse of drugs or alcohol (House & Thompson, 1988).

TABLE 36–1. *Transplant Epidemiology*

	No. Transplanted	No. Waiting	Survival 1 Year	Survival 5 Years	Cost (in thousands of dollars)
Heart[a]	1689	1763	85	75–80	100–150
Liver[a]	2191	1216	73	68–70	135–280
Pancreas[a]	418	472	89	40–50	30–40
Bone marrow					
Allogenic[b]	2526	NA	[e]	[e]	150–200
Autogenic[c]	1500				

NA = not available.
[a]Data obtained from United Network for organ sharing for 1989.
[b]Data obtained from Statistical Center of the International Bone Marrow Transplant Registry (IBMTR), Medical College of Wisconsin, Milwaukee. Analyses have not been reviewed or approved by the Advisory Committee of the IBMTR.
[c]Estimate obtained from Dr. James Armitage, Omaha, Nebraska.
[d]Information obtained from Department of Health & Human Services.
[e]Survival rate highly variable depending on presenting illness.

The issue of alcoholism is a particularly difficult one. Nearly 10% of Americans are affected by this disorder; and of these individuals, 15% (nearly two million people) develop alcoholic cirrhosis (Beresford, Turcotte, Marion, et al., 1990; Schenker, 1984). Patients with alcohol-related end-stage liver disease (ARESLD) represent about 50% of patients with end-stage liver disease (ESLD) (Moss & Siegler, 1991). Transplantation of such patients is an emotionally charged issue with attitudes running the gamut from those who vigorously support transplanting such patients to those who believe that these individuals have willfully destroyed their own organ and should not be trusted to take care of a new, transplanted organ (Cohen, Benjamin, and the Ethics and Social Impact Committee of the Transplant and Health Policy Center, Ann Arbor, Mich., 1991; Moss & Siegler, 1991). Centers willing to transplant such patients often require a period of abstinence ranging from 1 month to 2 years. The abstinence criterion is often an informal one and does not reflect the knowledge available from current alcohol research (Beresford, Turcotte, Marion, et al., 1990). In one court case, Allen v. Mansour (1986), an alcoholic patient successfully sued the state Medicaid board for refusing to pay for a liver transplant. At issue was his failure to meet the 2–year preoperative abstinence requirement, which was thought to be longer than the natural course of his illness. In settling in the patient's favor it was thought that experts in the field of alcoholism had not been consulted and that standards based on current knowledge of alcoholism had not been established. At issue was not the patient's suitability but, rather, the criteria used when making the suitability determination. Such cases highlight the importance of applying contemporary knowledge when establishing selection criteria.

THE DONOR

Living Related Donor

Bone marrow and kidneys comprise the majority of organs transplanted from living related donors. Transplantation of segments of pancreas and liver from living related donors has also been done because of the scarcity of available cadaveric organs.

Although many factors go into the decision to donate, it is often a spontaneous one with altruistic motives. Most donors have no difficulty deciding to donate and are gratified by the experience (Simmons, Hickey, Kjellstrand, et al., 1971). Aside from the anxiety a surgical procedure engenders, most potential donors give little thought to risks such as pain, infection, or death. Recipients often feel a sense of debt to the donor that they cannot repay. For some recipients a special bond or closeness may develop, whereas others may experience guilt. The latter is more likely if there has been a history of conflict between the two or reluctance to donate on the part of the donor. In fact, some recipients prefer a cadaveric transplant as a way to avoid feelings of guilt, indebtedness, and retribution (Castelnuovo-Tedesco, 1978).

Bone marrow donors often feel like the last hope for the recipient. A survey of adult bone marrow donors using the Profile of Mood States (POMS), the Simmons Scale, and an investigator-constructed questionnaire found 10 to 20% of donors experiencing psychiatric difficulties related to the transplant (Wolcott, Wellisch, Fawzy, et al., 1986). Often this reaction relates to the donor feeling regret or guilt if the recipient develops graft versus host disease (GVHD), if the graft fails to take, or if the patient has a recurrence (e.g., of leukemia) (Futterman & Wellisch, 1990).

Technically, transplantation of pancreas segments from living donors is more difficult than is transplantation of cadaveric organs, although the rejection rate is lower (Sutherland, Goetz, & Najarian, 1984). Psychological reactions of donors are similar to those of kidney donors, discussed in Chapter 27. Because pancreas transplants are generally confined to diabetics, donors must be screened carefully as to their risk for developing diabetes.

Donation of a liver lobe, a technique still in the experimental stage, was developed in response to the limited availability of livers for small children, a group where waiting list mortality ranges from 14 to 24% (Busuttil, 1991). Advantages of lobe transplants, usually from a parent, include availability and HLA matching. The most important disadvantage is the risk of death for the donor, although to date this event has not occurred. An important issue for the donor is guilt if the transplanted lobe fails or rejects. The risk of death by donating a lobe is unknown, although one can imagine the impact on the family if both the donor and patient died in this process. An alternative for increasing the number of organs available for transplantation is graft reduction. In this procedure the size of the cadaveric liver is surgically reduced at the time of transplant (Moreno, Garcia, Loinaz, et al., 1991). Advantages include preservation of vasculature not available with a living related donation, potential for donation to two recipients, and no risk to the donor.

Cadaver

Patients must be brain-dead before organs can be "harvested" for transplant, a process that may take several days. Generally, family members of a brain-dead patient feel positively about organ donation and believe the brain-dead patient can continue to live through someone else or that the untimely death may have served a beneficial purpose. Problems arise if the family wishes to know the name of the recipient or how the recipient is progressing. Similarly, recipients are often curious about the donor: wondering if they will take on characteristics of the donor, wishing to thank the family, or experiencing survivor guilt. Frequently, recipients do become aware of the donor's identity, often through the media.

Usually, little attention is paid to the staff who are taking care of the brain-dead patient, although they must often deal with the grief of the family as well as keep the donor "healthy" until the time of harvesting. Support and education as well as a chance to ventilate these feelings may decrease some of these difficulties (Sophie, Salloway, Sorack, et al., 1983).

PREOPERATIVE EVALUATION

The preoperative evaluation (see Table 36-2) can serve as a baseline for comparison should future problems arise. It identifies existing psychiatric problems and past psychiatric problems (especially those precipitated by stresses such as past surgeries), identifies means by which patients and their families cope with stress, provides a chance to find out about support systems, and provides a chance to educate the patient and the family about the psychiatric aspects of transplantation.

Meeting with patients and family preoperatively provides knowledge of family dynamics that can lead to the identification of potential future problems. Patients who have become totally dependent on a spouse may have difficulty becoming more independent after a successful transplant. Where the illness has been longstanding, a sense of stability may have developed that is upset when the sick member changes roles (Riether & Stoudemire, 1987). (See also Chapter 2 on family therapy for chronic medical illness).

Denial is a common first response when the patient is told of the need for a transplant. Rejection of the procedure as too extreme is often followed by fear, anger, and depression. This attitude may lead to a failure to follow through with the evaluation process. Doctor shopping, searching for alternative treatments, bargaining, and magical thinking are not uncommon. Often the question of "Why me?" is asked at this point along with an attempt to reverse the process causing the organ failure. The alcoholic may stop drinking, hoping the

TABLE 36–2. *Preoperative Evaluation*

1. Patient profile including relationships, educational history, work history, legal history
2. Organ failure: cause, complications, course, compliance with treatment
3. Means of coping with illness, past and present
4. Expectation of surgery, including fantasy material
5. Supports, including family, friends, church, employment
6. Existing psychiatric difficulties and treatment plan
7. Past psychiatric history
8. Family psychiatric history
9. Substance use/abuse history
10. Mental status examination; may include neuropsychological testing
11. Understanding of proposed procedure and ability to provide informed consent

alcoholic cirrhosis will reverse itself. When it does not occur, a sense of resignation and pursuit of transplant may ensue as the only chance for survival. A worsening physical condition, disability, or the onset of encephalopathy hastens this process. At this point, anxiety about meeting the selection criteria, as well as death anxiety, may be prominent (House & Thompson, 1988; Kuhn, Davis, & Lippman, 1988). Postponement of appointments, refusal to undergo diagnostic tests, failure to follow through with treatment recommendations, and anger toward the transplant team may reflect ambivalence about the surgery and must be thoroughly evaluated (Watts, Freeman, McGriffin, et al., 1984).

Compliance is imperative and is reflective of how the patient views the illness and the patient's adaptation to it. Moreover, how the patient has coped with this and other illnesses gives an idea of how he or she might cope with the transplant. Unrealistic expectations about the surgery will set the patient and the family up for disappointment (Dubovsky & Penn, 1980). Patients with adequate social supports tend to do better when faced with setbacks.

Initially, a baseline mental status examination must be performed. Formal neuropsychological testing can further assess cognitive changes and psychiatric status. (Neuropsychologic testing is discussed in Chapter 18.)

Computerized tomography (CT) and magnetic resonance imaging (MRI) may uncover structural etiologies of some psychiatric conditions. Use of CT or MRI scans along with neuropsychological testing may help in prognosticating whether a cognitive disorder will reverse with transplantation; patients without gross structural brain changes have a better chance of full cognitive recovery.

Identification and treatment of existing psychiatric conditions is also an important function of the preoperative evaluation. The prevalence of adjustment disorders, especially with depression or anxious mood, is high in these patients (Mai, 1986; Watts, Freeman, McGriffin, et al., 1984). The true prevalence of major depression, however, is difficult to assess owing to the overlap of biologic symptoms of depression with symptoms of severe medical illness (see Chapter 4). The reported prevalence of depression in transplant patients ranges from 16% to 60% (Kuhn, Davis, & Lippman, 1988; Trzepacz, Maue, Coffman, et al., 1987). The prevalence of delirium ranges between 3% and 50% (House, Dubovsky, & Penn, 1983; Trzepacz, Brenner, Coffman, et al., 1988). Anxiety is common and may affect as many as 39% of patients (Mai, 1986). As mentioned, drug and alcohol abuse need to be identified and treated. Patients awaiting transplants with acute cognitive changes need a full workup for organic etiologies that include withdrawal states, subdural hematomas, toxic effects to the brain, and strokes.

WAITING PERIOD

Once accepted, patients often describe the waiting period for a donor organ to become available as being the most stressful part of the transplant. Among heart transplant candidates, 47% experienced significant psychiatric distress during this time (Kuhn, Brennan, Lacefield, et al., 1990). Factors such as blood type, tissue type, body size, the presence of antibodies, and the unavailability or organs may prolong the wait. During this time, patients may experience guilt as the availability of the organ they need requires the death of another person. With prolonged waits the patient often experiences fear and a sense of helplessness that an organ will not become available in time or that his medical condition will be such that he will realize little benefit from the transplant. A sense of abandonment by the transplant team can be lessened by periodic follow-up visits or telephone contacts for those who live away from the transplant center. Involvement with transplant support groups can be particularly helpful (Levenson & Olbrisch, 1987). Continuous assessment of anxiety and depression allows initiation of psychotropic medication and psychotherapy when indicated.

POSTOPERATIVE PSYCHIATRIC PROBLEMS

A sense of euphoria and well-being often follows successful transplantation. It may be based on relief over having survived the operation, on satisfaction with having a functioning organ, or on side effects of corticosteroids given for immunosuppression. This emotional state may be short-lived if complications develop. Acute changes in mental status in the days following transplantation occur in 50 to 70% of patients (Dubovsky & Penn, 1980; Freeman, Volks, & Sokol, 1988).

Delirium

Delirium is the most common psychiatric problem both pre- and posttransplant. Its pretransplant incidence may be as high as 50% (Trzepacz, Brenner, Coffman, et al., 1988); and its posttransplant incidence ranges between 25% and 50% (House, Dubovsky, & Penn, 1983; Watts, Freeman, McGriffin, et al., 1984). In this population, drug toxicity, infection, especially viral and mycotic, and rejection of the transplanted organ are probably the most common causes of the delirium. Other causes include failure of another organ, electrolyte imbalance, hemorrhage, sleep disturbance, central nervous system (CNS) hemorrhage, and lung emboli and infarction.

Many of the medications used for transplantation may cause delirium (see Table 36-3). Drugs such as cyclosporine and corticosteroids are given in high doses immediately posttransplant and are usually tapered over a short period. Cyclosporine has a great deal of interpatient variability in pharmacokinetics and fluctuating bioavailability, and its metabolism is affected by several drugs (including fluoxetine) and by the status of the liver (Rodighiero, 1989). Most programs measure trough levels using one of several assays because of the extensive interpatient and intrapatient variabilities in peak time and concentrations. The therapeutic range varies according to type of transplant. Kidney transplants have the lowest requirement followed by heart and by liver with the highest requirement. Delirium is most commonly seen at high doses, although it has been reported with therapeutic doses (Lucey, Kolars, Merion, et al., 1990). Reduction of the daily dose usually corrects this problem.

Reports of corticosteroid-induced psychiatric disorders in medically ill patients range from 2% to 50% (Wolkowitz, Rubinow, Doran, et al., 1990). The incidence of psychiatric and behavioral problems increases as the dosage increases, especially beyond 40 mg/day (Boston Collaborative Drug Surveillance Program, 1972; Lewis & Smith, 1983). When these agents are suspected of being the cause of the delirium, it may only be necessary to provide supportive care while dosages of these drugs are lowered. Because transplant patients are immunosuppressed, the risk of infection is increased and

TABLE 36–3. *Medications Used for Transplantation*

Drug	$t_{1/2}$	Dosage Form	Neuropsychiatric Effects	Comments
Cyclosporine	8–24	PO, IV, IM	Anxiety, delirium Hallucinations, seizures, tremor, parasthesias, hirsutism	Nephrotoxic, hepatotoxic. Cimetidine, ketoconazole, fluoxetine, erythromycin, and metoclopramide increase serum level. Phenobarbital, phenytoin, and rifampin lower serum level. Verapamil may potentiate. Methods of assay are controversial. IM erratically absorbed.
FK506	12–24	PO, IV	Anxiety, delirium, insomnia, tremor, parasthesias	Experimental; used mainly for liver transplantation.
Muromonab-CD3 (OKT3)	Variable	IV	Aseptic meningitis, seizures, tremors, malaise, photophobia	Monoclonal antibody. Treatment of rejection episodes. Blocks T cell functions. Serum levels available.
Corticosteroids	3–4 1–2	PO IV	Delirium, euphoria, depression, mania, insomnia, tremors, irritability	Phenobarbital may decrease serum level. May increase metabolism of drugs by liver.
Azathioprine	0.10–0.25	PO, IV, IM	Meningitis-type syndrome (rare)	May cause bone marrow suppression. Use for transplantation not common.
Methotrexate	8–15	PO, IV	Headache, fever, motor dysfunction, severe demyelinating leukoencephalopathy	Leukoencephalopathy may occur at any time.
Cyclophosphamide	4 8	PO IV	Transient dizziness, nausea, vomiting	
Antibiotics	8–32	PO, IV, IM	Delirium, hallucinations, fear of impending death (usually at high doses)	Half-time depends on excretion mode.
Antiviral drugs	8–32	PO, IV, topical	Delirium, irritability, hallucinations	
Antifungal drugs	0.5–12.0	PO, IV, topical	Depression, hallucinations	

may precipitate delirium. All types of bacterial, viral, mycotic, and protozoal infections must be included in the differential diagnosis. Agents used to combat these infections can also precipitate delirium (see Table 36-3).

Treatment of delirium includes correction of the underlying medical or surgical problem, behavioral management including one-to-one nursing, orientation techniques, and the use of restraints and pharmacotherapy (see Chapter 19). Pharmacologically, high potency neuroleptics such as haloperidol should be considered. Haloperidol may be given either intravenously or intramuscularly with minimal extrapyramidal side effects and is safe, from a cardiovascular standpoint (see Chapters 9, 12 and 23). Cyclosporine and FK506, however, can cause a fine tremor that must be differentiated from extrapyramidal effects of neuroleptics (Trzepacz, Levenson, & Tringali, 1991).

Because of their anticholinergic effects, use of low-potency phenothiazines such as chlorpromazine should be limited so as not to worsen the delirium. In addition, low-potency phenothiazines can cause hypotension.

Benzodiazepines can be helpful in providing sedation, but because they can worsen delirium they should be used cautiously, with preference given to those having a shorter half-life. Benzodiazepines metabolized by conjugation (lorazepam, oxazepam, or the hypnotic temazepam) should be used for liver transplant patients if there is any question as to current hepatic function. Psychopharmacologic principles for using psychotropics in the medical-surgical patient are discussed in detail in Chapter 9.

Because delirium is frightening to the family as well as the patient, efforts should be made to educate the family regarding the delirium, enlisting its help in calming the patient when appropriate. Preoperative education about delirium may make it less frightening for both patient and family when it does occur. (See also Chapter 19.)

Depression

The reported incidence of pretransplant depression ranges from 16% to 60% (Kuhn, Davis, & Lippman, 1988; Trzepacz, Maue, Coffman, et al., 1987). Posttransplant depression may occur in as many as 80% of patients (House, Dubovsky, & Penn, 1983). Posttransplant depression can be precipitated by the onset of medical or surgical complications and from disillusionment concerning the potential outcome of the transplant.

Organic factors that need assessment include drug toxicity (see Table 36-3), infection, metabolic disturbance, graft rejection, and organ failure. Infections such as cytomegalic virus (CMV) infection can present with depression.

Psychological factors may include fear of rejection, fear of losing supports such as the hospital, inability to achieve expectations such as returning to the level of health prior to organ failure, or changes in body habitus. Before initiating antidepressant medication, a full assessment of organic factors must be made.

In general, antidepressants should not be started when depressive symptoms are secondary to a specific organic cause, and treatment for that cause is underway. If a patient continues to be depressed despite full medical assessment and remediation of known organic problems, antidepressant therapy should be reconsidered. In most instances patients can be treated with customary dosages, although serum levels should be followed for those drugs for which they are meaningful (Trzepacz, Levenson, & Tringali, 1991). When treating patients with antidepressants, caution must be exercised, as the risk for suicide has been reported to be greater than that of the general population (Gulledge, Buszta, & Montague, 1983; Washer, Schroter, & Starzl, 1983).

Tricyclic antidepressants are commonly used in the transplant population. Starting doses should be low and increased every 3 to 5 days to a therapeutic level as the patient tolerates. The dietary and drug interaction restrictions required by monamine oxidase inhibitors (MAOIs) make them more difficult to manage as antidepressants for this particular population. In addition, the combination of meperidine and MAOIs can precipitate a hypertensive crisis.

Fluoxetine has been little used to date because of its long half-life and the lack of meaningful, validated serum levels. However, the availability of a liquid form makes it possible to initiate it at a lower dose, and it is likely to see increased use in the future. The increased incidence of seizures associated with bupropion make it less desirable to use in conjunction with cyclosporine, which may also precipitate seizure activity (de Groen, 1989).

For those patients too debilitated to tolerate traditional antidepressant therapy, such as those in an intensive care unit (ICU), methylphenidate and dextroamphetamine may be considered (Kaufmann, Cassem, Murray, et al., 1984; Masand, Pickett, & Murray, 1991). These drugs are relatively safe when given in certain doses: 10 to 60 mg methylphenidate per day or 5 to 30 mg dextroamphetamine per day. Insomnia and agitation are the main side effects. Tachycardia and hypertension are unusual at these doses, even in heart transplant patients. They can precipitate manic episodes in patients with histories of bipolar disorder.

Most patients experience episodes of depression that

do not require pharmacologic treatment as they attempt to adjust to the demands of being a transplant patient. Medical and surgical complications and setbacks often result in depressions that are situational and transitory in nature. Such setbacks challenge the patients' fantasies that they will have an uneventful course—and hence make real the hazards of transplantation. Death of a fellow patient on the unit can precipitate a depression, so it is frequently dealt with by denial. Depression may also be precipitated by the first rejection episode, when the uncertainty of life is again experienced. For some, the realization that they will not regain their previous state of health, social status, financial status, or job position may result in discouragement and depression.

Body image or physical changes such as those induced by corticosteroids, the weight gain, hirsuitism, and hand tremor produced by cyclosporine, and surgical scarring are a concern for many patients, especially the young. Body image is a unique concern among transplant patients in that a foreign body part is added—in contrast to most surgical procedures in which there may be a change in or loss of a body part. This reaction may be affected by the degree of importance the patient imparts to the organ, such as the heart being the seat of the soul (Basch, 1973). House, Dubovsky, and Penn (1983) noted that questions of body image were voiced earlier post-operatively by renal transplant patients than by liver transplant patients. Having a backup system, dialysis, allows them more freedom to invest time and energy in their appearance. In contrast, liver transplant patients were more concerned about whether the transplant would succeed.

Patients need to know that such reactions are common. Brief, supportive psychotherapy, including grief work focusing on losses, can be helpful. Group meetings can also help patients learn that their concerns are shared and give them a chance to learn how others cope with similar problems. Such interventions can be supplemented with pharmacologic treatment when adjustment disorders are persistent and disabling despite psychosocial treatment.

Anxiety

Along with depression, anxiety in the form of adjustment disorders, generalized anxiety, and panic attacks are common. Its prevalence may range from 14% to 39% in the transplant population (Mai, 1986; Trzepacz, Maue, Coffman, et al., 1987). Drug toxicity, especially from cyclosporine (see Table 36-3), cardiac arrhythmias, hyperthyroidism, hypoglycemia, hypocalcemia, and hypoxia can precipitate anxiety and must be assessed before initiating anxiolytic therapy. Medical and surgical setbacks, long waits for a donor, waiting for laboratory reports and biopsies, and transitions such as leaving the hospital are also precipitants.

Because of their effectiveness and the paucity of significant drug interactions, benzodiazepines are safe drugs to use in transplant patients. Unless sedation is being sought, relatively low doses are recommended. The short half-lives of conjugated drugs such as lorazepam and oxazepam allow for rapid regulation according to the patient's clinical condition. Lorazepam also offers the convenience of intramuscular, intravenous, and even sublingual administration. Midazolam can be given intramuscularly or intravenously and has a rapid onset of action. For procedure-associated anxiety, midazolam can be useful for its sedating and amnestic effect. Alprazolam or clonazepam may be used for treatment of panic attacks, although abrupt discontinuation should be avoided (Trzepacz, Levenson, & Tringali, 1991).

Buspirone is a nonbenzodiazepine anxiolytic that has received little attention in the transplant literature. Potential advantages include a lack of respiratory suppression and no sedative effects. Because of its extensive first-pass-metabolism, initial dosage should be low in liver transplant patients.

Anxiety is a common response to such uncertainties as a lack of control, awaiting a donor, or fear of the unknown such as the result of a biopsy. Supportive psychotherapy and adjunctive therapies such as biofeedback, imaging, and relaxation training offer nonpharmacologic treatment and may be useful for helping patients regain a sense of control. Support groups can help by identifying possible coping mechanisms and by allowing patients to know they are not alone with their concerns.

Psychosis

The treatment of psychosis, including mania, requires rapid intervention to minimize the risk to the patient and transplant through noncompliance. Often these patients must be treated on the surgical unit but require close behavioral supervision and frequent medication monitoring and dosage adjustment. Psychosis can be precipitated by drug toxicity (see Table 36-3), infection, or electrolyte imbalance; it may represent an exacerbation of an existing psychotic condition; or it may be a manifestation of a secondary mental disorder. Drug-induced psychotic reactions are common. Immediately after transplant such reactions may be secondary to the high doses of immunosuppressants, which are quickly tapered during this period. Steroids, in particular, may precipitate episodes of mania. When initiating neuroleptics, low-dose, high-potency agents such as haloperidol are preferable. To avoid worsening a possible organic component, the more-anticholinergic agents such as chlorpromazine should be avoided.

Lithium is a safe drug to use to control mania but requires frequent monitoring. Because it is excreted by the kidney, renal function must be monitored closely, as must electrolyte levels. Cyclosporine has been found to increase reabsorption of lithium at the proximal tubules (Vincent, Wenting, Schalekamp, et al., 1987). For those patients unresponsive to lithium, carbamazepine or valproate should be considered. However, both are associated with hepatic toxicity (Trzepacz, Levenson, Tringali, 1991).

Some patients present with an exaggerated sense of optimism, euphoria, or agitation immediately postoperatively. This hypomanic presentation may, in part, be a response to prednisone. Such patients often respond to limit-setting and behavioral interventions. Moreover, this reaction is often short-lived, and antimanic medications are rarely needed. Clonazepam might be considered if insomnia were a major problem.

Patients with a more paranoid presentation may respond in a more threatening or aggressive manner. Delirium and cyclosporine toxicity must be considered in the differential diagnosis (Craven, 1991). Consistency in caregivers, limiting the number of people presenting information, and allowing some distance may give the patient more of a sense of control. Reassurance and explanation to the patient about what is happening can also be of help. Staff members must be cautioned as to the possibility of violence and be aware of the appropriate actions and medications to be used if needed. The risk of psychosis is greatest when the patient has included the transplanted organ or the immunosuppressant medications in their paranoid delusional system.

Alcohol Abuse

Despite careful screening and pretransplant counseling, some alcoholic patients suffer a relapse of alcohol abuse after transplantation. This possibility might lead to the consideration of using disulfiram as an aid to abstinence. In this connection, it is important to know that the liquid preparation of cyclosporine contains alcohol and cannot be combined with disulfiram. Gelatin capsules of cyclosporine, however, do not contain alcohol and can be safely given safely with disulfiram.

HEART

Few organs arouse as much emotion as the heart. Lunde (1969) noted that the heart is not only an organ absolutely essential for life but is often viewed as the center for love, loyalty, and other emotions. Heart disease affects nearly 20% of the U.S. population (Kannel & Thom, 1985). Of these patients, nearly 1700 (see Table 36-1) received a heart transplant during 1989. Because of the scarcity of donors, 20% of candidates die before a donor heart becomes available (Surman, 1989).

Heart transplantation is indicated for severe heart failure. The most common causes are coronary artery disease, viral cardiomyopathy, idiopathic cardiomyopathy, hypertensive cardiomyopathy, alcoholic cardiomyopathy, and congenital malformations. Possible contraindications include severe peripheral vascular disease, diabetes mellitus, systemic lupus erythematosus, amyloidosis, or other irreversible organ diseases (Schroeder & Hunt, 1987). Candidates for heart transplants are often in class IV heart failure. As such, they may manifest fatigue, insomnia, shortness of breath, anorexia, and cognitive deficits, especially as the ejection fraction drops below 20% (Trzepacz, Levenson & Tringali, 1991). As their condition worsens, assist devices including pacemakers, balloon pumps, and, in rare instances, mechanical hearts may be required. With each comes the risk of neuropsychiatric and psychological complications including seizures, strokes, delirium, anxiety, panic disorders, and depression. Continued deterioration in health prior to transplant increases morbidity and significantly delays the patient's return to work and to other activities (Caine, Sharples, English, et al., 1990).

Cardiac transplantation, like other types of cardiac surgery, is associated with a high incidence of neuropsychiatric complications. The incidence of postcardiotomy delirium may be as high as 32% for major cardiac surgeries, beginning 2 to 5 days after surgery (Smith & Dimsdale, 1989). In a retrospective study by Hotson and Pedley (1976), 54% of patients with heart transplantation had at least one neurologic complication and 20% died of this complication. Vascular complications accounted for 60% of cerebral lesions in a postmortem study by Montero and Martinez (1986). Chronic hypotension preoperatively followed by episodes of hypertension postoperatively may impede cerebral autoregulation, resulting in infarcts, emboli, and hemorrhages (Montero & Martinez, 1986). These problems may present as a psychiatric condition, ranging from a mood disorder to psychosis, and the first task of treatment is to exclude a neurologic or infectious component.

Depression and anxiety are common occurrences both before and after heart transplant. Anxiety and depression in the absence of a psychiatric history seem related to the severity of the cardiac illness (Kuhn, Davis, & Lippman, 1988). Advanced cardiac dysfunction can precipitate symptoms of depression that disappear after a successful transplant. Preoperative symptoms do not predict postoperative psychiatric complications (Kuhn, Davis, & Lippman, 1988; Mai, 1986).

Tricyclic antidepressants must be used carefully before transplant to avoid adversely affecting cardiac conduction, cardiac output, and blood pressure (Trzepacz, Levenson, & Tringali, 1991). Posttransplant, patients may not experience the changes in heart rate that normally accompany pain or a change in affect because the heart is denervated. Tricyclics such as nortriptyline and desipramine have the advantage that they have established therapeutic serum level ranges, as well as having a relatively low incidence of cardiac side effects (Kay, Bienenfeld, Slomowitz, et al., 1991; Roose, Glassman, Giardina, et al., 1991). Nortriptyline has been associated with increased QRS intervals and heart rate, but it does not significantly alter other hemodynamic variables (Shapiro, 1991). Maprotiline and trazodone are also reported to be less cardiotoxic (Watts, Freeman, McGriffin, et al., 1984). Dosages should be titrated and determined by clinical response. (See also Chapters 9 and 23.)

Psychodynamic Factors

The incorporation of a new heart as a part of self may be more difficult for some because of the emotion attached to it. The physical sensation of the beating heart may serve as a reminder of the transplanted organ. Some recipients experience "survival guilt" that their continuing to live was dependent on the death of someone else. Most are curious about the donor and wonder if they themselves will take on characteristics of the donor's personality as a result of the transplant. Others become regressed to the point of fearing their new heart will be reclaimed by the donor (Castelnuovo-Tedesco, 1970). In the heart transplant patient, denial is the most common coping mechanism (Mai, 1986). Some refer to their heart as a worn-out part, others as just a pump; and others become excessively protective of the organ (Tabler & Frierson, 1990). These attitudes seem to help patients distance themselves from the reality of transplantation and help in the integration process. Fears of annhilation, mutilation, and regression may interfere with the patient's ability to incorporate the transplanted heart as a part of self. Exploring patients' dreams and fantasies about the transplant help to identify some of these concerns.

Some personality traits may become more pronounced postoperatively, such as obsessive-compulsive behavior in which a patient may fear rejection if medication is not given precisely on time. Patients with a more dependent style may become more demanding and clinging toward the transplant team and family. Such reactions are more likely around rejection episodes or at the time of discharge from the hospital.

Quality of Life

Despite the many complications with which heart transplant patients are faced, most believe their quality of life is improved with transplantation. O'Brien, Buxton, and Ferguson (1987) in Great Britain suggested that 90 to 92% of heart transplant patients were free from quality of life restrictions as soon as 3 months after surgery. Similar studies in the United States found 80 to 89% of heart transplant recipients reporting good to excellent quality of life despite experiencing periodic side effects from immunosuppressant therapy (Lough, Lindsey, Shinn, et al., 1985). Other concerns commonly expressed include role relationships, sexual function, and activity-exercise patterns (Hyler, Corley, & McMahon, 1985). Concern about the ability to function sexually affects up to 48% of patients (Tabler & Frierson, 1990). Common problems include decreased libido, ejaculatory problems, impotence, and anorgasmia. Patients with difficulty incorporating the new heart are uncertain as to whether their heart can tolerate the stress of sexual activity. Similarly, spouses who are used to protecting the patient before transplant may avoid intercourse after transplant as an expression of concern about sexual activity on the new heart. Evaluation and treatment of sexual dysfunction may help improve the quality of life of patients after the transplant. (See also Chapter 15.)

LIVER

Liver transplantation is indicated for severe liver failure. Common indications include primary biliary cirrhosis, primary sclerosing cholangitis, chronic active hepatitis, inborn metabolic errors such as α_1-antitrypsin deficiency and Wilson's disease, and congenital malformations such as biliary atresia. As mentioned, transplantation for alcoholic cirrhosis is controversial, as is transplantation for cancer. Patients with tumors confined to the liver and no evidence of metastasis are considered at some centers. However, most such tumors recur, and the results have been disappointing.

Liver failure results in complex physiologic problems, including loss of muscle mass and compromised kidney, heart, pulmonary, and cognitive function. Cognitive dysfunction ranges from subtle deficits to hepatic coma. In addition, patients experience coagulopathies, variceal bleeding, jaundice, poor nutrition, and debilitating fluid retention.

Liver transplants are thought to have a lower incidence of rejection than other solid organ transplants (Williams & Sabesin, 1990). Others (Starzl, Demetris, & Van Thiel, 1989) have reported that histopathologic evidence of rejection can be found in two-thirds of patients. Unique to liver transplantation is the liver's ability to regenerate and regulate its size according to the need of the patient. This point is particularly important in children, who may receive a lobe of an adult liver, rather than the entire organ.

Hepatic encephalopathy is common prior to transplant. CT or MRI scans of the head in patients with advanced liver disease but with no overt signs of encephalopathy may reveal the presence of cortical and subcortical atrophy and edema (Moore, Dun, Crawford, et al., 1989). It can also be seen after transplant. On neuropsychological testing, the severity of impairment correlated with the magnitude of these morphologic changes (Bernthal, Hays, Tarter, et al., 1987). Moreover, successful transplantation does not completely reverse this impairment. At present, no guidelines exist for predicting from CT or MRI findings whether cognitive impairment resolves after transplantation. This impairment may be secondary to the neurologic sequelae of the existing brain pathology or the neurotoxic effects of cyclosporine (Tarter, Switla, Arria, et al., 1990). The prevalence of delirium before transplant has been noted to range from 16% to 30% (Adams, Ponsford, Gurson et al., 1987; Trzepacz, Brenner, Coffman, et al., 1988), whereas during the posttransplant period the incidence may run as high as 50% (House, Dubovsky, & Penn, 1983).

Graft failure and neurotoxicity secondary to cyclosporine must be considered when assessing delirium. Generally, liver transplants require higher doses of cyclosprine, and 25 to 35% may have some degree of neurotoxicity. Persistent drowsiness or somnolence may be early indicators of a neurotoxic complication (Craven, 1991). Factors associated with elevated levels of cyclosporine include high doses of prednisone, hypertension, hypomagnesemia, and hypocholesterolemia (de Groen, 1989). Cyclosporine is metabolized in the liver, and any impairment increases the cyclosporine levels. Cyclosporine has been associated with seizures in 12 to 25% of patients. Other factors contributing to seizures are metabolic encephalopathy and cerebral infarcts (Plevak, Southorn, Narr, et al., 1989). Phenytoin can be used to control cyclosporine-induced seizures. Barbiturates can increase the metabolism of cyclosporine, as can phenytoin. Carbamazepine has a potential for liver toxicity, and its use for cyclosporine-induced seizures require frequent monitoring of blood levels and liver functions. (In fact, due to its potential hepatotoxicity the unavailability of a parenteral form, carbamazepine is rarely used for this purpose.) Following an episode of toxicity, the cyclosporine may need to be withheld and restarted slowly once the levels have dropped.

The high immunosuppressant requirement makes infection a continual problem. Cytomegalovirus is the most common viral infection and the most common cause of postoperative hepatitis (Starzl, Demetris, & Van Thiel, 1989).

Liver transplant patients experience psychological issues that are common to other transplant patients. Pretransplant, however, the physical changes caused by their illness are often more evident than those seen with heart, pancreas, and bone marrow transplants. Distended abdomens, jaundice, and large ecchymotic skin changes are common. Posttransplant, the patients are particularly concerned with organ rejection and dying, like the heart transplant patients. Unlike kidney or pancreas transplants, there is no backup should the liver reject short of retransplantation (House, Dubovsky, & Penn, 1983). Therefore complications and setbacks often precipitate episodes of increased anxiety and depression. For those patients whose liver failure has been due to hepatitis, the fear of recurrence is ever present.

Posttransplant concerns include changes in appearance secondary to medication as well as the surgical scarring on the abdomen, slowness of the recovery process, and difficulty internalizing a new liver. The positive thinking that is present pretransplant may give way to fears and anxieties that were not acceptable prior to the operation (Heyink, Tymstra, Slooff, et al., 1990).

BONE MARROW

Bone marrow is a fundamental organ, the psychological impact of which can be seen in the expression: "I feel it to the marrow of my bones." Bone marrow transplant (BMT) is indicated for malignant as well as nonmalignant hematologic disorders. Malignant disorders include preleukemia, acute lymphoblastic leukemia, acute myelogenous leukemia, chronic myelogenous leukemia, and non-Hodgkin's lymphoma. Nonmalignant disorders include severe aplastic anemia, severe combined immunodeficiency syndrome, thalessemia major, and Fanconi's anemia.

Unlike solid organ transplant in which the goal is to replace a diseased organ and suppress the recipient's immune system, with BMT the goal is to eliminate the recipient's immune response and the disease response.

Bone marrow transplants are either autogenic or allogenic.

With autogenic BMT, the patient's own marrow is destroyed by chemotherapy or irradiation after marrow has first been aspirated and cryopreserved. The aspirated marrow is treated to remove tumor cells, then reinfused at a later date. Its advantage is avoiding graft versus host disease (GVHD); the risk arises from incomplete removal of tumor cells.

With allogenic bone marrow transplants the patient's own marrow is destroyed and donor marrow infused. The advantage is the marrow is without tumor; the disadvantages are the limited availability of HLA-matched donors and the possibility of GVHD. Additional risks for both methods include failure or delayed engrafting of the marrow and damage to other organ systems by the chemicals or radiation given for marrow ablation. Both allogenic and autogenic bone marrow transplants appear to have similar risks for psychosocial complications (Jenkins, Linington, & Whitaker, 1991).

Patients with cancer have increased psychiatric morbidity (see Chapter 24), and transplant may be viewed as their last chance for survival. Brown and Kelly (1976) identified eight stages in the BMT process (see Table 36-4). During the immunosuppression phase, patients are treated with high doses of cytotoxic agents such as *methotrexate* or *cyclophosphamide* or total body irradiation (or both methods are used) to destroy the recipient's hematopoietic tissue. Cytotoxic agents vary in degree of neurotoxicity, with methotrexate being the most likely to cause CNS impairment. Neurotoxic effects include confusion, delirium, and leukoencephalopathy.

Leukoencephalopathy is seen in about 7% of patients with acute lymphoblastic leukemia who have undergone cranial or intrathecal chemotherapy before preparation for bone marrow transplant. Symptoms include slurred speech, lethargy, seizures, confusion, and dementia. Often these symptoms show up within the first month after transplant, although they may show up much later. Leukoencephalopathy is a degenerative process for which there is no effective treatment (Klingmann, 1988).

TABLE 36–4. *Bone Marrow Transplant Stages*

Stage
Stage 1—Decision to accept treatment
Stage 2—Preparation
Stage 3– Immunosuppression
Stage 4—Transplantation of bone marrow
Stage 5—Waiting for graft rejection or take
Stage 6—Graft versus host disease (GVHD)
Stage 7—Predischarge
Stage 8—Adaptation

Total body irradiation is associated with an acute confusional state and decreased concentration; in children, long-term cognitive deficits may result (Jenkins, Linington, & Whittaker, 1991). Long-term effects of total body irradiation include primary ovarian failure, azospermia, and delayed development of secondary sexual characteristics in children. This period of aplasia leaves the patient susceptible to opportunistic infection, particularly *Pneumocystis carinii* and cytomegalic virus, which is a leading cause of death immediately after transplant.

The leading cause of death posttransplant for autogenic BMT is organ toxicity secondary to medications involving the lungs, liver, and heart. For allogenic BMT the leading cause of death is GVHD followed by infection. Pulmonary infection occurs in up to 50% of patients, with a mortality as high as 50% (Sullivan, Meyers, & Flournoy, 1986). To decrease this risk, patients are isolated in laminar flow units and treated with antibiotics. They may spend 6 weeks or more in these units during which they may experience environmental isolation, disorientation, depression, anxiety, and delirium.

Regression in response to pain, anxiety, and stress are common and may precipitate behavioral problems involving a wide range of transference and countertransference reactions (Futterman & Wellisch, 1990). To regain control, patients may threaten to leave the isolation unit. During this phase relaxation techniques may be helpful (Brown & Kelly, 1976). The physical restrictions of the BMT unit and the BMT process tend to generate dependence on the staff. The isolation period may bring up ambivalence about the transplant by both patient and spouse. In contrast to the initial enthusiasm about a possible "cure," families may become frightened of possibility of death once they see how ill the patient has become. Some respond to this sense of helplessness by becoming overwhelmed and depressed, whereas others respond in a more aggressive way, trying to assume control over decisions (Futterman & Wellisch, 1990).

During the first 10 days after transplant little may happen, after which complications may begin to develop. Complications may include failure of the transplant to engraft, infection, leukemic relapse, and GVHD. GVHD in these patients can be especially devastating. Because the patient is immunosuppressed, the T lymphocytes may first attack the patient's skin, liver, and gastrointestinal tract as well as causing pancytopenia. Initial treatment may begin with high-dose steroids. A combination of methotrexate and cyclosporine may reduce the severity of GVHD (Storb, Deeg, Whitehead, et al., 1986).

Prolonged isolation, anemia, and GVHD can precipitate depression. At some point 40 to 50% of BMT patients meet criteria for major depression (Jenkins, Linington, & Whittaker, 1991). During this time families feel helpless, and the donor may feel guilty about having caused this reaction (Brown & Kelly, 1976). In general, the use of tricyclic antidepressants in these patients is safe, although there is a possible remote risk of agranulocytosis. In patients with GVHD involving the liver, lower doses of tricyclic antidepressants may be necessitated, as they are hepatically metabolized.

Chronic GVHD affects 30 to 50% of long-term survivors and may appear 3 to 6 months after allogenic BMT. These patients may experience chronic scleroderma, liver disease, infections, and eye and mouth disorders. Currently, immunotoxin is being used to treat GVHD that fails to respond to steroid therapy (Wikle, Coyle, & Shapiro, 1990). Complications with autogenic BMT are similar but occur less often. Personal control is often a central issue and may be seen in conflicts with nursing staff and family over privacy and treatment plans. Death of a patient on the unit also elicits survivor guilt.

PANCREAS

Pancreas transplantation is indicated for treatment of insulin-requiring diabetes and is still considered experimental. The endocrine pancreas may be transplanted as a vascularized graft or as free isolated islets. Islet cell transplantation has met with limited success because of (1) the difficulty of isolating islets from cadaver donors and (2) subsequent rejection.

Whole-organ transplants offer normal homeostatic control that cannot be obtained through insulin injections alone. The goal of transplantation is to provide euglycemia and prevent or reverse nephropathy, neuropathy, and retinopathy. Improvements in these complications have been noted if they are caught at an early stage (Bently, Jung, & Garrison, 1990; Soon-Shiong & Lanza, 1990). Whole-organ transplantation is most often done in combination with kidney transplantation for patients with end-stage nephropathy. A functioning pancreas transplant may prevent development of diabetic changes in the transplanted kidney (Bohman, Tyden, Wilčzek, et al., 1985). Patients with severe neurovascular complications are not ideal candidates because of the risk for cerebral vascular accidents. Problems with whole-organ transplant include venous thrombosis and lack of markers for the early diagnosis of rejection. Both prednisone and cyclosporine may have an unfavorable effect on glucose homeostasis.

Diabetic patients have had to learn to adapt to chronic illness, loss of independence, personal, social, and dietary restrictions, and in most cases organ failure, especially of the kidney and eye. Their quality of life decreases substantially with the onset of retinopathy, neuropathy, and nephropathy. Therefore the timing of transplantation becomes the crucial factor. This factor is made difficult, as there is no clear marker to indicate which diabetics will go on to develop severe secondary complications (Bentley, Jung, & Garrison, 1990). Patients with functioning grafts have viewed themselves as happier (93%) and healthier (93%) than before transplant (Zehrer & Gross, 1990). Most thought that the maintenance demands of their diabetes before the transplant had been greater than the demands of the transplant (Zehrer & Gross, 1990). Most pancreas transplants are done in conjunction with renal transplants. The psychological and psychiatric complications are similar to those described for renal transplants. (See Chapter 27.)

REFERENCES

ADAMS DH, PONSFORD S, GURSON B, ET AL. (1987). Neurological complications following liver transplantation. Lancet 1:949–951.

ALLEN V. MANSOUR (1986). 86-73429 (District Court for the Eastern District of Michigan, Southern Division).

BASCH SH (1973). The intrapsychic integration of a new organ: A clinical study of kidney transplantation. Psychoanal Q 42:364–384.

BENTLY FR, JUNG S, & GARRISON RN (1990). Neuropathy and psychosocial adjustment after pancreas transplant in diabetics. Transplant Proc 22:691–695.

BERESFORD TP, TURCOTTE JG, MERION R, ET AL. (1990). A rational approach to liver transplantation for the alcoholic patient. Psychosomatics 31:241–254.

BERNTHAL P, HAYS A, TARTER RE, ET AL. (1987). Cerebral CT scan abnormalities in cholestatic and hepatocellular disease and their relationship to neuropsychologic test performance. Hepatology 7:107–114.

BOHMAN SO, TYDEN G, WILCZEK H, ET AL. (1985). Prevention of kidney graft diabetic nephropathy by pancreas transplantation in man. Diabetes 34:306–308.

Boston Collaborative Drug Surveillance Program (1972). Acute adverse reactions to prednisone in relation to dosage. Clin Pharmacol Ther 13:694–698.

BROWN HN, & KELLY MJ (1976). Stages of bone marrow transplantation: a psychiatric perspective. Psychosom Med 386:439–446.

BUSUTTIL RW (1991). Living-related liver donation. Transplant Proc 23:43–45.

CAINE N, SHARPLES LD, ENGLISH TA, ET AL. (1990). Prospective study comparing quality of life before and after heart transplantation. Transplant Proc 2:1437–1439.

CASTELNUOVO-TEDESCO P (1970). Psychoanalytic considerations in a case of cardiac transplantation. In S ARIETI (Ed.), The world biennial of psychiatry and psychotherapy (Vol 1; Ch. 15, pp. 336–352). New York: Basic Books.

CASTELNUOVO-TEDESCO P (1978). Ego vicissitudes in response to replacement or loss of body parts. Psychoanal Q 47:381–497.

CHRISTOPHERSON LK, & LUNDE DT (1971). Selection of cardiac transplant recipients and their subsequent psychological adjustment. In P CASTELNUOVO-TEDESCO (Ed.), Psychiatric aspects of organ transplantation (Ch. 4, pp. 36–45). Orlando, FL: Grune & Stratton.

COHEN C, BENJAMIN M, and the Ethics and Social Impact Committee of the Transplant and Health Policy Center, Ann Arbor, Mich. (1991). Alcoholics and liver transplantation. JAMA 265:1299–1301.

CRAVEN JL (1991). Cyclosporine-associated organic mental disorders in liver transplant recipients. Psychosomatics 32:92–102.

DE GROEN PC (1989). Cyclosporine: a review and its specific use in liver transplantation. Mayo Clin Proc 64:680–689.

DUBOVSKY SL, & PENN I (1980). Psychiatric considerations in renal transplant surgery. Psychosomatics 21:481–491.

FREEMAN AM, VOLKS DG, & SOKOL RS (1988). Cardiac transplantation: clinical correlates of psychiatric outcome. Psychosomatics 29:47–54.

FUTTERMAN AD, & WELLISCH DK (1990). Psychodynamic themes of bone marrow transplantation. Hematol Oncol Clin North Am 4:699–709.

GULLEDGE AD, BUSZTA KC, & MONTAGUE DK (1983). Psychological aspect of renal transplantation. Urol Clin North Am 10:327–335.

HEYINK J, TYMSTRA T, SLOOFF MJH, ET AL. (1990). Liver transplantation—psychological problems following the operation. Transplantation 49:1018–1019.

HOTSON JR, & PEDLEY TA (1976). The neurological complications of cardiac transplantation. Brain 99:673–694.

HOUSE RM, DUBOVSKY SL, & PENN I (1983). Psychiatric aspects of hepatic transplantation. Transplantation 36:146–150.

HOUSE RM, & THOMPSON TL (1988). Psychiatric aspects of organ transplantation. JAMA 260:535–539.

HYLER BJ, CORLEY MC, & MCMAHON D (1985). Role of nursing in a support group for heart transplantation recipients and their families. Heart Transplant 4:453–456.

JENKINS PL, LININGTON A, & WHITTAKER JA (1991). A retrospective study of psychosocial morbidity in bone marrow transplant recipients. Psychosomatics 32:65–71.

KANNEL WB, & THOM TJ (1985). Incidence, prevalence, and mortality of cardiovascular diseases. In JW HURST, RB LOGUE, CE RACKLEY, ET AL. (Eds.), The heart, arteries, and veins (6th ed., Ch. 34, pp. 557–565). New York: McGraw-Hill.

KAUFMANN MW, CASSEM N, MURRAY G, ET AL. (1984). The use of methylphenidate in depressed patients after cardiac surgery. J Clin Psychiatry 45:82–84.

KAY J, BIENENFELD D, SLOMOWITZ M, ET AL. (1991). Use of tricyclic antidepressants in recipients of heart transplants. Psychosomatics 32:165–170.

KLINGMANN HG (1988). Central nervous system. In HJ DEEG, HG KLINGMANN, & GL PHILLIPS (Eds.), Guide to bone marrow transplantation (Ch. 8, pp. 140–148). Berlin: Springer-Verlag.

KUHN WF, BRENNAN AF, LACEFIELD PK, ET AL. (1990). Psychiatric distress during stages of the heart transplant protocol. J Heart Transplant 9:25–29.

KUHN WF, DAVIS MH, & LIPPMAN SB (1988). Emotional adjustment to cardiac transplantation. Gen Hosp Psychiatry 10:108–113.

LEVENSON JL, & OLBRISCH ME (1987). Stress of waiting or a donor organ. Psychosomatics 28:399–403.

LEWIS OA, & SMITH RE (1983). Steroid-induced psychiatric syndromes. J Affect Disord 5:319–332.

LOUGH ME, LINDSEY AM, SHINN JA, ET AL. (1985). Life satisfaction following heart transplantation. Heart Transplant 4:446–449.

LUCEY MR, KOLARS JC, MERION RM, ET AL. (1990). Cyclosporin toxicity at therapeutic blood levels and cytochrome P450 IIIA. Lancet 335:11–15.

LUNDE D (1969). Psychiatric complications of heart transplants. Am J Psychiatry 126:369–373.

MAI FM (1986). Graft and donor denial in heart transplant recipients. Am J Psychiatry 143:1159–1161.

MASAND P, PICKETT P, & MURRAY GB (1991). Psychostimulants for secondary depression in medical illness. Psychosomatics 32:203–208.

MERRIKIN KJ, & OVERCAST TD (1985). Patient selection for heart transplantation: when is discrimination choice discrimination? J Health Polit Policy Law 10:7–32.

MONTERO CG, & MARTINEZ AJ (1986). Neuropathology of heart transplantation: 23 cases. Neurology 36:1149–1154.

MOORE JW, DUN KAA, CRAWFORD H, ET AL. (1989). Neuropsychological deficits & morphological MRI brain scan abnormalities in apparently healthy non-encephalopathic patients with cirrhosis. J Hepatol 9:319–325.

MORENO E, GARCIA I, LOINAZ IG, ET AL. (1991). Reduced-size liver transplantation in children and adults. Transplant Proc 23:1953.

MOSS AH, & SIEGLER M (1991). Should alcoholics compete equally for liver transplantation? JAMA 265:1295–1298.

O'BRIEN I, BUXTON MJ, & FERGUSON BA (1987). Measuring the effectiveness of heart transplant programmes: Quality of life data and their relationship to survival analysis. J Chronic Dis 4:1375–1585.

PETERS JC, ALPERT M, BEITMAN BD, ET AL. (1990). Panic disorder associated with permanent pacemaker implantation. Psychosomatics 31:345–347.

PLEVAK OJ, SOUTHORN PA, NARR BJ, ET AL. (1989). Intensive-care unit experience in the Mayo liver transplantation program: the first 100 cases. Mayo Clin Proc 64:433–445.

RIETHER AM, & STOUDEMIRE A (1987). Surgery and trauma. In A STOUDEMIRE & BS FOGEL (Eds.). Principles of Medical Psychiatry (Ch. 19, pp. 423–449). Orlando, FL: Grune & Stratton.

RODIGHIERO V (1989). Therapeutic drug monitoring of cyclosporin. Clin Pharmacokinet 16:27–37.

ROOSE SP, GLASSMAN AH, GIARDINA EG, ET AL. (1987). Tricyclic antidepressants in depressed patients with cardiac conduction disease. Arch Gen Psychiatry 44:273–275.

SCHENKER S (1984). Alcoholic liver disease: evaluation of natural history and prognostic factors. Hepatology 4:36–43.

SCHROEDER JS & HUNT S (1987). Cardiac transplantation update 1987. JAMA 258:3142–3145.

SHAPIRO PA (1991). Nortriptyline treatment of depressed cardiac transplant recipients. Am J Psychiatry 48:371–373.

SIMMONS RG, HICKEY K, KJELLSTRAND CM, ET AL. (1971). Donors and non-donors: The role of the family and the physician in kidney transplantation. In CASTELNUOVA-TEDESCO P (Ed.). Psychiatric aspects of organ transplantation (pp. 102–115). New York: Grune & Stratton.

SMITH LW, & DIMSDALE JE (1989). Postcardiotomy delirium: conclusions after 25 years? Am J Psychiatry 146:452–458.

SOON-SHIONG P, & LANZA RP (1990). Pancreas and islet-cell transplantation: potential cure for diabetes. Postgrad Med 8:133–140.

SOPHIE LR, SALLOWAY JC, SORACK G, ET AL. (1983). Intensive care nurses' perception of cadaver organ procurement. Heart Lung 12:261–267.

STARZL TE, DEMETRIS AJ, & VAN THIEL D (1989). Liver transplantation. N Engl J Med 321:1092–1099.

STORB R, DEEG HJ, WHITEHEAD J, ET AL. (1986). Methotrexate and cyclosporine compared with cyclosporine alone for prophylaxis of acute graft versus heart disease after marrow transplantation for leukemia. N Engl J Med 314:729–735.

SULLIVAN KM (1986). Acute and chronic graft-versus-heart disease in man. Int J Cell Cloning 4:42–93.

SULLIVAN KM, MEYERS JD, FLOURNOY N, ET AL. (1986). Early and late interstitial pneumonia following human bone marrow transplantation. Int J Cell Cloning 4:107–121.

SURMAN OS (1987). Hemodialysis and renal transplantation. In TP HACKETT & NH CASSEM (Eds.), Massachusetts General handbook of general hospital psychiatry (Ch. 19, pp. 380–402). St. Louis: Mosby.

SURMAN OS (1989). Psychiatric aspects of organ transplantation. Am J Psychiatry 146:972–982.

SUTHERLAND DER, GOETZ FC, & NAJARIAN JS (1984). Pancreas transplants from related donors. Transplantation 38:625–633.

TABLER JB, & FRIERSON RL (1990). Sexual concerns after heart transplantation. J Heart Transplant 9:397–403.

TARTER RE, SWITLA J, ARRIA A, ET AL. (1990). Subclinical hepatic encephalopathy. Transplantation 50:632–637.

TRZEPACZ PT, BRENNER R, COFFMAN G, ET AL. (1988). Delirium in liver transplantation candidates: discriminant analysis of multiple test variables. Biol Psychiatry 24:3–14.

TRZEPACZ PT, LEVENSON JL, & TRINGALI RA (1991). Psychopharmacology and neuropsychiatric syndromes in organ transplantation. Gen Hosp Psychiatry 13:233–245.

TRZEPACZ PT, MAUE FR, COFFMAN G, ET AL. (1987). Neuropsychiatric assessment of liver transplantation candidates: delirium and other psychiatric disorders. Int J Psychiatry Med 16:101–111.

VINCENT HH, WENTING GJ, SCHALEKAMP MADH, ET AL. (1987). Impaired fractional excretion of lithium: a very early marker of cyclosporine nephrotoxicity. Transplant Proc 19:4147–4148.

WASHER G, SCHROTER G, & STARZL TE (1983). Causes of death after kidney transplantation. JAMA 250:49–58.

WATTS D, FREEMAN AM, MCGRIFFIN D, ET AL. (1984). Psychiatric aspects of cardiac transplantation. Heart Transplant 3:243–247.

WIKLE T, COYLE K, & SHAPIRO D (1990). Bone marrow transplant. Am J Nurs 90:48–56.

WILLIAMS JW, & SABESIN SM (1990). Liver transplantation. Postgrad Med 87:191–207.

WOLCOTT DL, WELLISCH DK, FAWZY FI, ET AL. (1986). Psychological adjustment of adult bone marrow transplant donors where recipient survives. Transplantation 41:484–488.

WOLKOWITZ OM, RUBINOW D, DORAN AR, ET AL. (1990). Prednisone effects on neurochemistry and behavior. Arch Gen Psychiatry 47:963–968.

ZEHRER CL, & GROSS CR (1990). Quality of life after pancreas transplant. Diab Care 13:539–540.

37 Otolaryngology

HAROLD BRONHEIM, M.D., JAMES J. STRAIN, M.D., AND HUGH F. BILLER, M.D.

Ear, nose, and throat (ENT) disorders are common, and surgery for them is performed in hospital settings throughout the United States. The appropriate psychiatric evaluation and treatment of this patient population has been limited partly because of the difficulties in communication encountered in those patients with laryngectomy or tracheostomy. Although otolaryngology patients undergo severe stress, it occurs just at the time that they are least able to verbalize and communicate their basic needs. Therefore because of the limitations in communication that can occur postoperatively, it is important that preoperative psychiatric evaluation be done in high risk patients as often as possible.

The psychiatric consultant therefore confronts two important dilemmas in the approach to this patient group: diagnostic and therapeutic. Diagnostically, the consultant must delineate the overlapping boundaries between normal adaptation and psychiatric morbidity, while challenged by the severe limitations of verbal communication.

In this chapter, the special attention is given to issues of speech, disfigurement, and adaptation, followed by a discussion of the common psychiatric syndromes and basic treatment approaches that may be used in the care of these patients. Much of the material discussed in this chapter is based on our extensive review of the subject found elsewhere (Bronheim, Strain, & Biller, 1991a,b).

SPECIAL ISSUES IN OTOLARYNGOLOGY

Extensive head and neck surgery may lead to complicated postoperative courses with alterations in basic physical functions such as breathing, vision, or hearing. Facial contours may be deformed and occlusal relations disrupted. Chewing and swallowing may be grossly distorted. Surgery frequently involves tracheostomy, which may be temporary, or permanent if the patient undergoes total laryngectomy or suffers from extensive tracheal stenosis. Postoperatively, the patient must not only come to terms with cancer and surgery but also live with a changed, disfigured self with marked impairments in speech (Bronheim, Strain, Biller, et al., 1989; Lucente, Strain, & Wyatt, 1987; O'Hara, Ghonem, Hinrichs, et al., 1989; Shapiro & Kornfeld, 1987; Strain, 1985; Surman, 1986).

When approaching this patient population, the psychiatrist must be prepared to deal cohesively with the separate issues of (1) cancer, surgery and life-threatening illness; (2) disfigurement; (3) difficulties of verbalization; and (4) long-term physical and psychological adaptation.

Surgery

Although ENT cancers comprise only 5 to 6% of all cancers, patients with head and neck disease undergo an onslaught of frightening alterations that go well beyond that experienced by most other surgical patients (American Cancer Society, 1988). The multiple organs and functions encompassed within the small region of the face and neck give rise to an enormous variety of physical diseases, pathologies, and dysfunctions. The most common cancer is *squamous cell carcinoma*, which usually requires, in addition to surgery, adjunctive treatment with radiation or chemotherapy (Urken & Biller, 1988).

Multiple stressors face the ENT surgery patient, including cancer, fear of death, mutilation, and pain. However, because of its location on the face or in the head and neck, the mutilation resulting from surgery is often experienced as an assault on the total self (Greenacre, 1958). It is more difficult for ENT patients to isolate the disease and distance themselves defensively, especially when there are multiple alterations in sensory and motor functions postoperatively. Anxiety persists as the patients' concerns remain focused on the new significant physical derangements.

Advances in otolaryngology techniques have permitted more extensive resections of cancer and nononcologic disease. Older individuals with more extensive disease are now undergoing surgery impossible only a few years ago, and some patients undergo multiple surgical revisions, which can "stress" well-compensated individuals beyond endurable emotional limits. As a result, the incidence of adjustment disorder increases during hospitalization with extended length of stay, and psychiatric consultation is requested on average much later than that which occurs in the usual consultation setting (Bronheim, Strain, Biller, et al., 1989).

Disfigurement

Head and neck surgery is the most disfiguring of all surgery, with oral mandibular surgery considered to be the most stressful (Dropkin, Malgady, Scott, et al., 1983). Disfigurement may involve loss of an eye, shifting the nose across the midline, facial and jaw asymmetry, a new orifice for the tracheostomy site, asymmetry of the neck by either surgical dissection or skin and muscle grafts, and multiple scars.

Patients complain of having their "throats cut" or of an inability to recognize themselves. They may have difficulty looking at their reflection in mirrors or elsewhere. They feel like grotesque monsters and notice people staring at them or abruptly turning away. Head and neck patients hide in their rooms and fear going out in public. Disfigurement has an extraordinary impact on all visitors who are unfamiliar with the ENT hospital setting, including the medical staff. A perceptibly disturbed reaction of the psychiatric consultant to the patient can have a profound effect on the success or failure of the therapeutic relationship.

Speech

Speech is the most powerful, immediate, and versatile means of communication and is especially important during periods of acute stress. The inability to adequately regulate speech, which occurs as a result of laryngeal and oral maxillary surgery or tracheostomy, leads to a sense of urgency, increased tension and frustration, and finally helplessness and withdrawal (Hagglund & Heikki, 1980). Because of difficulties in speech, verbalization is limited in quantity and quality and of necessity is focused on the immediate issues of physical needs and discomforts. The expression of feelings, which are more subtle and complex, is compromised.

Consequently, otolaryngology patients are unable to verbalize sufficiently to discharge affects such as anxiety, frustration, sadness, or anger and are unable to communicate feelings of hopelessness or the wishes to end the current suffering one way or another. As a result, they commonly "act out" on the unit when they become frustrated, and are more likely to throw objects and strike out at staff (Edgecomb, 1984). The psychiatrist, by facilitating the process of verbalization, enables the process of integrating the new body image and personal identity and at the same time initiates the important discharge of affect with its concomitant stress reduction.

Adaptation

Although most otolaryngology patients finally manage a successful adaptation, they may undergo a stormy, prolonged, difficult course (West, 1974, 1977). The postoperative period is complicated by alterations that can lead to chronic physical deficits that may persist even after discharge from the hospital. A number of factors have been reported to be correlated with long-term postoperative adaptation (Anderson, 1987; Meyers, Aarons, Suzuki, et al., 1980; Mumford, Schlesinger, Glass, et al., 1982; Natvig, 1983a; Schleifer, Macara-Hinson, Coyle, et al., 1989; West, 1974). Preoperatively, the more patients are informed of what physical alterations they can expect immediately after surgery, the more rapidly they are likely to engage in physical rehabilitation (Natvig, 1983a). The unprepared patient almost certainly requires more time and attention to cope with the shock of the new physical alterations. Postoperatively, physical rehabilitation and speech therapy have been found to improve long-term outcome. After discharge, however, individuals with persistent major defects can suffer long-term psychiatric disturbances with depression and social withdrawal, especially if the defect involves speech (West, 1974).

The individuals who are particularly vulnerable are those with limited coping mechanisms as a result of either preexistent ego deficits or cognitive dysfunction associated with cognitive impairment disorders. In addition, individuals with poor family, social, or religious supports are also at risk for psychological dysfunction (Dhooper, 1985; Natvig, 1983a). Early psychiatric examination and intervention can improve long-term adaptation through the treatment of existing comorbid psychiatric disorders and relief of psychological dysfunction. Optimal long-term rehabilitation depends on efforts initiated immediately postoperatively, which include modulation of affect, recovery of physical function, and improved social and work relationships (Anderson, 1987; Mumford, Schlesinger, Glass, et al., 1982). Nevertheless, the distinction between normal and abnormal adaptation as well as the delineation of specific psychiatric syndromes postoperatively remains problematic. Although some patients make good physical adaptation to tracheostomy from the standpoint of speech, breathing, and physical endurance, psychologically they may not. Patients may become more withdrawn and less active than prior to surgery because of feelings of embarrasment about "disgusting" secretions or drooling.

Reconstruction and Cosmesis

The reconstructive surgery described earlier is undertaken mainly for the purpose of rapidly restoring fundamental functions such as chewing, swallowing, and breathing or simply covering an open surgical wound. From a psychiatric standpoint, the main issues revolve around the patient's realistic preparedness for surgery and the possible consequences of further disfigurement or surgical failure. Usually an attempt is made to preserve speech even at the risk of sacrificing swallowing (due to occasional difficulties with aspiration). Nevertheless, some individuals cannot tolerate an inability to ingest food orally (even when adequately nourished with gastric tube feedings) and need to have further reconstructive surgery to physically separate the airway from the orohypopharyngeal space. In so doing, speech is sacrificed in order to preserve swallowing. A psychiatric assessment of such individuals usually reveals either dementia with poor impulse control or dependent personality with a history of alcoholism.

Reconstructive surgery may also be undertaken to repair a preexisting physical deformity or simply to achieve enhanced cosmetic appearance. Reconstructive surgery in obviously deformed young children generally leads to an improvement in interpersonal behavior and cognitive development (Belfer, 1982; Kalick, 1982). Individuals who seek surgery for purely cosmetic reasons are found to be more anxious as a group and may suffer from poor self-image. Nevertheless, for many, cosmetic surgery, especially rhinoplasty in young women, leads to an enhanced sense of well-being with a corresponding increase in social activity (Marcus, 1984).

Psychiatric evaluation of individuals undergoing cosmetic or reconstructive surgery therefore must focus on the nature of the patients' body image and their expectations as to how they will be affected by the change. Unrealistic expectations are more likely to lead to postoperative disappointment and maladjustment. Other individuals at risk include those who insist on undergoing extensive or life-threatening surgical interventions for relatively benign deformities, males (versus females), and those undergoing surgery on features that have prominent psychological meanings that have existed for many years. At lower risk are females undergoing cosmetic surgery, especially on features that have rapidly changed due to aging and that have outpaced the changes in the inner body image (Goin & Goin, 1981). "Refreshing" the face or neck of a woman of age 50 who feels that she is looking old and unattractive is seldom as complicated psychologically as, for example, doing a rhinoplasty on a young man who feels his lack of success professionally or socially is due mainly to a slightly prominent nose. Psychiatric aspects of plastic and cosmetic surgery are also discussed in Chapter 34.

PSYCHIATRIC SYNDROMES

Utilizing criteria derived from DSM-III-R, the most frequent psychiatric syndromes found in consultation in the head and neck surgical setting are listed in Table 37-1. Adjustment disorders comprise the most common axis I disorder. Substance abuse and cognitive impairment disorders are also frequently diagnosed.

Major Mood Disorders

The diagnosis of depression and the delineation of psychiatric syndromes in the medically ill is difficult, and there is a paucity of rigorous clinical studies that have been made (Kathol, Noyes, Williams, et al., 1990b; Mason & Frosch, 1989; Strain, 1992; Strain, Snyder, & Fulop, 1992). Diagnosis of depression in the medically ill differs by as much as 13% depending on the diagnostic system, and it is frequently falsely positive (Kathol, Mutgi, Williams, et al., 1990a). The Hamilton Rating Scale for depression depends heavily on vegetative symptoms for the diagnosis and is likely to be overinclusive of physically ill patients (Hamilton, 1960). The Beck Depression Inventory, which emphasizes cognitive factors can yield false-positive results in patients with delirium, which is also common in medically ill populations (Beck, Ward, & Mendelson, 1961). Fur-

TABLE 37–1. *DSM-III* Diagnosis in ENT Patients*: The Mount Sinai Hospital 1980–1986

Diagnosis	%	No.
Axis I (*n* = 139)		
No disorder	23.0	32
Major affective disorder	10.1	14
Anxiety disorder	1.4	2
Somatoform/factitious	0.0	0
Psychotic	1.4	2
Organic (Secondary)	12.9	18
Substance abuse	18.0	25
Adjustment disorders	36.0	50
Axis II: personality disorders (*n* = 139)		
No disorder	54.0	75
Definite personality disorder	18.0	25
Deferred or unknown	28.1	39

From Bronheim, Strain, & Biller (1991a). With permission.

* The patients were consulted in the context of a liaison intervention. The cohort represents a mix of standard consultations plus others identified by or to the liaison psychiatrist. The cohort does not represent a screening sample.

thermore, psychological cognitive and vegetative symptoms commonly remit with the alleviation of the medical illness or improvement postoperatively (Kathol, Noyes, Williams, et al., 1990b). Proposals have been made for the substitution of criteria in *DSM-III* for the evaluation of depression in the medically ill, and they have been reported to yield a correct classification of 96% of the patients (Endicott, 1984; Rapp & Vrana, 1989).

The diagnosis of major depression therefore is complicated in the ENT patient population. Derangements in speech, appetite, swallowing, breathing, and sleeping and the direct effects of cancer obscure the diagnosis (Holland, Rowland, & Plumb, 1977). The vegetative symptoms are unreliable and need to be replaced with another domain that is ideational or behavioral, which is less liable to distortion by physical illness. When, in addition to depressed mood, the patient becomes overly needy, pessimistic, angry, brooding, or hopeless over a protracted period, the diagnosis of major depression should be made and treatment with antidepressants initiated (Endicott, 1984). Diagnosis of depression in the medically ill is discussed in Chapter 4.

Adjustment Disorder

Given the extreme to catastrophic nature of the stressors in the otolaryngology setting, it is not surprising that adjustment disorders are common. Although patients may suffer distress clinically judged to be in excess of the expected reaction to the physical deficits, the magnitude of psychological impairment may be subthreshold for a major mood disorder. Most patients with adjustment disorders make adequate psychological adaptation over time.

Anxiety-Panic

The vulnerable patient who is unprepared and subjected to the trauma of multiple surgeries of complications may develop a sustained anxiety state with intermittent panic and hyperarousal alternating with numbness and emotional exhaustion. On the other hand, the patient with a tracheostomy who appears to be acutely panicked may be primarily suffering discomfort due to intermittent airway obstruction, bronchospasm, and hypoxemia. Before making a diagnosis of panic disorder, the patient's airway and oxygenation status should be carefully evaluated (Basawaraj, Rifkin, Seshagiris, et al., 1990).

Substance Abuse Disorders and Cognitive Impairment Disorders

Because head and neck tumors are associated with long-term alcohol and tobacco use, a large proportion of otolaryngology patients are male, and many have alcohol-related dementia. In the head and neck surgical setting, the postoperative course is commonly complicated by alcohol and nicotine withdrawal, which can also mimic panic disorder. Because of the effects of alcohol use, underlying dementia may first present on the ENT service as postoperative delirium. Individuals with compromised cognitive capacity due to Alzheimer's disease are more susceptible to delirium occurring as a result of anesthesia, infection, anemia, hypoxemia, or other metabolic factors. Alcohol withdrawal must be considered in every patient with postoperative delirium. When patients undergo prolonged reconstructive surgery, the operative report should be reviewed for episodes of hypotension or a consequential significantly reduced hematocrit. A neurologic examination needs to be performed to exclude a new cerebral vascular accident. The medical evaluation should exclude infection of the surgical wound, or pneumonia from aspiration. Obviously, medication effects must also be suspected in a patient whose delirium persists over an extended period.

All patients in this setting with postoperative delirium who have a history of alcoholism should have a dementia workup while on the surgical service because an undetected cognitive impairment disorder may have a profound effect on recovery and long-term rehabilitation. The assessment of dementia on the ENT service is particularly difficult when the patient has a tracheostomy (Strain, Fulop, Lebovits, et al., 1988). Confused patients mouth long responses even when they cannot be understood and may refuse to write in response to direct requests to do so. They may also be unable to simply cover the tracheostomy tube opening with their finger in order to phonate. These maneuvers are steps that cognitively intact individuals attempt to learn in order to provide as much comprehensible communication as possible. Patients with significant cognitive impairment disorders can neither learn these basic maneuvers nor recognize that they cannot be understood by others and consequently persist in their ineffectual efforts.

THERAPEUTIC APPROACH

The treatment of head and neck patients to some extent poses an even greater challenge to the consulting psychiatrist than diagnosis. Treatment, as elsewhere in the medical and surgical setting, is tailored to the individual needs of the patient (e.g., antidepressants for major depression, supportive therapy for adjustment reactions). All too often, however, psychiatric consultants unfamiliar with the otolaryngology setting overprescribe medication when supportive psychotherapy may be the treatment of choice. Combination therapy or the sequential application of a variety of therapies is often necessary. A brief discussion of the most common therapeutic modalities utilized and how they may be specially modified in the head and neck setting follows.

Psychotherapy

Because of the profound disfigurement, the first reaction of many psychiatrists is to retreat from the situation as soon as possible (Buckley, 1986; Spikes & Holland, 1975). Patients, however, are often intensely sensitive to the effects of their appearance on others and may respond to the psychiatrist's behavior in kind. Patients observe from the beginning whether the consulting psychiatrist is sufficiently comfortable to enter into a meaningful ongoing therapeutic relationship.

By practice and history, psychiatry has adopted a receptive role when examining patients by trying to listen empathically while encouraging the patient to speak. Obviously, this method does not proceed smoothly on the ENT service, and the psychiatrist must be prepared to assist the patient in communication through active participation. The psychiatrist must, for the purpose of efficiency, supply phrases to fill in the gap of feelings probably intended but not fully expressed among the patient's short written phrases and facial and hand gestures. The psychiatrist must often try to express for patients that which they may not able to say for themselves. Direct inquiries should be made about changed appearances and feelings of disfigurement as well as the frustration of not being able to speak for themselves.

The more consulting psychiatrists know about the expected course of treatment for the patient the more helpful they can be to the patient in alleviating anxiety. Otolaryngology patients frequently have discussions with anyone knowledgeable about their condition and so benefit from a review of the events and emotions leading up to the present state. The psychiatrist may be the only one who enters into a deeper exploration of patients' fears and thoughts, and so is uniquely positioned to correct cognitive distortions and unfounded fantasies.

In this active participating manner, the psychiatrist builds an oratory of affect that communicates understanding to the patient and, as noted earlier, in some instances the psychiatrist literally speaks for the patient. As a result, therapeutic interviews tend to be of shorter duration, with frequent contacts (Groves & Kuchorski, 1986).

Drug Therapy

Pharmacologic therapy frequently is necessary and is the treatment of choice in a number of situations. In general, difficulties are seldom encountered with enteral administration. The gastrointestinal tract usually is functional, and placement of a nasogastric tube or percutaneous gastric tube ensures access.

Depression in ENT patients has been treated successfully with both monoamine oxidase inhibitors (MAOIs) and tricyclic antidepressants (Fernandez & Adams, 1986). Greater attention must be given to drug toxicity and delirium in the medically ill population (Popkin, Callies, & MacKenzie, 1985; Rodin & Voshart, 1986); therefore, of the tricyclics, nortriptyline and desipramine, whose serum levels can be carefully monitored, may be preferred. More recently, fluoxetine, sertraline and bupropion have been available, however, specific studies using the ENT surgical population have not been reported. Although many individuals diagnosed with depression remit with improvement in their physical condition alone, some studies of depression in medical and surgical settings indicate that antidepressant treatment is superior to no treatment in effecting long-term rehabilitation and return to work (Rifkin, Reardon, Siris, et al., 1985; Schleifer, Macara-Hinson, Coyle, et al., 1989).

The acute control of mania is a problem seldom encountered in this setting, when antimanic therapy is needed, there is no contraindication to any of the major antimanic agents, and lithium, carbamazepine, and valproate can be used. However, because of volume shifts and concomitant administration of other medications in extended surgical reconstructive cases, serum levels of antimanic agents should be checked postoperatively and dosages adjusted as necessary to provide therapeutic levels while avoiding toxicity.

Anxiety is common in this setting because of alcohol and nicotine withdrawal and intermittent airway obstructions. Frequent airway suctioning is the primary treatment for the latter problem, but benzodiazepines should be employed in adequate doses to increase the patient's comfort and tolerance of the tracheostomy, and to block agitation due to withdrawal. In the absence of severe chronic obstructive pulmonary disease, respiratory depression seldom is a problem and in general does not represent an impediment to adequate treatment.

In general, alcohol withdrawal is best managed with a long-acting benzodiazepine, such as diazepam (Valium) or chlordiazepoxide hydrochloride (Librium). However, because of the increased prevalence of cognitive impairment disorders in this patient population, drug-induced deliria occur frequently and the benzodiazepines should be tapered rapidly. Alprazolam and triazolam should be avoided because of the problems of rebound anxiety and the greater likelihood of nighttime confusion or amnesia, particularly with triazolam.

General anxiety frequently appears postoperatively, and currently lorazepam is the preferred medication. Its relatively short half-life and its suitability for parenteral administration are advantages.

Buspirone is usually not used perioperatively because of the 10 to 14 days necessary to obtain full anxiolysis. However, it may be used in the outpatient setting postoperatively if anxiety persists, especially in patients with unresectable malignancies.

The appearance of agitation postoperatively can be dramatic and dangerous for the patient in the surgical intensive care unit (ICU), especially if the patients attempt to pull out intravenous lines or extubate themselves. Usually the cause of delirium is unclear, and it may be due to a combination of factors, including accumulation of anesthesia, alcohol or nicotine withdrawal, fever, sepsis, hypoxemia, rapid fluid shifts, anemia, metabolic changes, and stroke—all in addition to preexisting cognitive impairment disorders. Combinations of neuroleptics, benzodiazepines, and narcotics are administered intravenously, including haloperidol, lorazepam, and fentanyl citrate or morphine sulfate, all of which should be tapered as rapidly as possible. If the patient seems to be mostly paranoid or psychotic, haloperidol should be used. If alcohol withdrawal is thought to predominate (by history or physical examination), more lorazepam should be administered. If the patient appears more discomfited by the endotracheal or tracheostomy tube, however, the narcotics should be titrated up to the point of sedation. The decision about which combination of medications to be used and in what dosage is usually made on an empiric clinical basis. Akathisia as a result of neuroleptic administration may be mistaken for agitation. The patient should be monitored for extrapyramidal symptoms and if akasthisia is suspected, propranolol may be administered (if the patient does not have lung disease or CHF). Psychopharmacology in the medically ill is also discussed in Chapters 9 and 12.

Pain is seldom psychogenic in nature in otolaryngology and, when present, usually reflects pathology, infection, or invasive tumor. Adequate treatment regarding both dosage and frequency of narcotic analgesics is indicated, with substance dependence seldom presenting a problem. More commonly, patients suffer excessively from insufficient dosing. The occasional substance-abusing patient who stridently demands narcotics at all times can be managed with methadone in fixed doses that can be slowly tapered over 7 to 10 days when no longer needed. Usually 20 to 30 mg of liquid methadone given in divided doses three times daily suffices. (See also Chapter 16.)

Behavioral Team Approach

Multiple functional disabilities occur in otolaryngology as a direct consequence of the surgery for the underlying disease. As a result, many individuals are overwhelmed and withdraw during the immediate postoperative period. A behaviorally oriented team involving the surgeon, the occupational, speech, and physical therapists, the nursing staff, and the psychiatrist work in conjunction to identify the patients at risk. The team creates a therapeutic milieu that is behaviorally active, rewarding the patient for progressive behaviors instead of passively accepting somatic preoccupation and regression. In so doing, the team actively prepares the patient for a resumption of life postoperatively. With the team approach, the social worker reviews with the family the preparations that will be necessary in the home for the transition of the patient from the hospital to the home environment (Dropkin, 1989; Natvig, 1983b; Petrucci & Harwick, 1984). Family members are directly affected by the alterations in the head and neck patient, and family therapy with multiple family groups is beneficial for supporting family members who are overwhelmed, and for facilitating in-hospital teaching of tracheostomy care (Dhooper, 1985; Fiegenbaum, 1981; Vogtsberger, Harris, & Mattox, 1985).

SPECIAL TOPICS

Aids to Communication

In order to communicate effectively with the head and neck surgical patient, especially those with laryngectomies, the psychiatrist must effect a close relationship despite all the physical obstacles to rapid communication. At first, nonverbal efforts on the part of the psychiatrist communicates to the patient a willingness to search for deeper thoughts and feelings. Helping to arrange the patients so that they can write more comfortably and then adopting a receptive pose allows the patients to unhurriedly express a wide range of emotions. Letter boards may be used initially for the critically ill or for those who cannot physically write.

Working closely with the speech therapist, a combination of phonation devices and techniques are utilized. What they all have in common is an effort to create a vibrating column of air or tone that, when funneled through the mouth, can be articulated into speech (Case, 1984; Lerman, 1991).

Esophageal, pharyngeal, and buccal speech involve production of a vibrating column of air between fissure spaces at various levels of the oropharynx to create speech. Mechanical and electronic devices produce the vibrating column artificially. The Tokyo artificial larynx, one of the simplest, has a vibrating reed that is placed over the tracheostomy. With exhalation, sound is funneled by connecting tube to the mouth for articulation. It has the advantage of being useful almost immediately postoperatively. The Cooper-Rand electronic device produces a tone electronically, which is again fed by tube intraorally. Its advantage is freedom from involving the tracheostomy and respiration. The most commonly employed communication aids involve transcervical sound production that may be generated by vibrating devices positioned carefully in the neck. The most common devices are the Western Electric 5C and Serrox electronic larynx. Vibration is transmitted to the vocal tract through the neck and then articulated into speech. Their major advantage is wider acceptance and freedom from unsightly oral tubes, and they are currently the communication aids most commonly preferred. Advanced technologic devices to assist with communication in ICU patients unable to speak are discussed in Chapter 12.

After evaluations of speech intelligibility in laryngectomy patients, the Tokyo aid has been rated the highest with regard to intelligibility, with esophageal speech having been judged the lowest. The mechanical devices, although most preferred, produce a voice that sounds the most artificial.

The presence of a tracheostomy does not necessarily imply the absence of the larynx or vocal cords. Tracheostomy may be necessary either temporarily postoperatively or in situations of chronic airway narrowing or obstruction. Speech may be produced by covering the tracheostomy opening and forcing air up through the vocal cords. A simple one-way valve is helpful to some individuals, allowing inspiration through the tracheostomy and expiration through the pharynx. Simple adjustments of the size of the tracheostomy or the addition of the one-way valves can make significant changes in overall speech production and therefore in the patient's acceptance of the tracheostomy.

Group therapy can be of great benefit to the laryngectomy patient and family. A network of self-help groups do exist, and patients often benefit from postoperative visits with former patients who can demonstrate the use of the communication devices. Knowing others who have undergone the same radical changes in physical and social function can be supportive during adjustment to autonomous function and alaryngeal speech.

Hearing Loss and Deafness

Approximately 6% of the population have significant bilateral hearing impairment, of whom approximately 45% are over age 65 (Bailey, Pappas, Graham, et al., 1976). Moreover, 25% of people ages 65 to 74 and 50% of those 75 years and older experience hearing difficulties (Anderson & Meyerhoff, 1986). Deafness is a major cause of speech retardation and associated behavioral impairments (Kaplan & Sadock, 1985). In the elderly, hearing loss is thought to be associated with late-onset paranoid disorders even if it is due to conductive middle ear causes and not to the sensorineural defects of aging (Eastwood, Corbin, Reed, et al., 1985; Savoy, Lazarus, Jarvik, et al., 1991). In children referred for a wide variety of school problems, 8% have had abnormal hearing activity, 30% abnormal speech discrimination in noise, and approximately 60% abnormal auditory memory (Oberklaid, Harris, & Keir, 1989). Therefore complete audiometric assessment is warranted as part of the evaluation of all children with learning difficulties or developmental language disorders and in select elderly patients with paranoid features. If hearing loss is suspected (as when, for example, the patient has difficulty comprehending whispered speech), full audiologic examination is indicated, which includes physical examination of the external ear canal and tympanic membrane and radiographic evaluation of the middle and inner ear. More precise functional assessment includes the Weber and Rinne tests, utilizing tuning forks to delineate conductive versus sensorineural deficits, and audiometry for both pure tones and speech.

Hearing sensitivity is measured by frequencies of 250 to 8000 cycles/second (Hertz). A hearing threshold of 15 decibels (dB) or less is considered normal. An average hearing threshold of 30 dB or less in the 500- to 3000-Hz zone is sufficient for ordinary needs, although amplification may be necessary if work or social activities require even better hearing.

Although pure tone measurements provide significant information, speech discrimination is equally important for the assessment of hearing deficits. Monosyllabic words can be presented at loud levels, and the percentage of words correctly identified is noted. Scores lower than 70 to 80% indicate significant difficulties in understanding speech (Anderson & Meyerhoff, 1986).

The appropriate management of the hearing-impaired requires continuous attention directed at effective communication. The psychiatrist must remember to speak loudly, slowly, and clearly, preferably with the patient facing the lips and face of the speaker. Augmenting speech with physical, facial, and hand gestures adds to overall comprehension and communication. If the patient is severely deaf, sound must be substituted for by sight—writing, lip reading, or sign language. Rehabilitation requires adequate instruction in these modalities in addition to a hearing aid (Anderson & Meyerhoff, 1986; Becher & Cohn, 1982). A functioning prosthetic hearing aid is indicated if deafness cannot be relieved or improved by other means such as surgery. Only with acoustic agnosia (sensory aphasia) is a hearing aid completely useless (Becher, Naumann, & Pfaltz, 1989, p. 153). For all other situations, a multitude of devices currently exist and should be appropriately fitted by an audiologist.

Complete deafness is rare, and those with almost total deafness usually have remnants that can be utilized by high performance hearing aids. Young children can be tested with electric response audiometry and impedance audiometry, which have now displaced many formerly used methods in pediatric audiology (Becher, Naumann, & Pfaltz, 1989, p. 59). The deaf child should be educated in separate schools where sign language is now commonly being replaced by articulated speech (Becher, Naumann, & Pfaltz, 1989, pp. 155–159).

In select cases of children and adults, implantation of electrodes into the inner ear (cochlear implant) has led to remarkable enhancement of acoustic perception. Although the device may lead to greater awareness of nonvisual stimuli, such as traffic sounds and the presence of others around them, patients do not necessarily obtain fine auditory discrimination of speech. As a result of the implant, though, individuals may develop enhanced communication and lip-reading skills. To date, psychological testing has not revealed in either adults or children those individuals most suitable for the procedure functionally, or those most likely to benefit from the implant psychologically and cognitively. The psychiatrist's role is to help prepare the patient and family for what they can generally expect postoperatively in order to avert the occurrence of depression as a result of unrealistically high expectations (McKenna, 1986; Tiber, 1985).

Resistance to the use of hearing aids is common and not well understood. Resistance may involve physical factors, including the discomfort of poor-fitting ear molds, leading to ear canal infections, erosions, and wax collection. Others individuals are supersensitive to noise amplification and dislike it (Becher & Cohn, 1982). In general, those individuals who are highly motivated and physically and socially active, accept hearing aids well. Others who are unmotivated, socially inactive, physically immobile, depressed, and withdrawn do not use the device. Cognitive impairment or confusion as to the proper use of the device also impedes the use of a hearing aid. Still others, as a result of growing dependence and rigidity with aging, resist using a hearing aid as a means to effect a power struggle or avoid communication with those on whom they depend (Savoy, Lazarus, & Jarvik, 1991, p. 523). They stubbornly resist all efforts to make communication with them easier for all those who would care to try. Behavioral approaches may be the best strategy in these situations.

Tinnitus

A hearing disorder closely related to hearing loss is tinnitus. Both hearing loss and tinnitus increase with age; and as the threshold for hearing increases, so does the severity of tinnitus. Approximately 2% of the population report significant tinnitus, resulting in some disability (Coles, 1985).

The etiology of tinnitus is unclear and may involve a combination of factors including spontaneous firing of the auditory nerve, denervation hypersensitivity with neural loss, or central excitation as with phantom limb pain (Slater & Terry, 1987, pp. 51–53). Patients who suffer from tinnitus may be highly sensitive to noise, as was Beethoven, who suffered recurrent bouts that worsened every winter (Slater & Terry, 1987, p. 96). A host of medical conditions are associated with tinnitus (e.g., Meniere's disease, diabetes, thyroid disorders, anemia) as are some drugs, particularly caffeine, nicotine, aspirin, and some antibiotics (Slater & Terry, 1987, pp. 143–155). It is of special significance that MAOIs and tricyclic antidepressants, as well as buspirone, also have been associated with tinnitus.

A significantly higher prevalence of psychiatric conditions are found in tinnitus sufferers. There is both a higher premorbid history of depression as well as a higher comorbid incidence of depression, with a lifetime prevalence of major depression as high as 78% (Sullivan, Katon, Dolice, et al., 1988). The patients most disabled by tinnitus are most disturbed by associated symptoms such as sleep disorder, poor concentration, and withdrawal from social relations, all of which are primary symptoms of depression.

The treatment of tinnitus patients is multimodal. Hearing aids that amplify external sounds or produce sounds of a specific tone are commonly prescribed. Clearly, antidepressants are indicated in individuals with associated major depression. Although the underlying sounds may not be eliminated by antidepressant treatment, the degree of disability and perceived suffering can be markedly improved. Biofeedback and relaxation training have also been beneficially employed (Slater & Terry, 1987, pp. 167–200).

Spastic Dysphonia

Spastic dysphonia for many years was considered a conversion disorder. More careful examination of these patients reveals a differentiation from other conversion disorders by the gradual nature of the disorder's evolution and its subsequent long-term persistence. Also, its occurrence immediately after a specific psychosocial stressor has not been substantiated. Physical examination of a conversion disorder reveals no organic basis, whereas the vocal cords are usually observed to be in spasm in spastic dysphonia. Spastic dysphonia seldom responds dramatically to psychiatric interventions with sodium amytal interview, psychotherapy, or benzodiazepines, whereas conversion disorder frequently does. Even though anxiety is generally known to exacerbate the symptoms of spastic dysphonia, psychological studies of these patients have failed to confirm the presence of somatization and conversion (Ginsberg, Wallach, Strain, et al., 1988).

There is growing evidence that spastic dysphonia is a disorder more of neurologic origin, with features that overlap with other dystonias caused possibly by brainstem dysfunction. Sectioning of the recurrent laryngeal nerve leads to dramatic improvement immediately postoperatively. Controversy exists, however, as to the benefits of recurrent laryngeal nerve section, in that improvement may not persist on long-term follow-up (Aronson & DeSanto, 1983; Dedo & Izdebski, 1983). Electromyographically guided injections of botulinum toxin have been used successfully for spastic dysphonia and are currently recommended as the treatment of choice for this disorder (Jankovic & Brin, 1991). However, patients may require repeat injections every 4 months, and the procedure may be complicated by aspiration. Nevertheless, anxiety or psychiatric dysfunction may persist after either medical or surgical intervention, and treatment aimed at either reducing anxiety or stress management may still be necessary.

CONCLUSION

The head and neck surgical service is a unique setting for the psychiatrist. The psychiatrist may encounter patients with profound alterations in appearance and function that are disturbing to observe. The psychiatrist must participate actively with the patient sometimes in unpleasant surroundings and must be familiar with surgical sequelae, tracheostomy, communication, and hearing devices, as well as other prosthetic and cosmetic options. Moreover, the psychiatrist must rely on a variety of skills, diagnostic as well as biologic and psychological techniques, to facilitate the best adaptation of these patients to the overwhelming effects of surgery.

Because of the frequent alterations in speech, detection of patients at risk is limited. Therefore even more than is indicated in other medical and surgical settings, a liaison psychiatric approach in conjunction with the otolaryngology team in a screening rather than a referral mode is most successful. The continual presence of the liaison psychiatrist and the consequential enhanced awareness of the surgical staff leads to the identification and referral of vulnerable or subclinically symptomatic patients.

Lastly, a review of the psychiatric literature in otolaryngology reveals a paucity of material. More rigorous study of diagnosis, intervention, and outcome is necessary to determine which patients undergoing the multiple stressors of head and neck surgery do so successfully and who do not.

REFERENCES

American Cancer Society (1988). Cancer facts and figures. New York: American Cancer Society.

Anderson EA (1987). Preoperative preparation for cardiac surgery facilitates recovery, reduces psychological distress, and reduces the incidence of acute postoperative hypertension. J Consult Clin Psychol 55:513–520.

Anderson RG, & Meyerhoff WL (1986). Otologic disorders. In E Calkins, P Davis, AB Ford (Eds.), The practice of geriatrics (Ch. 25). Philadelphia: WB Saunders.

Aronson AE, & DeSanto LW (1983). Adductor spastic dysphonia: three years after recurrent laryngeal nerve resection. Laryngoscope 93:1–8.

Bailey HAT, Pappas JJ, Graham S, et al. (1976). Total hearing rehabilitation. Arch Otolaryngol 162:323.

Basawaraj K, Rifkin R, Seshagiris D, et al. (1990). The prevalence of anciety disorders in patients with chronic obstructive pulmonary disease. Am J Psychiatry 147:200–201.

Becher RL, & Cohn ES (1982). Problems of the eye and ear. In RL Becher, JR Burton, & PD Ziere (Eds.), Principles of ambulatory medicine. Baltimore: Williams & Wilkins.

Becher W, Naumann HH, & Pfaltz CR (1989). Ear, nose, and throat disorders. Stuttgart: Thieme.

Beck AT, Ward CH, & Mendelson M (1961). An inventory for measuring depression. Arch Gen Psychiatry 4:561–571.

Belfer ML, Harrison AM, Pillemer FC, et al. (1982). Appearance and the influence of reconstructive surgery on body image. Clin Plast Surg 9:307–315.

Bronheim H, Strain JJ, & Biller HF (1991a). Psychiatric aspects of head and neck surgery. Part I. New surgical techniques and psychiatric consequences. Gen Hosp Psychiatry 13:165–176.

Bronheim H, Strain JJ, & Biller HF (1991b). Psychiatric aspects of head and neck surgery. Part II. Body image and psychiatric interventions. Gen Hosp Psychiatry 13:225–232.

BRONHEIM H, STRAIN JJ, BILLER HF, ET AL. (1989). Psychiatric consultation on an otolaryngology service. Gen Hosp Psychiatry 11:95–102.

BUCKLEY P (1986). Supportive psychotherapy: a neglected treatment. Psychiatr Ann 16:515–533.

CASE JL (1984). Clinical management of voice disorders (Ch. 7). Rockville, MD: Aspen.

COLES RRA (1985). Epidemiology of tinnitus. 1. Prevalence. J Laryngol Otol 9(suppl):7–15.

DEDO H, & IZDEBSKI K (1983). Intermediate results of 306 recurrent nerve sections for spastic dysphonia. Laryngoscope 93:9–16.

DHOOPER SS (1985). Social work with laryngectomies. Health Soc Work 10:217–227.

DROPKIN MJ (1989). Coping with disfigurement and dysfunction after head and neck surgery: a conceptual framework. Semin Oncol Nurs 5:213–219.

DROPKIN MS, MALGADY RG, SCOTT DW, ET AL. (1983). Scaling of disfigurement and dysfunction in post operative head and neck patients. Head Neck Surg 16:559–570.

EASTWOOD MR, CORBIN S, REED M, ET AL. (1985). Acquired hearing loss and psychiatric illness: an estimate of prevalence and comorbidity in a geriatric setting. Br J Psychiatry 147:552.

EDGECOMBE RM (1984). Models of communication. The differentiation of somatic and verbal expression. Psychoanalytic Study of the Child XXXIX (pp. 137–154). New Haven: Yale University Press.

ENDICOTT J (1984). Measurement of depression in patients with cancer. Cancer 53:2243–2249.

FERNANDEZ F, & ADAMS F (1986). Methylphenidate treatment of patients with head and neck cancer. Head Neck Surg 8:296–300.

FIEGENBAUM W (1981). A social training program for clients with facial disfigurations: a contribution to rehabilitation of cancer patients. Int J Rehabil Res 4:501–509.

GINSBERG BI, WALLACH J, STRAIN JJ, ET AL. (1988). Spastic dysphonia: toward defining the proper psychiatric role in the treatment of a disorder of unclear etiology. Gen Hosp Psychiatry 10:132–137.

GOIN JM, & GOIN MK (1981). Changing the body. The psychological effects of plastic surgery (p. 5). Baltimore: Williams & Wilkins.

GREENACRE P (1958). Early physical determinants in the development of the sense of identity. J Am Psychoanal Assoc 6:612–627.

GROVES JE, & KUCHORSKI A (1986). Brief psychotherapy. In TP HACKETT & N CASSEM (Eds.), MGH handbook of General Hospital Psychiatry (pp. 309–331). Littleton, MA: PSG.

HAGGLUND T, & HEIKKI P (1980). The inner space of the body image. Psychoanal Q 49:256–283.

HAMILTON M (1960). A rating scale for depression. J Neurol Neurosurg Psychiatry 23:56–62.

HOLLAND JC, ROWLAND J, & PLUMB M (1977). Psychological aspects of anorexia in cancer patients. Cancer Res 37:2425–2428.

JANKOVIC J, & BRIN MF (1991). Review article: therapeutic uses of botulinum toxin. N Eng J Med 324:1186–1194.

KALICK SM (1982). Clinician, social scientist and body image: collaboration and future prospects. Clin Plast Surg 9:379–385.

KAPLAN HI, & SADOCK BJ (Eds.) (1985). Comprehensive textbook of psychiatry IV (4th ed., p. 140). Baltimore: Williams & Wilkins.

KATHOL RG, MUTGI A, WILLIAMS J, ET AL. (1990a). Diagnosis of major depression in cancer patients according to four sets of criteria. Am J Psychiatry 147:1021–1024.

KATHOL RG, NOYES R, WILLIAMS J, ET AL. (1990b). Diagnosing depression in patients with medical illness. Psychosomatics 31:434–440.

LERMAN JW (1991). The artificial larynx. In SJ SALMAN & KH MOUNT (Eds.), Alaryngeal speech rehabilitation (Ch. 2). Austin, TX: ProEd.

LUCENTE F, STRAIN JJ, & WYATT DA (1987). Psychological problems of the patient with head and neck cancer. In SE THAWLEY & WR PANJE (Eds.), Comprehensive management of head and neck tumors (Ch. 5). Philadelphia: WB Saunders.

MARCUS P (1984). Psychological aspects of cosmetic rhinoplasty. Br J Plast Surg 37:313–318.

MASON B, & FROSCH WA (1989). Secondary depressions. In Treatments of psychiatric disorders: a task force report of the American Psychiatric Association (pp. 1898–1924). Washington, DC: American Psychiatric Association.

MCKENNA L (1986). The psychological assessment of cochlear implant patients. Br J Audiol 20:29–34.

MEYERS AD, AARONS B, SUZUKI B, ET AL. (1980). Sexual behavior following laryngectomy. Ear Nose Throat J 59(8):35–39.

MUMFORD E, SCHLESINGER HJ, GLASS GV, ET AL. (1982). The effects of psychological intervention on recovery from surgery and heart attacks: an analysis of the literature. Am J Public Health 72:141–151.

NATVIG K (1983a). Laryngectomies in Norway, study no. 1: social, personal and behavioral factors related to present mastery of the laryngectomy event. J Otolaryngol 12:155–162.

NATVIG K (1983b). Laryngectomies in Norway, study no. 2: preoperative counselling and post operative training evaluated by the patients and their spouses. J Otolaryngol 12:249–254.

OBERKLAID F, HARRIS C, & KEIR E (1989). Auditory dysfunction in children with school problems. Clin Pediatr 28:397–403.

PETRUCCI RJ, & HARWICK RD (1984). Role of the psychologist on a radical head and neck surgical service team. Professional Psychol Res Pract 15:538–543.

O'HARA MW, GHONEM MM, HINRICHS JV, ET AL. (1989). Psychological consequences of surgery. Psychosom Med 51:356–370.

POPKIN MK, CALLIES AL, & MACKENZIE TB (1985). The outcome of antidepressant use in the medically ill. Arch Gen Psychiatry 42:1160–1163.

RAPP SR, & VRANA S (1989). Substituting for somatic symptoms in the diagnosis of depression in elderly male medical patients. Am J Psychiatry 146:1197–1200.

RIFKIN A, REARDON G, SIRIS S, ET AL. (1985). Trimipramine in physical illness with depression. J Clin Psychiatry 46:4–8.

RODIN G, & VOSHART K (1986). Depression in the medically ill: an overview. Am J Psychiatry 143:696–705.

SAVOY J, LAZARUS LW, & JARVIK LF (Eds.) (1991). Comprehensive review of geriatric psychiatry. Washington, DC: American Psychiatric Press.

SCHLEIFER SJ, MACARA-HINSON MM, COYLE DA, ET AL. (1989). The nature and course of depression following myocardial infarction. Arch Intern Med 149:1785–1789.

SHAPIRO PA, & KORNFELD DS (1987). Psychiatric aspects of head and neck cancer surgery. Psychiatr Clin North Am 10:87–100.

SLATER R, & TERRY M (1987). Tinnitus: a guide to sufferers and professionals. London: Croem Helm.

SPIKES J, & HOLLAND J (1975). The physician's response to the dying patient. In JJ STRAIN & S GROSSMAN (Eds.), Psychiatric care of the medically ill (Ch. 11, pp. 138–148). Orlando, FL: Appleton-Century-Crofts.

STRAIN JJ (1985). The surgical patient. In R MICHELS & JO CAVENAR (Eds.), Psychiatry (Vol. 2; Ch. 121, pp. 1–11). Philadelphia: Lippincott.

STRAIN JJ (1991). Diagnostic considerations in the medical setting. Psychiatr Clin North Am 4:287–300.

STRAIN JJ, FULOP G, LEBOVITS A, ET AL. (1988). Screening devices for diminished cognitive capacity. Gen Hosp Psychiatry 10:16–23.

STRAIN JJ, SNYDER SS, & FULOP G (1992). Mood disorder and medical illness. In A Tasman (Ed.), Review in psychiatry (Vol. II; 23:453–476).

Sullivan M, Katon W, Dolice R, et al. (1988). Disabling tinnitus: association with affective disorder. Gen Hosp Psychiatry 10:285–291.

Surman OS (1986). The surgical patient. In TP Hacket & NH Cassem (Eds.), MGH handbook of general hospital psychiatry (Ch. 5, pp. 69–83). Littleton, MA: PSG Publishing.

Tiber N (1985). A psychological evaluation of cochlear implants in children. Ear Hear 6(3):48–51.

Urken ML, & Biller HF (1988). Management of early vocal cord carcinoma. Oncology 2:41–45.

Vogtsberger KW, Harris LL, & Mattox DE (1985). Group psychotherapy for head and neck cancer patients. Laryngoscope 95:585–587.

West DW (1974). Adaptation to surgically induced facial disfigurement among cancer patients. Dissert Abstr Int 34:442.

West DW (1977). Social adaptation patterns among cancer patients with facial disfigurements resulting from surgery. Arch Phys Med Rehabil 58:473–479.

38 Ophthalmology

WILLIAM V. GOOD, M.D.

Vision occupies a central position in human physical and mental functions. Visual acuity, the appreciation of fine visual detail, permits reading and recognition of subtle facial expressions. Acuity is determined by the cone cell population. Peripheral vision detects motion and shades of light; it is determined by rod cells. The eye is a complex optical system. Clear media, appropriate correction of refractive error, and intact visual pathways also are necessary for good vision.

Causes of vision failure can be categorized anatomically, epidemiologically, and developmentally. In adults in the Western world, the most common causes of vision failure are senile macular degeneration, glaucoma, and cataracts (Kahn & Moorhead, 1973). Effective treatments are available for glaucoma and cataracts but not for senile macular degeneration. In children, congenital cataracts, retinopathy of prematurity, and optic nerve hypoplasia are leading causes of blindness. Congenital cataracts are treatable (Beller, Hoyt, & Marg, 1981) and there is a promising new treatment for retinopathy of prematurity (Cryotherapy for Retinopathy Group, 1990). In the future, congenital birth defects will rank as a leading cause of childhood blindness.

Adults with acute vision loss undergo a grieving process similar to that for other types of loss. Chronic (i.e., long-standing) vision loss runs the risk of causing social isolation and depression. Even highly motivated visually impaired patients sometimes feel isolated, as traveling, reading, watching television, and looking at the emotional expressions of others all require vision. Vision loss exacerbates cognitive impairment due to neurologic or medical illness, or developmental disability. The best intervention for isolation is to minimize the impact of vision loss by providing low-vision aids, reading machines, seeing-eye dogs, Braille instruction, and companionship. If depression interferes with this program, it should be treated.

VISION LOSS IN CHILDREN

The reactions of children to vision loss assume two forms. Children demonstrate neurobehavioral responses to vision failure. This category of behavior is best interpreted as behavioral adaptation to poor vision. Second, personality development may be affected. Social and cognitive development are involved in this category.

Neurobehavioral responses to vision loss by children are often misinterpreted (by parents and teachers) as consciously determined. It is of interest that many behaviors are so specific to disease states that they can be used to topographically localize the lesion in the central nervous system (CNS) (Good & Hoyt, 1989). Bilateral, early vision loss due to disease anterior to the optic chiasm causes nystagmus (see Table 38-1). Children can reduce nystagmus by rotating their eyes to a so-called null zone. The result is a child who habitually turns his head to see more clearly. Holding objects close to the eyes also dampens nystagmus and improves the quality of vision.

Bilateral retinal disease (e.g., retinopathy of prematurity, cytomegalovirus retinitis, Leber's congenital amaurosis), when severe and early onset, induces eye poking in children (Jan, Freeman, McCormick, et al., 1983). The cause of the symptom presumably is the child's trying to induce phosphenes (visual sparks) by putting pressure on the eye. Periocular fat atrophy results from constant pressure on the globes, and such children develop a "sunken" eye appearance. The poking is commonly misinterpreted as autistic or "naughty."

Cortical blindness (termed cortical vision impairment, or CVI) causes variable visual performance in children (Jan, Greenveld, Sykanada, et al., 1987), perhaps due to slow recovery that occurs in these children, or to "swiss cheese" visual field defects that allow better visual function if eyes and head are in line with an object of visual interest. Subcortical vision, also termed blind sight (Scheider, 1969), allows for subconscious navigation even when neuroimaging scans demonstrate total ablation of the visual cortex (Campion, Latto, & Smith, 1983). Cortical color vision has bilateral representation. Thus it is preserved in all but the most severe disease states, which accounts for the interest that CVI children show for bright colors.

Development is affected in some visually impaired

TABLE 38–1. *Topographic Localization of Lesion Based on Behavior with Childhood Eye Disease*

Disease	Symptoms
Bilateral anterior visual pathways (1 year of age)	Nystagmus Head turn Holding objects of interest close to eyes
Bilateral retinal disease (e.g., ROP)	Eye poking
Cortical blindness	Variable visual performance Preference of colored object Staring at sun or bright lights Blind sight

children. Fraiberg (1977) studied children with vision impairment without other handicaps and found the following: Milestones for bonding and attachment are similar to sighted infants; mobility is delayed (average onset at 18 to 24 months); confusion of "I" and "you" pronouns occurs during the second and third years. Groenveld (1990) has described the pitfalls in developmental assessments of visually impaired children.

Attempts to link specific psychiatric conditions to specific ophthalmologic diagnoses have been made, for example, Leber's congenital amaurosis to infantile autism (Rogers & Newhart-Larson, 1989). Most authorities agree that "autistic-like" behavior occurs with increased frequency in any congenitally blinding condition (Parmelee, Cutsforth, & Jackson 1958). A balanced view would hold that vision impairment lowers the threshold for acquiring developmental delays and problems with social interaction. Individual and family vulnerabilities also play an important role.

More than 80 conditions cause a combination of vision and hearing impairment. Such children require special attention in terms of rehabilitation. In many cases hearing loss follows vision loss. Physicians can anticipate this situation when an accurate diagnosis has been made and prepare the child with early Braille training, which much more successful if undertaken before hearing impairment develops.

Two aspects of psychological reactions to vision loss are unique to children. First is the vulnerable child syndrome. The family's reaction to the child's illness may be exaggerated, based on the manner in which the physician presents the diagnosis and prognosis. A family may overreact to the diagnosis of vision loss and treat their child as though he were more generally vulnerable. Separation disorders can ensue. Second, the grieving process is exacerbated at many developmental stages as parents (and child) compare their child's progress with normally sighted counterparts. This recurrent grieving process also has been described for childhood disabilities.

SPECIFIC DISTURBANCES OF VISION

Visual functioning can be altered by the existence of visual *hallucinations*. In general, visual hallucinations are far more likely to be caused by organic brain disease than by mental illness. The new onset of visual hallucinations should prompt, at the minimum, a careful review of systems and physical examination.

Visual hallucinations may be more common in the visually impaired, but a change in character of frequency of hallucinations should prompt a careful evaluation for primary or secondary mental disorders. Cogan categorized visual hallucinations as either ictal or release, based on the presence of a stimulating cerebral cortex lesion (ictal) or reduced cortical stimulation (release) (Cogan, 1973). Ictal hallucinations are usually stereotyped, intermittent, and not associated with visual field defects during the interictal phase. Other seizure activity may also be present. Release hallucinations are usually continuous, intrude on the patient's activities (Lessell & Kylstra, 1988), and accompany visual field or acuity loss.

Palinopsia is characterized by persistence of an image after removal of the stimulus; i.e., a visual hallucination occurs that is modeled after the most recent visual experience. The persistent image usually lasts several minutes. Other disorders of perception also can occur, including size and shape distortion. An evolving hemianopic visual field defect can often be detected. The usual cause of palinopsia is a nondominant parietoccipital lesion (Bender & Feldman, 1968). Trazodone can rarely cause palinopsia (Hughes & Lessell, 1990). Symptoms remit when treatment is curtailed.

The *perception that the world is tilting* is caused by an infarct of the lateral medulla, the so-called Wallenberg syndrome (Hornston, 1974). Other neurologic symptoms can include dysphagia, ipsilateral face analgesia and contralateral body analgesia, ipsilateral ataxia, and Horner syndrome. *Upside-down reversal of vision* is usually a transient visual disturbance resulting from posterior circulation ischemia (Pamir, Ozer, Siva, et al., 1990).

REHABILITATION OF THE VISUALLY IMPAIRED

Disability caused by vision loss is defined according to an individual's life style. A pilot may be disabled with

20/25 vision, whereas a sedentary person might do well with vision as poor as 20/100. Psychiatrists and ophthalmologists must appreciate that a reduction in life style capability matters more than the specific disease entity and visual acuity in most cases.

Most of the United States apply vision criteria for obtaining a driver's license. On average, at least 20/40 vision in one eye (or both eyes together) and 30 degrees of visual field are required. Exceptions and restrictions can be made at a physician's recommendations. Loss of driver's license due to vision impairment occurs frequently in older people and can cause family upheaval. When an aging relative is unable to provide self-transportation, he or she must assume a more dependent role in the family. In turn, his or her family may become acutely aware of their relative's loss of functioning. Depressive feelings may follow.

Vision disability also occurs with hemianopias. Reading and driving may be affected. Children cope better than adults, because they learn compensatory strategies (Meienberg, Zangemeister, Rosenberg, et al., 1981).

Monocular blindness limits stereopsis (depth perception). A period of adjustment occurs with acquired monocular vision loss, but it should be emphasized that most suffer little restriction of activities.

Cataracts and Psychological Functioning

Ideally, rehabilitation involves correction of the underlying problem with full restitution of vision. Such is the case in most instances of cataracts. Modern cataract surgery has evolved into a rapid, safe, effective procedure. Modern lens implants for the management of aphakia have eliminated the need for postoperative thick spectacles and contact lenses. The patient's dissatisfaction with reduced vision is the usual indication for surgery. Restoration of good visual functioning improves life style, and there may be an accompanying improvement in mood and outlook on life. On the other hand, the existence of senile cataract is associated with increased mortality, irrespective of underlying specific systemic disease (e.g., diabetes mellitus). Cataracts presumably occur in patients with nonspecific systemic factors that affect overall health status.

When treatment cannot improve vision, rehabilitation takes on an entirely different meaning. The patient must learn to adapt to the vision problem and cope with the disability.

Adaptations include the use of other senses to help compensate for vision failure. Bar and stand magnifiers enlarge print and are useful in some cases. Closed-circuit television (CCTV), a reading machine, provides maximal enlargement of print. Because of the cumbersome procedure of using CCTV, however, many patients fail to find it beneficial. Braille is useful for patients with very low vision.

Mobility training helps patients with low vision. The use of a cane requires skill; rehabilitation centers often provide advice on its use. Guide dogs are also helpful. Dogs provide guidance in direction for outdoor activities, and they make excellent companions for isolated or single patients.

Many organizations exist to help visually impaired patients. Lighthouses for the Blind and the American Society for the Prevention of Blindness are organizations that provide direction to services for the visually impaired.

Impediments to Rehabilitation: Countertransference Reactions

There are many possible explanations for the symbolic role that eyes play in human psychological development. For example, the ocular surface (corneal epithelium) is rich in pain receptors and one of the most sensitive areas of the body. The thought or reality of an eye injury can thus elicit a sympathetic response in us. Eyes have assumed a symbolic significance perhaps because of their vulnerability. Thus one view holds that (fear of) eye injuries represents displaced castration anxiety. Less debatable is the fact that human interaction and communication often involves eye contact. So much of our understanding of the environment is based on vision.

No wonder, then, that an eye injury in a patient can elicit a countertransference response in a caregiver. The provider may respond by directing all of his or her attention to the eye injury, to the possible exclusion of other concomitant injuries. In one extreme case seen by me, a conspicuous eye injury completely distracted emergency room staff from a concomitant gunshot wound to the head!

On the other hand, care providers may be repulsed by eye injuries and fail to provide basic ophthalmic care for this reason. It is common for the ophthalmologist to be summoned to the emergency room for an "eye injury" only to find that emergency physicians have failed to measure visual acuity or to place a shield across an injured eye.

The above examples are general reactions to eye injuries. Countertransference and transference reactions to patients' situations occur regularly; characteristics of these reactions are determined by the personalities involved. It is my impression, though, that such reactions are more common with eye injuries than with most other types of wounds.

OPHTHALMOLOGY IN NEUROPSYCHIATRIC DIFFERENTIAL DIAGNOSIS: FUNCTIONAL VISION LOSS

To the ophthalmologist functional vision loss means complaint of vision failure in the absence of physical findings. The ophthalmologist's primary goal is to prove that vision is better than described. A secondary goal is to define the etiology of the functional loss.

Functional complaints can take many forms. Complaints of loss of acuity in one or both eyes are common. The physical examination must focus on objective findings to support the complaint. If the physical examination is normal, including refraction, the possibility of amblyopia must be considered. Amblyopia is failure of development of vision during childhood, caused by strabismus, occlusion (e.g., cataract, vitreous hemorrhage), or anisometropia (unequal refraction). In patients with amblyopia, the physical examination is normal, except for the presence of strabismus (may be subtle) or anisometropia.

A normal physical examination and no history supporting a diagnosis of amblyopia suggests functional loss. A number of methods for demonstrating better vision than claimed can be used, including rapidly changing Snellen letters and sizes to trick the patient into seeing several lines better than claimed, secretly fogging the good eye behind the phoropter (refraction instrument) in an attempt to get the patient to see better than claimed with the poor eye; stereo acuity testing (which requires good vision in both eyes for good performance); or handwriting sampling. A truly blind person usually signs his signature legibly. A functional blindness patient does not. Functionally ill patients follow their own eye movements in a mirror; i.e., they observe and visually fixate on their eyes as reflected in the mirror. Blind patients do not. Functional patients ambulate well; blind patients trip or bump objects.

Functional complaints of loss of visual field also are common. Tunnel vision is particularly prevalent among patients with conversion disorder. It can be demonstrated on automated perimetry, but the functional nature of the field loss is best demonstrated at the tanget screen. Here the examiner uses an absurdly large target to demonstrate that the visual field loss is dynamic; i.e., it worsens as the examination progresses.

The medical differential diagnosis of tunnel vision includes end-stage glaucoma, retinitis pigmentosa, and bilateral occipital lobe lesions with macular sparing. Vision failure with minimal physical findings also can occur with cone dystrophy. A careful eye examination should diagnose these conditions. By far, the most common cause of tunnel vision is psychiatric disturbance.

The psychiatric differential diagnosis includes malingering, sociopathy, hysteria, and depression. Child abuse should be suspected in childhood cases. The ophthalmologist should take the time to consider this differential diagnosis once functional loss is diagnosed.

Multiple sclerosis often is associated with an affective disturbance (depression or euphoria). In some individuals, a peripheral retinal phlebitis occurs (Arnold, Pepose, Hepler, et al., 1984). Anterior uveitis causing eye pain, redness, and photophobia has also been reported (Lim, Tessler, & Goodwin, 1991).

Neurologic complaints include unilateral (rarely bilateral) vision loss or diplopia. Unilateral or asymmetric vision loss causes an afferent pupil defect (Marcus-Gunn pupil). The pupil in the eye with poor vision reacts consensually, but when the same light is promptly shined directly at it, the pupil actually dilates. Beware that loss of vision due to multiple sclerosis (MS) does not preclude a measurement of 20/20, as there are many aspects of vision other than acuity that can cause symptoms and be damaged in MS (e.g., contrast sensitivity, color vision) (Good, Berg, Muci-Mendoza, et al., 1992).

Visual field loss (e.g., hemianopia) seldom accompanies MS. When the psychiatrist suspects MS, the ophthalmologist may aid in the diagnosis by performing a visual evoked potential test, which may be abnormal even in patients who note no vision problems. The psychiatrist can check for occult vision loss by examining color vision. The patient is asked to regard a bright red object with each eye separately and to compare color sensation. The perception of diminished color saturation in an eye suggests an optic nerve lesion.

The psychiatrist should be aware that debate exists concerning the indications for magnetic resonance imaging (MRI) to confirm MS in optic neuritis patients without other neurologic symptoms. For example, should the adult with unilateral optic neuritis be scanned to find a second demyelinating lesion? The psychological impact of diagnosing MS can be profound, and there is no current treatment for MS that is appropriate for such cases. The psychiatrist and neuroophthalmologist should consult with each other and their patient before endeavoring to diagnose MS.

Vogt-Koyanagi-Harada syndrome consists of uveitis, vitiligo, alopecia, and hearing loss. Vision loss can be profound and some individuals experience alterations in mental functions.

The acquired immunodeficiency diseases (AIDS) causes a constellation of neurologic and psychiatric syndromes (see Chapter 32). AIDS also causes a host of ophthalmologic diseases (see Table 38-2).

Cytomegalovirus (CMV) retinitis is the most dreaded and common ocular complication, occurring in about 25% of AIDS victims (Holland, Gottlieb, Yee, et al., 1982). Vision loss can be the initial manifestation of AIDS. CMV retinitis is usually unilateral initially but is soon followed by disease in the fellow eye. The retinitis occurs in the distribution of retinal vessels in a characteristic "cottage cheese and ketchup" configuration. An optic neuritis can also occur. Treatment with ganciclovir or foscarnet may arrest the disease. (Henderley, Freeman, Causey, et al., 1987).

Herpes zoster ophthalmitis is also common and may occur in advance of the full-blown AIDS syndrome. Keratitis and uveitis can occur. Dissemination to other organs is more likely in immunocompromised patients. Treatment with acyclovir is helpful.

What is a realistic role for psychiatrists in the management of eye disease in AIDS? First, they should be aware of vision-threatening diseases in AIDS and refer all patients for a baseline eye examination. Second, all AIDS patients should be questioned regularly (monthly) as to the presence of eye symptoms. It is not enough to ask: "Have you noticed any change in your vision?" Good acuity in one eye may mask problems in the fellow eye. It is most appropriate to ask the patient to check acuity in each eye separately. Any change should prompt an immediate ophthalmologic consultation. Visual field screening is of limited value in AIDS patients.

Metabolic diseases can affect both mental functioning and vision. In diabetes mellitus, hypoglycemia and ketoacidosis can cause profound alterations in consciousness, mood, and cognition. Ocular complications of diabetes mellitus are cataract, glaucoma, and retinopathy.

TABLE 38–2. *AIDS and the Eye*

Optic site	Disorder
Lids and cornea	Herpes zoster and herpes simplex keratitis Microbial keratitis Keratitis sicca Kaposi's sarcoma
Anterior segment	Iritis Recurrent syphilitic iritis
Posterior segment	CMV retinitis Herpes zoster uveitis and retinitis Fungal retinitis (*Candida, Cryptococcus*) Optic neuritis *Nocardia* Toxoplasmosis Idiopathic uveal effusion syndrome
Brain	Cortical blindness

Galactosemia causes developmental delay and hepatospleomegaly. The cause is a defect in galactose-1-phosphate uridyl transferase. The eye is affected by a lamellar-type cataract. Prompt removal of the cataract can save vision from irreversible amblyopia.

Wilson's disease is caused by a defect in copper metabolism. Low levels of copper-transporting protein accompany low serum and high tissue copper levels. Psychiatrists should be aware that mental status changes mimicking neurosis and psychosis can precede other physical signs. Eventually, liver failure and neurologic deterioration occurs. Descemet's membrane in the cornea stains with copper. This so-called Kaiser-Fleischer ring is an early sign of Wilson's disease. A subcapsular sunflower cataract occurs rarely. Treatment by chelation is beneficial.

Other neurodegenerative and metabolic conditions have associated eye findings. The interested reader is referred to Hoyt and Good (1991).

Retinitis pigmentosa is a disorder of photoreceptor cell function in the retina. The exact defect is unknown. Classically, rod cells are affected first (night vision), followed by cone cell degeneration (visual acuity). Usher syndrome consists of congenital deafness followed by retinitis pigmentosa. With type I Usher syndrome, psychosis may accompany hearing and vision loss (Fishman, Kumar, Joseph, et al., 1983).

Many conditions that cause vasculitis also affect the eyes. In some cases vasculitis involves the CNS and affects mental and neurologic functions (also see Chapter 33). Giant cell arteritis (temporal arteritis) causes jaw claudication, anorexia, and depression (Frohman, 1991). Eye findings include amaurosis fugax, ischemic optic neuropathy, and diplopia. Prompt recognition of this syndrome is essential. An elevation in the erythrocyte sedimentation rate aids in diagnosis. Vision loss in one eye often is followed by loss of vision in the fellow eye if treatment with prednisone is not initiated promptly. Table 38-3 lists eye findings that may be seen with conditions that cause vasculitis.

TRAUMA

Eye trauma occurs in settings that are of interest to the psychiatrist. In young children, poor supervision, child abuse, and low socioeconomic status are risk factors. Boys are affected more commonly than girls. In adolescents sports-related injuries are paramount (Strahlman, Elman, Daub, et al., 1990). The shaken baby syndrome causes

TABLE 38–3. *Vasculitis and the Eye*

Sjögren's syndrome
Sclerokeratitis
Diplopia
Retina arteritis and phlebitis
Uveitis
Papilledema
Amaurosis fugax
Ischemic optic neuritis
Optic neuritis

retinal hemorrhages, vitreous hemorrhage, and the risk of vision loss (Lambert, Johnson, & Hoyt, 1986). Neurologic damage is usually present. The cause is child abuse. A baby is shaken so hard the shearing force plus increased venous pressure cause retinal blood vessel rupture.

Signs suggesting child abuse include unexplained retinal hemorrhages. The differential diagnosis is limited to blood dyscrasias and trauma of another cause. Eyelid ecchymosis (black eye), hyphema, dislocated lens, and retinal detachment usually are caused by trauma. Eye wall lacerations and unilateral optic atrophy are also often traumatic in nature. The history should explore the possibility of child abuse in these cases, and of course other somatic stigmata of abuse should be sought (e.g., fractures and bruises). In adults alcoholism and aggressive behavior may play a role in the etiology of eye trauma.

Trauma can cause lid ecchymoses. Hyphemas, the result of blunt trauma, are defined as blood in the anterior chamber of the eye. Ruptured globes (eye wall lacerations) are the most serious and vision-threatening injury. Anterior globe lacerations carry the best prognosis; posterior lacerations usually damage the retina and carry a poor visual prognosis.

Suicide attempts occasionally cause serious eye injuries. A gun pointed at the side of the head, when discharged, can cause considerable damage to the orbital bones and eye. Retina sclopeteria (rupture of the choroid and retina caused by a shock wave) occurs after a bullet ricochets off the globe. When a gun is discharged in the mouth, one or both eyes may be lost if the patient survives. Suicide precautions and management of the considerable countertransference issues in these cases is paramount.

Management of eye trauma should include treatment aimed at psychosocial antecedents. Suspected child abuse must be reported. Many patients grow depressed after loss of an eye and require psychiatric intervention.

AUTOENUCLEATION

Because of the eye's symbolic position in mental functioning, it is sometimes chosen for self-mutilation. Patients who attempt autoenucleation (removal of their own eye) are virtually always decompensated schizophrenics. The eye is tethered in the orbit by extraocular muscles, facial attachments, and the optic nerve. Nevertheless, there are reports of people successfully removing an eye (Krauss, Yee, & Foos, 1984). The instrument for removal is usually fingers. The patient may be left with a temporal hemianopia in the surviving eye owing to traction on the optic chiasm.

Surgery to preserve vision is the first order of business. If the eye is only partially enucleated, the patient may decline signing the operative permit for repair. Emergency administrative intervention is then necessary. After surgery, psychiatric consultation should focus on the nature of the stress in the patient's life and his or her delusions. Delusions are often biblical in nature, or Oedipally based. "If thine eye offends thee, pluck it out."

Psychodynamically, most patients who attempt to injure their own eye have the perception of an Oedipal conflict. Sexual feelings or feelings of competition arouse great anxiety and superego concern. Patients follow their own superego directive to hurt themselves. The eye is chosen presumably as a displacement from autocastration.

Some patients are intensely relieved if enucleation was successful. However, it is no guarantee that they will not attempt the same injury to the other eye as pressure to act on delusions increases. Some are so driven to autoenucleation that there is hardly time to intervene psychiatrically. Restraints and constant vigilance are mandatory while rapid tranquilization is attempted. In some cases, emergency electroconvulsive therapy (ECT) is required.

PSYCHIATRIC IMPLICATIONS OF OCULAR TREATMENT

Topical eye medication is absorbed into the systemic circulation via the nasolacrimal duct and nasal mucosa, and so systemic side effects can occur. Topical beta-blockers used to treat glaucoma, for example, can cause depression. Theoretically, patients with a personal or family history of mood disorder might be more susceptible, but this theory has not been proved.

The potential for topical beta-blocker treatment to induce depression parallels that known for systemic beta-blockers (Avorn, Everett & Weiss, 1980). Concurrently administered systemic and topical beta-blockers could have an additive effect (Chamberlain, 1989). Other CNS effects of beta-blockers include hallucinations, lethargy, and confusion. Impotence also occurs (Berggren, 1990).

The mood disorder caused by beta-blockers is best characterized as a secondary mood disorder. Symptoms range from mild dysphoria and anergia to suicidal intentions and attempts. Patients prescribed topical beta-blockers should be warned of neurologic side effects. At follow-up, they should be asked specifically about changes in mental state, as they may not associate mood disturbance with eye drops. Depressed glaucoma patients on topical beta-blockers should have their topical beta-blocker withheld for 2 months to determine its relation to the depression. If necessary, pilocarpine or epinephrine-like compounds (or both) can be substituted for beta-blockers. Laser trabeculoplasty or surgery offer alternatives to medication in the occasional patient intolerant of medical treatment.

Topical anticholinergics are used in ophthalmology to provide surgical exposure and visibility (due to mydriasis) and to paralyze accommodation for accurate refraction. In adults, tropicamide is usually employed in the office setting. Tropicamide seldom causes psychiatric side effects. In children, stronger and potentially psychoactive agents such as cyclopentolate, scopolamine, and atropine are used. These agents are often used preoperatively in adults as well.

Systemic side effects from topically administered anticholinergic agents are similar to systemically administered preparations. Diminished salivation, increased heart rate, and a gastrointestinal motility disturbance may accompany a range of changes in mental status. Children can develop marked changes in activity level, ranging from drowsiness to a transient attention deficit disorder with marked hyperactivity. Transient delirium can occur as well. Regrettably, these side effects are often overlooked in young patients.

PSYCHIATRIC COMPLICATIONS OF OPHTHALMOLOGIC SURGERY

Acetazolamide is often prescribed for glaucoma in patients who respond inadequately to topical preparations. As many as 40% have to stop the medication due to side effects. Psychiatric side effects of depression, restlessness, fatigue, and diminished libido can occur (Epstein & Grant, 1977; Wallace & Fraunfelder, 1979).

Some elderly patients become overtly confused, disoriented, agitated, and even paranoid at the time of eye surgery. Contributing factors are preoperative anticholinergic eye drops used to dilate the pupil, dementia, and anxiety. Anticholinergic drops blur vision by paralyzing accommodation, and drapes used in ophthalmic "local" surgery obscure vision as well. Anxiolytic and narcotic agents may further compromise mental functioning. Escalating confusion, fear, and agitation often occur at the time of surgery. The physician is then faced with two critical decisions. (1) Should local anesthetic surgery be abandoned in preference for general anesthesia, usually at a later date? General anesthesia may be most appropriate for some patients. (2) Should more sedating medicine be used? The risk here is that more medication may cause a more pronounced paradoxical anxiety response.

The best management in this situation lies in its anticipation. Because the most potent preoperative risk factors are cognitive impairment and anxiety, screening for both should be part of preoperative assessment. An anesthesiologist attending a local anesthetic case often can control unwanted psychotropic side effects by administering a deeper anesthetic. Demented patients probably do better with general anesthesia. Patients often master an anticipated situation with education and guidance. In the end, the ophthalmologist should be prepared to cancel and postpone a case if necessary.

Ophthalmic surgery may be partially or entirely cosmetic. Blepharoplasty, wound repair, and even strabismus surgery offers the patient the opportunity for a more desirable appearance. The preoperative assessment for such patients should include at least an informal assessment of the patient's mental status and understanding of the goals and limitations of surgery. If a mental disorder precludes the patient from grasping the nature and goals of surgery, elective surgery should be postponed until the patient is able to fully comprehend the treatment.

Several questions may help to ferret out the potentially problematic patient. What does the patient hope to gain from the surgery? Why has the patient chosen the particular surgeon? In other words, is the patient's view of the doctor realistic? Or does the patient place undue esteem on the doctor? Is the patient's self-image appropriate? Distortion of body perception (i.e., body dysmorphic disorder) is a relative contraindication to cosmetic eye surgery. (See Chapter 34.)

OPHTHALMIC COMPLICATIONS OF PSYCHOTROPIC MEDICATIONS

Those who prescribe psychotropic medications should be aware of ocular side effects that occur in some predisposed patients. The first such important side effect is that of acute angle closure glaucoma precipitated by medications with anticholinergic preparations (e.g., tricyclic antidepressants and neuroleptics).

Glaucoma is categorized as narrow angle or open angle based on the mechanism of increased intraocular pressure. With open angle glaucoma, eye pressure increases irrespective of pupil size and position. The glaucoma process is usually insidious. Conversely, narrow angle glaucoma occurs in a predisposed eye when pupil size changes and causes blockage of the egress of aqueous humor from the posterior to the anterior eye chamber. Eye pressure usually increases suddenly, causing eye pain and redness. The patient may complain of seeing haloes around lights. This symptom is caused by corneal edema in angle closure glaucoma. Some patients have nausea and vomiting. Narrow angle patients are more likely hyperopic, Asian, and older. The diagnosis can be suspected if the patient shows a shallow inferior chamber on the so-called penlight test in which the depth of the anterior chamber is estimated by shining a light transversely across the eye. A deep anterior chamber places the patient at low risk for narrow angle glaucoma. Many patients know they suffer from glaucoma, but confirmation of narrow angle glaucoma often requires communication with the treating ophthalmologist.

Antidepressant medications can increase pupil size via anticholinergic side effects (Fraunfelder, 1976). Amitriptyline and doxepin have the greatest anticholinergic potency (Snyder & Yamamura, 1977) among antidepressants. Among the neuroleptics, chlorpromazine and thioridazine have strong anticholinergic properties.

Patients started on anticholinergic medication should be warned of symptoms of narrow angle glaucoma (Lieberman & Stoudemire, 1987). A history of any type of glaucoma is an indication for ophthalmologic consultation before implementing treatment, because even patients with open angle glaucoma can have a modest increase in intraocular pressure. This increase may be due to changes in uveoscleral outflow of aqueous, a mechanism of drainage that accounts for perhaps 10% of flow from the eye. In most cases systemically administered anticholinergic medication is tolerated by the glaucoma patient.

Thioridazine causes a retinal pigment epithelium disturbance that can produce blindness at doses greater than 800 mg/day (Grant, 1974). At doses between 600 and 800 mg, eye problems occur rarely (Heshe, Engelstoft, & Kirk, 1961). There are no reported cases of permanent visual loss at doses under 600 mg/day. Accumulation of the drug in the pigment epithelium layer can lead to continued visual loss even after the drug is stopped (Davidorff, 1973). Conversely, vision may improve even though the ocular fundus shows progressive atrophy (Marmor, 1990).

Patients treated with more than 600 mg of thioridazine per day should have regular eye examinations. Most ophthalmologists recommend examinations at 6-month intervals. An electroretinogram, visual field test, and fluorescein angiogram help detect early toxic changes, if suspected. Otherwise, the ophthalmologist should screen the patient with a routine examination with attention directed to the funduscopic examination. Clinicians are advised to avoid doses higher than 800 mg of thioridazine per day if possible and to warn patients at treatment initiation of the possibility of ocular side effects.

Chlorpromazine can cause an anterior cataract. The cataract is distinctive but rarely visually significant. Chlorpromazine probably does not affect the retina.

Lithium toxicity causes a variety of ocular motor disturbances, the most common of which is gaze-evoked nystagmus (Tesio, Porta, & Messa, 1987). Disorders of saccadic (fast) eye movements (Apte & Langston, 1983), oculogyric crisis (Sandyk, 1984), and opsoclonia (Cohen & Cohen, 1974) have also been reported. Downbeating nystagmus occurs occasionally. A clinicopathologic study showed that lithium stained the nuclei propositus hypoglossi in a patient with downbeating nystagmus (Corbett, 1989).

Oculogyric crisis also occurs with neuroleptic use. Haloperidol and piperazine phenothiazines are most likely to cause this problem (Leigh, Foley, Remler, et al., 1987), which usually can be managed with anticholinergic medication.

Blurred vision occurs occasionally in patients treated with anticholinergic drugs, which impair accommodation. The hyperope may experience blurring with distance and near vision. Others (emmetrope, myopes) could encounter blurred reading vision. A discrepancy between distance and near acuity (e.g., 20/20 distance, 20/100 at 14 inches) suggests the diagnosis. Treatment may include reducing dosage or waiting; the accommodative tone may improve with time. Pupilloconstrictor eye drops may be helpful for some patients.

ADVERSE DRUG INTERACTIONS

Tricyclic antidepressants (TCAs) act by blocking the reuptake of catecholamines at presynaptic nerve terminals. The effect of epinephrine can be prolonged in patients taking TCAs (Boakes, Laurence, & Teoh, 1973) but not tetracyclic antidepressants. Topical epinephrine for glaucoma management achieves some systemic absorption. To date, the potential systemic interaction with TCAs is theoretic only.

A potential interaction between ophthalmic beta-blockers used to treat glaucoma (timolol, betaxolol) and phenothiazines also exists. When systemic propranolol and chlorpromazine or thioridazine are administered together, blood levels of the beta-blocker and neuroleptic become elevated, which may be due to competition for metabolism in the liver (Gerber, Cantor, & Brater, 1990). Because of its systemic absorption, timolol has well-known adverse side effects (asthma). Although there are no reports to date of adverse interactions with topical beta-blocker treatment, some effects of elevated neuroleptic blood levels occur after prolonged treatment and they may not have been noted yet.

Calcium channel blocking agents find occasional use in the treatment of psychiatric disorders. When used concurrently with ophthalmic beta-blockers, they may cause conduction defects, heart failure, and hypotension in patient with impaired cardiovascular function (Pringle & MacEwen, 1987).

REFERENCES

APTE SN, & LANGSTON JW (1983). Permanent neurologic defects due to lithium toxicity. Ann Neurol 13:453–455.

ARNOLD AC, PEPOSE JS, HEPLER RS, ET AL. (1984). Retinal periphlebitis and retinitis in multiple sclerosis. I. Pathologic characteristics. Ophthalmology 91:255–262.

AVORN J, EVERETT DE, & WEISS S (1980). Increased antidepressant use in patients prescribed beta blockers. JAMA 244:2263.

BELLER R, HOYT CS, & MARG E (1981). Monocular cataracts: good visual results with neonatal surgery. Am J Ophthalmol 91:559–565.

BENDER M, & FELDMAN M (1968). Sabin A.J. Palinopsia. Brain 91:321–338.

BERGGREN L (1990). Pharmacological and clinical aspects of glaucoma therapy. Acta Ophthalmol 68:497–507.

BOAKES AJ, LAURENCE DR, TEOH PC (1973). Interactions between sympathomimetic and antidepressant agents in man. Br J Med 1:311–315.

CAMPION J, LATTO R, & SMITH YM (1983). Is blindsight an effect of scatter red light, spared cortex, and near threshold vision? Behav Brain Sci 6:423–486.

CHAMBERLAIN TJ (1989). Myocardial infraction after ophthalmic betaxolol. N Engl J Med 321:1342.

COGAN DG (1973). Visual hallucinations as release phenomena. Von Graefes Arch Klin Exp Ophthalmol 188:139–150.

COHEN WJ, & COHEN NH (1974). Lithium carbonate, haloperidol, and irreversible brain damage. JAMA 230:1283–1287.

CORBETT JJ (1989). Downbeating nystagmus and other ocular motor defects caused by lithium toxicity. Neurology 39:481–485.

Cryotherapy for Retinopathy of Prematurity Cooperative Group (1990): Multicenter trial of cryotherapy for retinopathy of prematurity. Arch Ophthalmol 108:195–204.

DAVIDORFF FH (1973). Thioridazine pigmentary retinopathy. Arch Ophthalmol 90:251–255.

EPSTEIN DL, & GRANT WM (1977). Carbonic anhydrase inhibitor side effects. Arch Ophthalmol 95:1378–1382.

FISHMAN GA, KUMAR A, JOSEPH ME, ET AL. (1983). Usher's syndrome: ophthalmic and neuro-otologic findings suggesting genetic heterogenecity. Arch Ophthalmol 101:1366–1374.

FRAIBERG S (1977). Insights from the blind: comparative studies of blind and sighted infants. New York: Basic Books.

FRAUNFELDER FT (1976). Drug induced ocular side effects and drug interactions (pp. 60–63). Philadelphia: Lea & Febiger.

FROHMAN L (1991). Neuro-ophthalmic manifestations of vasculitis. In Proceedings of North American Neuroophthalmology Society, Park City, Utah.

GERBER SL, CANTOR LB, & BRATER DC (1990). Systemic drug interaction with topical glaucoma medications. Surv Ophthalmol 35:205–218.

GOOD WV, BERG BO, MUCI-MENDOZA R, ET AL. (1992). Optic neuritis in children with poor recovery of vision. Aust NZ J Ophthalmol. In press.

GOOD WV, & HOYT CS (1989). Neurobehavioral adaptations of visually impaired children. Int Ophthalmol Clin 29:57–60.

GRANT WM (1974). Toxicology of the eye (pp. 1005–1008). Springfield, IL: Charles C Thomas.

GROENVELD M (1990). The dilemma of assessing the visually impaired child. Dev Med Child Neurol 32:1105–1113.

HENDERLEY DE, FREEMAN WR, CAUSEY DM, ET AL. (1987). Cytomegalovirus and response to therapy with gancyclovir. Ophthalmology 94:425–432.

HESHE J, ENGELSTOFT FH, & KIRK L (1961). Retinal injury developing under thioridazine treatment. Nord Psykiatr J 15:442–447.

HOLLAND GN, GOTTLIEB MS, YEE RD, ET AL. (1982). Ocular disorders associated with a new series acquired cellular immunodeficiency syndrome. Am J Ophthalmol 93:393–402.

HORNSTEN G (1974) Wallenberger's syndrome. I. General symptomatology with special reference to visual disturbances and imbalance. Acta Neurol Scand 50:434–446.

HOYT CS, & GOOD WV (1991). Pediatric eye disease. In A RUDOLPH (Ed.), Textbook of pediatrics.

HUGHES MS, & LESSELL S (1990). Trazadone-induced palinopsia. Arch Ophthalmol 108:399–400.

JAN JE, FREEMAN RD, MCCORMICK AQ, ET AL. (1983). Eye pressing by visually impaired children. Dev Med Child Neurol 25:755–762.

JAN JE, GROENVELD M, SYKANADA AM, ET AL. (1987). Behavioral characteristics of children with permanent cortical visual impairment. Dev Med Child Neurol 29:571–576.

KAHN HA, & MOORHEAD HB (1973) Statistics on blindness in the model reporting area, 1969–1970. U.S. Department of Health, Education and Welfare Publ. No. (NIH) 73-427. Washington, DC: U.S. Government Printing Office.

KRAUSS HR, YEE RD, & FOOS RY (1984). Autoenucleation. Surv Ophthalmol 29:179–187.

LAMBERT SR, JOHNSON TE, & HOYT CS (1986). Optic nerve sheath and retinal hemorrhage associated with the shaken baby syndrome. Arch Ophthalmol 104:1509–1512.

LEIGH RI, FOLEY JM, REMLER BF, ET AL. (1987). Oculogyric crisis: a syndrome of thought disorder and ocular deviation. Ann Neurol 22:13–17.

LESSELL S, & KYLSTRA J (1988) Exercise-induced visual hallucinations. J Clin Neuroophthalmol 8:81–83.

LIEBERMAN E, & STOUDEMIRE A (1987). Use of tricyclic antidepressant in patients with glaucoma. Psychomatics 28:145–148.

LIM JI, TESSLER HH, & GOODWIN JA (1991). Anterior granulomatous uveitis in patients with multiple sclerosis. Ophthalmology 98:142–145.

MARMOR MF (1990). Is thioridazine retinopathy progressive relationship of pigmentary changes to visual function. Br J Ophthalmol 74:738–742.

MEIENBERG O, ZANGEMEISTER WH, ROSENBERG M, ET AL. (1981). Saccadic eye movement strategies in patients with homanymous hemianopia. Ann Neurol 9:537–544.

PAMIR NM, OZAR AF, SIVA A, ET AL. (1990) "Upside down" reversal of vision after third ventriculostomy. J Clin Neuroophthalmol 10:271–272.

PARMELEE AH, CUTSFORTH MG, & JACKSON CL (1958). Mental development of children with blindness due to retrolental fibroplasia. Am J Dis Child 96:641–654.

PRINGLE SD, & MACEWEN CJ (1987). Severe bradycardia due to interaction of timolol eye drops and verapamil. BMJ 294:155–156.

ROGERS SJ, & NEWHART-LARSON S (1989). Characteristics of infantile autism in five children with Leber's congenital amaurosis. Dev Med Child Neurol 31:598–608.

SANDYK R (1984). Oculogyric crisis induced by lithium carbonate. Eur Neurol 23:92–94.

SCHEIDER GE (1969). Two visual systems: brain mechanisms for localization and discrimination are dissociated by tectal and cortical lesions. Science 163:895–902.

SNYDER SH, & YAMAMURA HI (1977). Antidepressants and the muscarinic actylcholine receptor. Arch Gen Psychiatry 34:236–239.

STRAHLMAN E, ELMAN M, DAUB E ET AL. (1990). Causes of pediatric eye injuries. Arch Ophthalmol 108:603–608.

TESIO L, PORTA GL, & MESSA E (1987). Cerebellar syndrome in lithium poisoning: a case of partial recovery. J Neurol Neurosurg Psychiatry 50:235.

WALLACE TR, & FRAUNFELDER FT (1979). Decreased libido—a side effect of carbonic anhydrase inhibitor. Ann Ophthalmol 11:1563–1566.

39 Diagnostic assessment and treatment of impotence

DANIEL J. BLAKE, M.D., Ph.D., AND JONATHAN P. JAROW, M.D.

Although impotence is undoubtedly a common complaint, valid and reliable epidemiologic data concerning it are lacking. An important reason for this lack is that the diagnosis is difficult to verify independently from the patient's complaint of erectile failure. Perhaps nearly all men experience failure of erection from time to time, and many of them make their anxieties known to physicians who then become suspicious of the medications the man is taking, problems in his marriage, or other worries. Levine (1989) stated that an estimated one man in ten over age 21 endures or has endured erectile dysfunction at one time or another. Considering all men who may have had at least one episode of erectile failure, the percentage is probably higher. For the United States in 1985, impotence accounted for more than 400,000 outpatient visits to physicians and more than 30,000 hospital admissions (Krane, Goldstein, & De Tejada, 1989).

When men who were identified as medical patients were asked about erectile dysfunction, more than one-third (34%) complained that they were impotent (Slag, Morley, Elson, et al., 1983). When Slag et al. examined 188 consenting impotent men (from a larger group of men identifying themselves as impotent) they found that 14% of the cases were presumably psychogenic and 79% of the men suffered from a variety of medical maladies that were believed the cause of the erectile failure; 7% of cases were idiopathic.

In a 1984 review of the psychiatric aspects of impotence, there was no clear agreement among investigators about the nature, the importance, or even the existence of causal relations between psychopathologic syndromes (e.g., major depression or schizophrenia) and erectile dysfunction as such (Blake, 1984). Little has changed since that review. Roose, Glassman, Walsh, and associates (1982) showed, in a study using nocturnal penile tumescence measurements, that the capacity for nocturnal erections impaired during severe depression was reestablished during recovery from depression. This finding was supported by Thase, Reynolds, and Jennings (1988a,b). In another study by Thase, Reynolds, Glanz, and colleagues (1987), however, erectile tumescence was present in depressive illness but the total duration of nocturnal tumescence was decreased. We know of no other quantitative studies concerning the possible relation of erectile insufficiency and psychiatric diagnosis. Although little is known of a relation between psychiatric syndromes and erectile failure, there has long been a belief that neurotic or "psychogenic" mechanisms frequently underlie impotence. This issue is discussed later.

ERECTILE MECHANISMS

The penis is comprised of three cylindical structures: the paired corpora cavernosa, or erectile bodies, and the corpus spongiosum, which contains the urethra. The cavernosal bodies of the penis are filled with interconnected lacunar spaces or sinuses, forming a three-dimensional maze. An erection begins with relaxation of trabecular smooth muscle around the lacunae; smooth muscle relaxation of the helicine arteries permits rapid filling of the lacunar space with blood. As pressure builds in the lacunar space the trabecular walls of the lacunar space press against the surrounding tunica albuginea, compressing the plexus of subtunical venules (see Figure 39-1). This step, in turn, diminishes venous outflow from the cavernosa. The pressure in the cavernosa thus builds, and the penis becomes rigid. Detumescence occurs with the contraction of penile smooth muscle within the sinusoids, allowing rapid drainage of the previously compressed venules and impeding arterial filling. The arterial blood supply to the penis originates from the internal pudendal artery, which gives rise to the three pairs of penile arteries: dorsal arteries, ventral (or spongiosa) arteries, and cavernosal arteries.

Erections may be elicited by local tactile reflex mechanisms, involving spinal reflex pathways, and by the brain (psychogenic erection). For reflexogenic erections

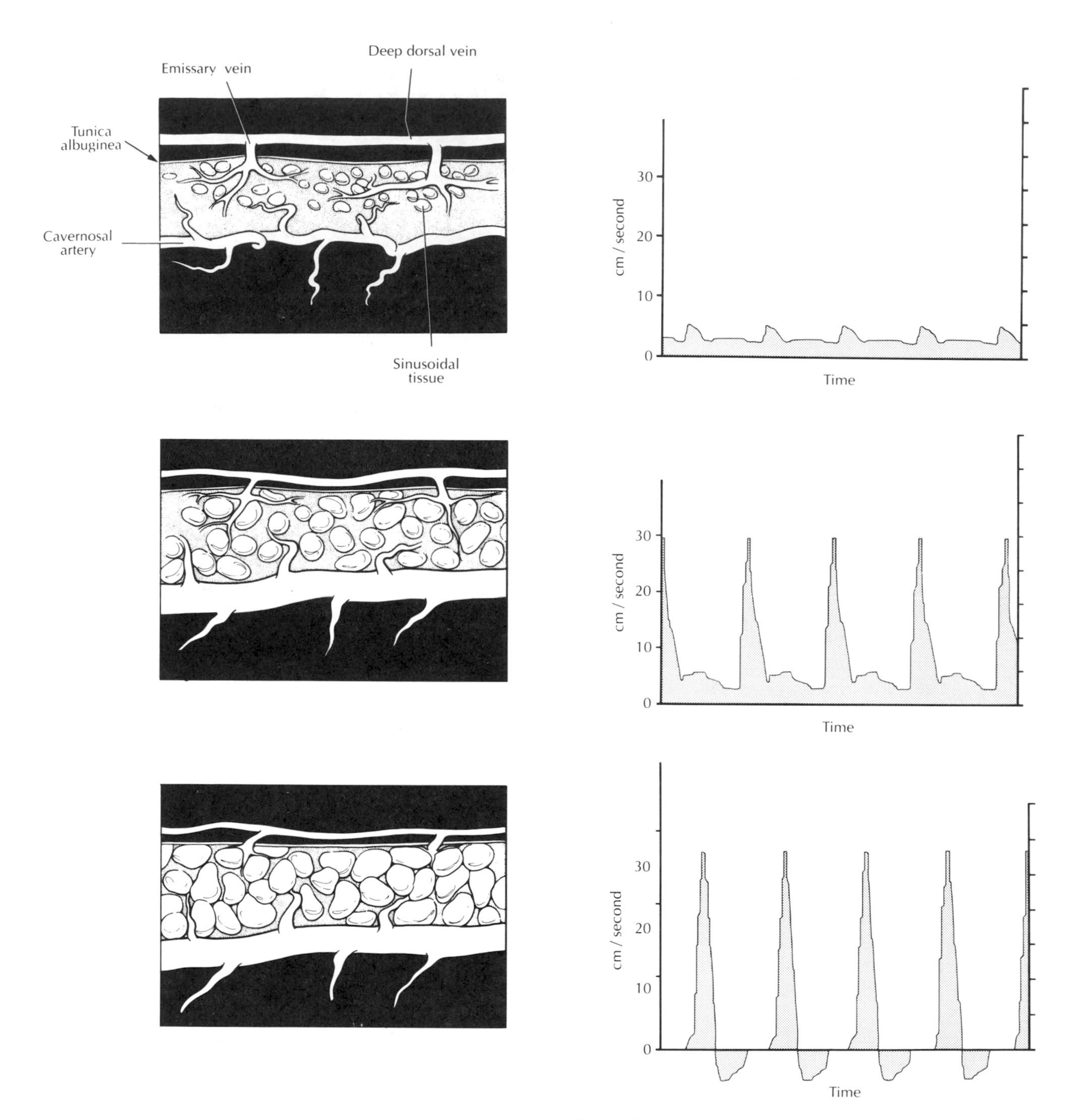

FIG. 39–1. Corporal bodies and cavernosal artery blood flow during the phases of an erection. In the flaccid state the sinusoidal smooth muscle is contracted with high resistance to cavernosal arterial blood flow. As the smooth muscle relaxes, blood flow increases and the sinusoidal spaces fill with blood. During a full erection, the engorged sinusoids compress the emissary veins, thereby increasing intracorporeal pressure (note reversal of diastolic blood flow) to produce adequate rigidity for sexual intercourse.

somatosensory impulses from the penis travel via the dorsal penile nerve to the pudendal nerve and thence to the sacral cord. Efferent fibers to the penis travel in the parasympathetic outflow from the sacral cord, giving rise to the pelvic nerve and then the cavernosal nerve. The principal staging area for brain transmission for psychogenic erection to the spinal cord is believed to be the medial anterior hypothalamus (Groat & Steers, 1988; Hart & Leedy, 1985). The hypothalamus receives projections from many parts of the forebrain and projects to the nuclei of both sympathetic and parasympathetic connections to the penis.

CAUSES OF ERECTILE FAILURE

When a complaint of impotence is encountered it is helpful to consider the complaint from the standpoint of: (1) local causes; (2) segmental causes; (3) systemic causes; and (4) psychological causes.

Local Pathology

An important form of vasculogenic impotence consists in excessive venous drainage through the subtunical venules. In effect, failure of the venoocclusive mechanism "short circuits" otherwise adequate perfusion of the corpora cavernosa. Corporal venoocclusive failure may reflect inadequate relaxation of trabecular smooth muscle, which in turn may reflect anxiety or damage to parasympathetic innervation (see also below). Venoocclusive failure may also reflect loss of fibroelastic compliance of the trabeculae, so adequate compression of the subtunical venules against the tunica albuginea cannot occur. Such loss of fibroelastic compliance can occur as a feature of aging (Cerami, et al., 1987; Krane, Goldstein, & De Tejada, 1989) and possibly hypercholestrolemia. Priapism, penile trauma, and penile surgery also can adversely affect fibroelastic compliance. Michal and Ruzbarsky (1980) showed that there was significant narrowing of the penile artery in men over age 38, associated with obstruction of arterioles. These findings were more prominent among diabetic patients. Ischemic changes within the corpora may also affect fibroelastic compliance.

Peyronie's disease of the penis usually presents with penile curvature, painful intercourse, or the presence of a hard knot on the dorsal penis. The source of offense is a fibrous plaque involving the tunica albuginea, usually on the dorsal surface of the corpora. Most cases of Peyronie's plaque occur during the fourth and fifth decades of life (Krane, 1986). Impotence in men with Peyronie's disease is usually caused by venoocclusive dysfunction.

Segmental Pathology

It is at the segmental level that most causes of erection insufficiency or failure probably are caused. Perhaps the most common cause of impotence is arterial occlusive disease proximal to the penis (Herman, Adar, & Rubinstein, 1978; Kaiser & Korenman, 1988). It involves occlusive disease in the pelvic arterial supply to the penis, which may derive from pelvic trauma, pelvic irradiation, or, most commonly, atherosclerosis.

Another important regional cause of erectile insufficiency is the complications of diabetes mellitus. A major contributor to diabetic impotence is vascular pathology, and vascular occlusive disease may be the most important factor in diabetic erectile insufficiency (Herman, Adar, & Rubenstein, 1978; Lehman & Jacobs, 1983). Perhaps the next most important cause of erectile insufficiency in diabetes is segmental neuropathy. Several investigators have found evidence of a neurologic basis for some cases of diabetic impotence. Ellenberg's (1971) clinical studies suggested a neurologic basis for diabetic erectile failure. Kolodny, Kahn, Goldstein, et al. (1973) compared a group of impotent diabetic men with a group of nonimpotent diabetic men and was able to differentiate them only on the parameter of greater peripheral neuropathy in the impotent group. Farming, Glocer, Fox, and associates (1974) found from a postmortem study of five impotent diabetic men that there was thickening of the corpus cavernosal nerve associated with areas of beading and vacuolation, not found in control material. Lin and Bradley (1985) studied dorsal penile nerve conduction in insulin-dependent diabetics and found it to be slower than in controls.

A less common but well-known segmental cause of erectile failure is trauma: blunt, penetrating, or iatrogenic. Pelvic fractures after motor vehicle accidents may result in either neurogenic or vasculogenic impotence. Pelvic surgery for either prostatic or rectal disease may injure the nerves responsible for erection.

Systemic Pathology

Systemic causes of erectile dysfunction are, perforce, complex. Perhaps the most investigated are endocrinologic problems. Most systemic illness that affects erectile function probably involves endocrinopathy as a final common pathway, although intercurrent neurologic vascular damage may also be implicated. The importance of adequate levels of androgens for libido is well established, but the role of these hormones in the mechanics of erection is unclear (Bancroft & Wu, 1983; Kwan, Greenleaf, Mann, et al., 1983). Although androgen receptor sites have been demonstrated in those areas of the central nervous system (CNS) believed to be involved with erection, erection can still be achieved by men with castration levels of plasma testosterone (Bancroft & Wu, 1983). The association of hyperprolactinemia with erectile dysfunction is well known. Loss of sexual arousal and impotence are symptoms associated with hyperprolactinemia, such as can occur with chronic renal failure, use of neuroleptic drugs, and prolactin-secreting pituitary adenoma (Carter, Tyson, Tolis, et al., 1978; Franks, Jacob, Martin, et al., 1978; Krane, Goldstein, & De Tejada, 1989; Spark, White, & Connolly, 1980). Thyroid disease is occasionally associated with erectile insufficiency. Both hyperthyroidism and hypothyroidism may be associated with erectile dysfunction, with the former most often involving loss of arousal and the latter with loss of erection and arousal (Krane, Goldstein, & De Tejada, 1989).

Erectile dysfunction also can occur as a side effect of drugs. Although there are numerous tables in texts and articles of pharmacologic agents that are associated with impotence, the evidence in the case of nearly every drug is either anecdotal or uncontrolled. Quantitative and reliable measures of erectile function, such as nocturnal penile tumescence and penile buckling pressure, have not been systematically used in studies of drug effects on erection. Moreover, mechanisms by which drugs may affect the erectile process have not been elucidated, although there is considerable presumptive evidence. Drugs can interfere with androgenic receptor activity as has been noted with the H2 receptor blocker cimetidine; or they can raise prolactin levels because of dopamine blockade, as with the neuroleptics. Central serotonergic and noradrenergic pathways associated presumably with sexual function can be disrupted by various antidepressant agents, particularly the tricyclic compounds. Most of the antihypertensive agents have been implicated in male (as well as female) sexual dysfunction at one time or another. The adverse effects of antihypertensive drugs may be mediated centrally via neurotransmitter systems involved in sexual functioning, or peripherally with autonomic effects (methyldopa, guanethidine, clonidine, beta-blockers). Vasodilators, such as hydralazine, also have been implicated in impotence. However, it is important to note that none of these drugs typically produce impotence in young men with normal erectile function, and that most patients receiving these drugs have underlying systemic medical illnesses that are associated with impotence.

Psychological Causes

In the past it was reported that as many as 90% of cases of impotence were of psychological origin (Hastings, 1963; Stafford-Clark, 1954). In point of fact, the percentage of cases known to be of psychological origin or of any other etiology is unknown for the reasons discussed at the outset of this chapter. Apart from the epidemiologic difficulties encountered when addressing this question, there is all manner of erectile unreliability or insufficiency, of both psychological and nonpsychological origins, that would not reasonably be called impotence. Some men lose their erection on certain occasions but not others. This circumstance could reflect, for example, an arterial steal syndrome (Wagner & Green, 1981) or a setting-dependent psychological phenomenon. Although we do not know what percentage of cases of erectile dysfunction ultimately are psychogenic, it is likely that emotional factors figure prominently in many cases.

Clinical formulation of psychogenic sexual dysfunction, including erectile dysfunction, still depends today a good deal on ideologic factors or "schools of thought." For example, in past decades it was fashionable to accumulate and interpret psychological data according to the method and theory of psychoanalysis, with its emphasis on the critical stages of psychosexual development. In more recent years, since the late 1960s, there has been emphasis on "here and now" sexual meanings and issues, principally performance anxiety, which arises in the immediate sexual setting and destroys the mental attitude of sexual desire that is necessary for erection. There has also been attention given to the relation of sexual dysfunction and clinical psychiatric diagnosis as discussed above. Another approach to psychogenic sexual disorders related to the concept of performance anxiety has been that of behavioral theory and conditioning (Birk, 1980; Ginsberg, Frosch, & Shapiro, 1972; Wolpe & Lazarus, 1966). Masters and Johnson (1970) thought that performance failure was conditioned by the demand for sexual performance placed by the culture on men and exercised directly through the family, peer group, and partner. Birk (1980) believed that sexual dysfunctions were learned aberrations of the normal sexual response cycle. Wolpe and Lazarus (1966) thought that men acquired sexual inhibitions and fears through a variety of malconditioning experiences. Finally, there is an old but pertinent idea that a man may be impotent because of his basic constitutional makeup. Ferenczi (1916) believed that certain men suffered from what he termed "congenital sexual inferiority." Cooper (1969), in a discussion of male sexual inadequacy, appeared to favor the clinical concept of constitutional impotence. He suspected a combined genetic and developmental basis.

It is probably fair to say that no one school of thought concerning psychogenic erectile failure is valid at the expense of others. All throw light on some area of the problem. Certainly the persistence of early wishes and fears, as described by psychoanalytic observers, probably sets the stage for performance anxiety in certain vulnerable men. This vulnerability with attendant maladaptive wishes and fears was brought about in part by the stressful conditioning experiences of these men. Kaplan (1974) suggested that some men may have erectile failure in response to stress because of a specific psychosomatic vulnerability. She speculated that this vulnerability involves experiential and constitutional factors.

DIAGNOSIS OF ERECTILE DISORDERS

Some writers advocate a rigorous diagnostic regimen. However, because patients do not die of impotence, many clinicians utilize a goal-oriented approach to the evaluation of organic impotence in order to limit the expense and risk exposed to the patient. In recent years there have emerged a number of techniques and procedures valuable in the diagnosis and treatment of erectile dysfunction (see Table 39-1). The diagnostic evaluation begins with routine and psychosexual anamnesis followed by physical examination. Many clinicians advocate the use of psychological testing, such as the Minnesota Multiphasic Personality Index (MMPI), but these tests are of uncertain, perhaps doubtful, value given the individuality that characterizes cases of erectile difficulty.

TABLE 39–1. *Diagnostic Tests Utilized for Impotence Evaluation*

Test	Accuracy	Cost
Endocrine		
Testosterone	**	*
Free testosterone	****	**
Prolactin	****	*
Blood glucose	***	*
Nocturnal testing		
Snap gauge	*	*
RigaScan	**	**
Visual sexual stimulation	***	**
NPT	****	****
Neurological testing		
Biothesiometry	*	*
Nerve conduction	**	**
Vascular testing	**	*
Pharmacologic erection test		
Penile-brachial index	*	*
Duplex ultrasonography	***	**
Arteriography	****	****
Dynamic infusion	**	***
Cavernosometry/ cavernosography		

* = least; **** = most.

A variety of specific physiologic tests may then be employed to assess erectile function, such as induction of artificial erection with intracavernosal injection of vasoactive agents, determination of the penile/brachial blood pressure ratio, cavernometric studies, and nocturnal penile tumescence (NPT) measurement. In addition, androgen and gonadotropin blood levels may be measured, as may local and regional blood flow. Neurologic function is often assessed with the bulbocavernosus reflex latency time measurement. Many clinicians prefer to assess the erection with a provocative visual stimulus such as an erotic videotape. This procedure is known as visual sexual stimulation (VSS), and it can rule out most nonpsychological causes of erectile failure (Ruutu, Virtanen, Bjorn, et al., 1988).

Biomedical Evaluation

In clinics where thorough evaluation of erectile complaints is performed and pursued, history, physical examination, and routine blood studies are followed by NPT determinations, often for three consecutive nights. This procedure is followed in an effort to determine whether the patient has psychogenic versus organic erectile dysfunction. There are many reports and studies concerning the use of NPT for the evaluation of erectile dysfunction, and they have been reviewed by Ware (1987). The basic strategy with NPT monitoring is to observe erectile behavior directly and quantitatively by taking advantage of the phenomenon of periodic penile erection during sleep, usually rapid-eye-movement (REM) sleep. As simple and as powerful as this idea seems, there are several potential pitfalls. One is that the standard procedure of measuring circumferential expansion using a strain gauge as an index of tumescence may not reveal rigidity. Ware and others have also studied and recommended the use of penile buckling force measurements to assess rigidity. It is also important to take into account age-related changes and sleep quality in the penile tumescence–sleep cycle relation, as well as the effects of possible concurrent depressive illness. An important mistake to be avoided is concluding that NPT studies can differentiate psychological from organic impotence with 100% accuracy. This point cannot be determined without reliance on an independent gold standard of diagnosis, which does not exist. There is no certainty that psychological mechanisms cannot affect central erectile mechanisms during sleep in some persons so that these persons would appear "organic" on NPT and thus be false-positive organic cases. Studies have demonstrated altered sleep patterns in depressed patients that produce misleading NPT results. A major drawback of NPT analysis is its high cost. Usually two or more nights spent in the sleep laboratory are required. Low cost, convenient alternatives to NPT, such as snap gauges, are available but are much less reliable. There is a system for in-the-home NPT using strain gauges and a computer pack for monitoring. The advantage of this system (RigiScan) is the reduced cost. However, the interpretation of borderline studies is difficult because of the lack of electroencephalographic (EEG) monitoring, and buckling force is not measured with this system.

If NPT or visual provocation studies disclose normal erections, especially if these are confirmed visually by the patient using wake-up or photographic confirmation, further quantitative efforts to diagnose impotence usually cease. Some urologists, however, may want to produce an erection in the patient using an injected vasoactive compound (see below) at this point. If tumescence measurements increase suspicion of a local, segmental, or systemic cause, specific neurologic, vascular, and endrocrinologic studies are then obtained.

Penile and pelvic neurologic studies are not routinely performed unless a neurologic disorder is suspected. Many patients with neurogenic impotence have overt signs on history and physical examination (i.e., spinal cord injury) . Unfortunately, it is not currently possible to directly assess the functional status of the cavernosal autonomic efferent nerves. Instead, neurologic testing focuses on the afferent somatic arc. Often vibratory sensory loss on the penis is investigated first using biothesiometry and, if found, bulbocavernosal reflex latency is measured in search of evidence of sacral outflow neuropathology.

Hypogonadism is assessed through repeated serum testosterone determinations, looking for consistently low levels. Whether a low testosterone level is primary or secondary is determined by measuring luteinizing hormone (LH) levels: If LH is low, secondary hypogonadism is suggested; if LH is high, primary testicular failure is suspected. It is important to differentiate between these two disorders, as they are managed differently. Prolactin levels are also measured, as are thyroid hormones, because these hormones can affect the hypothalamus-pituitary-testicular axis. In addition, all impotent patients should be screened for occult diabetes mellitus with a fasting blood glucose determination.

Assessment of the vascular sufficiency of the penis, its arterial supply and venous drainage, have already been touched on above. There are a number of clinical measurements available, and selected ones are chosen according to convenience, accuracy, expense, and degree of invasiveness. Virag (1982) first showed that intracorporeal injection of a vasoactive agent can produce an erection. Stimulation of an erection with papaverine or other vasoactive substances including phentolamine and prostaglandin E1 have since become established techniques for diagnosing erectile failure. Although most urologists agree that a full erectile response to papaverine excludes the presence of a significant vascular lesion, an impaired response does not diagnose one. Men with NPT-documented psychogenic impotence sometimes fail to obtain a full erection in response to papaverine (Buvat, Lemaire, Marcolin, et al., 1987). Failure may be due to anxiety associated with the test situation, which then leads to increased adrenergic tone or possibly cavernosal leakage. Erectile failure with papaverine therefore requires further evaluation before one can establish a diagnosis of either venous or arterial insufficiency. However, the pharmacologic erection test is useful as a screening procedure to identify patients for further testing.

There are numerous studies available to assess the arterial supply of the corpora cavernosa. The simplest test is the blood pressure measurement of the flaccid penis using a Doppler stethoscope. Comparison of penile blood pressure to that of the upper extremity produces an index that has been helpful for identifying vascular occlusive disease. This test is accurate when positive. However, there are significant false negatives, as pressure rather than flow is measured, the artery being tested is not responsible for erections, and the penis is flaccid at the time of measurement. Duplex ultrasonography can provide a functional evaluation of the cavernosal arteries following pharmacologic stimulation. This relatively noninvasive study is highly accurate. Internal pudendal arteriography is the gold standard diagnostic study for evaluation of the penile vasculature. However, this test reveals vascular anatomy and does not assess function. In addition, the test is both costly and invasive.

Venogenic impotence is diagnosed using dynamic infusion cavernosometry and cavernosography (DICC). Saline infusion of the penis is used to assess the veno-occlusive mechanism. The flow rate required to produce and maintain an erection as well as the rate of decay of intracorporeal pressure following discontinuation of saline infusion are measured. Studies have shown that the accuracy of this diagnostic study is significantly improved when performed using pharmacologic stimulation with vasoactive agents. However, because of the lack of controlled studies and a gold standard the sensitivity and specificity of cavernosometry is unknown. Cavernosography is performed using dilute nonionic contrast and provides an anatomic evaluation of the incompetent venous pathways.

Psychological Evaluation

Psychological evaluation of erectile dysfunction may be employed at any point during the impotence evaluation. Many centers perform a screening psychological evaluation at the onset, whereas others may not utilize it until a diagnosis of psychogenic impotence is suspected on the basis of specific testing. Usually a psychological diagnostic interview is undertaken. It may be fairly brief and informal, performed by a consulting or "team" psychiatrist or psychologist in the urology clinic; or it may be more prolonged and formal (Beutler, Ware, & Karacan, 1978; Karacan, Salis, & Williams, 1978). The diagnostic interview can never be conclusive, but in expert hands it can generate a case argument for or against a psychological basis for the erectile dysfunction, whether it is perceived in terms of performance anxiety, major psychiatric disturbance, or psychodynamic issues. Beutler, Ware, and Karacan (1978) described ten indicators that suggest psychogenic impotence. They include sudden onset of impotence, presence of early morning erections, better function with some partners than others, obsession with the penis, marital discord, diminished libido, performance pressure, discordant sexual object choice, sexual misinformation, and the long-standing absence of a sexual relationship.

Although psychological and personality testing are commonly part of the evaluation of erectile dysfunction, the test instruments cannot yield an answer to the question of psychogenic orgin. A number of years ago Beutler, Karacan, Anch, and colleagues (1975) examined the utility of the MMPI in discriminating a psychogenic group from a nonpsychogenic group of impotent men, divided using NPT criteria for organic erectile failure. Although they found that men who scored more than 60 on the MF scale or scored above 70 on any scale were statistically the most like to by psychogenic by NPT criteria, they could not thereby prove that a psychological cause was present. Later, several investigators were unable to reproduce the Beutler group results (Marshall, Surridge, & Delva, 1980; Robiner, Godec, Cass, et al., 1982; Staples, Fischer, Shapiro, et al., 1980).

RECOMMENDED EVALUATION APPROACH

There is no specific algorithm for the evaluation of all impotent patients. Instead, the evaluation should be tailored to the characteristics of each individual patient, the orientation of the treating physician, and availability of diagnostic and therapeutic modalities. Our own approach is to base the evaluation on the patient's own desires for therapy after performing routine screening diagnostic procedures. All patients should undergo an initial evaluation with a complete history, physical examination, and testosterone, prolactin, and blood glucose assays. The first step in management of the impotent patient is to differentiate between psychogenic and organic impotence. Although nonsurgical forms of therapy can be used in men with psychogenic impotence, sex therapy is more likely to provide a specific and long-lasting result.

Most patients with psychogenic impotence can be identified upon history alone. The sudden onset of intermittent erectile failure is consistent with a psychogenic etiology. In addition, the presence of full and long-lasting erections nocturnally or with masturbation implies a psychogenic etiology for those occasions of erectile failure. Finally, the presence of full erections at the time of orgasm is also consistent with psychogenic impotence. Some form of nocturnal monitoring may be used to confirm the presence of full nocturnal erections, which supports a diagnosis of psychogenic impotence.

Approximately 3% of impotent men attending a urologic clinic have an endocrine abnormality. Hypogonadism is the most common endocrinopathy and is usually primary. Hyperprolactinemia can present with either decreased libido or impotence. Elevations of serum prolactin over 50 mg/ml is consistent with prolactin-secreting pituitary tumors. Either computed tomography (CT) or magnetic resonance imaging (MRI) can be used to evaluate a patient for pituitary tumors. Patients with a low serum testosterone level should be retested, as testosterone is secreted in an episodic fashion. It is not unusual for the repeat level to be within normal limits. The serum LH level is used to differentiate between primary and secondary hypogonadism which may be managed differently. A fasting blood glucose level may detect occult diabetes mellitus in the impotent patient. More extensive endocrine testing, including thyroid function studies, may be performed in select patients.

In the patient with suspected organic impotence and normal endocrine studies, the next study is a pharmacologic erection test. Various drugs and dosages have been utilized, but a standard test dose used in our center is 10 μg of prostaglandin E1. Development of a full erection is consistent with either neurogenic or psychogenic impotence. Patients with mild vascular disease may also obtain a full erection. Patients who do not respond most likely have vasculogenic impotence. However, there may be a significant "clinical effect" with inhibition of response due to endogenous catecholamines associated with anxiety. Therefore it is worthwhile to repeat tests showing no response in a setting more comfortable to the patient.

Unfortunately, an ideal test for neurologic disorders does not exist. Occult abnormalities of the somatic-sensory pathways can be ruled out with either biothesiometry, pudendal evoked response testing, or bulbocavernosal reflex latency. However, most patients with neurogenic impotence can be identified from the initial history and physical examination.

A full vascular evaluation is indicated in those patients who do not respond to pharmacologic stimulation and are interested in vascular surgery. Arterial function can be assessed noninvasively with duplex ultrasonography or invasively with pudendal arteriography. Venous occlusive function is evaluated using DICC in patients who have normal arterial function and inadequate response to vasoactive agents.

TREATMENT OF ERECTILE DYSFUNCTION

Treatment options are usually elected as a function of the believed cause of the erectile dysfunction, but this picture is changing. Self-injection therapy, for example, may be elected for idiopathic (presumably psychogenic) erectile failure as well as for that due to neurologic damage. Psychological factors should definitely be evaluated before considering surgical therapy. As a rule, however, treatment decisions often reflect the outcome of diagnostic studies. The therapeutic options available to treat impotent patients are listed in Table 39-2. (See also Chapters 15 and 40.)

Psychological Treatment

Until the late 1950s erectile failure was thought usually to be psychogenic and was treated by relatively simple office interventions (Menninger, 1935; Stafford-Clark, 1954), psychoanalysis (Bergler, 1945; Ferenczi, 1916), or psychoanalytic psychotherapy (Fenichel, 1945). The principal assumption of treatment was that the erectile failure was the consequence of intrapsychic conflict associated with key unconscious wishes. The strategy of treatment was that of the classic (exploratory, uncovering) psychoanalytic approach for the most part. Psychoanalytic formulation and treatment of impotence is no longer common but is by no means anachronistic.

TABLE 39–2. *Therapeutic Modalities Utilized for Impotent Patients*

Treatment	Cost	Risk	Outcome
Sex therapy	***		***
Hormonal therapy[a]	**	*	***
Oral agents	*	*	*
Pharmacologic erection therapy	***	**	***
Vacuum erection device	*	*	***
Penile implant	****	***	****
Vascular surgery[b]	*****	**	?

* = least; ***** = most.

[a]Outcome based on results when treatment is performed for a specific endocrine abnormality.

[b]Procedures are still under investigational development.

During the late 1950s and 1960s systematic desensitization with reciprocal inhibition was introduced into the psychological therapy of sexual disorders (Wolpe & Lazarus, 1966). Systematic desensitization is a behavior therapy that focuses on the patient's anxiety about sexual performance. The technique of the therapy involves construction of a "densensitization hierarchy" of mental images or scenes by the patient and therapist. This hierarchy spans from non-anxiety-provoking images of a relaxed state of the patient on the one end to the image of maximal anxiety, provoking coital activity on the other. The patient is asked to picture each of these progressively more sexual scenes in his mind during the course of therapy. The associated feelings of anxiety are then countered by "reciprocal inhibition." Reciprocal inhibition usually takes the form of counterconditioning by means of pleasurable imagery or muscle relaxation (or both). It may be facilitated by hypnosis (Glick, 1975) or a short-acting barbiturate (Brady, 1966).

Wolpe and Lazarus (1966) and later Masters and Johnson (1970) developed in vivo desensitization in the treatment of psychosexual disorders. In vivo desensitization is founded on the same behavioral concepts as systematic desensitization but differs from the latter in the direct (rather than imagined) use of sexual arousal and pleasure as competitive responses to anxiety. Masters and Johnson were able to use sexual or "sensate" pleasure as a desensitizing competitive response by using the technique of prohibition. By prohibiting coitus and intromission, worry and anxiety about the performance of these actions was reduced markedly so that sensate pleasure could be indulged unfettered. The treatment process, which came to be known as sex therapy, evolves over a number of sessions between the man with erectile failure and his partner. The two engage in sensate focus exercises that gradually and increasingly involve the genitals. Coitus is proscribed until the end of the therapy, at which time it is hoped that sexual performance anxiety will have been mastered and faith in erectile stability restored to the patient.

Masters and Johnson (1970) reported a high success rate with sex therapy, about 70% with cases of erectile unreliability. Success rates have varied widely in studies and reports since that of Masters and Johnson, from 39% to 98% (LoPiccolo & Stock, 1986; Perreault, Wright, & Mathieu, 1979). Sex therapy can be performed on a short-term daily basis (Masters & Johnson, 1970) or on a weekly or biweekly basis over a longer period (Kaplan, 1974; Schmidt & Lucas, 1976).

Hormonal Therapy

Androgen therapy has been the most common form of empiric therapy used to treat impotence. As a rule, only men with an abnormally low serum testosterone level have a sustained positive response to androgen replacement therapy. Many eugonadal men experience an initial and temporary improvement that may be due to a placebo or libido effect. Free testosterone levels have been shown to be a better predictor of patient response to this form of therapy. Hormonal replacement therapy is achieved with intramuscular injection of 200 mg of testosterone esters every 2–4 weeks. Oral androgens should not be used because of their erratic absorption and potential liver toxicity. Testosterone replacement therapy is contraindicated in men with prostate cancer. Patients with secondary hypogonadism (low LH level) should be evaluated for a functional or nonfunctional pituitary tumor. These patients may be treated with either parenteral testosterone, human chorionic gonadotropin, or antiestrogens such as clomiphene citrate. Benign hyperprolactinemia is treated with bromocriptine. Surgery and radiation therapy do not always reduce prolactin levels to normal on their own. Medical therapy has been shown to be the most successful and is often effective as the only treatment even for macroadenomas. Testosterone therapy alone is not enough to restore sexual function in these patients even though their testosterone levels are often low.

Intracavernosal Injection

Physiologic erection mechanisms can be mimicked by the intracavernosal injection of vasoactive drugs. Injection of such agents can relax smooth muscle tone. Most commonly, the direct smooth muscle relaxant papaverine is injected into the cavernosum. With some patients the papaverine is supplemented by an alpha-adrenergic receptor blocking agent such as phenoxybenzamine or phentolamine. Intracavernosal injection of prostaglandin E1, a smooth muscle relaxant, has been shown to result in erection (Stackl, Hasun, & Marberger, 1988).

Intracavernosal injection therapy enjoys the advantage of complete reversibility and high efficacy in cases of both organic and psychogenic erectile failure. The procedure is a specific treatment of neurogenic impotence but is often effective in patients with vasculogenic impotence. Successful use of intracavernosal injection with psychogenic patients was described by Turner, Althof, and Levine (1989). This group described a high efficacy rate among men willing to continue with the treatment; 60% had dropped out of self-injection therapy by 6 months, however. A high drop-out rate from injection therapy was reported by Althof, Turner, Levin, et al. (1989), who studied the phenomenon with interest given its efficacy and self-administration capability. They found a 46% cumulative drop-out rate among a series of patients and partners. Patients were found to reject treatment because they were unable to accept the idea of self-injection. Others dropped out of treatment because they were disappointed in its efficacy.

Intracavernosal injection therapy begins under the supervision of the physician, who titrates to the lowest effective dose to produce erection. In this way, the patient learns the technique and becomes accustomed to the procedure. Erectogenic agents are selected and injected directly into the corporal bodies via a fine needle (27 to 30 gauge) usually using an insulin syringe. The erection that results usually persists 30 to 60 minutes.

Complications from self-injection therapy include painless fibrotic nodules, prolonged erection, and local infection (Zentgraf, Baccouche, & Junemann, 1988). Infection has been rare, but fibrotic nodules in the cavernosa have been common in some studies. Animal studies suggest that Prostaglandin E1 is less likely to induce fibrosis than other agents. Prolonged erection has been rare after the dose titration period. When prolonged erection does occur, no more than 4 hours should be allowed to pass before medical attention is sought. Intracavernosal injection of epinephrine or congeners is the treatment for prolonged erection.

Penile Prostheses

Penile implantation of erectile prostheses has been popular since 1973, when Small and Carrion developed a semirigid paired device for this purpose (Small, Carrion, & Gordon, 1975). Shortly afterward, Scott, Bradley, and Timm (1973) developed an inflatable prosthesis. The inflatable device was associated with frequent failure and complications, and the semirigid rods were thus preferred by many urologists. During the 1980s there were numerous refinements to the inflatable devices and the development of other types of prosthesis. Implantation is a safe and popular solution to erectile failure; and by 1989 more than 25,000 penile prostheses were being surgically implanted in the United States each year (Krane, Goldstein, & De Tejada, 1989). These devices have a simple mission—to effect an erection suitable to intromission—which has been accomplished with minimal or no effect on ejaculation and orgasm. Ninety percent of men receiving these devices are more or less satisfied with them (Blake, McCartney, Fried, et al., 1983; Gregory & Purcell, 1987; Mallory, Wein, & Carpiniello, 1986).

Because surgery is involved with risk of postoperative complications and incomplete reversibility to the preoperative status quo, urologists have been concerned to avoid prosthesis implantation in patients with psychogenic erectile disturbances or in emotionally unpredictable persons. Men believed to have psychogenic erectile failure but who have been deemed stable and have not been helped by psychological therapy or other biologic, nonsurgical measures are often considered acceptable candidates. As discussed above, screening for the unstable or psychosocially inappropriate candidate may be simple or elaborate. There is a consensus that it is important to identify the patient, and hopefully the partner, who seeks a broad solution to relationship failure by means of the "prosthesis cure." It is not clear to what extent this need is valid, but urologists who believe it to exist usually ask psychiatrists and psychologists to assess the patient's psychological and social agenda.

Complications of penile implantation include, individually or in combination: infection, extrusion of the device, or mechanical failure (in those with moving parts). Reoperation for complications was initially of notable frequency (10 to 44%), but these rates have fallen in recent years with improvements in design and manufacture (Furlow, Goldwasser, & Gundian, 1988; Kaufman, Lindner, & Raz, 1982; Kessler, 1980). (See also Chapters 15 and 40.)

External Vacuum Devices

There are available several devices that operate in the treatment of impotence by creating a negative pressure to produce an erection. These devices typically consist of a plastic cylinder connected by a tube to a hand-held vacuum pump that is used to fill the cavernosal sinuses by means of a negative pressure. Elastic construction bands are then placed around the base of the penis to impede drainage of blood and maintain the erection. The resulting erection is rigid only beyond the constrictor bands, so that it pivots on a "floppy" neck. There tends to be congestion of extracorporeal penile tissues, resulting in a larger circumference than found is in normal erections (Turner, Althof, & Levine, 1990). Tumescence of 30 minutes or more is afforded before penile temperature drops significantly. Vacuum devices are said to produce erections sufficient for intromission 73 to 100% of the time (Nadig, Ware, & Blumoff, 1986; Turner, Althof, & Levine, 1990; Witherington, 1989). Satisfaction with these devices has been generally high, despite a certain degree of hassle in producing the erections.

There are relatively few complications of vacuum-generated erection. They include initial minor pain and discomfort, occasional ecchymoses, and petechiae. Men with blood clotting disorders or who are taking anticoagulants should not use vacuum devices.

Yohimbine

Yohimbine is an indolic alkaloid derived from the bark of the *Coryunanthe yohimbine* plant. The drug was historically believed to have aphrodisiac properties and was marketed until 1973 in combination with methyltestosterone and strychnine (Afrodex). Margolis, Prieto, Stein, et al. (1971) summarized the use of Afrodex in 10,000 men with erectile and arousal difficulties and reported good to excellent results in 80% of cases. Later, the efficacy of Afrodex was challenged, and the drug compound ultimately removed from the market. Sonda, Mazo, and Chancellor (1990) investigated yohimbine in a controlled study of 40 men complaining of impotence. A statistically significant improvement with the drug was shown, but the actual percentage improved was relatively low, 33%. A problem with the study was the heterogeneity of the sample, which represented various specific sexual problems.

Vascular Surgery

Vascular surgery represents one of the newest treatment options for impotence. The aim of vascular surgery is to correct a specific vascular abnormality, either inadequate inflow or venous leak. Aortoiliac reconstruction and endarterectomy were largely unsuccessful owing perhaps in part to either unrecognized corporal veno-occlusive disease or distal arterial occlusive disease (Krane, Goldstein, & DeTejada, 1989). Later, microvascular bypass procedures were performed on men with arteriosclerosis, but they also had poor results; most anastomoses were not patent at 1 year.

The most successful revascularization procedures are those performed on young men with isolated proximal segmental occlusions of the pudendal artery typically following pelvic injury. New blood flow is brought to the penis via the inferior epigastric artery or sapheous vein graft, which is anastomosed to the dorsal artery using microsurgical technique. Current revascularization success rates appear to be improved over the past but vary widely from less than one-third of attempts to more than 80% (Krane, Goldstein, & DeTejada, 1989; Sharlip, 1988). Revascularization appears to be most successful in young patients with discrete arterial occlusive lesions secondary to trauma.

Patients with venoocclusive dysfunction are candidates for venous ligation surgery. Various techniques have been proposed, including complete excision of the deep dorsal vein and plication of the crura of the corpora cavernosa. However, the long-term success rate resulting from these procedures is less than 50%. Until there is significant improvement in our methods of diagnosis and treatment, vascular surgery for the treatment of impotence should be considered investigational.

REFERENCES

Althof SE, Turner LA, Levin SB, et al. (1989). Why do so many people drop out from auto-injection therapy for impotence. J Sex Mar Ther 15:121–129.

Bancroft J, & Wu FC (1983). Changes in erectile responsiveness during androgen replacement therapy. Arch Sex Behav 12:59–66.

Bergler E (1945). A short genetic surgery of psychic impotence. II. Psychiatr Q 19:657–767.

Beutler LE, Karacan I, Anch AM, et al (1975). MMPI and MIT discriminators of biogenic and psychogenic impotence. J Consult Clin Psychol 43:899–903.

Beutler LE, Ware C, & Karacan I (1978). Psychological assessment of the sexually impotent male. In RL Williams & I Karacan (Eds.), Sleep disorders: diagnosis and treatment (pp. 388–394). New York: John Wiley & Sons.

Birk L (1980). Shifting gears in treating psychogenic sexual dysfunctions: Medical assessment, sex therapy, psychotherapy and couple therapy. Psychiatr Clin North Am 3:153–172.

Blake DJ (1984). Issues in the psychiatric diagnosis and management of impotence. Psychiatr Med 2:109–130.

Blake DJ, McCartney C, Fried FA, et al. (1983). Psychiatric assessment of the penile implant recipient: preliminary study. Urology 21:252–256.

Brady JP (1966). Brevital relaxation treatment of frigidity. Behav Res Ther 4:71–77.

Buvat J, Lemaire A, Marcolin G, et al. (1987). Intracavernaous injection of papaverine. World J Urol 5:150–155.

Carter JN, Tyson JE, Tolis G, et al. (1978). Prolactin-secreting tumors and hypogonadism in 22 men. N Engl J Med 299:847–852.

Cerami A, Vlassara H, Brownlee M, et al. (1987). Glucose and aging. Sci Am 256:90–96.

Cooper AJ (1969). Factors in male sexual inadequacy: a review. J Nerv Ment Dis 149:337–359.

Dickinson IK, & Pryor JP (1989). Pharmacocaverometry: a modified papaverine test. Br J Urol 63:539–545.

Ellenberg M (1971). Impotence in diabetes: the neurological factor. Ann Intern Med 75:213–219.

Farming I, Glocer L, Fox D, et al. (1974). Impotence and diabetes: histologic studies of the autonomic nervous fibers of the corpora cavernosa in impotent diabetic males. Diabetes 23:971–975.

Fenichel O (1945). The psychoanalytic theory of neurosis. New York: WW Norton.

Ferenczi S (1916). Sex and psychoanalysis. Boston: Gorham Press.

Franks S, Jacob HS, Martin N, et al. (1978). Hyperprolactinaemia and impotence. Clin Endocrinol (Oxf) 8:277–287.

Furlow WL, Goldwasser B, & Gundian JC (1988). Implantation of model AMS 700 penile prosthesis: long term results. J Urol 139:741–742.

Ginsberg GL, Frosch WA, & Shapiro T (1972). The new impotence. Arch Gen Psychiatry 26:218–220.

Glick BS (1975). Desensitization therapy in impotence and frigidity: review of the literature and report of a case. Am J Psychiatry 132:169–171.

Goldstein I (1987). Vasculogenic impotence: its diagnosis and treatment. In R deVere White (Ed.), Problems in urology: sexual dysfunction (pp. 547–563). Philadelphia: Lippincott.

Gregory JG, & Purcell MH (1987). Scott's inflatable penile prosthesis: evaluation of mechanical survival in the series 700 model. J Urol 676–677.

Groat WC, & Steers WD (1988). Neuroanatomy and neurophysiology of penile erection. In EA Tanagho, TF Lue, & RD McClure (Eds.), Contempary management of impotence and infertility (pp. 3–27). Baltimore: Williams & Wilkins.

Hart BL, & Leedy MG (1985). Neurological basis of male sexual behavior: a comparative analysis. In N Adler, D Pfaff, & RW Gow (Eds.), Reproduction. (Vol 7; pp. 373–415). Handbook of behavioral neurobiology. New York: Plenum Press.

Hastings DW (1963). Impotence and frigidity. Boston: Little, Brown.

Herman A, Adar R, & Rubinstein Z (1978). Vascular lesions associated with impotence in diabetic and nondiabetic arterial occlusive disease. Diabetes 27:975–981.

Kaiser FE, & Korenman SG (1988). Impotence in diabetic men. Am J Med 85(suppl 5A): 147–152.

Kaplan HS (1974). The new sex therapy. Montreal: Brunner-Mazel.

Karacan I, Salis PF, & Williams RL (1978). The role of the sleep laboratory in the diagnosis and treatment of impotence. In RL Williams & I Karacan (Eds.), Sleep disorders: diagnosis and treatment (pp. 353–382). New York: John Wiley & Sons.

Kaufman JJ, Lindner A, & Raz S (1982). Complications of penile prosthesis surgery for impotence. J Urol 128:1192–1194.

Kessler R (1980). Surgical experience with the inflatable penile prosthesis. J Urol 124:611–612.

Kolodny RC, Kahn CB, Goldstein HH, et al. (1973). Sexual dysfunction in diabetic men. Diabetes 23:306.

Krane KJ (1986). Sexual function and dysfunction. In PC Walsh, RF Gittes, AD Perlmutter, et al. (Eds.), Campbell's urology (Vol 1; Ch. 13, pp. 700–735). Philadelphia: WB Saunders.

Krane RJ, Goldstein I, & Saenz De Tejada I (1989). Impotence. N Engl J Med 321:1648–1659.

Kwan M, Greenleaf WJ, Mann J, et al. (1983). The nature of androgen action on male sexuality: a combined laboratory-self report study of hypogonadal men. J Clin Endocrinol Metab 57:557–562.

Lehman TP, & Jacobs JA (1983). Etiology of diabetic impotence. J Urol 129:291–294.

Levine L (1989). Erectile dysfunction: causes, diagnosis and treatment. Comp Ther 15:54–58.

Lin JT, & Bradley WE (1985). Penile neuropathy in insulin dependent diabetes mellitus. J Urol 133:213–215.

Lin SN, Lin RS, Yu PC, et al. (1989). Diagnosis of vasculogenic impotence: combination of penil xenon-133 washout and paparerine tests. Urology 34:28–32.

Lo Piccolo J, & Stock WE (1986). Treatment of sexual dysfunction. J Consult Clin Psychol 54:158–167.

Lue TF, Hricak H, Marich KW, et al. (1985). Vasculogenic impotence evaluated by high resolution ultrasonography and pulsed doppler spectrum analysis. Radiology 155:777–781.

Mallory TR, Wein AJ, & Carpiniello VL (1986). Reliability of AMS M700 inflatable penile prosthesis. Urology 28:385–387.

Margolis R, Prieto P, Stein L, et al. (1971). Statistical summary of 10,000 male cases using afrodex in treatment of impotence. Curr Ther Res 13:616–622.

Marshall P, Surridge D, & Delva N (1980). Differentiation of organic and psychogenic impotence on the basis of MMPI decision rules. J Consult Clin Psychol 48:407–408.

Masters WH, & Johnson VE (1970). Human sexual inadequacy. Boston: Little, Brown.

Menninger K (1935). Impotence and frigidity from the standpoint of psychoanalysis. J Urol 34:166–188.

Michal V, & Ruzbarsky V (1980). Histological changes in the penile arterial bed with aging and diabetes. In AG Zornigatti & G Ross (Eds.), Vasculogenic impotence; proceedings of the first international conference on corpus cavernosum revascularization (pp. 42–48). Springfield, IL: Charles C Thomas.

Nadig PW, Ware JC, & Blumoff R (1986). Noninvasive device to produce and maintain on erection-like state. Urology 27:127–128.

Perreault R, Wright J, & Mathieu M (1979). The directive sex therapies in psychiatric outpatient setting. Can J Psychiatry 24:47–54.

Robiner WN, Godec CJ, Cass AS, et al. (1982). The role of the Minnesota Multiphasic Personality Inventory in the evaluation of erectile dysfunction. J Urol 128:487–488.

Roose SP, Glassman AM, Walsh BJ, et al. (1982). Reversible loss of nocturnal penile tumescence during depression: a preliminary study. Neuropsychobiology 8:284–288.

Ruutu ML, Virtanen JM, Bjorn L, et al. (1988). Which examinations are useful in erectile failure? Scand J Urol Nephrol (suppl) 110:251–255.

Schmidt CW, & Lucas J (1976). The short term, intermittent, conjoint treatment of sexual disorders. In JK Meyer (Ed.), Clinical management of sexual disorders (pp. 130–147). Baltimore: Williams & Wilkins.

Scott FB, Bradley WE, & Timm GW (1973). Management of erectile impotence: use of implantable inflatable prosthesis. Urology 2:80–82.

Sharlip ID (1988). Treatment of arteriogenic impotence by penile revascularization. In Proceedings of the sixth biennial international symposium for corpus cavernosum revascularization and third biennial world meeting on impotence (p. 133). Boston.

Slag MF, Morley JE, Elson MK, et al. (1983). Impotence in medical clinic outpatients. JAMA 249:1736–1740.

Small MP, Carrion HM, & Gordon JA (1975). Small-Carrion penile prosthesis: new implant for management of impotence. Urology 2:80–82.

Sonda LP, Mazo R, & Chancellor MB (1990). The role of yohimbine for the treatment of erectile impotence. J Sex Mar Ther 16:15–21.

Spark RF, White RA, & Connolly PB (1980). Impotence is not always psychogenic: newer insights into hypothalamic-pituitary-gonadal dysfunction. JAMA 243:750–755.

Stackl W, Hasun R, & Marberger M (1988). Intracavernous injection of prostaglandin E1 in impotent men. J Urol 140:66–68.

Stafford-Clark D (1954). The etiology and treatment of impotence. Practitioner 172:397–404.

Staples RB, Fischer M, Shapiro M, et al. (1980). A reevaluation of MMPI discriminators of biogenic and psychogenic impotence. J Consult Clin Psychol 48:543–545.

Thase ME, Reynolds CF III, Glanz LM, et al. (1987). Nocturnal penile tumescence in depressed men. Am J Psychiatry 144:89–92.

Thase ME, Reynolds CF III, & Jennings JR (1988a). Nocturnal penile tumescence is diminished in depressed men. Soc Biol Psychiatry 24:33–46.

Thase ME, Reynolds CF III, & Jennings JR (1988b). Diagnostic performance of nocturnal penile tumescence studies in healthy, dysfunctional (impotent), and depressed men. Psychiatr Res 26:79–87.

Turner LA, Althof SE, & Levine SB (1989). Self-injection of papaverine and phentolamine in the treatment of psychogenic impotence. J Sex Mar Ther 15:163–175.

Turner LA, Althof SE, & Levine SB (1990). Treating erectile dysfunction with external vacuum devices: impact upon sexual, psychosocial and marital functioning. J Urol 144:79–82.

Virag R (1982). Intracavernous injection of papaverine for erectile failure [Letter]. Lancet 2:938.

Wagner G, & Green R (1981). Impotence. New York: Plenum Press.

Ware JC (1987). Evaluation of impotence. Psychiatry Clin North Am 10:675–686.

Wespes E, Delcour C, Struyven J, et al. (1984). Cavernometry-cavernography: its role in organic impotence. Eur Urol 10:229–232.

Witherington R (1989). Vacuum constriction device for management of erectile impotence. J Urol 141:320–322.

Wolpe J, & Lazarus AA (1966). Behavior therapy techniques. Oxford: Pergamon Press.

Zentgraf M, Baccouche M, & Junemann KP (1988). Diagnosis and therapy of erectile dysfunction using papaverine and phentolamine. Urol Int 43:65–75.

40 Sexual aspects of rehabilitation after spinal cord injury

DONALD R. BODNER, M.D.

The annual incidence of spinal cord injury is estimated between 30 and 50 per million and may actually be higher, as patients who die shortly after injury and those with minimal residual weakness often are not included (Fine, Kuhlemeier, DeVivo, et al., 1979; Gehrig & Michaelis, 1968; Kraus, Franti, Riggins, et al., 1975; Young, Burns, Bowen, et al., 1982). In the United States there are approximately 300,000 people with spinal cord injury, and an additional 14,000 to 15,000 new injuries occur each year. It has been estimated that 1.5 billion dollars are spent per year in the United States to treat patients with spinal cord injury. The most frequent age at which spinal cord injury occurs is 19, with the mean age being 26 years.

Spinal cord injuries occur approximately four times more frequently in males than females (Stover, Fine, & Kennedy, 1986). Such an injury often changes a person's ability to walk and to control the bowels and bladder; and it usually alters a man's ability to achieve erections or ejaculate. It requires the affected individual not only to cope with physical disability but also to accommodate to a new sexual self. Psychiatric participation on the rehabilitative team is necessary if spinal cord injured patients are to be optimally rehabilitated and integrated back into society.

HISTORICAL BACKGROUND

The earliest documented mention of spinal cord injury was in the Edwin Smith Surgical Papyrus transcribed during the seventeenth century and thought to originate approximately 3000 BC (Smith, 1930). During Egyptian times, spinal cord injury was considered to be a fatal disease that was not treatable. Hippocrates first described the complications of spinal cord injury, including the loss of control of the bowel and bladder (Adams, 1886).

There was little progress in the clinical management of the spinal cord injured (SCI) patient for many years. Despite the achievements in understanding neuroanatomy and neurophysiology, during World War I most spinal cord injured patients died of urosepsis (Claude & Lhermitte, 1914; Holmes, 1915). The emphasis at that time was on neurologic assessment, not treatment, as the prognosis of this injury was considered so dismal.

During World War II, great strides were made in the understanding and management of the SCI patient. It was learned that catheterizing the flaccid neurogenic bladder caused by spinal shock immediately reduced mortality due to urosepsis from 100% to 50%. Sir Ludwig Guttman, a German immigrant in the United Kingdom, established a multidisciplinary unit to deal with spinal cord injury and is considered the father of modern spinal cord injury treatment. In the United States during this same period, Donald Munro in the private sector and Ernest Bors through the Veterans Administration were early pioneers in establishing regional units for the treatment of spinal cord injury. The Veterans Administration should be credited for developing the concept and creating regional spinal cord injury units in the United States.

Great progress was made in the urologic management of the SCI patient during the 1970s and 1980s. Renal failure was once the leading cause of death in this group of patients, but with improved understanding and management of the neurologic bladder it is no longer so (Dietrick & Russi, 1958; Freed, Bakst, & Barrie, 1966; Masters & Johnson, 1966; Mesard, Carmody, Mannarino, et al., 1978). With a more normal life expectancy, emphasis has shifted to improving the quality of life. The primary goals of current urologic management of patients with spinal cord injury are to preserve kidney function, prevent bladder and kidney infections, and avoid catheters whenever possible. With greater success in accomplishing these goals, sexual concerns have emerged as an important part of rehabilitation.

Sexual education, a need of all SCI patients, is best accomplished by a team approach, including a psychologist or psychiatrist (or both) and a urologist familiar with spinal cord injury. Instruction should be commenced shortly after injury and continued throughout the rehabilitation period. Because spinal cord injury usually involves young adults, sexuality is important to SCI victims. Many have not yet married nor found a regular partner. The ability to achieve erections, ejaculate, and achieve orgasm is important to most men, and the ability to achieve orgasm and pregnancy is important to most women. Education should include spouses and other central family members who provide support to the patient.

NEUROLOGIC PATHWAYS AND SEXUAL FUNCTION

In order for erections to occur, there must be adequate arterial inflow to the penis without abnormal venous outflow, thereby creating a steady state that permits a firm erection. In addition, these complex vascular changes that permit erection occur in a tightly controlled endocrine and neurologic milieu. (See Chapter 39.)

Erections can be reflexogenic or psychological (Newman & Northrup, 1981). Reflex erections occur as a result of neural transmission from pelvic parasympathetic nerves. These fibers, originating in the second through fourth sacral (S2–S4) roots, also provide parasympathetic enervation to the urinary bladder and are responsible for bladder emptying when stimulated. Direct stimulation of the penis by rubbing the glans or shaft results in reflex erections with the sacral reflex arc (S2–S4) is intact. Sensory stimuli travel via the dorsal nerve of the penis, which joins the pudendal nerve, and then travel to the spinal cord. Efferent enervation to the penis travels via the pelvic nerve, which enters the penis as the cavernosal nerve. Psychogenic erections, or those erections created by thoughts, are speculated to originate from cerebral impulses that travel down the lateral columns of the spinal cord primarily via the thoracolumbar portion of the cord and perhaps via the sacral roots as well (see Table 40-1) (Newman & Northup, 1981).

TABLE 40–1. *Classification of Erections*

Reflex erections
Produced by direct stimulation of the penis
Mediated by the parasympathetic nervous system
Require the S2–S4 nerve roots (sacral reflex arc) be intact
Occur independently of erotic stimuli or thoughts
Are generally preserved with spinal cord injury above the L2 vertebral level (S2–S4 nerve roots)
Psychogenic erections
Produced by thoughts and erotic stimuli
Mediated by the sympathetic nervous system
Occur independently of direct stimulation of the penis
Require the thoracolumbar nerve roots be intact
Often lost with injury to the thoracic and cervical cord

Ejaculation if mediated via the sympathetic nervous system and the thoracolumbar portion of the spinal cord. It includes emission, bladder neck closure, and antegrade ejaculation. Although the neurologic pathways for vaginal lubrication and orgasm in the female are less well described, the sacral parasympathetics are thought to be responsible for reflex lubrication of the vagina, and thoracolumbar sympathetics and sacral parasympathetics are probably responsible for psychogenic lubrication (Griffith & Trieschmann, 1975). Smooth muscle contraction of the uterus, fallopian tubes, and paraurethral glands—the corollary to emission in the male—should be mediated by the thoracolumbar sympathetics. Contraction of the striated pelvic floor muscles, the analogue of antegrade ejaculation is mediated through the parasympathetics (S2–S4) and the somatic efferents (Griffith & Trieschmann, 1975).

EFFECT OF SPINAL CORD INJURY ON SEXUAL FUNCTION

The ability to achieve erections and to ejaculate has been studied in patients with spinal cord injury. In 1955 Talbot reported that only 20% (40 of 186) men with spinal cord injury were able to have intercourse. He did not specify the level or completeness of injury. Bors and Comarr (1960) carefully questioned 529 patients with spinal cord injury and evaluated the patient's ability to obtain reflex or psychogenic erections and his ability to ejaculate relative to the level and completeness of injury. They differentiated between upper motor neuron and lower motor neuron lesions. In respect to bladder and sexual function, upper motor neuron lesions refer to an injury to the spinal cord above the sacral reflex arc. The S2–S4 nerve roots are at the mid lumbar vertebral level in an adult. Upper motor neuron injuries produce spasticity below the level of the injury. The bladder contracts reflexly, and lower extremity reflexes are hyperactive.

Lower motor neuron lesions refer to direct injuries to the sacral reflex arc. These lesions require injury to the cauda equina or conus medularis. The bladder is then flaccid, and anal tone and external urethral sphincter tone are diminished or absent. Patients with complete or incomplete upper motor neuron lesions in Bors and Comarr's study were generally able to obtain reflex erections, but the erections were often not satisfactory for coitus, whereas only a small percentage of patients with complete lower motor neuron lesions were noted to obtain any erection, and they were generally psychogenic in nature (Bors & Comarr, 1960).

Comarr (1970) subsequently studied 150 patients with spinal cord injury in which 115 had upper motor neuron lesions and 35 had lower motor neuron lesions. Ninety-two percent of the patients with upper motor neuron lesions were able to obtain erections by external stimulation or manipulation, and 94% obtained spontaneous erections; only 22% were capable of achieving psychogenic erections. Patients with lower motor neuron lesions had a much lower potential for achieving any erection. None of the 35 patients with such injury were able to obtain reflex erections, only 11% achieved spontaneous erections, and 26% noted psychogenic erections. Sixty-two percent of the patients in this series attempted coitus, and 43% considered themselves successful. Of the patients with upper motor neuron lesions 9% could ejaculate (only 1% if the injury was complete), whereas 17% of patients with lower motor neuron lesions could ejaculate. Although one must avoid making definitive predictions of ultimate sexual capability based on the level and completeness of injury alone, Table 40-2 relates the level of spinal cord injury to the residual sexual function that can generally be expected from damage in each area.

The ability to obtain erections is lost during the period of spinal shock. The length of time after injury required for erections to return varies from several days to 6 weeks with incomplete injuries to as long as a year or more. Irreversible treatment (i.e., surgical intervention) for sexual dysfunction should not be performed during the initial rehabilitation period, as further improvement in erections can occur with time.

Sexual education, however, is an important component of overall rehabilitation and should be started shortly after injury. When the period of spinal cord shock has worn off, the patient realizes that lifelong spinal cord dysfunction is a reality. Much clinical attention is directed to bladder management, prevention of pressure sores, and ambulation. Issues of sexuality and sexual function should be addressed at this early time after injury as well. It should be remembered that women with spinal cord injury can still achieve pregnancies even though amenorrhea can occur for 3 to 9 months after spinal cord injury (Francois & Maury, 1987).

INITIAL MANAGEMENT

During the initial rehabilitation following spinal cord injury, a patient must learn to live and function with the disability. It entails learning how to get around, which now may require the use of a wheelchair. Bladder emptying is often accomplished with intermittent catheterization, an external condom catheter, or occasionally an indwelling catheter. Bowel function can also be altered, and many SCI patients require a bowel program to provide regular emptying.

Adjustment to spinal cord injury is best facilitated through a multidisciplinary approach. Central to the team are a psychiatrist or psychologist and a urologist familiar with spinal cord injury. Education as to what can be expected with erection, ejaculation, and the ability to have menstrual periods and achieve pregnancies should be addressed during this period of rehabilitation. The patient must adjust to the injury itself. The transformation must then occur from hospital patient to individual capable of functioning in society. People with spinal cord injuries must first be comfortable with their own self-image before being capable of interacting as a spouse or significant other. Family members and significant others must be made an integral part of the rehabilitation process. Spinal cord injury affects the entire family, and each member must adjust.

Following spinal cord injury, patients are hospitalized for 3 to 6 months or more for initial rehabilitation. As a patient progresses during this hospitalization, he or she often goes home on pass during the weekends to facilitate adjustment to returning to society. Patients are encouraged to experiment with sex with their partner during these times if they are so inclined. Practical points include ensuring that the bladder is empty prior to having intercourse, thereby avoiding the embarrassing reflex loss of urine during intercourse. Patients on intermittent self-catheterization should catheterize the bladder immediately before intercourse. Suprapubic tapping is often helpful to elicit a bladder contraction when intermittent catheterization is not used. For patients with indwelling catheters, the catheter can be folded upon itself and covered with a condom prior to intercourse. Alternatively, some patients wish to remove a catheter prior to intercourse and replace it afterward. If a patient has poor leg or arm function, pelvic thrusting may be difficult, and that person should consider assuming the "down" position and have his or her partner on top. It should be remembered that male patients may obtain reflex erections that are short-lived and may be improved by direct stimulation of the penis. Continuous stimulation may be necessary to maintain the erection.

A young SCI patient who does not have a regular

TABLE 40–2. *Relation of Level of Injury to Sexual Function*

Effect	Comments
Cauda equina/conus injury	
Flaccid bladder	Injury to the S2–S4 nerve roots
Loss of external urethral tone	
Loss of anal tone	
Bulbocavernous reflex absent	
Males	
Usually no reflex erections	
Rare psychogenic erection	
Occasional ejaculation	
Females	
Vaginal secretion often absent	
Patients generally fertile	
Lumbar injury	
Reflex bladder	Injury above the S2–S4 roots.
Anal and external urethral tone present	
Males	
Reflex erections can occur.	
Psychogenic erections can occur.	
Ejaculation is rare.	
Females	
Vaginal secretions present as part of sacral/genital reflex.	
Fertility is preserved.	
Thoracic/cervical injury	
Reflex bladder	
Anal sphincter tone present	
External urethral sphincter often spastic	
Males	
Reflex erections predominate (usually of short duration).	
Psychogenic erections are generally absent.	
Ejaculation occurs occasionally.	
Females	
Vaginal secretions are present as part of genital reflex.	
Fertility is preserved.	
Sensation of labor pain is absent.	

sexual partner should be counseled about the treatment options for erectile dysfunction after spinal cord injury so he is aware of the treatment alternatives. He should be encouraged to meet and establish friendships with young women as if he had no spinal cord injury. We do not, however, recommend the use of sexual surrogates for this purpose although opinions regarding this issue vary among clinicians.

PATIENT EVALUATION

Although SCI patients may have problems obtaining or sustaining erections on a neurologic basis, psychological factors may contribute significantly to the erectile dysfunction (see Table 40-3). For this reason, patients with spinal cord injury complaining of difficulty with erections should be evaluated by both a psychiatrist or psychologist and a urologist.

A careful history including a sexual history should be obtained from the patient and his or her partner.

TABLE 40–3. *Psychogenic Versus Organic Factors in SCI Sexual Dysfunction*

Psychogenic Factors	Organic (Neurologic)
Sexual problems prior to injury	No erections
Depression	Short-lived erections
Normal bowel and bladder control	Loss of bowel and bladder control
Preserved rectal and external urethral tone	Absent perineal sensation
Bulbocavernous reflex present	Bulbocavernous reflex present depending on level of injury

The male patient's ability to achieve reflex and psychogenic erections should be ascertained. Did the patient have erectile difficulties prior to the injury? Is the sexual problem more a matter of discovering a position that is more functional after the spinal cord injury, such as having the male supine? As associated injuries can occur with spinal cord injury, the nature of the other injuries and any prior surgical intervention should be elicited. A quadriplegic patient with a severe pelvic fracture may not get reflex erections and may have a flaccid bladder on the basis of the pelvic fracture, not the spinal cord injury. The patient and the partner's psychological adjustment to the injury should be noted. A history of past medical problems along with a list of current medications is important, as many of the commonly used medications for SCI patients may adversely affect sexual function (see Table 40-4). However, because these medications are important for the overall well-being and function of the patient, a change or elimination of the drug may not be possible.

Physical examination should include a careful neurologic assessment in which the level and completeness of the injury is documented. Rectal tone and the bulbocavernous reflex when present are helpful in establishing the integrity of the sacral reflex arc. Perineal sensation should be tested along with reflexes and sensation in the lower extremity (see Table 40-5). It should be noted that if a male patient has preservation of pinprick sensation in the penile, scrotal, and perianal dermatomes, he can anticipate a normal return of psychogenic erections, ejaculation, and orgasm. In females, the ability to achieve orgasm is dependent on the status of sensation of the erogenous zones, especially the breasts, clitoris, and labia.

TABLE 40–4. *Common Medications Used in SCI Patients*

Drug	Purpose	Side Effect
Oxybutynin	Bladder relaxant	Dry mouth; impotence
Probanthine	Bladder relaxant	Dry mouth; impotence
Baclofen	Muscle relaxant	Drowsiness; impotence
Valium	Muscle relaxant	Drowsiness
Phenoxybenzamine		Hypotension and inhibition of ejaculation
Prazosin	Relax sphincter	
Terazosin		

RESTORATION OF ERECTIONS

Impotence is defined as the inability to obtain and sustain an erection adequate for penetration and satisfactory completion of the sexual act. Although most SCI patients are capable of obtaining an erection, the erection is often short-lived and may be unpredictable (Bors & Comarr, 1960). Notwithstanding, many patients with spinal cord injury do not require any specific treatment for erectile dysfunction. Being aware of the existing treatment options is sufficient for many patients. Patients are instructed in the treatment options available for erectile dysfunction during the initial rehabilitation hospitalization.

In the past, erectile problems in SCI men were corrected by use of external appliances placed over the penis to provide increased firmness or surgical implantation of a penile prosthesis. However, because of the lack of sensation in the penis and the neurogenic bladder associated with intermittent bladder infections, SCI patients treated with penile prostheses for erectile dysfunction were noted to have a higher incidence of complication, secondary to erosion or infection of the prosthesis, ultimately necessitating its removal (Collins & Hackler, 1988; Rossier & Fam, 1984). Results have been better when the penile prosthesis is implanted for the purpose of keeping an external condom catheter on the penis rather than to improve sexual function (Iwatsubo, Tanaka, Takahashi, et al., 1986; Steidle & Mulcahy, 1989). It may be due in part to better patient selection in these series and the fact that no pressure is being applied to the glans penis where the tip of the prosthesis is located when the external condom catheter is used. With better means of treating erectile dysfunction less invasively, penile prostheses are reserved for patients who require help in keeping condom catheters on the penis and those who are not satisfied with the

TABLE 40–5. *Physical Examination Findings After Spinal Shock*

Parameter	Injury Level		
	Spinal Shock	Cauda Equina	Thoracic/Cervical
Rectal tone	Absent/decreased	Absent/decreased	Increased
Bulbocavernous reflex	Absent/decreased	Absent/decreased	Present
Perineal sensation	Absent/decreased	Absent/decreased	Absent
Reflex penile erections	Absent/decreased	Absent/decreased	Present
Spontaneous bladder contractions	Absent	Absent	Present

Note: Spinal shock generally lasts from several days to approximately 6 weeks after spinal cord injury.

less invasive treatment alternatives for erectile dysfunction.

There is currently no evidence to support the use of yohimbine or other psychotropic agents to improve erections in the SCI patient. As other noninvasive options are readily available and have proved effective, we generally proceed directly with one of them as described below. Further studies are needed to determine if there is a role for psychotropic agents in this patient population.

INTRACORPOREAL INJECTION OF VASOACTIVE MEDICATIONS

In 1977 Michal and colleagues reported that accidental injection of papaverine, a smooth muscle relaxant, into the corpora cavernosa of the penis during a revascularization procedure produced an erection (Michal, Kramer & Pospichal, 1977). Virag (1982) and subsequently Brindley (1983) reported that intracorporeal injection of papaverine produced an erection adequate for intercourse. In 1985 Virag reported that sustained erections lasting up to 18 hours occurred in 42 of 227 patients given 80 mg of papaverine intracorporally, with 15 of the patients requiring aspiration of blood from the corpora for detumescence. Zorgniotti and LeFleur (1985) were the first to report intracorporeal self-injection therapy for the treatment of impotence, and they noted successful erections for intercourse in 59 of 62 patients. The dosage of medication injected was 30 mg papaverine combined with 0.5 mg of phentolamine, an alpha-blocking agent. One patient had a sustained erection for more than 24 hours requiring aspiration of the corpora. Gasser and associates compared the effect of intracorporeal injection of papaverine with saline to exclude the placebo effect; they reported achieving good erections in 82% of patients receiving intracorporeal papaverine versus no erections produced by injecting saline alone (Gasser, Roach, Larsen, et al., 1987).

Sidi and colleagues evaluated the effectiveness of intracorporeal injection of papaverine and phentolamine in 100 patients with impotence. All 17 patients with known neurologic impotence obtained a good erection with low doses of the medication, and 4 of the 17 had sustained erections (Sidi, Cameron, Duffy, et al., 1986). Patients with vasculogenic impotence responded less effectively to injection therapy. Neurologically impotent patients appear to respond best to intracorporeal injection therapy but are most prone to sustained erections. When sustained erections last more than 4 to 6 hours, the corpora should be aspirated of old blood; and if the penis does not detumesce, a dilute alpha-stimulant (phenylephrine, metaraminol, or epinephrine) should be injected intracorporally while carefully monitoring vital signs until detumescence is achieved.

Bodner et al. reported their experience with intracavernous injection of vasoactive medications for erection in men with spinal cord injury (Bodner, Lindan, Leffler, et al., 1987). Of 20 patients, 19 (95%) achieved satisfactory erections, and sustained erections lasting more than 4 hours occurred in three patients (15%) on six occasions. Wyndaele and associates used intracavernous injection therapy in 12 SCI patients and emphasized the need to begin with low doses of medication to prevent the complication of priapism (Wyndaele, de Meyer, de Sy, et al., 1986). Longer follow-up (minimum 2 years) has been reported in two studies specifically evaluating injection therapy in SCI patients (Bodner, Frost, & Leffler, 1992; Lloyd & Richards, 1990). In both of these series, the dropout rate was approximately 40% at 1 year, and the injection frequency was approximately once every 2 weeks. Penile plaque was noted in only 1 of 50 patients on chronic injection therapy, and the treatment was effective in 45 of the 50 patients (Bodner, Frost, & Leffler, 1992). With the exception of priapism, to which SCI patients appear susceptible, long-term complications of plaque formation occurred less frequently than was reported in non-SCI injectors (Levine, Althof, Turner, et al., 1989). The incidence of plaque formation in non-SCI injectors was approximately 30% at 1 year and was directly related to the frequency and dosage of the injection (Lakin, Montague, Vanderbrug Medendorp, et al., 1990; Levine, Althof, Turner, et al., 1989).

Intracavernous injections with vasoactive medications can effectively restore erections in SCI men. Patients must be followed closely because of the tendency for sustained erections to occur. Although SCI patients are at increased risk for sustained erection after intracorporeal injection of vasoactive medications, the incidence of priapism has not been correlated to the level of spinal injury. The dosage of medication required to produce an erection is much lower in SCI patients than other individuals with impotence. High doses of the medication are more prone to produce the sustained erections. Initial injections should use very low doses of the medication in this patient population; the dosage is then slowly titrated up by serial injections performed on a weekly basis until an adequate erection for intercourse, lasting less than 4 hours, is achieved.

The technique is shown in Figure 40-1. Patients are encouraged to alternate the side of the penis injected to minimize the risk of plaque formation. The incidence of long-term complications appears to be lower in this patient population than in patients with nonneurologic

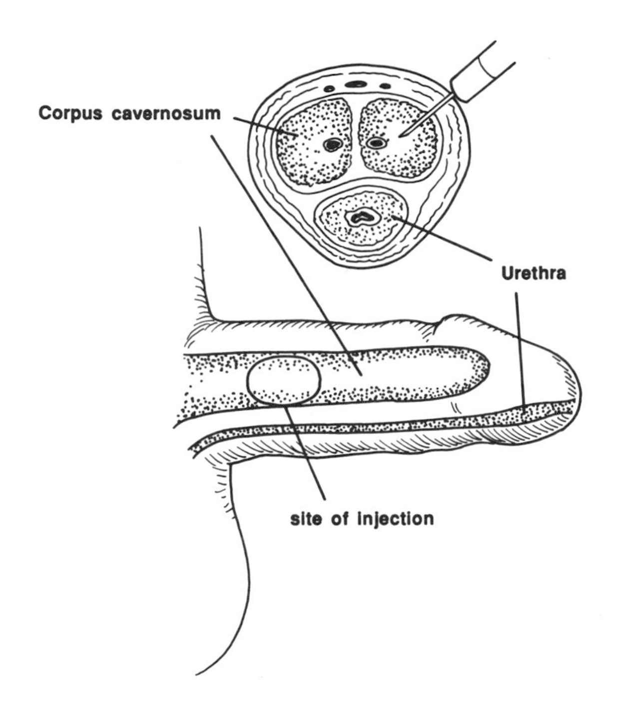

FIG. 40–1. Site of injection. The needle is inserted into the corpora cavernosa at the base of the penis.

impotence, which may be related to the fact that on average the dosage is lower and the frequency of injection is less in this group. We recommend that patients not inject the penis more than two times per week in order to minimize complications.

It has been postulated that the incidence of priapism and plaque formation may be less using prostaglandin E1 because of this agent is metabolized more rapidly (Earle, Keogh, Wisniewski, et al., 1990). The benefits and long-term roles of prostaglandin E1 are unknown and need to be determined (see Table 40-6).

Whereas intracorporeal injection of vasoactive medications can restore erections, and the quality of the erection produced by this technique is satisfactory, the medication has no effect in terms of producing ejaculation or orgasm. Most patients with spinal cord injury do not ejaculate, and restoring the erection does not change this situation. However, spouse satisfaction is acceptable with this method of treatment. It is of note that patients with spinal cord injury may ejaculate without sensation of orgasm. Others may note an altered feeling during ejaculation. (See also Chapter 39.)

VACUUM DEVICES

Vacuum devices for creating erections are not new. Lederer (1917) had a surgical patent for such a device as early as 1917. There has been interest in evaluating these devices with reports of success in their use (see Figure 40-2). Nadig and colleagues reported on 35 men using a vacuum device with a constriction band and noted that 91% of the men achieved a sufficient erection for intercourse and 80% of the men were satisfied. Although blood flow to the penis was noted to decrease while the constricting band was in place and the temperature of the penis dropped an average of 0.96°C over 30 minutes, most patients did not report it to be a problem (Nadig, Ware, & Blumoff, 1986). Turner et al. evaluated 36 non-SCI patients who were using one such device, the Erecaid, and found that 7 (19%) dropped out of the program within 6 months. The most common complaint in this series was blocked ejaculate, occurring in 39% of the men who regularly used the device (Turner, Althof, Levine, et al., 1990). Some of the newer rings may cause a lower incidence of blocked ejaculate.

There has been limited experience with vacuum devices used to restore potency in SCI patients. Lloyd and associates reported on 13 SCI men using the Erecaid and noted that 12 of the 13 (90%) achieved satisfactory erections (Lloyd, Toth, & Perkash, 1989). The Synergist system consists of a custom-fitted silicone sheath with a thin collar at its base. Negative pressure causes the collar to constrict against the base of the penis, increasing rigidity. The device is worn as a condom during intercourse. Zasler and Katz (1989) reported on their experience using the Synergist vacuum device in 20 SCI men. Of the 18 patients who completed the study protocol, all were able to have intercourse, and 88% of the men and 78% of the women were satisfied with the erection. As with intracorporeal injection therapy, the external suction devices have been effective in producing erections but do not change the patient's own ability

TABLE 40–6. *Vasoactive Medications Used for Erections*

Drug	Mechanism of Action	Comment
Papaverine	Smooth muscle relaxant	Frequently used; contraindicated by Lilly for injection into the penis
Phentolamine	Alpha-blocker	Synergistic with papaverine
Phenoxybenzamine	Alpha-blocker	Used in Europe; painful injection; slow onset
Prostaglandin E1	Vasodilator	Expensive, less penile plaque
Nitroglycerine	Vasodilator	Local application to penis can cause headache

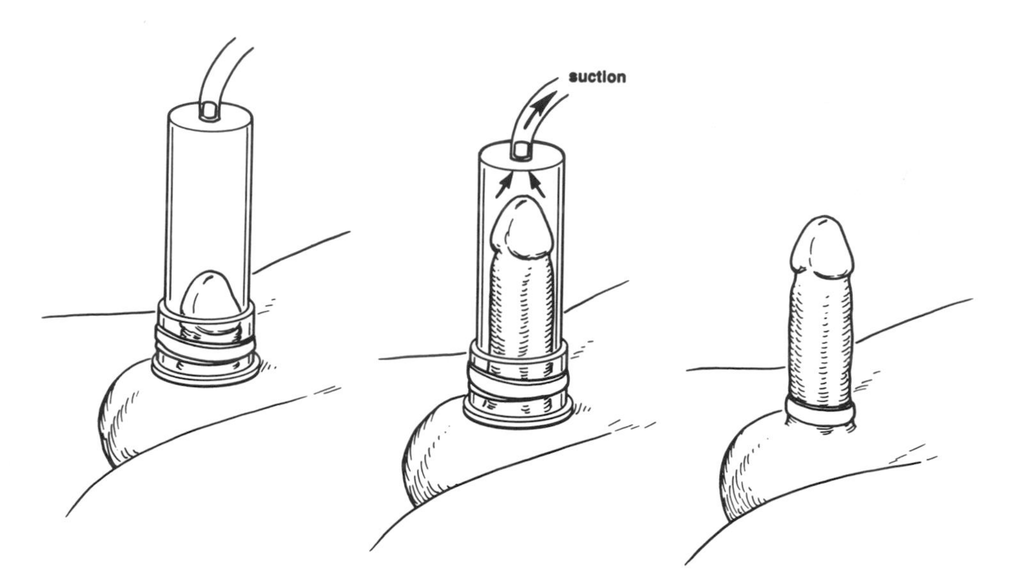

FIG. 40–2. Suction erection devices are placed about the penis, and a watertight seal is achieved against the abdomen. The vacuum is applied, creating the erection, and then a rubber band is left at the base of the penis to maintain the erection for as long as 30 minutes.

to ejaculate or achieve orgasm. Some patients with normal ejaculation may have retarded ejaculation due to the constricting band; and if it is a problem, newer bands are available that are not constricting about the urethra. (See also Chapter 39.)

FERTILITY

Ejaculation is more consistently lost than erection after spinal cord injury and more difficult to restore. Only about 10% of patients with complete spinal cord injury can ejaculate (Bors & Comarr, 1960; Francois & Maury, 1987). The frequency of ejaculation increases to approximately 30% when the lesion is incomplete and to approximately 70% when it is an incomplete lower motor neuron injury (Bors & Comarr, 1960; Ver Voot, 1987). If a patient is interested in fathering a child, he is encouraged to determine if he can obtain an ejaculate under any circumstance. If a specimen is produced, it is sent to the laboratory for routine semen analysis to confirm that sperm is present and to determine the count, motility, and morphology. If a specimen cannot be obtained, the patient is advised to attempt to produce a specimen with the use of a vibrator applied as direct stimulation to the penis (Francois & Maury, 1987). If a specimen can be produced, it is then inseminated into the patient's partner. When no specimen can be obtained by any of these techniques, early promising results have been obtained by intrarectal electroejaculation and artificial insemination (Bennett, Seager, Vasher, et al., 1988; Brindley, 1986). This technique was developed in animals in captivity to ensure their procreation. The technique involves placing a probe in the rectum and electrically stimulating the region of the prostate. Ejaculation usually occurs within several minutes and a semen specimen obtained.

The major problem with the semen obtained by electroejaculation is the poor motility of the sperm. The question often asked by patients is whether they should have electroejaculation performed shortly after injury and the semen frozen for later use as the sperm count may decline with time. Although studies are limited, it appears that chronic urinary tract infections adversely affect semen quality; hence prevention of infections tends to preserve semen quality for some time after spinal cord injury. Sperm are lost during the freezing-thawing process, so patients are generally advised to wait for electrical ejaculation until they are ready to use the specimen. Although problems still exist with sperm motility following electroejaculation, pregnancies have been achieved. When electroejaculation is unsuccessful, consideration should be given to donor semen specimens or adoption. In either of these settings, both partners should be well counseled by the a psychiatrist or psychologist.

It should be remembered that spinal cord injury generally does not cause infertility in women. However, pregnancy may be complicated by chronic urinary tract infections and management of the neurogenic bladder.

An addition problem is the fact that SCI women often do not have labor sensations, and stillbirths can occur because medical help is not sought during labor. SCI pregnant women are often admitted to the hospital during the last month of pregnancy and closely observed. Labor should not be induced in patients with upper thoracic and cervical spinal cord injury, as autonomic dysreflexia (unopposed sympathetic discharge resulting in severe hypertension) may occur.

SUMMARY

Spinal cord injury is a devastating injury that forces the affected person to make lifelong adjustments. Great progress has been made in the medical management of these patients, and life expectancy now approaches age-matched controls. The multidisciplinary approach to management of these patients is essential to their successful reintegration into society. The role of the psychiatrist or psychologist is central, beginning at the time of injury and continuing indefinitely. Patients must first cope with the injury and understand what secondary changes may occur in their functional ability. The patient should experiment to determine how sexual function can be optimized after the injury. When erections are not satisfactory for the patient, further assessment is indicated. Restoring penile erections can effectively be accomplished by vacuum devices or intracorporeal injection of vasoactive medications. Only rarely are penile prostheses required. When under treatment for sexual dysfunction, patients should be followed as closely by their psychotherapist for support and adjustment as they are followed by the urologist for potential complications from injection therapy. Pregnancy is possible in the SCI woman, and patients should be advised of this fact. Unique problems of pregnancy in SCI women, including urinary tract infections, autonomic dysreflexia, and the fact that labor often cannot be felt, can best be addressed through a SCI clinic familiar with these problems. For the man who does not ejaculate and desires to father a child, the future is promising with electroejaculation.

REFERENCES

Adams F (1886). The genuine works of Hippocrates II (pp. 568–654). New York: William Wood.

Bennett CJ, Seager SW, Vasher EA, et al (1988). Sexual dysfunction and electroejaculation in men with spinal cord injury: review. J Urol 139:453–456.

Bodner DR, Frost F, & Leffler E (1992). The role of intracorporeal injection of vasoactive medications for restoration of erection in the SCI male—a three year follow up. Paraplegia 30:118–120.

Bodner DR, Lindan R, Leffler E, et al. (1987). The application of intracavernous injection of vasoactive medications for erection in men with spinal cord injury. J Urol 138:310–311.

Bors E, & Comarr AE (1960). Neurological disturbances of sexual function with special reference to 529 patients with spinal cord injury. Urol Surv 10:191–222.

Brindley GS (1983). Cavernosal alpha-blockade: a new technique for investigating and treating erectile impotence. Br J Psychiatry 143:332–337.

Brindley GS (1986). Sexual and reproductive problems of paraplegic men. In JR Clarke (Ed.), Oxford Reviews and Reproductive Biology (pp. 214–222). Oxford: Oxford University Press.

Claude H, & Lhermitte J (1914). Etude clinique et anatomopathologique de la commotion muldullaire direct par projectile de guerre. Ann Med (Paris) 2:479–506.

Collins KP, & Hackler RH (1988). Complications of penile prosthesis in the spinal cord injury population. J Urol 140:984–985.

Comarr AE (1970). Sexual function among patients with spinal cord injury. Urol Int 25:134–168.

Dietrick RB, & Russi S (1958). Tabulation and review of autopsy findings in fifty-five paraplegics. JAMA 166:41–44.

Earle CM, Keogh EJ, Wisniewski ZS, et al. (1990). Prostaglandin E1 therapy for impotence, comparison with papaverine. J Urol 143:57–59.

Fine PR, Kuhlemeier KV, DeVivo, MJ, et al. (1979). Spinal cord injury: an epidemiologic perspective. Paraplegia 17:237–250.

Francois N, & Maury M (1987). Sexual aspects in paraplegic patients. Paraplegia 25:289–292.

Freed MM, Bakst HJ, & Barrie DL (1966). Life expectancy, survival rates, and causes of death in civilian patients with spinal cord trauma. Arch Phys Med Rehabil 47:457–463.

Gasser TC, Roach RM, Larsen EH, et al. (1987). Intracavernous self injection with phentolamine and papaverine for the treatment of impotence. J Urol 137:678–680.

Gehrig R, & Michaelis LS (1968). Statistics of acute paraplegia and tetraplegia on a national scale. Paraplegia 5:93–95.

Griffith ER, & Trieschmann RB (1975). Sexual functioning in women with spinal cord injury. Arch Phys Med Rehabil 56:18–21.

Holmes G (1915). The Goulstonian lectures on spinal injuries of warfare. BMJ 2:769–774.

Iwatsubo E, Tanaka M, Takahashi K, et al. (1986). Noninflatable penile prosthesis for the treatment of urinary incontinence and sexual disability of patients with spinal cord injury. Paraplegia 24:307–310.

Kraus JF, Franti CE, Riggins RS, et al. (1975). Incidence of traumatic spinal cord lesions. J Chronic Dis 28:471–492.

Lakin MM, Montague DR, Vanderbrug Medendorp S., et al. (1990). Intracavernous injection therapy: analysis of results and complications. J Urol 143:1138–1141.

Lederer O (May 8, 1917). Surgical device, U.S. patent number 1,255,341.

Levine SB, Althof SE, Turner LA, et al. (1989). Side effects of self-administration of intracavernous papaverine and phentolamine for the treatment of impotence. J Urol 141:54–57.

Lloyd KL, & Richards JS (1990). Intracorporeal injections in spinal cord injury: two year follow up (p. 68). Am Spinal Cord Injury Assoc Abstracts Dig.

Lloyd EE, Toth LL, & Perkash I (1989). Vacuum tumescence: an option for spinal cord injured males with erectile dysfunction SCI Nurs 6:25–28.

Masters WH, Johnson VE (1966). Human sexual response. Boston: Little, Brown.

Mesard L, Carmody A, Mannarino E, et al. (1978). Survival after spinal cord trauma. Arch Neurol 35:78–83.

MICHAL V, KRAMER R, & POSPICHAL J (1977). Arterial epigastrico cavernous anastomosis for the treatment of sexual impotence. World J Surg 1:515–520.

NADIG PW, WARE JC, & BLUMOFF R (1986). Noninvasive device to produce and maintain an erection-like state. Urology 27:126–131.

NEWMAN HF, & NORTHUP JD (1981). Mechanism of human penile erection: an overview. Urology 27:399–408.

ROSSIER AB, & FAM BA (1984). Indication and results of semirigid penile prostheses in spinal cord injury patients: long term follow up. J Urol 131:59–62.

SIDI AA, CAMERON JS, DUFFY LM, ET AL. (1986). Intracavernous drug-induced erections in the management of male erectile dysfunctions: experience with 100 patients. J Urol 135:704–706.

SMITH E (1930). The Edwin Smith Surgical Papyrus (pp. 323–332). (Translated from the Hieroglyphics by J. H. Breasted). Chicago: University of Chicago Press.

STEIDLE CP, & MULCAHY JJ (1989). Erosion of penile prosthesis: a complication of urethral catheterization. J Urol 142:736–739.

STOVER SL, FINE PR, & KENNEDY EJ (Eds.) (1986). Spinal cord injury: the facts and figures. Birmingham: University of Alabama.

TALBOT HS (1955). The sexual function in paraplegia. J Urol 73:91–100.

TURNER LA, ALTHOF SE, LEVINE SB, ET AL. (1990). Treating erectile dysfunction with external vacuum devices: impact upon sexual, psychological and marital functioning. J Urol 144:79–82.

VER VOOT SM (1987). Ejaculatory stimulation in spinal cord injured men. Urology 29:282–289.

VIRAG R (1982). Intracavernous injection of papaverine for erectile failure [Letter]. Lancet 2:938.

VIRAG R (1985). About pharmacologically induced prolonged erection [Letter]. Lancet 2:519.

WYNDAELE JJ, DE MEYER JM, DE SY WA, ET AL. (1986). Intracavernous injection of vasoactive drugs: one alternative for treating impotence in spinal cord injury patients. Paraplegia 24:271–275.

YOUNG JS, BURNS PE, BOWEN AM, ET AL. (1982). Spinal Cord Injury Statistics: Experience of Regional Model Spinal Cord Injury Systems. Phoenix: Good Samaritan Medical Center.

ZASLER ND, & KATZ PG (1989). Synergist erection system in the management of impotence secondary to spinal cord injury. Arch Phys Med Rehabil 70:712–716.

ZORGNIOTTI AW, & LEFLEUR RS (1985). Auto-injection of the corpus cavernosum with a vasoactive drug combination for vasculogenic impotence. J Urol 133:39–41.

Note: Information on reference books, monographs and videotapes for physician and patient education in spinal cord injury can be obtained from the Eastern Paralyzed Veterans of America, 75-20 Astoria Boulevard, Jackson Heights, NY 11370-1177.

VI Behavioral Medicine

41 Behavioral therapy strategies with medical patients

DRUE H. BARRETT, Ph.D., GENE G. ABEL, M.D.,
JOANNE L. ROULEAU, Ph.D., AND BARRY J. COYNE, Ph.D.

This chapter presents basic principles of behavioral therapy and illustrates how selected behavioral strategies may be applied to the treatment of medical patients. A comprehensive treatment of behavioral therapy in the medical population is beyond the scope of this chapter, so the discussion is limited to the most common disorders in the medical setting in which behaviorally oriented techniques are useful as part of the overall treatment plan.

DEFINITIONS

The definition of behavioral therapy currently endorsed by the Association for the Advancement of Behavior Therapy and approved by the American Psychiatric Association refers to a general approach rather than specialized techniques or a unified theoretic frame (Franks, Wilson, Kendall, et al., 1990). Behavioral therapy is distinguished from other approaches to treating human behavioral problems by using principles based on research in experimental and social psychology that are systematically applied and simultaneously evaluated.

Masters, Burish, Hallon, and colleagues (1987) have summarized the major assumptions on which behavioral therapy is based. The first assumption is that the focus of the intervention should be on behavior rather than on some presumed underlying cause. It should be cautioned, however, that despite common misconceptions of behavioral therapy the term "behavior" is interpreted as including both covert behaviors (such as emotional responding and cognitions) and overt behaviors. The second major assumption is that maladaptive behaviors are acquired through the same learning process as adaptive behaviors and thus can be modified by interventions grounded in learning principles.

The field of behavioral medicine is in part an application of behavioral therapy theory and principles to the management of disorders within the medical population. Pomerleau and Brady (1979) defined behavioral medicine as "the clinical use of techniques derived from the experimental analysis of behavior—behavior therapy and behavior modification—for the evaluation, prevention, or treatment of physical disease or physiological dysfunction . . . [and] . . . the conduct of research contributing to the functional strategies and understanding of behavior associated with medical disorders in health care." In general, however, strictly defining behavioral medicine as the application of behavioral therapy to the medical population is too strict. Behavioral medicine is more broadly defined as "the field concerned with the development of behavioral science knowledge and techniques relevant to the understanding of physical health and illness and the application of this knowledge and techniques to prevention, diagnosis and rehabilitation" (Schwartz & Weiss, 1978).

This chapter focuses primarily on specific behavioral therapy techniques as a guide for choosing which type of medical patients and conditions may be responsive to a behavioral approach. (In many circumstances, however, behavioral techniques can and should be integrated with individual psychotherapy, family therapy, or pharmacologic strategies.) For further reading and more detailed accounts of behavioral techniques, readers are referred to comprehensive texts on behavioral medicine and behavioral therapy, found in the reference list.

BEHAVIORAL THERAPY APPROACH

According to Kazdin (1978), the behavioral therapy approach to treatment can be summarized in seven rules. First, therapist and patient must agree on the specific, objective goals of treatment. Second, selected strategies for treatment and assessment must match these treatment goals. Third, treatment strategies must work directly to reach criterion or target behaviors required by the specific treatment goals. Fourth, the therapist must accept the patient's problems and goals in the patient's own terms instead of inferring some underlying disposition thought to indirectly influence the patient's behavior. Fifth, current conditions rather than historical determinants should shape the course of treatment. Sixth, specifying treatment in objective terms should be used so as to facilitate replication. Finally, the therapist should rely on basic research in behavioral psychology as the primary source for specific therapeutic techniques.

BASIC TECHNIQUES AND APPLICATION TO MEDICAL SETTINGS

Illness behavior, if more strongly reinforced than wellness behavior, may persist long after its organic cause had disappeared. Behavioral approaches have been used successfully with patients suffering from a variety of conditions for which organic causes could not be found, as well as with patients having difficulty coping with a persistent medical illness or physical disability. The following behavioral therapy methods have potential applications with medical patients.

Systematic Desensitization

Developed by Joseph Wolpe (1958, 1969), *systematic desensitization* (a form of counterconditioning) involves substituting an emotional response that is appropriate or adaptive to a given situation for one that is inappropriate or maladaptive. *Counterconditioning* describes any procedure using learning principles to train the patient to substitute one type of response for another. Specifically, systematic desensitization targets maladaptive anxiety through the substitution of relaxation for anxiety.

Systematic desensitization is most effective and appropriate for decreasing irrational anxiety reactions. Masters, Burish, Hallon, et al. (1987) defined anxiety as irrational when "there is evidence that the patient has sufficient skills for coping with whatever it is he or she fears, but habitually avoids the target situation, or if that is impossible, performs below his or her actual level of skill."

The method of systematic desensitization involves first training the patient in deep muscle relaxation procedures (described later in this chapter). Next, a hierarchy related to the patient's irrational anxiety is constructed. The hierarchy is a series of anxiety-eliciting scenes or situations ranked according to degree of severity. Beginning with the least anxiety-arousing items, these scenes are imagined by the patient while in a state of relaxation so that repeated presentations fail to elicit anxiety. The patient is then asked to approach the feared stimuli in vivo to further eradicate anxiety responses.

Modeling

Following the work of Bandura (1969, 1971), modeling procedures have been applied extensively in behavioral therapy. In its basic form, *modeling* is simple. The patient is exposed to one or more other individuals who demonstrate appropriate behaviors to be learned by the patient. The model(s) may be present (live) or filmed (symbolic). The modeling procedure should include exposure to the stimuli and situations that surround the model's behavior, so that not only the target behavior but also its appropriateness to relevant stimuli can be demonstrated. Ideally, the adoption of specific modeled behaviors should also provoke changes in collateral behaviors (thoughts and feelings) that accompany the target overt behavior. Many modeling techniques involve immediate participation of the patient, who performs the modeled behavior after its demonstration by the model. Verbal reinforcement often follows the patient's performance. Modeling is often used to teach patients appropriate assertion behaviors (in contrast to angry outbursts) or to accomplish complicated diagnostic procedures.

Contingency Management

With operant conditioning, reinforcement is said to occur when the contingent application or removal of a stimulus results in an increase in the frequency of behavior preceding it. There are two major categories of reinforcers, positive and negative. A *positive reinforcer* is any stimulus event whose contingent application increases the likelihood of the behavior preceding it, whereas a *negative reinforcer* is any event whose contingent withdrawal increases the rate of the behavior preceding it (Skinner, 1969). Care must be taken that the reinforcer is available only after the display of appropriate behavior. Usually, it is the close pairing in time between any action and its consequences that establishes the link between a behavior and its contingent reinforcement. With behavioral therapy, a systematic strategy of presenting and withdrawing contingent rewards and punishments is called *contingency management*.

Before establishing a contingency management program, the frequency of the target behavior prior to the initiation of treatment (baseline) must be determined. Once treatment begins, the frequency of the target behavior indicates the relative effectiveness of the behavior modification program. Finally, after termination of treatment, follow-up observations regarding the frequency of the target behavior should be recorded to verify the continued beneficial effect of treatment.

With all behavioral techniques, it is essential for the therapist to identify the particular antecedent stimuli, situations, and conditions that are consistently associated with the patient's inappropriate behavior, as well as those conditions that consistently follow the inappropriate behavior as maintenance factors or reinforcers. Prior to initiation of a contingency management program, the relative power of each reinforcer in the patient's environment should also be determined. There are different types of reinforcer, such as material reinforcers (candy, tokens, money), social reinforcers (smiles, praise, physical contact), and activity reinforcers (phone calls, walking privileges). An observant therapist or nursing staff can quickly discover those conditions for each patient that supply effective contingent reinforcement (McFarlane, Bellissimo, & Upton, 1982). The incentive value and power of particular stimuli to act as reinforcers depends on the individual's learning history; thus contingency management strategies must be individualized.

Once powerful reinforcers are identified, the therapist must establish when contingent reinforcement is to be delivered. Immediate reinforcement seems best; usually the longer the time between the completion of a behavior and the delivery of a reinforcer, the less effective is the reinforcer (Ware & Terrell, 1961). When the patient is expected to acquire an entire sequence of desired behaviors, reinforcement should be administered after each component of the sequence as well as at its end. As therapy progresses, delaying reinforcement until the entire sequence is completed becomes more cost-effective of the therapist's time. Patients do not have to be consciously aware of reinforcement contingencies for learning to occur. However, when a reinforcer follows a long sequence of behaviors, or when reinforcement is delayed following completion of a behavior, its effectiveness can be increased by clearly spelling out the contingency.

Adherence to medical treatment regimens in particular can be viewed as a function of specific situational factors (Shelton & Levy, 1981). The more the therapist can directly influence the patient's environment, the more effective is contingency management.

CASE STUDY: Contingency Management Program for Treatment of Anorexia Nervosa

Evaluation. Behavioral analysis of a 57-year-old anorexic woman suggested that the antecedent of the patient's weight loss was a lack of social interactions after she moved to a new state, with resultant loss of social contacts, and her husband's frequent absence on business trips. The patient's weight loss was rewarded by the extensive attention she received from her family and medical staff while in the cachectic state (Miller, 1983). The patient's history of noncompliance with traditional psychotherapy coupled with her marked hostility and anger when interviewed suggested that any discussion of why she failed to eat would be unproductive (Goldfarb, Dykens, & Gerrard, 1985; Halmi, 1974).

Behavioral Intervention. Because the patient's refusal to eat and resulting weakness had placed her in an environment where hospital staff could observe her behavior, contingency management was selected as the major treatment intervention. The first step of treatment was to identify and measure objectively the patient's target behavior. In this case weight gain was selected as the target behavior, and it was measured by having the patient's weight measured by the nurse-recorded analytic scale each morning after voiding, with the patient wearing the same minimal hospital garment. The next step was to determine possible reinforcers for this patient. (Most positive reinforcers are generally identified by their high frequency in the patient's environment or, in this case, the hospital environment.) The therapist and staff observed the patient frequently talking on the telephone, watching television, smoking about 30 cigarettes per day, and enjoying talks with other patients on her hospital floor. Finally, the behavioral criterion required to earn reinforcement must be established. In this case the criterion consisted of a 0.25-pound weight gain from the patient's previous highest weight. The following treatment strategy was promptly implemented.

1. Protocol for medical staff
   a. Discontinue patient's television, phone, and visiting privileges, and limit her cigarette intake to 15 per day.
   b. Confine patient to her room.

   c. Weigh patient each morning at 6:00 a.m., after voiding and in her hospital gown.
2. Contingency rules
   a. If patient gains 0.25 pound above previous highest weight, she enters step 1 of reinforcement program (see below).
   b. If patient sustains three consecutive days of step 1, she enters step 2 of reinforcement program.
   c. If patient sustains three consecutive days of step 2, she enters step 3 of reinforcement program.
3. Reinforcement program
   Step 1. Access to TV and 30 cigarettes per day.
   Step 2. Step 1 plus use of the telephone.
   Step 3. Steps 1 and 2 plus visiting privileges and freedom to leave her room.

After implementation of this contingency management program, the patient (and eventually her husband) gave impassioned pleas for exceptions to the contingency program or attempted to justify its termination so she would not be upset. These maneuvers by the patient and her family were met consistently with the unimpassioned response that: "This is the method we feel will be most effective in allowing you to leave the hospital as rapidly as possible." As was expected, the busy hospital staff, with constantly changing shifts, initially needed repeated clarification and reassurance about the criteria for the various steps of reinforcement.

A major advantage of the program, given the hospital setting, was its simplicity. After an initial 7 to 10 days of attempting to disrupt the contingency program, the patient learned that the staff were steadfast in their resolve to carry out the program and were consistent in rewarding her for weight gain. She steadily began gaining weight and within 4 weeks was discharged after gaining 20 pounds.

With contingency management programs, various artifacts can creep into the treatment that make it less than perfect. In this case, the patient began her treatment while receiving intravenous infusions and hyperalimentation, factors that produced either sudden weight loss or weight gain unrelated to the patient's eating behavior. No attempt was made to rectify the system for these contaminants, as such adjustments would be exhausting to the therapist, staff, and probably the patient herself. By setting the criteria for reinforcement at 0.25 pound above the patient's previous highest weight, the contingency system was still effective irrespective of the artifacts produced by the intravenous infusions and hyperalimentation.

Another important consideration is the generalization of the treatment once the patient leaves the hospital. In this case, will the patient continue to gain weight after discharge? For this reason, it is important to train family members in reinforcement procedures, especially instruction on how their own responses may contribute to the patient's maladaptive eating pattern. More extensive detailed discussions of the behavioral treatment of anorexia nervosa may be found in Bruch (1973) and Mizes (1985).

Relaxation Procedures

One of the most common forms of behavioral therapy is relaxation training. *Progressive relaxation* (Jacobson, 1938) is a method by which the patient is taught to tense and then relax all parts of the body systematically while concentrating on the sensations produced by relaxation and learning to progressively produce deeper relaxation. Somewhat similar methods include autogenic training (Schultz & Luthe, 1969), which consists of concentrated imagery of warmth and heaviness, and transcendental meditation, which involves deep breathing and repetition of a single word (a mantra).

Biofeedback has been used separately or in combination with other relaxation techniques (biofeedback-assisted relaxation) to help the patient learn muscular relaxation by providing immediate information or feedback from instruments that measure the electrical activity of muscles. Often the patient's knowledge that he or she is succeeding serves as a reward to facilitate further success. As the following case history illustrates, biofeedback may also have more subtle effects, such as helping patients to select successful coping strategies and to increase their sense of self-control.

CASE STUDY: Muscle Relaxation and Biofeedback for Treatment of Type A Behavior

Evaluation. A 57-year-old man was referred by the cardiology department because of persistent chest pain and depressive features following a myocardial infarction and coronary artery bypass. Clinical observation of the patient's behavior revealed the characteristics of the type A personality pattern. He exuded dominance and control over others, maintained an authoritarian attitude regarding nearly all issues, and displayed rapid speech with frequent hostile, angry comments. He reported that he worked 7 days a week and that a typical day off began Friday night at 7:00 p.m. when he would leave work, go home, take his boat to the lake, fish all night long, and then go back to work Saturday morning, feeling that he had enjoyed a day's relaxation and should now get back to work.

When his authoritarian, controlling personality style was not well accepted by others, he experienced marked anxiety, which was exacerbated by an increase in frequency of chest pain. With each onset of pain, his attitude was to ignore it and continue working, which inevitably led to persistent chest pain that would plague him throughout the night.

Behavioral Intervention. The patient was asked to monitor the frequency of symptoms (including chest pain and type A behaviors) and to identify attitudes and emotions associated with them. This homework allowed an ongoing daily appraisal by the patient of the various components of his anxiety, chest pain, and type A behavior. In addition to treatment with antidepressants for his depression, the patient was taught progressive muscle relaxation. This nonpharmacological intervention assisted the patient in controlling anxiety and taught him how to become more aware of the early situational antecedents that provoked his anxiety, such as not getting his way or being unable to control situations. To further assist in the patient's control of anxiety, he underwent muscle biofeedback assessment and treatment.

Biofeedback involves recording one or more of the patient's physiologic responses and displaying this information to the patient. The patient is then instructed on the association between emotional responding and physiologic responses (Burish, 1981). Biofeedback can be used as both an assessment tool and a treatment strategy. With this patient, biofeedback was first used to evaluate whether specific situations related to his type A personality pattern would provoke an elevated electromyographic (EMG) response and subsequent chest pain. In this case, in order to evaluate physiologic provocative responses objectively, the therapist presented the patient with brief, 2-minute audiotaped scenes specifically targeting type A personality stressors, such as hearing the boss lavishly praise a younger competitor, being criticized for not working hard enough, and being continuously interrupted when trying to talk (Abel, Blanchard, Barlow, et al., 1975). For many patients, an elevated EMG provides an objective assessment of anxiety. In this case, elevation of the EMG occurred as the patient recalled being angry with people or recalled becoming anxious by attempting to be perfect in various work situations.

The medical patient is frequently reluctant to conceptualize emotional factors as contributing to physical symptoms and often finds it exceedingly difficult to trust the opinion of a psychiatrist or psychologist, possibly because of prevalent negative attitudes toward such professionals in our culture. Psychophysiologic assessment, by contrast, provides a means of dispelling this distrust by presenting the patient with powerful evidence that demonstrates a connection between emotional factors and the body's physiologic reactivity.

This particular patient initially was skeptical that his anger and anxiety could affect his body's physiologic responding. However, during one session early in his biofeedback training, he had marked difficulty controlling the EMG tensions within his normal range. Puzzled by the elevated EMG, the biofeedback technician questioned the patient about his thoughts. The patient reported that he had been thinking about another hospitalized patient with whom he had exchanged hostile words the day before. The patient was instructed to ignore those thoughts for the moment and to proceed with biofeedback by attempting to relax. After the session, he was shown the EMG printout that demonstrated the elevated tension associated with his recall of his anger and the dramatic drop in EMG elevation that accompanied diverting his attention from anger to relaxation. Here was objective evidence that made it impossible for the patient to ignore the relation between anger and his body's physiologic responding. He then dedicated himself to learning how to control his anger and anxiety and to use both the muscle relaxation and biofeedback procedures to assist in therapy.

Effective treatment in a therapy session alone, however, is of limited value. Gains within treatment sessions must extend or generalize to the patient's natural environment so the patient is able to control his physiologic responses without mechanical assistance (Young & Blanchard, 1980). As his training continued, the physiologic recordings demonstrated a significant reduction in the patient's overall EMG activity when receiving biofeedback that generalized to when he was sitting quietly without biofeedback. In addition, he reported that when awakened by angina attacks during the night, he successfully applied the techniques he had learned during biofeedback training to control his anxiety, which eventually resulted in elimination of his chest pain episodes.

The patient was not "cured" by biofeedback alone. Antidepressants, assertiveness training, family therapy, progressive muscle relaxation, awareness of the antecedents of chest pain and anxiety, education to alter the type A personality pattern (Friedman & Rosenman, 1974), and biofeedback were all components of a total treatment program directed at his symptomatology. In this case, however, biofeedback motivated the patient to remain in treatment by providing objective evidence that emotional factors contributed to his physical illness.

A variety of biofeedback techniques are available. The most commonly used modalities include EMG, skin temperature, electrodermal response (EDR), and electroencephalography (EEG) (Bellissimo, 1981; Green, 1984; Tunks & Bellissimo, 1991). EMG measures muscle activity and is generally used in conjunction with relaxation training. It is the primary technique used for treating tension headache, bruxism, chronic pain, and fecal or urinary incontinence. Skin temperature monitoring is also useful with relaxation training and for treatment of vascular diseases including migraine headache, Raynaud's disease, and hypertension. EDR measures skin conductance and resistance and is generally used in relaxation training and systematic desensitization procedures. EEG measures brain wave activity and has been found to be helpful for treating chronic pain, insomnia, and hyperactivity, and for increasing attention control.

Another biofeedback modality that is being used more commonly is the pneumograph. Respiration sensors across the chest and abdomen measure breathing patterns. This technique is typically used to treat hyperventilation, anxiety, and panic attacks. It has also been used with patients who have breathing difficulties related to medical conditions such as chronic obstructive pulmonary disease (COPD) and asthma. Comprehensive guides to the use of biofeedback and its clinical applications can be found in Basmajian (1989) and Schwartz (1987).

Procedures to Increase the Effectiveness of a Behavioral Treatment Approach

Assuming that the behavioral therapist has adopted an effective treatment approach, what sort of evidence should be used to convince the patient that behavior X is responsible for problem Y? In some cases, to enhance the potential impact of a treatment procedure as well as to maximize the patient's compliance, a demonstration of the relation between the patient's behavior and a symptom can be highly effective.

An effective means of assessing a patient's problem is direct observation of the patient's motor behavior during symptom development. The closer the therapist can come to witnessing symptoms, the better the patient's and therapist's understanding of what strategies may help treat the patient's symptom complex. In some cases, symptoms can be provoked, as in the following example of a patient with hyperventilation syndrome.

CASE STUDY: Behavioral Diagnostic Technique for Treatment of Hyperventilation

Evaluation. A 38-year-old woman with below-average intellectual functioning reported a 2-year history of "spells." The symptoms she described appeared to be panic attacks with hyperventilation (Lum, 1975, 1981). To evaluate and treat this patient's symptoms, the therapist told the patient that she would undergo a breathing test but did not inform her that the test was likely to provoke her symptoms. This information was withheld in order to control any demand characteristics of the test and, more importantly, to ensure that any symptoms that were provoked were similar to those she had been having during her "spells." After medical clearance, the patient was taught to breathe at a rate of 15 breaths per minute as indicated by the therapist raising and lowering a hand to signal when the patient should inhale and exhale. The patient was verbally reinforced for following the therapist's pacing of the patient's breathing. At this point in the test the patient reported no symptoms. Her paced breathing was then rapidly accelerated to 30 breaths per minute, which was to be maintained until she became symptomatic (or for 5 minutes without symptoms). In this patient's case, within 1.5 minutes she developed marked flushing of the face and numbness of the hands; and within 2.5 minutes she developed the tightness of her fingers and toes that typically occurred with her "spells." Two minutes into the provocative test she also became less attentive to the examiner and began reporting her typical chest pain, along with tachycardia, marked shortness of breath, and sensations that she was going to pass out. The patient was asked to describe each of these symptoms and to report if she had ever experienced them before and under what conditions. In each case she reported the symptoms as identical to those that developed during her "spells." By 3.5 minutes into the provocative test, she asked to stop because she feared her symptoms would get out of control.

Behavioral Intervention. The therapist then reduced her pace of breathing to 15 breaths per minute and instructed her to place her right hand over her chest and her left hand over her abdomen and to breathe, not with chest excursion but with abdominal excursion (diaphragmatic breathing). The patient was also instructed to breathe through her nose or through pursed lips in order to cut down on the exchange of air, thereby decreasing carbon dioxide expulsion. As the patient's breathing was paced at this slower rate, there was a dramatic reduction of the symptoms described above.

The patient was then asked why her symptoms had decreased. She reported that it must have something to do with her breathing, as rapid breathing had brought on the symptoms she usually had with her "spells" and slower breathing eliminated those symptoms. At this point, the therapist reiterated the relation between the patient's symptoms and her breathing.

Subsequent treatment included training the patient in appropriate abdominal breathing while lying supine, sitting, standing, and lastly walking and exerting herself. The importance of the latter element is that many patients with hyperventilation syndrome precipitate their symptoms during the course of their work, when exertion leads to tachycardia and tachypnea, which triggers the anxiety–hyperventilation cycle (Moreyra, McGough, Hosler, et al., 1982).

If the patient's panic and hyperventilation symptoms are not responsive to breathing training, additional use of pharmacologic treatment may be needed. Our clinical experience has found use of a sedating tricyclic antidepressant (such as nortriptyline) to be most useful.

Shaping Procedures

Many patients are required to perform specific medical skills. The diabetic's inability to check his own blood glucose, the obese patient's inability to increase his physical activities to promote weight loss, the man with erectile dysfunction not being able to touch his partner sexually without focusing on his erection response—all are examples in which the patient may need to acquire skills in order to deal with a specific medical problem. Often the physician fails to appreciate that seemingly simple assignments involve complex skills that must be learned one small step at a time.

One behavioral technique effective in teaching patient's new behaviors is *shaping*, also referred to as successive approximation or response differentiation (Bandura, 1969). If a behavior is not emitted by an individual, it cannot be reinforced. Shaping involves the systematic application of reinforcement of successive approximations to a goal behavior. Initially, behaviors already present in the individual's response repertoire that have some similarity to the desired behavior are reinforced. This phase is followed by reinforcement of closer and closer approximations to the target behavior. The following case demonstrates the acquisition of sequential behaviors through shaping procedures.

CASE STUDY: Shaping Treatment to Increase Compliance with Physical Rehabilitation in a Patient with "Drop Attacks"

Evaluation. A 62-year-old widow was referred by the neurology department for treatment of possible depression. She had been an energetic, self-sufficient, responsible woman working 5 days per week and enjoying a full social life until 2 years prior to her referral, when she developed "drop attacks," episodes in which without warning she would suddenly lose control of her skeletal muscles and collapse. Falling without muscle tone, she was unable to prepare for the impact of the fall and in 2 years had suffered numerous lacerations, a skull fracture, various fractured ribs, and a fractured vertebra. Extensive neurologic workups and attempts at drug intervention had not reduced the frequency of these attacks. Her neurologist referred the patient when during her most recent 4-week hospitalization the patient refused to ambulate even with a walker.

When interviewed, the patient was attentive, engaging, and could easily identify the etiology of her noncompliance. Her last 2 years were filled with drop attack episodes. Falling to the ground, even if she sustained no fractures in the fall, was terrifying. Because she had not risked ambulating in over a month, the patient anticipated she would inevitably become an invalid and have to be cared for by others. These thoughts led her to be depressed.

Behavioral Intervention. After institution of an antidepressant medication, a behavioral analysis was conducted to assess the patient's fears and her skills at ambulation. Despite staff on each side to assist her in moving from bedside to her walker, the patient became anxious and fearful. Although merely standing at her walker was anxiety-provoking for the patient, she reported that her most terrifying fear was the prospect of having to get up from the floor. When lowered to the floor by staff (as if she had suffered a drop attack), in proximity to a chair for climbing support, the patient immediately became so frightened and confused that she gave up any attempt to raise herself.

An operant approach was instituted that targeted increasing the patient's use of the walker and training her to stand up from the floor. While on the neurology service, the patient received attention from staff regardless of whether she was attempting to ambulate, attempting to rise from the floor, or simply lying in bed. On the Behavioral Medicine Unit, attention from the staff was made contingent on the patient practicing walker ambulation and getting up from the floor, with minimal attention given to her while she was lying in bed.

The patient received social reinforcement in the form of praise and encouragement from staff during each training session when she practiced her needed skills. Each complex skill, however, was divided into small, separate components. Behaviors that are problematic for a patient must be divided into small, manageable steps so that reinforcement can be given frequently for even minor attempts at mastering each component of a complex behavior (such as using a walker or rising from the floor). Training the patient to stand up included instructed her to wait on the floor until she was no longer frightened before mentally organizing the steps of getting up, to identify furniture that would be supportive for rising, to turn furniture into position while still lying on the floor, to move her legs under her body so that her limbs could work in unison to "climb up" the furniture, to pause halfway to rest so she had adequate energy to complete the full climb, to climb up the furniture until standing erect, and to rest while standing before attempting to ambulate. She received separate training sessions on each component of this complex task, and after several days of training the patient regained her ability to walk.

Dividing a complex skill into manageable components presents the opportunity for frequent reinforcement and allows the patient to gain confidence without having to complete an entire complex task. This successive approximation, or shaping, strategy has frequent application in skills training of the medical patient. Once a skill is acquired, successful performance of a complex task improves mood and self-esteem, as it did in this case.

Extinction Procedures

Some medical patients do not have problems of behavioral deficits that require the training of new behaviors but instead perform dysfunctional behaviors that need to decrease in frequency. Ideally, it is best if appropriate behavior can be strengthened in a patient's repertoire because sometimes it weakens or eliminates the excessive dysfunctional behavior. At times, however, it is not possible, and more invasive, generally aversive procedures must be taught to the patient.

A variety of behavioral techniques have been developed to reduce or eliminate the frequency of problem behaviors (Masters, Burish, Hallon, et al., 1987). There are seven basic *response eliminating*, or *extinction*, procedures: extinction by contingency management; graduated extinction; covert extinction; negative practice; stimulus satiation; implosive therapy; and flooding. The following sections illustrate two of these approaches.

Flooding or Response Prevention

The therapeutic technique of *flooding* has its roots in laboratory experiments on the establishment and extinction of avoidance responses (Solomon, 1964). Flooding, also referred to as response prevention, involves exposing an individual to anxiety-provoking stimuli while preventing the occurrence of avoidance behaviors. In practice, patients are exposed to either imaginal or actual feared stimuli without chance of escape while they experience high levels of anxiety.

A variety of phobic symptoms can be treated by exposing patients to their fears under conditions from which they cannot escape. The result of such an exposure technique is prompt reduction of anxiety to the feared environment or situation. From a behavioral perspective, fear development and avoidance behaviors have been explained by Mowrer's (1960) two-factor theory of fear learning. This theory suggests that fear responses are acquired through classic conditioning in which anxiety is associated with a specific situation or behavior. Over time and as a result of generalization, this anxiety becomes associated not only with that specific environment but with environments similar to the original anxiety-provoking situation. Fear responses are thought to be maintained through an operant conditioning process. In an attempt to avoid the anxiety associated with the feared situation, the individual escapes from or avoids the situation. Escaping from the feared situation is momentarily associated with anxiety reduction. Because anxiety reduction is a strong reinforcer, further avoidance of the feared situation is maintained by the constant reinforcement that follows avoidance or escape from the feared situation. The fear-avoidance pattern thus becomes chronic.

CASE STUDY: Anxiety Reduction Through Flooding for Treatment of "Shy Bladder"

Evaluation. A 33-year-old man was referred from the urology service because of a "shy bladder." The patient recalled always being highly anxious when urinating in public restrooms. Throughout grade school and high school, this fear led him to reduce his fluid intake prior to going to school in order to avoid urinating there. His fears about urinating in public facilities continued after graduation and eventually extended to his work site. During the 12 years prior to referral, he had never urinated in any of the restrooms of the office building in which he worked. The patient was seeking treatment at this time because of a recently initiated drug awareness program in which employees were asked to submit urine samples for drug screening. He had never been able to provide physicians with a urine sample on demand.

Behavioral Intervention. After initiating a behavioral analysis of the patient's phobia through interview and self-report forms, the therapist walked with the patient into various restrooms to identify the precise characteristics and surroundings that provoked the patient's anxiety. To further quantify his fear response, the patient was asked to rate his anxiety before, during, and after entrance into each restroom. As is traditionally the case with such individuals, restrooms with frequent use, open stalls, and especially urinals with no dividers were most provocative of the patient's anxiety and inability to urinate (Abel & Blendinger, 1986; Anderson, 1977). The patient was next asked to "water load" by drinking 16 ounces of soda 1.5 hours prior to each future therapy session. With the resultant high urges to urinate, he was accompanied by the therapist to an anxiety-provoking public restroom where he was asked to imagine his usual thoughts about entering such a feared situation. He was instructed to enter the restroom and approach the urinal, unzip his pants, and carry out all of the physical and cognitive steps of urination except actually voiding. This procedure was repeated at several public restrooms with those characteristics that especially provoked his fears.

After 4 to 5 days of entering restrooms in proximity to the therapist's office, the patient was then asked to carry out the same procedure at work and in social surroundings to bring about generalization of the treatment. To the patient's surprise, his fears began to decrease drastically, and as the therapist expected on a number of such occasions the patient was unable to comply with instructions not to void: His fear under such circumstances was reduced; and, as a result of his water loading, his urges to urinate were high. He subsequently voided on a number of such occasions, which gave him considerable encouragement and further reduced his fear (Wolpe, 1958).

As a result of the treatment, the patient's life changed markedly. In the past the patient had avoided work and social situations that might necessitate his urinating in another's presence. He began socializing, attending parties and mixing more with work colleagues as his social limitation was now markedly reduced.

For the final step of his treatment, the patient had to overcome his fear of urinating on demand for laboratory testing. During the course of the treatment program, the therapist took the patient to a hospital laboratory and had him void in a restroom used for collecting urine samples. With his urinary urges high and anxiety lowered, the patient was overjoyed by his ability to void under this formerly anxiety-provoking situation. His follow-up treatments require him to maintain his gains by frequent use of new restrooms.

In vivo exposure with subsequent extinction of avoidance behavior is an exceeding powerful treatment with wide applicability to a variety of fears.

Covert Extinction

One method of decreasing high frequency behaviors is to associate them with negative events or consequences (Abel, Becker, Cunningham-Rathner, et al., 1984). *Covert extinction* involves a variety of procedures in which the consequences of behavior are imagined. One such procedure, covert sensitization, involves the imaginal pairing of a problem behavior with images of aversive or negative consequences (Cautela, 1966). This procedure involves three steps: (1) imagining the thoughts, images, and feelings antecedent to dysfunctional behavior; (2) imagining negative (usually social) consequences that could result from this behavior; and (3) pairing these two cognitions together in a repeated cycle over a 15-minute session so the antecedents begin to prompt cognitions of the negative, aversive consequences. In this manner the early antecedents are extinguished, and the subsequent dysfunctional behavior does not follow.

The next clinical case presents an example of covert sensitization used to extinguish bulimic behavior by asking the patient to cognitively pair the antecedents of binge behavior to its possible negative consequences.

CASE STUDY: Covert Sensitization for Treatment of Bulimia

Evaluation. A 29-year-old married woman was referred by her internist for treatment of bulimia. Her life was organized around binge–purge episodes. Working a different shift than her husband, she would leave work after an afternoon of planning her binge menu, which often included half-gallons of ice cream, dozens of cookies and doughnuts, pastries, and potato chips. Locking her apartment door, she would attack the food and relish its consumption, knowing that she could eat anything she wanted because vomiting was to follow. Futile attempts to conceal her binge–vomit episodes and pleas by her family to give up her bulimia eventually overwhelmed the patient.

Behavioral Intervention. Behavioral analysis revealed that most of the patient's bulimic episodes occurred during the late afternoon, evening, and early morning—periods when her husband was not at home and her 3- and 4-year-old daughters were with sitters or asleep. The patient's binge–purge episodes did not occur on the spur of the moment, however, but began hours earlier as she ruminated about binge–purging and planned her menu (Fairburn, 1982).

In this case there were a number of negative social consequences that might result from bulimia. For example, her two children were becoming aware of her binge–purge episodes. The patient feared that they might acquire an interest in binge–vomiting, simply modeling that behavior after her. She also feared that her husband might leave her on realizing that she had not ceased her binge–purge behavior as promised.

To ensure compliance with the treatment, the patient was asked to verbalize and tape-record her 15-minute covert sensitization scenes (Abel, Becker, Cunningham-Rathner, et al., 1984), during which the potential adverse consequences of her behavior were described in grim detail. After listening to the audiotapes, the therapist critiqued them and then taught the patient, first, how to make the aversive scenes even more aversive and, second, how to promote frequent pairings or associations between the antecedents of the binge–purge behavior and the possible social consequences that were most fearful to her. Such feedback by the therapist was not only instructional, it also provided the patient with attention and social reinforcement (DiMatteo & DiNicola, 1982). Once the patient had acquired the skills to develop such scenes within the therapy setting, she practiced developing covert sensitization tapes while at home. Finally, she was asked to use this symptom reduction technique in her natural environment whenever she had urges to ruminate about binge–purging. Following several weeks of covert sensitization, the patient's binges ceased.

Bringing It All Together: A Comprehensive Behavioral Approach

A behavioral medicine strategy typically involves utilizing a combination of behavioral techniques. The necessity of utilizing a comprehensive approach is typical in patients with chronic pain syndromes. During the late 1960s a behavioral framework was proposed by Fordyce and his colleagues to explain and treat chronic pain patients (Fordyce, Fowler, & DeLateur, 1968; Fordyce, Fowler, Lehmann, et al., 1968). According to this framework, reports of chronic pain are viewed as a set of behaviors that are influenced by environmental consequences in the same manner as other behaviors. Typical pain behaviors include grimacing, guarded movement, slow response time, limping, pain statements, and analgesic drug seeking (Gatchel, Baum, & Krantz, 1989). With an operant approach pain is not viewed simply as a neurophysiologic event but is seen as involving psychological, environmental, and biologic variables (Fordyce, 1976; Turk, Meichenbaum, & Genest, 1983).

Fordyce (1976) distinguished between two types of pain behavior: *respondent* pain and *operant* pain. Respondent pain behaviors occur as a response to antecedent stimuli (i.e., tissue damage) in a reflexive manner. Operant pain behaviors may be elicited by antecedent stimuli, but they are also influenced by environmental consequences that follow the patient's pain behaviors. The frequency of operant pain behaviors can be increased in three ways: (1) *positive reinforcement*, such as attention, sympathy and concern, financial compensation, or administration of pain medications; (2) *negative reinforcement*, or the removal of aversive stimuli, such as the avoidance of work or household chores; and (3) extinction, or *nonreinforcement*, of well behaviors.

The goals of a behaviorally oriented pain treatment program are to: (1) reduce the frequency of pain behaviors; (2) increase the patient's activity levels, physical abilities, and other well behaviors; (3) eliminate reliance on nonessential pain medications; and (4) reduce utilization of medical care for the relief of pain. In order to reach these goals a combination of the behavior techniques discussed in this chapter are required. The following case demonstrates the integration of several behavioral strategies.

CASE STUDY: Comprehensive Behavioral Treatment of Chronic Pain

Evaluation. A 23-year-old man was referred by his orthopedic surgeon because of persistent, severe back pain unresponsive to previous treatment. The patient had been in excellent health until 3 months prior to his most recent hospitalization. At this time the patient developed localized low back pain in the lumbar area. His orthopedist treated him conservatively with traction and narcotic analgesics for approximately 2 weeks, and he eventually underwent disk removal at L3–L4, L4–L5, and L5–S1 levels. He did well after surgery and was discharged.

The day after the patient's discharge he had the onset of severe pain and back spasms near the incision site and severe constipation. He was rehospitalized and evaluated by a gastroenterologist who determined that the gastrointestinal complaints were secondary to excessive narcotic use. The patient was treated conservatively, discharged a second time, only to be rehospitalized 2 days later, again due to complaints of pain and constipation. The patient also developed symptoms of depression, including loss of appetite, terminal sleep disturbance, crying spells, depressed mood, ruminations regarding pain, and suicidal ideation.

The medical unit staff found the patient nearly impossible to treat because of his numerous pain behaviors and hystrionic personality. He was demanding and manipulative of the staff, became hostile and angry if asked to move, and talked about his pain in dramatic terms, "It's like having a railroad tie driven into me with a sledge hammer." What would normally be routine care took at least an hour to perform with this patient.

Because of the patient's depressive symptoms and the nursing staff's irritation with the patient, the Behavioral Medicine Unit was consulted during this third hospitalization. It was determined that there was an organic basis to the patient's pain, and he was eventually diagnosed with a *Pseudomonas* infection related to the previous lumbar laminectomy. It was treated with intravenous antibiotic therapy. However, in addition to the organic component of his pain, there was found to be a strong operant component maintaining his pain behavior. The patient's mother had retired early from her job, anticipating that her son would be chronically ill and that she would need to move in with him and his wife to take care of her son on an outpatient basis. The patient's wife was constantly at his bedside when she was not required to be at work. The patient simply had to point with his head in the direction of his water, and his wife would jump up to provide fluids while adjusting the straw so her husband did not have to move in order to drink.

Treatment

Treatment involved in a multimodal approach including use of antidepressant medication to deal with the depressive symptomatology and to raise the patient's pain threshold, a contingency management system in which staff and family members were instructed to provide social reinforcement when the patient made attempts to be more independent in caring for himself and instruction in relaxation and pain reduction procedures utilizing biofeedback.

Additionally, the patient was started on a "step program" in order to gradually reinforce increasing activity levels. This program consisted of having the patient monitor pain levels on a 10-point scale, with 0 indicating no pain, 5 indicating pain that required some form of intervention, and 10 indicating severe pain. The patient was placed on a pedometer to measure his activity level in the form of number of steps he walked. A 5-step system was established in which the patient's maximum permitted activity level (increasing in increments of 500 steps) was determined by his pain level. The patient was required to have five consecutive days with no pain report greater than level 5 in order to proceed up to the next activity level. Additionally, if the patient experienced any pain greater than a level 5, he was required to move down one step to a lower activity level.

The rationale for this step program was that many patients with chronic pain attempt to overextend on the days their pain is temporarily reduced so as to "catch up" on their activity levels. The step program ensures that the patient maintain low levels of pain over time before attempting greater levels of activity. The use of the pedometer provides a somewhat objective means for the patient and staff to quantify the extent of the patient's activity rather than relying on less accurate estimates of activity levels.

During the step program the patient was placed on a fixed interval schedule of analgesics, delivered in a vehicle so as to negate the patient asking for more pain medication when he felt greater discomfort. This fixed interval schedule minimizes that possibility.

During the course of the patient's hospitalization, his analgesics were slowly discontinued; he mobilized with increasing ease; and by the time of discharge he was ambulating with a brace. Two months after discharge he was off all analgesics, was ambulating 6 hours per day, and was making arrangements to return to work. (Multimodal approaches to the treatment of chronic pain are also discussed in Chapter 17.)

SPECIFIC CHARACTERISTICS OF PATIENTS WHO BENEFIT FROM BEHAVIORAL THERAPY

The characteristics of some patients' symptoms make them more amenable to behavioral therapy. The factors listed below should help guide the clinician when making the decision whether to pursue a behavioral approach.

Observable Behavior

For a behavioral therapy program to be implemented, the patient, the clinician, or others in the patient's environment must be able to know when contingent intervention could be implemented. For example, if nursing personnel were to institute a treatment contingent on a patient's increased anxiety, a behavioral program would be difficult to follow if the nurses could not observe outward signs of the patient's anxiety. Observable components of the symptom must be identified. In this case, observable indications of anxiety might include hand-wringing, pacing, or tachycardia. These signs would provide observable behavior to carry out a contingency program.

Fluctuation of the Symptom

If observers are to respond to a behavior, it should have specific antecedents that influence whether the symptom occurs more frequently or with greater intensity. The behavioral therapist must evaluate whether fluctuations of the symptom suggest the presence of specific antecedents that worsen the patient's condition. For example, if a patient seeks treatment because of hypertension but shows no fluctuation of the symptom, behavioral treatment is more difficult. If the patient's blood pressure fluctuates, however, it increases the likelihood that behavioral analysis can identify antecedents that aggravate the symptom.

Social Environment That Can Observe Behavior and Respond to It Consistently

Treatment interventions necessitate a response from the social environment in which the behavioral excess or deficit occurs. In some situations patients themselves can respond after the occurrence of symptomatic behaviors, but this pattern generally requires attentive patients who are anxious to reduce their symptoms. When patients are less motivated to reduce their symptoms or are incapable of organizing their thoughts and actions following the occurrence of the symptom, other people in the patient's environment must be available to implement treatment interventions. This responsive environment does not mean that the treatment personnel must be highly trained in behavioral therapy. It does, however, require that they be observant of the patient's symptoms and able to respond with consistency.

Thus in many cases the designated "patient" is not the identified patient but his or her family members. Training family members in the implementation of behavioral interventions, such as reinforcing well behavior rather than illness behavior, is essential to ensure the generalization and maintenance of treatment gains.

Frequent Occurrence of the Symptom

If the patient is to learn a new response or if different consequences are to be taught to the patient following the occurrence of the symptom, such learning requires repetition if the new response is to be acquired. Repetition of the consequences of the behavior necessitates a fairly high frequency of the symptom. If a diabetic's hypoglycemic episodes occurred once every 4 months, a behavioral treatment would probably be ineffective, as implementation of the treatment would occur so infrequently. On the other hand, if the patient's blood glucose levels were monitored with frequent assays and treatment implemented not only after clinical episodes of hypoglycemia but after hypoglycemic episodes identified by blood chemistries, it would increase the frequency of implementing the treatment.

Adequate Time for Treatment Intervention

Learning requires repetition. Behavioral treatment interventions therefore require an adequate passage of time so that patients can repetitively come into contact with the consequences of their behavior. When clinical realities, however, require immediate intervention, behavioral interventions may be contraindicated. The acutely suicidal patient threatening to end his or her life necessitates psychiatric hospitalization for protection. Such suicidal patients can be treated with behavioral approaches for the cognitive distortions that increase their suicidal ideation, but this treatment takes time. Clinical necessity to protect a patient's life would warrant a prompt protective hospitalization rather than a less immediate cognitive treatment intervention.

CHOOSING A COMPETENT BEHAVIOR THERAPIST

Competent behavior therapists are trained not only in the techniques of behavior therapy but also in the theory behind these techniques. Many clinical psychology programs offer specialization in behavior analysis and behavior therapy. There are several national behaviorally oriented professional organizations that serve as good referral resources for competent behavior therapists. They include the Association for the Advancement of Behavior Therapy (AABT), the Association for Behavior Analysis (ABA), and the Society of Behavioral Medicine (SBM). These organizations offer membership directories that list the location and qualifications of members. The addresses and phone numbers of these organizations are listed below, under Further Information.

SUMMARY

Rather than attempt to define a specific psychiatric diagnosis in need of treatment, behavioral approaches recognize the variability of symptoms over time and environments, and therefore attempt to break down all problematic behavior as either behavioral excesses or behavioral deficits. In this chapter, we have presented behavioral therapy strategies for the assessment, treatment, and compliance evaluations for several behavioral medicine problems. No brief discussion of behavioral therapy can allude to all of the possible treatment interventions, as designing effective strategies is limited only by the therapist's imagination (Miller, 1983).

FURTHER INFORMATION

Readers interested in obtaining information about the most recent developments in Behavioral Medicine are advised to write two organizations that actively promote advances in this field:

Ms. Judith Woodward, The Society of Behavioral Medicine, 103 South Adams Street, Rockville, MD 20850. The society publishes both the *Annals of Behavioral Medicine* and the *Behavioral Medicine Abstracts*.

Dr. Andrew Baum, The Academy of Behavioral Medicine, Department of Medical Psychology, Uniformed Services University of the Health Sciences, 4301 Jones Bridge Road, Bethesda, MD 20814-4799. The Academy publishes, through Academic Press, *Perspectives on Behavioral Medicine*, a yearly volume of the annual proceedings of the Academy.

Readers interest in obtaining information about behavior therapy and theory may contact:

The Association for the Advancement of Behavior Therapy, 15 West 36 Street, New York, NY 10018; (212) 279-7970.

The Association for Behavior Analysis, Department of Psychology, Western Michigan University, Kalamazoo, MI 49008-3899; (616) 387-4494.

REFERENCES

ABEL GG, BECKER JV, CUNNINGHAM-RATHNER J, ET AL. (1984). Treatment manual: the treatment of child molesters. Unpublished monograph, available from the first author at the Behavioral Medicine Institute of Atlanta, 3280 Howell Mill Rd, NW, Suite T–30. Atlanta, GA 30327.

ABEL GG, BLANCHARD EB, BARLOW DH, ET AL. (1975). Identifying specific erotic cues in sexual deviation by audio-taped description. J Appl Behav Anal 8:247–260.

ABEL GG, & BLENDINGER D (1986). Behavioral urology. In JO CAVENAR (Ed.), Psychiatry (Vol. II; Ch. 98, pp. 1–11). Philadelphia: Lippincott.

ANDERSON LT (1977). Desensitization in vivo for men unable to urinate in a public facility. J Behav Ther Exp Psychiatry 8:105–106.

BANDURA A (1969). Principles of behavior modification. New York: Holt.

BANDURA A (1971). Psychotherapy based on modeling principles. In AE BERGIN & SL GARFIELD (Eds.), Handbook of psychotherapy and behavior change (Ch. 17, pp. 653–708). New York: John Wiley & Sons.

BASMAJIAN JV (Ed.) (1989). Biofeedback: principles and practice for clinicians (3rd ed.). Baltimore: Williams & Wilkins.

BELLISSIMO A (1981). Biofeedback in medicine. Mod Med Can 36:630–633.

BRUCH H (1973). Eating disorders: obesity, anorexia nervosa, and the person within. New York: Basic Books.

BURISH TG (1981). EMG biofeedback in the treatment of stress-related disorders. In CK PROKOP & LA BRADLEY (Eds.), Medical psychology: contributions to behavioral medicine (Ch. 21, pp. 395–421). Orlando, FL: Academic Press.

CAUTELA JR (1966). Treatment of compulsive behavior by covert sensitization. Psychol Res 16:33–41.

DIMATTEO MR, & DINICOLA DD (1982). Achieving patient compliance: the psychology of the medical practitioner's role. New York: Pergamon Press.

FAIRBURN CG (1982). Binge eating and its management. Br J Psychiatry 141:631–633.

FORDYCE WE (1976). Behavioral methods for chronic pain and illness. St. Louis: Mosby.

FORDYCE WE, FOWLER RS, & DELATEUR BJ (1968). An application of behavior modification technique to a problem of chronic pain. Behav Res Ther 6:105–107.

FORDYCE WE, FOWLER RS, LEHMANN JF, ET AL. (1968). Some implications of learning in problems of chronic pain. J Chronic Dis 21:179–190.

FRANKS CM, WILSON GT, KENDALL PC, ET AL. (1990). Review of behavior therapy. Vol. 12. Theory and practice. New York: Guilford Press.

FRIEDMAN M, & ROSENMAN RH (1974). Type A behavior and your heart. New York: Knopf.

GATCHEL RJ, BAUM A, & KRANTZ DS (1989). An introduction to health psychology (2nd ed.). New York: Random House.

GOLDFARB LA, DYKENS EM, & GERRARD M (1985). The Goldfarb fear of fat scale. J Pers Assess 49:329–332.

GREEN J (1984). Biofeedback training: a client information paper. Wheat Ridge, CO: Association for Applied Psychophysiology and Biofeedback.

HALMI KA (1974). Anorexia nervosa: demographic and clinical features in 94 cases. Psychosom Med 36:18–25.

JACOBSON E (1938). Progressive relaxation. Chicago: University of Chicago Press.

KAZDIN AE (1978). Behavior therapy: evolution and expansion. Counsel Psychol 3:34–37.

LUM LC (1975). Hyperventilation, the tip of the iceberg. J Psychosom Res 19:375.

LUM LC (1981). Hyperventilation and anxiety state. J R Soc Med 74:1–4.

MASTERS JC, BURISH TG, HALLON SD, ET AL. (1987). Behavioral therapy: techniques and empirical findings. New York: Harcourt, Brace, Javonovich.

MCFARLANE AH, BELLISSIMO A, & UPTON E (1982). Atypical anorexia nervosa: treatment and management on a behavioral medicine unit. Psychiatr J Univ Ottawa 7:158–162.

MILLER NE (1983). Behavioral medicine: symbiosis between laboratory and clinic. Annu Rev Psychol 34:1–31.

MIZES JS (1985). Bulimia: a review of its symptomatology and treatment. Adv Behav Res Ther 7:91–142.

MOREYRA AE, MCGOUGH WE, HOSLER M, ET AL. (1982). Clinical improvement of patients with chest pain and normal coronary arteries after coronary arteriography: the possible role of hyperventilation syndrome. Int J Cardiol 2:306–308.

MOWRER OH (1960). Learning theory and behavior. New York: John Wiley & Sons.

POMERLEAU OF, & BRADY JP (Eds.) (1979). Behavioral medicine: theory and practice. Baltimore: Williams & Wilkins.

SCHULTZ JH, & LUTHE W (1969). Autogenic therapy. Orlando, FL: Grune & Stratton.

SCHWARTZ MS (1987). Biofeedback: a practioner's guide. New York: Guilford Press.

SCHWARTZ GE, & WEISS SM (1978). Yale conference on behavioral medicine: a proposed definition and statement of goals. J Behav Med 1:3–12.

SHELTON JL, & LEVY RL (Eds.) (1981). Behavioral assignments and treatment compliance: a handbook of clinical strategies. Champaign, IL: Research Press.

SKINNER BF (1969). Contingenices of reinforcement: a theoretical analysis. Norwalk, CT: Appleton.

SOLOMON RL (1964). Punishment. Am Psychol 19:239–253.

TUNKS E, & BELLISSIMO A (1991). Behavioral medicine: concepts and procedures. New York: Pergamon Press.

TURK DC, MEICHENBAUM D, & GENEST M (1983). Pain and behavioral medicine: a cognitive-behavioral perspective. New York: Guilford Press.

WARE R, & TERRELL G (1961). Effects of delayed reinforcement on associative and incentive factors. Child Dev 32:789–793.

WOLPE J (1958). Psychotherapy by reciprocal inhibition. Stanford: Stanford University Press.

WOLPE J (1969). The practice of behavior therapy. Oxford: Pergamon.

YOUNG LD, & BLANCHARD EB (1980). Medical applications of biofeedback training—a selective review. In S RACHMAN (Ed.), Contributions to medical psychology (Vol. 2; Ch. 9, pp. 215–254). New York: Pergamon Press.

42 Nicotine dependence: assessment and management

RICHARD A. BROWN, Ph.D., MICHAEL G. GOLDSTEIN, M.D.,
RAYMOND NIAURA, Ph.D., KAREN M. EMMONS, Ph.D.,
AND DAVID B. ABRAMS, Ph.D.

Nicotine dependence is an area of interest and importance to psychiatrists and psychologists in behavioral medicine as well as other physicians who have become increasingly involved in research and treatment of tobacco use disorders. The stand taken by The Joint Commission for Accreditation of Health Care Organizations (JCAHO, 1991) to move toward the elimination of smoking in all hospitals has added impetus to physician efforts to provide smoking cessation treatment to patients. Other significant developments include the identification of nicotine as a powerful psychoactive drug with high addictive potential (U.S. Department of Health and Human Services [U.S. DHHS], 1988), the development of several promising new pharmacologic agents to treat nicotine dependence (Jarvik & Henningfield, 1988), the association of nicotine dependence with other psychiatric disorders (Davis, Faust, & Ordentlich, 1984; Hughes, Hatsukami, Mitchell, et al., 1986), and the long-standing awareness of the health hazards of smoking (U.S. DHHS, 1985, 1989) as well as the growing concerns about the hazards of passive smoke exposure (Emmons, Abrams, Marshall, et al., 1992).

This chapter offers guidelines for the diagnosis and treatment of nicotine dependence from a biopsychosocial or behavioral medicine perspective. The chapter begins with an overview of the scope of the nicotine dependence problem. A discussion follows of the characteristics of nicotine that have led to its identification as a psychoactive drug, the criteria for diagnosing and assessing nicotine dependence, and the metabolic consequences of cigarette smoking. The relation of nicotine dependence to other psychiatric disorders, including other substance use disorders and depression, is examined; and principles and strategies for effectively treating nicotine dependence are discussed. Both pharmacologic and nonpharmacologic treatment strategies are described. This chapter expands on our earler review of this topic (Goldstein, Niaura, & Abrams, 1991) and includes material that is covered in greater detail in other reviews (Abrams, Emmons, Niaura, et al., 1991; Brown & Emmons, 1991; Schwartz, 1987). Behavioral interventions for smoking cessation have been given additional emphasis.

SCOPE OF THE PROBLEM

Despite a steady decline in the prevalence of cigarette smoking since the late 1960s, cigarette smoking remains the most important preventable contributor to premature death, disability, and unnecessary expense in the United States (U.S. DHHS, 1989). Currently, about 50 million adults, or about 28% of the U.S. population, smoke cigarettes—down from 38% in 1976 (U.S. DHHS, 1988). The total number of adults who have quit smoking increased from approximately 30 million in 1976 to 40 million in 1988 (U.S. DHHS, 1988). However, because the proportion of "ever smokers" who are now former smokers is increasing at a yearly rate of less than 1%, smoking prevalence is likely to remain high well beyond the year 2000 (Pierce, Fiore, Novotny, et al., 1989).

Several factors contribute to the relatively slow decline in smoking prevalence. Though more than 90% of smokers surveyed say that they want to quit smoking (Gallup Opinion Index, 1974), only a fraction (15%) are actively attempting to do so, either on their own or with the help of some form of aid or program (Abrams & Biener, in press; Prochaska & DiClemente, 1983). High relapse rates also contribute to the slow overall rate of change of smoking prevalence. Almost 80% of smokers who quit relapse within a year after quitting (Hunt & Bespalec, 1974; Schwartz, 1987). Moreover, relapse appears to be common during the earliest stages of attempted abstinence. As many as 30% of smokers who quit without assistance relapse within 72 hours (Garvey, 1988). Highly dependent smokers have greater difficulty quitting (Pinto, Abrams, Monti, et al., 1987). Though formal treatment for nicotine dependence produces slightly higher 1-year abstinence rates of about 20 to 30% (Schwartz, 1987), there is considerable room for improvement (Abrams & Wilson, 1986).

As the prevalence of smoking in the general population declines, the percentage of "hard core" smokers may be increasing; a phenomenon that could slow the rate of decline of smoking (DiFranza & Guerrera, 1989; Pierce, Fiore, Novotny, et al., 1989). A growing body of evidence, reviewed below, suggests that development of physical dependence on nicotine in chronic smokers serves to maintain smoking and promote smoking relapse. Thus nicotine dependence may be a central characteristic of a growing group of smokers, one that erodes the motivation to quit smoking and promotes relapse when abstinence is attempted.

Although biologic aspects of nicotine dependence are important and receive considerable emphasis in this chapter, psychological, behavioral, and environmental factors also play a crucial role in the dependence process (Abrams, Emmons, Niaura, et al., 1991; Brown & Emmons, 1991; Lichtenstein & Brown, 1982; Marlatt & Gordon, 1985; Niaura, Goldstein & Abrams, 1991; Schacter, 1979; Wills & Shiffman, 1985). Thus an integrated biopsychosocial approach to the treatment of nicotine dependence is generally more effective than either behavioral or pharmacologic treatment alone, especially for the heavily addicted smoker or the smoker who has failed to remain abstinent after multiple attempts to quit (Abrams & Wilson, 1986; Goldstein, Niaura, Abrams, et al., 1989; U.S. DHHS, 1988). The high frequency of psychiatric comorbidity also increases the need for an integrated approach to treatment.

Recognition of the importance of a biopsychosocial approach has led researchers to increase their efforts to identify effective, multidimensional treatments for smoking cessation. These methods are described in the sections on treatment, below.

NICOTINE AS A PSYCHOTROPIC DRUG

Pharmacokinetics and Pharmacodynamics of Nicotine

Nicotine is readily absorbed from tobacco smoke in the lungs and from smokeless tobacco in the mouth and nose (U.S. DHHS, 1988). Nicotine rapidly accumulates in the brain after cigarette smoking; maximum brain concentrations are reached within 1 minute (U.S. DHHS, 1988). The rapid accumulation of nicotine in the brain combined with nicotine's effects on brain activity and function provide optimal reinforcement for the development of drug dependence (U.S. DHHS, 1988). Acute and chronic tolerance develop to many effects of nicotine (U.S. DHHS, 1988). The development of chronic tolerance contributes to an increase in cigarette consumption, as individuals smoke more to obtain desired effects of nicotine (Henningfield, Miyasato, & Janinske, 1985; U.S. DHHS, 1988).

Nicotine is a powerful pharmacologic agent with a wide variety of stimulant and depressant effects involving the central and peripheral nervous, cardiovascular, endocrine, and other systems (U.S. DHHS, 1988). Peripheral nervous system effects include skeletal muscle relaxation (U.S. DHHS, 1988). Central effects include electrocortical activation and increases in brain serotonin, endogenous opioid peptides, pituitary hormones, catecholamines, and vasopressin (Pomerleau & Pomerleau, 1984, 1987; U.S. DHHS, 1988). The rewarding properties of nicotine may be related to nicotine's stimulatory effects on dopaminergic pathways in the mesolimbic system. (Clarke, 1989; Fuxe, Anderson, & Eneroth, 1987; Pert & Clarke, 1987; U.S. DHHS, 1988).

Nicotine has been shown to increase attention, memory, and learning in smokers, especially when the environment exerts a relatively low demand and stimulation is most desirable (e.g., while performing clerical work or long-distance driving) (U.S. DHHS, 1988). Nicotine also has anxiolytic and antinociceptive effects (U.S. DHHS, 1988). Evidence from both surveys and experimental studies suggest that smokers smoke more during stressful situations or in situations involving negative mood, and that nicotine consumption is associated with decreases in negative affect (U.S. DHHS, 1988). High doses of nicotine can produce cocaine-like stimulant effects (Henningfield, Miyasato, & Jasinske, 1985).

Nicotine effects vary greatly among individuals and within the same individual over time (U.S. DHHS, 1988). Differences depend on the setting in which the smoking occurs, the current state of the individual (e.g., whether he or she is smoking, nicotine-deprived, tolerant, stressed, nonstressed) and individual differences in dependence, genetics, learning history, and psychosocial coping skills (Abrams, Emmons, Niaura, et al., 1991; Pomerleau & Pomerleau, 1987; U.S. DHHS, 1988). The implications of these differences for treatment are described below.

Addictive Properties of Nicotine

The characteristics and patterns of chronic nicotine use have much in common with the use of other psychotropic drugs (Henningfield, 1984; Jarvik & Henningfield, 1988; U.S. DHHS, 1988). As with other psychotropic drugs that produce addiction, humans will self-administer nicotine in the laboratory to reproduce desired effects and respond to nicotine as a positive reinforcer (Henningfield, 1984; Jarvik & Henningfield, 1988; U.S. DHHS, 1988). Patterns of relapse to smoking after smoking cessation are quite similar to the relapse patterns noted after treatment for other forms of drug abuse and dependence (Henningfield, 1984; Hunt, Barnett, & Branch, 1971; U.S. DHHS, 1988). Nicotine also can produce the physiologic changes associated with physical dependence, including tolerance and withdrawal states (Henningfield, 1986; Hughes, Higgins, & Hatsukami, 1990; U.S. DHHS, 1988). (See Nicotine Dependence, below, for more details.)

Nicotine meets the criteria for classification as an addicting drug: (1) the presence of highly controlled and compulsive patterns of drug taking; (2) the presence of psychoactive effects that contribute to use; and (3) the drug's capability of functioning as a reinforcer that can strengthen behavior, leading to further drug ingestion. (Henningfield, 1984, 1986; U.S. DHHS, 1988). Moreover, nicotine use has additional properties that are associated with drug addiction: stereotypic patterns of use, use despite harmful effects, relapse following abstinence, recurrent drug cravings, development of acute tolerance and physical dependence, and pleasant or euphoric effects (Henningfield, 1984, 1986; Hughes, Higgins, & Hatsukami, 1990; U.S. DHHS, 1988).

NICOTINE DEPENDENCE

Evidence has converged from several sources to suggest that nicotine is an addicting or dependence-producing drug. There are significant individual differences in degree of dependence, and the consensus is that there is a continuum of dependence from mild to severe among the population of smokers (Shiffman, 1989). The nicotine withdrawal syndrome is now fairly well characterized (Hughes & Hatsukami, 1986; Hughes, Higgins, & Hatsukami, 1990). The syndrome includes craving or urges to use nicotine; irritability, frustration, or anger; anxiety; difficulty concentrating; restlessness; decreased heart rate; and increased appetite or weight gain.

The signs and symptoms of the nicotine withdrawal syndrome can appear within 2 hours after the last use of tobacco, usually peak between 24 and 48 hours after cessation, and last from a few days to a few weeks (Hughes, Higgins, & Hatsukami, 1990; Hughes, Gust, Skoog, et al., 1991). However, there is a great deal of individual variation in both the symptom pattern and time course of withdrawal symptoms. Whereas a history of previous withdrawal symptoms is important to the diagnosis of nicotine dependence, the patient's future experiences during attempted abstinence cannot always be predicted.

Another aspect of nicotine dependence relates to the notion of nicotine regulation. Some evidence suggests that smokers consciously or unconsciously compensate and adjust their smoking patterns to maintain their usual plasma levels of nicotine (Moss & Prue, 1982; Russell, 1987; U.S. DHHS, 1988).

METABOLIC CONSEQUENCES OF CIGARETTE SMOKING

Attention has focused on the relations among smoking, nicotine, and body weight as potential mediators of smoking initiation, maintenance, and relapse. There is substantial evidence of an inverse relation between cigarette smoking and body weight (U.S. DHHS, 1988) and evidence to suggest that some people, especially women, smoke to prevent weight gain (Klesges & Klesges, 1988; Russell & Epstein, 1988; U.S. DHHS, 1988). Moreover, weight gain accompanying smoking cessation may trigger a relapse to smoking (Klesges & Klesges, 1988; U.S. DHHS, 1988), although some studies have found positive correlations between weight gain during abstinence and 1-year abstinence rates (Hall, Ginsberg, & Jones, 1986; Hughes, Gust, Skoog, et al., 1991). The mechanisms that may underlie the relation between smoking and body weight include a nicotine-induced increase in energy use (Benowitz, 1988; U.S. DHHS, 1988) and nicotine-induced suppression of food intake, especially carbohydrates (Benowitz, 1988; Grunberg, 1985; Jarvik, 1987; U.S. DHHS, 1988).

Smoking is known to accelerate the metabolism of many drugs (Benowitz, 1988, 1989). Table 42-1 lists drugs whose metabolism is significantly accelerated by smoking. It should be noted that the list includes several psychoactive drugs (i.e., imipramine, clomipramine, clorazepate, oxazepam, desmethyldiazepam, and pentazocine). Because metabolic effects of cigarette smoking are complex, it is difficult to generalize about the effects of cigarette smoking or nicotine use on entire classes of drugs (Benowitz, 1988, 1989). For example, though cigarette smoking increases the metabolism of imipramine, it has minimal or no effect on the metab-

TABLE 42–1. *Drugs Whose Metabolism Is Accelerated by Cigarette Smoking*

Antipyrine
Caffeine
Clomipramine
Clorazepate
Desmethyldiazepam
Imipramine
Lidocaine
Oxazepam
Pentazocine
Phenacetin
Propranolol
Theophylline

olism of nortriptyline (Benowitz, 1988, 1989). Moreover, although smoking accelerates the metabolism of such benzodiazepines as clorazepate, oxazepam, and desmethyldiazepam (an important metabolite of diazepam), it does not significantly alter the metabolism of chlordiazepoxide and lorazepam (Benowitz, 1988, 1989). Smoking may also increase the metabolism of some neuroleptics, though this area has not been carefully studied (Benowitz, 1989). Phenytoin levels are not significantly altered by smoking, but the effects of smoking on the metabolism of other anticonvulsants has also not been well studied (Benowitz, 1988, 1989).

The effects of smoking on the metabolism of drugs are likely due to the effects of nicotine as well as other constituents of cigarette smoking, such as polyaromatic hydrocarbons. Though it is likely that low-nicotine, low-tar cigarettes have less effect on drug metabolism, this question has not been studied. In addition to metabolic effects, nicotine use also results in pharmacodynamic interactions that may potentiate or interfere with the effects of other drugs. For example, nicotine's stimulatory effects may counteract the sedative effects of phenothiazines and benzodiazepines (Benowitz 1988, 1989).

The clinical importance of the effects of smoking on drug metabolism and pharmacodynamics is unclear. However, because smoking may increase metabolism and decrease the effectiveness of several psychoactive drugs, higher dosages of these drugs may be needed in patients who smoke (Bansil, Hymowitz, Keller, et al, 1989). Moreover, when smokers taking psychotropic drugs quit smoking or significantly reduce their smoking rate, drug toxicity may occur. As psychiatric inpatient units move toward becoming smoke-free environments, psychiatric patients may be forced to stop smoking. In this setting, the interactions between smoking and psychotropic drug levels and effects become even more salient. Because of the complex metabolic and pharmacodynamic effects of smoking, we recommend a low threshold for obtaining plasma levels of psychotropics in patients who smoke when there is evidence for drug toxicity or if the patient fails to respond to psychoactive drug treatment.

CIGARETTE SMOKING AND PSYCHIATRIC DISORDERS

A number of studies have established that psychiatric patients, especially those with psychoses or with other substance-related disorders are much more likely to smoke than the general population or even nonpsychiatric patients. Table 42-2 summarizes some of this literature.

Some concern has been focused on the association of smoking with depression and other substance abuse in both general and psychiatric populations, and on the implications of these associations for treatment. Some studies have demonstrated significant relations between self-reported depression and the frequency of smoking. For example, among the general population of adults in the United States, smokers with major depressive disorder (Glassman, Helzer, Covey, et al., 1990) or depressive symptoms (Anda, Williamson, Escobedo, et al., 1990) are less likely to quit smoking than nondepressed smokers, and smokers report higher levels of depression than nonsmokers (Frederick, Frerichs, & Clark, 1988; Leon, Kolotkin, & Korgeski, 1979; Waal-Manning & de Hamel, 1978). In a 9-year longitudinal study, self-reported depression during adolescence predicted subsequent frequency and duration of smoking during young adulthood (Kandel & Davies, 1986).

Whether or not depression contributes to smoking initiation, the presence of a mood disorder may play a role in precipitating a relapse to smoking after attempted abstinence. Among smokers undergoing cessation treatment, depressed mood at pretreatment predicts failure to achieve abstinence during treatment (Rausch, Nichinson, Lamke, et al., 1990); depressive symptoms following initial cessation predict subsequent relapse to smoking (West, Hajek, & Belcher, 1989); and a lifetime history of major depression predicts the prevalence and intensity of depressive symptoms upon initial cessation (Covey, Glassman, & Stetner, 1990; Glassman, Covey, & Stetner, 1989a; Hall, Munoz, & Reus, 1990) and subsequent relapse to smoking (Glassman, Stetner, Walsh, et al., 1988; Hall, Munoz, & Reus, 1990, 1991). For example, Glassman, Stetner, Walsh, and colleagues (1988) found that patients with a history of depression had significantly poorer smoking abstinence rates after treatment (33%) than did patients

TABLE 42–2. *Smoking Prevalence in Psychiatric Patients and General U.S. Populations*

Population	Rate (%)	References
Psychiatric inpatients		
VA inpatients	83	O'Farrell, Connors, & Upper (1983)
Alcoholics	>80	Battjes (1988); Maletzky & Klotter (1974); Walton (1972)
Drug abusers	>85	Burling & Ziff (1988); Rounsaville, Kosten, Weissman, et al. (1985)
Psychiatric outpatients		
Schizophrenia	88	Hughes, Hatsukami, Mitchel, et al. (1986) U.S. DHHS (1988)
Mania	7	
Other	45–49	
General U.S. population	28	

without a history of depression (57%), independent of the effects of gender or treatment. Moreover, none of the patients was depressed on entering the study. There is also preliminary evidence that pretreatment with antidepressants may prevent the development of abstinence-related depressive symptoms in smokers with a history of a depressive disorder (Glassman, Covey, & Stetner, 1989a).

Drug and polydrug abusing populations also have high smoking prevalence. Smoking rates over 85% have been noted in alcoholics (Battjes, 1988; Istvan & Matarazzo, 1984), opiate addicts (Rounsaville, Kosten, Weissman, et al., 1985), and polydrug users (Burling & Ziff, 1988). Alcohol use and abuse and possibly abuse of other drugs are also associated with difficulty giving up smoking. Drinking alcohol is reported to be a significant precipitant of smoking relapse in the general population (Shiffman, 1986). Moreover, alcoholics are considerably less likely to be successful than nonalcoholics in their attempts to quit smoking (DiFranza & Guerrera, 1990). However, at least half of alcoholics express a strong desire to attend a smoking treatment program (Kozlowski, Wilkinson, Skinner, et al., 1989), and many have made multiple attempts to quit smoking (Bobo, Gilchrist, Schilling, et al., 1987). Smoking may also be used as a means of resisting cravings for alcohol or other drugs in abstinent alcoholics. The commonalities, differences, and interactions among tobacco and other substances of abuse are gaining strong research attention (Abrams, Rohsenow, Niaura, et al., 1992; Sobell, Sobell, Kozlowski, et al., 1990).

The higher prevalence of smoking among patients with psychiatric disorders, relative to the general population, has important implications for treatment of both psychiatric disorders and nicotine dependence. A careful assessment of patients who seek help to stop smoking is warranted to uncover possible psychiatric comorbidity, as prolonged abstinence may not be possible until the psychiatric disorder is treated.

GENERAL CONSIDERATIONS FOR TREATMENT

States of Change

When developing a treatment plan, one must take into account a number of important psychological and social factors. Before attempting to intervene with smokers to provide treatment for nicotine dependence, it is essential to determine whether they are seriously considering quitting smoking. Only a fraction of smokers are ready to take action either on their own or with the help of some form of aid or program (Prochaska & DiClemente, 1983).

Prochaska and DiClemente (1986) found that smokers move through several *stages of change* during the process of quitting smoking: (1) *precontemplation*, a stage of unawareness or denial of smoking as a problem; (2) *contemplation*, a stage of ambivalence, when pros and cons for quitting are weighed without definite commitment to taking action; (3) *preparation*, a stage in which individuals are intending to change in the near future and are taking some small but significant steps toward action (e.g., cutting down, delaying their first cigarette in the morning); (4) *action*, characterized by serious attempts to quit smoking; and (5) *maintenance*, a stage in which individuals work to continue a healthier life style and to modify their environments and their experiences to prevent relapse (see Figure 42-1). Individuals are likely to benefit from different interventions matched to their stage of change (Prochaska & DiClemente, 1986).

Prochaska and Goldstein (1991) suggested that physicians often become frustrated and demoralized when their focus is exclusively on getting patients to quit smoking without consideration of the patient's stage of change. The authors provided detailed discussion of how physicians can easily assess stage of change and adopt a stage-matched, patient-centered approach with smokers in their practice. For example, precontempla-

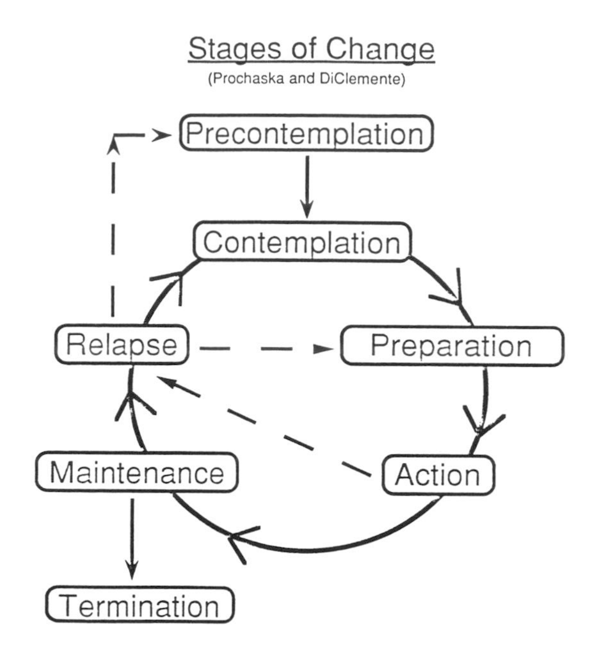

FIG. 42–1. Stages of change. (From Prochaska & DiClemente, 1983.)

tors tend to evaluate the pros of smoking as outweighing the cons. Personalized health feedback may increase these patients' awareness of the hazards (cons) of smoking, thus moving them closer to the contemplation stage. Likewise, smokers in contemplation can be encouraged to think about the benefits of quitting and to take small steps toward action, which may move them closer to the action stage. Overall, physicians can increase their effectiveness with patients who smoke by matching their intervention to the stage of readiness for change of the individual, rather than by uniformly intervening to promote immediate cessation with all smokers.

Role of the Psychiatrist

Because smoking is highly prevalent among patients with psychiatric disorders, general psychiatrists are likely to encounter many nicotine-dependent patients. Many of these patients are likely to be in the precontemplation or contemplation stage of change and are resistant to being treated for nicotine dependence. In contrast, psychiatrists who develop expertise in the treatment of nicotine dependence tend to see self-referred patients and those referred by primary care physicians specifically for treatment of this disorder.

The discussion of nicotine-dependence treatment strategies that follows focuses primarily on the treatment of smokers who are actively trying to quit smoking and therefore are in the preparation and action stages of change. A comprehensive review of individual and public health approaches to smoking cessation may be found in Abrams, Emmons, Niaura, et al. (1991). In addition, self-help strategies (Schwartz, 1987), brief or minimal physician-delivered interventions (Goldstein, Guise, Ruggiero, et al., 1990; Kottke, Battista, De Friesse, et al., 1988; Ockene, 1987; Orleans, 1985; Prochaska & Goldstein, 1991; Stokes & Rigotti, 1988), and worksite and community intervention (Abrams & Biener, 1989; Schwartz, 1987; U.S. DHHS, 1985) are not discussed in detail in this chapter.

The change in the guidelines of The Joint Commission for Accreditation of Health Care Organizations (JCAHO, 1991) that requires smoking restrictions among hospitalized patients accentuates the need for nicotine dependence interventions with hospitalized smokers. The creation of a smokefree hospital is likely to increase the demand for interventions to manage nicotine withdrawal while affording an opportunity to provide cessation-oriented messages and treatments for inpatients who smoke. Orleans, Rotberg, Quade, et al. (1990) found that 27% of physician-referred smokers (hospital inpatients and outpatients) were abstinent 6 months after receiving minimal-contact, behavioral smoking cessation treatment within a hospital quit-smoking consult service. Taylor, Houston-Miller, Killen, et al. (1990) reported that 61% of post-myocardial infarction patients randomly assigned to receive a nurse-managed, brief behavioral intervention for smoking cessation were abstinent at 1-year follow-up compared to 32% of those receiving usual care. Kristeller and Ockene (1987) described a strategy for assessment and treatment of smoking on a general hospital psychiatry consultation service using a model derived from behavioral medicine.

Although the emphasis in this section has been on underlying biologic mechanisms, as stated previously, multiple psychological, behavioral, and environmental factors contribute to the nicotine dependence process (Abrams, Emmons, Niaura, et al., 1991; Brown & Emmons, 1991; Lichtenstein & Brown, 1982; Marlatt & Gordon, 1985; Niaura, Goldstein, & Abrams, 1991; Schacter, 1979; Wills & Shiffman, 1985). Thus a combined biopsychosocial approach to treatment is generally more effective than either behavioral or pharmacologic treatment alone (Brown & Emmons, 1991; Goldstein, Niaura, Abrams, et al., 1989; U.S. DHHS, 1988).

PHARMACOLOGIC TREATMENT

Pharmacologic treatment for nicotine dependence can be characterized using the same typology that has been developed for treating other forms of drug dependence (Jarvik & Henningfield, 1988; U.S. DHHS, 1988). The four categories of pharmacologic treatment, based on the pharmacologic strategy employed, are listed in Table 42-3. Each of these strategies is described in some detail, using specific pharmacologic agents as examples.

Nicotine Replacement or Substitution Therapy

The principle of replacement therapy is to provide the patient with a more manageable and safer form of the drug to ameliorate withdrawal symptoms and allow the patient to gradually discontinue use of the drug (Jarvik & Henningfield, 1988; U.S. DHHS, 1988). Replacement therapy also provides an opportunity to develop strategies to deal with behavioral or learned components of the drug dependence while controlling the physiologic "need" for the drug. Moreover, if the method of administration of the replacement drug is sufficiently different than the method associated with the development of drug dependence, the learned associations between cues associated with drug administration and the drug's physiologic effect can be broken. Several nicotine replacement strategies are discussed below.

Nicotine Resin Complex

Also known as nicotine polacrilex (Nicorette), or nicotine gum, nicotine resin complex is effective in attenuating nicotine withdrawal symptoms (Fagerstrom, 1988: Lam, Sacks, Sze, et al., 1987; U.S. DHHS 1988). Several controlled trials have found that nicotine resin complex is more effective than placebo in promoting abstinence from smoking when it is combined with a formal treatment program (Fagerstrom, 1988; Goldstein & Niaura, 1991; Grabowski & Hall, 1985; Lam, Sacks, Sze, et al., 1987; Lichtenstein, 1986; Schwartz, 1987; U.S. DHHS, 1988).

However, nicotine resin complex appears to have limited effectiveness in producing long-term abstinence when the subject is provided with little or no behavioral counseling, especially as it is delivered in physicians' offices (Fagerstrom, 1988; Goldstein & Niaura, 1991; Hughes, Gust, Keenan, et al., 1989; Schwartz, 1987; U.S. DHHS, 1988). Thus formal cognitive-behavioral relapse prevention training is important to the effectiveness of nicotine resin complex. This statement is supported by: (1) research that associates nicotine resin complex use with significantly increased rates of short-term, but not long-term, abstinence (Hall, Tunstall, Rugg, et al., 1985; Harackiewicz, Blair, Sansone, et al., 1988; Hughes, Gust, Keenan, et al., 1989; U.S. DHHS, 1988); (2) a study that demonstrated improved outcome when relapse prevention skills training was added to nicotine resin complex treatment (Goldstein, Niaura, Abrams, et al., 1989); and (3) a critical review of a group of studies concluding that the addition of psychological therapy to nicotine gum treatment increases long-term quit rates (Hughes, 1991). Evidence suggests that nicotine resin complex is most effective for smokers who are heavily dependent on nicotine (Fagerstrom & Melin, 1985; Jarvik & Schneider, 1984; Tonnesen, Fryd, Hansen, et al., 1988; Tonnesen, Fryd, Hansen, et al., 1988; U.S. DHHS, 1988). There is also some evidence that the use of nicotine resin complex during smoking cessation limits postcessation weight gain (Klesges, Meyers, Klesges, et al., 1989).

Several issues related to the use of nicotine resin complex remain unresolved. The 4-mg dose, currently not available in the United States, may be more effective for the heavily dependent smoker (Kornitzer, Kittel, Dramaix, et al., 1987; Tonnesen, Fryd, Hansen, et al., 1988a; Tonnesen, Fryd, Hansen, et al., 1988b), but this possibility has not been adequately tested. Theoretically, a fixed schedule of schedule of nicotine gum administration might be more likely to be effective in reducing withdrawal symptoms and aiding abstinence in the highly dependent smoker. However, there is presently only suggestive evidence to support this hypothesis (Fortmann, Killen, Telch, et al., 1988; Goldstein, Niaura, Abrams, et al., 1989).

The properties of nicotine resin complex require that patients be provided with specific instructions about proper use. These instructions are outlined in Table 42-4. Side effects of nicotine resin complex are common but usually well tolerated, especially when patients receive instructions regarding proper use. Hiccups, nausea, anorexia, oral soreness, jaw soreness, and gastrointestinal distress are the most frequent side effects (Fortmann, Killen, Telch, et al., 1988). Individuals who

TABLE 42–3. *Pharmacologic Treatments for Nicotine Dependence*

Nicotine replacement or substitution
Nicotine resin complex
Nicotine transdermal patch
Nasal nicotine solution
Nicotine aerosol rods
Lobeline
Stimulants
Nonspecific pharmacotherapy
Clonidine
Antidepressants
Doxepin
Fluoxetine
Buspirone
Blockade therapy
Mecamylamine
Deterrent therapy
Silver acetate

TABLE 42–4. *Instructions for Patients Using Nicotine Resin Complex (Seven S's)*

Stop smoking cigarettes—do *not* smoke and chew
Substitute gum for cigarettes when the urge to smoke occurs—but do not wait until the urge gets too strong
Slowly chew—only a few chews, then "park" the gum
Several pieces per day—about one piece for every two cigarettes
Stay on the gum for 2 to 3 months and *gradually* decrease over time
Stay away from acidic beverages (e.g., coffee, juice) before and during gum chewing
Stop using the gum

Adapted from the treatment protocol developed for the Waterloo Smoking Project, JA Best and D Wilson, principal investigators, Waterloo University and McMaster University, National Cancer Institute, Smoking, Tobacco and Cancer Program, 1988.

utilize nicotine resin complex to stop smoking may become physically dependent on the gum and experience withdrawal when they abstain from chewing it (Hughes, Hatsukami, & Skoog, 1986; West & Russell, 1985b). However, only 6 to 9% of subjects who receive nicotine resin complex in specialty clinics continue to use the gum 1 year after treatment (West & Russell, 1986). Among those who successfully abstain from cigarettes, 34 to 54% use the gum for more than 6 months (Hughes, Hatsukami, & Skoog, 1986), and about 25% continue to use gum after 1 year (Hajek, Jackson, & Belcher, 1988). These individuals may require slow tapering of the gum use over a longer period or the use of other pharmacologic treatments to treat withdrawal after discontinuation of nicotine gum.

Nicotine Transdermal Patches

The nicotine transdermal patch, which delivers nicotine through the skin, is a promising new nicotine delivery system. Research has demonstrated that nicotine patches reduce nicotine withdrawal symptoms significantly more than placebo patches (Daughton, Heatley, Prendergast, et al., 1991; Rose, 1991; Transdermal Nicotine Study Group, 1991), though their effect on craving is inconsistent (Rose, 1991). The efficacy of nicotine patches has been demonstrated in a number of placebo-controlled double-blind trials (Abelin, Muller, Buehler, et al., 1989; Buchkremer, Bents, Horstmann, et al., 1989, Daughton, Heatley, Prendergast, et al., 1991; Tonnesen, Norregaard, Simonsen, et al., 1991; Transdermal Nicotine Study Group, 1991).

When assessing the emerging literature on nicotine patches, limitations in research methodology should be noted. Though the results of several published studies have demonstrated that nicotine patches are effective in enhancing short-term smoking cessation rates when used with or without behavioral treatment (Abelin, Muller, Buehler, et al., 1989; Buchkremer, Bents, Horstmann, et al., 1989; Hurt, Laugher, Offord, et al., 1990), only three studies, (Daughton, Heatley, Prendergast, et al., 1991; Tonnesen, Norregaard, Simonsen, et al., 1991; Transdermal Nicotine Study Group, 1991) have reported longer-term abstinence rates. All of these studies used carbon monoxide measurements to validate nonsmoking status, a technique that may miss smoking that occurred 24 hours before the assessment. Also, it should be noted that, except for one study (Daughton, Heatley, Prendergast, et al., 1991), nicotine patches were administered for at least 2 months in research trials.

No published study has directly compared nicotine resin complex with the nicotine patch, with or without cognitive-behavioral treatment. The patch provides several potential advantages over nicotine resin complex: administration only once per day; usefulness in patients with oral or temporomandibular disease or dental appliances; production of stable continuous nicotine blood levels over a 16- to 24-hour period (depending on the preparation used); the absence of problems associated with use of nicotine resin complex that are due to improper chewing technique (e.g., oral and gastrointestinal side effects; poor absorption of nicotine); and the potential for enhanced adherence. However some of the advantages of gum that are not present in the patch include the ability to use a behavioral substitute for a cigarette when the patient is craving (chewing gum is an oral activity) and the cognitive value of knowing that an alternative to cigarettes is available. Much more research is needed to determine (1) the optimal methods of administration (dose, formulation, duration, schedule for tapering); (2) the best way to combine nicotine patches with cognitive-behavioral interventions; and (3) individual differences that might predict successful outcomes.

Side effects associated with nicotine transdermal patches are relatively common, but well tolerated (Abelin, Muller, Buehler, et al., 1989; Daughton, Heatley, Prendergast, et al., 1991, Rose, 1991; Tonnesen, Norregaard, Simonsen, et al., 1991; Transdermal Nicotine Study Group, 1991). The most common side effect is skin irritation, which varies in incidence depending on the preparation used. In the Transdermal Nicotine Study Group (1991) study, almost 50% of the patients receiving active treatment reported transient itching or burning at skin sites, but only 14% had definite erythema. The frequency of dropouts due to skin reactions in recently published trials was low, ranging from 1% (Transdermal Nicotine Study Group, 1991) to 5% (Abelin, Muller, Buehler, et al., 1989). Insomnia and disturbing dreams occurred more commonly during active treatment with 24-hour transdermal patches than with placebo, an effect that may be attributed to overnight maintenance of nicotine levels (Transdermal Nicotine Study Group, 1991).

During late 1991, two transdermal patches were approved by the Food and Drug Administration (FDA): Habitrol (marketed by Ciba-Geigy), and Nicoderm (marketed by Marion Merrell Dow). Both are available in preparations that deliver approximately 7, 14, or 21 mg over 24 hours. Nicoderm, tested in the Transdermal Nicotine Study Group trial (1991), uses rate-control membrane technology to deliver nicotine, whereas Habitrol, utilized in the Abelin et al. (Abelin, Muller, Buehler, et al., 1989) study, uses a copolymer solution of nicotine dispersed in a pad of viscose and cotton. The 21-mg patches are the recommended starting dose for most patients. Based on the results of clinical trials, package inserts for these products recommend 4 to 6 weeks of treatment with the 21-mg dose, followed by separate 2- to 4-week courses with the 14- and 7-mg patches. The insert also suggests that the 14-mg patch be used as the starting dose for patients who have active coronary disease, weigh less than 100 pounds, and smoke less than half a pack of cigarettes a day. A 1-month supply of the 21-mg patches costs approximately $120. At least two other pharmaceutical companies have sought FDA approval for their transdermal nicotine delivery systems. Because of the advantages, it is likely that nicotine transdermal patches will become a popular pharmacologic agent for treating nicotine dependence.

Other Nicotine Delivery Systems and Lobeline

Several other systems for delivering nicotine are currently being developed or tested. Nicotine nasal solution showed promise in reducing withdrawal and aiding abstinence, but nasal irritation and embarrassment regarding its use has limited further development (Russell, Jarvis, Feyerabend, et al., 1983; U.S. DHHS, 1988). An aerosol preparation developed by Perkins and coworkers (Perkins, Epstein, Stiller, et al., 1986) is well tolerated and can provide excellent control for research purposes over dose delivery. Aerosolized rods, developed by the tobacco industry as smokeless cigarettes, can also deliver nicotine in significant doses, but only when puffed intensively (Russell, Jarvis, Sutherland, et al., 1987; U.S. DHHS, 1988). Sublingual nicotine tablets have also been used a research tool (Jarvik & Henningfield, 1988). However, nicotine aerosols, rods, and sublingual tablets have not yet been shown to be effective aids to smoking cessation. Lobeline, a weak nicotine receptor agonist that is found in many over-the-counter aids for quitting smoking, is of unproved efficacy for the treatment of tobacco dependence (Jarvik & Henningfield, 1988; Schwartz, 1987).

Other Pharmacotherapy

In addition to nicotine replacement therapy, several other pharmacologic strategies have been developed to treat manisfestations of the nicotine withdrawal syndrome. Medications intended to symptomatically reduce withdrawal discomfort (e.g., sedatives, anticholinergics, sympathomimetics) have generally been found to be ineffective compared with placebos (Jarvik & Henningfield, 1988; U.S. DHHS, 1988). In some cases, provision of sedatives or stimulants has actually led to increases in smoking (Jarvik & Henningfield, 1988). Moreover, there is a risk in prescribing potentially addictive drugs to treat nicotine dependence (Jarvik & Henningfield, 1988).

More recently efforts to manage withdrawal have focused on preventing or attenuating the entire withdrawal syndrome, rather than simply treating the symptoms. The role of clonidine and antidepressants in the treatment of nicotine dependence is discussed in the following sections.

Clonidine

Clonidine is a centrally acting antihypertensive agent that has been shown to be effective in preventing symptoms associated with opiate withdrawal (Gold, Pottash, & Sweeney, 1980; Washton & Resnick, 1981) and alcohol withdrawal (Baumgartner & Rowen, 1987; Manhem, Nilsson, Moberg, et al., 1985; Wilkins, Jenkins, & Steiner, 1983). Clonidine has also been shown to attenuate the nicotine withdrawal syndrome (Glassman, Jackson, Walsh, et al., 1984; Glassman, Stetner, & Raizman, 1986; Ornish, Zizkook, & McAdams, 1988). Clonidine is thought to reduce withdrawal symptoms by inhibiting activity of the locus ceruleus, which regulates noradrenergic activity (Aghajanian, 1978; Foote, Bloom, & Aston-Jones, 1983; Glassman, Stetner & Raizman, 1986; Issac, 1980). Noradrenergic overactivity is believed to be responsible for some of the symptoms of nicotine withdrawal, especially anxiety and craving (Glassman, Stetner, & Raizman, 1986).

Clonidine has also shown promise as an adjunct to smoking cessation treatment (Glassman, Stetner, Walsh, et al., 1988; Wei & Young, 1988). Glassman and colleagues (1988) demonstrated in a randomized, double-blind, placebo-controlled trial that treatment with oral clonidine significantly improved 6-month smoking-cessation treatment outcomes compared with placebo (27% and 5% smoking abstinence rates, respectively).

Clonidine's effect on 4-week outcomes in this study was significant for women, but not for men. Wei and Young (1988) reported that clonidine was significantly more effective after 4.5 months of treatment (57% abstinence rate) than either diazepam (37%) or placebo (37%) in a double-blind trial conducted in China.

Because all patients in the Glassman, Stetner, Walsh, et al. (1988) and Wei and Young (1988) studies also underwent behavioral treatment for smoking cessation, one cannot conclude from these results that clonidine used alone is effective as a treatment for smoking cessation. Indeed, when clonidine was dispensed without behavioral treatment in another study, a significant benefit over placebo was found at 1 week, but not at 4, 8, or 12 weeks, after the initiation of treatment (Davison, Kaplan, Fintel, et al., 1988). Like nicotine gum, it appears that clonidine may have limited effectiveness in producing long-term abstinence unless delivered in combination with cognitive-behavioral treatment.

The use of a transdermal preparation of clonidine may further increase its efficacy (Covey & Glassman, 1991), as this delivery system provides therapeutic levels at a steady state for several days with fewer side effects than the oral preparation (Ornish, Zisook, & McAdams, 1988; Weber & Drayer, 1984). Though transdermal clonidine effectively attenuates nicotine withdrawal symptoms (Ornish, Zisook, & McAdams, 1988), its efficacy as a smoking-cessation treatment has not yet been adequately tested.

When used to prevent nicotine withdrawal, the usual dose of oral or transdermal clonidine is 0.1 to 0.3 mg/day. Treatment should be continued for at least 2 weeks, and the dose should be tapered before discontinuation. The dose should be increased in increments (0.1 mg/day), and blood pressure should be measured in sitting and standing positions after each increase in dose. Patients should be told to watch for dizziness, lightheadedness, or excessive sedation; and they should be told to be cautious when using other sedatives (e.g., alcohol, over-the-counter cold remedies) or when they must change position rapidly.

TABLE 42–5. *Evaluation of Patients with Nicotine Dependence*

- Assessing the level of nicotine dependence
  - Smoking history
    - Nicotine intake
      - Smoking rate
      - Nicotine content of brand
    - Self-reported smoking pattern
      - Self-reported severity and pattern of withdrawal
      - Time to first cigarette in the morning
  - Measures of nicotine intake
    - Nicotine or cotinine assays
    - Alveolar carbon monoxide level
  - Fagerstrom Tolerance Questionnaire
- Assessing reasons for smoking
  - Self-monitoring of smoking
    - Cues or triggers associated with smoking
- Assessment of co-morbidity
  - Current or past psychiatric disorder
    - Schizophrenia
    - Mood disorder
    - Anxiety disorder
  - Other substance-related disorder
  - Obesity/history of weight gain with cessation
  - Psychoactive drug use
  - Obesity/history of weight gain with cessation

Antidepressants

The theoretic basis for considering antidepressants as a treatment for nicotine dependence comes from several sources. First, as noted previously, a history of depression is associated with increased difficulty with smoking cessation (Glassman, Stetner, Walsh, et al., 1988; Glassman, Covey, & Stetner, 1989a). Second, depressive symptoms commonly occur during nicotine withdrawal (Hughes, Higgins, & Hatsukami, 1990) (see Table 42-5), especially when there is a history of depression (Glassman, Covey, & Stetner, 1989a). Third, nicotine appears to be a powerful regulator of affect (U.S. DHHS, 1988). Finally, antidepressants have been useful for ameliorating the withdrawal state associated with cocaine, a psychoactive drug that shares stimulant properties with nicotine (Gawin & Kleber, 1987).

At present, doxepin hydrochloride is the only antidepressant that has been shown to be effective in attenuating withdrawal symptoms after smoking cessation (Edwards, Murphy, Downs, et al., 1988). Pilot data suggest that doxepin also has some benefit as an adjunct to smoking cessation (Edwards, Downs, et al., 1989; Edwards, Murphy, et al., 1989). We currently recommend that antidepressants be considered as an adjunct to smoking cessation treatment when (1) there is a history of a major depressive episode or dysthymia; (2) previous smoking cessation attempts were associated with prominent depressive symptoms, especially if they persisted beyond the first 2 weeks of abstinence; and (3) previous attempts at cessation using behavioral interventions and other pharmacologic agents have failed. When treating nicotine dependence, typical antidepressant doses should be utilized. Treatment should be continued for at least 2 weeks, or longer if depressive symptoms develop.

Buspirone and Other Anxiolytics

Buspirone, a novel anxiolytic with presynaptic dopaminergic activity, reduced craving for cigarettes, minimized withdrawal anxiety and fatigue, and led to reduced smoking in an open, uncontrolled pilot study (Gawin, Compton, & Byck, 1989). Buspirone might prove to be especially useful as an adjunct to treatment in patients who have a coexisting anxiety disorder, especially generalized anxiety disorder, or in patients who develop pronounced or persistent anxiety symptoms after smoking cessation. Typical antianxiety doses of buspirone should be used when treating nicotine dependence. Buspirone should be continued for at least 2 weeks and until anxiety symptoms are controlled. Patients with coexisting panic disorder and obsessive-compulsive disorder may also benefit from adjunctive treatment with an antidepressant with proved efficacy for these disorders (e.g., imipramine and fluoxetine, respectively). The use of benzodiazepines for a patient with a coexisting anxiety disorder is not recommended because they have no proven efficacy (Schwartz, 1987; Wei & Young, 1988).

Blockade Therapy

The goal of blockade therapy is to reduce or eliminate any rewarding pharmacologic effects that would occur if an individual resumed use of the drug after becoming abstinent (Jarvik & Henningfield, 1988; U.S. DHHS, 1988). Mecamylamine, a nicotine antagonist that acts both peripherally and centrally, has been shown to block the nicotine-mediated reinforcing consequences of cigarette smoking (Stolerman, 1986; U.S. DHHS, 1988). However, it has not been tested in a clinical trial as an aid to maintain abstinence, and its use is limited by its ganglionic blocking activity, which produces several unwanted side effects (U.S. DHHS, 1988).

Deterrent Therapy

Deterrent therapy is based on the idea that pretreatment with an agent that transforms nicotine use from a rewarding to an adverse experience would decrease relapse (Jarvik & Henningfield, 1988; U.S. DHHS, 1988). Disulfiram treatment of alcoholism is the pharmacologic analogy for this form of treatment. Presently, silver acetate preparations are available as deterrent treatments for smoking. Sulfide salts, which are distasteful, are produced whenever sulfides, present in tobacco smoke, come in contact with silver acetate residue in the mouth (Jarvik & Henningfield, 1988; U.S. DHHS, 1988).

Silver acetate gum, a currently available preparation, has to be chewed upon awakening and then several times throughout the day. Its efficacy has been tested in a double-blind, placebo-controlled trial (Malcolm, Currey, Mitchell, et al., 1986). At the end of 3 weeks of treatment, active silver acetate was significantly more effective in achieving abstinence than placebo (11% versus 4%). However, at 4-month follow-up, differences between active silver acetate and placebo gum were no longer significant (Malcolm, Currey, Mitchell, et al., 1986).

COGNITIVE-BEHAVIORAL TREATMENTS

Cognitive-behavioral treatment strategies are essential to the long-term success of the treatment of nicotine dependence (Abrams & Wilson, 1986; Brown & Emmons, 1991; U.S. DHHS, 1988). Several reviews have comprehensively describe cognitive-behavioral and other nonpharmacologic interventions for treating smokers (e.g., Abrams, Emmons, Niaura, et al., 1991; Brown & Emmons, 1991; Goldstein, Guise, Ruggiero, et al., 1990; Schwartz, 1987; U.S. DHHS, 1988).

In the following sections, we provide an overview of several cognitive-behavioral (and a few other types of nonpharmacologic) interventions listed in Table 42-6. Most of the interventions described below can be combined to meet the needs of the patient.

The research literature as well as our own clinical experience lead us to recommend a multicomponent approach (see further discussion of multicomponent approaches below). It is useful to think of three interrelated phases of multicomponent cessation programs:

TABLE 42–6. *Behavioral Treatments for Nicotine Dependence*

- Preparation
  - Motivation-commitment training
  - Health information
  - Reasons for quitting and for smoking
  - Contingency contracting
  - Target quit date
  - Self-monitoring
- Quitting
  - Self-management
    - Alter/avoid smoking cues
    - Substitute alternative behaviors
    - Stimulus control techniques
    - Relaxation techniques
  - Aversion strategies
    - Satiation
    - Rapid smoking
  - Nicotine fading
- Maintenance
  - Coping skills training
  - Social support
  - Cue exposure
  - Exercise
  - Coping with negative affect/depression
  - Avoiding weight gain

preparation, quitting, and maintenance. They are interrelated in the sense that they overlap in time, and some principles and methods are useful during more than one phase. The discussion of each phase includes a brief summary of the research literature on various strategies and methods as well as our own recommendations for clinical applications.

Preparation

Some patients are ready to quit immediately, whereas others do better with a "preparation" period prior to quitting, the length of which may vary according to individual needs. There are three key objectives for the preparation period. First, patients should review and strengthen their motivation and commitment to quitting. Second, there should be a clearly established target quit date that allows patients the time to "mentally prepare" and to develop the coping strategies needed to quit smoking. Third, patients should self-monitor their smoking behavior (keep a daily diary) in order to establish baseline levels and to begin to learn about smoking signals and consequences.

Motivation-Commitment Training

Motivation is a critical factor that determines whether a person will be successful in the effort to quit smoking. The preparation stage offers the opportunity to help smokers resolve their ambivalence and strengthen their commitment to quitting successfully.

The development of new approaches to help smokers increase their motivation or readiness to quit smoking is needed. Brown, Niles, and Emmons (1991) piloted a motivational intervention for smokers in the precontemplation and contemplation stages of change in which smokers made experimental changes in their smoking behavior, gained realistic expectations about the process of quitting, and contracted for current behavioral changes that might promote future cessation. Significant reductions in daily smoking rate and daily nicotine yield were produced as a result of the intervention and were maintained over a 6-month follow-up. Significant increases in self-efficacy about quitting were evidenced from pre- to posttreatment but were not maintained at the 6-month follow-up, as none of the ten subjects attempted cessation following the intervention. At 18-month follow-up, one subject had quit smoking and there had been eight other quit attempts, compared to 16 total quit attempts in all the years prior to the intervention.

Health Information

Whereas "scare tactics" are likely to be nonproductive, concern about the health risks of smoking motivates people to attempt cessation (Lichtenstein & Brown, 1980). The challenge is to move smokers from general acceptance ("Cigarette smoking is dangerous to health") to personalized acceptance ("Cigarette smoking is dangerous to *my* health") (Fishbein, 1977).

We recommend providing personalized health risk information to smokers whenever possible. However, health risk information should be coupled with specific directives as to what the person can do to change her or his smoking behavior. A concurrent focus on the health *benefits* of quitting may be more important to the prospective quitter than the fear of negative health consequences from continued smoking.

Reasons for Quitting and for Smoking

Another approach to enhancing motivation involves having participants write down their specific reasons for wanting to stop smoking and for wanting to continue smoking, as well as the short-term and long-term consequences of each. Identifying reasons to continue smoking may seem contradictory, but this method makes it possible to identify the likely barriers to quitting so they may be addressed.

Contingency Contracting

Increased commitment to abstinence may be achieved by establishing an agreement where monetary consequences are provided for smoking or not smoking. Contingency contracts have generally been effective in producing short-term abstinence, but relapse is common once the contract is withdrawn (U.S. DHHS, 1988).

Target Quit Date

It is often useful for patients to establish a target quit date. It gives them a specific date toward which to work and should allow sufficient time for the acquisition of coping skills needed to maintain cessation and prevent relapse. The time prior to quit date may also be used to reduce nicotine consumption by changing brands or by reducing the number of cigarettes prior to quitting completely. However, reductions below 50% of baseline rate or about 12 cigarettes a day (whichever is lower) are likely to increase the reinforcing value of the remaining cigarettes and thus may be counterproductive (Levinson, Shapiro, Schwartz, et al., 1971).

Self-Monitoring

Keeping a written record of the number of cigarettes serves to establish baseline data, to increase knowledge about the factors cueing and maintaining one's smoking habit, and to track progress throughout the cessation attempt. Self-monitoring is likely to result in a self-recorded rate at least several cigarettes per day less than the "real" baseline.

A preprinted card or sheet that can be attached to the cigarette pack facilitates self-monitoring. Smokers are typically instructed to self-record each cigarette *prior* to smoking it and to record the time of day, the situation in which the cigarette was smoked (e.g., with coffee), and associated mood (e.g., tense, relaxed). The situational notations allow for a functional anaylsis of smoking episodes, revealing the environmental influences that trigger smoking. These events or triggers that are associated with smoking need to be delineated, as they may become future high-risk situations for relapse requiring effective coping responses (Marlatt & Gordon, 1985).

Quitting

An exhaustive review of smoking intervention methods derived from social learning theory is beyond the scope of this chapter. The interested reader should consult two comprehensive reviews of the treatment literature (Schwartz, 1987; U.S. DHHS, 1988). We devote our attention here to the three major social learning-based approaches that are typically included in multicomponent programs: self-management, aversion, and nicotine fading.

Self-Management

Self-management (sometimes termed stimulus control) refers to those strategies intended to rearrange environmental cues that trigger smoking or alter the consequences of smoking. The assumption is that, as a learned behavior, smoking is strengthened through its association with environmental events, as well as by the immediate positive consequences of smoking. Smokers intervene actively in their natural environment to break up the smoking behavior chain (situation → urge → smoke) by utilizing one of two general strategies: (1) alter or totally avoid the controlling cues whenever possible; and (2) substitute an alternative behavior to replace smoking cigarettes in those cue situations that cannot be altered or avoided. Because smoking cues often precipitate relapse (Abrams, Monti, Pinto, et al., 1987; Abrams, Emmons, Niaura, et al., 1991; Marlatt & Gordon, 1985), it is important for patients to anticipate these potential high-risk situations and to develop effective coping strategies (Abrams, Monti, Pinto, et al., 1987; Shiffman, 1985). Details regarding the use of specific stimulus control techniques may be found elsewhere (Lichtenstein & Brown, 1980).

Aversion Strategies

Three aversive conditioning strategies have been used in the treatment of cigarette smoking: electric shock, imaginal stimuli (e.g., covert sensitization), and cigarette smoke itself. We limit our discussion to methods involving smoke aversion, as they have been utilized most frequently and have enjoyed the greatest measure of succcss.

Satiation. Satiation involves having smokers double or triple their usual smoking rate for a specified number of days just prior to quitting. Clinical trials of satiation with at least 1 year follow-up have produced quit rates ranging from 18% to 63%, with a median of 34.5% abstinence (Schwartz, 1987).

Rapid Smoking. With rapid smoking, the smoker is instructed to puff rapidly—every 6 to 8 seconds—and to continue puffing until it is no longer bearable. Standard treatment involves in two or three trials at six to eight treatment sessions. Results are better when rapid smoking is combined with other procedures (Schwartz, 1987). Such studics, with at lcast 1 ycar follow-up, have yielded cessation rates ranging from 7% to 52%, with a median of 30.5% abstinence (Schwartz, 1987). However, the best multicomponent programs involving rapid smoking have attained quit rates of about 50% at 6- and 12-month follow-ups (Best, Owen, & Trentadue, 1978; Brandon, Zelman, & Baker, 1987; Halls, Rugg, Tunstall, et al., 1984).

Practical drawbacks exist to the use of rapid smoking, as the procedure greatly intensifies the naturally harmful effects of smoking. Cardiovascular complications remain the major concern, as cardiac irregularities on electrocardiogram (ECG) readings have been reported but have not led to any significant clinical symptoms (Lichtenstein & Brown, 1982). However, suggested procedures for medical screening and selection of patients should be followed (Lichtenstein & Glasgow, 1977). Given these limitations, we recommend that rapid smoking and satiation be considered only after non-aversive alternatives have proved unsuccessful and only in an appropriate medical setting where emergency procedures are readily available.

Nicotine Fading

Nicotine fading, a nonaversive gradual nicotine reduction procedure, represents an alternative to pharmacologic treatment to manage nicotine withdrawal symptoms (Foxx & Brown, 1979). Smokers work toward a target quit date by switching brands to progressively lower nicotine content cigarettes over a period of several weeks. In this way, withdrawal symptoms tend to be experienced more diffusely over a gradual period and are less intense at the quit date, when total abstinence is attempted. Nicotine and tar yield can also be plotted each day to provide patients with positive feedback regarding their efforts, which may lead to an increased sense of self-efficacy about quitting.

One drawback of nicotine fading is that it is not applicable for smokers who are already on the lowest nicotine content brands. Also, it should be noted that smokers may compensate by smoking more cigarettes or changing the topography of their smoking behavior during the fading process. However, patients can be cautioned about this possibility and advised to keep such changes to a minimum.

Clinical trials of nicotine fading with at least 1 year follow-up have yielded quit rates ranging from 7% to 46%, with a median of 26% abstinence (Schwartz, 1987). The best multicomponent programs involving nicotine fading have produced quit rates of 40% or better at long-term follow-up (Foxx & Brown, 1979; Lando & McGovern, 1985). These success rates are among the highest for nonaversive procedures (Glasgow & Lichtenstein, 1987). Nicotine fading represents an acceptable treatment option to most smokers (Lando, 1986), and we recommend its use prior to consideration of aversive strategies.

Maintenance

Because most treated smokers do quit initially but resume smoking within several months of treatment termination (Hunt & Bespalec, 1974), maintenance is a critical issue in smoking cessation. We focus here on several of the more frequently employed maintenance strategies and on two more recent approaches that appear promising.

Coping Skills

The coping skills approach assumes that the individual lacks the behavioral and cognitive skills necessary to become a permanent nonsmoker. Relapse prevention theory (Marlatt & Gordon, 1980, 1985) proposes that the ability to cope with "high-risk" situations determines an individual's probability of maintaining abstinence. Successful coping in high-risk situations leads to an increased sense of self-efficacy (Bandura, 1977), but failure to cope initiates a chain of events in which diminished self-efficacy may lead to a slip and perhaps to a full-blown relapse. In these instances participants are taught to avoid self-defeating attributions and resulting negative emotional reactions (i.e., the abstinence violation effect) that promote continued smoking (Marlatt & Gordon, 1980, 1985).

A model relapse prevention program consists of five components: (1) identification of high-risk situations; (2) coping rehearsal; (3) avoiding the abstinence violation effect; (4) lifestyle balance; and (5) self-rewards (Brown & Lichtenstein, 1980). In two trials of this program in combination with nicotine fading, one yielded 46% abstinence at 6 months (Brown & Lichtenstein, 1980) and the other 19% abstinence at 1 year (Brown, Lichtenstein, McIntyre, et al., 1984).

Glasgow and Lichtenstein (1987) reviewed a number of other studies that found no effect for maintenance procedures based on relapse prevention theory, although there is evidence that less-dependent smokers are more likely to benefit from this approach. The relapse prevention model continues to focus attention on the central role of coping in maintaining long-term abstinence and will continue to stimulate research on relapse and its determinants.

Social Support

The role of family and peer influences in successful smoking cessation has been well documented (Colletti & Brownell, 1982). Although evidence suggests that social support influences outcome after smoking cessation attempts (Schwartz, 1987; U.S. DHHS, 1988), interventions to increase patients' social support during treatment for nicotine dependence have met with mixed results (Lichtenstein, Glasgow, & Abrams, 1986; U.S. DHHS, 1988). Nevertheless, inclusion of spouses in treatment and teaching them how to be supportive of patients' attempts to quit may be helpful.

Cue Exposure

Several studies of active smokers have found that smoking-related stimuli can elicit increases in peripheral vasoconstriction, as well as increases in blood pressure and urges to smoke (Abrams, Emmons, Niaura, et al., 1991; Niaura, Rohsenow, Binkoff, et al., 1988). Abrams and colleagues (Abrams, Monti, Carey, et al., 1988) found that increases in urge, anxiety, and heart rate in response to a smoking confederate were greater for smokers who had relapsed than for never-smoked control subjects. Successful quitters had intermediate levels of reactivity. At present, there is evidence to suggest that substance-use cues are present in a percentage of smoking relapses (Abrams, Emmons, Niaura, et al., 1991). Thus in vivo or imaginal exposure to physical smoking cues may become an important adjunct to relapse prevention programs for some smokers.

Exercise

There has been considerable interest in the role that physical activity may play in preventing relapse. Because vigorous exercise is incompatible with simultaneous smoking, exercise can serve as a substitute behavior following cessation. Exercise may also be a good alternative to "restrained eating" or dieting for individuals who are concerned about postcessation weight gain. In addition, exercise may moderate mood changes such as depression and anxiety, and therefore it may serve to attenuate nicotine withdrawal (Abrams, Monti, Pinto, et al., 1987; Shiffman, 1982, 1984). One study provided preliminary evidence for the role of exercise training as a behavioral relapse prevention strategy (Marcus, Albrecht, Niaura, et al., 1991).

Other Clinical Concerns

Negative Affect. Some writers have proposed that affect regulation is the driving force behind cigarette smoking and addiction (Carmody, 1989; Leventhal & Cleary, 1980; Tompkins, 1966). If, in fact, smokers use nicotine to regulate affect, poor outcomes in smoking cessation may be due, in part, to deficiencies in smokers' coping skills for managing affect without the use of cigarettes. The acquisition of skills for coping with depression could help prospective quitters to: (1) prevent the onset of depressive symptoms upon cessation from smoking; and (2) modify or improve their affect in potential negative affect situations, which are the most common precipitants of relapse (Bliss, Garvey, Heinold, et al., 1989; Brandon, Tiffany, Obremski, et al., 1990; O'Connell & Shiffman, 1988; Shiffman, 1982). Because the efficacy of cognitive-behavioral interventions for depression is well established (Hoberman & Lewinsohn, 1985; Hollon & Najavits, 1988), it may be advisable to integrate training in these skills as part of an integrated smoking cessation treatment program.

Weight Gain. Smoking cessation is associated with weight gain of 6.4 pounds and increases in caloric intake of 200 to 350 calories per day (Klesges, Meyers, Klesges, et al., 1989; Rodin, 1987). The issue of cessation-related weight gain clearly has implications for treatment outcome. Fear of weight gain may be a barrier to quitting for some individuals (Klesges, Brown, Pascale, et al., 1987). When the evaluation identifies weight gain as an issue, behavioral treatment interventions that focus on weight management are likely to increase the effectiveness of treatment.

Although some investigators have postulated that weight gain among recent quitters may cause relapse, two studies have found that postcessation weight gain predicted abstinence rather than recidivism (Hall, Ginsberg, & Jones, 1986; Streator, Sargent, & Ward, 1989). Hall and colleagues (1986) hypothesized that individuals in their study who did not gain weight may have actively deprived themselves of food, thus increasing the reinforcing properties of smoking, as well as the probability of relapse. An important aspect of smoking cessation treatment for such individuals may be to learn adaptive, alternative means of coping that minimizes their sense of deprivation. In addition, it may also be prudent to focus on preoccupations and concerns about weight gain. A complete description of behavioral treatment interventions for weight management is beyond the scope of this chapter but may be found in reviews of Abrams (1984) and Brownell (1986) and in Chapter 43 of this text.

Nonbehavioral Treatments

Hypnosis

Hypnosis, when applied to smoking cessation, produces only modest results when used alone (Schwartz, 1987). It appears to be most effective when hypnosis is provided by a skilled hypnotherapist, the patient is highly susceptible to hypnotic induction, there are several hours of treatment over several sessions, the relationship with the therapist is intense, hypnotic suggestions are personalized, and there is either adjunctive counseling or follow-up (Holroyd, 1980; Schwartz, 1987). Brown and Fromm (1987) have provided a comprehensive description of the use of hypnosis as an adjunct to other behavioral interventions for the treatment of nicotine dependence. In general, the efficacy of hypnosis therapy has not been proved in randomized clinical trials in comparison with cognitive-behavioral and pharmacologic therapies.

Acupuncture

Acupuncture is an ancient Chinese science that utilizes needles or staples to treat various conditions, including smoking. The needles or staples are placed in specific locations, or "points," that are thought to provide connections with the body's regulatory systems (Schwartz, 1987). The ear is a common site for needle or staple insertion because it is believed to connect to the "neurovegetative system," which controls appetitive behavior (Schwartz, 1987). Schwartz (1987), in a review, found no convincing evidence that acupuncture either relieves withdrawal symptoms after smoking cessation or significantly improves smoking-cessation outcome when used alone. However, it may have a positive placebo effect (Schwartz, 1987). See Choy, Purnell, and Jaffe (1978) for a description of the technique of auricular acupuncture.

Multicomponent Programs and Groups

Programs that include multiple components have success rates that are generally greater than programs utilizing only one or two strategies (Abrams & Wilson, 1986; Brown & Emmons, 1991; Lichtenstein, 1986; Schwartz, 1987; U.S. DHHS, 1988). However, the relative strengths and weaknesses of various combinations of behavioral and other nonpharmacologic interventions remains unresolved. In general, it appears that programs that combine strategies to help the patient to quit (e.g., nicotine fading and stimulus control) with strategies to help the patient maintain abstinence (e.g., relapse-prevention training and social support) have the best outcomes (Schwartz, 1987; Tiffany, Martin, & Baker, 1986; U.S. DHHS, 1988). However, including more strategies in a given program may overwhelm subjects and reduce adherence to treatment (Lichtenstein & Brown, 1982, U.S. DHHS, 1988).

Group treatment settings have frequently been used to deliver multicomponent treatment. This approach takes advantage of the social support component of group treatment, is less costly than individual treatment, and facilitates the involvement of multiple therapists who may have overlapping but distinct therapeutic skills (e.g., a psychiatrist skilled in pharmacologic interventions and a psychologist skilled in behavioral interventions). However, we are not aware of any studies that have compared the effectiveness of group versus individual nicotine dependence treatment that controlled for program content and therapist contact time. A description of group treatment technique is beyond the scope of this chapter. A manual for group treatment combining coping skills training with nicotine resin complex (Goldstein, Niaura, Abrams, et al., 1989) is available from the authors; contact Michael G. Goldstein, M.D., The Miriam Hospital, 164 Summit Avenue, Providence, RI 02906.

Table 42-7 provides a summary of follow-up abstinence rates according to different intervention methods for studies published during the years 1959 to 1985. The table provides only a general guide to the efficacy of various treatment modalities. The variability in results reflects methodologic differences among the trials as well as varying definitions of abstinence. Moreover, many of these studies used self-reported abstinence rather than relying on biochemical measures. Therefore the table should be interpreted with caution. Investigators interested in the effectiveness of a particular treatment should review the individual studies and evaluate results based on the quality and rigor of the experimental design and methods employed. The table also does not provide information concerning the efficacy of treatments tested after 1985. However, such studies concerning the effects of nicotine replacement, other medications, and physician interventions are reviewed elsewhere in this chapter.

STEP-BY-STEP APPROACH TO TREATMENT

This section describes a step-by-step approach to treatment of nicotine dependence for the general psychiatrist in an outpatient clinic or office setting. Though we have oversimplified the approach to increase clarity, we believe this strategy is clinically useful. It is anticipated that the treatment would be provided in 6 to 10 weekly individual sessions, each lasting 30 to 50 minutes. After an initial assessment, individual treatment sessions would include cognitive-behavioral interventions, as well as psychotherapy aimed at associated psychiatric disorders, if present. If successful smoking cessation is achieved, monthly 15-minute follow-up sessions are suggested for the next 6 to 12 months.

Step 1: Assessment

The first step in the treatment process is a careful assessment of the severity of nicotine dependence, psychiatric comorbidity, and other psychological, behavioral, and sociocultural factors contributing to nicotine use. A reasonable first step in evaluation is to obtain a detailed smoking history. It is useful to inquire about previous attempts to quit in order to document the duration and intensity of withdrawal symptoms and reasons for resuming smoking. Symptoms may also have been apparent when the patient switched to a low-tar, low-nicotine cigarette or after the patient stopped using smokeless tobacco products or nicotine gum. It may also be useful to assess the intensity of withdrawal symptoms during a prescribed period of abstinence of

TABLE 42–7. *Summary of 1-Year Follow-up Abstinence Rates of Smoking Cessation Trials by Method (1959–1985)*

Intervention Method	No.	Range	Median
Self-help	7	12–33	18
Educational	12	15–55	25
Group	31	5–71	28
Medication*	12	6–50	18.5
Nicotine resin complex	9	8–38	11
Nicotine resin complex and behavioral treatment	11	12–49	29
Hypnosis—individual	8	13–68	19.5
Hypnosis—group	2	14–88	—
Acupuncture	6	8–32	27
Physician advice or counseling	12	3–13	6
Physician intervention—more than counseling	10	13–38	22.5
Physician intervention			
Pulmonary patients	6	25–76	31.5
Cardiac patients	16	11–73	43
Rapid smoking	6	6–40	21
Rapid smoking and other procedures	10	7–52	30.5
Satiation smoking	12	18–63	34.5
Regular-paced aversive smoking	3	20–39	26
Nicotine fading	16	7–46	26
Contingency contracting	4	14–38	27
Multiple programs	17	6–76	40

Adapted from Schwartz (1987). See original report for complete explanations of methods and results.
*Medications included lobeline hydrochloride, meprobamate, amphetamine silver acetate, methylphenidate, diazepam, and hydroxyzine.

1 to 3 days. Nicotine withdrawal symptoms can be monitored using one of many available scales (Hughes & Hatsukami, 1986; Shiffman, 1979). Simultaneous monitoring of depressive symptoms may also be useful, particularly in patients with a history of depression.

Measures of nicotine intake may provide some index of the degree of nicotine dependence. They include self-reported smoking rate and nicotine content of cigarettes, biochemical assays of nicotine and its metabolites, and alveolar carbon monoxide levels (an indirect measure). Cotinine, a metabolite of nicotine, can be measured in saliva, blood, or urine and provides a reasonable reflection of nicotine intake (Abrams, Follick, Biener, et al., 1987b). Due to the relatively high cost of analysis and a turnaround time of up to several weeks, however, cotinine analysis is not a feasible method for determining nicotine intake for most clinicians. A less expensive and more convenient alternative to assess nicotine exposure is analysis of expired alveolar carbon monoxide. Two devices to measure concentrations of carbon monoxide, the EC50 (Vitalograph, Inc., Lenexa, Kansas) and the Ecolyzer 2000 (National Draeger, Pittsburgh, Pennsylvania), are readily available and relatively inexpensive. Studies have pointed to the value of carbon monoxide measurement as a predictor of the severity of nicotine withdrawal symptoms (West & Russell, 1985a).

Perhaps the most common measure of self-reported nicotine dependence in current use is the Fagerstrom Tolerance Questionnaire (FTQ) (Fagerstrom, 1978). The FTQ is a seven-item self-administered form that identifies behaviors thought to reflect nicotine dependence (e.g., high smoking rate, nicotine level of cigarette brand, and smoking when ill or soon after awakening). Time to first cigarette in the morning, a specific question on the FTQ, is an indirect measure of withdrawal following overnight abstinence and appears to be the better predictor of outcome after attempted smoking cessation than the total FTQ score (Lichtenstein & Mermelstein, 1986). Smoking a cigarette within 20 to 30 minutes after awakening suggests a high level of nicotine dependence.

In addition to assessing the patient's level of nicotine dependence, it is important to determine whether there is evidence of psychiatric comorbidity. A present or past history of major psychiatric disorder is likely to make smoking cessation more difficult (DiFranza & Guerrera, 1990; Glassman, Stetner, Walsh, et al., 1988). Moreover, smoking cessation may precipitate a relapse of an underlying psychiatric disorder, especially depression (Glassman, Covey, & Stetner, 1989b). Identification of psychiatric comorbidity facilitates monitoring and early treatment of exacerbations or relapses of psychiatric disorders.

After assessment, patients can be characterized as having high or low nicotine dependence, with or without associated psychiatric comorbidity, yielding four groups (see Figure 42-2).

Step 2: Initial Treatment

Initial treatment involves choosing a treatment plan that is matched to the specific needs of the patient. All pa-

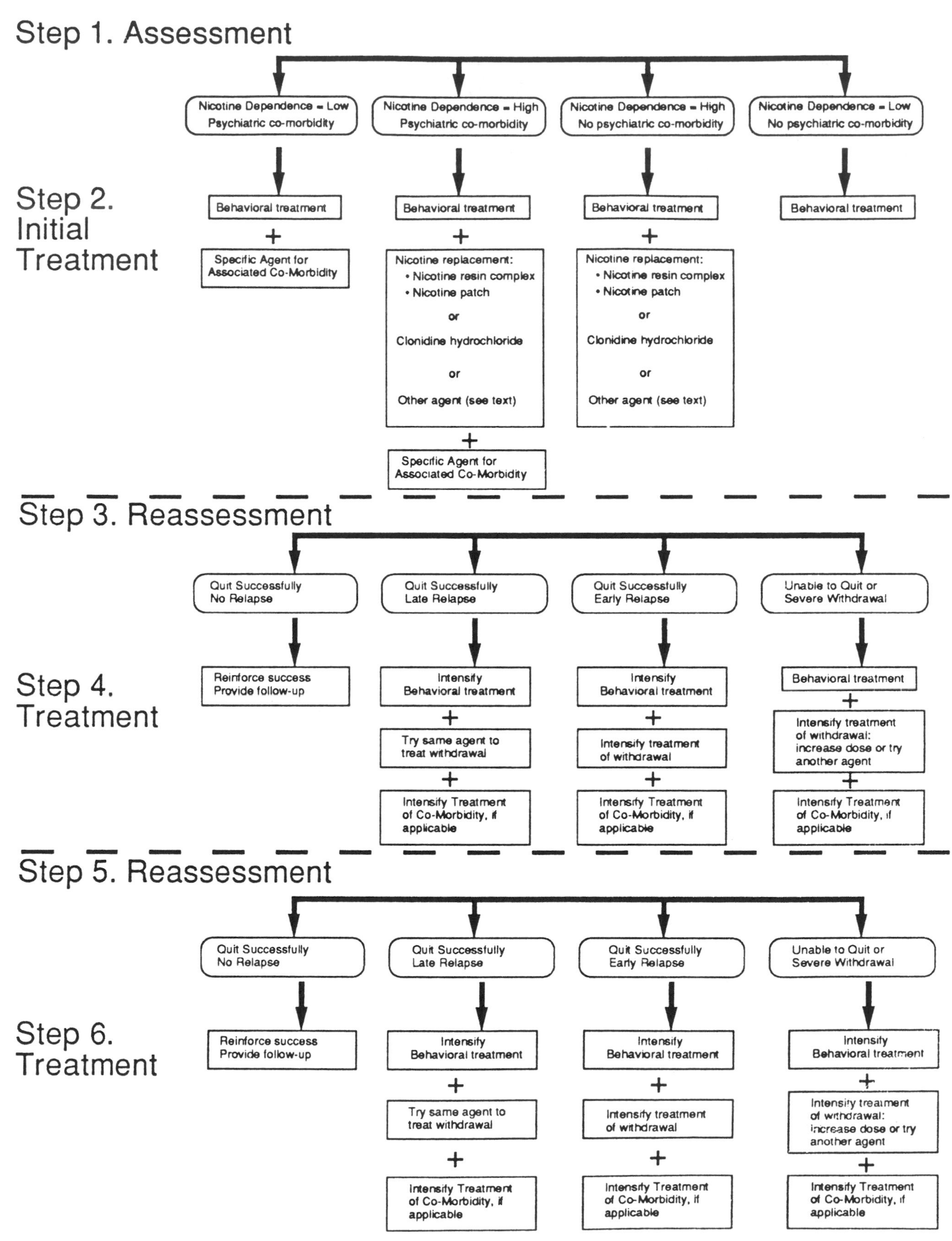

FIG. 42–2. Step-by-step approach to the treatment of nicotine dependence.

tients should be offered cognitive-behavioral treatment, as studies have demonstrated that pharmacologic treatments have limited efficacy when used alone (Abrams, Emmons, Niaura, et al., 1991; Brown & Emmons, 1991; Hughes, Gust, Keenan, et al., 1989; Schwartz, 1987; U.S. DHHS, 1988). If the evaluation identifies high levels of nicotine dependence, we recommend adding a pharmacologic intervention to help the patient manage withdrawal symptoms and craving. Response to pharmacologic agents during previous attempts to quit helps to guide the choice of a specific pharmacologic agent. Nicotine resin complex is usually our first choice if the patient has not previously tried nicotine resin complex or used it incorrectly, if there are no contraindications to its use (i.e., a recent myocardial infarction, pregnancy, or significant dental or gum disease), and if there is no associated psychiatric comorbidity (Goldstein, Niaura & Abrams, 1991). However, the nicotine transdermal patch now also warrants consideration. As noted previously, the patch provides several potential advantages over nicotine resin complex, including its ease of administration and the potential for enhanced compliance and adherence. Clonidine hydrochloride should be considered as an adjunct to cognitive-behavioral treatment for those highly dependent patients who do not want to undergo a several-week course of nicotine replacement.

If the highly dependent patient has failed to quit smoking while *correctly* using nicotine resin complex or nicotine transdermal patches during previous quit attempts, a prudent alternative would be to refer the patient to a specialized behavioral medicine program that can provide both pharmacologic and cognitive-behavioral treatment modalities, such as that described below. Staff in this type of program have the specialized training and expertise to provide more intensive cognitive-behavioral treatment and to tailor treatment to the individual needs of the patient, which will likely maximize the chances for success.

If psychiatric comorbidity is uncovered by the assessment, we recommend choosing a behavioral treatment *plus* a pharmacologic agent that specifically addresses the associated psychiatric disorder. When a patient is currently being treated with a psychoactive agent for an associated psychiatric disorder, it may be necessary to adjust the dose of the agent in anticipation of an exacerbation or recurrence of the underlying disorder. In addition, a change in metabolism or pharmacodynamics may result from smoking cessation. If the patient is not currently on a psychoactive agent to treat an associated psychiatric disorder, adding such an agent should be considered, especially if psychiatric symptoms have developed during previous attempts to quit smoking.

If both high nicotine dependence and psychiatric comorbidity are present, the patient may require separate agents to treat withdrawal and comorbidity. However, as discussed in the section on pharmacologic treatment above, several agents used to treat anxiety and depression have also shown some promise as treatments of withdrawal. If studies demonstrate the effectiveness of these agents as treatments of withdrawal, it may be possible to use these agents (e.g., doxepin, buspirone) as the sole pharmacologic treatment for patients with both high dependence and associated psychiatric comorbidity. Patients with neither high levels of nicotine dependence nor psychiatric morbidity might initially be treated with cognitive-behavioral treatments alone.

Step 3: Reassessment

The third step is a reassessment based on the results of the initial treatment. Subjects can be characterized along a continuum that reflects the difficulty they have had with quitting smoking and remaining abstinent.

Step 4: Treatment

The fourth step involves adjusting treatment to match the problems that emerge from the initial treatment attempt. Subjects who have been unable to remain abstinent during the first 48 hours after attempting to quit smoking may be suffering from severe withdrawal or experiencing an acute exacerbation of their underlying psychiatric disorder. For these patients, pharmacologic treatment of withdrawal should be initiated or, if already in place, intensified either by using higher doses of the initial agent or using another drug. More intensive treatment of the associated psychiatric disorder may also be needed.

Patients who have quit successfully but relapse within the first 2 weeks after quitting have different needs. This group of patients may have experienced prolonged withdrawal symptoms, subacute exacerbation of an associated psychiatric condition, or trouble coping with cues for smoking or the consequences of not smoking. They are likely to require adjustment of the pharmacologic intervention used to treat withdrawal symptoms as well as intensification of the behavioral interventions to address the triggers of relapse. Again, more intensive treatment of comorbidity may be required.

Those patients who relapse to smoking after more than 2 weeks of abstinence probably do not require more intensive pharmacologic treatment of withdrawal but do require more help to deal with the behavioral and psychological aspects of smoking. They also may require more intensive treatment of associated psychiatric disorders, if these disorders were exacerbated during abstinence from smoking. Weight gain is a common reason for late relapse and may need to be specifically addressed.

Finally, those patients who quit successfully and remain abstinent throughout treatment benefit from follow-up visits to allow the practitioner to reinforce the patients' success and monitor mental state for several months after smoking cessation. Medication, if used, requires monitoring and tapering as well. Nicotine gum or nicotine transdermal patches should be tapered over several weeks and discontinued after about 3 months. It is unclear how long clonidine and other agents should be prescribed, but we prefer to limit use to 4 weeks if the patient is successful in quitting. Treatment of associated or emergent psychiatric disorders may require longer treatment and follow-up. Relapse prevention strategies should be reviewed during these visits to increase the patient's self-efficacy and confidence.

Step 5: Further Reassessment

The fifth step is another reassessment based on the results of subsequent treatment. Subjects can again be characterized along a continuum that reflects the difficulty they have had with quitting smoking and remaining abstinent.

Step 6: Treatment Adjustments

The sixth step involves adjusting treatment to match the problems that emerge from the subsequent treatment attempt. The only difference between the fourth and sixth steps is that we recommend intensifying behavioral treatment for all subjects who have not been able to remain abstinent. The fifth and sixth steps are repeated until the patient successfully remains abstinent.

REFERRAL TO A SPECIALIZED PROGRAM

Patients with high levels of nicotine dependence who remain refractory to office-based treatment including pharmacologic agents may require referral to a specialized behavioral medicine program that can provide *both* pharmacologic and intensive behavioral treatment modalities. The mode of treatment in such a program ranges from brief, guided self-help to extended individual or group intervention, with or without medication. Behavioral treatment is provided individually or in groups. The choice of group versus individual treatment is primarily based on the patient's preferences. Group treatment is less costly but also less flexible and less intensive. Groups are particularly useful for those patients who lack a social network that is supportive of nonsmoking.

Psychiatrists in general practice may wish to develop a relationship with a local behavioral medicine program where patients can be sent for more intensive behavioral treatment while the psychiatrist continues to prescribe adjunctive pharmacologic agents.

RESOURCES

Resources have been developed to assist physicians and hospitals in providing interventions to their patients who smoke, including (1) educational materials for patients in the form of pamphlets, self-help manuals, audiotapes, and videotapes; (2) materials to facilitate the identification and tracking of smokers; (3) manuals and kits designed to assist physicians in their office practice to develop an integrated approach to smoking-cessation interventions; and (4) materials providing policy development and implementation strategies for hospitals intending to become smokefree, for nurses to assist inpatients to quit smoking, and for inpatient smokers to help them "survive" a smokefree hospitalization. Excellent manuals or kits are available from the National Cancer Institute (NCI) (1988, 1989) and the American Academy of Family Physicians (1987).

The NCI manual is particularly helpful for primary-care physicians in office practice. It provides specific guidelines for developing an office-based smoking-cessation program. Though this manual is designed for primary care physicians, it is also likely to be useful for psychiatrists wishing to provide only minimal behavioral interventions to their patients who smoke.

Among the other specific resources that are available, the self-help manuals developed by the American Lung Association (ALA) (1987; Davis, Faust, & Ordentlich, 1984) are particularly useful as adjuncts to treatment. The ALA has also developed videotapes for patients. The American Cancer Society (1988), the American Heart Association (1984), and NCI also have developed educational materials for patients. For those seeking formal training in smoking-cessation counseling, the American Society of Addiction Medicine holds a 3-day workshop on treating nicotine dependence each fall. Books written to help the physician to learn behavioral counseling strategies are also available (Russell, 1986).

With hospitals going smokefree (JCAHO, 1991), the materials developed to assist hospitals in negotiating the process of organizational change (American Hospital Association, undated), nurses in working with inpatient smokers (U.S. DHHS, 1990), and the smokers themselves in coping with abstinence during hospitalization (Herman, 1989) should be useful.

Information on how to order physician, hospital, and patient resources/materials can be found in the reference section of this chapter.

Acknowledgments

The authors express their appreciation to Linda Moreau, Barbara Doll, Lori Krawetz, and Mark Morgenstern for their capable secretarial assistance. We are also indebted to Suzanne Sales for her invaluable assistance in proofreading and editing this chapter. A special thanks is also due to Barrie Guise, Ph.D., for her helpful contribution to the chapter.

The work on this chapter was supported in part by the National Institute on Drug Abuse, grant DA05623 to M.G.G.; the National Cancer Institute, grant CA38309 to D.B.A.; the National Cancer Institute Cancer Prevention Research Consortium, grant PO1 CA50087 to James O. Prochaska, D.B.A., Wayne Velicer, and M.G.G.; and the National Heart Lung and Blood Institute, grant HL32318 to R.S.N.

REFERENCES

Abelin T, Muller P, Buehler A, et al. (1989). Controlled trial of transdermal nicotine patch in tobacco withdrawal. Lancet 1:7–10.

Abrams DB (1984). Current status and clinical developments in the behavioral treatment of obesity. In CM Franks (Ed.), New directions in behavior therapy (pp. 21–55). New York: Haworth Press.

Abrams DB, & Biener L (1989). Organizational approaches to worksite smoking cessation. Paper presented at the annual meeting of the Society of Behavioral Medicine, New York.

Abrams DB, & Biener L. Motivational characteristics of smokers at the worksite: a public health challenge. Prev Med. In press.

Abrams DB, Emmons KM, Niaura RS, et al. (1991). Tobacco dependence: an integration of individual and public health perspectives. In PE Nathan, JW Langenbucher, BS McCrady, et al. (Eds.), Annual review of addictions research and treatment (Vol 1). New York: Pergamon Press.

Abrams DB, Follick MJ, Biener L, et al. (1987). Saliva cotinine as a measure of smoking status in field settings. Am J Public Health 77:846–848.

Abrams DB, Monti PM, Carey KB, et al. (1988). Reactivity to smoking cues and relapse: two studies of discriminant validity. Behav Res Ther 26:223–225.

Abrams DB, Monti PM, Pinto RP, et al. (1987). Psychological stress and coping in smokers who relapse or quit. Health Psychol 6:289–303.

Abrams DB, Rohsenow DJ, Niaura RS, et al. (1992). Smoking and treatment outcome for alcoholics: effects on coping skills, urge to drink and drinking rates. Behav Ther 23:283–297.

Abrams DB, & Wilson GT (1986). Clinical advances in treatment of smoking and alcohol addiction. In AJ Frances & RE Hales (Eds.), The American Psychiatric Association: annual review, psychiatric update (Vol 5; Ch. 28, pp. 606–626). Washington, DC: American Psychiatric Press.

Aghajanian GK (1978). Tolerance of locus coeruleus neurons to morphine and suppression of withdrawal response by clonidine. Nature 276:186–188.

American Academy of Family Physicians (1987). AAFP stop smoking kit. Kansas City, MO: American Academy of Family Physicians (1-800-274-2237).

American Cancer Society (1988). Smart move. Atlanta, GA: American Cancer Society (404-320-3333).

American Heart Association (1984). Calling it quits. Dallas, TX: American Heart Association (214-373-6300).

American Hospital Association (undated). Smoking and hospitals are a bad match (Cat. No. 166901). Chicago: American Hospital Association (1-800-AHA-2626).

American Lung Association (1987). Freedom from smoking for you and your family. New York: American Lung Association (212-315-8700).

Anda RF, Williamson DF, Escobedo LG, et al. (1990). Depression and the dynamics of smoking. JAMA 264:1541–1545.

Bandura A (1977). Self-efficacy: toward a unifying theory of behavioral change. Psychol Rev 84:191–215.

Bansil RK, Hymowitz N, Keller S, et al. (1989). Cigarette smoking and neuroleptics. Presented at the Annual Meeting of the American Psychiatric Association, San Francisco.

Battjes RJ (1988). Smoking as an issue in alcohol and drug abuse treatment. Addict Behav 13:225–230.

Baumgartner GR, & Rowen RC (1987). Clonidine vs chlordiazepoxide in the management of acute alcohol withdrawal syndrome. Arch Intern Med 147:1223–1226.

Benowitz NL (1988). Pharmacologic aspects of cigarette smoking and nicotine addiction. N Engl J Med 319:1318–1330.

Benowitz NL (1989). Nicotine interactions with psychiatric medications. Presented at the Annual Meeting of the American Psychiatric Association, San Francisco.

Best JA, Owen LE, & Trentadue, L (1978). Comparison of satiation and rapid smoking in self-managed smoking cessation. Addict Behav 3:71–78.

Bliss RE, Garvey AJ, Heinold JW, et al. (1989). The influence of situation and coping on relapse crisis outcomes after smoking cessation. J Consult Clin Psychol 57:443–449.

Bobo JK, Gilchrist LD, Schilling RF II, et al. (1987). Cigarette smoking cessation attempts by recovering alcoholics. Addict Behav 12:209–215.

Brandon TH, Tiffany ST, Obremski KM, et al. (1990). Postcessation cigarette use: the process of relapse. Addict Behav 15:105–114.

Brandon TH, Zelman DC, & Baker TB (1987). Effects of maintenance sessions on smoking relapse: delaying the inevitable? J Consult Clin Psychol 55:780–782.

Brown D, & Fromm E (1987). Hypnosis and behavioral medicine. Hilsdale, NJ: Lawrence Erlbaum & Associates.

Brown RA, & Emmons KA (1991). Behavioral treatment of cigarette dependence. In JA Cocores (Ed.), The clinical management of nicotine dependence (Ch. 9, pp. 97–118). New York: Springer-Verlag.

Brown RA, & Lichtenstein E (September 1980). Effects of a cognitive-behavioral relapse prevention program for smokers. Paper presented at the annual meeting of the American Psychological Association, Montreal.

Brown RA, Lichtenstein E, McIntyre KO, et al. (1984). Effects of nicotine fading and relapse prevention on smoking cessation. J Consult Clin Psychol 52:307–308.

Brown RA, Niles BL, & Emmons KM (November 1991). "Thinking about quitting": a motivational intervention for smoking cessation. Paper presented at the annual meeting of the Association for Advancement of Behavior Therapy, New York.

Brownell KD (1986). Public health approaches to obesity and its management. Annu Rev Public Health 7:521–533.

Buchkremer G, Bents H, Horstmann M, et al. (1989). Combination of behavioral smoking cessation with transdermal nicotine substitution. Addict Behav 14:229–238.

Burling TA, & Ziff DC (1988). Tobacco smoking: a comparison between alcohol and drug abuse inpatients. Addict Behav 13:185–190.

Carmody TP (1989). Affect regulation, nicotine addiction, and smoking cessation. J Psychoactive Drugs 21:331–342.

Choy DSJ, Purnell F, & Jaffe R (1978). Auricular acupuncture for cessation of smoking. In JL Schwartz (Ed.), Progress in smoking cessation. Proceedings of the international conference on smoking cessation (pp. 329–334). New York: American Cancer Society.

Clarke PBS (1989). Nicotine receptors and mechanisms in the brain. Presented at the Annual Meeting of the American Psychiatric Association, San Francisco.

Colletti G, & Brownell K (1982). The physical and emotional benefits of social support: applications to obesity, smoking, and alcoholism. In M Hersen, R Eisler, & P Miller (Eds.), Progress in behavior modification (Vol 13). Orlando, FL: Academic Press.

Covey LS, & Glassman AH (1991). A meta-analysis of double-blind placebo-controlled trials of clonidine for smoking cessation. Br J Addict 86:991–998.

Covey LS, Glassman AH, & Stetner F (1990). Depression and depressive symptoms in smoking cessation. Compr Psychiatry 31:350–354.

Daughton DM, Heatley SA, Prendergast JJ, et al. (1991). Effect of transdermal nicotine delivery as an adjunct to low-intervention smoking cessation therapy. Arch Intern Med 151:749–752.

Davis AL, Faust R, & Ordentlich M (1984). Self-help smoking cessation and maintenance programs: a comparative study with 12 month follow-up by the American Lung Association. Am J Public Health 74:1212–1217.

Davison R, Kaplan K, Fintel D, et al. (1988). The effect of clonidine on the cessation of cigarette smoking. Clin Pharmacol Ther 44:265–267.

DiFranza JR, & Guerrera MP (1989). Hard core smokers [letter]. JAMA 261:2634.

DiFranza JR, & Guerrera MP (1990). Alcoholism and smoking. J Stud Alcohol 51:130–135.

Edwards NB, Downs AD, Murphy JK, et al. (1989). Brief doxepin therapy for smoking cessation. Presented at the Annual Meeting of the American Psychiatric Association, San Francisco.

Edwards NB, Murphy JK, Downs AD, et al. (1989). Doxepin as an adjunct to smoking cessation: a double-blind pilot study. Am J Psychiatry 146:373–376.

Edwards NB, Murphy JK, Downs AD, et al. (1988). Antidepressants and nicotine withdrawal symptoms. Presented at the Annual Meeting of the American Psychiatric Association, Montreal.

Emmons KM, Abrams DB, Marshall RJ, et al. (1992). Exposure to environmental tobacco smoke in naturalistic settings. Am J Public Health 82:24–28.

Fagerstrom KO (1978). Measuring degree of physical dependence to tobacco smoking with reference to individualization of treatment. Addict Behav 3:235–241.

Fagerstrom KO (1988). Efficacy of nicotine chewing gum: a review. In O Pomerleau & CS Pomerleau (Eds.), Nicotine replacement: a critical evaluation. New York: Alan R. Liss.

Fagerstrom KO, & Melin B (1985). Nicotine chewing gum in smoking cessation: efficiency, nicotine dependence, therapy duration and clinical recommendations. In J Grabowski & SM Hall (Eds.), Pharmacologic adjuncts to smoking cessation. NIDA Research Monograph 53. DHHS, Public Health Service, Alcohol, Drug Abuse, and Mental Health Administration, DHHS Publ. No. (ADM) 85-1333 (pp. 102–109). Rockville, MD.

Fishbein M (May 1977). Consumer beliefs and behavior with respect to cigarette smoking: a critical analysis of the public literature. In: Federal Trade Commission, report to Congress: pursuant to the Public Health Cigarette Smoking Act, for the year 1976. Washington, DC: Federal Trade Commission.

Foote SL, Bloom FE, & Aston-Jones G (1983). Nucleus locus ceruleus: new evidence of anatomical and physiological specificity. Physiol Rev 63:844–914.

Fortmann SP, Killen JD, Telch MJ, et al. (1988). Minimal contact treatment for smoking cessation: a placebo controlled trial of nicotine polacrilex and self-directed relapse prevention: initial results of the Stanford Stop Smoking Project. JAMA 260:1575–1580.

Foxx RM, & Brown RA (1979). Nicotine fading and self-monitoring for cigarette abstinence or controlled smoking. J Appl Behav Anal 12:111–125.

Frederick T, Frerichs RR, & Clark VA (1988). Personal health habits and symptoms of depression at the community level. Prev Med 17:173–182.

Fuxe K, Anderson K, & Eneroth P (1987). Effects of nicotine and exposure to cigarette smoke on discrete dopamine and noradrenaline nerve terminal systems of the telcephalon and diencephalon of the rat: relationship to reward mechanisms and distribution of nicotine binding sites in brain. In WR Martin, GR Van Loon, ET Iwamoto, et al. (Eds.), Tobacco smoking and nicotine: a neurobiological approach (pp. 225–262). New York: Plenum Press.

Gallup Opinion Index (June 1974). No. 108, pp. 20–21.

Garvey AJ (1988). Factors related to relapse after smoking cessation. Presented at the Annual Meeting of the Society of Behavioral Medicine, Boston.

Gawin FH, Compton M, & Byck R (1989). Buspirone reduces smoking. Arch Gen Psychiatry 46:288–289.

Gawin FH, & Kleber H (1987). Issues in cocaine abuse treatment research. In S Fisher, A Raskin, & EH Uhlenhuth (Eds.), Cocaine: clinical and biobehavioral aspects (pp. 174–192). New York: Oxford University Press.

Glasgow RE, & Lichtenstein E (1987). Long-term effects of behavioral smoking cessation interventions. Behav Ther 18:297–324.

Glassman AH, Covey LS, & Stetner F (1989a). Smoking cessation, depression and antidepressants. Presented at the Annual Meeting of the American Psychiatric Association, San Francisco.

Glassman AH, Covey LS, & Stetner F (1989b). Smoking, smoking cessation, and depression. Presented at the Annual Meeting of the American Psychiatric Association, San Francisco.

Glassman AH, Helzer JE, Covey LS, et al. (1990). Smoking, smoking cessation and major depression. JAMA 264:1546–1549.

Glassman AH, Jackson WK, Walsh BT, et al. (1984). Cigarette craving, smoking withdrawal, and clonidine. Science 226:864–866.

Glassman AH, Stetner F, & Raizman P (1986). Clonidine and cigarette smoking withdrawal. In JK Ockene (Ed.), The pharmacologic treatment of tobacco dependence: proceedings of the World Congress, November 4–5, 1985 (pp. 174–180). Cambridge, MA: Institute for the Study of Smoking Behavior and Policy.

Glassman AH, Stetner F, Walsh BT, et al. (1988). Heavy smokers, smoking cessation and clonidine. JAMA 259:2863–2866.

Gold MS, Pottash AC, & Sweeney DR (1980). Opiate withdrawal using clonidine. JAMA 243:343–346.

GOLDSTEIN MG, GUISE BJ, RUGGIERO L, ET AL. (1990). Behavioral medicine strategies for medical patients. In A STOUDEMIRE (Ed.), Clinical psychiatry for medical students. Philadelphia: Lippincott.

GOLDSTEIN MG, & NIAURA R (1991). Nicotine gum. In JA COCORES (Ed.), The clinical management of nicotine dependence (Ch. 15, pp. 181–195). New York: Springer-Verlag.

GOLDSTEIN MG, NIAURA R, & ABRAMS DB (1991). Pharmacological and behavioral treatment of nicotine dependence: nicotine as a drug of abuse. In A STOUDEMIRE & BS FOGEL (Eds.), Medical psychiatric practice (Vol 1; Ch. 16, pp. 541–596). Washington, DC: American Psychiatric Press.

GOLDSTEIN MG, NIAURA R, ABRAMS DB, ET AL. (1989). Effects of behavioral skills training and schedule of nicotine gum administration on smoking cessation. Am J Psychiatry 146:56–60.

GRABOWSKI J, & HALL SM (1985). Tobacco use, treatment strategies, and pharmacological adjuncts: an overview. In J GRABOWSKI & SM HALL (Eds.), Pharmacologic adjuncts in smoking cessation. NIDA Research Monograph 53. DHHS Publ. No. (ADM) 85-1333 (pp. 1–14). Rockville, MD: DHHS; Public Health Service; Alcohol, Drug Abuse, and Mental Health Administration.

GRUNBERG NE (1985). Nicotine, cigarette smoking, and body weight. Br J Addict 80:369–377.

HAJEK P, JACKSON P, & BELCHER M (1988). Long-term use of nicotine chewing gum. JAMA 260:1593–1596.

HALL SM, GINSBERG D, & JONES RT (1986). Smoking cessation and weight gain. J Consult Clin Psychol 54:342–346.

HALL SM, MUNOZ R, & REUS V (June 1990). Smoking cessation, depression and dysphoria. Paper presented at the annual scientific meeting of the Committee on the Problems of Drug Dependence, Richmond, VA.

HALL SM, MUNOZ R, & REUS V (1991). Depression and smoking treatment: a clinical trial of an affect regulation treatment. Paper presented at the annual scientific meeting of the Committee on the Problems of Drug Dependence, Puerto Rico.

HALL SM, TUNSTALL C, RUGG D, ET AL. (1985). Nicotine gum and behavioral treatment in smoking cessation. J Consult Clin Psychol 53:256–258.

HARACKIEWICZ JM, BLAIR LW, SANSONE C, ET AL. (1988). Nicotine gum and self-help manuals in smoking cessation: an evaluation in a medical context. Addict Behav 13:319–390.

HENNINGFIELD JE (1984). Pharmacologic basis and treatment of cigarette smoking. J Clin Psychiatry 45(12, Sec 2):24–34.

HENNINGFIELD JE (1986). How tobacco produces drug dependence. In JK OCKENE (Ed.), The pharmacologic treatment of tobacco dependence: proceedings of the World Congress, November 4–5, 1985 (pp. 19–31). Cambridge, MA: Institute for the Study of Smoking Behavior and Policy.

HENNINGFIELD JE, MIYASATO K, & JASINSKI DR (1985). Abuse liability and pharmacodynamic characteristics of intravenous and inhaled nicotine. J Pharmacol Exp Ther 234:1–12.

HERMAN S (1989). Surviving without cigarettes: a guide for patients in "smoke-free" hospitals. Durham, NC: Duke University Medical Center, Smokers' Consultation Service (919-684-6576).

HOBERMAN HM, & LEWINSOHN PM (1985). The behavioral treatment of depression. In E BECKHAM & W LEBER (Eds.), Handbook of depression. Homewood, IL: Dorsey Press.

HOLLON SD, & NAJAVITS L (1988). Review of empirical studies on cognitive therapy. In AJ FRANCES & R HALES (Eds.), Review of psychiatry (Vol 7). Washington, DC: American Psychiatric Press.

HOLROYD J (1980). Hypnosis treatment for smoking: an evaluative review. Int J Clin Exp Hypn 28:341–367.

HUGHES JR (1991). Combined psychological and nicotine gum treatment for smoking: a critical review. J Substance Abuse 3:337–350.

HUGHES JR, GUST SW, KEENAN RM, ET AL. (1989). Nicotine vs placebo gum in general medical practice. JAMA 261:1300–1305.

HUGHES JR, GUST SW, SKOOG K, ET AL. (1991). Symptoms of tobacco withdrawal: A replication and extension. Arch Gen Psychiatry 48:52–59.

HUGHES JR, & HATSUKAMI D (1986). Signs and symptoms of tobacco withdrawal. Arch Gen Psychiatry 43:289–294.

HUGHES JR, HATSUKAMI DK, MITCHELL JE, ET AL. (1986). Prevalence of smoking among psychiatric outpatients. Am J Psychiatry 143:993–997.

HUGHES JR, HATSUKAMI DK, & SKOOG KP (1986). Physical dependence on nicotine in gum: a placebo substitution trial. JAMA 255:3277–3279.

HUGHES JR, HIGGINS ST, & HATSUKAMI DK (1990). Effects of abstinence from tobacco. In LT KOZLOWSKI, HM ANNIS, HD CAPPELL, ET AL. (Eds.), Recent advances in alcohol and drug problems (Ch. 10, pp. 317–398). New York: Plenum Press.

HUNT WA, BARNETT LW, & BRANCH LG (1971). Relapse rates in addiction programs. J Clin Psychol 27:455–456.

HUNT WA, & BESPALEC DA (1974). An evaluation of current methods of modifying smoking behavior. J Clin Psychol 30:431–438.

HURT RD, LAUGHER GG, OFFORD KP, ET AL. (1990). Nicotine-replacement therapy with use of a transdermal nicotine patch: a randomized double-blind placebo-controlled trial. Mayo Clin Proc 65:1529–1537.

ISSAC L (1980). Clonidine in the central nervous system: site and mechanism of hypotensive action. J Cardiovasc Pharmacol 2:S5–S19.

ISTVAN J, & MATARAZZO J (1984). Tobacco, alcohol, and caffeine use: a review of their interrelationship. Psychol Bull 95:301–326.

JARVIK ME (1987). Does smoking decrease eating and eating increase smoking? In WR MARTIN, GR VAN LOON, ET IWAMOTO, ET AL. (Eds.), Tobacco, smoking and nicotine: a neurobiological approach (pp. 389–400). New York: Plenum Press.

JARVIK ME, & HENNINGFIELD JE (1988). Pharmacologic treatment of tobacco dependence. Pharmacol Biochem Behav 30:279–294.

JARVIK ME, & SCHNEIDER NG (1984). Degree of addiction and the effectiveness of nicotine gum therapy for smoking. Am J Psychiatry 141:790–791.

Joint Commission on Accreditation of Health Care Organizations (1991). AMH smoking policy revised. Joint Commission Perspectives 2:13.

KANDEL DB, & DAVIES M (1986). Adult sequelae of adolescent depressive symptoms. Arch Gen Psychiatry 43:255–262.

KLESGES RC, BROWN K, PASCALE RW, ET AL. (1987). Factors associated with participation, attrition, and outcome in a smoking cessation program at the workplace. Health Psychol 7:575–589.

KLESGES RC, & KLESGES LM (1988). Cigarette smoking as a dietary strategy in a university population. Int J Eating Disord 7:413–419.

KLESGES RC, MEYERS AW, KLESGES LM, ET AL. (1989). Smoking, body weight and their effects on smoking behavior: a comprehensive review of the literature. Psychol Bull 106:204–230.

KORNITZER M, KITTEL F, DRAMAIX M, ET AL. (1987). A double-blind study of 2 mg vs 4 mg nicotine chewing gum in an industrial setting. J Psychosom Res 31:171–176.

KOTTKE TE, BATTISTA RN, DEFRIESSE GH, ET AL. (1988). Attributes of successful smoking cessation interventions in medical practice. JAMA 259:2882–2889.

KOZLOWSKI LT, WILKINSON A, SKINNER W, ET AL. (1989). Comparing tobacco cigarette dependence with other drug dependencies: greater or equal 'difficulty quitting' and 'urges to use', but less 'pleasure' from cigarettes. JAMA 261:898–901.

KRISTELLER JL, & OCKENE JK (1987). Assessment and treatment of smoking on a consultation service. In R MICHELS & JO CAVENAR JR (Eds.), Psychiatry (Ch. 113, pp. 1–13). Philadelphia: Lippincott.

LAM W, SACKS HS, SZE PC, ET AL. (1987). Meta-analysis of randomized controlled trials of nicotine chewing-gum. Lancet 2:27–30.

LANDO HA (1986). Long-term modification of chronic smoking behavior: a paradigmatic approach. Bull Soc Psychol Addict Behav 5:5–17.

LANDO HA, & MCGOVERN PG (1985). Nicotine fading as a nonaversive alternative in a broad-spectrum treatment for eliminating smoking. Addict Behav 10:153–161.

LEON GR, KOLOTKIN R, & KORGESKI G (1979). MacAndrew addiction scale and other MMPI characteristics associated with obesity, anorexia and smoking behavior. Addict Behav 4:401–407.

LEVENTHAL H, & CLEARY PD (1980). The smoking problem: a review of the research and theory in behavioral risk modification. Psychol Bull 88:370–405.

LEVINSON BL, SHAPIRO D, SCHWARTZ GE, ET AL. (1971). Smoking elimination by gradual reduction. Behav Ther 2:477–487.

LICHTENSTEIN E (1986). Clinic based cessation strategies. In JK OCKENE (Ed.), The pharmacologic treatment of tobacco dependence: proceedings of the World Congress, November 4–5, 1985 (pp. 205–217). Cambridge, MA: Institute for the Study of Smoking Behavior and Policy.

LICHTENSTEIN E, & BROWN RA (1980). Smoking cessation methods: review and recommendations. In WR MILLER (Ed.), The addictive behaviors: treatment of alcoholism, drug abuse, smoking, and obesity (Ch. 4, pp. 169–206). Oxford: Pergammon Press.

LICHTENSTEIN E, & BROWN RA (1982). Current trends in the modification of cigarette dependence. In AS BELLACK, M HERSEN, & AE KAZDIN (Eds.), International handbook of behavior modification and therapy (Ch. 19, pp. 575–611). New York: Plenum.

LICHTENSTEIN E, & GLASGOW RE (1977). Rapid smoking: side effects and safeguards. J Consult Clin Psychol 45:815–821.

LICHTENSTEIN E, GLASGOW RE, & ABRAMS DB (1986). Social support and smoking cessation: in search of effective interventions. Behav Ther 17:607–619.

LICHTENSTEIN E, & MERMELSTEIN RJ (1986). Some methodological caution in the use of the tolerance questionnaire. Addict Behav 11:439–442.

MALCOLM R, CURREY HS, MITCHELL MA, ET AL. (1986). Silver acetate as a deterrent to smoking. Chest 90:107–111.

MALETZKY BM, & KLOTTER J (1974). Smoking and alcoholism. Am J Psychiatry 131:445–447.

MANHEM P, NILSSON LH, MOBERG AL, ET AL. (1985). Alcohol withdrawal: effects of clonidine treatment on sympathetic activity, the renin-aldosterone system, and clinical symptoms. Alcohol Clin Exp Res 9:238–243.

MARCUS BH, ALBRECHT AE, NIAURA RS, ET AL. (1991). Usefulness of physical exercise for maintaining smoking cessation in women. Am J Cardiol 68:406–407.

MARLATT GA, & GORDON JR (1980). Determinants of relapse: implications for the maintenance of behavior change. In PO DAVIDSON & SM DAVIDSON (Eds.), Behavioral medicine: changing health lifestyles. New York: Bruner/Mazel.

MARLATT GA, & GORDON JR (Eds.) (1985). Relapse prevention. New York: Guilford.

MOSS RA, & PRUE DM (1982). Research on nicotine regulation. Behav Ther 13:31–46.

National Cancer Institute (1988). Clearing the air: how to quit smoking and stay quit for keeps. Bethesda: National Cancer Institute, Office of Cancer Communications (1-800-4-CANCER).

National Cancer Institute (1989). How to help your patients stop smoking: a National Cancer Institute manual for physicians (NIH Publ No 90-3064). Washington, DC: US Government Printing Office (or National Cancer Institute, Office of Cancer Communications, 1-800-4-CANCER).

NIAURA R, GOLDSTEIN M, & ABRAMS D (1991). A bioinformational systems perspective on tobacco dependence. Br J Addict 86:593–597.

NIAURA RS, ROHSENOW DJ, BINKOFF JA, ET AL. (1988). Relevance of cue reactivity to understanding alcohol and smoking relapse. J Abnorm Psychol 97:133–152.

O'CONNELL KA, & SHIFFMAN S (1988). Negative affect smoking and smoking relapse. J Substance Abuse 1:25–33.

OCKENE JK (1987). Physician-delivered interventions for smoking cessation: strategies for increasing effectiveness. Prev Med 16:723–737.

O'FARRELL TJ, CONNORS GJ, & UPPER D (1983). Addictive behaviors among hospitalized psychiatric patients. Addict Behav 18:329–333.

ORLEANS CT (1985). Understanding and promoting smoking cessation: overview and guidelines for physician intervention. Annu Rev Med 36:51–61.

ORLEANS CT, ROTBERG HL, QUADE D, ET AL. (1990). A hospital quit-smoking consult service: clinical report and intervention guidelines. Prev Med 19:198–212.

ORNISH SA, ZISOOK S, & MCADAMS LA (1988). Effects of transdermal clonidine treatment on withdrawal symptoms associated with cigarette smoking: a randomized controlled trial. Arch Intern Med 148:2027–2031.

PERKINS KA, EPSTEIN LH, STILLER R, ET AL. (1986). An aerosol spray alternative to cigarette smoking in the study of the behavioral and physiological effects of nicotine. Behav Res Methods Instrum Comput 18:420–426.

PERT A, & CLARKE PBS (1987). Nicotine modulation of dopaminergic transmission: functional implications. In WR MARTIN, GR VAN LOON, ET IWAMOTO, ET AL. (Eds.), Tobacco, smoking and nicotine: a neurobiological approach (pp. 169–189). New York: Plenum Press.

PIERCE JP, FIORE MC, NOVOTNY TE, ET AL. (1989). Trends in smoking in the United States: projections to the year 2000. JAMA 261:61–65.

PINTO R, ABRAMS DB, MONTI PM, ET AL. (1987). Nicotine dependence and the likelihood of quitting smoking. Addict Behav 12:371–374.

POMERLEAU OF, & POMERLEAU CS (1984). Neuroregulators and the reinforcement of smoking: towards a biobehavioral explanation. Neurosci Biochem Rev 8:503–513.

POMERLEAU OF, & POMERLEAU CS (1987). A biobehavioral view of substance abuse and addiction. J Drug Issues 17:111–131.

PROCHASKA JO, & DICLEMENTE CC (1983). Stages and processes of self-change of smoking: toward an integrative model of change. J Consult Clin Psychol 51:390–395.

PROCHASKA JO, & DICLEMENTE CC (1986). Towards a comprehensive model of change. In WR MILLER & N HEATHER (Eds.), Treating addictive disorders: processes of change (pp. 3–27). New York: Plenum Press.

PROCHASKA JO, & GOLDSTEIN MG (1991). Process of smoking cessation: implications for clinicians. In JM SAMET & DB COULTAS (Eds.), Clin Chest Med 12:727–735.

RAUSCH JL, NICHINSON B, LAMKE C, ET AL. (1990). Influence of negative affect on smoking cessation treatment outcome: a pilot study. Br J Addict 85:929–933.

RODIN J (1987). Weight change following smoking cessation: the role of food intake and exercise. Addict Behav 12:303–317.

ROSE JE (1991). Transdermal nicotine and nasal nicotine administration as smoking-cessation treatments. In JA COCORES (Ed.), The clinical management of nicotine dependence (Ch. 16, 196–207). New York: Springer-Verlag.

ROUNSAVILLE BJ, KOSTEN TR, WEISSMAN MM, ET AL. (1985). Evaluating and treating depressive disorders in opiate addicts. DHHS Publ. No. (ADM) 85-1406. Washington, DC: US Government Printing Office.

Russell MAH (1987). Nicotine intake and its regulation by smokers. In WR Martin, GR Van Loon, ET Iwamoto, et al. (Eds.), Tobacco, smoking and nicotine: a neurobiological approach (pp. 25–50). New York: Plenum Press.

Russell MAH, Jarvis MJ, Feyerabend C, et al. (1983). Nasal nicotine solution: a potential aid to giving up smoking? BMJ 286:683–684.

Russell MAH, Jarvis MJ, Sutherland G, et al. (1987). Nicotine replacement in smoking cessation: absorption of nicotine vapor from smoke-free cigarettes. JAMA 257:3262–3265.

Russell ML (1986). Behavioral counseling in medicine: strategies for modifying at-risk behavior. New York: Oxford University Press.

Russell PO, & Epstein LH (1988). Smoking. In EA Blechman & KD Brownell (Eds.), Handbook of behavioral medicine for women (pp. 369–383). New York: Pergamon Press.

Schacter S (1979). Pharmacological and psychological determinants of smoking. Ann Intern Med 88:104–114.

Schwartz JL (1987). Review and evaluation of smoking cessation methods: the United States and Canada, 1978–1985. Bethesda: National Institutes of Health.

Shiffman SM (1979). The tobacco withdrawal syndrome. In KA Krasnegor (Ed.), Cigarette smoking as a dependence process. NIDA Research Monograph 23. DHHS Publication No. (ADM) 84-800 (Ch. 13, pp. 158–185). Washington, DC: U.S. Government Printing Office.

Shiffman S (1982). Relapse following smoking cessation: a situational analysis. J Consult Clin Psychol 50:71–86.

Shiffman S (1984). Coping with temptations to smoke. J Consult Clin Psychol 52:261–267.

Shiffman S (1985). Coping with temptations to smoke. In TA Wills & S Shiffman (Eds.), Coping and substance use (pp. 223–242). Orlando, FL: Academic Press.

Shiffman S (1986). A cluster-analytic typology of smoking relapse episodes. Addict Behav 11:295–307.

Shiffman S (1989). Tobacco "chippers"—individual differences in tobacco dependence. Psychopharmacology 97:539–547.

Sobell LC, Sobell MO, Kozlowski L, et al. (1990). Alcohol or tobacco research versus alcohol and tobacco research. Br J Addict 85:263–269.

Stokes J, & Rigotti NA (1988). The health consequences of cigarette smoking and the internist's role in smoking cessation. Adv Intern Med 33:431–460.

Stolerman IP (1986). Could nicotine antagonists be used in smoking cessation? Br J Addict 81:47–53.

Streator JA, Sargent RC, & Ward DC (1989). A study of factors associated with weight change in women who attempt smoking cessation. Addict Behav 14:523–530.

Taylor CB, Houston-Miller N, Killen JD, et al. (1990). Smoking cessation after acute myocardial infarction: effects of a nurse-managed intervention. Ann Intern Med 113:118–123.

Tiffany ST, Martin EM, & Baker TB (1986). Treatments for cigarette smoking: an evaluation of the contributions of aversion and counseling procedures. Behav Res Ther 24:437–452.

Tompkins S (1966). Psychological model for smoking behavior. Am J Public Health 56:17–20.

Tonnesen P, Fryd V, Hansen M, et al. (1988a). Effect of nicotine chewing gum in combination with group counseling on the cessation of smoking. N Engl J Med 318:15–18.

Tonnesen P, Fryd V, Hansen M, et al. (1988b). Two and four mg nicotine chewing gum in combination with group counseling in smoking cessation: an open, randomized, controlled trial with a 22 month follow-up. Addict Behav 13:17–27.

Tonnesen P, Norregaard J, Simonsen K, et al. (1991). A double-blind trial of a 16-hour transdermal nicotine patch in smoking cessation. N Engl J Med 325:311–315.

Transdermal Nicotine Study Group (1991). Transdermal nicotine for smoking cessation: six-month results from two multicenter controlled clinical trials. JAMA 266:3133–3138.

U.S. Department of Health and Human Services (1985). The health consequence of smoking: cancer and chronic lung disease in the workplace: a report of the Surgeon General. Rockville, MD: U.S. Department of Health and Human Services, Public Health Service, Office on Smoking and Health.

U.S. Department of Health and Human Services (1988). The health consequences of smoking: nicotine addiction: a report of the Surgeon General. Rockville, MD: U.S. Department of Health and Human Services, Public Health Service, Office on Smoking and Health.

U.S. Department of Health and Human Services (1989). Reducing the health consequences of smoking: 25 years of progress: a report of the Surgeon General. Rockville, MD: U.S. Department of Health and Human Services, Public Health Service, Office on Smoking and Health.

U.S. Department of Health and Human Services (October 1990). Nurses: help your patients stop smoking. NIH Publ. No. 90-2962. Washington, DC: Department of Health and Human Services, Public Health Service, National Institutes of Health.

Waal-Manning HJ, & de Hamel FA (1978). Smoking habit and psychometric scores: a community study. NZ Med J 88:188–191.

Walton RG (1972). Smoking and alcoholism: a brief report. Am J Psychiatry 128:1455–1456.

Washton AM, & Resnick RB (1981). Clonidine in opiate withdrawal: review and appraisal of clinical findings. Pharmacotherapy 1:140–146.

Weber MA, & Drayer JIM (1984). Clinical experience with rate-controlled delivery of antihypertensive therapy by a transdermal system. Am Heart J 108:231–236.

Wei H, & Young D (1988). Effect of clonidine on cigarette cessation and in the alleviation of withdrawal symptoms. Br J Addict 83:1221–1226.

West RJ, Hajek P, & Belcher M (1989). Severity of withdrawal symptoms as a predictor of outcome of an attempt to quit smoking. Psychol Med 19:981–985.

West RJ, & Russell MAH (1985a). Pre-abstinence smoke intake and smoking motivation as predictors of severity of cigarette withdrawal symptoms. Psychopharmacology 87:334–336.

West RJ, & Russell MAH (1985b). Effects of withdrawal from long-term nicotine gum use. Psychol Med 15:891–893.

West RJ, & Russell MAH (1986). Dependence on nicotine chewing gum. JAMA 256:3214–3215.

Wilkins AJ, Jenkins WJ, & Steiner JA (1983). Efficacy of clonidine in treatment of alcohol withdrawal state. Psychopharmacology 81:78–80.

Wills TA, & Shiffman S (1985). Coping and substance use: A conceptual framework. In TA Wills & S Shiffman (Eds.), Coping and substance use (Ch. 1, pp. 3–24). Orlando, FL: Academic Press.

43 Assessment, classification, and treatment of obesity: behavioral medicine perspective

MATTHEW M. CLARK, Ph. D., LAURIE RUGGIERO, Ph. D., VINCENT PERA, JR., M. D., MICHAEL G. GOLDSTEIN, M. D., AND DAVID B. ABRAMS, M. D

Obesity, an excess of body fat, is a common and significant health hazard (Peterson, Rothschild, Weinberg, et al., 1988). Obesity is associated with hypertension, diabetes, hypercholesterolemia, respiratory failure, hypoventilation syndrome, gallbladder disease, certain forms of cancer, arthritis, gout, and sleep apnea (Bray, 1985; National Institutes of Health Consensus Development Panel, 1985). Obesity is a risk factor for cardiovascular disease in both men (Hubert, Feinleib, McNamara, et al., 1983) and women (Manson, Colditz, Stampfer, et al., 1990). In particular, moderate to severe obesity is associated with an increasing prevalence of medical illness (Blackburn & Kanders, 1987).

This chapter focuses on adult obesity, although a literature exists on the special needs and treatment of childhood and adolescent obesity (Epstein & Squires, 1988).

Approximately 25 to 50% of the adult populations in affluent societies are obese (Roberts, Savage, Coward, et al., 1988). In the United States estimates of the prevalence of obesity in adults ranges from 15% to 50% (Bray, 1976; Van Itallie, 1979). As many as 60 million Americans are obese and over 30 billion dollars are spent annually on weight loss measures (Atkinson, 1990).

The assessment and treatment of obesity is not clear-cut. Enormous commercial opportunities have led to serious concerns about inappropriate and expensive treatments. Sociocultural pressures for thinness, equated with success and happiness, are embedded in our norms and in the mass media and marketing of numerous products from dishwashing liquid to children's toys and dolls (Abrams, 1991; Abrams, Emmons, Niaura, et al., 1991; Garner, Rockert, Olmstead, et al., 1985; Wachtel, 1989). Questions have been raised about whether the medical and scientific community has been unduly influenced by these pressures and societal norms (Garner & Wooley, 1991).

Cultural pressures influence individual behaviors and attitudes regarding food and eating. Many eating disorders, from anorexia and bulimia nervosa to binge eating among the obese, are driven in part by the cultural imperative to "have the best body" or to be perfect. Especially troublesome is the individual expectation that weight reduction is a panacea for dysphoria: poor self-esteem, marital and interpersonal conflicts, mood disturbance, sexual conflicts or posttraumatic disorders such as sexual abuse during early adolescence or childhood (Silverstein, Perdue, Peterson, et al., 1986). Obesity is therefore a complex condition with genetic-biologic, cognitive-behavioral, and sociocultural-environmental factors implicated in its etiology. Caution therefore should be exercised to ensure that all of the known factors are considered and that scientific evidence is not too strongly influenced by the *zeitgeist* of cultural and commercial values of our modern Western industrial society.

In our opinion, obesity is a condition that responds best to an interdisciplinary approach to assessment and treatment. Increasingly, expert consensus is that the more severe forms of obesity should be treated in a clinic using a team of specialists (Wadden, Foster, Letizia, et al., 1990). At a minimum, the team should be comprised of a psychologist or psychiatrist specializing in behavioral medicine, including the psychology of eating behaviors and eating disorders, an internist, an exercise physiologist, and a nutritionist. This chapter offers a systematic approach to the assessment and the treatment of obesity such as might be found within such an interdisciplinary clinic.

Assessment of Obesity

Direct and Indirect Measurement of Body Fat

The only direct way of measuring excess body fat is at autopsy. There is no consensus on the standard for measurement of body fat in a living individual (Kraemer, Berkowitz, & Hammer, 1990). The most accurate method of assessing body fat in a living individual is by hydrostatic weighing (Wadden, 1985). However, this procedure is complicated and unavailable to most practitioners. Other laboratory measures include ultrasonography, computed tomography (CT), magnetic resonance imaging (MRI), and electric impedance. These laboratory methods are expensive, time-consuming, and cumbersome (Marshall, Hazlett, Spady, et al., 1990). Their value in predicting health outcomes is largely unknown in comparison to more practical measures of obesity.

Practical Measurement of Body Fat

Common measurements include body fat distribution, waist/hip ratio, weight in pounds, Metropolitan relative weight, body mass index, skinfold thickness, and body circumference. The two most practical for the clinician are the Metropolitan relative weight and the body mass index.

The most widely used weight standards in America are those provided by the Metropolitan Life Insurance Company. The tables were developed in 1959, were revised in 1983 (see Table 43-1) and are based on gender, height, and body frame (Andres, Elahi, Tobin, et al., 1985; MLIC, 1959, 1984). The 1959 tables were used in the Framingham Heart Study. Metropolitan Relative Weight (MRW) is based on the midpoint of the desirable weight range for medium frame for the appropriate sex and height of the subject.

Body mass index (BMI), defined as a subject's weight in kilograms divided by the square of the height in meters (kg/m^2), is an acceptable and practical measure of obesity (Kraemer, Berkowitz, & Hammer, 1990). Blackburn and Kanders (1987) reviewed obesity treatment research and concluded that a change of 5 units on the BMI scale improves heart function, blood pressure, glucose tolerance, sleep disorders, respiratory function, lipid profile, and requirements for medication in 90% of obese patients who are in treatment studies.

TABLE 43–1. *1983 Metropolitan Height and Weight Tables*

Height				
Feet	Inches	Small Frame	Medium Frame	Large Frame
Men				
5	2	128–134	131–141	138–150
5	3	130–136	133–143	140–153
5	4	132–138	135–145	142–156
5	5	134–140	137–148	144–160
5	6	136–142	139–151	146–164
5	7	138–145	142–154	149–168
5	8	140–148	145–157	152–172
5	9	142–151	148–160	155–176
5	10	144–154	151–163	158–180
5	11	146–157	154–166	161–184
6	0	149–160	157–170	164–188
6	1	152–164	160–174	168–192
6	2	155–168	164–178	172–197
6	3	158–172	167–182	176–202
6	4	162–176	171–187	181–207
Women				
4	10	102–111	109–121	118–131
4	11	103–113	111–123	120–134
5	0	104–115	113–126	122–137
5	1	106–118	115–129	125–140
5	2	108–121	118–132	128–143
5	3	111–124	121–135	131–147
5	4	114–127	124–138	134–151
5	5	117–130	127–141	137–155
5	6	120–133	130–144	140–159
5	7	123–136	133–147	143–163
5	8	126–139	136–150	146–167
5	9	129–142	139–153	149–170
5	10	132–145	142–156	152–173
5	11	135–148	145–159	155–176
6	0	138–151	148–162	158–179

Source of basic data 1979 Build Study Society of Actuaries and Association of Life Insurance Medical Directors of America 1980. From *Metropolitan Life Insurance Statistical Bulletin*, 1983, p. 2. Reprinted with permission.

Kraemer, Berkowitz, and Hammer (1990) concluded that BMI was a better predictor of mortality in the Framingham Study than were the Metropolitan Life Tables. These authors stated that BMI is arguably the best pragmatic measure available to the clinician owing to its accessibility, reliability, and validity. The National Institutes of Health consensus conference on obesity (1985) also recommended use of BMI as the standard for the measurement of obesity. Kraemer and colleagues (1990) cited evidence that the BMI should be in the range of 21 to 27. Blackburn and Kanders (1987) recommended achieving a BMI of less than 30.

Classification of Obesity

Stunkard (1984) proposed a classification system for obesity characterized by body weight as follows: mild obesity for those 20 to 40% overweight, moderate obesity for those 41 to 100% overweight, and severe obesity for those more than 100% overweight. This classification system is used in this chapter when reviewing treatment recommendations.

CAUSES OF OBESITY

Genetic Factors

The probability of an individual being obese is related to the presence of obesity in that individual's family (Charney, Goodman, McBride, et al., 1976; Wadden, 1985). These findings, however, do not provide a direct assessment of the importance of genetic factors, as these children experienced both genetic and environmental influences. Several studies were designed to sort out the relation between genetic and environmental factors in the development of obesity. Stunkard and colleagues reported that the concordance rate for BMI was twice as high for monozygotic twins as for dizygotic twins (Stunkard, Foch, & Hrubec, 1986); that identical twins are more similar than fraternal twins on BMI regardless of whether they were reared together or apart (Stunkard, Harris, Pedersen, et al., 1990); and that there is a significant relation between the weight class of adult adoptees and the BMI of their biologic parents, but there is no relation between the weight class of adult adoptees and their adoptive parents' BMI (Stunkard, Sorenson, Harris, et al., 1986). Stunkard, Harris, Pederson, and colleagues (1990), after reviewing the literature, concluded that genetic factors may account for as much as 70% of the variance in BMI and that combined environmental factors may account for about 30% of the variance. Regardless of the accuracy of these estimates, it is clear that genetic factors play an important role in the etiology of obesity.

In a study examining the potential contribution of reduced energy expenditure to the development of obesity, Ravussin and colleagues reported that reduced 24-hour energy expenditure, measured in a respiratory chamber, was correlated to weight gain over a 2-year period and there was familial effect on 24-hour energy expenditure (Ravussin, Lillioja, Knowler, et al., 1988). These findings provide additional support to a physiologic influence on obesity and suggest that energy efficiency may be inherited and that high metabolic efficiency may be a risk factor for the development of obesity.

Energy Balance

An individual's weight status is a function of the amount of calories consumed (intake) and the amount of calories burned (output). Positive energy balance exists when more calories are consumed than are burned, and this imbalance may result in obesity if the imbalance is maintained over a period of time (Poehlman, Tremblay, Despres, et al., 1986). Small imbalances in this equation can lead to the development of obesity. We gain 1 pound of fat whenever we consume 3500 calories more than we burn. Overeating 100 calories per day results in a 1 pound weight gain after 5 weeks, or about a 10 pound weight gain over a year, or a 100 pound weight gain over a decade. To underscore how easily it can happen, consider that two cookies, one banana, or 6 oz of juice are examples of food selections that contain about 100 calories.

Energy Intake

Studies of caloric intake differences between lean versus obese subjects have reported conflicting results (Kromhout, 1983). Researchers have examined both total caloric amount and caloric content (fat versus protein versus carbohydrate). Dreon and colleagues (Dreon, Frey-Hewitt, Ellsworth, et al., 1988) reported that total caloric intake was not related to body fat but that percent body fat correlated positively with fat intake and negatively with carbohydrate and plant protein intake. Fat content therefore appears to be a more important factor than total caloric intake in the development or maintenance of obesity.

Energy Expenditure

Obese individuals may be less physically active than lean individuals. In a study involving naturalistic observation of stairs usage versus escalator usage, Brownell, Stunkard, and Albaum (1980) reported that obese individuals used the stairs less frequently than nonobese individuals. However, in a literature review of physical activity level in obesity, Bray (1990) concluded that obese individuals may be either as active or less physically active than lean individuals, and that research has failed to demonstrate consistent findings on this question.

Energy Intake and Energy Expenditure

Obese individuals may exercise less and consume a greater number of calories from fat than do lean individuals (Romieu, Willett, Stampfer, et al., 1988). However, whether these findings are a cause or a consequence of obesity remains to be answered. Physiologically, obese individuals may be metabolically inefficient and thus predisposed to accumulate body fat. Therefore, as noted previously, genetic, biologic, psychological, and socioenvironmental factors, including energy intake, availability of specific food types, and activity level, interact to determine whether a given individual will become obese. After reviewing genetic and environmental influences on obesity, Stunkard and colleagues concluded that genetics determine the presence of obesity, and environmental, psychological, and sociocultural factors determine the severity of obesity.

Psychological Factors Influencing Obesity

Maladaptive eating behaviors and psychological factors probably are both causes and consequences of obesity in a cycle that is difficult to tease apart. Comorbid psychological factors, such as eating disorders, may influence the weight status of a significant subgroup (up to 30%) of obese individuals who present for treatment (Agras, 1991). However, the prevalence of eating disorders in the entire obese population (a community or population-based epidemiologic sample) probably is lower than that reported in treatment samples (Wilson, Nonas, & Rosenblum, 1992). This section reviews the role that psychological factors play in the etiology of obesity.

In the past, investigators proposed psychodynamic explanations (Wise & Gordon, 1977) and general personality traits (McReynolds, 1982; Moore & Rodin, 1986) as causes of obesity. Neither of these theories has enjoyed strong empiric support, but some of these ideas may simply be difficult to research using traditional scientific methods. Obese children do not score differently than normal weight peers on general psychological measures (Sallade, 1973), and studies of obese adults have reported similar results (Halmi, Long, Stunkard, et al., 1980; Holland, Masling, & Copley, 1970; Weinberg, Mendelson, & Stunkard, 1961).

Investigators have proposed that either an eating disorder or specific types of psychopathology may be important psychological factors in a subgroup of the obese in treatment settings (Agras, 1991). Binge eating is defined as a rapid consumption of a large amount of food during a discrete period of time. Obese bingers (proposed binge eating disorder) appear to differ from obese nonbingers in terms of cognitions, psychopathology, and treatment outcome (Wing & Marcus, 1992). Estimates of binge eating in an obese population seeking treatment for weight control range from 20% to 55% (Telch, Agras, Rossiter, et al., 1990); and as binge severity increases so does psychological disturbance (Kolotkin, Revis, Kirkley, et al., 1987). Marcus and colleagues (Marcus, Wing, Ewing, et al., 1990) reported that 60% of obese binge eaters seeking treatment for weight control met criteria for psychiatric comorbidity compared to a 28% comorbidity rate in obese nonbingers. Kolotkin and colleagues (Kolotkin, Revis, Kirkley, et al., 1987) found that severity of binge episodes was associated with increased psychological disturbance on the Minnesota Multiphasic Personality Inventory (MMPI) (Dahlstrom, Welsh, & Dahlstrom, 1972). Obese binge eaters undergoing weight reduction treatment lose less weight than nonbinge eaters (Keefe, Wychogrod, Weinberger, et al., 1984), and binge eaters are more likely to drop out of treatment than are nonbinge eaters (Marcus, Wing, & Hopkins, 1988).

Some have suggested that obese individuals are unsuccessful dieters. It is estimated that 44% of women and 25% of men in the United States are trying to lose weight (Blackburn, Wilson, Kanders, et al., 1989). Marlatt and Gordon (1985) have proposed that when individuals are trying to abstain from a substance and then use that substance they experience a disruption in their self-control that leads to further substance use. They have called this phenomenon the abstinence violation effect. Obese individuals who are dieting may be more likely to overeat after experiencing a "slip." Therefore obese individuals may consume more calories after unsuccessfully restricting their intake. A cycle of restriction, violation, and disinhibition may ensue, causing the individual to become more obese.

Moreover, a subset of obese individuals meet the criteria for bulimia nervosa, as defined by *DSM-III-R* (APA, 1987). In addition to binge eating, these individuals also (1) experience a feeling of lack of control over eating behavior during binge episodes; (2) regularly engage in other potentially abusive activities to prevent weight gain (i.e., self-induced vomiting, use of laxatives or diuretics, strict dieting or fasting, or vigorous exercise); and (3) exhibit persistent overconcern with body shape and weight (APA, 1987). It has been estimated that 5% of obese persons in clinical populations, primarily women, meet the diagnostic criteria for bulimia nervosa (Stunkard, 1989). A larger percentage of obese individuals in clinical settings, approximately 10%, suffer from the night-eating syndrome, characterized by morning anorexia, evening hyperphagia, and insomnia (Stunkard, 1989). Patients with this syndrome also have carbohydrate craving and depressive symptoms and may have a depressive syndrome mediated by serotonin (Stunkard, 1989; Wurtman, 1988).

SHOULD OBESITY BE TREATED?

For most individuals, obesity is a health risk factor that warrants treatment. However, prior to discussing treatment recommendations, we consider a review that questions whether obesity should be treated. In this selective review, Garner and Wooley (1991) questioned the appropriateness of treatment of obesity because of the generally poor results of most interventions in achieving long-term success. They argued that the stigmatization of obesity, overstatement of the health risks associated with obesity, and the pervasive influence of the weight loss industry have combined to maintain public demand and have slowed the integration of scientific findings to the contrary (Garner & Wooley, 1991).

They also pointed out that because weight loss is difficult to achieve alternatives to improving the life style of the obese should be considered, such as changing other health risk factors (e.g., smoking, dietary fat/fiber, salt consumption, body image and self-esteem concerns, counseling or psychotherapy, and increasing physical activity, among others). Indeed, a major study (Blair, Kohl, Paffenbarger, et al., 1989) of all-cause mortality in 10,224 men found that even moderate exercise "had a (tremendous) beneficial health effect in the obese."

One can also consider the possible iatrogenic effects of obesity treatment, such as the danger of weight cycling (Ernsberger & Nelson, 1988, Stunkard & Penick, 1979); adverse effects on acute high density lipoprotein (HDL) to total cholesterol changes (Follick, Abrams, Smith, et al., 1984), psychological risks associated with repeated failure to maintain weight loss (Wadden, Stunkard, & Liebschutz, 1988), creation or exacerbation of binge-eating and other eating disorders (Foreyt, 1987; Telch, Agras, & Rossiter, 1988), and a tendency to stop exercising once weight relapse has occurred (Garner & Wooley, 1991).

In our opinion, the main controversy in this area concerns at what level of obesity and in what subpopulations of the obese should treatment be recommended, rather than whether obesity should be treated at all. In individuals who are only 5 to 20% overweight, it is important to consider whether the individual would benefit more from participation in a weight loss program or from making other life style changes (e.g., increasing self-esteem or activity level). There is a positive and increasingly strong association between the degree of severity of obesity, other risk factors (e.g., hypertension, type II diabetes) and disease morbidity and mortality (Jeffrey, 1991). Many patients do benefit from treatment, and some maintain their weight loss. Thus we recommend that patients with moderate or severe obesity, or those with mild obesity plus other risk factors, be offered a trial of formal treatment. However, relative benefits and risks associated with attempted treatment versus nontreatment of obesity have not been adequately researched at this time, so clinicians should be appropriately skeptical when making treatment recommendations.

TREATMENT OF OBESITY

The National Institutes of Health (1985) recommended weight reduction for individuals with a body weight in excess of 20% of desirable weight; and Blackburn and Kanders (1987) recommended achieving a BMI of less than 30. We recommend that obesity evaluation and treatment be conducted within a behavioral medicine framework or in an equivalent clinical program. Program components should include, at least, medical supervision, cognitive-behavior therapy, nutritional education, and exercise instruction. The intake evaluation should include an interdisciplinary case conference.

Recommended treatment for obesity frequently has been based on its severity: mild (20 to 40% overweight); moderate (41 to 100% overweight); or severe (more than 100% overweight). Table 43-2 displays the recommended treatments for each class of obese patients.

Mild Obesity

Balanced-Deficit Diet

A balanced-deficit diet provides nutritionally balanced daily meals with a caloric intake level below the individual's caloric expenditure level. The only component that varies over time is the caloric level, which depends primarily on the person's current weight level; the nu-

TABLE 43–2. *Recommended Treatment Modalities by Obesity Classification*

Treatment	Mild	Moderate	Severe
Nutrition education	Yes	Yes	Yes
BDD	Yes	Yes[a]	Yes[b]
VLCD	No	Yes[a]	Yes[c]
Behavior therapy	Yes	Yes	Yes
Exercise	Yes	Yes	Yes
Medication	No	?No	?No
Surgery	No	No	Yes

BDD = balanced deficit diet; VLCD = very-low-calorie diet.
[a]Choice of specific diet should be based on level of obesity within a moderate range, as well as on other patient–treatment matching factors.
[b]Used with certain patients such as high risk.
[c]Dietary treatment of choice.

tritional content of the diet remains essentially the same. The level of caloric deficit should be determined in consultation with a dietitian and based on established dietary guidelines. Decreasing daily caloric intake by 500 Kcal below that necessary for weight maintenance generally yields a 1 lb weight loss per week. It is generally recommended that mildly obese individuals lose 1.0 to 1.5 lb per week (Brownell & Wadden, 1991).

Behavioral Approaches

A behavioral approach, also referred to as cognitive-behavior therapy or behavior modification, represents one of the most widely used approaches for weight loss (Brownell & Wadden, 1986). It is, in fact, the central core around which all the other components are built (Abrams, 1983; Wadden & Brownell, 1986). This approach has its conceptual origin in learning theory and in cognitive-behavioral social learning theory (Abrams, 1983; Bandura, 1986; Brownell & Wadden, 1986).

The basic premise of the behavioral approach is that, in order to lose weight and keep it off, individuals must gradually replace current maladaptive eating and exercise behaviors with new healthy eating and exercise habits that can be incorporated into their life styles and maintained indefinitely. Usual components of a behavioral program include self-monitoring (keeping a daily record) of food or calorie intake (or both), stimulus control procedures (i.e., rearranging the environment in order to eliminate food cues), and goal setting and self-reinforcement to help shape new behaviors (e.g., slower eating). Programs that are more comprehensive include other components such as cognitive restructuring (e.g., changing maladaptive thinking regarding dieting), an emphasis on a gradual increase and then maintenance of physical activity, assertiveness training focused on weight loss issues (e.g., food refusal), facilitation of social support for weight loss efforts, and relapse prevention training (Foreyt, 1987). Abrams (1983) reviewed the clinical techniques that form the core behavior modification approach, and Brownell and Wadden (1986) described a weight loss program that included five primary components: behavioral techniques, exercise, cognitive or attitude change, social support, and nutritional education.

Nutritional Education

The importance of including a nutritional education component in all treatment programs is underscored by research suggesting that the high fat content of the American diet may be an important contributor to the high prevalence of obesity in the country (Wadden & Brownell, 1984). An ideal nutritionally balanced diet consists of approximately 50 to 55% of calories from carbohydrates, no more than 30% from fat, and 15 to 20% from protein (Wadden, Foster, Letizia, et al., 1990). Such a diet should not be viewed as a "diet" that one "goes on and off" but, instead, as a life style change that can be maintained forever (Kayman, Bruvold, & Stern, 1990).

Increasing Physical Activity Level

Physical activity involves structured exercise and life style activity. A typical exercise prescription for weight management is a minimum of 30 minutes of aerobic exercise at least three times per week. Exercise is important to weight loss in many ways: Exercise increases energy expenditure and basal metabolism (Thompson, Jarvie, Lahey, et al., 1982), suppresses appetite (Epstein, Masek, & Marshall, 1978), counteracts some of the deleterious effects of obesity (e.g., improves cardiac efficiency) (Dowdy, Cureton, DuVal, et al., 1985), and minimizes the loss of lean tissue during weight loss (Pavlou, Steffee, Lerman, et al., 1985). Furthermore, exercise has been identified as one of the few predictors of success in weight loss programs (Brownell, 1984).

Exercise is also important in the long-term maintenance of weight loss. In one study, individuals who were treated with diet plus exercise demonstrated maintenance of weight loss for 18 months while those treated with diet alone regained weight to baseline levels (Pavlou, Krey, & Steffee, 1989).

Diet plus increased life style activity is as effective as diet plus programmed exercise in the short term and more effective over the long term (Epstein, Wing, Koeske, et al., 1982). It is therefore recommended that obese individuals be encouraged not only to increase their energy expenditure through regular structured exercise, (e.g., walking, swimming, aerobics) but also to increase their level of life style activity (e.g., using stairs rather than elevators, walking to the corner store instead of driving). Patients should keep up exercise regardless of their response to obesity treatment (Blair, Kohl, Paffenbarger, et al., 1989; Garner & Wooley, 1991).

Treatment Efficacy

To date, more than 150 studies have demonstrated the efficacy of behavioral weight loss programs in treating individuals with mild and moderate obesity (Brownell & Wadden, 1986). Studies on the efficacy of comprehensive behavioral approaches have indicated that such programs yield steady weight losses of 1 to 2 lb per week during treatment (Brownell, 1982). Behavioral programs that last at least 15 weeks and result in weight losses of at least 20 lb may be sufficient for mildly obese individuals. It is noteworthy that the short-term effectiveness of behavioral approaches has improved since the early 1970s (Abrams, 1983; Brownell & Wadden, 1986). Combination programs (e.g., behavior therapy plus exercise) produce better results than unimodal approaches (Dahlkoetter, Callahan, & Lanton, 1979). The average weight loss at 1-year followup is approximately 10 to 11 lb (Brownell, 1982).

Factors associated with maintenance of weight loss include (1) consistent exercise after treatment; (2) restriction of caloric intake after treatment; (3) participation in further weight loss programs or further use of diets after treatment; (4) regular weighing after treatment; (5) client-therapist contact after treatment; (6) problem-solving training during treatment; and (7) cognitive relapse prevention training (Westover & Lanyon, 1990). A focus on techniques unique to maintenance of weight loss (rather than techniques used for initial weight loss) can increase the chance of maintaining losses after treatment (Abrams & Follick, 1983). Maintenance-related components included gradual fading out of contact over 2 months rather than abrupt termination, relapse prevention training using a "planned minirelapse homework assignment," problem-solving training, and regular weighing.

Moderate Obesity

The behavior modification approach is the core foundation of all levels of treatment (mild, moderate, severe). However, individuals who are moderately obese and therefore need to lose 50 to 100 lb usually require more aggressive caloric restriction treatment than is offered by a balanced-deficit diet (Brownell & Wadden, 1991). Conventional balanced-deficit diets produce weight loss that is too slow, whereas total fasting (abstinence from all caloric intake) produces large weight losses but is associated with significant risks such as nitrogen loss, ketonuria, hyperuricemia, reduction in lean body mass, electrolyte depletion, and glycogen depletion (Runcie & Thompson, 1970).

Very-low-calorie diets (VLCDs) are defined as diets containing less than 800 Kcal per day (Life Sciences Research Office, 1979). The latest formulations of these diets are much safer than the original products developed during the 1970s, but they are still associated with some risks and side effects. They allow rapid weight loss with acceptable medical risks, by providing sufficient protein to maximize fat loss and minimize the loss of lean body mass (Wadden, Stunkard, & Brownell, 1983). VLCDs reduce daily intake to 400 to 800 calories per day and provide an average weight loss of 2 to 5 lb per week in women and 3 to 6 lb per week in men (Matz, 1987; Wadden, Stunkard, & Brownell, 1983).

In contrast to the "liquid protein diets" (LPDs) used during the 1970s, VLCDs that provide high quality protein are considered safe when used appropriately and under close medical supervision (see Medical Evaluation section). In more than 10,000 cases reported in one study (Patton, Amatruda, Biddle, et al., 1981), no diet-related fatalities were identified and no increase in cardiac abnormalities was found with 24-hour Holter monitoring. Prior to participating in a VLCD, patients must undergo a thorough medical evaluation by a physician trained in administering VLCDs (Wadden, Van Itallie, & Blackburn, 1990).

There are currently two primary types of VLCDs available: the protein-sparing modified fast and the liquid protein diet (Wadden, Stunkard, Brownell, et al., 1985). The food version of protein-sparing modified fast provides 1.5 g of protein per kilogram of ideal body weight (70 to 100 g). Protein is obtained from lean meat, fish, and fowl (Blackburn, Bistrian, & Flatt, 1975; Flatt & Blackburn, 1974). Dietary intake must be supplemented with vitamins and minerals, especially calcium, potassium, and sodium (Wadden, Stunkard, Brownell, et al., 1985). Liquid protein diets are commercially prepared and provide 33 to 70 g of protein, 30 to 45 g of carbohydrate, and approximately 2 g of fat. Furthermore, patients on VLCDs are required to consume a minimum of 2 liters of noncaloric fluid per day. Three to six weeks of "refeeding" is provided following use of a VLCD. Refeeding involves the gradual reintroduction of conventional foods.

As mentioned previously, it is recommended that VLCDs be used in a clinical setting that provides the complete interdisciplinary team approach including a behavioral specialist, internist, dietitian, and exercise physiologist (Wadden, Van Itallie, & Blackburn, 1990). Such a comprehensive behavioral medicine approach ensures inclusion of ongoing attention to psychological and eating behavior issues, medical monitoring, nutritional education, exercise instruction, and cognitive-behavior therapy techniques for weight loss and relapse prevention. Research has demonstrated that VLCDs generally are safe when medically supervised and used for periods of 16 weeks or less (Wadden, Van Itallie, & Blackburn, 1990).

Although most patients can lose large amounts of weight using VLCDs alone, maintenance of weight loss is a problem with their use (Genuth, Vertes, & Hazelton, 1978; Wadden, Stunkard, & Liebschutz, 1988). A controlled trial (Wadden & Stunkard, 1986) that compared VLCD alone, behavior therapy alone, and the combination of these approaches found average 1-year follow-up weight losses of 4.7 kg (±7.3), 9.5 kg (±6.7), and 12.9 kg (±9.3), respectively. Furthermore, 29% of the subjects in the combined-treatment group maintained weight losses within 2 kg of their end-of-treatment weight compared with 0% in the VLCD alone group and 44% in the behavior therapy alone group. Overall, programs that have used a combination of VLCD and behavioral treatment have shown promise for both weight loss and maintenance (Genuth, Castro, & Vertes, 1974; Wadden & Stunkard 1986; Wadden, Stunkard, & Liebschutz, 1988). Given the short-term efficacy of comprehensive interdisciplinary programs that combine behavior therapy and VLCDs, and the promising maintenance of change in at least 50% of the patients over 1 to 3 years, this integrated approach is the treatment of choice for moderately obese patients.

Severe Obesity

Treatment for severe obesity is controversial, mainly regarding the role of surgical procedures. There is general agreement, however, that the behavioral approaches described above should be included as a component of a comprehensive approach to managing severe obesity, as individuals must make long-term changes in behavior and life style in order to maintain weight loss. Furthermore, VLCDs may be useful with severely obese individuals, particularly at the lower end of the severe obesity category.

Surgery is the last resort treatment for severely obese individuals who have been treated unsuccessfully with conventional approaches (Mason, 1987). Because 1 to 5% of patients die during the surgery, other avenues of weight loss should be given a fair trial before turning to surgery (Halmi, 1980). In our opinion, at least 6 months to 1 year of individual and intense treatment should be carried out before surgery is considered. However, pressing health concerns or impatience on the part of the patient make it difficult to adopt this ideal recommendation. Patients generally lose large amounts of weight (75 to 100 lb) during the year following surgery and many maintain most of the weight loss for up to 3 years (Halmi, 1980).

Jejunoileal bypass is effective, but it has been associated with severe medical complications, such as chronic, severe diarrhea and related electrolyte losses. It has been supplanted by the gastric bypass and "gastric stapling" procedures (Kral, 1988; Stunkard, Sorenson, Hanis, et al., 1986). The gastric stapling procedure involves significantly reducing the size of the stomach by inserting staples or sutures. The remaining stomach volume is approximately 15 to 30 ml and therefore significantly limits a patient's food intake (Kral, 1988; Stunkard, Sorenson, Harris, et al., 1986). A variation of gastric stapling, the Vertical Banded Gastroplasty (VBG), adds a strip of plastic mesh to reinforce the gastric outlet and prevent stretching (Mason, 1987). With the gastric bypass procedure, in addition to producing a gastric pouch, a gastrojejunostomy is created, connecting the gastric pouch to a segment of upper jejunum. The remainder of the stomach drains into a segment of small bowel that is attached to the distal jejunum, creating a Roux-en-Y anastomosis (Hall, Watts, O'Brien, et al., 1990; Kral, 1988) (see Figure 43-1). Vomiting is the primary complication of bypass surgery and VBG, and it generally results when patients eat more than their stomachs can hold. Follow-up research on individuals with gastric bypasses has indicated that they are at high risk of developing deficiencies in vitamin B_{12} and iron and, less commonly, calcium and thiamine (Amaral, Thompson, Caldwell, et al., 1985; Halverson, 1987).

Research on long-term follow-up of gastric bypass patients indicates that, in general, patients tend to continue to lose weight for up to 24 months, after which many experience weight gain (Flickinger, Sinar, & Swanson, 1987). Using a definition of morbid obesity as at least 100 lb over ideal weight, 94% of subjects in one study were no longer morbidly obese at 2 years after surgery (Flickinger, Sinar, & Swanson, 1987). At the 5-year follow-up point, the mean percent of original weight of subjects was 66%, and the mean percent of ideal body weight was 142% compared with 214% preoperatively. Hall, Watts, O'Brien, and colleagues (1990) performed a randomized trial of three gastric restriction operations (VBG, gastrogastrostomy, and gastric bypass) and found that gastric bypass significantly outperformed the other procedures. At a 3-year followup, 67% of patients with gastric bypass had

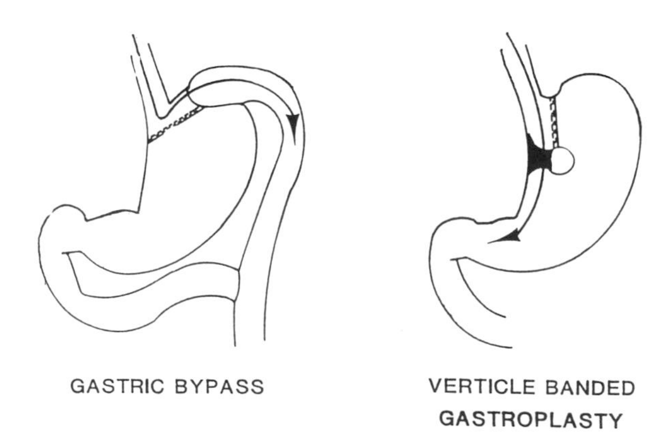

FIG. 43–1. Gastric surgery. Figures redrawn from Kral (1989) and Pories (1989).

maintained a 50% loss of excess weight, compared to 17% and 48% for gastrogastrostomy and VBG, respectively (Hall, Watts, O'Brien, et al., 1990).

The Roux-en-Y gastric bypass is the preferred procedure in centers specializing in obesity surgery because of its relative efficacy and safety (Hall, Watts, O'Brien, et al., 1990; Kral, 1989; Walters, Pories, Swanson, et al., 1991). Though the few studies noted above provide some preliminary positive data on the outcome of surgery for severe obesity, they must be interpreted with caution. These reports come from centers where surgery for obesity is a major focus, and the results may not reflect the experience of the average surgical treatment program. Moreover, research comparing surgery with nonsurgical treatment of severe obesity is needed.

Severely obese individuals often experience negative emotional responses when using conventional weight loss approaches such as diets, medication, and fasting (Halmi, Stunkard, & Mason, 1980). Therefore it is important to note that treatment using surgery with this population has been found to be beneficial from a psychological perspective as well. The first study to report this finding (Solow, Silberfarb, & Swift, 1974) noted improvements in mood, self-esteem, interpersonal and vocational effectiveness, body image, and activity level, as well as decreased use of denial. In another study (Halmi, Stunkard, & Mason, 1980) that assessed psychological factors using a structured interview, more than one-half of subjects experienced postoperative improvements in depression, anxiety, irritability, self-confidence, and body image. Saltzstein and Gutman (1980) also found that 50% of their sample had improved postoperative psychological status measured by MMPI (Dahlstrom, Walch, & Dahlstrom, 1972) clinical scale scores. Gastric bypass, as well as the now defunct intestinal bypass surgery, also leads to beneficial changes in satiety, food preference, and eating patterns consistent with a lowering of the set point about which body weight is regulated (Stunkard, 1989; Stunkard, Stinnett, & Smoller, 1986). These changes in psychological status, however, are usually measured shortly after surgery or within the first 2 years, when the surgery appears to be maximally effective. The psychological consequences of weight gain in those patients who relapse 3 to 5 years after surgery have generally not been evaluated and are unknown.

Pharmacotherapy

The role of medications to treat obesity is an area of considerable interest and research (Nauss-Karol & Sullivan, 1988; Weintraub & Bray, 1989) (see Table 43-3). Existing medications are mostly appetite-suppressing drugs that act on the catecholamine neurotransmitters of the central nervous system (CNS). These drugs are generally derivatives of phenylethylamine and act as CNS stimulants (Nauss-Karol & Sullivan, 1988; Weintraub & Bray, 1989). In addition to suppressing appetite, they may increase heart rate and blood pressure; cause irritability, nervousness, and insomnia; cause diarrhea, nausea, and abdominal distress; and produce an elevation in plasma levels of free fatty acids, glycerol, or both (Nauss-Karol & Sullivan, 1988).

The efficacy of CNS stimulants was reviewed in 1975 by the Food and Drug Administration (FDA) which found that efficacy was comparable among all such drugs (Nauss-Karol & Sullivan, 1988). In general, these agents produced an average weight loss of 0.6 lb per week more than that achieved with placebo (Nauss-Karol & Sullivan, 1988). However, weight gain occurs after withdrawal of these agents, especially if pharmacologic treatment is not combined with behavioral treatment (Craighead, Stunkard, & O'Brien, 1981). There is also evidence that the addition of drug treatment to behavioral treatment may actually provoke relapse and weight gain after cessation of the drug because subjects attribute success to the drug and not to their own efforts at using alternative behavioral strategies (Craighead, 1984; Rodin, Elias, Silberstein, et al., 1988). Most importantly, many of these CNS stimulants have the potential to produce tolerance, withdrawal, and dependence, especially those agents classified as schedule II by the Drug Enforcement Agency (see Table 43-3).

Other CNS medications used to treat obesity include drugs that affect the neurotransmitter serotonin. Serotonin is known to affect a number of physiologic drives, including food intake (Weintraub & Bray, 1989).

Fenfluramine increases the amount of serotonin released and decreases postsynaptic uptake of serotonin in the CNS (Rowland & Carlton, 1986). Fenfluramine, unlike other derivatives of phenylethylamine, has few stimulant effects and little or no risk of habituation (Rowland & Carlton, 1986; Weintraub & Bray, 1989). Though it is effective as an appetite suppressant, fenfluramine may produce a number of adverse effects, including gastrointestinal complaints (especially diarrhea), drowsiness, dizziness, potentiation of antihypertensives, pulmonary hypertension, dry mouth, sleep disturbances, headache, impotence, and loss of libido (Editorial, 1991; Rogers & Blundell, 1979). Depression also may occur after discontinuation (Editorial, 1991).

The D-isomer of fenfluramine, dexfenfluramine, was developed and tested in trials in Europe. Dexfenfluramine is twice as potent an appetite suppressant as fenfluramine with many fewer side effects (Editorial, 1991). Clinical trials have demonstrated dexfenfluramine's ef-

TABLE 43–3. *Pharmacologic Agents for Treating Obesity*

Generic Name	Trade Names	Comments
Appetite Suppressing agents		
Schedule II		
Amphetamine	Dexedrine, Obetrol	Schedule II drugs are most likely to be abused. There is no justification for their use to treat obesity
Methamphetamine	Desoxyn	
Phenmetrazine	Preludin	
Schedule III		
Benzphetamine	Didrex	Schedule III drugs have low abuse potential, but have no advantage over Schedule IV drugs
Phendimetrazine	Plegine, Obalan	
Schedule IV		
Diethylpropion	Tenuate, Tepanil	Overall, schedule IV drugs have the lowest abuse potential, but diethylpropion has substantial reinforcing properties. Fenfluramine, dexfenfluramine, and fluoxetine are serotinergic. Dexfenfluramine, a potent isomer, is not yet available in the U.S.
Fenfluramine	Pondimin	
Dexfenfluramine		
Fluoxetine	Mazanor, Sanorex	
Mazindol	Fastin, Lonamin	
Phenteramine		
Over-the-counter		
Phenylpropanolamine	Acutrim	No reinforcing properties. Effectiveness is similar to prescription anorectics.
Thermogenic Drugs		
Ephedrine		Efficacy unproved.
Thyroid hormone		Weight loss is associated with loss of fat-free mass
Beta-agonists		Increase metabolism and lipolysis. Promising but still experimental. Selective agents that increase metabolism without affecting cardiopulmonary receptors or lean body mass are being studied

Data from Nauss-Karol and Sullivan (1988) and Weintraub and Bray (1989).

fectiveness when combined with dietary intervention or counseling in both short-term studies (Enzi, Crepaldi, Inelman, et al., 1988; Finer, Craddock, Lavielle, et al., 1985; Goodall, Oxtoby, Richards, et al., 1988) and long-term studies (Guy-Grand, Apfelbaum, Crepaldi, et al., 1989). Of special interest is that one of these studies suggested the effectiveness of dexfenfluramine in a group of obese schizophrenic patients taking neuroleptics (Goodall, Oxtoby, Richards, et al., 1988).

The FDA has labeled appetite-suppressing drugs as a "short term" (a few weeks) adjunct to the treatment of obesity (Bray, 1991). There is considerable controversy as to whether these drugs should ever be used chronically to treat obesity. Whereas some authors advise that they should never be used on a chronic long-term basis (Nauss-Karol & Sullivan, 1988), others suggest that those agents that have little abuse potential (e.g., class IV agents and phenylpropanolamine) should be considered for chronic administration to some obese patients (Bray, 1991; Weintraub & Bray, 1989). Weintraub and Bray (1989) argued that the weight gain that usually follows the cessation of appetite-suppressing drug therapy is an expected consequence of having a "disease" such as obesity. Clinicians, they assert, should expect drug treatment to help control obesity, not "cure" it. They draw parallels between the use of these drugs to treat obesity and the use of lipid-lowering agents to treat hypercholesterolemia (Bray, 1991; Weintraub & Bray, 1989). In both cases, cessation of drug therapy usually leads to reversal of the treatment effect. However, Weintraub and Bray also noted that there are few clinical trials that demonstrate the effectiveness and safety of the chronic use of these agents. Until such evidence is available, and because of the potential of these agents to produce toxicity or dependence, we cannot recommend their routine or chronic use.

Fluoxetine is currently used primarily as an antidepressant. Its potential efficacy as a pharmacotherapeutic agent for obesity was noted in clinical studies of depressed patients, who gained less weight during fluoxetine treatment than patients treated with other classes of antidepressants (Nauss-Karol & Sullivan, 1988). Fluoxetine, as well as other serotonergic antidepressants and dexfenfluramine, may be especially useful for limiting carbohydrate craving and binge eating (Editorial, 1991; Nauss-Karol & Sullivan, 1988).

There are a number of other groups of medications currently under investigation for their potential benefit in the treatment of obesity. Thermogenic drugs such as ephedrine have shown promise (Nauss-Karol & Sullivan, 1988; Weintraub & Bray, 1989). Ephedrine is a synthetic adrenergic agent with both alpha- and beta-agonist properties that can increase energy expenditure (Malchow-Moller, Larsen, Hey, et al., 1985). Other drugs that enhance thermogenesis include thyroid hormone and beta-agonists, which increase basal metabolic expenditure (Nauss-Karol & Sullivan, 1988; Weintraub & Bray, 1989), and growth hormone, which reduces protein loss during low calorie dieting (Bray, 1976). The efficacy of these agents for treatment of obesity is not established, though trials of beta-agonists have shown promise (Connacher, Jung, & Mitchell, 1988; Weintraub & Bray, 1989). Beta-agonists also increase lipolysis in human adipose tissue, especially in sympathetically innervated brown adipose tissue (Nauss-Karol & Sullivan, 1988; Weintraub & Bray, 1989). The identification of a subclass of beta-adrenergic receptors (b3) responsible for control of thermogenesis is likely to lead to new pharmacologic agents in the near future (Bray, 1976).

Finally, research is being conducted to identify drugs that might prove to be effective for managing obesity through their direct effects on the gastrointestinal tract and lipid metabolism. These drugs include cholecystokinin, chlorocitrate, inhibitors of dietary carbohydrate absorption, inhibitors of dietary lipid absorption, and inhibitors of lipid synthesis (Nauss-Karol & Sullivan, 1988).

In summary, though some individuals may benefit from the addition of pharmacotherapy to cognitive-behavioral treatments of obesity, the multifactorial etiology of obesity makes it unlikely that a focus on a unidimensional factor, such as a single drug, can become the cure for this disorder. Pharmacologic treatment must be linked with behavioral and environmental interventions that address the biobehavioral, socioeconomic, and sociocultural factors contributing to the development and maintenance of obesity (Abrams, 1991).

PATIENT EVALUATION

When deciding when and how to intervene to treat obesity, several dimensions of patient characteristics must be evaluated. These factors are (1) physiological and medical findings; (2) activity level; and (3) psychological functioning, eating habits, and sociocultural factors. The following is a description of a comprehensive evaluation.

Medical Evaluation

Patients with moderate or severe obesity require a medical evaluation; we recommend it also for patients with mild obesity. The evaluation should include laboratory studies and a resting electrocardiogram. It should address appropriateness of obesity treatment at all, screen for endogenous etiologies of obesity (e.g., hypothyroidism, Cushing's syndrome), determine the patient's ability to participate in an exercise program, screen for medical contraindications to active weight loss, and screen for illnesses or conditions (e.g., type II diabetes) that may require more intensive medical and behavioral monitoring during active weight loss.

A thorough *medical history* should include assessment of cardiac risk factors and family history of obesity. The review of systems should be expanded to screen for endogenous causes of obesity and illnesses that predispose to obesity. Examples include hypothyroidism, Cushing's syndrome, hypothalamic syndromes, polycystic ovary syndrome, pituitary dysfunction, pseudohypoparathyroidism, hypogonadism, diabetes mellitus, insulinoma, and hyperinsulinemia. Patients should be questioned regarding the presence of symptoms that are associated with these disorders, such as headache, difficulty hearing, visual disturbance, alteration of facial appearance, slowed speech, lethargy, easy fatigability, easy bruising, increased appetite, increased sleep requirement, constipation, cold intolerance, frequent skin infections, vaginitis, polydipsia, polyuria, polyphagia, galactorrhea, diminished libido, menorrhagia, and altered menses. Additional emphasis also should be placed on screening for depression, gallbladder disease, hyperlipidemia, sleep apnea, obesity-hypoventilation syndrome, and cardiopulmonary symptoms indicative of coronary artery disease.

Moreover, the patient should be questioned regarding the use of medications, as there are many commonly prescribed drugs associated with weight gain. A partial list of the medications associated with weight gain includes corticosteroids, tricyclic antidepressants, medroxyprogesterone, phenothiazines, cyproheptadine, and lithium (see Table 43-4).

Assessment of smoking status is also useful because of the relation between nicotine and body weight. There is an inverse relation between cigarette smoking and body weight (U.S. DHHS, 1988) and evidence to suggest that some people, especially women, smoke to prevent weight gain (Klesges & Klesges, 1988; Russell & Epstein, 1988; U.S. DHHS, 1988). Moreover, weight gain frequently accompanies smoking cessation (Klesges & Klesges, 1988; U.S. DHHS, 1988; Williamson, Madams, Anda, et al., 1991). Findings from a longitudinal epidemiologic study of smokers and former smokers found that the mean weight gain attributable to the cessation of smoking, as adjusted for a number of relevant covariates, was 2.8 kg in men and 3.8 kg in women (Williamson, Madams, Anda, et al., 1991). Major weight gain (> 13 kg) occurred in 9.8% of the men and 13.4% of the women who quit smoking (Williamson, Madams, Anda, et al., 1991).

TABLE 43–4. *Medications Associated with Weight Gain*

Agents Most Commonly Associated With Weight Gain	Potential Alternatives
Antidepressants	
Heterocyclics	
Amitriptyline	Fluoxetine
Nortriptyline	Bupropion
Imipramine	Protriptyline
Trazodone	Desipramine
MAO inhibitors	
Phenelzine	Tranylcypromine Isocarboxazid
Lithium	Clonazepam Verapamil
Neuroleptics	
Phenothiazines	Molindone Haloperidol Thiothixene
Corticosteroids	
Hydrocortisone	Dexamethasone
Prednisone	Triamcinolone
Corticosterone	
Fluocortisol	
Estrogens and progestins	
Oral contraceptives	Oral contraceptives with low estrogen content
Medroxyprogesterone	
Antihistamines	
Cyproheptadine	Other antihistamines

The mechanisms that may underlie the relation between smoking and body weight include a nicotine-induced increase in metabolic rate (Benowitz, 1988; U.S. DHHS, 1988) and nicotine-induced suppression of food intake, especially carbohydrates (Benowitz, 1988; Grunberg, 1985; Jarvik, 1987; U.S. DHHS, 1988). The latter mechanism may be especially important in women, as both animal and human studies have found that female smokers who quit increase their dietary intake significantly more than male smokers (U.S. DHHS, 1988).

Because of smoking cessation's effects on metabolic rate, eating behavior, and weight, obese individuals may experience increased difficulty losing weight after they quit smoking. Thus when treating the obese smoker, it is important to address the role that nicotine may be playing in regulating weight. Because of the serious health risks associated with smoking, health care providers must advise obese patients to stop smoking. Obese smokers who want to quit smoking should accept weight control interventions as an essential component of their smoking cessation treatment program. The reader is referred to Chapter 42 for recommendations regarding smoking cessation treatment when weight gain is an issue.

Choice of Psychopharmacologic Agents in the Obese Patient

Psychopharmacologic agents are well represented on the list of drugs associated with weight gain. Weight gain during treatment with these agents may be an important reason for nonadherence to psychopharmacologic treatment of psychiatric disorders (Cantu & Korek, 1988; Garland, Remick, & Zis, 1988; Noyes, Garvey, Cook, et al., 1989). Tricyclic antidepressants stimulate appetite and craving for carbohydrates or sweets (Garland, Remick, & Zis, 1988; Stein, Stein, & Linn, 1985), decrease resting metabolic rate (Fernstrom, 1989), and may produce a dose-dependent continuous weight gain of 0.57 to 1.37 kg per month of treatment (Garland, Remick, & Zis, 1988). Clinical experience suggests that those agents with prominent antihistaminic effects (e.g., amitriptyline, imipramine, trazodone) may be more likely to produce these effects (Guggenheim & Uzogara, 1981). Antidepressant agents with stimulant properties, such as fluoxetine and bupropion, which usually do not produce weight gain, are the agents of choice for treating depression in patients with preexisting obesity. Monoamine oxidase inhibitors (MAOIs), especially phenelzine, also are associated with weight gain. Tranylcypromine is not as likely to produce weight gain as phenelzine and can be substituted (after a washout period) if obesity or weight gain is a problem for patients taking phenelzine. Lithium maintenance therapy stimulates weight gains over 10 kg in 20% of patients (Garland, Remick, & Zis, 1988). Purported mechanisms include insulin-like actions on carbohydrate and fat metabolism, polydipsia, and sodium retention (Garland, Remick, & Zis, 1988). Weight gain and increased appetite have been associated with use of carbamazepine (Lampl, Eshel, Rapaport, et al., 1991). If obesity or weight gain is a problem during lithium or carbamazepine maintenance therapy, an alternative for the prophylaxis of bipolar disorder should be tried (Prien & Gelenberg, 1989). Although all of the effective antimanic drugs can cause weight gain in some, patients may do substantially better on one than on another.

Neuroleptics, especially phenothiazines, frequently are associated with weight gain and obesity (Goodall, Oxtoby, Richards, et al., 1988; Medical Letter, 1991). If possible, nonphenothiazines should be substituted for phenothiazines when weight gain or obesity is present in patients who require neuroleptics. Molindone is much less likely to produce weight gain than other neuroleptics and is the agent of choice when moderate or severe obesity is present.

Screening for Genetic Syndromes

A group of rare genetic syndromes associated with obesity must be considered when evaluating the obese patient. This group includes the Prader-Willi syndrome, Lawrence-Moon syndrome, Bardet-Biedl syndrome, Allstrom-Hallgren syndrome, Carpenter syndrome, and Cohen syndrome. Many of these syndromes are associated with mental retardation, short stature, hypogonadism, abnormal facies, and other physical abnormalities. These syndromes are rare and often are diagnosed early in life. A comparison of clinical findings commonly seen with these syndromes is provided in Table 43-5.

Physical Examination of the Obese Patient

The medical evaluation of the obese patient requires a thorough physical examination (adapted in part from Bray, 1989). Emphasis should be placed on the portions of the examination that might reveal abnormal physical findings associated with the illnesses mentioned earlier. Examples of skin abnormalities include hyperpigmentation, parchment skin, ecchymoses, hirsutism, telangiectasia, and purplish abdominal striae. Hair thickness, texture, and distribution must be assessed and female frontal baldness, sparse eyebrows, and low pitched, raspy voice noted. Other facial features to look for include moon facies, periorbital edema, and facial puffiness. Visual field testing, visual and hearing acuity testing, and fundoscopic examinations must be done. A careful neck and thyroid gland examination also is required. Abdominal examination must assess for ascites. Gynecologic evaluation should look for clitoral enlargement suggesting excess androgen effect. Neurologic evaluation must look for focal and general weakness, focal and general sensory deficits, and neuromuscular irritability (e.g., Chvostek's and Trousseau's signs). The patient's weight, height, and distribution of obesity should be documented and the level of obesity (expressed as body mass index or percentage above ideal body weight) calculated at this time.

Laboratory Evaluation of the Obese Patient

Initial laboratory studies for all moderately and severely obese patients should include the following: fasting serum glucose, serum electrolytes, blood urea nitrogen and creatinine. Thyroid studies, including total thyroxine (T_4), triiodothyronine (T_3), and thyroid-stimulating hormone (TSH), are required. Liver function and the question of gallbladder disease is assessed with serum studies, including alkaline phosphatase, bilirubin, trans-

TABLE 43–5. *Clinical Findings with Rare Genetic Syndromes Associated with Obesity*

Parameter	Allstrom-Hallgren	Bardet-Biedl	Carpenter	Cohen	Lawrence-Moon	Prader-Willi
Obesity	Truncal; onset age 2–10 years	Generalized; onset age 1–2 years	Truncal and gluteal; onset age 2–5 years	Truncal; onset age 4–6 years	Generalized; onset age 1–2 years	Generalized; onset age 1–3 years
Stature	Normal to short	Normal to short	Normal	Short to tall	Normal to short	Short
Craniofacies	No specific abnormal characteristics	No specific abnormal characteristics	Apparent exophthalmos acrocephaly, flat nasal ridge, high arched palate, low set ears, retrognathism	High arched palate, small jaw, high nasal bridge, open mouth	No specific abnormal characteristics	Almond-shaped eyes, narrow bifrontal diameter, high arched palate, V-shaped mouth
Limbs	No abnormality	Polydactyly	Polydactyly; syndactyly	Narrow hands and feet	Polydactyly (rare)	Small hands and feet
Mental status	Normal intelligence	Mild mental retardation	Mild mental retardation	Mild mental retardation	Mild mental retardation	Mild to moderate mental retardation
Other findings	Hypogonadism (males only)	Hypogonadism	Hypogonadism	Hypogonadism	Spastic paraplegia	Hypotonia
	Diabetes mellitis	Congenital heart disease		Hypotonia	Retinopathy	Hypogonadism
	Blindness			Dysplastic ears		Enamel hypoplaria
	Sensory nerve deafness	Nephropathy				

Adapted from Bray (1989).

aminases (SGOT, SGPT), lactate dehydrogenase (LDH), and GGT. If serum bilirubin or liver enzymes are elevated, a right upper quadrant abdominal ultrasound scan is obtained next. It is done to rule out cholelithiasis and obstructive liver disease. Serum calcium and phosphorus are obtained to screen for parathyroid dysfunction. Serum protein and urinalysis also are obtained to screen for kidney dysfunction. Hematologic status is assessed through a complete blood count with differential and platelet estimate. Lipid status is assessed through a lipid profile (see Figure 43-2).

It is not recommended that additional endocrinologic testing be done routinely during the initial phase of obesity evaluation. Data obtained from the history, review of systems, physical examination, and initial laboratory testing allow the physician to determine if additional endocrinologic testing is warranted. For example, physical findings such as hirsutism accompanied by complaints of headache, irregular menses, and weight gain may imply ovarian dysfunction. Further studies of the pituitary-ovarian axis should then be pursued.

Should the evaluation generate findings suspicious for an endogenous cause of obesity, it is recommended that active weight loss be postponed while further workup is pursued. Once the condition is treated or controlled, it then may be appropriate to pursue active weight loss.

Matching Obese Patients to Appropriate Treatment

Patients who have no obesity treatment-related medical issues or psychological problems (e.g., eating disorder, affective disorder) should be placed in the appropriate treatment modality as a function of degree of obesity. It is generally recommended that patients with mild obesity (20 to 40% overweight) be treated in a group program with a balanced deficit diet in conjunction with exercise, nutritional education, and cognitive-behavior therapy. Those patients with moderate (41 to 100% overweight) and severe (>100% overweight) obesity are best treated with VLCD administered by a properly trained physician in conjunction with group cognitive-behavior therapy, nutritional education, and exercise instruction. Surgical intervention should be considered for severely obese individuals who have been unsuccessful with a multidisciplinary VLCD approach (Brownell & Wadden, 1991). A patient considering surgery should have a thorough presurgical evaluation and should undergo a waiting period of at least 1 month before a final decision is made. Once the treatment decision is made, both pre- and postsurgery treatment should include individual cognitive-behavioral treatment for 10 to 12 weeks followed by a postsurgery support group with other patients who have had surgery and who could be further along in their postsurgical experiences.

Very-low-calorie diets are safe and appropriate treatments for most patients when administered with proper medical monitoring. (Wadden, Van Itallie, & Blackburn, 1990). The VLCD is an appropriate choice for patients with serious medical problems. Absolute medical contraindications to the use of VLCDs include a myocardial infarction within 3 months, a bleeding ulcer, active thrombophlebitis, high dose steroid medication use (prednisone > 20 mg/day), a prolonged QT interval, and a cerebral vascular accident within 3 to 6 months. The routine medical history, physical examination, and laboratory studies outlined above should identify these contraindications.

The initial evaluation may identify medical problems that are not contraindications to the use of VLCD but represent high risk for intercurrent problems during VLCD treatment. These factors include unstable angina, significant left ventricular cardiac dysfunction or arrhythmia, moderate to severe insulin-dependent diabetes, a history of a cerebral vascular accident (more than 6 months prior to treatment), a history of cholelithiasis, chronic pulmonary disease requiring moderate-dose steroid medication, chronic arthritic conditions requiring moderate-dose steroid medication or high-dose nonsteroidal antiinflammatory medications, and psychiatric illnesses such as major depression or bipolar disorder.

Because obesity has a severe negative impact on such conditions, the benefit gained from weight loss through the VLCD may outweigh the risk. Increased risk from the presence of the above conditions can often be minimized through vigilant medical monitoring of these patients by a qualified physician. Laboratory testing and physician visits may be required more frequently during active weight loss. Treatment of insulin-requiring type II diabetics with VLCD requires a highly coordinated effort between patient and physician. The patient must have home glucose monitoring capability and maintain daily contact with the physician during the initial weeks of the VLCD in order to regulate insulin dosage adequately.

Patients not medically appropriate for VLCD liquid diets may do well on modified versions of these diets. A modified version allows the patient one small meal of solid food per day. The remainder of the daily nutritional requirement is provided by a liquid protein supplement, which allows the patient to receive all daily nutritional requirements while remaining on a VLCD (≤800 cal/day) but not on a total liquid diet. The introduction of one meal per day can minimize some of the metabolic changes seen with total liquid fasting,

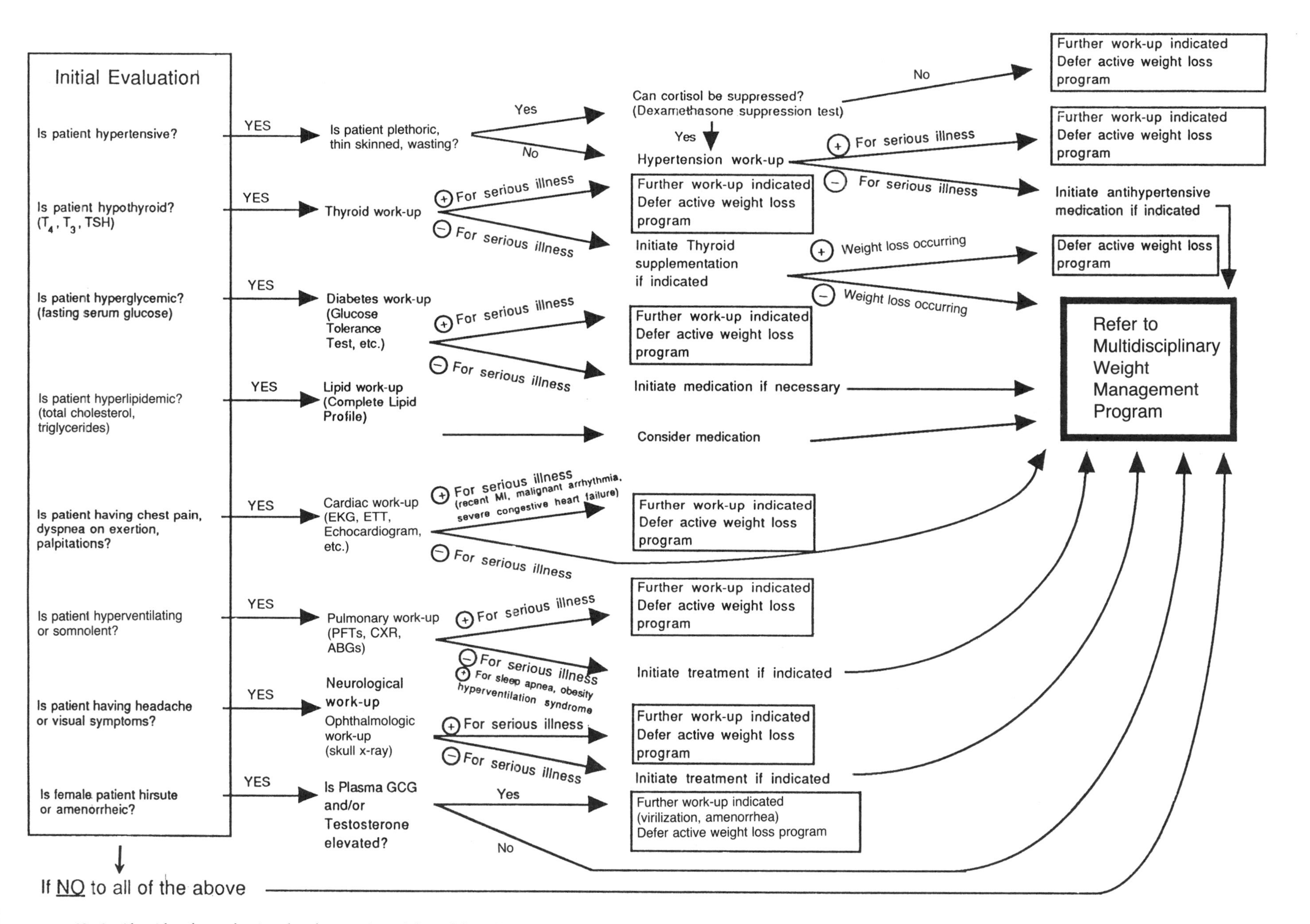

FIG. 43–2. Algorithm for evaluating the obese patient. Adapted from Bray (1989).

lessen the degree of intravascular fluid depletion seen with the liquid VLCD, and cause less of an increase in the workload for the liver and kidney. This modified approach is more appropriate for patients receiving moderate doses of steroids or moderate to high doses of nonsteroidal antiinflammatory medications, patients with moderate to severe cardiac dysfunction (requiring diuretics), and patients with moderate liver or kidney dysfunction or recent strokes. The issue with stroke patients is avoiding dehydration, which could lead to hypercoagulability.

The balanced deficit diet is the recommended modality for those patients with mild obesity. Patients on diuretics, antiarrhythmics such as digoxin and quinidine, theophylline preparations, lithium, thyroid supplementation, and anticoagulation medications require more frequent medical monitoring and laboratory studies while on a balanced-deficit diet.

Exercise Evaluation

The medical evaluation should also determine each patient's capability to exercise. Aerobic exercise is effective in maximizing weight loss and preventing lowering of the metabolic rate (Craighead & Blum, 1989; Pavlou, Krey, & Steffee, 1989). Generally, an aerobic level of activity is defined by achieving a heart rate of 70% of maximal heart rate. The medical evaluation serves to identify patients with cardiac risk factors and those who may be experiencing symptoms that may be of cardiac origin (i.e., chest pain, palpitations, dyspnea). These patients should undergo maximal exercise tolerance testing to assess the cardiac safety of vigorous exercise. Low cardiac risk patients optionally may undergo low level exercise testing (70% of estimated maximal heart rate) as a baseline assessment of exercise tolerance to help plan the exercise prescription (American College of Sports Medicine, 1991).

An exercise prescription is then generated from this information to allow the patient to safely participate in an exercise program. An exercise prescription should be designed to assist patients in initiating an exercise program that is safe, enhances their weight loss, and increases the probability of weight maintenance. The prescription should include type, intensity, duration, frequency, and progression of aerobic activity. Although patients may start at a low level of aerobic activity, they should progress to 15 to 60 minutes of aerobic activity three to five days per week (American College of Sports Medicine, 1991; Craighead & Blum, 1989; Pavlou, Krey, & Steffee, 1989).

Psychological or Psychiatric Evaluation

Commercial weight loss programs frequently minimize the importance of psychological assessment (Fitzgibbon & Kirschenbaum, 1991). Yet many obese patients have psychiatric comorbidity or psychosocial factors that warrant individualized treatment. Therefore obese patients should be screened for the presence of a comorbid psychiatric disorder. Special attention should be placed on assessing the presence of an eating disorder, substance abuse or dependence, body image disturbance, depression, anxiety, binge eating, history of sexual abuse, or a personality disorder. It has been suggested that up to 30% of obese patients in weight control programs have eating disorders or Axis II personality disorders (Agras, 1991; Wing & Marcus, 1992). Patients with an active substance abuse problem or eating disorder are best treated by being offered specific services for these problems. Patients with other comorbid psychiatric disorders may be appropriate for participation in a weight management program, provided they are in simultaneous treatment for their comorbid psychiatric disorder, and that their current level of functioning would not prohibit their ability to adhere to weight management treatment.

To assist the clinical interview, patients should complete weight and diet history questionnaires that inquire about weight history, binge episodes, and purging. We recommend the Beck Depression Inventory (Beck, Rush, Shaw, et al., 1979) for assessing depressive symptomatology and the Eating Inventory (Stunkard & Messick, 1985) for assessing dietary disinhibition, hunger, and cognitive restraint. These measures take about 20 minutes for patients to complete and are quickly scored. Other psychological inventories [(e.g., Eating Disorders Inventory (1984) and the MMPI (Dahlstrom, Welch, & Dahlstrom, 1972)] may be useful for evaluating some individuals. These measures complement the social history, psychiatric history, and mental status examination. In our experience, patients seeking treatment for obesity may either minimize their psychiatric difficulties owing to a lack of insight or may conceal their known difficulties to prevent being "rejected" for obesity treatment.

In addition to conducting a standard psychosocial assessment, we recommend that clinicians inquire about previous weight loss attempts, current eating habits, motivation for weight loss, body image, positive aspects of their obesity (secondary gain), binge eating, weight goal, expectations, and willingness to make long-term life style changes. To assess binge eating (proposed binge eating disorder) (APA, 1991), patients should be asked about the occurrence of binge episodes (rapid consumption of a large amount of food), frequency of binge episodes, history of binge eating, cognition during a binge (perceived loss of control of their eating), physical distress following a binge, negative affect or cognition following a binge episode. The Binge Eating Scale (Gormally, Black, Daston, et al., 1982) can be used to assist in the evaluation of binge episodes. To assess body image disturbance (persistent overconcern with body shape and weight), patients can be asked about how important is their weight in their self-evaluation (self-esteem); are there activities they avoid because of how they feel about their weight; do they avoid reflections (mirrors or windows); when were they last filmed or photographed; and how frequently do they think about their weight (Wilson, Nonas, & Rosenblum, 1992).

If a patient has a history of having been sexually abused, the trauma may or may not be related to the patient's current weight status. For such patients, clinicians should additionally attempt to assess how likely it is that weight loss will (1) trigger memories of sexual abuse (based on a return to a lower weight, which coincides with the weight at which the patient was abused); (2) cause problematic disruptions in their relationships secondary to potential increased sexual attention or demands from others; or (3) lead to feelings of vulnerability. These areas are difficult to assess but may affect treatment; therefore we recommend that clinicians be trained in this area or maintain a close referral relationship with a program and clinicians who specialize in sexual abuse issues.

Clinicians should also evaluate the patient's appropriateness for group treatment. Current level of functioning, past group experiences, and personality disorders may be contraindications to group treatment. When patients are already in treatment with another mental health professional, the current therapist should be contacted to verify assessment information, provide additional clinical information, and coordinate care.

Patient–Treatment Matching: Summary

Upon completion of the evaluation, physiologic findings, weight status, physical activity level, eating habits, binge eating, body image, psychological functioning, and any psychiatric comorbidity should all be considered when matching patient to treatment (Brownell, 1986; Brownell & Wadden, 1991). The challenge is to determine which type of treatment is most appropriate and cost-effective for a particular patient. Because group treatment is least expensive, it should be recommended except when life style factors (i.e., work schedule) or a comorbid psychiatric disorder (i.e., borderline personality disorder) make the patient inappropriate for group treatment. Patient–treatment matching is summarized in Figure 43-3.

Example of Patient–Treatment Matching in Special Populations: Individuals with Type II Diabetes

Diabetes mellitus is a major health problem, and obesity has been implicated in type II diabetes (Wing, Epstein, Paternostro-Bayles, et al., 1988). Approximately 75% of individuals with type II diabetes are obese. Obesity may be associated with: (1) the onset of type II diabetes; (2) insulin resistance; and (3) a high risk of diabetes-related vascular disease (Bierman & Brunzell, 1978; Knowler, Pettitt, Savage, et al., 1981). Weight loss is considered the treatment of choice for obese individuals with type II diabetes, as weight loss reduces hyperglycemia, increases insulin sensitivity, and improves coronary heart disease risk factors (Henry & Gumbiner, 1991; Wing, Marcus, Epstein, et al., 1991). Either caloric restriction or weight loss itself may be the mediating mechanism for these health improvements.

The interdisciplinary approach of behavioral medicine is ideally suited to address the problem of treating type II diabetes (Abrams, Steinberg, Follick, et al., 1986). Indeed, the complex etiology of type II diabetes can include insulin dysregulation and disturbances of other neuroactive pathways (Rodin, 1985). These disturbances can result in greater behavioral excesses or deficits in areas such as stress reactivity, affect regulation, and hence greater disturbances in eating behavior (Rodin, 1985; Rodin, Wack, Ferrannini, et al., 1985). In part because of their special needs, type II diabetic patients should be referred to a comprehensive multidisciplinary weight management program. Research studies indicate that weight loss in this group is greater following a comprehensive behavioral program than after psychotherapy, nutritional education alone, or standard medical care (Wing, Epstein, Norwalk, et al., 1985). To underscore the importance of a comprehensive behavioral medicine program, level of physical activity was related to weight loss and improvements in glycosylated hemoglobin at a 1-year follow up of obese patients with type II diabetes (Wing, Epstein, Paternostro-Bayles, et al., 1988).

It may be more difficult for obese patients with type II diabetes to lose weight and maintain their weight loss than it is for nondiabetic patients. After reviewing the literature, it is unclear whether this difference is accounted for by: (1) design error; (2) actual differences in biobehavioral self-regulatory mechanisms (Rodin,

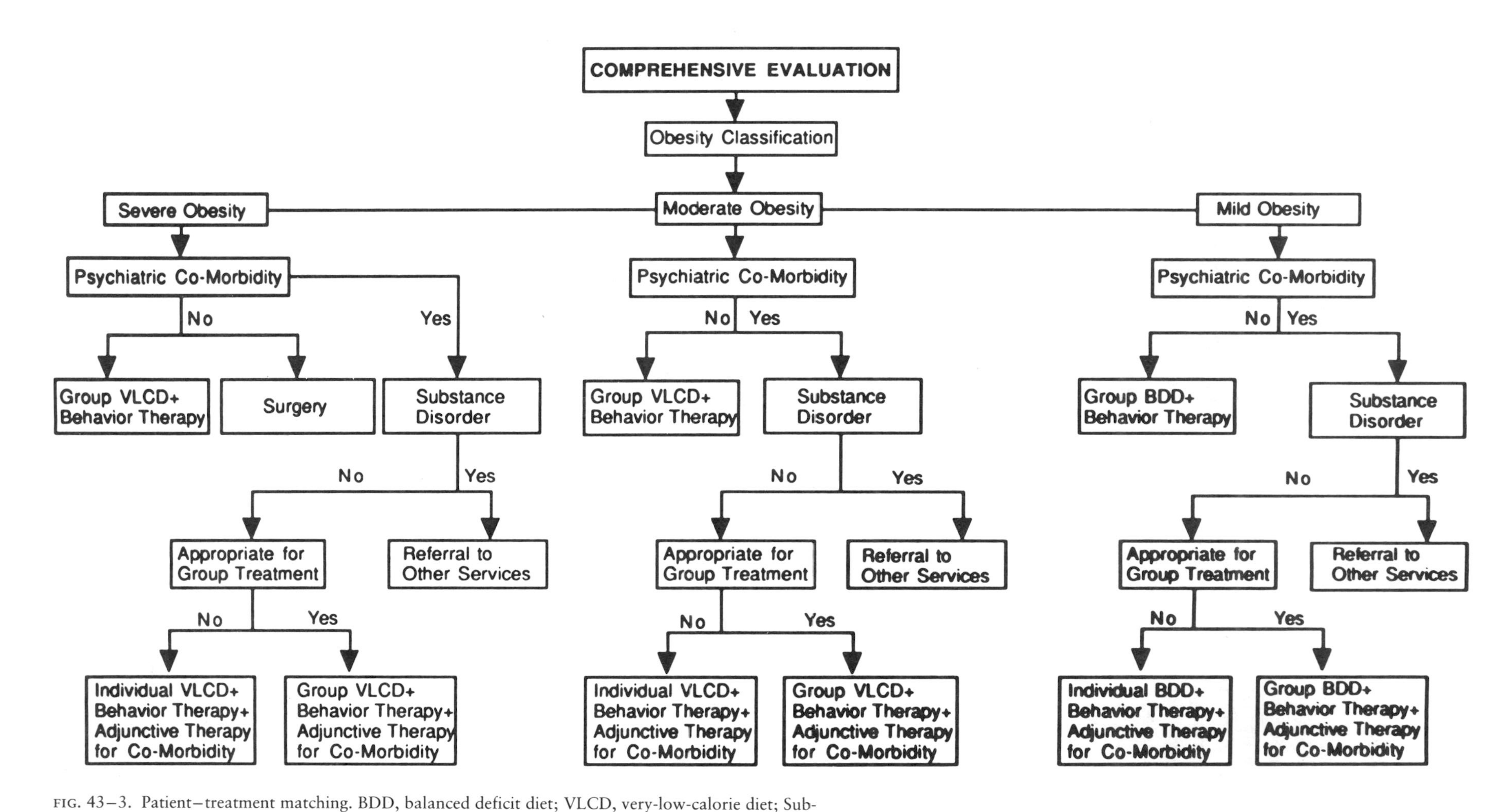

FIG. 43–3. Patient–treatment matching. BDD, balanced deficit diet; VLCD, very-low-calorie diet; Substance Disorder, alcohol, drug, or eating disorder.

1985); or (3) patients with type II diabetes being less successful at caloric restriction than nondiabetic patients (Wing, Marcus, Epstein, et al., 1987).

Research has demonstrated the safety and efficacy of a VLCD for treatment of type II diabetes (Amatruda, Richeson, Weile, et al., 1988; Henry, Scheaffer, & Olefsky, 1985; Wing, Marcus, Salata, et al., 1991). Wing, Marcus, Salata, and colleagues (1991) reported that subjects who were randomly assigned an 8-week VLCD program demonstrated greater reductions in glycosylated hemoglobin (HbA_1) and in fasting blood glucose at a 1-year follow up compared to subjects who were assigned to a balanced-deficit diet program despite similar weight losses at the 1-year follow up. The authors stated that the improved glycemic control following a VLCD may be due to increased insulin secretion, but further research is needed. Pories, Caro, Flickinger, and colleagues (1987) reported that 4 months after gastric bypass surgery, 86 of 88 morbidly obese patients with non-insulin-dependent diabetes became euglycemic with no diabetic medication or special diet. It is unclear at this time whether treatment or maintenance treatment for obese patients with type II diabetes should differ from that of others. Further research is needed, but the studies described above indicate that this question may be an important one to consider for patient-treatment matching.

Consequences of Weight Cycling: Fact or Fiction

It has been proposed that repeated weight cycling, (i.e., weight loss followed by weight regain) may be harmful. Brownell, Greenwood, Stellar, and colleagues (1986) studied the effects of weight cycling in obese rats. The authors reported that it took the obese rats twice as long (21 versus 46 days) to lose the same amount of weight after regaining weight, despite the same calorie intake during both restrictions (diets). Steen, Oppliger, and Brownell (1988) then studied the resting metabolic rate of 27 high school wrestlers. Wrestlers who reported frequent weight cycles had lower resting metabolic rates than noncyclers. These data suggested that weight cycling may be harmful. Other studies with rats, however, have not demonstrated a weight cycling effect (Cleary, 1986).

A prospective study of 12 weight-cycling collegiate wrestlers (Melby, Schmidt, & Corrigan, 1990) reported that numerous cycles of weight loss and weight regain did not lower metabolic rate. In another study weight cycling was not associated with changes in metabolic control or medication requirements of 327 non-insulin-dependent diabetic men (Schotte, Cohen, & Singh, 1990). Additionally, Wadden, Foster, Letizia, and colleagues (1990) reported that, at a 48-week follow up the metabolic rate of 18 obese women who lost weight either through a balanced-deficit diet (1200 Kcal/day) or a very-low-calorie diet (420 Kcal/day) was not reduced more than was expected by the subject's reduction in body weight. Furthermore, Jebb, Goldberg, Coward, and colleagues (1991) reported that three consecutive cycles of 2 weeks of dieting (455 Kcal/day) followed by 4 weeks of ad libitum eating did not decrease absolute basal metabolic rate significantly in 11 obese women. The above findings suggest that there is no long-term negative effect of dieting on metabolic rate.

The potential consequences of weight cycling deserve serious attention at this time. Rodin and colleagues (Rodin, Radke-Sharpe, Rebuffe-Serive, et al., 1990) reported, in a study of 87 women, that self-reported repeated weight cycling promoted abdominal adiposity. Because abdominal adiposity may be an independent risk factor for vascular disease (Rodin, Radke-Sharpe, Rebuffe-Serive, et al., 1990), weight cycling may increase long-term health risks. More recently, Lissner and colleagues (Lissner, Odell, D'Agostino, et al., 1991) examined the variability in body weight of Framingham Heart Study subjects. Fluctuations in body weight were associated with a higher risk of coronary heart disease and death. In summary, research thus far on weight cycling is subject to several methodologic limitations and has yielded conflicting results. More prospective and longitudinal studies using large sample sizes are needed to clarify potential negative consequences of weight cycling.

MAINTENANCE

As previously discussed, most patients experience weight regain at follow up (Hovell, Koch, Hofstetter, et al., 1988; Wadden, Stunkard, & Liebschutz, 1988). Because maintenance of weight loss is problematic, all patients should be encouraged to learn weight maintenance skills. Surveying people who have either maintained their weight loss or who have regained weight, Kayman and colleagues (Kayman, Bruvold, & Stern, 1990) reported that weight maintainers exercised regularly, directly confronted their problems, and had established new eating habits consistent with their overall life style. In contrast, weight regainers were less likely to exercise regularly, used avoidance strategies such as eating or sleeping to manage stress, and continued to eat special diet foods, often preparing special separate meals for themselves. Involvement of family members may improve maintenance. Wing and colleagues (Wing, Marcus, Epstein, et al., 1991) reported that women treated in a together condition with their spouse lost more weight and maintained more of their weight loss than women treated in an alone condition. Men, however, did better when treated alone.

Maintenance treatment, designed to prevent relapse, has proved effective in preventing weight regain (Perri, Nezu, Patti, et al., 1989). Maintenance treatment should consist of a weekly weigh-in, exercise instruction, dietary education, social support, and relapse prevention training (Perri, McAllister, Gange, et al., 1988). Involvement of the patient's primary physician may enhance maintenance, as physician attention has a positive effect on whether patients lose weight (Levy & Williamson, 1988). Some patients require some form of support for the remainder of their lives because there are genetic, biologic, psychological, and environmental factors that may undermine the individual's resolve and their newly acquired coping skills. Family and work site "systems," changes in food availability, and norms can be used to help maintain healthy life styles (Abrams, 1991). Continued attention to comorbid psychiatric disorders during the maintenance phase is important as either ongoing or new psychiatric difficulties may complicate a patient's ability to maintain their life style changes.

EVALUATION OF OBESITY TREATMENT PROGRAMS

When deciding to which program to refer a patient, we recommend that clinicians assess the patient's psychological or psychiatric status and medical condition, and then evaluate the available weight management programs. Based on patient-treatment matching guidelines (see Figure 43-3) clinicians can assess which type of program (e.g., VLCD versus balanced-deficit diet, group versus individual) is recommended. When evaluating the programs available to the patient, we recommend that clinicians use the following criteria: (1) Is there medical supervision? (2) Is there an interdisciplinary treatment program involving a registered dietitian, an exercise physiologist, a psychologist or psychiatrist, and an internist all of whom specialize in obesity treatment? (3) What are the credentials, qualifications, and experience of the staff? (4) What are the program's short- and long-term results? (5) Are individual and group programs available? (6) Is maintenance treatment provided? (7) What percentage of patients stay in maintenance treatment for 1 year? (ADA Reports, 1990; Michigan Health Council, 1990; Wadden, Foster, Letizia, et al., 1990) Clinicians who elect not to participate in an interdisciplinary treatment program can assist their patients in managing their weight by assessing the patient's readiness to incorporate necessary life style changes, providing appropriate referrals, communicating with the weight management program to coordinate care, and supporting the patient's involvement in maintenance treatment.

SUMMARY, CONCLUSIONS, AND FUTURE DIRECTIONS

It is our belief that obesity is a complex biopsychosocial phenomenon. Because of the complexity of this problem and the recidivism following treatment, we recommend that obese patients undergo an in-depth interdisciplinary evaluation to match them appropriately to comprehensive treatment. Physiologic findings, weight status, activity level, eating habits, comorbid psychiatric or characterologic factors, and psychological functioning should all be integrated into assessment and treatment planning. Treatment should combine caloric restriction, medical monitoring, cognitive-behavioral therapy, exercise instruction, and nutritional education. After treatment, patients should be encouraged to enroll in a structured maintenance program, as relapse after treatment remains a significant problem.

REFERENCES

ABRAMS DB (1983). Developments in the behavioral treatment of obesity. In CM FRANKS (Ed.), New developments in behavior therapy. New York: Hawthorn Press.

ABRAMS DB (1991). Conceptual models to integrate individual and public health interventions: the example of the workplace. In M Henderson (Chair), Proceedings of the international conference on promoting dietary change in communities (pp. 170–190). Seattle: The Fred Hutchinson Cancer Research Center.

ABRAMS DB, EMMONS KM, NIAURA RS, ET AL. (1991). Tobacco dependence: an integration of individual and public health perspectives. In PE NATHAN, JW LANGENBUCHER, BS MCCRADY, ET AL. (Eds.), Annual review of addictions treatment and research (Vol 1; pp. 391–436). New York: Pergamon Press.

ABRAMS DB, & FOLLICK MJ (1983). Behavioral weight loss intervention at the worksite: feasibility and maintenance. J Consult Clin Psychol 51:226–233.

ABRAMS DB, STEINBERG JL, FOLLICK MA, ET AL. (1986). Obesity and type II diabetes: behavioral medicine's contribution. Behav Med Abstr 7:1–4.

ADA (American Dietetic Association) Reports (1990). Position of the American Dietetic Association: very-low-calorie weight loss diets. J Am Diet Assoc 90:722–726.

AGRAS WS (1991). Overweight people are not all the same: how types and differences should guide interventions. Paper presented at the annual convention of the Society of Behavioral Medicine, Washington, DC.

AMARAL JF, THOMPSON WR, CALDWELL MD, ET AL. (1985). Prospective hematologic evaluation of gastric exclusion surgery for morbid obesity. Ann Surg 201:186.

AMATRUDA JM, RICHESON JF, WEILE SL, ET AL. (1988). The safety and efficacy of a controlled low-energy (very-low-calorie) diet in the treatment of non-insulin-dependent diabetes and obesity. Arch Intern Med 148:873–877.

American College of Sports Medicine (1991). Guidelines for exercise testing and prescription (4th ed.). Philadelphia: Lea & Febiger.

American Psychiatric Association (1987). Diagnostic and statistical manual of mental disorders (3rd ed., revised). Washington, DC: American Psychiatric Press.

American Psychiatric Association (1991). Task Force on DSM-IV. Washington, DC: American Psychiatric Press.

ANDRES R, ELAHI D, TOBIN JD, ET AL. (1985). Impact of age on weight goals. Ann Intern Med 103:1030–1033.

ATKINSON R (1990). Cost effectiveness of the treatment of obesity. Paper presented at a conference on treatment of the patient with medically significant obesity, Atlanta.

BANDURA A (1986). Social foundations of thought and action: a cognitive theory. Englewood Cliffs, NJ: Prentice-Hall.

BECK AT, RUSH AJ, SHAW BF, ET AL. (1979). Cognitive therapy of depression. New York: Guildford Press.

BENOWITZ NL (1988). Pharmacologic aspects of cigarette smoking and nicotine addiction. N Engl J Med 319:1318–1330.

BIERMAN EL, & BRUNZELL JD (1978). Interrelation of atherosclerosis, abnormal lipid metabolism and diabetes mellitus. In HM KATZEN & RJ MAHLER (Eds.), Diabetes, obesity, and vascular disease: metabolic and molecular interrelationships (pp. 187–210). New York: John Wiley & Sons.

BLACKBURN GL, BISTRIAN BR, & FLATT JP (1975). Role of a protein sparing-modified fast in a comprehensive program of weight reduction. In A HOWARD (Ed.), Recent advances in weight reduction (pp. 279–281). London: Newman Publishing.

BLACKBURN GL, & KANDERS BS (1987). Medical evaluation and treatment of the obese patient with cardiovascular disease. Am J Cardiol 60:55G–58G.

BLACKBURN GL, WILSON GT, KANDERS BS, ET AL. (1989). Weight cycling: the experience of human dieters. Am J Clin Nutr 49:1105–1109.

BLAIR SN, KOHL HW III, PAFFENBARGER RS, ET AL. (1989). Physical fitness and all-cause mortality: a prospective study of healthy men and women. JAMA 262:2395–2401.

BRAY GA (1976). The obese patient. Philadelphia: WB Saunders.

BRAY GA (1985). Complications of obesity. Ann Intern Med 103:1052–1062.

BRAY GA (1989). Obesity: basic aspects and clinical application. Med Clin North Am 73:1–269.

BRAY GA (1990). Exercise and obesity. In C BOUCHARD, RJ SHEPERD, T STEPHENS, ET AL. (Eds.), Exercise, fitness, and health (Ch. 41, pp. 497–515). Champaign, IL: Human Kinetics Books.

BRAY GA (1991). Barriers to the treatment of obesity. Ann Intern Med 115:152–153.

BROWNELL KD (1982). Obesity: understanding and treating a serious, prevalent, and refractory disorder. J Consult Clin Psychol 50:820–840.

BROWNELL KD (1984). Behavioral, psychological, and environmental predictors of obesity and success at weight reduction. Int J Obes 8:543–550.

BROWNELL KD (1986). Public health approaches to obesity and its management. Annu Rev Public Health 7:521–533.

BROWNELL KD, GREENWOOD MRC, STELLAR E, ET AL. (1986). The effects of repeated cycles of weight loss and regain in rats. Physiol Behav 38:459–464.

BROWNELL KD, MARLATT GA, LICHTENSTEIN E, ET AL. (1986). Understanding and preventing relapse. Am Psychol 41:765–782.

BROWNELL KD, STUNKARD AJ, & ALBAUM JM (1980). Evaluation and modification of exercise patterns in the natural environment. Am J Psychiatry 137:1540–1545.

BROWNELL KD, & WADDEN TA (1986). Behavior therapy for obesity: modern approaches and better results. In KD BROWNELL & JP FOREYT (Eds.), Handbook of eating disorders (Ch. 9, pp. 180–197). New York: Basic Books.

BROWNELL KD, & WADDEN TA (1991). The heterogeneity of obesity: fitting treatments to individuals. Behav Ther 22:153–177.

CANTU TG, & KOREK JS (1988). Monoamine oxidase inhibitors and weight gain. Drug Intell Clin Pharm 22:755–759.

CHARNEY E, GOODMAN HC, MCBRIDE M, ET AL. (1976). Childhood antecedents of adult obesity: do chubby infants become obese adults? N Engl J Med 295:6–9.

CLEARY MP (1986). Response of adult lean and obese female Zucker rats to intermittent food restriction/refeeding. J Nutr 116:1489–1499.

CONNACHER AA, JUNG RT, & MITCHELL PEG (1988). Clinical research: weight loss in obese subjects on a restricted diet given BRL 26830A, a new atypical beta adrenoceptor agonist. BMJ 296:1217–1220.

CRAIGHEAD L (1984). Sequencing of behavioral therapy and pharmacotherapy for obesity. J Consult Clin Psychol 52:190–199.

CRAIGHEAD LM, & BLUM MD (1989). Supervised exercise in behavioral treatment for moderate obesity. Behav Ther 20:49–59.

CRAIGHEAD LM, STUNKARD AJ, & O'BRIEN R (1981). Behavior therapy and pharmacotherapy of obesity. Arch Gen Psychiatry 38:763–768.

DAHLKOETTER J, CALLAHAN EJ, & LANTON J (1979). Obesity and the unbalanced energy equation: exercise versus eating habit change. J Consult Clin Psychol 47:898–905.

DAHLSTROM WG, WELCH GS, & DAHLSTROM LE (1972). An MMPI handbook. Vol. 1. Clinical interpretation (2nd ed.). Minneapolis: University of Minnesota Press.

DOWDY DB, CURETON KJ, DUVAL HP, ET AL. (1985). Effects of aerobic dance on physical work capacity, cardiovascular function, and body composition of middle-aged women. Res Q Exerc Sport 56:227–233.

DREON DM, FREY-HEWITT B, ELLSWORTH N, ET AL. (1988). Dietary fat: carbohydrate ratio and obesity in middle-aged men. Am J Clin Nutr 47:995–1000.

Eating Disorders Inventory–GARNER DM & OLMSTED MP (1984). Manual for eating disorder inventory. Psychological Assessment Resources, Odessa, FL.

Editorial (1991). Dexfenfluramine. Lancet 337:1315–1316.

ENZI G, CREPALDI G, INELMAN EM, ET AL. (1988). Efficacy and safety of dexfenfluramine in obese patients: a multicentre study. Clin Neuropharmacol 11(suppl 1):S173–178.

EPSTEIN LH, MASEK B, & MARSHALL W (1978). Prelunch exercise increases lunchtime caloric intake. Behav Therapist 1:15.

EPSTEIN LH, & SQUIRES S (1988). The stoplight diet for children: an eight-week program for parents and children. Boston: Little, Brown.

EPSTEIN LH, WING RR, KOESKE R, ET AL. (1982). A comparison of lifestyle change and programmed aerobic exercise on weight and fitness changes in obese children. Behav Ther 13:651–665.

ERNSBERGER P, & NELSON DO (1988). Refeeding hypertension in dietary obesity. Am J Physiol 254:R47–R55.

FERNSTROM MH (1989). Depression, antidepressants, and body weight change. Ann NY Acad Sci 575:31–40.

FINER N, CRADDOCK D, LAVIELLE R, ET AL. (1985). Dexfenfluramine in the treatment of refractory obesity. Curr Ther Res 38:847–854.

FITZGIBBON ML, & KIRSCHENBAUM DS (1991). Distressed binge eaters as a distant subgroup among obese individuals. Addict Behav 16:441–451.

FLATT JP, & BLACKBURN GL (1974). The metabolic regulatory system: implications for protein-sparing therapies during caloric deprivation and disease. Am J Clin Nutr 27:175–187.

FLICKINGER EG, SINAR DR, & SWANSON M (1987). Gastric bypass. Gastro Clin North Am 16:2.

FOLLICK MJ, ABRAMS DB, SMITH TW, ET AL. (1984). Behavioral intervention for weight loss: acute versus long-term effects on HDL and LDL cholesterol levels. Arch Intern Med 144:1571–1574.

FOREYT JP (1987). Issues in the assessment and treatment of obesity. J Consult Clin Psychol 55:677–684.

GARLAND EJ, REMICK RA, & ZIS AP (1988). Weight gain with antidepressants and lithium. J Clin Psychopharmacol 8:323–330.

GARNER DM, ROCKERT W, OLMSTED MP, ET AL. (1985). Psychoeducational principles in the treatment of bulimia and anorexia nervosa. In DM GARNER & PE GARFINKEL (Eds.), Handbook of psychotherapy for anorexia nervosa and bulimia (pp. 513–572). New York: Gilford.

GARNER DM, & WOOLEY SC (1991). Confronting the failure of behavioral and dietary treatments for obesity. Clin Psychol Rev 11:729–780.

GENUTH SM, CASTRO JH, & VERTES V (1974). Weight reduction in obesity by outpatient semistarvation. JAMA 230:987–991.

GENUTH SM, VERTES V, & HAZELTON I (1978). Supplemented fasting in the treatment of obesity. In G BRAY (Ed.), Recent advances in obesity research (Vol 2; pp. 370–378). London: Newman.

GOODALL E, OXTOBY C, RICHARDS R, ET AL. (1988). A clinical trial of the efficacy and acceptability of d-fenfluramine in the treatment of neuroleptic-induced obesity. Br J Psychiatry 153:208–213.

GORMALLY J, BLACK S, DASTON S, ET AL. (1982). The assessment of binge eating severity among obese persons. Addict Behav 7:47–55.

GRUNBERG NE (1985). Nicotine, cigarette smoking, and body weight. Br J Addict 80:369–377.

GUGGENHEIM FG, & UZOGARA EO (1981). Understanding and treating obesity. In TC Manschreck (Ed.), Psychiatric medicine update (pp. 129–145). New York: Elsevier.

GUY-GRAND B, APFELBAUM M, CREPALDI G, ET AL. (1989). International trial of long-term dexfenfluramine in obesity. Lancet 2:1142–1145.

HALL JC, WATTS JM, O'BRIEN PE, ET AL. (1990). Gastric surgery for morbid obesity: the Adelaide study. Ann Surg 211:419–427.

HALMI KA (1980). Gastric bypass for massive obesity. In AJ Stunkard (Ed.), Obesity (pp. 388–394). Philadelphia: WB Saunders.

HALMI KA, LONG M, STUNKARD AJ, ET AL. (1980). Psychiatric diagnosis of morbidly obese gastric bypass patients. Am J Psychiatry 137:470–472.

HALMI KA, STUNKARD AJ, & MASON EE (1980). Emotional responses to weight reduction by three methods: diet, jejunoileal bypass, and gastric bypass. Am J Clin Nutr 33:446–451.

HALVERSON JD (1987). Vitamin and mineral deficiencies following obesity surgery. Gastroenterol Clin North Am 16:307.

HENRY RR, & GUMBINER B (1991). Benefits and limitations of very-low-calorie diet therapy in obese NIDDM. Diab Care 14:802–823.

HENRY R, SCHEAFFER L, & OLEFSKY JM (1985). Glycemic effects of intensive caloric restriction and isocaloric refeeding in noninsulin-dependent diabetes mellitus. J Clin Endocrinol Metab 61:917–925.

HOLLAND J, MASLING J, & COPLEY D (1970). Mental illness in lower class normal, obese and hyper-obese women. Psychosom Med 32:351–357.

HOVELL MF, KOCH A, HOFSTETTER R, ET AL. (1988). Long-term weight maintenance: assessment of a behavioral and supplemented fasting regimen. Am J Public Health 78:663–666.

HUBERT HB, FEINLEIB M, MCNAMARA PM, ET AL. (1983). Obesity as an independent risk factor for cardiovascular disease: a 26-year follow-up of participants in the Framingham Heart Study. Circulation 67:968–977.

JARVIK ME (1987). Does smoking decrease eating and eating increase smoking? In WR MARTIN, GR VAN LOON, ET IWAMOTO, ET AL (Eds.), Tobacco, smoking and nicotine: a neurobiological approach (pp. 389–400). New York: Plenum Press.

JEBB SA, GOLDBERG GR, COWARD WA, ET AL. (1991). Effects of weight cycling caused by intermittent dieting on metabolic rate and body composition in obese women. Int J Obes 15:367–374.

JEFFREY RW (1991). Biobehavioral influences on diet, obesity and weight control strategies in women. Paper presented at the conference on women, behavior and cardiovascular disease sponsored by the National Heart, Lung, and Blood Institute. Chevy Chase, MD.

KAYMAN S, BRUVOLD W, & STERN JS (1990). Maintenance and relapse after weight loss in women: behavioral aspects. Am J Clin Nutr 52:800–807.

KEEFE PH, WYCHOGROD D, WEINBERGER E, ET AL. (1984). Binge eating and outcome of behavioral treatment of obesity: a preliminary report. Behav Res Ther 22:319–321.

KLESGES RC, & KLESGES LM (1988). Cigarette smoking as a dietary strategy in a university population. Int J Eating Disord 7:413–419.

KNOWLER WC, PETTITT DJ, SAVAGE PJ, ET AL. (1981). Diabetes incidence in Pima Indians: contributions of obesity and parental diabetes. Am J Epidemiol 113:144–156.

KOLOTKIN RL, REVIS ES, KIRKLEY BG, ET AL. (1987). Binge eating in obesity: associated MMPI characteristics. J Consult Clin Psychol 55:872–876.

KRAEMER HC, BERKOWITZ RI, & HAMMER LD (1990). Methodological difficulties in studies of obesity. I. Measurement issues. Ann Behav Med 12:112–118.

KRAL J (1988). Surgery for obesity. In RT FRANKLE & M YOUNG (Eds.), Obesity and weight control (Ch. 15, pp. 297–313). Rockville, MD: Aspen Publishers.

KRAL JG (1989). Surgical treatment of obesity. Med Clin North Am 73:251–264.

KROMHOUT D (1983). Energy and macronutrient intake in lean and obese middle-aged men (the Zutphen Study). Am J Clin Nutr 37:295–299.

LAMPL Y, ESHEL Y, RAPAPORT A, ET AL. (1991). Weight gain, increased appetite, and excessive food intake induced by carbamazepine. Clin Neuropharmacol 14:251–255.

LEVY BT, & WILLIAMSON PS (1988). Patient perceptions and weight loss of obese adults. J Fam Pract 27:285–290.

Life Sciences Research Office (1979). Research needs in the management of obesity by severe caloric restriction. In Federation of American societies for experimental biology. Contract No. FDA 223-75-2090. Washington, DC: U.S. Government Printing Office.

LISSNER L, ODELL P, D'AGOSTINO RB, ET AL. (1991). Variability of body weight and health outcomes in the Framingham population. N Engl J Med 324:1839–1844.

MALCHOW-MOLLER A, LARSEN S, HEY H, ET AL. (1985). Ephedrine as an anorectic: the story of the Elsinore pill. Int J Obes 9:347–353.

MANSON JE, COLDITZ GA, STAMPFER MJ, ET AL. (1990). A prospective study of obesity and risk of coronary heart disease in women. N Engl J Med 322:882–889.

MARCUS MD, WING RR, EWING L, ET AL. (1990). Psychiatric disorders among obese binge eaters. Int J Eating Disord 9:69–77.

MARCUS MD, WING RR, & HOPKINS J (1988). Obese binge eaters: affect, cognitions, and response to behavioral weight control. J Consult Clin Psychol 56:433–439.

MARLATT GA, & GORDON JR (1985). Relapse prevention: maintenance strategies in the treatment of addictive behaviors. New York: Guilford Press.

MARSHALL JD, HAZLETT CB, SPADY DW, ET AL. (1990). Comparison of convenient indicators of obesity. Am J Clin Nutr 51:22–28.

MASON EE (1987). Morbid obesity: use of vertical banded gastroplasty. Surg Clin North Am 67:521–537.

MATZ R (1987). Obesity: an eclectic review. Hosp Pract 22:152A–152C, 152F–152J.

MCREYNOLDS WT (1982). Toward a psychology of obesity: review of research on the role of personality and level of adjustments. Int J Eating Disord 2:37–57.

Medical Letter (1991). Drugs for psychiatric disorders. Med Lett 33:43–50.

MELBY CL, SCHMIDT WD, & CORRIGAN D (1990). Resting metabolic rate in weight-cycling collegiate wrestlers compared with physically active, noncycling control subjects. Am J Clin Nutr 52:409–414.

Metropolitan Life Insurance Company (1959). New weight standards for men and women. Stat Bull Metropolitan Life Insurance Company 40:1–4.

Metropolitan Life Insurance Company (1984). 1983 Metropolitan height and weight tables. Stat Bull Metropolitan Life Insurance Company 64:2–9.

Michigan Health Council (1990). Final report: task force to establish weight loss guidelines; recommendations for adult weight loss programs in Michigan. In K PETERSMARCK (Ed.), Toward safe weight loss (pp. 1–23). East Lansing: Michigan Health Council.

MOORE RS, & RODIN J (1986). The influence of psychological variables in obesity. In KD BROWNELL & JP FOREYT (Eds.), Handbook of eating disorders (Ch. 5, pp. 99–121). New York: Basic Books.

National Institutes of Health Consensus Development Panel on the Health Implications of Obesity (1985). Health implications of obesity: National Institutes of Health consensus development conference statement. Ann Intern Med 103:1073–1077.

NAUSS-KAROL C, & SULLIVAN AC (1988). Pharmacological approaches to the treatment of obesity. In RT FRANKLE & M YOUNG (Eds.), Obesity and weight control (Ch. 14, pp. 275–296). Rockville, MD: Aspen Publishers.

NOYES R, GARVEY MJ, COOK BL, ET AL. (1989). Problems with tricyclic antidepressant use in patients with panic disorder or agoraphobia: results of a naturalistic follow-up study. J Clin Psychiatry 50:163–169.

PATTON ML, AMATRUDA JM, BIDDLE TL, ET AL. (1981). Prevention of life threatening cardiac arrhythmias associated with a modified fast by dietary supplementation with trace metals and fatty acids. Clin Res 29:663A.

PAVLOU KN, KREY S, & STEFFEE WP (1989). Exercise adjunct to weight loss and maintenance in moderately obese subjects. Am J Clin Nutr 49:1115–1123.

PAVLOU KN, STEFFEE WP, LERMAN RH, ET AL. (1985). Effects of dieting and exercise on lean body mass, oxygen intake, and strength. Med Sci Sports Exerc 17:466–471.

PERRI MG, MCALLISTER DA, GANGE JJ, ET AL. (1988). Effect of four maintenance programs on the long-term management of obesity. J Consult Clin Psychol 57:450–452.

PERRI MG, NEZU AM, PATTI ET, ET AL. (1989). Effect of length of treatment on weight loss. J Consult Clin Psychol 57:450–452.

PETERSON HR, ROTHSCHILD M, WEINBERG CR, ET AL. (1988). Body fat and the activity of the autonomic nervous system. N Engl J Med 318:1077–1083.

POEHLMAN ET, TREMBLAY A, DESPRES JP, ET AL. (1986). Genotype-controlled changes in body composition and fat morphology following overfeeding in twins. Am J Clin Nutr 43:723–731.

PORIES WJ (1989). Progress report: the treatment of morbid obesity. Paper presented at a conference on treatment of obesity, Boston.

PORIES WJ, CARO JF, FLICKINGER EG, ET AL. (1987). The control of diabetes mellitus (NIDDM) in the morbidly obese with the Greenville gastric bypass. Ann Surg 206:316–323.

PRIEN RF, & GELENBERG AJ (1989). Alternatives to lithium for preventive treatment of bipolar disorder. Am J Psychiatry 146:840–848.

RAVUSSIN E, LILLIOJA S, KNOWLER WC, ET AL. (1988). Reduced rate of energy expenditure as a risk factor for body-weight gain. N Engl J Med 318:467–472.

ROBERTS SB, SAVAGE J, COWARD WA, ET AL. (1988). Energy expenditure and intake in infants born to lean and overweight mothers. N Engl J Med 318:461–466.

RODIN J (1985). Insulin levels, hunger and food intake: an example of feedback loops and body weight regulation. Health Psychol 4:1–18.

RODIN J, ELIAS M, SILBERSTEIN LR, ET AL. (1988). Combined behavioral and pharmacologic treatment for obesity: predictors of successful weight maintenance. J Consult Clin Psychol 56:399–404.

RODIN J, RADKE-SHARPE N, REBUFFE-SCRIVE M, ET AL. (1990). Weight cycling and fat distribution. Int J Obes 14:303–310.

RODIN J, WACK J, FERRANNINI E, ET AL. (1985). Effects of insulin and glucose on feeding behavior. Metabolism 34:826–831.

ROGERS PJ, & BLUNDELL JE (1979). Effect of anorexic drugs on food intake and the microstructure of eating in human subjects. Psychopharmacology 66:159–165.

ROMIEU I, WILLETT WC, STAMPFER MJ, ET AL. (1988). Energy intake and other determinants of relative weight. Am J Clin Nutr 47:406–412.

ROWLAND NE, & CARLTON J (1986). Neurobiology of an anorectic drug: fenfluramine. Prog Neurobiol 27:16–62.

RUNCIE J, & THOMPSON TS (1970). Prolonged starvation—a dangerous procedure? BMJ 3:432–435.

RUSSELL PO, & EPSTEIN LH (1988). Smoking. In EA BLECHMAN & KD BROWNELL (Eds.), Handbook of behavioral medicine for women (Ch. 28, pp. 369–383). New York: Pergamon Press.

SALLADE J (1973). A comparison of the psychological adjustment of obese vs. nonobese children. J Psychosom Res 7:89–96.

SALTZSTEIN EC, & GUTMAN MC (1980). Gastric bypass for morbid obesity: pre-operative and post-operative psychological evaluation of patients. Arch Surg 115:21–28.

SCHOTTE DE, COHEN E, & SINGH SP (1990). Effects of weight cycling on metabolic control in male outpatients with non-insulin-dependent diabetes mellitus. Health Psychol 9:599–605.

SILVERSTEIN B, PERDUE L, PETERSON B, ET AL. (1986). Possible causes of the thin standard of bodily attractiveness for women. Int J Eating Disord 5:907–916.

SOLOW C, SLIBERFARB PM, & SWIFT K (1974). Psychosocial effects of intestinal bypass surgery for severe obesity. N Engl J Med 290:300–304.

STEEN S, OPPLIGER RA, & BROWNELL KD (1988). Metabolic effects of repeated weight loss and regain in adolescent wrestlers. JAMA 260:47–50.

STEIN EM, STEIN S, & LINN MW (1985). Geriatric sweet tooth: a problem with tricyclics. J Am Geriatr Soc 33:687–692.

STUNKARD AJ (1984). The current status of treatment of obesity in adults. In AJ STUNKARD & E STELLER (Eds.), Eating and its disorders (Ch. 13, pp. 157–174). New York: Raven Press.

STUNKARD AJ (1989). Obesity. In R MICHELS, AM COOPER, SB GUZE, ET AL. (Eds.), Psychiatry (Ch. 103, pp. 1–13). Philadelphia: Lippencott.

STUNKARD AJ, FOCH TT, & HRUBEC Z (1986). A twin study of human obesity. JAMA 256:51–54.

STUNKARD AJ, HARRIS JR, PEDERSEN NL, ET AL. (1990). The body-mass index of twins who have been reared apart. N Engl J Med 322:1483–1487.

Stunkard AJ, & Messick S (1985). The three-factor eating questionnaire to measure dietary restraint, disinhibition and hunger. J Psychosom Res 29:71–83.

Stunkard AJ, & Penick SB (1979). Behavior modification in the treatment of obesity: the problem of maintaining weight loss. Arch Gen Psychiatry 36:801–806.

Stunkard AJ, Sorenson TIA, Hanis C, et al. (1986). An adoption study of human obesity. N Engl J Med 314:193–198.

Stunkard AJ, Stinnett JL, & Smoller JW (1986). Psychological and social aspects of the surgical treatment of obesity. Am J Psychiatry 143:417–429.

Telch CF, Agras WS, & Rossiter EM (1988). Binge eating increases with increasing adiposity. Int J Eating Disord 7:115–119.

Telch CF, Agras WS, Rossiter EM, et al. (1990). Group cognitive-behavioral treatment for the nonpurging bulimic: an initial evaluation. J Consult Clin Psychol 58:629–635.

Thompson JK, Jarvie GJ, Lahey BB, et al. (1982). Exercise and obesity: etiology, physiology, and intervention. Psychol Bull 91:55–79.

U.S. Department of Health and Human Services (1988). The health consequence of smoking: nicotine addiction: a report of the Surgeon General. Rockville, MD: US Department of Health and Human Services, Public Health Service, Office on Smoking and Health.

Van Itallie TB (1979). Obesity: adverse effects on health and longevity. Am J Clin Nutr 32:2723–2733.

Wachtel PL (1989). The poverty of affluence: a psychological portrait of the American way of life. Philadelphia: New Society Publishers.

Wadden TA (1985). Treatment of obesity in adults: a clinical perspective. In PA Keller & LG Ritt (Eds.), Innovations in clinical practice: a source book (Vol IV; Ch. 5, pp. 127–152). Sarasota, FL: Professional Resource Exchange.

Wadden TA, & Brownell KD (1984). The alteration of eating and nutrition habits and health populations. In JB Matarazzo, N Miller, SM Weiss, et al. (Eds.), Behavioral health: a handbook of health education and disease prevention (Ch. 39, pp. 608–631). New York: John Wiley & Sons.

Wadden TA, Foster GD, Letizia KA, et al. (1990). Long-term effects of dieting on resting metabolic rate in obese outpatients. JAMA 264:707–711.

Wadden TA, & Stunkard AJ (1986). Controlled trial of very low calorie diet, behavior therapy, and their combination in the treatment of obesity. J Consult Clin Psychol 54:482–488.

Wadden TA, Stunkard AJ, & Brownell KD (1983). Very low calorie diets: their efficacy, safety and future. Ann Intern Med 99:675–684.

Wadden TA, Stunkard AJ, & Liebschutz J (1988). Three-year follow-up of the treatment of obesity by very low calorie diet, behavior therapy, and their combination. J Consult Clin Psychol 56:925–928.

Wadden TA, Stunkard AJ, Brownell KD, et al. (1985). A comparison of two very-low-calorie diets: protein-sparing modified fast versus protein formula liquid diet. Am J Clin Nutr 40:533–539.

Wadden TA, Van Itallie TB, & Blackburn GL (1990). Responsible and irresponsible use of very-low-calorie diets in the treatment of obesity. JAMA 263:83–85.

Walters GS, Pories WJ, Swanson MS, et al. (1991). Long term studies of mental health after the Greenville gastric bypass operation. Am J Surg 161:154–158.

Weinberg N, Mendelson M, & Stunkard A (1961). Failure to find distinctive personality features in a group of obese men. Am J Psychiatry 117:1035–1037.

Weintraub MT, & Bray GA (1989). Drug treatment of obesity. Med Clin North Am 73:237–249.

Westover SA, & Lanyon RI (1990). The maintenance of weight loss after behavioral treatment: a review. Behav Modif 14:123–137.

Williamson DF, Madams J, Anda RF, et al. (1991). Smoking cessation and severity of weight gain in a national cohort. N Engl J Med 324:739–745.

Wilson GT, Nonas CA, & Rosenblum G (1992). Assessment of binge-eating in obese patients. Int J Eating Dis. In press.

Wing RR, Epstein LH, Norwalk MP, et al. (1985). Behavior change, weight loss and physiological improvements in type II diabetic patients. J Consult Clin Psychol 53:111–122.

Wing RR, Epstein LH, Paternostro-Bayles M, et al. (1988). Exercise in a behavioral weight control programme for obese patients with type 2 (non-insulin-dependent) diabetes. Diabetologia 31:902–909.

Wing RR, & Marcus MD (1992, in press). Binge eating among obese individuals. Curr Opin Psychiatry.

Wing RR, Marcus MD, Epstein LH, et al. (1987). Type II diabetic subjects lose less weight than their overweight nondiabetic spouses. Diab Care 10:563–566.

Wing RR, Marcus MD, Epstein LH, et al. (1991). A family based approach to the treatment of obese type II diabetic patients. J Consult Clin Psychol 59:156–162.

Wing RR, Marcus MD, Salata R, et al. (1991). Effects of a very- low-calorie diet on long-term glycemic control in obese type 2 diabetic subjects. Arch Intern Med 151:1334–1340.

Wise TN, & Gordon J (1977). Sexual functioning in the hyperobese. Obes/Bariatric Med 6:84–87.

Wurtman JJ (1988). Carbohydrate craving, mood changes and obesity. J Clin Psychiatry 49:8(suppl)37–39.

VII Medical-Legal Issues

44 Medical-legal issues

RODNEY J. S. DEATON, M.D., J.D.,
CHRISTOPHER C. COLENDA, M.D, M.P.H., AND
HAROLD J. BURSZTAJN, M.D.

Today, with the increasing legal regulation of medical practice, consulting psychiatrists often find themselves confronting legal questions as well as clinical questions as they conduct their evaluations. Even though most hospitals now retain attorneys to answer specific legal questions, a working knowledge of the most common legal issues arising in a medical-surgical setting can aid consulting psychiatrists in several ways.

First, whenever a legal issue arises during a consultation, clinicians can structure their evaluations so they may provide the medical team, hospital administrators, and attorneys with psychiatric information relevant to the particular legal issue at hand. Second, with general knowledge about medical-legal issues, consulting psychiatrists are better able to distinguish exactly which questions are legal and which are clinical. Moreover, in cases requiring careful risk management, consulting psychiatrists know when to avail themselves, if possible, of a consultation with a forensically trained psychiatrist—or perhaps a clinically astute hospital attorney or risk management officer.

In a medical-surgical setting, legal issues often arise in the context of crisis and uncertainty (Bursztajn, Feinbloom, Hamm, et al., 1990). In such situations the answer to a precise legal question may not solve underlying conflicts. Consulting psychiatrists should be on the lookout for these conflicts so they can apply their clinical expertise to try to resolve the conflicts as quickly and as satisfactorily as possible. With the resolution of the conflicts, the legal problem may, in fact, disappear.

This chapter focuses on the major legal issues likely to be encountered by consulting psychiatrists. When discussing each issue, we focus not only on legal doctrine but also on the clinical assessments and interpersonal dialogues clinicians should carry out, if necessary, to provide as complete and as useful an evaluation as possible. Because of space limitations, this chapter focuses only on issues of American law.

COMPETENCY

Competency is a legal term (Appelbaum & Grisso, 1988; Mahler & Perry, 1988). As such, it encompasses a familiar meaning of the term, that being the idea of capacity: the cognitive, emotional, and behavioral abilities a person must possess to be able to take certain actions. In addition, though, the term encompasses the powers an individual has, granted and protected by law, to exercise those capacities.

All adults are presumed to be competent, i.e., to have the psychological capacity and the legal power to undertake legally significant transactions, such as to execute a will or a contract. Similarly, all adults are presumed competent to decide what they will allow to happen to their bodies, such as whether to allow a physician to perform a medical procedure.

Only a judge can declare an individual to be incompetent, by means of a formal declaration issued pursuant to a formal judicial proceeding. During such a proceeding, the judge hears evidence about the individual's capacity to make the legal decisions in question from those representing the possibly incompetent individual and those questioning his or her competence. Then, after both sides have presented their cases, the judge is to make an independent assessment about the patient's capacities based on all the evidence presented. The judge then decides what powers, if any, the individual should be allowed to retain in order to exercise those capacities. If the judge declares the individual to be incompetent, the judge appoints for the individual a guardian or conservator to take the particular legal action(s) at issue in the case. The powers granted to the guardian may be broad, or they may be limited to making a particular decision, such as deciding whether a particular treatment should be instituted.

Even though competency is a legal decision, to be decided by a judge alone, consulting psychiatrists often are asked to assess a patient's "competency." Such a request is usually a request to assess a patient's psychological capacity to make a particular decision. Often, primary treatment teams request these consultations for older patients with cognitive disorders—and usually only when the patients refuse treatment (Golinger & Federoff, 1989; Myers & Barrett, 1986; Strain, Taintor, Gise, et al., 1985). Unfortunately—from both risk management and ethical standpoints—treatment staffs rarely, if ever, question a patient's competency when the patient and everyone surrounding the patient is in agreement with the action to be taken (Mahler, Perry, & Miller, 1990). Clinicians may not even recognize the more subtle forms of incapacity (Gutheil & Bursztajn, 1986).

In fact, without some disagreement no one may question the patient's capacity at all. Urgent requests for psychiatric consultation arise when a medical crisis has precipitated a family crisis manifested by ambivalence, intense affect, or accusations toward the treatment team, or has provoked intense disagreements over treatment options among family members or between family members and the treatment team (Mahler & Perry, 1988). Thus, as mentioned before, the psychiatrist often must make an accurate assessment of the patient's capacities and needs and then help all participants in the treatment decision to communicate more effectively with one another to bring about informed, fruitful decisions.

Finally, the consulting psychiatrist must remember that "competency" and "committability" are two different concepts legally and clinically. As is discussed below, "committed" persons—those required to remain in a facility for treatment for reasons of dangerousness to self or others—are still legally "competent" to make treatment decisions until they are adjudicated "incompetent" to do so. Similarly, from a clinical standpoint, persons may be dangerous to themselves or others yet still have the capacity to assess the risks and benefits of certain medical procedures. As a practical matter, many "committed" individuals do indeed lack the capacity to make informed choices about their treatment, especially their psychiatric treatment. Nevertheless, the consulting psychiatrist should never assume that these patients do lack that capacity and so must perform the following assessment on every patient, regardless of whether the patient is legally committed to the facility.

Assessment of Competency

A person with an adequate capacity to undertake a particular action should, at a minimum, be able to understand the basic nature of the proposed action in addition to the consequences, both good and bad, that may result if the proposed action is taken and if it is not taken. The patient need not understand the technical details of the action or its consequences yet should be able to articulate the most salient aspects of the action in layman's terms. The degree of understanding required may vary, depending on the particular action, usually a medical treatment, being considered (Mahler & Perry, 1988).

From a more complete psychological standpoint, patients should have the following basic capabilities: (1) an awareness of the nature of their present situation and insight as to the level of their impairment, if any exists; (2) a factual understanding of the issues with which they must deal; and (3) the ability to manipulate information rationally in order to reach a decision on these issues (Gutheil & Appelbaum, 1982).

To perform an evaluation, the psychiatrist should review the patient's medical history and current medical condition, looking especially for a history of changes in orientation or cognition. The clinician should then perform a thorough clinical interview of the patient, including a complete mental status examination, looking for evidence of impaired cognition or of a specific psychiatric disorder.

Although the psychiatrist should endeavor to make a diagnosis at this point, such a diagnosis does not end the inquiry: even persons suffering from severe schizophrenia may have the capacity to understand the nature and consequences of medical procedures. More importantly, even persons who have cognitive deficits—even permanent ones—may still have the capacity to understand the nature and consequences of certain procedures. A person diagnosed as having dementia is not thereby "incompetent" in the legal or the psychological sense. Each case is context dependent, and the clinician must go through all the above steps with even the most cognitively impaired patients to determine if they may indeed have the capacity to understand the nature and consequences of the particular procedure being proposed for them.

In order to perform the more complete psychological assessment, the psychiatrist should carry out the evaluation as a dialogue, not as a one-sided series of questions by the clinician followed by answers by the patient. The psychiatrist should encourage the patient to ask questions and to volunteer thoughts and feelings about the particular action to be taken. By doing so, the psychiatrist not only gains information about the thought processes and thought content of the patient but also may discover that the patient lacks information about the procedure or is refusing to undergo a procedure because of personal values or purposes he or she may wish to accomplish. Furthermore, this dialogue gives the psychiatrist an ability to assess the patient's capacity to form an alliance with the examiner, an often useful predictor of the patient's ability to engage in a course of treatment. In another context, one of the authors has termed this latter capacity "interpersonal competency" (Bursztajn & Hamm, 1982).

In order to understand the patient's current life situation, the psychiatrist should then speak with members of the treatment team and, with the patient's permission, family members. With each individual, the psychiatrist should try to learn that person's beliefs about the action to be taken and about the goals the person believes should be accomplished. Such conversations may promote further dialogue between the patient and others involved in the decision, so a mutually acceptable solution can be reached without need for further legal proceedings.

In the end, after thorough discussion by all parties, the issue of the patient's capacity to take the particular legal action still may be relevant. If so, the consulting psychiatrist may render an opinion on it. (See also Chapter 45.)

CONSENT

As a rule, consulting psychiatrists are most often asked to evaluate a patient's capacity to give (or refuse) consent for a specific treatment. As a legal doctrine, consent to treatment has undergone changes since the 1960s. Formerly, if a physician merely obtained a patient's consent to perform a procedure, that consent was sufficient to insulate the physician from a claim of battery, which is a tort (injury) defined as a harmful, offensive touching (Mohr v. Williams, 1905; Restatement (Second) of Torts § 18).

By the 1960s, however, courts began to require doctors not only to tell their patients that a procedure would occur but also to explain to patients the reasons for doing the procedure and the nature and consequences of it. The courts framed the requirement as a duty to obtain "informed consent," a duty that all physicians were declared to owe to their patients before undertaking any treatment or procedure. If a doctor fails to obtain such consent from a patient and if, as a result, the patient undergoes a treatment and then suffers physical injury, the doctor may then be liable to the patient in negligence, another type of tort (Canterbury v. Spence, 1972; Natanson v. Kline, 1960). Although courts have differed as to the scope of information that needs to be communicated, almost all states now require doctors to obtain, in at least some form, the informed consent of their patients before undertaking any medical treatment on them.

If the consulting psychiatrist is asked to determine if a patient has the capacity to give informed consent, the psychiatrist should evaluate if the patient understands the nature of the treatment proposed, the risks and benefits of the treatment, the nature of alternative treatments with their risks and benefits, and, importantly, the risks and benefits of no treatment. Again, such an evaluation should be conducted as a dialogue, with the patient free to ask questions and, hopefully, develop a strong alliance with the psychiatrist (Gutheil, Bursztajn, & Brodsky, 1984). If the patient is refusing to give consent, the psychiatrist should try to understand the reasons given for the refusal, looking for evidence of a psychiatric illness that could impair the patient's judgment, such as depression or dementia. Furthermore, the psychiatrist should try to ascertain whether any consent given is voluntary, free from any undue coercion by either the patient's family or the staff (Appelbaum & Grisso, 1988; Fogel, Mills, & Landen, 1986).

Unfortunately, the laws governing who may or may not have the power to give informed consent to medical treatment are not uniformly straightforward. In some situations, they are clear: Every competent individual who has the capacity to give informed consent to treatment has the legal right to give consent or to refuse it, even if that means the patient signs out of the hospital against the advice of the treatment team (Mills, 1985). Moreover, for incompetent individuals who have had guardians appointed to make treatment decisions for them, their guardians have similar powers. The law is far less clear, however, in the case of the patient who, though legally competent, lacks the capacity to give or to refuse consent for the medical treatment, such as the patient who is delirious or comatose.

In emergency situations in which lifesaving measures must be instituted, physicians may institute treatment (and necessary diagnostic procedures) immediately, without fear of incurring legal liability. In such cases, courts imply that the individual would consent to the procedures and declare that such emergencies constitute exceptions to the usual requirement of informed consent.

Under this exception many if not most patients suffering from an acute delirium may be treated, even if the patient has not expressly given consent for the treatment, as acute delirium does usually signal a medical emergency, requiring immediate diagnosis and if possible immediate treatment (Fogel, Mills, & Landen, 1986). If necessary to complete diagnosis and acute treatment, appropriate psychotropic medication may be administered to calm the agitated patient. If pharmacologic treatment is contraindicated or unsuccessful, appropriate physical restraints may be applied until diagnosis and acute treatment can be completed. If family is available to give consent in any of these situations, though, it is wise for clinicians to discuss the procedures with them (Overman & Stoudemire, 1988). (See Chapter 45.)

Once the emergency has passed—or without such an emergency—no long-term course of treatment can, as a matter of law in many states, be initiated without a patient's informed consent. One possible solution may be available: The psychiatrist and treatment team should ask friends or family members whether, while fully capacitated, the patient executed an advanced directive such as a durable power of attorney, thereby giving to a particular third party the authority to consent to treatment. (This document is explained in the next section.) If so, that person may give informed consent for the treatment.

Without such an advanced directive, further nonemergent treatment should not be instituted without the advice of the facility's general counsel, who should be acquainted with the legal options available in the particular jurisdiction. Of course, as a practical matter, medical teams do institute such treatment all the time, often depending on the consent of family members to do so (Appelbaum & Roth, 1984). As a matter of law, however, these teams are open to future liability for failure to obtain informed consent should the patient suffer harm as a result of the treatment (Areen, 1987).

Some states have in fact allowed family members to make these decisions by enacting health proxy laws that automatically empower next of kin or other individuals to give necessary consent for an incapacitated individual (Areen, 1987). Without such laws, however, no treatment may be authorized without some type of judicial intervention, whether that be a court order or a declaration of incompetence and the appointment of a guardian. Such a procedure may be time-consuming; many jurisdictions, however, have procedures to expedite hearings in medical cases.

In short, for many cases in which psychiatrists are called routinely to assess capacity to consent, such as individuals with dementia who have not been adjudicated incompetent, a team cannot, in a legal sense, safely institute treatment without judicial authorization. Psychiatrists' roles in these difficult situations are most important. They can explain these issues to the medical team in the context of the patient's particular needs and capacities; they can direct the team to consult legal counsel as quickly as possible; and then, if necessary, they can aid lawyers and judges to make informed decisions that lead to proper treatment for the patient. Moreover, should extensive legal procedures be necessary, they also can serve as a source of support for families and team members during the sometimes complex, often frustrating course of the proceedings. (See also Chapter 45.)

Advanced Directives and the "Right to Die"

In a related area—the morally and emotionally complex area of the withdrawal or withholding of treatment for severely ill, incapacitated individuals—psychiatrists also may be asked to take several roles. For example, because some patients try to plan for such occurrences by executing an advanced directive for treatment, the psychiatrist may be asked to evaluate the patient's capacity to execute such a document. The two most common advanced directive documents are the living will and the durable power of attorney (Overman & Stoudemire, 1988; Siner, 1989).

Living wills are documents, recognized in an increasing number of jurisdictions, that allow competent persons (in this situation, called testators) to leave instructions for future caregivers about what treatment measures they do or do not wish to have performed on them should they ever lack the capacity to give consent for a procedure. Living wills may take various forms, depending on the personal desires of the testator and, often, on state law requirements. Yet most involve some refusal by the testator of certain medical procedures, and most often leave instructions about the use of life-support systems should certain medical conditions be met. In some states, patients can execute living wills only upon receiving a diagnosis of a terminal illness (Areen, 1987).

A power of attorney is a document, executed according to the laws of a particular jurisdiction, that enables a third party to take legal actions on behalf of the person executing the document. The ordinary power of attorney is not valid once the person executing the document becomes incapacitated. A *durable* power of attorney, on the other hand, does remain valid, even when the one giving the power lacks capacity to make the decision himself. Durable powers of attorney ordinarily must be authorized by statute; and an increasing number of states are passing such statutes. Furthermore, in many of these states, the person executing the instrument can give the holder of the durable power the authority to make medical treatment decisions, even including the decision to withdraw treatment.

As a clinical matter, when evaluating a person who plans to execute one of the advanced directives the consulting psychiatrist should perform the usual evaluation for capacity, probing the understanding and the reasons behind any treatment requests or refusals by the patient. The psychiatrist should then document his or her clinical impressions, so that they are available to future decision makers should any dispute arise as to the validity of the directive.

As a legal matter, even though the case law interpreting these instruments remains in some flux, some legal authorities believe that a medical team that follows the directives outlined in these documents will not suffer liability for doing so. In fact, the team may face legal liability if it refuses to do so (Weir & Gostin, 1990). On the other hand, as a practical matter, a medical team that blindly follows such directives when the patient's competence has been questioned, or when there are signs that the patient had a change of heart subsequent to the signing, may still face litigation by distraught family members. In such situations, further consultation by a competent psychiatrist—ideally, one who is forensically trained—is highly advisable.

Most people, however, have not executed one of these advanced directives. Consequently, if the need arises to consider withdrawing life-supportive treatment from an incapacitated individual who has not left a directive, other legal procedures must be followed. Many states now have laws defining the procedures to be taken in these cases, loosely termed "right to die" cases (Areen, 1987; Weir & Gostin, 1990). Some states allow family members and medical personnel to decide what to do, generally free from the oversight of courts. Other states require judicial hearings similar to these in guardianship proceedings.

The United States Supreme Court, in its only decision to date on this issue, did not set any uniform standard procedure for the states to follow, instead allowing states to make their own rules (Cruzan v. Director, Missouri Department of Health, 1990; Orentlicher, 1990). To know what to do in particular cases, medical teams and consulting psychiatrists must work closely with family, knowledgeable attorneys, and, if possible, hospital ethics committees. Yet should a particular patient reveal to a consulting psychiatrist his or her wishes as to any advanced directive, that psychiatrist should document those wishes carefully and clearly in the patient's record. Had such a situation occurred in the case of Ms. Cruzan, with clear documentation of her wishes in her medical record, the Missouri Supreme Court possibly would have allowed her family, as her guardians, to carry out those wishes.

In these situations, psychiatrists may be asked to evaluate a surrogate decision maker's capacity to make a treatment decision. They also may be asked to assist in conflict resolution if the treatment team and the surrogate(s) are at odds over the decision of withdrawing or withholding medical support. Still, as families become aware of the seriousness of the patient's condition and begin to accept the inevitable outcome, more often than not they accept the recommendation to limit care (Luce, 1990). In these cases, the psychiatrist's role then may be to offer support, perhaps to perform crisis family therapy, and to facilitate communication between families and physicians over resolving particular problems. (See also Chapter 45.)

Consent for Psychiatric Treatment

The laws governing the giving of consent for psychiatric treatment are essentially the same as those governing consent for medical treatment. Competent, fully capacitated individuals have the right to give or refuse consent for psychiatric treatment. Individuals who have been declared incompetent to give consent for psychiatric treatment may be treated only with the approval of the guardian appointed for that purpose. (The rights of legally competent, yet psychiatrically incapacitated, patients are discussed in the next section.)

As a practical matter, however, these situations do pose different clinical and legal issues for consulting psychiatrists, as here they often are the doctors implementing the treatment. Consequently, they often are not only evaluating the patient's capacity for giving informed consent but also are seeking informed consent from the patient for treatment.

Nevertheless, the clinical evaluation to be done is the same: The clinician must engage the patient in a dialogue about the risks and benefits of the treatment advocated, as well as the risks and benefits of alternative treatments and of no treatment. In doing so, the clinician is well advised to discuss the long-term effects of the proposed treatment, such as tardive dyskinesia from neuroleptics, because as a legal matter some patients have collected sizable damages from psychiatrists for the inappropriate use of these drugs (Wettstein, 1983). Should psychiatric treatment continue over a long period, psychiatrists should periodically reassess the need for the treatments chosen, monitoring the effects of the treatments, especially if neuroleptics are being used, and discussing with their patients the need for such treatment (Mills, 1985).

Refusal of Psychiatric Treatment

In many states, specific laws govern the psychiatric treatment of legally competent, yet psychiatrically incapacitated, individuals—those with treatable psychiatric disorders who are refusing treatment for that disorder. "Psychiatric" disorders in this context generally means, to use older nomenclature, "functional" disorders (e.g., schizophrenia and mood disorder) compared to "organic" or cognitive disorders (e.g., delirium and dementia). Consulting psychiatrists must have a working knowledge of the laws in their jurisdiction governing the "right to refuse treatment" (Gutheil & Appelbaum, 1982), as well as the laws governing civil commitment (Fogel, Mills, & Landen, 1986).

As with cases of medical treatment, when the patient is in immediate danger of harming himself or others psychiatrists may institute emergency treatment, usually neuroleptics, without the patient's consent and without immediate resort to review procedures (Rogers v. Commissioner of the Department of Mental Health, 1983). Beyond these limited emergency situations, states vary as to the procedures psychiatrists must follow in order to be able to treat these individuals (Appelbaum, 1988). Some states require a full judicial hearing, leading to the appointment of a specialized guardian for the purposes of psychiatric treatment. Other states allow more informal review of the treatment decision by disinterested psychiatrists. Still others allow the physician to treat a patient unless the patient personally requests judicial oversight. In its only major decision in this area, a case involving a prisoner's refusal of antipsychotic medications, the United States Supreme Court did not promulgate uniform standards for the country (Washington v. Harper, 1990). Thus consulting psychiatrists must still be familiar with the procedures followed in their own state so they can advise medical staff accordingly.

CONFIDENTIALITY

As an ethical precept, physicians have adhered to the value of confidentiality at least since the time of Hippocrates. Today this ethical value is also embodied in legal principle: every physician owes a duty of confidentiality to every patient. Except in limited circumstances, the physician cannot volunteer information about a patient to a third party without the patient's permission (Gutheil & Appelbaum, 1982).

Generally speaking, most patients have consented to the psychiatrist's sharing of information with members of the hospital staff, if not expressly then implicitly, by consenting to the consultation, knowing that it was ordered by a member of the staff. Still, the psychiatrist should share only such information as is necessary for the treatment team to understand the patient's psychiatric condition and needs, so they can assist in any treatment planned.

On the other hand, a patient's agreement to a consultation does not imply consent for a psychiatrist to share information with family members. If consulting psychiatrists plan to interview other family members, they should ask the patient's permission to share any confidential information with them, making a note of such consent in the chart. If the patient refuses consent, the psychiatrist should explore the reasons for the refusal and, if necessary, explain why a good evaluation and treatment plan may depend on the communication. If the patient persists in refusal, the psychiatrist should not reveal the information to the family but should notify the referral source of the refusal and explain any limitations it places on the completeness of the evaluation.

Except for a medical emergency involving the patient, such as a transfer to another facility for more intensive care, the psychiatrist must receive the patient's express permission to release information to third parties outside the hospital or clinic. Again, in an emergency, the law implies that a patient would consent to the release of information that would help third parties carry out treatment. As when sharing information with the treatment team, the psychiatrist should share only information essential to the immediate treatment of the patient.

One situation merits special comment: the release of information to third party payors and managed care companies. As a legal matter, most patients have, as part of their insurance agreement, given explicit consent for these organizations to receive information about their diagnosis and treatment. Yet psychiatrists should interpret this consent narrowly. Certainly, the psychiatrist should not release any information that could not be discussed openly with the patient or guardian. Even then, the information released should be only relevant information that is necessary to determine the need for and the scope of any treatments given. If the companies demand more information, the consulting psychiatrist should consider discussing the situation directly with the patient or guardian to determine what steps should be taken to guarantee appropriate confidentiality.

Confidentiality and the Duty to Warn or Protect

Ever since the California Supreme Court decided that psychiatrists in that state were under a duty to protect third parties from the threatened actions of their patients (Tarasoff v. Regents of the University of California, 1976), psychiatrists throughout the United States have wondered to what extent they must take similar actions. As a legal matter, states vary in their requirements. Some states have no such requirements, and in recent years some states have been narrowing the scope of information that must be reported in these situations (Appelbaum, Zonana, Bonnie, et al., 1989; Beck, 1988).

Clinically, the issue in all these situations is generally the same: The patient announces to some individual that he plans to take some action that would lead to harm to a third party. When consulting psychiatrists interview such patients, in addition to performing a thorough mental status examination, they should explore in detail the patient's contemplated action, endeavoring to assess the patient's thoughts, feelings, motivations, and readiness to act. Furthermore, the clinician should explore with the patient the consequences of the patient's proposed action, trying if possible to form an alliance with the part of the patient that feels ambivalent about the potential victim and that does not wish to harm the victim (Wulsin, Bursztajn, & Gutheil, 1983). Such an approach may help the patient gain self-control, thereby obviating a need for further intervention.

Nevertheless, there are times when such an approach is not successful or, for various situational reasons, cannot or should not be followed. If the patient is threatening physical violence against a third party and the psychiatrist believes that the patient is capable of carrying out those threats in the near future, the psychiatrist should discuss the concern with the medical team. If appropriate, the psychiatrist should institute some type of treatment, possibly civilly committing the patient should the patient meet the state's commitment criteria. If this step cannot be taken, in conjunction with the medical team and legal counsel the psychiatrist should consider notifying some third party that can take appropriate action, whether it be local police authorities, the potential victim, family members, or someone else. Because of the variation in legal requirements among states, no general legal rule can be given (Appelbaum, Zonana, Bonnie, et al., 1989). Above all, the psychiatrist should try to take the action that best preserves whatever alliance has been established with the patient and involves the least possible breach of confidentiality—yet that also protects the endangered third party (Wulsin, Bursztajn, & Gutheil, 1983).

MALPRACTICE

Since the so-called "malpractice crises" of the 1970s and 1980s, psychiatrists, like all physicians, have been concerned about how to avoid actions that could lead to a lawsuit. Consulting psychiatrists may indeed face risks of malpractice liability. To understand these risks correctly, however, they need to understand the legal concept of malpractice and how malpractice is proved at a trial.

Like the failure to obtain informed consent, malpractice is one type of negligence, a tort defined as a harm that occurs when one person breaches a duty of care owed to another person (Restatement [Second] of Torts § 281, 1965). Generally, whenever a person takes any action that conceivably could lead to someone's harm, that person is under a duty to perform the action in, as the law puts it, a reasonable manner. If the person performs the action in an unreasonable manner, and another person is harmed as a result, the first person is said to be negligent and thus liable for the ensuing damages.

The meaning of "reasonable" is more carefully defined for professionals such as physicians, who perform services for other individuals. For physicians, to perform reasonably is to perform consistent with standards of good care, defined as the care and performance that would be given under similar circumstances by a reasonably knowledgeable and skilled physician with training similar to that of the doctor being evaluated. As a rule now, the standard reflects the reasonable practices of physicians throughout the United States, a national standard, not just the practices of physicians from the region of the doctor being evaluated (Restatement [Second] of Torts § 299A, 1965).

In any case, legal decision makers, whether they be a judge or a jury, learn what the appropriate standard of care is by hearing testimony from expert witnesses who define that standard for them and then give them an opinion as to whether the particular professional's actions met that standard. These expert witnesses are doctors, some of whom testify in court for a living, others of whom testify only in certain cases. For almost all cases, it is these doctors who testify in courts, not judges, who are setting the standards by which, as a legal matter, fellow doctors will be judged.

The latter point is important for all physicians, including consulting psychiatrists, to grasp. Judges and juries do not determine what is good medical care. Doctors who testify in court determine what is good medical care. Judges and juries determine only which testifying doctor is most persuasive and credible when there is a difference of opinion in a particular case.

Given the way the system works, consulting psychiatrists can take several actions to reduce the risk of malpractice liability. Primarily, they must strive to be knowledgeable about current practices in psychiatry, as defined in the leading journals and texts of the profession, and then conform their practices to those standards. Those standards are the ones to which physician expert witnesses refer in their evaluations.

Psychiatrists practicing in medical settings should be well versed in the medical conditions that can cause mental or behavioral symptoms, and in the medical side effects of psychotropic drugs, particularly severe ones such as the neuroleptic malignant syndrome. They should look for medical conditions that may have been missed by the patient's referring doctor. They should keep in mind that patients may suffer from more than one psychiatric disorder, paying particular attention to possible comorbid conditions such as alcohol or drug abuse.

Furthermore, medical psychiatrists should prescribe only medications the indications and effects of which they understand fully. They should discharge patients only when they are stable enough, both medically and psychologically, to continue treatment at home. They should understand the time courses and prognoses of the illnesses they treat.

In short, the expert witness evaluates whether the consulting psychiatrist performed in a manner consistent with the practice of a "reasonable" (i.e., competent and careful) medical psychiatrist. Moreover, should the psychiatrist also assume responsibility for the concurrent medical care of the patient, the psychiatrist's treatments will be compared not only to those of the "reasonable" medical psychiatrist but also to those of the "reasonable" internist, family practitioner, or other appropriate medical specialist, depending on the type of medical treatment the psychiatrist chooses.

Beyond having information about treatments, medical psychiatrists also must be able to justify, in a risk-benefit format, the reasons for choosing a particular diagnosis or treatment when they apply this knowledge to particular cases. From a risk management standpoint, they then must write these reasons down in the patient's record, using phrases such as the following:

> The patient appears to have X diagnosis because. . . .
> Treatment will be instituted so that. . . .
> This treatment has the following risks. . . .; nonetheless the risks are outweighed by the following benefits. . . .
> The patient has been informed of the above treatment plan, as well as of the following treatment options. . . .
> As evidenced by the following, the patient has the capacity to give informed consent. . . .

By taking this step the psychiatrist demonstrates to later readers of the record, specifically expert witnesses, that not only did he or she possess the reasonable practitioner's knowledge of the patient's condition, but that his or her knowledge was applied reasonably in the particular case and a competent patient's informed consent was obtained before proceeding with treatment.

If the psychiatrist has followed the recommendations made in this chapter by conducting each evaluation as a dialogue, one in which the patient is respected as an active participant, who not only can give information but can ask questions and then give even more pertinent information based on those questions, the physician almost certainly has gained the maximum amount of information possible to make accurate assessments and institute well-reasoned, beneficial treatment plans. In addition, by promoting such a dialogue, the psychiatrist has established a solid alliance with the patient, one that can weather the occasional adverse outcome that occurs (Gutheil, Bursztajn, & Brodsky, 1984). Furthermore, the patient knows that the doctor who has treated her with such respect before can be relied on to continue to treat her, despite the adverse outcome, with similar expertise and respect. A solid alliance is not a guarantee that the patient will not later file a malpractice suit, but it remains the best tool of a psychiatrist not only to conduct good patient care but to carry out successful risk management.

OTHER ISSUES

Sexual Misconduct and Other Boundary Violations

Sexual contact between clinician and patient is always ethically wrong. Depending on the clinical facts and state law, it also may be grounds for a civil or criminal action (Appelbaum, 1990). It seems to be prevalent across all clinical specialties. The consulting psychiatrist who becomes aware of such misconduct by any member of the clinical team has a duty to take steps to ensure that it stops. Furthermore, proper supervision of clinical staff includes education about ubiquity of transference and the wrongfulness of the acting-out of countertransference responses. Psychiatrists who fail to provide such education—even in instances in which they are not aware of any sexual acting-out by any member of the clinical team—may face liability for negligently failing to monitor those individuals under their supervision (Bursztajn, 1990).

If psychiatrists perform physical examinations, they may wish to take reasonable steps to provide proper chaperoning of the examination, tailoring the exact extent of such chaperoning to the particular patient's psychological state. Not every examination requires a chaperon, but psychiatrists should use good clinical judgment to determine which situations may call for one. For example, they may wish to have a chaperon present when examining a suspicious, frightened patient or one who is known to be considering litigation against the psychiatrist or another clinician.

Physical and Chemical Restraints

It is well known that staffs on general medical-surgical services use both physical and chemical restraints to prevent patients from harming themselves and to allow staff to continue providing treatment to combative, agitated patients (Appelbaum & Roth, 1984). Consulting psychiatrists may be asked questions on the use of behavioral interventions and psychopharmacologic agents to control combative patients and thus should be familiar with some of the legal issues involved in their use.

From a legal standpoint, most cases and regulations in this area have not involved the use of restraints in acute care medical facilities but, rather, their use in chronic care and psychiatric facilities (Johnson, 1990). The current state of the law is not clear. Technically, the use of restraints, either physical or chemical, could give rise to tort claims by patients of battery and false imprisonment. On the other hand, caretakers are also under a duty to provide quality care for patients and to prevent them from harming themselves or others. Thus treatment teams probably are privileged to use these techniques to a limited extent, but only to accomplish treatment and protection goals, not to punish or to avoid proper supervision of patients.

Many states have promulgated detailed guidelines for the use of restraints in nursing homes and psychiatric facilities, and the federal government has promulgated guidelines for the use of restraints in facilities such as nursing homes (Johnson, 1990). As a general rule, these guidelines have required that restraints be used only for short periods and only for the purposes of protecting the patient or furthering treatment; they should be instituted and monitored only under the supervision of a physician. Even if not required by regulations within their states, staffs at acute medical facilities should follow similar procedures, using restraints only when necessary, carefully documenting the need for their use, the alternative measures attempted, and the frequent reassessments of their need. Certainly if it appears that the need for restraints, even if only intermittent, may be necessary for any extended period, psychiatrists and medical staff should seek guidance from hospital counsel as to what procedures to follow.

As a clinical matter, consultation with a psychiatrist may be critical in helping staff avoid the use of restraints by suggesting behavioral and environmental interventions that may be applied. If the psychiatrist and staff determine that such interventions may not sufficiently protect the patient to allow further medical treatment, the psychiatrist can make certain that, if psychopharmacologic agents are used to calm the agitated patient, appropriate agents are used, considering the age and medical status of the patient (Fogel, Mills, & Landen, 1986). For any patient who is treated for any period with neuroleptic agents, consulting psychiatrists should document the risk-benefit rationale of such a treatment plan. The presence of extrapyramidal movements and tardive dyskinesia should be noted; and the patient's capacity to give or to withhold informed consent should be documented.

Medical-Legal Issues involving Persons with AIDS

As the acquired immunodeficiency syndrome (AIDS) epidemic spreads, psychiatrists are increasingly asked to consult on patients infected with the human immunodeficiency virus (HIV). The legal implications of these consultations are only beginning to be sorted out (Gostin, 1990; Orr, 1989). As a rule, the major issues involved in these evaluations are the same as those discussed earlier in this chapter. A few points, however, are worth noting.

The HIV virus not only attacks a patient's immune system; it also infects the central nervous system, potentially causing a variety of neuropsychiatric conditions, most of which can temporarily or permanently compromise an individual's capacity to reason and make informed judgments. Hence the infection may impair an individual's capacity to make decisions about treatments or to execute wills, contracts, or other legal documents (Gostin, 1990). Furthermore, when these questions arise, it usually follows that deep, divisive rifts occur within families or among loved ones. Consequently, a psychiatrist's role in these situations may include not only a systematic evaluation of the patient's cognitive and decision making capacities but also conscientious counseling and even family therapy.

Confidentiality issues also are important in these cases. As a rule, the consulting psychiatrist is not the first physician to learn of a patient's HIV infection, obviating many of the problems involved with reports to public health officials. Still, during their consultations they may learn that the patient is actively involved in sexual or needle-sharing behavior with partners who have not been informed of the patient's HIV seropositive status. It remains an open legal question whether the psychiatrist is under a Tarasoff-like duty to warn the partners of the patient's status (Brennan, 1989; Kermani & Weiss, 1989; Melton, 1988). Some states have legislated that individuals such as spouses, funeral directors, emergency workers, sexual assault victims, laboratory workers, and others may or must be notified of an individual's HIV status should they have had risk-related contact with the individual (Gostin, 1989). As a clinical matter, psychiatrists should follow the guidelines outlined in the earlier discussion on the duty to warn that advocate exploring with the patient his or her reasons for not notifying the third party at risk. If the individual plans to persist in the dangerous behavior, however, the psychiatrist should discuss with hospital counsel what steps should be taken in conformity with the laws of the jurisdiction.

REFERENCES

APPELBAUM PS (1988). The right to refuse treatment with antipsychotic medications: retrospect and prospect. Am J Psychiatry 145:413–419.

APPELBAUM PS (1990). Statuses regulating patient-therapist sex. Hosp Community Psychiatry 41:15–16.

APPELBAUM PS, & GRISSO T (1988). Assessing patients' capacities to consent to treatment. N Engl J Med 319:1635–1638.

APPELBAUM PS, & ROTH LH (1984). Involuntary treatment in medicine and psychiatry. Am J Psychiatry 141:202–205.

APPELBAUM PS, ZONANA H, BONNIE R, ET AL. (1989). Statutory approaches to limiting psychiatrists' liability for their patients' violent acts. Am J Psychiatry 146:821–828.

AREEN J (1987). The legal status of consent obtained from families of adult patients to withhold or withdraw treatment. JAMA 258:229–235.

BECK JC (1988). The therapist's legal duty when the patient may be violent. Psychiatr Clin North Am 11:665–679.

BRENNAN TA (1989). AIDS and the limits of confidentiality: the physician's duty to warn contacts of seropositive individuals. J Gen Intern Med 4:242–246.

BURSZTAJN HJ (1990). Supervisory responsibility for prevention of supervisee-patient sexual contact. Presented at the Massachusetts Psychiatric Society, Newton, Massachusetts.

BURSZTAJN HJ, FEINBLOOM RI, HAMM RM, ET AL. (1990). Medical choices, medical chances: how patients, families, and physicians can cope with uncertainty. New York: Routledge, Chapman & Hall.

BURSZTAJN HJ, & HAMM RM (1982). The clinical utility of utility assessment. Med Decis Making 2:161–165.

Canterbury v. Spence (1972). 464 F2d 772 (D C Cir), cert. den., 409 US 1064.

Cruzan v. Director, Missouri Department of Health (1990). 110 S Ct 2841.

FOGEL BS, MILLS MJ, & LANDEN JE (1986). Legal aspects of the treatment of delirium. Hosp Community Psychiatry 37:154–158.

GOLINGER RC, & FEDOROFF JP (1989). Characteristics of patients referred to psychiatrists for competency evaluations. Psychosomatics 30:296–299.

GOSTIN LO (1989). The politics of AIDS: compulsory state powers, public health and civil liberties. Ohio State Law J 49:1017–1058.

GOSTIN LO (1990). The AIDS Litigation Project: a national review of court and human rights commissions. Part I. The social impact of AIDS. JAMA 263:1961–1970.

GUTHEIL TG, & APPELBAUM PS (1982). Clinical handbook of psychiatry and the law. New York: McGraw-Hill.

GUTHEIL TG, & BURSZTAJN HJ (1986). Clinician's guidelines for assessing and presenting subtle forms of patient incompetence in legal settings. Am J Psychiatry 143:1020–1023.

GUTHEIL TG, BURSZTAJN HJ, & BRODSKY A (1984). Malpractice prevention through the sharing of uncertainty: informed consent and the therapeutic alliance. N Engl J Med 311:49–51.

JOHNSON S (1990). The fear of liability and the use of restraints in nursing homes. Law Med Health Care 18:263–273.

KERMANI EJ, & WEISS BA (1989). AIDS and confidentiality: the legal concept and its application in psychotherapy. Am J Psychother 43:25–31.

LUCE JM (1990). Ethical principles in critical care. JAMA 263:696–700.

MAHLER J, & PERRY S (1988). Assessing competency in the physically ill: guidelines for psychiatric consultants. Hosp Community Psychiatry 39:856–861.

MAHLER JC, PERRY S, & MILLER F (1990). Psychiatric evaluation of competency in physically ill patients who refuse treatment. Hosp Community Psychiatry 41:1140–1141.

MELTON GB (1988). Ethical and legal issues in AIDS-related practice. Am Psychol 43: 941–947.

MILLS MJ (1985). Legal issues in psychiatric treatment. Psychiatr Med 2:245–261.

Mohr v. Williams (Minn 1905). 104 N W 12.

MYERS BG, & BARRETT CL (1986). Competency issues in referrals to a consultation liaison service. Psychosomatics 27:782–789.

Natanson v. Kline (1960). 186 Kan 393, 350 P2d 1093.

ORENTLICHER D (1990). The right to die after Cruzan. JAMA 264:2444–2446.

ORR A (1989). Legal AIDS: implications of AIDS and HIV for British and American law. J Med Ethics 15:61–67.

OVERMAN W, & STOUDEMIRE A (1988). Guidelines for legal and financial counseling of Alzheimer's disease patients and their families. Am J Psychiatry 145:1495–1500.

Restatement (Second) of Torts §§ 18, 281, 299A (1965).

Rogers v. Commissioner of the Department of Mental Health (1983). 390 Mass 489, 458 NE2d 308.

SINER DA (1989). Advance directives in emergency medicine: medical, legal and ethical implications. Ann Emerg Med 18:1364–1369.

STRAIN JJ, TAINTOR Z, GISE LH, (1985). Informed consent: mandating the consultation. Gen Hosp Psychiatry 7:228–233.

Tarasoff v. Regents of the University of California (1974/1976). 13 Cal 3d 177, 529 P2d 553, 118 Cal Rptr 129 (1974); modified on rehearing, 17 Cal 3d 425, 551 P2d 334, 131 Cal Rptr 14. (1976).

Washington v. Harper (1990). 110 S Ct 1028.

WEIR RF, & GOSTIN L (1990). Decisions to abate life-sustaining treatment for nonautonomous patients: ethical standards and legal liability for physicians after Cruzan. JAMA 264:1846–1853.

WETTSTEIN RM (1983). Tardive dyskinesia and malpractice. Behav Sci Law 1:85–107.

WULSIN LR, BURSZTAJN HJ, & GUTHEIL TG (1983). Unexpected clinical features of the Tarasoff decision: the therapeutic alliance and the "duty to warn." Am J Psychiatry 140:601–603.

45 Medical legal issues of dementia

WILLIAM H. OVERMAN, J. D.

As more of the population survives to old age, legal issues involving the neuropsychiatric disorders of later life will occur with ever-increasing frequency. This chapter addresses some of these issues and describes the elements of legal planning for the aging patient at risk for dementia or displaying early signs of cognitive impairment.

PRESENT THE DIAGNOSIS

When the diagnosis is dementia, it is of the utmost importance for the clinician to thoroughly explain the diagnosis, although some degree of care must be taken when deciding what the patient is told directly about the diagnosis and prognosis. Some clinicians advocate a blunt approach, whereas others "titrate" the amount of information given the patient. One must at least inform the patient of any significant cognitive problems that have been detected on examination and how they might progress over time, especially to the extent that the individual's ability to make decisions may be impaired. It is important for the clinician to keep in mind that as the patient's competency diminishes, so does the number of planning options that are available.

Care should be taken not to offer false hope to the patient or family, as it may facilitate denial and cause delays in making critical legal and financial arrangements for the future. One can point out that common sense and prudence dictate "preparing for the worst" so the patient and the family can cope with any eventuality. Thus it is critical that the patients and their families be advised to seek professional advice from their attorney and accountant while the patient is still in the early stages of the illness. In this way the clinician plays a crucial role in helping the client and family understand the need for professional legal and financial planning (Overman & Stoudemire, 1988).

This chapter contains revised materials that originally appeared in Overman W, & Stoudemire A. Guidelines for legal and financial counseling of Alzheimer's disease patients and their families. *American Journal of Psychiatry* 145:1495–1500, 1988.

PLANNING WITH A PURPOSE

The primary purpose of planning for the aging patient is to preserve autonomy in later life. Proper planning also provides protection for assets and income, minimizes the risk of economic exploitation, and provides the means for all professions to provide the services the patient needs. In the legal area, proper advance planning means at a minimum that four foundation legal documents need to be put in place: a Living Will; a Durable Power of Attorney for Health Care; a Durable Financial Power of Attorney; and a Last Will and Testament. Essential for the patient's planning in the legal, financial, and medical areas, they enable the physician to practice medicine with a minimum of legal and financial complications. These documents can only be validly executed by competent people; thus it is crucial to complete them before the patient with a progressive dementia becomes incompetent.

Properly prepared documents provide protections for the patient, for the physician, and for health care facilities involved in the patient's care. Without proper safeguards and controls, however, they can also become a great source of danger and abuse. Thus the need for experienced professional advice and guidance becomes paramount.

CONSENT; INFORMED CONSENT; COMPETENCY

Issues of competency may be among the first to arise when assessing patients with dementia, as these issues impact directly on the legal ability of the patient to execute legal documents and to make medical decisions. In this regard it is important to note that the legal definition of the term "competency" differs, depending on whether the issue at hand is competency to execute legal documents or is the capacity to make medical decisions.

Competency and the Doctrine of Informed Consent

Competency assessment in the area of determining capacity to make legal decisions is based on the legal doctrine of consent. Derived from English common law, it gives patients the right to make their own decisions about (1) whether and (2) how they receive medical care. This doctrine of consent is composed of two closely related rules: (1) the rule of consent, which requires that the physician obtain the patient's permission before rendering a treatment; and (2) the rule of informed consent, which requires that the patient's permission be based on a reasonable or appropriate explanation of the proposed treatment by the physician (Meretsky v. Ellenby, 1979; Pugsley v. Privette, 1980; Schloendorff v. Society, 1914). Contrary to the beliefs of many, physicians do not automatically have unrestricted authority or the right to treat a patient without the patient's permission.

For a patient to be able to exercise the right of consent, he or she must first understand from the physician the exact nature of the diagnosis, the prognosis, and what course of treatment is to be expected. The physician must inform the patient to such an extent that a reasonably informed decision can be made (Natanson v. Kline, 1960; Salgo v. Board, 1957). The courts have found that patients must also be advised of the risks they run if treatment is declined. This corollary of the doctrine of informed consent is often referred to as the requirement of "informed refusal" (Crisher v. Spak,1983; Truman v. Thomas, 1980). These doctrines have been codified in all states, and local law should be consulted to ascertain the legal standards specified.

Consent can be considered to be informed only if the patient (1) is competent, (2) has been provided with sufficient information to reach an informed decision, and (3) has not been coerced. The consent must be given voluntarily (Kaimowitz v. Michigan, 1973/1974; Macdonald, Meyer, & Essig, 1986; Mills & Daniels, 1987).

The essence of the first criterion for informed consent, the matter of competency, is "that a particular patient at a particular time have the capacity to make a meaningful choice" about a particular matter at hand (Macdonald, Meyer, & Essig, 1986). In the context of informed consent, "competency" as a term usually denotes "capacity" or, perhaps more accurately, "decisional capacity." The term "capacity" means the quantum of legal capacity necessary to make decisions on one's own behalf.

The determination of competency is a legal issue decided by the courts. Thus a physician may perform a mental status examination and render an opinion regarding the ability of a patient to make a reasoned judgment about a particular matter, but the actual decision about competency is made by the court (Appelbaum & Roth, 1981; Macdonald, Mayer, & Essig, 1986; Mills, 1985; Roth, Meisel, & Lidz, 1977). Different legal standards of competency apply in different situations (e.g., the degree of competency needed to make a valid will is different from the degree needed to enter into a valid contract; both of these areas are different from the degree of competency needed to execute a valid power of attorney or to be able to give effective consent for medical treatment). The legal assessment of competency is made on the extent to which a person possesses (1) a set of values and goals; (2) an ability to understand information and to communicate; (3) an ability to reason and deliberate about the choices and their consequences.

Several other aspects of assessing competency should be addressed, as outlined by Mills and Daniels (1987). Does the patient have a clear choice with respect to the proposed diagnostic or treatment procedures? Is the choice the patient has made reasonable? Is the choice rational? Does the patient possess the ability to make a reasonable decision? Is the patient able to comprehend and understand the relevant information required to make a rational choice? Most courts apply a reasonability standard when deciding how much information a physician is required to release to a patient. This standard is sometimes called the "materiality rule." The physician is required to disclose information that a reasonable person in the patient's position would consider necessary for deciding whether to undergo a proposed treatment (Canterbury v. Spence, 1972; Cobbs v. Grant, 1972; Wilkinson v. Vesey, 1972).

Although legally the determination of competency is made final by a court decision, unless the question of competency is brought to the court it is usually the physician who determines "competency" on a day-to-day basis. The law provides legal standards, yet looks to the physician to apply the standards. Thus it is incumbent on the physician to understand the legal standards and how they apply in variant situations. Given that the law recognizes that a particular person's level of capacity may vary, and given the current emphasis on personal autonomy and rights of self-determination, it thus becomes critical that the physician properly and adequately determine competency and adequately document that determination. In this regard, with the legislative attention being given nationally to advance medical directives and informed consent, it appears that the documentation of the determination of competency by the physician and the patient's grant of consent to treatment must appear in the patient's chart and medical file. If the physician believes that there is a question of competency that the physician cannot resolve, the final decision must be made by the appropriate court in a legal proceeding.

It is clear that the doctrine of informed consent has resurfaced as a legal-medical issue. This doctrine of informed consent has resurfaced as a "patient's rights" approach that has been disseminated throughout the health care field. In 1973 the American Hospital Association published a Statement on a Patient's Bill of Rights. During the 1980s this approach gained momentum as state legislatures around the country passed laws creating "Bills of Rights" for hospital and long term care facility patients. At the national level, the Nursing Home Reform Act (Omnibus Budget Reconciliation Act, 1987) includes a mandatory Patient Bill of Rights, and in 1990 Congress passed the Patient Self-Determination Act (Omnibus Budget Reconciliation Act, 1990), which focuses on patient rights in the area of medical decision making. (See also Chapter 44.)

Incompetency

Issues of consent and competency quickly arise regarding patients with Alzheimer's disease and other dementias in which competency is eventually lost as the disease progresses. For this reason, reassessment of the patient's competency, done at the time of any medical, legal, or financial decisions, is usually necessary before proceeding with specific recommendations regarding the affairs of the patient.

In this regard, legal standards of competency should be compared with the general criteria of assessing legal incompetency. Although these criteria can be applied for other purposes (such as finding a person incompetent to have executed a valid contract), they appear in the most detail as legal standards for guardianship purposes. A typical current statutory definition of legal "incompetency" is "one who, by reason of mental illness, drunkenness, drug addiction, or old age is incapable of caring for themselves or providing for their family, or is liable to dissipate their property or become the victim of designing persons" (Brakel, Parry, & Weiner, 1986). The Uniform Probate Code uses the term "incapacity" rather than "incompetency" and defines an incapacitated person as "any person who is impaired by reason of mental illness, mental deficiency, physical illness or disability, advanced age, chronic use of drugs, chronic intoxication, or other cause (except minority) to the extent that he lacks sufficient understanding or capacity to make or communicate responsible decisions concerning his person" (Uniform Probate Code, 1969).

In guardianship proceedings, the Court makes two findings when finding someone to be "incompetent": (1) the person suffers from a condition affecting mental capacity; and (2) certain functional disabilities result from the condition. The common element in traditional state definitions of incompetent persons is an inability to perform essential life functions resulting from a lack of mental capacity. Generally there is no requirement of absolute inability or incapacity to function but, rather, only an inability to manage affairs "adequately" or to make "responsible" decisions. Details of competency assessment, including procedures for obtaining informed consent when patients are adjudged not to be competent, may be found elsewhere (American Psychiatric Association Task Report, 1986; Auerbach & Banja, in press, 1993; Mills & Daniels, 1987; Nolan, 1984; Roth, 1982).

Emergencies and Implied Consent

In certain emergency situations, medical decisions may need to be made for the mentally incompetent dementia patient. If emergency treatment is needed, the doctrine of implied consent may be invoked (Mills & Daniels, 1987; Fogel, Mills & Landen, 1986). This doctrine allows the physician to evaluate and treat the emergency condition, even against the apparent will of the patient, and is an exception to the general rule requiring informed consent. Even so, this exception is circumscribed by the legal requirement that if the patient is unconscious or otherwise unable to consent, the physician has a duty to attempt to contact the patient's family or guardian. In a South Dakota case (Dewes v. Indian Health Service, 1980), the court refused to permit the physician to rely on the emergency exception because the unconscious patient's parents were in the next room and could easily have been asked to consent.

Incompetency in Nonemergencies

If the medical situation is not an emergency, several strategies may be considered in the event of a patient's incompetency (Mills & Daniels, 1987; Fogel, Mills, & Landen, 1986). First, a second opinion about the care of the patient may be obtained. Simultaneously, legal consultation from the hospital may be secured to guide further decision making and legal proceedings if necessary, particularly for any matter in which life support systems or the use of resuscitation procedures is involved. Although in most states the spouse may consent for the patient in emergency situations if the patient is incompetent, the same may not be true in nonemergency situations.

Parents, siblings, children, and common-law spouses may or may not legally give consent for incompetent adult patients in medical matters, depending on state law. This area of consent to treatment for the incompetent patient is one in which the law is currently evolving from day to day. A number of states have revised their laws to provide for such consent, but many states do not have such laws. Formal legal proceedings may be advisable to designate a guardian or determine who gives consent for the incompetent patient. This point is especially true in any situation involving the possible use of extreme resuscitation procedures or life support systems. Emergency guardianship arrangements are usually available in most states and generally can be obtained within 5 days to a week. In addition, some states now have laws that provide for "emergency emergency" guardianships, or "medical emergency" guardianships, which generally can be obtained within 24 to 48 hours. If no family member is available, the court can appoint a guardian. Local law must be consulted. In the absence of such laws, the rule of implied consent may apply even though the medical situation is not a true "emergency," but professional advice should be sought if this question arises.

GUARDIANSHIP OF ADULTS

Guardianship (and in some states, conservatorship) has long been recognized as a means for providing a legal way for someone to make decisions for an incapacitated person. The guardian or conservator is appointed to handle the affairs of the ward. The relationship between guardian or conservator and ward is an artificial legal relationship similar in some ways to that of parent and child. It is a fiduciary relationship in which the law demands that the guardian or conservator show the highest degree of responsibility and care when handling the ward's affairs. As terms, "guardianship" can include authority to make decisions both as to the person and the property of the incapacitated person; "conservatorship," however, usually includes only the authority to handle property. Local law should be consulted for the definition and use of these terms. The remaining discussion in this section of the chapter materials applies to both guardianships and conservatorships.

Guardianship should be considered only after other alternatives have been reviewed when planning with the aging or elderly patient. Guardianship proceedings strip away basic rights. They are difficult to understand and manage. Although intended as protective measures, they usually are not well supervised by the courts and too frequently become a means of abuse of the elderly patient. In addition, the procedural rules and the traditional legal definitions of competency, incompetency, and incapacity used in guardianship proceedings have become subject to increasing criticism in recent years.

Guardianship laws treat the guardian and ward as total strangers, regardless of the "actual" relationship. Although the guardian receives authority to handle the ward's affairs, a great deal of responsibility is placed on the guardian, who may easily find it to be more of a burden than expected. Court orders may be needed before taking actions such as making major medical decisions. The guardian is accountable to the Probate Court and must report and account for every action. Guardianship does not automatically terminate (except by death of the ward). If something were to happen to the guardian or the guardian wished to resign, a proceeding would have to be filed in the Probate Court. More importantly to the ward, the guardian's authority to make a care placement of the ward is not subject to court review and could easily lead to inappropriate institutionalization.

On the positive side, the ward receives certain legal protections designed to help prevent the ward's being abused by the guardian. The accountability of the guardian to the court may provide some protection to the ward. Where the guardian handles the ward's property (like a conservator), the guardian must file a bond. Guardian's bonds have become difficult to obtain and expensive. Generally only the ward's income can be used without obtaining a court order and then only for maintenance of the ward and for certain lawful expenses. The guardian cannot sell any of the ward's property without an order from the Probate Court. Each year the guardian must file an annual accounting, informing the Court of what was received, what was spent, and the remaining balance.

There are three general types of guardianship that may be considered, depending on the needs of the particular ward. Most often thought of is guardianship of person and property or "plenary" guardianship, where the guardian handles both the ward's personal affairs and property. However, it is also possible to preserve some of the ward's autonomy by using, instead, a guardianship of person only or property only. With a guardianship of the person, the guardian handles only the ward's personal needs. With a guardianship of the property, the guardian handles only the ward's property.

To apply for guardianship, the procedures set by law must be strictly followed. Guardianship proceedings are technical, complex, and expensive. Across the United States many of the Probate Judges are not attorneys and find it difficult to handle these technical proceedings involving the most basic personal civil rights. In states with more contemporary mental health laws, the application for permanent guardianship generally takes 5 to 7 weeks. Where the proposed ward is in grave circumstances, emergency guardianship proceedings are usually available.

Generally emergency guardianships terminate automatically after a time specified in the law has elapsed from the appointment of the emergency guardian unless an earlier date is specified by the Probate Court. In many states the specified duration of the emergency guardianship is either 30 or 45 days. As to permanent guardianships, either the guardian or ward can file a petition to terminate the guardianship. Usually the procedure is similar to that of the guardianship application. Permanent guardianships terminate automatically upon the ward's death.

GUARDIANSHIP REFORM

The basic concepts of guardianship are coming under increased scrutiny. Physicians are recognizing that guardianships are not a "cure-all." Of particular concern are the ethical issues of preservation of autonomy, protection of the right of self-determination, and protection of allegedly incapacitated persons from unnecessary losses of liberty (Hommel, Wang, & Bergman, 1990; Wang, Burns, & Hommel, 1990). In many existing guardianship laws, including the Uniform Probate Code, one of the statutory grounds for applying for guardianship is the "advanced age" of the proposed ward (Uniform Probate Code, 1969). Such deficiencies in guardianship laws frequently lead to guardianships that are inappropriately applied for and inappropriately granted. This situation is viewed as a major form of abuse of the elderly.

Recognition of the deficiencies in guardianship laws around the United States has led to discussion of the establishment of "diversion projects." Court personnel are trained to intercept and interview people who seek to file guardianship applications. The purpose of the interview is to ascertain if alternatives, short of guardianship, might be available that would meet the needs of the allegedly incapacitated person (Hommel, Wang, & Bergman, 1990).

This recognition has also led to legislative efforts of reform at national and state levels. Several versions of proposed national guardianship bills have been introduced within the past several years in the U.S. House of Representatives and the U.S. Senate. In July 1988 there was a National Guardianship Symposium that published recommendations for guardianship reform (ABA Commission, 1989).

In 1989 comprehensive new guardianship laws were passed in several states that reflect two major trends in guardianship reform (Fla Stat, 1989; Mich Comp Laws Ann, 1989; NM Stat Ann, 1989; ND Cent Code, 1989). The first is a change in the definition of incapacity, focusing on specific functional abilities rather than the traditional definitions, which provide no meaningful indication of a person's ability to function independently. The second is a clear preference for limited guardianships, where the only rights removed from the ward are those that the ward does not have the functional ability to exercise. Interest in similar guardianship reform is being seen at the state level as well (Wang, Burns, & Hommel, 1990). In addition, preference for the use of durable powers of attorney is clearly being evidenced in the area of adult protective services legislation.

GUARDIANSHIP AVOIDANCE; ALTERNATIVES

Each alternative to guardianship has its own benefits and limitations that must be carefully considered, including the potential for misuse or abuse. In the area of legal and financial planning for the aging client, there are several alternatives.

Joint Ownership of Property

Perhaps a person needs help only with personal business activities such as making bank deposits, writing checks, or paying monthly bills. Adding a trusted relative or friend to their bank account can help. This other person could then make deposits, write checks, and pay bills. However, consideration must be given to the fact that this other person could remove all the money on his or her own signature. Also, if the joint bank account has right of survivorship, it automatically becomes the property of the other person at the death of the original owner, outside the control of the original owner's will. This potential result should be considered carefully.

A more restricted alternative may be that of giving the trusted person signature authority by way of a bank power of attorney. The person appointed as power of attorney on the account does not receive any ownership interest, and no right of survivorship is created. This arrangement allows them to handle the account, but only under the directions of the owner, a safeguard that can help avoid abuse and can be safer than creating a joint account with right of survivorship.

Transferring real estate to create joint ownership can

be particularly problematic. The person added to the title becomes part owner of the property, and the original owner is no longer able to sell, rent, or obtain a loan on the property without the added person's permission. If there is right of survivorship, the original owner's will does not control the disposition of the property. The person added automatically inherits the property at the death of the original owner.

If there is no right of survivorship, the interest given in the property does not pass to the added person at the death of the original owner, unless it is so directed in the will of the original owner. If the added person dies before the original owner, their share of the property goes to their estate, not back to the original owner. The original owner may find himself being forced out of his home because of problems in the estate of the added person. In addition, the person added has access to the property during the lifetime of the original owner. Yet the original owner may intend for the person added to receive the property only upon the original owner's death. Either type of joint ownership clearly can create unanticipated problems.

Gifting

Gifting property to another may serve to protect and preserve the property where the giver faces chronic illness or the prospects of long-term care. The person making the gift can enjoy the giving or see the beneficial results of the gift. However, a host of ethical issues may arise. Once the property has been given away, the giver has lost all control over the gifted property. If an unforeseen emergency arises and the giver wants or needs the property returned, the giver has no legal claim to the property. Legal action could be attempted in some circumstances but would be expensive, time-consuming, and of uncertain outcome.

Direct Deposit and Representative Payee Arrangements

Perhaps a person is receiving Social Security, Railroad Retirement, or Veterans Administration benefits, and needs help only with receiving or handling these funds. The agency paying the benefits can arrange to deposit the benefits directly into the person's bank account, thereby avoiding the possibility that another person may somehow gain access to the check. Alternatively, if an appropriate application is made, the agency could appoint a person as representative payee to receive and handle the funds.

The beneficiary can terminate the arrangement at any time by proving to the agency that they can properly handle his or her own benefits. Frequently a doctor's statement showing capability is sufficient. The relationship can also be terminated by showing to the paying agency that the representative payee does not have the best interests of the beneficiary in mind or has misused the benefits.

Trust Agreements

Another frequently used alternative for business and property management is the Trust Agreement. A trust is a much more formal vehicle than a power of attorney and provides more protection than a power of attorney. A trust is frequently used where there are family problems, as part of a general estate plan, and whenever tax planning is an issue. Because the subject of trusts is complex and technical, only general comments are made here. Experienced professional financial and legal advice is essential when considering any form of a trust arrangement. As with a power of attorney, a Trust Agreement must be drafted with the appropriate safeguards and controls to avoid potential misuse and abuse.

A Trust Agreement signed during life is called an Intervivos or Living Trust. In addition to signing a Trust Agreement or Declaration of Trust (the document creating the trust), the person must also "fund" the trust by transferring property into it. The person who creates the trust (the donor) actually transfers ownership of the property to the trust. The property in the trust (the corpus) is then handled by the person or persons (the trustees) named by the donor. Creating a trust is like founding a corporation or creating an artificial person. The tax laws and regulations that apply to trusts tend to be complicated and highly technical.

Among the benefits of setting up a trust is the ability of the donor to choose the trustee or trustees. In the Trust Agreement the donor specifies exactly how the trust is to be handled and the powers and authorities given to the trustee or trustees, and specifies who is to receive the benefits of the trust (the beneficiary or beneficiaries). The donor can be both the trustee and a beneficiary and can provide for a successor trustee to take over if the donor becomes incapacitated.

A trust can be revocable (the donor keeps the right to terminate the trust) or irrevocable (the donor cannot terminate the trust). The trust can end at the death of the donor or continue after the death of the donor, as the donor wishes. If the trust terminates at the death of the donor, the trust property can then either go to the donor's estate to be handled through the will or be distributed outside the donor's estate according to instructions contained in the Trust Agreement. This document can be an important estate and gift tax planning device.

Trusts are popular vehicles when considering issues of planning for advancing age and planning for the potential need for long-term care. During the early 1980s, trusts gained popularity as a device for qualifying the donor for governmental financial assistance such as Medicaid. Trusts set up for this purpose became known as "Medicaid Qualifying Trusts." However, the increasing use of trusts for this purpose became considered as abusive, and the Medicaid law was changed in a way that virtually ended the use of the Medicaid Qualifying Trust. Today the establishment of a trust does not have the benefit of qualifying the donor for financial assistance, such as Medicaid, unless specific provisions are included in the Trust Agreement that most people find unacceptable. Under current law, the existence of a trust can prevent a person from qualifying for Medicaid assistance, which is yet another reason to consider a trust arrangement with great care and only with proper professional advice.

FOUR FOUNDATION LEGAL DOCUMENTS

Considering and implementing any of the alternatives discussed above is not effective without implementation of the four basic legal planning documents: a Living Will, a Durable Power of Attorney for Health Care, a Durable Financial Power of Attorney, and a Will. These documents are intended to serve as valuable advance planning tools and to protect and preserve a person's right to live and die with dignity.

Without proper safeguards and controls, however, these documents can also become a means of abuse. Although forms may be available and at first glance appear to be less expensive and easier to use, they can create as many problems as they solve. Legal documents should always be tailored to the individual client's need, should state the client's particular instructions and personal desires, and should provide the necessary controls and safeguards. An attorney who is qualified and experienced in preparing such documents is able to identify legal issues and help the person executing the document understand these issues and clarify personal preferences and desires. The attorney is then able to put these preferences and desires into understandable and legally enforceable language.

Professionally drafted documents may cost more up front but ultimately save time, money, and grief for all concerned, especially when crisis hits. This point is particularly true of the Durable Power of Attorney for Health Care. This document is the single most dangerous document the client will ever execute. It literally gives to someone the authority to make life-and-death decisions for the patient if the time comes that the patient cannot. Thus the decision may well be made at a time when the client does not have the capacity to control the decision-making process. Providing for this kind of authority and its exercise is too important to trust to a form. In addition, the life-and-death decision will likely be made in a care facility setting. It is important, therefore, that the advance directive be properly prepared to address the complex and technical, and often conflicting, legal interests and issues that arise in such settings.

Durability

As two of the foundation legal planning documents are powers of attorney, it is important to note that under English Common Law (the law on which the laws of the United States are founded), a power of attorney was automatically revoked if the person who signed it later became physically or mentally incapacitated. A "durable" power of attorney is one in which the specific statement is made that the document "survives" such later incapacity. If the "durability" statement is not included in the document, in many jurisdictions the old Common Law could apply, and the power of attorney would be automatically revoked when the person who signed it later became incapacitated. For these reasons, the national standard is that all powers of attorney specifically be labeled as "durable" and include the "durability" language. It is important to note that as long as the patient is not incapacitated, he or she may override or revoke a durable power of attorney at any time. (See also Chapter 44.)

Living Will

The issue of withholding life supports is one of the most difficult ethical and legal issues our society faces. In recent decades most people have taken the position that uncontrolled use of life support systems is an abuse of personal autonomy. They have demanded recognition of a personal "right to die" with dignity. The Living Will represents the first effort of the legal community to meet this demand.

The Living Will may not be available in every state, although many states have recognized this document by law. If available in their state, the client and family should be advised of this important health-care-related document. The document varies from state to state in terms of how it may be drafted. In most states that have laws recognizing the Living Will, those laws specify the language required to have a legal Living Will. Local law must be consulted.

In most states the Living Will has a dual purpose. One purpose is to express a person's desire not to be kept alive by artificial means. The other is to protect the physician to whom it is presented, as it is usually the treating physician who, by law, makes the decision to withhold or discontinue a life-sustaining treatment. For the patient's protection, most Living Will laws specify the prognosis the physician must make before utilizing the document. Terms usually mentioned include terminal illness, intolerable pain, permanent coma, persistent vegetative state, and "no quality of life." The physician must respect the patient's wishes as set forth in the Living Will or face severe legal penalties. Although required to consult with the patient's family, neither the physician nor the family can go against the desires stated in the Living Will. The Living Will does not give the physician unrestricted authority to "pull the plug"; specific criteria regarding prognosis must always be satisfied.

Given the nature of the issues with which it deals, one problem has been that the name of this document is misleading and has caused some confusion. In addition, the legal definition of the term "Living Will" varies among states. In some, Living Wills are not allowed, provision being made instead for what is usually called a "Natural Death Directive." Despite the difference in name, the concept is similar. Where the term "Living Will" is used, in some states the term refers to a document such as the one described above. In others, the term refers to a document known as the Durable Power of Attorney for Health Care; and in yet others this same term refers to a document that is a combination of both the Living Will and the Durable Power of Attorney for Health Care. Local law again should be consulted. If a patient executes a Living Will and then moves to another state, the document must be reviewed to be sure that it meets local legal criteria.

There is a rising misperception that the Living Will may be losing some importance. Many state laws limit the use of the Living Will to situations involving a "terminal illness" or "terminal condition." Furthermore, many states prohibit the use of a Living Will to disconnect nutrition-hydration devices. Thus there are situations in which the Living Will may not apply. For example, many diseases that make patients candidates for life support systems are not defined medically as "terminal," which is an inherent weakness in the document. Also, depending on local legal definition, Living Wills may not deal with the broader issues of medical treatment and health care. Because of these perceived limitations on the Living Will, in some ways the document was obsolete before it received wide legislative recognition.

In an effort to correct this problem, legislatures have begun to recognize by specific law the Durable Power of Attorney for Health Care. This document is much more comprehensive than the Living Will. Although it may therefore seem to make the Living Will unnecessary, the Living Will is still an important document. The Living Will deals with specific desires of the client with regard to life support systems and creates clear evidence of their desires regarding the use or nonuse of such systems. This point is of great importance in light of the decision of the U.S. Supreme Court in the Cruzan case (Cruzan v. Director, 1990). In this case the Court ruled that the State of Missouri was legally authorized to require "clear and convincing evidence" of a person's desires regarding withdrawal of life supports. A Living Will constitutes such evidence.

Durable Power of Attorney for Health Care

Many states do not have laws that give one family member, even the spouse, the legal right (with the possible exception of emergency situations) to make medical or health care decisions for an incapacitated family member. Once competency has been lost, guardianship may become the only alternative. The consensus among experts is that guardianships are particularly ineffective in the area of medical decision making. Guardians do not have the legal authority to make many medical decisions, particularly life-and-death decisions, without obtaining a court order. Courts find it difficult to deal with the legal and ethical issues that arise when guardians seek court orders allowing them to make medical decisions. In some states, guardians cannot make life-and-death decisions even with a court order.

The traditional legal tool when planning for incapacity is the power of attorney. Hence this document was targeted by the legal profession as the appropriate tool when providing for surrogate medical decision making. Medical decision making powers began to appear in powers of attorney. As these powers were developed and became more technical, the legal profession began separating and segregating medical powers into a separate document, the Durable Power of Attorney for Health Care. Many states now recognize the Durable Power of Attorney for Health Care by law. For example, the Georgia Legislature recognized its importance by passing a statute specifically authorizing its use, effective July 1, 1990 (Official Code of Ga Ann, 1990). New York has also recognized this document, called in that state a "Health Care Proxy" (NY Session Laws, 1990).

A medical power of attorney helps individuals retain their basic civil rights, including the "right to die with dignity" recognized in the Cruzan case. While still competent, patients can make their own decisions and can change or revoke the documents; and they can choose the person or persons they want to make medical decisions for them when they themselves cannot. Provisions can be made for an alternate to serve if the first person cannot. Specific directions can be given in black and white on how decisions are to be made. Personal desires and beliefs can be stated concerning medical and health care preferences, including what types of treatment the patient wants or does not want provided, e.g., withholding, connecting, and disconnecting life support systems. Thus this document is especially valuable when used in combination with a Living Will.

Several points must be emphasized with regard to implementation of the medical power of attorney, especially with regard to dementia patients. At the time the document is executed, the person executing it needs a basic comprehension of the purpose of the document, must understand that he or she is executing the document, and must understand who is being appointed as the agent. The legal test is comprehension *at the moment of executing the document.* The legality of the document is not affected if the person executing it later forgets he has executed the document. It is important to remember that even though a patient has executed a medical power of attorney and appointed another person to make medical decisions in case of incapacity, the patient makes his or her own decisions until such time as the physician determines that the patient has lost the capacity to decide. The patient maintains the right to make all decisions within his or her decisional capacity. Thus the patient may not lose the capacity to make *all* medical decisions and may retain, or regain temporarily or from time to time, the capacity to make some or all medical decisions.

Finally, it is crucially important to understand that the standard most generally applied by the law in the area of surrogate medical decision making is that of "substituted judgment." That is, a standard requiring that the agent appointed in the medical power of attorney make medical decisions *as the patient would make them*, not as the agent would make them. In this regard, proper drafting of the medical power of attorney becomes paramount, as the national standard is that the document must specifically reflect the personal beliefs, desires, and value system of the patient.

Cruzan Case

In its decision in the Cruzan case on June 25, 1990 (Cruzan v. Director, 1990), the U.S. Supreme Court underlined the importance of medical powers of attorney. Nancy Cruzan was in a skilled care facility, being kept alive by feeding tubes. She was in a "persistent vegetative state" according to an attending neurosurgeon, who testified that she exhibited motor reflexes but evidenced no indications of significant cognitive function. She was then in her early thirties and had been in this condition with no improvement since 1986. Being advised that their daughter would never improve and feeling that she was actually "dead," Nancy Cruzan's parents went to court to get an order to have the tubes removed. Although this order was granted in the lower court, on appeal the Missouri Supreme Court turned them down. This court ruled that where a patient is incapacitated and can not make medical decisions (specifically as to life support systems), no other person has the legal authority to make such decisions for the patient, regardless of that person's relationship to the patient or the patient's condition. The court further held that oral statements about life support systems made to others by a patient are not legally binding. On this basis the Court went on to deny the Cruzan's request to have their daughter's feeding tube removed, finding that they had not sufficiently proved Nancy Cruzan's desires concerning life supports. They had not provided "clear and convincing" evidence. The parents took their case to the U.S. Supreme Court.

It was the first time in history that the U.S. Supreme Court agreed to face the issues raised in the Cruzan case. Is there a "right to die"? If a person does not have the capacity to make a decision regarding that right, can a decision be made for them? If so, how and by whom?

Chief Justice Rehnquist wrote the majority opinion issued by the U.S. Supreme Court in this case, finding that a competent person has a constitutional right to refuse unwanted medical treatment. This finding was based on the protection of each person's liberty interest found in the Fourteenth Amendment to the Constitution. Part of this protected liberty interest is the right of each person to "self-determination;" that is, the right to decide what will or will not be done to them. In a separate opinion, Justice Brennan stated specifically that "Nancy Cruzan is entitled to choose to die with dignity" (Cruzan v. Director, 1990, p. 4926). However, what about persons such as Nancy Cruzan who are not competent to make their own medical decisions? The answer provided to this question by the Court was that *each individual state must set its own standards for decisions regarding life support systems for incapacitated persons.* The appropriate mechanisms would be legislation or court decision. It is interesting to note that the Court, as dicta, specifically indicated that advance directives (such as the medical power of attorney) would be an appropriate way to resolve this question and avoid the problem.

This U.S. Supreme Court decision is having the effect in many states of severely limiting any authority of families to make decisions such as designating an incapacitated family member as a "no code" or "do not resuscitate," as well as limiting any authority for family members to make decisions regarding the use, withholding, or disconnection of life support systems. In other words, it may well be that if the patient wants someone to be able to make medical decisions for him when he cannot, he must put it in writing with a Living Will and a Durable Medical Power of Attorney. Without these documents, the Cruzan decision implies that some states may make it impossible even for a guardian to obtain legal authority to make decisions regarding life support systems.

The U.S. Congress responded to this perceived problem on November 5, 1990, with the passage of the Patient Self-Determination Act (Omnibus Budget Reconciliation Act, 1990). This Act requires health care providers who participate in the Medicare and Medicaid programs to provide written information about the availability and use of advance directives. They are required to provide this information to patients, families, and the community. The Act is effective with respect to services furnished on or after December 1, 1991.

Following the Cruzan decision, Nancy Cruzan's parents went back to the lower court in Missouri. On November 1, 1990, the lower court reviewed all the evidence from the first hearing and heard additional evidence from all parties concerned. It also considered the decision of the U.S. Supreme Court and a number of amicus briefs filed by organizations such as the International Anti-Euthanasia Task Force, the National Legal Center for the Medically Dependent and Disabled, Inc., and the National Academy of Elder Law Attorneys. On December 14, 1990, the court issued an order finding "by clear and convincing evidence" that Nancy Cruzan, if mentally able, would intend to terminate her nutrition and hydration (Cruzan v. Mouton, 1989). The court also found that Mr. and Mrs. Cruzan, as co-guardians for their daughter, were authorized to have the nutrition and hydration systems removed. It was done, and Nancy Cruzan died shortly thereafter.

Durable Power of Attorney

The Durable Power of Attorney for finance is the most frequently used alternative to guardianship or conservatorship. The Durable Financial Power of Attorney deals exclusively with matters of business and finance. Basic civil rights are retained; and while still competent, the person makes his or her own decisions and can revoke the document. Such individuals may choose who will handle their business and make financial decisions for them if they become incapacitated. The patient may specify particular desires as to how decisions are to be made, and specific checks and balances can be included to help avoid potential financial abuse. Provisions can be made for one or more alternates to serve if the first person cannot or chooses not to serve.

Last Will and Testament

A will is an essential planning instrument for all stages of life, not just for the aged or wealthy. In the will, the maker decides who will handle his or her estate and how it will be distributed. There are certain legal requirements to have a valid will. The specifics of these requirements vary somewhat from state to state, so local law should be consulted. Generally, the testator must be of a certain age, of sound mind, understand that he or she is making the Last Will and Testament, and be clearly identified in the document. Even victims of Alzheimer's disease or other dementia can generally execute a valid will if they have the requisite legal capacity.

Careful thought must be given to the distribution of the estate. In some states it is not necessary to leave anything to family members, including spouse or children. Local law should be reviewed in this regard. If any beneficiaries are minors or incapacitated (such as an adult beneficiary who suffers from Alzheimer's disease or dementia or is otherwise a "special needs" person), special arrangements may be needed for someone to receive and hold their share and protect them from potential financial abuses. Making such arrangements can avoid the need for guardianship. Also, the bequest may prevent the beneficiary from qualification for government benefits. If a married couple make wills, they are likely to be what are known as "mirror" wills. Each spouse leaves everything to the other, with a provision that if the other spouse does not survive everything goes to their children. Without careful review, the potential result could be that a healthy spouse dies first, leaving everything to an ill spouse who is in a care facility. The ill spouse has been qualified for Medicaid, and the inheritance increases the spouse's resources beyond Medicaid limits. The survivor now no longer qualifies for Medicaid assistance and may well have to have a guardian appointed to handle these assets. The assets must be spent on the costs of the survivor's care, and the desire of the couple to see that their children receive their estate is totally defeated.

Regardless of one's personal feelings about disposition of assets in order to qualify for assistance, it must also be remembered that Medicaid does not pay for everything. An important objective of asset management should be that some assets remain in the hands of trustworthy persons who will use them to pay for the expenses not covered by the Medicaid program. This provision helps to improve the quality of the patient's life during the final years, which is especially important in light of current and anticipated cutbacks in Medicaid programs.

MEDICAID

One rapidly growing area of special need for our older population is that of health care, particularly long-term care. It was first reflected at the federal level in 1965 when the Medicare and Medicaid programs were legislated by amendment to Title XIX of the Social Security Act. Medicaid (42 U.S.C. §§ 1396, et seq.) was created as a welfare program to provide medical assistance to low income individuals and families. As such, it represents the first step toward a national health care policy.

During the 1970s and 1980s there was an explosive increase in recognition of the continued expansion of our older population and in the concern over issues of public health care and the drastic increases in the cost of health care. In 1988 it resulted in passage of the first major amendment to the Medicare and Medicaid programs: *The Medicare Catastrophic Coverage Act of 1988* (MCCA) [P. L. 100-360, Section 303, July 1, 1988]. MCCA was seen by some as another step toward a national health care policy. However, the basic federal philosophy continued to be to create only a limited public policy and to let the private insurance industry handle the bulk of health care coverage for most of the population, including the older population.

After passage, MCCA generated some of the strongest public controversy in decades, primarily over the Medicare surcharges which were intended to finance the expanded Medicare coverage. This controversy prompted the repeal of MCCA in November 1989. Given the intense national concern over issues of health care for our aging population, there was considerable disagreement between the House and the Senate over the repeal of MCCA. There was strong sentiment in the Senate to save the Medicaid portions of MCCA, as these portions were not tied to the Medicare surcharges. The Senate was successful, leaving repealed the Medicare portions that had started the legislative process.

In brief, the Medicaid program is intended to provide assistance with medical expenses to persons whose income is not sufficient to meet their medical needs. Thus it is at heart a poverty program. To be eligible for Medicaid, people must qualify in two areas. First, they must fit within one of the specified eligibility categories: aged (over 65); blind; disabled; or the caretaker of minor children. Second, they must meet certain financial requirements as to income (which varies from state to state) and resources (universally limited to $2,000 for the Medicaid applicant). Once all eligibility criteria are met, the person qualifies for Medicaid assistance. Payment for approved Medicaid services is made directly to the service provider.

One of the difficulties with the Medicaid program is that there is voluminous law, regulation, and administrative policy, at both the federal and the state level, that controls the program. Another more problematic difficulty is that there are significant variations from state to state in program eligibility and coverage. Medicaid is a federal- and state-financed program and is administered locally by each state. Although a basic framework is provided by federal law, each state may choose from a variety of coverage options within that framework. As an example of the degree of variation in state Medicaid programs, California went so far as to elect not to participate directly in the Medicaid program, instead creating its own program called "Medi-Cal." MCCA further complicated state-to-state variations in Medicaid programs by providing new options in program coverage and eligibility criteria.

There are several basic categories of people eligible for the Medicaid program. The availability of each varies from state to state so the local program should be consulted. Medicaid is for most older adults the only source of government assistance in paying for long-term nursing home care. To qualify, the patient must be certified as needing a nursing home level of care, and meet the specified financial eligibility criteria in the areas of income and assets. This category is often referred to as "nursing home Medicaid."

Spousal Impoverishment

"Spousal impoverishment" as a term and as a concept was also created by the passage of MCCA in 1988 (Medicare Catastrophic Coverage Act, 1988). Prior to MCCA, the Medicaid eligibility rules often resulted in the impoverishment not only of the institutionalized Medicaid applicant but also their at-home or "community" spouse. MCCA was intended to help avoid impoverishment of the community spouse, i.e., to avoid "spousal impoverishment." The concept was to provide financial protections designed to ensure that the community spouse would retain a minimal amount of assets and receive an acceptable minimal level of income so as to allow him or her to continue to live in the community without the need of public assistance. The MCCA spousal impoverishment rules apply to all Medicaid applicants institutionalized on or after September 30, 1989.

Income Protection

Under MCCA the states were required to establish a "protected" amount of income for the community spouse ("spousal income allowance"), to be set within a specified range, from an amount tied to the federal poverty level for a family of two, to a maximum of $1,500 per month. If less than the allowed maximum was chosen, specific provision was made for calculation of a spousal "minimum monthly maintenance needs allowance" and for an additional "excess shelter allowance." The figure chosen for the spousal income allowance is indexed to increases in the cost of living during subsequent years (as evidenced by the Consumer Price Index), effective January 1 of each year. Local law should be consulted to ascertain what amount was chosen by the individual state.

Resource Protection

MCCA required the states to allow the community spouse to retain a "protected" amount of assets ("spousal resource allowance") within a range of not less than $12,000, nor more than a cap of $60,000 (excluding the home, as discussed below). Whatever amount was adopted, it is indexed to increases in the cost of living (as evidenced by the Consumer Price Index) in subsequent years. This increase is also effective on January 1 of each year. Local law should be consulted to ascertain what amount was chosen by the individual state. The community spouse resource allowance is in addition to the $2,000 resource allowance permitted the Medicaid applicant.

"Snapshot"

MCCA provides a formal mechanism for officially applying the rules for treatment of marital resources and determining financial eligibility for Medicaid assistance. This mechanism is the "assessment" procedure, also called the "snapshot." Prior to MCCA, the assessment was made as of the date the Medicaid application was filed. Under MCCA, the "snapshot" is a compilation of marital resources as they were on the date of admission of the applicant to the nursing home, regardless of the date of application. It is another major change from the pre-MCCA rules. The purpose of the snapshot is to determine how the resource rules apply—how much a couple will have to spend down for Medicaid eligibility.

Transfer of Resources

Usually the first strategy thought of by most couples for reducing excess resources is that of transferring or giving away resources. Although sometimes used in Medicaid eligibility planning, this strategy should be used as a "last resort." Other available alternatives can provide more direct value to the client without causing them to give away their hard-earned estate to other people. In addition, for Medicaid purposes transfers are penalized.

MCCA had a major impact on the transfer penalty provisions of the Medicaid eligibility rules. The workings of the MCCA rules on transfer of assets are therefore important to consider, particularly in light of the changed transfer penalty rules. Subject to certain limited exceptions, the MCCA transfer penalty rules apply to all transfers for less than full consideration made by either spouse to a third person. A formula is provided by MCCA for calculating transfer penalties, the penalty being a period of ineligibility for Medicaid assistance from the date of each transfer. Regardless of the amount transferred, the maximum penalty period is capped at 30 months.

When considering assets, the principal residence usually becomes the subject of major consideration. It is usually the single asset the client most wants to protect. Particularly for the older client, the home is a symbol of autonomy and independence. The preference of Congress for continued home ownership is clearly evidenced throughout MCCA. The homeplace and contiguous land are exempt assets, so long as a spouse or dependent child lives there or the applicant has an intention to return home. Also, a number of specific exceptions to the transfer rules are provided in MCCA to protect the home. However, these exceptions are particularly subject to variation in application from state to state, and it is of extreme importance to consult local law in this regard.

Medicaid Qualifying Trusts

Although a thorough discussion of Medicaid Qualifying Trusts is beyond the scope of these materials, there are some basics that are well to keep in mind. With the rising concern over escalating health care costs and fear of the need for catastrophic long-term care, intervivos trusts became popular during the late 1970s and early 1980s as a vehicle for obtaining Medicaid eligibility.

Simply put, potential applicants would put their assets into a trust, wait 2 years (to avoid the Two Year Transfer Rule now replaced by the MCCA Thirty Month Transfer Penalty Rule), and, *voilà*, they qualified. They no longer owned assets, the trust owned them. A trust set up for this purpose became known as a "Medicaid Qualifying Trust" (MQT). When a national news magazine ran an article explaining why everyone should have such a trust, Congress responded sharply in 1986 with an "antitrust" amendment to the Medicaid law (sometimes call the "antidiscretion rule"). This amendment created one of the major "cons" to MQTs by requiring that assets in the trust be "locked away" by prohibiting the trustees from paying out any money and property as would disqualify the beneficiary/applicant from Medicaid benefits. This rule severely limits the usefulness of the trust. What if the beneficiary needs extra money for an emergency expense? What if the Medicaid program changes or ceases to exist in the future?

CONCLUSION

The increasing complexity of legal, financial, medical, and ethical issues has extensively affected the medical and mental health professions as well as the health care industry. The legal and ethical aspects of long-recognized issues, such as consent to treatment, informed consent, withholding treatment, and assessment of competency, have once again become emergent issues. Particular attention is being focused on the many problematic issues and needs related to increasing needs for long-term care. In this regard, increasing attention is being given to recognition of the fact that if attorneys focus too narrowly on the financial issues and alternatives for Medicaid qualification planning, it severely limits the client's options and may well do more harm than good in the long run.

No one profession can give competent advice on all these issues. Each issue has more than one facet. For example, what may appear best financially is not always best emotionally. Approaching an issue exclusively from the financial perspective also may lead to a solution that has a negative legal or even medical result. The client must work with several professionals, ideally as an interdisciplinary team.

Consideration must not be limited to the immediate problem or situation. Client needs, family needs, and the relations between them must be identified. Clear, realistic goals must be established to meet these needs. Effective planning, which demands open communication, is adaptive planning that changes with life.

REFERENCES

ABA Commission on the Mentally Disabled and Commission of Legal Problems of the Elderly (1989). Guardianship: an agenda for reform.

American Hospital Association statement on a patient's bill of rights (1973). Hospitals 47:41.

American Psychiatric Association Task Force Report 23 (1986). An overview of legal issues in geriatric psychiatry. Washington, DC: APA.

Auerbach VS, & Banja JD (in press) (1993). Competency determinations: Issues for the mental health professional. In A Stoudemire, & BS Fogel (Eds.), Medical-psychiatric practice (Vol. 2). Washington, DC: American Psychiatric Press, Inc.

Appelbaum PS, & Roth LH (1981). Clinical issues in the assessment of competency. Am J Psychiatry 138:1462–1467.

Brakel SH, Parry J, & Weiner BA (Eds.) (1986). The mentally disabled and the law (3rd ed.). Chicago: American Bar Foundation.

Canterbury v. Spence (1972). 464 F 2d 772, DC Cir Ct, cert denied, 409 US 1064.

Cobbs v. Grant (1972). 8 Cal 3d 229, 502 P 2d 1, 104 Cal Rptr 505.

Crisher v. Spak (1983). 122 Misc 2d 355, 471 NYS 2d 741.

Nancy Beth Cruzan, Lester L. and Joyce Cruzan, petitioners v. David D. Mouton, esq., and Thad C. McCanse, esq., Co-Guardian ad litems, Estate No. CV384-9P, Cir Ct, Jasper County, Missouri (1989).

Nancy Beth Cruzan, by her Parents and Co-Guardians, Lester L. Cruzan, et ux., Petitioners v. Director, Missouri Department of Health, et al. (No. 88-1503, June 26, 1990; 58 LW 4916).

Dewes v. Indian Health Service (1980). 504 F Supp 203, SD.

Fla Stat 744.101 et seq. (1989) (effective Oct 1, 1989); Mich Comp Laws Ann 700.441 et seq. (1989 Supp) (effective March 30, 1989); NM Stat Ann 45-5-301 et seq. (1989) (effective June 16, 1989); ND Cent Code 30.1-26-01 et seq. (1989 Supp) (effective July 1, 1990).

Hommel PA, Wang L, Bergman JA (1990). Trends in guardianship reform: implications for the medical and legal professions. Law Med Health Care 18:213–226.

Fogel, B, Mills MJ, & Landen JE (1986). Legal aspects of the treatment of delirium. Hosp Comm Psychiatry 37:154–158.

Kaimowitz v. Michigan Department of Mental Health Div No 73-91434-AW, Cir Ct, Wayne County, Mich (1973). Abstracted in 13 Criminal Law Rep 2452. Reprinted: Brooks AD (1974). Law, psychiatry and the mental health system. New York: Little, Brown.

Macdonald MG, Meyer KC, & Essig B (1986). Health care law: a practical guide. New York: Matthew Bender.

Medicare Catastrophic Coverage Act (1988). Public Law 100- 360, §303(a). 42 U.S.C. §1396a.

Meretsky v. Ellenby (1979). 370 So 2d 1222, Fla Dist Ct App.

Mills MJ (1985). Legal issues in psychiatric treatment. Psychiatr Med 2:245–261.

Mills MF, & Daniels ML (1987). Medical-legal issues. In A Stoudemire & BF Fogel (Eds.), Principles of medical psychiatry. Orlando, FL: Grune & Stratton.

Natanson v. Kline (1960). 186 Kan 393, 350 P 2d 1093, clarified at 187 Kan 186, 354 P 2d 670.

Nolan BS (1984). Functional evaluation of the elderly in guardianship proceedings. Law Med Health Care 12:210–218.

NY Session Laws, Chapt 752:1540 (1990).

Official Code of Ga Ann 31-36-1 (1990).

Omnibus Budget Reconciliation Act, Public Law 100-203 (1987).

Omnibus Budget Reconciliation Act, §§ 4206 and 4751 (1990).

Overman WH, & Stoudemire A (1988). Guidelines for legal and

financial counseling of Alzheimer's disease patients and their families. Am J Psychiatry 145:1495–1500.

Pugsley v. Privett (1980). 220 VA 892, 263 SE 2d 69.

ROTH LH (1982). Competency to consent to or refuse treatment. In L GRINSPOON (Ed.), Psychiatry 1982: The American Psychiatric Association Annual Review. Washington, DC: American Psychiatric Press.

ROTH LH, MEISEL A, & LIDZ CW (1977). Tests of competency to consent to treatment. Am J Psychiatry 134:279–284.

Salgo v. Leland Stanford Jr University Board of Trustees (1957). 154 Cal App 2d 560, 317 P 2d 170, 1st Dist.

Schloendorff v. Society of New York Hospital (1914). 211 NY 125, 105 NE 92.

Truman v. Thomas (1980). 27 Cal 3d 285, 611 P 2d 902, 165 Cal Rptr 308.

Uniform Probate Code, Section 5-101(1) (August 1969). Adopted by National Conference of Commissioners on Uniform Laws, American Bar Association.

WANG L, BURNS AM, & HOMMEL PA (1990). Trends in guardianship reform: roles and responsibilities of legal advocates. Clearinghouse Rev 24:561–569.

Wilkinson v. Vesey (1972). 110 RI 606, 295 A2d 676.

Index

Note: Page numbers followed by t refer to tables, page numbers followed by f refer to figures.